Atmospheric and Oceanic Fluid Dynamics

Fundamentals and Large-Scale Circulation

Second Edition

The atmosphere and ocean are two of the most important components of the climate system, and fluid dynamics is central to our understanding of both. This book provides a unified and comprehensive treatment of the field that blends classical results with modern interpretations. It takes the reader seamlessly from the basics to the frontiers of knowledge, from the equations of motion to modern theories of the general circulation of the atmosphere and ocean. These concepts are illustrated throughout the book with observations and numerical examples. As well as updating existing chapters, this second, full-colour edition includes new chapters on tropical dynamics, El Niño, the stratosphere and gravity waves. Supplementary resources are provided online, including figures from the book and problem sets, making this new edition an ideal resource for students and scientists in the atmospheric, oceanic and climate sciences, as well as in applied mathematics and engineering.

Geoffrey K. Vallis is a professor of applied mathematics at the University of Exeter, UK. Prior to taking up his position there, he taught for many years at Princeton University in the USA. He has carried out research in the atmospheric sciences, oceanography and the planetary sciences, and has published over 100 peer-reviewed journal articles. He is the recipient of various prizes and awards, including the Adrian Gill Prize (Royal Meteorological Society) in 2014, and the Stanislaw M. Ulam Distinguished Scholar award (Los Alamos National Laboratory) in 2013.

"Vallis writes explanations as clear as tropical ocean waters, bringing fresh new light to complex concepts. This expanded text will be immediately useful both for graduate students and seasoned researchers in the field."

Dargan M. W. Frierson, *University of Washington*

"In 2006, Vallis' first edition of AOFD offered the atmospheric and oceanic sciences community a truly great book, marking a milestone in our discipline. Well, Vallis has done it again! This second edition of AOFD represents the pinnacle of a maturing discipline. It is **The Great Book** of the field, and it will remain so for a generation or longer. AOFD-2 dives deep into atmospheric and oceanic fluid dynamics, spanning a wealth of topics while offering the reader lucid, pedagogical, and thorough presentations across a universe of knowledge. There are really three books here: one focused on geophysical fluid dynamic fundamentals; a second on atmospheric dynamics; and a third on oceanic dynamics. Each part offers new material relative to the first edition, as well as the reworking of earlier presentations to enhance pedagogy and update understanding based on recent research. ... the reader is privileged to receive a unified presentation from a master scientific writer whose pedagogy is unmatched in the discipline. This book is a truly grand achievement. It will be well used by fluid dynamicists, oceanographers, atmospheric scientists, applied mathematicians, and physicists for decades to come. Each sentence, paragraph, section, chapter, and figure, are thoughtful and erudite, providing the reader with insights and rigor needed to truly capture the physical and mathematical essence of each topic."

Stephen M. Griffies, *Geophysical Fluid Dynamics Laboratory and Princeton University*

"Vallis speaks my language. He successfully weaves together fundamental theory, physical intuition, and observed phenomena to tell the story of geophysical fluid behavior at local and global scales. This multi-pronged approach makes this an ideal text for both beginners and experts alike - there is something for everyone. This is why it is the book I use for my class, the book I recommend to incoming graduate students (no matter their background) and the book I go to first when I need clarity on GFD topics. The first edition of this book has been my go-to text since it was first published ... With the new edition, we now get an even more comprehensive view of how the fundamental processes that dictate the evolution of our atmosphere and oceans drive the complex phenomena we observe."

Elizabeth A. Barnes, *Colorado State University*

"The first edition ... provided an exceptionally valuable introduction to the dynamical theory of the large-scale circulation of the atmosphere and ocean ... This second edition is a further major achievement ... It includes significant new material on the atmosphere and on the ocean, presented in two separate later sections of the book, but building carefully and clearly on the 'unified' material in the first part of the book ... The second edition will be an exceptionally valuable resource for those designing advanced-level courses, for the students taking those courses and for researchers, many of whom will surely be stimulated by the clear presentation of existing theory to identify what such theory does not explain and where progress is needed."

Peter Haynes, *University of Cambridge*

"This second edition is even more comprehensive than the first. It now covers subjects such as the derivation of the first law of thermodynamics, the fundamental physics involved in the meridional overturning of the ocean, and equatorial oceanography. The book concentrates on the fundamentals of each subject, with sufficient motivation to make the exposition clear. For good reason, the first edition is now the standard text for courses in oceanography, and this will clearly continue with this second edition, helping all of us, not just students, to clarify our understanding of this field."

Trevor J. McDougall, *University of New South Wales*

"Researchers looking for an informative and coherent treatment of the dynamics of the atmosphere and ocean, starting at a fundamental level, and proceeding to advanced topics, will find that this book is a truly superb resource. The book is particularly notable for its even-handed treatment of the ocean and the atmosphere and its synthetic discussion of observations, numerics and analytic methods."

William R. Young, *Scripps Institution of Oceanography*

Atmospheric and Oceanic Fluid Dynamics

Fundamentals and Large-Scale Circulation

Second Edition

Geoffrey K. Vallis
University of Exeter

CAMBRIDGE
UNIVERSITY PRESS

CAMBRIDGE
UNIVERSITY PRESS

University Printing House, Cambridge CB2 8BS, United Kingdom

One Liberty Plaza, 20th Floor, New York, NY 10006, USA

477 Williamstown Road, Port Melbourne, VIC 3207, Australia

314-321, 3rd Floor, Plot 3, Splendor Forum, Jasola District Centre, New Delhi - 110025, India

79 Anson Road, #06-04/06, Singapore 079906

Cambridge University Press is part of the University of Cambridge.

It furthers the University's mission by disseminating knowledge in the pursuit of education, learning and research at the highest international levels of excellence.

www.cambridge.org
Information on this title: www.cambridge.org/9781107065505
DOI: 10.1017/9781107588417

First published 2017
Reprinted 2019

A catalogue record for this publication is available from the British Library

ISBN 978-1-107-06550-5 Hardback

The Falconer

Turning and turning in the widening gyre
W. B. Yeats.

Soaring and diving in the spiraling gyre
The freed falcon divines
Earth's air and water.

B. A. Wingate, 2017.

Contents

Preface xiii
Notation xvii

PART I FUNDAMENTALS OF GEOPHYSICAL FLUID DYNAMICS 1

1 Equations of Motion 3
1.1 Time Derivatives for Fluids 3
1.2 The Mass Continuity Equation 7
1.3 The Momentum Equation 11
1.4 The Equation of State 13
1.5 Thermodynamic Relations 14
1.6 Thermodynamic Equations for Fluids 21
1.7 Thermodynamics of Seawater 30
1.8 Sound Waves 40
1.9 Compressible and Incompressible Flow 41
1.10 The Energy Budget 42
1.11 An Introduction to nondimensionalization and Scaling 46
Appendix A: Thermodynamics of an Ideal gas from the Gibbs function 47
Appendix B: The First Law of Thermodynamics for Fluids 49

2 Effects of Rotation and Stratification 55
2.1 Equations of Motion in a Rotating Frame 55
2.2 Equations of Motion in Spherical Coordinates 59
2.3 Cartesian Approximations: The Tangent Plane 69
2.4 The Boussinesq Approximation 70
2.5 The Anelastic Approximation 75
2.6 Pressure and other Vertical Coordinates 79
2.7 Scaling for Hydrostatic Balance 83
2.8 Geostrophic and Thermal Wind Balance 87
2.9 Gradient Wind Balance 94
2.10 Static Instability and the Parcel Method 97
Appendix A: Asymptotic Derivation of the Boussinesq Equations 101

3 Shallow Water Systems 105
3.1 Dynamics of a Single Shallow Layer of Fluid 105
3.2 Reduced Gravity Equations 110
3.3 Multi-Layer Shallow Water Equations 112
3.4 From Continuous Stratification to Shallow Water 114
3.5 Geostrophic Balance and Thermal Wind 118
3.6 Form Stress 119
3.7 Conservation Properties of Shallow Water Systems 120
3.8 Shallow Water Waves 123
3.9 Geostrophic Adjustment 127
3.10 Isentropic Coordinates 134
3.11 Available Potential Energy 137

4 Vorticity and Potential Vorticity 143
4.1 Vorticity and Circulation 143
4.2 The Vorticity Equation 145
4.3 Vorticity and Circulation Theorems 147
4.4 Vorticity Equation in a Rotating Frame 153
4.5 Potential Vorticity Conservation 156
4.6 Potential Vorticity in the Shallow Water System 162
4.7 Potential Vorticity in Approximate, Stratified Models 163
4.8 The Impermeability of Isentropes to Potential Vorticity 165

5 Geostrophic Theory 171
5.1 Geostrophic Scaling 171
5.2 The Planetary-Geostrophic Equations 176
5.3 The Shallow Water Quasi-Geostrophic Equations 180
5.4 The Continuously Stratified Quasi-Geostrophic System 187
5.5 Quasi-Geostrophy and Ertel Potential Vorticity 195
5.6 Energetics of Quasi-Geostrophy 198
5.7 The Ekman Layer 201

PART II WAVES, INSTABILITIES AND TURBULENCE 213

6 Wave Fundamentals 215
6.1 Fundamentals and Formalities 215
6.2 Group Velocity 220
6.3 Ray Theory 224
6.4 Rossby Waves 226
6.5 Rossby Waves in Stratified Quasi-Geostrophic Flow 231
6.6 Energy Propagation and Reflection of Rossby Waves 234
6.7 Group Velocity, Revisited 240
6.8 Energy Propagation of Poincaré Waves 244
Appendix A: The WKB Approximation for Linear Waves 247

7 Gravity Waves 251
7.1 Surface Gravity Waves 251
7.2 Shallow Water Waves on Fluid Interfaces 257
7.3 Internal Waves in a Continuously Stratified Fluid 259
7.4 Internal Wave Reflection 268

	7.5	Internal Waves in a Fluid with Varying Stratification	271
	7.6	Internal Waves in a Rotating Frame of Reference	276
	7.7	Topographic Generation of Internal Waves	283
	7.8	Acoustic-Gravity Waves in an Ideal Gas	293
8	**Linear Dynamics at Low Latitudes**		**297**
	8.1	Co-existence of Rossby and Gravity Waves	298
	8.2	Waves on the Equatorial Beta Plane	303
	8.3	Ray Tracing and Equatorial Trapping	314
	8.4	Forced-Dissipative Wavelike Flow	316
	8.5	Forced, Steady Flow: the Matsuno–Gill Problem	321
		Appendix A: Nondimensionalization and Parabolic Cylinder Functions	330
		Appendix B: Mathematical Relations in the Matsuno–Gill Problem	333
9	**Barotropic and Baroclinic Instability**		**335**
	9.1	Kelvin–Helmholtz Instability	335
	9.2	Instability of Parallel Shear Flow	337
	9.3	Necessary Conditions for Instability	345
	9.4	Baroclinic Instability	347
	9.5	The Eady Problem	351
	9.6	Two-Layer Baroclinic Instability	356
	9.7	A Kinematic View of Baroclinic Instability	363
	9.8	The Energetics of Linear Baroclinic Instability	367
	9.9	Beta, Shear and Stratification in a Continuous Model	369
10	**Waves, Mean-Flows, and their Interaction**		**379**
	10.1	Quasi-Geostrophic Wave–Mean-Flow Interaction	380
	10.2	The Eliassen–Palm Flux	383
	10.3	The Transformed Eulerian Mean	387
	10.4	The Non-Acceleration Result	394
	10.5	Influence of Eddies on the Mean-Flow in the Eady Problem	399
	10.6	Necessary Conditions for Instability	403
	10.7	Necessary Conditions for Instability: Use of Pseudoenergy	406
11	**Basics of Incompressible Turbulence**		**413**
	11.1	The Fundamental Problem of Turbulence	413
	11.2	The Kolmogorov Theory	416
	11.3	Two-dimensional Turbulence	423
	11.4	Predictability of Turbulence	433
	11.5	Spectra of Passive Tracers	437
12	**Geostrophic Turbulence and Baroclinic Eddies**		**445**
	12.1	Differential Rotation in Two-dimensional Turbulence	445
	12.2	Stratified Geostrophic Turbulence	454
	12.3	A Scaling Theory for Geostrophic Turbulence	460
	12.4	Phenomenology of Baroclinic Eddies in the Atmosphere and Ocean	464

13 Turbulent Diffusion and Eddy Transport 473

13.1 Diffusive Transport 473
13.2 Turbulent Diffusion 475
13.3 Two-Particle Diffusivity 480
13.4 Mixing Length Theory 484
13.5 Homogenization of a Scalar that is Advected and Diffused 487
13.6 Diffusive Fluxes and Skew Fluxes 490
13.7 Eddy Diffusion in the Atmosphere and Ocean 493
13.8 Thickness and Potential Vorticity Diffusion 502

PART III LARGE-SCALE ATMOSPHERIC CIRCULATION 509

14 The Overturning Circulation: Hadley and Ferrel Cells 511

14.1 Basic Features of the Atmosphere 511
14.2 A Steady Model of the Hadley Cell 516
14.3 A Shallow Water Model of the Hadley Cell 524
14.4 Asymmetry Around the Equator 525
14.5 Eddy Effects on the Hadley Cell 528
14.6 Non-local Eddy Effects and Numerical Results 532
14.7 The Ferrel Cell 534

15 Zonally-Averaged Mid-Latitude Atmospheric Circulation 539

15.1 Surface Westerlies and the Maintenance of a Barotropic Jet 540
15.2 Layered Models of the Mid-Latitude Circulation 549
15.3 Eddy Fluxes and an Example of a Closed Model 562
15.4 A Stratified Model and the Real Atmosphere 566
15.5 Tropopause Height and the Stratification of the Troposphere 572
15.6 A Model for both Stratification and Tropopause Height 579
Appendix A: TEM for the Primitive Equations in Spherical Coordinates 581

16 Planetary Waves and Zonal Asymmetries 585

16.1 Rossby Wave Propagation in a Slowly Varying Medium 585
16.2 Horizontal Propagation of Rossby Waves 588
16.3 Critical Lines and Critical Layers 594
16.4 A WKB Wave–Mean-Flow Problem for Rossby Waves 598
16.5 Vertical Propagation of Rossby waves 599
16.6 Vertical Propagation of Rossby Waves in Shear 606
16.7 Forced and Stationary Rossby Waves 609
16.8 Effects of Thermal Forcing 615
16.9 Wave Propagation Using Ray Theory 621

17 The Stratosphere 627

17.1 A Descriptive Overview 627
17.2 Waves in the Stratosphere 634
17.3 Wave Momentum Transport and Deposition 639
17.4 Phenomenology of the Residual Overturning Circulation 642
17.5 Dynamics of the Residual Overturning Circulation 644
17.6 The Quasi-Biennial Oscillation 652
17.7 Variability and Extra-Tropical Wave–Mean-Flow Interaction 663

18 Water Vapour and the Tropical Atmosphere 673
18.1 A Moist Ideal Gas 673
18.2 The Distribution of Relative Humidity 680
18.3 Atmospheric Convection 691
18.4 Convection in a Moist Atmosphere 695
18.5 Radiative Equilibrium 700
18.6 Radiative-Convective Equilibrium 703
18.7 Vertically-Constrained Equations of Motion for Large Scales 708
18.8 Scaling and Balanced Dynamics for Large-Scale Flow in the Tropics 711
18.9 Scaling and Balance for Large-Scale Flow with Diabatic Sources 714
18.10 Convectively Coupled Gravity Waves and the MJO 717
Appendix A: Moist Thermodynamics from the Gibbs Function 720
Appendix B: Equations of Radiative Transfer 724
Appendix C: Analytic Approximation of Tropopause Height 725

PART IV LARGE-SCALE OCEANIC CIRCULATION 729

19 Wind-Driven Gyres 731
19.1 The Depth Integrated Wind-Driven Circulation 733
19.2 Using Viscosity Instead of Drag 740
19.3 Zonal Boundary Layers 744
19.4 The Nonlinear Problem 745
19.5 Inertial Solutions 747
19.6 Topographic Effects on Western Boundary Currents 753

20 Structure of the Upper Ocean 761
20.1 Vertical Structure of the Wind-Driven Circulation 761
20.2 A Model with Continuous Stratification 767
20.3 Observations of Potential Vorticity 770
20.4 The Main Thermocline 774
20.5 Scaling and Simple Dynamics of the Main Thermocline 776
20.6 The Internal Thermocline 779
20.7 The Ventilated Thermocline 785
Appendix A: Miscellaneous Relationships in a Layered Model 796

21 The Meridional Overturning Circulation and the ACC 801
21.1 Sideways Convection 802
21.2 The Maintenance of Sideways Convection 808
21.3 Simple Box Models 813
21.4 A Laboratory Model of the Abyssal Circulation 818
21.5 A Model for Oceanic Abyssal Flow 821
21.6 A Model of Deep Wind-Driven Overturning 829
21.7 The Antarctic Circumpolar Current 836
21.8 A Dynamical Model of the Residual Overturning Circulation 845
21.9 A Model of the Interhemispheric Circulation 853

22 Equatorial Circulation and El Niño **861**

22.1	Observational Preliminaries	861
22.2	Dynamical Preliminaries	862
22.3	A Local Model of the Equatorial Undercurrent	865
22.4	An Ideal Fluid Model of the Equatorial Undercurrent	876
22.5	An Introduction to El Niño and the Southern Oscillation	886
22.6	The Walker Circulation	891
22.7	The Oceanic Response	893
22.8	Coupled Models and Unstable Interactions	895
22.9	Simple Conceptual and Numerical Models of ENSO	898
22.10	Numerical Solutions of the Shallow Water Equations	902
	Appendix A: Derivation of a Delayed-Oscillator Model	904

References **909**

Index **936**

In the main text, sections that are more advanced or that contain material that is peripheral to the main narrative are marked with a black diamond, ♦. Sections that contain material that is still not settled or that describe active areas of research are marked with a dagger, †.

The sea obscured by vapour that ... slid in one mighty mass along the sea shore ... The distant country, overhung by straggling clouds that sailed over it, appeared like the darker clouds, seen at a great distance apparently motionless, while the nearer ones pass quickly over them, driven by the lower winds. I never saw such a union of earth, sky and sea. The clouds beneath our feet spread to the water, and the clouds of the sky almost joined them.

Dorothy Wordsworth, *Alfoxden Journal*, 1798.

Preface

THIS IS A BOOK ON THE DYNAMICS OF THE ATMOSPHERE AND OCEAN, with an emphasis on the fundamentals and on the large-scale circulation. By 'large-scale' I mean scales between that of the weather (a few hundred kilometres in the atmosphere and a few tens of kilometres in the ocean, which indeed has its own weather) and the global scale. My focus is our own planet Earth, for that is where we live, but the principles and methodology used should be appropriate for the study of the atmospheres and oceans of other planets. And even if we stay at home, I try to take the reader on a journey — from the most basic and classical material to the frontiers of knowledge.

The book is written at a level appropriate for advanced undergraduate or graduate courses; it is also meant to be a useful reference for researchers, and some aspects of the book have the nature of a research monograph. Prior knowledge of fluid dynamics is helpful but is not a requirement, for the fluid equations are introduced in the first chapter. Similarly, some knowledge of thermodynamics will ease the reader's path, but is not essential. On the other hand, the reader is assumed to have a working knowledge of vector calculus and to know what a partial differential equation is.

Atmospheric and oceanic fluid dynamics (AOFD) is both pure and applied. It is a pure because it involves some of the most fundamental and unsolved problems in fluid dynamics — problems in turbulence and wave–mean flow interaction, in chaos and predictability, and in the general circulation itself. It is applied because the climate and weather so profoundly affect the human condition, and the practice of weather forecasting is a notable example of a successful (yes it is) applied science. The field is also broad, encompassing such subjects as the general circulation, instabilities, gyres, boundary layers, waves, convection and turbulence. My goal in this book is to present a coherent selection of these topics so that the reader will gain a solid grounding in the fundamentals, motivated by and with an appreciation for the problems of the real world. The book is primary a theoretical one, but in a number of places observations are used to illustrate the dynamics and define the problems. And I have tried to lead the reader to the edge of active areas of research — for a book that limits itself to what is absolutely settled would, I think, be rather dry, a quality best reserved for martinis and humour.

This book is a major revision of one that first appeared in 2006, and about half of the material here is new or rewritten. Some of the changes were motivated by the fact that half of the planet lies equatorward of 30°, that about 20% of the mass of the atmosphere lies above 10 km, and that it rains. Overall, I have tried to encompass the most important aspects of the dynamics and circulation of both atmosphere and ocean, for the similarities and differences between the two are so instructive that even if one's interest is solely in one there is much to be learned by studying the other. Where the subject matter verges on areas of research, I have focused on ideas and topics that, I think, will be of lasting value rather than of current fashion; however, this choice is undoubtedly influenced by my own interests and expertise, not to mention my prejudices.

How to Use this Book

This book is a long one, but it might help to think of it as four short ones with a common theme. These books are on the fundamentals of geophysical fluid dynamics (GFD); waves, instabilities and turbulence; atmospheric circulation; and ocean circulation. Each short book forms a Part in this book, and each might form the basis for a course of about a term or a semester, although there is more material in each than can comfortably be covered unless the course hums along at a torrid pace. The ordering of the topics follows a logical sequence, but is not the order that the book need be read. For example, Chapter 1 covers the basic equations of motion for a fluid, including the requisite thermodynamics. The thermodynamics of seawater belongs to and is included in that chapter, but this material is quite advanced and is not needed to understand many of the later chapters, and may be skipped on a first reading. Typically, sections that are more advanced or that are peripheral to the main narrative are marked with a black diamond, ✦, and sections that contain matters that are not settled or that describe active areas of research are marked with a dagger, †, but there is some overlap and arbitrariness in the markings. In some of these areas the book may be thought of as a entrance to the original literature and it gives my own view on the subject.

Putting the various topics together in one book will, I hope, emphasize their coherence as part of a single field. Still, there are many paths through the woods and the following notes may help steer the reader, although I do not wish to be prescriptive and experienced instructors will chart their own course. In a nutshell, the material of Part I forms the basis of all that follows, and subsequent Parts may then be read in any order if the reader is willing to cross refer, often to Part II. With care, one may also construct a single course that combines aspects of Parts II, III and/or IV.

A basic GFD course

A first course in GFD might cover much of Part I and, in many cases, Rossby waves and possibly baroclinic instability from Part II. If the students have already had a course in fluid dynamics then much of Chapter 1 can be skipped. Some of the thermodynamics may in any case be omitted on a first reading, and a basic course might encompass the following:

For students with no fluid mechanics:

Equations of motion, Sections 1.1–1.6, omitting starred sections.
Compressibility, Sections 1.8 and 1.9.
Energetics, Section 1.10 (optional).

Then, for all students:

Rotational effects and Boussinesq equations, Sections 2.1–2.4.
Pressure coordinates, Section 2.6 (optional, mainly for meteorologists).
Hydrostatic and geostrophic balance, Sections 2.7 and 2.8.
Static instability, Section 2.10.

Shallow water equations and geostrophic adjustment: Sections 3.1, 3.2, 3.5, 3.7–3.9.

Vorticity, Sections 4.1–4.4 (for students with no fluids background).
Potential vorticity, some or all of Sections 4.5 and 4.6.

Geostrophic theory, Sections 5.1–5.5.
Ekman layers, Section 5.7.

Rossby waves, Sections 6.4 and 6.5, and possibly Section 16.5 on vertical propagation.
Jet formation, Section 15.1.
Baroclinic instability, Sections 9.5 and/or 9.6.

One option is to use only the Boussinesq equations (and later the shallow water equations) and the beta plane, eschewing pressure co-ordinates and sphericity, coming back to them later for atmospheric applications. This option also makes the derivation of quasi-geostrophy a little easier.

But meteorologists may well prefer to introduce these topics from the outset, and use pressure coordinates and not the Boussinesq approximation. Many first courses will end after Rossby waves, although a rapidly-moving course for well-prepared students might cover some aspects of baroclinic instability. If waves are being covered separately, then an alternative is to cover some aspect of the general circulation of the atmosphere or ocean, for example the Stommel model of ocean gyres in Section 19.1, which is a pedagogical delight.

Subsequent reading

The other three short books, or Parts to this book, are loosely based on more advanced courses that I have taught at various times, and taken together are an attempt to pull together classical ideas and recent developments in the field into a coherent whole. Each Part is more-or-less self-contained, although some aspects of Parts III and IV build on Part II, especially the sections on Rossby waves and baroclinic instability. An understanding of the essentials of geostrophic turbulence and turbulent diffusion will also help. Provided these sections and the material in Part I is mastered the remaining Parts may be read in any order if the reader is willing to flip pages occasionally and consult other sections as needed. I won't suggest any specific syllabus to follow apart from the contents of the book itself, but let me make a few remarks.

Part II: Waves, instabilities and turbulence

Rossby waves, gravity waves and baroclinic instability could form the first half of a second course on GFD, with wave–mean-flow interaction and/or turbulence the second half. Any one of these topics could be a full course, especially if supplemented from other sources. WKB methods are very useful in a variety of practical problems and can readily be taught in an elementary way. If needs be equatorial waves (Chapter 8) could be treated separately, for it is a little mathematical, and perhaps folded into a course covering equatorial circulation more generally. One might also construct a course based around geostrophic turbulence and the mid-latitude circulation of the atmosphere and/or ocean.

Part III: Atmospheric circulation

My focus here is on the large scale, and on the fundamental aspects of the general circulation — how wide is the Hadley Cell? why do the surface winds blow eastward in mid-latitudes? why is there a tropopause and why is it 10 km high and not 5 or 50 km? and so forth. A basic knowledge of GFD, such as is in the sample course above, is a pre-requisite; other sections of Part II may be needed but can be consulted as necessary. Chapters 14 and 15 form a pair and should be read consecutively, and Chapter 16 is in some ways a continuation of Chapter 5 on waves. There are numerous ways to navigate through these chapters — one might omit the stratosphere chapter, or it could form the basis for an entire course if other literature were added. The same applies to Chapter 18 on the tropics, and here especially the material cuts close to the bone of what is known and other researchers might write a very different chapter.

Part IV: Oceanic circulation

Many of the above comments on atmospheric circulation apply here, as my goal is to describe and explain the fundamental dynamical processes determining the large-scale structure and circulation of the ocean — why go gyres have western boundary currents? what determines the structure of the thermocline? why does El Niño occur? As for Part III, a basic knowledge of GFD, such as is in the sample course above, is a pre-requisite. Chapter 19 on wind-driven gyres is the foundation of much that follows, and Chapter 20 flows naturally from it. The chapters on the meridional overturning circulation and El Niño delve into active areas of research, and again others might have written differently. Both here and in my discussion of atmospheric circulation I take the position of a *practical theoretician;* such a person seeks theories or explanations of phenomena, but they should be relevant to the world about us.

Miscellany and Acknowledgements

The book was written in LaTeX using Minion fonts for text and Minion Math from Typoma for equations. I fear that the references at the end disproportionately represent articles written in English and by British and North American authors, for that is the community I have mainly socialized with. The references are simply those with which I happen to most familiar — but this is a lazy choice and I apologize for it. Student exercises, various codes, and all of the figures, may be downloaded from the CUP website or my own website, which is best obtained using a search engine.

Many, many, colleagues and students have helped in the writing of this book, both by offering constructive suggestions and by gently and not so gently pointing out errors and misconceptions. Many thanks to all of you! I acknowledge your input in the endnotes after each chapter, although I am afraid that I have omitted many of you. If you have additional input or find mistakes, please email me. I would also like to thank three of my predecessor authors. The pioneering books by Joe Pedlosky and Adrian Gill paved the way and if those books had not existed I would have had neither the knowledge nor the courage to write mine. And the book by Rick Salmon taught me that careful arguments can be couched in plain language, and that an easy-to-read style can at the same time be clear and precise. I have tried to follow that example.

As with the first edition, this book ultimately owes its existence to my own hubris and selfishness: hubris to think that others might wish to read what I have written, and selfishness because the enjoyable task of writing such a book masquerades as work.

NOTATION

Variables are normally set in italics, constants (e.g, π, i) in roman (i.e., upright), differential operators in roman, vectors in bold, and tensors in bold sans serif. Thus, vector variables are in bold italics, vector constants (e.g., unit vectors) in bold upright, and tensor variables are in bold slanting sans serif. A subscript denotes a derivative only if the subscript is a coordinate, such as x, y, z or t, or when so denoted in the text. A subscript 0 generally denotes a reference value (e.g., ρ_0). The components of a vector are denoted by superscripts. If a fraction contains only two terms in the denominator then brackets are not always used; thus $1/2\pi = 1/(2\pi) \neq \pi/2$.

The lists below contain only the more important variables or instances of ambiguous notation, in quasi-alphabetical order, first of Roman characters and then of mainly Greek characters and operators. Distinct meanings are separated with a semi-colon.

Variable	Description
a	Radius of Earth.
b	Buoyancy, $-g\delta\rho/\rho_0$ or $g\delta\theta/\tilde{\theta}$.
B	Planck function, often σT^4.
$\boldsymbol{c}_g$	Group velocity, (c_g^x, c_g^y, c_g^z).
c_p	Phase speed; heat capacity at constant pressure.
c_v	Heat capacity at constant volume.
c_s	Sound speed.
f, f_0	Coriolis parameter, and its reference value.
$\boldsymbol{g}, g$	Vector acceleration due to gravity, magnitude of $\boldsymbol{g}$.
g	Gibbs function.
$\mathbf{i}, \mathbf{j}, \mathbf{k}$	Unit vectors in (x, y, z) directions.
i; i	An integer index; square root of minus one.
I	Internal energy.
$\boldsymbol{k}$	Wave vector, with components (k, l, m) or (k^x, k^y, k^z).
k_d	Wave number corresponding to deformation radius.
L_d	Deformation radius.
L, H	Horizontal length scale, vertical (height) scale.
m	Angular momentum about the Earth's axis of rotation.
N	Buoyancy, or Brunt–Väisälä, frequency.
p, p_R	Pressure, and a reference value of pressure.
Pr	Prandtl ratio, f_0/N.
q	Quasi-geostrophic potential vorticity; water vapour specific humidity.
Q	Potential vorticity (in particular Ertel PV).
$\dot{Q}$	Rate of heating.
Ra	Rayleigh number.
Re ; Re	Real part of expression; Reynolds number, UL/ν.
Ro	Rossby number, U/fL.
S	Salinity; source term on right-hand side of an evolution equation.
$S_o, \boldsymbol{S}_o$	Solenoidal term, solenoidal vector.
T	Temperature; scaling value for time.
t	Time.
$\boldsymbol{u}$	Two-dimensional (horizontal) velocity, (u, v).
$\boldsymbol{v}$	Three-dimensional velocity, (u, v, w).
w	Vertical velocity; water vapour mixing ratio.
x, y, z	Cartesian coordinates, usually in zonal, meridional and vertical directions.
Z	Log-pressure, $-H \log p/p_R$; scaling for z.

Variable	Description
$\mathcal{A}$	Wave activity.
α	Inverse density, or specific volume; aspect ratio.
β; β^*	Rate of change of f with latitude, $\partial f/\partial y$; $\beta^* = \beta - u_{yy}$
$\beta_T, \beta_S, \beta_p$	Coefficient of expansion with respect to temperature, salinity and pressure, respectively.
ϵ	Generic small parameter (epsilon).
ε	Cascade or dissipation rate of energy (varepsilon).
η	Specific entropy; perturbation height; enstrophy cascade or dissipation rate.
$\boldsymbol{\mathcal{F}}$	Eliassen Palm flux, $(\mathcal{F}^y, \mathcal{F}^z)$.
γ	The ratio c_p/c_v; Vorticity gradient, e.g., $\beta - u_{yy}$.
Γ	Lapse rate (sometimes subscripted, e.g., Γ_z, but here this does not denote a differential).
κ	Diffusivity; the ratio R/c_p.
$\mathcal{K}$	Kolmogorov or Kolmogorov-like constant.
Λ	Shear, e.g., $\partial U/\partial z$.
μ	Viscosity; chemical potential.
ν	Kinematic viscosity, μ/ρ.
v	Meridional component of velocity.
$\mathcal{P}$	Pseudomomentum.
ϕ	Pressure divided by density, p/ρ.
φ	Passive tracer.
Φ	Geopotential, usually gz; scaling value of ϕ.
Π	Exner function, $\Pi = c_pT/\theta = c_p(p/p_R)^{R/c_p}$; an enthalpy-like quantity.
$\boldsymbol{\omega}$	Vorticity.
$\Omega, \boldsymbol{\Omega}$	Rotation rate of Earth and associated vector.
ψ	Streamfunction.
ρ	Density.
ρ_θ	Potential density.
σ	Layer thickness, $\partial z/\partial\theta$; Prandtl number ν/κ; measure of density, $\rho - 1000$.
$\boldsymbol{\tau}$	Stress vector, often wind stress.
$\tilde{\boldsymbol{\tau}}$	Kinematic stress, $\boldsymbol{\tau}/\rho$.
τ	Zonal component or magnitude of wind stress; eddy turnover time; optical depth.
θ; Θ	Potential temperature; generic thermodynamic variable, often conservative temperature.
ϑ, λ	Latitude, longitude.
ζ	Vertical component of vorticity.
$\left(\dfrac{\partial a}{\partial b}\right)_c$	Derivative of a with respect to b at constant c.
$\left.\dfrac{\partial a}{\partial b}\right\|_c$	Derivative of a with respect to b evaluated at $b = c$.
∇_a	Gradient operator at constant value of coordinate a. Thus, $\nabla_z = \mathbf{i}\partial_x + \mathbf{j}\partial_y$.
$\nabla_a\cdot$	Divergence operator at constant value of coordinate a. Thus, $\nabla_z\cdot = (\mathbf{i}\partial_x + \mathbf{j}\partial_y)\cdot$.
$\nabla^\perp$	Perpendicular gradient, $\nabla^\perp\phi \equiv \mathbf{k}\times\nabla\phi$.
curl_z	Vertical component of $\nabla\times$ operator, $\mathrm{curl}_z\boldsymbol{A} = \mathbf{k}\cdot\nabla\times\boldsymbol{A} = \partial_x A^y - \partial_y A^x$.
$\dfrac{\mathrm{D}}{\mathrm{D}t}$	Material derivative (generic).
$\dfrac{\mathrm{D}_g}{\mathrm{D}t}$	Material derivative using geostrophic velocity, for example $\partial/\partial t + \boldsymbol{u}_g\cdot\nabla$.

Part I

FUNDAMENTALS OF GEOPHYSICAL FLUID DYNAMICS

To begin at the beginning:
It is spring, moonless night in the small town,
starless and bible-black, the cobblestreets silent
and the hunched, courters'-and-rabbits' wood limping invisible down
to the sloeblack, slow, black, crowblack, fishingboat-bobbing sea.

Dylan Thomas, *Under Milk Wood*, 1954.

CHAPTER 1

Equations of Motion

HAVING NOTHING BUT A BLANK SLATE, we begin by establishing the governing equations of motion for a fluid, with particular attention to the fluids of Earth's atmosphere and ocean. These equations determine how a fluid flows and evolves when forces are applied to it, or when it is heated or cooled, and so involve both dynamics and thermodynamics. And because the equations of motion are nonlinear the two become intertwined and at times inseparable.

1.1 TIME DERIVATIVES FOR FLUIDS

The equations of motion of fluid mechanics differ from those of rigid-body mechanics because fluids form a continuum, and because fluids flow and deform. Thus, even though the same relatively simple physical laws (Newton's laws and the laws of thermodynamics) govern both solid and fluid media, the expression of these laws differs between the two. To determine the equations of motion for fluids we must clearly establish what the time derivative of some property of a fluid actually means, and that is the subject of this section.

1.1.1 Field and Material Viewpoints

In solid-body mechanics one is normally concerned with the position and momentum of identifiable objects — the angular velocity of a spinning top or the motions of the planets around the Sun are two well-worn examples. The position and velocity of a particular object are then computed as a function of time by formulating equations of the form

$$\frac{\mathrm{d}x_i}{\mathrm{d}t} = F(\{x_i\}, t), \tag{1.1}$$

where $\{x_i\}$ is the set of positions and velocities of all the interacting objects and the operator F on the right-hand side is formulated using Newton's laws of motion. For example, two massive point objects interacting via their gravitational field obey

$$\frac{\mathrm{d}\boldsymbol{r}_i}{\mathrm{d}t} = \boldsymbol{v}_i, \qquad \frac{\mathrm{d}\boldsymbol{v}_i}{\mathrm{d}t} = \frac{Gm_j}{(\boldsymbol{r}_i - \boldsymbol{r}_j)^2}\hat{\boldsymbol{r}}_{i,j}, \qquad i = 1, 2;\; j = 3 - i. \tag{1.2}$$

We thereby predict the positions, $\boldsymbol{r}_i$, and velocities, $\boldsymbol{v}_i$, of the objects given their masses, m_i, and the gravitational constant G, and where $\hat{\boldsymbol{r}}_{i,j}$ is a unit vector directed from $\boldsymbol{r}_i$ to $\boldsymbol{r}_j$.

In fluid dynamics such a procedure would lead to an analysis of fluid motions in terms of the positions and momenta of different fluid parcels, each identified by some label, which might simply be their position at an initial time. We call this a *material* point of view, because we are concerned with identifiable pieces of material; it is also sometimes called a *Lagrangian* view, after J.-L. Lagrange. The procedure is perfectly acceptable in principle, and if followed would provide a complete description of the fluid dynamical system. However, from a practical point of view it is much more than we need, and it would be extremely complicated to implement. Instead, for most problems we would like to know what the values of velocity, density and so on are at *fixed points* in space as time passes. (A weather forecast we might care about tells us how warm it will be where we live and, if we are given that, we do not particularly care where a fluid parcel comes from, or where it subsequently goes.) Since the fluid is a continuum, this knowledge is equivalent to knowing how the fields of the dynamical variables evolve in space and time, and this is often known as the *field* or *Eulerian* viewpoint, after L. Euler.[1] Thus, whereas in the material view we consider the time evolution of identifiable fluid elements, in the field view we consider the time evolution of the fluid field from a particular frame of reference. That is, we seek evolution equations of the general form

$$\frac{\partial}{\partial t}\varphi(x, y, z, t) = \mathcal{G}(\varphi, x, y, z, t), \tag{1.3}$$

where the field $\varphi(x, y, z, t)$ represents all the dynamical variables (velocity, density, temperature, etc.) and $\mathcal{G}$ is some operator to be determined from Newton's laws of motion and appropriate thermodynamic laws.

Although the field viewpoint will often turn out to be the most practically useful, the material description is invaluable both in deriving the equations and in the subsequent insight it frequently provides. This is because the important quantities from a fundamental point of view are often those which are associated with a given fluid element: it is these which directly enter Newton's laws of motion and the thermodynamic equations. It is thus important to have a relationship between the rate of change of quantities associated with a given fluid element and the local rate of change of a field. The material or advective derivative provides this relationship.

1.1.2 The Material Derivative of a Fluid Property

A *fluid element* is an infinitesimal, indivisible, piece of fluid — effectively a very small fluid parcel of fixed mass. The *material derivative* is the rate of change of a property (such as temperature or momentum) of a particular fluid element or finite mass of fluid; that is, it is the total time derivative of a property of a piece of fluid. It is also known as the 'substantive derivative' (the derivative associated with a parcel of fluid substance), the 'advective derivative' (because the fluid property is being advected), the 'convective derivative' (convection is a slightly old-fashioned name for advection, still used in some fields), or the 'Lagrangian derivative' (after Lagrange).

Let us suppose that a fluid is characterized by a given velocity field $\boldsymbol{v}(\boldsymbol{x}, t)$, which determines its velocity throughout. Let us also suppose that the fluid has another property φ, and let us seek an expression for the rate of change of φ of a fluid element. Since φ is changing in time and in space we use the chain rule,

$$\delta\varphi = \frac{\partial\varphi}{\partial t}\delta t + \frac{\partial\varphi}{\partial x}\delta x + \frac{\partial\varphi}{\partial y}\delta y + \frac{\partial\varphi}{\partial z}\delta z = \frac{\partial\varphi}{\partial t}\delta t + \delta\boldsymbol{x}\cdot\nabla\varphi. \tag{1.4}$$

This is true in general for any δt, δx, etc. The total time derivative is then

$$\frac{\mathrm{d}\varphi}{\mathrm{d}t} = \frac{\partial\varphi}{\partial t} + \frac{\mathrm{d}\boldsymbol{x}}{\mathrm{d}t}\cdot\nabla\varphi. \tag{1.5}$$

If this equation is to represent a material derivative we must identify the time derivative in the second term on the right-hand side with the rate of change of position of a fluid element, namely its velocity. Hence, the material derivative of the property φ is

$$\frac{d\varphi}{dt} = \frac{\partial\varphi}{\partial t} + \boldsymbol{v} \cdot \nabla\varphi. \tag{1.6}$$

The right-hand side expresses the material derivative in terms of the local rate of change of φ plus a contribution arising from the spatial variation of φ, experienced only as the fluid parcel moves. Because the material derivative is so common, and to distinguish it from other derivatives, we denote it by the operator D/Dt. Thus, the material derivative of the field φ is

$$\frac{D\varphi}{Dt} = \frac{\partial\varphi}{\partial t} + (\boldsymbol{v} \cdot \nabla)\varphi. \tag{1.7}$$

The brackets in the last term of this equation are helpful in reminding us that $(\boldsymbol{v} \cdot \nabla)$ is an operator acting on φ. The operator $\partial/\partial t + (\boldsymbol{v} \cdot \nabla)$ is the *Eulerian representation of the Lagrangian derivative as applied to a field.* We use the notation D/Dt rather generally for Lagrangian derivatives, but the operator may take a different form when applied to other objects, such as a fluid volume.

Material derivative of vector field

The material derivative may act on a vector field $\boldsymbol{b}$, in which case

$$\frac{D\boldsymbol{b}}{Dt} = \frac{\partial\boldsymbol{b}}{\partial t} + (\boldsymbol{v} \cdot \nabla)\boldsymbol{b}. \tag{1.8}$$

In Cartesian coordinates this is

$$\frac{D\boldsymbol{b}}{Dt} = \frac{\partial\boldsymbol{b}}{\partial t} + u\frac{\partial\boldsymbol{b}}{\partial x} + v\frac{\partial\boldsymbol{b}}{\partial y} + w\frac{\partial\boldsymbol{b}}{\partial z}, \tag{1.9}$$

and for a particular component of $\boldsymbol{b}$, b^x say,

$$\frac{Db^x}{Dt} = \frac{\partial b^x}{\partial t} + u\frac{\partial b^x}{\partial x} + v\frac{\partial b^x}{\partial y} + w\frac{\partial b^x}{\partial z}, \tag{1.10}$$

and similarly for b^y and b^z. In Cartesian tensor notation the expression becomes

$$\frac{Db_i}{Dt} = \frac{\partial b_i}{\partial t} + v_j\frac{\partial b_i}{\partial x_j} = \frac{\partial b_i}{\partial t} + v_j\partial_j b_i, \tag{1.11}$$

where the subscripts denote the Cartesian components, repeated indices are summed, and $\partial_j b_i \equiv \partial b_i/\partial x_j$. In coordinate systems other than Cartesian the advective derivative of a vector is not simply the sum of the advective derivative of its components, because the coordinate vectors themselves change direction with position; this will be important when we deal with spherical coordinates. Finally, we remark that the advective derivative of the position of a fluid element, $\boldsymbol{r}$ say, is its velocity, and this may easily be checked by explicitly evaluating $D\boldsymbol{r}/Dt$.

1.1.3 Material Derivative of a Volume

The volume that a given, unchanging, mass of fluid occupies is deformed and advected by the fluid motion, and there is no reason why it should remain constant. Rather, the volume will change as a result of the movement of each element of its bounding material surface, and in particular will change if there is a non-zero normal component of the velocity at the fluid surface. That is, if the volume of some fluid is $\int dV$, then

$$\frac{D}{Dt}\int_V dV = \int_S \boldsymbol{v} \cdot d\boldsymbol{S}, \tag{1.12}$$

where the subscript V indicates that the integral is a definite integral over some finite volume V, although the limits of the integral will be functions of time if the volume is changing. The integral on the right-hand side is over the closed surface, S, bounding the volume. Although intuitively apparent (to some), this expression may be derived more formally using Leibniz's formula for the rate of change of an integral whose limits are changing. Using the divergence theorem on the right-hand side, (1.12) becomes

$$\frac{\mathrm{D}}{\mathrm{D}t}\int_V \mathrm{d}V = \int_V \nabla\cdot\boldsymbol{v}\,\mathrm{d}V. \tag{1.13}$$

The rate of change of the volume of an infinitesimal fluid element of volume ΔV is obtained by taking the limit of this expression as the volume tends to zero, giving

$$\lim_{\Delta V\to 0}\frac{1}{\Delta V}\frac{\mathrm{D}\Delta V}{\mathrm{D}t} = \nabla\cdot\boldsymbol{v}. \tag{1.14}$$

We will often write such expressions informally as

$$\frac{\mathrm{D}\Delta V}{\mathrm{D}t} = \Delta V\nabla\cdot\boldsymbol{v}, \tag{1.15}$$

with the limit implied.

Consider now the material derivative of some fluid property, ξ say, multiplied by the volume of a fluid element, ΔV. Such a derivative arises when ξ is the amount per unit volume of ξ-substance — the mass density or the amount of a dye per unit volume, for example. Then we have

$$\frac{\mathrm{D}}{\mathrm{D}t}(\xi\Delta V) = \xi\frac{\mathrm{D}\Delta V}{\mathrm{D}t} + \Delta V\frac{\mathrm{D}\xi}{\mathrm{D}t}. \tag{1.16}$$

Using (1.15) this becomes

$$\frac{\mathrm{D}}{\mathrm{D}t}(\xi\Delta V) = \Delta V\left(\xi\nabla\cdot\boldsymbol{v} + \frac{\mathrm{D}\xi}{\mathrm{D}t}\right), \tag{1.17}$$

and the analogous result for a finite fluid volume is just

$$\frac{\mathrm{D}}{\mathrm{D}t}\int_V \xi\,\mathrm{d}V = \int_V\left(\xi\nabla\cdot\boldsymbol{v} + \frac{\mathrm{D}\xi}{\mathrm{D}t}\right)\mathrm{d}V. \tag{1.18}$$

This expression is to be contrasted with the Eulerian derivative for which the volume, and so the limits of integration, are fixed and we have

$$\frac{\mathrm{d}}{\mathrm{d}t}\int_V \xi\,\mathrm{d}V = \int_V \frac{\partial\xi}{\partial t}\,\mathrm{d}V. \tag{1.19}$$

Now consider the material derivative of a fluid property φ multiplied by the mass of a fluid element, $\rho\Delta V$, where ρ is the fluid density. Such a derivative arises when φ is the amount of φ-substance per unit mass (note, for example, that the momentum of a fluid element is $\rho\boldsymbol{v}\Delta V$). The material derivative of $\varphi\rho\Delta V$ is given by

$$\frac{\mathrm{D}}{\mathrm{D}t}(\varphi\rho\Delta V) = \rho\Delta V\frac{\mathrm{D}\varphi}{\mathrm{D}t} + \varphi\frac{\mathrm{D}}{\mathrm{D}t}(\rho\Delta V). \tag{1.20}$$

But $\rho\Delta V$ is just the mass of the fluid element, and that is constant — that is how a fluid element is defined. Thus the second term on the right-hand side vanishes and

$$\frac{\mathrm{D}}{\mathrm{D}t}(\varphi\rho\Delta V) = \rho\Delta V\frac{\mathrm{D}\varphi}{\mathrm{D}t} \quad\text{and}\quad \frac{\mathrm{D}}{\mathrm{D}t}\int_V \varphi\rho\,\mathrm{d}V = \int_V \rho\frac{\mathrm{D}\varphi}{\mathrm{D}t}\,\mathrm{d}V, \tag{1.21a,b}$$

Material and Eulerian Derivatives

The material derivatives of a scalar (φ) and a vector ($\boldsymbol{b}$) field are given by:

$$\frac{\mathrm{D}\varphi}{\mathrm{D}t} = \frac{\partial \varphi}{\partial t} + \boldsymbol{v}\cdot\nabla\varphi, \qquad \frac{\mathrm{D}\boldsymbol{b}}{\mathrm{D}t} = \frac{\partial \boldsymbol{b}}{\partial t} + (\boldsymbol{v}\cdot\nabla)\boldsymbol{b}. \tag{D.1}$$

Various material derivatives of integrals are:

$$\frac{\mathrm{D}}{\mathrm{D}t}\int_V \varphi\,\mathrm{d}V = \int_V \left(\frac{\mathrm{D}\varphi}{\mathrm{D}t} + \varphi\nabla\cdot\boldsymbol{v}\right)\mathrm{d}V = \int_V \left(\frac{\partial\varphi}{\partial t} + \nabla\cdot(\varphi\boldsymbol{v})\right)\mathrm{d}V, \tag{D.2}$$

$$\frac{\mathrm{D}}{\mathrm{D}t}\int_V \mathrm{d}V = \int_V \nabla\cdot\boldsymbol{v}\,\mathrm{d}V, \tag{D.3}$$

$$\frac{\mathrm{D}}{\mathrm{D}t}\int_V \rho\varphi\,\mathrm{d}V = \int_V \rho\frac{\mathrm{D}\varphi}{\mathrm{D}t}\,\mathrm{d}V. \tag{D.4}$$

These formulae also hold if φ is a vector. The Eulerian derivative of an integral is:

$$\frac{\mathrm{d}}{\mathrm{d}t}\int_V \varphi\,\mathrm{d}V = \int_V \frac{\partial\varphi}{\partial t}\,\mathrm{d}V, \tag{D.5}$$

so that

$$\frac{\mathrm{d}}{\mathrm{d}t}\int_V \mathrm{d}V = 0 \quad \text{and} \quad \frac{\mathrm{d}}{\mathrm{d}t}\int_V \rho\varphi\,\mathrm{d}V = \int_V \frac{\partial\rho\varphi}{\partial t}\,\mathrm{d}V. \tag{D.6}$$

where (1.21b) applies to a finite volume. That expression may also be derived more formally using Leibniz's formula for the material derivative of an integral, and the result also holds when φ is a vector. The result is quite different from the corresponding Eulerian derivative, in which the volume is kept fixed; in that case we have:

$$\frac{\mathrm{d}}{\mathrm{d}t}\int_V \varphi\rho\,\mathrm{d}V = \int_V \frac{\partial}{\partial t}(\varphi\rho)\,\mathrm{d}V. \tag{1.22}$$

Various material and Eulerian derivatives are summarized in the shaded box above.

1.2 THE MASS CONTINUITY EQUATION

In classical mechanics mass is absolutely conserved and in solid-body mechanics we normally do not need an explicit equation of mass conservation. However, in fluid mechanics fluid flows into and away from regions, and fluid density may change, and an equation that explicitly accounts for the flow of mass is one of the equations of motion of the fluid.

1.2.1 An Eulerian Derivation

We will first derive the mass conservation equation from an Eulerian point of view; that is to say, our reference frame is fixed in space and the fluid flows through it.

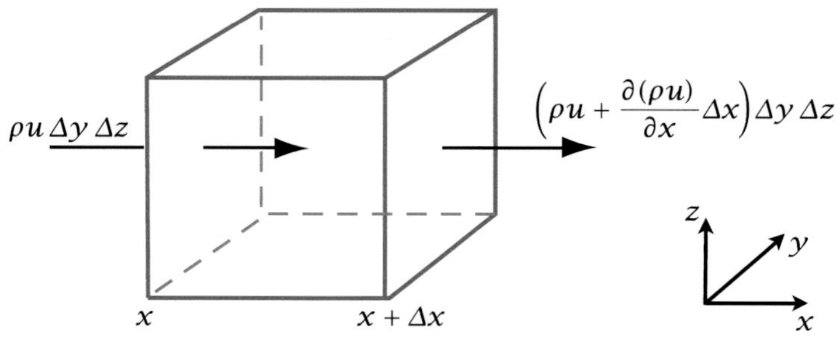

Fig. 1.1 Mass conservation in an Eulerian cuboid control volume. The mass convergence, $-\partial(\rho u)/\partial x$ (plus contributions from the y and z directions), must be balanced by a density increase, $\partial\rho/\partial t$.

Cartesian derivation

Consider an infinitesimal, rectangular cuboid, control volume, $\Delta V = \Delta x \Delta y \Delta z$ that is fixed in space, as in Fig. 1.1. Fluid moves into or out of the volume through its surface, including through its faces in the y–z plane of area $\Delta A = \Delta y \Delta z$ at coordinates x and $x + \Delta x$. The accumulation of fluid within the control volume due to motion in the x-direction is evidently

$$\Delta y \Delta z[(\rho u)(x, y, z) - (\rho u)(x + \Delta x, y, z)] = -\left.\frac{\partial(\rho u)}{\partial x}\right|_{x,y,z} \Delta x\,\Delta y\,\Delta z. \tag{1.23}$$

To this must be added the effects of motion in the y- and z-directions, namely

$$-\left[\frac{\partial(\rho v)}{\partial y} + \frac{\partial(\rho w)}{\partial z}\right]\Delta x\,\Delta y\,\Delta z. \tag{1.24}$$

This net accumulation of fluid must be accompanied by a corresponding increase of fluid mass within the control volume. This is

$$\frac{\partial}{\partial t}(\text{density} \times \text{volume}) = \Delta x\,\Delta y\,\Delta z \frac{\partial \rho}{\partial t}, \tag{1.25}$$

because the volume is constant. Thus, because mass is conserved, (1.23), (1.24) and (1.25) give

$$\Delta x\,\Delta y\,\Delta z\left[\frac{\partial \rho}{\partial t} + \frac{\partial(\rho u)}{\partial x} + \frac{\partial(\rho v)}{\partial y} + \frac{\partial(\rho w)}{\partial z}\right] = 0. \tag{1.26}$$

The quantity in square brackets must be zero and we therefore have

$$\frac{\partial \rho}{\partial t} + \nabla \cdot (\rho \boldsymbol{v}) = 0. \tag{1.27}$$

This is called the *mass continuity equation* for it recognizes the continuous nature of the mass field in a fluid. There is no diffusion term in (1.27), no term like $\kappa\nabla^2\rho$. This is because mass is transported by the macroscopic movement of molecules; even if this motion appears diffusion-like any net macroscopic molecular motion constitutes, by definition, a velocity field.

Vector derivation

Consider an arbitrary control volume V bounded by a surface S, fixed in space, with by convention the direction of $\mathbf{S}$ being toward the outside of V, as in Fig. 1.2. The rate of fluid loss due to flow through the closed surface S is then given by

$$\text{fluid loss} = \int_S \rho \boldsymbol{v} \cdot \mathrm{d}\mathbf{S} = \int_V \nabla \cdot (\rho \boldsymbol{v})\, \mathrm{d}V, \tag{1.28}$$

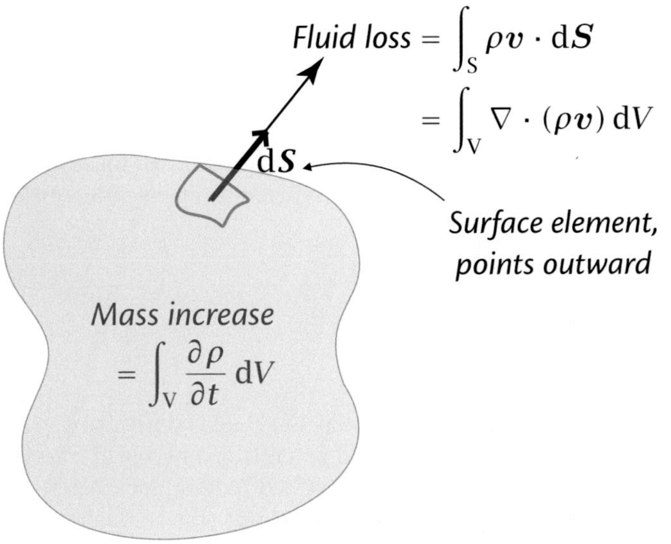

Fig. 1.2 Mass conservation in an arbitrary Eulerian control volume V bounded by a surface S. The mass increase, $\int_V (\partial\rho/\partial t)\,\mathrm{d}V$ is equal to the mass flowing into the volume, $-\int_S (\rho\boldsymbol{v})\cdot \mathrm{d}\boldsymbol{S} = -\int_V \nabla\cdot(\rho\boldsymbol{v})\,\mathrm{d}V$.

using the divergence theorem.

This must be balanced by a change in the mass M of the fluid within the control volume, which, since its volume is fixed, implies a density change. That is

$$\text{fluid loss} = -\frac{\mathrm{d}M}{\mathrm{d}t} = -\frac{\mathrm{d}}{\mathrm{d}t}\int_V \rho \,\mathrm{d}V = -\int_V \frac{\partial \rho}{\partial t}\,\mathrm{d}V. \tag{1.29}$$

Equating (1.28) and (1.29) yields

$$\int_V \left[\frac{\partial \rho}{\partial t} + \nabla\cdot(\rho\boldsymbol{v})\right] \mathrm{d}V = 0. \tag{1.30}$$

Because the volume is arbitrary, the integrand must vanish and we recover (1.27).

1.2.2 Mass Continuity via the Material Derivative

We now derive the mass continuity equation (1.27) from a material perspective. This is the most fundamental approach of all since the principle of mass conservation states simply that the mass of a given element of fluid is, by definition of the element, constant. Thus, consider a small mass of fluid of density ρ and volume ΔV. Then conservation of mass may be represented by

$$\frac{\mathrm{D}}{\mathrm{D}t}(\rho\Delta V) = 0. \tag{1.31}$$

Both the density and the volume of the parcel may change, so

$$\Delta V \frac{\mathrm{D}\rho}{\mathrm{D}t} + \rho \frac{\mathrm{D}\Delta V}{\mathrm{D}t} = \Delta V \left(\frac{\mathrm{D}\rho}{\mathrm{D}t} + \rho\nabla\cdot\boldsymbol{v}\right) = 0, \tag{1.32}$$

where the second expression follows using (1.15). Since the volume element is arbitrary, the term in brackets must vanish and

$$\frac{\mathrm{D}\rho}{\mathrm{D}t} + \rho\nabla\cdot\boldsymbol{v} = 0. \tag{1.33}$$

After expansion of the first term this becomes identical to (1.27). This result may be derived more formally by rewriting (1.31) as the integral expression

$$\frac{\mathrm{D}}{\mathrm{D}t}\int_V \rho\,\mathrm{d}V = 0. \tag{1.34}$$

Expanding the derivative using (1.18) gives

$$\frac{\mathrm{D}}{\mathrm{D}t}\int_V \rho\,\mathrm{d}V = \int_V \left(\frac{\mathrm{D}\rho}{\mathrm{D}t} + \rho\nabla\cdot\boldsymbol{v}\right)\mathrm{d}V = 0. \tag{1.35}$$

Because the volume over which the integral is taken is arbitrary the integrand itself must vanish and we recover (1.33). Summarizing, equivalent partial differential equations representing conservation of mass are:

$$\frac{\mathrm{D}\rho}{\mathrm{D}t} + \rho\nabla\cdot\boldsymbol{v} = 0, \qquad \frac{\partial\rho}{\partial t} + \nabla\cdot(\rho\boldsymbol{v}) = 0. \tag{1.36a,b}$$

1.2.3 A General Continuity Equation

The derivation of a continuity equation for a general scalar property of a fluid is similar to that for density, except that there may be an external source or sink, and potentially a means of transferring the property from one location to another differently than by fluid motion, for example by diffusion. If ξ is the amount of some property of the fluid per unit volume (the volume concentration, sometimes simply called the concentration), and if the net effect per unit volume of all non-conservative processes is denoted by $Q_{[v,\xi]}$, then the continuity equation for concentration may be written:

$$\frac{\mathrm{D}}{\mathrm{D}t}(\xi\Delta V) = Q_{[v,\xi]}\Delta V. \tag{1.37}$$

Expanding the left-hand side and using (1.15) we obtain

$$\frac{\mathrm{D}\xi}{\mathrm{D}t} + \xi\nabla\cdot\boldsymbol{v} = Q_{[v,\xi]}, \qquad \text{or} \qquad \frac{\partial\xi}{\partial t} + \nabla\cdot(\xi\boldsymbol{v}) = Q_{[v,\xi]}. \tag{1.38}$$

If we are interested in a tracer that is normally measured per unit mass of fluid (which is typical when considering thermodynamic quantities) then the conservation equation would be written

$$\frac{\mathrm{D}}{\mathrm{D}t}(\varphi\rho\Delta V) = Q_{[m,\varphi]}\rho\Delta V, \tag{1.39}$$

where φ is the tracer mixing ratio or mass concentration — that is, the amount of tracer per unit fluid mass — and $Q_{[m,\varphi]}$ represents non-conservative sources of φ per unit mass. Then, since $\rho\Delta V$ is constant we obtain

$$\frac{\mathrm{D}\varphi}{\mathrm{D}t} = Q_{[m,\varphi]} \qquad \text{or} \qquad \frac{\partial(\rho\varphi)}{\partial t} + \nabla\cdot(\rho\varphi\boldsymbol{v}) = \rho Q_{[m,\varphi]}, \tag{1.40}$$

using the mass continuity equation, (1.36), to obtain the equation on the right. The source term $Q_{[m,\varphi]}$ is evidently equal to the rate of change of φ of a fluid element. When this is so, we often write it simply as $\dot{\varphi}$, so that

$$\frac{\mathrm{D}\varphi}{\mathrm{D}t} = \dot{\varphi}. \tag{1.41}$$

A tracer obeying (1.41) with $\dot{\varphi} = 0$ is said to be *materially conserved.* If a tracer is materially conserved except for the effects of sources or sinks, or diffusion terms, then it is sometimes (if rather loosely) said to be an 'adiabatically conserved' variable, although adiabatic properly means with no heat exchange. If those sources and sinks are in the form of the divergence of a flux with φ satisfying $\rho\mathrm{D}\varphi/\mathrm{D}t = \nabla\cdot\boldsymbol{F}_\varphi$ or equivalently, using the mass continuity equation, $\partial(\rho\varphi)/\partial t + \nabla\cdot(\rho\boldsymbol{v}\varphi) = \nabla\cdot\boldsymbol{F}_\varphi$, then φ is said to be a *conservative* variable because, with no flux boundary conditions, $\int \rho\varphi\,\mathrm{d}V$ = constant. Although momentum as a whole is conserved, momentum is not a materially conserved variable, as we are about to see.

1.3 THE MOMENTUM EQUATION

The momentum equation is a partial differential equation that describes how the velocity or momentum of a fluid responds to internal and imposed forces. We will derive it using material methods, with a very heuristic treatment of the terms representing pressure and viscous forces.

1.3.1 Advection

Let $\boldsymbol{m}(x, y, z, t)$ be the momentum-density field (momentum per unit volume) of the fluid. Thus, $\boldsymbol{m} = \rho\boldsymbol{v}$ and the total momentum of a volume of fluid is given by the volume integral $\int_V \boldsymbol{m}\,\mathrm{d}V$. Now, for a fluid the rate of change of a momentum of an identifiable fluid mass is given by the material derivative, and by Newton's second law this is equal to the force acting on it. Thus,

$$\frac{\mathrm{D}}{\mathrm{D}t}\int_V \rho\boldsymbol{v}\,\mathrm{d}V = \int_V \boldsymbol{F}\,\mathrm{d}V, \tag{1.42}$$

where $\boldsymbol{F}$ is the force per unit volume. Now, using (1.21b) (with φ replaced by $\boldsymbol{v}$) to transform the left-hand side of (1.42), we obtain

$$\int_V \left(\rho\frac{\mathrm{D}\boldsymbol{v}}{\mathrm{D}t} - \boldsymbol{F}\right)\mathrm{d}V = 0. \tag{1.43}$$

Because the volume is arbitrary the integrand itself must vanish and we obtain

$$\rho\frac{\mathrm{D}\boldsymbol{v}}{\mathrm{D}t} = \boldsymbol{F} \quad \text{or} \quad \frac{\partial\boldsymbol{v}}{\partial t} + (\boldsymbol{v}\cdot\nabla)\boldsymbol{v} = \frac{\boldsymbol{F}}{\rho}, \tag{1.44a,b}$$

having used (1.8) to expand the material derivative.

We have thus obtained an expression for how a fluid accelerates if subject to known forces. These forces are, however, not all external to the fluid itself: a stress arises from the direct contact between one fluid parcel and another, giving rise to pressure and viscous forces, sometimes referred to as *contact* forces. Because a complete treatment of these forces would be very lengthy, and is available elsewhere, we treat them informally and intuitively.

1.3.2 Pressure and Viscous Forces

Pressure

Within or at the boundary of a fluid the pressure is the normal force per unit area due to the collective action of molecular motion. Thus

$$\mathrm{d}\widehat{\boldsymbol{F}}_p = -p\,\mathrm{d}\boldsymbol{S}, \tag{1.45}$$

where p is the pressure, $\widehat{\boldsymbol{F}}_p$ is the pressure force and $\mathrm{d}\boldsymbol{S}$ an infinitesimal surface element. If we grant ourselves this intuitive notion, it is a simple matter to assess the influence of pressure on a fluid, for the pressure force on a volume of fluid is the integral of the pressure over its boundary and so

$$\widehat{\boldsymbol{F}}_p = -\int_S p\,\mathrm{d}\boldsymbol{S}. \tag{1.46}$$

The minus sign arises because the pressure force is directed inwards, whereas $\boldsymbol{S}$ is a vector normal to the surface and directed outwards. Applying a form of the divergence theorem to the right-hand side gives

$$\widehat{\boldsymbol{F}}_p = -\int_V \nabla p\,\mathrm{d}V, \tag{1.47}$$

where the volume V is bounded by the surface S. The pressure force per unit volume, $\boldsymbol{F}_p$, is therefore just $-\nabla p$, and the force per unit mass is $\nabla p/\rho$. The force is evidently non-zero only if the pressure varies in space and for this reason it is more properly known as the *pressure-gradient* force.

Table 1.1 Experimental values of viscosity for air, water and mercury at room temperature and pressure.

	$\mu\,(\mathrm{kg\,m^{-1}\,s^{-1}})$	$\nu\,(\mathrm{m^2\,s^{-1}})$
Air	1.8×10^{-5}	1.5×10^{-5}
Water	1.1×10^{-3}	1.1×10^{-6}
Mercury	1.6×10^{-3}	1.2×10^{-7}

Viscosity

Viscosity, like pressure, is a force due to the internal motion of molecules. The effects of viscosity are apparent in many situations — the flow of treacle or volcanic lava are obvious examples. In other situations, for example large-scale flow in the atmosphere, viscous effects are negligible. However, for a constant density fluid viscosity is the *only* way that energy may be removed from the fluid, so that if energy is being added in some way viscosity must ultimately become important if the fluid is to reach an equilibrium in which energy input equals energy dissipation. When tea is stirred in a cup, it is viscous effects that cause the fluid to eventually stop spinning after we have removed our spoon.

A number of textbooks show that, for most Newtonian fluids, the viscous force per unit volume is approximately equal to $\mu\nabla^2\boldsymbol{v}$, where μ is the viscosity. Obtaining this expression involves making an incompressibility assumption and is not exact, but it is in fact a good approximation for most liquids and gases. With this term and the pressure-gradient force the momentum equation becomes,

$$\frac{\partial \boldsymbol{v}}{\partial t} + (\boldsymbol{v}\cdot\nabla)\boldsymbol{v} = -\frac{1}{\rho}\nabla p + \nu\nabla^2\boldsymbol{v} + \boldsymbol{F}_b, \tag{1.48}$$

where $\nu \equiv \mu/\rho$ is the *kinematic viscosity* and $\boldsymbol{F}_b$ represents external body forces (per unit mass) such as gravity, $\boldsymbol{g}$. Equation (1.48) is sometimes called the Navier–Stokes equation. For gases, dimensional arguments suggest that the magnitude of ν should be given by

$$\nu \sim \text{mean free path} \times \text{mean molecular velocity}, \tag{1.49}$$

which for a typical molecular velocity of $300\ \mathrm{m\,s^{-1}}$ and a mean free path of 7×10^{-8} m gives the not unreasonable estimate of $2.1\times10^{-5}\ \mathrm{m^2\,s^{-1}}$, within a factor of two of the experimental value (Table 1.1). Interestingly, the kinematic viscosity is smaller for water and mercury than it is for air.

1.3.3 Hydrostatic Balance

The vertical component — the component parallel to the gravitational force, $\boldsymbol{g}$ — of the momentum equation is

$$\frac{\mathrm{D}w}{\mathrm{D}t} = -\frac{1}{\rho}\frac{\partial p}{\partial z} - g, \tag{1.50}$$

where w is the vertical component of the velocity and $\boldsymbol{g} = -g\mathbf{k}$. If the fluid is static the gravitational term is balanced by the pressure term and we have

$$\frac{\partial p}{\partial z} = -\rho g, \tag{1.51}$$

and this relation is known as *hydrostatic balance,* or hydrostasy. It is clear in this case that the pressure at a point is given by the weight of the fluid above it, provided $p = 0$ at the top of the fluid. It might also appear that (1.51) would be a good *approximation* to (1.50) provided that vertical accelerations, $\mathrm{D}w/\mathrm{D}t$, are small compared to gravity, which is nearly always the case in the atmosphere and ocean. While this statement is true if we need only a reasonable approximate value of the pressure at a point or in a column, the satisfaction of this condition is *not* sufficient to ensure that (1.51) provides an accurate enough pressure to determine the horizontal pressure gradients responsible for producing motion. We return to this point in Section 2.7.

1.4 THE EQUATION OF STATE

In three dimensions the momentum and continuity equations provide four equations, but contain five unknowns — three components of velocity, density and pressure. Obviously other equations are needed, and an *equation of state* is an expression that diagnostically relates the various thermodynamic variables to each other. The *conventional* equation of state, or the *thermal* equation of state, is an expression that relates temperature, pressure, composition (the mass fraction of the various constituents) and density, and we may write it, rather generally, as

$$p = p(\rho, T, \varphi_n), \tag{1.52}$$

where φ_n is mass fraction of the nth constituent. An equation of this form is not the most fundamental equation of state from a thermodynamic perspective, an issue we visit later, but it connects readily measurable quantities.

For an ideal gas (and the air in the Earth's atmosphere is very close to ideal) the thermal equation of state is

$$p = \rho R T, \tag{1.53}$$

where R is the gas constant for the gas in question and T is temperature. R is a specific constant, and is related to the universal gas constant R^* by $R = R^*/\overline{\mu}$, where $\overline{\mu}$ is the mean molar mass (molecular weight in kg/mol) of the constituents of the gas. Equivalently, $R = n_m k_B$, where k_B is Boltzmann's constant and n_m is the number of molecules per unit mass, so that R is proportional to the number of molecules contained in a unit mass. Since $R^* = 8.314\,\mathrm{J\,mol^{-1}\,K^{-1}}$ and, for dry air, $\mu = 29.0 \times 10^{-3}\,\mathrm{kg\,mol^{-1}}$ we obtain $R = 287\,\mathrm{J\,kg^{-1}\,K^{-1}}$. Air has virtually constant composition except for variations in water vapour; these variations make the gas constant, R, in the equation of state for air a weak function of the water vapour content but for now we regard R as a constant.

For a liquid such as seawater no simple expression akin to (1.53) is easily derivable, and semi-empirical equations are usually resorted to. For water in a laboratory setting a reasonable approximation of the equation of state is $\rho = \rho_0[1-\beta_T(T-T_0)]$, where β_T is a thermal expansion coefficient and ρ_0 and T_0 are constants. In the ocean the density is also significantly affected by pressure and dissolved salts: seawater is a solution of many ions in water — chloride ($\approx$ 1.9% by weight), sodium (1%), sulfate (0.26%), magnesium (0.13%) and so on, with a total average concentration of about 35‰ (ppt, or parts per thousand). The ratios of the fractions of these salts are almost constant throughout the ocean, and their total concentration may be parameterized by a single measure, the *salinity*, S.[2] Given this, the density of seawater is a function of three variables — pressure, temperature, and salinity — and we may write the conventional equation of state as

$$\rho = \rho(T, S, p) \qquad \text{or} \qquad \alpha = \alpha(T, S, p), \tag{1.54}$$

where $\alpha = 1/\rho$ is the specific volume, or inverse density. For small variations around a reference value we have

$$\mathrm{d}\alpha = \left(\frac{\partial \alpha}{\partial T}\right)_{S,p} \mathrm{d}T + \left(\frac{\partial \alpha}{\partial S}\right)_{T,p} \mathrm{d}S + \left(\frac{\partial \alpha}{\partial p}\right)_{T,S} \mathrm{d}p = \alpha(\beta_T\,\mathrm{d}T - \beta_S\,\mathrm{d}S - \beta_p\,\mathrm{d}p), \tag{1.55}$$

where the rightmost expression serves to define the thermal expansion coefficient β_T, the saline contraction coefficient β_S, and the compressibility coefficient (or inverse bulk modulus) β_p. In general these quantities are not constants, but for small variations around a reference state they may be treated as such and we have

$$\alpha = \alpha_0\left[1 + \beta_T(T - T_0) - \beta_S(S - S_0) - \beta_p(p - p_0)\right]. \tag{1.56}$$

Typical values of these parameters, with variations typically encountered through the ocean, are: $\beta_T \approx 2(\pm 1.5) \times 10^{-4}\,\mathrm{K}^{-1}$, $\beta_S \approx 7.6(\pm 0.2) \times 10^{-4}\,\mathrm{ppt}^{-1}$, $\beta_p \approx 4.4(\pm 0.5) \times 10^{-10}\,\mathrm{Pa}^{-1}$. The value of β_p is also related to the speed of sound, c_s, by $\beta_p = \alpha_0/c_s^2$. Since the variations around the mean density are small, (1.56) implies that

$$\rho = \rho_0 \left[1 - \beta_T(T - T_0) + \beta_S(S - S_0) + \beta_p(p - p_0)\right]. \tag{1.57}$$

In the ocean the pressure term leads to larger density changes than either the salinity or temperature terms but it is not normally as important for the dynamics, because pressure is largely determined by the hydrostatic pressure giving a large vertical density gradient. It is the lateral variations in density that are often more important for the dynamics and these are affected just as much by the saline and temperature terms.

A linear equation of state for seawater is emphatically *not* accurate enough for quantitative oceanography; as mentioned the β parameters in (1.56) themselves vary with pressure, temperature and (more weakly) salinity so introducing nonlinearities to the equation. The most important of these are captured by an equation of state of the form

$$\alpha = \alpha_0 \left[1 + \beta_T(1 + \gamma^* p)(T - T_0) + \frac{\beta_T^*}{2}(T - T_0)^2 - \beta_S(S - S_0) - \beta_p(p - p_0)\right]. \tag{1.58}$$

The starred constants β_T^* and γ^* capture the leading nonlinearities: γ^* is the *thermobaric parameter,* which determines the extent to which the thermal expansion depends on pressure, and β_T^* is the second thermal expansion coefficient.[3] Even this equation of state has some quantitative deficiencies and more complicated empirical formulae are often used if very high accuracy is needed. The variation of density of seawater with temperature, salinity and pressure is illustrated in Fig. 1.3, with more discussion in Section 1.7.2.

Clearly, the equation of state introduces, in general, a sixth unknown, temperature, and we will have to introduce another physical principle — one coming from thermodynamics — to obtain a complete set of equations. However, if the equation of state were such that it linked only density and pressure, without introducing another variable, then the equations would be complete; the simplest case of all is a constant density fluid for which the equation of state is just ρ = constant. A fluid for which the density is a function of pressure alone is called a *barotropic fluid* or a *homentropic fluid*; otherwise, it is a *baroclinic fluid.* (In this context, 'barotropic' is a shortening of the original phrase 'auto-barotropic'.) Equations of state of the form $p = C\rho^\gamma$, where γ is a constant, are called 'polytropic'.

1.5 THERMODYNAMIC RELATIONS

In this section we review a few aspects of thermodynamics. We provide neither a complete nor an a priori development of the subject; rather, we focus on aspects that are particularly relevant to fluid dynamics, and that are needed to derive a 'thermodynamic equation' for fluids. Readers whose interest is solely in an ideal gas or a simple Boussinesq fluid may skim this section and then refer to it later as needed.

1.5.1 A Few Fundamentals

A fundamental postulate of thermodynamics is that the internal energy of a system in equilibrium is a function of its extensive properties: volume, entropy, and the mass of its various constituents. Extensive means that the property value is proportional to the amount of material present, in contrast to an intensive property such as temperature. For our purposes it is more convenient to divide all of these quantities by the mass of fluid present, so expressing the internal energy per unit mass (or the specific internal energy) I, as a function of the specific volume $\alpha = \rho^{-1}$, the specific entropy

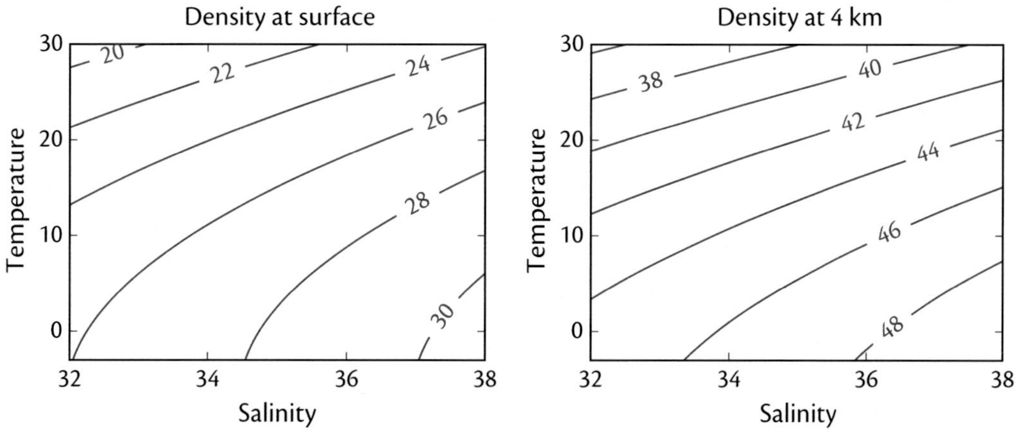

Fig. 1.3 Contours of density as a function of temperature and salinity for seawater. Contour labels are (density – 1000) kg m^{-3}. Left panel: at sea-level ($p = 10^5$ Pa, or 1000 mb). Right panel: at $p = 4\times10^7$ Pa (about 4 km depth). In both cases the contours are slightly convex, so that if two parcels at the same density but different temperatures and salinities are mixed, the resulting parcel is of higher density. (The average temperature is not exactly conserved on mixing, but it very nearly is.)

η, and the mass fractions of its various components. (However, multiplying by the mass does not turn an intensive variable into a properly extensive one.) Our interest is in two-component fluids (dry air and water vapour, or water and salinity) so that we may parameterize the composition by a single parameter, S. (We follow conventional thermodynamical notation as much as possible, except that we use I instead of u for internal energy, since u is a fluid velocity, and η instead of S for entropy, since S is salinity.) We can also write entropy in terms of internal energy, density and salinity, and thus we have

$$I = I(\alpha, \eta, S) \qquad \text{or} \qquad \eta = \eta(I, \alpha, S). \tag{1.59a,b}$$

Given the functional forms on the right-hand sides, either of these expressions constitutes a complete description of the macroscopic state of a system in equilibrium, and we call either of them the *fundamental equation of state*. The thermal equation of state can be derived from (1.59), but not vice versa. The first differential of (1.59a) gives, formally,

$$\mathrm{d}I = \left(\frac{\partial I}{\partial \eta}\right)_{\alpha,S} \mathrm{d}\eta + \left(\frac{\partial I}{\partial \alpha}\right)_{\eta,S} \mathrm{d}\alpha + \left(\frac{\partial I}{\partial S}\right)_{\alpha,\eta} \mathrm{d}S. \tag{1.60}$$

We will now ascribe physical meaning to these differentials.

Conservation of energy states that the internal energy of a body may change because of work done by or on it, or because of a heat input, or because of a change in its chemical composition. We write this as

$$\mathrm{d}I = \text{đ}Q + \text{đ}W + \text{đ}C, \tag{1.61}$$

where đW is the work done on the body, đQ is the heat input to the body, and đC is the change in internal energy caused by a change in its chemical composition (e.g., its salinity, or water vapour content), sometimes called the 'chemical work'. The infinitesimal quantities on the right-hand side (denoted with a đ) are so-called imperfect or inexact differentials:[4] Q, W and C are not functions of the state of a body, and the internal energy cannot be regarded as the sum of a 'heat' and a 'work'. We should think of heat and work as having meaning only as fluxes of energy, or rates of energy input, and not as amounts of energy; their sum changes the internal energy of a body, which *is* a function of its state. Equation (1.61) is sometimes called 'the first law of thermodynamics' (discussed more in Appendix B at the end of the chapter). We first consider the causes of variations of the quantities on the right-hand side, and then make a connection to (1.60).

Heat input: In an infinitesimal quasi-static or reversible process, if an amount of heat $đQ$ (per unit mass) is externally supplied then the specific entropy η will change according to

$$T\,\mathrm{d}\eta = đQ. \tag{1.62}$$

The entropy is a function of the state of a body and is, by definition, an adiabatic invariant. As we are dealing with the amount of a quantity per unit mass, η is the specific entropy, although we will often refer to it just as the entropy. We may regard (1.61) as defining the heat input, $đQ$, by way of a statement of conservation of energy, and (1.62) then says that there is a function of state, the entropy, that changes by an amount equal to the heat input divided by the temperature.

Work done: The work done on a body during a reversible process is equal to the pressure times its change in volume, and the work is positive if the volume change is negative. Thus if an infinitesimal amount of work $đW$ (per unit mass) is applied to a body then its thermodynamic state will change according to

$$-p\,\mathrm{d}\alpha = đW, \tag{1.63}$$

where $\alpha = 1/\rho$ is the specific volume of the fluid and p is the pressure.

Composition: The chemical work, which produces a change in internal energy due to a small change in composition, $\mathrm{d}S$, is given by

$$đC = \mu\,\mathrm{d}S, \tag{1.64}$$

where μ is the *chemical potential* of the solution. In the ocean, salinity is the compositional variable and changes arise through precipitation and evaporation at the surface and molecular diffusion. When salinity changes the internal energy of a fluid parcel changes by (1.64), but this change is usually small compared to other changes in internal energy. In practice, the most important effect of salinity is that it changes the density of seawater. In the atmosphere the composition of a parcel of air varies mainly according to the amount of water vapour present. Since water vapour and dry air have different chemical potentials these variations cause changes in internal energy, but in the absence of phase-changes the changes are small. An important compositional effect does arise when condensation or evaporation occurs, for then energy is released (or required), as discussed in Chapter 18.

Collecting equations (1.61)–(1.64) together we have

$$\mathrm{d}I - T\,\mathrm{d}\eta = -p\,\mathrm{d}\alpha + \mu\,\mathrm{d}S. \tag{1.65}$$

We refer to this as *the fundamental thermodynamic relation*. The fundamental equation of state, (1.59), describes the properties of a particular fluid, and the fundamental relation, (1.65), is associated with conservation of energy. Much of classical thermodynamics follows from these two expressions.

1.5.2 Thermodynamic Potentials and Maxwell Relations

Given the fundamental thermodynamic relation, various other 'thermodynamic potentials' and relations between variables can be derived that prove extremely useful. The thermodynamic potentials are, like internal energy and entropy, functions of the state but they have different natural variables by which they are expressed.

If we begin with the internal energy itself, then from (1.60) and (1.65) it follows that

$$T = \left(\frac{\partial I}{\partial \eta}\right)_{\alpha,S}, \qquad p = -\left(\frac{\partial I}{\partial \alpha}\right)_{\eta,S}, \qquad \mu = \left(\frac{\partial I}{\partial S}\right)_{\eta,\alpha}. \tag{1.66a,b,c}$$

These may be regarded as the defining relations for these variables; because of the connection between (1.61) and (1.65) these are not just formal definitions, and the pressure and temperature so defined are indeed related to our intuitive concepts of these variables and to the motion of the fluid molecules. If we write

$$\mathrm{d}\eta = \frac{1}{T}\,\mathrm{d}I + \frac{p}{T}\,\mathrm{d}\alpha - \frac{\mu}{T}\,\mathrm{d}S, \tag{1.67}$$

it is also clear that

$$p = T\left(\frac{\partial \eta}{\partial \alpha}\right)_{I,S}, \qquad T^{-1} = \left(\frac{\partial \eta}{\partial I}\right)_{\alpha,S}, \qquad \mu = -T\left(\frac{\partial \eta}{\partial S}\right)_{I,\alpha}. \tag{1.68a,b,c}$$

We also see that I and α (and S) are the natural variables for entropy.

Because the right-hand side of (1.65) is equal to an exact differential, the second derivatives are independent of the order of differentiation. That is,

$$\frac{\partial^2 I}{\partial \eta\,\partial \alpha} = \frac{\partial^2 I}{\partial \alpha\,\partial \eta}, \tag{1.69}$$

and therefore, using (1.66)

$$\left(\frac{\partial T}{\partial \alpha}\right)_{\eta} = -\left(\frac{\partial p}{\partial \eta}\right)_{\alpha}. \tag{1.70}$$

This is one of the *Maxwell relations,* which are a collection of four similar relations that follow directly from the fundamental thermodynamic relation (1.65) and simple relations between second derivatives. (Additional Maxwell-like relations exist if we consider chemical effects.) To derive the other Maxwell relations we will introduce thermodynamic potentials enthalpy, h, the Gibbs function, g, and the free energy, f. These are all closely related to the internal energy and they are all extensive functions (and then denoted with an uppercase letter), but for fluid-dynamical purposes it is convenient to divide them by the mass and use their specific forms, denoted with a lowercase letter (with the exception of I itself, the specific internal energy).

Define the *enthalpy* of a fluid by

$$h \equiv I + p\alpha, \tag{1.71}$$

and then (1.65) becomes

$$\mathrm{d}h = T\,\mathrm{d}\eta + \alpha\,\mathrm{d}p + \mu\,\mathrm{d}S. \tag{1.72}$$

Evidently, the natural variables for enthalpy are entropy and pressure so that, in general,

$$\mathrm{d}h = \left(\frac{\partial h}{\partial \eta}\right)_{p,S}\mathrm{d}\eta + \left(\frac{\partial h}{\partial p}\right)_{\eta,S}\mathrm{d}p + \left(\frac{\partial h}{\partial S}\right)_{\eta,p}\mathrm{d}S. \tag{1.73}$$

Comparing the last two equations we have

$$T = \left(\frac{\partial h}{\partial \eta}\right)_{p,S}, \qquad \alpha = \left(\frac{\partial h}{\partial p}\right)_{\eta,S}, \qquad \mu = \left(\frac{\partial h}{\partial S}\right)_{\eta,p}. \tag{1.74}$$

Noting that

$$\frac{\partial^2 h}{\partial \eta \, \partial p} = \frac{\partial^2 h}{\partial p \, \partial \eta} \tag{1.75}$$

we evidently must have

$$\left(\frac{\partial T}{\partial p}\right)_\eta = \left(\frac{\partial \alpha}{\partial \eta}\right)_p, \tag{1.76}$$

and this is our second Maxwell relation.

To obtain the third, we write

$$\mathrm{d}I = T\,\mathrm{d}\eta - p\,\mathrm{d}\alpha + \mu\,\mathrm{d}S = \mathrm{d}(T\eta) - \eta\,\mathrm{d}T - \mathrm{d}(p\alpha) + \alpha\,\mathrm{d}p + \mu\,\mathrm{d}S, \tag{1.77}$$

or

$$\mathrm{d}g = -\eta\,\mathrm{d}T + \alpha\,\mathrm{d}p + \mu\,\mathrm{d}S \quad \text{where} \quad g \equiv I - T\eta + p\alpha = h - TS. \tag{1.78a,b}$$

The quantity g is the *Gibbs function,* also known as the 'Gibbs free energy' or 'Gibbs potential'. (We use g for gravity and g for the specific Gibbs function.) Now, formally, we have,

$$\mathrm{d}g = \left(\frac{\partial g}{\partial T}\right)_{p,S} \mathrm{d}T + \left(\frac{\partial g}{\partial p}\right)_{T,S} \mathrm{d}p + \left(\frac{\partial g}{\partial S}\right)_{T,p} \mathrm{d}S. \tag{1.79}$$

Comparing the last two equations we see that

$$\eta = -\left(\frac{\partial g}{\partial T}\right)_{p,S}, \qquad \alpha = \left(\frac{\partial g}{\partial p}\right)_{T,S}, \qquad \mu = \left(\frac{\partial g}{\partial S}\right)_{T,p}. \tag{1.80}$$

Furthermore, because

$$\frac{\partial^2 g}{\partial p \, \partial T} = \frac{\partial^2 g}{\partial T \, \partial p} \tag{1.81}$$

we have our third Maxwell equation,

$$\left(\frac{\partial \eta}{\partial p}\right)_T = -\left(\frac{\partial \alpha}{\partial T}\right)_p. \tag{1.82}$$

The Gibbs function is unique among the thermodynamic potentials in that its natural variables, T and p, are intensive quantities.

The fourth Maxwell equation makes use of the specific free energy or Helmholtz function, f, where

$$f \equiv I - T\eta, \qquad \text{and} \qquad \mathrm{d}f = -\eta\,\mathrm{d}T - p\,\mathrm{d}\alpha + \mu\,\mathrm{d}S, \tag{1.83}$$

giving

$$\left(\frac{\partial \eta}{\partial \alpha}\right)_T = \left(\frac{\partial p}{\partial T}\right)_\alpha, \tag{1.84}$$

and all four of these Maxwell equations are summarized in the box on the next page. All of them follow from the fundamental thermodynamic relation, (1.65), which is the real silver hammer of thermodynamics. The fundamental relation can also be written in the following form connecting internal energy, enthalpy and entropy,

$$\mathrm{d}I + p\,\mathrm{d}\alpha = \mathrm{d}h - \alpha\,\mathrm{d}p = T\mathrm{d}\eta + \mu\,\mathrm{d}S, \tag{1.85}$$

which turns out to be useful for fluid dynamical applications.

Thermodynamic Functions and Maxwell Relations

The four canonical Maxwell relations, with their associated potentials, are

Internal energy:

$$I = I,$$

$$\mathrm{d}I = T\,\mathrm{d}\eta - p\,\mathrm{d}\alpha + \mu\,\mathrm{d}S,$$

$$\left(\frac{\partial T}{\partial \alpha}\right)_{\eta,S} = -\left(\frac{\partial p}{\partial \eta}\right)_{\alpha,S}.$$

Enthalpy:

$$h = I + p\alpha,$$

$$\mathrm{d}h = T\,\mathrm{d}\eta + \alpha\,\mathrm{d}p + \mu\,\mathrm{d}S,$$

$$\left(\frac{\partial T}{\partial p}\right)_{\eta,S} = \left(\frac{\partial \alpha}{\partial \eta}\right)_{p,S}.$$

Gibbs function:

$$g = I - T\eta + p\alpha,$$

$$\mathrm{d}g = -\eta\,\mathrm{d}T + \alpha\,\mathrm{d}p + \mu\,\mathrm{d}S,$$

$$\left(\frac{\partial \eta}{\partial p}\right)_{T,S} = -\left(\frac{\partial \alpha}{\partial T}\right)_{p,S}.$$

Helmholtz free energy:

$$f = I - T\eta,$$

$$\mathrm{d}f = -\eta\,\mathrm{d}T - p\,\mathrm{d}\alpha + \mu\,\mathrm{d}S,$$

$$\left(\frac{\partial \eta}{\partial \alpha}\right)_{T,S} = \left(\frac{\partial p}{\partial T}\right)_{\alpha,S}.$$

Meaning of the state functions

Internal energy, enthalpy, the Gibbs function, the free energy and the entropy are all state functions from which other thermodynamic quantities can be derived, but with different meanings and uses. The utility of these quantities will become apparent as we proceed, but here is a brief summary.

The internal energy of a body is the total energy within a body, excluding the kinetic energy and the potential energy due to external fields like gravity. It is an invariant if the volume is fixed and there is no heating or chemical change to the body (this is the first law). The other state functions are related to internal energy by Legendre transformations, but they are not necessarily equal to the energy that a body contains. The enthalpy is the Legendre transformation of the internal energy (a function of entropy, density and composition) to a function of entropy, pressure and composition, and it is important in fluids because the energy transfer between a fluid parcel and its environment is associated with a flux of enthalpy, not internal energy. If a parcel is adiabatically displaced its change in potential energy will be balanced by a change in enthalpy, for it is enthalpy that accounts for the work done by pressure forces. When two adjacent parcels at the same pressure mix, their total enthalpy is conserved.

The Gibbs function is useful because it is constant for systems at constant temperature, pressure and composition. Its natural variables, temperature, pressure and composition are all measurable and for that reason it finds use as the fundamental state function from which all other thermodynamic variables may be derived. The Helmholtz free energy is also sometimes used as a fundamental state variable, and is useful for systems at constant temperature, density and composition for then it is constant. For small isothermal and isohaline changes, the increase of free energy is equal to the work done on the system. Free energy is not in fact commonly used in the atmospheric or oceanic sciences. Finally, whole books have been written on entropy (which is not a Legendre transformation of the internal energy and is not a thermodynamic potential in the same sense as the others). Suffice it to say here that entropy is the state function that responds directly to heating and it is a measure of the disorder of a system.[5]

Fundamental equation of state

The fundamental equation of state (1.59) gives complete information about a fluid in thermodynamic equilibrium, and given this we can obtain expressions for the temperature, pressure and chemical potential using (1.66). These expressions are also equations of state; however, each of them, taken individually, contains less information than the fundamental equation because a derivative has been taken. Equivalent to the fundamental equation of state are, using (1.72), an expression for the enthalpy as a function of its natural variables pressure, entropy and composition, or, using (1.78), the Gibbs function as a function of pressure, temperature and composition. Of these, the Gibbs function is particularly useful in practice because the pressure, temperature and composition may all be measured in the laboratory. Given the fundamental equation of state, the thermodynamic state of a body is fully specified by a knowledge of any two of p, ρ, T, η and I, plus its composition. The thermal equation of state, (1.52), is obtained by using (1.59a) to eliminate entropy from (1.66a) and (1.66b).

One simple fundamental equation of state is to take the internal energy to be a function of density and not entropy; that is, $I = I(\alpha)$. Bodies with such a property are called *homentropic.* Using (1.66), temperature and chemical potential have no role in the fluid dynamics and the density is a function of pressure alone — the defining property of a barotropic fluid.[6] Neither water nor air is, in general, homentropic but under some circumstances the flow may be adiabatic and $p = p(\rho)$.

In an ideal gas the molecules do not interact except by elastic collisions, and the volume of the molecules is presumed to be negligible compared to the total volume they occupy. The internal energy of the gas then depends only on temperature, and not on density. A *simple* ideal gas, also called a *perfect* gas (although nomenclature in the literature varies), is an ideal gas for which the heat capacity is constant, so that

$$I = cT, \tag{1.86}$$

where c is a constant. Using this and the conventional ideal gas equation, $p = \rho RT$ (where R is also constant), along with the fundamental thermodynamic relation (1.65), we can infer the fundamental equation of state; however, we will defer that until we discuss potential temperature in Section 1.6.1 — if curious, look ahead to page 25. A general ideal gas also obeys $p = \rho RT$, but has heat capacities that may be a function of temperature (but only of temperature, if composition is fixed), but in this book we only deal with simple ideal gases.

Internal energy and specific heats

We can obtain some useful relations between the internal energy and specific heat capacities, and some useful estimates of their values, by some simple manipulations of the fundamental thermodynamic relation. Assuming that the composition of the fluid is constant, (1.65) is

$$T\,\mathrm{d}\eta = \mathrm{d}I + p\,\mathrm{d}\alpha, \tag{1.87}$$

so that, taking I to be a function of α and T,

$$T\,\mathrm{d}\eta = \left(\frac{\partial I}{\partial T}\right)_{\alpha,S} \mathrm{d}T + \left[\left(\frac{\partial I}{\partial \alpha}\right)_{T,S} + p\right] \mathrm{d}\alpha. \tag{1.88}$$

From this, we see that the heat capacity at constant volume (i.e., constant α) c_v is given by

$$c_v \equiv T\left(\frac{\partial \eta}{\partial T}\right)_{\alpha,S} = \left(\frac{\partial I}{\partial T}\right)_{\alpha,S}. \tag{1.89}$$

Thus, c in (1.86) is equal to c_v.

Similarly, using (1.72) we have

$$T\,\mathrm{d}\eta = \mathrm{d}h - \alpha\,\mathrm{d}p = \left(\frac{\partial h}{\partial T}\right)_{p,S} \mathrm{d}T + \left[\left(\frac{\partial h}{\partial p}\right)_{T,S} - \alpha\right] \mathrm{d}p. \tag{1.90}$$

The heat capacity at constant pressure, c_p, is then given by

$$c_p \equiv T\left(\frac{\partial \eta}{\partial T}\right)_{p,S} = \left(\frac{\partial h}{\partial T}\right)_{p,S}. \tag{1.91}$$

For an ideal gas $h = I + RT = T(c_v + R)$. But $c_p = (\partial h/\partial T)_p$, and hence $c_p = c_v + R$. For future use we define $\gamma \equiv c_p/c_v$ and $\kappa \equiv R/c_p$, and $(\gamma - 1)/\gamma = \kappa$. Statistical mechanics tells us that for a simple ideal gas the internal energy is equal to $kT/2$ per molecule, or $RT/2$ per unit mass, for each excited degree of freedom, where k is the Boltzmann constant and R the gas constant. The diatomic molecules N_2 and O_2 that form most of our atmosphere have two rotational and three translational degrees of freedom, so that $I \approx 5RT/2$, and so $c_v \approx 5R/2$ and $c_p \approx 7R/2$, both being constants. These are in fact very good approximations to the measured values for the Earth's atmosphere, and give $c_v \approx 714\,\mathrm{J\,kg^{-1}K^{-1}}$ and $c_p \approx 1000\,\mathrm{J\,kg^{-1}K^{-1}}$ (whereas c_p is measured to be $1003\,\mathrm{J\,kg^{-1}K^{-1}}$). The internal energy is simply c_vT and the enthalpy is c_pT. For a liquid, especially one like seawater that contains dissolved salts, no such simple relations are possible: the heat capacities are functions of the state of the fluid, and the internal energy is a function of pressure (or density) as well as temperature and composition.

1.6 THERMODYNAMIC EQUATIONS FOR FLUIDS

The thermodynamic relations — for example (1.65) — apply to identifiable bodies or systems; thus, the heat input affects the fluid parcel to which it is applied, and we can apply the material derivative to the above thermodynamic relations to obtain equations of motion for a moving fluid. In doing so we make two assumptions:

(i) That locally the fluid is in thermodynamic equilibrium. This means that, although the thermodynamic quantities like temperature, pressure and density vary in space and time, locally they are related by the thermodynamic relations such as the equation of state and Maxwell's relations.

(ii) That macroscopic fluid motions are reversible and so not entropy producing. Thus, such effects as the viscous dissipation of energy, radiation, and conduction may produce entropy whereas the macroscopic fluid motion itself does not.

The first point requires that the temperature variation on the macroscopic scales must be slow enough that there can exist a volume that is small compared to the scale of macroscopic variations, so that temperature is effectively constant within it, but that is also sufficiently large to contain enough molecules so that macroscopic variables such as temperature have a proper meaning.

Now from (1.61), conservation of energy for an infinitesimal fluid parcel may be written as

$$\mathrm{d}I = -p\,\mathrm{d}\alpha + đQ_E, \tag{1.92}$$

where $p\,\mathrm{d}\alpha$ is the work done by the parcel and $đQ_E$ is the total energy input to the parcel with contributions from heating and changes in composition. Given the first assumption above, we may form the material derivative of (1.92) to obtain

$$\frac{\mathrm{D}I}{\mathrm{D}t} + p\frac{\mathrm{D}\alpha}{\mathrm{D}t} = \dot{Q}_E, \tag{1.93}$$

where $\dot{Q}_E$ is the rate of total energy input, per unit mass, with (as for $đQ_E$) possible contributions from thermal fluxes (including radiative heating, thermal diffusion and heat generated by viscous damping) and fluxes of composition, and note that $\dot{Q}_E$ does not include any mechanical effects. (In general, both the diffusion of heat and composition depend on the gradients of both temperature and composition, although the thermal flux is largely determined by the temperature gradient, and the compositional flux by the gradient of composition.)

Using the mass continuity equation in the form $\mathrm{D}\alpha/\mathrm{D}t = \alpha\nabla\cdot\boldsymbol{v}$, we write (1.93) as

$$\frac{\mathrm{D}I}{\mathrm{D}t} + p\alpha\nabla\cdot\boldsymbol{v} = \dot{Q}_E. \tag{1.94}$$

This is the *internal energy equation* for a fluid. Internal energy is not a conservative variable because of the compression term involving $\nabla\cdot\boldsymbol{v}$. The internal energy equation may also be written in terms of enthalpy, and using (1.71) and (1.93) we obtain the equivalent equation,

$$\frac{\mathrm{D}h}{\mathrm{D}t} - \alpha\frac{\mathrm{D}p}{\mathrm{D}t} = \dot{Q}_E. \tag{1.95}$$

Because $\dot{Q}_E$ contains, in general, energy fluxes due to changes in composition, we need to know that composition. The composition of a fluid parcel is carried with it as it moves, and changes only if there are non-conservative sources and sinks, such as diffusive fluxes. Thus, and analogously to (1.41), the evolution of composition is determined by

$$\frac{\mathrm{D}S}{\mathrm{D}t} = \dot{S}, \tag{1.96}$$

where $\dot{S}$ represents all the non-conservative terms.

Rather than use an internal energy equation, we may use the fundamental thermodynamic relation to infer an evolution equation for entropy. Thus, forming the material derivative from (1.65) and using (1.94) and (1.96), we obtain the *entropy equation*

$$\frac{\mathrm{D}\eta}{\mathrm{D}t} = \frac{1}{T}\dot{Q}_E - \frac{\mu}{T}\dot{S} \equiv \frac{1}{T}\dot{Q}, \tag{1.97}$$

where $\dot{Q}$ is the heating rate per unit mass. This equation is simply the material derivative of (1.62), along with assumption *(ii)* above. The heating of a fluid parcel generally needs to be *derived,* since it involves the viscous dissipation of energy as well as radiative and diffusive fluxes. The derivation is the topic of Appendix A at the chapter end, and from now on we assume the heating and energy input are known quantities.

The entropy equation is not independent of the internal energy equation, but is connected via the thermodynamic relations of Sections 1.5.1 and 1.5.2, and by the equation of state. If we use the internal energy equation we can in principle then calculate the entropy using the equation of state in the form $\eta = \eta(I, \alpha, S)$, or if we use the entropy equation we can then calculate the internal energy using $I = I(\eta, \alpha, S)$. Indeed, the internal energy equation and the entropy equation are both commonly referred to as 'the thermodynamic equation'. The heating term, $\dot{Q}$, is, however, not always easy to accurately determine in practice — it is affected by gradients of composition as well as viscosity and radiative fluxes — and the use of the internal energy equation may be more straightforward than the use of an entropy equation, especially in multi-component fluids. On the other hand, internal energy is affected by the $\nabla\cdot\boldsymbol{v}$ term in (1.94) and in a liquid this is small but non-zero and that may cause difficulties. See Section 1.7.3 for more discussion.

In any case, given evolution equations for composition and internal energy or entropy, and given the fundamental equation of state, we have, in principle, a complete set of equations for a fluid, as summarized in the shaded box on the facing page. Let us now look at the special, but very important, case of a dry ideal gas.

Fundamental Equations of Motion of a Fluid

The following equations constitute, in principle, a complete set of equations for a fluid heated at a rate $\dot{Q}$ and whose composition, S, changes at a rate $\dot{S}$.

Evolution equations for velocity, density and composition

$$\frac{\mathrm{D}\boldsymbol{v}}{\mathrm{D}t} = -\frac{1}{\rho}\nabla p + \nu\nabla^2\boldsymbol{v} + \boldsymbol{F}, \qquad \frac{\mathrm{D}\rho}{\mathrm{D}t} + \rho\nabla\cdot\boldsymbol{v} = 0, \qquad \frac{\mathrm{D}S}{\mathrm{D}t} = \dot{S}, \tag{F.1}$$

where $\boldsymbol{F}$ is some body force per unit mass, such as gravity.

Internal energy equation or entropy equation (the 'thermodynamic' equation)

$$\frac{\mathrm{D}I}{\mathrm{D}t} + \frac{p}{\rho}\nabla\cdot\boldsymbol{v} = \dot{Q}_E \qquad \text{or} \qquad \frac{\mathrm{D}\eta}{\mathrm{D}t} = \frac{1}{T}\dot{Q}, \tag{F.2}$$

where $\dot{Q}$ is the heating and $\dot{Q}_E = \dot{Q} + \mu\dot{S}$ is the total rate of energy input.

Fundamental equation of state for internal energy, I, entropy, η, or Gibbs function, g

$$I = I(\rho, \eta, S), \quad \eta = \eta(\rho, I, S) \quad \text{or} \quad g = g(T, p, S). \tag{F.3}$$

Diagnostic equations for temperature and pressure

$$T = \left(\frac{\partial I}{\partial \eta}\right)_{\alpha,S} \qquad \text{or} \qquad T^{-1} = \left(\frac{\partial \eta}{\partial I}\right)_{\alpha,S}. \tag{F.4}$$

$$p = -\left(\frac{\partial I}{\partial \alpha}\right)_{\eta,S} = \rho^2\left(\frac{\partial I}{\partial \rho}\right)_{\eta,S} \qquad \text{or} \qquad p = T\left(\frac{\partial \eta}{\partial \alpha}\right)_{I,S}. \tag{F.5}$$

The actual method of solution of these equations will depend on the equation of state. For example, for an ideal gas $T = I/c_v$ and the thermal equation of state, $p = \rho RT$, may be used to infer pressure.

The equations describing fluid motion are called the *Euler equations* if the viscous term is omitted, and the *Navier–Stokes equations* if viscosity is included.[7] These appellations are often taken to mean only the momentum and mass conservation equations, and the Euler equations are sometimes taken to be the equations for a fluid of constant density.

1.6.1 Thermodynamic Equation for an Ideal Gas

For a dry ideal gas the internal energy is a function of temperature only and $\mathrm{d}I = c_v\,\mathrm{d}T$. The first law of thermodynamics becomes

$$đQ = c_v\,\mathrm{d}T + p\,\mathrm{d}\alpha, \qquad \text{or} \qquad đQ = c_p\,\mathrm{d}T - \alpha\,\mathrm{d}p, \tag{1.98a,b}$$

where the second expression is derived using $\alpha = RT/p$ and $c_p - c_v = R$. Forming the material derivative of the above gives two forms of the internal energy equation:

$$c_v\frac{\mathrm{D}T}{\mathrm{D}t} + p\frac{\mathrm{D}\alpha}{\mathrm{D}t} = \dot{Q} \qquad \text{or} \qquad c_p\frac{\mathrm{D}T}{\mathrm{D}t} - \frac{RT}{p}\frac{\mathrm{D}p}{\mathrm{D}t} = \dot{Q}. \tag{1.99a,b}$$

Using the mass continuity equation, (1.99a) may be written as

$$c_v \frac{DT}{Dt} + p\alpha \nabla \cdot \boldsymbol{v} = \dot{Q}. \tag{1.100}$$

This is one of the most common and useful forms of the thermodynamic equation for the atmosphere. A less common but equivalent form arises if we use the ideal gas equation to eliminate T in favour of p, giving

$$\frac{Dp}{Dt} + \gamma p \nabla \cdot \boldsymbol{v} = \dot{Q}\frac{\rho R}{c_v}. \tag{1.101}$$

The Earth's atmosphere also contains water vapour with mixing ratio q (commonly referred to as specific humidity), and an evolution equation for it takes the form

$$\frac{Dq}{Dt} = \dot{q}, \tag{1.102}$$

where $\dot{q}$ represents the effects of condensation and evaporation. The main thermodynamic effects of water vapour occur when it condenses and latent heat is released; this heating appears in the thermodynamic equation, with $\dot{Q} = -L\dot{q}$, where L is the latent heat of condensation, as discussed in Chapter 18.

Potential temperature, potential density and entropy

We can use entropy instead of temperature for our thermodynamic equation, and this corresponds to using (1.97) instead of (1.94). However, it is common in meteorology to express entropy in terms of a temperature-like quantity, potential temperature, which has a more intuitive appeal to some.[8] Seawater also has a potential temperature variable, but with a different form.

We begin with the observation that when a fluid parcel changes pressure adiabatically, it will expand or contract and, using (1.98b) with $đQ = 0$, its temperature change is determined by $c_p\,dT = \alpha\,dp$. This temperature change is plainly not caused by heating, but we may construct a temperature-like quantity that changes *only* if diabatic effects are present; specifically *potential temperature*, θ, is defined to be the temperature that a fluid would have if moved adiabatically and at constant composition to some reference pressure (usually taken to be 1000 hPa, which is close to the pressure at the Earth's surface). Thus, in adiabatic flow the potential temperature obeys $D\theta/Dt = 0$.

In order to relate θ to the other thermodynamic variables we use (1.98b) and the equation of state for an ideal gas to write the fundamental thermodynamic relation as

$$d\eta = c_p\, d\ln T - R\, d\ln p. \tag{1.103}$$

If we move a parcel adiabatically ($d\eta = 0$) from p to p_R the temperature changes, by definition, from T to θ, and (1.103) gives

$$\int_T^\theta c_p d\ln T - \int_p^{p_R} R\, d\ln p = 0. \tag{1.104}$$

For constant c_p and R this equation may be solved to give

$$\theta = T\left(\frac{p_R}{p}\right)^\kappa, \tag{1.105}$$

where p_R is the reference pressure and $\kappa \equiv R/c_p$. (Another derivation of this result is given in Appendix A.) It follows from (1.103) and (1.105) that potential temperature is related to entropy by

$$d\eta = c_p \, d\ln\theta, \tag{1.106}$$

and, if c_p is constant, which it nearly is for Earth's atmosphere,

$$\eta = c_p \ln\theta. \tag{1.107}$$

Forming the material derivative of (1.107) and using (1.97) we obtain

$$c_p \frac{D\theta}{Dt} = \frac{\theta}{T}\dot{Q}, \tag{1.108}$$

with θ given by (1.105). Equations (1.100), (1.101) and (1.108) are all equivalent forms of the thermodynamic equation for an ideal gas.

The *potential density*, ρ_θ, is the density that a fluid parcel would have if moved adiabatically and at constant composition to a reference pressure, p_R. If the equation of state is written as $\rho = f(p, T)$ then the potential density is just $\rho_\theta = f(p_R, \theta)$ and for an ideal gas we therefore have

$$\rho_\theta = \frac{p_R}{R\theta} = \rho\left(\frac{p_R}{p}\right)^{1/\gamma}. \tag{1.109}$$

Finally, for later use, a little manipulation of the equation of state for an ideal gas, for small variations around a reference state, reveals that

$$\frac{\delta\theta}{\theta} = \frac{\delta T}{T} - \kappa\frac{\delta p}{p} = \frac{1}{\gamma}\frac{\delta p}{p} - \frac{\delta\rho}{\rho}. \tag{1.110}$$

The fundamental equation of state for an ideal gas

Equations (1.107) and (1.105) are closely related to the fundamental equation of state, and using $I = c_v T$ and the equation of state $p = \rho RT$, we can express the entropy explicitly in terms of the density and the internal energy. Similarly, we may derive an expression for Gibbs function as a function of pressure and temperature, and we find

$$\eta = c_v \ln I - R\ln\rho + A, \qquad g = c_p T(1 - \ln T) + RT\ln p + BT + C, \tag{1.111a,b}$$

where A, B and C are constants. Either of these two expressions may be regarded as *the fundamental equation of state for a simple ideal gas*. They could in fact be used to define a simple ideal gas (although we have not motivated that approach) and if we were to *begin* with either of them we could derive all the other thermodynamic quantities of interest. For example, using (1.68a) and (1.111a) we immediately recover $p = \rho RT$, and using (1.68b) we obtain $T = c_v/I$. Similarly, from the Gibbs function we obtain $\alpha = (\partial g/\partial p)_T = RT/p$, and the entropy satisfies $\eta = -(\partial g/\partial T)_p$. We provide a more complete set of derivations from the Gibbs function in Appendix A on page 47 and, for moist air, the appendix on page 720.

1.6.2 ✦ Other Forms of the Thermodynamic Equation

For a liquid such as seawater no simple exact equation of state exists and writing down a useful thermodynamic equation is not easy. Thus, although (1.97) holds in general, we need to be able to evaluate the heating and we need an expression relating entropy to the other thermodynamic variables — that is, an equation of state. For quantitative modelling and observational work such an equation of state must be quite accurate and we come back to this in Section 1.7. However, we can gain understanding — for both liquids and gases — by beginning with the entropy equation and making simplifications to it, as follows.

Forms of the Thermodynamic Equation

General form

$$\frac{DI}{Dt} + p\frac{D\alpha}{Dt} = \dot{Q}_E \qquad \text{or} \qquad \frac{DI}{Dt} + p\alpha\nabla\cdot\boldsymbol{v} = \dot{Q}_E, \tag{T.1a,b}$$

where I is the internal energy and $\dot{Q}_E$ is the rate of energy input, per unit mass. This may be written in terms of enthalpy, h, or entropy, η

$$\frac{Dh}{Dt} - \alpha\frac{Dp}{Dt} = \dot{Q}_E, \qquad T\frac{D\eta}{Dt} = \dot{Q}_E - \mu\dot{S} = \dot{Q}, \tag{T.2a,b}$$

where $\dot{Q}$ is the heating rate and $\dot{S}$ the rate of change of composition. For a fluid parcel of constant composition, $c_p \,\mathrm{d}\ln\theta = \mathrm{d}\eta$ and (T.2b) may be written as a potential temperature equation.

Ideal gas

For an ideal gas $\mathrm{d}I = c_v \mathrm{d}T$, $\mathrm{d}h = c_p\mathrm{d}T$, $\mathrm{d}\eta = c_p \mathrm{d}\ln\theta$ and the adiabatic thermodynamic equation may be written in the following equivalent, exact, forms:

$$\begin{aligned} c_p\frac{DT}{Dt} - \alpha\frac{Dp}{Dt} &= 0, & \frac{Dp}{Dt} + \gamma p\nabla\cdot\boldsymbol{v} &= 0, \\ c_v\frac{DT}{Dt} + p\alpha\nabla\cdot\boldsymbol{v} &= 0, & \frac{D\theta}{Dt} &= 0, \end{aligned} \tag{T.3}$$

where $\theta = T(p_R/p)^\kappa$, and energy or heating terms in various forms appear on the right-hand sides as needed. The two expressions on the second line are usually the most useful in modelling and theoretical work in meteorology.

I. Thermodynamic equation using pressure and density

If we regard η as a function of pressure and density, and salinity S where appropriate, we obtain

$$\begin{aligned} T\,\mathrm{d}\eta &= T\left(\frac{\partial\eta}{\partial\rho}\right)_{p,S}\mathrm{d}\rho + T\left(\frac{\partial\eta}{\partial p}\right)_{\rho,S}\mathrm{d}p + T\left(\frac{\partial\eta}{\partial S}\right)_{\rho,p}\mathrm{d}S \\ &= T\left(\frac{\partial\eta}{\partial\rho}\right)_{p,S}\mathrm{d}\rho - T\left(\frac{\partial\eta}{\partial\rho}\right)_{p,S}\left(\frac{\partial\rho}{\partial p}\right)_{\eta,S}\mathrm{d}p + T\left(\frac{\partial\eta}{\partial S}\right)_{\rho,p}\mathrm{d}S. \end{aligned} \tag{1.112}$$

Forming the material derivative, and using (1.97) and (1.96), we obtain for a moving fluid

$$T\left(\frac{\partial\eta}{\partial\rho}\right)_{p,S}\frac{D\rho}{Dt} - T\left(\frac{\partial\eta}{\partial\rho}\right)_{p,S}\left(\frac{\partial\rho}{\partial p}\right)_{\eta,S}\frac{Dp}{Dt} = \dot{Q} - T\left(\frac{\partial\eta}{\partial S}\right)_{\rho,p}\dot{S}. \tag{1.113}$$

But $(\partial p/\partial\rho)_{\eta,S} = c_s^2$ where c_s is the speed of sound (see Section 1.8). This is a measurable quantity in a fluid, and often nearly constant, and so useful to keep in an equation. The thermodynamic equation may then be written in the form

$$\frac{D\rho}{Dt} - \frac{1}{c_s^2}\frac{Dp}{Dt} = Q_{[\rho,p]}, \tag{1.114}$$

where $Q_{[\rho,p]} \equiv (\partial\rho/\partial\eta)_{p,S}\dot{Q}/T + (\partial\rho/\partial S)_{\eta,p}\dot{S}$ represents the effects of entropy and salinity source terms. This form of the thermodynamic equation is valid for both liquids and gases.

Approximations using pressure and density

The speed of sound in a fluid is related to its compressibility — the less compressible the fluid, the greater the sound speed. In a liquid, sound speed is often sufficiently high that the second term in (1.114) can be neglected, and the thermodynamic equation takes the simple form:

$$\frac{D\rho}{Dt} = Q_{[\rho,p]}. \tag{1.115}$$

The above equation is a very good approximation for many laboratory fluids. It is a *thermodynamic* equation, arising from the principle of conservation of energy for a liquid; it is a very different equation from the mass conservation equation, which for compressible fluids is also an evolution equation for density.

In the ocean the enormous pressures resulting from columns of seawater kilometres deep mean that although the second term in (1.114) may be small, it is not negligible, and a better approximation results if we suppose that the pressure is given by the weight of the fluid above it — the hydrostatic approximation. In this case $dp = -\rho g\, dz$ and (1.114) becomes

$$\frac{D\rho}{Dt} + \frac{\rho g}{c_s^2}\frac{Dz}{Dt} = Q_{[\rho,p]}. \tag{1.116}$$

In the second term the height field varies much more than the density field, so a good approximation is to replace ρ by a constant, ρ_0, in this term only. Taking the speed of sound also to be constant gives

$$\frac{D}{Dt}\left(\rho + \frac{\rho_0 z}{H_\rho}\right) = Q_{[\rho,p]}, \qquad \text{where} \qquad H_\rho = c_s^2/g. \tag{1.117a,b}$$

H_ρ is the *density scale height* of the ocean. In water, $c_s \approx 1500\,\mathrm{m\,s^{-1}}$ so that $H_\rho \approx 200\,\mathrm{km}$. The quantity in brackets on the left-hand side of (1.117a) is (in this approximation) the *potential density*, this being the density that a parcel would have if moved adiabatically and with constant composition to the reference height $z = 0$. The adiabatic lapse rate of density is the rate at which the density of a parcel changes when undergoing an adiabatic displacement. From (1.117) it is approximately

$$-\left(\frac{\partial\rho}{\partial z}\right)_\eta \approx \frac{\rho_0 g}{c_s^2} \approx 5\,(\mathrm{kg\,m^{-3}})/\mathrm{km}, \tag{1.118}$$

so that if a parcel is moved adiabatically from the surface to the deep ocean (5 km depth, say) its density will increase by about $25\,\mathrm{kg\,m^{-3}}$, a fractional change of about 1/40 or 2.5%.

II. Thermodynamic equation using pressure and temperature

Taking entropy to be a function of pressure and temperature (and salinity if appropriate), we have

$$\begin{aligned} T\,d\eta &= T\left(\frac{\partial\eta}{\partial T}\right)_{p,S} dT + T\left(\frac{\partial\eta}{\partial p}\right)_{T,S} dp + T\left(\frac{\partial\eta}{\partial S}\right)_{T,p} dS \\ &= c_p\,dT + T\left(\frac{\partial\eta}{\partial p}\right)_{T,S} dp + T\left(\frac{\partial\eta}{\partial S}\right)_{T,p} dS. \end{aligned} \tag{1.119}$$

For a moving fluid, and using (1.97) and (1.96), this implies,

$$\frac{DT}{Dt} + \frac{T}{c_p}\left(\frac{\partial\eta}{\partial p}\right)_{T,S}\frac{Dp}{Dt} = Q_{[T,p]}, \tag{1.120}$$

where $Q_{[T,p]} \equiv \dot{Q}/c_p - Tc_p^{-1}\dot{S}(\partial\eta/\partial S)$. Now substitute the Maxwell relation (1.82) in the form

$$\left(\frac{\partial \eta}{\partial p}\right)_T = \frac{1}{\rho^2}\left(\frac{\partial \rho}{\partial T}\right)_p \tag{1.121}$$

to give

$$\frac{DT}{Dt} + \frac{T}{c_p\rho^2}\left(\frac{\partial \rho}{\partial T}\right)_p \frac{Dp}{Dt} = Q_{[T,p]}, \quad \text{or} \quad \frac{DT}{Dt} - \frac{T}{c_p}\left(\frac{\partial \alpha}{\partial T}\right)_p \frac{Dp}{Dt} = Q_{[T,p]}. \tag{1.122a,b}$$

The density and temperature are related through a coefficient of thermal expansion β_T where

$$\left(\frac{\partial \rho}{\partial T}\right)_p = -\beta_T \rho. \tag{1.123}$$

Equation (1.122) then becomes

$$\frac{DT}{Dt} - \frac{\beta_T T}{c_p \rho}\frac{Dp}{Dt} = Q_{[T,p]}\,. \tag{1.124}$$

This form of the thermodynamic equation is valid for both liquids and gases. In an ideal gas we have $\beta_T = 1/T$, whereas in a liquid β_T is usually quite small.

Approximations using pressure and temperature

In the hydrostatic approximation we suppose that the pressure in (1.124) varies according only to the weight of the fluid above it. Then $dp = -\rho g\, dz$ and (1.124) becomes

$$\frac{1}{T}\frac{DT}{Dt} + \frac{\beta_T g}{c_p}\frac{Dz}{Dt} = \frac{Q_{[T,p]}}{T}. \tag{1.125}$$

For an ideal gas we have $\beta_T = 1/T$, whence, if c_p is constant,

$$\frac{D}{Dt}\left(c_p T + gz\right) = c_p Q_{[T,p]}. \tag{1.126}$$

The quantity $c_p T + gz$ is known as the dry static energy and we will encounter it again throughout the book. The above equation is closely related to the potential temperature form of the thermodynamic equation, since in hydrostatic balance a little manipulation reveals that

$$\frac{T}{\theta}\frac{\partial \theta}{\partial z} = \frac{\partial T}{\partial z} + \frac{g}{c_p}. \tag{1.127}$$

The quantity $T + gz/c_p$ is a height form of potential temperature for an ideal gas in hydrostatic balance, being the temperature that a fluid at a level z and temperature T would have if moved adiabatically to a reference level of $z = 0$.

If β_T is constant, which is a fair approximation for many liquids, then for small variations of temperature around the value T_0, (1.125) simplifies to

$$\frac{D}{Dt}\left(T + \frac{T_0 z}{H_T}\right) = Q_{[T,p]}, \quad \text{where} \quad H_T = \frac{c_p}{\beta_T g}. \tag{1.128a,b}$$

The quantity H_T is the *temperature scale height* of the fluid, and with the oceanic values $\beta_T \approx 2\times10^{-4}\,\mathrm{K}^{-1}$ and $c_p \approx 4\times10^3\,\mathrm{J\,kg^{-1}\,K^{-1}}$ we obtain $H_T \approx 2000\,\mathrm{km}$. The field $T + T_0 z/H_T$ is a height form of potential temperature for liquids; that is,

$$\theta \approx T + \frac{\beta_T g T_0}{c_p} z. \tag{1.129}$$

The temperature changes because of the work done by or on the fluid parcel as it expands or is compressed. In seawater the expansion coefficient β_T and c_p are functions of pressure and (1.129) is not good enough if high accuracy is required, whereas in a laboratory setting we can often simply neglect the term involving β_T.

The adiabatic lapse rate of temperature is the rate at which the temperature of a parcel changes in the vertical when undergoing an adiabatic displacement. From (1.125) it is

$$\Gamma_z \equiv -\left(\frac{\partial T}{\partial z}\right)_\eta = \frac{T g \beta_T}{c_p}. \tag{1.130}$$

In general Γ_z is a function of temperature, salinity and pressure, but it is a calculable quantity if β_T is known and, with the oceanic values above, it is approximately $0.15\,\mathrm{K\,km^{-1}}$. Equation (1.130) is not accurate enough for quantitative oceanography because the expansion coefficient is a function of pressure; nor is it a good measure of stability, because of the effects of salt. In a dry atmosphere the ideal gas relationship gives $\beta_T = 1/T$ and so

$$\Gamma_z = \frac{g}{c_p}, \tag{1.131}$$

which is approximately $10\,\mathrm{K\,km^{-1}}$. The only approximation involved in deriving this is the use of the hydrostatic relationship.

It is noteworthy that the scale heights given by (1.117b) and (1.128b) differ so much. The first is due to the pressure compressibility of seawater [and so related to c_s^2, or β_p in (1.57)] whereas the second is due to the change of density with temperature [β_T in (1.57)], and is the distance over which the difference between temperature and potential temperature changes by an amount equal to the temperature itself (i.e., by about 273 K). The two heights differ so much because the value of the thermal expansion coefficient is not directly related to the pressure compressibility — for example, fresh water at 4° C has zero thermal expansivity, and so would have an infinite temperature scale height, but its pressure compressibility differs little from water at 20° C.

III. Thermodynamic equation using density and temperature

Taking entropy to be a function of density and temperature (and salinity if appropriate) we have

$$\begin{aligned} T\,\mathrm{d}\eta &= T\left(\frac{\partial \eta}{\partial T}\right)_{\alpha,S} \mathrm{d}T + T\left(\frac{\partial \eta}{\partial \alpha}\right)_{T,S} \mathrm{d}\alpha + T\left(\frac{\partial \eta}{\partial S}\right)_{T,\alpha} \mathrm{d}S \\ &= c_v\,\mathrm{d}T + T\left(\frac{\partial \eta}{\partial \alpha}\right)_{T,S} \mathrm{d}\alpha + T\left(\frac{\partial \eta}{\partial S}\right)_{T,\alpha} \mathrm{d}S. \end{aligned} \tag{1.132}$$

For a moving fluid this implies,

$$\frac{\mathrm{D}T}{\mathrm{D}t} + \frac{T}{c_v}\left(\frac{\partial \eta}{\partial \alpha}\right)_{T,S} \frac{\mathrm{D}\alpha}{\mathrm{D}t} = Q_{[T,\alpha]}, \tag{1.133}$$

where $Q_{[T,\alpha]} \equiv \dot{Q}/c_v - T c_v^{-1} \dot{S}(\partial\eta/\partial S)$.

For an ideal gas, and using (1.84), which is one of Maxwell's relations, (1.133) may be written as

$$c_v \frac{DT}{Dt} + p\frac{D\alpha}{Dt} = \dot{Q}, \tag{1.134}$$

so recovering the internal energy equation (1.99a). On the other hand, for a liquid of nearly constant density, the second term on the left-hand side of (1.133) is small, and $c_p \approx c_v$, and we have to a first approximation $DT/Dt = Q_{[T,\alpha]}$.

Various forms of the thermodynamic equation are summarized in the box on page 26. For an ideal gas, (1.114) and (1.124) are exactly equivalent to (1.100) or (1.101), and numerical models of an ideal gas usually use either a prognostic equation for internal energy $c_v T$ or potential temperature. A discussion of an accurate thermodynamic equation for seawater is given below, with some summary remarks in the box on the next page.

1.7 ✦ THERMODYNAMICS OF SEAWATER

We now discuss the thermodynamics of liquids such as seawater in a little more detail. Readers whose interest is mainly in an ideal gas may skim this section. We begin with a phenomenological discussion of potential temperature and potential density followed by a more accurate treatment of the equation of state.

1.7.1 Potential Temperature, Potential Density and Entropy

Potential temperature and entropy

The potential temperature is defined as the temperature that a parcel would have if moved adiabatically and at constant composition to a given reference pressure p_R, often taken as 10^5 Pa (or 1000 hPa, or 1000 mb, approximately the pressure at the sea-surface). Thus it may be calculated, at least in principle, by the integral

$$\theta(S, T, p, p_R) = T + \int_p^{p_R} \Gamma'_{ad}(S, T, p')\,dp', \tag{1.135}$$

where $\Gamma'_{ad} = (\partial T/\partial p)_{\eta,S}$. Such integrals may be hard to calculate, and if we know the equation of state in the form $\eta = \eta(S, T, p)$ then we can calculate potential temperature more directly because potential temperature must satisfy

$$\eta(S, T, p) = \eta(S, \theta, p_R). \tag{1.136}$$

Solving this equation for θ gives, in principle, $\theta = \theta(\eta, S, p_R) = \theta(T, S, p)$, and examples will be given in (1.152) and in Appendix A. Potential temperature is not a materially conserved variable in the presence of salinity changes.

For a parcel of constant composition, changes in entropy are directly related to changes in potential temperature because, from the right-hand side of (1.136),

$$d\eta = \left(\frac{\partial \eta(S, \theta, p_R)}{\partial \theta}\right)_S d\theta. \tag{1.137}$$

Thus, if a fluid parcel moves adiabatically and at constant composition then $d\eta = 0$ and $d\theta = 0$. Furthermore, if we express entropy as a function of temperature and pressure then

$$T\,d\eta = T\left(\frac{\partial \eta}{\partial T}\right)_{p,S} dT + T\left(\frac{\partial \eta}{\partial p}\right)_{T,S} dp + T\left(\frac{\partial \eta}{\partial S}\right)_{p,T} dS = c_p\,dT + T\left(\frac{\partial \eta}{\partial p}\right)_{T,S} dp + T\left(\frac{\partial \eta}{\partial S}\right)_{p,T} dS, \tag{1.138}$$

Thermodynamics of Liquids

Liquids, unlike ideal gases, do not have a simple equation of state and this has ramifications for the thermodynamic equation. For seawater, a very accurate, albeit equally complex, equation of state is given by TEOS-10. The simpler expression for the Gibbs function, (1.146), is a good approximation in many circumstances and using it we can derive the thermal equation of state and some useful forms of the thermodynamic equation.

Thermal equation of state

A very useful expression for many purposes is given by

$$\alpha = \alpha_0 \left[1 + \beta_T(1 + \gamma^* p)(T - T_0) + \frac{\beta_T^*}{2}(T - T_0)^2 - \beta_S(S - S_0) - \beta_p(p - p_0) \right]. \quad \text{(TL.1)}$$

For laboratory situations where pressure variations are not large a useful approximation is

$$\alpha = \alpha_0 \left[1 + \beta_T(T - T_0) - \beta_S(S - S_0) \right]. \quad \text{(TL.2)}$$

Thermodynamic equation

Using entropy or potential enthalpy, as discussed in Section 1.7.3, as a primary thermodynamic variable is the best way to proceed if high accuracy is required. For idealized or laboratory work we can make further approximations, as follows.

Entropy evolution:
Using the hydrostatic approximation and simplifying (1.148) gives

$$\frac{D\eta}{Dt} = 0, \qquad \eta = c_{p0} \ln \frac{T}{T_0}\left[1 + \beta_S^*(S - S_0) \right] + gz\beta_T. \quad \text{(TL.3)}$$

Given entropy and salinity we can infer temperature and, using (TL.1), density.

Potential temperature and potential density:
Potential temperature or potential density are could also be used as thermodynamic variables. Accurate expressions can be derived, but approximate expressions are

$$\frac{D\theta}{Dt} = 0, \qquad \theta = T + \frac{\beta_T g T_0 z}{c_{p0}}, \quad \text{(TL.4)}$$

$$\text{or} \quad \frac{D\rho_\theta}{Dt} = 0, \qquad \rho_\theta = \rho + \frac{\rho_0 g z}{c_s^2}. \quad \text{(TL.5)}$$

Given θ or ρ_θ, as well as salinity, we then infer density using an appropriate equation of state as needed.

using the definition of c_p. If we evaluate this expression at the reference pressure, where $T = \theta$ and $dp = 0$, and consider isohaline changes with $dS = 0$, then we have $\theta \, d\eta = c_p(p_R, \theta, S) \, d\theta$, and therefore

$$d\eta = c_p(p_R, \theta, S)\frac{d\theta}{\theta}, \quad (1.139)$$

and $\mathrm{d}\eta/\mathrm{d}\theta = c_p(p_R, \theta, S)/\theta$. In the special case of constant c_p integration yields

$$\eta = c_p \ln\theta + \text{constant}, \tag{1.140}$$

as for a simple ideal gas — see (1.107). Given (1.139), the thermodynamic equation can be written

$$c_p \frac{\mathrm{D}\theta}{\mathrm{D}t} = \frac{\theta}{T}\dot{Q}, \tag{1.141}$$

where the right-hand side represents heating. We can use (1.141) as a thermodynamic equation instead of (1.97), although we must now relate θ instead of η to the other state variables via an equation of state.

The notion of potential temperature is useful because it is connected to the actual temperature, with which we are familiar; roughly speaking, potential temperature is temperature plus a correction for the effects of thermal expansion. Entropy, on the other hand, may seem alien and unnecessarily exotic. However, the use of potential temperature brings no true simplifications to the equations of motion beyond those already afforded by the use of entropy as a thermodynamic variable.

Potential density

Potential density, ρ_θ, is the density that a parcel would have if it were moved adiabatically and with fixed composition to a given reference pressure, p_R, that is often, but not always, taken as 10^5 Pa, or 1 bar. If the equation of state is of the form $\rho = \rho(S, T, p)$ then by definition we have

$$\rho_\theta = \rho(S, \theta, p_R). \tag{1.142}$$

For a parcel moving adiabatically and at fixed salinity its potential density is therefore conserved, and it is the vertical gradient of potential density that provides the appropriate measure of stability (as we find in Section 2.10.1). Because density of seawater is nearly constant we can obtain an approximate expression for potential density by Taylor-expanding the density around the potential density at the reference level at which $T = \theta$ and $p = p_R$. At first order we then have

$$\rho(S, \theta, p) \approx \rho(S, \theta, p_R) + (p - p_R)\left(\frac{\partial \rho}{\partial p}\right)_{S,\theta}. \tag{1.143}$$

The first term on the right hand side is, by definition, the potential density and the derivative in the second term is the inverse of the square of speed of sound, evaluated at the reference level, and so

$$\rho_\theta \approx \rho - \frac{1}{c_s^2}(p - p_R) \approx \rho + \frac{\rho_0 g z}{c_s^2}. \tag{1.144}$$

To obtain the right-most expression we use hydrostatic balance and take $p = -\rho_0 g z$ and $p_R = 0$ (at $z = 0$), giving the same expression as occurs in (1.117).

Because the density of seawater is nearly constant, it is common in oceanography to subtract the amount 1000 kg m^{-3} before quoting its value; then, depending on whether we are referring to *in situ* density or the potential density the results are called σ_T ('sigma-tee') or σ_θ ('sigma-theta') respectively. Thus,

$$\sigma_T = \rho(p, T, S) - 1000, \qquad \sigma_\theta = \rho(p_R, \theta, S) - 1000. \tag{1.145a,b}$$

Instead of the subscript θ, a number can be used to denote the level to which potential density is referenced. Thus, σ_0 is the potential density referenced to the surface and σ_2 is the potential density referenced to 200 bars of pressure, or about 2 kilometres depth.

Parameter	Description	Value
ρ_0	Reference density	1.027×10^3 kg m^{-3}
α_0	Reference specific volume	9.738×10^{-4} m^3 kg^{-1}
T_0	Reference temperature	283K
S_0	Reference salinity	35 ppt = 35 g kg^{-1}
c_{s0}	Reference sound speed	1490 m s^{-1}
β_T	First thermal expansion coefficient	1.67×10^{-4} K^{-1}
β_T^*	Second thermal expansion coefficient	1.00×10^{-5} K^{-2}
β_S	Haline contraction coefficient	0.78×10^{-3} ppt^{-1}
β_p	Compressibility coefficient ($= \alpha_0/c_{s0}^2$)	4.39×10^{-10} m s^2 kg^{-1}
γ^*	Thermobaric parameter ($\approx \gamma'^*$)	1.1×10^{-8} Pa^{-1}
c_{p0}	Specific heat capacity at constant pressure	3986 J kg^{-1} K^{-1}
β_S^*	Haline heat capacity coefficient	1.5×10^{-3} ppt^{-1}

Table 1.2 Various thermodynamic and equation-of-state parameters appropriate for the seawater equations of state (1.58) and (1.146). The unit ppt (or ‰) is parts per thousand by weight, or g/kg.

We cannot use $p_R = 0$ everywhere and maintain accuracy. Thus, for a parcel near 2 km, σ_2 is more relevant than σ_0. The 'neutral density' (or quasi-neutral density) is a semi-empirical way to avoid this reference-level difficulty. Neutral density is, by construction, a quantity such that the buoyancy force is locally perpendicular to its iso-surfaces, so that a parcel that is displaced adiabatically along a neutral density iso-surface will remain neutrally buoyant.[9] Neutral density is not a thermodynamic state variable and because of form of the seawater equation of state there is no continuous, unique field of neutral density extending through the ocean. Wherever it appears in figures in this book it can be assumed that potential density would look similar.

1.7.2 Equation of State for Seawater

Oceanographers go to great lengths to obtain an accurate equation of state and other physical properties of seawater, and we noted in Section 1.4 that seawater has some nonlinear properties that, although small, are nevertheless important. We need to be able to calculate these properties, and in this section we illustrate how the thermodynamic variables for seawater — the conventional equation of state, an expression for potential temperature, and so on — can be obtained directly from the fundamental equation of state. Writing the fundamental equation in the form $I = I(\eta, S, \alpha)$ is not practically useful, because the variables are not easily measured in the laboratory. However, if we cast the fundamental equation in terms of the Gibbs function, $g = I - T\eta + p\alpha$, then $\mathrm{d}g = -\eta\,\mathrm{d}T + \alpha\,\mathrm{d}p + \mu\,\mathrm{d}S$ and the independent variables are the familiar and measurable (T, S, p). A similar, but simpler, procedure is carried out for an ideal gas in Appendix A.

A Gibbs function that reproduces the properties of seawater with high accuracy is very complicated, but we can write down a Gibbs function that, although slightly less accurate, captures the most important properties with a certain degree of economy and transparency. The expression is[10]

$$
\begin{aligned}
g = g_0 - \eta_0(T - T_0) + \mu_0(S - S_0) - c_{p0}T\left[\ln(T/T_0) - 1\right]\left[1 + \beta_S^*(S - S_0)\right] \\
+ \alpha_0(p - p_0)\Big[1 + \beta_T(T - T_0) - \beta_S(S - S_0) - \frac{\beta_P}{2}(p - p_0) \\
+ \frac{\beta_T\gamma^*}{2}(p - p_0)(T - T_0) + \frac{\beta_T^*}{2}(T - T_0)^2\Big].
\end{aligned}
\tag{1.146}
$$

In this equation the variables are g, T, S and p. The parameters (which, as in (1.58), all have subscripts or stars, with the starred parameters giving rise to nonlinear effects) are all constants that could in principle be determined in the laboratory with the help of the derived quantities like heat capacity, and their approximate values are given in Table 1.2. We will take $p_0 = 0$ and $\beta_p = \alpha_0/c_{s0}^2$, where c_{s0} is a reference sound speed. Equation (1.146) is in fact quite accurate for most oceanographic situations, and from it we may derive the following quantities of interest:

- The conventional or thermal equation of state, $\alpha = (\partial g/\partial p)_{T,S}$:

$$\alpha = \alpha_0 \left[1 + \beta_T(1 + \gamma^* p)(T - T_0) + \frac{\beta_T^*}{2}(T - T_0)^2 - \beta_S(S - S_0) - \beta_p p \right]. \tag{1.147}$$

- The entropy, $\eta = -(\partial g/\partial T)_{p,S}$:

$$\eta = \eta_0 + c_{p0} \ln \frac{T}{T_0} \left[1 + \beta_S^*(S - S_0) \right] - \alpha_0 p \left[\beta_T + \beta_T \gamma^* \frac{p}{2} + \beta_T^*(T - T_0) \right]. \tag{1.148}$$

  For temperatures in the range 0°–30° Celsius, entropy increases linearly with temperature to within a few percent.

- The heat capacity, $c_p = T(\partial \eta/\partial T)_{p,S}$:

$$c_p = c_{p0} \left[1 + \beta_S^*(S - S_0) \right] - \alpha_0 p \beta_T^* T. \tag{1.149}$$

  This is to a first approximation constant, varying mildly with salinity and more weakly with temperature and pressure.

- The thermal expansion coefficient, $\widehat{\beta}_T = \alpha^{-1}(\partial \alpha/\partial T)_{S,p}$:

$$\widehat{\beta}_T = (\alpha_0/\alpha) \left[\beta_T + \beta_T \gamma^* p + \beta_T^*(T - T_0) \right], \tag{1.150}$$

  where α is given by (1.147).

- The adiabatic lapse rate, $\Gamma = (\partial T/\partial p)_{\eta,S}$. Using (1.138) gives

$$\Gamma = \left(\frac{\partial T}{\partial p} \right)_{\eta,S} = -\frac{T}{c_p} \left(\frac{\partial \eta}{\partial p} \right)_{T,S} = \frac{T}{c_p} \alpha_0 [\beta_T(1 + \gamma^* p) + \beta_T^*(T - T_0)]. \tag{1.151}$$

  where c_p is given by (1.149).

Potential temperature and potential density, revisited

An expression for the potential temperature, θ, may be obtained by solving (1.136) for θ. In general, such an equation must be solved numerically, but for our equation of state, using (1.148) and taking $p_R = 0$, we find

$$\theta = T \exp \left\{ -\frac{\alpha_0 \beta_T p}{c_p'} \left[1 + \frac{1}{2} \gamma^* p + \frac{\beta_T^*}{\beta_T}(T - T_0) \right] \right\}, \tag{1.152}$$

where $c_p' = c_{p0} \left[1 + \beta_S^*(S - S_0) \right]$. Equation (1.152) is a relationship between T, θ and p analogous to (1.105) for an ideal gas. The exponent itself is small, the second and third terms in square brackets are small compared to unity, the deviations of both T and θ from T_0 are also presumed to be small,

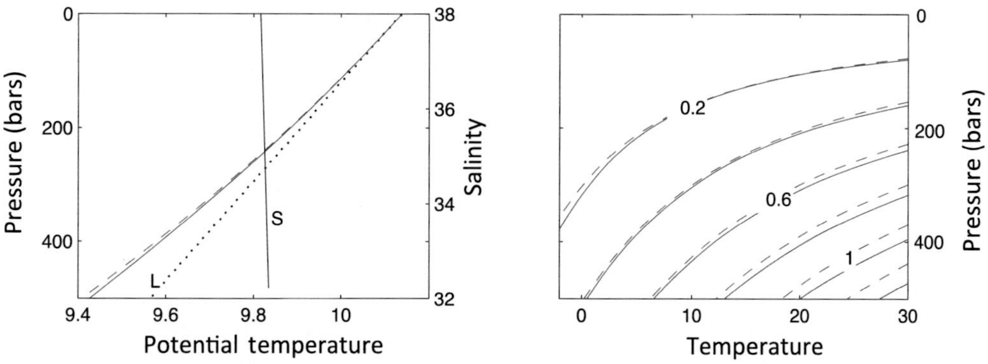

Fig. 1.4 Examples of the variation of potential temperature of seawater with pressure, temperature and salinity. Left panel: the sloping lines show potential temperature as a function of pressure at fixed salinity (S = 35 psu) and temperature (10° C). The solid line is computed using an accurate, empirical equation of state, the almost-coincident dashed line uses the simpler expression (1.152) and the dotted line (labelled L) uses the linear expression (1.153c). The near vertical solid line, labelled S, shows the variation of potential temperature with salinity at fixed temperature and pressure. Right panel: Contours of the difference between temperature and potential temperature, $(T - \theta)$ in the pressure–temperature plane, for S = 35 psu. The dashed lines use (1.152) and the solid lines use an accurate empirical formula. (Note: 100 bars of pressure (10^7 Pa or 10 MPa) is approximately 1 km depth.)

and c_p' is nearly constant. Taking advantage of all of this enables (1.152) to be rewritten, with increasing levels of approximation, as

$$T' \approx \frac{T_0\alpha_0\beta_T}{c_{p0}} p\left(1 + \frac{1}{2}\gamma^* p + T_0\frac{\alpha_0\beta_T^*}{c_{p0}} p\right) + \theta'\left(1 + T_0\frac{\alpha_0\beta_T^*}{c_{p0}} p\right) \tag{1.153a}$$

$$\approx \frac{T_0\alpha_0\beta_T}{c_{p0}} p\left(1 + \frac{1}{2}\gamma^* p\right) + \theta'\left(1 + T_0\frac{\alpha_0\beta_T^*}{c_{p0}} p\right) \tag{1.153b}$$

$$\approx \frac{T_0\alpha_0\beta_T}{c_{p0}} p + \theta', \tag{1.153c}$$

where $T' = T - T_0$ and $\theta' = \theta - T_0$. The last of the three, (1.153c), holds for a linear equation of state, and is useful for calculating approximate differences between temperature and potential temperature; making use of the hydrostatic approximation reveals that it is essentially the same as (1.129). Plots of the difference between temperature and potential temperature, using both a highly accurate empirical equation of state and using the simplified equation, (1.152), are given in Fig. 1.4, and some examples of the density variation of seawater are given in Fig. 1.5.

To obtain an equation of state that gives density in terms of potential temperature, pressure and salinity we use (1.153b) in the equation of state, (1.147), to give

$$\alpha \approx \alpha_0\left[1 - \frac{\alpha_0 p}{c_{s0}'^2} + \beta_T(1 + \gamma'^* p)\theta' + \frac{1}{2}\beta_T^*\theta'^2 - \beta_S(S - S_0)\right], \tag{1.154}$$

where $\gamma'^* = \gamma^* + T_0\beta_T^*\alpha_0/c_{p0}$ and $c_{s0}^{-2} = c_{s0}'^{-2} - \beta_T^2 T_0/c_p$. The parameters γ^* and γ'^* differ by a few percent, and c_s^2 and $c_s'^2$ differ by a few parts in a thousand, and we may neglect the differences. We may further approximate (1.154) by using the hydrostatic pressure instead of the actual pressure; thus, letting $p = -g(z - z_0)/\alpha_0$ where z_0 is the nominal value of z at which $p = 0$, we obtain

$$\alpha \approx \alpha_0\left[1 + \frac{g(z - z_0)}{c_{s0}^2} + \beta_T\left(1 - \gamma^*\frac{g(z - z_0)}{\alpha_0}\right)\theta' + \frac{\beta_T^*}{2}\theta'^2 - \beta_S(S - S_0)\right]. \tag{1.155}$$

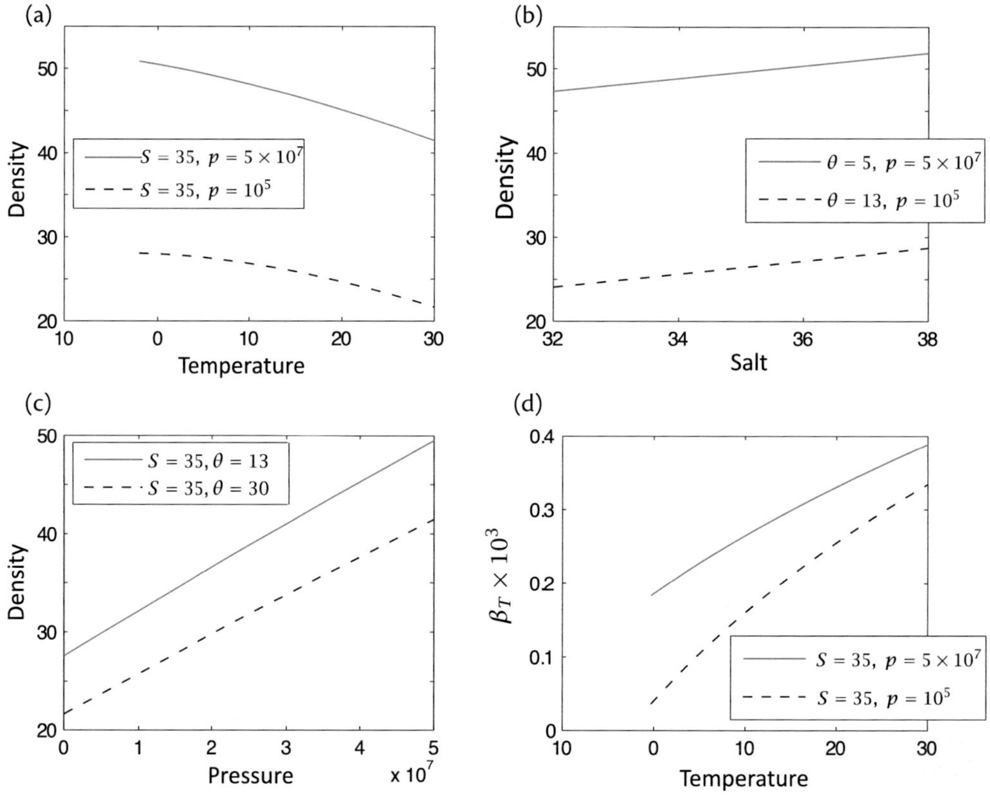

Fig. 1.5 Examples of the variation of density of seawater, $(\rho - 1000)\,\mathrm{kg\,m^{-3}}$. (a) With potential temperature (° C); (b) with salt (g/kg); and (c) with pressure (Pa), for seawater. Panel (d), shows the thermal expansion coefficient, $\widehat{\beta}_T = -\rho^{-1}(\partial\rho/\partial T)_{p,S}\,\mathrm{K}^{-1}$, for each of the two curves in panel (a).

Using z instead of p in the equation of state entails a slight loss of accuracy, but is necessary to ensure that the Boussinesq equations maintain good conservation properties, as will be discussed in Sections 2.4 and 4.7.1.

Given an expression for density in terms of potential temperature we can write down an expression for potential density, ρ_θ, since by definition it is just the density at a reference pressure, p_R. From (1.147) we have, assuming density variations are small,

$$\rho_\theta \approx \rho_0\left[1 + \frac{\alpha_0 p_R}{c_{s0}^2} - \beta_T(1+\gamma^* p_R)\theta' - \frac{1}{2}\beta_T^*\theta'^2 + \beta_S(S - S_0)\right], \tag{1.156a}$$

$$\approx \rho + (p_R - p)\left(\frac{1}{c_{s0}^2} - \beta_T\gamma^*\rho_0\theta'\right). \tag{1.156b}$$

This expression may be compared to the more approximate one, (1.144). The second term in large brackets in (1.156b) is quite small and is a manifestation of the thermobaric effect, that the compressibility of seawater is a weak function of temperature.

1.7.3 Potential Enthalpy as a Thermodynamic Variable

As discussed in Section 1.6, there are various forms of thermodynamic equation that are, in principle, equivalent. However, they are not equivalent in practice, even if the fundamental equation of state of is known. Thus, for example, for a nearly incompressible fluid like seawater the velocity divergence is small but non-zero and numerically integrating the internal energy equation, (1.94), may be awkward and inaccurate. An ideal choice might be some conservative variable that obeys

$$\rho\frac{\mathrm{D}\chi}{\mathrm{D}t} = \nabla\cdot\boldsymbol{F}_\chi \qquad \text{or equivalently} \qquad \frac{\partial}{\partial t}(\rho\chi) + \nabla\cdot(\rho\boldsymbol{v}\chi) = \nabla\cdot\boldsymbol{F}_\chi, \tag{1.157}$$

where χ is a thermodynamical variable from which all the other variables can be inferred using the fundamental equation of state, and $\boldsymbol{F}_\chi$ represents molecular and radiative fluxes. Unfortunately, there is no variable that exactly has an advective left-hand side and a right-hand side that is the divergence of a flux — but some are nearly so, as we discuss.

Entropy

An obvious choice for a primary thermodynamic variable is entropy, an advected variable satisfying

$$\frac{\mathrm{D}\eta}{\mathrm{D}t} = \frac{\dot{Q}}{T}, \tag{1.158}$$

where $\dot{Q}$ is the heating. The main difficulties with an entropic approach are the determination of the heating and the accurate treatment of that heating once determined. The right-hand side does not have a conservative form even in the case which the heating is due solely to diffusive molecular fluxes and radiation. A further, albeit minor, complication is that heating is affected by irreversible molecular fluxes of salt in the interior. Further difficulties arise in using (1.158) in a turbulent ocean, because heating is affected not only by thermal effects but also by freshwater fluxes at the ocean surface and unresolved fluxes of salinity in the ocean interior. Care must taken to ensure that parameterized fluxes of temperature and salinity do produce (and not reduce) entropy.

Potential temperature, θ, provides no fundamental advantage over entropy, because θ is a function of both entropy and salinity and still requires a computation of heating. Indeed, because $\mathrm{d}\eta \approx c_p \mathrm{d}\ln\theta$ the evolution equation for potential temperature properly involves the heat capacity, c_p, which is a function (albeit a weak one) of both salinity and pressure. Having said all this, the above problems are not in practice large ones, and entropy and (more commonly) potential temperature have been used very successfully in quantitative ocean models.

Potential enthalpy

An alternative to entropy is to construct a near-conservative variable related to enthalpy. Consider first the fundamental thermodynamic relation, (1.85), in the form

$$\frac{\mathrm{D}I}{\mathrm{D}t} + p\frac{\mathrm{D}\alpha}{\mathrm{D}t} = T\frac{\mathrm{D}\eta}{\mathrm{D}t} + \mu\frac{\mathrm{D}S}{\mathrm{D}t} = \frac{\mathrm{D}h}{\mathrm{D}t} - \alpha\frac{\mathrm{D}p}{\mathrm{D}t} = \dot{Q}_E, \tag{1.159}$$

where $\dot{Q}_E = \dot{Q} + \mu\dot{S}$ is the total rate of non-mechanical energy input per unit mass, $\dot{Q}$ is the heating and $\dot{S}$ represents saline sources and sinks. Now, it is usually easier and more accurate to determine energy input, $\dot{Q}_E$, than the heating, $\dot{Q}$, because $\dot{Q}_E$ is very nearly equal to the divergence of an energy flux. That is to say, $\rho\dot{Q}_E \approx \nabla\cdot\boldsymbol{F}_E$ where $\boldsymbol{F}_E$ is an energy flux, with the small difference arising from the heating due to viscous dissipation of kinetic energy — see Appendix B. This property suggests the use of internal energy or enthalpy as a primary thermodynamical variable, but neither are conservative quantities because of the $\mathrm{D}\alpha/\mathrm{D}t$ and $\mathrm{D}p/\mathrm{D}t$ terms in (1.159). However we can form a quantity, *potential enthalpy,* that very nearly is conservative, as follows.[11]

The potential enthalpy, h^0, is defined to be the enthalpy that a fluid parcel has if taken at constant composition and entropy from its current location to a fixed reference pressure, p_R. Thus,

$$h^0(S,\eta,p_R) \equiv h(S,\eta,p) + \int_p^{p_R}\left(\frac{\partial h}{\partial p}\right)_{S,\eta}\mathrm{d}p' = h(S,\eta,p) + \int_p^{p_R}\alpha\,\mathrm{d}p', \tag{1.160}$$

using (1.74). That is, $h = h^0 + h^d$ where

$$h^d(\eta, p, S) = -\int_p^{p_R} \left(\frac{\partial h}{\partial p}\right)_{S,\eta} \mathrm{d}p' = -\int_p^{p_R} \alpha\, \mathrm{d}p' \tag{1.161}$$

is the 'dynamic enthalpy'. The material derivative of the dynamic enthalpy is given by

$$\frac{\mathrm{D}h^d}{\mathrm{D}t} = \frac{\mathrm{D}\eta}{\mathrm{D}t}\frac{\partial h^d}{\partial \eta} + \frac{\mathrm{D}S}{\mathrm{D}t}\frac{\partial h^d}{\partial S} + \alpha\frac{\mathrm{D}p}{\mathrm{D}t}. \tag{1.162}$$

To obtain this expression we have used $(\partial h^d/\partial p)_{\eta,S} = (\partial h/\partial p)_{\eta,S} = \alpha$, because h^0 is not a function of pressure. Using (1.162) and (1.160) we obtain

$$\frac{\mathrm{D}h^0}{\mathrm{D}t} = \frac{\mathrm{D}h}{\mathrm{D}t} - \frac{\mathrm{D}h^d}{\mathrm{D}t} = \frac{\mathrm{D}h}{\mathrm{D}t} - \left[\alpha\frac{\mathrm{D}p}{\mathrm{D}t} + \frac{\mathrm{D}\eta}{\mathrm{D}t}\frac{\partial h^d}{\partial \eta} + \frac{\mathrm{D}S}{\mathrm{D}t}\frac{\partial h^d}{\partial S}\right], \tag{1.163}$$

and using the fundamental thermodynamic relation, (1.159), we can write this equation as

$$\frac{\mathrm{D}h^0}{\mathrm{D}t} = \dot{Q}_E - \frac{\partial h^d}{\partial \eta}\dot{\eta} - \frac{\partial h^d}{\partial S}\dot{S}. \tag{1.164}$$

The second and third terms on the right-hand side are much smaller than the other terms and can be neglected in most oceanographic applications. To see this, realize that an approximate size of the first term on the right-hand side is $c_p \mathrm{d}\theta/\mathrm{d}t$, whereas the approximate size of the second term is $(\partial h^d/\partial \eta) \times (c_p\theta^{-1}\mathrm{d}\theta/\mathrm{d}t)$. That is, the second term is smaller than the first by the factor

$$\gamma = \frac{1}{\theta}\frac{\partial h^d}{\partial \eta} = \frac{1}{c_p}\frac{\partial h_d}{\partial \theta} \sim \frac{1}{c_p}\frac{\partial \alpha}{\partial \theta}\Delta p, \tag{1.165}$$

using (1.161), where $\Delta p = p - p_R \le 4 \times 10^7$ Pa. Putting in values from the seawater equation of state, (1.154), we find

$$\gamma \sim \frac{1}{c_p}\frac{\partial \alpha}{\partial \theta}\Delta p \approx \frac{\beta_T \Delta p}{\rho_0 c_p} \approx \frac{1.7 \times 10^{-4}\;\; 4 \times 10^7}{10^3\;\; 4 \times 10^3} \approx 1.7 \times 10^{-3}, \tag{1.166}$$

with a similarly small value for the saline term, where the smallness of these terms ultimately stems from the near incompressibility of seawater. These terms may actually be smaller than (1.166) suggests if we choose p_R appropriately. Most of the energy flux into the ocean occurs at the ocean surface and if we choose $p_R = 0$ the fluxes affect enthalpy and potential enthalpy in the same way.

Given the above arguments, an accurate and computable thermodynamic equation for seawater is (1.164) with the last two terms neglected and the source term in flux form, namely

$$\rho\frac{\mathrm{D}h^0}{\mathrm{D}t} = \nabla \cdot \boldsymbol{F}_E \qquad \text{or} \qquad \frac{\partial}{\partial t}(\rho h^0) + \nabla \cdot (\rho \boldsymbol{v} h^0) = \nabla \cdot \boldsymbol{F}_E. \tag{1.167}$$

Given that the right-hand side is a flux divergence, in steady state the interior fluxes of potential enthalpy are (in this approximation) in exact balance with the energy fluxes at the ocean boundaries. The integral of ρh^0 is thus a sensible measure of the total non-kinetic energy, or 'heat content', of the ocean because it responds almost exactly to energy fluxes at the ocean surface.[12]

Why does internal energy not have these same advantages since, from (1.94), it too is conservative apart from a small compression term? The reason is that internal energy can be changed by processes that are entirely internal to the ocean, whereas enthalpy cannot. When two adjacent fluid parcels mix, the total enthalpy is conserved because of the form of the thermodynamic equation, $\rho Dh/Dt - Dp/Dt = \rho Q_E$. If ρQ_E is the divergence of a flux then the total enthalpy ($\int \rho h \, dV$) is conserved on mixing (because the parcels are at the same pressure), but this conservative property is not shared by internal energy, because of the compression term, $pD\alpha/Dt$, in the internal energy equation. Indeed, when two parcels at different temperatures and salinities are mixed the density of the resulting parcel is higher than the average of the two (see Fig. 1.3); this effect is called cabbeling and the contraction leads to a small increase in internal energy. (The differences in the conservation of enthalpy and internal energy arise from the differences in the equation of motion for enthalpy and internal energy and not directly from the equation of state.) Thus, internal energy is not a good measure of the 'heat content' of the ocean and, furthermore, in practice the very small compression term would be hard to treat accurately. In contrast, internal energy often *is* used as a primary thermodynamic variable in ideal-gas atmospheric models, where $I = c_v T$.

Thus, in short, the practical advantages of potential enthalpy are three-fold: (i) Potential enthalpy is nearly conservative, as in (1.167); (ii) As a consequence, potential enthalpy is itself a useful measure of the non-kinetic energy of the ocean; (iii) Energy flux is a little more easily and more accurately computed than heating. Property (ii) is not shared by internal energy, since that is not conserved when parcels mix in the interior. In practice, potential enthalpy is almost proportional to potential temperature, as we now see.

Conservative temperature and an expression for potential enthalpy

To make a connection to the perhaps more familiar potential temperature it is convenient to define *conservative temperature*, Θ, by

$$\Theta \equiv \frac{h^0(\eta, p_R, S)}{c_p^0}, \tag{1.168}$$

where $c_p^0 = 3991.87\,\mathrm{J\,kg^{-1}\,K^{-1}}$ is the average heat capacity at the ocean surface and $p_R = 0$. (Potential enthalpy can have any reference pressure; conservative temperature has, by definition $p_R = 0$.) As we see below, conservative temperature is then very similar to potential temperature.

The enthalpy is given in terms of the Gibbs function by

$$h = g + T\eta = g - T\left(\frac{\partial g}{\partial T}\right)_{p,S}. \tag{1.169}$$

The potential enthalpy is therefore given by

$$h^0(S, T, p) = h(S, \theta, p_R) = g(S, \theta, p_R) - T\frac{\partial}{\partial T}g(S, \theta, p_R), \tag{1.170}$$

where θ is the potential temperature referenced to $p = p_R$ (and $p_R = 0$) and the last term is the derivative of the Gibbs function evaluated at $T = \theta$ and $p = p_R$. We can evaluate the derivative using the seawater equation of state, (1.146), giving, to within a constant factor,

$$\begin{aligned} h(T, p, S) = {}& \mu_0 S + c_{p0} T\left[1 + \beta_S^*(S - S_0)\right] \\ & + \alpha_0 p\left[1 - \beta_T T_0 - \beta_S(S - S_0) - \frac{\beta_p p}{2} - \frac{\beta_T \gamma^*}{2} p T_0 + \frac{\beta_T^*}{2}(T_0^2 - T^2)\right], \end{aligned} \tag{1.171}$$

and

$$h^0(\theta, 0, S) = h(\theta, 0, S) = \theta c_{p0}\left[1 + \beta_S^*(S - S_0)\right] + \mu_0 S. \tag{1.172}$$

The last term on the right-hand side of the above equation, $\mu_0 S$, is over two orders of magnitude smaller than the first. Also, from (1.149), the factor multiplying θ in (1.172) is just the heat capacity at $p = 0$, namely c_p^0, which varies only by few percent over the ocean. Thus, using the definition of conservative temperature given in (1.168), we have to a good approximation

$$c_p^0 \Theta \equiv h^0(\theta, 0, S) \approx c_p^0 \theta. \tag{1.173}$$

That is, to this approximation, conservative temperature equals potential temperature. Finally, if we are to use potential enthalpy (or entropy) as a thermodynamic variable we must manipulate the equation of state to obtain variables such as pressure or density from it, analogous to (1.154).

1.8 SOUND WAVES

Full of sound and fury, signifying nothing.
William Shakespeare, *Macbeth*, c. 1606.

We now consider, rather briefly, one of the most common phenomena in fluid dynamics, yet one which is in most circumstances relatively unimportant for geophysical fluid dynamics — sound waves. Their unimportance stems from the fact that the pressure disturbances produced by sound waves are a tiny fraction of the ambient pressure and are too small to affect the circulation. For example, the ambient surface pressure in the atmosphere is about 10^5 Pa and variations due to large-scale weather phenomena are about 10^3 Pa or larger, whereas sound waves of 70 dB (i.e., a loud conversation) produce pressure variations of about 0.06 Pa. (To convert, dBs = $20\log_{10}(\Delta p/p_r)$ where Δp is the pressure change in Pascals and $p_r = 2 \times 10^{-5}$.)

The smallness of the disturbance produced by sound waves justifies a linearization of the equations of motion and we do so about a spatially uniform basic state that is a time-independent solution to the equations of motion. Thus, we write $\boldsymbol{v} = \boldsymbol{v}_0 + \boldsymbol{v}'$, $\rho = \rho_0 + \rho'$ (where a subscript 0 denotes a basic state and a prime denotes a perturbation) and so on, substitute in the equations of motion, and neglect terms involving products of primed quantities. By choice of our reference frame we will simplify matters further by setting $\boldsymbol{v}_0 = 0$. The linearized momentum and mass conservation equations are then, respectively,

$$\rho_0 \frac{\partial \boldsymbol{v}'}{\partial t} = -\nabla p', \qquad \frac{\partial \rho'}{\partial t} = -\rho_0 \nabla \cdot \boldsymbol{v}'. \tag{1.174a,b}$$

(These linear equations do not in themselves determine the magnitude of the disturbance.) Now, sound waves are largely adiabatic. Thus,

$$\frac{\mathrm{d}p}{\mathrm{d}t} = \left(\frac{\partial p}{\partial \rho}\right)_\eta \frac{\mathrm{d}\rho}{\mathrm{d}t}, \tag{1.175}$$

where $(\partial p/\partial \rho)_\eta$ is the derivative at constant entropy, whose particular form is given by the equation of state for the fluid at hand. Then, from (1.174) and (1.175) we obtain a single equation for pressure,

$$\frac{\partial^2 p'}{\partial t^2} = c_s^2 \nabla^2 p', \tag{1.176}$$

where $c_s^2 = (\partial p/\partial \rho)_\eta$. Equation (1.176) is the classical wave equation; solutions propagate at a speed c_s, which may be identified as the speed of sound. For adiabatic flow in an ideal gas, manipulation of the equation of state leads to $p = C\rho^\gamma$, where $\gamma = c_p/c_v$, whence $c_s^2 = \gamma p/\rho = \gamma RT$. Values of γ typically range from 5/3 for a monatomic gas to 7/5 for a diatomic gas, and so for air, which is almost entirely diatomic, we find $c_s \approx 350\,\mathrm{m\,s^{-1}}$ at 300 K. In seawater no such theoretical approximation is easily available, but measurements show that $c_s \approx 1500\,\mathrm{m\,s^{-1}}$.

1.9 COMPRESSIBLE AND INCOMPRESSIBLE FLOW

Although there may be no fluids of truly constant density, in many cases the density of a fluid will vary so little that it is a very good approximation to consider the density effectively constant. The fluid is then said to be *incompressible.* (Some sources take incompressible to mean that density is unaffected by pressure. We take it to mean that density is also unaffected by temperature and composition.) For example, in the Earth's oceans the density varies by less than 5% (usually much less) even though the pressure at the ocean bottom is several hundred times that at the surface. We first consider how the mass continuity equation simplifies when density is truly constant, and then consider conditions under which treating density as constant is a good approximation.

1.9.1 Constant Density Fluids

If a fluid is strictly of constant density then the mass continuity equation, $\mathrm{D}\rho/\mathrm{D}t + \rho\nabla\cdot\boldsymbol{v} = 0$, simplifies to

$$\nabla\cdot\boldsymbol{v} = 0. \tag{1.177}$$

The *prognostic* equation (1.36) has become a *diagnostic* equation (1.177), or a constraint to be satisfied by the velocity. The volume of each material fluid element is therefore constant; to see this recall that conservation of mass is $\mathrm{D}(\rho\Delta V)/\mathrm{D}t = 0$ and if ρ is constant this becomes $\mathrm{D}\Delta V/\mathrm{D}t = 0$, whence (1.177) is recovered because $\mathrm{D}\Delta V/\mathrm{D}t = \Delta V\nabla\cdot\boldsymbol{v}$.

1.9.2 Incompressible Flows

In reality no fluid is truly incompressible and for (1.177) to approximately hold we just require that

$$\left|\frac{\mathrm{D}\rho}{\mathrm{D}t}\right| \ll \rho\left(\left|\frac{\partial u}{\partial x}\right| + \left|\frac{\partial v}{\partial y}\right| + \left|\frac{\partial w}{\partial z}\right|\right); \tag{1.178}$$

that is, the material derivative of density is much smaller than the individual terms constituting the divergence. As a working definition we say that *in an incompressible fluid, density changes are so small that they have a negligible effect on the mass balance.* We do not need to assume that the densities of differing fluid elements are similar to each other, but in the ocean (and in most liquids) variations in density, $\delta\rho$, are in fact everywhere small compared to the mean density, ρ_0. A sufficient condition for incompressibility, then, is that

$$\frac{\delta\rho}{\rho_0} \ll 1. \tag{1.179}$$

The fact that $\nabla\cdot\boldsymbol{v} = 0$ does *not* imply that we may independently use $\mathrm{D}\rho/\mathrm{D}t = 0$. Indeed for a liquid with an equation of state $\rho = \rho_0(1 - \beta_T(T - T_0))$ and a thermodynamic equation $c_p DT/Dt = \dot{Q}$ we have

$$\frac{\mathrm{D}\rho}{\mathrm{D}t} = -\frac{\beta_T\rho_0}{c_p}\dot{Q}. \tag{1.180}$$

Furthermore, incompressibility does not necessarily imply the neglect of density variations in the momentum equation — it is only in the mass continuity equation that density variations are neglected, as will become apparent in our discussion of the Boussinesq equations in chapter 2.

Conditions for incompressibility

The conditions under which incompressibility is a good approximation to the full mass continuity equation depend not only on the physical nature of the fluid but also on the flow itself. The condition that density is largely unaffected by pressure gives one necessary condition for the legitimate use of (1.177), as follows. First assume adiabatic flow, and omit the gravitational term. Then

$$\frac{\mathrm{d}p}{\mathrm{d}t} = \left(\frac{\partial p}{\partial \rho}\right)_\eta \frac{\mathrm{d}\rho}{\mathrm{d}t} = c_s^2 \frac{\mathrm{d}\rho}{\mathrm{d}t}, \tag{1.181}$$

so that the density and pressure variations of a fluid parcel are related by

$$\delta p \sim c_s^2 \delta \rho. \tag{1.182}$$

From the momentum equation we estimate

$$\frac{U^2}{L} \sim \frac{1}{L}\frac{\delta p}{\rho_0}, \tag{1.183}$$

where U and L are typical velocities and lengths and where ρ_0 is a representative value of the density. Using (1.182) and (1.183) gives $U^2 \sim c_s^2 \delta\rho/\rho_0$. The incompressibility condition (1.179) then becomes

$$\frac{U^2}{c_s^2} \ll 1. \tag{1.184}$$

Thus, for a flow to be incompressible the fluid velocities must be less than the speed of sound; that is, the Mach number, $M \equiv U/c_s$, must be small.

In the Earth's atmosphere it is apparent that density changes with height. To estimate how much density does change, let us first assume hydrostatic balance and an ideal gas, so that $\partial p/\partial z = -\rho g$. If we also assume that atmosphere is isothermal then

$$\frac{\partial p}{\partial z} = \left(\frac{\partial p}{\partial \rho}\right)_T \frac{\partial \rho}{\partial z} = RT_0 \frac{\partial \rho}{\partial z}. \tag{1.185}$$

Using hydrostasy and (1.185) gives

$$\rho = \rho_0 \exp(-z/H_\rho), \tag{1.186}$$

where $H_\rho = RT_0/g$ is the (density) *scale height* of the atmosphere. (It is also the pressure scale height here.) It is easy to see that density changes are negligible only if we concern ourselves with motion less than the scale height, so this is another necessary condition for incompressibility.

In the atmosphere, although the Mach number is small for most flows, vertical displacements often exceed the scale height and the flow cannot then be considered incompressible. In the ocean, density changes from all causes are small and in most circumstances the ocean may be considered to contain an incompressible fluid.

1.10 THE ENERGY BUDGET

The total energy of a fluid includes the kinetic, potential and internal energies. Both fluid flow and pressure forces will, in general, move energy from place to place, but we nevertheless expect, even demand, energy to be conserved in an enclosed volume. Is it?

1.10.1 Constant Density Fluid

For a constant density fluid the momentum equation and the mass continuity equation $\nabla \cdot \boldsymbol{v} = 0$, are sufficient to completely determine the evolution of a system. The momentum equation is

$$\frac{\mathrm{D}\boldsymbol{v}}{\mathrm{D}t} = -\nabla(\phi + \Phi) + \nu\nabla^2\boldsymbol{v}, \tag{1.187}$$

where $\phi = p/\rho_0$ and Φ is the potential for any conservative force per unit mass (e.g., gz for a uniform gravitational field). We can rewrite the advective term on the left-hand side using the identity,

$$(\boldsymbol{v} \cdot \nabla)\boldsymbol{v} = -\boldsymbol{v} \times \boldsymbol{\omega} + \nabla(\boldsymbol{v}^2/2), \tag{1.188}$$

where $\boldsymbol{\omega} \equiv \nabla \times \boldsymbol{v}$ is the *vorticity*, discussed more in later chapters. Then, omitting viscosity, we have

$$\frac{\partial \boldsymbol{v}}{\partial t} + \boldsymbol{\omega} \times \boldsymbol{v} = -\nabla B, \tag{1.189}$$

where $B = (\phi + \Phi + \boldsymbol{v}^2/2)$ is the *Bernoulli function* for constant density flow. Consider for a moment steady flows ($\partial \boldsymbol{v}/\partial t = 0$). Streamlines are, by definition, parallel to $\boldsymbol{v}$ everywhere, and the vector $\boldsymbol{v} \times \boldsymbol{\omega}$ is everywhere orthogonal to the streamlines, so that taking the dot product of the steady version of (1.189) with $\boldsymbol{v}$ gives $\boldsymbol{v} \cdot \nabla B = 0$. That is, for steady flows the Bernoulli function is constant along a streamline, and $\mathrm{D}B/\mathrm{D}t = 0$.

Reverting to the time-varying case, take the dot product of (1.189) with $\boldsymbol{v}$ and include the density to yield

$$\frac{1}{2}\frac{\partial \rho_0 \boldsymbol{v}^2}{\partial t} + \rho_0 \boldsymbol{v} \cdot (\boldsymbol{\omega} \times \boldsymbol{v}) = -\rho_0 \boldsymbol{v} \cdot \nabla B. \tag{1.190}$$

The second term on the left-hand side vanishes identically. Defining the kinetic energy density K, or energy per unit volume, by $K = \rho_0 \boldsymbol{v}^2/2$, (1.190) becomes an expression for the rate of change of K,

$$\frac{\partial K}{\partial t} + \nabla \cdot (\rho_0 \boldsymbol{v} B) = 0. \tag{1.191}$$

Because Φ is time-independent this may be written

$$\frac{\partial E}{\partial t} + \nabla \cdot (\rho_0 \boldsymbol{v} B) = 0, \tag{1.192}$$

where $E = K + \rho_0 \Phi$ is the total energy density (i.e, the total energy per unit volume). This has the form of a general conservation equation in which a local change in a quantity is balanced by the divergence of its flux. However, the energy flux, $\rho_0 \boldsymbol{v} B = \rho_0 \boldsymbol{v}(\boldsymbol{v}^2/2 + \Phi + \phi)$, is *not* simply the velocity times the energy density $\rho_0(\boldsymbol{v}^2/2 + \Phi)$; there is an additional term, $\boldsymbol{v} p$, that represents the energy transfer occurring when work is done by the fluid against the pressure force.

Now consider a volume through which there is no mass flux, for example a domain bounded by rigid walls. The rate of change of energy within that volume is then given by the integral of (1.192),

$$\frac{\mathrm{d}}{\mathrm{d}t}\int_V E \,\mathrm{d}V = -\int_V \nabla \cdot (\rho_0 \boldsymbol{v} B)\,\mathrm{d}V = -\int_S \rho_0 B \boldsymbol{v} \cdot \mathrm{d}\boldsymbol{S} = 0, \tag{1.193}$$

using the divergence theorem. Thus, the total energy within the volume is conserved. The total kinetic energy is also conserved, total gravitational potential energy is equal to $\int_V \rho_0 g z \,\mathrm{d}V$, and this is a constant, unaffected by a rearrangement of the fluid. Thus, in a constant density fluid there is no exchange between kinetic energy and potential energy.

1.10.2 Variable Density Fluids

We start with the inviscid momentum equation with a time-independent potential Φ,

$$\rho \frac{\mathrm{D}\boldsymbol{v}}{\mathrm{D}t} = -\nabla p - \rho \nabla \Phi, \tag{1.194}$$

and take its dot product with $\boldsymbol{v}$ to obtain an equation for the evolution of kinetic energy,

$$\frac{1}{2}\rho \frac{\mathrm{D}\boldsymbol{v}^2}{\mathrm{D}t} = -\boldsymbol{v} \cdot \nabla p - \rho \boldsymbol{v} \cdot \nabla \Phi = -\nabla \cdot (p\boldsymbol{v}) + p \nabla \cdot \boldsymbol{v} - \rho \boldsymbol{v} \cdot \nabla \Phi. \tag{1.195}$$

The internal energy equation for adiabatic flow is

$$\rho\frac{\mathrm{D}I}{\mathrm{D}t} = -p\nabla\cdot\boldsymbol{v}. \tag{1.196}$$

Finally, and somewhat trivially, the potential energy density obeys

$$\rho\frac{\mathrm{D}\Phi}{\mathrm{D}t} = \rho\boldsymbol{v}\cdot\nabla\Phi. \tag{1.197}$$

Adding (1.195), (1.196) and (1.197) we obtain

$$\rho\frac{\mathrm{D}}{\mathrm{D}t}\left(\frac{1}{2}\boldsymbol{v}^2 + I + \Phi\right) = -\nabla\cdot(p\boldsymbol{v}), \tag{1.198}$$

which, on expanding the material derivative and using the mass conservation equation, becomes

$$\frac{\partial}{\partial t}\left[\rho\left(\frac{1}{2}\boldsymbol{v}^2 + I + \Phi\right)\right] + \nabla\cdot\left[\rho\boldsymbol{v}\left(\frac{1}{2}\boldsymbol{v}^2 + I + \Phi + p/\rho\right)\right] = 0. \tag{1.199}$$

This may be written

$$\frac{\partial E}{\partial t} + \nabla\cdot\left[\boldsymbol{v}(E+p)\right] = 0, \tag{1.200}$$

where $E = \rho(\boldsymbol{v}^2/2+I+\Phi)$ is the total energy per unit volume of the fluid. This is the energy equation for an unforced, inviscid and adiabatic, compressible fluid. The energy flux term vanishes when integrated over a closed domain with rigid boundaries, implying that the total energy is conserved. However, there can be an exchange of energy between kinetic, potential and internal components. It is the divergent term, $\nabla\cdot\boldsymbol{v}$, that connects the kinetic energy equation, (1.195), and the internal energy equation, (1.196). In an incompressible fluid this term is absent, and the internal energy is divorced from the other components of energy. This consideration will be important when we consider the Boussinesq equations in Section 2.4. Note finally that the flux of energy, $\boldsymbol{F}_E = \boldsymbol{v}(E+p)$ is not equal to the velocity times the energy; rather, energy is also transferred by pressure. We may write the energy flux as

$$\boldsymbol{F}_E = \rho\boldsymbol{v}\left(\frac{\boldsymbol{v}^2}{2} + \Phi + h\right), \tag{1.201}$$

where $h = I + p/\rho$ is the enthalpy. That is, the local rate of change of energy is effected by the fluxes of kinetic and potential energy and *enthalpy*, not internal energy.

Bernoulli's theorem

The quantity

$$B = \left(E + \frac{p}{\rho}\right) = \left(\frac{1}{2}\boldsymbol{v}^2 + I + \Phi + p\alpha\right) = \left(\frac{1}{2}\boldsymbol{v}^2 + h + \Phi\right), \tag{1.202}$$

is the general form of the Bernoulli function, equal to the sum of the kinetic energy, the potential energy and the enthalpy. Equation (1.200) may be written as

$$\frac{\partial E}{\partial t} + \nabla\cdot(\rho\boldsymbol{v}B) = 0. \tag{1.203}$$

This equation may also be written $\partial(\rho B)/\partial t + \nabla\cdot(\rho\boldsymbol{v}B) = \partial p/\partial t$, so obviously the Bernoulli function itself is not conserved, even for adiabatic flow. For steady flow $\nabla\cdot(\rho\boldsymbol{v}) = 0$, and the $\partial/\partial t$ terms vanish so that (1.203) may be written $\boldsymbol{v}\cdot\nabla B = 0$, or even $\mathrm{D}B/\mathrm{D}t = 0$. The Bernoulli function is then a constant along streamlines, a result commonly known as Bernoulli's theorem.[13] For adiabatic flow at constant composition we also have $\mathrm{D}\theta/\mathrm{D}t = 0$. Thus, steady flow is both along surfaces of constant θ and along surfaces of constant B, and the vector

$$\boldsymbol{l} = \nabla\theta \times \nabla B \tag{1.204}$$

is parallel to streamlines. A related result for unsteady flow is given in Section 4.8.

Viscous effects

We might expect that viscosity will always act to reduce the kinetic energy of a flow, and we will demonstrate this for a constant density fluid. Retaining the viscous term in (1.187), the energy equation becomes

$$\frac{\mathrm{d}\widehat{E}}{\mathrm{d}t} \equiv \frac{\mathrm{d}}{\mathrm{d}t}\int_V E\,\mathrm{d}V = \mu\int_V \boldsymbol{v}\cdot\nabla^2\boldsymbol{v}\,\mathrm{d}V. \tag{1.205}$$

The right-hand side is negative definite. To see this we use the vector identity

$$\nabla\times(\nabla\times\boldsymbol{v}) = \nabla(\nabla\cdot\boldsymbol{v}) - \nabla^2\boldsymbol{v}, \tag{1.206}$$

and because $\nabla\cdot\boldsymbol{v} = 0$ we have $\nabla^2\boldsymbol{v} = -\nabla\times\boldsymbol{\omega}$, where $\boldsymbol{\omega} \equiv \nabla\times\boldsymbol{v}$. Thus,

$$\frac{\mathrm{d}\widehat{E}}{\mathrm{d}t} = -\mu\int_V \boldsymbol{v}\cdot(\nabla\times\boldsymbol{\omega})\,\mathrm{d}V = -\mu\int_V \boldsymbol{\omega}\cdot(\nabla\times\boldsymbol{v})\,\mathrm{d}V = -\mu\int_V \boldsymbol{\omega}^2\,\mathrm{d}V, \tag{1.207}$$

after integrating by parts, providing $\boldsymbol{v}\times\boldsymbol{\omega}$ vanishes at the boundary. Thus, viscosity acts to extract kinetic energy from the flow. The loss of kinetic energy reappears as an irreversible warming of the fluid (called 'Joule heating'), and the total energy of the fluid is conserved, but this effect plays no role in a constant density fluid. The heating is normally locally small, at least in the Earth's ocean and atmosphere, but it is needed to preserve total energy.

1.10.3 Enthalpy, Static Energy and Energy Flux

We saw above that it is the quantity

$$B = \text{kinetic energy} + \text{potential energy} + \text{enthalpy}, \tag{1.208}$$

that, when multiplied by $\rho\boldsymbol{v}$, fluxes the energy. However, the conserved energy contains only the kinetic and potential energies plus the internal energy. The difference between B and energy is the fluid dynamical analogue of the thermodynamical principle that it is the flux of *enthalpy* that changes the energy of a system, because this accounts for work done by the pressure.

For an example, consider an ideal gas in a uniform gravitational field for which $I = c_v T$ and $p\alpha = RT$ so that $B = c_v T + RT + \mathrm{KE} + gz = c_p T + \mathrm{KE} + gz$. The sum of the enthalpy and the potential energy,

$$h_d^* \equiv c_p T + gz, \tag{1.209}$$

is known as the *dry static energy*. Dry static energy is not an integral conserved quantity nor is it a measure of the total energy itself. To see the importance of enthalpy fluxes, consider an adiabatic rearrangement of a fluid with $\mathrm{d}\eta = 0$. From (1.87) we then have $\mathrm{d}I = -p\,\mathrm{d}\alpha$, meaning that the internal energy changes because of work done. At the same time, from (1.72) with $\mathrm{d}\eta = 0$ we also have $\mathrm{d}h = \alpha\,\mathrm{d}p$. For a fluid in hydrostatic balance, (1.51), we have that $\mathrm{d}p = -g\,\mathrm{d}z/\alpha$ and thus

$$\mathrm{d}(h + gz) = 0. \tag{1.210}$$

The change in potential energy of a parcel arises from a change in the enthalpy, not the internal energy, because of the work that must be done by the pressure force to move the parcel.

The form of the conservation law (1.210) is different from the total energy conservation previously derived. Equation (1.200) is an *integral* conservation law, whereas (1.210) is a *parcel* conservation law, which we might write in fluid dynamical form as

$$\frac{\mathrm{D}}{\mathrm{D}t}(h + gz) = 0. \tag{1.211}$$

Equation (1.211) is a form of the thermodynamic equation for adiabatic changes to an ideal gas in hydrostatic balance, similar to (1.126) or (1.141). The equivalence arises because changes in h_d^* and θ are related by $\mathrm{d}h_d^* = c_p\,\mathrm{d}T + g\,\mathrm{d}z = c_p(T/\theta)\,\mathrm{d}\theta$, as in (1.127).

Potential enthalpy (see Section 1.7.3 for an oceanographic discussion) is the enthalpy that a parcel would have if adiabatically moved to a reference pressure, and since the enthalpy for an ideal gas is c_pT, the potential enthalpy is just $c_p\theta$. The dry static energy, $c_pT + gz$, is the enthalpy that a parcel at a height z and temperature T would have if moved adiabatically and hydrostatically to $z = 0$, and is thus a special form of potential enthalpy (and sometimes called 'generalized enthalpy'). Energy and enthalpy both play major roles in the chapters ahead, but let's round off this chapter with an introduction to scaling in a pure fluid-dynamical setting.

1.11 AN INTRODUCTION TO NONDIMENSIONALIZATION AND SCALING

The units we use to measure length, velocity and so on are irrelevant to the dynamics and it is useful to express the equations of motion in terms of 'nondimensional' variables, by which we mean expressing every variable as the ratio of its value to some reference value. We choose the reference as a natural one for a given flow so that, as far as possible, the nondimensional variables are order-unity quantities, and doing this is called *scaling the equations.* There is no reference that is universally appropriate, and much of the art of fluid dynamics lies in choosing sensible scaling factors for the problem at hand. We introduce the methodology here with a simple example.

1.11.1 The Reynolds Number

Consider the constant-density momentum equation in Cartesian coordinates. If a typical velocity is U, a typical length is L, a typical time scale is T, and a typical value of the pressure deviation is Φ, then the approximate sizes of the various terms in the momentum equation are given by

$$\frac{\partial \boldsymbol{v}}{\partial t} + (\boldsymbol{v}\cdot\nabla)\boldsymbol{v} = -\nabla\phi + \nu\nabla^2\boldsymbol{v}, \tag{1.212a}$$

$$\frac{U}{T} \quad \frac{U^2}{L} \sim \frac{\Phi}{L} \quad \nu\frac{U}{L^2}. \tag{1.212b}$$

The ratio of the inertial terms to the viscous terms is $(U^2/L)/(\nu U/L^2) = UL/\nu$, and this is the *Reynolds number.*[14] More formally, we can nondimensionalize the momentum equation by writing

$$\widehat{\boldsymbol{v}} = \frac{\boldsymbol{v}}{U}, \qquad \widehat{\boldsymbol{x}} = \frac{\boldsymbol{x}}{L}, \qquad \widehat{t} = \frac{t}{T}, \qquad \widehat{\phi} = \frac{\phi}{\Phi}, \tag{1.213}$$

where the terms with hats on are *nondimensional* values of the variables and the capitalized quantities are known as *scaling values*, and these are the approximate magnitudes of the variables. We now choose the scaling values so that the nondimensional variables are of order unity, or $\widehat{u} = \mathcal{O}(1)$. Thus, for example, we choose U so that $u = \mathcal{O}(U)$, where this notation should be taken to mean that the magnitude of the variable u is of order U, or that $u \sim U$, and we say that 'u scales like U'.

Because there are no external forces in this problem, appropriate scaling values for time and pressure are

$$T = \frac{L}{U}, \qquad \Phi = U^2. \tag{1.214}$$

Substituting (1.213) and (1.214) into the momentum equation gives

$$\frac{U^2}{L}\left[\frac{\partial \hat{\boldsymbol{v}}}{\partial \hat{t}} + (\hat{\boldsymbol{v}}\cdot\nabla)\hat{\boldsymbol{v}}\right] = -\frac{U^2}{L}\nabla\hat{\phi} + \frac{\nu U}{L^2}\nabla^2\hat{\boldsymbol{v}}, \tag{1.215}$$

where we use the convention that when ∇ operates on a nondimensional variable it is a nondimensional operator. Equation (1.215) then simplifies to

$$\frac{\partial \hat{\boldsymbol{v}}}{\partial \hat{t}} + (\hat{\boldsymbol{v}}\cdot\nabla)\hat{\boldsymbol{v}} = -\nabla\hat{\phi} + \frac{1}{Re}\nabla^2\hat{\boldsymbol{v}}, \qquad \text{where} \qquad Re \equiv \frac{UL}{\nu}. \tag{1.216a,b}$$

The parameter Re is, as before, the Reynolds number. If we have chosen our length and velocity scales sensibly — that is, if we have scaled them properly — each variable in (1.216a) is order unity, with the viscous term being multiplied by $1/Re$. There are two important conclusions:

(i) The ratio of the importance of the inertial terms to the viscous terms is given by the Reynolds number, defined above. In the absence of other forces, such as those due to gravity and rotation, the Reynolds number is the only nondimensional parameter explicitly appearing in the momentum equation. Hence its value, along with the boundary conditions and geometry, controls the behaviour of the system.

(ii) More generally, by scaling the equations of motion appropriately the parameters determining the behaviour of the system become explicit. *Scaling the equations is intelligent nondimensionalization.*

Nondimensionalizing the equations does not, however, absolve the investigator from the responsibility of producing dimensionally correct equations. One should regard 'nondimensional' equations as dimensional equations in units appropriate for the problem at hand.

APPENDIX A: THERMODYNAMICS OF AN IDEAL GAS FROM THE GIBBS FUNCTION

All the thermodynamic quantities of interest for a simple ideal gas, sometimes called a perfect gas, may be derived in a straightforward way from the fundamental equation of state. (An analogous treatment for moist air is given in Appendix A of Chapter 18.) To show this we begin with the specific Gibbs function, $g = I - T\eta + p\alpha$, which for an ideal gas is given by

$$g = \mathcal{C}_p T(1 - \ln T) + \mathcal{R}T\ln p + BT + C, \tag{1.217}$$

where $\mathcal{C}_p$, $\mathcal{R}$, B and C are constants (and our notation anticipates what $\mathcal{C}_p$ and $\mathcal{R}$ really are). The procedure below is especially useful when the Gibbs function is more complex than (1.217), with the main difficulty then lying in obtaining the Gibbs function in the first instance.

Density and the thermal equation of state

From the fundamental relation involving the Gibbs function, (1.78a), specific volume is given by

$$\alpha = \left(\frac{\partial g}{\partial p}\right)_T = \frac{\mathcal{R}T}{p}, \qquad \text{or} \qquad p = \rho\mathcal{R}T. \tag{1.218}$$

Thus, R, the gas constant used elsewhere in this chapter, is equal to $\mathcal{R}$.

Entropy

$$\eta = -\left(\frac{\partial g}{\partial T}\right)_p = \mathcal{C}_p \ln T - \mathcal{R} \ln p - B \tag{1.219}$$

Using (1.218) and (1.221) in (1.219), the entropy may be expressed in terms of I and α, giving $\eta = (\mathcal{C}_p - R) \ln I - \mathcal{R} \ln \rho + \text{constant}$, as in (1.111).

The internal energy

Using (1.218) and (1.219) in the definition of the Gibbs function, $g = I - T\eta + p\alpha$, gives

$$I = g + T\eta - p\alpha = g - T\left(\frac{\partial g}{\partial T}\right)_p - p\left(\frac{\partial g}{\partial p}\right)_T, \tag{1.220}$$

and so

$$I = (\mathcal{C}_p - \mathcal{R})T + C = \mathcal{C}_v T + C, \tag{1.221}$$

where $\mathcal{C}_v \equiv \mathcal{C}_p - \mathcal{R}$. Equation (1.221) also suggests we may sensibly take $C = 0$, but no physical result in classical mechanics depends on this choice.

Heat capacities

Let c_p be the heat capacity at constant pressure. We have

$$c_p \equiv T\left(\frac{\partial \eta}{\partial T}\right)_p = \mathcal{C}_p. \tag{1.222}$$

To obtain the heat capacity at constant volume, c_v, first rewrite the entropy using the thermal equation of state as

$$\eta = (\mathcal{C}_p \ln T - \mathcal{R} \ln T + \mathcal{R} \ln \alpha) + \text{constant}. \tag{1.223}$$

We then have

$$c_v \equiv T\left(\frac{\partial \eta}{\partial T}\right)_\alpha = \mathcal{C}_p - \mathcal{R}. \tag{1.224}$$

Adiabatic lapse rate

The adiabatic lapse rate, Γ_p, is the rate of change of temperature with pressure at constant entropy. Thus

$$\Gamma_p \equiv \left(\frac{\partial T}{\partial p}\right)_\eta, \tag{1.225}$$

Now, from (1.138) we have

$$\left(\frac{\partial T}{\partial p}\right)_\eta = -\frac{T}{\mathcal{C}_p}\left(\frac{\partial \eta}{\partial p}\right)_T, \tag{1.226}$$

and using (1.219) to evaluate the right-hand side gives

$$\Gamma_p = -\frac{T}{\mathcal{C}_p}\left(\frac{\partial \eta}{\partial p}\right)_T = \frac{T}{\mathcal{C}_p}\frac{\mathcal{R}}{p} = \frac{1}{\mathcal{C}_p \rho}, \tag{1.227}$$

using the thermal equation of state. It is common to write this expression in terms of a rate of change of temperature with respect to height by using the hydrostatic approximation, $\mathrm{d}p = -\rho g\, \mathrm{d}z$, whence

$$\Gamma_z \equiv -\left(\frac{\partial T}{\partial z}\right)_\eta = \frac{g}{\mathcal{C}_p}. \tag{1.228}$$

The physical significance of this quantity is explored in Chapter 2.

Potential temperature

As in (1.136), potential temperature, θ, satisfies

$$\eta(T, p) = \eta(\theta, p_R). \tag{1.229}$$

That is to say, θ is the temperature that a parcel will have if moved at constant entropy from a pressure p to a reference pressure p_R. Using (1.219) gives

$$\mathcal{C}_p \ln T - \mathcal{R} \ln p = \mathcal{C}_p \ln \theta - \mathcal{R} \ln p_R, \qquad \text{or} \qquad \ln(T/\theta)^{\mathcal{C}_p} = \ln(p/p_R)^{\mathcal{R}}. \tag{1.230}$$

Re-arranging gives

$$\theta = T\left(\frac{p_R}{p}\right)^{\mathcal{R}/\mathcal{C}_p}. \tag{1.231}$$

We can derive the same result by noting that, by definition, potential temperature satisfies

$$\theta \equiv T(p_R) = T(p) + \int_p^{p_R} \left(\frac{\partial T}{\partial p'}\right)_\eta \mathrm{d}p' = T(p) + \int_p^{p_R} \frac{\mathcal{R}T}{\mathcal{C}_p p'} \mathrm{d}p', \tag{1.232}$$

where the rightmost expression uses (1.227). It is easy to verify that the solution to this integral equation is $T = \theta(p/p_R)^{\mathcal{R}/\mathcal{C}_p}$, although solving the equation *ab initio* is a little more difficult. Finally, in an ideal gas potential temperature can be related to entropy using (1.219) and (1.231), giving $\eta = \mathcal{C}_p \ln \theta + \text{constant}$.

Enthalpy and potential enthalpy

Enthalpy is related to the Gibbs function by

$$h = g + T\eta = g - T\left(\frac{\partial g}{\partial T}\right)_p = \mathcal{C}_p T. \tag{1.233}$$

Potential enthalpy, h^0, is the enthalpy that a parcel would have if moved adiabatically to a reference pressure. It therefore satisfies an equation similar to (1.229), namely $\eta(h, p) = \eta(h^0, p_R)$. Since $h = \mathcal{C}_p T$ we immediately obtain

$$h^0 = \mathcal{C}_p \theta. \tag{1.234}$$

That is, the potential enthalpy is equal to the potential temperature times the heat capacity at constant pressure.

APPENDIX B: THE FIRST LAW OF THERMODYNAMICS FOR FLUIDS

In its usual form the first law states that changes in the internal energy of a body are equal to the sum of heat supplied, the work done, and changes due to composition (the chemical work), and in a reversible process the entropy of a body then changes according to the heat supplied. However, the heating is not known a priori and the first law may be regarded as a definition of heating by way of energy conservation; heating should then be considered a derived quantity, as we noted in our discussion of (1.61) and (1.62). The real problem is to determine what the heating actually is, and this is not wholly trivial for fluids that also have mechanical energy and viscosity, as well as thermal and compositional diffusion. The first law is not a statement of total energy conservation, since it does not involve kinetic energy.

To obtain an unambiguous prescription for the heating — and hence a useful thermodynamic equation — we *begin* with total energy conservation and work backwards to obtain the heating, and since energy conservation is the more fundamental physical law this procedure is natural.[15]

Effectively, we subtract off the evolution of mechanical energy from an equation for total energy conservation, and the remainder is the thermodynamic equation. Equivalently (and this is how we proceed below) we demand consistency between total energy conservation and an energy equation derived in a forward fashion using a thermodynamic equation representing the first law but with the heating unspecified, and thereby deduce an explicit expression for that heating. Underlying this procedure is the fundamental notion that the work done and the generation of mechanical energy are less ambiguous than heating because they can be measured and/or arise through well-defined forces in the momentum equation. We will assume that energy and energy fluxes are knowable or calculable, but we will not discuss the questions of what energy fundamentally is or why it is conserved.

For reference, we first write down the fundamental thermodynamic relation, (1.85), and its inputs,

$$\frac{DI}{Dt} + p\frac{D\alpha}{Dt} = T\frac{D\eta}{Dt} + \mu\frac{DS}{Dt} = \frac{Dh}{Dt} - \alpha\frac{Dp}{Dt} = \dot{Q}_E, \tag{1.235}$$

where $\dot{Q}_E = \dot{Q} + \mu\dot{S}$ accounts for the total energy input to a fluid parcel from both heating, $\dot{Q}$, and compositional changes, $\mu\dot{S}$. The inclusion of the term $\dot{Q}_E$ connects the above equation to the first law, but the first law is not useful until we know what the heating is.

B.1 Single Component Fluid

Consider the energetics of a fluid as in Section 1.10.2, with two additional effects: the fluid is viscous, and there is an additional energy source in the fluid, for example radiation or thermal conduction, that does not appear in the momentum equation. We write the momentum equation, (1.194), in Cartesian tensor notation (where repeated indices are summed) as

$$\rho\frac{Dv_i}{Dt} = -\frac{\partial p}{\partial x_i} - \rho\frac{\partial \Phi}{\partial x_i} + \mu_v\frac{\partial^2 v_i}{\partial x_j \partial x_j}, \tag{1.236}$$

where μ_v is the coefficient of viscosity and Φ is a time-independent external potential field, such as gz. The form of the viscous term is exact only for incompressible fluids with constant viscosity but it is usually an excellent approximation in the atmosphere and ocean, because the Mach number is small and the scale on which dissipation occurs is very much smaller than the density scale height. (In any case those restrictions can be relaxed.) We also write the entropy equation as

$$T\frac{D\eta}{Dt} = \dot{Q}, \tag{1.237}$$

where $\dot{Q}$ is the heating, whose form we do not yet know. In a single component fluid we can use the fundamental thermodynamic relation, (1.235), to write (1.237) as an internal energy equation,

$$\rho\frac{DI}{Dt} + p\nabla\cdot\boldsymbol{v} = \rho\dot{Q}, \tag{1.238}$$

also having used $D\alpha/Dt = \alpha\nabla\cdot\boldsymbol{v}$.

We obtain a kinetic energy equation by multiplying (1.236) by v_i to give

$$\frac{1}{2}\rho\frac{Dv_i^2}{Dt} = -\partial_i(pv_i) + p\partial_i v_i - \rho v_i\partial_i\Phi + \mu_v v_i\partial_j(\partial_j v_i), \tag{1.239}$$

where $\partial_i \equiv \partial/\partial x_i$. The viscous term may be written as

$$\mu_v v_i\partial_j(\partial_j v_i) = \mu_v\left[\partial_j(v_i\partial_j v_i) - (\partial_j v_i)^2\right]. \tag{1.240}$$

The first term on the right-hand side is the divergence of a flux, and so is energy conserving, and the second term is negative definite, representing kinetic energy dissipation.

We now proceed, just as in Section 1.10.2, to obtain a total energy equation from (1.238) and (1.239) and the result is

$$\frac{\partial}{\partial t}\left[\rho\left(\frac{1}{2}v_i^2 + I + \Phi\right)\right] + \partial_i\left[\rho v_i\left(\frac{1}{2}v_j^2 + I + \Phi + p/\rho\right)\right] = \mu_v\left[\partial_j(v_i\partial_j v_i) - (\partial_j v_i)^2\right] + \rho\dot{Q}. \quad (1.241)$$

Now, the general form of the energy conservation law for a fluid takes the form

$$\frac{\partial}{\partial t}\left[\rho\left(\frac{1}{2}v_i^2 + I + \Phi\right)\right] + \partial_i\left[\rho v_i\left(\frac{1}{2}v_j^2 + I + \Phi + p/\rho\right)\right] = \partial_j\left[\mu_v v_i\partial_j v_i + F_{Ej}\right], \quad (1.242)$$

where F_{Ej} (or $\boldsymbol{F}_E$, and $\partial_j F_{Ej} = \nabla\cdot\boldsymbol{F}_E$) is the total energy flux due to radiation, conduction and any other effects — the important point being that the right-hand side of (1.242) must be the divergence of a flux in order to guarantee energy conservation. We have not used the first law to obtain (1.242), just the fundamental thermodynamic relation, the momentum equation and energy conservation. The above two equations are consistent only if the heating term has the form $\rho\dot{Q} = \partial_j F_{Ej} + \mu_v(\partial_j v_i)^2$, and the internal energy and entropy equations are then

$$\rho\frac{DI}{Dt} + p\nabla\cdot\boldsymbol{v} = \nabla\cdot\boldsymbol{F}_E + \mu_v(\partial_j v_i)^2, \qquad \rho T\frac{D\eta}{Dt} = \nabla\cdot\boldsymbol{F}_E + \mu_v(\partial_j v_i)^2. \quad (1.243a,b)$$

Evidently, the 'heating' of a fluid is given by the sum of the energy fluxes and a positive definite term due to viscous dissipation. Either of the above equivalent equations may be considered to be statements of the first law with explicit expressions for energy input and heating, and either of them then provides a useful predictive thermodynamic equation for a fluid.

B.2 Multi-Component Fluids

Consider now a two component fluid, such as dry air and water vapour or water and salinity. We refer to the second component as concentration and we assume it obeys

$$\rho\frac{DS}{Dt} = \nabla\cdot\boldsymbol{F}_S. \quad (1.244)$$

The only difference from the previous derivation is that, using the fundamental relation, the internal energy equation is now

$$\rho\frac{DI}{Dt} + p\nabla\cdot\boldsymbol{v} = \dot{Q} + \mu\nabla\cdot\boldsymbol{F}_S, \quad (1.245)$$

where μ is the chemical potential and the second term on the right-hand side accounts for the effects of concentration fluxes on internal energy (i.e., the chemical work). Proceeding to calculate the energy equation as before, we find that (1.241) then has an additional term $\mu\nabla\cdot\boldsymbol{F}_S$ on the right-hand side. This is consistent with (1.242) only if the internal energy and entropy equations obey

$$\rho\frac{DI}{Dt} + p\nabla\cdot\boldsymbol{v} = \nabla\cdot\boldsymbol{F}_E + \mu_v(\partial_j v_i)^2, \qquad \rho T\frac{D\eta}{Dt} = \nabla\cdot\boldsymbol{F}_E - \mu\nabla\cdot\boldsymbol{F}_S + \mu_v(\partial_j v_i)^2, \quad (1.246a,b)$$

where the energy flux, $\boldsymbol{F}_E$, now includes the effects of any fluxes of composition. Again, either of the above two equations is a statement of the first law, and the right-hand side of (1.246b) is the heating. Additional terms may appear on the right-hand sides of the above equations if there are additional source or sink terms, for example a source of concentration.

Equation (1.246) differs from the single component case by the addition of a concentration flux. However, in a two component fluid the diffusion of temperature and of concentration are affected by gradients of both temperature and concentration, so that the heat flux itself differs from the single component case. Thermodynamics provides constraints on these fluxes, but the reader must look elsewhere to learn about them.[16] Finally, in both oceanography and meteorology the viscous heating term is small, at least on Earth, but it must be included if energy balance is desired — for example if incoming solar radiation is to balance outgoing infrared radiation.

Notes

1 Joseph-Louis Lagrange (1736–1813) was a Franco-Italian, born and raised in Turin who then lived and worked mainly in Germany and France. He made notable contributions in analysis, number theory and mechanics and was recognized as one of the greatest mathematicians of the eighteenth century. He laid the foundations of the calculus of variations (to wit, the 'Lagrange multiplier') and first formulated the principle of least action, and his treatise *Mécanique Analytique* (1788) provides a unified analytic framework (it contains no diagrams, a feature emulated in Whittaker's *Treatise on Analytical Dynamics*, 1927) for all Newtonian mechanics.

Leonard Euler (1707–1783), a Swiss mathematician who lived and worked for extended periods in Berlin and St. Petersburg, made important contributions in many areas of mathematics and mechanics, including the analytical treatment of algebra, the theory of equations, calculus, number theory and classical mechanics. He was the first to establish the form of the equations of motion of fluid mechanics, writing down both the field description of fluids *and* what we now call the material or advective derivative.

Truesdell (1954) points out that 'Eulerian' and 'Lagrangian', especially the latter, are inappropriate eponyms. The Eulerian description was introduced by d'Alembert in 1749 and generalized by Euler in 1752, and the so-called Lagrangian description was introduced by Euler in 1759. (It is sometimes said that advances in mathematics are named after the next person to discover them after Euler — the Coriolis effect is another example.) The modern confusion evidently stems from a monograph by Dirichlet in 1860 that credits Euler in 1757 and Lagrange in 1788 for the respective methods.

Clifford Truesdell (1919–2000) was a remarkable figure himself, known both for his own contributions to many areas of continuum mechanics and for his scholarly investigations on the history of mathematics and science. He also had a trenchant and at times pungent writing style. Ball & James (2002) provide a biography.

2 Salinity is a mass fraction and thus is nondimensional, but it is commonly referred to in units of g/kg. For many years the measure of salinity of seawater that was used in oceanography was based on electrical conductivity and referred to as 'practical salinity', S_P, since this was (and still is) more easily measured. In thermodynamical calculations practical salinity is now largely dropped in favour of the true salinity, generally referred to as absolute salinity and denoted S_A. Differences between practical and absolute salinity are small but not negligible (Millero *et al.* 2008, IOC *et al.* 2010).

3 See also de Szoeke (2004). Nycander & Roquet (2015) and Roquet *et al.* (2015) show that this equation of state can, in fact, be used to give a quantitatively accurate simulation of the ocean.

4 The use of inexact differentials in thermodynamics is questionable for they do not have a straightforward mathematical foundation. Their use can be avoided and, were this a rigorous treatise on thermodynamics, probably should be avoided, but here they are useful artifacts. Reif (1965) and Callen (1985) both make use of them but Truesdell (1969) is particularly scathing on the matter.

5 It is said that the early students of ideas related to entropy were unusually prone to suicide, Ludwig Boltzmann being a tragic example. Thankfully there are many counter-examples, such as William Thompson (Lord Kelvin). He did foundational work early in his career on thermodynamics and among other achievements put forward a formulation of the second law. Neither this nor his much less successful later work seems to have caused him too much distress, and he lived for 83 years. Perhaps Truesdell (1969) gets it right when he says that entropy gives 'intense headaches to those who have studied thermodynamics'.

6 Because the word barotropic has other meanings — sometimes it is just taken to mean the vertical average — it might be better to always refer to fluids for which density is a function only of pressure as homentropic. Unfortunately the current usage is deeply ingrained and to insist on homentropic would be tilting at windmills.

7 Claude-Louis-Marie-Henri Navier (1785–1836) was a French civil engineer, professor at the École Polytechnique and later at the École des Ponts et Chaussée. He was an expert in road and bridge building (he developed the theory of suspension bridges) and, relatedly, made lasting theoretical contributions to the theory of elasticity, being the first to publish a set of general equations for the dynamics of an elastic solid. In fluid mechanics, he laid down the now-called *Navier–Stokes equations*, including the viscous terms, in 1822.

George Gabriel Stokes (1819–1903). Irish born (in Skreen, County Sligo), he was a professor of mathematics at Cambridge from 1849 until his retirement. As well as having a role in the development of fluid mechanics, especially through his considerations of viscous effects, Stokes worked on the dynamics of elasticity, fluorescence, the wave theory of light, and was (perhaps rather ill-advisedly in hindsight) a proponent of the idea of an ether permeating all space.

8 Potential temperature was known to William Thomson in 1857.

9 Jackett & McDougall (1997), extending McDougall (1987). A similar quantity was described by Eden & Willebrand (1999). de Szoeke (2000), Nycander (2011) and Tailleux (2016) provide more discussion.

10 Building from de Szoeke (2004) and with input from W. R. Young. See also Fofonoff (1959) and Warren (2006) for some historical background. A very accurate semi-empirical formula for the Gibbs function is given by Feistel (2008). Using this as a basis, seawater equations of state are now available in the form of the TEOS-10 standard (IOC *et al.* 2010) and from Roquet *et al.* (2015), and these fit laboratory measurements close to the accuracy of the measurements themselves.

11 Potential enthalpy was introduced to oceanography by McDougall (2003) and its use is advocated in IOC *et al.* (2010). The advantages and disadvantages of various thermodynamic variables, including entropy and potential enthalpy, are discussed there and in Graham & McDougall (2013). Useful discussion is also to be found in Warren (1999), Young (2010) and Nycander (2011). I am very grateful to T. McDougall for discussions on these and other thermodynamic matters.

12 Referring to the 'heat content' of a fluid borders on dangerous language, because heat itself is a type of energy transfer, like work, and not a state variable. On the other hand, heat content has an intuitive appeal and, provided it is properly understood, conveys a useful meaning in oceanography, since the compression work done on the ocean is small (as water is almost incompressible) and kinetic energy is small compared to internal and potential energy.

13 Bernoulli's theorem was developed mainly by Daniel Bernoulli (1700–1782). It was based on earlier work on the conservation of energy that Daniel had done with his father, Johann Bernoulli (1667–1748), and so perhaps should be known as Bernoullis' theorem. The two men fell out when Daniel was a young man, reputedly because of Johann's jealousy of Daniel's abilities, and subsequently had a very strained relationship. The Bernoulli family produced several (at least eight) talented mathematicians over three generations in the seventeenth and eighteenth centuries, and is often regarded as the most mathematically distinguished family of all time.

14 Osborne Reynolds (1842–1912) was an Irish born (Belfast) physicist who was professor of engineering at Manchester University from 1868–1905. His early work was in electricity and magnetism, but he is now most famous for his work in hydrodynamics. The 'Reynolds number', which determines the ratio of inertial to viscous forces, and the 'Reynolds stress', which is the stress on the mean flow due to the fluctuating components, are both named after him. He was also one of the first scientists to think about the concept of group velocity.

15 See also Landau & Lifshitz (1987) and IOC *et al.* (2010). For an interesting and somewhat idiosyncratic view of the first law applied to the ocean, read Warren (2006).

16 Onsager (1931), Salmon (1998). For example, in a single component fluid, total entropy increases if the heat flux is proportional to a downgradient temperature flux.

Further Reading

General fluid dynamics

There are numerous books on hydrodynamics, an early one being

Lamb, H., 1932. *Hydrodynamics.*
Lamb's book is a classic in the field, although now too dated to make it useful as an introduction.

Two somewhat more modern references, at a fairly advanced level, are

Batchelor, G. K., 1967. *An Introduction to Fluid Dynamics.*

Landau, L. D. & Lifshitz, E. M., 1987. *Fluid Mechanics.*
These two books both contain a detailed derivation of the equations of motion, including viscous and pressure forces.

At a more elementary level we have

Kundu, P., Cohen, I. &. Dowling, D., 2015. *Fluid Mechanics.*
This book is written at the advanced undergraduate/beginning graduate level, is easier-going than Batchelor or Landau & Lifshitz, and contains material on geophysical fluid dynamics.

For the connoisseur, a more specialized treatment is

Truesdell, C., 1954. *The Kinematics of Vorticity.*
Written in Truesdell's inimitable style, this book discusses many aspects of vorticity with numerous historical references. Truesdell's books are all gems in their own way.

Thermodynamics

There are many books on thermodynamics, and two that I have found particularly useful are

Reif, F., 1965. *Fundamentals of Statistical and Thermal Physics.*

Callen, H. B., 1985. *Thermodynamics and an Introduction to Thermostatistics.*
Reif's book has become something of a classic, and Callen provides an axiomatic approach that will be an antidote for those who feel that thermodynamic reasoning is mysterious or even circular.

For the subtopic of atmospheric thermodynamics see the further reading section at the end of Chapter 14.

Geophysical fluid dynamics

Gill, A. E., 1982. *Atmosphere–Ocean Dynamics.*
A richly textured book, especially strong on equatorial dynamics and gravity wave motion.

Pedlosky, J., 1987. *Geophysical Fluid Dynamics.*
A primary reference for flow at low Rossby number. Although the book requires some effort, there is a handsome pay-off for those who study it closely.

Holton, J. R. & Hakim, G., 2012. *An Introduction to Dynamical Meteorology.*
A very well-known textbook at the undergraduate/beginning graduate level.

Salmon, R., 1998. *Lectures on Geophysical Fluid Dynamics.*
Covers the fundamentals as well as Hamiltonian fluid dynamics, geostrophic turbulence and oceanic circulation.

The heavens themselves, the planets, and this centre
Observe degree, priority, and place,
Insisture, course, proportion, season, form,
Office, and custom, in all line of order.
And therefore is the glorious planet Sol
In noble eminence enthroned and sphered.
William Shakespeare, *Troilus and Cressida*, c. 1602.

Eppur si muove. (And yet it does move.)
Galileo Galilei, apocryphal, 1633.

CHAPTER 2

Effects of Rotation and Stratification

THE ATMOSPHERE AND OCEAN are shallow layers of fluid on a sphere, 'shallow' because their thickness is much less than their horizontal extent. Their motion is strongly influenced by two effects: rotation and stratification, the latter meaning that there is a mean vertical gradient of (potential) density that is often large compared with the horizontal gradient. Here we consider how the equations of motion are affected by these effects. First, we consider some elementary effects of rotation on a fluid and derive the Coriolis and centrifugal forces, and write down the equations of motion appropriate for motion on a sphere. Then we discuss some approximations to the equations of motion that are appropriate for large-scale flow in the ocean and atmosphere, in particular the hydrostatic and geostrophic approximations, and finally we look at the possible static instability of stratified flows.

2.1 EQUATIONS OF MOTION IN A ROTATING FRAME

Newton's second law of motion, that the acceleration of a body is proportional to the imposed force divided by the body's mass, applies in so-called inertial frames of reference; that is, frames that are stationary or moving only with a constant rectilinear velocity relative to the distant galaxies. Now Earth spins round its own axis with a period of almost 24 hours (23h 56m, the difference due to Earth's rotation around the Sun) and so the surface of the Earth manifestly is not an inertial frame. Nevertheless, it is very convenient to describe the flow relative to Earth's surface (which in fact is moving at speeds of up to a few hundreds of metres per second), rather than in some inertial frame.[1] This necessitates recasting the equations into a form appropriate in a rotating frame of reference, and that is the subject of this section.

2.1.1 Rate of Change of a Vector

Consider first a vector $\boldsymbol{C}$ of constant length rotating relative to an inertial frame at a constant angular velocity $\boldsymbol{\Omega}$. Then, in a frame rotating with that same angular velocity it appears stationary and constant. If in a small interval of time δt the vector $\boldsymbol{C}$ rotates through a small angle $\delta\lambda$ then the change in $\boldsymbol{C}$, as perceived in the inertial frame, is given by (see Fig. 2.1)

$$\delta \boldsymbol{C} = |\boldsymbol{C}| \cos\vartheta \, \delta\lambda \, \boldsymbol{m}, \tag{2.1}$$

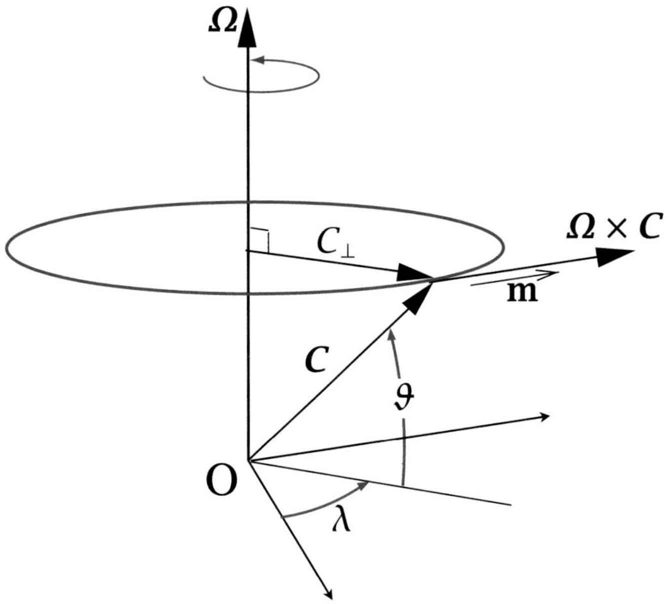

Fig. 2.1 A vector $\boldsymbol{C}$ rotating at an angular velocity $\boldsymbol{\Omega}$. It appears to be a constant vector in the rotating frame, whereas in the inertial frame it evolves according to $(\mathrm{d}\boldsymbol{C}/\mathrm{d}t)_I = \boldsymbol{\Omega} \times \boldsymbol{C}$.

where the vector $\boldsymbol{m}$ is the unit vector in the direction of change of $\boldsymbol{C}$, which is perpendicular to both $\boldsymbol{C}$ and $\boldsymbol{\Omega}$. But the rate of change of the angle λ is just, by definition, the angular velocity so that $\delta\lambda = |\boldsymbol{\Omega}|\delta t$ and

$$\delta\boldsymbol{C} = |\boldsymbol{C}||\boldsymbol{\Omega}| \sin\hat{\vartheta}\, \boldsymbol{m}\, \delta t = \boldsymbol{\Omega} \times \boldsymbol{C}\, \delta t, \tag{2.2}$$

using the definition of the vector cross product, where $\hat{\vartheta} = (\pi/2 - \vartheta)$ is the angle between $\boldsymbol{\Omega}$ and $\boldsymbol{C}$. Thus

$$\left(\frac{\mathrm{d}\boldsymbol{C}}{\mathrm{d}t}\right)_I = \boldsymbol{\Omega} \times \boldsymbol{C}, \tag{2.3}$$

where the left-hand side is the rate of change of $\boldsymbol{C}$ as perceived in the inertial frame.

Now consider a vector $\boldsymbol{B}$ that changes in the inertial frame. In a small time δt the change in $\boldsymbol{B}$ as seen in the rotating frame is related to the change seen in the inertial frame by

$$(\delta\boldsymbol{B})_I = (\delta\boldsymbol{B})_R + (\delta\boldsymbol{B})_{rot}, \tag{2.4}$$

where the terms are, respectively, the change seen in the inertial frame, the change due to the vector itself changing as measured in the rotating frame, and the change due to the rotation. Using (2.2) $(\delta\boldsymbol{B})_{rot} = \boldsymbol{\Omega} \times \boldsymbol{B}\,\delta t$, and so the rates of change of the vector $\boldsymbol{B}$ in the inertial and rotating frames are related by

$$\left(\frac{\mathrm{d}\boldsymbol{B}}{\mathrm{d}t}\right)_I = \left(\frac{\mathrm{d}\boldsymbol{B}}{\mathrm{d}t}\right)_R + \boldsymbol{\Omega} \times \boldsymbol{B}. \tag{2.5}$$

This relation applies to a vector $\boldsymbol{B}$ that, as measured at any one time, is the same in both inertial and rotating frames.

2.1.2 Velocity and Acceleration in a Rotating Frame

The velocity of a body is not measured to be the same in the inertial and rotating frames, so care must be taken when applying (2.5) to velocity. First apply (2.5) to $\boldsymbol{r}$, the position of a particle to obtain

$$\left(\frac{\mathrm{d}\boldsymbol{r}}{\mathrm{d}t}\right)_I = \left(\frac{\mathrm{d}\boldsymbol{r}}{\mathrm{d}t}\right)_R + \boldsymbol{\Omega} \times \boldsymbol{r} \tag{2.6}$$

or

$$\boldsymbol{v}_I = \boldsymbol{v}_R + \boldsymbol{\Omega} \times \boldsymbol{r}. \tag{2.7}$$

We refer to $\boldsymbol{v}_R$ and $\boldsymbol{v}_I$ as the relative and inertial velocity, respectively, and (2.7) relates the two. Apply (2.5) again, this time to the velocity $\boldsymbol{v}_R$ to give

$$\left(\frac{\mathrm{d}\boldsymbol{v}_R}{\mathrm{d}t}\right)_I = \left(\frac{\mathrm{d}\boldsymbol{v}_R}{\mathrm{d}t}\right)_R + \boldsymbol{\Omega} \times \boldsymbol{v}_R, \tag{2.8}$$

or, using (2.7)

$$\left(\frac{\mathrm{d}}{\mathrm{d}t}(\boldsymbol{v}_I - \boldsymbol{\Omega} \times \boldsymbol{r})\right)_I = \left(\frac{\mathrm{d}\boldsymbol{v}_R}{\mathrm{d}t}\right)_R + \boldsymbol{\Omega} \times \boldsymbol{v}_R, \tag{2.9}$$

or

$$\left(\frac{\mathrm{d}\boldsymbol{v}_I}{\mathrm{d}t}\right)_I = \left(\frac{\mathrm{d}\boldsymbol{v}_R}{\mathrm{d}t}\right)_R + \boldsymbol{\Omega} \times \boldsymbol{v}_R + \frac{\mathrm{d}\boldsymbol{\Omega}}{\mathrm{d}t} \times \boldsymbol{r} + \boldsymbol{\Omega} \times \left(\frac{\mathrm{d}\boldsymbol{r}}{\mathrm{d}t}\right)_I. \tag{2.10}$$

Then, noting that

$$\left(\frac{\mathrm{d}\boldsymbol{r}}{\mathrm{d}t}\right)_I = \left(\frac{\mathrm{d}\boldsymbol{r}}{\mathrm{d}t}\right)_R + \boldsymbol{\Omega} \times \boldsymbol{r} = (\boldsymbol{v}_R + \boldsymbol{\Omega} \times \boldsymbol{r}), \tag{2.11}$$

and assuming that the rate of rotation is constant, (2.10) becomes

$$\left(\frac{\mathrm{d}\boldsymbol{v}_R}{\mathrm{d}t}\right)_R = \left(\frac{\mathrm{d}\boldsymbol{v}_I}{\mathrm{d}t}\right)_I - 2\boldsymbol{\Omega} \times \boldsymbol{v}_R - \boldsymbol{\Omega} \times (\boldsymbol{\Omega} \times \boldsymbol{r}). \tag{2.12}$$

This equation may be interpreted as follows. The term on the left-hand side is the rate of change of the relative velocity as measured in the rotating frame. The first term on the right-hand side is the rate of change of the inertial velocity as measured in the inertial frame (the inertial acceleration, which is, by Newton's second law, equal to the force on a fluid parcel divided by its mass). The second and third terms on the right-hand side (including the minus signs) are the *Coriolis force* and the *centrifugal force* per unit mass. Neither of these is a true force — they may be thought of as quasi-forces (i.e., 'as if' forces); that is, when a body is observed from a rotating frame it behaves as if unseen forces are present that affect its motion. If (2.12) is written, as is common, with the terms $+2\boldsymbol{\Omega} \times \boldsymbol{v}_r$ and $+\boldsymbol{\Omega} \times (\boldsymbol{\Omega} \times \boldsymbol{r})$ on the left-hand side then these terms should be referred to as the Coriolis and centrifugal *accelerations.*[2]

Centrifugal force

If $\boldsymbol{r}_\perp$ is the perpendicular distance from the axis of rotation (see Fig. 2.1 and substitute $\boldsymbol{r}$ for $\mathbf{C}$), then, because $\boldsymbol{\Omega}$ is perpendicular to $\boldsymbol{r}_\perp$, $\boldsymbol{\Omega} \times \boldsymbol{r} = \boldsymbol{\Omega} \times \boldsymbol{r}_\perp$. Then, using the vector identity $\boldsymbol{\Omega} \times (\boldsymbol{\Omega} \times \boldsymbol{r}_\perp) = (\boldsymbol{\Omega} \cdot \boldsymbol{r}_\perp)\boldsymbol{\Omega} - (\boldsymbol{\Omega} \cdot \boldsymbol{\Omega})\boldsymbol{r}_\perp$ and noting that the first term is zero, we see that the centrifugal force per unit mass is just given by

$$\boldsymbol{F}_{ce} = -\boldsymbol{\Omega} \times (\boldsymbol{\Omega} \times \boldsymbol{r}) = \Omega^2 \boldsymbol{r}_\perp. \tag{2.13}$$

This may usefully be written as the gradient of a scalar potential,

$$\boldsymbol{F}_{ce} = -\nabla \Phi_{ce}, \tag{2.14}$$

where $\Phi_{ce} = -(\Omega^2 r_\perp^2)/2 = -(\boldsymbol{\Omega} \times \boldsymbol{r}_\perp)^2/2$.

Coriolis force

The Coriolis force per unit mass is given by

$$\boldsymbol{F}_{Co} = -2\boldsymbol{\Omega} \times \boldsymbol{v}_R. \tag{2.15}$$

It plays a central role in much of geophysical fluid dynamics and will be considered extensively later on. For now, we just note three basic properties:

(i) There is no Coriolis force on bodies that are stationary in the rotating frame.
(ii) The Coriolis force acts to deflect moving bodies at right angles to their direction of travel.
(iii) The Coriolis force does no work on a body because it is perpendicular to the velocity, and so $\boldsymbol{v}_R \cdot (\boldsymbol{\Omega} \times \boldsymbol{v}_R) = 0$.

2.1.3 Momentum Equation in a Rotating Frame

Since (2.12) simply relates the accelerations of a particle in the inertial and rotating frames, then in the rotating frame of reference the momentum equation may be written

$$\frac{\mathrm{D}\boldsymbol{v}}{\mathrm{D}t} + 2\boldsymbol{\Omega} \times \boldsymbol{v} = -\frac{1}{\rho}\nabla p - \nabla\Phi, \tag{2.16}$$

incorporating the centrifugal term into the potential, Φ. We have dropped the subscript R; henceforth, unless we need to be explicit (as in the next section), all velocities without a subscript will be considered to be relative to the rotating frame.

2.1.4 Mass and Tracer Conservation in a Rotating frame

Let φ be a scalar field that, in the inertial frame, obeys

$$\frac{\mathrm{D}\varphi}{\mathrm{D}t} + \varphi\nabla \cdot \boldsymbol{v}_I = 0. \tag{2.17}$$

Now, observers in both the rotating and inertial frame measure the same value of φ. Further, $\mathrm{D}\varphi/\mathrm{D}t$ is simply the rate of change of φ associated with a material parcel, and therefore is reference frame invariant. Thus, without further ado, we write

$$\left(\frac{\mathrm{D}\varphi}{\mathrm{D}t}\right)_R = \left(\frac{\mathrm{D}\varphi}{\mathrm{D}t}\right)_I, \tag{2.18}$$

where $(\mathrm{D}\varphi/\mathrm{D}t)_R = (\partial\varphi/\partial t)_R + \boldsymbol{v}_R \cdot \nabla\varphi$ and $(\mathrm{D}\varphi/\mathrm{D}t)_I = (\partial\varphi/\partial t)_I + \boldsymbol{v}_I \cdot \nabla\varphi$, and the local temporal derivatives $(\partial\varphi/\partial t)_R$ and $(\partial\varphi/\partial t)_I$ are evaluated at fixed locations in the rotating and inertial frames, respectively.

Further, using (2.7), we have that

$$\nabla \cdot \boldsymbol{v}_I = \nabla \cdot (\boldsymbol{v}_R + \boldsymbol{\Omega} \times \boldsymbol{r}) = \nabla \cdot \boldsymbol{v}_R, \tag{2.19}$$

since $\nabla \cdot (\boldsymbol{\Omega} \times \boldsymbol{r}) = 0$. Thus, using (2.18) and (2.19), (2.17) is equivalent to

$$\frac{\mathrm{D}\varphi}{\mathrm{D}t} + \varphi\nabla \cdot \boldsymbol{v}_R = 0, \tag{2.20}$$

where all observables are measured in the *rotating* frame. Thus, the equation for the evolution of a scalar whose measured value is the same in rotating and inertial frames is unaltered by the presence of rotation. In particular, the mass conservation equation is unaltered by the presence of rotation.

Although we have taken (2.18) as true a priori, the individual components of the material derivative differ in the rotating and inertial frames. In particular

$$\left(\frac{\partial \varphi}{\partial t}\right)_I = \left(\frac{\partial \varphi}{\partial t}\right)_R - (\boldsymbol{\Omega} \times \boldsymbol{r}) \cdot \nabla \varphi, \tag{2.21}$$

because $\boldsymbol{\Omega} \times \boldsymbol{r}$ is the velocity, in the inertial frame, of a uniformly rotating body. Similarly,

$$\boldsymbol{v}_I \cdot \nabla \varphi = (\boldsymbol{v}_R + \boldsymbol{\Omega} \times \boldsymbol{r}) \cdot \nabla \varphi. \tag{2.22}$$

Adding the last two equations reprises and confirms (2.18).

2.2 EQUATIONS OF MOTION IN SPHERICAL COORDINATES

The Earth is very nearly spherical and it might appear obvious that we should cast our equations in spherical coordinates. Although this does turn out to be true, the presence of a centrifugal force causes some complications that we should first discuss. The reader who is willing ab initio to treat the Earth as a perfect sphere and to neglect the horizontal component of the centrifugal force may skip the next section.

2.2.1 ♦ The Centrifugal Force and Spherical Coordinates

The centrifugal force is a potential force, like gravity, and so we may therefore define an 'effective gravity' equal to the sum of the true, or Newtonian, gravity and the centrifugal force. The Newtonian gravitational force is directed approximately toward the centre of the Earth, with small deviations due mainly to the Earth's oblateness. The line of action of the effective gravity will in general differ slightly from this, and therefore have a component in the 'horizontal' plane, that is the plane perpendicular to the radial direction. The magnitude of the centrifugal force is $\Omega^2 r_\perp$, and so the effective gravity is given by

$$\boldsymbol{g} \equiv \boldsymbol{g}_{\mathit{eff}} = \boldsymbol{g}_{grav} + \Omega^2 \boldsymbol{r}_\perp, \tag{2.23}$$

where $\boldsymbol{g}_{grav}$ is the Newtonian gravitational force due to the gravitational attraction of the Earth and $\boldsymbol{r}_\perp$ is normal to the rotation vector (in the direction $\boldsymbol{C}$ in Fig. 2.2), with $r_\perp = r\cos\vartheta$. Both gravity and centrifugal force are potential forces and therefore we may define the *geopotential*, Φ, such that

$$\boldsymbol{g} = -\nabla \Phi. \tag{2.24}$$

Surfaces of constant Φ are not quite spherical because $r_\perp$, and hence the centrifugal force, vary with latitude (Fig. 2.2); this has certain ramifications, as we now discuss.

The components of the centrifugal force parallel and perpendicular to the radial direction are $\Omega^2 r \cos^2\vartheta$ and $\Omega^2 r \cos\vartheta \sin\vartheta$. Newtonian gravity is much larger than either of these, and at the Earth's surface the ratio of centrifugal to gravitational terms is approximately, and no more than,

$$\alpha \approx \frac{\Omega^2 a}{g} \approx \frac{(7.27 \times 10^{-5})^2 \times 6.4 \times 10^6}{9.8} \approx 3 \times 10^{-3}. \tag{2.25}$$

(At the equator and pole the horizontal component of the centrifugal force is zero and the effective gravity is aligned with Newtonian gravity.) The angle between $\boldsymbol{g}$ and the line to the centre of the Earth is given by a similar expression and so is also small, typically around 3×10^{-3} radians. However, the horizontal component of the centrifugal force is still large compared to the Coriolis force, the ratio of their magnitudes in mid-latitudes being given by

$$\frac{\text{horizontal centrifugal force}}{\text{Coriolis force}} \approx \frac{\Omega^2 a \cos\vartheta \sin\vartheta}{2\Omega |u|} \approx \frac{\Omega a}{4|u|} \approx 10, \tag{2.26}$$

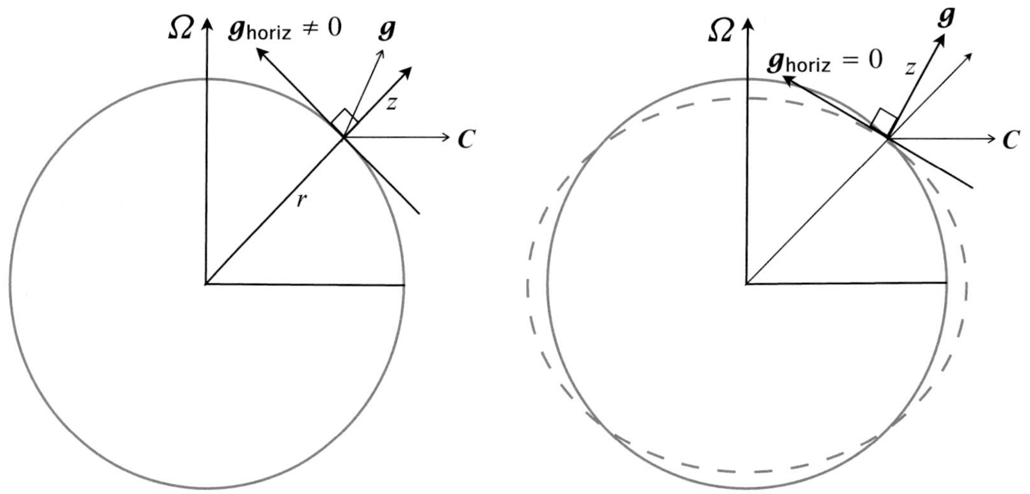

Fig. 2.2 Left: directions of forces and coordinates in true spherical geometry. g is the effective gravity (including the centrifugal force, C) and its horizontal component is evidently non-zero. Right: a modified coordinate system, in which the vertical direction is defined by the direction of g, and so the horizontal component of g is identically zero. The dashed line schematically indicates a surface of constant geopotential. The differences between the direction of g and the direction of the radial coordinate, and between the sphere and the geopotential surface, are much exaggerated and in reality are similar to the thickness of the lines themselves.

using $u = 10\,\mathrm{m\,s^{-1}}$. The centrifugal term therefore dominates over the Coriolis term, and is largely balanced by a pressure gradient force. Thus, if we adhered to true spherical coordinates, both the horizontal and radial components of the momentum equation would be dominated by a static balance between a pressure gradient and gravity or centrifugal terms. Although in principle there is nothing wrong with writing the equations this way, it obscures the dynamical balances involving the Coriolis force and pressure that determine the large-scale horizontal flow.

A way around this problem is to use the direction of the geopotential force to *define* the vertical direction, and then for all geometric purposes to regard the surfaces of constant Φ as if they were true spheres.[3] The horizontal component of effective gravity is then identically zero, and we have traded a potentially large dynamical error for a very small geometric error. In fact, over time, the Earth has developed an equatorial bulge to compensate for and neutralize the centrifugal force, so that the effective gravity does act in a direction virtually normal to the Earth's surface; that is, the surface of the Earth is an oblate spheroid of nearly constant geopotential. The geopotential Φ is then a function of the vertical coordinate alone, and for many purposes we can just take $\Phi = gz$; that is, the direction normal to geopotential surfaces, the local vertical, is, in this approximation, taken to be the direction of increasing r in spherical coordinates. It is because the oblateness is very small (the polar diameter is about 12 714 km, whereas the equatorial diameter is about 12 756 km) that using spherical coordinates is a very accurate way to map the spheroid. If the angle between effective gravity and a natural direction of the coordinate system were not small then more heroic measures would be called for.

If the solid Earth did not bulge at the equator, the *behaviour* of the atmosphere and ocean would differ significantly from that of the present system. For example, the surface of the ocean is, necessarily, very nearly a geopotential surface; if the solid Earth were exactly spherical then the ocean would perforce become much deeper at low latitudes and the ocean basins would dry out completely at high latitudes. We could still choose to use the spherical coordinate system discussed above to describe the dynamics, but the shape of the surface of the solid Earth would have to be represented by a topography, with the topographic height increasing monotonically polewards nearly everywhere.

2.2.2 Some Identities in Spherical Coordinates

The location of a point is given by the coordinates (λ, ϑ, r) where λ is the angular distance eastwards (i.e., longitude), ϑ is angular distance polewards (i.e., latitude) and r is the radial distance from the centre of the Earth — see Fig. 2.3. (In some other fields of study co-latitude is used as a spherical coordinate.) If a is the radius of the Earth, then we also define $z = r - a$. At a given location we may also define the Cartesian increments $(\delta x, \delta y, \delta z) = (r\cos\vartheta\delta\lambda, r\delta\vartheta, \delta r)$.

For a scalar quantity ϕ the material derivative in spherical coordinates is

$$\frac{D\phi}{Dt} = \frac{\partial \phi}{\partial t} + \frac{u}{r\cos\vartheta}\frac{\partial \phi}{\partial \lambda} + \frac{v}{r}\frac{\partial \phi}{\partial \vartheta} + w\frac{\partial \phi}{\partial r}, \tag{2.27}$$

where the velocity components corresponding to the coordinates (λ, ϑ, r) are

$$(u, v, w) \equiv \left(r\cos\vartheta\frac{D\lambda}{Dt}, r\frac{D\vartheta}{Dt}, \frac{Dr}{Dt} \right). \tag{2.28}$$

That is, u is the zonal velocity, v is the meridional velocity and w is the vertical velocity. If we define $(\mathbf{i}, \mathbf{j}, \mathbf{k})$ to be the unit vectors in the direction of increasing (λ, ϑ, r) then

$$\boldsymbol{v} = \mathbf{i}u + \mathbf{j}v + \mathbf{k}w. \tag{2.29}$$

Note also that $Dr/Dt = Dz/Dt$.

The divergence of a vector $\boldsymbol{B} = \mathbf{i}B^\lambda + \mathbf{j}B^\vartheta + \mathbf{k}B^r$ is

$$\nabla\cdot\boldsymbol{B} = \frac{1}{\cos\vartheta}\left[\frac{1}{r}\frac{\partial B^\lambda}{\partial \lambda} + \frac{1}{r}\frac{\partial}{\partial \vartheta}(B^\vartheta\cos\vartheta) + \frac{\cos\vartheta}{r^2}\frac{\partial}{\partial r}(r^2 B^r)\right]. \tag{2.30}$$

The vector gradient of a scalar is:

$$\nabla\phi = \mathbf{i}\frac{1}{r\cos\vartheta}\frac{\partial \phi}{\partial \lambda} + \mathbf{j}\frac{1}{r}\frac{\partial \phi}{\partial \vartheta} + \mathbf{k}\frac{\partial \phi}{\partial r}. \tag{2.31}$$

The Laplacian of a scalar is:

$$\nabla^2\phi \equiv \nabla\cdot\nabla\phi = \frac{1}{r^2\cos\vartheta}\left[\frac{1}{\cos\vartheta}\frac{\partial^2 \phi}{\partial \lambda^2} + \frac{\partial}{\partial \vartheta}\left(\cos\vartheta\frac{\partial \phi}{\partial \vartheta}\right) + \cos\vartheta\frac{\partial}{\partial r}\left(r^2\frac{\partial \phi}{\partial r}\right)\right]. \tag{2.32}$$

The curl of a vector is:

$$\operatorname{curl}\boldsymbol{B} = \nabla\times\boldsymbol{B} = \frac{1}{r^2\cos\vartheta}\begin{vmatrix} \mathbf{i}\,r\cos\vartheta & \mathbf{j}\,r & \mathbf{k} \\ \partial/\partial\lambda & \partial/\partial\vartheta & \partial/\partial r \\ B^\lambda r\cos\vartheta & B^\vartheta r & B^r \end{vmatrix}. \tag{2.33}$$

The vector Laplacian $\nabla^2\boldsymbol{B}$ (used for example when calculating viscous terms in the momentum equation) may be obtained from the vector identity:

$$\nabla^2\boldsymbol{B} = \nabla(\nabla\cdot\boldsymbol{B}) - \nabla\times(\nabla\times\boldsymbol{B}). \tag{2.34}$$

Only in Cartesian coordinates does this take the simple form:

$$\nabla^2\boldsymbol{B} = \frac{\partial^2\boldsymbol{B}}{\partial x^2} + \frac{\partial^2\boldsymbol{B}}{\partial y^2} + \frac{\partial^2\boldsymbol{B}}{\partial z^2}. \tag{2.35}$$

The expansion in spherical coordinates is of itself, to most eyes, rather uninformative.

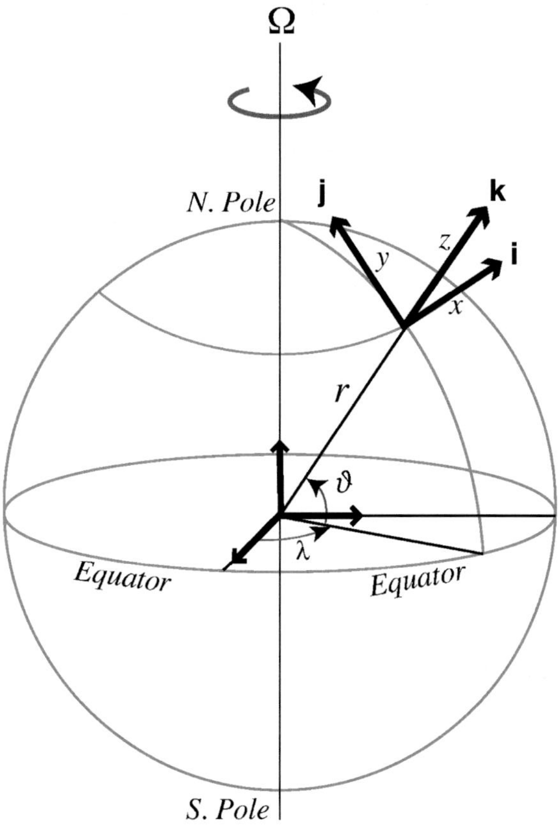

Fig. 2.3 The spherical coordinate system. The orthogonal unit vectors **i**, **j** and **k** point in the direction of increasing longitude λ, latitude ϑ, and altitude z. Locally, one may apply a Cartesian system with variables x, y and z measuring distances along **i**, **j** and **k**.

Rate of change of unit vectors

In spherical coordinates the defining unit vectors are **i**, the unit vector pointing eastwards, parallel to a line of latitude; **j** is the unit vector pointing polewards, parallel to a meridian; and **k**, the unit vector pointing radially outward. The directions of these vectors change with location, and in fact this is the case in nearly all coordinate systems, with the notable exception of the Cartesian one, and thus their material derivative is not zero. One way to evaluate this is to consider geometrically how the coordinate axes change with position. Another way, and the way that we shall proceed, is to first obtain the effective rotation rate $\boldsymbol{\Omega}_{flow}$, relative to the Earth, of a unit vector as it moves with the flow, and then apply (2.3). Specifically, let the fluid velocity be $\boldsymbol{v} = (u, v, w)$. The meridional component, v, produces a displacement $r\delta\vartheta = v\delta t$, and this gives rise to a local effective vector rotation rate around the local zonal axis of $-(v/r)\mathbf{i}$, the minus sign arising because a displacement in the direction of the north pole is produced by negative rotational displacement around the **i** axis. Similarly, the zonal component, u, produces a displacement $\delta\lambda r\cos\vartheta = u\delta t$ and so an effective rotation rate, about the Earth's rotation axis, of $u/(r\cos\vartheta)$. Now, a rotation around the Earth's rotation axis may be written as (see Fig. 2.4)

$$\boldsymbol{\Omega} = \Omega(\mathbf{j}\cos\vartheta + \mathbf{k}\sin\vartheta). \tag{2.36}$$

If the scalar rotation rate is not Ω but is $u/(r\cos\vartheta)$, then the vector rotation rate is

$$\frac{u}{r\cos\vartheta}(\mathbf{j}\cos\vartheta + \mathbf{k}\sin\vartheta) = \mathbf{j}\frac{u}{r} + \mathbf{k}\frac{u\tan\vartheta}{r}. \tag{2.37}$$

Thus, the total rotation rate of a vector that moves with the flow is

$$\boldsymbol{\Omega}_{flow} = -\mathbf{i}\frac{v}{r} + \mathbf{j}\frac{u}{r} + \mathbf{k}\frac{u\tan\vartheta}{r}. \tag{2.38}$$

Applying (2.3) to (2.38), we find

$$\frac{\mathrm{D}\mathbf{i}}{\mathrm{D}t} = \boldsymbol{\Omega}_{flow} \times \mathbf{i} = \frac{u}{r\cos\vartheta}(\mathbf{j}\sin\vartheta - \mathbf{k}\cos\vartheta), \tag{2.39a}$$

$$\frac{\mathrm{D}\mathbf{j}}{\mathrm{D}t} = \boldsymbol{\Omega}_{flow} \times \mathbf{j} = -\mathbf{i}\frac{u}{r}\tan\vartheta - \mathbf{k}\frac{v}{r}, \tag{2.39b}$$

$$\frac{\mathrm{D}\mathbf{k}}{\mathrm{D}t} = \boldsymbol{\Omega}_{flow} \times \mathbf{k} = \mathbf{i}\frac{u}{r} + \mathbf{j}\frac{v}{r}. \tag{2.39c}$$

2.2.3 Equations of Motion

Mass conservation and thermodynamic equation

The mass conservation equation, (1.36a), expanded in spherical co-ordinates, is

$$\frac{\partial\rho}{\partial t} + \frac{u}{r\cos\vartheta}\frac{\partial\rho}{\partial\lambda} + \frac{v}{r}\frac{\partial\rho}{\partial\vartheta} + w\frac{\partial\rho}{\partial r} + \frac{\rho}{r\cos\vartheta}\left[\frac{\partial u}{\partial\lambda} + \frac{\partial}{\partial\vartheta}(v\cos\vartheta) + \frac{1}{r}\frac{\partial}{\partial r}(wr^2\cos\vartheta)\right] = 0. \tag{2.40}$$

Equivalently, using the form (1.36b), this is

$$\frac{\partial\rho}{\partial t} + \frac{1}{r\cos\vartheta}\frac{\partial(u\rho)}{\partial\lambda} + \frac{1}{r\cos\vartheta}\frac{\partial}{\partial\vartheta}(v\rho\cos\vartheta) + \frac{1}{r^2}\frac{\partial}{\partial r}(r^2 w\rho) = 0. \tag{2.41}$$

The thermodynamic equation, (1.108), is a tracer advection equation. Thus, using (2.27), its (adiabatic) spherical coordinate form is

$$\frac{\mathrm{D}\theta}{\mathrm{D}t} = \frac{\partial\theta}{\partial t} + \frac{u}{r\cos\vartheta}\frac{\partial\theta}{\partial\lambda} + \frac{v}{r}\frac{\partial\theta}{\partial\vartheta} + w\frac{\partial\theta}{\partial r} = 0, \tag{2.42}$$

and similarly for tracers such as water vapour or salt.

Momentum equation

Recall that the inviscid momentum equation is:

$$\frac{\mathrm{D}\boldsymbol{v}}{\mathrm{D}t} + 2\boldsymbol{\Omega} \times \boldsymbol{v} = -\frac{1}{\rho}\nabla p - \nabla\Phi, \tag{2.43}$$

where Φ is the geopotential. In spherical coordinates the directions of the coordinate axes change with position and so the component expansion of (2.43) is

$$\frac{\mathrm{D}\boldsymbol{v}}{\mathrm{D}t} = \frac{\mathrm{D}u}{\mathrm{D}t}\mathbf{i} + \frac{\mathrm{D}v}{\mathrm{D}t}\mathbf{j} + \frac{\mathrm{D}w}{\mathrm{D}t}\mathbf{k} + u\frac{\mathrm{D}\mathbf{i}}{\mathrm{D}t} + v\frac{\mathrm{D}\mathbf{j}}{\mathrm{D}t} + w\frac{\mathrm{D}\mathbf{k}}{\mathrm{D}t} \tag{2.44a}$$

$$= \frac{\mathrm{D}u}{\mathrm{D}t}\mathbf{i} + \frac{\mathrm{D}v}{\mathrm{D}t}\mathbf{j} + \frac{\mathrm{D}w}{\mathrm{D}t}\mathbf{k} + \boldsymbol{\Omega}_{flow} \times \boldsymbol{v}, \tag{2.44b}$$

using (2.39). Using either (2.44a) and the expressions for the rates of change of the unit vectors given in (2.39), or (2.44b) and the expression for $\boldsymbol{\Omega}_{flow}$ given in (2.38), (2.44) becomes

$$\frac{\mathrm{D}\boldsymbol{v}}{\mathrm{D}t} = \mathbf{i}\left(\frac{\mathrm{D}u}{\mathrm{D}t} - \frac{uv\tan\vartheta}{r} + \frac{uw}{r}\right) + \mathbf{j}\left(\frac{\mathrm{D}v}{\mathrm{D}t} + \frac{u^2\tan\vartheta}{r} + \frac{vw}{r}\right) + \mathbf{k}\left(\frac{\mathrm{D}w}{\mathrm{D}t} - \frac{u^2+v^2}{r}\right). \tag{2.45}$$

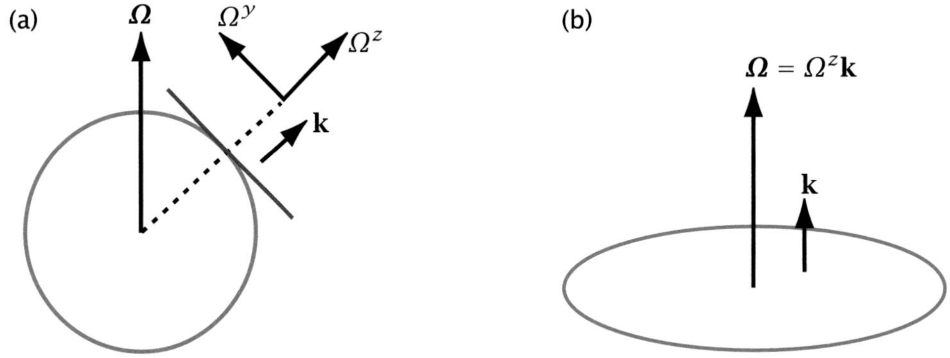

Fig. 2.4 (a) On the sphere the rotation vector $\boldsymbol{\Omega}$ can be decomposed into two components, one in the local vertical and one in the local horizontal, pointing toward the pole. That is, $\boldsymbol{\Omega} = \Omega_y \mathbf{j} + \Omega_z \mathbf{k}$ where $\Omega_y = \Omega \cos\vartheta$ and $\Omega_z = \Omega \sin\vartheta$. In geophysical fluid dynamics, the rotation vector in the local vertical is often the more important component in the horizontal momentum equations. On a rotating disk, (b), the rotation vector $\boldsymbol{\Omega}$ is parallel to the local vertical $\mathbf{k}$.

Using the definition of a vector cross product the Coriolis term is:

$$2\boldsymbol{\Omega} \times \boldsymbol{v} = \begin{vmatrix} \mathbf{i} & \mathbf{j} & \mathbf{k} \\ 0 & 2\Omega\cos\vartheta & 2\Omega\sin\vartheta \\ u & v & w \end{vmatrix}$$
$$= \mathbf{i}\,(2\Omega w\cos\vartheta - 2\Omega v\sin\vartheta) + \mathbf{j}\,2\Omega u\sin\vartheta - \mathbf{k}\,2\Omega u\cos\vartheta. \tag{2.46}$$

Using (2.45) and (2.46), and the gradient operator given by (2.31), the momentum equation (2.43) becomes:

$$\frac{\mathrm{D}u}{\mathrm{D}t} - \left(2\Omega + \frac{u}{r\cos\vartheta}\right)(v\sin\vartheta - w\cos\vartheta) = -\frac{1}{\rho r\cos\vartheta}\frac{\partial p}{\partial\lambda}, \tag{2.47a}$$

$$\frac{\mathrm{D}v}{\mathrm{D}t} + \frac{wv}{r} + \left(2\Omega + \frac{u}{r\cos\vartheta}\right)u\sin\vartheta = -\frac{1}{\rho r}\frac{\partial p}{\partial\vartheta}, \tag{2.47b}$$

$$\frac{\mathrm{D}w}{\mathrm{D}t} - \frac{u^2 + v^2}{r} - 2\Omega u\cos\vartheta = -\frac{1}{\rho}\frac{\partial p}{\partial r} - g. \tag{2.47c}$$

The terms involving Ω are called Coriolis terms, and the quadratic terms on the left-hand sides involving $1/r$ are often called metric terms.

2.2.4 The Primitive Equations

The so-called *primitive equations* of motion are simplifications of the above equations frequently used in atmospheric and oceanic modelling.[4] Three related approximations are involved:

(i) *The hydrostatic approximation.* In the vertical momentum equation the gravitational term is assumed to be balanced by the pressure gradient term, so that

$$\frac{\partial p}{\partial z} = -\rho g. \tag{2.48}$$

The advection of vertical velocity, the Coriolis terms, and the metric term $(u^2 + v^2)/r$ are all neglected.

(ii) *The shallow-fluid approximation.* We write $r = a + z$ where the constant a is the radius of the Earth and z increases in the radial direction. The coordinate r is then replaced by a except where it is used as the differentiating argument. Thus, for example,

$$\frac{1}{r^2}\frac{\partial(r^2 w)}{\partial r} \to \frac{\partial w}{\partial z}. \tag{2.49}$$

(iii) *The traditional approximation.* Coriolis terms in the horizontal momentum equations involving the vertical velocity, and the still smaller metric terms uw/r and vw/r, are neglected.

The second and third of these approximations should be taken, or not, together, the underlying reason being that they both relate to the presumed small aspect ratio of the motion, so the approximations succeed or fail together. If we make one approximation but not the other then we are being asymptotically inconsistent, and angular momentum and energy conservation are not assured.[5] The hydrostatic approximation also depends on the small aspect ratio of the flow, but in a slightly different way. For large-scale flow in the terrestrial atmosphere and ocean all three approximations are in fact very accurate approximations. We defer a more complete treatment until Section 2.7, in part because a treatment of the hydrostatic approximation is done most easily in the context of the Boussinesq equations, derived in Section 2.4.

Making these approximations, the momentum equations for a shallow layer are

$$\frac{\mathrm{D}u}{\mathrm{D}t} - 2\Omega\sin\vartheta v - \frac{uv}{a}\tan\vartheta = -\frac{1}{a\rho\cos\vartheta}\frac{\partial p}{\partial\lambda}, \tag{2.50a}$$

$$\frac{\mathrm{D}v}{\mathrm{D}t} + 2\Omega\sin\vartheta u + \frac{u^2\tan\vartheta}{a} = -\frac{1}{\rho a}\frac{\partial p}{\partial\vartheta}, \tag{2.50b}$$

$$0 = -\frac{1}{\rho}\frac{\partial p}{\partial z} - g, \tag{2.50c}$$

where

$$\frac{\mathrm{D}}{\mathrm{D}t} = \left(\frac{\partial}{\partial t} + \frac{u}{a\cos\vartheta}\frac{\partial}{\partial\lambda} + \frac{v}{a}\frac{\partial}{\partial\vartheta} + w\frac{\partial}{\partial z}\right). \tag{2.51}$$

We note the ubiquity of the factor $2\Omega\sin\vartheta$, and take the opportunity to define the *Coriolis parameter*, $f \equiv 2\Omega\sin\vartheta$. The associated mass conservation equation for a shallow fluid layer is:

$$\frac{\partial\rho}{\partial t} + \frac{u}{a\cos\vartheta}\frac{\partial\rho}{\partial\lambda} + \frac{v}{a}\frac{\partial\rho}{\partial\vartheta} + w\frac{\partial\rho}{\partial z} + \rho\left[\frac{1}{a\cos\vartheta}\frac{\partial u}{\partial\lambda} + \frac{1}{a\cos\vartheta}\frac{\partial}{\partial\vartheta}(v\cos\vartheta) + \frac{\partial w}{\partial z}\right] = 0, \tag{2.52}$$

or equivalently,

$$\frac{\partial\rho}{\partial t} + \frac{1}{a\cos\vartheta}\frac{\partial(u\rho)}{\partial\lambda} + \frac{1}{a\cos\vartheta}\frac{\partial}{\partial\vartheta}(v\rho\cos\vartheta) + \frac{\partial(w\rho)}{\partial z} = 0. \tag{2.53}$$

2.2.5 Primitive Equations in Vector Form

The primitive equations on a sphere may be written in a compact vector form provided we make a slight reinterpretation of the material derivative of the coordinate axes. Instead of (2.39) we take the material derivative of the unit vectors to be

$$\frac{\mathrm{D}\mathbf{i}}{\mathrm{D}t} = \widetilde{\boldsymbol{\Omega}}_{flow} \times \mathbf{i} = \mathbf{j}\frac{u\tan\vartheta}{a}, \tag{2.54a}$$

$$\frac{\mathrm{D}\mathbf{j}}{\mathrm{D}t} = \widetilde{\boldsymbol{\Omega}}_{flow} \times \mathbf{j} = -\mathbf{i}\frac{u\tan\vartheta}{a}, \tag{2.54b}$$

where $\widetilde{\boldsymbol{\Omega}}_{flow} = \mathbf{k} u \tan\vartheta / a$, which is the vertical component of (2.38) with r replaced by a. Given (2.54), the primitive equations (2.50a) and (2.50b) may be written as

$$\frac{\mathrm{D}\boldsymbol{u}}{\mathrm{D}t} + \boldsymbol{f} \times \boldsymbol{u} = -\frac{1}{\rho}\nabla_z p, \tag{2.55}$$

where $\boldsymbol{u} = u\mathbf{i} + v\mathbf{j} + 0\,\mathbf{k}$ is the horizontal velocity, $\nabla_z p = [(a\cos\vartheta)^{-1}\partial p/\partial\lambda,\ a^{-1}\partial p/\partial\vartheta]$ is the gradient operator at constant z, and $\boldsymbol{f} = f\mathbf{k} = 2\Omega\sin\vartheta\mathbf{k}$. In (2.55) the material derivative of the horizontal velocity is given by

$$\frac{\mathrm{D}\boldsymbol{u}}{\mathrm{D}t} = \mathbf{i}\frac{\mathrm{D}u}{\mathrm{D}t} + \mathbf{j}\frac{\mathrm{D}v}{\mathrm{D}t} + u\frac{\mathrm{D}\mathbf{i}}{\mathrm{D}t} + v\frac{\mathrm{D}\mathbf{j}}{\mathrm{D}t}. \tag{2.56}$$

The advection of the horizontal wind $\boldsymbol{u}$ is still by the three-dimensional velocity $\boldsymbol{v}$.

The vertical momentum equation is the hydrostatic equation, (2.50c), and the mass conservation equation is

$$\frac{\mathrm{D}\rho}{\mathrm{D}t} + \rho\nabla\cdot\boldsymbol{v} = 0 \qquad \text{or} \qquad \frac{\partial\rho}{\partial t} + \nabla\cdot(\rho\boldsymbol{v}) = 0, \tag{2.57}$$

where $\mathrm{D}/\mathrm{D}t$ is given by (2.51), and the second expression is written out in full in (2.53).

2.2.6 The Vector Invariant Form of the Momentum Equation

The 'vector invariant' form of the momentum equation is so-called because it appears to take the same form in all coordinate systems — there is no advective derivative of the coordinate system to worry about. With the aid of the identity $(\boldsymbol{v}\cdot\nabla)\boldsymbol{v} = -\boldsymbol{v}\times\boldsymbol{\omega} + \nabla(v^2/2)$, where $\boldsymbol{\omega} \equiv \nabla\times\boldsymbol{v}$ is the relative vorticity (which we explore at greater length in Chapter 4) the three-dimensional momentum equation, (2.16), may be written:

$$\frac{\partial\boldsymbol{v}}{\partial t} + (2\boldsymbol{\Omega} + \boldsymbol{\omega})\times\boldsymbol{v} = -\frac{1}{\rho}\nabla p - \frac{1}{2}\nabla\boldsymbol{v}^2 + \boldsymbol{g}, \tag{2.58}$$

and this is the vector invariant momentum equation. In spherical coordinates the relative vorticity is given by:

$$\boldsymbol{\omega} = \nabla\times\boldsymbol{v} = \frac{1}{r^2\cos\vartheta}\begin{vmatrix} \mathbf{i}\,r\cos\vartheta & \mathbf{j}\,r & \mathbf{k} \\ \partial/\partial\lambda & \partial/\partial\vartheta & \partial/\partial r \\ ur\cos\vartheta & rv & w \end{vmatrix} \tag{2.59}$$

$$= \mathbf{i}\frac{1}{r}\left(\frac{\partial w}{\partial\vartheta} - \frac{\partial(rv)}{\partial r}\right) - \mathbf{j}\frac{1}{r\cos\vartheta}\left(\frac{\partial w}{\partial\lambda} - \frac{\partial}{\partial r}(ur\cos\vartheta)\right) + \mathbf{k}\frac{1}{r\cos\vartheta}\left(\frac{\partial v}{\partial\lambda} - \frac{\partial}{\partial\vartheta}(u\cos\vartheta)\right).$$

We can write the horizontal momentum equations of the primitive equations in a similar way. Making the traditional and shallow fluid approximations, the horizontal components of (2.58) become

$$\frac{\partial\boldsymbol{u}}{\partial t} + (\boldsymbol{f} + \mathbf{k}\zeta)\times\boldsymbol{u} + w\frac{\partial\boldsymbol{u}}{\partial z} = -\frac{1}{\rho}\nabla_z p - \frac{1}{2}\nabla\boldsymbol{u}^2, \tag{2.60}$$

where $\boldsymbol{u} = (u, v, 0)$, $\boldsymbol{f} = \mathbf{k}\,2\Omega\sin\vartheta$ and ∇_z is the horizontal gradient operator (the gradient at a constant value of z). Using (2.59), ζ is given by

$$\zeta = \frac{1}{a\cos\vartheta}\frac{\partial v}{\partial\lambda} - \frac{1}{a\cos\vartheta}\frac{\partial}{\partial\vartheta}(u\cos\vartheta) = \frac{1}{a\cos\vartheta}\frac{\partial v}{\partial\lambda} - \frac{1}{a}\frac{\partial u}{\partial\vartheta} + \frac{u}{a}\tan\vartheta. \tag{2.61}$$

The separate components of the momentum equation are given by:

$$\frac{\partial u}{\partial t} - (f + \zeta)v + w\frac{\partial u}{\partial z} = -\frac{1}{a\cos\vartheta}\left(\frac{1}{\rho}\frac{\partial p}{\partial\lambda} + \frac{1}{2}\frac{\partial\boldsymbol{u}^2}{\partial\lambda}\right), \tag{2.62}$$

and

$$\frac{\partial v}{\partial t} + (f + \zeta)u + w\frac{\partial v}{\partial z} = -\frac{1}{a}\left(\frac{1}{\rho}\frac{\partial p}{\partial \vartheta} + \frac{1}{2}\frac{\partial \boldsymbol{u}^2}{\partial \vartheta}\right). \tag{2.63}$$

2.2.7 Angular Momentum

The zonal momentum equation can be usefully expressed as a statement about axial angular momentum; that is, angular momentum about the rotation axis. The zonal angular momentum per unit mass is the component of angular momentum in the direction of the axis of rotation and it is given by, without making any shallow atmosphere approximation,

$$m = (u + \Omega r \cos\vartheta) r \cos\vartheta. \tag{2.64}$$

The evolution equation for this quantity follows from the zonal momentum equation and has the simple form

$$\frac{\mathrm{D}m}{\mathrm{D}t} = -\frac{1}{\rho}\frac{\partial p}{\partial \lambda}, \tag{2.65}$$

where the material derivative is

$$\frac{\mathrm{D}}{\mathrm{D}t} = \frac{\partial}{\partial t} + \frac{u}{r\cos\vartheta}\frac{\partial}{\partial \lambda} + \frac{v}{r}\frac{\partial}{\partial \vartheta} + w\frac{\partial}{\partial r}. \tag{2.66}$$

Using the mass continuity equation, (2.65) can be written as

$$\frac{\mathrm{D}\rho m}{\mathrm{D}t} + \rho m \nabla\cdot\boldsymbol{v} = -\frac{\partial p}{\partial \lambda} \tag{2.67}$$

or

$$\frac{\partial \rho m}{\partial t} + \frac{1}{r\cos\vartheta}\frac{\partial(\rho u m)}{\partial \lambda} + \frac{1}{r\cos\vartheta}\frac{\partial}{\partial \vartheta}(\rho v m\cos\vartheta) + \frac{1}{r^2}\frac{\partial}{\partial r}(\rho m w r^2) = -\frac{\partial p}{\partial \lambda}. \tag{2.68}$$

This is an angular momentum conservation equation.

If the fluid is confined to a shallow layer near the surface of a sphere, then we may replace r, the radial coordinate, by a, the radius of the sphere, in the definition of m, and we define $\widetilde{m} \equiv (u + \Omega a\cos\vartheta)a\cos\vartheta$. Then (2.65) is replaced by

$$\frac{\mathrm{D}\widetilde{m}}{\mathrm{D}t} = -\frac{1}{\rho}\frac{\partial p}{\partial \lambda}, \tag{2.69}$$

where now

$$\frac{\mathrm{D}}{\mathrm{D}t} = \frac{\partial}{\partial t} + \frac{u}{a\cos\vartheta}\frac{\partial}{\partial \lambda} + \frac{v}{a}\frac{\partial}{\partial \vartheta} + w\frac{\partial}{\partial z}. \tag{2.70}$$

In the shallow fluid approximation (2.68) becomes

$$\frac{\partial \rho m}{\partial t} + \frac{1}{a\cos\vartheta}\frac{\partial(\rho u m)}{\partial \lambda} + \frac{1}{a\cos\vartheta}\frac{\partial}{\partial \vartheta}(\rho v m\cos\vartheta) + \frac{\partial}{\partial z}(\rho m w) = -\frac{\partial p}{\partial \lambda}, \tag{2.71}$$

which is an angular momentum conservation equation for a shallow atmosphere.

♦ *From angular momentum to the spherical component equations*

An alternative way of deriving the three components of the momentum equation in spherical polar coordinates is to *begin* with (2.65) and the principle of conservation of energy. That is, we take the equations for conservation of angular momentum and energy as true a priori and demand that the forms of the momentum equation be constructed to satisfy these. Expanding the material derivative in (2.65), noting that $\mathrm{D}r/\mathrm{D}t = w$ and $\mathrm{D}\cos\vartheta/\mathrm{D}t = -(v/r)\sin\vartheta$, immediately gives (2.47a). Multiplication by u then yields

$$u\frac{\mathrm{D}u}{\mathrm{D}t} - 2\Omega uv\sin\vartheta + 2\Omega uw\cos\vartheta - \frac{u^2 v\tan\vartheta}{r} + \frac{u^2 w}{r} = -\frac{u}{\rho r\cos\vartheta}\frac{\partial p}{\partial\lambda}. \tag{2.72}$$

Now suppose that the meridional and vertical momentum equations are of the form

$$\frac{\mathrm{D}v}{\mathrm{D}t} + \text{Coriolis and metric terms} = -\frac{1}{\rho r}\frac{\partial p}{\partial\vartheta}, \tag{2.73a}$$

$$\frac{\mathrm{D}w}{\mathrm{D}t} + \text{Coriolis and metric terms} = -\frac{1}{\rho}\frac{\partial p}{\partial r}, \tag{2.73b}$$

but that we do not know what form the Coriolis and metric terms take. To determine that form, construct the kinetic energy equation by multiplying (2.73) by v and w, respectively. Now, the metric terms must vanish when we sum the resulting equations along with (2.72), so that (2.73a) must contain the Coriolis term $2\Omega u\sin\vartheta$ as well as the metric term $u^2\tan\vartheta/r$, and (2.73b) must contain the term $-2\Omega u\cos\phi$ as well as the metric term u^2/r. But if (2.73b) contains the term u^2/r it must also contain the term v^2/r by isotropy, and therefore (2.73a) must also contain the term vw/r. In this way, (2.47) is precisely reproduced, although the sceptic might argue that the uniqueness of the form has not been demonstrated.

A particular advantage of this approach arises in determining the appropriate momentum equations that conserve angular momentum and energy in the shallow-fluid approximation. We begin with (2.69) and expand to obtain (2.50a). Multiplying by u gives

$$u\frac{\mathrm{D}u}{\mathrm{D}t} - 2\Omega uv\sin\vartheta - \frac{u^2 v\tan\vartheta}{a} = -\frac{u}{\rho a\cos\vartheta}\frac{\partial p}{\partial\lambda}. \tag{2.74}$$

To ensure energy conservation, the meridional momentum equation must contain the Coriolis term $2\Omega u\sin\vartheta$ and the metric term $u^2\tan\vartheta/a$, but the vertical momentum equation must have neither of the metric terms appearing in (2.47c). Thus we deduce the following equations:

$$\frac{\mathrm{D}u}{\mathrm{D}t} - \left(2\Omega\sin\vartheta + \frac{u\tan\vartheta}{a}\right)v = -\frac{1}{\rho a\cos\vartheta}\frac{\partial p}{\partial\lambda}, \tag{2.75a}$$

$$\frac{\mathrm{D}v}{\mathrm{D}t} + \left(2\Omega\sin\vartheta + \frac{u\tan\vartheta}{a}\right)u = -\frac{1}{\rho a}\frac{\partial p}{\partial\vartheta}, \tag{2.75b}$$

$$\frac{\mathrm{D}w}{\mathrm{D}t} = -\frac{1}{\rho}\frac{\partial p}{\partial r} - g. \tag{2.75c}$$

This equation set, when used in conjunction with the thermodynamic and mass continuity equations, conserves appropriate forms of angular momentum and energy. In the hydrostatic approximation the material derivative of w in (2.75c) is *additionally* neglected. Thus, the hydrostatic approximation is mathematically and physically consistent with the shallow-fluid approximation, but it is an additional approximation with slightly different requirements that one may choose, rather than being required, to make. From an asymptotic perspective, the difference lies in the small parameter necessary for either approximation to hold, namely:

$$\text{shallow fluid and traditional approximations:} \qquad \gamma \equiv \frac{H}{a} \ll 1, \tag{2.76a}$$

$$\text{small aspect ratio for hydrostatic approximation:} \qquad \alpha \equiv \frac{H}{L} \ll 1, \tag{2.76b}$$

where L is the horizontal scale of the motion and a is the radius of the Earth. For hemispheric or global scale phenomena $L \sim a$ and the two approximations coincide. (Requirement (2.76b) for the hydrostatic approximation will be derived in Section 2.7.)

2.3 CARTESIAN APPROXIMATIONS: THE TANGENT PLANE

2.3.1 The f-plane

Although the rotation of the Earth is central for many dynamical phenomena, the sphericity of the Earth is not always so. This is especially true for phenomena on a scale somewhat smaller than global where the use of spherical coordinates becomes awkward, and it is more convenient to use a locally Cartesian representation of the equations. Referring to the red line in Fig. 2.4 we will define a plane tangent to the surface of the Earth at a latitude ϑ_0, and then use a Cartesian coordinate system (x, y, z) to describe motion on that plane. For small excursions on the plane, $(x, y, z) \approx (a\lambda \cos\vartheta_0, a(\vartheta - \vartheta_0), z)$. Consistently, the velocity is $\boldsymbol{v} = (u, v, w)$, so that u,v and w are the components of the velocity *in the tangent plane,* in approximately in the east–west, north–south and vertical directions, respectively.

The momentum equations for flow in this plane are then

$$\frac{\partial u}{\partial t} + (\boldsymbol{v}\cdot\nabla)u + 2(\Omega^y w - \Omega^z v) = -\frac{1}{\rho}\frac{\partial p}{\partial x}, \tag{2.77a}$$

$$\frac{\partial v}{\partial t} + (\boldsymbol{v}\cdot\nabla)v + 2(\Omega^z u - \Omega^x w) = -\frac{1}{\rho}\frac{\partial p}{\partial y}, \tag{2.77b}$$

$$\frac{\partial w}{\partial t} + (\boldsymbol{v}\cdot\nabla)w + 2(\Omega^x v - \Omega^y u) = -\frac{1}{\rho}\frac{\partial p}{\partial z} - g, \tag{2.77c}$$

where the rotation vector $\boldsymbol{\Omega} = \Omega^x\mathbf{i} + \Omega^y\mathbf{j} + \Omega^z\mathbf{k}$ and $\Omega^x = 0$, $\Omega^y = \Omega\cos\vartheta_0$ and $\Omega^z = \Omega\sin\vartheta_0$. If we make the traditional approximation, and so ignore the components of $\boldsymbol{\Omega}$ not in the direction of the local vertical, then the above equations become

$$\frac{\mathrm{D}u}{\mathrm{D}t} - f_0 v = -\frac{1}{\rho}\frac{\partial p}{\partial x}, \qquad \frac{\mathrm{D}v}{\mathrm{D}t} + f_0 u = -\frac{1}{\rho}\frac{\partial p}{\partial y}, \qquad \frac{\mathrm{D}w}{\mathrm{D}t} = -\frac{1}{\rho}\frac{\partial p}{\partial z} - g, \tag{2.78a,b,c}$$

where $f_0 = 2\Omega^z = 2\Omega\sin\vartheta_0$. Defining the horizontal velocity vector $\boldsymbol{u} = (u, v, 0)$, the first two equations may be written as

$$\frac{\mathrm{D}\boldsymbol{u}}{\mathrm{D}t} + \boldsymbol{f}_0 \times \boldsymbol{u} = -\frac{1}{\rho}\nabla_z p, \tag{2.79}$$

where $\mathrm{D}\boldsymbol{u}/\mathrm{D}t = \partial\boldsymbol{u}/\partial t + \boldsymbol{v}\cdot\nabla\boldsymbol{u}$, $\boldsymbol{f}_0 = 2\Omega\sin\vartheta_0\mathbf{k} = f_0\mathbf{k}$, and $\mathbf{k}$ is the direction perpendicular to the plane. These equations are, evidently, exactly the same as the momentum equations in a system in which the rotation vector is aligned with the local vertical, as illustrated in panel (b) of Fig. 2.4. They will describe flow on the surface of a rotating sphere to a good approximation provided the flow is of limited latitudinal extent so that the effects of sphericity are unimportant; we have made what is known as the *f-plane*. We may in addition make the hydrostatic approximation, in which case (2.78c) becomes the familiar $\partial p/\partial z = -\rho g$.

2.3.2 The Beta-plane Approximation

The magnitude of the vertical component of rotation varies with latitude, and this has important dynamical consequences. We can approximate this effect by allowing the effective rotation vector to vary. Thus, noting that, for small variations in latitude,

$$f = 2\Omega\sin\vartheta \approx 2\Omega\sin\vartheta_0 + 2\Omega(\vartheta - \vartheta_0)\cos\vartheta_0, \tag{2.80}$$

then on the tangent plane we may mimic this by allowing the Coriolis parameter to vary as

$$f = f_0 + \beta y, \tag{2.81}$$

where $f_0 = 2\Omega \sin \vartheta_0$ and $\beta = \partial f/\partial y = (2\Omega \cos \vartheta_0)/a$. This important approximation is known as the *beta-plane*, or *β-plane*, approximation; it captures the the most important *dynamical* effects of sphericity, without the complicating *geometric* effects, which are not essential to describe many phenomena. The momentum equations (2.78) are unaltered except that f_0 is replaced by $f_0 + \beta y$ to represent a varying Coriolis parameter. Thus, sphericity combined with rotation is dynamically equivalent to a *differentially rotating* system. For future reference, we write down the β-plane horizontal momentum equations:

$$\frac{\mathrm{D}\boldsymbol{u}}{\mathrm{D}t} + \boldsymbol{f} \times \boldsymbol{u} = -\frac{1}{\rho}\nabla_z p, \tag{2.82}$$

where $\boldsymbol{f} = (f_0 + \beta y)\hat{\mathbf{k}}$. In component form this equation becomes

$$\frac{\mathrm{D}u}{\mathrm{D}t} - fv = -\frac{1}{\rho}\frac{\partial p}{\partial x}, \qquad \frac{\mathrm{D}v}{\mathrm{D}t} + fu = -\frac{1}{\rho}\frac{\partial p}{\partial y}. \tag{2.83a,b}$$

The mass conservation, thermodynamic and hydrostatic equations in the β-plane approximation are the same as the usual Cartesian, f-plane, forms of those equations.

2.4 EQUATIONS FOR A STRATIFIED OCEAN: THE BOUSSINESQ APPROXIMATION

The density variations in the ocean are quite small compared to the mean density, and we may exploit this to derive somewhat simpler but still quite accurate equations of motion. Let us first examine how much density does vary in the ocean.

2.4.1 Variation of Density in the Ocean

The variations of density in the ocean are due to three effects: the compression of water by pressure (which we denote as $\Delta_p\rho$), the thermal expansion of water if its temperature changes ($\Delta_T\rho$), and the haline contraction if its salinity changes ($\Delta_S\rho$). How big are these? An appropriate equation of state to approximately evaluate these effects is the linear one

$$\rho = \rho_0\left[1 - \beta_T(T - T_0) + \beta_S(S - S_0) + \beta_p p\right], \tag{2.84}$$

where $\beta_T \approx 2 \times 10^{-4}\,\mathrm{K}^{-1}$, $\beta_S \approx 10^{-3}\,\mathrm{g/kg}^{-1}$ and $\beta_p = 1/(\rho_0 c_s^2) \approx 4.4 \times 10^{-10}\,\mathrm{Pa}^{-1}$ with $c_s \approx 1500\,\mathrm{m\,s}^{-1}$ (see Table 1.2 on page 33). The three effects may then be evaluated as follows:

Pressure compressibility. We have $\Delta_p\rho \approx \Delta p/c_s^2 \approx \rho_0 gH/c_s^2$ where H is the depth we evaluate the pressure change (quite accurately) using the hydrostatic approximation. Thus,

$$\frac{|\Delta_p\rho|}{\rho_0} \approx \frac{gH}{c_s^2} \sim 4 \times 10^{-2}, \tag{2.85}$$

with $H = 8\,\mathrm{km}$ and $c_s^2/g \approx 200\,\mathrm{km}$. The latter quantity is the density scale height of the ocean. Thus, the pressure at the bottom of the ocean, enormous as it is, is insufficient to compress the water enough to make a significant change in its density. Changes in density due to dynamical variations of pressure are small if the Mach number is small, and this is also usually the case.

Thermal expansion. We have $\Delta_T\rho \approx -\beta_T\rho_0\Delta T$ and therefore

$$\frac{|\Delta_T\rho|}{\rho_0} \approx \beta_T\Delta T \sim 4 \times 10^{-3} \tag{2.86}$$

with $\Delta T = 20\,\mathrm{K}$. Evidently we would require temperature differences of order β_T^{-1}, or 5000 K to obtain order one variations in density.

Saline contraction. We have $\Delta_S\rho \approx \beta_S\rho_0\Delta S$ and therefore

$$\frac{|\Delta_S\rho|}{\rho_0} \approx \beta_S\Delta S \sim 1.5 \times 10^{-3}, \tag{2.87}$$

with $\Delta S = 5g\,\mathrm{kg}^{-1}$. The fractional change in the density of seawater due to salinity variations is thus also very small.

Evidently, fractional density changes in the ocean are very small due to the above effects.

2.4.2 The Boussinesq Equations

The *Boussinesq equations* are a set of equations that exploit the smallness of density variations in liquids.[6] An asymptotic derivation is given in Appendix A (page 101) but in what follows we are more heuristic. To set notation we write

$$\begin{aligned} \rho &= \rho_0 + \delta\rho(x, y, z, t) & \text{(2.88a)}\\ &= \rho_0 + \widehat{\rho}(z) + \rho'(x, y, z, t) & \text{(2.88b)}\\ &= \widetilde{\rho}(z) + \rho'(x, y, z, t), & \text{(2.88c)} \end{aligned}$$

where ρ_0 is a constant and we assume that

$$|\widehat{\rho}|, |\rho'|, |\delta\rho| \ll \rho_0. \tag{2.89}$$

We need not assume that $|\rho'| \ll |\widehat{\rho}|$, but this is often the case in the ocean. The horizontal gradients (i.e., gradients at constant z, ∇_z) satisfy $\nabla_z\rho = \nabla_z\rho' = \nabla_z\delta\rho$. To obtain the Boussinesq equations we will just use (2.88a), but (2.88c) will be useful for the anelastic equations considered later.

Associated with the reference density is a reference pressure that is defined to be in hydrostatic balance with it. That is,

$$p = p_0(z) + \delta p(x, y, z, t), \tag{2.90}$$

where $|\delta p| \ll p_0$ and

$$\frac{\mathrm{d}p_0}{\mathrm{d}z} \equiv -g\rho_0. \tag{2.91a,b}$$

Momentum equations

Letting $\rho = \rho_0 + \delta\rho$ the momentum equation can be written, without approximation, as

$$(\rho_0 + \delta\rho)\left(\frac{\mathrm{D}\boldsymbol{v}}{\mathrm{D}t} + 2\boldsymbol{\Omega}\times\boldsymbol{v}\right) = -\nabla\delta p - \frac{\partial p_0}{\partial z}\mathbf{k} - g(\rho_0 + \delta\rho)\mathbf{k}, \tag{2.92}$$

and using (2.91) this becomes, again without approximation,

$$(\rho_0 + \delta\rho)\left(\frac{\mathrm{D}\boldsymbol{v}}{\mathrm{D}t} + 2\boldsymbol{\Omega}\times\boldsymbol{v}\right) = -\nabla\delta p - g\delta\rho\mathbf{k}. \tag{2.93}$$

If $\delta\rho/\rho_0 \ll 1$ then we may neglect the $\delta\rho$ term on the left-hand side and the above equation becomes

$$\frac{\mathrm{D}\boldsymbol{v}}{\mathrm{D}t} + 2\boldsymbol{\Omega}\times\boldsymbol{v} = -\nabla\phi + b\mathbf{k}, \tag{2.94}$$

where $\phi = \delta p/\rho_0$ and $b = -g\,\delta\rho/\rho_0$ is the *buoyancy*. We should not and do not neglect the term $g\,\delta\rho$, for there is no reason to believe it to be small: $\delta\rho$ may be small, but g is big! Equation (2.94) is the momentum equation in the Boussinesq approximation, and it is common to say that the

Boussinesq approximation ignores all variations of density of a fluid in the momentum equation, except when associated with the gravitational term.

For most large-scale motions in the ocean the *deviation* pressure and density fields are also approximately in hydrostatic balance, and in that case the vertical component of (2.94) becomes

$$\frac{\partial \phi}{\partial z} = b. \tag{2.95}$$

A condition for (2.95) to hold is that vertical accelerations are small *compared to* $g\,\delta\rho/\rho_0$, *and not compared to the acceleration due to gravity itself.* For more discussion of this point, see Section 2.7.

Mass continuity

The unapproximated mass continuity equation is

$$\frac{\mathrm{D}\delta\rho}{\mathrm{D}t} + (\rho_0 + \delta\rho)\nabla\cdot\boldsymbol{v} = 0. \tag{2.96}$$

Provided that time scales advectively — that is to say that $\mathrm{D}/\mathrm{D}t$ scales in the same way as $\boldsymbol{v}\cdot\nabla$ — then we may approximate this equation by

$$\nabla\cdot\boldsymbol{v} = 0, \tag{2.97}$$

which is the same as that for a constant density fluid. This *absolutely does not* allow one to go back and use (2.96) to say that $D\delta\rho/Dt = 0$; the evolution of density is given by the thermodynamic equation in conjunction with an equation of state, and this should not be confused with the mass conservation equation. Note also that in eliminating the time-derivative of density we eliminate the possibility of sound waves.

Thermodynamic equation and equation of state

The Boussinesq equations are closed by the addition of an equation of state, a thermodynamic equation and, as appropriate, a salinity equation. Neglecting salinity for the moment, a useful starting point is to write the thermodynamic equation, (1.114), as

$$\frac{\mathrm{D}\rho}{\mathrm{D}t} - \frac{1}{c_s^2}\frac{\mathrm{D}p}{\mathrm{D}t} = \frac{\dot{Q}}{(\partial\eta/\partial\rho)_p T} \approx -\dot{Q}\left(\frac{\rho_0\beta_T}{c_p}\right) \tag{2.98}$$

using $(\partial\eta/\partial\rho)_p = (\partial\eta/\partial T)_p(\partial T/\partial\rho)_p \approx -c_p/(T\rho_0\beta_T)$. Given the expansions (2.88a) and (2.90a), (2.98) can be written to a good approximation as

$$\frac{\mathrm{D}\delta\rho}{\mathrm{D}t} - \frac{1}{c_s^2}\frac{\mathrm{D}p_0}{\mathrm{D}t} = -\dot{Q}\left(\frac{\rho_0\beta_T}{c_p}\right), \tag{2.99}$$

or, using (2.91a),

$$\frac{\mathrm{D}}{\mathrm{D}t}\left(\delta\rho + \frac{\rho_0 g}{c_s^2}z\right) = -\dot{Q}\left(\frac{\rho_0\beta_T}{c_p}\right). \tag{2.100}$$

The term in brackets on left-hand side is the potential density, as in (1.117). The severest approximation to this is to neglect the second term there, and noting that $b = -g\delta\rho/\rho_0$ we obtain

$$\frac{\mathrm{D}b}{\mathrm{D}t} = \dot{b}, \tag{2.101}$$

where $\dot{b} = g\beta_T\dot{Q}/c_p$. The momentum equation (2.94), mass continuity equation (2.97) and thermodynamic equation (2.101) then form a closed set, called the *simple Boussinesq equations.*

In the ocean the compressibility effect can be important and it is convenient to write the thermodynamic equation as

$$\frac{\mathrm{D}b_\sigma}{\mathrm{D}t} = \dot{b}_\sigma, \tag{2.102}$$

where b_σ is the potential buoyancy given by

$$b_\sigma \equiv -g\frac{\delta\rho_\theta}{\rho_0} = -\frac{g}{\rho_0}\left(\delta\rho + \frac{\rho_0 g z}{c_s^2}\right) = b - g\frac{z}{H_\rho}, \tag{2.103}$$

where $H_\rho = c_s^2/g$. Buoyancy itself is obtained from b_σ by the 'equation of state', $b = b_\sigma + gz/H_\rho$.

In many applications we may need to use a still more accurate equation of state. In that case (and see Section 1.7.3) we replace (2.101) by the thermodynamic equations

$$\frac{\mathrm{D}\Theta}{\mathrm{D}t} = \dot{\Theta}, \qquad \frac{\mathrm{D}S}{\mathrm{D}t} = \dot{S}, \tag{2.104a,b}$$

where Θ is an appropriate thermodynamic state variable, such as potential enthalpy or entropy, S is salinity, and an equation of state then gives the buoyancy. The equation of state has the general form $b = b(\Theta, S, p)$, but to be consistent with the level of approximation in the other Boussinesq equations we replace p by the hydrostatic pressure calculated with the reference density, that is by $-\rho_0 g z$, and the equation of state then takes the general form

$$b = b(\Theta, S, z). \tag{2.105}$$

An example of (2.105) is (1.155), taken with the definition of buoyancy $b = -g\delta\rho/\rho_0$. The closed set of equations (2.94), (2.97), (2.104) and (2.105) are sometimes called the general Boussinesq equations, or, in oceanographic contexts, the seawater Boussinesq equations. Using an accurate equation of state and the Boussinesq approximation is the procedure used in many comprehensive ocean general circulation models. The Boussinesq equations, which with the hydrostatic and traditional approximations are often considered to be the oceanic primitive equations, are summarized in the shaded box on the following page.

♦ *Mean stratification and the buoyancy frequency*

The processes that cause density to vary in the vertical often differ from those that cause it to vary in the horizontal. For this reason it is sometimes useful to write $\rho = \rho_0 + \tilde{\rho}(z) + \rho'(x, y, z, t)$ and define $\tilde{b}(z) \equiv -g\tilde{\rho}/\rho_0$ and $b' \equiv -g\rho'/\rho_0$. Using the hydrostatic equation to evaluate pressure, the thermodynamic equation (2.98) becomes, to a good approximation,

$$\frac{\mathrm{D}b'}{\mathrm{D}t} + N^2 w = 0, \tag{2.106}$$

where $\mathrm{D}/\mathrm{D}t$ remains a three-dimensional operator and

$$N^2(z) = \left(\frac{\mathrm{d}\tilde{b}}{\mathrm{d}z} - \frac{g^2}{c_s^2}\right) = \frac{\mathrm{d}\tilde{b}_\sigma}{\mathrm{d}z}, \tag{2.107}$$

where $\tilde{b}_\sigma = \tilde{b} - gz/H_\rho$. The quantity N^2 is a measure of the mean stratification of the fluid, and is equal to the vertical gradient of the mean potential buoyancy. N is known as the buoyancy frequency, something we return to in Section 2.10. Equations (2.106) and (2.107) also hold in the simple Boussinesq equations, but with $c_s^2 = \infty$.

Summary of Boussinesq Equations

The simple Boussinesq equations are, for an inviscid fluid:

momentum equations: $$\frac{\mathrm{D}\boldsymbol{v}}{\mathrm{D}t} + \boldsymbol{f} \times \boldsymbol{v} = -\nabla\phi + b\mathbf{k}, \tag{B.1}$$

mass conservation: $$\nabla \cdot \boldsymbol{v} = 0, \tag{B.2}$$

buoyancy equation: $$\frac{\mathrm{D}b}{\mathrm{D}t} = \dot{b}. \tag{B.3}$$

A more general form replaces the buoyancy equation by:

thermodynamic equation: $$\frac{\mathrm{D}\Theta}{\mathrm{D}t} = \dot{\Theta}, \tag{B.4}$$

salinity equation: $$\frac{\mathrm{D}S}{\mathrm{D}t} = \dot{S}, \tag{B.5}$$

equation of state: $$b = b(\Theta, S, z). \tag{B.6}$$

An equation of state of the form $b = b(\Theta, S, \phi)$ is not asymptotically correct and good conservation properties are not assured.

2.4.3 Energetics of the Boussinesq System

In a uniform gravitational field but with no other forcing or dissipation, we write the simple Boussinesq equations as

$$\frac{\mathrm{D}\boldsymbol{v}}{\mathrm{D}t} + 2\boldsymbol{\Omega} \times \boldsymbol{v} = b\mathbf{k} - \nabla\phi, \qquad \nabla \cdot \boldsymbol{v} = 0, \qquad \frac{\mathrm{D}b}{\mathrm{D}t} = 0. \tag{2.108a,b,c}$$

From (2.108a) and (2.108b) the kinetic energy density evolution (cf. Section 1.10) is given by

$$\frac{1}{2}\frac{\mathrm{D}v^2}{\mathrm{D}t} = bw - \nabla \cdot (\phi\boldsymbol{v}), \tag{2.109}$$

where the constant reference density ρ_0 is omitted. Let us now define the potential $\Phi \equiv -z$, so that $\nabla\Phi = -\mathbf{k}$ and

$$\frac{\mathrm{D}\Phi}{\mathrm{D}t} = \nabla \cdot (\boldsymbol{v}\Phi) = -w, \tag{2.110}$$

and using this and (2.108c) gives

$$\frac{\mathrm{D}}{\mathrm{D}t}(b\Phi) = -wb. \tag{2.111}$$

Adding (2.111) to (2.109) and expanding the material derivative gives

$$\frac{\partial}{\partial t}\left(\frac{1}{2}\boldsymbol{v}^2 + b\Phi\right) + \nabla \cdot \left[\boldsymbol{v}\left(\frac{1}{2}\boldsymbol{v}^2 + b\Phi + \phi\right)\right] = 0. \tag{2.112}$$

This constitutes an energy equation for the Boussinesq system, and may be compared to (1.199). The energy density (divided by ρ_0) is just $\boldsymbol{v}^2/2 + b\Phi$. What does the term $b\Phi$ represent? Its integral, multiplied by ρ_0, is the potential energy of the flow minus that of the basic state, or $\int g(\rho - \rho_0)z\,\mathrm{d}z$. If there were a heating term on the right-hand side of (2.108c) this would directly provide a source of potential energy, rather than internal energy as in the compressible system. Because the fluid is incompressible, there is no conversion from kinetic and potential energy into internal energy.

✦ *Energetics with a general equation of state*

Now consider the energetics of the general Boussinesq equations. Suppose first that we allow the equation of state to be a function of pressure; the equations of motion are then (2.108) except that (2.108c) is replaced by

$$\frac{\mathrm{D}\Theta}{\mathrm{D}t} = 0, \qquad \frac{\mathrm{D}S}{\mathrm{D}t} = 0, \qquad b = b(\Theta, S, \phi). \tag{2.113a,b,c}$$

where Θ is some conservative thermodynamic variable and S is salinity. A little algebraic experimentation will reveal that no energy conservation law of the form (2.112) generally exists for this system! The problem arises because, by requiring the fluid to be incompressible, we eliminate the proper conversion of internal energy to kinetic energy. However, if we use the approximation $b = b(\Theta, S, z)$, the system does conserve an energy, as we now show.[7]

Define the potential, Π, as the integral of b at constant potential temperature and salinity, namely

$$\Pi(\Theta, S, z) \equiv -\int_a^z b\,\mathrm{d}z', \tag{2.114}$$

where a is a constant, so that $\partial \Pi/\partial z = -b$. (The quantity Π is related to the dynamic enthalpy of Section 1.7.3.) Taking the material derivative of the left-hand side gives

$$\frac{\mathrm{D}\Pi}{\mathrm{D}t} = \left(\frac{\partial \Pi}{\partial \Theta}\right)_{S,z}\frac{\mathrm{D}\Theta}{\mathrm{D}t} + \left(\frac{\partial \Pi}{\partial S}\right)_{\Theta,z}\frac{\mathrm{D}S}{\mathrm{D}t} + \left(\frac{\partial \Pi}{\partial z}\right)_{\Theta,S}\frac{\mathrm{D}z}{\mathrm{D}t} = -bw, \tag{2.115}$$

using (2.113a,b). Combining (2.115) and (2.109) gives

$$\frac{\partial}{\partial t}\left(\frac{1}{2}\boldsymbol{v}^2 + \Pi\right) + \nabla\cdot\left[\boldsymbol{v}\left(\frac{1}{2}\boldsymbol{v}^2 + \Pi + \phi\right)\right] = 0. \tag{2.116}$$

Thus, energetic consistency is maintained with an arbitrary equation of state, provided that the buoyancy (or density) is taken as a function of z and not pressure — as Appendix A indicates is the proper thing to do.

2.5 EQUATIONS FOR A STRATIFIED ATMOSPHERE: THE ANELASTIC APPROXIMATION

2.5.1 Preliminaries

In the atmosphere the density varies significantly, especially in the vertical. However, deviations of both ρ and p from a statically balanced state are often quite small, and the relative vertical variation of potential temperature is also small. We can usefully exploit these observations to give a somewhat simplified set of equations, useful both for theoretical and numerical analyses because sound waves are eliminated by way of an 'anelastic' approximation.[8] To begin we set

$$\rho = \widetilde{\rho}(z) + \delta\rho(x, y, z, t), \qquad p = \widetilde{p}(z) + \delta p(x, y, z, t), \tag{2.117a,b}$$

where we assume that $|\delta\rho| \ll |\widetilde{\rho}|$ and we define $\widetilde{p}$ such that

$$\frac{\partial \widetilde{p}}{\partial z} \equiv -g\widetilde{\rho}(z). \tag{2.118}$$

The notation is similar to that for the Boussinesq case except that, importantly, the density basic state is now a (given) function of the vertical coordinate. As with the Boussinesq case, the idea is to ignore dynamic variations of density (i.e., of $\delta\rho$) except where associated with gravity. First recall a couple of ideal gas relationships involving potential temperature, θ. If we define $s = \log\theta$ (so that s is entropy divided by c_p) then

$$s = \log\theta = \log T - \frac{R}{c_p}\log p = \frac{1}{\gamma}\log p - \log\rho, \tag{2.119}$$

where $\gamma = c_p/c_v$, implying

$$\delta s = \frac{\delta\theta}{\theta} = \frac{1}{\gamma}\frac{\delta p}{p} - \frac{\delta\rho}{\rho} \approx \frac{1}{\gamma}\frac{\delta p}{\tilde{p}} - \frac{\delta\rho}{\tilde{\rho}}. \tag{2.120}$$

Further, if $\tilde{s} \equiv \gamma^{-1}\log\tilde{p} - \log\tilde{\rho}$ then

$$\frac{\mathrm{d}\tilde{s}}{\mathrm{d}z} = \frac{1}{\gamma\tilde{p}}\frac{\mathrm{d}\tilde{p}}{\mathrm{d}z} - \frac{1}{\tilde{\rho}}\frac{\mathrm{d}\tilde{\rho}}{\mathrm{d}z} = -\frac{g\tilde{\rho}}{\gamma\tilde{p}} - \frac{1}{\tilde{\rho}}\frac{\mathrm{d}\tilde{\rho}}{\mathrm{d}z}. \tag{2.121}$$

In the atmosphere, the left-hand side is, typically, much smaller than either of the two terms on the right-hand side.

2.5.2 The Momentum Equation

The exact inviscid horizontal momentum equation is

$$(\tilde{\rho} + \delta\rho)\left(\frac{\mathrm{D}\boldsymbol{u}}{\mathrm{D}t} + \boldsymbol{f} \times \boldsymbol{u}\right) = -\nabla_z \delta p. \tag{2.122}$$

Neglecting $\delta\rho$ where it appears with $\tilde{\rho}$ leads to

$$\frac{\mathrm{D}\boldsymbol{u}}{\mathrm{D}t} + \boldsymbol{f} \times \boldsymbol{u} = -\nabla_z \phi, \tag{2.123}$$

where $\phi = \delta p/\tilde{\rho}$, and this is similar to the corresponding equation in the Boussinesq approximation.

The vertical component of the inviscid momentum equation is, without approximation,

$$(\tilde{\rho} + \delta\rho)\frac{\mathrm{D}w}{\mathrm{D}t} = -\frac{\partial\tilde{p}}{\partial z} - \frac{\partial\delta p}{\partial z} - g\tilde{\rho} - g\delta\rho = -\frac{\partial\delta p}{\partial z} - g\delta\rho, \tag{2.124}$$

using (2.118). Neglecting $\delta\rho$ on the left-hand side we obtain

$$\frac{\mathrm{D}w}{\mathrm{D}t} = -\frac{1}{\tilde{\rho}}\frac{\partial\delta p}{\partial z} - g\frac{\delta\rho}{\tilde{\rho}} = -\frac{\partial}{\partial z}\left(\frac{\delta p}{\tilde{\rho}}\right) - \frac{\delta p}{\tilde{\rho}^2}\frac{\partial\tilde{\rho}}{\partial z} - g\frac{\delta\rho}{\tilde{\rho}}. \tag{2.125}$$

This is not a useful form for a gaseous atmosphere, since the variation of the mean density cannot be ignored. However, we may eliminate $\delta\rho$ in favour of δs using (2.120) to give

$$\frac{\mathrm{D}w}{\mathrm{D}t} = g\,\delta s - \frac{\partial}{\partial z}\left(\frac{\delta p}{\tilde{\rho}}\right) - \frac{g}{\gamma}\frac{\delta p}{\tilde{p}} - \frac{\delta p}{\tilde{\rho}^2}\frac{\partial\tilde{\rho}}{\partial z}, \tag{2.126}$$

and using (2.121) gives

$$\frac{\mathrm{D}w}{\mathrm{D}t} = g\,\delta s - \frac{\partial}{\partial z}\left(\frac{\delta p}{\tilde{\rho}}\right) + \frac{\mathrm{d}\tilde{s}}{\mathrm{d}z}\frac{\delta p}{\tilde{\rho}}. \tag{2.127}$$

What have these manipulations gained us? Two things:

(i) The gravitational term now involves δs rather than $\delta\rho$ which enables a more direct connection with the thermodynamic equation.

(ii) The potential temperature scale height ($\sim$ 100 km) in the atmosphere is much larger than the density scale height ($\sim$ 10 km), and so the last term in (2.127) is small.

The second item thus suggests that we choose our reference state to be one of constant potential temperature. The term $\mathrm{d}\tilde{s}/\mathrm{d}z$ then vanishes and the vertical momentum equation becomes

$$\frac{\mathrm{D}w}{\mathrm{D}t} = g\,\delta s - \frac{\partial \phi}{\partial z}, \tag{2.128}$$

where $\delta s = \delta\theta/\theta_0$, where θ_0 is a constant. If we define a buoyancy by $b_a \equiv g\delta s = g\delta\theta/\theta_0$, then (2.123) and (2.128) have the same form as the Boussinesq momentum equations, but with a slightly different definition of buoyancy.

2.5.3 Mass Conservation

Using (2.117a) the mass conservation equation may be written, without approximation, as

$$\frac{\partial \delta\rho}{\partial t} + \nabla\cdot[(\tilde{\rho}+\delta\rho)\boldsymbol{v}] = 0. \tag{2.129}$$

We neglect $\delta\rho$ where it appears with $\tilde{\rho}$ in the divergence term. Further, the local time derivative will be small if time itself is scaled advectively (i.e., $T \sim L/U$ and sound waves do not dominate), giving

$$\nabla\cdot\boldsymbol{u} + \frac{1}{\tilde{\rho}}\frac{\partial}{\partial z}(\tilde{\rho}w) = 0. \tag{2.130}$$

It is here that the eponymous anelastic approximation arises: the elastic compressibility of the fluid is neglected, and this serves to eliminate sound waves. For reference, in spherical coordinates the equation is

$$\frac{1}{a\cos\vartheta}\frac{\partial u}{\partial\lambda} + \frac{1}{a\cos\vartheta}\frac{\partial}{\partial\vartheta}(v\cos\vartheta) + \frac{1}{\tilde{\rho}}\frac{\partial(w\tilde{\rho})}{\partial z} = 0. \tag{2.131}$$

In an ideal gas, the choice of constant potential temperature determines how the reference density $\tilde{\rho}$ varies with height. In some circumstances it is convenient to let $\tilde{\rho}$ be a constant, ρ_0 (effectively choosing a different equation of state), in which case the anelastic equations become identical to the Boussinesq equations, albeit with the buoyancy interpreted in terms of potential temperature in the former and density in the latter.

2.5.4 Thermodynamic Equation

The thermodynamic equation for an ideal gas may be written

$$\frac{\mathrm{D}\ln\theta}{\mathrm{D}t} = \frac{\dot{Q}}{Tc_p}. \tag{2.132}$$

In the anelastic equations, $\theta = \theta_0 + \delta\theta$, where θ_0 is constant, and the thermodynamic equation is

$$\frac{\mathrm{D}\delta s}{\mathrm{D}t} = \frac{\tilde{\theta}}{Tc_p}\dot{Q}. \tag{2.133}$$

Summarizing, the complete set of anelastic equations, with rotation but with no dissipation or diabatic terms, is

$$\frac{\mathrm{D}\boldsymbol{v}}{\mathrm{D}t} + 2\boldsymbol{\Omega}\times\boldsymbol{v} = \mathbf{k}b_a - \nabla\phi, \qquad \frac{\mathrm{D}b_a}{\mathrm{D}t} = 0, \qquad \nabla\cdot(\tilde{\rho}\boldsymbol{v}) = 0, \tag{2.134a,b,c}$$

where $b_a = g\delta s = g\delta\theta/\theta_0$. The anelastic equations are sometimes called the 'weak Boussinesq equations', with the original incompressible set then called the 'strong Boussinesq equations'.

The main difference between the anelastic and Boussinesq sets is in the mass continuity equation, and when $\widetilde{\rho} = \rho_0 =$ constant the two equation sets are identical. However, whereas the Boussinesq approximation is a very good one for ocean dynamics, the anelastic approximation is less so for large-scale atmosphere flow: the constancy of the reference potential temperature state is not a particularly good approximation, and the deviations in density from its reference profile are not especially small, leading to inaccuracies in the momentum equation. Nevertheless, the anelastic equations have been used very productively in limited area 'large-eddy simulations' where one does not wish to make the hydrostatic approximation but where sound waves are unimportant.[9] The equations also provide a good jumping-off point for theoretical studies and for the still simpler models of Chapter 5.

2.5.5 ♦ Energetics of the Anelastic Equations

Conservation of energy follows in much the same way as for the Boussinesq equations, except that $\widetilde{\rho}$ enters. Take the dot product of (2.134a) with $\widetilde{\rho}\boldsymbol{v}$ to obtain

$$\widetilde{\rho}\frac{\mathrm{D}}{\mathrm{D}t}\left(\frac{1}{2}\boldsymbol{v}^2\right) = -\nabla\cdot(\phi\widetilde{\rho}\boldsymbol{v}) + b_a\widetilde{\rho}w. \tag{2.135}$$

Now, define a potential $\Phi(z)$ such that $\nabla\Phi = -\mathbf{k}$, and so

$$\widetilde{\rho}\frac{\mathrm{D}\Phi}{\mathrm{D}t} = -w\widetilde{\rho}. \tag{2.136}$$

Combining this with the thermodynamic equation (2.134b) gives

$$\widetilde{\rho}\frac{\mathrm{D}(b_a\Phi)}{\mathrm{D}t} = -wb_a\widetilde{\rho}. \tag{2.137}$$

Adding this to (2.135) gives

$$\widetilde{\rho}\frac{\mathrm{D}}{\mathrm{D}t}\left(\frac{1}{2}\boldsymbol{v}^2 + b_a\Phi\right) = -\nabla\cdot(\phi\widetilde{\rho}\boldsymbol{v}), \tag{2.138}$$

or, expanding the material derivative,

$$\frac{\partial}{\partial t}\left[\widetilde{\rho}\left(\frac{1}{2}\boldsymbol{v}^2 + b_a\Phi\right)\right] + \nabla\cdot\left[\widetilde{\rho}\boldsymbol{v}\left(\frac{1}{2}\boldsymbol{v}^2 + b_a\Phi + \phi\right)\right] = 0. \tag{2.139}$$

This equation has the form

$$\frac{\partial E}{\partial t} + \nabla\cdot\left[\boldsymbol{v}(E + \widetilde{\rho}\phi)\right] = 0, \tag{2.140}$$

where $E = \widetilde{\rho}(\boldsymbol{v}^2/2 + b_a\Phi)$ is the energy density of the flow. This is a consistent energetic equation for the system, and when integrated over a closed domain the total energy is evidently conserved. The total energy density comprises the kinetic energy and a term $\widetilde{\rho}b_a\Phi$, which is analogous to the potential energy of a simple Boussinesq system. However, it is not exactly equal to potential energy because b_a is the buoyancy based on potential temperature, not density; rather, the term combines contributions from both the internal energy and the potential energy into an enthalpy-like quantity.

2.6 PRESSURE AND OTHER VERTICAL COORDINATES

Although using z as a vertical coordinate is a natural choice given our Cartesian worldview, it is not the only option, nor is it always the most useful one. Any variable that has a one-to-one correspondence with z in the vertical, so any variable that varies monotonically with z, could be used; pressure and, more surprisingly, entropy, are common choices. In the atmosphere pressure almost always falls monotonically with height, and using it instead of z provides a useful simplification of the mass conservation and geostrophic relations, as well as a more direct connection with observations, which are often taken at fixed values of pressure. (In the ocean pressure coordinates are essentially almost the same as height coordinates because density is almost constant.) Entropy seems an exotic vertical coordinate, but it is very useful in adiabatic flow and we consider it in Chapter 3.

2.6.1 General Relations

First consider a general vertical coordinate, ξ. Any variable Ψ that is a function of the coordinates (x, y, z, t) may be expressed instead in terms of (x, y, ξ, t) by considering ξ to be a function of the independent variables (x, y, z, t). Derivatives with respect to z and ξ are related by

$$\frac{\partial \Psi}{\partial \xi} = \frac{\partial \Psi}{\partial z}\frac{\partial z}{\partial \xi} \qquad \text{and} \qquad \frac{\partial \Psi}{\partial z} = \frac{\partial \Psi}{\partial \xi}\frac{\partial \xi}{\partial z}. \tag{2.141a,b}$$

Horizontal derivatives in the two coordinate systems are related by the chain rule,

$$\left(\frac{\partial \Psi}{\partial x}\right)_\xi = \left(\frac{\partial \Psi}{\partial x}\right)_z + \left(\frac{\partial z}{\partial x}\right)_\xi \frac{\partial \Psi}{\partial z}, \tag{2.142}$$

and similarly for time.

The material derivative in ξ coordinates may be derived by transforming the original expression in z coordinates using the chain rule, but because (x, y, ξ, t) are independent coordinates, and noting that the 'vertical velocity' in ξ coordinates is just $\dot{\xi}$ (i.e., $\mathrm{D}\xi/\mathrm{D}t$, just as the vertical velocity in z coordinates is $w = \mathrm{D}z/\mathrm{D}t$), we can write down

$$\frac{\mathrm{D}\Psi}{\mathrm{D}t} = \left(\frac{\partial \Psi}{\partial t}\right)_{x,y,\xi} + \boldsymbol{u}\cdot\nabla_\xi \Psi + \dot{\xi}\frac{\partial \Psi}{\partial \xi}, \tag{2.143}$$

where ∇_ξ is the gradient operator at constant ξ. The operator $\mathrm{D}/\mathrm{D}t$ is the same in z or ξ coordinates because it is the total derivative of some property of a fluid parcel, and this is independent of the coordinate system. However, the individual terms within it will differ between coordinate systems.

2.6.2 Pressure Coordinates

In pressure coordinates the analogue of the vertical velocity is $\omega \equiv Dp/Dt$, and the advective derivative itself is given by

$$\frac{\mathrm{D}}{\mathrm{D}t} = \frac{\partial}{\partial t} + \boldsymbol{u}\cdot\nabla_p + \omega\frac{\partial}{\partial p}. \tag{2.144}$$

Note, though, that the advective derivative is the same operator as it is in height coordinates, since it is just the total derivative of a given fluid parcel; it is just written with different coordinates.

To obtain an expression for the pressure force, now let $\xi = p$ in (2.142) and apply the relationship to p itself to give

$$0 = \left(\frac{\partial p}{\partial x}\right)_z + \left(\frac{\partial z}{\partial x}\right)_p \frac{\partial p}{\partial z}, \tag{2.145}$$

which, using the hydrostatic relationship, gives

$$\left(\frac{\partial p}{\partial x}\right)_z = \rho\left(\frac{\partial \Phi}{\partial x}\right)_p, \tag{2.146}$$

where $\Phi = gz$ is the *geopotential*. Thus, the horizontal pressure force in the momentum equations is

$$\frac{1}{\rho}\nabla_z p = \nabla_p \Phi, \tag{2.147}$$

where the subscripts on the gradient operator indicate that the horizontal derivatives are taken at constant z or constant p. The horizontal momentum equation thus becomes

$$\frac{\mathrm{D}\boldsymbol{u}}{\mathrm{D}t} + \boldsymbol{f} \times \boldsymbol{u} = -\nabla_p \Phi, \tag{2.148}$$

where $\mathrm{D}/\mathrm{D}t$ is given by (2.144). The hydrostatic equation in height coordinates is $\partial p/\partial z = -\rho g$ and in pressure coordinates this becomes

$$\frac{\partial \Phi}{\partial p} = -\alpha \qquad \text{or} \qquad \frac{\partial \Phi}{\partial p} = -\frac{p}{RT}. \tag{2.149}$$

The mass continuity equation simplifies attractively in pressure coordinates, if the hydrostatic approximation is used. Recall that the mass conservation equation can be derived from the material form

$$\frac{\mathrm{D}}{\mathrm{D}t}(\rho\,\delta V) = 0, \tag{2.150}$$

where $\delta V = \delta x\,\delta y\,\delta z$ is a volume element. But by the hydrostatic relationship $\rho\delta z = -(1/g)\delta p$ and thus

$$\frac{\mathrm{D}}{\mathrm{D}t}(\delta x\,\delta y\,\delta p) = 0. \tag{2.151}$$

This is completely analogous to the expression for the material conservation of volume in an incompressible fluid, (1.15). Thus, without further ado, we write the mass conservation in pressure coordinates as

$$\nabla_p \cdot \boldsymbol{u} + \frac{\partial \omega}{\partial p} = 0, \tag{2.152}$$

where the horizontal derivative is taken at constant pressure.

The (adiabatic) thermodynamic equation is still $\mathrm{D}\theta/\mathrm{D}t = 0$, and θ may be related to pressure and temperature using its definition and the ideal gas equation to complete the equation set. However, because the hydrostatic equation is written in terms of temperature and not potential temperature it is convenient to write the thermodynamic equation accordingly. To do this we begin with the thermodynamic equation in the form of (1.99b), namely $c_p\mathrm{D}T/\mathrm{D}t - \alpha\mathrm{D}p/\mathrm{D}t = 0$. Since $\omega \equiv \mathrm{D}p/\mathrm{D}t$ this equation is simply

$$c_p\frac{\mathrm{D}T}{\mathrm{D}t} - \frac{RT}{p}\omega = 0, \tag{2.153}$$

which is an appropriate thermodynamic equation in pressure coordinates. It is sometimes useful to write this as

$$\frac{\partial T}{\partial t} + u\frac{\partial T}{\partial x} + v\frac{\partial T}{\partial y} - \omega S_p = 0, \qquad \text{where} \qquad S_p = \frac{\kappa T}{p} - \frac{\partial T}{\partial p} = -\frac{T}{\theta}\frac{\partial \theta}{\partial p}, \tag{2.154a,b}$$

Equations of Motion in Pressure and Log-pressure Coordinates

The adiabatic, inviscid primitive equations in pressure coordinates are:

$$\frac{D\boldsymbol{u}}{Dt} + \boldsymbol{f} \times \boldsymbol{u} = -\nabla_p \Phi, \tag{P.1}$$

$$\frac{\partial \Phi}{\partial p} = \frac{-RT}{p}, \tag{P.2}$$

$$\nabla_p \cdot \boldsymbol{u} + \frac{\partial \omega}{\partial p} = 0, \tag{P.3}$$

$$c_p \frac{DT}{Dt} - \frac{RT}{p}\omega = 0 \quad \text{or} \quad \frac{\partial T}{\partial t} + u\frac{\partial T}{\partial x} + v\frac{\partial T}{\partial y} - \omega S_p = 0. \tag{P.4}$$

where $S_p = \kappa T/p - \partial T/\partial p$ and $\kappa = R/c_p$. The above equations are, respectively, the horizontal momentum equation, the hydrostatic equation, the mass continuity equation and the thermodynamic equation. Using hydrostasy and the ideal gas relation, the thermodynamic equation may also be written as

$$\frac{\partial T}{\partial t} + u\frac{\partial T}{\partial x} + v\frac{\partial T}{\partial y} + \omega\frac{\partial s}{\partial p} = 0, \tag{P.5}$$

where $s = T + gz/c_p$ is the dry static energy divided by c_p.

The corresponding equations in log-pressure coordinates are

$$\frac{D\boldsymbol{u}}{Dt} + \boldsymbol{f} \times \boldsymbol{u} = -\nabla_Z \Phi, \tag{P.6}$$

$$\frac{\partial \Phi}{\partial Z} = \frac{RT}{H}, \tag{P.7}$$

$$\nabla_Z \cdot \boldsymbol{u} + \frac{1}{\rho_R}\frac{\partial \rho_R W}{\partial z} = 0, \tag{P.8}$$

$$c_p \frac{DT}{Dt} + W\frac{RT}{H} = 0 \quad \text{or} \quad \frac{\partial T}{\partial t} + u\frac{\partial T}{\partial x} + v\frac{\partial T}{\partial y} + WS_Z = 0. \tag{P.9}$$

where $\rho_R = \rho_0 \exp(-Z/H)$ and $S_Z = \kappa T/H + \partial T/\partial Z$. The thermodynamic equation may also be written as

$$\frac{\partial}{\partial t}\frac{\partial \Phi}{\partial Z} + u\frac{\partial}{\partial x}\frac{\partial \Phi}{\partial z} + v\frac{\partial}{\partial y}\frac{\partial \Phi}{\partial z} + WN_*^2 = 0, \tag{P.10}$$

where $N_*^2 = (R/H)S_Z$.

having used the ideal gas equation and the definition of potential temperature, with $\kappa = R/c_p$. Evidently, S_p is an appropriate measure of static stability in pressure coordinates and it is closely related to the buoyancy frequency N, as we see in the next subsection.

The main practical difficulty with the pressure-coordinate equations is the lower boundary condition. Using

$$w \equiv \frac{Dz}{Dt} = \frac{\partial z}{\partial t} + \boldsymbol{u} \cdot \nabla_p z + \omega\frac{\partial z}{\partial p}, \tag{2.155}$$

and (2.149), the boundary condition of $w = 0$ at $z = z_s$ becomes

$$\frac{\partial \Phi}{\partial t} + \boldsymbol{u} \cdot \nabla_p \Phi - \alpha \omega = 0, \tag{2.156}$$

at $p(x, y, z_s, t)$. In theoretical studies, it is common to assume that the lower boundary is in fact a constant pressure surface and simply assume that $\omega = 0$, or that $\omega = -\alpha^{-1}\partial\Phi/\partial t$. For more realistic studies the fact that the level $z = 0$ is not a coordinate surface must be properly considered. For this reason, and especially if the lower boundary is uneven because of the presence of topography, so-called *sigma coordinates* are sometimes used, in which the vertical coordinate is chosen so that the lower boundary is itself a coordinate surface. Sigma coordinates may use height itself as a vertical measure (typical in oceanic applications) or use pressure (typical in atmospheric applications). In the latter case the vertical coordinate is $\sigma = p/p_s$ where $p_s(x, y, t)$ is the surface pressure. The difficulty of applying (2.156) is replaced by a prognostic equation for the surface pressure, derived from the mass conservation equation.

Interestingly, the pressure coordinate equations (collected together in the shaded box on the previous page) are isomorphic to the hydrostatic, salt-free general Boussinesq equations (see the shaded box on page 74) with $z \Leftrightarrow -p$, $w \Leftrightarrow -\omega$, $\phi \Leftrightarrow \Phi$, $b \Leftrightarrow \alpha$, $\Theta \Leftrightarrow \theta$ and an equation of state $b = b(\Theta, z) \Leftrightarrow \alpha = \alpha(\theta, p)$ (and in an ideal gas $\alpha = (\theta R/p_R)(p_R/p)^{1/\gamma}$). The dynamics of one system can often therefore be expected to have an analogue in the other.

2.6.3 Log-pressure Coordinates

A variant of pressure coordinates arises by using *log-pressure* coordinates, in which the vertical coordinate is $Z = -H \ln(p/p_R)$ where p_R is a reference pressure (say 1000 mb) and H is a constant (for example a scale height RT_0/g where T_0 is a constant) so that Z has units of length. (Uppercase letters are conventionally used for some variables in log-pressure coordinates, and these are not to be confused with scaling parameters.) The 'vertical velocity' for the system is now

$$W \equiv \frac{\mathrm{D}Z}{\mathrm{D}t}, \tag{2.157}$$

and the advective derivative is

$$\frac{\mathrm{D}}{\mathrm{D}t} \equiv \frac{\partial}{\partial t} + \boldsymbol{u} \cdot \nabla_p + W\frac{\partial}{\partial Z}. \tag{2.158}$$

The horizontal momentum equation is unaltered from (2.148), although we use (2.158) to evaluate the advective derivative. It is straightforward to show that the hydrostatic equation becomes

$$\frac{\partial \Phi}{\partial Z} = \frac{RT}{H}. \tag{2.159}$$

The mass continuity equation (2.152) becomes

$$\frac{\partial u}{\partial x} + \frac{\partial v}{\partial y} + \frac{\partial W}{\partial Z} - \frac{W}{H} = 0, \tag{2.160}$$

which may be written as

$$\nabla_Z \cdot \boldsymbol{u} + \frac{1}{\rho_R}\frac{\partial(\rho_R W)}{\partial z} = 0, \tag{2.161}$$

where $\nabla_Z \cdot$ is the divergence at constant Z and $\rho_R = \rho_0 \exp(-Z/H)$, so giving a form similar to the mass conservation equation in the anelastic equations. (The value of the constant ρ_0 may be set to one.)

As with pressure coordinates, it is convenient to write the thermodynamic equation in terms of temperature and not potential temperature, and in an analogous procedure to the one leading to (2.153) we obtain

$$c_p \frac{\mathrm{D}T}{\mathrm{D}t} + W\frac{RT}{H} = 0. \tag{2.162}$$

This equation may be written as

$$\frac{\partial T}{\partial t} + u\frac{\partial T}{\partial x} + v\frac{\partial T}{\partial y} + WS_Z = 0, \tag{2.163}$$

where

$$S_Z = \frac{\kappa T}{H} + \frac{\partial T}{\partial Z}, \tag{2.164}$$

and we may note that $S_Z = S_p p/H$. Using the hydrostatic equation we may write (2.163) as

$$\frac{\partial}{\partial t}\frac{\partial \Phi}{\partial Z} + u\frac{\partial}{\partial x}\frac{\partial \Phi}{\partial z} + v\frac{\partial}{\partial y}\frac{\partial \Phi}{\partial z} + WN_*^2 = 0, \tag{2.165}$$

where $N_*^2 = (R/H)S_Z$. The quantity N_* is not exactly equal to the square of the buoyancy frequency as normally defined (for an ideal gas $N^2 = (g/\theta)\partial\theta/\partial z$), but the two can be shown to be related by $N_*/N = p/(\rho g H) = RT/gH$, and are equal for an isothermal atmosphere.[10] Integrating the hydrostatic equation between two pressure levels gives, with $\Phi = gz$,

$$z(p_2) - z(p_1) = -\frac{R}{g}\int_{p_1}^{p_2} T \,\mathrm{d}\ln p. \tag{2.166}$$

Thus, the thickness of the layer is proportional to the average temperature of the layer, and at constant temperature the geometric height increases linearly with the logarithm of pressure. At a temperature of 240 K (280 K) the scale height, RT/g, is about 7 km (8.2 km). A useful rule of thumb for Earth's atmosphere (and one that holds at 240 K) is that geometric height increases by about 16 km for each factor of ten decrease in pressure, and pressures of 1000 hPa, 100 hPa, 10 hPa roughly correspond to heights of 0, 16 km 32 km and so on.

2.7 SCALING FOR HYDROSTATIC BALANCE

We first encountered hydrostatic balance in Section 1.3.3; we now look in more detail at the conditions required for it to hold. Along with geostrophic balance, considered in the next section, it is one of the most fundamental balances in geophysical fluid dynamics. The corresponding states, hydrostasy and geostrophy, are not exactly realized, but their approximate satisfaction has profound consequences on the behaviour of the atmosphere and ocean.

2.7.1 Preliminaries

Consider the relative sizes of terms in (2.77c):

$$\frac{W}{T} + \frac{UW}{L} + \frac{W^2}{H} + \Omega U \sim \left|\frac{1}{\rho}\frac{\partial p}{\partial z}\right| + g. \tag{2.167}$$

For most large-scale motion in the atmosphere and ocean the terms on the right-hand side are orders of magnitude larger than those on the left, and therefore must be approximately equal. Explicitly, suppose $W \sim 1\,\mathrm{cm\,s^{-1}}$, $L \sim 10^5\,\mathrm{m}$, $H \sim 10^3\,\mathrm{m}$, $U \sim 10\,\mathrm{m\,s^{-1}}$, $T = L/U$. Then by substituting into (2.167) it seems that the pressure term is the only one which could balance the gravitational term, and we are led to approximate (2.77c) by,

$$\frac{\partial p}{\partial z} = -\rho g. \tag{2.168}$$

This equation, which is a vertical momentum equation, is known as *hydrostatic balance*.

However, (2.168) is not always a useful equation! Let us suppose that the density is a constant, ρ_0. We can then write the pressure as

$$p(x, y, z, t) = p_0(z) + p'(x, y, z, t), \qquad \text{where} \qquad \frac{\partial p_0}{\partial z} \equiv -\rho_0 g. \tag{2.169}$$

That is, p_0 and ρ_0 are in hydrostatic balance. On the f-plane, the inviscid vertical momentum equation becomes, without approximation,

$$\frac{\mathrm{D}w}{\mathrm{D}t} = -\frac{1}{\rho_0}\frac{\partial p'}{\partial z}. \tag{2.170}$$

Thus, *for constant density fluids the gravitational term has no dynamical effect:* there is no buoyancy force, and the pressure term in the horizontal momentum equations can be replaced by p'. Hydrostatic balance, and in particular (2.169), is not a useful vertical momentum equation in this case. If the fluid is stratified, we should therefore subtract off the hydrostatic pressure associated with the mean density before we can determine whether hydrostasy is a useful *dynamical* approximation, accurate enough to determine the horizontal pressure gradients. This is automatic in the Boussinesq equations, where the vertical momentum equation is

$$\frac{\mathrm{D}w}{\mathrm{D}t} = -\frac{\partial \phi}{\partial z} + b, \tag{2.171}$$

and the hydrostatic balance of the basic state is already subtracted out. In the more general equation,

$$\frac{\mathrm{D}w}{\mathrm{D}t} = -\frac{1}{\rho}\frac{\partial p}{\partial z} - g, \tag{2.172}$$

we need to compare the advective term on the left-hand side with the pressure variations arising from horizontal flow in order to determine whether hydrostasy is an appropriate vertical momentum equation. Nevertheless, if we only need to determine the pressure for use in an equation of state then we simply need to compare the sizes of the dynamical terms in (2.77c) with g itself, in order to determine whether a hydrostatic approximation will suffice.

2.7.2 Scaling and the Aspect Ratio

In a Boussinesq fluid we write the horizontal and vertical momentum equations as

$$\frac{\mathrm{D}\boldsymbol{u}}{\mathrm{D}t} + \boldsymbol{f} \times \boldsymbol{u} = -\nabla_z \phi, \qquad \frac{\mathrm{D}w}{\mathrm{D}t} = -\frac{\partial \phi}{\partial z} + b. \tag{2.173a,b}$$

With $\boldsymbol{f} = 0$, (2.173a) implies the scaling

$$\phi \sim U^2. \tag{2.174}$$

If we use mass conservation, $\nabla_z \cdot \boldsymbol{u} + \partial w/\partial z = 0$, to scale vertical velocity then

$$w \sim W = \frac{H}{L}U = \alpha U, \tag{2.175}$$

where $\alpha \equiv H/L$ is the aspect ratio. The advective terms in the vertical momentum equation all scale as

$$\frac{Dw}{Dt} \sim \frac{UW}{L} = \frac{U^2H}{L^2}. \tag{2.176}$$

Using (2.174) and (2.176) the ratio of the advective term to the pressure gradient term in the vertical momentum equations then scales as

$$\frac{|Dw/Dt|}{|\partial\phi/\partial z|} \sim \frac{U^2H/L^2}{U^2/H} \sim \left(\frac{H}{L}\right)^2. \tag{2.177}$$

Thus, the condition for hydrostasy, that $|Dw/Dt|/|\partial\phi/\partial z| \ll 1$, is:

$$\alpha^2 \equiv \left(\frac{H}{L}\right)^2 \ll 1. \tag{2.178}$$

The advective term in the vertical momentum may then be neglected. Thus, *hydrostatic balance arises from a small aspect ratio approximation.*

We can obtain the same result more formally by nondimensionalizing the momentum equations. Using uppercase symbols to denote scaling values we write

$$\begin{aligned} (x, y) &= L(\hat{x}, \hat{y}), \qquad z = H\hat{z}, \qquad \boldsymbol{u} = U\hat{\boldsymbol{u}}, \qquad w = W\hat{w} = \frac{HU}{L}\hat{w}, \\ t &= T;\hat{t} = \frac{L}{U}\hat{t}, \qquad \phi = \Phi\hat{\phi} = U^2\hat{\phi}, \qquad b = B\hat{b} = \frac{U^2}{H}\hat{b}, \end{aligned} \tag{2.179}$$

where the hatted variables are nondimensional and the scaling for w is suggested by the mass conservation equation, $\nabla_z \cdot \boldsymbol{u} + \partial w/\partial z = 0$. Substituting (2.179) into (2.173) (with $f = 0$) gives us the nondimensional equations

$$\frac{D\hat{\boldsymbol{u}}}{D\hat{t}} = -\nabla\hat{\phi}, \qquad \alpha^2\frac{D\hat{w}}{D\hat{t}} = -\frac{\partial\hat{\phi}}{\partial\hat{z}} + \hat{b}, \tag{2.180a,b}$$

where $D/D\hat{t} = \partial/\partial\hat{t} + \hat{u}\partial/\partial\hat{x} + \hat{v}\partial/\partial\hat{y} + \hat{w}\partial/\partial\hat{z}$ and we use the convention that when ∇ operates on nondimensional quantities the operator itself is nondimensional. From (2.180b) it is clear that hydrostatic balance pertains when $\alpha^2 \ll 1$.

2.7.3 ♦ Effects of Stratification on Hydrostatic Balance

To include the effects of stratification we need to involve the thermodynamic equation, so let us first write down the complete set of non-rotating dimensional equations:

$$\frac{D\boldsymbol{u}}{Dt} = -\nabla_z\phi, \qquad \frac{Dw}{Dt} = -\frac{\partial\phi}{\partial z} + b', \tag{2.181a,b}$$

$$\frac{Db'}{Dt} + wN^2 = 0, \qquad \nabla\cdot\boldsymbol{v} = 0. \tag{2.182a,b}$$

We have written, without approximation, $b = b'(x, y, z, t) + \tilde{b}(z)$, with $N^2 = d\tilde{b}/dz$; this separation is useful because the horizontal and vertical buoyancy variations may scale in different ways, and often N^2 may be regarded as given. (We have also redefined ϕ by subtracting off a static component in hydrostatic balance with $\tilde{b}$.) We nondimensionalize (2.182) by first writing

$$\begin{aligned} (x, y) &= L(\hat{x}, \hat{y}), \qquad z = H\hat{z}, \qquad \boldsymbol{u} = U\hat{\boldsymbol{u}}, \quad w = W\hat{w} = \epsilon\frac{HU}{L}\hat{w}, \\ t &= T\hat{t} = \frac{L}{U}\hat{t}, \qquad \phi = U^2\hat{\phi}, \qquad b' = \Delta b\hat{b}' = \frac{U^2}{H}\hat{b}', \qquad N^2 = \overline{N}^2\hat{N}^2, \end{aligned} \tag{2.183}$$

where ϵ is, for the moment, undetermined, $\overline{N}$ is a representative constant value of the buoyancy frequency and Δb scales only the horizontal buoyancy variations. Substituting (2.183) into (2.181) and (2.182) gives

$$\frac{\mathrm{D}\hat{\boldsymbol{u}}}{\mathrm{D}\hat{t}} = -\nabla_z \hat{\phi}, \qquad \epsilon\alpha^2 \frac{\mathrm{D}\widehat{w}}{\mathrm{D}t} = -\frac{\partial \hat{\phi}}{\partial \hat{z}} + \hat{b}' \tag{2.184a,b}$$

$$\frac{U^2}{\overline{N}^2 H^2} \frac{\mathrm{D}\hat{b}'}{\mathrm{D}\hat{t}} + \epsilon \widehat{w} \widehat{N}^2 = 0, \qquad \nabla \cdot \hat{\boldsymbol{u}} + \epsilon \frac{\partial \widehat{w}}{\partial \hat{z}} = 0. \tag{2.185a,b}$$

where now $\mathrm{D}/\mathrm{D}\hat{t} = \partial/\partial\hat{t} + \hat{\boldsymbol{u}} \cdot \nabla_z + \epsilon\widehat{w}\partial/\partial\hat{z}$. To obtain a non-trivial balance in (2.185a) we choose $\epsilon = U^2/(\overline{N}^2 H^2) \equiv Fr^2$, where Fr is the *Froude number,* a measure of the stratification of the flow. A strong stratification corresponds to a small Froude number. From (2.183), the vertical velocity then scales as

$$W = \frac{Fr^2 UH}{L} \tag{2.186}$$

and if the flow is highly stratified the vertical velocity will be even smaller than a pure aspect ratio scaling might suggest. (There must, therefore, be some cancellation in horizontal divergence in the mass continuity equation; that is, $|\nabla_z \cdot \boldsymbol{u}| \ll U/L$.) With this choice of ϵ the nondimensional Boussinesq equations may be written:

$$\frac{\mathrm{D}\hat{\boldsymbol{u}}}{\mathrm{D}\hat{t}} = -\nabla_z \hat{\phi}, \qquad Fr^2\alpha^2 \frac{\mathrm{D}\widehat{w}}{\mathrm{D}\hat{t}} = -\frac{\partial \hat{\phi}}{\partial \hat{z}} + \hat{b}', \tag{2.187a,b}$$

$$\frac{\mathrm{D}\hat{b}'}{\mathrm{D}\hat{t}} + \widehat{w}\widehat{N}^2 = 0, \qquad \nabla \cdot \hat{\boldsymbol{u}} + Fr^2 \frac{\partial \widehat{w}}{\partial \hat{z}} = 0. \tag{2.188a,b}$$

The nondimensional parameters in the system are the aspect ratio and the Froude number (in addition to $\widehat{N}$, but by construction this is just an order one function of z). From (2.187b) the condition for hydrostatic balance to hold is evidently that

$$Fr^2\alpha^2 \ll 1, \tag{2.189}$$

so generalizing the aspect ratio condition (2.178) to a stratified fluid. Because Fr is a measure of stratification, (2.189) formalizes our intuitive expectation that the more stratified a fluid the more vertical motion is suppressed and therefore the more likely hydrostatic balance is to hold. Equation (2.189) is equivalent to

$$\frac{U^2}{\overline{N}^2 H^2} \frac{H^2}{L^2} = \frac{U^2}{L^2 \overline{N}^2} \ll 1, \tag{2.190}$$

and in a hydrostatic model the condition is always, by construction, satisfied.

Why bother with any of this scaling? Why not just say that hydrostatic balance holds when $|Dw/Dt| \ll |\partial\phi/\partial z|$? One reason is that we do not have a good idea of the value of w from direct measurements, and it may change significantly in different oceanic and atmospheric parameter regimes. On the other hand the Froude number and the aspect ratio are familiar nondimensional parameters with a wide applicability in other contexts, and which we can control in a laboratory setting or estimate in the ocean or atmosphere. Still, when equations are scaled, ascertaining which parameters are to be regarded as given and which should be derived is often a choice, rather than being set a priori.

2.7.4 Hydrostasy in the Ocean and Atmosphere

Is the hydrostatic approximation in fact a good one in the ocean and atmosphere?

In the ocean

For the large-scale ocean circulation, let $N \sim 10^{-2}\,\mathrm{s}^{-1}$, $U \sim 0.1\,\mathrm{m\,s}^{-1}$ and $H \sim 1\,\mathrm{km}$. Then $Fr = U/(NH) \sim 10^{-2} \ll 1$. Thus, $Fr^2\alpha^2 \ll 1$ even for unit aspect-ratio motion. In fact, for larger scale flow the aspect ratio is also small; for basin-scale flow $L \sim 10^6\,\mathrm{m}$ and $Fr^2\alpha^2 \sim 0.01^2{\times}0.001^2 = 10^{-10}$ and hydrostatic balance is an extremely good approximation.

For intense convection, for example in the Labrador Sea, the hydrostatic approximation may be less appropriate, because the intense descending plumes may have an aspect ratio (H/L) of one or greater and the stratification is very weak. The hydrostatic condition then often becomes the requirement that the Froude number is small. Representative orders of magnitude are $U \sim W \sim 0.1\,\mathrm{m\,s}^{-1}$, $H \sim 1\,\mathrm{km}$ and $N \sim 10^{-3}\,\mathrm{s}^{-1}$ to $10^{-4}\,\mathrm{s}^{-1}$. For these values Fr ranges between 0.1 and 1, and at the upper end of this range hydrostatic balance is violated.

In the atmosphere

Over much of the troposphere $N \sim 10^{-2}\,\mathrm{s}^{-1}$ so that with $U = 10\,\mathrm{m\,s}^{-1}$ and $H = 1\,\mathrm{km}$ we find $Fr \sim 1$. Hydrostasy is then maintained because the aspect ratio H/L is much less than unity. For larger scale synoptic activity a larger vertical scale is appropriate, and with $H = 10\,\mathrm{km}$ both the Froude number and the aspect ratio are much smaller than one; indeed with $L = 1000\,\mathrm{km}$ we find $Fr^2\alpha^2 \sim 0.1^2 \times 0.01^2 = 10^{-6}$ and the flow is hydrostatic to a very good approximation indeed. However, for smaller scale atmospheric motions associated with fronts and, especially, convection, there can be little expectation that hydrostatic balance will be a good approximation.

For large-scale flows in both atmosphere and ocean, the conceptual and practical simplifications afforded by the hydrostatic approximation can hardly be overemphasized.

2.8 GEOSTROPHIC AND THERMAL WIND BALANCE

We now consider the dominant dynamical balance in the horizontal components of the momentum equation. In the horizontal plane (meaning along geopotential surfaces) we find that the Coriolis term is much larger than the advective terms and the dominant balance is between it and the horizontal pressure force. This balance is called *geostrophic balance,* and it occurs when the Rossby number is small, as we now investigate.

2.8.1 The Rossby Number

The *Rossby number* characterizes the importance of rotation in a fluid.[11] It is, essentially, the ratio of the magnitude of the relative acceleration to the Coriolis acceleration, and it is of fundamental importance in geophysical fluid dynamics. It arises from a simple scaling of the horizontal momentum equation, namely

$$\frac{\partial \boldsymbol{u}}{\partial t} + (\boldsymbol{v}\cdot\nabla)\boldsymbol{u} + \boldsymbol{f}\times\boldsymbol{u} = -\frac{1}{\rho}\nabla_z p, \tag{2.191a}$$

$$U^2/L \qquad fU, \tag{2.191b}$$

where U is the approximate magnitude of the horizontal velocity and L is a typical length scale over which that velocity varies. (We assume that $W/H \lesssim U/L$, so that vertical advection does not dominate the advection.) The ratio of the sizes of the advective and Coriolis terms is defined to be the Rossby number,

$$Ro \equiv \frac{U}{fL}. \tag{2.192}$$

If the Rossby number is small then rotation effects are important and, as the values in Table 2.1 indicate, this is the case for large-scale flow in both ocean and atmosphere.

Variable	Scaling symbol	Meaning	Atmos. value	Ocean value
(x, y)	L	Horizontal length scale	10^6 m	10^5 m
t	T	Time scale	1 day (10^5 s)	10 days (10^6 s)
(u, v)	U	Horizontal velocity	$10\ \mathrm{m\,s^{-1}}$	$0.1\ \mathrm{m\,s^{-1}}$
	Ro	Rossby number, U/fL	0.1	0.01

Table 2.1 Scales of large-scale flow in atmosphere and ocean. The choices given are representative of large-scale mid-latitude eddying motion in both systems.

Another intuitive way to think about the Rossby number is in terms of time scales. The Rossby number based on a time scale is

$$Ro_T \equiv \frac{1}{fT}, \tag{2.193}$$

where T is a time scale associated with the dynamics at hand. If the time scale is an advective one, meaning that $T \sim L/U$, then this definition is equivalent to (2.192). Now, $f = 2\Omega \sin\vartheta$, where Ω is the angular velocity of the rotating frame and equal to $2\pi/T_p$ where T_p is the period of rotation (24 hours). Thus,

$$Ro_T = \frac{T_p}{4\pi T \sin\vartheta} = \frac{T_i}{T}, \tag{2.194}$$

where $T_i = 1/f$ is the 'inertial time scale', about three hours in mid-latitudes. Thus, for phenomena with time scales much longer than this, such as the motion of the Gulf Stream or a mid-latitude atmospheric weather system, the effects of the Earth's rotation can be expected to be important, whereas a short-lived phenomenon, such as a cumulus cloud or tornado, may be oblivious to such rotation.

2.8.2 Geostrophic Balance

If the Rossby number is sufficiently small in (2.191a) then the rotation term will dominate the nonlinear advection term, and if the time period of the motion scales advectively then the rotation term also dominates the local time derivative. The only term that can then balance the rotation term is the pressure term, and therefore we must have

$$\boldsymbol{f} \times \boldsymbol{u} \approx -\frac{1}{\rho}\nabla_z p, \tag{2.195}$$

or, in Cartesian component form

$$fu \approx -\frac{1}{\rho}\frac{\partial p}{\partial y}, \qquad fv \approx \frac{1}{\rho}\frac{\partial p}{\partial x}. \tag{2.196}$$

This balance is known as *geostrophic balance*, and its consequences are profound, giving geophysical fluid dynamics a special place in the broader field of fluid dynamics. We *define* the geostrophic velocity by

$$fu_g \equiv -\frac{1}{\rho}\frac{\partial p}{\partial y}, \qquad fv_g \equiv \frac{1}{\rho}\frac{\partial p}{\partial x}, \tag{2.197}$$

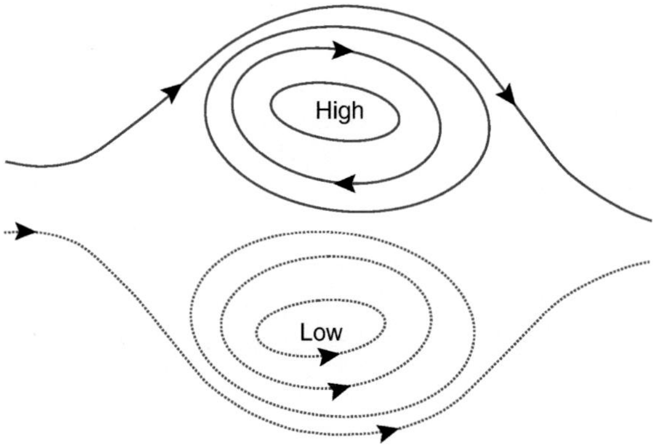

Fig. 2.5 Geostrophic flow with a positive value of the Coriolis parameter f. Flow is parallel to the lines of constant pressure (isobars). Cyclonic flow is anticlockwise around a low pressure region and anticyclonic flow is clockwise around a high. If f were negative, as in the Southern Hemisphere, (anti)cyclonic flow would be (anti)clockwise.

and for low Rossby number flow $u \approx u_g$ and $v \approx v_g$. In spherical coordinates the geostrophic velocity is

$$fu_g = -\frac{1}{\rho a}\frac{\partial p}{\partial \vartheta}, \qquad fv_g = \frac{1}{a\rho \cos\vartheta}\frac{\partial p}{\partial \lambda}, \tag{2.198}$$

where $f = 2\Omega \sin\vartheta$. Geostrophic balance has a number of immediate ramifications:

- Geostrophic flow is parallel to lines of constant pressure (isobars). If $f > 0$ the flow is anticlockwise round a region of low pressure and clockwise around a region of high pressure (see Fig. 2.5).

- If the Coriolis force is constant and if the density does not vary in the horizontal the geostrophic flow is horizontally non-divergent and

$$\nabla_z \cdot \boldsymbol{u}_g = \frac{\partial u_g}{\partial x} + \frac{\partial v_g}{\partial y} = 0. \tag{2.199}$$

  We may define the *geostrophic streamfunction*, ψ, by

$$\psi \equiv \frac{p}{f_0 \rho_0}, \qquad \text{whence} \qquad u_g = -\frac{\partial \psi}{\partial y}, \quad v_g = \frac{\partial \psi}{\partial x}. \tag{2.200}$$

  The vertical component of vorticity, ζ, is then given by

$$\zeta = \mathbf{k} \cdot \nabla \times \boldsymbol{v} = \frac{\partial v}{\partial x} - \frac{\partial u}{\partial y} = \nabla_z^2 \psi. \tag{2.201}$$

- If the Coriolis parameter is not constant, then cross-differentiating (2.197) gives, for constant density geostrophic flow,

$$v_g \frac{\partial f}{\partial y} + f \nabla_z \cdot \boldsymbol{u}_g = 0. \tag{2.202}$$

  Using the mass continuity equation, $\nabla_z \cdot \boldsymbol{u}_g = -\partial w/\partial z$, we then obtain

$$\beta v_g = f \frac{\partial w}{\partial z}, \tag{2.203}$$

  where $\beta \equiv \partial f/\partial y = 2\Omega \cos\vartheta/a$. This geostrophic vorticity balance is sometimes known as 'Sverdrup balance', although that expression is better restricted to the case when the vertical velocity comes from a wind stress, as considered in Chapter 19.

2.8.3 Taylor–Proudman Effect

If $\beta = 0$, then (2.203) implies that the vertical velocity is not a function of height. In fact, in that case none of the components of velocity vary with height if density is also constant. To show this, in the limit of zero Rossby number we first write the three-dimensional momentum equation as

$$\boldsymbol{f}_0 \times \boldsymbol{v} = -\nabla\phi - \nabla\chi, \tag{2.204}$$

where $\boldsymbol{f}_0 = 2\boldsymbol{\Omega} = 2\Omega\mathbf{k}$, $\phi = p/\rho_0$, and $\nabla\chi$ represents other potential forces. If $\chi = gz$ then the vertical component of this equation represents hydrostatic balance, and the horizontal components represent geostrophic balance. On taking the curl of this equation, the terms on the right-hand side vanish and the left-hand side becomes

$$(\boldsymbol{f}_0 \cdot \nabla)\boldsymbol{v} - \boldsymbol{f}_0 \nabla \cdot \boldsymbol{v} - (\boldsymbol{v} \cdot \nabla)\boldsymbol{f}_0 + \boldsymbol{v}\nabla \cdot \boldsymbol{f}_0 = 0. \tag{2.205}$$

But $\nabla \cdot \boldsymbol{v} = 0$ by mass conservation, and because $\boldsymbol{f}_0$ is constant both $\nabla \cdot \boldsymbol{f}_0$ and $(\boldsymbol{v} \cdot \nabla)\boldsymbol{f}_0$ vanish. Equation (2.205) thus reduces to

$$(\boldsymbol{f}_0 \cdot \nabla)\boldsymbol{v} = 0, \tag{2.206}$$

which, since $\boldsymbol{f}_0 = f_0\mathbf{k}$, implies $f_0 \partial\boldsymbol{v}/\partial z = 0$, and in particular we have

$$\frac{\partial u}{\partial z} = 0, \qquad \frac{\partial v}{\partial z} = 0, \qquad \frac{\partial w}{\partial z} = 0. \tag{2.207}$$

All three components of velocity are uniform along the axis of rotation.

A different presentation of this argument proceeds as follows. If the flow is exactly in geostrophic and hydrostatic balance then

$$v = \frac{1}{f_0}\frac{\partial \phi}{\partial x}, \qquad u = -\frac{1}{f_0}\frac{\partial \phi}{\partial y}, \qquad \frac{\partial \phi}{\partial z} = -g. \tag{2.208a,b,c}$$

Differentiating (2.208a,b) with respect to z, and using (2.208c) yields

$$\frac{\partial v}{\partial z} = \frac{-1}{f_0}\frac{\partial g}{\partial x} = 0, \qquad \frac{\partial u}{\partial z} = \frac{1}{f_0}\frac{\partial g}{\partial y} = 0. \tag{2.209}$$

Noting that the geostrophic velocities are horizontally non-divergent ($\nabla_z \cdot \boldsymbol{u} = 0$), and using mass continuity then gives $\partial w/\partial z = 0$, as before.

If there is a solid horizontal boundary anywhere in the fluid, for example at the surface, then $w = 0$ at that surface and thus $w = 0$ everywhere. Hence the motion occurs in planes that lie perpendicular to the axis of rotation, and the flow is effectively two dimensional. This result is known as the *Taylor–Proudman effect,* namely that for constant density flow in geostrophic and hydrostatic balance the vertical derivatives of the horizontal and the vertical velocities are zero.[12] At zero Rossby number, if the vertical velocity is zero somewhere in the flow then it is zero everywhere in that vertical column; furthermore, the horizontal flow has no vertical shear, and the fluid moves like a slab. The effects of rotation have provided a *stiffening* of the fluid in the vertical.

In neither the atmosphere nor the ocean do we observe precisely such vertically coherent flow, mainly because of the effects of stratification. However, it is typical of geophysical fluid dynamics that the assumptions underlying a derivation are not fully satisfied, yet there are manifestations of it in real flow. For example, one might have naïvely expected, because $\partial w/\partial z = -\nabla_z \cdot \boldsymbol{u}$, that the scales of the various variables would be related by $W/H \sim U/L$. However, if the flow is rapidly rotating we expect that the horizontal flow will be in near geostrophic balance and therefore nearly divergence free; thus $\nabla_z \cdot \boldsymbol{u} \ll U/L$, and $W \ll HU/L$.

2.8.4 Thermal Wind Balance

Thermal wind balance arises by combining the geostrophic and hydrostatic approximations, and this is most easily done in the context of the anelastic (or Boussinesq) equations, or in pressure coordinates. For the anelastic equations, geostrophic balance may be written

$$-fv_g = -\frac{\partial \phi}{\partial x} = -\frac{1}{a\cos\vartheta}\frac{\partial \phi}{\partial \lambda}, \qquad fu_g = -\frac{\partial \phi}{\partial y} = -\frac{1}{a}\frac{\partial \phi}{\partial \vartheta}. \tag{2.210a,b}$$

Combining these relations with hydrostatic balance, $\partial\phi/\partial z = b$, gives

$$-f\frac{\partial v_g}{\partial z} = -\frac{\partial b}{\partial x} = -\frac{1}{a\cos\lambda}\frac{\partial b}{\partial \lambda}, \qquad f\frac{\partial u_g}{\partial z} = -\frac{\partial b}{\partial y} = -\frac{1}{a}\frac{\partial b}{\partial \vartheta}. \tag{2.211a,b}$$

These equations represent *thermal wind balance*, and the vertical derivative of the geostrophic wind is the 'thermal wind'. Equation (2.211) may be written in terms of the zonal angular momentum as

$$\frac{\partial m_g}{\partial z} = -\frac{a}{2\Omega\tan\vartheta}\frac{\partial b}{\partial y}, \tag{2.212}$$

where $m_g = (u_g + \Omega a\cos\vartheta)a\cos\vartheta$. Potentially more accurate than geostrophic balance is the so-called gradient wind balance, discussed more in Section 2.9, which retains a centrifugal term in the momentum equation. The meridional momentum equation (2.50b) becomes

$$2u\Omega\sin\vartheta + \frac{u^2}{a}\tan\vartheta \approx -\frac{\partial \phi}{\partial y} = -\frac{1}{a}\frac{\partial \phi}{\partial \vartheta}. \tag{2.213}$$

For large-scale flow this only differs significantly from geostrophic balance very close to the equator. Taking the vertical derivative of (2.213) and using the hydrostatic relation $\partial\phi/\partial z = b$ gives a modified thermal wind relation, and this may be put in the simple form

$$\frac{\partial m^2}{\partial z} = -\frac{a^3\cos^3\vartheta}{\sin\vartheta}\frac{\partial b}{\partial y}, \tag{2.214}$$

where $m = (u + \Omega a\cos\vartheta)a\cos\vartheta$ is the annular angular momentum.

If the density or buoyancy is constant then there is no shear and (2.211) or (2.214) give the Taylor–Proudman result. But suppose that the temperature falls in the poleward direction. Then thermal wind balance implies that the (eastward) wind will increase with height — just as is observed in the atmosphere! In general, a vertical shear of the horizontal wind is associated with a horizontal temperature gradient, and this is one of the most simple and far-reaching effects in geophysical fluid dynamics. The underlying physical mechanism is illustrated in Fig. 2.6.

Geostrophic and thermal wind balance in pressure coordinates

In pressure coordinates geostrophic balance is just

$$\boldsymbol{f}\times\boldsymbol{u}_g = -\nabla_p\Phi, \tag{2.215}$$

where Φ is the geopotential and ∇_p is the gradient operator taken at constant pressure. If f is constant, it follows from (2.215) that the geostrophic wind is non-divergent on pressure surfaces.

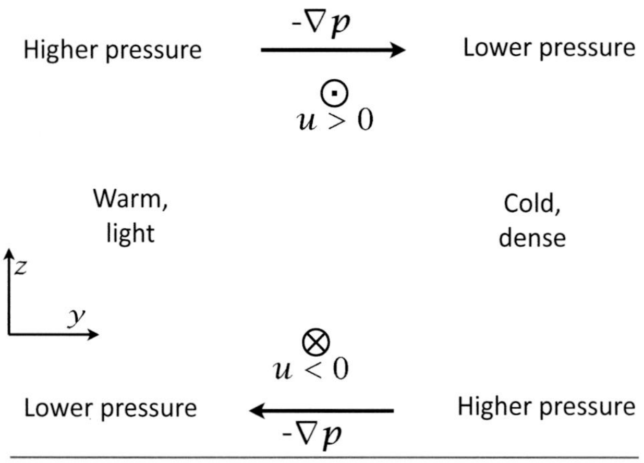

Fig. 2.6 The mechanism of thermal wind. A cold fluid is denser than a warm fluid, so by hydrostasy the vertical pressure gradient is greater where the fluid is cold. Thus, pressure gradients form as shown, where 'higher' and 'lower' mean relative to the average at that height. The horizontal pressure gradients are balanced by the Coriolis force, producing (for $f > 0$) the horizontal winds shown. Only the wind *shear* is given by the thermal wind.

Taking the vertical derivative of (2.215) (that is, its derivative with respect to p) and using the hydrostatic equation, $\partial\Phi/\partial p = -\alpha$, gives the thermal wind equation

$$\boldsymbol{f} \times \frac{\partial \boldsymbol{u}_g}{\partial p} = \nabla_p \alpha = \frac{R}{p}\nabla_p T, \tag{2.216}$$

where the last equality follows using the ideal gas equation and because the horizontal derivative is at constant pressure. In component form this is

$$-f\frac{\partial v_g}{\partial p} = \frac{R}{p}\frac{\partial T}{\partial x}, \qquad f\frac{\partial u_g}{\partial p} = \frac{R}{p}\frac{\partial T}{\partial y}. \tag{2.217}$$

In log-pressure coordinates, with $Z = -H\ln(p/p_R)$, thermal wind is

$$\boldsymbol{f} \times \frac{\partial \boldsymbol{u}_g}{\partial Z} = -\frac{R}{H}\nabla_Z T. \tag{2.218}$$

The effect in all these cases is the same: a horizontal temperature gradient, or a temperature gradient along an isobaric surface, is accompanied by a vertical shear of the horizontal wind.

2.8.5 ♦ Vertical Velocity and Hydrostatic Balance

Scaling for vertical velocity

If the Coriolis parameter is constant then flows that are in geostrophic balance have zero horizontal divergence ($\nabla_x \cdot \boldsymbol{u} = 0$) and zero vertical velocity. We can therefore expect that any flow with small Rossby number will have a correspondingly small vertical velocity. Let us make this statement more precise using the rotating Boussinesq equations, (2.173) with constant Coriolis parameter. Let $\boldsymbol{u} = \boldsymbol{u}_g + \boldsymbol{u}_a$ where the geostrophic flow satisfies $\boldsymbol{f}_0 \times \boldsymbol{u}_g = -\nabla\phi$. The horizontal momentum equation, with corresponding scales for each term, then becomes

$$\frac{\partial \boldsymbol{u}}{\partial t} + \boldsymbol{u}\cdot\nabla\boldsymbol{u} + w\frac{\partial \boldsymbol{u}}{\partial z} + \boldsymbol{f}_0 \times \boldsymbol{u}_a = 0, \tag{2.219}$$

$$\frac{U^2}{L} \qquad \frac{U^2}{L} \qquad \frac{WU}{H} \qquad f_0 U_a. \tag{2.220}$$

This equation suggests a scaling for the ageostrophic flow of

$$U_a = \frac{U}{f_0 L}U = Ro\, U. \tag{2.221}$$

That is, the ageostrophic flow is Rossby number smaller (at least) than the geostrophic flow. To obtain a scaling for the vertical velocity we look to the mass continuity equation written in the form

$$\frac{\partial w}{\partial z} = -\nabla \cdot \boldsymbol{u}_a, \tag{2.222}$$

since only the ageostrophic flow has a divergence. Equations (2.221) and (2.222) suggest the scaling

$$W = Ro \frac{HU}{L}. \tag{2.223}$$

That is, the vertical velocity is order Rossby number smaller than an estimate based purely on the mass continuity equation would suggest.

If the Coriolis parameter is not constant then the geostrophic flow itself is divergent and this induces a vertical velocity, as in (2.203). The scaling for vertical velocity is now

$$W = \frac{\beta}{f} HU = Ro_\beta \frac{HU}{L}, \tag{2.224}$$

where $Ro_\beta = \beta L/f$ is the *beta Rossby number*. It is less than one for all flows except those with a truly global scale.

Scaling for hydrostatic balance

Let us nondimensionalize the rotating Boussinesq equations, (2.173), by writing

$$\begin{gathered}(x, y) = L(\hat{x}, \hat{y}), \qquad z = H\hat{z}, \qquad \boldsymbol{u} = U\hat{\boldsymbol{u}}, \qquad t = T\hat{t} = \frac{L}{U}\hat{t}, \qquad f = f_0\hat{f}, \\ w = \frac{\epsilon HU}{L}\hat{w}, \qquad \phi = \Phi\hat{\phi} = f_0 UL\hat{\phi}, \qquad b = B\hat{b} = \frac{f_0 UL}{H}\hat{b}.\end{gathered} \tag{2.225}$$

These relations are almost the same as (2.179), except for the factor of ϵ in the scaling of w. If the Coriolis parameter is constant or nearly so then, from (2.223), $\epsilon = Ro$, whereas if the Coriolis parameter varies then $\epsilon = Ro_\beta$, as in (2.223). The scaling for ϕ and b' are suggested by geostrophic and thermal wind balance with f_0 a representative value of f. Substituting these values into (2.173) we obtain the scaled momentum equations:

$$Ro \frac{\mathrm{D}\hat{\boldsymbol{u}}}{\mathrm{D}\hat{t}} + \hat{\boldsymbol{f}} \times \hat{\boldsymbol{u}} = -\nabla\hat{\phi}, \qquad Ro\, \epsilon\alpha^2 \frac{\mathrm{D}\hat{w}}{\mathrm{D}\hat{t}} = -\frac{\partial\hat{\phi}}{\partial\hat{z}} - \hat{b}, \tag{2.226a,b}$$

where $\mathrm{D}/\mathrm{D}\hat{t} = \partial/\partial\hat{t} + \hat{\boldsymbol{u}} \cdot \nabla_z + \epsilon\hat{w}\partial/\partial\hat{z}$. There are two notable aspects to these equations. First and most obviously, when $Ro \ll 1$, (2.226a) reduces to geostrophic balance, $\boldsymbol{f} \times \boldsymbol{u} \approx -\nabla\hat{\phi}$. Second, the material derivative in (2.226b) is multiplied by three nondimensional parameters, and we can understand the appearance of each as follows:

(i) The aspect ratio dependence (α^2) arises in the same way as for non-rotating flows — that is, because of the presence of w and z in the vertical momentum equation as opposed to (u, v) and (x, y) in the horizontal equations.

(ii) The Rossby number dependence (Ro) arises because in rotating flow the pressure gradient is balanced by the Coriolis force, and the advective terms are Rossby-number smaller.

(iii) The factor ϵ arises because in rotating flow w is smaller than u by ϵ times the aspect ratio. The factor may be the Rossby number itself, or the beta Rossby number.

The factor $Ro\, \epsilon\alpha^2$ is very small for large-scale flow; the reader is invited to calculate representative values. Evidently, a rapidly rotating fluid is more likely to be in hydrostatic balance than a non-rotating fluid, other conditions being equal. The combined effects of rotation and stratification are, not surprisingly, quite subtle and we leave that topic for Chapter 5.

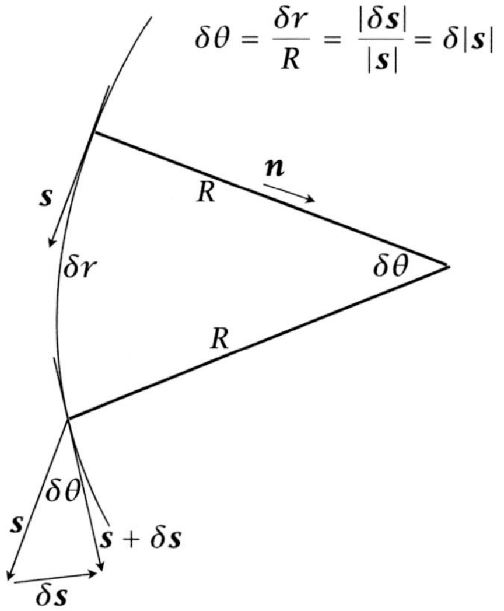

Fig. 2.7 A parcel tracing a curved path with radius of curvature R, moving a small distance δr and through a small angle $\delta\theta$. The parcel may experience centrifugal forces along $\boldsymbol{n}$ as well as Coriolis forces due to Earth's rotation.

2.9 ✦ GRADIENT WIND BALANCE

If a flow follows a curved path then our intuition suggests that it will experience a centrifugal force of some kind in addition to the Coriolis force. We can easily imagine that a parcel then experiences a three-way balance, between Coriolis, centrifugal and pressure forces, and this balance is called *gradient wind balance.* It is a more general, and more accurate, balance than geostrophic balance but it is not always as useful. To illustrate it we will keep matters simple and consider purely horizontal flow of constant density. We first introduce the notion of *natural coordinates* and show how gradient wind balance emerges straightforwardly.[13] We then discuss how gradient wind balance arises in the Eulerian equations of motion in a fixed coordinate system.

2.9.1 Natural Coordinates

For most purposes a coordinate system that is fixed in space, or rotating coincidentally with the Earth, is the most practically useful. However, it is sometimes useful to use a coordinate system that is moving with the local flow. To that end, and restricting our attention to horizontal flow, consider a parcel of fluid moving with velocity $\boldsymbol{u}$. We define a *natural coordinate system* by the set of unit vectors $\boldsymbol{s}$, $\boldsymbol{n}$ and $\mathbf{k}$, where $\boldsymbol{s}$ is the unit vector tangential to the flow, $\boldsymbol{n}$ is the horizontal unit vector normal to the flow and $\mathbf{k}$ is the unit vector in the vertical. Apart from $\mathbf{k}$, all these vectors evolve with the flow. Our goal is to split the horizontal momentum equation into components parallel to and normal to the direction of the local flow and thereby to discern the force balances in either direction. (Readers who trust their intuition may skip ahead to (2.231), which they may find obvious.)

If U is the speed of the parcel then $\boldsymbol{u} = U\boldsymbol{s}$ and so

$$\frac{\mathrm{d}\boldsymbol{u}}{\mathrm{d}t} = \boldsymbol{s}\frac{\mathrm{d}U}{\mathrm{d}t} + U\frac{\mathrm{d}\boldsymbol{s}}{\mathrm{d}t}. \tag{2.227}$$

Furthermore, if the flow of a parcel follows the curve $r(t)$ — that is, r gives the distance moved by the parcel — then $U = \mathrm{d}r/\mathrm{d}t$. To obtain a useful expression for $\mathrm{d}\boldsymbol{s}/\mathrm{d}t$ we note that, as in Fig. 2.7,

$$\theta \equiv \frac{\delta r}{R} = \frac{|\delta \boldsymbol{s}|}{|\boldsymbol{s}|} = |\delta \boldsymbol{s}|, \tag{2.228}$$

where R is the *radius of curvature* and $|\boldsymbol{s}| = 1$. [The radius of curvature may be evaluated geometrically as follows. Draw a line tangent to the path at some point, and draw a line perpendicular to the tangent through that point. An infinitesimal distance along the curve, construct another perpendicular line in the same manner. The two perpendicular lines meet at the center of curvature, and the distance along one of the perpendicular lines from the center of curvature to the curve itself is the radius of curvature. By convention, the radius of curvature is positive (negative) if the parcel is curving to the left (right).]

The change of $\boldsymbol{s}$ is directed along $\boldsymbol{n}$, and so from (2.228) we infer that

$$\frac{\mathrm{d}\boldsymbol{s}}{\mathrm{d}r} = \frac{\boldsymbol{n}}{R}. \tag{2.229}$$

Thus, the rate of change of $\boldsymbol{s}$ is given by

$$\frac{\mathrm{d}\boldsymbol{s}}{\mathrm{d}t} = \frac{\mathrm{d}\boldsymbol{s}}{\mathrm{d}r}\frac{\mathrm{d}r}{\mathrm{d}t} = \frac{\boldsymbol{n}}{R}U, \tag{2.230}$$

and using this expression in (2.227), the acceleration of the fluid parcel is given by

$$\frac{\mathrm{d}\boldsymbol{u}}{\mathrm{d}t} = \boldsymbol{s}\frac{\mathrm{d}U}{\mathrm{d}t} + \boldsymbol{n}\frac{U^2}{R}. \tag{2.231}$$

The two terms on the right-hand side may be interpreted as being, respectively, the change in speed of the parcel along its path and the centripetal acceleration owing to the curvature of the path (if the path is a straight line then R is infinite and the centripetal acceleration is zero).

2.9.2 Application to Fluids

Now consider the application of the above to a fluid in a rotating frame of reference. The total time derivatives of (2.231) may be replaced by material derivatives, and additional pressure and Coriolis forces act on the flow. The pressure-gradient force, $-\nabla\phi$, has components along and perpendicular to the flow so that

$$-\nabla\phi = -\left(\frac{\partial\phi}{\partial r}\boldsymbol{s} + \frac{\partial\phi}{\partial n}\boldsymbol{n}\right). \tag{2.232}$$

Given our convention, $\partial\phi/\partial n$ is positive if pressure increases to the left of the trajectory of the flow. The Coriolis force, $-\boldsymbol{f}\times\boldsymbol{u}$, has only a component perpendicular to the flow so that

$$-\boldsymbol{f}\times\boldsymbol{u} = -fU\boldsymbol{n}. \tag{2.233}$$

Because $\boldsymbol{n}$ is directed to the left of the flow, for positive f the Coriolis force is directed to the right of the flow. Using (2.231), (2.232) and (2.233), and neglecting friction, we can write down the components of the momentum equation parallel and perpendicular to the flow, namely

$$\frac{\mathrm{D}U}{\mathrm{D}t} = -\frac{\partial\phi}{\partial r}, \qquad \frac{U^2}{R} + fU = -\frac{\partial\phi}{\partial n}. \tag{2.234a,b}$$

These equations tell us, at least in principle, how a fluid field will evolve from some initial conditions. The radius of curvature can only be determined from the instantaneous velocity field for steady flow: if the flow is unsteady then the radius of curvature for a parcel depends on the pressure field and differs from the radius of curvature of a streamline.

Gradient wind balance, cyclostrophic balance and inertial flow

Gradient flow (or, as it is commonly called, the gradient wind) is the flow that satisfies (2.234b). We may always define a gradient wind at a given point by using this equation, and the only forces that are neglected are frictional ones. Given the pressure field ϕ we may use (2.234b) to calculate the gradient wind using the quadratic formula. Gradient wind balance is thus a better approximation to the real flow than is geostrophic balance (which omits the centrifugal term). The gradient wind will not in general be a steady solution to the equations of motion — the flow will accelerate and the pressure field will change.

If the flow is in a straight line then the radius of curvature is infinite and (2.234b) reduces to $fU = -\partial\phi/\partial n$, which is geostrophic balance. For this to hold exactly the flow need not be steady, just in a straight line. It will be a good approximation to the real flow when $fU \gg U^2/R$ or $U/fR \ll 1$, taking all quantities as positive, which is similar to a condition of low Rossby number. In contrast, cyclostrophic flow occurs when the quantity U/fR is large and the Coriolis force may be neglected, giving $U^2/R = -\partial\phi/\partial n$.

Inertial flow arises when the pressure gradient vanishes and (2.234b) reduces to $U/R + f = 0$. Now, if the pressure gradient vanishes then, from (2.234a) the speed of fluid parcels is constant. Thus, if f is constant the fluid parcels execute circles, known as inertia circles, of radius U/f and period $2\pi/f = 1\,\text{day}/(2\sin\vartheta)$ where ϑ is latitude. Such a motion is evidently not truly inertial, for inertial motion is in a straight line and at a constant speed; other forces — gravitational and centrifugal — must act to keep the flow in the horizontal plane. Inertial motion is rarely a good approximation to flow in the atmosphere or ocean, although occasionally parcels are observed to trace approximations to inertia circles, especially in the ocean.

2.9.3 Gradient Wind Balance in the Two-dimensional Eulerian Equations

Let us now consider gradient wind balance in the Eulerian equations of motion. The basic ideas can be exposed by considering unforced constant-density two dimensional flow in a rotating reference frame, for which the equations of motion are

$$\frac{\mathrm{D}u}{\mathrm{D}t} - fv = -\frac{\partial\phi}{\partial x}, \qquad \frac{\mathrm{D}v}{\mathrm{D}t} + fu = -\frac{\partial\phi}{\partial y}, \qquad \frac{\partial u}{\partial x} + \frac{\partial v}{\partial y} = 0. \tag{2.235a,b,c}$$

We may introduce a stream function ψ such that $u = -\partial\psi/\partial y$ and $v = \partial\psi/\partial x$. If the flow is geostrophic then $f \times (u, v) = (-\partial\phi/\partial y, \partial\phi/\partial x)$; however, unless the Coriolis parameter f is constant the streamfunction is not proportional to the pressure field. The vorticity (considered further in Chapter 4), which is defined to be the curl of the velocity, is given by

$$\boldsymbol{\omega} \equiv \nabla \times \boldsymbol{u} = \mathbf{k}\zeta, \qquad \text{where} \qquad \zeta = \left(\frac{\partial v}{\partial x} - \frac{\partial u}{\partial y}\right) = \frac{\partial^2\psi}{\partial x^2} + \frac{\partial^2\psi}{\partial y^2} \equiv \nabla_z^2\psi. \tag{2.236}$$

Let us form the evolution equations for vorticity and divergence by taking the curl and divergence of (2.235a,b). After a little algebra the vorticity equation is found to be

$$\frac{\partial\zeta}{\partial t} + \boldsymbol{u}\cdot\nabla(\zeta + f) = 0, \qquad \text{or equivalently} \qquad \frac{\partial\zeta}{\partial t} + J(\psi, \zeta + f) = 0, \tag{2.237}$$

where $J(a, b) \equiv (\partial a/\partial x)(\partial b/\partial y) - (\partial a/\partial y)(\partial b/\partial x)$ is the *Jacobian*. If $f = f(y)$ then $J(\psi, f) = \boldsymbol{u}\cdot\nabla f = \beta v$ where $\beta = \partial f/\partial y$. Because the velocity is divergence-free the vorticity equation above is closed; that is, it may be integrated without the need to solve any other evolution equations, and in particular without the need to solve for the pressure field. (In general the vorticity equation alone is not closed.)

If we take the divergence of (2.235a,b) then, using (2.235c) the time derivative disappears and again after a little algebra we obtain the divergence equation,

$$2J(u, v) + \nabla\cdot(f\nabla\psi) = \nabla^2\phi. \tag{2.238}$$

In this context, (2.238) may be regarded as an equation for pressure, ϕ, given the velocity or the streamfunction. Equation (2.238) is commonly referred to as the *gradient wind balance relation*, and it is analogous to (2.234b) in that it generalizes and is more accurate than geostrophic balance. Equation (2.238) is *exact* for two-dimensional, incompressible, inviscid and unforced flow, even when time-dependent. If the Rossby number is small then the second term on the left-hand side of (2.238) dominates the first, and $\nabla\cdot(f\nabla\psi)\approx\nabla^2\phi$. This is equivalent to geostrophic balance, and if f is constant the streamfunction and pressure field are proportional to each other.

In the more general case the two-dimensional divergence is not zero; that is $\partial_t(\partial_x u+\partial_y v)\neq 0$. However, if the Rossby number is small and if f is nearly constant, then the geostrophic relation implies that the two-dimensional divergence will be small compared to the two-dimensional vorticity. In this case, gradient wind balance will be a good *approximation* to the flow, and a somewhat better approximation, if not more useful, than geostrophic balance alone.

2.10 STATIC INSTABILITY AND THE PARCEL METHOD

In this and the next couple of sections we consider how a fluid might oscillate if it were perturbed away from a resting state. Our focus is on vertical displacements, and the restoring force is gravity. Given that, the simplest way to approach the problem is to consider from first principles the pressure and gravitational forces on a displaced parcel, as in Fig. 2.8. To this end, consider a fluid initially at rest in a constant gravitational field, and therefore in hydrostatic balance. Suppose that a small parcel of the fluid is adiabatically displaced upwards by the small distance δz, without altering the overall pressure field; that is, the fluid parcel instantly assumes the pressure of its environment. If after the displacement the parcel is lighter than its environment, it will accelerate upwards, because the upward pressure gradient force is now greater than the downward gravity force on the parcel; that is, the parcel is *buoyant* (a manifestation of Archimedes' principle) and the fluid is *statically unstable*. If on the other hand the fluid parcel finds itself heavier than its surroundings, the downward gravitational force will be greater than the upward pressure force and the fluid will sink back towards its original position and an oscillatory motion will develop. Such an equilibrium is *statically stable*. Using such simple parcel arguments we will now develop criteria for the stability of the environmental profile.

2.10.1 Stability and the Profile of Potential Density

Consider the case of a stationary fluid whose density varies with altitude. We denote this background state with a tilde, as in $\widetilde{\rho}(z)$. We then displace a fluid parcel adiabatically a small distance from z to δz, as in Fig. 2.8. In such a displacement it is the *potential density* ρ_θ (not the actual density) that is materially conserved, because potential density takes into account the effects of pressure compressibility. Let us also use the pressure at level $z+\delta z$ as the reference level, where potential density equals in situ density.

The parcel at z takes on the potential density of its environment so that $\rho_\theta(z)=\widetilde{\rho}_\theta(z)$ and it preserves this as it rises, so that $\rho_\theta(z+\delta z)=\rho_\theta(z)$. But since $z+\delta z$ is the reference level, the *in situ* density of the displaced parcel, $\rho(z+\delta z)$, is equal to its potential density $\rho_\theta(z+\delta z)$, which is equal to $\widetilde{\rho}_\theta(z)$. Thus, at $z+\delta z$ the environment has in situ density equal to $\widetilde{\rho}_\theta(z+\delta z)$ and the parcel has in situ density equal to $\widetilde{\rho}_\theta(z)$. Putting all this together in a single equation, the difference between the parcel density and the environmental density, $\delta\rho$, is given by

$$\begin{aligned}\delta\rho&=\rho(z+\delta z)-\widetilde{\rho}(z+\delta z)=\rho_\theta(z+\delta z)-\widetilde{\rho}_\theta(z+\delta z)\\&=\rho_\theta(z)-\widetilde{\rho}_\theta(z+\delta z)=\widetilde{\rho}_\theta(z)-\widetilde{\rho}_\theta(z+\delta z).\end{aligned}\tag{2.239}$$

Thus, for small δz,

$$\delta\rho=-\frac{\partial\widetilde{\rho}_\theta}{\partial z}\delta z,\tag{2.240}$$

where the derivative on the right-hand side is the environmental gradient of potential density. If the right-hand side is positive, the parcel is heavier than its surroundings and the displacement is

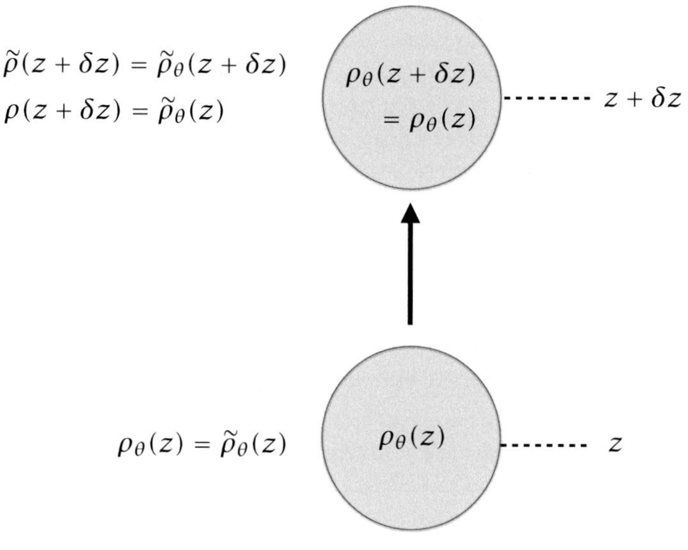

Fig. 2.8 A parcel is adiabatically displaced upward from level z to $z + \delta z$. A tilde denotes the value in the environment, and variables without tildes are those in the parcel.

The parcel preserves its potential density, ρ_θ, which it takes from the environment at level z. If $z+\delta z$ is the reference level, the potential density there is equal to the actual density. The parcel's stability is determined by the difference between its density and the environmental density, as in (2.239). If the difference is positive the displacement is stable, and if negative the displacement is unstable.

stable. That is, *the stability of a parcel of fluid is determined by the gradient of the locally-referenced potential density.*

The conditions for stability are

$$\text{Stability}: \quad \frac{\partial \widetilde{\rho}_\theta}{\partial z} < 0, \qquad \text{Instability}: \quad \frac{\partial \widetilde{\rho}_\theta}{\partial z} > 0. \tag{2.241a,b}$$

The equation of motion of the fluid parcel is then given by a direct application of Newton's second law, that the mass times the acceleration is given by the force acting on the parcel. The force is the above-derived buoyancy force so we have

$$\frac{\partial^2 \delta z}{\partial t^2} = \frac{g}{\rho}\left(\frac{\partial \widetilde{\rho}_\theta}{\partial z}\right)\delta z = -N^2 \delta z, \tag{2.242}$$

where, noting that $\rho(z) = \widetilde{\rho}_\theta(z)$ to within $O(\delta z)$,

$$N^2 = -\frac{g}{\widetilde{\rho}_\theta}\left(\frac{\partial \widetilde{\rho}_\theta}{\partial z}\right). \tag{2.243}$$

A parcel that is displaced in a stably stratified fluid will thus oscillate at the *buoyancy frequency* N, proportional to the vertical gradient of potential density. (The buoyancy frequency is also known as the Brunt–Väisälä frequency, after its discovers.) The above expression for the buoyancy frequency is a general one, true in both liquids and gases in a constant gravitational field. The quantity $\widetilde{\rho}_\theta$ is the *locally-referenced* potential density of the environment. The reference level turns out not to be important for the atmosphere, but it is for the ocean: parcels at the same level with the same in situ density may have different potential densities if their salinity differs. In contrast, for fresh water in a laboratory setting potential density is virtually equal to in situ density.

2.10.2 A Dry Ideal-gas Atmosphere

Buoyancy frequency

In the atmosphere potential density is related to potential temperature by $\rho_\theta = p_R/(\theta R)$, where p_R is the reference level for potential temperature. Using this expression in (2.243) gives

$$N^2 = \frac{g}{\tilde{\theta}}\left(\frac{\partial \tilde{\theta}}{\partial z}\right), \tag{2.244}$$

where $\tilde{\theta}$ is the environmental potential temperature. The reference value p_R does not appear, and we are free to choose this value arbitrarily — the surface pressure is a common choice. The conditions for stability, (2.241), then correspond to $N^2 > 0$ for stability and $N^2 < 0$ for instability. On average the atmosphere is stable and in the troposphere (the lowest several kilometres of the atmosphere) the average N is about $0.01\,\mathrm{s}^{-1}$, with a corresponding period, $(2\pi/N)$, of about 10 minutes. In the stratosphere (which lies above the troposphere) N is a few times higher than this.

Dry adiabatic lapse rate

The negative of the rate of change of the (real) temperature in the vertical is known as the *temperature lapse rate*, or often just the lapse rate, and denoted Γ. The lapse rate corresponding to $\partial\theta/\partial z = 0$ is called the *dry adiabatic lapse rate* and denoted Γ_d. Using $\theta = T(p_0/p)^{R/c_p}$ and $\partial p/\partial z = -\rho g$ we find that the lapse rate and the potential temperature lapse rate are related by

$$\frac{T}{\theta}\frac{\partial \theta}{\partial z} = \frac{\partial T}{\partial z} + \frac{g}{c_p}, \tag{2.245}$$

so that the dry adiabatic lapse rate is given by

$$\Gamma_d = \frac{g}{c_p}, \tag{2.246}$$

as we derived in (1.131). The conditions for static stability corresponding to (2.241) are thus:

$$\begin{aligned} \text{Stability}: &\quad \frac{\partial \tilde{\theta}}{\partial z} > 0, \quad \text{or} \quad -\frac{\partial \tilde{T}}{\partial z} < \Gamma_d. \\ \text{Instability}: &\quad \frac{\partial \tilde{\theta}}{\partial z} < 0, \quad \text{or} \quad -\frac{\partial \tilde{T}}{\partial z} > \Gamma_d. \end{aligned} \tag{2.247a,b}$$

The observed lapse rate (look ahead to Fig. 15.25) is often less than $7\,\mathrm{K\,km^{-1}}$ (corresponding to a buoyancy frequency of about $10^{-2}\,\mathrm{s}^{-1}$) whereas a dry adiabatic lapse rate is about $10\,\mathrm{K\,km}^{-1}$. Why the discrepancy? Why is the atmosphere so apparently stable? One reason is that in mid-latitudes heat is transferred upwards by in large-scale weather systems that keep the atmosphere stable even in the absence of convection. A second reason is that the atmosphere contains water vapour and a column of air that contains water may be unstable even if its lapse rate is less than dry adiabatic. If a moist parcel rises then, as it enters a cooler environment, water vapour may condense releasing more heat and leading to more ascent; that is, a moist atmosphere may be unstable when a dry atmosphere is stable. We defer more discussion to Chapters 15 and 18.

2.10.3 A Liquid Ocean

No simple, accurate, analytic expression is available for computing static stability in the ocean. If the ocean had no salt, then the potential density referenced to the surface would generally be a measure of the sign of stability of a fluid column, if not of the buoyancy frequency. However, in the presence of salinity, the surface-referenced potential density is not necessarily even a measure of the sign of stability, because the coefficients of compressibility β_T and β_S vary in different ways with pressure. To see this, suppose two neighbouring fluid elements at the surface have the same potential density, but different salinities and temperatures, and displace them both adiabatically to the deep ocean. Although their potential densities referenced to the surface are still equal, we can say little about their actual densities, and hence their stability relative to each other, without doing a detailed calculation because they will each have been compressed by different amounts. It is the profile of the *locally-referenced* potential density that determines the stability.

A useful expression for stability arises by noting that in an adiabatic displacement

$$\delta\rho_\theta = \delta\rho - \frac{1}{c_s^2}\delta p = 0. \tag{2.248}$$

If the fluid is hydrostatic $\delta p = -\rho g \delta z$, so that if a parcel is displaced adiabatically its density changes according to

$$\left(\frac{\partial\rho}{\partial z}\right)_{\rho_\theta} = -\frac{\rho g}{c_s^2}. \tag{2.249}$$

If a parcel is displaced a distance δz upwards then the density difference between it and its new surroundings is

$$\delta\rho = -\left[\left(\frac{\partial\rho}{\partial z}\right)_{\rho_\theta} - \left(\frac{\partial\tilde{\rho}}{\partial z}\right)\right]\delta z = \left[\frac{\rho g}{c_s^2} + \left(\frac{\partial\tilde{\rho}}{\partial z}\right)\right]\delta z, \tag{2.250}$$

where the tilde again denotes the environmental field. It follows that the stratification is given by

$$N^2 = -g\left[\frac{g}{c_s^2} + \frac{1}{\tilde{\rho}}\left(\frac{\partial\tilde{\rho}}{\partial z}\right)\right]. \tag{2.251}$$

This expression holds for both liquids and gases, and it is proportional to the vertical gradient of potential density. For ideal gases it is the same as (2.244), as a little algebra will show, using $c_s^2 = \gamma p/\rho$. In seawater the expression may be compared to the gradient of (1.144). The factor of g/c_s^2 is small but not negligible; it is a slightly destabilising factor in the sense that a density profile with an in situ density that increases with depth is not necessarily stable. In liquids, a good approximation is to use a reference value ρ_0 for the undifferentiated density in the denominator, whence (2.251) becomes equal to the Boussinesq expression (2.107). On average the ocean is statically stable, with typical values of N in the upper ocean being about 0.01 s^{-1}, falling to 0.001 s^{-1} in the more homogeneous abyssal ocean. These frequencies correspond to periods of about 10 and 100 minutes, respectively.

2.10.4 Gravity Waves and Convection Using the Equations of Motion

The parcel approach to oscillations and stability, while simple and direct, seems divorced from the fluid-dynamical equations of motion. To remedy this, we now use the equations of motion for a stratified Boussinesq fluid to analyze the motion resulting from a small disturbance. Our treatment here is brief and introductory, with a fuller treatment given in Chapter 7.

Consider a Boussinesq fluid, initially at rest, in which the buoyancy varies linearly with height. Linearizing the equations of motion about this basic state gives the linear momentum equations,

$$\frac{\partial u'}{\partial t} = -\frac{\partial\phi'}{\partial x}, \qquad \frac{\partial w'}{\partial t} = -\frac{\partial\phi'}{\partial z} + b', \tag{2.252a,b}$$

the mass continuity and thermodynamic equations,

$$\frac{\partial u'}{\partial x} + \frac{\partial w'}{\partial z} = 0, \qquad \frac{\partial b'}{\partial t} + w' N^2 = 0, \tag{2.253a,b}$$

where $N^2 = \mathrm{d}\tilde{b}/\mathrm{d}z$ is the basic state buoyancy profile, and we assume that the flow is a function only of x and z. A little algebra reduces the above equations to a single one for w',

$$\left[\left(\frac{\partial^2}{\partial x^2} + \frac{\partial^2}{\partial z^2} \right) \frac{\partial^2}{\partial t^2} + N^2 \frac{\partial^2}{\partial x^2} \right] w' = 0. \tag{2.254}$$

Seeking solutions of the form $w' = \mathrm{Re}\, W \exp[\mathrm{i}(kx + mz - \omega t)]$ yields the dispersion relationship for gravity waves:

$$\omega^2 = \frac{k^2 N^2}{k^2 + m^2}. \tag{2.255}$$

The frequency (look ahead to Fig. 7.2) is always less than N, approaching N for small horizontal scales, $k \gg m$.

Consider two special cases. First, if we neglect pressure perturbations, as in the parcel argument, then the two equations,

$$\frac{\partial w'}{\partial t} = b', \qquad \frac{\partial b'}{\partial t} + w' N^2 = 0, \tag{2.256}$$

form a closed set and give $\omega^2 = N^2$, as in the parcel argument. Second, if we make the hydrostatic approximation and omit $\partial w'/\partial t$ in (2.252b) then the dispersion relation becomes $\omega^2 = k^2 N^2/m^2$. The frequency then grows, artifactually, without bound as the horizontal scale becomes smaller.

If the basic state density increases with height then $N^2 < 0$ and we expect this state to be unstable. Indeed, the disturbance grows exponentially according to $\exp(\sigma t)$ where $\sigma = \mathrm{i}\omega = \pm k\widetilde{N}/(k^2 + m^2)^{1/2}$, and where $\widetilde{N}^2 = -N^2$. We have reproduced the result previously obtained by parcel theory, namely that if the basic state density (or more generally potential density) increases with height the flow is unstable. Most convective activity in the ocean and atmosphere is, in the end, related to an instability of this form.

APPENDIX A: ASYMPTOTIC DERIVATION OF THE BOUSSINESQ EQUATIONS

The Boussinesq equations are those equations that are appropriate when the density variations are very small but gravitational effects are large, and here we provide an asymptotic derivation. Two key results are that the velocity field is divergence-free, and the buoyancy should be taken as a function of z and not p in the equation of state. The first result follows from fact that density variations are presumptively small, and the second follows because the lowest order balance in the vertical momentum equation is $\partial p_0/\partial z = -\rho_0 g$, whence $p_0 = -\rho_0 g z$, and p_0 and not p should be used in the equation of state at lowest order. The following derivation, which assumes familiarity with elementary asymptotics (or which can be taken as a gentle introduction to asymptotics) mainly just formalizes these results.

Let us suppose that the density varies like $\rho(x, y, z, t) = \rho_0 + \delta\rho(x, y, z, t)$, where ρ_0 is a constant and $|\delta\rho| \ll \rho_0$. Specifically, let $\epsilon\rho_0$ be a typical magnitude for $\delta\rho$ where $\epsilon \ll 1$ so that

$$\delta\rho = (\epsilon\rho_0)\delta\widehat{\rho} \qquad \text{and} \qquad \rho = \rho_0(1 + \epsilon\delta\widehat{\rho}) \tag{2.257}$$

where a hat denotes a nondimensional quantity and $\delta\widehat{\rho}$ is an $\mathcal{O}(1)$ quantity.

The dimensional vertical momentum equation, omitting rotation and viscosity for simplicity, is

$$(\rho_0 + \delta\rho)\frac{\mathrm{D}w}{\mathrm{D}t} = -\frac{\partial p}{\partial z} - (\rho_0 + \delta\rho)g. \tag{2.258}$$

Now, g is 'big' and variations in density are 'small' and so $g\delta\rho$ is taken to be the same approximate size as the advection term $\rho_0 \mathrm{D}w/\mathrm{D}t$ on the left-hand side. The term $\rho_0 g$ must then be balanced by the pressure gradient. Also, there is no necessary difference between vertical and horizontal scales and velocities. With these points in mind we nondimensionalize with the following scales:

$$(u, v, w) = U(\hat{u}, \hat{v}, \hat{w}), \quad (x, y, z) = L(\hat{x}, \hat{y}, \hat{z}), \quad t = \frac{L}{U}\hat{t}, \quad p = \rho_0 \frac{U^2}{\epsilon}\hat{p}, \quad g = \frac{U^2}{\epsilon L}\hat{g}, \tag{2.259}$$

where the hatted quantities are nondimensional and are presumptively $\mathcal{O}(1)$. Equation (2.258) becomes

$$(1 + \epsilon\delta\hat{\rho})\frac{\mathrm{D}\hat{w}}{\mathrm{D}\hat{t}} = -\frac{1}{\epsilon}\frac{\partial \hat{p}}{\partial \hat{z}} - \frac{1}{\epsilon}(1 + \epsilon\delta\hat{\rho})\hat{g}. \tag{2.260}$$

We now take ϵ as an asymptotic ordering parameter and expand the nondimensional fields as series in ϵ. Then, with subscripts denoting the asymptotic order (for nondimensional quantities only), we have

$$\delta\hat{\rho} = \delta\hat{\rho}_0 + \epsilon\delta\hat{\rho}_1 + \epsilon^2\delta\hat{\rho}_2 \dots, \qquad \hat{p} = \hat{p}_0 + \epsilon\hat{p}_1 + \epsilon^2\hat{p}_2 \dots, \qquad \hat{w} = \hat{w}_0 + \epsilon\hat{w}_1 + \epsilon^2\hat{w}_2 \dots, \tag{2.261}$$

and similarly for $\hat{u}$ and $\hat{v}$. If we substitute the above series into (2.260) and equate terms with the same power of ϵ, the first two orders are

$$\frac{\partial \hat{p}_0}{\partial \hat{z}} = -\hat{g}, \qquad \frac{\mathrm{D}\hat{w}_0}{\mathrm{D}\hat{t}} = -\frac{\partial \hat{p}_1}{\partial \hat{z}} - \delta\hat{\rho}_0 g. \tag{2.262a,b}$$

Evidently the leading order pressure, $\hat{p}_0$, is hydrostatic. We may now revert to dimensional variables and (2.262a) gives $p_0(z) = -\rho_0 g z$, and (2.262b) becomes

$$\frac{\mathrm{D}w}{\mathrm{D}t} = -\frac{1}{\rho_0}\frac{\partial p'}{\partial z} - \frac{\delta\rho}{\rho_0}g \qquad \text{or} \qquad \frac{\mathrm{D}w}{\mathrm{D}t} = -\frac{\partial \phi}{\partial z} + b, \tag{2.263}$$

where p' is the deviation from the hydrostatic pressure, $\phi = p'/\rho_0$ and $b = -g\delta\rho/\rho_0$.

In the horizontal momentum equations only the perturbation pressure p' appears since p_0 is a function of z only. Furthermore, the density must be taken to be the constant ρ_0 (since this is $1/\epsilon$ larger than $\delta\rho$). Then, if $\boldsymbol{u}$ is the horizontal velocity, (u, v), and $\boldsymbol{v}$ is the three-dimensional one, (u, v, w), we obtain

$$\frac{\partial \boldsymbol{u}}{\partial t} + \boldsymbol{v} \cdot \nabla \boldsymbol{u} = -\frac{1}{\rho_0}\nabla_z p' \qquad \text{or} \qquad \frac{\mathrm{D}\boldsymbol{u}}{\mathrm{D}t} = -\nabla_z \phi, \tag{2.264}$$

where the gradients (i.e., ∇_z) on the right-hand side are horizontal, at constant z. Equations (2.263) and (2.264) combine to give

$$\frac{\mathrm{D}\boldsymbol{v}}{\mathrm{D}t} = -\nabla\phi + b\mathbf{k}, \tag{2.265}$$

where $\mathbf{k}$ is the vertical unit vector and frictional and rotational terms may be added as needed.

The mass continuity equation is

$$\frac{\mathrm{D}}{\mathrm{D}t}(\rho_0 + \delta\rho) + (\rho_0 + \delta\rho)\nabla \cdot \boldsymbol{v} = 0. \tag{2.266}$$

Since $\mathrm{D}\rho_0/\mathrm{D}t = 0$ and ρ_0 is $1/\epsilon$ larger than $\delta\rho$, we see without further ado that the lowest order mass conservation equation is that of an incompressible, divergence-free fluid, namely $\nabla \cdot \boldsymbol{v} = 0$.

Boussinesq thermodynamics

The Boussinesq equations are completed with a thermal equation of state, an equation of evolution for the composition (e.g., S, the salinity), and an equation that provides a thermodynamic state variable Θ (e.g., entropy or internal energy). The thermal equation of state gives the density in terms of pressure, composition and the thermodynamic variable and so is of the form $b = b(p, \Theta, S)$, meaning b is a function of (p, Θ, S). However, to be consistent with the derivation of (2.263) we should use the lowest order variables on the right-hand side of (2.267), and specifically *we should use p_0 and not p itself.* The equation of state becomes

$$b = b(p_0, \Theta, S) \qquad \text{or} \qquad b = b(z, \Theta, S), \tag{2.267}$$

and an example is (1.155); that equation gives the inverse density, α, in terms of z, potential temperature and salinity, and from which an expression for b immediately follows.

Evolution of the thermodynamic variable is obtained from the first law, and the discussions of Sections 1.6 and 1.7.3 generally apply. The internal energy equation is, with no external source or diffusion,

$$\frac{\mathrm{D}I}{\mathrm{D}t} + p\alpha\nabla\cdot\boldsymbol{v} = 0. \tag{2.268}$$

The lowest order velocities are divergence-free and at lowest order we thus have $\mathrm{D}I/\mathrm{D}t = 0$. (To determine the actual magnitudes of the terms, we can obtain the internal energy from the Gibbs function using $I = g + \eta T - p\alpha = g - T\partial g/\partial T - p\partial g/\partial p$, and with the seawater Gibbs function, (1.146), we obtain $I = c_{p0}T$ + smaller terms. Referring to the values in table 1.2, $c_{p0}\Delta T$ is roughly comparable to $p\alpha$, so that $p\alpha\nabla\cdot\boldsymbol{v}$ is indeed much smaller than $\mathrm{D}I/\mathrm{D}t$.) If we evolved I, we would then need an equation of state of the form $b = b(z, I, S)$ to complete the system.

However, as for the full (non-Boussinesq) system it is often advantageous to evolve potential enthalpy, or potential temperature or entropy, rather than internal energy. We then obtain buoyancy using an accurate equation of state but with the hydrostatic pressure instead of the full pressure. For example, and for idealized or laboratory work with fresh water, a thermodynamic equation and an approximate equation of state might use (1.128) and a simplified form of (1.155) giving

$$\frac{\mathrm{D}\theta}{\mathrm{D}t} = 0 \qquad \text{and} \qquad b = g\frac{\delta\alpha}{\alpha_0} = g\left[\frac{gz}{c_s^2} + \beta_T(\theta - \theta_0)\right], \tag{2.269}$$

where θ is potential temperature, θ_0 is a constant reference value and c_s is the speed of sound. The potential temperature is related to the actual temperature by $\theta = T + \beta_T g\theta_0 z/c_p$, but T is not needed to evolve the system. The term in gz/c_s^2 is often small enough to be neglected, in which case the thermodynamic equation becomes an evolution of buoyancy itself, $\mathrm{D}b/\mathrm{D}t = 0$.

Notes

1 The geocentric view was slowly and contentiously being replaced by the Copernican or heliocentric view during Shakespeare's lifetime, with the upheaval of matters thought settled and stable. Galileo, whose telescopes helped confirm the heliocentric view, was born in the same year as Shakespeare, 1564. In the geocentric view Earth's surface is an inertial frame and there is no Coriolis force.

2 The distinction between Coriolis force and acceleration is not always made in the literature, even after noting that the force is considered as a force per unit mass. For a fluid in geostrophic balance, one might either say that there is a balance between the pressure force and the Coriolis force, with no net acceleration, or that the pressure force produces a Coriolis acceleration. The descriptions are equivalent, because of Newton's second law, but should not be conflated.

The Coriolis effect is named after Gaspard Gustave de Coriolis (1792–1843), who discussed the eponymous force in the context of rotating mechanical systems (Coriolis 1832, 1835), but Euler was aware of the effect almost a century before. Persson (1998) provides a historical account.

3 Phillips (1973). A related discussion can be found in Stommel & Moore (1989).

4 Phillips (1966) and White (2002, 2003) form a pleasing set of review articles that synthesize the various forms and approximations of the equations of motion. In the early days of numerical modelling the primitive equations were indeed the most primitive — i.e., the least filtered — equations that could practically be integrated numerically. Associated with increasing computer power there is a tendency for comprehensive numerical models to use non-hydrostatic equations of motion that do not make the shallow-fluid or traditional approximations, and it is conceivable that the meaning of the word 'primitive' may evolve to accommodate them.

5 It is nevertheless possible to derive dynamically consistent equations for a shallow atmosphere that do not make the traditional approximation (Tort & Dubos 2014, Dellar 2011). See White *et al.* (2005) for a related discussion.

6 The Boussinesq approximation is named for Boussinesq (1903), although similar approximations were used earlier by Oberbeck (1879, 1888). Spiegel & Veronis (1960) give a physically based derivation for an ideal gas, and Mihaljan (1962) and Gray & Giorgini (1976) provide more systematic derivations that include the effects of viscosity and diffusion and discussions of the energetics.

7 Young (2010).

8 Various versions of anelastic and pseudo-incompressible equations exist — see Batchelor (1953a), Ogura & Phillips (1962), Gough (1969), Gilman & Glatzmaier (1981), Lipps & Hemler (1982), and Durran (1989), although not all have potential vorticity and energy conservation laws (Bannon 1995, 1996; Scinocca & Shepherd 1992). The system we derive is most similar to that of Ogura & Phillips (1962) and unpublished notes by J. S. A. Green. The connection between the Boussinesq and anelastic equations is discussed by Lilly (1996) and Ingersoll (2005), and the extension to a complex equation of state and inclusion of moisture is discussed by Pauluis (2008).

9 A numerical model that explicitly includes sound waves must take very small timesteps in order to maintain numerical stability, in particular to satisfy the Courant–Friedrichs–Lewy (CFL) criterion. An alternative is to use implicit timestepping that effectively lets the numerics filter the sound waves. If we make the hydrostatic approximation then all sound waves except those that propagate horizontally are eliminated, and there is little numerical need to make the anelastic approximation.

10 Gill (1982) provides a longer discussion. Not all authors differentiate between N and N_*.

11 The Rossby number is named for C.-G. Rossby (see endnote 2 on page 211), but it was also used by Kibel (1940) and is sometimes called the Kibel or Rossby–Kibel number. The notion of geostrophic balance and so, implicitly, that of a small Rossby number, predates both Rossby and Kibel.

12 After Taylor (1921b) and Proudman (1916). The Taylor–Proudman effect is sometimes called the Taylor–Proudman 'theorem', but it is more usefully thought of as a physical effect, with manifestations even when the conditions for its satisfaction are not precisely met. In fact, Hough (1897) seems to have been aware of the effect well before Taylor and Proudman.

13 This discussion owes much to that in Holton (1992). Inertial motion is discussed by Durran (1993).

Jim Holton (1938–2004) made many contributions to atmospheric dynamics over the course of a distinguished career spent almost entirely at the University of Washington in Seattle. In his early career he elucidated, with Richard Lindzen of MIT, the essential mechanism of the quasi-biennial oscillation, or QBO, and he continued to make important contributions to wave–mean-flow interaction, stratosphere-troposphere interaction and stratospheric dynamics more generally throughout his career. He is also known, both to scientists and students, for his popular textbook *An Introduction to Dynamical Meteorology.*

Since all models are wrong the scientist cannot obtain a 'correct' one by excessive elaboration ... he should seek an economical description of natural phenomena.

George E. Box, *Science and Statistics*, 1976.

The sciences do not try to explain ... they mainly make models ... a mathematical construct the justification [of which] is that it is expected to work.

John von Neumann, *Methods in the Physical Sciences*, 1955.

CHAPTER 3

Shallow Water Systems

CONVENTIONALLY, 'THE' SHALLOW WATER EQUATIONS describe a thin layer of constant density fluid in hydrostatic balance, rotating or not, bounded from below by a rigid surface and from above by a free surface, above which we suppose is another fluid of negligible inertia. Such a configuration can be generalized to multiple layers of immiscible fluids of different densities lying one on top of another, forming a stably-stratified 'stacked shallow water' system, which in many ways behaves like a continuously stratified fluid. These types of systems are the main subject of this chapter. We also introduce the notion of available potential energy, which involves thinking about a continuously stratified system as if it were a stacked shallow water system.

The single-layer model is one of the simplest useful models in geophysical fluid dynamics because it allows for a consideration of the effects of rotation in a simple framework without the complicating effects of stratification. A model with just two layers is not only a simple model of a stratified fluid, it is a surprisingly good model of many phenomena in the ocean and atmosphere. Such models are more than just pedagogical tools — we will find that there is a close physical and mathematical analogy between the shallow water equations and a description of the continuously stratified ocean or atmosphere written in isopycnal or isentropic coordinates, with a meaning beyond a coincidental similarity in the equations. Let us begin with the single-layer case.

3.1 DYNAMICS OF A SINGLE SHALLOW LAYER OF FLUID

Shallow water dynamics apply, by definition, to a fluid layer of constant density in which the horizontal scale of the flow is much greater than the layer depth. The fluid motion is fully determined by the momentum and mass continuity equations, and because of the assumed small aspect ratio the hydrostatic approximation is well satisfied, and we invoke this from the outset. Consider, then, fluid in a container above which is another fluid of negligible density (and therefore negligible inertia) relative to the fluid of interest, as illustrated in Fig. 3.1. Our notation is that $\boldsymbol{v} = u\mathbf{i} + v\mathbf{j} + w\mathbf{k}$ is the three-dimensional velocity and $\boldsymbol{u} = u\mathbf{i} + v\mathbf{j}$ is the horizontal velocity. $h(x, y)$ is the thickness of the liquid column, H is its mean height, and η is the height of the free surface. In a flat-bottomed container $\eta = h$, whereas in general $h = \eta - \eta_b$, where η_b is the height of the floor of the container.

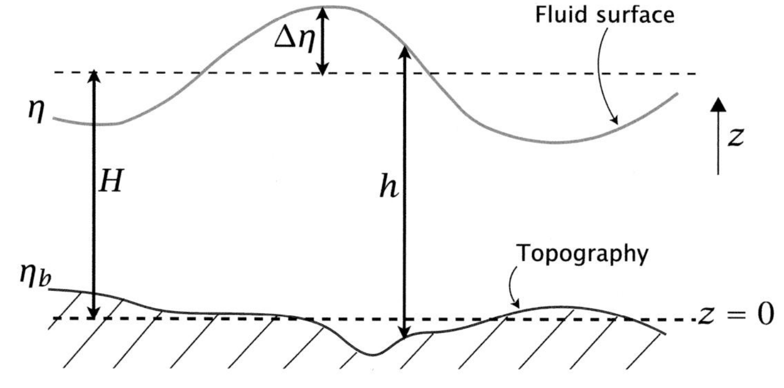

Fig. 3.1 A shallow water system. h is the thickness of a water column, H its mean thickness, η the height of the free surface and η_b is the height of the lower, rigid, surface above some arbitrary origin, typically chosen such that the average of η_b is zero. $\Delta\eta$ is the deviation free surface height, so we have $\eta = \eta_b + h = H + \Delta\eta$.

3.1.1 Momentum Equations

The vertical momentum equation is just the hydrostatic equation,

$$\frac{\partial p}{\partial z} = -\rho_0 g, \tag{3.1}$$

and, because density is assumed constant, we may integrate this to

$$p(x, y, z, t) = -\rho_0 g z + p_o. \tag{3.2}$$

At the top of the fluid, $z = \eta$, the pressure is determined by the weight of the overlying fluid and this is assumed to be negligible. Thus, $p = 0$ at $z = \eta$, giving

$$p(x, y, z, t) = \rho_0 g(\eta(x, y, t) - z). \tag{3.3}$$

The consequence of this is that the horizontal gradient of pressure is independent of height. That is

$$\nabla_z p = \rho_0 g \nabla_z \eta, \tag{3.4}$$

where

$$\nabla_z = \mathbf{i}\frac{\partial}{\partial x} + \mathbf{j}\frac{\partial}{\partial y} \tag{3.5}$$

is the gradient operator at constant z. (In the rest of this chapter we will drop the subscript z unless that causes ambiguity. The three-dimensional gradient operator will be denoted by ∇_3. We will also mostly use Cartesian coordinates, but the shallow water equations may certainly be applied over a spherical planet — 'Laplace's tidal equations' are essentially the shallow water equations on a sphere.) The horizontal momentum equations therefore become

$$\frac{\mathrm{D}\boldsymbol{u}}{\mathrm{D}t} = -\frac{1}{\rho_0}\nabla p = -g\nabla\eta. \tag{3.6}$$

The right-hand side of this equation is independent of the vertical coordinate z. Thus, if the flow is initially independent of z, it must stay so. (This z-independence is unrelated to that arising from the rapid rotation necessary for the Taylor–Proudman effect.) The velocities u and v are functions of x, y and t only, and the horizontal momentum equation is therefore

$$\frac{\mathrm{D}\boldsymbol{u}}{\mathrm{D}t} = \frac{\partial \boldsymbol{u}}{\partial t} + u\frac{\partial \boldsymbol{u}}{\partial x} + v\frac{\partial \boldsymbol{u}}{\partial y} = -g\nabla\eta. \tag{3.7}$$

That the horizontal velocity is independent of z is a consequence of the hydrostatic equation, which ensures that the horizontal pressure gradient is independent of height. (Another starting point

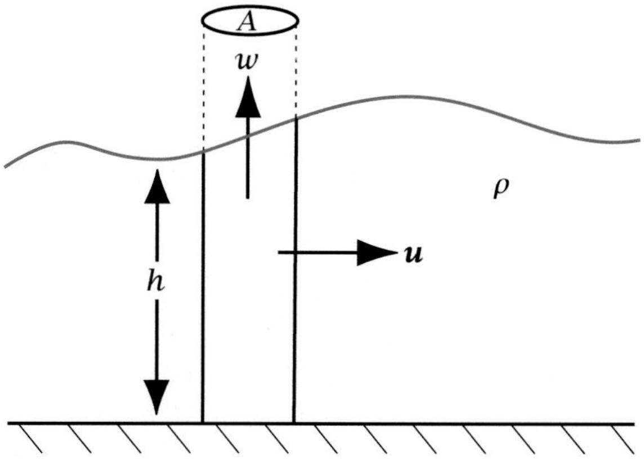

Fig. 3.2 The mass budget for a column of area A in a shallow water system. The fluid leaving the column is $\oint \rho_0 h\boldsymbol{u} \cdot \boldsymbol{n}\,\mathrm{d}l$ where $\boldsymbol{n}$ is the unit vector normal to the boundary of the fluid column. There is a non-zero vertical velocity at the top of the column if the mass convergence into the column is non-zero.

would be to take this independence of the horizontal motion with height as the *definition* of shallow water flow. In real physical situations such independence does not hold exactly — for example, friction at the bottom may induce a vertical dependence of the flow in a boundary layer.) In the presence of rotation, (3.7) easily generalizes to

$$\frac{\mathrm{D}\boldsymbol{u}}{\mathrm{D}t} + \boldsymbol{f} \times \boldsymbol{u} = -g\nabla\eta, \tag{3.8}$$

where $\boldsymbol{f} = f\mathbf{k}$. Just as with the primitive equations, f may be constant or may vary with latitude, so that on a spherical planet $f = 2\Omega \sin\vartheta$ and on the β-plane $f = f_0 + \beta y$.

3.1.2 Mass Continuity Equation

From first principles

The mass contained in a fluid column of height h and cross-sectional area A is given by $\int_A \rho_0 h\,\mathrm{d}\boldsymbol{A}$ (see Fig. 3.2). If there is a net flux of fluid across the column boundary (by advection) then this must be balanced by a net increase in the mass in A, and therefore a net increase in the height of the water column. The mass convergence into the column is given by

$$F_m = \text{mass flux in} = -\int_S \rho_0 \boldsymbol{u} \cdot \mathrm{d}\mathbf{S}, \tag{3.9}$$

where S is the area of the vertical boundary of the column. The surface area of the column is composed of elements of area $h\boldsymbol{n}\,\delta l$, where δl is a line element circumscribing the column and $\boldsymbol{n}$ is a unit vector perpendicular to the boundary, pointing outwards. Thus (3.9) becomes

$$F_m = -\oint \rho_0 h\boldsymbol{u} \cdot \boldsymbol{n}\,\mathrm{d}l. \tag{3.10}$$

Using the divergence theorem in two dimensions, (3.10) simplifies to

$$F_m = -\int_A \nabla \cdot (\rho_0 \boldsymbol{u} h)\,\mathrm{d}A, \tag{3.11}$$

where the integral is over the cross-sectional area of the fluid column (looking down from above). This is balanced by the local increase in height of the water column, given by

$$F_m = \frac{\mathrm{d}}{\mathrm{d}t}\int \rho_0\,\mathrm{d}V = \frac{\mathrm{d}}{\mathrm{d}t}\int_A \rho_0 h\,\mathrm{d}A = \int_A \rho_0 \frac{\partial h}{\partial t}\,\mathrm{d}A. \tag{3.12}$$

Because ρ_0 is constant, the balance between (3.11) and (3.12) leads to

$$\int_A \left[\frac{\partial h}{\partial t} + \nabla \cdot (\boldsymbol{u}h) \right] \mathrm{d}A = 0, \tag{3.13}$$

and because the area is arbitrary the integrand itself must vanish, whence,

$$\frac{\partial h}{\partial t} + \nabla \cdot (\boldsymbol{u}h) = 0 \qquad \text{or} \qquad \frac{\mathrm{D}h}{\mathrm{D}t} + h\nabla \cdot \boldsymbol{u} = 0. \tag{3.14a,b}$$

This derivation holds whether or not the lower surface is flat. If it is, then $h = \eta$, and if not $h = \eta - \eta_b$. Equations (3.8) and (3.14) form a complete set, summarized in the shaded box on the facing page.

From the 3D mass conservation equation

Since the fluid is incompressible, the three-dimensional mass continuity equation is just $\nabla \cdot \boldsymbol{v} = 0$. Writing this out in component form

$$\frac{\partial w}{\partial z} = -\left(\frac{\partial u}{\partial x} + \frac{\partial v}{\partial y} \right) = -\nabla \cdot \boldsymbol{u}. \tag{3.15}$$

Integrate this from the bottom of the fluid ($z = \eta_b$) to the top ($z = \eta$), noting that the right-hand side is independent of z, to give

$$w(\eta) - w(\eta_b) = -h\nabla \cdot \boldsymbol{u}. \tag{3.16}$$

At the top the vertical velocity is the material derivative of the position of a particular fluid element. But the position of the fluid at the top is just η, and therefore (see Fig. 3.2)

$$w(\eta) = \frac{\mathrm{D}\eta}{\mathrm{D}t}. \tag{3.17a}$$

At the bottom of the fluid we have similarly

$$w(\eta_b) = \frac{\mathrm{D}\eta_b}{\mathrm{D}t}, \tag{3.17b}$$

where, apart from earthquakes and the like, $\partial \eta_b / \partial t = 0$. Using (3.17a,b), (3.16) becomes

$$\frac{\mathrm{D}}{\mathrm{D}t}(\eta - \eta_b) + h\nabla \cdot \boldsymbol{u} = 0 \tag{3.18}$$

or, as in (3.14b),

$$\frac{\mathrm{D}h}{\mathrm{D}t} + h\nabla \cdot \boldsymbol{u} = 0. \tag{3.19}$$

3.1.3 A Rigid Lid

The case where the *upper* surface is held flat by the imposition of a rigid lid is sometimes of interest. The ocean suggests one such example, since the bathymetry at the bottom of the ocean provides much larger variations in fluid thickness than do the small variations in the height of the ocean surface. If we suppose that the upper surface is at a constant height H, then from (3.14a) with $\partial h / \partial t = 0$ the mass conservation equation is

$$\nabla_h \cdot (\boldsymbol{u}h_b) = 0, \tag{3.20}$$

The Shallow Water Equations

For a single-layer fluid, and including the Coriolis term, the inviscid shallow water equations are

$$\text{momentum:} \qquad \frac{D\boldsymbol{u}}{Dt} + \boldsymbol{f} \times \boldsymbol{u} = -g\nabla\eta. \tag{SW.1}$$

$$\text{mass continuity:} \qquad \frac{Dh}{Dt} + h\nabla\cdot\boldsymbol{u} = 0 \quad \text{or} \quad \frac{\partial h}{\partial t} + \nabla\cdot(h\boldsymbol{u}) = 0, \tag{SW.2}$$

where $\boldsymbol{u}$ is the horizontal velocity, h is the total fluid thickness, η is the height of the upper free surface and η_b is the height of the lower surface (the bottom topography). Thus,

$$h(x, y, t) = \eta(x, y, t) - \eta_b(x, y) \tag{SW.3}$$

The material derivative is

$$\frac{D}{Dt} = \frac{\partial}{\partial t} + \boldsymbol{u}\cdot\nabla = \frac{\partial}{\partial t} + u\frac{\partial}{\partial x} + v\frac{\partial}{\partial y}, \tag{SW.4}$$

with the rightmost expression holding in Cartesian coordinates.

where $h_b = H - \eta_b$. Note that (3.20) allows us to define an incompressible *mass-transport velocity*, $\boldsymbol{U} \equiv h_b\boldsymbol{u}$.

Although the upper surface is flat, the pressure there is no longer constant because a force must be provided by the rigid lid to keep the surface flat. The horizontal momentum equation is

$$\frac{D\boldsymbol{u}}{Dt} = -\frac{1}{\rho_0}\nabla p_{lid}, \tag{3.21}$$

where p_{lid} is the pressure at the lid, and the complete equations of motion are then (3.20) and (3.21).[1] If the lower surface is flat, the two-dimensional flow itself is divergence-free, and the equations reduce to the two-dimensional incompressible Euler equations.

3.1.4 Stretching and the Vertical Velocity

Because the horizontal velocity is depth independent, the vertical velocity plays no role in advection. However, w is certainly not zero for then the free surface would be unable to move up or down, but because of the vertical independence of the horizontal flow w does have a simple vertical structure; to determine this we write the mass conservation equation as

$$\frac{\partial w}{\partial z} = -\nabla\cdot\boldsymbol{u}, \tag{3.22}$$

and integrate upwards from the bottom to give

$$w = w_b - (\nabla\cdot\boldsymbol{u})(z - \eta_b). \tag{3.23}$$

Thus, the vertical velocity is a linear function of height. Equation (3.23) can be written as

$$\frac{Dz}{Dt} = \frac{D\eta_b}{Dt} - (\nabla\cdot\boldsymbol{u})(z - \eta_b), \tag{3.24}$$

and at the upper surface $w = \mathrm{D}\eta/\mathrm{D}t$ so that here we have

$$\frac{\mathrm{D}\eta}{\mathrm{D}t} = \frac{\mathrm{D}\eta_b}{\mathrm{D}t} - (\nabla \cdot \boldsymbol{u})(\eta - \eta_b). \tag{3.25}$$

Eliminating the divergence term from the last two equations gives

$$\frac{\mathrm{D}}{\mathrm{D}t}(z - \eta_b) = \frac{z - \eta_b}{\eta - \eta_b}\frac{\mathrm{D}}{\mathrm{D}t}(\eta - \eta_b), \tag{3.26}$$

which in turn gives

$$\frac{\mathrm{D}}{\mathrm{D}t}\left(\frac{z - \eta_b}{\eta - \eta_b}\right) = \frac{\mathrm{D}}{\mathrm{D}t}\left(\frac{z - \eta_b}{h}\right) = 0. \tag{3.27}$$

This means that the ratio of the height of a fluid parcel above the floor to the total depth of the column is fixed; that is, the fluid stretches uniformly in a column, and this is a kinematic property of the shallow water system.

3.1.5 Analogy with Compressible Flow

The shallow water equations (3.8) and (3.14) are analogous to the compressible gas dynamic equations in two dimensions, namely

$$\frac{\mathrm{D}\boldsymbol{u}}{\mathrm{D}t} = -\frac{1}{\rho}\nabla p \tag{3.28}$$

and

$$\frac{\partial \rho}{\partial t} + \nabla \cdot (\boldsymbol{u}\rho) = 0, \tag{3.29}$$

along with an equation of state which we take to be $p = f(\rho)$. The mass conservation equations (3.14) and (3.29) are identical, with the replacement $\rho \leftrightarrow h$. If $p = C\rho^\gamma$, then (3.28) becomes

$$\frac{\mathrm{D}\boldsymbol{u}}{\mathrm{D}t} = -\frac{1}{\rho}\frac{\mathrm{d}p}{\mathrm{d}\rho}\nabla\rho = -C\gamma\rho^{\gamma-2}\nabla\rho. \tag{3.30}$$

If $\gamma = 2$ then the momentum equations (3.8) and (3.30) become equivalent, with $\rho \leftrightarrow h$ and $C\gamma \leftrightarrow g$. In an ideal gas $\gamma = c_p/c_v$ and values typically are in fact less than 2 (in air $\gamma \approx 7/5$); however, if the equations are linearized, then the analogy is exact for all values of γ, for then (3.30) becomes $\partial \boldsymbol{v}'/\partial t = -\rho_0^{-1}c_s^2\nabla\rho'$ where $c_s^2 = \mathrm{d}p/\mathrm{d}\rho$, and the linearized shallow water momentum equation is $\partial \boldsymbol{u}'/\partial t = -H^{-1}(gH)\nabla h'$, so that $\rho_0 \leftrightarrow H$ and $c_s^2 \leftrightarrow gH$. The sound waves of a compressible fluid are then analogous to shallow water waves, which are considered in Section 3.8.

3.2 REDUCED GRAVITY EQUATIONS

Consider now a single shallow moving layer of fluid on top of a deep, quiescent fluid layer (Fig. 3.3), and beneath a fluid of negligible inertia. This configuration is often used as a model of the upper ocean: the upper layer represents flow in perhaps the upper few hundred metres of the ocean, the lower layer being the near-stagnant abyss. If we turn the model upside-down we have a perhaps slightly less realistic model of the atmosphere: the lower layer represents motion in the troposphere above which lies an inactive stratosphere. The equations of motion are virtually the same in both cases.

3.2.1 Pressure Gradient in the Active Layer

We will derive the equations for the oceanic case (active layer on top) in two cases, which differ slightly in the assumption made about the upper surface.

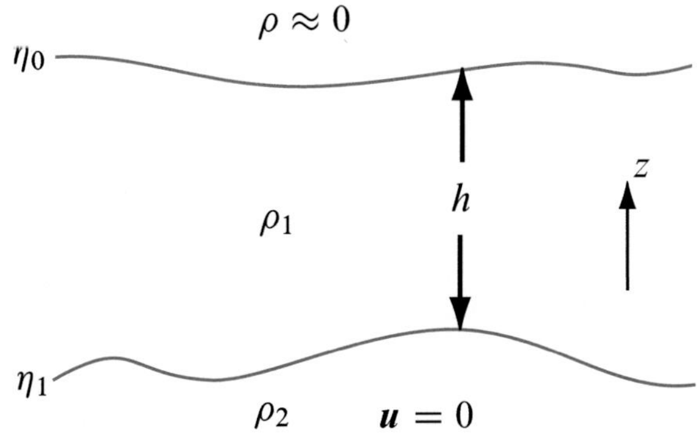

Fig. 3.3 The reduced gravity shallow water system. An active layer lies over a deep, denser, quiescent layer. In a common variation the upper surface is held flat by a rigid lid, and $\eta_0 = 0$.

I Free upper surface

The pressure in the upper layer is given by integrating the hydrostatic equation down from the upper surface. Thus, at a height z in the upper layer

$$p_1(z) = g\rho_1(\eta_0 - z), \tag{3.31}$$

where η_0 is the height of the upper surface. Hence, everywhere in the upper layer,

$$\frac{1}{\rho_1}\nabla p_1 = g\nabla\eta_0, \tag{3.32}$$

and the momentum equation is

$$\frac{\mathrm{D}\boldsymbol{u}}{\mathrm{D}t} + \boldsymbol{f} \times \boldsymbol{u} = -g\nabla\eta_0. \tag{3.33}$$

In the lower layer the pressure is also given by the weight of the fluid above it. Thus, at some level z in the lower layer,

$$p_2(z) = \rho_1 g(\eta_0 - \eta_1) + \rho_2 g(\eta_1 - z). \tag{3.34}$$

But if this layer is motionless the horizontal pressure gradient in it is zero and therefore

$$\rho_1 g\eta_0 = -\rho_1 g'\eta_1 + \text{constant}, \tag{3.35}$$

where $g' = g(\rho_2 - \rho_1)/\rho_1$ is the *reduced gravity*, and normally $(\rho_2 - \rho_1)/\rho \ll 1$ and $g' \ll g$. The momentum equation becomes

$$\frac{\mathrm{D}\boldsymbol{u}}{\mathrm{D}t} + \boldsymbol{f} \times \boldsymbol{u} = g'\nabla\eta_1. \tag{3.36}$$

The equations are completed by the usual mass conservation equation,

$$\frac{\mathrm{D}h}{\mathrm{D}t} + h\nabla\cdot\boldsymbol{u} = 0, \tag{3.37}$$

where $h = \eta_0 - \eta_1$. Because $g \gg g'$, (3.35) shows that surface displacements are *much smaller* than the displacements at the interior interface. We see this in the real ocean where the mean interior isopycnal displacements may be several tens of metres but variations in the mean height of ocean surface are of the order of centimetres.

II The rigid lid approximation

The smallness of the upper surface displacement suggests that we will make little error if we impose a *rigid lid* at the top of the fluid. Displacements are no longer allowed, but the lid will in general

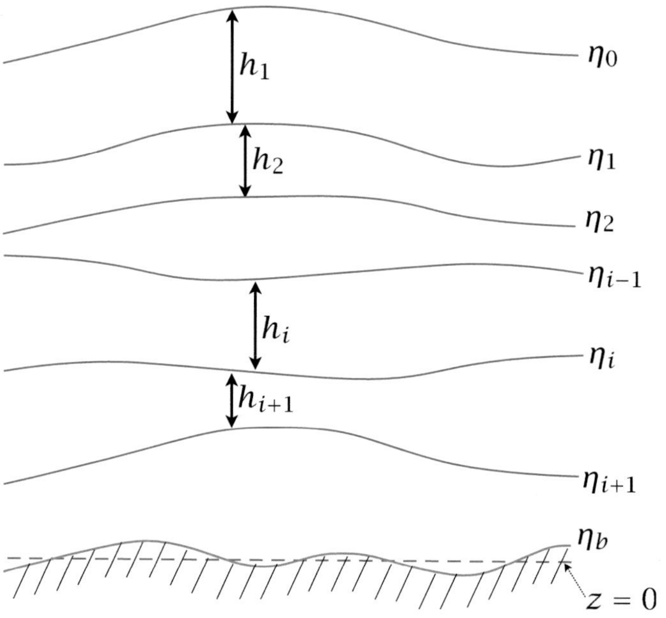

Fig. 3.4 The multi-layer shallow water system. The layers are numbered from the top down. The coordinates of the interfaces are denoted by η, and the layer thicknesses by h, so that $h_i = \eta_{i-1} - \eta_i$.

impart a pressure force to the fluid. Suppose that this is $P(x, y, t)$, then the horizontal pressure gradient in the upper layer is simply

$$\nabla p_1 = \nabla P. \tag{3.38}$$

The pressure in the lower layer is again given by hydrostasy, and is

$$p_2 = -\rho_1 g \eta_1 + \rho_2 g(\eta_1 - z) + P = \rho_1 g h - \rho_2 g(h + z) + P, \tag{3.39}$$

so that

$$\nabla p_2 = -g(\rho_2 - \rho_1)\nabla h + \nabla P. \tag{3.40}$$

Then if $\nabla p_2 = 0$ (because the lower layer is stationary) we have $g(\rho_2 - \rho_1)\nabla h = \nabla P$, and the momentum equation for the upper layer is just

$$\frac{\mathrm{D}\boldsymbol{u}}{\mathrm{D}t} + \boldsymbol{f} \times \boldsymbol{u} = -g'\nabla h, \tag{3.41}$$

where $g' = g(\rho_2 - \rho_1)/\rho_1$. These equations differ from the usual shallow water equations only in the use of a reduced gravity g' in place of g itself. It is the density *difference* between the two layers that is important. Similarly, if we take a shallow water system, with the moving layer on the bottom, and we suppose that overlying it is a stationary fluid of finite density, then we would easily find that the fluid equations for the moving layer are the same as if the fluid on top had zero inertia, except that g would be replaced by an appropriate reduced gravity.

3.3 MULTI-LAYER SHALLOW WATER EQUATIONS

We now consider the dynamics of multiple layers of fluid stacked on top of each other. This is a crude representation of continuous stratification, but it turns out to be a powerful model of many geophysically interesting phenomena as well as being physically realizable in the laboratory. The pressure is continuous across the interface, but the density jumps discontinuously and this allows the horizontal velocity to have a corresponding discontinuity. The set up is illustrated in Fig. 3.4.

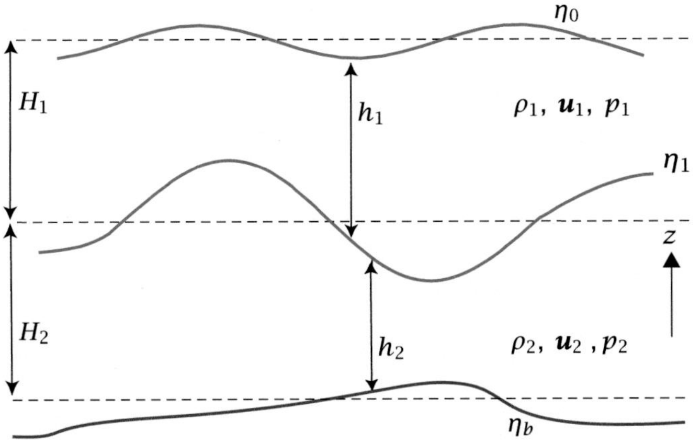

Fig. 3.5 The two-layer shallow water system. A fluid of density ρ_1 lies over a denser fluid of density ρ_2. In the reduced gravity case the lower layer is arbitrarily thick and is assumed stationary and so has no horizontal pressure gradient. In the 'rigid-lid' approximation the top surface displacement is neglected, but there is then a non-zero pressure gradient induced by the lid.

In each layer pressure is given by the hydrostatic approximation, and so anywhere in the interior we can find the pressure by integrating down from the top. Thus, at a height z in the first layer we have

$$p_1 = \rho_1 g(\eta_0 - z), \tag{3.42}$$

and in the second layer,

$$p_2 = \rho_1 g(\eta_0 - \eta_1) + \rho_2 g(\eta_1 - z) = \rho_1 g\eta_0 + \rho_1 g_1'\eta_1 - \rho_2 g z, \tag{3.43}$$

where $g_1' = g(\rho_2 - \rho_1)/\rho_1$, and so on. The term involving z is irrelevant for the dynamics, because only the horizontal derivative enters the equation of motion. Omitting this term, for the nth layer the dynamical pressure is given by the sum from the top down:

$$p_n = \rho_1 \sum_{i=0}^{n-1} g_i'\eta_i, \tag{3.44}$$

where $g_i' = g(\rho_{i+1} - \rho_i)/\rho_1$ (and $g_0 = g$). The interface displacements may be expressed in terms of the layer thicknesses by summing from the bottom up:

$$\eta_n = \eta_b + \sum_{i=n+1}^{i=N} h_i. \tag{3.45}$$

The momentum equation for each layer may then be written, in general,

$$\frac{\mathrm{D}\boldsymbol{u}_n}{\mathrm{D}t} + \boldsymbol{f} \times \boldsymbol{u}_n = -\frac{1}{\rho_n}\nabla p_n, \tag{3.46}$$

where the pressure is given by (3.44) and in terms of the layer depths using (3.46). If we make the Boussinesq approximation then ρ_n on the right-hand side of (3.46) is replaced by ρ_1.

Finally, the mass conservation equation for each layer has the same form as the single-layer case, and is

$$\frac{\mathrm{D}h_n}{\mathrm{D}t} + h_n\nabla\cdot\boldsymbol{u}_n = 0. \tag{3.47}$$

The two- and three-layer cases

The two-layer model (Fig. 3.5) is the simplest model to capture the effects of stratification. Evaluating the pressures using (3.44) and (3.45) we find:

$$p_1 = \rho_1 g\eta_0 = \rho_1 g(h_1 + h_2 + \eta_b), \tag{3.48a}$$

$$p_2 = \rho_1[g\eta_0 + g_1'\eta_1] = \rho_1\left[g(h_1 + h_2 + \eta_b) + g_1'(h_2 + \eta_b)\right]. \tag{3.48b}$$

The momentum equations for the two layers are then

$$\frac{\mathrm{D}\boldsymbol{u}_1}{\mathrm{D}t} + \boldsymbol{f} \times \boldsymbol{u}_1 = -g\nabla\eta_0 = -g\nabla(h_1 + h_2 + \eta_b), \tag{3.49a}$$

and in the bottom layer

$$\begin{aligned}\frac{\mathrm{D}\boldsymbol{u}_2}{\mathrm{D}t} + \boldsymbol{f} \times \boldsymbol{u}_2 &= -\frac{\rho_1}{\rho_2}\left(g\nabla\eta_0 + g_1'\nabla\eta_1\right)\\ &= -\frac{\rho_1}{\rho_2}\left[g\nabla(\eta_b + h_1 + h_2) + g_1'\nabla(h_2 + \eta_b)\right].\end{aligned} \tag{3.49b}$$

In the Boussinesq approximation ρ_1/ρ_2 is replaced by unity.

In a three-layer model the dynamical pressures are found to be

$$p_1 = \rho_1 g h, \tag{3.50a}$$

$$p_2 = \rho_1\left[gh + g_1'(h_2 + h_3 + \eta_b)\right], \tag{3.50b}$$

$$p_3 = \rho_1\left[gh + g_1'(h_2 + h_3 + \eta_b) + g_2'(h_3 + \eta_b)\right], \tag{3.50c}$$

where $h = \eta_0 = \eta_b + h_1 + h_2 + h_3$ and $g_2' = g(\rho_3 - \rho_2)/\rho_1$. More layers can obviously be added in a systematic fashion.

3.3.1 Reduced-gravity Multi-layer Equation

As with a single active layer, we may envision multiple layers of fluid overlying a deeper stationary layer. This is a useful model of the stratified upper ocean overlying a nearly stationary and nearly unstratified abyss. Indeed we use such a model to study the 'ventilated thermocline' in Chapter 20 and a detailed treatment may be found there. If we suppose there is a lid at the top, then the model is almost the same as that of the previous section. However, now the horizontal pressure gradient in the lowest model layer is zero, and so we may obtain the pressures in all the active layers by integrating the hydrostatic equation upwards from this layer. Suppose we have N moving layers, then the reader may verify that the dynamic pressure in the nth layer is given by

$$p_n = -\sum_{i=n}^{i=N} \rho_1 g_i' \eta_i, \tag{3.51}$$

where as before $g_i' = g(\rho_{i+1} - \rho_i)/\rho_1$. If we have a lid at the top, and take $\eta_0 = 0$, then the interface displacements are related to the layer thicknesses by

$$\eta_n = -\sum_{i=1}^{i=n} h_i. \tag{3.52}$$

From these expressions the momentum equation in each layer is easily constructed.

3.4 ✦ FROM CONTINUOUS STRATIFICATION TO SHALLOW WATER

In this section we show that the *continuously stratified* equations have a close correspondence to the shallow water equations, without breaking the fluid into discrete layers of differing densities. In particular, if the continuous equations are linearized and the flow is stably stratified, then each vertical mode of the continuous equations has the same form as the shallow water equations, with the modes being distinguished by the phase speed of the associated gravity waves.[2]

3.4.1 Vertical Normal Modes of the Linear Equations

We begin with a hydrostatic Boussinesq system, linearized about a state of rest and with fixed stratification, $N(z)$, noting that a similar derivation can be applied to an ideal gas using pressure coordinates. The equations are

$$\frac{\partial u}{\partial t} - fv = -\frac{\partial \phi}{\partial x}, \qquad \frac{\partial v}{\partial t} + fu = -\frac{\partial \phi}{\partial y}, \qquad 0 = -\frac{\partial \phi}{\partial z} + b, \tag{3.53a,b,c}$$

$$\nabla \cdot \boldsymbol{u} + \frac{\partial w}{\partial z} = 0, \qquad \frac{\partial b}{\partial t} + wN^2 = 0. \tag{3.53c,d}$$

The first line above contains the u and v momentum equations and the hydrostatic equation, and the second line contains the mass continuity equation and the buoyancy or thermodynamic equation, with the ∇ operator being purely horizontal (or at constant pressure), and we will take $N^2 > 0$. We assume a lid at the bottom and top of the domain. Including a free surface at the top, as appropriate for an ocean, is a slight extension. Including a 'leaky' tropopause with a stratosphere above is a more major extension.

The difficulty with these equations is that there are five independent variables in three spatial coordinates so that even the linear problems are algebraically complex, especially when f is variable. The equations are more general than is needed, because it is often observed that the vertical structure of solutions is relatively simple, especially in linear problems. A solution is to project the vertical structure onto appropriate eigenfunctions, and then to retain a very small number — often only one — of these eigenfunctions.

To determine what those eigenfunctions should be, we first combine the hydrostatic and buoyancy equations to give

$$\frac{\partial}{\partial t}\left(\frac{\phi_z}{N^2}\right) + w = 0. \tag{3.54}$$

Differentiating with respect to z and using the mass continuity equation gives

$$\frac{\partial}{\partial t}\left(\frac{\phi_z}{N^2}\right)_z - \nabla \cdot \boldsymbol{u} = 0. \tag{3.55}$$

It is this equation that motivates our choice of basis functions: we choose to expand the pressure and horizontal components of velocity in terms of an eigenfunction that satisfies the following Sturm–Liouville problem:

$$\frac{\mathrm{d}}{\mathrm{d}z}\left(\frac{1}{N^2}\frac{\mathrm{d}C_m}{\mathrm{d}z}\right) + \frac{1}{c_m^2}C_m = 0, \qquad \frac{\mathrm{d}}{\mathrm{d}z}C_m(0) = \frac{\mathrm{d}}{\mathrm{d}z}C_m(-H) = 0. \tag{3.56}$$

The eigenfunctions C_m are orthogonal in the sense that

$$\int_{-H}^{0} C_m C_n \,\mathrm{d}z = \frac{c_m^2}{g}\delta_{mn}, \tag{3.57}$$

where $\delta_{mn} = 0$ unless $m = n$, in which case it equals one. The normalization is by convention and the factor of g makes the functions C_m nondimensional. There are an infinite number of eigenvalues, c_m, namely $c_0, c_1, c_2 \ldots$, normally arranged in descending order of size, and for each there is a corresponding eigenfunction C_m. The pressure and horizontal velocity components are then expressed as

$$[u, v, \phi] = \sum_0^\infty [u_m(x,y,t), v_m(x,y,t), \phi_m(x,y,t)]\, C_m(z). \tag{3.58}$$

The benefit of this procedure is that the z-derivatives in the equations of motion are replaced by multiplications, and in particular (3.55) becomes

$$\frac{\partial \phi_m}{\partial t} + c_m^2 \nabla \cdot \boldsymbol{u}_m = 0 \quad \text{or} \quad \frac{\partial \eta_m^*}{\partial t} + H_m \nabla \cdot \boldsymbol{u}_m = 0, \tag{3.59a,b}$$

where $\eta^* \equiv \phi/g$. The quantity $H_m = c_m^2/g$ is the *equivalent depth* associated with the eigenmode. Equations (3.59) are evidently of the same form as the familiar linear mass continuity equation in the shallow water equations, namely

$$\frac{\partial \hat{\eta}}{\partial t} + c^2 \nabla \cdot \boldsymbol{u} = 0 \qquad \text{or} \qquad \frac{\partial \eta}{\partial t} + H \nabla \cdot \boldsymbol{u} = 0, \tag{3.60a,b}$$

where $c = \sqrt{gH}$ and $\hat{\eta} = g\eta$.

The horizontal momentum equations are simply,

$$\frac{\partial u_m}{\partial t} - f v_m = -\frac{\partial \phi_m}{\partial x}, \qquad \frac{\partial v_m}{\partial t} + f u_m = -\frac{\partial \phi_m}{\partial y}. \tag{3.61a,b}$$

Equations (3.59) and (3.61) are a closed set, once we have calculated the equivalent depth H_m for each mode. If there is a forcing in the momentum equation then the transformed forcing appears on the right-hand sides of (3.61). If there is a source in the buoyancy equation then a corresponding term appears on the right-hand side of (3.59), analogous to a mass source term in the shallow water equations. Note that a thermodynamic source affects $\partial\phi/\partial z$ and not ϕ itself.

Eigenfunctions for buoyancy and vertical velocity

The vertical velocity and the buoyancy do not satisfy the same boundary conditions and so should not be expanded in the same way. Rather, we let

$$\left[w, \frac{b}{N^2}\right] = \sum_0^\infty \left[w_m(x,y,t), \hat{b}_m(x,y,t)\right] S_m(z), \tag{3.62}$$

where the eigenfunctions satisfy

$$\frac{1}{N^2}\frac{\mathrm{d}^2 S_m}{\mathrm{d}z^2} + \frac{1}{c_m^2} S_m = 0, \qquad S_m(0) = S_m(-H) = 0, \tag{3.63a,b}$$

where $S_m = 0$ if $N = 0$, and we may use the orthonormalization,

$$\int_{-H}^0 N^2 S_m S_n \,\mathrm{d}z = g\delta_{mn}. \tag{3.64}$$

The functions S_m and C_m are related by

$$C_m = \frac{c_m^2}{g}\frac{\mathrm{d}S_m}{\mathrm{d}z}, \qquad N^2 S_m = -g\frac{\mathrm{d}C_m}{\mathrm{d}z}, \tag{3.65}$$

and it is these relationships that motivate the form of (3.63). The vertical velocity may be evaluated from the mass continuity equation, $\partial w/\partial z = -\nabla \cdot \boldsymbol{u}$, which becomes

$$w_m \frac{\mathrm{d}S_m}{\mathrm{d}z} = -C_m \nabla \cdot \boldsymbol{u}_m \qquad \Longrightarrow \qquad w_m = -\frac{c_m^2}{g}\nabla \cdot \boldsymbol{u}_m. \tag{3.66a,b}$$

Buoyancy is obtained from (3.53c) which, using (3.65), gives $\hat{b}_m = -\phi_m/g$.

3.4.2 Examples and Approximations

The values of c_m can be computed by solving the eigenvalue problem for the given stratification, although in general this must be carried out numerically. Consider, though, the simplest case in which N is constant, which is a reasonable approximation for the troposphere, less so for the ocean. The normal modes are sines and cosines, and for $m = 1, 2 \dots$ we have

$$C_m(z) = A_m \cos\frac{m\pi z}{H}, \qquad S_m(z) = B_m \sin\frac{m\pi z}{H}, \qquad c_m = \frac{NH}{m\pi}, \tag{3.67}$$

where, for $m > 0$, $A_m = c_m/\sqrt{gH/2}$ and $B_m = \sqrt{2g/HN^2}$. The equivalent depth is given by

$$H_m = \frac{N^2H^2}{gm^2\pi^2} = \frac{g'H}{gm^2\pi^2}, \tag{3.68}$$

where $g' \equiv HN^2$ and for a Boussinesq fluid $g' = (gH/\rho_0)\partial\rho/\partial z$. Using (3.67) we see that $c_m = \sqrt{gH_m} = \sqrt{g'H}/m\pi$, and note the factors of π are significant in these expressions. The mode with $m = 0$ is a special one and is called the *barotropic mode* with

$$C_0 = A_0/2, \qquad c_0^2 = gH. \tag{3.69}$$

The above expressions allow us to estimate equivalent depths and phase speeds for the atmosphere and ocean, with some caveats. For the atmosphere we should properly take into account its compressibility and a leaky tropopause, but proceeding nevertheless let us take $H = 10\,\mathrm{km}$ and $N = 10^{-2}\,\mathrm{s^{-1}}$ (a typical tropospheric value), whence

$$c_0 \approx 300\,\mathrm{m\,s^{-1}},\ c_1 \approx 30\,\mathrm{m\,s^{-1}},\ c_2 \approx 15\,\mathrm{m\,s^{-1}} \quad \text{and} \quad H_1 \approx 100\,\mathrm{m},\ H_2 \approx 25\,\mathrm{m}. \tag{3.70}$$

These equivalent depths are much smaller than the actual depth of the atmosphere, a fact that transcends our approximations and that greatly affects the properties of atmospheric gravity waves, as we discover in later chapters. The best fits to observations of internal gravity waves in the atmosphere are often in fact made with an equivalent depth of 50 m or less and a speed of about $20\,\mathrm{m\,s^{-1}}$.

The oceanic stratification is in fact not constant, but decreases significantly below the thermocline, which is about 1 km thick. We might proceed by simply using values appropriate for the thermocline in the above, and if we take $N = 10^{-2}\,\mathrm{s^{-1}}$ and $H = 1\,\mathrm{km}$ we find, using (3.67) and (3.68),

$$c_0 \approx 200\,\mathrm{m\,s^{-1}},\ c_1 \approx 3\,\mathrm{m\,s^{-1}},\ c_2 \approx 1.5\,\mathrm{m\,s^{-1}} \quad \text{and} \quad H_1 \approx 1\,\mathrm{m},\ H_2 \approx 0.25\,\mathrm{m}. \tag{3.71}$$

The speed c_0 is (as for the atmosphere) vastly larger than any parcel speed in the ocean. In contrast, the equivalent depths are very small, but this just reflects the smallness of the density variations in the ocean and the fact that H_m is proportional to g'/g.

If the oceanic stratification varies reasonably slowly we can use WKB methods (page 247) to good effect to better evaluate the eigenvalues and eigenfunctions.[3] Roughly speaking NH is replaced by $\int N\,\mathrm{d}z$ in (3.67), and the WKB solution, for $m \geq 1$, is

$$S_m \sim S_0 \sin\left(\frac{1}{c_m}\int_{-H}^{z} N(z)\,\mathrm{d}z\right), \qquad C_m \sim \left(\frac{c_m N S_0}{g}\right)\cos\left(\frac{1}{c_m}\int_{-H}^{z} N(z)\,\mathrm{d}z\right),$$
$$c_m \approx \frac{1}{m\pi}\int_{-H}^{0} N\,\mathrm{d}z, \tag{3.72a,b,c}$$

where $S_0 = (c_m/N)^{1/2}$. Using (3.72c) still gives values of c_1 of around 2–3 $\mathrm{m\,s^{-1}}$ over the ocean gyres, less in equatorial regions, providing some post facto justification for using $H = 1\,\mathrm{km}$ previously. The eigenfunctions, (3.72a,b), are 'stretched' sines and cosines, with local wavenumbers

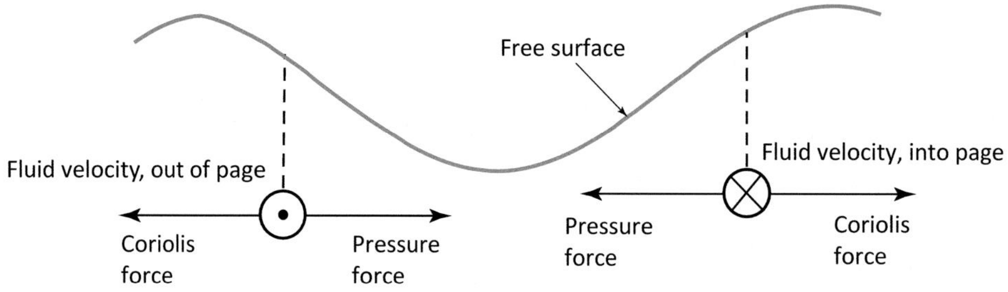

Fig. 3.6 Geostrophic flow in a shallow water system, with a positive value of the Coriolis parameter f, as in the Northern Hemisphere. The pressure force is directed down the gradient of the height field, and this can be balanced by the Coriolis force if the fluid velocity is at right angles to it. If f were negative, the geostrophic flow would be reversed.

proportional to $N(z)$ and so varying more rapidly in the upper ocean than at depth (look ahead to Fig. 12.12). The vertical velocity eigenfunctions, S_m, have a smaller amplitude in the upper ocean but the pressure and horizontal velocity amplitudes are larger.

For the remainder of this chapter we will use the shallow water equations in their conventional form, for if there *is* a region where density changes rapidly in the vertical then the layered equations are quite natural, and allow for the incorporation of nonlinearities more easily.

3.5 GEOSTROPHIC BALANCE AND THERMAL WIND

We now turn our attention to the *dynamics* of shallow water systems, beginning with the effects of rotation. Geostrophic balance occurs in the shallow water equations, just as in the continuously stratified equations, when the Rossby number U/fL is small and the Coriolis term dominates the advective terms in the momentum equation. In the single-layer shallow water equations the geostrophic flow is:

$$\boldsymbol{f} \times \boldsymbol{u}_g = -g\nabla\eta. \tag{3.73}$$

Thus, the geostrophic velocity is proportional to the slope of the surface, as sketched in Fig. 3.6. (For the rest of this section we drop the subscript g, and take all velocities to be geostrophic.)

In both the single-layer and multi-layer cases, the slope of an interfacial surface is directly related to the difference in pressure gradient on either side and so, by geostrophic balance, to the shear of the flow. This is the shallow water analogue of the thermal wind relation. To obtain an expression for this, consider the interface, η, between two layers labelled 1 and 2. The pressure in two layers is given by the hydrostatic relation and so,

$$p_1 = A(x,y) - \rho_1 g z \qquad \text{(at some } z \text{ in layer 1)}, \tag{3.74a}$$

$$\begin{aligned} p_2 &= A(x,y) - \rho_1 g\eta + \rho_2 g(\eta - z) \\ &= A(x,y) + \rho_1 g_1'\eta - \rho_2 g z \qquad \text{(at some } z \text{ in layer 2)}, \end{aligned} \tag{3.74b}$$

where $A(x, y)$ is a function of integration. Thus we find

$$\frac{1}{\rho_1}\nabla(p_1 - p_2) = -g_1'\nabla\eta. \tag{3.75}$$

If the flow is geostrophically balanced and Boussinesq then, in each layer, the velocity obeys

$$f\boldsymbol{u}_i = \frac{1}{\rho_1}\mathbf{k} \times \nabla p_i\,. \tag{3.76}$$

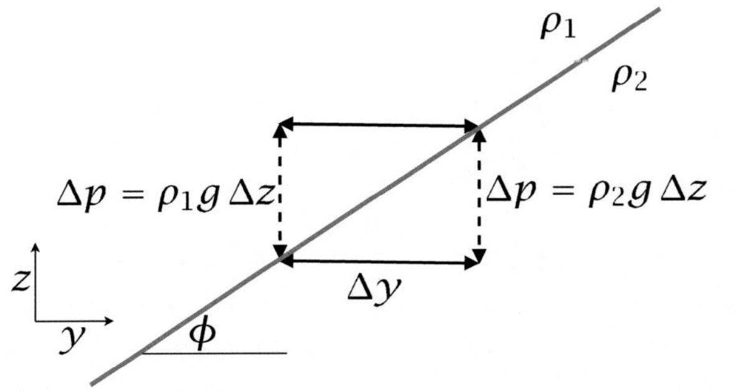

Fig. 3.7 Margules' relation: using hydrostasy, the difference in the horizontal pressure gradient between the upper and the lower layer is given by $-g'\rho_1 s$, where $s = \tan\phi = \Delta z/\Delta y$ is the interface slope and $g' = g(\rho_2 - \rho_1)/\rho_1$. Geostrophic balance then gives $f(u_1 - u_2) = g's$, which is a special case of (3.78).

Using (3.75) then gives

$$f(\boldsymbol{u}_1 - \boldsymbol{u}_2) = -\mathbf{k} \times g_1' \nabla \eta, \tag{3.77}$$

or in general

$$f(\boldsymbol{u}_n - \boldsymbol{u}_{n+1}) = -\mathbf{k} \times g_n' \nabla \eta. \tag{3.78}$$

This is the thermal wind equation for the shallow water system. It applies at any interface, and it implies *the shear is proportional to the interface slope,* a result known as the Margules relation[4] (Fig. 3.7).

Suppose that we represent the atmosphere by two layers of fluid; a meridionally decreasing temperature may then be represented by an interface that slopes upwards toward the pole. Then, in either hemisphere, we have

$$u_1 - u_2 = \frac{g_1'}{f}\frac{\partial \eta}{\partial y} > 0, \tag{3.79}$$

and the temperature gradient is associated with a positive shear.

3.6 FORM STRESS

When the interface between two layers varies with position — that is, when it is wavy — the layers exert a pressure force on each other. Similarly, if the bottom of the fluid is not flat then the topography and the bottom layer will in general exert forces on each other. This kind of force (normally arising as a force per unit area) is known as *form stress,* and it is an important means whereby momentum can be added to or extracted from a flow.[5] Consider a layer confined between two interfaces, $\eta_1(x, y)$ and $\eta_2(x, y)$. Then over some zonal interval L the average zonal pressure force on that fluid layer is given by

$$F_p = -\frac{1}{L}\int_{x_1}^{x_2}\int_{\eta_2}^{\eta_1} \frac{\partial p}{\partial x}\,\mathrm{d}x\,\mathrm{d}z. \tag{3.80}$$

Integrating by parts first in z and then in x, and noting that by hydrostasy $\partial p/\partial z$ does not depend on horizontal position within the layer, we obtain

$$F_p = -\frac{1}{L}\int_{x_1}^{x_2}\left[\frac{\partial p}{\partial x} z\right]_{\eta_2}^{\eta_1} \mathrm{d}x = -\overline{\eta_1 \frac{\partial p_1}{\partial x}} + \overline{\eta_2 \frac{\partial p_2}{\partial x}} = +\overline{p_1 \frac{\partial \eta_1}{\partial x}} - \overline{p_2 \frac{\partial \eta_2}{\partial x}}, \tag{3.81}$$

where p_1 is the pressure at η_1, and similarly for p_2, and to obtain the second line we suppose that the integral is around a closed path, such as a circle of latitude, and the average is denoted with an overbar. These terms represent the transfer of momentum from one layer to the next, and at a particular interface, i, we may define the form stress, τ_i, by

$$\tau_i \equiv \overline{p_i \frac{\partial \eta_i}{\partial x}} = -\overline{\eta_i \frac{\partial p_i}{\partial x}}. \tag{3.82}$$

The form stress is a force per unit area and its vertical derivative, $\partial\tau/\partial z$, is the force (per unit volume) on the fluid. Form stress is a particularly important means for the vertical transfer of momentum and its ultimate removal in an eddying fluid, and is one of the main mechanisms whereby the wind stress at the top of the ocean is communicated to the ocean bottom. At the fluid bottom the form stress is $\overline{p\partial_x\eta_b}$, where η_b is the bottom topography, and this is proportional to the momentum exchange with the solid Earth. This is a significant mechanism for the ultimate removal of momentum in the ocean, especially in the Antarctic Circumpolar Current where it is likely to be much larger than bottom (or Ekman) drag arising from small-scale turbulence and friction. In the two-layer, flat-bottomed case the only form stress occurring is that at the interface, and the momentum transfer between the layers is just $\overline{p_1\partial\eta_1/\partial x}$ or $-\overline{\eta_1\partial p_1/\partial x}$; then, the force on each layer due to the other is equal and opposite, as we would expect from momentum conservation. (Form stress is discussed more in an oceanographic context in Sections 19.6.3 and 21.7.2.)

For flows in geostrophic balance, the form stress is related to the meridional heat flux. The pressure gradient and velocity are related by $\rho f v' = \partial p'/\partial x$ and the interfacial displacement is proportional to the temperature perturbation, b' — in fact one may show that $\eta' \approx -b'/(\partial\overline{b}/\partial z)$. Thus $-\overline{\eta'\partial p'/\partial x} \propto \overline{v'b'}$, a correspondence that will recur when we consider the *Eliassen–Palm flux* in Chapter 10.

3.7 CONSERVATION PROPERTIES OF SHALLOW WATER SYSTEMS

There are two common types of conservation property in fluids: (i) material invariants; and (ii) integral invariants. Material invariance occurs when a property (φ say) is conserved on each fluid element, and so obeys the equation $\mathrm{D}\varphi/\mathrm{D}t = 0$. An integral invariant is one that is conserved after an integration over some, usually closed, volume; energy is an example.

3.7.1 Potential Vorticity: a Material Invariant

The vorticity of a fluid (considered at greater length in chapter 4), denoted $\boldsymbol{\omega}$, is defined to be the curl of the velocity field. Let us also define the shallow water vorticity, $\boldsymbol{\omega}^*$, as the curl of the horizontal velocity. We therefore have:

$$\boldsymbol{\omega} \equiv \nabla \times \boldsymbol{v}, \qquad \boldsymbol{\omega}^* \equiv \nabla \times \boldsymbol{u}. \tag{3.83}$$

Because $\partial u/\partial z = \partial v/\partial z = 0$, only the vertical component of $\boldsymbol{\omega}^*$ is non-zero and

$$\boldsymbol{\omega}^* = \mathbf{k}\left(\frac{\partial v}{\partial x} - \frac{\partial u}{\partial y}\right) = \mathbf{k}\zeta. \tag{3.84}$$

Considering first the non-rotating case, we use the vector identity

$$(\boldsymbol{u}\cdot\nabla)\boldsymbol{u} = \frac{1}{2}\nabla(\boldsymbol{u}\cdot\boldsymbol{u}) - \boldsymbol{u}\times(\nabla\times\boldsymbol{u}), \tag{3.85}$$

to write the momentum equation, (3.8) with $f = 0$, as

$$\frac{\partial \boldsymbol{u}}{\partial t} + \boldsymbol{\omega}^* \times \boldsymbol{u} = -\nabla\left(g\eta + \frac{1}{2}\boldsymbol{u}^2\right). \tag{3.86}$$

To obtain an evolution equation for the vorticity we take the curl of (3.86), and make use of the vector identity

$$\begin{aligned}\nabla\times(\boldsymbol{\omega}^*\times\boldsymbol{u}) &= (\boldsymbol{u}\cdot\nabla)\boldsymbol{\omega}^* - (\boldsymbol{\omega}^*\cdot\nabla)\boldsymbol{u} + \boldsymbol{\omega}^*\nabla\cdot\boldsymbol{u} - \boldsymbol{u}\nabla\cdot\boldsymbol{\omega}^* \\ &= (\boldsymbol{u}\cdot\nabla)\boldsymbol{\omega}^* + \boldsymbol{\omega}^*\nabla\cdot\boldsymbol{u},\end{aligned} \tag{3.87}$$

using the fact that $\nabla \cdot \boldsymbol{\omega}^*$ is the divergence of a curl and therefore zero, and $(\boldsymbol{\omega}^* \cdot \nabla)\boldsymbol{u} = 0$ because $\boldsymbol{\omega}^*$ is perpendicular to the surface in which $\boldsymbol{u}$ varies. Taking the curl of (3.86) gives

$$\frac{\partial \zeta}{\partial t} + (\boldsymbol{u} \cdot \nabla)\zeta = -\zeta \nabla \cdot \boldsymbol{u}, \tag{3.88}$$

where $\zeta = \mathbf{k} \cdot \boldsymbol{\omega}^*$. Now, the mass conservation equation may be written as

$$-\zeta \nabla \cdot \boldsymbol{u} = \frac{\zeta}{h}\frac{\mathrm{D}h}{\mathrm{D}t}, \tag{3.89}$$

and using this (3.88) becomes

$$\frac{\mathrm{D}\zeta}{\mathrm{D}t} = \frac{\zeta}{h}\frac{\mathrm{D}h}{\mathrm{D}t}, \tag{3.90}$$

which simplifies to

$$\frac{\mathrm{D}Q}{\mathrm{D}t} = 0 \qquad \text{where} \qquad Q = \left(\frac{\zeta}{h}\right). \tag{3.91}$$

The important quantity Q is known as the *potential vorticity*, and (3.91) is the potential vorticity equation. We re-derive this conservation law in a different way in Section 4.6.

Because Q is conserved on parcels, then so is any function of Q; that is, $F(Q)$ is a material invariant, where F is any function. To see this algebraically, multiply (3.91) by $F'(Q)$, the derivative of F with respect to Q, giving

$$F'(Q)\frac{\mathrm{D}Q}{\mathrm{D}t} = \frac{\mathrm{D}}{\mathrm{D}t}F(Q) = 0. \tag{3.92}$$

Since F is arbitrary there are an infinite number of material invariants corresponding to different choices of F.

Effects of rotation

In a rotating frame of reference, the shallow water momentum equation is

$$\frac{\mathrm{D}\boldsymbol{u}}{\mathrm{D}t} + \boldsymbol{f} \times \boldsymbol{u} = -g\nabla\eta, \tag{3.93}$$

where (as before) $\boldsymbol{f} = f\mathbf{k}$. This may be written in vector invariant form as

$$\frac{\partial \boldsymbol{u}}{\partial t} + (\boldsymbol{\omega}^* + \boldsymbol{f}) \times \boldsymbol{u} = -\nabla\left(g\eta + \frac{1}{2}\boldsymbol{u}^2\right), \tag{3.94}$$

and taking the curl of this gives the vorticity equation

$$\frac{\partial \zeta}{\partial t} + (\boldsymbol{u} \cdot \nabla)(\zeta + f) = -(f + \zeta)\nabla \cdot \boldsymbol{u}. \tag{3.95}$$

This is the same as the shallow water vorticity equation in a non-rotating frame, save that ζ is replaced by $\zeta + f$, the reason for this being that f is the vorticity that the fluid has by virtue of the background rotation. Thus, (3.95) is simply the equation of motion for the total or absolute vorticity, $\boldsymbol{\omega}_a = \boldsymbol{\omega}^* + \boldsymbol{f} = (\zeta + f)\mathbf{k}$.

The potential vorticity equation in the rotating case follows, much as in the non-rotating case, by combining (3.95) with the mass conservation equation, giving

$$\frac{\mathrm{D}}{\mathrm{D}t}\left(\frac{\zeta + f}{h}\right) = 0. \tag{3.96}$$

That is, the potential vorticity in a rotating shallow system is given by $Q = (\zeta + f)/h$ and is a material invariant. (The same symbol, Q, is commonly used for many of the manifestations of potential vorticity.)

Vorticity and circulation

Although vorticity itself is not a material invariant, its integral over a horizontal material area is invariant. To demonstrate this in the non-rotating case, consider the integral

$$C = \int_A \zeta \, dA = \int_A Qh \, dA, \tag{3.97}$$

over a surface A, the cross-sectional area of a column of height h (as in Fig. 3.2). Taking the material derivative of this gives

$$\frac{DC}{Dt} = \int_A \frac{DQ}{Dt} h \, dA + \int_A Q \frac{D}{Dt}(h \, dA). \tag{3.98}$$

On the right-hand side the first term is zero, by (3.91), and the second term is just the derivative of the volume of a column of fluid of constant density and so it too is zero. Thus,

$$\frac{DC}{Dt} = \frac{D}{Dt} \int_A \zeta \, dA = 0. \tag{3.99}$$

Thus, the integral of the vorticity over some cross-sectional area of the fluid is unchanging, although both the vorticity and area of the fluid may individually change. Using Stokes' theorem, it may be written as

$$\frac{DC}{Dt} = \frac{D}{Dt} \oint \boldsymbol{u} \cdot d\boldsymbol{l}, \tag{3.100}$$

where the line integral is around the boundary of A. This is an example of Kelvin's circulation theorem, which we shall meet again in a more general form in Chapter 4, where we also consider the rotating case.

A slight generalization of (3.99) is possible. Consider the integral $I = \int F(Q)h \, dA$ where again F is any differentiable function of its argument. It is clear that

$$\frac{D}{Dt} \int_A F(Q)h \, dA = 0. \tag{3.101}$$

If the area of integration in (3.86) or (3.101) is the whole domain (enclosed by frictionless walls, for example) then it is clear that the integral of $hF(Q)$ is a constant, including as a special case the integral of ζ.

3.7.2 Energy Conservation: an Integral Invariant

Since we have made various simplifications in deriving the shallow water system, it is not self-evident that energy should be conserved, or indeed what form the energy takes. The kinetic energy density (KE), meaning the kinetic energy per unit area, is $\rho_0 h \boldsymbol{u}^2/2$. The potential energy density of the fluid is

$$\text{PE} = \int_0^h \rho_0 g z \, dz = \frac{1}{2}\rho_0 g h^2. \tag{3.102}$$

The factor ρ_0 appears in both kinetic and potential energies and, because it is a constant, we will omit it. For algebraic simplicity we also assume the bottom is flat, at $z = 0$.

Using the mass conservation equation (3.14b) we obtain an equation for the evolution of potential energy density, namely

$$\frac{D}{Dt}\frac{gh^2}{2} + gh^2 \nabla \cdot \boldsymbol{u} = 0 \tag{3.103a}$$

or

$$\frac{\partial}{\partial t}\frac{gh^2}{2} + \nabla \cdot \left(\boldsymbol{u} \frac{gh^2}{2} \right) + \frac{gh^2}{2} \nabla \cdot \boldsymbol{u} = 0. \tag{3.103b}$$

From the momentum and mass continuity equations we obtain an equation for the evolution of kinetic energy density, namely

$$\frac{\mathrm{D}}{\mathrm{D}t}\frac{h\boldsymbol{u}^2}{2} + \frac{\boldsymbol{u}^2 h}{2}\nabla\cdot\boldsymbol{u} = -g\boldsymbol{u}\cdot\nabla\frac{h^2}{2} \tag{3.104a}$$

or

$$\frac{\partial}{\partial t}\frac{h\boldsymbol{u}^2}{2} + \nabla\cdot\left(\boldsymbol{u}\frac{h\boldsymbol{u}^2}{2}\right) + g\boldsymbol{u}\cdot\nabla\frac{h^2}{2} = 0. \tag{3.104b}$$

Adding (3.103b) and (3.104b) we obtain

$$\frac{\partial}{\partial t}\frac{1}{2}\left(h\boldsymbol{u}^2 + gh^2\right) + \nabla\cdot\left[\frac{1}{2}\boldsymbol{u}\left(gh^2 + h\boldsymbol{u}^2 + gh^2\right)\right] = 0, \tag{3.105}$$

or

$$\frac{\partial E}{\partial t} + \nabla\cdot\boldsymbol{F} = 0, \tag{3.106}$$

where $E = \text{KE} + \text{PE} = (h\boldsymbol{u}^2 + gh^2)/2$ is the density of the total energy and $\boldsymbol{F} = \boldsymbol{u}(h\boldsymbol{u}^2/2 + gh^2)$ is the energy flux. If the fluid is confined to a domain bounded by rigid walls, on which the normal component of velocity vanishes, then on integrating (3.105) over that area and using Gauss's theorem, the total energy is seen to be conserved; that is

$$\frac{\mathrm{d}\hat{E}}{\mathrm{d}t} = \frac{1}{2}\frac{\mathrm{d}}{\mathrm{d}t}\int_A \left(h\boldsymbol{u}^2 + gh^2\right)\mathrm{d}A = 0. \tag{3.107}$$

Such an energy principle also holds in the case with bottom topography. Just as we found in the case for a compressible fluid in Chapter 2, the energy flux in (3.106) is not just the energy density multiplied by the velocity; it contains an additional term $g\boldsymbol{u}h^2/2$, and this represents the energy transfer occurring when the fluid does work against the pressure force.

3.8 SHALLOW WATER WAVES

Let us now look at the gravity waves that occur in shallow water. To isolate the essence we will consider waves in a single fluid layer, with a flat bottom and a free upper surface, in which gravity provides the sole restoring force.

3.8.1 Non-rotating Shallow Water Waves

Given a flat bottom the fluid thickness is equal to the free surface displacement (Fig. 3.1), and taking the basic state of the fluid to be at rest we let

$$h(x, y, t) = H + h'(x, y, t) = H + \eta'(x, y, t), \tag{3.108a}$$

$$\boldsymbol{u}(x, y, t) = \boldsymbol{u}'(x, y, t). \tag{3.108b}$$

The mass conservation equation, (3.14b), then becomes

$$\frac{\partial \eta'}{\partial t} + (H + \eta')\nabla\cdot\boldsymbol{u}' + \boldsymbol{u}'\cdot\nabla\eta' = 0, \tag{3.109}$$

and neglecting squares of small quantities this yields the linear equation

$$\frac{\partial \eta'}{\partial t} + H\nabla\cdot\boldsymbol{u}' = 0. \tag{3.110}$$

Similarly, linearizing the momentum equation, (3.8) with $\boldsymbol{f} = 0$, yields

$$\frac{\partial \boldsymbol{u}'}{\partial t} = -g\nabla\eta'. \tag{3.111}$$

Eliminating velocity by differentiating (3.110) with respect to time and taking the divergence of (3.111) leads to

$$\frac{\partial^2 \eta'}{\partial t^2} - gH\nabla^2\eta' = 0, \tag{3.112}$$

which may be recognized as a wave equation. We can find the dispersion relationship for this by substituting the trial solution

$$\eta' = \mathrm{Re}\,\tilde{\eta}\mathrm{e}^{\mathrm{i}(\boldsymbol{k}\cdot\boldsymbol{x}-\omega t)}, \tag{3.113}$$

where $\tilde{\eta}$ is a complex constant, $\boldsymbol{k} = \mathbf{i}k + \mathbf{j}l$ is the horizontal wavenumber and Re indicates that the real part of the solution should be taken. If, for simplicity, we restrict attention to the one-dimensional problem, with no variation in the y-direction, then substituting into (3.112) leads to the dispersion relationship

$$\omega = \pm ck, \tag{3.114}$$

where $c = \sqrt{gH}$; that is, the wave speed is proportional to the square root of the mean fluid depth and is independent of the wavenumber — the waves are dispersionless. The general solution is a superposition of all such waves, with the amplitudes of each wave (or Fourier component) being determined by the Fourier decomposition of the initial conditions.

Because the waves are dispersionless, the general solution can be written as

$$\eta'(x,t) = \frac{1}{2}\left[F(x-ct) + F(x+ct)\right], \tag{3.115}$$

where $F(x)$ is the height field at $t = 0$. From this, it is easy to see that the shape of an initial disturbance is preserved as it propagates both to the right and to the left at speed c.

3.8.2 Rotating Shallow Water (Poincaré) Waves

We now consider the effects of rotation on shallow water waves. Linearizing the rotating, flat-bottomed f-plane shallow water equations, (SW.1) and (SW.2) on page 109, about a state of rest we obtain

$$\frac{\partial u'}{\partial t} - f_0 v' = -g\frac{\partial \eta'}{\partial x}, \qquad \frac{\partial v'}{\partial t} + f_0 u' = -g\frac{\partial \eta'}{\partial y}, \qquad \frac{\partial \eta'}{\partial t} + H\left(\frac{\partial u'}{\partial x} + \frac{\partial v'}{\partial y}\right) = 0. \tag{3.116a,b,c}$$

To obtain a dispersion relationship we let

$$(u, v, \eta) = (\tilde{u}, \tilde{v}, \tilde{\eta})\mathrm{e}^{\mathrm{i}(\boldsymbol{k}\cdot\boldsymbol{x}-\omega t)}, \tag{3.117}$$

and substitute into (3.116), giving

$$\begin{pmatrix} -\mathrm{i}\,\omega & -f_0 & \mathrm{i}\,gk \\ f_0 & -\mathrm{i}\,\omega & \mathrm{i}\,gl \\ \mathrm{i}\,Hk & \mathrm{i}\,Hl & -\mathrm{i}\,\omega \end{pmatrix} \begin{pmatrix} \tilde{u} \\ \tilde{v} \\ \tilde{\eta} \end{pmatrix} = 0. \tag{3.118}$$

This homogeneous equation has non-trivial solutions only if the determinant of the matrix vanishes, and that condition gives

$$\omega(\omega^2 - f_0^2 - c^2K^2) = 0, \tag{3.119}$$

where $K^2 = k^2 + l^2$ and $c^2 = gH$. There are two classes of solution to (3.119). The first is simply $\omega = 0$, i.e., time-independent flow corresponding to geostrophic balance in (3.116). Because

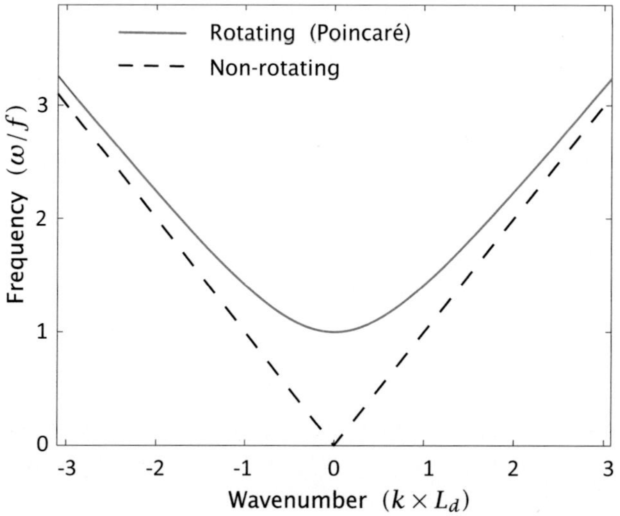

Fig. 3.8 Dispersion relation for Poincaré waves and non-rotating shallow water waves. Frequency is scaled by the Coriolis frequency f, and wavenumber by the inverse deformation radius $\sqrt{gH}/f$. For small wavenumbers the frequency of the Poincaré waves is approximately f, and for high wavenumbers it asymptotes to that of non-rotating waves.

geostrophic balance gives a divergence-free velocity field for a constant Coriolis parameter the equations are satisfied by a time-independent solution. (If the Coriolis parameter varies in space then the $\omega = 0$ solution morphs into a non-trivial dispersion relation for *Rossby waves,* considered in Chapter 5.) The second set of solutions gives the dispersion relation

$$\omega^2 = f_0^2 + c^2(k^2 + l^2), \tag{3.120}$$

or

$$\omega^2 = f_0^2 + gH(k^2 + l^2). \tag{3.121}$$

The corresponding waves are known as *Poincaré* waves,[6] and the dispersion relationship is illustrated in Fig. 3.8. Note that the frequency is always greater than the Coriolis frequency f_0. There are two interesting limits:

(i) The short wave limit. If

$$K^2 \gg \frac{f_0^2}{gH}, \tag{3.122}$$

where $K^2 = k^2 + l^2$, then the dispersion relationship reduces to that of the non-rotating case (3.114). This condition is equivalent to requiring that the wavelength be much shorter than the *deformation radius,* $L_d \equiv \sqrt{gH}/f$. Specifically, if $l = 0$ and $\lambda = 2\pi/k$ is the wavelength, the condition is

$$\lambda^2 \ll L_d^2(2\pi)^2. \tag{3.123}$$

The numerical factor of $(2\pi)^2$ is more than an order of magnitude, so care must be taken when deciding if the condition is satisfied in particular cases. Furthermore, the wavelength must still be longer than the depth of the fluid, otherwise the shallow water condition is not met.

(ii) The long wave limit. If

$$K^2 \ll \frac{f_0^2}{gH}, \tag{3.124}$$

that is if the wavelength is much longer than the deformation radius L_d, then the dispersion relationship is

$$\omega = f_0. \tag{3.125}$$

These are known as *inertial oscillations*. The equations of motion giving rise to them are

$$\frac{\partial u'}{\partial t} - f_0 v' = 0, \qquad \frac{\partial v'}{\partial t} + f_0 u' = 0, \tag{3.126}$$

which are equivalent to material equations for free particles in a rotating frame, unconstrained by pressure forces, namely

$$\frac{\mathrm{d}^2 x}{\mathrm{d}t^2} - f_0 v = 0, \qquad \frac{\mathrm{d}^2 y}{\mathrm{d}t^2} + f_0 u = 0. \tag{3.127}$$

3.8.3 Kelvin Waves

The Kelvin wave is a particular type of gravity wave that exists in the presence of both rotation and a lateral boundary. Suppose there is a solid boundary at $y = 0$; clearly harmonic solutions in the y-direction are not allowable, as these would not satisfy the condition of no normal flow at the boundary. Do any wave-like solutions exist? The affirmative answer to this question was provided by W. Thomson and the associated waves are now eponymously known as *Kelvin waves*.[7] We begin with the linearized shallow water equations, namely

$$\frac{\partial u'}{\partial t} - f_0 v' = -g\frac{\partial \eta'}{\partial x}, \qquad \frac{\partial v'}{\partial t} + f_0 u' = -g\frac{\partial \eta'}{\partial y}, \qquad \frac{\partial \eta'}{\partial t} + H\left(\frac{\partial u'}{\partial x} + \frac{\partial v'}{\partial y}\right) = 0. \tag{3.128a,b,c}$$

The fact that $v' = 0$ at $y = 0$ suggests that we look for a solution with $v' = 0$ everywhere, whence these equations become

$$\frac{\partial u'}{\partial t} = -g\frac{\partial \eta'}{\partial x}, \qquad f_0 u' = -g\frac{\partial \eta'}{\partial y}, \qquad \frac{\partial \eta'}{\partial t} + H\frac{\partial u'}{\partial x} = 0. \tag{3.129a,b,c}$$

Equations (3.129a) and (3.129c) lead to the standard wave equation

$$\frac{\partial^2 u'}{\partial t^2} = c^2 \frac{\partial^2 u'}{\partial x^2}, \tag{3.130}$$

where $c = \sqrt{gH}$, the usual wave speed of shallow water waves. The solution of (3.130) is

$$u' = F_1(x + ct, y) + F_2(x - ct, y), \tag{3.131}$$

with corresponding surface displacement

$$\eta' = \sqrt{H/g}\left[-F_1(x + ct, y) + F_2(x - ct, y)\right]. \tag{3.132}$$

The solution represents the superposition of two waves, one (F_1) travelling in the negative x-direction, and the other in the positive x-direction. To obtain the y dependence of these functions we use (3.129b) which gives

$$\frac{\partial F_1}{\partial y} = \frac{f_0}{\sqrt{gH}} F_1, \qquad \frac{\partial F_2}{\partial y} = -\frac{f_0}{\sqrt{gH}} F_2, \tag{3.133}$$

with solutions

$$F_1 = F(x + ct)\mathrm{e}^{y/L_d}, \qquad F_2 = G(x - ct)\mathrm{e}^{-y/L_d}, \tag{3.134}$$

where $L_d = \sqrt{gH}/f_0$ is the radius of deformation. If we consider flow in the half-plane in which $y > 0$, then for positive f_0 the solution F_1 grows exponentially away from the wall, and so fails to satisfy the condition of boundedness at infinity. It thus must be eliminated, leaving the general solution

$$u' = e^{-y/L_d} G(x - ct), \qquad v' = 0, \qquad \eta' = \sqrt{H/g}\, e^{-y/L_d} G(x - ct). \tag{3.135a,b,c}$$

These are Kelvin waves, and they decay exponentially away from the boundary. In general, for f_0 positive the boundary is to the right of an observer moving with the wave. Given a constant Coriolis parameter, we could equally well have obtained a solution on a meridional wall, in which case we would find that the wave again moves such that the wall is to the right of the wave direction. (This is obvious once it is realized that f-plane dynamics are isotropic in x and y.) Thus, in the Northern Hemisphere the wave moves anticlockwise round a basin, and conversely in the Southern Hemisphere, and in both hemispheres the direction is cyclonic.

3.9 GEOSTROPHIC ADJUSTMENT

We noted in Chapter 2 that the large-scale, extratropical circulation of the atmosphere is in near-geostrophic balance. Why is this? Why should the Rossby number be small? Arguably, the magnitude of the velocity in the atmosphere and ocean is ultimately given by the strength of the forcing, and so ultimately by the differential heating between pole and equator (although even this argument is not satisfactory, since the forcing mainly determines the energy throughput, not directly the energy itself, and the forcing is itself dependent on the atmosphere's response). But even supposing that the velocity magnitudes are given, there is no a-priori guarantee that the forcing or the dynamics will produce length scales that are such that the Rossby number is small. However, there is in fact a powerful and ubiquitous process whereby a fluid in an initially unbalanced state naturally evolves toward a state of geostrophic balance, namely *geostrophic adjustment*. This process occurs quite generally in rotating fluids, whether stratified or not. To pose the problem in a simple form we consider the free evolution of a single shallow layer of fluid whose initial state is manifestly unbalanced, and we suppose that surface displacements are small so that the evolution of the system is described by the linearized shallow equations of motion. These are

$$\frac{\partial \boldsymbol{u}}{\partial t} + \boldsymbol{f} \times \boldsymbol{u} = -g\nabla\eta, \qquad \frac{\partial \eta}{\partial t} + H\nabla\cdot\boldsymbol{u} = 0, \tag{3.136a,b}$$

where η is the free surface displacement and H is the mean fluid depth, and we omit the primes on the linearized variables.

3.9.1 Non-rotating Flow

We consider first the non-rotating problem set, with little loss of generality, in one dimension. We suppose that initially the fluid is at rest but with a simple discontinuity in the height field so that

$$\eta(x, t = 0) = \begin{cases} +\eta_0 & x < 0 \\ -\eta_0 & x > 0, \end{cases} \tag{3.137}$$

and $u(x, t = 0) = 0$ everywhere. We can realize these initial conditions physically by separating two fluid masses of different depths by a thin dividing wall, and then quickly removing the wall. What is the subsequent evolution of the fluid? The general solution to the linear problem is given by (3.115) where the functional form is determined by the initial conditions so that here

$$F(x) = \eta(x, t = 0) = -\eta_0 \operatorname{sgn}(x). \tag{3.138}$$

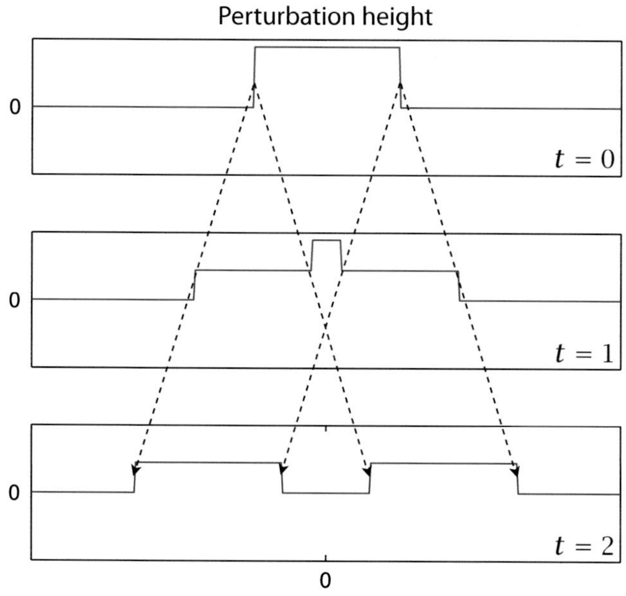

Fig. 3.9 The time development of an initial 'top hat' height disturbance, with zero initial velocity, in non-rotating flow. Fronts propagate in both directions, and the velocity is non-zero between fronts, but ultimately the disturbances are radiated away to infinity, and the fluid is left at rest with zero perturbation height.

Equation (3.115) states that this initial pattern is propagated to the right and to the left. That is, two discontinuities in fluid height move to the right and left at a speed $c = \sqrt{gH}$. Specifically, the solution is

$$\eta(x,t) = -\frac{1}{2}\eta_0[\mathrm{sgn}(x+ct) + \mathrm{sgn}(x-ct)]. \tag{3.139}$$

The initial conditions may be much more complex than a simple front, but, because the waves are dispersionless, the solution is still simply a sum of the translation of those initial conditions to the right and to the left at speed c. The velocity field in this class of problem is obtained from

$$\frac{\partial u}{\partial t} = -g\frac{\partial \eta}{\partial x}, \tag{3.140}$$

which gives, using (3.115),

$$u = -\frac{g}{2c}[F(x+ct) - F(x-ct)]. \tag{3.141}$$

Consider the case with initial conditions given by (3.137). At a given location, away from the initial disturbance, the fluid remains at rest and undisturbed until the front arrives. After the front has passed, the fluid surface is again undisturbed and the velocity is uniform and non-zero. Specifically:

$$\eta = \begin{cases} -\eta_0 \mathrm{sgn}(x) \\ 0 \end{cases} \qquad u = \begin{cases} 0 & |x| > ct \\ (\eta_0 g/c) & |x| < ct. \end{cases} \tag{3.142}$$

The solution with 'top-hat' initial conditions in the height field, and zero initial velocity, is a superposition of two discontinuities similar to (3.142) and is illustrated in Fig. 3.9. Two fronts propagate in either direction from each discontinuity and, in this case, the final velocity, as well as the fluid displacement, is zero after all the fronts have passed. That is, the disturbance is radiated completely away.

3.9.2 Rotating Flow

Rotation makes a profound difference to the adjustment problem of the shallow water system, because a steady, adjusted, solution can exist with non-zero gradients in the height field — the associated pressure gradients being balanced by the Coriolis force — and potential vorticity conservation provides a powerful constraint on the fluid evolution.[8] In a rotating shallow fluid that conservation is represented by

$$\frac{\partial Q}{\partial t} + \boldsymbol{u} \cdot \nabla Q = 0, \tag{3.143}$$

where $Q = (\zeta + f)/h$. In the linear case with constant Coriolis parameter, (3.143) becomes

$$\frac{\partial q}{\partial t} = 0, \qquad q = \left(\zeta - f_0 \frac{\eta}{H}\right). \tag{3.144}$$

This equation may be obtained either from the linearized velocity and mass conservation equations, (3.136), or from (3.143) directly. In the latter case, we write

$$Q = \frac{\zeta + f_0}{H + \eta} \approx \frac{1}{H}(\zeta + f_0)\left(1 - \frac{\eta}{H}\right) \approx \frac{1}{H}\left(f_0 + \zeta - f_0 \frac{\eta}{H}\right) = \frac{f_0}{H} + \frac{q}{H}, \tag{3.145}$$

having used $f_0 \gg |\zeta|$ and $H \gg |\eta|$. The term f_0/H is a constant and so dynamically unimportant, as is the H^{-1} factor multiplying q. Further, the advective term $\boldsymbol{u} \cdot \nabla Q$ becomes $\boldsymbol{u} \cdot \nabla q$, and this is second order in perturbed quantities and so is neglected. Thus, making these approximations, (3.143) reduces to (3.144). The potential vorticity field is therefore fixed in space! Of course, this was also true in the non-rotating case where the fluid is initially at rest. Then $q = \zeta = 0$ and the fluid remains irrotational throughout the subsequent evolution of the flow. However, this is rather a weak constraint on the subsequent evolution of the fluid; it does nothing, for example, to prevent the conversion of all the potential energy to kinetic energy. In the rotating case the potential vorticity is non-zero, and potential vorticity conservation and geostrophic balance are all we need to infer the final steady state, assuming it exists, without solving for the details of the flow evolution, as we now see.

With an initial condition for the height field given by (3.137), the initial potential vorticity is given by

$$q(x, y) = \begin{cases} -f_0 \eta_0 / H & x < 0 \\ f_0 \eta_0 / H & x > 0, \end{cases} \tag{3.146}$$

and this remains unchanged throughout the adjustment process. The final steady state is then the solution of the equations

$$\zeta - f_0 \frac{\eta}{H} = q(x, y), \qquad f_0 u = -g \frac{\partial \eta}{\partial y}, \qquad f_0 v = g \frac{\partial \eta}{\partial x}, \tag{3.147a,b,c}$$

where $\zeta = \partial v/\partial x - \partial u/\partial y$. Because the Coriolis parameter is constant, the velocity field is horizontally non-divergent and we may define a streamfunction $\psi = g\eta/f_0$. Equations (3.147) then reduce to

$$\left(\nabla^2 - \frac{1}{L_d^2}\right)\psi = q(x, y), \tag{3.148}$$

where $L_d = \sqrt{gH}/f_0$ is known as the *Rossby radius of deformation* or often just the 'deformation radius' or the 'Rossby radius'. It is a naturally occurring length scale in problems involving both rotation and gravity, and arises in a slightly different form in stratified fluids.

The initial conditions (3.146) admit of a nice analytic solution, for the flow will remain uniform in y, and (3.148) reduces to

$$\frac{\partial^2 \psi}{\partial x^2} - \frac{1}{L_d^2}\psi = \frac{f_0 \eta_0}{H}\operatorname{sgn}(x). \tag{3.149}$$

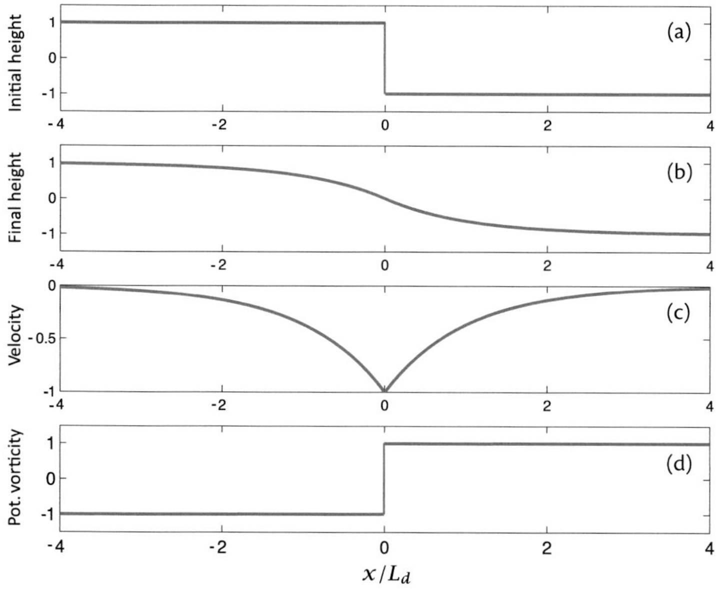

Fig. 3.10 Solutions of a linear geostrophic adjustment problem. (a) Initial height field, given by (3.137) with $\eta_0 = 1$. (b) Equilibrium (final) height field, η given by (3.150) and $\eta = f_0\psi/g$. (c) Equilibrium geostrophic velocity, normal to the gradient of height field, given by (3.151). (d) Potential vorticity, given by (3.146), and this does not evolve.

The distance, x is nondimensionalized by the deformation radius L_d and the velocity by $\eta_0(g/f_0L_d)$. Changes to the initial state occur within $\mathcal{O}(L_d)$ of the initial discontinuity.

We solve this separately for $x > 0$ and $x < 0$ and then match the solutions and their first derivatives at $x = 0$, also imposing the condition that the velocity decays to zero as $x \to \pm\infty$. The solution is

$$\psi = \begin{cases} -(g\eta_0/f_0)(1 - e^{-x/L_d}) & x > 0 \\ +(g\eta_0/f_0)(1 - e^{x/L_d}) & x < 0. \end{cases} \tag{3.150}$$

The velocity field associated with this is obtained from (3.147b,c), and is

$$u = 0, \qquad v = -\frac{g\eta_0}{f_0L_d} e^{-|x|/L_d}. \tag{3.151}$$

The velocity is perpendicular to the slope of the free surface, and a jet forms along the initial discontinuity, as illustrated in Fig. 3.10.

The important point of this problem is that the variations in the height and field are not radiated away to infinity, as in the non-rotating problem. Rather, potential vorticity conservation constrains the influence of the adjustment to within a deformation radius (we see now why this name is appropriate) of the initial disturbance. This property is a general one in geostrophic adjustment — it also arises if the initial condition consists of a velocity jump.

A snapshot of the time evolution of flow, obtained by a numerical integration of the shallow water equations for both rotating and non-rotating flow, is illustrated in Fig. 3.11. The initial conditions are a jump in the height field, as in Fig. 3.10. Fronts propagate away at a speed $\sqrt{gH} = 1$ in both cases, but in the rotating flow they leave behind a geostrophically balanced state with a non-zero meridional velocity.

3.9.3 ♦ Energetics of Adjustment

How much of the initial potential energy of the flow is lost to infinity by gravity wave radiation, and how much is converted to kinetic energy? The linear equations (3.136) lead to

$$\frac{1}{2}\frac{\partial}{\partial t}(H\boldsymbol{u}^2 + g\eta^2) + gH\nabla\cdot(\boldsymbol{u}\eta) = 0, \tag{3.152}$$

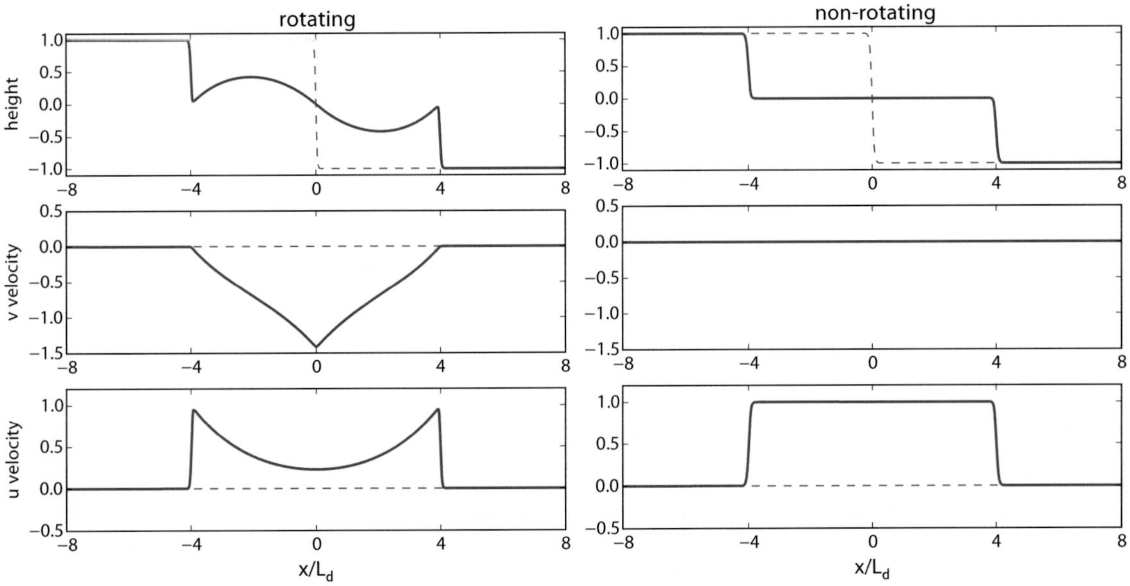

Fig. 3.11 The solutions of the shallow water equations obtained by numerically integrating the equations of motion with and without rotation. The panels show snapshots of the state of the fluid (solid lines) soon after being released from a stationary initial state (red dashed lines) with a height discontinuity. The rotating flow is evolving toward an end state similar to Fig. 3.10 whereas the non-rotating flow will eventually become stationary. In the non-rotating case L_d is defined using the rotating parameters.[9]

so that energy conservation holds in the form

$$E = \frac{1}{2}\int (H\boldsymbol{u}^2 + g\eta^2)\,\mathrm{d}\boldsymbol{x}, \qquad \frac{\mathrm{d}E}{\mathrm{d}t} = 0, \tag{3.153}$$

provided the integral of the divergence term vanishes, as it normally will in a closed domain. The fluid has a non-zero potential energy, $(1/2)\int_{-\infty}^{\infty} g\eta^2\,\mathrm{d}x$, if there are variations in fluid height, and with the initial conditions (3.137) the initial potential energy is

$$PE_I = \int_0^{\infty} g\eta_0^2\,\mathrm{d}x. \tag{3.154}$$

This is nominally infinite if the fluid has no boundaries, and the initial potential energy density is $g\eta_0^2/2$ everywhere.

In the non-rotating case, and with initial conditions (3.137), after the front has passed, the potential energy density is zero and the kinetic energy density is $Hu^2/2 = g\eta_0^2/2$, using (3.142) and $c^2 = gH$. Thus, all the potential energy is locally converted to kinetic energy as the front passes, and eventually the kinetic energy is distributed uniformly along the line. In the case illustrated in Fig. 3.9, the potential energy and kinetic energy are both radiated away from the initial disturbance. (Note that although we can superpose the solutions from different initial conditions, we cannot superpose their potential and kinetic energies.) The general point is that the evolution of the disturbance is not confined to its initial location.

In contrast, in the rotating case the conversion from potential to kinetic energy *is largely confined to within a deformation radius of the initial disturbance,* and at locations far from the initial disturbance the initial state is essentially unaltered. The conservation of potential vorticity has prevented the complete conversion of potential energy to kinetic energy, a result that is not sensitive to the precise form of the initial conditions.

In fact, in the rotating case, some of the initial potential energy is converted to kinetic energy, some remains as potential energy and some is lost to infinity; let us calculate these amounts. The final potential energy, after adjustment, is, using (3.150),

$$PE_F = \frac{1}{2} g\eta_0^2 \left[\int_0^{\infty} \left(1 - e^{-x/L_d}\right)^2 dx + \int_{-\infty}^{0} \left(1 - e^{x/L_d}\right)^2 dx \right]. \tag{3.155}$$

This is nominally infinite, but the change in potential energy is finite and is given by

$$PE_I - PE_F = g\eta_0^2 \int_0^{\infty} (2e^{-x/L_d} - e^{-2x/L_d})\, dx = \frac{3}{2} g\eta_0^2 L_d. \tag{3.156}$$

The initial kinetic energy is zero, because the fluid is at rest, and its final value is, using (3.151),

$$KE_F = \frac{1}{2} H \int \boldsymbol{u}^2\, dx = H \left(\frac{g\eta_0}{fL_d} \right)^2 \int_0^{\infty} e^{-2x/L_d}\, dx = \frac{g\eta_0^2 L_d}{2}. \tag{3.157}$$

Thus one-third of the difference between the initial and final potential energies is converted to kinetic energy, and this is trapped within a distance of the order of a deformation radius of the disturbance; the remainder, an amount $gL_d\eta_0^2$ is radiated away and lost to infinity. In any finite region surrounding the initial discontinuity the final energy is less than the initial energy.

3.9.4 ♦ General Initial Conditions

Because of the linearity of the (linear) adjustment problem a spectral viewpoint is useful, in which the fields are represented as the sum or integral of *non-interacting* Fourier modes. For example, suppose that the height field of the initial disturbance is a two-dimensional field given by

$$\eta(0) = \iint \tilde{\eta}_{k,l}(0) e^{i(kx+ly)}\, dk\, dl, \tag{3.158}$$

where the Fourier coefficients $\tilde{\eta}_{k,l}(0)$ are given, and the initial velocity field is zero. Then the initial (and final) potential vorticity field is given by

$$q = -\frac{f_0}{H} \iint \tilde{\eta}_{k,l}(0) e^{i(kx+ly)}\, dk\, dl. \tag{3.159}$$

To obtain an expression for the final height and velocity fields, we express the potential vorticity field as

$$q = \iint \tilde{q}_{k,l}\, dk\, dl. \tag{3.160}$$

The potential vorticity field does not evolve, and it is related to the initial height field by

$$\tilde{q}_{k,l} = -\frac{f_0}{H} \eta_{k,l}(0). \tag{3.161}$$

In the final, geostrophically balanced state, the potential vorticity is related to the height field by

$$q = \frac{g}{f_0} \nabla^2 \eta - \frac{f_0}{H} \eta \qquad \text{and} \qquad \tilde{q}_{k,l} = \left(-\frac{g}{f_0} K^2 - \frac{f_0}{H} \right) \tilde{\eta}_{k,l}, \tag{3.162a,b}$$

where $K^2 = k^2 + l^2$. Using (3.161) and (3.162), the Fourier components of the final height field satisfy

$$\left(-\frac{g}{f_0} K^2 - \frac{f_0}{H} \right) \tilde{\eta}_{k,l} = -\frac{f_0}{H} \tilde{\eta}_{k,l}(0) \tag{3.163}$$

or

$$\widetilde{\eta}_{k,l} = \frac{\widetilde{\eta}_{k,l}(0)}{K^2 L_d^2 + 1}. \tag{3.164}$$

In physical space the final height field is just the spectral integral of this, namely

$$\eta = \iint \widetilde{\eta}_{k,l} \mathrm{e}^{\mathrm{i}(kx+ly)} \,\mathrm{d}k\,\mathrm{d}l = \iint \frac{\widetilde{\eta}_{k,l}(0)\mathrm{e}^{\mathrm{i}(kx+ly)}}{K^2 L_d^2 + 1} \,\mathrm{d}k\,\mathrm{d}l. \tag{3.165}$$

We see that at large scales ($K^2 L_d^2 \ll 1$) $\eta_{k,l}$ is almost unchanged from its initial state; the velocity field, which is then determined by geostrophic balance, thus adjusts to the pre-existing height field. At large scales most of the energy in geostrophically balanced flow is potential energy; thus, it is energetically easier for the velocity to change to come into balance with the height field than vice versa. At small scales, however, the final height field has much less variability than it did initially.

Conversely, at small scales the height field adjusts to the velocity field. To see this, let us suppose that the initial conditions contain vorticity but have zero height displacement. Specifically, if the initial vorticity is $\nabla^2\psi(0)$, where $\psi(0)$ is the initial streamfunction, then it is straightforward to show that the final streamfunction is given by

$$\psi = \iint \widetilde{\psi}_{k,l} \mathrm{e}^{\mathrm{i}(kx+ly)} \,\mathrm{d}k\,\mathrm{d}l = \iint \frac{K^2 L_d^2 \widetilde{\psi}_{k,l}(0)\mathrm{e}^{\mathrm{i}(kx+ly)}}{K^2 L_d^2 + 1} \,\mathrm{d}k\,\mathrm{d}l. \tag{3.166}$$

The final height field is then obtained from this, via geostrophic balance, by $\eta = (f_0/g)\psi$. Evidently, for small scales ($K^2 L_d^2 \gg 1$) the streamfunction, and hence the vortical component of the velocity field, are almost unaltered from their initial values. On the other hand, at large scales the final streamfunction has much less variability than it does initially, and so the height field is largely governed by whatever variation it (and not the velocity field) had initially. In general, the final state is a superposition of the states given by (3.165) and (3.166). The divergent component of the initial velocity field does not affect the final state because it has no potential vorticity, and so all of the associated energy is eventually lost to infinity.

Finally, we remark that just as in the problem with a discontinuous initial height profile, the change in total energy during adjustment is negative — this can be seen from the form of the integrals above, although we leave the specifics as a problem to the reader. That is, some of the initial potential and kinetic energy is lost to infinity, but some is trapped by the potential vorticity constraint.

3.9.5 A Variational Perspective

In the non-rotating problem, all of the initial potential energy is eventually radiated away to infinity. In the rotating problem, the final state contains both potential and kinetic energy. Why is the energy not all radiated away to infinity? It is because potential vorticity conservation on parcels prevents all of the energy being dispersed. This suggests that it may be informative to think of the geostrophic adjustment problem as a *variational problem:* we seek to minimize the energy consistent with the conservation of potential vorticity. We stay in the linear approximation in which, because the advection of potential vorticity is neglected, potential vorticity remains constant at each point.

The energy of the flow is given by the sum of potential and kinetic energies, namely

$$\text{energy} = \int (H\boldsymbol{u}^2 + g\eta^2)\,\mathrm{d}A, \tag{3.167}$$

(where $\mathrm{d}A \equiv \mathrm{d}x\,\mathrm{d}y$) and the potential vorticity field is

$$q = \zeta - f_0\frac{\eta}{H} = (v_x - u_y) - f_0\frac{\eta}{H}, \tag{3.168}$$

where the subscripts x, y denote derivatives. The problem is then to extremize the energy subject to potential vorticity conservation. This is a constrained problem in the calculus of variations, sometimes called an *isoperimetric* problem because of its origins in maximizing the area of a surface for a given perimeter.[10] The mathematical problem is to extremize the integral

$$I = \int \left\{ H(u^2 + v^2) + g\eta^2 + \lambda(x, y)\big[(v_x - u_y) - f_0\eta/H\big] \right\} \mathrm{d}A, \tag{3.169}$$

where $\lambda(x, y)$ is a Lagrange multiplier, undetermined at this stage. It is a function of space: if it were a constant, the integral would merely extremize energy subject to a given integral of potential vorticity, and rearrangements of potential vorticity (which here we wish to disallow) would leave the integral unaltered.

As there are three independent variables there are three Euler–Lagrange equations that must be solved in order to minimize I. These are

$$\frac{\partial L}{\partial \eta} - \frac{\partial}{\partial x}\frac{\partial L}{\partial \eta_x} - \frac{\partial}{\partial y}\frac{\partial L}{\partial \eta_y} = 0,$$
$$\frac{\partial L}{\partial u} - \frac{\partial}{\partial x}\frac{\partial L}{\partial u_x} - \frac{\partial}{\partial y}\frac{\partial L}{\partial u_y} = 0, \qquad \frac{\partial L}{\partial v} - \frac{\partial}{\partial x}\frac{\partial L}{\partial v_x} - \frac{\partial}{\partial y}\frac{\partial L}{\partial v_y} = 0, \tag{3.170}$$

where L is the integrand on the right-hand side of (3.169). Substituting the expression for L into (3.170) gives, after a little algebra,

$$2g\eta - \frac{\lambda f_0}{H} = 0, \qquad 2Hu + \frac{\partial \lambda}{\partial y} = 0, \qquad 2Hv - \frac{\partial \lambda}{\partial x} = 0, \tag{3.171}$$

and then eliminating λ gives the simple relationships

$$u = -\frac{g}{f_0}\frac{\partial \eta}{\partial y}, \qquad v = \frac{g}{f_0}\frac{\partial \eta}{\partial x}, \tag{3.172}$$

which are the equations of geostrophic balance. Thus, in the linear approximation, *geostrophic balance is the minimum energy state for a given field of potential vorticity.*

3.10 ISENTROPIC COORDINATES

We now return to the continuously stratified primitive equations, and consider the use of potential density as a vertical coordinate. In practice this means using potential temperature in the atmosphere and (for simple equations of state) buoyancy in the ocean; such coordinate systems are generically called *isentropic coordinates,* and sometimes *isopycnal coordinates* if density is used. This may seem an odd thing to do but for adiabatic flow the resulting equations of motion have an attractive form that aids the interpretation of large-scale flow. The thermodynamic equation becomes a statement for the conservation of the mass of fluid with a given value of potential density and, because the flow of both the atmosphere and the ocean is largely along isentropic surfaces, the momentum and vorticity equations have a quasi-two-dimensional form.

The particular choice of vertical coordinate is determined by the form of the thermodynamic equation in the equation-set at hand; thus, if the thermodynamic equation is $\mathrm{D}\theta/\mathrm{D}t = \dot{\theta}$, we transform the equations from (x, y, z) coordinates to (x, y, θ) coordinates. The material derivative in this coordinate system is

$$\frac{\mathrm{D}}{\mathrm{D}t} = \frac{\partial}{\partial t} + u\left(\frac{\partial}{\partial x}\right)_\theta + v\left(\frac{\partial}{\partial y}\right)_\theta + \frac{\mathrm{D}\theta}{\mathrm{D}t}\frac{\partial}{\partial \theta} = \frac{\partial}{\partial t} + \boldsymbol{u}\cdot\nabla_\theta + \dot{\theta}\frac{\partial}{\partial \theta}, \tag{3.173}$$

where the last term on the right-hand side is zero for adiabatic flow.

3.10.1 A Hydrostatic Boussinesq Fluid

In the simple Boussinesq equations (see the table on page 74) the buoyancy is the relevant thermodynamic variable. With hydrostatic balance the horizontal and vertical momentum equations are, in height coordinates,

$$\frac{\mathrm{D}\boldsymbol{u}}{\mathrm{D}t} + \boldsymbol{f} \times \boldsymbol{u} = -\nabla\phi, \qquad b = \frac{\partial \phi}{\partial z}, \tag{3.174}$$

where b is the buoyancy, the variable analogous to the potential temperature θ of an ideal gas. The thermodynamic equation is

$$\frac{\mathrm{D}b}{\mathrm{D}t} = \dot{b}, \tag{3.175}$$

and because $b = -g\delta\rho/\rho_0$, isentropic coordinates are analogous to isopycnal coordinates.

Using (2.142) the horizontal pressure gradient may be transformed to isentropic coordinates:

$$\left(\frac{\partial \phi}{\partial x}\right)_z = \left(\frac{\partial \phi}{\partial x}\right)_b - \left(\frac{\partial z}{\partial x}\right)_b \frac{\partial \phi}{\partial z} = \left(\frac{\partial \phi}{\partial x}\right)_b - b\left(\frac{\partial z}{\partial x}\right)_b = \left(\frac{\partial M}{\partial x}\right)_b, \tag{3.176}$$

where

$$M \equiv \phi - zb. \tag{3.177}$$

Thus, the horizontal momentum equation becomes

$$\frac{\mathrm{D}\boldsymbol{u}}{\mathrm{D}t} + \boldsymbol{f} \times \boldsymbol{u} = -\nabla_b M. \tag{3.178}$$

where the material derivative is given by (3.173), with b replacing θ. Using (3.177) the hydrostatic equation becomes

$$\frac{\partial M}{\partial b} = -z. \tag{3.179}$$

The mass continuity equation may be derived by noting that for a Boussinesq fluid the mass element may be written as

$$\delta m = \rho_0 \frac{\partial z}{\partial b} \delta b\, \delta x\, \delta y. \tag{3.180}$$

The mass continuity equation, $\mathrm{D}\delta m/\mathrm{D}t = 0$, becomes

$$\frac{\mathrm{D}}{\mathrm{D}t}\frac{\partial z}{\partial b} + \frac{\partial z}{\partial b}\nabla_3 \cdot \boldsymbol{v} = 0, \tag{3.181}$$

where $\nabla_3 \cdot \boldsymbol{v} = \nabla_b \cdot \boldsymbol{u} + \partial \dot{b}/\partial b$ is the three-dimensional derivative of the velocity in isentropic coordinates. Equation (3.181) may thus be written

$$\frac{\mathrm{D}\sigma}{\mathrm{D}t} + \sigma \nabla_b \cdot \boldsymbol{u} = -\sigma \frac{\partial \dot{b}}{\partial b}, \tag{3.182}$$

where $\sigma \equiv \partial z/\partial b$ is a measure of the thickness between two isentropic surfaces and the material derivative is given by (3.173) with θ replaced by b. Equations (3.178), (3.179) and (3.182) comprise a closed set, with dependent variables $\boldsymbol{u}$, M and z in the space of independent variables x, y and b.

3.10.2 A Hydrostatic Ideal Gas

Deriving the equations of motion for this system requires a little more work than in the Boussinesq case but the idea is the same. For an ideal gas in hydrostatic balance we have, using (1.110),

$$\frac{\delta\theta}{\theta} = \frac{\delta T}{T} - \kappa\frac{\delta p}{p} = \frac{\delta T}{T} + \frac{\delta\Phi}{c_p T} = \frac{1}{c_p T}\delta M, \tag{3.183}$$

where $\delta\Phi = g\delta z$ and $M \equiv c_p T + \Phi$ is the 'Montgomery potential', equal to the dry static energy. (We use some of the same symbols as in the Boussinesq case to facilitate comparison, but their meanings are slightly different.) From this

$$\frac{\partial M}{\partial\theta} = \Pi, \tag{3.184}$$

where $\Pi \equiv c_p T/\theta = c_p(p/p_R)^{R/c_p}$ is the 'Exner function'. Equation (3.184) represents the hydrostatic relation in isentropic coordinates. Note also that $M = \theta\Pi + \Phi$.

To obtain an appropriate form for the horizontal pressure gradient force first note that, in the usual height coordinates, it is given by

$$\frac{1}{\rho}\nabla_z p = \theta\nabla_z\Pi, \tag{3.185}$$

where $\Pi = c_p T/\theta$. Using (2.142) gives

$$\theta\nabla_z\Pi = \theta\nabla_\theta\Pi - \frac{\theta}{g}\frac{\partial\Pi}{\partial z}\nabla_\theta\Phi. \tag{3.186}$$

Then, using the definition of Π and the hydrostatic approximation to help evaluate the vertical derivative, we obtain

$$\frac{1}{\rho}\nabla_z p = c_p\nabla_\theta T + \nabla_\theta\Phi = \nabla_\theta M. \tag{3.187}$$

Thus, the horizontal momentum equation is

$$\frac{\mathrm{D}\boldsymbol{u}}{\mathrm{D}t} + \boldsymbol{f}\times\boldsymbol{u} = -\nabla_\theta M. \tag{3.188}$$

Much as in the Boussinesq case, the mass continuity equation may be derived by noting that the mass element may be written as

$$\delta m = -\frac{1}{g}\frac{\partial p}{\partial\theta}\delta\theta\,\delta x\,\delta y. \tag{3.189}$$

The mass continuity equation, $\mathrm{D}\delta m/\mathrm{D}t = 0$, becomes

$$\frac{\mathrm{D}}{\mathrm{D}t}\frac{\partial p}{\partial\theta} + \frac{\partial p}{\partial\theta}\nabla_3\cdot\boldsymbol{v} = 0 \quad\text{or}\quad \frac{\mathrm{D}\sigma}{\mathrm{D}t} + \sigma\nabla_\theta\cdot\boldsymbol{u} = -\sigma\frac{\partial\dot\theta}{\partial\theta}, \tag{3.190a,b}$$

where now $\sigma \equiv \partial p/\partial\theta$ is a measure of the (pressure) thickness between two isentropic surfaces. Equations (3.184), (3.188) and (3.190b) form a closed set, analogous to (3.179), (3.178) and (3.182).

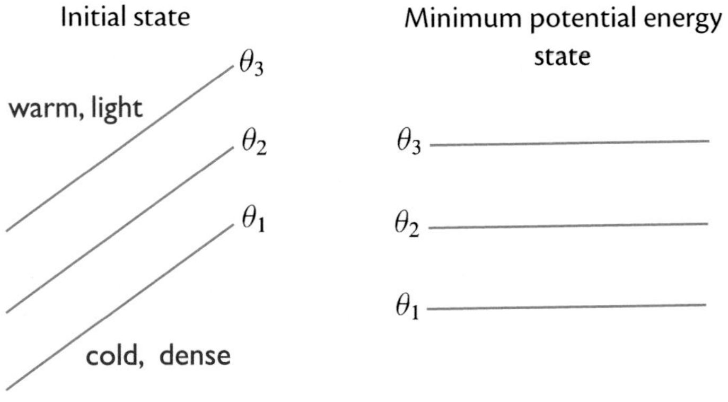

Fig. 3.12 If a stably stratified initial state with sloping isentropes (left) is adiabatically rearranged then the state of minimum potential energy has flat isentropes, as on the right, but the amount of fluid contained between each isentropic surface is unchanged. The difference between the potential energies of the two states is the *available potential energy*.

3.10.3 ✦ Analogy to Shallow Water Equations

The equations of motion in isentropic coordinates have an analogy with the shallow water equations, and we may think of the shallow water equations as a finite-difference representation of the primitive equations written in isentropic coordinates, or think of the latter as the continuous limit of the shallow water equations as the number of layers increases. For example, consider a two-isentropic-level representation of (3.184), (3.188) and (3.190), in which the lower boundary is an isentrope. A natural finite differencing gives

$$-M_1 = \Pi_0 \Delta\theta_0, \qquad M_1 - M_2 = \Pi_1 \Delta\theta_1, \tag{3.191a,b}$$

where the $\Delta\theta$s are constants, and the momentum equations for each layer become

$$\frac{\mathrm{D}\boldsymbol{u}_1}{\mathrm{D}t} + \boldsymbol{f} \times \boldsymbol{u}_1 = -\Delta\theta_0 \nabla \Pi_0, \qquad \frac{\mathrm{D}\boldsymbol{u}_2}{\mathrm{D}t} + \boldsymbol{f} \times \boldsymbol{u}_2 = -\Delta\theta_0 \nabla \Pi_0 - \Delta\theta_1 \nabla \Pi_1. \tag{3.192}$$

Together with the mass continuity equation for each level these are similar to the two-layer shallow water equations (3.49).

3.11 AVAILABLE POTENTIAL ENERGY

We now revisit the issue of the internal and potential energy in stratified flow, motivated by the following remarks. In adiabatic, inviscid flow the total amount of energy is conserved, and there are conversions between internal energy, potential energy and kinetic energy. In an ideal gas the potential energy and the internal energy of a column extending throughout the atmosphere are in a constant ratio to each other — their sum is called the total potential energy. In a simple Boussinesq fluid, energetic conversions involve only the potential and kinetic energy, and not the internal energy. Yet, plainly, in neither a Boussinesq fluid nor an ideal gas can *all* the total potential energy in a fluid be converted to kinetic energy, for then all of the fluid would be adjacent to the ground and the fluid would have no thickness. Given a state of the atmosphere or ocean, how much of its total potential energy is available for conversion to kinetic energy? In particular, because energy is conserved only in adiabatic flow, we may usefully ask: how much potential energy is available for conversion to kinetic energy under an adiabatic rearrangement of fluid parcels?

Suppose that at any given time the flow is stably stratified, but that the isentropes (or more generally the surfaces of constant potential density) are sloping, as in Fig. 3.12. The potential energy of the system would be reduced if the isentropes were flattened, for then heavier fluid would be moved to lower altitudes, with lighter fluid replacing it at higher altitudes. In an adiabatic rearrangement the amount of fluid between the isentropes would remain constant, and a state with flat isentropes (meaning parallel to the geopotential surfaces) evidently constitutes a state of minimum total potential energy. The difference between the total potential energy of the fluid and the

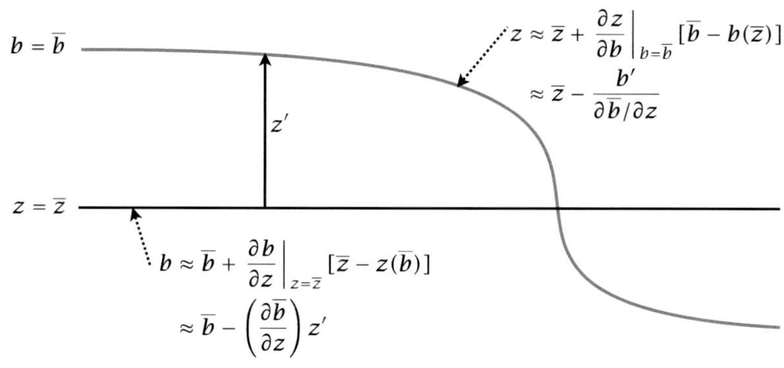

Fig. 3.13 An isopycnal surface, $b = \overline{b}$, and the constant height surface, $z = \overline{z}$, where $\overline{z}$ is the height of the isopycnal surface after a rearrangement to a minimum potential energy state, equal to the average height of the isopycnal surface. The values of z on the isopycnal surface, and of b on the constant height surface, can be obtained by the Taylor expansions shown. For an ideal gas in pressure coordinates, replace z by p and b by θ.

total potential energy after an adiabatic rearrangement to a state in which the isentropic surfaces are flat is called the *available potential energy,* or APE.[11]

3.11.1 A Boussinesq Fluid

The potential energy of a column of a Boussinesq fluid of unit area is given by

$$P = -\int_0^H bz\,\mathrm{d}z = -\int_0^H \frac{b}{2}\,\mathrm{d}z^2. \tag{3.193}$$

and the potential energy of the entire fluid is given by the horizontal integral of this. The minimum potential energy of the fluid arises after an adiabatic rearrangement in which the isopycnals are flattened, and the resulting buoyancy is only a function of z. The available potential energy is then the difference between the energy of the initial state and of this minimum state, and to obtain an approximate expression for this we first integrate (3.193) by parts to give

$$P = \frac{1}{2}\int_0^{b_m} z^2\,\mathrm{d}b - \left[\frac{bz^2}{2}\right]_0^H = \frac{1}{2}\int_0^{b_m} z^2\,\mathrm{d}b - \frac{b_m H^2}{2}, \tag{3.194}$$

where b_m is the maximum value of b in the domain, and we may formally take the upper boundary to have this value of b without affecting the final result. The minimum potential energy state arises when z is a function only of b, $z = Z(b)$ say. Because mass is conserved in the rearrangement, Z is equal to the horizontally averaged value of z on a given isopycnal surface, $\overline{z}$, and the surfaces $\overline{z}$ and $\overline{b}$ thus define each other completely. The average available potential energy, per unit area, is then given by

$$\text{APE} = \frac{1}{2}\int_0^{b_m} (\overline{z^2} - \overline{z}^2)\,\mathrm{d}b = \frac{1}{2}\int_0^{b_m} \overline{z'^2}\,\mathrm{d}b, \tag{3.195}$$

where $z = \overline{z} + z'$; that is, z' is the height variation of an isopycnal surface, and the last term on the right-hand side of (3.194) has cancelled with an identical term in the expression for the potential energy of the re-arranged state. The available potential energy is thus proportional to the integral of the variance of the altitude of such a surface, and it is a positive-definite quantity. To obtain an expression in z-coordinates, we express the height variations on an isopycnal surface in terms of buoyancy variations on a surface of constant height by Taylor-expanding the height about its value on the isopycnal surface. Referring to Fig. 3.13 this gives

$$z(\overline{b}) = \overline{z} + \left.\frac{\partial z}{\partial b}\right|_{b=\overline{b}} [\overline{b} - b(\overline{z})] = \overline{z} - \left.\frac{\partial z}{\partial b}\right|_{b=\overline{b}} b', \tag{3.196}$$

where $b' = b(\overline{z}) - \overline{b}$ is corresponding buoyancy perturbation on the $\overline{z}$ surface and $\overline{b}$ is the average value of b on the $\overline{z}$ surface. Furthermore, $\partial z/\partial b|_{z=\overline{b}} \approx \partial \overline{z}/\partial b \approx (\partial \overline{b}/\partial z)^{-1}$, and (3.196) thus becomes

$$z' = z(\overline{b}) - \overline{z} \approx -b'\left(\frac{\partial \overline{z}}{\partial b}\right) \approx -\frac{b'}{(\partial \overline{b}/\partial z)}, \tag{3.197}$$

where $z' = z(b) - \overline{z}$ is the height perturbation of the isopycnal surface, from its average value. Using (3.197) in (3.195) we obtain an expression for the APE per unit area, to wit

$$\text{APE} \approx \frac{1}{2}\int_0^H \frac{\overline{b'^2}}{\partial \overline{b}/\partial z}\,\mathrm{d}z. \tag{3.198}$$

The total APE of the fluid is the horizontal integral of the above, and so is proportional to the variance of the buoyancy on a height surface. We emphasize that APE is not defined for a single column of fluid, for it depends on the variations of buoyancy over a horizontal surface. Note too that the derivation neglects the effects of topography; this, and the use of a basic-state stratification, effectively restrict the use of (3.198) to a single ocean basin, and even for that the approximations used limit the accuracy of the expressions.

3.11.2 An Ideal Gas

The expression for the APE for an ideal gas is obtained, *mutatis mutandis,* in the same way as for a Boussinesq fluid and the trusting reader may skip directly to (3.206). The internal energy of an ideal gas column of unit area is given by

$$I = \int_0^\infty c_v T\rho\,\mathrm{d}z = \int_0^{p_s} \frac{c_v}{g} T\,\mathrm{d}p, \tag{3.199}$$

where p_s is the surface pressure, and the corresponding potential energy is given by

$$P = \int_0^\infty \rho g z\,\mathrm{d}z = \int_0^{p_s} z\,\mathrm{d}p = \int_0^\infty p\,\mathrm{d}z = \int_0^{p_s} \frac{R}{g} T\,\mathrm{d}p. \tag{3.200}$$

In (3.199) we use hydrostasy, and in (3.200) the equalities make successive use of hydrostasy, an integration by parts, hydrostasy and the ideal gas relation. Thus, the total potential energy (TPE) is given by

$$\text{TPE} \equiv I + P = \frac{c_p}{g}\int_0^{p_s} T\,\mathrm{d}p. \tag{3.201}$$

Using the ideal gas equation of state we can write this as

$$\text{TPE} = \frac{c_p}{g}\int_0^{p_s}\left(\frac{p}{p_s}\right)^{\kappa}\theta\,\mathrm{d}p = \frac{c_p p_s}{g(1+\kappa)}\int_0^\infty \left(\frac{p}{p_s}\right)^{\kappa+1}\mathrm{d}\theta, \tag{3.202}$$

after an integration by parts. (We omit a term proportional to $p_s\theta_s$ that arises in the integration by parts, because it cancels in a similar fashion to the boundary term in the Boussinesq derivation; or take $\theta_s = 0$.) The total potential energy of the entire fluid is equal to a horizontal integral of (3.202). The minimum total potential energy arises when the pressure in (3.202) is a function only of θ, $p = P(\theta)$, where by conservation of mass P is the average value of the original pressure on the isentropic surface, $P = \overline{p}$. The average available potential energy per unit area is then given by the difference between the initial state and this minimum, namely

$$\text{APE} = \frac{c_p p_s}{g(1+\kappa)}\int_0^\infty \left[\overline{\left(\frac{p}{p_s}\right)^{\kappa+1}} - \left(\frac{\overline{p}}{p_s}\right)^{\kappa+1}\right]\mathrm{d}\theta, \tag{3.203}$$

which is a positive-definite quantity. A useful approximation to this is obtained by expressing the right-hand side in terms of the variance of the potential temperature on a pressure surface. We first use the binomial expansion to expand $p^{\kappa+1} = (\overline{p} + p')^{\kappa+1}$. Neglecting third- and higher-order terms (3.203) becomes

$$\text{APE} = \frac{R\overline{p}_s}{2g} \int_0^\infty \left(\frac{\overline{p}}{\overline{p}_s}\right)^{\kappa+1} \overline{\left(\frac{p'}{\overline{p}}\right)^2} \, d\theta. \tag{3.204}$$

The variable $p' = p(\theta) - \overline{p}$ is a pressure perturbation on an isentropic surface, and is related to the potential temperature perturbation on an isobaric surface by [cf. (3.197)]

$$p' \approx -\theta' \frac{\partial \overline{p}}{\partial \theta} \approx -\frac{\theta'}{\partial \overline{\theta}/\partial p}, \tag{3.205}$$

where $\theta' = \theta(p) - \theta(\overline{p})$ is the potential temperature perturbation on the $\overline{p}$ surface. Using (3.205) in (3.204) we finally obtain

$$\text{APE} = \frac{R\overline{p}_s^{-\kappa}}{2} \int_0^{p_s} p^{\kappa-1} \left(-g\frac{\partial \overline{\theta}}{\partial p}\right)^{-1} \overline{\theta'^2} \, dp. \tag{3.206}$$

The APE is thus proportional to the variance of the potential temperature on the pressure surface or, from (3.204), proportional to the variance of the pressure on an isentropic surface.

3.11.3 Use and Interpretation

The potential energy of a fluid is reduced when the dynamics acts to flatten the isentropes. Consider, for example, Earth's atmosphere, with isentropes sloping upwards toward the pole (as in the left panel of Fig. 3.12 with the pole on the right). Flattening these isentropes amounts to a sinking of dense air and a rising of light air, and this reduction of potential energy leads to a corresponding production of kinetic energy. Thus, if the dynamics is such as to reduce the temperature gradient between equator and pole by flattening the isentropes then APE is converted to KE by that process. A statistically steady state is achieved because the heating from the Sun continually acts to restore the horizontal temperature gradient between equator and pole, thus replenishing the pool of APE, and to this extent the large-scale atmospheric circulation acts like a heat engine.

It is a useful exercise to calculate the total potential energy, the available potential energy and the kinetic energy of atmosphere and the ocean. One finds

$$\text{TPE} \gg \text{APE} > \text{KE} \tag{3.207}$$

with, very approximately, TPE ~ 100 APE and APE ~ 10 KE. The first inequality should not surprise us (as it was this that led us to define APE in the first instance), but the second inequality is not obvious (and in fact the ratio is larger in the ocean). It is related to the fact that the instabilities of the atmosphere and ocean occur at a scale smaller than the size of the domain, and are unable to release all the potential energy that might be available. Understanding this more fully is the topic of Chapters 9 and 12.

Notes

1 The algorithm to solve these equations numerically differs from that of the free-surface shallow water equations because the mass conservation equation can no longer be stepped forward in time. Rather, an elliptic equation for p_{lid} must be derived by eliminating time derivatives between (3.21) using (3.20), and this is then solved at each timestep.

2 This correspondence was known to Matsuno (1966). Gill & Clarke (1974), McCreary (1985) and others also provide derivations of various kinds.

3 Chelton *et al.* (1998), who also provide maps of the first deformation radius and related quantities for the world's oceans.

4 After Margules (1903). Margules sought to relate the energy of fronts to their slope. In this same paper the notion of available potential energy arose.

5 'Form stress' is an expression derived from 'form drag', an expression commonly used in aerodynamics. In aerodynamics, form drag is the force due to the pressure difference between the front and rear of an object, or any other 'form', moving through a fluid. Aerodynamic form drag may, albeit uncommonly, also include frictional effects between the wind and the surface itself.

6 (Jules) Henri Poincaré (1854–1912) was a prodigious French mathematician, physicist and philosopher, certainly one of the greatest mathematicians living at the turn of the twentieth century. He is remembered for his original work in (among other things) algebra, topology, dynamical systems and celestial mechanics, obtaining many results in what would be called nonlinear dynamics and chaos when these fields re-emerged some 60 years later — the notion of 'sensitive dependence on initial conditions', for example, is present in his work. He obtained a number of the results of special relativity independently of Einstein, and worked on the theory of rotating fluids — hence the Poincaré waves of this chapter. He also wrote extensively and successfully for the general public on the meaning, importance and philosophy of science. Among other things he discussed whether scientific knowledge was an arbitrary convention, a notion that remains discussed and controversial to this day. (His answer: 'convention', in part, yes; 'arbitrary', no.) He was a proponent of the role of intuition in mathematical and scientific progress, and did not believe that mathematics could ever be wholly reduced to pure logic.

7 Thomson (1869). William Thompson later became Lord Kelvin.

8 As was considered by Rossby (1938).

9 The code (available from the author's web site) will also reproduce Fig. 3.9.

10 An introduction to variational problems may be found in Weinstock (1952) and a number of other textbooks. Applications to many traditional problems in mechanics are discussed by Lanczos (1970).

11 Margules (1903) introduced the concept of potential energy that is available for conversion to kinetic energy, Lorenz (1955) clarified its meaning and derived useful, approximate formulae for its computation, and there has since been a host of papers on the subject. Thus, for example, Shepherd (1993) showed that the APE is just the non-kinetic part of the pseudoenergy, Huang (1998) looked at some of the limitations of the approximate expressions in an oceanic context, and on the atmospheric side Pauluis (2007) looked at the effects of moisture.

In addition to his formulation of available potential energy, Edward Lorenz (1917–2008) made enormous contributions to the atmospheric sciences over the course of a long career spent almost entirely at MIT. He is perhaps most famous for being one of the modern founders of chaos theory as it emerged in the 1960s, and his paper *Deterministic non-periodic flow,* published in the *Journal of the Atmospheric Sciences* in 1963, was not only a watershed in meteorology but it changed the way we think about irregular systems — see also the endnotes on page 443. Lorenz also introduced the idea of empirical orthogonal functions to meteorology and wrote with clarity and insight about the atmospheric general circulation in a monograph in 1967.

In common with a number of meteorologists of his generation (Eric Eady was another), Lorenz was first educated in mathematics and then, because of World War II, was trained as a weather forecaster. After the war he moved to MIT in 1946 for a PhD, and in 1953 was hired on the MIT faculty, apparently on Jule Charney's recommendation, and stayed there for the rest of his career. An avid hiker who often spent summers in the mountains at NCAR in Boulder, he was a quiet, modest man with nothing to be modest about. See Palmer (2009) and Emanuel (2011) for longer biographical memoirs.

Strange as it may sound, the power of mathematics rests on its evasion of all unnecessary thought and on its wonderful saving of mental operations.
Ernst Mach (1838–1916), quoted in Bell (1937).

CHAPTER 4

Vorticity and Potential Vorticity

VORTICITY AND POTENTIAL VORTICITY both play a central role in geophysical fluid dynamics, especially in the dynamics of the large scale circulation. In this chapter we define and discuss these quantities and deduce some of their dynamical properties and effects. Along the way we will come across *Kelvin's circulation theorem,* one of the most fundamental conservation laws in all of fluid mechanics, to which the conservation of potential vorticity is intimately tied.

4.1 VORTICITY AND CIRCULATION

4.1.1 Preliminaries

Vorticity, $\boldsymbol{\omega}$, is defined to be the curl of velocity and so is given by

$$\boldsymbol{\omega} \equiv \nabla \times \boldsymbol{v}. \tag{4.1}$$

Circulation, C, is defined to be the integral of velocity around a closed fluid loop and so is given by

$$C \equiv \oint \boldsymbol{v} \cdot \mathrm{d}\boldsymbol{r} = \int_S \boldsymbol{\omega} \cdot \mathrm{d}\boldsymbol{S}, \tag{4.2}$$

where the second expression uses Stokes' theorem and S is any surface bounded by the loop. The circulation around the path is equal to the integral of the normal component of vorticity over *any* surface bounded by that path. The circulation is not a field like vorticity and velocity; rather, we think of the circulation around a particular material line of finite length, and so its value generally depends on the path chosen. If δS is an infinitesimal surface element whose normal points in the direction of the unit vector $\hat{\boldsymbol{n}}$, then

$$\hat{\boldsymbol{n}} \cdot (\nabla \times \boldsymbol{v}) = \frac{1}{\delta S} \oint_{\delta r} \boldsymbol{v} \cdot \mathrm{d}\boldsymbol{r}, \tag{4.3}$$

where the line integral is around the infinitesimal area. Thus at a point the component of vorticity in the direction of $\boldsymbol{n}$ is proportional to the circulation around the surrounding infinitesimal fluid element, divided by the elemental area bounded by the path of the integral. A heuristic test for the presence of vorticity is to imagine a small paddle wheel in the flow; the paddle wheel acts as a 'circulation-meter', and rotates if the vorticity is non-zero. Vorticity might seem to be similar to angular momentum, in that it is a measure of spin. However, unlike angular momentum, *the value of vorticity at a point does not depend on the particular choice of an axis of rotation;* indeed, the definition of vorticity makes no reference at all to an axis of rotation or to a coordinate system. Rather, vorticity is a measure of the *local* spin of a fluid element.

4.1.2 Simple Axisymmetric Examples

Consider axisymmetric motion in two dimensions, so that the flow is confined to a plane. We use cylindrical coordinates (r, ϕ, z), where z is the direction perpendicular to the plane, with velocity components (u^r, u^ϕ, u^z). For axisymmetric flow $u^z = u^r = 0$ but $u^\phi \neq 0$. The following two examples are quite instructive. (A third example that the reader may wish to consider is solid body rotation on a sphere, which has a vorticity gradient in latitude.)

Rigid body motion

For a body in rigid body rotation, the velocity distribution is given by

$$u^\phi = \Omega r, \tag{4.4}$$

where Ω is the angular velocity of the fluid and r is the distance from the axis of rotation. Associated with this rotation is a vorticity given by

$$\boldsymbol{\omega} = \nabla \times \boldsymbol{v} = \omega^z \mathbf{k}, \tag{4.5}$$

where

$$\omega^z = \frac{1}{r}\frac{\partial}{\partial r}(r u^\phi) = \frac{1}{r}\frac{\partial}{\partial r}(r^2 \Omega) = 2\Omega. \tag{4.6}$$

The vorticity of a fluid in solid body rotation is thus twice the angular velocity of the fluid about the axis of rotation, and is pointed in a direction orthogonal to the plane of rotation.

The 'vr' vortex

This vortex is so-called because the tangential velocity (historically denoted by 'v' in this context) is such that the product vr is constant. In our notation we would have

$$u^\phi = \frac{K}{r}, \tag{4.7}$$

where K is a constant determining the vortex strength. Evaluating the z-component of vorticity gives

$$\omega^z = \frac{1}{r}\frac{\partial}{\partial r}(r u^\phi) = \frac{1}{r}\frac{\partial}{\partial r}\left(r\frac{K}{r}\right) = 0, \tag{4.8}$$

except where $r = 0$, at which the expression is singular and the vorticity is infinite. Our paddle wheel rotates when placed at the vortex center, but, less obviously, does not if placed elsewhere.

The circulation around a circle that encloses the origin is given by

$$C = \oint \frac{K}{r} r \,\mathrm{d}\phi = 2\pi K. \tag{4.9}$$

This does not depend on the radius, and so it is true even as the radius tends to zero. Since the vorticity is the circulation divided by the area, the vorticity at the origin must be infinite. Consider now an integration path that does *not* enclose the origin, for example the contour A–B–C–D–A in Fig. 4.1. Over the segments A–B and C–D the velocity is orthogonal to the contour, and so the contribution is zero. Over B–C and D–A we have

$$C_{BC} = \frac{K}{r_2}\phi r_2 = K\phi, \qquad C_{DA} = -\frac{K}{r_1}\phi r_1 = -K\phi. \tag{4.10}$$

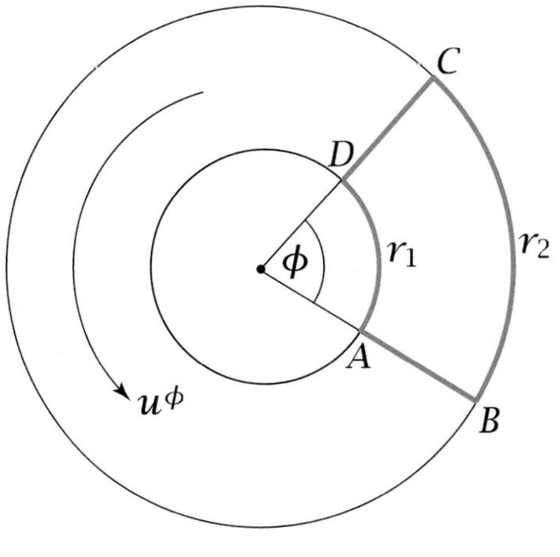

Fig. 4.1 Evaluation of circulation in the axisymmetric vr vortex. The circulation around the path A–B–C–D is zero. This result does not depend on the radii r_1 or r_2 or the angle ϕ, and the circulation around any infinitesimal path not enclosing the origin is zero. Thus the vorticity is zero everywhere except at the origin.

Adding these two expressions we see that the net circulation around the contour C_{ABCDA} is zero. If we shrink the integration path to an infinitesimal size then, within the path, by Stokes' theorem, the vorticity is zero. We can of course place the path anywhere we wish, except surrounding the origin, and obtain this result. Thus the vorticity is everywhere zero, except at the origin.

4.2 THE VORTICITY EQUATION

Using the vector identity $\boldsymbol{v} \times (\nabla \times \boldsymbol{v}) = \nabla(\boldsymbol{v} \cdot \boldsymbol{v})/2 - (\boldsymbol{v} \cdot \nabla)\boldsymbol{v}$, we write the momentum equation as

$$\frac{\partial \boldsymbol{v}}{\partial t} + \boldsymbol{\omega} \times \boldsymbol{v} = -\frac{1}{\rho}\nabla p - \frac{1}{2}\nabla v^2 + \boldsymbol{F}, \tag{4.11}$$

where $\boldsymbol{F}$ represents viscous and body forces. Taking the curl of (4.11) gives the vorticity equation

$$\frac{\partial \boldsymbol{\omega}}{\partial t} + \nabla \times (\boldsymbol{\omega} \times \boldsymbol{v}) = \frac{1}{\rho^2}(\nabla \rho \times \nabla p) + \nabla \times \boldsymbol{F}. \tag{4.12}$$

Now, the vector identity

$$\nabla \times (\boldsymbol{a} \times \boldsymbol{b}) = (\boldsymbol{b} \cdot \nabla)\boldsymbol{a} - (\boldsymbol{a} \cdot \nabla)\boldsymbol{b} + \boldsymbol{a}\nabla \cdot \boldsymbol{b} - \boldsymbol{b}\nabla \cdot \boldsymbol{a}, \tag{4.13}$$

implies that the second term on the left-hand side of (4.12) may be written as

$$\nabla \times (\boldsymbol{\omega} \times \boldsymbol{v}) = (\boldsymbol{v} \cdot \nabla)\boldsymbol{\omega} - (\boldsymbol{\omega} \cdot \nabla)\boldsymbol{v} + \boldsymbol{\omega}\nabla \cdot \boldsymbol{v} - \boldsymbol{v}\nabla \cdot \boldsymbol{\omega}. \tag{4.14}$$

Because vorticity is the curl of velocity its divergence vanishes, and so (4.12) becomes

$$\frac{\partial \boldsymbol{\omega}}{\partial t} + (\boldsymbol{v} \cdot \nabla)\boldsymbol{\omega} = (\boldsymbol{\omega} \cdot \nabla)\boldsymbol{v} - \boldsymbol{\omega}\nabla \cdot \boldsymbol{v} + \frac{1}{\rho^2}(\nabla \rho \times \nabla p) + \nabla \times \boldsymbol{F}. \tag{4.15}$$

The divergence term may be eliminated with the aid of the mass-conservation equation to give

$$\frac{\mathrm{D}\tilde{\boldsymbol{\omega}}}{\mathrm{D}t} = (\tilde{\boldsymbol{\omega}} \cdot \nabla)\boldsymbol{v} + \frac{1}{\rho^3}(\nabla \rho \times \nabla p) + \frac{1}{\rho}\nabla \times \boldsymbol{F}, \tag{4.16}$$

where $\tilde{\boldsymbol{\omega}} \equiv \boldsymbol{\omega}/\rho$. We will set $\boldsymbol{F} = 0$ in most of what follows.

The third term on the right-hand side of (4.15), as well as the second term on the right-hand side of (4.16), is variously called the *baroclinic* term, the *non-homentropic* term, or the *solenoidal* term. (A solenoidal vector has no divergence, hence the name.) The solenoidal vector, $\boldsymbol{S}_o$, is defined by

$$\boldsymbol{S}_o \equiv \frac{1}{\rho^2}\nabla\rho \times \nabla p = -\nabla\alpha \times \nabla p. \tag{4.17}$$

A solenoid is a tube directed perpendicular to both $\nabla\alpha$ and ∇p, with elements of length proportional to $\nabla p \times \nabla\alpha$. If the isolines of p and α are parallel to each other, then solenoids do not exist. This occurs when the density is a function only of pressure, for then

$$\nabla\rho \times \nabla p = \nabla\rho \times \nabla\rho \frac{\mathrm{d}p}{\mathrm{d}\rho} = 0. \tag{4.18}$$

The solenoidal vector may also be written

$$\boldsymbol{S}_o = -\nabla\eta \times \nabla T. \tag{4.19}$$

This follows most easily by first writing the momentum equation in the form $\partial\boldsymbol{v}/\partial t + \boldsymbol{\omega} \times \boldsymbol{v} = T\nabla\eta - \nabla B$, and taking its curl. Evidently the solenoidal term vanishes if: (i) isolines of pressure and density are parallel; (ii) isolines of temperature and entropy are parallel; (iii) density, entropy, temperature or pressure are constant. A *barotropic* fluid has by definition $\rho = \rho(p)$ and therefore no solenoids. A *baroclinic* fluid is one for which ∇p is not parallel to $\nabla\rho$. From (4.16) we see that the baroclinic term must be balanced by terms involving velocity or its tendency and therefore, in general, *a baroclinic fluid is a moving fluid,* even in the presence of viscosity.

For a barotropic fluid the vorticity equation takes the simple form,

$$\frac{\mathrm{D}\widetilde{\boldsymbol{\omega}}}{\mathrm{D}t} = (\widetilde{\boldsymbol{\omega}} \cdot \nabla)\boldsymbol{v}. \tag{4.20}$$

If the fluid is also incompressible, meaning that $\nabla \cdot \boldsymbol{v} = 0$, then we have the even simpler form,

$$\frac{\mathrm{D}\boldsymbol{\omega}}{\mathrm{D}t} = (\boldsymbol{\omega} \cdot \nabla)\boldsymbol{v}. \tag{4.21}$$

When expanded into components, the terms on the right-hand side of (4.20) or (4.21) can be divided into 'stretching' and 'tipping' (or 'tilting') terms, and we return to that in Section 4.3.1.

An integral conservation property

Consider a single Cartesian component in (4.15). Then, using superscripts to denote components,

$$\begin{aligned}\frac{\partial\omega^x}{\partial t} &= -\boldsymbol{v} \cdot \nabla\omega^x - \omega^x\nabla \cdot \boldsymbol{v} + (\boldsymbol{\omega} \cdot \nabla)v^x + S_o^x \\ &= -\nabla \cdot (\boldsymbol{v}\omega^x) + \nabla \cdot (\boldsymbol{\omega}v^x) + S_o^x,\end{aligned} \tag{4.22}$$

where S_o^x is the (x-component of the) solenoidal term. Equation (4.22) may be written as

$$\frac{\partial\omega^x}{\partial t} + \nabla \cdot (\boldsymbol{v}\omega^x - \boldsymbol{\omega}v^x) = S_o^x, \tag{4.23}$$

and this implies the Cartesian tensor form of the vorticity equation, namely

$$\frac{\partial\omega_i}{\partial t} + \frac{\partial}{\partial x_j}(v_j\omega_i - v_i\omega_j) = S_{oi}, \tag{4.24}$$

with summation over repeated indices. The tendency of the components of vorticity is thus given by the solenoidal term plus the divergence of a vector field, and if the solenoidal term vanishes the volume integrated vorticity can only be altered by boundary effects. However, in both the atmosphere and the ocean the solenoidal term *is* important, although we will see in Section 4.5 that a useful conservation law for a scalar quantity can still be obtained.

4.2.1 Two-dimensional Flow

In two-dimensional flow the fluid is confined to a surface, and independent of the dimension normal to that surface. In the simplest case in Cartesian geometry the flow is on a flat plane, and the velocity normal to the plane and the rate of change of any quantity normal to that plane are zero. Let the normal direction be the z-direction; the fluid velocity in the plane, $\boldsymbol{u}$, is $\boldsymbol{u} = u\mathbf{i} + v\mathbf{j}$, and the velocity normal to the plane, w, is zero. Only one component of vorticity is non-zero and this is

$$\boldsymbol{\omega} = \mathbf{k}\left(\frac{\partial v}{\partial x} - \frac{\partial u}{\partial y}\right). \tag{4.25}$$

That is, in two-dimensional flow the vorticity is perpendicular to the velocity. We let $\zeta \equiv \omega^z = \boldsymbol{\omega}\cdot\mathbf{k}$. Both the stretching and tilting terms vanish in two-dimensional flow, and the two-dimensional vorticity equation becomes, for incompressible flow,

$$\frac{\mathrm{D}\zeta}{\mathrm{D}t} = 0, \tag{4.26}$$

where $\mathrm{D}\zeta/\mathrm{D}t = \partial\zeta/\partial t + \boldsymbol{u}\cdot\nabla\zeta$. That is, in two-dimensional flow vorticity is conserved following the fluid elements; each material parcel of fluid keeps its value of vorticity even as it is being advected around. Furthermore, specification of the vorticity completely determines the flow field. To see this, we use the incompressibility condition to define a streamfunction ψ such that

$$u = -\frac{\partial \psi}{\partial y}, \qquad v = \frac{\partial \psi}{\partial x}, \qquad \zeta = \nabla^2\psi. \tag{4.27a,b,c}$$

Given the vorticity, the Poisson equation (4.27c) can be solved for the streamfunction and the velocity fields obtained through (4.27a,b), and this process is called 'inverting the vorticity'.

Numerical integration of (4.26) is then a process of timestepping plus inversion. The vorticity equation may then be written as an advection equation for vorticity,

$$\frac{\partial \zeta}{\partial t} + \boldsymbol{u}\cdot\nabla\zeta = 0, \tag{4.28}$$

in conjunction with (4.27). The vorticity is stepped forward one timestep using a finite-difference representation of (4.28), and the vorticity inverted to obtain a velocity using (4.27).

Two-dimensional flow is not restricted to a Cartesian plane — it exists on the surface of a sphere for example. In that case the velocity normal to the spherical surface (the 'vertical velocity') vanishes, and the equations are naturally expressed in spherical coordinates. Nevertheless, vorticity (absolute vorticity if the sphere is rotating) is still conserved on parcels as they move over the spherical surface.

4.3 VORTICITY AND CIRCULATION THEOREMS

4.3.1 The 'Frozen-in' Property of Vorticity

Let us first consider some simple topological properties of the vorticity field and its evolution. We define a *vortex line* to be a line drawn through the fluid which is everywhere in the direction of the local vorticity. This definition is analogous to that of a streamline, which is everywhere in the direction of the local velocity. A *vortex tube* is formed by the collection of vortex lines passing through a closed curve (Fig. 4.2). A *material line* is just a line that connects material fluid elements. Suppose we draw a vortex line through the fluid; such a line obviously connects fluid elements and therefore defines a coincident material line. As the fluid moves the material line deforms, and the vortex line also evolves in a manner determined by the equations of motion. A remarkable

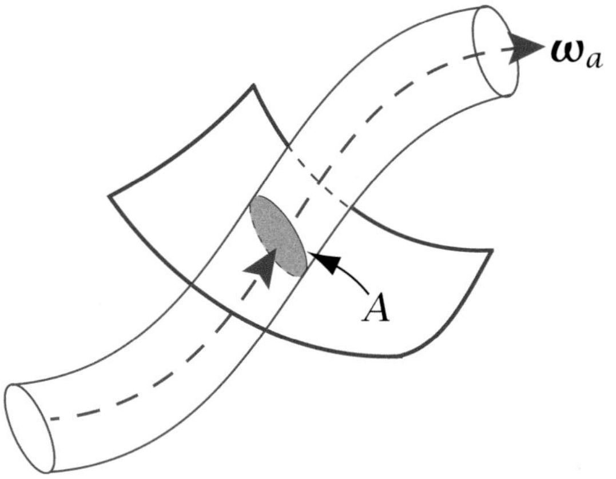

Fig. 4.2 A vortex tube passing through a material sheet. The circulation is the integral of the velocity around the boundary of A, and is equal to the integral of the normal component of vorticity over A.

property of vorticity is that, for an unforced and inviscid barotropic fluid, the flow evolution is such that a vortex line remains coincident with the material line that it was initially associated with. Put another way, a vortex line always contains the same material elements — the vorticity is 'frozen' or 'glued' to the material fluid.[1]

To prove this we consider how an infinitesimal material line element $\delta \boldsymbol{l}$ evolves, $\delta \boldsymbol{l}$ being the infinitesimal material element connecting $\boldsymbol{l}$ with $\boldsymbol{l} + \delta \boldsymbol{l}$. The rate of change of $\delta \boldsymbol{l}$ following the flow is given by

$$\frac{\mathrm{D}\delta \boldsymbol{l}}{\mathrm{D}t} = \frac{1}{\delta t}\left[\delta \boldsymbol{l}(t + \delta t) - \delta \boldsymbol{l}(t)\right], \tag{4.29}$$

which follows from the definition of the material derivative in the limit $\delta t \to 0$. From the Taylor expansion of $\delta \boldsymbol{l}(t)$ and the definition of velocity it is also apparent that

$$\delta \boldsymbol{l}(t + \delta t) = \boldsymbol{l}(t) + \delta \boldsymbol{l}(t) + (\boldsymbol{v} + \delta \boldsymbol{v})\delta t - (\boldsymbol{l}(t) + \boldsymbol{v}\delta t) = \delta \boldsymbol{l} + \delta \boldsymbol{v}\, \delta t, \tag{4.30}$$

as illustrated in Fig. 4.3. Substituting (4.30) into (4.29) gives $\mathrm{D}\delta \boldsymbol{l}/\mathrm{D}t = \delta \boldsymbol{v}$, as expected, and because $\delta \boldsymbol{v} = (\delta \boldsymbol{l} \cdot \nabla)\boldsymbol{v}$ we obtain

$$\frac{\mathrm{D}\delta \boldsymbol{l}}{\mathrm{D}t} = (\delta \boldsymbol{l} \cdot \nabla)\boldsymbol{v}. \tag{4.31}$$

Comparing this with (4.16), we see that vorticity evolves in the same way as a line element in an unforced barotropic fluid. To see what this means, at some initial time we can define an infinitesimal material line element parallel to the vorticity at that location, that is,

$$\delta \boldsymbol{l}(\boldsymbol{x}, t = 0) = A\boldsymbol{\omega}(\boldsymbol{x}, t = 0), \tag{4.32}$$

where A is a constant. Then, for all subsequent times the magnitude of the vorticity of that fluid element, even as it moves to a new location $\boldsymbol{x}'$, remains proportional to the length of the fluid element at that point and is oriented in the same way; that is $\boldsymbol{\omega}(\boldsymbol{x}', t) = A^{-1}\delta \boldsymbol{l}(\boldsymbol{x}', t)$.

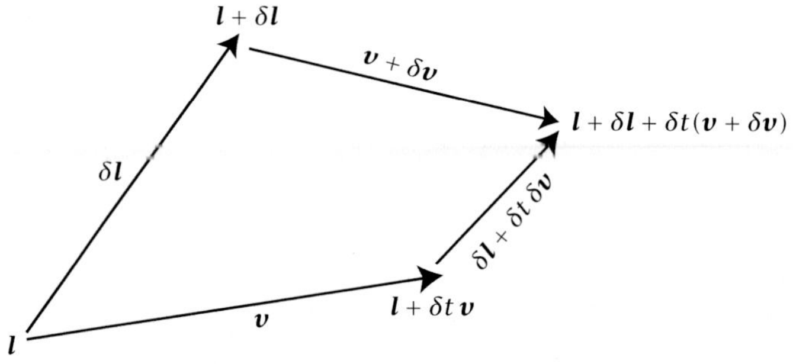

Fig. 4.3 Evolution of an infinitesimal material line $\delta \boldsymbol{l}$ from time t to time $t + \delta t$. It can be seen from the diagram that $\mathrm{D}\delta \boldsymbol{l}/\mathrm{D}t = \delta \boldsymbol{v}$.

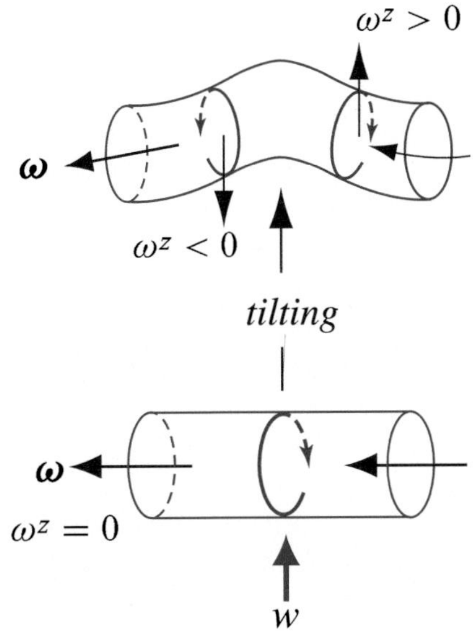

Fig. 4.4 The tilting of vorticity. Suppose that the vorticity, $\boldsymbol{\omega}$ is initially directed horizontally, as in the lower figure, so that ω^z, its vertical component, is zero. The material lines, and therefore also the vortex lines, are tilted by the positive vertical velocity w thus creating a non-zero vertically oriented vorticity. This mechanism is important in creating vertical vorticity in the atmospheric boundary layer, and is connected to the β-effect in large-scale flow.

To verify this result in a different way note that a vortex line element is determined by the condition $\delta \boldsymbol{l} = A\boldsymbol{\omega}$ which implies $\boldsymbol{\omega} \times \delta \boldsymbol{l} = 0$. Now, for any line element we have that

$$\frac{\mathrm{D}}{\mathrm{D}t}(\boldsymbol{\omega} \times \delta \boldsymbol{l}) = \frac{\mathrm{D}\boldsymbol{\omega}}{\mathrm{D}t} \times \delta \boldsymbol{l} - \frac{\mathrm{D}\delta \boldsymbol{l}}{\mathrm{D}t} \times \boldsymbol{\omega}. \tag{4.33}$$

We also have that

$$\frac{\mathrm{D}\delta \boldsymbol{l}}{\mathrm{D}t} = \delta \boldsymbol{v} = (\delta \boldsymbol{l} \cdot \nabla)\boldsymbol{v} \quad \text{and} \quad \frac{\mathrm{D}\boldsymbol{\omega}}{\mathrm{D}t} = (\boldsymbol{\omega} \cdot \nabla)\boldsymbol{v}. \tag{4.34}$$

If the line element is initially a vortex line element then, at $t = 0$, $\delta \boldsymbol{l} = A\boldsymbol{\omega}$ and, using (4.34), the right-hand side of (4.33) vanishes. Thus, the *tendency* of $\boldsymbol{\omega} \times \delta \boldsymbol{l}$ is zero, and the vortex line continues to be a material line.

Stretching and tilting

The terms on the right-hand side of (4.20) or (4.21) may be interpreted in terms of 'stretching' and 'tipping' (or 'tilting'). Consider a single Cartesian component of (4.21),

$$\frac{\mathrm{D}\omega^x}{\mathrm{D}t} = \omega^x \frac{\partial u}{\partial x} + \omega^y \frac{\partial u}{\partial y} + \omega^z \frac{\partial u}{\partial z}. \tag{4.35}$$

The second and third terms on the right-hand side are the tilting or tipping terms because they involve changes in the orientation of the vorticity vector. They tell us that vorticity in the x-direction may be generated from vorticity in the y- and z-directions if the advection acts to tilt the material lines. Because vorticity is tied to these lines, vorticity oriented in one direction becomes oriented in another, as in Fig. 4.4.

The first term on the right-hand side of (4.35) is the stretching term, and it acts to intensify the x-component of vorticity if the velocity is increasing in the x-direction — that is, if the material lines are being stretched (Fig. 4.5). The effect arises because a vortex line is tied to a material line, and therefore vorticity is amplified in proportion to the stretching of the material line aligned with it. This effect is important in tornadoes, to give one example. If the fluid is incompressible, stretching of a fluid mass in one direction must be accompanied by convergence in another, and this leads to the conservation of circulation, as we now discuss.

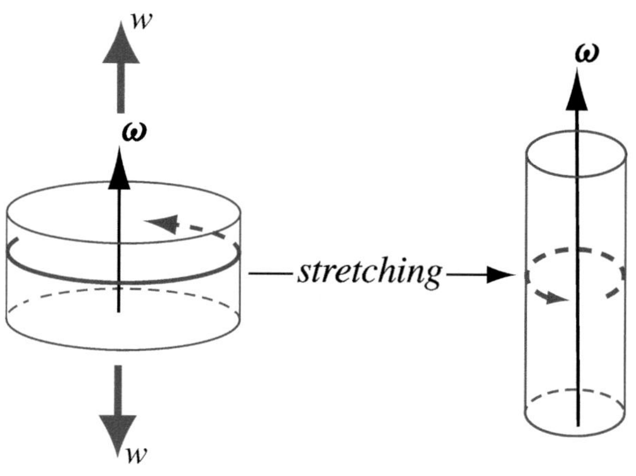

Fig. 4.5 A vertical velocity, w, stretches the cylinder. Vorticity is tied to material lines and so is amplified in the direction of the stretching. However, because the volume of fluid is conserved, the end surfaces shrink, the material lines through the cylinder ends converge and the integral of vorticity over a material surface (the circulation) remains constant.

4.3.2 Kelvin's Circulation Theorem

Kelvin's circulation theorem states that under certain circumstances the circulation around a material fluid parcel is conserved; that is, the circulation is conserved 'moving with the flow'.[2] The primary restrictions are that body forces are conservative (i.e., they are representable as potential forces, and therefore that the flow be inviscid), and that the fluid is barotropic with $\rho = \rho(p)$. Of these, the latter is more restrictive for geophysical fluids. The circulation in the theorem is defined with respect to an inertial frame of reference; specifically, the velocity in (4.39) is the velocity relative to an inertial frame. To prove the theorem, we begin with the inviscid momentum equation,

$$\frac{\mathrm{D}\boldsymbol{v}}{\mathrm{D}t} = -\frac{1}{\rho}\nabla p - \nabla\Phi, \tag{4.36}$$

where $\nabla\Phi$ represents the conservative body forces on the system. Applying the material derivative to the circulation, (4.2), gives

$$\begin{aligned}\frac{\mathrm{D}C}{\mathrm{D}t} = \frac{\mathrm{D}}{\mathrm{D}t}\oint \boldsymbol{v}\cdot\mathrm{d}\boldsymbol{r} &= \oint\left(\frac{\mathrm{D}\boldsymbol{v}}{\mathrm{D}t}\cdot\mathrm{d}\boldsymbol{r} + \boldsymbol{v}\cdot\mathrm{d}\boldsymbol{v}\right)\\ &= \oint\left[\left(-\frac{1}{\rho}\nabla p - \nabla\Phi\right)\cdot\mathrm{d}\boldsymbol{r} + \boldsymbol{v}\cdot\mathrm{d}\boldsymbol{v}\right]\\ &= \oint -\frac{1}{\rho}\nabla p\cdot\mathrm{d}\boldsymbol{r},\end{aligned} \tag{4.37}$$

using (4.36) and $D(\mathrm{d}\boldsymbol{r})/Dt = \mathrm{d}\boldsymbol{v}$, where $\mathrm{d}\boldsymbol{r}$ is the line element and with the line integration being over a closed, material, circuit. The second and third terms on the second line vanish separately, because they are exact differentials integrated around a closed loop. The term on the last line vanishes if the density is constant or, more generally, if the density is a function of pressure alone, in which case ∇p is parallel to $\nabla\rho$. To see this, note that

$$\oint \frac{1}{\rho}\nabla p\cdot\mathrm{d}\boldsymbol{r} = \int_S \nabla\times\left(\frac{\nabla p}{\rho}\right)\cdot\mathrm{d}\boldsymbol{S} = \int_S \frac{-\nabla\rho\times\nabla p}{\rho^2}\cdot\mathrm{d}\boldsymbol{S}, \tag{4.38}$$

using Stokes' theorem where S is any surface bounded by the path of the line integral. The integral evidently vanishes identically if p is a function of ρ alone. The right-most expression above is the integral of the solenoidal vector, and if it is zero (4.37) becomes

$$\frac{\mathrm{D}}{\mathrm{D}t}\oint \boldsymbol{v}\cdot\mathrm{d}\boldsymbol{r} = 0. \tag{4.39}$$

This is Kelvin's circulation theorem. In words, *the circulation around a material loop is invariant for a barotropic fluid that is subject only to conservative forces.* Using Stokes' theorem, the circulation theorem may also be written as

$$\frac{\mathrm{D}}{\mathrm{D}t}\int_S \boldsymbol{\omega}\cdot \mathrm{d}\boldsymbol{S} = 0. \tag{4.40}$$

That is, the area integral of the normal component of vorticity across any material surface is constant, under the same conditions. This form is both natural and useful, and it arises because of the way vorticity is tied to material fluid elements. Kelvin's circulation theorem is the one conservation law that is unique to fluids. Unlike, say, the conservation of energy, it has no analogue in solid body mechanics. Potential vorticity conservation, which we come to later on, is an extension of circulation conservation.

Stretching and circulation

Let us informally consider how vortex stretching and mass conservation work together to give the circulation theorem. Let the fluid be incompressible so that the volume of a fluid mass is constant, and consider a surface normal to a vortex tube, as in Fig. 4.5. Let the volume of a small material box around the surface be δV, the length of the material lines be δl and the surface area be δA. Then

$$\delta V = \delta l\,\delta A. \tag{4.41}$$

Because of the frozen-in property, the vorticity passing through the surface is proportional to the length of the material lines. That is, $\omega \propto \delta l$, and

$$\delta V \propto \omega\,\delta A. \tag{4.42}$$

The right-hand side is just the circulation around the surface. Now, if the corresponding material tube is stretched δl increases, but the volume, δV, remains constant by mass conservation. Thus, the circulation given by the right-hand side of (4.42) also remains constant. In other words, because of the frozen-in property vorticity is amplified by the stretching, but the vortex lines get closer together in such a way that the product $\omega\,\delta A$ remains constant and circulation is conserved.

4.3.3 Baroclinic Flow and the Solenoidal Term

In baroclinic flow, the circulation is not generally conserved, and from (4.37) we have

$$\frac{\mathrm{D}C}{\mathrm{D}t} = -\oint \frac{\nabla p}{\rho}\cdot \mathrm{d}\boldsymbol{r} = -\oint \frac{\mathrm{d}p}{\rho}, \tag{4.43}$$

and this is called the baroclinic circulation theorem.[3] Recalling the fundamental thermodynamic relation $T\,\mathrm{d}\eta = \mathrm{d}I + p\,\mathrm{d}\alpha$, where $\alpha = \rho^{-1}$, we have

$$\alpha\,\mathrm{d}p = \mathrm{d}(p\alpha) - T\,\mathrm{d}\eta + \mathrm{d}I, \tag{4.44}$$

and the first and last terms on the right-hand side will vanish upon integration around a circuit. The solenoidal term on the right-hand side of (4.43) may therefore be written as

$$S_o \equiv -\oint \alpha\,\mathrm{d}p = \oint T\,\mathrm{d}\eta = -\oint \eta\,\mathrm{d}T = -R\oint T\,\mathrm{d}\log p, \tag{4.45}$$

where the last equality holds only for an ideal gas. Using Stokes' theorem, S_o can also be written as

$$S_o = -\int_S \nabla\alpha\times\nabla p\cdot \mathrm{d}\boldsymbol{S} = -\int_S \left(\frac{\partial \alpha}{\partial T}\right)_p \nabla T\times\nabla p\cdot \mathrm{d}\boldsymbol{S} = \int_S \nabla T\times\nabla\eta\cdot \mathrm{d}\boldsymbol{S}. \tag{4.46}$$

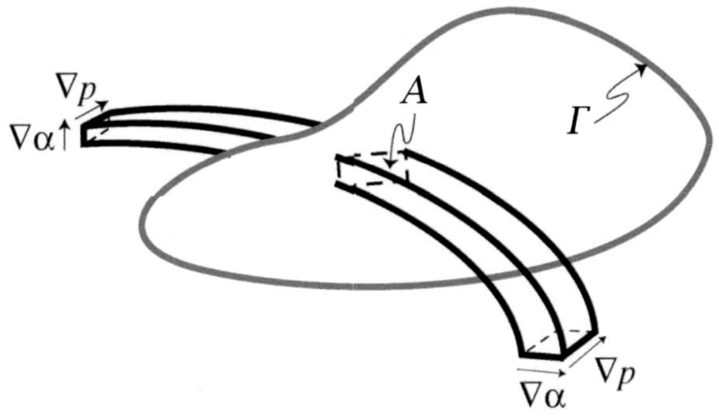

Fig. 4.6 Solenoids and the circulation theorem. Solenoids are tubes perpendicular to both $\nabla\alpha$ and ∇p, and they have a non-zero cross-sectional area if isolines of α and p do not coincide. The rate of change of circulation over a material surface is given by the sum of all the solenoidal areas crossing the surface. If $\nabla\alpha \times \nabla p = 0$ there are no solenoids.

The rate of change of the circulation across a surface depends on the existence of this solenoidal term (Fig. 4.6). However, even if the solenoidal vector is in non-zero, circulation *is* conserved if the material path is in a surface of constant entropy, η, *and if* $\mathrm{D}\eta/\mathrm{D}t = 0$. The solenoidal term then vanishes and, because $\mathrm{D}\eta/\mathrm{D}t = 0$, entropy remains constant on that same material loop as it evolves. This result gives rise to the conservation of potential vorticity, discussed in Section 4.5.

4.3.4 Circulation in a Rotating Frame

The absolute and relative velocities are related by $\boldsymbol{v}_a = \boldsymbol{v}_r + \boldsymbol{\Omega} \times \boldsymbol{r}$, so that in a rotating frame the rate of change of circulation is given by

$$\frac{\mathrm{D}}{\mathrm{D}t}\oint(\boldsymbol{v}_r + \boldsymbol{\Omega}\times\boldsymbol{r})\cdot\mathrm{d}\boldsymbol{r} = \oint\left[\left(\frac{\mathrm{D}\boldsymbol{v}_r}{\mathrm{D}t} + \boldsymbol{\Omega}\times\boldsymbol{v}_r\right)\cdot\mathrm{d}\boldsymbol{r} + (\boldsymbol{v}_r + \boldsymbol{\Omega}\times\boldsymbol{r})\cdot\mathrm{d}\boldsymbol{v}_r\right]. \tag{4.47}$$

But $\oint \boldsymbol{v}_r \cdot \mathrm{d}\boldsymbol{v}_r = 0$ and, integrating by parts,

$$\begin{aligned}\oint(\boldsymbol{\Omega}\times\boldsymbol{r})\cdot\mathrm{d}\boldsymbol{v}_r &= \oint\left\{\mathrm{d}[(\boldsymbol{\Omega}\times\boldsymbol{r})\cdot\boldsymbol{v}_r] - (\boldsymbol{\Omega}\times\mathrm{d}\boldsymbol{r})\cdot\boldsymbol{v}_r\right\}\\ &= \oint\left\{\mathrm{d}[(\boldsymbol{\Omega}\times\boldsymbol{r})\cdot\boldsymbol{v}_r] + (\boldsymbol{\Omega}\times\boldsymbol{v}_r)\cdot\mathrm{d}\boldsymbol{r}\right\}.\end{aligned} \tag{4.48}$$

The first term on the right-hand side is zero and so (4.47) becomes

$$\frac{\mathrm{D}}{\mathrm{D}t}\oint(\boldsymbol{v}_r + \boldsymbol{\Omega}\times\boldsymbol{r})\cdot\mathrm{d}\boldsymbol{r} = \oint\left(\frac{\mathrm{D}\boldsymbol{v}_r}{\mathrm{D}t} + 2\boldsymbol{\Omega}\times\boldsymbol{v}_r\right)\cdot\mathrm{d}\boldsymbol{r} = -\oint\frac{\mathrm{d}p}{\rho}, \tag{4.49}$$

where the second equality uses the momentum equation. The last term vanishes if the fluid is barotropic, and if so the circulation theorem is, unsurprisingly,

$$\frac{\mathrm{D}}{\mathrm{D}t}\oint(\boldsymbol{v}_r + \boldsymbol{\Omega}\times\boldsymbol{r})\cdot\mathrm{d}\boldsymbol{r} = 0, \qquad \text{or} \qquad \frac{\mathrm{D}}{\mathrm{D}t}\int_S(\boldsymbol{\omega}_r + 2\boldsymbol{\Omega})\cdot\mathrm{d}\boldsymbol{S} = 0, \tag{4.50a,b}$$

where the second equation uses Stokes' theorem and we have used $\nabla\times(\boldsymbol{\Omega}\times\boldsymbol{r}) = 2\boldsymbol{\Omega}$, and where $\boldsymbol{\omega}_r = \nabla\times\boldsymbol{v}_r$ is the *relative vorticity*.[4]

4.3.5 The Circulation Theorem for Hydrostatic Flow

Kelvin's circulation theorem holds for hydrostatic flow, with a slightly different form. For simplicity we restrict attention to the f-plane, and start with the hydrostatic momentum equations,

$$\frac{\mathrm{D}\boldsymbol{u}_r}{\mathrm{D}t} + 2\boldsymbol{\Omega}\times\boldsymbol{u}_r = -\frac{1}{\rho}\nabla_z p, \qquad 0 = -\frac{1}{\rho}\frac{\partial p}{\partial z} - \nabla\Phi, \tag{4.51a,b}$$

where $\Phi = gz$ is the gravitational potential and $\boldsymbol{\Omega} = \Omega\mathbf{k}$. The advecting field is three-dimensional, and in particular we still have $\mathrm{D}\delta\boldsymbol{r}/\mathrm{D}t = \delta\boldsymbol{v} = (\delta\boldsymbol{r}\cdot\nabla)\boldsymbol{v}$. Thus, using (4.51) we have

$$\begin{aligned}\frac{\mathrm{D}}{\mathrm{D}t}\oint(\boldsymbol{u}_r + \boldsymbol{\Omega}\times\boldsymbol{r})\cdot\mathrm{d}\boldsymbol{r} &= \oint\left[\left(\frac{\mathrm{D}\boldsymbol{u}_r}{\mathrm{D}t} + \boldsymbol{\Omega}\times\boldsymbol{v}_r\right)\cdot\mathrm{d}\boldsymbol{r} + (\boldsymbol{u}_r + \boldsymbol{\Omega}\times\boldsymbol{r})\cdot\mathrm{d}\boldsymbol{v}_r\right] \\ &= \oint\left(\frac{\mathrm{D}\boldsymbol{u}_r}{\mathrm{D}t} + 2\boldsymbol{\Omega}\times\boldsymbol{u}_r\right)\cdot\mathrm{d}\boldsymbol{r} \\ &= \oint\left(-\frac{1}{\rho}\nabla p - \nabla\Phi\right)\cdot\mathrm{d}\boldsymbol{r}, \end{aligned} \tag{4.52}$$

as with (4.49), having used $\boldsymbol{\Omega}\times\boldsymbol{v}_r = \boldsymbol{\Omega}\times\boldsymbol{u}_r$, and where the gradient operator ∇ is three-dimensional. The last term on the right-hand side vanishes because it is the integral of the gradient of a potential around a closed path. The first term vanishes if the fluid is barotropic, so that the circulation theorem is

$$\frac{\mathrm{D}}{\mathrm{D}t}\oint(\boldsymbol{u}_r + \boldsymbol{\Omega}\times\boldsymbol{r})\cdot\mathrm{d}\boldsymbol{r} = 0. \tag{4.53}$$

Using Stokes' theorem we have the equivalent form

$$\frac{\mathrm{D}}{\mathrm{D}t}\int_S(\boldsymbol{\omega}_{hy} + 2\boldsymbol{\Omega})\cdot\mathrm{d}\boldsymbol{S} = 0, \tag{4.54}$$

where the subscript '*hy*' denotes hydrostatic and, in Cartesian coordinates,

$$\boldsymbol{\omega}_{hy} = \nabla\times\boldsymbol{u}_r = -\mathbf{i}\frac{\partial v_r}{\partial z} + \mathbf{j}\frac{\partial u_r}{\partial z} + \mathbf{k}\left(\frac{\partial v_r}{\partial x} - \frac{\partial u_r}{\partial y}\right). \tag{4.55}$$

4.4 VORTICITY EQUATION IN A ROTATING FRAME

Perhaps the easiest way to derive the vorticity equation appropriate for a rotating reference frame is to begin with the momentum equation in the form

$$\frac{\partial\boldsymbol{v}_r}{\partial t} + (2\boldsymbol{\Omega} + \boldsymbol{\omega}_r)\times\boldsymbol{v}_r = -\frac{1}{\rho}\nabla p - \nabla\left(\Phi + \frac{1}{2}v_r^2\right), \tag{4.56}$$

where the potential Φ contains the gravitational and centrifugal forces. Take the curl of this and use the identity (4.13), which here implies

$$\nabla\times[(2\boldsymbol{\Omega} + \boldsymbol{\omega}_r)\times\boldsymbol{v}_r] = (2\boldsymbol{\Omega} + \boldsymbol{\omega}_r)\nabla\cdot\boldsymbol{v}_r + (\boldsymbol{v}_r\cdot\nabla)(2\boldsymbol{\Omega} + \boldsymbol{\omega}_r) - [(2\boldsymbol{\Omega} + \boldsymbol{\omega}_r)\cdot\nabla]\boldsymbol{v}_r, \tag{4.57}$$

(noting that $\nabla\cdot(2\boldsymbol{\Omega} + \boldsymbol{\omega}) = 0$), to give the vorticity equation

$$\frac{\mathrm{D}\boldsymbol{\omega}_r}{\mathrm{D}t} = [(2\boldsymbol{\Omega} + \boldsymbol{\omega}_r)\cdot\nabla]\,\boldsymbol{v} - (2\boldsymbol{\Omega} + \boldsymbol{\omega}_r)\nabla\cdot\boldsymbol{v}_r + \frac{1}{\rho^2}(\nabla\rho\times\nabla p). \tag{4.58}$$

If the rotation rate, $\boldsymbol{\Omega}$, is a constant then $\mathrm{D}\boldsymbol{\omega}_r/\mathrm{D}t = \mathrm{D}\boldsymbol{\omega}_a/\mathrm{D}t$ where $\boldsymbol{\omega}_a = 2\boldsymbol{\Omega} + \boldsymbol{\omega}_r$ is the absolute vorticity. The only difference between the vorticity equation in the rotating and inertial frames of reference is in the presence of the solid-body vorticity $2\boldsymbol{\Omega}$ on the right-hand side. The second term on the right-hand side may be folded into the material derivative using mass continuity, and after a little manipulation (4.58) becomes

$$\frac{\mathrm{D}}{\mathrm{D}t}\left(\frac{\boldsymbol{\omega}_a}{\rho}\right) = \frac{1}{\rho}(2\boldsymbol{\Omega} + \boldsymbol{\omega}_r)\cdot\nabla\boldsymbol{v}_r + \frac{1}{\rho^3}(\nabla\rho\times\nabla p). \tag{4.59}$$

However, note that it is the absolute vorticity, $\boldsymbol{\omega}_a$, that now appears on the left-hand side. If ρ is constant, $\boldsymbol{\omega}_a$ may be replaced by $\boldsymbol{\omega}_r$.

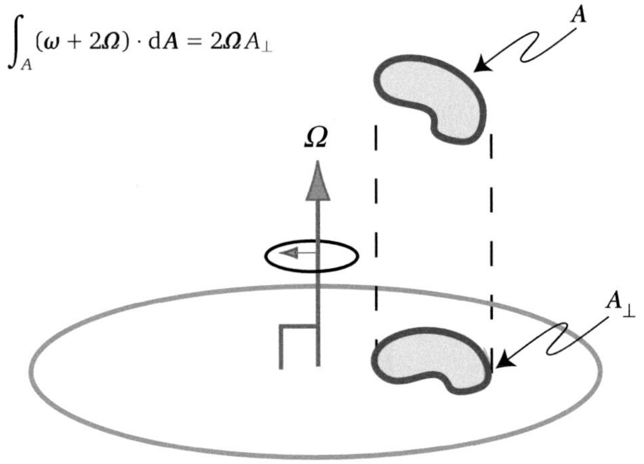

Fig. 4.7 The projection of a material circuit on to the equatorial plane. If a fluid element moves poleward, keeping its orientation to the local vertical fixed (i.e., it stays horizontal), then the area of its projection on to the equatorial plane increases. If its total (absolute) circulation is to be maintained, then the vertical component of its relative vorticity must diminish; that is, $\int_A (\boldsymbol{\omega} + 2\boldsymbol{\Omega}) \cdot \mathrm{d}\boldsymbol{A} = \int_A (\zeta + f)\,\mathrm{d}A = \text{constant}$. Thus, the β term in $\mathrm{D}(\zeta + f)/\mathrm{D}t = \mathrm{D}\zeta/\mathrm{D}t + \beta v = 0$ arises from the *tilting* of a parcel relative to the axis of rotation as it moves meridionally.

4.4.1 The Circulation Theorem and the Beta Effect

What are the implications of the circulation theorem on a rotating, spherical planet? Let us define relative circulation over some material loop as

$$C_r \equiv \oint \boldsymbol{v}_r \cdot \mathrm{d}\boldsymbol{r}. \tag{4.60}$$

Because $\boldsymbol{v}_r = \boldsymbol{v}_a - \boldsymbol{\Omega} \times \boldsymbol{r}$ (where $\boldsymbol{r}$ is the distance from the axis of rotation), we use Stokes' theorem to give

$$C_r = C_a - \int 2\boldsymbol{\Omega} \cdot \mathrm{d}\boldsymbol{S} = C_a - 2\Omega A_\perp, \tag{4.61}$$

where C_a is the total or absolute circulation and $A_\perp$ is the area enclosed by the projection of the material circuit on to the plane normal to the rotation vector; that is, on to the equatorial plane (Fig. 4.7). If the solenoidal term is zero, then the circulation theorem, (4.50), may be written as

$$\frac{\mathrm{D}}{\mathrm{D}t}(C_r + 2\Omega A_\perp) = 0. \tag{4.62}$$

Thus, the relative circulation around a circuit changes if the orientation of the plane changes; that is, if the area of its projection on to the equatorial plane changes. In large scale dynamics the most common cause of this is when a fluid parcel changes its latitude. For example, consider the flow of a two-dimensional, infinitesimal, horizontal (i.e., tangent to the the radial vector), constant-density fluid parcel at a latitude ϑ with area A, so that the projection of its area on to the equatorial plane is $A_\perp = A \sin\vartheta$ and $C_r = \zeta_r A$. If the fluid surface moves, but remains horizontal, its area is preserved (because it is incompressible) and directly from (4.62) its relative vorticity changes as

$$\frac{\mathrm{D}\zeta_r}{\mathrm{D}t} = -\frac{2\Omega}{A}\frac{\mathrm{D}A_\perp}{\mathrm{D}t} = -2\Omega\frac{\mathrm{D}}{\mathrm{D}t}\sin\vartheta = -v_r\frac{2\Omega\cos\vartheta}{a} = -\beta v_r, \tag{4.63}$$

where

$$\beta \equiv \frac{\mathrm{d}f}{\mathrm{d}y} = \frac{2\Omega}{a}\cos\vartheta. \tag{4.64}$$

The means by which the vertical component of the relative vorticity of a parcel changes by virtue of its latitudinal displacement is known as the *beta effect*, or the *β-effect*. It is a manifestation of the tilting term in the vorticity equation, and it is often the most important means by which relative vorticity does change in large-scale flow. The β-effect arises in the full vorticity equation, as we now see.

4.4.2 The Vertical Component of the Vorticity Equation

In large-scale dynamics, the most important, although not the largest, component of the vorticity is often the vertical one, because this contains much of the information about the horizontal flow. We can obtain an explicit expression for its evolution by taking the vertical component of (4.58), although care must be taken because the unit vectors (**i**, **j**, **k**) are functions of position (see Section 2.2).

An alternative derivation begins with the horizontal momentum equations,

$$\frac{\partial u}{\partial t} - v(\zeta + f) + w\frac{\partial u}{\partial z} = -\frac{1}{\rho}\frac{\partial p}{\partial x} - \frac{1}{2}\frac{\partial}{\partial x}(u^2 + v^2) + F^x \tag{4.65a}$$

$$\frac{\partial v}{\partial t} + u(\zeta + f) + w\frac{\partial v}{\partial z} = -\frac{1}{\rho}\frac{\partial p}{\partial y} - \frac{1}{2}\frac{\partial}{\partial y}(u^2 + v^2) + F^y, \tag{4.65b}$$

where in this section we again drop the subscript r on variables measured in the rotating frame. Cross-differentiating gives, after a little algebra,

$$\begin{aligned}\frac{\mathrm{D}}{\mathrm{D}t}(\zeta + f) = -(\zeta + f)\left(\frac{\partial u}{\partial x} + \frac{\partial v}{\partial y}\right) + \left(\frac{\partial u}{\partial z}\frac{\partial w}{\partial y} - \frac{\partial v}{\partial z}\frac{\partial w}{\partial x}\right) \\ + \frac{1}{\rho^2}\left(\frac{\partial \rho}{\partial x}\frac{\partial p}{\partial y} - \frac{\partial \rho}{\partial y}\frac{\partial p}{\partial x}\right) + \left(\frac{\partial F^y}{\partial x} - \frac{\partial F^x}{\partial y}\right).\end{aligned} \tag{4.66}$$

We interpret the various terms as follows:

$\mathrm{D}\zeta/\mathrm{D}t = \partial\zeta/\partial t + \boldsymbol{v}\cdot\nabla\zeta$. The material derivative of the vertical component of the vorticity.

$\mathrm{D}f/\mathrm{D}t = v\partial f/\partial y = v\beta$. The β-effect. The vorticity is affected by the meridional motion of the fluid, so that, apart from the terms on the right-hand side, $(\zeta + f)$ is conserved on parcels. Because the Coriolis parameter changes with latitude this is like saying that the system has differential rotation. This effect is precisely that due to the change in orientation of fluid surfaces with latitude, as discussed in Section 4.4.1 and illustrated Fig. 4.7.

$-(\zeta + f)(\partial u/\partial x + \partial v/\partial y)$. The divergence term, which gives rise to vortex stretching. In an incompressible fluid this may be written $(\zeta + f)\partial w/\partial z$, so that vorticity is amplified if the vertical velocity increases with height, so stretching the material lines and the vorticity.

$(\partial u/\partial z)(\partial w/\partial y) - (\partial v/\partial z)(\partial w/\partial x)$. The tilting term, whereby a vertical component of vorticity may be generated by a vertical velocity acting on a horizontal vorticity. See Fig. 4.4.

$\rho^{-2}\left[(\partial\rho/\partial x)(\partial p/\partial y) - (\partial\rho/\partial y)(\partial p/\partial x)\right] = \rho^{-2}J(\rho, p)$. The solenoidal term, also called the non-homentropic or baroclinic term, arising when isosurfaces of pressure and density are not parallel.

$(\partial F^y/\partial x - \partial F^x/\partial y)$. The forcing and friction term. If the only contribution is from molecular viscosity then this term is $\nu\nabla^2\zeta$.

Two-dimensional and shallow water vorticity equations

In an inviscid two-dimensional incompressible flow, all of the terms on the right-hand side of (4.66) vanish and we have the simple equation

$$\frac{\mathrm{D}(\zeta + f)}{\mathrm{D}t} = 0, \tag{4.67}$$

implying that the absolute vorticity, $\zeta_a \equiv \zeta + f$, is materially conserved. If f is a constant, then (4.67) reduces to (4.28), and background rotation plays no role. If f varies linearly with y, so that $f = f_0 + \beta y$, then (4.67) becomes

$$\frac{\partial \zeta}{\partial t} + \boldsymbol{u} \cdot \nabla \zeta + \beta v = 0, \tag{4.68}$$

which is known as the two-dimensional β-plane vorticity equation.

For inviscid shallow water flow, we can show that (see Chapter 3)

$$\frac{\mathrm{D}(\zeta + f)}{\mathrm{D}t} = -(\zeta + f)\left(\frac{\partial u}{\partial x} + \frac{\partial v}{\partial y}\right). \tag{4.69}$$

In this equation the vanishing of the tilting term is perhaps the only aspect which is not immediately apparent, but this succumbs to a little thought.

4.5 POTENTIAL VORTICITY CONSERVATION

Too much of a good thing is wonderful.
Mae West (1892–1990).

Although Kelvin's circulation theorem is a general statement about vorticity conservation, in its original form it is not always a practically useful statement for two reasons. First, it is not a statement about a *field*, such as vorticity itself. Second, it is not satisfied for baroclinic flow, such as is found in the atmosphere and ocean. (Non-conservative forces such as viscosity also lead to circulation non-conservation, but this applies to virtually all conservation laws and does not diminish them.) It turns out that it is possible to derive a beautiful conservation law that overcomes both of these failings and one that, furthermore, is extraordinarily useful in geophysical fluid dynamics. This is the conservation of *potential vorticity* (PV) introduced first by Rossby and then in a more general form by Ertel.[5] The idea is that we can use a scalar field that is being advected by the flow to keep track of, or to take care of, the evolution of fluid elements. For a baroclinic fluid this scalar field must be chosen in a special way (it must be a function of the density and pressure alone), but there is no restriction to a barotropic fluid. Then using the scalar evolution equation in conjunction with the vorticity equation gives us a scalar conservation equation. In the next few subsections we derive the equation for potential vorticity conservation in a number of superficially different ways — different explications but the same explanation.[6]

4.5.1 PV Conservation from the Circulation Theorem

Barotropic fluids

Let us begin with the simple case of a barotropic fluid. For an infinitesimal volume we write Kelvin's theorem as

$$\frac{\mathrm{D}}{\mathrm{D}t}\left[(\boldsymbol{\omega}_a \cdot \boldsymbol{n})\delta A\right] = 0, \tag{4.70}$$

where $\boldsymbol{n}$ is a unit vector normal to an infinitesimal surface δA. Now consider a volume bounded by two isosurfaces of values χ and $\chi + \delta\chi$, where χ is any materially conserved tracer, thus satisfying $\mathrm{D}\chi/\mathrm{D}t = 0$, so that δA initially lies in an isosurface of χ (see Fig. 4.8). Since $\boldsymbol{n} = \nabla\chi/|\nabla\chi|$ and the infinitesimal volume $\delta V = \delta h\,\delta A$, where δh is the separation between the two surfaces, we have

$$\boldsymbol{\omega}_a \cdot \boldsymbol{n}\,\delta A = \boldsymbol{\omega}_a \cdot \frac{\nabla\chi}{|\nabla\chi|}\frac{\delta V}{\delta h}. \tag{4.71}$$

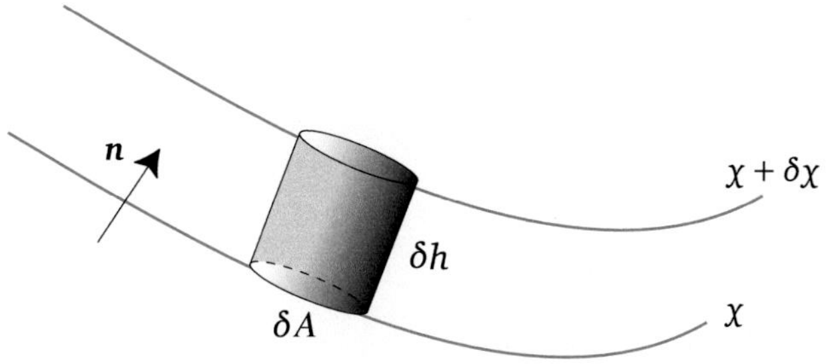

Fig. 4.8 An infinitesimal fluid element, bounded by two isosurfaces of the conserved tracer χ. As $\mathrm{D}\chi/\mathrm{D}t = 0$, then $\mathrm{D}\delta\chi/\mathrm{D}t = 0$.

Now, the value of δh may be obtained from

$$\delta\chi = \delta\boldsymbol{x} \cdot \nabla\chi = \delta h |\nabla\chi|, \tag{4.72}$$

and using this in (4.70) we obtain

$$\frac{\mathrm{D}}{\mathrm{D}t}\left[\frac{(\boldsymbol{\omega}_a \cdot \nabla\chi)\delta V}{\delta\chi}\right] = 0. \tag{4.73}$$

Since χ is conserved on material elements then so is $\delta\chi$ and it may be taken out of the differentiation. The mass of the volume element $\rho\,\delta V$ is also conserved, so that (4.73) becomes

$$\frac{\rho\delta V}{\delta\chi}\frac{\mathrm{D}}{\mathrm{D}t}\left(\frac{\boldsymbol{\omega}_a}{\rho} \cdot \nabla\chi\right) = 0 \tag{4.74}$$

or

$$\frac{\mathrm{D}}{\mathrm{D}t}(\widetilde{\boldsymbol{\omega}}_a \cdot \nabla\chi) = 0, \tag{4.75}$$

where $\widetilde{\boldsymbol{\omega}}_a = \boldsymbol{\omega}_a/\rho$. Equation (4.75) is a statement of potential vorticity conservation for a barotropic fluid. The field χ may be chosen arbitrarily, provided that it is materially conserved.

The general case

For a baroclinic fluid the above derivation fails simply because the statement of the conservation of circulation, (4.70) is not, in general, true: there are solenoidal terms on the right-hand side and from (4.43) and (4.45) we have

$$\frac{\mathrm{D}}{\mathrm{D}t}[(\boldsymbol{\omega}_a \cdot \boldsymbol{n})\delta A] = \boldsymbol{S}_o \cdot \boldsymbol{n}\,\delta A, \qquad \boldsymbol{S}_o = -\nabla\alpha \times \nabla p = -\nabla\eta \times \nabla T. \tag{4.76a,b}$$

However, the right-hand side of (4.76a) may be annihilated by choosing the circuit around which we evaluate the circulation to be such that the solenoidal term is identically zero. Given the form of $\boldsymbol{S}_o$, this occurs if the values of any of p, ρ, η, T are constant on that circuit; that is, if $\chi = p, \rho, \eta$ or T. But the derivation also demands that χ be a materially conserved quantity, which usually restricts the choice of χ to be η (or potential temperature), or to be ρ itself if the thermodynamic equation is $\mathrm{D}\rho/\mathrm{D}t = 0$. Thus, the conservation of potential vorticity for inviscid, adiabatic flow is

$$\frac{\mathrm{D}}{\mathrm{D}t}(\widetilde{\boldsymbol{\omega}}_a \cdot \nabla\theta) = 0, \tag{4.77}$$

where $\mathrm{D}\theta/\mathrm{D}t = 0$. For diabatic flow source terms appear on the right-hand side, and we derive these later on. A summary of this derivation is provided by Fig. 4.9.

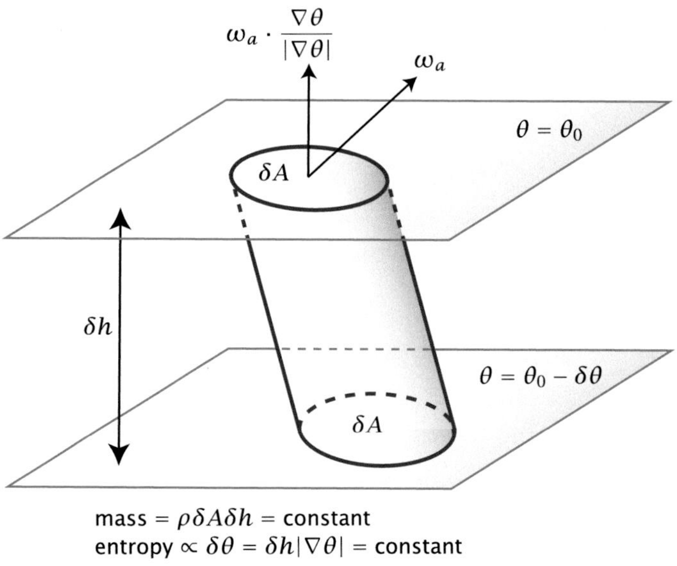

Fig. 4.9 Geometry of potential vorticity conservation. The circulation equation is $D[(\boldsymbol{\omega}_a \cdot \boldsymbol{n})\delta A]/Dt = \boldsymbol{S}_o \cdot \boldsymbol{n}\delta A$, where $\boldsymbol{S}_o \propto \nabla\theta \times \nabla T$. We choose $\boldsymbol{n} = \nabla\theta/|\nabla\theta|$, where θ is materially conserved, to annihilate the solenoidal term on the right-hand side, and we note that $\delta A = \delta V/\delta h$, where δV is the volume of the cylinder, and the height of the column is $\delta h = \delta\theta/|\nabla\theta|$. The circulation is $C \equiv \boldsymbol{\omega}_a \cdot \boldsymbol{n}\delta A = \boldsymbol{\omega}_a \cdot (\nabla\theta/|\nabla\theta|)(\delta V/\delta h) = [\rho^{-1}\boldsymbol{\omega}_a \cdot \nabla\theta](\delta M/\delta\theta)$, where $\delta M = \rho\delta V$ is the mass of the cylinder. As δM and $\delta\theta$ are materially conserved, so is the potential vorticity $\rho^{-1}\boldsymbol{\omega}_a \cdot \nabla\theta$.

4.5.2 PV Conservation from the Frozen-in Property

In this section we show that potential vorticity conservation is a consequence of the frozen-in property of vorticity. This is not surprising, because the circulation theorem itself has a similar origin. Thus, this derivation is not independent of the derivation in the previous section, just a re-expression of it. We first consider the case in which the solenoidal term vanishes from the outset.

Barotropic fluids

If χ is a materially conserved tracer then the difference in χ between two infinitesimally close fluid elements is also conserved and

$$\frac{D}{Dt}(\chi_1 - \chi_2) = \frac{D\delta\chi}{Dt} = 0. \tag{4.78}$$

But $\delta\chi = \nabla\chi \cdot \delta\boldsymbol{l}$, where $\delta\boldsymbol{l}$ is the infinitesimal vector connecting the two fluid elements. Thus

$$\frac{D}{Dt}(\nabla\chi \cdot \delta\boldsymbol{l}) = 0. \tag{4.79}$$

However, as the line element and the vorticity (divided by density) obey the same equation, we can replace the line element by vorticity (divided by density) in (4.79) to obtain again

$$\frac{D}{Dt}\left(\frac{\nabla\chi \cdot \boldsymbol{\omega}_a}{\rho}\right) = 0. \tag{4.80}$$

That is, the potential vorticity, $Q = (\tilde{\boldsymbol{\omega}}_a \cdot \nabla\chi)$ is a material invariant, where χ is any scalar quantity that satisfies $D\chi/Dt = 0$.

Baroclinic fluids

In baroclinic fluids we cannot casually substitute the vorticity for that of a line element in (4.79) because of the presence of the solenoidal term, and in any case a little more detail would not be amiss. From (4.79) we obtain

$$\delta\boldsymbol{l} \cdot \frac{D\nabla\chi}{Dt} + \nabla\chi \cdot \frac{D\delta\boldsymbol{l}}{Dt} = 0, \tag{4.81}$$

or, using (4.31),

$$\delta \boldsymbol{l} \cdot \frac{\mathrm{D}\nabla\chi}{\mathrm{D}t} + \nabla\chi \cdot [(\delta \boldsymbol{l} \cdot \nabla)\boldsymbol{v}] = 0. \tag{4.82}$$

Now, let us choose $\delta \boldsymbol{l}$ to correspond to a vortex line, so that at the initial time $\delta \boldsymbol{l} = \epsilon \widetilde{\boldsymbol{\omega}}_a$. (Note that in this case the association of $\delta \boldsymbol{l}$ with a vortex line can only be made instantaneously, and we cannot set $\mathrm{D}\delta \boldsymbol{l}/\mathrm{D}t \propto \mathrm{D}\boldsymbol{\omega}_a/\mathrm{D}t$.) Then,

$$\widetilde{\boldsymbol{\omega}}_a \cdot \frac{\mathrm{D}\nabla\chi}{\mathrm{D}t} + \nabla\chi \cdot [(\widetilde{\boldsymbol{\omega}}_a \cdot \nabla)\boldsymbol{v}] = 0, \tag{4.83}$$

or, using the vorticity equation (4.16),

$$\widetilde{\boldsymbol{\omega}}_a \cdot \frac{\mathrm{D}\nabla\chi}{\mathrm{D}t} + \nabla\chi \cdot \left(\frac{\mathrm{D}\widetilde{\boldsymbol{\omega}}_a}{\mathrm{D}t} - \frac{1}{\rho^3}\nabla\rho \times \nabla p \right) = 0. \tag{4.84}$$

This may be written as

$$\frac{\mathrm{D}}{\mathrm{D}t}\widetilde{\boldsymbol{\omega}}_a \cdot \nabla\chi = \frac{1}{\rho^3}\nabla\chi \cdot (\nabla\rho \times \nabla p). \tag{4.85}$$

The term on the right-hand side is, in general, non-zero for an arbitrary choice of scalar, but it will evidently vanish if ∇p, $\nabla\rho$ and $\nabla\chi$ are coplanar. If χ is any function of p and ρ this will be satisfied, but χ must also be a materially conserved scalar. If, as for an ideal gas, $\rho = \rho(\eta, p)$ (or $\eta = \eta(p, \rho)$) where η is the entropy (which is materially conserved), and if χ is a function of entropy η alone, then χ satisfies both conditions. Explicitly, the solenoidal term vanishes because

$$\nabla\chi \cdot (\nabla\rho \times \nabla p) = \frac{\mathrm{d}\chi}{\mathrm{d}\eta}\nabla\eta \cdot \left[\left(\frac{\partial\rho}{\partial p}\nabla p + \frac{\partial\rho}{\partial\eta}\nabla\eta \right) \times \nabla p \right] = 0. \tag{4.86}$$

Thus, provided χ satisfies the two conditions

$$\frac{\mathrm{D}\chi}{\mathrm{D}t} = 0 \quad \text{and} \quad \chi = \chi(p, \rho), \tag{4.87}$$

then (4.85) becomes

$$\frac{\mathrm{D}}{\mathrm{D}t}\left(\frac{\boldsymbol{\omega}_a \cdot \nabla\chi}{\rho} \right) = 0. \tag{4.88}$$

The natural choice for χ is potential temperature, whence

$$\frac{\mathrm{D}}{\mathrm{D}t}\left(\frac{\boldsymbol{\omega}_a \cdot \nabla\theta}{\rho} \right) = 0. \tag{4.89}$$

The presence of a density term in the denominator is not necessary for incompressible flows (i.e., if $\nabla \cdot \boldsymbol{v} = 0$).

4.5.3 PV Conservation: an Algebraic Derivation

Finally, we give an algebraic derivation of potential vorticity conservation. We will take the opportunity to include frictional and, diabatic processes, although these may also be included in the derivations above.[7] We begin with the frictional vorticity equation in the form

$$\frac{\mathrm{D}\widetilde{\boldsymbol{\omega}}_a}{\mathrm{D}t} = (\widetilde{\boldsymbol{\omega}}_a \cdot \nabla)\boldsymbol{v} + \frac{1}{\rho^3}(\nabla\rho \times \nabla p) + \frac{1}{\rho}(\nabla \times \boldsymbol{F}), \tag{4.90}$$

where $\boldsymbol{F}$ represents any non-conservative force term on the right-hand side of the momentum equation (i.e., $\mathrm{D}\boldsymbol{v}/\mathrm{D}t = -\rho^{-1}\nabla p + \boldsymbol{F}$). We have also the equation for our materially conserved scalar χ,

$$\frac{\mathrm{D}\chi}{\mathrm{D}t} = \dot{\chi}, \tag{4.91}$$

where $\dot{\chi}$ represents any sources and sinks of χ. Now

$$(\tilde{\boldsymbol{\omega}}_a \cdot \nabla)\frac{\mathrm{D}\chi}{\mathrm{D}t} = \tilde{\boldsymbol{\omega}}_a \cdot \frac{\mathrm{D}\nabla\chi}{\mathrm{D}t} + [(\tilde{\boldsymbol{\omega}}_a \cdot \nabla)\boldsymbol{v}] \cdot \nabla\chi, \tag{4.92}$$

which may be obtained just by expanding the left-hand side. Thus, using (4.91),

$$\tilde{\boldsymbol{\omega}}_a \cdot \frac{\mathrm{D}\nabla\chi}{\mathrm{D}t} = (\tilde{\boldsymbol{\omega}}_a \cdot \nabla)\,\dot{\chi} - [(\tilde{\boldsymbol{\omega}}_a \cdot \nabla)\boldsymbol{v}] \cdot \nabla\chi. \tag{4.93}$$

Now take the dot product of (4.90) with $\nabla\chi$:

$$\nabla\chi \cdot \frac{\mathrm{D}\tilde{\boldsymbol{\omega}}_a}{\mathrm{D}t} = \nabla\chi \cdot [(\tilde{\boldsymbol{\omega}}_a \cdot \nabla)\boldsymbol{v}] + \nabla\chi \cdot \left[\frac{1}{\rho^3}(\nabla\rho \times \nabla p)\right] + \nabla\chi \cdot \left[\frac{1}{\rho}(\nabla \times \boldsymbol{F})\right]. \tag{4.94}$$

The sum of the last two equations yields

$$\frac{\mathrm{D}}{\mathrm{D}t}(\tilde{\boldsymbol{\omega}}_a \cdot \nabla\chi) = \tilde{\boldsymbol{\omega}}_a \cdot \nabla\dot{\chi} + \nabla\chi \cdot \left[\frac{1}{\rho^3}(\nabla\rho \times \nabla p)\right] + \frac{\nabla\chi}{\rho} \cdot (\nabla \times \boldsymbol{F}). \tag{4.95}$$

This equation reprises (4.85), but with the addition of frictional and diabatic terms. As before, the solenoidal term is annihilated if we choose $\chi = \theta(p, \rho)$, so giving the evolution equation for potential vorticity in the presence of forcing and diabatic terms, namely

$$\frac{\mathrm{D}}{\mathrm{D}t}(\tilde{\boldsymbol{\omega}}_a \cdot \nabla\theta) = \tilde{\boldsymbol{\omega}}_a \cdot \nabla\dot{\theta} + \frac{\nabla\theta}{\rho} \cdot (\nabla \times \boldsymbol{F}). \tag{4.96}$$

4.5.4 Effects of Salinity and Moisture

For seawater the equation of state may be written as

$$\theta = \theta(\rho, p, S), \tag{4.97}$$

where θ is the potential temperature and S is the salinity. In the absence of diabatic terms and saline diffusion the potential temperature is a materially conserved quantity. However, because of the presence of salinity, potential temperature cannot be used to annihilate the solenoidal term; that is

$$\nabla\theta \cdot (\nabla\rho \times \nabla p) = \left(\frac{\partial\theta}{\partial S}\right)_{p,\rho} \nabla S \cdot (\nabla\rho \times \nabla p) \neq 0. \tag{4.98}$$

Strictly speaking then, *there is no potential vorticity conservation principle for seawater.* However, such a blunt statement overemphasizes the non-conservation of potential vorticity because the saline effect is small. In fact, we can derive an approximate potential vorticity conservation law, as follows.[0]

Suppose that we use potential density to try to annihilate the solenoidal term. Potential density is adiabatically conserved but, like θ, it is a function of salinity so that

$$\nabla\rho_\theta \cdot (\nabla\rho \times \nabla p) = \left(\frac{\partial\rho_\theta}{\partial S}\right)_{p,\rho} \nabla S \cdot (\nabla\rho \times \nabla p) \neq 0. \tag{4.99}$$

Now, potential density may be written as function of salinity and potential temperature (or entropy) with no pressure dependence and therefore we can rewrite the above expression as

$$(\nabla \rho \times \nabla p) \cdot \nabla \rho_\theta = (\nabla \rho_\theta \times \nabla \rho) \cdot \nabla p \tag{4.100a}$$

$$= \left[\left(\frac{\partial \rho_\theta}{\partial S} \nabla S + \frac{\partial \rho_\theta}{\partial \theta} \nabla \theta \right) \times \left(\frac{\partial \rho}{\partial S} \nabla S + \frac{\partial \rho}{\partial \theta} \nabla \theta + \frac{\partial \rho}{\partial p} \nabla p \right) \right] \cdot \nabla p \tag{4.100b}$$

$$= \left[\frac{\partial \rho_\theta}{\partial S} \nabla S \times \frac{\partial \rho}{\partial \theta} \nabla \theta + \frac{\partial \rho_\theta}{\partial \theta} \nabla \theta \times \frac{\partial \rho}{\partial S} \nabla S \right] \cdot \nabla p \tag{4.100c}$$

$$= \left[\frac{\partial \rho_\theta}{\partial S} \frac{\partial \rho}{\partial \theta} - \frac{\partial \rho_\theta}{\partial \theta} \frac{\partial \rho}{\partial S} \right] (\nabla S \times \nabla \theta) \cdot \nabla p, \tag{4.100d}$$

If the term in square brackets in (4.100d) is zero then potential vorticity is conserved, and this is the case if the density and potential density are related by

$$\rho(S, \theta, p) = \rho_\theta(S, \theta) + F(p), \tag{4.101}$$

where F is some function of p; the result also follows directly using (4.101) and (4.100a). Equation (4.101) does not exactly hold because the compressibility of seawater is not in fact just a function of pressure (this is the thermobaric effect). However, as can be seen from (1.156b), the equation holds to a good approximation, a result related to the fact that the speed of sound in seawater is nearly constant. Thus, to this approximation, potential vorticity is adiabatically conserved in seawater if potential density is used as the scalar variable. The derivation does not care whether density itself is a function of salinity; rather, it asks that the difference between density and potential density is a function only of pressure.

Similarly, in a moist atmosphere there is, strictly, no conservation of a conventional potential vorticity because potential temperature is a function of density, pressure and water vapour, although the moisture dependence is usually weak. Condensational heating also provides a diabatic source term that provides a source of potential vorticity. These effects may accounted for in part by using a virtual potential temperature in the definition of potential vorticity, and including the contribution of liquid water to the entropy.[9] In any case, and as with seawater, the compositional effects are fairly small, especially in mid-latitudes, and the dynamics of potential vorticity conservation play a central role in the large-scale dynamics of both atmosphere and ocean.

4.5.5 Effects of Rotation, and Summary Remarks

In a rotating frame the potential vorticity conservation equation is obtained simply by replacing $\boldsymbol{\omega}_a$ by $\boldsymbol{\omega} + 2\boldsymbol{\Omega}$, where $\boldsymbol{\Omega}$ is the rotation rate of the rotating frame. The operator $\mathrm{D}/\mathrm{D}t$ is reference-frame invariant, and so may be evaluated using the usual formulae with velocities measured in the rotating frame.

We have generally referred to the quantity $\boldsymbol{\omega}_a \cdot \nabla\theta/\rho$ as the potential vorticity; however, this form (often referred to as the Ertel or Rossby–Ertel potential vorticity) is not unique. If θ is a materially conserved variable, then so is $g(\theta)$ where g is any function, so that $\boldsymbol{\omega}_a \cdot \nabla g(\theta)/\rho$ is also a potential vorticity. In the atmosphere θ itself is in fact commonly used, whereas in the ocean potential density is the more appropriate scalar, with $f\partial\rho_\theta/\partial z$ being a common approximation for low Rossby number flows.

The conservation of potential vorticity has profound consequences in fluid dynamics, especially in a rotating, stratified fluid. The non-conservative terms are often small, and large-scale flow in both the ocean and the atmosphere is well characterized by conservation of potential vorticity. Such conservation is a very powerful constraint on the flow, and indeed it turns out that potential vorticity is usually a more useful quantity for baroclinic, or non-homentropic, fluids than for barotropic fluids, because the required use of a special conserved scalar imparts additional information; in barotropic fluids potential vorticity has little more power than vorticity itself.

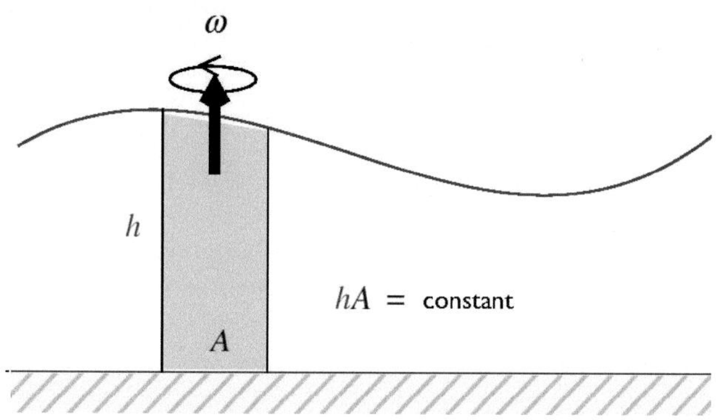

Fig. 4.10 The volume of a column of fluid, hA, is conserved. Furthermore, the vorticity is tied to material lines so that ζA is also a material invariant, where $\zeta = \boldsymbol{\omega} \cdot \mathbf{k}$ is the vertical component of the vorticity. From this, ζ/h must be materially conserved, or $\mathrm{D}(\zeta/h)/\mathrm{D}t = 0$, which is the conservation of potential vorticity in a shallow water system. With rotation this generalizes to $\mathrm{D}[(\zeta + f)/h]/\mathrm{D}t = 0$.

4.6 ✦ POTENTIAL VORTICITY IN THE SHALLOW WATER SYSTEM

In Chapter 3 we derived potential vorticity conservation by direct manipulation of the shallow water equations. In this short section we show that shallow water potential vorticity is also derivable from the conservation of circulation. Specifically, we will begin with the three-dimensional form of Kelvin's theorem, and then make the small aspect ratio assumption (which is the key assumption underlying shallow water dynamics), and thereby recover shallow water potential vorticity conservation (see also Fig. 4.10).

We begin with

$$\frac{\mathrm{D}}{\mathrm{D}t}(\boldsymbol{\omega}_3 \cdot \delta\mathbf{S}) = 0, \tag{4.102}$$

where $\boldsymbol{\omega}_3$ is the curl of the three-dimensional velocity and $\delta\mathbf{S} = \boldsymbol{n}\,\delta S$ is an arbitrary infinitesimal vector surface element, with $\boldsymbol{n}$ being a unit vector pointing in the direction normal to the surface. If we separate the vorticity and the surface element into vertical and horizontal components we can write (4.102) as

$$\frac{\mathrm{D}}{\mathrm{D}t}\left[(\zeta + f)\delta A + \boldsymbol{\omega}_h \cdot \delta\mathbf{S}_h\right] = 0, \tag{4.103}$$

where $\boldsymbol{\omega}_h$ and $\delta\mathbf{S}_h$ are the horizontally directed components of the vorticity and the surface element, and $\delta A = \mathbf{k}\,\delta\mathbf{S}$ is the area of a horizontal cross-section of a fluid column. In Cartesian form the horizontal component of the vorticity is

$$\boldsymbol{\omega}_h = \mathbf{i}\left(\frac{\partial w}{\partial y} - \frac{\partial v}{\partial z}\right) - \mathbf{j}\left(\frac{\partial w}{\partial x} - \frac{\partial u}{\partial z}\right) = \mathbf{i}\frac{\partial w}{\partial y} - \mathbf{j}\frac{\partial w}{\partial x}, \tag{4.104}$$

where vertical derivatives of the horizontal velocity are zero by virtue of the nature of the shallow water system. Now, the vertical velocity in the shallow water system is smaller than the horizontal velocity by the order of the aspect ratio — the ratio of the fluid depth to the horizontal scale of the motion. Furthermore, the size of the horizontally directed surface element is also smaller than the vertically-directed component by the aspect ratio; that is,

$$|\boldsymbol{\omega}_h| \sim \alpha|\zeta| \qquad \text{and} \qquad |\delta\mathbf{S}_h| \sim \alpha|\delta A|, \tag{4.105}$$

where $\alpha = H/L$ is the aspect ratio. Thus $\boldsymbol{\omega}_h \cdot \delta\mathbf{S}_h$ is smaller than the term $\zeta\delta A$ by the aspect number squared, and in the small aspect ratio approximation should be neglected. Kelvin's circulation theorem, (4.103), becomes

$$\frac{\mathrm{D}}{\mathrm{D}t}[(\zeta + f)\delta A] = 0 \qquad \text{or} \qquad \frac{\mathrm{D}}{\mathrm{D}t}\left[\frac{(\zeta + f)}{h}h\,\delta A\right] = 0, \tag{4.106a,b}$$

where h is the depth of the fluid column. But $h\,\delta A$ is the volume of the fluid column, and this is constant. Thus, (4.106b) gives, as in (3.96),

$$\frac{\mathrm{D}}{\mathrm{D}t}\left(\frac{\zeta + f}{h}\right) = 0, \tag{4.107}$$

where, because horizontal velocities are independent of the vertical coordinate, the advection is purely horizontal.

4.7 POTENTIAL VORTICITY IN APPROXIMATE, STRATIFIED MODELS

If approximate models of stratified flow (Boussinesq, hydrostatic and so on) are to be useful then they should conserve an appropriate form of potential vorticity, and we consider a few such cases.

4.7.1 The Boussinesq Equations

A Boussinesq fluid is incompressible; that is, the volume of a fluid element is conserved and the flow is divergence-free, with $\nabla\cdot\boldsymbol{v} = 0$. The equation for vorticity itself is then isomorphic to that for a line element. However, the Boussinesq equations are not barotropic — $\nabla\rho$ is not parallel to ∇p — and although the pressure gradient term $\nabla\phi$ disappears on taking its curl (or equivalently disappears on integration around a closed path) the buoyancy term $\mathbf{k}b$ does not, and it is this term that prevents Kelvin's circulation theorem from holding. Specifically, the evolution of circulation in the Boussinesq equations obeys

$$\frac{\mathrm{D}}{\mathrm{D}t}[(\boldsymbol{\omega}_a\cdot\boldsymbol{n})\delta A] = (\nabla\times b\mathbf{k})\cdot\boldsymbol{n}\,\delta A, \tag{4.108}$$

where here, as in (4.70), $\boldsymbol{n}$ is a unit vector orthogonal to an infinitesimal surface element of area δA. The right-hand side is annihilated if we choose $\boldsymbol{n}$ to be parallel to ∇b, because $\nabla b\cdot\nabla\times(b\mathbf{k}) = 0$. In the simple Boussinesq equations the thermodynamic equation is

$$\frac{\mathrm{D}b}{\mathrm{D}t} = 0, \tag{4.109}$$

and potential vorticity conservation is therefore (with $\boldsymbol{\omega}_a = \boldsymbol{\omega} + 2\boldsymbol{\Omega}$)

$$\frac{\mathrm{D}Q}{\mathrm{D}t} = 0, \qquad Q = (\boldsymbol{\omega} + 2\boldsymbol{\Omega})\cdot\nabla b. \tag{4.110a,b}$$

Expanding (4.110b) in Cartesian coordinates with $2\boldsymbol{\Omega} = f\mathbf{k}$ we obtain:

$$Q = (v_x - u_y)b_z + (w_y - v_z)b_x + (u_z - w_x)b_y + fb_z. \tag{4.111}$$

In the general Boussinesq equations b itself is not materially conserved. We cannot expect to obtain a conservation law if salinity is present, but if the equation of state and the thermodynamic equation are:

$$b = b(\theta, z), \qquad \frac{\mathrm{D}\theta}{\mathrm{D}t} = 0, \tag{4.112}$$

then potential vorticity conservation follows, because taking $\boldsymbol{n}$ to be parallel to $\nabla\theta$ will cause the right-hand side of (4.108) to vanish; that is,

$$\nabla\theta\cdot\nabla\times(b\mathbf{k}) = \left(\frac{\partial\theta}{\partial z}\nabla z + \frac{\partial\theta}{\partial b}\nabla b\right)\cdot\nabla\times(b\mathbf{k}) = 0. \tag{4.113}$$

The materially conserved potential vorticity in the Boussinesq approximation, Q_B, is thus

$$Q_B = \boldsymbol{\omega}_a\cdot\nabla\theta. \tag{4.114}$$

Note that if the equation of state is $b = b(\theta, \phi)$, where ϕ is the pressure, then potential vorticity is not conserved because then, in general, $\nabla\phi\cdot\nabla\times(b\mathbf{k}) \neq 0$.

4.7.2 The Hydrostatic Equations

Making the hydrostatic approximation has no effect on whether or not the circulation theorem is satisfied. Thus, in a baroclinic hydrostatic fluid we have

$$\frac{D}{Dt}\int(\boldsymbol{\omega}_{hy} + 2\boldsymbol{\Omega})\cdot d\mathbf{S} = -\int \nabla\alpha\times\nabla p\cdot d\mathbf{S}, \tag{4.115}$$

where, from (4.55) $\boldsymbol{\omega}_{hy} = \nabla\times\boldsymbol{u} = -\mathbf{i}v_z + \mathbf{j}u_z + \mathbf{k}(v_x - u_y)$, but the gradient operator and material derivative are fully three-dimensional. Derivation of potential vorticity conservation then proceeds, as in Section 4.5.1, by choosing the circuit over which the circulation is calculated to be such that the right-hand side vanishes; that is, to be such that the solenoidal term is annihilated. Precisely as before, this occurs if the circuit is barotropic, and without further ado we write

$$\frac{DQ_{hy}}{Dt} = \frac{D}{Dt}\left[\frac{(\boldsymbol{\omega}_{hy} + 2\boldsymbol{\Omega})\cdot\nabla\theta}{\rho}\right] = 0. \tag{4.116}$$

Expanding the expression for Q_{hy} in Cartesian coordinates gives

$$Q_{hy} = \frac{1}{\rho}\left[(v_x - u_y)\theta_z - v_z\theta_x + u_z\theta_y + 2\Omega\theta_z\right]. \tag{4.117}$$

In spherical coordinates the hydrostatic approximation is usually accompanied by the traditional approximation and the expanded expression for a conserved potential vorticity is more complicated. It can still be derived from Kelvin's theorem, but this is left as an exercise for the reader.

4.7.3 Potential Vorticity on Isentropic Surfaces

If we begin with the primitive equations in isentropic coordinates then potential vorticity conservation follows quite simply. Cross-differentiating the horizontal momentum equations (3.178) gives the vorticity equation

$$\frac{D}{Dt}(\zeta + f) + (\zeta + f)\nabla_\theta\cdot\boldsymbol{u} = 0, \tag{4.118}$$

where $D/Dt = \partial/\partial t + \boldsymbol{u}\cdot\nabla_\theta$. The thermodynamic equation is

$$\frac{D\sigma}{Dt} + \sigma\nabla\cdot\boldsymbol{u} = 0, \tag{4.119}$$

where $\sigma = \partial z/\partial b$ (Boussinesq) or $\partial p/\partial\theta$ (ideal gas) is the thickness of an isopycnal layer. Eliminating the divergence between (4.118) and (4.119) gives

$$\frac{DQ_{IS}}{Dt} = 0, \quad \text{where} \quad Q_{IS} = \left(\frac{\zeta + f}{\sigma}\right). \tag{4.120}$$

The derivation, and the result, are precisely the same as with the shallow water equations (Sections 3.7.1 and 4.6).

A connection between isentropic and height coordinates

The hydrostatic potential vorticity written in height coordinates may be transformed into a form that reveals its intimate connection with isentropic surfaces. Let us make the Boussinesq approximation for which the hydrostatic potential vorticity is, with no rotation,

$$Q_{hy} = (v_x - u_y)b_z - v_z b_x + u_z b_y, \tag{4.121}$$

where b is the buoyancy. We can write this as

$$Q_{hy} = b_z \left[\left(v_x - v_z \frac{b_x}{b_z} \right) - \left(u_y - u_z \frac{b_y}{b_z} \right) \right]. \tag{4.122}$$

But the terms in the inner brackets are just the horizontal velocity derivatives at constant b. To see this, note that

$$\left(\frac{\partial v}{\partial x}\right)_b = \left(\frac{\partial v}{\partial x}\right)_z + \frac{\partial v}{\partial z}\left(\frac{\partial z}{\partial x}\right)_b = \left(\frac{\partial v}{\partial x}\right)_z - \frac{\partial v}{\partial z}\left(\frac{\partial b}{\partial x}\right)_z \Big/ \frac{\partial b}{\partial z}, \tag{4.123}$$

with a similar expression for $(\partial u/\partial y)_b$. (These relationships follow from standard rules of partial differentiation. Derivatives with respect to z are taken at constant x and y.) Thus, we obtain

$$Q_{hy} = \frac{\partial b}{\partial z}\left[\left(\frac{\partial v}{\partial x}\right)_b - \left(\frac{\partial u}{\partial y}\right)_b\right] = \frac{\partial b}{\partial z}\zeta_b. \tag{4.124}$$

Thus, potential vorticity is simply the horizontal vorticity evaluated on a surface of constant buoyancy, multiplied by the vertical derivative of buoyancy. An analogous derivation, with a similar result, proceeds for the ideal gas equations, with potential temperature replacing buoyancy.

4.8 ♦ THE IMPERMEABILITY OF ISENTROPES TO POTENTIAL VORTICITY

An interesting property of isentropic surfaces is that they are 'impermeable' to potential vorticity, meaning that the mass integral of potential vorticity ($\int Q\rho \, \mathrm{d}V$) over a volume bounded by an isentropic surface remains constant, even in the presence of diabatic sources, provided the surfaces do not intersect a non-isentropic surface such as the ground.[10] This may seem surprising, especially because unlike most conservation laws the result does not require adiabatic flow, and for that reason it leads to interesting interpretations of a number of phenomena. However, impermeability is a consequence of the definition of potential vorticity rather than the equations of motion, and in that sense it is a kinematic and not dynamical property.

To derive the result we define $s \equiv \rho Q = \nabla \cdot (\theta \boldsymbol{\omega}_a)$, where $\boldsymbol{\omega}_a$ is the absolute vorticity, and integrate over some volume V to give

$$I = \int_V s \, \mathrm{d}V = \int_V \nabla \cdot (\theta \boldsymbol{\omega}_a) \, \mathrm{d}V = \int_S \theta \boldsymbol{\omega}_a \cdot \mathrm{d}\mathbf{S}, \tag{4.125}$$

using the divergence theorem, where S is the surface surrounding the volume V. If this is an isentropic surface then we have

$$I = \theta \int_S \boldsymbol{\omega}_a \cdot \mathrm{d}\mathbf{S} = \theta \int_V \nabla \cdot \boldsymbol{\omega}_a \, \mathrm{d}V = 0, \tag{4.126}$$

again using the divergence theorem. That is, over a volume wholly enclosed by a single isentropic surface the integral of s vanishes. If the volume is bounded by more than one isentropic surface none of which intersect the surface, for example by concentric spheres of different radii as in Fig. 4.11(a), the result still holds. The quantity s is called 'potential vorticity concentration', or 'PV concentration'. The integral of s over a volume is akin to the total amount of a conserved material property, such as salt content, and so may be called 'PV substance'. That is, the PV concentration is the amount of potential vorticity substance per unit volume and

$$\text{PV substance} = \int s \, \mathrm{d}V = \int \rho Q \, \mathrm{d}V. \tag{4.127}$$

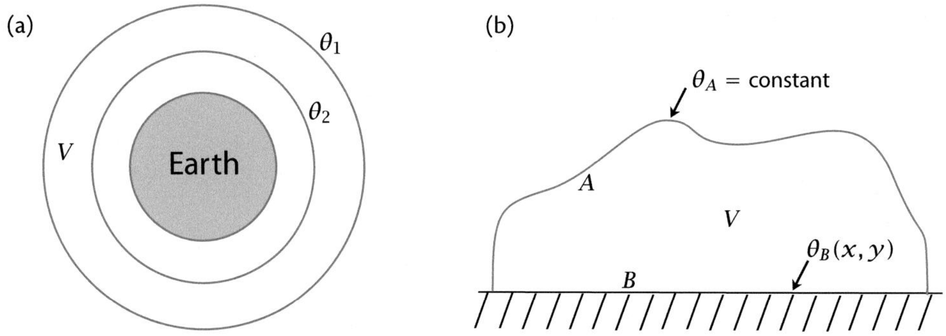

Fig. 4.11 (a) Two isentropic surfaces that do not intersect the ground. The integral of PV concentration over the volume between them, V, is zero, even if there is heating and the contours move. (b) An isentropic surface, A, intersects the ground, B, thus enclosing a volume V. The rate of change of PV concentration over the volume is given by an integral over B.

Suppose now that the fluid volume is enclosed by an isentrope that intersects the ground, as in Fig. 4.11(b). Let A denote the isentropic surface, B denote the ground, θ_A the constant value of θ on the isentrope, and $\theta_B(x, y, t)$ the non-constant value of θ on the ground. The integral of s over the volume is then

$$I = \int_V \nabla\cdot(\theta\boldsymbol{\omega}_a)\,\mathrm{d}V = \theta_A \int_A \boldsymbol{\omega}_a \cdot \mathrm{d}\boldsymbol{S} + \int_B \theta_B \boldsymbol{\omega}_a \cdot \mathrm{d}\boldsymbol{S}$$
$$= \theta_A \int_{A+B} \boldsymbol{\omega}_a \cdot \mathrm{d}\boldsymbol{S} + \int_B (\theta_B - \theta_A)\boldsymbol{\omega}_a \cdot \mathrm{d}\boldsymbol{S} \tag{4.128}$$
$$= \int_B (\theta_B - \theta_A)\boldsymbol{\omega}_a \cdot \mathrm{d}\boldsymbol{S}.$$

The first term on the second line vanishes after using the divergence theorem. Thus, the value of I, and hence its rate of change, is a function *only of an integral over the surface B,* and the PV flux there must be calculated using the full equations of motion. However, we do not need to be concerned with any flux of PV concentration through the isentropic surface; put another way, the PV substance in a volume can change only when isentropes enclosing the volume intersect a boundary such as the Earth's surface.

4.8.1 Interpretation and Application

Motion of the isentropic surface

How can the above results hold in the presence of heating? The isentropic surfaces must move in such a way that the total amount of PV concentration contained between them nevertheless stays fixed, and we now demonstrate this explicitly. The potential vorticity equation may be written

$$\frac{\partial Q}{\partial t} + \boldsymbol{v}\cdot\nabla Q = S_Q, \tag{4.129}$$

where, from (4.96), $S_Q = (\boldsymbol{\omega}_a/\rho)\cdot\nabla\dot{\theta} + \nabla\theta\cdot(\nabla\times\boldsymbol{F})/\rho$. Using mass continuity this may be written as

$$\frac{\partial s}{\partial t} + \nabla\cdot\boldsymbol{J} = 0, \tag{4.130}$$

where $\boldsymbol{J} \equiv \rho\boldsymbol{v}Q + \boldsymbol{N}$ and $\nabla\cdot\boldsymbol{N} = -\rho S_Q$. Written in this way, the quantity $\boldsymbol{J}/s$ is a notional velocity, $\boldsymbol{v}_Q$ say, and s satisfies

$$\frac{\partial s}{\partial t} + \nabla\cdot(\boldsymbol{v}_Q s) = 0. \tag{4.131}$$

That is, s evolves as if it were being fluxed by the velocity $\boldsymbol{v}_Q$. The concentration of a chemical tracer χ (i.e., χ is the amount of tracer per unit volume) obeys a similar equation, to wit

$$\frac{\partial \chi}{\partial t} + \nabla \cdot (\boldsymbol{v}\chi) = 0. \tag{4.132}$$

However, whereas (4.132) implies that $\mathrm{D}(\chi/\rho)/\mathrm{D}t = 0$, (4.131) does not imply that $\partial Q/\partial t + \boldsymbol{v}_Q \cdot \nabla Q = 0$ because $\partial \rho/\partial t + \nabla \cdot (\rho \boldsymbol{v}_Q) \neq 0$.

Now, the impermeability result tells us that there can be no notional velocity across an isentropic surface. How can this be satisfied by the equations of motion? We write the right-hand side of (4.129) as

$$\rho S_Q = \nabla \cdot (\dot{\theta}\boldsymbol{\omega}_a + \theta \nabla \times \boldsymbol{F}) = \nabla \cdot (\dot{\theta}\boldsymbol{\omega}_a + \boldsymbol{F} \times \nabla\theta). \tag{4.133}$$

Thus, $\boldsymbol{N} = -\dot{\theta}\boldsymbol{\omega}_a - \boldsymbol{F} \times \nabla\theta$ and we may write the $\boldsymbol{J}$ vector as

$$\boldsymbol{J} = \rho \boldsymbol{v} Q - \dot{\theta}\boldsymbol{\omega}_a - \boldsymbol{F} \times \nabla\theta = \rho Q(\boldsymbol{v}_\perp + \boldsymbol{v}_\parallel) - \dot{\theta}\boldsymbol{\omega}_\parallel - \boldsymbol{F} \times \nabla\theta, \tag{4.134}$$

where, making use of the thermodynamic equation,

$$\boldsymbol{v}_\parallel = \boldsymbol{v} - \frac{\boldsymbol{v} \cdot \nabla\theta}{|\nabla\theta|^2}\nabla\theta, \qquad \boldsymbol{v}_\perp = -\frac{\partial\theta/\partial t}{|\nabla\theta|^2}\nabla\theta, \tag{4.135a}$$

$$\boldsymbol{\omega}_\parallel = \boldsymbol{\omega}_a - \frac{\boldsymbol{\omega}_a \cdot \nabla\theta}{|\nabla\theta|^2}\nabla\theta = \boldsymbol{\omega}_a - \frac{Q\rho}{|\nabla\theta|^2}\nabla\theta. \tag{4.135b}$$

The subscripts '⊥' and '‖' denote components perpendicular and parallel to the local isentropic surface, and $\boldsymbol{v}_\perp$ is the velocity of the isentropic surface normal to itself. Equation (4.134) may be verified by using (4.135) and $\mathrm{D}\theta/\mathrm{D}t = \dot{\theta}$.

The 'parallel' terms in (4.135) are all vectors parallel to the local isentropic surface, and therefore do not lead to any flux of PV concentration across that surface. Furthermore, the term $\rho Q \boldsymbol{v}_\perp$ is ρQ multiplied by the normal velocity of the surface. That is to say, the notional velocity associated with the flux normal to the isentropic surface is equal to the normal velocity of the isentropic surface itself, and so it too provides no flux of PV concentration across that surface (even though there may well be a mass flux across the surface). Put simply, the isentropic surface always moves in such a way as to ensure that there is no flux of PV concentration across it. In our proof of the impermeability result in the previous subsection we used the fact that the potential vorticity multiplied by the density is the divergence of a vector. In the demonstration above we used the fact that the terms *forcing* potential vorticity are the divergence of a vector.

† *Dynamical choices of PV flux and a connection to Bernoulli's theorem*

If we add a non-divergent vector to the flux, $\boldsymbol{J}$, then it has no effect on the evolution of s. This gauge invariance means that the notional velocity, $v_Q = \boldsymbol{J}/(\rho Q)$ is similarly non-unique, although it does not mean that there are not dynamical choices for it that are more appropriate in given circumstances. To explore this, let us obtain a general expression for $\boldsymbol{J}$ by starting with the definition of s, so that

$$\begin{aligned} \frac{\partial s}{\partial t} &= \nabla\theta \cdot \frac{\partial \boldsymbol{\omega}_a}{\partial t} + \boldsymbol{\omega}_a \cdot \nabla \frac{\partial \theta}{\partial t} \\ &= \nabla\theta \cdot \nabla \times \frac{\partial \boldsymbol{v}}{\partial t} + \nabla \cdot \left(\boldsymbol{\omega}_a \frac{\partial \theta}{\partial t}\right) = -\nabla \cdot \boldsymbol{J}', \end{aligned} \tag{4.136}$$

where

$$\boldsymbol{J}' = \nabla\theta \times \frac{\partial \boldsymbol{v}}{\partial t} - \frac{\partial \theta}{\partial t}\boldsymbol{\omega}_a + \nabla\phi \times \nabla\chi. \tag{4.137}$$

The last term in this expression is an arbitrary divergence-free vector. If we choose $\phi = \theta$ and $\chi = B$, where B is the Bernoulli function given by $B = I + \boldsymbol{v}^2/2 + p/\rho$ where I is the internal energy per unit mass, then

$$\boldsymbol{J}' = \nabla\theta \times \left(\nabla B + \frac{\partial \boldsymbol{v}}{\partial t}\right) - \boldsymbol{\omega}_a(\dot{\theta} - \boldsymbol{v}\cdot\nabla\theta), \tag{4.138}$$

having used the thermodynamic equation $\mathrm{D}\theta/\mathrm{D}t = \dot{\theta}$. Now, the momentum equation may be written, without approximation, in the form

$$\frac{\partial \boldsymbol{v}}{\partial t} = -\boldsymbol{\omega}_a \times \boldsymbol{v} + T\nabla\eta + \boldsymbol{F} - \nabla B, \tag{4.139}$$

where η is the specific entropy ($\mathrm{d}\eta = c_p \,\mathrm{d}\ln\theta$). Using (4.138) and (4.139) gives

$$\boldsymbol{J}' = \rho Q \boldsymbol{v} - \dot{\theta}\boldsymbol{\omega}_a + \nabla\theta \times \boldsymbol{F}, \tag{4.140}$$

which is the same as (4.134). Furthermore, using (4.137) for steady flow,

$$\boldsymbol{J} = \nabla\theta \times \nabla B. \tag{4.141}$$

That is, the flux of potential vorticity (in this gauge) is aligned with the intersection of θ- and B-surfaces. For steady *inviscid and adiabatic* flow the Bernoulli function is constant along streamlines; that is, surfaces of constant Bernoulli function are aligned with streamlines, and, because θ is materially conserved, streamlines are formed at intersecting θ- and B-surfaces, as in (1.204). In the presence of forcing, this property is replaced by (4.141), and the flux of PV concentration is along such intersections.

This choice of gauge leading to (4.140) is physical in that it reduces to the true advective flux $\boldsymbol{v}\rho Q$ for unforced, adiabatic flow, but it is not a unique choice, nor is it mandated by the dynamics. Choosing $\chi = 0$ leads to the flux

$$\boldsymbol{J}_1 = \rho Q \boldsymbol{v} - \dot{\theta}\boldsymbol{\omega}_a + \nabla\theta \times (\boldsymbol{F} - \nabla B), \tag{4.142}$$

and using (4.137) this vanishes for steady flow, which is a potentially useful property.

Summary remarks

The impermeability result is kinematic, but can provide an interesting point of view and useful diagnostic tool.[11] We make the following summary remarks:

- There can be no net transport of potential vorticity across an isentropic surface, and the total amount of potential vorticity in a volume wholly enclosed by isentropic surfaces is zero. Thus, and with hindsight trivially, the amount of potential vorticity contained between two isentropes isolated from the Earth's surface in the Northern Hemisphere is the negative of the corresponding amount in the Southern Hemisphere.
- Potential vorticity flux lines (i.e., lines everywhere parallel to $\boldsymbol{J}$) can either close in on themselves or begin and end at boundaries (e.g., the ground or the ocean surface). However, $\boldsymbol{J}$ may change its character. Thus, for example, at the base of the oceanic mixed layer $\boldsymbol{J}$ may change from being a diabatic flux above to an adiabatic advective flux below. There may be a similar change in character at the atmospheric tropopause.
- The flux vector $\boldsymbol{J}$ is defined only to within the curl of a vector. Thus the vector $\boldsymbol{J}' = \boldsymbol{J} + \nabla\times\boldsymbol{A}$, where $\boldsymbol{A}$ is arbitrary, is as valid as is $\boldsymbol{J}$ in the above derivations.

Notes

1 The frozen-in property — that vortex lines are material lines — was derived by Helmholtz (1858) and is sometimes called Helmholtz's theorem.

2 The theorem originates with William Thomson (1824–1907), who became Lord Kelvin in 1892. The circulation theorem was published in Thomson (1869) and is a conservation law that is unique to a fluid: unlike, for example, the conservation of energy, it has no analogue in solid-body mechanics. Thomson was born in Belfast but spent most of his life in Scotland, becoming a professor at the University of Glasgow in 1846 (at the age of 22!) and staying there for 53 years. A prolific and creative scientist, he made a lasting impact on both fluid dynamics and thermodynamics — among other achievements he proposed an absolute temperature scale and a formulation of the second law of thermodynamics. Later in life he turned to engineering and was one of the proponents of a telegraph cable under the Atlantic. To his credit he also had some grand failures — his estimates of the age of Earth and how long oxygen would last in the atmosphere were both wrong by orders of magnitude.

3 Silberstein (1896) proved that 'the necessary and sufficient condition for the generation of vortical flow...influenced only by conservative forces ...is that the surface of constant pressure and surface of constant density...intersect', as we derived in Section 4.2, and this leads to (4.43). Bjerknes (1898a,b) explicitly put this into the form of a circulation theorem and applied it to problems of meteorological and oceanographic importance (see Thorpe *et al.* 2003), and the theorem is sometimes called the Bjerknes theorem or the Bjerknes–Silberstein theorem.

Vilhelm Bjerknes (1862–1951) was a physicist and hydrodynamicist who in 1917 moved to the University of Bergen as founding head of the Bergen Geophysical Institute. Here he did what was probably his most influential work in meteorology, setting up and contributing to the 'Bergen School of Meteorology'. Among other things he and his colleagues were among the first to consider, as a practical proposition, the use of numerical methods — initial data in conjunction with the fluid equations of motion — to forecast the state of the atmosphere, based on earlier work describing how that task might be done (Abbe 1901, Bjerknes 1904). Inaccurate initial velocity fields compounded with the shear complexity of the effort ultimately defeated them, but the effort was continued (also unsuccessfully) by L. F. Richardson (Richardson 1922), before J. Charney, R. Fjørtoft and J. Von Neumann eventually made what may be regarded as the first successful numerical forecast (Charney *et al.* 1950). Their success can be attributed to the used of a simplified, filtered, set of equations and the use of an electronic computer.

4 The result (4.50) was given by Poincaré (1893) although it is sometimes attributed to Bjerknes (1902).

5 The first derivation of the PV conservation law was given for the hydrostatic shallow water equations by Rossby (1936), with a generalization to the stratified case, via the use of isentropic coordinates, in Rossby (1938) and Rossby (1940). In the 1936 paper Rossby noted — his Eq. (75) — that a fluid column satisfies $f + \zeta = cD$, where c is a constant and D is the thickness of a fluid column; equivalently, $(f + \zeta)/D$ is a material invariant. The expression 'potential vorticity' was introduced in Rossby (1940), as follows: '*This quantity, which may be called the potential vorticity, represents the vorticity the air column would have it it were brought, isopycnally or isentropically, to a standard latitude (f_0) and stretched or shrunk vertically to a standard depth D_0 or weight Δ_0.*' (Rossby's italics.) That is,

$$\text{potential vorticity} = \zeta_0 = \left(\frac{\zeta + f}{D}\right) D_0 - f_0, \tag{4.143}$$

which follows from his Eq. (11), and this is the sense he uses it in that paper. However, potential vorticity has come to mean the quantity $(\zeta + f)/D$, which of course does not have the dimensions of vorticity. We use it in this latter, now conventional, sense throughout this book. Ironically, quasi-geostrophic potential vorticity as usually defined does have the dimensions of vorticity.

The expression for potential vorticity in a non-hydrostatic, continuously stratified fluid was given by Ertel (1942a), and its relationship to circulation was given by Ertel (1942b). It is now commonly known as the *Ertel potential vorticity*, or the *Rossby–Ertel* potential vorticity. Interestingly, in Rossby

(1940) we find the Fermat-like comment 'It is possible to derive corresponding results for an atmosphere in which the potential temperature varies continuously with elevation.... The generalized treatment will be presented in another place.' Given his prior (1938) derivation of the stratified quantity in isentropic coordinates, he must not have regarded his own derivations as very general. Opinions differ as to whether Rossby's and Ertel's derivations were independent, and Cressman (1996) remarks that the origin of the concept of potential vorticity is a 'delicate one that has aroused some passion in private correspondences'. In fact, Ertel visited MIT in autumn 1937 and presumably talked to Rossby and became aware of his work. It seems almost certain that Ertel knew of Rossby's shallow water and isentropic theorems, but it is also clear that Ertel subsequently provided a significant generalization, most likely independently. Rossby and Ertel apparently remained on good terms, but further collaboration was stymied by World War II. They later published a pair of short joint papers, one in German and the other in English, describing related conservation theorems (Ertel & Rossby 1949a,b). English translations of a number of Ertel's papers are to be found in Schubert *et al.* (2004). I thank Roger Samelson for enlightening me about the history of Rossby and Ertel.

6 Native French speakers may be confused by the difference. In English, an explication is a particular way of performing an analysis or presenting an explanation, so there can be different explications of the same mechanism.

7 Truesdell (1951, 1954) and Obukhov (1962) were early explorers of the consequences of heating and friction on potential vorticity. The work of F. P. Bretherton, R. E. Dickinson and J. S. A. Green (e.g., Bretherton 1966a, Dickinson 1969, Green 1970) helped bring potential vorticity ideas further into the mainstream of GFD.

8 Many thanks to Stephen Griffies for pointing out this argument, and for many other comments and discussions on matters related to this book. A study of saline effects on potential vorticity is to be found in Straub (1999).

9 Schubert *et al.* (2001) provide more discussion of these matters. They derive a 'moist PV' that is an extension of the dry Ertel PV to moist atmospheres and that has invertibility and impermeability properties.

10 Haynes & McIntyre (1987, 1990). See also Danielsen (1990), Schär (1993), who obtained the result (4.141), Bretherton & Schär (1993) and Davies-Jones (2003).

11 See, for example, McIntyre & Norton (1990) and Marshall & Nurser (1992). The latter use $\boldsymbol{J}$ vectors to study the creation and transport of potential vorticity in the oceanic thermocline.

A little inaccuracy sometimes saves a ton of explanation.
H. M. Munro (Saki), *The Square Egg*, 1924.

Every decoding is another encoding.
David Lodge, in the voice of Morris Zapp, *Small World*, 1984.

CHAPTER 5

Geostrophic Theory

LARGE-SCALE FLOW IN THE OCEAN AND THE ATMOSPHERE is characterized by an approximate balance in the vertical direction between the pressure gradient and gravity (hydrostatic balance), and in the horizontal direction between the pressure gradient and the Coriolis force (geostrophic balance). In this chapter we exploit these balances to simplify the Navier–Stokes equations and thereby obtain various sets of simplified 'geostrophic equations'. Depending on the precise nature of the assumptions we make, we are led to the *quasi-geostrophic* (QG) system for horizontal scales similar to that on which most synoptic activity takes place and, for very large-scale motion, to the *planetary-geostrophic* (PG) set of equations. By eliminating unwanted or unimportant modes of motion, in particular sound waves and gravity waves, and by building in the important balances between flow fields, these filtered equation sets allow the investigator to better focus on a particular class of phenomena and to potentially achieve a deeper understanding than might otherwise be possible.[1]

Simplifying the equations in this way relies first on scaling the equations. The idea is that we *choose* the scales we wish to describe, typically either on some a-priori basis or by using observations as a guide. We then attempt to derive a set of equations that is simpler than the original set but that consistently describes motion of the chosen scale. An asymptotic method is one way to achieve this, for it systematically tells us which terms we can drop and which we should keep. The combined approach — scaling plus asymptotics — has proven enormously useful, but we should always remember two things: (i) that scaling is a choice; (ii) that the approach does not explain the existence of particular scales of motion, it just describes the motion that might occur on such scales. We have already employed this general approach in deriving the hydrostatic primitive equations, but now we go further.

5.1 GEOSTROPHIC SCALING

5.1.1 Scaling in the Shallow Water Equations

Postponing the complications that come with stratification, we begin with the shallow water equations. With the odd exception, we will denote the scales of variables by capital letters; thus, if L is a typical length scale of the motion we wish to describe, and U is a typical velocity scale, and assuming the scales are horizontally isotropic, we write

$$\begin{aligned} (x,y) &\sim L \quad \text{or} \quad (x,y) = \mathcal{O}(L) \\ (u,v) &\sim U \quad \text{or} \quad (u,v) = \mathcal{O}(U), \end{aligned} \tag{5.1}$$

and similarly for other variables. We may then nondimensionalize the variables by writing

$$(x,y) = L(\hat{x},\hat{y}), \qquad (u,v) = U(\hat{u},\hat{v}), \tag{5.2}$$

where the hatted variables are nondimensional and, by supposition, are $\mathcal{O}(1)$. The various terms in the momentum equation then scale as:

$$\frac{\partial \boldsymbol{u}}{\partial t} + \boldsymbol{u}\cdot\nabla\boldsymbol{u} + \boldsymbol{f}\times\boldsymbol{u} = -g\nabla\eta, \tag{5.3a}$$

$$\frac{U}{T} \qquad \frac{U^2}{L} \qquad fU \sim g\frac{\mathcal{H}}{L}, \tag{5.3b}$$

where the ∇ operator acts in the x–y plane and $\mathcal{H}$ is the amplitude of the variations in the surface displacement. (We use η to denote the height of the free surface above some arbitrary reference level, as in Fig. 3.1. Thus, $\eta = H + \Delta\eta$, where $\Delta\eta$ denotes the variation of η about its mean position.)

The ratio of the advective term to the rotational term in the momentum equation (5.3) is $(U^2/L)/(fU) = U/fL$; this is the Rossby number,[2] first encountered in Chapter 2. Using values typical of the large-scale circulation (e.g., from Table 2.1) we find that $Ro \approx 0.1$ for the atmosphere and $Ro \approx 0.01$ for the ocean: small in both cases. If we are interested in motion that has the advective time scale $T = L/U$ then we scale time by L/U so that

$$t = \frac{L}{U}\hat{t}, \tag{5.4}$$

and the local time derivative and the advective term then both scale as U^2/L, and both are smaller than the rotation term by a factor of the order of the Rossby number. Then, either the Coriolis term is the dominant term in the equation, in which case we have a state of no motion with $-fv = 0$, or else the Coriolis force is balanced by the pressure force, and the dominant balance is

$$-fv = -g\frac{\partial\eta}{\partial x}, \tag{5.5}$$

namely *geostrophic balance*, as encountered in Chapter 2. If we make this non-trivial choice, then the equation informs us that variations in η (i.e., $\Delta\eta$) scale according to

$$\Delta\eta \sim \mathcal{H} = \frac{fUL}{g}. \tag{5.6}$$

We can also write $\mathcal{H}$ as

$$\mathcal{H} = Ro\,\frac{f^2L^2}{g} = Ro\,H\frac{L^2}{L_d^2}, \tag{5.7}$$

where $L_d = \sqrt{gH}/f$ is the deformation radius and H is the mean depth of the fluid. The variations in fluid height thus scale as

$$\frac{\Delta\eta}{H} \sim Ro\,\frac{L^2}{L_d^2}, \tag{5.8}$$

and the height of the fluid may be written as

$$\eta = H\left(1 + Ro\,\frac{L^2}{L_d^2}\hat{\eta}\right) \qquad \text{and} \qquad \Delta\eta = Ro\,\frac{L^2}{L_d^2}H\hat{\eta}, \tag{5.9}$$

where $\hat{\eta}$ is the $\mathcal{O}(1)$ nondimensional value of the surface height deviation.

Nondimensional momentum equation

If we use (5.9) to scale height variations, (5.2) to scale lengths and velocities, and (5.4) to scale time, then the momentum equation (5.3) becomes

$$Ro\left[\frac{\partial \hat{\boldsymbol{u}}}{\partial \hat{t}} + (\hat{\boldsymbol{u}} \cdot \nabla)\hat{\boldsymbol{u}}\right] + \hat{\boldsymbol{f}} \times \hat{\boldsymbol{u}} = -\nabla\hat{\eta}, \tag{5.10}$$

where $\hat{\boldsymbol{f}} = \mathbf{k}\hat{f} = \mathbf{k}f/f_0$, where f_0 is a representative value of the Coriolis parameter. (If f is a constant, then $\hat{f} = 1$, but it is informative to explicitly write $\hat{f}$ in the equations. Also, where the operator ∇ operates on a nondimensional variable then the differentials are taken with respect to the nondimensional variables $\hat{x}$, $\hat{y}$.) All the variables in (5.10) will be assumed to be of order unity, and the Rossby number multiplying the local time derivative and the advective terms indicates the smallness of those terms. By construction, the dominant balance in this equation is the geostrophic balance between the last two terms.

Nondimensional mass continuity (height) equation

The (dimensional) mass continuity equation can be written as

$$\frac{1}{H}\frac{\mathrm{D}\eta}{\mathrm{D}t} + \left(1 + \frac{\Delta\eta}{H}\right)\nabla \cdot \boldsymbol{u} = 0. \tag{5.11}$$

Using (5.2), (5.4) and (5.9) this equation may be written

$$Ro\left(\frac{L}{L_d}\right)^2 \frac{\mathrm{D}\hat{\eta}}{\mathrm{D}\hat{t}} + \left[1 + Ro\left(\frac{L}{L_d}\right)^2 \hat{\eta}\right] \nabla \cdot \hat{\boldsymbol{u}} = 0. \tag{5.12}$$

Equations (5.10) and (5.12) are the nondimensional versions of the full shallow water equations of motion. Evidently, some terms in the equations of motion are small and may be eliminated with little loss of accuracy, and the way this is done will depend on the size of the second nondimensional parameter, $(L/L_d)^2$, which we come to shortly.

Froude and Burger numbers

The Froude number may be generally defined as the ratio of a fluid particle speed to a wave speed. In a shallow water system this gives

$$Fr \equiv \frac{U}{\sqrt{gH}} = \frac{U}{f_0 L_d} = Ro\frac{L}{L_d}. \tag{5.13}$$

The Burger number[3] is a useful measure of the scale of motion of the fluid, relative to the deformation radius, and may be defined by

$$Bu \equiv \left(\frac{L_d}{L}\right)^2 = \frac{gH}{f_0^2 L^2} = \left(\frac{Ro}{Fr}\right)^2. \tag{5.14}$$

It is also useful to define the parameter $F \equiv Bu^{-1}$, which is like the square of a Froude number but uses the rotational speed fL instead of U in the numerator.

5.1.2 Geostrophic Scaling in the Stratified Equations

We now apply the same scaling ideas, *mutatis mutandis,* to the stratified primitive equations. We use the hydrostatic anelastic equations, which we write as

$$\frac{\mathrm{D}\boldsymbol{u}}{\mathrm{D}t} + \boldsymbol{f} \times \boldsymbol{u} = -\nabla_z \phi, \tag{5.15a}$$

$$\frac{\partial \phi}{\partial z} = b, \tag{5.15b}$$

$$\frac{\mathrm{D}b}{\mathrm{D}t} = 0, \tag{5.15c}$$

$$\nabla \cdot (\tilde{\rho} \boldsymbol{v}) = 0, \tag{5.15d}$$

where b is the buoyancy and $\tilde{\rho}$ is a reference density profile. Anticipating that the average stratification may not scale in the same way as the deviation from it, let us separate out the contribution of the advection of a reference stratification in (5.15c) by writing

$$b = \tilde{b}(z) + b'(x, y, z, t). \tag{5.16}$$

The thermodynamic equation then becomes

$$\frac{\mathrm{D}b'}{\mathrm{D}t} + N^2 w = 0, \tag{5.17}$$

where $N^2 \equiv \partial \tilde{b}/\partial z$ (and the advective derivative is still three-dimensional). We then let $\phi = \tilde{\phi}(z) + \phi'$, where $\tilde{\phi}$ is hydrostatically balanced by $\tilde{b}$, and the hydrostatic equation becomes

$$\frac{\partial \phi'}{\partial z} = b'. \tag{5.18}$$

Equations (5.17) and (5.18) replace (5.15c) and (5.15b), and ϕ' is used in (5.15a).

Nondimensional equations

We scale the basic variables by supposing that

$$(x, y) \sim L, \quad (u, v) \sim U, \quad t \sim \frac{L}{U}, \quad z \sim H, \quad f \sim f_0, \quad N \sim N_0 \tag{5.19}$$

where the scaling variables (capitalized, except for f_0) are chosen to be such that the nondimensional variables have magnitudes of the order of unity, and the parameters N_0 and f_0 are representative values of N and f. The scales chosen are such that the Rossby number is small; that is $Ro = U/(f_0 L) \ll 1$. In the momentum equation the pressure term then balances the Coriolis force,

$$|\boldsymbol{f} \times \boldsymbol{u}| \sim |\nabla \phi'|, \tag{5.20}$$

and so the pressure scales as

$$\phi' \sim \Phi = f_o U L. \tag{5.21}$$

Using the hydrostatic relation, (5.21) implies that the buoyancy scales as

$$b' \sim B = \frac{f_0 U L}{H}, \tag{5.22}$$

and from this we obtain

$$\frac{(\partial b'/\partial z)}{N^2} \sim Ro \frac{L^2}{L_d^2}, \tag{5.23}$$

where $L_d = N_0 H/f_0$ is the deformation radius in the continuously stratified fluid, analogous to the quantity $\sqrt{gH}/f_0$ in the shallow water system, and we use the same symbol, L_d, for both. In the continuously stratified system, *if the scale of motion is the same as or smaller than the deformation radius, and the Rossby number is small, then the variations in stratification are small.* The choice of scale is the key difference between the planetary-geostrophic and quasi-geostrophic equations.

Finally, we will nondimensionalize the vertical velocity by using the mass conservation equation,

$$\frac{1}{\tilde{\rho}}\frac{\partial \tilde{\rho} w}{\partial z} = -\left(\frac{\partial u}{\partial x} + \frac{\partial v}{\partial y}\right), \tag{5.24}$$

and we suppose that this implies

$$w \sim W = \frac{UH}{L}. \tag{5.25}$$

This is a naïve scaling for rotating flow: if the Coriolis parameter is nearly constant the geostrophic velocity is nearly horizontally non-divergent and the right-hand side of (5.24) is small, and $W \ll UH/L$. We might then estimate w by cross-differentiating geostrophic balance (with $\tilde{\rho}$ constant for simplicity) to obtain the linear geostrophic vorticity equation and corresponding scaling:

$$\beta v \approx f\frac{\partial w}{\partial z}, \qquad w \sim W = \frac{\beta U H}{f_0}. \tag{5.26a,b}$$

However, rather than using (5.26b) from the outset, we will use (5.25) and let the asymptotics guide us to a proper scaling in the fullness of time. Note that if variations in the Coriolis parameter are large and $\beta \sim f_0/L$, then (5.26b) is the same as (5.25).

Given the scalings above (using (5.25) for w) we nondimensionalize by setting

$$\begin{gathered}(\hat{x}, \hat{y}) = L^{-1}(x, y), \qquad \hat{z} = H^{-1}z, \quad (\hat{u}, \hat{v}) = U^{-1}(u, v), \quad \hat{t} = \frac{U}{L}t, \\ \widehat{w} = \frac{L}{UH}w, \quad \widehat{f} = \frac{f}{f_0}, \quad \widehat{N} = \frac{N}{N_0}, \quad \widehat{\phi} = \frac{\phi'}{f_0 U L}, \quad \hat{b} = \frac{H}{f_0 U L}b',\end{gathered} \tag{5.27}$$

where the hatted variables are nondimensional. The horizontal momentum and hydrostatic equations then become

$$Ro\frac{\mathrm{D}\hat{\boldsymbol{u}}}{\mathrm{D}\hat{t}} + \widehat{\boldsymbol{f}} \times \hat{\boldsymbol{u}} = -\nabla\widehat{\phi}, \tag{5.28}$$

and

$$\frac{\partial\widehat{\phi}}{\partial\hat{z}} = \hat{b}. \tag{5.29}$$

The nondimensional mass conservation equation is simply

$$\frac{1}{\tilde{\rho}}\nabla\cdot(\tilde{\rho}\hat{\boldsymbol{v}}) = \left(\frac{\partial\hat{u}}{\partial\hat{x}} + \frac{\partial\hat{v}}{\partial\hat{y}} + \frac{1}{\tilde{\rho}}\frac{\partial\tilde{\rho}\widehat{w}}{\partial\hat{z}}\right) = 0, \tag{5.30}$$

and the nondimensional thermodynamic equation is

$$\frac{f_0 U L}{H}\frac{U}{L}\frac{\mathrm{D}\hat{b}}{\mathrm{D}\hat{t}} + \widehat{N}^2 N_0^2\frac{HU}{L}\widehat{w} = 0, \tag{5.31}$$

or

$$Ro\frac{\mathrm{D}\hat{b}}{\mathrm{D}\hat{t}} + \left(\frac{L_d}{L}\right)^2\widehat{N}^2\widehat{w} = 0. \tag{5.32}$$

The nondimensional primitive equations are summarized in the box on the following page.

Nondimensional Primitive Equations

Horizontal momentum: $$Ro\frac{\mathrm{D}\hat{\boldsymbol{u}}}{\mathrm{D}\hat{t}} + \widehat{\boldsymbol{f}} \times \hat{\boldsymbol{u}} = -\nabla\widehat{\phi} \qquad \text{(PE.1)}$$

Hydrostatic: $$\frac{\partial\widehat{\phi}}{\partial\widehat{z}} = \hat{b} \qquad \text{(PE.2)}$$

Mass continuity: $$\left(\frac{\partial\hat{u}}{\partial\hat{x}} + \frac{\partial\hat{v}}{\partial\hat{y}} + \frac{1}{\widetilde{\rho}}\frac{\partial\widetilde{\rho}\widehat{w}}{\partial\widehat{z}}\right) = 0 \qquad \text{(PE.3)}$$

Thermodynamic: $$Ro\frac{\mathrm{D}\hat{b}}{\mathrm{D}\hat{t}} + \left(\frac{L_d}{L}\right)^2 \widehat{N}^2\widehat{w} = 0 \qquad \text{(PE.4)}$$

These equations are written for the anelastic equations in a rotating frame of reference. The Boussinesq equations result if we take $\widetilde{\rho} = 1$. The equations in pressure coordinates also have a similar form — see Section 2.6.2.

5.2 THE PLANETARY-GEOSTROPHIC EQUATIONS

We now use the low Rossby number scalings above to derive equation sets that are simpler than the original, 'primitive', ones. The planetary-geostrophic equations are probably the simplest such set of equations, and we derive these equations first for the shallow water equations, and then for the stratified primitive equations.

5.2.1 Using the Shallow Water Equations

Informal derivation

The advection and time derivative terms in the momentum equation (5.10) are order Rossby number smaller than the Coriolis and pressure terms (the term in square brackets is multiplied by Ro), and therefore let us neglect them. The momentum equation straightforwardly becomes

$$\widehat{\boldsymbol{f}} \times \hat{\boldsymbol{u}} = -\nabla\widehat{\eta}. \tag{5.33}$$

The mass conservation equation (5.12), contains two nondimensional parameters, $Ro = U/(f_0 L)$ (the Rossby number), and $F = (L/L_d)^2$ (the ratio of the length scale of the motion to the deformation scale; $F = Bu^{-1}$) and we must make a choice as to the relationship between these two numbers. We will choose

$$F\,Ro = \mathcal{O}(1), \tag{5.34}$$

which implies

$$L^2 \gg L_d^2 \qquad \text{or equivalently} \qquad F \gg 1, \quad Bu \ll 1. \tag{5.35}$$

That is to say, we suppose that the scales of motion are much larger than the deformation scale. Given this choice, all the terms in the mass conservation equation, (5.12), are of roughly the same size, and we retain them all. Thus, the shallow water planetary geostrophic equations are the full mass continuity equation along with geostrophic balance and a geometric relationship between the height field and the fluid thickness, and in dimensional form these are:

$$\frac{\mathrm{D}h}{\mathrm{D}t} + h\nabla\cdot\boldsymbol{u} = 0, \qquad \boldsymbol{f}\times\boldsymbol{u} = -g\nabla\eta, \qquad \eta = h + \eta_b\,. \tag{5.36a,b,c}$$

We emphasize that *the planetary-geostrophic equations are only valid for scales of motion much larger than the deformation radius.* The height variations are then as large as the mean height field itself; that is, using (5.8), $\Delta\eta/H = \mathcal{O}(1)$.

Formal derivation

We make the following assumptions:

(i) The Rossby number is small. $Ro = U/f_0 L \ll 1$.

(ii) The scale of the motion is significantly larger than the deformation scale. That is, (5.34) holds or equivalently

$$F = Bu^{-1} = \left(\frac{L}{L_d}\right)^2 \gg 1 \tag{5.37}$$

and in particular

$$F\,Ro = \mathcal{O}(1). \tag{5.38}$$

(iii) Time scales advectively, so that $T = L/U$.

The idea is now to expand the nondimensional velocity and height fields in an asymptotic series with the Rossby number as the small parameter, substitute into the equations of motion and derive a simpler set of equations. It is a nearly trivial exercise in this instance, and so illustrates the methodology well. The expansions are

$$\widehat{\boldsymbol{u}} = \widehat{\boldsymbol{u}}_0 + Ro\,\widehat{\boldsymbol{u}}_1 + Ro^2\,\widehat{\boldsymbol{u}}_2 + \dots \qquad \text{and} \qquad \widehat{\eta} = \widehat{\eta}_0 + Ro\,\widehat{\eta}_1 + Ro^2\,\widehat{\eta}_2 + \dots. \tag{5.39a,b}$$

Substituting (5.39) into the momentum equation then gives

$$Ro\left[\frac{\partial \widehat{\boldsymbol{u}}_0}{\partial \widehat{t}} + \widehat{\boldsymbol{u}}_0\cdot\nabla\widehat{\boldsymbol{u}}_0 + \widehat{\boldsymbol{f}}\times\widehat{\boldsymbol{u}}_1\right] + \widehat{\boldsymbol{f}}\times\widehat{\boldsymbol{u}}_0 = -\nabla\widehat{\eta}_0 - Ro\left[\nabla\widehat{\eta}_1\right] + \mathcal{O}(Ro^2). \tag{5.40}$$

The Rossby number is an asymptotic ordering parameter; thus, the sum of all the terms at any particular order in Rossby number must vanish. At lowest order we obtain the simple expression

$$\widehat{\boldsymbol{f}}\times\widehat{\boldsymbol{u}}_0 = -\nabla\widehat{\eta}_0. \tag{5.41}$$

Note that although f_0 is a representative value of f, we have made no assumptions about the constancy of f. In particular, f is allowed to vary by an order one amount, provided that it does not become so small that the Rossby number $U/(fL)$ is not small.

The appropriate height (mass conservation) equation is similarly obtained by substituting (5.39) into the shallow water mass conservation equation. Because $F\,Ro = \mathcal{O}(1)$ at lowest order we simply retain all the terms in the equation to give

$$F\,Ro\left[\frac{\partial\widehat{\eta}_0}{\partial t} + \widehat{\boldsymbol{u}}_0\cdot\nabla\widehat{\eta}_0\right] + \left[1 + F\,Ro\,\widehat{\eta}\right]\nabla\cdot\widehat{\boldsymbol{u}}_0 = 0. \tag{5.42}$$

Equations (5.41) and (5.42) are a closed set, namely the nondimensional planetary-geostrophic equations. The dimensional forms of these equations are just (5.36).

Variation of the Coriolis parameter

Suppose then that f is a constant (f_0). Then, from the curl of (5.41), $\nabla\cdot\boldsymbol{u}_0 = 0$. This means that we can define a streamfunction for the flow and, from geostrophic balance, the height field is just that streamfunction. That is, in dimensional form,

$$\psi = \frac{g}{f_0}\eta, \qquad \boldsymbol{u} = \mathbf{k}\times\nabla\psi, \tag{5.43a,b}$$

and (5.42) becomes, in dimensional form,

$$\frac{\partial \eta}{\partial t} + \boldsymbol{u} \cdot \nabla \eta = 0 \qquad \text{or} \qquad \frac{\partial \eta}{\partial t} + J(\psi, \eta) = 0, \tag{5.44}$$

where $J(a, b) \equiv a_x b_y - a_y b_x$. But since $\eta \propto \psi$ the advective term is proportional to $J(\psi, \psi)$, which is zero. Thus, the flow does not evolve at this order. The planetary-geostrophic equations are *uninteresting* if the scale of the motion is such that the Coriolis parameter is not variable. On Earth, the scale of motion on which this parameter regime exists is rather limited, since the planetary-geostrophic equations require that the scale of motion also be larger than the deformation radius. In the Earth's atmosphere, any scale that is larger than the deformation radius will be such that the Coriolis parameter varies significantly over it, and we do not encounter this parameter regime. On the other hand, in the Earth's ocean the deformation radius is relatively small and there exists a small parameter regime (called the frontal geostrophic regime) that has scales larger than the deformation radius but smaller than that on which the Coriolis parameter varies.

Potential vorticity

The shallow water PG equations may be written as an evolution equation for an appropriate potential vorticity. A little manipulation reveals that (5.36) are equivalent to:

$$\frac{\mathrm{D}Q}{\mathrm{D}t} = 0, \qquad Q = \frac{f}{h}, \quad \boldsymbol{f} \times \boldsymbol{u} = -g\nabla\eta, \quad \eta = h + \eta_b\,. \tag{5.45}$$

Thus, potential vorticity is a material invariant in the approximate equation set, just as it is in the full equations. The other variables — the free surface height and the velocity — are diagnosed from it, a process known as *potential vorticity inversion*. In the planetary geostrophic approximation, the inversion proceeds using the approximate form f/h rather than the full potential vorticity, $(f + \zeta)/h$. Thus, in a strict sense, we do not approximate potential vorticity, because this is the evolving variable. Rather, we approximate the inversion relations from which we derive the height and velocity fields. The simplest way of all to derive the shallow water PG equations is to *begin* with the conservation of potential vorticity, and to note that at small Rossby number the expression $(\zeta + f)/h$ may be approximated by f/h. Then, noting in addition that the flow is geostrophic, (5.45) immediately emerges. *Every* approximate set of equations that we derive in this chapter may be expressed as the evolution of potential vorticity, with the other fields being obtained diagnostically from it.

5.2.2 Planetary-Geostrophic Equations for Stratified Flow

To explore the stratified system we will use the inviscid and adiabatic Boussinesq equations of motion with the hydrostatic approximation. The derivation carries through easily enough using the anelastic or pressure-coordinate equations, but as the PG equations have more oceanographic than atmospheric importance, using the incompressible equations is quite appropriate.

Simplifying the equations

The nondimensional equations we begin with are (5.28)–(5.32). As in the shallow water case we expand these in a series in the Rossby number, so that:

$$\hat{u} = \hat{u}_0 + Ro\,\hat{u}_1 + Ro^2\,\hat{u}_2 + \dots, \qquad \hat{b} = \hat{b}_0 + Ro\,\hat{b}_1 + Ro^2\,\hat{b}_2 + \dots, \tag{5.46}$$

and similarly for $\hat{v}$, $\hat{w}$ and $\hat{\phi}$. Substituting into the nondimensional equations of motion (on page 176) and equating powers of Ro gives the lowest-order momentum, hydrostatic, and mass conservation equations:

$$\hat{\boldsymbol{f}} \times \hat{\boldsymbol{u}}_0 = -\nabla\hat{\phi}_0, \qquad \frac{\partial \hat{\phi}_0}{\partial \hat{z}} = \hat{b}_0, \qquad \nabla \cdot \hat{\boldsymbol{v}}_0 = 0. \tag{5.47a,b,c}$$

If we also assume that $L_d/L = \mathcal{O}(1)$, then the thermodynamic equation (5.32) becomes

$$\left(\frac{L_d}{L}\right)^2 \hat{N}^2 \hat{w}_0 = 0. \tag{5.48}$$

Of course we have neglected any diabatic terms in this equation, which would in general provide a non-zero right-hand side. Nevertheless, this is not a useful equation, because the set of the equations we have derived, (5.47) and (5.48), can no longer evolve: all the time derivatives have been scaled away! Thus, although instructive, these equations are not very useful. If instead we assume that the scale of motion is much larger than the deformation scale then the other terms in the thermodynamic equation will become equally important. Thus, we suppose that $L_d^2 \ll L^2$ or, more formally, that $L^2 = \mathcal{O}(Ro^{-1})L_d^2$, and then all the terms in the thermodynamic equation are retained. A closed set of equations is then given by (5.47) and the thermodynamic equation (5.32).

Dimensional equations

Restoring the dimensions, dropping the asymptotic subscripts, and allowing for the possibility of a source term, denoted by $S_{[b']}$, in the thermodynamic equation, the *planetary-geostrophic* equations of motion are:

$$\begin{gathered} \frac{\mathrm{D}b'}{\mathrm{D}t} + wN^2 = S_{[b']}, \\ \boldsymbol{f} \times \boldsymbol{u} = -\nabla\phi', \qquad \frac{\partial \phi'}{\partial z} = b', \qquad \nabla \cdot \boldsymbol{v} = 0. \end{gathered} \tag{5.49}$$

The thermodynamic equation may also be written simply as

$$\frac{\mathrm{D}b}{\mathrm{D}t} = \dot{b}, \tag{5.50}$$

where b now represents the total stratification. The relevant pressure, ϕ, is then the pressure that is in hydrostatic balance with b, so that geostrophic and hydrostatic balance are most usefully written as

$$\boldsymbol{f} \times \boldsymbol{u} = -\nabla\phi, \qquad \frac{\partial \phi}{\partial z} = b. \tag{5.51a,b}$$

Potential vorticity

Manipulation of (5.49) reveals that we can equivalently write the equations as an evolution equation for potential vorticity. Thus, the evolution equations may be written as

$$\frac{\mathrm{D}Q}{\mathrm{D}t} = \dot{Q}, \qquad Q = f\frac{\partial b}{\partial z}, \tag{5.52}$$

where $\dot{Q} = f\partial\dot{b}/\partial z$, and the inversion — i.e., the diagnosis of velocity, pressure and buoyancy — is carried out using the hydrostatic, geostrophic and mass conservation equations.

Applicability to the ocean and atmosphere

In the atmosphere a typical deformation radius NH/f is about 1000 km. The constraint that the scale of motion be significantly larger than the deformation radius is thus hard to satisfy, since one quickly runs out of room on a planet whose equator-to-pole distance is 10 000 km. Only the largest planetary waves can satisfy the planetary-geostrophic scaling in the atmosphere and we should then also write the equations in spherical coordinates. In the ocean the deformation radius is about 100 km, so there is lots of room for the planetary-geostrophic equations to hold, and much of the theory of the large-scale structure of the ocean involves these equations.

5.3 THE SHALLOW WATER QUASI-GEOSTROPHIC EQUATIONS

We now derive a set of geostrophic equations that is valid (unlike the PG equations) when the horizontal scale of motion is similar to that of the deformation radius. These equations are called the *quasi-geostrophic* equations, and are perhaps the most widely used set of equations for theoretical studies of the atmosphere and ocean. The specific assumptions we make are as follows:

(i) The Rossby number is small, so that the flow is in near-geostrophic balance.

(ii) The scale of the motion is not significantly larger than the deformation scale. Specifically, we shall require that

$$Ro\left(\frac{L}{L_d}\right)^2 = \mathcal{O}(Ro). \tag{5.53}$$

For the shallow water equations, this assumption implies, using (5.9), that the variations in fluid depth are small compared to its total depth. For the continuously stratified system it implies, using (5.23), that the variations in stratification are small compared to the background stratification.

(iii) Variations in the Coriolis parameter are small; that is, $|\beta L| \ll |f_0|$ where L is the length scale of the motion.

(iv) Time scales advectively; that is, the scaling for time is given by $T = L/U$.

The second and third of these differ from the planetary-geostrophic counterparts: we make the second assumption because we wish to explore a different parameter regime, and we then find that the third assumption is necessary to avoid the rather trivial state of $\beta v = 0$ (as we discuss more below). All of the assumptions are the same whether we consider the shallow water equations or a continuously stratified flow, and in this section we consider the former.

5.3.1 Single-layer Shallow Water Quasi-Geostrophic Equations

The algorithm is, again, to expand the variables $\widehat{u}, \widehat{v}, \widehat{\eta}$ in an asymptotic series with the Rossby number as the small parameter, substitute into the equations of motion, and derive a simpler set of equations. Thus we let

$$\widehat{u} = \widehat{u}_o + Ro\,\widehat{u}_1 + Ro^2\,\widehat{u}_2 + \dots, \qquad \widehat{v} = \widehat{v}_o + Ro\,\widehat{v}_1 + Ro^2\,\widehat{v}_2 + \dots, \tag{5.54a}$$

$$\widehat{\eta} = \widehat{\eta}_0 + Ro\,\widehat{\eta}_1 + Ro^2\,\widehat{\eta}_2 \dots. \tag{5.54b}$$

We recognize the smallness of β compared to f_0/L by letting $\beta = \widehat{\beta}U/L^2$, where $\widehat{\beta}$ is assumed to be a parameter of order unity. Then the expression $f = f_0 + \beta y$ becomes

$$\widehat{f} = f/f_0 = \widehat{f}_0 + Ro\,\widehat{\beta}\widehat{y}, \tag{5.55}$$

where $\widehat{f}_0$ is the nondimensional value of f_0; its value is unity, but it is helpful to denote it explicitly. Substitute (5.54) into the nondimensional momentum equation (5.10), and equate powers of Ro. At lowest order we obtain

$$\widehat{f}_0 \widehat{u}_0 = -\frac{\partial \widehat{\eta}_0}{\partial \widehat{y}}, \qquad \widehat{f}_0 \widehat{v}_0 = \frac{\partial \widehat{\eta}_0}{\partial \widehat{x}}. \tag{5.56}$$

Cross-differentiating gives

$$\nabla \cdot \widehat{\boldsymbol{u}}_0 = 0, \tag{5.57}$$

where, as always, when ∇ operates on a nondimensional variable, the derivatives are taken with respect to the nondimensional coordinates. Evidently the velocity field is divergence-free, with this property arising from the momentum equation rather than the mass conservation equation.

The mass conservation equation is also, at lowest order, $\nabla \cdot \widehat{\boldsymbol{u}}_0 = 0$, and at next order we have

$$F \frac{\partial \widehat{\eta}_0}{\partial \widehat{t}} + F \widehat{\boldsymbol{u}}_0 \cdot \nabla \widehat{\eta}_0 + \nabla \cdot \widehat{\boldsymbol{u}}_1 = 0. \tag{5.58}$$

This equation is not closed, because the evolution of the zeroth-order term involves evaluation of a first-order quantity. For closure, we go to the next order in the momentum equation,

$$\frac{\partial \widehat{\boldsymbol{u}}_0}{\partial \widehat{t}} + (\widehat{\boldsymbol{u}}_0 \cdot \nabla)\widehat{\boldsymbol{u}}_0 + \widehat{\beta} \widehat{y} \mathbf{k} \times \widehat{\boldsymbol{u}}_0 + \widehat{f}_0 \mathbf{k} \times \widehat{\boldsymbol{u}}_1 = -\nabla \widehat{\eta}_1, \tag{5.59}$$

and take its curl to give the vorticity equation:

$$\frac{\partial \widehat{\zeta}_0}{\partial \widehat{t}} + (\widehat{\boldsymbol{u}}_0 \cdot \nabla)(\widehat{\zeta}_0 + \widehat{\beta} \widehat{y}) = -\widehat{f}_0 \nabla \cdot \widehat{\boldsymbol{u}}_1. \tag{5.60}$$

The term on the right-hand side is the *vortex stretching* term. Only vortex stretching by the background or planetary vorticity is present, because the vortex stretching by the relative vorticity is smaller by a factor of the Rossby number. Equation (5.60) is also not closed; however, we may use (5.58) to eliminate the divergence term to give

$$\frac{\partial \widehat{\zeta}_0}{\partial \widehat{t}} + (\widehat{\boldsymbol{u}}_0 \cdot \nabla)(\widehat{\zeta}_0 + \widehat{\beta} \widehat{y}) = \widehat{f}_0 \left(F \frac{\partial \widehat{\eta}_0}{\partial \widehat{t}} + F \widehat{\boldsymbol{u}}_0 \cdot \nabla \widehat{\eta}_0 \right), \tag{5.61}$$

or

$$\frac{\partial}{\partial \widehat{t}}(\widehat{\zeta}_0 - \widehat{f}_0 F \widehat{\eta}_0) + (\widehat{\boldsymbol{u}}_0 \cdot \nabla)(\widehat{\zeta}_0 + \widehat{\beta} \widehat{y} - F \widehat{f}_0 \widehat{\eta}_0) = 0. \tag{5.62}$$

The final step is to note that the lowest-order vorticity and height fields are related through geostrophic balance, so that using (5.56) we can write

$$\widehat{u}_0 = -\frac{\partial \widehat{\psi}_0}{\partial \widehat{y}}, \qquad \widehat{v}_0 = \frac{\partial \widehat{\psi}_0}{\partial \widehat{x}}, \qquad \widehat{\zeta}_0 = \nabla^2 \widehat{\psi}_0, \tag{5.63}$$

where $\widehat{\psi}_0 = \widehat{\eta}_0 / \widehat{f}_0$ is the streamfunction. Equation (5.62) can thus be written as

$$\frac{\partial}{\partial \widehat{t}}(\nabla^2 \widehat{\psi}_0 - \widehat{f}_0^2 F \widehat{\psi}_0) + (\widehat{\boldsymbol{u}}_0 \cdot \nabla)(\widehat{\zeta}_0 + \widehat{\beta} \widehat{y} - \widehat{f}_0^2 F \widehat{\psi}_0) = 0 \tag{5.64}$$

or

$$\frac{\mathrm{D}_0}{\mathrm{D}\widehat{t}}(\nabla^2 \widehat{\psi}_0 + \widehat{\beta} \widehat{y} - \widehat{f}_0^2 F \widehat{\psi}_0) = 0, \tag{5.65}$$

where the subscript '0' on the material derivative indicates that the lowest order velocity, the geostrophic velocity, is the advecting velocity. Restoring the dimensions, (5.65) becomes

$$\frac{\mathrm{D}}{\mathrm{D}t}\left(\nabla^2 \psi + \beta y - \frac{1}{L_d^2} \psi \right) = 0, \tag{5.66}$$

where $\psi = (g/f_0)\eta$, $L_d^2 = gH/f_0^2$, and the advective derivative is

$$\frac{\mathrm{D}\,\cdot}{\mathrm{D}t} = \frac{\partial\,\cdot}{\partial t} + u_g\frac{\partial\,\cdot}{\partial x} + v_g\frac{\partial\,\cdot}{\partial y} = \frac{\partial\,\cdot}{\partial t} - \frac{\partial\psi}{\partial y}\frac{\partial\,\cdot}{\partial x} + \frac{\partial\psi}{\partial x}\frac{\partial\,\cdot}{\partial y} = \frac{\partial\,\cdot}{\partial t} + J(\psi,\cdot). \tag{5.67}$$

Another form of (5.66) is

$$\frac{\mathrm{D}}{\mathrm{D}t}\left(\zeta + \beta y - \frac{f_0}{H}\eta\right) = 0, \tag{5.68}$$

with $\zeta = (g/f_0)\nabla^2\eta$. Equations (5.66) and (5.68) are forms of the shallow water quasi-geostrophic potential vorticity equation. The quantity

$$q \equiv \zeta + \beta y - \frac{f_0}{H}\eta = \nabla^2\psi + \beta y - \frac{1}{L_d^2}\psi \tag{5.69}$$

is the *shallow water quasi-geostrophic potential vorticity.*

Connection to shallow water potential vorticity

The quantity q given by (5.69) is an approximation (except for dynamically unimportant constant additive and multiplicative factors) to the shallow water potential vorticity. To see the truth of this statement, begin with the expression for the shallow water potential vorticity,

$$Q = \frac{f + \zeta}{h}. \tag{5.70}$$

Now let $h = H(1 + \eta'/H)$, where η' is the perturbation of the free-surface height, and assume that η'/H is small to obtain

$$Q = \frac{f + \zeta}{H(1 + \eta'/H)} \approx \frac{1}{H}(f + \zeta)\left(1 - \frac{\eta'}{H}\right) \approx \frac{1}{H}\left(f_0 + \beta y + \zeta - f_0\frac{\eta'}{H}\right). \tag{5.71}$$

Because f_0/H is a constant it has no effect in the evolution equation, and the quantity given by

$$q = \beta y + \zeta - f_0\frac{\eta'}{H} \tag{5.72}$$

is materially conserved. Using geostrophic balance we have $\zeta = \nabla^2\psi$ and $\eta' = f_0\psi/g$ so that (5.72) is identical to (5.69). Only the variation in η is important in (5.68) or (5.69).

The approximations needed to go from (5.70) to (5.72) are the same as those used in our earlier, more long-winded, derivation of the quasi-geostrophic equations. That is, we assumed that f itself is nearly constant, and that f_0 is much larger than ζ, equivalent to a low Rossby number assumption. It was also necessary to assume that $H \gg \eta'$ to enable the expansion of the height field which, using assumption *(ii)* on page 180, is equivalent to requiring that the scale of motion not be significantly larger than the deformation scale. The derivation is completed by noting that the advection of the potential vorticity should be by the geostrophic velocity alone, and we recover (5.66) or (5.68).

Two interesting limits

There are two interesting limits to the quasi-geostrophic potential vorticity equation which, taking $\beta = 0$ for simplicity, are as follows:

(i) *Motion on scales much smaller than the deformation radius.* That is, $L \ll L_d$ and thus $Bu \gg 1$ or $F \ll 1$. Then (5.66) becomes

$$\frac{\partial \zeta}{\partial t} + J(\psi, \zeta) = 0, \tag{5.73}$$

where $\zeta = \nabla^2 \psi$ and $J(\psi, \zeta) = \psi_x \zeta_y - \psi_y \zeta_x$. Thus, the motion obeys the two-dimensional vorticity equation. Physically, on small length scales the deviations in the height field are very small and may be neglected.

(ii) *Motion on scales much larger than the deformation radius.* Although scales are not allowed to become so large that $Ro(L/L_d)^2$ is of order unity, we may, a posteriori, still have $L \gg L_d$, whence the potential vorticity equation, (5.66), becomes

$$\frac{\partial \psi}{\partial t} + J(\psi, \psi) = 0 \qquad \text{or} \qquad \frac{\partial \eta}{\partial t} + J(\psi, \eta) = 0, \tag{5.74}$$

because $\psi = g\eta/f_0$. The Jacobian term evidently vanishes. Thus, one is left with a trivial equation that implies there is no advective evolution of the height field. There is nothing wrong with our reasoning; the mathematics has indeed pointed out a limit interesting in its uninterestingness. From a physical point of view, however, such a lack of motion is likely to be rare, because on such large scales the Coriolis parameter varies considerably, and we are led to the planetary-geostrophic equations.

In practice, often the most severe restriction of quasi-geostrophy is that variations in layer thickness are small: what does this have to do with geostrophy? If we scale η assuming geostrophic balance then $\eta \sim fUL/g$ and $\eta/H \sim Ro(L/L_d)^2$. Thus, if Ro is to remain small, η/H can only be of order one if $(L/L_d)^2 \gg 1$. That is, the height variations must occur on a large scale, or we are led to a scaling inconsistency. Put another way, *if there are order-one height variations over a length scale of less than or of the order of the deformation scale, the Rossby number will not be small.* Large height variations are allowed if the scale of motion is large, but this contingency is described by the planetary-geostrophic equations.

Another flow regime

Although perhaps of little terrestrial interest, we can imagine a regime in which the Coriolis parameter varies fully, but the scale of motion remains no larger than the deformation radius. This parameter regime is not quasi-geostrophic, but it gives an interesting result. Because $\eta'/H \sim Ro(L/L_d)^2$ deviations of the height field are at least of order Rossby number smaller than the reference height and $|\eta'| \ll H$. The dominant balance in the height equation is then

$$H\nabla \cdot \boldsymbol{u} = 0, \tag{5.75}$$

presuming that time still scales advectively. This zero horizontal divergence must remain consistent with geostrophic balance,

$$\boldsymbol{f} \times \boldsymbol{u} = -g\nabla\eta, \tag{5.76}$$

where now f is a fully variable Coriolis parameter. Taking the curl of (that is, cross-differentiating) (5.76) gives

$$\beta v + f\nabla \cdot \boldsymbol{u} = 0, \tag{5.77}$$

whence, using (5.75), $v = 0$, and the flow is purely zonal. Although not at all useful as an evolution equation, this illustrates the constraining effect that differential rotation has on meridional velocity. This effect may be the cause of the banded, highly zonal flow on some of the giant planets, and we will revisit this issue in our discussion of geostrophic turbulence.

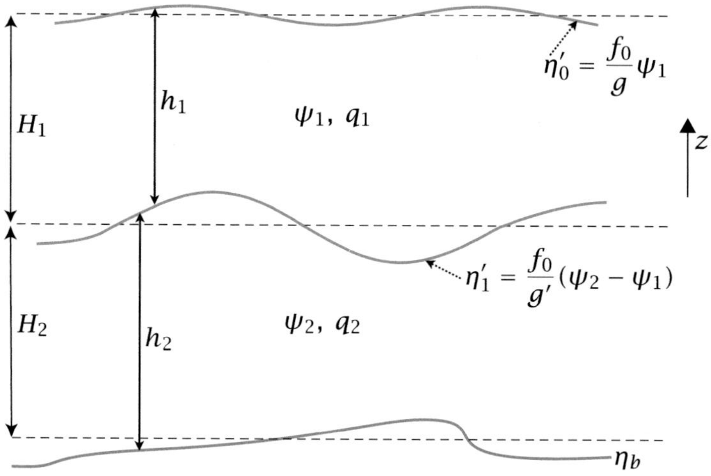

Fig. 5.1 A quasi-geostrophic fluid system consisting of two immiscible fluids of different density. The quantities η' are the interface displacements from the resting basic state, denoted with dashed lines, with η_b being the bottom topography.

5.3.2 Two-layer and Multi-layer Quasi-Geostrophic Systems

Just as for the one-layer case, the multi-layer shallow water equations simplify to a corresponding quasi-geostrophic system in appropriate circumstances. The assumptions are virtually the same as before, although we assume that the variation in the thickness of *each* layer is small compared to its mean thickness. The basic fluid system for a two-layer case is sketched in Fig. 5.1 (and see also Fig. 3.5), and for the multi-layer case in Fig. 5.2.

Let us proceed directly from the potential vorticity equation for each layer. We will also stay in dimensional variables, foregoing a strict asymptotic approach for the sake of informality and insight, and use the Boussinesq approximation. For each layer the potential vorticity equation is just

$$\frac{\mathrm{D}Q_i}{\mathrm{D}t} = 0, \qquad Q_i = \frac{\zeta_i + f}{h_i}. \tag{5.78}$$

Let $h_i = H_i + h_i'$ where $|h_i'| \ll H_i$. The potential vorticity then becomes

$$Q_i \approx \frac{1}{H_i}(\zeta_i + f)\left(1 - \frac{h_i'}{H_i}\right) \qquad \text{— variations in layer thickness are small,} \tag{5.79a}$$

$$\approx \frac{1}{H_i}\left(f + \zeta_i - f\frac{h_i'}{H_i}\right) \qquad \text{— the Rossby number is small,} \tag{5.79b}$$

$$\approx \frac{1}{H_i}\left(f + \zeta_i - f_0\frac{h_i'}{H_i}\right) \qquad \text{— variations in Coriolis parameter are small.} \tag{5.79c}$$

Now, because Q appears in the equations only as an advected quantity, it is only the *variations* in the Coriolis parameter that are important in the first term on the right-hand side of (5.79c), and given this all three terms are of the same approximate magnitude. Then, because mean layer thicknesses are constant, we can define the quasi-geostrophic potential vorticity in each layer by

$$q_i = \left(\beta y + \zeta_i - f_0\frac{h_i'}{H_i}\right), \tag{5.80}$$

and this will evolve according to $\mathrm{D}q_i/\mathrm{D}t = 0$, where the advective derivative is by the geostrophic wind. As in the one-layer case, the quasi-geostrophic potential vorticity has different dimensions from the full shallow water potential vorticity.

Two-layer model

To obtain a closed set of equations we must obtain an advecting field from the potential vorticity. We use geostrophic balance to do this, and neglecting the advective derivative in (3.49) gives

$$f_0 \times \boldsymbol{u}_1 = -g\nabla\eta_0 = -g\nabla(h_1' + h_2' + \eta_b), \tag{5.81a}$$
$$f_0 \times \boldsymbol{u}_2 = -g\nabla\eta_0 - g'\nabla\eta_1 = -g\nabla(h_1' + h_2' + \eta_b) - g'\nabla(h_2' + \eta_b), \tag{5.81b}$$

where $g' = g(\rho_2 - \rho_1)/\rho_1$ and η_b is the height of any bottom topography, and, because variations in the Coriolis parameter are presumptively small, we use a constant value of f (i.e., f_0) on the left-hand side. For each layer there is therefore a streamfunction, given by

$$\psi_1 = \frac{g}{f_0}(h_1' + h_2' + \eta_b), \qquad \psi_2 = \frac{g}{f_0}(h_1' + h_2' + \eta_b) + \frac{g'}{f_0}(h_2' + \eta_b), \tag{5.82a,b}$$

and these two equations may be manipulated to give

$$h_1' = \frac{f_0}{g'}(\psi_1 - \psi_2) + \frac{f_0}{g}\psi_1, \qquad h_2' = \frac{f_0}{g'}(\psi_2 - \psi_1) - \eta_b. \tag{5.83a,b}$$

We note as an aside that the interface displacements are given by

$$\eta_0' = \frac{f_0}{g}\psi_1, \qquad \eta_1' = \frac{f_0}{g'}(\psi_2 - \psi_1). \tag{5.84a,b}$$

Using (5.80) and (5.83) the quasi-geostrophic potential vorticity for each layer becomes

$$\begin{aligned} q_1 &= \beta y + \nabla^2\psi_1 + \frac{f_0^2}{g'H_1}(\psi_2 - \psi_1) - \frac{f_0^2}{gH_1}\psi_1, \\ q_2 &= \beta y + \nabla^2\psi_2 + \frac{f_0^2}{g'H_2}(\psi_1 - \psi_2) + f_0\frac{\eta_b}{H_2}. \end{aligned} \tag{5.85a,b}$$

In the rigid-lid approximation the last term in (5.85a) is neglected. The potential vorticity in each layer is advected by the geostrophic velocity, so that the evolution equation for each layer is just

$$\frac{\partial q_i}{\partial t} + J(\psi_i, q_i) = 0, \qquad i = 1, 2. \tag{5.86}$$

Multi-layer model

A multi-layer quasi-geostrophic model may be constructed by a straightforward extension of the above two-layer procedure (see Fig. 5.2). The quasi-geostrophic potential vorticity for each layer is still given by (5.80). The pressure field in each layer can be expressed in terms of the thickness of each layer using (3.44) and (3.45), and by geostrophic balance the pressure is proportional to the streamfunction, ψ_i, for each layer. Carrying out these steps we obtain, after a little algebra, the following expression for the quasi-geostrophic potential vorticity of an interior layer, in the Boussinesq approximation:

$$q_i = \beta y + \nabla^2\psi_i + \frac{f_0^2}{H_i}\left(\frac{\psi_{i-1} - \psi_i}{g_{i-1}'} - \frac{\psi_i - \psi_{i+1}}{g_i'}\right), \tag{5.87}$$

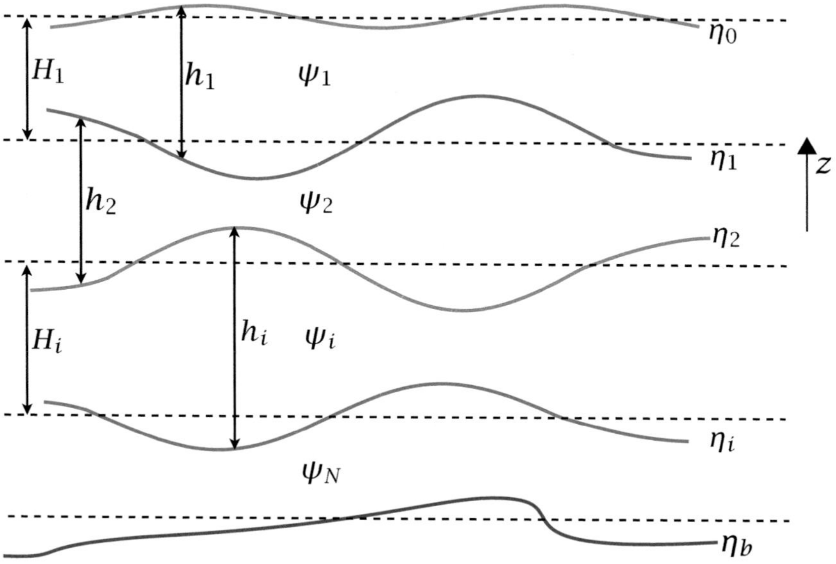

Fig. 5.2 A multi-layer quasi-geostrophic fluid system. Layers are numbered from the top down, i denotes a general interior layer and N denotes the bottom layer.

and for the top and bottom layers,

$$q_1 = \beta y + \nabla^2\psi_1 + \frac{f_0^2}{H_1}\left(\frac{\psi_2 - \psi_1}{g_1'}\right) - \frac{f_0^2}{gH_1}\psi_1, \tag{5.88a}$$

$$q_N = \beta y + \nabla^2\psi_N + \frac{f_0^2}{H_N}\left(\frac{\psi_{N-1} - \psi_N}{g_{N-1}'}\right) + \frac{f_0}{H_N}\eta_b. \tag{5.88b}$$

In these equations H_i is the basic-state thickness of the ith layer, and $g_i' = g(\rho_{i+1} - \rho_i)/\rho_1$. In each layer the evolution equation is (5.86), now for $i = 1 \dots N$. The displacements of each interface are given, similarly to (5.84), by

$$\eta_0' = \frac{f_0}{g}\psi_1, \qquad \eta_i' = \frac{f_0}{g_i'}(\psi_{i+1} - \psi_i). \tag{5.89a,b}$$

5.3.3 † Non-asymptotic and Intermediate Models

The form of the derivation of the previous section suggests that we might be able to improve on the accuracy and the range of applicability of the quasi-geostrophic equations, whilst still filtering gravity waves. For example, a seemingly improved set of geostrophic evolution equations might be

$$\frac{\partial q_i}{\partial t} + \boldsymbol{u}_i \cdot \nabla q_i = 0, \tag{5.90}$$

with

$$q_i = \frac{f + \zeta_i}{h_i}, \qquad \zeta_i = \frac{\partial v_i}{\partial x} - \frac{\partial u_i}{\partial y}, \tag{5.91a,b}$$

and with the velocities given by geostrophic balance, and therefore a function of the layer depths. Thus, the vorticity, height and velocity fields may all be inverted from potential vorticity. Note that the inversion does not involve the linearization of potential vorticity about a resting state — compare (5.91a) with (5.80)] — and we might also choose to keep the full variation of the Coriolis parameter in (5.81). Thus, the model consisting of (5.90) and (5.91) contains both the planetary geostrophic and quasi-geostrophic equations. However, the informality of the derivation hides the fact that this is not an asymptotically consistent set of equations: it mixes asymptotic orders in the same equation, and good conservation properties are not assured. The set above does not, in fact, exactly conserve energy. Models that are either more accurate or more general than the quasi-geostrophic or planetary-geostrophic equations yet that still filter gravity waves are called 'intermediate models'.[4]

A model that is derived asymptotically will, in general, maintain the conservation properties of the original set. To see this, albeit in a rather abstract way, suppose that the original equations (e.g., the primitive equations) may be written in nondimensional form, as

$$\frac{\partial \varphi}{\partial t} = F(\varphi, \epsilon), \tag{5.92}$$

where φ is a set of variables, F is some operator and ϵ is a small parameter, such as the Rossby number. Suppose also that this set of equations has various invariants (such as energy and potential vorticity) that hold for any value of ϵ. The asymptotically derived lowest-order model (such as quasi-geostrophy) is simply a version of this equation set valid in the limit $\epsilon = 0$, and therefore it will preserve the invariants of the original set. These invariants may seem to have a different form in the simplified set: for example, in deriving the hydrostatic primitive equations from the Navier–Stokes equations the small parameter is the aspect ratio, and this multiplies the vertical velocity. Thus, in the limit of zero aspect ratio, and therefore in the primitive equations, the kinetic energy component of the energy invariant has contributions only from the horizontal velocity. In other cases, some invariants may be reduced to trivialities in the simplified set. On the other hand, there is nothing to preclude new invariants emerging that hold only in the limit $\epsilon = 0$, and enstrophy (considered later in this chapter) is one example.

5.4 THE CONTINUOUSLY STRATIFIED QUASI-GEOSTROPHIC SYSTEM

We now consider the quasi-geostrophic equations for the continuously stratified hydrostatic system. The primitive equations of motion are given by (5.15), and we extract the mean stratification so that the thermodynamic equation is given by (5.17). We also stay on the β-plane for simplicity. Readers who wish for a briefer, more informal derivation may peruse the box on page 193; however, it is important to realize that there is a systematic asymptotic derivation of the quasi-geostrophic equations, for it is this that ensures that the resulting equations have good conservation properties, as explained above.

5.4.1 Scaling and Assumptions

The scaling assumptions we make are just those we made for the shallow water system on page 180, with a deformation radius now given by $L_d = NH/f_0$. The nondimensionalization and scaling are initially precisely that of Section 5.1.2, and we obtain the following nondimensional equations:

horizontal momentum: $$Ro \frac{\mathrm{D}\widehat{\boldsymbol{u}}}{\mathrm{D}\widehat{t}} + \widehat{\boldsymbol{f}} \times \widehat{\boldsymbol{u}} = -\nabla_z \widehat{\phi}, \tag{5.93}$$

hydrostatic: $$\frac{\partial \widehat{\phi}}{\partial \widehat{z}} = \widehat{b}, \tag{5.94}$$

mass continuity: $$\frac{\partial \widehat{u}}{\partial \widehat{x}} + \frac{\partial \widehat{v}}{\partial \widehat{y}} + \frac{1}{\widetilde{\rho}} \frac{\partial \widetilde{\rho} \widehat{w}}{\partial \widehat{z}} = 0, \tag{5.95}$$

thermodynamic: $$Ro \frac{\mathrm{D}\widehat{b}}{\mathrm{D}\widehat{t}} + \left(\frac{L_d}{L}\right)^2 \widehat{N}^2 \widehat{w} = 0. \tag{5.96}$$

In Cartesian coordinates we may express the Coriolis parameter as

$$\boldsymbol{f} = \boldsymbol{f}_0 + \beta y \mathbf{k}, \tag{5.97}$$

where $\boldsymbol{f}_0 = f_0 \mathbf{k}$. The variation of the Coriolis parameter is assumed to be small (this is a key difference between the quasi-geostrophic system and the planetary-geostrophic system), and in particular we shall assume that βy is approximately the size of the relative vorticity, and so is much smaller than f_0 itself.[5] Thus,

$$\beta y \sim \frac{U}{L}, \qquad \beta \sim \frac{U}{L^2}, \tag{5.98}$$

and so we define an $\mathcal{O}(1)$ nondimensional beta parameter by

$$\widehat{\beta} = \frac{\beta L^2}{U} = \frac{\beta L}{Ro\, f_0}. \tag{5.99}$$

From this it follows that if $f = f_0 + \beta y$, the corresponding nondimensional version is

$$\widehat{f} = \widehat{f}_0 + Ro\, \widehat{\beta} \widehat{y}. \tag{5.100}$$

where $\widehat{f} = f/f_0$ and $\widehat{f}_0 = f_0/f_0 = 1$.

5.4.2 Asymptotics

We now expand the nondimensional dependent variables in an asymptotic series in Rossby number, and write

$$\widehat{\boldsymbol{u}} = \widehat{\boldsymbol{u}}_0 + Ro\, \widehat{\boldsymbol{u}}_1 + \dots, \qquad \widehat{\phi} = \widehat{\phi}_0 + Ro\, \widehat{\phi}_1 + \dots, \qquad \hat{b} = \hat{b}_0 + Ro\, \hat{b}_1 + \dots. \tag{5.101}$$

Substituting these into the equations of motion, the lowest-order momentum equation is simply geostrophic balance,

$$\widehat{\boldsymbol{f}}_0 \times \widehat{\boldsymbol{u}}_0 = -\nabla \widehat{\phi}_0, \tag{5.102}$$

with a *constant* value of the Coriolis parameter. (Here and for the rest of this chapter we drop the subscript z from the ∇ operator.) From (5.102) it is evident that

$$\nabla \cdot \widehat{\boldsymbol{u}}_0 = 0. \tag{5.103}$$

Thus, the horizontal flow is, to leading order, non-divergent; this is a consequence of geostrophic balance, and is *not* a mass conservation equation. Using (5.103) in the mass conservation equation, (5.95), gives

$$\frac{\partial}{\partial \widehat{z}}(\widetilde{\rho} \widehat{w}_0) = 0, \tag{5.104}$$

which implies that if w_0 is zero somewhere (e.g., at a solid surface) then w_0 is zero everywhere (essentially the Taylor–Proudman effect). A physical way of saying this is that the scaling estimate $W = UH/L$ is an overestimate of the size of the vertical velocity, because even though $\partial w/\partial z \approx -\nabla\cdot\boldsymbol{u}$, the horizontal divergence of the geostrophic flow is small if f is nearly constant and $|\nabla\cdot\boldsymbol{u}| \ll U/L$. We might have anticipated this from the outset, and scaled w differently, perhaps using the geostrophic vorticity balance estimate, $w \sim \beta U H/f_0 = Ro\, UH/L$, as the scaling factor for w, but there is no a-priori guarantee that this would be correct.

At next order the momentum equation is

$$\frac{\mathrm{D}_0 \widehat{\boldsymbol{u}}_0}{\mathrm{D}\widehat{t}} + \widehat{\beta}\widehat{y}\mathbf{k} \times \widehat{\boldsymbol{u}}_0 + \widehat{\boldsymbol{f}} \times \widehat{\boldsymbol{u}}_1 = -\nabla \widehat{\phi}_1, \tag{5.105}$$

where $D_0/Dt = \partial/\partial\hat{t} + (\hat{\boldsymbol{u}}_0 \cdot \nabla)$, and the next order mass conservation equation is

$$\nabla_z \cdot (\tilde{\rho}\hat{\boldsymbol{u}}_1) + \frac{\partial}{\partial z}(\tilde{\rho}\widehat{w}_1) = 0. \tag{5.106}$$

From (5.96), the lowest-order thermodynamic equation is just

$$\left(\frac{L_d}{L}\right)^2 \widehat{N}^2 \widehat{w}_0 = 0, \tag{5.107}$$

provided that, as we have assumed, the scales of motion are not sufficiently large that $Ro(L/L_d)^2 = \mathcal{O}(1)$. (This is a key difference between quasi-geostrophy and planetary geostrophy.) At next order we obtain an evolution equation for the buoyancy, and this is

$$\frac{\mathrm{D}_0\hat{b}_0}{\mathrm{D}\hat{t}} + \widehat{w}_1\widehat{N}^2\left(\frac{L_d}{L}\right)^2 = 0. \tag{5.108}$$

The potential vorticity equation

To obtain a single evolution equation for lowest-order quantities we eliminate w_1 between the thermodynamic and momentum equations. Cross-differentiating the first-order momentum equation (5.105) gives the vorticity equation,

$$\frac{\partial\widehat{\zeta}_0}{\partial\hat{t}} + (\hat{\boldsymbol{u}}_0 \cdot \nabla)\widehat{\zeta}_0 + \hat{v}_0\widehat{\beta} = -\widehat{f}_0\nabla_z \cdot \hat{u}_1. \tag{5.109}$$

(In dimensional terms, the divergence on the right-hand side is small, but is multiplied by the large term f_0, and their product is of the same order as the terms on the left-hand side.) Using the mass conservation equation (5.106), (5.109) becomes

$$\frac{\mathrm{D}_0}{\mathrm{D}\hat{t}}(\zeta_0 + \widehat{f}) = \frac{\widehat{f}_0}{\tilde{\rho}}\frac{\partial}{\partial z}(w_1\tilde{\rho}). \tag{5.110}$$

Combining (5.110) and (5.108) gives

$$\frac{\mathrm{D}_0}{\mathrm{D}\hat{t}}(\zeta_0 + \widehat{f}) = -\frac{\widehat{f}_0}{\tilde{\rho}}\frac{\partial}{\partial\hat{z}}\left[\frac{\mathrm{D}_0}{\mathrm{D}\hat{t}}(F\tilde{\rho}\hat{b}_0)\right], \tag{5.111}$$

where $F \equiv (L/\widehat{N}L_d)^2$. The right-hand side of this equation is

$$\frac{\partial}{\partial\hat{z}}\left(\frac{\mathrm{D}_0\hat{b}_0}{\mathrm{D}\hat{t}}\right) = \frac{\mathrm{D}_0}{\mathrm{D}\hat{t}}\left(\frac{\partial\hat{b}_0}{\partial\hat{z}}\right) + \frac{\partial\hat{\boldsymbol{u}}_0}{\partial\hat{z}} \cdot \nabla\hat{b}_0. \tag{5.112}$$

The second term on the right-hand side vanishes identically using the thermal wind equation

$$\mathbf{k} \times \frac{\partial\hat{\boldsymbol{u}}_0}{\partial\hat{z}} = -\frac{1}{\widehat{f}_0}\nabla\hat{b}_0, \tag{5.113}$$

and so (5.111) becomes

$$\frac{\mathrm{D}_0}{\mathrm{D}\hat{t}}\left[\widehat{\zeta}_0 + \widehat{f} + \frac{\widehat{f}_0}{\tilde{\rho}}\frac{\partial}{\partial\hat{z}}\left(\tilde{\rho}F\hat{b}_0\right)\right] = 0, \tag{5.114}$$

or, after using the hydrostatic equation,

$$\frac{\mathrm{D}_0}{\mathrm{D}\hat{t}}\left[\widehat{\zeta}_0 + \widehat{f} + \frac{\widehat{f}_0}{\tilde{\rho}}\frac{\partial}{\partial\hat{z}}\left(\tilde{\rho}F\frac{\partial\widehat{\phi}_0}{\partial z}\right)\right] = 0. \tag{5.115}$$

Since the lowest-order horizontal velocity is divergence-free, we can define a streamfunction $\widehat{\psi}$ such that

$$\widehat{u}_0 = -\frac{\partial \widehat{\psi}}{\partial \widehat{y}}, \qquad \widehat{v}_0 = \frac{\partial \widehat{\psi}}{\partial \widehat{x}}, \tag{5.116}$$

where also, using (5.102), $\phi_0 = \widehat{f}_0 \widehat{\psi}$. The vorticity is then given by $\widehat{\zeta}_0 = \nabla^2 \widehat{\psi}$ and (5.115) becomes a single equation in a single unknown:

$$\frac{\mathrm{D}_0}{\mathrm{D}\widehat{t}} \left[\nabla^2 \widehat{\psi} + \widehat{\beta} \widehat{y} + \frac{\widehat{f}_0^2}{\widetilde{\rho}} \frac{\partial}{\partial \widehat{z}} \left(\widetilde{\rho} F \frac{\partial \widehat{\psi}}{\partial \widehat{z}} \right) \right] = 0, \tag{5.117}$$

where the material derivative is evaluated using $\widehat{\boldsymbol{u}}_0 = \mathbf{k} \times \nabla \widehat{\psi}$. This is the nondimensional form of the quasi-geostrophic potential vorticity equation, one of the most important equations in dynamical meteorology and oceanography. In deriving it we have reduced the Navier–Stokes equations, which are six coupled nonlinear partial differential equations in six unknowns (u, v, w, T, p, ρ) to a single (albeit nonlinear) first-order partial differential equation in a single unknown.[6]

Dimensional equations

The dimensional version of the quasi-geostrophic potential vorticity equation may be written as

$$\frac{\mathrm{D}q}{\mathrm{D}t} = 0, \qquad q = \nabla^2 \psi + f + \frac{f_0^2}{\widetilde{\rho}} \frac{\partial}{\partial z} \left(\frac{\widetilde{\rho}}{N^2} \frac{\partial \psi}{\partial z} \right), \tag{5.118a,b}$$

where only the variable part of f (e.g., βy) is relevant in the second term on the right-hand side of the expression for q. The quantity q is known as the *quasi-geostrophic potential vorticity.* It is analogous to the exact (Ertel) potential vorticity (see Section 5.5 for more about this), and it is conserved when advected by the *horizontal* geostrophic flow. All the other dynamical variables may be obtained from potential vorticity as follows:

(i) Streamfunction, using (5.118b).

(ii) Velocity: $\boldsymbol{u} = \mathbf{k} \times \nabla \psi$ $[\equiv \nabla^{\perp} \psi = -\nabla \times (\mathbf{k}\psi)]$.

(iii) Relative vorticity: $\zeta = \nabla^2 \psi$.

(iv) Perturbation pressure: $\phi = f_0 \psi$.

(v) Perturbation buoyancy: $b' = f_0 \partial \psi / \partial z$.

The length scale, $L_d = NH/f_0$, emerges naturally from the quasi-geostrophic dynamics. It is the scale at which buoyancy and relative vorticity effects contribute equally to the potential vorticity, and is called the *deformation radius;* it is analogous to the quantity $\sqrt{gH}/f_0$ arising in shallow water theory. In the upper ocean, with $N \approx 10^{-2}\,\mathrm{s}^{-1}$, $H \approx 10^3\,\mathrm{m}$ and $f_0 \approx 10^{-4}\,\mathrm{s}^{-1}$, then $L_d \approx 100\,\mathrm{km}$. At high latitudes the ocean is much less stratified and f is somewhat larger, and the deformation radius may be as little as 30 km (see Fig. 12.13 on page 469, where the deformation radius is defined slightly differently). In the atmosphere, with $N \approx 10^{-2}\,\mathrm{s}^{-1}$, $H \approx 10^4\,\mathrm{m}$, then $L_d \approx 1000\,\mathrm{km}$. It is this order of magnitude difference in the deformation scales that accounts for a great deal of the quantitative difference in the dynamics of the ocean and the atmosphere. If we take the limit $L_d \to \infty$ then the stratified quasi-geostrophic equations reduce to

$$\frac{\mathrm{D}q}{\mathrm{D}t} = 0, \qquad q = \nabla^2 \psi + f. \tag{5.119}$$

This is the two-dimensional vorticity equation, identical to (4.67). The high stratification of this limit has suppressed all vertical motion, and variations in the flow become confined to the horizontal plane. Finally, we note that it is typical in quasi-geostrophic applications to omit the prime on the buoyancy perturbations, and write $b = f_0 \partial\psi/\partial z$; however, we will keep the prime in this chapter.

5.4.3 Buoyancy Advection at the Surface

The solution of the elliptic equation in (5.118) requires vertical boundary conditions on ψ at the ground and at the top of the atmosphere, and these are given by use of the thermodynamic equation. For a flat, slippery, rigid surface the vertical velocity is zero so that the thermodynamic equation may be written as

$$\frac{\mathrm{D}b'}{\mathrm{D}t} = 0, \qquad b' = f_0 \frac{\partial \psi}{\partial z}. \tag{5.120}$$

We apply this at the ground and at the tropopause, treating the latter as a lid on the lower atmosphere. In the presence of friction and topography the vertical velocity is not zero, but is given by

$$w = r\nabla^2\psi + \boldsymbol{u}\cdot\nabla\eta_b, \tag{5.121}$$

where the first term represents Ekman friction (with the constant r proportional to the thickness of the Ekman layer) and the second term represents topographic forcing. The boundary condition becomes

$$\frac{\partial}{\partial t}\left(f_0\frac{\partial\psi}{\partial z}\right) + \boldsymbol{u}\cdot\nabla\left(f_0\frac{\partial\psi}{\partial z} + N^2\eta_b\right) + N^2 r\nabla^2\psi = 0, \tag{5.122}$$

where all the fields are evaluated at $z = 0$ or at $z = H$, the height of the lid. Thus, the quasi-geostrophic system is characterized by the horizontal advection of potential vorticity in the interior and the advection of buoyancy at the boundary. Instead of a lid at the top, then in a compressible fluid such as the atmosphere we may suppose that all disturbances tend to zero as $z \to \infty$.

✦ *A potential vorticity sheet at the boundary*

Rather than regarding buoyancy advection as providing the boundary condition, it is sometimes useful to think of there being a very thin sheet of potential vorticity just above the ground and another just below the lid, specifically with a vertical distribution proportional to $\delta(z-\epsilon)$ or $\delta(z - H + \epsilon)$, where ϵ is small. The boundary condition (5.120) or (5.122) can be replaced by this, along with the condition that there are no variations of buoyancy at the boundary and $\partial\psi/\partial z = 0$ at $z = 0$ and $z = H$.[7]

To see this, we first note that the differential of a step function is a delta function. Thus, a discontinuity in $\partial\psi/\partial z$ at a level $z = z_1$ is equivalent to a delta function in potential vorticity there:

$$q(z_1) = \left[\frac{f_0^2}{N^2}\frac{\partial\psi}{\partial z}\right]_{z_1-}^{z_1+} \delta(z - z_1). \tag{5.123}$$

Now, suppose that the lower boundary condition, given by (5.120), has some arbitrary distribution of buoyancy on it. We can replace this condition by the simpler condition $\partial\psi/\partial z = 0$ at $z = 0$, provided we also add to our definition of potential vorticity a term given by (5.123) with $z_1 = \epsilon$. This term is then advected by the horizontal flow, as are the other contributions. A buoyancy source at the boundary must similarly be treated as a sheet of potential vorticity source in the interior. Any flow with buoyancy variations over a horizontal boundary is thus equivalent to a flow with uniform buoyancy at the boundary, but with a spike in potential vorticity adjacent to the boundary. This approach brings notational and conceptual advantages, in that now everything is expressed in terms of potential vorticity and its advection. However, in practice there may be less to be gained, because the boundary terms must still be included in any particular calculation that is to be performed.

5.4.4 Vertical Velocity and the Omega Equation

The vertical velocity is not needed in order to evolve the quasi-geostrophic equations. However, it is not zero and a relatively simple recipe can be found that is of practical use in diagnosing the vertical velocity in weather charts. When deriving the potential vorticity equation, we eliminated vertical velocity from the vorticity equation and thermodynamic equations to give a single evolution equation. Here our approach is complementary: we begin with the same two equations, but eliminate the time derivatives. We will proceed using dimensional variables and write the vorticity and thermodynamic equations as

$$\frac{\partial \zeta}{\partial t} + J(\psi, \zeta) = \frac{f_0}{\widetilde{\rho}} \frac{\partial (\widetilde{\rho} w)}{\partial z} + \mathcal{Z}, \qquad \frac{\partial b}{\partial z} + J(\psi, b) + wN^2 = Q, \tag{5.124a,b}$$

where $b = f_0 \partial\psi/\partial z$ and $\zeta = \nabla^2\psi$, $\mathcal{Z}$ and Q are friction and heating terms that we can leave unspecified, and $\widetilde{\rho}$ is a reference density profile. If we take $f_0\partial/\partial z$ of the first equation and ∇^2 of the second we can eliminate time derivatives to find

$$\frac{\partial}{\partial z}\left[\frac{f_0^2}{\widetilde{\rho}} \frac{\partial (\widetilde{\rho} w)}{\partial z}\right] + N^2\nabla^2 w = f_0 \frac{\partial}{\partial z}[J(\psi, \zeta)] - \nabla^2[J(\psi, b)] - f_0 \frac{\partial \mathcal{Z}}{\partial z} + \nabla^2 Q. \tag{5.125}$$

The equation is called the *omega equation* because omega (ω) is the vertical velocity in pressure coordinates, which was where the equation first appeared. It is an elliptic equation for w, and is in fact a Poisson equation if $\widetilde{\rho}$ is a constant. It may be easily solved by numerical methods, given the state of the flow at any given time. However, there is rarely a need to solve it exactly, for there is no need to calculate w to step forward the equations. Rather, the equation finds use as an interpretive guide for meteorologists: in the thermodynamic equation both heating itself and warm advection will tend to produce vertical motion, as will the vertical differential of vorticity advection.

5.4.5 Quasi-Geostrophy in Pressure Coordinates

The derivation of the quasi-geostrophic system in pressure coordinates is very similar to that in height coordinates, with the main difference coming at the boundaries, and we give only the results. The starting point is the primitive equations in pressure coordinates, (P.1) on page 81. In pressure coordinates, the quasi-geostrophic potential vorticity is found to be

$$q = f + \nabla^2\psi + \frac{\partial}{\partial p}\left(\frac{f_0^2}{S^2}\frac{\partial \psi}{\partial p}\right), \tag{5.126}$$

where $\psi = \Phi / f_0$ is the streamfunction and Φ the geopotential, and

$$S^2 \equiv -\frac{R}{p}\left(\frac{p}{p_R}\right)^{\kappa} \frac{\mathrm{d}\widetilde{\theta}}{\mathrm{d}p} = -\frac{1}{\rho\theta}\frac{\mathrm{d}\widetilde{\theta}}{\mathrm{d}p}, \tag{5.127}$$

where $\widetilde{\theta}$ is a reference profile and a function of pressure only. In log-pressure coordinates, with $Z = -H \ln p$, the potential vorticity may be written as

$$q = f + \nabla^2\psi + \frac{1}{\rho_*}\frac{\partial}{\partial Z}\left(\frac{\rho_* f_0^2}{N_Z^2}\frac{\partial \psi}{\partial Z}\right), \tag{5.128}$$

where

$$N_Z^2 = S^2\left(\frac{p}{H}\right)^2 = -\left(\frac{R}{H}\right)\left(\frac{p}{p_R}\right)^{\kappa}\frac{\mathrm{d}\widetilde{\theta}}{\mathrm{d}Z} \tag{5.129}$$

Informal Derivation of Stratified QG Equations

We will use the Boussinesq equations, but very similar derivations could be given using the anelastic equations or pressure coordinates. The first ingredient is the vertical component of the vorticity equation, (4.66); in the Boussinesq version there is no baroclinic term and we have

$$\frac{D_3}{Dt}(\zeta + f) = -(\zeta + f)\left(\frac{\partial u}{\partial x} + \frac{\partial v}{\partial y}\right) + \left(\frac{\partial u}{\partial z}\frac{\partial w}{\partial y} - \frac{\partial v}{\partial z}\frac{\partial w}{\partial x}\right). \tag{QG.1}$$

We now apply the assumptions on page 180. The advection and the vorticity on the left-hand side are geostrophic, but we keep the horizontal divergence (which is small) on the right-hand side where it is multiplied by the big term f. Furthermore, because f is nearly constant we replace it with f_0 except where it is differentiated. The second term (tilting) on the right-hand side is smaller than the advection terms on the left-hand side by the ratio $[UW/(HL)]/[U^2/L^2] = [W/H]/[U/L] \ll 1$, because w is small ($\partial w/\partial z$ equals the divergence of the ageostrophic velocity). We therefore neglect it, and given all this (QG.1) becomes

$$\frac{D_g}{Dt}(\zeta_g + f) = -f_0\left(\frac{\partial u}{\partial x} + \frac{\partial v}{\partial y}\right) = f_0\frac{\partial w}{\partial z}, \tag{QG.2}$$

where the second equality uses mass continuity and $D_g/Dt = \partial/\partial t + \boldsymbol{u}_g \cdot \nabla$.

The second ingredient is the three-dimensional thermodynamic equation,

$$\frac{D_3 b}{Dt} = 0. \tag{QG.3}$$

The stratification is assumed to be nearly constant, so we write $b = \tilde{b}(z) + b'(x, y, z, t)$, where $\tilde{b}$ is the basic state buoyancy. Furthermore, because w is small it only advects the basic state, and with $N^2 = \partial\tilde{b}/\partial z$ (QG.3) becomes

$$\frac{D_g b'}{Dt} + wN^2 = 0. \tag{QG.4}$$

Hydrostatic and geostrophic wind balance enable us to write the geostrophic velocity, vorticity, and buoyancy in terms of streamfunction ψ [$= p/(f_0\rho_0)$]:

$$\boldsymbol{u}_g = \mathbf{k} \times \nabla\psi, \qquad \zeta_g = \nabla^2\psi, \qquad b' = f_0\partial\psi/\partial z. \tag{QG.5}$$

The quasi-geostrophic potential vorticity equation is obtained by eliminating w between (QG.2) and (QG.4), and this gives

$$\frac{D_g q}{Dt} = 0, \qquad q = \zeta_g + f + \frac{\partial}{\partial z}\left(\frac{f_0 b'}{N^2}\right). \tag{QG.6}$$

This equation is the Boussinesq version of (5.118), and using (QG.5) it may be expressed entirely in terms of the streamfunction, with $D_g \cdot /Dt = \partial/\partial t + J(\psi, \cdot)$. The vertical boundary conditions, at $z = 0$ and $z = H$ say, are given by (QG.4) with $w = 0$, with straightforward generalizations if topography or friction are present.

is the buoyancy frequency and $\rho_* = \exp(-z/H)$. Temperature and potential temperature are related to the streamfunction by

$$T = -\frac{f_0 p}{R}\frac{\partial \psi}{\partial p} = \frac{H f_0}{R}\frac{\partial \psi}{\partial Z}, \tag{5.130a}$$

$$\theta = -\left(\frac{p_R}{p}\right)^{\kappa}\left(\frac{f_0 p}{R}\right)\frac{\partial \psi}{\partial p} = \left(\frac{p_R}{p}\right)^{\kappa}\left(\frac{H f_0}{R}\right)\frac{\partial \psi}{\partial Z}. \tag{5.130b}$$

In pressure or log-pressure coordinates, potential vorticity is advected along isobaric surfaces, analogously to the horizontal advection in height coordinates.

The surface boundary condition again is derived from the thermodynamic equation. In log-pressure coordinates this is

$$\frac{\mathrm{D}}{\mathrm{D}t}\left(\frac{\partial \psi}{\partial Z}\right) + \frac{N_Z^2}{f_0} W = 0, \tag{5.131}$$

where $W = \mathrm{D}Z/\mathrm{D}t$. This is not the real vertical velocity, w, but it is related to it by

$$w = \frac{f_0}{g}\frac{\partial \psi}{\partial t} + \frac{RT}{gH} W. \tag{5.132}$$

Thus, choosing $H = RT(0)/g$, we have, at $Z = 0$,

$$\frac{\partial}{\partial t}\left(\frac{\partial \psi}{\partial Z} - \frac{N_Z^2}{g}\psi\right) + \boldsymbol{u}\cdot\nabla\frac{\partial \psi}{\partial Z} = -\frac{N^2}{f_0} w, \tag{5.133}$$

where

$$w = \boldsymbol{u}\cdot\nabla\eta_b + r\nabla^2\psi. \tag{5.134}$$

This differs from the expression in height coordinates only by the second term in the local time derivative. In applications where accuracy is not the main issue the simpler boundary condition $\mathrm{D}(\partial_Z\psi)/\mathrm{D}t = 0$ is sometimes used. Finally, we remark that in pressure coordinates, the equivalent to vertical velocity, $\partial p/\partial t$, is denoted ω (omega), but it need not be evaluated to solve the equations.

5.4.6 The Two-level Quasi-Geostrophic System

The quasi-geostrophic system has, in general, continuous variation in the vertical direction (and horizontal, of course). By finite-differencing the continuous equations we can obtain a *multi-level* model, and a crude but important special case of this is the *two-level* model, also known as the Phillips model.[8] To obtain the equations of motion one way to proceed is to take a crude finite difference of the continuous relation between potential vorticity and streamfunction given in (5.118b). In the Boussinesq case (or in pressure coordinates, with a slight reinterpretation of the meaning of the symbols) the continuous expression for potential vorticity is

$$q = \zeta + f + \frac{\partial}{\partial z}\left(\frac{f_0 b'}{N^2}\right), \tag{5.135}$$

where $b' = f_0\partial\psi/\partial z$. In the case with a flat bottom and rigid lid at the top (and incorporating topography is an easy extension) the boundary condition of $w = 0$ is satisfied by $\mathrm{D}\partial_z\psi/\mathrm{D}t = 0$ at the top and bottom. An obvious finite-differencing of (5.135) in the vertical direction (see Fig. 5.3) then gives

$$q_1 = \zeta_1 + f + \frac{2f_0^2}{N^2 H_1 H}(\psi_2 - \psi_1), \qquad q_2 = \zeta_2 + f + \frac{2f_0^2}{N^2 H_2 H}(\psi_1 - \psi_2). \tag{5.136}$$

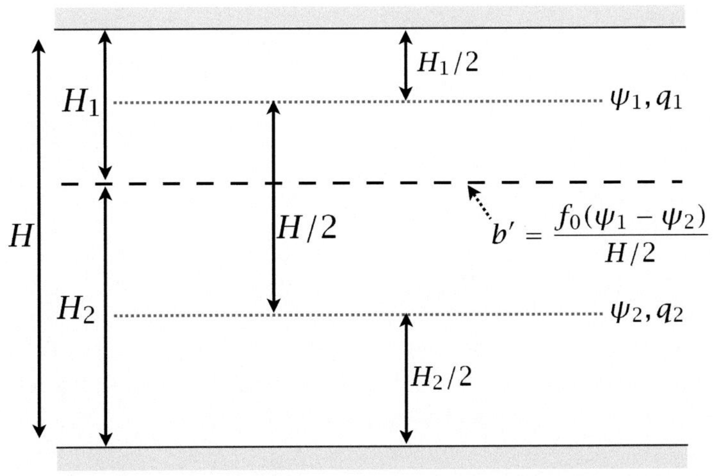

Fig. 5.3 A two-level quasi-geostrophic system with a flat bottom and rigid lid at which $w = 0$.

In atmospheric problems it is common to choose $H_1 = H_2$, whereas in oceanic problems we might choose to have a thinner upper layer, representing the flow above the main thermocline. Note that the boundary conditions of $w = 0$ at the top and bottom are already taken care of in (5.136): *they are incorporated into the definition of the potential vorticity* — a finite-difference analogue of the delta-function construction of Section 5.4.3. At each level the potential vorticity is advected by the streamfunction so that the evolution equation for each level is:

$$\frac{\mathrm{D}q_i}{\mathrm{D}t} = \frac{\partial q_i}{\partial t} + \boldsymbol{u}_i \cdot q_i = \frac{\partial q_i}{\partial t} + J(\psi_i, q_i) = 0, \qquad i = 1, 2. \tag{5.137}$$

Models with more than two levels can be constructed by extending the finite-differencing procedure in a natural way.

Connection to the layered system

The two-level expressions, (5.136), have an obvious similarity to the *two-layer* expressions, (5.85). Noting that $N^2 = \partial\hat{b}/\partial z$ and that $b = -g\delta\rho/\rho_0$ it is natural to let

$$N^2 = -\frac{g}{\rho_0}\frac{\rho_1 - \rho_2}{H/2} = \frac{g'}{H/2}. \tag{5.138}$$

With this identification we find that (5.136) becomes

$$q_1 = \zeta_1 + f + \frac{f_0^2}{g' H_1}(\psi_2 - \psi_1), \qquad q_2 = \zeta_2 + f + \frac{f_0^2}{g' H_2}(\psi_1 - \psi_2). \tag{5.139}$$

These expressions are identical to (5.85) in the flat-bottomed, rigid lid case. Similarly, a multi-layered system with n layers is equivalent to a finite-difference representation with n levels. It should be said, though, that in the pantheon of quasi-geostrophic models the two-level and two-layer models hold distinguished places.

5.5 ✦ QUASI-GEOSTROPHY AND ERTEL POTENTIAL VORTICITY

When using the shallow water equations, quasi-geostrophic theory could be naturally developed beginning with the expression for potential vorticity. Is such an approach possible for the stratified primitive equations? The answer is yes, but with complications.

5.5.1 ✦ Using Height Coordinates

Noting the general expression, (4.117), for potential vorticity in a hydrostatic fluid, the potential vorticity in the Boussinesq hydrostatic equations is given by

$$Q = \left[(v_x - u_y)b_z - v_z b_x + u_z b_y + f b_z\right], \tag{5.140}$$

where the x, y, z subscripts denote derivatives. Without approximation, we write the stratification as $b = \tilde{b}(z) + b'(x, y, z, t)$, and (5.140) becomes

$$Q = [f_0 N^2] + [(\beta y + \zeta)N^2 + f_0 b'_z] + [(\beta y + \zeta)b'_z - (v_z b'_x - u_z b'_y)], \tag{5.141}$$

where, under quasi-geostrophic scaling, the terms in square brackets are in decreasing order of size. Neglecting the third term, and taking the velocity and buoyancy fields to be in geostrophic and thermal wind balance, we can write the potential vorticity as $Q \approx \widetilde{Q} + Q'$, where $\widetilde{Q} = f_0 N^2$ and

$$Q' = (\beta y + \zeta)N^2 + f_0 b'_z = (\beta y + \nabla^2\psi)N^2 + f_0^2 \frac{\partial^2 \psi}{\partial z^2}. \tag{5.142}$$

The potential vorticity evolution equation is then

$$\frac{DQ'}{Dt} + w\frac{\partial \widetilde{Q}}{\partial z} = 0. \tag{5.143}$$

The vertical advection is important only in advecting the basic state potential vorticity $\widetilde{Q}$ and so, neglecting $w\partial Q'/\partial z$ and dividing by N^2, (5.143) becomes

$$\frac{\partial q_*}{\partial t} + \boldsymbol{u}_g \cdot \nabla q_* + \frac{w}{N^2}\frac{\partial \widetilde{Q}}{\partial z} = 0, \tag{5.144}$$

where $\widehat{q}$ is

$$q_* = (\beta y + \zeta) + \frac{f_0}{N^2} b'_z. \tag{5.145}$$

This is the approximation to the (perturbation) Ertel potential vorticity in the quasi-geostrophic limit. However, it is not the same as the expression for the quasi-geostrophic potential vorticity, (5.118b) and, furthermore, (5.144) involves a vertical advection. (Thus, we might refer to the expression in (5.118) as the 'quasi-geostrophic pseudopotential vorticity', but the prefix 'quasi-geostrophic' alone normally suffices.) We can derive (5.118) by eliminating w between (5.144) and the quasi-geostrophic thermodynamic equation $\partial b'/\partial t + \boldsymbol{u}_g \cdot \nabla b' + w\partial\tilde{b}/\partial z = 0$.

5.5.2 Using Isentropic Coordinates

An illuminating and somewhat simpler path from Ertel potential vorticity to the quasi-geostrophic equations goes by way of isentropic coordinates.[9] We begin with the isentropic expression for the Ertel potential vorticity of an ideal gas,

$$Q = \frac{f + \zeta}{\sigma}, \tag{5.146}$$

where $\sigma = -\partial p/\partial\theta$ is the thickness density (which we will just call the thickness), and in adiabatic flow the potential vorticity is advected along isopycnals. We now employ quasi-geostrophic scaling to derive an approximate equation set from this. First, assume that variations in thickness are small compared with the reference state, so that

$$\sigma = \tilde{\sigma}(\theta) + \sigma', \qquad |\sigma'| \ll |\sigma|, \tag{5.147}$$

and similarly for pressure and density. Assuming also that the variations in the Coriolis parameter are small, then on the β–plane (5.146) becomes

$$Q \approx \left[\frac{f_0}{\widetilde{\sigma}}\right] + \left[\frac{1}{\widetilde{\sigma}}(\zeta + \beta y) - \frac{f_0}{\widetilde{\sigma}}\frac{\sigma'}{\widetilde{\sigma}}\right]. \tag{5.148}$$

We now use geostrophic and hydrostatic balance to express the terms on the right-hand side in terms of a single variable, noting that the first term does not vary along isentropic surfaces. Hydrostatic balance is

$$\frac{\partial M}{\partial \theta} = \Pi, \tag{5.149}$$

where $M = c_p T + gz$ and $\Pi = c_p(p/p_R)^\kappa$. Writing $M = \widetilde{M}(\theta) + M'$ and $\Pi = \widetilde{\Pi}(\theta) + \Pi'$, where $\widetilde{M}$ and $\widetilde{\Pi}$ are hydrostatically balanced reference profiles, we obtain

$$\frac{\partial M'}{\partial \theta} = \Pi' \approx \frac{\mathrm{d}\widetilde{\Pi}}{\mathrm{d}p}p' = \frac{1}{\theta\widetilde{\rho}}p', \tag{5.150}$$

where the last equality follows using the equation of state for an ideal gas and $\widetilde{\rho}$ is a reference profile. The perturbation thickness field may then be written as

$$\sigma' = -\frac{\partial}{\partial \theta}\left(\widetilde{\rho}\theta\frac{\partial M'}{\partial \theta}\right). \tag{5.151}$$

Geostrophic balance is $\boldsymbol{f}_0 \times \boldsymbol{u} = -\nabla_\theta M'$ where the velocity, and the horizontal derivatives, are along isentropic surfaces. This enables us to define a flow streamfunction by

$$\psi \equiv \frac{M'}{f_0}, \tag{5.152}$$

and we can then write all the variables in terms of ψ:

$$u = -\left(\frac{\partial \psi}{\partial y}\right)_\theta, \qquad v = \left(\frac{\partial \psi}{\partial x}\right)_\theta, \qquad \zeta = \nabla_\theta^2 \psi, \qquad \sigma' = -f_0\frac{\partial}{\partial \theta}\left(\widetilde{\rho}\theta\frac{\partial \psi'}{\partial \theta}\right). \tag{5.153}$$

Using (5.148), (5.152) and (5.153), the quasi-geostrophic system in isentropic coordinates may be written

$$\frac{\mathrm{D}q}{\mathrm{D}t} = 0, \qquad q = f + \nabla_\theta^2 \psi + \frac{f_0^2}{\widetilde{\sigma}}\frac{\partial}{\partial \theta}\left(\widetilde{\rho}\theta\frac{\partial \psi}{\partial \theta}\right), \tag{5.154a,b}$$

where the advection of potential vorticity is by the geostrophically balanced flow, along isentropes. The variable q is an approximation to the second term in square brackets in (5.148), multiplied by $\widetilde{\sigma}$.

Projection back to physical-space coordinates

We can recover the height or pressure coordinate quasi-geostrophic systems by projecting (5.154) on to the appropriate coordinate. This is straightforward because, by assumption, the isentropes in a quasi-geostrophic system are nearly flat. Recall that, from (2.142), a transformation between vertical coordinates may be effected by

$$\left.\frac{\partial}{\partial x}\right|_\theta = \left.\frac{\partial}{\partial x}\right|_p + \left.\frac{\partial p}{\partial x}\right|_\theta \frac{\partial}{\partial p}, \tag{5.155}$$

but the second term is $\mathcal{O}(Ro)$ smaller than the first one because, under quasi-geostrophic scaling, isentropic slopes are small. Thus $\nabla_\theta^2\psi$ in (5.154b) may be replaced by $\nabla_p^2\psi$ or $\nabla_z^2\psi$. The vortex stretching term in (5.154) becomes, in pressure coordinates,

$$\frac{f_0^2}{\tilde{\sigma}}\frac{\partial}{\partial\theta}\left(\tilde{p}\theta\frac{\partial\psi}{\partial\theta}\right) \approx \frac{f_0^2}{\tilde{\sigma}}\frac{\mathrm{d}\tilde{p}}{\mathrm{d}\theta}\frac{\partial}{\partial p}\left(\tilde{p}\theta\frac{\mathrm{d}\tilde{p}}{\mathrm{d}\theta}\frac{\partial\psi}{\partial p}\right) = \frac{\partial}{\partial p}\left(\frac{f_0^2}{S^2}\frac{\partial\psi}{\partial p}\right), \tag{5.156}$$

where S^2 is given by (5.127). The expression for the quasi-geostrophic potential vorticity in isentropic coordinates is thus approximately equal to the quasi-geostrophic potential vorticity in pressure coordinates. This near-equality holds because the isentropic expression, (5.154b), does not contain a component proportional to the mean stratification: the second square-bracketed term on the right-hand side of (5.148) is the only dynamically relevant one, and its evolution along isentropes is mirrored by the evolution along isobaric surfaces of quasi-geostrophic potential vorticity in pressure coordinates.

5.6 ♦ ENERGETICS OF QUASI-GEOSTROPHY

If the quasi-geostrophic set of equations is to represent a real fluid system in a physically meaningful way then it should have a consistent set of energetics. In particular, the total energy should be conserved, and there should be analogues of kinetic and potential energy and conversion between the two. We now show that such energetic properties do hold, using the Boussinesq set as an example.

Let us write the governing equations as a potential vorticity equation in the interior,

$$\frac{\mathrm{D}}{\mathrm{D}t}\left[\nabla^2\psi + \frac{\partial}{\partial z}\left(\frac{f_0^2}{N^2}\frac{\partial\psi}{\partial z}\right)\right] + \beta\frac{\partial\psi}{\partial x} = 0, \qquad 0 < z < 1, \tag{5.157}$$

and buoyancy advection at the boundary,

$$\frac{\mathrm{D}}{\mathrm{D}t}\left(\frac{\partial\psi}{\partial z}\right) = 0, \qquad z = 0, 1. \tag{5.158}$$

For lateral boundary conditions we may assume that ψ = constant, or impose periodic conditions. If we multiply (5.157) by $-\psi$ and integrate over the domain, using the boundary conditions, we easily find

$$\frac{\mathrm{d}\widehat{E}}{\mathrm{d}t} = 0, \qquad \widehat{E} = \frac{1}{2}\int_V \left[(\nabla\psi)^2 + \frac{f_0^2}{N^2}\left(\frac{\partial\psi}{\partial z}\right)^2\right]\mathrm{d}V. \tag{5.159a,b}$$

The term involving β makes no direct contribution to the energy budget. Equation (5.159) is the fundamental energy equation for quasi-geostrophic motion, and it states that in the absence of viscous or diabatic terms the total energy is conserved. The two terms in (5.159b) can be identified as the kinetic energy (KE) and available potential energy (APE) of the flow, where

$$\mathrm{KE} = \frac{1}{2}\int_V (\nabla\psi)^2\,\mathrm{d}V, \qquad \mathrm{APE} = \frac{1}{2}\int_V \frac{f_0^2}{N^2}\left(\frac{\partial\psi}{\partial z}\right)^2\mathrm{d}V. \tag{5.160a,b}$$

The available potential energy may also be written as

$$\mathrm{APE} = \frac{1}{2}\int_V \frac{H^2}{L_d^2}\left(\frac{\partial\psi}{\partial z}\right)^2\mathrm{d}V, \tag{5.161}$$

where L_d is the deformation radius NH/f_0 and we may choose H such that $z \sim H$. At some scale L the ratio of the kinetic energy to the potential energy is thus, roughly,

$$\frac{\mathrm{KE}}{\mathrm{APE}} \sim \frac{L_d^2}{L^2}. \tag{5.162}$$

For scales much larger than L_d the potential energy dominates the kinetic energy, and contrariwise.

5.6.1 Conversion Between APE and KE

Let us return to the vorticity and thermodynamic equations,

$$\frac{\mathrm{D}\zeta}{\mathrm{D}t} = f\frac{\partial w}{\partial z}, \qquad \frac{\mathrm{D}b'}{\mathrm{D}t} + N^2 w = 0 \tag{5.163a,b}$$

where $\zeta = \nabla^2\psi$, and $b' = f_0\partial\psi/\partial z$. From (5.163a) we form a kinetic energy equation, namely

$$\frac{1}{2}\frac{\mathrm{d}}{\mathrm{d}t}\int_V (\nabla\psi)^2\,\mathrm{d}V = -\int_V f_0\frac{\partial w}{\partial z}\psi\,\mathrm{d}V = \int_V f_0 w\frac{\partial\psi}{\partial z}\,\mathrm{d}V. \tag{5.164}$$

From (5.163b) we form a potential energy equation, namely

$$\frac{\mathrm{d}}{\mathrm{d}t}\frac{1}{2}\int_V \frac{f_0^2}{N^2}\left(\frac{\partial\psi}{\partial z}\right)^2\,\mathrm{d}V = -\int_V f_0 w\frac{\partial\psi}{\partial z}\,\mathrm{d}V. \tag{5.165}$$

Thus, the *conversion* from APE to KE is represented by

$$\frac{\mathrm{d}}{\mathrm{d}t}\mathrm{KE} = -\frac{\mathrm{d}}{\mathrm{d}t}\mathrm{APE} = \int_v f_0 w\frac{\partial\psi}{\partial z}\,\mathrm{d}V. \tag{5.166}$$

Because the buoyancy is proportional to $\partial\psi/\partial z$, when warm fluid rises there is a correlation between w and $\partial\psi/\partial z$ and APE is converted to KE. Whether such a phenomenon occurs depends of course on the dynamics of the flow; however, such a conversion *is*, in fact, a common feature of geophysical flows, and in particular of baroclinic instability, as we shall see in Chapter 9.

5.6.2 Energetics of Two-layer Flows

Two-layer or two-level flows are an important special case. For layers of equal thickness let us write the evolution equations as

$$\frac{\mathrm{D}}{\mathrm{D}t}\left[\nabla^2\psi_1 - \frac{1}{2}k_d^2(\psi_1 - \psi_2)\right] + \beta\frac{\partial\psi_1}{\partial x} = 0, \tag{5.167a}$$

$$\frac{\mathrm{D}}{\mathrm{D}t}\left[\nabla^2\psi_2 + \frac{1}{2}k_d^2(\psi_1 - \psi_2)\right] + \beta\frac{\partial\psi_2}{\partial x} = 0, \tag{5.167b}$$

where $k_d^2/2 = (2f_0/NH)^2$. On multiplying these two equations by $-\psi_1$ and $-\psi_2$, respectively, and integrating over the horizontal domain, the advective term in the material derivatives and the beta term all vanish, and we obtain

$$\int_A \left[\frac{\mathrm{d}}{\mathrm{d}t}\frac{1}{2}(\nabla\psi_1)^2 + \frac{1}{2}k_d^2\psi_1\frac{\mathrm{d}}{\mathrm{d}t}(\psi_1 - \psi_2)\right]\,\mathrm{d}A = 0, \tag{5.168a}$$

$$\int_A \left[\frac{\mathrm{d}}{\mathrm{d}t}\frac{1}{2}(\nabla\psi_2)^2 - \frac{1}{2}k_d^2\psi_2\frac{\mathrm{d}}{\mathrm{d}t}(\psi_1 - \psi_2)\right]\,\mathrm{d}A = 0. \tag{5.168b}$$

Adding these gives

$$\frac{\mathrm{d}}{\mathrm{d}t}\int_A \left[\frac{1}{2}(\nabla\psi_1)^2 + \frac{1}{2}(\nabla\psi_2)^2 + \frac{k_d^2}{4}(\psi_1 - \psi_2)^2\right]\,\mathrm{d}A = 0. \tag{5.169}$$

This is the energy conservation statement for the two layer model. The first two terms represent the kinetic energy and the last term represents the available potential energy.

Energy in the baroclinic and barotropic modes

A useful partitioning of the energy is between the energy in the barotropic and baroclinic modes. The barotropic streamfunction, $\overline{\psi}$, is the vertically averaged streamfunction and the baroclinic mode is the difference between the streamfunctions in the two layers. That is, for equal layer thicknesses,

$$\overline{\psi} \equiv \frac{1}{2}(\psi_1 + \psi_2), \qquad \tau \equiv \frac{1}{2}(\psi_1 - \psi_2). \tag{5.170}$$

Substituting (5.170) into (5.169) reveals that

$$\frac{\mathrm{d}}{\mathrm{d}t}\int_A \left[(\nabla\overline{\psi})^2 + (\nabla\tau)^2 + k_d^2\tau^2\right] \mathrm{d}A = 0. \tag{5.171}$$

The energy density in the barotropic mode is thus just $(\nabla\overline{\psi})^2$, and that in the baroclinic mode is $(\nabla\tau)^2 + k_d^2\tau^2$. This partitioning will prove particularly useful when we consider baroclinic turbulence in Chapter 12.

5.6.3 Enstrophy Conservation

Potential vorticity is advected only by the horizontal flow, and thus it is materially conserved on the horizontal surface at every height and

$$\frac{\mathrm{D}q}{\mathrm{D}t} = \frac{\partial q}{\partial t} + \boldsymbol{u}\cdot\nabla q = 0. \tag{5.172}$$

Furthermore, the advecting flow is divergence-free so that $\boldsymbol{u}\cdot\nabla q = \nabla\cdot(\boldsymbol{u}q)$. Thus, on multiplying (5.172) by q and integrating over a horizontal domain A we obtain

$$\frac{\mathrm{d}\widehat{Z}}{\mathrm{d}t} = 0, \qquad \widehat{Z} = \frac{1}{2}\int_A q^2\,\mathrm{d}A. \tag{5.173}$$

The result holds in an enclosed domain, with no-normal flow boundary conditions, or in a channel with periodic boundary conditions in x and no-normal flow conditions in y. The quantity $\widehat{Z}$ is known as the *enstrophy*, and it is conserved at each height as well as, naturally, over the entire volume. (In a doubly-periodic domain, only the relative enstrophy, $\int \zeta^2\,\mathrm{d}A$, is conserved.)

The enstrophy is just one of an infinity of invariants in quasi-geostrophic flow. Because the potential vorticity of a fluid element is conserved, *any* function of the potential vorticity must be a material invariant and we can immediately write

$$\frac{\mathrm{D}}{\mathrm{D}t}F(q) = 0. \tag{5.174}$$

To verify that this is true, simply note that (5.174) implies that $(\mathrm{d}F/\mathrm{d}q)\mathrm{D}q/\mathrm{D}t = 0$, which is true by virtue of (5.172). (However, by virtue of the material advection, the function $F(q)$ need not be differentiable in order for (5.174) to hold.) Each of the material invariants corresponding to different choices of $F(q)$ has a corresponding integral invariant; that is,

$$\frac{\mathrm{d}}{\mathrm{d}t}\int_A F(q)\,\mathrm{d}A = 0, \tag{5.175}$$

with the boundary conditions as before. The enstrophy invariant corresponds to choosing $F(q) = q^2$; it plays a particularly important role because, like energy, it is a quadratic invariant, and its presence profoundly alters the behaviour of two-dimensional and quasi-geostrophic flow compared to three-dimensional flow (see Section 11.3).

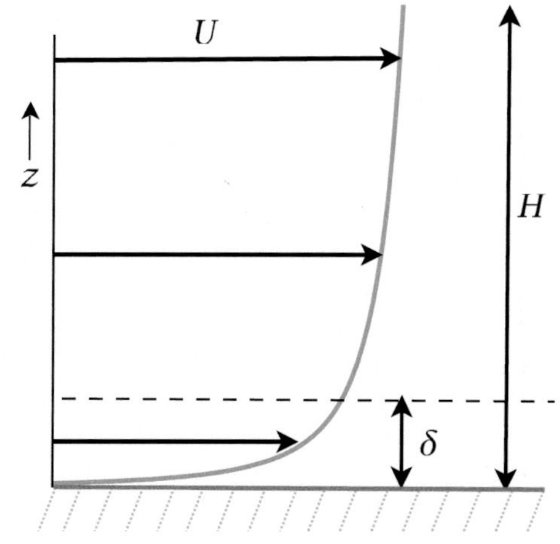

Fig. 5.4 An idealized boundary layer. The values of a field, such as velocity, U, may vary rapidly in a boundary in order to satisfy the boundary conditions at a rigid surface. The parameter δ is a measure of the boundary layer thickness, H is a typical scale of variation away from the boundary, and typically a boundary layer has $\delta \ll H$.

5.7 THE EKMAN LAYER

In the final topic of this chapter we consider the effects of friction. The fluid fields in the interior of a domain are often set by different physical processes from those occurring at a boundary, and consequently often change rapidly in a thin *boundary layer*, as in Fig. 5.4. Such boundary layers nearly always involve one or both of viscosity and diffusion, because these appear in the terms of highest differential order in the equations of motion, and so are responsible for the number and type of boundary conditions that the equations must satisfy — for example, the presence of molecular viscosity leads to the condition that the tangential flow, as well as the normal flow, must vanish at a rigid surface.

In many boundary layers in non-rotating flow the dominant balance in the momentum equation is between the advective and viscous terms. In some contrast, in large-scale atmospheric and oceanic flow the effects of rotation are large, and this results in a boundary layer, known as the *Ekman layer*, in which the dominant balance is between Coriolis and frictional or stress terms.[10] Now, the direct effects of molecular viscosity and diffusion are nearly always negligible at distances more than a few millimetres away from a solid boundary, but it is inconceivable that the entire boundary layer between the free atmosphere (or free ocean) and the surface is only a few millimetres thick. Rather, in practice a balance occurs between the Coriolis terms and the forces due to the stress generated by small-scale turbulent motion, and this gives rise to a boundary layer that has a typical depth of a few tens to several hundreds of metres. Because the stress arises from the turbulence we cannot with confidence determine its precise form; thus, we should try to determine what general properties Ekman layers may have that are *independent* of the precise form of the friction, and these properties turn out to be integral ones such as the total mass flux in the Ekman layer.

The atmospheric Ekman layer occurs near the ground, and the stress at the ground itself is due to the surface wind (and its vertical variation). In the ocean the main Ekman layer is near the surface, and the stress at the ocean surface is largely due to the presence of the overlying wind. There is also a weak Ekman layer at the bottom of the ocean, analogous to the atmospheric Ekman layer. To analyze all these layers, let us assume that:

- The Ekman layer is Boussinesq. This is a very good assumption for the ocean, and a reasonable one for the atmosphere if the boundary layer is not too deep.
- The Ekman layer has a finite depth that is less than the total depth of the fluid, this depth being given by the level at which the frictional stresses essentially vanish. Within the Ekman layer, frictional terms are important, whereas geostrophic balance holds beyond it.
- The nonlinear and time-dependent terms in the equations of motion are negligible, hydro-

static balance holds in the vertical, and buoyancy is constant, not varying in the horizontal.

- As needed, the friction can be parameterized by a viscous term of the form $\rho_0^{-1}\partial\boldsymbol{\tau}/\partial z = A\,\partial^2\boldsymbol{u}/\partial z^2$, where A is constant and $\boldsymbol{\tau}$ is the stress. (In general, stress is a tensor, τ_{ij}, with an associated force given by $F_i = \partial\tau_{ij}/\partial x_j$, summing over the repeated index. It is common in geophysical fluid dynamics that the vertical derivative dominates, and in this case the force is $\boldsymbol{F} = \partial\boldsymbol{\tau}/\partial z$. We still use the word stress for $\boldsymbol{\tau}$, but it now refers to a vector whose derivative in a particular direction (z in this case) is the force on a fluid.) In laboratory settings A may be the molecular viscosity, whereas in the atmosphere and ocean it is a so-called *eddy viscosity.* In turbulent flows momentum is transferred by the near-random motion of small parcels of fluid and, by analogy with the motion of molecules that produces a molecular viscosity, the associated stress is approximately given by using a turbulent or eddy viscosity that may be orders of magnitude larger than the molecular one.

5.7.1 Equations of Motion and Scaling

Frictional–geostrophic balance in the horizontal momentum equation is:

$$\boldsymbol{f} \times \boldsymbol{u} = -\nabla_z \phi + \frac{\partial \widetilde{\boldsymbol{\tau}}}{\partial z}, \tag{5.176}$$

where $\widetilde{\boldsymbol{\tau}} \equiv \boldsymbol{\tau}/\rho_0$ is the kinematic stress and $\boldsymbol{f} = f\mathbf{k}$, where the Coriolis parameter f is allowed to vary with latitude. If we model the stress with an eddy viscosity, (5.176) becomes

$$\boldsymbol{f} \times \boldsymbol{u} = -\nabla_z \phi + A\frac{\partial^2 \boldsymbol{u}}{\partial z^2}. \tag{5.177}$$

The vertical momentum equation is $\partial\phi/\partial z = b$, i.e., hydrostatic balance, and, because buoyancy is constant, we may without loss of generality write this as

$$\frac{\partial \phi}{\partial z} = 0. \tag{5.178}$$

The equation set is completed by the mass continuity equation, $\nabla \cdot \boldsymbol{v} = 0$.

The Ekman number

We nondimensionalize the equations by setting

$$(u, v) = U(\widehat{u}, \widehat{v}), \quad (x, y) = L(\widehat{x}, \widehat{y}), \quad f = f_0\widehat{f}, \quad z = H\widehat{z}, \quad \phi = \Phi\widehat{\phi}, \tag{5.179}$$

where hatted variables are nondimensional. H is a scaling for the height, and at this stage we will suppose it to be some height scale in the free atmosphere or ocean, not the height of the Ekman layer itself. Geostrophic balance suggests that $\Phi = f_0 UL$. Substituting (5.179) into (5.177) we obtain

$$\widehat{\boldsymbol{f}} \times \widehat{\boldsymbol{u}} = -\widehat{\nabla}\widehat{\phi} + Ek\frac{\partial^2 \widehat{\boldsymbol{u}}}{\partial \widehat{z}^2}, \tag{5.180}$$

where the parameter

$$Ek \equiv \left(\frac{A}{f_0 H^2}\right), \tag{5.181}$$

is the *Ekman number*, and it determines the importance of frictional terms in the horizontal momentum equation. If $Ek \ll 1$ then the friction is small in the flow interior where $\widehat{z} = \mathcal{O}(1)$. However, the friction term cannot necessarily be neglected in the boundary layer because it is of the highest differential order in the equation, and so determines the boundary conditions; if Ek is small the vertical scales become small and the second term on the right-hand side of (5.180) remains finite. The case when this term is simply omitted from the equation is therefore a *singular limit,* meaning that it differs from the case with $Ek \to 0$. If $Ek \geq 1$ friction is important everywhere, but it is usually the case that Ek is small for atmospheric and oceanic large-scale flow, and the interior flow is very nearly geostrophic. (In part this is because A itself is only large near a rigid surface where the presence of a shear creates turbulence and a significant eddy viscosity.)

Momentum balance in the Ekman layer

For definiteness, suppose the fluid lies above a rigid surface at $z = 0$. Sufficiently far away from the boundary the velocity field is known, and we suppose this flow to be in geostrophic balance. We then write the velocity field and the pressure field as the sum of the interior geostrophic part, plus a boundary layer correction:

$$\widehat{\boldsymbol{u}} = \widehat{\boldsymbol{u}}_g + \widehat{\boldsymbol{u}}_E, \qquad \widehat{\phi} = \widehat{\phi}_g + \widehat{\phi}_E, \tag{5.182}$$

where the Ekman layer corrections, denoted with a subscript E, are negligible away from the boundary layer. Now, in the fluid interior we have, by hydrostatic balance, $\partial\widehat{\phi}_g/\partial\widehat{z} = 0$. In the boundary layer we still have $\partial\widehat{\phi}_g/\partial\widehat{z} = 0$ so that, to satisfy hydrostasy, $\partial\widehat{\phi}_E/\partial\widehat{z} = 0$. But because $\widehat{\phi}_E$ vanishes away from the boundary we have $\widehat{\phi}_E = 0$ everywhere. Thus, *there is no boundary layer in the pressure field.* Note that this is a much stronger result than saying that pressure is continuous, which is nearly always true in fluids; rather, it is a special result for Ekman layers.

Using (5.182) with $\widehat{\phi}_E = 0$, the dimensional horizontal momentum equation (5.176) becomes, in the Ekman layer,

$$\boldsymbol{f} \times \boldsymbol{u}_E = \frac{\partial \widetilde{\boldsymbol{\tau}}}{\partial z}. \tag{5.183}$$

The dominant force balance in the Ekman layer is thus between the Coriolis force and the friction. We can determine the thickness of the Ekman layer if we model the stress with an eddy viscosity so that

$$\boldsymbol{f} \times \boldsymbol{u}_E = A\frac{\partial^2 \boldsymbol{u}_E}{\partial z^2} \qquad \text{or} \qquad \widehat{\boldsymbol{f}} \times \widehat{\boldsymbol{u}}_E = Ek\frac{\partial^2 \widehat{\boldsymbol{u}}_E}{\partial \widehat{z}^2}, \tag{5.184a,b}$$

where the second equation is nondimensional. It is evident that (5.184b) can only be satisfied if $\widehat{z} \neq \mathcal{O}(1)$, implying that H is not a proper scaling for z in the boundary layer. Rather, if the vertical scale in the Ekman layer is $\widehat{\delta}$ (meaning $\widehat{z} \sim \widehat{\delta}$) we must have $\widehat{\delta} \sim Ek^{1/2}$. In dimensional terms this means the thickness of the Ekman layer is

$$\delta = H\widehat{\delta} = HEk^{1/2} \tag{5.185}$$

or, using (5.181),

$$\delta = \left(\frac{A}{f_0}\right)^{1/2}. \tag{5.186}$$

This estimate also emerges directly from (5.184a). Note that (5.185) can be written as

$$Ek = \left(\frac{\delta}{H}\right)^2. \tag{5.187}$$

That is, the Ekman number is equal to the square of the ratio of the depth of the Ekman layer to an interior depth scale of the fluid motion. In laboratory flows where A is the molecular viscosity we can thus estimate the Ekman layer thickness, and if we know the eddy viscosity of the ocean or atmosphere we can estimate their respective Ekman layer thicknesses. We can invert this argument and obtain an estimate of A if we know the Ekman layer depth. In the atmosphere, deviations from geostrophic balance are very small above 1 km, and using this gives $A \approx 10^2\,\mathrm{m^2\,s^{-1}}$. In the ocean Ekman depths are often 50 m or less, and eddy viscosities are about $0.1\,\mathrm{m^2\,s^{-1}}$.

5.7.2 Integral Properties of the Ekman Layer

What can we deduce about the Ekman layer without specifying the detailed form of the frictional term? Using dimensional notation we recall frictional–geostrophic balance,

$$\boldsymbol{f} \times \boldsymbol{u} = -\nabla\phi + \frac{1}{\rho_0}\frac{\partial \boldsymbol{\tau}}{\partial z}, \tag{5.188}$$

where $\boldsymbol{\tau}$ is zero at the edge of the Ekman layer. In the Ekman layer itself we have

$$\boldsymbol{f} \times \boldsymbol{u}_E = \frac{1}{\rho_0}\frac{\partial \boldsymbol{\tau}}{\partial z}. \tag{5.189}$$

Consider either a top or bottom Ekman layer, and integrate over its thickness. From (5.189) we obtain

$$\boldsymbol{f} \times \boldsymbol{M}_E = \boldsymbol{\tau}_T - \boldsymbol{\tau}_B, \qquad \text{where} \qquad \boldsymbol{M}_E = \int_{Ek} \rho_0 \boldsymbol{u}_E \,\mathrm{d}z. \tag{5.190}$$

Here $\boldsymbol{M}_E$ is the ageostrophic mass transport in the Ekman layer and $\boldsymbol{\tau}_T$ and $\boldsymbol{\tau}_B$ are the respective stresses at the top and the bottom of the Ekman layer at hand. The stress at the top (bottom) will be zero in a bottom (top) Ekman layer and therefore, from (5.190),

$$\begin{aligned} &\text{top Ekman layer:} && \boldsymbol{M}_E = -\frac{1}{f}\mathbf{k} \times \boldsymbol{\tau}_T, \\ &\text{bottom Ekman layer:} && \boldsymbol{M}_E = \frac{1}{f}\mathbf{k} \times \boldsymbol{\tau}_B. \end{aligned} \tag{5.191a,b}$$

The transport is thus at right angles to the stress at the surface, and proportional to the magnitude of the stress. These properties have a simple physical explanation: integrated over the depth of the Ekman layer the surface stress must be balanced by the Coriolis force, which in turn acts at right angles to the mass transport. A consequence of (5.191) is that the mass transports in adjacent oceanic and atmospheric Ekman layers are equal and opposite, because the stress is continuous across the ocean–atmosphere interface. Equation (5.191a) is particularly useful in the ocean, where the stress at the surface is primarily due to the wind, and is largely independent of the interior oceanic flow. In the atmosphere, the surface stress mainly arises as a result of the interior atmospheric flow, and to calculate it we need to parameterize the stress in terms of the flow.

Finally, we obtain an expression for the vertical velocity induced by an Ekman layer. The mass conservation equation is

$$\frac{\partial u}{\partial x} + \frac{\partial v}{\partial y} + \frac{\partial w}{\partial z} = 0. \tag{5.192}$$

Integrating this over an Ekman layer gives

$$\frac{1}{\rho_0}\nabla \cdot \boldsymbol{M}_{To} = -(w_T - w_B), \tag{5.193}$$

where $\boldsymbol{M}_{To}$ is the total (Ekman plus geostrophic) mass transport in the Ekman layer,

$$\boldsymbol{M}_{To} = \int_{Ek} \rho_0 \boldsymbol{u} \,\mathrm{d}z = \int_{Ek} \rho_0 (\boldsymbol{u}_g + \boldsymbol{u}_E) \,\mathrm{d}z \equiv \boldsymbol{M}_g + \boldsymbol{M}_E, \tag{5.194}$$

and w_T and w_B are the vertical velocities at the top and bottom of the Ekman layer; the former (latter) is zero in a top (bottom) Ekman layer. Equations (5.194) and (5.190) give

$$\mathbf{k} \times (\boldsymbol{M}_{To} - \boldsymbol{M}_g) = \frac{1}{f}(\boldsymbol{\tau}_T - \boldsymbol{\tau}_B). \tag{5.195}$$

Taking the curl of this (i.e., cross-differentiating) gives

$$\nabla \cdot (\boldsymbol{M}_{To} - \boldsymbol{M}_g) = \mathrm{curl}_z[(\boldsymbol{\tau}_T - \boldsymbol{\tau}_B)/f], \tag{5.196}$$

where the curl_z operator on a vector $\boldsymbol{A}$ is defined by $\mathrm{curl}_z \boldsymbol{A} \equiv \partial_x A_y - \partial_y A_x$. Using (5.193) we obtain, for top and bottom Ekman layers respectively,

$$w_B = \frac{1}{\rho_0}\left(\mathrm{curl}_z \frac{\boldsymbol{\tau}_T}{f} + \nabla \cdot \boldsymbol{M}_g\right), \qquad w_T = \frac{1}{\rho_0}\left(\mathrm{curl}_z \frac{\boldsymbol{\tau}_B}{f} - \nabla \cdot \boldsymbol{M}_g\right), \tag{5.197a,b}$$

where $\nabla \cdot \boldsymbol{M}_g = -(\beta/f)\boldsymbol{M}_g \cdot \mathbf{j}$ is the divergence of the geostrophic transport in the Ekman layer, and this is often small compared to the other terms in these equations. Thus, friction induces a vertical velocity at the edge of the Ekman layer, proportional to the curl of the stress at the surface, and this is perhaps the most used result in Ekman layer theory. Numerical models sometimes do not have the vertical resolution to explicitly resolve an Ekman layer, and (5.197) provides a means of *parameterizing* the layer in terms of resolved or known fields. This is useful for the top Ekman layer in the ocean, where the stress can be regarded as a function of the overlying wind.

5.7.3 Explicit Solutions I: a Bottom Boundary Layer

We now assume that the frictional terms can be parameterized as an eddy viscosity and calculate the explicit form of the solution in the boundary layer. The frictional–geostrophic balance may be written as

$$\boldsymbol{f} \times (\boldsymbol{u} - \boldsymbol{u}_g) = A\frac{\partial^2 \boldsymbol{u}}{\partial z^2}, \tag{5.198a}$$

where

$$f(u_g, v_g) = \left(-\frac{\partial \phi}{\partial y}, \frac{\partial \phi}{\partial x}\right). \tag{5.198b}$$

We continue to assume there are no horizontal gradients of temperature, so that, via thermal wind, $\partial u_g/\partial z = \partial v_g/\partial z = 0$.

Boundary conditions and solution

Appropriate boundary conditions for a bottom Ekman layer are

$$\text{at } z = 0: \quad u = 0, \quad v = 0 \quad \text{(the no slip condition)} \tag{5.199a}$$

$$\text{as } z \to \infty: \quad u = u_g, \quad v = v_g \quad \text{(a geostrophic interior).} \tag{5.199b}$$

Let us seek solutions to (5.198a) of the form

$$u = u_g + A_0 e^{\alpha z}, \qquad v = v_g + B_0 e^{\alpha z}, \tag{5.200}$$

where A_0 and B_0 are constants. Substituting into (5.198a) gives two homogeneous algebraic equations

$$A_0 f - B_0 A\alpha^2 = 0, \qquad -A_0 A\alpha^2 - B_0 f = 0. \tag{5.201a,b}$$

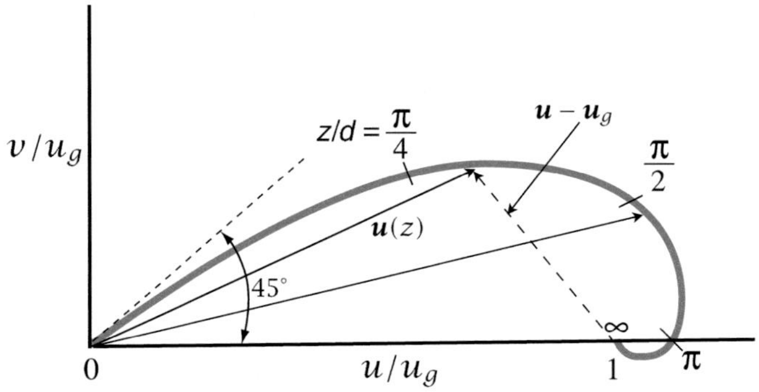

Fig. 5.5 The idealized Ekman layer solution in the lower atmosphere, plotted as a hodograph of the wind components: the arrows show the velocity vectors at particular heights, and the curve traces out the continuous variation of the velocity. The values on the curve are of the nondimensional variable z/d, where $d = (2A/f)^{1/2}$, and v_g is chosen to be zero.

For non-trivial solutions the solvability condition $\alpha^4 = -f^2/A^2$ must hold, from which we find $\alpha = \pm(1 \pm \mathrm{i})\sqrt{f/2A}$. Using the boundary conditions we then obtain the solution

$$u = u_g - \mathrm{e}^{-z/d}\left[u_g \cos(z/d) + v_g \sin(z/d)\right], \tag{5.202a}$$

$$v = v_g + \mathrm{e}^{-z/d}\left[u_g \sin(z/d) - v_g \cos(z/d)\right], \tag{5.202b}$$

where $d = \sqrt{2A/f}$ is, within a constant factor, the depth of the Ekman layer obtained from scaling considerations. The solution decays exponentially from the surface with this e-folding scale, so that d is a good measure of the Ekman layer thickness. Note that the boundary layer correction depends on the interior flow, since the boundary layer serves to bring the flow to zero at the surface.

To illustrate the solution, suppose that the pressure force is directed in the y-direction (northwards), so that the geostrophic current is eastwards. Then the solution, the now-famous *Ekman spiral*, is plotted in Figs. 5.5 and 5.6. The wind falls to zero at the surface, and its direction just above the surface is northeastwards; that is, it is rotated by 45° to the left of its direction in the free atmosphere. Although this result is independent of the value of the frictional coefficients, it is dependent on the form of the friction chosen. The force balance in the Ekman layer is between the Coriolis force, the stress, and the pressure force. At the surface the Coriolis force is zero, and the balance is entirely between the northward pressure force and the southward stress force.

Transport, force balance and vertical velocity

The cross-isobaric flow is given by (for $v_g = 0$)

$$V = \int_0^\infty v \,\mathrm{d}z = \int_0^\infty u_g \mathrm{e}^{-z/d} \sin(z/d)\,\mathrm{d}z = \frac{u_g d}{2}. \tag{5.203}$$

For positive f, this is to the left of the geostrophic flow — that is, down the pressure gradient. In the general case ($v_g \neq 0$) we obtain

$$V = \int_0^\infty (v - v_g)\,\mathrm{d}z = \frac{d}{2}(u_g - v_g). \tag{5.204}$$

Similarly, the additional zonal transport produced by frictional effects is, for $v_g = 0$,

$$U = \int_0^\infty (u - u_g)\,\mathrm{d}z = -\int_0^\infty \mathrm{e}^{-z/d} \sin(z/d)\,\mathrm{d}z = -\frac{u_g d}{2}, \tag{5.205}$$

and in the general case

$$U = \int_0^\infty (u - u_g)\,\mathrm{d}z = -\frac{d}{2}(u_g + v_g). \tag{5.206}$$

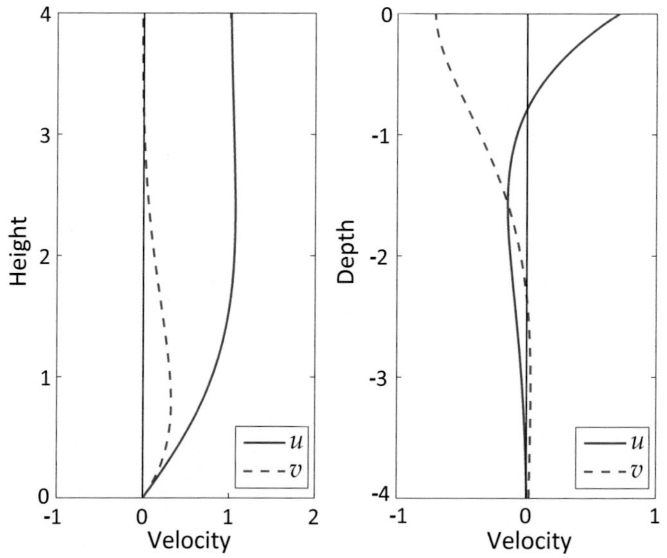

Fig. 5.6 Solutions for a bottom Ekman layer with a given flow in the fluid interior (left), and for a top Ekman layer with a given surface stress (right), both with $d = 1$. On the left we have $u_g = 1, v_g = 0$. On the right we have $u_g = v_g = 0$, $\widetilde{\tau}_y = 0$ and $\sqrt{2}\widetilde{\tau}_x/(fd) = 1$.

Thus, the total transport caused by frictional forces is

$$\boldsymbol{M}_E = \frac{\rho_0 d}{2}\left[-\mathbf{i}(u_g + v_g) + \mathbf{j}(u_g - v_g)\right]. \tag{5.207}$$

The total stress at the bottom surface $z = 0$ induced by frictional forces is

$$\widetilde{\boldsymbol{\tau}}_B = A\frac{\partial \boldsymbol{u}}{\partial z}\bigg|_{z=0} = \frac{A}{d}\left[\mathbf{i}(u_g - v_g) + \mathbf{j}(u_g + v_g)\right], \tag{5.208}$$

using the solution (5.202). Thus, using (5.207), (5.208) and $d^2 = 2A/f$, we see that the total frictionally induced transport in the Ekman layer is related to the stress at the surface by $\boldsymbol{M}_E = (\mathbf{k} \times \boldsymbol{\tau}_B)/f$, reprising the result of the previous more general analysis, (5.197). From (5.208), the stress is at an angle of 45° to the left of the velocity at the surface. (However, this result is not generally true for all forms of stress.) These properties are illustrated in Fig. 5.7.

The vertical velocity at the top of the Ekman layer, w_E, is obtained using (5.207) or (5.208). If f is constant we obtain

$$w_E = -\frac{1}{\rho_0}\nabla \cdot \boldsymbol{M}_E = \frac{1}{f_0}\mathrm{curl}_z \widetilde{\boldsymbol{\tau}}_B = \frac{d}{2}\zeta_g, \tag{5.209}$$

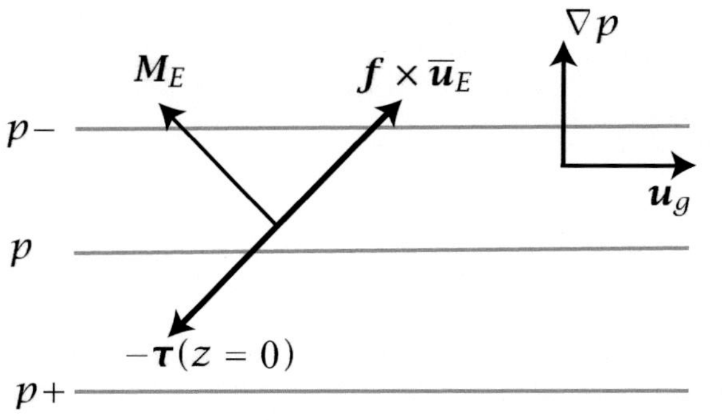

Fig. 5.7 An Ekman layer generated from an eastward geostrophic flow above with associated pressure levels as shown (blue lines). An overbar denotes a vertical integral over the Ekman layer, so that $-\boldsymbol{f} \times \overline{\boldsymbol{u}}_E$ is the Coriolis force on the vertically integrated Ekman velocity. $\boldsymbol{M}_E$ is the frictionally induced boundary layer transport, and $\boldsymbol{\tau}$ is the stress.

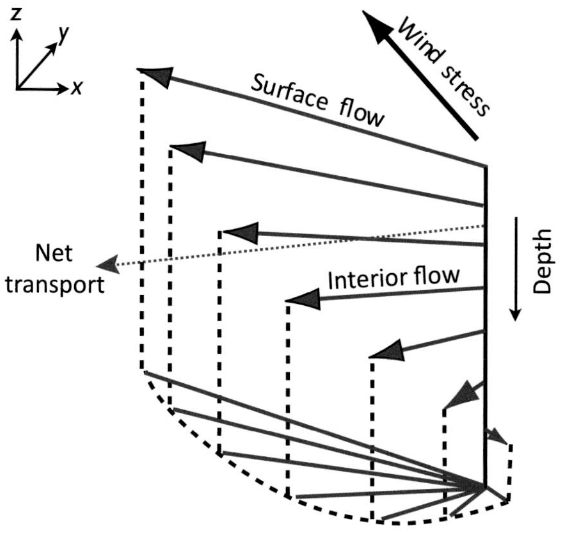

Fig. 5.8 An idealized Ekman spiral in a southern hemisphere ocean, driven by an imposed wind stress. A northern hemisphere spiral would be the reflection of this in the vertical plane. Such a clean spiral is rarely observed in the real ocean. The net transport is at right angles to the wind, independent of the detailed form of the friction. The angle of the surface flow is 45° to the wind only for a Newtonian viscosity.

where ζ_g is the vorticity of the geostrophic flow. Thus, the vertical velocity at the top of the Ekman layer, which arises because of the frictionally-induced divergence of the cross-isobaric flow in the Ekman layer, is proportional to the geostrophic vorticity in the free fluid and is proportional to the Ekman layer height $\sqrt{2A/f_0}$.

Another bottom boundary condition

In the analysis above we assumed a *no slip* condition at the surface, namely that the velocity tangential to the surface vanishes. This is formally appropriate if A is a molecular viscosity, but in a turbulent flow, where A is to be interpreted as an eddy viscosity, the flow close to the surface may be far from zero. Then, unless we wish to explicitly calculate the flow in an additional very thin viscous boundary layer the no-slip condition may be inappropriate. An alternative, slightly more general boundary condition is to suppose that the stress at the surface is given by

$$\boldsymbol{\tau} = \rho_0 C \boldsymbol{u}, \tag{5.210}$$

where C is a constant. The surface boundary condition is then

$$A\frac{\partial \boldsymbol{u}}{\partial z} = C\boldsymbol{u}. \tag{5.211}$$

If C is infinite we recover the no-slip condition. If $C = 0$, we have a condition of no stress at the surface, also known as a *free slip* condition. For intermediate values of C the boundary condition is known as a 'mixed condition'. Evaluating the solution in these cases is left as an exercise for the reader.

5.7.4 Explicit Solutions II: the Upper Ocean

Boundary conditions and solution

The wind provides a stress on the upper ocean, and the Ekman layer serves to communicate this to the oceanic interior. Appropriate boundary conditions are thus:

$$\text{at } z = 0: \quad A\frac{\partial u}{\partial z} = \widetilde{\tau}^x, \quad A\frac{\partial v}{\partial z} = \widetilde{\tau}^y, \quad \text{(a given surface stress)} \tag{5.212a}$$

$$\text{as } z \to -\infty: \quad u = u_g, \quad v = v_g, \quad \text{(a geostrophic interior)} \tag{5.212b}$$

where $\widetilde{\boldsymbol{\tau}}$ is the given (kinematic) wind stress at the surface. Solutions to (5.198a) with (5.212) are found by the same methods as before, and are

$$u = u_g + \frac{\sqrt{2}}{fd} e^{z/d} \left[\widetilde{\tau}^x \cos(z/d - \pi/4) - \widetilde{\tau}^y \sin(z/d - \pi/4) \right], \tag{5.213}$$

and

$$v = v_g + \frac{\sqrt{2}}{fd} e^{z/d} \left[\widetilde{\tau}^x \sin(z/d - \pi/4) + \widetilde{\tau}^y \cos(z/d - \pi/4) \right]. \tag{5.214}$$

Note that the boundary layer correction depends only on the imposed surface stress, and not the interior flow itself. This is a consequence of the type of boundary conditions chosen, for in the absence of an imposed stress the boundary layer correction is zero — the interior flow already satisfies the gradient boundary condition at the top surface. Similarly to the bottom boundary layer, the velocity vectors of the solution trace a diminishing spiral as they descend into the interior (see Fig. 5.8, which is drawn for the Southern Hemisphere).

Transport, surface flow and vertical velocity

The transport induced by the surface stress is obtained by integrating (5.213) and (5.214) from the surface to $-\infty$. We explicitly find

$$U = \int_{-\infty}^{0} (u - u_g)\, dz = \frac{\widetilde{\tau}^y}{f}, \qquad V = \int_{-\infty}^{0} (v - v_g)\, dz = -\frac{\widetilde{\tau}^x}{f}, \tag{5.215}$$

which indicates that the ageostrophic transport is perpendicular to the wind stress, as noted previously from more general considerations. Suppose that the surface wind is eastward; in this case $\widetilde{\tau}^y = 0$ and the solutions immediately give

$$u(0) - u_g = \widetilde{\tau}^x / fd, \qquad v(0) - v_g = -\widetilde{\tau}^x / fd. \tag{5.216}$$

Therefore the magnitudes of the frictional flow in the x and y directions are equal to each other, and the ageostrophic flow is 45° to the right (for $f > 0$) of the wind. This result depends on the form of the frictional parameterization, but not on the size of the viscosity.

At the edge of the Ekman layer the vertical velocity is given by (5.197), and so is proportional to the curl of the wind stress. (The second term on the right-hand side of (5.197) is the vertical velocity due to the divergence of the geostrophic flow, and is usually much smaller than the first term.) The production of a vertical velocity at the edge of the Ekman layer is one of the most important effects of the layer, especially with regard to the large-scale circulation, for it provides an efficient means whereby surface fluxes are communicated to the interior flow (see Fig. 5.9).

5.7.5 † Observations of the Ekman Layer

Ekman layers — and in particular the Ekman spiral — are generally quite hard to observe, in either the ocean or atmosphere, both because of the presence of phenomena that are not included in the theory and because of the technical difficulties of actually measuring the vector velocity profile, especially in the ocean. Ekman-layer theory does not take into account the effects of stratification or of inertial and gravity waves (Section 2.10.4 and Chapter 7), nor does it account for the effects of convection or buoyancy-driven turbulence. If gravity waves are present, the instantaneous flow will be non-geostrophic and so time-averaging will be required to extract the geostrophic flow. If strong convection is present, the simple eddy-viscosity parameterizations used to derive the Ekman spiral will be rendered invalid, and the spiral Ekman profile cannot be expected to be observed in either atmosphere or ocean.

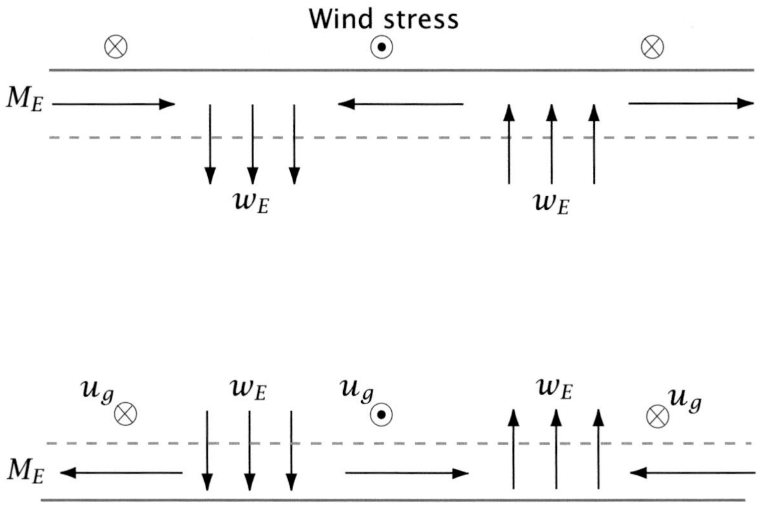

Fig. 5.9 Upper and lower Ekman layers. The upper Ekman layer in the ocean is primarily driven by an imposed wind stress, whereas the lower Ekman layer in the atmosphere or ocean largely results from the interaction of interior geostrophic velocity and a rigid lower surface. The upper part of the figure shows the vertical Ekman 'pumping' velocities that result from the given wind stress, and the lower part of the figure shows the Ekman pumping velocities given the interior geostrophic flow.

In the atmosphere, in convectively neutral cases, the Ekman profile can sometimes be qualitatively discerned. In convectively unstable situations the Ekman profile is generally not observed, but the flow is nevertheless cross-isobaric, from high pressure to low, consistent with the theory. (For most purposes, it is in any case the integral properties of the Ekman layer that is most important.) In the ocean, from about 1980 onwards improved instruments have made it possible to observe the vector current with depth, and to average that current and correlate it with the overlying wind, and a number of observations generally consistent with Ekman dynamics have emerged.[11] There are some differences between observations and theory, and these can be ascribed to the effects of stratification (which causes a shallowing and flattening of the spiral), and to the interaction of the Ekman spiral with turbulence (and the inadequacy of the eddy-diffusivity parameterization). In spite of these differences, Ekman layer theory remains a remarkable and enduring foundation of geophysical fluid dynamics.

5.7.6 ✦ Frictional Parameterization of the Ekman Layer

Suppose that the free atmosphere is described by the quasi-geostrophic vorticity equation,

$$\frac{\mathrm{D}\zeta_g}{\mathrm{D}t} = f_0 \frac{\partial w}{\partial z}, \tag{5.217}$$

where ζ_g is the geostrophic relative vorticity. Let us further model the atmosphere as a single homogeneous layer of thickness H lying above an Ekman layer of thickness $d \ll H$. If the vertical velocity is negligible at the top of the layer (at $z = H + d$) the equation of motion becomes

$$\frac{\mathrm{D}\zeta_g}{\mathrm{D}t} = \frac{f_0[w(H+d) - w(d)]}{H} = -\frac{f_0 d}{2H}\zeta_g, \tag{5.218}$$

using (5.209). This equation shows that the Ekman layer acts as a *linear drag* on the interior flow, with a drag coefficient r equal to $f_0 d/2H$ and with associated time scale T_{Ek} given by

$$T_{Ek} = \frac{2H}{f_0 d} = \frac{2H}{\sqrt{2f_0 A}}. \tag{5.219}$$

In the oceanic case the corresponding vorticity equation for the interior flow is

$$\frac{\mathrm{D}\zeta_g}{\mathrm{D}t} = \frac{1}{H}\mathrm{curl}_z \tau_s, \tag{5.220}$$

where τ_s is the surface stress. The surface stress thus acts as if it were a body force on the interior flow, and neither the Coriolis parameter nor the depth of the Ekman layer explicitly appear in this formula.

The Ekman layer is a very efficient way of communicating surface stresses to the interior. To see this, suppose that eddy mixing were the sole mechanism of transferring stress from the surface to the fluid interior, and there were no Ekman layer. Then the time scale of spindown of the fluid would be given by using

$$\frac{\mathrm{d}\zeta}{\mathrm{d}t} = A\frac{\partial^2 \zeta}{\partial z^2}, \tag{5.221}$$

implying a turbulent spin-down time, T_{turb}, of

$$T_{turb} \sim \frac{H^2}{A}, \tag{5.222}$$

where H is the depth over which we require a spin-down. This is much longer than the spin-down of a fluid that has an Ekman layer, for we have

$$\frac{T_{turb}}{T_{Ek}} = \frac{(H^2/A)}{(2H/f_0 d)} = \frac{H}{d} \gg 1, \tag{5.223}$$

using $d = \sqrt{2A/f_0}$. The effects of friction are enhanced because of the presence of a secondary circulation confined to the Ekman layers (as in Fig. 5.9) in which the vertical scales are much smaller than those in the fluid interior and so where viscous effects become significant; these frictional stresses are then communicated to the fluid interior via the induced vertical velocities at the edge of the Ekman layers.

Notes

1 The phrase 'quasi-geostrophic' seems to have been introduced by Durst & Sutcliffe (1938) and the concept used in Sutcliffe's development theory of baroclinic systems (Sutcliffe 1939, 1947). The first systematic derivation of the quasi-geostrophic equations based on scaling theory was given by Charney (1948). The planetary-geostrophic equations were used by Robinson & Stommel (1959) and Welander (1959) in studies of the thermocline (and were first known as the 'thermocline equations'), and were put in the context of other approximate equation sets by Phillips (1963).

2 Carl-Gustav Rossby (1898–1957) played a dominant role in the development of dynamical meteorology in the early and middle parts of the twentieth century, and his work permeates all aspects of dynamical meteorology today. Perhaps the most fundamental nondimensional number in rotating fluid dynamics, the Rossby number, is named for him, as is the perhaps most fundamental wave, the Rossby wave. He also discovered the conservation of potential vorticity (later generalized by Ertel) and contributed important ideas to atmospheric turbulence and the theory of air masses. Swedish born, he studied first with V. Bjerknes before taking a position in Stockholm in 1922 with

the Swedish Meteorological Hydrologic Service and receiving a 'Licentiat' from the University of Stockholm in 1925. Shortly thereafter he moved to the United States, joining the Government Weather Bureau, a precursor of NOAA's National Weather Service. In 1928 he moved to MIT, playing an important role in developing the meteorology department there, while still maintaining connections with the Weather Bureau. In 1940 he moved to the University of Chicago, where he similarly helped develop meteorology there. In 1947 he became Director of the newly formed Institute of Meteorology in Stockholm, and subsequently divided his time between there and the United States. Thus, as well as his scientific contributions, he played an influential role in the institutional development of the field.

3 Burger (1958).

4 Numerical integrations of the potential vorticity equation using (5.91), and performing the inversion without linearizing potential vorticity, do in fact indicate improved accuracy over either the quasi-geostrophic or planetary-geostrophic equations (Mundt *et al.* 1997). In a similar vein, McIntyre & Norton (2000) show how useful potential vorticity inversion can be, and Allen *et al.* (1990a,b) demonstrate the high accuracy of certain intermediate models. Certainly, asymptotic correctness should not be the only criterion used in constructing a filtered model, because the parameter range in which the model is useful may be too limited. Note that there is a difference between extending the parameter range in which a filtered model is useful, as in the inversion of (5.91), and going to higher asymptotic order accuracy in a given parameter regime, as in Allen (1993) and Warn *et al.* (1995). Using Hamiltonian mechanics it is possible to derive equations that both span different asymptotic regimes and that have good conservation properties (Salmon 1983, Allen *et al.* 2002).

5 There is a difference between the *dynamical* demands of the quasi-geostrophic system in requiring β to be small, and the *geometric* demands of the Cartesian geometry. On Earth the two demands are similar in practice. But without dynamical inconsistency we may imagine a Cartesian system in which $\beta y \sim f$, and indeed this is common in idealized, planetary geostrophic models of the large-scale ocean circulation.

6 The atmospheric and oceanic sciences are sometimes thought of as not being 'beautiful' in the same way as some branches of theoretical physics. Yet surely quasi-geostrophic theory, and the quasi-geostrophic potential vorticity equation, are quite beautiful, combining austerity of description and richness of behaviour.

7 Bretherton (1966b). Schneider *et al.* (2003) look at the non-QG extension. The equivalence between boundary conditions and delta-function sources is a common feature of elliptic and similar problems, and is analogous to the generation of electromagnetic fields by point charges. It is sometimes exploited in the numerical solution of elliptic equations, both as a simple way to include non-homogeneous boundary conditions and, using the so-called capacitance matrix method, to solve problems in irregular domains (e.g., Hockney 1970).

8 Phillips (1954, 1956) used a two-level model for instability studies and to construct a simple general circulation model of the atmosphere.

9 Charney & Stern (1962). See also Berrisford *et al.* (1993) and Vallis (1996).

10 After Ekman (1905). The problem is said to have been posed to V. W. Ekman (1874–1954), a student of Vilhelm Bjerknes, by Fridtjof Nansen, the polar explorer and statesman, who wanted to understand the motion of pack ice and of his ship, the *Fram*, embedded in the ice.

11 For oceanic observations see Davis *et al.* (1981), Price *et al.* (1987), Rudnick & Weller (1993). For the atmosphere see, e.g., Nicholls (1985).

Part II

WAVES, INSTABILITIES AND TURBULENCE

The waves broke and spread their waters swiftly over the shore. One after another they massed themselves and fell; the spray tossed itself back with the energy of their fall. The waves were steeped deep-blue save for a pattern of diamond-pointed light on their backs which rippled as the backs of great horses ripple with muscles as they move. The waves fell; withdrew and fell again.

Virginia Woolf, *The Waves*, 1931.

CHAPTER 6

Wave Fundamentals

WAVES ARE EVERYWHERE: on the sea-shore, on piano wires, in football stadiums, and filling the space between the distant stars and Earth. This chapter provides an introduction to their properties, paying particular attention to a wave that is especially important to the large scale flow in both ocean and atmosphere — the Rossby wave. We start with an elementary introduction to wave kinematics, discussing such basic concepts as phase speed and group velocity. Then, beginning with Section 6.4, we discuss the dynamics of Rossby waves, and this part may be considered to be the natural follow-on from the geostrophic theory of the previous chapter. Finally, in Section 6.7, we return to group velocity in a more general way and illustrate the results using Poincaré waves, with more applications to gravity and Rossby waves in later chapters.

The reason for such an ordering of topics is that wave kinematics without a dynamical example is jejune and dry, yet understanding wave dynamics of any sort is hardly possible without appreciating at least some of its formal structure, and readers should flip pages back and forth through the chapter as needed. Those readers who wish to cut to the chase may skip the first few sections and begin at Section 6.4, referring back as needed. (Many of the key elementary results are summarized in the shaded box on page 218.) Other readers may wish to skip the sections on Rossby waves altogether and, after absorbing the sections on the wave theory move on to Chapter 7 on gravity waves, returning to Rossby waves (or not) later on. The Rossby wave and gravity wave discussions are largely independent of each other, although they both require that the reader is familiar with such ideas as group velocity and phase speed. Rossby waves and gravity waves can co-exist and close to the equator the two kinds of waves become more intertwined; we deal with the ensuing waves in Chapter 8. We also extend our discussion of Rossby waves in a global atmospheric context in Chapter 16.

6.1 FUNDAMENTALS AND FORMALITIES

6.1.1 Definitions and Kinematics

A wave may be more easily recognized than defined. Loosely speaking, a wave is a propagating disturbance that has a characteristic relationship between its frequency and size, and a linear wave may be defined as a disturbance that satisfies a *dispersion relation.* (Nonlinear waves exist, but the curious reader must look elsewhere to learn about them.[1]) In order to see what all this means, and what a dispersion relation is, suppose that a disturbance, $\psi(\boldsymbol{x}, t)$ (where ψ might be velocity, streamfunction, pressure, etc.), satisfies the equation

$$L(\psi) = 0, \tag{6.1}$$

where L is a linear operator, typically a polynomial in time and space derivatives; one example is $L(\psi) = \partial\nabla^2\psi/\partial t + \beta\partial\psi/\partial x$. If (6.1) has constant coefficients (if β is constant in this example) then harmonic solutions may often be found that are a superposition of *plane waves,* each of which satisfy

$$\psi = \mathrm{Re}\,\widetilde{\psi}\mathrm{e}^{\mathrm{i}\theta(\boldsymbol{x},t)} = \mathrm{Re}\,\widetilde{\psi}\mathrm{e}^{\mathrm{i}(\boldsymbol{k}\cdot\boldsymbol{x}-\omega t)}, \tag{6.2}$$

where $\widetilde{\psi}$ is a complex constant, θ is the phase, ω is the wave frequency and $\boldsymbol{k}$ is the vector wavenumber (k, l, m) (also written as (k^x, k^y, k^z) or, in subscript notation, k_i). The prefix Re denotes the real part of the expression, but we will drop it if there is no ambiguity.

Earlier, we said that waves are characterized by having a particular relationship between the frequency and wavevector known as the *dispersion relation.* This is an equation of the form

$$\omega = \Omega(\boldsymbol{k}), \tag{6.3}$$

where $\Omega(\boldsymbol{k})$, or $\Omega(k_i)$, and meaning $\Omega(k, l, m)$, is some function determined by the form of L in (6.1) and which thus depends on the particular type of wave — the function is different for sound waves, light waves and the Rossby waves and gravity waves we will encounter in this book (peek ahead to (6.60) and (7.56), and there is more discussion in Section 6.1.3). Unless it is necessary to explicitly distinguish the function Ω from the frequency ω, we will often write $\omega = \omega(\boldsymbol{k})$.

If the medium in which the waves are propagating is inhomogeneous then (6.1) will probably not have constant coefficients (for example, β may vary with y). Nevertheless, if the medium is varying sufficiently slowly, wave solutions may often still be found with the general form

$$\psi(\boldsymbol{x}, t) = a(\boldsymbol{x}, t)\mathrm{e}^{\mathrm{i}\theta(\boldsymbol{x},t)}, \tag{6.4}$$

where the amplitude $a(\boldsymbol{x}, t)$ varies slowly compared to the phase, θ. The frequency and wavenumber are then defined by

$$\boldsymbol{k} \equiv \nabla\theta, \qquad \omega \equiv -\frac{\partial\theta}{\partial t}. \tag{6.5}$$

The example of (6.2) is clearly just a special case of this. Equation (6.5) implies the formal relation between $\boldsymbol{k}$ and ω:

$$\frac{\partial\boldsymbol{k}}{\partial t} + \nabla\omega = 0. \tag{6.6}$$

The WKB method, described in Appendix A at the end of this chapter, is one way of finding such solutions.

6.1.2 Wave Propagation and Phase Speed

A common property of waves is that they propagate through space with some velocity, which in special cases might be zero. Waves in fluids may carry energy and momentum but not normally, at least to a first approximation, fluid parcels themselves. Further, it turns out that the speed at which properties like energy are transported (the group speed) may be different from the speed at which the wave crests themselves move (the phase speed). Let's try to understand this statement, beginning with the phase speed. A summary of key results is given on page 218.

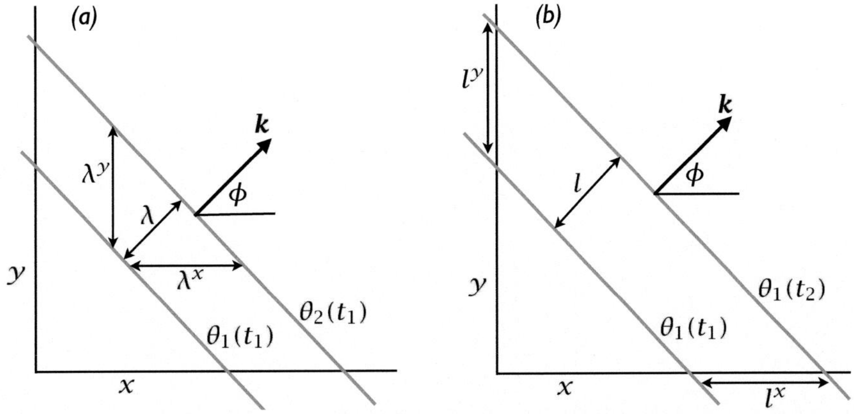

Fig. 6.1 The propagation of a two-dimensional wave. (a) Two lines of constant phase (e.g., two wavecrests) at a time t_1. The wave is propagating in the direction $\boldsymbol{k}$ with wavelength λ. (b) The same line of constant phase at two successive times. The phase speed is the speed of advancement of the wavecrest in the direction of travel, and so $c_p = l/(t_2 - t_1)$. The phase speed in the x-direction is the speed of propagation of the wavecrest along the x-axis, and $c_p^x = l^x/(t_2 - t_1) = c_p/\cos\phi$.

Phase speed

Consider the propagation of monochromatic plane waves, for that is all that is needed to introduce the phase speed. Given (6.2) a wave will propagate in the direction of $\boldsymbol{k}$ (Fig. 6.1). At a given instant and location we can align our coordinate axis along this direction, and we write $\boldsymbol{k}\cdot\boldsymbol{x} = Kx^*$, where x^* increases in the direction of $\boldsymbol{k}$ and $K^2 = |\boldsymbol{k}|^2$ is the magnitude of the wavenumber. With this, we can write (6.2) as

$$\psi = \mathrm{Re}\,\widetilde{\psi}\mathrm{e}^{\mathrm{i}(Kx^* - \omega t)} = \mathrm{Re}\,\widetilde{\psi}\mathrm{e}^{\mathrm{i}K(x^* - ct)}, \tag{6.7}$$

where $c = \omega/K$. From this equation it is evident that the phase of the wave propagates at the speed c in the direction of $\boldsymbol{k}$, and we define the *phase speed* by

$$c_p \equiv \frac{\omega}{K}. \tag{6.8}$$

The wavelength of the wave, λ, is the distance between two wavecrests — that is, the distance between two locations along the line of travel whose phase differs by 2π — and evidently this is given by

$$\lambda = \frac{2\pi}{K}. \tag{6.9}$$

In (for simplicity) a two-dimensional wave, and referring to Fig. 6.1, the wavelength and wave vectors in the x- and y-directions are given by,

$$\lambda^x = \frac{\lambda}{\cos\phi}, \quad \lambda^y = \frac{\lambda}{\sin\phi}, \qquad k^x = K\cos\phi, \quad k^y = K\sin\phi. \tag{6.10}$$

In general, lines of constant phase intersect both the coordinate axes and propagate along them. The speed of propagation along these axes is given by

$$c_p^x = c_p\frac{l^x}{l} = \frac{c_p}{\cos\phi} = c_p\frac{K}{k^x} = \frac{\omega}{k^x}, \qquad c_p^y = c_p\frac{l^y}{l} = \frac{c_p}{\sin\phi} = c_p\frac{K}{k^y} = \frac{\omega}{k^y}, \tag{6.11}$$

using (6.8) and (6.10), and again referring to Fig. 6.1 for notation. The speed of phase propagation along any one of the axes is in general *larger* than the phase speed in the primary direction of

Wave Fundamentals

- A wave is a propagating disturbance that has a characteristic relationship between its frequency and size, known as the dispersion relation. Waves typically arise as solutions to a linear problem of the form
$$L(\psi) = 0, \tag{WF.1}$$
where L is, commonly, a linear operator in space and time. Two examples are
$$\frac{\partial^2 \psi}{\partial t^2} - c^2 \nabla^2 \psi = 0 \qquad \text{and} \qquad \frac{\partial}{\partial t}\nabla^2 \psi + \beta \frac{\partial \psi}{\partial x} = 0. \tag{WF.2}$$
The first example is so common in all areas of physics it is sometimes called 'the' wave equation. The second example gives rise to Rossby waves.

- Solutions to the governing equation are often sought in the form of plane waves that have the form
$$\psi = \mathrm{Re}\, A e^{\mathrm{i}(\boldsymbol{k}\cdot\boldsymbol{x} - \omega t)}, \tag{WF.3}$$
where A is the wave amplitude, $\boldsymbol{k} = (k, l, m)$ is the wavevector, and ω is the frequency.

- The dispersion relation connects the frequency and wavevector through an equation of the form $\omega = \Omega(\boldsymbol{k})$ where Ω is some function. The relation is normally derived by substituting a trial solution like (WF.3) into the governing equation (WF.1). For the examples of (WF.2) we obtain $\omega = c^2 K^2$ and $\omega = -\beta k / K^2$ where $K^2 = k^2 + l^2 + m^2$ or, in two dimensions, $K^2 = k^2 + l^2$.

- The phase speed is the speed at which the wave crests move. In the direction of propagation and in the x, y and z directions the phase speeds are given by, respectively,
$$c_p = \frac{\omega}{K}, \qquad c_p^x = \frac{\omega}{k}, \qquad c_p^y = \frac{\omega}{l}, \qquad c_p^z = \frac{\omega}{m}, \tag{WF.4}$$
where $K = 2\pi/\lambda$ and λ is the wavelength. The wave crests have both a speed (c_p) and a direction of propagation (the direction of $\boldsymbol{k}$), like a vector, but the components defined in (WF.4) are not the components of that vector.

- The group velocity is the velocity at which a wave packet or wave group moves. It is a vector and is given by
$$\boldsymbol{c}_g = \frac{\partial \omega}{\partial \boldsymbol{k}} \quad \text{with components} \quad c_g^x = \frac{\partial \omega}{\partial k}, \qquad c_g^y = \frac{\partial \omega}{\partial l}, \qquad c_g^z = \frac{\partial \omega}{\partial m}. \tag{WF.5}$$
Many physical quantities of interest are transported at the group velocity.

- If the coefficients of the wave equation are not constant (for example if the medium is inhomogeneous) then, if the coefficients are only slowly varying, approximate solutions may sometimes be found in the form
$$\psi = \mathrm{Re}\, A(\boldsymbol{x}, t) e^{\mathrm{i}\theta(\boldsymbol{x}, t)}, \tag{WF.6}$$
where the amplitude A is also slowly varying and the local wavenumber and frequency are related to the phase, θ, by $\boldsymbol{k} = \nabla\theta$ and $\omega = -\partial\theta/\partial t$. The dispersion relation is then a *local* one of the form $\omega = \Omega(\boldsymbol{k}; x, t)$.

the wave. The phase speeds are clearly *not* components of a vector: for example, $c_p^x \neq c_p \cos\phi$. Analogously, the wavevector $\boldsymbol{k}$ is a true vector, whereas the wavelength λ is not.

To summarize, the phase speed and its components are given by

$$c_p = \frac{\omega}{K}, \qquad c_p^x = \frac{\omega}{k^x}, \quad c_p^y = \frac{\omega}{k^y}. \tag{6.12}$$

Phase velocity

Although it is not particularly useful, there is a way of defining a phase speed so that it is a true vector, and which might then be called phase velocity. We define the phase velocity to be the velocity that has the magnitude of the phase speed and the direction in which wave crests are propagating; that is

$$\boldsymbol{c}_p \equiv \frac{\omega}{K}\frac{\boldsymbol{k}}{|K|} = c_p \frac{\boldsymbol{k}}{|K|}, \tag{6.13}$$

where $\boldsymbol{k}/|K|$ is the unit vector in the direction of wave-crest propagation. The components of the phase velocity in the x- and y-directions are then given by

$$c_p^x = c_p \cos\phi, \qquad c_p^y = c_p \sin\phi. \tag{6.14}$$

Defined this way, the quantity given by (6.13) is a vector velocity. However, the components in the x- and y-directions are manifestly not the speed at which wave crests propagate in those directions, and it is a misnomer to call these quantities phase speeds. Still, it can be helpful to ascribe a direction to the phase speed and the quantity given by (6.13) may then be useful.

6.1.3 The Dispersion Relation

The above description is kinematic, in that it applies to almost any disturbance that has a wavevector and a frequency. The particular *dynamics* of a wave are determined by the relationship between the wavevector and the frequency; that is, by the *dispersion relation.* Once the dispersion relation is known a great many of the properties of the wave follow in a more-or-less straightforward manner. Picking up from (6.3), the dispersion relation is a functional relationship between the frequency and the wavevector of the general form

$$\omega = \Omega(\boldsymbol{k}). \tag{6.15}$$

Perhaps the simplest example of a linear operator that gives rise to waves is the one-dimensional equation

$$\frac{\partial \psi}{\partial t} + c\frac{\partial \psi}{\partial x} = 0. \tag{6.16}$$

Substituting a trial solution of the form $\psi = \mathrm{Re}\, A e^{\mathrm{i}(kx-\omega t)}$ we obtain $(-\mathrm{i}\omega + c\mathrm{i}k)A = 0$, giving the dispersion relation

$$\omega = ck. \tag{6.17}$$

The phase speed of this wave is $c_p = \omega/k = c$. A few other examples of governing equations, dispersion relations and phase speeds are:

$$\frac{\partial \psi}{\partial t} + \boldsymbol{c}\cdot\nabla\psi = 0, \qquad \omega = \boldsymbol{c}\cdot\boldsymbol{k}, \qquad c_p = |\boldsymbol{c}|\cos\theta, \qquad c_p^x = \frac{\boldsymbol{c}\cdot\boldsymbol{k}}{k}, \qquad c_p^y = \frac{\boldsymbol{c}\cdot\boldsymbol{k}}{l}, \tag{6.18a}$$

$$\frac{\partial^2 \psi}{\partial t^2} - c^2\nabla^2\psi = 0, \qquad \omega^2 = c^2K^2, \qquad c_p = \pm c, \qquad c_p^x = \pm\frac{cK}{k}, \qquad c_p^y = \pm\frac{cK}{l}, \tag{6.18b}$$

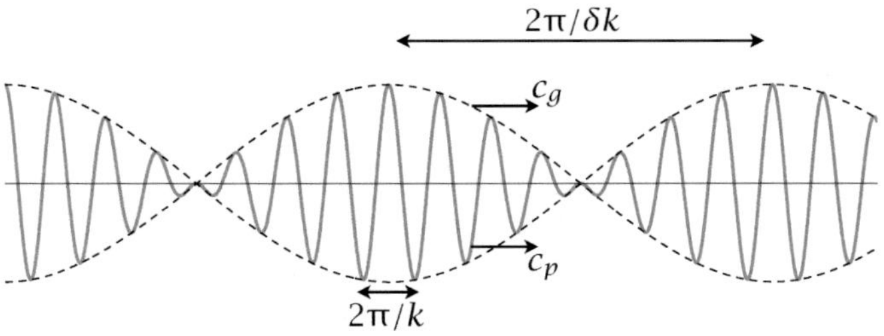

Fig. 6.2 Superposition of two sinusoidal waves with wavenumbers k and $k+\delta k$, producing a wave (solid line) that is modulated by a slowly varying wave envelope or packet (dashed).

The envelope moves at the group velocity, $c_g = \partial\omega/\partial k$, and the phase moves at the group speed, $c_p = \omega/k$.

$$\frac{\partial}{\partial t}\nabla^2\psi + \beta\frac{\partial\psi}{\partial x} = 0, \qquad \omega = \frac{-\beta k}{K^2}, \qquad c_p = \frac{\omega}{K}, \qquad c_p^x = -\frac{\beta}{K^2}, \qquad c_p^y = -\frac{\beta k/l}{K^2}, \tag{6.18c}$$

where $K^2 = k^2 + l^2$ and θ is the angle between $\boldsymbol{c}$ and $\boldsymbol{k}$, and the examples are all two-dimensional, with variation in x and y only.

A wave is said to be *nondispersive* if the phase speed is independent of the wavelength. This condition is satisfied for the simple example (6.16) but is manifestly not satisfied for (6.18c), and these waves (Rossby waves, in fact) are *dispersive*. Waves of different wavelengths then travel at different speeds so that a group of waves will spread out — disperse — even if the medium is homogeneous. When a wave is dispersive there is another characteristic speed at which the waves propagate, the group velocity, and we come to this shortly.

Most media are inhomogeneous, but if the medium varies sufficiently slowly in space and time — and in particular if the variations are slow compared to the wavelength and period — we may still have a *local* dispersion relation between frequency and wavevector,

$$\omega = \Omega(\boldsymbol{k}; \boldsymbol{x}, t), \tag{6.19}$$

where x and t are slowly varying parameters. We'll resume our discussion of this topic in Section 6.3, but before that we must introduce the group velocity.

6.2 GROUP VELOCITY

Information and energy do not, in general, propagate at the phase speed. Rather, most quantities of interest propagate at the *group velocity*, a quantity of enormous importance in wave theory.[2] Roughly speaking, group velocity is the velocity at which a packet or a group of waves will travel, whereas the individual wave crests travel at the phase speed. To introduce the idea we will consider the superposition of plane waves, noting that a truly monochromatic plane wave already fills all space uniformly so that there can be no propagation of energy from place to place.

6.2.1 Superposition of Two Waves

Consider the linear superposition of two waves. Limiting attention to the one-dimensional case for simplicity, consider a disturbance represented by

$$\psi = \mathrm{Re}\,\widetilde{\psi}(\mathrm{e}^{\mathrm{i}(k_1x-\omega_1t)} + \mathrm{e}^{\mathrm{i}(k_2x-\omega_2t)}). \tag{6.20}$$

Let us further suppose that the two waves have similar wavenumbers and frequency, and, in particular, that $k_1 = k + \Delta k$ and $k_2 = k - \Delta k$, and $\omega_1 = \omega + \Delta\omega$ and $\omega_2 = \omega - \Delta\omega$. With this, (6.20) becomes

$$\begin{aligned}\psi &= \mathrm{Re}\,\widetilde{\psi}\mathrm{e}^{\mathrm{i}(kx-\omega t)}[\mathrm{e}^{\mathrm{i}(\Delta kx-\Delta\omega t)} + \mathrm{e}^{-\mathrm{i}(\Delta kx-\Delta\omega t)}] \\ &= 2\,\mathrm{Re}\,\widetilde{\psi}\mathrm{e}^{\mathrm{i}(kx-\omega t)}\cos(\Delta k\,x - \Delta\omega\,t).\end{aligned} \tag{6.21}$$

The resulting disturbance, illustrated in Fig. 6.2 has two aspects: a rapidly varying component, with wavenumber k and frequency ω, and a more slowly varying envelope, with wavenumber Δk and frequency $\Delta\omega$. The envelope modulates the fast oscillation, and moves with velocity $\Delta\omega/\Delta k$; in the limit $\Delta k \to 0$ and $\Delta\omega \to 0$ this is the *group velocity*, $c_g = \partial\omega/\partial k$. Group velocity is equal to the phase speed, ω/k, only when the frequency is a linear function of wavenumber. The energy in the disturbance must move at the group velocity — note that the node of the envelope moves at the speed of the envelope and no energy can cross the node. These concepts generalize to more than one dimension, and if the wavenumber is the three-dimensional vector $\boldsymbol{k} = (k, l, m)$ then the three-dimensional envelope propagates at the group velocity given by

$$\boldsymbol{c}_g = \frac{\partial\omega}{\partial\boldsymbol{k}} \equiv \left(\frac{\partial\omega}{\partial k}, \frac{\partial\omega}{\partial l}, \frac{\partial\omega}{\partial m}\right). \tag{6.22}$$

The group velocity is also written as $\boldsymbol{c}_g = \nabla_{\boldsymbol{k}}\omega$ or, in subscript notation, $c_{gi} = \partial\Omega/\partial k_i$, with the subscript i denoting the component of a vector.

6.2.2 Superposition of Many Waves

Now consider a slight extension of the above arguments to the case in which many waves are excited. The essential assumption of this derivation is that the wavenumber distribution is sufficiently narrow so that the dispersion relation can be approximated as

$$\omega(k) \approx \omega(k_0) + \left.\frac{d\omega}{dk}\right|_{k_0} (k - k_0). \tag{6.23}$$

We treat only the one-dimensional case but the three-dimensional generalization is possible.

A superposition of plane waves, each satisfying some dispersion relation, can be represented by the Fourier integral

$$\psi(x, t) = \int_{-\infty}^{\infty} \widetilde{A}(k) e^{i(kx-\omega t)} \, dk. \tag{6.24a}$$

The function $\widetilde{A}(k)$ is given by the initial conditions:

$$\widetilde{A}(k) = \frac{1}{2\pi}\int_{-\infty}^{\infty} \psi(x, 0) e^{-ikx} \, dx. \tag{6.24b}$$

Note that if the waves are dispersionless and $\omega = ck$ where c is a constant, then

$$\psi(x, t) = \int_{-\infty}^{+\infty} \widetilde{A}(k) e^{ik(x-ct)} \, dk = \psi(x - ct, 0), \tag{6.25}$$

by comparison with (6.24a) at $t = 0$. That is, the initial condition simply translates at a speed c, with no change in structure.

Now consider the case for which the disturbance is a *wave packet* — essentially a nearly plane wave or superposition of waves that is confined to a finite region of space. We will consider a case with the initial condition

$$\psi(x, 0) = a(x) e^{ik_0 x}, \tag{6.26}$$

where $a(x, t)$, rather like the envelope in Fig. 6.3, modulates the amplitude of the wave on a scale much longer than that of the wavelength $2\pi/k_0$, and more slowly than the wave period. That is,

$$\frac{1}{a}\frac{\partial a}{\partial x} \ll k_0, \qquad \frac{1}{a}\frac{\partial a}{\partial t} \ll k_0 c, \tag{6.27a,b}$$

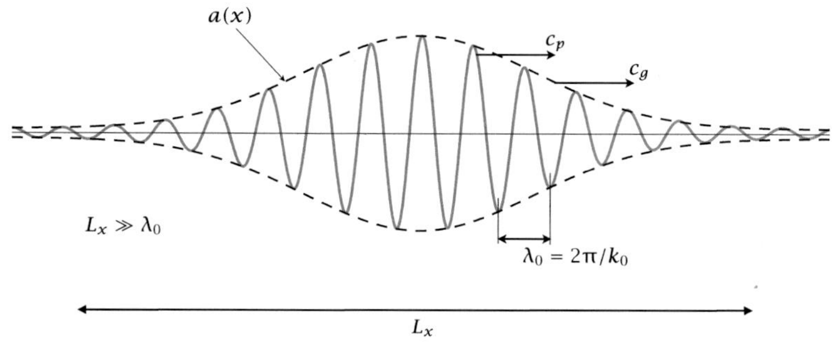

Fig. 6.3 A wave packet. The envelope, $a(x)$, has a scale, L_x, much larger than the wavelength, λ_0, of the wave embedded inside. The envelope moves at the group velocity, c_g, and the phase of the waves at the phase speed, c_p.

and the disturbance is essentially a slowly modulated plane wave. We suppose that $a(x,0)$ is peaked around some value x_0 and is very small if $|x-x_0| \gg k_0^{-1}$; that is, $a(x,0)$ is small if we are sufficiently many wavelengths of the plane wave away from the peak, as is the case in Fig. 6.3. We would like to know how such a packet evolves.

We can express the envelope as a Fourier integral by first writing the initial conditions as a Fourier integral,

$$\psi(x,0) = \int_{-\infty}^{\infty} \widetilde{A}(k) e^{ikx} \, dk \quad \text{where} \quad \widetilde{A}(k) = \frac{1}{2\pi} \int_{-\infty}^{+\infty} \psi(x,0) e^{-ikx} \, dx, \tag{6.28a,b}$$

so that, using (6.26),

$$\widetilde{A}(k) = \frac{1}{2\pi} \int_{-\infty}^{+\infty} a(x,0) e^{i(k_0-k)x} \, dx \quad \text{and} \quad a(x) = \int_{-\infty}^{\infty} \widetilde{A}(k) e^{i(k-k_0)x} \, dk. \tag{6.29a,b}$$

We still haven't made much progress beyond (6.24). To do so, we note first that $a(x)$ is confined in space, so that to a good approximation the limits of the integral in (6.29a) can be made finite, $\pm L$ say, provided $L \gg k_0^{-1}$. We then note that when $(k_0 - k)$ is large the integrand in (6.29a) oscillates rapidly; successive intervals in x therefore cancel each other and make a small net contribution to the integral. Thus, the integral is dominated by values of k near k_0, and $\widetilde{A}(k)$ is peaked near k_0. (The finite spatial extent of $a(x,0)$ is needed for this argument.)

We can now evaluate how the wave packet evolves. Beginning with (6.24a) we have

$$\psi(x,t) = \int_{-\infty}^{\infty} \widetilde{A}(k) \exp\{i(kx - \omega(k)t)\} \, dk \tag{6.30a}$$

$$\approx \int_{-\infty}^{\infty} \widetilde{A}(k) \exp\left\{ i[k_0 x - \omega(k_0)t] + i(k-k_0)x - i(k-k_0) \left.\frac{\partial \omega}{\partial k}\right|_{k=k_0} t \right\} dk, \tag{6.30b}$$

having expanded $\omega(k)$ in a Taylor series about k_0, using (6.23). We thus have

$$\begin{aligned} \psi(x,t) &= \exp\{i[k_0 x - \omega(k_0)t]\} \int \int_{-\infty}^{\infty} \widetilde{A}(k) \exp\left\{ i(k-k_0)\left[x - \left.\frac{\partial \omega}{\partial k}\right|_{k=k_0} t \right]\right\} dk \\ &= \exp\{i[k_0 x - \omega(k_0)t]\}\, a(x - c_g t), \end{aligned} \tag{6.31}$$

using (6.29b) and where $c_g = \partial\omega/\partial k$ evaluated at $k = k_0$. That is to say, the envelope $a(x,t)$ moves at the group velocity, keeping its initial shape.

The group velocity has a meaning beyond that implied by the derivation above: it turns out to be a quite general property of waves that energy (and certain other quadratic properties) propagate at the group velocity. This is to be expected, at least in the presence of coherent wave packets, because there is no energy outside the wave envelope so the energy must propagate with the envelope.

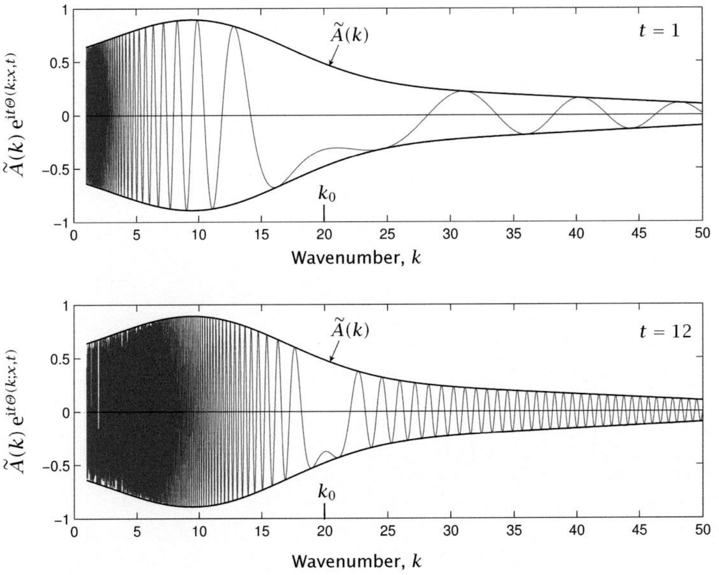

Fig. 6.4 The integrand of (6.32), namely the function that when integrated over wavenumber gives the wave amplitude at a particular x and t.

The example shown is for a Rossby wave with $\omega = -\beta/k$, with $\beta = 400$ and $x/t = 1$, and hence $k_0 = 20$, for two times $t = 1$ and $t = 12$. (The amplitude of the envelope, $\tilde{A}(k)$ is arbitrary.) At the later time the oscillations are much more rapid in k, so that the contribution is more peaked from wavenumbers near to k_0.

6.2.3 ✦ The Method of Stationary Phase

We will now relax the assumption that wavenumbers are confined to a narrow band but (since there is no free lunch) we confine ourselves to seeking solutions at large t; that is, we will be seeking a description of waves far from their source. Consider a disturbance of the general form

$$\psi(x,t) = \int_{-\infty}^{\infty} \tilde{A}(k) e^{i[kx-\omega(k)t]}\, dk = \int_{-\infty}^{\infty} \tilde{A}(k) e^{i\Theta(k;x,t)t}\, dk, \tag{6.32}$$

where $\Theta(k;x,t) \equiv kx/t - \omega(k)$. (Here we regard Θ as a function of k with parameters x and t; we will sometimes just write $\Theta(k)$ with $\Theta'(k) = \partial\Theta/\partial k$.) Now, a standard result in mathematics (known as the 'Riemann–Lebesgue lemma') states that

$$I = \lim_{t\to\infty} \int_{-\infty}^{\infty} f(k) e^{ikt}\, dk = 0, \tag{6.33}$$

provided that $f(k)$ is integrable and $\int_{-\infty}^{\infty} f(k)\, dk$ is finite. Intuitively, as t increases the oscillations in the integral increase and become much faster than any variation in $f(k)$; successive oscillations thus cancel and the integral becomes very small.

Looking at (6.32), with $\tilde{A}$ playing the role of $f(k)$, the integral will be small if Θ is everywhere varying with k. However, if there is a region where Θ does not vary with k — that is, if there is a region where the phase is stationary and $\partial\Theta/\partial k = 0$ — then there will be a contribution to the integral from that region. Thus, for large t, an observer will predominantly see waves for which $\Theta'(k) = 0$ and so, using the definition of Θ, for which

$$\frac{x}{t} = \frac{\partial\omega}{\partial k}. \tag{6.34}$$

In other words, at some space-time location (x,t) the waves that dominate are those whose group velocity $\partial\omega/\partial k$ is x/t. An example is plotted in Fig. 6.4 with a dispersion relation $\omega = -\beta/k$; the wavenumber that dominates, k_0 say, is thus given by solving $\beta/k_0^2 = x/t$, which for $x/t = 1$ and $\beta = 400$ gives $k_0 = 20$.

We may actually approximately calculate the contribution to $\psi(x,t)$ from waves moving with the group velocity. Let us expand $\Theta(k)$ around the point, k_0, where $\Theta'(k_0) = 0$. We obtain

$$\psi(x,t) = \int_{-\infty}^{\infty} \widetilde{A}(k) \exp\left\{it\left[\Theta(k_0) + (k-k_0)\Theta'(k_0) + \tfrac{1}{2}(k-k_0)^2\Theta''(k_0)\ldots\right]\right\} \, dk \tag{6.35}$$

The higher order terms are small because $k - k_0$ is presumed small (for if it is large the integral vanishes), and the term involving $\Theta'(k_0)$ is zero. The integral becomes

$$\psi(x,t) = \widetilde{A}(k_0)\mathrm{e}^{\mathrm{i}\Theta(k_0)} \int_{-\infty}^{\infty} \exp\left\{\mathrm{i}t\tfrac{1}{2}(k-k_0)^2\Theta''(k_0)\right\} \, dk. \tag{6.36}$$

We thus have to evaluate a Gaussian, and because $\int_{-\infty}^{\infty} \mathrm{e}^{-cx^2}\,dx = \sqrt{\pi/c}$ we obtain

$$\psi(x,t) \approx \widetilde{A}(k_0)\mathrm{e}^{\mathrm{i}\Theta(k_0)t}\left[\frac{-2\pi}{(\mathrm{i}t\theta''(k_0))}\right]^{1/2} = \widetilde{A}(k_0)\left[\frac{2\mathrm{i}\pi}{(t\theta''(k_0))}\right]^{1/2} \mathrm{e}^{\mathrm{i}(k_0x-\omega(k_0)t)}. \tag{6.37}$$

The solution is therefore a plane wave, with wavenumber k_0 and frequency $\omega(k_0)$, slowly modulated by an envelope determined by the form of $\Theta(k_0; x, t)$, where k_0 is the wavenumber such that $x/t = c_g = \partial\omega/\partial k|_{k=k_0}$.

6.3 RAY THEORY

Most waves propagate in a medium that is inhomogeneous; for example, in the Earth's atmosphere and ocean the stratification varies with altitude and the Coriolis parameter varies with latitude. In these cases it can be hard to obtain the solution of a wave problem by Fourier methods, even approximately. Nonetheless, the idea of signals propagating at the group velocity is a robust one, and we can often obtain some of the information we want — and in particular the trajectory of a wave — using a recipe known as *ray theory*.[3]

6.3.1 Introduction

In an inhomogeneous medium let us suppose that the solution to a particular wave problem is

$$\psi(\boldsymbol{x},t) = a(\boldsymbol{x},t)\mathrm{e}^{\mathrm{i}\theta(\boldsymbol{x},t)}, \tag{6.38}$$

where a is the wave amplitude and θ the phase, and a varies slowly in a sense we will make more precise shortly. The local wavenumber and frequency are defined by,

$$k_i \equiv \frac{\partial\theta}{\partial x_i}, \qquad \omega \equiv -\frac{\partial\theta}{\partial t}, \tag{6.39}$$

where the first expression is equivalent to $\boldsymbol{k} \equiv \nabla\theta$ and so $\nabla\times\boldsymbol{k} = 0$. We suppose that the amplitude a varies slowly over a wavelength and a period; that is $|\Delta a|/|a|$ is small over the length $1/k$ and the period $1/\omega$ or

$$\frac{|\partial a/\partial x|}{a} \ll |k|, \qquad \frac{|\partial a/\partial t|}{a} \ll \omega, \tag{6.40}$$

and similarly in the other directions. We will assume that the wavenumber and frequency as defined by (6.39) are the same as those that would arise if the medium were homogeneous and a were a constant. Thus, we may obtain a local dispersion relation from the governing equation by keeping the spatially (and possibly temporally) varying parameters fixed and obtain

$$\omega = \Omega(k_i; x_i, t), \tag{6.41}$$

and then allow x_i and t to vary, albeit slowly.

Let us now consider how the wavevector and frequency might change with position and time. It follows from their definitions above that the wavenumber and frequency are related by

$$\frac{\partial k_i}{\partial t} + \frac{\partial \omega}{\partial x_i} = 0, \tag{6.42}$$

where we use a subscript notation for vectors and repeated indices are summed. Using (6.42) and (6.41) gives

$$\frac{\partial k_i}{\partial t} + \frac{\partial \Omega}{\partial x_i} + \frac{\partial \Omega}{\partial k_j}\frac{\partial k_j}{\partial x_i} = 0 \qquad \text{or} \qquad \frac{\partial k_i}{\partial t} + \frac{\partial \Omega}{\partial x_i} + \frac{\partial \Omega}{\partial k_j}\frac{\partial k_i}{\partial x_j} = 0, \tag{6.43a,b}$$

where to get (6.43b) we use $\partial k_j/\partial x_i = \partial k_i/\partial x_j$, true because $\boldsymbol{k}$ has no curl. Equation (6.43b) may be written as

$$\frac{\partial \boldsymbol{k}}{\partial t} + \boldsymbol{c}_g \cdot \nabla \boldsymbol{k} = -\nabla \Omega, \tag{6.44}$$

where

$$\boldsymbol{c}_g = \frac{\partial \Omega}{\partial \boldsymbol{k}} = \left(\frac{\partial \Omega}{\partial k}, \frac{\partial \Omega}{\partial l}, \frac{\partial \Omega}{\partial m} \right) \tag{6.45}$$

is, once more, the group velocity. The left-hand side of (6.44) is similar to an advective derivative, but uses the group velocity. Evidently, if the dispersion relation for frequency is not an explicit function of space then *the wavevector is propagated at the group velocity.*

The frequency is, in general, a function of space, wavenumber and time, and from the dispersion relation, (6.41), its variation is governed by

$$\frac{\partial \omega}{\partial t} = \frac{\partial \Omega}{\partial t} + \frac{\partial \Omega}{\partial k_i}\frac{\partial k_i}{\partial t} = \frac{\partial \Omega}{\partial t} - \frac{\partial \Omega}{\partial k_i}\frac{\partial \omega}{\partial x_i}, \tag{6.46}$$

using (6.42). Using the definition of group velocity, we may write (6.46) as

$$\frac{\partial \omega}{\partial t} + \boldsymbol{c}_g \cdot \nabla \omega = \frac{\partial \Omega}{\partial t}. \tag{6.47}$$

As with (6.44) the left-hand side is like an advective derivative, but uses a group velocity. Thus, if the dispersion relation is not a function of time, the frequency also propagates at the group velocity.

Motivated by (6.44) and (6.47) we define a *ray* as the trajectory traced by the group velocity, and we see that if the function Ω is not an explicit function of space or time, then *both the wavevector and the frequency are constant along a ray.*

6.3.2 Ray Theory in Practice

What use is ray theory? The idea is that we may use (6.44) and (6.47) to track a group of waves from one location to another without solving the full wave equations of motion. We can then sometimes solve interesting problems using ordinary differential equations (ODEs) rather than partial differential equations (PDEs).

Suppose that the initial conditions consist of a group of waves at a position x_0, for which the amplitude and wavenumber vary only slowly with position. We also suppose that we know the dispersion relation for the waves at hand; that is, we know the functional form of $\Omega(k; x, t)$. Now, the total derivative following the group velocity is given by

$$\frac{\mathrm{d}}{\mathrm{d}t} = \frac{\partial}{\partial t} + \boldsymbol{c}_g \cdot \nabla, \tag{6.48}$$

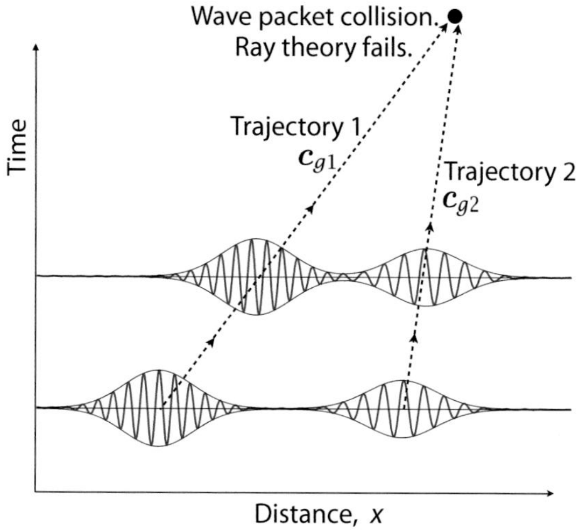

Fig. 6.5 Idealised trajectory of two wavepackets, each with a different wavelength and moving with a different group velocity, as might be calculated using ray theory. If the wave packets collide ray theory must fail. Ray theory gives only the trajectory of the wave packet, not the detailed structure of the waves within a packet.

so that (6.44) and (6.47) may be written as

$$\frac{\mathrm{d}\boldsymbol{k}}{\mathrm{d}t} = -\nabla\Omega, \qquad \frac{\mathrm{d}\omega}{\mathrm{d}t} = -\frac{\partial\Omega}{\partial t}. \tag{6.49a,b}$$

These are ordinary differential equations for wavevector and frequency, solvable provided we know the right-hand sides — that is, provided we know the space and time location at which the dispersion relation [i.e., $\Omega(k; x, t)$] is to be evaluated. But the location *is* known because it is moving with the group velocity and so

$$\frac{\mathrm{d}\boldsymbol{x}}{\mathrm{d}t} = \boldsymbol{c}_g, \tag{6.49c}$$

where $\boldsymbol{c}_g = \partial\Omega/\partial\boldsymbol{k}|_{\boldsymbol{x},t}$ (i.e., $c_{gi} = \partial\Omega/\partial k_i|_{\boldsymbol{x},t}$). The set (6.49a) is a triplet of ordinary differential equations for the wavevector, frequency and position of a wave group. The equations may be solved, albeit sometimes numerically, to give the trajectory of a wave packet or collection of wave packets as schematically illustrated in Fig. 6.5. Of course, if the medium or the wavepacket amplitude is not slowly varying ray theory will fail, and this will perforce happen if two wave packets collide.

The evolution of the amplitude of the wave packet is not given by ray theory. However, the evolution of a related quantity — the wave activity — may be calculated if the group velocity is known. In Section 6.7 we will find that the wave activity, $\mathcal{A}$, satisfies $\partial\mathcal{A}/\partial t + \nabla\cdot(\boldsymbol{c}_g\mathcal{A}) = 0$; that is, the flux of wave activity is along a ray. Another way to calculate the evolution of a wave and its amplitude in a varying medium is to use 'WKB theory' — see Appendix A. Before all that we shift gears and turn our attention to a specific wave, the Rossby wave, but the reader whose interest is more in the general properties of waves may skip forward to Section 6.7.

6.4 ROSSBY WAVES

Rossby waves are perhaps the most important large-scale wave in the atmosphere and ocean, although gravity waves rival them in some contexts.[4] They are most easily described using the quasi-geostrophic equations, as follows.

6.4.1 The Linear Equation of Motion

The relevant equation of motion is the inviscid, adiabatic potential vorticity equation in the quasi-geostrophic system, as discussed in Chapter 5, namely

$$\frac{\partial q}{\partial t} + \boldsymbol{u}\cdot\nabla q = 0, \tag{6.50}$$

where $q(x, y, z, t)$ is the potential vorticity and $\boldsymbol{u}(x, y, z, t)$ is the horizontal velocity. The velocity is related to a streamfunction by $u = -\partial\psi/\partial y$, $v = \partial\psi/\partial x$ and the potential vorticity is some function of the streamfunction, which might differ from system to system. Two examples, one applying to a continuously stratified system and the second to a single layer system, are

$$q = f + \zeta + \frac{\partial}{\partial z}\left(S(z)\frac{\partial \psi}{\partial z}\right), \qquad q = \zeta + f - k_d^2\psi, \tag{6.51a,b}$$

where $S(z) = f_0^2/N^2$, $\zeta = \nabla^2\psi$ is the relative vorticity and $k_d = 1/L_d$ is the inverse radius of deformation for a shallow water system. (Note that definitions of k_d and L_d can vary, typically by factors of 2, π, etc.) Boundary conditions may be needed to form a complete system.

We now *linearize* (6.50); that is, we suppose that the flow consists of a time-independent component (the 'basic state') plus a perturbation, with the perturbation being small compared with the mean flow. The basic state must satisfy the time-independent equation of motion, and it is common and useful to linearize about a zonal flow, $\overline{u}(y, z)$. The basic state is then purely a function of y and so we write

$$q = \overline{q}(y, z) + q'(x, y, t), \qquad \psi = \overline{\psi}(y, z) + \psi'(x, y, z, t) \tag{6.52}$$

with a similar notation for the other variables. Note that $\overline{u} = -\partial\overline{\psi}/\partial y$ and $\overline{v} = 0$. Substituting into (6.50) gives, without approximation,

$$\frac{\partial q'}{\partial t} + \overline{\boldsymbol{u}}\cdot\nabla\overline{q} + \overline{\boldsymbol{u}}\cdot\nabla q' + \boldsymbol{u}'\cdot\nabla\overline{q} + \boldsymbol{u}'\cdot\nabla q' = 0. \tag{6.53}$$

The primed quantities are presumptively small so we neglect terms involving their products. Further, we are assuming that we are linearizing about a state that is a solution of the equations of motion, so that $\overline{\boldsymbol{u}}\cdot\nabla\overline{q} = 0$. Finally, since $\overline{v} = 0$ and $\partial\overline{q}/\partial x = 0$ we obtain

$$\frac{\partial q'}{\partial t} + \overline{u}\frac{\partial q'}{\partial x} + v'\frac{\partial \overline{q}}{\partial y} = 0. \tag{6.54}$$

This equation or one very similar appears very commonly in studies of Rossby waves. To proceed, let us consider the simple example of waves in a single layer.

6.4.2 Waves in a Single Layer

Consider a system obeying (6.50) and (6.51b). The equation could be written in spherical coordinates with $f = 2\Omega\sin\vartheta$, but the dynamics are more easily illustrated on a Cartesian β-plane for which $f = f_0 + \beta y$, and since f_0 is a constant it does not appear in our subsequent derivations.

Infinite deformation radius

If the scale of motion is much less than the deformation scale then we make the approximation that $k_d = 0$ and the equation of motion may be written as

$$\frac{\partial \zeta}{\partial t} + \boldsymbol{u}\cdot\nabla\zeta + \beta v = 0. \tag{6.55}$$

We linearize about a constant zonal flow, $\overline{u}$, by writing

$$\psi = \overline{\psi}(y) + \psi'(x, y, t), \tag{6.56}$$

where $\overline{\psi} = -\overline{u}y$. Substituting(6.56) into (6.55) and neglecting the nonlinear terms involving products of ψ' gives

$$\frac{\partial}{\partial t}\nabla^2\psi' + \overline{u}\frac{\partial \nabla^2\psi'}{\partial x} + \beta\frac{\partial \psi'}{\partial x} = 0. \tag{6.57}$$

This equation is just a single-layer version of (6.54), with $\partial\overline{q}/\partial y = \beta$, $q' = \nabla^2\psi'$ and $v' = \partial\psi'/\partial x$.

The coefficients in (6.57) are not functions of y or z; this is not a requirement for wave motion to exist but it does enable solutions to be found more easily. Let us seek solutions in the form of a plane wave, namely

$$\psi' = \mathrm{Re}\,\widetilde{\psi}\mathrm{e}^{\mathrm{i}(kx+ly-\omega t)}, \tag{6.58}$$

where $\widetilde{\psi}$ is a complex constant. Solutions of this form are valid in a domain with doubly-periodic boundary conditions; solutions in a channel can be obtained using a meridional variation of $\sin ly$, with no essential changes to the dynamics. The amplitude of the oscillation is given by $\widetilde{\psi}$ and the phase by $kx + ly - \omega t$, where k and l are the x- and y-wavenumbers and ω is the frequency of the oscillation.

Substituting (6.58) into (6.57) yields

$$[(-\omega + \overline{u}k)(-K^2) + \beta k]\widetilde{\psi} = 0, \tag{6.59}$$

where $K^2 = k^2 + l^2$. For non-trivial solutions the above equation implies

$$\omega = \overline{u}k - \frac{\beta k}{K^2}, \tag{6.60}$$

and this is the *dispersion relation* for barotropic Rossby waves. Evidently the velocity U Doppler shifts the frequency by the amount Uk. The components of the phase speed and group velocity are given by, respectively,

$$c_p^x \equiv \frac{\omega}{k} = \overline{u} - \frac{\beta}{K^2}, \qquad c_p^y \equiv \frac{\omega}{l} = \overline{u}\frac{k}{l} - \frac{\beta k}{K^2 l}, \tag{6.61a,b}$$

and

$$c_g^x \equiv \frac{\partial\omega}{\partial k} = \overline{u} + \frac{\beta(k^2 - l^2)}{(k^2 + l^2)^2}, \qquad c_g^y \equiv \frac{\partial\omega}{\partial l} = \frac{2\beta kl}{(k^2 + l^2)^2}. \tag{6.62a,b}$$

The phase speed in the absence of a mean flow is *westward*, with waves of longer wavelengths travelling more quickly, and the eastward current speed required to hold the waves of a particular wavenumber stationary (i.e., $c_p^x = 0$) is $U = \beta/K^2$. The background flow $\overline{u}$ evidently just provides a uniform shift to the phase speed, and (in this case) can be transformed away by a change of coordinate. The x-component of the group velocity may also be written as the sum of the phase speed plus a positive quantity, namely

$$c_g^x = c_p^x + \frac{2\beta k^2}{(k^2 + l^2)^2}. \tag{6.63}$$

This means that the zonal group velocity for Rossby wave packets moves eastward faster than its zonal phase speed. A stationary wave ($c_p^x = 0$) can only propagate eastward, and this has implications for the 'downstream development' of Rossby wave packets.[5]

Finite deformation radius

For a finite deformation radius the basic state $\Psi = -\overline{u}y$ is still a solution of the original equations of motion, but the potential vorticity corresponding to this state is $q = \overline{u}yk_d^2 + \beta y$ and its gradient is $\nabla q = (\beta + \overline{u}k_d^2)\mathbf{j}$. The linearized equation of motion is thus

$$\left(\frac{\partial}{\partial t} + \overline{u}\frac{\partial}{\partial x}\right)(\nabla^2\psi' - \psi' k_d^2) + (\beta + \overline{u}k_d^2)\frac{\partial \psi'}{\partial x} = 0. \tag{6.64}$$

Substituting $\psi' = \widetilde{\psi}e^{i(kx+ly-\omega t)}$ we obtain the dispersion relation,

$$\omega = \frac{k(\overline{u}K^2 - \beta)}{K^2 + k_d^2} = \overline{u}k - k\frac{\beta + \overline{u}k_d^2}{K^2 + k_d^2}. \tag{6.65}$$

The corresponding components of phase speed and group velocity are

$$c_p^x = \overline{u} - \frac{\beta + \overline{u}k_d^2}{K^2 + k_d^2} = \frac{\overline{u}K^2 - \beta}{K^2 + k_d^2}, \qquad c_p^y = \overline{u}\frac{k}{l} - \frac{k}{l}\left(\frac{\overline{u}K^2 - \beta}{K^2 + k_d^2}\right), \tag{6.66a,b}$$

and

$$c_g^x = \overline{u} + \frac{(\beta + \overline{u}k_d^2)(k^2 - l^2 - k_d^2)}{(k^2 + l^2 + k_d^2)^2}, \qquad c_g^y = \frac{2kl(\beta + \overline{u}k_d^2)}{(k^2 + l^2 + k_d^2)^2}. \tag{6.67a,b}$$

The uniform velocity field now no longer provides just a simple Doppler shift of the frequency, nor a uniform addition to the phase speed. From (6.66a) the waves are stationary when $K^2 = \beta/\overline{u} \equiv K_s^2$; that is, the current speed required to hold waves of a particular wavenumber stationary is $\overline{u} = \beta/K^2$. However, this is *not* simply the magnitude of the phase speed of waves of that wavenumber in the absence of a current — this is given by

$$c_p^x = \frac{-\beta}{K_s^2 + k_d^2} = \frac{-\overline{u}}{1 + k_d^2/K_s^2} \neq -\overline{u}. \tag{6.68}$$

Why is there a difference? It is because the current does not just provide a uniform translation, but, if k_d is non-zero, it also modifies the basic potential vorticity gradient. The basic state height field η_0 is sloping; that is $\eta_0 = -(f_0/g)\overline{u}y$, and the ambient potential vorticity field increases with y and $q = (\beta + Uk_d^2)y$. Thus, the basic state defines a preferred frame of reference, and the problem is not Galilean invariant.[6]

We also note that, from (6.67a), the group velocity is negative (westward) if the x-wavenumber is sufficiently small compared to the y-wavenumber or the deformation wavenumber. That is, said a little loosely, *long waves move information westward and short waves move information eastward*, and this is a common property of Rossby waves. The x-component of the phase speed, on the other hand, is always westward relative to the mean flow.

6.4.3 The Mechanism of Rossby Waves

The fundamental mechanism underlying Rossby waves may be understood as follows. Consider a material line of stationary fluid parcels along a line of constant latitude, and suppose that some disturbance causes their displacement to the line marked $\eta(t = 0)$ in Fig. 6.6. In the displacement, the potential vorticity of the fluid parcels is conserved, and in the simplest case of barotropic flow on the β-plane the potential vorticity is the absolute vorticity, $\beta y + \zeta$. Thus, in either hemisphere, a northward displacement leads to the production of negative relative vorticity and a southward displacement leads to the production of positive relative vorticity. The relative vorticity gives rise

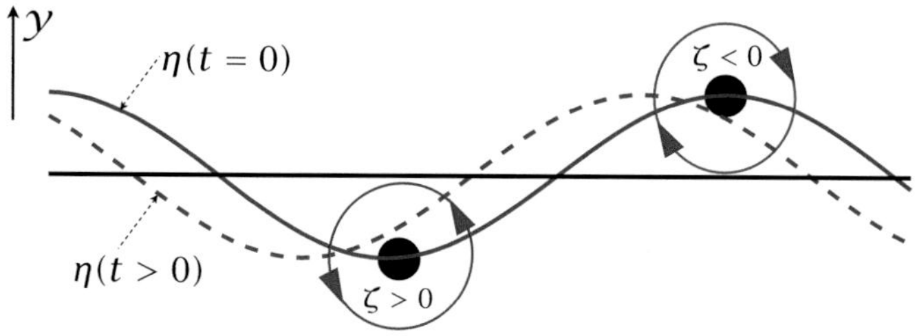

Fig. 6.6 A two-dimensional (x–y) Rossby wave. An initial disturbance displaces a material line at constant latitude (the straight horizontal line) to the solid line marked $\eta(t=0)$. Conservation of potential vorticity, $\beta y + \zeta$, leads to the production of relative vorticity, ζ, as shown. The associated velocity field (arrows on the circles) then advects the fluid parcels, and the material line evolves into the dashed line with the phase propagating westward.

to a velocity field which, in turn, advects the parcels in the material line in the manner shown, and the wave propagates westwards.

In more complicated situations, such as flow in two layers, considered below, or in a continuously stratified fluid, the mechanism is essentially the same. A displaced fluid parcel carries with it its potential vorticity and, in the presence of a potential vorticity gradient in the basic state, a potential vorticity anomaly is produced. The potential vorticity anomaly produces a velocity field (an example of potential vorticity inversion) which further displaces the fluid parcels, leading to the formation of a Rossby wave. The vital ingredient is a basic state potential vorticity gradient, such as that provided by the change of the Coriolis parameter with latitude.

6.4.4 Rossby Waves in Two Layers

Now consider the dynamics of the two-layer model, linearized about a state of rest. The two, coupled, linear equations describing the motion in each layer are

$$\frac{\partial}{\partial t}\left[\nabla^2\psi_1' + F_1(\psi_2' - \psi_1')\right] + \beta\frac{\partial\psi_1'}{\partial x} = 0, \tag{6.69a}$$

$$\frac{\partial}{\partial t}\left[\nabla^2\psi_2' + F_2(\psi_1' - \psi_2')\right] + \beta\frac{\partial\psi_2'}{\partial x} = 0, \tag{6.69b}$$

where $F_1 = f_0^2/g'H_1$ and $F_2 = f_0^2/g'H_2$. By inspection (6.69) may be transformed into two uncoupled equations: the first is obtained by multiplying (6.69a) by F_2 and (6.69b) by F_1 and adding, and the second is the difference of (6.69a) and (6.69b). Then, defining

$$\overline{\psi} = \frac{F_1\psi_2' + F_2\psi_1'}{F_1 + F_2}, \qquad \tau = \frac{1}{2}(\psi_1' - \psi_2'), \tag{6.70a,b}$$

(think 'τ for temperature'), (6.69) become

$$\frac{\partial}{\partial t}\nabla^2\overline{\psi} + \beta\frac{\partial\overline{\psi}}{\partial x} = 0, \tag{6.71a}$$

$$\frac{\partial}{\partial t}\left[(\nabla^2 - k_d^2)\tau\right] + \beta\frac{\partial\tau}{\partial x} = 0, \tag{6.71b}$$

where now $k_d = (F_1 + F_2)^{1/2}$. The internal radius of deformation for this problem is the inverse of this, namely

$$L_d = k_d^{-1} = \frac{1}{f_0}\left(\frac{g'H_1H_2}{H_1 + H_2}\right)^{1/2}. \tag{6.72}$$

The variables $\overline{\psi}$ and τ are the *normal modes* for the two-layer model, as they oscillate independently of each other. (For the continuous equations the analogous modes are the eigenfunctions of $\partial_z[(f_0^2/N^2)\partial_z\phi] = \lambda^2\phi$.) The equation for $\overline{\psi}$, *the barotropic mode,* is identical to that of the single-layer, rigid-lid model, namely (6.57) with $U = 0$, and its dispersion relation is just

$$\omega = -\frac{\beta k}{K^2}. \tag{6.73}$$

The barotropic mode corresponds to synchronous, depth-independent, motion in the two layers, with no undulations in the dividing interface.

The displacement of the interface is given by $2f_0\tau/g'$ and so is proportional to the amplitude of τ, the *baroclinic mode.* The dispersion relation for the baroclinic mode is

$$\omega = -\frac{\beta k}{K^2 + k_d^2}. \tag{6.74}$$

The mass transport associated with this mode is identically zero, since from (6.70) we have

$$\psi_1 = \overline{\psi} + \frac{2F_1\tau}{F_1 + F_2}, \qquad \psi_2 = \overline{\psi} - \frac{2F_2\tau}{F_1 + F_2}, \tag{6.75a,b}$$

and this implies

$$H_1\psi_1 + H_2\psi_2 = (H_1 + H_2)\overline{\psi}. \tag{6.76}$$

The left-hand side is proportional to the total mass transport, which is evidently associated with the barotropic mode.

The dispersion relation and associated group and phase velocities are plotted in Fig. 6.7. The x-component of the phase speed, ω/k, is negative (westwards) for both baroclinic and barotropic Rossby waves. The group velocity of the barotropic waves is always positive (eastwards), but the group velocity of long baroclinic waves may be negative (westwards). For very short waves, $k^2 \gg k_d^2$, the baroclinic and barotropic velocities coincide and their phase and group velocities are equal and opposite. With a deformation radius of 50 km, typical for the mid-latitude ocean, a non-dimensional frequency of unity in the figure corresponds to a dimensional frequency of $5\times10^{-7}\,\mathrm{s}^{-1}$ or a period of about 100 days. In an atmosphere with a deformation radius of 1000 km a nondimensional frequency of unity corresponds to $1\times10^{-5}\,\mathrm{s}^{-1}$ or a period of about 7 days. nondimensional velocities of unity correspond to respective dimensional velocities of about $0.25\,\mathrm{m\,s^{-1}}$ (ocean) and $10\,\mathrm{m\,s^{-1}}$ (atmosphere).

The deformation radius only affects the baroclinic mode. For scales much smaller than the deformation radius, $K^2 \gg k_d^2$, we see from (6.71b) that the baroclinic mode obeys the same equation as the barotropic mode so that

$$\frac{\partial}{\partial t}\nabla^2\tau + \beta\frac{\partial\tau}{\partial x} = 0. \tag{6.77}$$

Using this and (6.71a) implies that

$$\frac{\partial}{\partial t}\nabla^2\psi_i + \beta\frac{\partial\psi_i}{\partial x} = 0, \qquad i = 1, 2. \tag{6.78}$$

That is to say, the two layers themselves are uncoupled from each other. At the other extreme, for very long baroclinic waves the relative vorticity is unimportant.

6.5 ROSSBY WAVES IN STRATIFIED QUASI-GEOSTROPHIC FLOW

6.5.1 Preliminaries

Let us now consider the dynamics of linear waves in stratified quasi-geostrophic flow on a β-plane, with a resting basic state. (In Chapter 16 we explore the role of Rossby waves in a more realistic

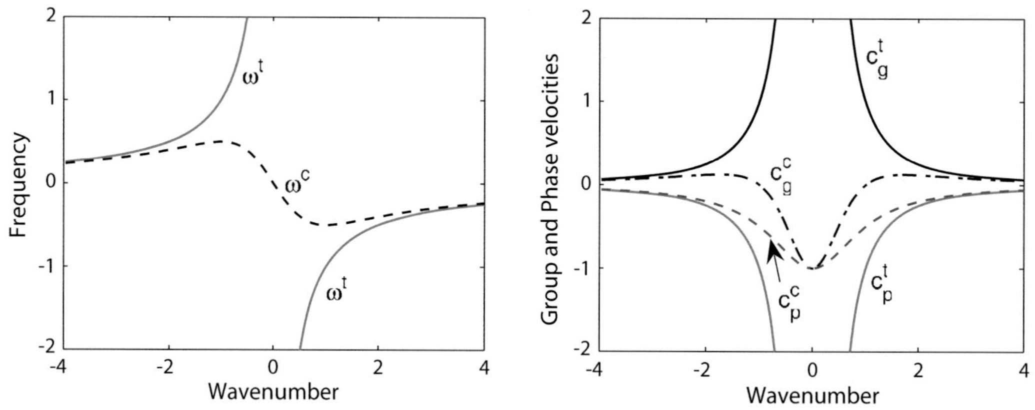

Fig. 6.7 Left: the dispersion relation for barotropic (ω^t, solid line) and baroclinic (ω^c, dashed line) Rossby waves in the two-layer model, calculated using (6.73) and (6.74) with $k^y = 0$, plotted for both positive and negative zonal wavenumbers and frequencies. The wavenumber is nondimensionalized by k_d, and the frequency is nondimensionalized by β/k_d. Right: the corresponding zonal group and phase velocities, $c_g = \partial\omega/\partial k^x$ and $c_p = \omega/k^x$, with superscript 't' or 'c' for the barotropic or baroclinic mode, respectively. The velocities are nondimensionalized by β/k_d^2.

setting.) The interior flow is governed by the potential vorticity equation, (5.118), and linearizing this about a state of rest gives

$$\frac{\partial}{\partial t}\left[\nabla^2\psi' + \frac{1}{\widetilde{\rho}(z)}\frac{\partial}{\partial z}\left(\widetilde{\rho}(z)F(z)\frac{\partial\psi'}{\partial z}\right)\right] + \beta\frac{\partial\psi'}{\partial x} = 0, \tag{6.79}$$

where $\widetilde{\rho}$ is the density profile of the basic state and $F(z) = f_0^2/N^2$. (F is the square of the inverse Prandtl ratio, N/f_0.) In the Boussinesq approximation $\widetilde{\rho} = \rho_0$, i.e., a constant. The vertical boundary conditions are determined by the thermodynamic equation, (5.120). If the boundaries are flat, rigid, slippery surfaces then $w = 0$ at the boundaries and if there is no surface buoyancy gradient the linearized thermodynamic equation is

$$\frac{\partial}{\partial t}\left(\frac{\partial\psi'}{\partial z}\right) = 0. \tag{6.80}$$

We apply this at the ground and, with somewhat less justification, at the tropopause: we assume the higher static stability of the stratosphere inhibits vertical motion. If the ground is not flat or if friction provides a vertical velocity by way of an Ekman layer, the boundary condition must be modified, but we will stay with the simplest case and apply (6.80) at $z = 0$ and $z = H$.

6.5.2 Wave Motion

As in the single-layer case, we seek solutions of the form

$$\psi' = \mathrm{Re}\,\widetilde{\psi}(z)\mathrm{e}^{\mathrm{i}(kx+ly-\omega t)}, \tag{6.81}$$

where $\psi(z)$ will determine the vertical structure of the waves. The case of a sphere is more complicated but introduces no truly new physical phenomena.

Substituting (6.81) into (6.79) gives

$$\omega\left[-K^2\widetilde{\psi}(z) + \frac{1}{\widetilde{\rho}}\frac{\mathrm{d}}{\mathrm{d}z}\left(\widetilde{\rho}F(z)\frac{\mathrm{d}\widetilde{\psi}}{\mathrm{d}z}\right)\right] - \beta k\widetilde{\psi}(z) = 0. \tag{6.82}$$

Now, let us suppose that $\widetilde{\psi}$ satisfies

$$\frac{1}{\widetilde{\rho}}\frac{\mathrm{d}}{\mathrm{d}z}\left(\widetilde{\rho}F(z)\frac{\mathrm{d}\widetilde{\psi}}{\mathrm{d}z}\right) = -\Gamma\widetilde{\psi}, \tag{6.83}$$

where Γ is some constant (in fact it is an eigenvalue, as discussed below). Equation (6.82) becomes

$$-\,\omega\left[K^2 + \Gamma\right]\widetilde{\psi} - \beta k\widetilde{\psi} = 0, \tag{6.84}$$

and the dispersion relation follows, namely

$$\omega = -\frac{\beta k}{K^2 + \Gamma}. \tag{6.85}$$

The value of Γ is obtained by solving the eigenvalue problem, (6.83), for the vertical structure; the boundary conditions, derived from (6.80), are $\partial\widetilde{\psi}/\partial z = 0$ at $z = 0$ and $z = H$. The resulting eigenvalues, Γ are proportional to the inverse of the squares of the deformation radii for the problem and the eigenfunctions are the vertical structure functions.

A simple example

Consider the case in which $F(z)$ and $\widetilde{\rho}$ are constant, and in which the domain is confined between two rigid surfaces at $z = 0$ and $z = H$. Then the eigenvalue problem for the vertical structure is

$$F\frac{\mathrm{d}^2\widetilde{\psi}}{\mathrm{d}z^2} = -\Gamma\widetilde{\psi}, \tag{6.86a}$$

with boundary conditions of

$$\frac{\mathrm{d}\widetilde{\psi}}{\mathrm{d}z} = 0, \qquad \text{at } z = 0, H. \tag{6.86b}$$

There is a sequence of solutions to this, namely

$$\widetilde{\psi}_n(z) = \cos(n\pi z/H), \qquad n = 1, 2\ldots, \tag{6.87}$$

with corresponding eigenvalues

$$\Gamma_n = n^2\frac{F\pi^2}{H^2} = (n\pi)^2\left(\frac{f_0}{NH}\right)^2, \qquad n = 1, 2\ldots. \tag{6.88}$$

Equation (6.88) may be used to define the deformation radii for this problem, namely

$$L_n \equiv \frac{1}{\sqrt{\Gamma_n}} = \frac{NH}{n\pi f_0}. \tag{6.89}$$

The first deformation radius is the same as the expression obtained by dimensional analysis, namely NH/f_0, (or NH/f if f varies) except for a factor of π. (Definitions of the deformation radii both with and without the factor of π are common in the literature, and neither is obviously more correct. In the latter case, the first deformation radius in a problem with uniform stratification is given by NH/f_0, equal to $\pi/\sqrt{\Gamma_1}$.) In addition to these baroclinic modes, the case with $n = 0$, that is with $\widetilde{\psi} = 1$, is also a solution of (6.86) for any $F(z)$.

Using (6.85) and (6.88) the dispersion relation becomes

$$\omega = -\frac{\beta k}{K^2 + (n\pi)^2(f_0/NH)^2}, \qquad n = 0, 1, 2\ldots, \tag{6.90}$$

and the horizontal wavenumbers k and l are also quantized in a finite domain. The dynamics of the barotropic mode are independent of height and independent of the stratification of the basic state, and so these Rossby waves are *identical* with the Rossby waves in a homogeneous fluid contained between two flat rigid surfaces. The structure of the baroclinic modes, which in general depends on the structure of the stratification, becomes increasingly complex as the vertical wavenumber n increases. This increasing complexity naturally leads to a certain delicacy, making it rare that they can be unambiguously identified in nature. The eigenproblem for a realistic atmospheric profile is further complicated because of the lack of a rigid lid at the top of the atmosphere.[7]

6.6 ENERGY PROPAGATION AND REFLECTION OF ROSSBY WAVES

We now consider how energy is fluxed in Rossby waves. To keep matters algebraically simple we consider waves in a single layer and without a mean flow, but we allow for a finite radius of deformation. To remind ourselves, the dynamics are governed by the evolution of potential vorticity and the linearized evolution equation is

$$\frac{\partial}{\partial t}\left(\nabla^2 - k_d^2\right)\psi + \beta\frac{\partial \psi}{\partial x} = 0. \tag{6.91}$$

The dispersion relation follows in the usual way and is

$$\omega = \frac{-k\beta}{K^2 + k_d^2}, \tag{6.92}$$

which is a simplification of (6.65), and the group velocities are

$$c_g^x = \frac{\beta(k^2 - l^2 - k_d^2)}{(K^2 + k_d^2)^2}, \qquad c_g^y = \frac{2\beta kl}{(K^2 + k_d^2)^2}, \tag{6.93a,b}$$

which are simplifications of (6.67), and as usual $K^2 = k^2 + l^2$.

To obtain an energy equation multiply (6.91) by $-\psi$ to obtain, after a couple of lines of algebra,

$$\frac{1}{2}\frac{\partial}{\partial t}\left((\nabla\psi)^2 + k_d^2\psi^2\right) - \nabla\cdot\left(\psi\nabla\frac{\partial\psi}{\partial t} + \mathbf{i}\frac{\beta}{2}\psi^2\right) = 0, \tag{6.94}$$

where $\mathbf{i}$ is the unit vector in the x-direction. The first group of terms are the energy itself, or more strictly the energy density. (An energy density is an energy per unit mass or per unit volume, depending on the context.) The term $(\nabla\psi)^2/2 = (u^2 + v^2)/2$ is the kinetic energy and $k_d^2\psi^2/2$ is the potential energy, proportional to the displacement of the free surface, squared. The second term is the energy flux, so that we may write

$$\frac{\partial E}{\partial t} + \nabla\cdot\boldsymbol{F} = 0, \tag{6.95}$$

where $E = (\nabla\psi)^2/2 + k_d^2\psi^2$ and $\boldsymbol{F} = -(\psi\nabla\partial\psi/\partial t + \mathbf{i}\beta\psi^2)$. We haven't yet used the fact that the disturbance has a dispersion relation, and if we do so we may expect, following the derivations of Section 6.2, that the energy moves at the group velocity. Let us now demonstrate this explicitly.

We assume a solution of the form

$$\psi = A(x)\cos(\boldsymbol{k}\cdot\boldsymbol{x} - \omega t) = A(x)\cos(kx + ly - \omega t) \tag{6.96}$$

where $A(x)$ is assumed to vary slowly compared to the nearly plane wave. (Note that $\boldsymbol{k}$ is the wave vector, to be distinguished from $\mathbf{k}$, the unit vector in the z-direction.) The kinetic energy in a wave is given by

$$\text{KE} = \frac{A^2}{2}\left(\psi_x^2 + \psi_y^2\right), \tag{6.97}$$

Essentials of Rossby Waves

- Rossby waves owe their existence to a gradient of potential vorticity in the fluid. If a fluid parcel is displaced, it conserves its potential vorticity and so its relative vorticity will in general change. The relative vorticity creates a velocity field that displaces neighbouring parcels, whose relative vorticity changes and so on.
- A common source of a potential vorticity gradient is differential rotation, or the β-effect and the associated Rossby waves are called *planetary waves*. In the presence of non-zero β the ambient potential vorticity increases northward and the phase of the Rossby waves propagates westward. In general, Rossby waves propagate pseudo-westwards, meaning to the left of the direction of increasing potential vorticity.
- A common equation of motion for Rossby waves is

$$\frac{\partial q'}{\partial t} + \overline{u}\frac{\partial q'}{\partial x} + v'\frac{\partial \overline{q}}{\partial y} = 0, \tag{RW.1}$$

  with an overbar denoting the basic state and a prime a perturbation. In the case of a single layer of fluid with no mean flow this equation becomes

$$\frac{\partial}{\partial t}(\nabla^2 + k_d^2)\psi' + \beta\frac{\partial \psi'}{\partial x} = 0, \tag{RW.2}$$

  with dispersion relation

$$\omega = \frac{-\beta k}{k^2 + l^2 + k_d^2}. \tag{RW.3}$$

- In the absence of a mean flow (i.e., $\overline{u} = 0$), the phase speed in the zonal direction ($c_p^x = \omega/k$) is always negative, or westward, and is larger for large waves. For (RW.3) the components of the group velocity are given by

$$c_g^x = \frac{\beta(k^2 - l^2 - k_d^2)}{(k^2 + l^2 + k_d^2)^2}, \qquad c_g^y = \frac{2\beta kl}{(k^2 + l^2 + k_d^2)^2}. \tag{RW.4}$$

  The group velocity is westward if the zonal wavenumber is sufficiently small, and eastward if the zonal wavenumber is sufficiently large.
- Rossby waves exist in stratified fluids, and have a similar dispersion relation to (RW.3) with an appropriate vertical wavenumber appearing in place of the inverse deformation radius, k_d.
- The reflection of such Rossby waves at a wall is specular, meaning that the group velocity of the reflected wave makes the same angle with the wall as the group velocity of the incident wave. The energy flux of the reflected wave is equal and opposite to that of the incoming wave in the direction normal to the wall.

so that, averaged over a wave period,

$$\overline{\mathrm{KE}} = \frac{A^2}{2}(k^2 + l^2)\frac{\omega}{2\pi}\int_0^{2\pi/\omega} \sin^2(\boldsymbol{k}\cdot\boldsymbol{x} - \omega t)\,\mathrm{d}t. \tag{6.98}$$

The time-averaging produces a factor of one half, and applying a similar procedure to the potential energy we obtain

$$\overline{\mathrm{KE}} = \frac{A^2}{4}(k^2 + l^2), \qquad \overline{\mathrm{PE}} = \frac{A^2}{4}k_d^2, \tag{6.99}$$

so that the average total energy is

$$\overline{E} = \frac{A^2}{4}(K^2 + k_d^2), \tag{6.100}$$

where $K^2 = k^2 + l^2$.

The flux, $\boldsymbol{F}$, is given by

$$\boldsymbol{F} = -\left(\psi\nabla\frac{\partial\psi}{\partial t} + \mathbf{i}\frac{\beta}{2}\psi^2\right) = -A^2\cos^2(\boldsymbol{k}\cdot\boldsymbol{x} - \omega t)\left(\boldsymbol{k}\omega - \mathbf{i}\frac{\beta}{2}\right), \tag{6.101}$$

so that evidently the energy flux has a component in the direction of the wavevector, $\boldsymbol{k}$, and a component in the x-direction. Averaging over a wave period straightforwardly gives us additional factors of one half:

$$\overline{\boldsymbol{F}} = -\frac{A^2}{2}\left(\boldsymbol{k}\omega + \mathbf{i}\frac{\beta}{2}\right). \tag{6.102}$$

We now use the dispersion relation $\omega = -\beta k/(K^2 + k_d^2)$ to eliminate the frequency, giving

$$\overline{\boldsymbol{F}} = \frac{A^2\beta}{2}\left(\boldsymbol{k}\frac{k}{K^2 + k_d^2} - \mathbf{i}\frac{1}{2}\right), \tag{6.103}$$

and writing this in component form we obtain

$$\overline{\boldsymbol{F}} = \frac{A^2\beta}{4}\left[\mathbf{i}\left(\frac{k^2 - l^2 - k_d^2}{K^2 + k_d^2}\right) + \mathbf{j}\left(\frac{2kl}{K^2 + k_d^2}\right)\right]. \tag{6.104}$$

Comparison of (6.104) with (6.93) and (6.100) reveals that

$$\overline{\boldsymbol{F}} = \mathbf{c}_g\overline{E} \tag{6.105}$$

so that the energy propagation equation (6.95), when averaged over a wave, becomes

$$\frac{\partial\overline{E}}{\partial t} + \nabla\cdot\mathbf{c}_g\overline{E} = 0. \tag{6.106}$$

It is interesting that the variation of A plays no role in the above manipulations, so that the derivation appears to go through if the amplitude $A(\boldsymbol{x}, t)$ is in fact a constant and the wave is a single plane wave. This seems hard to reconcile with our previous discussion, in which we noted that the group velocity was the velocity of a wave *packet* involving a superposition of plane waves. Indeed, the derivative of the frequency with respect to wavenumber means little if there is only one wavenumber. In fact there is nothing wrong with the above derivation if A is a constant and only a single plane wave is present. The resolution of the paradox arises by noting that a plane wave fills all of space and time; in this case there is no convergence of the energy flux and the energy propagation equation is trivially true.

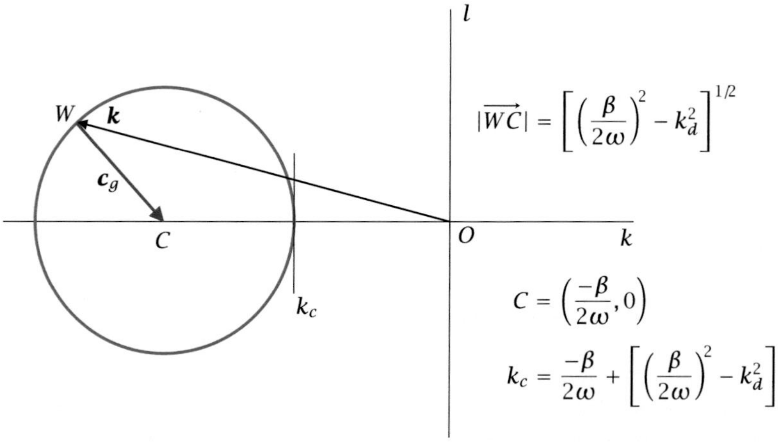

Fig. 6.8 The energy propagation diagram for Rossby waves. The wavevectors of a given frequency all lie in a circle of radius $[(\beta/2\omega)^2 - k_d^2]^{1/2}$, centred at the point C. The closest distance of the circle to the origin is k_c, and if the deformation radius is infinite then $k_c = 0$ and the circle touches the origin. For a given wavenumber $\boldsymbol{k}$, the group velocity is along the line directed from W to C.

6.6.1 ✦ Rossby Wave Reflection

We now consider how Rossby waves might be reflected from a solid boundary. The topic has an obvious oceanographic relevance, for the reflection of Rossby waves turns out to be one way of interpreting why intense oceanic boundary currents form on the western sides of ocean basins, not the east. Rossby waves also reflect off the western boundary in equatorial regions during the El Niño phenomenon. There is also an atmospheric relevance, for meridionally propagating Rossby waves may effectively be reflected as they approach a 'turning latitude' where the meridional wavenumber goes to zero, as considered in Chapter 16. As a preliminary we give a useful graphic interpretation of Rossby wave propagation.[8]

The energy propagation diagram

The dispersion relation for Rossby waves, $\omega = -\beta k/(k^2 + l^2 + k_d^2)$, may be rewritten as

$$(k + \beta/2\omega)^2 + l^2 = (\beta/2\omega)^2 - k_d^2. \tag{6.107}$$

For constant ω this equation is the parametric representation of a circle, meaning that the wavevector (k, l) must lie on a circle centred at the point $(-\beta/2\omega, 0)$ and with radius $[(\beta/2\omega)^2 - k_d^2]^{1/2}$, as illustrated in Fig. 6.8. If k_d is zero the circle touches the origin, and if it is non-zero the distance of the closest point to the circle, k_c say, is given by $k_c = -\beta/2\omega + [(\beta/2\omega)^2 - k_d^2]^{1/2}$. For low frequencies, specifically if $\omega \ll \beta/2k$, then $k_c \approx -\omega k_d^2/\beta$. The radius of the circle is a positive real number only when $\omega < \beta/2k_d$. This is the maximum frequency possible, and it occurs when $l = 0$ and $k = k_d$ and when $c_g^x = c_g^y = 0$.

The group velocity, and hence the energy flux, can be visualized graphically from Fig. 6.8. By direct manipulation of the expressions for group velocity and frequency (Equations (RW.3) and (RW.4) on page 235) we find that

$$c_g^x = \frac{-2\omega}{K^2 + k_d^2}\left(k + \frac{\beta}{2\omega}\right), \qquad c_g^y = \frac{-2\omega}{K^2 + k_d^2}\, l. \tag{6.108a,b}$$

(To check this, it is easiest to begin with the right-hand sides and use the dispersion relation for ω.) Now, since the centre of the circle of wavevectors is at the position $(-\beta/2\omega, 0)$, and referring to

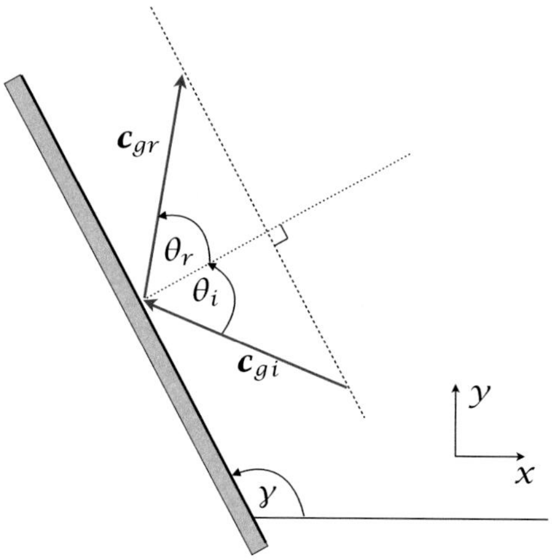

Fig. 6.9 The reflection of a Rossby wave at a western wall, in physical space. A Rossby wave with a westward group velocity impinges at an angle θ_i to a wall, inducing a reflected wave moving eastward at an angle θ_r. The reflection is specular, with $\theta_r = \theta_i$, and energy conserving, with $|\boldsymbol{c}_{gr}| = |\boldsymbol{c}_{gi}|$ — see text and Fig. 6.10.

Fig. 6.8, we have

$$\boldsymbol{c}_g = \frac{2\omega}{K^2 + k_d^2}\boldsymbol{R} \tag{6.109}$$

where $\boldsymbol{R} = \overrightarrow{WC}$ is the vector directed from W to C, that is from the end of the wavevector itself to the centre of the circle around which all the wavevectors lie.

Equation (6.109) and Fig. 6.8 allow for a useful visualization of the energy and phase. The phase propagates in the direction of the wave vector, and for Rossby waves this is always westward. The group velocity is in the direction of the wave vector to the centre of the circle, and this can be either eastward (if $k^2 > l^2 + k_d^2$) or westward ($k^2 < l^2 + k_d^2$). Interestingly, the velocity vector is normal to the wave vector. To see this, consider a purely westward propagating wave for which $l = 0$. Then $v = \partial\psi/\partial x = \mathrm{i}k\widetilde{\psi}$ and $u = -\partial\psi/\partial y = -\mathrm{i}l\widetilde{\psi} = 0$. We now see how some of these properties can help us understand the reflection of Rossby waves.

Reflection at a wall

Consider Rossby waves incident on a wall making an angle γ with the x-axis, and suppose that somehow these waves are reflected back into the fluid interior. This is a reasonable expectation, for the wall cannot normally simply absorb all the wave energy, and if reflection does occur it will have the following two properties:

(i) The incident and reflected wave will have the same wavenumber component along the wall.

(ii) The incident and reflected wave will have the same frequency.

To understand these properties, first consider the case in which the wall is oriented meridionally along the y-axis with $\gamma = 90°$. For our immediate concerns there is loss of generality in this choice, because we may simply choose coordinates so that y is parallel to the wall and the β-effect, which differentiates x from y, does not enter the initial argument. The incident and reflected waves are

$$\psi_i(x, y, t) = A_i \exp\left[\mathrm{i}\left(k_i x + l_i y - \omega_i t\right)\right], \qquad \psi_r(x, y, t) = A_r \exp\left[\mathrm{i}\left(k_r x + l_r y - \omega_r t\right)\right], \tag{6.110}$$

with subscripts i and r denoting incident and reflected. At the wall, which we take to be at $x = 0$, the normal velocity $u = -\partial\psi/\partial y$ must be zero, so that

$$A_i l_i \exp\left[\mathrm{i}\left(l_i y - \omega_i t\right)\right] + A_r l_r \exp\left[\mathrm{i}\left(l_r y - \omega_r t\right)\right] = 0. \tag{6.111}$$

For this equation to hold for all y and all time then we must have

$$l_r = l_i, \qquad \omega_r = \omega_i. \tag{6.112}$$

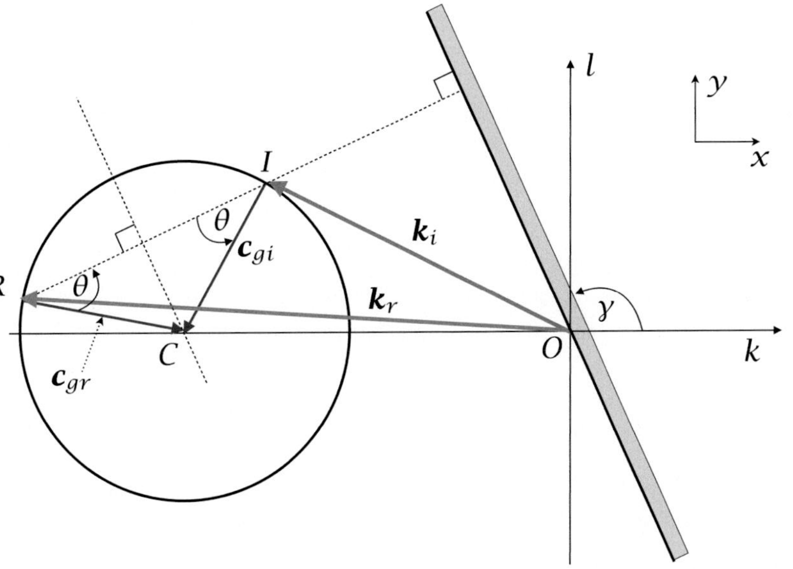

Fig. 6.10 Graphical representation of the reflection of a Rossby wave at a western wall, in spectral space. The incident wave has wavevector $\boldsymbol{k}_i$, ending at point I. Construct the wavevector circle of constant frequency through point I with radius $(\beta/2\omega)^2 - k_d^2)^{1/2}$ and centre $C = (-\beta/2\omega, 0)$; the group velocity vector then lies along $\overrightarrow{IC}$ and is directed westward. The reflected wave has a wavevector $\boldsymbol{k}_r$ such that its projection on the wall is equal to that of $\boldsymbol{k}_i$, and this fixes the point R. The group velocity of the reflected wave then lies along $\overrightarrow{RC}$, and it can be seen that $\boldsymbol{c}_{gr}$ makes the same angle to the wall as does $\boldsymbol{c}_{gi}$, except that it is directed eastward. The reflection is therefore specular and is such that the energy flux directed away from the wall is equal to the energy flux directed toward the wall.

This result is independent of the detailed dynamics of the waves, requiring only that the velocity is determined from a streamfunction. If we consider Rossby-wave dynamics specifically, the x- and y-coordinates are not arbitrary and the y-axis cannot be taken to be aligned with the wall; however, the underlying result still holds, meaning that the *projection* of the incident wavevector, $\boldsymbol{k}_i$ on the wall must equal the *projection* of the reflected wavevector, $\boldsymbol{k}_r$. The magnitude of the wavevector (the wavenumber) is not in general conserved by reflection. Finally, given these results and using (6.111) we see that the incident and reflected amplitudes are related by

$$A_r = -A_i. \tag{6.113}$$

Now let's delve a little deeper into the problem.

Generally, when we consider a wave to be incident on a wall, we are supposing that the *group velocity* is directed toward the wall. Suppose that a wave of given frequency, ω, and wavevector, $\boldsymbol{k}_i$, and with westward group velocity is incident on a predominantly western wall, as in Fig. 6.9. (Similar reasoning, *mutatis mutandis,* can be applied to a wave incident on an eastern wall.) Let us suppose that incident wave, $\boldsymbol{k}_i$ lies at the point I on the wavenumber circle, and the group velocity is found by drawing a line from I to the centre of the circle, C (so $\boldsymbol{c}_{gi} \propto \overrightarrow{IC}$), and in this case the vector is directed westward.

The projection of the $\boldsymbol{k}_i$ must be equal to the projection of the reflected wave vector, $\boldsymbol{k}_r$, and both wavevectors must lie in the same wavenumber circle, centred at $-\beta/2\omega$, because the frequencies of the two waves are the same. We may then graphically determine the wavevector of the reflected wave using the construction of Fig. 6.10. (This is a figure in spectral space and the position of the wall does *not* imply that it is an eastern boundary.) Given the wavevector, the group velocity of the reflected wave follows by drawing a line from the wavevector to the centre of the circle (the line

$\overrightarrow{RC}$). We see from the figure that the reflected group velocity is directed eastward and that it forms the same angle to the wall as does the incident wave; that is, the reflection is *specular.* Since the amplitudes of the incoming and reflected wave are the same, the components of the energy flux perpendicular to the wall are equal and opposite. Furthermore, we can see from the figure that the wavenumber of the reflected wave has a larger magnitude than that of the incident wave. For waves reflecting off an eastern boundary, the reverse is true. Put simply, at a western boundary incident long waves are reflected as short waves, whereas at an eastern boundary incident short waves are reflected as long waves.

Quantitatively solving for the wavenumbers of the reflected wave is a little tedious in the case when the wall is at angle, but easy enough if the wall is along the y-axis. We know the frequency, ω, and the y-wavenumber, l, so that the x-wavenumber may be deduced from the dispersion relation

$$\omega = \frac{-\beta k_i}{k_i^2 + l^2 + k_d^2} = \frac{-\beta k_r}{k_r^2 + l^2 + k_d^2}. \tag{6.114}$$

We obtain

$$k_i = \frac{-\beta}{2\omega} + \sqrt{\left(\frac{\beta}{2\omega}\right)^2 - (l^2 + k_d^2)}, \qquad k_r = \frac{-\beta}{2\omega} - \sqrt{\left(\frac{\beta}{2\omega}\right)^2 - (l^2 + k_d^2)}. \tag{6.115a,b}$$

The signs of the square-root terms are chosen for reflection at a western boundary, for which, as we noted, the reflected wave has a larger (absolute) wavenumber than the incident wave. For reflection at an eastern boundary we reverse the signs.

Oceanographic relevance

The behaviour of Rossby waves at lateral boundaries is not surprisingly of some oceanographic importance. Suppose that Rossby waves are generated in the middle of the ocean, for example by the wind or by some fluid dynamical instability in the ocean. Shorter waves will tend to propagate eastward, and be reflected back at the eastern boundary as long waves, and long waves will tend to propagate westward, being reflected back as short waves.

Reflection at the western boundary is complicated by friction and by other waves. In mid-latitudes the reflection at a western boundary generates Rossby waves that have a short *zonal* length scale (the meridional scale of the reflected wave is the same as the incident wave if the wall is meridional), which means that their *meridional* velocity is large. Now, if the zonal wavenumber is much larger than both the meridional wavenumber l and the inverse deformation radius k_d then, using either (6.62) or (6.67) the group velocity in the x-direction is given by $c_g^x = \overline{u} + \beta/k^2$, where $\overline{u}$ is the zonal mean flow. If the mean flow is westward, so that $\overline{u}$ is negative, then very short waves will be unable to escape from the boundary; specifically, if $k > \sqrt{-\beta/U}$ then the waves will be trapped in a western boundary layer. Even with no mean flow, the short zonal length scale means that frictional effects will be large.

In some circumstances Rossby waves incident on a boundary also have the option of generating coastal Kelvin waves. Also, at a western boundary at the equator Rossby waves may also reflecting back as equatorial Kelvin waves, which we introduce in the next chapter. This effect may be particularly important in the dynamics of El Niño, as we discuss in Section 22.7.

6.7 ✦ GROUP VELOCITY, REVISITED

We now return to a more general discussion of group velocity. Our goal is to show that the group velocity arises in fairly general ways, not just from methods stemming from Fourier analysis or from ray theory. We first give a simple and direct derivation of group velocity that is valid in the simple but important special case of a homogeneous medium.Then, in Section 6.7.2, we give a rather general derivation of the *group velocity property,* namely that conserved quantities that are quadratic in the wave amplitude — that is, *wave activities* — are transported at the group velocity.

6.7.1 Group Velocity in Homogeneous Media

Consider waves propagating in a medium in which the wave equation has the form

$$L(\psi) = \Lambda\left[\frac{\partial}{\partial t}, \frac{\partial}{\partial x}\right]\psi(x,t) = 0, \tag{6.116}$$

where Λ is a polynomial operator in the space and time derivatives, with constant coefficients, and its arguments are in square brackets. For simplicity we restrict attention to waves in one dimension, and a simple example is $\Lambda = \partial(\partial_{xx})/\partial t + \beta\partial/\partial x$ so that $L(\psi) = \partial(\partial_{xx}\psi)/\partial t + \beta\partial\psi/\partial x$. We will seek a solution of the form

$$\psi(x,t) = A(x,t)\mathrm{e}^{\mathrm{i}\theta(x,t)}, \tag{6.117}$$

where θ is the phase of the disturbance and $A(x,t)$ is the slowly varying amplitude, so that the solution has the form of a wave packet. The phase is such that $k = \partial\theta/\partial x$ and $\omega = -\partial\theta/\partial t$, and the slowly varying nature of the envelope $A(x,t)$ is formalized by demanding that

$$\frac{1}{A}\frac{\partial A}{\partial x} \ll k, \qquad \frac{1}{A}\frac{\partial A}{\partial t} \ll \omega. \tag{6.118}$$

The space and time derivatives of ψ are then given by

$$\frac{\partial\psi}{\partial x} = \left(\frac{\partial A}{\partial x} + \mathrm{i}A\frac{\partial\theta}{\partial x}\right)\mathrm{e}^{\mathrm{i}\theta} = \left(\frac{\partial A}{\partial x} + \mathrm{i}Ak\right)\mathrm{e}^{\mathrm{i}\theta}, \tag{6.119a}$$

$$\frac{\partial\psi}{\partial t} = \left(\frac{\partial A}{\partial t} + \mathrm{i}A\frac{\partial\theta}{\partial t}\right)\mathrm{e}^{\mathrm{i}\theta} = \left(\frac{\partial A}{\partial t} - \mathrm{i}A\omega\right)\mathrm{e}^{\mathrm{i}\theta}, \tag{6.119b}$$

so that the wave equation becomes

$$\Lambda\psi = \Lambda\left[\frac{\partial}{\partial t} - \mathrm{i}\omega, \frac{\partial}{\partial x} + \mathrm{i}k\right]A = 0. \tag{6.120}$$

Noting that the space and time derivative of A are small compared to k and ω we expand the polynomial in a Taylor series about (ω, k) to obtain

$$\Lambda[-\mathrm{i}\omega, \mathrm{i}k]A + \frac{\partial\Lambda}{\partial(-\mathrm{i}\omega)}\frac{\partial A}{\partial t} + \frac{\partial\Lambda}{\partial(\mathrm{i}k)}\frac{\partial A}{\partial x} \approx 0. \tag{6.121}$$

Now, the dispersion relation for plane waves is $\Lambda[-\mathrm{i}\omega, \mathrm{i}k] = 0$. Taking this to be satisfied the first term in (6.121) vanishes giving

$$\frac{\partial A}{\partial t} - \frac{\partial\Lambda/\partial k}{\partial\Lambda/\partial\omega}\frac{\partial A}{\partial x} = \frac{\partial A}{\partial t} + \frac{\partial\omega}{\partial k}\frac{\partial A}{\partial x} = 0, \tag{6.122}$$

having used $(\partial\Lambda/\partial k)/(\partial\Lambda/\partial\omega) = -(\partial\omega/\partial k)_\Lambda$. Then, since $c_g \equiv \partial\omega/\partial k$, we have

$$\frac{\partial A}{\partial t} + c_g\frac{\partial A}{\partial x} = 0, \tag{6.123}$$

meaning that *the envelope moves at the group velocity.*

6.7.2 ♦ The Group Velocity Property

In Section 6.6 we found that, when averaged over the phase, the energy of a Rossby wave, $\overline{E}$, obeys

$$\frac{\partial \overline{E}}{\partial t} + \nabla \cdot c_g \overline{E} = 0. \tag{6.124}$$

This equation tells us that the energy flux is equal to $c_g \overline{E}$, and it turns out to be a very general feature of waves. It is called the *group velocity property,* and it is not restricted to Rossby waves, or energy, or homogeneous media; it holds for almost any conserved quantity that is quadratic in the wave amplitude, and we now demonstrate this in a more general way.[9] A quantity that is quadratic and conserved is known as a *wave activity.* (The corresponding local quantity, such as the wave activity per unit volume, is strictly called the wave activity *density,* but also often just wave activity.) The group velocity property is useful because if we can determine $\boldsymbol{c}_g$ then we know straight away how wave activities propagate. Energy itself is sometimes a wave activity but often is not: in a growing baroclinic wave energy is drawn from the background state and is not conserved. However, we will see in Chapter 10 that even in a growing baroclinic disturbance it is possible to define a conserved wave activity.

The formal procedure

The derivation, which is rather formal, will hold for waves and wave activities that satisfy the following three assumptions:

(i) The wave activity, A, and flux, $\boldsymbol{F}$, obey the general conservation relation

$$\frac{\partial A}{\partial t} + \nabla \cdot \boldsymbol{F} = 0. \tag{6.125}$$

(ii) Both the wave activity and the flux are quadratic functions of the wave amplitude.

(iii) The waves themselves are of the general form

$$\psi = \widetilde{\psi} e^{i\theta(\boldsymbol{x},t)} + \text{c.c.}, \qquad \theta = \boldsymbol{k} \cdot \boldsymbol{x} - \omega t, \qquad \omega = \omega(\boldsymbol{k}), \tag{6.126a,b,c}$$

where (6.126c) is the dispersion relation, and ψ is any wave field. We will carry out the derivation in the case in which $\widetilde{\psi}$ is treated as a constant in space, but the treatment applies when the amplitude varies slowly over a wavelength (that is, when the wave is 'WKB-able').

The wave activity and flux in (6.125) are not unique, as (6.125) is unaffected by the following transformation,

$$A \to A + \nabla \cdot \boldsymbol{C}, \qquad \boldsymbol{F} \to \boldsymbol{F} - \frac{\partial \boldsymbol{C}}{\partial t}, \tag{6.127}$$

where $\boldsymbol{C}$ is any vector, as well as the addition of any non-divergent vector to $\boldsymbol{F}$ or a constant to A. The conditions above, in particular that the wave activity and flux are quadratic functions, remove the ambiguity and provide a flux that is related to the activity by the group velocity; that is, by $\boldsymbol{F} = \boldsymbol{c}_g A$, as we see below.

To proceed, from assumption *(ii)*, the wave activity must have the general form

$$A = b + a e^{2i(\boldsymbol{k}\cdot\boldsymbol{x}-\omega t)} + a^* e^{-2i(\boldsymbol{k}\cdot\boldsymbol{x}-\omega t)}, \tag{6.128a}$$

where the asterisk, *, denotes complex conjugacy, and b is a real constant and a is a complex constant. For example, suppose that $A = \psi^2$ and $\psi = c e^{i(\boldsymbol{k}\cdot\boldsymbol{x}-\omega t)} + c^* e^{-i(\boldsymbol{k}\cdot\boldsymbol{x}-\omega t)}$, then we find that (6.128a) is satisfied with $a = c^2$ and $b = 2cc^*$. Similarly, the flux has the general form

$$\boldsymbol{F} = \boldsymbol{g} + \boldsymbol{f} e^{2i(\boldsymbol{k}\cdot\boldsymbol{x}-\omega t)} + \boldsymbol{f}^* e^{-2i(\boldsymbol{k}\cdot\boldsymbol{x}-\omega t)}. \tag{6.128b}$$

where $\boldsymbol{g}$ is a real constant vector (not gravity) and $\boldsymbol{f}$ is a complex constant vector. The mean activity and mean flux are obtained by averaging over a cycle; the oscillating terms vanish on integration and therefore the wave activity and flux are given by

$$\overline{A} = b, \qquad \overline{\boldsymbol{F}} = \boldsymbol{g}, \tag{6.129}$$

where the overbar denotes the mean.

Now consider a wave with a slightly different phase, $\theta + \mathrm{i}\,\delta\theta$, where $\delta\theta$ is small compared with θ. Thus, we formally replace $\boldsymbol{k}$ by $\boldsymbol{k} + \mathrm{i}\,\boldsymbol{\delta k}$ and ω by $\omega + \mathrm{i}\,\delta\omega$ where, to satisfy the dispersion relation, we have

$$\omega + \mathrm{i}\,\delta\omega = \omega(\boldsymbol{k} + \mathrm{i}\,\boldsymbol{\delta k}) \approx \omega(\boldsymbol{k}) + \mathrm{i}\,\boldsymbol{\delta k}\cdot\frac{\partial\omega}{\partial\boldsymbol{k}}, \tag{6.130}$$

and therefore

$$\delta\omega = \boldsymbol{\delta k}\cdot\frac{\partial\omega}{\partial\boldsymbol{k}} = \boldsymbol{\delta k}\cdot\boldsymbol{c}_g, \tag{6.131}$$

where $\boldsymbol{c}_g \equiv \partial\omega/\partial\boldsymbol{k}$ is the group velocity.

The new wave has the general form

$$\psi' = (\widetilde{\psi} + \delta\widetilde{\psi})\mathrm{e}^{\mathrm{i}(\boldsymbol{k}\cdot\boldsymbol{x}-\omega t)}\mathrm{e}^{-\boldsymbol{\delta k}\cdot\boldsymbol{x}+\delta\omega t} + \text{c.c.}, \tag{6.132}$$

and, analogously to (6.128), the associated wave activity and flux have the forms:

$$A' = \left[b + \delta b + (a+\delta a)\mathrm{e}^{2\mathrm{i}(\boldsymbol{k}\cdot\boldsymbol{x}-\omega t)} + (a^* + \delta a^*)\mathrm{e}^{-2\mathrm{i}(\boldsymbol{k}\cdot\boldsymbol{x}-\omega t)}\right]\mathrm{e}^{-2\boldsymbol{\delta k}\cdot\boldsymbol{x}+2\delta\omega t} \tag{6.133a}$$

$$\boldsymbol{F}' = \left[\boldsymbol{g} + \delta\boldsymbol{g} + (\boldsymbol{f}+\delta\boldsymbol{f})\mathrm{e}^{2\mathrm{i}(\boldsymbol{k}\cdot\boldsymbol{x}-\omega t)} + (\boldsymbol{f}^* + \delta\boldsymbol{f}^*)\mathrm{e}^{-2\mathrm{i}(\boldsymbol{k}\cdot\boldsymbol{x}-\omega t)}\right]\mathrm{e}^{-2\boldsymbol{\delta k}\cdot\boldsymbol{x}+2\delta\omega t}, \tag{6.133b}$$

where the δ quantities are small. If we now demand that A' and $\boldsymbol{F}'$ satisfy assumption (i), then substituting (6.133) into (6.125) gives, on averaging over the phase of a wave and after a little algebra,

$$(\boldsymbol{g} + \delta\boldsymbol{g})\cdot\boldsymbol{\delta k} = (b + \delta b)\delta\omega, \tag{6.134}$$

and therefore at first order in δ quantities, $\boldsymbol{g}\cdot\boldsymbol{\delta k} = b\delta\omega$. Using (6.131) and (6.129) we obtain

$$\boldsymbol{c}_g = \frac{\boldsymbol{g}}{b} = \frac{\overline{\boldsymbol{F}}}{\overline{A}} \qquad \text{or} \qquad \overline{\boldsymbol{F}} = \boldsymbol{c}_g\overline{A}. \tag{6.135}$$

Using this the conservation law, (6.125), becomes

$$\frac{\partial\overline{A}}{\partial t} + \nabla\cdot(\boldsymbol{c}_g\overline{A}) = 0. \tag{6.136}$$

Thus, for waves satisfying our three assumptions, the flux velocity — that is, the propagation velocity of the wave activity — is equal to the group velocity. Henceforth, we will normally denote $\overline{A}$ by $\mathcal{A}$, and $\overline{\boldsymbol{F}}$ by $\boldsymbol{\mathcal{F}}$.

A remark: In this section and in Section 6.1 we have seen various derivations of the group velocity, some more kinematic, some more physical, some more general. Why are there different ways to derive it? Which is the 'real' derivation? Perhaps the answer is that quantity $\partial\omega/\partial k$ is a fundamental velocity for a wave. If, for example, the energy propagates at this speed then the wave envelope must move at this speed, and vice versa. The karma of mathematics means that the derivations, even if stylistically very different, must give the same answer.

6.8 ENERGY PROPAGATION OF POINCARÉ WAVES

In the final section of this chapter we discuss the energetics of Poincaré waves (first encountered in Section 3.8.2) and show explicitly that the energy propagation occurs at the group velocity. We begin with the one-dimensional problem as this shows the essential aspects and the algebra is a little simpler.

6.8.1 Energetics in One Dimension

The one-dimensional (i.e., no variations in the y-direction), inviscid linear shallow-water equations on the f-plane, linearized about a state of rest, are

$$\frac{\partial u}{\partial t} - f_0 v = -g\frac{\partial h}{\partial x}, \qquad \frac{\partial v}{\partial t} + f_0 u = 0, \qquad \frac{\partial \eta}{\partial t} = -H\frac{\partial u}{\partial x}. \tag{6.137a,b,c}$$

To obtain the dispersion relation we differentiate the first equation with respect to t and substitute from the second and third to obtain

$$\frac{\partial^2 u}{\partial t^2} - Hg\frac{\partial u}{\partial x} + f_0^2 u = 0, \tag{6.138}$$

whence, assuming solutions of the form $u = \operatorname{Re}\tilde{u}e^{i(kx-\omega t)}$, we obtain the dispersion relation,

$$\omega^2 = f_0^2 + Hgk^2, \tag{6.139}$$

similar to (3.121) on page 125. Differentiating with respect to k gives $2\omega\,\partial\omega/\partial k = 2kHg$ or

$$c_g = \frac{Hg}{c_p}, \tag{6.140}$$

where $c_g = \partial\omega/\partial k$ and $c_p = \omega/k$. Using (6.139) and (6.140) the ratio of the group and phase velocities is found to be

$$\frac{c_g}{c_p} = \frac{L_d^2 k^2}{1 + L_d^2 k^2}, \tag{6.141}$$

where $L_d = \sqrt{gH}/f$ is the deformation radius. This ratio is always less than unity, tending to zero in the long-wave limit ($kL_d \ll 1$) and to unity for short waves ($kL_d \gg 1$).

The energy equations are obtained by multiplying the three equations of (6.137) by u, v and η respectively, and adding, to give

$$\frac{\partial E}{\partial t} + \frac{\partial F}{\partial x} = 0, \tag{6.142a}$$

where

$$E = \frac{1}{2}(Hu^2 + Hv^2 + g\eta^2), \qquad F = gHu\eta, \tag{6.142b}$$

are the energy density and the energy flux, respectively. Here in the linear approximation the energy is transported only by the pressure term, whereas in the full nonlinear equations there is also an advective transport.

The group velocity property for Poincaré waves

To specialize to the case of propagating *waves* we need to average over a wavelength and use the phase relationships between u, v and η implied by the equations of motion. Writing $u = \operatorname{Re}\tilde{u}e^{i(kx-\omega t)}$, and similarly for v and η, we have

$$\tilde{v} = -\mathrm{i}f\frac{\tilde{\eta}}{Hk}, \qquad \tilde{u} = \omega\frac{\tilde{\eta}}{Hk}. \tag{6.143a,b}$$

The kinetic energy, averaged over a wavelength, is then

$$\mathrm{KE} = \frac{1}{2}H(\overline{u^2} + \overline{v^2}) = \frac{1}{4}(\omega^2 + f^2)\frac{\tilde{\eta}^2}{Hk^2} = \frac{1}{4}\frac{\omega^2 + f^2}{\omega^2 - f^2}g\tilde{\eta}^2 \tag{6.144}$$

using (6.143) and the dispersion relation, with the extra factor of one half arising from the averaging over a wavelength. Similarly, the potential energy of the wave is

$$\mathrm{PE} = \frac{1}{2}g\overline{\eta^2} = \frac{1}{4}g\tilde{\eta}^2. \tag{6.145}$$

Thus, the ratio of kinetic to potential energy is just

$$\frac{\mathrm{KE}}{\mathrm{PE}} = \frac{\omega^2 + f^2}{\omega^2 - f^2} = 1 + \frac{2}{k^2 L_d^2}, \tag{6.146}$$

using the dispersion relation, and where $L_d = \sqrt{gH/f}$ is the deformation radius. Thus, the kinetic energy is always *greater* than the potential energy (there is no equipartition in this problem), with the ratio approaching unity for small scales (large k).

The total energy (kinetic plus potential) is then

$$\mathrm{KE} + \mathrm{PE} = \frac{1}{4}\left(\frac{\omega^2 + f^2}{\omega^2 - f^2} + 1\right)g\tilde{\eta}^2 = \frac{1}{2}\frac{\omega^2}{k^2 H}\tilde{\eta}^2 = \frac{1}{2}\frac{c_p^2}{H}\tilde{\eta}^2, \tag{6.147}$$

again using the dispersion relation. The energy flux, F, averaged over a wavelength, is

$$F = gH\overline{u\eta} = \frac{1}{2}\frac{g\omega}{k}\tilde{\eta}^2 = \frac{1}{2}gc_p\tilde{\eta}^2. \tag{6.148}$$

From (6.147) an (6.148) the flux and the energy are evidently related by

$$F = \frac{Hg}{c_p}E = c_g E, \tag{6.149}$$

using (6.140). That is, the energy flux is equal to the group velocity times the energy itself. Note that in this problem there is no flux in the y direction, because v and η are exactly out of phase from (6.143a).

6.8.2 ✦ Energetics in Two Dimensions

The derivations of the preceding section carry through, *mutatis mutandis,* in the full two-dimensional case. We will give only the key results and allow the reader to fill in the algebra.

As derived in Section 3.8.2 the dispersion relation is

$$\omega^2 = f_0^2 + gH(k^2 + l^2). \tag{6.150}$$

The relation between the components of the group velocity and the phase speed is very similar to the one-dimensional case, and in particular we have

$$c_g^x = \frac{\partial\omega}{\partial k} = gH\frac{k}{\omega} = \frac{gH}{c_p^x}, \qquad c_g^y = \frac{\partial\omega}{\partial l} = gH\frac{l}{\omega} = \frac{gH}{c_p^y}. \tag{6.151}$$

The magnitude of the group velocity is $c_g \equiv |\mathbf{c}_g| = (c_g^{x2} + c_g^{y2})^{1/2}$. The magnitude of the phase speed, in the direction of travel of the wave crests, is $c_p = \omega/(k^2 + l^2)^{1/2}$ (note that in general this is *smaller* than the phase speed in either the x or y directions, ω/k or ω/l). Thus, we have

$$c_g^2 = (gH)^2 \frac{k^2 + l^2}{\omega^2} = \frac{(gH)^2}{c_p^2}, \qquad \mathbf{c}_g = \left(\frac{gH}{c_p K}\right)\mathbf{k}, \tag{6.152}$$

which is analogous to (6.140). The ratio of the magnitudes of the group and phase velocities is, analogously to (6.141),

$$\frac{c_g}{c_p} = \frac{gH}{c_p^2} = \frac{L_d^2 K^2}{1 + L_d^2 K^2}, \tag{6.153}$$

where $K^2 = k^2 + l^2$. As in the one-dimensional case the group velocity is large for short waves, in which rotation plays no role, and small for long waves.

The energy equation is found to be

$$\frac{\partial E}{\partial t} + \nabla \cdot \boldsymbol{F} = 0, \tag{6.154a}$$

with

$$E = \frac{1}{2}(Hu^2 + Hv^2 + g\eta^2), \qquad F = gH(u\mathbf{i} + v\mathbf{j})\eta. \tag{6.154b}$$

From the equations of motion the phase relations between the fields are

$$\widetilde{v} = \frac{\omega l - ikf}{HK^2}\widetilde{\eta}, \qquad \widetilde{u} = \frac{\omega k - ilf}{HK^2}\widetilde{\eta}, \tag{6.155}$$

so that the kinetic energy is given by, similarly to (6.144),

$$\mathrm{KE} = \frac{1}{2}H(\overline{u^2} + \overline{v^2}) = \frac{1}{4}(\omega^2 + f^2)\frac{\widetilde{\eta}^2}{Hk^2} = \frac{1}{4}\frac{\omega^2 + f^2}{\omega^2 - f^2}g\widetilde{\eta}^2, \tag{6.156}$$

and the potential energy by

$$\mathrm{PE} = \frac{1}{2}g\overline{\eta^2} = \frac{1}{4}g\widetilde{\eta}^2. \tag{6.157}$$

The ratio of the kinetic and potential energies is given by

$$\frac{\mathrm{KE}}{\mathrm{PE}} = \frac{\omega^2 + f^2}{\omega^2 - f^2} = 1 + \frac{2}{K^2 L_d^2}. \tag{6.158}$$

The total (kinetic plus potential) energy is given by

$$E = \mathrm{KE} + \mathrm{PE} = \frac{1}{4}\left(\frac{\omega^2 + f^2}{\omega^2 - f^2} + 1\right)g\widetilde{\eta}^2 = \frac{1}{2}\frac{\omega^2}{K^2 H}\widetilde{\eta}^2 = \frac{1}{2}\frac{c_p^2}{H}\widetilde{\eta}^2, \tag{6.159}$$

The energy flux, $\boldsymbol{F}$, averaged over a wavelength, is

$$\boldsymbol{F} = gH\overline{u\eta} = \frac{1}{2}\frac{g\omega}{k^2 + l^2}\widetilde{\eta}^2\mathbf{k} = \frac{1}{2}\frac{g\omega}{K^2}\widetilde{\eta}^2\mathbf{k}, \tag{6.160}$$

using (6.155) and where $\boldsymbol{k} = k\mathbf{i} + l\mathbf{j}$ is the wavevector of the wave.

From (6.159) and (6.160), and using (6.152), the flux and the energy are related by

$$\boldsymbol{F} = \mathbf{c}_g E. \tag{6.161}$$

That is, the energy flux is equal to the group velocity times the energy itself.

APPENDIX A: THE WKB APPROXIMATION FOR LINEAR WAVES

The WKB method (after Wentzel, Kramers and Brillouin, the last people to discover the technique[10]) is a way of finding approximate solutions to certain linear differential equations in which the term with the highest derivative is multiplied by a small parameter. The theory for such equations is quite extensive but our interests are modest, being mainly in dispersive waves, and WKB theory can be used to find approximate solutions in cases in which the coefficients of the wave equation vary slowly in space or time. Consider an equation of the form

$$\frac{d^2\xi}{dz^2} + m^2(z)\xi = 0. \tag{6.162}$$

Such an equation commonly arises in wave problems. If m^2 is positive the equation has wavelike solutions, and if m is constant the solution has the harmonic form

$$\xi = \text{Re}\, A_0 e^{imz}, \tag{6.163}$$

where A_0 is a complex constant. If m varies only slowly with z — meaning that the variations in m only occur on a scale much longer than $1/m$ — one might reasonably expect that the harmonic solution above would provide a decent first approximation; that is, we expect the solution to locally look like a plane wave with local wavenumber $m(z)$. However, we might also expect that the solution would not be *exactly* of the form $\exp(im(z)z)$, because the phase of ξ is $\theta(z) = mz$, so that $d\theta/dz = m + z dm/dz \neq m$. Thus, in (6.163) m is not the wavenumber unless m is constant.

The condition that variations in m, or in the wavelength $\lambda \sim m^{-1}$, occur only slowly may be variously expressed as

$$\lambda \left|\frac{\partial \lambda}{\partial z}\right| \ll \lambda \quad \text{or} \quad \left|\frac{\partial m^{-1}}{\partial z}\right| \ll 1 \quad \text{or} \quad \left|\frac{\partial m}{\partial z}\right| \ll m^2. \tag{6.164a,b,c}$$

This condition will generally be satisfied if variations in the background state, or in the medium, occur on a scale much longer than the wavelength. Let us first find a solution by way of a perturbation expansion.

A.1 Solution by Perturbation Expansion

To explicitly recognize the rapid oscillations of the wave compared to its slow variations in amplitude we rescale the coordinate z with a small parameter ϵ. Thus, we let $\hat{z} = \epsilon z$ where $\epsilon \ll 1$ (ϵ may be similar to the nondimensional parameter $|d\lambda/dz|$) and the new variable $\hat{z}$ then varies by $\mathcal{O}(1)$ over the scale on which m varies. Equation (6.162) becomes

$$\epsilon^2 \frac{d^2\xi}{d\hat{z}^2} + m^2(\hat{z})\xi = 0, \tag{6.165}$$

and we may now suppose that all variables are $\mathcal{O}(1)$. If m were constant the solution would be of the form $\xi = A\exp(m\hat{z}/\epsilon)$ and this suggests that we look for a solution to (6.165) of the form

$$\xi(z) = e^{g(\hat{z})/\epsilon}, \tag{6.166}$$

where $g(\hat{z})$ is some as yet unknown function. We then have, with primes denoting derivatives,

$$\xi' = \frac{1}{\epsilon} g' e^{g/\epsilon}, \qquad \xi'' = \left(\frac{1}{\epsilon^2} g'^2 + \frac{1}{\epsilon} g''\right) e^{g/\epsilon}. \tag{6.167a,b}$$

Using these expressions in (6.165) yields

$$\epsilon g'' + g'^2 + m^2 = 0, \tag{6.168}$$

and if we let $g = \int h \, \mathrm{d}\hat{z}$ we obtain

$$\epsilon \frac{\mathrm{d}h}{\mathrm{d}\hat{z}} + h^2 + m^2 = 0. \tag{6.169}$$

To obtain a solution of this equation we expand h in powers of the small parameter ϵ,

$$h(\hat{z};\epsilon) = h_0(\hat{z}) + \epsilon h_1(\hat{z}) + \epsilon^2 h_2(\hat{z}) + \cdots. \tag{6.170}$$

Substituting this in (6.169) and setting successive powers of ϵ to zero gives, at first and second order,

$$h_0^2 + m^2 = 0, \qquad 2h_0 h_1 + \frac{\mathrm{d}h_0}{\mathrm{d}\hat{z}} = 0. \tag{6.171a,b}$$

The solutions of these equations are

$$h_0 = \pm \mathrm{i}m, \qquad h_1(\hat{z}) = -\frac{1}{2}\frac{\mathrm{d}}{\mathrm{d}\hat{z}} \ln \frac{m(\hat{z})}{m_0}, \tag{6.172a,b}$$

where m_0 is a constant. Now, ignoring higher-order terms, (6.166) may be written in terms of h_0 and h_1 as

$$\xi(\hat{z}) = \exp\left(\int h_0 \, \mathrm{d}\hat{z}/\epsilon\right) \exp\left(\int h_1 \, \mathrm{d}\hat{z}\right), \tag{6.173}$$

and, using (6.172) and with z in place of $\hat{z}$, we obtain

$$\xi(z) = A_0 m^{-1/2} \exp\left(\pm \mathrm{i} \int m \, \mathrm{d}z\right), \tag{6.174}$$

where A_0 is a constant, and this is the WKB solution to (6.162). Explicitly, the solution is

$$\xi(z) = A_0 m^{-1/2} \exp\left(\mathrm{i}\int m \, \mathrm{d}z\right) + A_0^* m^{-1/2} \exp\left(-\mathrm{i}\int m \, \mathrm{d}z\right), \tag{6.175}$$

or, in terms of real quantities,

$$\xi(z) = B_0 m^{-1/2} \cos\left(\int m \, \mathrm{d}z\right) + C_0 m^{-1/2} \sin\left(\int m \, \mathrm{d}z\right), \tag{6.176}$$

where B_0, and C_0 are real constants.

A property of (6.174) is that the derivative of the phase is just m; that is, m is indeed the local wavenumber. A crucial aspect of the derivation is that m varies slowly, so that there is a small parameter, ϵ, in the problem. Having said this, WKB theory can often provide qualitative guidance even when there is little scale separation between the variation of the background state and the wavelength. Asymptotics often works when it does not have to.

A.2 Alternate Derivation

A quick and informative, but less systematic, way to obtain the same result is to seek solutions of (6.162) in the form

$$\xi = A(z) \mathrm{e}^{\mathrm{i}\theta(z)}, \tag{6.177}$$

where $A(z)$ and $\theta(z)$ are both presumptively real. Using (6.177) in (6.162) yields

$$\mathrm{i}\left[2\frac{\mathrm{d}A}{\mathrm{d}z}\frac{\mathrm{d}\theta}{\mathrm{d}z} + A\frac{\mathrm{d}^2\theta}{\mathrm{d}z^2}\right] + \left[A\left(\frac{\mathrm{d}\theta}{\mathrm{d}z}\right)^2 - \frac{\mathrm{d}^2 A}{\mathrm{d}z^2} - m^2 A\right] = 0. \tag{6.178}$$

The terms in square brackets must each be zero. The WKB approximation is to assume that the amplitude varies sufficiently slowly that $|A^{-1}\mathrm{d}^2A/\mathrm{d}z^2| \ll m^2$, and hence that the term involving $\mathrm{d}^2A/\mathrm{d}z^2$ may be neglected. The real and imaginary parts of (6.178) become

$$\left(\frac{\mathrm{d}\theta}{\mathrm{d}z}\right)^2 = m^2, \qquad 2\frac{\mathrm{d}A}{\mathrm{d}z}\frac{\mathrm{d}\theta}{\mathrm{d}z} + A\frac{\mathrm{d}^2\theta}{\mathrm{d}z^2} = 0. \tag{6.179a,b}$$

These two equations are very similar to (6.171). The solution of the first one is

$$\theta = \pm \int m \,\mathrm{d}z, \tag{6.180}$$

and substituting this into (6.179b) gives

$$2\frac{\mathrm{d}A}{\mathrm{d}z}m + A\frac{\mathrm{d}m}{\mathrm{d}z} = 0, \qquad \text{with solution} \qquad A = A_0 m^{-1/2}. \tag{6.181a,b}$$

Using (6.180) and (6.181b) in (6.177) recovers (6.174). Using (6.179a) and the real part of (6.178) we see that the condition for the validity of the approximation is that

$$\left|A^{-1}\frac{\mathrm{d}^2A}{\mathrm{d}z^2}\right| \ll m^2, \quad \text{which using (6.181b) is} \quad \left|\frac{1}{m^{-1/2}}\frac{\mathrm{d}^2m^{-1/2}}{\mathrm{d}z^2}\right| \ll m^2. \tag{6.182a,b}$$

Equation (6.164) expresses a similar condition to (6.182b).

Notes

1 For example, *Linear and Nonlinear Waves* by G. B. Whitham, *Nonlinear Dispersive Waves* by M. J. Ablowitz, or Boyd (1980).

2 Group velocity seems to have been first articulated in about 1841 by the Irish mathematician and physicist William Rowan Hamilton (1806–1865), who is also remembered for his formulation of 'Hamiltonian mechanics'. Hamilton may have been motivated by optics, and it was George Stokes, Osborne Reynolds and John Strutt (better known as Lord Rayleigh) who further developed and generalized the idea in a more hydrodynamic context.

3 More detailed treatments of ray theory and related matters are given by Whitham (1974), Lighthill (1978) and LeBlond & Mysak (1980).

4 What are now called Rossby waves were probably first discovered in a theoretical context by Hough (1897, 1898). He considered the linear shallow water equations on a sphere (i.e., Laplace's tidal equations) expanding the solution in powers of the sine of latitude, and obtained two classes of waves: long, rotationally modified, gravity waves and a balanced wave dependent on variations in Coriolis parameter. However, his work was mainly aimed at understanding ocean tides and it was not until the topic was revisited by Rossby (1939) that the meteorological relevance was appreciated. Rossby used the beta-plane approximation in Cartesian co-ordinates, and the simplicity of the presentation along with the meteorological context led to the work attracting significant notice.

5 Chang & Orlanski (1994), Orlanski & Sheldon (1995). Edmund Chang made number of very helpful remarks to me on waves more generally. I would also like to thank Emma Howard for comments on this chapter and Chapter 18.

6 The non-Doppler effect also even in models in height coordinates. See White (1977).

7 See Chapman & Lindzen (1970).

8 As in Longuet-Higgins (1964).

9 The form of this derivation was originally given by Hayes (1977) in the context of wave energy. Vanneste & Shepherd (1998) provide further discussion, in particular of the uniqueness or otherwise of the wave activity density and flux.

10 A description of the WKB method, also called the JWKB method and sometimes the geometrical optics approximation, can be found in many books on perturbation methods, for example Simmonds & Mann (1998), Holmes (2013) and Bender & Orszag (1978). Developments in perturbation theory and multiple scale analysis have generalized the method to the extent that 'WKB' does not appear as a separate topic, or in the index, in the well-known book by Kevorkian & Cole (2011). Wentzel, Kramers and Brillouin separately (or at least in different articles) presented the technique in 1926 as a way to find approximate solutions of the Schrödinger equation. Harold Jeffreys, a mathematical geophysicist, had proposed a similar technique in 1924, and Rayleigh in 1912 had already addressed some aspects of the theory. A mathematical treatment of the topic was in fact given by Joseph Liouville and George Green in 1837, with even earlier relevant work by Francesco Carlini, an Italian astronomer and director of an observatory in Milan, in 1817. The story thus affirms the hypothesis that methods are named after the last people to discover them...

Further Reading

Waves

Bühler, O. 2009. *Waves and Mean Flows.*
A modern, advanced, discussion of waves, mean flows and their interaction, including the transformed Eulerian mean, the generalized Lagrangian mean, and more.

LeBlond, P. H. & Mysak, L. A. 1980. *Waves in the Ocean.*
A comprehensive review of the dynamics of many kinds of waves in the ocean.

Pedlosky, J., 2003. *Waves in the Ocean and Atmosphere: Introduction to Wave Dynamics.*
A compact, informal introduction to the main waves to be found in the atmosphere and ocean, in the style of lecture notes.

Instabilities

The following two books cover most forms of hydrodynamic instability.

Chandrasekhar, S., 1961. *Hydrodynamic and Hydromagnetic Stability.*
This book has become an enduring classic, but has no discussion of baroclinic instability at all.

Drazin P. & Reid, W. H., 1981. *Hydrodynamic Stability.*
A standard text on hydrodynamic instability theory. It has straightforward and extensive discussions of most of the standard cases, but alas only a brief treatment of baroclinic instability.

Turbulence

There are numerous books on turbulence, but one place to start is

Tennekes, H. & Lumley, J., 1972. *A First Course in Turbulence.*
The book remains a classic introduction to the subject.

Another book that has stood the test of time is

Monin, A. S. & Yaglom, A. M., 1971. *Statistical Fluid Mechanics.*
The two volumes are encyclopædic in content and contain a wealth of information, especially on turbulent diffusion. They are not nearly as daunting as they seem.

Two more modern general introductions and references are

Davidson, P. D., 2015. *Turbulence: An Introduction for Scientists and Engineers.*

Pope, S. B., 2000. *Turbulent Flows.*
Both of these are written at about the graduate student level.

Two books containing review and synthesis articles covering a range of topics related to jets and turbulence on both large and small scales, in atmospheres and oceans, are

Galperin, B. & Read, P. L. Eds., 2017. *Zonal Jets: Phenomenology, Genesis, Physics.*

Baumert, H. Z., Simpson, J. & Sündermann, J. Eds., 2005. *Marine Turbulence: Theories, Observations and Models.*

It mounts at sea, a concave wall,
Down-ribbed with shine,
And pushes forward, building tall,
Its steep incline.
Thom Gunn, *From the Wave*, 1971.

CHAPTER 7

Gravity Waves

GRAVITY WAVES ARE, UNSURPRISINGLY, waves in a fluid in which gravity provides the restoring force. (Gravitational waves are a prediction of general relativity theory.) For gravity to have an effect the fluid density must vary, and thus the waves must either exist at a fluid interface or stratification must be present — and a fluid interface is just an abrupt form of stratification. It is thus common to think of gravity waves as being either internal waves or surface waves: the former being in the interior of a fluid where the density changes may be continuous, and the latter at a fluid interface. Naturally enough the two waves have many similarities — indeed surface waves (also called interfacial waves) are a limiting form of internal waves, when the density variations in the vertical become discontinuous. We considered such interfacial waves in the hydrostatic, shallow water case in Chapter 3, and we will first extend that to the nonhydrostatic case. We then consider internal waves in the continuously stratified equations and that constitutes the bulk of the chapter.[1]

In most of the chapter we will restrict attention to the Boussinesq equations, because in making the incompressibility approximation sound waves are eliminated, greatly simplifying the treatment. In the atmosphere the Boussinesq equations are not a quantitatively good approximation except for motions of a small vertical extent; the anelastic equations improve matters in allowing for a vertical variation of the basic state density, an effect important when considering the vertical propagation of gravity waves high into the atmosphere. Nevertheless, no truly new types of waves are introduced in this way, and so we leave the details to the original literature. If, on the other hand, the fluid is truly compressible then sound waves make themselves heard, and we consider the algebraically complex case of *acoustic-gravity* waves at the end of this chapter. We begin with the simpler case of surface gravity waves atop a constant density fluid.

7.1 SURFACE GRAVITY WAVES

Let us consider an incompressible fluid with a free surface and a flat bottom that obeys the three-dimensional momentum and mass continuity equations, namely

$$\frac{\mathrm{D}\boldsymbol{v}}{\mathrm{D}t} = -\nabla_3\phi - g\mathbf{k}, \qquad \nabla\cdot\boldsymbol{v} = 0, \tag{7.1}$$

using our standard notation where $\phi = p/\rho_0$. The free surface at the top of the fluid is at $z = \eta(x, y, t)$, the mean position of the free surface is at $z = 0$ and the bottom of the fluid, assumed flat, is at $z = -H$ — refer to Fig. 3.1 on on page 106. In this chapter we use a subscript 3 on ∇ to denote a three-dimensional operator, and the unsubscripted operator is horizontal.

In a state of rest the pressure, ϕ_0 say, is given by hydrostatic balance and so $\phi_0 = -gz$. If we write $\phi = -gz + \phi'$ the momentum equation becomes, without approximation,

$$\frac{\mathrm{D}\boldsymbol{v}}{\mathrm{D}t} = -\nabla_3\phi'. \tag{7.2}$$

Linearizing the equations of motion about such a resting state straightforwardly yields

$$\frac{\partial \boldsymbol{v}'}{\partial t} = -\nabla_3\phi', \qquad \nabla_3 \cdot \boldsymbol{v}' = 0, \tag{7.3a,b}$$

where a prime denotes a perturbation quantity in the usual way. We now proceed by expressing the problem solely in terms of pressure. (An equivalent alternative would be to use a velocity potential, ξ say, such that $\boldsymbol{v} = \nabla_3\xi$, which is possible because, from (7.3a) the flow is irrotational.) Taking the divergence of (7.3a) and using (7.3b) gives us Laplace's equation for the pressure, namely

$$\nabla_3^2\phi' = 0. \tag{7.4}$$

This equation has no explicit time dependence, but the boundary conditions are time dependent and that is how we will obtain the dispersion relation.

7.1.1 Boundary Conditions

Since (7.4) is an equation for pressure we seek boundary conditions on pressure. At the bottom of the fluid ($z = -H$) the condition that $w = 0$ may be turned into a condition on pressure using (7.3a), namely that

$$\frac{\partial \phi'}{\partial z} = 0 \qquad \text{at } z = -H. \tag{7.5}$$

At the top surface, $z = \eta$, the pressure must equal that of the atmosphere above. We will take this to be a constant, and in particular zero, so that $\phi = 0$ at $z = \eta$. Now, the perturbation pressure is given by $\phi = -gz + \phi'$, so that at $z = \eta$ we obtain

$$\phi' = g\eta. \tag{7.6}$$

A second boundary condition at the top is the kinematic condition that a fluid parcel in the free surface must remain within the fluid, and therefore that (with full nonlinearity)

$$\frac{\mathrm{D}}{\mathrm{D}t}(z - \eta) = 0. \tag{7.7}$$

If we linearize this equation and use the definition of w we obtain $w' = \partial\eta/\partial t$ at $z = \eta$, which using (7.6) becomes $w' = g^{-1}\partial\phi'/\partial t$. Using the vertical component of the momentum equation, (7.3a), we obtain the pressure boundary condition

$$\frac{1}{g}\frac{\partial^2\phi'}{\partial t^2} = -\frac{\partial \phi'}{\partial z} \qquad \text{at } z = \eta. \tag{7.8}$$

The value of η is in fact unknown without solving the problem itself, and in the general (nonlinear) case we have to solve the whole problem in a self-consistent fashion. However, in the linear problem η is presumptively small (we are linearizing the free surface about $z = 0$) and we will apply this boundary condition at $z = 0$ rather than at $z = \eta$, for the error will only be second order.

Having established the equations and the boundary conditions, and noting that we will be dealing exclusively with linear equations in the rest of this section (and for most of this chapter), we'll now drop the primes on perturbation quantities unless needed.

7.1.2 Wave Solutions

We now seek solutions to (7.4) in the form

$$\phi = \mathrm{Re}\,\Phi(z)\exp(\mathrm{i}[\boldsymbol{k}\cdot\boldsymbol{x} - \omega t]) \tag{7.9}$$

where $\boldsymbol{x} = \mathbf{i}x + \mathbf{j}y$ and $\boldsymbol{k} = \mathbf{i}k + \mathbf{j}l$ and Re denotes that the real part is to be taken, a notation that we drop unless it causes ambiguity. Equation (7.4) becomes

$$\frac{\mathrm{d}^2\Phi}{\mathrm{d}z^2} - K^2\Phi = 0, \tag{7.10}$$

where $K^2 = k^2 + l^2$ and the boundary conditions are that $\mathrm{d}\Phi/\mathrm{d}z = 0$ at $z = -H$ and $\mathrm{d}^2\Phi/\mathrm{d}z^2 = -g\mathrm{d}\Phi/\mathrm{d}z$ at $z = 0$. The bottom boundary condition is satisfied by a solution of the form

$$\Phi = A\cosh K(z + H). \tag{7.11}$$

Substituting into the top boundary condition, (7.8) at $z = 0$, we obtain

$$-\omega^2\cosh KH = gK\sinh KH = 0, \tag{7.12}$$

or

$$\omega = \pm\sqrt{gK\tanh KH}. \tag{7.13}$$

This is the dispersion relation for surface gravity waves. The corresponding phase speed is given by

$$c_p = \frac{\omega}{K} = \pm\sqrt{gH}\left(\frac{\tanh KH}{KH}\right)^{1/2}. \tag{7.14}$$

Using (7.9) and (7.11) the full solution for the pressure field is

$$\phi = \mathrm{Re}\,\Phi_0\cosh K(z + H)\exp(\mathrm{i}[\boldsymbol{k}\cdot\boldsymbol{x} - \omega t]) \tag{7.15}$$

with ω given by (7.13) and the amplitude Φ_0 being set by the initial conditions. It is convenient to write the amplitude Φ_0 in terms of the amplitude of the free surface elevation, η_0, using the upper boundary condition that $\phi = g\eta$, so that $\eta_0 = \Phi_0/g$. The other field variables may be found from (7.3a) and are given by

$$u = \eta_0\frac{k}{\omega}gC\cosh K(z + H), \quad v = \eta_0\frac{l}{\omega}gC\cosh K(z + H), \quad w = -i\eta_0\frac{K}{\omega}gC\sinh K(z + H), \tag{7.16a,b,c}$$

where $C = \exp(\mathrm{i}[\boldsymbol{k}\cdot\boldsymbol{x} - \omega t])/\cosh KH$, and only the real parts of each expression should be taken. Thus, if we take η_0 to be real then u and v vary like $\cos(\boldsymbol{k}\cdot\boldsymbol{x} - \omega t)$ and w varies as $\sin(\boldsymbol{k}\cdot\boldsymbol{x} - \omega t)$, and this is what we will assume.

7.1.3 Properties of the Solution

First, from (7.13) we see that for each wavevector amplitude there are two waves propagating in opposite directions, with a frequency and phase speed that depend only on the wavelength K and not the orientation of the wave vector. Second, the waves are *dispersive*. That is, similary to Rossby waves but unlike light waves in a vacuum or shallow water waves, the phase speed is different for waves of different wavelengths. Since the frequency is a function only of K (and not of k or l individually) the group velocity is parallel to the wave vector itself and is given by

$$\boldsymbol{c}_g = \nabla_{\boldsymbol{k}}\omega = \frac{\partial\omega}{\partial K}\frac{\boldsymbol{k}}{K}, \tag{7.17}$$

where $\mathbf{k} = k\mathbf{i} + l\mathbf{j}$ and $\mathbf{k}/K$ is the unit vector in the direction of propagation. Using the dispersion relation $\omega^2 = gK \tanh KH$ we obtain

$$2\omega \frac{\partial \omega}{\partial K} = g\left(\tanh KH + \frac{KH}{\cosh^2 KH}\right), \tag{7.18}$$

so that

$$\boldsymbol{c}_g = \frac{g}{2c_p K}\left(\tanh KH + \frac{KH}{\cosh^2 KH}\right)\mathbf{k}, \tag{7.19}$$

and, now using (7.14), the ratio of the group speed (i.e., the magnitude of the group velocity) to the phase speed is given by

$$\frac{c_g}{c_p} = \frac{1}{2}\left(1 + \frac{2KH}{\sinh 2KH}\right), \tag{7.20}$$

having used the relation $2\sinh x \cosh x = \sinh 2x$.

We note two important limiting cases:

(i) The long wavelength or shallow water limit, $KH \ll 1$. It is this limit that is relevant to large-scale flow in the ocean and atmosphere. In this limit the wavelength is much greater than the depth of the fluid and the dispersion relation (7.13) reduces to $\omega = K\sqrt{gH}$ (since for small x, $\tanh x \to x$) and $c_p = c_g = \sqrt{gH}$, and the waves are nondispersive. This result is apparent from (7.20) in the limit of $KH \ll 1$. As expected, this is the same dispersion relation as was previously derived *ab initio* for shallow water waves in Chapter 3. This limit is appropriate as water waves approach the shore and start feeling the bottom, and for long waves such as tides and tsunamis.

The pressure field in this limit is given, using (7.15),

$$\phi = \eta_0 g \exp(\mathrm{i}[\boldsymbol{k}\cdot\boldsymbol{x} - \omega t]). \tag{7.21}$$

This is the *perturbation* pressure associated with the wave, and evidently it does not depend on depth. The total pressure at a given point in the fluid is given by the static pressure plus perturbation pressure and this is, including the density ρ_0,

$$p = -\rho_0 g z + \rho_0 \phi = \rho_0 g(\eta - z). \tag{7.22}$$

Evidently, the pressure in the shallow water limit is hydrostatic. If $1/k > 20H$ the error in this approximation is less than 3%.

(ii) The short wavelength or deep water limit, $KH \gg 1$. For large KH, $\tanh KH \to 1$ so that the dispersion relation becomes $\omega^2 = gK$ and $c_p^2 = g/K$. These waves are dispersive, with long waves travelling faster than short waves. A familiar manifestation of this arises when a rock is thrown into a pool; initially, waves of all wavelengths are excited (for the initial disturbance is like a delta function), but the long waves propagate away faster than the short waves and reach distant objects first. The group speed in this case is given by

$$c_g = \frac{\partial \omega}{\partial k} = \frac{g}{2\omega} = \frac{1}{2}\sqrt{\frac{g}{K}} = \frac{c_p}{2}. \tag{7.23}$$

This result is also apparent from (7.20) in the limit of short waves, $KH \gg 1$, and it has an interesting consequence for wave packets. Consider a packet of short waves moving in the positive x-direction; the envelope moves with the group speed and the individual crests move with the phase speed, so that individual crests enter the packet from the rear and travel through the packet, exiting at the front.

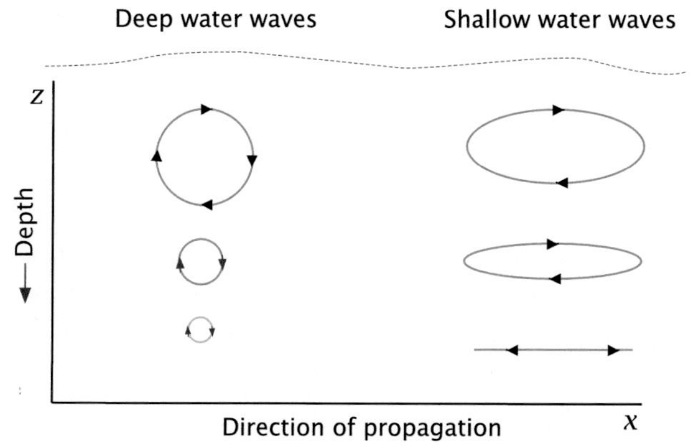

Fig. 7.1 Parcel motion for deep and shallow water waves. The motion is circular for deep water waves, with an amplitude that decreases exponentially with depth. The motion is elliptical for shallow water waves, but the horizontal excursion is independent of depth and the vertical excursion decays linearly with depth.

Parcel motion

The trajectories of water parcels are rather interesting in water waves. It turns out that in deep water the parcels make circular orbits with an amplitude diminishing with depth, whereas shallow water waves trace elliptic paths, as illustrated in Fig. 7.1 and as we now explain.

We obtain the parcel excursions using the expressions for velocity (7.16), taking $v = 0$ without loss of generality. For shallow water waves ($KH \ll 1$) u is depth independent and the velocity and the excursion in the x direction, which we denote as X, are given by

$$u = \eta_0 \frac{kg}{\omega} \cos(kx - \omega t), \qquad X = -\eta_0 \frac{gk}{\omega^2} \sin(kx - \omega t), \tag{7.24a}$$

and this is independent of z. The excursion in the z direction, Z, is given by

$$w = \eta_0 \frac{k^2}{\omega}(z + H) \sin(kx - \omega t), \qquad Z = \eta_0 \frac{gk^2}{\omega^2}(z + H) \cos(kx - \omega t), \tag{7.24b}$$

where $\omega = k\sqrt{gH}$. We see that $Z = \eta_0$ at $z = 0$, as expected. The above expressions for X and Z are, at some fixed location x and z, parametric representations of an ellipse. As z varies the horizontal amplitude of the ellipses remains constant whereas the vertical amplitude decreases linearly from the top $z = 0$ to a zero amplitude at the bottom, $z = -H$. The vertical amplitude is also generally much less than the horizontal amplitude, by the ratio

$$\frac{|Z|}{|X|} = \frac{|w|}{|u|} \sim kH \ll 1. \tag{7.25}$$

That is, the fluid motion is mostly horizontal.

In the deep water limit, $kH \gg 1$, the horizontal and vertical velocities and excursions are given by

$$u = \eta_0 \frac{kg}{\omega} \exp kz \cos(kx - \omega t), \qquad X = -\eta_0 \frac{kg}{\omega^2} \exp kz \sin(kx - \omega t), \tag{7.26a}$$

$$w = \eta_0 \frac{kg}{\omega} \exp kz \sin(kx - \omega t), \qquad Z = \eta_0 \frac{kg}{\omega^2} \exp kz \cos(kx - \omega t), \tag{7.26b}$$

where $\omega^2 = gk$, and again we have that $Z = \eta$ at $z = 0$. The expressions for X and Z, having the same amplitude, are now parametric representations of circles whose amplitudes diminish exponentially with depth. Evidently, all the dynamical variables decrease exponentially with depth, with an e-folding scale of the wavelength itself. The wave field cannot feel the bottom of the fluid container and all the expressions become independent of the water depth H.

♦ *Energy propagation*

For our final discussion on this topic we look at the energy propagation of surface waves. The kinetic energy per unit horizontal area is given by

$$\mathrm{KE} = \int_{-H}^{0} \frac{1}{2}\rho_0 \boldsymbol{v}^2 \,\mathrm{d}z. \tag{7.27}$$

The upper limit on the integration is taken to be $z = 0$, rather than $z = \eta$, because using the latter would lead to a term of order $\eta \boldsymbol{v}^2$, which is third order in perturbation quantities. The potential energy per unit horizontal area is

$$\mathrm{PE} = \int_{-H}^{\eta} \rho_0 g z \,\mathrm{d}z = \frac{\rho_0 g}{2}(\eta^2 - H^2). \tag{7.28}$$

The integral now must be over the complete depth of the fluid in order to calculate the potential energy to quadratic order. The term in H^2 is a constant and so is largely irrelevant to the problem of energy propagation. Also, since ρ_0 is a constant we will set its value to unity.

The kinetic energy equation is obtained by taking the dot product of the linearized momentum equation, (7.3a) with $\boldsymbol{v}$ and integrating over the depth of the fluid to give

$$\int_{-H}^{0} \mathrm{d}z \left[\frac{\partial}{\partial t}\frac{\boldsymbol{v}^2}{2} + \nabla\cdot(\boldsymbol{u}\phi) + \frac{\partial w\phi}{\partial z} \right] = 0, \tag{7.29}$$

noting that $\boldsymbol{v} = \boldsymbol{u} + w\mathbf{k}$ and $\nabla\cdot\boldsymbol{v} = 0$. The boundary conditions on w are that $w = 0$ at $Z = -H$ and $w = \partial\eta/\partial t$ at $z = 0$. Further, at $z = 0$ $\phi = g\eta$, and using these results (7.29) becomes

$$\int_{-H}^{0} \mathrm{d}z \left[\frac{\partial}{\partial t}\frac{\boldsymbol{v}^2}{2} + \nabla\cdot(\boldsymbol{u}\phi) \right] + g\frac{\partial}{\partial t}\frac{\eta^2}{2} = 0, \tag{7.30}$$

which, using (7.27) and (7.28), is just

$$\frac{\partial}{\partial t}(\mathrm{KE} + \mathrm{PE}) + \nabla\cdot\boldsymbol{F} = 0, \tag{7.31}$$

where $\boldsymbol{F} = \int_{-H}^{z} \boldsymbol{u}\phi \,\mathrm{d}z$ is the energy flux, a vector with only horizontal components. (Thus, the divergence term in (7.31) is just a horizontal divergence.)

Equation (7.31) is an energy conservation equation for the linearized equations. It is fairly general at the moment, for we have not specialized to the case of *wave* motion. Let's do that now, by using the properties of the waves derived above and averaging over a wave period. Without loss of generality we'll assume the waves are propagating in the x direction so that $v = 0$ and $K = k$; nevertheless, the calculation is rather algebraic and the trusting reader may skim it.

The kinetic energy averaged over a wave period, $\overline{\mathrm{KE}}$ is given by

$$\begin{aligned}\overline{\mathrm{KE}} &= \frac{\omega}{2\pi}\int \mathrm{d}t \left(\int \frac{1}{2}\boldsymbol{v}^2 \,\mathrm{d}z \right) \\ &= \frac{k^2\eta_0^2 g^2}{2\omega^2\cosh^2 kH}\frac{\omega}{2\pi}\int \mathrm{d}t \int \mathrm{d}z \left[\cosh^2 k(z+H)\cos^2(kx-\omega t) + \sinh^2 k(z+H)\sin^2(kx-\omega t)\right].\end{aligned} \tag{7.32}$$

In this expression the time integrals range from 0 to $2\pi/\omega$ and the vertical integrals range from $-H$ to 0. The time averages of $\sin^2$ and $\cos^2$ produce a factor of $1/2$, and noting that $\cosh^2 x + \sinh^2 x = \cosh 2x$ we obtain

$$\overline{\mathrm{KE}} = \frac{k^2\eta_0^2 g^2}{2\omega^2\cosh^2 kH}\frac{1}{2}\frac{\sinh(2kH)}{2k}. \tag{7.33}$$

Using the dispersion relation $\omega^2 = gk \tanh kH$ we finally obtain the simple expression

$$\overline{\mathrm{KE}} = \frac{g\eta_0^2}{4}. \tag{7.34}$$

The perturbation potential energy is given by

$$\overline{\mathrm{PE}} = \frac{\omega}{2\pi}\int \frac{1}{2} g\eta^2 \,\mathrm{d}t = \frac{g\eta_0^2}{2}\frac{\omega}{2\pi}\int \cos^2(kx - \omega t)\,\mathrm{d}t = \frac{g\eta_0^2}{4}. \tag{7.35}$$

Evidently, from (7.34) and (7.35) there is equipartitioning of energy time-averaged potential and kinetic energy components. Such equipartitioning is not, however, a universal property of wave motion.

The time averaged energy flux, which is in the x direction, is given by

$$\overline{F} = \frac{\omega}{2\pi}\int \mathrm{d}t \int u'\phi' \,\mathrm{d}z. \tag{7.36}$$

Using the wave expressions (7.15) we obtain, after a couple of lines of algebra,

$$\overline{F} = \frac{1}{2}\eta_0^2 \frac{g^2}{2c}\frac{1}{\cosh^2 kH}\left[\frac{\sinh 2kH}{2k} + H\right]. \tag{7.37}$$

Using (7.19) and the fact that $\sinh 2hK = 2\sinh kH \cosh kH$ we obtain

$$\overline{F} = \frac{\eta_0^2 g}{2} c_g = (\overline{\mathrm{KE}} + \overline{\mathrm{PE}}) c_g. \tag{7.38}$$

Thus, using (7.34), (7.35) and (7.38), and generalizing the direction of propagation, we have that

$$\frac{\partial \overline{E}}{\partial t} + \nabla \cdot \mathbf{c}_g \overline{E} = 0, \tag{7.39}$$

where $\overline{E} = \overline{\mathrm{KE}} + \overline{\mathrm{PE}}$. Thus, the flux of energy is equal to the energy times the group velocity, or equivalently the energy in the wave propagates with the group velocity. As we established in Chapter 6, this property is a rather general one for wave motion and it is satisfying to see how it applies to surface waves.

7.2 SHALLOW WATER WAVES ON FLUID INTERFACES

Let us now generalize our treatment of surface gravity waves to those waves that exist on the interface between *two* moving fluids of different densities. The ensuing waves are a simple model of gravity waves that exist in the interior of the atmosphere and, perhaps especially, the ocean, in which we idealize the continuous stratification of the real fluid by supposing that the fluid comprises two (or conceivably more) layers of immiscible fluids of different densities stacked on top of each other. We will consider only the hydrostatic case in which case the layers form a 'stacked shallow water' system. We further limit ourselves to two moving layers; an extension to multiple layers is conceptually if not algebraically straightforward, but it soon becomes easier to treat the continuously stratified case, which we do in later sections.

7.2.1 Equations of Motion

Consider a two-layer shallow water model as illustrated in Fig. 3.5 on page 113. From Section 3.3 the nonlinear momentum equations are, for the upper layer,

$$\frac{\mathrm{D}\boldsymbol{u}_1}{\mathrm{D}t} + \boldsymbol{f} \times \boldsymbol{u}_1 = -g\nabla\eta_0, \tag{7.40a}$$

and in the lower layer

$$\frac{\mathrm{D}\boldsymbol{u}_2}{\mathrm{D}t} + \boldsymbol{f} \times \boldsymbol{u}_2 = -\frac{\rho_1}{\rho_2}\left(g\nabla\eta_0 + g_1'\nabla\eta_1\right). \tag{7.40b}$$

where $g_1' = g(\rho_2 - \rho_1)/\rho_1$ (we will henceforth drop the subscript 1 and denote this as g'), and in the Boussinesq case we take $\rho_1/\rho_2 = 1$. We will only consider the non-rotating case, and after linearization about a resting state we have for the upper and lower layers respectively

$$\frac{\partial \boldsymbol{u}_1'}{\partial t} = -g\nabla\eta_0', \qquad \frac{\partial \boldsymbol{u}_2'}{\partial t} = -g\nabla\eta_0' - g'\nabla\eta_1'. \tag{7.41}$$

The equations of motion are completed by the mass continuity equations for each layer, namely

$$\frac{\mathrm{D}}{\mathrm{D}t}(\eta_0 - \eta_1) + h_1\nabla\cdot\boldsymbol{u}_1 = 0 \quad\Longrightarrow\quad \frac{\partial}{\partial t}(\eta_0' - \eta_1') + H_1\nabla\cdot\boldsymbol{u}_1' = 0 \tag{7.42a,b}$$

and

$$\frac{\mathrm{D}\eta_1}{\mathrm{D}t} + h_2\nabla\cdot\boldsymbol{u}_2 = 0 \quad\Longrightarrow\quad \frac{\partial\eta_1'}{\partial t} + H_2\nabla\cdot\boldsymbol{u}_2' = 0, \tag{7.43a,b}$$

where the two rightmost expressions follow after linearization and we assume that the bottom is flat; that is $\eta_b = 0$. Henceforth we will also omit any primes on the perturbed quantities.

7.2.2 Dispersion Relation

We first eliminate the velocity from (7.41a) and (7.42b) to give

$$\frac{\partial^2}{\partial t^2}(\eta_0 - \eta_1) - gH_1\nabla^2\eta_0 = 0, \tag{7.44}$$

and similarly for the lower layer:

$$\frac{\partial^2\eta_1}{\partial t^2} - H_2(g\nabla^2\eta_0 + g'\nabla^2\eta_1) = 0. \tag{7.45}$$

Equations (7.44), and (7.45) form a complete set and in the usual fashion we may look for solutions of the form $\eta_i = \mathrm{Re}\,\tilde{\eta}_i \exp[\mathrm{i}(\boldsymbol{k}\cdot\boldsymbol{x} - \omega t)]$. We obtain

$$(\omega^2 - gH_1K^2)\tilde{\eta}_0 - \omega^2\tilde{\eta}_1 = 0, \tag{7.46a}$$

$$-gH_2K^2\tilde{\eta}_0 + (\omega^2 - g'H_2K^2)\tilde{\eta}_1 = 0, \tag{7.46b}$$

where $K^2 = k^2 + l^2$. For these equations to have non-trivial solutions we must have

$$(\omega^2 - gH_1K^2)(\omega^2 - g'H_2K^2) - \omega^2 gH_2K^2 = 0, \tag{7.47}$$

which, for small $g'/g \ll 1$ gives, after a couple of lines of algebra,

$$\omega^2 = \frac{1}{2}K^2gH \pm \frac{1}{2}K^2gH\sqrt{1 - 4\frac{g'}{g}\frac{H_1H_2}{H^2}} \approx \frac{1}{2}K^2gH \pm \frac{1}{2}K^2gH\left(1 - 2\frac{g'}{g}\frac{H_1H_2}{H^2}\right), \tag{7.48}$$

where $H = H_1 + H_2$. If $g' = 0$ we recover the familiar single-layer dispersion relation, $\omega = K\sqrt{gH}$ (as well as $\omega = 0$). In the more general case there are two distinct modes:

(*i*) A fast mode with phase speed given by

$$c_p^2 = \left(\frac{\omega}{k}\right)^2 = gH\left(1 - \frac{g'}{g}\frac{H_1H_2}{H^2},\right), \tag{7.49}$$

where, for simplicity (and, in fact, without loss of generality, since it amounts only to an alignment of our coordinate system), we take $l = 0$. Using (7.46a) we then find that

$$\frac{\eta_0}{\eta_1} \approx \frac{H}{H_2}. \tag{7.50}$$

That is, since $H > H_2$, the displacement of the upper surface is larger than that of the lower. This mode is sometimes called the 'barotropic' mode, for the oscillations are vertically coherent (the phase on the interior surface is the same as that at the surface), and virtually the same oscillation would exist even in the absence of a density jump in the interior.

(*ii*) A slower mode with phase speed given by

$$c_p^2 \approx g'\frac{H_1H_2}{H}, \tag{7.51}$$

and vertical structure

$$\frac{\eta_0}{\eta_1} \approx \frac{g'H_2}{gH} \ll 1. \tag{7.52}$$

In this case the displacement of the upper surface is smaller than the interior displacement by the ratio of g' to g; in the ocean, where density differences are small, the ratio might well be of order 1/100. Furthermore, the internal displacement is *out of phase* with that at the surface. Often, in oceanic situations the interface may be taken as representing the thermocline, in which case $H_2 \gg H_1$ (i.e., the abyss has a greater depth than the thermocline) and $H \approx H_2$. In this case $c_p^2 \approx g'H_1$, and internal waves on the thermocline behave rather like surface waves, but with a weaker restoring force (and consequently a larger amplitude) because the density difference between the two layers of seawater is much smaller than the density difference between the seawater and air above it.

7.3 INTERNAL WAVES IN A CONTINUOUSLY STRATIFIED FLUID

We now turn our attention to *internal gravity waves,* namely waves that are internal to a given fluid and that owe their existence to the restoring force of gravity. Interfacial waves are, of course, a model of internal waves with a discontinuous jump in density within the fluid. Surface waves might even be thought of as internal waves if one supposes that part of the fluid has zero density, although this stretches the definition of the word internal somewhat. In this section we will consider the simplest and most fundamental case, that of internal waves in a Boussinesq fluid with constant stratification and no background rotation.

Reprising and extending the material of Section 2.10.4, let us consider a continuously stratified Boussinesq fluid, initially at rest, in which the background buoyancy varies only with height and so the buoyancy frequency, N, is a function only of z. Linearizing the equations of motion about this basic state gives the linear momentum equations,

$$\frac{\partial \boldsymbol{u}'}{\partial t} = -\nabla\phi', \qquad \frac{\partial w'}{\partial t} = -\frac{\partial \phi'}{\partial z} + b', \tag{7.53a,b}$$

the mass continuity and thermodynamic equations,

$$\frac{\partial u'}{\partial x} + \frac{\partial v'}{\partial y} + \frac{\partial w'}{\partial z} = 0, \qquad \frac{\partial b'}{\partial t} + w'N^2 = 0. \tag{7.53c,d}$$

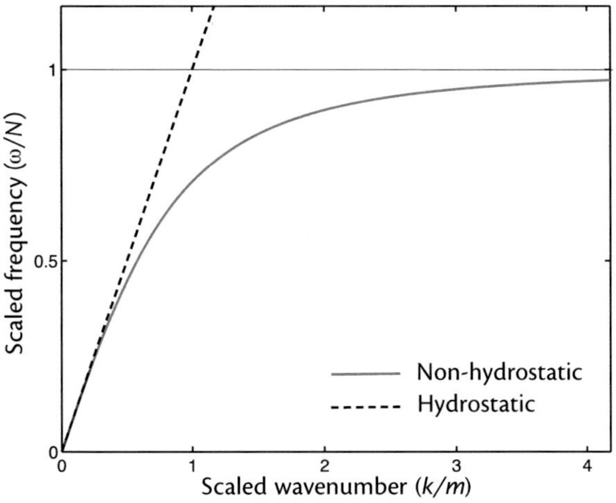

Fig. 7.2 Scaled frequency, ω/N, plotted as a function of scaled horizontal wavenumber, k/m, using the full dispersion relation of (7.56) with $l = 0$ (solid line, asymptoting to unit value for large k/m), and with the hydrostatic dispersion relation (7.60) (dashed line, tending to ∞ for large k/m).

Our notation is such that $\boldsymbol{u} \equiv u\mathbf{i}+v\mathbf{j}$, $\boldsymbol{v} \equiv u\mathbf{i}+v\mathbf{j}+w\mathbf{k}$, and the gradient operator is two-dimensional unless noted. Thus, $\nabla \equiv \mathbf{i}\,\partial_x + \mathbf{j}\,\partial_y$ and $\nabla_3 \equiv \mathbf{i}\,\partial_x + \mathbf{j}\,\partial_y + \mathbf{k}\partial_z$.

A little algebra gives a single equation for w',

$$\left[\frac{\partial^2}{\partial t^2}\left(\nabla^2 + \frac{\partial^2}{\partial z^2}\right) + N^2\nabla^2\right] w' = 0. \tag{7.54}$$

This equation is evidently *not* isotropic. If N^2 is a constant — that is, if the background buoyancy varies linearly with z — then the coefficients of each term are constant, and we may then seek solutions of the form

$$w' = \mathrm{Re}\ \widetilde{w}\mathrm{e}^{\mathrm{i}(kx+ly+mz-\omega t)}, \tag{7.55}$$

where Re denotes the real part, a denotion that will frequently be dropped unless ambiguity arises, and other variables oscillate in a similar fashion. Using (7.55) in (7.54) yields the dispersion relation:

$$\omega^2 = \frac{(k^2+l^2)N^2}{k^2+l^2+m^2} = \frac{K^2N^2}{K_3^2}, \tag{7.56}$$

where $K^2 = k^2 + l^2$ and $K_3^2 = k^2 + l^2 + m^2$. The frequency (see Fig. 7.2) is thus always less than N, approaching N for small horizontal scales, $K^2 \gg m^2$. If we neglect pressure perturbations, as in the parcel argument, then the two equations,

$$\frac{\partial w'}{\partial t} = b', \qquad \frac{\partial b'}{\partial t} + w'N^2 = 0, \tag{7.57}$$

form a closed set, and give $\omega^2 = N^2$.

If the basic state density increases with height then $N^2 < 0$ and we expect this state to be unstable. Indeed, the disturbance grows exponentially according to $\exp(\sigma t)$ where

$$\sigma = \mathrm{i}\omega = \pm\frac{K\widetilde{N}}{K_3}, \tag{7.58}$$

where $\widetilde{N}^2 \equiv -N^2$ and $K_3 = \sqrt{K_3^2}$. Most convective activity in the ocean and atmosphere is, ultimately, related to an instability of this form, although of course there are many complicating issues — water vapour in the atmosphere, salt in the ocean, the effects of rotation and so forth.

7.3.1 Hydrostatic Internal Waves

Let us now suppose that the fluid satisfies the hydrostatic Boussinesq equations, and for simplicity assume that $l = 0$. The linearized two-dimensional equations of motion become

$$\frac{\partial \boldsymbol{u}'}{\partial t} = -\nabla\phi', \qquad 0 = -\frac{\partial \phi'}{\partial z} + b', \tag{7.59a}$$

$$\frac{\partial u'}{\partial x} + \frac{\partial v'}{\partial y} + \frac{\partial w'}{\partial z} = 0, \qquad \frac{\partial b'}{\partial t} + w'N^2 = 0, \tag{7.59b}$$

where these are the horizontal and vertical momentum equations, the mass continuity equation and the thermodynamic equation respectively. A little algebra gives the dispersion relation,

$$\omega^2 = \frac{(k^2 + l^2)N^2}{m^2}. \tag{7.60}$$

The frequency and, if N^2 is negative, the growth rate, are unbounded as $K^2/m^2 \to \infty$, and the hydrostatic approximation thus has quite unphysical behaviour for small horizontal scales. Many numerical models of the large-scale circulation in the atmosphere and ocean do make the hydrostatic approximation. In these models convection must be *parameterized;* otherwise, it would simply occur at the smallest scale available, namely the size of the numerical grid, and this type of unphysical behaviour should be avoided. In nonhydrostatic models convection must also be parameterized if the horizontal resolution of the model is too coarse to properly resolve the convective scales.

7.3.2 Some Properties of Internal Waves

Internal waves have a number of interesting and counter-intuitive properties — let's discuss them.

The dispersion relation

We can write the dispersion relation, (7.56), as

$$\omega = \pm N \cos\vartheta, \tag{7.61}$$

where $\cos^2\vartheta = K^2/(K^2 + m^2)$ so that ϑ is the angle between the three-dimensional wave-vector, $\mathbf{k} = k\mathbf{i} + l\mathbf{j} + m\mathbf{k}$, and the horizontal. The frequency is evidently a function only of N and ϑ, and, if this is given, the frequency is not a function of wavelength. This has some interesting consequences for wave reflection, as we see below.

We can also write the dispersion relation, (7.56), as

$$\frac{\omega^2}{N^2 - \omega^2} = \frac{K^2}{m^2}. \tag{7.62}$$

Thus, and consistently with our first point, given the wave frequency the ratio of the vertical to the horizontal wavenumber is fixed.

Polarization relations

The oscillations of pressure, velocity and buoyancy are, naturally, connected, and we can obtain the relations between them with some simple manipulations. If the pressure field is oscillating like $\phi' = \widetilde{\phi}\exp[\mathrm{i}(\boldsymbol{k}\cdot\boldsymbol{x} - \omega t)] = \widetilde{\phi}\exp[\mathrm{i}(kx + ly + mz - \omega t)]$ then, using (7.53a), the horizontal velocity components have the phases

$$(\widetilde{u}, \widetilde{v}) = (k, l)\,\omega^{-1}\widetilde{\phi}. \tag{7.63a}$$

As the frequency is real, the velocities are in phase with the pressure. A little algebra also reveals that the buoyancy perturbation is related to the pressure perturbation by

$$\tilde{b} = \frac{\mathrm{i}mN^2}{N^2 - \omega^2}\tilde{\phi} = \frac{\mathrm{i}N^2K^2}{m\omega^2}\tilde{\phi} = \frac{\mathrm{i}K_3^2}{m}\tilde{\phi}, \tag{7.63b}$$

using the dispersion relation, so that the buoyancy and pressure perturbations are $\pi/2$ out of phase.

The vertical velocity is related to the pressure perturbation by

$$\tilde{w} = \frac{-\omega m}{N^2 - \omega^2}\tilde{\phi} = \frac{-K^2}{m\omega}\tilde{\phi}, \tag{7.63c}$$

where the second expression uses (7.62). The vertical velocity is in phase with the pressure perturbation, and for regions of positive m (and so with upward phase propagation) regions of high relative pressure are associated with downward fluid motion.

The pressure, buoyancy and velocity fields are all real fields and we can write the above phase relationships in terms of sines and cosines as follows:

$$\phi = \Phi_0 \cos(kx + ly + mz - \omega t), \tag{7.64a}$$

$$(u, v) = (k, l)\frac{\Phi_0}{\omega}\cos(kx + ly + mz - \omega t), \tag{7.64b}$$

$$w = \left(\frac{-\omega m}{N^2 - \omega^2} = \frac{-K^2}{m\omega}\right)\Phi_0 \cos(kx + ly + mz - \omega t). \tag{7.64c}$$

$$b = \left(\frac{mN^2}{N^2 - \omega^2} = \frac{N^2K^2}{m\omega^2}\right)\Phi_0 \sin(kx + ly + mz - \omega t), \tag{7.64d}$$

where Φ_0 is a constant. We might equally well have chosen ϕ to have a sine dependence in (7.64a); nothing of substance differs, but (7.64b,c,d) should be changed appropriately. The relations between pressure, buoyancy and velocity in (7.63) and (7.64) are known as *polarization relations.*

Relation between wave vector and velocity

On multiplying (7.64b) and (7.64c) by (k, l) and m, respectively, we see that

$$\boldsymbol{k} \cdot \tilde{\boldsymbol{v}} = 0, \tag{7.65}$$

where $\boldsymbol{k}$ and $\tilde{\boldsymbol{v}}$ are three-dimensional vectors. This means that, at any instant, the wave vector is perpendicular to the velocity vector, and the velocity is therefore aligned *along* the direction of the troughs and crests, along which there is no pressure gradient. If the wave vector is purely horizontal (i.e., $m = 0$), then the motion is purely vertical and $\omega = N$.

The vertical and horizontal velocities are related to the wavenumbers. If (for simplicity, and with no loss of generality) the motion is in the x–y plane with $v = l = 0$, then it is a corollary of (7.65) that

$$\frac{\tilde{u}}{\tilde{w}} = -\frac{m}{k}. \tag{7.66}$$

Furthermore, from (7.55) with $l = 0$, at any given instant all of the perturbation quantities in the wave are constant along the lines $kx + mz$ = constant. Thus, *all fluid parcel motions are parallel to the wave fronts.* Now, since the wave frequency is related to the background buoyancy frequency by $\omega = \pm N \cos\vartheta$, it follows that the fluid parcels oscillate along lines that are at an angle $\vartheta = \cos^{-1}(\omega/N)$ to the vertical. The polarization relations and the group and phase velocities are illustrated in Fig. 7.3. Let us now discuss the wave properties in a little more detail.

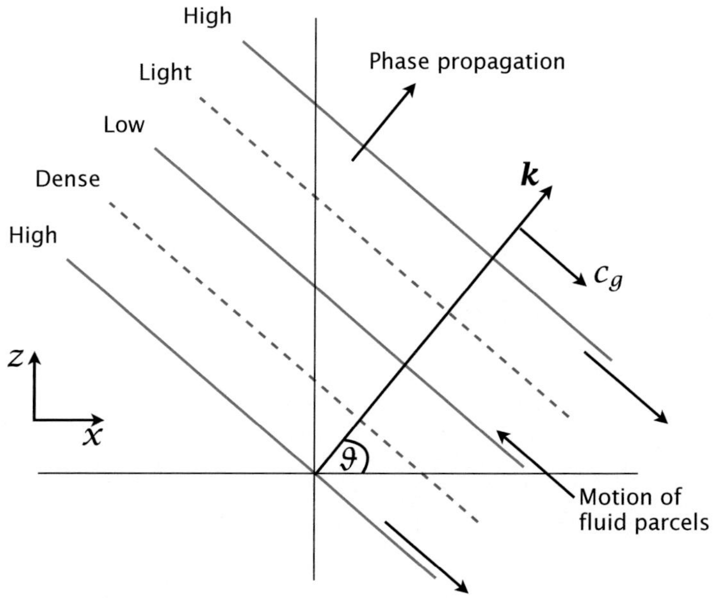

Fig. 7.3 An internal wave propagating in the direction $\boldsymbol{k}$. Both k and m are positive for the wave shown. The solid lines show crests and troughs of constant pressure, and the dashed lines the corresponding crests and troughs of buoyancy (or density). The motion of the fluid parcels is along the lines of constant phase, as shown, and is parallel to the group velocity and perpendicular to the phase speed.

7.3.3 A Parcel Argument and Physical Interpretation

Let us consider first the dispersion relation itself and try to derive it more physically, or at least heuristically. Let us suppose there is a wave propagating in the (x, z) plane at some angle ϑ to the horizontal, with fluid parcels moving parallel to the troughs and crests, as in Fig. 7.3. In general the restoring force on a parcel is due to both the pressure gradient and gravity, but along the crests there is no pressure gradient. Referring to Fig. 7.4, for a total displacement Δs the restoring force, F_{res}, in the direction of the particle displacement is

$$F_{res} = g\cos\vartheta \times \Delta\rho = g\cos\vartheta \times \frac{\partial\rho}{\partial z}\Delta z = g\cos\vartheta \times \frac{\partial\rho}{\partial z}\Delta s\cos\vartheta = \rho_0\frac{\partial b}{\partial z}\cos^2\vartheta\,\Delta s, \tag{7.67}$$

noting that $\Delta z = \cos\vartheta\,\Delta s$. The equation of motion of a parcel moving along a trough or crest is therefore

$$\rho_0\frac{\mathrm{d}^2\Delta s}{\mathrm{d}t^2} = -\rho_0 N^2\cos^2\vartheta\,\Delta s, \tag{7.68}$$

which implies a frequency $\omega = N\cos\vartheta$, as in (7.61). One of the $\cos\vartheta$ factors in (7.68) comes from the fact that the parcel displacement is at an angle to the direction of gravity, and the other comes from the fact that the restoring force that a parcel experiences is proportional to $N\cos\vartheta$. (The reader may also wish to refer ahead to Fig. 7.14 and Section 7.6.1 for a similar argument.)

Now consider the wave illustrated in Fig. 7.3. For this wave both k and m are positive, and the frequency is assumed positive by convention to avoid duplicative solutions. The slanting solid and dashed lines are lines of constant phase, and from (7.63b) the buoyancy and pressure are 1/4 of a wavelength out of phase. When k and m are both positive the extrema in the buoyancy field lag the extrema in the vertical velocity by $\pi/2$, as illustrated. The perturbation velocities are zero along the lines of extreme buoyancy. This follows because the velocities are in phase with the pressure, which as we noted is out of phase with the buoyancy.

Given the direction of the fluid parcel displacement in Fig. 7.3, the direction of the phase propagation c_p up and to the right may be deduced from the following argument. Buoyancy perturbations arise because of vertical advection of the background stratification, $w'\partial b_0/\partial z = w'N^2$. A local maximum in rising motion, and therefore a tendency to increase the fluid density, is present

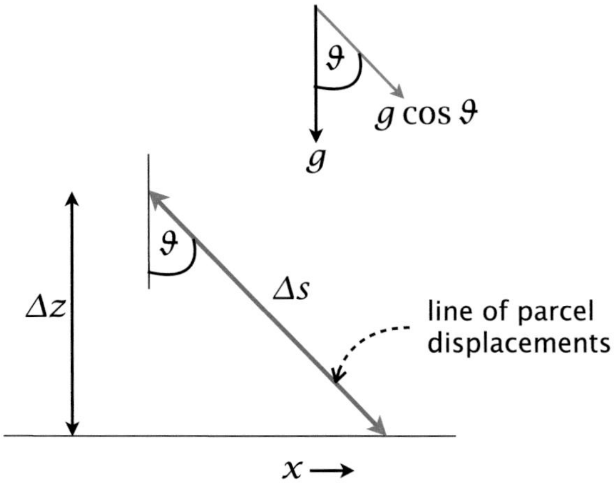

Fig. 7.4 Parcel displacements and associated forces in an internal gravity wave in which the parcel displacements are occurring at an angle ϑ to the vertical, as in Fig. 7.3.

along the 'Low' line 1/4 wavelength upward and to the right of the 'Dense' phase line. Thus, the density of fluid along the 'Low' phase line increases and the 'Dense' phase line moves upward and to the right. If the fluid parcel motion were reversed the pattern of 'High–Dense–Low–Light–High' in Fig. 7.3 would remain the same. However, the downward fluid motion along the 'Low' line would cause the fluid to lose density, and so the phase lines would propagate downward and to the left. Evidently, the wave fronts, or the lines of constant phase, move at right angles to the fluid-parcel trajectories. In the figure we see that the group velocity is denoted as being at right angles to the phase speed, so let's discuss this.

7.3.4 Group Velocity and Phase Speed

As we noted above, the frequency of internal waves is given by $\omega = N\cos\vartheta$, where ϑ is the angle the wave vector makes with the horizontal. This means that the surfaces of constant frequency are *cones*, as illustrated in Fig. 7.5.

To evaluate phase and group velocities in a useful way it is convenient to use spherical polar coordinates, as in Fig. 7.6, in which

$$k = K_3\cos\vartheta\cos\lambda, \qquad l = K_3\cos\vartheta\sin\lambda, \qquad m = K_3\sin\vartheta, \tag{7.69}$$

so that $\boldsymbol{k} = K_3(\cos\vartheta\cos\lambda, \cos\vartheta\sin\lambda, \sin\vartheta)$. The angles are ϑ, the angle of the wave vector with the horizontal and λ, which determines the orientation in the horizontal plane. (The notation is similar to the spherical coordinates of Chapter 2 — see Fig. 2.3 — although here ϑ is the angle with the horizontal, not the angle with the equatorial plane.) We also note that

$$\sin^2\vartheta = \frac{m^2}{k^2+l^2+m^2}, \qquad \cos^2\vartheta = \frac{K^2}{K_3^2} = \frac{k^2+l^2}{k^2+l^2+m^2}, \qquad \tan\lambda = \frac{l}{k}. \tag{7.70}$$

In many problems we can align the direction of the wave propagation with the x-axis and take $l = 0$ and $\tan\lambda = 0$.

The phase speed of the internal waves in the direction of the wave vector (sometimes referred to as the phase velocity) is given by

$$c_p = \frac{\omega}{K_3} = \frac{N}{K_3}\cos\vartheta = \frac{NK}{K_3^2}. \tag{7.71}$$

The phase speeds (as conventionally-defined) in the x, y and z directions are

$$c_p^x \equiv \frac{\omega}{k} = \frac{N}{k}\cos\vartheta, \qquad c_p^y \equiv \frac{\omega}{l} = \frac{N}{l}\cos\vartheta, \qquad c_p^z \equiv \frac{\omega}{m} = \frac{N}{m}\cos\vartheta. \tag{7.72a,b,c}$$

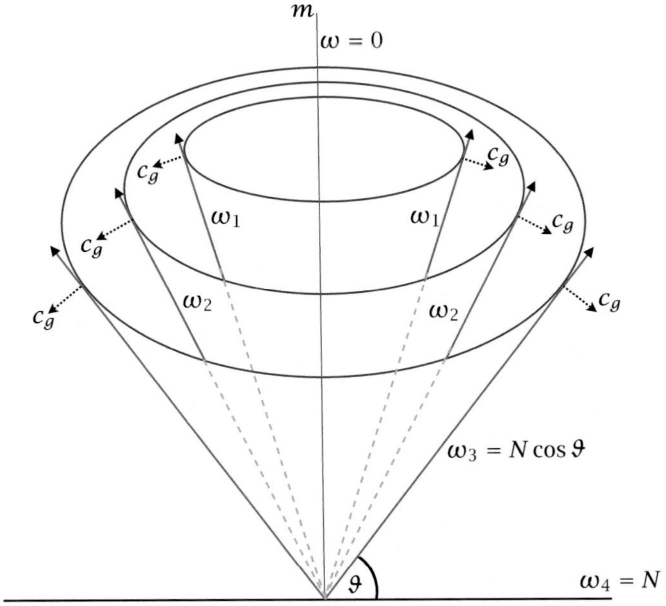

Fig. 7.5 Internal wave cones. The surfaces of constant frequency are cones, defined by the surface that has a constant angle to the horizontal. The wave vector, and so the phase velocity, point along the cone away from the origin, and the frequency of any wave with a wave vector in the cone is $N \cos \vartheta$. The group velocity is at right angles to the cone and pointed in the direction of increasing frequency, as indicated by the arrows on the dotted lines. In the vertical direction the phase speed and group velocity have opposite signs.

As noted in Section 6.1.2, these quantities are the speed of propagation of the wave crests in the respective directions. In general, each speed is *larger* than the phase speed in the direction perpendicular to the wave crests (that is, in the direction of the wave vector), but no information is transmitted at these speeds.

The group velocity is given by

$$\boldsymbol{c}_g = \left(\frac{\partial \omega}{\partial k}, \frac{\partial \omega}{\partial l}, \frac{\partial \omega}{\partial m} \right). \tag{7.73}$$

Using (7.56) we find

$$c_g^x = \frac{\partial \omega}{\partial k} = \frac{Nm}{K_3^2} \frac{km}{KK_3} = \left(\frac{N}{K_3} \sin \vartheta \right) \cos \lambda \sin \vartheta, \tag{7.74a}$$

$$c_g^y = \frac{\partial \omega}{\partial l} = \frac{Nm}{K_3^2} \frac{lm}{KK_3} = \left(\frac{N}{K_3} \sin \vartheta \right) \sin \lambda \sin \vartheta, \tag{7.74b}$$

$$c_g^z = \frac{\partial \omega}{\partial m} = -\frac{Nm}{K_3^2} \frac{K}{K_3} = -\left(\frac{N}{K_3} \sin \vartheta \right) \cos \vartheta. \tag{7.74c}$$

The magnitude of the group velocity is evidently

$$|\boldsymbol{c}_g| = \frac{N}{K_3} \sin \vartheta, \tag{7.75}$$

and the group velocity vector is directed at an angle ϑ to the vertical, as in Fig. 7.5. This angle is perpendicular to the cone itself; that is, the group velocity is perpendicular to the wave vector, as may be verified by taking the dot product of (7.69) and (7.74) which gives

$$\boldsymbol{k} \cdot \boldsymbol{c}_g = 0. \tag{7.76}$$

The group velocity is therefore parallel to the motion of the fluid parcels, as illustrated in Fig. 7.3. Furthermore, because energy propagates with the group velocity, and the latter is *parallel* to lines

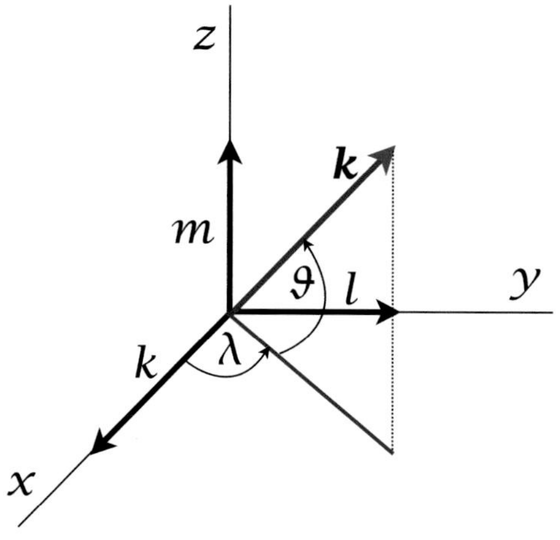

Fig. 7.6 The spherical coordinates used to describe internal waves, as in (7.69). The angle ϑ is the angle of the wave vector with the horizontal, and λ determines the orientation in the horizontal plane. The wave vector $\boldsymbol{k}$ is given by $\boldsymbol{k} = (k, l, m)$, these being the wavenumbers in the direction of increasing (x, y, z), respectively.

of constant phase, energy propagates perpendicular to the direction of phase propagation — very different from the case of acoustic waves or even shallow water waves. In the vertical direction we see from (7.72c) and (7.74c) that

$$\frac{\omega}{m}\frac{\partial \omega}{\partial m} = -\frac{N^2}{K_3^2}\cos^2\vartheta < 0. \tag{7.77}$$

That is, the phase speed and the group velocity have opposite signs, meaning that if the wave crests move downward the group moves upward!

Effect of a mean flow

Suppose that there is a mean flow, U, in the x-direction, as is common in both atmosphere and ocean. The dispersion relation, (7.56), simply becomes

$$(\omega - Uk)^2 = \frac{K^2N^2}{K^2 + m^2}. \tag{7.78}$$

The frequency is Doppler shifted, as expected, but the upward propagation of waves is affected in an interesting way. From (7.78) we find that the vertical component of the group velocity may be written as

$$\frac{\partial \omega}{\partial m} = \frac{-m(\omega - Uk)}{K^2 + m^2} = \frac{-mk(c - U)}{K^2 + m^2}, \tag{7.79}$$

where $c = \omega/k$ is the phase speed in the x-direction. If U is not constant but is varying slowly with z then (7.79) still holds, although m itself will also vary slowly with z. The point to note is that the group velocity goes to zero at the location where $U = c$, that is at a critical line, and the wave stalls. Of course m itself may become large near a critical line (as we consider in more detail in Section 17.3). In this case — which is essentially the hydrostatic one, with $m^2 \gg K^2$ — we obtain

$$\frac{\partial \omega}{\partial m} = \frac{-k(c - U)}{m} = \frac{-k^2(c - U)^2}{KN}. \tag{7.80}$$

The physical consequence of group velocity going to zero as the wave approaches a critical line is that any dissipation that may be present has more time to act. That is, we can expect a wave to be preferentially dissipated near a critical line, giving up its momentum to the mean flow and its energy to create mixing — the former being important in the atmosphere (for this is the mechanism producing the quasi-biennial oscillation) and the latter in the ocean.

7.3.5 Energetics of Internal Waves

In this section we explore the energetics of internal waves, and we first show that the linearized equations conserve a sensible form of energy. Linearized equations do not, of course, automatically conserve energy even if the original nonlinear equations from which they derive do: an unstable wave will draw energy from the background state and grow in amplitude, as we will see in Chapter 9 on baroclinic instability.

Energy conservation

From (7.53a,b) we obtain an equation for the evolution of kinetic energy, namely

$$\frac{\partial}{\partial t}\left(\frac{\boldsymbol{v}'^2}{2}\right) = b'w' - \nabla_3 \cdot (\phi' \boldsymbol{v}'), \tag{7.81}$$

where $\boldsymbol{v}'^2 = u'^2 + v'^2 + w'^2$, and from (7.53c,d) we obtain

$$\frac{1}{N^2}\frac{\partial}{\partial t}\frac{b'^2}{2} + w'b' = 0. \tag{7.82}$$

Adding the above two equations gives

$$\frac{\partial}{\partial t}\frac{1}{2}\left(\boldsymbol{v}'^2 + \frac{b'^2}{N^2}\right) + \nabla_3 \cdot (\phi' \boldsymbol{v}') = 0. \tag{7.83}$$

This is the linear version of the energy conservation equation for Boussinesq flow, as in (2.112) on page 74.

Two differences are apparent: (i) the transport of energy is only by way of the pressure term and the advective transport is absent, as expected in a linear model; (ii) the potential energy term bz of the linear model is replaced by b'^2/N^2. It is less obvious why this should be so. However, the quantity

$$A = \frac{1}{2}\int \frac{\overline{b'^2}}{\partial \bar{b}/\partial z}\,\mathrm{d}z\,\mathrm{d}A = \frac{1}{2}\int \frac{\overline{b'^2}}{N^2}\,\mathrm{d}z\,\mathrm{d}A \tag{7.84}$$

is just the *available potential energy* (APE) of a Boussinesq fluid in which the isopycnal surfaces vary only slightly from a stable, purely horizontal, resting state, and it is only the APE that participates in the linear system.

If we integrate (7.83) over a volume such that the normal component of the velocity vanishes at the boundaries (for example, we integrate over a volume enclosed by rigid walls), then the divergence term vanishes and we obtain the integral conservation statement:

$$\widehat{E} = \frac{1}{2}\int \left(\boldsymbol{v}'^2 + \frac{b'^2}{N^2}\right) \mathrm{d}V, \qquad \frac{\mathrm{d}\widehat{E}}{\mathrm{d}t} = 0. \tag{7.85}$$

The quantity $\widehat{E}$ is an example of a *wave activity:* a conserved quantity that is quadratic in wave amplitude. This conservation statement (7.83) is true whether or not the basic state is stably stratified; that is, whether or not N^2 is positive. However, (7.85) only provides a bound on growing perturbations if N^2 is positive, in which case all the terms that constitute $\widehat{E}$ are positive definite. If $N^2 < 0$ then both $\boldsymbol{v}'^2$ and b'^2 can grow without bound even as $\widehat{E}$ itself remains constant.

Consider now the energy in a *wave*, and we will denote by $\overline{E}$ the energy density, meaning the mean perturbation energy per unit volume, averaged over a wavelength. Thus

$$2\overline{E} = \overline{\boldsymbol{v}'^2} + \frac{\overline{b'^2}}{N^2}. \tag{7.86}$$

If we use the polarization relations of Section 7.3.2 then the kinetic and potential energy densities may be written in terms of the pressure amplitude as

$$2\overline{KE} = \left(\frac{k^2}{\omega^2} + \frac{l^2}{\omega^2} + \frac{(k^2+l^2)^2}{m^2\omega^2}\right)|\widetilde{\phi}|^2 = \frac{K^2K_3^2}{m^2\omega^2}|\widetilde{\phi}|^2, \tag{7.87a}$$

$$2\overline{PE} = \frac{N^2K^4}{m^2\omega^4} = \frac{K^2K_3^2}{m^2\omega^2}|\widetilde{\phi}|^2, \tag{7.87b}$$

using also the dispersion relation, $\omega^2K_3^2 = K^2N^2$. Thus, there is *equipartition* between the kinetic and potential energies, a common feature of waves in non-rotating systems (although not a universal feature of waves). The total energy density is thus

$$\overline{E} = \frac{K^2K_3^2}{m^2\omega^2}|\widetilde{\phi}|^2 = \frac{K_3^2}{K^2}|\widetilde{w}|^2 = \frac{|\widetilde{w}|^2}{\cos^2\vartheta}, \tag{7.88}$$

using (7.64c), where $\widetilde{w}$ is the amplitude of the vertical component of the velocity perturbation.

Energy propagation and the group velocity property

In Section 6.7 we derived, from rather general considerations, the 'group velocity property' for wave activity. We showed that if a wave activity, $\mathcal{A}$, and its flux, $\boldsymbol{\mathcal{F}}$ obeyed a conservation law of the form $\partial\mathcal{A}/\partial t + \nabla\cdot\boldsymbol{\mathcal{F}} = 0$, and if the wave activity and its flux were both quadratic functions of the wave amplitude, then the flux is related to the wave activity by $\boldsymbol{\mathcal{F}} = \mathbf{c}_g\mathcal{A}$. The internal wave energy density and its flux do have these properties — see (7.83) — so we should expect the group velocity property to hold, and we now demonstrate that explicitly, albeit briefly.

The energy flux vector for internal waves is $\boldsymbol{\mathcal{F}} = \overline{\phi'\boldsymbol{v}'}$ and using (7.63a) and (7.63c) this is

$$\boldsymbol{\mathcal{F}} = \left(\frac{k}{\omega}, \frac{l}{\omega}, -\frac{K^2}{m\omega}\right)|\widetilde{\phi}|^2. \tag{7.89}$$

Using (7.74) and (7.88) the group velocity times the energy density is

$$c_g^x \times \overline{E} = \left[\frac{Nm^2}{K_3^3}\frac{k}{K}\right] \times \left[\frac{K^2K_3^2}{m^2\omega^2}|\widetilde{\phi}|^2\right] = \frac{k}{\omega}|\widetilde{\phi}|^2, \tag{7.90a}$$

$$c_g^y \times \overline{E} = \left[\frac{Nm^2}{K_3^3}\frac{l}{K}\right] \times \left[\frac{K^2K_3^2}{m^2\omega^2}|\widetilde{\phi}|^2\right] = \frac{l}{\omega}|\widetilde{\phi}|^2, \tag{7.90b}$$

$$c_g^z \times \overline{E} = \left[\frac{NmK}{K_3^3}\right] \times \left[\frac{K^2K_3^2}{m^2\omega^2}|\widetilde{\phi}|^2\right] = -\frac{K^2}{m\omega}|\widetilde{\phi}|^2, \tag{7.90c}$$

which evidently is the same as (7.89), completing our demonstration.

7.4 ✦ INTERNAL WAVE REFLECTION

Suppose a propagating internal wave encounters a solid boundary — sloping topography, for example. The boundary effectively acts as a source of waves and so the original wave is reflected in some fashion. However, because of the nature of the dispersion relation for internal waves the reflection occurs in a rather peculiar way, as we now discuss.

For algebraic simplicity let us initially suppose that the wave is propagating in the x–z plane, and the equation of mass continuity $\partial_x u + \partial_z w = 0$ is then satisfied by introducing a streamfunction ψ such that

$$u = -\frac{\partial\psi}{\partial z}, \qquad w = \frac{\partial\psi}{\partial x}. \tag{7.91}$$

If the incident wave is denoted ψ_1 and the reflected wave ψ_2 then the total wave field is

$$\psi = \widetilde{\psi}_1 \exp\{\mathrm{i}(k_1 x + m_1 z - \omega_1 t)\} + \widetilde{\psi}_2 \exp\{\mathrm{i}(k_2 x + m_2 z - \omega_2 t)\}, \tag{7.92}$$

where as usual a tilde denotes a complex wave amplitude and the real part of the expression is implied. The total streamfunction must be constant *at the boundary* — in fact without loss of generality we may suppose that $\psi = 0$ at the boundary — and this can only be achieved if

$$k_1 x + m_1 z - \omega_1 t = k_2 x + m_2 z - \omega_2 t \tag{7.93}$$

for all t and for all x and z along the boundary. This implies that

$$\omega_1 = \omega_2 \tag{7.94}$$

and

$$k_1 x + m_1 z_b(x) = k_2 x + m_2 z_b(x), \tag{7.95}$$

where $z_b(x)$ parameterizes the height of the reflecting boundary. We can view this another way: suppose that the boundary slopes at an angle γ to the horizontal, as in Fig. 7.7 or Fig. 7.8. We then have $z_b = x \tan\gamma$ and a unit vector along the boundary satisfies $\boldsymbol{j}_\gamma = \mathbf{i}\cos\gamma + \mathbf{j}\sin\gamma$. Equation (7.92) may be written as

$$\psi = \widetilde{\psi}_1 \exp\{\mathrm{i}\,[(k_1 + m_1 \tan\gamma)x - \omega_1 t]\} + \widetilde{\psi}_2 \exp\{\mathrm{i}\,[(k_2 + m_1 \tan\gamma)x - \omega_2 t]\}, \tag{7.96}$$

from which the wavenumber condition that must be satisfied is

$$k_1 + m_1 \tan\gamma = k_2 + m_2 \tan\gamma \tag{7.97}$$

or, and as may also be seen from (7.95),

$$\boldsymbol{k}_1 \cdot \boldsymbol{j} = \boldsymbol{k}_2 \cdot \boldsymbol{j}. \tag{7.98}$$

This means that the components of the wave vector parallel to the boundary for the incoming and outgoing wave are equal to each other. This, and the conservation of frequency expressed by (7.94), are *general* results about linear wave reflection; they apply to light waves, for example. However, the dispersion relation of internal waves gives rise to rather unintuitive and decidedly non-specular properties of reflection.

7.4.1 Properties of Internal Wave Reflection

Suppose an internal wave is incident on a solid boundary, sloping at an angle γ to the horizontal, as in Fig. 7.7 or Fig. 7.8. The incident and reflected waves must satisfy the following conditions:

(i) The frequency of the reflected wave is equal to that of the incident wave. Because the frequency is given by $\omega = N\cos\vartheta$, the angle of the reflected wave *with respect to the horizontal* is equal to that of the incident wave.

(ii) The components of the wave vector along the slope of the reflected wave and incident wave are equal.

(iii) The group velocity of the reflected wave must be directed away from the slope.

We did not derive the third of these conditions, but the reflected wave must carry energy and information away from the slope, and these are carried by the group velocity. Similarly, a wave incident on a boundary is one in which the group velocity is directed toward the slope.

Consider a wave approaching a slope as in Fig. 7.7, such that the incoming wave vector makes an angle of ϑ_1 with the horizontal, and the boundary slope is γ. The condition (7.98) states that the

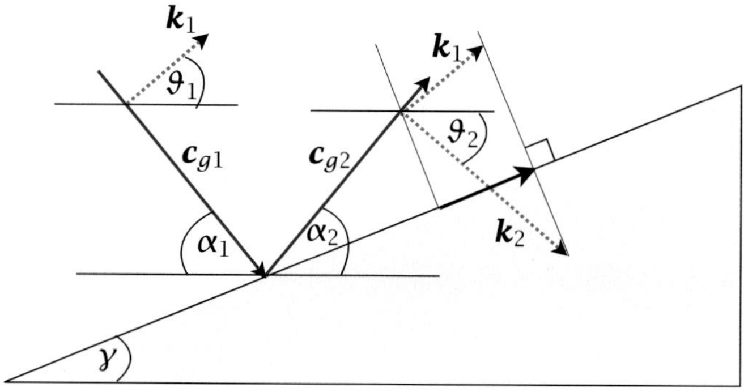

Fig. 7.7 Internal wave reflection from a shallow sloping boundary. The incoming wave vector, $\boldsymbol{k}_1$, makes an angle ϑ_1 with the horizontal, and the incoming group velocity, $\boldsymbol{c}_{g1}$ makes an angle $\alpha_1 = \pi/2 - \vartheta_1$.
The group velocity of the reflected wave, $\boldsymbol{c}_{g2}$ is directed away from the slope, and to satisfy the frequency condition $\alpha_2 = \alpha_1$. The projection along the slope of the reflected wave vector, $\boldsymbol{k}_2$ must be equal to that of the incoming wave vector (the projection is the short thick arrow along the slope), and so the magnitude of the reflected wave vector is larger than that of the incoming wave.

projections along the boundary of the the incoming and outgoing wave vectors are equal to each other, and so

$$\kappa_1 \cos(\vartheta - \gamma) = \kappa_2 \cos(\vartheta + \gamma), \tag{7.99}$$

where κ_1 and κ_2 are the magnitudes of the incoming and reflected wave vectors and $\vartheta = \vartheta_1 = \vartheta_2$, because the outgoing wave makes the same angle with the horizontal as does the incoming wave. The group velocity is perpendicular to the wave vector and makes an angle $\alpha = \pi/2 - \vartheta$ to the horizontal, and in terms of this (7.99) may be written, provided $\alpha > \gamma$,

$$\kappa_1 \sin(\alpha + \gamma) = \kappa_2 \sin(\alpha - \gamma). \tag{7.100}$$

For a sufficiently steep boundary slope we may have $\alpha < \gamma$, and in this case the wave will be back reflected down the slope, as in Fig. 7.8. A little geometry reveals that the condition (7.100) should be replaced by

$$\kappa_1 \sin(\alpha + \gamma) = \kappa_2 \sin(\gamma - \alpha). \tag{7.101}$$

The case with $\alpha = \gamma$ is plainly a critical one. In this case the group velocity of the reflected wave is directed along the slope, and the wave vector is perpendicular to the slope. The magnitude of the reflected wave vector is infinite; that is, the waves have zero wavelength, and so would in reality be subject to viscous dissipation and diffusion. Reflection of internal waves is in fact an important mechanism leading to mixing in the ocean.

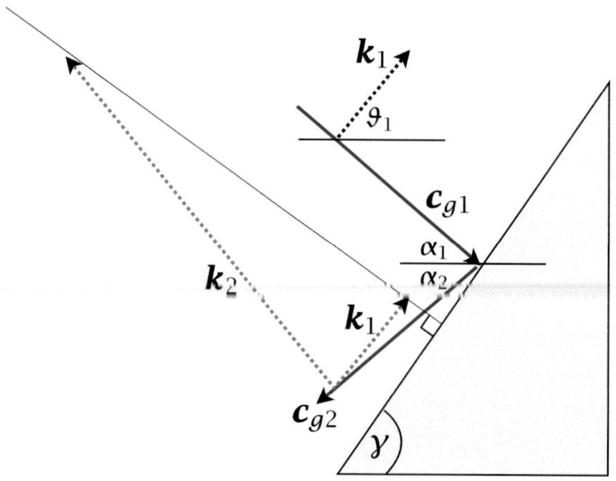

Fig. 7.8 As for Fig. 7.7, but now showing reflection from a steep slope. The wave is back-reflected down the slope, and in this example the magnitude of the reflected wave is again larger than that of the incoming wave.

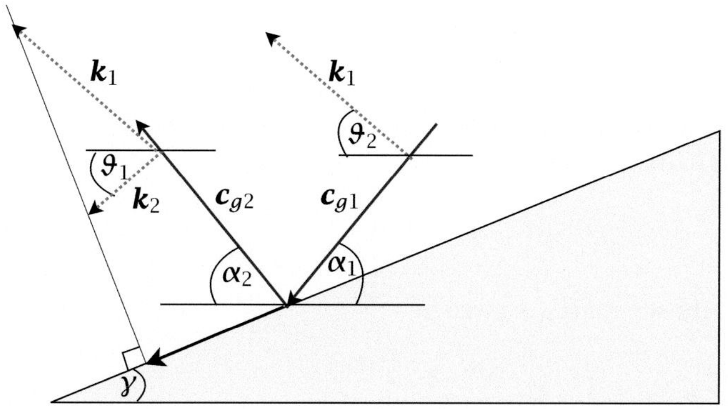

Fig. 7.9 As for Fig. 7.7, but now showing the production of a reflected wave with a longer wavelength than the incident wave. The wavevector of the reflected wave is more nearly parallel to the sloping boundary than is the wave vector of the incident wave.

The reflected wave need not, of course, always have a wavenumber that is higher than that of the incident wave: it is a matter of whether the incoming wave vector is more nearly aligned with the slope of the boundary than is the reflected wave, and if it is the reflected wave will have a higher wavenumber, and contrariwise. An example of reflection producing a longer wave is illustrated in Fig. 7.9. Still, the process whereby waves are reflected to produce waves of a shorter wavelength that are then dissipated is an irreversible one, and the net effect of many quasi-random wave reflections is likely to be the dissipation of short waves.

Finally, one might ask why the reflected wave could not simply be back along the track of the incident wave — for example, why could we not have $\boldsymbol{c}_{g1} = -\boldsymbol{c}_{g2}$? If this were so then we would have $\boldsymbol{k}_2 = -\boldsymbol{k}_1$, and it would be impossible for the two wave vectors to project equally on the sloping boundary.

7.5 ✦ INTERNAL WAVES IN A FLUID WITH VARYING STRATIFICATION

In most realistic situations the stratification N^2 is not constant. In the ocean the stratification is largest in the upper ocean (in the 'pycnocline') diminishing with depth in the weakly stratified abyss. In the atmosphere the stratification tends to be fairly constant in the troposphere but increases fairly abruptly as we pass into the stratosphere. In such circumstances the wave equation (7.54) no longer has constant coefficients and we cannot easily obtain wavelike solutions. However, if the stratification varies *slowly* in the vertical direction, meaning that its variations occur on a larger space scale than the vertical wavelength, while remaining constant in the horizontal direction, then we expect the solution to look locally like plane waves and we can obtain approximate solutions. This is the territory of the WKB approximation, as described in Appendix A of Chapter 6, and we will employ this technology.

7.5.1 Obtaining a WKB Solution

No assumptions are made about the uniformity of N when deriving (7.54), so the equation of motion is again

$$\left[\frac{\partial^2}{\partial t^2}\left(\nabla^2 + \frac{\partial^2}{\partial z^2}\right) + N^2\nabla^2\right] w' = 0, \tag{7.102}$$

where $N^2 = N^2(z)$. Let us seek solutions in the form

$$w' = \operatorname{Re} W(z) \mathrm{e}^{i(kx+ly-\omega t)}, \tag{7.103}$$

whence we obtain

$$\frac{\mathrm{d}^2 W}{\mathrm{d}z^2} + m^2(z) W = 0, \tag{7.104}$$

where

$$m^2 \equiv \frac{(N^2 - \omega^2)K^2}{\omega^2} . \tag{7.105}$$

This is closely related to the dispersion relation for the gravity waves, with m being the vertical wavenumber. The WKB solution to (7.104) is

$$W(z) = Cm^{-1/2} \exp\left(\pm i \int^z m \, dz'\right), \tag{7.106}$$

where C is a constant. The phase of the wave, $\theta(z)$, is given by

$$\theta(z) = \int^z m \, dz' = \int^z \pm K \left(\frac{N^2 - \omega^2}{\omega^2}\right)^{1/2} dz'. \tag{7.107}$$

Since $m = d\theta/dz$, locally the flow behaves like a plane wave with vertical wavenumber m and with amplitude varying as $m^{-1/2}$. Re-arranging (7.105) we obtain the dispersion relation in a familiar form,

$$\omega^2 = \frac{N^2 K^2}{K^2 + m^2} = N^2 \cos^2 \vartheta(z), \tag{7.108}$$

where $\cos^2 \vartheta = K^2/(K^2 + m^2)$. Given this, we can interpret (7.105) as giving the vertical wavenumber in a medium in which the stratification is varying and the frequency and horizontal wavenumber are known. We see that N, ϑ and m are functions of z, but ω is not, because the medium is time independent (cf., the discussion in Section 6.3).

7.5.2 Properties of the Solution

The WKB solution above is *almost* that of a plane wave with slowly varying wavenumber. Thus, it seems that the solution (7.106) might be further approximated as

$$w \approx Cm^{-1/2} \exp\left(\pm i m(z) z\right), \tag{7.109}$$

where $m(z)$ is given by (7.105). The accuracy of this solution increases as the variation of m diminishes, and in many circumstances (7.109) may be used to infer the qualitative behaviour of a wave. Nonetheless, it is an integral that appears in the phase in the solution (7.106), so the solution is not truly local.

From (7.106) the amplitude varies with height as $m^{-1/2}$, so that if the stratification (N^2) increases m will increase and the amplitude will decrease. Here we have derived this result directly by solving the wave equations of motion, but the result is a consequence of the conservation of energy in internal waves: energy here is a 'wave activity' — namely a conserved quantity, quadratic in the wave amplitude — in this problem. As discussed in Section 7.3.5, the vertical component of the energy flux, F^z, is $c_g^z \overline{E}$, where $\overline{E}$ is the energy density and c_g^z is the vertical component of the group velocity, and for a wave propagating vertically this energy flux must be constant. Now, manipulating (7.74c) and using (7.88) we have

$$c_g^z = -\frac{\omega m}{K_3^2}, \qquad \overline{E} = \left(\frac{W}{\cos \vartheta}\right)^2, \tag{7.110a,b}$$

so that

$$F^z = c_g^z \overline{E} = -\frac{W^2 \omega m}{K^2} = \text{constant}. \tag{7.111}$$

Thus, because the horizontal wavenumber K is preserved (since there are no inhomogeneities in the horizontal) and the frequency is constant (because the medium itself is not time varying), we must have $W \propto m^{-1/2}$, as in (7.106).

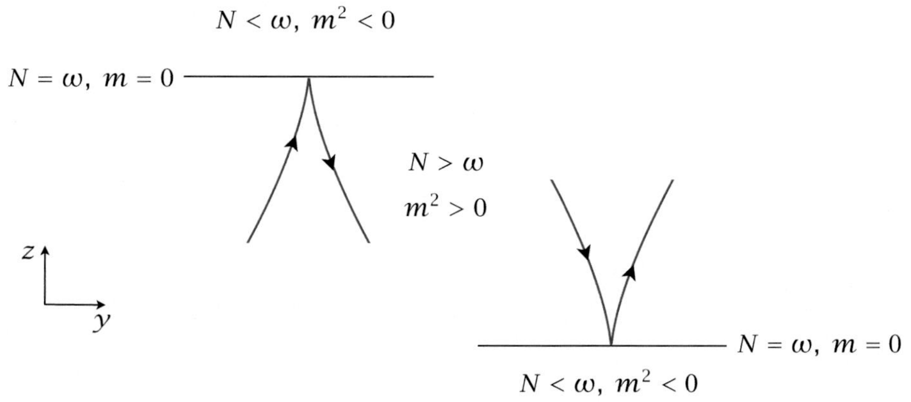

Fig. 7.10 Trajectories of internal waves approaching a turning height where $N = \omega$. The trajectory makes a cusp, as given by (7.115). If a region of high stratification is sandwiched between two regions of lower stratification then the waves may be vertically confined to a waveguide.

7.5.3 Wave Trajectories and an Idealized Example

Rays

As we discussed in Section 6.3, a wave packet will follow a *ray,* where a ray is a trajectory following the group velocity. Restricting attention to two dimensions and using (7.74) the horizontal and vertical components of the group velocity are (for $l > 0$),

$$c_g^y = \frac{Nm^2}{(l^2 + m^2)^{3/2}}, \qquad c_g^z = \frac{-Nlm}{(l^2 + m^2)^{3/2}}. \tag{7.112a,b}$$

The path of a ray may thus be parameterized by the expression

$$\frac{\mathrm{d}z}{\mathrm{d}y} = \frac{c_g^z}{c_g^y} = -\frac{l}{m} = \frac{-\omega}{\sqrt{N^2 - \omega^2}}, \tag{7.113}$$

where the rightmost expression follows from the dispersion relation (7.56) with $k = 0$. The above expressions hold even when N varies in the vertical. Now, for there to be vertical propagation the vertical wavenumber must be positive and the wave frequency must be less than N. Suppose a wave is generated in a strongly stratified region and propagates vertically to a more weakly stratified region (with smaller N). The vertical wavenumber m becomes smaller and smaller, both the vertical and horizontal components of the group velocity tend to zero and the wave packet will stall. However, c_g^y goes to zero faster than c_g^z and the ray path turns toward the region of lower stratification.

This behaviour may be interpreted in terms of the dispersion relation $\omega = N\cos\vartheta$, where $\vartheta = \cos^{-1}[l^2/(l^2 + m^2)]$ is the angle between the three-dimensional wavevector and the horizontal (see Section 7.3.2). If N decreases as we move vertically then ϑ must decrease until we reach the maximum value of $\cos\vartheta = 1$ and the wave vector is purely horizontal. The group velocity is perpendicular to the wave vector and so is then purely vertical. The wave cannot propagate into the region in which $N^2 < \omega^2$ for then m is imaginary and the disturbance will decay. Rather, the wave will tend to reflect, and the region where $N = \omega$ is often called a turning level. The trajectory can be obtained analytically in the region of the turning level as follows. Suppose that $N = \omega$ at $z = z^*$ so that, expanding N^2 around that point, we have $N^2(z) \approx N^2(z^*) + (z - z^*)\mathrm{d}N^2(z^*)/\mathrm{d}z$. Equation (7.113) becomes

$$\frac{\mathrm{d}z}{\mathrm{d}y} = \frac{-\omega}{\sqrt{(z - z^*)\mathrm{d}N^2/\mathrm{d}z}}, \tag{7.114}$$

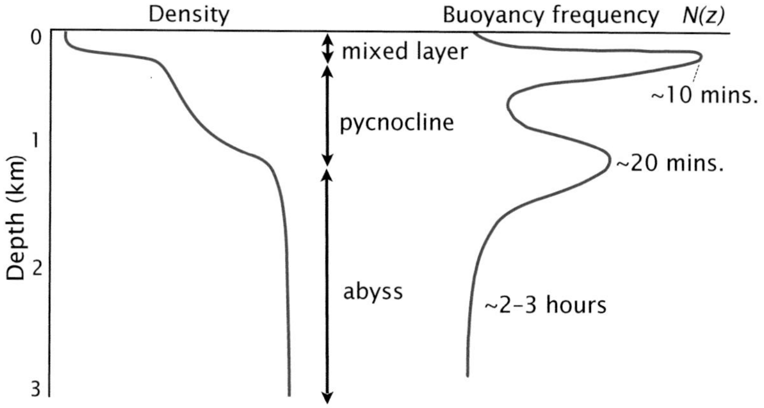

Fig. 7.11 Sketch of an ocean density profile, left, and buoyancy frequency, right, labelled with the approximate period. The pycnocline is sandwiched between two weakly stratified regions. The double peak in the buoyancy frequency is exaggerated, but the pycnocline is generally the region of highest frequency internal waves.

which, upon integrating, yields

$$z - z^* = \frac{\omega(y^* - y)^{2/3}}{\sqrt{\mathrm{d}N^2/\mathrm{d}z}}. \tag{7.115}$$

This cusp-like trajectory is illustrated in Fig. 7.10.

An idealized oceanic waveguide

The stratification of the ocean is decidedly nonuniform in the vertical, as schematically illustrated in Fig. 7.11. The density is almost uniform in a layer at the top of the ocean about 50–100 m deep known as the mixed layer. The density then increases fairly rapidly over a region 500–1000 m deep known as the pycnocline, and is then fairly uniform in the abyss. The weak stratification in the abyss and in the mixed layer will inhibit the propagation of internal waves generated in the thermocline. For example, consider a wave of frequency ω propagating downwards from the oceanic thermocline with and into the weakly stratified abyss. As soon as $N(z) < \omega$ the vertical wavenumber becomes imaginary and the disturbance will vary like $e^{\pm mz}$. On physical grounds we must choose the solution that evanesces with depth. Similar behaviour will occur for a wave propagating up from the thermocline into the weakly stratified mixed layer. Thus, waves are trapped in a region where $N^2 > \omega^2$, and this region forms a *wave guide*, as sketched in Fig. 7.12. Similar dynamics are described again in an atmospheric context below.

The profile of N^2 is a simple exponential and the corresponding value of m^2 is calculated using (7.105) with $K = \omega = 1$ (the values are nondimensional). The value of m goes to zero near the top and the bottom of the domain, as illustrated. The corresponding group velocities are illustrated in Fig. 7.13, and can be seen to be purely vertical at the two turning heights. The amplitude of a wave becomes very large near the turning heights, but the wave itself need not break because its energy is constant and its vertical wavelength is very large. Rather, the wave will be reflected (following the trajectory illustrated in Fig. 7.10), and the wave is confined in the waveguide.

7.5.4 Atmospheric Considerations

The atmosphere differs from the ocean in many ways, but for the purposes of internal waves two of these are particularly important: (i) the density diminishes in the vertical and so the Boussinesq approximation is not valid, except for small vertical displacements; (ii) there is no upper surface, so we must consider radiation conditions for large z, or require that the solutions remain bounded for $z \to \infty$, rather than conventional boundary conditions.

There are two commonly-used ways to deal with density variations — through the use of pressure coordinates or the anelastic equations. We will use pressure coordinates in Chapter 17, but

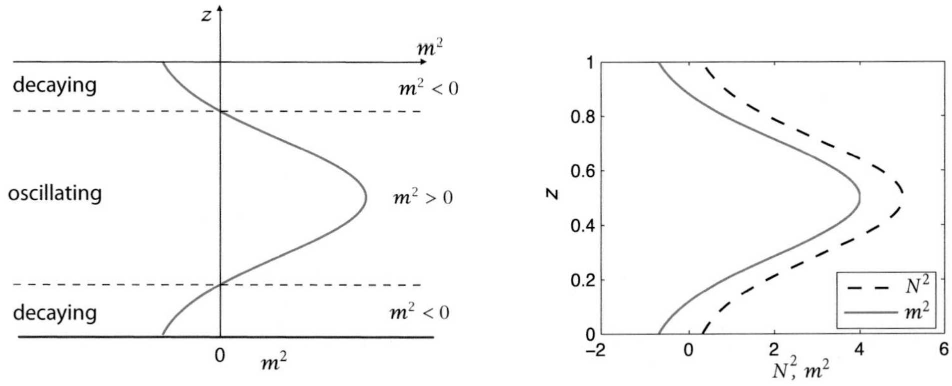

Fig. 7.12 An oceanic wave guide. The left panel shows m^2 and the regions of oscillation and decay, and the right panel also shows the value of N^2 from which m^2 is calculated using (7.105). Waves generated in the central region will propagate before evanescing in the region of negative m^2, so confining the waves to the central wave guide.

here we briefly consider the anelastic equations. These (see Section 2.5) differ from the Boussinesq primarily in the mass continuity equation, which becomes

$$\frac{\partial u}{\partial x} + \frac{\partial v}{\partial y} + \frac{1}{\rho_0}\frac{\partial}{\partial z}(w\rho_0) = 0, \tag{7.116}$$

where $\rho_0 = \rho_0(z)$ is a specified profile of density. Using (7.116) instead of (7.53c) gives the equation of motion

$$\frac{\partial^2}{\partial t^2}\left(\nabla^2 w' + \frac{\partial}{\partial z}\frac{1}{\rho_0}\frac{\partial \rho_0 w'}{\partial z}\right) + N^2\nabla^2 w' = 0, \tag{7.117}$$

in place of (7.54). Because ρ_0 is a function of z we cannot find plane wave solutions without additional approximation — for example unless we assume that ρ_0 changes only slowly with z. For this reason the Boussinesq approximation is often imposed from the outset in theoretical work, even for the atmosphere; the approximation is quantitatively poor, but the qualitative character of the waves is captured.

The second factor (the lack of an upper surface) becomes an issue when considering gravity waves propagating high into the atmosphere, a phenomenon we look at in Chapter 17 and in Section 7.7, where we consider the generation of internal waves by flow over topography. To finish this section off, let us consider an atmospheric waveguide. The dynamics are very similar to those of the oceanic waveguide discussed above but, for the sake of variety, we will treat it in a slightly different way.

An atmospheric waveguide

We suppose the atmosphere to be a semi-infinite region from the ground at $z = 0$ to infinity. If N^2 is constant then solutions, as in the bounded case, vary sinusoidally in z, for example $w' \sim \sin mz$, where m is the vertical wavenumber. These solutions remain bounded as $z \to \infty$, although they do not decay. If N varies, then other possibilities exist. Suppose that a region of small stratification, N_1 overlies a region of larger stratification, N_2; that is

$$N = \begin{cases} N_1 & z > H, \\ N_2 & 0 < z < H, \end{cases} \tag{7.118}$$

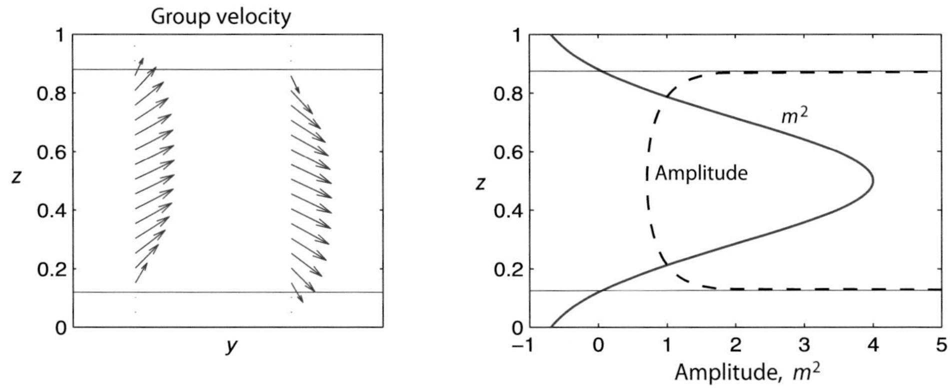

Fig. 7.13 Left panel: Group velocity vectors for upward and downward propagating gravity waves in a stratification illustrated in Fig. 7.12, calculated using (7.112). Right panel: The values of m^2 and the amplitude of the wave, the latter varying as $m^{-1/2}$. The thin horizontal lines in both panels indicate the height at which $m^2 = 0$.

where $N_2 > N_1$. (This is *not* a model of the stratosphere overlying the troposphere, because the stratosphere is highly stratified. If anything, it is a model of the mesosphere overlying the stratosphere and troposphere.) The frequency in the two regions must be the same and if $\omega < N_1 < N_2$ then

$$\omega^2 = \frac{N_1^2}{K^2 + m_1^2} = \frac{N_2^2}{K^2 + m_2^2}, \tag{7.119}$$

whence

$$m_1 = m_2 \left(\frac{N_1^2 - \omega^2}{N_2^2 - \omega^2} \right)^{1/2}. \tag{7.120}$$

In contrast, if $N_1 < \omega < N_2$ then wave-like solutions are not allowed in the upper region, because the frequency must always be less than the local value of N. Rather, solutions in the upper region evanesce according to

$$w_1' = \widetilde{w}_1 \mathrm{e}^{-\mu z} \mathrm{e}^{\mathrm{i}(kx+ly-\omega t)}, \tag{7.121}$$

where

$$\mu^2 = \frac{\omega^2 - N_1^2}{\omega^2} K^2. \tag{7.122}$$

The solutions still vary sinusoidally in the lower layer, according to

$$w_2' = \widetilde{w}_2 \sin m_1 z \mathrm{e}^{\mathrm{i}(kx+ly-\omega t)}, \tag{7.123}$$

where m now takes on only discrete values in order to satisfy the boundary conditions that w and ϕ are continuous $z = H$, and that w vanishes at $z = 0$.

7.6 INTERNAL WAVES IN A ROTATING FRAME OF REFERENCE

In the presence of both a Coriolis force and stratification a displaced fluid will feel two restoring forces — one due to gravity and the other to rotation. The first gives rise to gravity waves, as we have discussed, and the second to inertial waves. When the two forces both occur the resulting waves are called *inertia-gravity waves*. The algebra describing them can be complicated so we begin with a simple parcel argument to lay bare the basic dynamics; refer back to Section 7.3.3 as needed.

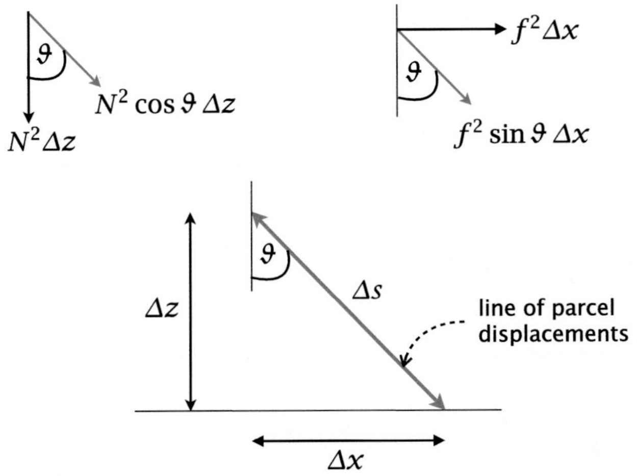

Fig. 7.14 Parcel displacements and associated forces in an inertia-gravity wave in which the parcel displacements are occurring at an angle ϑ to the vertical. Both Coriolis and buoyancy forces are present, and $\Delta s = \Delta z / \cos\vartheta = \Delta x / \sin\vartheta$.

7.6.1 A Parcel Argument

Consider a parcel that is displaced along a slantwise path in the x–z plane, as shown in Fig. 7.14, with a horizontal displacement of Δx and a vertical displacement of Δz. Let us suppose that the fluid is Boussinesq and that there is a stable and uniform stratification given by $N^2 = -g\rho_0^{-1}\partial\rho_0/\partial z = \partial b/\partial z$. Referring to (7.67) as needed, the component of the restoring buoyancy force, F_b say, in the direction of the parcel oscillation is given by (7.67),

$$F_b = -N^2 \cos\vartheta\, \Delta z = -N^2 \cos^2\vartheta\, \Delta s. \tag{7.124}$$

The parcel will also experience a restoring Coriolis force, F_C, and the component of this in the direction of the parcel displacement is

$$F_C = -f^2 \sin\vartheta\, \Delta x = -f^2 \sin^2\vartheta\, \Delta s. \tag{7.125}$$

Here, and for the rest of the chapter, we denote the Coriolis parameter by f. It should be regarded as a constant in any given problem (so there are no Rossby waves), but its value varies with latitude. Using (7.124) and (7.125) the (Lagrangian) equation of motion for a displaced parcel is

$$\frac{\mathrm{d}^2 \Delta s}{\mathrm{d}t^2} = -(N^2 \cos^2\vartheta + f^2 \sin^2\vartheta)\Delta s, \tag{7.126}$$

and hence the frequency is given by

$$\omega^2 = N^2 \cos^2\vartheta + f^2 \sin^2\vartheta. \tag{7.127}$$

Now, nearly everywhere in both atmosphere and ocean, $N^2 > f^2$. From (7.127) we then see that the frequency lies in the interval $N^2 > \omega^2 > f^2$. (To see this, put $N = f$ or $f = N$ in (7.127), and use $\sin^2\vartheta + \cos^2\vartheta = 1$.) If the parcel displacements approach the vertical then the Coriolis force diminishes and $\omega \to N$, and similarly $\omega \to f$ as the displacements become horizontal. The ensuing waves are then pure inertial waves.

We can write (7.127) in terms of wavenumbers since, for motion in the x-z plane,

$$\cos^2\vartheta = \frac{k^2}{k^2 + m^2}, \qquad \sin^2\vartheta = \frac{m^2}{k^2 + m^2}, \tag{7.128}$$

where k and m are the horizontal and vertical wavenumbers and $l = 0$. The dispersion relation becomes

$$\omega^2 = \frac{N^2 k^2 + f^2 m^2}{k^2 + m^2}. \tag{7.129}$$

Let's now move on to a discussion using the linearized equations of motion.

7.6.2 Equations of Motion

In a rotating frame of reference, specifically on an f-plane, the linearized equations of motion are the momentum equations

$$\frac{\partial \boldsymbol{u}'}{\partial t} + \boldsymbol{f}_0 \times \boldsymbol{u}' = -\nabla\phi', \qquad \frac{\partial w'}{\partial t} = -\frac{\partial \phi'}{\partial z} + b', \tag{7.130a,b}$$

and the mass continuity and thermodynamic equations,

$$\frac{\partial u'}{\partial x} + \frac{\partial v'}{\partial y} + \frac{\partial w'}{\partial z} = 0, \qquad \frac{\partial b'}{\partial t} + w'N^2 = 0. \tag{7.130c,d}$$

These are similar to (7.53), with the addition of a Coriolis term in the horizontal momentum equations.

To obtain a single equation for w' we take the horizontal divergence of (7.130a) and use the continuity equation to give

$$\frac{\partial}{\partial t}\left(\frac{\partial w'}{\partial z}\right) + f\zeta' = \nabla^2\phi', \tag{7.131}$$

where $\zeta' \equiv (\partial v'/\partial x - \partial u'/\partial y)$ is the vertical component of the vorticity. We may obtain an evolution equation for that vorticity by taking the curl of (7.130a), giving

$$\frac{\partial \zeta'}{\partial t} = f\frac{\partial w'}{\partial z}. \tag{7.132}$$

Eliminating vorticity between these equations gives

$$\left(\frac{\partial^2}{\partial t^2} + f^2\right)\frac{\partial w'}{\partial z} = \frac{\partial}{\partial t}\nabla^2\phi'. \tag{7.133}$$

We may obtain another equation linking pressure and vertical velocity by eliminating the buoyancy between (7.130b) and (7.130d), so giving

$$\frac{\partial^2 w'}{\partial t^2} + N^2 w' = -\frac{\partial}{\partial t}\frac{\partial \phi'}{\partial z}. \tag{7.134}$$

Eliminating ϕ' between (7.133) and (7.134) gives a single equation for w' analogous to (7.54), namely

$$\left[\frac{\partial^2}{\partial t^2}\left(\nabla^2 + \frac{\partial^2}{\partial z^2}\right) + f^2\frac{\partial^2}{\partial z^2} + N^2\nabla^2\right] w' = 0. \tag{7.135}$$

If we assume a time dependence of the form $w' = \widehat{w}\mathrm{e}^{-\mathrm{i}\omega t}$, this equation may be written in the sometimes useful form,

$$\frac{\partial^2 \widehat{w}}{\partial z^2} = \left(\frac{N^2 - \omega^2}{\omega^2 - f^2}\right)\nabla^2\widehat{w}. \tag{7.136}$$

7.6.3 Dispersion Relation

Assuming wave solutions to (7.135) of the form $w' = \tilde{w} \exp[\mathrm{i}(kx + ly + mz - \omega t)]$ we readily obtain the dispersion relation

$$\omega^2 = \frac{f^2 m^2 + (k^2 + l^2) N^2}{k^2 + l^2 + m^2}, \tag{7.137}$$

which is a minor generalization of (7.129). We can also write the dispersion relation as

$$\omega^2 = f^2 \sin^2 \vartheta + N^2 \cos^2 \vartheta, \tag{7.138}$$

or

$$\omega^2 = f^2 + (N^2 - f^2) \cos^2 \vartheta, \qquad \text{or} \qquad \omega^2 = N^2 - (N^2 - f^2) \sin^2 \vartheta, \tag{7.139}$$

where ϑ is the angle of the wavevector with the horizontal. The frequency therefore lies between N and f. The waves satisfying (7.137) are called inertia-gravity waves and are analogous to surface gravity waves in a rotating frame — that is, Poincaré waves — discussed in Section 3.8.2.

In many atmospheric and oceanic situations $f \ll N$ (in fact typically $N/f \sim 100$, the main exception being weakly stratified near-surface mixed layers and the deep abyss in the ocean) and $f < \omega < N$. From (7.138) the frequency is dependent only on the angle the wavevector makes with the horizontal, and the surfaces of constant frequency again form cones in wavenumber space, although depending on the values of f and ω the frequency does not necessarily decrease monotonically with ϑ as in the non-rotating case. For reference, the group velocity is

$$c_g^x = \left[\frac{N^2 - f^2}{\omega K_3^4} Km\right] \frac{km}{K} = \left[\frac{N^2 - f^2}{\omega K_3} \cos\vartheta \sin\vartheta\right] \cos\lambda \sin\vartheta, \tag{7.140a}$$

$$c_g^y = \left[\frac{N^2 - f^2}{\omega K_3^4} Km\right] \frac{lm}{K} = \left[\frac{N^2 - f^2}{\omega K_3} \cos\vartheta \sin\vartheta\right] \sin\lambda \sin\vartheta, \tag{7.140b}$$

$$c_g^z = -\left[\frac{N^2 - f^2}{\omega K_3^4} Km\right] K = -\left[\frac{N^2 - f^2}{\omega K_3} \cos\vartheta \sin\vartheta\right] \cos\vartheta, \tag{7.140c}$$

using (7.69), and where $K_3^4 \equiv (k^2 + l^2 + m^2)^2$. These expressions reduce to (7.74) if $f = 0$, in which case $\omega = N \cos\vartheta$. Notice that the directional factors — the sin and cos terms outside of the square brackets — are the same as those in (7.74). Thus, the group velocity is, as in the non-rotating case, at an angle ϑ to the vertical, or $\alpha = \pi/2 - \vartheta$ to the horizontal. The magnitude of the group velocity is now given by

$$|\mathbf{c}_g| = \frac{N^2 - f^2}{\omega K_3^3} Km = \frac{N^2 - f^2}{\omega K_3} \cos\vartheta \sin\vartheta. \tag{7.141}$$

There are a few notable limits:

1. A purely horizontal wave vector. In this case $m = 0$ and $\omega = N$. The waves are then unaffected by the Earth's rotation. This is because the Coriolis force is (in the f-plane approximation) due to the product of the Coriolis parameter and the horizontal component of the velocity. If the wave vector is horizontal, the fluid velocities are purely vertical and so the Coriolis force vanishes.

2. A purely vertical wave vector. In this case $\omega = f$, the fluid velocities are horizontal and the fluid parcels do not feel the stratification. The oscillations are then known as *inertial waves,* although they are not inertial in the sense of there being no implied force in an inertial frame of reference. This case and the previous one show that when $\omega = N$ or $\omega = f$ the group velocity is zero.

3. In the limit $N \to 0$ we have pure inertial waves with a frequency $0 < \omega < f$, and specifically $\omega = f \sin\vartheta$. Similarly, as $f \to 0$ we have pure internal waves, as discussed previously, with $\omega = N\cos\vartheta$.
4. The hydrostatic limit, which we discuss below.

The hydrostatic limit

Hydrostasy occurs in the limit of large horizontal scales, $k, l \ll m$. If we therefore neglect k^2 and l^2 where they appear with m^2 in (7.137) we obtain

$$\omega^2 = f^2 + N^2\frac{k^2 + l^2}{m^2} = f^2 + N^2\cos^2\vartheta, \tag{7.142}$$

where the rightmost expression arises from (7.138) if we take

$$\sin^2\vartheta = \frac{m^2}{k^2 + l^2 + m^2} \to 1, \qquad \cos^2\vartheta = \frac{K^2}{k^2 + m^2} \to \frac{K^2}{m^2} \ll 1, \tag{7.143}$$

with $K^2 = k^2 + l^2$.

If we make the hydrostatic approximation from the outset in the rotating, linearized, equations of motion then we have

$$\frac{\partial u'}{\partial t} - fv = -\frac{\partial \phi'}{\partial x}, \qquad \frac{\partial v'}{\partial t} + fu = -\frac{\partial \phi'}{\partial y}, \qquad 0 = -\frac{\partial \phi'}{\partial z} + b', \tag{7.144a}$$

$$\frac{\partial u'}{\partial x} + \frac{\partial v'}{\partial y} + \frac{\partial w'}{\partial z} = 0, \qquad \frac{\partial b'}{\partial t} + w'N^2 = 0. \tag{7.144b}$$

This reduces to the single equation

$$\left[\frac{\partial^2}{\partial t^2}\frac{\partial^2}{\partial z^2} + f^2\frac{\partial^2}{\partial z^2} + N^2\nabla^2\right] w' = 0, \tag{7.145}$$

and corresponding dispersion relation

$$\omega^2 = \frac{f^2m^2 + K^2N^2}{m^2} = f^2 + N^2\frac{K^2}{m^2}, \tag{7.146}$$

so recovering (7.142). This limit is sometimes known as the *rapidly rotating regime.* The Coriolis parameter f now appears in isolation, and simply provides inertial oscillations that are independent of the wavenumber and the stratification.

Another way to think about the small aspect ratio limit follows if we define

$$\alpha' \equiv \frac{\text{vertical scale}}{\text{horizontal scale}} = \frac{K}{m} = \frac{1}{\tan\vartheta} \ll 1. \tag{7.147}$$

From the nonhydrostatic dispersion relation, (7.137), a line or two of algebra gives

$$\alpha'^2 = \left(\frac{\omega^2 - f^2}{N^2 - \omega^2}\right). \tag{7.148}$$

The hydrostatic limit requires that this aspect ratio is small, and therefore that $N^2 \gg \omega^2$. Put differently, low frequencies will tend to have a small aspect ratio and be hydrostatic. Using (7.140) and (7.139) we find that the ratio of the vertical to the horizontal group velocities scales as the aspect ratio, that is

$$\frac{c_g^z}{c_g^h} = \alpha' = \left(\frac{\omega^2 - f^2}{N^2 - \omega^2}\right)^{1/2}. \tag{7.149}$$

where $c_g^h = (c_g^{x2} + c_g^{y2})^{1/2}$, and the above ratio is small in the hydrostatic limit. We return to this limit in Section 17.2 on gravity waves in the stratosphere.

7.6.4 Polarization Relations

Just as in the non-rotating case, we can derive phase relations between the various fields, useful if we are trying to identify internal waves from observations. As for all waves in an incompressible fluid, the condition $\nabla_3 \cdot \boldsymbol{v} = 0$ gives

$$\boldsymbol{k} \cdot \boldsymbol{v}' = 0, \tag{7.150}$$

so that the fluid motion is in the plane that is perpendicular to the wave vector. The derivations of the other polarization relations are left as exercises for the reader, and the relations are found to be

$$\widetilde{u} = \frac{k\omega + \mathrm{i}lf}{\omega^2 - f^2}\widetilde{\phi}, \qquad \widetilde{v} = \frac{l\omega - \mathrm{i}kf}{\omega^2 - f^2}\widetilde{\phi}, \tag{7.151a,b}$$

which should be compared with (7.63a). We also have a relation between buoyancy and pressure,

$$\widetilde{b} = \frac{\mathrm{i}mN^2}{N^2 - \omega^2}\widetilde{\phi}, \tag{7.152}$$

and one between vertical velocity and pressure,

$$\widetilde{w} = \frac{-m\omega}{N^2 - \omega^2}\widetilde{\phi} = \frac{-\omega K_3^2}{(N^2 - f^2)m}\widetilde{\phi}, \tag{7.153}$$

with the second equality following with use of the dispersion relation.

7.6.5 Geostrophic Motion and Vortical Modes

If we seek *steady* solutions to (7.130), the equations of motion become

$$-fv = -\frac{\partial \phi'}{\partial x}, \qquad fu = -\frac{\partial \phi'}{\partial y}, \qquad 0 = -\frac{\partial \phi'}{\partial z} + b', \tag{7.154a,b}$$

and

$$\frac{\partial u'}{\partial x} + \frac{\partial v'}{\partial y} + \frac{\partial w'}{\partial z} = 0, \qquad w'N^2 = 0. \tag{7.155a,b}$$

These are the equations of geostrophic and hydrostatic balance, with zero vertical velocity. What can we say about this solution? If instead of eliminating pressure between (7.133) and (7.134) we eliminate vertical velocity we obtain

$$\frac{\partial}{\partial t}\left[\frac{\partial^2}{\partial t^2}\left(\nabla^2 + \frac{\partial^2}{\partial z^2}\right) + f^2\frac{\partial^2}{\partial z^2} + N^2\nabla^2\right]\phi' = 0, \tag{7.156}$$

which is similar to (7.135), except for the extra time derivative, which allows for the possibility of a solution with $\omega = 0$. If $\omega \neq 0$ then

$$\left[\frac{\partial^2}{\partial t^2}\left(\nabla^2 + \frac{\partial^2}{\partial z^2}\right) + f^2\frac{\partial^2}{\partial z^2} + N^2\nabla^2\right]\phi' = 0, \tag{7.157}$$

and the dispersion relation is given by (7.137). If $\omega = 0$, then the quantity in square brackets in (7.156) may not be a function of time; that is

$$\left[\frac{\partial^2}{\partial t^2}\left(\nabla^2 + \frac{\partial^2}{\partial z^2}\right) + f^2\frac{\partial^2}{\partial z^2} + N^2\nabla^2\right]\phi' = \chi(x, y, z), \tag{7.158}$$

where χ is a function of space, but not time, and so determined by the initial conditions of ϕ'. When $\omega \neq 0$, then $\chi = 0$. What is χ? We shall see that it is nothing but the potential vorticity of the flow!

Potential vorticity

Recall the vorticity equation and the buoyancy equation, namely

$$\frac{\partial \zeta'}{\partial t} = f\frac{\partial w'}{\partial z}, \qquad \frac{\partial b'}{\partial t} + w'N^2 = 0. \tag{7.159a,b}$$

If we eliminate w' from these equations we obtain

$$\frac{\partial q}{\partial t} = 0, \qquad \text{where} \qquad q = \left[\zeta' + f\frac{\partial}{\partial z}\left(\frac{b'}{N^2}\right)\right] \tag{7.160a,b}$$

and q is the potential vorticity for this problem. In general, for adiabatic flow, potential vorticity is conserved on fluid parcels and $\mathrm{D}Q/\mathrm{D}t = 0$ where for a Boussinesq fluid $Q = \omega_a \cdot \nabla b$. There are two differences between this general case and ours; first, because we have linearized the dynamics the advective term is omitted, and $\partial q/\partial t = 0$. Second, q is not exactly the same as Q, but it is an approximation to it valid when the stratification is dominated by its background value, N^2. Very informally, we have then, for constant N,

$$Q = (\boldsymbol{\omega} + \boldsymbol{f}_0)\cdot\nabla b \approx (\zeta + f)\left(N^2 + \frac{\partial b'}{\partial z}\right) \approx fN^2 + f\frac{\partial b'}{\partial z} + \zeta N^2 = N^2\left[f + \zeta + f\frac{\partial}{\partial z}\left(\frac{b'}{N^2}\right)\right]. \tag{7.161}$$

The first term on the right-hand side of this expression, fN^2, is a constant and so dynamically unimportant, and the remaining terms are equal to q as given by (7.160b).

Another way to see that (7.160b) is the potential vorticity is to note that the displacement of an isentropic surface, η say, is related to the change in buoyancy by

$$\eta \approx -\frac{b'}{\partial \bar{b}/\partial z} = -\frac{b'}{N^2}, \tag{7.162}$$

as illustrated in Fig. 3.13 on page 138. The thickness of an isentropic layer is the difference between the heights of two neighbouring isentropic surfaces, and so is given by

$$h = -\frac{b_1'}{N^2} + \frac{b_2'}{N^2} \approx -H\frac{\partial}{\partial z}\left(\frac{b'}{N^2}\right), \tag{7.163}$$

where H is the mean separation between the surfaces, or the mean thickness. Thus, the expression (7.160b) may be written

$$q = \left[\zeta' - \frac{fh}{H}\right], \tag{7.164}$$

which is the shallow water expression for the potential vorticity of a fluid layer, linearized about a mean thickness H and a state of rest (with $|\zeta'| \ll f$).

Let us now relate q to χ, and we do this by expressing ζ' and b' in terms of ϕ' and w'. From (7.131) and (7.130b) respectively we have,

$$f\zeta' = \nabla^2\phi' - w'_{zt}, \tag{7.165a}$$

$$\frac{f^2}{N^2}b'_z = \frac{f^2}{N^2}w_{zt} + \frac{f^2}{N^2}\phi'_{zz}, \tag{7.165b}$$

using subscripts to denote derivatives. Thus, f times the potential vorticity is

$$fq = \nabla^2\phi' + \frac{f^2}{N^2}\phi'_{zz} + \frac{f^2}{N^2}w_{zt} - w_{zt}. \tag{7.166}$$

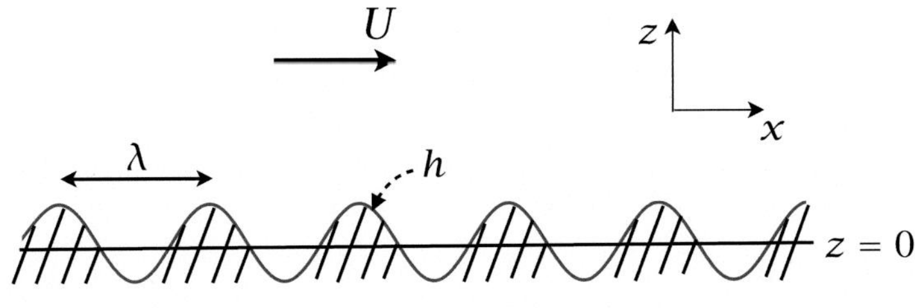

Fig. 7.15 Uniform flow, U, in the x-direction flowing over sinusoidal topography, h. The vertical co-ordinate is stretched, and in reality $|h| \ll \lambda$.

We now use (7.134) to express the second w_{zt} term in terms of ϕ', giving

$$fq = \nabla^2\phi' + \frac{f^2}{N^2}\phi'_{zz} + \frac{f^2}{N^2}w'_{zt} + \frac{1}{N^2}\left(w'_{zttt} + \phi'_{zztt}\right), \tag{7.167}$$

and we then use (7.133) to eliminate w', giving

$$fq = \nabla^2\phi' + \frac{f^2}{N^2}\phi'_{zz} + \frac{1}{N^2}\left(\nabla^2\phi'_{tt} + \phi'_{zztt}\right), \tag{7.168}$$

or, re-arranging,

$$fq = \frac{1}{N^2}\left[\frac{\partial^2}{\partial t^2}\left(\nabla^2\phi' + \frac{\partial^2\phi'}{\partial z^2}\right) + N^2\nabla^2\phi' + f^2\frac{\partial^2\phi'}{\partial z^2}\right]. \tag{7.169}$$

Comparing this with (7.158), we can see that

$$\chi = fN^2q. \tag{7.170}$$

That is to say, the conserved quantity for motions with $\omega = 0$ is nothing but a constant multiple of the potential vorticity. When $\omega \neq 0$, then χ and hence the potential vorticity are zero. In other words, *oscillating linear gravity waves, even in a rotating reference frame, have zero potential vorticity.* This is an important result, because large-scale balanced dynamics is characterized by the advection of potential vorticity, so that (in the linear approximation at least) internal waves play no direct role in the potential vorticity budget. However, they *do* play an important role in transporting and dissipating energy, as we will see later on.

7.7 TOPOGRAPHIC GENERATION OF INTERNAL WAVES

How are internal waves generated? One way that is important in both the ocean and atmosphere is by way of a horizontal flow, such as a mean wind or, in the ocean, a tide or a mesoscale eddy, passing over a topographic feature. This forces the fluid to move up and/or down, so generating an internal wave, commonly known as a mountain wave. In this section we illustrate the mechanism with simple examples of steady, uniform, flow with constant stratification over idealized topography.[2]

7.7.1 Sinusoidal Mountain Waves

For simplicity we ignore the effects of the Earth's rotation and pose the problem in two dimensions, x and z, using the Boussinesq approximation. Our goal is to calculate the response to a steady, uniform flow of magnitude U over a sinusoidally varying boundary $h = \tilde{h}\cos kx$ at $z = 0$, as in Fig. 7.15 with $k = 2\pi/\lambda$. The topographic variations are assumed small, so allowing the dynamics to be linearized, which will enable an arbitrarily shaped boundary to be considered by appropriately summing over Fourier modes.

Equation of motion and dispersion relation

The momentum equations, the buoyancy equation, and the mass continuity equation are, respectively,

$$\left(\frac{\partial}{\partial t} + U\frac{\partial}{\partial x}\right)u' = -\frac{\partial \phi}{\partial x}, \qquad \left(\frac{\partial}{\partial t} + U\frac{\partial}{\partial x}\right)w' = -\frac{\partial \phi}{\partial z} + b', \tag{7.171a,b}$$

$$\left(\frac{\partial}{\partial t} + U\frac{\partial}{\partial x}\right)u' + w'N^2 = 0, \qquad \frac{\partial u'}{\partial x} + \frac{\partial w'}{\partial z} = 0. \tag{7.171c,d}$$

We henceforth drop the primes on the perturbation quantities.

The dispersion relation is obtained by noting that the equations are the same is with no mean flow, save that $\partial/\partial t$ is replaced by $\partial/\partial t + U\partial/\partial x$. Thus, similar to (7.78), the dispersion relation is

$$(\omega - Uk)^2 = \frac{k^2N^2}{k^2 + m^2} \qquad \text{or} \qquad \omega = Uk \pm \frac{k^2N^2}{(k^2 + m^2)^{1/2}}. \tag{7.172a,b}$$

The horizontal and vertical components of the group velocity are then given by

$$c_g^z = \pm\frac{Nmk}{(k^2 + m^2)^{3/2}}, \qquad c_g^x = U \pm \frac{Nm^2}{(k^2 + m^2)^{3/2}}. \tag{7.173a,b}$$

Steady waves have $\omega = 0$ and if $U > 0$ such waves are associated with the negative root in (7.172b), the positive root in (7.173a) and the negative root in (7.173b). Energy will propagate upwards, and away from the mountain, if c_g^z is positive.

The solution

If we are looking for steady solutions we can follow the recipe of Section 7.5 with $U\partial/\partial x$ replacing the time derivative. By analogy to (7.54) we find a single equation for w namely

$$\left[U\frac{\partial^2}{\partial x^2}\left(\frac{\partial^2}{\partial x^2} + \frac{\partial^2}{\partial z^2}\right) + N^2\frac{\partial^2}{\partial x^2}\right]w = 0. \tag{7.174}$$

We take the topography to have the form $h = \mathrm{Re}\, h_0 \exp(\mathrm{i}kx)$ and its presence is felt via a lower boundary condition, $w = U\partial h/\partial x = \mathrm{Re}\, Uh_0\mathrm{i}k \exp(\mathrm{i}kx)$ at $z = 0$. We thus seek solutions to (7.173) of the form

$$w = \mathrm{Re}\, Uh_0\mathrm{i}k\mathrm{e}^{\mathrm{i}(kx+mz)}. \tag{7.175}$$

The harmonic dependence in z is valid only because we take N to be constant. The value of m is given by the dispersion relation (7.172) with $\omega = 0$, which gives

$$m^2 = \left(\frac{N}{U}\right)^2 - k^2. \tag{7.176}$$

Equation (7.176) is just the dispersion relation for internal gravity waves, but here we are using it to determine the vertical wavenumber since the frequency is given (it is zero). We see that m^2 may be negative, and so m imaginary, if $N^2 < k^2U^2$, so evidently there will be a qualitative difference between short waves, with $k > (N/U)^2$, and long waves, with $k < (N/U)^2$. In the atmosphere we might take $U \sim 10\,\mathrm{m\,s^{-1}}$ and $N \sim 10^{-2}\,\mathrm{s^{-1}}$, in which case $U/N \sim 1\,\mathrm{km}$. In the ocean below the thermocline we might take $U \sim 0.1\,\mathrm{m\,s^{-1}}$ and $N \sim 10^{-3}\,\mathrm{s^{-1}}$, whence $U/N \sim 100\,\mathrm{m}$.

A useful interpretation of (7.176) arises if the mean flow is constant and we consider the problem in the frame of reference of that mean flow. In this case the topography has the form $h = \mathrm{Re}\, h_0 \exp \mathrm{i}k(x - Ut)$, and the solution in the moving frame is

$$w = \mathrm{Re}\, Uh_0\mathrm{i}k\mathrm{e}^{\mathrm{i};(kx+mz+Ukt)}. \tag{7.177}$$

That is, the fluid now oscillates with a frequency $\omega = -Uk$. The condition for propagation, $N^2 < k^2U^2$ is just the same as $\omega^2 < N^2$, which is just the condition for gravity waves to oscillate in a stratified fluid.

Given the solution for w we can use the polarization relations of section 7.3.2 to obtain the solutions for perturbation horizontal velocity and pressure. One way to proceed is to pose the problem in the moving frame, with $\omega = -Uk$, and directly use (7.64). Then back in the stationary frame we obtain the solutions

$$w = w_0 e^{i(kx+mz)} = iUkh_0 e^{imz} e^{ikx}, \tag{7.178a}$$

$$u = u_0 e^{i(kx+mz)} = -imUh_0 e^{imz} e^{ikx}, \tag{7.178b}$$

$$\phi = \phi_0 e^{i(kx+mz)} = imU^2 h_0 e^{imz} e^{ikx}, \tag{7.178c}$$

where m is given by (7.176). Let us see what the solutions mean, and if and how waves propagate.

7.7.2 Energy Propagation

The direction of energy propagation is given by the group velocity. For steady waves ($\omega = 0$) we have, using (7.173a),

$$c_g^z = \frac{Nkm}{(k^2 + m^2)^{3/2}} = \frac{Ukm}{k^2 + m^2}. \tag{7.179a,b}$$

This means that for positive U an upward group velocity, and hence upward energy propagation, occur when k and m have the same sign. This property also may be deduced by evaluating the vertical energy flux, $\overline{w\phi}$ using (7.178), with appropriate care to take the real part of the fields. The phase speed, $c_p^z = \omega/m$, is zero in the stationary frame and it is $-Uk/m$ in the translating frame.

Short, trapped waves

If the undulations on the boundary are sufficiently short then $k^2 > (N/U)^2$ and m^2 is negative and m is pure imaginary. Writing $m = is$, so that $s^2 = k^2 - (N/U)^2$, the solutions have the form

$$w = \mathrm{Re}\, w_0 e^{ikx - sz}. \tag{7.180}$$

We must choose the solution with $s > 0$ in order that the solution decays away from the mountain, and internal waves are not propagated into the interior. (If there were a rigid lid or a density discontinuity at the top of the fluid, as at the top of the ocean, then the possibility of reflection would arise and we would seek to satisfy the upper boundary condition with a combination of decaying and amplifying modes.) The above result is entirely consistent with the dispersion relation for internal waves, namely $\omega = N\cos\vartheta$: because $\cos\vartheta < 1$ the frequency ω must be less than N so that if the forcing frequency is higher than N no internal waves will be generated.

Because the waves are trapped waves we do not expect energy to propagate away from the mountains. To verify this, from the polarization relation (7.178) we have

$$w = \frac{k}{mU}\phi = \frac{-ik}{sU}\phi. \tag{7.181}$$

The pressure and the vertical velocity are therefore out of phase by $\pi/2$, and the vertical energy flux, $\overline{w\phi}$ (see (7.83) and Section 7.3.5) is identically zero. This is consistent with the fact that the energy flux is in the direction of the group velocity; the group velocity is given by (7.179a) and for an imaginary m the real part is zero. A solution to the problem in the short wave limit is shown in top panels of Fig. 7.16 with $s = 1$ ($m = i$).

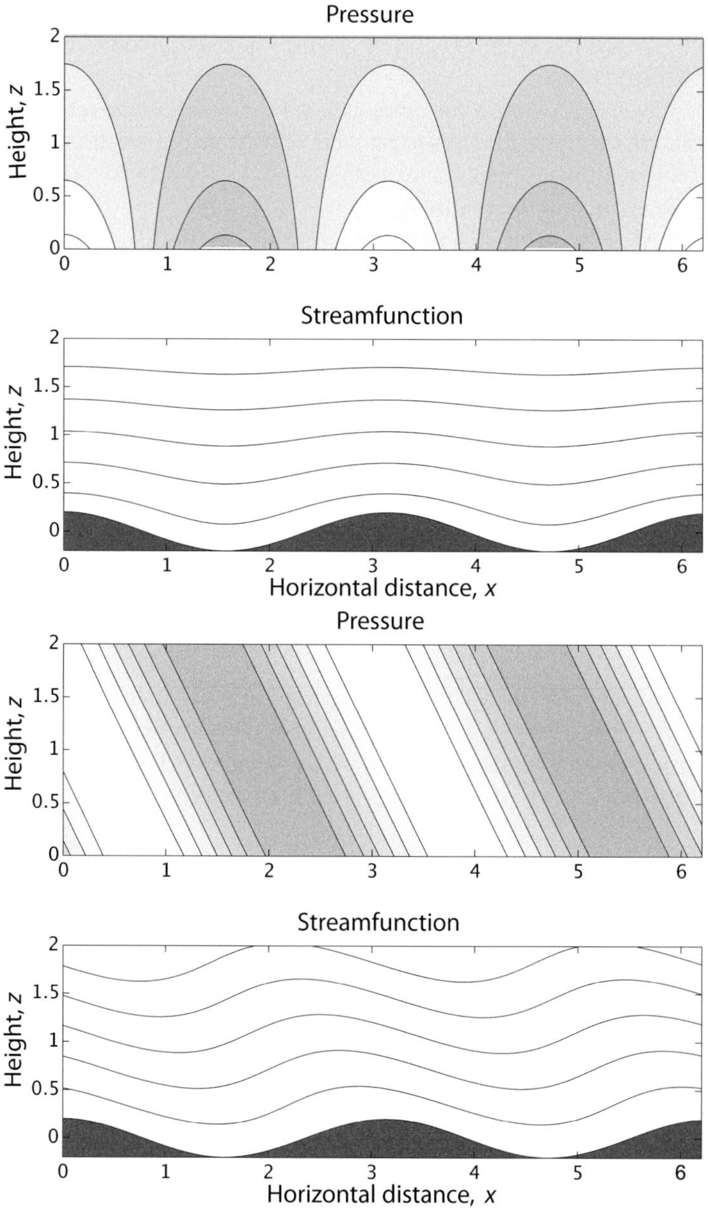

Fig. 7.16 Top two panels. Solutions for the flow over a sinusoidal ridge, using (7.178), in the short wave limit ($Uk > N$) and with $m = \mathrm{i}$. The top panel shows the pressure, with darker gray indicating higher pressure. The second panel shows contours of the total streamfunction, $\psi - Uz$, with flow coming in from the left, and the topography itself (solid). The perturbation amplitude decreases exponentially with height.

Bottom two panels. The same as above, but in the long wave limit ($Uk < N$) with $m = 1$. The pressure is high on the windward side of the topography, and phase lines tilt upstream with height for both pressure and streamfunction.

Long, propagating waves

Suppose now that $k^2 < (N/U)^2$ so that $\omega^2 < N^2$. From (7.176) m is now real and the solution has propagating waves of the form

$$w = w_0 \mathrm{e}^{\mathrm{i}(kx+mz)}, \qquad m^2 = \left(\frac{N}{U}\right)^2 - k^2. \tag{7.182}$$

Vertical propagation is occurring because the forcing frequency is less than the buoyancy frequency. The angle at which fluid parcel oscillations occur is then slanted off the vertical at an angle ϑ such

Topographically Generated Gravity Waves (Mountain Waves)

- In both atmosphere and ocean an important mechanism for the generation of gravity waves is flow over bottom topography, and the ensuing waves are sometimes called mountain waves. A canonical case is that of a uniform flow over a sinusoidal topography, with constant stratification. If the flow is in the x-direction and there is no y-variation then the boundary condition is

$$w(x, z = 0) = U\frac{\partial h}{\partial x} = -\mathrm{i}Uk\tilde{h}. \tag{MW.1}$$

Solutions of the problem may be found in the form $w(x, z, t) = w_0 \exp[i(kx + mz - \omega t)]$, where the boundary condition at $z = 0$ is given by (MW.1), the frequency is given by the internal wave dispersion relation, and the other dynamical fields are obtained using the polarization relations.

- One way to easily solve the problem is to transform into a frame moving with the background flow, U. The topography then appears to oscillate with a frequency $-Uk$, and this in turn becomes the frequency of the gravity waves.
- Propagating gravity waves can only be supported if the frequency is less than N, meaning that $Uk < N$. That is, the waves must be sufficiently long and therefore the topography must be of sufficiently large scale.
- When propagating waves exist, energy is propagated upward away from the topography. The topography also exerts a drag on the background flow.
- If the waves are too short they are evanescent, decaying exponentially with height. That is, they are trapped near the topography
- In the presence of rotation the wave frequency must lie between the buoyancy frequency N and the inertial frequency f. If the flow is constant, waves can thus radiate upward if

$$f < Uk < N. \tag{MW.2}$$

Thus, both very long waves and very short waves are evanescent.

that the forcing frequency is equal to the natural frequency of oscillations at that angle, namely

$$\vartheta = \cos^{-1}\left(\frac{Uk}{N}\right). \tag{7.183}$$

The angle ϑ is also the angle between the wavevector $\boldsymbol{k}$ and the horizontal, as in (7.61), because the wavevector is at right angles to the parcel oscillations. If $Uk = N$ then the fluid parcel oscillations are vertical and, using (7.176), $m = 0$. Thus, although the group velocity is directed vertically, parallel to the fluid parcel oscillations, its magnitude is zero, from (7.179).

Our intuition suggests that if there is vertical propagation there must be an upwards energy flux, since the energy source is at the ground. Let's confirm this. Using the polarization relations (7.178a,c) we obtain

$$w_0 = \frac{k}{mU}\phi_0. \tag{7.184}$$

and the energy flux in the vertical direction is, from (7.89)

$$F^z = \frac{k}{2mU}|\phi_0|^2 = \frac{mU}{2k}|w_0|^2, \tag{7.185}$$

which is evidently non-zero. This energy flux must be upward, away from the source (the topography), and this determines the sign of m that must be chosen by the solution. Specifically, for positive U, the group velocity must be positive so from (7.179) m must be positive. If U were negative the sign of m would be negative, and if $m = 0$ there is no vertical energy propagation.

Because energy is propagating upward and away from the topography there must be a drag at the lower boundary. The stress at the boundary, τ, is the rate at which horizontal momentum is transported upwards and so is given by

$$\tau = -\rho_0 \overline{uw}, \tag{7.186}$$

where the overbar denotes averaging over a wavelength and ρ_0 is the density, which is constant. From (7.178)

$$u_0 = -\mathrm{i}mUh_0, \qquad w_0 = \mathrm{i}Ukh_0, \tag{7.187}$$

so that

$$\widetilde{\tau} \equiv \tau/\rho_0 = -\overline{uw} = \frac{1}{2}kmU^2h_0^2, \tag{7.188}$$

where the factor of 1/2 comes from the averaging, and we take the product $u_0 w_0^*$ where w_0^* is the complex conjugate of w_0. The sign of the stress depends on the sign of m, and thus on the sign of U. For positive U, m is positive and so the stress is positive at the surface.

Solutions for flow over topography in the long wave limit and $m = 1$ are shown in the lower panels Fig. 7.16. The flow is coming in from the left, and the phase lines evidently tilt upstream with height. Lines of constant phase follow $kx + mz$ = constant, and in the solution shown both k and m are positive ($k = m = 1$). Thus, the lines slope back at a slope $x/z = -m/k$, and energy propagates up and to the left. The phase propagation is actually downward in this example, as the reader may confirm. The pressure is high on the upstream side of the mountain, and this provides a drag on the flow — a topographic form drag.

7.7.3 Flow over an Isolated Ridge

Most mountains are of course not perfect sinusoids, but we can construct a solution for any given topography using a superposition of Fourier modes. In this section we will illustrate the solution for a mountain consisting of a single ridge; the actual solution must usually be obtained numerically, and we will sketch the method and show some results.

Sketch of the methodology

The methodology to compute a solution is as follows. Consider a topographic profile, $h(x)$, and let us suppose that it is periodic in x over some distance L. Such a profile can (nearly always) be decomposed into a sum of Fourier coefficients, meaning that we can write

$$h(x) = \sum_k \widetilde{h}_k \mathrm{e}^{\mathrm{i}kx}, \tag{7.189}$$

where $\widetilde{h}_k$ are the Fourier coefficients. We can obtain the set of $\widetilde{h}_k$ by multiplying (7.189) by $\mathrm{e}^{-\mathrm{i}kx}$ and integrating over the domain from $x = 0$ to $x = L$, a procedure known as taking the discrete Fourier transform of $h(x)$, and there are standard computer algorithms for doing this efficiently. Once we have obtained the values of $\widetilde{h}_k$ we essentially solve the problem separately for each k in precisely the same manner as we did in the previous section. For *each* k there will be a vertical wavenumber

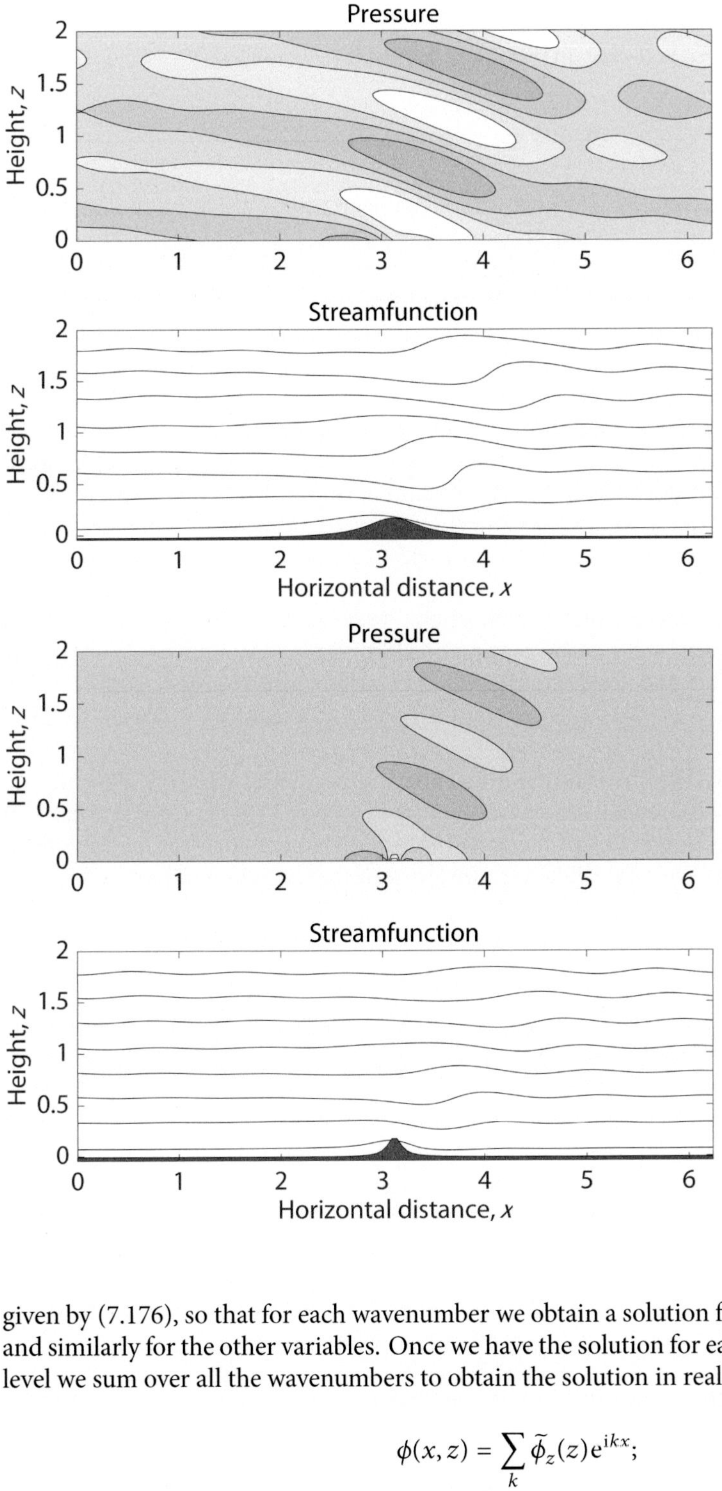

Fig. 7.17 Upper two panels. Solutions for the flow over a bell-shaped ridge (7.191), with $a^2 = 4U^2/N^2$. High pressure is shaded darker, and the flow comes in from the left.

Lower two panels. Same as above, but now for a narrow ridge, of the same height but with $a^2 = U^2/4N^2$.

(The dark shading for the mountain is graphically superimposed on the solution, and the streamfunction does not actually intersect the topography.)

given by (7.176), so that for each wavenumber we obtain a solution for pressure of the form $\widetilde{\phi}_k(z)$, and similarly for the other variables. Once we have the solution for each wavenumber, then at each level we sum over all the wavenumbers to obtain the solution in real space; that is, we evaluate

$$\phi(x, z) = \sum_k \widetilde{\phi}_z(z) \mathrm{e}^{\mathrm{i}kx}; \tag{7.190}$$

that is to say, we take the inverse discrete Fourier transform.

A witches' brew

For specificity let us consider the bell-shaped topographic profile

$$h(x) = \frac{h_0 a^2}{a^2 + x^2}, \tag{7.191}$$

sometimes called the Witch of Agnesi.[3] (Results with a Gaussian profile are quite similar.) Such a profile is composed of *many* (in fact an infinite number of) Fourier coefficients of differing amplitudes. If the profile is narrow (meaning a is small, in a sense made clearer below) then there will be a great many significant coefficients at high wavenumbers. In fact, in the limiting case of an infinitely thin ridge (a delta function) all wavenumbers are present with equal weight, so there are certainly more large wavenumbers than small wavenumbers. However, if a is large, then the contributing wavenumbers will predominantly be small.

In the problem of flow over topography the natural horizontal scale is U/N. If $a \gg U/N$ then the dominant wavenumbers are small and the solution will consist of waves propagating upward with little loss of amplitude and phase lines tilting upstream, as illustrated in the upper two panels of Fig. 7.17. (The influence of the mountain is downstream, but the solutions are obtained on a finite, periodic domain and some influence returns upstream.) In the case of a narrow ridge, as illustrated in the lower two panels of Fig. 7.17, the perturbation is largely trapped near to the mountain and the perturbation fields largely decay exponentially with height. Nevertheless, because the ridge *does* contain some small wavenumbers, some weak, propagating large-scale disturbances are generated. The fluid acts as a low-pass filter, and the perturbation aloft consists only of large scales.

A hydrostatic solution

If the ridge is sufficiently wide then the solution is essentially hydrostatic, with little dependence of the vertical structure on the horizontal wavenumber; that is, using (7.176) at large scales, $m^2 \approx (N/U)^2$. Furthermore, the x-component of the group velocity is zero, which can be seen from (7.173b) using $m = N/U$ (since m is positive for upward wave propagation and $m \gg k$ by hydrostasy). Explicitly we have

$$c_g^x = U - \frac{Nm^2}{(k^2 + m^2)^{3/2}} \approx U - \frac{N}{m} = U - U = 0. \tag{7.192}$$

Thus, the disturbance appears directly over the mountain, with no downstream propagation, as in Fig. 7.18. The pattern therefore repeats itself in the vertical at intervals of $2\pi/m$ or $2\pi U/N$, with neither a downnstream nor upstream influence. This solution is in fact the most atmospherically relevant one, for it is that produced by atmospheric flow over mountains that have horizontal scales larger than a few kilometres (i.e. $a \gg U/N$). It is also oceanographically relevant for flow over features of greater than a few kilometers, except possibly in regions of very weak stratification and large abyssal currents. Scales that are much larger are filtered by the Coriolis effect, as we now see.

7.7.4 Effects of Rotation

General considerations

We now briefly consider the effects of a Coriolis force on mountain waves. The problem is in many ways similar to the non-rotating case but the dispersion relation and so the criteria for upward propagation differ accordingly, for it transpires that the Coriolis effect filters the waves at very large scales. To proceed, we first note that the steady flow must be in geostrophic balance, so that if the flow is zonal there is a background meridional pressure gradient that satisfies

$$fU = -\frac{\partial \Phi}{\partial y}, \tag{7.193}$$

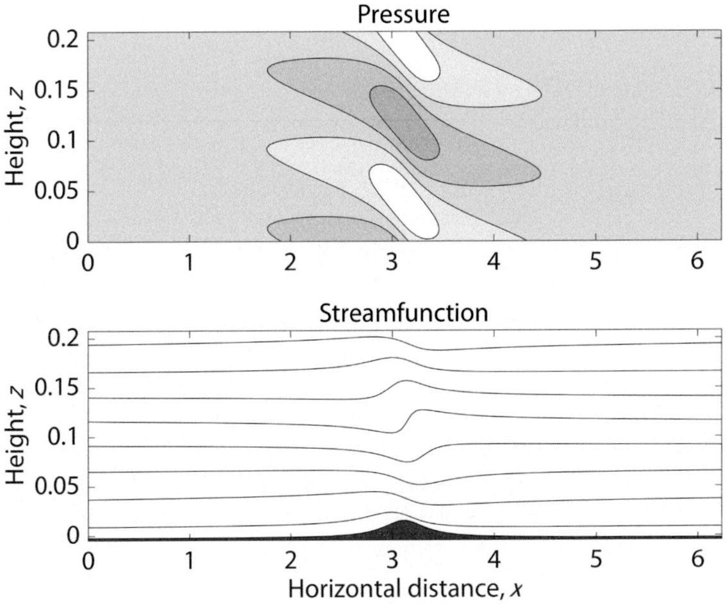

Fig. 7.18 Similar to the upper two panels of Fig. 7.17, but with a smaller U. Thus, the ridge is of the same height and width, but the flow is in a hydrostatic regime with $a^2 = 10U^2/N^2$. This is a very relevant regime for flow over mountains in the atmosphere.

The group velocity is almost purely vertical and the disturbance is confined to a region directly above the mountain. The vertical extent of the domain shown contains about one wavenumber, and the streamfunction is a mirror image of the mountain at about $z = 0.1$.

where as before we take f to be constant in any given problem. The main difference arises from the fact that the waves now obey the dispersion relation with rotation, namely (7.137) or, restricting attention to the x–z plane,

$$\omega^2 = \frac{f^2 m^2 + k^2 N^2}{k^2 + m^2}. \tag{7.194}$$

As in the non-rotating case we can obtain an expression for the vertical wavenumber in steady flow by letting $\omega = -Uk$, whence

$$m^2 = \frac{k^2(N^2 - U^2k^2)}{U^2k^2 - f^2}. \tag{7.195}$$

Evanescent solutions arise when m is imaginary and, as before, such solutions arise for small scales for which $k > N/U$. However, from (7.195), evanescent solutions also arise for very large scales for which $k < f/U$. Propagating waves thus exist in the wavenumber and lengthscale ($L = 2\pi/k$) intervals

$$\frac{N}{U} > k > \frac{f}{U} \qquad \text{or} \qquad \frac{U}{N} < \frac{L}{2\pi} < \frac{U}{f}, \tag{7.196}$$

and these waves have frequencies between N and f. In an atmosphere with $U = 10\,\mathrm{m\,s^{-1}}$ and $f = 10^{-4}\,\mathrm{s^{-1}}$ the large scale at which evanescence reappears is $L = 2\pi U/f \approx 600\,\mathrm{km}$, which of course is not very large at all relative to global scales (and still smaller if we take $U = 5\,\mathrm{m\,s^{-1}}$). Thus, upward propagating gravity waves exist between scales of a few kilometres and several hundred kilometres. For the deep ocean, let us take $N = 10^{-3}\,\mathrm{s^{-1}}$, $f = 10^{-4}\,\mathrm{s^{-1}}$ and $U = 1\,\mathrm{cm\,s^{-1}}$. Then, very roughly, propagating waves exist between scales of a few tens of meters to a few hundred metres, and the hydrostatic regime (which requires $L \gg U/N$ and $L < 2\pi U/f$)) may be quite limited.

Wave solutions and energy propagation

Obtaining a wave solution in the rotating case follows a similar path to the non-rotating case. In the resting frame vertical velocity satisfies the boundary condition $w = U\partial h/\partial x$, and in the moving frame $w = \partial h/\partial t$. If U is constant we may use the polarization relations of Section 7.6.4, with $l = 0$ and $\omega = -Uk$, to obtain relations analogous to (7.178), and we find

$$w = \widetilde{w}(z)\mathrm{e}^{\mathrm{i}kx} = w_0 \mathrm{e}^{\mathrm{i}(kx+mz)} = \mathrm{i}Ukh_0 \mathrm{e}^{\mathrm{i}mz}\mathrm{e}^{\mathrm{i}kx}, \tag{7.197a}$$

$$u = \widetilde{u}(z)\mathrm{e}^{\mathrm{i}kx} = u_0 \mathrm{e}^{\mathrm{i}(kx+mz)} = -\mathrm{i}mUh_0 \mathrm{e}^{\mathrm{i}mz}\mathrm{e}^{\mathrm{i}kx}, \tag{7.197b}$$

$$\phi = \widetilde{\phi}(z)\mathrm{e}^{\mathrm{i}kx} = \phi_0 \mathrm{e}^{\mathrm{i}(kx+mz)} = \frac{\mathrm{i}m(U^2k^2 - f^2)}{k^2} h_0 \mathrm{e}^{\mathrm{i}mz}\mathrm{e}^{\mathrm{i}kx}, \tag{7.197c}$$

$$v = \widetilde{v}(z)\mathrm{e}^{\mathrm{i}kx} = v_0 \mathrm{e}^{\mathrm{i}(kx+mz)} = -\mathrm{i}f\frac{m}{k} h_0 \mathrm{e}^{\mathrm{i}mz}\mathrm{e}^{\mathrm{i}kx}, \tag{7.197d}$$

Of these, the expressions for w and u are no different from the non-rotating case, because w is set by the same boundary condition and u is given by mass continuity, $\partial u/\partial x + \partial w/\partial z = 0$, in both rotating and non-rotating cases. However, the solution now produces a meridional velocity, (7.197d), even when there is no variation in the topography in the y-direction, and to obtain it we use (7.151) with $l = 0$, giving $\widetilde{v} = -\mathrm{i}\widetilde{u}f\omega = \mathrm{i}\widetilde{u}f/Uk$.

As in the non-rotating case, when there are propagating waves there is high pressure on the windward (upstream) side of the topography and low pressure on the leeward side, and the phase lines tilt upstream with height. The drag on the flow is equal to the rate of upward momentum transport and using (7.197a,b) we obtain

$$\overline{uw} = -\frac{1}{2}kmU^2h_0^2 < 0. \tag{7.198}$$

There are some subtleties associated with the horizontal force across a wavy boundary that we shall not go into.[4] In any case, as in the non-rotating case a momentum flux divergence will only arise in the free atmosphere if dissipation occurs, for example if the waves break and/or if viscous effects become important. The vertical flux of energy density is given by

$$\overline{\phi w} = \frac{1}{2}U\frac{m}{k}h_0^2(U^2k^2 - f^2) > 0. \tag{7.199}$$

If m and k have the same sign then energy propagates away from the mountain, which as the reader may verify is consistent with the group velocity being directed upward.

Atmospheric and oceanic parameters

Are evanescent or propagating gravity waves more likely to be excited for typical atmospheric and oceanic parameters? Consider the atmosphere with a surface flow of $U = 10\,\mathrm{m\,s^{-1}}$ and $N = 10^{-2}\,\mathrm{s^{-1}}$. The critical wavenumber separating evanescent and propagating waves is then $k = N/U = 2 \times 10^{-3}$, corresponding to a wavelength of about 6000 m. Topographic features like the Rockies, Andes and Himalayas certainly contain such large wavelengths and so we can expect them to excite upward propagating gravity waves. For larger horizontal scales the flow is hydrostatic and the influence is felt directly above the mountain, as in Fig. 7.18. At still larger scales the waves are inhibited by the Coriolis effect, which causes evanescence for waves with a horizontal wavelength larger than a few hundred kilometres.

In the ocean the abyssal stratification is quite weak, typically with $N \sim 10^{-3}\,\mathrm{s^{-1}}$, and the velocities are also weak compared to those of the upper ocean, although they can be of order $1\,\mathrm{cm\,s^{-1}}$ in eddying regions. With $N = 10^{-3}$ and $U = 1\,\mathrm{cm\,s^{-1}}$ we find a critical wavelength of about 60 m. The ocean bathymetry obviously has many scales larger than this (and for smaller values of U the critical scales are correspondingly smaller) meaning that it is relatively easy for abyssal flow to generate gravity waves that propagate upward into the ocean interior, and as in the atmospheric case in many circumstances the flow will be hydrostatic and the influence will be directly above the bathymetry. However, as noted earlier, the Coriolis parameter will inhibit gravity waves of too large a wavelength — for $U = 1\,\text{cm s}^{-1}$ the horizontal scales must be less than $2\pi U/f$ or about 600 m. The upper ocean is much more greatly stratified, with $N \approx 10^{-2}\,\text{s}^{-1}$. Gravity waves are no longer generated by flow over topography but by the stirring effects of winds making a turbulent mixed layer. The forcing frequency must still be less than N in order to efficiently generate gravity waves, and using a velocity of $10\,\text{cm s}^{-1}$ we might heuristically estimate that propagating gravity waves can be generated with scales of tens of metres, with the Coriolis cut off now occurring at about 6 km. One message of these very rough calculations is that, in both atmosphere and ocean, gravity waves tend to have a smaller horizontal scale than Rossby waves.

7.8 ♦ ACOUSTIC-GRAVITY WAVES IN AN IDEAL GAS

In the final section of this chapter we consider wave motion in a stratified, *compressible* fluid such as the Earth's atmosphere. The stratification allows gravity waves to exist, and the compressibility allows sound waves to exist. The resulting problem is, not surprisingly, complicated and arcane and to make it as tractable as possible we will specialize to the case of an isothermal, stationary atmosphere and ignore the effects of rotation and sphericity. The results are not without interest, both in themselves and in illustrating the importance of simplifying the equations of motion from the outset, for example by making the Boussinesq or hydrostatic approximation, in order to isolate phenomena of interest.

In what follows we denote the unperturbed state with a subscript 0 and the perturbed state with a prime ($'$), and we omit some of the algebraic details. Because it is at rest, the basic state is in hydrostatic balance,

$$\frac{\partial p_0}{\partial z} = -\rho_0(z) g. \tag{7.200}$$

Ignoring variations in the y-direction for algebraic simplicity (and without loss of generality, in fact) the linearized equations of motion are:

$$u \text{ momentum:} \qquad \rho_0 \frac{\partial u'}{\partial t} = -\frac{\partial p'}{\partial x}, \tag{7.201a}$$

$$w \text{ momentum:} \qquad \rho_0 \frac{\partial w'}{\partial t} = -\frac{\partial p'}{\partial z} - \rho' g, \tag{7.201b}$$

$$\text{mass conservation:} \qquad \frac{\partial \rho'}{\partial t} + w' \frac{\partial \rho_0}{\partial z} = -\rho_0 \left(\frac{\partial u'}{\partial x} + \frac{\partial w'}{\partial z} \right), \tag{7.201c}$$

$$\text{thermodynamic:} \qquad \frac{\partial p'}{\partial t} - w' \frac{p_0}{H} = -\gamma p_0 \left(\frac{\partial u'}{\partial x} + \frac{\partial w'}{\partial z} \right). \tag{7.201d}$$

where $\gamma = c_p/c_v = 1/(1-\kappa)$. For an isothermal basic state we have $p_0 = \rho_0 R T_0$ where T_0 is a constant, so that $\rho_0 = \rho_s e^{-z/H}$ and $p_0 = p_s e^{-z/H}$ where $H = RT_0/g$. The thermodynamic equation, (7.201d), is the linear form of (1.101) on page 24, which has the equation of state for an ideal gas built-in, and (7.201a,b,c,d) thus form a complete set with variables u', w', ρ' and p'.

Differentiating (7.201a) with respect to time and using (7.201d) leads to

$$\left(\frac{\partial^2}{\partial t^2} - c_s^2 \frac{\partial^2}{\partial x^2} \right) u' = c_s^2 \left(\frac{\partial}{\partial z} - \frac{1}{\gamma H} \right) \frac{\partial}{\partial x} w', \tag{7.202a}$$

where $c_s^2 = \gamma p_0/\rho_0$ is the square of the speed of sound (equal to $\partial p/\partial \rho$ or $\gamma R T_0$). Similarly, differentiating (7.201b) with respect to time and using (7.201c) and (7.201d) leads to

$$\left(\frac{\partial^2}{\partial t^2} - c_s^2 \left[\frac{\partial^2}{\partial z^2} - \frac{1}{H} \frac{\partial}{\partial z} \right] \right) w' = c_s^2 \left(\frac{\partial}{\partial z} - \frac{\kappa}{H} \right) \frac{\partial u'}{\partial x}. \tag{7.202b}$$

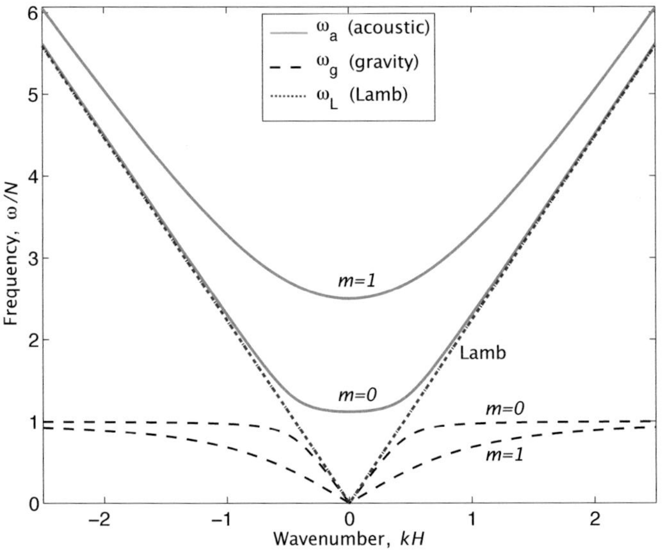

Fig. 7.19 Dispersion diagram for acoustic-gravity waves in an isothermal atmosphere, calculated using (7.208). The frequency is in units of buoyancy frequency N, and the wavenumbers are nondimensionalized by the inverse of the scale height, H. Solid curves indicate acoustic waves, whose frequency is always higher than that of the corresponding Lamb wave (i.e., ck), and of the base acoustic frequency $\approx 1.12N$. The dashed curves are internal gravity waves, whose frequency asymptotes to N at small horizontal scales.

Equations (7.202a) and (7.202b) combine to give, after some cancellation,

$$\frac{\partial^4 w'}{\partial t^4} - c_s^2 \frac{\partial^2}{\partial t^2}\left(\frac{\partial^2}{\partial x^2} + \frac{\partial^2}{\partial z^2} - \frac{1}{H}\frac{\partial}{\partial z}\right) w' - c_s^2 \frac{\kappa g}{H}\frac{\partial^2 w'}{\partial x^2} = 0. \tag{7.203}$$

If we set $w' = W(x,z,t)e^{z/(2H)}$, so that $W = (\rho_0/\rho_s)^{1/2} w'$, then the term with the single z-derivative is eliminated, giving

$$\frac{\partial^4 W}{\partial t^4} - c_s^2 \frac{\partial^2}{\partial t^2}\left(\frac{\partial^2}{\partial x^2} + \frac{\partial^2}{\partial z^2} - \frac{1}{4H^2}\right) W - c_s^2 \frac{\kappa g}{H}\frac{\partial^2 W}{\partial x^2} = 0. \tag{7.204}$$

Although superficially complicated, this equation has constant coefficients and we may seek wave-like solutions of the form

$$W = \mathrm{Re}\, \widetilde{W} e^{i(kx+mz-\omega t)}, \tag{7.205}$$

where $\widetilde{W}$ is the complex wave amplitude. Using (7.205) in (7.204) leads to the dispersion relation for acoustic-gravity waves, namely

$$\omega^4 - c_s^2 \omega^2 \left(k^2 + m^2 + \frac{1}{4H^2}\right) + c_s^2 N^2 k^2 = 0, \tag{7.206}$$

with solution

$$\omega^2 = \frac{1}{2} c_s^2 K^2 \left[1 \pm \left(1 - \frac{4N^2k^2}{c_s^2 K^4}\right)^{1/2}\right], \tag{7.207}$$

where $K^2 = k^2 + m^2 + 1/(4H^2)$. (The reader may investigate as to whether the factor $[1 - 4N^2k^2/(c_s^2K^4)]$ is positive.) For an isothermal, ideal gas atmosphere $4N^2H^2/c_s^2 \approx 0.8$ and so this may be written

$$\frac{\omega^2}{N^2} \approx 2.5\widehat{K}^2 \left[1 \pm \left(1 - \frac{0.8\widehat{k}^2}{\widehat{K}^4}\right)^{1/2}\right], \tag{7.208}$$

where $\widehat{K}^2 = \widehat{k}^2 + \widehat{m}^2 + 1/4$, and $(\widehat{k}, \widehat{m}) = (kH, mH)$.

7.8.1 Interpretation

Acoustic and gravity waves

There are two branches of roots in (7.207), corresponding to acoustic waves (using the plus sign in the dispersion relation) and internal gravity waves (using the minus sign). These (and the Lamb wave, described below) are plotted in Fig. 7.19. If $4N^2k^2/c_s^2K^4 \ll 1$ then the two sets of waves are well separated. From (7.208) this is satisfied when

$$\frac{4\kappa}{\gamma}(kH)^2 \approx 0.8(kH)^2 \ll \left[(kH)^2 + (mH)^2 + \frac{1}{4}\right]^2; \tag{7.209}$$

that is, when *either* $mH \gg 1$ or $kH \gg 1$. The two roots of the dispersion relation are then

$$\omega_a^2 \approx c_s^2 K^2 = c_s^2\left(k^2 + m^2 + \frac{1}{4H^2}\right) \tag{7.210}$$

and

$$\omega_g^2 \approx \frac{N^2k^2}{k^2 + m^2 + 1/(4H^2)}, \tag{7.211}$$

corresponding to acoustic and gravity waves, respectively. The acoustic waves owe their existence to the presence of compressibility in the fluid, and they have no counterpart in the Boussinesq system. On the other hand, the internal gravity waves are just modified forms of those found in the Boussinesq system, and if we take the limit $(kH, mH) \to \infty$ then the gravity wave branch reduces to $\omega_g^2 = N^2k^2/(k^2 + m^2)$, which is the dispersion relationship for gravity waves in the Boussinesq approximation. We may consider this to be the limit of infinite scale height or (equivalently) the case in which wavelengths of the internal waves are sufficiently small that the fluid is essentially incompressible.

Vertical structure

Recall that $w' = W(x, z, t)e^{z/(2H)}$ and, by inspection of (7.202), u' has the same vertical structure. That is,

$$w' \propto e^{z/(2H)}, \qquad u' \propto e^{z/(2H)}, \tag{7.212}$$

and the amplitude of the velocity field of the internal waves increases with height. The pressure and density perturbation amplitudes fall off with height, varying like

$$p' \propto e^{-z/(2H)}, \qquad \rho' \propto e^{-z/(2H)}. \tag{7.213}$$

The kinetic energy of the perturbation, $\rho_0(u'^2 + w'^2)$ is *constant* with height, because $\rho_0 = \rho_s e^{-z/H}$.

Hydrostatic approximation and Lamb waves

Equations (7.202) also admit to a solution with $w' = 0$. We then have

$$\left(\frac{\partial^2}{\partial t^2} - c_s^2\frac{\partial^2}{\partial x^2}\right)u' = 0 \qquad \text{and} \qquad \left(\frac{\partial}{\partial z} - \frac{\kappa}{H}\right)\frac{\partial u'}{\partial x} = 0, \tag{7.214}$$

and these have solutions of the form

$$u' = \mathrm{Re}\,\widetilde{U}e^{\kappa z/H}e^{i(kx-\omega t)}, \qquad \omega = ck, \tag{7.215}$$

where $\widetilde{U}$ is the wave amplitude. These are horizontally propagating sound waves, known as *Lamb waves* after the hydrodynamicist Horace Lamb. Their velocity perturbation amplitude increases with height, but the pressure perturbation falls with height; that is

$$u' \propto e^{\kappa z/H} \approx e^{2z/(7H)}, \qquad p' \propto e^{(\kappa-1)z/H} \approx e^{-5z/(7H)}. \tag{7.216}$$

Their kinetic energy density, $\rho_0 u'^2$, varies as

$$\mathrm{KE} \propto \mathrm{e}^{-z/H+2\kappa z/H} = \mathrm{e}^{(2R-c_p)z/(c_p H))]} = \mathrm{e}^{(R-c_v)z/(c_p H)} \approx \mathrm{e}^{-3z/(7H)}, \tag{7.217}$$

for an ideal gas. (In a simple ideal gas, $c_v = nR/2$ where n is the number of excited degrees of freedom, 5 for a diatomic molecule.) The kinetic energy density thus falls away exponentially from the surface, and in this sense Lamb waves are an example of edge waves or surface-trapped waves.

Now consider the case in which we make the hydrostatic approximation ab initio, but without restricting the perturbation to have $w' = 0$. The linearized equations are identical to (7.201), except that (7.201b) is replaced by

$$\frac{\partial p'}{\partial z} = -\rho' g. \tag{7.218}$$

The consequence of this is that first term $(\partial^2 w'/\partial t^2)$ in (7.202b) disappears, as do the first two terms in (7.203) (i.e., the terms $\partial^4 w'/\partial t^4 - c^2(\partial^2/\partial t^2)(\partial^2 w'/\partial x^2)$). It is a simple matter to show that the dispersion relation is then

$$\omega^2 = \frac{N^2 k^2}{m^2 + 1/(4H^2)}. \tag{7.219}$$

These are long gravity waves, and may be compared with the corresponding Boussinesq result (7.60). Again, the frequency increases without bound as the horizontal wavelength diminishes. The Lamb wave, of course, still exists in the hydrostatic model, because (7.214) is still a valid solution. Thus, horizontally propagating sound waves still exist in hydrostatic (primitive equation) models, but vertically propagating sound waves do not — essentially because the term $\partial w/\partial t$ is absent from the vertical momentum equation.

Notes

1 The book by Sutherland (2010) treats internal waves in some detail. I would like to thank S. Legg for encouraging me to add material on gravity waves and for her unpublished lecture notes.

2 See Durran (1990) for more discussion and, for a review, Durran (2015). I am also grateful to Dale Durran for many comments and corrections on the chapters on waves.

3 A treatment of this rather canonical profile was given by Queney (1948). The profile is named for Maria Agnesi, 1718–1799, an Italian mathematician and later a theologian, who had discussed the properties of the curve, as had Pierre de Fermat and Guido Grandi (a professor at the University of Pisa) somewhat earlier. The term 'witch' (in 'Witch of Agnesi') seems to be a mistranslation, inadvertent or otherwise, from Italian of versiera, which refers to a curve and not an adversary of God or a she-devil (which would be avversiera). Maria Agnesi may have been the first woman appointed to a professorship of mathematics, at the University of Bologna. The curve is similar to a Cauchy distribution and to a Lorentzian, and has some useful Fourier-transform properties that enable an analytic approach, although our solution is numerical.

4 Jones (1967) and Bretherton (1969) show that it is the quantity $\overline{\rho(u - f\eta)w}$, where η is the particle displacement parallel to the y-axis, that represents the force across a wavy boundary and that is constant with height, with an extra form drag arising because of the Coriolis force.

Changes in latitude, changes in attitude,
Nothing remains quite the same.
Jimmy Buffett, *Changes in Latitude*, 1997.

CHAPTER 8

Linear Dynamics at Low Latitudes

At low latitudes the atmosphere and ocean take on rather different characters than in mid-latitudes, and this chapter is our first taste of that. The tasting will be rather anodyne and mathematical, focusing on the linear dynamics of wave motion. We won't get into the real *phenomenology* of low latitudes: the tropical atmosphere with its humidity, its convection, and its towering cumulonimbus clouds, or the equatorial ocean with its undercurrents and countercurrents. And most certainly we don't get into low latitude atmosphere-ocean interaction and the wonderful phenomenon of El Niño — these all come later. Rather, this chapter is really just about the linear geophysical fluid dynamics of the shallow water equations at low latitudes, when the beta effect is important and the flow is not completely geostrophically balanced. Still, let us not be too deprecatory about these dynamics — they are important both in their own right and as prerequisites for these more complex phenomena that we encounter later.[1]

Wave motion at low latitudes can be more complicated than its mid-latitude counterpart. In mid-latitudes there is a fairly clear separation in the time and space scales between balanced and unbalanced motion, and it is useful to recognize this by explicitly filtering out gravity wave motion and considering purely balanced motion, using for example the quasi-geostrophic equations. In equatorial regions, where the Coriolis parameter can become very small and is zero at the equator, the Rossby number may be order unity or larger and such a separation is less useful. However, even as f becomes small, β becomes large and Rossby waves remain important but the frequency separation between Rossby and gravity waves is smaller. The reader may then readily imagine the complications arising even from linear wave problems in equatorial regions and determining the dispersion relation for combined Rossby and gravity waves in a continuously stratified fluid is an algebraically complex task. The task is much simplified by posing the problem in the context of the shallow water equations, and these arise both as a physical model (e.g., of the thermocline in the ocean) or via a modal expansion, as in Section 3.4.1.

Before diving into the details, ask why do we talk about the 'tropical' atmosphere but the 'equatorial' ocean? It is because an essential demarcation in the atmosphere lies at the edge of the Hadley Cell, at about 25°–30° latitude, and the dynamics are rather different poleward and equatorward of this edge. In some contrast, the dynamics of the ocean do not change their essential character until we approach quite close to the equator. At 10° latitude ocean dynamics has many of the characteristics of the mid-latitudes — the Rossby number is still very small, for example. Only when we get to within a very few degrees of the equator do the dynamics change in a qualitative way.

8.1 CO-EXISTENCE OF ROSSBY AND GRAVITY WAVES

To see how Rossby waves and gravity waves co-exist, consider the linear, single layer, rotating shallow water equations,

$$\frac{\partial u}{\partial t} - fv = -\frac{\partial \phi}{\partial x}, \qquad \frac{\partial v}{\partial t} + fu = -\frac{\partial \phi}{\partial y}, \qquad \frac{\partial \phi}{\partial t} + c^2\left(\frac{\partial u}{\partial x} + \frac{\partial v}{\partial y}\right) = 0, \qquad (8.1a,b,c)$$

where, in terms of the familiar shallow water variables, $\phi = g'\eta$ and $c^2 = g'H$, where ϕ is the kinematic pressure, η is the free surface height, H is the reference depth of the fluid and g' is the reduced gravity. After some manipulation (described in Section 8.2), but with no additional approximation, these equations reduce to a single equation for v, namely

$$\frac{1}{c^2}\frac{\partial^3 v}{\partial t^3} + \frac{f^2}{c^2}\frac{\partial v}{\partial t} - \frac{\partial}{\partial t}\nabla^2 v - \beta\frac{\partial v}{\partial x} = 0. \qquad (8.2)$$

On the beta plane the Coriolis parameter is given by $f = f_0 + \beta y$; thus, (8.2) has a non-constant coefficient, entailing considerable algebraic difficulties. We will address these difficulties in Section 8.2, but for now let us suppose that both β and f are constants in (8.2). Effectively, we are assuming that Coriolis parameter, f, is constant except where differentiated, an approximation common in a mid-latitude setting in the quasi-geostrophic equations.[2] This approximation provides a useful introduction to the more complex problem.

Equation (8.2) then has constant coefficients and we may look for plane wave solutions of the form $v = \widetilde{v}\exp[\mathrm{i}(\boldsymbol{k}\cdot\boldsymbol{x} - \omega t)]$, whence

$$\frac{\omega^2 - f_0^2}{c^2} - (k^2 + l^2) - \frac{\beta k}{\omega} = 0. \qquad (8.3a)$$

This is a cubic equation in ω, as expected given (8.1). Written differently it becomes

$$\left(k + \frac{\beta}{2\omega}\right)^2 + l^2 = \left(\frac{\beta}{2\omega}\right)^2 + \frac{\omega^2 - f_0^2}{c^2}, \qquad (8.3b)$$

which may be compared to (6.107). Noting that $k_d^2 = f_0^2/g'H = f_0^2/c^2$, the two equations are identical except for the appearance of a term involving frequency in the last term on the right-hand side of (8.3b). The wave propagation diagram is illustrated in Fig. 8.1. The wave vectors at a given frequency all lie on a circle centred at $(-\beta/2\omega, 0)$ and with radius R given by

$$R = \left[\left(\frac{\beta}{2\omega}\right)^2 + \frac{\omega^2 - f_0^2}{c^2}\right]^{1/2}, \qquad (8.4)$$

and the radius must be positive in order for the waves to exist. In the low frequency case the diagram is essentially the same as that shown in Fig. 6.8, but is quantitatively significantly different in the high frequency case. These limiting cases are discussed further in Section 8.1.1 below.

To plot the full dispersion relation it is useful to nondimensionalize using the following scales for time (T), distance (L) and velocity (U):

$$T = f_0^{-1}, \qquad L = L_d = k_d^{-1} = c/f_0, \qquad U = L/T = c. \qquad (8.5a,b,c)$$

Denoting nondimensional quantities with a hat we then have

$$\omega = \widehat{\omega} f_0, \qquad (k, l) = (\widehat{k}, \widehat{l})k_d, \qquad \beta = \widehat{\beta}\frac{f_0^2}{c} = \widehat{\beta}\frac{f_0}{L_d} = \widehat{\beta} f_0 k_d. \qquad (8.6)$$

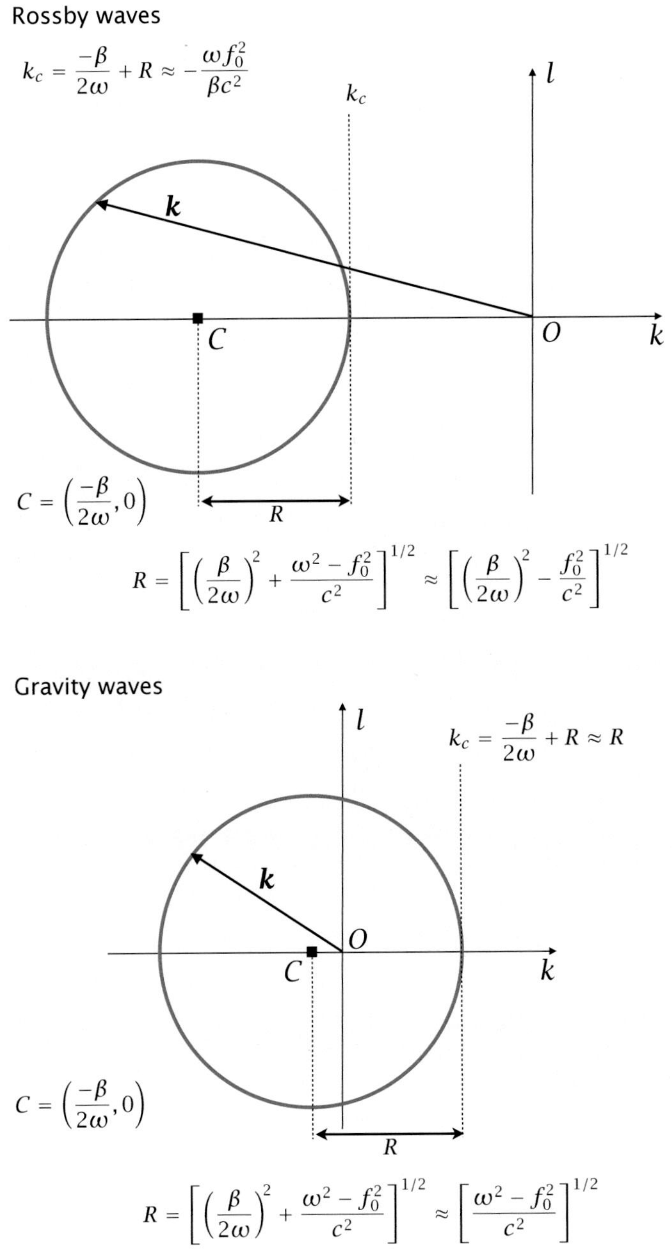
Rossby waves
$k_c = \frac{-\beta}{2\omega} + R \approx -\frac{\omega f_0^2}{\beta c^2}$
l
k_c
$\boldsymbol{k}$
C
O
k
$C = \left(\frac{-\beta}{2\omega}, 0\right)$
R
$R = \left[\left(\frac{\beta}{2\omega}\right)^2 + \frac{\omega^2 - f_0^2}{c^2}\right]^{1/2} \approx \left[\left(\frac{\beta}{2\omega}\right)^2 - \frac{f_0^2}{c^2}\right]^{1/2}$
Gravity waves
l
$k_c = \frac{-\beta}{2\omega} + R \approx R$
$\boldsymbol{k}$
O
C
k
$C = \left(\frac{-\beta}{2\omega}, 0\right)$
R
$R = \left[\left(\frac{\beta}{2\omega}\right)^2 + \frac{\omega^2 - f_0^2}{c^2}\right]^{1/2} \approx \left[\frac{\omega^2 - f_0^2}{c^2}\right]^{1/2}$

Fig. 8.1 Wave propagation diagrams for Rossby-gravity waves, obtained using (8.3). The top figure shows the diagram in the low frequency, Rossby wave limit, and the bottom figure shows the high frequency, gravity wave limit.

In each case the the locus of wavenumbers for a given frequency is a circle centred at $C = (-\beta/2\omega, 0)$ with a radius R given by (8.4), but the approximate expressions differ significantly at high and low frequency.

The dispersion relation (8.3) may then be written as

$$\hat{\omega}^2 - 1 - (\hat{k}^2 + \hat{l}^2) - \hat{\beta}\frac{\hat{k}}{\hat{\omega}} = 0. \tag{8.7}$$

We may expect that two of the roots correspond to gravity waves and the third to Rossby waves. The only parameter in the dispersion relation is $\hat{\beta} = \beta c / f_0^2 = \beta L_d / f_0$. In the atmosphere a representative value for L_d is 1000 km, whence $\hat{\beta} = 0.1$. In the ocean $L_d \sim 100$ km, whence $\hat{\beta} = 0.01$. If we allow ourselves to consider 'external' Rossby waves (which are of some oceanographic relevance) then $c = \sqrt{gH} = 200\,\mathrm{m\,s^{-1}}$ and $L_d = 2000$ km, whence $\hat{\beta} = 0.2$.

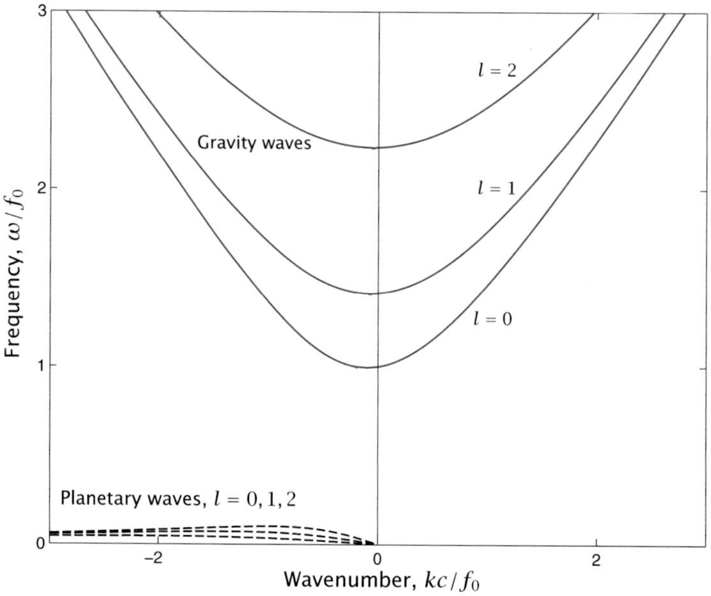

Fig. 8.2 Dispersion relation for Rossby-gravity waves, obtained from (8.8) with $\widehat{\beta} = 0.2$ for three values of l. There is a frequency gap between the Rossby or planetary waves and the gravity waves. For the stratified mid-latitude atmosphere or ocean the frequency gap is in reality even larger.

To actually obtain a solution we regard the equation as a quadratic in k and solve in terms of the frequency, giving

$$\widehat{k} = -\frac{\widehat{\beta}}{2\widehat{\omega}} \pm \frac{1}{2}\left[\frac{\widehat{\beta}^2}{\widehat{\omega}^2} + 4(\widehat{\omega}^2 - \widehat{l}^2 - 1)\right]^{1/2}. \tag{8.8}$$

The solutions are plotted in Fig. 8.2, with $\widehat{\beta} = 0.2$, and we see that the waves fall into two groups, labelled gravity waves and planetary waves in the figure. The gap between the two groups of waves is in fact still larger if a smaller (and generally more relevant) value of $\widehat{\beta}$ is used. To interpret all this let us consider some limiting cases.

8.1.1 Special Cases and Properties of the Waves

We now consider a few special cases of the dispersion relation.

(i) Constant Coriolis parameter

If $\beta = 0$ then the dispersion relation becomes

$$\omega\left[\omega^2 - f_0^2 - (k^2 + l^2)c^2\right] = 0, \tag{8.9}$$

with the roots

$$\omega = 0, \qquad \omega^2 = f_0^2 + c^2(k^2 + l^2). \tag{8.10a,b}$$

The root $\omega = 0$ corresponds to geostrophic motion (and, since $\beta = 0$, Rossby waves are absent), with the other root corresponding to Poincaré waves, considered in Chapter 3. The frequency is higher than the inertial frequency f; that is, $\omega^2 > f_0^2$.

(ii) High frequency waves

If we take the limit of $\omega \gg f_0$ then (8.3a) gives

$$\frac{\omega^2}{c^2} - (k^2 + l^2) - \frac{\beta k}{\omega} = 0. \tag{8.11}$$

Rossby and Gravity Waves

- Generically speaking, Rossby-gravity waves are waves that arise under the combined effects of a potential vorticity gradient and stratification. Sometimes the definition is restricted to a wave on a single branch of the dispersion curve connecting Rossby and gravity waves. In mid-latitudes on Earth Rossby waves and gravity waves are well separated and have distinct physical mechanisms.

- The simplest setting in which such waves occur is the linearized shallow water equations which may be written as a single equation for v, namely

$$\frac{1}{c^2}\frac{\partial^3 v}{\partial t^3} + \frac{f^2}{c^2}\frac{\partial v}{\partial t} - \frac{\partial}{\partial t}\nabla^2 v - \beta\frac{\partial v}{\partial x} = 0. \tag{RG.1}$$

- If we take both f and β to be constants then the equation above admits of plane-wave solutions with dispersion relation

$$\omega^2 - \frac{\beta k c^2}{\omega} = f_0^2 + c^2(k^2 + l^2). \tag{RG.2}$$

- In Earth's atmosphere and ocean it is common, especially in mid-latitudes, for there to be a frequency separation between two classes of solution. To a good approximation, high frequency waves satisfy

$$\omega^2 = f_0^2 + c^2(k^2 + l^2). \tag{RG.3}$$

These are inertio-gravity waves, also known as Poincaré waves. The low frequency waves satisfy

$$\omega = \frac{-\beta k c^2}{f_0^2 + c^2(k^2 + l^2)} = \frac{-\beta k}{k_d^2 + k^2 + l^2}, \tag{RG.4}$$

where $k_d^2 = f_0^2/c^2$, and these are called Rossby waves or planetary waves.

- Rossby-gravity waves also exist in the stratified equations. Solutions may be found by decomposing the vertical structure into a series of orthogonal modes, and a sequence of shallow water equations for each mode results, with a different c for each mode. Solutions may also be found if f is allowed to vary in (RG.1), at the price of some algebraic complexity, as in Section 8.2.

To be physically realistic we should also now eliminate the β term, because if $\omega \gg f_0$ then, from geometric considerations on a sphere, $k^2 \gg \beta k/\omega$. Thus, the dispersion relation is simply $\omega^2 = c^2(k^2 + l^2)$. These waves are just gravity waves uninfluenced by rotation, and are a special case of Poincaré waves.

(iii) Low frequency waves

Consider the limit of $\omega \ll f_0$. The dispersion relation reduces to

$$\omega = \frac{-\beta k}{k^2 + l^2 + k_d^2}. \tag{8.12}$$

This is just the dispersion relation for quasi-geostrophic Rossby waves as previously obtained — see (6.65) or (6.92). In this limit, the requirement that the radius of the circle be positive becomes

$$\omega^2 < \frac{\beta^2}{4k_d^2}. \tag{8.13}$$

That is to say, the Rossby waves have a maximum frequency, and directly from (8.12) this occurs when $k = k_d$ and $l = 0$.

The frequency gap

The maximum frequency of Rossby waves is usually much less than the frequency of the Poincaré waves: the lowest frequency of the Poincaré waves is f_0 and the highest frequency of the Rossby waves is $\beta/2k_d$. Thus,

$$\frac{\text{Low gravity wave frequency}}{\text{High Rossby wave frequency}} = \frac{f_0}{\beta/2k_d} = \frac{f_0^2}{2\beta c}. \tag{8.14}$$

If $f_0 = 10^{-4}\,\mathrm{s}^{-1}$, $\beta = 10^{-11}\,\mathrm{m}^{-1}\,\mathrm{s}^{-1}$ and $k_d = 1/100\,\mathrm{km}^{-1}$ (a representative oceanic baroclinic deformation radius) then $f_0/(\beta/2k_d) = 200$. If $L_d = 1000\,\mathrm{km}$ (an atmospheric baroclinic radius) then the ratio is 20. If we use a barotropic deformation radius of $L_d = 2000\,\mathrm{km}$ then the ratio is 10. Evidently, for most mid-latitude applications there is a large gap between the Rossby wave frequency and the gravity wave frequency. Because of this frequency gap, to a good approximation Fig. 8.2 may be obtained by separately plotting (8.10b) for the gravity waves, and (8.12) for the Rossby or planetary waves. The differences between these and the exact results become smaller as $\widehat{\beta}$ gets smaller, virtually indistinguishable in the plots shown.

Finally, we remark that a 'Rossby-gravity wave' is sometimes defined to be the wave on a single branch of the dispersion curve that connects Rossby waves and gravity waves across a range of wavenumbers. The equatorial beta plane does support such a wave — the 'Yanai wave' that will be derived in Section 8.2 and shown in Fig. 8.6, and this wave bridges the frequency gap near the equator. However, in the mid-latitude system above there is no such wave; rather, there are what are essentially separate Rossby waves and gravity waves.

8.1.2 Planetary-Geostrophic Rossby waves

A good approximation for the large-scale ocean circulation involves ignoring the time-derivatives and nonlinear terms in the momentum equation, allowing evolution only to occur in the thermodynamic equation. This is the planetary-geostrophic approximation, introduced in Section 5.2, and it is interesting to see to what extent that system supports Rossby waves.[3] It is easiest just to begin with the linear shallow water equations themselves, and omitting time derivatives in the momentum equation gives

$$-fv = -\frac{\partial\phi}{\partial x}, \qquad fu = -\frac{\partial\phi}{\partial y}, \tag{8.15a,b}$$

$$\frac{\partial\phi}{\partial t} + c^2\left(\frac{\partial u}{\partial x} + \frac{\partial v}{\partial y}\right) = 0. \tag{8.15c}$$

From these equations we straightforwardly obtain

$$\frac{\partial\phi}{\partial t} - \frac{c^2\beta}{f^2}\frac{\partial\phi}{\partial x} = 0. \tag{8.16}$$

Again we will treat both f and β as constants so that we may look for solutions in the form $\phi = \widetilde{\phi}\exp[\mathrm{i}(\boldsymbol{k}\cdot x - \omega t)]$. The ensuing dispersion relation is

$$\omega = -\frac{c^2\beta}{f_0^2}k = -\frac{\beta k}{k_d^2}, \tag{8.17}$$

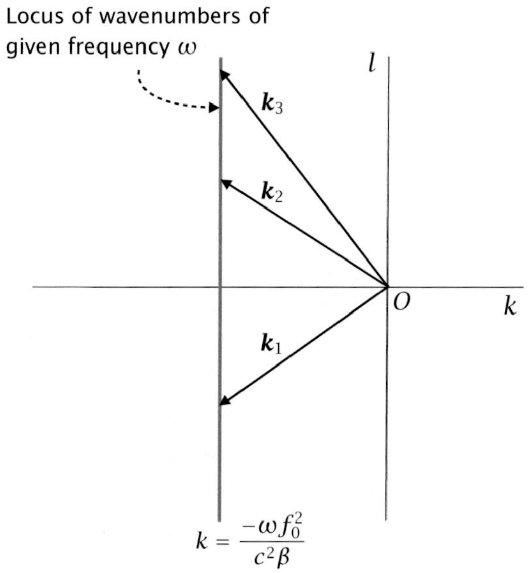

Fig. 8.3 The locus of points on planetary-geostrophic Rossby waves. Waves of a given frequency all have the same x-wavenumber, given by (8.17).

which is a limiting case of (8.12) with $k^2, l^2 \ll k_d^2$. The waves are a form of Rossby waves with phase and group speeds given by

$$c_p = -\frac{c^2\beta}{f_0^2}, \qquad c_g^x = -\frac{c^2\beta}{f_0^2}. \tag{8.18}$$

That is, the waves are non-dispersive and propagate westward. Equation (8.16) has the general solution $\phi = G(x + \beta c^2/f^2 t)$, where G is any function, so an initial disturbance will just propagate westward at a speed given by (8.18), without any change in form.

Note finally that the locus of wavenumbers in k–l space is no longer a circle, as it is for the usual Rossby waves. Rather, since the frequency does not depend on the y-wavenumber, the locus is a straight line, parallel to the y-axis, as in Fig. 8.3. Waves of a given frequency all have the same x-wavenumber, given by $k = -\omega f_0^2/(c^2\beta) = -\omega k_d^2/\beta$, as shown in Fig. 8.3.

Physical mechanism

Because the waves *are* a form of Rossby wave their physical mechanism is related to that discussed in Section 6.4.3, but with an important difference: relative vorticity is no longer important, but the flow divergence is. Thus, consider flow round a region of high pressure, as illustrated in Fig. 8.4. If the pressure is circularly symmetric as shown, the flow to the south of H in the left-hand sketch, and to the south of L in the right-hand sketch, is larger than that to the north. Hence, in the left sketch the flow converges at W and diverges at E, and the flow pattern moves westward. In the flow depicted in the right sketch the low pressure propagates westward in a similar fashion.

8.2 WAVES ON THE EQUATORIAL BETA PLANE

We now discuss the properties of shallow water waves at low latitudes, allowing the Coriolis parameter to properly vary in all terms, albeit using the β-plane approximation.[4] Thus, Taylor-expanding the Coriolis parameter around a latitude ϑ_0 we obtain

$$f = 2\Omega \sin\vartheta \approx 2\Omega \sin\vartheta_0 + 2\Omega(\vartheta - \vartheta_0)\cos\vartheta_0 = f_0 + \beta y, \tag{8.19}$$

where $f_0 = 2\Omega \sin\vartheta_0$, $\beta = 2\Omega\cos\vartheta_0/a$ and $y = a(\vartheta - \vartheta_0)$ where a is the radius of the Earth. For motions at low latitudes we take $\vartheta_0 = 0$, giving the *equatorial beta-plane approximation* in which

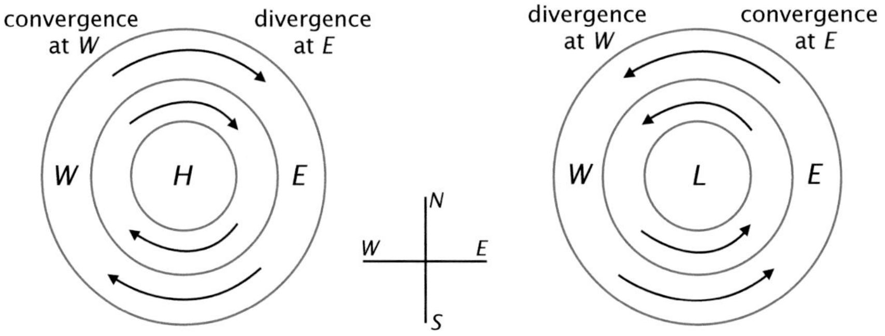

Fig. 8.4 The westward propagation of planetary-geostrophic Rossby waves. The circular lines are isobars centred around high and low pressure centres. Because of the variation of the Coriolis force, the mass flux between two isobars is greater to the south of a pressure centre than it is to the north. Hence, in the left-hand sketch there is convergence to the west of the high pressure and the pattern propagates westward. Similarly, if the pressure centre is a low, as in the right-hand sketch, there is divergence to the west of the pressure centre and the pattern still propagates westward.

$\sin\vartheta \approx \vartheta$, $\cos\vartheta \approx 1$ and $f = 2\Omega\vartheta = \beta y$. The linearized momentum and mass conservation equations are then

$$\frac{\partial u}{\partial t} - fv = -\frac{\partial \phi}{\partial x}, \qquad \frac{\partial v}{\partial t} + fu = -\frac{\partial \phi}{\partial y}, \qquad \frac{\partial \phi}{\partial t} + c^2\left(\frac{\partial u}{\partial x} + \frac{\partial v}{\partial y}\right) = 0. \tag{8.20a,b,c}$$

These are exactly the same as (8.1) except that now $f = \beta y$. There are just two dimensional parameters in the above equations, c and β, and from these we may form the scales

$$T_{eq} = (c\beta)^{-1/2}, \qquad L_{eq} = \left(\frac{c}{\beta}\right)^{1/2}. \tag{8.21a,b}$$

These are the fundamental time and length scales for equatorial dynamics (although some definitions differ by a factor of $\sqrt{2}$, as discussed later). The length scale L_{eq} is known as the equatorial radius of deformation. If we regard the above shallow water equations as coming from a modal decomposition of the primitive equations, as in Section 3.4, then there is a wave speed, $c_m = \sqrt{gH_m}$ and deformation radius for each mode. For the first baroclinic mode and for the atmosphere, if $c_1 = 25\,\mathrm{m\,s^{-1}}$ and $\beta = 2.3\times10^{-11}\,\mathrm{m^{-1}\,s^{-1}}$, then $L_{eq} \approx 1000\,\mathrm{km}$ and $T_{eq} \approx 0.5$ days; for the equatorial ocean with $c = 2\,\mathrm{m\,s^{-1}}$ then $L_{eq} \approx 300\,\mathrm{km}$ and $T_{eq} \approx 1.7$ days.

Cross-differentiating (8.20a) and (8.20b) and using (8.20c) to eliminate the divergence we may also derive the linearized potential vorticity equation, namely

$$\frac{\partial}{\partial t}\left(\zeta - \frac{f\phi}{c^2}\right) + \beta v = 0. \tag{8.22}$$

This is the same as the familiar linearized potential vorticity equation on the f-plane, with the addition of the term $\mathrm{D}f/\mathrm{D}t = \beta v$. Equation (8.22) is not independent of (8.20) but it will be convenient to use it sometimes.

To obtain a single equation for a single unknown, operate on (8.20a) with $(f/c^2)\partial_t$, on (8.20b) with $(1/c^2)\partial_{tt}$, on (8.20c) with $(1/c^2)\partial_{ty}$) and on (8.22) with ∂_x. Using subscripts to denote derivatives the resulting equations are

$$\frac{f}{c^2}u_{tt} - \frac{f^2}{c^2}v_t = -\frac{f}{c^2}\phi_{xt}, \qquad \frac{1}{c^2}v_{ttt} + \frac{f}{c^2}u_{tt} = -\frac{1}{c^2}\phi_{ytt}, \tag{8.23a,b}$$

$$\frac{1}{c^2}\phi_{tty} + (u_{xyt} + v_{yyt}) = 0, \qquad v_{xxt} - u_{xyt} - \frac{f}{c^2}\phi_{xt} + \beta v_x = 0. \tag{8.23c,d}$$

These equations linearly combine (a - (b+c+d)) to give a single equation for v, namely

$$\frac{1}{c^2}\frac{\partial^3 v}{\partial t^3} + \frac{f^2}{c^2}\frac{\partial v}{\partial t} - \frac{\partial}{\partial t}\left(\frac{\partial^2 v}{\partial y^2} + \frac{\partial^2 v}{\partial x^2}\right) - \beta\frac{\partial v}{\partial x} = 0. \tag{8.24}$$

This equation is third order and has non-constant coefficients, and is thus somewhat complicated. Before proceeding, note one common approximation, sometimes called the *longwave approximation*. If zonal scales are much greater than meridional scales then we expect the zonal wind to be in geostrophic balance with the meridional pressure gradient. In this case we replace (8.20b) by

$$fu = -\frac{\partial \phi}{\partial y}, \tag{8.25}$$

and (8.22) and (8.23b,d) are modified accordingly. Then, in instead of (8.24), we obtain

$$\frac{f^2}{c^2}\frac{\partial v}{\partial t} - \frac{\partial}{\partial t}\left(\frac{\partial^2 v}{\partial y^2}\right) - \beta\frac{\partial v}{\partial x} = 0. \tag{8.26}$$

This equation is first order in time and the dispersion relation may be obtained reasonably straightforwardly. This approximation is particularly useful in the forced-dissipative problem as we will see in Section 8.4. In the free problem the dispersion equation can in fact be obtained easily enough in the general case, that is from (8.24), allowing us to make the longwave approximation at a later stage.

8.2.1 Dispersion Relations

In this section we explore the properties of (8.24), in particular obtaining a dispersion relation. The coefficients of (8.24) vary in the meridional direction but are constant in the zonal direction. We thus search for solutions in the form of a plane wave in the zonal direction only and we let

$$v = \tilde{v}(y)e^{i(kx-\omega t)}, \tag{8.27}$$

and assume boundary conditions of $\tilde{v}(y) \to 0$ as $y \to \pm\infty$. Substituting (8.27) into (8.24) gives

$$\frac{d^2\tilde{v}}{dy^2} + \left(\frac{\omega^2}{c^2} - k^2 - \frac{\beta k}{\omega} - \frac{\beta^2 y^2}{c^2}\right)\tilde{v} = 0. \tag{8.28}$$

Given the velocity, c, and the presence of the beta effect there is a rather obvious way to nondimensionalize the equations. However, it turns out that by introducing an additional factor of $\sqrt{2}$ into the scaling, the mathematics of one of the problems that we address later is simplified. At the risk of discussing a trivial difference, let's do both — the confident and impatient reader may choose one and skim the other.

Nondimensionalization I

Let us scale time and distance with the quantities $T_{eq} = (c\beta)^{-1/2}$, $L_{eq} = (c/\beta)^{1/2}$ as in (8.21). The nondimensional frequency, lengthscale and wavenumber are then given by

$$\hat{\omega} = \frac{\omega}{(\beta c)^{1/2}}, \qquad \hat{y} = y\left(\frac{\beta}{c}\right)^{1/2}, \qquad \hat{k} = k\left(\frac{c}{\beta}\right)^{1/2}. \tag{8.29}$$

If we take $\delta\rho/\rho_0 = 0.002$, $H = 100$ m and $\beta = 2\Omega/a = 2.3 \times 10^{-11}\,\mathrm{m^{-1}\,s^{-1}}$ we find

$$g' \approx 0.02\,\mathrm{m\,s^{-2}}, \quad c \approx 1.4\,\mathrm{m\,s^{-1}}, \quad L_{eq} \approx 250\,\mathrm{km}, \quad T_\beta = 1.7 \times 10^5\,\mathrm{s} \approx 2\,\mathrm{days}. \tag{8.30}$$

The mid-latitude shallow-water deformation radius, L_d is usually defined as $L_d = c/f$ which differs from (8.21b) most notably in the power of f. However, if in the mid-latitude expression we take $f = \beta y$, as if near the equator, and $y = L_d$, then $L_d = c/(\beta L_d)$, which is the same as (8.21b).

Substituting (8.29) into (8.28) gives the slightly simpler-looking equation

$$\frac{\mathrm{d}^2\widehat{v}}{\mathrm{d}\widehat{y}^2} + \left(\widehat{\omega}^2 - \widehat{k}^2 - \frac{\widehat{k}}{\widehat{\omega}} - \widehat{y}^2\right)\widehat{v} = 0. \tag{8.31}$$

This equation may be put into a standard form[5] by writing $v(\widehat{y}) = \Psi(\widehat{y})\exp(-\widehat{y}^2/2)$, whence (8.31) becomes

$$\frac{\mathrm{d}^2\Psi}{\mathrm{d}\widehat{y}^2} - 2\widehat{y}\frac{\mathrm{d}\Psi}{\mathrm{d}\widehat{y}} + \lambda\Psi = 0, \tag{8.32}$$

where $\lambda = \widehat{\omega}^2 - \widehat{k}^2 - \widehat{k}/\widehat{\omega} - 1$. Equation (8.32) is known as *Hermite's equation,* and it is an eigenvalue equation, with solutions if and only if $\lambda = 2m$, for $m = 0, 1, 2, \ldots$. The solutions are Hermite polynomials, $\Psi(\widehat{y}) = H_m(\widehat{y})$, where the first few polynomials are given by

$$\begin{aligned} &H_0 = 1, \qquad H_1 = 2\widehat{y}, \qquad H_2 = 4\widehat{y}^2 - 2, \\ &H_3 = 8\widehat{y}^3 - 12\widehat{y}, \qquad H_4 = 16\widehat{y}^4 - 48\widehat{y}^2 + 12. \end{aligned} \tag{8.33}$$

A Hermite polynomial is even or odd when m is even or odd, respectively; that is $H_m(-\widehat{y}) = (-1)^m H_m(\widehat{y})$. Note that we are using Hermite polynomials to describe the v field, so that mirror symmetry across the equator occurs when m is odd. The v field is then odd, but the u and ϕ fields are even, as shown in Appendix A to this chapter.

Hermite polynomials multiplied by a Gaussian are a form of *parabolic cylinder function,*

$$V_m(y) = H_m(y)\exp(-y^2/2). \tag{8.34}$$

These functions are also orthogonal in the interval $[-\infty, +\infty]$; that is

$$\int_{-\infty}^{\infty} V_n V_m \,\mathrm{d}y = \int_{-\infty}^{\infty} H_n(y) H_m(y)\exp(-y^2)\,\mathrm{d}y = \sqrt{\pi}2^n n!\,\delta_{nm}, \tag{8.35}$$

Appendix A gives additional details. Given the Hermite solution for Ψ, the solutions for v are given by

$$v(\widehat{y}) = V_m(\widehat{y}) = H_m(\widehat{y})\mathrm{e}^{-\widehat{y}^2/2}, \qquad m = 0,\, 1,\, 2\ldots, \tag{8.36}$$

and so decay exponentially as $\widehat{y} \to \pm\infty$ (as we require) with a decay scale of the equatorial deformation radius $\sqrt{c/\beta}$. The functions V_m are plotted in Fig. 8.5 for $m = 0$ to 3.

The dispersion relation follows from the quantization condition $\lambda = 2m$, which implies

$$\widehat{\omega}^2 - \widehat{k}^2 - \frac{\widehat{k}}{\widehat{\omega}} = 2m + 1, \tag{8.37a}$$

or, using (8.29), the dimensional form,

$$\omega^2 - c^2k^2 - \beta\frac{kc^2}{\omega} = (2m+1)\beta c., \tag{8.37b}$$

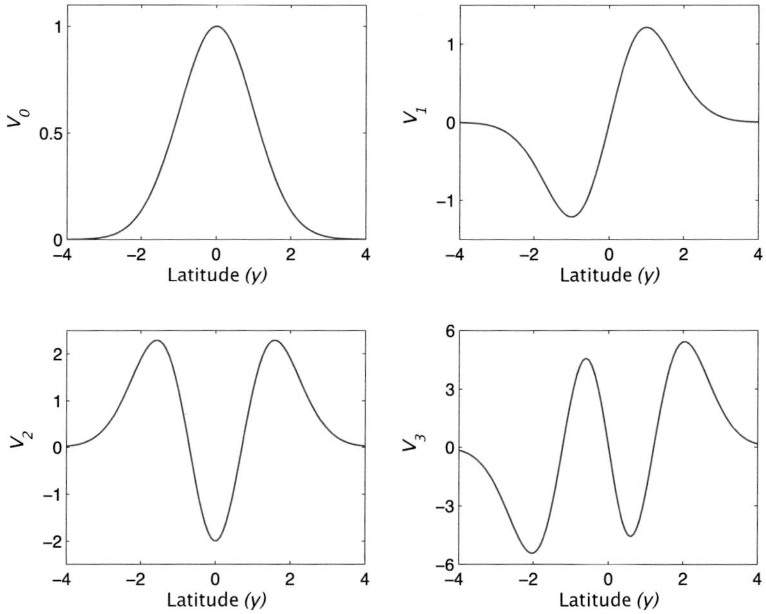

Fig. 8.5 Latitudinal variation of the wave amplitudes, the parabolic cylinder functions $V_m(y)$, given by (8.34), for $m = 0, 1, 2, 3$. The parameter m is analogous to a meridional wavenumber. The parabolic cylinder functions given by (8.43) have a similar but not identical form.

This is a cubic equation in ω, and although a solution is possible, it is easier to solve the quadratic equation for the wavenumber in terms of the frequency giving

$$\widehat{k} = -\frac{1}{2\widehat{\omega}} \pm \frac{1}{2}\left[\left(\frac{1}{\widehat{\omega}} - 2\widehat{\omega}\right)^2 - 8m\right]^{1/2}, \tag{8.37c}$$

or, in dimensional form,

$$k = -\frac{\beta}{2\omega} \pm \frac{1}{2}\left[\left(\frac{\beta}{\omega} - \frac{2\omega}{c}\right)^2 - \frac{8m\beta}{c}\right]^{1/2}. \tag{8.37d}$$

Equations (8.37) are forms of the dispersion relation for the shallow water equations on an equatorial beta-plane. Before exploring their properties we nondimensionalize in a different way.

Nondimensionalization II

We now scale time and distance with the quantities

$$T_{eq} = (2c\beta)^{-1/2}, \qquad L_{eq} = (c/2\beta)^{1/2}. \tag{8.38a,b}$$

Velocity is still nondimensionalized by c. The nondimensional version of (8.28) becomes

$$\frac{\mathrm{d}^2\widetilde{v}}{\mathrm{d}\widehat{y}^2} + \left(\widehat{\omega}^2 - \widehat{k}^2 - \frac{\widehat{k}}{2\widehat{\omega}} - \frac{\widehat{y}^2}{4}\right)\widetilde{v} = 0, \tag{8.39}$$

which may be compared with (8.31). We now make the substitution

$$\widetilde{v}(\widehat{y}) = \Phi \exp(-\widehat{y}^2/4), \tag{8.40}$$

which leads to

$$\frac{\mathrm{d}^2\Phi}{\mathrm{d}\widehat{y}^2} - \widehat{y}\frac{\mathrm{d}\Phi}{\mathrm{d}\widehat{y}} + \gamma\Phi = 0, \tag{8.41}$$

where $\gamma = \widehat{\omega}^2 - \widehat{k}^2 - \widehat{k}/2\widehat{\omega} - 1/2$. Equation (8.41) could be transformed into (8.32) by changing to the independent variable $y' = \widehat{y}/\sqrt{2}$, and the dispersion relation then follows in the same way. More directly, solutions of (8.41) are given by the modified Hermite polynomials $\Phi(\widehat{y}) = G_m(\widehat{y})$ where

$$(G_0, G_1, G_2, G_3, G_4) = (1, \widehat{y}, \widehat{y}^2 - 1, \widehat{y}^3 - 3\widehat{y}, \widehat{y}^4 - 6\widehat{y}^2 + 3). \tag{8.42}$$

These are also known as the probabilists' Hermite polynomials, with (8.33) or (8.151) being the physicists' Hermite polynomials, reflecting historical use; the two sets of polynomials are connected by $H_n(y) = 2^{n/2} G_n(y\sqrt{2})$. The corresponding parabolic cylinder functions are given by

$$D_n(\widehat{y}) = G_n(\widehat{y}) \exp(-\widehat{y}^2/4), \tag{8.43}$$

and these functions are solutions of (8.39). The orthonormality condition on the modified polynomials is that

$$\int_{-\infty}^{\infty} D_n(y) D_m(y) \, dy = \int_{-\infty}^{\infty} G_n(y) G_m(y) \exp(-y^2/2) \, dy = \sqrt{2\pi}\, n! \; \delta_{nm}, \tag{8.44}$$

which may be compared to (8.35). The quantization condition on γ is that $\gamma = m$, where $m = 0, 1, 2, \ldots$. Thus, the nondimensional dispersion relation is

$$\widehat{\omega}^2 - \widehat{k}^2 - \frac{\widehat{k}}{2\widehat{\omega}} - \frac{1}{2} = m, \tag{8.45}$$

and restoring the dimensions using (8.38) gives (8.37b). Later on, when dealing with the steady, forced-dissipative problem, the use of the probabilists' polynomials is a little more convenient.

8.2.2 Limiting and Special Cases

To further explore the wave case we stay with our first nondimensionalization, namely (8.21), and with the goal of figuring out what's going on we'll consider various special cases of the dispersion relations (8.37). It is convenient to first partition the waves by frequency, and consider separately high frequency gravity waves and low frequency planetary waves. We need do this only for the case $m \geq 1$ because the $m = 0$ case (mixed Rossby-gravity waves) may be treated exactly. Then finally we look at the so-called $m = -1$ case, namely Kelvin waves.

High and low frequency waves

1. *High frequency waves.* The term $\beta k c^2/\omega$ in (8.37) is small and may be neglected. The dispersion relation becomes

$$\widehat{\omega}^2 = \widehat{k}^2 + 2m + 1 \qquad \text{or} \qquad \omega^2 = c^2 k^2 + \beta c(2m + 1). \tag{8.46a,b}$$

   This dispersion relation is similar to that of mid-latitude Poincaré waves, with βc replacing f_0^2: recall the form of (3.121), namely $\omega^2 = c^2(k^2 + l^2) + f_0^2$. Waves satisfying (8.46) are thus sometimes called equatorially trapped Poincaré waves or equatorially trapped gravity waves.

   The approximation requires that $\omega \gg \beta/|k|$, and is somewhat inaccurate for small k: note that (8.46) is symmetric around $k = 0$, whereas the full dispersion relation, plotted in Fig. 8.6, is offset. (Formally, the limit is valid for $\widehat{k} \to \infty$, $\widehat{\omega} \to \infty$ and $\widehat{k}/\omega =$ constant.)

   For finite m the limiting case at high wavenumbers is just $\widehat{\omega} = \pm\widehat{k}$, or, in dimensional form, $\omega = \pm ck$. This is just the dispersion relation for familiar conventional shallow water gravity waves, unaffected by rotation and the β-effect. However, in the rotating case the waves are trapped at the equator and propagate only in the zonal direction, albeit both eastward and westward.

2. *Low frequency waves.* For low frequency waves we neglect the term involving ω^2 in (8.37) and the dispersion relation becomes

$$\widehat{\omega} = \frac{-\widehat{k}}{2m + 1 + \widehat{k}^2}, \qquad \omega = \frac{-\beta k}{(2m+1)\beta/c + k^2}, \tag{8.47}$$

nondimensionally and dimensionally, respectively. This is recognizable as the dispersion relation for a zonally propagating Rossby wave with large x-wavenumber, and these waves are called equatorially trapped Rossby waves, or equatorially trapped planetary waves. We may further consider two limits of these waves, as follows.

(a) *Short, low frequency waves, with* $\widehat{k} \to \infty$, $\widehat{\omega} \to 0$. The dispersion relation becomes

$$\widehat{\omega} = -\frac{1}{\widehat{k}}, \qquad \omega = -\frac{\beta}{k}. \tag{8.48}$$

The phase speed and group velocity in this limit are given by, dimensionally,

$$c_p = -\frac{\beta}{k^2}, \qquad c_g = \frac{\beta}{k^2}. \tag{8.49}$$

Thus, the phase speed is westward but the group velocity, and so the direction of energy propagation, is eastward.

(b) *Long low frequency waves, with* $\widehat{k} \to 0$, $\widehat{\omega} \to 0$. The dispersion relation (8.37) becomes, in nondimensional and dimensional form,

$$\widehat{\omega} = \frac{-\widehat{k}}{2m+1}, \qquad \omega = \frac{-ck}{2m+1}. \tag{8.50}$$

These represent westward propagating waves whose speed is given by $c/(2m+1)$. For $m = 1$ (the smallest allowable value for planetary waves) they have one-third the speed of a non-rotating gravity wave or of a Kelvin wave (discussed below). However, these waves propagate only westward, and they match with the westward propagating planetary waves derived above as wavenumber increases. They are conveniently nondispersive, and are also important near western boundaries where they superpose to create western boundary currents. The longwave approximation may be made from the outset, and is equivalent to assuming that the zonal flow is in geostrophic balance; that is, (8.20b) is replaced by $fu = -g'\partial\eta/\partial y$. Then, instead of solving (8.24) we solve (8.26). The only difference is in the value of λ in (8.32) — we find $\lambda = -\widehat{k}/\widehat{\omega} - 1$ — and so (8.50) immediately emerges. Short waves are filtered out of the system. This approximation will turn out to be particularly important when we consider the steady problem in Section 8.5.

There is a distinct gap in frequencies between the minimum frequency of the gravity waves, given by (8.46), and the maximum frequency of the planetary waves, given by (8.49) also with m small. The minimum gravity wave frequency occurs when $k = 0$ and is $\omega_{gmin}^2 = \beta c(2m+1)$. From (8.47) the maximum planetary wave frequency occurs when $k^2 = (2m+1)\beta/c$ and gives $\omega_{pmax}^2 = \beta c/[4(2m+1)]$. The ratio of these two frequencies is

$$\frac{\omega_{gmin}}{\omega_{pmax}} = 2(2m+1), \tag{8.51}$$

giving a value of six for $m = 1$ and two for $m = 0$ (a case we consider more below). Note that this ratio is *independent* of the values of the physical parameters β and c. Although the gap is distinct, it is not as large as the corresponding gap at mid-latitudes, which may be an order of magnitude or more.

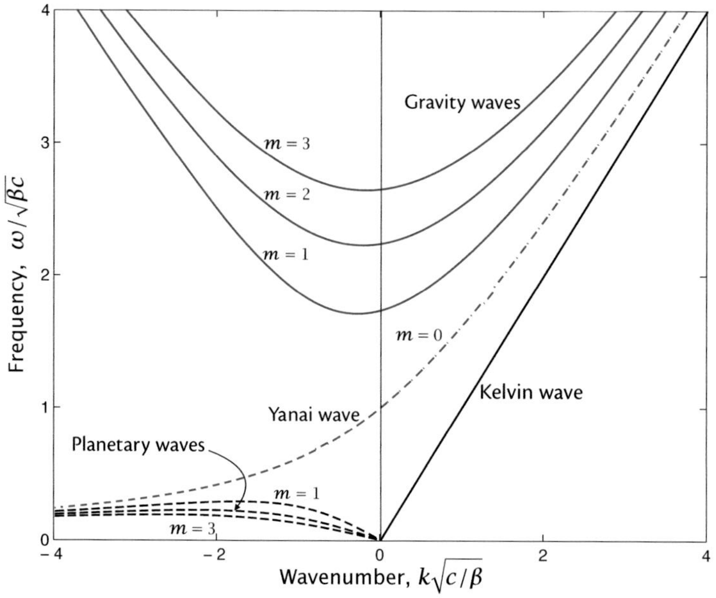

Fig. 8.6 Dispersion relation for equatorial waves, as given by (8.37), for $m = 0, 1, 2, 3$. The upper group of curves are gravity waves, given approximately by (8.46). The lower group with $k < 0$ are westward propagating planetary waves, given approximately by (8.47). Also shown are the Yanai wave with $m = 0$, satisfying (8.54), and the eastward propagating Kelvin wave (the '$m = -1$' wave) satisfying $\omega = ck$ for $k \geq 0$.

Special values of m

In addition to the limiting cases at low and high frequency, there are two other cases in which we can readily solve the dispersion relation, namely when $m = 0$ and the Kelvin wave case, as follows:

1. *The case with* $m = 0$. The resulting waves are known as *Yanai waves*,[6] or *Rossby-gravity* waves, since they span the two types of waves. They are antisymmetric across the equator. From (8.37a) the dispersion relation simplifies to

$$\widehat{k} = -\widehat{\omega} \qquad \text{or} \qquad \widehat{k} = -\frac{1}{\widehat{\omega}} + \widehat{\omega}. \tag{8.52a,b}$$

or dimensionally

$$k = -\frac{\omega}{c}, \qquad k = -\frac{\beta}{\omega} + \frac{\omega}{c}. \tag{8.53a,b}$$

The case $k = -\omega/c$ is non-physical, for it represents a gravity wave moving westward. Such a wave grows without bound as $|y|$ increases away from the equator, as we demonstrate explicitly in the discussion on Kelvin waves below. The physically realizable case, (8.53b), has the explicit dispersion relation

$$\omega = \frac{kc}{2} \pm \frac{1}{2}\sqrt{k^2c^2 + 4\beta c}. \tag{8.54}$$

Again it is useful to consider various limiting cases:

- $k = 0$. In this case (8.54) gives $\omega = \sqrt{\beta c}$ and there is a balance between the two terms on the right-hand side of (8.53b). Note that in Fig. 8.6 the Yanai wave at $k = 0$ intercepts the ordinate at a value of nondimensional frequency of 1.
- $k \to +\infty$. In this case $\omega = ck$, with a balance between the left-hand side and the second term on the right-side of (8.53b). Evidently, this corresponds to eastward propagating gravity waves.
- $k \to -\infty$. In this case, because ω must be positive, we have $\omega = -\beta/k$, and a balance between the left-hand side and the first term on the right-side of (8.53b). The waves are westward propagating Rossby or planetary waves.

Yanai waves, therefore, are mixed Rossby-gravity waves: the phase of the Rossby wave propagates westward (like all Rossby waves) and has a low frequency, and the gravity wave propagates eastward (and only eastward, unlike conventional gravity waves). The group velocity of Yanai waves is positive in all cases, being given by, from (8.53b),

$$c_g^x \equiv \frac{\partial \omega}{\partial k} = \frac{\omega^2 c}{\beta c + \omega^2}. \tag{8.55}$$

The group velocity of the full problem is, from (8.37),

$$c_g^x = \frac{c^2 \omega(\beta + 2\omega k)}{2\omega^3 + \beta k c^2}. \tag{8.56}$$

This may be positive or negative, and vanishes when $\omega = -\beta/2k$.

2. *Kelvin waves, or the 'm = −1' case.* (This section may be considered to be an extension of Section 3.8.3.) In general, Hermite's equation, (8.32), has solutions when m is a positive integer or zero. However, there is a class of waves that happens to satisfy the dispersion relation (8.37) with $m = -1$, namely equatorial Kelvin waves. These waves have identically zero meridional velocity and so their equations of motion are

$$\frac{\partial u}{\partial t} = -g'\frac{\partial \eta}{\partial x}, \qquad fu = -g'\frac{\partial \eta}{\partial y}, \qquad \frac{\partial \eta}{\partial t} + H\frac{\partial u}{\partial x} = 0, \tag{8.57a,b,c}$$

where $f = \beta y$. The zonal velocity is in geostrophic balance with the meridional pressure gradient, and (8.57a) and (8.57c) give the classic wave equation,

$$\frac{\partial^2 u}{\partial t^2} - c^2\frac{\partial^2 u}{\partial x^2} = 0, \tag{8.58}$$

where $c = \sqrt{g'H}$ as before, and so the dispersion relation $\omega = \pm ck$. This is, in fact, a solution of (8.37) with $m = -1$, as may easily be checked.

The solution to (8.58), and the corresponding solution for η, is

$$u = F_1(x + ct, y) + F_2(x - ct, y), \qquad \eta = \left(\frac{H}{g'}\right)^{1/2}\left[-F_1(x + ct, y) + F_2(x - ct, y)\right], \tag{8.59}$$

where F_1 and F_2 are arbitrary functions, representing waves travelling westwards and eastwards, respectively. We obtain the y-dependence of these functions by using (8.57b) giving

$$\beta y F_1 = c\frac{\partial F_1}{\partial y}, \qquad \beta y F_2 = -c\frac{\partial F_2}{\partial y}. \tag{8.60}$$

The solutions of these equations are

$$F_1 = F(x + ct)\exp[y^2/(2L_{eq}^2)], \qquad F_2 = G(x - ct)\exp[-y^2/(2L_{eq}^2)], \tag{8.61a,b}$$

where F and G are the amplitudes at $y = 0$. Evidently, F_1 increases without bound away from the equator, and so this solution must be eliminated. The complete solution is thus:

$$u = G(x - ct)\exp[-y^2/(2L_{eq}^2)], \qquad \eta = \frac{H}{c}u, \qquad v = 0, \tag{8.62}$$

with dispersion relation

$$\omega = ck. \tag{8.63}$$

These waves are equatorially trapped Kelvin waves. They propagate eastward only, without dispersion, and their amplitude decays away from the equator in precisely the same way as the other equatorial waves considered above, and in a slightly different way from the Kelvin waves on the f-plane given by (3.135).

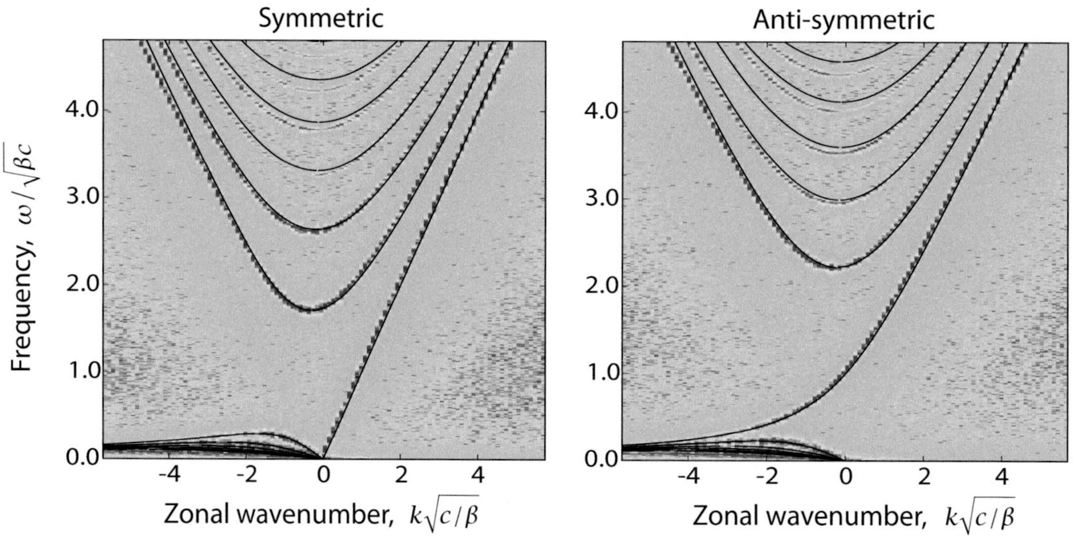

Fig. 8.7 Power spectra from a numerical simulation of the shallow water equations on the sphere(colour shading, with red the most intense), with the analytic dispersion relation for equatorial Rossby and gravity waves overlaid (solid black lines, as in Fig. 8.6). The left panel shows the symmetric component, obtained by adding Northern and Southern Hemispheres and with only the odd values of m plotted analytically, and the right panel plots the antisymmetric component and the even values of m.

8.2.3 A Numerical Illustration

After the many mathematical manipulations above, the reader, if like the author, may well be sceptical that such waves do actually exist, especially on a sphere where Kelvin waves are not exactly realizable. To assuage this doubt, Fig. 8.7 shows the power spectrum from a numerical simulation of the nonlinear shallow water equations over the full sphere. The height field is initialized with small random perturbations everywhere and the system allowed to freely evolve, with no damping except for that residing in the numerical scheme.[7] The figure shows the resulting power spectrum, over a region from 15° S to 15° N, from the near-statistical equilibrium state that emerges. The equatorial waves emerge with beautiful clarity above the noise, with only small deviations for the highest modes due to resolution issues with the numerics that slow the waves. To see a simulation showing Rossby and Kelvin waves in physical space look ahead to Fig. 22.18. The real world is never quite so limpid, but Fig. 18.23 and Fig. 18.24 will suggest that the waves are not merely figments of the imathination.

8.2.4 Why do Kelvin Waves have a Preferred Direction of Travel?

Both equatorial and coastal Kelvin waves have a preferred direction of travel: equatorial Kelvin waves move eastward and, consistently, coastal Kelvin waves travel such that they have a wall to their right in the Northern Hemisphere and to their left in the Southern Hemisphere. Why?

Consider the linear zonal momentum and mass continuity equations,

$$\frac{\partial u}{\partial t} = -g'\frac{\partial \eta}{\partial x}, \qquad \frac{\partial \eta}{\partial t} = -H\frac{\partial u}{\partial x}. \tag{8.64}$$

Looking for wavelike solutions of the form $(u, h) = (\tilde{u}, \tilde{\eta})e^{i(kx-\omega t)}$ we obtain $\tilde{u} = g'\tilde{\eta}/c$ and $c\tilde{\eta} = H\tilde{u}$. This means that under the crests of fluid (i.e., positive values of η) u has the same sign as c; the parcels of fluid are moving in the same direction as the phase of the wave. This property is also apparent if one considers how the fluid must move in order that the troughs and crests progress in

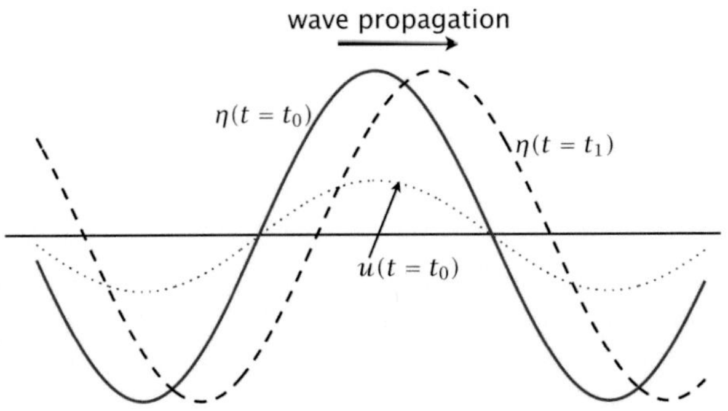

Fig. 8.8 A shallow water gravity wave, showing the fluid interface at an initial and later time $\eta(t_0)$ and $\eta(t_1)$, and the fluid velocity at the initial time, $u(t_0)$. The fluid flow is in the same direction as the phase speed (positive in this example) under the fluid crests, and is in the opposite direction under the troughs.

a a particular direction, as illustrated in Fig. 8.8. This property holds for shallow water waves quite generally, and is not restricted to Kelvin waves.

Now we add rotation and restrict attention to Kelvin waves. In the direction perpendicular to the direction of travel of the wave, the flow is in geostrophic balance:

$$fu = -g' \frac{\partial \eta}{\partial y}. \tag{8.65}$$

Consider the flow under a fluid crest in an equatorial Kelvin wave, as illustrated in Fig. 8.9. The pressure gradient force is directed away from the equator and, if the wave is travelling eastward the pressure force can balanced by the Coriolis force directed toward the equator. Under a trough the fluid is flowing in the opposite direction to the wave itself, and both the pressure gradient force and the Coriolis force are reversed and geostrophic balance still holds. If the wave were to travel westwards, no such balances could be achieved.

Very similar reasoning holds for coastal Kelvin waves, with a cross-wave pressure gradient supported by a wall. Geostrophic balance can now be maintained only if the wall is to the right of the direction of travel in the Northern Hemisphere (where $f > 0$) and to the left in the Southern Hemisphere (where $f < 0$).

8.2.5 Potential Vorticity Dynamics of Equatorial Rossby Waves

The Rossby waves and Rossby-gravity waves derived above are rather similar to their mid- latitude counterparts, which can be derived from a balanced potential vorticity equation without involving unbalanced dynamics at all. Can we do something similar for equatorial Rossby waves? The answer is yes, although the method is a little ad hoc.[8] Kelvin waves and inertia-gravity waves are filtered out, but Rossby waves and Rossby-gravity waves are reproduced in a way that transparently illuminates their dynamics.

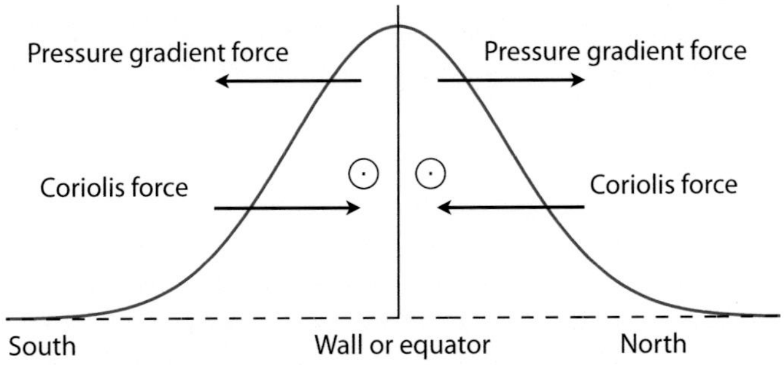

Fig. 8.9 Balance of forces across a Kelvin wave. The solid line is the fluid surface and the phase speed is directed out of the page. Beneath a crest the fluid flow is in the direction of the phase speed and produces Coriolis forces as shown, so balancing the pressure gradient forces. If the wave were travelling in the opposite direction no such geostrophic balance could be achieved.

Let us begin with the unforced linearized potential vorticity equation, which, to remind ourselves, is

$$\frac{\partial}{\partial t}\left(\zeta - \frac{f\phi}{c^2}\right) + \beta v = 0, \tag{8.66}$$

where, as before, $f = \beta y$. Let us now suppose that the divergence is small and the flow close to geostrophic balance so that the velocity, vorticity and height fields can all be written in terms of a streamfunction,

$$u = -\frac{\partial \psi}{\partial y}, \qquad v = \frac{\partial \psi}{\partial x}, \qquad \zeta = \nabla^2\psi, \qquad \phi = f\psi. \tag{8.67}$$

This is similar to what is done in the quasi-geostrophic approximation, except that here the Coriolis parameter is allowed to vary, with $f = \beta y$. Equation (8.67) is best regarded as an ansatz — an approximation or assumption made for convenience — for it has not been rigorously justified.

Using (8.67) in (8.66) gives

$$\frac{\partial}{\partial t}\left(\nabla^2\psi - \frac{f^2\psi}{c^2}\right) + \beta\frac{\partial \psi}{\partial x} = 0. \tag{8.68}$$

We can seek wavelike solutions of this in the form

$$\psi = \widetilde{\psi}(y)\mathrm{e}^{\mathrm{i}(kx-\omega t)}, \tag{8.69}$$

and (8.68) becomes

$$\frac{\mathrm{d}^2\widetilde{\psi}}{\mathrm{d}y^2} - \left(k^2 + \frac{\beta k}{\omega} + \frac{\beta^2 y^2}{c^2}\right)\widetilde{\psi} = 0. \tag{8.70}$$

This is almost the same as (8.28) except for the replacement of $\widetilde{v}$ by $\widetilde{\psi}$ and the absence of the ω^2 term in the bracketed expression. Since meridional velocity is just $\partial\psi/\partial x \propto k\psi$ the meridional velocity obeys the same equation as $\widetilde{\psi}$, and the absence of ω^2 arises because we are in the low-frequency limit. We thus simply repeat the development following (8.28) and obtain a dispersion relation similar to (8.37b) but without the ω^2 term, to wit

$$\omega = \frac{-\beta k}{(2m+1)\beta/c + k^2}. \tag{8.71}$$

This is the same as the dispersion relation for low frequency waves discussed in Section 8.2.2. The balanced system (8.68) thus *exactly* reproduces the Rossby waves and Rossby-gravity waves in the low frequency limit. We are not able to recover the behaviour of Kelvin waves by this methodology because such waves are essentially non-balanced: in the meridional direction the Coriolis force balances the height field, as in (8.65), but in the zonal direction there is a balance between the zonal acceleration and the pressure gradient.

8.3 RAY TRACING AND EQUATORIAL TRAPPING

We have seen that equatorial waves are trapped near the equator. What then happens to a wave that initially propagates in a direction away from the equator? The waves must either change their character completely, or be refracted back toward the equator. The former can only happen if there exists a class of mid-latitude waves with similar frequency and wavenumber; otherwise no such waves can be excited and the waves must, if they are not absorbed, bend back if energy is to be conserved. Let us explore this using some ideas from ray theory, as discussed in Section 6.3.

8.3.1 Dispersion Relation and Ray Equations

Consider again the wave equation of motion for the meridional velocity, (8.24). We seek solutions of the form $v = \tilde{v}(y)e^{i(kx-\omega t)}$, giving

$$\frac{d^2\tilde{v}}{dy^2} + \left(\frac{\omega^2}{c^2} - k^2 - \frac{\beta k}{\omega} - \frac{\beta^2 y^2}{c^2}\right)\tilde{v} = 0. \tag{8.72}$$

If the term in brackets is positive then sinusoidal-like solutions in y are possible, but if the term is negative, which will occur for y larger than some critical value y_c, then the physically realizable solutions decay exponentially with y; that is, wavelike solutions are trapped between two critical latitudes. Using the dispersion relation (8.37), Equation (8.72) becomes

$$\frac{d^2\tilde{v}}{dy^2} + \left(\frac{(2m+1)\beta}{c} - \frac{\beta^2 y^2}{c^2}\right)\tilde{v} = 0, \tag{8.73}$$

and therefore the critical latitudes are given by

$$y_c = \pm\left(\frac{\omega^2}{\beta^2} - \frac{c^2k^2}{\beta^2} - \frac{c^2k}{\beta\omega}\right)^{1/2} = \left((2m+1)\frac{c}{\beta}\right)^{1/2}, \tag{8.74}$$

For $k = 0$, and so for meridionally propagating waves, the critical latitudes are given by $y_c = \omega/\beta$, and at the critical latitude $\omega = f$. The waves are therefore trapped within their *inertial latitudes*, the latitudes at which their frequency is f. For larger k the critical latitudes are correspondingly smaller.

To explore this phenomenon using ray theory we assume that the medium is varying sufficiently slowly that it is possible to find wavelike solutions with spatially varying y-wavenumbers. We write (8.72) as

$$\frac{d^2\tilde{v}}{dy^2} + l^2(y)\tilde{v} = 0, \tag{8.75}$$

$$l^2(y) = \frac{\omega^2}{c^2} - k^2 - \frac{\beta k}{\omega} - \frac{\beta^2 y^2}{c^2} = \frac{\beta^2}{c^2}(y_c^2 - y^2)m \tag{8.76}$$

where $l(y)$ is assumed to vary slowly in the WKB sense (see Appendix A on page 247). The WKB solution is

$$\tilde{v}(y) = l^{-1/2}\exp\left(\pm i\int l\,dy\right). \tag{8.77}$$

The trajectory of the waves is determined by the ray paths — paths that are parallel to the direction of the group velocity — so that their trajectory, $x(t)$, $y(t)$ is given by

$$\frac{dx}{dt} = c_g^x, \qquad \frac{dy}{dt} = c_g^y \quad \text{and} \quad \frac{dy}{dx} = \frac{c_g^y}{c_g^x}. \tag{8.78}$$

Using the dispersion relation (8.76) gives

$$\frac{\partial\omega}{\partial l} = \frac{2\omega^2 lc^2}{2\omega^3 + \beta kc^2}, \tag{8.79}$$

and using this and (8.56) gives the slope of the ray in the x–y plane,

$$\frac{dy}{dx} = \frac{c_g^y}{c_g^x} = \frac{l}{k + \beta/(2\omega)}. \tag{8.80}$$

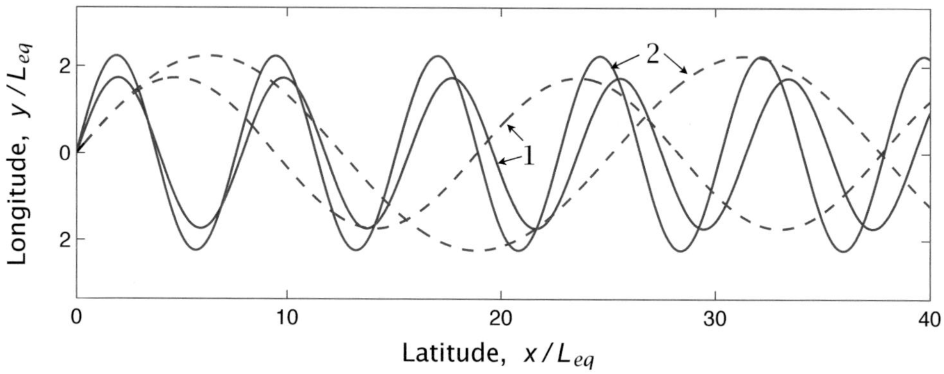

Fig. 8.10 Rays in the equatorial waveguide calculated using (8.83). The dashed lines show planetary wave trajectories and the solid lines are gravity wave trajectories, with $m = 1, 2$ (numbers marked on the graph) and $\widehat{k} = 1$. The turning latitude for each wave is $(2m+1)^{1/2} L_{eq}$, where $L_{eq} = \sqrt{c/\beta}$.

Using the expression for l given by (8.76) we can write this in terms of y instead of l, so that

$$\frac{\mathrm{d}y}{\mathrm{d}x} = \frac{\beta(y_c^2 - y^2)^{1/2}}{kc + \beta c/(2\omega)}. \tag{8.81}$$

Using the standard result that

$$\int \frac{\mathrm{d}y}{(y_c^2 - y^2)^{1/2}} = \sin^{-1}\frac{y}{y_c}, \tag{8.82}$$

we finally obtain

$$y = y_c \sin\left[\frac{\beta x}{ck + \beta c/(2\omega)}\right], \qquad \widehat{y} = (2m+1)^{1/2} \sin\left[\frac{\widehat{x}}{\widehat{k} + 1/(2\widehat{\omega})}\right]. \tag{8.83}$$

where the second expression is the nondimensional form. The ray path is therefore a sinusoid moving along the equator; the waves are confined to a *waveguide* centred at the equator and with a polewards extent of $y = \pm y_c$, as in Fig. 8.10. Equation (8.83) holds for both planetary and gravity waves, and for the latter the term $\beta c/(2\omega)$ may be neglected.

8.4 ✦ FORCED-DISSIPATIVE WAVELIKE FLOW

We now consider linear equatorial dynamics in the presence of forcing and damping. (A somewhat simpler treatment of this issue is given in Section 22.7 and some readers may wish to delay considering the problem until then. The important special case of the forced, steady problem is treated in Section 8.5, and it is also possible to skip to that section and refer back here as needed.) If forcing is present then damping is needed so that a steady state can be reached, and the simplest form is a linear drag. From a physical perspective the presence of such a drag is the most unsatisfactory aspect of our treatment, for it has no real physical justification especially as, for mathematical reasons, the drag must be the same for momentum and height (implying a frictional spindown time equal to a radiative spindown time). Nevertheless, unresolved small scale processes often do act as some form of damping and a linear damping is the simplest form. We consider the full problem initially and then special cases.[9]

The dimensional linear forced-dissipative equations of motion are

$$\frac{\partial u}{\partial t} + \alpha u - fv + \frac{\partial \phi}{\partial x} = F^x, \qquad \frac{\partial v}{\partial t} + \alpha v + fu + \frac{\partial \phi}{\partial y} = F^y, \tag{8.84a,b}$$

$$\frac{\partial \phi}{\partial t} + \alpha\phi + c^2\left(\frac{\partial u}{\partial x} + \frac{\partial v}{\partial y}\right) = -Q, \tag{8.84c}$$

where F^x and F^y are the x and y components of the imposed forces, Q is a thermal or mass source and α is a damping coefficient, assumed the same for all three variables. If we interpret $\boldsymbol{F} = (F^x, F^y)$ as wind stress, $\boldsymbol{\tau}$, acting on a layer of fluid we might make the association of $\boldsymbol{F} = \boldsymbol{\tau}/\rho_0 H$. Still, for now we will treat this system simply as a problem in geophysical fluid dynamics. The potential vorticity equation corresponding to (8.84), obtained by cross-differentiating (8.84a) and (8.84b), is

$$\left[\frac{\partial}{\partial t} + \alpha\right]\left(\zeta - \frac{f}{c^2}\phi\right) + \beta v = \mathrm{curl}_z \boldsymbol{F} + \frac{fQ}{c^2}. \tag{8.85}$$

In much the same way as we derived (8.24) we can derive a single partial differential equation for v, namely

$$\begin{aligned}
\frac{1}{c^2}\left[\frac{\partial}{\partial t} + \alpha\right]^3 v + \frac{f^2}{c^2}\left[\frac{\partial}{\partial t} + \alpha\right] v - \left[\frac{\partial}{\partial t} + \alpha\right]\left(\frac{\partial^2 v}{\partial y^2} + \frac{\partial^2 v}{\partial x^2}\right) - \beta\frac{\partial v}{\partial x} \\
= \frac{1}{c^2}\left[\frac{\partial}{\partial t} + \alpha\right]\frac{\partial Q}{\partial y} - \frac{f}{c^2}\frac{\partial Q}{\partial x} \\
+ \frac{1}{c^2}\left[\frac{\partial}{\partial t} + \alpha\right]^2 F^y - \frac{f}{c^2}\left[\frac{\partial}{\partial t} + \alpha\right] F^x - \frac{\partial}{\partial x}\left(\frac{\partial F^y}{\partial x} - \frac{\partial F^x}{\partial y}\right).
\end{aligned} \tag{8.86}$$

The left-hand side is a minor variation of that of (8.24). This equation is obviously very complicated and perhaps not very attractive. Although the equation might be solved by similar methods to those used on (8.24) (or solved numerically) we will proceed in a slightly more informative way, with two differences:

(i) We consider only special cases of (8.84). For example, we simplify (8.84b) to geostrophic balance, $fu = -\partial\phi/\partial y$, and in Section 8.5 we will pay particular attention to the steady version of the equations.

(ii) We will change variables from (u, v, ϕ) to a set denoted (q, r, v), defined below, that allow an easier connection to be made between v and the variables u and ϕ.

8.4.1 Mathematical Development

For algebraic convenience we introduce the following linear combinations of u and ϕ,

$$q \equiv \frac{\phi}{c} + u, \qquad r \equiv \frac{\phi}{c} - u. \tag{8.87}$$

Note that u and ϕ have the same natural symmetry across the equator, with both symmetric unless forcing deems otherwise, whereas v tends to be antisymmetric. The u-momentum and the height (ϕ) equations may be written as equations for q and r, namely

$$\left(\frac{\partial}{\partial t} + \alpha\right) q + c\frac{\partial q}{\partial x} + c\frac{\partial v}{\partial y} - fv = F^x - Q/c, \tag{8.88a}$$

$$\left(\frac{\partial r}{\partial t} + \alpha\right) r - c\frac{\partial r}{\partial x} + c\frac{\partial v}{\partial y} + fv = -F^x - Q/c, \tag{8.88b}$$

and the v-momentum equation becomes

$$\left(\frac{\partial}{\partial t} + \alpha\right) v + \frac{f}{2}(q - r) = -\frac{c}{2}\frac{\partial}{\partial y}(q + r) + F^y. \tag{8.88c}$$

Nondimensionalization

We scale velocity by c and time and distance by

$$T_{eq} = (c2\beta)^{-1/2}, \qquad L_{eq} = (c/2\beta)^{1/2}. \tag{8.89a,b}$$

The nondimensional equations of motion are then

$$\left(\frac{\partial}{\partial t} + \alpha\right) q + \frac{\partial q}{\partial x} + \frac{\partial v}{\partial y} - \frac{1}{2} y v = F^x - Q, \tag{8.90a}$$

$$\left(\frac{\partial}{\partial t} + \alpha\right) r - \frac{\partial r}{\partial x} + \frac{\partial v}{\partial y} + \frac{1}{2} y v = -F^x - Q, \tag{8.90b}$$

$$\left(\frac{\partial}{\partial t} + \alpha\right) v + \frac{y}{4}(q - r) = -\frac{1}{2}\frac{\partial}{\partial y}(q + r) + F^y. \tag{8.90c}$$

To avoid clutter here we do not use special notation for nondimensional variables.

The solutions of these equations may be expressed in terms of parabolic cylinder functions, $D_n(y)$. That is, we seek solutions of the form

$$(v, q, r) = \sum_{n=0}^{\infty} \left(v_n(x,t), q_n(x,t), r_n(x,t)\right) D_n(y). \tag{8.91}$$

with the forcing terms expanded in a similar fashion. The parabolic cylinder functions themselves have the form

$$(D_0, D_1, D_2, D_3) = (1, y, y^2 - 1, y^3 - 3y) \exp(-y^2/4), \tag{8.92}$$

and so on. The polynomial terms are just the modified Hermite polynomials $G_n(y)$ given by (8.42). The parabolic cylinder functions obey the ladder properties that

$$\frac{dD_n}{dy} + \frac{1}{2} y D_n = n D_{n-1}, \qquad \frac{dD_n}{dy} - \frac{1}{2} y D_n = -D_{n+1}. \tag{8.93a,b}$$

If we substitute (8.91) into (8.90) we obtain ordinary differential equations for the amplitudes. From the q equation, (8.90a), we obtain, after a little algebra,

$$\left(\frac{\partial}{\partial t} + \alpha\right) q_0 + \frac{\partial q_0}{\partial x} = F_0^x - Q_0, \tag{8.94a}$$

$$\left(\frac{\partial}{\partial t} + \alpha\right) q_{n+1} + \frac{\partial q_{n+1}}{\partial x} - v_n = F_{n+1}^x - Q_{n+1}, \qquad n = 0, 1, 2, 3, \ldots. \tag{8.94b}$$

From the r equation, (8.90b), we find

$$\left(\frac{\partial}{\partial t} + \alpha\right) r_{n-1} - \frac{\partial r_{n-1}}{\partial x} + n v_n = -\left(F_{n-1}^x + Q_{n-1}\right), \qquad n = 1, 2, 3, \ldots, \tag{8.95}$$

and from the v equation, (8.90c), we find

$$\left(\frac{\partial}{\partial t} + \alpha\right) v_0 + \frac{q_1}{2} = F_0^y, \tag{8.96a}$$

$$\left(\frac{\partial}{\partial t} + \alpha\right) v_n + \frac{(n+1)}{2} q_{n+1} - \frac{r_{n-1}}{2} = F_n^y, \qquad n = 1, 2, 3, \ldots. \tag{8.96b}$$

Finally, we note without derivation that these equations may be combined to give

$$\left[\frac{\partial}{\partial t} + \alpha\right]^3 v_n + \left[\frac{\partial}{\partial t} + \alpha\right]\left((2n+1)v_n - \frac{\partial^2 v_n}{\partial x^2}\right) - \frac{\partial v_n}{\partial x} = G, \tag{8.97}$$

where G is a combination of the various forcing terms. This equation can be most easily derived by substituting (8.91) into the nondimensional form of (8.86).

In principle, the above equations provide a means of solving the problem for almost any forcing. The equations have constant coefficients and may be solved by a superposition of harmonic functions in the x-direction, in conjunction with the variation in the y-direction given by the parabolic cylinder functions. In general this procedure would be somewhat tedious and uninformative. Thus, and to avoid being asphyxiated by an avalanche of algebra, we will consider some special cases. Enthusiasts may continue with the general development by themselves. (We might also note that modern geophysical fluid dynamics has advanced by way of using numerical methods to find solutions to complicated equations, in conjunction with using analytic methods to find solutions of simplified cases or to find general relations.) We will discuss two particularly important problems later, in Sections 8.5 and 22.7,and for the rest of this section we content ourselves with some general comments about forced waves.[10]

8.4.2 ✦ Forced Waves

Consider the problem of forced waves in which we retain some of the forcing terms but neglect the damping. Our purpose is to show what kinds of waves might be excited and to help interpret (8.94)–(8.97). When $\alpha = 0$ Equation (8.97) becomes, in dimensional form,

$$\frac{\partial}{\partial t}\left[\frac{1}{c^2}\frac{\partial^2 v_n}{\partial t^2} - \frac{\partial^2 v_n}{\partial x^2} + (2n+1)\frac{\beta}{c}v_n\right] - \beta\frac{\partial v_n}{\partial x} = G. \tag{8.98}$$

From (8.98) the dispersion relation for free waves, that is, (8.37), follows if we let $G = 0$ and seek harmonic solutions of the form $\exp(\mathrm{i}kx - \mathrm{i}\omega t)$.

Consider now a forcing, $\widetilde{F}$ say, that projects only onto the zeroth order parabolic function, D_0, which is a constant. Equation (8.94a) becomes, in dimensional form,

$$\left(\frac{\partial}{\partial t} + c\frac{\partial}{\partial x}\right)q_0 = \widetilde{G}. \tag{8.99}$$

The free solutions of this are Kelvin waves propagating eastwards at speeds $c = \omega/k$ for each k that might be excited; that is $q_0 = \mathrm{Re}\, C\exp[\mathrm{i}k(x - ct)]$, where C is a constant. Suppose that the forcing is harmonic in x and time,

$$\widetilde{F} = \mathrm{Re}\, A\,\{\exp[\mathrm{i}(k_1 x - \omega_1 t)] + \exp[\mathrm{i}(k_1 x + \omega_1 t)]\} = A\,[\cos(k_1 x - \omega_1 t) + \cos(k_1 x + \omega_1 t)]\,, \tag{8.100}$$

and A is real. The solution to (8.99) with this forcing is given by

$$q_0 = -\frac{A\sin(k_1 x - \omega_1 t)}{\omega_1 - ck_1} + \frac{A\sin(k_1 x + \omega_1 t)}{\omega_1 + ck_1}. \tag{8.101}$$

All the parameters in the above equation, c, k_1, ω_1, are positive. If the forcing is just one harmonic then, in general, $c \neq \omega_1/k_1$. However, if the forcing is a superposition of many harmonics then there may be one that is in resonance with the free mode, and this wave, an eastward propagating Kelvin wave represented by an expression like the first term on the right-hand side of (8.101), will be preferentially excited. Similar considerations apply to other waves too; that is, the forcing will excite waves that most resemble the forcing and can resonate with it. Sometimes, a forcing will resemble a delta function in both space and time: for example, a sudden and localized burst of wind over the ocean because of intense storm activity, will give rise to a forcing that contains nearly all space and time scales, since a Dirac delta function has equal representation of all Fourier modes. In this case, both eastward propagating Kelvin waves and westward propagating planetary waves will be excited, and to look at some of these it is useful to make a longwave approximation, as we now discuss.

Planetary waves, revisited

In the planetary wave, or longwave, approximation the highest time derivative in (8.98) is omitted, leaving

$$\frac{\partial}{\partial t}\left[\frac{\partial^2 v_n}{\partial x^2} - (2n+1)\frac{\beta}{c}v_n\right] + \beta\frac{\partial v_n}{\partial x} = G. \tag{8.102}$$

If $G = 0$ this equation gives the dispersion relation

$$\omega = \frac{-\beta k}{(2n+1)\beta/c + k^2}, \tag{8.103}$$

as in (8.47). Planetary waves will be excited when the forcing itself has a low frequency.

The longwave approximation, revisited

Many situations in low latitudes are characterized by having a longer zonal scale than meridional scale; thus, $|\partial\phi/\partial y| \gg |\partial\phi/\partial x|$. When this is the case, geostrophic balance will hold to a good approximation for the zonal flow even in the presence of forcing and dissipation, but not for the meridional flow, and to a good approximation the meridional momentum equation (8.84b) may be replaced by

$$fu = -\frac{\partial\phi}{\partial y}. \tag{8.104}$$

In this limit (8.98) simplifies to

$$\frac{\partial}{\partial t}\left[(2n+1)\frac{\beta}{c}v_n\right] - \beta\frac{\partial v_n}{\partial x} = G, \tag{8.105}$$

from which, with $G = 0$, the dispersion relation,

$$\omega = \frac{-kc}{2n+1}, \tag{8.106}$$

immediately follows. This equation is the small k limit of (8.103) and the wave is non-dispersive. When $n = -1$ we have eastward propagating Kelvin waves and when $n \geq 0$ we have westward propagating long Rossby waves.

The amplitude equations, (8.94)–(8.96) then simplify as follows, also taking $\alpha = 0$. The q equations become

$$\frac{\partial q_0}{\partial t} + \frac{\partial q_0}{\partial x} = F_0^x - Q_0, \tag{8.107a}$$

$$\frac{\partial q_{n+1}}{\partial t} + \frac{\partial q_{n+1}}{\partial x} - v_n = F_{n+1}^x - Q_{n+1}, \qquad n = 0, 1, 2, 3, \ldots. \tag{8.107b}$$

The r equation becomes

$$\frac{\partial r_{n-1}}{\partial t} - \frac{\partial r_{n-1}}{\partial x} + nv_n = -(F_{n-1}^x + Q_{n-1}), \qquad n = 1, 2, 3, \ldots, \tag{8.108}$$

and from the v equation (geostrophic balance) we find

$$q_1 = 0, \tag{8.109a}$$

$$(n+1)q_{n+1} = r_{n-1}, \qquad n = 1, 2, 3, \ldots \tag{8.109b}$$

If we use (8.109b) to eliminate r_{n-1} in (8.108), and then use (8.107b) to eliminate v_n we obtain

$$(2n+1)\frac{\partial q_{n+1}}{\partial t} - \frac{\partial q_{n+1}}{\partial x} = n\,(F^x_{n+1} - Q_{n+1}) - (F^x_{n-1} + Q_{n-1})\,. \tag{8.110}$$

The above set of equations provide, in principle, a means for studying the response of the system to an imposed forcing, such as winds blowing over the ocean or a diabatic source in the atmosphere. Having neglected dissipation, wavelike solutions of constant amplitude will be found only if the forcing is oscillatory rather than steady. Solutions are found by solving the first-order wave equations (8.107a) and (8.110) for q_n, and then using (8.109b) to obtain r_n. A simple expression for v_n results if we add (8.107b) and (8.110).

Waves and adjustment

The wave described by (8.107a) is a Kelvin wave, moving eastwards with nondimensional speed unity, or dimensional speed c. The speed also follows from the dispersion relation, $\omega = -kc/(2n+1)$, with $n = -1$. In contrast, the waves described by (8.110) are westwards propagating, long, low frequency planetary waves. In dimensional form (8.110) becomes

$$(2n+1)\frac{\partial q_{n+1}}{\partial t} - c\frac{\partial q_{n+1}}{\partial x} = n\,(F^x_{n+1} - Q_{n+1}) - (F^x_{n-1} + Q_{n-1})\,, \tag{8.111}$$

and hence the waves have a speed $-c/(2n+1)$, just as in (8.50). There are no short low frequency waves in this approximation.

As we noted above, an arbitrary forcing will in general excite both gravity waves and planetary waves and the initial flow will be out of geostrophic balance. In the mid-latitude case (Section 3.9) the gravity waves radiate to infinity (in the idealized problem) leaving behind an adjusted flow in geostrophic balance, determined by potential vorticity conservation. The process of adjustment is less efficient at low latitudes, because the waves are trapped between their inertial latitudes (Section 8.3) and in the absence of dissipation the fluid will oscillate endlessly. In the zonal direction both planetary and Kelvin waves propagate. A gravity wave front moves away more quickly, with the eventual adjustment occurring by way of planetary waves.

Let us now turn our attention to a rather concrete problem, that of the steady response to a localised thermal forcing.

8.5 FORCED, STEADY FLOW: THE MATSUNO–GILL PROBLEM

We now consider the forced, steady version of the equatorial wave problem; that is to say, we seek steady solutions of (8.84), but with a mechanical or thermal forcing on the right-hand side.[11] Because of its importance to the tropical circulation of the atmosphere this problem has become somewhat iconic and some readers may be tempted to begin reading this chapter here. However, the problem is really just the forced, steady version of the wave problems studied in Sections 8.2 and 8.4, and the reader should have at least a passing familiarity with that material before proceeding. Those readers who have followed the previous sections closely will find the material that follows, namely the *Matsuno–Gill* problem, a pleasant stroll in the park.

8.5.1 Mathematical Development

We begin with (8.84) and make two additional simplifications. First, that the flow is steady and second that the zonal wind is in geostrophic balance with the meridional pressure gradient. This 'semi-geostrophic' approximation is similar to the longwave approximation discussed in previous sections, and requires that αv is smaller than fu. The equations of motion become

$$\alpha u - fv + \frac{\partial \phi}{\partial x} = F^x, \qquad fu + \frac{\partial \phi}{\partial y} = 0, \qquad \alpha\phi + c^2\left(\frac{\partial u}{\partial x} + \frac{\partial v}{\partial y}\right) = -Q. \tag{8.112a,b,c}$$

From these equations we may derive a single equation for v, namely

$$\frac{f^2}{c^2}\alpha v - \alpha\frac{\partial^2 v}{\partial y^2} - \beta\frac{\partial v}{\partial x} = \frac{\alpha}{c^2}\frac{\partial Q}{\partial y} - \frac{f}{c^2}\frac{\partial Q}{\partial x} - \frac{f}{c^2}\alpha F^x + \frac{\partial^2 F^x}{\partial x\,\partial y}. \tag{8.113}$$

This is just a simplification of (8.86) appropriate for a steady system with the zonal wind in geostrophic balance, obtained by omitting all the time derivatives, the term involving α^3, and the F^y term on the right-hand side. We nondimensionalize all the variables using the time and length scales

$$T_{eq} = (2c\beta)^{-1/2}, \qquad L_{eq} = \left(\frac{c}{2\beta}\right)^{1/2}, \tag{8.114a,b}$$

so that the various dimensional and nondimensional (hatted) variables are related by

$$(u,v) = c\,(\widehat{u},\widehat{v}), \qquad \phi = c^2\widehat{\phi}, \qquad (x,y) = L_{eq}(\widehat{x},\widehat{y}),$$
$$\alpha = \frac{\widehat{\alpha}}{T_{eq}}, \qquad Q = \frac{c^3}{L_{eq}}\widehat{Q}, \qquad F^x = \frac{c^2}{L_{eq}}\widehat{F}^x, \qquad \widehat{f} = \frac{1}{2}\widehat{y}, \qquad \widehat{\beta} = \frac{1}{2}. \tag{8.115}$$

Unless specifically noted the variables from now on are nondimensional and we drop the hats. The equations of motion, (8.116), become

$$\alpha u - \frac{y}{2}v + \frac{\partial\phi}{\partial x} = F^x, \qquad \frac{y}{2}u + \frac{\partial\phi}{\partial y} = 0, \qquad \alpha\phi + \left(\frac{\partial u}{\partial x} + \frac{\partial v}{\partial y}\right) = -Q. \tag{8.116a,b,c}$$

and the v equation becomes

$$\frac{y^2}{4}\alpha v - \alpha\frac{\partial^2 v}{\partial y^2} - \frac{1}{2}\frac{\partial v}{\partial x} = \alpha\frac{\partial Q}{\partial y} - \frac{y}{2}\frac{\partial Q}{\partial x} - \frac{\alpha y}{2}\alpha F^x + \frac{\partial^2 F^x}{\partial x\partial y}. \tag{8.117}$$

As before when dealing with wave-like problems it is convenient to change variables to p and q where

$$q = \phi + u, \qquad r = \phi - u. \tag{8.118a,b}$$

The equations of motion, (8.116), become

$$\alpha q + \frac{\partial q}{\partial x} + \frac{\partial v}{\partial y} - \frac{1}{2}yv = F^x - Q, \qquad \alpha r - \frac{\partial r}{\partial x} + \frac{\partial v}{\partial y} + \frac{1}{2}yv = -F^x - Q,$$
$$\frac{y}{4}(q-r) + \frac{1}{2}\frac{\partial}{\partial y}(q+r) = 0. \tag{8.119a,b,c}$$

These are special cases of (8.90), the first two equations being combinations of the u-momentum and pressure equations and the last one being the v-momentum equation (zonal geostrophic balance).

As in the general treatment given earlier we expand the variables and the forcing in terms of parabolic cylinder functions. Thus, for example,

$$Q(x) = \sum_{n=0}^{\infty}{}' Q_n(x)D_n(y), \tag{8.120}$$

and similarly for the other variables. The resulting ordinary differential equations are special cases of (8.94)–(8.97), specifically

$$\alpha q_0 + \frac{\partial q_0}{\partial x} = F_0^x - Q_0, \tag{8.121a}$$

$$\alpha q_{n+1} + \frac{\partial q_{n+1}}{\partial x} - v_n = F^x_{n+1} - Q_{n+1}, \qquad n = 0, 1, 2, 3, \ldots \tag{8.121b}$$

$$\alpha r_{n-1} - \frac{\partial r_{n-1}}{\partial x} + n v_n = -\left(F^x_{n-1} + Q_{n-1}\right), \qquad n = 1, 2, 3, \ldots, \tag{8.122}$$

$$q_1 = 0, \tag{8.123a}$$

$$(n+1) q_{n+1} - r_{n-1} = 0, \qquad n = 1, 2, 3, \ldots, \tag{8.123b}$$

Using (8.121b), (8.122) and (8.123b) we obtain

$$\alpha(2n+1) q_{n+1} - \frac{\partial q_{n+1}}{\partial x} = n\left(F^x_{n+1} - Q_{n+1}\right) - \left(F^x_{n-1} + Q_{n-1}\right) \qquad n = 1, 2, 3, \ldots. \tag{8.124}$$

Finally, although we shall not use it, the v equation (8.113) becomes

$$\alpha\left((2n+1) v_n - \frac{\partial^2 v_n}{\partial x^2}\right) - \frac{\partial v_n}{\partial x} = G, \tag{8.125}$$

where G represents the various forcing terms.

As in the wavelike case, the above equations provide, at least in principle, a means of solving for the response for any particular forcing. The procedure is to project the forcing onto parabolic cylinder functions, and then solve the amplitude equations (8.121)–(8.123) for the zonal dependence, and then finally to reconstruct the solutions using the $q_n(x)$, $r_n(x)$ and $v_n(x)$ and the parabolic cylinder functions. Naturally enough, this is easier said than done and we will go through the procedure in detail for just one important case.

8.5.2 Symmetric Heating

A canonical case is that in which the system is forced by a heating that is confined in both the x- and y-directions, and is symmetric across the equator. Confinement in the y-direction may be achieved by supposing that the heating projects solely onto the first parabolic function, so that

$$Q(x) = Q_0(x) D_0(y) = G(x) \exp(-y^2/4), \tag{8.126}$$

and confinement in the x-direction may be achieved by supposing that the heating is of the form

$$G(x) = \begin{cases} A \cos kx & |x| < L \\ 0 & |x| > L, \end{cases} \tag{8.127}$$

where $k = \pi/2L$. This may seem an odd form to choose, but the harmonic variation for $|x| < L$ enables an analytic solution to be found in that region, and the absence of any forcing at all in the far field enables solutions to be found there in the form of decaying wavelike disturbances. Although this problem is clearly a special case the qualitative form of the solution transcends its precise details.

Kelvin wave contribution

We noted in Section 8.4.2 that the equation for q_0 represents an eastwards propagating Kelvin wave, and this holds in the damped case also. That is to say, there will be a non-zero solution of (8.121) only in the forced region and eastward of it, where it will be progressively damped. Using this insight we can easily derive the solution in all three regions. First, for $X < -L$, we have

$$q_0 = 0, \qquad x < -L. \tag{8.128a}$$

In the forcing region we have to solve (8.121) with a boundary condition of $q_0 = 0$ at $x = -L$. The solution is

$$q_0 = \frac{-A}{\alpha^2 + k^2} \left\{\alpha \cos kx + k \left[\sin kx + e^{-\alpha(x+L)}\right]\right\} \qquad |x| < L. \tag{8.128b}$$

For $x > L$ we solve (8.121), but with a right-hand side equal to zero, with a boundary condition at $x = L$ given by (8.128b), namely $q_0 = -Ak(\alpha^2 + k^2)^{-1}[1 + \exp(-2\alpha L)]$. The solution is

$$q_0 = \frac{-Ak}{\alpha^2 + k^2} \left(1 + e^{-2\alpha L}\right) e^{\alpha(L-x)}, \qquad x > L. \tag{8.128c}$$

Because the motion is a decaying Kelvin wave $v = 0$ and the nondimensional u and ϕ fields are equal to each other, with $r = 0$. Thus, from (8.118) and (8.128),

$$u = \phi = \frac{1}{2} q_0(x) \exp(-y^2/4), \qquad v = 0. \tag{8.129}$$

This does not mean r_0 is zero; rather, it is associated with the planetary wave solution discussed below. The vertical velocity may be reconstructed from

$$w = -\left(\frac{\partial u}{\partial x} + \frac{\partial v}{\partial y}\right) = \alpha\phi + Q, \tag{8.130}$$

whence

$$w = \frac{1}{2} \left[\alpha q_0(x) + Q_0(x)\right] \exp(-y^2/4). \tag{8.131}$$

We now complete the solution by finding a planetary wave contribution.

Planetary wave contribution

We now find the solution associated with q_2 and r_0. From (8.124) we have

$$\frac{dq_2}{dx} - 3\alpha q_2 = Q_0. \tag{8.132a}$$

From (8.123b) and (8.121b) we have

$$r_0 = 2q_2, \qquad v_1 = \alpha q_2 + \frac{dq_2}{dx}. \tag{8.132b,c}$$

These are planetary waves propagating westwards at a dimensional speed of $c/(2n+1) = c/3$. Thus, no signal is transmitted eastwards and we can find a solution to the above equations in an analogous fashion to the way we found a solution for the Kelvin wave problem. After just a little algebra, the solution is found to be:

$$q_2 = 0, \qquad x > L, \tag{8.133a}$$

$$q_2 = \frac{A}{(3\alpha)^2 + k^2} \left[-3\alpha \cos kx + k\left(\sin kx - e^{3\alpha(x-L)}\right)\right], \qquad |x| < L, \tag{8.133b}$$

$$q_2 = \frac{-Ak}{(3\alpha)^2 + k^2} \left[1 + e^{-6\alpha L}\right] e^{3\alpha(x+L)}, \qquad x < -L. \tag{8.133c}$$

The corresponding solutions for the pressure and velocity fields are

$$u = \frac{e^{-y^2/4}}{2} q_2(x)(y^2 - 3), \qquad v = y e^{-y^2/4} \left[Q_0(x) + 4\alpha q_2(x)\right], \tag{8.134a,b}$$

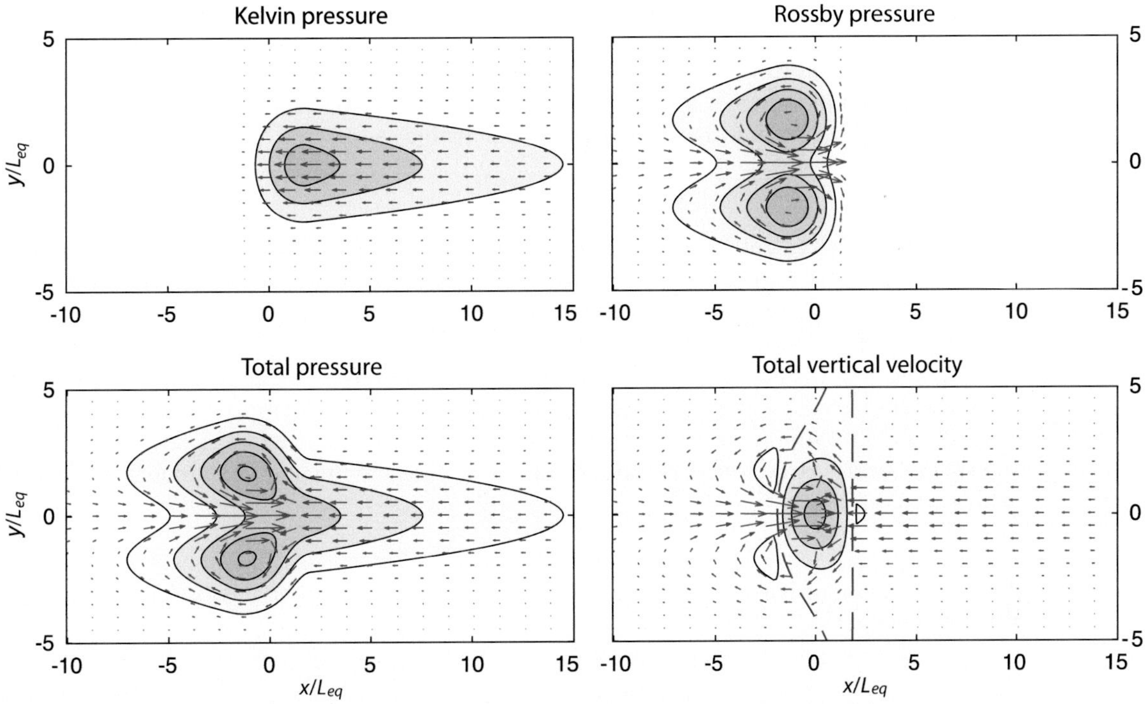

Fig. 8.11 Nondimensional solutions of the Matsuno–Gill model, with heating close to the origin and given by (8.127) with $L = 2$ and $\alpha = 0.1$. The shaded contours show the fields as indicated, and the arrows show the associated horizontal velocities. The 'Kelvin' and 'Rossby' designations indicate that just the Kelvin wave or Rossby (planetary) wave contributions are plotted as given by (8.129)–(8.131) and (8.134)–(8.135), respectively. For the pressure fields the contour interval is 0.3 and all fields are negative (so dark shading is low pressure) with the zero contour omitted. For vertical velocity the contour interval is 0.3 beginning at -0.1, and so is -0.1, 0.2, 0.5..., with an additional zero contour (red dashed) with upward motion within it.

$$\phi = \frac{e^{-y^2/4}}{2} q_2(x)(1+y^2), \qquad w = \frac{e^{-y^2/4}}{2} \left[Q_0(x) + \alpha q_2(x)(1+y^2)\right]. \tag{8.135a,b}$$

The solutions appear complicated (they are complicated!), although still amenable to interpretation. But first we combine the Kelvin and planetary wave contributions and restore the dimensions, to give

$$u = \frac{c}{2}\left[q_0(x) + q_2(x)(2\beta y^2/c - 3)\right] e^{-\beta y^2/2c}, \tag{8.136a}$$

$$v = cy\Big[Q_0(x) + (4\alpha/c)q_2(x)\Big] e^{-\beta y^2/2c}, \tag{8.136b}$$

$$\phi = \frac{c^2}{2}\left[q_0(x) + q_2(x)(2\beta y^2/c + 1)\right] e^{-\beta y^2/2c}, \tag{8.136c}$$

$$w = w_0 \frac{e^{-\beta y^2/2c}}{2}\left[2Q_0(x) + \alpha q_0(x) + \alpha q_2(x)(1 + 2\beta y^2/c)\right], \tag{8.136d}$$

where $w_0 = (2c\beta)^{1/2}H$, and q_0, q_2 and Q_0 are still nondimensional functions. The nondimensional forms are recovered by setting $w_0 = c = 1$ and $\beta = 1/2$. The solutions above are specific to the form of the forcing function we chose. However, a similar methodology could in principle be applied to forcing of any form, including forcing in the momentum equations, and, because the equations are linear, the solutions could be superposed. The solution above represents the physically important case of a localized heating, and the gross structure of the far field is largely independent of the details of the forcing: there is a rapidly decaying disturbance west and polewards of the forcing and a more slowly decaying disturbance east of the forcing close to the equator (Fig. 8.11).

Interpretation

Let's now try to figure out what's going on. A solution is illustrated in Fig. 8.11. The heating is confined to a region from $-2 < x < 2$ and exponentially falls away from the equator with an e-folding distance of 2, more-or-less corresponding to the shaded region of vertical velocity in the lower right panel, as intuitively expected and discussed more below.

Consider first the flow in the forcing region. Here the vertical velocity is positive, with the associated horizontal convergence being that of the zonal flow: the meridional flow is polewards, *away* from the maximum of the heating. To understand this, consider the limit $\alpha \to 0$. From (8.136) the vertical velocity field coincides with the heating and the (nondimensional) meridional velocity is given by

$$v = yQ_0 \exp(-y^2/4) = yw. \tag{8.137}$$

Thus, vertical motion is associated with poleward motion. To understand this, consider the inviscid vorticity equation

$$\beta v + f\nabla \cdot \boldsymbol{u} = 0, \qquad \text{or} \qquad \beta v = fw, \tag{8.138a,b}$$

which in nondimensional form is

$$v + y\nabla \cdot \boldsymbol{u} = 0, \qquad \text{or} \qquad v = yw. \tag{8.139a,b}$$

Evidently, (8.137) and (8.139b) are equivalent. Another way to think about this is to note that the rising motion in the region of the forcing causes vortex stretching, as discussed in Chapter 4, and hence the generation of cyclonic vertical vorticity and a polewards migration. From the perspective of potential vorticity, then to the extent that the flow is adiabatic the quantity $(f + \zeta)/h$ is conserved following the flow. The heating increases the value of h (the stretching), so that $f + \zeta$ also tends to increase in magnitude. The flow finds it easier to migrate polewards to increase its value of f than to increase its relative vorticity alone, for the latter would require more energy. If we interpret these equations as the lower layer of a two-layer system, then the flow in the lower layer is away from the source, and toward the source in the upper layer.

Consider now the flow to the west of the heating, associated with $q_2(x)$. The disturbance here is produced by a decaying westwards propagating Rossby wave — a form of 'Rossby plume' that we will also encounter in Chapter 19 (see Fig. 19.14 on page 754 and the associated discussion). The vertical velocity is negative, and the horizontal velocity is almost geostrophically balanced: the pressure perturbation is negative everywhere, and so circulating cyclonically around the centres of low pressure just to the west of heating. The flow converges to the equator, producing an eastward flow along the equator, converging in the heating zone. We may be tempted to interpret this in terms of the inviscid vorticity equation, as we did in the forcing region. This would suggest that, away from the forcing region, because the flow is divergent ($\nabla \cdot \boldsymbol{u} > 0,\ w < 0$) then from (8.138) the meridional velocity should be toward the equator in both hemispheres. However, this explanation is at best qualitative, because the vorticity equation above is not exactly satisfied by the solution (8.138), because non-zero solutions away from the forcing region depend entirely on the presence of dissipation.

The flow east of the forcing region motion is induced by an eastward propagating Kelvin wave, or more precisely the steady, eastward-decaying analogue of such a wave. Evidently, from Fig. 8.11, the pressure field extends further east of the source than west of the source, and this is because

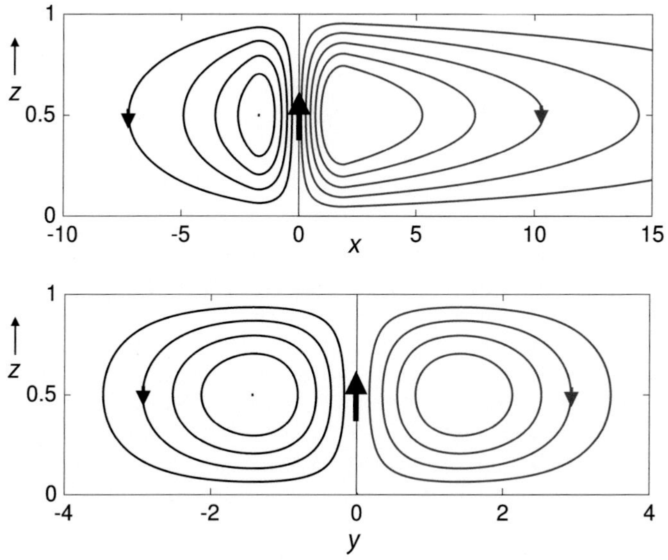

Fig. 8.12 The zonal overturning streamfunction of the meridionally averaged flow (top), and the meridional overturning streamfunction of the zonally-averaged flow (bottom) in the Matsuno–Gill problem with symmetric heating with a maximum at $y = 0$ and $x = 0$, the same as in Fig. 8.11.

There is rising motion around the location of the heating, sinking elsewhere. The contour interval in the top plot is about four times that of the bottom plot; that is, the Walker circulation here is stronger than the Hadley circulation.

Kelvin waves decay more slowly than Rossby waves. Keeping both the time derivative and the damping, the unforced Kelvin wave satisfies, from (8.94),

$$\left[\alpha + \frac{\partial}{\partial t}\right] q_0 + \frac{\partial q_0}{\partial x} = 0, \tag{8.140}$$

whereas the unforced Rossby wave satisfies, from (8.110) and (8.124) for $n = 1$,

$$3\left[\alpha + \frac{\partial}{\partial t}\right] q_1 - \frac{\partial q_1}{\partial x} = 0. \tag{8.141}$$

Thus, the effective damping rate of the Rossby wave is three times that of the Kelvin wave. Put another way, the Kelvin wave travels three times as fast as the Rossby wave so that if the damping rate, α, is the same the influence of the Kelvin wave spreads three times further east. The horizontal velocity in the Kelvin wave is purely zonal, and near the surface it is directed toward the heating source.

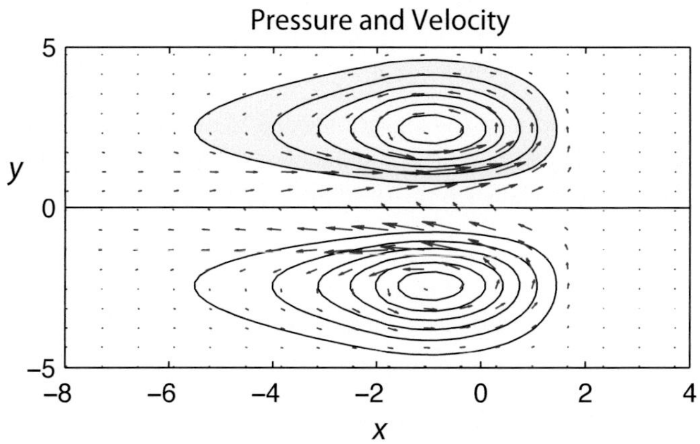

Fig. 8.13 Pressure (contours) and horizontal velocity (arrows) in the Matsuno–Gill problem with an antisymmetric heating given by (8.143) and nondimensional decay factor $\alpha = 0.1$. The heating is in the Northern Hemisphere generating a low pressure region (shaded) with inflow and ascent, and cooling is in the Southern Hemisphere. The contour interval is 0.3 and the zero contour is along $y = 0$.

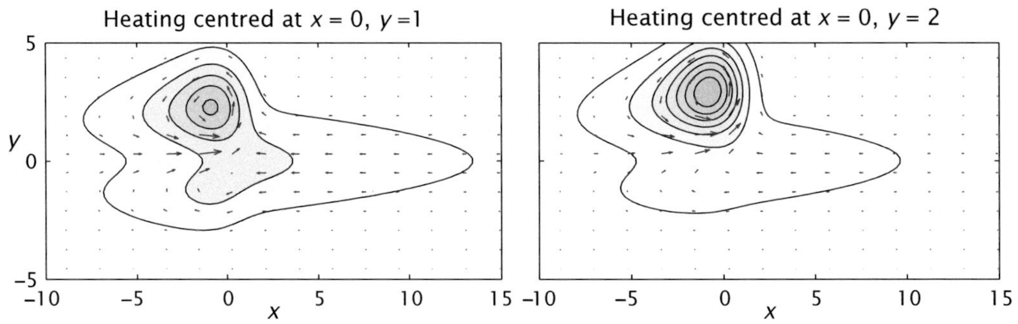

Fig. 8.14 Pressure (contours) and horizontal velocity (arrows) in the Matsuno–Gill model with the heating centred off the equator, as labelled, but otherwise similar to that of Fig. 8.11. As the heating moves to higher latitudes the Kelvin wave response weakens but the magnitude of the local response increases (the contour interval is the same in both panels).

Vertical structure

The zonal structure of the solution is a coarse representation of the Walker circulation in the equatorial Pacific. Here, the sea-surface temperature is high in the west, near Indonesia, and low in the east, near South America, because of the upwelling that brings deep, cold water to the surface. This distribution of sea-surface temperature effectively provides a heating in the western Pacific and induces westward winds along the equator, enhancing the westward trade winds that already exist as part of the general circulation. The overturning circulation in the zonal and meridional planes is illustrated in Fig. 8.12. This solution is obtained by supposing that the fields represent the first vertical mode, as discussed in Section 3.4. If the stratification is uniform then the modes are just sines and cosines and so we have

$$(u, v, \phi) = (\widetilde{u}, \widetilde{v}, \widetilde{\phi}) \cos(\pi z/D), \qquad w = \widetilde{w} \sin(\pi z/D). \tag{8.142}$$

Now the modal form of the mass continuity equation, (3.66), is $\widetilde{w} = -(c^2/g)\nabla \cdot \widetilde{\boldsymbol{u}}$. If this is to be consistent with the usual form of $\partial w/\partial z = -\nabla\cdot\boldsymbol{u}$ then we make the association $\pi/D = g/c^2 = 1/H_1^*$, where H_1^* is the equivalent depth of the first mode. Given this vertical structure, we integrate the solutions meridionally so enabling a streamfunction to be defined (because $v = 0$ as $y \to \pm\infty$, so $\partial\overline{u}^y/\partial x + \partial\overline{w}^y/\partial z = 0$, with the overbar denoting meridional integration). The expressions for the meridional integrals are given in Appendix B to this chapter, with the streamfunctions given by (8.175) and (8.175).

8.5.3 Antisymmetric Forcing

An analytic solution with asymmetric forcing may be obtained by using a forcing of the form

$$Q(x, y) = Q_1(x)D_1(y) = y \cos kx \exp(-y^2/4), \tag{8.143}$$

using the same form of zonal localization as before. The algebra needed to obtain a solution is somewhat tedious but straightforward, of a very similar nature to that described above. One finds that there are, again, two parts to the response. One part corresponds to a long planetary wave with $n = 0$ and using (8.121)–(8.123) we find

$$q_1 = 0, \qquad v_0 = Q_1. \tag{8.144}$$

There is no response outside the forcing region because long mixed waves have zero propagation velocity. The other part of the solution is obtained, again using (8.121)–(8.123), from

$$v_2 = \frac{\mathrm{d}q_3}{\mathrm{d}x} + \alpha q_3, \qquad r_1 = 3q_3, \qquad \frac{\mathrm{d}q_3}{\mathrm{d}x} - 5\alpha q_3 = Q_1. \tag{8.145a,b,c}$$

The solution of these equations is left as an exercise for the reader and is illustrated in Fig. 8.13. The solutions are zero east of the forcing region because there is no long wave so propagating. West of the forcing region there is eastward inflow into the heating region in the Nortern Hemisphere (which is being heated), as well as a tendency for poleward flow for the reasons described earlier. Thus, there is a cyclone with upward motion somewhat west of the main heating region, and a corresponding anti-cyclone in the cooled region, as illustrated in Fig. 8.13. The zonally averaged solutions (not shown) resemble an asymmetric Hadley Cell, with the air rising in the Northern (summer) Hemisphere, moving southwards aloft into the winter hemisphere before sinking.

8.5.4 Other Forcings

The solution to more general forcings can be constructed by using other forcing coefficients, or a superposition of forcing coefficients, and many solutions of interest to the tropical atmosphere and ocean may be so constructed. Solutions may also be constructed (sometimes more easily) numerically, either by time-stepping the linear shallow water equations to equilibrium or by solving the elliptic equation (8.113) using standard techniques.[12] We will present the solutions to two such cases: (i) a heating source centred off the equator in the Northern Hemisphere; and (ii) a line source of heating, either centred at the equator or just north of it, mimicking the Inter-Tropical Convergence Zone (ITCZ).

Heating off the equator

Solutions for heating off the equator may be constructed by adding the solutions for antisymmetric and symmetric heating presented above. In Fig. 8.14, we present a solution that has heating of a very similar form to that of the symmetric heating shown in Fig. 8.11, but centred off the equator at $y = 1$ and $y = 2$. The pattern is dominated by a low pressure region just to the west of the heating, with convergence and upward motion within it, and an eastward inflow between the equator and the centre of the heating. In the solution with the heating centred at $y = 1$ there is also a response east of the heating region, largest at the equator, produced by the eastward propagating, damped Kelvin wave. As the heating moves further from the equator (in the right panel of Fig. 8.14), the pressure response becomes stronger but the flow around the heating is in near geostrophic balance.

A line of heating

Finally, let us consider the solutions when the heating is independent of x, and the solutions themselves are then independent of x. Two such solutions are presented in Fig. 8.15 and in Fig. 8.16, for a line of heating at the equator and at $y = 1$. As we noted above, these solutions might be thought of as rather idealized versions of the ITCZ (although in the real ITCZ the location of the convective region is determined as part of the solution for the overall flow).

Consider first the solution with heating at the equator. A low pressure region develops over the heating and the flow converges there, producing equatorward and westward 'trade winds' and consequent upward motion at the equator, with the zonal velocity rapidly decreasing actually at the equator. Now consider what happens when the heating is off-equator, noting that the real ITCZ is generally situated a little north of the equator, especially in the Pacific Ocean. A low pressure region is formed along the line of the heating and the meridional velocity converges sharply there, with more inflow coming from the equatorial side of the line of heating (as can be seen in the right-hand panels of both Fig. 8.15 and Fig. 8.16). As regards the zonal velocity, there is an *eastward* jet along the line of the heating, with westward flow to either side. That is to say, there is a splitting of the westward trades caused by the line of sharp heating.

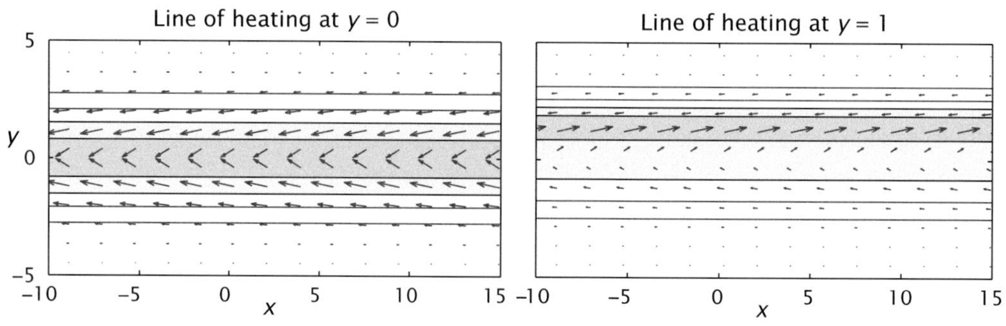

Fig. 8.15 As for Fig. 8.14 but with a line of heating at the equator (left panel) and at $y = 1$ (right panel). The heating generates a region of low pressure (shaded) where the flow converges. In the right panel the meridional velocity is larger on the equatorward side of the line than on the poleward side. See also Fig. 8.16.

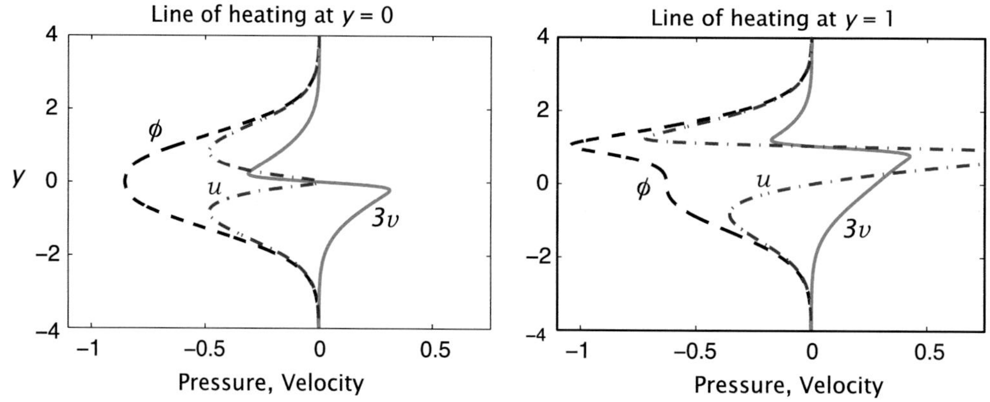

Fig. 8.16 As for Fig. 8.15, but showing line plots of pressure, ϕ, zonal velocity, u and three times (for presentational purposes) the meridional velocity v. Left panel is for a line of heating at the equator and the right panel for heating at $y = 1$. The heating creates a region of low pressure where the flow converges. Note that in the right panel the meridional velocity is larger on the equatorward side of the line than on the poleward side.

APPENDIX A: NONDIMENSIONALIZATION AND PARABOLIC CYLINDER FUNCTIONS

This appendix provides a brief discussion of the nondimensionalization used to derive the various dispersion relations in this chapter and some of the properties of the associated Hermite polynomials and parabolic cylinder functions. We do not provide proofs or detailed derivations.[13]

In discussions of equatorial waves and their steady counterparts, one of two slightly different nondimensionalizations is often employed. They lead to the use of parabolic cylinder functions in two slightly different forms; they are essentially equivalent but one may be more convenient than the other depending on the setting. For definiteness, we begin with (8.28), namely

$$\frac{\mathrm{d}^2\tilde{v}}{\mathrm{d}y^2} + \left(\frac{\omega^2}{c^2} - k^2 - \frac{\beta k}{\omega} - \frac{\beta^2 y^2}{c^2}\right)\tilde{v} = 0. \tag{8.146}$$

If we nondimensionalize time and distance using

$$T_{eq} = (c\beta)^{-1/2}, \qquad L_{eq} = (c/\beta)^{1/2}, \tag{8.147a,b}$$

we obtain

$$\frac{d^2 v}{d\widehat{y}^2} + \left(\widehat{\omega}^2 - \widehat{k}^2 - \frac{\widehat{k}}{\widehat{\omega}} - \widehat{y}^2\right) v = 0. \tag{8.148}$$

The substitution

$$v(\widehat{y}) = \Psi(\widehat{y}) e^{-\widehat{y}^2/2}, \tag{8.149}$$

leads to

$$\frac{d^2\Psi}{d\widehat{y}^2} - 2\widehat{y}\frac{d\Psi}{d\widehat{y}} + \lambda\Psi = 0, \tag{8.150}$$

where $\lambda = \widehat{\omega}^2 - \widehat{k}^2 - \widehat{k}/\widehat{\omega} - 1$. This is Hermite's equation with solutions if and only if $\lambda = 2m$ for $m = 0, 1, 2, \ldots$, and it is this quantization condition that gives the dispersion relation. The solutions are Hermite polynomials; that is, $\Psi(\widehat{y}) = H_m(\widehat{y})$, where

$$(H_0, H_1, H_2, H_3, H_4) = (1,\ 2\widehat{y},\ 4\widehat{y}^2 - 2,\ 8\widehat{y}^3 - 12\widehat{y},\ 16\widehat{y}^4 - 48\widehat{y}^2 + 12). \tag{8.151}$$

The Hermite polynomial multiplied by a Gaussian is a form of parabolic cylinder function, $V_m(y)$; that is

$$V_m(y) = H_m(y)\exp(-y^2/2). \tag{8.152}$$

The function $V_m(y)$ satisfies

$$\frac{d^2 V_m}{dy^2} + (2m + 1 - y^2)V_m = 0. \tag{8.153}$$

It is often useful to include the normalization coefficient in the definition of the cylinder function; that is, if

$$P_m = \frac{V_m}{\sqrt{2^m m!\,\sqrt{\pi}}}, \qquad \text{then} \qquad \int_{-\infty}^{\infty} P_m P_n = \delta_{mn}. \tag{8.154a,b}$$

As may be verified by direct manipulation, these forms of parabolic cylinder functions obey certain recurrence relations, namely

$$\frac{dP_m}{dy} = -\frac{(m+1)^{1/2}}{\sqrt{2}}P_{m+1} + \frac{m^{1/2}}{\sqrt{2}}P_{m-1} \quad \text{and} \quad yP_m = \frac{m^{1/2}}{\sqrt{2}}P_{m-1} + \frac{(m+1)^{1/2}}{\sqrt{2}}P_{m+1}, \tag{8.155a,b}$$

or equivalently

$$\frac{dP_m}{dy} + yP_m = (2m)^{1/2}P_{m-1} \qquad \text{and} \qquad \frac{dP_m}{dy} - yP_m = -\sqrt{2}(m+1)^{1/2}P_{m+1}. \tag{8.156}$$

When $m = 0$ the recurrence relations are

$$\frac{dP_0}{dy} = \frac{-1}{\sqrt{2}}P_1 \qquad \text{and} \qquad yP_0 = \frac{1}{\sqrt{2}}P_1. \tag{8.157a,b}$$

Had we developed the forced-dissipative problem using this form of cylinder functions these relations would have been used instead of (8.93).

The above relations may be used in conjunction with (8.20) and (8.36) to obtain the relations between u, v and ϕ in the equatorial wave problem. Using (8.20) the v and u, and the v and ϕ, fields are related by

$$\frac{\partial^2 u}{\partial t^2} - c^2\frac{\partial^2 u}{\partial x^2} = \beta y\frac{\partial v}{\partial t} + c^2\frac{\partial^2 v}{\partial x\,\partial y}, \qquad \frac{\partial^2 \phi}{\partial x^2} - c^{-2}\frac{\partial^2 \phi}{\partial t^2} = \beta y\frac{\partial v}{\partial x} + \frac{\partial^2 v}{\partial y\,\partial t}, \tag{8.158}$$

which in nondimensional form have $\beta = c = 1$. Suppose that the v field is a single Hermite mode of unit amplitude, meaning that

$$\hat{v}(x, y, t) = \tilde{v}(y)\mathrm{e}^{\mathrm{i}(kx-\omega t)} = P_m(y)\mathrm{e}^{\mathrm{i}(kx-\omega t)}. \tag{8.159}$$

If $u(x, y, t) = \tilde{u}(y)\mathrm{e}^{\mathrm{i}(kx-\omega t)}$ then, using (8.158a) we obtain

$$\tilde{u} = \frac{-\mathrm{i}}{(k^2 - \omega^2)}\left[\hat{y}\omega P_m - k\frac{\mathrm{d}P_m}{\mathrm{d}\hat{y}}\right] = \frac{\mathrm{i}}{\sqrt{2}}\left[\frac{m^{1/2}P_{m-1}}{k+\omega} - \frac{(m+1)^{1/2}P_{m+1}}{k-\omega}\right], \tag{8.160}$$

where the rightmost expression uses the recurrence relations (8.155). Evidently, if m and $\tilde{v}$ are odd (even) then $\tilde{u}$ is an even (odd) function of y. The height field $\tilde{\phi}$ may similarly be related to P_m using (8.158b), giving

$$\tilde{\phi} = \frac{-\mathrm{i}}{k^2 - \omega^2/c^2}\left[ykP_m + \mathrm{i}\omega\frac{\mathrm{d}P_m}{\mathrm{d}y}\right] = \frac{-\mathrm{i}}{\sqrt{2}}\left[\frac{m^{1/2}P_{m-1}}{k+\omega} + \frac{(m+1)^{1/2}P_{m+1}}{k-\omega}\right]. \tag{8.161}$$

Thus, for a given m, ϕ has the same symmetry across the equator as does u, the opposite of v.

Implications

As discussed in the main text, the gravest mode in y is the Kelvin wave, which has $v = 0$, and u and ϕ fields centred on the equator that decay away exponentially in y^2, as for P_0. The Yanai, or mixed Rossby-gravity mode, has $m = 0$ and the v field is even around the equator, meaning it is an antisymmetric mode. Using the recurrence relation (8.157) we see that this mode only generates the P_1 mode in u and ϕ. The gravest Rossby mode has $m = 1$, and from (8.160) and (8.161) we see that this generates $m = 0$ and $m = 2$ modes in u and ϕ. Thus, a symmetric disturbance centred at the equator will in general generate an eastward propagating Kelvin mode and a Rossby mode with both an equatorial signal and off equatorial modes with a structure similar to those of the stationary pattern in Fig. 8.11, and as we will see again in Fig. 22.18 when we study El Niño.

Other parabolic cylinder functions

The other commonly used form of parabolic cylinder functions, denoted $D_n(y)$, are the modified Hermite polynomials (8.42) multiplied by a Gaussian; that is

$$D_n(y) = G_n(y)\exp(-y^2/4), \tag{8.162}$$

and these functions are solutions of (8.39) which arises when we use the nondimensionalization

$$T_{eq} = (2c\beta)^{-1/2}, \qquad L_{eq} = (c/2\beta)^{1/2}. \tag{8.163a,b}$$

These parabolic cylinder functions satisfy

$$\frac{\mathrm{d}^2 D_m}{\mathrm{d}y^2} + \frac{1}{2}(2m + 1 - \frac{1}{2}y^2)D_m = 0, \tag{8.164}$$

which is sometimes called the Weber differential equation. The functions have the property that

$$\frac{\mathrm{d}D_n}{\mathrm{d}y} + \frac{1}{2}yD_n = nD_{n-1}, \qquad \frac{\mathrm{d}D_n}{\mathrm{d}y} - \frac{1}{2}yD_n = -D_{n+1}. \tag{8.165a,b}$$

The above two equations may be combined to give (8.164), and by subtracting them we see that

$$D_{n+1} - yD_n + nD_{n-1} = 0. \tag{8.166}$$

The form of these particular ladder operators makes these parabolic cylinder functions convenient in our development of the forced, steady (i.e., Matsuno–Gill) problem, although the use of (8.156) would be equivalent.

APPENDIX B: MATHEMATICAL RELATIONS IN THE MATSUNO–GILL PROBLEM

Here we provide various zonal and meridional integrals of the solutions given in Section 8.5.2. The zonal integral of the forcing is given by

$$I = \int_{-\infty}^{\infty} Q_0(x)\,\mathrm{d}x = \int_{-L}^{L} \cos(kx)\,\mathrm{d}x = \frac{4L}{\pi}, \tag{8.167}$$

using $k = \pi/2L$. The zonal integrals of the various q, r and v fields are given as follows. Using (8.121) with $F_0 = 0$ we see that

$$\int_{-\infty}^{\infty} \alpha q_0\,\mathrm{d}x = -\big[q_0\big]_{-\infty}^{\infty} - \int_{-\infty}^{\infty} Q_0(x)\,\mathrm{d}x = -I. \tag{8.168}$$

Using (8.132) we obtain similar results for q_2, r_0 and v_1, to wit

$$\int_{-\infty}^{\infty} \left(q_0, q_2, r_0, v_1\right)\,\mathrm{d}x = \left(-1, -\frac{1}{3}, -\frac{2}{3}, -\frac{\alpha}{3}\right)\frac{I}{\alpha}. \tag{8.169}$$

The zonally integrated pressure and velocity fields are obtained using (8.167), (8.169) and the non-dimensional form of (8.136), giving

$$\int_{-\infty}^{\infty} (u, v, w, \phi)\,\mathrm{d}x = \left(\frac{-y^2}{6\alpha}, \frac{-y}{3}, \frac{2-y^2}{6}, \frac{-4-y^2}{6\alpha}\right)\left(\frac{4L}{\pi}\right)\exp(-y^2/4). \tag{8.170}$$

The meridional integrals of the velocity fields may also be calculated. To do this we first note the integrals

$$\int_{-\infty}^{\infty} (1, y, y^2)\exp(-y^2/4)\,\mathrm{d}y = (2, 0, 4)\sqrt{\pi}. \tag{8.171}$$

The first of these is a standard result, the second follows from considerations of symmetry and the third follows on integration by parts. Using (8.171) and the nondimensional form of (8.136) we obtain

$$\int_{-\infty}^{\infty} u\,\mathrm{d}y = \sqrt{\pi}\left[q_0(x) - q_2(x)\right], \qquad \int_{-\infty}^{\infty} v\,\mathrm{d}y = 0, \tag{8.172a,b}$$

$$\int_{-\infty}^{\infty} w\,\mathrm{d}y = \sqrt{\pi}\left[\alpha q_0(x) + 3\alpha q_2(x) + 2Q_0(x)\right], \qquad \int_{-\infty}^{\infty} \phi\,\mathrm{d}y = \sqrt{\pi}\left[q_0(x) + 3q_2(x)\right]. \tag{8.172c,d}$$

Equations (8.170) and (8.172) are useful because they allow us to define streamfunctions for the overturning circulation in the zonal and meridional plane, respectively. From (8.170) and (8.172), and using (8.121) and (8.132a), we find that

$$\overline{w}^x + \frac{\partial \overline{v}^x}{\partial y} = 0, \qquad \overline{w}^y + \frac{\partial \overline{u}^y}{\partial x} = 0, \tag{8.173}$$

with the overbar denoting a zonal or meridional average, as indicated. These results are to be expected from the mass continuity equation, $w = -(\partial_x u + \partial_y v)$, on zonal and meridional integration, respectively, but the fact that the solutions show it so explicitly is a demonstration of the karma of mathematics.

A streamfunction may be constructed by supposing that, in a fluid of depth H, the horizontal and vertical velocities vary as

$$(u, v) = (\tilde{u}, \tilde{v})\cos(\pi z/H), \qquad w = \tilde{w}\sin(\pi z/H). \tag{8.174a,b}$$

Using (8.170) the streamfunction in the meridional plane, Ψ_M is given by

$$\Psi_M(y, z) = \frac{IH}{\pi}\frac{-y}{3}\exp(-y^2/4)\sin \pi z/H. \tag{8.175}$$

Using (8.172) the streamfunction in the zonal plane, Ψ_Z, is given by

$$\Psi_Z(x,z) = \frac{\sqrt{\pi}H}{\pi}\left[q_0(x) - q_2(x)\right]\sin \pi z/H. \tag{8.176}$$

Notes

1 My thanks to Jacob Wenegrat for many comments on this chapter and others, and to Peter Gent for various comments on equatorial dynamics.

2 Drawing from unpublished lecture notes of M. Hendershott in Chapman *et al.* (1989). For more, see Paldor *et al.* (2007) and Heifetz & Caballero (2014).

3 Waves of this type were deduced by Bjerknes (1937).

4 The first complete treatment of this problem seems to have been given by Matsuno (1966), with special cases to be found in Stern (1963) and Bretherton (1964). An analysis of the rotating linear shallow water equations on the sphere (as opposed to β-plane) was given by Longuet-Higgins (1968) with Paldor & Sigalov (2011) providing an extension to any rotating, smooth surface. A review of equatorial waves in an oceanic context was provided by McCreary (1985).

5 Standard forms are in the eye of the beholder.

6 After M. Yanai. See Yanai & Maruyama (1966).

7 A semi-implicit, semi-Lagrangian scheme. I am grateful to James Penn for the simulation and the figure.

8 Verkley & van der Velde (2010).

9 For more on this type of problem see Gill & Clarke (1974).

10 Readers who wish to study the forced problem in more detail might start with Lighthill (1969), McCreary (1981) or Clarke (2008).

11 This problem was considered by Matsuno (1966) and revisited by Gill (1980) in the context of understanding the response of the tropical atmosphere to diabatic heating. It is now commonly referred to as the *Matsuno–Gill* problem. The treatment given here is similar to that of Gill.

Adrian Gill (1937–1986) was an Australian who spent his career in the U.K., first at Cambridge University and then, all too briefly, at Oxford as part of the U.K. Meteorological Office. He is known both for his marvellous book (*Atmosphere–Ocean Dynamics*, 1982) and for his insightful work on, to name but a few topics, equatorial dynamics, internal waves, the Antarctic Circumpolar Current and adjustment processes. Gill is admired for his scientific style, for he was somehow able to distil complex problems to an austere essence that he was often able to solve analytically. He also communicated clearly and concisely, and is said to have had a rather understated sense of humour that those close to him greatly appreciated. The field suffered an untimely loss when he died, of natural causes, decades before his time.

12 A code that solves the elliptic problem using Fourier transforms and a tridiagonal inversion was graciously provided by Chris Bretherton and Adam Sobel. Timestepping can also be a particularly simple way to obtain some solutions, as noted by Matthew Barlow. Both numerical and (where possible) analytic methods were used to obtain the solutions shown.

13 For more information about Hermite polynomials and parabolic cylinder functions see, for example, Jeffreys & Jeffreys (1946), Abramowitz & Stegun (1965) or mathematical software such as Maple or Python.

Come with rain, O loud Southwester!
Bring the singer, bring the nester;
Give the buried flower a dream;
Make the settled snowbank steam.
Robert Frost, *To the Thawing Wind*, 1915.

CHAPTER 9

Barotropic and Baroclinic Instability

WHAT HYDRODYNAMIC STATES OCCUR IN NATURE? If we take as given the applicability of the Navier–Stokes equations, any flow must be a solution of these equations, subject to the relevant initial and boundary conditions. Most of the flows we experience are of course *time-dependent* solutions, not steady solutions. Why should this be? There are many steady solutions to the equations of motion — certain purely zonal flows, for example. However, steady solutions do not abound in nature because, in order to persist, they must be stable to those small perturbations that inevitably arise. Indeed, all the steady solutions that are known for the large-scale flow in the Earth's atmosphere and ocean have been found to be unstable. It is such instability that makes the subject an interesting one.

There are a myriad forms of hydrodynamic instability, but our focus in this chapter is on barotropic and baroclinic instability. *Baroclinic instability* (and we will define the term more precisely later on) is an instability that arises in rotating, stratified fluids that are subject to a horizontal temperature gradient. It is the instability that gives rise to the large- and mesoscale motion in the atmosphere and ocean — it produces atmospheric weather systems, for example — and so is, perhaps, the form of hydrodynamic instability that most affects the human condition. *Barotropic instability* is an instability that arises because of the shear in a flow, and may occur in fluids of constant density. It is important to us for two reasons: first, it is important in its own right as an instability mechanisms for jets and vortices, and as an important process in both two- and three-dimensional turbulence; second, many problems in barotropic and baroclinic instability are formally and dynamically similar, so that the solutions and insight we obtain in the often simpler problems in barotropic instability may be useful in the baroclinic problem.

9.1 KELVIN–HELMHOLTZ INSTABILITY

To introduce the issue, we first consider, rather informally, perhaps the simplest physically interesting instance of a fluid-dynamical instability — that of a constant-density flow with a shear perpendicular to the fluid's mean velocity, this being an example of a *Kelvin–Helmholtz instability*,[1] and of a barotropic instability. (More generally, Kelvin–Helmholtz instability involves fluid with varying density.) Let us consider two fluid masses of equal density, with a common interface at $y = 0$, moving with velocities $-U$ and $+U$ in the x-direction, respectively (Fig. 9.1). There is no variation in the basic flow in the z-direction (normal to the page), and we will assume this is also true for the instability (these restrictions are not essential). This flow is clearly a solution of

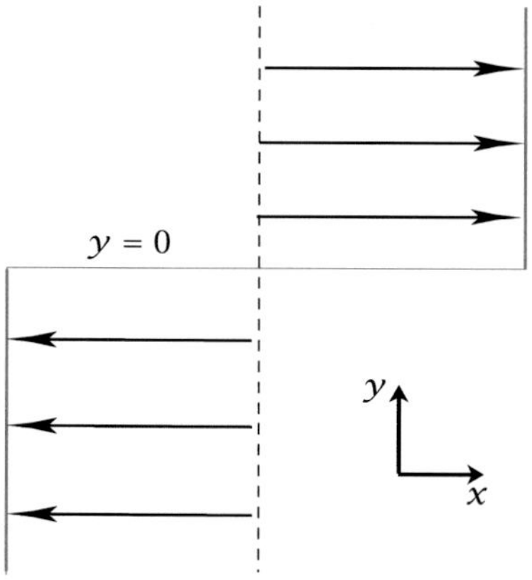

Fig. 9.1 A simple basic state giving rise to shear-flow instability. The velocity profile is discontinuous and the density is uniform.

the Euler equations. What happens if the flow is perturbed slightly? If the perturbation is initially small then even if it grows we can, for small times after the onset of instability, neglect the nonlinear interactions in the governing equations because these are the squares of small quantities. The equations determining the evolution of the initial perturbation are then the Euler equations linearized about the steady solution. Thus, denoting perturbation quantities with a prime and basic state variables with capital letters, for $y > 0$ the perturbation satisfies

$$\frac{\partial \boldsymbol{u}'}{\partial t} + U\frac{\partial \boldsymbol{u}'}{\partial x} = -\nabla p', \qquad \nabla\cdot\boldsymbol{u}' = 0, \tag{9.1a,b}$$

and a similar equation holds for $y < 0$, but with U replaced by $-U$. Given periodic boundary conditions in the x-direction, we may seek solutions of the form

$$\phi'(x,y,t) = \mathrm{Re}\sum_k \widetilde{\phi}_k(y)\exp[ik(x-ct)], \tag{9.2}$$

where ϕ is any field variable (e.g., pressure or velocity), and Re denotes that only the real part should be taken. (Typically we use tildes over variables to denote Fourier-like modes, and we will often omit the marker 'Re'.) Because (9.1a) is linear, the Fourier modes do not interact and we may confine attention to just one. Taking the divergence of (9.1a), the left-hand side vanishes and the pressure satisfies Laplace's equation

$$\nabla^2 p' = 0. \tag{9.3}$$

This has solutions in the form

$$p' = \begin{cases} \mathrm{Re}\,\widetilde{p}_1 \mathrm{e}^{ikx-ky}\mathrm{e}^{\sigma t} & y > 0, \\ \mathrm{Re}\,\widetilde{p}_2 \mathrm{e}^{ikx+ky}\mathrm{e}^{\sigma t} & y < 0, \end{cases} \tag{9.4}$$

where, anticipating the possibility of growing solutions, we have written the time variation in terms of a growth rate, $\sigma = -ikc$. In general σ is complex: if it has a positive real component, the amplitude of the perturbation will grow and there is an instability; if σ has a non-zero imaginary component, then there will be oscillatory motion, and there may be both oscillatory motion *and* an instability. To obtain the dispersion relationship, we consider the y-component of (9.1a), namely (for $y > 0$)

$$\frac{\partial v_1'}{\partial t} + U\frac{\partial v_1'}{\partial x} = -\frac{\partial p_1'}{\partial y}. \tag{9.5}$$

Substituting a solution of the form $v_1' = \widetilde{v}_1\exp(ikx + \sigma t)$ yields, with (9.4),

$$(\sigma + ikU)\widetilde{v}_1 = k\widetilde{p}_1. \tag{9.6}$$

But the velocity normal to the interface is, at the interface, nothing but the rate of change of the position of the interface itself; that is, at $y = +0$

$$v_1 = \frac{\partial \eta'}{\partial t} + U\frac{\partial \eta'}{\partial x}, \tag{9.7}$$

or

$$\tilde{v}_1 = (\sigma + \mathrm{i}kU)\tilde{\eta}, \tag{9.8}$$

where η' is the displacement of the interface from its equilibrium position. Using this in (9.6) gives

$$(\sigma + \mathrm{i}kU)^2\tilde{\eta} = k\tilde{p}_1. \tag{9.9}$$

The above few equations pertain to motion on the $y > 0$ side of the interface. Similar reasoning on the other side gives (at $y = -0$)

$$(\sigma - \mathrm{i}kU)^2\tilde{\eta} = -k\tilde{p}_2. \tag{9.10}$$

But at the interface $p_1 = p_2$, because pressure must be continuous. The dispersion relationship then emerges from (9.9) and (9.10), giving

$$\sigma^2 = k^2U^2. \tag{9.11}$$

This equation has two roots, one of which is positive. Thus, the amplitude of the perturbation grows exponentially, like $\mathrm{e}^{\sigma t}$, and the flow is *unstable*. The instability itself can be seen in the natural world when billow clouds appear wrapped up into spirals: the clouds are acting as tracers of fluid flow, and are a manifestation of the instability at finite amplitude, as seen later in Fig. 9.6.

9.2 INSTABILITY OF PARALLEL SHEAR FLOW

We now consider a little more systematically the instability of *parallel shear flows,* such as are illustrated in Fig. 9.2. This is a classic problem in hydrodynamic stability theory, and there are two particular reasons for our own interest:

(i) The instability is an example of *barotropic instability,* which abounds in the ocean and atmosphere. Roughly speaking, barotropic instability arises when a flow is unstable by virtue of its horizontal shear, with gravitational and buoyancy effects being secondary.

(ii) The instability is in many ways analogous to *baroclinic instability,* which is the main instability giving rise to weather systems in the atmosphere and similar phenomena in the ocean.

We will restrict attention to two-dimensional, incompressible flow; this illustrates the physical mechanisms in the most transparent way, in part because it allows for the introduction of a streamfunction and the automatic satisfaction of the mass continuity equation. In fact, for parallel two-dimensional shear flows the most unstable disturbances are two-dimensional ones.[2]

The vorticity equation for incompressible two-dimensional flow is just

$$\frac{\mathrm{D}\zeta}{\mathrm{D}t} = 0. \tag{9.12}$$

We suppose the basic state to be a parallel flow in the x-direction that may vary in the y-direction. That is

$$\overline{\boldsymbol{u}} = U(y)\mathbf{i}. \tag{9.13}$$

The linearized vorticity equation is then

$$\frac{\partial \zeta'}{\partial t} + U\frac{\partial \zeta'}{\partial x} + v'\frac{\partial Z}{\partial y} = 0, \tag{9.14}$$

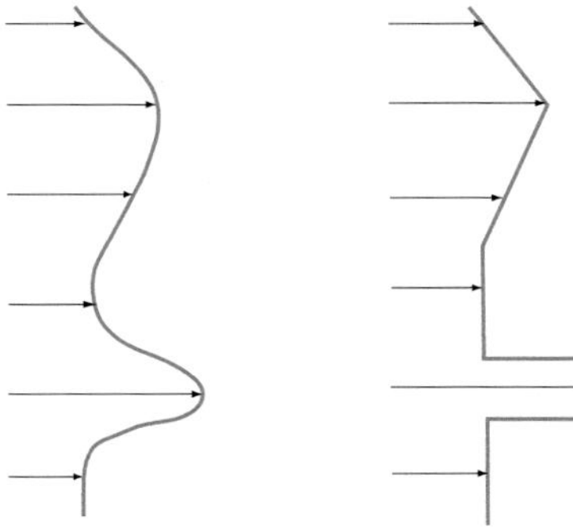

Fig. 9.2 Left: example of a smooth velocity profile, in which both the velocity and the vorticity are continuous. Right: example of a piecewise continuous profile, in which the velocity and vorticity may have finite discontinuities.

where $\mathcal{Z} = -\partial_y U$. Because the mass continuity equation has the simple form $\partial u'/\partial x + \partial v'/\partial y = 0$, we may introduce a streamfunction ψ such that $u' = -\partial\psi'/\partial y$, $v' = \partial\psi'/\partial x$ and $\zeta' = \nabla^2\psi'$. The linear vorticity equation becomes

$$\frac{\partial \nabla^2\psi'}{\partial t} + U\frac{\partial \nabla^2\psi'}{\partial x} + \frac{\partial \mathcal{Z}}{\partial y}\frac{\partial \psi'}{\partial x} = 0. \tag{9.15}$$

The coefficients of the x-derivatives are not themselves functions of x; thus, we may seek solutions that are harmonic functions (sines and cosines) in the x-direction, but the y dependence must remain arbitrary at this stage and we write

$$\psi' = \mathrm{Re}\,\widetilde{\psi}(y)\mathrm{e}^{\mathrm{i}k(x-ct)}. \tag{9.16}$$

The full solution is a superposition of all wavenumbers, but since the problem is linear the waves do not interact and it suffices to consider them separately. If c is purely real then c is the phase speed of the wave; if c has a positive imaginary component then the wave will grow exponentially and is thus *unstable.*

From (9.16) we have

$$u' = \widetilde{u}(y)\mathrm{e}^{\mathrm{i}k(x-ct)} = -\widetilde{\psi}_y\mathrm{e}^{\mathrm{i}k(x-ct)}, \tag{9.17a}$$

$$v' = \widetilde{v}(y)\mathrm{e}^{\mathrm{i}k(x-ct)} = \mathrm{i}k\widetilde{\psi}\mathrm{e}^{\mathrm{i}k(x-ct)}, \tag{9.17b}$$

$$\zeta' = \widetilde{\zeta}(y)\mathrm{e}^{\mathrm{i}k(x-ct)} = (-k^2\widetilde{\psi} + \widetilde{\psi}_{yy})\mathrm{e}^{\mathrm{i}k(x-ct)}, \tag{9.17c}$$

where the y subscript denotes a derivative. Using (9.17) in (9.14) gives

$$(U-c)(\widetilde{\psi}_{yy} - k^2\widetilde{\psi}) - U_{yy}\widetilde{\psi} = 0, \tag{9.18}$$

which is known as *Rayleigh's equation.*[3] It is the linear vorticity equation for disturbances to parallel shear flow, and in the presence of a β-effect it generalizes slightly to

$$(U-c)(\widetilde{\psi}_{yy} - k^2\widetilde{\psi}) + (\beta - U_{yy})\widetilde{\psi} = 0, \tag{9.19}$$

which is known as the Rayleigh–Kuo equation.

9.2.1 Piecewise Linear Flows

Although Rayleigh's equation is linear and has a simple form, it is nevertheless quite difficult to analytically solve for an arbitrary smoothly varying profile. It is simpler to consider *piecewise linear* flows, in which U_y is a constant over some interval, with U or U_y changing abruptly to another value at a line of discontinuity, as illustrated in Fig. 9.2. The curvature, U_{yy} is accounted for through the satisfaction of matching conditions, analogous to boundary conditions, at the lines of discontinuity (as in Section 9.1), and solutions in each interval are then exponential functions.

Jump or matching conditions

The idea, then, is to solve the linearized vorticity equation separately in the continuous intervals in which vorticity is constant, matching the solution with that in the adjacent regions. The matching conditions arise from two physical conditions:

(i) That normal stress should be continuous across the interface. For an inviscid fluid this implies that pressure be continuous.

(ii) That the normal velocity of the fluid on either side of the interface should be consistent with the motion of the interface itself.

Let us consider the implications of these two conditions.

(i) Continuity of pressure
The linearized momentum equation in the direction along the interface is:

$$\frac{\partial u'}{\partial t} + U\frac{\partial u'}{\partial x} + v'\frac{\partial U}{\partial y} = -\frac{\partial p'}{\partial x}. \tag{9.20}$$

For normal modes, $u' = -\widetilde{\psi}_y e^{ik(x-ct)}$, $v' = ik\widetilde{\psi}e^{ik(x-ct)}$ and $p' = \widetilde{p}e^{ik(x-ct)}$, and (9.20) becomes

$$ik(U-c)\widetilde{\psi}_y - ik\widetilde{\psi}U_y = -ik\widetilde{p}. \tag{9.21}$$

Because pressure is continuous across the interface we have the first *matching* or *jump condition,*

$$\Delta[(U-c)\widetilde{\psi}_y - \widetilde{\psi}U_y] = 0, \tag{9.22}$$

where the operator Δ denotes the difference in the values of the argument (in square brackets) across the interface. That is, the quantity $(U-c)\widetilde{\psi}_y - \widetilde{\psi}U_y$ is continuous.

We can obtain this condition directly from Rayleigh's equation, (9.19), written in the form

$$[(U-c)\widetilde{\psi}_y - U_y\widetilde{\psi}]_y + [\beta - k^2(U-c)]\widetilde{\psi} = 0. \tag{9.23}$$

Integrating across the interface gives (9.22).

(ii) Material interface condition
At the interface, the normal velocity v is given by the kinematic condition

$$v = \frac{D\eta}{Dt}, \tag{9.24}$$

where η is the interface displacement. The linear version of (9.24) is

$$\frac{\partial \eta'}{\partial t} + U\frac{\partial \eta'}{\partial x} = \frac{\partial \psi'}{\partial x}. \tag{9.25}$$

If the fluid itself is continuous then this equation must hold at either side of the interface, giving two equations and their normal-mode counterparts, namely,

$$\frac{\partial \eta'}{\partial t} + U_1\frac{\partial \eta'}{\partial x} = \frac{\partial \psi_1'}{\partial x} \quad\longrightarrow\quad (U_1 - c)\widetilde{\eta} = \widetilde{\psi}_1, \tag{9.26}$$

$$\frac{\partial \eta'}{\partial t} + U_2 \frac{\partial \eta'}{\partial x} = \frac{\partial \psi_2'}{\partial x} \quad \longrightarrow \quad (U_2 - c)\tilde{\eta} = \tilde{\psi}_2. \tag{9.27}$$

Material continuity at the interface thus gives the second jump condition

$$\Delta\left[\frac{\tilde{\psi}}{U - c}\right] = 0. \tag{9.28}$$

That is, $\tilde{\psi}/(U - c)$ is continuous at the interface. Note that if U is continuous across the interface the condition becomes one of continuity of the normal velocity.

9.2.2 Kelvin–Helmholtz Instability, Revisited

We now use Rayleigh's equation and the jump conditions to consider the situation illustrated in Fig. 9.1; that is, vorticity is everywhere zero except in a thin sheet at $y = 0$. On either side of the interface, Rayleigh's equation is simply

$$(U_i - c)(\partial_{yy}\tilde{\psi}_i - k^2\tilde{\psi}_i) = 0, \qquad i = 1, 2 \tag{9.29}$$

or, assuming that $U_i \neq c$, $\partial_{yy}\tilde{\psi}_i - k^2\tilde{\psi}_i = 0$. (This is just Laplace's equation, coming from $\nabla^2\psi' = \zeta'$, with $\zeta' = 0$ everywhere except at the interface.) Solutions of this that decay away on either side of the interface are

$$y > 0: \qquad \tilde{\psi}_1 = \Psi_1 e^{-ky}, \tag{9.30a}$$

$$y < 0: \qquad \tilde{\psi}_2 = \Psi_2 e^{ky}, \tag{9.30b}$$

where Ψ_1 and Ψ_2 are constants. The boundary condition (9.22) gives

$$(U_1 - c)(-k)\Psi_1 = (U_2 - c)(k)\Psi_2, \tag{9.31}$$

and (9.28) gives

$$\frac{\Psi_1}{(U_1 - c)} = \frac{\Psi_2}{(U_2 - c)}. \tag{9.32}$$

The last two equations combine to give $(U_1 - c)^2 = -(U_2 - c)^2$, which, supposing that $U = U_1 = -U_2$ gives $c^2 = -U^2$. Thus, since U is purely real, $c = \pm iU$, and the disturbance grows exponentially as $\exp(kU_1t)$, just as we obtained in Section 9.1. All wavelengths are unstable, and indeed the shorter the wavelength the greater the instability. In reality, viscosity will damp the smallest waves, although the presence of viscosity would also mean that the initial profile is not an exact, steady solution of the equations of motion.

9.2.3 Edge Waves

We now consider a case sketched in Fig. 9.3 in which the velocity is continuous, but the vorticity is discontinuous. Since on either side of the interface $U_{yy} = 0$, Rayleigh's equation is just

$$(U(y) - c)(\tilde{\psi}_{yy} - k^2\tilde{\psi}) = 0. \tag{9.33}$$

Provided $c \neq U$ this has solutions,

$$\tilde{\psi} = \begin{cases} \Phi_1 e^{-ky} & y > 0 \\ \Phi_2 e^{ky} & y < 0. \end{cases} \tag{9.34}$$

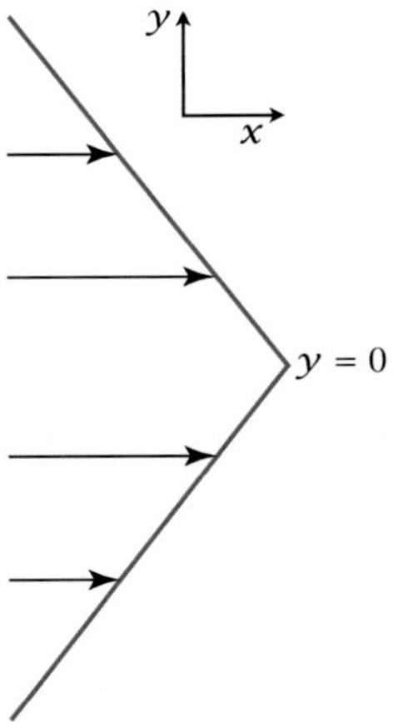

Fig. 9.3 Velocity profile of a point jet, in which vorticity is concentrated at a point. Although the vorticity is discontinuous, a small perturbation gives rise only to *edge waves* centred at $y = 0$, and so the jet is stable.

The value of c is found by applying the jump conditions (9.22) and (9.28) at $y = 0$. Using (9.34) these give

$$-k(U_0 - c)\Phi_1 - \Phi_1 \partial_y U_1 = k(U_0 - c)\Phi_2 - \Phi_2 \partial_y U_2 \qquad (9.35a)$$

$$\Phi_1 = \Phi_2, \qquad (9.35b)$$

where $U_1(y)$ and $U_2(y)$ are the velocities on either side of the interface, and both are equal to U_0 at the interface. (In Fig. 9.3 we illustrate the case with $U_1 = -Ay$ and $U_2 = Ay$, where A is a positive constant.) After a line of algebra the above equations give

$$c = U_0 + \frac{\partial_y U_1 - \partial_y U_2}{2k}. \qquad (9.36)$$

This is the dispersion relationship for *edge waves* that propagate along the interface with speed equal to the sum of the fluid speed and a factor proportional to the difference in the vorticity between the two layers. No matter what the shear is on either side of the interface, the phase speed is purely real and there is no instability. Equation (9.36) is analogous to the Rossby wave dispersion relation $c = U_0 - \beta/K^2$, and reflects a similarity in the physics — β is a planetary vorticity gradient, which in (9.36) is collapsed to a front and represented by the difference $\partial_y U_1 - \partial_y U_2 = -(Z_1 - Z_2)$, where Z_1 and Z_2 are the basic-state vorticities on either side of the interface. One might even say that such edge waves *are* Rossby waves of a very simple form.

9.2.4 Interacting Edge Waves Producing Instability

Now we consider a slightly more complicated case in which edge waves may interact giving rise, as we shall see, to an instability. The physical situation is illustrated in Fig. 9.4. We consider the simplest case, that of a shear layer (which we denote as region 2) sandwiched between two semi-infinite layers, regions 1 and 3, as in the left-hand panel of the figure. Thus, the basic state is

$$y > a: \qquad U = U_1 = U_0 \ \text{(a constant)}, \qquad (9.37a)$$

$$-a < y < a: \qquad U = U_2 = \frac{U_0}{a} y, \qquad (9.37b)$$

$$y < -a: \qquad U = U_3 = -U_0. \qquad (9.37c)$$

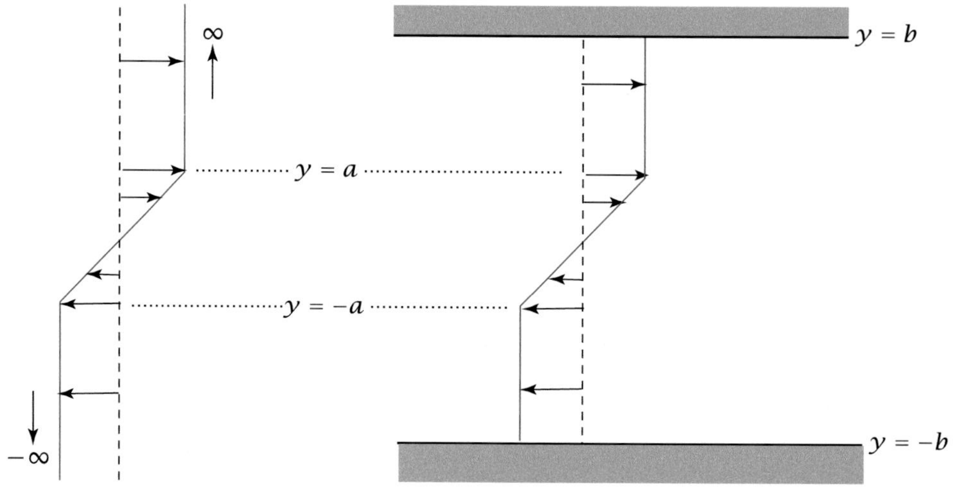

Fig. 9.4 Barotropically unstable velocity profiles. In the simplest case, on the left, a region of shear is sandwiched between two infinite regions of constant velocity. The edge waves at $y = \pm a$ interact to produce an instability. If $a = 0$, then the situation corresponds to that of Fig. 9.1, giving Kelvin–Helmholtz instability. In the case on the right, the flow is bounded at $y = \pm b$. The flow is unstable provided that b is sufficiently larger than a. If $b = a$ (plane Couette flow) the flow is stable to infinitesimal disturbances.

We assume a solution of Rayleigh's equation of the form:

$$y > a: \qquad \widetilde{\psi}_1 = A\mathrm{e}^{-k(y-a)}, \tag{9.38a}$$

$$-a < y < a: \qquad \widetilde{\psi}_2 = B\mathrm{e}^{k(y-a)} + C\mathrm{e}^{-k(y+a)}, \tag{9.38b}$$

$$y < -a: \qquad \widetilde{\psi}_3 = D\mathrm{e}^{k(y+a)}. \tag{9.38c}$$

These particular forms all decay away from the interfaces at the edges of domains in which the assumed solutions apply, as edge waves must do. Applying the jump conditions (9.22) and (9.28) at the interfaces at $y = a$ and $y = -a$ gives the following relations between the coefficients:

$$-A[(U_0 - c)k] = B\left[(U_0 - c)k - \frac{U_0}{a}\right] - C\mathrm{e}^{-2ka}\left[\frac{U_0}{a} + (U_0 - c)k\right], \tag{9.39a}$$

$$A = B + C\mathrm{e}^{-2ka}, \tag{9.39b}$$

$$D[(U_0 + c)k] = B\mathrm{e}^{-2ka}\left[(U_0 + c)k + \frac{U_0}{a}\right] + C\left[\frac{U_0}{a} - (U_0 + c)k\right], \tag{9.39c}$$

$$D = B\mathrm{e}^{-2ka} + C. \tag{9.39d}$$

These are a set of four homogeneous equations, with the unknown parameters A, B, C and D, which may be written in the form of a matrix equation,

$$\begin{pmatrix} k(U_0 - c) & k(U_0 - c) - U_0/a & -\mathrm{e}^{-2ka}[U_0/a + k(U_0 - c)] & 0 \\ 1 & -1 & -\mathrm{e}^{-2ka} & 0 \\ 0 & -\mathrm{e}^{-2ka}[k(U_0 + c) + U_0/a] & k(U_0 + c) - (U_0/a) & k(U_0 + c) \\ 0 & \mathrm{e}^{-2ka} & 1 & -1 \end{pmatrix} \begin{pmatrix} A \\ B \\ C \\ D \end{pmatrix} = 0. \tag{9.40}$$

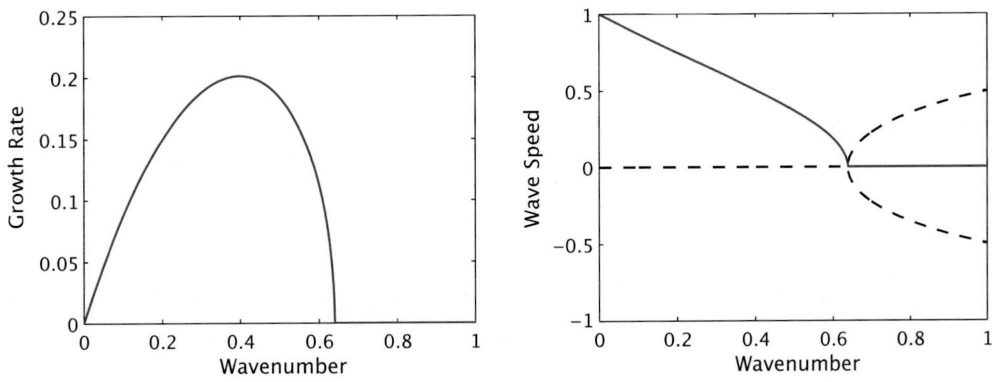

Fig. 9.5 Left: Growth rate ($\sigma = kc_i$) calculated from (9.41) with c nondimensionalized by U_0 and k nondimensionalized by $1/a$ (equivalent to setting $a = U_0 = 1$). Right: Real (c_r, dashed) and imaginary (c_i, solid) wave speeds. The flow is unstable for $k < 0.63$, with the maximum instability occurring at $k = 0.39$.

For non-trivial solutions the determinant of the matrix must be zero, and solving the ensuing equation gives, after some algebra, the dispersion relationship[4]

$$c^2 = \left(\frac{U_0}{2ka}\right)^2 \left[(1 - 2ka)^2 - \mathrm{e}^{-4ka}\right], \tag{9.41}$$

and this is plotted in Fig. 9.5. The flow is unstable for sufficiently long wavelengths, since then the right-hand side of (9.41) is negative. The critical wavenumber below which instability occurs is found by solving $(1 - 2ka)^2 = \mathrm{e}^{-4ka}$, which gives instability for $ka < 0.63293$. A numerical solution of the initial value problem is illustrated in Figs. 9.6 and 9.7. Here, the initial perturbation is small and random, containing components at all wavenumbers. All the modes in the unstable range grow exponentially, and the pattern is soon dominated by the mode that grows fastest — a horizontal wavenumber of 3 in this problem. Eventually, the perturbation grows sufficiently that the linear equations are no longer valid and, as is seen in the second column of Fig. 9.6, vortices form and pinch off. The vortices interact and the flow develops into two-dimensional turbulence, as considered in Chapter 11.

The mechanism of the instability — an informal view

(A similar mechanism is discussed in Section 9.7, and the reader may wish to read the two descriptions in tandem.) We have seen that an edge wave in isolation is stable, with the instability arising when two edge waves have sufficient cross-stream extent that they can interact with each other. This occurs for sufficiently long wavelengths because the cross-stream decay scale is proportional to the along-stream wavelength — hence the high-wavenumber cut-off. To see the mechanism of the instability transparently, let us first suppose that the interfaces are, in fact, sufficiently far away that the edge waves at each interface do not interact. Using (9.36) the edge waves at $y = -a$ and $y = +a$ have dispersion relationships

$$c_{+a} = U_0 - \frac{U_0/a}{2k}, \qquad c_{-a} = -U_0 + \frac{U_0/a}{2k}. \tag{9.42a,b}$$

If the two waves are to interact these phase speeds must be equal, giving the condition

$$c = 0, \qquad k = 1/(2a). \tag{9.43a,b}$$

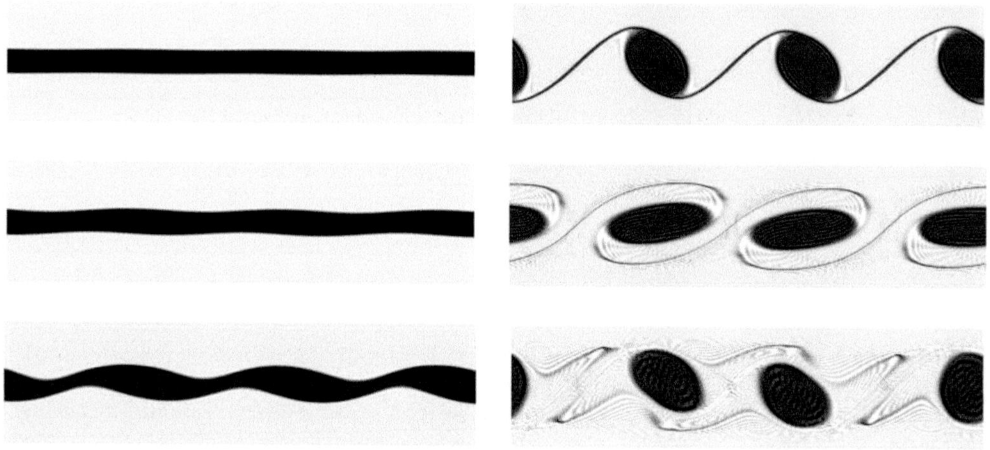

Fig. 9.6 A sequence of plots of the vorticity, at equal time intervals, from a numerical solution of the nonlinear vorticity equation (9.12), with initial conditions as in Fig. 9.4 with $a = 0.1$, plus a very small random perturbation. Time increases first down the left column and then down the right column. The solution is obtained in a rectangular (4×1) domain, with periodic conditions in the x-direction and slippery walls at $y = (0, 1)$. The maximum linear instability occurs for a wavelength of 1.57, which for a domain of length 4 corresponds to a wavenumber of 2.55. Since the periodic domain quantizes the allowable wavenumbers, the maximum instability is at wavenumber 3, and this is what emerges. Only in the first two or three frames is the linear approximation valid.

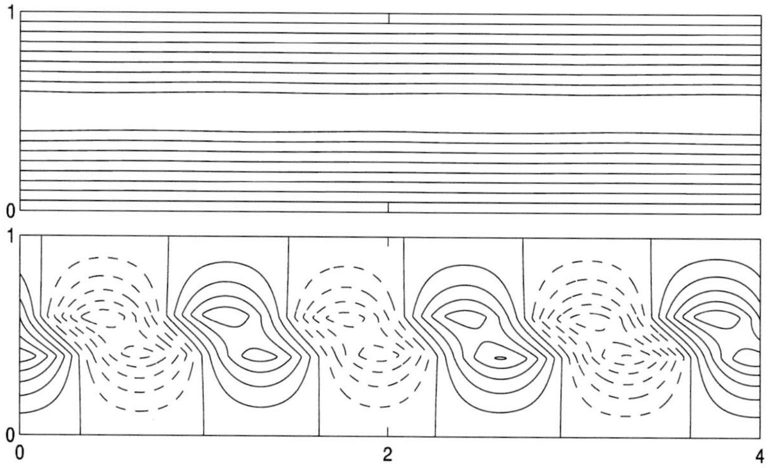

Fig. 9.7 The total streamfunction (top panel) and the perturbation streamfunction from the same numerical calculation as in Fig. 9.6, at a time corresponding to the second frame. Positive values are solid lines, and negative values are dashed. The perturbation pattern leans into the shear, and grows exponentially in place.

That is, the waves are stationary, and their wavelength is proportional to the separation of the two edges. In fact, (9.43) approximately characterizes the conditions at the critical wavenumber $k = 0.63/a$ (see Fig. 9.5). In the region of the shear the two waves have the form

$$\psi_{+a} = \mathrm{Re}\,\widetilde{\psi}_{+a}(t)\mathrm{e}^{k(y-a)}\mathrm{e}^{\mathrm{i}\phi}\mathrm{e}^{ikx}, \qquad \psi_{-a} = \mathrm{Re}\,\widetilde{\psi}_{-a}(t)\mathrm{e}^{-k(y+a)}\mathrm{e}^{ikx}, \tag{9.44a,b}$$

where ϕ is the phase shift between the waves; in the case of pure edge waves we have $\widetilde{\psi}_{\pm a} = A_{\pm a}\mathrm{e}^{-ikct}$ where we may take $A_{\pm a}$ to be real.

Now consider how the wave generated at $y = -a$ might affect the wave at $y = +a$ and vice versa. The contribution of ψ_{-a} to the acceleration of ψ_{+a} is given by applying the x-momentum equation, (9.20), at $y = +a$. Thus we take the kinematic solutions, (9.44), and use them in a dynamical equation, the momentum equation, to calculate the ensuing acceleration. At $y = +a$ we heuristically write

$$\frac{\partial u'_{+a}}{\partial t} = -[v'_{+a}(+a) + v'_{-a}(+a)]\frac{\mathrm{d}U}{\mathrm{d}y} \tag{9.45}$$

with a similar expression at $y = -a$, and omitting the pressure terms. Here $v'_{-a}(+a)$ denotes the value of v at $y = +a$ due to the edge wave generated at $-a$. It is the second term on the right-hand side that is necessary for any potential instability, as the first term gives only neutral edge waves. If the spatial dependence of the waves is given by (9.44), then at $y = +a$ we have, omitting the contribution from the edge waves,

$$\frac{\partial u'_{+a}}{\partial t} = -v'_{-a}(+a)\frac{\mathrm{d}U}{\mathrm{d}y} \qquad \Rightarrow \qquad -k\mathrm{e}^{\mathrm{i}\phi}\mathrm{e}^{ikx}\frac{\mathrm{d}\widetilde{\psi}_{+a}}{\mathrm{d}t} = -ik\mathrm{e}^{ikx}\mathrm{e}^{-2ka}\frac{\partial U}{\partial y}\widetilde{\psi}_{-a}. \tag{9.46}$$

The real part of this equation is: $\cos(kx + \phi)\mathrm{d}\widetilde{\psi}_{+a}/\mathrm{d}t = -\sin kx\,\mathrm{e}^{-2ka}\mathrm{d}U/\mathrm{d}y$. If the edge waves have the appropriate phase with respect to each other then the two edge waves can feed back on each other and couple to form a single growing mode. In particular if $\phi = \pi/2$ then the evolution equations for ψ_{+a} and ψ_{-a} become

$$\frac{\mathrm{d}\widetilde{\psi}_{+a}}{\mathrm{d}t} = \mathrm{e}^{-2ka}\frac{\mathrm{d}U}{\mathrm{d}y}\widetilde{\psi}_{-a}, \qquad \frac{\mathrm{d}\widetilde{\psi}_{-a}}{\mathrm{d}t} = \mathrm{e}^{-2ka}\frac{\mathrm{d}U}{\mathrm{d}y}\widetilde{\psi}_{+a}, \tag{9.47a,b}$$

which give exponential growth. When $\phi = \pi/2$ the wave at $y = +a$ lags the wave at $y = -a$, and *the unstable perturbation tilts into the shear;* this property of the instability is seen in the full solution, Fig. 9.7.

9.3 NECESSARY CONDITIONS FOR INSTABILITY

9.3.1 Rayleigh's Criterion

For simple profiles it may be possible to calculate, or even intuit, the instability properties, but for continuous profiles of $U(y)$ this is often impossible and it would be nice to have some general guidelines as to when a profile might be unstable. To this end, we derive a couple of *necessary* conditions for instability, or *sufficient* conditions for stability, that will at least tell us if a flow *might* be unstable. We begin by writing Rayleigh's equation, (9.19), as

$$\widetilde{\psi}_{yy} - k^2\widetilde{\psi} + \frac{\beta - U_{yy}}{U - c}\widetilde{\psi} = 0. \tag{9.48}$$

Multiply by $\widetilde{\psi}^*$ (the complex conjugate of $\widetilde{\psi}$) and integrate over the domain of interest. After integrating the first term by parts we obtain

$$\int_{y_1}^{y_2}\left(\left|\frac{\partial\widetilde{\psi}}{\partial y}\right|^2 + k^2|\widetilde{\psi}|^2\right)\mathrm{d}y - \int_{y_1}^{y_2}\frac{\beta - U_{yy}}{U - c}|\widetilde{\psi}|^2\,\mathrm{d}y = 0, \tag{9.49}$$

assuming that $\widetilde{\psi}$ vanishes at the boundaries. (The limits to the integral may be infinite, in which case it is assumed that $\widetilde{\psi}$ decays to zero as $|y|$ approaches ∞.) The only variable in this expression that is complex is c, and thus the first integral is real. The imaginary component of the second integral is

$$c_i \int \frac{\beta - U_{yy}}{|U - c|^2} |\widetilde{\psi}|^2 \, \mathrm{d}y = 0. \tag{9.50}$$

Thus, *either* c_i vanishes or the integral does. For there to be an instability c_i must be non-zero. The eigenvalues of Rayleigh's equation come in pairs, for each decaying mode (negative c_i) there is a corresponding growing mode (positive c_i). If c_i is to be non-zero, the integrand of (9.50) must be zero and therefore:

> *A necessary condition for instability is that the expression*
> $$\beta - U_{yy}$$
> *change sign somewhere in the domain.*

Equivalently, a sufficient criterion for stability is that $\beta - U_{yy}$ does not vanish in the domain interior. This condition is known as Rayleigh's inflection-point criterion, or when $\beta \neq 0$, the Rayleigh–Kuo inflection point criterion.[5]

A more general derivation

Consider again the vorticity equation, linearized about a parallel shear flow (cf. (9.14) with a β-term),

$$\frac{\partial \zeta}{\partial t} + U \frac{\partial \zeta}{\partial x} + v \left(\frac{\partial Z}{\partial y} + \beta \right) = 0, \tag{9.51}$$

(dropping the primes on the perturbation quantities). Multiply by ζ and divide by $\beta + Z_y$ to obtain

$$\frac{\partial}{\partial t} \left(\frac{\zeta^2}{\beta + Z_y} \right) + \frac{U}{\beta + Z_y} \frac{\partial \zeta^2}{\partial x} + 2v\zeta = 0, \tag{9.52}$$

and then integrate with respect to x to give

$$\frac{\partial}{\partial t} \int \left(\frac{\zeta^2}{\beta + Z_y} \right) \mathrm{d}x = -2 \int v\zeta \, \mathrm{d}x. \tag{9.53}$$

Now, using $\nabla \cdot \boldsymbol{u} = 0$, the vorticity flux may be written as

$$v\zeta = -\frac{\partial}{\partial y}(uv) + \frac{1}{2} \frac{\partial}{\partial x}(v^2 - u^2). \tag{9.54}$$

That is, *the flux of vorticity is the divergence of some quantity.* Its integral therefore vanishes provided there are no contributions from the boundary, and integrating (9.53) with respect to y gives

$$\frac{\mathrm{d}}{\mathrm{d}t} \int \left(\frac{\zeta^2}{\beta + Z_y} \right) \mathrm{d}x \, \mathrm{d}y = 0. \tag{9.55}$$

If there is to be an instability ζ must grow, but the integral is identically zero. These two conditions can only be simultaneously satisfied if $\beta + Z_y$, or equivalently $\beta - U_{yy}$, is zero somewhere in the domain.

This derivation shows that the inflection-point criterion applies even if disturbances are not of normal-mode form. The quantity $\zeta^2/(\beta + Z_y)$ is an example of a *wave-activity density* — a wave activity being a conserved quantity, quadratic in the amplitude of the wave. Such quantities play an important role in instabilities, and we consider then further in Chapter 10.

9.3.2 Fjørtoft's Criterion

Another necessary condition for instability was obtained by Fjørtoft.[6] In this section we will derive his condition for normal-mode disturbances, and provide a more general derivation in Section 10.7. From the real part of (9.49) we find

$$\int_{y_1}^{y_2} (\beta - U_{yy}) \frac{(U - c_r)}{|U - c|^2} |\widetilde{\psi}|^2 \, dy = \int_{y_1}^{y_2} \left| \frac{\partial \widetilde{\psi}}{\partial y} \right|^2 + k^2 |\widetilde{\psi}|^2 \, dy > 0. \tag{9.56}$$

Now, from (9.50), we know that for an instability we must have

$$\int_{y_1}^{y_2} \frac{\beta - U_{yy}}{|U - c|^2} |\widetilde{\psi}|^2 \, dy = 0. \tag{9.57}$$

Using this equation and (9.56) we see that, for an instability,

$$\int_{y_1}^{y_2} (\beta - U_{yy}) \frac{(U - U_s)}{|U - c|^2} |\widetilde{\psi}|^2 \, dy > 0, \tag{9.58}$$

where U_s is *any* real constant. It is most useful to choose this constant to be the value of $U(y)$ at which $\beta - U_{yy}$ vanishes. This leads directly to the criterion:

> *A necessary condition for instability is that the expression*
> $$(\beta - U_{yy})(U - U_s),$$
> *where U_s is the value of $U(y)$ at which $\beta - U_{yy}$ vanishes, be positive somewhere in the domain.*

This criterion is satisfied if the magnitude of the vorticity has an extremum inside the domain, and not at the boundary or at infinity (Fig. 9.8). Why choose U_s in the manner we did? Suppose we chose U_s to have a large positive value, so that $U - U_s$ is negative everywhere. Then (9.58) just implies that $\beta - U_{yy}$ must be negative somewhere, and this is already known from Rayleigh's criterion. If we choose U_s to be large and negative, we simply find that $\beta - U_{yy}$ must be positive somewhere. The most stringent criterion is obtained by choosing U_s to be the value of $U(y)$ at which $\beta - U_{yy}$ vanishes.

Interestingly, the β-effect can be either stabilizing or destabilizing: It can stabilize the middle two profiles of Fig. 9.8, because if it is large enough $\beta - U_{yy}$ will be one-signed. However, the β-effect will destabilize a westward point jet, $U(y) = -(1 - |y|)$ (the negative of the jet in Fig. 9.3), because $\beta - U_{yy}$ is negative at $y = 0$ and positive elsewhere. An eastward point jet is stable, with or without β. Finally, we emphasize that both Fjørtoft's and Rayleigh's criteria are necessary conditions for instability, and examples exist that do satisfy their criterion, yet which are stable to infinitesimal perturbations.

9.4 BAROCLINIC INSTABILITY

Baroclinic instability is a hydrodynamic instability that occurs in stably stratified, rotating fluids, and it is ubiquitous in the Earth's atmosphere and oceans, and almost certainly occurs in other planetary atmospheres. It gives rise to weather, and so is perhaps the form of hydrodynamic instability that affects us most, and that certainly we talk about most.

9.4.1 A Physical Picture

We will first draw a picture of baroclinic instability as a form of 'sloping convection' in which the fluid, although statically stable, is able to release available potential energy when parcels move along

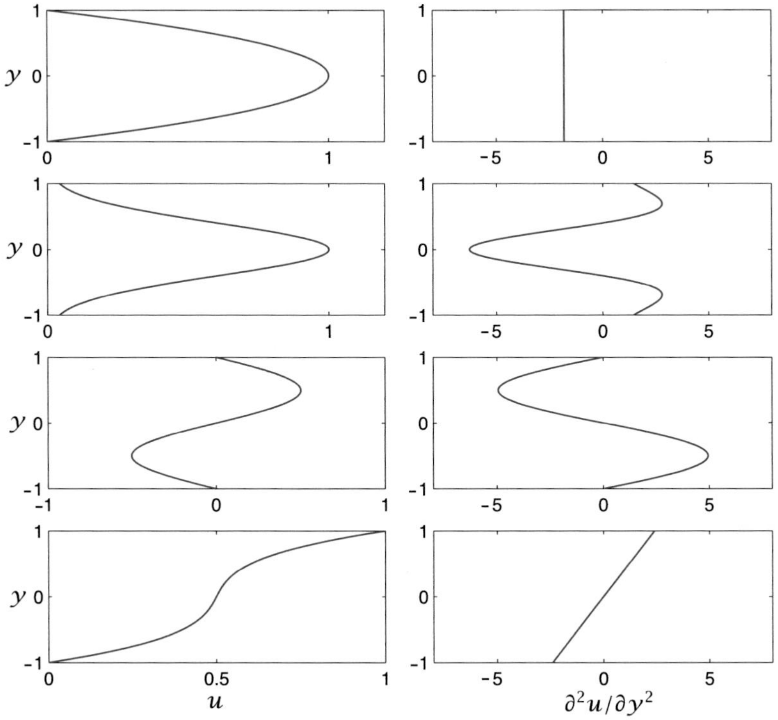

Fig. 9.8 Example parallel velocity profiles (left column) and their second derivatives (right column). From the top: Poiseuille flow ($u = 1 - y^2$); a Gaussian jet; a sinusoidal profile; a polynomial profile.

By Rayleigh's criterion, the top profile is stable, whereas the lower three are potentially unstable. However, the bottom profile is stable by Fjørtoft's criterion (note that the vorticity maxima are at the boundaries). If the β-effect were present and large enough it would stabilize the middle two profiles.

a sloping path. To this end, let us first ask: what is the basic state that is baroclinically unstable? In a stably stratified fluid potential density decreases with height; we can also easily imagine a state in which the basic state temperature decreases, and the potential density increases, polewards. (We will couch most of our discussion in terms of the Boussinesq equations, and henceforth drop the qualifier 'potential' from density.) Can we construct a steady solution from these two conditions? The answer is yes, provided the fluid is also rotating; rotation is necessary because the meridional temperature gradient generally implies a meridional pressure gradient; there is nothing to balance this in the absence of rotation, and a fluid parcel would therefore accelerate. In a rotating fluid this pressure gradient can be balanced by the Coriolis force and a steady solution can be maintained even in the absence of viscosity. Consider a stably stratified Boussinesq fluid in geostrophic and hydrostatic balance on an f-plane, with buoyancy decreasing uniformly polewards. Then $fu = -\partial\phi/\partial y$ and $\partial\phi/\partial z = b$, where $b = -g\delta\rho/\rho_0$ is the buoyancy. These together give the thermal wind relation, $\partial u/\partial z = \partial b/\partial y$. If there is no variation of these fields in the zonal direction, then, for *any* variation of b with y, this is a steady solution to the primitive equations of motion, with $v = w = 0$.

The density structure corresponding to a uniform increase of density in the meridional direction is illustrated in Fig. 9.9. Is this structure stable to perturbations? The answer is no, although the perturbations must be a little special. Suppose the particle at 'A' is displaced upwards; then, since the fluid is (by assumption) stably stratified it will be denser than its surroundings and hence experience a restoring force, and similarly if displaced downwards. Suppose, however, we interchange the two parcels at positions 'A' and 'B'. Parcel 'A' finds itself surrounded by parcels of higher density that itself, and it is therefore buoyant; it is also higher than where it started. Parcel 'B' is negatively buoyant, and at a lower altitude than where is started. Thus, overall, the centre of gravity of the fluid has been lowered, and so its overall potential energy lowered. This loss in potential energy (PE) of the basic state must be accompanied by a gain in kinetic energy of the perturbation. Thus, the perturbation amplifies and converts potential energy into kinetic energy.

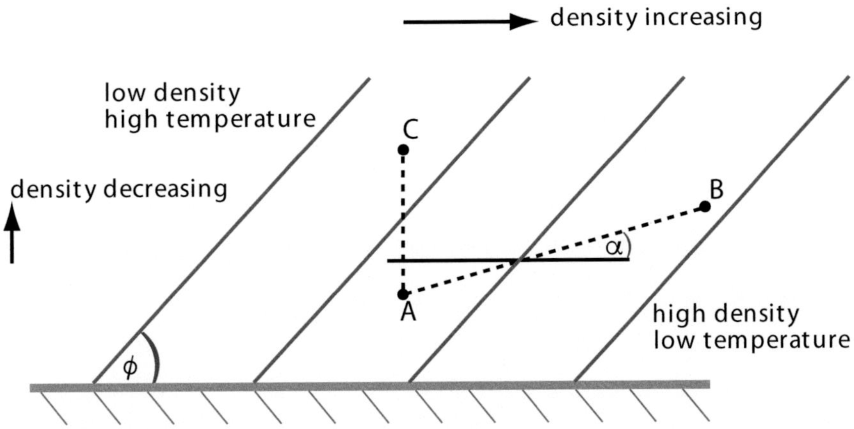

Fig. 9.9 A steady basic state giving rise to baroclinic instability. Potential density decreases upwards and equatorwards, and the associated horizontal pressure gradient is balanced by the Coriolis force. Parcel 'A' is heavier than 'C', and so statically stable, but it is lighter than 'B'. Hence, if 'A' and 'B' are interchanged there is a release of potential energy.

The loss of potential energy is easily calculated. Since

$$\mathrm{PE} = \int \rho g \, \mathrm{d}z, \tag{9.59}$$

the change in potential energy due to the interchange is

$$\Delta\mathrm{PE} = g(\rho_A z_A + \rho_B z_B - \rho_A z_B - \rho_B z_A) = g(z_A - z_B)(\rho_A - \rho_B) = g\,\Delta\rho\,\Delta z. \tag{9.60}$$

If both $\rho_B > \rho_A$ and $z_B > z_A$ then the initial potential energy is larger than the final one, energy is released and the state is unstable. If the slope of the isopycnals is ϕ [so that $\phi = -(\partial_y \rho)/(\partial_z \rho)$] and the slope of the displacements is α, then for a displacement of horizontal distance L the change in potential energy is given by

$$\Delta\mathrm{PE} = g\Delta\rho\Delta z = g\left(L\frac{\partial\rho}{\partial y} + L\alpha\frac{\partial\rho}{\partial z}\right)\alpha L = gL^2\alpha\frac{\partial\rho}{\partial y}\left(1 - \frac{\alpha}{\phi}\right), \tag{9.61}$$

if α and ϕ are small. If $0 < \alpha < \phi$ then energy is released by the perturbation, and it is maximized when $\alpha = \phi/2$. For the atmosphere the actual slope of the isotherms is about 10^{-3}, so that the slope and potential parcel trajectories are indeed shallow.

Although intuitively appealing, the thermodynamic arguments presented in this section pay no attention to satisfying the dynamical constraints of the equations of motion, and we now turn our attention to that.

9.4.2 Linearized Quasi-Geostrophic Equations

To explore the dynamics of baroclinic instability we use the quasi-geostrophic equations, specifically a potential vorticity equation for the fluid interior and a buoyancy or temperature equation at two vertical boundaries, one representing the ground and the other the tropopause. (The tropopause is the boundary between the troposphere and stratosphere at about 10 km; it is not a true rigid surface, but the higher static stability of the stratosphere inhibits vertical motion. We discuss the effects of that in Section 9.9.) For a Boussinesq fluid, the potential vorticity equation is

$$\frac{\partial q}{\partial t} + \boldsymbol{u} \cdot \nabla q = 0, \qquad 0 < z < H,$$
$$q = \nabla^2 \psi + \beta y + \frac{\partial}{\partial z}\left(F \frac{\partial \psi}{\partial z}\right), \tag{9.62}$$

where $F = f_0^2/N^2$, and the buoyancy equation, with $w = 0$, is

$$\frac{\partial b}{\partial t} + \boldsymbol{u} \cdot \nabla b = 0, \qquad z = 0, H,$$
$$b = f_0 \frac{\partial \psi}{\partial z}. \tag{9.63}$$

A solution of these equations is a purely zonal flow, $\boldsymbol{u} = U(y, z)\mathbf{i}$ with a corresponding temperature field given by thermal wind balance. The potential vorticity of this basic state is

$$Q = \beta y - \frac{\partial U}{\partial y} + \frac{\partial}{\partial z} F \frac{\partial \Psi}{\partial z} = \beta y + \frac{\partial^2 \Psi}{\partial y^2} + \frac{\partial}{\partial z} F \frac{\partial \Psi}{\partial z}, \tag{9.64}$$

where Ψ is the streamfunction of the basic state, related to U by $U = -\partial \Psi / \partial y$. Linearizing (9.62) about this zonal flow gives the potential vorticity equation for the interior,

$$\frac{\partial q'}{\partial t} + U \frac{\partial q'}{\partial x} + v' \frac{\partial Q}{\partial y} = 0, \qquad 0 < z < H, \tag{9.65}$$

where $q' = \nabla^2 \psi' + \partial_z (F \partial_z \psi')$ and $v' = \partial_x \psi'$. Similarly, the linearized buoyancy equation at the boundary is

$$\frac{\partial b'}{\partial t} + U \frac{\partial b'}{\partial x} + v' \frac{\partial B}{\partial y} = 0, \qquad z = 0, H, \tag{9.66}$$

where $b' = f_0 \partial_z \psi'$ and $\partial_y B = \partial_y (f_0 \partial_z \Psi) = -f_0 \partial U / \partial z$.

Just as for the barotropic problem, a standard way of proceeding is to seek normal-mode solutions. Since the coefficients of (9.65) and (9.66) are functions of y and z, but not of x, we seek solutions of the form

$$\psi'(x, y, z, t) = \mathrm{Re}\, \widetilde{\psi}(y, z) \mathrm{e}^{ik(x - ct)}, \tag{9.67}$$

and similarly for the derived quantities u', v', b' and q'. In particular,

$$\widetilde{q} = \frac{\partial^2 \widetilde{\psi}}{\partial y^2} + \frac{\partial}{\partial z} F \frac{\partial \widetilde{\psi}}{\partial z} - k^2 \widetilde{\psi}. \tag{9.68}$$

Using (9.68) and (9.67) in (9.65) and (9.66) gives, with subscripts y and z denoting derivatives,

$$(U - c)\left(\widetilde{\psi}_{yy} + (F\widetilde{\psi}_z)_z - k^2 \widetilde{\psi}\right) + Q_y \widetilde{\psi} = 0 \qquad 0 < z < H, \tag{9.69a}$$

$$(U - c)\widetilde{\psi}_z - U_z \widetilde{\psi} = 0 \qquad z = 0,\ H. \tag{9.69b}$$

These equations are analogous to Rayleigh's equations for parallel shear flow, and illustrate the similarity between the baroclinic instability problem and that of a parallel shear flow.

9.4.3 Necessary Conditions for Baroclinic Instability

Necessary conditions for instability may be obtained (as for parallel shear flows) by multiplying (9.69) by $\widetilde{\psi}^*$ and integrating over the domain. Integrating by parts, we first note that

$$\int_{y_1}^{y_2} \widetilde{\psi}^* \widetilde{\psi}_{yy} \, dy = \left[\widetilde{\psi}^* \widetilde{\psi}_y\right]_{y_1}^{y_2} - \int_{y_1}^{y_2} |\widetilde{\psi}_y|^2 \, dy. \tag{9.70}$$

If the integral is performed between two quiescent latitudes, or $\psi = 0$ at the meridional boundaries, then the first term on the right-hand side vanishes. Similarly,

$$\begin{aligned}\int_0^H \widetilde{\psi}^* (F\widetilde{\psi}_z)_z \, dz &= [F\widetilde{\psi}^* \widetilde{\psi}_z]_0^H - \int_0^H F|\widetilde{\psi}_z|^2 \, dz \\ &= \left[\frac{FU_z|\widetilde{\psi}|^2}{(U-c)}\right]_0^H - \int_0^H F|\widetilde{\psi}_z|^2 \, dz,\end{aligned} \tag{9.71}$$

using (9.69b). Now, multiply (9.69a) by $\widetilde{\psi}^*$ and integrate over y and z, and use (9.70) and (9.71) to obtain

$$\int_0^H \int_{y_1}^{y_2} |\psi_y|^2 + F|\widetilde{\psi}_z|^2 + k^2|\widetilde{\psi}|^2] \, dy \, dz - \int_{y_1}^{y_2} \left\{ \int_0^H \frac{Q_y}{U-c} |\widetilde{\psi}|^2 \, dz + \left[\frac{FU_z|\widetilde{\psi}|^2}{U-c}\right]_0^H \right\} dy = 0. \tag{9.72}$$

The first term is purely real whereas the second term is complex. The imaginary component of the second term must be zero and therefore

$$-c_i \int_{y_1}^{y_2} \left\{ \int_0^H \frac{Q_y}{|U-c|^2} |\widetilde{\psi}|^2 \, dz + \left[\frac{FU_z|\widetilde{\psi}|^2}{|U-c|^2}\right]_0^H \right\} dy = 0. \tag{9.73}$$

If there is to be instability c_i must be non-zero and the integrand must therefore vanish. This gives the *Charney–Stern–Pedlosky* (CSP) necessary condition for instability,[7] namely that one of the following criteria must be satisfied:

(i) Q_y changes sign in the interior.
(ii) Q_y is the opposite sign to U_z at the upper boundary, $z = H$.
(iii) Q_y is the same sign as U_z at the lower boundary, $z = 0$.
(iv) U_z is the same sign at the upper and lower boundaries, a condition that differs from *(ii)* or *(iii)* if $Q_y = 0$.

In the Earth's mid-latitude atmosphere, Q_y is often dominated by β, and is positive everywhere, as, frequently, is the shear. The instability criterion is then normally satisfied through *(iii)*: that is, both Q_y and $U_z(0)$ are positive. A more general, and in some ways simpler, derivation that does not rely on normal-mode disturbances is given in Section 10.7.2.

9.5 THE EADY PROBLEM

We now proceed to explicitly calculate the stability properties of a particular configuration that has become known as the *Eady problem*. This was one of the first two mathematical descriptions of baroclinic instability, the other being the *Charney problem*.[8] The two problems were formulated independently, each being the (largely unsupervised) PhD thesis of its author, and although the Charney problem is in some respects more complete (for example in allowing a β-effect), the Eady problem displays the instability in a more transparent form. The Charney problem in its entirety is also quite mathematically opaque, and so we first consider the Eady problem.[9] The β-effect can be incorporated relatively simply in the two-layer model of the next section, and in Section 9.9.1 we look at some aspects of the Charney problem approximately. To begin, let us make the following simplifying assumptions:

(i) The motion is on the f-plane ($\beta = 0$). This assumption, although not particularly realistic for the Earth's atmosphere, greatly simplifies the analysis.

(ii) The fluid is uniformly stratified; that is, N^2 is a constant. This is a decent approximation for the atmosphere below the tropopause, but less so for the ocean where the stratification varies considerably, being much larger in the upper ocean.

(iii) The basic state has uniform shear; that is, $U_0(z) = \Lambda z = Uz/H$, where Λ is the (constant) shear and U is the zonal velocity at $z = H$, where H is the domain depth. This profile is more appropriate for the atmosphere than the ocean — below the thermocline the ocean is relatively quiescent and the shear is small.

(iv) The motion is contained between two rigid, flat horizontal surfaces. In the atmosphere this corresponds to the ground and a 'lid' at a constant-height tropopause.

Although, apart from *(i)*, these assumptions are more appropriate for the atmosphere than the ocean, the same qualitative nature of baroclinic instability carries through to the ocean.

9.5.1 The Linearized Problem

With a basic state streamfunction of $\Psi = -\Lambda z y$, the basic state potential vorticity, Q, is

$$Q = \nabla^2\Psi + \frac{H^2}{L_d^2}\frac{\partial}{\partial z}\left(\frac{\partial \Psi}{\partial z}\right) = 0. \tag{9.74}$$

The fact that $Q = 0$ makes the Eady problem a special case, albeit an illuminating one. The linearized potential vorticity equation is

$$\left(\frac{\partial}{\partial t} + \Lambda z\frac{\partial}{\partial x}\right)\left(\nabla^2\psi' + \frac{H^2}{L_d^2}\frac{\partial^2\psi'}{\partial z^2}\right) = 0. \tag{9.75}$$

This equation has no x-dependent coefficients and in a periodic channel we may seek solutions of the form $\psi'(x, y, z, t) = \mathrm{Re}\,\widetilde{\psi}(y, z)\mathrm{e}^{\mathrm{i}k(x-ct)}$, yielding

$$(\Lambda z - c)\left(\frac{\partial^2\widetilde{\psi}}{\partial y^2} + \frac{H^2}{L_d^2}\frac{\partial^2\widetilde{\psi}}{\partial z^2} - k^2\widetilde{\psi}\right) = 0. \tag{9.76}$$

This equation is (9.69a) applied to the Eady problem.

Boundary conditions

There are two sets of boundary conditions to satisfy, the vertical boundary conditions at $z = 0$ and $z = H$ and the lateral boundary conditions. In the horizontal plane we may either consider the flow to in a channel, periodic in x and confined between two meridional walls, or, with a slightly greater degree of idealization but with little change to the essential dynamics, we may suppose that the domain is doubly-periodic. Either case is dealt with easily enough by the choice of geometric basis function; we choose a channel of width L and impose $\psi = 0$ at $y = +L/2$ and $y = -L/2$ and, to satisfy this, seek solutions of the form $\Psi = \Phi(z)\sin ly$ or, using (9.67)

$$\psi'(x, y, z, t) = \mathrm{Re}\,\Phi(z)\sin ly\,\mathrm{e}^{\mathrm{i}k(x-ct)}, \tag{9.77}$$

where $l = n\pi/L$, with n being a positive integer.

The vertical boundary conditions are that $w = 0$ at $z = 0$ and $z = H$. We follow the procedure of Section 9.4.2 and from (9.66) we obtain

$$\left(\frac{\partial}{\partial t} + \Lambda z\frac{\partial}{\partial x}\right)\frac{\partial\psi'}{\partial z} - \Lambda\frac{\partial\psi'}{\partial x} = 0, \qquad \text{at } z = 0, H. \tag{9.78}$$

Solutions

Substituting (9.77) into (9.76) gives the interior potential vorticity equation

$$(\Lambda z - c)\left[\frac{H^2}{L_d^2}\frac{\partial^2\Phi}{\partial z^2} - (k^2 + l^2)\Phi\right] = 0, \tag{9.79}$$

and substituting (9.77) into (9.78) gives, at $z = 0$ and $z = H$,

$$c\frac{\mathrm{d}\Phi}{\mathrm{d}z} + \Lambda\Phi = 0 \qquad \text{and} \qquad (c - \Lambda H)\frac{\mathrm{d}\Phi}{\mathrm{d}z} + \Lambda\Phi = 0. \tag{9.80a,b}$$

These are equivalent to (9.69b) applied to the Eady problem. If $\Lambda z \neq c$ then (9.79) becomes[10]

$$H^2\frac{\mathrm{d}^2\Phi}{\mathrm{d}z^2} - \mu^2\Phi = 0, \tag{9.81}$$

where $\mu^2 = L_d^2(k^2 + l^2)$. The nondimensional parameter μ is a horizontal wavenumber, scaled by the inverse of the Rossby radius of deformation. Solutions of (9.81) are

$$\Phi(z) = A\cosh\mu\hat{z} + B\sinh\mu\hat{z}, \tag{9.82}$$

where $\hat{z} = z/H$; thus, μ determines the vertical structure of the solution. The boundary conditions (9.80) are satisfied if

$$\begin{gathered} A\,[\Lambda H] + B\,[\mu c] = 0, \\ A\,[(c - \Lambda H)\mu\sinh\mu + \Lambda H\cosh\mu] + B\,[(c - \Lambda H)\mu\cosh\mu + \Lambda H\sinh\mu] = 0. \end{gathered} \tag{9.83}$$

Equations (9.83) are two coupled homogeneous equations in the two unknowns A and B. Nontrivial solutions will only exist if the determinant of their coefficients (the terms in square brackets) vanishes, and this leads to

$$c^2 - Uc + U^2(\mu^{-1}\coth\mu - \mu^{-2}) = 0, \tag{9.84}$$

where $U \equiv \Lambda H$ and $\coth\mu = \cosh\mu/\sinh\mu$. The solution of (9.84) is

$$c = \frac{U}{2} \pm \frac{U}{\mu}\left[\left(\frac{\mu}{2} - \coth\frac{\mu}{2}\right)\left(\frac{\mu}{2} - \tanh\frac{\mu}{2}\right)\right]^{1/2}. \tag{9.85}$$

The waves, being proportional to $\exp(-\mathrm{i}kct)$, will grow exponentially if c has an imaginary part. Since $\mu/2 > \tanh(\mu/2)$ for all μ, for an instability we require that

$$\frac{\mu}{2} < \coth\frac{\mu}{2}, \tag{9.86}$$

which is satisfied when $\mu < \mu_c$ where $\mu_c = 2.399$. The growth rates of the instabilities themselves are given by the imaginary part of (9.85), multiplied by the x-wavenumber; that is

$$\sigma = kc_i = k\frac{U}{\mu}\left[\left(\coth\frac{\mu}{2} - \frac{\mu}{2}\right)\left(\frac{\mu}{2} - \tanh\frac{\mu}{2}\right)\right]^{1/2}. \tag{9.87}$$

These solutions suggest a natural nondimensionalization: scale length by L_d, height by H and time by $L_d/U = L_d/(H\Lambda)$. The growth rate scales as the inverse of the time scaling and so by U/L_d. The timescale is also usefully written as

$$T_E = \frac{L_d}{U} = \frac{NH}{f_0 U} = \frac{N}{f_0\Lambda} = \frac{1}{Fr f_0} = \frac{\sqrt{Ri}}{f_0}, \tag{9.88}$$

where $Fr = U/(NH)$ and $Ri = N^2/\Lambda^2$ are the Froude and Richardson numbers for this problem.

From (9.87) we can (with a little work) determine that the maximum growth rate occurs when $\mu = \mu_m = 1.61$. For any given x-wavenumber, the most unstable wavenumber has the gravest meridional scale, which here is $n = 1$, and we may further consider a wide channel so that $l^2 \ll k^2$. The maximum growth rate, σ_E, is then given by

$$\sigma_E = \frac{0.31U}{L_d} = \frac{0.31\Lambda H}{L_d} = \frac{0.31\Lambda f}{N}, \tag{9.89}$$

and this is known as the *Eady growth rate*. We have removed the subscript 0 from the Coriolis parameter here. Although f is taken as constant in the quasi-geostrophic derivation, we might wish to calculate the Eady growth rate at various locations around the globe, in which case we should use the local value of the Coriolis parameter and deformation radius. Evidently, the growth rate is proportional to the shear times the Prandtl ratio, f/N. The associate phase speed is the real part of c and is given by $c_r = 0.5U$.

For small l the unstable x-wavenumbers and corresponding wavelengths occur for

$$k < k_c = \frac{\mu_c}{L_d} = \frac{2.4}{L_d}, \qquad \lambda > \lambda_c = \frac{2\pi L_d}{\mu_c} = 2.6L_d. \tag{9.90a,b}$$

The wavenumber and wavelength at which the instability is greatest are:

$$k_m = \frac{1.6}{L_d}, \qquad \lambda_m = \frac{2\pi L_d}{\mu_m} = 3.9L_d. \tag{9.91a,b}$$

These properties are illustrated in the left-hand panels of Fig. 9.10 and in Fig. 9.11.

Given c, we may use (9.83) to determine the vertical structure of the Eady wave and this is, to within an arbitrary constant factor,

$$\Phi(\hat{z}) = \cosh \mu\hat{z} - \frac{U}{\mu c}\sinh \mu\hat{z} = \left(\cosh \mu\hat{z} - \frac{Uc_r \sinh \mu\hat{z}}{\mu|c^2|} + \frac{\mathrm{i}Uc_i \sinh \mu\hat{z}}{\mu|c^2|}\right). \tag{9.92}$$

The wave therefore has a phase, $\theta(z)$, given by

$$\theta(\hat{z}) = \tan^{-1}\left(\frac{Uc_i \sinh \mu\hat{z}}{\mu|c^2| \cosh \mu\hat{z} - Uc_r \sinh \mu\hat{z}}\right). \tag{9.93}$$

The phase and amplitude of the Eady waves are plotted in the right panels of Fig. 9.10, and their overall structure in Fig. 9.12, where we see the unstable wave tilting into the shear.

9.5.2 Atmospheric and Oceanic Parameters

To get a qualitative sense of the nature of the instability we choose some typical parameters, as follows.

For the atmosphere

Let us choose

$$H \sim 10\,\mathrm{km}, \qquad U \sim 10\,\mathrm{m\,s^{-1}}, \qquad N \sim 10^{-2}\,\mathrm{s^{-1}}. \tag{9.94}$$

We then obtain:

$$\text{deformation radius:} \quad L_d = \frac{NH}{f} \approx \frac{10^{-2}\,10^4}{10^{-4}} \approx 1000\,\mathrm{km}, \tag{9.95}$$

$$\text{scale of maximum instability:} \quad L_{\max} \approx 3.9L_d \approx 4000\,\mathrm{km}, \tag{9.96}$$

$$\text{growth rate:} \quad \sigma \approx 0.3\frac{U}{L_d} \approx \frac{0.3\times 10}{10^6}\,\mathrm{s^{-1}} \approx 0.26\,\mathrm{day^{-1}}. \tag{9.97}$$

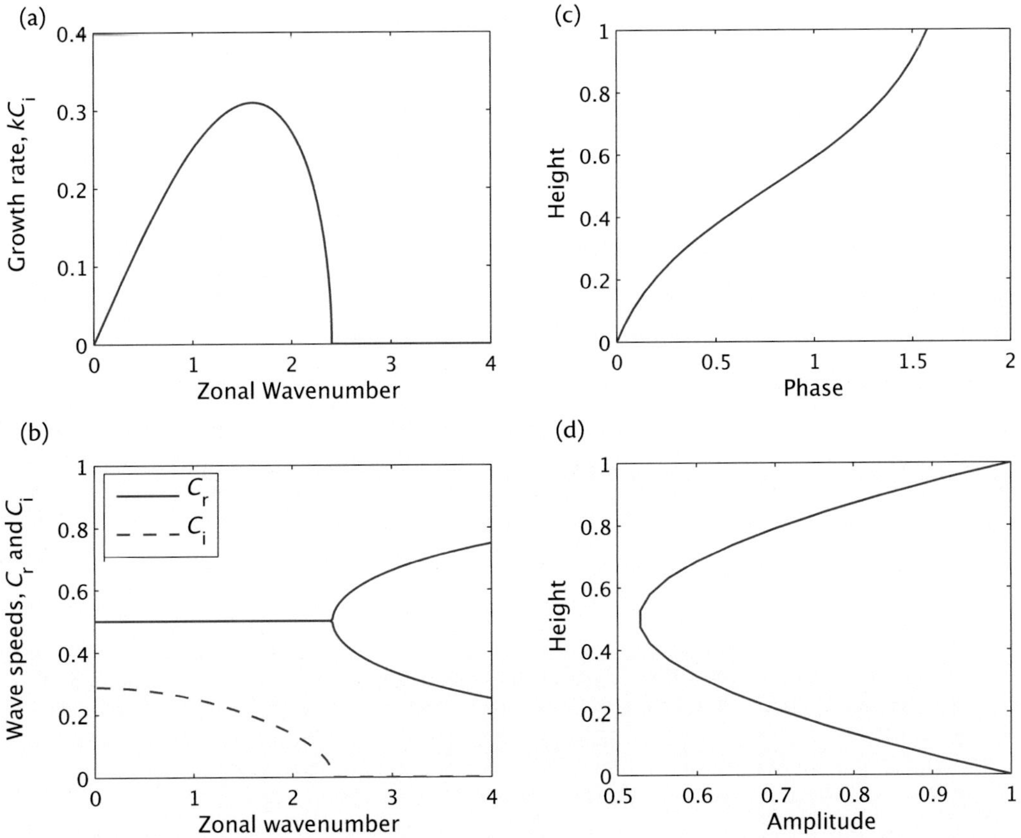

Fig. 9.10 Solution of the Eady problem, in nondimensional units. (a) The growth rate, kc_i as a function of scaled wavenumber μ, from (9.87) with $\Lambda = H = 1$ and for the gravest meridional mode. (b) The real (solid) and imaginary (dashed) wave speeds of those modes, as a function of horizontal wavenumber. (c) The phase of the single most unstable mode as a function of height. (d) The amplitude of that mode as a function of height. To obtain dimensional values, multiply the growth rate by $\Lambda H/L_d$ and the wavenumber by $1/L_d$.

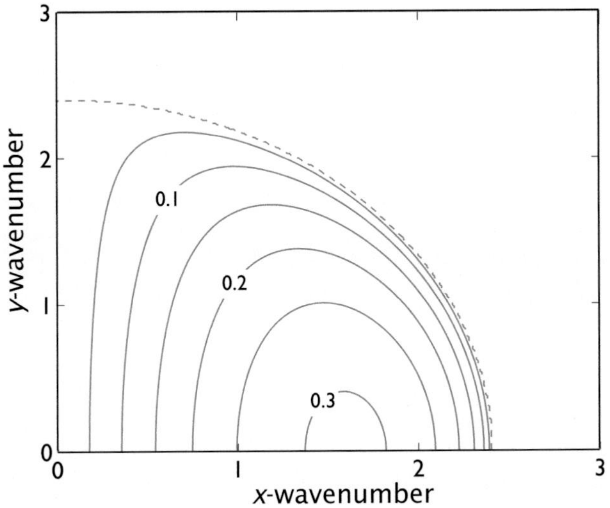

Fig. 9.11 Contours of the growth rate, σ, in the Eady problem, in the k–l plane using (9.87), nondimensionalized as in Fig. 9.10. The growth rate peaks near the deformation scale, and for any given zonal wavenumber the most unstable wavenumber is that with the gravest meridional scale.

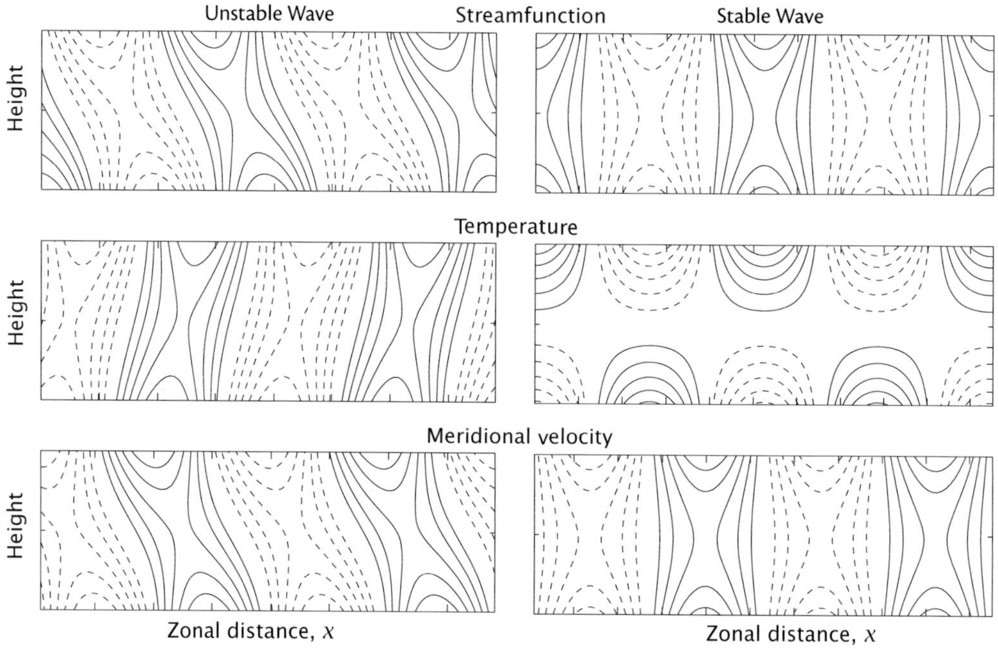

Fig. 9.12 Left column: vertical structure of the most unstable Eady mode. Top: contours of streamfunction. Middle: temperature, $\partial\psi/\partial z$. Bottom: meridional velocity, $\partial\psi/\partial x$. Negative contours are dashed, and two complete wavelengths are present in the horizontal direction. Polewards flowing (positive v) air is generally warmer than equatorwards flowing air. Right column: the same, but now for a wave just beyond the short-wave cut-off.

For the ocean

For the main thermocline in the ocean let us choose

$$H \sim 1\,\text{km}, \qquad U \approx 0.1\,\text{m}\,\text{s}^{-1}, \qquad N \sim 10^{-2}\,\text{s}^{-1}. \tag{9.98}$$

We then obtain:

$$\text{deformation radius:} \qquad L_d = \frac{NH}{f} \approx \frac{10^{-2} \times 1000}{10^{-4}} = 100\,\text{km}, \tag{9.99}$$

$$\text{scale of maximum instability:} \qquad L_{\text{max}} \approx 3.9\,L_d \approx 400\,\text{km}, \tag{9.100}$$

$$\text{growth rate:} \qquad \sigma \approx 0.3\frac{U}{L_d} \approx \frac{0.3 \times 0.1}{10^5}\,\text{s}^{-1} \approx 0.026\,\text{day}^{-1}. \tag{9.101}$$

In the ocean, the Eady problem is not quantitatively applicable because of the non-uniformity of the stratification. Nevertheless, the above estimates give a qualitative sense of the scale and growth rate of the instability relative to the corresponding values in the atmosphere. A summary of the main points of the Eady problem is given in the shaded box on the facing page.

9.6 TWO-LAYER BAROCLINIC INSTABILITY

The eigenfunctions displaying the largest growth rates in the Eady problem have a relatively simple vertical structure. This suggests that an even simpler mathematical model of baroclinic instability might be constructed in which the vertical structure is a priori restricted to a very simple form, namely the two-layer or two-level quasi-geostrophic (QG) model of Sections 5.3.2 and 5.4.6. This

Some Results in Baroclinic Instability

Eady Problem

- The length scales of the instability are characterized by the deformation scale. The most unstable scale has a wavelength about four times the deformation radius L_d, where $L_d = NH/f$.
- The growth rate of the instability is approximately

$$\sigma_E \approx \frac{0.3U}{L_d} = 0.3\Lambda\frac{f}{N}. \tag{B.1}$$

  That is, it is proportional to the shear scaled by the Prandtl ratio f/N. The value σ_E is known as the *Eady growth rate.*

- The most unstable waves for a given zonal scale are those with the gravest meridional scale.
- There is a *short-wave cutoff* beyond which (i.e., at higher wavenumber than) there is no instability. This occurs near the deformation radius.
- The instability relies on an interaction between waves at the upper and lower boundaries. If either boundary is removed, the instability dies. This point is considered in Section 9.7.
- The two-layer (Phillips) problem with zero beta captures many of the results of the Eady model.

Effects of beta

- The beta effect allows the instability to grow by the interaction of edge waves at the surface with Rossby waves in the interior, not just with edge waves at the top. Potential vorticity changes sign because of an interaction between the surface temperature gradient and the interior potential vorticity gradient. Thus, not all unstable modes are deep.
- There is a long-wave cut-off to the main instability branch. At scales larger than this the instabilities are slowly growing, and absent in the two-layer (Phillips) problem.
- In the continuously stratified problem there is no short wave cut-off, but these modes are slowly growing and in the two-layer model they are absent.
- In the two-layer model with beta, there is a minimum shear, Λ_c, for instability given by

$$\Lambda_c = \frac{\beta H}{2}\frac{N^2}{f^2}. \tag{B.2}$$

  This shear does not arise in the continuous problem, although it may be a useful criterion for the onset of rapidly growing deep modes.

- The above differences between the two-layer problem and the continuously stratified one arise because in the former all modes are deep and so there can be no interaction between edge waves and shallow Rossby waves.

instability problem is often called the 'Phillips problem'.[11] One notable advantage over the Eady model is that it is possible to include the β-effect in a simple way.

9.6.1 Posing the Problem

For two layers or two levels of equal thickness, we write the potential vorticity equations in the dimensional form,

$$\frac{\mathrm{D}}{\mathrm{D}t}\left[\zeta_i + \beta y + \frac{k_d^2}{2}(\psi_j - \psi_i)\right] = 0, \qquad i = 1,2, \quad j = 3 - i, \tag{9.102}$$

where, using two-level notation for definiteness,

$$\frac{k_d^2}{2} = \left(\frac{2f_0}{NH}\right)^2 \quad \rightarrow \quad k_d = \frac{\sqrt{8}}{L_d}, \tag{9.103}$$

where H is the total depth of the domain, as in the Eady problem. The basic state we choose is

$$\Psi_1 = -U_1 y, \qquad \Psi_2 = -U_2 y = +U_1 y. \tag{9.104}$$

It is possible to choose $U_2 = -U_1$ without loss of generality because there is no topography and the system is Galilean invariant. The basic state potential vorticity gradient is then given by

$$Q_1 = \beta y + k_d^2 U y, \qquad Q_2 = \beta y - k_d^2 U y, \tag{9.105}$$

where $U = U_1$. (Note that U differs by a constant multiplicative factor from the U in the Eady problem.) Even in the absence of β there is a non-zero potential vorticity gradient. Why should this be different from the Eady problem? — after all, the shear is uniform in both problems. The difference arises from the vertical boundary conditions. In the standard layered formulation the temperature gradient at the boundary is absorbed into the definition of the potential vorticity in the interior. This results in a non-zero interior potential vorticity gradient at the two levels adjacent to the boundary (the only layers in the two-layer problem), but with isothermal boundary conditions $\mathrm{D}(\partial\psi/\partial z)/\mathrm{D}t = 0$. In the Eady problem we have a zero interior gradient of potential vorticity but a temperature gradient at the boundary. The two formulations are physically equivalent — a finite-difference example of the Bretherton boundary layer, encountered in Section 5.4.3.

The linearized potential vorticity equation is, for each layer,

$$\frac{\partial q_i'}{\partial t} + U_i \frac{\partial q_i'}{\partial x} + v_i' \frac{\partial Q_i}{\partial y} = 0, \qquad i = 1,2, \tag{9.106}$$

or, more explicitly,

$$\left(\frac{\partial}{\partial t} + U\frac{\partial}{\partial x}\right)\left[\nabla^2\psi_1' + \frac{k_d^2}{2}(\psi_2' - \psi_1')\right] + \frac{\partial \psi_1'}{\partial x}(\beta + k_d^2 U) = 0, \tag{9.107a}$$

$$\left(\frac{\partial}{\partial t} - U\frac{\partial}{\partial x}\right)\left[\nabla^2\psi_2' + \frac{k_d^2}{2}(\psi_1' - \psi_2')\right] + \frac{\partial \psi_2'}{\partial x}(\beta - k_d^2 U) = 0. \tag{9.107b}$$

For simplicity we will set the problem in a square, doubly periodic domain, and so seek solutions of the form,

$$\psi_i' = \mathrm{Re}\,\widetilde{\psi}_i \mathrm{e}^{\mathrm{i}(kx+ly-\omega t)} = \mathrm{Re}\,\widetilde{\psi}_i \mathrm{e}^{\mathrm{i}k(x-ct)}\mathrm{e}^{\mathrm{i}ly}, \qquad i = 1,2. \tag{9.108}$$

Here, k and l are the x- and y-wavenumbers, and $(k,l) = (2\pi/L)(m,n)$, where L is the size of the domain, and m and n are integers. The constant $\widetilde{\psi}_i$ is the complex amplitude.

9.6.2 The Solution

Substituting (9.108) into (9.107) we obtain

$$[ik(U-c)]\left[-K^2\widetilde{\psi}_1 + k_d^2(\widetilde{\psi}_2 - \widetilde{\psi}_1)/2\right] + ik\widetilde{\psi}_1(\beta + k_d^2 U) = 0, \tag{9.109a}$$

$$[-ik(U+c)]\left[-K^2\widetilde{\psi}_2 + k_d^2(\widetilde{\psi}_1 - \widetilde{\psi}_2)/2\right] + ik\widetilde{\psi}_2(\beta - k_d^2 U) = 0, \tag{9.109b}$$

where $K^2 = k^2 + l^2$. Re-arranging these two equations gives

$$\left[(U-c)(k_d^2/2 + K^2) - (\beta + k_d^2 U)\right]\widetilde{\psi}_1 - \left[k_d^2(U-c)/2\right]\widetilde{\psi}_2 = 0, \tag{9.110a}$$

$$-\left[k_d^2(U+c)/2\right]\widetilde{\psi}_1 + \left[(U+c)(k_d^2/2 + K^2) + (\beta - k_d^2 U)\right]\widetilde{\psi}_2 = 0. \tag{9.110b}$$

These equations are of the form

$$[A]\widetilde{\psi}_1 + [B]\widetilde{\psi}_2 = 0, \qquad [C]\widetilde{\psi}_1 + [D]\widetilde{\psi}_2 = 0, \tag{9.111}$$

where A, B, C, D correspond to the terms in square brackets in (9.110). For non-trivial solutions the determinant of coefficients must be zero, that is $AD - BC = 0$. This gives a quadratic equation in c and solving this we obtain

$$c = -\frac{\beta}{K^2 + k_d^2}\left\{1 + \frac{k_d^2}{2K^2} \pm \frac{k_d^2}{2K^2}\left[1 + \frac{4K^4(K^4 - k_d^4)}{k_\beta^4 k_d^4}\right]^{1/2}\right\}, \tag{9.112}$$

where $K^4 = (k^2 + l^2)^2$ and $k_\beta = \sqrt{\beta/U}$ (its inverse is known as the Kuo scale). We may nondimensionalize this equation using the deformation radius L_d as the length scale and the shear velocity U as the velocity scale.[12] Then, denoting nondimensional parameters with hats, we have

$$k = \frac{\widehat{k}}{L_d}, \qquad c = \widehat{c}U, \qquad t = \frac{L_d}{U}\widehat{t}, \tag{9.113}$$

and the nondimensional form of (9.112) is just

$$\widehat{c} = -\frac{\widehat{k}_\beta^2}{\widehat{K}^2 + \widehat{k}_d^2}\left\{1 + \frac{\widehat{k}_d^2}{2\widehat{K}^2} \pm \frac{\widehat{k}_d^2}{2\widehat{K}^2}\left[1 + \frac{4\widehat{K}^4(\widehat{K}^4 - \widehat{k}_d^4)}{\widehat{k}_\beta^4 \widehat{k}_d^4}\right]^{1/2}\right\}, \tag{9.114}$$

where $\widehat{k}_\beta = k_\beta L_d$ and $\widehat{k}_d = \sqrt{8}$, as in (9.103). The nondimensional parameter

$$\gamma = \frac{1}{4}\widehat{k}_\beta^2 = \frac{\beta L_d^2}{4U}, \tag{9.115}$$

is often useful as a measure of the importance of β; it is proportional to the square of the ratio of the deformation radius to the Kuo scale $\sqrt{U/\beta}$. (It is the two-layer version of the 'Charney–Green number' considered more in Section 9.9.1.) Let us look at two special cases first, before considering the general solution to these equations.

I. *Zero shear, non-zero β*

If there is no shear (i.e., $U = 0$) then (9.110a) and (9.110b) are identical and two roots of the equation give the purely real phase speeds c,

$$c = -\frac{\beta}{K^2} \qquad \text{and} \qquad c = -\frac{\beta}{K^2 + k_d^2}. \tag{9.116}$$

The first of these is the dispersion relationship for Rossby waves in a purely barotropic flow, and corresponds to the eigenfunction $\widetilde{\psi}_1 = \widetilde{\psi}_2$. The second solution corresponds to the baroclinic eigenfunction $\widetilde{\psi}_1 + \widetilde{\psi}_2 = 0$.

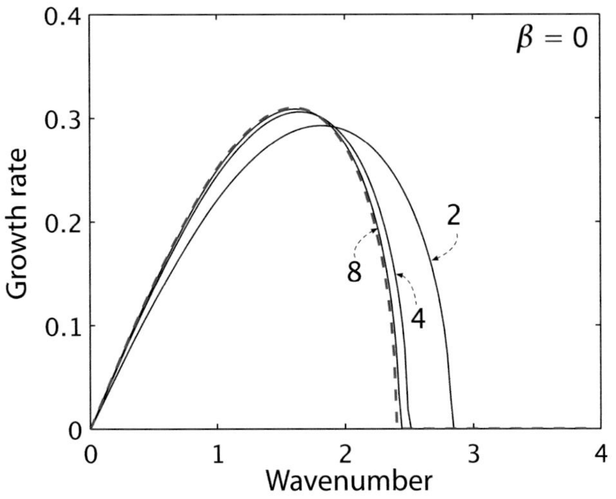

Fig. 9.13 Baroclinic growth rate as calculated with two, four and eight vertical levels (solid lines, as labelled), and in the continuous case (red dashed line), all with $\beta = 0$. The two-level result is the analytic result of (9.117), and the continuous result is the analytic result of the Eady problem. The four- and eight-level results were obtained numerically, and are almost the same as the result from the Eady problem.

II. Zero β, non-zero shear

If $\beta = 0$, then (9.110) yields, after a little algebra,

$$c = \pm U \left(\frac{K^2 - k_d^2}{K^2 + k_d^2} \right)^{1/2} \quad \text{or} \quad \sigma = Uk \left(\frac{k_d^2 - K^2}{K^2 + k_d^2} \right)^{1/2}, \tag{9.117}$$

where $\sigma = -i\omega$ is the growth rate. These expressions are similar to those in the Eady problem. Indeed, as we increase the number of levels (using a numerical method to perform the calculation) the growth rate converges to that of the Eady problem (Fig. 9.13). We note the following:

- There is an instability for *all* values of U.
- There is a high-wavenumber cut-off, at a scale proportional to the radius of deformation. For the two-layer model, if $K > k_d = 2.82/L_d$ there is no growth. For the Eady problem, the high wavenumber cut-off occurs at $2.4/L_d$.
- There is no low wavenumber cut-off.
- For any given k, the highest growth rate occurs for $l = 0$. In the two-layer model, from (9.117), for $l = 0$ the maximum growth rate occurs when $k = 0.634k_d = 1.79/L_d$. For the Eady problem, the maximum growth rate occurs at $1.61/L_d$.

Solution in the general case: non-zero shear and non-zero β

Using (9.114), the growth rate and wave speeds as function of wavenumber are plotted in Fig. 9.14. We observe that there still appears to be a high-wavenumber cut-off and, for $\beta = 0$, there is a low-wavenumber cut-off. A little analysis elucidates the origin of these features.

The neutral curve

For instability, there must be an imaginary component to the phase speed in (9.112); that is, we require

$$k_\beta^4 k_d^4 + 4K^4(K^4 - k_d^4) < 0. \tag{9.118}$$

This is a quadratic equation in K^4 for the value of K, K_c say, at which the growth rate is zero. Solving, we find

$$K_c^4 = \frac{1}{2} k_d^4 \left(1 \pm \sqrt{1 - k_\beta^4 / k_d^4} \right), \tag{9.119}$$

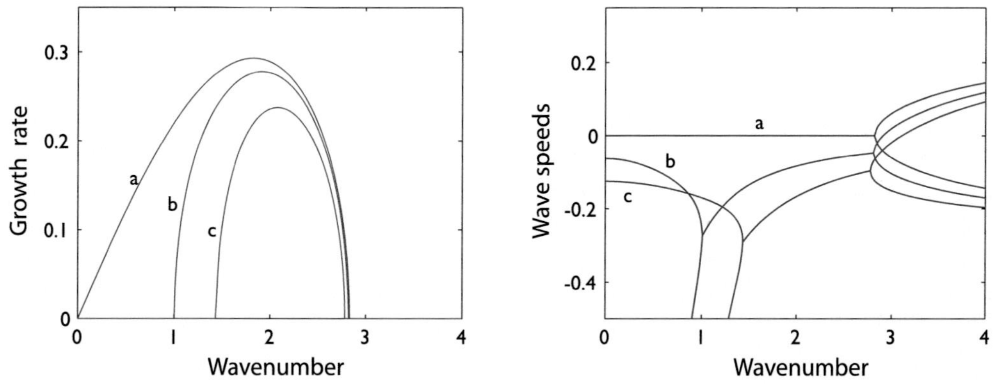

Fig. 9.14 Growth rates and wave speeds for the two-layer baroclinic instability problem, from (9.114), with three (nondimensional) values of β as labelled: a, $\gamma = 0$ ($\hat{k}_\beta = 0$); b, $\gamma = 0.5$ ($\hat{k}_\beta = \sqrt{2}$); c, $\gamma = 1$ ($\hat{k}_\beta = 2$). As β increases, so does the low-wavenumber cut-off to instability, but the high-wavenumber cut-off is little changed. The solutions are obtained from (9.114), with $\hat{k}_d = \sqrt{8}$ and $U_1 = -U_2 = 1/4$.

and this is plotted in Fig. 9.15. From (9.118) useful approximate expressions can be obtained for the critical shear as a function of wavenumber in the limits of small K and $K \approx k_d$, and these are left as exercises for the reader.

Minimum shear for instability

From (9.118), instability arises when $\beta^2 k_d{}^4/U^2 < 4K^4(k_d{}^4 - K^4)$. The maximum value of the right-hand side of this expression arises when $K^4 = \hat{k}_d^4/2$; thus, instability arises only when

$$\frac{\beta^2 k_d^4}{U^2} < 4\frac{k_d^4}{2}\frac{k_d^4}{2} \qquad \text{or} \qquad \kappa_\beta < k_d. \tag{9.120}$$

That is, *instability only arises if the deformation radius is sufficiently smaller than the Kuo scale.* The critical velocity difference required for instability is then

$$U_1 - U_2 > U_c = \frac{2\beta}{k_d^2} = \frac{1}{4}\beta L_d^2, \tag{9.121}$$

recalling our notation that $U_1 - U_2 = 2U$ and that $k_d^2 = 8/L_d^2$, as in (9.103). The critical shear for instability is

$$\Lambda_c = \frac{\beta H}{2}\frac{N^2}{f^2}, \tag{9.122a,b}$$

where the shear Λ is defined by $(U_1 - U_2)/(0.5H)$, where H is the total depth of the domain.

In any given quasi-geostrophic calculation f is held constant, but if we wish to see how the critical shear varies with latitude we vary f accordingly. Figure 9.16 sketches how this critical shear might vary with latitude in the atmosphere and ocean, allowing f to vary. If the shear is just the critical value, the instability occurs at $k = 2^{-1/4}k_d = 0.84k_d = 2.37/L_d$. As the shear increases, the wavenumber at which the growth rate is maximum decreases slightly (see Fig. 9.15), and for a sufficiently large shear the β-effect is negligible and the wavenumber of maximum instability is, as we saw earlier, $0.634\,k_d$ or $1.79/L_d$

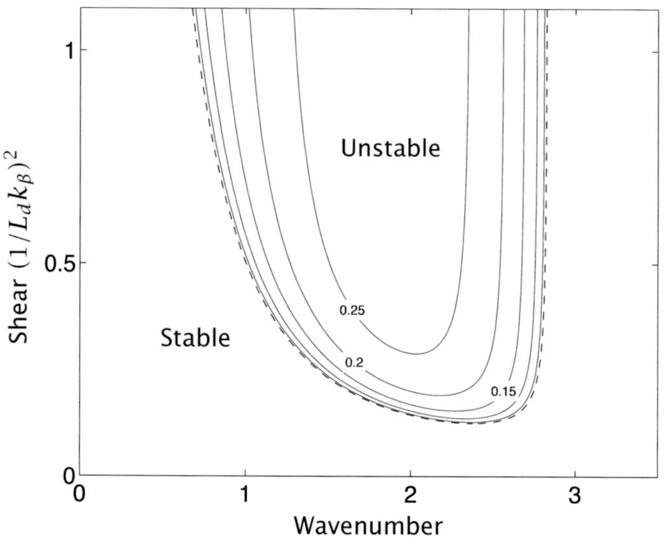

Fig. 9.15 Contours of growth rate in the two-layer baroclinic instability problem. The dashed line is the neutral stability curve, (9.119), and the other curves are contours of growth rates obtained from (9.114). The wavenumber is scaled by $1/L_d$ (i.e., by $k_d/\sqrt{8}$) and growth rates are scaled by the inverse of the Eady time scale (i.e., by U/L_d). Thus, for $L_d = 1000\,\text{km}$ and $U = 10\,\text{m}\,\text{s}^{-1}$, a nondimensional growth rate of 0.25 corresponds to a dimensional growth rate of $0.25 \times 10^{-5}\,\text{s}^{-1} = 0.216\,\text{day}^{-1}$.

Note the relationship of the minimum shear to the basic state potential vorticity gradient in the respective layers. In the upper and lower layers the potential vorticity gradients are given by, respectively,

$$\frac{\partial Q_1}{\partial y} = \beta + k_d^2 U, \qquad \frac{\partial Q_2}{\partial y} = \beta - k_d^2 U. \tag{9.123a,b}$$

Thus, the requirement for instability is exactly that which causes the potential vorticity gradient to change sign somewhere in the domain, in this case becoming negative in the lower layer. This is an example of the general rule that potential vorticity (suitably generalized to include the surface boundary conditions) must change sign somewhere in order for there to be an instability.

High-wavenumber cut-off

Instability can only arise when, from (9.118),

$$4K^4(k_d^4 - K^4) > k_\beta^4 k_d^4, \tag{9.124}$$

so that a necessary condition for instability is

$$k_d^2 > K^2. \tag{9.125}$$

Thus, waves shorter than the deformation radius are always stable, no matter what the value of β. We also see from Fig. 9.14 and Fig. 9.15 that the high wavenumber cut-off in fact varies little with β if $k_d \gg k_\beta$. The critical shear required for instability approaches infinity as K approaches k_d.

Low-wavenumber cut-off

Suppose that $K \ll k_d$. Then (9.118) simplifies to $k_\beta^4 < 4K^4$. That is, for instability we require

$$K^2 > \frac{1}{2}k_\beta^2 = \frac{\beta}{2U}. \tag{9.126}$$

Thus, using (9.125) and (9.126) the unstable waves lie approximately in the interval $\beta/(\sqrt{2}U) < K < k_d$.

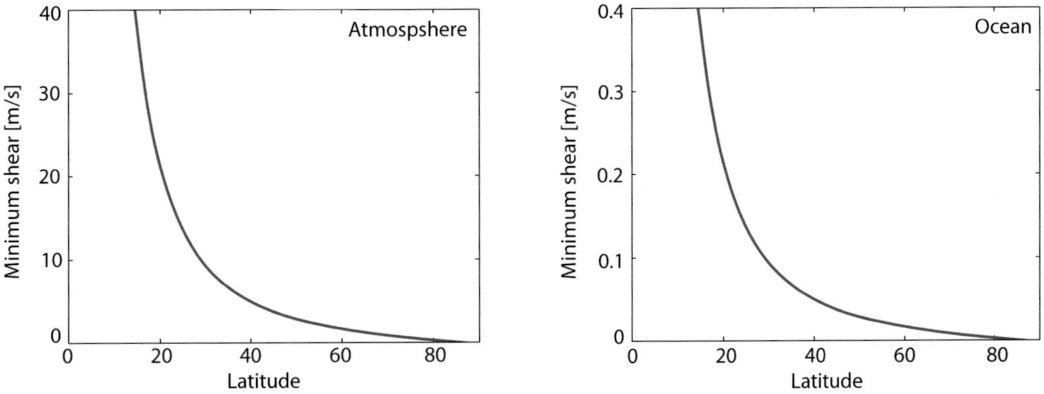

Fig. 9.16 The minimum shear (the velocity difference $U_1 - U_2$) required for baroclinic instability in a two-layer model, calculated using (9.121), i.e., $U_{min} = \beta L_d^2/4$ where $\beta = 2\Omega a^{-1} \cos\vartheta$ and $L_d = NH/f$, with $f = 2\Omega \sin\vartheta$. The left panel uses $H = 10\,\mathrm{km}$ and $N = 10^{-2}\,\mathrm{s}^{-1}$, and the right panel uses parameters representative of the main thermocline, $H = 1\,\mathrm{km}$ and $N = 10^{-2}\,\mathrm{s}^{-1}$. The results are not quantitatively accurate, but the implications that the minimum shear is much less for the ocean, and that in both the atmosphere and the ocean the shear increases rapidly at low latitudes, are robust.

9.7 A KINEMATIC VIEW OF BAROCLINIC INSTABILITY

In this section we take a more intuitive look at baroclinic instability, trying to understand the mechanism without treating the problem in full generality or exactness. We will do this by way of a semi-kinematic argument that shows how the waves in each layer of a two-layer model, or the waves on the top and bottom boundaries in the Eady model, can constructively interact to produce a growing instability. It is kinematic in the sense that we initially treat the waves independently, and only subsequently allow them to interact — but it is this dynamical interaction that gives the instability. We first revisit the two-layer model and simplify it to its bare essentials.

9.7.1 The Two-layer Model

A simple dynamical model

We first re-derive the instability ab initio from the equations of motion written in terms of the baroclinic streamfunction τ and the barotropic streamfunction ψ where

$$\tau \equiv \frac{1}{2}(\psi_1 - \psi_2), \qquad \psi \equiv \frac{1}{2}(\psi_1 + \psi_2). \tag{9.127}$$

We linearize about a sheared basic state of zero barotropic velocity and with $\beta = 0$. Thus, with $\psi = 0 + \psi'$ and $\tau = -Uy + \tau'$ the linearized equations of motion, equivalent to (9.107) with $\beta = 0$, are

$$\frac{\partial}{\partial t}\nabla^2\psi' = -U\frac{\partial}{\partial x}\nabla^2\tau', \tag{9.128a}$$

$$\frac{\partial}{\partial t}(\nabla^2 - k_d^2)\tau' = -U\frac{\partial}{\partial x}(\nabla^2 + k_d^2)\psi'. \tag{9.128b}$$

Neglecting the y-dependence for simplicity, we may seek solutions of the form $(\psi', \tau') = \mathrm{Re}\,(\widetilde{\psi}, \widetilde{\tau})\,\exp[\mathrm{i}k(x - ct)]$, where $c = \mathrm{i}\,c_i + c_r$, giving

$$c\widetilde{\psi} - U\widetilde{\tau} = 0, \tag{9.129a}$$

$$c(K^2 + k_d^2)\widetilde{\tau} - U(K^2 - k_d^2)\widetilde{\psi} = 0. \tag{9.129b}$$

These equations have non-trivial solutions if the determinant of the matrix of coefficients of $\widetilde{\tau}$ and $\widetilde{\psi}$ is zero, giving the quadratic equation $c^2(K^2 + k_d^2) - U^2(K^2 - k_d^2) = 0$. Solving this gives, reprising (9.117),

$$c = \pm U \left(\frac{K^2 - k_d^2}{K^2 + k_d^2} \right)^{1/2}. \tag{9.130}$$

Instabilities occur for $K^2 < k_d^2$, for which $c_r = 0$; that is, the wave speed is purely imaginary. From (9.129) unstable modes have

$$\widetilde{\tau} = \mathrm{i}\frac{c_i}{U}\widetilde{\psi} = \mathrm{e}^{\mathrm{i}\pi/2}\frac{c_i}{U}\widetilde{\psi}. \tag{9.131}$$

That is, τ *lags* ψ by 90° for a growing wave ($c_i > 0$). Similarly, τ *leads* ψ by 90° for a decaying wave. Now, the temperature is proportional to τ, and in the two-level model is advected by the vertically averaged perturbation meridional velocity, v say (with Fourier amplitude $\widetilde{v}$), where $v = \partial\psi/\partial x$. Thus, for growing or decaying waves,

$$\widetilde{v} = \widetilde{\tau}\frac{kU}{c_i}, \tag{9.132}$$

and the meridional velocity is exactly *in phase* with the temperature for growing modes, and is *out of phase* with the temperature for decaying modes. That is, for unstable modes, poleward flow is correlated with high temperatures, and for decaying modes poleward flow is correlated with low temperatures. For neutral waves, $\widetilde{\tau} = c_r\widetilde{\psi}/U$ and so $\widetilde{v} = \mathrm{i}k\widetilde{\tau}U/c_r$, and the meridional velocity and temperature are $\pi/2$ out of phase. Thus, to summarize:

- growing waves transport heat (or buoyancy) polewards;
- decaying waves transport heat equatorwards;
- neutral waves do not transport heat.

Further simplifications to the two-layer model

First consider (9.128) for waves much larger than the deformation radius, $K^2 \ll k_d^2$; we obtain

$$\frac{\partial}{\partial t}\nabla^2\psi = -U\frac{\partial}{\partial x}\nabla^2\tau, \qquad \frac{\partial}{\partial t}\tau = U\frac{\partial}{\partial x}\psi. \tag{9.133a,b}$$

for which we obtain, either directly or from (9.130), $c = \pm\mathrm{i}U$; that is, the flow is unstable. To see the mechanism, suppose that the initial perturbation is barotropic and sinusoidal in x, with no y variation. Polewards flowing fluid (i.e., $\partial\psi/\partial x > 0$) will, by (9.133b), generate a positive τ, and the baroclinic flow will be out of phase with the barotropic flow. Then, by (9.133a), the advection of τ by the mean shear produces growth of ψ that is in phase with the original disturbance. Contrast this case with that for very small disturbances, for which $K^2 \gg k_d^2$, and (9.128) becomes

$$\frac{\partial}{\partial t}\nabla^2\psi = -U\frac{\partial}{\partial x}\nabla^2\tau, \qquad \frac{\partial}{\partial t}\nabla^2\tau = -U\frac{\partial}{\partial x}\nabla^2\psi, \tag{9.134a,b}$$

or, in terms of the equations for each layer,

$$\frac{\partial}{\partial t}\nabla^2\psi_1 = -U\frac{\partial}{\partial x}\nabla^2\psi_1, \qquad \frac{\partial}{\partial t}\nabla^2\psi_2 = +U\frac{\partial}{\partial x}\nabla^2\psi_2. \tag{9.135a,b}$$

That is, the layers are completely decoupled and no instability can arise. Motivated by this, consider waves that propagate independently in each layer on the potential vorticity gradient caused by β

(if non-zero) and shear. Thus, in (9.107) we keep the potential vorticity gradients but neglect k_d^2 where it appears alongside ∇^2 and find

$$\left(\frac{\partial}{\partial t} + U\frac{\partial}{\partial x}\right)\nabla^2\psi_1' + \frac{\partial \psi_1'}{\partial x}\frac{\partial Q_1}{\partial y} = 0, \tag{9.136a}$$

$$\left(\frac{\partial}{\partial t} - U\frac{\partial}{\partial x}\right)\nabla^2\psi_2' + \frac{\partial \psi_2'}{\partial x}\frac{\partial Q_2}{\partial y} = 0, \tag{9.136b}$$

where $\partial Q_1/\partial y = \beta + k_d^2 U$ and $\partial Q_2/\partial y = \beta - k_d^2 U$. Seeking solutions of the form (9.108), the phase speeds of the associated waves are

$$c_1 = U - \frac{\partial_y Q_1}{K^2}, \qquad c_2 = -U - \frac{\partial_y Q_2}{K^2}. \tag{9.137a,b}$$

In the upper layer the phase speed is a combination of an eastward advection and a fast westward wave propagation due to a strong potential vorticity gradient. In the lower layer the phase speed is a combination of a westward advection and a slow eastward wave propagation due to the weak potential vorticity gradient. The two phase speeds are, in general, not equal, but they would need to be so if they were to combine to cause an instability. From (9.137) this occurs when $K^2 = k_d^2$ and $c_1 = c_2 = -\beta/k_d^2$. These conditions are just those occurring at the high-wavenumber cut-off to instability in the two-level model. At higher wavenumbers, the waves are unable to synchronize, whereas at lower wavenumbers they may become inextricably coupled.

Let us suppose that the phase of the wave in the upper layer lags (i.e., is westward of) that in the lower layer, as illustrated in the top panel Fig. 9.17. The lower panel shows the temperature field, $\tau = (\psi_1 - \psi_2)/2$, and the average meridional velocity, $v = \partial_x(\psi_1 + \psi_2)/2$. In this configuration, the temperature field is *in phase* with the meridional velocity, meaning that warm fluid is advected polewards. Now, let us allow the waves in the two layers to interact by adding one dynamical equation, the thermodynamic equation, which in its simplest form is

$$\frac{\partial \tau}{\partial t} = -v\frac{\partial \overline{\tau}}{\partial y} = vU, \tag{9.138}$$

where $\overline{\tau}$ is proportional to the basic state temperature field. The temperature field, τ, grows in proportion to v, which is proportional to τ if the waves tilt westwards with height, and an instability results. This dynamical mechanism is just that which is compactly described by (9.133), and an important consequence is that baroclinic waves transfer energy polewards. It is a straightforward matter to show that if the streamfunction tilts eastwards with height, v is out of phase with τ and the waves decay. A similar description applies to the Eady problem, as we now see.

9.7.2 Interacting Edge Waves in the Eady Problem

We now explore how the edge waves at the top and bottom surfaces in the Eady problem give rise to an instability, by way of a semi-kinematic description that is very similar to that of the barotropic problem described on page 343. Let us first consider the case in which the bottom and top surfaces are essentially uncoupled. Instead of solutions of (9.81) that have the structure (9.82) (which satisfies both boundary conditions), consider solutions that *separately* satisfy the bottom and top boundary conditions and that decay into the interior. These are

$$\psi_B = \mathrm{Re}\, A_B e^{ik(x-c_T t)} e^{-\mu z/H}, \qquad \psi_T = \mathrm{Re}\, A_T e^{i\phi} e^{ik(x-c_B t)} e^{\mu(z-H)/H}, \tag{9.139a,b}$$

for the bottom and top surfaces, respectively, and ϕ is the phase shift, with A_B and A_T being real constants. The boundary conditions (9.80) then determine the phase speeds of the two systems and we find

$$c_B = \frac{\Lambda H}{\mu}, \qquad c_T = \Lambda H\left(1 - \frac{1}{\mu}\right). \tag{9.140a,b}$$

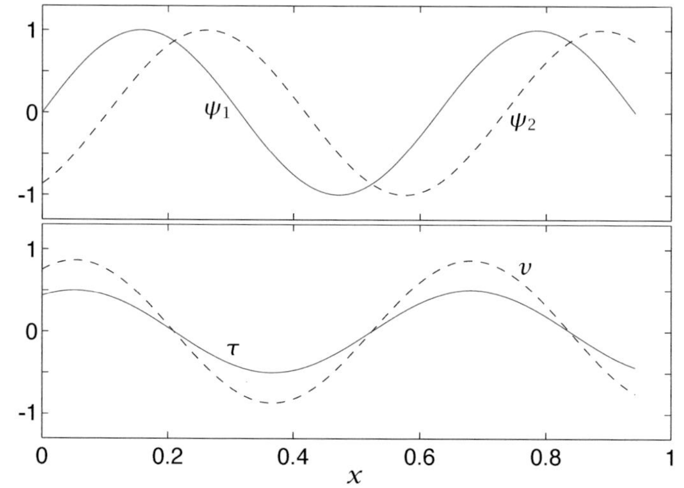

Fig. 9.17 Baroclinically unstable waves in a two layer model. The streamfunction is shown in the top panel, ψ_1 for the top layer and ψ_2 for the bottom layer. Given the westward tilt shown, the temperature, τ, and meridional velocity, υ (bottom panel) are in phase, and the instability grows.

These are the phase speeds of *edge waves* in the Eady problem; they are real and in general they are unequal. It must therefore be the *interaction* of the waves on the upper and lower boundaries that is necessary for instability, because the unstable wave has but a single phase speed. This interaction can occur when their phase speeds are equal, and from (9.140) this occurs when $\mu = 2$, giving

$$k = \frac{2}{L_d} \quad \text{and} \quad c = \frac{\Lambda H}{2}. \tag{9.141a,b}$$

This phase speed is just that of the flow at mid-level, and at the critical wavenumber in the full Eady problem, $k_c = 2.4/L_d$, from (9.90), the phase speed is purely real and equal to that of (9.141b) — see Fig. 9.10. Thus, (9.141) approximately characterizes the critical wavenumber in the full problem.

To turn this kinematic description into a dynamical instability, suppose that the two rigid surfaces are close enough so that the waves can interact, but still far enough so that their structure is approximately given by (9.139). (Note that if μ is too large, the waves decay rapidly away from the edges and will not intersect.) Specifically, let the buoyancy perturbation at a given boundary be advected by the total meridional velocity perturbation, including that arising from the perturbation at the other boundary, so that at the top and bottom boundaries

$$\frac{\partial b'_T}{\partial t} = -(v'_B + v'_T)\frac{\partial \overline{b}_T}{\partial y}, \qquad \frac{\partial b'_B}{\partial t} = -(v'_B + v'_T)\frac{\partial \overline{b}_B}{\partial y}. \tag{9.142}$$

The waves will reinforce each other if v'_T is in phase with b'_B at the lower boundary, and if v'_B is in phase with b'_T at the upper boundary. Now, using (9.139), the velocity and buoyancy associated with the edge waves are given by

$$b_B = -\mathrm{Re}\; kNA_B \mathrm{e}^{-\mu z/H} \mathrm{e}^{ikx}, \qquad b_T = \mathrm{Re}\; kNA_T \mathrm{e}^{\mathrm{i}\phi} \mathrm{e}^{\mu(z-H)/H} \mathrm{e}^{ikx}, \tag{9.143a}$$

$$v_B = \mathrm{Re}\; ikA_B \mathrm{e}^{-\mu z/H} \mathrm{e}^{ikx}, \qquad v_T = \mathrm{Re}\; ikA_T \mathrm{e}^{\mathrm{i}\phi} \mathrm{e}^{\mu(z-H)/H} \mathrm{e}^{ikx}. \tag{9.143b}$$

The fields b_B and v_T, and b_T and v_B, will be positively correlated if $0 < \phi < \pi$, and will be exactly in phase if $\phi = \pi/2$, and this case is illustrated in Fig. 9.18. Just as in the two-layer case, this phase corresponds to a westward tilt with height, and it is this, in conjunction with geostrophic and hydrostatic balance, that allows warm fluid to move polewards and available potential energy to be released. From (9.142), the perturbation will grow and an instability will result. The analogy between baroclinic instability and barotropic instability should be evident from the similarity of this description and that of Section 9.2.4, with z in the baroclinic problem playing the role of y in

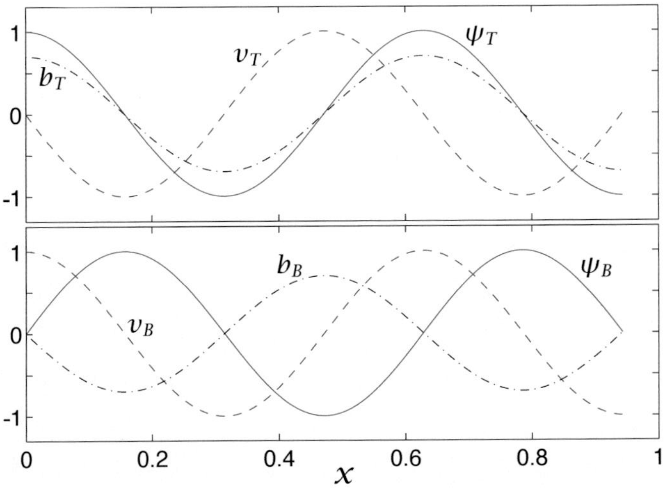

Fig. 9.18 Interacting edge waves in the Eady model. The upper panel shows waves on the top surface, and the lower panel shows waves on the bottom. If the streamfunction tilts westwards with height, then the temperature on the top (bottom) is correlated with the meridional velocity on the bottom (top), the waves can reinforce each other. See also Fig. 9.12.

the barotropic problem, and b the role of u; note that (9.142) is almost identical to (9.45). However, the analogy of the two problems in full is not perfect because the boundary condition that $w = 0$ does not have an exact correspondence in the barotropic problem. More importantly, the nonlinear development of the baroclinic problem, discussed in Chapter 12, is generally three-dimensional, which need not be the case in the barotropic problem.

9.8 ✦ THE ENERGETICS OF LINEAR BAROCLINIC INSTABILITY

In baroclinic instability, warm parcels move polewards and cold parcels move equatorwards. This motion draws on the available potential energy of the mean state, because warm light parcels move upwards and cold dense parcels move downwards and the height of the mean centre of gravity of the fluid falls, and the loss of potential energy is converted to kinetic energy of the perturbation. However, because the instability is growing, the energy of the perturbation is of course not conserved, and both the kinetic energy and the available potential energy of the perturbation will grow. However, we still expect a conversion of potential energy to kinetic energy, and the purpose of this section is to demonstrate that explicitly. For simplicity, we restrict attention to the flat-bottomed two-level model with $\beta = 0$.

As in Section 5.6, the energy may be partitioned into kinetic energy and available potential energy. In a three-dimensional quasi-geostrophic flow the kinetic energy is given by, in general,

$$\mathrm{KE} = \frac{1}{2}\int (\nabla\psi)^2 \,\mathrm{d}V, \tag{9.144}$$

which, in the case of the two-layer model becomes

$$\mathrm{KE} = \frac{1}{2}\int (\nabla\psi_1)^2 + (\nabla\psi_2)^2 \,\mathrm{d}A = \int (\nabla\psi)^2 + (\nabla\tau)^2 \,\mathrm{d}A. \tag{9.145}$$

Restricting attention to a single Fourier mode this becomes

$$\mathrm{KE} = k^2\widetilde{\psi}^2 + k^2\widetilde{\tau}^2. \tag{9.146}$$

The available potential energy in the continuous case is given by

$$\mathrm{APE} = \frac{1}{2}\int \left(\frac{f_0}{N}\right)^2 \left(\frac{\partial\psi}{\partial z}\right)^2 \mathrm{d}V. \tag{9.147}$$

For a single Fourier mode in a two-layer model this becomes

$$\text{APE} = k_d^2 \widetilde{\tau}^2. \tag{9.148}$$

Now, the nonlinear vorticity equations for each level are

$$\frac{\partial}{\partial t}\nabla^2\psi_1 + J(\psi_1, \nabla^2\psi_1) = -2\frac{f_0 w}{H}, \tag{9.149a}$$

$$\frac{\partial}{\partial t}\nabla^2\psi_2 + J(\psi_2, \nabla^2\psi_2) = 2\frac{f_0 w}{H}, \tag{9.149b}$$

where w is the vertical velocity between the levels. (These equations are the two-level analogues of the continuous vorticity equation, with the right-hand sides being finite-difference versions of $f_0 \partial w/\partial z$.) Multiplying the two equations of (9.149) by $-\psi_1$ and $-\psi_2$, respectively, and adding we readily find

$$\frac{\mathrm{d}}{\mathrm{d}t}\text{KE} = \frac{4f_0}{H}\int w\tau \,\mathrm{d}A. \tag{9.150a}$$

For a single Fourier mode this becomes

$$\frac{\mathrm{d}}{\mathrm{d}t}\text{KE} = \text{Re}\,\frac{4f_0}{H}\widetilde{w}\widetilde{\tau}^*, \tag{9.150b}$$

where $w = \widetilde{w}\exp[\mathrm{i}(kx - ct)] + \text{c.c.}$, and the asterisk denotes a complex conjugate.

The continuous thermodynamic equation is

$$\frac{\mathrm{D}b}{\mathrm{D}t} + wN^2 = 0. \tag{9.151}$$

Using $b = f_0 \partial\psi/\partial z$ and finite-differencing (with $\partial\psi/\partial z \to (\psi_1 - \psi_2)/(H/2) = 4\tau/H$), we obtain the two-level thermodynamic equation:

$$\frac{\partial\tau}{\partial t} + J(\psi, \tau) + \frac{wN^2 H}{4f_0} = 0. \tag{9.152}$$

The change of available potential energy is obtained from this by multiplying by $k_d^2\tau$ and integrating, giving

$$\int\left(\frac{1}{2}\frac{\mathrm{d}}{\mathrm{d}t}k_d^2\tau^2 + \tau w\frac{2f_0}{H}\right)\mathrm{d}A = 0, \tag{9.153}$$

or

$$\frac{\mathrm{d}}{\mathrm{d}t}\text{APE} = -\frac{4f_0}{H}\int w\tau\,\mathrm{d}A, \tag{9.154a}$$

or, for a single Fourier mode,

$$\frac{\mathrm{d}}{\mathrm{d}t}\text{APE} = -\text{Re}\,\frac{4f_0}{H}\widetilde{w}\,\widetilde{\tau}^*. \tag{9.154b}$$

From (9.150) and (9.154) it is clear that in the nonlinear equations the sum of the kinetic energy and the available potential energy is conserved.

We now specialize by obtaining w from the linear baroclinic instability problem. Using this in (9.150) and (9.154) will give us the conversion between kinetic energy and potential energy in the growing baroclinic wave. It is important to realize that the total energy of the disturbance will not be conserved — both the potential and kinetic energy are growing, exponentially in this problem, because they are extracting energy from the mean state. To calculate w we use the linearized thermodynamic equation. From (9.152) this is

$$\frac{\partial\tau}{\partial t} - U\frac{\partial\psi}{\partial x} + \frac{HwN^2}{4f_0} = 0, \tag{9.155}$$

omitting the primes on perturbation quantities. For a single Fourier mode, this gives

$$\frac{HN^2}{4f_0}\widetilde{w} = ik(c\widetilde{\tau} + U\widetilde{\psi}). \tag{9.156}$$

But, from (9.129), $c\widetilde{\psi} = U\widetilde{\tau}$ in two-layer f-plane baroclinic instability and so

$$\frac{HN^2}{4f_0}\widetilde{w} = ikc\widetilde{\tau}\left(1 + \frac{U^2}{c^2}\right) = ikc\widetilde{\tau}\left(\frac{2K^2}{K^2 - k_d^2}\right), \tag{9.157}$$

using (9.130). For stable waves, $K^2 > k_d^2$ and $c = c_r$ and in that case the vertical velocity is $\pi/2$ out of phase with the temperature, and there is no conversion of APE to KE. For unstable waves $c = \mathrm{i}\, c_i$ and $K^2 < k_d^2$, and the vertical velocity is in phase with the temperature; that is, *warm air is rising and there is a conversion of APE to KE*. To see this more formally, recall that the conversion from APE to KE is given by $4\widetilde{w}\widetilde{\tau}^* f_0/H$. Thus, using (9.157),

$$\frac{\mathrm{d}}{\mathrm{d}t}(\text{APE} \to \text{KE}) = \operatorname{Re} 2ikck_d^2\left(\frac{2K^2}{K^2 - k_d^2}\right)\widetilde{\tau}^2, \tag{9.158}$$

using also the definition of k_d given in (9.103). If the wave is growing, then $K^2 < k_d^2$ and $c = \mathrm{i}c_i$ and the right-hand side is real and positive. For neutral waves, if $c = c_r$ the right-hand side of (9.158) is pure imaginary, and so the conversion is zero. This completes our demonstration that baroclinic instability converts potential energy into kinetic energy.

9.9 ♦ BETA, SHEAR AND STRATIFICATION IN A CONTINUOUS MODEL

The two-layer model of Section 9.6 indicates that β has a number of important effects on baroclinic instability. Do these carry over to the continuously stratified case? The answer by and large is yes, but with some important qualifications that generally concern weak or shallow instabilities. In particular, we will find that there is no short-wave cut-off in the continuous model with non-zero beta, and that the instability determines its own depth scale. We will illustrate these properties first by way of scaling arguments and then by way of numerical calculations.[13]

9.9.1 Scaling Arguments and Estimates

With finite density scale height and non-zero β, the quasi-geostrophic potential vorticity equation, linearized about a mean zonal velocity $U(z)$, is

$$\left(\frac{\partial}{\partial t} + U\frac{\partial}{\partial x}\right)q' + \frac{\partial \psi'}{\partial x}\frac{\partial Q}{\partial y} = 0, \tag{9.159}$$

where

$$q' = \nabla^2\psi' + \frac{f_0^2}{\rho_R}\frac{\partial}{\partial z}\left(\frac{\rho_R}{N^2}\frac{\partial \psi'}{\partial z}\right), \qquad \frac{\partial Q}{\partial y} = \beta - \frac{f_0^2}{\rho_R}\frac{\partial}{\partial z}\left(\frac{\rho_R}{N^2}\frac{\partial U}{\partial z}\right), \tag{9.160a,b}$$

and ρ_R is a specified density profile. If we assume that $U = \Lambda z$ where Λ is constant and that N is constant, and let $H_\rho^{-1} = -\rho_R^{-1}\partial\rho_R/\partial z$, then

$$\frac{\partial Q}{\partial y} = \beta + \frac{f_0^2\Lambda}{N^2 H_\rho} = \beta(1 + \alpha), \quad \text{where} \quad \alpha = \left(\frac{f_0^2\Lambda}{\beta N^2 H_\rho}\right). \tag{9.161}$$

The boundary conditions on (9.159) are

$$\left(\frac{\partial}{\partial t}+U\frac{\partial}{\partial x}\right)\frac{\partial \psi'}{\partial z}-\frac{\partial \psi'}{\partial x}\frac{\partial U}{\partial z}=0, \qquad \text{at } z=0, \tag{9.162}$$

and that $\psi \to 0$ as $z \to \infty$. The problem we have defined essentially constitutes the Charney problem. We can reduce this to the Eady problem by setting $\beta = 0$ and $H_\rho = \infty$, and providing a lid that is some finite height above the ground.

As in the Eady problem, we seek solutions of the form

$$\psi = \mathrm{Re}\,\widetilde{\psi}(z)\mathrm{e}^{\mathrm{i}(kx+ly-kct)}, \tag{9.163}$$

and substituting into (9.159) gives

$$\left(\frac{f_0^2}{H_\rho^2 N^2}\right)\left(H_\rho^2\frac{\mathrm{d}^2\widetilde{\psi}}{\mathrm{d}z^2}-H_\rho\frac{\mathrm{d}\widetilde{\psi}}{\mathrm{d}z}\right)-\left(K^2-\frac{\beta+\Lambda f_0^2/(N^2H_\rho)}{\Lambda z-c}\right)\widetilde{\psi}=0. \tag{9.164}$$

The Boussinesq version of this expression, for a fluid contained between two horizontal surfaces, is obtained by letting $H_\rho = \infty$, giving

$$\left(\frac{f_0^2}{N^2}\right)\frac{\mathrm{d}^2\widetilde{\psi}}{\mathrm{d}z^2}-\left(K^2-\frac{\beta}{\Lambda z-c}\right)\widetilde{\psi}=0. \tag{9.165}$$

It seems natural to nondimensionalize (9.164) using:

$$z = H_\rho\hat{z}, \qquad c = \Lambda H_\rho\hat{c}, \qquad K=\left(\frac{f_0}{NH_\rho}\right)\widehat{K}, \tag{9.166}$$

whence the equation becomes

$$\frac{\mathrm{d}^2\widetilde{\psi}}{\mathrm{d}\hat{z}^2}-\frac{\mathrm{d}\widetilde{\psi}}{\mathrm{d}\hat{z}}-\left(\widehat{K}^2-\frac{\gamma+1}{\hat{z}-\hat{c}}\right)\widetilde{\psi}=0, \tag{9.167}$$

where

$$\gamma=\alpha^{-1}=\frac{\beta N^2 H_\rho}{f_0^2\Lambda}=\frac{\beta L_d^2}{H_\rho\Lambda}=\frac{H_\rho}{h}, \tag{9.168}$$

where $h \equiv \Lambda f_0^2/(\beta N^2)$. The nondimensional parameter γ is known as the Charney–Green number.[14] The Boussinesq version, (9.165), may be nondimensionalized using H_D in place of H_ρ, where H_D is the depth of the fluid between two rigid surfaces. In that case

$$\frac{\mathrm{d}^2\widetilde{\psi}}{\mathrm{d}\hat{z}^2}-\left(\widehat{K}^2-\frac{\gamma}{\hat{z}-\hat{c}}\right)\widetilde{\psi}=0, \tag{9.169}$$

where here the nondimensional variables are scaled with H_D.

Now, suppose that γ is large, for example if β or the static stability are large or the shear is weak. Equation (9.167) admits no non-trivial balance, suggesting that we rescale the variables using h instead of H_ρ as the vertical scale in (9.166). The rescaled version of (9.167) is then

$$\frac{\mathrm{d}^2\widetilde{\psi}}{\mathrm{d}\hat{z}^2}-\frac{1}{\gamma}\frac{\mathrm{d}\widetilde{\psi}}{\mathrm{d}\hat{z}}-\left(\widehat{K}^2-\frac{1+\gamma^{-1}}{\hat{z}-\hat{c}}\right)\widetilde{\psi}=0, \tag{9.170}$$

or, approximately,

$$\frac{\mathrm{d}^2\widetilde{\psi}}{\mathrm{d}\hat{z}^2}-\left(\widehat{K}^2-\frac{1}{\hat{z}-\hat{c}}\right)\widetilde{\psi}=0. \tag{9.171}$$

This is exactly the same equation as results from a similar rescaling of the Boussinesq system, (9.169), as we might have expected because now the dynamical vertical scale, h, is much smaller than the scale height H_ρ (or H_D) and the system is essentially Boussinesq. Thus, noting that (9.171) has the same nondimensional form as (9.169) save that γ is replaced by unity, and that (9.171) with $\gamma = 1$ must produce the same scales and growth rates as in the Eady problem, we may deduce that:

(i) the wavelength of the instability is $\mathcal{O}(Nh/f_0)$;

(ii) the growth rate of the instability is $\mathcal{O}(Kc) = \mathcal{O}(f_0\Lambda/N)$;

(iii) the vertical scale of the instability is $\mathcal{O}(h) = \mathcal{O}(f_0^2\Lambda/(\beta N^2))$.

These are the same as for the Eady problem, except with the dynamical height h replacing the geometric or scale height H_D. Effectively, the dynamics has determined its own vertical scale, h, which is much less than the scale height or geometric height, producing 'shallow modes'.

In the limit $\gamma \ll 1$ (strong shear, weak β), the Boussinesq and compressible problems differ. The Boussinesq problem reduces to the Eady problem, considered previously, whereas (9.167) becomes, approximately,

$$\frac{\mathrm{d}^2\widetilde{\psi}}{\mathrm{d}\widehat{z}^2} - \frac{\mathrm{d}\widetilde{\psi}}{\mathrm{d}\widehat{z}} - \left(\widehat{K}^2 - \frac{1}{\widehat{z} - \widehat{c}}\right)\widetilde{\psi} = 0, \tag{9.172}$$

and in this limit the appropriate vertical scale is the density scale height H_ρ. Because $H_\rho \gg h$ these are 'deep modes', occupying the entire vertical extent of the domain.

The scale h does not arise in the two-level model, but there is a connection between it and the critical shear for instability in the two-level model. The condition $\gamma \ll 1$, or $h \gg H$, may be written as

$$H\Lambda \gg \beta\left(\frac{NH}{f_0}\right)^2. \tag{9.173}$$

Compare this with the necessary condition for instability in a two-level model, (9.122), namely

$$(U_1 - U_2) > \beta\left(\frac{NH_\Delta}{f_0}\right)^2, \tag{9.174}$$

where H_Δ is the vertical distance between the two levels. Thus, essentially the same condition governs the onset of instability in the two-level model as governs the production of deep modes in the continuous model. This correspondence is a natural one, because in the two-level model *all* modes are 'deep', and the model fails (as it should) to capture the shallow modes of the continuous system. For similar reasons, there is a high-wavenumber cut-off in the two-level model: in the continuous model these modes are shallow and so cannot be captured by two-level dynamics. Somewhat counter-intuitively, for these modes the β-effect must be important, even though the modes have small horizontal scale: when $\beta = 0$ the instability arises via an interaction between edge waves at the top and bottom of the domain, whereas the shallow instability arises via an interaction of the edge waves at the surface with Rossby waves just above the surface.

9.9.2 Some Numerical Calculations

Adding β to the Eady model

Our first step is to add the β-effect to the Eady problem.[15] That is, we suppose a Boussinesq fluid with uniform stratification, that the shear is zonal and constant and that the entire problem is sandwiched between two rigid surfaces. Growth rates and phase speeds of such an instability calculation are illustrated in Fig. 9.19 and the vertical structure is shown in Fig. 9.20. As in the two-layer problem, there is a low-wavenumber cut-off to the main instability, although there is now an additional weak instability at very large-scales. These so-called *Green modes* have no counterpart in the two-layer model — they are deep, slowly growing modes that will be dominated by faster growing

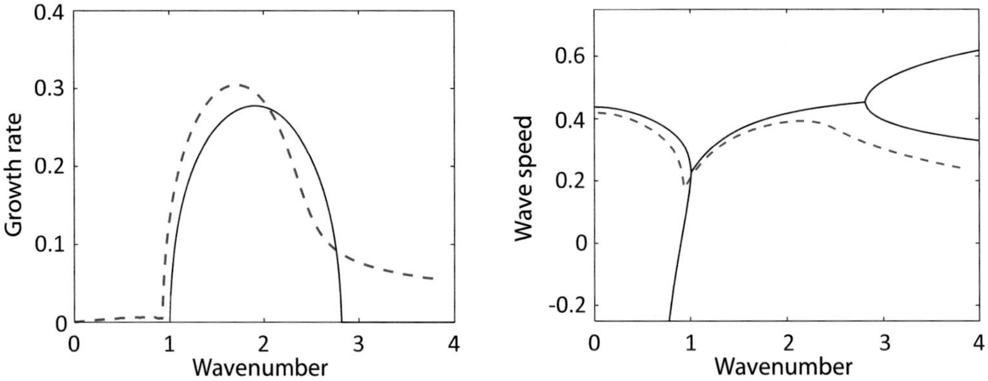

Fig. 9.19 Growth rates and wave speeds for the two-layer (solid) and continuous (dashed) models, with the same values of the Charney–Green number, γ, and uniform shear and stratification. (In the two-layer case $\gamma = \beta L_d^2/[2(U_1 - U_2)] = 0.5$, and in the continuous case $\gamma = \beta L_d^2/(H\Lambda) = 0.5$.) In the continuous case only the wave speed associated with the unstable mode is shown. In the two-layer case there are two real wave speeds which coalesce in the unstable region.

modes in most real situations. (Also, the fact that the Green modes have a scale much larger than the deformation scale suggests a degree of caution in the accuracy of the quasi-geostrophic calculation.) At high wavenumbers there is no cut-off to the instability in the continuous problem in the case of non-zero beta; the high-wavenumber modes are shallow and unstable via an interaction between edge waves at the lower boundary and Rossby waves in the lower atmosphere, and so have no counterpart in either the two-layer problem (where the modes are deep) or the Eady problem (which has no Rossby waves).

Effects of non-uniform shear and stratification

If the shear or stratification is non-uniform an analytic treatment is, even in problems without β, usually impossible and the resulting equations must be solved numerically. However, if we restrict attention to a *discontinuity* in the shear or the stratification, then the resulting problem is very similar to the problem with rigid boundaries, and this property provides some justification for using the Eady problem to model instabilities in the Earth's atmosphere: in the troposphere the stratification is (approximately) constant, and the rapid increase in stratification in the stratosphere can be approximated by a lid at the tropopause. Heuristically, we can see this from the form of the thermodynamic equation, namely

$$\frac{Db}{Dt} + N^2 w = 0. \tag{9.175}$$

If N^2 is high this suggests w will be small, and a lid is the limiting case of this. The oceanic problem is rather more involved, because although both the stratification and the shear are concentrated in the upper ocean, they vary relatively smoothly; furthermore, the shear is high where the stratification is high, and the two have opposing effects.

To go one step further, consider the Boussinesq potential vorticity equation, linearized about a zonally uniform state $\Psi(y, z)$, with a rigid surface at $z = 0$. The normal-mode evolution equations are similar to (9.69), namely

$$(U - c)\left[\frac{\partial^2}{\partial y^2} - k^2 + \frac{\partial}{\partial z}\left(F\frac{\partial}{\partial z}\right)\right]\widetilde{\psi} + \frac{\partial Q}{\partial y}\widetilde{\psi} = 0, \qquad z > 0, \tag{9.176a}$$

$$(U - c)\frac{\partial \widetilde{\psi}}{\partial z} - \frac{\partial U}{\partial z}\widetilde{\psi} = 0, \qquad \text{at } z = 0, \tag{9.176b}$$

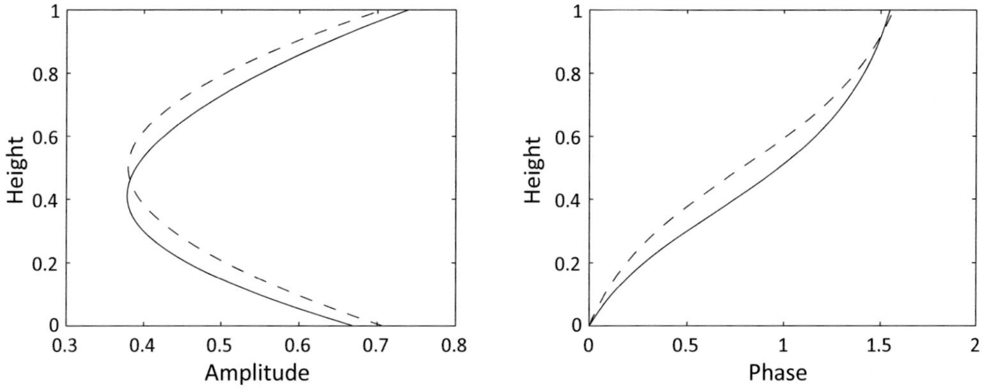

Fig. 9.20 Vertical structure of the most unstable modes in a continuously stratified instability calculation with $\beta = 0$ (dashed lines, the Eady problem) and $\beta \neq 0$ (solid lines, as in the continuous problem in Fig. 9.19). The effect of beta is to depress the height of the maximum amplitude of the instability.

where $\partial_y Q = \beta - \partial_{yy}U - \partial_z(F\partial_z U)$. Now suppose that there is a discontinuity in the shear and/or the stratification in the interior of the fluid, at some level $z = z_c$. Integrating (9.176a) across the discontinuity, noting that $\widetilde{\psi}$ is continuous in z, gives

$$(U - c)\left[F\frac{\partial\widetilde{\psi}}{\partial z}\right]_{z_c-}^{z_c+} - \widetilde{\psi}(y, z_c)\left[F\frac{\partial U}{\partial z}\right]_{z_c-}^{z_c+} = 0. \tag{9.177}$$

which has similar form to (9.176b). This construction is evocative of the equivalence of a delta-function sheet of potential vorticity at a rigid boundary, except that now a discontinuity in the potential vorticity in the *interior* has a similarity with a rigid boundary.

We can illustrate the effects of an interior discontinuity that crudely represents the tropopause by numerically solving the linear eigenvalue problem. We pose the problem on the f-plane, in a horizontally doubly-periodic domain, with no horizontal variation of shear, and between two horizontal rigid lids. The eigenvalue problem is defined by (9.69), and the numerical procedure then solves for the complex eigenvalue c and eigenfunction $\widetilde{\psi}(z)$; various results are illustrated in Fig. 9.21. To parse this rather complex figure, first look at the solid curves in all the panels. These arise when the problem is solved with a uniform shear and a uniform stratification, with a lid at $z = 0$ and $z = 1$, hence simply giving the Eady problem. The familiar growth rates and vertical structure of the solution are given by the solid curves in panels (b), (c) and (d), and these are just the same as in Fig. 9.10. The various dotted and dashed curves show the results when the lid at $z = 1$ is replaced by a stratosphere stretching from $1 < z < 2$ either with high stratification, zero shear, or both, and in all of these cases the stratosphere acts qualitatively in the same way as a rigid lid. The vertical structure of the solution in the troposphere is, in all cases, quite similar, and the amplitude decays rapidly above the idealized tropopause, consistent with the almost uniform phase of the disturbance illustrated in panel (d) — recall that a tilting of the disturbance with height is necessary for instability. It is these properties that make the Eady problem of more general applicability to the Earth's atmosphere than might be first thought: the high stratification above the tropopause and consequent decay of the instability are mimicked by the imposition of a rigid lid.

In the ocean, the stratification is highest in its upper regions (e.g., in the main thermocline) where the shear is also strongest, and numerical calculations of the structure and growth rate of idealized profiles are illustrated in Fig. 9.22.[16] The solid curve shows the Eady problem, and the various dashed curves show the phase speeds, growth rates and phase with combinations of the profiles illustrated in panel (a). Much of the ocean is characterized by having both a higher shear

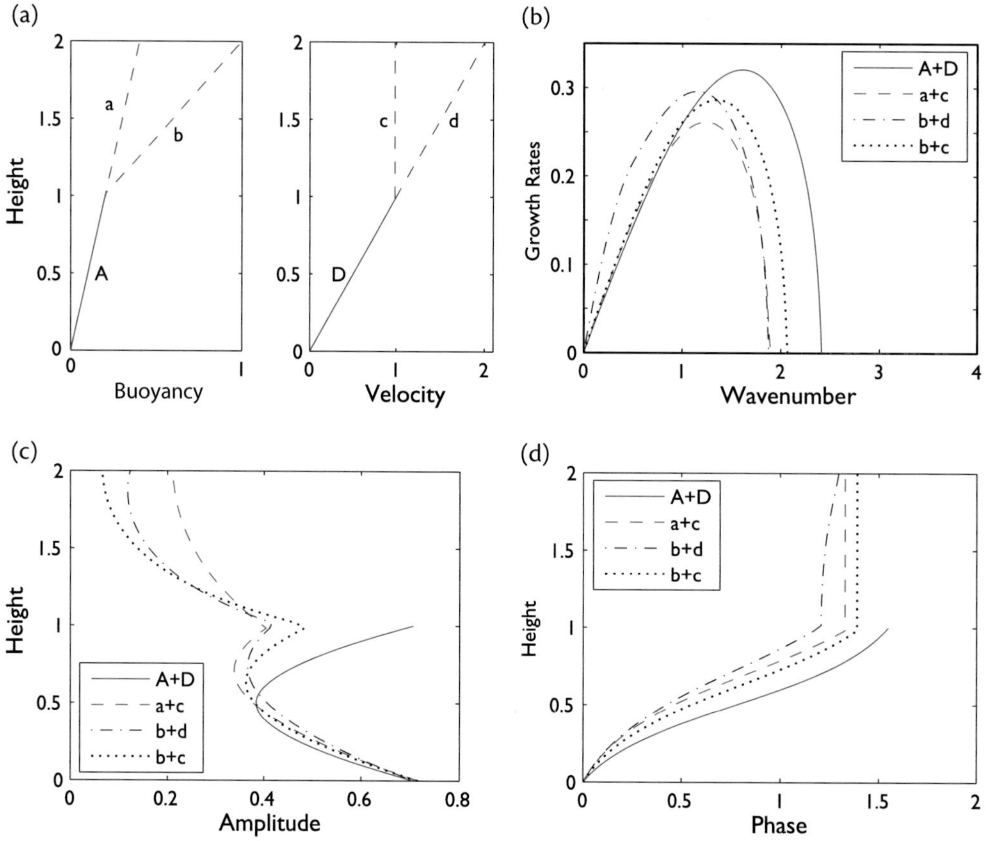

Fig. 9.21 The effect of a stratosphere on baroclinic instability: (a) the given profiles of shear and stratification; (b) the growth rate of the instabilities; (c) the amplitude of the most unstable mode as a function of height; (d) the phase of the most unstable mode. The instability problem is solved numerically with various profiles of stratification and shear. In each profile, in the idealized troposphere ($z < 1$) the shear and stratification are uniform and the same in each case. We consider four idealized stratospheres ($z \geq 1$): (1) A lid at $z = 1$, i.e., no stratosphere, giving the Eady problem itself (profiles A+D, solid lines); (2) stratospheric stratification as for the troposphere, but with zero shear (profiles a+c, dashed); (3) stratospheric shear as for troposphere, but with stratification (N^2) four times the tropospheric value (b+d, dot-dashed); (4) zero shear and high stratification in the stratosphere (b+c, dotted). In the troposphere the amplitude and structure of the instability are similar in all cases, illustrating the similarity of a rigid lid and abrupt changes in shear or stratification. Either a high stratification or a low shear (or both) will result in weak stratospheric instability.

and a higher stratification in the upper 1 km or so, and this case is shown with the dotted line in Fig. 9.22. The amplitude of the instability is also largely confined to the upper ocean, and unlike the Eady problem it does not arise through the interaction of edge waves at the top and bottom: the potential vorticity changes sign because of the interior variations due to the non-uniform shear, mainly in the upper ocean. Consistently, the phase of the baroclinic waves is nearly constant in the lower ocean in those cases in which the shear is confined to the upper ocean. The real ocean is still more complicated, because the most unstable regions near intense western boundary currents are often also barotropically unstable, and the mean flow itself may be meridionally directed. Nevertheless, the result that linear baroclinic instability is primarily an *upper ocean* phenomenon is quite robust. However, we will find in Chapter 12 that the nonlinear evolution of baroclinic instability leads to eddies throughout the water column.

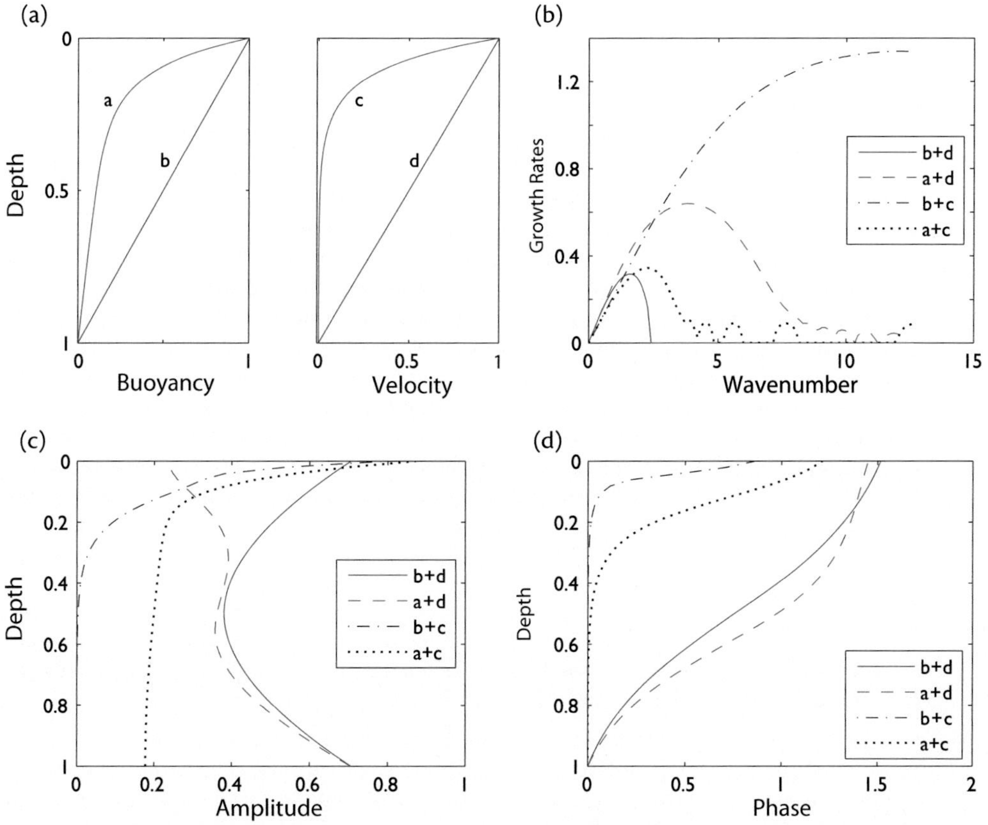

Fig. 9.22 The baroclinic instability in an idealized ocean, with four different profiles of shear or stratification. The panels are: (a) the profiles of velocity and buoyancy (and so N^2) used; (b) the growth rates of the various cases; (c) the vertical structure of the amplitude of the most unstable models; (d) the phase in the vertical of the most unstable modes. The instability is calculated numerically with four combinations of shear and stratification: (1) uniform stratification and shear i.e., the Eady problem, (profiles b+d, solid lines); (2) uniform shear, upper-ocean enhanced stratification (a+d, dashed); (3) uniform stratification, upper ocean enhanced shear (b+c, dot-dashed); (4) both stratification and shear enhanced in the upper ocean (a+c, dotted). Case 2 (a+d, dashed) is really more like an atmosphere with a stratosphere (see Fig. 9.21), and the amplitude of the disturbance falls off, rather unrealistically, in the upper ocean. Case 4 (a+c, dotted) is the most oceanographically relevant.

Notes

1 Thomson (1871), Helmholtz (1868). Thomson later became known as Lord Kelvin.

2 This is Squire's theorem, which states that for every three-dimensional disturbance to a plane-parallel flow there exists a more unstable two-dimensional one (Squire 1933). This means there is no need to consider three-dimensional effects to determine whether such a flow is unstable.

3 Rayleigh (1880). John Strutt (1842–1919) became 3rd Baron Rayleigh on the death of his father in 1873 and, in a testament to the enduring British class system, is almost universally known as Lord Rayleigh. He made major contributions in many areas of physics, among them fluid mechanics (including the theory of sound and instability theory), the analysis of the composition of gases (leading to the discovery of argon), and scattering theory.

4 First obtained by Rayleigh (1894). I thank Adrian Matthews for comments and pointing out an error in an earlier derivation. For more applications of this kind of argument see Harnik & Heifetz (2007) and references therein.

5 Rayleigh (1880) and, for the case with β, Kuo (1949).

6 Fjørtoft (1950).

7 Charney & Stern (1962), Pedlosky (1964).

8 Eady (1949), Charney (1947). Eric Eady (1915–1966) is best remembered today as the author of the iconic 'Eady model' of baroclinic instability, which describes the fundamental hydrodynamic instability mechanism that gives rise to weather systems. After an undergraduate education in mathematics he joined the UK Meteorological Office in 1937, becoming a forecaster and upper air analyst, in which capacity he served throughout the war. In 1946 he joined the Department of Mathematics at Imperial College, presenting his PhD thesis in 1948 on 'The theory of development in dynamical meteorology', subsequently summarized in *Tellus* (Eady 1949). This work, with a skilled combination of austerity and relevance, provides a mathematical description of the essential aspects of cyclone development that stands to this day as a canonical model in the field. It also includes, rather obliquely, a derivation of the stratified quasi-geostrophic equations, albeit in a special form. The impact of the work was immediate and it led to visits to Bergen (in 1947 with J. Bjerknes), Stockholm (in 1952 with C.-G. Rossby) and Princeton (in 1953 with J. von Neumann and Charney). Eady followed his baroclinic instability work with prescient discussions of the general circulation of the atmosphere (Eady 1950, Eady & Sawyer 1951, Eady 1954). A perfectionist who sought to understand it all, Eady's subsequent published output was small and he later turned his attention to fundamental problems in other areas of fluid mechanics, the dynamics of the Sun and the Earth's interior, and biochemistry. In his younger days he was a lively personality but he became increasingly divorced from normal scientific and human discourse, and finally took his own life. There is little published about him, save for the obituary by Charnock *et al.* (1966). Elsewehere, Charnock said 'Talking with Eric Eady was one of the pleasures of life'.

Jule Charney (1917–1981) played a defining role in dynamical meteorology in the second half of the twentieth century. He made seminal contributions in many areas including: the theory of baroclinic instability (Charney 1947); a systematic scaling theory for large-scale atmospheric motions and the derivation of the quasi-geostrophic equations (Charney 1948); a theory of stationary waves in the atmosphere (Charney & Eliassen 1949); the demonstration of the feasibility of numerical weather forecasts (Charney *et al.* 1950); planetary wave propagation into the stratosphere (Charney & Drazin 1961); a criterion for baroclinic instability (Charney & Stern 1962); a theory for hurricane growth (Charney & Eliassen 1964); and the concept of geostrophic turbulence (Charney 1971). His PhD from UCLA in 1946 was entitled 'Dynamics of long waves in a baroclinic westerly current' and this became his well-known 1947 paper. After this he spent a year at Chicago and another at Oslo, and in 1948 joined the Institute of Advanced Study in Princeton where he stayed until 1956 (and where Eady visited for a while). He spent most of his subsequent career at MIT, interspersed with many visits to Europe, especially Norway. For a more complete picture of Charney, see Lindzen *et al.* (1990) and a brief biography by N. Phillips, available at http://www.nap.edu/readingroom/books/biomems/jcharney.html.

9 The solution to Eady problem follows relatively straightforwardly using the quasi-geostrophic approximation from the outset. But when Eady was formulating his problem the quasi-geostrophic equations were not known, except perhaps to Charney. Eady (1949) began with something more akin to the primitive equations, and introduced an essentially quasi-geostrophic assumption late in his derivation, and independently derived a linear version of the quasi-geostrophic potential vorticity equation.

10 If c is real (and the waves are neutral), then there exists the possibility that $\Lambda z - c = 0$, and the equation for Φ is

$$\frac{\mathrm{d}^2\Phi}{\mathrm{d}z^2} - \mu^2\Phi = C\delta(z - z_c), \qquad z_c = c/\Lambda, \tag{9.178}$$

where C is a constant. Because z_c is continuous in the interval $[0, 1]$ so is c, and these solutions have a continuous spectrum of eigenvalues. The associated eigenfunctions provide formal completeness to the normal modes, enabling any function to be represented as their superposition.

11 After Phillips (1954).

12 Our nondimensionalization of the two-layer system is such as to be in correspondence with that for the continuous system. Thus we choose H to be the total depth of the domain. This choice produces growth rates and wavenumbers that are equivalent to those in the Eady problem.

13 Green (1960) and Branscome (1983). Lindzen & Farrell (1980) also provide an approximate calculation of growth rates in the Charney problem.

14 After Charney (1947), in whose problem it appears, and Green (1960), who appreciated its importance.

15 Our numerical procedure is to assume a wave-like solution in the horizontal direction of the form $\widetilde{\psi} \exp[\mathrm{i}(kx + ly - \omega t)]$, and to finite difference the equations in the vertical direction. The resulting eigenvalue equations are solved by standard matrix methods, for each horizontal wavenumber. See Smith & Vallis (1998) and Green (1960).

16 Gill *et al.* (1974) and Robinson & McWilliams (1974) were among the first to look at baroclinic instability in the ocean.

Below, a myriad, myriad waves hastening, lifting up their necks,
Tending in ceaseless flow toward the track of the ship,
Waves of the ocean bubbling and gurgling, blithely prying,
Waves, undulating waves, liquid, uneven, emulous waves,
Toward that whirling current, laughing and buoyant.
Walt Whitman, *After the Sea-Ship*, in *Leaves of Grass*, 1881.

CHAPTER 10

Waves, Mean-Flows, and their Interaction

WAVE–MEAN-FLOW INTERACTION is concerned with how some mean-flow, perhaps a time or zonal average, interacts with a wave-like departure from that mean, and this chapter provides an elementary introduction to this topic. It is 'elementary' because our derivations and discussion are obtained by straightforward manipulations of the equations of motion in the simplest case that illustrates the relevant principle. It is implicit in what we do that it is a sensible thing to decompose the fields into a mean plus some departure, and one case when this is so is when the departure is of small amplitude. Departures from the mean — generically called *eddies* — are in reality not always small; for example, in the mid-latitude troposphere the eddies are often of similar amplitude to the mean-flow, and Chapters 12 and 13 explore this from the standpoint of turbulence. However, in this chapter we will assume that eddies are indeed of small amplitude, and, in particular, that eddy–mean-flow interaction is larger than eddy–eddy interaction.

A *wave* is an eddy that satisfies, at least approximately, a dispersion relation. It is the presence of such a dispersion relation that enables a number of results to be obtained that would otherwise be out of our reach. It is implicit in defining waves this way that they are generally of small amplitude, for it is this that allows the equations of motion to be sensibly linearized and a dispersion relation to be obtained (although some waves have finite amplitude and still satisfy a dispersion relation), and the interaction with the mean-flow calculated. The qualitative nature of such interaction can then provide insights into the finite-amplitude problem, and one goal of wave–mean-flow theory is to provide a way of qualitatively understanding more realistic situations, and to suggest diagnostics that might be used to analyze both observations and numerical solutions of the nonlinear problem. In this chapter we will largely concern ourselves with a *zonal* mean, since this is the simplest and often most useful case because of the presence of simple boundary conditions. We will also be mainly concerned with quasi-geostrophic dynamics on a β-plane (and hence Rossby waves), using Boussinesq dynamics, since here the concepts are most clearly illustrated. Thus, in this chapter the reader will find an introduction to such matters as the 'transformed Eulerian mean', the 'Eliassen–Palm flux' and the 'non-acceleration result', and later in the chapter we look at how some related ideas can be used to prove stability of a flow without invoking normal modes. Impatient readers who are anxious for real examples may wish to first look at Chapters 15 and 17 and then come back to this chapter as needed.

10.1 QUASI-GEOSTROPHIC WAVE–MEAN-FLOW INTERACTION

10.1.1 Preliminaries

To fix our dynamical system and notation, we write down the Boussinesq quasi-geostrophic potential vorticity equation

$$\frac{\partial q}{\partial t} + J(\psi, q) = D, \tag{10.1}$$

where D represents any non-conservative terms and the potential vorticity in a Boussinesq system is

$$q = \beta y + \zeta + \frac{\partial}{\partial z}\left(\frac{f_0}{N^2} b\right), \tag{10.2}$$

where ζ is the relative vorticity and b is the buoyancy perturbation from the background state characterized by N^2. (In an ideal gas $q = \beta y + \zeta + (f_0/\rho_R)\partial_z(\rho_R b/N^2)$, where ρ_R is a specified density profile, and most of our derivations can be extended to that case.) We will refer to lines of constant b as isentropes. In terms of the streamfunction, the variables are

$$\zeta = \nabla^2\psi, \qquad b = f_0\frac{\partial \psi}{\partial z}, \qquad q = \beta y + \left[\nabla^2 + \frac{\partial}{\partial z}\left(\frac{f_0^2}{N^2}\frac{\partial}{\partial z}\right)\right]\psi. \tag{10.3}$$

where $\nabla^2 \equiv (\partial_x^2 + \partial_y^2)$. The potential vorticity equation holds in the fluid interior; the boundary conditions on (10.3) are provided by the thermodynamic equation

$$\frac{\partial b}{\partial t} + J(\psi, b) + wN^2 = H, \tag{10.4}$$

where H represents heating terms. The vertical velocity at the boundary, w, is zero in the absence of topography and Ekman friction, and if H is also zero the boundary condition is just

$$\frac{\partial b}{\partial t} + J(\psi, b) = 0. \tag{10.5}$$

Equations (10.1) and (10.5) are the evolution equations for the system and if both D and H are zero they conserve both the total energy, $\widehat{E}$ and the total enstrophy, $\widehat{Z}$:

$$\begin{aligned} \frac{\mathrm{d}\widehat{E}}{\mathrm{d}t} &= 0, \qquad \widehat{E} = \frac{1}{2}\int_V (\nabla\psi)^2 + \frac{f_0^2}{N^2}\left(\frac{\partial \psi}{\partial z}\right)^2 \mathrm{d}V, \\ \frac{\mathrm{d}\widehat{Z}}{\mathrm{d}t} &= 0, \qquad \widehat{Z} = \frac{1}{2}\int_V q^2 \,\mathrm{d}V, \end{aligned} \tag{10.6}$$

where V is a volume bounded by surfaces at which the normal velocity is zero, or that has periodic boundary conditions. The enstrophy is also conserved layerwise; that is, the horizontal integral of q^2 is conserved at every level.

10.1.2 Potential Vorticity Flux in the Linear Equations

Let us decompose the fields into a mean (to be denoted with an overbar) plus a perturbation (denoted with a prime), and let us suppose the perturbation fields are of small amplitude. (In linear problems, such as those considered in Chapter 9, we decomposed the flow into a 'basic state' plus a perturbation, with the basic state fixed in time. Our approach here is similar, but soon we will allow the mean state to evolve.) The linearized quasi-geostrophic potential vorticity equation is then

$$\frac{\partial q'}{\partial t} + \overline{u}\frac{\partial q'}{\partial x} + u'\frac{\partial \overline{q}}{\partial x} + \overline{v}\frac{\partial q'}{\partial y} + v'\frac{\partial \overline{q}}{\partial y} = D', \tag{10.7}$$

where D' represents eddy forcing and dissipation and, in terms of streamfunction,

$$(u'(x,y,z,t), v'(x,y,z,t)) = \left(-\frac{\partial \psi'}{\partial y}, \frac{\partial \psi'}{\partial x}\right), \tag{10.8a}$$

$$q'(x,y,z,t) = \nabla^2 \psi' + \frac{\partial}{\partial z}\left(\frac{f_0^2}{N^2}\frac{\partial \psi'}{\partial z}\right). \tag{10.8b}$$

If the mean is a zonal mean then $\partial \overline{q}/\partial x = 0$ and $\overline{v} = 0$ (because v is purely geostrophic) and (10.7) simplifies to

$$\frac{\partial q'}{\partial t} + \overline{u}\frac{\partial q'}{\partial x} + v'\frac{\partial \overline{q}}{\partial y} = D', \tag{10.9}$$

where

$$\overline{q} = \beta y - \frac{\partial \overline{u}}{\partial y} + \frac{\partial}{\partial z}\left(\frac{f_0}{N^2}\overline{b}\right), \quad \text{and} \quad \frac{\partial \overline{q}}{\partial y} = \beta - \frac{\partial^2 \overline{u}}{\partial y^2} - \frac{\partial}{\partial z}\left(\frac{f_0^2}{N^2}\frac{\partial \overline{u}}{\partial z}\right). \tag{10.10a,b}$$

using thermal wind, $f_0 \partial \overline{u}/\partial z = -\partial b/\partial y$.

Multiplying by q' and zonally averaging gives the enstrophy equation:

$$\frac{1}{2}\frac{\partial}{\partial t}\overline{q'^2} = -\overline{v'q'}\frac{\partial \overline{q}}{\partial y} + \overline{D'q'}. \tag{10.11}$$

The quantity $\overline{v'q'}$ is the meridional flux of potential vorticity; this is downgradient (by definition) when the first term on the right-hand side is positive (i.e., $\overline{v'q'}\partial \overline{q}/\partial y < 0$), and it then acts to increase the variance of the perturbation. (This occurs, for example, when the flux is diffusive so that $\overline{v'q'} = -\kappa \partial \overline{q}/\partial y$, where κ may vary but is everywhere positive.) This argument may be inverted: for inviscid flow ($D = 0$), if the waves are growing, as for example in the canonical models of baroclinic instability discussed in Chapter 9, then *the potential vorticity flux is downgradient.*

If the second term on the right-hand side of (10.11) is negative, as it will be if D' is a dissipative process (e.g., if $D' = A\nabla^2 q'$ or if $D' = -rq'$, where A and r are positive) then a statistical balance can be achieved between enstrophy production via downgradient transport, and dissipation. If the waves are steady (by which we mean statistically steady, neither growing nor decaying in amplitude) and conservative (i.e., $D' = 0$) then we must have

$$\overline{v'q'} = 0. \tag{10.12}$$

Similar results follow for the buoyancy at the boundary; we start by linearizing the thermodynamic equation (10.5) to give

$$\frac{\partial b'}{\partial t} + \overline{u}\frac{\partial b'}{\partial x} + v'\frac{\partial \overline{b}}{\partial y} = H', \tag{10.13}$$

where H' is a diabatic source term. Multiplying (10.13) by b' and averaging gives

$$\frac{1}{2}\frac{\partial}{\partial t}\overline{b'^2} = -\overline{v'b'}\frac{\partial \overline{b}}{\partial y} + \overline{H'b'}. \tag{10.14}$$

Thus growing adiabatic waves have a downgradient flux of buoyancy at the boundary. In the Eady problem there is no interior gradient of basic-state potential vorticity and all the terms in (10.11) are zero, but the perturbation grows at the boundary. If the waves are steady and adiabatic then, analogously to (10.12),

$$\overline{v'b'} = 0. \tag{10.15}$$

The boundary conditions and fluxes may be absorbed into the interior definition of potential vorticity and its fluxes by way of the delta-function boundary layer construction, described in Section 5.4.3. In models with discrete vertical layers or a finite number of levels it is common practice to absorb the boundary conditions into the definition of potential vorticity at top and bottom.

10.1.3 Wave–Mean-Flow Interaction

In linear problems we usually suppose that the mean-flow is fixed and that the zonal mean terms, $\overline{u}$ and $\overline{q}$ in (10.9), are functions only of y and z. However, in reality we might expect that the mean-flow would change because of momentum and heat flux convergences arising from the eddy–eddy interactions. To calculate these changes we begin with the potential vorticity equation (10.1) and, in the usual way, express the variables as a zonal mean plus an eddy term and obtain

$$\frac{\partial \overline{q}}{\partial t} + \nabla \cdot (\overline{\boldsymbol{u}}\,\overline{q}) + \nabla \cdot \overline{(\boldsymbol{u}'q')} = \overline{D}. \tag{10.16}$$

Now, since the mean-flow is a zonal mean, and $\overline{v} = 0$, the first term is zero and the mean-flow evolves according to

$$\frac{\partial \overline{q}}{\partial t} + \frac{\partial}{\partial y}\overline{v'q'} = \overline{D}. \tag{10.17}$$

Similarly, at the boundary the mean buoyancy evolution equation is

$$\frac{\partial \overline{b}}{\partial t} + \frac{\partial}{\partial y}\overline{v'b'} = \overline{H}. \tag{10.18}$$

To obtain $\overline{u}$ from $\overline{q}$ and $\overline{b}$ we use thermal wind balance to define a streamfunction Ψ. That is, since

$$f_0\frac{\partial \overline{u}}{\partial z} = -\frac{\partial \overline{b}}{\partial y}, \quad \text{then} \quad \left(\overline{u}, \frac{1}{f_0}\overline{b}\right) = \left(-\frac{\partial \Psi}{\partial y}, \frac{\partial \Psi}{\partial z}\right) \tag{10.19a,b}$$

whence, using (10.10a), the potential vorticity is

$$\overline{q}(y,z,t) - \beta y = \frac{\partial}{\partial z}\left(\frac{f_0^2}{N^2}\frac{\partial \Psi}{\partial z}\right) + \frac{\partial^2 \Psi}{\partial y^2}. \tag{10.20}$$

If $\overline{q}$ is known in the interior from (10.18), and $\overline{b}$ (i.e., $f_0\partial\Psi/\partial z$) is known at the boundaries, then $\overline{u}$ and $\overline{b}$ in the interior may be obtained using (10.20) and (10.19b). The equations are also summarized in the grey box on page 390.

To close the system we suppose that the eddy terms themselves evolve according to (10.9) and (10.13). If in those equations we were to include the eddy–eddy interaction terms we would simply recover the full system, so in neglecting those terms we have constructed an eddy–mean-flow system, commonly called a *wave–mean-flow* system because by eliminating the nonlinear terms in the perturbation equation the eddies will often be wavelike. Non quasi-geostrophic wave–mean-flow systems may be constructed in a similar fashion: for example, we could construct a system using the primitive equations with separate equations for eddy and zonal-mean temperature and velocity fields, and an example involving gravity waves is given in Chapter 17.

It is important to realise that such systems do differ from linear ones. In constructing linear systems we posit that the eddy terms are small compared to the mean-flow and thus neglect the eddy–eddy interaction terms and keep the mean-flow fixed. In a wave–mean-flow problem we similarly suppose the eddy terms are small, and we neglect eddy–eddy interaction terms where they produce another eddy, because the terms involving the mean-flow are larger. However, in the mean-flow equation, (10.16), there are no mean-flow terms that are larger, so we keep the eddy–eddy terms and allow the mean-flow to evolve. Such a justification is hardly a rigorous one, since if the eddy terms are small then the effects on the mean-flow will be small, and so one might suppose that the mean-flow should be held fixed. The wave–mean-flow equations really can only be justified on a case-by-case basis with a detailed examination of the size of the terms and the rate at which they evolve, and that is the subject of weakly nonlinear theory. Another justification for wave–mean-flow problems is that they lead to insight into the behaviour of the full system.

We now consider some more properties of the waves themselves — how they propagate and what they conserve — beginning with a discussion of the potential vorticity flux and its relative, the Eliassen–Palm flux

10.2 THE ELIASSEN–PALM FLUX

The eddy flux of potential vorticity may be expressed in terms of vorticity and buoyancy fluxes as

$$v'q' = v'\zeta' + f_0 v' \frac{\partial}{\partial z}\left(\frac{b'}{N^2}\right). \tag{10.21}$$

The second term on the right-hand side can be written as

$$\begin{aligned} f_0 v' \frac{\partial}{\partial z}\left(\frac{b'}{N^2}\right) &= f_0 \frac{\partial}{\partial z}\left(\frac{v'b'}{N^2}\right) - f_0 \frac{\partial v'}{\partial z}\frac{b'}{N^2} \\ &= f_0 \frac{\partial}{\partial z}\left(\frac{v'b'}{N^2}\right) - f_0 \frac{\partial}{\partial x}\left(\frac{\partial \psi'}{\partial z}\right)\frac{b'}{N^2} \\ &= f_0 \frac{\partial}{\partial z}\left(\frac{v'b'}{N^2}\right) - \frac{f_0^2}{2N^2}\frac{\partial}{\partial x}\left(\frac{\partial \psi'}{\partial z}\right)^2, \end{aligned} \tag{10.22}$$

using $b' = f_0 \partial\psi'/\partial z$.

Similarly, the flux of relative vorticity can be written

$$v'\zeta' = -\frac{\partial}{\partial y}(u'v') + \frac{1}{2}\frac{\partial}{\partial x}(v'^2 - u'^2), \tag{10.23}$$

Using (10.22) and (10.23), (10.21) becomes

$$v'q' = -\frac{\partial}{\partial y}(u'v') + \frac{\partial}{\partial z}\left(\frac{f_0}{N^2}v'b'\right) + \frac{1}{2}\frac{\partial}{\partial x}\left((v'^2 - u'^2) - \frac{b'^2}{N^2}\right). \tag{10.24}$$

Thus the meridional potential vorticity flux, in the quasi-geostrophic approximation, can be written as the divergence of a vector: $v'q' = \nabla \cdot \boldsymbol{\mathcal{E}}$ where

$$\boldsymbol{\mathcal{E}} \equiv \frac{1}{2}\left((v'^2 - u'^2) - \frac{b'^2}{N^2}\right)\mathbf{i} - (u'v')\mathbf{j} + \left(\frac{f_0}{N^2}v'b'\right)\mathbf{k}. \tag{10.25}$$

A particularly useful form of this arises after zonally averaging, for then (10.24) becomes

$$\overline{v'q'} = -\frac{\partial}{\partial y}\overline{u'v'} + \frac{\partial}{\partial z}\left(\frac{f_0}{N^2}\overline{v'b'}\right). \tag{10.26}$$

The vector defined by

$$\boldsymbol{\mathcal{F}} \equiv -\overline{u'v'}\,\mathbf{j} + \frac{f_0}{N^2}\overline{v'b'}\,\mathbf{k} \tag{10.27}$$

is called the (quasi-geostrophic) *Eliassen–Palm (EP) flux*,[1] and its divergence, given by (10.26), gives the poleward flux of potential vorticity:

$$\overline{v'q'} = \nabla_x \cdot \boldsymbol{\mathcal{F}}, \tag{10.28}$$

where $\nabla_x \cdot \equiv (\partial/\partial y, \partial/\partial z)\cdot$ is the divergence in the meridional plane. Unless the meaning is unclear, the subscript x on the meridional divergence will be dropped.

10.2.1 The Eliassen–Palm Relation

On dividing by $\partial\overline{q}/\partial y$ and using (10.28), the enstrophy equation (10.11) becomes

$$\frac{\partial \mathcal{A}}{\partial t} + \nabla \cdot \mathcal{F} = \mathcal{D}, \qquad (10.29a)$$

where

$$\mathcal{A} = \frac{\overline{q'^2}}{2\partial\overline{q}/\partial y}, \qquad \mathcal{D} = \frac{\overline{D'q'}}{\partial\overline{q}/\partial y}, \qquad (10.29b)$$

and $\mathcal{F}$ is given by (10.27). Equation (10.29a) is known as the *Eliassen–Palm relation,* and it is a conservation law (when $\mathcal{D} = 0$) for the *wave activity density* $\mathcal{A}$. (We also encountered wave activities in Section 6.7.2). The conservation law is exact (in the linear approximation) if the mean-flow is constant in time; it will be a good approximation if $\partial\overline{q}/\partial y$ varies slowly compared to the variation of $\overline{q'^2}$. In this instance $\mathcal{A}$ is the *pseudomomentum density*, $\mathcal{P}$, but other kinds of wave activity density exist — the pseudoenergy density for example, which we will encounter later.

If we integrate (10.29a) over a meridional area A bounded by walls where the eddy activity vanishes, and if $\mathcal{D} = 0$, we obtain

$$\frac{\mathrm{d}}{\mathrm{d}t}\int_A \mathcal{A}\,\mathrm{d}A = 0. \qquad (10.30)$$

The integral is a wave activity — a quantity that is quadratic in the amplitude of the perturbation and that is conserved in the absence of forcing and dissipation. If there is no ambiguity we will drop the word density and also refer to $\mathcal{A}$ and $\mathcal{P}$ as wave activities. ('Wave action' is related to wave activity, but specifically means energy divided by the frequency; it is also conserved in many problems.) Note that neither the perturbation energy nor the perturbation enstrophy are wave activities of the linearized equations, because there can be an exchange of energy or enstrophy between mean and perturbation — indeed, this is how a perturbation grows in baroclinic or barotropic instability! This is already evident from (10.11), or in general take (10.7) with $D' = 0$ and multiply by q' to give the enstrophy equation,

$$\frac{1}{2}\frac{\partial q'^2}{\partial t} + \frac{1}{2}\overline{\boldsymbol{u}} \cdot \nabla q'^2 + \boldsymbol{u}'q' \cdot \nabla\overline{q} = 0, \qquad (10.31)$$

where here the overbar is an average (although it need not be a zonal average). Integrating this over a volume V gives

$$\frac{\mathrm{d}\hat{Z}'}{\mathrm{d}t} \equiv \frac{\mathrm{d}}{\mathrm{d}t}\int_V \frac{1}{2}q'^2\,\mathrm{d}V = -\int_V \boldsymbol{u}'q' \cdot \nabla\overline{q}\,\mathrm{d}V. \qquad (10.32)$$

The right-hand side does not, in general, vanish and so $\hat{Z}'$ is not in general conserved.

10.2.2 The Group Velocity Property for Rossby Waves

The vector $\mathcal{F}$ describes how the wave activity propagates. We noted in Chapter 6 that in the case in which the disturbance is composed of plane or almost plane waves that satisfy a dispersion relation, then $\mathcal{F} = \boldsymbol{c}_g\mathcal{A}$, where $\boldsymbol{c}_g$ is the group velocity and (10.29a) becomes

$$\frac{\partial \mathcal{A}}{\partial t} + \nabla \cdot (\mathcal{A}\boldsymbol{c}_g) = 0. \qquad (10.33)$$

This is a useful property, because if we can diagnose $\boldsymbol{c}_g$ from observations we can use (10.29a) to determine how wave activity density propagates. Let us demonstrate this explicitly for the pseudomomentum in Rossby waves.

The Boussinesq quasi-geostrophic equation on the β-plane, linearized around a uniform zonal flow and with constant static stability, is

$$\frac{\partial q'}{\partial t} + \overline{u}\frac{\partial q'}{\partial x} + v'\frac{\partial \overline{q}}{\partial y} = 0, \tag{10.34}$$

where $q' = [\nabla^2 + (f_0^2/N^2)\partial^2/\partial z^2]\psi'$ and, if $\overline{u}$ is constant, $\partial\overline{q}/\partial y = \beta$. Thus we have

$$\left(\frac{\partial}{\partial t} + \overline{u}\frac{\partial}{\partial x}\right)\left[\nabla^2\psi' + \frac{\partial}{\partial z}\left(\frac{f_0^2}{N^2}\frac{\partial \psi'}{\partial z}\right)\right] + \beta\frac{\partial \psi'}{\partial x} = 0. \tag{10.35}$$

Seeking solutions of the form

$$\psi' = \mathrm{Re}\,\widetilde{\psi}e^{\mathrm{i}(kx+ly+mz-\omega t)}, \tag{10.36}$$

we find the dispersion relation,

$$\omega = \overline{u}k - \frac{\beta k}{\kappa^2}, \tag{10.37}$$

where $\kappa^2 = (k^2 + l^2 + m^2 f_0^2/N^2)$, and the group velocity components:

$$c_g^y = \frac{2\beta kl}{\kappa^4}, \qquad c_g^z = \frac{2\beta km f_0^2/N^2}{\kappa^4}. \tag{10.38}$$

Also, if $u' = \mathrm{Re}\,\widetilde{u}\exp[\mathrm{i}(kx + ly + mz - \omega t)]$, and similarly for the other fields, then

$$\widetilde{u} = -\mathrm{Re}\ \mathrm{i}l\widetilde{\psi}, \qquad \widetilde{v} = \mathrm{Re}\ \mathrm{i}k\widetilde{\psi}, \qquad \widetilde{b} = \mathrm{Re}\,\mathrm{i}mf_0\widetilde{\psi}, \qquad \widetilde{q} = -\mathrm{Re}\,\kappa^2\widetilde{\psi}. \tag{10.39}$$

The wave activity density is then

$$\mathcal{A} = \frac{1}{2}\frac{\overline{q'^2}}{\beta} = \frac{\kappa^4}{4\beta}|\widetilde{\psi}^2|, \tag{10.40}$$

where the additional factor of 2 in the denominator arises from the averaging. Using (10.39) the EP flux, (10.27), is

$$\mathcal{F}^y = -\overline{u'v'} = \frac{1}{2}kl|\widetilde{\psi}^2|, \qquad \mathcal{F}^z = \frac{f_0}{N^2}\overline{v'b'} = \frac{f_0^2}{2N^2}km|\widetilde{\psi}^2|. \tag{10.41}$$

Using (10.38), (10.40) and (10.41) we obtain

$$\boldsymbol{\mathcal{F}} = (\mathcal{F}^y, \mathcal{F}^z) = \mathbf{c}_g\mathcal{A}. \tag{10.42}$$

If the properties of the medium are slowly varying, so that a (spatially varying) group velocity can still be defined, then this is a useful expression to estimate how the wave activity propagates in the atmosphere and in numerical simulations.

10.2.3 ✦ The Orthogonality of Modes

It is a direct consequence of the conservation of wave activity that disturbance modes are orthogonal in the 'wave activity norm', defined later on, and thus are a useful measure of the amplitude of a particular mode.[2] To explore this, we start with the linearized potential vorticity equation,

$$\frac{\partial q'}{\partial t} + \overline{u}\frac{\partial q'}{\partial x} + v'\frac{\partial \overline{q}}{\partial y} = 0. \tag{10.43}$$

Let us formally seek solutions of the form $\psi' = \mathrm{Re}\,\Psi \exp(ikx)$ where Ψ is the sum of *modes*,

$$\Psi = \sum_n \widetilde{\psi}_n(y,z)\mathrm{e}^{-ikc_n t}, \tag{10.44}$$

where n is an identifier of the modes. The modes satisfy

$$(\overline{u}\Delta_k^2 + \overline{q}_y)\widetilde{\psi}_n = c_n \Delta_k^2 \widetilde{\psi}_n, \tag{10.45}$$

where

$$\Delta_k^2 = \frac{\partial^2}{\partial y^2} + \frac{\partial}{\partial z}\left(\frac{f_0^2}{N^2}\frac{\partial}{\partial z}\right) - k^2. \tag{10.46}$$

The upper and lower boundary conditions (at $z = 0, -H$) are given by the thermodynamic equation

$$\frac{\partial b'}{\partial t} + \overline{u}\frac{\partial b'}{\partial x} + v'\frac{\partial \overline{b}}{\partial y} = 0, \tag{10.47}$$

and if we simplify further by supposing $\partial\overline{u}/\partial z = 0$ then the boundary condition becomes

$$\frac{\partial \psi_z'}{\partial t} + \overline{u}\frac{\partial \psi_z'}{\partial x} = 0. \tag{10.48}$$

There are no meridional buoyancy fluxes at the boundary. If N^2 is a constant (a simplifying but not essential assumption) then we can let $\widetilde{\psi}_n(y,z) = \psi_n(y)\cos pz$, with $p = j\pi/H$ where j is an integer and the mode n now labels only the meridional modes. The corresponding potential vorticity modes are given by

$$q_n = \Delta_{k,m}^2 \psi_n, \qquad \Delta_{k,m}^2 = \frac{\partial^2}{\partial y^2} - \frac{f_0^2}{N^2}m^2 - k^2, \tag{10.49}$$

and the boundary conditions are then built in to any solution we construct from (10.45) and (10.49).[3] We may then consider a single zonal and a single vertical wavenumber. (If there is no horizontal variation of the shear, the meridional modes are harmonic functions, for example $\psi_n \propto \sin(n\pi y/L)$ for a channel of width L.)

For a given basic state we may imagine solving (10.45), numerically or analytically, and determining the modes. However, these modes are not orthogonal in the sense of either energy or enstrophy. That is, denoting the inner product by

$$\langle a, b\rangle \equiv \frac{1}{2L}\int_L ab\,\mathrm{d}y, \tag{10.50}$$

then, in general,

$$I_E = \langle \psi_n, q_m\rangle \neq 0, \qquad I_Z = \langle q_n, q_m\rangle \neq 0, \tag{10.51a,b}$$

for $n \neq m$, where $q_n = \Delta_{k,p}^2\psi_n$. Perturbation energy and enstrophy are thus not wave activities of the linearized equations, and it is not meaningful to talk about the energy or enstrophy of a particular mode. However, by the same token we may expect orthogonality in the wave activity norm. To prove this and understand what it means, suppose that at $t = 0$ the disturbance consists of two modes, n and m, so that at a later time $q = (q_n\mathrm{e}^{-ikc_n t} + q_m\mathrm{e}^{-ikc_m t} + \text{c.c.})$, where $c_m \neq c_n$ and we assume that both are real. The wave activity is

$$P \equiv \int \mathcal{A}\,\mathrm{d}y\,\mathrm{d}z = \left\langle q_n, q_m^*/\overline{q}_y\right\rangle \mathrm{e}^{-ik(c_n-c_m)t} + \left\langle q_m, q_m^*/\overline{q}_y\right\rangle + \left\langle q_n, q_n^*/\overline{q}_y\right\rangle + \text{c.c.} \tag{10.52}$$

The second and third terms on the right-hand side are the wave activities of each mode, and these are constants (to see this, consider the case when the disturbance is just a single mode). Now, because $\mathrm{d}P/\mathrm{d}t = 0$ the first term must vanish if $c_n \neq c_m$, implying the modes are orthogonal and, in particular,

$$\operatorname{Re} \int \frac{1}{\overline{q}_y} q_n q_m^* \,\mathrm{d}y = 0, \tag{10.53}$$

for $n \neq m$. The inner product weighted by $1/\overline{q}_y$ defines the wave activity norm. Orthogonality is a useful result, for it means that the wave activity is a proper measure of the amplitude of a given mode unlike, for example, energy. The conservation of wave activity will lead to a particularly straightforward derivation of the necessary conditions for stability, given in Section 10.6.

10.3 THE TRANSFORMED EULERIAN MEAN

The so-called *transformed Eulerian mean,* or TEM, is a transformation of the equations of motion that provides a useful framework for discussing eddy effects under a wide range of conditions.[4] It is useful because, as we shall see, it is equivalent to a very natural form of averaging the equations that serves to eliminate eddy fluxes in the thermodynamic equation and collect them together, in a simple form, in the momentum equation, and in so doing it highlights the role of potential vorticity fluxes. The TEM also provides a natural separation between diabatic and adiabatic effects or between advective and diffusive fluxes and, in the case in which the flow is adiabatic, a pleasing simplification of the equations. In later chapters we will use the TEM to better understand the mid-latitude troposphere and the dynamics of the Antarctic Circumpolar Current, and as a framework for the parameterization of eddy fluxes. Of course, there being no free lunch, the TEM brings with it its own difficulties, and in particular the implementation of boundary conditions can cause difficulties, especially in the actual numerical integration of the equations.

10.3.1 Quasi-Geostrophic Form

For simplicity we will use the Boussinesq equations on the beta-plane. The zonally-averaged Eulerian mean equations for the zonally-averaged zonal velocity and buoyancy may then be written as (see Section 2.2.6)

$$\frac{\partial \overline{u}}{\partial t} - (f + \overline{\zeta})\overline{v} + \overline{w}\frac{\partial \overline{u}}{\partial z} = -\frac{\partial}{\partial y}\overline{u'v'} - \frac{\partial}{\partial z}\overline{u'w'} + \overline{F}, \tag{10.54a}$$

$$\frac{\partial \overline{b}}{\partial t} + \overline{v}\frac{\partial \overline{b}}{\partial y} + \overline{w}\frac{\partial \overline{b}}{\partial z} = -\frac{\partial}{\partial y}\overline{v'b'} - \frac{\partial}{\partial z}\overline{w'b'} + \overline{S}, \tag{10.54b}$$

where $\overline{F}$ and $\overline{S}$ represent frictional and heating terms, respectively, and the meridional velocity, $\overline{v}$, is purely ageostrophic. Using quasi-geostrophic scaling we neglect the vertical eddy flux divergences and all ageostrophic velocities except when multiplied by f_0 or N^2. The above equations then become

$$\frac{\partial \overline{u}}{\partial t} = f_0\overline{v} - \frac{\partial}{\partial y}\overline{u'v'} + \overline{F}, \tag{10.55a}$$

$$\frac{\partial \overline{b}}{\partial t} = -N^2\overline{w} - \frac{\partial}{\partial y}\overline{v'b'} + \overline{S}. \tag{10.55b}$$

These two equations are connected by the thermal wind relation,

$$f_0\frac{\partial \overline{u}}{\partial z} = -\frac{\partial \overline{b}}{\partial y}, \tag{10.56}$$

which is a combination of the geostrophic v-momentum equation ($f_0\overline{u} = -\partial\overline{\phi}/\partial y$) and hydrostasy ($\partial\overline{\phi}/\partial z = \overline{b}$). One less than ideal aspect of (10.55) is that in the extratropics the dominant balance is usually between the first two terms on the right-hand sides of each equation, even in time-dependent cases. Thus, the Coriolis force closely balances the divergence of the eddy momentum fluxes, and the advection of the mean stratification ($N^2 w$, or 'adiabatic cooling') often balances the divergence of eddy heat flux, with heating being a small residual. This may lead to an under-estimation of the importance of diabatic heating, as this is ultimately responsible for the mean meridional circulation. Furthermore, the link between $\overline{u}$ and $\overline{b}$ via thermal wind dynamically couples buoyancy and momentum, and obscures the understanding of how the eddy fluxes influence these fields — is it through the eddy heat fluxes or momentum fluxes, or some combination?

To address this issue we combine the terms $N^2 w$ and the eddy flux in (10.55b) into a single total or *residual* (so recognizing the cancellation between the mean and eddy terms) heat transport term that in a steady state is balanced by the diabatic term $\overline{S}$. To do this, we first note that because $\overline{v}$ and $\overline{w}$ are related by mass conservation we can define a mean meridional streamfunction ψ_m such that

$$(\overline{v}, \overline{w}) = \left(-\frac{\partial \psi_m}{\partial z}, \frac{\partial \psi_m}{\partial y}\right). \tag{10.57}$$

The velocities then satisfy $\partial\overline{v}/\partial y + \partial\overline{w}/\partial z = 0$ automatically. If we define a *residual streamfunction* by

$$\psi^* \equiv \psi_m + \frac{1}{N^2}\overline{v'b'}, \tag{10.58a}$$

the components of the *residual mean meridional circulation* are then given by

$$(\overline{v}^*, \overline{w}^*) = \left(-\frac{\partial \psi^*}{\partial z}, \frac{\partial \psi^*}{\partial y}\right), \tag{10.58b}$$

and

$$\overline{v}^* = \overline{v} - \frac{\partial}{\partial z}\left(\frac{1}{N^2}\overline{v'b'}\right), \qquad \overline{w}^* = \overline{w} + \frac{\partial}{\partial y}\left(\frac{1}{N^2}\overline{v'b'}\right). \tag{10.59}$$

Note that by construction, the residual overturning circulation satisfies

$$\frac{\partial \overline{v}^*}{\partial y} + \frac{\partial \overline{w}^*}{\partial z} = 0. \tag{10.60}$$

Substituting (10.59) into (10.55a) and (10.55b) the zonal momentum and buoyancy equations then take the simple forms

$$\begin{aligned}\frac{\partial \overline{u}}{\partial t} &= f_0\overline{v}^* + \overline{v'q'} + \overline{F}, \\ \frac{\partial \overline{b}}{\partial t} &= -N^2\overline{w}^* + \overline{S},\end{aligned} \tag{10.61a,b}$$

which are known as the (quasi-geostrophic) *transformed Eulerian mean equations,* or TEM equations. The potential vorticity flux, $\overline{v'q'}$, is given in terms of the heat and vorticity fluxes by (10.26), and is equal to the divergence of the Eliassen–Palm flux as in (10.28).

The TEM equations make it apparent that we may consider the potential vorticity fluxes, rather than the separate contributions of the vorticity and heat fluxes, to force the circulation. If we know the potential vorticity flux as well as $\overline{F}$ and $\overline{S}$, then (10.60) and (10.61), along with thermal wind balance

$$f_0\frac{\partial \overline{u}}{\partial z} = -\frac{\partial \overline{b}}{\partial y}, \tag{10.62}$$

form a complete set. The meridional overturning circulation is obtained by eliminating time derivatives from (10.61) using (10.62), giving

$$f_0^2 \frac{\partial^2 \psi^*}{\partial z^2} + N^2 \frac{\partial^2 \psi^*}{\partial y^2} = f_0 \frac{\partial}{\partial z} \overline{v'q'} + f_0 \frac{\partial \overline{F}}{\partial z} + \frac{\partial \overline{S}}{\partial y}. \tag{10.63}$$

Thus, the residual or net overturning circulation is driven by the (vertical derivative of the) potential vorticity fluxes and the diabatic terms — 'driven' in the sense that if we know those terms we can calculate the overturning circulation, although of course the fluxes themselves depend on the circulation. Note that this equation applies at every instant, even if the equations are not in a steady state.

Use of the equations in TEM form is particularly useful when the eddy potential vorticity flux arises from wave activity, for example from Rossby waves. The potential vorticity flux is the convergence of the EP flux $\boldsymbol{\mathcal{F}}$, as in (10.28), and if the eddies satisfy a dispersion relation the components of the EP flux are equal to the group velocity multiplied by the wave activity density $\mathcal{A}$, as in (10.42). Thus, knowing the group velocity tells us a great deal about how momentum is transported by waves. We'll use the TEM to deduce the mean-flow acceleration in Sections 10.4, 10.5 and, in particular, in Section 17.3.

Connection to potential vorticity and wave–mean-flow interaction

If we cross-differentiate (10.61) then, after using the residual mass continuity equation (10.60), we recover the zonally-averaged potential vorticity equation, namely

$$\frac{\partial \overline{q}}{\partial t} = -\frac{\partial}{\partial y} \overline{v'q'} - \frac{\partial \overline{F}}{\partial y}, \qquad \text{where} \qquad \overline{q}(y,t) = \frac{\partial}{\partial z}\left(\frac{f_0}{N^2}\overline{b}\right) - \frac{\partial \overline{u}}{\partial y}, \tag{10.64a,b}$$

which is essentially the same as (10.18) and (10.20), noting that we may add βy to the definition of zonally-averaged potential vorticity with no effect.

The corresponding equation for the evolution of eddy potential vorticity is, in its inviscid form,

$$\left(\frac{\partial}{\partial t} + \overline{u}(y,t)\frac{\partial}{\partial x}\right) q' + v' \frac{\partial \overline{q}}{\partial y} = 0, \tag{10.65}$$

as in (10.7). Equations (10.64) and (10.65) are a closed set of quasi-linear equations, and we have recovered the wave–mean-flow system described in Section 10.1.3.

10.3.2 The TEM in Isentropic Coordinates

The residual circulation has an illuminating interpretation if we think of the fluid as comprising multiple layers of shallow water, or equivalently if we cast the problem in isentropic coordinates (Section 3.10). Using the notation of a shallow water system, the momentum and mass conservation equation can be written as

$$\frac{\partial u}{\partial t} + \boldsymbol{u} \cdot \nabla u - f v = F, \qquad \frac{\partial h}{\partial t} + \nabla \cdot (h\boldsymbol{u}) = S. \tag{10.66a,b}$$

The quantity h is the thickness — the separation between two isentropic surfaces — and S is a thickness source term. (The field h plays the same role as σ in Section 3.10.) With quasi-geostrophic scaling, so that variations in Coriolis parameter and layer thickness are small, zonally averaging in a conventional way gives

$$\frac{\partial \overline{u}}{\partial t} - f_0 \overline{v} = \overline{v'\zeta'} + \overline{F}, \qquad \frac{\partial \overline{h}}{\partial t} + H \frac{\partial \overline{v}}{\partial y} = -\frac{\partial}{\partial y}\overline{v'h'} + \overline{S}. \tag{10.67a,b}$$

Quasi-Geostrophic Wave–Mean-Flow Interaction

The inviscid and unforced Boussinesq quasi-geostrophic set of wave–mean-flow equations is

$$\frac{\partial q'}{\partial t} + \overline{u}\frac{\partial q'}{\partial x} + v'\frac{\partial \overline{q}}{\partial y} = 0, \tag{WMF.1a}$$

$$\frac{\partial \overline{q}}{\partial t} + \frac{\partial}{\partial y}\overline{v'q'} = 0, \tag{WMF.1b}$$

along with similar equations as needed for buoyancy at the boundary (see main text). The eddy terms are

$$q' = \left[\nabla^2 + \frac{\partial}{\partial z}\left(\frac{f_0^2}{N^2}\frac{\partial}{\partial z}\right)\right]\psi', \qquad (u', v') = \left(-\frac{\partial \psi'}{\partial y}, \frac{\partial \psi'}{\partial x}\right). \tag{WMF.2a,b}$$

The mean-flow terms are

$$\overline{q}(y,t) = \beta y - \frac{\partial \overline{u}}{\partial y} + \frac{\partial}{\partial z}\left(\frac{f_0}{N^2}\overline{b}\right), \tag{WMF.3}$$

and

$$\frac{\partial \overline{q}}{\partial y} = \beta - \frac{\partial^2 \overline{u}}{\partial y^2} - \frac{\partial}{\partial z}\left(\frac{f_0}{N^2}\frac{\partial \overline{b}}{\partial y}\right) = \beta - \frac{\partial^2 \overline{u}}{\partial y^2} - \frac{\partial}{\partial z}\left(\frac{f_0^2}{N^2}\frac{\partial \overline{u}}{\partial z}\right), \tag{WMF.4}$$

using thermal wind. To solve for the mean-flow we may define a streamfunction Ψ such that

$$\left(\overline{u}, \frac{1}{f_0}\overline{b}\right) = \left(-\frac{\partial \Psi}{\partial y}, \frac{\partial \Psi}{\partial z}\right), \tag{WMF.5}$$

whence

$$\overline{q}(y,t) - \beta y = \frac{\partial}{\partial z}\left(\frac{f_0^2}{N^2}\frac{\partial \Psi}{\partial z}\right) + \frac{\partial^2 \Psi}{\partial y^2}. \tag{WMF.6}$$

Given $\overline{q}$ from (WMF.1b) we solve (WMF.6) to give $\overline{u}$ and $\overline{b}$. Equivalently, we may derive a single equation for the zonal wind by differentiating (WMF.1b) with respect to y and, using (WMF.4), we obtain

$$\left[\frac{\partial^2}{\partial y^2} + \frac{\partial}{\partial z}\left(\frac{f_0^2}{N^2}\frac{\partial}{\partial z}\right)\right]\frac{\partial \overline{u}}{\partial t} = \frac{\partial^2}{\partial y^2}\overline{v'q'}. \tag{WMF.7}$$

The evolution of the mean-flow may also usefully be written in TEM form as

$$\frac{\partial \overline{u}}{\partial t} - f_0\overline{v}^* + \overline{v'q'} = 0, \tag{WMF.8a}$$

$$\frac{\partial \overline{b}}{\partial t} + N^2\overline{w}^* = 0, \tag{WMF.8b}$$

where $\overline{v}^*$ and $\overline{w}^*$ are found by solving the elliptic equation (10.63), and the value of $\partial\overline{q}/\partial y$, for use in (WMF.1a), is obtained using (WMF.4).

The overbars in these equations denote averages taken along isentropes — i.e., they are averages for a given layer — but are otherwise conventional, and the meridional velocity is purely ageostrophic. By analogy with (10.59), we define the residual circulation by

$$\overline{v}^* \equiv \overline{v} + \frac{1}{H}\overline{v'h'}, \tag{10.68}$$

where H is the mean thickness of the layer. Using (10.68) in (10.67) gives

$$\frac{\partial \overline{u}}{\partial t} - f_0 \overline{v}^* = \overline{v'q'} + \overline{F}, \qquad \frac{\partial \overline{h}}{\partial t} + H\frac{\partial \overline{v}^*}{\partial y} = \overline{S}, \tag{10.69a,b}$$

where

$$\overline{v'q'} = \overline{v'\zeta'} - \frac{f_0}{H}\overline{v'h'}, \tag{10.70}$$

is the meridional potential vorticity flux in a shallow water system. From (10.68) we see that the residual velocity is a measure of the *total meridional thickness flux*, eddy plus mean, in an isentropic layer. This is often a more useful quantity than the Eulerian velocity $\overline{v}$ because it is generally the former, not the latter, that is constrained by the external forcing. What we have done, of course, is to effectively use a thickness-weighted mean in (10.66b); to see this, define the thickness-weighted mean by

$$\overline{v}_* \equiv \frac{\overline{hv}}{\overline{h}}. \tag{10.71}$$

(We use $\overline{v}_*$ to denote a thickness- or mass-weighted mean, and $\overline{v}^*$ to denote a residual velocity; the quantities are closely related, as we will see.) From (10.71) we have

$$\overline{v}_* = \overline{v} + \frac{1}{\overline{h}}\overline{v'h'}, \tag{10.72}$$

then the zonal average of (10.66b) is just

$$\frac{\partial \overline{h}}{\partial t} + \frac{\partial}{\partial y}(\overline{h}\overline{v}_*) = \overline{S}, \tag{10.73}$$

which is the same as (10.69b) if we take $H = \overline{h}$. Similarly, if we use the thickness weighted velocity (10.72) in the momentum equation (10.67a) we obtain (10.69a).

Evidently, if the mass-weighted meridional velocity is used in the momentum and thickness equations then the eddy mass flux does not enter the equations explicitly: the only eddy flux in (10.69) is that of potential vorticity. That is, in isentropic coordinates the equations in TEM form are equivalent to the equations that arise from a particular form of averaging — thickness weighted averaging — rather than the conventional Eulerian averaging. A similar correspondence occurs in height coordinates, as we now see.

10.3.3 Connection between the Residual and Thickness-weighted Circulation

It is evident from the above arguments that, in a shallow water system or in isentropic coordinates, the residual velocity is a measure of the total (i.e., mean plus eddy) thickness transport. In height coordinates, the definition of residual velocity, (10.58), does not lend itself so easily to such an interpretation. However, the residual velocity in height coordinates is, in fact, also a measure of the total thickness transport, or equivalently of the mass transport between two isentropic surfaces, as we now discover. Specifically, we show that averaging the total transport in isentropic layers is equivalent to the mass transport evaluated by the TEM formalism in height coordinates, and

Aspects of the TEM Formulation

Properties and features

- The residual mean circulation is equivalent to the total mass-weighted (eddy plus Eulerian mean) circulation, and it is this circulation that is driven by the diabatic forcing.
- There are no explicit eddy fluxes in the buoyancy budget; the only eddy term is the flux of potential vorticity, and this is the divergence of the Eliassen–Palm flux; that is $\overline{v'q'} = \nabla_x \cdot \boldsymbol{\mathcal{F}}$.
- The residual circulation, $\overline{v}^*$, becomes part of the solution, just as $\overline{v}$ is part of the solution in an Eulerian mean formulation.

But note

- The TEM formulation does not solve the parameterization problem, and eddy fluxes are still present in the equations.
- The theory and practice are well developed for a zonal average, but less so for three-dimensional, non-zonal flow. This is because the geometry enforces simple boundary conditions in the zonal mean case.[6]
- The boundary conditions on the residual circulation are neither necessarily simple nor easily determined; for example, at a horizontal boundary $\overline{w}^*$ is not zero if there are horizontal buoyancy fluxes.

Examples of the use of the TEM and its relatives in the general circulation of the atmosphere and ocean arise in Sections 15.2, 15.4, 17.3, 17.7 and 21.7.

specifically that the thickness-weighted mean, $\overline{v}_*$, is equivalent to the residual velocity, $\overline{v}^*$, in height coordinates. Our demonstration is for a Boussinesq system, but the extension to a compressible gas is reasonably straightforward.[5]

Consider two isentropic surfaces, η_1 and η_2 with mean positions $\overline{\eta}_1$ and $\overline{\eta}_2$, as in Fig. 10.1. (We use z to denote the vertical coordinate, and η to denote the location of isentropic surfaces.) The meridional transport between these surfaces is given by

$$T = \int_{\eta_2}^{\eta_1} v \,\mathrm{d}z. \tag{10.74}$$

If the velocity does not vary with height within the layer (and in the limit of layer thickness going to zero this is the case) then $T = vh$ where $h = \eta_1 - \eta_2$ is the thickness of the isentropic layer. The zonally-averaged transport is then given by

$$\overline{T} = \frac{1}{L}\int_L T \,\mathrm{d}x = \frac{1}{L}\int_L \left(\int_{\eta_2}^{\eta_1} v \,\mathrm{d}z \right) \mathrm{d}x = \overline{\int_{\eta_2}^{\eta_1} v \,\mathrm{d}z} = \overline{vh} = \overline{v}\overline{h} + \overline{v'h'}, \tag{10.75}$$

with obvious notation, and with an overbar denoting a zonal average. Letting the distance between isentropes shrink to zero this result allows us to write

$$\overline{v}_* \equiv \frac{\overline{v\sigma}^b}{\overline{\sigma}} = \overline{v}^b + \frac{\overline{v'\sigma'}^b}{\overline{\sigma}}, \tag{10.76}$$

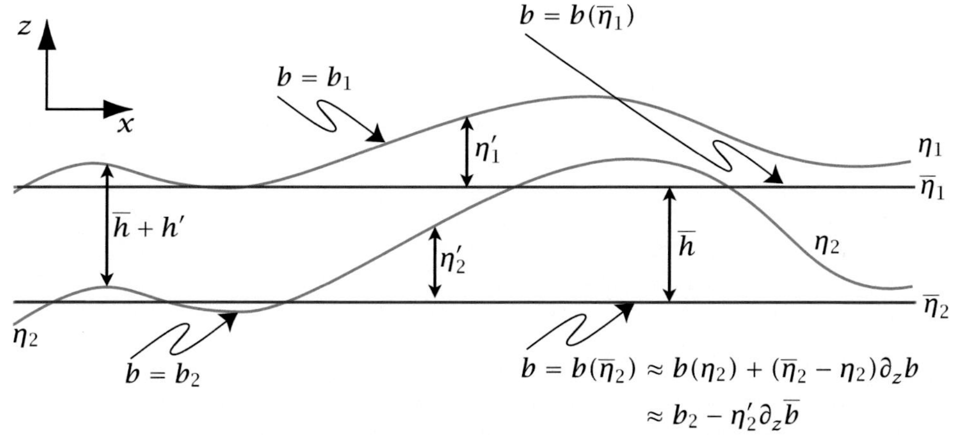

Fig. 10.1 Two isentropic surfaces, η_1 and η_2, and their mean positions, $\overline{\eta}_1$ and $\overline{\eta}_2$. The departure of an isentrope from its mean position is proportional to the temperature perturbation at the mean position of the isentrope, and the variations in thickness (h') of the isentropic layer are proportional to the vertical derivative of this.

where $\overline{(\cdot)}^b$ denotes an average along an isentrope and $\overline{\sigma} = \overline{\partial z/\partial b}$ is the thickness density, a measure of the thickness between two isentropes. Equation (10.76) is analogous to (10.72), for a continuously stratified system. The averaged quantity $\overline{v}_*$ is not proportional to the average of the velocity at constant height, or even to the average along an isentrope; rather, it is the *thickness-weighted* zonal average of the velocity *between* two isentropic surfaces, Δb apart, of mean separation proportional to $\overline{\sigma}\Delta b$. Our goal is to express this transport in terms of Eulerian-averaged quantities, at a constant height z.

Let us first connect an average along an isentrope of some variable χ to its average at constant height by writing, for small isentropic displacements,

$$\overline{\chi}^b = \overline{\chi(z + \eta')}^z \approx \overline{\chi(z) + \eta'\partial\chi/\partial z}^z, \tag{10.77}$$

where the superscript explicitly denotes how the zonal average is taken, and η' is the displacement of the isentrope from its mean position. This can be expressed in terms of the temperature perturbation at the location of the mean isentrope by Taylor-expanding b around its value on that mean isentrope. That is,

$$b(\eta) = b(\overline{\eta}) + \left(\frac{\partial b}{\partial z}\right)_{z=\overline{\eta}} (\eta - \overline{\eta}) + \dots, \tag{10.78}$$

where $\overline{\eta} = \overline{\eta}(z)$, giving

$$\eta' \approx \frac{-b'}{\partial_z b(\overline{\eta})} \approx -\frac{b'}{\partial_z \overline{b}^z}, \tag{10.79}$$

where $\eta' = \eta - \overline{\eta}$ and $b' = b(\overline{\eta}) - b(\eta)$. Using (10.79) in (10.77) (and omitting the superscript z on $\partial_z \overline{b}$) we obtain, with $\chi = v$,

$$\overline{v}^b = \overline{v}^z - \frac{\overline{b'\partial_z v'}^z}{\partial_z \overline{b}}. \tag{10.80}$$

Note that if v is in thermal wind balance with b then the second term vanishes identically, but we will not invoke this.

We now transform the second term on the right-hand side of (10.76) to an average at constant z. The variations in thickness of an isothermal layer are given by

$$\sigma' \approx \overline{\sigma}\frac{\partial \eta'}{\partial z} = -\overline{\sigma}\frac{\partial}{\partial z}\left(\frac{b'}{\partial_z \overline{b}}\right), \tag{10.81}$$

using (10.79). Thus, neglecting terms that are third-order in amplitude,

$$\overline{v'\sigma'}^b = -\overline{\sigma}\overline{v'\frac{\partial}{\partial z}\left(\frac{b'}{\partial_z \overline{b}}\right)}^z. \tag{10.82}$$

Using both (10.80) and (10.82), (10.76) becomes

$$\overline{v}_* = \overline{v}^z - \frac{\overline{b'\partial_z v'}^z}{\partial_z \overline{b}} - \overline{v'\frac{\partial}{\partial z}\left(\frac{b'}{\partial_z \overline{b}}\right)}^z = \overline{v}^z - \frac{\partial}{\partial z}\overline{\left(\frac{v'b'}{\partial_z \overline{b}}\right)}^z. \tag{10.83}$$

The right-hand side of the last equation is the TEM form of the residual velocity; thus, we have shown that

$$\overline{v}_* \equiv \frac{\overline{v\sigma}}{\overline{\sigma}} = \overline{v}^b + \frac{\overline{v'\sigma'}^b}{\overline{\sigma}} \approx \overline{v}^z - \frac{\partial}{\partial z}\left(\frac{\overline{v'b'}^z}{\partial_z \overline{b}}\right) \equiv \overline{v}^*. \tag{10.84}$$

We see the equivalence of the thickness-weighted mean velocity on the left-hand side and the residual velocity on the right-hand side. In the quasi-geostrophic limit $N^2 = \partial_z \overline{b}$ and $\overline{\sigma}$ is a reference thickness.

10.4 THE NON-ACCELERATION RESULT

We now consider further the interpretation and application of the potential vorticity flux and its relatives, using a quasi-geostrophic framework. We first derive an important result in wave–mean-flow dynamics, the non-acceleration condition.[7] This result shows that under certain conditions, to be made precise below, waves have no net effect on the mean-flow, an important and somewhat counter-intuitive result.

10.4.1 A Derivation from the Potential Vorticity Equation

Consider how the potential vorticity fluxes affect the mean fields. The unforced and inviscid zonally-averaged potential vorticity equation is

$$\frac{\partial \overline{q}}{\partial t} + \frac{\partial \overline{v'q'}}{\partial y} = 0. \tag{10.85}$$

Now, in quasi-geostrophic theory the geostrophically balanced velocity and buoyancy can be determined from the potential vorticity via an elliptic equation, and in particular

$$\overline{q} - \beta y = \frac{\partial^2 \overline{\psi}}{\partial y^2} + \frac{\partial}{\partial z}\left(\frac{f_0^2}{N^2}\frac{\partial \overline{\psi}}{\partial z}\right), \tag{10.86}$$

where $\overline{\psi}$ is such that $(\overline{u}, \overline{b}/f_0) = (-\partial\overline{\psi}/\partial y, \partial\overline{\psi}/\partial z)$. Differentiating (10.85) with respect to y we obtain

$$\left[\frac{\partial^2}{\partial y^2} + \frac{\partial}{\partial z}\left(\frac{f_0^2}{N^2}\frac{\partial}{\partial z}\right)\right]\frac{\partial \overline{u}}{\partial t} = (\nabla \cdot \boldsymbol{\mathcal{F}})_{yy}, \tag{10.87}$$

where $\nabla\cdot\boldsymbol{\mathcal{F}} = \overline{v'q'}$ is the divergence of the EP flux (in the y–z plane, i.e., $\nabla_x\cdot\boldsymbol{\mathcal{F}}$). This is determined using the wave activity equation for pseudomomentum which, reprising (10.29a), is

$$\frac{\partial \mathcal{P}}{\partial t} + \nabla\cdot\boldsymbol{\mathcal{F}} = \mathcal{D}, \tag{10.88}$$

now using $\mathcal{P}$ for the wave activity since we are specifically talking about pseudomomentum. If the waves are statistically steady (i.e., $\partial\mathcal{P}/\partial t = 0$) and have no dissipation ($\mathcal{D} = 0$) then evidently $\nabla\cdot\boldsymbol{\mathcal{F}} = 0$. If there is no acceleration at the boundaries then the solution of (10.87) is

$$\frac{\partial \overline{u}}{\partial t} = 0. \tag{10.89}$$

This is a *non-acceleration result.* That is to say, under certain conditions the tendency of the mean fields, and in particular of the zonally-averaged zonal flow, are independent of the waves. To be explicit, those conditions are:

(i) The waves are steady (so that, using the wave activity equation $\mathcal{P}$ does not vary).

(ii) The waves are conservative; that is, $\mathcal{D} = 0$ in (10.29a). Given this and item *(i)*, the Eliassen–Palm relation implies that $\nabla\cdot\boldsymbol{\mathcal{F}} = 0$ and the potential vorticity flux is zero.

(iii) The waves are of small amplitude (all of our analysis has neglected terms that are cubic in perturbation amplitude).

(iv) The waves do not affect the boundary conditions (so there are no boundary contributions to the acceleration).

Given the way we have derived it, the result does not seem too surprising; however, it can be powerful and counter-intuitive, for it means that steady waves (i.e., those whose amplitude does not vary) do not affect the zonal flow. However, they *do* affect the Eulerian meridional overturning circulation, and the relative vorticity flux may also be non-zero. In fact, the non-acceleration theorem is telling us that the changes in the vorticity flux are exactly compensated for by changes in the meridional circulation, and there is no net effect on the zonally-averaged zonal flow. It is *irreversibility*, often manifested by the breaking of waves, that leads to permanent changes in the mean-flow.

The derivation of this result by way of the momentum equation, which one might expect to be more natural, is rather awkward because one must consider momentum and buoyancy fluxes separately. Furthermore, the zonally-averaged meridional circulation comes into play: for example, the meridional velocity, $\overline{v}$, is small because it is purely ageostrophic, but it is not zero and we cannot neglect it because it is multiplied by the Coriolis parameter, which is large. Thus, the eddy vorticity fluxes can affect both the meridional circulation and the acceleration of the zonal mean-flow, and it might seem impossible to disentangle the two effects without completely solving the equations of motion. Nevertheless, we *can* proceed by way of the momentum and buoyancy equations if we use the transformed Eulerian mean and this provides a useful alternate derivation, as follows.

10.4.2 Using TEM to Give the Non-Acceleration Result

We may use the TEM formalism to obtain the non-acceleration result. The explanation is largely equivalent to that given above, but the explication may be useful.

A two-dimensional case

Consider two-dimensional incompressible flow on the β-plane, for which there is no buoyancy flux. The linearized vorticity equation is

$$\frac{\partial \zeta'}{\partial t} + \overline{u}\frac{\partial \zeta'}{\partial x} + v'\frac{\partial \overline{\zeta}}{\partial y} = D', \tag{10.90}$$

from which we derive, analogously to (10.29a), the Eliassen–Palm relation

$$\frac{\partial \mathcal{P}}{\partial t} + \frac{\partial \mathcal{F}}{\partial y} = \mathcal{D}, \tag{10.91}$$

where $\mathcal{F} = -\overline{u'v'}$, $\mathcal{D}$ represents non-conservative forces, and

$$\mathcal{P} = \frac{\overline{\zeta'^2}}{2\partial_y\overline{\zeta}} = \frac{1}{2}\overline{\eta'^2}\frac{\partial\overline{\zeta}}{\partial y}. \tag{10.92}$$

The quantity $\eta' \equiv -\zeta'/\partial_y\overline{\zeta}$ is proportional to the meridional particle displacement in a disturbance. Now consider the x-momentum equation

$$\frac{\partial u}{\partial t} = -\frac{\partial u^2}{\partial x} - \frac{\partial uv}{\partial y} - \frac{\partial \phi}{\partial x} + fv. \tag{10.93}$$

Zonally averaging, noting that $\overline{v} = 0$, gives

$$\frac{\partial \overline{u}}{\partial t} = -\frac{\partial \overline{uv}}{\partial y} = \overline{v'\zeta'} = \frac{\partial \mathcal{F}}{\partial y}. \tag{10.94}$$

Finally, combining (10.91) and (10.94) gives

$$\frac{\partial}{\partial t}(\overline{u} + \mathcal{P}) = \mathcal{D}. \tag{10.95}$$

In the absence of non-conservative terms (i.e., if $\mathcal{D} = 0$) the quantity $\overline{u} + \mathcal{P}$ is constant.[8] Further, if the waves are steady and conservative then $\mathcal{P}$ is constant and, therefore, so is $\overline{u}$. This is the non-acceleration result.

The stratified case

In the stratified case we can use the TEM form of the momentum equation to derive a similar result. The unforced zonally-averaged zonal momentum equation can be written as

$$\frac{\partial \overline{u}}{\partial t} - f_0\overline{v}^* = \nabla \cdot \boldsymbol{\mathcal{F}}, \tag{10.96}$$

and using the Eliassen–Palm relation, (10.29a), this may be written as

$$\frac{\partial}{\partial t}(\overline{u} + \mathcal{P}) - f_0\overline{v}^* = \mathcal{D}, \tag{10.97}$$

and so again $\mathcal{P}$ is related to the momentum of the flow. If, furthermore, the waves are steady ($\partial\mathcal{P}/\partial t = 0$) and conservative ($\mathcal{D} = 0$), then $\partial\overline{u}/\partial t - f_0\overline{v}^* = 0$. However, under these same conditions the residual circulation will also be zero. This is because the residual meridional circulation $(\overline{v}^*, \overline{w}^*)$ arises via the necessity to keep the temperature and velocity fields in thermal wind balance, and is thus determined by an elliptic equation, namely (10.63). If the waves are steady and adiabatic then, since $\overline{v'q'} = 0$, the right-hand side of the equation is zero and it becomes

$$f_0^2\frac{\partial^2\psi^*}{\partial z^2} + N^2\frac{\partial^2\psi^*}{\partial y^2} = 0. \tag{10.98}$$

If $\psi^* = 0$ at the boundaries, then the unique solution of this is $\psi^* = 0$ everywhere. At the meridional boundaries we may certainly suppose that ψ^* vanishes if these are quiescent latitudes, and at the horizontal boundaries the buoyancy flux will vanish if the waves there are steady, because from (10.14) we have

$$\overline{v'b'}\frac{\partial \overline{b}}{\partial y} = -\frac{1}{2}\frac{\partial}{\partial t}\overline{b'^2} = 0. \tag{10.99}$$

Under these circumstances, then, the residual meridional circulation vanishes in the interior and, from (10.96), the mean-flow is steady, thus reprising the non-acceleration result.

Compare (10.96) with the momentum equation in conventional Eulerian form, namely

$$\frac{\partial \overline{u}}{\partial t} - f_0\overline{v} = \overline{v'\zeta'}. \tag{10.100}$$

There is no reason that the vorticity flux should vanish when waves are present, even if they are steady. However, such a flux is (under non-acceleration conditions) precisely compensated by the meridional circulation $f_0\overline{v}$, something that is hard to infer or intuit directly from (10.100); even when non-acceleration conditions do not apply there will be a significant cancellation between the Coriolis and eddy terms. The difficulty boils down to the fact that, in contrast to $\overline{v'q'}$, $\overline{v'\zeta'}$ is not the flux of a wave activity.

Unlike the proof of the non-acceleration result given in Section 10.4.1, the above argument does not use the invertibility property of potential vorticity directly, suggesting an extension to the primitive equations, and the reader may pursue that elsewhere.[9] Various results regarding the TEM and non-acceleration are summarized in the shaded box on the following page.

10.4.3 The EP Flux and Form Drag

It may seem a little magical that the zonal flow is driven by the Eliassen–Palm flux via (10.96). The poleward vorticity flux is clearly related to the momentum flux convergence, but why should a poleward buoyancy flux affect the momentum? The TEM form of the momentum equation may be written as

$$\frac{\partial \overline{u}}{\partial t} = \frac{\partial}{\partial z}\left(\frac{f_0}{N^2}\overline{v'b'}\right) + F_m, \tag{10.101}$$

where $F_m = \overline{v'\zeta'} + f_0\overline{v}^*$ represents forces from the momentum flux and Coriolis force. The first term on the right-hand side certainly does not look like a force; however, it turns out to be directly proportional to the *form drag* between isentropic layers. Recall from Section 3.6 that the form drag, τ_d, at an interface between two layers of shallow water is

$$\tau_d = -\overline{\eta'\frac{\partial p'}{\partial x}}, \tag{10.102}$$

where η is the interfacial displacement. But from (10.79) $\eta' = -b'/N^2$ and with this and geostrophic balance we have

$$\tau_d = \frac{\rho_0 f_0}{N^2}\overline{v'b'}. \tag{10.103}$$

Thus, the vertical component of the EP flux (i.e., the meridional buoyancy flux) is in fact a real stress acting on a fluid layer and equal to the momentum flux caused by the wavy interface. The net momentum convergence into an infinitesimal layer of mean thickness $\overline{h}$ is then (cf. (3.81)),

$$F_d = \overline{h}\frac{\partial \tau_d}{\partial z} = \overline{h}\rho_0 f_0\frac{\partial}{\partial z}\left(\frac{\overline{v'b'}}{N^2}\right), \tag{10.104}$$

and a layer of mean thickness $\overline{h}$ is accelerated according to

$$\frac{\partial \overline{u}}{\partial t} = f_0\frac{\partial}{\partial z}\left(\frac{\overline{v'b'}}{\partial_z\overline{b}}\right) + F_m. \tag{10.105}$$

TEM, Residual Velocities and Non-Acceleration

For a Boussinesq quasi-geostrophic system, the TEM form of the unforced momentum equation and the thermodynamic equation are:

$$\frac{\partial \overline{u}}{\partial t} - f_0 \overline{v}^* = \nabla \cdot \boldsymbol{\mathcal{F}}, \qquad \frac{\partial \overline{b}}{\partial t} + \overline{w}^* N^2 = \overline{S}, \tag{T.1}$$

where $N^2 = \partial \overline{b}_0 / \partial z$, $\overline{S}$ represents diabatic effects, $\boldsymbol{\mathcal{F}}$ is the Eliassen–Palm (EP) flux and its divergence is the potential vorticity flux; thus, $\nabla \cdot \boldsymbol{\mathcal{F}} = \nabla_x \cdot \boldsymbol{\mathcal{F}} = \overline{v'q'}$. The residual velocities are

$$\overline{v}^* = \overline{v} - \frac{\partial}{\partial z}\left(\frac{1}{N^2}\overline{v'b'}\right), \qquad \overline{w}^* = \overline{w} + \frac{\partial}{\partial y}\left(\frac{1}{N^2}\overline{v'b'}\right). \tag{T.2}$$

Spherical coordinate and ideal gas versions of these take a similar form. We may define a meridional overturning streamfunction such that $(\overline{v}^*, \overline{w}^*) = (-\partial \psi^* / \partial z, \partial \psi^* / \partial y)$, and using thermal wind to eliminate time-derivatives in (T.1) we obtain

$$f_0^2 \frac{\partial^2 \psi^*}{\partial z^2} + N^2 \frac{\partial^2 \psi^*}{\partial y^2} = f_0 \frac{\partial}{\partial z}\overline{v'q'} + \frac{\partial \overline{S}}{\partial y}. \tag{T.3}$$

The manipulations (given in the main text) that lead to the above equations may seem formal, in that they simply transform the momentum and thermodynamic equations from one form to another. However, the resulting equations have two potential advantages over the untransfomed ones:

(i) The residual meridional velocity is approximately equal to the average thickness-weighted velocity between two neighbouring isentropic surfaces, and so is a measure of the total (Eulerian mean plus eddy) meridional transport of thickness or buoyancy.

(ii) The EP flux is directly related to certain conservation properties of waves. The divergence of the EP flux is the meridional flux of potential vorticity:

$$\boldsymbol{\mathcal{F}} = -(\overline{u'v'})\,\mathbf{j} + \left(\frac{f_0}{N^2}\overline{v'b'}\right)\mathbf{k}, \qquad \nabla \cdot \boldsymbol{\mathcal{F}} = \overline{v'q'}. \tag{T.4}$$

Furthermore, the EP flux satisfies, to second order in wave amplitude,

$$\frac{\partial \mathcal{P}}{\partial t} + \nabla \cdot \boldsymbol{\mathcal{F}} = \mathcal{D}, \qquad \text{where} \quad \mathcal{P} = \frac{\overline{q'^2}}{2\partial \overline{q}/\partial y}, \qquad \mathcal{D} = \frac{\overline{D'q'}}{\partial \overline{q}/\partial y}. \tag{T.5}$$

The quantity $\mathcal{P}$ is a *wave activity density*, specifically the *pseudomomentum*, and $\mathcal{D}$ is its dissipation. For nearly plane waves, $\mathcal{P}$ and $\boldsymbol{\mathcal{F}}$ are connected by the *group velocity property*,

$$\boldsymbol{\mathcal{F}} = (\mathcal{F}^y, \mathcal{F}^z) = \mathbf{c}_g \mathcal{P}, \tag{T.6}$$

where $\mathbf{c}_g$ is the group velocity of the waves. If the waves are steady ($\partial \mathcal{P}/\partial t = 0$) and dissipationless ($\mathcal{D} = 0$) then $\nabla \cdot \boldsymbol{\mathcal{F}} = 0$ and using (T.1) and (T.3) there is no wave-induced acceleration of the mean-flow; this is the 'non-acceleration' result. Commonly there is enstrophy dissipation, or wave-breaking, and $\nabla \cdot \boldsymbol{\mathcal{F}} < 0$; such *wave drag* leads to flow deceleration and/or a poleward residual meridional velocity.

The appearance of the buoyancy flux is really a consequence of the way we have chosen to average the equations: obtaining (10.105) involved averaging the forces over an isentropic layer, and given this it can only be the residual circulation that contributes to the Coriolis force. One might say that the vertical component of the EP flux is a force in drag, masquerading as a buoyancy flux.

10.5 ✦ INFLUENCE OF EDDIES ON THE MEAN-FLOW IN THE EADY PROBLEM

We now consider the eddy fluxes in the Eady problem, and, in particular, how these might feed back on to the mean-flow. Because of the simplicity of the setting the problem can be fully solved in both the Eulerian or residual frameworks and it is therefore a very instructive, albeit algebraically complex, example.[10]

10.5.1 Formulation

Let us first distinguish between the basic flow, the zonal mean fields, and the perturbation. The basic flow is the flow around which the equations of motion are linearized; this flow is unstable, and the perturbations, assumed to be small, grow exponentially with time. Because the perturbations are formally always small they do not affect the basic flow, but they do produce changes in the zonal mean velocity and buoyancy fields. In Eulerian form this is represented by,

$$\frac{\partial \overline{u}}{\partial t} = f_0\overline{v} - \frac{\partial \overline{u'v'}}{\partial y}, \qquad \frac{\partial \overline{b}}{\partial t} = -N^2\overline{w} - \frac{\partial \overline{b'v'}}{\partial y}, \tag{10.106}$$

and the TEM version of these equations is

$$\frac{\partial \overline{u}}{\partial t} = f_0\overline{v}^* + \overline{v'q'}, \qquad \frac{\partial \overline{b}}{\partial t} = -N^2\overline{w}^*, \tag{10.107}$$

where in the Eady problem $\partial_y(\overline{u'v'})$ and $\overline{v'q'}$ are both zero. We can calculate the perturbation quantities from the solution to the Eady problem (e.g., calculate $\overline{v'b'}$) and thus infer the structure of the mean-flow tendencies $\partial\overline{u}/\partial t$ and $\partial\overline{b}/\partial t$ and the meridional circulation, $(\overline{v}, \overline{w})$ or $(\overline{v}^*, \overline{w}^*)$. All of these fields are perturbation quantities and all are exponentially growing, and so in reality they will eventually have a finite effect on the pre-existing zonal flow, but in the Eady problem, or any similar linear problem, such rectification is assumed to be small and is neglected.

Using the thermal wind relation, $f_0\partial_z\overline{u} = -\partial_y\overline{b}$ to eliminate time derivatives in (10.106) gives an equation for the meridional streamfunction ψ_E, namely,

$$\frac{L^2}{L_d^2}\frac{\partial^2\psi_E}{\partial z^2} + \frac{\partial^2\psi_E}{\partial y^2} = -\frac{1}{N^2}\frac{\partial^2\overline{b'v'}}{\partial y^2}, \tag{10.108}$$

where $(\overline{v}, \overline{w}) = (-\partial\psi_E/\partial z, \partial\psi_E/\partial y)$ and we have nondimensionalized z with D and y with L. The boundary conditions are that $\psi_E = 0$ at $y = 0, L$ and $z = 0, D$. Similarly, and analogously to (10.63), we obtain an equation for the residual streamfunction, ψ^*, namely

$$\frac{L^2}{L_d^2}\frac{\partial^2\psi^*}{\partial z^2} + \frac{\partial^2\psi^*}{\partial y^2} = 0, \tag{10.109}$$

where now the boundary conditions are that $N^2\overline{w}^* = \partial\overline{v'b'}/\partial y$ at the upper and lower boundaries, and $\overline{v} = 0$ at the lateral boundaries. In terms of the residual streamfunction this is

$$\psi^* = \frac{1}{N^2}\overline{v'b'}, \text{ at } z = 0,\ 1, \qquad \psi^* = 0, \text{ at } y = 0,\ 1. \tag{10.110}$$

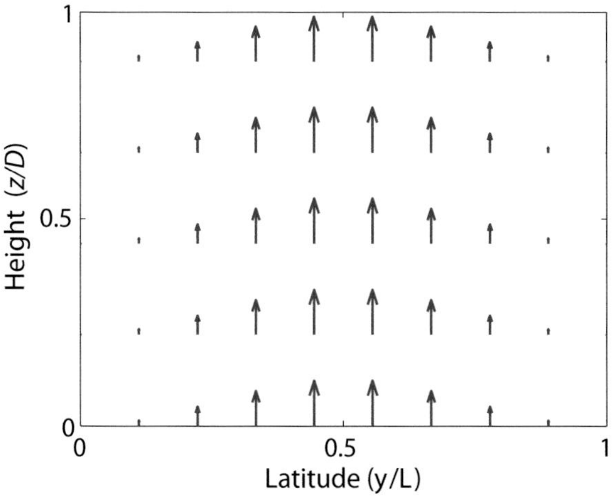

Fig. 10.2 The Eliassen–Palm vector in the Eady problem. It is directed purely vertically, .

The residual and overturning circulations are related by (10.58a), and (10.108) and (10.109) are, at one level, simply different representations of the same problem, connected by a simple mathematical transformation. However, the residual streamfunction better represents the total transport of the fluid. Equation (10.109) is particularly simple, because of the absence of potential vorticity fluxes in the interior, and it is apparent that the residual circulation is driven by boundary sources. We care only about the spatial structure of the right-hand sides of (10.108) and of the boundary conditions of (10.110). The former is given by

$$-\frac{\partial^2 \overline{b'v'}}{\partial y^2} \propto -\frac{\partial^2}{\partial y^2}\sin^2 ly = -2l^2\cos 2ly. \tag{10.111}$$

The eddy heat fluxes in the Eady problem are independent of height, as may be calculated explicitly from the solutions of Chapter 9. In fact, the result follows without detailed calculation, by first noting that the eddy potential vorticity flux is zero because the basic state has zero QG potential vorticity and therefore none may be generated. Further, because the basic state does not vary in y there can be no momentum flux convergence in the y-direction, and so the momentum flux itself is zero if it is zero on the boundary. Thus [using for example (10.27) and (10.28)] the eddy heat flux is independent of height and the EP vectors are directed purely vertically (Fig. 10.2).

The boundary conditions for the residual circulation are

$$\psi^*(y,0) = \psi^*(y,1) \propto \sin^2 ly. \tag{10.112}$$

10.5.2 ♦ Solution

The solutions to (10.108) and (10.109) may be obtained either analytically or numerically. In a domain $0 < y < 1$ and $0 < z < 1$ the residual streamfunction for $l = \pi$ is given by:

$$\begin{aligned} \psi^* &= \sum_{n=1}^{\infty} A_n \sin[(2n-1)ly]\frac{\cosh[L_d\pi(2n-1)(z-0.5)/L]}{\cosh[L_d\pi(2n-1)/2L]}, \\ A_n &= \frac{2}{\pi(2n-1)} - \frac{1}{\pi(2n-1)-2l} - \frac{1}{\pi(2n-1)+2l}. \end{aligned} \tag{10.113}$$

The solution is obtained by first projecting the boundary conditions (proportional to $\sin^2 ly$, or $(1-\cos 2ly)/2$) on to the eigenfunctions of the horizontal part of the Laplacian (i.e., sine functions), and this gives the coefficients of A_n. The vertical structure is then obtained by solving $(L/L_d)^2\partial_z^2\psi^* =$

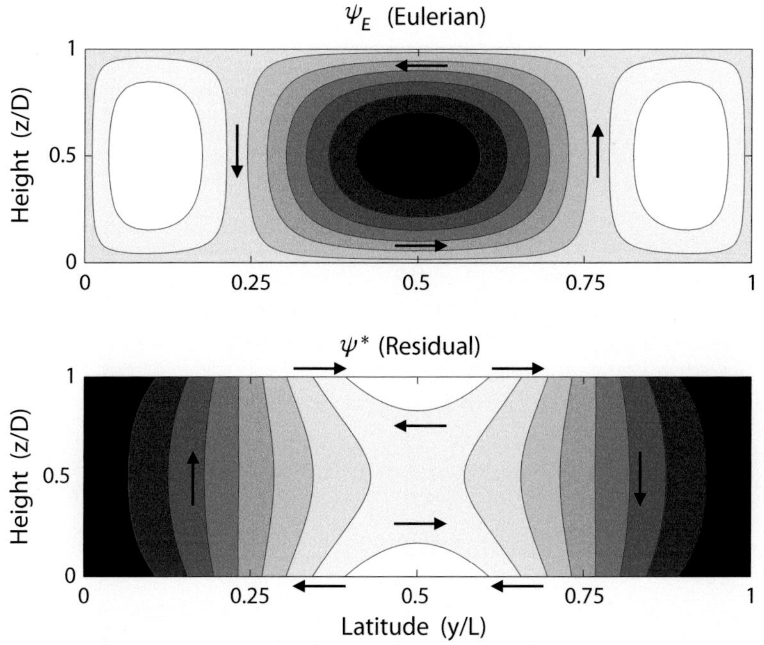

Fig. 10.3 The Eulerian streamfunction (top) and the residual streamfunction for the Eady problem, calculated using (10.108) and (10.109), with $L^2/L_d^2 = 9$.

$-\partial_y^2\psi^*$, which gives the cosh functions. The series converges very quickly, and the first term in the series captures the dominant structure of the solution, essentially because, for $l = \pi$, $\sin ly$ is not unlike $\sin^2 ly$ on the interval $[0, 1]$.

The Eulerian circulation is obtained from the residual circulation using (10.58a), and so by the addition of a field independent of z and proportional to $\sin^2 ly$. The resulting structure is dominated by this and the first term of (10.113) (proportional to $\sin ly$) and, noting that the circulation is symmetric about $z = 0.5$, we obtain a circulation dominated by a single cell, with equatorward motion aloft and poleward motion near the surface (Fig. 10.3). The heat flux convergence in high latitudes is leading to mean rising motion, with the precise shape of the streamfunction determined by the boundary conditions. Although this is true, the heat flux arises *because* of the motion of fluid parcels, so it may be a little misleading to infer, as one might from the Eulerian streamfunction, that the heat flux *causes* the individual parcels to rise or sink in this fashion. The residual streamfunction is a better indicator of the total mass transport and, perhaps as one might intuitively expect, these show parcels rising in the low latitudes and sinking in high latitudes, providing a tendency to flatten the isopycnals and to reduce the meridional temperature gradient.

The residual circulation also shows fluid entering or leaving the domain at the boundary — what does this represent? Suppose that instead of solving the continuous problem we had posed the problem in a finite number of layers (and we explicitly consider the two-layer problem below). As the number of layers increases the solution to the linear baroclinic instability problem approaches that of the Eady problem (e.g., Fig. 9.13); however, as we saw in Section 10.3, the residual circulation is closed in the layered model, and the sum over all the layers of the meridional transport vanishes. Now, in the layered model the vertical boundary conditions are built in to the representation by way of a redefinition of the potential vorticity of the top and bottom layers, so that, in the layered version of the Eady problem there appears to be a potential vorticity gradient in these two layers, instead of a buoyancy gradient at the boundary. The residual circulation is then closed by a return flow that occurs only in the top and bottom layers, and as the number of layers increases this flow is confined to a thinner and thinner layer, and to a delta-function in the continuous limit. To indicate this we have placed arrows just above and below the domain in Fig. 10.3. (This equivalence between boundary conditions and delta-function sources is the same

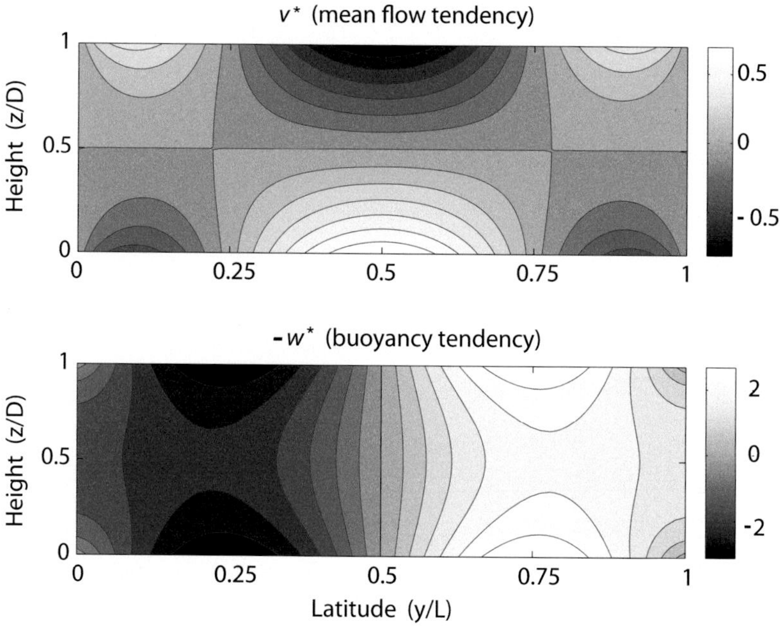

Fig. 10.4 The tendency of the zonal mean-flow ($\partial\overline{u}/\partial t$) and the buoyancy ($\partial\overline{b}/\partial t$) for the Eady problem. Lighter (darker) shading means a positive (negative) tendency, but the units themselves are arbitrary.

as that giving rise to the delta-function boundary layer of Section 5.4.3.)

The effect on the mean-flow is inferred directly from the residual circulation: the mean-flow acceleration is proportional to $\overline{v}^*$ and the buoyancy tendency is proportional to $-\overline{w}^*$, and these are plotted in Figs. 10.4 and 10.5. Because there is no momentum flux convergence in the problem the zonal flow tendency is entirely baroclinic — its vertical integral is zero — and over most of the domain is such as to reduce the mean shear. Consistently (using thermal wind) the buoyancy tendency is such as to reduce the meridional temperature gradient; that is, the instabilities act to transport heat polewards and so reduce the instability of the mean-flow.

10.5.3 The Two-level Problem

The residual circulation and mean-flow tendencies can also be calculated for the two-level (Phillips) problem, with the β-effect. The potential vorticity fluxes in each layer are non-zero and the mean-flow equations are, for $i = 1, 2$,

$$\frac{\partial \overline{u}_i}{\partial t} = f_0\overline{v}_i^* + \overline{v_i' q_i'}, \qquad \frac{\partial \overline{b}}{\partial t} = -N^2\overline{w}^*. \tag{10.114}$$

The vertical velocity and buoyancy are evaluated at mid-depth, and the thermal wind equation is $\overline{u}_1 - \overline{u}_2 = -(H/2)\partial_y\overline{b}$ where H is the total depth of the fluid and, by mass conservation, $\overline{v}_1^* = -\overline{v}_2^*$. If we define a residual streamfunction ψ^* such that

$$\overline{v}_1^* = -\overline{v}_2^* = \psi^*, \qquad \overline{w}^* = \frac{\partial \psi^*}{\partial y}, \tag{10.115}$$

then eliminating time derivatives in (10.114) gives an equation for the residual streamfunction,

$$\frac{\partial^2\psi^*}{\partial y^2} - \frac{k_d^2}{2}\psi^* = \frac{2f_0L^2}{N^2H}(\overline{v_1'q_1'} - \overline{v_2'q_2'}), \tag{10.116}$$

where $k_d^2/2 = [2f_0/(NH)]^2$, and we have nondimensionalized vertical scales by D and horizontal scales by L. As in the Eady problem, it is only the spatial structures of the terms on the right-hand

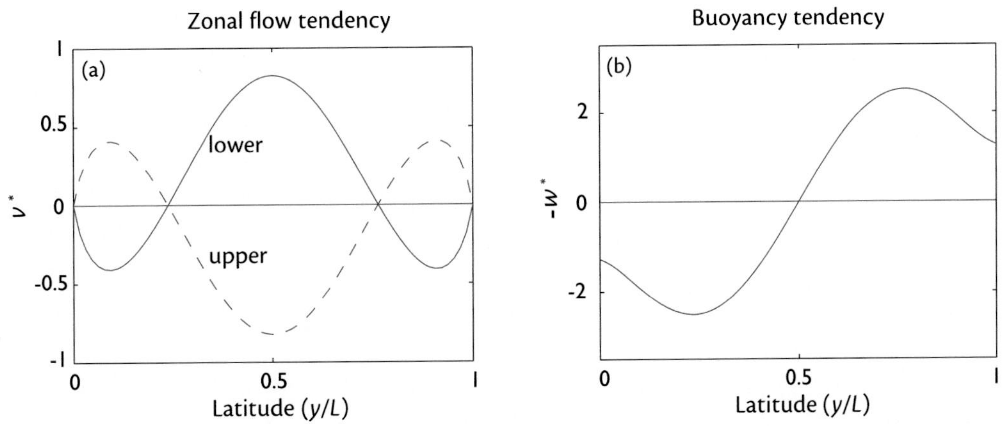

Fig. 10.5 (a) The tendency of the zonal mean-flow ($\partial\overline{u}/\partial t$) just below the upper lid (dashed) and just above the surface (solid) in the Eady problem. The vertically integrated tendency is zero. (b) The vertically averaged buoyancy tendency.

side that are relevant, and these may be calculated from the solutions to the two-level instability problem. The main difference from the Eady problem is that the interior potential vorticity fluxes are non-zero, even in the case with $\beta = 0$: effectively, the boundary fluxes of the Eady problem are absorbed into the potential vorticity fluxes of the two layers. Solving for the residual circulation and interpreting the mean-flow tendencies is left as an exercise for the reader.

10.6 ✦ NECESSARY CONDITIONS FOR INSTABILITY

As we noted in Chapter 9, necessary conditions for instability, or sufficient conditions for stability, can be very useful because when satisfied they obviate the need to perform a detailed calculation. In the remainder of this chapter we use the conservation of wave activities — pseudomomentum and pseudoenergy — to derive such conditions. In sections 9.3 and 9.4.3 we derived such conditions assuming the instability to be of normal-mode form. Here we give derivations that are both more general and, in some ways, simpler; they utilize the fact that the potential vorticity flux may be written as a divergence of a vector and therefore vanishes when integrated over a domain, aside from possible boundary contributions.

10.6.1 Stability Conditions from Pseudomomentum Conservation

Consider the perturbation enstrophy equation,

$$\frac{1}{2}\frac{\partial}{\partial t}\overline{q'^2} = -\frac{\partial\overline{q}}{\partial y}\nabla_x \cdot \boldsymbol{\mathcal{F}}, \tag{10.117}$$

where $\boldsymbol{\mathcal{F}}$ is the Eliassen–Palm flux given by (10.27), the overbar is a zonal mean and the divergence is in the y–z plane. Dividing by $\partial\overline{q}/\partial y$ and integrating over a domain A which is such that the Eliassen–Palm flux vanishes at the boundaries gives the pseudomomentum conservation law,

$$\int_A \frac{\partial}{\partial t}\left(\frac{\overline{q'^2}}{\partial_y\overline{q}}\right) \mathrm{d}y\,\mathrm{d}z = 0. \tag{10.118}$$

Equation (10.118) implies that, in the *norm* $\left[q'^2/\partial_y\overline{q}\right]$, the perturbation cannot grow unless $\partial\overline{q}/\partial y$ changes sign somewhere in the domain, or at the boundaries. This result does not depend upon the instability being of normal-mode form. The simplest result of all occurs in a barotropic problem with no vertical variation. Then $\partial\overline{q}/\partial y = \partial/\partial\overline{\zeta}_a y = \beta - \partial^2\overline{u}/\partial y^2$, and demanding that this must change sign for an instability reprises the inflection point (Rayleigh–Kuo) condition. In the more general case, if $\partial\overline{q}/\partial y$ changes sign along a vertical line then the instability is called a baroclinic instability, and if it changes sign along a horizontal line the instability is barotropic — these may be taken as the definitions of those terms. A mixed instability has a change of sign along both horizontal and vertical lines.

10.6.2 Inclusion of Boundary Terms

Suppose now that the flow is contained between two flat boundaries, at $z = 0$ and $z = H$. The relevant equations of motion are the potential vorticity evolution in the interior, supplemented by the thermodynamic equation at the boundary. For unforced and inviscid flow these give (cf. (10.11) and (10.14)),

$$\frac{\partial}{\partial t}\left(\frac{1}{2}\frac{\overline{q'^2}}{\partial_y\overline{q}}\right) = -\overline{v'q'}, \qquad 0 < z < H, \tag{10.119}$$

and

$$\frac{\partial}{\partial t}\left(\frac{1}{2}\frac{\overline{b'^2}}{\partial_y\overline{b}}\right) = -\overline{v'b'}, \qquad z = 0,\ H. \tag{10.120}$$

The poleward flux of potential vorticity is

$$\overline{v'q'} = -\frac{\partial}{\partial y}\overline{u'v'} + \frac{\partial}{\partial z}\left(\frac{f_0}{N^2}\overline{v'b'}\right), \tag{10.121}$$

and integrating this expression with respect to both y and z gives

$$\int_A \overline{v'q'}\,\mathrm{d}y\,\mathrm{d}z = \left[\frac{f_0}{N^2}\overline{v'b'}\right]_0^H, \tag{10.122}$$

assuming that the meridional boundaries are at quiescent latitudes. Integrating (10.119) over y and z, and using (10.122) gives

$$\frac{\partial}{\partial t}\iint\frac{1}{2}\frac{\overline{q'^2}}{\partial_y\overline{q}}\,\mathrm{d}y\,\mathrm{d}z = -\left[\frac{f_0}{N^2}\overline{v'b'}\right]_0^H. \tag{10.123}$$

Using (10.120) to eliminate $\overline{v'b'}$ finally gives

$$\frac{\partial}{\partial t}\left\{\iint\frac{1}{2}\frac{\overline{q'^2}}{\partial_y\overline{q}}\,\mathrm{d}y\,\mathrm{d}z - \int\left[\frac{1}{2}\frac{f_0}{N^2}\frac{\overline{b'^2}}{\partial_y\overline{b}}\right]_0^H \mathrm{d}y\right\} = 0. \tag{10.124}$$

If this expression is positive or negative definite the perturbation cannot grow and therefore the basic state is stable. Stability thus depends on the meridional gradient of potential vorticity in the interior, and the meridional gradient of buoyancy at the boundary. If $\partial\overline{q}/\partial y$ changes sign in the interior, or $\partial\overline{b}/\partial y$ changes sign at the boundary, we have the potential for instability. If these are both one signed, then various possibilities exist, and using the thermal wind relation ($f_0\partial\overline{u}/\partial z = -\partial\overline{b}/\partial y$) we obtain the following.

I. A stable case:

$$\frac{\partial\overline{q}}{\partial y} > 0 \text{ and } \left.\frac{\partial u}{\partial z}\right|_{z=0} < 0 \text{ and } \left.\frac{\partial u}{\partial z}\right|_{z=H} > 0 \implies \text{stability}. \tag{10.125}$$

Stability also ensues if all inequalities are switched.

II. *Instability via interior–surface interactions:*

$$\frac{\partial \overline{q}}{\partial y} > 0 \text{ and } \left.\frac{\partial u}{\partial z}\right|_{z=0} > 0 \text{ or } \left.\frac{\partial u}{\partial z}\right|_{z=H} < 0 \implies \text{potential instability.} \tag{10.126}$$

The condition $\partial q/\partial y > 0$ and $(\partial u/\partial z)_{z=0} > 0$ is the most common criterion for instability that is met in the atmosphere. In the troposphere we can sometimes ignore contributions of the buoyancy fluxes at the tropopause ($z = H$), and stability is then determined by the interior potential vorticity gradient and the surface buoyancy gradient. Similarly, in the ocean contributions from the ocean floor are normally very small.

III. *Instability via edge wave interaction:*

$$\left.\frac{\partial u}{\partial z}\right|_{z=0} > 0 \text{ and } \left.\frac{\partial u}{\partial z}\right|_{z=H} > 0 \implies \text{potential instability.} \tag{10.127}$$

(And similarly, with both inequalities switched.) Such an instability may occur where the troposphere acts like a lid, as for example in the Eady problem. If $\partial \overline{q}/\partial y = 0$ and there is no lid at $z = H$ (e.g., the Eady problem with no lid) then the instability disappears.

One consequence of the upper boundary condition is that it provides a condition on the depth of the disturbance. In the Eady problem the evolution of the system is determined by temperature evolution at the surface,

$$\frac{\mathrm{D}b}{\mathrm{D}t} = 0 \qquad \text{at} \quad z = 0,\ H, \tag{10.128}$$

(where $b = f_0 \partial\psi/\partial z$) and zero potential vorticity in the interior, which implies that

$$\nabla^2\psi + k_d^2 H^2 \frac{\partial^2 \psi}{\partial z^2} = 0, \qquad 0 < z < H, \tag{10.129}$$

where $k_d = f_0/(HN)$. Assuming a solution of the form $b \sim \sin kx$ then the Poisson equation (10.129) becomes

$$H^2 k_d^2 \frac{\partial^2 \psi}{\partial z^2} = k^2 \psi, \tag{10.130}$$

with solutions $\psi = A\exp(-\alpha z) + B\exp(\alpha z)$, where $\alpha^2 = k^2 N^2/f_0^2$. The scale height of the disturbance is thus

$$h \sim \frac{f_0 L}{2\pi N}. \tag{10.131}$$

where $L \sim 2\pi/k$ is the horizontal scale of the disturbance. If the upper boundary is higher than this, it cannot interact strongly with the surface, because the disturbances at either boundary decay before reaching the other. Put another way, if the structure of the disturbance is such that it is shallower than H, the presence of the upper boundary is not felt. In the Eady problem, we know that the upper boundary must be important, because it is only by its presence that the flow can be unstable. Thus, all unstable modes in the Eady problem must be 'deep' in this sense, which can be verified by direct calculation. This condition gives rise to a physical interpretation of the high-wavenumber cut-off: if L is too small, the modes are too shallow to span the full depth of the fluid, and from (10.131) the condition for stability is thus

$$L < L_c = 2\pi \frac{NH}{f_0} \qquad \text{or} \qquad K > K_c = \frac{f_0}{NH} = L_d^{-1}, \tag{10.132}$$

where L_c and K_c are the critical length scales and wavenumbers. Wavenumbers larger than the reciprocal of the deformation radius are stable in the Eady problem. If β is non-zero, this condition does not apply, because the necessary condition for instability can be satisfied by a combination of a surface temperature gradient and an interior gradient of potential vorticity provided by β, as in condition (*II.*) in Section 10.6.2. Thus, we may expect that, if $\beta \neq 0$, higher wavenumbers ($k > k_d$) may be unstable but if so they will be shallow, and this may be confirmed by explicit calculation (see Figs. 9.12 and 9.19). In the two-level model shallow modes are, by construction, not allowed so that high wavenumbers will be stable, with or without beta.

10.7 ✦ NECESSARY CONDITIONS FOR INSTABILITY: USE OF PSEUDOENERGY

In this section we derive another necessary condition for instability, sometimes called an 'Arnold condition', that is based on the conservation properties of energy and enstrophy. Such conditions can be derived more generally by variational methods, and these lead to somewhat stronger results (in particular, nonlinear results that do not require the perturbation to be small) but our derivations will be elementary and direct.[11]

10.7.1 Two-dimensional Flow

First consider inviscid, incompressible two-dimensional flow governed by the equation of motion

$$\frac{\partial q}{\partial t} + J(\psi, q) = 0, \tag{10.133}$$

where $q = \zeta + f = \nabla^2\psi + f$ is the absolute vorticity and ψ is the streamfunction. In a steady state, the streamfunction and the potential vorticity are functions of each other so that

$$q = Q(\Psi) \qquad \text{and} \qquad \psi = \Psi(Q), \tag{10.134}$$

where Q is a differentiable but otherwise arbitrary function of its argument, and Ψ its functional inverse. Equation (10.133) is then

$$\frac{\partial q}{\partial t} = -\frac{dQ}{d\psi} J(\Psi, \Psi) = 0, \tag{10.135}$$

and all steady solutions are of the form (10.134). We shall prove that if $d\Psi/dQ > 0$ then the flow is stable, in a sense to be made explicit below. Consider the evolution of perturbations about such a steady state, so that

$$q = Q + q', \qquad \psi = \Psi + \psi', \tag{10.136}$$

and we suppose that the perturbation vanishes at the domain boundary or that the boundary conditions are periodic. The potential vorticity perturbation satisfies, in the linear approximation,

$$\frac{\partial q'}{\partial t} + J(\psi', Q) + J(\Psi, q') = 0. \tag{10.137}$$

Now, because potential vorticity is conserved on parcels, any function of potential vorticity is also materially conserved, and in particular

$$\frac{D\Psi(q)}{Dt} = \frac{\partial \Psi}{\partial t} + J(\psi, \Psi) = 0. \tag{10.138}$$

Linearizing this using (10.136) gives

$$\frac{d\Psi}{dQ}\frac{\partial q'}{\partial t} + J(\psi', \Psi) + J\left(\Psi, \frac{d\Psi}{dQ} q'\right) = 0. \tag{10.139}$$

We now form an energy equation from (10.137) by multiplying by $-\psi'$ and integrating over the domain. Integrating the first term by parts we find

$$\frac{\mathrm{d}}{\mathrm{d}t}\int \frac{1}{2}(\nabla\psi')^2\,\mathrm{d}A = \int \psi' J(\Psi, q')\,\mathrm{d}A. \tag{10.140}$$

Similarly, from (10.139) we obtain

$$\frac{\mathrm{d}}{\mathrm{d}t}\int \frac{1}{2}\frac{\mathrm{d}\Psi}{\mathrm{d}Q}q'^2\,\mathrm{d}A = -\int\left[q' J(\psi', \Psi) + q' J\left(\Psi, \frac{\mathrm{d}\Psi}{\mathrm{d}Q}q'\right)\right]\mathrm{d}A. \tag{10.141}$$

The second term in square brackets vanishes. This follows using the property of Jacobians, obtained by integrating by parts, that

$$\langle aJ(b,c)\rangle = \langle bJ(c,a)\rangle = \langle cJ(a,b)\rangle = -\langle cJ(b,a)\rangle, \tag{10.142}$$

where the angle brackets denote horizontal integration. Using this we have

$$\begin{aligned}\left\langle q' J\left(\Psi, \frac{\mathrm{d}\Psi}{\mathrm{d}Q}q'\right)\right\rangle = -\left\langle \frac{\mathrm{d}\Psi}{\mathrm{d}Q}q' J(\Psi, q')\right\rangle &= -\frac{1}{2}\left\langle \frac{\mathrm{d}\Psi}{\mathrm{d}Q}J(\Psi, q'^2)\right\rangle \\ &= -\frac{1}{2}\left\langle q'^2 J\left(\frac{\mathrm{d}\Psi}{\mathrm{d}Q}, \Psi\right)\right\rangle = 0.\end{aligned} \tag{10.143}$$

Adding (10.140) and (10.141) the remaining nonlinear terms cancel and we obtain the conservation law,

$$\frac{\mathrm{d}\widehat{H}}{\mathrm{d}t} = 0, \qquad \text{where} \qquad \widehat{H} = \frac{1}{2}\int\left[(\nabla\psi')^2 + \frac{\mathrm{d}\Psi}{\mathrm{d}Q}q'^2\right]\mathrm{d}A. \tag{10.144}$$

The quantity $\widehat{H}$ is known as the *pseudoenergy* of the disturbance and because it is a conserved quantity, quadratic in the wave amplitude, it is (like pseudomomentum) a wave activity. Its conservation holds whether the disturbance is growing, decaying or neutral.

If $\mathrm{d}\Psi/\mathrm{d}Q$ is positive everywhere the pseudoenergy is a positive-definite quantity, and the growth of the disturbance is then largely prevented and the basic state is said to be *stable in the sense of Liapunov.* This means that the magnitude of the perturbation, as measured by some norm, is bounded by its initial magnitude. In the case here we define the norm

$$||\psi||^2 \equiv \int\left[(\nabla\psi)^2 + \frac{\mathrm{d}\Psi}{\mathrm{d}Q}(\nabla^2\psi)^2\right]\mathrm{d}A, \tag{10.145}$$

so that

$$||\psi'(t)||^2 = ||\psi'(0)||^2. \tag{10.146}$$

If $\mathrm{d}\Psi/\mathrm{d}Q > 0$ then, although the energy of the disturbance can grow, its final amplitude is bounded by the initial value of the pseudoenergy, because if perturbation energy is to grow perturbation enstrophy must shrink but it cannot shrink past zero. Normal-mode instability, in which modes grow exponentially, is completely precluded.

If the pseudoenergy is *negative definite* then stability is also assured, but this is a less common situation for it demands that $\mathrm{d}\Psi/\mathrm{d}Q$ be sufficiently negative so that the (negative of the) enstrophy contribution is always larger than the energy contribution, and this can usually only be satisfied in a sufficiently small domain. To see this, suppose that $q' = \nabla^2\psi'$, and that in the domain under consideration the Laplacian operator has eigenvalues $-k^2$, where

$$\nabla^2\psi' = -k^2\psi' \tag{10.147}$$

and the smallest eigenvalue, by magnitude, is k_0^2. Then, using Poincaré's inequality,

$$\int (\nabla^2 \psi')^2 \, \mathrm{d}A \geq k_0^2 \int (\nabla \psi')^2 \, \mathrm{d}A, \tag{10.148}$$

a sufficient condition to make $\widehat{H}$ negative definite is that

$$\frac{\mathrm{d}\Psi}{\mathrm{d}Q} < -\frac{1}{k_0^2}. \tag{10.149}$$

As the domain gets bigger, k_0 diminishes and this condition becomes harder to satisfy.[12]

Parallel shear flow and Fjørtoft's condition

Consider the stability of a zonal flow (i.e., a flow in the x-direction), that varies only with y. The flow stability condition is then

$$\frac{\mathrm{d}\Psi}{\mathrm{d}Q} = \frac{\mathrm{d}\Psi/\mathrm{d}y}{\mathrm{d}Q/\mathrm{d}y} = -\frac{U - U_s}{\beta - U_{yy}} > 0, \tag{10.150}$$

where U_s is a constant, representing an arbitrary, constant, zonal flow. The last equality follows because the problem is Galilean invariant, and we are therefore at liberty to choose U_s arbitrarily. To connect this with Fjørtoft's condition (Chapter 9) multiply the top and bottom by $(\beta - U_{yy})$, whence we see that a sufficient condition for stability is that $(U - U_s)(\beta - U_{yy})$ is everywhere negative. The derivation here, unlike our earlier one in Section 9.3.2, makes it clear that the condition does not apply only to normal-mode instabilities.

10.7.2 ♦ Stratified Quasi-Geostrophic Flow

The extension of the pseudoenergy arguments to quasi-geostrophic flow is mostly straightforward, but with a complication from the vertical boundary conditions at the surface and at an upper boundary, and the trusting reader may wish to skip straight to the results, (10.155)–(10.157).[13] For definiteness, we consider Boussinesq, β-plane quasi-geostrophic flow confined between flat rigid surfaces at $z = 0$ and $z = H$. The interior flow is governed by the familiar potential vorticity equation $\mathrm{D}q/\mathrm{D}t = 0$ and the buoyancy equation $\mathrm{D}b/\mathrm{D}t = 0$ at the two boundaries, where

$$q = \nabla^2 \psi + \beta y + \frac{\partial}{\partial z}\left(S(z) \frac{\partial \psi}{\partial z} \right), \qquad b = f_0 \frac{\partial \psi}{\partial z}, \tag{10.151}$$

and $S(z) = f_0^2/N^2$ is positive. The basic state $(\psi = \Psi, q = Q, b = B_1, B_2)$ satisfies

$$\begin{gathered} \psi = \Psi(Q), \quad 0 < z < H, \\ \psi = \Psi_1(B_1), \quad z = 0 \qquad \text{and} \qquad \psi = \Psi_2(B_2), \quad z = H. \end{gathered} \tag{10.152}$$

Analogously to the barotropic case, we obtain the equations of motion for the interior perturbation

$$\frac{\partial q'}{\partial t} + J(\psi', Q) + J(\Psi, q') = 0, \tag{10.153a}$$

$$\frac{\mathrm{d}\Psi}{\mathrm{d}Q}\frac{\partial q'}{\partial t} + J(\psi', \Psi) + J\left(\Psi, \frac{\mathrm{d}\Psi}{\mathrm{d}Q} q'\right) = 0, \tag{10.153b}$$

and at the two boundaries

$$\frac{\partial b'}{\partial t} + J(\psi', B_i) + J(\Psi_i, b') = 0, \tag{10.154a}$$

$$\frac{d\Psi_i}{dB_i}\frac{\partial b'}{\partial t} + J(\psi', \Psi_i) + J\left(\Psi_i, \frac{d\Psi_i}{dB_i} b'\right) = 0, \qquad (10.154b)$$

for $i = 1, 2$. (By $d\Psi_i/dB_i$ we mean the derivative of Ψ_i with respect to its argument, evaluated at B_i.) From these equations, we form the pseudoenergy by multiplying (10.153a) by $-\psi'$, (10.153b) by q', and (10.154a) by ψ', (10.154b) by b'. After some manipulation we obtain the pseudoenergy conservation law:

$$\frac{d\widehat{H}}{dt} = 0, \qquad \text{where} \qquad \widehat{H} = \mathcal{E} + \mathcal{Z} + \mathcal{B}_1 + \mathcal{B}_2, \qquad (10.155)$$

and

$$\mathcal{E} = \frac{1}{2}\left\{(\nabla\psi')^2 + S\left(\frac{\partial\psi'}{\partial z}\right)^2\right\}, \qquad \mathcal{Z} = \frac{1}{2}\left\{\frac{d\Psi}{dQ}q'^2\right\},$$
$$\mathcal{B}_1 = \frac{1}{2}\left\langle\frac{S(0)}{f_0}\frac{d\Psi_1}{dB_1}b'(0)^2\right\rangle, \qquad \mathcal{B}_2 = -\frac{1}{2}\left\langle\frac{S(H)}{f_0}\frac{d\Psi_2}{dB_2}b'(H)^2\right\rangle. \qquad (10.156)$$

where the curly brackets denote a three-dimensional integration over the fluid interior, and the angle brackets denote a horizontal integration over the boundary surfaces at 0 and H. The pseudoenergy $\widehat{H}$ is positive-definite, and therefore stability is assured in that norm, if all of the following conditions are satisfied:

$$\frac{d\Psi}{dQ} > 0, \qquad \frac{1}{f_0}\frac{d\Psi_1}{dB_1} > 0, \qquad \frac{1}{f_0}\frac{d\Psi_2}{dB_2} < 0. \qquad (10.157)$$

If the flow is compressible, the potential vorticity is $q = \nabla^2\psi + \beta y + \rho_R^{-1}\partial_z(\rho_R S \partial_z \psi)$, where $\rho_R = \rho_R(z)$, but the final stability conditions are unaltered. If the upper boundary is then removed to infinity where $\rho_R(z) = 0$ then only the lower boundary condition contributes to (10.157). In the layered form of the quasi-geostrophic equations the vertical boundary conditions are built in to the definitions of potential vorticity in the top and bottom layers. In this case, a sufficient condition for stability is that $d\Psi/dQ > 0$ in each layer. Indeed, an alternative derivation of (10.155)–(10.157) would be to incorporate the boundary conditions on buoyancy into the definition of potential vorticity by the delta-function construction of Section 5.4.3.

Zonal shear flow

Consider now zonally uniform zonal flows, such as might give rise to baroclinic instability in a channel. The fields are then functions of y and z only, and the sufficient conditions for stability are:

$$\begin{aligned}
\frac{d\Psi}{dQ} &= \frac{\partial\Psi/\partial y}{\partial Q/\partial y} = -\frac{U}{dQ/dy} > 0,\\
\frac{d\Psi_1}{dB_1} &= \frac{d\Psi_1/dy}{dB_1/dy} = \frac{U(0)}{dU(0)/dz} > 0,\\
\frac{d\Psi_2}{dB_2} &= \frac{d\Psi_2/dy}{dB_2/dy} = \frac{U(H)}{dU(H)/dz} < 0,
\end{aligned} \qquad (10.158)$$

using the thermal wind relation, and setting $f_0 = 1$ (its value is irrelevant). These results generalize Fjørtoft's condition to the stratified case,[14] and as in that case we are at liberty to add a uniform zonal flow to all the velocities.

10.7.3 ♦ Applications to Baroclinic Instability

We may use the stability conditions derived above to provide a few more results about baroclinic instability, including an alternative derivation of the minimum shear criterion in two-layer flow, and a derivation of the high-wavenumber cut-off to instability. In what follows we do not derive any new criteria; rather, the derivations make it apparent that the criteria are not restricted to perturbations of normal-mode form.

Minimum shear in two-layer flow

We consider two layers of equal depth, on a flat-bottomed β-plane with basic state

$$\Psi_1 = -U_1 y, \qquad \Psi_2 = -U_2 y \tag{10.159a}$$

$$Q_1 = \beta y - \frac{k_d^2}{2}(U_2 - U_1)y, \qquad Q_2 = \beta y - \frac{k_d^2}{2}(U_1 - U_2)y. \tag{10.159b}$$

This state is characterized by $Q_i = \gamma_i \Psi_i$ where

$$\gamma_1 = -\frac{(\beta + k_d^2\widehat{U})}{(\overline{U} + \widehat{U})}, \qquad \gamma_2 = -\frac{(\beta - k_d^2\widehat{U})}{(\overline{U} - \widehat{U})}, \tag{10.160}$$

with $\overline{U} = (U_1 + U_2)/2$ and $\widehat{U} = (U_1 - U_2)/2$. The barotropic flow does not affect the stability properties, so without loss of generality we may choose $\overline{U} < -\widehat{U}$, and this makes $\gamma_1 > 0$. Then γ_2 is also positive if $\beta > k_d^2\widehat{U}/2$. Thus, a sufficient condition for stability is that

$$\widehat{U} < \frac{\beta}{k_d^2}, \tag{10.161}$$

as obtained in Chapter 9. However, we now see that the stability condition does not apply only to normal-mode instabilities.[15]

Use of pseudomomentum conservation provides an alternative derivation of the same result. The flow will also be stable if in both layers $\partial Q/\partial y > 0$, for then the conserved pseudomomentum will be positive definite. If $U_1 > U_2$ then, from (10.159) $\mathrm{d}Q_1/\mathrm{d}y > 0$. The flow will be stable if $\mathrm{d}Q_2/\mathrm{d}y > 0$, and this gives

$$\widehat{U} = \frac{1}{2}(U_1 - U_2) < \frac{\beta}{k_d^2}, \tag{10.162}$$

as in (10.161).

The high-wavenumber cut-off in two-layer baroclinic instability

We can use a pseudoenergy argument to show that there is a high-wavenumber cut-off to two-layer baroclinic instability, with the basic state (10.159). The conserved pseudoenergy analogous to (10.155) and (10.156) is readily found to be

$$\widehat{H} = \left\langle (\nabla\psi_1')^2 + (\nabla\psi_2')^2 + \frac{1}{2}k_d^2(\psi_1' - \psi_2')^2 + \frac{q_1'^2}{\gamma_1} + \frac{q_2'^2}{\gamma_2} \right\rangle = 0. \tag{10.163}$$

Let us choose (without loss of generality) the barotropic flow to be $\overline{U} = \beta/k_d^2$. We then have $\gamma_1 = \gamma_2 = -1/k_d^2$, and the pseudoenergy is then just the actual energy minus k_d^{-2} times the total enstrophy. If we define $\psi = (\psi_1' + \psi_2')/2$ and $\tau = (\psi_1' - \psi_2')/2$ then, using (12.41a) and (12.44), (10.163) may be expressed as

$$\widehat{H} = \left\langle (\nabla\psi)^2 + (\nabla\tau)^2 + k_d^2\tau^2 - k_d^{-2}\left\{(\nabla^2\psi)^2 + [(\nabla^2 - k_d^2)\tau]^2\right\}\right\rangle. \tag{10.164}$$

Now, let us express the fields as Fourier sums,

$$(\tau, \psi) = \sum_{k,l} (\widetilde{\tau}_{k,l}, \widetilde{\psi}_{k,l}) \mathrm{e}^{\mathrm{i}(kx+ly)}. \tag{10.165}$$

(This expression assumes a doubly-periodic domain; essentially the same end-result is obtained in a channel.) The pseudoenergy may then be written as

$$\widehat{H} = \sum_{k,l} \left[K^2 \widetilde{\psi}_{k,l}^2 (k_d^2 - K^2) + K'^2 \widetilde{\tau}_{k,l}^2 (k_d^2 - K'^2) \right], \tag{10.166}$$

where $K^2 = k^2 + l^2$ and $K'^2 = K^2 + k_d^2$. If the deformation radius is sufficiently large (or the domain sufficiently small) that $K^2 > k_d^2$, then the pseudoenergy is *negative-definite*, so the flow is stable, no matter what the shear may be. Such a situation might arise on a planet whose circumference was less than the deformation radius, or in a small ocean basin. In the linear problem, in which perturbation modes do not interact, horizontal wavenumbers with $k^2 > k_d^2$ are stable and there is thus a high-wavenumber cut-off to instability, as was found in Chapter 9 by direct calculation.

Notes

1 After Eliassen & Palm (1961).

2 Andrews & McIntyre (1976), Ripa (1981) and Held (1985).

3 These restrictions on the basic state are not necessary to prove orthogonality, but they make the algebra simpler. Also, we pay no attention here to the nature of the eigenvalues of (10.45), which, in general, consist of both a discrete and a continuous spectrum. See Farrell (1984) and McIntyre & Shepherd (1987).

4 The TEM was introduced by Andrews & McIntyre (1976, 1978) and Boyd (1976). A precursor is the paper of Riehl & Fultz (1957), who noted the shortcomings of zonal averaging in uncovering the meaning of indirect cells in laboratory experiments, and by extension the atmosphere.

5 The main result of this subsection was originally obtained by McIntosh & McDougall (1996). I thank A. Plumb for a discussion about the derivation given here. It is in fact possible to write exact TEM-like equations wholly in terms of the thickness-weighted averaged quantities and without taking a zonal average. The literature is extensive and at times hard to follow, but the reader may usefully look to de Szoeke & Bennett (1993) and Young (2012), who show that thickness-weighted averaged equations may be derived that are identical to the unweighted equations except for the appearance, in the horizontal momentum equations, of an eddy forcing by the divergence of three-dimensional Eliassen–Palm vectors. The divergence of these EP vectors is related to the eddy flux of the full potential vorticity, in an analogous manner to (but more general than) the quasi-geostrophic result.

6 This problem can be worked around in some cases (Plumb 1990, Greatbatch 1998).

7 Non-acceleration arguments have a long history, with contributions from Charney & Drazin (1961), Eliassen & Palm (1961), Holton (1974) and, in particular, Boyd (1976) and Andrews & McIntyre (1978). Dunkerton (1980) reviews and provides examples. Non-acceleration is now so prevalent in the literature that it could be written nonacceleration.

8 Conservation laws of this ilk, their connection to the underlying symmetries of the basic state and (relatedly) their finite-amplitude extension, are discussed by McIntyre & Shepherd (1987) and Shepherd (1990). Conservation of momentum is related to the translational invariance of the medium whereas conservation of $\mathcal{P}$ is related to the translational invariance of the basic state, and hence the appellation 'pseudomomentum'.

9 See Andrews & McIntyre (1978) and Young (2012).

10 Steve Garner and Raffaele Ferrari both provided very helpful input to this section. Shepherd (1983) considers the two-layer problem.

11 The original papers are Arnold (1965, 1966), with many results being developed by Holm *et al.* (1985).

12 The stability criterion is sometimes referred to as 'Arnold's second condition'. More discussion is given in Holm *et al.* (1985) and McIntyre & Shepherd (1987).

13 Blumen (1968), but the method we use is more direct.

14 Pedlosky (1964) derived these conditions by a normal-mode approach.

15 Pierini & Vulpiani (1981) and Vallis (1985) further consider the finite-amplitude case.

Shootings of water thread down the slope
of the huge green stone —
The white eddy-rose that blossom'd up
against the stream in the scollop,
by fits and starts, obstinate in resurrection —
It is the life *that we live.*
Samuel Taylor Coleridge, *The Eddy-Rose*, adapted from notebook, 1799.

CHAPTER 11

Basics of Incompressible Turbulence

TURBULENCE IS HIGH REYNOLDS NUMBER FLUID FLOW, dominated by nonlinearity, containing both spatial and temporal disorder. No definition is perfect, and it is hard to disentangle a definition from a property, but this statement captures the essential aspects. A turbulent flow has eddies with a spectrum of sizes between some upper and lower bounds, the former often determined by the forcing scale or the domain scale, and the latter usually by viscosity. The individual eddies come and go, and are inherently unpredictable. Rather like life, turbulent flows are endlessly fascinating and not a little frustrating.[1]

The circulation of the atmosphere and ocean is, *inter alia,* the motion of a forced-dissipative fluid subject to various constraints such as rotation and stratification. The larger scales are orders of magnitude larger than the dissipation scale (the scale at which molecular viscosity becomes important) and at many if not all scales the motion is highly nonlinear and quite unpredictable. Thus, we can justifiably say that the atmosphere and ocean are turbulent fluids. We are not simply talking about the small-scale flows traditionally regarded as turbulent; rather, our main focus will be the large-scale flows associated with baroclinic instability and greatly influenced by rotation and stratification, a kind of turbulence known as *geostrophic turbulence.* However, before discussing turbulence in the atmosphere and ocean, in this chapter we consider from a fairly elementary standpoint the basic theory of two- and three-dimensional turbulence, and in particular the theory of inertial ranges. We do not provide a comprehensive discussion of turbulence; rather, we provide an introduction to those aspects of most interest or relevance to the dynamical oceanographer or meteorologist. In the next chapter we consider the effects of rotation and stratification, and after that we look at turbulent diffusion.

11.1 THE FUNDAMENTAL PROBLEM OF TURBULENCE

Turbulence is a difficult subject because it is nonlinear, and because, and relatedly, there are interactions between scales of motion. Let us first see what difficulties these bring, beginning with the closure problem itself.

11.1.1 The Closure Problem

Although in a turbulent flow it may be virtually impossible to predict the detailed motion of each eddy, the statistical properties — time averages for example — are not necessarily changing and we might like to predict such averages. Effectively, we accept that we cannot predict the weather, but we can try to predict the climate. Even though we know the equations that determine the system, this task proves to be very difficult because the equations are nonlinear, and we come up against the *closure problem*. To see what this is, let us decompose the velocity field into mean and fluctuating components,

$$\boldsymbol{v} = \overline{\boldsymbol{v}} + \boldsymbol{v}'. \tag{11.1}$$

Here $\overline{\boldsymbol{v}}$ is the mean velocity field, and $\boldsymbol{v}'$ is the deviation from that mean. The mean might be a time average, in which case $\overline{\boldsymbol{v}}$ is a function only of space and not of time, or it might be a time mean over a finite period (e.g., a season if we are dealing with the weather), or it might be some form of ensemble mean. The average of the deviation is, by definition, zero; that is $\overline{\boldsymbol{v}'} = 0$. The idea is to substitute (11.1) into the momentum equation and try to obtain a closed equation for the mean quantity $\overline{\boldsymbol{v}}$. Rather than dealing with the full Navier–Stokes equations, let us carry out this program for a model nonlinear system that obeys

$$\frac{\mathrm{d}u}{\mathrm{d}t} + uu + ru = 0, \tag{11.2}$$

where r is a constant. The average of this equation is:

$$\frac{\mathrm{d}\overline{u}}{\mathrm{d}t} + \overline{uu} + r\overline{u} = 0. \tag{11.3}$$

The value of the term $\overline{uu}$ (i.e., $\overline{u^2}$) is not deducible simply by knowing $\overline{u}$, since it involves correlations between eddy quantities, namely $\overline{u'u'}$. That is, $\overline{uu} = \overline{u}\,\overline{u} + \overline{u'u'} \neq \overline{u}\,\overline{u}$. We can go to the next order to try (vainly!) to obtain an equation for $\overline{u}\,\overline{u}$. First multiply (11.2) by u to obtain an equation for u^2, and then average it to yield

$$\frac{1}{2}\frac{\mathrm{d}\overline{u^2}}{\mathrm{d}t} + \overline{uuu} + r\overline{u^2} = 0. \tag{11.4}$$

This equation contains the undetermined cubic term $\overline{uuu}$. An equation determining this would contain a quartic term, and so on in an unclosed hierarchy. Many methods of closing the hierarchy make assumptions about the relationship of $(n+1)$th order terms to nth order terms, for example by supposing that

$$\overline{uuuu} = \alpha\overline{uu}\,\overline{uu} + \beta\overline{uuu}, \tag{11.5}$$

where α and β are some parameters, and closures set in physical space or in spectral space (i.e., acting on the Fourier transformed variables) have both been proposed. If we know that the variables are distributed normally then such closures can sometimes be made exact, but this is not generally the case in turbulence and all closures that have been proposed so far are, at best, approximations.

This same closure problem arises in the Navier–Stokes equations. If density is constant (say $\rho = 1$) the x-momentum equation for an averaged flow is

$$\frac{\partial \overline{u}}{\partial t} + (\overline{\boldsymbol{v}} \cdot \nabla)\overline{u} = -\frac{\partial \overline{p}}{\partial x} - \nabla \cdot \overline{\boldsymbol{v}'u'}. \tag{11.6}$$

Written out in full in Cartesian coordinates, the last term is

$$\nabla \cdot \overline{\boldsymbol{v}'u'} = \frac{\partial}{\partial x}\overline{u'u'} + \frac{\partial}{\partial y}\overline{u'v'} + \frac{\partial}{\partial z}\overline{u'w'}. \tag{11.7}$$

These terms, and the similar ones in the y- and z-momentum equations, represent the effects of eddies on the mean flow and are known as *Reynolds stress* terms. The 'closure problem' of turbulence may be thought of as finding a representation of such Reynolds stress terms in terms of mean flow quantities. Nobody has been able to close the system, in any useful way, without introducing physical assumptions not directly deducible from the equations of motion themselves. Indeed, not only has the problem not been solved, it is not clear that in general a useful closed-form solution actually exists.

11.1.2 Triad Interactions in Turbulence

The nonlinear term in the equations of motion not only leads to difficulties in closing the equations, but it leads to interactions among different length scales, and in this section we write the equations of motion in a form that makes this explicit. For algebraic simplicity we will restrict our attention to two-dimensional flows, but very similar considerations also apply in three dimensions, and the details of the algebra following are not of themselves important to subsequent sections.[2]

The equation of motion for an incompressible fluid in two dimensions — see for example (4.67) or (5.119) — may be written as

$$\frac{\partial \zeta}{\partial t} + J(\psi, \zeta) = F + \nu\nabla^2\zeta, \qquad \zeta = \nabla^2\psi. \tag{11.8}$$

We include a forcing and viscous term but no Coriolis term. Let us suppose that the fluid is contained in a square, doubly-periodic domain of side L, and let us expand the streamfunction and vorticity in Fourier series so that, with a tilde denoting a Fourier coefficient,

$$\psi(x,y,t) = \sum_{\boldsymbol{k}} \widetilde{\psi}(\boldsymbol{k},t)\mathrm{e}^{\mathrm{i}\boldsymbol{k}\cdot\boldsymbol{x}}, \qquad \zeta(x,y,t) = \sum_{\boldsymbol{k}} \widetilde{\zeta}(\boldsymbol{k},t)\mathrm{e}^{\mathrm{i}\boldsymbol{k}\cdot\boldsymbol{x}}, \tag{11.9}$$

where $\boldsymbol{k} = \mathbf{i}k^x + \mathbf{j}k^y$, $\widetilde{\zeta} = -k^2\widetilde{\psi}$ where $k^2 = k^{x2} + k^{y2}$ and, to ensure that ψ is real, $\widetilde{\psi}(k^x, k^y, t) = \widetilde{\psi}^*(-k^x, -k^y, t)$, where * denotes the complex conjugate, and this property is known as conjugate symmetry. The summations are over all positive and negative x- and y-wavenumbers, and $\widetilde{\psi}(\boldsymbol{k}, t)$ is shorthand for $\widetilde{\psi}(k^x, k^y, t)$. Substituting (11.9) in (11.8) gives, with (for the moment) F and ν both zero,

$$\begin{aligned}\frac{\partial}{\partial t}\sum_{\boldsymbol{k}} \widetilde{\zeta}(\boldsymbol{k},t)\mathrm{e}^{\mathrm{i}\boldsymbol{k}\cdot\boldsymbol{x}} &= \sum_{\boldsymbol{p}} p^x\widetilde{\psi}(\boldsymbol{p},t)\mathrm{e}^{\mathrm{i}\boldsymbol{p}\cdot\boldsymbol{x}} \times \sum_{\boldsymbol{q}} q^y\widetilde{\zeta}(\boldsymbol{q},t)\mathrm{e}^{\mathrm{i}\boldsymbol{q}\cdot\boldsymbol{x}} \\ &\quad - \sum_{\boldsymbol{p}} p^y\widetilde{\psi}(\boldsymbol{p},t)\mathrm{e}^{\mathrm{i}\boldsymbol{p}\cdot\boldsymbol{x}} \times \sum_{\boldsymbol{q}} q^x\widetilde{\zeta}(\boldsymbol{q},t)\mathrm{e}^{\mathrm{i}\boldsymbol{q}\cdot\boldsymbol{x}},\end{aligned} \tag{11.10}$$

where $\boldsymbol{p}$ and $\boldsymbol{q}$ are, like $\boldsymbol{k}$, horizontal wave vectors. We may obtain an evolution equation for the wavevector $\boldsymbol{k}$ by multiplying (11.10) by $\exp(-\mathrm{i}\boldsymbol{k}\cdot\boldsymbol{x})$ and integrating over the domain, and using the fact that the Fourier modes are orthogonal; that is

$$\int \mathrm{e}^{\mathrm{i}\boldsymbol{p}\cdot\boldsymbol{x}}\mathrm{e}^{\mathrm{i}\boldsymbol{q}\cdot\boldsymbol{x}}\,\mathrm{d}A = L^2\delta(\boldsymbol{p}+\boldsymbol{q}), \tag{11.11}$$

where $\delta(\boldsymbol{p}+\boldsymbol{q})$ equals unity if $\boldsymbol{p} = -\boldsymbol{q}$ and is zero otherwise. Using this, (11.10) becomes, restoring the forcing and dissipation terms,

$$\frac{\partial}{\partial t}\widetilde{\psi}(\boldsymbol{k},t) = \sum_{\boldsymbol{p},\boldsymbol{q}} A(\boldsymbol{k},\boldsymbol{p},\boldsymbol{q})\widetilde{\psi}(\boldsymbol{p},t)\widetilde{\psi}(\boldsymbol{q},t) + \widetilde{F}(\boldsymbol{k}) - \nu k^2\widetilde{\psi}(\mathbf{k},t), \tag{11.12}$$

where $A(\boldsymbol{k},\boldsymbol{p},\boldsymbol{q}) = (q^2/k^2)(p^xq^y - p^yq^x)\delta(\boldsymbol{p}+\boldsymbol{q}-\mathbf{k})$ is an 'interaction coefficient', and the summation is over all $\boldsymbol{p}$ and $\boldsymbol{q}$; however, only those wavevector triads with $\boldsymbol{p}+\boldsymbol{q} = \mathbf{k}$ make a non-zero contribution, because of the presence of the delta function.

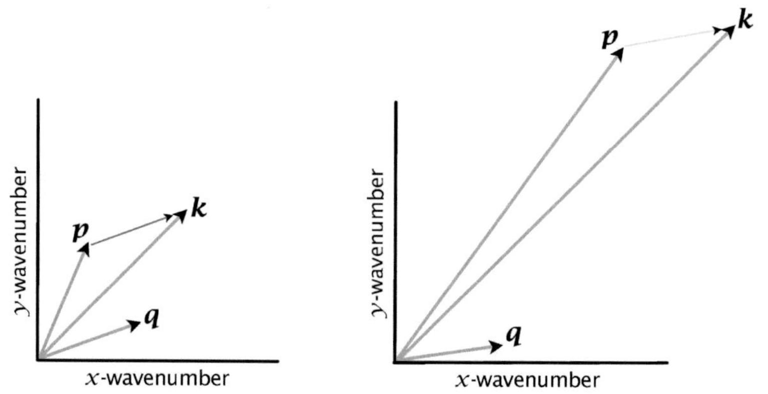

Fig. 11.1 Two interacting triads, each with $\boldsymbol{k} = \boldsymbol{p} + \boldsymbol{q}$. On the left, a local triad with $k \sim p \sim q$. On the right, a non-locpwal triad with $k \sim p \gg q$.

Consider, then, a fluid in which just two Fourier modes are initially excited, with wavevectors $\boldsymbol{p}$ and $\boldsymbol{q}$, along with their conjugate-symmetric partners at $-\boldsymbol{p}$ and $-\boldsymbol{q}$. These modes interact, obeying (11.12), to generate third and fourth wavenumbers, $\boldsymbol{k} = \boldsymbol{p} + \boldsymbol{q}$ and $\boldsymbol{m} = \boldsymbol{p} - \boldsymbol{q}$ (along with their conjugate-symmetric partners). These four wavenumbers can interact among themselves to generate several additional wavenumbers, $\boldsymbol{k} + \boldsymbol{p}$, $\boldsymbol{k} + \boldsymbol{m}$ and so on, so potentially filling out the entire spectrum of wavenumbers. The individual interactions are called *triad interactions,* and it is by way of such interactions that energy is transferred between scales in turbulent flows, in both two and three dimensions. The dissipation term does not lead to interactions between modes with different wavevectors; rather, it acts like a drag on each Fourier mode, with a coefficient that increases with wavenumber and therefore that preferentially affects small scales.

The selection rule for triad interactions — that $\boldsymbol{k} = \boldsymbol{p} + \boldsymbol{q}$ — does not restrict the scales of these interacting wavevectors, and the types of triad interactions fall between two extremes:

(i) local interactions, in which $k \sim p \sim q$;

(ii) non-local interactions, in which $k \sim p \gg q$.

These two kinds of triads are schematically illustrated in Fig. 11.1. Without a detailed analysis of the solutions of the equations of motion — an analysis that is in general impossible for fully-developed turbulence — we cannot say with certainty whether one kind of triad interaction dominates. The theory of Kolmogorov considered below, and its two-dimensional analogue, assume that it is the *local* triads that are most important in transferring energy; this is a reasonable assumption because, from the perspective of a small eddy, large eddies appear as a nearly-uniform flow and so translate the small eddies around without distortion and thus without transferring energy between scales.

11.2 THE KOLMOGOROV THEORY

The foundation of many theories of turbulence is the spectral theory of Kolmogorov.[3] This theory does not close the equations as explicit a manner as (11.5), but it does provide a prediction for the energy spectrum of a turbulent flow (i.e., how much energy is present at a particular spatial scale) and it does this by suggesting a relationship between the energy spectrum (a second-order quantity in velocity) and the spectral energy flux (a third-order quantity).

11.2.1 The Physical Picture

Consider high Reynolds number (Re) incompressible flow that is being maintained by some external force. Then the evolution of a system that has $\rho = 1$ is governed by

$$\frac{\partial \boldsymbol{v}}{\partial t} + (\boldsymbol{v} \cdot \nabla)\boldsymbol{v} = -\nabla p + \boldsymbol{F} + \nu \nabla^2 \boldsymbol{v} \tag{11.13}$$

and

$$\nabla \cdot \boldsymbol{v} = 0. \tag{11.14}$$

Here, $\boldsymbol{F}$ is some force we apply to maintain fluid motion — for example, we stir the fluid with a spoon. (One might argue that such stirring is not a force, like gravity, but a continuous changing of the boundary conditions. Having noted this, we treat it as a force.) A simple scale analysis of these equations seems to indicate that the ratio of the size of the inertial terms on the left-hand side to the viscous term is the Reynolds number VL/ν, where V and L are velocity and length scales. To be explicit let us consider the ocean, and take $V = 0.1\,\mathrm{m\,s^{-1}}$, $L = 1000\,\mathrm{km}$ and $\nu = 10^{-6}\,\mathrm{m^2\,s^{-1}}$. Then $Re = VL/\nu \approx 10^{11}$, and it seems that we can neglect the viscous term on the right-hand side of (11.13). But this can lead to a paradox, as if the fluid is being forced this forcing is likely to put energy into the fluid. To see this, we obtain the energy budget for (11.13) by multiplying by $\boldsymbol{v}$ and integrating over a domain. If there is no flow into or out of our domain, the inertial terms in the momentum equation conserve energy and, recalling Section 1.10, the energy equation is

$$\frac{\mathrm{d}\hat{E}}{\mathrm{d}t} = \frac{\mathrm{d}}{\mathrm{d}t}\int \frac{1}{2}\boldsymbol{v}^2\,\mathrm{d}V = \int \left(\boldsymbol{F}\cdot\boldsymbol{v} + \nu\boldsymbol{v}\cdot\nabla^2\boldsymbol{v}\right)\mathrm{d}V = \int \left(\boldsymbol{F}\cdot\boldsymbol{v} - \nu\boldsymbol{\omega}^2\right)\mathrm{d}V, \tag{11.15}$$

where $\hat{E}$ is the total energy. If we neglect the viscous term we are led to an inconsistency, since the forcing term is a source of energy ($\overline{\boldsymbol{F}\cdot\boldsymbol{v}} > 0$), because a force will normally, on average, produce a velocity that is correlated with the force itself. Without viscosity, energy keeps on increasing.

What is amiss? It is true that for motion with a 1000 km length scale and a velocity of a few centimetres per second we can neglect viscosity when considering the balance of forces in the momentum equation. But this does not mean that there is no motion at much smaller length scales — indeed we seem to be led to the inescapable conclusion that there must be some motion at smaller scales in order to remove energy. Scale analysis of the momentum equation suggests that viscous terms will be comparable with the inertial terms at a scale L_ν where the Reynolds number based on that scale is of order unity, giving

$$L_\nu \sim \frac{\nu}{V}. \tag{11.16}$$

This is a very small scale for geophysical flows, of order millimetres or less. Where and how are such small scales generated? Boundaries are one important region; if there is high Reynolds number flow above a solid boundary, for example the wind above the ground, then viscosity *must* become important in bringing the velocity to zero in order that it can satisfy the no-slip condition at the surface, as illustrated in Fig. 5.4.

Motion on very small scales may also be generated in the fluid interior. How might this happen? Suppose the forcing acts only at large scales, and its direct action is to set up some correspondingly large-scale flow, composed of eddies and shear flows and such-like. Then typically there will be an instability in the flow, and a smaller eddy will grow: initially, the large-scale flow may be treated as an unchanging shear flow, and the disturbance while small will obey linear equations of motion similar to those applicable in an idealized Kelvin–Helmholtz instability. This instability clearly must draw from the large scale quasi-stationary flow, and it will eventually saturate at some finite amplitude. Although it has grown in intensity, it is still typically smaller than the large scale flow that fostered it (remember how the growth rate of the shear instability gets larger as the wavelength of the perturbation decreased). As it reaches finite amplitude, the perturbation itself may become unstable, and smaller eddies will feed off its energy and grow, and so on.[4] Vortex stretching plays an important role in all this, stretching line elements and creating eddies, and energy, at small scales. The general picture that emerges is of a large-scale flow that is unstable to eddies somewhat smaller in scale. These eddies grow, and develop still smaller eddies, and energy is transferred to smaller and smaller scales in a cascade-like process, sketched in Fig. 11.2. Finally, eddies are generated that are sufficiently small that they feel the effects of viscosity, and energy is drained away. Thus, there is a flux of kinetic energy from the large to the small scales, where it is dissipated into heat.

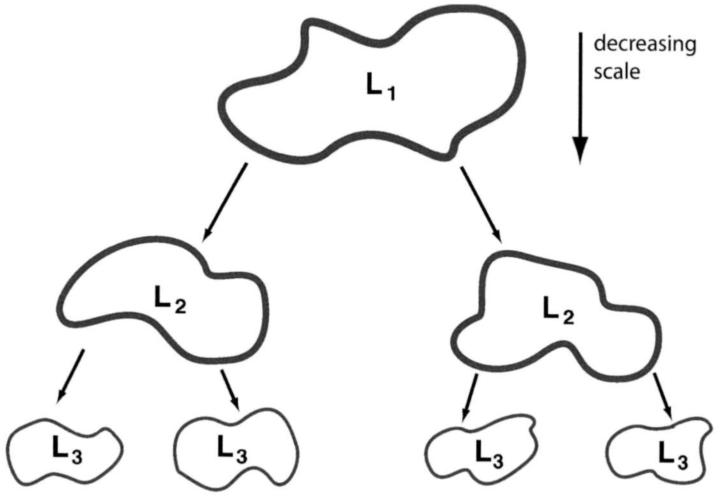

Fig. 11.2 The passage of energy to smaller scales: eddies at large scale break up into ones at smaller scale, thereby transferring energy to smaller scales. (The eddies in reality are embedded within each other.)

If the passage occurs between eddies of similar sizes (i.e., if it is spectrally local) the transfer is said to be a cascade.

11.2.2 Inertial-range Theory

Given the above picture it becomes possible to predict what the energy spectrum is. Let us suppose that the flow is statistically isotropic (i.e., the same in all directions) and homogeneous (i.e., the same everywhere; all isotropic flows are homogeneous, but not vice versa). Homogeneity precludes the presence of solid boundaries but can be achieved in a periodic domain, and the finite domain puts an upper limit, sometimes called the outer scale, on the size of eddies.

If we decompose the velocity field into Fourier components, then in a finite domain we may write

$$u(x, y, z, t) = \sum_{\boldsymbol{k}} \tilde{u}(\boldsymbol{k}, t) e^{i(k^x x + k^y y + k^z z)}, \tag{11.17}$$

where $\tilde{u}$ is the Fourier transformed field of u, with similar identities for v and w, and $\boldsymbol{k} = (k^x, k^y, k^z)$. The sum is a triple sum over all wavenumbers (k^x, k^y, k^z), and in a finite domain these wavenumbers are quantized. Finally, to ensure that u is real we require that $\tilde{u}(-\boldsymbol{k}) = \tilde{u}^*(\boldsymbol{k})$, where the asterisk denotes the complex conjugate. Using Parseval's theorem (and assuming density is unity, as we shall throughout this chapter) the energy in the fluid is given by

$$\frac{1}{V}\int_V E \, dV = \frac{1}{2V}\int_V \left(u^2 + v^2 + w^2\right) dV = \frac{1}{2}\sum_{\boldsymbol{k}} \left(|\tilde{u}|^2 + |\tilde{v}|^2 + |\tilde{w}|^2\right) \equiv \sum_{\boldsymbol{k}} \mathcal{E}_{\boldsymbol{k}}, \tag{11.18}$$

where E is the energy density per unit mass, V is the volume of the domain, and the last equality serves to define the discrete energy spectrum $\mathcal{E}_{\boldsymbol{k}}$. We will now assume that the turbulence is isotropic, and that the domain is sufficiently large that the sums in the above equations may be replaced by integrals. We may then write

$$\overline{E} = \frac{1}{V}\hat{E} = \frac{1}{2V}\int_V \boldsymbol{v}^2 \, dV = \int \mathcal{E}(k) \, dk, \tag{11.19}$$

where $\overline{E}$ is the average energy, $\hat{E}$ is the total energy and $\mathcal{E}(k)$ is the energy spectral density, or the energy spectrum, so that $\mathcal{E}(k)\delta k$ is the energy in the small wavenumber interval δk. Because of the assumed isotropy, the energy is a function only of the scalar wavenumber k, where $k^2 = k^{x2} + k^{y2} + k^{z2}$. The units of $\mathcal{E}(k)$ are L^3/T^2 and the units of $\overline{E}$ are L^2/T^2.

We now suppose that the fluid is stirred at large scales and, via the nonlinear terms in the momentum equation, that this energy is transferred to small scales where it is dissipated by viscosity. The key assumption is to suppose that, if the forcing scale is sufficiently larger than the dissipation

Dimensions and the Kolmogorov Spectrum

Quantity	**Dimension**
Wavenumber, k	1/L
Energy per unit mass, E	$U^2 = L^2/T^2$
Energy spectrum, $\mathcal{E}(k)$	$EL = L^3/T^2$
Energy flux, ε	$E/T = L^2/T^3$

If $\mathcal{E} = f(\varepsilon, k)$ then the only dimensionally consistent relation for the energy spectrum is

$$\mathcal{E} = \mathcal{K}\varepsilon^{2/3}k^{-5/3},$$

where $\mathcal{K}$ is a dimensionless constant.

scale, there exists a range of scales that is intermediate between the large scale and the dissipation scale and where neither forcing nor dissipation are explicitly important to the dynamics. This assumption, known as the *locality hypothesis,* depends on the nonlinear transfer of energy being sufficiently local (in spectral space). This intermediate range is known as the *inertial range,* because the inertial terms and not forcing or dissipation must dominate in the momentum balance. If the rate of energy input per unit volume by stirring is equal to ε, then if we are in a steady state there must be a flux of energy from large to small scales that is also equal to ε, and an energy dissipation rate, also ε.

Now, we have no general theory for the energy spectrum of a turbulent fluid, but we might suppose it takes the general form

$$\mathcal{E}(k) = g(\varepsilon, k, k_0, k_\nu), \tag{11.20}$$

where the right-hand side denotes a function of the spectral energy flux or cascade rate ε, the wavenumber k, the forcing wavenumber k_0 and the wavenumber at which dissipation acts, k_ν (and $k_\nu \sim L_\nu^{-1}$). The function g will of course depend on the particular nature of the forcing. Now, the locality hypothesis essentially says that at some scale within the inertial range the flux of energy to smaller scales depends only on processes occurring at or near that scale. That is to say, the energy flux is only a function of $\mathcal{E}$ and k, or equivalently that the energy spectrum can be a function *only* of the energy flux ε and the wavenumber itself. From a physical point of view, as energy cascades to smaller scales the details of the forcing are forgotten but the effects of viscosity are not yet apparent, and the energy spectrum takes the form,

$$\mathcal{E}(k) = g(\varepsilon, k). \tag{11.21}$$

The function g is assumed to be *universal,* the same for every turbulent flow.

Let us now use dimensional analysis to give us the form of the function $g(\varepsilon, k)$ (see the shaded box above). In (11.21), the left-hand side has dimensions L^3/T^2; the factor T^{-2} can only be balanced by $\varepsilon^{2/3}$ because k has no time dependence; that is, (11.21), and its dimensions, must take the form

$$\mathcal{E}(k) = \varepsilon^{2/3}g(k), \tag{11.22a}$$

$$\frac{L^3}{T^2} \sim \frac{L^{4/3}}{T^2}g(k), \tag{11.22b}$$

where $g(k)$ is some function. Evidently $g(k)$ must have dimensions $L^{5/3}$, and the functional rela-

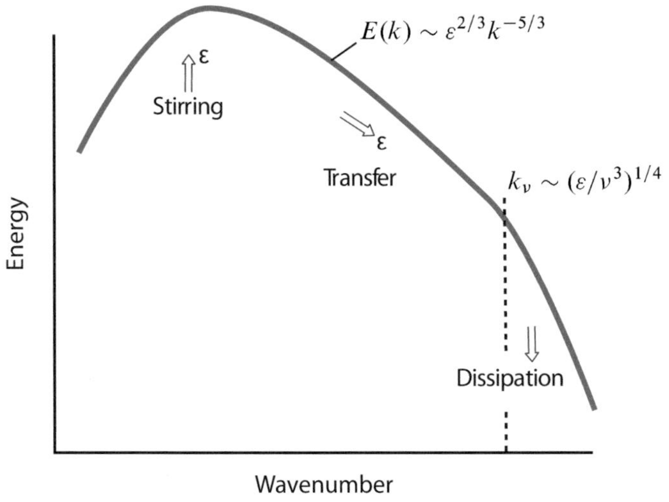

Fig. 11.3 The energy spectrum in three-dimensional turbulence, in the theory of Kolmogorov. Energy is supplied at some rate ε; it is cascaded to small scales, where it is ultimately dissipated by viscosity. There is no systematic energy transfer to scales larger than the forcing scale, so here the energy falls off.

tionship we must have, if the physical assumptions are right, is

$$\mathcal{E}(k) = \mathcal{K}\varepsilon^{2/3}k^{-5/3}. \tag{11.23}$$

This is the famous 'Kolmogorov -5/3 spectrum', enshrined as one of the cornerstones of turbulence theory. It is sketched in Fig. 11.3, and some experimental results are shown in Fig. 11.4. The parameter $\mathcal{K}$ is a dimensionless constant, undetermined by this theory; it is known as Kolmogorov's constant and experimentally its value is found to be about 1.5.[5]

An equivalent, and revealing, way to derive this result is to first define an eddy turnover time τ_k, which is the time taken for a parcel with velocity v_k to move a distance $1/k$, v_k being the velocity associated with the (inverse) scale k. On dimensional considerations $v_k = [\mathcal{E}(k)k]^{1/2}$ so that

$$\tau_k = \left[k^3\mathcal{E}(k)\right]^{-1/2}. \tag{11.24}$$

Kolmogorov's assumptions are then equivalent to setting

$$\varepsilon \sim \frac{v_k^2}{\tau_k} = \frac{k\mathcal{E}(k)}{\tau_k}. \tag{11.25}$$

If we demand that ε be a constant then (11.24) and (11.25) yield (11.23).

The viscous scale and energy dissipation

At some small length scale we should expect viscosity to become important and the scaling theory we have just set up will fail. What is that scale? In the inertial range friction is unimportant because the time scales on which it acts are too long for it be important and dynamical effects dominate. In the momentum equation the viscous term is $\nu\nabla^2 u$ so that a viscous or dissipation time scale at a scale k^{-1}, τ_k^ν, is

$$\tau_k^\nu \sim \frac{1}{k^2\nu}, \tag{11.26}$$

so that the viscous time scale decreases with scale. The eddy turnover time, τ_k — that is, the inertial time scale — in the Kolmogorov spectrum is

$$\tau_k = \varepsilon^{-1/3}k^{-2/3}. \tag{11.27}$$

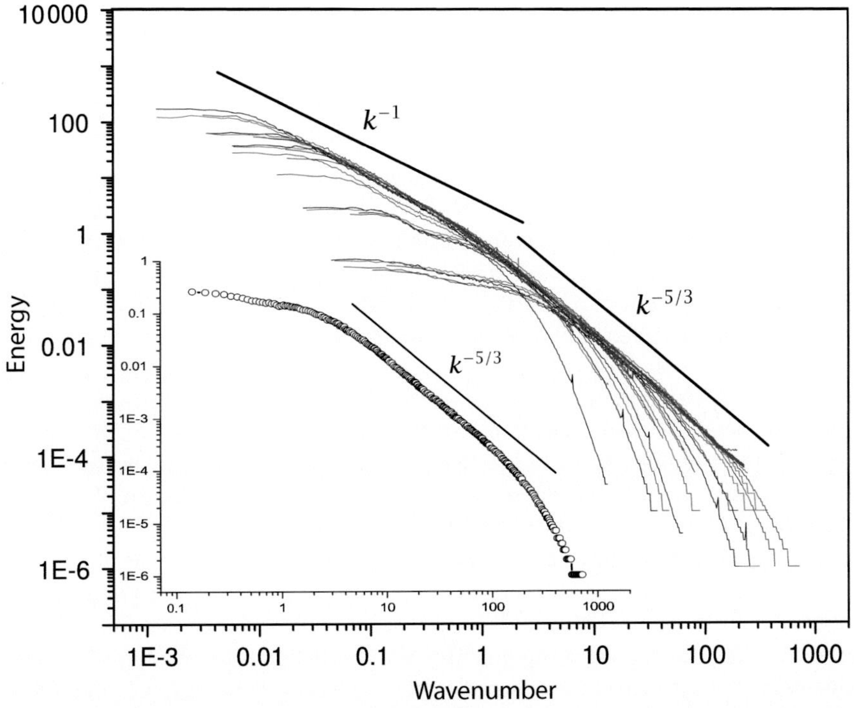

Fig. 11.4 The energy spectrum of 3D turbulence measured in some experiments at the Princeton Superpipe facility.[6] The outer plot shows the spectra from a large number of experiments at different Reynolds numbers up to 10^6, with the magnitude of their spectra appropriately rescaled. Smaller scales show a good −5/3 spectrum, whereas at larger scales the eddies feel the effects of the pipe wall and the spectra are a little shallower. The inner plot shows the spectrum in the centre of the pipe in a single experiment at $Re \approx 10^6$.

The wavenumber at which dissipation becomes important is then given by equating the above two time scales, yielding the dissipation wavenumber, k_ν and the associated length scale, L_ν,

$$k_\nu \sim \left(\frac{\varepsilon}{\nu^3}\right)^{1/4}, \qquad L_\nu \sim \left(\frac{\nu^3}{\varepsilon}\right)^{1/4}. \tag{11.28a,b}$$

L_ν is called the *Kolmogorov scale*. It is the *only* quantity which can be created from the quantities ν and ε that has the dimensions of length. (It is the same as the scale given by (11.16) provided that in that expression V is the velocity magnitude at the Kolmogorov scale.) Thus, for $L \gg L_\nu$, $\tau_k \ll \tau_k^\nu$ and inertial effects dominate. For $L \ll L_\nu$, $\tau_k^\nu \ll \tau_k$ and frictional effects dominate. In fact for length scales smaller than the dissipation scale, (11.27) is inaccurate; the energy spectrum falls off more rapidly than $k^{-5/3}$ and the inertial time scale falls off less rapidly than (11.27) implies, and dissipation dominates even more.

Given the dissipation scale, let us estimate the average energy dissipation rate, $\mathrm{d}/\mathrm{d}T\overline{E}$. This is given by

$$\frac{\mathrm{d}}{\mathrm{d}t}\overline{E} = \frac{1}{V}\int \nu \boldsymbol{v} \cdot \nabla^2 \boldsymbol{v} \,\mathrm{d}V. \tag{11.29}$$

The length at which dissipation acts is the Kolmogorov scale and, noting that $v_k^2 \sim \varepsilon^{2/3}k^{-2/3}$ and using (11.28a), the average energy dissipation rate scales as

$$\mathrm{d}/\mathrm{d}t\overline{E} \sim \nu k_\nu^2 v_{k_\nu}^2 \sim \nu k_\nu^2 \frac{\varepsilon^{2/3}}{k_\nu^{2/3}} \sim \varepsilon. \tag{11.30}$$

That is, the energy dissipation rate is equal to the energy cascade rate. On the one hand this seems sensible, but on the other hand it is *independent of the viscosity.* In particular, in the limit of viscosity tending to zero the energy dissipation remains finite! Surely the energy dissipation rate must go to zero if viscosity goes to zero? To see that this is not the case, consider that energy is input at some large scales, and the magnitude of the stirring largely determines the energy input and cascade rate. The scale at which viscous effects then become important is determined by the viscous scale, L_ν, given by (11.28b). *As viscosity tends to zero L_ν becomes smaller in just such a way as to preserve the constancy of the energy dissipation.* This is one of the most important results in three-dimensional turbulence. Now, we established in Section 1.10 that the Euler equations (i.e., the fluid equations with the viscous term omitted from the outset) do conserve energy. This means that the Euler equations are a *singular limit* of the Navier–Stokes equations: the behaviour of the Navier–Stokes equations as viscosity tends to zero is different from the behaviour resulting from simply omitting the viscous term from the equations ab initio.

How big is L_ν in the atmosphere? A crude estimate, perhaps wrong by an order of magnitude, comes from noting that ε has units of U^3/L, and that at length scales of the order of 100 m in the atmospheric boundary layer (where there might be a three-dimensional energy cascade to small scales) velocity fluctuations are of the order of 1 cm s^{-1}, giving $\varepsilon \approx 10^{-8}\,\mathrm{m^2\,s^{-3}}$. Using (11.28b) we then find the dissipation scale to be of the order of a millimetre or so. In the ocean the dissipation scale is also of the order of millimetres. Various inertial range properties, in both three and two dimensions, are summarized in the shaded box on the facing page.

Degrees of freedom

How many degrees of freedom does a turbulent fluid like the atmosphere potentially have? We might estimate this number, N say, by the expression

$$N \sim \left(\frac{L}{L_\nu}\right)^3, \tag{11.31}$$

where L is the length scale of the energy-containing eddies at the large scale. If we take $L = 1000$ km and $L_\nu = 1$ mm this gives about 10^{27}! On a rather more general basis, we can obtain an expression for N using (11.28b), to give

$$N \sim L^3\left(\frac{\varepsilon}{\nu^3}\right)^{3/4}, \tag{11.32}$$

or, using $\varepsilon \sim U^3/L$,

$$N \sim \left(\frac{UL}{\nu}\right)^{9/4} = Re^{9/4}, \tag{11.33}$$

where Re is the Reynolds number based on the large-scale flow. For typical large-scale atmospheric flows with $U \sim 10\,\mathrm{m\,s^{-1}}$, $L \sim 10^6$ m and $\nu = 10^{-5}\,\mathrm{m^2\,s^{-1}}$, $Re \sim 10^{12}$ and again $N \sim 10^{27}$. Obviously, this number is very approximate, but nevertheless the number of potential degrees of freedom in the atmosphere is truly enormous, greater than Avogadro's number. Thus trying to model the turbulent atmosphere explicitly is akin to trying to model the gas in a room by following the motion of each individual molecule, and it seems unnecessary. How *should* we model it? That, in a nutshell, is the (unsolved) problem of turbulence.

Inertial Range Properties in 3D and 2D Turbulence

For reference, a few inertial range properties are listed below, omitting non-dimensional constants.

	3D energy range	2D enstrophy range	
Energy spectrum	$\varepsilon^{2/3}k^{-5/3}$	$\eta^{2/3}k^{-3}$	(T.1)
Turnover time	$\varepsilon^{-1/3}k^{-2/3}$	$\eta^{-1/3}$	(T.2)
Viscous scale, L_ν	$(\nu^3/\varepsilon)^{1/4}$	$(\nu^3/\eta)^{1/6}$	(T.3)
Passive tracer spectrum	$\chi\varepsilon^{-1/3}k^{-5/3}$	$\chi\eta^{-1/3}k^{-1}$	(T.4)

In these expressions:

ν = viscosity, k = wavenumber, ε = energy cascade rate,
η = enstrophy cascade rate, χ = tracer variance cascade rate.

11.2.3 A Final Note on our Assumptions

The assumptions of homogeneity and isotropy that are made in the Kolmogorov theory are ansatzes, in that we make them because we want to have a tractable model of turbulence (and certainly we can conceive of an experiment in which turbulence is for most practical purposes homogeneous and isotropic). The essential *physical* assumptions are: (i) that there exists an inertial range in which the energy flux is constant; and (ii) that the energy is cascaded from large to small scales in a series of small steps, as the energy spectra will then be determined by spectrally local quantities. The second assumption is the locality assumption and without it we could have, instead of (11.23),

$$\mathcal{E}(k) = C\varepsilon^{2/3}k^{-5/3}g(k/k_0)h(k/k_\nu), \tag{11.34}$$

where g and h are unknown functions. We essentially postulate that there exists a range of intermediate wavenumbers over which $g(k/k_0) = h(k/k_\nu) = 1$.

The first, and less obvious, assumption might be called the *non-intermittency* assumption, and it demands that rare events (in time or space) with large amplitudes do not dominate the energy flux or the dissipation rate. If they were to do so, then the flux would fluctuate strongly, the turbulent statistics would not be completely characterized by ε and Kolmogorov's theory would not be exactly right. Note that in the theory ε is the mean energy cascade rate, and $\varepsilon^{2/3}$ is the two-thirds power of the mean, which is not equal to the mean of the two-thirds power. In fact, in high Reynolds turbulence the $-5/3$ spectra is often observed to a fairly high degree of accuracy (e.g., as in Fig. 11.4), although the higher-order statistics (e.g., higher-order structure functions) predicted by the theory are often found to be in error, and it is generally believed that Kolmogorov's theory is not exact.[7]

11.3 TWO-DIMENSIONAL TURBULENCE

Two-dimensional turbulence behaves in a profoundly different way from three-dimensional turbulence, largely because of the presence of another quadratic invariant, the enstrophy (defined below;

see also Section 5.6.3). In two dimensions, the vorticity equation for incompressible flow is:

$$\frac{\partial \zeta}{\partial t} + \boldsymbol{u} \cdot \nabla \zeta = F + \nu \nabla^2 \zeta, \tag{11.35}$$

where $\boldsymbol{u} = u\boldsymbol{i} + v\boldsymbol{j}$ and $\zeta = \boldsymbol{k} \cdot \nabla \times \boldsymbol{u}$ and F is a stirring term. In terms of a streamfunction, $u = -\partial\psi/\partial y$, $v = \partial\psi/\partial x$, and $\zeta = \nabla^2\psi$, and (11.35) may be written as

$$\frac{\partial \nabla^2 \psi}{\partial t} + J(\psi, \nabla^2\psi) = F + \nu\nabla^4\psi. \tag{11.36}$$

We obtain an energy equation by multiplying by $-\psi$ and integrating over the domain, and an enstrophy equation by multiplying by ζ and integrating. When $F = \nu = 0$ we find

$$\widehat{E} = \frac{1}{2}\int_A (u^2 + v^2)\,\mathrm{d}A = \frac{1}{2}\int_A (\nabla\psi)^2\,\mathrm{d}A, \qquad \frac{\mathrm{d}\widehat{E}}{\mathrm{d}t} = 0, \tag{11.37a}$$

$$\widehat{Z} = \frac{1}{2}\int_A \zeta^2\,\mathrm{d}A = \frac{1}{2}\int_A (\nabla^2\psi)^2\,\mathrm{d}A, \qquad \frac{\mathrm{d}\widehat{Z}}{\mathrm{d}t} = 0, \tag{11.37b}$$

where the integral is over a finite area with either no-normal flow or periodic boundary conditions. The quantity $\widehat{E}$ is the energy, and $\widehat{Z}$ is known as the *enstrophy.* The enstrophy invariant arises because the vortex stretching term, so important in three-dimensional turbulence, vanishes identically in two dimensions. In fact, because vorticity is conserved on parcels it is clear that the integral of *any* function of vorticity is zero when integrated over A; that is, from (11.35)

$$\frac{\mathrm{D}g(\zeta)}{\mathrm{D}t} = 0 \qquad \text{and} \qquad \frac{\mathrm{d}}{\mathrm{d}t}\int_A g(\zeta)\,\mathrm{d}A = 0, \tag{11.38}$$

where $g(\zeta)$ is an arbitrary function. Of this infinity of conservation properties, enstrophy conservation (with $g(\zeta) = \zeta^2$) in particular has been found to have enormous consequences to the flow of energy between scales, as we will soon discover.[8]

11.3.1 Energy and Enstrophy Transfer

In three-dimensional turbulence we posited that energy is cascaded to small scales via vortex stretching. In two dimensions that mechanism is absent, and there is reason to expect energy to be transferred to *larger* scales. This counter-intuitive behaviour arises from the twin integral constraints of energy and enstrophy conservation, and the following three arguments illustrate why this should be so.

I. Vorticity elongation

Consider a band or a patch of vorticity, as in Fig. 11.5, in a nearly inviscid fluid. The vorticity of each element of fluid is conserved as the fluid moves. Now, we should expect that the quasi-random motion of the fluid will act to elongate the band but, as its area must be preserved, the band narrows and so vorticity gradients will increase. This is equivalent to the enstrophy moving to smaller scales. Now, the energy in the fluid is

$$\widehat{E} = -\frac{1}{2}\int \psi\zeta\,\mathrm{d}A, \tag{11.39}$$

where the streamfunction is obtained by solving the Poisson equation $\nabla^2\psi = \zeta$. If the vorticity is locally elongated primarily only in one direction (as it must be to preserve area), the integration involved in solving the Poisson equation will lead to the scale of the streamfunction becoming larger in the direction of stretching, but virtually no smaller in the perpendicular direction. Because stretching occurs, on average, in all directions, the overall scale of the streamfunction will increase in all directions, and the cascade of enstrophy to small scales will be accompanied by a transfer of energy to large scales.

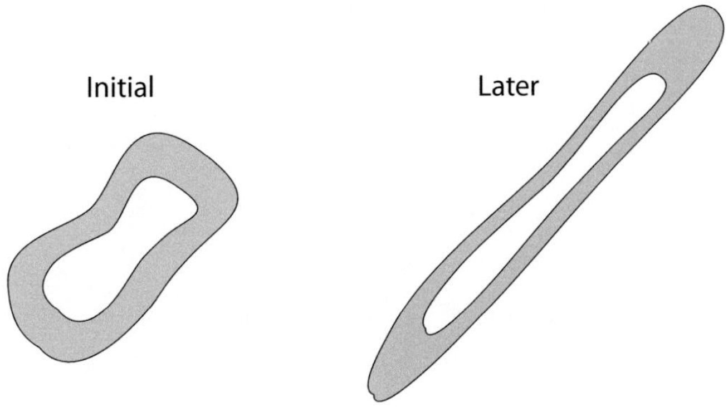

Fig. 11.5 In two-dimensional incompressible flow, a band of fluid is elongated, but its area is preserved. Elongation is followed by folding and more elongation, producing filaments as in Fig. 11.8. As vorticity is tied to fluid parcels, the values of the vorticity in the shaded area (and in the hole) are maintained; thus, vorticity gradients increase and the enstrophy is thereby, on average, moved to smaller scales.

II. An energy-enstrophy conservation argument

A moment's thought will reveal that the distribution of energy and enstrophy in wavenumber space are respectively analogous to the distribution of mass and moment of inertia of a lever, with wavenumber playing the role of distance from the fulcrum. Any re-arrangement of mass such that its distribution also becomes wider must be such that the centre of mass moves toward the fulcrum. Thus, analogously, any rearrangement of a flow that preserves both energy and enstrophy, and that causes the distribution to spread out in wavenumber space, will tend to move energy to small wavenumbers and enstrophy to large. To prove this we begin with expressions for the average energy and enstrophy:

$$\overline{E} = \int \mathcal{E}(k)\,\mathrm{d}k, \qquad \overline{Z} = \int \mathcal{Z}(k)\,\mathrm{d}k = \int k^2\mathcal{E}(k)\,\mathrm{d}k, \tag{11.40}$$

where $\mathcal{E}(k)$ and $\mathcal{Z}(k)$ are the energy and enstrophy spectra. A wavenumber characterizing the spectral location of the energy is the centroid,

$$k_e = \frac{\int k\mathcal{E}(k)\,\mathrm{d}k}{\int \mathcal{E}(k)\,\mathrm{d}k}, \tag{11.41}$$

and, for simplicity, we normalize units so that the denominator is unity. The spreading out of the energy distribution is formalized by setting

$$I \equiv \int (k-k_e)^2\mathcal{E}(k)\,\mathrm{d}k, \qquad \frac{\mathrm{d}I}{\mathrm{d}t} > 0. \tag{11.42}$$

Here, I measures the width of the energy distribution, and this is assumed to increase. Expanding out the integral gives

$$\begin{aligned} I &= \int k^2\mathcal{E}(k)\,\mathrm{d}k - 2k_e\int k\mathcal{E}(k)\,\mathrm{d}k + k_e^2\int \mathcal{E}(k)\,\mathrm{d}k \\ &= \int k^2\mathcal{E}(k)\,\mathrm{d}k - k_e^2\int \mathcal{E}(k)\,\mathrm{d}k, \end{aligned} \tag{11.43}$$

where the last equation follows because $k_e = \int k\mathcal{E}(k)\,\mathrm{d}k$ is, from (11.41), the energy-weighted centroid. Because both energy and enstrophy are conserved, (11.43) gives

$$\frac{\mathrm{d}k_e^2}{\mathrm{d}t} = -\frac{1}{\overline{E}}\frac{\mathrm{d}I}{\mathrm{d}t} < 0. \tag{11.44}$$

Thus, the centroid of the distribution moves to smaller wavenumbers and to larger scales (see Fig. 11.6).

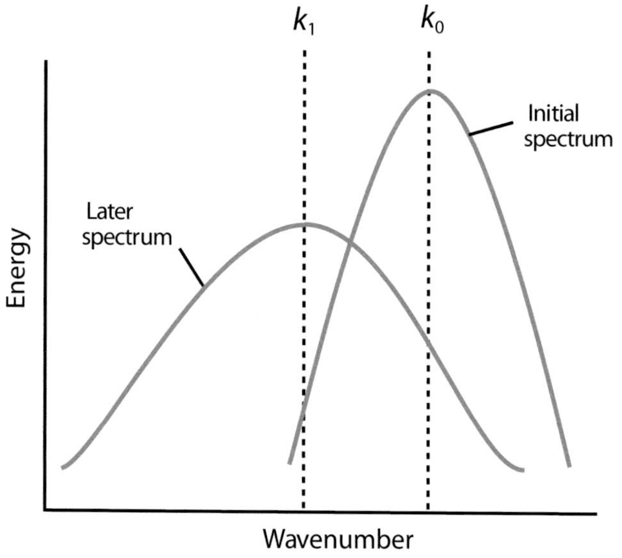

Fig. 11.6 In two-dimensional flow, the centroid of the energy spectrum will move to large scales (smaller wavenumber) provided that the width of the distribution increases — as can be expected in a nonlinear, eddying flow.

An appropriately defined measure of the centre of the enstrophy distribution, on the other hand, moves to higher wavenumbers. The demonstration follows easily if we work with the inverse wavenumber, which is a direct measure of length. Let $q = 1/k$ and assume that the enstrophy distribution spreads out by nonlinear interactions, so that, analogously to (11.42),

$$J \equiv \int (q - q_e)^2 \mathcal{Y}(q)\,\mathrm{d}q, \qquad \frac{\mathrm{d}J}{\mathrm{d}t} > 0, \tag{11.45}$$

where $\mathcal{Y}(q)$ is such that the enstrophy is $\int \mathcal{Y}(q)\mathrm{d}q$ and

$$q_e = \frac{\int q\mathcal{Y}(q)\,\mathrm{d}q}{\int \mathcal{Y}(q)\,\mathrm{d}q}. \tag{11.46}$$

Expanding the integrand in (11.45) and using (11.46) gives

$$J = \int q^2\mathcal{Y}(q)\,\mathrm{d}q - q_e^2 \int \mathcal{Y}(q)\,\mathrm{d}q. \tag{11.47}$$

But $\int q^2\mathcal{Y}(q)\,\mathrm{d}q$ is conserved, because this is the energy. Thus,

$$\frac{\mathrm{d}J}{\mathrm{d}t} = -\frac{\mathrm{d}}{\mathrm{d}t}q_e^2 \int \mathcal{Y}(q)\,\mathrm{d}q, \tag{11.48}$$

whence

$$\frac{\mathrm{d}q_e^2}{\mathrm{d}t} = -\frac{1}{\overline{Z}}\frac{\mathrm{d}J}{\mathrm{d}t} < 0. \tag{11.49}$$

Thus, the length scale characterizing the enstrophy distribution gets smaller, and the corresponding wavenumber gets larger.

III. A similarity argument

Consider an initial value problem, in which a fluid with some initial distribution of energy is allowed to freely evolve, unencumbered by boundaries. We note two aspects of the problem:

(i) there is no externally imposed length scale (because of the way the problem is posed);

(ii) the energy is conserved (this being an assumption).

It is the second condition that limits the argument to two dimensions, for in three dimensions energy is quickly cascaded to small scales and dissipated, but let us here posit that this does not occur. These two assumptions are then sufficient to infer the general direction of transfer of energy, using a rather general similarity argument. To begin, write the energy per unit mass of the fluid as

$$\overline{E} = U^2 = \int \mathcal{E}(k,t)\,\mathrm{d}k, \tag{11.50}$$

where $\mathcal{E}(k,t)$ is the energy spectrum and U is proportional to the square root of the total energy and has units of velocity. On dimensional considerations we could write

$$\mathcal{E}(k,t) = U^2 L \widehat{\mathcal{E}}(\widehat{k},\widehat{t}), \tag{11.51}$$

where $\widehat{\mathcal{E}}$, and its arguments, are nondimensional quantities, and L is some length scale. However, if, over time, the initial conditions are forgotten then there is no length scale in the problem and the only parameters available to determine the energy spectrum are energy, time, and wavenumber, that is U, t and k. A little thought reveals that the most general form for the energy spectrum is then

$$\mathcal{E}(k,t) = U^3 t \widehat{\mathcal{E}} = U^3 t g(Ukt), \tag{11.52}$$

where g is an arbitrary function of its arguments. The argument of g is the only nondimensional grouping of U, t and k, and $U^3 t$ provides the proper dimensions for $\mathcal{E}$. Conservation of energy now implies that the integral

$$I = \int_0^\infty t g(Ukt)\,\mathrm{d}k \tag{11.53}$$

is *not* a function of time. Defining $\vartheta = Ukt$, this requirement is met if

$$\int_0^\infty g(\vartheta)\,\mathrm{d}\vartheta = \text{constant}. \tag{11.54}$$

Now, the spectrum is a function of k only through the combination $\vartheta = Ukt$. Thus, as time proceeds features in the spectrum move to smaller k. Suppose, for example, that the energy is initially peaked at some wavenumber k_p; the product tk_p is preserved, so k_p must diminish with time and the energy must move to larger scales. Similarly, the energy weighted mean wavenumber, k_e, moves to smaller wavenumbers, or larger scales. To see this explicitly, we have

$$k_e = \frac{\int k\mathcal{E}\,\mathrm{d}k}{\int \mathcal{E}\,\mathrm{d}k} = \frac{\int k\mathcal{E}\,\mathrm{d}k}{U^2} = \int kUt\, g(Ukt)\,\mathrm{d}k = \int \frac{\vartheta g(\vartheta)}{Ut}\,\mathrm{d}\vartheta = \frac{C}{Ut}, \tag{11.55}$$

where all the integrals are over the interval $(0,\infty)$ and $C = \int \vartheta g(\vartheta)\,\mathrm{d}\vartheta$ is a constant. Thus, the wavenumber centroid of the energy distribution decreases with time, and the characteristic scale of the flow, $1/k_e$, increases with time. Interestingly, the enstrophy does not explicitly enter this argument, and in general it is not conserved; rather, it is the requirement that energy be conserved that limits the argument to two dimensions; if we accept ab initio that energy *is* conserved, it must be transferred to larger scales.[9]

11.3.2 Inertial Ranges in Two-dimensional Turbulence

If, unlike the case in three dimensions, energy is transferred to larger scales in inviscid, nonlinear, two-dimensional flow then we might expect the inertial ranges of two-dimensional turbulence to be quite different from their three-dimensional counterparts. But before looking in detail at the inertial ranges themselves, let us establish a couple of general properties of forced-dissipative flow in two dimensions.

Some properties of forced-dissipative flow

We will first show that, unlike the case in three dimensions, energy dissipation goes to zero as the Reynolds number rises. In the absence of forcing terms, the total dissipation of energy is, from (11.35)),

$$\frac{\mathrm{d}\widehat{E}}{\mathrm{d}t} = -\nu \int \zeta^2 \,\mathrm{d}A. \tag{11.56}$$

Energy dissipation can only remain finite as $\nu \to 0$ if vorticity becomes infinite. However, this cannot happen because vorticity is conserved on parcels except for the action of viscosity, meaning that $\mathrm{D}\zeta/\mathrm{D}t = \nu\nabla^2\zeta$. However, the viscous term can only *reduce* the value of vorticity on a parcel, and so vorticity can never become infinite if it is not so initially, and therefore using (11.56) energy dissipation goes to zero with ν. (In three dimensions vorticity becomes infinite as viscosity goes to zero because of the effect of vortex stretching.) This conservation of energy is related to the fact that energy is trapped at large scales, even in forced-dissipative flow. On the other hand, enstrophy is transferred to small scales and therefore we expect it to be dissipated at large wavenumbers, even as the Reynolds number becomes very large.

We can show that energy is trapped at large scales in forced-dissipative two-dimensional flow (in a sense that will be made explicit) by the following argument.[10] Suppose that the forcing of the fluid is confined to a particular scale, characterized by the wavenumber k_f, and that dissipation is effected by a linear drag and a small viscosity. The equation of motion is

$$\frac{\partial \zeta}{\partial t} + J(\psi, \zeta) = F - r\zeta + \nu\nabla^2\zeta, \tag{11.57}$$

where F is the stirring and r and ν are positive constants. This leads to the following energy and enstrophy equations:

$$\frac{\mathrm{d}\widehat{E}}{\mathrm{d}t} = -2r\widehat{E} - \int \psi F \,\mathrm{d}A - \int \nu\zeta^2 \,\mathrm{d}A \approx -2r\widehat{E} - \int \psi F \,\mathrm{d}A, \tag{11.58a}$$

$$\frac{\mathrm{d}\widehat{Z}}{\mathrm{d}t} = -2r\widehat{Z} + \int \zeta F \,\mathrm{d}A - D_Z \approx -2r\widehat{Z} - k_f^2 \int \psi F \,\mathrm{d}A - D_Z, \tag{11.58b}$$

where $D_Z = \int \nu(\nabla\zeta)^2 \,\mathrm{d}A$ is the enstrophy dissipation, which is positive. To obtain the right-most expressions, in (11.58a) we assume there is no dissipation of energy by the viscous term, and in (11.58b) we assume that the forcing is confined to wavenumbers near k_f. Consider a statistically steady state and also write $\widehat{E} = \int \mathcal{E}(k)\,\mathrm{d}k$ and $\widehat{Z} = \int k^2\mathcal{E}(k)\,\mathrm{d}k$, where the integrations are over all wavenumbers. If we then eliminate the integral involving ψF between (11.58a) and (11.58b) we obtain

$$\int k^2\mathcal{E}(k)\,\mathrm{d}k + \frac{D_Z}{2r} = \int k_f^2\mathcal{E}(k)\,\mathrm{d}k, \tag{11.59}$$

Now, from the obvious inequality $\int (k - k_e)^2\mathcal{E}(k)\,\mathrm{d}k \geq 0$, where k_e is the energy centroid defined in (11.41), we obtain

$$\int \left(k^2 - k_e^2\right) \mathcal{E}(k)\,\mathrm{d}k \geq 0. \tag{11.60}$$

Combining (11.59) and (11.60) gives

$$\int \left(k_f^2 - k_e^2\right) \mathcal{E}(k)\,\mathrm{d}k \geq \frac{D_Z}{2r} > 0. \tag{11.61}$$

Thus, in a statistically steady state, k_f is larger than k_e, and the energy containing scale, as characterized by k_e^{-1}, must be larger than the forcing scale k_f^{-1}. This demonstration (like argument II in Section 11.3.1) relies both on the conservation of energy and enstrophy by the nonlinear terms and on the particular relationship between the energy and enstrophy spectra.

This result, as well as the arguments of Section 11.3.1, suggest that in a forced-dissipative two-dimensional fluid, energy is transferred to larger scales and enstrophy is transferred to small scales. To obtain a statistically steady state some friction, such as the Rayleigh drag of (11.57), is necessary to remove energy at large scales, and enstrophy must be removed at small scales, but if the forcing scale is sufficiently well separated in spectral space from such frictional effects then two inertial ranges may form — an *energy inertial range* carrying energy to larger scales, and an *enstrophy inertial range* carrying enstrophy to small scales (Fig. 11.7). These ranges are analogous to the three-dimensional inertial range of Section 11.2, and similar conditions must apply if the ranges are to be truly inertial — in particular we must assume spectral locality of the energy or enstrophy transfer. Given that, we can then calculate their properties, as follows.

The enstrophy inertial range

In the enstrophy inertial range the enstrophy cascade rate η, equal to the rate at which enstrophy is supplied by stirring, is assumed constant. By analogy with (11.25) we may assume that this rate is given by

$$\eta \sim \frac{k^3 \mathcal{E}(k)}{\tau_k}. \tag{11.62}$$

With τ_k (still) given by (11.24) we obtain

$$\mathcal{E}(k) = \mathcal{K}_\eta \eta^{2/3} k^{-3}, \tag{11.63}$$

where $\mathcal{K}_\eta$ is, we presume, a universal constant, analogous to the Kolmogorov constant of (11.23). This, and various other properties in two- and three-dimensional turbulence, are summarized in the shaded box on page 423.

The velocity and time at a particular wavenumber then scale as

$$v_k \sim \eta^{1/3} k^{-1}, \qquad t_k \sim l_k / v_k \sim 1/(k v_k) \sim \eta^{-1/3}. \tag{11.64a,b}$$

We may also obtain (11.64) by substituting (11.63) into (11.24). Thus, *the eddy turnover time in the enstrophy range of two-dimensional turbulence is length-scale invariant.* The appropriate viscous scale is given by equating the inertial and viscous terms in (11.35). Using (11.64a) we obtain, analogously to (11.28a), the viscous wavenumber

$$k_\nu \sim \left(\frac{\eta^{1/3}}{\nu} \right)^{1/2}. \tag{11.65}$$

The enstrophy dissipation, analogously to (11.30) goes to a finite limit given by

$$\frac{\mathrm{d}}{\mathrm{d}t}\overline{Z} = \nu \int_A \zeta \nabla^2 \zeta \, \mathrm{d}A \sim \nu k_\nu^4 v_{k_\nu}^2 \sim \eta, \tag{11.66}$$

using (11.64a) and (11.65). Thus, the enstrophy dissipation in two-dimensional turbulence is (at least according to this theory) independent of the viscosity.

Energy inertial range

The energy inertial range of two-dimensional turbulence is quite similar to that of three-dimensional turbulence, except in one major respect: the energy flows from smaller to larger scales! Because the atmosphere and ocean behave in some ways as two-dimensional fluids, this has profound consequences on their behaviour, and is something we return to in the next chapter. The upscale

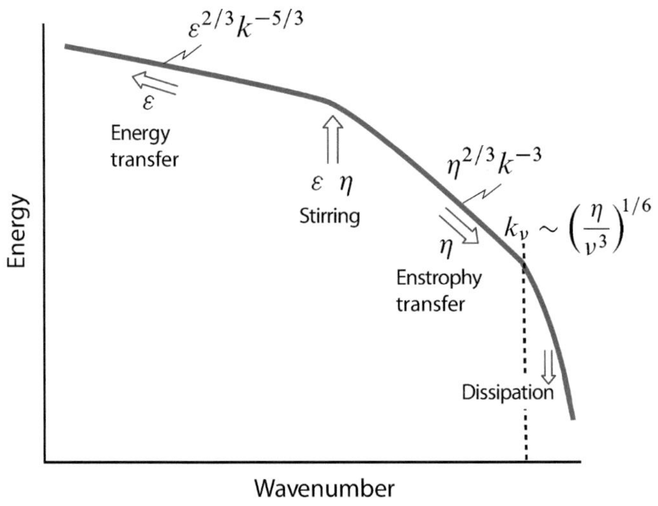

Fig. 11.7 The energy spectrum of two-dimensional turbulence. (Compare with Fig. 11.3.) Energy supplied at some rate ε is transferred to large scales, whereas enstrophy supplied at some rate η is transferred to small scales, where it may be dissipated by viscosity. If the forcing is localized at a scale k_f^{-1} then $\eta \approx k_f^2\varepsilon$.

energy flow is known as the *inverse cascade,* and the associated energy spectrum is, as in the three-dimensional case,

$$\mathcal{E}(k) = \mathcal{K}_\varepsilon \varepsilon^{2/3} k^{-5/3}, \tag{11.67}$$

where $\mathcal{K}_\varepsilon$ is a nondimensional constant — sometimes called the Kolmogorov–Kraichnan constant, and not necessarily equal to $\mathcal{K}$ in (11.23) — and ε is the rate of energy transfer to larger scales. Of course we now need a mechanism to remove energy at large scales, otherwise it will pile up at the scale of the domain and a statistical steady state will not be achieved. Introducing a linear drag, $-r\zeta$, into the vorticity equation, as in (11.57), is one means of removing energy, and such a term may be physically justified by appeal to Ekman layer theory (Section 5.7). Although such a term appears to be scale invariant, its effects will be felt primarily at large scales because at smaller scales the time scale of the turbulence is much shorter than that of the friction, and we may estimate the scale at which the drag becomes important by equating the two time scales. The turbulent timescale is given by (11.27), and equating this to the frictional time scale r^{-1} gives $r^{-1} = \varepsilon^{-1/3}k_r^{-2/3}$, or

$$k_r = \left(\frac{r^3}{\varepsilon}\right)^{1/2} \quad \text{or} \quad L_r = \left(\frac{\varepsilon}{r^3}\right)^{1/2}, \tag{11.68}$$

where k_r is the frictional wavenumber, and frictional effects are important at scales *larger* than the frictional scale L_r.

11.3.3 † More Phenomenology

The phenomenology of two-dimensional turbulence is not quite as settled as the above arguments imply. Note, for example, that time scale (11.64b) is independent of length scale, whereas in three-dimensional turbulence the time scale decreases with length scale, which seems more physical and more conducive to spectrally local interactions. A useful heuristic measure of this locality is given by estimating the contributions to the straining rate, $S(k)$, from motions at all scales larger than k^{-1}. The strain rate scales like the shear, so that an estimate of the total strain rate is given by

$$S(k) = \left[\int_{k_0}^{k} \mathcal{E}(p)p^2 \,\mathrm{d}p\right]^{1/2}, \tag{11.69}$$

where k_0 is the wavenumber of the largest scale present. The contributions to the integrand from a given wavenumber octave are given by

$$\int_p^{2p} \mathcal{E}(p')p'^3 \,\mathrm{d}\log p' \sim \mathcal{E}(p)p^3. \tag{11.70}$$

In three dimensions, use of the −5/3 spectrum indicates that the contributions from each octave below a given wavenumber k increase with wavenumber, being a maximum close to k, and this is a posteriori consistent with the locality hypothesis. However, in two-dimensional turbulence with a −3 spectrum each octave makes the same contribution. That is to say, the contributions to the strain rate at a given wavenumber, as defined by (11.69), are not spectrally local. This does not prove that the enstrophy transfer is spectrally non-local, but nor does it build confidence in the theory.

Dimensionally the strain rate is the inverse of a time, and if this is a spectrally non-local quantity then, instead of (11.24), we might use the inverse of the strain rate as an eddy turnover time giving

$$\tau_k = \left[\int_{k_0}^{k} p^2 \mathcal{E}(p)\,\mathrm{d}p\right]^{-1/2}. \tag{11.71}$$

This has the advantage over (11.24) in that it is a non-increasing function of wavenumber, whereas if the spectrum is steeper than k^{-3}, (11.24) implies a time scale increasing with wavenumber. Using this in (11.62) gives a prediction for the enstrophy inertial range, namely

$$\mathcal{E}(k) = \mathcal{K}_\eta \eta^{2/3}\left[\log(k/k_0)\right]^{-1/3} k^{-3}, \tag{11.72}$$

which is similar to (11.63) except for a logarithmic correction. This expression is, of course, spectrally non-local, in contradiction to our original assumption: it has arisen by noting the spectral locality inherent in (11.69), and proposing a reasonable, although ad hoc, solution.

The discussion above suggests that the phenomenology of the forward enstrophy cascade is on the verge of being internally inconsistent, and that the k^{-3} spectral slope might be the shallowest limit that is likely to be actually achieved in nature or in any particular computer simulation or laboratory experiment rather than a robust, universal slope. To see this, suppose the detailed fluid dynamics strives in some way to produce a slope shallower than k^{-3}; then, using (11.70), the strain is local and the shallow slope is forbidden by the Kolmogorovian scaling results. However, if the dynamics organizes itself into structures with a slope steeper than k^{-3} the strain is quite non-local. The fundamental assumption of Kolmogorov scaling is not satisfied, and there is no internal inconsistency. The theory then simply does not apply, and a slope steeper than k^{-3} is not theoretically inconsistent.

There are two other potential issues with the theory of two-dimensional turbulence. One is that enstrophy is only one of an infinity of invariants of inviscid two-dimensional flow, and the theory takes no account of the presence of others. The second is that, as in three-dimensional turbulence, if there is strong intermittency the flow cannot be fully characterized by single enstrophy and energy cascade rates. In two-dimensional turbulence the main form of intermittency may be the formation of coherent vortices, discussed more below. In spite of these problems, the notions of a forward transfer of enstrophy and an inverse transfer of energy are quite robust, and have considerable numerical support. Indeed, the realization that in two-dimensional turbulence energy is transferred to larger scales was arguably one of the most important developments in fluid mechanics in the second half of the twentieth century, with important ramifications in rotating and stratified *three-dimensional* fluids.[11]

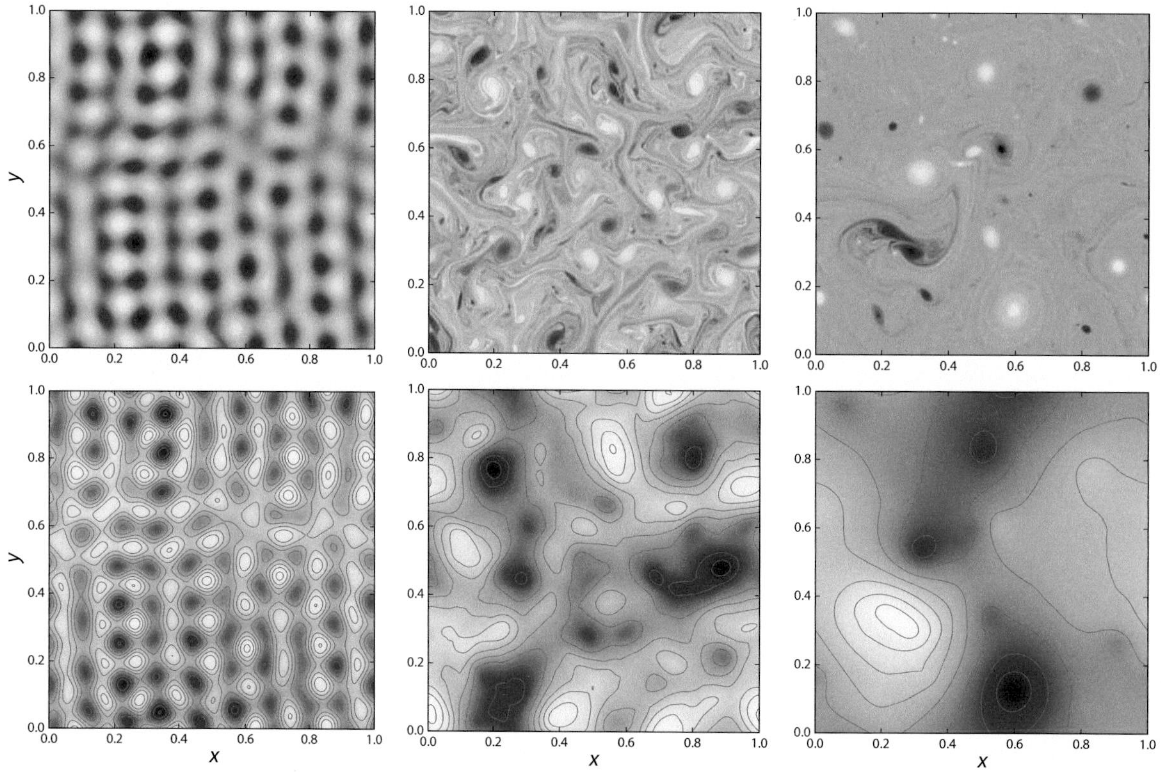

Fig. 11.8 Nearly-free evolution of vorticity (top) and streamfunction (bottom) in a doubly-periodic domain obeying the two-dimensional vorticity equation with no forcing but with a weak viscous term, in a numerical simulation with 512^2 equivalent grid points. Time proceeds from left to right. The initial conditions have just a few non-zero Fourier modes (around wavenumber 9) with randomly generated phases. Kelvin–Helmholtz instability leads to vortex formation and roll-up (as in Fig. 9.6), and like-signed vortices merge (with an example in the top right panel) ultimately leading to a state of just two oppositely-signed vortices. Between the vortices, enstrophy cascades to smaller scales. The scale of the streamfunction grows larger, reflecting the transfer of energy to larger scales.

11.3.4 Numerical Illustrations

Numerical simulations nicely illustrate both the classical phenomenology and its shortcomings. In the simulations shown in Fig. 11.8 the vorticity field is initialized quasi-randomly, with little structure in the initial field, and with only a few non-zero Fourier components; the flow then freely evolves, save for the effects of a weak viscosity.

Vortices soon form, and between them enstrophy is cascaded to small scales where it is dissipated, producing a flat and nearly featureless landscape. The energy cascade to larger scales is reflected in the streamfunction field (bottom row of Fig. 11.8), the length scale of which slowly grows larger with time. The vortices themselves form through a roll-up mechanism, similar to that illustrated in Fig. 9.6, and their presence provides problems to the phenomenology. Because circular vortices are nearly exact, stable solutions of the inviscid equations they can 'store' enstrophy, disrupting the relationship between enstrophy flux and enstrophy itself that is assumed in the Kolmogorov–Kraichnan phenomenology. When vortices merge, the enstrophy dissipation rate increases rapidly for a short period of time, so providing a source of intermittency.

Nevertheless, some forced-dissipative numerical simulations suggest that the presence of vortices may be confined to scales close to that of the forcing, and if the resolution is sufficiently high

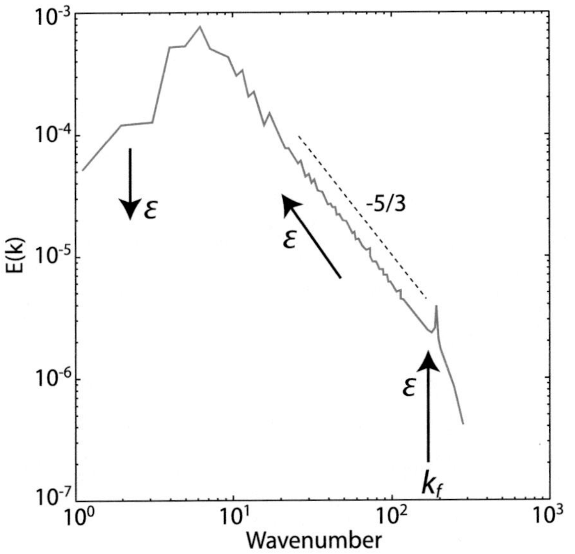

Fig. 11.9 The energy spectrum in a numerical simulation of forced-dissipative two-dimensional turbulence. The fluid is stirred at wavenumber k_f and dissipated at large scales with a linear drag, and there is a $k^{-5/3}$ spectrum at intermediate scales. The arrows indicate the direction of the energy flux, ε.[12]

then the $-5/3$ inverse cascade and -3 forward enstrophy cascade appear. Certainly, if the forcing is spectrally localized, then a well-defined $-5/3$ spectrum robustly forms, as illustrated in Fig. 11.9. The forward k^{-3} spectrum is typically more delicate, being influenced by the presence of coherent vortices, but it does arise in some numerical simulations when the resolution is sufficiently high.[13] The atmosphere itself is not observed to have an inverse $-5/3$ spectrum at large scales; indeed there is no well-defined inverse energy cascade in the sense described above. The atmosphere does have an approximate -3 cascade (see Fig. 12.9 in the next chapter), but whether we should attribute this to a classical forward enstrophy cascade is not settled.

11.4 PREDICTABILITY OF TURBULENCE

> *Small differences in the initial conditions may produce very great ones in the final phenomenon... Prediction becomes impossible... A tenth of a degree more or less at any given point, and the cyclone will burst here and not there, and extend its ravages over districts it might otherwise have spared.*
>
> Henri Poincaré, *Science and Method*, 1908.

Forecasting the weather is hard. That this is so stems from the fact that the atmosphere is chaotic, and chaotic systems are unpredictable virtually by definition. However, the atmosphere (and turbulence in general) is certainly not a *low-dimensional* chaotic system (meaning a system with only a few degrees of freedom), and the connection between atmospheric unpredictability and the so-called 'sensitive dependence on initial conditions' of low-dimensional systems is not as straightforward as it might seem. In this section we expand on and clarify these issues, beginning with an informal discussion of a few aspects of low-dimensional dynamical systems.

11.4.1 Low-dimensional Chaos and Unpredictability

Chaos, or temporal disorder leading to effective indeterminism, is a ubiquitous property of nonlinear dynamical systems. Much of this was known to Poincaré, but in its modern reincarnation it stems in part from the 'Lorenz equations'.[14] These are a set of three coupled nonlinear ordinary differential equations, originally derived by way of a rather ad hoc truncation of the fluid equations governing a two-dimensional convective system: the streamfunction of a convective role is written as $\psi(x, z, t) = X(t) \sin kx \sin \pi z$, and the temperature perturbation as $\theta(x, z, t) =$

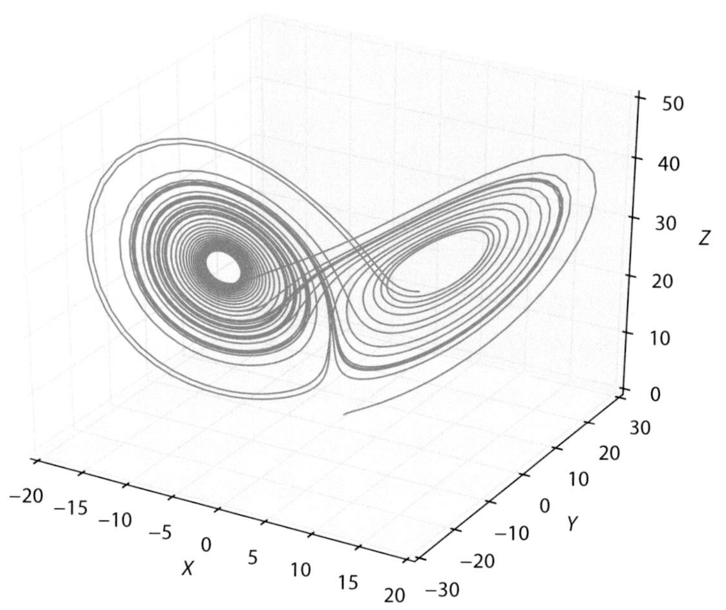

Fig. 11.10 A solution of the Lorenz equations, with $\sigma = 10$, $r = 28$ and $b = 8/3$. The plot shows a trajectory of a solution in phase space, with initial conditions $X = Y = Z = 1$.

$Y(t)\cos kx \sin \pi z + Z(t)\sin 2\pi z$, where x and z are the horizontal and vertical coordinates in physical space, k is a wavenumber, and X, Y, Z are amplitudes (not coordinates). Thus, X represents the rotational speed of the convection roll, Y the temperature difference horizontally across the roll, and Z the deviation temperature from a background vertical stratification. The resulting equations are, in notation standard for them,

$$\frac{dX}{dt} = \sigma(Y - X), \qquad \frac{dY}{dt} = rX - Y - XZ, \qquad \frac{dZ}{dt} = XY - bZ, \tag{11.73a,b,c}$$

where the parameters are: σ, the Prandtl number; r, proportional to the Rayleigh number; and b, a wavenumber dependent dissipation coefficient. The behaviour of the system varies with the parameters, and a well studied set uses $\sigma = 10$, $r = 28$ and $b = 8/3$. A typical solution of the system, a 'flow', is given in Fig. 11.10, and evidently the behaviour is quite complex. It is aperiodic, and the frequency spectrum (not shown) is quite broad. Now, suppose that at any given instant the flow is perturbed slightly. Or to put it another way, suppose that we are trying to predict the future behaviour of the system by integrating the equations of motion but that we have inaccurate knowledge about the system at some particular time. We find that the evolution of the original flow and that of the perturbed flow diverge from each other, and after a little while the two systems are completely different (Fig. 11.11). Because we can never expect to have completely accurate information about the state of the system, the system is thus unpredictable; that is, the details of the evolution depend sensitively on the initial conditions, and small errors in the initial conditions grow to finite amplitude. Let us make two other points:

(i) The time taken for the trajectories to diverge depends on the magnitude of the initial perturbation. Small perturbations grow exponentially at first, and at any given point in the trajectory, the smaller the perturbation the longer the predictability.

(ii) Once the perturbation has reached finite amplitude, the predictability time — the time for the error to become as large as the solution itself — will typically be of the order of the characteristic advective time of the system, the time for a convective roll to overturn.

Deterministic unpredictability is in fact a common feature of nonlinear dynamical systems, and may be taken as an informal definition of a chaotic system.

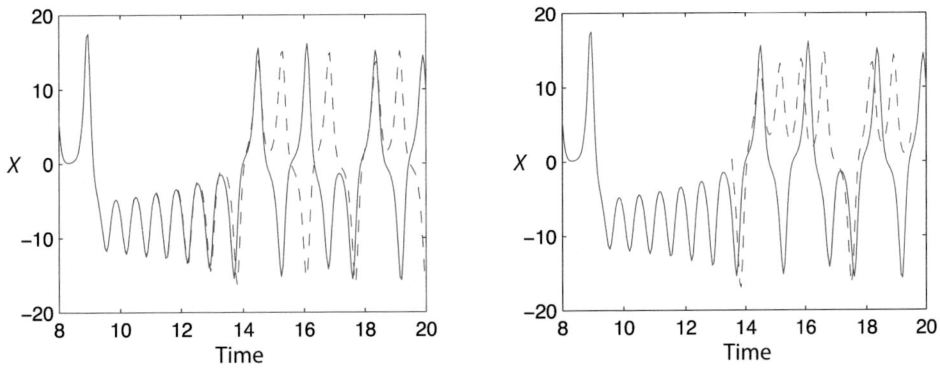

Fig. 11.11 Examples of the evolution of the variable X in the Lorenz model subject to a small perturbation at time 10 (left panel) and time 13.5 (right panel). The original and perturbed systems are the solid and dashed lines, respectively.

11.4.2 ♦ Predictability of a Turbulent Flow

A turbulent flow will be unpredictable if turbulence is chaotic. Because it is our common experience that turbulence is both spatially and temporally disordered, it seems, perhaps with the benefit of hindsight, that turbulence must be chaotic and unpredictable. However, this has not always been evident, and it relies on the non-trivial recognition that chaotic systems exist.[15]

Presuming that a turbulent flow *is* unpredictable, can we estimate its predictability time, namely the time taken for an initially small error to completely contaminate the system? Let us first note that turbulent flows in general contain multiple scales of motion, and let us suppose that the initial error is confined to small scales. The predictability time of the system may be taken as the time taken for the error to contaminate all scales of motion. There are two possible routes that the error may take in affecting the larger scales. In the first we suppose, following classical turbulence phenomenology, that errors on a small scale will mostly contaminate the motion on the next larger scale (in a logarithmic sense), and that this contamination occurs on the local eddy turnover time. Eddies on this larger scale then grow and affect the next larger scale, and the error field is so cascaded upscale via local triad interactions finally reaching the largest scales of the fluid. This mechanism does not rely on there being an inverse cascade of energy — it is only the error, or the contamination, that is cascaded upscale. In the second route we suppose that errors occurring on the small scale immediately contaminate the largest scales, with an initial error equal to the amplitude of the small scale, and that the large-scale error then grows exponentially.

I. Error growth via a local cascade

Let us suppose that the error is initially confined to some small scale characterized by the (inverse of) the wavenumber k_1, as determined by the resolution of our observing network. For modes at that scale the error may be considered finite rather than infinitesimal, and it will saturate and contaminate the next largest scale in a time scale comparable to the eddy turnover time at that scale. Thus, in general, errors initially confined to a scale k will contaminate the scale $2k$ after a time τ_k, with τ_k given by (11.24). The total time taken for errors to propagate from the small scale k_1 to the largest scale k_0 is then given by

$$T = \int_{k_0}^{k_1} \tau_k \,\mathrm{d}(\ln k) = \int_{k_0}^{k_1} [k^3 \mathcal{E}(k)]^{-1/2} \,\mathrm{d}(\ln k), \tag{11.74}$$

treating the wavenumber spectrum as continuous. The logarithmic integral arises because the cascade proceeds logarithmically — error cascades from k to $2k$ in a time τ_k. For an energy spectrum of the form $E = Ak^{-n}$ this becomes

$$T = \frac{2}{A^{1/2}(n-3)}\left[k^{(n-3)/2}\right]_{k_0}^{k_1}, \tag{11.75}$$

for $n \neq 3$, and $T = A^{-1/2}\ln(k_1/k_0)$ for $n = 3$. If in two-dimensional turbulence we have $n = 3$ and $A = \eta^{2/3}$, and if in three-dimensional turbulence we have $n = 5/3$ and $A = \varepsilon^{2/3}$, then the respective predictability times are given by

$$T_{2d} \sim \eta^{-1/3}\ln(k_1/k_0), \qquad T_{3d} \sim \varepsilon^{-1/3}k_0^{-2/3}. \tag{11.76a,b}$$

As $k_1 \to \infty$, that is as the initial error is confined to smaller and smaller scales, predictability time grows larger for two-dimensional turbulence (and for $n \geq 3$ in general), but remains finite for three-dimensional turbulence.

II. Error growth via a direct interaction

Let us now assume that the small scale error directly affects the large scales, where the error then grows exponentially until it saturates. That is, if φ is a measure of the amplitude of the large-scale error then

$$\varphi \sim \varphi_0 \exp(\sigma t), \tag{11.77}$$

where σ is the inverse of the eddy turnover time at the large scale, and φ_0 is the amplitude of its initial error and this, we assume, is equal to the amplitude of the motion at the poorly-observed small scales at wavenumber k_1. In the enstrophy cascade of two-dimensional turbulence the eddy turnover time is given by $\tau_k \sim \eta^{-1/3}$ and $A \sim \eta^{2/3}$, and so we take

$$\sigma = \eta^{1/3} = A^{1/2}, \qquad \varphi_0 = Ak_1^{-n}, \tag{11.78}$$

where $n = 3$. The time, T'_{2d}, needed for the error to saturate the large scales, k_0, is then approximately given by the solution of

$$Ak_0^{-n} = Ak_1^{-n}\exp\left(A^{1/2}T'_{2d}\right), \tag{11.79}$$

giving

$$T'_{2d} \sim \eta^{-1/3}\ln(k_1/k_0). \tag{11.80}$$

In three-dimensional turbulence the eddy turnover time is given by $\tau_k \sim \varepsilon^{-1/3}k^{-2/3}$ and $A \sim \varepsilon^{2/3}$, and so we take

$$\sigma = k_0^{2/3}A^{1/2}, \qquad \varphi_0 = Ak_1^{-n}, \tag{11.81}$$

where $n = 5/3$. The time, T'_{3d}, needed for an error to saturate the large scale is then approximately given by the solution of

$$Ak_0^{-n} = Ak_1^{-n}\exp\left(k_0^{2/3}A^{1/2}T'_{3d}\right), \tag{11.82}$$

giving

$$T'_{3d} \sim k_0^{-2/3}\varepsilon^{-1/3}\ln(k_1/k_0). \tag{11.83}$$

The estimates (11.80) and (11.83) are to be compared with (11.76). For two-dimensional turbulence, the estimates are equal (reflecting the scale independence of the eddy turnover time), whereas for three-dimensional turbulence the estimate from (11.76) is much shorter than that from (11.83) if $k_1 \gg k_0$, meaning that the local cascade mechanism of error growth will dominate.

11.4.3 Implications and Weather Predictability

In two-dimensional flow the predictability time, (11.76a), increases without bound as the scale of the initial error decreases. This is consistent with what has been rigorously proven about the two-dimensional Navier–Stokes equations, namely that provided the initial conditions are sufficiently smooth, the solutions have a continuous dependence on the initial conditions, and a change in solution at some later time may be bounded by reducing the magnitude of the change in the initial conditions. This does not mean that two-dimensional flow is in practice necessarily predictable: a small error or small amount of noise in the system will still render a flow truly unpredictable sometime in the future, but we can put off that time indefinitely if we know the initial conditions well enough.

In three dimensions, with a spectrum of $k^{-5/3}$, the predictability-time estimate from (11.76b) is not dependent on the scale of the initial error. Thus, even if the initial error is confined to smaller and smaller scales, the predictability time is bounded. The time it takes for such errors to spread to the largest scales is simply a few large eddy turnover times, essentially because the eddy turnover times of the small scales are so small. For such a fluid, there is no unique error doubling time, because the error growth rate is a function of scale.

In the troposphere the large-scale flow behaves more like a two-dimensional fluid than a three-dimensional fluid and from scales from a few hundred to a few thousand kilometres it has, roughly, a k^{-3} spectrum (look ahead to Fig. 12.9). If this spectrum extended indefinitely to small scales the predictability time would be correspondingly large, but at scales smaller than about 100 km or so, the atmosphere starts to behave more three dimensionally and the predictability time cannot be significantly extended by making observations at still finer scales. That is, the effective limit to predictability is governed by the horizontal scale at which the atmosphere turns three dimensional. (Hypothetically, we might be able to increase the predictability time if we could observe scales well into the viscous regime where the spectrum steepens again, but this is not a practicable proposition.) Putting in the numbers gives a predictability limit of about 12 days (but there is at least a factor of two uncertainty in such a calculation), and small perturbations that are impossible to observe will change the course of the large-scale weather systems on this time scale. The 'butterfly effect' has its origins in this argument: a butterfly flapping its wings is, so it goes, able to change the course of the weather a week or so later.[16]

As regards our attempts to predict weather, as the atmosphere becomes observed more and more accurately, the initial error will become concentrated at smaller and smaller scales, eventually reaching the scale at which the atmosphere ceases to behave as a quasi-two-dimensional fluid and where its spectrum flattens. The initial error growth rate will then increase, indicating the unavoidable onset of diminishing returns in adding resolution to our observing systems and models. Unlike the situation in a low-order chaotic system, the growth rate of errors in a turbulent flow is not, in general, exponential (except for a pure −3 spectrum) even for a small initial error. This is because the initial error will never be properly infinitesimal, in that a given error will nearly always project on to some scale at finite amplitude.[17]

11.5 ✦ SPECTRA OF PASSIVE TRACERS

In fluid dynamics we are often concerned with the transport of tracers by turbulent flows. An *active tracer* is one that affects the flow itself (potential vorticity and salt are examples), whereas a *passive tracer* does not affect the flow field (a neutrally buoyant dye, for example), and its dynamics are often simpler. In some circumstances an active tracer can be considered to be approximately passive; the atmospheric temperature field at very large scales turns out to be one example, and its transport is obviously a key determinant of our climate. Let us therefore spend some time discussing the dynamics of a passive tracer; in this section we will consider their spectra, and in Chapter 13 we will consider their diffusive transport.

We consider a tracer that obeys

$$\frac{D\varphi}{Dt} = F_{[\varphi]} + \kappa \nabla^2 \varphi, \tag{11.84}$$

where $F_{[\varphi]}$ is the stirring of the dye and κ is its diffusivity, and κ in general differs from the kinematic molecular viscosity ν. If φ is temperature, the ratio of viscosity to diffusivity is called the *Prandtl number* and denoted σ, so that $\sigma \equiv \nu/\kappa$. If φ is a passive tracer, the ratio is sometimes called the *Schmidt number*, but we shall call it the Prandtl number in all cases. We assume that the tracer variance is created at some well-defined scale k_0, and that κ is sufficiently small that dissipation only occurs at very small scales. (Dissipation only reduces the tracer *variance*, not the amount of tracer itself.) The turbulent flow will generically tend to stretch patches of dye into elongated filaments, in much the same way as vorticity in two-dimensional turbulence is filamented — Fig. 11.5 applies just as well to a passive tracer in either two or three dimensions as it does to vorticity in two dimensions. Thus we expect a transfer of tracer variance from large to small scales. If the dye is stirred at a rate χ then, by analogy with our treatment of the cascade of energy, we posit that

$$\mathcal{K}_\chi \chi \propto \frac{\mathcal{P}(k)k}{\tau_k}, \tag{11.85}$$

where $\mathcal{P}(k)$ is the spectrum of the tracer, k is the wavenumber, τ_k is an eddy time scale and $\mathcal{K}_\chi$ is a constant, not necessarily the same constant in all cases. (In the rest of the section, Kolmogorov-like constants will be denoted by $\mathcal{K}$, differentiated by miscellaneous superscripts or subscripts.) We will also assume that τ_k is given by

$$\tau_k = [k^3 \mathcal{E}(k)]^{-1/2}. \tag{11.86}$$

Suppose that the turbulent spectrum is given by $\mathcal{E}(k) = Ak^{-n}$. Using (11.86), (11.85) becomes

$$\mathcal{K}_\chi \chi = \frac{\mathcal{P}(k)k}{[Ak^{3-n}]^{-1/2}}, \tag{11.87}$$

and

$$\mathcal{P}(k) = \mathcal{K}_\chi A^{-1/2} \chi k^{(n-5)/2}. \tag{11.88}$$

We see that the steeper the energy spectrum the shallower the tracer spectrum. If the energy spectrum is steeper than −3 then (11.86) may not be a good estimate of the eddy turnover time, and we use instead

$$\tau_k = \left[\int_{k_0}^{k} p^2 \mathcal{E}(p) \, dp \right]^{-1/2}, \tag{11.89}$$

where k_0 is the low-wavenumber limit of the spectrum. If the energy spectrum is shallower than −3, then the integrand is dominated by the contributions from high wavenumbers and (11.89) effectively reduces to (11.86). If the energy spectrum is steeper than −3, then the integrand is dominated by contributions from low wavenumbers. For $k \gg k_0$ we can approximate the integral by $[k_0^3 E(k_0)]^{-1/2}$, that is the eddy-turnover time at large scales, τ_{k_0}, given by (11.86). The tracer spectrum then becomes

$$\mathcal{P}(k) = \mathcal{K}'_\chi \chi \tau_{k_0} k^{-1}, \tag{11.90}$$

where $\mathcal{K}'_\chi$ is a constant. In all these cases the tracer cascade is to smaller scales even if, as may happen in two-dimensional turbulence, energy is cascading to larger scales.

The scale at which diffusion becomes important is given by equating the turbulent time scale τ_k to the diffusive time scale $(\kappa k^2)^{-1}$. This is independent of the flux of tracer, χ, essentially because the equation for the tracer is linear. Determination of expressions for these scales in two and three dimensions is left as a problem for the reader.

11.5.1 Examples of Tracer Spectra

Energy inertial range flow in three dimensions

Consider a range of wavenumbers over which neither viscosity nor diffusivity directly influence the turbulent motion and the tracer. Then, in (11.88), $A = \mathcal{K}\varepsilon^{2/3}$ where ε is the rate of energy transfer to small scales, $\mathcal{K}$ is the Kolmogorov constant, and $n = 5/3$. The tracer spectrum becomes[18]

$$\mathcal{P}(k) = \mathcal{K}_\chi^{3d}\varepsilon^{-1/3}\chi k^{-5/3}, \tag{11.91}$$

where $\mathcal{K}_\chi^{3d}$ is a (putatively universal) constant. It is interesting that the $-5/3$ exponent appears in both the energy spectrum and the passive tracer spectrum. Using (11.86), this is the only spectral slope for which this occurs. Experiments show that this range does, at least approximately, exist with a value of $\mathcal{K}_\chi^{3d}$ of about 0.5–0.6 in three dimensions.

Inverse energy-cascade range in two-dimensional turbulence

Suppose that the energy injection occurs at a smaller scale than the tracer injection, so that there exists a range of wavenumbers over which energy is cascading to larger scales while tracer variance is simultaneously cascading to smaller scales. The tracer spectrum is then

$$\mathcal{P}(k) = \mathcal{K}_\chi^{2d}\varepsilon^{-1/3}\chi k^{-5/3}, \tag{11.92}$$

the same as (11.91), although ε is now the energy cascade rate to larger scales and the constant $\mathcal{K}_\chi^{2d}$ does not necessarily equal $\mathcal{K}_\chi^{3d}$.

Enstrophy inertial range in two-dimensional turbulence

In the forward enstrophy inertial range the eddy time scale is $\tau_k = \eta^{-1/3}$ (assuming of course that the classical phenomenology holds). Directly from (11.85) the corresponding tracer spectrum is then

$$\mathcal{P}(k) = \mathcal{K}_\chi^{2d*}\eta^{-1/3}\chi k^{-1}. \tag{11.93}$$

The passive tracer spectrum now has the same slope as the spectrum of vorticity variance (i.e., the enstrophy spectrum), which is perhaps reassuring since the tracer and vorticity obey similar equations in two dimensions.

The viscous-advective range of large Prandtl number flow

If the Prandtl number $\sigma = \nu/\kappa \gg 1$ (and in seawater $\sigma \approx 7$) then there may exist a range of wavenumbers in which viscosity is important, but not tracer diffusion. The energy spectrum is then very steep, and (11.90) will apply. The straining then comes from wavenumbers near the viscous scale, so that for three dimensional flow the appropriate k_0 to use in (11.90) is the viscous wavenumber, and $k_0 = k_\nu = (\varepsilon/\nu^3)^{1/4}$. The dynamical time scale at this wavenumber is given by

$$\tau_{k_\nu} = \left(\frac{\nu}{\varepsilon}\right)^{1/2}, \tag{11.94}$$

and using this and (11.90) the tracer spectrum in this viscous-advective range becomes

$$\mathcal{P}(k) = \mathcal{K}_B'\left(\frac{\nu}{\varepsilon}\right)^{1/2}\chi k^{-1}. \tag{11.95}$$

This spectral form applies for $k_\nu < k < k_\kappa$, where k_κ is the wavenumber at which diffusion becomes important, found by equating the eddy turnover time given by (11.94) with the diffusive time scale $(\kappa k^2)^{-1}$. This gives

$$k_\kappa = \left(\frac{\varepsilon}{\nu\kappa^2}\right)^{1/4}, \tag{11.96}$$

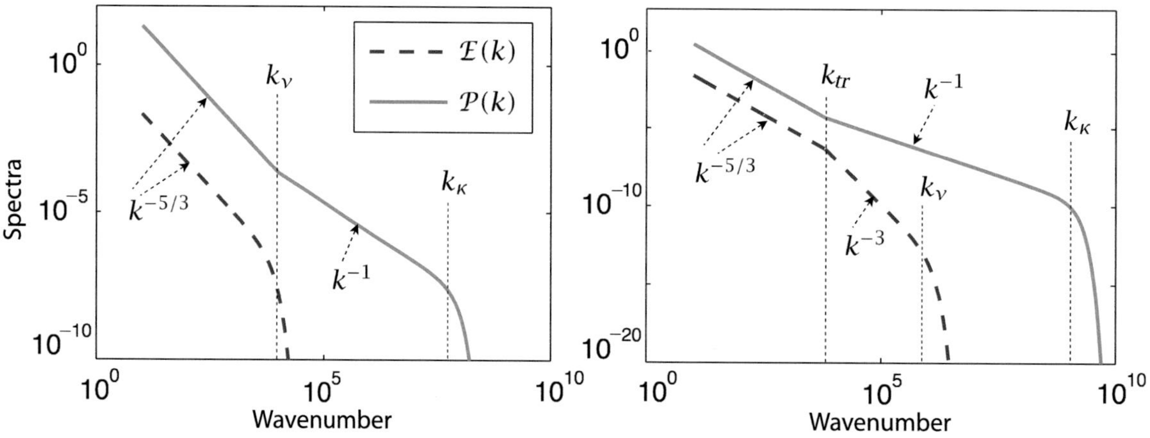

Fig. 11.12 The energy spectra, $\mathcal{E}(k)$ and passive tracer spectra $\mathcal{P}(k)$ in large Prandtl number three-dimensional (left) and two-dimensional (right) turbulence. In three dimensions $\mathcal{P}(k)$ is given by (11.91) for $k < k_\nu$ and by (11.100) for $k > k_\nu$. In two dimensions, if k_{tr} marks the transition between a $k^{-5/3}$ inverse energy cascade and a k^{-1} forward enstrophy cascade, then $\mathcal{P}(k)$ is given by (11.92) for $k < k_{tr}$ and by (11.101) for $k > k_{tr}$. In both two and three dimensions the tracer spectra fall off rapidly for $k > k_\kappa$.

and k_κ is known as the Batchelor wavenumber (and its inverse is the Batchelor scale). Beyond k_κ, the diffusive flux is not constant and the tracer spectrum can be expected to decay as wavenumber increases. A heuristic way to calculate the spectrum in the diffusive range is to first note that the flux of the tracer is not constant but diminishes according to

$$\frac{\mathrm{d}\chi'(k)}{\mathrm{d}k} = -2\kappa k^2 \mathcal{P}(k), \tag{11.97}$$

where χ' is the wavenumber-dependent rate of tracer transfer. Let us still assume that χ' and $\mathcal{P}(k)$ are related by an equation of the form (11.85), where now τ_k is a constant, given by (11.94). Thus,

$$\mathcal{K}_B \chi' = \frac{\mathcal{P}(k)k}{\tau_{k_\kappa}} = \frac{\mathcal{P}(k)k}{(\nu/\varepsilon)^{1/2}}, \tag{11.98}$$

where $\mathcal{K}_B$ is a constant. Using (11.97) and (11.98) we obtain

$$\frac{\mathrm{d}\chi'}{\mathrm{d}k} = -2\mathcal{K}_B \kappa k \left(\frac{\nu}{\varepsilon}\right)^{1/2} \chi'. \tag{11.99}$$

Solving this, using $\chi' = \chi$ (where χ is a constant) for small k, gives

$$\mathcal{P}(k) = \mathcal{K}_B \left(\frac{\nu}{\varepsilon}\right)^{1/2} \chi k^{-1} \exp\left[-\mathcal{K}_B (k/k_\kappa)^2\right]. \tag{11.100}$$

This reduces to (11.95) if $k \ll k_\kappa$, and is known as the Batchelor spectrum.[19] Its high-wavenumber part $k \gg k_\kappa$ is known as the viscous-diffusive subrange. The spectrum, and its two-dimensional analogue, are illustrated in Fig. 11.12.

In two dimensions the viscous-advective range occurs for wavenumbers greater than $k_\nu = (\eta/\nu^3)^{1/6}$. The appropriate time scale within this subrange is $\eta^{-1/3}$, which gives a spectrum with precisely the same form as (11.93). At sufficiently high wavenumbers tracer diffusion becomes important, with the diffusive scale now given by equating the eddy turnover time $\eta^{-1/3}$ with the viscous time scale $(\kappa k^2)^{-1}$. This gives the diffusive wavenumber, analogous to (11.96), of $k_\kappa = (\eta/\kappa^3)^{1/6}$. Using (11.99) and the procedure above we then obtain an expression for the spectrum in the region $k > k_\nu$, that is a two-dimensional analogue of (11.100), namely

$$\mathcal{P}(k) = \mathcal{K}'_B \eta^{-1/3} \chi k^{-1} \exp\left[-\mathcal{K}'_B (k/k_\kappa)^2\right]. \tag{11.101}$$

For $k \ll k_\kappa$ this reduces to (11.93), possibly with a different value of the Kolmogorov-like constant.

† *The inertial-diffusive range of small Prandtl number flow*

For small Prandtl number ($\nu/\kappa \ll 1$) the energy inertial range may coexist with a range over which tracer variance is being dissipated, giving us the so-called inertial-diffusive range. The tracer will begin to be dissipated at a wavenumber obtained by equating a dynamical eddy turnover time with a diffusive time, and this gives a diffusive wavenumber

$$k'_\kappa = \begin{cases} (\varepsilon/\kappa^3)^{1/4} & \text{in three dimensions,} \\ (\eta/\kappa^3)^{1/6} & \text{in two dimensions.} \end{cases} \tag{11.102}$$

Beyond the diffusive wavenumber the flux of the tracer is no longer constant but diminishes according to (11.97).

Given a non-constant flux and an eddy-turnover time that varies with wavenumber there is no self-evidently correct way to proceed. One way is to assume that χ and $\mathcal{P}(k)$ are related by $\mathcal{K}''_\chi \chi = \mathcal{P}(k)k/\tau_k$ (as in (11.98), but with a potentially different proportionality constant) and with τ_k given by (11.86); that is, $\tau_k = \varepsilon^{-1/3}k^{-2/3}$ in three-dimensional turbulence. Using this in (11.97) leads to

$$\mathcal{P}(k) = \mathcal{K}''_\chi \chi \varepsilon^{-1/3} k^{-5/3} \exp[-(\mathcal{K}''_\chi 3/2)(k/k'_\kappa)^{4/3}], \tag{11.103}$$

where χ is the tracer flux at the beginning of the tracer dissipation range. (A similar expression emerges in two-dimensional turbulence.) However, given such a steep spectrum an argument based on spectral locality is likely to be suspect. Another argument posits a particular relationship between the tracer spectrum and energy spectrum in the inertial-diffusive range, and this leads to

$$\mathcal{P}(k) = \frac{\mathcal{K}''_B}{3} \chi_0 \varepsilon^{2/3} \kappa^{-3} k^{-17/3} = \mathcal{K}''_B \chi_0 \varepsilon^{-1/3} k^{-5/3} g(k/k_\kappa), \tag{11.104}$$

where $g(\alpha) = \alpha^{-4}/3$ and $\mathcal{K}''_B$ is a constant.[20]

Notes

1 Horace Lamb has been quoted as saying that when he died and went to Heaven he hoped for enlightenment on two things, quantum electrodynamics and turbulence, although he was only optimistic about the former. Werner Heisenberg expressed a similar sentiment. But Heaven may be the wrong place to seek such enlightenment, for turbulence is the invention of the Devil, put on Earth to torment us.

2 The algebra of the three-dimensional case is more complicated because of the pressure term and because the momentum equation is a vector equation. Nevertheless, in incompressible flow we can take the divergence of the momentum equation to obtain an elliptic equation for pressure of the form $\nabla^2 p = Q(\boldsymbol{v})$, where the right-hand side is a quadratic function of velocity and its derivatives, Fourier transform this and then proceed much as in the two-dimensional case.

3 A. N. Kolmogorov (1903–1987) was a Russian mathematician and theoretical physicist, who made seminal contributions to turbulence (with famous papers in 1941 and in 1962), to probability and statistics, and to classical mechanics (e.g., the Kolmogorov–Arnold–Moser theorem). Yaglom (1994) provides more details on both the man and his scientific contributions.

4 The process has been encapsulated in the following ditty by Lewis Fry (L.F.) Richardson (1881–1953), his own summary of Richardson (1920):

Big whorls have little whorls, that feed on their velocity;
And little whorls have lesser whorls, and so on to viscosity.

The verse follows a well-known one by the mathematician Augustus De Morgan (in *A Budget of Paradoxes* 1872), *'Great fleas have little fleas upon their backs to bite 'em...'*, which in turn is a parody on a poem by Jonathan Swift. Richardson himself was a British scientist best known as the person who (following earlier work by Cleveland Abbe in 1901 and Vilhelm Bjerknes in 1904) envisioned weather forecasting in its current form — that is, numerical weather prediction. However, as described in his 1922 book, instead of an electronic computer performing the calculations, he imagined, perhaps fancifully, a hall full of people performing calculations in unison all directed by a conductor at the front. His first numerical forecast, calculated by hand, was wildly inaccurate because he failed to initialize his atmosphere properly and because his timestep was too long and did not satisfy the CFL condition, and unrealistic gravity waves dominated the solution. However, it was a prescient and important effort. He also worked on turbulence, and seems to have envisioned the turbulent cascade prior to Kolmogorov (to wit the verse above), and the 'Richardson number', a measure of fluid stratification, is named after him. He also made contributions to the theory of war and was known as a pacifist — he was a conscientious objector and drove ambulances in the first World War, and resigned from the UK Meteorological Office because it became part of the Air Ministry.

5 Kolmogorov (1941) obtained the result in a slightly different way, using distances in real space rather than wavenumber and deriving the equivalent result for the longitudinal structure function, $D(r) \equiv \langle [\boldsymbol{u}_l(\boldsymbol{x}+\boldsymbol{r}) - \boldsymbol{u}_l(\boldsymbol{x})]^2 \rangle \sim r^{2/3}$. It was Obukhov (1941) who gave an argument in spectral space and first wrote down that $E(k) \sim k^{-5/3}$. Kolmogorov's argument is regarded as more general and hence the 5/3 spectrum is usually named for him, but sometimes it is called the 'Kolmogorov–Obukhov' 5/3 spectrum.

6 Results kindly provided by R. Zhao.

7 The first observations confirming the -5/3 predictions were from a tidal channel (Grant *et al.* 1962). These results were initially presented at a turbulence conference in Marseille in 1961, ironically at the same time as Kolmogorov presented a modification of his original theory that incorporated a local mean dissipation rate, to try to take intermittency into account, recognizing that his first theory was incomplete (Kolmogorov 1962). It is said to have been L. D. Landau who pointed out the consequences of intermittency to Kolmogorov, soon after the K41 theory first appeared.

8 Two early papers on two-dimensional turbulence are those of Lee (1951) and Fjørtoft (1953), the former noting the incompatibility of the material conservation of vorticity with Kolmogorov's energy inertial range, and the latter recognizing the two-dimensional nature of large-scale atmospheric motion. Batchelor (1953b) noted the tendency of energy to concentrate in small wavenumbers as a consequence of energy and enstrophy conservation. The theory was developed by Kraichnan (1967), who predicted the spectral slopes of the two-dimensional cascades, Leith (1968) and Batchelor (1969). Lilly (1969) performed some early numerical integrations. For reviews see Kraichnan & Montgomery (1980), Danilov & Gurarie (2001) and Boffetta & Ecke (2012).

9 This similarity argument is due to Batchelor (1969), and its validity was explored by Bartello & Warn (1996) using numerical simulations of decaying two-dimensional turbulence. They found that the similarity hypothesis is not quantitatively accurate, and in particular that higher-order moments of the vorticity do not obey the predictions of the theory. This failure may be ascribed to the fact that in two-dimensional flow vorticity is conserved on parcels, and in the presence of coherent vortices this is an effective constraint that is not included in the theory. In flow with a finite deformation radius coherent vortices are found to be less important and an analogous similarity hypothesis appears to work better (Iwayama *et al.* 2002).

10 Arbic *et al.* (2007), Colin de Verdière (1980). See also Scott (2001).

11 For example, Peltier & Stuhne (2002) and Smith & Waleffe (1999).

12 Adapted from Maltrud & Vallis (1991).

13 For numerical simulations illustrating these and other properties of two-dimensional turbulence see, among others, McWilliams (1984), Maltrud & Vallis (1991), Oetzel & Vallis (1997), Lindborg & Alvelius (2000) and Smith *et al.* (2002). Also look at Jupiter through a telescope! A statistical-mechanical argument that in two-dimensional turbulence there is tendency for a small number of vortices to form was given by Onsager (1949).

14 The Lorenz equations were written down by Lorenz (1963), based on some earlier work of Saltzman (1962), and inspired a veritable industry of study. The field of chaos, or more generally nonlinear dynamics, has grown enormously since then, prompted also by work in mathematics occurring at about the same time, and its development is sometimes regarded as one of the true revolutions of science in the twentieth century. Aubin & Dahan Dalmedico (2002) write a history. The correspondence of the Lorenz equations to a real fluid system is tenuous, but the importance of the properties they demonstrate transcends this; we regard the equations simply as an example of a chaotic system with some fluid relevance. The equations and variations about them have reappeared in studies of, among other things, lasers, dynamos, chemical reactions, mechanical waterwheels and El Niño.

15 That a turbulent flow is, *inter alia*, chaotic and unpredictable follows from the work of Lorenz (1963), Ruelle & Takens (1971) and others who showed that fluid turbulence was generically a consequence of a small number of bifurcations as some controlling parameter (such as the Reynolds number) is changed. Prior to this, turbulence was sometimes thought, following Landau (1944), to be a large collection of periodic motions with incommensurate frequencies that would have complex and non-repeating but presumably predictable behaviour. Notwithstanding the Landau picture, it seems to have been known, well before the development of nonlinear dynamical systems theory in the 1960s and 1970s, that the weather was inherently unpredictable. Poincaré was perhaps the first to properly understand this at the turn of the twentieth century, although in the review of a book by the thermodynamicist P. Duhem, W. S. Franklin wrote in 1898 that 'An infinitesimal cause produces a finite effect. Long range detailed weather prediction is therefore impossible... the flight of a grasshopper in Montana may turn a storm aside from Philadelphia to New York!'

Weather forecasters themselves long seem to have intuited that the atmosphere was intrinsically, and not just practically, difficult to forecast. In the 1941 novel *Storm* by G. R. Stewart (1895–1980, a professor of English at UC Berkeley) we find a forecaster recalling his old professor's saying that 'A Chinaman sneezing in Shen-si may set men to shoveling snow in New York City'. (*Storm* is also notable because it used female names for intense storms, a practice that became common among forecasters in World War II and that was used by the US Weather Service from 1953–1978, after which gender equity obtained.) In a more academic setting, the predictability problem is mentioned in the book by Godske *et al.* (1957), much of which was written in the 1930s and 1940s, and Thompson (1957) and Novikov (1959) studied the unpredictability of atmospheric flows from the perspective of turbulence, evidently unaware of, or at least uninfluenced by, either Poincaré or Landau. Phillip Thompson himself was in the Joint Numerical Weather Prediction Unit of the US government in the 1950s, whose task was to numerically produce weather forecasts, and this practical experience undoubtedly confronted and guided his theoretical thinking. These various strands came together and were clarified by the dynamical systems viewpoint coupled with the view of the atmosphere as a geostrophically turbulent fluid, and this led to the viewpoint we describe here, and to estimates of the limit to predictability of the atmospheric weather of about two weeks, although at any given time the predictability may certainly be shorter or longer than that.

16 The more technical phrase 'sensitive dependence on initial conditions' is a paraphrase of one of Poincaré, but the catchier one 'butterfly effect' is more recent. It seems to have first appeared in a meteorological context in Smagorinsky (1969), where we find 'Would the flutter of a butterfly's wings ultimately amplify to the point where the numerical simulation departed from reality...? If not the flutter of the butterfly's wings, the disturbance might be the result of instrumental errors in the initial conditions...' The phrase became more well known following a lecture by Lorenz to the American Association for the Advancement of Science (AAAS) in 1972 entitled 'Predictability: does the flap of a butterfly's wings in Brazil set off a tornado in Texas?' The shape of the Lorenz attractor in phase space also resembles a butterfly (Fig. 11.10), and the two (not unrelated) phenomena are sometimes conflated. The death of a butterfly also changes the course of history in a science fiction story by Ray Bradbury, *A Sound of Thunder* (1952). See also endnote 15 above and the brief history

by Hilborn (2004).

17 A related point is that the predictability of a turbulent system is not well characterized by its spectrum of Lyapunov exponents: in a turbulent system in three dimensions the largest Lyapunov exponent is likely to be associated with very small scales of motion, and the error growth associated with this effectively saturates at small scales.

18 First derived by Obukhov (1949). See also Corrsin (1951).

19 Batchelor (1959) also suggests that the constant $\mathcal{K}_B$ in (11.100) should have the value 2. There is some observational support for the k^{-1} viscous-advective range in the temperature spectra of the ocean, one of the first measurements being that of Grant *et al.* (1968). Aside from their intrinsic interest, the viscous and diffusive scales are used in microstructure theory and measurements that lead to estimates of the ocean's energy dissipation rate (Gregg 1998, Stips 2005).

20 Batchelor (1959). There is some numerical support for the $-17/3$ spectrum using a large-eddy simulation (LES) model (Chasnov 1991). See also O'Gorman & Pullin (2005).

O, wonder!
How many goodly creatures are there here!
How beauteous mankind is! O brave new world,
That has such people in't.
William Shakespeare, *The Tempest*, c. 1611.

CHAPTER 12

Geostrophic Turbulence and Baroclinic Eddies

GEOSTROPHIC TURBULENCE is turbulence in flows that are stably stratified and close to geostrophic balance. Like most aspects of turbulence the subject is difficult, and a real 'solution' — meaning an accurate, informative statement about average states, without computation of the detailed evolution — may be out of our reach, and may not exist. Ironically, it is sometimes easier to say something interesting about geostrophic turbulence than about incompressible isotropic two- or three-dimensional turbulence. In the latter class of problems there is nothing else to understand other than the problem of turbulence itself, and this is like climbing a smooth marble wall. On the other hand, rotation and stratification give one something else to grasp, a thorn though it may be, and it becomes possible to address geophysically interesting phenomena without having to solve the whole turbulence problem. Furthermore, in inhomogeneous geostrophic turbulence, asking questions about the *mean fields* is meaningful and useful, whereas this is uninteresting in isotropic turbulence.

The subject of geostrophic turbulence is a wondrous one, giving rise to phenomena that are both beautiful and important — the jets and eddies on Jupiter and the weather on Earth are but two examples. The subject is not restricted to quasi-geostrophic flow, and the large scale turbulence of Earth's ocean and atmosphere is sometimes simply called 'macro-turbulence'. Nevertheless, the quasi-geostrophic equations describe the main effects: they retain advective nonlinearity in the vorticity equation, and they capture the constraining effects of rotation and stratification that are so important in geophysical flows in a simple and direct way; for these reasons the quasi-geostrophic equations will be our main tool. Let us consider the effects of rotation first, then stratification.

12.1 DIFFERENTIAL ROTATION IN TWO-DIMENSIONAL TURBULENCE

In the limit of motion of a scale much shorter than the deformation radius, and with no topography, the quasi-geostrophic potential vorticity equation, (5.118), reduces to the two-dimensional equation,

$$\frac{\mathrm{D}q}{\mathrm{D}t} = 0, \tag{12.1}$$

where $q = \zeta + f$. This is the perhaps the simplest equation with which to study the effects of rotation on turbulence. Suppose first that the Coriolis parameter is constant, so that $f = f_0$. Then (12.1) becomes simply the two-dimensional vorticity equation

$$\frac{D\zeta}{Dt} = 0. \tag{12.2}$$

Thus constant rotation has *no* effect on purely two-dimensional motion. Flow that is already two-dimensional — flow on a soap film, for example — is unaffected by rotation. (In the ocean and atmosphere, or in a rotating tank, it is of course the effects of rotation that lead to the flow being quasi-two dimensional in the first instance.)

Suppose, though, that the Coriolis parameter is variable, as in $f = f_0 + \beta y$. Then we have

$$\frac{D}{Dt}(\zeta + \beta y) = 0 \qquad \text{or} \qquad \frac{D\zeta}{Dt} + \beta v = 0. \tag{12.3a,b}$$

If the asymptotically dominant term in these equations is the one involving β then we must have $v = 0$ at lowest order. That is, if β is very large, then the meridional flow v must be correspondingly small to ensure that the equation can balance, and this argument holds in the presence of forcing and dissipation. Any flow must then be predominantly *zonal*. A closely related argument uses the conservation of angular momentum as follows. A ring of fluid encircling the Earth at a velocity u has an angular momentum per unit mass $a \cos\vartheta(u + \Omega a \cos\vartheta)$, where ϑ is the latitude and a is the radius of the Earth. Moving this ring of air polewards while conserving its angular momentum requires that its zonal velocity and hence energy must increase, so unless there is a source for that energy the flow is constrained to remain zonal.

In the following sections we look at the mechanism of jet formation in a little more detail. We first give some fairly general scaling arguments, and then proceed in two complementary ways — first with an argument couched in spectral space using the language of turbulent cascades, and second with an argument in physical space. In Section 15.1 we come back to the problem from a third perspective, one particularly appropriate for Earth's atmosphere.

12.1.1 The Wave–Turbulence Cross-over

Scaling

Let us now consider how turbulent flow might interact with Rossby waves. We write (12.1) in full as

$$\frac{\partial\zeta}{\partial t} + \boldsymbol{u}\cdot\nabla\zeta + \beta v = 0. \tag{12.4}$$

If $\zeta \sim U/L$ and if $t \sim T$ then the respective terms in this equation scale as

$$\frac{U}{LT} \qquad \frac{U^2}{L^2} \qquad \beta U. \tag{12.5}$$

The way that time scales (i.e., advectively or with a Rossby wave frequency scaling) is determined by which of the other two terms dominates, and this in turn is scale dependent. For large scales the β-term is dominant, and at smaller scales the advective term is dominant. The cross-over scale, denoted L_R, is called the *Rhines scale* and is given by[1]

$$L_R \sim \left(\frac{U}{\beta}\right)^{1/2}. \tag{12.6}$$

The U in (12.6) should be interpreted as the root-mean-square velocity at the energy containing scales, not a mean or translational velocity. We refer to the specific scale $\sqrt{U/\beta}$ as the Rhines scale,

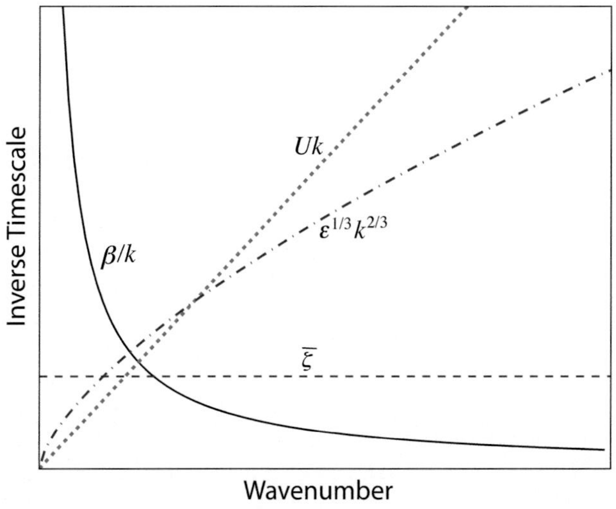

Fig. 12.1 Inverse timescales, or frequencies, in wavenumber space in beta-plane turbulence. The solid curve is the frequency of Rossby waves, proportional to β/k, and the other three curves are various estimates of the inverse turbulence time scale. These are the turbulent eddy transfer rate, proportional to $\varepsilon^{1/3}k^{2/3}$ in a $k^{-5/3}$ spectrum; the estimate Uk where U is a root-mean-squared velocity; and the mean vorticity, which is constant.

Where the Rossby wave frequency is larger (smaller) than the turbulent frequency, i.e., at large (small) scales, Rossby waves (turbulence) dominate the dynamics.

L_R, and more general scales involving a balance between nonlinearity and β as the β-scale, denoted L_β.

This is not a unique way to arrive at a β-scale, since we have chosen the length scale that connects vorticity to velocity to also be the β-scale, and it is not obvious that this should be so. If the two scales are different, the three terms in (12.4) scale as

$$\frac{\mathcal{Z}}{T} \qquad \frac{U\mathcal{Z}}{L} \qquad \beta U, \tag{12.7}$$

respectively, where $\mathcal{Z}$ is the scaling for vorticity; that is, $\zeta = \mathcal{O}(\mathcal{Z})$. Equating the second and third terms gives the scale

$$L_{\beta\mathcal{Z}} = \frac{\mathcal{Z}}{\beta}. \tag{12.8}$$

Although (12.6) and (12.8) differ in detail, both indicate that at some *large* scale Rossby waves are likely to dominate whereas at small scales advection and turbulence dominate.

Another heuristic way to derive (12.6) is by a direct consideration of time scales. Ignoring anisotropy, Rossby wave frequency is β/k and an inverse advective time scale is Uk, where k is the wavenumber. Equating these two gives an equation for the Rhines wavenumber

$$k_R \sim \left(\frac{\beta}{U}\right)^{1/2}. \tag{12.9}$$

This equation is the inverse of (12.6), but factors of order unity (e.g., π) cannot be revealed by simple scaling arguments such as these. We denote the Rhines wavenumber as k_R and a more generic cross-over wavenumber as k_β (and below we also introduce k_ε). The cross-over between waves and turbulence is reasonably sharp, as indicated in Fig. 12.1.

Turbulent phenomenology

We now examine wave–turbulence cross-over using the phenomenology of two-dimensional turbulence. We will suppose that the fluid is stirred at some well-defined scale k_f, producing an energy input ε. Then (assuming no energy is lost to smaller scales) energy cascades to large scales at that same rate. At some scale, the β-term in the vorticity equation will start to make its presence felt. By analogy with the procedure for finding the viscous dissipation scale in turbulence, we can find the scale at which linear Rossby waves dominate by equating the inverse of the turbulent eddy turnover time to the Rossby wave frequency. The eddy-turnover time is

$$\tau_k = \varepsilon^{-1/3} k^{-2/3}, \tag{12.10}$$

and equating this to the inverse Rossby wave frequency k/β gives estimates for the β-wavenumber and its inverse, the β-scale, namely

$$k_\varepsilon = \left(\frac{\beta^3}{\varepsilon}\right)^{1/5}, \qquad L_\varepsilon = \left(\frac{\varepsilon}{\beta^3}\right)^{1/5}. \tag{12.11a,b}$$

We denote these scales with a subscript ε because of the appearance of the energy cascade rate. In a real fluid these expressions are harder to evaluate than (12.9), since it is generally much easier to measure velocities than energy transfer rates, or even vorticity. On the other hand, (12.11) is more satisfactory from the point of view of turbulence theory because ε may be determined by processes largely independent of β, whereas the magnitude of the eddies at the energy containing scales is likely to be a function of β. We also remark that the scale given by (12.11b) is not necessarily the energy-containing scale, and may in principle differ considerably from the scale given by (12.9). This is because the inverse cascade is not necessarily *halted* at the scale (12.11b) — this is just the scale at which Rossby waves become important. Energy may continue to cascade to larger scales, albeit anisotropically as discussed below, and so the energy containing scale may be larger.

12.1.2 Generation of Zonal Flows and Jets

None of the effects discussed so far takes into account the anisotropy inherent in Rossby waves, and such anisotropy can give rise to predominantly zonal flows and jets. To understand this, let us first note that energy transfer will be relatively inefficient at those scales where linear Rossby waves dominate the dynamics. But the wave–turbulence boundary is not isotropic; the Rossby wave frequency is quite anisotropic, being given by

$$\omega = -\frac{\beta k^x}{k^{x2} + k^{y2}}. \tag{12.12}$$

If, albeit a little crudely, we suppose that the turbulent part of the flow remains isotropic, the wave–turbulence boundary is then given by equating the inverse of (12.10) with (12.12) and is a solution of

$$\varepsilon^{1/3} k^{2/3} = \frac{\beta k^x}{k^2}, \tag{12.13}$$

where k is the isotropic wavenumber. Solving this gives expressions for the x- and y-wavenumber components of the wave–turbulence boundary, namely

$$k_\varepsilon^x = \left(\frac{\beta^3}{\varepsilon}\right)^{1/5} \cos^{8/5}\theta, \qquad k_\varepsilon^y = \left(\frac{\beta^3}{\varepsilon}\right)^{1/5} \sin\theta \cos^{3/5}\theta, \tag{12.14}$$

where the polar coordinate is parameterized by the angle $\theta = \tan^{-1}(k^y/k^x)$. This odd-looking formula is illustrated in Fig. 12.2, and it defines the anisotropic wave-turbulence boundary. (Slight variations on this theme are produced by using different expressions for the turbulence time scale.)

What occurs physically? The region inside the dumbbell shapes in Fig. 12.2 is dominated by Rossby waves, where the natural frequency of the oscillation is *higher* than the turbulent frequency. If the flow is stirred at a wavenumber higher than this the energy will cascade to larger scales, but because of the frequency mismatch the turbulent flow will be unable to efficiently excite modes within the dumbbell. Nevertheless, there is still a natural tendency of the energy to seek the gravest

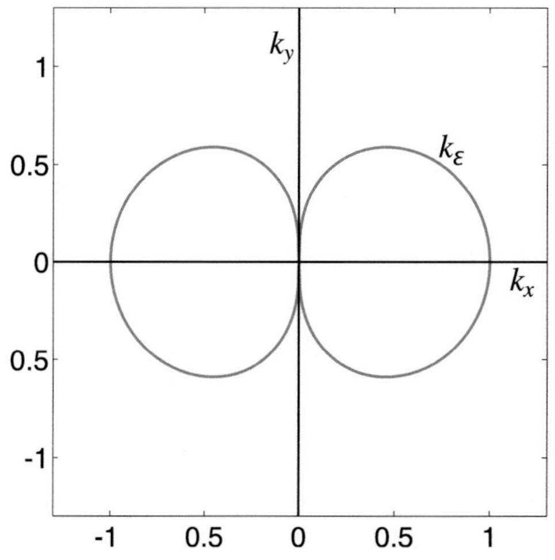

Fig. 12.2 The anisotropic wave–turbulence boundary k_ε, in wave-vector space calculated by equating the turbulent eddy transfer rate, proportional to $k^{2/3}$ in a $k^{-5/3}$ spectrum, to the Rossby wave frequency $\beta k^x/k^2$, as in (12.14).The wavenumbers are scaled such that $\beta^3/\varepsilon = 1$.

Within the dumbbell Rossby waves dominate and energy transfer is inhibited. The inverse cascade plus Rossby waves thus leads to a generation of zonal flow.

mode, and it will do this by cascading toward the $k^x = 0$ axis; that is, toward zonal flow. Thus, the combination of Rossby waves and turbulence will lead to the formation of zonal flow and, potentially, zonal jets.[3]

Figure 12.3 illustrates this mechanism; it shows the freely evolving (unforced, inviscid) energy spectrum in a simulation on a β–plane, with an initially isotropic spectrum. The energy implodes, cascading to larger scales but avoiding the region inside the dumbbell and piling up at $k^x = 0$. In physical space the flow organizes itself into zonally elongated structures and jets, in both freely-decaying and forced-dissipative simulations (Fig. 12.4 and Fig. 12.7).

† *Joint effect of beta and friction*

The β term does not remove energy from a fluid. Thus, if energy is being added to a fluid at some small scales, and the energy is cascading to larger scales, then the β-effect does not of itself halt the inverse cascade, it merely deflects the cascade such that the flow becomes more zonal. Suppose that the fluid obeys the barotropic vorticity equation,

$$\frac{\partial \zeta}{\partial t} + J(\psi, \zeta) + \beta \frac{\partial \psi}{\partial x} = F - r\zeta + \nu \nabla^2 \zeta, \tag{12.15}$$

where the viscosity, ν, is small and acts only to remove enstrophy, and not energy, at very small scales. The forcing, F, supplies energy at a rate ε and this is cascaded upscale and removed by the linear drag term $-r\zeta$, where the drag coefficient r is a constant. If the friction is sufficiently large then the energy is removed before it feels the effect of β at the scale (11.68), namely $L_r = (\varepsilon/r^3)^{1/2}$. However, if friction is small, such that $L_\varepsilon = (\varepsilon/\beta^3)^{1/5}$ is smaller than L_r, then the cascade feels the beta effect before it feels frictional effects, and L_r is then unlikely to be a relevant parameter.

If the forcing scale is small and an inverse cascade exists, the turbulence is characterized by three parameters, β, ε and r. There is only one way to make a nondimensional parameter from these and this is

$$\gamma = \beta^2 \varepsilon r^{-5}, \tag{12.16}$$

along with powers of γ. There is no unique way to make a length scale from these parameters, but various scales arise phenomenologically. One is L_ε, noted above, and another arises from energetic considerations. If we form an energy equation from (12.15) by multiplying by $-\psi$ and spatially integrating (and neglecting viscosity) we find the energy balance

$$\varepsilon = -\frac{1}{A}\int_A \psi F \, \mathrm{d}A = \frac{r}{A}\int_A (\nabla\psi)^2 \, \mathrm{d}A = 2r\overline{E}, \tag{12.17}$$

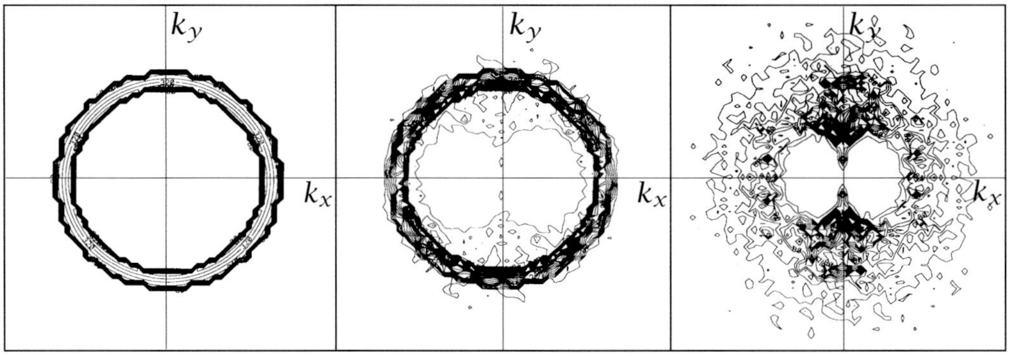

Fig. 12.3 Evolution of the energy spectrum in a freely evolving two-dimensional simulation on the β-plane. The panels show contours of energy in wavenumber (k_x, k_y) space at successive times. The initial spectrum is isotropic. The energy 'implodes', but its passage to large scales is impeded by the β-effect, and the second and third panels show the spectrum at later times, illustrating the dumbbell predicted by (12.14) and Fig. 12.2.[2]

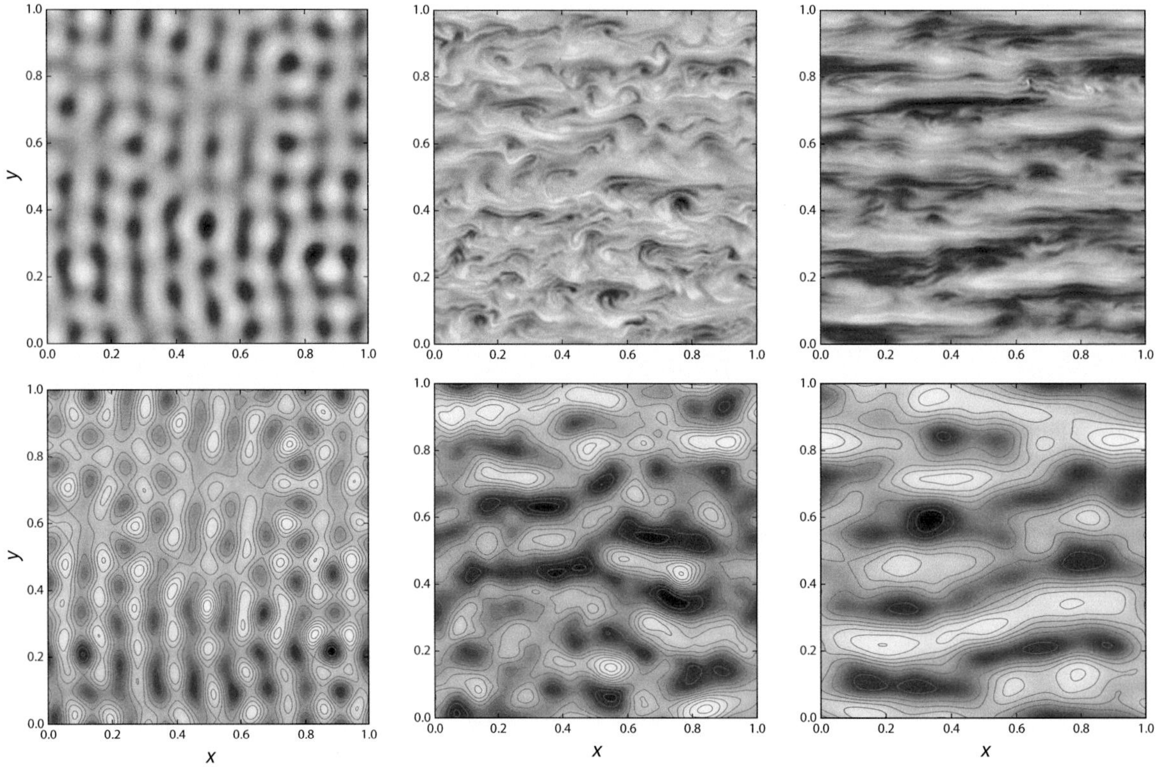

Fig. 12.4 Evolution of vorticity (top) and streamfunction (bottom) in a doubly periodic domain on the β-plane, obeying (12.4) with the addition of a weak viscous term on the right-hand side. The initial conditions are the same as for Fig. 11.8, and time proceeds from left to right. Compared to Fig. 11.8, vortex formation is inhibited and there is a tendency toward zonal flow.

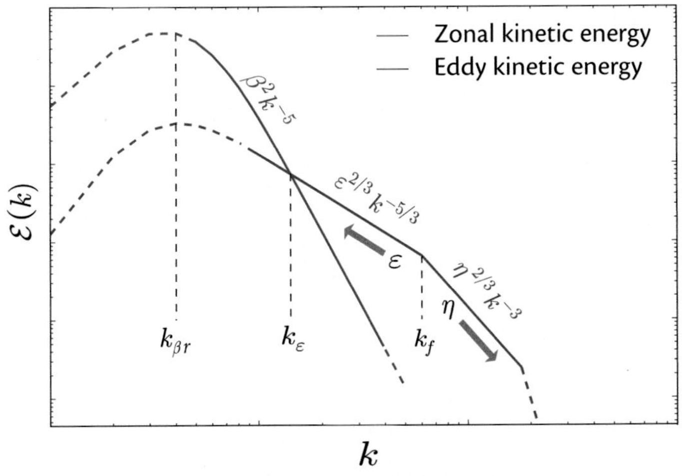

Fig. 12.5 Putative kinetic spectra, $\mathcal{E}(k)$, in beta-plane turbulence. Energy injected at wavenumber k_f cascades to larger scales until it reaches the turbulent beta scale, k_ε. The energy in the zonal modes then forms a zonostrophic k^{-5} spectrum which extends as far as the frictional Rhines scale, $k_{\beta r}$.

where $\overline{E}$ is the average energy of the fluid per unit mass. This expression holds whether or not β is non-zero. Using (12.6) with $U = \sqrt{2\overline{E}}$, where $\overline{E}$ is obtained from (12.17), we obtain the scale

$$L_{\beta r} = \left(\frac{\varepsilon}{r\beta^2}\right)^{1/4}, \tag{12.18}$$

which may be thought of as a prediction for the Rhines scale, L_R. The ratio of the scales $L_{\beta r}$ to L_ε is

$$\frac{L_{\beta r}}{L_\varepsilon} = \left(\frac{\beta^2\varepsilon}{r^5}\right)^{1/20} = \gamma^{1/20}. \tag{12.19}$$

What do the two scales, L_ε and $L_{\beta r}$, represent? The first is the scale at which the β-effect is first felt by the inverse cascade, and it is relevant for large-scale meridional mixing, and so for the meridional heat transport in the atmosphere. If friction is small (i.e., if $\gamma \gg 1$) the inverse energy transfer will continue past this scale, with the energy largely being at very small values of k_x, and (12.18) then characterizes the largest scale reached by the inverse cascade, and so the meridional scale of the jets. In this region one might hypothesize that the spectrum is determined by β and k alone, and if so it is given by

$$\mathcal{E}(k) = \mathcal{C}_z\beta^2 k^{-5}, \tag{12.20}$$

where $\mathcal{C}_z$ is a constant, and most of the energy is in the zonal flow. Only the extent (and not the amplitude) of this region is then determined by the forcing strength. Turbulence of this character is called *zonostrophic,* and is illustrated in Fig. 12.5. However, we might also expect the drag coefficient, r, to be a factor in at least part of spectral slope because, with such a steep slope, the timescale of the flow diminishes as the wavenumber falls, so there is no scale, L_{fric} say, such that friction acts only at scales larger than it.[4] But rather than pursuing more arguments in spectral space, let us turn to a physical space argument.

12.1.3 Potential Vorticity Staircases

Let us now consider a complimentary approach in physical space, with the essential idea being the following. Suppose that a turbulent flow exists on a background state that has a potential vorticity gradient, such as might be caused by the beta-effect. The eddies will seek to homogenize the potential vorticity but, in general, they will be unable to do so completely. Instead, the flow becomes partially homogenized, with regions of homogeneous potential vorticity separated by jumps, as

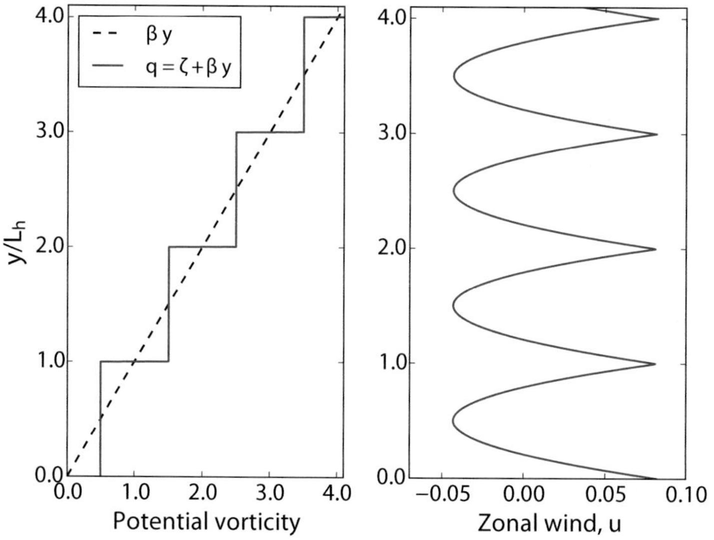

Fig. 12.6 An idealized potential vorticity staircase on a β-plane with $\beta = 1$. The left panel shows homogeneous regions of potential vorticity of meridional extent L_h separated by jumps, as calculated using (12.24). The right-hand panel shows the corresponding zonal flow, calculated using (12.28) with u_0 chosen to be such that the average zonal flow is zero.

illustrated in Fig. 12.6. Such a structure is called a *potential vorticity staircase,* and the staircase implies the existence of zonal jets.[5]

The homogenization of a scalar is discussed further in Section 13.5, but it can be imagined that, in a turbulent flow, the gradients of a scalar that is both advected and diffused will become smeared out as much as possible: diffusion will dissipate any extrema and advection cannot recreate them. Potential vorticity is a scalar, so we might expect it too to become homogenized, but it is not passive and therefore the process of homogenization will affect the flow itself, preventing complete homogenization. One reason for that may be the boundary conditions, another that the process of homogenization requires more energy than the flow contains. To see this, consider a freely-evolving flow on a beta-plane obeying the barotropic vorticity equation

$$\frac{\partial \zeta}{\partial t} + J(\psi, \zeta + \beta y) = \nu \nabla^2 \zeta, \tag{12.21}$$

where ν is sufficiently small that energy is well conserved over the timescales of interest. If potential vorticity, $q = \zeta + \beta y$, is to be homogenized over some meridional scale L then, in the homogenized region, $\zeta \approx -\beta y$. If the flow is predominantly zonal then this gives the estimate $U \sim \beta y^2$ and the energy in the region is, very approximately,

$$\frac{1}{2} \int u^2 \,\mathrm{d}x\,\mathrm{d}y \approx \int (\beta y^2)^2 \,\mathrm{d}x\,\mathrm{d}y, \tag{12.22}$$

giving the estimate $U^2 L \sim \beta^2 L^5$. If we suppose that the initial root-mean square velocity is also U, and that the flow becomes predominantly zonal, then solving the above estimate for L suggests that the potential vorticity will become homogenized over the scale

$$L_h \sim \left(\frac{U}{\beta}\right)^{1/2} \tag{12.23}$$

This is the same as the Rhines scale given in (12.6), as it has to be by dimensional analysis, although now it is supposed that much of the energy lies in the zonal flow. Although the scale is the same, this way of looking at the problem adds something to the physical picture — *an asymmetry between eastward and westward flow,* as we now discuss.

In the idealized staircase of Fig. 12.6 potential vorticity is piecewise continuous and given by

$$q = \zeta + \beta y = q_0, \qquad 0 < y < L_h, \tag{12.24a}$$

$$q = \zeta + \beta y = q_0 + \beta L_h \equiv q_1, \qquad L_h < y < 2L_h, \quad \text{and so on.} \tag{12.24b}$$

In any one of the homogenized regions the flow is given by solving

$$\frac{\partial u}{\partial y} = \beta y - q_n \quad \text{giving} \quad u = \frac{1}{2}\beta y^2 - q_n y + \text{constant}, \tag{12.25}$$

where $q_n = \beta L_h/2 + n\beta L_h$. The constant may be determined by requiring continuity of u across the regions, and we then obtain

$$u = \frac{1}{2}\beta(y - L_h/2)^2 + u_0, \qquad 0 < y < L_h, \tag{12.26}$$

$$u = \frac{1}{2}\beta(y - 3L_h/2)^2 + u_0, \qquad L_h < y < 2L_h, \tag{12.27}$$

or in general,

$$u = \frac{1}{2}\beta\left(y - (n - 1/2)L_h\right)^2 + u_0, \qquad (n-1)L_h < y < nL_h, \qquad n = 1, 2, 3\ldots, \tag{12.28}$$

where u_0 is a constant. Relative to u_0, the flow is weakly westward in the homogenized regions, whereas in the transition region the flow has a sharp eastward peak, as shown in Fig. 12.6. The spectrum of such a flow is k^{-4}, similar to the k^{-5} spectrum of zonostrophic flow.

Finally, let us make a few remarks. First, the potential vorticity gradient does not change sign, so the flow is (just) barotropically stable — one is supposing that the eddies arise by some external mechanism (e.g., some external stirring, or baroclinic instability) and they are creating the jet, not vice versa. Second, the staircase structure implies that potential vorticity is (obviously) well mixed in the regions of constant potential vorticity, but that there is no mixing across the jumps. The jumps are known as *mixing barriers* and, once formed, are found to be very persistent in numerical simulations. The edge of the stratospheric polar vortex, and the boundaries of ocean gyres, may also be examples. A fluid parcel is energetically incapable of jumping across the barrier, and the barrier's existence makes clear that parameterizing turbulent mixing as a downgradient flux of a conserved quantity cannot always work, at least with a constant diffusivity, since here the gradient is strong but the mixing is weak.

Numerical simulations do reproduce the staircase in some, but not all, circumstances (Fig. 12.7). The simulations suggest that there needs to be a good separation between the turbulent beta scale and the Rhines scales, in particular with $L_R/L_\epsilon \gg 1$, and that this is best achieved with weak friction and/or forcing that is at large scales, perhaps even larger than L_ϵ. However, the subject continues to evolve and the reader should consult the primary literature to learn more.

12.1.4 † Zonal Jets — Other Mechanisms and Final Remarks

In the above we have described how zonal jets may be created from two distinct points of view — one in spectral space and the other in physical space. We described the spectral approach in terms of a cascade — a local energy transfer from one wavenumber to another. However, in many circumstances the energy transfer from eddies to zonal flow may be much more direct, with no spectral intermediaries. That is, a zonally asymmetric flow may be unstable and, even in the absence of eddy-eddy interactions, produce a zonal jet. The interaction is then deemed 'quasi-linear' and the triad interaction is highly nonlocal, as in the right-hand panel of Fig. 11.1. There are still other ways in which we may describe jet formation, some described in section 15.1. It is the parameters of the problem at hand that will determine which mechanism is dominant in a particular situation, although a full understanding of their similarities and differences remains tenebrous.[7]

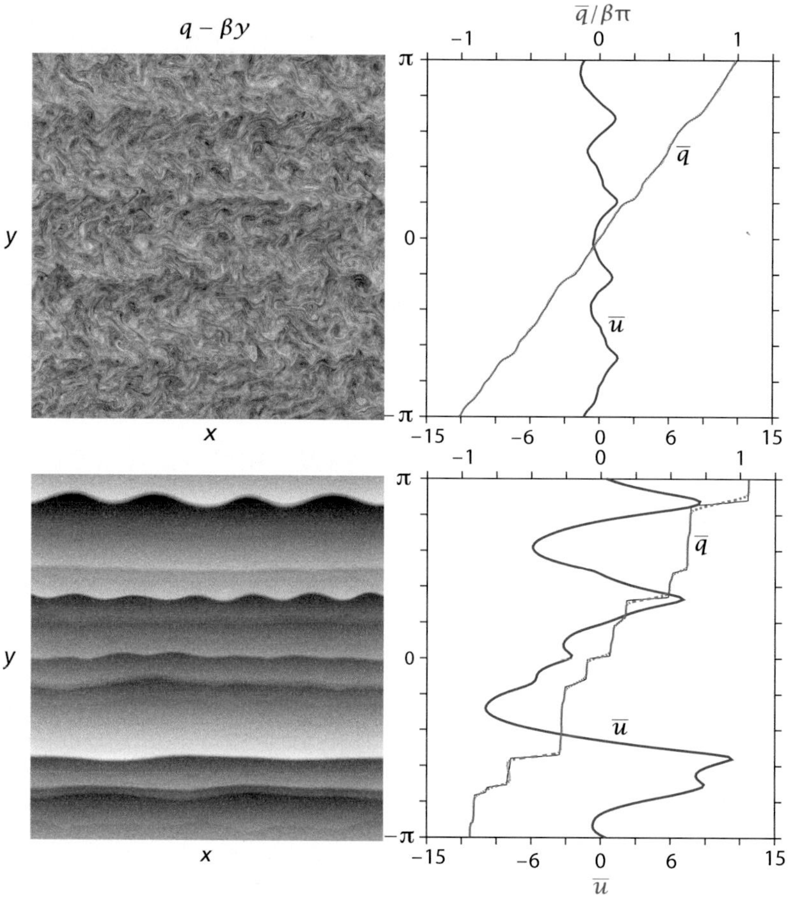

Fig. 12.7 Simulations of two-dimensional turbulence on a beta-plane.[6]

The left panels show snapshots of the relative vorticity ζ, and right panels show zonally-averaged potential vorticity and zonal wind. The top panels show an integration with $L_R/L_\varepsilon \approx 3$ and the bottom panels have $L_R/L_\varepsilon \approx 11$.

12.2 STRATIFIED GEOSTROPHIC TURBULENCE

12.2.1 An Analogue to Two-dimensional Flow

Now let us consider stratified effects in a simple setting, using the quasi-geostrophic equations with constant Coriolis parameter and constant stratification.[8] The (dimensional) unforced and inviscid governing equation may then be written as

$$\frac{\mathrm{D}q}{\mathrm{D}t} = 0, \qquad q = \nabla^2\psi + Pr^2\frac{\partial^2\psi}{\partial z^2}, \tag{12.29a}$$

where $Pr = f_0/N$ is the *Prandtl ratio* (and Pr/H is the inverse of the deformation radius) and $\mathrm{D}/\mathrm{D}t = \partial/\partial t + \boldsymbol{u}\cdot\nabla$ is the two-dimensional material derivative. The vertical boundary conditions are

$$\frac{\mathrm{D}}{\mathrm{D}t}\left(\frac{\partial\psi}{\partial z}\right) = 0, \qquad \text{at } z = 0, H. \tag{12.29b}$$

These equations are analogous to the equations of motion for purely two-dimensional flow. In particular, with periodic lateral boundary conditions, or conditions of no-normal flow, there are two quadratic invariants of the motion, the energy and the enstrophy, which are obtained by multiplying (12.29a) by $-\psi$ and q and integrating over the domain, as in Chapter 5. The conserved energy is

$$\frac{\mathrm{d}\hat{E}}{\mathrm{d}t} = 0, \qquad \hat{E} = \frac{1}{2}\int_V \left[(\nabla\psi)^2 + Pr^2\left(\frac{\partial\psi}{\partial z}\right)^2\right]\mathrm{d}V, \tag{12.30}$$

where the integral is over a *three-dimensional* domain. The enstrophy is conserved at each vertical level, and of course the volume integral is also conserved, namely

$$\frac{\mathrm{d}\widehat{Z}}{\mathrm{d}t} = 0, \qquad \widehat{Z} = \frac{1}{2}\int_V q^2 \,\mathrm{d}V = \frac{1}{2}\int_V \left[\nabla^2\psi + Pr^2\left(\frac{\partial^2\psi}{\partial z^2}\right)\right]^2 \mathrm{d}V. \tag{12.31}$$

The analogy with two-dimensional flow is even more transparent if we further rescale the vertical coordinate by $1/Pr$, and so let $z' = z/Pr$. Then the energy and enstrophy invariants are:

$$\widehat{E} = \int (\nabla_3\psi)^2 \,\mathrm{d}V, \qquad \widehat{Z} = \int q^2 \,\mathrm{d}V = \int (\nabla_3^2\psi)^2 \,\mathrm{d}V, \tag{12.32}$$

where $\nabla_3 = \mathbf{i}\partial/\partial x + \mathbf{j}\partial/\partial y + \mathbf{k}\partial/\partial z'$. The invariants then have almost the same form as the two-dimensional invariants, but with a three-dimensional Laplacian operator instead of a two-dimensional one.

Given these invariants, we should expect that any dynamical behaviour that occurs in the two-dimensional equations *that depends solely on the energy/enstrophy constraints* should have an analogue in quasi-geostrophic flow. In particular, the transfer of energy to large-scales and enstrophy to small scales will also occur in quasi-geostrophic flow with, in so far as these transfers are effected by a local cascade, corresponding spectra of $k^{-5/3}$ and k^{-3}. However, in the quasi-geostrophic case, it is the *three-dimensional* wavenumber that is relevant, with the vertical component scaled by the Prandtl ratio. As a consequence, the energy cascade to larger horizontal scales is generally accompanied by a cascade to larger vertical scales — a *barotropization* of the flow. Still, the analogy between two-dimensional and quasi-geostrophic cascades should not be taken too far, because in the latter the potential vorticity is advected only by the horizontal flow. Thus, the dynamics of quasi-geostrophic turbulence will *not* in general be isotropic in three-dimensional wavenumber. To examine these dynamics more fully we first turn to a simpler model, that of two-layer flow.

12.2.2 Two-layer Geostrophic Turbulence

Let us consider flow in two layers of equal depth, governed by the quasi-geostrophic equations with (for now) $\beta = 0$, namely

$$\frac{\partial q_i}{\partial t} + J(\psi_i, q_i) = 0, \qquad i = 1, 2, \tag{12.33}$$

where

$$q_1 = \nabla^2\psi_1 + \frac{1}{2}k_d^2(\psi_2 - \psi_1), \qquad q_2 = \nabla^2\psi_2 + \frac{1}{2}k_d^2(\psi_1 - \psi_2), \tag{12.34a}$$

$$J(a,b) = \frac{\partial a}{\partial x}\frac{\partial b}{\partial y} - \frac{\partial a}{\partial y}\frac{\partial b}{\partial x}, \qquad \frac{1}{2}k_d^2 = \frac{2f_0^2}{g'H} \equiv \frac{4f_0^2}{N^2H^2}. \tag{12.34b}$$

The wavenumber k_d is inversely proportional to the baroclinic radius of deformation, and the two equivalent expressions given are appropriate in a layered model and a level model, respectively. The equations conserve the total energy,

$$\frac{\mathrm{d}\widehat{E}}{\mathrm{d}t} = 0, \qquad \widehat{E} = \frac{1}{2}\int_A \left[(\nabla\psi_1)^2 + (\nabla\psi_2)^2 + \frac{1}{2}k_d^2(\psi_1 - \psi_2)^2\right] \mathrm{d}A, \tag{12.35}$$

and the enstrophy in each layer

$$\frac{\mathrm{d}\widehat{Z}_1}{\mathrm{d}t} = 0, \qquad \widehat{Z}_1 = \int_A q_1^2 \,\mathrm{d}A, \tag{12.36a}$$

$$\frac{d\widehat{Z}_2}{dt} = 0, \qquad \widehat{Z}_2 = \int_A q_2^2 \, dA. \tag{12.36b}$$

The first two terms in the energy expression, (12.35), represent the kinetic energy, and the last term is the available potential energy, proportional to the variance of temperature.

Baroclinic and barotropic decomposition

Define the barotropic and baroclinic streamfunctions by

$$\psi \equiv \frac{1}{2}(\psi_1 + \psi_2), \qquad \tau \equiv \frac{1}{2}(\psi_1 - \psi_2). \tag{12.37}$$

Then the potential vorticities for each layer may be written as

$$q_1 = \nabla^2\psi + (\nabla^2 - k_d^2)\tau, \qquad q_2 = \nabla^2\psi - (\nabla^2 - k_d^2)\tau, \tag{12.38a,b}$$

and the equations of motion may be rewritten as evolution equations for ψ and τ as follows:

$$\frac{\partial}{\partial t}\nabla^2\psi + J(\psi, \nabla^2\psi) + J(\tau, (\nabla^2 - k_d^2)\tau) = 0, \tag{12.39a}$$

$$\frac{\partial}{\partial t}(\nabla^2 - k_d^2)\tau + J(\tau, \nabla^2\psi) + J(\psi, (\nabla^2 - k_d^2)\tau) = 0. \tag{12.39b}$$

We note the following about the nonlinear interactions in (12.39):

(i) ψ and τ are like vertical modes. That is, ψ is the barotropic mode with a 'vertical wavenumber', k^z, of zero, and τ is a baroclinic mode with an effective vertical wavenumber of one.

(ii) Just as purely two-dimensional turbulence can be considered to be a plethora of interacting triads, whose two-dimensional vector wavenumbers sum to zero, it is clear from (12.39b) that geostrophic turbulence may be considered to be similarly composed of a sum of interacting triads, although now there are two distinct kinds, namely

$$(i)\quad (\psi, \psi) \to \psi, \qquad (ii)\quad (\tau, \tau) \to \psi \quad \text{or} \quad (\psi, \tau) \to \tau. \tag{12.40}$$

The first kind is a *barotropic triad,* for it involves only the barotropic mode. The other two are examples of a *baroclinic triad.* If a barotropic mode has a vertical wavenumber of zero, and a baroclinic mode has a vertical wavenumber of plus or minus one, then the three vertical wavenumbers of the triad interactions must sum to zero. There is no triad that involves only the baroclinic mode, as we may see from the form of (12.39). (If the layers are of unequal depths, then purely baroclinic triads do exist.)

(iii) Wherever the Laplacian operator acts on τ, it is accompanied by $-k_d^2$. That is, it is *as if* the effective horizontal wavenumber (squared) of τ is shifted, so that $k^2 \to k^2 + k_d^2$.

Conservation properties

Multiplying (12.39a) by ψ and (12.39b) by τ and horizontally integrating over the domain of area A, assuming once again that this is either periodic or has solid walls, gives

$$\widehat{T} = \int_A (\nabla\psi)^2 \, dA, \qquad \frac{d\widehat{T}}{dt} = \int_A \psi J(\tau, (\nabla^2 - k_d^2)\tau) \, dA, \tag{12.41a}$$

$$\widehat{C} = \int_A [(\nabla\tau)^2 + k_d^2\tau^2] \, dA, \qquad \frac{d\widehat{C}}{dt} = \int_A \tau J(\psi, (\nabla^2 - k_d^2)\tau) \, dA. \tag{12.41b}$$

Here, $\widehat{T}$ is the energy associated with the barotropic flow and $\widehat{C}$ is the energy of the baroclinic flow. An integration by parts shows that

$$\int_A \psi J(\tau, (\nabla^2 - k_d^2)\tau)\,\mathrm{d}A = -\int_A \tau J(\psi, (\nabla^2 - k_d^2)\tau)\,\mathrm{d}A, \tag{12.42}$$

and therefore

$$\frac{\mathrm{d}\widehat{E}}{\mathrm{d}t} = \frac{\mathrm{d}}{\mathrm{d}t}(\widehat{T} + \widehat{C}) = 0. \tag{12.43}$$

That is, total energy is conserved.

An enstrophy invariant is obtained by multiplying (12.39a) by $\nabla^2\psi$ and (12.39b) by $(\nabla^2 - k_d^2)\tau$ and integrating over the domain and adding the two expressions. The result is

$$\frac{\mathrm{d}\widehat{Z}}{\mathrm{d}t} = 0, \qquad \widehat{Z} = \int_A (\nabla^2\psi)^2 + \left[(\nabla^2 - k_d^2)\tau\right]^2 \mathrm{d}A. \tag{12.44}$$

This also follows from (12.36).

Just as for two-dimensional turbulence, we may define the spectra of the energy and enstrophy. Then, with obvious notation, for the energy we have

$$\frac{1}{A}\widehat{T} = \int \mathcal{T}(k)\,\mathrm{d}k \quad \text{and} \quad \frac{1}{A}\widehat{C} = \int \mathcal{C}(k)\,\mathrm{d}k. \tag{12.45}$$

The enstrophy spectrum $\mathcal{Z}(k)$ is related to the energy spectra by

$$\frac{1}{A}\widehat{Z} = \int \mathcal{Z}(k)\,\mathrm{d}k = \int \left[k^2\mathcal{T}(k) + (k^2 + k_d^2)\mathcal{C}(k)\right]\,\mathrm{d}k, \tag{12.46}$$

which is analogous to the relationship between energy and enstrophy in two-dimensional flow. We thus begin to suspect that the phenomenology of two-layer turbulence is closely related to, but richer than, that of two-dimensional turbulence.

12.2.3 Phenomenology of Two-layer Turbulence

In this section we explore the phenomenology of two-layer geostrophic turbulence, by examining how the equations of motion simplify if we suppose that the nonlinear interactions occur at particular scales, and by looking at the forms of the triad interactions.

Triad interactions and scales of interactions

Two types of triad interactions are possible:

Barotropic triads. An interaction that is purely barotropic (i.e., as if $\tau = 0$) conserves $\widehat{T}$, the barotropic energy, and the associated enstrophy $\int k^2\mathcal{T}(k)\,\mathrm{d}k$, and a barotropic triad behaves as purely two-dimensional flow. Explicitly, the conserved quantities are

energy:
$$\frac{\mathrm{d}}{\mathrm{d}t}\left[\mathcal{T}(k) + \mathcal{T}(p) + \mathcal{T}(q)\right] = 0, \tag{12.47a}$$

enstrophy:
$$\frac{\mathrm{d}}{\mathrm{d}t}\left[k^2\mathcal{T}(k) + p^2\mathcal{T}(p) + q^2\mathcal{T}(q)\right] = 0. \tag{12.47b}$$

Baroclinic triads. Baroclinic triads involve two baroclinic wavenumbers (say p, q) interacting with a barotropic wavenumber (say k). The energy and enstrophy conservation laws for this triad are

energy:
$$\frac{\mathrm{d}}{\mathrm{d}t}\left[\mathcal{T}(k) + \mathcal{C}(p) + \mathcal{C}(q)\right] = 0, \tag{12.48a}$$

enstrophy:
$$\frac{\mathrm{d}}{\mathrm{d}t}\left[k^2\mathcal{T}(k) + (p^2 + k_d^2)\mathcal{C}(p) + (q^2 + k_d^2)\mathcal{C}(q)\right] = 0. \tag{12.48b}$$

To the extent that the triad interactions are local, they may be divided into three general types:

(i) interactions only involving scales much smaller than the deformation radius;

(ii) interactions only involving scales much larger than the deformation radius;

(iii) interactions involving scales comparable to the deformation radius.

We consider each in turn.

(i) Interactions at small scales

For small scales the potential vorticity of each layer is given by

$$q_i = \nabla^2\psi_i + \frac{1}{2}k_d^2(\psi_j - \psi_i) \approx \nabla^2\psi_i\,, \tag{12.49}$$

where $i = 1, 2$ and $j = 3 - i$. Thus, each layer is decoupled from the other and obeys the equations of purely two-dimensional turbulence. Enstrophy is cascaded to small scales and, were there to be an energy source at small scales, energy would be transferred upscale until it reached a scale comparable to the deformation scale. As regards triad interactions, interactions at small scales correspond to the case $(p, q) \gg k_d$ in (12.48). If we neglect k_d^2 then we see that a baroclinic triad behaves like a barotropic triad, for (12.48) has the same form as (12.47). In reality, the small scales of a continuously stratified flow may not be representable by a two-layer model, because in a continuously stratified quasi-geostrophic model the enstrophy cascade occurs in *three-dimensional* wavenumber space. Thus, as the horizontal scales become smaller, so does the vertical scale and higher deformation radii (e.g., $n > 1$ in (6.89)) may play a role in reality.

(ii) Interactions at large scales

At scales much larger than the deformation radius we take $|\nabla^2| \sim k^2 \ll k_d^2$, and neglect terms in (12.39) that involve ∇^2 if they appear alongside terms involving k_d^2. Noting that $J(\tau, k_d^2\tau) = 0$, and assuming that $|J(\tau, \nabla^2\psi)| \sim |J(\psi, \nabla^2\tau)|$, we obtain

$$\frac{\partial}{\partial t}\nabla^2\psi + J(\psi, \nabla^2\psi) = -J(\tau, \nabla^2\tau), \qquad \frac{\partial \tau}{\partial t} + J(\psi, \tau) = 0. \tag{12.50a,b}$$

Thus, the baroclinic streamfunction obeys the equation of a passive tracer, although because of the term on the right-hand side of (12.50a) it is not truly passive. Nevertheless, and given the discussion of passive tracers in Section 11.5, these equations suggest that the variance of the baroclinic streamfunction, and thus the energy in the baroclinic mode, will be transferred to smaller scales.

In terms of triad interactions, this case corresponds to $(p, q, k) \ll k_d$. The energy and enstrophy conservation triad interaction laws, (12.48), collapse to

$$\frac{\mathrm{d}}{\mathrm{d}t}\left[\mathcal{C}(p) + \mathcal{C}(q)\right] = 0. \tag{12.51}$$

That is to say, energy is conserved among the baroclinic modes alone, with the barotropic mode k mediating the interaction. Consistent with the analysis in physical space, there is no constraint preventing the transfer of baroclinic energy to smaller scales, and no production of barotropic energy at $k \ll k_d$.

(iii) Interactions at scales comparable to the deformation radius

In this case there is, in general, no simplification of the equations in physical space, or of the conservation laws governing triad interactions, and both baroclinic and barotropic modes are important. However, we can write the triad interaction conservation laws in an instructive form if we define the pseudowavenumber k' by $k'^2 \equiv k^2 + k_d^2$ for a baroclinic mode and $k'^2 \equiv k^2$ for a barotropic mode, and similarly for p' and q'. Then (12.47) and (12.48) can be written as

$$\frac{\mathrm{d}}{\mathrm{d}t}\left[\mathcal{E}(k) + \mathcal{E}(p) + \mathcal{E}(q)\right] = 0, \tag{12.52a}$$

$$\frac{\mathrm{d}}{\mathrm{d}t}\left[k'^2\mathcal{E}(k) + p'^2\mathcal{E}(p) + q'^2\mathcal{E}(q)\right] = 0, \tag{12.52b}$$

where $\mathcal{E}(k)$ is the energy (barotropic or baroclinic) of the particular mode. These are formally identical with the conservation laws for purely two-dimensional flow and so we expect energy to seek the gravest mode (i.e., smallest pseudowavenumber). Since the gravest mode has $k_d = 0$ this implies a tendency of energy to be transferred into the barotropic mode — a *barotropization* of the flow.

Baroclinic instability in the classic two-layer problem concerns the instability of a flow with vertical but no horizontal shear. This is like a triad interaction for which $p \ll (k, q, k_d)$, where p is the wavenumber of a baroclinic mode, and thus $k^2 \sim q^2$. The triad conservation laws then become

$$\frac{\mathrm{d}}{\mathrm{d}t}\left[\mathcal{T}(k) + \mathcal{C}(p) + \mathcal{C}(q)\right] = 0, \tag{12.53a}$$

$$\frac{\mathrm{d}}{\mathrm{d}t}\left[k^2\mathcal{T}(k) + k_d^2\mathcal{C}(p) + (q^2 + k_d^2)\mathcal{C}(q)\right] = 0. \tag{12.53b}$$

From these two equations, and with $k^2 \approx q^2$, we derive

$$q^2\dot{\mathcal{C}}(q) = (k_d^2 - k^2)\dot{\mathcal{T}}(k). \tag{12.54}$$

Baroclinic instability requires that both $\dot{\mathcal{C}}(q)$ and $\dot{\mathcal{T}}(k)$ be positive. This can occur only if

$$k^2 < k_d^2. \tag{12.55}$$

Thus, there is a *high-wavenumber cut-off* for baroclinic instability, as we previously found in Section 9.6 by direct calculation. We can see that the cut-off arises by virtue of energy and enstrophy conservation, and is not dependent on linearizing the equations and looking for exponentially growing normal-mode instabilities.

For small scales, we previously noted that each layer is decoupled from the other and that enstrophy, but not energy, may cascade to smaller scales. Now, baroclinic instability (of the large-scale flow) occurs at scales comparable to or larger than the deformation radius. Thus, the energy that is extracted from the mean flow is essentially trapped at scales larger than the deformation scale: it cannot be transferred to smaller scales in a cascade; rather, it is transferred to barotropic flow at scales comparable to the deformation radius, from which it cascades upwards to larger barotropic scales.

Summary of two-layer phenomenology

Putting together the considerations above leads to the following picture of geostrophic turbulence in a two-layer system (Fig. 12.8). At large horizontal scales we imagine some source of baroclinic energy, which in the atmosphere might be the differential heating between pole and equator, or in the ocean might be the wind and surface heat fluxes. Baroclinic instability effects a non-local transfer of energy to the deformation scale, where both baroclinic and barotropic modes are excited. From here there is an enstrophy cascade in each layer to smaller and smaller scales, until eventually the wavenumber is large enough (denoted by k_{3D} in Fig. 12.8) that non-geostrophic effects become important and enstrophy is scattered by three-dimensional effects, and dissipated. At scales larger than the deformation radius, there is an inverse barotropic cascade of energy to larger scales, where

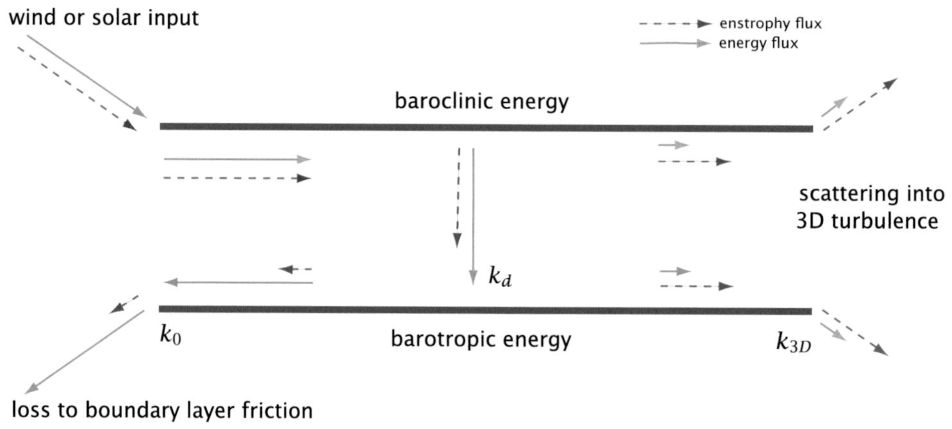

Fig. 12.8 Schema of idealized two-layer baroclinic turbulence.[9] The horizontal axis represents horizontal wavenumber, and the vertical variation is decomposed into two vertical modes — the barotropic and first baroclinic. Energy transfer is shown by solid arrows and enstrophy transfer by dashed arrows.
Large-scale forcing maintains the available potential energy, and so provides energy to the baroclinic mode at very large scales. Energy is transferred to smaller baroclinic scales, and then into barotropic energy at horizontal scales comparable to and larger than the deformation radius (this is baroclinic instability), and then transferred to larger barotropic scales in an upscale cascade. The entire process of baroclinic instability and energy transfer may be thought of as a generalized inverse cascade in which the energy passes to smaller pseudowavenumber $k'^2 \equiv k^2 + k_d^2$.

it is modified by the β-effect and halted by the effects of friction, with the latter being responsible for the dissipation of the large-scale barotropic energy.

There are two aspects of this phenomenology that prevent it from being a quantitative theory of baroclinic eddies in the atmosphere or ocean:

(i) Even as a model of two-layer quasi-geostrophic turbulence the assumptions are questionable. For example, we have largely neglected the effects of friction and we have assumed the baroclinic mode to be passive at large scales. Numerical simulations tend to show a significant transfer of energy from the baroclinic mode to the barotropic mode at scales larger than the deformation radius.

(ii) The ideas do not perfectly apply to either the atmosphere or ocean. In the latter, the turbulence is quite inhomogeneous except perhaps in the Antarctic Circumpolar Current. In the atmosphere, observations indicate that the deformation radius is almost as large as the Rhines scale, and is only a factor-of-a-few smaller than the pole–equator scale, leaving little room for isotropic inverse cascade to fully develop. On the other hand, the atmosphere does display k^{-3} spectra at scales somewhat smaller than the deformation radius (Fig. 12.9), and it may be associated with a forward cascade of enstrophy.[10]

12.3 † A SCALING THEORY FOR GEOSTROPHIC TURBULENCE

Let us now build on the phenomenological model of the previous section and construct a quantitative theory of two-layer forced-dissipative geostrophic turbulence.[12] Although, as we have noted, the underlying model is imperfect, some of its qualitative properties transcend its limitations and are quite revealing about the real system. We will consider a system in which the basic state is a purely zonal flow, with constant vertical shear and no horizontal variation; our goal is to predict the amplitude and the scale of the eddies that result from the baroclinic instability of this flow. The

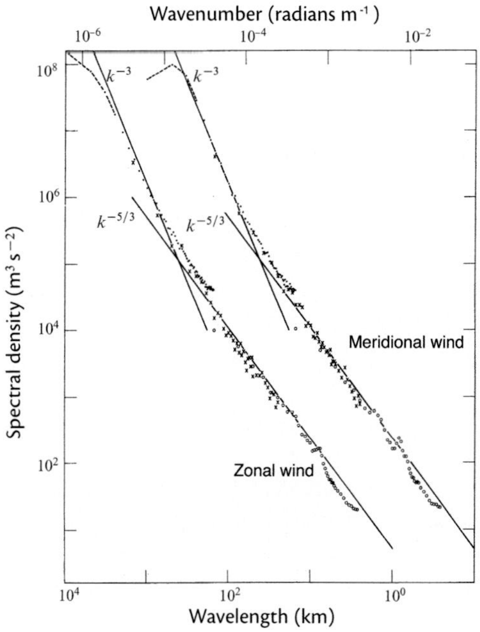

Fig. 12.9 Energy spectra of the zonal and meridional wind near the tropopause, from thousands of commercial aircraft measurements between 1975 and 1979. The meridional spectrum is shifted one decade to the right.

The -3 spectrum may well be associated with a forward enstrophy cascade, but the origin of the $-5/3$ spectrum at smaller scales is not definitively known.[11]

assumptions that the mean flow and the stratification are constants are quite severe and thus the model, even if correct within its own terms of reference, is certainly not a complete theory of the baroclinic eddies in mid-latitude flow.

12.3.1 Preliminaries

We shall suppose that, as in Fig. 12.8, baroclinic instability at large scales leads to a transfer of energy to the barotropic mode at a scale comparable to the deformation radius, followed by an inverse cascade of energy within the barotropic mode, and the energy is finally dissipated in the Ekman layer. At scales smaller than the deformation radius the layers are largely uncoupled, and in each layer there is an enstrophy cascade to small scales. The equations of motion describing all of this are (12.39), but we will explicitly recognize the effect of a shear flow by replacing τ by $\tau - Uy$, where U is constant, this being equivalent to supposing there is a constant shear in the flow. The equations of motion, (12.39), become

$$\frac{\partial}{\partial t}\nabla^2\psi + J(\psi, \nabla^2\psi) + J(\tau, (\nabla^2 - k_d^2)\tau) + U\frac{\partial}{\partial x}\nabla^2\tau = D_\psi, \tag{12.56a}$$

$$\frac{\partial}{\partial t}(\nabla^2 - k_d^2)\tau + J(\tau, \nabla^2\psi) + J(\psi, (\nabla^2 - k_d^2)\tau) + U\frac{\partial}{\partial x}(\nabla^2\psi + k_d^2\psi) = D_\tau. \tag{12.56b}$$

These are similar to the equations used for studying two-layer baroclinic instability in Chapter 9, but we now retain the nonlinear terms and include dissipation, represented by D.

For scales much larger than the deformation radius (12.56) become, analogously to (12.50),

$$\frac{\partial}{\partial t}\nabla^2\psi + J(\psi, \nabla^2\psi) = -J(\tau, \nabla^2\tau) - U\frac{\partial}{\partial x}\nabla^2\tau + D_\psi, \tag{12.57}$$

$$\frac{\partial \tau}{\partial t} + J(\psi, \tau) = U\frac{\partial \psi}{\partial x} - k_d^{-2}D_\tau]. \tag{12.58}$$

The equation for ψ is just the barotropic vorticity equation, 'forced' via its interaction with the baroclinic mode, namely the terms on the right-hand side of (12.57). The equation for the baroclinic streamfunction is the same as the equation for a passive scalar, except for the forcing term $U\partial\psi/\partial x$.

12.3.2 Scaling Properties

In the barotropic equation, we may argue that in the energy-containing scales, $k^2 \ll k_d^2$, the magnitude of the barotropic streamfunction is much larger than that of the baroclinic streamfunction; that is, $|\psi| \gg |\tau|$, as follows. We may reasonably suppose that the forcing of the barotropic vorticity equation occurs at wavenumbers close to k_d, as in baroclinic instability. At larger scales the barotropic streamfunction obeys the two-dimensional vorticity equation, and we may expect an energy cascade to large scales with energy spectrum given by

$$\mathcal{E}_\psi(k) = \mathcal{K}_1 \varepsilon^{2/3} k^{-5/3}, \tag{12.59}$$

where $\mathcal{K}_1$ is a constant and ε is the as yet undetermined energy flux through the system. We may suppose that this cascade holds for wavenumbers $k_0 < k \ll k_d$, where the wavenumber k_0 is the halting scale of the inverse cascade, determined by one or more of: friction, the β-effect, and the domain size. Now, in this same wavenumber regime the baroclinic streamfunction is being advected as a passive tracer — it is being stirred by ψ. Thus, any baroclinic energy that is put in at large scales by the interaction with the mean flow (via the term proportional to $U\psi_x$) will be cascaded to smaller scales. Thus, we expect the baroclinic energy spectrum to be that of a passive tracer whose variance is cascading to smaller scales in a forward cascade, with a spectrum given by (cf. (11.92))

$$\mathcal{E}_\tau(k) = \mathcal{K}_2 \varepsilon_\tau \varepsilon^{-1/3} k^{-5/3}, \tag{12.60}$$

where $\mathcal{K}_2$ is a constant, ε_τ is the transfer rate of baroclinic energy and ε is the same quantity appearing in (12.59). Now, because energy is not lost to small scales we have $\varepsilon_\tau = \varepsilon$, both being equal to the energy flux in the system. Thus, using (12.59) and (12.60), the energy in the barotropic and baroclinic modes is comparable at sufficiently large scales. Since the energy density in the former is $(\nabla\psi)^2$ and in the latter $(\nabla\tau)^2 + k_d^2\tau^2 \sim k_d^2\tau^2$, the magnitude of ψ must then be much larger than that of τ, and specifically

$$|\psi| \sim \frac{k_d|\tau|}{k_0} \gg |\tau|. \tag{12.61}$$

Let us write the baroclinic equation (12.58) in the form

$$\frac{\partial \tau}{\partial t} + J(\psi, \tau - Uy) = 0, \tag{12.62}$$

which is the equation for a passive tracer (τ) in a mean gradient (U), stirred by the flow (ψ), and we omit dissipation. Because there is a large scale separation (in fact, an infinite one) between the scale of the mean gradient of τ (i.e., the scale of variations of U) and the scale of its fluctuations, we can write

$$\tau \sim l' \frac{\partial \overline{\tau}}{\partial y} = -l'U, \tag{12.63}$$

where l' is the scale of the fluctuation. Thus, at the scale k_0^{-1}, the magnitude of τ, and the associated baroclinic velocity v_τ, are given by

$$\tau \sim \frac{U}{k_0}, \qquad v_\tau \sim U. \tag{12.64}$$

At this scale, and using (12.61), the magnitude of the barotropic streamfunction and its associated velocity are given by

$$\psi \sim \frac{k_d U}{k_0^2}, \qquad v_\psi \sim \frac{k_d U}{k_0}. \tag{12.65a,b}$$

This is an important result, regardless of the other details in the calculation. In particular, (12.65b) implies that the barotropic velocity, v_ψ, scales like the mean velocity U multiplied by the ratio of the length scale of the eddies to the deformation radius. In the Earth's atmosphere, this ratio is an $\mathcal{O}(1)$ number; in the ocean it is somewhat larger.

How much energy flows through the system? The mean shear is the ultimate source of energy, and for simplicity this shear is kept constant in time, analogous to an infinite heat bath supplying energy to a smaller system without its own temperature changing. The conversion of energy from the mean shear to the eddy flow is given by multiplying (12.58) by $k_d^2\tau$ and integrating over the domain. This gives an expression for the rate of increase of the available potential energy of the system, namely

$$\frac{\mathrm{d}}{\mathrm{d}t}\mathrm{APE} = \frac{\mathrm{d}}{\mathrm{d}t}\int_A k_d^2\tau^2\,\mathrm{d}A = \int_A 2Uk_d^2\tau\frac{\partial\psi}{\partial x}\,\mathrm{d}A. \tag{12.66}$$

(Note that the energy input to the system equals the poleward heat flux.) From this we estimate the average energy flux as

$$\varepsilon = 2Uk_d^2\overline{\psi_x\tau} \sim \frac{U^3k_d^3}{k_0^2}. \tag{12.67}$$

The correlation between ψ_x and τ cannot be determined by this argument. This aside, we have produced a physically based closure for the flux of energy through the system in terms only of the mean shear, the halting scale k_0 (discussed below) and the deformation scale k_d.

Finally, we calculate the eddy diffusivity (considered at greater length in chapter 13),

$$\kappa \equiv -\frac{\overline{v_\psi\tau}}{\partial_y\overline{\tau}} = \kappa \sim \frac{k_dU}{k_0^2}, \tag{12.68}$$

using (12.64) and (12.65). If the mixing velocity is the barotropic stirring velocity, this result implies a mixing length of k_0^{-1} and that the eddy diffusivity is just the magnitude of the barotropic streamfunction at the energy-containing scales.

12.3.3 The Halting Scale and the β Effect

Let us suppose that, as discussed in Section 12.1, the β effect provides a barrier for the inverse cascade at the scale (12.11), so that $k_0 = k_\varepsilon \sim (\beta^3/\varepsilon)^{1/5}$. Using (12.67) we find that the stopping scale is given by

$$k_0 = k_\varepsilon \sim \frac{\beta}{Uk_d}, \tag{12.69}$$

and using (12.68) and (12.69) we obtain for the energy flux and the eddy diffusivity,

$$\varepsilon \sim \frac{U^5k_d^5}{\beta^2}, \qquad \kappa \sim \frac{U^3k_d^3}{\beta^2}. \tag{12.70}$$

The magnitudes of the eddies themselves are easily given using (12.65) and (12.64), whence

$$\tau \sim \frac{U^2k_d}{\beta}, \quad v_\tau \sim U, \qquad \psi \sim \frac{U^3k_d^3}{\beta^2}, \quad v_\psi \sim \frac{U^2k_d^2}{\beta}. \tag{12.71}$$

Evidently, in this model (in which the mean shear and deformation radius are fixed), the eddies become more energetic with decreasing β, and the eddy amplitudes and poleward heat flux increase very rapidly with the mean shear, more so than in a model in which the energy-containing scale is fixed. This is because as β decreases, the inverse cascade can extend to larger scales, thereby increasing the overall energy of the flow. Similarly, as U increases not only does the eddy amplitude increase as a direct consequence, as in (12.64) and (12.65), but also k_ε falls (see (12.69)), and these effects combine to give a rapid increase of the eddy magnitudes with U.

Frictional effects

Whether β is present or not, friction is necessary to ultimately remove the energy flowing through the system, as well as to remove enstrophy at small scales. Friction provides another mechanism for halting the inverse cascade, and the simplest case is that of a linear drag representing Ekman friction, in which case we write

$$D_\psi = -r\nabla^2\psi. \tag{12.72}$$

and the stopping wavenumber for a given ε is given, as in (11.68), by $k_r = (r^3/\varepsilon)^{1/2}$. However, a little algebra shows that the use of this in (12.67) fails to give a result for ε. From a physical perspective, a linear drag that is weak enough to allow an inverse cascade to form is, *ipso facto,* too weak to equilibrate the flow. A friction that becomes larger at larger scales, for example an 'inverse Laplacian', has no such problems. More physically, a nonlinear drag, proportional to the square of the amplitude of the flow, also leads to a well-posed problem with the cascade halting at a well-defined scale. Finally, we should point out that in neither the atmosphere nor ocean is there an extended inverse cascade, because the deformation scale and the beta scale are not asymptotically well separated (although the β-effect does not prevent an inverse cascade to large zonal scales).

12.4 † PHENOMENOLOGY OF BAROCLINIC EDDIES IN THE ATMOSPHERE AND OCEAN

In the remaining sections of this chapter we take a more informal approach, illustrated by numerical experiments, to the problem of baroclinic eddies in the atmosphere and ocean. We draw from our treatment of geostrophic turbulence but by being a little less formal we are able to travel farther, for we spend less time looking at the map (but with a concomitant danger that we lose our way).

12.4.1 The Magnitude and Scale of Baroclinic Eddies

How big, in both amplitude and scale, do baroclinic eddies become? Suppose that the time-mean flow is given, and that it is baroclinically unstable. Eddies will grow, initially according to the linear theory of Chapter 9, but they cannot and do not continue to amplify: they ultimately equilibrate, and this by way of nonlinear mechanisms. The eddies will extract energy from the mean flow, but at the same time the available energy of the mean flow is being replenished by external forcing (i.e., the maintenance of an equator–pole temperature gradient by radiative forcing in the atmosphere, and wind and buoyancy forcing at the surface in the ocean). Thus, we cannot a priori determine the amplitude of baroclinic eddies by simply assuming that all of the available potential energy in the mean flow is converted to eddying motion. To close the problem we find we need to make three, not necessarily independent, assumptions:

(i) an assumption about the magnitude of the baroclinic eddies;

(ii) an assumption relating eddy kinetic energy to eddy available potential energy;

(iii) an assumption about the horizontal scale of the eddies.

Baroclinic eddies extract available potential energy (APE) from the mean flow, and it is reasonable to suppose that an eddy of horizontal scale L_e can extract, as an upper bound, the APE of the mean flow contained within that scale. The APE is proportional to the variation of the buoyancy field so that

$$(\Delta b')^2 \sim |\Delta\overline{b}|^2 \sim L_e^2|\nabla\overline{b}|^2, \tag{12.73}$$

where Δb is the variation in the buoyancy over the horizontal scale L_e. (For simplicity we stay with the Boussinesq equations, and $b = -g\delta\rho/\rho_0$. We could extend the arguments to an ideal-gas atmosphere with $b = g\delta\theta/\theta_0$.) Equivalently, we might simply write

$$b' \sim L_e|\nabla\overline{b}|, \tag{12.74}$$

which arises from a mixing-length approach. Supposing that the temperature gradient is mainly in the y-direction then, using thermal wind, we have

$$b' \sim L_e f \frac{\partial \overline{u}}{\partial z} \qquad \text{and} \qquad v'_\tau \sim \overline{u}, \tag{12.75a,b}$$

where v'_τ is an estimate of the shear (multiplied by the depth scale) of the eddying flow. (These estimates are the same as (12.64), with $\overline{u}$ replacing U.)

Our second assumption is to relate the barotropic eddy kinetic energy to the eddy available potential energy, and the most straightforward one to make is that there is a rough equipartition between the two. This assumption is reasonable because in the baroclinic lifecycle (or baroclinic inverse cascade) energy is continuously transferred from eddy available potential energy to eddy kinetic energy, and the assumption is then equivalent to supposing that the relevant eddy magnitude is always proportional to this rate of transfer. Thus we assume $v_\psi^2 \sim (b'/N)^2$ or

$$v'_\psi \sim \frac{b'}{N}. \tag{12.76}$$

Finally, the scale of the eddies is determined by the extent to which the eddies might grow through nonlinear interactions. As we discussed earlier, possibilities for this scale include the deformation radius itself (if the inverse cascade is weak) or the Rhines scale (if the inverse cascade is slowed by the β-effect), or even the domain scale if neither of these applies.

Some consequences

The simple manipulations above have some very interesting consequences. Using (12.75) and (12.76) we find

$$v'_\psi \sim \frac{f L_e}{NH}\overline{u} = \frac{L_e}{L_d}\overline{u}, \tag{12.77}$$

where $L_d = NH/f_0$ is the deformation radius and $\overline{u}$ is the amplitude of the mean baroclinic velocity, that is the mean shear multiplied by the height scale. This important relationship relates the magnitude of the eddy kinetic energy to that of the mean. In the atmosphere the scale of the motion is not much larger than the deformation radius (which is about 1000 km) and the eddy and mean kinetic energies are, consistently, comparable to each other. In the ocean the deformation radius (about 50 km over large areas) is significantly smaller than the scale of mesoscale eddies (which is more like 200 km or more), and observations consistently reveal that the eddy kinetic energy is an order of magnitude larger than the mean kinetic energy.[13]

One other important and slightly counter-intuitive result concerns the timescale of eddies. From (12.77) we have

$$T_e \sim \frac{L_e}{v'_\psi} \sim \frac{L_d}{\overline{u}} \equiv T_E, \tag{12.78}$$

where T_E is the Eady time scale. That is, the eddy time scale (at the scale of the largest eddies) is independent of the process that ultimately determines the spatial scale of those eddies; if the eddy length scale increases somehow, perhaps because friction or β are decreased, the velocity scale increases in proportion.

12.4.2 Baroclinic Eddies and their Lifecycle in the Atmosphere

Amplitude and scale

We saw in Section 9.9.2 that baroclinic instability in the atmosphere occurs predominantly in the troposphere, i.e., in the lowest 10 km or so of the atmosphere, with the higher stratification of the

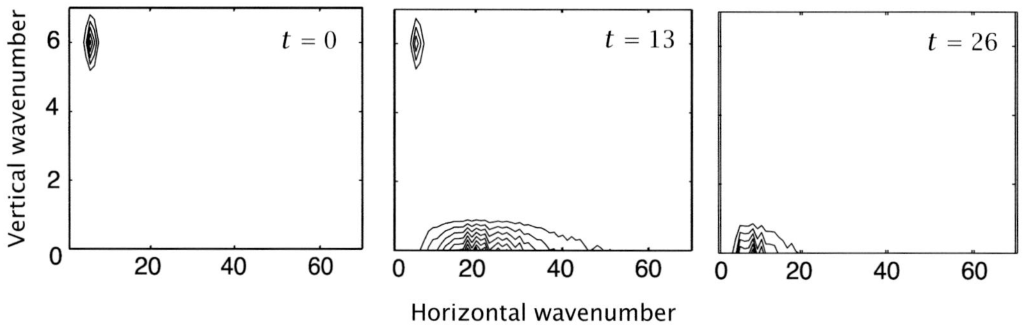

Fig. 12.10 A numerical simulation of an idealized baroclinic lifecycle, showing contours of energy in spectral space at successive times. Initially, there is baroclinic energy at low horizontal wavenumber, as in a large-scale shear. Baroclinic instability transfers this energy to barotropic flow at the scale of the deformation radius, and this is followed by a barotropic inverse cascade to large scales. Most of the transfer to the barotropic mode in fact occurs quite quickly, between times 11 and 14, but the ensuing barotropic inverse cascade is slower. The entire process may be thought of as a generalized inverse cascade. The stratification (N^2) is uniform, and the first deformation radius is at about wavenumber 15. Times are in units of the eddy turnover time.[15]

eponymous stratosphere inhibiting instability. In the mid-latitude troposphere the vertical shear and the stratification are relatively uniform and fairly simple models, such as the two-layer model or the Eady model (with the addition of the β-effect) are reasonable first-order models.

The mean pole–equator temperature gradient is about 40 K and the deformation radius NH/f is about 1000 km. The Rhines scale, $\sqrt{U/\beta}$ is a little larger than the deformation radius, being perhaps 2000 km, and is similar to the width of the main mid-latitude baroclinic zone which lies between about 40° and 65°, in either hemisphere. Given these, and especially given that the maximum wavelength for instability occurs at scales somewhat larger than the deformation radius, there is little prospect of an extended upscale cascade, and for this reason the Earth's atmosphere has comparable eddy kinetic and mean kinetic energies.[14]

The baroclinic lifecycle

The baroclinic lifecycle of geostrophic turbulence, sketched schematically in Fig. 12.8, can be nicely illustrated by way of numerical initial value problems, and we describe two such. The first is extremely idealized: take a doubly-periodic quasi-geostrophic model on the f-plane, initialize it with baroclinic energy at large horizontal scales, and then let the flow freely evolve. Figure 12.10 shows the results. The flow, initially concentrated in high vertical wavenumbers to best illustrate the energy transfer, is baroclinically unstable, and energy is transferred to barotropic flow at wavenumbers close to the first radius of deformation, here at about wavenumber 15. Energy then slowly cascades back to large scales in a predominantly barotropic inverse cascade, piling up at the largest scales much as in decaying, two-dimensional turbulence. Nearly all of the initial baroclinic energy is converted to barotropic, eddy kinetic energy and, even without any surface friction, the flow evolves to a baroclinically stable state. Couched in these terms, it is easy to see that the baroclinic lifecycle is a form of baroclinic inverse cascade, with an energy transfer to large total wavenumber, K_{tot}, that is made up of contributions from both horizontal and vertical wavenumbers:

$$K_{tot}^2 = K_h^2 + k_d^2 m^2, \tag{12.79}$$

where m is the vertical wavenumber and K_h is the horizontal wavenumber. As we noted earlier, the twin constraints of energy and enstrophy conservation prevent the excitation of horizontal scales with very large horizontal wavenumbers, and so the lifecycle proceeds through wavenumbers at the deformation scale.

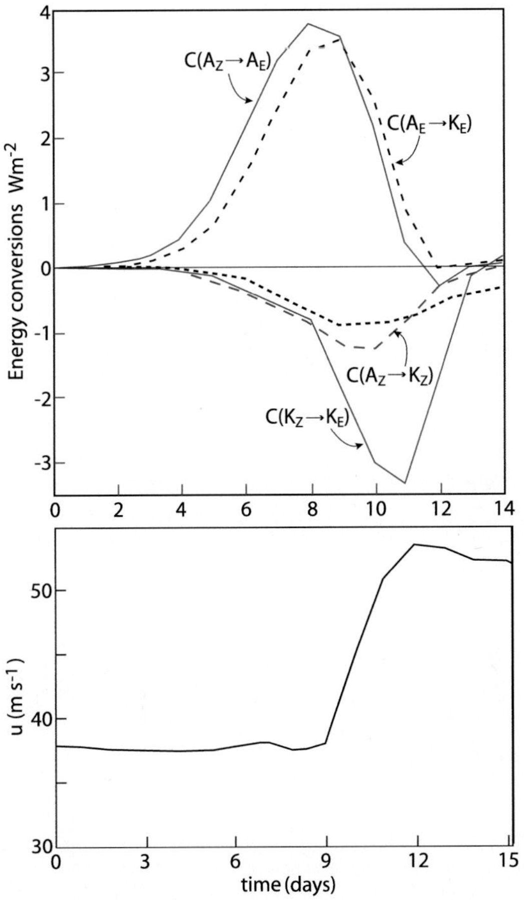

Fig. 12.11 Top: energy conversion and dissipation processes in a numerical simulation of an idealized atmospheric baroclinic lifecycle, simulated with a GCM. Bottom: evolution of the maximum zonal-mean velocity. Top: energy conversions. A_Z and A_E are zonal and eddy available potential energies (APEs), and K_Z and K_E are the corresponding kinetic energies, and $C(\cdot \to \cdot)$ represents the corresponding conversion.

Initially baroclinic processes dominate, with conversions from zonal to eddy APE, $C(A_Z \to A_E)$, and then eddy APE to eddy kinetic energy, $C(A_E \to K_E)$, followed by the barotropic conversion of eddy kinetic to zonal kinetic energy, a negative value of $C(K_Z \to K_E)$. The latter process is reflected in the *increase* of the maximum zonal-mean velocity at about day 10, shown in the lower panel.[16]

The results of the second, and more realistic, initial value problem are illustrated in Fig. 12.11. Here, the atmospheric primitive equations on a sphere are integrated forward, beginning from a baroclinically unstable zonal flow, plus a small-amplitude disturbance at zonal wavenumber 6. The disturbance grows rapidly through baroclinic instability, accompanied by a conversion of energy initially from the zonal mean potential energy to eddy available potential energy (EAPE), and then from EAPE to eddy kinetic energy (EKE), and finally from EKE to zonal kinetic energy (ZKE). The last stage of this roughly corresponds to the barotropic inverse cascade of quasi-geostrophic theory, and because of the presence of a β-effect the flow becomes organized into a zonal jet. The parameters in the Earth's atmosphere are such that there is only one such jet, and in the lower panel of Fig. 12.11 we see its amplitude increase quickly from days 10 to 12, associated with the conversion of EKE to ZKE. (In other lifecycle experiments, the end state is found to be more akin to a barotropic vortex or, in meteorological parlance, a 'cut-off cyclone'.[17]) A commonality between all the lifecycles is the evolution toward a generally barotropic flow, with variable degrees of zonality in the final state, depending on the importance of the β-effect.

Of course, the real atmosphere is never in the zonally uniform state that is used in idealized baroclinic instability or lifecycle studies. Rather, at any given time, finite-amplitude eddies exist and these provide a finite amplitude perturbation to the baroclinically unstable zonal flow. For this reason we rarely, if ever, see an exponentially growing normal mode. Furthermore, given any instantaneous atmospheric state, zonally symmetric or otherwise, the fastest growing (linear) instability is not necessarily exponential but may be 'non-modal', with a secular or linear growth that, over some finite time period and in some given norm, is much more rapid than exponential.[18]

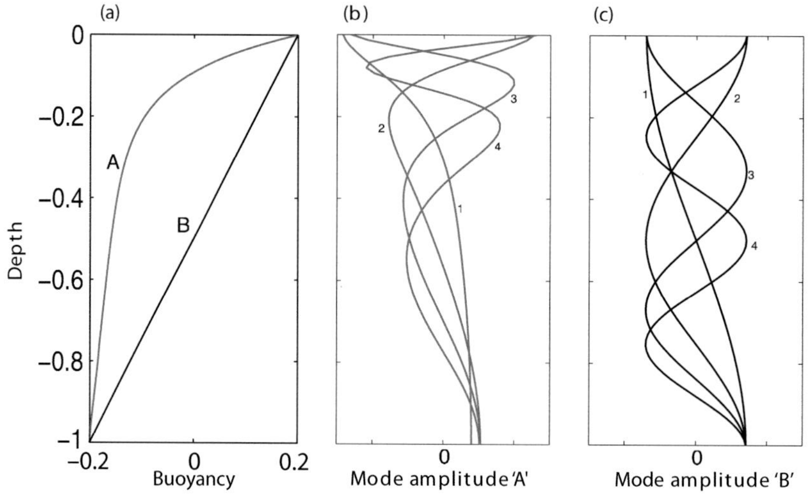

Fig. 12.12 (a) Two buoyancy profiles, $-g\delta\rho/\rho_0$, one being a fairly realistic oceanic case with enhanced stratification in the upper ocean (profile A), and the other with uniform stratification (profile B). (b) and (c) The first four baroclinic modes [eigenfunctions of (12.80)] for A and B. With profile B the eigenmodes are cosines, whereas in profile A they have a larger amplitude and shorter local wavelength in the upper ocean. The number of zero crossings is equal to the mode number.

12.4.3 Baroclinic Eddies and their Lifecycle in the Ocean

Basic ideas

Baroclinic instability was first developed as a theory for mid-latitude instabilities in the atmosphere and the original problems were set in a zonally re-entrant channel. The ocean, apart from the Antarctic Circumpolar Current (ACC), is not zonally re-entrant. However, it is driven by buoyancy and wind-forcing at the surface, and these combine to produce a region of enhanced stratification and associated shear in the ocean in the upper 500–1000 m or so — that is, in the 'thermocline' — as discussed more fully in Chapter 20. The associated sloping isopycnals constitute a pool of available potential energy, and so the ocean is potentially baroclinically unstable. Satellite observations indicate that baroclinic eddies are in fact almost ubiquitous in the mid- and high-latitude oceans, especially in and around intense western boundary currents, such as the Gulf Stream, and the ACC. The ocean is, literally, a sea of eddies.[19]

In addition to the geometry, the main differences between the oceanic and atmospheric problems are twofold:

(i) In the ocean, the shear and the stratification are not uniform between two rigid lids, nor even uniform between one rigid lid and a structure like the tropopause. Instead, both stratification and shear are largest in the upper ocean, decaying into a quiescent and nearly unstratified abyss.

(ii) In the ocean, the first radius of deformation is much smaller than the scale of the large-scale flow; that is, of the gyres or the large-scale overturning circulation. On dimensional grounds we have, using a height scale and stratification representative of the upper ocean, $L_d \sim NH/f \sim 10^{-2} \times 10^3/10^{-4} = 100\,\text{km}$.

A consequence of the enhanced shear in the upper ocean is that the amplitude of the growing waves is also largely concentrated in the upper ocean, as we saw in Fig. 9.22. Regarding the stratification, in quasi-geostrophic theory we may, as in Section 6.5.2, define the deformation radii by

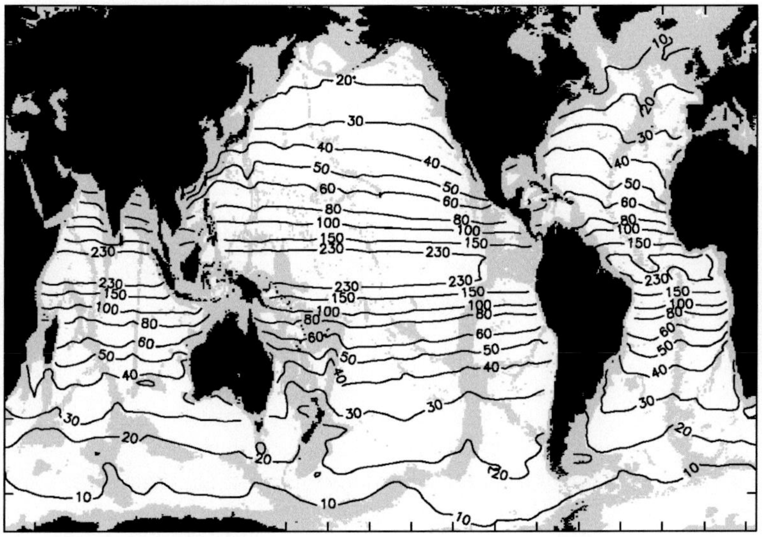

Fig. 12.13 The oceanic first deformation radius L_d in kilometres, calculated using the observed stratification and an equation similar to (12.80). Near equatorial regions are excluded, and regions of ocean shallower than 3500 m are shaded.

Variations in Coriolis parameter are responsible for much of large-scale variability, although weak stratification also reduces the deformation radius at high latitudes.[21]

solution of the eigenvalue problem

$$\frac{\partial}{\partial z}\frac{f^2}{N^2}\frac{\partial \phi_n}{\partial z} + \Gamma_n \phi_n = 0, \qquad \text{with} \qquad \frac{\partial \phi_n}{\partial z} = 0 \quad \text{at } z = 0,\, H. \tag{12.80a,b}$$

The successive eigenvalues, Γ_n, are related to the successive deformation radii, L_n, by $L_n^2 = 1/\Gamma_n$. If the stratification (N^2) is uniform the resulting eigenfunctions are cosines with corresponding eigenvalues and deformation radii given by

$$\Gamma_n = n^2 \frac{f^2 \pi^2}{N^2 H^2}, \qquad L_n = \frac{1}{\sqrt{\Gamma_n}} = \frac{NH}{n\pi f_0}. \tag{12.81a,b}$$

If the stratification is non-uniform, we must in most cases solve the eigenproblem numerically (or by WKB methods, see Section 3.4.2), and the results of one such calculation are given in Fig. 12.12. The case with uniform stratification reproduces cosine modes, whereas in the more realistic case the modes tend to have highest amplitude in the upper ocean, where the stratification is strongest — a result that is typical of oceanic profiles.[20]

The results of a calculation of the first deformation radius using observed oceanic profiles are given in Fig. 12.13, and values of 50–100 km emerge in mid-latitudes.[22] We may therefore expect oceanic baroclinic instability to occur on a scale much smaller than that in the atmosphere, and much smaller than the scale of an ocean basin. (However, the scale of baroclinic instability will typically be larger than the first deformation radius, L_1, shown in Fig. 12.13, because of two compounding effects. First, in uniform stratification the first deformation radius, L_1, as given by (12.81b) is a factor of π smaller than the simple definition NH/f. Second, in simple baroclinic instability problems like the Eady problem the wavelength of maximum instability is a few times NH/f. Thus, the wavelength of maximum instability may be an order of magnitude larger than L_1.)

Eddy amplitudes and scales

The consequences of this small deformation radius on the lifecycle and finite-amplitude equilibration of oceanic baroclinic eddies are far-reaching, one being that there is more scope for an inverse cascade than in the atmosphere, and indeed observations indicate that the horizontal scale of the

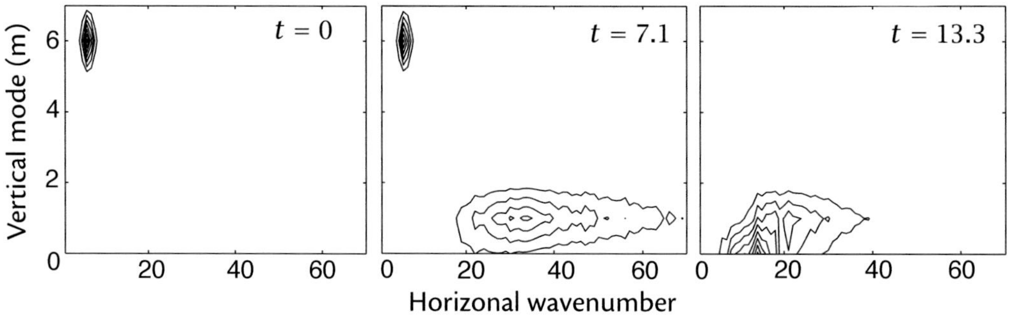

Fig. 12.14 Idealized baroclinic lifecycle, similar to that in Fig. 12.10, but with enhanced stratification of the basic state in the upper domain, representing the oceanic thermocline.

eddies is typically a few to several times larger than the local deformation radius itself. The situation is not clear cut, however, as some observations[23] indicate that the eddy size nevertheless scales with the local deformation radius, suggesting that the eddy scale may be set by the instability scale and not an inverse cascade.

In any case, suppose that an ocean eddy has a horizontal scale 200 km, and that it sits in the subtropical gyre where the mean temperature gradient is $10^{-5}\,\mathrm{K\,m^{-1}}$, that the mean shear and ensuing baroclinic activity are mainly confined to the upper 1000 m of the ocean, and that the deformation radius is 50 km. The temperature gradient corresponds to a temperature difference of about 20 K across 2000 km, a horizontal buoyancy gradient of about $2 \times 10^{-9}\,\mathrm{s^{-2}}$ [using the simple equation of state $\rho = \rho(1 - \beta_T \Delta T)$ where $\beta_T = 2 \times 10^{-4}\,\mathrm{K^{-1}}$] and a shear of about $2\,\mathrm{cm\,s^{-1}}$ over the upper 1 km of ocean. Then, using (12.77), we can estimate a typical eddy velocity scale as

$$v'_\psi \sim \frac{L_e}{L_d}\overline{u} \approx 4\overline{u} \approx 8\,\mathrm{cm\,s^{-1}}, \tag{12.82}$$

implying, as we noted earlier, an EKE that is an order of magnitude larger than the mean kinetic energy. Associated with this are typical temperature perturbations whose magnitude we can estimate using (12.74) or (12.75) as being about 2 K. These estimates are comparable to those observed in mid-ocean, with more energetic eddies forming near intense western boundary currents where gradients are large and barotropic instability also provides a source of energy for the eddies. There is least a factor-of-a-few uncertainty, but it is noteworthy that they are roughly comparable to the values observed.

Eddy lifecycles

The lifecycle of a mid-oceanic baroclinic eddy will differ from its atmospheric counterpart in two main respects:

(*i*) Baroclinic eddies may be advected by the mean flow into regions with quite different properties from where they initially formed.

(*ii*) The non-uniformity of the stratification affects the passage to barotropic flow.

Both of these can best be studied by numerical means. Regarding the first, eddies will often form in or near intense western boundary currents, but then will be advected by that current into the potentially less unstable open ocean before completing their lifecycle. Regarding the second, an oceanic analogue of the lifecycle illustrated in Fig. 12.10 is shown in Fig. 12.14. The main difference between this case and the atmospheric one is that baroclinic instability initially leads to the transfer of energy to vertical mode one, followed by a transfer to larger horizontal scales in the barotropic mode, as illustrated schematically in Fig. 12.15.[24] If the energy is initially solely in the

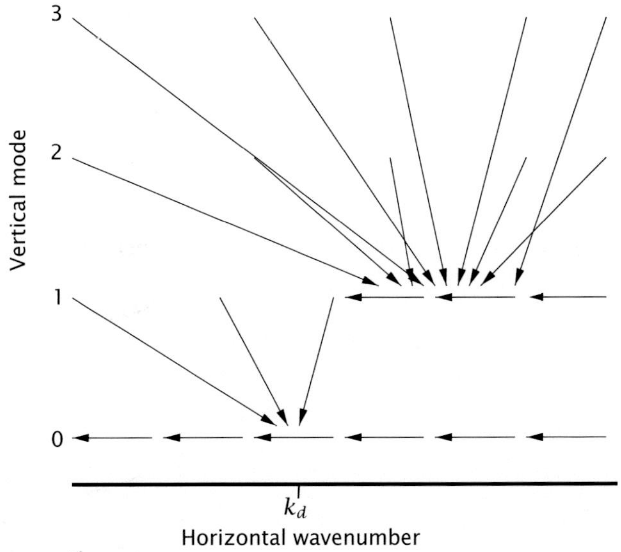

Fig. 12.15 Energy transfer paths as a function of vertical mode and horizontal wavenumber, in a fluid with an oceanic stratification; i.e., with a thermocline. Vertical mode 0 is the barotropic mode.

first baroclinic mode the cycle is more similar to the atmospheric one, but higher baroclinic modes may be more readily excited in the ocean than the atmosphere.

Notes

1 Rhines (1975). See also Holloway & Hendershott (1977) and Vallis & Maltrud (1993).

2 Adapted from Vallis & Maltrud (1993).

3 The mechanism described here follows Vallis & Maltrud (1993). Interactions between Rossby waves will also give rise to zonal flow, as described by Newell (1969) and Rhines (1975). Wave–mean-flow interactions provide a direct route to the production of zonal flows, as discussed in Chapter 15. Williams (1978) was one of the first numerical simulations to show the production of jets.

4 More discussion on these matters is variously given by Smith *et al.* (2002), Danilov & Gryanik (2002), Galperin *et al.* (2006, 2010), Sukoriansky *et al.* (2007), Scott & Dritschel (2012), Chai (2016) and others.

5 A PV staircase was proposed for the jets of Jupiter by Marcus (1993) and further discussed by Peltier & Stuhne (2002), and a review is to be found in Dritschel & McIntyre (2008). Staircases in turbulent flow can and have been found more generally, in particular in stratified flow in which an initially smooth density gradient may break down into steps and layers when stirred (Phillips 1972, Ruddick *et al.* 1989).

6 Adapted from Scott & Dritschel (2012).

7 Farrell & Ioannou (1995, 2008) and Srinivasan & Young (2012) describe a direct pathway to zonal jets that does not require eddy–eddy interactions. Tobias & Marston (2013) describe the mechanisms in a more general context at the price of some analytic accessibility, discussing the circumstances in which the jet formation proceeds by way of a cascade or a direct interaction with the zonal flow.

8 Quasi-geostrophic turbulence was introduced by Charney (1971). Salmon (1980) and Rhines (1977) provided much of the two-layer phenomenology. Various laboratory experiments are discussed by Read (2001).

9 Adapted from Salmon (1980).

10 Lindborg (1999) concluded that the data were consistent with a forward cascade of enstrophy between wavelengths of a few thousand kilometres and a few hundred kilometres. Boer & Shepherd

(1983) and Shepherd (1987) also found a k^{-3} spectrum at similar scales, noting the importance of interactions involving stationary waves. At scales smaller than than 100 km or so the spectrum is shallower than -3, and more like $-5/3$. This may be due to non-geostrophic effects, for example a forward cascade of energy associated with gravity wave breaking, or it may be due to a two-dimensional inverse cascade of energy with an energy source at very small scales associated with convection, or it may be due to effects associated with 'surface quasi-geostrophic' dynamics.

11 Adapted from Gage & Nastrom (1986).

12 Larichev & Held (1995) and Held & Larichev (1996). See Spall (2000) for an oceanic extension, and Thompson & Young (2006) for evidence that the theory fails to account for coherent structures.

13 Related arguments concerning eddy magnitudes were given by Gill *et al.* (1974). Atmospheric energetics, and atmospheric observations in general, are described by Peixoto & Oort (1992). For the oceanic case, see Wyrtki *et al.* (1976), Richardson (1983), and Stammer (1997).

14 This does not address the issue as to *why* the Rhines scale and deformation radius are similar. See also Chapter 15.

15 Modified from Smith & Vallis (2001).

16 Adapted from Simmons & Hoskins (1978).

17 See Thorncroft *et al.* (1993). These authors identify two classes of lifecycles, which they call LC1 and LC2, which have differing degrees of decay of eddy kinetic energy in the later parts of the lifecycle, and with LC2 producing cut-off cyclones. Initial conditions and spatial inhomogeneities, including the horizontal shear of the flow and the presence of critical layers, play an important role in guiding the location of the wave breaking, and hence the final state that is reached.

18 The theory of this as applied to the atmosphere has been developed by Farrell (1984), Farrell & Ioannou (1996) and other papers by these authors. Exponential growth is the exception, not the rule, in baroclinic instability in the real world, both because of nonlinear effects and non-normal instabilities. Indeed, if the basic flow is oscillatory, baroclinic instability can arise even if the basic flow is never unstable via the CSP criterion.

19 This realization came to fruition as a result of the bilateral US–USSR POLYMODE project in the 1970s. See Robinson (1984).

20 For example, Kundu *et al.* (1975).

21 From Chelton *et al.* (1998).

22 The eigenproblem actually solved was

$$\partial^2\phi/\partial z^2 + (N^2(z)/c^2)\phi = 0, \qquad \phi = 0 \quad \text{at } z = 0, H, \tag{12.83}$$

where H is the ocean depth and N is the observed buoyancy frequency. The deformation radius is given by $L_d = c/f$ where c is the first eigenvalue and f is the latitudinally varying Coriolis parameter.

23 Stammer (1997).

24 Fu & Flierl (1980) and Smith & Vallis (2001) examined this issue in more detail, both analytically and numerically. Figure 12.15 is adapted from these papers.

The smallest eddies are almost numberless, and large things are rotated only by large eddies and not by small ones, and small things are turned by small eddies and large.
Leonardo da Vinci, describing *turbolenza* in a sketch book, c. 1500.

CHAPTER 13

Turbulent Diffusion and Eddy Transport

THE TRANSPORT OF FLUID PROPERTIES BY UNSTEADY MOTION — that is, the way in which the properties of a fluid may be carried from one location to another by waves and turbulence — is one of the most important topics in geophysical fluid dynamics. It may be the dominant transport in a fluid, greatly exceeding that of the mean flow — in the atmosphere, for example, heat is transferred polewards primarily by the action of unsteady weather systems, not by the much weaker time-mean flow. However, we are often not interested in the details of the turbulent eddies and hence we might seek to *parameterize* the turbulent transport in terms of the mean flow; unfortunately, no general theory exists for such transport, for indeed such a theory would amount to a theory of turbulence. In the absence of this, we focus our attention in this chapter on the theory (such as it is) and practice of *turbulent diffusion*. In models of turbulent diffusion, the turbulent transport is generally related to the gradient of the mean flow, and it is the simplicity of the resulting expressions that has led to their wide adoption in areas as different as turbulent pipe flow, atmospheric boundary layer transport and large-scale ocean modelling. Diffusive models are, or aim to be, rational, simple and tractable — a blend of heuristic reasoning and elementary mathematics, the latter needed to ensure that certain basic requirements (conservation laws, for example) of a physical process are captured by a parameterization. However, just like other turbulent closures, they rely on physical assumptions that cannot be rigorously justified. In the first part of the chapter we consider turbulent diffusion from a general standpoint, and then specialize our discussion to geofluids, and in particular to large-scale transport by baroclinic eddies. Those readers with some prior knowledge of turbulent diffusion may choose to skip ahead to Section 13.6.

13.1 DIFFUSIVE TRANSPORT

We begin with a brief discussion of the diffusion equation itself, to wit

$$\frac{\partial \varphi}{\partial t} = \kappa \nabla^2 \varphi, \tag{13.1}$$

where κ is a constant, positive, scalar diffusivity and the tracer φ is a scalar field. We expect that an initially concentrated blob of tracer would spread out — it would diffuse — and thus small parcels of tracer are transported. How quickly does this occur, or, put another way, is there an effective diffusive transport velocity?

If the rate of spreading becomes independent of the initial conditions then, purely from dimensional considerations, the spreading can depend only on the diffusivity and time itself and we can write

$$\overline{X^2} = \alpha\kappa t, \tag{13.2}$$

where $\overline{X^2}$ is the mean-square displacement, t is time and α is a nondimensional constant. Let us quantify this with an explicit calculation. If φ is interpreted as the density of markers of fluid parcels, then the mean-square displacement of the markers is given by (in three dimensions)

$$\overline{X^2} = \frac{\int_0^\infty r^2 \varphi r^2 \,\mathrm{d}r}{\int_0^\infty \varphi r^2 \,\mathrm{d}r}, \tag{13.3}$$

where the denominator, the total amount of tracer present, is a constant and we have assumed a spherically symmetric distribution of tracer. Using (13.1) we find

$$\frac{\mathrm{d}}{\mathrm{d}t}\int_0^\infty r^2 \varphi r^2 \,\mathrm{d}r = \kappa \int_0^\infty r^2 \frac{1}{r^2}\frac{\partial}{\partial r}\left(r^2 \frac{\partial \varphi}{\partial r}\right) r^2 \,\mathrm{d}r = 6\kappa \int_0^\infty \varphi r^2 \,\mathrm{d}r, \tag{13.4}$$

after a couple of integrations by parts. Thus

$$\frac{\mathrm{d}}{\mathrm{d}t}\overline{X^2} = 6\kappa, \tag{13.5}$$

and because κ is a constant we have the important result that

$$\overline{X^2} = 6\kappa t. \tag{13.6}$$

In two dimensions the equivalent calculation begins with

$$\overline{X^2} = \frac{\int_0^\infty r^2 \varphi r \,\mathrm{d}r}{\int_0^\infty \varphi r \,\mathrm{d}r} \tag{13.7}$$

and using the diffusion equation we find

$$\frac{\mathrm{d}}{\mathrm{d}t}\int_0^\infty r^2 \varphi r \,\mathrm{d}r = \kappa \int_0^\infty r^2 \frac{1}{r}\frac{\partial}{\partial r}\left(r \frac{\partial \varphi}{\partial r}\right) r \,\mathrm{d}r = 4\kappa \int_0^\infty \varphi r \,\mathrm{d}r. \tag{13.8}$$

Thus we obtain

$$\overline{X^2} = 4\kappa t. \tag{13.9}$$

Finally, in one dimension (i.e., spreading along a line) it is easy to show that

$$\overline{X^2} = 2\kappa t. \tag{13.10}$$

Thus, in both three and two dimensions, *the spread of a diffused scalar increases with the half power of time.*

13.1.1 An Explicit Example

We gain a little more intuition about what the above calculations mean by considering the case in which the initial tracer distribution is a delta function at the origin. If the total amount of tracer is unity, then in three dimensions at subsequent times the tracer is given by the distribution

$$\varphi(r, t) = \frac{1}{8(\pi\kappa t)^{3/2}} \exp(-r^2/4\kappa t), \tag{13.11}$$

as may be checked by substitution back into the equation of motion. The distribution clearly broadens with time, and the mean-square distance from the origin is given by

$$\overline{X^2} = \int_0^\infty \frac{4\pi r^2}{8(\pi\kappa t)^{3/2}} \exp(-r^2/4\kappa t)\,\mathrm{d}r = 6\kappa t, \tag{13.12}$$

as in (13.6). The important point is that the mean distance travelled by a particle during a time interval t is proportional to the square root of that time interval. This is, of course, redolent of a random walk (see the shaded box on the following page), which brings us to the subject of turbulent diffusion.

13.2 TURBULENT DIFFUSION

Fluids differ from solids in that they can transport properties by advection — thus, heat is primarily transferred polewards in the atmosphere by means of air movement and not by molecular diffusion. Turbulent fluid motion differs from laminar fluid motion in that such advective transport may be greatly enhanced by the seemingly random motion of the fluid, the net transport being much larger than that which would be effected by the time-mean fluid motion alone. Indeed, to continue the atmospheric example, away from the tropics the poleward transport of heat in the atmosphere is largely effected by way of the (large-scale) turbulent transfer of heat in mid-latitude weather systems. Of course, such transfer *is* simply by advection, and if we could explicitly calculate the motion of the fluid parcels we could explicitly calculate the transport. However, turbulent transport is both very complicated and very sensitive to the initial conditions, so that any hope of performing such a calculation exactly in a real situation is often a forlorn one.

Turbulent transport is most important in *inhomogeneous* situations, because it is the divergence of the transport that is important and the mean divergence is non-zero only if there is inhomogeneity. The theories of Chapters 11 and 12 do not lend themselves to an easy extension to inhomogeneous flow, and we turn to a slightly more empirical approach.[1]

13.2.1 Simple Theory

Let us consider how fluid markers are transported in a statistically steady, homogeneous, turbulent flow. The markers are introduced at the origin $x = y = z = 0$ at $t = 0$; we may create an ensemble of such markers by performing many such tracer release experiments on different realizations of the turbulent flow, but with each flow having the same statistical properties. The question is, what is the average rate of dispersion of a single particle of fluid?

The displacement of a marker at a time t is given by

$$\boldsymbol{X}(t) = \int_0^t \boldsymbol{V}(t')\,\mathrm{d}t', \tag{13.13}$$

where $\boldsymbol{V}$ is the velocity of the fluid parcel — a material velocity. (We will use uppercase variables to denote material ('Lagrangian') quantities.) The mean-square displacement is

$$\overline{X^2(t)} = \int_0^t \mathrm{d}t_1 \int_0^t \overline{\boldsymbol{V}(t_1)\cdot\boldsymbol{V}(t_2)}\,\mathrm{d}t_2, \tag{13.14}$$

where the overbar denotes an ensemble average, and thus $\overline{\boldsymbol{V}(t_1)\cdot\boldsymbol{V}(t_2)}$ is a measure of the velocity correlation between the velocities of the fluid parcels at times t_1 and t_2. That is,

$$\overline{\boldsymbol{V}(t_1)\cdot\boldsymbol{V}(t_2)} = \overline{v^2}R(t_2 - t_1) = \overline{v^2}R(\tau), \tag{13.15}$$

A Random Walk

Here we give an elementary derivation of the most basic result in random walk theory, the relationship of the mean-square displacement to the number of steps taken.[2] A loose analogy is that of drunkards staggering randomly from here to there, with no correlation between their successive steps. After any number of steps, the *mean* displacement of the drunkards is zero, but we expect their *root-mean-square* displacement to increase: this is because drunkards independently thrown out of the same bar will generally wander off in different directions (which is why the mean displacement is zero), but after some time most of them will indeed end up some distance away.

For simplicity consider steps, $\boldsymbol{s}_n$, each with random orientation but equal magnitude, s. The displacement after n steps is related to the displacement after $n-1$ steps by

$$\boldsymbol{D}_n = \boldsymbol{D}_{n-1} + \boldsymbol{s}_n, \tag{R.1}$$

so that the amplitude of $\boldsymbol{D}_n$, namely D_n, is given by

$$\begin{aligned} D_n^2 &= (\boldsymbol{D}_{n-1} + \boldsymbol{s}_n) \cdot (\boldsymbol{D}_{n-1} + \boldsymbol{s}_n) \\ &= D_{n-1}^2 + s^2 + 2\boldsymbol{D}_{n-1} \cdot \boldsymbol{s}_n. \end{aligned} \tag{R.2}$$

Taking an ensemble average over many realizations gives

$$\overline{D_n^2} = \overline{D_{n-1}^2} + s^2, \tag{R.3}$$

having used $\overline{\boldsymbol{D}_{n-1} \cdot \boldsymbol{s}_n} = 0$, because each step is random.

Now, $D_0 = 0$, so that $\overline{D_1^2} = s^2$, $\overline{D_2^2} = 2s^2$ and so on. Thus, using (R.3) to proceed inductively, we have

$$\overline{D_n^2} = ns^2, \tag{R.4}$$

or

$$\overline{D_n^2}^{1/2} = \sqrt{n}s. \tag{R.5}$$

Thus, in a random walk the *root-mean-square displacement increases with the half-power of the number of steps taken.* More work is required to calculate the distribution of the random walkers, but it may be shown that in the limit of infinitesimally small steps the random walk becomes a Wiener process and the distribution becomes Gaussian, as in (13.11) (with the exact form depending on the dimensionality of the problem), indicating a diffusive process. Diffusion may be thought of as a continuous random walk, with the displacement proportional to the square root of time.

where $R(t_2 - t_1)$ is the velocity correlation function and, because the turbulence is statistically steady, this depends only on the time difference $\tau = t_2 - t_1$. Furthermore, $R(-\tau) = R(\tau)$. Thus,

$$\overline{X^2(t)} = \int_0^t \mathrm{d}t_1\, \overline{v^2} \int_0^t R(t_2 - t_1)\, \mathrm{d}t_2 = \int_0^t \mathrm{d}\hat{t}\, \overline{v^2} \int_{-\hat{t}}^{t-\hat{t}} R(\tau)\, \mathrm{d}\tau, \tag{13.16}$$

changing variables to τ and $\hat{t} = t_1$ (Fig. 13.1). We expect the velocity correlation function to fall monotonically from its initial value of unity to a value approaching zero as $\tau \to \infty$, as in Fig. 13.2,

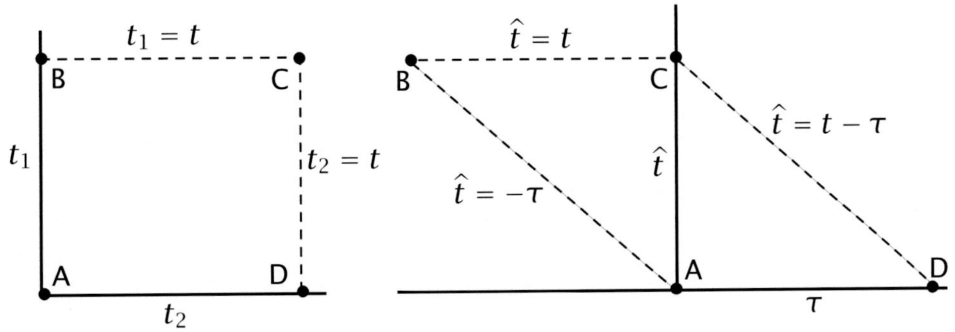

Fig. 13.1 Changes of time variables involved in (13.16) and (13.32). The original two-dimensional integral is over the rectangle ABCD. Defining $\tau = t_2 - t_1$ and $\hat{t} = t_1$, then the area is spanned by $[\hat{t} = (0, t),\ \tau = (-\hat{t}, t - \hat{t})]$ as in (13.16), or by $[\tau = (0, t),\ \hat{t} = (0, t - \tau)]$ (i.e., ACD) plus $[\tau = (-t, 0),\ \hat{t} = (-\tau, t)]$ (i.e., ABC) in (13.32).

and typically, there will be some characteristic time τ_{corr} that parameterizes the behaviour of the function. In general, we cannot obtain explicit general solutions without detailed knowledge of this correlation function, but there are two interesting limits:

(i) The short-time limit. For small times, i.e., for $t \ll \tau_{corr}$ (and so $t_1, t_2 \ll \tau_{corr}$) the correlation function will be approximately unity and so (13.16) becomes

$$\overline{X^2(t)} \approx \int_0^t \mathrm{d}\hat{t}\, \overline{v^2} \int_0^t \mathrm{d}\hat{t} = \overline{v^2} t^2. \tag{13.17}$$

Thus, the root-mean-square displacement increases linearly with time, and linearly with the root-mean-square velocity of the flow. For small times, the fluid parcel's behaviour is well correlated with that at the initial time, and so the displacement increases linearly in the direction it was initially going. Indeed, directly from (13.13) we have, for small times, $\boldsymbol{X}(t) \approx \boldsymbol{V}t$, which leads directly to (13.17).

(ii) The long-time limit. We are now concerned with the case $t \gg \tau_{corr}$. Because the correlation function falls with time, most of the contributions to the second integrand (involving $R(\tau)$) in (13.16) are from $\tau \leq \tau_{corr}$. Thus, without much loss in accuracy, we can replace the limits of integration by $-\infty$ and $+\infty$; that is

$$\overline{X^2(t)} \approx \int_0^t \mathrm{d}\hat{t}\, \overline{v^2} \int_{-\infty}^{\infty} R(\tau)\, \mathrm{d}\tau. \tag{13.18}$$

Assuming the second integral converges, it is just a number; in fact, noting that $R(\tau) = R(-\tau)$, we may use it to *define* the correlation time τ_{corr} by

$$\tau_{corr} \equiv \int_0^{\infty} R(\tau)\, \mathrm{d}\tau. \tag{13.19}$$

We then have the important result that

$$\overline{X^2(t)} \approx 2\overline{v^2} t \int_0^{\infty} R(\tau)\, \mathrm{d}\tau = 2\overline{v^2} \tau_{corr} t. \tag{13.20}$$

That is, for times that are long compared with the turbulence correlation time, the distance travelled by a fluid parcel in some time interval is proportional to the square-root of that time interval, just as for a diffusive process; this is because the fluid parcels are essentially undergoing random walks. Equation (13.20) connects two quite different fluid properties: the left-hand side tells us how tracers are dispersed in a turbulent flow, a material property, whereas the right-hand side can be evaluated from the Eulerian velocity field at different times. Both the left- and right-hand sides can be directly measured, by looking at the dispersion of a dye and by measuring the velocity at successive times.

We may define a *coefficient of turbulent diffusivity* by

$$K_{turb} = \tfrac{1}{3}\overline{v^2}\tau_{corr} , \tag{13.21}$$

and then we have the result that

$$\overline{X^2(t)} = 6K_{turb}t. \tag{13.22}$$

Comparison of (13.22) with (13.6) indicates that the transport of turbulent flow, under these conditions, is like a diffusive transport, with a coefficient of diffusivity given by (13.21). [Sometimes, the numerical factors are neglected, and a diffusivity is defined by the expressions

$$K_{turb} \equiv \frac{\mathrm{d}\overline{X^2(t)}}{\mathrm{d}t} \quad \text{or} \quad K_{turb} \equiv \frac{1}{2}\frac{\mathrm{d}\overline{X^2(t)}}{\mathrm{d}t}. \tag{13.23}$$

These lose the exact connection with a true diffusion coefficient, but usually the turbulent diffusivity can only be estimated, anyway.]

We may define a correlation length scale to be the approximate distance that a parcel moves, on average, in a material (i.e., 'Lagrangian') correlation time. Thus

$$l_{corr} \equiv v_{rms}\tau_{corr}, \tag{13.24}$$

where $v_{rms} = (\overline{v^2})^{1/2}$, whence

$$K_{turb} = \tfrac{1}{3}v_{rms}l_{corr} . \tag{13.25}$$

In most situations, the numerical coefficient (1/3 here) cannot be trusted because a real turbulent flow is unlikely to satisfy the restrictions of stationarity and homogeneity that we have imposed. Nevertheless, a relationship similar to (13.25) — that a turbulent diffusivity is proportional to an r.m.s. turbulent velocity and a correlation length scale, is the foundation for semi-empirical *mixing length* theories that we discuss in Section 13.4.

The simple relationships between the mean-square displacement, the Lagrangian time scale, the mean-square velocity and the eddy diffusivity allow the diffusivity to be computed from the statistics of particle trajectories. Thus, suppose that a cluster of floats is released into the ocean, or some balloons are released in the atmosphere. If neutrally buoyant, these instruments then essentially become labelled fluid particles, and one may compute K_{turb} directly from the dispersion of the cluster using (13.22). If it is possible to measure their root-mean-square velocity, then one may use (13.21) to estimate the diffusivity from this and the material correlation time scale.

13.2.2 ♦ An Anisotropic Generalization

We now consider the correlation between the different components of the displacement in anisotropic, but still homogeneous, flow.[3] The displacement of a fluid particle is given by (13.13), and this is a

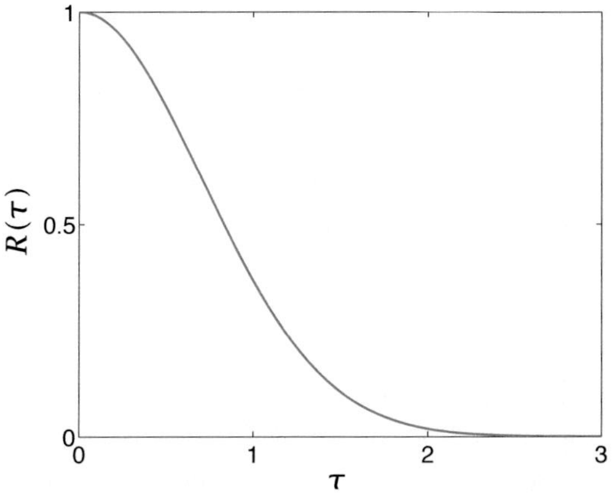

Fig. 13.2 Idealized velocity correlation function in turbulent flow with correlation time $\tau_{corr} = \mathcal{O}(1)$. For small times, $\tau \ll \tau_{corr}$, $R(\tau) \approx 1$. For large times, $\tau \gg \tau_{corr}$, $R(\tau) \ll 1$. We may define the correlation time by $\tau_{corr} = \int_0^\infty R(\tau)\,\mathrm{d}\tau$.

random vector. Thus, generalizing (13.14), we may define the fluid particle displacement covariance tensor by

$$D_{ij}(t) = \overline{X_i(t)X_j(t)} = \int_0^t \int_0^t \overline{V_i(t_1)V_j(t_2)}\,\mathrm{d}t_1\,\mathrm{d}t_2, \tag{13.26}$$

where the velocity denoted by V_i is the ith component of the velocity of a fluid element. For small times, $X_i(t) \approx v_i(\boldsymbol{a}, 0)t$, where $v_i(\boldsymbol{a}, 0)$ is the fluid velocity at the parcel's initial position, $\boldsymbol{a}$, and we obtain

$$D_{ij}(t) \approx \overline{v_i(\boldsymbol{a}, 0)v_j(\boldsymbol{a}, 0)}t^2. \tag{13.27}$$

If the flow is statistically steady and homogeneous the average of any quantity has no spatial or temporal dependence and so

$$D_{ij}(t) = A_{ij}t^2, \tag{13.28}$$

where the tensor $A_{ij} = \overline{v_i v_j}$ has constant entries, and this is a slight generalization of (13.17).

The velocity covariance of a fluid parcel at times t_1 and t_2 is, as before, a function only of the time difference $t_1 - t_2$ and so it must have the form

$$\overline{V_i(t_1)V_j(t_2)} = \left(\overline{v_i^2}\,\overline{v_j^2}\right)^{1/2} R_{ij}(t_2 - t_1). \tag{13.29}$$

Except in the case of isotropic flow $R_{ij}(\tau) \neq R_{ij}(-\tau)$, but we do have, in general,

$$R_{ij}(\tau) = R_{ji}(-\tau). \tag{13.30}$$

Now, to obtain a generalization of (13.20), we first use (13.29) in (13.26) to obtain

$$D_{ij}(t) = \left(\overline{v_i^2}\,\overline{v_j^2}\right)^{1/2} \int_0^t \int_0^t R_{ij}(t_2 - t_1)\,\mathrm{d}t_1\,\mathrm{d}t_2. \tag{13.31}$$

If we change variables to $\tau = t_2 - t_1$ and $\hat{t} = t_1$ (see Fig. 13.1) we obtain[4]

$$D_{ij}(t) = \left(\overline{v_i^2}\,\overline{v_j^2}\right)^{1/2} \left(\int_0^t \mathrm{d}\tau \int_0^{t-\tau} \mathrm{d}\hat{t}\,R_{ij}(\tau) + \int_{-t}^0 \mathrm{d}\tau \int_{-\tau}^t \mathrm{d}\hat{t}\,R_{ij}(\tau)\right), \tag{13.32}$$

and using (13.30) this becomes

$$D_{ij}(t) = 2\left(\overline{v_i^2}\,\overline{v_j^2}\right)^{1/2} \int_0^t \mathrm{d}\tau \int_0^{t-\tau} \mathrm{d}\hat{t}\,\widehat{R}_{ij}(\tau), \tag{13.33}$$

where $\widehat{R}_{ij} = (R_{ij} + R_{ji})/2$. This order of integration enables us to perform the integration over $\hat{t}$, giving

$$D_{ij}(t) = 2\left(\overline{v_i^2}\,\overline{v_j^2}\right)^{1/2} \int_0^t (t-\tau)\widehat{R}_{ij}(\tau)\,\mathrm{d}\tau. \tag{13.34}$$

For long times, i.e., for $t \gg \tau_{corr}$, the upper limit of the integration may be taken to be infinity, again because the contributions to the integrand from $\widehat{R}_{ij}(\tau)$ all come from small τ. Furthermore, we expect that for large t,

$$\int_0^\infty t\widehat{R}_{ij}(\tau)\,\mathrm{d}\tau \gg \int_0^\infty \tau\widehat{R}_{ij}(\tau)\,\mathrm{d}\tau, \tag{13.35}$$

because $R_{ij}(\tau)$ is only non-negligible for small τ, and $t \gg \tau$ in this range. Thus, we finally obtain a generalization of (13.20) for the displacement covariance of two components of the displacement, namely

$$D_{ij} = 2\left(\overline{v_i^2}\,\overline{v_j^2}\right)^{1/2} t \int_0^\infty \widehat{R}_{ij}(\tau)\,\mathrm{d}\tau. \tag{13.36}$$

The integral is a tensor with constant entries, analogous to the turbulent decorrelation time scale of (13.19). Then, with $\tau_{ij} \equiv \int_0^\infty \widehat{R}_{ij}(\tau)\,\mathrm{d}\tau$, the corresponding turbulent diffusivity is

$$K_{ij} = \tfrac{1}{3}\left(\overline{v_i^2}\,\overline{v_j^2}\right)^{1/2} \tau_{ij}. \tag{13.37}$$

13.2.3 Discussion

We have shown that, for sufficiently long times, the distance travelled by a fluid parcel in some time is proportional to the square root of that time, just as for a diffusive process and just as for a random walk. The motion of our fluid parcel is analogous to that of a dust particle undergoing Brownian motion — both are continually buffeted and undergo random walks as a result. Still, it may appear that the usefulness of our results is limited by the assumptions of stationarity and homogeneity — it is well-nigh impossible in nature to produce a statistically stationary, homogeneous turbulent flow, because statistical stationarity implies there must be an energy source and this, as well as the presence of boundaries, militates against homogeneity. However, we should not be so pessimistic, on two counts:

(i) Similar ideas may be directly applied to flows that are homogeneous in one direction, which is more easily achievable in nature.

(ii) Often, a flow will *not* be homogeneous in any direction. However, if the *statistics* of the eddy motion vary on a space scale that is longer than $v_{rms}\tau_{corr}$, then the eddy transport properties may be determined by a local theory. For example, the size of the eddy diffusivity is then determined by $D_t \sim v_{\rm rms} l_{corr}$ where the parameters, and hence the diffusion coefficient, vary, but only on a scale longer than the energy-containing scale.

The essential results of this section thus lie in (13.20), (13.21) and (13.25): that the dispersion of a fluid particle in a turbulent flow is *diffusive* in nature, and that the turbulent diffusivity is proportional to the product of the root-mean-square velocity and the correlation length.

13.3 TWO-PARTICLE DIFFUSIVITY

Let us now consider the problem of determining the mutual separation of two fluid parcels; the problem is relevant to geofluids because by tracking the separation of floats in the ocean, or balloons in the atmosphere, we can learn much about the nature of large-scale turbulence in those systems. The problem differs from the one-particle problem, because the separation of the particles itself will affect the rate of increase of the separation. In the one-particle problem in homogeneous flow, the position of the particular tagged fluid particle plays no direct role in determining its rate of spreading from its initial condition — any one position is the same as any other. But if two particles are close together, they may be swept away together by some large eddy, without affecting their mutual separation whereas two particles that are widely separated will undergo largely uncorrelated motion. Thus, we identify two regimes:

(i) A regime in which the separation of the particles is greater than the scale of the largest eddies. In this case, each particle is undergoing a random walk that is effectively uncorrelated with that of the other particle.

(ii) A regime in which the separation of the particles is less than the energy-containing scale of the motion. In this case, the eddies that contribute most to the two-particle separation are those that are comparable in scale to the separation itself.

If we attempt to apply the Taylor analysis ab initio we evidently have, by analogy to (13.13)

$$\boldsymbol{Y}(t) = \boldsymbol{X}_1(0) - \boldsymbol{X}_2(0) + \int_0^t [\boldsymbol{V}_1(t') - \boldsymbol{V}_2(t')]\,\mathrm{d}t', \tag{13.38}$$

and a mean-square separation of

$$\overline{\boldsymbol{Y}^2(t)} = \overline{(\boldsymbol{X}_1(0) - \boldsymbol{X}_2(0))^2} + \int_0^t \mathrm{d}t_1 \int_0^t [\boldsymbol{W}_1(t_1)\cdot\boldsymbol{W}_2(t_2)]\,\mathrm{d}t_2, \tag{13.39}$$

where $\boldsymbol{W}(t) = \boldsymbol{V}_1(t) - \boldsymbol{V}_2(t)$. However, it is now difficult to proceed much further. The problem is that we cannot write

$$\overline{\boldsymbol{W}(t_1)\cdot\boldsymbol{W}(t_2)} = \overline{w^2}R(t_2 - t_1), \tag{13.40}$$

because the correlation will depend on the initial separation of the particles as well as the time since then. Thus, the diffusivity itself will depend on both time and the initial particle separation, and the results analogous to those of the single-particle diffusivity cannot easily be recovered. However, we can make some progress by separately considering the two above-mentioned regimes.

13.3.1 Large Particle Separation

This case is analogous to the single-particle case. The particle separation is given by

$$\boldsymbol{Y}(t) = \boldsymbol{X}_1(t) - \boldsymbol{X}_2(t), \tag{13.41}$$

so the mean-square separation is

$$\overline{Y^2(t)} = \overline{X_1(t)^2} + \overline{X_2(t)^2} - 2\overline{\boldsymbol{X}_1(t)\cdot\boldsymbol{X}_2(t)}. \tag{13.42}$$

For long times, the last term is zero because the motion of the two particles is uncorrelated. Furthermore, each of the first two terms is given by (13.20) or (13.22), so that the mean separation varies as

$$\overline{Y^2(t)} = 4\overline{v^2}\tau_{corr}t \tag{13.43}$$

and the rate of separation, for large t, is given by

$$\frac{\mathrm{d}\overline{Y^2(t)}}{\mathrm{d}t} = 4\overline{v^2}\tau_{corr} = 12K_{turb}. \tag{13.44}$$

Thus, the relative diffusion is twice that of the single-particle process, in the limit that the particles are separated by an amount larger than the largest eddies.

13.3.2 Separation Within the Inertial Range

How do fluid parcels whose separation is at inertial scales behave relative to each other?[5] Suppose that two particles are tagged, and that their separation is greater than the viscous scale but smaller than the scales of the largest eddies — that is, the separation lies within the inertial range of the flow. Then, the rate of separation of the two particles can depend only on two quantities, the separation itself and properties of the inertial range, meaning (in three dimensions) the energy flux, ε, through the system. It cannot depend on the time, because this would imply that the subsequent rate of particle separation depends on the history of how the particles came to their current positions. Thus we can write

$$\frac{d\overline{L}^2}{dt} = g(\overline{L}, \varepsilon), \tag{13.45}$$

where $\overline{L} \equiv \overline{Y(t)^2}^{1/2}$. Dimensional analysis then gives

$$\frac{d\overline{L}^2}{dt} = A\varepsilon^{1/3}\overline{L}^{4/3}, \tag{13.46}$$

where A is a nondimensional constant, and this is known as 'Richardson's four-thirds law'. We can integrate (13.46) to give

$$\overline{L}^2 \sim \varepsilon t^3. \tag{13.47}$$

Another way of deriving (13.46) is to suppose that the separation obeys the diffusive law

$$\frac{d\overline{L}^2}{dt} = K_{turb}, \tag{13.48}$$

where K_{turb} is a turbulent diffusivity that is *a function of the separation itself*. This is because the farther apart the eddies are, the larger the scale of the eddies that can move the two particles independently, rather than just sweeping them along together. An estimate of the diffusivity is then

$$\mathcal{K}_{turb} \sim vl, \tag{13.49}$$

where v is the characteristic velocity of an eddy of scale l, and $l \sim \overline{L} = \overline{Y^2}^{1/2}$. Using the inertial range scaling $v \sim (l\varepsilon)^{1/3}$ this is

$$\mathcal{K}_{turb} \sim \varepsilon^{1/3}\overline{L}^{4/3}, \tag{13.50}$$

and so (13.48) becomes

$$\frac{d\overline{L}^2}{dt} \sim \varepsilon^{1/3}\overline{L}^{4/3}, \tag{13.51}$$

as before. Of course, dimensional consistency demands that we obtain the same result, but the derivation is intuitive and the estimate of the two-particle diffusivity (i.e., (13.50), that the eddy diffusivity governing the separation of two fluid parcels goes as the 4/3 power of their root-mean-square separation) is useful. If the particle separation is greater than the scale of the largest eddies in the system, l_{max}, then

$$K_{turb} \sim v(l_{max})l_{max} \sim \varepsilon^{1/3}l_{max}^{4/3} = \text{constant}. \tag{13.52}$$

The two-particle separation then proceeds as a conventional random walk or diffusive process, with the mean-square separation increasing linearly with time.

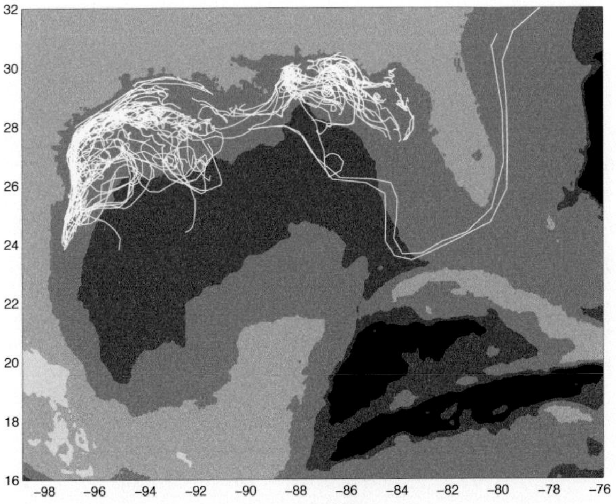

Fig. 13.3 Trajectories of surface drifters in the Gulf of Mexico, each truncated to produce paths of just 25 days. The drifters were released as part of 'SCULP' — the Surface CUrrent and Lagrangian drift Program.[6]

Diffusion in two-dimensional flow

In two dimensions the turbulent diffusivity will differ depending on whether the two-particle separation is in the energy inertial range or in the enstrophy inertial range. In the energy inertial range the scaling is the same as in the three-dimensional case, but in the enstrophy inertial range the rate of separation will depend on the enstrophy cascade rate, η, and the separation itself. Dimensional analysis then leads to

$$\frac{\mathrm{d}\overline{L}^2}{\mathrm{d}t} = B\eta^{1/3}\overline{L}^2, \tag{13.53}$$

where B is a nondimensional constant. This integrates to

$$\overline{L}^2 = \overline{L(0)}^2 \exp(B\eta^{1/3}t), \tag{13.54}$$

or $\overline{Y^2(t)} = \overline{Y^2(0)} \exp(B\eta^{1/3}t)$. Thus, the rate of separation is *exponential* in the enstrophy inertial range, a result unique to two-dimensional turbulence. Similarly, using $v \sim \eta^{1/3}l$, the turbulent diffusivity is given by

$$K_{turb} \sim \eta^{1/3}\overline{L}^2. \tag{13.55}$$

A geophysical example

The above ideas are well illustrated by analysing the trajectories of surface drifters in the Gulf of Mexico. The drifters are free-moving buoys which float about a half metre below the surface and which thus act as imperfect fluid markers — imperfect because they cannot follow the full three-dimensional motion of water parcels. Nevertheless, the motion at these scales can be expected to be quasi-geostrophic and nearly horizontal, so the associated error will be small. The drifters are tracked by satellite and their trajectories, proxies for the motion of fluid parcels, are shown in Fig. 13.3. The two-particle, or two-drifter, separation is illustrated in Fig. 13.4 and two regimes may be discerned. In the first, the pair separations grow approximately exponentially in time, with an e-folding time of 2 days, consistent with motion within an enstrophy inertial range using (13.54). The second regime is characterized by a power-law growth, proportional to $t^{2.2}$, somewhat slower than the t^3 separation expected for an energy inverse cascade using (13.47).[7] The boundary for the two regimes occurs at about 75 km, which is similar to the first deformation radius. No late-time diffusive regime (where the dispersion goes like t) is observed, suggesting that there exist long-time drifter correlations; these correlations arise because the separation of the drifters is never significantly larger than the energy-containing scale of the eddies themselves.

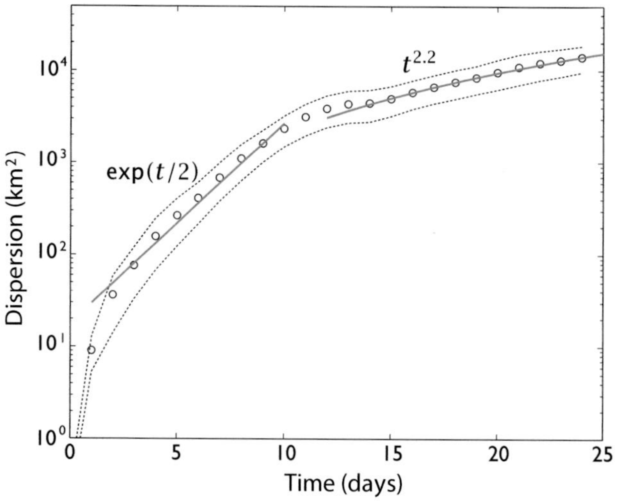

Fig. 13.4 Relative dispersion (the mean-square separation) for 140 drifter pairs as a function of time. The analysis utilizes all drifter pairs which come within 1 km of each other during their lifetimes.

In the atmosphere similar exponential separation of pairs of drifting balloons in the stratosphere at scales of less than 1000 km has been seen, consistent with an enstrophy inertial range. Evidence of a t^3 separation at larger scales, consistent with an energy inverse cascade, has been less forthcoming.[8]

13.4 MIXING LENGTH THEORY

The discussion of the previous two sections deals with the dispersion of marked fluid parcels. However, both for practical and fundamental reasons, we would like to be able to represent the turbulent transport of a fluid property in an Eulerian form. Thus, consider the equation for a conserved quantity φ in an incompressible turbulent flow:

$$\frac{\mathrm{D}\overline{\varphi}}{\mathrm{D}t} = -\nabla \cdot (\overline{\boldsymbol{v}'\varphi'}) + \kappa\nabla^2\overline{\varphi}, \tag{13.56}$$

where κ is the molecular diffusivity and the overbar denotes some kind of averaging or a filtering, so that $\overline{\varphi}$ represents only large scales. (We also adopt the convention that, unless noted, whenever the material derivative written as $\mathrm{D}/\mathrm{D}t$ is applied to an averaged field, the advection is by the averaged velocity only.) We expect the transport of φ to be enhanced by the turbulent flow and, as we saw in the previous sections, in some circumstances this transport will have a diffusive nature, completely overwhelming the molecular diffusivity. Let us consider this from an Eulerian angle, and by analogy with molecular mixing.

Given the mean distribution $\overline{\varphi}(x, y, z)$, let a fluid parcel be displaced from its mean position by a turbulent fluctuation. Suppose that the displaced parcel of fluid is able to carry its initial properties a distance l' before mixing with its surroundings. Then just prior to mixing with the environment the fluctuation of φ is given by, in the one-dimensional case,

$$\varphi' = -l'\frac{\partial\overline{\varphi}}{\partial x} - \frac{1}{2}l'^2\frac{\partial^2\overline{\varphi}}{\partial x^2} + O(l'^3). \tag{13.57}$$

If the mean gradient is varying on a space-scale, L, that is larger than the mixing length l', that is if

$$L \equiv \frac{|\partial\overline{\varphi}/\partial x|}{|\partial^2\overline{\varphi}/\partial x^2|} \gg l', \tag{13.58}$$

then we can neglect terms in l'^2 and higher. The turbulent flux of φ-stuff is then given by

$$F = \overline{u'\varphi'} = -\overline{u'l'}\frac{\partial\overline{\varphi}}{\partial x}. \tag{13.59}$$

In more than one dimension, we have

$$\boldsymbol{F} = F_i = -\overline{v_i'l_j'}\partial_j\overline{\varphi} = -K_{ij}\partial_j\overline{\varphi}, \tag{13.60}$$

with summation over repeated indices, where $K_{ij} \equiv \overline{v_i'l_j'}$. The quantity K_{ij} (which we also write as $\boldsymbol{K}$) is known as the eddy (or turbulent) diffusivity tensor. At high Reynolds number it is a property of the flow rather than the fluid itself but, supposing that it can somehow be determined, the equation for the mean value of φ becomes

$$\frac{\mathrm{D}\overline{\varphi}}{\mathrm{D}t} = \nabla\cdot(\boldsymbol{K}\nabla\overline{\varphi}) = \partial_i(K_{ij}\partial_j\overline{\varphi}), \tag{13.61}$$

neglecting molecular diffusion.

Suppose that there exists a coordinate system in which the displacements in one direction ($l'^{\widehat{x}}$, displacements in the $\widehat{x}$-direction) are not correlated with the fluctuating velocity in another, orthogonal, direction (v', the velocity in the $\widehat{y}$-direction) and for simplicity we restrict ourselves to two dimensions. Then, in that coordinate system $\boldsymbol{K}$ is symmetric and

$$\boldsymbol{K} = \begin{pmatrix} \overline{u'l'^{\widehat{x}}} & \overline{u'l'^{\widehat{y}}} \\ \overline{v'l'^{\widehat{x}}} & \overline{v'l'^{\widehat{y}}} \end{pmatrix} = \begin{pmatrix} \overline{u'l'^{\widehat{x}}} & 0 \\ 0 & \overline{v'l'^{\widehat{y}}} \end{pmatrix}. \tag{13.62}$$

The tensor may then, if needs be, be rotated to some other Cartesian coordinate system, but it will remain a symmetric tensor. In isotropic flow the two diagonal entries are equal and the equation of motion is,

$$\frac{\mathrm{D}\overline{\varphi}}{\mathrm{D}t} = \nabla\cdot(K\nabla\overline{\varphi}), \tag{13.63}$$

which is identical to the equation with molecular diffusion, save that the eddy diffusivity scalar, K, is different from the molecular diffusivity. To the extent, then, that K_{ij} is a symmetric tensor with constant entries, the turbulence acts like an enhanced diffusion. If the flow is homogeneous, then K does not vary spatially.

13.4.1 Requirements for Turbulent Diffusion

Turbulent diffusion evidently is a tractable and rational approach for parameterizing the effects of turbulent transport.[9] However, the premises required for the derivations above are not always satisfied and the derivation itself is rather heuristic, and turbulent diffusion is in no way a fundamental solution to the turbulence closure problem. Nonetheless, it can be an extremely useful parameterization in the appropriate circumstances, these being:

(i) There should be a scale separation between the mean gradient and the maximum mixing length, and the mixing length and decorrelation time scale should be well-defined.

(ii) The diffused property φ should be a materially conserved quantity, except for the effects of molecular diffusion.

(iii) The diffused property φ should be able to *mix* with its environment.

These are all largely self-evident from the derivation, but let us discuss items *(ii)* and *(iii)* a little more.

(ii) Material conservation of tracer

We assumed that a parcel of fluid carries its value of φ a distance, on average, equal to its mixing length before irreversibly mixing with its environment; this assumption is necessary in order that one may write $\varphi' = -l'\partial\overline{\varphi}/\partial x$. If φ is not materially-conserved over this scale other terms enter this formula. In particular, momentum is affected by the pressure force, and so is not normally a good candidate for turbulent diffusion. Potential vorticity is a better candidate, because it is a true material invariant, save for dissipative terms, and in large-scale geophysical flows potential vorticity also contains a great deal of the information about the flow. There is no ab initio requirement that the tracer be passive, and if it is not then its turbulent transport will affect the flow itself.

(iii) Tracer mixing and turbulent cascades

If a parcel cannot mix with its surroundings, then turbulent mixing cannot take place at all. Instead, we have what might be called turbulent stirring and if φ were, say, a dye then it would merely become threaded through the environment, producing streaks and swirls of colour rather than a truly mixed fluid. As another example, let φ be temperature and suppose that it has a mean gradient, so that temperature falls in the direction of increasing y. If a displaced parcel of fluid does not mix with or assume the value of its new environment at some stage, then there will be no correlation between the velocity producing the displacement and the value of the fluctuating quantity φ'. Suppose, for example, that an eddy causes parcels to be displaced from their mean positions. If a displaced parcel mixes with its surroundings, then a correlation will develop between v' and φ', and we would have $\overline{v'\varphi'} \neq 0$. However, if no mixing occurs, then the eddy simply recirculates with eddies retaining their initial values, and $\overline{v'\varphi'}$ is zero because of a lack of correlation between the two quantities. Thus, it is essential that there be a degree of *irreversibility* to the flow in order for turbulent diffusion to be appropriate.

Molecular diffusion is not the only process that enables an eddy to assume the value of its surroundings — a Newtonian or other relaxation back to a specified temperature may have much the same effect. Indeed, in the atmosphere a displaced parcel will be subject to a radiation field that acts qualitatively in this way. For example, suppose that the temperature equation is

$$\frac{\mathrm{D}T}{\mathrm{D}t} = -\lambda(T - T^*(y,z)), \tag{13.64}$$

where the right-hand side crudely represents radiative effects via a relaxation back to a specified profile. Then a displaced parcel will be subject to a radiative damping that is different from that at its initial position, and this will allow the parcel to take on the value of its surroundings, and so potentially enable turbulent diffusion to occur (provided λ is small so that T is approximately materially conserved).

For molecular diffusion to be the mechanism whereby a parcel mixes with its surroundings, the turbulence must create scales that are small enough for diffusion to act. This means that turbulence must create a cascade of φ-stuff to small scales. This is quite consistent with the notion that φ is a materially conserved quantity, because a scalar field φ that satisfies

$$\frac{\mathrm{D}\varphi}{\mathrm{D}t} = F + \kappa\nabla^2\varphi, \tag{13.65}$$

where F might represent a spectrally local source of variance of φ, is certainly cascaded to smaller scales. The presence of a molecular diffusion does not substantially affect the requirement that φ be conserved on parcels, because on scales comparable to the eddy mixing length the effect of molecular diffusion is negligible. (And if it were not, perhaps because κ was extremely large or because the turbulence was anaemic, we would not be particularly interested in the turbulent transport.)

13.4.2 A Macroscopic Perspective

Consider turbulent diffusion from a more macroscopic point of view, and in particular consider the transport of a nearly materially conserved tracer obeying

$$\frac{D\varphi}{Dt} = D, \tag{13.66}$$

where the advecting flow is incompressible and D is a dissipative process such that $\overline{D\varphi} \leq 0$ (a conventional harmonic diffusion has this property). By decomposing the fields into mean and eddy components in the usual way an equation for the evolution of the tracer variance can be straightforwardly derived, namely

$$\frac{1}{2}\frac{\partial}{\partial t}\overline{\varphi'^2} + \overline{\boldsymbol{v}'\varphi'} \cdot \nabla\overline{\varphi} + \frac{1}{2}\overline{\boldsymbol{v}} \cdot \nabla\overline{\varphi'^2} + \frac{1}{2}\nabla \cdot \overline{\boldsymbol{v}'\varphi'^2} = \overline{D'\varphi'}, \tag{13.67}$$

and where we may assume $\overline{D'\varphi'} < 0$. If the mean flow is small and if the third-order term may be neglected then in a statistically steady state we have

$$\overline{\boldsymbol{v}'\varphi'} \cdot \nabla\overline{\varphi} \approx \overline{D'\varphi'} < 0. \tag{13.68}$$

Therefore, on average, the flux of φ is downgradient in regions of dissipation, implying a positive average eddy diffusivity, and a balance is maintained between the downgradient flux of φ (which increases the variance) and dissipation. However, it should also be clear from (13.67) that if the turbulence is not statistically stationary, or if there is a mean flow, then downgradient transport cannot necessarily be expected. Indeed, the transport may be upgradient in regions where the eddy variance is falling, for then we may have the balance

$$\overline{\boldsymbol{v}'\varphi'} \cdot \nabla\overline{\varphi} \approx -\frac{1}{2}\frac{\partial}{\partial t}\overline{\varphi'^2} > 0. \tag{13.69}$$

13.5 HOMOGENIZATION OF A SCALAR THAT IS ADVECTED AND DIFFUSED

Let us now assume that the effects of turbulence on a tracer are indeed diffusive. An important consequence of this is that, in the absence of additional forcing, there can be no extreme values of the tracer in the interior of the fluid and, in some circumstances, the diffusion will *homogenize* values of the tracer in broad regions. In this section we demonstrate and explore these properties.

13.5.1 Non-existence of Extrema

Consider a tracer that obeys the equation

$$\frac{D\varphi}{Dt} = \nabla \cdot (\kappa\nabla\varphi) + S, \tag{13.70}$$

where $\kappa > 0$ and the advecting velocity is divergence-free. We now show that in regions where the source term, S, is zero there can be no interior extrema of φ if the flow is steady. The proof is in the form of a *reductio ad absurdum* argument — we first suppose there *is* an extrema of φ in the fluid, and show a contradiction.

Given an extremum, there will then be a surrounding surface (in three dimensions), or a surrounding contour (in two), connecting constant values of φ. For definiteness consider two-dimensional incompressible flow for which the steady flow satisfies

$$\nabla \cdot (\boldsymbol{u}\varphi) = \nabla \cdot (\kappa\nabla\varphi). \tag{13.71}$$

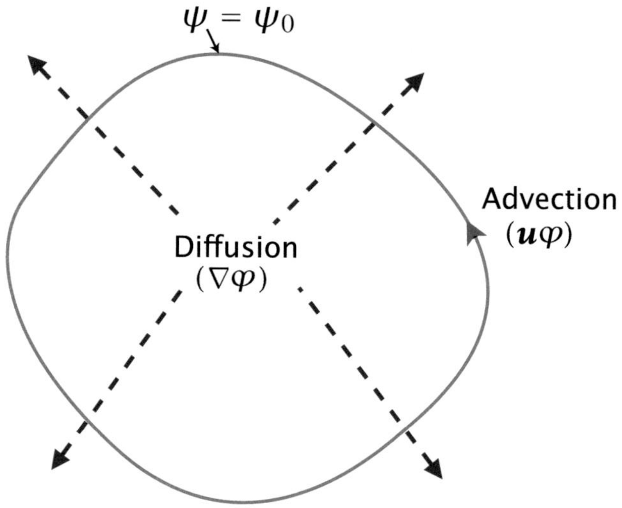

Fig. 13.5 If an extremum of a tracer φ exists in the fluid interior, then diffusion will provide a downgradient tracer flux. But over an area bounded by a streamline, or by an isoline of φ, the net advective flux is zero. Thus, the diffusion cannot be balanced by advection and so in a steady state no extrema can exist.

Integrating the left-hand side over the area, A, enclosed by the iso-line of φ, and applying the divergence theorem, gives

$$\iint_A \nabla\cdot(\boldsymbol{u}\varphi)\,\mathrm{d}A = \oint(\boldsymbol{u}\varphi)\cdot\boldsymbol{n}\,\mathrm{d}l = \varphi\oint\boldsymbol{u}\cdot\boldsymbol{n}\ \mathrm{d}l = \varphi\iint_A \nabla\cdot\boldsymbol{u}\,\mathrm{d}A = 0, \tag{13.72}$$

where $\boldsymbol{n}$ is a unit vector normal to the contour. (If we were to integrate over an area bounded by a velocity contour, then $\boldsymbol{u}\cdot\boldsymbol{n} = 0$ and the integral would similarly vanish.) But the integral of the right-hand side of (13.71) over the same area is non-zero; that is

$$\iint_A \nabla\cdot(\kappa\nabla\varphi)\,\mathrm{d}A = \oint\kappa\nabla\varphi\cdot\boldsymbol{n}\,\mathrm{d}l \neq 0, \tag{13.73}$$

if the integral surrounds an extremum. This is a contradiction for steady flow. Hence, there can be no isolated extrema of a conserved quantity in the interior of a fluid, if there is any diffusion at all. The result (which applies in two or three dimensions) is kinematic, in that φ can be any tracer at all, active or passive. The physical essence of the result is that the integrated effects of diffusion are non-zero surrounding an extremum, and cannot be balanced by advection. Thus, if the initial conditions contain an extremum, diffusion will smooth away the extremum until it no longer exists. This process, and the homogenization discussed below, are illustrated in Fig. 13.5.

13.5.2 Homogenization in Two-dimensional Flow

For two-dimensional flow we can obtain a still stronger result if we allow ourselves to make more assumptions about the strength and nature of the diffusion. The steady distribution of a scalar quantity being advected by an incompressible flow is governed by

$$J(\psi,\varphi) = \nabla\cdot(\kappa\nabla\varphi) + S, \tag{13.74}$$

where the terms on the right-hand side represent diffusion and source terms. Suppose that these terms are small, in the sense that the individual terms on the left-hand side nearly balance each other, so that

$$|J(\psi,\varphi)| \ll \frac{U|\varphi|}{L}. \tag{13.75}$$

This means we are in the high Peclet number limit ($P = UL/\kappa \gg 1$), and the dominance of advection suggests that any steady solution to (13.74) is of the form

$$\varphi = G(\psi) + \mathcal{O}(P^{-1}), \tag{13.76}$$

where G is (for the moment) any function of its argument. Thus, isolines of φ are nearly coincident with streamlines, and

$$\nabla\varphi \approx \nabla\psi \frac{\mathrm{d}\varphi}{\mathrm{d}\psi}. \tag{13.77}$$

On integrating (13.74) over the area, A, bounded by some closed streamline, $\psi = \psi_0$ say, the left-hand side vanishes and we obtain

$$0 = \iint_A S \,\mathrm{d}A + \oint_{\psi_0} \kappa \nabla\varphi \cdot \boldsymbol{n} \,\mathrm{d}l. \tag{13.78}$$

Using (13.77) then gives

$$\iint_A S \,\mathrm{d}A = -\oint_{\psi_0} \kappa \frac{\mathrm{d}\varphi}{\mathrm{d}\psi} \nabla\psi \cdot \boldsymbol{n} \,\mathrm{d}l. \tag{13.79}$$

Since $\mathrm{d}\varphi/\mathrm{d}\psi$ is constant along streamlines, and using $\boldsymbol{u} = \nabla^{\perp}\psi$, we have

$$\frac{\mathrm{d}\varphi}{\mathrm{d}\psi} = -\frac{\iint S \,\mathrm{d}A}{\oint_{\psi_0} \kappa \boldsymbol{u} \cdot \mathrm{d}\boldsymbol{l}}. \tag{13.80}$$

This relationship determines φ as a function of ψ — that is, it determines $G(\psi)$ — in terms of the forcing and dissipation acting on the fluid. If the fluid is both unforced and inviscid, then a steady solution obtains when φ is an *arbitrary* function of ψ. If the source term S is zero, but dissipation is non-zero then the denominator of (13.80) is non-zero and therefore φ must be uniform: φ has been *homogenized*. The homogenization result also follows if we choose to integrate over an area surrounded by an isoline of φ, φ_0 say, but we leave this as a problem for the reader.

Interpretation

The homogenization result applies to a statistically steady flow in which the eddy transport of φ-stuff by the eddying motion may be parameterized diffusively, and in which there is an approximate functional relationship between mean φ and mean ψ. The first of these assumptions we have discussed at length in previous sections. The second requires that the diffusion must not be too strong, so that locally the tracer is conserved on fluid parcels. In the steady state the tracer is then a function of the streamfunction, the same function everywhere within the closed region.

Given these assumptions, the dynamics giving rise to homogenization is transparent: integrating round a contour of ψ or φ the effect of the advective terms vanishes; the source (S) and the diffusion must balance each other, and if there is no source term there can be no tracer gradient. Put another way, the flow will circulate endlessly and steadily around the contours of ψ, which nearly coincide with contours of φ. Advection cannot alter the mean value of φ, so diffusion smooths out gradients within the closed contours, effectively *expelling* gradients of φ to the boundaries and forming a plateau of φ-values. Because extrema of φ are forbidden, the value of φ on the plateau cannot be a maximum or minimum: at the edge of the plateau the values of φ must fall somewhere, and rise somewhere else. The plateau can be a flat region etched out of a hillside, but a plateau on top of a butte is forbidden, for in that case diffusion would erode the butte down to the level of the surrounding land. Our derivation makes no distinction between a passive scalar like a dye and an active scalar, like potential vorticity. In reality, in the latter case the dynamics will further constrain the flow because the scalar distribution must be consistent with the velocity field that advects it, and this is particularly important in the dynamics of ocean gyres.

13.6 ♦ DIFFUSIVE FLUXES AND SKEW FLUXES

Thus far we have considered diffusion using a scalar diffusivity, and this is what is usually meant by diffusion. However, if only from a mathematical point of view, we may allow the diffusivity to be a tensor and this turns out to be very useful when considering the effects of baroclinic eddies. Having a tensor diffusivity allows for the possibility of *skew fluxes,* which are perpendicular to the gradient of the diffused quantity, unlike conventional diffusive fluxes, which are downgradient.

13.6.1 Symmetric and Antisymmetric Diffusivity Tensors

A tracer evolving freely save for the effects of molecular diffusivity, κ_m, obeys the equation

$$\frac{\mathrm{D}\varphi}{\mathrm{D}t} = \nabla \cdot (\kappa_m \nabla \varphi), \tag{13.81}$$

where κ_m is a positive scalar quantity. In the more general case we might have

$$\frac{\mathrm{D}\varphi}{\mathrm{D}t} = -\nabla \cdot \boldsymbol{F} = \nabla \cdot \mathsf{K} \nabla \varphi, \tag{13.82}$$

where K is (if φ is a scalar) a second-rank tensor and $\boldsymbol{F} = -\mathsf{K}\nabla\varphi$ is the diffusive flux of φ. The flux has a component across the isosurfaces of φ, called the diffusive flux, and a component along the iso-surfaces, called the skew flux We will see that these fluxes are associated with the symmetric and antisymmetric components of the diffusivity tensor, respectively, where

$$\mathsf{K} = \mathsf{S} + \mathsf{A}, \tag{13.83}$$

and, using component notation,

$$S_{mn} = \frac{1}{2}(K_{mn} + K_{nm}), \qquad A_{mn} = \frac{1}{2}(K_{mn} - K_{nm}). \tag{13.84}$$

The diagonal elements of the antisymmetric tensor are zero. The transport that is effected by these two tensors has different physical characteristics, as we now discuss.

Diffusion with the symmetric tensor

In the simplest case of all, with an isotropic medium, K is diagonal with equal entries,

$$\mathsf{K} = \mathsf{S} = \begin{pmatrix} \kappa & 0 & 0 \\ 0 & \kappa & 0 \\ 0 & 0 & \kappa \end{pmatrix}, \tag{13.85}$$

and we have the familiar $\boldsymbol{F} = -\kappa\nabla\varphi$, and (13.82) has the same form as (13.81). If κ is positive, then the flux is *downgradient*, meaning that

$$\boldsymbol{F} \cdot \nabla\varphi < 0, \tag{13.86}$$

even if κ is spatially non-uniform. Furthermore, such a diffusion is variance-dissipating; to see this, suppose we have the equation of motion

$$\frac{\mathrm{D}\varphi}{\mathrm{D}t} = \nabla \cdot (\kappa \nabla \varphi). \tag{13.87}$$

Multiplying by φ and integrating over the domain V gives

$$\frac{1}{2}\frac{\mathrm{d}}{\mathrm{d}t}\int_V \varphi^2 \,\mathrm{d}V = \int_V \boldsymbol{F} \cdot \nabla\varphi \,\mathrm{d}V = -\int_V \kappa(\nabla\varphi)^2 \,\mathrm{d}V \leq 0, \tag{13.88}$$

after an integration by parts and assuming that the normal derivative of φ vanishes at the boundaries; that is, there is no flux of φ-stuff through the boundary. However, diffusion does preserve the first moment of the field; that is

$$\frac{\mathrm{d}}{\mathrm{d}t}\int_V \varphi \,\mathrm{d}V = \int_V \nabla\cdot(\kappa\nabla\varphi)\,\mathrm{d}V = 0, \tag{13.89}$$

again assuming no flux through the boundaries.

The transport that is effected by the symmetric diffusion tensor is the diffusive flux, $\boldsymbol{F}_d$, where

$$\boldsymbol{F}_d = -\mathbf{S}\nabla\varphi = -S_{mn}\partial_n\varphi, \tag{13.90}$$

where we employ the common convention that repeated indices are summed. In general, the flux has a component that is parallel to the tracer gradient; that is, $\boldsymbol{F}_d\cdot\nabla\varphi \neq 0$. Suppose we have the simple equation of motion

$$\frac{\partial\varphi}{\partial t} = -\nabla\cdot\boldsymbol{F}_d = \nabla\cdot(\mathbf{S}\nabla\varphi). \tag{13.91}$$

This equation preserves the first moment of φ, provided there is no flux through the boundary. Tracer variance evolves according to

$$\frac{1}{2}\frac{\partial}{\partial t}\int_V \varphi^2\,\mathrm{d}V = \int_V \varphi\nabla\cdot(\mathbf{S}\nabla\varphi) = -\int_V (\mathbf{S}\nabla\varphi)\cdot\nabla\varphi\,\mathrm{d}V. \tag{13.92}$$

This can be shown to be negative or zero, provided that $\mathbf{S}$ is positive semi-definite, meaning that

$$\nabla\varphi\,\mathbf{S}\,\nabla\varphi = \partial_m\varphi\, S_{mn}\partial_n\varphi \geq 0. \tag{13.93}$$

The flux effected by such a diffusivity is then downgradient in the sense that

$$\boldsymbol{F}_d\cdot\nabla\varphi = -\mathbf{S}\nabla\varphi\cdot\nabla\varphi \leq 0. \tag{13.94}$$

The skew flux

The transport associated with the antisymmetric transport tensor is perpendicular to the gradient of φ, and so is neither upgradient nor downgradient. The flux is

$$\boldsymbol{F}_{sk} = -\mathbf{A}\nabla\varphi = -A_{mn}\partial_n\varphi, \tag{13.95}$$

and thus

$$\boldsymbol{F}_{sk}\cdot\nabla\varphi = -\mathbf{A}\nabla\varphi\cdot\nabla\varphi = -A_{mn}\partial_n\varphi\partial_m\varphi = 0, \tag{13.96}$$

where the final result follows because of the antisymmetry of $\mathbf{A}$ — the contraction of a symmetric tensor and an antisymmetric tensor is zero.[10] For this reason, the associated transport is known as a *skew flux* (a term applying in general to fluxes that are perpendicular to the tracer gradient) or a *skew diffusion* (when those fluxes are parameterized using an antisymmetric diffusivity). It follows from this that if a tracer obeys

$$\frac{\partial\varphi}{\partial t} = \nabla\cdot(\mathbf{A}\nabla\varphi), \tag{13.97}$$

then the tracer variance is conserved. This may be verified by multiplying this equation by φ and integrating by parts, assuming that the flux vanishes at the boundaries. That is, *a skew diffusion has no effect on the variance of the skew diffused variable.* One other familiar physical process shares these properties, and that is advection by a divergence-free flow. A skew diffusion is physically equivalent to such an advection in that the divergence of a skew diffusive flux is the same as the divergence of an appropriately chosen advective flux.

To explore this more, define an advective flux of a tracer φ to be a flux of the form

$$\boldsymbol{F}_{ad} \equiv \widetilde{\boldsymbol{v}}\varphi, \tag{13.98}$$

where $\widetilde{\boldsymbol{v}}$ is a divergence-free vector field. The divergence of the flux is just

$$\nabla \cdot \boldsymbol{F}_{ad} = \nabla \cdot (\widetilde{\boldsymbol{v}}\varphi) = \widetilde{\boldsymbol{v}} \cdot \nabla\varphi. \tag{13.99}$$

The field $\widetilde{\boldsymbol{v}}$ might be called a *pseudovelocity* or a *quasi-velocity* — it acts like a velocity but is not necessarily the velocity of any fluid particle. Because $\widetilde{\boldsymbol{v}}$ is divergence-free, we may define a vector streamfunction $\boldsymbol{\psi}$ such that

$$\widetilde{\boldsymbol{v}} = \nabla \times \boldsymbol{\psi} \qquad \text{or} \qquad \widetilde{v}_n = \epsilon_{lmn}\partial_l \psi_m. \tag{13.100}$$

The Levi–Civita symbol ϵ_{lmn} is such that $\epsilon_{123} = \epsilon_{231} = \epsilon_{312} = 1$, $\epsilon_{132} = \epsilon_{321} = \epsilon_{213} = -1$, and $\epsilon_{lmn} = 0$ for other combinations. The equivalence of the two expressions may be verified by expansion in Cartesian coordinates. The field $\boldsymbol{\psi}$ is not unique: the gradient of an arbitrary function may be added to it, this gradient vanishing upon taking the curl, resulting in the same velocity field. That is, if $\boldsymbol{\psi}' = \boldsymbol{\psi} + \nabla\gamma$, then $\widetilde{\boldsymbol{v}} = \nabla \times \boldsymbol{\psi} = \nabla \times \boldsymbol{\psi}'$. The scalar field γ is known as the *gauge,* and the freedom to choose it is the gauge freedom.

The advective flux $\boldsymbol{F}_{ad}$ is related to the skew flux $\boldsymbol{F}_{sk}$ by

$$\varphi\widetilde{\boldsymbol{v}} = \varphi\nabla \times \boldsymbol{\psi} = \nabla \times (\varphi\boldsymbol{\psi}) - \nabla\varphi \times \boldsymbol{\psi}, \tag{13.101}$$

or

$$\boldsymbol{F}_{ad} = \boldsymbol{F}_r + \boldsymbol{F}_{sk}, \tag{13.102}$$

where $\boldsymbol{F}_r = \nabla \times (\varphi\boldsymbol{\psi})$ is a rotational flux with no divergence, and

$$\boldsymbol{F}_{sk} = -\nabla\varphi \times \boldsymbol{\psi} \tag{13.103}$$

is a skew flux — 'skewed' because it is manifestly orthogonal to the gradient of φ, i.e., $\nabla\varphi \cdot \boldsymbol{F}_{sk} = 0$. Because $\nabla \cdot \boldsymbol{F}_{\mathrm{r}} = 0$ the divergence of the skew flux and advective flux are equal:

$$\nabla \cdot \boldsymbol{F}_{ad} = \nabla \cdot \boldsymbol{F}_{sk}. \tag{13.104}$$

However, the skew flux, $-\nabla\varphi \times \boldsymbol{\psi}$, and the advective flux, $\varphi\nabla \times \boldsymbol{\psi}$, may have, and in general do have, different magnitudes and directions; only their divergences are equal. If the divergences of the skew fluxes given by (13.95) and (13.103) are to be the same then $\boldsymbol{\psi}$ must be related to the antisymmetric tensor A. Using (13.95) we have

$$\begin{aligned}\nabla \cdot \boldsymbol{F}_{sk} &= -\partial_m(A_{mn}\partial_n\varphi) \\ &= -(\partial_n\varphi)(\partial_m A_{mn}) - [A_{mn}\partial_n\partial_m\varphi] \\ &= -\partial_n(\varphi\partial_m A_{mn}) + [\varphi\partial_n\partial_m A_{mn}],\end{aligned} \tag{13.105}$$

where the quantities in square brackets are zero as a consequence of the antisymmetry of A — a symmetric operator acting on an antisymmetric tensor is zero. But the skew flux divergence is equal to the advective flux divergence

$$\nabla \cdot \boldsymbol{F}_{sk} = \nabla \cdot \boldsymbol{F}_{ad} = \partial_n(\varphi\widetilde{v}_n), \tag{13.106}$$

so the associated skew velocity is related to the antisymmetric tensor

$$\widetilde{v}_n = -\partial_m A_{mn}, \tag{13.107}$$

and this is divergence-free because $\partial_n \partial_m A_{mn} = 0$. The streamfunction and the antisymmetric tensor are thus related.

Using (13.107) and (13.100), and just a little algebra, gives

$$A_{mn} = \epsilon_{mnp}\psi_p = \begin{pmatrix} 0 & \psi_3 & -\psi_2 \\ -\psi_3 & 0 & \psi_1 \\ \psi_2 & -\psi_1 & 0 \end{pmatrix}, \tag{13.108}$$

which provides an explicit connection between the antisymmetric tensor A_{mn} and the streamfunction for the skew velocity $\tilde{\boldsymbol{v}}$. Thus, to summarize:

- Any flux can be decomposed into a component across iso-surfaces of a scalar (the along-gradient or diffusive flux) and a component along isosurfaces and so perpendicular to the gradient (the skew flux).
- The along-gradient (usually downgradient) flux is effected by a diffusion using a symmetric diffusivity tensor.
- The skew flux is effected by a diffusion using an antisymmetric diffusivity tensor, and this is equivalent to an advection by some divergence-free velocity.
- The diffusive flux reduces tracer variance if the diffusivity is positive (in which case the diffusion is downgradient), whereas the skew diffusion has no effect on variance.

Let us now consider how all of this is relevant to the large-scale flow in the atmosphere and the ocean.

13.7 † EDDY DIFFUSION IN THE ATMOSPHERE AND OCEAN

We now, rather heuristically, discuss the transport of fluid properties by large-scale eddies typically generated by baroclinic instability — mesoscale eddies in the ocean, and weather systems in the atmosphere. The practical motivation is perhaps more oceanographic than atmospheric. Specifically, mesoscale and submesoscale eddies in the ocean cannot be easily resolved in numerical models of its large-scale circulation, especially those used for climate simulations involving integrations of the global ocean over decades and centuries. In such models, the effects of eddies must be *parameterized* in terms of properties of the mean flow.

In the end this problem will be solved for us by the increasing power of computers, as it largely has in atmospheric flows since numerical models of the general circulation already resolve most of the effects of baroclinic eddies. However, we then have to deal with the hardly less difficult problem of understanding those massive, turbulent, numerical integrations, and for that task a theory of turbulent transport is a *sine qua non.*

13.7.1 Preliminaries

Consider a tracer that obeys the advective-diffusive equation

$$\frac{\mathrm{D}\varphi}{\mathrm{D}t} = \nabla\cdot(\kappa_m \nabla\varphi). \tag{13.109}$$

If the advecting flow is divergence-free then the ensemble average or filtered flow obeys, neglecting the molecular diffusion,

$$\frac{\mathrm{D}\overline{\varphi}}{\mathrm{D}t} = -\nabla\cdot\overline{\boldsymbol{v}'\varphi'}, \tag{13.110}$$

where the right-hand side is the eddy transport (akin to Reynolds stresses). If we parameterize this transport by a diffusion then

$$\overline{\boldsymbol{v}'\varphi'} = -\boldsymbol{K}\nabla\overline{\varphi}, \tag{13.111}$$

where $\boldsymbol{K}$ is, in general, a second-rank tensor. If, say, the average is a zonal average then

$$\frac{\mathrm{D}\overline{\varphi}}{\mathrm{D}t} = -\frac{\partial \overline{v'\varphi'}}{\partial y} - \frac{\partial \overline{w'\varphi'}}{\partial z}. \tag{13.112}$$

If we are to employ a diffusive parameterization for the eddy terms in these equations, the issues that then arise fall into two general camps:

(i) the overall *magnitude* of the eddy diffusivity, possibly as a function of the mean flow;

(ii) the *structure* of the diffusivity tensor, and in particular the separate structure of its symmetric and antisymmetric parts.

13.7.2 Magnitude of the Eddy Diffusivity

The magnitude and scale of eddies were considered in Sections 12.3 and 12.4. Here we see how these give rise to corresponding estimates of the magnitude of an eddy diffusivity. If we restrict attention for the moment to the meridional transfer of tracer properties, then we might write

$$\overline{v'\varphi'} = -\kappa^{vy}\frac{\partial \overline{\varphi}}{\partial y} - \kappa^{vz}\frac{\partial \overline{\varphi}}{\partial z}, \tag{13.113}$$

where κ^{vy} and κ^{vz} are components of the eddy diffusivity tensor with obvious notation. These components have the dimensions of a length times a velocity and, to the extent that the diffusion represents the eddying motion we expect that κ^{vy} has an approximate magnitude of

$$\kappa^{vy} \sim v'l', \tag{13.114}$$

where v' is a typical magnitude of the horizontal eddy velocity, and l' is the *mixing length* of the eddies, generally taken to be a typical length scale of the eddies. Larger and more energetic eddies thus have a larger effect on the mean flow. We can estimate v' and l' in a number of reasonable ways depending on the flow conditions, and we consider a few such below.[11] The magnitude of the component κ^{vz} may then be estimated by making choices about the plane of parcel displacements, and this is considered in the next section.

Perhaps the simplest assumption to make follows from the fact that the eddies are a consequence of baroclinic instability, and so one might suppose that the eddy length scale is the scale of the instability — the first deformation radius. One might also suppose that the eddy velocity is of the same approximate magnitude as the mean flow, $\overline{u}$, thus giving

$$\kappa^{vy} \sim L_d\overline{u} = \frac{NH\overline{u}}{f}. \tag{13.115}$$

Another way of deriving this result is by noting that $\kappa^{vy} \sim l'^2/T_e$, where T_e is a characteristic eddy time scale. If T_e is the Eady time scale, $L_d/\overline{u}$, and if $l' \sim L_d$, we reproduce (13.115). Equation (13.115) may be written as

$$\kappa^{vy} \sim L_d\overline{u} \sim \frac{L_d^2 f}{\sqrt{Ri}} \sim L_d^2 Fr f, \tag{13.116}$$

where $Ri = N^2/\Lambda^2 = N^2H^2/\overline{u}^2$ and $Fr = U/(NH)$ are the Richardson and Froude numbers for this problem, respectively.

A little more generally, if there is a cascade to larger scales then the eddy scale, L_e say, may be larger than the deformation scale. Depending on circumstances, L_e might be the domain scale (if eddies grow to the size of the domain), the β-scale (if the β-effect halts the cascade), or some scale determined by frictional effects (possibly in conjunction with β). However, the arguments of

Section 12.3 suggest that the eddy time scale is the Eady time scale in all cases. We therefore have:

$$\text{eddy length scale} \sim L_e, \tag{13.117a}$$
$$\text{eddy time scale, } T_e \sim L_d/\overline{u}, \tag{13.117b}$$
$$\text{eddy velocity scale, } U_e \sim \overline{u}(L_e/L_d). \tag{13.117c}$$

These give the general estimate for the horizontal diffusivity of

$$\kappa^{vy} \sim \overline{u}\left(\frac{L_e^2}{L_d}\right). \tag{13.118}$$

The estimate (13.115) is a special case of this, with $L_e = L_d$; the two estimates will thus differ if the eddy scale is much larger than the deformation radius.

In the case in which the inverse cascade is modified by the Rossby waves we might (and neglecting friction effects; see Section 12.1.2) suppose that the eddy scale is the β-scale, (12.6), and we have

$$L_e \sim L_\beta = \left(\frac{U_e}{\beta}\right)^{1/2} = \frac{\overline{u}}{\beta L_d}, \tag{13.119}$$

using (13.117c). The eddy velocity scale is, using (13.117b),

$$U_e \sim \overline{u}\frac{L_e}{L_d} = \frac{\overline{u}^2}{\beta L_d^2}, \tag{13.120}$$

and combining (13.119) and (13.120) gives the estimate for the eddy diffusivity,

$$\kappa^{vy} \sim \frac{\overline{u}^3}{\beta^2 L_d^3}. \tag{13.121}$$

A similar estimate can be written in terms of the inverse energy cascade rate, ε, giving

$$\kappa^{vy} \sim \left(\frac{\varepsilon^3}{\beta^4}\right)^{1/5}. \tag{13.122}$$

This expression may be obtained purely by dimensional analysis, if it is assumed that the only factors determining κ are ε and β. The estimate may be useful if ϵ is known independently, for example by calculating the energy throughput in the system.

To summarize: the magnitude of any eddy diffusion may be estimated as the product of the velocity scale and the energy-containing length scale of the eddies. If we assume that the time scale is the Eady time scale we obtain (13.118), where L_e is undetermined. If the eddy scale is the β-scale, then (13.118) becomes (13.121). However, in neither the atmosphere nor the ocean is the β-scale significantly (e.g., an order of magnitude) larger than the deformation scale, but nor, complicating matters, does the inverse cascade necessarily halt at the β-scale (see Section 12.1.2). From an observational standpoint, the atmosphere has no $-5/3$ inverse cascade, although there is some evidence for one in some regions of ocean.[12] In other oceanic regions the eddies may be advected away from each other and away from the unstable zone, or dispersed by Rossby waves, before an inverse cascade can be organized, and energy will remain at the deformation scale. These arguments suggest that although we can make sensible estimates, we cannot determine with certainty what the magnitude of an eddy diffusivity should be, in either the atmosphere or ocean.

13.7.3 ♦ Use of the Symmetric Transport Tensor

Along-gradient transport, or diffusion, is transport by a symmetric transport tensor as in (13.90). The structure of the eddy diffusivity will determine, among other things, the surface along which transport occurs; for example, a diffusion of temperature might occur meridionally and/or vertically. To illustrate, let us consider transfer in a re-entrant channel with zonally homogeneous eddy statistics so that the averaging operator is the zonal average; the meridional and upward transport of a tracer φ are then given by

$$\overline{v'\varphi'} = -\kappa^{vy}\frac{\partial\overline{\varphi}}{\partial y} - \kappa^{vz}\frac{\partial\overline{\varphi}}{\partial z}, \tag{13.123}$$

$$\overline{w'\varphi'} = -\kappa^{wy}\frac{\partial\overline{\varphi}}{\partial y} - \kappa^{wz}\frac{\partial\overline{\varphi}}{\partial z}, \tag{13.124}$$

where $\kappa^{wy} = \kappa^{vz}$ by the posited symmetry. The relationship between the various transfer coefficients will be determined by the trajectories of the fluid parcels in the eddying motion. In the Cartesian y–z frame the transport tensor is not necessarily diagonal (i.e., κ^{vz} and κ^{wy} may be non-zero) but locally there is always a natural coordinate system in which the diffusivity tensor is diagonal. (A symmetric matrix may always be diagonalized by a suitable rotation of axes.) In that diagonal frame we can write

$$\mathbf{S}' = \kappa_s\begin{pmatrix}1 & 0\\ 0 & \alpha\end{pmatrix}, \tag{13.125}$$

where κ_s determines the overall size of the transfer coefficients (as estimated in the previous section), and α is the ratio of sizes of the components in the two orthogonal directions. Now, fluid displacements in large-scale baroclinic eddies are nearly, but not exactly, horizontal — they may be along isopycnals, for example, or at an angle between the horizontal and the isopycnals. We may argue that the coordinate system in which the tensor is diagonal is the coordinate system defined by the plane along which fluid displacements occur. This is sensible because the transfers along and orthogonal to the fluid paths are each a consequence of different physical phenomena, and so we may expect the transfer tensor to be diagonal in these coordinates.

Because eddy displacements are predominantly horizontal, the diagonal coordinate system has a small slope, s, at an angle θ with respect to the horizontal, where $s = \tan\theta \approx \theta \ll 1$. Furthermore, we expect the parameter α to be small (i.e., $\alpha \ll 1$), because this represents transfer in a direction orthogonal to the eddy fluid motion. We rotate the tensor $\mathbf{S}'$ through an angle θ to move into the usual y-z frame; that is

$$\mathbf{S} = \kappa_s\begin{pmatrix}\cos\theta & -\sin\theta\\ \sin\theta & \cos\theta\end{pmatrix}\begin{pmatrix}1 & 0\\ 0 & \alpha\end{pmatrix}\begin{pmatrix}\cos\theta & \sin\theta\\ -\sin\theta & \cos\theta\end{pmatrix} \tag{13.126a}$$

$$\approx \kappa_s\begin{pmatrix}1+s^2\alpha & s(1-\alpha)\\ s(1-\alpha) & s^2+\alpha\end{pmatrix} \qquad \text{(for small } s) \tag{13.126b}$$

$$\approx \kappa_s\begin{pmatrix}1 & s\\ s & s^2+\alpha\end{pmatrix} \qquad \text{(for small } s \text{ and small } \alpha). \tag{13.126c}$$

We can follow the same procedure in three dimensions. Then, if the eddy transport is isotropic in the plane of eddy displacements, the three-dimensional transport tensor is

$$\mathbf{S}' = \kappa_s\begin{pmatrix}1 & 0 & 0\\ 0 & 1 & 0\\ 0 & 0 & \alpha\end{pmatrix}, \tag{13.127}$$

and the slope of the motion is a two-dimensional vector $\boldsymbol{s} = (s^x, s^y)$, with the superscripts denoting components. If we rotate the transport tensor into physical space then we obtain, analogously to

(13.126),

$$\mathsf{S} = \kappa_s \begin{pmatrix} 1 + s^{y2} + \alpha s^{x2} & (\alpha - 1)s^x s^y & (1-\alpha)s^x \\ (\alpha - 1)s^x s^y & 1 + s^{x2} + \alpha s^{y2} & (1-\alpha)s^y \\ (1-\alpha)s^x & (1-\alpha)s^y & \alpha + s^2 \end{pmatrix} \tag{13.128a}$$

$$\approx \kappa_s \begin{pmatrix} 1 & 0 & s^x \\ 0 & 1 & s^y \\ s^x & s^y & \alpha + s^2 \end{pmatrix}, \tag{13.128b}$$

for small s and small α, where $s^2 = s^{x2} + s^{y2}$.

The plane of eddy displacements

We are now in a position to make heuristic choices about the transfer coefficients, and we will consider two bases for this:

I. *Using linear baroclinic instability theory.*[13] In a simple model of a growing baroclinic (Eady) wave, parcel trajectories that are along half the slope of the mean isopycnals are able to release the most potential energy. We thus suppose that $s = s_\rho/2$, where s_ρ is the isopycnal slope, and that $\alpha = 0$ in (13.126c) or (13.128b). In two dimensions this gives

$$\mathsf{S} = \kappa_s \begin{pmatrix} 1 & s_\rho/2 \\ s_\rho/2 & s_\rho^2/4 \end{pmatrix}, \tag{13.129}$$

and so

$$\overline{v'\varphi'} = -\kappa_s \left(\frac{\partial \overline{\varphi}}{\partial y} + \frac{1}{2} s_\rho \frac{\partial \overline{\varphi}}{\partial z} \right), \tag{13.130a}$$

$$\overline{w'\varphi'} = -\frac{1}{2}\kappa_s s_\rho \left(\frac{\partial \overline{\varphi}}{\partial y} + \frac{1}{2} s_\rho \frac{\partial \overline{\varphi}}{\partial z} \right). \tag{13.130b}$$

If the tracer φ is potential temperature (and not just a passive tracer) then (13.130), along with one of the estimates for the size of κ_s given in Section 13.7.2, constitutes a parameterization for the diffusive poleward and upward heat flux in the atmosphere.

II. *Flow along neutral surfaces.* If the fluid interior is adiabatic and steady, then fluid trajectories are along neutral surfaces; that is, along surfaces of potential density or potential temperature. One might therefore be inclined to assume that the eddy fluxes are aligned along the mean neutral surfaces and choose $s = s_\rho$. However, even in the adiabatic case, this is not always a good choice. From the adiabatic thermodynamic equation $\mathrm{D}b/\mathrm{D}t = 0$ we may derive the equation for the eddy buoyancy variance, namely

$$\frac{1}{2}\frac{\partial \overline{b'^2}}{\partial t} + \frac{1}{2}\overline{\boldsymbol{u}}\cdot\nabla_z \overline{b'^2} + \frac{1}{2}\overline{w}\frac{\partial \overline{b'^2}}{\partial z} + \overline{\boldsymbol{u}'b'}\cdot\nabla_z \overline{b} + \overline{w'b'}\frac{\partial \overline{b}}{\partial z} + \frac{1}{2}\nabla_z\cdot\overline{\boldsymbol{u}'b'^2} + \frac{1}{2}\frac{\partial}{\partial z}\overline{w'b'^2} = 0, \tag{13.131}$$

and specialize to the case of a zonally uniform basic state and small-amplitude wave. In that case

$$\frac{1}{2}\frac{\partial \overline{b'^2}}{\partial t} = -\overline{v'b'}\frac{\partial \overline{b}}{\partial y} - \overline{w'b'}\frac{\partial \overline{b}}{\partial z}. \tag{13.132}$$

If the wave is statistically steady then the left-hand side is zero and

$$\overline{\boldsymbol{v}'b'}\cdot\nabla_x \overline{b} = 0, \tag{13.133}$$

where the subscript x indicates that the vectors are in the meridional plane, with no variation in x. In this case there is indeed no along-gradient flux. However, if the wave is growing then $\overline{\boldsymbol{v}'b'} \cdot \nabla_x \overline{b} < 0$ and if $\partial \overline{b}/\partial y < 0$ and $\overline{v'b'} > 0$, as in the Northern Hemisphere, then

$$\frac{\overline{w'b'}}{\overline{v'b'}} > -\frac{\partial \overline{b}/\partial y}{\partial \overline{b}/\partial z}, \tag{13.134}$$

and so the mixing slope is *less* steep than the mean isopycnal slope, even though the flow may be adiabatic. Similarly, if the wave is decaying the mixing slope is steeper than that of the mean isopycnals. In an inhomogeneous flow, the advection by the mean flow in (13.131) plays a similar role to time dependence: the advection of eddy variance by the mean flow into a region of larger variance will give rise to a mixing slope that is less steep than the isopycnal slope, and conversely for a flow entering a region of less variance. Only for a statistically steady, adiabatic, linear wave field is the mixing slope guaranteed to be along the isopycnals.

Having said all this, let us suppose that the fluid trajectories are indeed along neutral surfaces. If there is no diffusion orthogonal to this then $\alpha = 0$, and the transport tensor is, in two or three dimensions respectively.

$$\mathsf{S} = \kappa_s \begin{pmatrix} 1 & s_\rho \\ s_\rho & s_\rho^2 \end{pmatrix}, \qquad \mathsf{S} = \kappa_s \begin{pmatrix} 1 & 0 & s_\rho^x \\ 0 & 1 & s_\rho^y \\ s_\rho^x & s_\rho^y & |s_\rho|^2 \end{pmatrix}. \tag{13.135}$$

In the two-dimensional case

$$\overline{v'\varphi'} = -\kappa_s \left(\frac{\partial \overline{\varphi}}{\partial y} + s_\rho \frac{\partial \overline{\varphi}}{\partial z} \right), \qquad \overline{w'\varphi'} = -\kappa_s s_\rho \left(\frac{\partial \overline{\varphi}}{\partial y} + s_\rho \frac{\partial \overline{\varphi}}{\partial z} \right). \tag{13.136a,b}$$

Suppose that φ is potential temperature θ, and that surfaces of potential temperature define neutral surfaces. Then plainly eddy motion along potential temperature surfaces does not transfer potential temperature, and the diffusion defined by (13.136) should have no effect. The equations themselves respect this, for then

$$s_\rho = -\frac{\partial_y \overline{\theta}}{\partial_z \overline{\theta}}, \tag{13.137}$$

and using this in (13.136) gives

$$\overline{v'\theta'} = 0, \qquad \overline{w'\theta'} = 0. \tag{13.138a,b}$$

There is no eddy transport at all, as expected.

Application to atmosphere and ocean

In the atmosphere, if we wished to parameterize the heat transporting effects of baroclinic eddies we might choose the mixing slope to be *shallower* than the isothermal slope. Heat may then be transported downgradient, from equator to pole, in a diabatic process. This is a reasonable choice because the eddy transport, in the atmosphere, is diabatic. This choice is *less* appropriate in the ocean, because the ocean interior is almost adiabatic. That is to say, fluid transport is almost along isopycnals, except for some rather small effects involving diapycnal diffusivity. Diffusing buoyancy along isopycnals has no effect at all. In the real ocean the presence of salinity means that the potential temperature, potential density and salinity surfaces are not parallel, and there will be eddy diffusion of θ and S (salinity) along neutral surfaces, but this fact does not help provide a parameterization for the heat flux by baroclinic eddies, because we cannot expect such a flux to depend for its existence on the presence of a second tracer, salinity. However, baroclinic eddies certainly *do* have an effect on the ocean structure, and if our ocean model does not resolve them we must parameterize them. For this, we turn to the antisymmetric transport tensor.

13.7.4 ♦ Use of the Antisymmetric Transport Tensor

The antisymmetric transport tensor gives rise to the skew flux, or the pseudoadvection. In two dimensions (one horizontal, one vertical) we can immediately write down its form, namely

$$\mathsf{A} = \begin{pmatrix} 0 & -\kappa'_a \\ \kappa'_a & 0 \end{pmatrix}, \tag{13.139}$$

where the transfer coefficient κ'_a, which may vary in space and time depending on the flow itself, determines the overall strength of the transport. In three dimensions we can write, by inspection,

$$\mathsf{A} = A_{ij} = \begin{pmatrix} 0 & 0 & -\kappa_a'^x \\ 0 & 0 & -\kappa_a'^y \\ \kappa_a'^x & \kappa_a'^y & 0 \end{pmatrix}, \tag{13.140}$$

where we have used our gauge freedom to choose $A_{21} = -A_{12} = 0$. Equation (13.140) preserves the form of (13.139) if one of the horizontal dimensions is absent — that is, if either row one and column one, or row two and column two, is eliminated. Our remaining choice is to determine the sign and magnitude of the transport coefficients.

An adiabatic, potential-energy diminishing, eddy transport scheme

A very useful parameterization for the transport of tracers in ocean models by baroclinic eddy fluxes, commonly known as the Gent–McWilliams or GM scheme,[14] can be constructed using the antisymmetric transport tensor. The satisfaction of two properties is the foundation of the scheme:

(i) Moments of the tracer should be preserved; in particular, the amount of fluid between two isopycnal surfaces should be preserved. This suggests the scheme should not diffuse buoyancy across constant-buoyancy surfaces.

(ii) The amount of available potential energy in the flow should be reduced, so mimicking the effects of baroclinic instability, which transfers available potential energy to kinetic energy.

The first of these is automatically satisfied by using an antisymmetric diffusivity tensor. The second property can be satisfied by choosing the transfer coefficients to be proportional to the slope of the isopycnals, in which case we may write (13.140) as

$$\mathsf{A} = \kappa_a \begin{pmatrix} 0 & 0 & -s^x \\ 0 & 0 & -s^y \\ s^x & s^y & 0 \end{pmatrix}, \tag{13.141}$$

where $\boldsymbol{s} = (s^x, s^y) = \nabla_\rho z = -\nabla_z \rho/(\partial\rho/\partial z)$ is the isopycnal slope (recall that s^x denotes a component of a vector, and s_x a derivative) and κ_a determines the overall magnitude of the diffusivity. In an ocean model separately carrying temperature and salinity fields, then (13.141) would be applied to each of these, with the isopycnal slope being determined using the equation of state. To more easily see what properties are implied by the transport, let us specialize to the salt-free case, with buoyancy, b, the only thermodynamic variable. The isopycnal slope is then $\boldsymbol{s} = -(b_x/b_z, b_y/b_z)$ and the horizontal eddy buoyancy transfer $\boldsymbol{F}_h = (F^x, F^y)$ is given by

$$\boldsymbol{F}_h = -\left(-\kappa_a \boldsymbol{s}\frac{\partial b}{\partial z}\right) = -\kappa_a\left(\frac{\partial b}{\partial x}, \frac{\partial b}{\partial y}\right) = -\kappa_a \nabla_z b, \tag{13.142a}$$

which for positive κ_a is the same as conventional downgradient diffusion.

The vertical transfer is given by

$$F^z = -\kappa_a\left(s^x\frac{\partial b}{\partial x} + s^y\frac{\partial b}{\partial y}\right) = \kappa_a s^2\frac{\partial b}{\partial z}, \tag{13.142b}$$

where $s^2 = \mathbf{s} \cdot \mathbf{s}$. This flux is *up* the vertical gradient; however, by construction, the total skew flux is neither upgradient nor downgradient.

The combination of the downgradient horizontal flux and the upgradient vertical flux acts to reduce the potential energy of the flow at the same time as preserving the volume of fluid within each density interval. The upgradient flux in the vertical is a consequence of the need to reduce the available potential energy: suppose warm light fluid overlays cold dense fluid in a statically stable configuration, then a downgradient vertical diffusion would raise the centre of gravity of the fluid, increasing its potential energy — just the opposite of the action of baroclinic instability. Thus, the sign on the vertical diffusivity must be negative and this, in combination with the structure of (13.141) (and so a positive horizontal diffusivity) allows both properties *(i)* and *(ii)* above to be satisfied. The parameterization does not preserve total energy; the loss of potential energy is not balanced by a corresponding gain of kinetic energy, rather it is assumed to be lost to dissipation. Finally, to determine the magnitude of the (skew) eddy diffusivity we may turn again to the phenomenological estimates of Section 13.7.2.

The eddy transport velocity

Applying (13.107) to (13.141) gives the eddy transport velocities,

$$\widetilde{\boldsymbol{u}} = -\frac{\partial}{\partial z}(\kappa_a \mathbf{s}), \qquad \widetilde{w} = \nabla_z \cdot (\kappa_a \mathbf{s}). \tag{13.143}$$

The streamfunction associated with $\boldsymbol{A}$ is found using (13.108) and (13.141) giving

$$\boldsymbol{\psi} = (-\kappa_a s^y, \kappa s^x, 0) = \mathbf{k} \times \kappa_a \mathbf{s}. \tag{13.144}$$

Two equivalent ways of implementing the GM parameterization are thus as a skew flux, as in (13.142), or as an advection by the pseudovelocities (13.143). The vanishing of the normal component of the velocity is equivalent to the vanishing of the normal component of the flux at the boundary, and ensures that the scheme conserves tracer moments. The advective flux of buoyancy is just

$$\boldsymbol{F}_{ad} = b\widetilde{\boldsymbol{v}} = b\nabla \times \boldsymbol{\psi} = b\nabla \times (\mathbf{k} \times \kappa_a \mathbf{s}), \tag{13.145}$$

whereas using (13.103) the skew flux is given by

$$\boldsymbol{F}_{sk} = -\nabla b \times \boldsymbol{\psi} = -\nabla b \times (\mathbf{k} \times \kappa_a \mathbf{s}). \tag{13.146}$$

Vector manipulation readily shows that the divergences of these two fluxes are equal.

13.7.5 Examples

Consider a situation with sloping isotherms (and with the density determined solely by temperature) as illustrated in Fig. 13.6. The vertical flux attempts to tighten the temperature distribution, whereas the horizontal flux, being downgradient, attempts to smooth out horizontal inhomogeneities. Taken together, their net effect is to preserve the amount of fluid between any two isotherms, but at the same time to rotate and flatten the isotherms, so reducing the available potential energy of the flow. This is different from a conventional downgradient diffusion. A purely horizontal diffusion would, in principle, act to equalize values at each level, and a three-dimensional downgradient diffusion would try to equalize all values. Thus, a skew flux behaves quite differently from the usual downgradient diffusion, which merely acts to reduce gradients without caring much about other fluid properties.

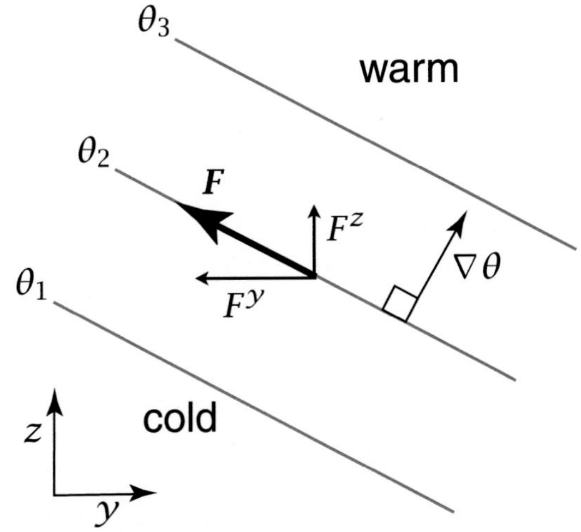

Fig. 13.6 The GM skew fluxes arising from sloping isotherms. The flux itself, $\boldsymbol{F}$, is parallel to the isotherms, with the horizontal flux being directed down the horizontal gradient but the vertical flux being upgradient. The effect of the vertical flux is to lower the centre of gravity of the fluid, and reduce the potential energy. The horizontal flux tries to make the temperature more uniform in the horizontal direction. The net effect of the skew flux is to *flatten* the isotherms.

To illustrate this consider a very simple example, that of a two-dimensional (y–z) fluid in which the initial density field is a 3×3 grid, with initial conditions

$$\rho_{init} = \begin{bmatrix} 2 & 1 & 1 \\ 3 & 2 & 1 \\ 3 & 3 & 2 \end{bmatrix}. \tag{13.147}$$

The isopycnals are sloping, much as in Fig. 13.6, and the flow is statically stable everywhere.

A purely horizontal diffusion would lead to, in the absence of other processes and with zero normal flux at the boundaries, a final state of

$$\rho_{hd} = \begin{bmatrix} 1.33 & 1.33 & 1.33 \\ 2 & 2 & 2 \\ 2.66 & 2.66 & 2.66 \end{bmatrix}, \tag{13.148}$$

and a full (vertical and horizontal) diffusion would give

$$\rho_{hvd} = \begin{bmatrix} 2 & 2 & 2 \\ 2 & 2 & 2 \\ 2 & 2 & 2 \end{bmatrix}. \tag{13.149}$$

Neither of the above two final states preserves the density census (i.e., its distribution) and both imply strong diabatic effects — the fluid has been *mixed*, and the density variance has been reduced.

In contrast, a skew diffusion or eddy-transport advection will rotate the density surfaces until the isopycnal slope is zero, at which point the value of the transfer coefficients becomes zero and the process stops. The final state is then

$$\rho_{GM} = \begin{bmatrix} 1 & 1 & 1 \\ 2 & 2 & 2 \\ 3 & 3 & 3 \end{bmatrix}. \tag{13.150}$$

This action both preserves the density census and reduces the available potential energy.

We can equally well interpret these effects in terms of eddy-transport velocities, so emphasizing that it is not the eddy flux itself that is important; rather, it is the flux divergence. If the slopes of Fig. 13.6 extended uniformly everywhere, then the associated fluxes would have zero divergence, and the eddy-induced velocities, given by (13.143), would be zero. On the other hand, consider

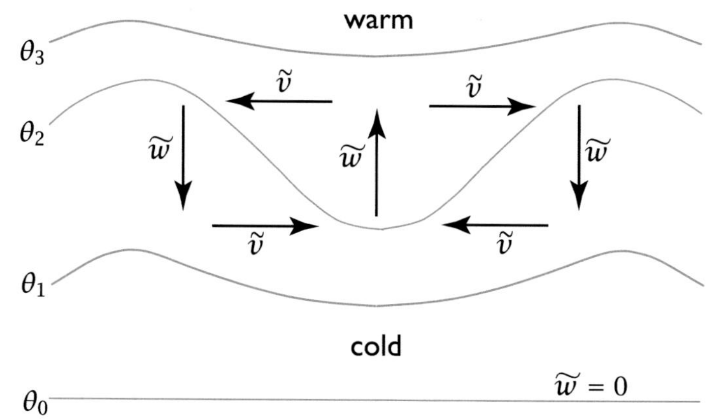

Fig. 13.7 The eddy-induced velocities in the Gent–McWilliams parameterization. The induced circulation attempts to flatten the sloping isopycnals. The induced vertical velocity, $\widetilde{w}$, is zero on flat isopycnals.

the case illustrated in Fig. 13.7, with variously sloping isotherms. For a constant value of the eddy diffusivity κ the slope of the isopycnals, $s = -(\partial\rho/\partial y)/(\partial\rho/\partial z)$, provides the streamfunction for the eddy-induced velocity:

$$\widetilde{v} = -\frac{\partial\psi}{\partial z} = -\frac{\partial}{\partial z}(\kappa_a s), \qquad \widetilde{w} = \frac{\partial\psi}{\partial y} = \frac{\partial}{\partial y}(\kappa_a s). \tag{13.151}$$

This induces the velocities illustrated in Fig. 13.7, which evidently serve to flatten the isopycnals.

Downgradient (symmetric) diffusion and skew diffusion would normally be used together, and the transport tensor will then have both symmetric and antisymmetric components. In ocean models the mixing of temperature and salinity is often chosen to be along isopycnal surfaces. On the other hand, in the atmosphere (especially in the troposphere) diabatic effects are quite important and the symmetric tensor may be chosen to represent cross-isothermal transport.

13.8 † THICKNESS AND POTENTIAL VORTICITY DIFFUSION

In the previous section, we considered the structure of the diffusivity tensor, and then chose the entries by physical reasoning to mimic the effects of baroclinic instability. An alternative approach is to choose, a priori, a quantity to be diffused downgradient (i.e., not skew diffused), and then to represent the effect in the equations of motion as commonly used. In this section we first explore the use of thickness as a 'diffusee', and then look at potential vorticity. Thickness is the vertical distance between two isotherms or isopycnals, and it is a candidate for diffusion because a downgradient thickness transfer within an isopycnal layer satisfies the following two conditions:

(i) The total mass contained between two isopycnals is preserved, provided there are no boundary fluxes, so the effect is adiabatic.

(ii) If there is no orography a thickness flux serves to flatten isopycnals, and hence to reduce the available potential energy of the flow, mimicking baroclinic instability.

These are similar to the two properties listed on page 499, and indeed we will find that thickness diffusion is very similar to the GM scheme. However, thickness is *not* a materially conserved quantity; thus, the arguments of Section 13.4 do not apply and turbulent diffusion of thickness is rather ad hoc. Let us explore all these issues further.

13.8.1 Equations for Thickness

Recalling the results of Section 3.10, in a Boussinesq fluid the 'thickness', or the distance between two surfaces of constant buoyancy, is given by

$$\text{thickness} = \int_{z(b_1)}^{z(b_2)} \mathrm{d}z = \int_{b_1}^{b_2} \frac{\partial z}{\partial b}\,\mathrm{d}b, \tag{13.152}$$

and thus we may define the thickness field (strictly a 'thickness density' field), $\sigma \equiv \partial z/\partial b$. The volume of fluid between two isopycnal surfaces is proportional to $\sigma \Delta A$, where ΔA is an infinitesimal area, and in the absence of diabatic processes this is conserved. Thus, we have $\mathrm{D}(\sigma\Delta A)/\mathrm{D}t = 0$ and using $\mathrm{D}\Delta A/\mathrm{D}t = \Delta A \nabla_b \cdot \boldsymbol{u}$ we obtain the equation of motion for thickness (cf. (3.182))

$$\frac{\mathrm{D}\sigma}{\mathrm{D}t} + \sigma\nabla_b \cdot \boldsymbol{u} = D_\sigma \qquad \text{or} \qquad \frac{\partial \sigma}{\partial t} + \nabla_b \cdot (\boldsymbol{u}\sigma) = D_\sigma, \tag{13.153}$$

now including a term D_σ to represent any diabatic terms. From (13.153) we obtain, after a little algebra, the variance equation

$$\frac{1}{2}\frac{\partial}{\partial t}\overline{\sigma'^2} + \overline{\boldsymbol{u}'\sigma'} \cdot \nabla_b \overline{\sigma} + \frac{1}{2}\overline{\boldsymbol{u}} \cdot \nabla_b \overline{\sigma'^2} + \frac{1}{2}\overline{\boldsymbol{u}' \cdot \nabla_b \sigma'^2} = -\overline{w'\sigma'} + \overline{D'_\sigma \sigma'}, \tag{13.154}$$

where we have written $w' \equiv (\sigma\nabla_b \cdot \boldsymbol{u})'$. This equation is to be compared with the corresponding equation for a conserved tracer, (13.67). If the mean flow is small and the third-order correlations may be neglected then (13.154) becomes, in a statistically steady state,

$$\overline{\boldsymbol{u}'\sigma'} \cdot \nabla_b \overline{\sigma} \approx -\overline{w'\sigma'} + \overline{D'_\sigma \sigma'}. \tag{13.155}$$

Unlike the case for a tracer that is materially conserved except for dissipative effects, the transport of thickness is not necessarily downgradient. However, in regions of baroclinic instability, where there is conversion of available potential energy to kinetic energy $\overline{w'\sigma'}$ is positive, thickness may be transferred downgradient, suggesting a diffusive parameterization.

The eddy-induced and residual velocities

Now, let us decompose these variables in the usual manner into a mean component, denoted with an overbar, and an eddy component, denoted with a prime. The averaged thickness equation is

$$\frac{\partial \overline{\sigma}}{\partial t} + \nabla_b \cdot \left(\overline{\sigma}\,\overline{\boldsymbol{u}} + \overline{\sigma'\boldsymbol{u}'}\right) = 0, \tag{13.156}$$

where $\overline{\sigma'\boldsymbol{u}'}$ is the eddy thickness flux. This equation may be written as

$$\frac{\partial \overline{\sigma}}{\partial t} + \nabla_b \cdot [(\overline{\boldsymbol{u}} + \widetilde{\boldsymbol{u}})\overline{\sigma}] = 0, \tag{13.157}$$

where

$$\widetilde{\boldsymbol{u}} \equiv \frac{\overline{\sigma'\boldsymbol{u}'}}{\overline{\sigma}} \tag{13.158}$$

is the 'eddy-induced velocity', sometimes referred to as the 'bolus velocity', so-called because the thickness flux is said to be evocative of a peristaltic transfer along a passage bounded by impermeable but elastic walls. The quantity

$$\overline{\boldsymbol{u}}^* = \overline{\boldsymbol{u}} + \widetilde{\boldsymbol{u}} \tag{13.159}$$

is the residual velocity we encountered in Chapter 10, and it accounts for the total transport of thickness, including both eddy and Eulerian means.

In adiabatic flow, the evolution of a materially conserved tracer τ is given by

$$\frac{\mathrm{D}}{\mathrm{D}t}(\tau\sigma\Delta A) = 0, \tag{13.160}$$

whence, because $\sigma\Delta A$ is a constant,

$$\frac{\partial \tau}{\partial t} + \boldsymbol{u}\cdot\nabla_b\tau = 0. \tag{13.161}$$

Combining this with the thickness equation, (13.153) with $D_\sigma = 0$, gives

$$\frac{\partial}{\partial t}(\sigma\tau) + \nabla_b\cdot(\sigma\boldsymbol{u}\tau) = 0, \tag{13.162}$$

which in turn leads to

$$\frac{\partial}{\partial t}(\overline{\sigma}\,\overline{\tau} + \overline{\sigma'\tau'}) + \nabla_b\cdot(\overline{\sigma}\,\overline{\boldsymbol{u}}\,\overline{\tau}) + \nabla_b\cdot\overline{\sigma'\boldsymbol{u}'}\,\overline{\tau} + \nabla_b\cdot[\overline{(\sigma\boldsymbol{u})'\tau'}] = 0, \tag{13.163}$$

or, using (13.156),

$$\frac{\partial\overline{\tau}}{\partial t} + \frac{1}{\overline{\sigma}}\frac{\partial}{\partial t}(\overline{\sigma'\tau'}) + \left[\overline{\boldsymbol{u}} + \frac{\overline{\sigma'\boldsymbol{u}'}}{\overline{\sigma}}\right]\cdot\nabla_b\overline{\tau} = -\frac{1}{\overline{\sigma}}\nabla_b\cdot[\overline{(\sigma\boldsymbol{u})'\tau'}]. \tag{13.164}$$

(To derive these, first let $\overline{\sigma\tau\boldsymbol{u}} = \overline{(\sigma\boldsymbol{u})'\tau'} + \overline{\sigma\boldsymbol{u}}\,\overline{\tau}$.) If we neglect the correlation between σ' and τ', then (13.164) has the form

$$\frac{\partial\overline{\tau}}{\partial t} + (\overline{\boldsymbol{u}} + \widetilde{\boldsymbol{u}})\cdot\nabla_b\tau = -\frac{1}{\overline{\sigma}}\nabla_b\cdot[\overline{(\sigma\boldsymbol{u})'\tau'}]. \tag{13.165}$$

Thus, the averaged tracer evolves as if it were advected by two velocity fields: the large-scale field itself, $\overline{\boldsymbol{u}}$, and the eddy-induced velocity $\widetilde{\boldsymbol{u}}$, their sum being the residual velocity. The term on the right-hand side of (13.165) is the divergence of the transport of the tracer along the isopycnals by the eddy transport $(\sigma\boldsymbol{u})'$. These equations are not yet closed because we don't know the eddy-induced velocity, $\widetilde{\boldsymbol{u}}$.

13.8.2 Diffusive Thickness Transport

A downgradient diffusion of thickness parameterizes the eddy transport velocity by

$$\widetilde{\boldsymbol{u}} \equiv \frac{\overline{\sigma'\boldsymbol{u}'}}{\overline{\sigma}} = -\frac{1}{\overline{\sigma}}\kappa\nabla_b\overline{\sigma}, \tag{13.166}$$

where κ is an eddy diffusivity. Similarly, we might parameterize the right-hand side of (13.164) by

$$-\frac{1}{\overline{\sigma}}\nabla_b\cdot[\overline{(\sigma\boldsymbol{u})'\tau'}] = \frac{1}{\overline{\sigma}}\nabla_b\cdot(\kappa\overline{\sigma}\nabla_b\overline{\tau}), \tag{13.167}$$

that is, as a diffusion of the tracer along isopycnals.

In height coordinates the eddy transport velocity will be a three-dimensional field, obtained by appropriately transforming $\widetilde{\boldsymbol{u}}$. We have

$$\widetilde{\boldsymbol{u}} \approx -\frac{1}{\sigma}\kappa\nabla_b\overline{\sigma} = -\kappa\frac{\partial b}{\partial z}\nabla_b\left(\frac{\partial\overline{z}}{\partial b}\right) = -\kappa\frac{\partial b}{\partial z}\frac{\partial\overline{s}}{\partial b} = -\kappa\frac{\partial\overline{s}}{\partial z}, \tag{13.168}$$

where the third equality uses $\nabla_b z = \mathbf{s}$, the isopycnal slope. The final result is not quite the same as (13.143), because the diffusivity is now outside the z-derivative. It is a subtle but important distinction, because it means that if κ varies the vertical velocity can no longer be obtained easily as a local function. That is to say, given (13.168), we no longer have $\widetilde{w} = \nabla_z \cdot \kappa\overline{\mathbf{s}}$ as in (13.143). Rather, $\widetilde{w}$ must be evaluated by a non-local integration of the mass conservation requirement, so that

$$\widetilde{w} = \int \nabla_z \cdot \left(\kappa \frac{\partial \overline{\mathbf{s}}}{\partial z}\right) \mathrm{d}z. \tag{13.169}$$

This result should not be disconcerting from a physical standpoint, because the baroclinic activity of eddies certainly involves vertical communication — recall the tendency toward barotropic flow in baroclinic lifecycles. From a computational standpoint, it is a little less convenient. A less satisfactory feature of thickness diffusion arises when the ocean floor is not flat; a strict thickness diffusion might then increase the slope of the isopycnals and increase the available potential energy, which would be an unwanted effect. Nevertheless, overall the GM scheme is evidently similar to a thickness diffusion.

13.8.3 † Potential Vorticity Diffusion

From a more fundamental perspective, potential vorticity, Q, is a better candidate for diffusion than thickness because it is a materially conserved quantity.[15] It is not the only variable that is materially conserved — potential temperature is also. However, potential temperature is advected by the *three-dimensional* velocity field and the vertical advection complicates matters, since any diffusion tensor certainly cannot be isotropic and is probably not symmetric. On the other hand, in isentropic coordinates the adiabatic potential vorticity advection occurs in the isentropic plane (and in quasi-geostrophic flow the advection is purely horizontal). Thus, only the two-dimensional diffusion need be considered, and the diffusion tensor will be much simplified. Near the upper and lower boundaries buoyancy may still be the appropriate field to diffuse, because $w = 0$ and buoyancy is conserved on parcels when advected by the horizontal flow. Horizontal diffusion of buoyancy is not an adiabatic parameterization, but diabatic effects do occur at the surface. These considerations suggest that downgradient potential vorticity diffusion on isentropic surfaces in the fluid interior, combined with downgradient buoyancy diffusion at the upper and lower boundaries, may be as rational a parameterization of eddy transfer effects as any simple diffusion scheme can be.

Actually implementing a potential vorticity diffusion in the equations of motion is not easily done, because when the equations of motion are written in conventional form the potential vorticity flux does not directly appear. Furthermore, if potential vorticity is diffused with a constant diffusivity, momentum is not conserved. The resolution of issues remains research task, and below we just make a couple of remarks.

Connection to thickness diffusion

Potential vorticity diffusion is closely connected to thickness diffusion, especially in an oceanic setting in which the scales of motion are much larger than the deformation radius and fluxes of relative vorticity are relatively unimportant. To see this, consider the expression for potential vorticity in isentropic coordinates, namely

$$Q = \frac{f + \zeta}{\sigma} \approx \frac{f}{\sigma}, \tag{13.170}$$

where the second expression holds under planetary-geostrophic scaling. The eddy flux of potential vorticity is then

$$\overline{\mathbf{u}'Q'} \approx -\frac{f}{\overline{\sigma}^2}\overline{\mathbf{u}'\sigma'}, \tag{13.171}$$

which is similar to thickness flux, and using (13.158) and (13.171) gives the bolus velocity

$$\widetilde{\boldsymbol{u}} = \frac{1}{\overline{\sigma}}\overline{\boldsymbol{u}'\sigma'} \approx -\frac{\overline{\sigma}}{f}\overline{\boldsymbol{u}'Q'}. \tag{13.172}$$

Now, the gradient of potential vorticity is approximately given by

$$\nabla_b\overline{Q} = \frac{1}{\overline{\sigma}}\nabla_b f - \frac{f}{\overline{\sigma}^2}\nabla_b\overline{\sigma} = \frac{\beta}{\overline{\sigma}}\mathbf{j} - \frac{f}{\overline{\sigma}^2}\nabla_b\overline{\sigma}, \tag{13.173}$$

so that potential vorticity diffusion is then

$$\overline{\boldsymbol{u}'Q'} = -\mathsf{K}\nabla_b\overline{Q} = -\mathsf{K}\left(\frac{\beta}{\overline{\sigma}}\mathbf{j} - \frac{f}{\overline{\sigma}^2}\nabla_b\overline{\sigma}\right). \tag{13.174}$$

Using this equation with (13.172) gives the bolus velocity

$$\widetilde{\boldsymbol{u}} = \mathsf{K}\left(\frac{\beta}{f}\mathbf{j} - \frac{1}{\overline{\sigma}}\nabla_b\overline{\sigma}\right). \tag{13.175}$$

This differs from (13.166) mainly in the existence of the term involving β on the right-hand side. The expression is singular at the equator, a consequence of ignoring the relative vorticity term in the expression for potential vorticity.

Other recipes for diffusing potential vorticity are possible, but it may fairly be said that, although potential vorticity diffusion has considerable theoretical appeal, no implementation in a comprehensive ocean model has shown practical advantages over simpler GM or thickness-diffusing schemes.

Using the transformed Eulerian mean

A natural framework to discuss how eddy fluxes interact with the mean flow is the transformed Eulerian mean (TEM), discussed in Chapter 10., and that framework may also be useful for the eddy parameterization problem, especially in idealized settings. The connection is not unexpected, given the connection between the residual velocity and the thickness-weighted mean velocity demonstrated in Section 10.3.3. We'll illustrate the use with a simple example.

Recall the TEM form of quasi-geostrophic zonally averaged momentum and thermodynamic equation,

$$\frac{\partial\overline{u}}{\partial t} - f_0\overline{v}^* = \overline{v'q'}, \tag{13.176a}$$

$$\frac{\partial\overline{b}}{\partial t} + N^2\overline{w}^* = 0, \tag{13.176b}$$

with no forcing or dissipation terms for simplicity. The eddy flux terms now explicitly appear *only* in the momentum equations, and the eddy flux on the right-hand side of (13.176a) is the potential vorticity flux. A potential vorticity flux parameterization is thus both natural and adiabatic. Having said that, the advecting velocities are the residual velocities, so that if the Eulerian velocity is required one must pass from $\overline{v}^*$ to $\overline{v}$ using an eddy-flux parameterization. Similar considerations apply to using the TEM in the primitive equations, but once more our reach has exceeded our grasp, and this is where the chapter stops.

Notes

1 A significant fraction of the theory of turbulent diffusion stems from G. I. (Geoffrey Ingram) Taylor (1886–1975), who made important contributions to both fluid and solid mechanics, in the former to meteorology, oceanography and aerodynamics. In addition to his work in turbulence, Taylor is known for his work on the theory of rotating fluids (the 'Taylor–Proudman' effect, for example) and on hydrodynamic stability (analysis of stability of Couette flow, for example), and for his clear and simple laboratory experiments. The main results of this section were first derived by Taylor (1921a). Ludwig Prandtl (1875–1953) was the other great pioneer of turbulent diffusion; he is also famous for his work in boundary-layer theory and aerodynamics.

2 A number of textbooks in both fluid dynamics and stochastic processes give more detail on this topic. Gardiner (1985) is one.

3 See also Monin & Yaglom (1971).

4 The way variables are changed between (13.31) and (13.32) could also have been used to derive (13.20), but in that case a simpler transformation sufficed.

5 This topic was first addressed empirically by Richardson (1926), although it was Obukhov (1941) who first theoretically obtained the '4/3 power law' describing how the eddy diffusivity varies with separation for parcels in the inertial range. Our treatment takes advantage of Kolmogorov scaling.

6 This figure and Fig. 13.4 were kindly provided by Joe LaCasce; see LaCasce & Ohlmann (2003).

7 In the open ocean, Ollitrault *et al.* (2005) do find float separation that increases with t^3, consistent with a $-5/3$ inverse cascade range.

8 Morel & Larcheveque (1974) and Er-El & Peskin (1981). Earlier dispersion calculations were made by Richardson (1926) who measured smoke spreading from chimneys, finding results that are consistent with a three-dimensional energy inertial range at small scales.

9 Turbulent diffusion is both widely used and widely criticized. If there is a scale-separation between a well-defined mean flow and the eddies then turbulent diffusion can be a very useful parameterization. However, this condition is often *not* satisfied, because it is unusual in fluid mechanics for the turbulent eddies to be significantly smaller than the mean flow. Baroclinic turbulence is something of an exception because there is a natural scale of the turbulence — the deformation radius — that is in general different from the scale of the mean flow, although even this scale separation may be lost if there is an inverse cascade or if the deformation scale is sufficiently large, as in the Earth's atmosphere. Furthermore, properly choosing what variable is to be diffused, and ensuring that various fluid conservation properties remain respected by the diffusion, remain difficult problems.

10 If S_{ij} and A_{ij} are symmetric and antisymmetric tensors respectively, then, summing over repeated indices, their contraction is $A_{ij}S_{ij} = -A_{ji}S_{ij} = -A_{ji}S_{ji} = -A_{ij}S_{ij}$, where the last equality follows because the indices are dummy. Thus, the contraction must equal zero.

11 Green (1970) and Stone (1972), in the context of the meridional transport of heat in the Earth's atmosphere, suggested that the magnitude of the turbulent diffusivity coefficients could be obtained by dynamical arguments using such things as baroclinic instability theory and the amount of available potential energy in the atmosphere, although their suggestions differ in such important details as the eddy mixing length. Other efforts have drawn on geostrophic turbulence theory, for example Larichev & Held (1995) and Smith & Vallis (2002).

12 Ollitrault *et al.* (2005).

13 Green (1970).

14 The Gent–McWilliams (GM) scheme originated in Gent & McWilliams (1990) and was much clarified by Gent *et al.* (1995). Previously, Plumb (1979) and Moffatt (1983) had noted the connection between symmetric and antisymmetric diffusivities and diffusive and advective fluxes, and Griffies (1998) explicitly showed how the GM bolus velocities are related to a skew flux and can be calculated using an antisymmetric diffusivity tensor, which led to notable improvements in the scheme. See McDougall (1998) and (Griffies 2004) for reviews and more discussion. Visbeck *et al.* (1997) suggested that the values of eddy diffusivities in the GM scheme might be determined by dynamical

arguments similar to those of Green (1970) and Stone (1972). In spite of the manifest imperfections of the GM scheme, to date (2017) no commonly implemented scheme has proven to be better, in practice, for ocean climate models.

15 Potential vorticity diffusion was suggested by Green (1970) as a parameterization for large-scale eddies in the atmosphere, and further explored and used in ocean contexts by Welander (1973), Marshall (1981), Rhines & Young (1982a), Tréguier *et al.* (1997), Greatbatch (1998) and others. Lee *et al.* (1997), Marshall *et al.* (1999), Drijfhout & Hazeleger (2001), and others, have explored numerically whether the eddy transfer of tracers in the ocean is in fact diffusive, and whether potential vorticity or thickness is a better quantity to diffuse. For other examples and methodologies see, for example, Killworth (1997), Smith & Vallis (2002), Ferrari *et al.* (2010) or Marshall *et al.* (2012).

Part III

LARGE-SCALE ATMOSPHERIC CIRCULATION

Blow the wind southerly, southerly, southerly,
Blow the wind south o'er the bonny blue sea;
Blow the wind southerly, southerly, southerly,
Blow bonnie breeze, my lover to me.
Traditional English folk song, *Blow the Wind Southerly*, c. 1834.

CHAPTER 14

The Overturning Circulation: Hadley and Ferrel Cells

THE LARGE-SCALE CIRCULATION OF THE ATMOSPHERE is normally taken to mean the flow on scales of the weather — several hundred or a thousand kilometres, say — to the global scale. The *general circulation* is virtually synonymous with the large-scale circulation, although the former is sometimes taken to be the time- or ensemble-averaged flow. Our goal in this and the next few chapters is understand this circulation and other properties of the atmosphere that accompany it — the temperature and moisture fields, for example. We might hope to answer the simple question, why do the winds blow as they do? In this chapter we focus on the dynamics of the Hadley Cell and, rather descriptively, on the mid-latitude overturning cell or the Ferrel Cell, moving to a more dynamical view of the extratropical zonally averaged circulation in Chapter 15.

The atmosphere is a terribly complex system, and we cannot hope to fully explain its motion as the analytic solution to a small set of equations. Rather, a full understanding of the atmosphere requires describing it in a consistent way on many levels simultaneously. One of these levels involves simulating the flow by numerically solving the governing equations of motion as completely as possible by using a comprehensive General Circulation Model (GCM). Such a simulation brings problems sof its own, for example understanding the simulation itself and discerning whether it is a good representation of reality, and so we shall concentrate on simpler, more conceptual models and the basic theory of the circulation. We begin this chapter with a brief observational overview of some of the large-scale features of the atmosphere, concentrating on the zonally-averaged fields.[1]

14.1 BASIC FEATURES OF THE ATMOSPHERE

14.1.1 The Radiative Equilibrium Distribution

A gross but informative measure characterizing the atmosphere, and the effects that dynamics have on it, is the pole-to-equator temperature distribution. The *radiative equilibrium* temperature is the hypothetical, three-dimensional, temperature field that would obtain if there were no atmospheric or oceanic motion, given the composition and radiative properties of the atmosphere and surface. The field is a function of the incoming solar radiation and the atmospheric composition, and its determination entails a complicated calculation, especially as the radiative properties of the atmosphere depend heavily on the amount of water vapour and cloudiness it contains. (The distribution

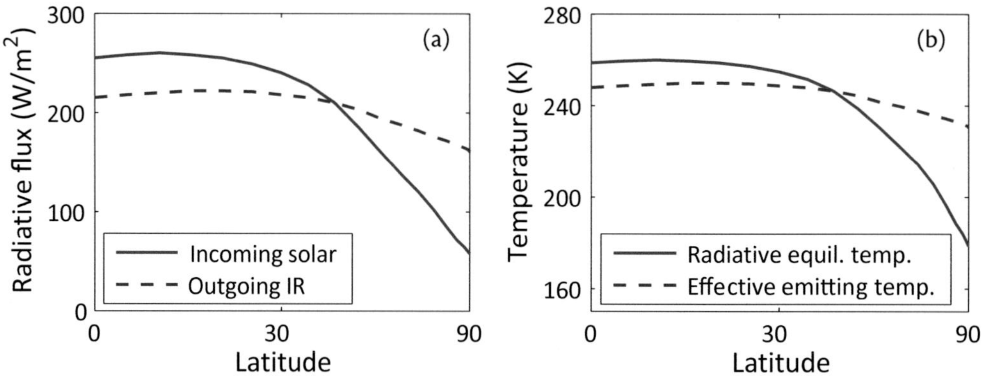

Fig. 14.1 (a) The (approximate) observed net average incoming solar radiation and outgoing infrared radiation at the top of the atmosphere, as a function of latitude (plotted on a sine scale). (b) The temperatures associated with these fluxes, calculated using $T = (R/\sigma)^{1/4}$, where R is the solar flux for the radiative equilibrium temperature and where R is the infrared flux for the effective emitting temperature. Thus, the solid line is an approximate radiative equilibrium temperature

of absorbers is usually taken to be that which obtains in the observed, moving, atmosphere, in order that the differences between the calculated radiative equilibrium temperature and the observed temperature are due to fluid motion.)

A much simpler calculation that illustrates the essence of the situation is to first note that at the top of the atmosphere the globally averaged incoming solar radiation is balanced by the outgoing infrared radiation. If there is no lateral transport of energy in the atmosphere or ocean then *at each latitude* the incoming solar radiation will be balanced by the outgoing infrared radiation, and if we parameterize the latter using a single latitudinally-dependent temperature we will obtain a crude radiative-equilibrium temperature (the 'radiative emitting temperature') for the atmospheric column at each latitude. Specifically, a black body subject to a net incoming radiation of S (watts per square metre) has a radiative-equilibrium temperature T_{rad} given by $\sigma T_{rad}^4 = S$, this being Stefan's law with Stefan–Boltzmann constant $\sigma = 5.67 \times 10^{-8}\,\mathrm{W\,m^{-2}\,K^{-4}}$. Thus, for the Earth, we have, at each latitude,

$$\sigma T_{rad}^4 = S(\vartheta)(1-\alpha), \tag{14.1}$$

where α is the albedo of the Earth and $S(\vartheta)$ is the incoming solar radiation at the top of the atmosphere, and its solution is shown in Fig. 14.1. The solid lines in the two panels show the net solar radiation and the solution to (14.1), T_{rad}; the dashed lines show the observed outgoing infrared radiative flux, I, and the effective emitting temperature associated with it, $(I/\sigma)^{1/4}$. The emitting temperature does not quantitatively characterize that temperature at the Earth's surface, nor at any single level in the atmosphere, because the atmosphere is not a black body and the outgoing radiation originates from multiple levels. Nevertheless, the qualitative point is evident: the radiative equilibrium temperature has a much stronger pole-to-equator gradient than does the effective emitting temperature, indicating that there is a poleward transport of heat in the atmosphere–ocean system. More detailed calculations indicate that the atmosphere is further from its radiative equilibrium in winter than summer, indicating a larger heat transport. The transport occurs because poleward moving air tends to have a higher static energy ($c_pT + gz$ for dry air; in addition there is some energy transport associated with water vapour evaporation and condensation) than the equatorward moving air, most of this movement being associated with the large-scale circulation. The radiative forcing thus seeks to maintain a pole-to-equator temperature gradient, and the ensuing circulation seeks to reduce this gradient.

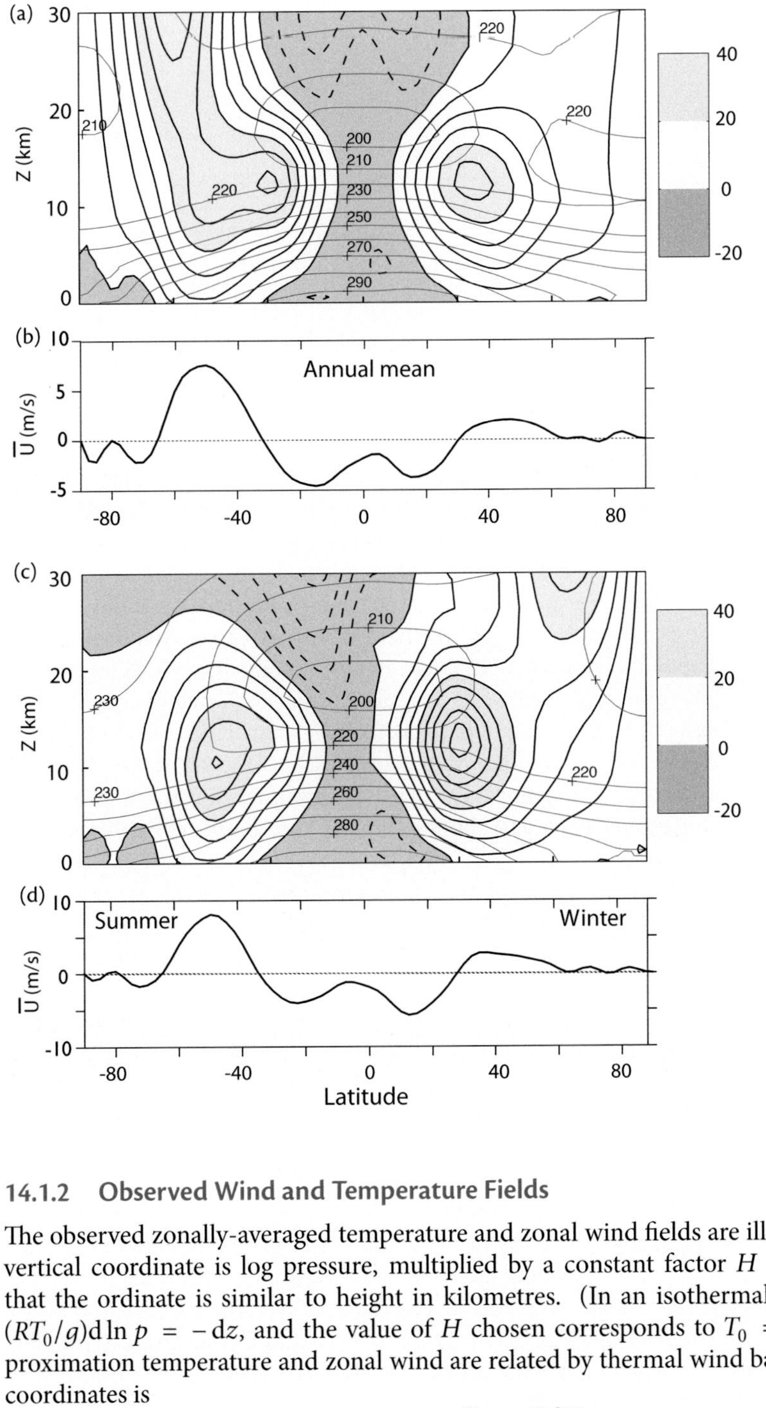

Fig. 14.2 (a) Annual mean, zonally-averaged zonal wind (heavy contours and shading) and the zonally-averaged temperature (red, thinner contours).

(b) Annual mean, zonally averaged zonal winds at the surface.

(c) and (d) Same as (a) and (b), except for northern hemisphere winter (December–January–February, or DJF).

The wind contours are at intervals of $5\,\mathrm{m\,s^{-1}}$ with shading for eastward winds above $20\,\mathrm{m\,s^{-1}}$ and for all westward winds, and the temperature contours are labelled. The ordinate of (a) and (c) is $Z = -H \log(p/p_R)$, where $H = 7.5$ km and $p_R = 100$ hPa.

14.1.2 Observed Wind and Temperature Fields

The observed zonally-averaged temperature and zonal wind fields are illustrated in Fig. 14.2. The vertical coordinate is log pressure, multiplied by a constant factor $H = RT_0/g = 7.5\,\mathrm{km}$, so that the ordinate is similar to height in kilometres. (In an isothermal hydrostatic atmosphere $(RT_0/g)\mathrm{d}\ln p = -\mathrm{d}z$, and the value of H chosen corresponds to $T_0 = 256\,\mathrm{K}$.) To a good approximation temperature and zonal wind are related by thermal wind balance, which in pressure coordinates is

$$f\frac{\partial u}{\partial p} = \frac{R}{p}\frac{\partial T}{\partial y}. \tag{14.2}$$

In the lowest several kilometres of the atmosphere temperature falls almost monotonically with latitude and height, and this region is called the *troposphere* (look ahead to Fig. 15.25). The temperature in the lower troposphere in fact varies more rapidly with latitude than does the effective emitting temperature, T_E, the latter being more characteristic of the temperature in the mid-to-

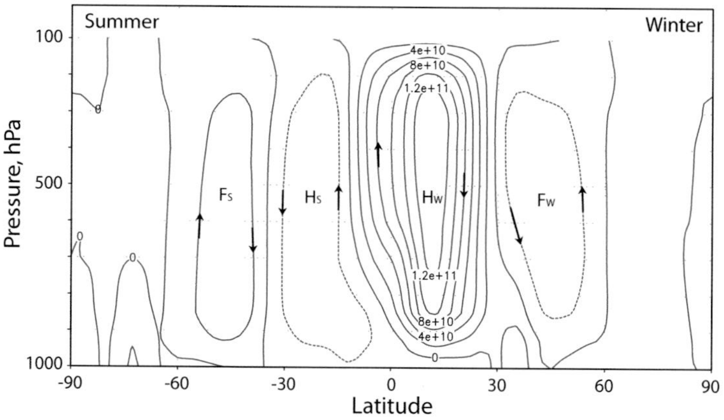

Fig. 14.3 The observed meridional overturning circulation (MOC) of the atmosphere ($kg\,s^{-1}$) averaged over December–January–February. Note the direct *Hadley Cells,* particularly strong in winter (H_W and H_S, in winter and summer respectively) with rising motion near the equator, descending motion in the subtropics, and the weaker, indirect, *Ferrel Cells* (F_W and F_S) at mid-latitudes.

upper troposphere. The meridional temperature gradient is much larger in winter than summer, because in winter high latitudes receive virtually no direct heating from the Sun. The gradient is also strongest at the edge of the subtropics, and here it is associated with a zonal jet, particularly strong in winter. There is no need to 'drive' this wind with any kind of convergent momentum fluxes: given the temperature, the flow is a consequence of thermal wind balance, and to the extent that the upper troposphere is relatively frictionless there is no need to maintain it against dissipation. Of course just as the radiative-equilibrium temperature gradient is much larger than that observed, so the zonal wind shear associated with it is much larger than that observed. Thus, the overall effect of the atmospheric and oceanic circulation, and in particular of the turbulent circulation of the mid-latitude atmosphere, is to *reduce* the amplitude of the vertical shear of the eastward flow by way of a poleward heat transport. Observations indicate that about two-thirds of this transport is effected by the atmosphere, and about a third by the ocean, rather more in low latitudes.[2]

Above the troposphere is the *stratosphere,* and here temperature typically increases with height. The boundary between the two regions is called the *tropopause,* and this varies in height from about 16 km in the tropics to about 8 km in polar regions. We consider the maintenance of this stratification in Section 15.5.

The surface winds typically have, going from the equator to the pole, an E–W–E (easterly–westerly–easterly) pattern, although the polar easterlies are weak and barely present in the Northern Hemisphere. (Meteorologists use 'westerly' to denote winds from the west, that is eastward winds; similarly 'easterlies' are westward winds.) In a given hemisphere, the surface winds are stronger in winter than summer, and they are also consistently stronger in the Southern Hemisphere than in the Northern Hemisphere, because in the former the surface drag is weaker because of the relative lack of continental land masses and topography. The surface winds are *not* explained by thermal wind balance. Indeed, unlike the upper level winds, they must be maintained against the dissipating effects of friction, and this implies a momentum convergence into regions of surface westerlies and a divergence into regions of surface easterlies. Typically, the maxima in the eastward surface winds are in mid-latitudes and somewhat poleward of the subtropical maxima in the upper-level westerlies and at latitudes where the zonal flow is a little more constant with height. The mechanisms of the momentum transport in the mid-latitudes and the maintenance of the surface westerly winds are the topics of section 15.1.

14.1.3 Meridional Overturning Circulation

The observed (Eulerian) zonally-averaged meridional overturning circulation (MOC) is shown in Fig. 14.3. The figure shows a streamfunction, Ψ for the vertical and meridional velocities such that,

Some Features of the Large-scale Atmospheric Circulation

From Figures 14.1–14.3 we see or infer the following:

1. A pole–equator temperature gradient that is much smaller than the radiative equilibrium gradient.
2. A troposphere, in which temperature generally falls with height, above which lies the stratosphere, in which temperature increases with height. The two regions are separated by a tropopause, which varies in height from about 16 km at the equator to about 6 km at the pole.
3. A monotonically decreasing temperature from equator to pole in the troposphere, but a weakening and sometimes reversal of this above the tropopause.
4. A westerly (i.e., eastward) tropospheric jet. The time and zonally-averaged jet is a maximum at the edge or just poleward of the subtropics, where it is associated with a strong meridional temperature gradient. In mid-latitudes the jet has a stronger barotropic component.
5. An E–W–E (easterlies–westerlies–easterlies) surface wind distribution. The latitude of the maximum in the surface westerlies is in mid-latitudes, where the zonally-averaged flow is more barotropic. The surface easterlies at high latitudes are very weak and seasonal, barely showing on an annual average.

in the pressure coordinates used in the figure,

$$\frac{\partial \Psi}{\partial y} = -\overline{\omega}, \qquad \frac{\partial \Psi}{\partial p} = \overline{v}, \tag{14.3}$$

where the overbar indicates a zonal average. In each hemisphere there is rising motion near the equator and sinking in the subtropics, and this circulation is known as the *Hadley Cell*.[3] The Hadley Cell is a thermally direct cell (i.e., the warmer fluid rises, the colder fluid sinks), much stronger in the winter hemisphere, and extends to about 25–30°. In mid-latitudes the sense of the overturning circulation is apparently reversed, with rising motion in the high-mid-latitudes, at around 60° and sinking in the subtropics, and this is known as the *Ferrel Cell*. However, as with most pictures of averaged streamlines in unsteady flow, this gives a misleading impression as to the actual material flow of parcels of air because of the presence of eddying motion, and we discuss this in the next chapter. At low latitudes the circulation is more nearly zonally symmetric and the picture does give a qualitatively correct representation of the actual flow. At high latitudes there is again a thermally direct cell (although it is weak and not always present), and thus the atmosphere is often referred to as having a three-celled structure.

14.1.4 Summary

Some of the main features of the zonally-averaged circulation are summarized in the shaded box above. We emphasize that the zonally-averaged circulation is not synonymous with a zonally symmetric circulation, and the mid-latitude circulation is highly asymmetric. Any model of the mid-latitudes that did not take into account the zonal asymmetries in the circulation — of which the weather is the main manifestation — would be seriously in error. This was first explicitly realized in the 1920s, and taking into account such asymmetries is the main task of the dynamical

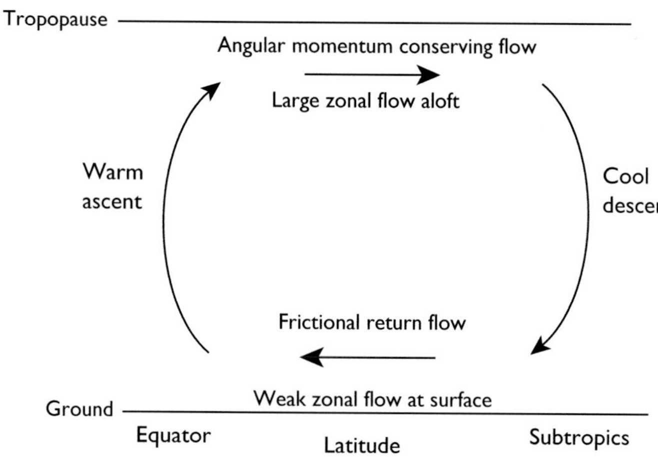

Fig. 14.4 A simple model of the Hadley Cell. Rising air near the equator moves poleward near the tropopause, descending in the subtropics and returning.

The poleward moving air conserves its angular momentum, leading to a shear of the zonal wind that increases away from the equator. By thermal wind the temperature of the air falls as it moves poleward, and to satisfy the thermodynamic budget it sinks in the subtropics.

meteorology of the mid-latitudes, and is the subject of the next chapter. The large-scale tropical circulation of the atmosphere is to a much larger degree zonally symmetric, and although monsoonal circulations and the Walker circulation (a cell with rising air in the Western Pacific and descending motion in the Eastern Pacific) are zonally asymmetric, they are relatively weaker than typical mid-latitude weather systems. Indeed the boundary between the tropics and mid-latitude may be usefully defined by the latitude at which such zonal asymmetries become dynamically important on the large scale and this boundary, at about 25°–30° on average, roughly coinciding with the edge of the Hadley Cell. We begin our dynamical description with a study of the low-latitude zonally symmetric atmospheric circulation.

14.2 A STEADY MODEL OF THE HADLEY CELL

Ceci n'est pas une pipe.
René Magritte. Title of painting, 1929.

14.2.1 Assumptions

Let us try to construct a zonally symmetric model of the Hadley Cell, recognizing that such a model is likely applicable mainly to the tropical atmosphere, this being more zonally symmetric than the mid-latitudes.[4] We suppose that heating is maximum at the equator, and our intuitive picture, drawing on the observed flow of Fig. 14.3, is of air rising at the equator and moving poleward at some height H, descending at some latitude ϑ_H, and returning equatorward near the surface. We will make three major assumptions:

(i) that the circulation is steady;

(ii) that the poleward moving air conserves its axial angular momentum, whereas the zonal flow associated with the near-surface, equatorward moving flow is frictionally retarded and weak;

(iii) that the circulation is in thermal wind balance.

We also assume the model is symmetric about the equator (an assumption we relax in Section 14.4). These are all reasonable assumptions, but they cannot be rigorously justified; in other words, we are constructing a *model* of the Hadley Cell, schematically illustrated in Fig. 14.4. The model defines a limiting case — steady, inviscid, zonally-symmetric flow — that cannot be expected to describe the atmosphere quantitatively, but that can be analysed fairly completely. Another limiting case, in which eddies play a significant role, is described in Section 14.5. The real atmosphere may defy such simple characterizations, but the two limiting cases provide useful benchmarks of understanding.

14.2.2 Dynamics

We now try to determine the strength and poleward extent of the Hadley circulation in our steady model. For simplicity we work with a Boussinesq atmosphere, but this is not an essential aspect. We first derive the conditions under which conservation of angular momentum will hold, and then determine the consequences of that.

The zonally-averaged zonal momentum equation may be easily derived from (2.50a) and/or (2.62) and in the absence of friction it is

$$\frac{\partial \overline{u}}{\partial t} - (f + \overline{\zeta})\overline{v} + \overline{w}\frac{\partial \overline{u}}{\partial z} = -\frac{1}{a\cos^2\vartheta}\frac{\partial}{\partial \vartheta}(\cos^2\vartheta\,\overline{u'v'}) - \frac{\partial \overline{u'w'}}{\partial z}, \tag{14.4}$$

where $\overline{\zeta} = -(a\cos\vartheta)^{-1}\partial_\vartheta(\overline{u}\cos\vartheta)$ and the overbars represent zonal averages. If we neglect the vertical advection and the eddy terms on the right-hand side, then a steady solution, if it exists, obeys

$$(f + \overline{\zeta})\overline{v} = 0. \tag{14.5}$$

Presuming that the meridional flow $\overline{v}$ is non-zero (an issue we address in Section 14.2.8) then $f + \overline{\zeta} = 0$, or equivalently

$$2\Omega\sin\vartheta = -\frac{1}{a}\frac{\partial \overline{u}}{\partial \vartheta} - \frac{\overline{u}\tan\vartheta}{a}. \tag{14.6}$$

At the equator we shall assume that $\overline{u} = 0$, because here parcels have risen from the surface where, by assumption, the flow is weak. Equation (14.6) then has a solution of

$$\overline{u} = \Omega a\frac{\sin^2\vartheta}{\cos\vartheta} \equiv U_M. \tag{14.7}$$

This gives the zonal velocity of the poleward moving air in the upper branch of the (model) Hadley Cell, above the frictional boundary layer. We can derive (14.7) directly from the conservation of axial angular momentum, m, of a parcel of air at a latitude ϑ. In the shallow atmosphere approximation we have (cf. (2.64) and equations following)

$$\overline{m} = (\overline{u} + \Omega a\cos\vartheta)a\cos\vartheta, \tag{14.8}$$

and if $\overline{u} = 0$ at $\vartheta = 0$ and if $\overline{m}$ is conserved on a poleward moving parcel, then (14.8) leads to (14.7). It also may be directly checked that

$$f + \overline{\zeta} = -\frac{1}{a^2\cos\vartheta}\frac{\partial \overline{m}}{\partial \vartheta}. \tag{14.9}$$

We have thus shown that, if eddy fluxes and frictional effects are negligible, the poleward flow will conserve its angular momentum, the result of which, by (14.7), is that the magnitude of the zonal flow in the Earth's rotating frame will increase with latitude (see Fig. 14.5). (Also, given the absence of eddies our model is zonally symmetric and we shall drop the overbars over the variables.)

If (14.7) gives the zonal velocity in the upper branch of the Hadley Cell, and that in the lower branch is close to zero, then the thermal wind equation can be used to infer the vertically averaged temperature. Although the geostrophic wind relation is not valid at the equator (a more accurate balance is the gradient wind balance, $fu + u^2\tan\vartheta/a = -a^{-1}\partial\phi/\partial\vartheta$) the zonal wind is in fact geostrophically balanced until very close to the equator, and at the equator itself the horizontal temperature gradient in our model vanishes, because of the assumed interhemispheric symmetry. Thus, conventional thermal wind balance suffices for our purposes, and this is

$$2\Omega\sin\vartheta\frac{\partial u}{\partial z} = -\frac{1}{a}\frac{\partial b}{\partial \vartheta}, \tag{14.10}$$

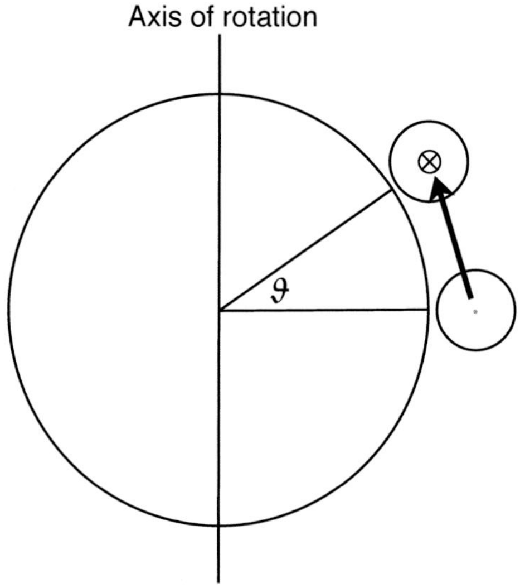

Fig. 14.5 If a ring of air at the equator moves poleward it moves closer to the axis of rotation. If the parcels in the ring conserve their angular momentum their zonal velocity must increase; thus, if $m = (\overline{u} + \Omega a \cos\vartheta) a \cos\vartheta$ is preserved and $\overline{u} = 0$ at $\vartheta = 0$ we recover (14.7).

where $b = g\,\delta\theta/\theta_0$ is the buoyancy and $\delta\theta$ is the deviation of potential temperature from a constant reference value θ_0. (Be reminded that θ is potential temperature, whereas ϑ is latitude.) Vertically integrating from the ground to the height H where the outflow occurs and substituting (14.7) for u yields

$$\frac{1}{a\theta_0}\frac{\partial\theta}{\partial\vartheta} = -\frac{2\Omega^2 a}{gH}\frac{\sin^3\vartheta}{\cos\vartheta}, \tag{14.11}$$

where $\theta = H^{-1}\int_0^H \delta\theta\,\mathrm{d}z$ is the vertically averaged potential temperature. If the latitudinal extent of the Hadley Cell is not too great we can make the small-angle approximation, and replace $\sin\vartheta$ by ϑ and $\cos\vartheta$ by one, then integrating (14.11) gives

$$\theta = \theta(0) - \frac{\theta_0\Omega^2 y^4}{2gHa^2}, \tag{14.12}$$

where $y = a\vartheta$ and $\theta(0)$ is the potential temperature at the equator, as yet unknown. Away from the equator, the zonal velocity given by (14.7) increases rapidly poleward and the temperature correspondingly drops. How far poleward is this solution valid? And what determines the value of the integration constant $\theta(0)$? To answer these questions we turn to thermodynamics.

14.2.3 Thermodynamics

In the above discussion, the temperature field is slaved to the momentum field in that it seems to follow passively from the dynamics of the momentum equation. Nevertheless, the thermodynamic equation must still be satisfied. Let us assume that the thermodynamic forcing can be represented by a Newtonian cooling to some specified radiative equilibrium temperature, θ_E; this is a severe simplification, especially in equatorial regions where the release of heat by condensation is important. The thermodynamic equation is then

$$\frac{\mathrm{D}\theta}{\mathrm{D}t} = \frac{\theta_E - \theta}{\tau}, \tag{14.13}$$

where τ is a relaxation time scale, perhaps a few weeks. Let us suppose that θ_E falls monotonically from the equator to the pole, and that it increases linearly with height, and a simple representation of this is

$$\frac{\theta_E(\vartheta, z)}{\theta_0} = 1 - \frac{2}{3}\Delta_H P_2(\sin\vartheta) + \Delta_V\left(\frac{z}{H} - \frac{1}{2}\right), \tag{14.14}$$

where Δ_H and Δ_V are nondimensional constants that determine the fractional temperature difference between the equator and the pole, and the ground and the top of the fluid, respectively. P_2 is the second Legendre polynomial, and it is usually the leading term in the Taylor expansion of symmetric functions (symmetric around the equator) that decrease from pole to equator; it also integrates to zero over the sphere. $P_2(y) = (3y^2 - 1)/2$, so that in the small-angle approximation and at $z = H/2$, or for the vertically averaged field, we have

$$\frac{\theta_E}{\theta_0} = 1 + \frac{1}{3}\Delta_H - \Delta_H\left(\frac{y}{a}\right)^2 \quad \text{or} \quad \theta_E = \theta_{E0} - \Delta\theta\left(\frac{y}{a}\right)^2, \tag{14.15a,b}$$

where θ_{E0} is the equilibrium temperature at the equator, $\Delta\theta$ determines the equator–pole radiative-equilibrium temperature difference, and

$$\theta_{E0} = \theta_0(1 + \Delta_H/3), \qquad \Delta\theta = \theta_0\Delta_H. \tag{14.16}$$

Now, let us suppose that the solution (14.12) is valid between the equator and a latitude ϑ_H where $v = 0$, so that within this region the system is essentially closed. Conservation of potential temperature then requires that the solution (14.12) must satisfy

$$\int_0^{Y_H} \theta \, \mathrm{d}y = \int_0^{Y_H} \theta_E \, \mathrm{d}y, \tag{14.17}$$

where $Y_H = a\vartheta_H$ is as yet undetermined. Poleward of this, the solution is just $\theta = \theta_E$. Now, we may demand that the solution be continuous at $y = Y_H$ (without temperature continuity the thermal wind would be infinite) and so

$$\theta(Y_H) = \theta_E(Y_H). \tag{14.18}$$

The constraints (14.17) and (14.18) determine the values of the unknowns $\theta(0)$ and Y_H. A little algebra gives

$$Y_H = \left(\frac{5\Delta\theta g H}{3\Omega^2\theta_0}\right)^{1/2}, \tag{14.19}$$

and

$$\theta(0) = \theta_{E0} - \left(\frac{5\Delta\theta^2 g H}{18a^2\Omega^2\theta_0}\right). \tag{14.20}$$

A useful nondimensional number that parameterizes these solutions is

$$R \equiv \frac{gH\Delta\theta}{\theta_0\Omega^2 a^2} = \frac{gH\Delta_H}{\Omega^2 a^2}, \tag{14.21}$$

which is the square of the ratio of the speed of shallow water waves to the rotational velocity of the Earth, multiplied by the fractional temperature difference from equator to pole. Typical values for the Earth's atmosphere are a little less than 0.1. In terms of R we have

$$Y_H = a\left(\frac{5}{3}R\right)^{1/2}, \qquad \theta(0) = \theta_{E0} - \left(\frac{5}{18}R\right)\Delta\theta. \tag{14.22a,b}$$

The solution, (14.12) with $\theta(0)$ given by (14.22b) is plotted in Fig. 14.6. Perhaps the single most important aspect of the model is that it predicts that the Hadley Cell has a *finite* meridional extent, *even for an atmosphere that is completely zonally symmetric.* The baroclinic instability that does occur in mid-latitudes is not necessary for the Hadley Cell to terminate in the subtropics, although it may be an important factor, or even the determining factor, in the real world.

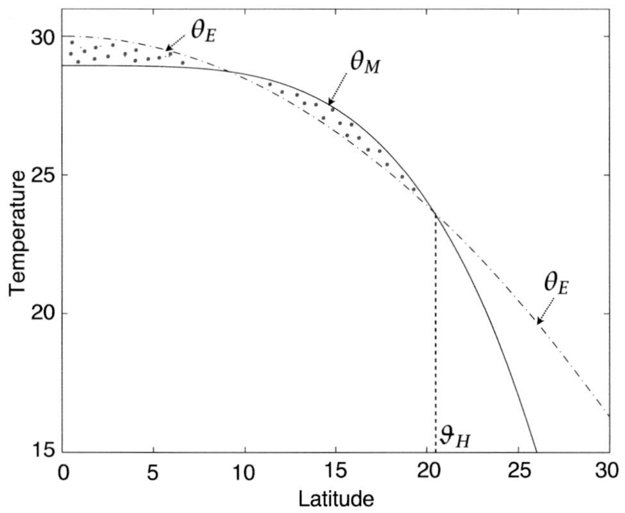

Fig. 14.6 The radiative equilibrium temperature (θ_E, dashed line) and the angular-momentum-conserving solution (θ_M, solid line) as a function of latitude. The two dotted regions have equal areas. The parameters are: $\theta_{EO} = 303\,\mathrm{K}$, $\Delta\theta = 50\,\mathrm{K}$, $\theta_0 = 300\,\mathrm{K}$, $\Omega = 7.272 \times 10^{-5}\,\mathrm{s}^{-1}$, $g = 9.81\,\mathrm{m\,s}^{-2}$, $H = 10\,\mathrm{km}$. These give $R = 0.076$ and $Y_H/a = 0.356$, corresponding to $\vartheta_H = 20.4°$.

14.2.4 Zonal Wind

The angular-momentum-conserving zonal wind is given by (14.7), which in the small-angle approximation becomes

$$U_M = \Omega \frac{y^2}{a}. \tag{14.23}$$

This relation holds for $y < Y_H$. The zonal wind corresponding to the radiative-equilibrium solution is given using thermal wind balance and (14.15b), which leads to

$$U_E = \Omega a R. \tag{14.24}$$

That the radiative-equilibrium zonal wind is a constant follows from our choice of the second Legendre function for the radiative equilibrium temperature and is not a fundamental result; nonetheless, for most reasonable choices of θ_E the corresponding zonal wind will vary much less than the angular-momentum-conserving wind (14.23). The winds are illustrated in Fig. 14.7. There is a discontinuity in the zonal wind at the edge of the Hadley Cell, and of the meridional temperature gradient, but not of the temperature itself.

14.2.5 Properties of the Solution

From (14.22) we can see that the model predicts that the latitudinal extent of the Hadley Cell is:

- proportional to the square root of the meridional radiative equilibrium temperature gradient: the stronger the gradient, the farther the circulation must extend to achieve thermodynamic balance via the equal-area construction in Fig. 14.6;
- proportional to the square root of the height of the outward flowing branch: the higher the outward flowing branch, the weaker the ensuing temperature gradient of the solution (via thermal wind balance), and so the further poleward the circulation must go;
- inversely proportional to the rotation rate Ω: the stronger the rotation rate, the stronger the angular-momentum-conserving wind, the stronger the ensuing temperature gradient and so the more compact the circulation.

These precise dependencies on particular powers of parameters are not especially significant in themselves, nor are they robust to changes in parameters. For example, were we to choose a meridional distribution of radiative equilibrium temperature different from (14.14) we might find different exponents in some of the solutions, although we would expect the same qualitative dependen-

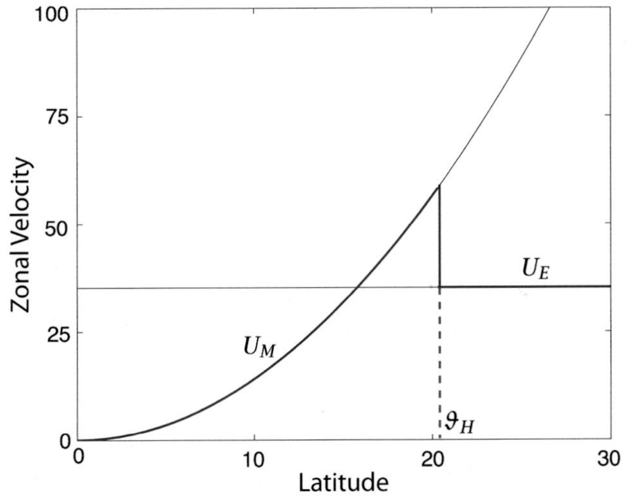

Fig. 14.7 The zonal wind corresponding to the radiative equilibrium temperature (U_E,) and the angular-momentum-conserving solution (U_M) as a function of latitude, given (14.23) and (14.24) respectively.

The parameters are the same as those of Fig. 14.6, and the radiative equilibrium wind, U_E is a constant, ΩaR. The actual zonal wind (in the model) follows the thick solid line: $u = U_m$ for $\vartheta < \vartheta_H$ ($y < Y_H$), and $u = U_E$ for $\vartheta > \vartheta_H$ ($y > Y_H$).

cies. However, the dependencies do provide predictions that may be tested with a numerical model. Also, as we have already noted, a key property of the model is that it predicts that the Hadley Cell has a finite meridional extent, even in the absence of mid-latitude baroclinic instability.

Another interesting property of the solutions is a discontinuity in the zonal wind. For tropical latitudes (i.e., $y < Y_H$), then $\overline{u} = U_M$ (the constant angular momentum solution), whereas for $y > Y_H$, $\overline{u} = U_E$ (the thermal wind associated with radiative equilibrium temperature θ_E). There is therefore a discontinuity of $\overline{u}$ at $y = Y_H$, because u is related to the meridional gradient of θ which changes discontinuously, even though θ itself is continuous. No such discontinuity is observed in the real world, although one may observe a baroclinic jet at the edge of the Hadley Cell.

14.2.6 Strength of the Circulation

We can make an estimate of the strength of the Hadley Cell by consideration of the thermodynamic equation at the equator, namely

$$w\frac{\partial \theta}{\partial z} \approx \frac{\theta_{E0} - \theta}{\tau}, \tag{14.25}$$

this being a balance between adiabatic cooling and radiative heating. If the static stability is determined largely by the forcing, and not by the meridional circulation itself, then $\theta_0^{-1}\partial\theta/\partial z \approx \Delta_V/H$, and (14.25) gives

$$w \approx \frac{H}{\theta_0 \Delta_V}\frac{\theta_{E0} - \theta}{\tau}. \tag{14.26}$$

Thus, the strength of the circulation is proportional to the distance of the solution from the radiative equilibrium temperature. The right-hand side of (14.25) can be evaluated from the solution itself, and from (14.22b) we have

$$\frac{\theta_{E0} - \theta}{\tau} = \frac{5R\Delta\theta}{18\tau}. \tag{14.27}$$

The vertical velocity is then given by

$$w \approx \frac{5R\Delta\theta H}{18\tau\Delta_V\theta_0} = \frac{5R\Delta_H H}{18\tau\Delta_V}. \tag{14.28}$$

Using mass continuity we can transform this into an estimate for the meridional velocity. Thus, if we let $(v/Y_H) \sim (w/H)$ and use (14.22), we obtain

$$v \sim \frac{R^{3/2} a \Delta_H}{\tau \Delta_V} \propto \frac{\Delta_H^{5/2}}{\Delta_V} \qquad \text{and} \qquad \Psi \sim vH \sim \frac{R^{3/2} a H \Delta_H}{\tau \Delta_V} \propto (\Delta\theta)^{5/2}, \tag{14.29}$$

where Ψ is the meridional overturning stream function Ψ, which evidently increases fairly rapidly as the gradient of the radiative equilibrium temperature increases. The characteristic overturning time of the circulation, τ_d is then

$$\tau_d = \frac{H}{w} \sim \frac{\tau \Delta_V}{R \Delta_H}. \tag{14.30}$$

We require $\tau_d/\tau \gg 1$ for the effects of the circulation on the static stability to be small and therefore $\Delta_V/(R\Delta_H) \gg 1$, or equivalently, using (14.16),

$$\theta_0 \Delta_V \gg R(\theta_{E0} - \theta_0). \tag{14.31}$$

If instead $\tau \gg \tau_d$, then the potential temperature would be nearly conserved as a parcel ascended in the rising branch of the Hadley Cell, and the static stability would be nearly neutral.

14.2.7 † Effects of Moisture

Suppose now that moisture is present, but that the Hadley Cell remains a self-contained system; that is, it neither imports nor exports moisture. We envision that water vapour joins the circulation by way of evaporation from a saturated surface into the equatorward, lower branch of the Hadley Cell, and that this water vapour then condenses in and near the upward branch of the cell. The latent heat released by condensation is exactly equal to the heat required to evaporate moisture from the surface, and no heat is lost or gained to the system. However, the heating *distribution* is changed from the dry case, becoming a strong function of the solution itself and likely to have a sharp maximum near the equator. Even if we were to try to parameterize the latent heat release by simply choosing a flow dependent radiative equilibrium temperature, the resulting problem would still be quite nonlinear and a general analytic solution seems out of our reach.[5]

Nevertheless, we may see quite easily the qualitative features of moisture, at least within the context of this model. The meridional distribution of temperature is still given by way of thermal wind balance with an angular-momentum-conserving zonal wind, and so is still given by (14.12). We may also assume that the meridional extent of the Hadley Cell is unaltered; that is, a solution exists with circulation confined to $\vartheta < \vartheta_H$ (although it may not be the unique solution). Then, if θ_E^* is the effective radiative equilibrium temperature of the moist solution, we have that $\theta_E^*(Y_H) = \theta_E(Y_H)$ and, in the small-angle approximation,

$$\int_0^{Y_H} \theta \, dy = \int_0^{Y_H} \theta_E^* \, dy = \int_0^{Y_H} \theta_E \, dy, \tag{14.32}$$

where the first equality holds because it defines the solution, and the second equality holds because moisture provides no net energy source. Because condensation will occur mainly in the upward branch of the Hadley Cell, θ_E^* will be peaked near the equator, as sketched in Fig. 14.8. This construction makes it clear that the main difference between the dry and moist solutions is that the latter has a more intense overturning circulation, because, from (14.25), the circulation increases with the temperature difference between the solution and the forcing temperature. Concomitantly, our intuition suggests that the upward branch of the moist Hadley circulation will become much narrower and more intense than the downward branch because of the enhanced efficiency of moist convection, and these expectations are generally confirmed by numerical integrations of the moist equations of motion.

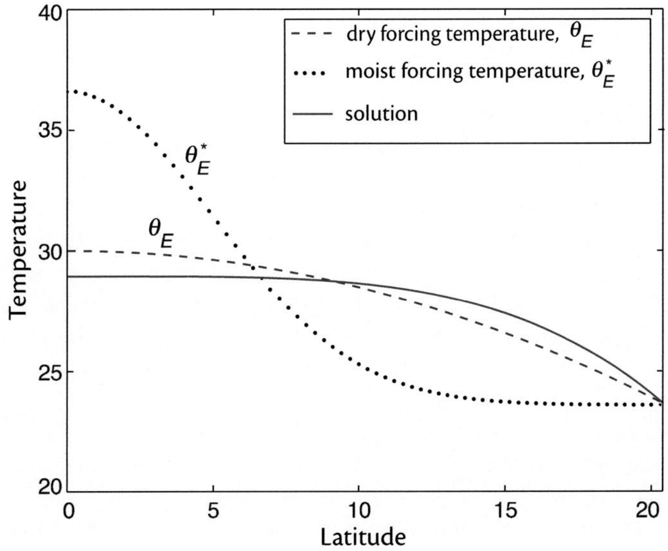

Fig. 14.8 Schematic of the effects of moisture on a model of the Hadley Cell. The temperature of the solution (solid line) is the same as that of a dry model, because this is determined from the angular-momentum-conserving wind. The heating distribution (as parameterized by a forcing temperature) is peaked near the equator in the moist case, leading to a more vigorous overturning circulation.

14.2.8 The Radiative Equilibrium Solution

Instead of a solution given by (14.12), could the temperature not simply be in radiative equilibrium everywhere? Such a state would have no meridional overturning circulation and the zonal velocity would be determined by thermal wind balance; that is,

$$v = 0, \qquad \theta = \theta_E, \qquad f\frac{u}{H} = -g\frac{\partial}{\partial y}\left(\frac{\theta_E}{\theta_0}\right). \tag{14.33}$$

To answer this question we consider the steady zonally symmetric zonal angular momentum equation with viscosity; that is, the zonally-averaged, viscous, steady, shallow atmosphere version of (2.68), namely

$$\frac{1}{a\cos\vartheta}\frac{\partial}{\partial\vartheta}(vm\cos\vartheta) + \frac{\partial(mw)}{\partial z} = \frac{\nu}{a\cos\vartheta}\frac{\partial}{\partial\vartheta}\left(\cos^2\vartheta\frac{\partial}{\partial\vartheta}\frac{u}{\cos\vartheta}\right) + \nu a\cos\vartheta\frac{\partial^2 u}{\partial z^2}, \tag{14.34}$$

where the variables vary only in the ϑ–z plane. The viscous term on the right-hand side arises from the expansion in spherical coordinates of the Laplacian. Note that it is angular *velocity*, not the angular momentum, that is diffused, because there is no diffusion of the angular momentum due to the Earth's rotation. However, to a very good approximation, the viscous term will be dominated by vertical derivatives and we may then write (14.34) as

$$\nabla_x \cdot (\boldsymbol{v}m) = \nu\frac{\partial^2 m}{\partial z^2}. \tag{14.35}$$

where $\nabla_x\cdot$ is the divergence in the meridional plane. The right-hand side now has a diffusive form, and in Section 13.5.1 we showed that variables obeying equations like this can have no extrema within the fluid. Thus, there can be no maximum or minimum of angular momentum in the interior of the fluid, a result known as Hide's theorem.[6] In effect, diffusion always acts to smooth away an isolated extremum, and this cannot be counterbalanced by advection. The result also implies that there can be no interior extrema in a statistically steady state if there is any zonally asymmetric eddy motion that transports angular momentum downgradient.

If the viscosity were so large that the viscous term was dominant in (14.34), and so with the horizontal term now important, then the fluid would evolve toward a state of solid body rotation, this being the fluid state with no internal stresses. In that case, there would be a maximum of angular momentum at the equator — a state of 'super-rotation'. (Related mechanisms have been proposed for the maintenance of super-rotation on Venus.[7])

Returning now to the question posed at the head of this section, suppose that the radiative equilibrium solution does hold. Then a radiative equilibrium temperature decreasing away from the equator more rapidly than the angular-momentum-conserving solution θ_M implies, using thermal wind balance, a maximum of m at the equator and above the surface, in violation of the no-extremum principle. Of course, we have derived the angular-momentum-conserving solution in the inviscid limit, in which the no-extrema principle does not apply. But any small viscosity will make the radiative equilibrium solution completely invalid, but potentially have only a small effect on the angular-momentum-conserving solution; that is, in the *limit* of small viscosity the angular-momentum-conserving solution can conceivably hold approximately, at least in the absence of boundary layers, whereas the radiative equilibrium solution cannot.

However, if the radiative equilibrium temperature varies more slowly with latitude than the temperature corresponding to the angular momentum conserving solution then a radiative equilibrium solution *can* obtain, without violating Hide's theorem. In particular, this is the case if $\theta_E \propto P_4(\sin\vartheta)$, where P_4 is the fourth Legendre polynomial, and so the possibility exists of two equilibrium solutions for the same forcing; however, P_4 is an unrealistically flat radiative equilibrium temperature for the Earth's atmosphere.

14.3 A SHALLOW WATER MODEL OF THE HADLEY CELL

Although expressed in the notation of the primitive equations, the model described above takes no account of any vertical structure in its stratification and is, *de facto,* a shallow water model. (We discuss how the primitive equations reduce to the shallow water equations in Sections 3.4 and 18.7.) Furthermore, the geometric aspects of sphericity play no essential role. Thus, we may transparently express the essence of the model by:

(i) explicitly using the shallow water equations instead of the stratified equations;

(ii) using the equatorial β-plane, with $f = f_0 + \beta y$ and $f_0 = 0$.

Let us therefore, if only as an exercise, construct a reduced-gravity model with an active upper layer overlying a stationary lower layer.

14.3.1 Momentum Balance

The inviscid zonal momentum equation of the upper layer is

$$\frac{Du}{Dt} - \beta y v = 0 \tag{14.36}$$

or

$$\frac{D}{Dt}\left(u - \frac{\beta y^2}{2}\right) = 0, \tag{14.37}$$

which is the β-plane analogue of the conservation of axial angular momentum. (In this section, all variables are zonally averaged, but we omit any notation denoting that.) From (14.37) we obtain the zonal wind as a function of latitude,

$$u = \frac{1}{2}\beta y^2 + A, \tag{14.38}$$

where A is a constant, which is zero if $u = 0$ at the equator, $y = 0$. The flow given by (14.38) is then analogous to the angular momentum conserving flow in the spherical model, (14.7). Because the lower layer is stationary, the analogue of thermal wind balance in the stratified model is just geostrophic balance, namely

$$fu = -g'\frac{\partial h}{\partial y}, \tag{14.39}$$

where h is the thickness of the active upper layer. Using (14.39) and $f = \beta y$ we obtain

$$g'\frac{\partial h}{\partial y} = -\frac{1}{2}\beta^2 y^3, \qquad \text{whence} \qquad h = -\frac{1}{8g'}\beta^2 y^4 + h(0), \tag{14.40a,b}$$

where $h(0)$ is the value of h at $y = 0$.

14.3.2 Thermodynamic Balance

The thermodynamic equation in the shallow water equations is just the mass conservation equation, which we write as

$$\frac{\mathrm{D}h}{\mathrm{D}t} = -\frac{1}{\tau}(h - h^*), \tag{14.41}$$

where the right-hand side represents heating — h^* is the field to which the height relaxes on a time scale τ. For illustrative purposes we will choose

$$h^* = h_0(1 - \alpha|y|). \tag{14.42}$$

(If we chose the more realistic quadratic dependence on y, the model would be more similar to that of the previous section.) To be in thermodynamic equilibrium we require that the right-hand side integrates to zero over the Hadley Cell; that is

$$\int_0^Y (h - h^*)\,\mathrm{d}y = 0, \tag{14.43}$$

where Y is the latitude of the poleward extent of the Hadley Cell, thus far unknown. Poleward of this, the height field is simply in equilibrium with the forcing — there is no meridional motion and $h = h^*$. Since the height field must be continuous, we require that

$$h(Y) = h^*(Y). \tag{14.44}$$

The two constraints (14.43) and (14.44) provide values of the unknowns $h(0)$ and Y, and give

$$Y = \left(\frac{5h_0\alpha g'}{\beta^2}\right)^{1/3}, \tag{14.45}$$

which is analogous to (14.19), as well as an expression for $h(0)$ that we leave as a problem for the reader. The qualitative dependence on the parameters is similar to that of the full model, although the latitudinal extent of the Hadley Cell is proportional to the cube root of the meridional thickness gradient α.

14.4 † ASYMMETRY AROUND THE EQUATOR

The Sun is overhead at the equator but two days out of the year, and in this section we investigate the effects that asymmetric heating has on the Hadley circulation. Observations indicate that except for the brief periods around the equinoxes, the circulation is dominated by a single cell with rising motion centred in the summer hemisphere, but extending well into the winter hemisphere. That is, as seen in Fig. 14.3, the 'winter cell' is broader and stronger than the 'summer cell', and it behoves us to try to explain this. We will stay in the framework of the inviscid angular-momentum model

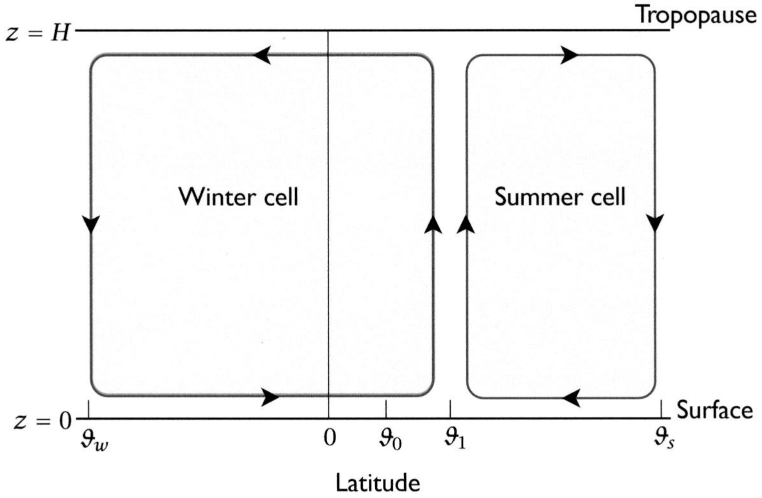

Fig. 14.9 A Hadley circulation model in which the heating is centred off the equator, at a latitude ϑ_0. The lower level convergence occurs at a latitude ϑ_1 that is not in general equal to ϑ_0. The resulting winter Hadley Cell is stronger and wider than the summer cell.

of Section 14.2, changing only the forcing field to represent the asymmetry and being a little more attentive to the details of spherical geometry.[8]

To represent an asymmetric heating we may choose a radiative equilibrium temperature of the form

$$\begin{aligned}\frac{\theta_E(\vartheta, z)}{\theta_0} &= 1 - \frac{2}{3}\Delta_H P_2(\sin\vartheta - \sin\vartheta_0) + \Delta_V\left(\frac{z}{H} - \frac{1}{2}\right)\\ &= 1 + \frac{\Delta_H}{3}\left[1 - 3(\sin\vartheta - \sin\vartheta_0)^2\right] + \Delta_V\left(\frac{z}{H} - \frac{1}{2}\right).\end{aligned} \tag{14.46}$$

This is similar to (14.14), but now the forcing temperature falls monotonically from a specified latitude ϑ_0. If $\vartheta_0 = 0$ the model is identical to the earlier one, but if not we envision a circulation as qualitatively sketched in Fig. 14.9, with rising motion off the equator at some latitude ϑ_1, extending into the winter hemisphere to a latitude ϑ_w, and into the summer hemisphere to ϑ_s. We will discover that, in general, $\vartheta_1 \neq \vartheta_0$ except when $\vartheta_0 = 0$. Following our procedure we used in the symmetric case as closely as possible, we then make the following assumptions:

(i) The flow is quasi-steady. That is, at any time of year the flow adjusts to a steady circulation on a time scale more rapid than that on which the solar zenith angle appreciably changes.

(ii) The flows in the upper branches conserve angular momentum, m. Further assuming that $u = 0$ at $\vartheta = \vartheta_1$ so that $m = \Omega a^2 \cos^2\vartheta_1$ we obtain

$$u(\vartheta) = \frac{\Omega a(\cos^2\vartheta_1 - \cos^2\vartheta)}{\cos\vartheta}. \tag{14.47}$$

Thus, we expect to see westward (negative) winds aloft at the equator. In the lower branches the zonal flow is assumed to be approximately zero, i.e., $u(0) \approx 0$.

(iii) The flow satisfies hydrostatic and gradient wind balance. The meridional momentum equation is then

$$fu + \frac{u^2\tan\vartheta}{a} = -\frac{1}{a}\frac{\partial\phi}{\partial\vartheta}, \tag{14.48}$$

and because the flow crosses the equator we cannot neglect the second term on the left-hand side. Combining this with hydrostatic balance ($\partial\phi/\partial z = g\theta/\theta_0$) leads to a generalized thermal wind balance, which may be written as

$$m\frac{\partial m}{\partial z} = -\frac{ga^2\cos^2\vartheta}{2\theta_0\tan\vartheta}\frac{\partial\theta}{\partial\vartheta}. \tag{14.49}$$

If the undifferentiated m is approximated by $\Omega a^2\cos^2\vartheta$, this reduces to conventional thermal wind balance, (14.10).

(iv) Potential temperature in each cell is conserved when integrated over the extent of the cell. Thus,

$$\int_{\vartheta_1}^{\vartheta_s}(\theta - \theta_E)\cos\vartheta\,\mathrm{d}\vartheta = 0, \qquad \int_{\vartheta_1}^{\vartheta_w}(\theta - \theta_E)\cos\vartheta\,\mathrm{d}\vartheta = 0, \tag{14.50}$$

for the summer and winter cells, respectively, where θ is the vertically averaged potential temperature.

(v) Potential temperature is continuous at the edge of each cell, so that

$$\theta(\vartheta_s) = \theta_E(\vartheta_s), \qquad \theta(\vartheta_w) = \theta_E(\vartheta_w), \tag{14.51}$$

and is also continuous at ϑ_1. This last condition must be explicitly imposed in the asymmetric model, whereas in the symmetric model it holds by symmetry. Now, recall from the symmetric model that the value of the temperature at the equator was determined by the integral constraint (14.17) and the continuity constraint (14.18). We have analogues of these in each hemisphere, namely (14.50) and (14.51), and thus, if ϑ_1 is set equal to ϑ_0 we cannot expect that they each would give the same temperature at ϑ_0. Thus, ϑ_1 must be a free parameter to be determined.

Given these assumptions, the solution may be calculated. Using thermal wind balance, (14.49), with $m(H) = \Omega a^2\cos^2\vartheta_1$ and $m(0) = \Omega a^2\cos^2\vartheta$ we find

$$-\frac{1}{\theta_0}\frac{\partial\theta}{\partial\vartheta} = \frac{\Omega^2a^2}{gH}\left(\frac{\sin\vartheta}{\cos^3\vartheta}\cos^4\vartheta_1 - \sin\vartheta\cos\vartheta\right), \tag{14.52}$$

which integrates to

$$\theta(\vartheta) - \theta(\vartheta_1) = -\frac{\theta_0\Omega^2a^2}{2gH}\frac{(\sin^2\vartheta - \sin^2\vartheta_1)^2}{\cos^2\vartheta}. \tag{14.53}$$

The value of ϑ_1, and the value of $\theta(\vartheta_1)$, are determined by the constraints (14.50) and (14.51). It is not in general possible to obtain a solution analytically, but one may be found numerically by an iterative procedure and one such is illustrated in Fig. 14.10. The zonal wind of the solution is always symmetric around the equator, because it is determined solely by angular momentum conservation. The temperature is therefore also symmetric, as (14.53) explicitly shows. However, the width of the solution in each hemisphere will, in general, be different.

Furthermore, because the strength of the circulation increases with difference between the temperature of the solution and the radiative equilibrium temperature, the circulation in the winter hemisphere will also be much stronger than that in the summer, a prediction that is consistent with the observations (see Fig. 14.3). More detailed calculations show that, because the strength of the model Hadley Cell increases nonlinearly with ϑ_0, the time-average strength of the Hadley Cell with seasonal forcing is stronger that that produced by annually averaged forcing. However, this does not appear to be a feature of either the observations or more complete numerical simulations, suggesting that an angular-momentum-conserving model has some deficiencies.[9]

The lack of consideration of zonal asymmetries and the lack of angular momentum conservation because of the effects of baroclinic eddies are issues that are shared with the steady model with

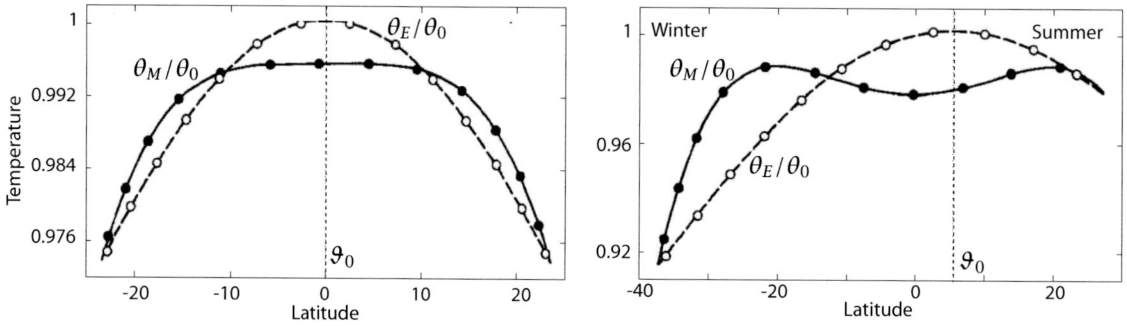

Fig. 14.10 Solutions of the Hadley Cell model with heating centred at the equator ($\vartheta_0 = 0°$, left) and off the equator ($\vartheta_0 = +6°$ N, right), with $\Delta_H = 1/6$. The dashed line is the radiative equilibrium temperature and the solid line is the angular-momentum-conserving solution. In the right-hand panel, $\vartheta_1 \approx +18°$, and the circulation is dominated by the cell extending from $+18°$ to $-36°$.[8]

hemispheric symmetry. A problem that is unique to the asymmetric model is the quasi-steady assumption, given the presence of a temporally progressing seasonal cycle. Because the latitude of the upward branch of the Hadley Cell varies with season, the value of the angular momentum entering the system also varies with time, and so a homogenized value of angular momentum is hard to achieve. Nonetheless, the overall picture that the model paints, with its qualitative explanation of the strengthened and extended winter Hadley Cell, is very useful, even if quantitatively flawed.

14.5 † EDDY EFFECTS ON THE HADLEY CELL

So far, we have ignored the effects of baroclinic eddies on the Hadley circulation although we have no reason to believe that their effects will be negligible. In fact, as the upper-level flow moves poleward the shear of the zonal wind increases, as described above, and at some point the flow will become baroclinically unstable. We first describe a simple model of this, before considering eddy fluxes more generally.[10]

14.5.1 A Hadley Cell Limited by Baroclinic Instability

Suppose that the flow moving poleward conserves its angular momentum, and for simplicity consider flow on a beta-plane. The flow is given by (14.7), which in the small angle approximation implies a shear, Λ_M, of

$$\Lambda_M \approx \frac{\Omega a \vartheta^2}{H} = \frac{\beta y^2}{2H}, \tag{14.54}$$

where H is the height of the outflow and the last two expressions hold in the small angle approximation. Now, in a quasi-geostrophic two-level model, the flow becomes unstable when the shear between upper and lower levels reaches a critical value, Λ_C, given by

$$\Lambda_C \equiv \frac{U_1 - U_2}{H/2} = \frac{1}{2H}\beta L_d^2, \tag{14.55}$$

where $L_d = NH/f$ is the baroclinic deformation radius, and on the sphere $\beta = 2\Omega\cos\phi/a$. Both β factor and the f hiding in L_d make Λ_C grow towards the equator. Equating (14.54) and (14.55) suggests that the angular-momentum conserving flow will become unstable at a latitude ϑ_C given by, in the small angle approximation,

$$\vartheta_C \approx \left(\frac{NH}{2\Omega a}\right)^{1/2}. \tag{14.56}$$

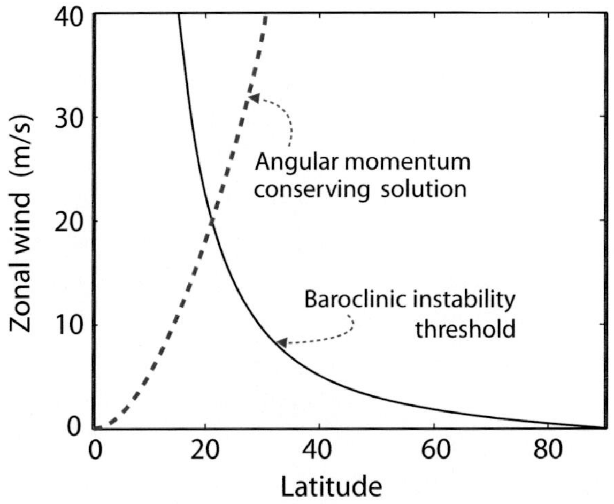

Fig. 14.11 Upper tropospheric zonal winds. The dashed curve shows the angular-momentum conserving wind, with $u = 0$ at the equator. The solid curve shows the threshold for baroclinic instability of the upper-level flow using a two-layer quasi-geostrophic calculation, (14.55), with f varying with latitude in the deformation radius.

The value of ϑ_C above should not be taken literally — the real atmosphere is not a two-level quasi-geostrophic model! But it does capture the essential truth that the angular momentum conserving solution will become baroclinically unstable at some latitude, as sketched in Fig. 14.11. It is a quantitative issue as to whether the Hadley flow becomes strongly unstable before it reaches its natural poleward extent. However, even if baroclinic instability itself is weak over much of the tropics, baroclinic instability further poleward will have an effect and lead to the non-conservation of angular momentum, as we now discuss.

14.5.2 Diagnostic Considerations

The zonally-averaged zonal momentum equation, (14.4), may be written as an equation for angular momentum, $\overline{m}$. Referring back to Section 2.2 if needs be, the equation may be written as

$$\begin{aligned}\frac{\partial \overline{m}}{\partial t} + \frac{1}{\cos\vartheta}\frac{\partial}{\partial y}(\overline{v}\,\overline{m}\cos\vartheta) + \frac{\partial}{\partial z}(\overline{w}\,\overline{m}) &= -\frac{1}{\cos\vartheta}\frac{\partial}{\partial y}(\overline{m'v'}\cos\vartheta) - \frac{\partial}{\partial z}(\overline{m'w'}) \\ &= -\frac{1}{\cos\vartheta}\frac{\partial}{\partial y}(\overline{u'v'}\,a\cos^2\vartheta) - \frac{\partial}{\partial z}(\overline{u'w'}\,a\cos\vartheta),\end{aligned} \tag{14.57}$$

where $\overline{m} = (\overline{u}+\Omega a\cos\vartheta)a\cos\vartheta$, $m' = u'a\cos\vartheta$, $y = a\vartheta$, and the vertical and meridional velocities are related by the mass continuity relation

$$\frac{1}{\cos\vartheta}\frac{\partial}{\partial y}(\overline{v}\cos\vartheta) + \frac{\partial \overline{w}}{\partial z} = 0. \tag{14.58}$$

In the angular-momentum-conserving model the eddy fluxes were neglected and (14.57) was approximated by the simple expression $\partial\overline{m}/\partial\vartheta = 0$, and by construction the Rossby number is $\mathcal{O}(1)$, because $\zeta = -f$.

The observed eddy heat and momentum fluxes are shown in Fig. 14.12. The eddy momentum flux is generally poleward, converging in the region of the mid-latitude surface westerlies. Its magnitude, and more particularly its meridional gradient, is as large or larger than the momentum flux associated with the mean flow. Neglecting vertical advection and vertical eddy fluxes, and using (14.58), (14.57) may be written as

$$\frac{\partial \overline{m}}{\partial t} + \overline{v}\frac{\partial \overline{m}}{\partial y} = -\frac{1}{\cos\vartheta}\frac{\partial}{\partial y}(\overline{u'v'}\,a\cos^2\vartheta). \tag{14.59}$$

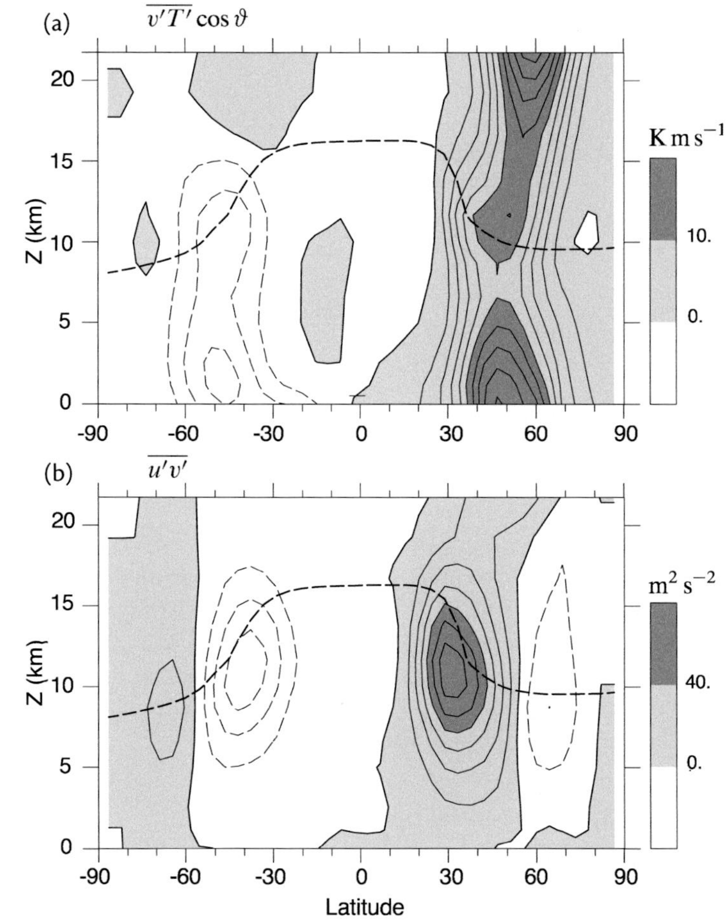

Fig. 14.12 (a) The average meridional eddy heat flux and (b) the eddy momentum flux in the northern hemisphere winter (DJF). The ordinate is log-pressure, with scale height H = 7.5 km. Positive fluxes are shaded, and the dashed line marks the thermal tropopause.

The eddy heat flux (contour interval 2 K m s^{-1}) is largely poleward and downgradient in both hemispheres. The eddy momentum flux (contour interval 10 m^2 s^{-2}) is upgradient and *converges* in mid-latitudes in the region of the mean jet, leading to eastward surface winds.[11]

The quantity $\overline{u'v'}$ increases to a maximum at about 30°, so the right-hand side is negative in the tropics and lower subtropics. Thus, if $\overline{v} > 0$ (as in the upper branch of the Northern Hemisphere Hadley Cell) and the flow is steady, the observed eddy fluxes are such as to cause the angular momentum of the zonal flow to *decrease* as it moves poleward, and the zonal velocity is lower than it would be in the absence of eddies. (In the Southern Hemisphere the signs of v and the eddy momentum flux are reversed, but the dynamics are equivalent.)

The eddy flux of heat will also affect the Hadley Cell, although in a different fashion. We see from Fig. 14.12 that the eddy flux of temperature is predominantly poleward, and therefore that eddies export heat from the subtropics to higher latitudes. Now, the zonally-averaged thermodynamic equation may be written

$$\frac{\partial \overline{b}}{\partial t} + \frac{1}{\cos\vartheta}\frac{\partial}{\partial y}(\overline{v}\overline{b}\cos\vartheta) + \frac{\partial}{\partial z}(\overline{w}\overline{b}) = -\frac{1}{\cos\vartheta}\frac{\partial}{\partial y}(\overline{v'b'}\cos\vartheta) - \frac{\partial}{\partial z}(\overline{w'b'}) + Q_b, \tag{14.60}$$

where Q_b represents the heating. After vertical averaging, the vertical advection terms vanish and the resulting equation is the thermodynamic equation implicitly used in the angular momentum-conserving model, with the addition of the meridional eddy flux on the right-hand side. A diverging eddy heat flux in the subtropics (as in Fig. 14.12) is evidently equivalent to increasing the meridional gradient of the radiative equilibrium temperature, and therefore will increase the intensity of the overturning circulation.

14.5.3 An Idealized Eddy-driven Model

Consider now the extreme case of an 'eddy-driven' Hadley Cell. (The driving for the Hadley Cell, and the atmospheric circulation in general, ultimately comes from the differential heating between equator and pole. Recognizing this, 'eddy driving' is a convenient way to refer to the role of eddies in producing a zonally-averaged circulation. See also endnote 2 on page 858.) The model is over-simple, but revealing. Neglecting vertical derivatives the zonally-averaged zonal momentum equation (14.4) may be written

$$\frac{\partial \overline{u}}{\partial t} - (f + \overline{\zeta})\overline{v} = -\frac{\partial}{\partial y}\overline{u'v'}. \tag{14.61}$$

using Cartesian geometry for simplicity. If the Rossby number is sufficiently low this becomes

$$\frac{\partial \overline{u}}{\partial t} - f\overline{v} = M, \tag{14.62}$$

where $M = -\partial_y(\overline{u'v'})$, . This approximation is not quantitatively accurate but it will highlight the role of the eddies. Note the contrast between this model and the angular-momentum-conserving model. In the latter we assumed $f + \overline{\zeta} \approx 0$, and $Ro = \mathcal{O}(1)$; now we are neglecting $\overline{\zeta}$ and assuming the Rossby number is small. At a similar level of approximation let us write the thermodynamic equation, (14.60), as

$$\frac{\partial \overline{b}}{\partial t} + N^2 w = J, \tag{14.63}$$

where $J = Q_b - \partial_y(\overline{v'b'})$ represents the diabatic terms and eddy forcing. We are assuming, as in quasi-geostrophic theory, that the mean stratification, N^2 is fixed, and now $\overline{b}$ represents only the (zonally averaged) deviations from this. The mass continuity equation allows us to define a meridional streamfunction Ψ; that is

$$\frac{\partial \overline{v}}{\partial y} + \frac{\partial \overline{w}}{\partial z} = 0 \qquad \text{allows} \qquad \overline{w} = \frac{\partial \Psi}{\partial y}, \quad \overline{v} = -\frac{\partial \Psi}{\partial z}. \tag{14.64a,b}$$

We may then use the thermal wind relation, $f\partial\overline{u}/\partial z = -\partial\overline{b}/\partial y$, to eliminate time derivatives in (14.62) and (14.63), giving[12]

$$f^2\frac{\partial^2 \Psi}{\partial z^2} + N^2\frac{\partial^2 \Psi}{\partial y^2} = f\frac{\partial M}{\partial z} + \frac{\partial J}{\partial y}. \tag{14.65}$$

This is a linear equation for the overturning streamfunction, one that holds even if the flow is not in a steady state, and a positive value of Ψ, in the Northern Hemisphere, corresponds to rising at the equator. The equation is equally (or in fact more) valid in mid-latitudes as in the tropics. We see that the overturning circulation is forced by eddy fluxes of heat and momentum, as well as heating and other terms that might appear on the right-hand sides of (14.62) and (14.63). If we rescale the vertical coordinate by the Prandtl ratio (i.e., let $z = z'f/N$) then (14.65) is a Poisson equation for the streamfunction. A few other germane points are as follows:

- The horizontal gradient of the thermodynamic forcing partially drives the circulation, and both the heating term and the horizontal eddy flux divergence act in the same sense. A thermodynamically forced overturning circulation, with warm fluid rising and cold fluid sinking, is called a 'direct cell'.

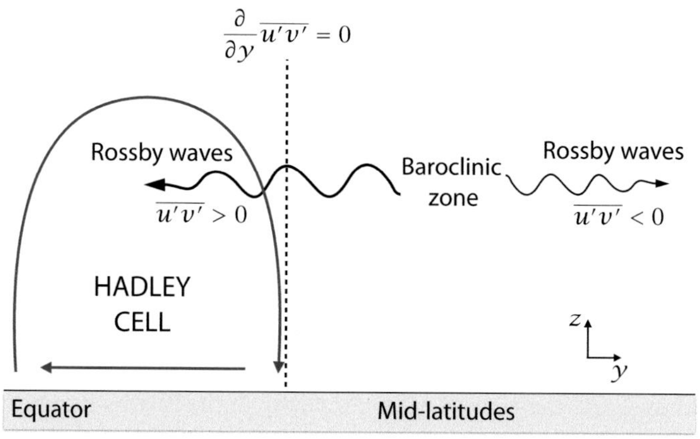

Fig. 14.13 Sketch of how eddy fluxes can affect the Hadley Cell even when the baroclinic zone is centred well poleward of the Cell.

Rossby waves are generated by baroclinic instability at mid-latitudes. Some propagate equatorward, and deposit westward momentum, $\partial(\overline{u'v'})/\partial y > 0$ inside the Hadley Cell. At some latitude the Rossby wave momentum flux is neither convergent nor divergent, $\partial(\overline{u'v'})/\partial y = 0$, corresponding to the edge of the Hadley cell.

- The vertical gradient of the horizontal eddy momentum divergence partially drives the circulation, and because $\partial_y(\overline{u'v'}) > 0$ over the tropics and subtropics (for the Northern Hemisphere, see Fig. 14.12) these fluxes intensify the circulation, weakening the zonal flow aloft terms and strengthening the overturning circulation. The Coriolis term fv is balanced by the eddy momentum flux convergence.
- If N is small, then the circulation becomes stronger if the other terms remain the same, because the air can circulate without transporting any heat.
- In winter, the increased strength of eddy momentum and buoyancy fluxes drives a stronger Hadley Cell. This constitutes a different mechanism from that given in Section 14.4 for the increased strength of the winter cell.

14.6 NON-LOCAL EDDY EFFECTS AND NUMERICAL RESULTS

14.6.1 † A Non-local Model

We saw above that eddy fluxes will tend to strengthen the Hadley Cell and weaken the zonal winds. These eddy fluxes can be important even if the main zone of baroclinic instability is well poleward of the Hadley Cell termination, because Rossby waves can propagate equatorward from the baroclinic zone into the subtropics (as sketched in Fig. 14.13). This propagation will be discussed more in Sections 15.1 and 16.2, but suffice it to say here that equatorward propagating Rossby waves produce a poleward momentum flux, $\overline{u'v'} > 0$ (we use Northern Hemisphere and Cartesian notation). Even if the main baroclinic activity occurs around, say, 45°, then the amplitude of the poleward eddy flux may reach its maximum value some distance equatorward of that latitude if the baroclinic zone itself extends further equatorward, and observations (Fig. 14.12) show that the flux is a maximum at about 30° (varying with season), diminishing equatorward of that, so that in the Hadley Cell $\partial_y(\overline{u'v'}) > 0$. The edge of the Hadley Cell is coincident with the latitude at which the poleward eddy flux divergence is zero, since the steady-state momentum equation is approximately

$$-(f + \overline{\zeta})\overline{v} = -\frac{\partial}{\partial y}\overline{u'v'}. \tag{14.66}$$

At the edge of the Hadley Cell we have $\overline{v} = 0$ and thus $\partial_y\left(\overline{u'v'}\right) = 0$. The model is a little simplistic because other terms in the momentum equation may then become important, but it is nevertheless instructive.

The latitude at which the right-hand side of (14.66) becomes small is not necessarily the same as the one where the poleward flow in the Hadley Cell becomes baroclinically unstable, although the two may be similar in practice. In fact, determining the latitudinal distribution of eddy fluxes is a difficult problem in wave–mean-flow interaction, since the eddy fluxes affect and are affected by the mean flow. As well as the generation of eddy fluxes by baroclinic instability, the absorption or dissipation of Rossby waves is an important factor, since in their absence the momentum flux would be approximately constant equatorward of the baroclinic zone, and the right-hand side of (14.66) would be zero over a range of latitudes. The absorption of Rossby waves is enhanced near critical latitudes (where the wave speed equals the mean fluid speed) and the wave activity and the momentum flux diminish equatorward of that, but the critical latitude itself need not correspond to the maximum of the eddy momentum fluxes. In the real world the critical latitude is not sharply defined and, although it is often equatorward of the edge of the Hadley Cell rather than coincident with it, Rossby waves may begin to dissipate well before reaching it, and so poleward of it.

The details of the dynamics determining the Hadley Cell edge are plainly rather complex, although a qualitative picture is simply described. The eddy momentum flux convergence, $\partial_y\overline{(u'v')}$, is negative in mid-latitudes and positive in low-latitudes, with a zero crossing ($\partial_y\overline{(u'v')} = 0$) near the edge of the Hadley Cell, as in Fig. 14.13. Consistent with this, observations (e.g., Fig. 14.12) show that the latitude of the eddy momentum flux maximum coincides with a rapid change in tropopause height, which itself is generally coincident with the edge of the Hadley Cell.

Evidently, the effects of baroclinic instability, and not just the effects of a local instability, can greatly influence the Hadley Cell, and to round out this section we illustrate that fact with some numerical simulations.

14.6.2 Numerical Solutions

Some illustrative results from two idealized numerical experiments with a GCM are shown in Figs. 14.14 and 14.15. The GCM has no explicit representation of moisture, except that the lapse rate is adjusted to a value close to the moist adiabatic lapse rate if it exceeds that value. In one experiment the model is constrained to produce an axisymmetric solution (left-hand panels of the figures), and the zonal wind produced by the model in the Hadley Cell outflow is fairly close to being angular-momentum-conserving. In a three-dimensional version of the model, in which baroclinic eddies are allowed to form, the zonal wind is significantly reduced from its angular-momentum-conserving value, and correspondingly the overturning circulation is much stronger (right-hand panels). Indeed, the strength of the Hadley Cell increases roughly linearly with the strength of the eddies in a sequence of numerical integrations similar to those shown, as suggested by (14.65). Qualitatively similar results are found in a model with no convective parameterization. In this case, the lapse rate is closer to neutral, N^2 is small, and the overturning circulation is generally stronger, as also expected from (14.65). The results generally indicate very strong eddy effects on the strength of the Hadley Cell, and the value of the zonal wind within it, although a dry model may overemphasize the importance of eddy effects, because the circulation in a zonally symmetric dry model is weaker than a similar moist model, as discussed in Section 14.2.7.

14.6.3 Final Remarks

Is the real Hadley circulation 'eddy-driven', as in Sections 14.5 and above, or is it a largely zonally symmetric structure constrained by angular momentum conservation, as in Section 14.2? And how does this balance vary with season?

Observations of the overturning flow in summer and winter provide a guide. Figure 14.16 shows the thickness-weighted transport overturning circulation in isentropic coordinates, and (as discussed in Chapter 10) this circulation includes both the Eulerian mean transport and the transport due to eddies. The winter cell (the cross-equatorial cell with the upward branch in the summer hemisphere) is strong and self-contained, with considerable recirculation most of which comes from its zonally symmetric component. The winter cell is quite distinct from the mid-latitude circulation, suggesting the dominance of axisymmetric dynamics, for if it were solely a response to eddy heat and momentum fluxes one might expect it to join more smoothly with the mid-latitude

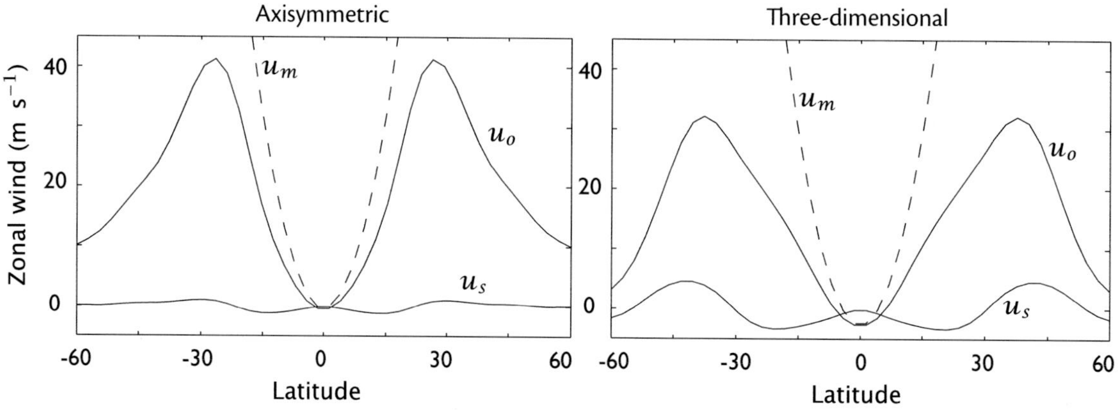

Fig. 14.14 The zonal wind in two numerical simulations. The right panel is from an idealized dry, three-dimensional atmospheric GCM, and the left panel is an axisymmetric version of the same model. Plotted are the zonal wind at the level of the Hadley Cell outflow, u_o; the surface wind, u_s; and the angular-momentum-conserving value, u_m.[13]

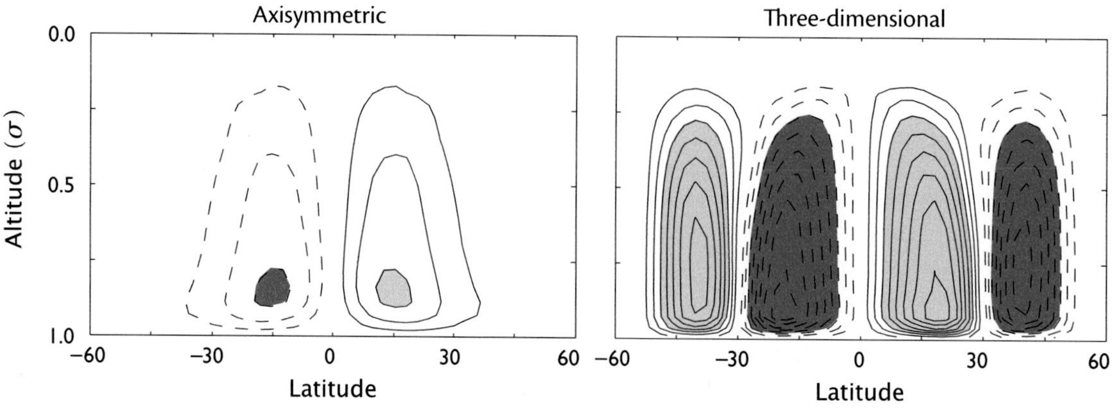

Fig. 14.15 As for Fig. 14.14, but now showing the streamfunction of the overturning circulation. 'Altitude' is $\sigma = p/p_s$, where p_s is surface pressure, and contour interval is 5 Sv (i.e., 5×10^9 kg s^{-1}). The effects of eddies may be exaggerated because the model is dry.

Ferrel Cell. The axisymmetric winter Hadley Cell is naturally stronger than its summer counterpart, even in a dry atmosphere, and the further effects of condensation and the concomitant concentration of the thermodynamic source may strengthen it further, giving the axisymmetric circulation a dominant role.

In summer, in contrast, there is virtually no recirculation within the Hadley Cell and it does not appear as a self-contained structure, suggestive of baroclinic eddy effects and/or a strong mid-latitude influence. And even without baroclinic eddies, zonally asymmetric circulations are important, for the Hadley Cell over India and South East Asia is intimately linked with monsoonal circulations. But tying the monsoon circulation into a theory of the Hadley Cell, and in particular into the transition from winter to summer dynamics, is, alas, a task for another day.

14.7 THE FERREL CELL

In this section we give a descriptive introduction to the Ferrel Cell, taking the eddy fluxes of heat and momentum to be given and viewing the circulation from a zonally averaged and Eulerian perspective. We investigate the associated dynamics in the next chapter.

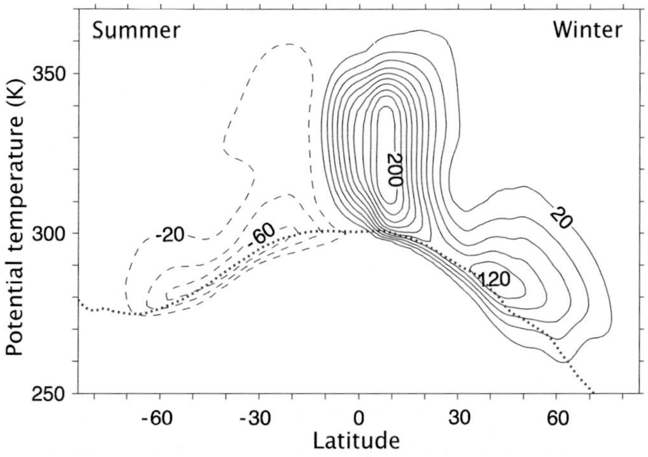

Fig. 14.16 The observed mass transport streamfunction in isentropic coordinates in northern hemisphere winter (DJF). The dotted line is the median surface temperature.

The return flow is nearly all in a layer near the surface, much of it at a lower temperature than the median surface temperature. Note the more vigorous circulation in the winter hemisphere.[14]

The Ferrel Cell is an indirect meridional overturning circulation in mid-latitudes (see Fig. 14.3) that is apparent in the zonally-averaged v and w fields, or the meridional overturning circulation defined by (14.3) or (14.64b). It is 'indirect' because cool air apparently rises in high latitudes, moves equatorward and sinks in the subtropics. Why should such a circulation exist? The answer, in short, is that it is there to balance the eddy momentum convergence of the mid-latitude eddies and it is effectively driven by those eddies. To see this, consider the zonally-averaged zonal momentum equation in mid-latitudes; at low Rossby number, and for steady flow this is just

$$-f\overline{v} = -\frac{1}{\cos^2\vartheta}\frac{\partial}{\partial\vartheta}(\cos^2\vartheta\,\overline{u'v'}) + \frac{1}{\rho}\frac{\partial\tau}{\partial z}. \tag{14.67}$$

This equation is a steady version of (14.62) with the addition of a frictional term $\partial\tau/\partial z$ on the right-hand side. At the surface we may approximate the stress by a drag, $\tau = r\overline{u}_s$, where r is a constant, with the stress falling away with height so that it is important only in the lowest kilometre or so of the atmosphere, in the atmospheric Ekman layer. Above this layer, the eddy momentum flux convergence is balanced by the Coriolis force on the meridional flow. In mid-latitudes (from about 30° to 70°) the eddy momentum flux divergence is negative in both hemispheres (Fig. 14.12) and therefore, from (14.67), the averaged meridional flow must be equatorward, as illustrated schematically in Fig. 14.17.

The flow cannot be equatorward everywhere, simply by mass continuity, and the return flow occurs largely in the Ekman layer, of depth d say. Here the eddy balance is between the Coriolis term and the frictional term, and integrating over this layer gives

$$-fV \approx -r\overline{u}_s, \tag{14.68}$$

where $V = \int_0^d \rho\overline{v}\,dz$ is the meridional transport in the boundary layer, above which the stress vanishes. The return flow is poleward (i.e., $V > 0$ in the Northern Hemisphere) producing an eastward Coriolis force. This can be balanced by a westward frictional force provided that the surface flow has an eastward component. In this picture, then, the mid-latitude eastward zonal flow at the surface is a proximate consequence of the poleward flowing surface branch of the Ferrel Cell, this poleward flow being required by mass continuity given the equatorward flow in the upper branch of the cell. In this way, the Ferrel Cell is responsible for bringing the mid-latitude eddy momentum flux convergence to the surface where it may be balanced by friction (refer again to Fig. 14.17).

A more direct way to see that the surface flow must be eastward, given the eddy momentum flux convergence, is to vertically integrate (14.67) from the surface to the top of the atmosphere.

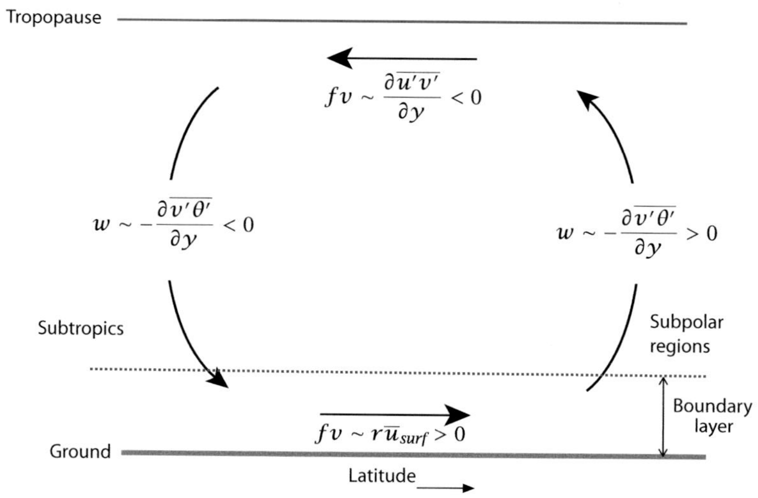

Fig. 14.17 The eddy-driven Ferrel Cell, from an Eulerian point of view.

Above the planetary boundary layer the mean flow is largely in balance with the eddy heat and momentum fluxes. The lower branch of the Ferrel Cell is largely confined to the boundary layer, where it is in a frictional–geostrophic balance.

By mass conservation, the Coriolis term vanishes (i.e., $\int_0^\infty f\rho\overline{v}\,\mathrm{d}z = 0$) and we obtain

$$\int_0^\infty \frac{1}{\cos^2\vartheta}\frac{\partial}{\partial\vartheta}(\cos^2\vartheta\,\overline{u'v'})\rho\,\mathrm{d}z = [\tau]_0^\infty = -r\overline{u}_s. \tag{14.69}$$

That is, the surface wind is proportional to the vertically integrated eddy momentum flux convergence. Because there *is* a momentum flux convergence, the left-hand side is negative and the surface winds are eastward.

The eddy heat flux also plays a role in the Ferrel Cell, for in a steady state we have, from (14.63)

$$w = \frac{1}{N^2}\left[Q_b - \frac{1}{\cos\vartheta}\frac{\partial(\overline{v'b'}\cos\vartheta)}{\partial y}\right], \tag{14.70}$$

and inspection of Fig. 14.12 shows that the observed eddy heat flux produces an overturning circulation in the same sense as the observed Ferrel Cell (again see Fig. 14.17).

Is the circulation produced by the heat fluxes *necessarily* the same as that produced by the momentum fluxes? In a non-steady state the effects of both heat and momentum fluxes on the Ferrel Cell are determined by (14.65) (an equation which applies more accurately at mid-latitudes than at low ones because of the low-Rossby number assumption), and there is no particular need for the heat and momentum fluxes to act in the same way. But in a steady state they must act to produce a consistent circulation. To see this, for simplicity let us take f and N^2 to be constant, let us suppose the fluid is incompressible and work in Cartesian coordinates. Take the y-derivative of (14.67) and the z-derivative of (14.70) and use the mass continuity equation. Noting that $\overline{v'\zeta'} = -\partial\overline{u'v'}/\partial y$ we obtain

$$\frac{\partial}{\partial y}\left(\overline{v'\zeta'} + \frac{f_0}{N^2}\frac{\partial\overline{v'b'}}{\partial z}\right) = \frac{\partial}{\partial z}\left(\frac{f_0}{N^2}Q[b]\right) + \frac{\partial}{\partial y}\left(\frac{1}{\rho_0}\frac{\partial\tau}{\partial z}\right). \tag{14.71}$$

The expression on the left-hand side is the divergence of the eddy flux of quasi-geostrophic potential vorticity! That the heat and momentum fluxes act to produce a consistent overturning circulation is thus equivalent to requiring that the terms in the quasi-geostrophic potential vorticity equation are in a steady-state balance. The eddy fluxes of heat and momentum evidently play a huge role in the mid-latitude circulation, and in the next chapter we examine the fluid dynamics giving rise to these eddy fluxes.

Notes

1 Many of the observations presented here are so-called *reanalyses*, prepared by the National Centers for Environmental Prediction (NCEP) and the European Centre for Medium-Range Weather Forecasts (ECMWF), described in Kalnay (1996) and Dee *et al.* (2011). Reanalysis products are syntheses of observations and model results and so are not wholly accurate representations of the atmosphere. However, especially in data-sparse regions of the globe and for poorly measured fields, they are likely to be more accurate representations of the atmosphere than could be achieved using only the raw data. Of course, this in turn means they contain biases introduced by the models. A reanalysis is a particular form of *state estimate*, which is the more general name given to similar products and is the name used in oceanography.

2 Trenberth & Caron (2001).

3 George Hadley (1685–1768) was a British meteorologist who formulated the first dynamical theory for the trade winds, presented in a paper (Hadley 1735) entitled 'Concerning the cause of the general trade winds.' At that time, trade winds referred to any large-scale prevailing wind, and not just tropical winds. The name 'trade' may be associated with the commercial (i.e., trade) exploitation of the wind by mariners on long ocean journeys, but trade also means (or at least meant) customary, and trade winds customarily blow in one direction. Relatedly, in Middle English the word trade means path or track — hence the phrase 'the wind blows trade', meaning the wind is on track. Hadley realized that in order to account for the zonal winds, the Earth's rotation makes it necessary for there also to be a meridional circulation. His vision was of air heated at low latitudes, cooled at high latitudes, giving rise to a single meridional cell between the equator and each pole. Although he thought of the cell as essentially filling the hemisphere, and he did not account for the instability of such a flow, it was nevertheless a foundational contribution to meteorology. The thermally direct cell in low latitudes is now named after him.

A three-celled circulation was proposed by William Ferrel (1817–1891), an American school teacher and meteorologist, and the middle of these cells is now named for him. His explanation of the cell (Ferrel 1856a) was not correct, hardly surprising because the eddy motion that drives the Ferrel Cell was not understood for another 100 years or so. Ferrel's ideas evolved to something more akin to a two-celled picture (Ferrel 1859), similar to that proposed by J. Thomson in 1857. The history of these ideas is discussed by Thomson (1892). Ferrel did however give the first essentially correct description of the role of the Coriolis force and the geostrophic wind in the general circulation (Ferrel 1858, a paper with a quite modern style), a key development in the history of geophysical fluid dynamics. Ferrel also contributed to tidal theory (Ferrel 1856b) and to ocean dynamics. (See http://www.history.noaa.gov/giants/ferrel2.html).

Although Hadley's single-celled viewpoint was superseded by the three-celled and two-celled structures, the modern view of the overturning circulation is, ironically, that of a single cell of 'residual circulation', which, although having distinct tropical and extratropical components, in some ways qualitatively resembles Hadley's original picture.

4 Schneider (1977) proposed an axially-symmetric, angular momentum conserving model of the Hadley Cell. Held & Hou (1980) developed the model we follow here.

5 Fang & Tung (1996) do find some analytic solutions in the presence of moisture.

6 After Hide (1969).

7 Gierasch (1975).

8 Largely following Lindzen & Hou (1988). The solutions of Fig. 14.10 are taken from that paper.

9 See Dima & Wallace (2003) for some relevant observations. They noted that the asymmetry of the Hadley Cell is affected by monsoonal circulations (which are, of course, not accounted for in the model presented here). Fang & Tung (1999) investigated the effects of time dependence and note that quasi-steadiness is not well satisfied, although this alone was unable to limit the nonlinear amplification effect. In reality, the effects of baroclinic eddies are important in Hadley Cell dynamics, as we discuss in Section 14.5.

10 Although our presentation does not follow the historical order, that the effects of baroclinic eddies

are likely to be important in modifying the Hadley circulation is in fact the traditional view, emerging in the second half of the 20th century. Lorenz (1967), for example, discusses the fact that an ideal, axisymmetric, Hadley circulation is likely to be baroclinically unstable, although the Hadley circulation envisioned was one that filled the hemisphere. More recently Kim & Lee (2001), Walker & Schneider (2005), Frierson *et al.* (2007), and many others, have discussed the effects of baroclinic eddies on the Hadley Cell.

11 Figure courtesy of M. Juckes, using an ECMWF reanalysis.

12 A simple generalization of (14.65) is to replace (14.62) by (14.59) and then to use the thermal wind equation in the form

$$\frac{f}{a\cos\vartheta}\frac{\partial\overline{m}}{\partial z} = -\frac{\partial\overline{b}}{\partial y}. \tag{14.72}$$

An equation very similar to, but a little more general than, (14.65) may then be derived. A still more general, usually elliptic equation for the overturning circulation may be derived from the zonally-averaged primitive equations, assuming only that the zonally-averaged zonal wind is in gradient wind balance with the pressure field (Vallis 1982).

13 The simulations, kindly performed by C. Walker, are similar to those in Walker & Schneider (2005).

14 Figure courtesy of T. Schneider, using an ECMWF reanalysis.

Further Reading

Atmospheric dynamics and circulation

Andrews, D. G., Holton, J. R. & Leovy, C. B., 1987. *Middle Atmosphere Dynamics.*
Discusses the theory and observations of the middle atmosphere in some detail, including Rossby waves and wave–mean-flow interaction.

Green, J. S. A., 1999. *Atmospheric Dynamics.*
A personal view of the subject, with numerous insights about how the atmosphere works.

Lorenz, E. N., 1967. *The Nature and Theory of the General Circulation of the Atmosphere.*
A classic monograph on the atmospheric general circulation.

Marshall, J. & Plumb, R. A., 2008. *Atmosphere, Ocean and Climate Dynamics: An Introductory Text.*
Discusses the dynamics and circulation of both atmosphere and ocean at an advanced undergraduate, beginning graduate level.

Peixoto, J. P. & Oort, A. H., 1992. *Physics of Climate.*
A descriptive but physically-based discussion of the general circulation, emphasizing observations.

Randall, D. 2015. *An Introduction to the Global Circulation of the Atmosphere.*
Discusses both the observations and mechanisms of the large-scale atmospheric circulation.

Schneider, T. & Sobel, A., 2007. *The Global Circulation of the Atmosphere: Phenomena, Theory, Challenges.*
Contains several useful review articles on the large-scale atmospheric circulation.

Atmospheric physics and thermodynamics

Ambaum, M., 2010. *Thermal Physics of the Atmosphere.*
Distinguished by treating the subject as a branch of classical thermodynamics, which indeed it is.

Bohren, C. F., & Albrecht, B., 1998. *Atmospheric Thermodynamics.*
A well-known, and rather lively, introduction to the field.

Cabellero, R. 2014. *Physics of the Atmosphere.*
A short introduction to the subject of its title.

Emanuel, K., 1994. *Atmospheric Convection.*
A comprehensive discussion of convection, progressing from basic theory to an advanced level.

Pierrehumbert, R. T., 2010. *Principles of Planetary Climate.*
Pierrehumbert's book is an ideal complement to the one you are currently looking at.

Wallace, J. M. & Hobbs, P. V., 2006. *Atmospheric Science: An Introductory Survey.*
An introduction to a broad range of topics in atmospheric sciences, for scientists at all levels.

Round and round and round she goes,
And where she stops, nobody knows.
Origin unknown, used in *The Original Amateur Hour,* radio and TV program, 1934–1970.

CHAPTER 15

Zonally-Averaged Mid-Latitude Atmospheric Circulation

THE FOCUS OF THIS CHAPTER is the zonally-averaged structure and circulation of the extratropical troposphere. Because of the presence of strong zonal asymmetries — in particular baroclinic eddies or, more simply, the weather — this circulation differs markedly from the zonally *symmetric* circulation that would exist if eddies did not develop at all. The angular-momentum-conserving model of the Hadley Cell discussed in Chapter 14 is an example of a zonally-symmetric circulation, but this may be quite different from the zonally-*averaged* circulation in a turbulent atmosphere. Let us explain more.

When studying some aspects of the large-scale ocean circulation, or the low-latitude atmospheric circulation, we can make a great deal of progress by treating the large-scale flow as if it were absolutely steady with the eddies having only a perturbative effect. However, this approach fails badly for the mid-latitude atmosphere: the large-scale mid-latitude circulation is intrinsically unsteady on the large-scale to the extent that the associated eddies essentially *are* the circulation. The eddies are also unpredictable and chaotic; that is to say, *the large-scale mid-latitude circulation of the atmosphere is a turbulent flow.* This turbulence involves large-scale, geostrophically and hydrostatically balanced flow — that is, it is geostrophic turbulence — and so has different properties than the smaller-scale, more nearly three- dimensional turbulence that may occur in boundary layers and the like. Further, this large-scale turbulence (sometimes called 'macro-turbulence') is neither fully-developed nor isotropic — it interacts with the Rossby waves that arise from the planetary rotation and might be regarded as a form of weak turbulence. Why is the large-scale flow turbulent? Why is it zonally asymmetric at all? There are two potential sources for zonal asymmetries:

(i) The zonal asymmetries that exist in the underlying boundary conditions and forcing: mountains, land–sea contrasts, the diurnal cycle, and so on.

(ii) Hydrodynamic instability: even if the surface and the forcing were exactly zonally symmetric, the corresponding zonally symmetric solutions of the equations of motion would have a large shear in the zonal wind and this might be baroclinically unstable to zonally asymmetric perturbations.

If the flow were not unstable, then we might expect that the zonal asymmetries of item *(i)* would give rise to corresponding, steady, zonal asymmetries in the resulting circulation, and this process

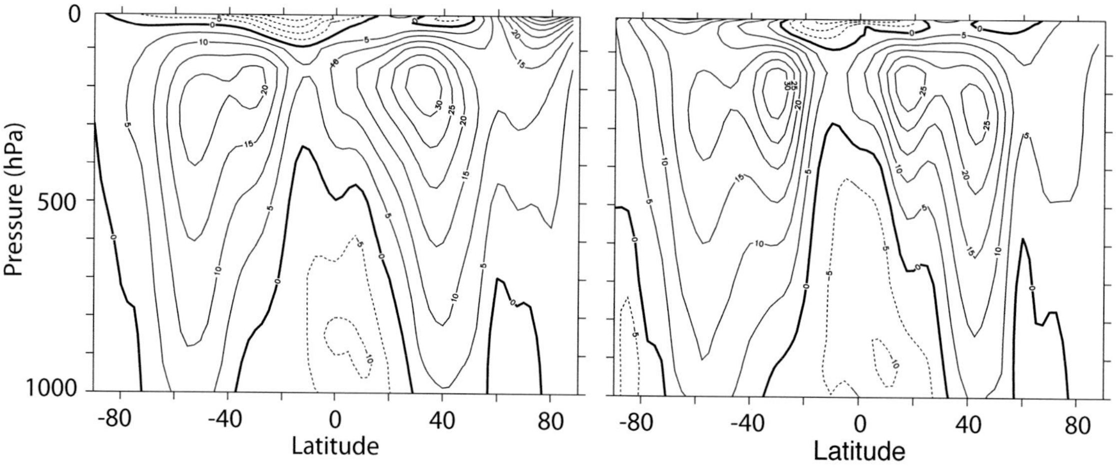

Fig. 15.1 The time averaged zonal wind at 150° W (in the mid-Pacific) in December–January–February (DJF, left), March–April–May (MAM, right). The contour interval is $5\,\mathrm{m\,s^{-1}}$. Note the double jet in each hemisphere, one in the subtropics and one in mid-latitudes. The subtropical jets are associated with a strong meridional temperature gradient, whereas the mid-latitude, eddy-driven, jets have a stronger barotropic component and are associated with westerly winds at the surface.

is discussed in the next chapter. However, the flow *is* unstable, primarily via baroclinic instability (as discussed in Chapter 9), and this leads to eddy growth and ultimately to geostrophic turbulence; this is, essentially, what gives rise to *weather*. The large-scale circulation is not, however, so turbulent that such things as Rossby waves cease to have meaning. Indeed, the most important nonlinear interactions are those involving Rossby waves and the zonally-averaged flow, and ideas of wave–mean-flow interaction go a long way in explaining the generation of the zonally-averaged flow. In particular, it is this interaction that produces the momentum convergence that gives rise to surface wind pattern.

In this chapter we focus on the effects of item *(ii)*, and try to understand the mid-latitude circulation of an atmosphere with zonally symmetric forcing and boundary conditions. We assume that the zonal asymmetries due to the boundary conditions do not *qualitatively* affect our arguments, and that the circulation of the atmosphere with a perfectly smooth surface would resemble that of the real circulation. We will present our arguments semi-independently of earlier chapters, and in particular we will develop some of the results of wave–mean flow interaction ab initio, so that a reader with just a little experience can start here and refer back to these chapters as needed. We begin with a discussion of the mechanisms that maintain the surface westerlies.[1]

15.1 SURFACE WESTERLIES AND THE MAINTENANCE OF A BAROTROPIC JET

15.1.1 Observations and Motivation

The atmosphere above the surface has a generally eastward flow, with a broad maximum about 10 km above the surface at around 40° in either hemisphere. But if we look a little more at the zonally average wind in Fig. 14.2(a) we see hints of there being two jets — one (the subtropical jet) at around 30°, and another somewhat poleward of this, especially apparent in the Southern Hemisphere. Such a jet is particularly noticeable in certain regions of the globe, when a zonal average is not taken, as in Fig. 15.1. The subtropical jet is associated with a strong meridional temperature gradient at the edge of the Hadley Cell, and is quite baroclinic. On the other hand, the mid-latitude jet (sometimes called the subpolar jet) is more barotropic (it has little vertical structure, with less shear than the subtropical jet) and lies above an eastward surface flow. This

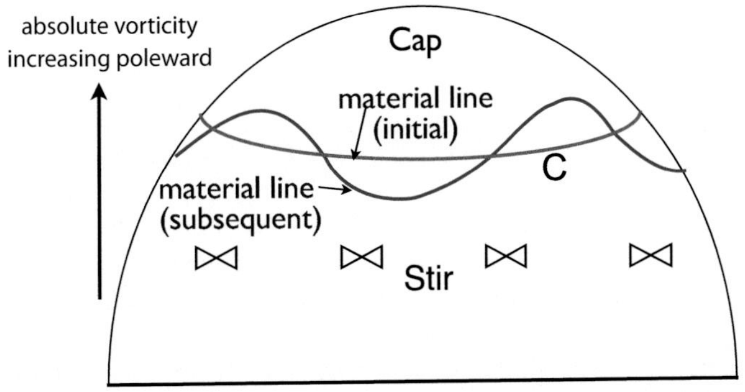

Fig. 15.2 Sketch of the effects of a mid-latitude disturbance on the circulation around the latitude line C.

If initially the absolute vorticity increases monotonically poleward, then the disturbance will bring fluid with lower absolute vorticity into the cap region. Then, using Stokes theorem, the velocity around the latitude line C will become more westward.

flow feels the effect of friction and so there must be a momentum *convergence* into this region, as is seen in Fig. 14.12. We will find that this momentum convergence occurs largely in transient eddies, and the jet is known as the *eddy-driven jet.* Although the eddies are a product of baroclinic instability, the essential mechanism of jet production is present in barotropic dynamics, so we first consider how an eastward jet can be maintained in a turbulent two-dimensional flow on the surface of a rotating sphere.

In barotropic turbulence, alternating east–west jets can be maintained if β is non-zero, as described in Section 12.1. However, that case was homogeneous, with no preferred latitude for a particular direction of jet, whereas in the atmosphere there appears to be but one mid-latitude jet, and although it meanders it certainly has a preferred average location. In the subsections that follow we give four explications as to how the jet is maintained; the first has a different flavour from the others, but they are all really just different perspectives on the same mechanism; they are all the same explanation.[2]

15.1.2 The Mechanism of Jet Production

I. The vorticity budget

Suppose that the absolute vorticity normal to the surface (i.e., $\zeta + 2\Omega\sin\vartheta$) increases monotonically poleward. (A sufficient condition for this is that the fluid is at rest.) By Stokes' theorem, the circulation around a line of latitude circumscribing the polar cap, I, is equal to the integral of the absolute vorticity over the cap. That is,

$$I_i = \int_{\text{cap}} \boldsymbol{\omega}_{ia} \cdot \mathrm{d}\boldsymbol{A} = \oint_C u_{ia}\,\mathrm{d}l = \oint_C (u_i + \Omega a\cos\vartheta)\,\mathrm{d}l, \tag{15.1}$$

where $\boldsymbol{\omega}_{ia}$ and u_{ia} are the initial absolute vorticity and velocity, respectively, u_i is the initial zonal velocity in the Earth's frame of reference, and the line integrals are around the line of latitude. For simplicity let us take $u_i = 0$ and suppose there is a disturbance equatorward of the polar cap, and that this results in a distortion of the material line around the latitude circle C (Fig. 15.2). Since we are supposing the source of the disturbance to be distant from the latitude of interest, then if we neglect viscosity the circulation along the material line is conserved, by Kelvin's circulation theorem. Thus, vorticity with a lower value is brought into the region of the polar cap — that is, the region poleward of the latitude line C. Using Stokes' theorem again the circulation around the latitude circle C must therefore fall; that is, denoting values after the disturbance with a subscript f,

$$I_f = \int_{\text{cap}} \boldsymbol{\omega}_{fa} \cdot \mathrm{d}\boldsymbol{A} < I_i \tag{15.2}$$

so that

$$\oint_C (u_f + \Omega a \cos\vartheta)\,\mathrm{d}l < \oint_C (u_i + \Omega a \cos\vartheta)\,\mathrm{d}l, \tag{15.3}$$

and

$$\overline{u}_f < \overline{u}_i, \tag{15.4}$$

with the overbar indicating a zonal average. Thus, there is a tendency to produce *westward* flow poleward of the disturbance. By a similar argument westward flow is also produced equatorward of the disturbance — to see this one might apply Kelvin's theorem over all of the globe south of the source of the disturbance (taking care to take the dot-product correctly between the direction of the vorticity vector and the direction normal to the surface). Finally, note that the overall situation is the same in the Southern Hemisphere. Thus, on the surface of a rotating sphere, external stirring will produce westward flow *away* from the region of the stirring.

Now suppose, furthermore, that the disturbance imparts no net angular momentum to the fluid. Then the integral of $ua\cos\vartheta$ over the entire hemisphere must be constant. But the fluid is accelerating westward away from the disturbance. Therefore, the fluid in the region of the disturbance must accelerate *eastward;* that is, angular momentum must converge into the stirred region, producing an eastward flow. This simple mechanism is the essence of the production of eastward eddy-driven jets in the atmosphere, and of the eastward surface winds in mid-latitudes. The stirring that here we have externally imposed comes, in reality, from baroclinic instability.

If the stirring subsides then the flow may reversibly go back to its initial condition, with a concomitant reversal of the momentum convergence that caused the zonal flow. Thus, we must have some form of dissipation and irreversibility in order to produce permanent changes, and in particular we need to irreversibly mix vorticity. (This result is closely related to the non-acceleration results of Chapter 10.) If the fluid is continuously mixed, then we also need a source that restores the absolute vorticity gradient, otherwise we will completely homogenize the vorticity over the hemisphere.

II. Rossby waves and momentum flux

We saw above that a mean gradient of vorticity is an essential ingredient in the mechanism whereby a mean flow is generated by stirring. Given such, we expect Rossby waves to be excited, and we now show how Rossby waves are intimately related to the momentum flux maintaining the mean flow.

If a stirring is present in mid-latitudes then we expect that Rossby waves will be generated there, propagate away and break and dissipate. To the extent that the waves are quasi-linear and do not interact, then just away from the source region each wave has the form

$$\psi = \mathrm{Re}\, C \mathrm{e}^{\mathrm{i}(kx+ly-\omega t)} = \mathrm{Re}\, C \mathrm{e}^{\mathrm{i}(kx+ly-kct)}, \tag{15.5}$$

where C is a constant, with dispersion relation

$$\omega = ck = \overline{u}k - \frac{\beta k}{k^2 + l^2} \equiv \omega_R, \tag{15.6}$$

provided that there is no meridional shear in the zonal flow. The meridional component of the group velocity is given by

$$c_g^y = \frac{\partial \omega}{\partial l} = \frac{2\beta k l}{(k^2 + l^2)^2}\,. \tag{15.7}$$

Now, the direction of the group velocity must be *away* from the source region; this is a radiation condition (discussed more in the next subsection), demanded by the requirement that Rossby waves transport energy *away* from the disturbance. Thus, northward of the source kl is positive

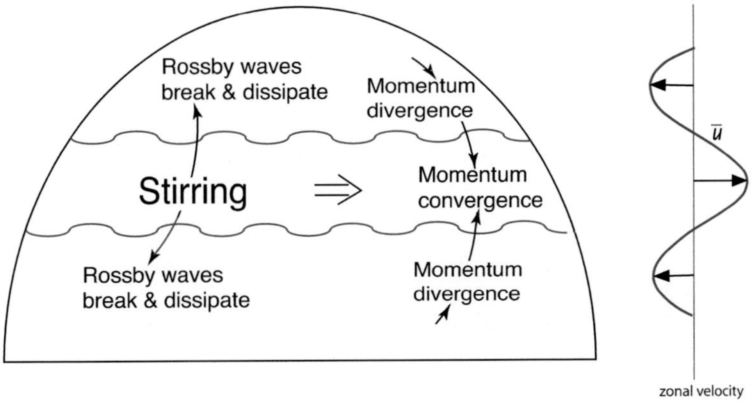

Fig. 15.3 Generation of zonal flow on a rotating sphere.
Stirring in mid-latitudes (by baroclinic eddies) generates Rossby waves that propagate away. Momentum converges in the region of stirring, producing eastward flow there and weaker westward flow on its flanks.

and southward of the source kl is negative. That the product kl can be positive or negative arises because for each k there are two possible values of l that satisfy the dispersion relation (15.6), namely

$$l = \pm\left(\frac{\beta}{\overline{u}-c} - k^2\right)^{1/2}, \tag{15.8}$$

assuming that the quantity in parentheses is positive.

The velocity variations associated with the Rossby waves are

$$u' = -\mathrm{Re}\, C\,\mathrm{i}l e^{\mathrm{i}(kx+ly-\omega t)}, \qquad v' = \mathrm{Re}\, C\,\mathrm{i}k e^{\mathrm{i}(kx+ly-\omega t)}, \tag{15.9a,b}$$

and the associated momentum flux is

$$\overline{u'v'} = -\frac{1}{2}C^2 kl. \tag{15.10}$$

Thus, given that the sign of kl is determined by the group velocity, northward of the source the momentum flux associated with the Rossby waves is southward (i.e., $\overline{u'v'}$ is negative), and southward of the source the momentum flux is northward (i.e., $\overline{u'v'}$ is positive). That is, the momentum flux associated with the Rossby waves is *toward* the source region. Momentum converges in the region of the stirring, producing net eastward flow there and westward flow to either side (Fig. 15.3).

Another way of describing the same effect is to note that if kl is positive then lines of constant phase ($kx + ly$ = constant) are tilted north-west/south-east, and the momentum flux associated with such a disturbance is negative ($\overline{u'v'} < 0$). Similarly, if kl is negative then the constant-phase lines are tilted north-east/south-west and the associated momentum flux is positive ($\overline{u'v'} > 0$). The net result is a convergence of momentum flux into the source region. In physical space this is reflected by having eddies that are 'bow-shaped', as in Fig. 15.4.

♦ *The radiation condition and Rayleigh friction*

Why is the group velocity directed away from the source region? It is because the energy flux travels at the group velocity, and the energy flux must be directed away from the source region; the reader comfortable with that statement may stop here. (See Section 6.7 for more on group velocity.) Another way to determine the direction of the group velocity is to employ a common trick in problems of wave propagation, that of adding a small amount of friction to the inviscid problem.[3] The solution of the ensuing problem in the limit of small friction will often make clear which solution is physically meaningful in the inviscid problem, and therefore which solution nature chooses. Consider the linear barotropic vorticity equation with linear friction,

$$\frac{\partial\zeta}{\partial t} + \beta\frac{\partial\psi}{\partial x} = -r\zeta, \tag{15.11}$$

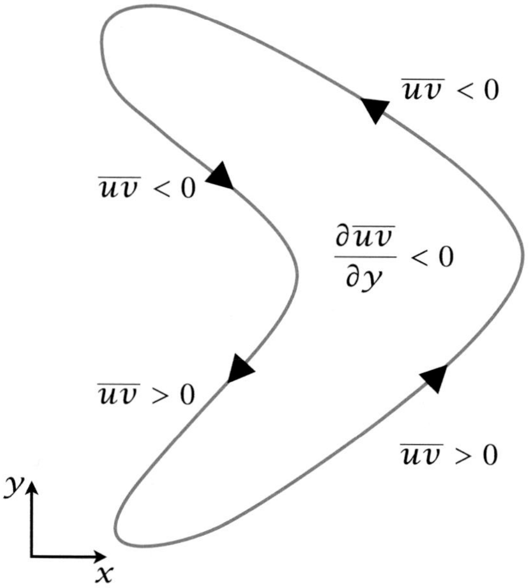

Fig. 15.4 The momentum transport in physical space, caused by the propagation of Rossby waves away from a source in mid-latitudes. The ensuing bow-shaped eddies are responsible for a convergence of momentum, as indicated. If the arrows were reversed the momentum transport would still have the same sign.

where r is a small friction coefficient. The dispersion relation is

$$\omega = -\frac{\beta k}{K^2} - \mathrm{i}r = \omega_R(k,l) - \mathrm{i}r, \tag{15.12}$$

where ω_R is defined by (15.6), with $\overline{u} = 0$, and so the wave decays with time. Now suppose a wave is generated in some region, and that it propagates meridionally away, decaying as it moves away. Then, instead of an imaginary frequency, we may suppose that the frequency is real and the y-wavenumber is imaginary. Specifically, we take $l = l_0 + l'$, where $l_0 = \pm[\beta/(\overline{u} - c) - k^2]^{1/2}$ for some zonal wavenumber k, as in (15.8), and $\omega = \omega_R(k, l_0)$. For small friction, we obtain l' by Taylor-expanding the dispersion relation around its inviscid value, $\omega_R(k, l_0)$, giving

$$\omega + \mathrm{i}r = \omega_R(k,l) \approx \omega_R(k,l_0) + \left.\frac{\partial \omega_R(k,l)}{\partial l}\right|_{l=l_0} l', \tag{15.13}$$

and therefore

$$l' = \frac{\mathrm{i}r}{c_g^y}, \tag{15.14}$$

where $c_g^y = \partial_l \omega_R(k,l)|_{l=l_0}$ is the y-component of the group velocity. The wavenumber is imaginary, so that the wave either grows or decays in the y-direction, and the wave solution obeys

$$\psi \approx \mathrm{Re}\, C \exp[\mathrm{i}(kx - \omega_R t)] \exp(\mathrm{i}l_0 y - ry/c_g^y). \tag{15.15}$$

We now demand that the solution decay away from the source, because any other choice is manifestly unphysical, even as we let r be as small as we please. Thus, with the source at $y = 0$, c_g^y must be positive for positive y and negative for negative y. In other words, the group velocity must be directed *away* from the source region, and so momentum flux converges on the source region.

III. The pseudomomentum budget

The kinematic relation between vorticity flux and momentum flux for non-divergent two-dimensional flow is

$$v\zeta = \frac{1}{2}\frac{\partial}{\partial x}\left(v^2 - u^2\right) - \frac{\partial}{\partial y}(uv). \tag{15.16}$$

After zonal averaging this gives

$$\overline{v'\zeta'} = -\frac{\partial \overline{u'v'}}{\partial y}, \tag{15.17}$$

noting that $\overline{v} = 0$ for two-dimensional incompressible (or geostrophic) flow. In spherical coordinates this expression becomes

$$\overline{v'\zeta'} \cos\vartheta = -\frac{1}{a\cos\vartheta}\frac{\partial}{\partial\vartheta}(\cos^2\vartheta\,\overline{u'v'}). \tag{15.18}$$

If either (15.17), or (15.18) are integrated with respect to y between two quiescent latitudes then their right-hand sides vanish. That is the zonally-averaged meridional vorticity flux vanishes when integrated over latitude.

Now, the barotropic zonal momentum equation is (for horizontally non-divergent flow)

$$\frac{\partial u}{\partial t} + \frac{\partial u^2}{\partial x} + \frac{\partial uv}{\partial y} - fv = -\frac{\partial \phi}{\partial x} + F_u - D_u, \tag{15.19}$$

where F_u and D_u represent the effects of any forcing and dissipation. Zonal averaging, with $\overline{v} = 0$, gives

$$\frac{\partial \overline{u}}{\partial t} = -\frac{\partial \overline{uv}}{\partial y} + \overline{F}_u - \overline{D}_u, \qquad \text{or} \qquad \frac{\partial \overline{u}}{\partial t} = \overline{v'\zeta'} + \overline{F}_u - \overline{D}_u. \tag{15.20}$$

using (15.17). Thus, the zonally-averaged wind is maintained by the zonally-averaged vorticity flux. On average there is little if any direct forcing of horizontal momentum and we may set $\overline{F}_u = 0$, and if the dissipation is parameterized by a linear drag (15.20) becomes

$$\frac{\partial \overline{u}}{\partial t} = \overline{v'\zeta'} - r\overline{u}, \tag{15.21}$$

where the constant r is an inverse frictional time scale.

Now consider the maintenance of this vorticity flux. The barotropic vorticity equation is

$$\frac{\partial \zeta}{\partial t} + \boldsymbol{u}\cdot\nabla\zeta + v\beta = F_\zeta - D_\zeta, \tag{15.22}$$

where F_ζ and D_ζ are forcing and dissipation of vorticity. Linearize about a mean zonal flow to give

$$\frac{\partial \zeta'}{\partial t} + \overline{u}\frac{\partial \zeta'}{\partial x} + \gamma v' = F'_\zeta - D'_\zeta, \tag{15.23}$$

where

$$\gamma = \beta - \frac{\partial^2 \overline{u}}{\partial y^2} \tag{15.24}$$

is the meridional gradient of absolute vorticity. (We use γ rather than β^* to denote this quantity because the argument may be extended to layered models, where $\gamma = \partial\overline{q}/\partial y$.) Now multiply (15.23) by ζ'/γ and zonally average, assuming that $\overline{u}_{yy}$ is small compared to β or varies only slowly, to form the pseudomomentum equation,

$$\frac{\partial \mathcal{P}}{\partial t} + \overline{v'\zeta'} = \frac{1}{\gamma}(\overline{\zeta' F'_\zeta} - \overline{\zeta' D'_\zeta}), \tag{15.25}$$

where

$$\mathcal{P} = \frac{1}{2\gamma}\overline{\zeta'^2} \tag{15.26}$$

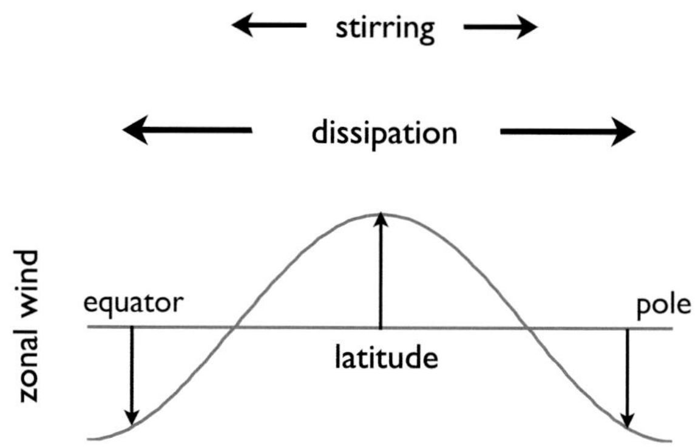

Fig. 15.5 Mean flow generation by a meridionally confined stirring. Because of Rossby wave propagation away from the source region, the distribution of pseudomomentum dissipation is broader than that of pseudomomentum forcing, and the sum of the two leads to the zonal wind distribution shown, with positive (eastward) values in the region of the stirring. See also Fig. 15.8.

is a wave activity density, equal to the pseudomomentum for this problem (see Section 10.2 and (10.29b) for related discussion). The parameter γ is positive if the average absolute vorticity increases monotonically northward, and this is usually the case in both Northern and Southern Hemispheres.

In the absence of forcing and dissipation, (15.21) and (15.25) imply an important relationship between the change of the mean flow and the pseudomomentum, namely

$$\frac{\partial \overline{u}}{\partial t} + \frac{\partial \mathcal{P}}{\partial t} = 0. \tag{15.27}$$

Now if for some reason $\mathcal{P}$ increases, perhaps because a wave enters an initially quiescent region because of stirring elsewhere, then mean flow must decrease. However, because the vorticity flux integrates to zero, the zonal flow cannot decrease everywhere. Thus, if the zonal flow decreases in regions away from the stirring, it must *increase* in the region of the stirring. In the presence of forcing and dissipation this mechanism can lead to the production of a statistically steady jet in the region of the forcing, since (15.21) and (15.25) combine to give

$$\frac{\partial \overline{u}}{\partial t} + \frac{\partial \mathcal{P}}{\partial t} = -r\overline{u} + \frac{1}{\gamma}\left(\overline{\zeta' F'_\zeta} - \overline{\zeta' D'_\zeta}\right), \tag{15.28}$$

and in a statistically steady state

$$r\overline{u} = \frac{1}{\gamma}\left(\overline{\zeta' F'_\zeta} - \overline{\zeta' D'_\zeta}\right). \tag{15.29}$$

The terms on the right-hand side represent the stirring and dissipation of vorticity, and integrated over latitude their sum will vanish, or otherwise the pseudomomentum budget cannot be in a steady state. However, let us suppose that forcing is confined to mid-latitudes. In the forcing region, the first term on the right-hand side of (15.29) will be larger than the second, and an eastward mean flow will be generated. Away from the direct influence of the forcing, the dissipation term will dominate and westward mean flows will be generated, as sketched in Fig. 15.5. Thus, *on a β-plane or on the surface of a rotating sphere an eastward mean zonal flow can be maintained by a vorticity stirring that imparts no net momentum to the fluid.* In general, stirring in the presence of a vorticity gradient will give rise to a mean flow, and on a spherical planet the vorticity gradient is provided by differential rotation.

It is crucial to the generation of a mean flow that the dissipation has a broader latitudinal distribution than the forcing: if all the dissipation occurred in the region of the forcing then from (15.29) no mean flow would be generated. This broadening arises via the action of Rossby waves that are generated in the forcing region and that propagate meridionally before dissipating, as described in the previous subsection, so allowing the generation of a mean flow.

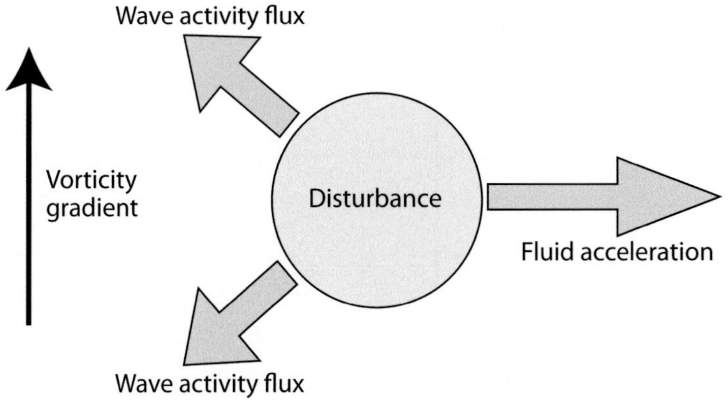

Fig. 15.6 If a region of fluid on the β-plane or on a rotating sphere is stirred, then Rossby waves propagate away from the disturbance, and this is the direction of the wave-activity flux vector. Thus, there is divergence of wave activity in the stirred region, and using (15.34) this produces an eastward acceleration.

IV. The Eliassen–Palm flux

The Eliassen–Palm (EP) flux (Section 10.2) provides a convenient framework for determining how waves affect the mean flow, and the barotropic case is a particularly simple and instructive example. In the unforced case, the zonally-averaged momentum equation may be written as

$$\frac{\partial \overline{u}}{\partial t} - f_0 \overline{v}^* = \nabla_x \cdot \boldsymbol{\mathcal{F}}, \tag{15.30}$$

where $\overline{v}^*$ is the residual meridional velocity and $\boldsymbol{\mathcal{F}}$ is the Eliassen–Palm flux, and $\nabla_x\cdot$ is the divergence in the meridional plane. In the barotropic case $\overline{v}^* = 0$ and

$$\boldsymbol{\mathcal{F}} = -\mathbf{j}\,\overline{u'v'}. \tag{15.31}$$

If the momentum flux is primarily the result of interacting nearly monochromatic waves, then the EP flux obeys the group velocity property (Section 6.7), namely that the flux of wave activity density is equal to the group velocity multiplied by the wave activity density. Thus,

$$\mathcal{F}^y \equiv \mathbf{j} \cdot \boldsymbol{\mathcal{F}} \approx c_g^y \mathcal{P}, \tag{15.32}$$

where $\mathcal{P}$ is the wave activity density, or pseudomomentum, given by

$$\mathcal{P} = \frac{\overline{\zeta'^2}}{2\overline{q}_y} = \frac{\overline{\zeta'^2}}{2\gamma}, \tag{15.33}$$

and, if $\gamma > 0$, $\mathcal{P}$ is a positive-definite quantity. The zonal momentum equation and the Eliassen–Palm relation (10.29a) become respectively

$$\frac{\partial \overline{u}}{\partial t} = \frac{\partial}{\partial y}(c_g^y \mathcal{P}), \qquad \frac{\partial \mathcal{P}}{\partial t} + \frac{\partial}{\partial y}(c_g^y \mathcal{P}) = 0, \tag{15.34a,b}$$

and so

$$\frac{\partial \overline{u}}{\partial t} = -\frac{\partial \mathcal{P}}{\partial t}, \tag{15.35}$$

as in (15.27).

Now suppose that we initiate a disturbance at some latitude, and then let the fluid evolve freely. The disturbance generates Rossby waves whose group velocity will be directed away from the region of disturbance, and from (15.34b) the wave activity density $\mathcal{P}$ will diminish in the region of

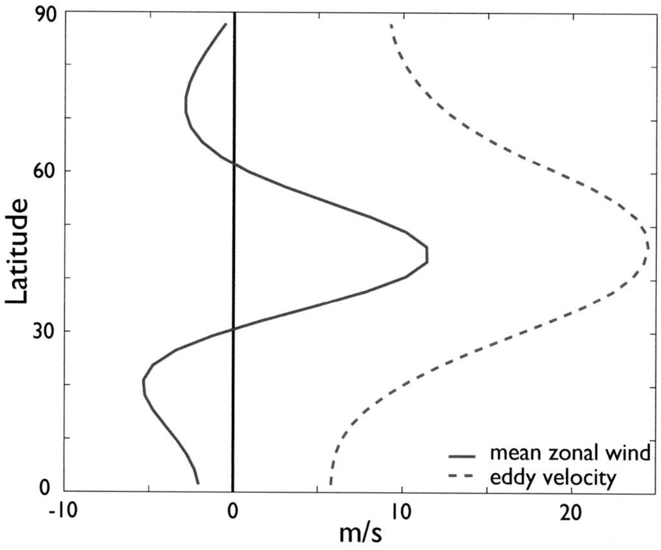

Fig. 15.7 The time and zonally-averaged wind (solid line) obtained by an integration of the barotropic vorticity equation (15.36) on the sphere.

The fluid is stirred in mid-latitudes by a random wavemaker that is statistically zonally uniform, acting around zonal wavenumber 8, and that supplies no net momentum. Momentum converges in the stirring region leading to an eastward jet with a westward flow to either side, and zero area-weighted spatially integrated velocity. The dashed line shows the r.m.s. (eddy) velocity created by the stirring.

the disturbance (and increase elsewhere). However, from (15.34a) the zonal velocity will *increase* in the region of the disturbance, and an eastward flow will be generated. That is, momentum converges in the region of the disturbance and an eastward jet is generated, as sketched in Fig. 15.6.

The EP flux argument, the pseudomomentum argument and the Rossby wave argument are just different expressions of the same physical process. Indeed, the result of (15.10) can be regarded as illustrating the group velocity property of the EP flux for barotropic Rossby waves. The vorticity budget argument is a little more general than these arguments, because it does not depend on linearization or small amplitude disturbances. None of these arguments requires that the flow be truly turbulent and, although they all involve nonlinear interactions, it is the presence of the β-effect — a linear term in the vorticity equation — that is crucial in the development of a mean flow.

15.1.3 A Numerical Example

We conclude from the above arguments that momentum will converge into a rapidly rotating flow that is stirred in a meridionally localized region. To illustrate this, we numerically integrate the barotropic vorticity equation on the sphere, with a meridionally localized stirring term; explicitly, the equation that is integrated is

$$\frac{\partial \zeta}{\partial t} + J(\psi, \zeta) + \beta \frac{\partial \psi}{\partial x} = -r\zeta + \kappa \nabla^4 \zeta + F_\zeta. \tag{15.36}$$

The first term on the right-hand side is a linear drag, parameterizing momentum loss in an Ekman layer. The second term removes enstrophy that has cascaded to small scales; it has a negligible impact at large scales. The forcing term F_ζ is a wavemaker confined to a zonal strip of about 15° meridional extent, centred at about 45° N, that is statistically zonally uniform and that spatially integrates to zero. Within that region it is a random stirring with a temporal decorrelation scale of a few days and a spatial decorrelation scale corresponding to about wavenumber 8, thus mimicking weather scales. Thus, it provides no net source of vorticity or momentum, but it is a source of pseudomomentum because $\overline{F_\zeta \zeta} > 0$.

The results of a numerical integration of (15.36) are illustrated in Figs. 15.7 and 15.8. An eastward jet forms in the vicinity of the forcing, with westward flow on either side. The pseudomomentum stirring and dissipation that produce this flow are shown in Fig. 15.8. As expected, the

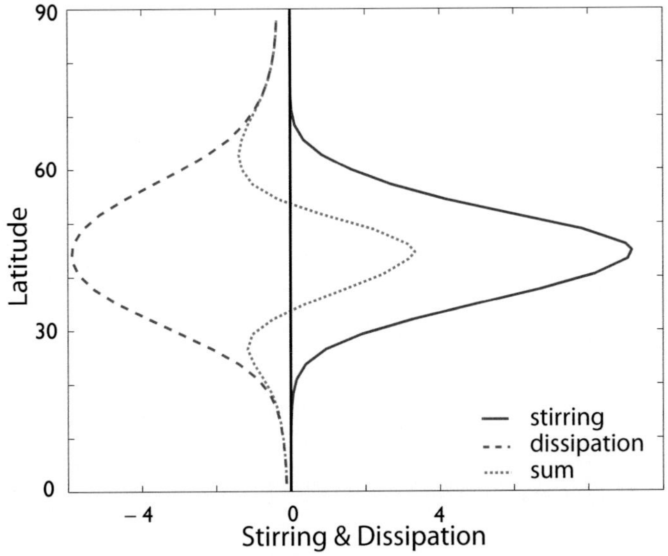

Fig. 15.8 The pseudomomentum stirring (solid line, $\overline{F_\zeta'\zeta'}$), dissipation (red dashed line, $\overline{D_\zeta'\zeta'}$) and their sum (dotted), for the same integration as Fig. 15.7.

Because Rossby waves propagate away from the stirred region before breaking, the distribution of dissipation is broader than the forcing, resulting in an eastward jet where the stirring is centred, with westward flow on either side.

dissipation has a broader distribution than the forcing, and the sum of the two (the dotted line) has the same meridional distribution as the zonal flow itself.

15.2 LAYERED MODELS OF THE MID-LATITUDE CIRCULATION

Let us now extend our barotropic model in the direction of increasing realism. So far we have shown that localized stirring can give rise to an eastward acceleration as Rossby waves propagate away from the disturbance. The source of the disturbance is baroclinic instability, and to incorporate that effects will necessitate some vertical structure into the problem; we do this by way of layered models of the circulation.[4] That is, we consider an atmosphere to consist of one or more isentropic layers, as described in Section 3.10. The equations describing such layers are virtually isomorphic to the shallow water equations and, for the sake of familiarity and simplicity, and with no loss of essential dynamics, we use the Boussinesq shallow water equations. We begin with a model of a single layer, with summaries for the impatient on page 552 and on page 556.

15.2.1 A Single-layer Model

We first consider a single layer obeying the shallow water equations. We further restrict the flow by supposing that it is constrained by two rigid surfaces: an upper flat lid and a lower, wavy (but stationary) surface (Fig. 15.9). We may imagine the fluid layer to crudely represent the upper troposphere, with the (given) lower wavy surface corresponding to the undulating mid-atmosphere interface of a two-layer model. (This section is in some ways an exercise, and too much realism should not be ascribed to the model.) Thus frictional effects are small in the momentum equation, and in particular there is no Ekman layer and no drag on the velocity field. However, there may be some dissipative effects in the vorticity equation, arising from the cascade of enstrophy to small scales. We also suppose that the Rossby number is small, that the variations in layer thickness are small compared to the mean layer thickness, and that variations in Coriolis parameter are small. Let the initial flow be a uniform zonal current, passing over the wavy lower boundary. The boundary is waviest in mid-latitudes, creating a disturbance from which Rossby waves emanate. Our questions are: (i) How does the wavy interface affect the mean zonal flow? (ii) What if any meridional circulation is induced?

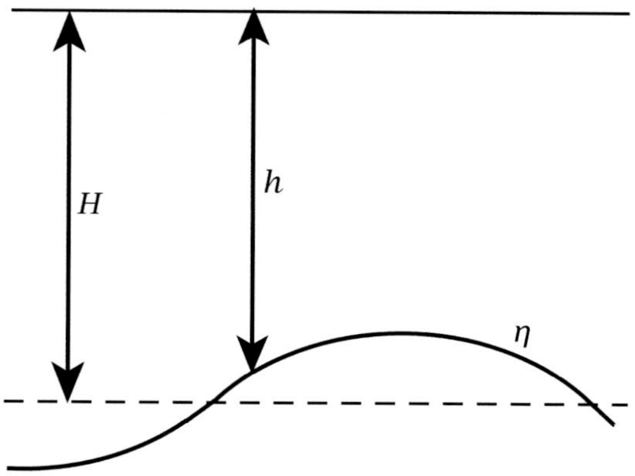

Fig. 15.9 A model atmosphere with an active layer of mean thickness H, local thickness h, and a variable lower surface of height displacement η, lying above a stationary layer with a slightly larger potential density.

Equations of motion

The zonal momentum equation for the layer may be written as

$$\frac{\partial u}{\partial t} - (f + \zeta)v = -\frac{\partial B}{\partial x}, \tag{15.37}$$

where $B = \phi + \boldsymbol{u}^2/2$ is the Bernoulli function for the problem and ϕ is the kinematic pressure, p/ρ_0. The zonal average of the equation is

$$\frac{\partial \overline{u}}{\partial t} - f\overline{v} = \overline{\zeta}\overline{v} + \overline{\zeta' v'}. \tag{15.38}$$

Note that $\overline{v}$ is wholly ageostrophic ($\overline{v}_g = \overline{\partial_x \psi} = 0$). Now, using $\partial u/\partial x + \partial v/\partial y = 0$, the vorticity flux is related to the momentum flux by

$$v\zeta = -\frac{\partial}{\partial y}(uv) + \frac{1}{2}\frac{\partial}{\partial x}(v^2 - u^2), \tag{15.39}$$

so that, under quasi-geostrophic scaling, (15.38) simplifies to

$$\frac{\partial \overline{u}}{\partial t} - f_0\overline{v} = \overline{\zeta' v'} = -\frac{\partial}{\partial y}\overline{u'v'}. \tag{15.40}$$

Although $\overline{v}$ is small and ageostrophic, mass conservation does not demand that it be zero, because the thickness of the layer is not constant — look ahead to (15.43). Thus, as $\overline{v}$ is multiplied by the large term f_0, the term $f_0\overline{v}$ term should be retained (whereas $\overline{\zeta}\overline{v}$ is dropped). If the flow is statistically steady and there are no sources or sinks of momentum (15.40) becomes

$$f_0\overline{v} = \frac{\partial}{\partial y}\overline{u'v'}. \tag{15.41}$$

To complete the model we use the zonally-averaged mass conservation equation, namely

$$\frac{\partial \overline{h}}{\partial t} + \frac{\partial}{\partial y}\overline{vh} = 0. \tag{15.42}$$

In the situation here $\partial \overline{h}/\partial t = 0$, because the flow is confined between two rigid surfaces, and so $\partial \overline{vh}/\partial y = 0$. If the mass flux vanishes somewhere, for example at a meridional boundary, it therefore vanishes everywhere and we have

$$\overline{v}\overline{h} + \overline{v'h'} = 0. \tag{15.43}$$

Using (15.41) and (15.43) gives

$$\frac{1}{f_0}\frac{\partial}{\partial y}\overline{u'v'} + \frac{\overline{v'h'}}{\overline{h}} = 0, \qquad \text{or} \qquad \overline{v'\zeta'} - f_0\frac{\overline{v'h'}}{\overline{h}} = 0. \tag{15.44a,b}$$

Because thickness variations are assumed to be small we may write this as

$$\overline{v'\zeta'} - f_0\frac{1}{H}\overline{v'h'} = 0, \tag{15.45}$$

where H is the reference thickness of the layer, which may be taken as its mean thickness. The left-hand side of (15.45) is actually just the potential vorticity flux for this problem — look ahead to (15.51). The potential vorticity equation for the layer is

$$\frac{\mathrm{D}Q}{\mathrm{D}t} = \frac{\mathrm{D}}{\mathrm{D}t}\left[\frac{\zeta + f}{h}\right] = 0, \tag{15.46}$$

where h is the fluid layer thickness. For small variations in layer thickness and Coriolis parameter this becomes

$$\frac{\mathrm{D}q}{\mathrm{D}t} = \frac{\partial q}{\partial t} + u\frac{\partial q}{\partial x} + v\frac{\partial q}{\partial y} = 0, \qquad q = \zeta + \beta y + f_0\frac{\eta}{H}, \tag{15.47a,b}$$

where $\eta = H - h$ is the height of the lower interface (Fig. 15.9) and this is a function of x and y but not, in this model, time. Using the horizontal non-divergence of the flow, the zonally-averaged potential vorticity equation is

$$\frac{\partial \overline{q}}{\partial t} = -\frac{\partial \overline{v}\,\overline{q}}{\partial y} - \frac{\partial \overline{v'q'}}{\partial y}. \tag{15.48}$$

The term involving $\overline{v}$ is very small, and omitting it and using (15.47a) we obtain the perturbation potential vorticity equation

$$\frac{\partial q'}{\partial t} + \overline{u}\frac{\partial q'}{\partial x} + v'\frac{\partial \overline{q}}{\partial y} = -D', \tag{15.49}$$

where we include a term, D', to represent dissipative processes. Multiplying by $q'/(\partial\overline{q}/\partial y)$ and zonally averaging we obtain the pseudomomentum equation for this system, namely

$$\frac{\partial \mathcal{P}}{\partial t} = \frac{\partial}{\partial t}\left(\frac{\overline{q'^2}}{2\gamma}\right) = -\overline{v'q'} - \frac{\overline{D'q'}}{\gamma}, \tag{15.50}$$

where $\gamma = \partial\overline{q}/\partial y$. This equation is the equivalent of (15.25), but now for the layered system. In a turbulent fluid we cannot, in general, demand that $D' = 0$, even as the viscosity goes to zero, because of the presence of an enstrophy flux to smaller scales and a concomitant dissipation. But in regions where D' *is* zero (where there is no wave breaking) then the potential vorticity flux must also be zero in a steady state. For our argument let us assume that, in fact, $D' = 0$.

Using (15.47b), the eddy potential vorticity flux is

$$\overline{v'q'} = \overline{v'\zeta'} + \frac{f_0}{H}\overline{v'\eta'} = \overline{v'\zeta'} - \frac{f_0}{H}\overline{v'h'}, \tag{15.51}$$

where η' is the topography and h' is the layer thickness perturbation. Using this in the zonal momentum equation (15.40) gives

$$\frac{\partial \overline{u}}{\partial t} = \overline{v'q'} + \frac{f_0}{H}\overline{v'h'} + f_0\overline{v}. \tag{15.52}$$

Informal Summary of the Single-layer Arguments

The zonally-averaged momentum equation is

$$\frac{\partial \overline{u}}{\partial t} - f_0 \overline{v} = \overline{v'\zeta'} = -\frac{\partial \overline{u'v'}}{\partial y}. \tag{SL.1}$$

A region that is the source of Rossby waves will generally be a region where there is momentum flux convergence, where $\partial \overline{u'v'}/\partial y < 0$. In this region $\overline{v}$ will be directed *equatorward* if $\overline{u}$ is steady, and this flow is the upper branch of the Ferrel Cell. To think about this in terms of potential vorticity, first define the residual meridional velocity by

$$\overline{v}^* = \frac{\overline{v'h'}}{\overline{h}} + \overline{v}. \tag{SL.2}$$

This is proportional to the total meridional mass flux in a layer, and is zero in this one-layer model. The momentum equation is then

$$\frac{\partial \overline{u}}{\partial t} = f_0 \overline{v}^* - \frac{f_0}{\overline{h}} \overline{v'h'} + \overline{v'\zeta'} \tag{SL.3a}$$

$$= f_0 \overline{v}^* + \overline{v'q'}. \tag{SL.3b}$$

using $\overline{v'q'} = \overline{v'\zeta'} - (f_0/\overline{h})\overline{v'h'}$, where q the is potential vorticity. The second term on the right-hand side of (SL.3a) is the *form drag* exerted by the topography on the flow, and in a steady state this balances the momentum flux convergence of the Rossby waves. Because of the presence of Rossby waves we expect $\overline{v'\zeta'} > 0$. If there is no dissipation then in steady flow $\overline{v'q'} = -f_0 \overline{v}^* = 0$ and the eddy mass flux is poleward (positive if $f_0 > 0$) and a meridional flow is generated as in Fig. 15.10.

We can infer the potential vorticity flux more directly using the pseudomomentum equation:

$$\frac{\partial \mathcal{P}}{\partial t} = \frac{\partial}{\partial t}\left(\frac{\overline{q'^2}}{2\gamma}\right) = -\overline{v'q'} - \frac{\overline{D'q'}}{\gamma}, \tag{SL.4}$$

where $\gamma = \partial \overline{q}/\partial y$. If dissipation is identically zero, then the potential vorticity flux is zero if the waves are steady. Then, using (SL.3b), there is no acceleration of the zonal flow — an example of the *non-acceleration theorem*.

More generally (and in the real atmosphere) there *will* be some dissipation away from the source region: Rossby waves will preferentially break in critical layers (near where $\overline{u} = c$) and/or more generally Rossby waves will interact producing an enstrophy cascade. These processes give $\overline{D'q'} > 0$ and (for $\gamma > 0$) a negative potential vorticity flux, $\overline{v'q'} < 0$. In these regions, a balance in the momentum equation (SL.3b) can be achieved either by balancing the PV flux with a friction term, as in the barotropic model of Section 15.1, or by a Coriolis force on a poleward residual meridional velocity. That is, $f_0 \overline{v}^* \approx -\overline{v'q'} > 0$, so generating a poleward residual flow.

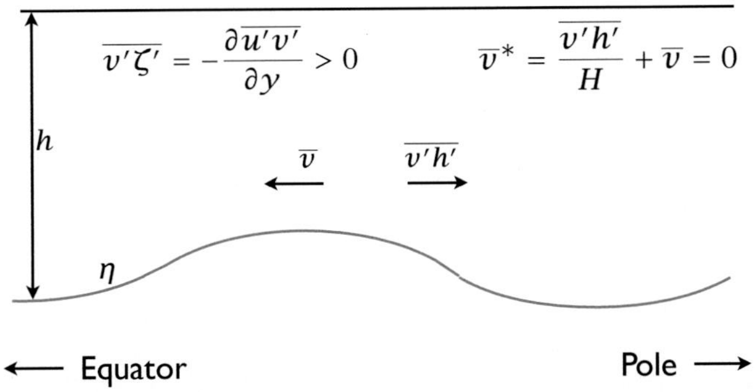

Fig. 15.10 Dynamics of a single layer, with no dissipation. The force on the active layer arises from the form drag exerted by the interface. Vorticity dynamics demands that this produce a converging eddy momentum flux ($\partial_y \overline{u'v'} < 0$), which in turn produces a poleward eddy mass flux ($\overline{v'h'} > 0$), and so an equatorward Eulerian flow.

But the last two terms on the right-hand side constitute the total mass flux, so we finally write

$$\frac{\partial \overline{u}}{\partial t} = \overline{v'q'} + f_0 \overline{v}^*, \qquad \overline{v}^* = \overline{v} + \frac{\overline{v'h'}}{H}. \tag{15.53a,b}$$

The quantity $\overline{v}^*$ is the *residual circulation* for this problem; it is proportional to the sum of mass flux from the mean flow and the eddies (see Section 10.3 for more discussion). Now, $\overline{v}^*$ is proportional to the total meridional mass flux and therefore here, because the flow is confined between rigid lids and if there are no sources or sinks of mass, $\overline{v}^* = 0$ everywhere (see (15.43) with $\overline{h} = H$).

Dynamics

When the flow passes over the wavy boundary, Rossby waves will, as in the barotropic case, cause momentum flux to converge in the generation region. If the flow is steady and dissipation-free then from the momentum equation

$$f_0 \overline{v} = \frac{\partial \overline{u'v'}}{\partial y}, \tag{15.54}$$

and, in regions of momentum flux convergence (i.e., where $\partial \overline{u'v'}/\partial y < 0$) *the mean meridional velocity is equatorward.* Thus, whereas frictional forces balance the vorticity flux in a constant-thickness barotropic model (because in that case $\overline{v} = 0$) in the free atmosphere a meridional circulation may be generated, and this is the basis of the equatorial flow in the upward branch of the Ferrel cell. However, this does not imply that the *total* mass flux is equatorward; in fact, for this single-layer model it must be zero, and therefore

$$\overline{v'h'} = -\overline{h}\overline{v} > 0. \tag{15.55}$$

That is, *the eddy mass flux is poleward,* balancing the equatorward mean flow. These balances are sketched in Fig. 15.10.

Another way to arrive at this result is to utilize potential vorticity fluxes directly. For steady, dissipation-free flow the pseudomomentum equation (15.50) reveals that the potential vorticity flux vanishes. Then using (15.53) and noting that $\overline{v}^* = 0$ we have $\partial \overline{u}/\partial t = 0$ — an example of the non-acceleration theorem that steady non-dissipative waves do not induce a change in the zonal momentum. Then, using (15.51) we find

$$\overline{v'\zeta'} = \frac{f_0}{H} \overline{v'h'}, \tag{15.56}$$

and using (15.43) we recover (15.54).

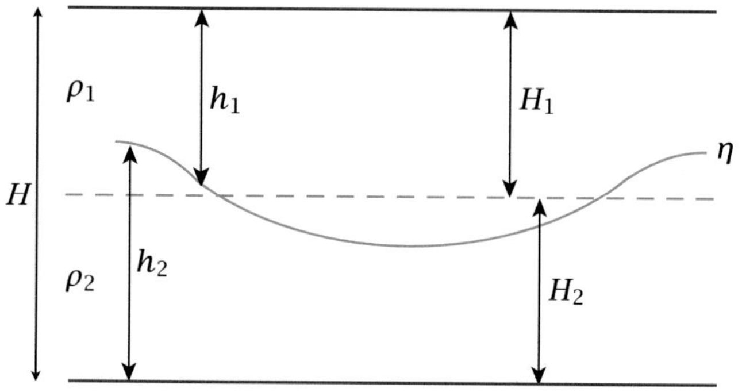

Fig. 15.11 An atmosphere with two homogeneous (or isentropic) layers of mean thickness H_1 and H_2, local thickness h_1 and h_2, and interface η, contained between two flat, rigid surfaces.

From the point of view of the momentum equation, the momentum flux convergence is balanced by the *form drag* caused by the flow over the wavy boundary. To see this we use (15.53b) to write the zonal momentum equation as

$$\frac{\partial \overline{u}}{\partial t} = f_0\overline{v} + \overline{v'\zeta'} = f_0\overline{v}^* - \frac{f_0}{H}\overline{v'h'} + \overline{v'\zeta'}, \tag{15.57}$$

where here $\overline{v}^*$ (but not $\overline{v}$) is zero. The term $-(f_0/H)\overline{v'h'}$ represents the force on the fluid layer coming from the wavy boundary — the form drag, as described in Section 3.6. Specifically, the average force per unit area exerted on the layer by the sloping surface is given by

$$F = -f_0\rho_0\overline{v'\eta'} = f_0\rho_o\overline{v'h'}, \tag{15.58}$$

and dividing by $\rho_0 H$ provides the acceleration on the active fluid layer. The atmosphere also exerts an equal and opposite force on the wavy surface, an effect we consider in the next section. A steady state is achieved, without dissipation, when the form drag is balanced by the eddy momentum flux convergence. From (15.51) or (15.53), this state is the same as the condition that the potential vorticity flux vanishes.

Final remarks on the one-layer model

A summary of the single-layer arguments is given in the shaded box on page 552. In the single-layer model, as in the barotropic model, the zonal flow is proximately driven by eddy fluxes of potential vorticity, and in the model the eddy fluxes must be zero if a steady state is to be achieved. *Vis-à-vis* the real atmosphere this is a little unrealistic, because from the pseudomomentum equation (15.50) we expect these fluxes to be negative, and there is then nothing to balance them in the momentum equation, (15.53), if $\overline{v}^* = 0$. In the real atmosphere, there are effectively sources and sinks in the mass conservation equation that arise from the thermodynamics that allow $\overline{v}^*$ to be non-zero; we then expect $\overline{v}^* > 0$, but to explore this requires a two-layer model, in which the single layer of the one-layer model will correspond to the upper layer of the two-layer model.

15.2.2 A Two-layer Model

We now consider a model with two active layers, constructing what is probably the simplest model that can capture the dynamics of the mid-latitude tropospheric general circulation without undue approximation. Indeed virtually all of the phenomenology that we associate with the circulation — a thermal wind, mid-latitude surface westerly winds, the Ferrel cell, breaking Rossby waves — is present. A three-layer model introduces no new physics, although a continuously stratified model does lead to some differences of interpretation. The physical model we have in mind is one of two isentropic layers of a compressible ideal gas, virtually equivalent to a two-layer shallow water model illustrated in Fig. 15.11, and our presentation will be in terms of the latter. The upper layer may be thought of as being forced by an undulating interface between the lower and upper layers, a crude representation of stratification. We continue to assume that quasi-geostrophic scaling holds; that is, the flow is in near geostrophic balance, variations in layer thickness are small compared to their mean thickness and variations in the Coriolis parameter are small. We also assume that the two fluid layers are held between two flat rigid lids — topography is an unnecessary complication at this stage.

Equations of motion

The equations of motion are those of a two-layer Boussinesq shallow water model confined between two rigid flat surfaces, and readers comfortable with these dynamics (see Sections 3.3 and 3.5) may quickly skip through this section, glancing at the boxed equations, and look at the summary on the next page. The momentum equations of each layer are

$$\frac{\mathrm{D}\boldsymbol{u}_1}{\mathrm{D}t} + \boldsymbol{f} \times \boldsymbol{u}_1 = -\nabla\phi_1, \tag{15.59a}$$

$$\frac{\mathrm{D}\boldsymbol{u}_2}{\mathrm{D}t} + \boldsymbol{f} \times \boldsymbol{u}_2 = -\nabla\phi_2 - r\boldsymbol{u}_2, \tag{15.59b}$$

where $\phi_1 = p_T/\rho_0$ and $\phi_2 = p_T/\rho_0 + g'\eta$, with p_T being the pressure at the lid at the top, η the interface displacement (see Fig. 15.11) and $g' = g(\rho_2 - \rho_1)/\rho_0$ the reduced gravity, and we may take $\rho_0 = \rho_1$. We have also included a simple representation of surface drag, $-r\boldsymbol{u}_2$, in the lowest layer, and r is a constant. We will use a constant value of the Coriolis parameter except where it is differentiated, and on zonal averaging the zonal components of (15.59) become

$$\frac{\partial \overline{u}_1}{\partial t} - f_0\overline{v}_1 = \overline{v_1'\zeta_1'} \tag{15.60a}$$

$$\frac{\partial \overline{u}_2}{\partial t} - f_0\overline{v}_2 = \overline{v_2'\zeta_2'} - r\overline{u}_2. \tag{15.60b}$$

Geostrophic balance in each layer implies

$$f_0\boldsymbol{u}_{g1} = \mathbf{k} \times \nabla\phi_1, \qquad f_0\boldsymbol{u}_{g2} = \mathbf{k} \times \nabla\phi_1 + g'\mathbf{k} \times \nabla\eta, \tag{15.61a,b}$$

where the subscript g denotes geostrophic. Subtracting one equation from the other gives

$$f_0(\boldsymbol{u}_1 - \boldsymbol{u}_2) = -g'\mathbf{k} \times \nabla\eta, \tag{15.62}$$

dropping the subscripts g on $\boldsymbol{u}$. This equation represents thermal wind balance (or the Margules relation) for this system. A temperature gradient thus corresponds to a slope of the interface height, with the interface sloping upwards toward lower temperatures, analogous to isentropes sloping up toward the pole in the real atmosphere.

The quasi-geostrophic potential vorticity for each layer is

$$q_i = \zeta_i + f - f_0\frac{h_i}{H_i}, \tag{15.63}$$

where H_i is the reference thickness of each layer, which we take to be its mean thickness. The potential vorticity flux in each layer is then

$$\overline{v_i'q_i'} = \overline{v_i'\zeta_i'} - \frac{f_0}{H_i}\overline{v_i'h_i'}. \tag{15.64}$$

Phenomenology of a Two-layer Mid-Latitude Atmosphere

A radiative forcing that heats low latitudes and cools high latitudes will lead to an isentropic interface that slopes upward with increasing latitude, and a poleward total mass flux in the upper layer and an equatorward flux in the lower layer. The interface implies a thermal wind shear between the two layers. Neglecting relative vorticity, the potential vorticity gradients in each layer are given by

$$\frac{\partial \overline{q}_1}{\partial y} = \beta - \frac{f_0}{H_1}\frac{\partial \overline{h}_1}{\partial y} > 0 \quad \text{and} \quad \frac{\partial \overline{q}_2}{\partial y} = \beta - \frac{f_0}{H_2}\frac{\partial \overline{h}_2}{\partial y} \lesssim 0. \tag{TL.1}$$

The gradient generally is large and positive in the upper layer and small and negative in the lower layer — the gradient must change sign if there is to be baroclinic instability as we assume to be the case. This baroclinic instability generates eddy fluxes that largely determine the surface winds and the meridional overturning circulation. The zonal momentum equation in each layer is

$$\frac{\partial \overline{u}_1}{\partial t} = f_0\overline{v}_1 + \overline{v_1'\zeta_1'} = f_0\overline{v}_1^* + \overline{v_1'q_1'}, \tag{TL.2a}$$

$$\frac{\partial \overline{u}_2}{\partial t} = f_0\overline{v}_2 + \overline{v_2'\zeta_2'} - r\overline{u}_2 = f_0\overline{v}_2^* + \overline{v_2'q_2'} - r\overline{u}_2, \tag{TL.2b}$$

where

$$\overline{v}_i^* = \overline{v}_i + \frac{\overline{v_i'h_i'}}{H_i} \tag{TL.3}$$

is the residual meridional flow. In steady state the potential vorticity flux will be equatorward in the upper layer and poleward in the lower layer. Because the mass flux in each layer is equal and opposite, the surface (i.e., lower-layer) wind is given by the vertical integral of the vorticity or potential vorticity fluxes, namely

$$rH_2\overline{u}_2 = H_1\overline{v_1'q_1'} + H_2\overline{v_2'q_2'} = H_1\overline{v_1'\zeta_1'} + H_2\overline{v_2'\zeta_2'}. \tag{TL.4}$$

The vorticity flux is positive in the upper layer and negative in the lower layer. However, because the potential vorticity gradient in the upper layer is large, this layer is more linear than the lower layer and Rossby waves are better able to transport momentum. The magnitude of the vorticity flux is thus larger in the upper layer than in the lower layer and, using (TL.4), the surface winds are positive (eastward) in the mid-latitude baroclinic zone (see Fig. 15.14).

To balance the upper-layer mid-latitude momentum flux convergence a meridional overturning circulation (a Ferrel cell) is generated. In a steady state $f_0\overline{v}_1 = -\overline{v_1'\zeta_1'}$ so that the zonally-averaged upper level flow is equatorward. However, the total mass flux in the upper level is poleward; thus, the equatorward meridional velocity in the upper branch of the Ferrel cell is the consequence of an Eulerian zonal average and does not correspond to a net equatorward mass transport.

In the real atmosphere, the equatorward residual flow occurs close to the surface. Thus, for more realism, we might think of the lower layer as representing a near-surface layer and choose $H_1 \gg H_2$, or even construct a three-layer model with a shallow near-surface layer and two interior layers.

Using this in (15.60) gives

$$\frac{\partial \overline{u}_1}{\partial t} = \overline{v_1' q_1'} + f_0 \overline{v}_1^*, \qquad \frac{\partial \overline{u}_2}{\partial t} = \overline{v_2' q_2'} + f_0 \overline{v}_2^* - r\overline{u}_2, \tag{15.65a,b}$$

where

$$\overline{v}_i^* = \overline{v}_i + \frac{\overline{v_i' h_i'}}{H_i} \tag{15.66}$$

is the meridional component of the residual velocity in each layer, proportional to the *total* meridional mass flux in each layer. These are the transformed Eulerian mean (TEM) forms of the equations, first encountered in Section 10.3.

In the barotropic model of Section 15.1 the mean meridional velocity vanished at every latitude, a consequence of mass conservation in a single layer between two rigid flat surfaces. In the single-layer model of Section 15.2.1 the mean meridional velocity was in general non-zero, but the total meridional mass flux (i.e., the meridional component of the residual velocity) was zero if the domain is bounded laterally by solid walls. In the two-layer model we will allow a transformation of mass from one layer to another, which is the equivalent of heating: a conversion of mass from the lower layer to the upper layer is heating, and conversely for cooling. Thus, heating at low latitudes and cooling at high latitudes leads to the interface sloping upwards toward the pole. In the two-layer model the constraint that mass conservation supplies is that, assuming a statistically steady state, the total poleward mass flux summed over both layers must vanish.

The mass conservation equation for each layer is

$$\frac{\partial h_i}{\partial t} + \nabla \cdot (h_i \boldsymbol{u}_i) = S_i, \tag{15.67}$$

where S_i is the mass source term and we may suppose that $S_1 + S_2 = 0$ everywhere. A zonal average gives

$$\frac{\partial \overline{h}_i}{\partial t} + \frac{\partial \overline{h_i v_i}}{\partial y} = \overline{S}_i, \tag{15.68}$$

or, setting $\overline{h}_i = H_i$ and using (15.66),

$$\frac{\partial \overline{h}_i}{\partial t} + H_i \frac{\partial \overline{v}_i^*}{\partial y} = \overline{S}_i. \tag{15.69}$$

The mass source term in these equations is equivalent to heating, and let us suppose that this is such as to provide heating at low latitudes and cooling at high ones. This is equivalent to conversion of an upper-layer mass to a lower-layer mass at high latitudes, and the reverse at low latitudes; such a conversion can only be balanced by a poleward mass flux in the upper layer and an equatorward mass flux in the lower layer (Fig. 15.12). That is to say, an Earth-like radiative forcing between equator and pole implies that *the total mass flux in the upper layer will be poleward.* This is the opposite of the mean meridional circulation of the Ferrel cell shown in Fig. 14.3! What's going on? Before we can answer that, let us manipulate the equations of motion and obtain a couple of useful preliminary results.

Manipulating the equations

Because the total depth of the fluid is fixed, the mass conservation equations in each layer, (15.67), may each be written as an equation for the interface displacement, namely

$$\frac{\partial \eta}{\partial t} + \nabla \cdot (\eta \boldsymbol{u}_1) = -S_1, \qquad \text{or} \qquad \frac{\partial \eta}{\partial t} + \nabla \cdot (\eta \boldsymbol{u}_2) = S_2, \tag{15.70a,b}$$

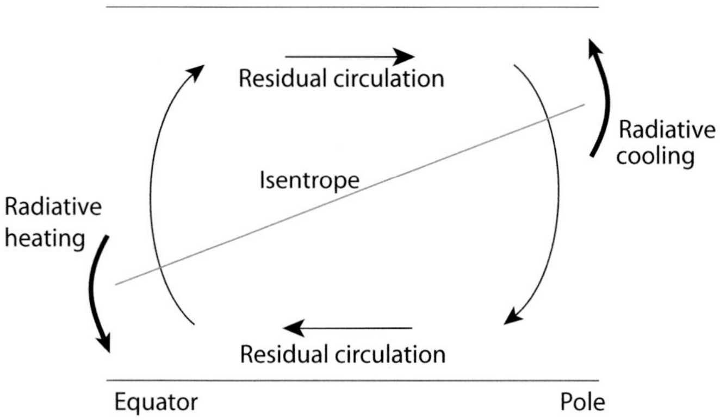

Fig. 15.12 Thermodynamics of a two-layer model, with an isentrope (or an interface between two layers) sloping up toward a cold pole, caused by cooling at high latitudes and heating at low.

The heating is balanced by a net mass flux — the meridional overturning circulation. In the tropics this circulation is the Hadley Cell, and is nearly all in the mean flow. In mid-latitudes the circulation is largely in the residual flow, and the Eulerian mean flow (the Ferrell Cell) is in the opposite sense.

where $\eta = H_1 - h_1 = h_2 - H_2$ and $h_1 + h_2 = H_1 + H_2$. Because of the thermal wind equation, (15.62), (15.70a) and (15.70b) are identical: $\boldsymbol{u}_1 \cdot \nabla\eta = \boldsymbol{u}_2 \cdot \nabla\eta$ and $S_1 = -S_2$. (If $S_1 \neq -S_2$ the flow would not remain balanced and the thermal wind equation could not be satisfied.) The zonally-averaged interface equation may be written as

$$\frac{\partial \overline{\eta}}{\partial t} - H_1 \frac{\partial \overline{v}_1^*}{\partial y} = \overline{S}, \qquad \text{or} \qquad \frac{\partial \overline{\eta}}{\partial t} + H_2 \frac{\partial \overline{v}_2^*}{\partial y} = \overline{S}, \tag{15.71}$$

where $\overline{S} = -\overline{S}_1 = +\overline{S}_2$, consistent with the mass conservation statement

$$H_1 \overline{v}_1^* + H_2 \overline{v}_2^* = 0, \tag{15.72}$$

which states that the vertically integrated total mass flux vanishes at each latitude.

Now, whereas (15.72) is a kinematic statement about the total mass flux, the dynamics provides a constraint on the *eddy* mass flux in each layer. Using the thermal wind relationship we have

$$f_0 \overline{(v_1' - v_2')\eta'} = g' \overline{\frac{\partial \eta'}{\partial x} \eta'} = 0. \tag{15.73}$$

Hence, if the upper and lower surfaces are both flat, we have that

$$\overline{v_1' h_1'} = -\overline{v_2' h_2'}, \tag{15.74}$$

and the eddy meridional mass fluxes in each layer are equal and opposite. If the bounding surfaces are not flat, we have

$$\overline{v_1' \eta_T'} - \overline{v_1' h_1'} = \overline{v_2' \eta_B'} + \overline{v_2' h_2'} \tag{15.75}$$

instead, where η_T and η_B are the topographies at the top and the bottom. Equations (15.74) and (15.75) are *dynamical* results, and not just kinematic ones; they are equivalent to noting that the form drag on one layer due to the interface displacement is equal and opposite to that on the other, namely

$$\overline{v_1' \eta'} = -[-\overline{v_2' \eta'}], \tag{15.76}$$

where the minus sign inside the square brackets arises because the interface displacement is into layer one but out of layer two.

Using (15.64) and (15.75) the eddy potential vorticity fluxes in the two layers are related by

$$H_1 \overline{v_1' q_1'} + H_2 \overline{v_2' q_2'} = H_1 \overline{v_1' \zeta_1'} + H_2 \overline{v_2' \zeta_2'} - f_0 \overline{v_1' \eta_T'} + f_0 \overline{v_2' \eta_B'}, \tag{15.77}$$

which is the layered version of the continuous result (see (10.26) on page 383),

$$\int_B^T \overline{v'q'}\,\mathrm{d}z = \int_B^T \overline{v'\zeta'}\,\mathrm{d}z + \frac{f_0}{N^2}\left[\overline{v'b'}\right]_B^T. \tag{15.78}$$

For flat upper and lower surfaces, and using $\overline{v_i\zeta_i} = -\partial\overline{u_iv_i}/\partial y$, (15.77) becomes

$$H_1\overline{v_1'q_1'} + H_2\overline{v_2'q_2'} = -H_1\frac{\partial}{\partial y}\overline{u_1'v_1'} - H_2\frac{\partial}{\partial y}\overline{u_2'v_2'}, \tag{15.79}$$

and integrating with respect to y between quiescent latitudes gives

$$\int\left[H_1\overline{v_1'q_1'} + H_2\overline{v_2'q_2'}\right]\mathrm{d}y = 0. \tag{15.80}$$

That is, the total meridional flux of potential vorticity must vanish. This is a consequence of the fact that the potential vorticity flux is the divergence of a vector field; in the continuous case

$$\overline{v'q'} = -\frac{\partial\overline{u'v'}}{\partial y} + f_0\frac{\partial}{\partial z}\frac{\overline{v'b'}}{N^2}, \tag{15.81}$$

which similarly vanishes when integrated over a volume if there are no boundary contributions.

15.2.3 Dynamics of the Two-layer Model

We now consider the climate, or the time averaged statistics, of our two-layer model. The equations of motion are (15.60) or (15.65), and (15.68) or (15.69). These equations are not closed because of the presence of eddy fluxes, and in this section we make some phenomenological and rather general arguments about how these behave in order to get a sense of the general circulation. In the next section we use a specific closure to address the same problem.

Let us summarize the physical situation. The two layers of our model are confined in the vertical direction between two flat, rigid surfaces, and they are meridionally confined between slippery walls at high and low latitudes (the 'pole' and 'equator'). The circulation is driven thermodynamically by heating at low latitudes and cooling at high ones, which translates to a conversion of layer 1 fluid to layer 2 fluid at high latitudes, and the converse at low latitudes (see Fig. 15.12). This sets up an interface that slopes upwards toward the pole and, by thermal wind, a shear. This situation is baroclinically unstable, and this sets up a field of eddies, most vigorous in mid-latitudes where the temperature gradient (or interface slope) is largest. Three fields encapsulate the dynamics — the lower-layer wind field, the meridional circulation, and the meridional temperature gradient, and our goal is to understand their qualitative structure. We note from the outset that the residual circulation is poleward in the upper layer, equatorward in the lower layer, and that this is a thermodynamic result, a consequence of heating at low latitudes and cooling at high latitudes.

From (15.65), the steady-state lower-layer wind is given by

$$rH_2\overline{u}_2 = H_1\overline{v_1'q_1'} + H_2\overline{v_2'q_2'} = H_1\overline{v_1'\zeta_1'} + H_2\overline{v_2'\zeta_2'}, \tag{15.82}$$

where the second equality uses (15.79). That is, *the lower-layer wind is determined by the vertical integral of either the vorticity flux or the potential vorticity flux.*

Neglecting contributions due to the mean horizontal shear (which are small if the *beta-Rossby number, $U/\beta L^2$*, is small) the potential vorticity gradient in each layer is given by

$$\frac{\partial\overline{q}_1}{\partial y} = \beta - \frac{f_0}{H_1}\frac{\partial\overline{h}_1}{\partial y} \gg 0 \qquad \text{and} \qquad \frac{\partial\overline{q}_2}{\partial y} = \beta - \frac{f_0}{H_2}\frac{\partial\overline{h}_2}{\partial y} \lesssim 0. \tag{15.83a,b}$$

In the upper layer $\partial\overline{h}_1/\partial y$ is negative so that the total potential vorticity gradient is positive and larger than β itself. In the lower layer $\partial\overline{h}_2/\partial y$ is positive and indeed if there is to be baroclinic instability it must be as large as β in order for $\partial\overline{q}/\partial y$ to change sign somewhere. Thus, although negative the potential vorticity gradient is much weaker in the lower layer. Thus, Rossby waves (meaning waves that exist because of a background gradient in potential vorticity) will propagate further in the upper layer, and this asymmetry is the key to the production of surface winds.

Now, the potential vorticity flux must be negative (and downgradient) in the upper layer, and there are various ways to see this. One is from the upper-layer momentum equation (15.65a) which in a steady state gives

$$\overline{v_1'q_1'} = -f_0\overline{v}_1^*. \tag{15.84}$$

Because $\overline{v}_1^*$ is poleward, $f_0\overline{v}_1^*$ is positive and the potential vorticity flux is negative in both Northern and Southern Hemispheres. Equivalently, in the upper layer the radiative forcing is increasing the potential vorticity gradient between the equator and the pole, so there must be an equatorward potential vorticity flux to compensate. Finally, the perturbation enstrophy or pseudomomentum equations tell us that in a steady state the potential vorticity flux is downgradient (also see Section 15.3.2). This is not an independent argument, since it merely says that the enstrophy budget may be balanced through a balance between production proportional to the potential vorticity gradient and the dissipation. For similar reasons we expect the potential vorticity flux to be positive (poleward) in the lower layer.

Now, (15.80) tells us that the latitudinally integrated potential vorticity flux is equal and opposite in the two layers. If the potential vorticity flux in the lower layer were everywhere equal and opposite to that in the upper layer, then using (15.82) there would be no surface wind, in contrast to the observations. In fact, the potential vorticity flux is more uniformly distributed in the upper layer, and this gives rise to the surface wind observed. Let us give a couple of perspectives (on the same argument) as to why this should be so. The argument centres around the fact that the potential vorticity gradient is stronger in the upper layer, as we can see from (15.83).

I. *Rossby waves and the vorticity flux*
The stronger potential vorticity gradient of the upper layer is better able to support linear Rossby waves than the lower layer. Thus, the vorticity flux in the region of Rossby-wave genesis in mid-latitudes will be large and positive in the upper layer, and small and negative in the lower layer. Thus, there will be more momentum convergence into the source region in the upper layer than in the lower layer, and the vertical integral of the vorticity flux will largely be dominated by that of the upper layer. This is positive in mid-latitudes and, to ensure that its latitudinal integral is zero, it is negative on either side. Using (15.82), a surface wind has the same pattern as the net vorticity flux, and so is eastward in the mid-latitude source region and westward on either side.

II. *Potential vorticity flux*
Rossby waves are generated in the region of baroclinic instability, at approximately the same latitude in both upper and lower layers. However, because the potential vorticity gradient is larger in the upper layer than in the lower layer, Rossby waves are able to propagate more efficiently and breaking and associated dissipation will tend to be further from the source region in the upper layer than in the lower layer. Now, the pseudomomentum equation for each layer is, similarly to (15.50) for the one-layer case,

$$\frac{\partial \mathcal{P}_i}{\partial t} = \frac{\partial}{\partial t}\left(\frac{\overline{q_i'^2}}{2\gamma_i}\right) = -\overline{v_i'q_i'} - \frac{\overline{D_i'q_i'}}{\gamma_i}, \qquad i = 1, 2, \tag{15.85}$$

where γ_i, the potential vorticity gradient, has opposite signs in each layer. In a statistically steady state, the region of strongest dissipation is the region where the potential vorticity

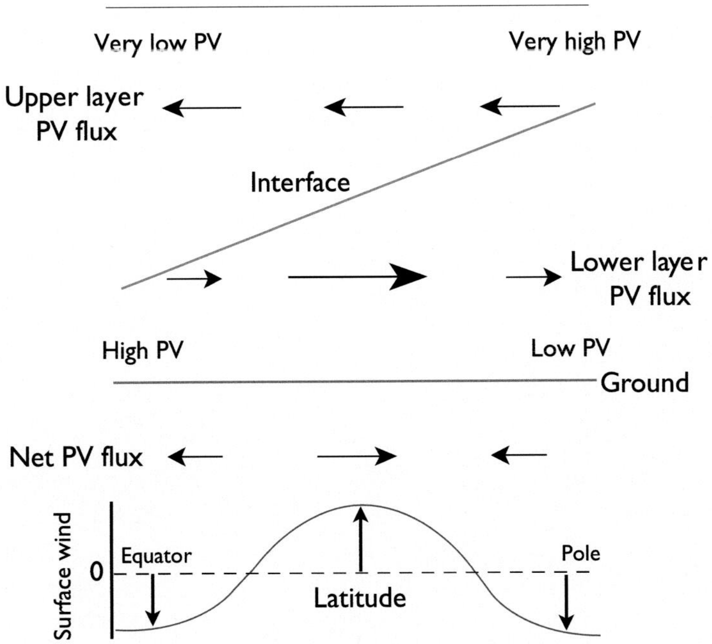

Fig. 15.13 The upper panel sketches the potential vorticity fluxes in each layer in a two-layer model. The surface wind is proportional to their vertical integral. The PV fluxes are negative (positive) in the upper (lower) layer, but are more uniformly distributed at upper levels.

The lower panel shows the net (vertically integrated) PV fluxes and the associated surface winds. In both panels the x-axis is latitude.

flux is largest. In the upper layer, Rossby-wave propagation allows the dissipation region to spread out from the source, whereas in the lower layer the dissipation region will be concentrated near the source. The distribution of the potential vorticity flux then becomes as illustrated in Fig. 15.13. The surface winds, being the vertical integral of the potential vorticity fluxes, are westerly in the baroclinic region and easterly to either side.

Momentum balance and the overturning circulation

From thermodynamic arguments we deduced that the residual circulation is direct, meaning that warm fluid rises at low latitudes, moves poleward aloft, and returns near the surface. At low latitudes where eddy effects are small the zonally-averaged Eulerian circulation circulates in the same way, giving us the Hadley Cell. In mid-latitudes, we may determine the Eulerian circulation from the momentum equation, (15.60). In the upper layer the balance is between the vorticity flux and the Coriolis term, namely

$$f_0\overline{v}_1 = -\overline{v_1'\zeta_1'} < 0. \tag{15.86}$$

That is, *the mean Eulerian flow is equatorward,* and this is the upper branch of the Ferrel cell. Note that the Eulerian circulation is in the opposite sense to the residual circulation.

In the lower layer the vorticity fluxes are weak and the balance is largely between the Coriolis force on the meridional wind and the frictional force on the zonal wind (as in Fig. 14.17). If the upper-layer flow is equatorward, the lower-layer flow must be poleward by mass conservation, and so the zonal wind is positive (eastward); that is

$$r\overline{u}_2 \approx f_0\overline{v}_2 = -\frac{H_1}{H_2}f_0\overline{v}_1 > 0, \tag{15.87a,b}$$

where the second equality follows by mass conservation of the Eulerian flow.

In terms of the TEM form of the equations, (15.65), the corresponding balances in the centre of the domain are

$$f_0\overline{v}_1^* = -\overline{v_1'q_1'} > 0 \tag{15.88a}$$

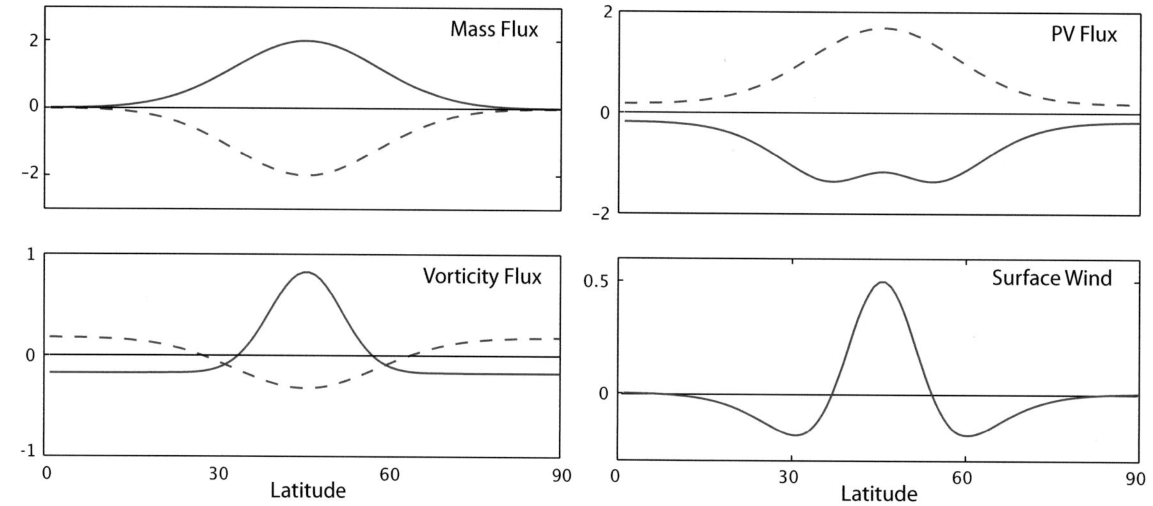

Fig. 15.14 Eddy fluxes in a two-layer model of an atmosphere with a single mid-latitude baroclinic zone. The upper-layer fluxes are solid lines and the lower-layer fluxes are dashed. The bottom-right panel shows the sum of the lower- and upper-layer vorticity fluxes (or, equivalently, the sum of the potential vorticity fluxes), which is proportional (when the surface friction is a linear drag) to the surface wind. The fluxes satisfy the various relationships and integral constraints of Section 15.2.2 but are otherwise idealized.

and

$$r\overline{u}_2 = f_0\overline{v}_2^* + \overline{v_2'q_2'} = -f_0\frac{H_1}{H_2}\overline{v}_1^* + \overline{v_2'q_2'} = \frac{H_1}{H_2}\overline{v_1'q_1'} + \overline{v_2'q_2'} > 0, \tag{15.88b}$$

using mass conservation and the fact that the lower-layer potential vorticity fluxes are larger than those of the upper layer. Illustrations of the dynamical balances of the two-layer model are given in Figs. 15.13 and 15.14.

15.3 † EDDY FLUXES AND AN EXAMPLE OF A CLOSED MODEL

The arguments above are heuristic and phenomenological and, although quite plausible, they are not wholly systematic. In this section we give a more axiomatic calculation by making certain closure assumptions that relate the eddy fluxes to the mean fields; specifically, we invoke the diffusion of potential vorticity, and then calculate the zonal winds and meridional circulation. The main purpose of this section is to explicitly show that if we do invoke a potential vorticity closure then a complete solution of the flow follows. However, too much credence should not be given to the particular closure we do invoke, for it cannot be rigorously justified.

15.3.1 Equations for a Closed Model

With quasi-geostrophic scaling, the equations of motion are the momentum equations written in residual form

$$\frac{\partial \overline{u}_1}{\partial t} = f_0\overline{v}_1^* + \overline{v_1'q_1'}, \tag{15.89a}$$

$$\frac{\partial \overline{u}_2}{\partial t} = f_0\overline{v}_2^* + \overline{v_2'q_2'} - r\overline{u}_2, \tag{15.89b}$$

and the mass conservation equation for each layer which may be written as an equation for the interface height,

$$\frac{\partial \overline{\eta}}{\partial t} - H_1 \frac{\partial \overline{v}_1^*}{\partial y} = S, \tag{15.90}$$

where $\eta = H_1 - h_1 = h_2 - H_2$, with the notation of the previous sections. This can be written in terms of $\overline{v}_2^*$ because

$$H_1 \overline{v}_1^* + H_2 \overline{v}_2^* = 0. \tag{15.91}$$

The velocities and thickness of the layers are related by the thermal wind relation

$$f_0(\overline{u}_1 - \overline{u}_2) = g' \frac{\partial \overline{\eta}}{\partial y} = -g' \frac{\partial \overline{h}_1}{\partial y}. \tag{15.92}$$

Using this to eliminate time derivatives between (15.89) and (15.90) reveals that the residual circulation satisfies

$$f_0^2 \frac{H}{H_2} \overline{v}_1^* - H_1 g' \frac{\partial^2 \overline{v}_1^*}{\partial y^2} = +g' \frac{\partial S}{\partial y} - f_0 \left(\overline{v_1' q_1'} - \overline{v_2' q_2'} \right) - f_0 r \overline{u}_2, \tag{15.93}$$

where $H = H_1 + H_2$. Thus, *the residual circulation is driven by the potential vorticity fluxes,* plus the diabatic terms. We may derive a similar expression for the Eulerian mean meridional flow, namely

$$f_0^2 \frac{H}{H_2} \overline{v}_1 - H_1 g' \frac{\partial^2 \overline{v}_1}{\partial y^2} = g' \frac{\partial S}{\partial y} + g' \frac{\partial^2}{\partial y^2} \overline{v_1' h_1'} - f_0 \left(\overline{v_1' \zeta_1'} - \overline{v_2' \zeta_2'} \right) - f_0 r \overline{u}_2. \tag{15.94}$$

However, the right-hand side now involves *both* the eddy vorticity fluxes and the eddy mass fluxes. The above equations illustrate the natural way in which the potential vorticity fluxes proximately 'drive' the extratropical atmosphere (see also the box on page 566).

Potential vorticity equation

A single prognostic equation for each layer is obtained by eliminating the residual circulation from (15.89) and (15.90), giving

$$\frac{\partial \overline{q}_1}{\partial t} = -\frac{\partial \overline{v_1' q_1'}}{\partial y} + \frac{f_0}{H_1} S, \tag{15.95a}$$

$$\frac{\partial \overline{q}_2}{\partial t} = -\frac{\partial \overline{v_2' q_2'}}{\partial y} - \frac{f_0}{H_2} S + r \frac{\partial \overline{u}_2}{\partial y}, \tag{15.95b}$$

where q_i are the quasi-geostrophic potential vorticities of each layer given by

$$\overline{q}_1 = -\frac{\partial \overline{u}_1}{\partial y} + f_0 \frac{\overline{\eta}}{H_1}, \qquad \overline{q}_2 = -\frac{\partial \overline{u}_2}{\partial y} - f_0 \frac{\overline{\eta}}{H_2}. \tag{15.96a,b}$$

Closure

If the potential vorticity fluxes can be expressed in terms of the mean fields then (15.95) is a closed set of equations. We can then solve for the potential vorticity in each layer and, using (15.93), for the residual circulation. One simple and rational closure is to assume that potential vorticity flux is transferred downgradient so that

$$\overline{v_i' q_i'} = -K_i \frac{\partial \overline{q}_i}{\partial y}, \tag{15.97}$$

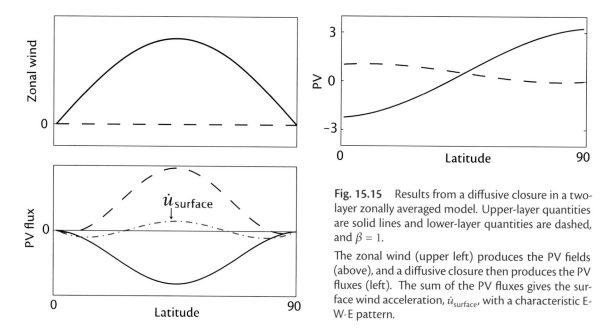

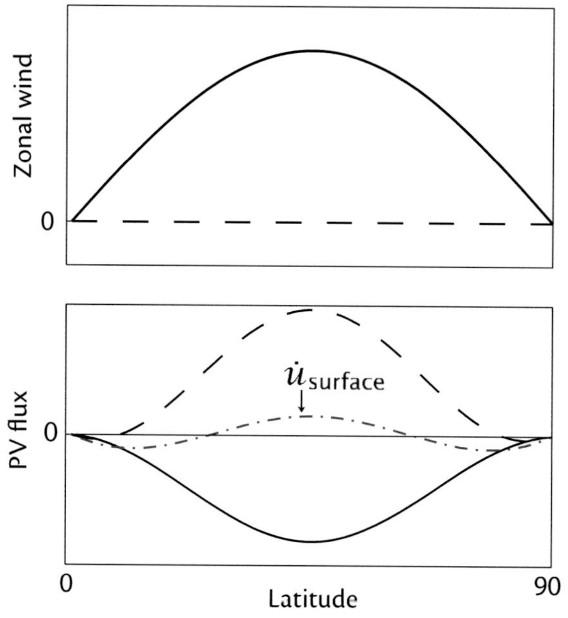

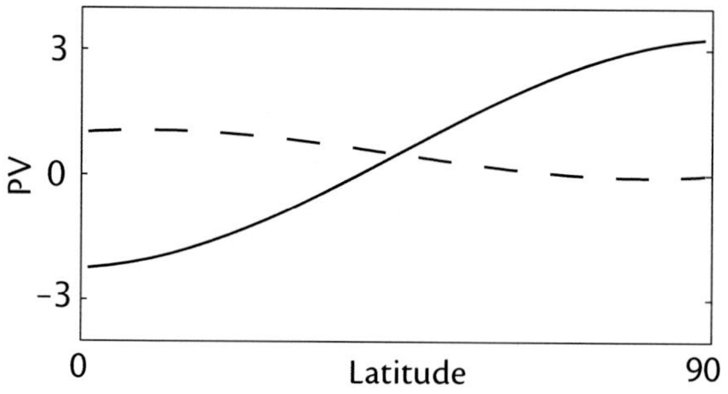

Fig. 15.15 Results from a diffusive closure in a two-layer zonally averaged model. Upper-layer quantities are solid lines and lower-layer quantities are dashed, and $\beta = 1$.

The zonal wind (upper left) produces the PV fields (above), and a diffusive closure then produces the PV fluxes (left). The sum of the PV fluxes gives the surface wind acceleration, $\dot{u}_{\text{surface}}$, with a characteristic E-W-E pattern.

where K_i is an eddy diffusivity, or transfer coefficient, which here is just a scalar quantity.[5] Note that the model demands a closure of the potential vorticity flux — not momentum, vorticity or the mass flux — and potential vorticity, being a materially conserved variable, is also that field for which a diffusive closure is most applicable.

Such a closure has all of the features and problems associated with diffusive closures discussed in Chapter 13, plus some of its own. One is that such a diffusive closure will not automatically respect the kinematic constraint that the volume integral of the potential vorticity flux must vanish, which for the two-layer model is expressed by (15.80). We may *choose* the vertical structure of the diffusivity in such a way that this constraint is satisfied, and in that case the model produces the results illustrated in Fig. 15.15.

The diffusive closure does indeed then produce potential vorticity fluxes similar to the observed westward–eastward–westward surface wind pattern, and a residual circulation of the same sense as in Fig. 15.12, and constitutes perhaps the simplest closed model of the zonally-averaged atmospheric circulation. Note that the surface wind is produced by the integral of the potential vorticity flux and, because the fluxes are quite different in the two layers, two layers are needed to produce a realistic pattern of surface wind without oversimplification, as well as to represent the meridional overturning and residual circulations. However, the model is a little ad hoc and the results depend on the structure of the transfer coefficients and the boundary conditions chosen.

15.3.2 ✦ Necessary Conditions for a Statistically Steady State

In linear baroclinic instability problems, a necessary condition for instability (the Charney–Stern–Pedlosky, or CSP, condition) is that the potential vorticity change sign in the interior of the fluid, or that the potential vorticity gradient in the interior has a particular sign with respect to the buoyancy gradient at horizontal bounding surfaces, as discussed in Chapters 9 and 10. These conditions do not apply in the statistical steady state of the forced-dissipative problem, but we may derive related conditions that do, although they are not completely general. We will focus on the interior condition and not the boundary conditions, as is appropriate in a layered model, but the argument could be extended to cover boundary issues explicitly.

The linear perturbation potential vorticity equation is

$$\frac{\partial q'}{\partial t} = -\overline{u}\frac{\partial q'}{\partial x} - v'\frac{\partial \overline{q}}{\partial y} - D', \tag{15.98}$$

where D' represents dissipative processes. From this we form the enstrophy equation

$$\frac{1}{2}\frac{\partial \overline{q'^2}}{\partial t} + \overline{D'q'} = -\overline{v'q'}\frac{\partial \overline{q}}{\partial y}, \tag{15.99}$$

where an overbar is a zonal average and we generally assume $\overline{D'q'} > 0$, as with a linear drag on q. In the standard linear problem we take $D = 0$ and then for growing waves ($\partial_t \overline{q'^2} > 0$) the right-hand side must be positive. But the integral of $\overline{v'q'}$ over latitude and height is zero, and thus $\overline{v'q'}$ must take both positive and negative signs. Hence, for growing waves, $\partial\overline{q}/\partial y$ must also take both positive and negative signs, and we recover the CSP condition that $\partial\overline{q}/\partial y$ must change sign for an instability. (We need not assume that the instabilities have normal form. A very similar argument was given in Section 10.6.)

In a statistically steady state the production of variance by the terms on the right-hand side is balanced by a cascade of variance to small scales. The quantity $\overline{v'q'}$ is non-zero, but its spatial integral still vanishes. We therefore have

$$\int \frac{\overline{D'q'}}{\partial\overline{q}/\partial y}\,\mathrm{d}y\,\mathrm{d}z = -\int \overline{v'q'}\,\mathrm{d}y\,\mathrm{d}z = 0. \tag{15.100}$$

Now, $\overline{D'q'} > 0$, and therefore to satisfy the equation $\partial\overline{q}/\partial y$ *must change sign somewhere.* That is, in a statistically steady state with dissipation, $\partial\overline{q}/\partial y$ must be both positive and negative somewhere in the domain. This is an analogue of the CSP result for wave–mean-flow interaction. Furthermore, if (as we have assumed) the left-hand side of (15.99) is positive everywhere, then eddy flux must be downgradient everywhere.

We can go a little further and obtain a result with the nonlinear terms. The zonally-averaged perturbation enstrophy equation is then

$$\frac{1}{2}\frac{\partial \overline{q'^2}}{\partial t} = -\overline{v'q'}\frac{\partial \overline{q}}{\partial y} - \frac{1}{2}\frac{\partial}{\partial y}\overline{v'q'^2} - \overline{D'q'}. \tag{15.101}$$

On integrating in y the third-order term vanishes and we obtain

$$\int \left(\frac{1}{2}\frac{\partial}{\partial t}\overline{q'^2} + \overline{D'q'}\right)\mathrm{d}y = -\int \overline{v'q'}\frac{\partial \overline{q}}{\partial y}\,\mathrm{d}y, \tag{15.102}$$

and so, if the left-hand side is positive, the flux must still be downgradient in the integrated sense that

$$\int \overline{v'q'}\frac{\partial \overline{q}}{\partial y}\,\mathrm{d}y < 0. \tag{15.103}$$

Suppose that the flux is *locally* downgradient, meaning that $\partial\overline{q}/\partial y$ and $\overline{v'q}$ have opposite signs, and in the nonlinear case this is an additional physical assumption. Then, because $\overline{v'q'}$ has both positive and negative values (because its integral is zero), then so must the potential vorticity gradient, $\partial\overline{q}/\partial y$. That is, *when dissipation is present and if the potential vorticity fluxes are downgradient, a statistically steady state can be maintained only if the potential vorticity gradient changes sign somewhere.* In the continuously stratified case, this condition is replaced by ones involving a combination of the interior potential vorticity gradient and the buoyancy gradient at the boundary, the conditions being the same as necessary conditions for instability.

Potential Vorticity Fluxes and the Extratropical Atmosphere

The extratropical circulation of the atmosphere is driven by the differential heating between equator and pole, mediated by fluxes of potential vorticity. Thus, in a layered model we have the following:

(i) Zonal winds. At each level the acceleration of the zonal winds is governed by the potential vorticity fluxes:

$$\frac{\partial \overline{u}_i}{\partial t} = \overline{v_i' q_i'} + f_0 \overline{v}_i^* + F_i, \tag{PV.1}$$

where $\overline{v}_i^*$ is the residual meridional flow and F_i represents friction.

(ii) Surface winds. In steady state, the surface winds are produced by the vertically integrated potential vorticity fluxes:

$$r H_s \overline{u}_s = \sum_i H_i \overline{v_i' q_i'}, \tag{PV.2}$$

where u_s is the surface wind, H_s the thickness of the lowest layer, and r is a frictional coefficient.

(iii) Meridional transport. The total (or residual) meridional transport is, proximately, forced by the potential vorticity fluxes. For example, in a two-layer model

$$f_0^2 \frac{H}{H_2} \overline{v}_1^* - H_1 g' \frac{\partial^2 \overline{v}_1^*}{\partial y^2} = +g' \frac{\partial S}{\partial y} - f_0 \left(\overline{v_1' q_1'} - \overline{v_2' q_2'} \right) - f_0 (F_1 - F_2), \tag{PV.3}$$

where S is proportional to the diabatic forcing, and this equation holds at all times. In a steady state the momentum equation gives simply

$$f_0 \overline{v}_i^* = -\overline{v_i' q_i'} - F_i. \tag{PV.4}$$

Above the surface layer friction is negligible and the meridional transport responds almost solely to the potential vorticity fluxes.

15.4 A STRATIFIED MODEL AND THE REAL ATMOSPHERE

In the previous section we introduced the effects of stratification by way of a two-layer model. Let us now discuss, albeit rather qualitatively, the dynamics of a continuously stratified model more relevant to the real atmosphere. These dynamics are generally similar to that of the two-layer model, although a number of differences in interpretation do arise. In particular, rather than the potential vorticity flux in the two layers, it is the potential vorticity flux in the interior and the buoyancy flux near the boundary that are the key aspects in producing the mean circulation.

15.4.1 Potential Vorticity and its Fluxes

The observed zonally-averaged potential vorticity field is shown in Fig. 15.16. Of interest to us is the fact that over most of the atmosphere, over most of the year, the potential vorticity gradient is monotonic, with the potential vorticity increasing northward. (The potential vorticity in the troposphere also increases moving northward along isentropes.) How, then, can the atmosphere be baroclinically unstable? It is because the surface buoyancy (or temperature) decreases poleward,

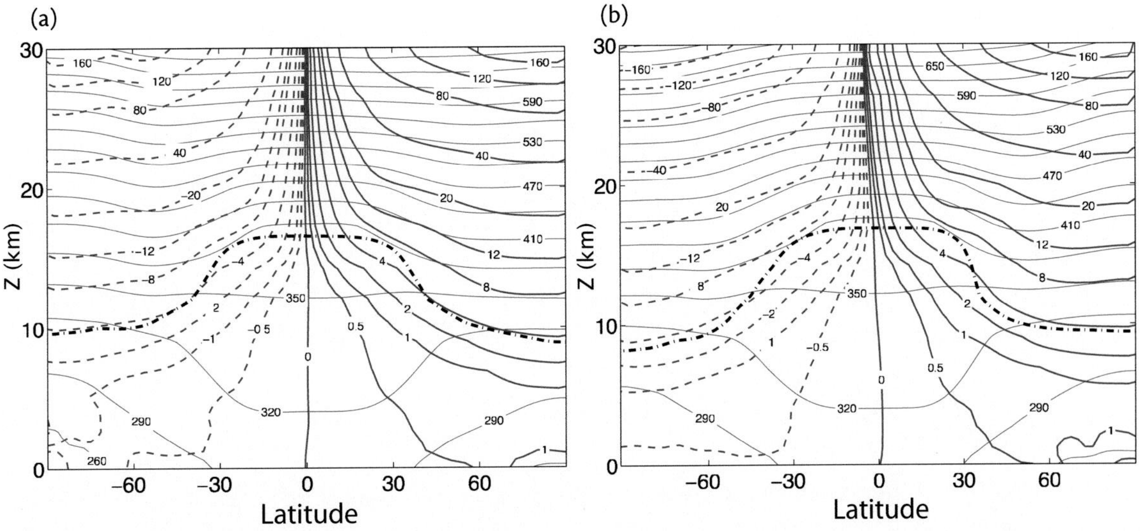

Fig. 15.16 The observed zonally-averaged Ertel potential vorticity distribution (dark solid and dashed lines, peaking up at the equator) and the potential temperature (lighter, red lines) for (a) annual mean, (b) December–January–February. Also shown is the position of the thermal tropopause (black, dot-dashed line). The potential vorticity is in 'PV units', 1 PVU $\equiv 1.0 \times 10^{-6}\,\mathrm{m^2\,K\,s^{-1}\,kg^{-1}}$, and has uneven contour intervals. The vertical coordinate is log pressure, with $Z = -H \log(p/p_R)$ km, where $p_R = 10^5$ Pa and H = 7.5 km.

and thus the atmosphere becomes unstable via the interaction of a surface edge wave with an interior Rossby wave (see the conditions on page 351 or page 404). This baroclinic instability may then excite Rossby waves which propagate meridionally, producing a momentum convergence and westward surface flow, and an associated meridional circulation or Ferrel Cell, much as described in Section 15.1. Let us explore these phenomena in a little more detail.

Surface winds

Consider the zonally-averaged, continuously stratified momentum equations with quasi-geostrophic scaling,

$$\frac{\partial \overline{u}}{\partial t} = \overline{v'\zeta'} + f_0\overline{v} + F = \overline{v'q'} + f_0\overline{v}^* + F, \tag{15.104}$$

where F represents frictional effects and the residual velocity $\overline{v}^*$ is given by

$$\overline{v}^* = -\frac{\partial \psi^*}{\partial z} = \overline{v} - \frac{\partial}{\partial z}\left(\frac{1}{N^2}\overline{v'b'}\right). \tag{15.105}$$

The friction is given by the vertical gradient of a stress, $F = \partial\tau/\partial z$, and at the surface we may parameterize the stress, following (5.210) on page 208, by $\tau = r\overline{u}$ where r is a constant. Then, vertically integrating (15.104) from the surface to the top of the atmosphere (where frictional stresses and the buoyancy flux both vanish) we find, in steady state,

$$r\overline{u}(0) = \left\langle \overline{v'\zeta'} \right\rangle = \left\langle \overline{v'q'} \right\rangle + \frac{f_0}{N^2}\overline{v'b'}(0), \tag{15.106}$$

where the angle brackets denote a vertical integral and (0) denotes surface values. Thus, the surface winds are determined, analogously to (15.82), by the vertically integrated relative vorticity fluxes, or equivalently by the integral of the interior potential vorticity fluxes and the buoyancy fluxes at

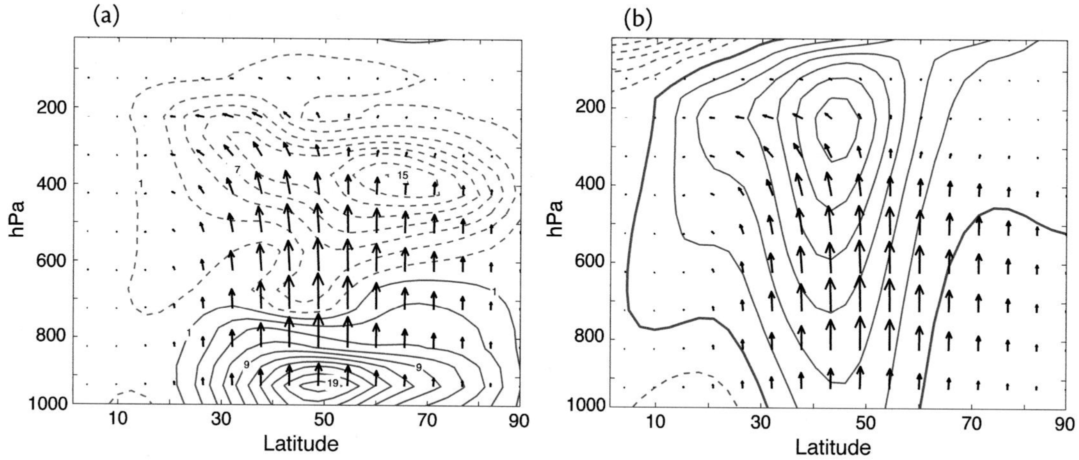

Fig. 15.17 The Eliassen–Palm flux in an idealized primitive equation of the atmosphere. (a) The EP flux (arrows) and its divergence (contours, with intervals of 2 m s^{-1}/day). The solid contours denote flux divergence, a positive PV flux, and eastward flow acceleration; the dashed contours denote flux convergence and deceleration. (b) The EP flux (arrows) and the time and zonally-averaged zonal wind (contours). See the Appendix A for details of plotting EP fluxes.

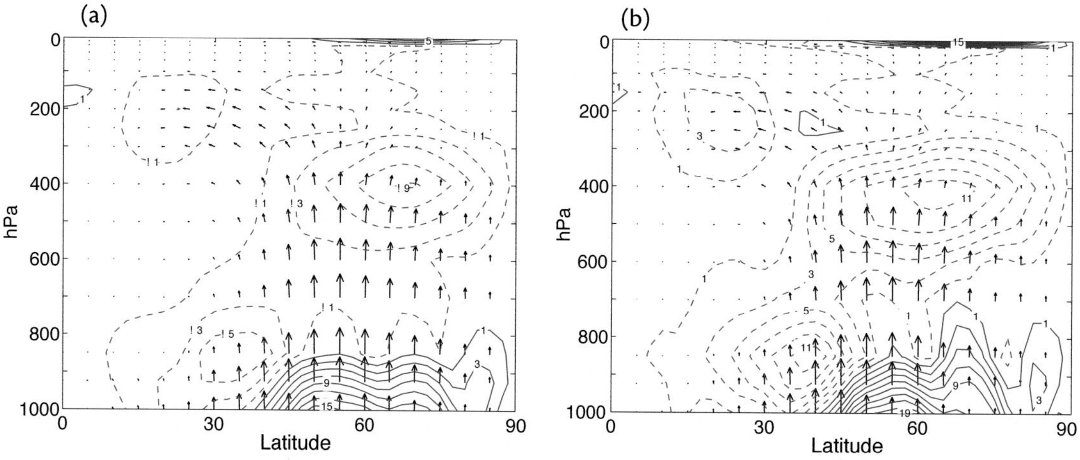

Fig. 15.18 The observed Eliassen–Palm flux (arrows) and its divergence (contours, with intervals of 2 m s^{-1}/day, zero contour omitted) in the Northern Hemisphere. Solid contours denote divergence, a positive (eastward) torque on the flow, and dashed contours denote convergence, a westward torque. (a) Annual mean, (b) DJF (December-January-February).

the surface. The advantage of the latter representation is that both potential vorticity and buoyancy are materially conserved variables and it may be easier to deduce some properties of their fluxes than of the fluxes of relative vorticity. Compared to the two-layer formulation, the interior fluxes are analogous to those of the upper layer whereas the surface fluxes are analogous to those of the lower layer, especially as the lower layer becomes thin.

Potential vorticity and Eliassen–Palm fluxes

As in Section 10.2, the quasi-geostrophic potential vorticity flux may be written as the divergence of the Eliassen–Palm (EP) vector,

$$\overline{v'q'} = \nabla_x \cdot \boldsymbol{\mathcal{F}}, \tag{15.107}$$

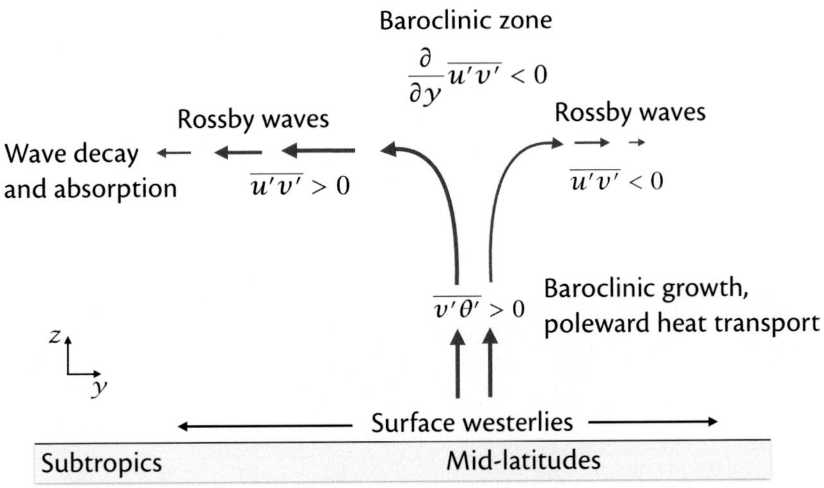

Fig. 15.19 Schematic of Eliassen-Palm fluxes in a baroclinic atmosphere. The vertical component, $\overline{v'\theta'}$ (or $\overline{v'b'}$) occurs primarily during the growing phase of the baroclinic lifecycle and corresponds to a poleward heat flux. The meridional fluxes in the upper atmosphere are associated with Rossby waves, and the vertical integral of the momentum flux divergence gives rise to surface westerlies.

where $\nabla_x\cdot \equiv \mathbf{j}\,\partial/\partial y + \mathbf{k}\,\partial/\partial z$ and

$$\boldsymbol{\mathcal{F}} \equiv -\overline{u'v'}\,\mathbf{j} + \frac{f_0}{N^2}\overline{v'b'}\,\mathbf{k}. \tag{15.108}$$

(See Appendix A to this chapter for primitive equation and spherical coordinate versions; in practice the quasi-geostrophic expression qualitatively captures the dominant terms in the primitive equation expressions.) The EP vector as obtained from an idealized primitive equation general circulation model integration is shown in Fig. 15.17, and the EP vector from observations is shown in Fig. 15.18, and both show qualitatively similar properties — a generally upwards-pointing vector in mid-latitude, veering equatorward aloft (and sometimes with some poleward propagation) as illustrated in Fig. 15.19.

The upward component represents the meridional transfer of heat, and this occurs during the growth phase of the baroclinic lifecycle and is qualitatively captured by linear models — for example, in the Eady problem the EP flux is directed purely vertically (Fig. 10.2), and this aspect resembles the vertical components of Figs. 15.17 and 15.18. But why should the average over a complete baroclinic lifecycle (which Fig. 15.19 schematically represents) even approximately resemble that of the growing phase of the baroclinic lifecycle? After all, the eddies must subsequently decay, and one might imagine that the fluxes would then reverse themselves. In fact, this is not the case: the baroclinic lifecycle is not reversible, because of two effects. First, there is transfer of baroclinic energy to barotropic modes (as described in Chapter 12) followed by a barotropic decay. Thus, over the complete cycle, there is no downwelling branch of EP fluxes that would correspond to equatorward heat transfer, and on average the poleward heat transfer (the $\overline{v'\theta'}$ branch) balances the net atmospheric heating. Second, there is an irreversible absorption and decay of the Rossby waves emanating from the baroclinic zone. These Rossby waves give rise to the lateral component of the EP flux, much as described in Section 15.1, although the wave activity is predominantly in the upper troposphere where the potential vorticity gradient is larger. This propagation is an irreversible process since the Rossby waves break and dissipate some distance from their source; this dissipation breaks the non-acceleration conditions and provides the mean flow acceleration and, consequentially, the observed zonal wind

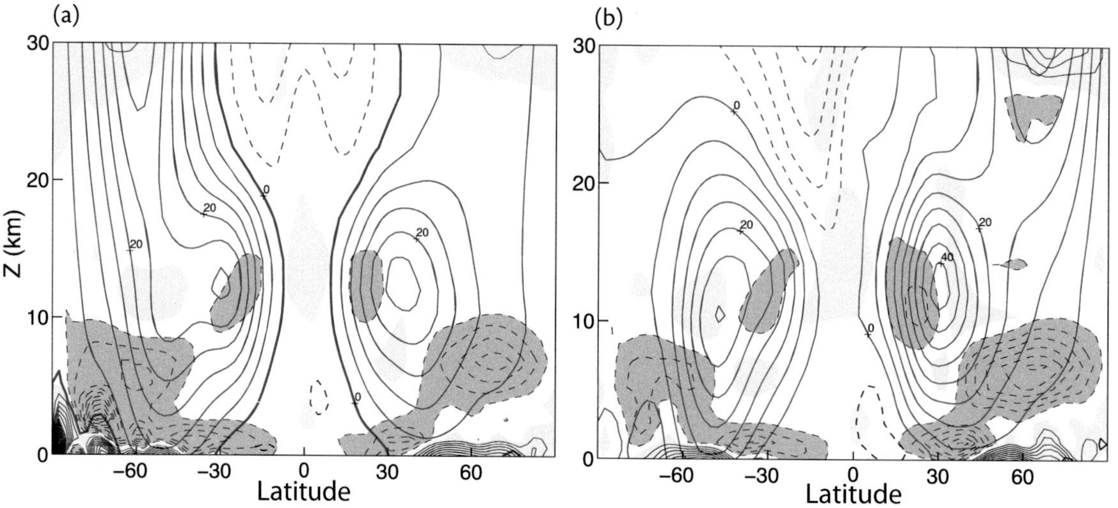

Fig. 15.20 The observed zonally-averaged zonal wind (thicker, red, contours, interval $5\,\mathrm{m\,s^{-1}}$), and the Eliassen–Palm flux divergence (contour interval $2\,\mathrm{m\,s^{-1}}$/day, zero contour omitted). Regions of positive EP flux divergence (eastward acceleration) are lightly shaded; regions less than $-2\,\mathrm{m\,s^{-1}}$/day are more darkly shaded. (a) Annual mean, (b) DJF (December–January–February).

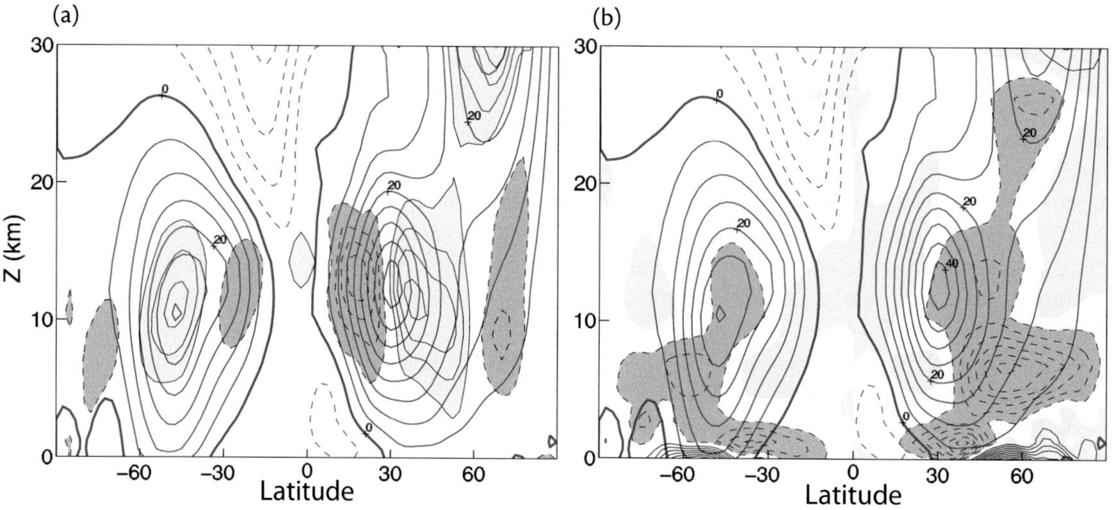

Fig. 15.21 The divergence of the two components of the EP flux (shaded), and the zonally-averaged zonal wind (thicker, red, contours) for DJF. (a) The momentum fluxes, $-\partial_y \overline{u'v'}$, contour interval is $1\,\mathrm{m\,s^{-1}/day^{-1}}$, light shaded for positive values > 1, dark shaded for negative values < -1. (b) The buoyancy flux, $f\partial_z(\overline{v'b'}/N^2)$, with contour interval and shading convention as in Fig. 15.20.

The divergence of the EP flux — that is, the potential vorticity flux — accelerates (or decelerates) the mean flow, as can be seen from (15.104) and Fig. 15.20. Broadly speaking, the EP flux decelerates the flow aloft (where it is balanced by the Coriolis force on the poleward residual flow) but provides an eastward acceleration at the surface (where it is largely balanced by friction). However, the two components of the flux (Fig. 15.21) have rather different effects on the mean flow. The horizontal component acts to extract momentum from the subtropics and deposit it in mid-latitudes, and so accelerate the flow producing a fairly barotropic eastward jet. (It is this component that gives rise to so-called negative viscosity, in which the eddies transfer momentum upgradient.) The vertical component of the EP flux arises from the meridional buoyancy flux and acts to reduce the intensity of the mid-latitude westerlies aloft, transferring momentum to the surface where it may be balanced by friction, and producing the surface westerlies. Two questions spring to mind:

(i) Why is the meridional wave-activity propagation predominantly in the upper atmosphere? That is to say, why do the EP vectors only veer laterally above about 400 hPa, and not in the lower atmosphere?

(ii) Why is the wave-activity propagation (the direction of the EP flux vectors) predominantly equatorward?

As regards item *(i)*, the propagation is mainly in the upper atmosphere because it is here that the potential vorticity gradient is strongest, as can be seen from Fig. 15.16. In the upper troposphere the beta effect is reinforced by the thermodynamic vortex stretching term, whereas the two effects partial cancel in the lower troposphere, an effect that is seen most clearly in a two-layer model (for example, Fig. 15.15), Thus, wave propagation is more efficient in the upper troposphere, whereas the lower troposphere is more nonlinear and so here the enstrophy cascade, and wavebreaking, occur locally and closer to the region of baroclinic instability itself. Regarding item *(ii)*, the proximate reason is that waves predominantly break on the equatorial side of the instability, and this in turn is for two possible reasons. One is that β increases towards the equator, so that linear propagation is more dominant. The other is that there is often a critical layer in the subtropics, where the speed of the waves equals that of the flow itself ($\overline{u} = c$), and here breaking can efficiently occur.

15.4.2 Overturning Circulation

The Eulerian overturning circulation (meaning the circulation from a conventional zonal average at constant height) in mid-latitudes is a single indirect cell, the Ferrel Cell, with rising motion at high latitudes and sinking in the subtropics (top panel of Fig. 15.22). The residual circulation is direct and, consistent with the theory of Section 10.3.3, resembles closely the thickness-weighted circulation. The observed circulation is shown in Fig. 15.22 with a schematic in Fig. 15.23. The main features are qualitatively captured by the two-layer dynamics of Section 15.2.2, but the continuously stratified case differs in some respects.

The main difference between the continuous and two-layer cases is that in the former the return flows — both the lower branch of the Ferrel Cell and the equatorial branch of the residual circulation — are not distributed over the lower troposphere, but are confined to a relatively thin layer. In the lower branch of the Ferrel Cell the dynamical balance is between friction and the Coriolis force on the meridional flow, so that its thickness is that of a turbulent Ekman layer and about a kilometre. To understand this better, let us take a quasi-geostrophic perspective. The mean potential vorticity gradient in the free atmosphere is nearly everywhere poleward and the potential vorticity flux is largely downgradient and equatorward. This means that here the residual circulation is largely poleward, satisfying the balance

$$f\overline{v}^* \approx -\overline{v'q'}. \tag{15.109}$$

In a multi-layer quasi-geostrophic model, with friction acting only in the lowest layer, the circulation is closed by return flow in the lowest layer; thus, as the number of layers increases the return flow is carried in an ever-thinner layer, this becoming a delta-function in the continuous limit, just as in the example of residual flow in the Eady problem (Section 10.5). In the real atmosphere, the return flow cannot be confined to a delta-function, but this argument suggests that it will occur close to the surface and this expectation is borne out in the lower panels of Fig. 15.22 and in Fig. 14.16. In fact, much of the equatorial return flow occurs in isentropic layers that have a potential temperature below the mean value at the surface — that is, in cold air outbreaks.

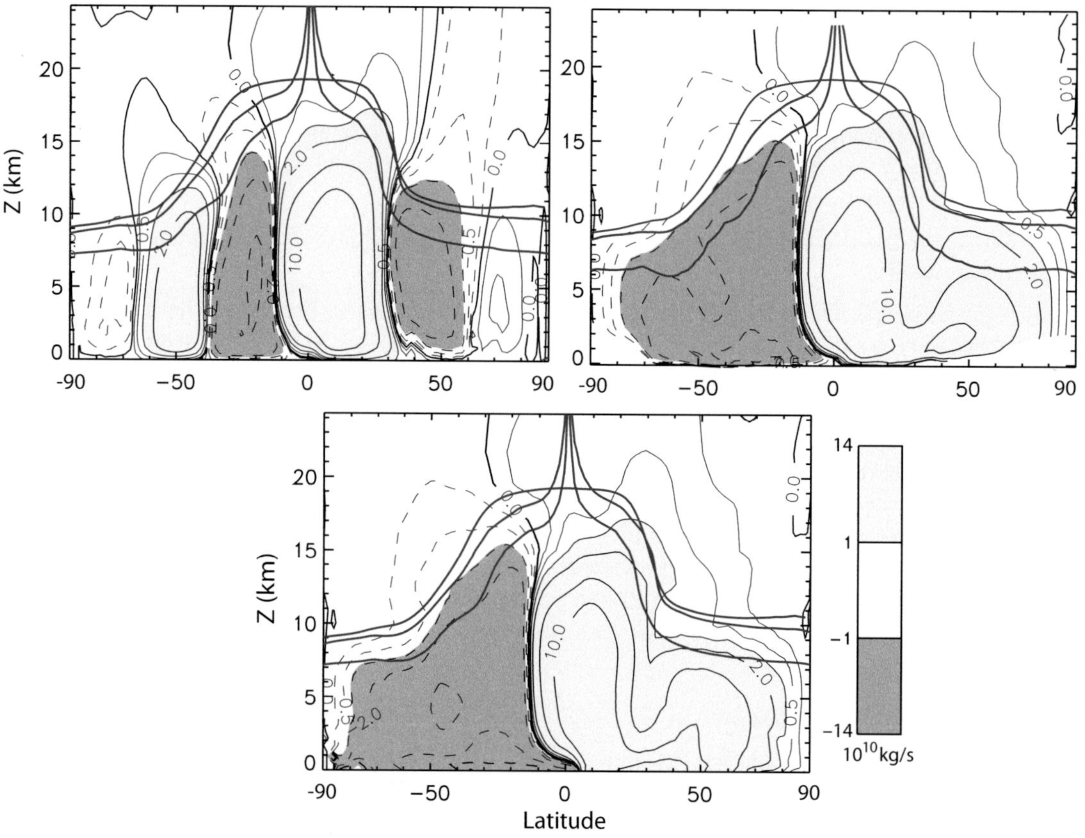

Fig. 15.22 Top left: The observed zonally-averaged, Eulerian-mean, streamfunction in Northern Hemisphere winter (DJF, 1994–1997). Negative contours are dashed, and values greater or less than $10^{10}\,\mathrm{kg\,s^{-1}}$ (10 Sv) are shaded, darker for negative values. The circulation is clockwise around the lighter shading. The three thick solid lines indicate various measures of the tropopause: the two that peak at the equator are isolines of potential vorticity, $Q = \pm 1.5, \pm 4$ PV units, and the flatter one is the thermal tropopause. Top right: The thickness-weighted meridional mass streamfunction. After calculation in isentropic coordinates, the streamfunction is projected back on to log-pressure coordinates for display. Bottom: the residual streamfunction calculated from the Eulerian circulation and the eddy fluxes.[6]

15.5 † TROPOPAUSE HEIGHT AND THE STRATIFICATION OF THE TROPOSPHERE

Let us now explore the physical processes that determine atmospheric stratification. The atmosphere may be divided by stratification into certain distinct regions, illustrated in Fig. 15.24 and Fig. 15.25. The left panel of Fig. 15.24 shows the so-called 'US standard atmosphere', a rough average temperature profile and a sometimes-useful standard, as well as actual observed values in the lower atmosphere. In the lower 10 km or so of the atmosphere we have the *troposphere*, a dynamically active region wherein most of the weather and the vast predominance of heat transport occurs. The troposphere is capped by the *tropopause*, above which lies the *stratosphere*, a region of stable stratification extending upwards to about 50 km. (Troposphere means 'turning sphere', appropriately so as within it dynamical overturning is prevalent. Stratosphere means 'layered sphere', and here there is much less vertical motion.) The stratosphere is capped by the *stratopause*, above which are the mesosphere, thermosphere and exosphere, regions of the upper atmosphere that do not concern us here. Our focus will be on the processes that determine the stratification of the lower atmosphere and the height of the tropopause.[7]

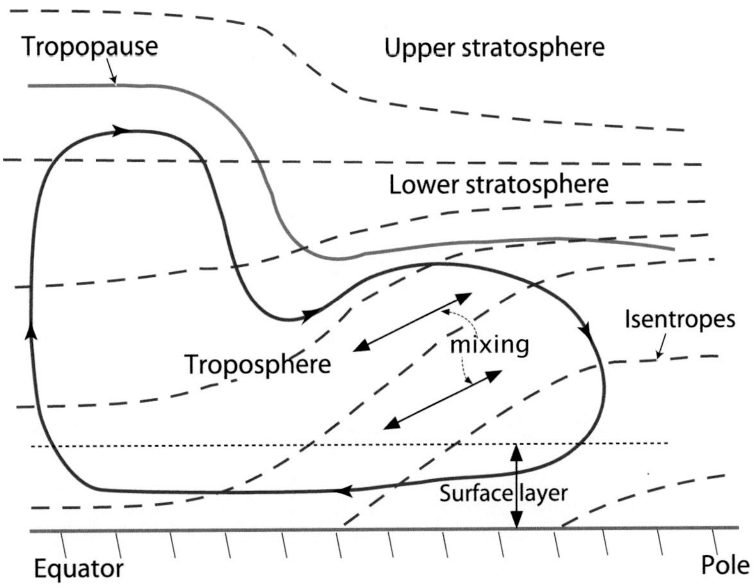

Fig. 15.23 Schematic of the stratification and residual overturning circulation in the lower atmosphere. The overturning circulation has two distinct parts, a tropical Hadley Cell where the large-scale flow is largely zonally-symmetric and a shallower extratropical cell where the heat and momentum transfer occurs in eddying motion. The equatorward return flow is mostly confined to a shallow surface layer. The lower stratosphere is ventilated by the troposphere along isentropic surfaces, whereas in the upper stratosphere isentropes do not intersect the tropopause. The tropopause is the boundary between the partially mixed troposphere and the near-radiative equilibrium stratosphere. Mixing tends to occur along slopes somewhat shallower than the isentropes.

In the troposphere temperature generally falls with height, whereas in the stratosphere it increases with height, and this gives rise to a thermal definition of the tropopause:[8] *The tropopause is the lowest level at which the lapse rate decreases to 2 K km^{-1} or less, provided also that the average lapse rate between this level and all higher levels within 2 km does not exceed 2 K km^{-1}*. At any particular time there might also be a second tropopause: if above the first tropopause the average lapse rate between any level and all higher levels within 1 km exceed 3 K km^{-1}, then a second (higher) tropopause is defined by that same criterion. Finally, such definitions are presumed not to apply if they are satisfied below 500 mb. As so defined, the thermal tropopause typically varies in height from about 16 km at low latitudes to about 8 km near the poles. These statements are a practical definition of the tropopause appropriate for today's climate on Earth — they would not hold on another planet or in a changed climate. Regardless, the tropopause is a distinct boundary separating two differently stratified regions, the troposphere and the stratosphere. The thermal tropopause is marked in Fig. 15.22 and, as we see there and in Fig. 15.16, in the extratropics it is almost parallel to isolines of potential vorticity, and sometimes an isoline of potential vorticity (say $Q = 3$ or 4 PV units) is used as a somewhat ad hoc definition of the extratropical tropopause.

Finally we note that the tropopause appears as a rather sharp feature when viewed instantaneously, although this sharpness is often blurred when time or spatial averages are taken. The solid line in Fig. 15.24, denoted 'tropopause-based average', shows the profile obtained when the tropopause height itself is taken as a common reference level, using data from individual radiosonde ascents over the United States.[9] The sharpness may indicate that the tropopause is acting as a mixing barrier for potential vorticity, separating the better-mixed troposphere and the more quiescent stratosphere.

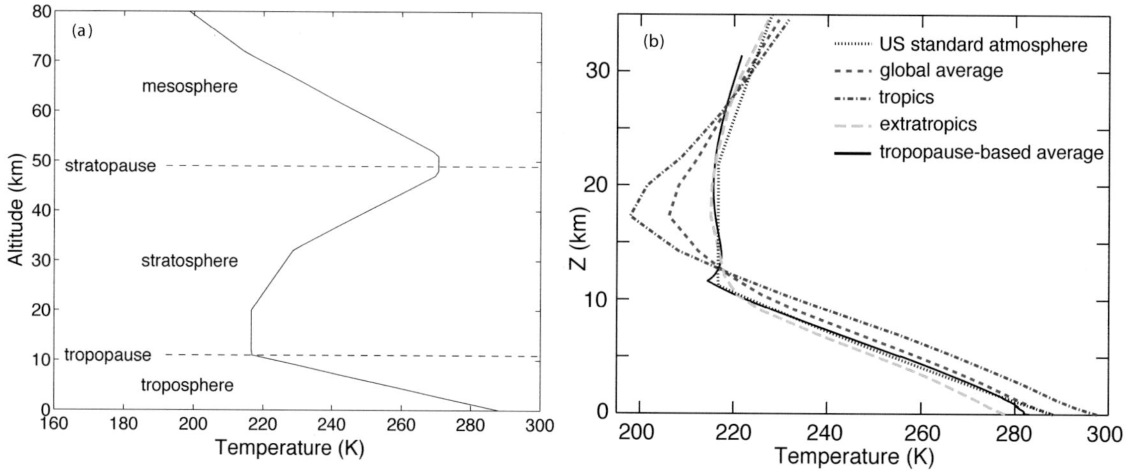

Fig. 15.24 (a) The temperature profile of the 'US standard atmosphere', marking the standard regions of the atmosphere below 80 km. In addition to the regions shown, the top of the mesosphere is marked by the mesopause, at about 80 km, above which lies the 'thermosphere', in which temperatures rise again into the 'exosphere', extending a few thousand kilometres and where the atmospheric temperature ceases to have a useful meaning. (b) Observed, annually averaged profiles of temperature in the atmosphere, where the ordinate is log-pressure. 'Tropics' is the average from 30° S to 30° N, and the extratropics is the average over the rest of the globe. The observations are from a reanalysis over 1958–2003 that extends upwards to about 35 km. See text for the meaning of 'tropopause-based average'.

15.5.1 Baroclinic Eddies and the Maintenance of Stratification

The atmosphere is largely heated at the surface – the ground is heated by the Sun and the ground heats the atmosphere. If the air near the surface becomes too warm it will become convectively unstable and the resulting convection (here meaning predominantly vertical convection occurring on relatively small scales) will transport heat upwards, stabilizing the temperature profile. If we look at the temperature profiles in Fig. 15.25 we see that in the tropics and subtropics the lapse rate is very close to the saturated adiabatic lapse rate, which is the lapse rate that is neutrally stable to convection in the presence of moisture (as we consider further in Chapter 18), suggesting that the stratification is indeed maintained by convection. But in mid- and high latitudes the lapse rate is quite stable with respect to convection, suggesting that some process other than convection is transporting heat upwards. The obvious candidate is baroclinic instability because that has the property of transporting heat both poleward and upward, both of which occur in the atmosphere, and if the vertical transport is efficient enough convection will then not be needed.

The baroclinic eddies do not extend infinitely upwards, and so we expect a boundary — a *tropopause* — between a dynamical troposphere and a stratosphere that is more nearly in radiative equilibrium. That is to say, the troposphere is that region in which a dynamical distribution of energy takes place; in the tropic this redistribution is by convection, and in mid-latitudes the redistribution is effected by baroclinic eddies. Given this picture, two questions present themselves:

(i) What determines the stratification (i.e., the lapse rate) of the mid-latitude troposphere?

(ii) What determines the height of the tropopause?

The full answer to these questions involves both radiation and dynamics. In the following sections, we will focus more on dynamics, and in Section 15.6 we will bring radiation into the mix. However, we won't discuss radiation itself until Chapter 18, so the reader who desires a full understanding will have to skip back and forth a little.

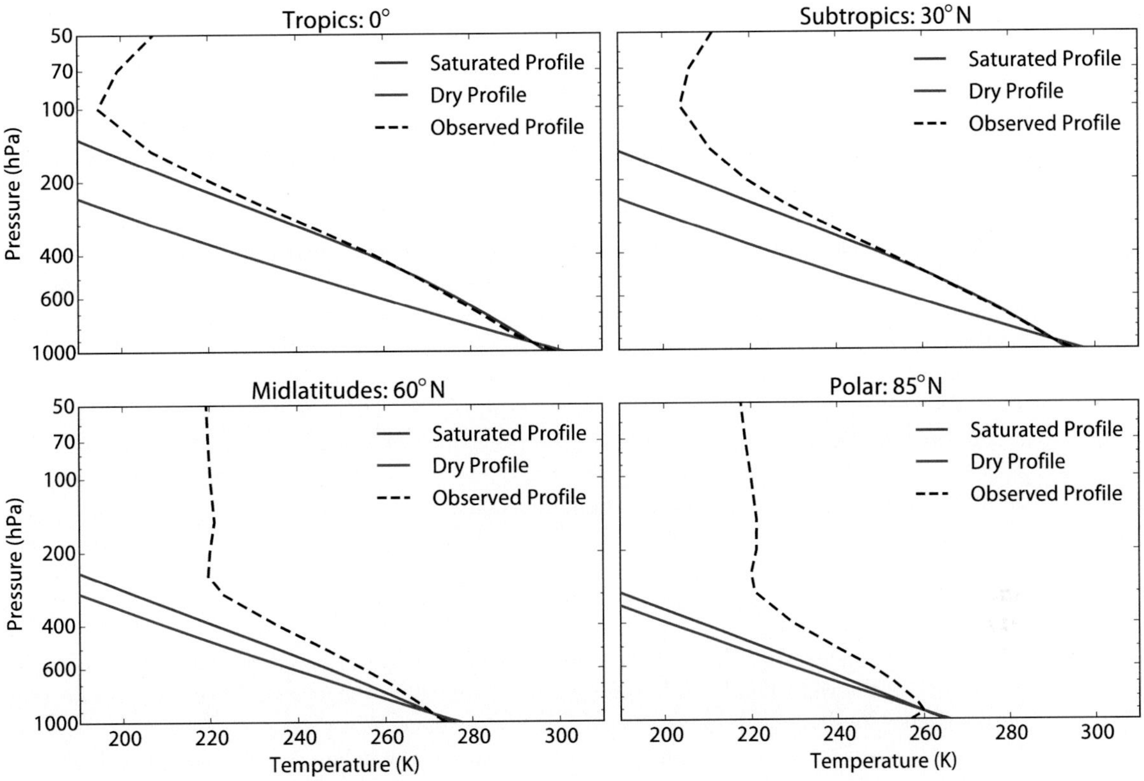

Fig. 15.25 The annually- and zonally-averaged observed (from reanalysis) temperature profiles at various latitudes, along with profiles constructed by integrating the saturated and dry adiabatic lapse rates, chosen to coincide with the observed temperature at 925 hPa, or about 750 m.

15.5.2 † Potential Vorticity Mixing and Baroclinic Adjustment

Baroclinic eddies grow from small beginnings to finite amplitude, and it is the finite amplitude eddies that stir the atmosphere and define the troposphere. However, they do not forget their linear properties, so let us remind ourselves of one.

A linear height scale

On the β-plane, linear baroclinic instability can produce a height scale that is different from the height of any pre-existing 'lid' or from the density scale height. This height scale is (Section 9.9.1)

$$h = \frac{\Lambda f^2}{\beta N^2}, \tag{15.110}$$

where $\Lambda = \partial \overline{u}/\partial z$. That is to say, if $h < H$, where H is the density scale height (or the height of some lid in a Boussinesq model), then the baroclinic eddies will extend upwards to a height h, and this will be the vertical extent of significant heat fluxes. Thus (one might argue), below h the thermal structure is determined by the dynamical effects of baroclinic instability, whereas above h the atmosphere is more nearly in radiative equilibrium. Using $\Lambda = (15\,\mathrm{m\,s^{-1}})/(10\,\mathrm{km})$, $\beta = 1.6\times10^{-11}\,\mathrm{s^{-1}\,m^{-1}}$, $f = 1\times10^{-4}\,\mathrm{s^{-1}}$ and $N = 10^{-2}\,\mathrm{s^{-1}}$ gives $h \approx 10\,\mathrm{km}$, which approximates the height of the tropopause in mid-latitudes. This is not a prediction because we have taken observed values for the stratification and shear. The amplitude of meridional heat transfer, which determines the meridional temperature gradient and so the shear Λ, is really determined by nonlinear effects.

Mixing

Let us now suppose that baroclinic eddies grow to finite amplitude and mix potential vorticity (see Section 13.5 for a general discussion of potential vorticity mixing). Such mixing will try to homogenize potential vorticity, or equivalently to expel potential vorticity gradients to a boundary, and if so the (extratropical) tropopause will occur at an isoline of potential vorticity, and be marked by a near-discontinuity in the potential vorticity distribution. Because $Q \approx (f/\rho)\partial\theta/\partial z$, the tropopause would also correspond to a discontinuity in stratification. These ideas are, at least qualitatively, in accord with observations: the potential vorticity distribution in the troposphere is somewhat more homogeneous than in the stratosphere — see Fig. 15.16, noting the unequal contour intervals of PV, and there is a near-discontinuity in stratification (by definition) at the tropopause. What are the effects of that mixing?

Potential vorticity mixing will occur only so far as needed in order to stabilize the mean flow. We also know that the meridional surface temperature gradient remains negative, so that if the flow is stabilized it must involve changes in the interior potential vorticity distribution. If horizontal scales are sufficiently large we may neglect the contribution of relative vorticity, and the quasi-geostrophic potential vorticity and its gradient become

$$q = \beta y + \frac{\partial}{\partial z}\left(\frac{f_0^2}{N^2}\frac{\partial \psi}{\partial z}\right) \qquad \text{and} \qquad \frac{\partial \overline{q}}{\partial y} = \beta - \frac{\partial}{\partial z}\left(\frac{f_0^2}{N^2}\Lambda\right). \tag{15.111}$$

We might hypothesize that the vertical extent to which mixing occurs is just sufficient to make the two terms on the right-hand side a similar size in order that the potential vorticity can become homogeneous, or that it can change sign and be just unstable. This gives $\beta \sim f_0^2\Lambda/(N^2 H_T)$, where H_T is the vertical extent of the instability and, we assume, the height of the tropopause; that is

$$H_T \sim \frac{f_0^2\Lambda}{N^2\beta}. \tag{15.112}$$

Put another way, in this model *the troposphere extends vertically as far as baroclinic waves can alter the potential vorticity from its planetary value.* (A similar depth scale occurs when evaluating the depth of the wind's influence in an ocean circulation model, Section 20.2.1.) If the lapse rate (and the shear) is known, this height determines the tropopause. Note the similarity of (15.112) to (15.110) — a similarity that is unsurprising given that we are constructing a height scale from a shear, f, β and N^2 using potential vorticity dynamics in both cases. We still do not have a prediction for H_T, because the stratification N^2 is unknown.

Baroclinic adjustment

Equilibration by potential vorticity mixing is closely related to a process known as *baroclinic adjustment,* by analogy with convective adjustment. The essential idea is that baroclinic eddies are sufficiently efficient that they can stabilize the mean flow by transferring heat poleward and upward until the necessary condition for instability (the Charney–Stern–Pedlosky condition, in so far as the flow is quasi-geostrophic and inviscid) is just satisfied, and the atmosphere is marginally supercritical to baroclinic instability. The adjustment might conceivably proceed predominantly by changes in the static stability, N^2, or predominantly by changes in the horizontal temperature gradient. The arguments for such an adjustment process are a little ad hoc, and the final state to which the atmosphere putatively adjusts is not well defined since there is no critical shear for instability in a continuously stratified model. In a two-level model the critical shear for instability is given by (9.121), which may be put in the form

$$\Lambda_{\text{crit}} = \frac{\beta N^2 H}{f_0^2}, \tag{15.113}$$

which is of the same form as (15.112). In a continuously stratified model, (15.113) may be regarded as a value of the shear above which the baroclinic eddies become deep (all eddies are deep in the two-level model) and transport heat efficiently.

Allowing the parameters f and β to vary with latitude, and using the thermal wind relation, (15.112) or (15.113) may be written in the form

$$H_T = -\left(\frac{\partial_y \theta}{\partial_z \theta}\right)\frac{f}{\beta} = s\frac{f}{\beta} \sim sa, \tag{15.114}$$

where $s = -(\partial_y \theta)/(\partial_z \theta)$ is the isentropic slope and a is the Earth's radius. This equation suggests that the isentropic slope is roughly such that isentropes extend from the surface in the subtropics to the tropopause at the poles, and this is more-or-less true in the present atmosphere (Fig. 15.16). That is to say, if adjustment-like arguments do hold, the isentropic slope will remain roughly constant even as other parameters change — for example, the horizontal temperature gradient may change with season but the stratification changes such as to keep s the same. The numerical and observational evidence for such an adjustment is mixed, although it is plausible that a weaker version may hold in which potential vorticity is imperfectly homogenized and (15.112) provides a plausible scaling, but not a precise prediction.[10]

Even if the above scalings were correct, they provide a closed prediction of neither the tropopause height nor stratification; they only provide a relation between the two. We will close the problem in Section 15.6, but first we consider another point of view.

15.5.3 † Extratropical Convection and the Ventilated Troposphere

A point of view that differs *qualitatively* from the potential vorticity one is to suppose that the mid-latitude tropospheric lapse rate is maintained by convection, the convection occurring predominantly in the warm sector of mature baroclinic waves.[11] In reality such convection will involve moisture, but first consider a dry atmosphere. In a given baroclinic zone the minimum potential temperature difference between the tropopause and the surface, $\Delta_z \theta = \theta_T - \theta_S$ is approximately zero: if the tropopause were colder than this the column would be convectively unstable, and the difference would become zero. The essential assumption that we make is that within a baroclinic zone there generally *does* exist a region that is convectively unstable, and that convection then ensues with sufficient efficiency to partially fill the troposphere with air with that surface value of potential temperature. The process differs from convection in the tropics because it is organized by baroclinic waves and, if we imagine a succession of baroclinic waves around a latitude band the mean value of $\Delta_z \theta$ will be approximately, we assume, its minimum (zero) plus a fraction of its standard deviation. The standard deviation in turn is a consequence of the pre-existing meridional temperature gradient and meridional advection across that gradient, and therefore

$$\text{standard deviation}(\Delta_z \theta) \propto \Delta_y \theta, \tag{15.115}$$

where the term on the right-hand side is the meridional temperature difference at the surface across the baroclinic zone. The mean potential temperature difference between the surface and tropopause is then simply proportional to the meridional temperature gradient at that latitude, with an undetermined constant of proportionality and so

$$\Delta_z \theta \propto \Delta_y \theta. \tag{15.116}$$

Finally, if moisture is present (as it is!) the potential temperature should be replaced by the equivalent potential temperature — the potential temperature achieved when all the water vapour in a parcel of air condenses and the latent heat of condensation is used to heat the parcel (Section 18.3.2).

The physical hypothesis is essentially that within a baroclinic wave the advection of warm air into a cold region *necessarily* leads to convection, and that this convection then efficiently fills the

The Stratification of the Troposphere and Stratosphere

- The troposphere is that region of the atmosphere in which a dynamical redistribution of heat occurs. The stratosphere above it is more nearly in radiative equilibrium (although in winter in the lower stratosphere, radiative equilibrium is not a very good approximation). The tropopause is the change in stratification between the two regions.

- The tropospheric lapse rate and the height of the tropopause are determined by a combination of dynamics and radiation. The tropopause temperature is, to a fair approximation, determined by the requirements of radiative balance and the height of the tropopause then follows if the lapse rate is known, and vice versa. Dynamical and radiative consistency leads to a state in which the tropopause height is the height to which dynamical effects extend and equal to the height needed for radiative balance.

- In Earth's atmosphere, there are two dynamical processes important for the maintenance of stratification:

  (i) Convection, especially moist convection. It is generally thought that moist convection plays the dominant role in determining tropical stratification, leading to a lapse rate that is approximately neutral to moist convection (discussed more in Chapter 18).

  (ii) Baroclinic eddies. In mid-latitudes baroclinic eddies transport heat poleward and upwards, so determining in the mid-latitude meridional temperature gradient and stratification. Various theories for this have been proposed, but none are accepted as having wide applicability and accuracy. In one incarnation ('baroclinic adjustment') baroclinic activity is so efficient that the atmosphere becomes only marginally supercritical to baroclinic instability, a process related to homogenization of potential vorticity. In a variation on this theme, potential vorticity and surface potential temperature are diffused downgradient, but potential vorticity is not necessarily homogenized. A complete theory would require estimates of the structure and magnitude of the eddy diffusivities.

  In the extratropics moist convection is less dominant, but convection in the warm sector of baroclinic eddies may act to produce a stratification that is related to the meridional temperature difference across a baroclinic zone.

- If we have a theory for the stratification produced by baroclinic eddies then we can use that theory in conjunction with radiative calculations to make predictions of the tropopause height, as in Section 15.6. For example, various dynamical arguments suggest the importance of the height scale $H \sim (f^2 \Lambda)/(N^2 \beta)$, where we associate H with the tropopause height. Radiative arguments also lead to a expression for tropopause height in terms of the stratification. Using the two together gives predictions for both tropopause height and the stratification.

- In the mid-stratosphere, ozone absorbs solar radiation and this gives rise to an increase in temperature with height from the lower- to the mid-stratosphere. This effect makes the stratosphere still more stable to baroclinic instability and may sharpen the tropopause, but it is not the root cause of the tropopause.

available volume, to the extent possible, with the warmest possible fluid. Oceanographers will find this a comfortable concept, for they are used to the notion of convection filling the domain with the densest available fluid (densest in the oceanic case because oceanic convection usually occurs from the top, with cold, dense water sinking). However, unlike the ocean in which the bottom of the container limits the volume of dense water that can be made, here it is the tropopause that provides the upper lid; the height of this is determined by the dynamics itself, in conjunction with the requirement of radiative balance. Thus, the baroclinic zone becomes, in oceanographic parlance, *ventilated* by the warmest air at the surface. However, the entire baroclinic zone does not completely fill with this warm air because the convection is maintained by a meridional temperature gradient and it is necessarily intermittent: baroclinic instability would shut off if the entire baroclinic zone were filled with homogeneous warm fluid, and the zone would then meridionally restratify. It is this continual maintenance of variance that leads to (15.116).

The ultimate consequence of these convection arguments is that the moist isentropic slope is proportional to that slope which would take an isentrope at the surface to the tropopause across a baroclinic zone. On Earth, this is similar to the potential vorticity mixing ideas, which suggest that the isentropic slope is proportional to the slope that goes from the ground to the tropopause over a horizontal scale f/β, the equator-to-pole scale, but the reasons are different.

15.6 † A MODEL FOR BOTH STRATIFICATION AND TROPOPAUSE HEIGHT

In the previous section we discussed various ideas regarding the effects of baroclinic eddies on the mid-latitude stratification. The discussion was incomplete on two grounds. First, the ideas themselves are heuristic. Second, even if true, they don't give us a complete picture; they just give us a relation between stratification and tropopause height. A second relation is needed, and this is provided by the radiative-dynamical arguments of Section 18.6 and Appendix C of Chapter 18, which the reader may now read with profit. In those sections we showed that, if the lapse rate in the lower atmosphere is assumed constant, it can extend upward only to a certain height in order to maintain an overall radiative balance — that is, in order that the outgoing longwave radiation equal the incoming solar radiation. Specifically, we obtained an approximate analytic expression for the tropopause height, H_T, (18.172), namely

$$8\Gamma H_T^2 - C H_T T_T - \tau_s H_a T_T = 0. \tag{15.117}$$

where $C = 2\log 2 \approx 1.39$. In this expression Γ is the lapse rate, $-\partial T/\partial z$, T_T is the temperature at the tropopause, τ_s is the surface optical depth and H_a is the scale height of the main infrared absorber. The radiative parameters are determined by the composition of the atmosphere and we regard them as given, and the temperature at the tropopause is, to a decent approximation, given by radiative balance, namely $2\sigma T_T^4 \approx S_{net}$ where S_{net} is the net incoming solar radiation. (If there is a lateral convergence of heat in the atmosphere then S_{net} should be modified appropriately. Alternatively, we regard (15.117) as a theory for the global mean tropopause height.) The main assumptions leading to (15.117) are that the atmosphere is radiatively grey in the infrared and that the lapse rate is uniform up to H_T.

In mid-latitudes, (15.117) does not provide a closed prediction for tropopause height because the lapse rate is not known (whereas in the tropics we might take the lapse rate to be determined by convection). However, if we combine (15.117) with the ideas about baroclinic transport we obtain a closed model. We do not have a good model of baroclinic transport, but for the purposes here let us suppose that the isentropic slope is such that (15.114) roughly holds. Noting that, in a hydrostatic atmosphere, $(T/\theta)\,\partial\theta/\partial z = \Gamma_d - \Gamma$ where Γ_d is the dry adiabatic lapse rate, we rewrite (15.114) as

$$\Gamma = \Gamma_d + \left(\frac{\partial_y T}{H_T}\right)\frac{f}{\beta}. \tag{15.118}$$

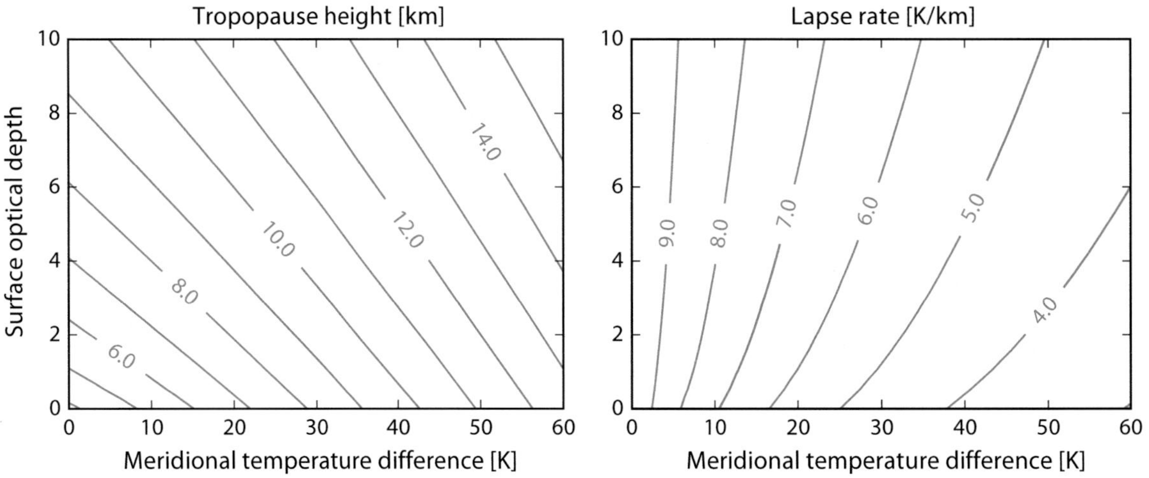

Fig. 15.26 Left: Contours of tropopause height as a function of the temperature difference between the subtropics and pole, and the optical depth. Right: Corresponding lapse rate, $-\partial T/\partial y$. The results are from a theoretical calculation using (15.120) and (15.118), here with $H_a = 2\,\text{km}$ and a tropopause temperature of 215 K.

If we take $H_T \approx 10\,\text{km}$ and $\partial T/\partial y \approx 40\,\text{K}/(7000\,\text{km})$ we obtain not-unreasonable values for Γ of around $4\,\text{K}\,\text{km}^{-1}$.

If we suppose that both (15.117) and (15.118) are true we obtain predictions for the height of the tropopause in mid-latitudes and the tropospheric stratification, as a function of the horizontal temperature gradient and the radiative properties of the atmosphere. In doing so we are saying the following, in rather general terms. *The height of the tropopause is the height to which baroclinic eddies extend, and it is also the height demanded by radiative balance.* The simultaneous satisfaction of these two conditions provides predictions for both the tropopause height and the lapse rate. The theory is not complete, because the lateral temperature gradient is not predicted, and this is a function of the efficiency of baroclinic eddies. It is also the case that we have chosen a particular closure for baroclinic eddies that may not have general applicability. If we had a better theory we would use it instead, and the reader is invited to explore other options.

15.6.1 Some Calculations

If we combine (15.118) and (15.117) we obtain

$$8\Gamma_d H_T^2 - H_T\left[C T_T - \left(\frac{8 f \partial_y T}{\beta}\right)\right] - \tau_s H_a T_T = 0, \tag{15.119}$$

the solution of which is

$$H_T = \frac{1}{16\Gamma_d}\left(A + \sqrt{A^2 + 32\Gamma_d \tau_s H_a T_T}\right), \tag{15.120}$$

where $A = C T_T - 8 f \partial_y T/\beta$. The solutions are plotted in Fig. 15.26, but it may help interpret these if we consider the optically thin and thick cases, namely

$$\text{Optically thick:} \quad \tau_s H_a \gg \frac{A^2}{32\Gamma_d T_T} \quad \text{whence} \quad H_T \approx \sqrt{\frac{T_T \tau_s H_a}{8\Gamma_d}}, \tag{15.121}$$

$$\text{Optically thin:} \quad \tau_s H_a \ll \frac{A^2}{32\Gamma_d T_T} \quad \text{whence} \quad H_T \approx \frac{C T_T - 8\partial_y T f/\beta}{8\Gamma_d}. \tag{15.122}$$

The optically thin case shows the effects of baroclinic eddies. If the horizontal heat transport is very strong and $\partial T/\partial y$ is small then the lapse rate will approach the dry adiabat and the tropopause height will diminish, and conversely for a weak heat transport. Solutions to the full problem show that behaviour, as well as a slight increase in height with optical depth, because with bigger optical depth the effective emitting level is higher, and the tropopause height tracks the emitting level.

The solutions are qualitatively reasonable, but we repeat some caveats. On the radiative side the atmosphere is not grey, although that is a quantitative rather than qualitative deficiency. On the dynamical side we have used an ansatz — that the isentropic slopes are fixed — that we believe not to be exactly true and even this is not a complete treatment of baroclinic activity, since it does not give us the horizontal temperature gradient. Notwithstanding these points, the calculation does capture a fundamental principle, that the tropopause height and stratification are determined by the joint effects of dynamics and radiation. Finally, the stratosphere is quite stably stratified and this strongly inhibits baroclinic instability. Thus, once formed by the mechanisms above, the tropopause provides a natural lid on baroclinic instability, keeping it tropospherically confined.

APPENDIX A: TEM FOR THE PRIMITIVE EQUATIONS IN SPHERICAL COORDINATES

In spherical and log-pressure coordinates let us define the residual streamfunction for the ideal-gas primitive equations by[12]

$$\psi^* \equiv \psi + \frac{\overline{v'\theta'}}{\partial_Z \overline{\theta}}. \tag{15.123}$$

Here, an overbar denotes a conventional (Eulerian) zonal average, ψ is the streamfunction of the zonally-averaged flow, $Z = -H \ln(p/p_R)$ where p_R is a reference pressure and H is a scale height, and $\rho_R = \rho_0 \exp(-Z/H)$ where ρ_0 is a constant. The associated transformed, or residual, velocities are:

$$\overline{v}^* = -\frac{1}{\rho_R}\frac{\partial}{\partial Z}(\psi^* \rho_R), \qquad \overline{w}^* = \frac{1}{a\cos\vartheta}\frac{\partial}{\partial\vartheta}(\psi^* \cos\vartheta), \tag{15.124}$$

with an equivalent expression for $\overline{v}$ and $\overline{w}$ in terms of ψ. (The notation for log-pressure coordinates follows Section 2.6.3 on page 82, except here we use a lowercase w for the vertical velocity.) If we write the equations of motion in terms of the residual velocities instead of the Eulerian velocities we obtain the 'transformed Eulerian mean', or TEM, equations (section 10.3). The TEM forms of the zonally-averaged thermodynamic and zonal momentum equations are:

$$\frac{\partial\overline{\theta}}{\partial t} + \frac{\overline{v}^*}{a}\frac{\partial\overline{\theta}}{\partial\vartheta} + \overline{w}^*\frac{\partial\overline{\theta}}{\partial Z} = \frac{1}{\rho_R}\frac{\partial G}{\partial z}, \tag{15.125a}$$

$$\frac{\partial\overline{u}}{\partial t} + \overline{v}^*\left(\frac{1}{a\cos\vartheta}\frac{\partial}{\partial\vartheta}(\overline{u}\cos\vartheta) - f\right) + \overline{w}^*\frac{\partial\overline{u}}{\partial Z} = \frac{1}{\rho_R\cos\vartheta}\nabla\cdot\boldsymbol{\mathcal{F}}. \tag{15.125b}$$

The transformed equations of motion are completed by the meridional momentum, mass continuity and hydrostatic equations:

$$\overline{u}\left(f + \frac{\overline{u}}{a}\tan\vartheta\right) = -\frac{1}{a}\frac{\partial\Phi}{\partial\vartheta} + \widehat{S}, \tag{15.126a}$$

$$\frac{1}{a\cos\vartheta}\frac{\partial}{\partial\vartheta}(\overline{v}^*\cos\vartheta) + \frac{1}{\rho_R}\frac{\partial}{\partial Z}(\rho_R\overline{w}^*) = 0, \tag{15.126b}$$

$$\frac{\partial\Phi}{\partial Z} = \frac{R\overline{T}}{H}T = \frac{R}{H\overline{\theta}}e^{-\kappa Z/H}. \tag{15.126c}$$

In (15.125b), $\boldsymbol{\mathcal{F}} = (\mathcal{F}^\vartheta, \mathcal{F}^Z)$ is the Eliassen–Palm flux, given by

$$\mathcal{F}^\vartheta = \rho_R \cos\vartheta \left[\overline{u}_Z \frac{\overline{v'\theta'}}{\partial_Z \overline{\theta}} - \overline{u'v'} \right], \tag{15.127a}$$

$$\mathcal{F}^Z = \rho_R \cos\vartheta \left[\left(f - \frac{\partial_\vartheta(\overline{u}\cos\vartheta)}{a\cos\vartheta} \right) \frac{\overline{v'\theta'}}{\partial_Z \overline{\theta}} - \overline{u'w'} \right], \tag{15.127b}$$

with

$$\nabla\cdot\boldsymbol{\mathcal{F}} = \frac{1}{a\cos\vartheta}\frac{\partial}{\partial\vartheta}(\mathcal{F}^\vartheta\cos\vartheta) + \frac{\partial}{\partial Z}\mathcal{F}^Z. \tag{15.127c}$$

In (15.125a)

$$G = \frac{\rho_R}{\partial_Z\overline{\theta}} \left(\overline{v'\theta'} \frac{1}{a}\frac{\partial\overline{\theta}}{\partial\vartheta} + \overline{w'\theta'}\frac{\partial\overline{\theta}}{\partial Z} \right), \tag{15.127d}$$

and $\widehat{S}$ in (15.126a) contains various, generally small, terms that lead to departures from gradient-wind balance between $\overline{u}$ and the geopotential Φ. Expressions very similar to the ones above also arise in pressure coordinates.

In many circumstances, the EP flux is well approximated by

$$\boldsymbol{\mathcal{F}} = \left(-\rho_R\cos\vartheta\,\overline{u'v'},\, f\rho_R\cos\vartheta\,\frac{\overline{v'\theta'}}{\partial_Z\overline{\theta}} \right), \tag{15.128}$$

in which case the zonal flow is accelerated by the EP flux according to

$$\frac{\partial\overline{u}}{\partial t} + \cdots = \frac{1}{a\cos^2\vartheta}\frac{\partial}{\partial\vartheta}\left(-\overline{u'v'}\,\cos^2\vartheta\right) + \frac{1}{\rho_R}\frac{\partial}{\partial Z}\left(\rho_R f\frac{\overline{v'\theta'}}{\partial_Z\overline{\theta}}\right). \tag{15.129}$$

With $f = f_0$ (15.128) becomes the quasi-geostrophic EP flux, and in this limit G is also neglected.

In the figures that show the EP vectors, the horizontal and vertical components of the EP flux are scaled by a (the Earth's radius) and by $H = 1000\,\text{hPa}$ (the pressure depth of the atmosphere), respectively. The scaling determines the direction of the arrows and makes it possible to see the divergence by eye, and which component dominates in producing that divergence. In the figures that show the EP flux divergence, we plot the right-hand side of (15.125b), namely the EP flux divergence divided by $\rho_R\cos\vartheta$, this being the quantity that directly contributes to the acceleration of the zonal flow.[13]

Notes

1 The modern view of the mid-latitude general circulation — of a largely zonally-asymmetric motion that provides the bulk of the meridional transport of heat and momentum in the extratropics — began to take form in the 1920s in papers by Defant (1921) and Jeffreys (1926). Defant regarded the mid-latitude circulation as turbulence on a large scale (albeit without realizing the important organizing effects of waves), and calculated the horizontal eddy-diffusivities using Prandtl-like mixing length arguments. Soon after, Jeffreys presciently wrote of 'the dynamical necessity for a continual exchange of air between high and low latitude' and that 'no general circulation of the atmosphere without cyclones is dynamically possible when friction is taken into account.' This point of view slowly gained ground, with, for example, Starr (1948) advocating the point of view that large-scale eddies were responsible for the bulk of the meridional transport of momentum in mid-latitudes, and Rossby (1949) eventually noting in a review article that 'One is forced to conclude that there no

longer exists a compelling reason to build the theory of the maintenance of the general circulation exclusively on *meridional* solenoidal circulations'.

Following this work came a pair of discussion papers by Eady (1950, 1954), that, setting the stage for the modern viewpoint, struggle with the turbulent transport of mid-latitude eddies and the maintenance of the surface currents — the importance of the enstrophy budget is discussed, for example, and Eady comes close to deriving wave activity conservation. Around that time Kuo (1951) discussed the maintenance of zonal flows by the mechanism of vorticity transfer in a state with a meridional background gradient, similar to mechanism I of Section 15.1.2. Another milestone was the influential monograph by Lorenz (1967) that synthesized the progress to that date, noting (in his last paragraph) that the cause of the poleward eddy momentum transport across mid-latitudes (and hence the cause of the surface eastward winds) had not at that time been rigorously explained. If perhaps not rigorous, we do now have a qualitative explanation of these dynamics by way of potential vorticity dynamics and the momentum transport in Rossby waves, as described in this chapter and in Chapter 10.

2 The mechanism producing a westerly jet in the atmosphere, and the associated surface westerlies, used to be referred to as 'negative viscosity' (Starr 1968), because it is associated with an upgradient transfer of momentum. The generation of a zonal flow by rearrangements of vorticity on a background state with a meridional gradient was noted by Kuo (1951). Dickinson (1969) and Thompson (1971, 1980) calculated the momentum transport by Rossby waves, with Thompson explicitly noting that the zonal momentum flux was in the opposite direction to the group velocity. There is thus a potential for upgradient transfer and mean flow generation, as experimentally verified by Whitehead (1975). These ideas were developed further by Green (1970), Rhines & Holland (1979) and others since.

3 This technique is noted by Lighthill (1965), who remarks that the idea goes back to Rayleigh.

4 Following Held (2000).

5 Models of the general circulation of this ilk were introduced by Green (1970). Dickinson (1969) also considered the potential vorticity transport in planetary waves.

6 Adapted from Juckes (2001).

7 Early evidence that the temperature increases above about 11 km came from the balloon measurements of Tesserenc De Bort (1902), who also suggested the names tropopause and stratosphere, and Assmann (1902). See Hoinka (1997) for a historical account. More recently, radiative and dynamical issues relevant to this topic are discussed by, among others, Stone (1972), Held (1982), Juckes (2000) and some of the articles in Schneider & Sobel (2007), and we draw on many of them.

8 Paraphrasing World Meteorological Organization (1957); see also Lewis (1991).

9 Thomas Birner calculated the tropopause-based averages. See also Birner *et al.* (2002) and Birner (2006).

10 In some but not all circumstances it seems that rotating fluids do seek to become marginally supercritical, in some sense, to baroclinic instability. The original suggestion ('baroclinic adjustment') was due to Stone (1978), and Stone & Nemet (1996) found that the isentropic slope of the real atmosphere does not vary strongly with season, even though the heat flux does, a result supportive of baroclinic adjustment ideas. Related to this (although their interpretation and reasoning were different) Schneider & Walker (2006) found that an idealized model atmosphere was only marginally supercritical over a broad parameter regime. However, other simulations have found examples of supercritical flows. Salmon (1980) and Vallis (1988b) found that quasi-geostrophic flow could be strongly supercritical, and Thuburn & Craig (1997) and Zurita-Gotor & Vallis (2009, 2011) found results that were not supportive of baroclinic adjustment in primitive equation models. In Earth's atmosphere linear calculations show that the mean atmospheric state, certainly in winter, is baroclinically unstable, with growth rates of about 0.2 day^{-1} or more (Valdes & Hoskins 1988), and the Eady growth rate is similarly positive over a large fraction of the mid-latitudes. See Zurita-Gotor & Lindzen (2007) for a review of some of these ideas. Jansen & Ferrari (2012, 2013) discussed and modelled the reasons for the various differences, and concluded that marginal criticality is not a

general property of a rotating stratified system. It is likely that much of Earth's ocean is supercritical. Nevertheless, the idea is one we use in Section 15.6.

11 Following Juckes (2000).

12 For more detail see Edmon *et al.* (1980) or Andrews (1987).

13 E. Gerber kindly constructed these figures.

Place me on Sunium's marbled steep,
Where nothing, save the waves and I,
May hear our mutual murmurs sweep.
George Gordon Byron (Lord Byron), *The Isles of Greece*, 1820.

CHAPTER 16

Planetary Waves and Zonal Asymmetries

PLANETARY WAVES ARE LARGE-SCALE ROSSBY WAVES in which the potential vorticity gradient is provided by differential rotation (i.e., the beta-effect). They are ubiquitous in Earth's atmosphere and almost certainly in other planetary atmospheres. They propagate horizontally over the two Poles, and they propagate vertically into the stratosphere and beyond. In the previous chapter we saw that it is the propagation of Rossby waves away from their mid-latitude source that gives rise to the mean eastward eddy-driven jet. In this chapter we will see that the dynamics of such waves also largely determines the large-scale *zonally asymmetric* circulation of the mid-latitude atmosphere. In the first few sections we discuss the properties and propagation of planetary waves themselves, and in many ways these sections are a continuation of Chapter 6. We then look more specifically at planetary waves forced by surface variations in topography and thermal properties, for it is these waves that give rise to the zonally asymmetric circulation.

In proceeding this way we are dividing our task of constructing a theory of the general circulation of the extratropical atmosphere into two. The first task (Chapters 14 and 15) was to understand the zonally averaged circulation and the transient zonal asymmetries by supposing that, to a first approximation, this circulation is qualitatively the same as it would be if the boundary conditions were zonally symmetric, with no mountains or land–sea contrasts. Given the statistically zonally symmetric circulation, the second task is to understand the zonally asymmetric circulation. We may do this by supposing that the latter is a perturbation on the former, and using a theory linearized about the zonally symmetric state. It is by no means obvious that such a procedure will be successful, for it depends on the nonlinear interactions among the zonal asymmetries being weak. We might make some a priori estimates that suggest that this might be the case, but the ultimate justification for the approach lies in its a posteriori success. In our discussion of stationary waves we will focus first on the response to orography at the lower boundary, and then consider thermodynamic forcing — arising, for example, from an inhomogeneous surface temperature field. Our focus throughout this chapter is the mid-latitudes.

16.1 ROSSBY WAVE PROPAGATION IN A SLOWLY VARYING MEDIUM

In Chapters 6 and 7 we looked at wave propagation using linearized equations of motion. We now focus and extend this discussion by looking at Rossby wave propagation in a medium in which the parameters (such as the zonal wind and the stratification) vary spatially — as occurs in the real atmosphere. If the parameters do vary then waves may propagate into a region in which they amplify, perhaps violating the initial assumption of linearity, so let us first look at what the conditions for linearity are.[1]

16.1.1 Linear Dynamics

If the linear equations are to be an accurate representation of the dynamics then the perturbation quantities need to be small compared to the background state, or at least the nonlinear terms must be small. In reality this is not always the case and indeed it may be that in course of propagation the waves amplify and may even *break.* Wave breaking is familiar to anyone who has been to the beach and watched water waves move toward the shore and crash in the 'surf zone' as the mean depth becomes too shallow to support laminar surface waves. Manifestly, the linear approximation breaks down at this point. More generally, wave breaking simply refers to an irreversible deformation of material surfaces, generally leading to dissipation. Since Rossby waves generally grow in amplitude as they propagate up (because density falls) we can expect Rossby wave breaking to occur somewhere in the atmosphere, but waves can also break as they propagate laterally, if and when they grow in size to such an extent that the nonlinear terms in the equations of motion become important.

To examine this consider the quasi-geostrophic potential vorticity equation,

$$\left(\frac{\partial}{\partial t}+\boldsymbol{u}\cdot\nabla\right)q=0,\qquad q=\beta y+\nabla^2\psi'+\frac{f_0^2}{\rho_R}\frac{\partial}{\partial z}\left(\frac{\rho_R}{N^2}\frac{\partial\psi}{\partial z}\right). \tag{16.1a,b}$$

The derivation of this equation was given in Chapter 5 and all the terms are defined there. In brief, q is the quasi-geostrophic potential vorticity and ψ the streamfunction, f_0 is the Coriolis parameter and ρ_R is a density profile, a function of z only. Breaking the above equation up into mean and perturbation quantities in the usual way we obtain

$$\left(\frac{\partial}{\partial t}+\overline{u}(y,z)\frac{\partial}{\partial x}\right)q'+v'\frac{\partial\overline{q}}{\partial y}=-\left(\frac{\partial u'q'}{\partial x}+\frac{\partial v'q'}{\partial y}\right). \tag{16.2}$$

In the linear approximation we neglect the terms on the right-hand side and, seeking wave-like solutions of the form $\psi=F(x-ct)$, we obtain

$$(\overline{u}-c)\frac{\partial q'}{\partial x}+v'\frac{\partial\overline{q}}{\partial y}=0. \tag{16.3}$$

For the linear approximation to be valid the terms in this equation must be larger than the nonlinear terms in (16.2), and this will be the case if

$$|\overline{u}-c|\gg|u'|\qquad\text{and}\qquad\left|\frac{\partial\overline{q}}{\partial y}\right|\gg\left|\frac{\partial q'}{\partial y}\right|. \tag{16.4a,b}$$

Although it is common to only treat the case in which $\overline{u}$ is a constant, we may also consider the case in which $\overline{u}$ varies slowly, either in latitude or height or both, and (16.3) then approximately holds locally. But if a wave propagates into a region in which $\overline{u}=c$ then the linear criterion *must* break down. Regions where $\overline{u}=c$ are called *critical lines, critical surfaces, critical heights* or *critical latitudes,* depending on context, and in many circumstances a *critical layer* of finite width will surround the critical line, in which frictional and/or nonlinear effects are important. The location of a critical surface does not depend on the frame of reference used to measure the velocities.

For reference we first write down a few results for the simplest case when $\partial\overline{q}/\partial y,\overline{u},N^2$ and ρ_R are all constant, referring to Section 6.5 as needed. We look for solutions of the form

$$\psi'=\mathrm{Re}\,\widetilde{\psi}e^{i(kx+ly+mz-\omega t)}, \tag{16.5}$$

and obtain the dispersion relation

$$\omega = \overline{u}k - \frac{k\beta}{k^2 + l^2 + Pr^2 m^2}, \tag{16.6}$$

where $Pr = f_0/N$ is the Prandtl ratio. The components of the group velocity are given by

$$c_g^x = \overline{u} + \frac{(k^2 - l^2 - Pr^2 m^2)\beta}{(k^2 + l^2 + Pr^2 m^2)^2}, \qquad c_g^y = \frac{2kl\beta}{(k^2 + l^2 + Pr^2 m^2)^2}, \qquad c_g^z = \frac{2kmPr^2\beta}{(k^2 + l^2 + Pr^2 m^2)^2}. \tag{16.7a,b,c}$$

16.1.2 Conditions for Wave Propagation

Suppose that the zonal wind varies slowly with latitude and height, but that, for simplicity, the density, ρ_R, is a constant. The equation of motion is

$$\left(\frac{\partial}{\partial t} + \overline{u}(y,z)\frac{\partial}{\partial x}\right)q' + v'\frac{\partial \overline{q}}{\partial y} = 0. \tag{16.8}$$

Because the coefficients of the equation are not constant we cannot assume harmonic solutions in the y and z directions; rather, we seek solutions of the form

$$\psi' = \widetilde{\psi}(y,z)\mathrm{e}^{\mathrm{i}k(x-ct)}. \tag{16.9}$$

If the parameters in (16.8) are varying slowly compared to the wavelength of the waves then a dispersion relation still exists (as discussed in Section 6.3), but the relation will be of the form $\omega = \Omega(\boldsymbol{k}; \boldsymbol{x}, t)$; where the function Ω varies slowly in space. Now, if the medium is not an explicit function of x or of time the x-wavenumber and the frequency will be a constant, and hence c is constant too, and we can use the dispersion relation to find what are effectively the other wavenumbers in the problem. Using (16.9) in (16.8) we find (with N^2 constant)

$$\frac{\partial^2 \widetilde{\psi}}{\partial y^2} + \frac{f_0^2}{N^2}\frac{\partial^2 \widetilde{\psi}}{\partial z^2} + n^2(y,z)\widetilde{\psi} = 0, \qquad \text{where} \qquad n^2(y,z) = \frac{\partial \overline{q}/\partial y}{\overline{u} - c} - k^2. \tag{16.10a,b}$$

Equation (16.10a) is similar to the Rayleigh or Rayleigh–Kuo equation encountered in Chapter 9, but now c is given and is not an eigenvalue; rather, the frequency is known and the dispersion relation gives the quantity n. The quantity n is the *refractive index* and it greatly affects how the waves propagate: solutions are wavelike when n^2 is positive and evanescent when n^2 is negative. To see this in a simple case, suppose there is no z-variation so that $\partial^2\widetilde{\psi}/\partial y^2 + n^2\widetilde{\psi} = 0$, whereas if n is constant and real we have harmonic solutions in the y-direction of the form $\exp(\mathrm{i}ny)$. If $n^2 < 0$ the solutions will evanesce. Waves tend to propagate toward regions of large n^2 and turn away from regions of negative n^2, as we will see in the examples to follow.

The value of n^2 will become very large if and as $\overline{u}$ approaches c from above and the waves, being very short, will tend to break. If $\overline{u}$ continues to diminish and becomes smaller than c then n^2 switches from being large and positive to large and negative. If n^2 diminishes because $\partial\overline{q}/\partial y$ diminishes then it will transition smoothly to a negative value. The location where $\overline{u} = c$ is called a critical surface (or line). The location where n^2 passes through zero is called a turning surface (or line).

The bounds on n^2 can be translated into bounds on the zonal phase speed c. Given a zonal wind $\overline{u}$, wave propagation requires that c is bounded by

$$\overline{u} - \frac{\partial \overline{q}/\partial y}{k^2 + \gamma^2} < c < \overline{u}. \tag{16.11}$$

At the upper bound (a critical surface) the wavelength is small and wave breaking is likely to occur. At the lower bound (a turning surface) the refractive index tends to zero and the wavelength tends to infinity. Waves will tend to propagate away from regions with a small n and be refracted toward regions of large n. The bounds can also be expressed in terms of the zonal velocity:

$$0 < \overline{u} - c < \frac{\partial \overline{q}/\partial y}{k^2 + \gamma^2}. \tag{16.12}$$

This form is useful when considering a situation in which the wave speed is given, for example by boundary conditions; Equation (16.12) then tells us under what configurations of zonal velocity wave propagation can occur. The lower bound corresponds to a critical surface and the upper bound to a turning surface.

It is algebraically complicated to continue our analysis in the three-dimensional case, so let us consider the cases in which the inhomogeneities in the medium occur separately in the horizontal and vertical. A summary of some key concepts is provided on page 590.

16.2 HORIZONTAL PROPAGATION OF ROSSBY WAVES

Consider the purely horizontal problem for which the linearized equation of motion is

$$\left(\frac{\partial}{\partial t} + \overline{u}(y)\frac{\partial}{\partial x}\right) q' + v' \frac{\partial \overline{q}}{\partial y} = 0, \tag{16.13}$$

where $q' = \nabla^2 \psi'$, $v' = \partial \psi'/\partial x$ and $\partial \overline{q}/\partial y = \beta - \overline{u}_{yy}$, which we will denote β^*. If $\overline{u}$ and β^* do not vary in space then we may obtain wavelike solutions in the usual way and obtain the dispersion relation

$$\omega \equiv ck = \overline{u}k - \frac{\partial \overline{q}/\partial y}{k^2 + l^2}, \tag{16.14}$$

where k and l are the x- and y-wavenumbers.

If the parameters do vary in the y-direction then we seek a solution $\psi' = \widetilde{\psi}(y)\exp[ik(x - ct)]$ and obtain, analogous to (16.10),

$$\frac{\partial^2 \widetilde{\psi}}{\partial y^2} + l^2(y)\widetilde{\psi} = 0, \quad \text{where} \quad l^2(y) = \frac{\beta^*}{\overline{u} - c} - k^2. \tag{16.15a,b}$$

If the parameter variation is sufficiently small, occurring on a spatial scale longer than the wavelength of the waves, then we may expect that the disturbance will propagate locally as a plane wave. The solution is then of WKB form (see Appendix A to Chapter 6), namely

$$\widetilde{\psi}(y) = A_0 l^{-1/2} \exp\left(\mathrm{i}\int l \, \mathrm{d}y\right), \tag{16.16}$$

where A_0 is a constant. The phase of the wave in the y-direction, θ, is evidently given by $\theta = \int l \, \mathrm{d}y$, so that the local wavenumber is given by $\mathrm{d}\theta/\mathrm{d}y = l$. The group velocity is calculated in the normal way using the dispersion relation (16.11) and we obtain

$$c_g^x = \overline{u} + \frac{(k^2 - l^2)\beta^*}{(k^2 + l^2)^2}, \qquad c_g^y = \frac{2kl\,\beta^*}{(k^2 + l^2)^2}, \tag{16.17a,b}$$

where l is given by (16.15b), with both quantities varying slowly in the y-direction.

16.2.1 Wave Amplitude

As a Rossby wave propagates its amplitude is not necessarily constant because, in the presence of a shear, the wave may exchange energy with the background state, and the WKB solution, (16.15), tells us that the variation goes like $l^{-1/2}(y)$. This variation can be understood from somewhat more general considerations. As discussed in Chapter 10 (specifically Section 10.2.1) an inviscid, adiabatic wave will conserve its wave activity, and specifically its pseudomomentum, meaning that

$$\frac{\partial \mathcal{P}}{\partial t} + \nabla \cdot \boldsymbol{\mathcal{F}} = 0, \tag{16.18}$$

where $\mathcal{P}$ is quadratic in the wave amplitude and $\boldsymbol{\mathcal{F}}$ is the flux of $\mathcal{P}$, and the two are related by the group velocity property $\boldsymbol{\mathcal{F}} = \mathbf{c}_g \mathcal{P}$. In the zonally-averaged case the pseudomomentum and flux for the stratified quasi-geostrophic equations are given by

$$\mathcal{P} = \frac{\overline{q'^2}}{2\beta^*}, \qquad \boldsymbol{\mathcal{F}} = -\overline{u'v'}\,\mathbf{j} + \frac{f_0}{N^2}\overline{v'b'}\,\mathbf{k}, \tag{16.19}$$

with $\boldsymbol{\mathcal{F}}$ being the Eliassen–Palm (EP) flux. If the waves are steady then $\nabla \cdot \boldsymbol{\mathcal{F}} = 0$, and in the two-dimensional case under consideration $b' = 0$ and $\partial \overline{u'v'}/\partial y = 0$. Thus $u'v' = kl|\widetilde{\psi}|^2 =$ constant, and since k is constant along a ray the amplitude of a wave varies like

$$|\widetilde{\psi}| = \frac{A_0}{\sqrt{l(y)}}, \tag{16.20}$$

as in the WKB solution. The energy of the wave then varies like

$$\text{Energy} = (k^2 + l^2)\frac{A_0^2}{l}. \tag{16.21}$$

16.2.2 Two Examples

To illustrate the above ideas in a concrete fashion we consider two examples, one with a turning line and one with a critical line. Very close to the turning line and critical line more detailed analysis is needed to obtain a complete solution, but we can obtain a sense of the behaviour with an elementary treatment.

Waves with a turning latitude

A turning line arises where $l = 0$ and it corresponds to the lower bound of c in (16.11). The line arises if the potential vorticity gradient diminishes to such an extent that $l^2 < 0$ and the waves then cease to propagate in the y-direction. This may happen even in unsheared flow as a wave propagates polewards and the magnitude of beta diminishes.

As a wave packet approaches a turning latitude then n goes to zero so the amplitude and the energy of the wave approach infinity. However, the wave will never reach the turning latitude because the meridional component of the group velocity is zero, as can be seen from the expressions for the group velocity, (16.17). As a wave approaches the turning latitude $c_g^x \to (\beta - \overline{u}_{yy})/k^2$ and $c_g^y \to 0$, so the group velocity is purely zonal and indeed, as $l \to 0$,

$$\frac{c_g^x - \overline{u}}{c_g^y} = \frac{k}{2l} \to \infty. \tag{16.22}$$

Because the meridional wavenumber is small the wavelength is large, so we do not expect the waves to break. Rather, we intuitively expect that a wave packet will turn — hence the eponym 'turning latitude' — and be reflected.

Rossby Wave Propagation in a Slowly Varying Medium

The linear equation of motion is, in terms of streamfunction,

$$\left(\frac{\partial}{\partial t}+\overline{u}(y,z)\frac{\partial}{\partial x}\right)\left[\nabla^2\psi'+\frac{f_0^2}{\rho_R}\frac{\partial}{\partial z}\left(\frac{\rho_R}{N^2}\frac{\partial\psi'}{\partial z}\right)\right]+\frac{\partial\psi'}{\partial x}\frac{\partial\overline{q}}{\partial y}=0. \qquad \text{(RP.1)}$$

We suppose that the parameters of the problem vary slowly in y and/or z but are uniform in x and t. The frequency and zonal wavenumber are therefore constant. We seek solutions of the form $\psi'=\widetilde{\psi}(y,z)e^{ik(x-ct)}$ and find (if, for simplicity, N^2 and ρ_R are constant)

$$\frac{\partial^2\widetilde{\psi}}{\partial y^2}+\frac{f_0^2}{N^2}\frac{\partial^2\widetilde{\psi}}{\partial z^2}+n^2(y,z)\widetilde{\psi}=0, \qquad \text{where} \qquad n^2(y,z)=\frac{\partial\overline{q}/\partial y}{\overline{u}-c}-k^2. \qquad \text{(RP.2)}$$

The value of n^2 must be positive in order that waves can propagate, and so waves cease to propagate when they encounter either

(i) *A turning surface,* where $n^2=0$, or
(ii) *A critical surface,* where $\overline{u}=c$ and n^2 becomes infinite.

For a given wave speed, the location of the turning surface, but not that of the critical surface, depends on wavenumber. The condition for wave propagation may be expressed as bounds on the zonal flow, to wit

$$0<\overline{u}-c<\frac{\partial\overline{q}/\partial y}{k^2}. \qquad \text{(RP.3)}$$

If the length scale over which the parameters of the problem vary is much longer than the wavelengths themselves we can expect the solution to look locally like a plane wave and a WKB analysis can be employed. In the purely horizontal problem we assume a solution of the form $\psi'=\widetilde{\psi}(y)e^{ik(x-ct)}$ and find

$$\frac{\partial^2\widetilde{\psi}}{\partial y^2}+l^2(y)\widetilde{\psi}=0, \qquad l^2(y)=\frac{\partial\overline{q}/\partial y}{\overline{u}-c}-k^2. \qquad \text{(RP.4)}$$

The WKB solution is of the form

$$\widetilde{\psi}(y)=Al^{-1/2}\exp\left(\pm i\int l\,dy\right). \qquad \text{(RP.5)}$$

Thus, $l(y)$ is the local y-wavenumber, and the amplitude of the solution varies like $l^{-1/2}$. However, the WKB condition fails at both a critical line and a turning line.

Approaching a critical line the amplitude of the wave diminishes (in the WKB approximation) because l is large. In the critical layer wave amplitude is in fact nearly constant but the vorticity becomes very large, and either nonlinearity and/or dissipation become important, and since wavelength is small the waves may break. At a turning line the amplitude and energy will both be large, but since the wavelength is long the waves will not necessarily break; rather, they are reflected.

A similar analysis may be employed for vertically propagating Rossby waves, and either a turning level or a critical level will prevent the upward propagation of waves into the stratosphere — this is the 'Charney–Drazin' condition, discussed in Section 16.5.3.

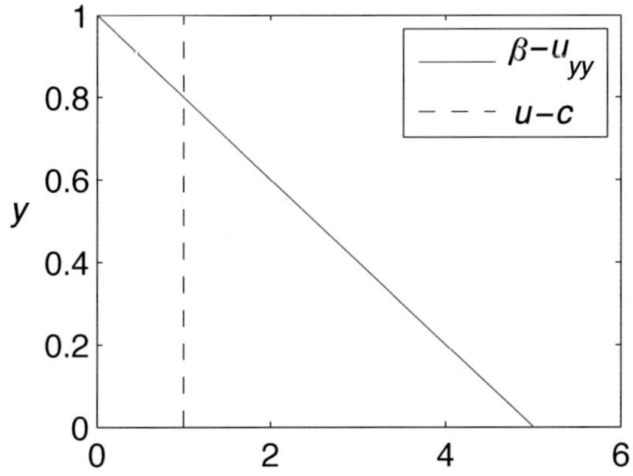

Fig. 16.1 Parameters for the first example considered in Section 16.2.2, with all variables nondimensional. The zonal flow is uniform with $u = 1$ and $c = 0$ (so that $\overline{u}_{yy} = 0$) and β diminishes linearly as y increases polewards as shown. With zonal wavenumber $k = 1$ there is a turning latitude at $y = 0.8$, and the wave properties are illustrated in Fig. 16.2.

To illustrate this, consider waves propagating in a background state that has no horizontal shear but with a beta effect that diminishes polewards. To be concrete suppose that $\beta = 5$ at $y = 0$, diminishing linearly to $\beta = 0$ at $y = 0$, and that $\overline{u} - c = 1$ everywhere. There is no critical line but depending on the x-wavenumber there may be a turning line, and if we choose $k = 1$ then the turning line occurs when $\beta = 1$ and so at $y = 0.8$. The turning latitude depends on the value of the x-wavenumber — if the zonal wavenumber is larger then waves will turn further south. The parameters are illustrated in Fig. 16.1.

For a given zonal wavenumber ($k = 1$ in this example) the value of l^2 is computed using (16.15b), and the components of the group velocity using (16.17), and these are illustrated in Fig. 16.2. We may choose either a positive or a negative value of l, corresponding to northward or southward oriented waves, and we illustrate both in the figure. The value of l^2 becomes zero at $y = 0.8$, and this corresponds to a turning latitude. The values of the wave amplitude and energy are computed using (16.20) and (16.21) (with an arbitrary amplitude at $y = 0$) and these both become infinite at the turning latitude.

What is happening physically? We may suppose that at some location in the domain there is a source of waves — baroclinic disturbances for example. Waves propagate away from the source (since the waves must carry energy away), and this determines the sign of n of any particular wave packet. The disturbance may in general consist of many zonal wavenumbers and many frequencies (or phase speeds, c), but the dispersion relation must be satisfied for each pair and this determines the meridional wavenumber via an equation like (16.15). As the wave packet propagates away from the source then, as we noted in Section 6.3 on ray theory, if the medium is zonally symmetric the x-wavenumber, k, is preserved. If the medium is not time-varying then the frequency, and therefore the wave speed c, are also preserved. We may approximately construct a ray by following the arrows in Fig. 16.2, and we see that a ray propagating polewards will bend eastward as it approaches the turning latitude Although its amplitude will become large it will not necessarily break because the wavelength is large; in fact, the packet may be reflected southward. We may heuristically construct a ray trajectory by drawing a line that is always parallel to the arrows marking the group velocity. Indeed, the entire procedure might be thought of as an Eulerian analogue of ray theory; rather than following a wave packet we just evaluate the field of group velocity, and if there is no explicit time dependence in the problem a ray follows the arrows.

The above argument suggests but does not demonstrate reflection at the turning latitude. The arrows of Fig. 16.2 are all zonal at the turning line and do not actually turn back. One might imagine using a WKB analysis but the WKB approximation fails in the vicinity of a turning latitude: the meridional wavenumber l tends to zero but $\mathrm{d}l/\mathrm{d}y$ does not, and the WKB condition (6.164)

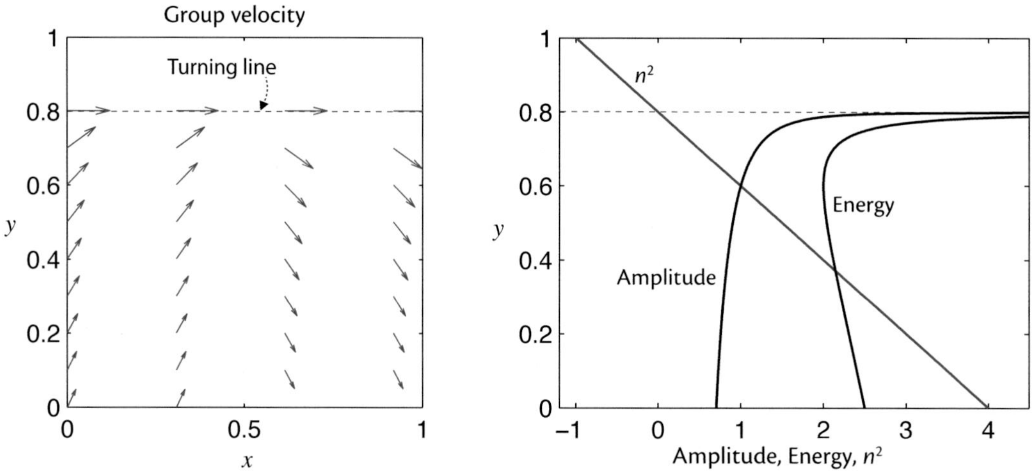

Fig. 16.2 Left: The group velocity evaluated using (16.17) for the parameters illustrated in Fig. 16.1, which give a turning latitude at $y = 0.8$. For $x < 0.5$ we choose positive values of n, and a northward group velocity, whereas for $x > 0.5$ we choose negative values of n. Right panel: Values of refractive index squared (n^2), the energy and the amplitude of a wave. n^2 is negative for $y > 0.8$. See text for more description.

cannot be satisfied (even though the wave equation itself is not singular). However, a momentum flux argument shows that the reflection is in fact perfect in the absence of dissipation. Suppose there is a wave source in mid- or low latitudes (say at $y = 0$ in Fig. 16.2) producing poleward propagating waves. Poleward of the turning line the waves evanesce and the zonally-averaged polewards momentum flux is zero at large y. However, this decay has occurred in the absence of friction, and therefore that momentum flux is zero *everywhere*. To see this with equations, away from forcing regions the inviscid barotropic pseudomomentum conservation equation, (16.18), becomes

$$\frac{\partial \mathcal{P}}{\partial t} + \overline{v'\zeta'} = 0, \qquad \text{or} \qquad \frac{\partial \mathcal{P}}{\partial t} - \frac{\partial}{\partial y}\overline{u'v'} = 0. \tag{16.23}$$

Thus, in a statistically steady state, $\overline{u'v'}$ is a constant, and that constant is zero if the flux is zero at large y. Since the forcing is producing a poleward propagating Rossby wave (with $\overline{u'v'} < 0$ in the Northern Hemisphere) there must be a reflected wave with $\overline{u'v'} > 0$, and that reflected wave must come from the vicinity of the turning line.

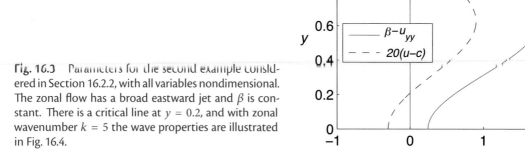

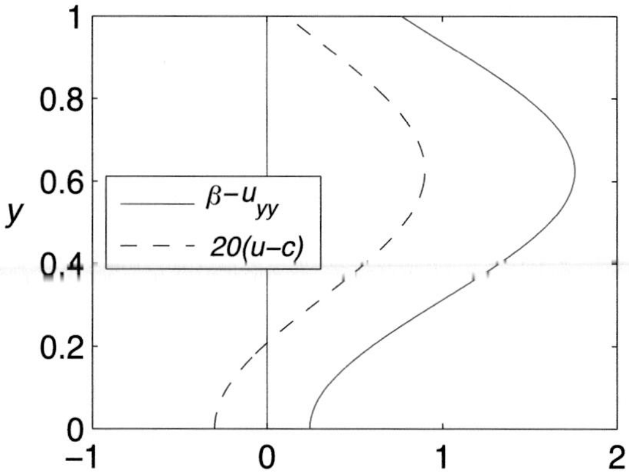

Fig. 16.3 Parameters for the second example considered in Section 16.2.2, with all variables nondimensional. The zonal flow has a broad eastward jet and β is constant. There is a critical line at $y = 0.2$, and with zonal wavenumber $k = 5$ the wave properties are illustrated in Fig. 16.4.

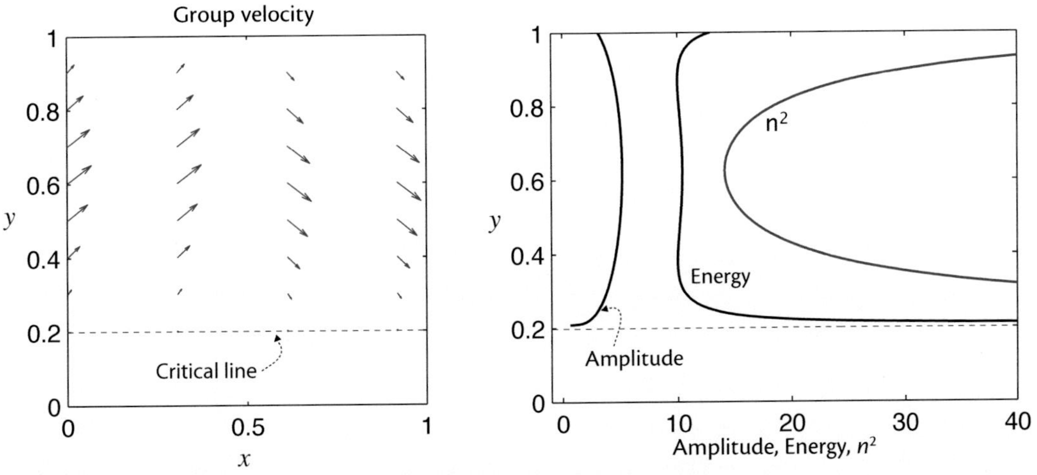

Fig. 16.4 Left: The group velocity evaluated using (16.17) for the parameters illustrated in Fig. 16.1, which give a critical line at $y = 0.2$. For $x < 0.5$ we choose positive values of n, and a northward group velocity, whereas for $x > 0.5$ we choose negative values of n. Right panel: Values of refractive index squared, the energy and the amplitude of a wave. The value of n^2 becomes infinite at the critical line and the linear theory breaks down in its vicinity. See text for more description.

Waves with a critical latitude

A critical line occurs when $\overline{u} = c$, corresponding to the upper bound of c in (16.11), and from (16.15) we see that at a critical line the meridional wavenumber approaches infinity. From (16.17) we see that both the x- and y-components of the group velocity are zero — a wave packet approaching a critical line just stops (at least according to ray theory). Specifically, as l becomes large

$$c_g^x - \overline{u} \to 0, \qquad c_g^y \to 0, \qquad \frac{c_g^x - \overline{u}}{c_g^y} \to -\frac{l}{k} \to -\infty. \tag{16.24}$$

From (16.20) the amplitude of the wave packet also approaches zero, but its energy approaches infinity. Since the wavelength is very small we expect the waves to *break* and deposit their momentum, and this situation commonly arises when Rossby waves excited in mid-latitudes propagate equatorward and encounter a critical latitude in the subtropics.

To illustrate this let us construct a background state that has an eastward jet in mid-latitudes becoming westward at low latitudes, with β constant chosen to be large enough so that $\beta - \overline{u}_{yy}$ is positive everywhere. (Specifically, we choose $\beta = 1$ and $\overline{u} = -0.03 \sin(8\pi y/5 + \pi/2) - 0.5$, but the precise form is not important.) If $c = 0$ then there is a critical line when $\overline{u}$ passes through zero, which in this example occurs at $x = 0.2$. (The value of $\overline{u} - c$ is small at $y = 1$, but no critical line is actually reached.) These parameters are illustrated in Fig. 16.3. We also choose $k = 5$, which results in a positive value for l^2 everywhere.

As in the previous example we compute the value of l^2 using (16.15b) and the components of the group velocity using (16.17), and these are illustrated in Fig. 16.4, with northward propagating waves shown for $x < 0.5$ and southward propagating waves for $x > 0.5$. The value of n^2 increases considerably at the northern and southern edges of the domain, and is actually infinite at the critical line at $y = 0.2$. Using (16.20) the amplitude of the wave diminishes as the critical line approaches, but the energy increases rapidly, suggesting that the linear approximation will break down. The waves (in the linear approximation) tend to stall before reaching the critical line, because both the x and the y components of the group velocity become very small. This is a little misleading because ray theory breaks down and a disturbance can reach the critical line, and in the region

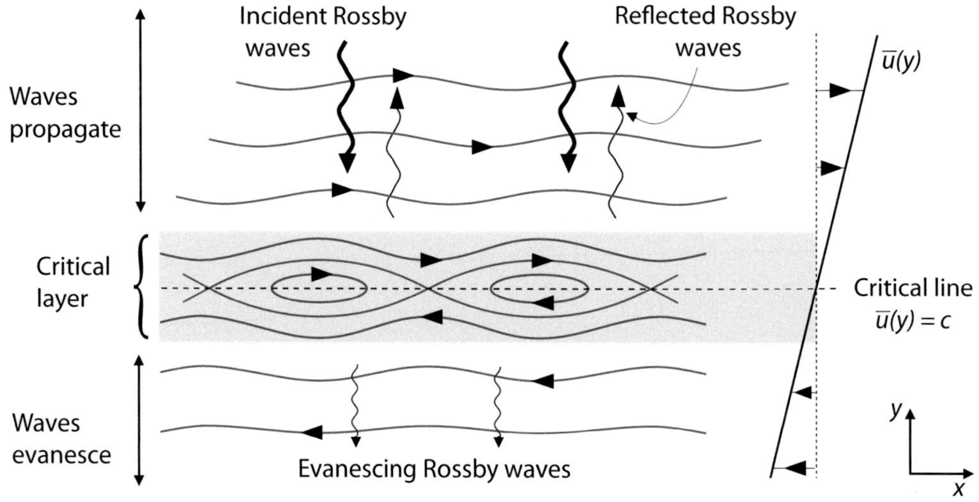

Fig. 16.5 Sketch of a Rossby-wave critical layer. Incident Rossby waves from mid-latitudes propagate in a horizontally sheared flow toward a critical line at which $\overline{u}(y) = c$. Surrounding the critical line is a critical layer, in which either nonlinear or frictional effects, or both, are important. If nonlinear effects are important then the critical layer may reflect, but there is still likely to be dissipation in the critical layer. Equatorward of the critical layer the waves evanesce.[3]

of the critical line — that is, in the critical layer — either frictional or nonlinear effects, or both, become important, as we see in the next section. The situation illustrated in this example is of particular relevance to the maintenance of the zonal wind structure in the troposphere: waves are generated in mid-latitudes and propagate equatorward, and as they approach a critical line in the subtropics they break, deposit westward momentum and retard the flow.

16.3 ✦ CRITICAL LINES AND CRITICAL LAYERS

We now look a little more closely at the behaviour of Rossby waves near a critical line and then, more briefly, at gravity wave behaviour. Critical layer theory is extensive and technical, and our discussion only scratches the surface and is mostly linear; readers wishing for more should go to the literature.[2] We use Northern Hemisphere conventions, envisioning waves propagating southward toward a subtropical critical line.

16.3.1 Preliminaries

Consider horizontally propagating Rossby waves obeying the linear barotropic vorticity equation on the beta-plane (vertically propagating waves may be considered using similar techniques). The equation of motion is

$$\left(\frac{\partial}{\partial t} + \overline{u}\frac{\partial}{\partial x}\right)\nabla^2\psi' + \beta^*\frac{\partial \psi'}{\partial x} = -r\nabla^2\psi', \tag{16.25}$$

where $\beta^* = \beta - \partial_y^2\overline{u}$. The parameter r is a drag coefficient that acts directly on the relative vorticity, and we shall assume that it is small compared to the Doppler-shifted frequency of the waves, except possibly near a critical line. It is not a particularly realistic form of dissipation but it is simple and captures the essential process. We seek solutions of the form

$$\psi'(x, y, t) = \widetilde{\psi}(y)e^{\mathrm{i}(k(x-ct))}. \tag{16.26}$$

Substituting into (16.25) we find, after a couple of lines of algebra, that $\widetilde{\psi}$ satisfies, analogously to (16.15),

$$\frac{d^2\widetilde{\psi}}{dy^2} + l^2(y)\widetilde{\psi} = 0, \quad \text{where} \quad l^2(y) = \frac{\beta^*}{\overline{u} - c - ir/k} - k^2. \tag{16.27a,b}$$

If the zonal wind has a lateral shear then β^* may vary with y, and thus so does l. If r is non-zero then l has an imaginary component so that the wave decays away from its source region, and as $\overline{u} \to c$ the decay will be particularly strong.

Suppose that the friction is zero. Near the critical line the y-wavenumber will be much larger than k and the streamfunction will obey an equation of the form

$$\frac{d^2\widetilde{\psi}}{dy^2} \approx -\frac{\beta^*}{\overline{u} - c}\widetilde{\psi} = -\frac{A}{y - y_c}\widetilde{\psi}, \tag{16.28}$$

where A is a constant, and by shifting the origin we will take $y_c = 0$. By inspection an approximate solution to this equation for small y is

$$\frac{d\widetilde{\psi}}{dy} = B \ln y, \qquad \widetilde{\psi} = B\left(y \ln y - y - \frac{1}{A}\right) \approx -\frac{B}{A}. \tag{16.29}$$

where B is a constant. (Equation (16.28) actually has Bessel function solutions.) The vorticity, $\widetilde{\psi}_{yy}$ then goes as $1/y$ and the velocity goes as $\ln y$ near the critical line. Both quantities blow up, *but the streamfunction itself does not.* The pseudomomentum also blows up as the critical layer. To see this, multiply (16.25) by ζ/β^* and zonally average to give

$$\frac{\partial \mathcal{P}}{\partial t} + \frac{\partial \mathcal{F}}{\partial y} = -\alpha \mathcal{P}, \tag{16.30}$$

where $\mathcal{P} = \overline{\zeta'^2}/2\beta^*$ is the pseudomomentum, $\partial \mathcal{F}/\partial y = -\partial_y(\overline{u'v'}) = \overline{v'\zeta'}$ is its flux divergence, and $\alpha = 2r$. If $\alpha = 0$ then $\mathcal{P}$ must blow up at the critical line because ζ does. Finally, the WKB approximation fails approaching a critical line: equations (16.27) and (16.28) tell us that

$$l^2 \sim \frac{1}{y} \qquad \text{and} \qquad \frac{dl}{dy} \sim \frac{1}{y^{3/2}}. \tag{16.31}$$

Thus, for small y, $dl/dy > l^2$ because $y^{-3/2} > y^{-1}$ for small y. The WKB condition that the wavenumber, l, varies more slowly than l^2 (requiring, as in (6.164), that $dl/dy \ll l^2$) is not satisfied.

Behaviour near a critical line

A detailed analysis is required to determine what does happen at the critical line, but in the linear problem we can make some useful headway. As in Fig. 16.5 we imagine there is a small region surrounding the critical line — the critical layer — in which either the nonlinear terms or the frictional terms, or both, are important. In the critical layer y derivatives are much larger than x derivatives so that $\zeta' \approx -\partial u'/\partial y = \partial^2\psi'/\partial y^2$. Furthermore, since $\overline{u} - c$ is small, a Taylor expansion gives

$$\overline{u} - c \approx y \left.\frac{\partial \overline{u}}{\partial y}\right|_{\overline{u}=c}, \tag{16.32}$$

and we will denote the derivative as $\partial_y \overline{u}_c$. Using (16.27) with $l^2 \gg k^2$ then gives

$$\widetilde{\zeta} = \frac{-\beta^*\widetilde{\psi}}{\overline{u} - c - ir/k} \approx \frac{-\beta^*\widetilde{\psi}}{y\partial_y\overline{u}_c - ir/k} = \frac{-\hat{\beta}\widetilde{\psi}}{y - i\hat{r}} = \frac{-\hat{\beta}(y + i\hat{r})\widetilde{\psi}}{y^2 + \hat{r}^2}, \tag{16.33}$$

where $\widehat{\beta} = \beta^*/\partial_y \overline{u}_c$ and $\hat{r} = r/(k\partial_y\overline{u}_c)$. The vorticity flux in the critical layer is given by

$$\overline{v'\zeta'} = \frac{1}{2}\mathrm{Re}\, \overline{\mathrm{i}k\widetilde{\psi}\tilde{\zeta}^*} = \frac{-k\widehat{\beta}\hat{r}}{2(y^2+\hat{r}^2)}|\widetilde{\psi}^2|. \tag{16.34}$$

The factor of 2 comes from the averaging, and only the part of $\tilde{\zeta}$ proportional to $\mathrm{i}\hat{r}$ contributes because of its phase relative to $\mathrm{i}\widetilde{\psi}$. There are two interesting aspects to (16.34):

(i) The vorticity flux is negative. (All the individual terms, including k, are positive.)

(ii) The stream function is almost constant in the critical layer, so that for small friction (16.34) is sharply peaked around $y = 0$. It tends to a delta-function as $\hat{r} \to 0$, because

$$\delta(y) = \frac{1}{\pi}\lim_{\hat{r}\to 0}\frac{\hat{r}}{y^2+\hat{r}^2}. \tag{16.35}$$

We can see that the thickness of the critical layer, δ_r is just $\hat{r}$.

The first property above tells us that the critical layer is dissipative. The second property tells us that the eddy momentum flux has a finite jump across the critical line. For small r we have

$$\left[\overline{u'v'}\right]_-^+ = -\int_-^+ \overline{\zeta'v'}\,\mathrm{d}y = \int_-^+ \frac{-k\widehat{\beta}\hat{r}}{2(y^2+\hat{r}^2)}|\widetilde{\psi}^2|\,\mathrm{d}y = -\int_-^+ \frac{1}{2}k\widehat{\beta}\pi|\widetilde{\psi}^2|\delta(y)\,\mathrm{d}y = -\frac{1}{2}k\widehat{\beta}\pi|\widetilde{\psi}^2|, \tag{16.36}$$

where the integrals are over the critical layer, and the momentum flux therefore diminishes across it. The last term on the right-hand side has no dependence on r, which means *there is finite absorption, even as friction tends to zero.* There will be no transmission through the critical layer, as we can see using a wave activity argument. For $y < 0$ (south of the critical line) the waves evanesce, but if (as we assume) friction there is negligible then in a statistically steady $\partial_y\overline{u'v'} = 0$. Thus $\overline{u'v'} = 0$ at the southern edge of the critical line, and therefore pseudomomentum must fall to zero across the critical line. Put simply, the eddy momentum fluxes are zero far from the critical layer (because waves evanesce), but the flux does not vary with y so it is zero at the southern edge of the critical layer. In fact, in such a layer there is also no reflection, and absorption is complete.

The zonal velocity itself also jumps across the layer. Since the streamfunction, and hence $v' = \partial\psi'/\partial x$, vary only weakly across the critical layer, $\zeta' \approx -\partial u'/\partial y$. The jump in velocity across the critical layer is then given by

$$[u']_-^+ = -\int_-^+ \zeta'\,\mathrm{d}y, \tag{16.37a,b}$$

which also may be calculated.

Nonlinear effects

We will say only a few words about this problem. First consider the critical layer thickness. Another way to evaluate the thickness in the linear problem is to directly note that the relative sizes of the advection term and the frictional term in the equation of motion is $y\partial_y\overline{u}_c/r$ so that an estimate for the frictional critical layer thickness is $\delta_r \sim r/(k\partial_y\overline{u}_c)$, as before. We can use a similar argument in the nonlinear case, and if we suppose a balance between the nonlinear terms and advection of the mean flow then the critical layer thickness, δ_{nl}, can be estimated from

$$y\frac{\partial u_c}{\partial y}\frac{\partial\zeta'}{\partial x} \sim v'\frac{\partial\zeta'}{\partial y} \sim k\psi'\frac{\zeta'}{\delta} \qquad \text{whence} \qquad \delta_{nl} \sim \left|\frac{v'}{k\partial_y\overline{u}_c}\right|^{1/2} \sim \left|\frac{\psi'}{\partial_y\overline{u}_c}\right|^{1/2}, \tag{16.38}$$

where v' is the meridional velocity near the critical layer. In Earth's atmosphere the nonlinear terms typically *are* important (and $\delta_{nl} > \delta_r$). Nevertheless, dissipation can and normally will occur in a nonlinear critical layer, either directly by the drag term in (16.25) or because the nonlinear interactions lead to wave breaking and a nonlinear cascade to dissipation within the layer itself, and that dissipation will retard the mean flow (i.e., make it more westward). This effect is important in the subtropics, where Rossby waves propagating equatorward from mid-latitudes in the upper troposphere encounter a critical level, and generate a critical layer. Rossby waves propagating upwards toward and into the stratosphere can also encounter a critical level, contributing to the QBO phenomenon described in Section 17.6.

If the critical layer is thin, then the total flow is a superposition of the background flow, $\overline{u}(y) \approx y\partial\overline{u}_c/\partial y$ and the disturbance field. The disturbance streamfunction, ψ' varies like $\exp(ikx)$, with no y variation because it is continuous across the critical line, so that the total streamfunction varies like

$$\psi \approx -\frac{1}{2}\frac{\partial \overline{u}_c}{\partial y} y^2 + \mathrm{Re}\,\psi'(0)\mathrm{e}^{ikx}. \tag{16.39}$$

This gives a pattern like that illustrated in Fig. 16.5, known as a Kelvin cat's eye pattern.

The dynamics of the pattern determines the value of $\int \overline{v'\zeta'}$ which in turn determines the value of $\overline{u'v'}$ at the edges, as in (16.36). If this value is reduced from the value of the incoming waves then the critical layer is reflecting, but determining whether this is so requires a more detailed analysis than we can provide here.

16.3.2 Internal Gravity Wave Critical Layers

Gravity wave critical layers have a rather different character than Rossby wave critical layers because of the nature of the dispersion relation. Referring back to (7.53) on page 259, we linearize the Boussinesq equations about a sheared mean flow $\overline{u}(z)$. Confining attention to two dimensions, x and z, then a little algebra results in the 'Taylor–Goldstein' equation,

$$\frac{\mathrm{d}^2\widetilde{\Psi}}{\mathrm{d}z^2} + \left[\frac{N^2}{(\overline{u}-c)^2} - \frac{\overline{u}_{zz}}{\overline{u}-c} - k^2\right]\widetilde{\Psi} = 0. \tag{16.40}$$

Here, $\widetilde{\Psi}$ is the amplitude of the streamfunction in the vertical plane, with $(u, w) = (-\partial\Psi/\partial z, \partial\Psi/\partial x)$. The presence of the first term in square brackets gives the equation a different nature than the Rossby problem, since if we omit the second term the equation has the form

$$\frac{\mathrm{d}^2\widetilde{\Psi}}{\mathrm{d}z^2} \sim -\frac{Ri\,\widetilde{\Psi}}{z^2}, \tag{16.41}$$

where $Ri = N^2/(\partial_z\overline{u}_c)^2$ is the Richardson number. This means that, if the flow is stable, the waves oscillate *extremely* rapidly as the critical line is approached — in fact an infinite number of times — and the group velocity also diminishes rapidly. (Solutions to (16.41) have the form $\widetilde{\psi} = z^\alpha$ where $\alpha = (1 \pm \sqrt{1-4Ri})/2$ and are oscillatory if $Ri > 1/4$, which is the condition for flow stability.) The slowdown gives dissipation more time to act, and commonly a wave will completely dissipate before the critical line is reached. If nonlinearity is allowed, the flow will break down into turbulence near the critical layer with a rapid cascade to dissipative scales.

The WKB approximation and notions of group velocity remain valid to a much greater degree than in the Rossby wave case. From (16.41) the vertical wavenumber, m, obeys

$$m^2 \sim \frac{1}{z^2}, \qquad \frac{\mathrm{d}m}{\mathrm{d}z} \sim \frac{1}{z^2} \sim m^2, \tag{16.42}$$

Thus, the WKB condition is neither obviously well satisfied nor badly violated, and we can expect group velocity and WKB theory to provide useful information much closer to the critical line than in the Rossby wave case. In particular, a Rossby wave will typically reach a critical layer even as the group velocity stalls, whereas a gravity wave may be absorbed before that. Aspects of the gravity wave analysis are continued in the discussion of the stratosphere in Section 17.3.3.

16.4 ♦ A WKB WAVE–MEAN-FLOW PROBLEM FOR ROSSBY WAVES

Let us now assume that the background properties do vary slowly and see how far we can get with a WKB approximation. As we saw above, WKB theory fails approaching a critical line, so we cannot determine what happens when $\overline{u} - c$ is very small, but the analysis is nevertheless instructive. If the friction is small and $r \ll k(\overline{u} - c)$, and if the meridional wavenumber l is larger than the zonal wavenumber k then l is given by

$$l^2(y) \approx \left[\frac{\beta^*(\overline{u} - c + \mathrm{i}r/k)}{(\overline{u} - c)^2 + r^2/k^2} \right] \approx \frac{\beta^*}{\overline{u} - c} \left[1 + \frac{\mathrm{i}r}{k(\overline{u} - c)} \right], \tag{16.43}$$

whence

$$l(y) \approx \left(\frac{\beta^*}{\overline{u} - c} \right)^{1/2} \left[1 + \frac{\mathrm{i}r}{2k(\overline{u} - c)} \right]. \tag{16.44}$$

The solution for the streamfunction is given by, in the WKB approximation,

$$\widetilde{\psi} = A l^{-1/2} \exp\left(\pm \mathrm{i} \int^y l \, \mathrm{d}y' \right), \tag{16.45}$$

just as in (16.16), but now the wave will decay as it moves away from its source and deposit momentum into the mean flow. Let us calculate this.

The momentum flux, F_k, associated with an x-wavenumber of k is given by

$$F_k(y) = \overline{u'v'} = -\mathrm{i}k \left(\psi \frac{\partial \psi^*}{\partial y} - \psi^* \frac{\partial \psi}{\partial y} \right), \tag{16.46}$$

and using (16.44) and (16.45) in (16.46) we obtain

$$F_k(y) = F_0 \exp\left(\int_0^y \frac{\pm r \beta^{*1/2}}{k(\overline{u} - c)^{3/2}} \, \mathrm{d}y' \right). \tag{16.47}$$

In deriving this expression we use the fact that the amplitude of $\widetilde{\psi}$ (i.e., $l^{-1/2}$) varies only slowly with y so that when calculating $\partial \widetilde{\psi}/\partial y$ the derivative of l may be ignored. In (16.47) F_0 is the value of the flux at $y = 0$ and the sign of the exponent must be chosen so that the group velocity is directed away from the wave source region. Clearly, if $r = 0$ then the momentum flux is constant.

The integrand in (16.47) is the attenuation rate of the wave and it has a straightforward physical interpretation. Using the real part of (16.44) in (16.17b), and assuming $|l| \gg |k|$, the meridional component of the group velocity is given by

$$c_g^y = \frac{2kl\beta^*}{(k^2 + l^2)^2} \approx \frac{2k\beta^*}{l^3} = \frac{2k(\overline{u} - c)^{3/2}}{\beta^{*1/2}}. \tag{16.48a,b}$$

Thus we have

$$\text{Wave attenuation rate} = \frac{r\beta^{*1/2}}{k(\overline{u} - c)^{3/2}} = \frac{2 \times \text{Dissipation rate } (2r)}{\text{Meridional group velocity } (c_g^y)}. \tag{16.49}$$

This result is of some generality in wave dynamics, and a simple interpretation is that as the group velocity diminishes the dissipation has more time to act and the wave is preferentially attenuated.

How does this attenuation affect the mean flow? The mean flow is subject to many waves and so obeys the equation

$$\frac{\partial \overline{u}}{\partial t} = -\sum_k \frac{\partial F_k}{\partial y} + \text{viscous terms}. \tag{16.50}$$

Because the amplitude varies only slowly compared to the phase, the amplitude of $\partial F_k/\partial y$ varies mainly with the attenuation rate (16.49). Consider a Rossby wave propagating away from some source region with a given frequency and x-wavenumber. Because k is negative a Rossby wave always carries westward (or negative) momentum with it. That is, F_k is always negative and increases (becomes more positive) as the wave is attenuated; that is to say, if $r \neq 0$ then $\partial F_k/\partial y$ is positive and from (16.50) the mean flow is accelerated *westward* as the wave dissipates. The dissipation, and attendant acceleration, will be particularly strong as the wave approaches a critical line where $\overline{u} = c$, although here the quantitative aspects of the analysis begin to fail.

The situation arises when Rossby waves, generated in mid-latitudes, propagate equatorward. As the waves enter the subtropics $\overline{u} - c$ becomes smaller and the waves dissipate, producing a westward force on the mean flow. Globally, momentum is conserved because there is an equal and opposite (and therefore eastward) wave force at the wave source producing an eastward eddy-driven jet, as discussed in the previous chapter.

Interpretation using wave activity

We can derive and interpret the above results by thinking about the propagation of wave activity, specifically the pseudomomentum given by (16.30). Referring as needed to the discussion in Sections 10.2.1 and 10.2.2, the flux obeys the group velocity property so that

$$\frac{\partial \mathcal{P}}{\partial t} + \frac{\partial}{\partial y}(\mathbf{c}_g \mathcal{P}) = -2r\mathcal{P}. \tag{16.51}$$

Let us suppose that the wave is in a statistical steady state and that the spatial variation of the group velocity occurs on a longer spatial scale than the variations in wave activity, consistent with the WKB assumption that group velocity varies slowly. We then have

$$c_g^y \frac{\partial \mathcal{P}}{\partial y} = -2r\mathcal{P}. \tag{16.52}$$

which integrates to give

$$\mathcal{P}(y) = \mathcal{P}_0 \exp\left(- \int^y \frac{2r}{c_g^y} \, \mathrm{d}y' \right). \tag{16.53}$$

That is, the attenuation rate of the wave activity is the dissipation rate of wave activity divided by the group velocity, as in (16.47) and (16.49). The wave-activity method of derivation suggests that this result is a general one, not restricted to Rossby waves, and indeed in Section 17.3.2 we will find that the attenuation rate of vertically propagating gravity waves is given by a similar expression.

The divergence of wave activity will lead to a force on the mean zonal flow, much as discussed in Section 15.1. For definiteness, suppose that waves propagate away from a mid-latitude source in the Northern Hemisphere. South of the source c_g^y is negative and north of the source c_g^y is positive. In either case, from (16.53) the wave activity density decreases away from the source and, with reference to (15.34a), the ensuing force on the mean flow is negative, or westward.

16.5 VERTICAL PROPAGATION OF ROSSBY WAVES

We now consider the vertical propagation of Rossby waves in a stratified atmosphere. The vertical propagation is important both because it must be taken into account to obtain an accurate picture of the tropospheric response to topographic and thermal forcing, and because it can excite motion in the stratosphere, as considered in Chapter 17. We will continue to use the stratified quasi-geostrophic equations, but we now allow the model to be compressible and semi-infinite, extending from $z = 0$ to $z = \infty$. It is simplest to first consider the problem slightly generally, without regard to boundary conditions; in Section 16.5.2 we will consider the lower boundary conditions and the requirements for waves to propagate vertically into the stratosphere. Our governing equation is the quasi-geostrophic potential vorticity equation, and with applications to the stratosphere in mind we will use log-pressure coordinates so that the equation of motion is

$$\frac{\partial q}{\partial t} + J(\psi, q) = 0, \qquad q = \nabla^2\psi + \beta y + \frac{f_0^2}{\rho_R}\frac{\partial}{\partial z}\left(\frac{\rho_R}{N^2}\frac{\partial \psi}{\partial z}\right), \tag{16.54}$$

where $z = H\ln(p/p_0)$ and $\rho_R = \rho_0 \mathrm{e}^{-z/H}$ with H being a specified density scale height, typically $RT(0)/g$.

16.5.1 Conditions for Wave Propagation

Let us linearize (16.54) about a zonal wind that depends only on z; that is, we let

$$\psi = -\overline{u}(z)y + \psi', \tag{16.55}$$

and obtain

$$\frac{\partial q'}{\partial t} + \overline{u}\frac{\partial q'}{\partial x} + v'\frac{\partial \overline{q}}{\partial y} = 0, \qquad \frac{\partial \overline{q}}{\partial y} = \beta - \frac{f_0^2}{\rho_R}\frac{\partial}{\partial z}\left(\frac{\rho_R}{N^2}\frac{\partial \overline{u}}{\partial z}\right), \tag{16.56}$$

or equivalently, in terms of streamfunction,

$$\left(\frac{\partial}{\partial t} + \overline{u}\frac{\partial}{\partial x}\right)\left[\nabla^2\psi' + \frac{f_0^2}{\rho_R}\frac{\partial}{\partial z}\left(\frac{\rho_R}{N^2}\frac{\partial \psi'}{\partial z}\right)\right] + \frac{\partial \psi'}{\partial x}\left[\beta - \frac{f_0^2}{\rho_R}\frac{\partial}{\partial z}\left(\frac{\rho_R}{N^2}\frac{\partial \overline{u}}{\partial z}\right)\right] = 0. \tag{16.57}$$

The first term in square brackets is the perturbation potential vorticity, q' and the second term equals $\partial\overline{q}/\partial y$. Seeking solutions of the form $\psi' = \mathrm{Re}\,\widetilde{\psi}(z)\exp[\mathrm{i}(kx + ly - kct)]$ gives

$$\left[\frac{f_0^2}{\rho_R}\frac{\partial}{\partial z}\left(\frac{\rho_R}{N^2}\frac{\partial \widetilde{\psi}}{\partial z}\right)\right] = \widetilde{\psi}\left(K^2 - \frac{\partial\overline{q}/\partial y}{\overline{u} - c}\right). \tag{16.58}$$

Let us simplify by assuming that both $\overline{u}$ and N^2 are constants so that $\partial\overline{q}/\partial y = \beta$. Equation (16.58) further simplifies if we define

$$\Phi(z) = \widetilde{\psi}(z)\left(\frac{\rho_R}{\rho_R(0)}\right)^{1/2} = \widetilde{\psi}(z)\mathrm{e}^{-z/2H} \tag{16.59}$$

whence we obtain

$$\frac{\mathrm{d}^2\Phi}{\mathrm{d}z^2} + m^2\Phi = 0, \quad \text{where} \quad m^2 = \frac{N^2}{f_0^2}\left(\frac{\beta}{\overline{u} - c} - K^2 - \gamma^2\right), \tag{16.60a,b}$$

where $\gamma^2 = f_0^2/(4N^2H^2) = 1/(2L_d)^2$ and where L_d is the deformation radius as sometimes defined (i.e., $L_d = NH/f_0$). The above equation has the same form as (16.10b). If the parameters on the right-hand side of (16.60b) are constant then so is m, and (16.60) has solutions of the form $\Phi(z) = \Phi_0 \mathrm{e}^{\mathrm{i}mz}$, so that the streamfunction itself varies as

$$\psi' = \mathrm{Re}\,\Phi_0 \exp\left[\mathrm{i}(kx + ly + mz - kct) + z/2H\right]. \tag{16.61}$$

In the (more realistic) case in which m varies with height then, if the variation is slow enough, the solution looks locally like a plane wave with m being a slowly varying vertical wavenumber, and WKB techniques may be used to find a solution, as we discuss further in Section 16.6. But

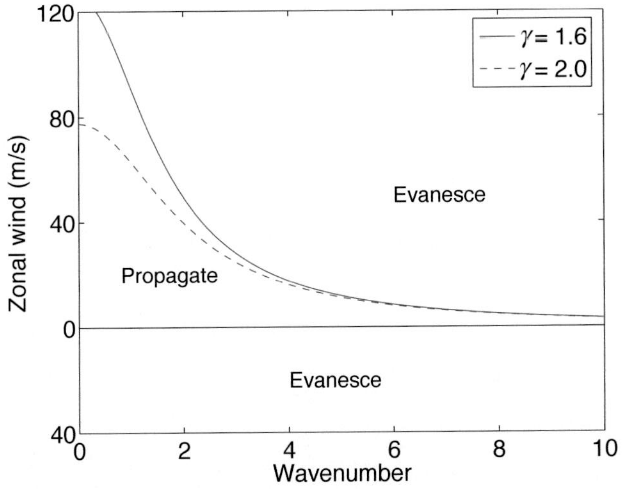

Fig. 16.6 The boundary between propagating waves and evanescent waves as a function of zonal wind and wavenumber, using (16.63), for a couple of values of γ. With $N = 2 \times 10^{-2}\,\mathrm{s}^{-1}$, $\gamma = 1.6$ ($\gamma = 2$) corresponds to a scale height of 7.0 km (5.5 km) and a deformation radius NH/f of 1400 km (1100 km).

even then essentially the same condition for propagation applies, namely that $m^2 > 0$ and, using (16.60b), this condition is satisfied if

$$0 < \overline{u} - c < \frac{\beta}{K^2 + \gamma^2}. \tag{16.62}$$

This condition is obviously similar to (16.12). For waves of some given frequency ($\omega = kc$) the above expression provides a condition on $\overline{u}$ for the vertical propagation of planetary waves. For stationary waves, $c = 0$, the condition becomes

$$0 < \overline{u} < \frac{\beta}{K^2 + \gamma^2}. \tag{16.63}$$

That is, in words, stationary, vertically oscillatory modes can exist only for zonal flows that are eastwards and that are less than the critical velocity $U_c = \beta/(K^2 + \gamma^2)$. This criterion, known as the *Charney–Drazin condition*,[4] is illustrated in Fig. 16.6 and we return to it in Section 16.5.3. The critical velocity for stationary waves evidently depends on the scale of the wave. For waves of a non-zero frequency the criterion is less severe, but stationary waves have a particular importance because they can be readily generated by surface topography.

Dispersion relation and group velocity

Noting that $\omega = ck$ and rearranging (16.60b) we obtain the dispersion relation for three-dimensional Rossby waves, namely

$$\omega = \overline{u}k - \frac{\beta k}{K^2 + \gamma^2 + m^2 f_0^2/N^2}. \tag{16.64}$$

The three components of the group velocity for these waves are then:

$$c_g^x = \overline{u} + \frac{\beta[k^2 - (l^2 + m^2 f_0^2/N^2 + \gamma^2)]}{(K^2 + m^2 f_0^2/N^2 + \gamma^2)^2}, \tag{16.65a}$$

$$c_g^y = \frac{2\beta kl}{(K^2 + m^2 f_0^2/N^2 + \gamma^2)^2}, \qquad c_g^z = \frac{2\beta km f_0^2/N^2}{(K^2 + m^2 f_0^2/N^2 + \gamma^2)^2}. \tag{16.65b,c}$$

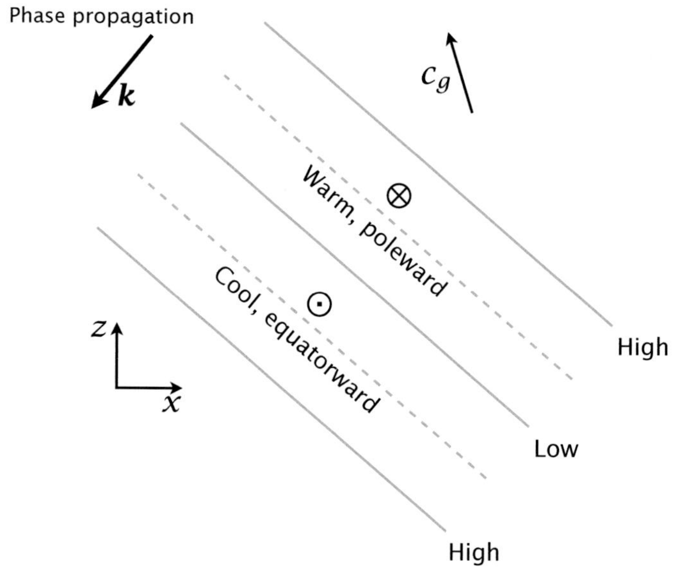

Fig. 16.7 East–west section of an upwardly propagating Rossby wave. The slanting lines are lines of constant phase and 'high' and 'low' refer to the pressure or streamfunction values. Both k and m are negative so the phase lines are oriented up and to the west. The phase propagates westward and downward, but the group velocity is upward.

The propagation in the horizontal is analogous to the propagation in a shallow water model, as in (6.66b); we see that higher baroclinic modes (bigger m) will have a more westward group velocity. The vertical group velocity is proportional to m, and for waves that propagate signals upward we must choose m to have the same sign as k so that c_g^z is positive. If there is no mean flow then the zonal wavenumber k is negative (in order that frequency is positive) and m must then also be negative. Energy then propagates upward but the phase propagates downward.

16.5.2 ✦ A Solution for Topographically-excited Waves

We now derive some explicit solutions for Rossby waves excited at a lower boundary by topography. Rossby waves may also be excited by thermal anomalies at the lower boundary, although in Earth's atmosphere their amplitude is somewhat smaller, and the treatment of such waves is left to the reader.[5] The lower boundary is obtained using the thermodynamic equation,

$$\frac{\partial}{\partial t}\left(\frac{\partial \psi}{\partial z}\right) + J\left(\psi, \frac{\partial \psi}{\partial z}\right) + \frac{N^2}{f_0} w = 0, \tag{16.66}$$

along with an equation for the vertical velocity, w, at the lower boundary. This is

$$w = \boldsymbol{u} \cdot \nabla h_b + r\zeta, \tag{16.67}$$

where the two terms respectively represent the kinematic contribution to vertical velocity due to flow over topography and the contribution from Ekman pumping, with r a constant, and the effects are taken to be additive. Linearizing the thermodynamic equation about the zonal flow and using (16.67) gives the boundary condition at $z = 0$,

$$\frac{\partial}{\partial t}\left(\frac{\partial \psi'}{\partial z}\right) + \overline{u}\frac{\partial}{\partial x}\frac{\partial \psi'}{\partial z} - v'\frac{\partial \overline{u}}{\partial z} = -\frac{N^2}{f_0}\left(\overline{u}\frac{\partial h_b}{\partial x} + r\nabla^2\psi'\right), \tag{16.68}$$

Solution

We look for solutions of (16.56) and (16.68) in the form

$$\psi' = \operatorname{Re}\widetilde{\psi}(z)\sin ly e^{ik(x-ct)} \qquad \text{with} \qquad h_b = \operatorname{Re}\widetilde{h}_b \sin ly e^{ikx}, \tag{16.69}$$

with $\widetilde{h}_b$ being purely real. Solutions must then satisfy

$$\left[\frac{f_0^2}{\rho_R}\frac{\partial}{\partial z}\left(\frac{\rho_R}{N^2}\frac{\partial\widetilde{\psi}}{\partial z}\right)\right] = \widetilde{\psi}\left(K^2 - \frac{\partial\overline{q}/\partial y}{\overline{u}-c}\right) \tag{16.70}$$

in the interior, and the boundary condition

$$(\overline{u}-c)\frac{\partial\widetilde{\psi}}{\partial z} - \widetilde{\psi}\frac{\partial\overline{u}}{\partial z} + \frac{irN^2K^2}{kf_0}\widetilde{\psi} = -\frac{N^2\overline{u}\widetilde{h}_b}{f_0}, \qquad \text{at } z=0, \tag{16.71}$$

as well as a radiation condition at plus infinity (and we must have that $\rho_0\widetilde{\psi}^2$ be finite). Let us simplify by considering the case of constant $\overline{u}$ and N^2 and with $r = 0$. As before we let $\Phi(z) = \widetilde{\psi}(z)\exp(-z/2H)$ and obtain the interior equation

$$\frac{\mathrm{d}^2\Phi}{\mathrm{d}z^2} + m^2\Phi = 0, \qquad \text{where} \quad m^2 = \frac{N^2}{f_0^2}\left(\frac{\beta}{\overline{u}-c} - K^2 - \gamma^2\right), \tag{16.72a,b}$$

and $\gamma^2 = f_0^2/(4N^2H^2) = 1/(2L_d)^2$, where L_d is the deformation radius, which is essentially the same as (16.60). The conditions for wave propagation, and in particular the Charney–Drazin condition, are just as previously derived. The surface boundary condition is now

$$(\overline{u}-c)\left(\frac{\mathrm{d}\Phi}{\mathrm{d}z} + \frac{\Phi}{2H}\right) = -\frac{N^2\overline{u}\widetilde{h}_b}{f_0}, \qquad \text{at } z=0. \tag{16.73}$$

This leads to an expression for the streamfunction amplitude, namely

$$\Phi_0 = \frac{N^2\widetilde{h}_b/f_0}{(\alpha, -im) - (2H)^{-1}}, \qquad \text{where} \qquad \alpha = +\frac{N}{f_0}\left(K^2 + \gamma^2 - \frac{\beta}{\overline{u}}\right)^{1/2}, \tag{16.74a,b}$$

and $(\alpha, -im)$ refers to the (trapped, oscillatory) case. A little algebra gives the solutions in the form

$$\psi'(x,y,z) = \mathrm{Re}\,\exp[\mathrm{i}(kx+mz) + z/2H]\sin ly\frac{f_0\widetilde{h}_b\left[im - (2H)^{-1}\right]}{K_s^2 - K^2}, \qquad m^2 > 0 \tag{16.75a}$$

$$\psi'(x,y,z) = \mathrm{Re}\,\exp[(2H)^{-1} - \alpha z + \mathrm{i}kx]\sin ly\frac{N^2\widetilde{h}_b}{f_0\left[\alpha - (2H)^{-1}\right]}, \qquad m^2 < 0 \tag{16.75b}$$

where $K_s^2 = \beta/\overline{u}$.

Resonance is possible when $\alpha = 1/(2H)$ or $K^2 = K_s^2$ and this condition obtains when barotropic Rossby waves are stationary. The wave resonates because the wave is a solution of the unforced, inviscid equations for the barotropic wave. If $K > K_s$ then $\alpha > 1/(2H)$ and the forced wave (i.e., the amplitude of ψ) decays with height with no phase variation. If $\alpha < 1/(2H)$ then $\widetilde{\psi}$ increases with height (although $\rho_R|\widetilde{\psi}|^2$ decreases with height), and this occurs when $(K_s^2 - \gamma^2)^{1/2} < K < K_s$. If $(K_s^2 - \gamma^2)^{1/2} > K$ then the amplitude of ϕ, (i.e., $\rho_R|\widetilde{\psi}|^2$) is independent of height; their vertical structure is oscillatory, like $\exp(imz)$.

16.5.3 Properties of the Solutions

Upward propagation and the Charney–Drazin condition

Let us return to the criterion for upward propagation given in (16.63) and illustrated in Fig. 16.6. One way to interpret this condition is to note that in a resting medium the Rossby wave frequency has a minimum value (and maximum absolute value), when $m = 0$, of

$$\omega = -\frac{\beta k}{K^2 + \gamma^2}. \tag{16.76}$$

Suppose that the waves are generated by bottom topography, and that $\overline{u}$ is uniform. In a frame moving with speed $\overline{u}$ our Rossby waves (stationary in the Earth's frame) have frequency $-\overline{u}k$, and this is the forcing frequency arising from the now-moving bottom topography. Thus, (16.63) is equivalent to saying that for oscillatory waves to exist *the forcing frequency must lie within the frequency range of vertically propagating Rossby waves.*

For westward flow, or for sufficiently strong eastward flow, the waves decay exponentially as $\Phi = \Phi_0 \exp(-\alpha z)$ where α is given by (16.74b). The critical velocity $u_c = (\beta/K^2 + \gamma^2)$ is a function of wavenumber, increasing with horizontal wavelength. Thus, for a given eastward flow long waves may penetrate vertically when short waves are trapped, an effect sometimes referred to as 'Charney–Drazin filtering'. There are three important consequences of this:

(i) Stratospheric motion is typically of larger horizontal scale than that of the troposphere, because Rossby waves tend to be excited first in the troposphere (by both baroclinic instability and flow over topography), but the shorter waves are trapped and only the longer ones reach the stratosphere.

(ii) The Rossby waves more commonly reach the stratosphere in the Northern Hemisphere than in the Southern Hemisphere. In both hemispheres the shorter, baroclinically forced Rossby waves are filtered, but the Northern Hemisphere also generates long Rossby waves through topographic interactions. The overturning circulation in the stratosphere (that is, the Brewer–Dobson circulation) is therefore stronger in the Northern Hemisphere.

(iii) Rossby waves find it hard to reach the stratosphere in summer, for then the stratospheric winds are often westwards (because the pole is warmer than the equator) and all waves are trapped in the troposphere. The eastward stratospheric winds that favour vertical penetration occur in the other three seasons, although very strong eastward winds can suppress penetration in mid-winter.

Other properties of the solution

Various other properties of the solution are described below, and a summary is given in the box on the next page.

Amplitudes and phases. The decaying solutions have no vertical phase variations (they are 'equivalent barotropic') and the streamfunction is exactly in phase or out of phase with the topography according as $K > K_s$ and $\alpha > (2H)^{-1}$, or $K < K_s$ and $\alpha < (2H)^{-1}$. In the latter case the amplitude of the streamfunction actually increases with height, but the energy, proportional to $\rho_R|\psi'^2|$ falls. The oscillatory solutions have (if there is no shear) constant energy with height but a shifting phase. The phase of the streamfunction at the surface may be in or out of phase with the topography, depending on m, but the potential temperature, $\partial\psi/\partial z$ is always out of phase with the topography. That is, positive values of h_b are associated with cool fluid parcels.

Vertical energy propagation. As noted, the energy propagates upwards for the oscillatory waves. This may be verified by calculating $\overline{p'w'}$ (the vertical component of the energy flux), where p' is the pressure perturbation, proportional to ψ', and w' is the vertical velocity perturbation. To this end, linearize the thermodynamic equation (16.66) to give

$$\frac{\partial}{\partial t}\left(\frac{\partial \psi'}{\partial z}\right) + \overline{u}\frac{\partial}{\partial x}\frac{\partial \psi'}{\partial z} - \frac{\partial \overline{u}}{\partial z}\frac{\partial \psi'}{\partial x} + \frac{N^2}{f_0}w' = 0. \tag{16.77}$$

Then, multiplying by ψ' and integrating by parts gives a balance between the second and fourth terms,

$$N^2\overline{\psi' w'} = \overline{u}\overline{b' v'}, \tag{16.78}$$

Upward Propagating Rossby Waves

In general Rossby waves satisfy

$$\psi' = \mathrm{Re}\, \Phi_0 \exp\left[\mathrm{i}(kx + ly + mz - kct) + z/2H\right], \tag{T.1}$$

where Φ_0 is a constant determined by the lower boundary conditions and

$$m = \pm \frac{N}{f_0}\left(\frac{\beta}{\overline{u}} - K^2 - \gamma^2\right)^{1/2}, \quad \text{with} \quad \gamma = \frac{f_0}{2NH}. \tag{T.2}$$

If $m^2 > 0$ the solutions are propagating, or radiating, waves in the vertical direction. If $m^2 < 0$ the energy of the solution, $|\rho_R \psi'^2|$, is vertically evanescent. The condition $m^2 > 0$ is equivalent to

$$0 < \overline{u} < \frac{\beta}{k^2 + l^2 + (f_0/2NH)^2}, \tag{T.3}$$

which is known as the Charney–Drazin condition. Vertical penetration is favoured when the winds are weakly eastwards, and the range of $\overline{u}$-values that allows this is larger for longer waves. Some other properties of the solution are:

- In order that the energy propagate upwards the vertical component of the group velocity must be positive, and hence k and m must have the same sign.
- The meridional heat flux is proportional to km, and thus upward propagation of waves is associated with a poleward heat transport.
- In an atmosphere in which density falls exponentially with height the amplitude of the streamfunction grows exponentially, so eventually nonlinear terms will become important. The waves may break, even in the absence of a critical layer.

where $b' = f_0 \partial\psi'/\partial z$ and $v' = \partial\psi'/\partial x$. Thus, the upward transfer of energy is proportional to the poleward heat flux. Evidently, the transfer of energy is upward when $km > 0$, and from (16.65), this corresponds to the condition that the vertical component of group velocity is positive, which has to be the case from general arguments. For Rossby waves $k < 0$ so that upward energy propagation requires $m < 0$ and therefore downward phase propagation.

Meridional heat transport. The meridional heat transport associated with a wave is

$$\rho_R \overline{v'b'} = \rho_R f_0 \overline{\frac{\partial\psi'}{\partial x}\frac{\partial\psi'}{\partial z}}. \tag{16.79}$$

For an oscillatory wave the heat flux is proportional to km. Now, the condition that energy is directed upward is that km is positive, for then c_g^z is positive. Thus, upward propagation is associated with a polewards heat flux. The meridional transport associated with a trapped solution is identically zero.

Form drag. If the waves propagate energy upwards, there must be a surface interaction to supply that energy. There is a force due to *form drag* (see Section 3.6) associated with this interaction given by

$$\text{form drag} = \overline{p' \frac{\partial h_b}{\partial x}}. \tag{16.80}$$

In the trapped case, the streamfunction is either exactly in or out of phase with the topography, so this interaction is zero. In the oscillatory case

$$\overline{\psi' \frac{\partial h_b}{\partial x}} = \frac{f_0 \widetilde{h}_b^2 km}{4(K_s^2 - K^2)}, \tag{16.81}$$

where the factor of 4 arises from the x and y averages of the squares of sines and cosines. The rate of doing work is $\overline{u}$ times (16.81).

16.6 ♦ VERTICAL PROPAGATION OF ROSSBY WAVES IN SHEAR

In the real atmosphere the zonal wind and the stratification change with height and there may be regions in which propagation occurs and regions where it does not, and in this section we illustrate that phenomenon with two examples. In one example the zonal wind increases sufficiently with height that wave propagation ceases because the wind is too strong, and in the other the zonal wind decreases aloft and becomes negative (westward), again causing wave propagation to cease. If the zonal wind and the stratification both vary sufficiently slowly with height — meaning that the scale of the variation is much greater than a vertical wavelength — then *locally* the solution will look like a plane wave and the analysis is straightforward, very similar to that performed in Section 7.5 (where we looked at internal waves with varying stratification) and Section 16.2.2 (where we looked at horizontally propagating Rossby waves).

For simplicity we consider Rossby waves in a flow with vertical shear but no horizontal shear, with constant stratification and constant density. With reference to Section 16.1.2, the equation of motion is

$$\left(\frac{\partial}{\partial t} + \overline{u}(z)\frac{\partial}{\partial x}\right) q' + \beta v' = 0. \tag{16.82}$$

We seek solutions of the form

$$\psi' = \widetilde{\psi}(z) e^{ik(x-ct)+ly}, \tag{16.83}$$

obtaining

$$\frac{\partial^2 \widetilde{\psi}}{\partial z^2} + m^2(z)\widetilde{\psi} = 0, \qquad \text{where} \qquad m^2(z) = \frac{N^2}{f_0^2}\left[\frac{\beta}{\overline{u} - c} - (k^2 + l^2)\right]. \tag{16.84a,b}$$

The WKB solution to this equation (see Appendix A to Chapter 6) is

$$\widetilde{\psi}(z) = Am^{-1/2} e^{i\int m\,dz}. \tag{16.85}$$

where A is a constant. The local vertical wavenumber is just m itself (for this is the derivative of the phase), and the amplitude varies like $m^{-1/2}$. This variation of amplitude is consistent with the conservation of wave activity, which in this case means that the Eliassen–Palm flux is constant. As there is no horizontal divergence in this problem, the constancy of $\boldsymbol{\mathcal{F}}$ in (16.19) implies $\partial_z \overline{v'b'} = 0$ and therefore

$$km|\widetilde{\psi}|^2 = \text{constant}. \tag{16.86}$$

Since the horizontal wavenumber is constant the dependence of the amplitude on $m^{-1/2}$ immediately follows. The energy of the wave is not constant unless there is no shear, since it may be extracted or given up to the mean flow.

As discussed in earlier sections, wave propagation requires than m^2 be positive. For stationary waves ($c = 0$) this gives the condition that $0 < \overline{u} < \beta/(k^2 + l^2)$ At the lower bound there is a critical line and $m^2 \to \infty$. At the upper bound $m^2 = 0$ and this is a turning line. Let us illustrate the behaviour approaching these regions with two examples.

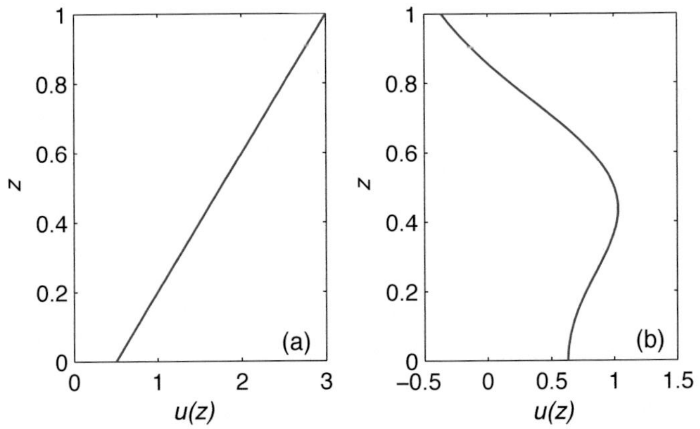

Fig. 16.8 Two profiles of nondimensional zonal wind used in the calculations illustrated in Fig. 16.9 and Fig. 16.10. (a) is a uniform shear that gives rise to a turning latitude, and (b) shows a profile in which the zonal wind diminishes to zero aloft, giving rise to a critical line.

16.6.1 Two Examples

In much the same way as we illustrated horizontal propagation in Section 16.2 we'll calculate the group velocity and wave amplitudes in two cases, one with a turning line and the other with a critical line. The same caveats apply — that the slowly-varying assumption fails at theses lines, and we cannot use WKB or ray theory to properly calculate reflection or absorption.

Waves with a turning line

Consider Rossby waves propagating in a background state in which the zonal wind increases uniformly with height, as in Fig. 16.8a, but in which all other parameters are constant. Specifically, we choose (nondimensional) values of $\beta = 5$, $k = l = 1$ and $c = 0$ (the reader may re-dimensionalize). We also scale the vertical coordinate so that $Pr = 1$. For the profile chosen m^2 is positive for $\overline{u} < 2.5$ and so for $0 < z < 0.8$, as shown in Fig. 16.9. For l fixed and m given by (16.84b) we calculate the group velocity using (16.7) and these are displayed in Fig. 16.9. We choose upwardly propagating waves (i.e. $m > 0$); in any physical situation the group velocity will be directed away from the source, and we are assuming this occurs at the surface. We also show equatorward moving waves for $y < 0.5$ and poleward moving waves for $y > 0.5$, but this is for illustrative purposes. The right-hand panel of the figure shows the value of m^2 diminishing with height, along with the vertical profiles of the amplitude (which goes like $m^{-1/2}$) and the energy (which goes like $(k^2 + l^2 + m^2)m$).

We see from Fig. 16.9 that the group velocity turns away from the turning line, and we can understand this from the ratio of the group velocities given in (16.7), namely

$$\frac{c_g^z}{c_g^y} = \frac{Pr^2 m}{l}, \tag{16.87}$$

where $Pr = f/N$. The group velocity is horizontal at the turning line. The amplitude of the waves is infinite, but the waves do not necessarily break because the vertical wavelength is very large. In fact, for the reasons given in the horizontally propagating case, we expect the waves to reflect.

Waves with a critical line

Now consider waves in a zonal wind that initially increases with height and then decreases and becomes negative, as illustrated in Fig. 16.8b. There is a critical line where $\overline{u}$ passes through zero, but all the other parameters are the same as in the previous example. The value of m^2 now generally increases with height, as illustrated in the right-hand panel of Fig. 16.10, becoming infinite at the critical line and negative above it. The amplitude of the wave, being proportional to $m^{-1/2}$ actually goes to zero at the critical line but the energy increases without bound (in the linear, inviscid approximation).

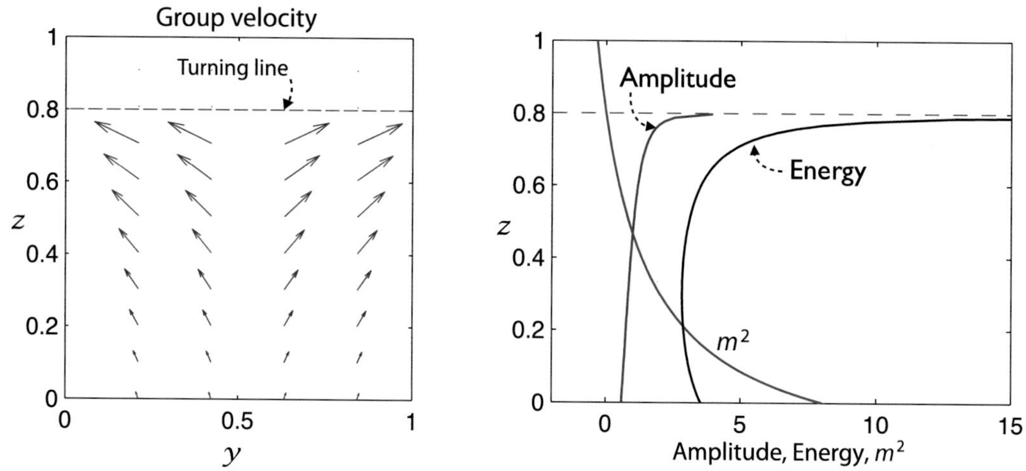

Fig. 16.9 Vertically propagating Rossby waves approaching a turning line. Left panel: group velocity vectors calculated using (16.7) for the parameters shown in Fig. 16.8a and with a source at $y = 0.5$. Right: profiles of m^2, wave amplitude and energy. The horizontal line at $z = 0.8$ marks a turning line: the group velocity turns away from it and the amplitude and energy both tend to infinity there.

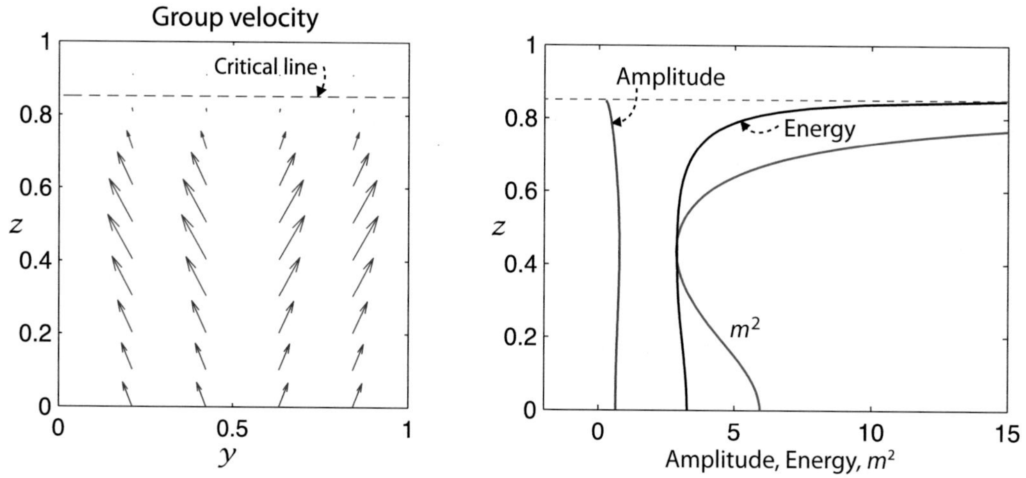

Fig. 16.10 Vertically propagating Rossby waves approaching a critical line. Left panel: group velocity vectors calculated using (16.7) for the parameters shown in Fig. 16.8b. Right: profiles of m^2, wave amplitude and energy. The horizontal line at $z \approx 0.85$ marks a critical line; the group velocity turns toward it but its amplitude diminishes as the critical line is approached.

The group velocity, shown in the left panel of Fig. 16.10, turns upward and toward the critical line and, from (16.87), is purely vertical at the critical line. The amplitude of the group velocity also diminishes, frictional and nonlinear effects become more important, and the notion of group velocity itself ceases to be meaningful close to the critical line where the WKB approximation breaks down.

As a final remark, we note that the condition that there be neither a critical line or a turning line is essentially a version of the Charney–Drazin criterion for wave propagation.

16.7 FORCED AND STATIONARY ROSSBY WAVES

We now turn our attention to understanding the large-scale zonally asymmetric circulation of the atmosphere, much of which is determined by the presence of stationary Rossby waves forced by topographic and thermal anomalies at the surface.[6]

16.7.1 A Simple One-layer Case

Many of the essential ideas can be illustrated by a one-layer quasi-geostrophic model, with potential vorticity equation

$$\frac{Dq}{Dt} = 0, \qquad q = \zeta + \beta y - \frac{f_0}{H}(\eta - h_b), \tag{16.88}$$

where H is the mean thickness of the layer, η is the height of the free surface, h_b is the bottom topography, and the velocity and vorticity are given by $\boldsymbol{u} = (g/f_0)\nabla^{\perp}\eta \equiv (g/f_0)\mathbf{k} \times \nabla\eta$ and $\zeta = (\partial v/\partial x - \partial u/\partial y) = (g/f_0)\nabla^2\eta$. Linearizing (16.88) about a flat-bottomed state with zonal flow $\overline{u}(y) = -(g/f_0)\partial\overline{\eta}/\partial y$ gives

$$\frac{\partial q'}{\partial t} + \overline{u}\frac{\partial q'}{\partial x} + v'\frac{\partial \overline{q}}{\partial y} = 0, \tag{16.89}$$

where $q' = \zeta' - (f_0/H)(\eta' - h_b)$ and $\partial\overline{q}/\partial y = \beta + \overline{u}/L_d^2$ with $L_d = \sqrt{gH}/f_0$, the radius of deformation. Equation (16.89) may be written, after the cancellation of a term proportional to $\overline{u}\partial\eta'/\partial x$, as

$$\frac{\partial}{\partial t}\left(\zeta' - \frac{\psi'}{L_d^2}\right) + \overline{u}\frac{\partial \zeta'}{\partial x} + \beta v' = -\overline{u}\frac{\partial \widehat{h}}{\partial x}, \tag{16.90}$$

where $\psi' = (g/f_0)\eta'$ and $\widehat{h} = h_b f_0/H = h_b g/(L_d^2 f_0)$.

The solution of this equation consists of the solution to the homogeneous problem (with the right-hand side equal to zero, as considered in section 6.4 on Rossby waves) and the particular solution. We proceed by decomposing the variables into their Fourier components

$$(\zeta', \psi', \widehat{h}) = \mathrm{Re}\,(\widetilde{\zeta}, \widetilde{\psi}, \widetilde{h}_b) \sin ly\, e^{ikx}, \tag{16.91}$$

where such a decomposition is appropriate for a channel, periodic in the x-direction and with no variation at the meridional boundaries, $y = (0, L)$. The full solution will be a superposition of such Fourier modes and, because the problem is linear, these modes do not interact. The free Rossby waves, the solution to the homogeneous problem, evolve according to

$$\psi = \mathrm{Re}\,\widetilde{\psi} \sin ly\, e^{i(kx - \omega t)}, \tag{16.92}$$

where ω is given by the dispersion relation,

$$\omega = k\overline{u} - \frac{k\partial_y\overline{q}}{K^2 + k_d^2} = \frac{k(\overline{u}K^2 - \beta)}{K^2 + k_d^2}, \tag{16.93a,b}$$

where $K^2 = k^2 + l^2$ and $k_d = 1/L_d$. Stationary waves occur at the wavenumbers for which $K = K_S \equiv \sqrt{\beta/\overline{u}}$. To the free waves we add the solution to the steady problem,

$$\overline{u}\frac{\partial \zeta'}{\partial x} + \beta v' = -\overline{u}\frac{\partial \widehat{h}}{\partial x}, \tag{16.94}$$

which is, using the notation of (16.91)

$$\widetilde{\psi} = \frac{\widetilde{h}_b}{(K^2 - K_s^2)}. \tag{16.95}$$

Now, $\widetilde{h}_b$ is a complex amplitude; thus, for $K > K_s$ the streamfunction response is *in phase* with the topography. For $K^2 \gg K_s^2$ the steady equation of motion is

$$\overline{u}\frac{\partial \zeta'}{\partial x} \approx -\overline{u}\frac{\partial \widehat{h}}{\partial x}, \tag{16.96}$$

and the topographic vorticity source is balanced by zonal advection of relative vorticity. For $K^2 < K_s^2$ the streamfunction response is *out of phase* with the topography, and the dominant balance for very large scales is between the meridional advection of planetary vorticity, $v\partial f/\partial y$ or βv, and the topographic source. For $K = K_s$ the response is infinite, with the stationary wave resonating with the topography. Now, any realistic topography can be expected to have contributions from *all* Fourier components. Thus, for *any* given zonal wind there will be a resonant wavenumber and an infinite response. This, of course, is not observed, and one reason is that the real system contains friction. The simplest way to include this is by adding a linear damping to the right-hand side of (16.90), giving

$$\frac{\partial}{\partial t}\left(\zeta' - \frac{\psi'}{L_d^2}\right) + \overline{u}\frac{\partial \zeta'}{\partial x} + \beta v' = -r\zeta' - \overline{u}\frac{\partial \widehat{h}}{\partial x}. \tag{16.97}$$

The free Rossby waves all decay monotonically to zero. However, the steady problem, obtained by omitting the first term on the left-hand side, has solutions

$$\widetilde{\psi} = \frac{\widetilde{h}_b}{(K^2 - K_s^2 - \mathrm{i}R)}, \tag{16.98}$$

where $R = (rK^2/\overline{u}k)$, and the singularity has been removed. The amplitude of the response is still a maximum for the stationary wave, and for this wave the phase of the response is shifted by $\pi/2$ with respect to the topography. The solution is shown in Fig. 16.11.[7] It is typical that for a mountain range whose Fourier composition contains all wavenumbers, there is a minimum in the streamfunction a little downstream of the mountain ridge.

16.7.2 Application to Earth's Atmosphere

With three parameters, I can fit an elephant.
William Thomson, Lord Kelvin (1824–1907).

Perhaps surprisingly, given the complexity of the real system and the simplicity of the model, when used with realistic topography a one-layer model can give reasonably realistic answers for the Earth's atmosphere (although we must always be careful of just fitting a model to observations). Thus, we calculate the stationary response to the Earth's topography using (16.97), using a reasonably realistic representation of the Earth's topography and, with qualification, the zonal wind. The zonal wind on the left-hand side of (16.97) is interpreted as the wind in the mid-troposphere, whereas the wind on the right-hand side is better interpreted as the surface wind, and so perhaps is about 0.4 times the mid-troposphere wind. Since the problem is linear, this amounts to tuning the amplitude of the response. The results, obtained using a rather crude representation of the Earth's topography, are plotted in Fig. 16.12. Also plotted is the observed time averaged response of the real atmosphere (the 500 hPa height field at 45° N). The agreement between model and observation is quite good, but this must be regarded as somewhat fortuitous if only because the other main source of the stationary wave field — thermal forcing — has been completely omitted from the calculation. Quantitative agreement is thus a consequence of the aforesaid tuning. Nevertheless, the calculation does suggest that the stationary, zonally asymmetric, features of the Earth's atmosphere arise via the interaction of the zonally symmetric wind field and the zonally asymmetric lower boundary, and that these may be calculated to a reasonable approximation with a linear model.

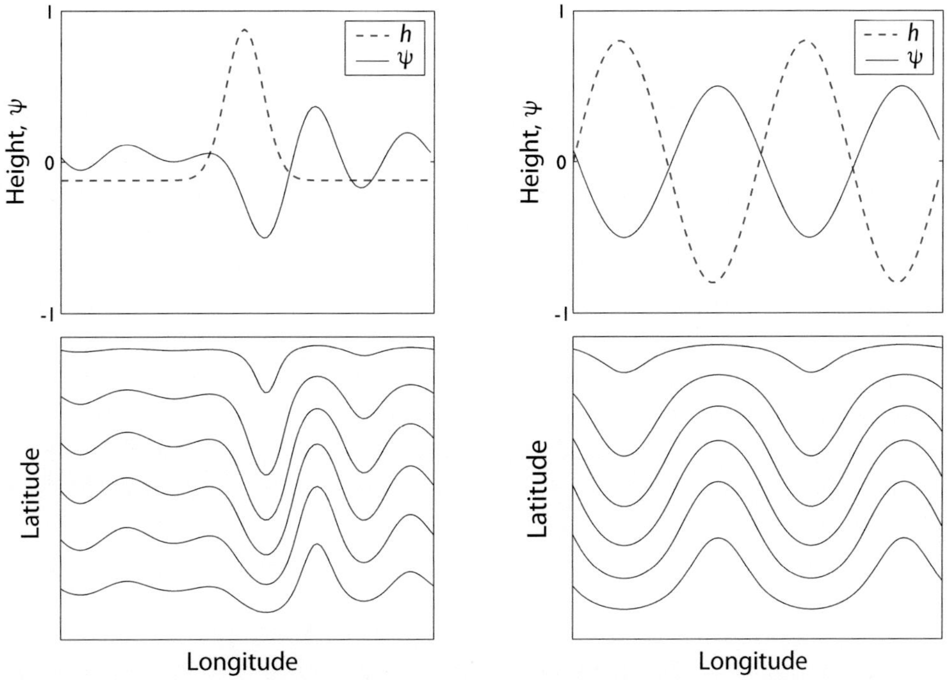

Fig. 16.11 The response to topographic forcing, i.e., the solution to the steady version of (16.97), for topography consisting of an isolated Gaussian ridge (left panels) and a pure sinusoid (right panels). The wavenumber of the stationary wave is about 4 and $r/(\overline{u}k) = 1$. The upper panels show the amplitude of the topography (dashed curve) and the perturbation streamfunction response (solid curve). The lower panels are contour plots of the streamfunction, including the mean flow. With the ridge, the response is dominated by the resonant wave and there is a streamfunction minimum, a 'trough', just downstream of the ridge. In the case on the right, the flow cannot resonate with the topography, which consists only of wavenumber 2, and the response is exactly out of phase with the topography.

16.7.3 ♦ One-dimensional Rossby Wave Trains

Although the Fourier analysis above gives exact (linear) results, it is not particularly revealing of the underlying dynamics. We see from Fig. 16.11 that the response to the Gaussian ridge is largely downstream of the ridge, and this suggests that it will be useful to consider the response as being due to Rossby *wavetrains* being excited by local features. This is also suggested by Fig. 16.13, which shows that the response to realistic topography is relatively local, and may be considered to arise from two relatively well-defined wavetrains, each of finite extent, one coming from the Rockies and the other from the Himalayas.

One way to analyse these wavetrains, and one which also brings up the concept of group velocity in a natural way, is to exploit (as in Section 15.1.2) a connection between changes in wavenumber and changes in frequency. Consider the linear barotropic vorticity equation in the form

$$\frac{\partial}{\partial t}(\zeta - k_d^2\psi) + \overline{u}\frac{\partial \zeta}{\partial x} + \beta\frac{\partial \psi}{\partial x} = -r\zeta, \tag{16.99}$$

where r is a frictional coefficient, which we presume to be small. Setting $k_d = 0$ for simplicity, the linear dispersion relation is

$$\omega = \overline{u}k - \frac{\beta k}{K^2} - \mathrm{i}r \equiv \omega_R(k, l) - \mathrm{i}r, \tag{16.100}$$

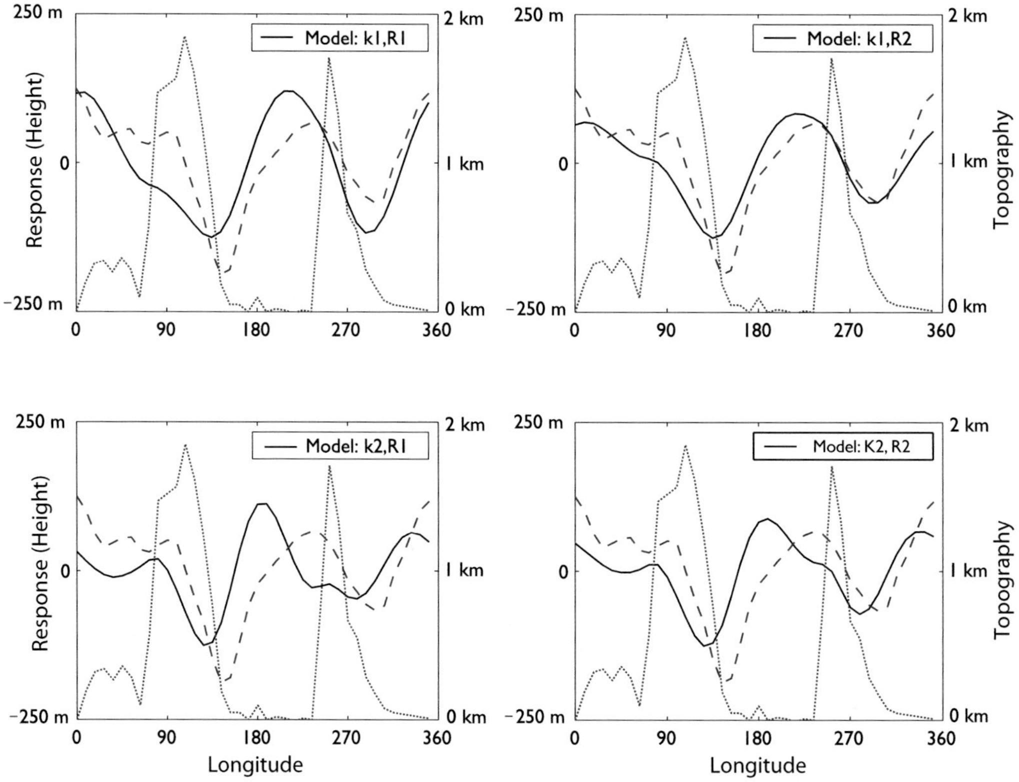

Fig. 16.12 Solutions of the Charney–Eliassen model. The solid lines are the steady solution of (16.97) using the Earth's topography at 45° N with two values of friction (R1 $\approx$ 6 days, R2 $\approx$ 3 days) and two values of resonant zonal wavenumber (2.5 for k1, 3.5 for k2), corresponding to zonal winds of approximately 17 and 13 m s^{-1}. The solutions are given in terms of height, η', where $\eta' = f_0\psi'/g$, with the scale on the left of each panel. The dashed line in each panel is the observed average height field at 500 hPa at 45° N in January. The dotted line is the topography used in the calculations, with the scale on the right of each panel.

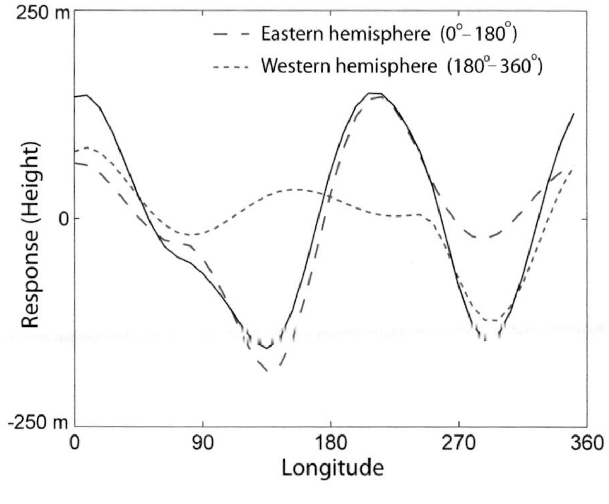

Fig. 16.13 The solution of the upper left-hand panel of Fig. 16.12 (solid line) and the solution divided into two contributions (dashed lines), one due to the topography only of the western hemisphere (i.e., with the topography in the east set to zero) and the other due to the topography only of the eastern hemisphere.

where $K^2 = k^2 + l^2$ and $\omega_R(k,l)$ is the inviscid dispersion relation for Rossby waves. Now, if there is a local source of the waves, for example an isolated mountain, we may expect to see a *spatial* attenuation of the wave as it moves away from the source. We may then regard the system as having a fixed, real frequency, but a changing, possibly complex, wavenumber. To determine this wavenumber for stationary waves (and so with $\omega = 0$), for small friction we expand the dispersion relation in a Taylor series about the inviscid value of ω_R at the real stationary wavenumber k_s, where $k_s = (K_s^2 - l^2)^{1/2}$ and $K_s = \sqrt{\beta/\overline{u}}$. This gives

$$\omega + \mathrm{i}r = \omega_R(k,l) \approx \omega_R(k_s,l) + \left.\frac{\partial \omega_R}{\partial k}\right|_{k=k_s} k' + \cdots . \tag{16.101}$$

Thus, $k' \approx \mathrm{i}r/c_g^x$, where c_g^x is the zonal component of the group velocity evaluated at a fixed position and at the stationary wavenumber; using (6.61b) this is given by

$$c_g^x = \left.\frac{\partial \omega_R}{\partial k}\right|_{k=k_s} = \frac{2\overline{u}k_s^2}{k_s^2 + l^2} . \tag{16.102}$$

The solution therefore decays away from a source at $x = 0$ according to

$$\psi \sim \exp(\mathrm{i}kx) = \exp\left[\mathrm{i}(k_s + k')x\right] \approx \exp(\mathrm{i}k_s x - rx/c_g^x), \tag{16.103}$$

and, because $c_g^x > 0$, the response is east of the source. The approximate solution for the streamfunction (denoted ψ_δ) of (16.97) in an infinite channel, with the topography being a δ-function mountain ridge at $x = x'$, and with all fields varying meridionally like $\sin ly$, is thus

$$\psi_\delta(x - x', y) \sim \begin{cases} 0 & x \le x' \\ -\dfrac{1}{k_s} \sin ly \sin[k_s(x - x')] \exp[-r(x - x')/c_g^x] & x \ge x' . \end{cases} \tag{16.104}$$

In the more general problem in which the topography is a general function of space, every location constitutes a separate source of wavetrains, and the complete (approximate) solution is given by the integral

$$\psi'(x,y) = \int_{-\infty}^{\infty} \widehat{h}(x)\psi_\delta(x - x', y)\,\mathrm{d}x' . \tag{16.105}$$

The field $\psi_\delta(x - x', y)$, is the *Green function* for the problem, sometimes denoted $G(x - x', y)$.

Example solutions calculated using both the Fourier and Green function methods are illustrated in Fig. 16.14. As in Fig. 16.11 there is a trough immediately downstream of the mountain, a result that holds for a broad range of parameters. In these solutions, the streamfunction decays almost completely in one circumnavigation of the channel, and thus, downstream of the mountain, both methods give virtually identical results. Such a correspondence will not hold if the wave can circumnavigate the globe with little attenuation, for then resonance will occur and the Green function method will be inaccurate; thus, whether the resonant picture or the wavetrain picture is more appropriate depends largely on the frictional parameter. A frictional time scale of about 10 days is often considered to approximately represent the Earth's atmosphere, in which case waves are only slightly damped on a global circumnavigation, and the Fourier picture is natural with the possibility of resonance. However, the smaller (more frictional) value of 5 days seems to give quantitatively better results in the barotropic problem, and the solution is more evocative of wavetrains. The larger friction may perform better because it is crudely parameterizing the meridional propagation and dispersion of Rossby waves that is neglected in the one-dimensional model.[8]

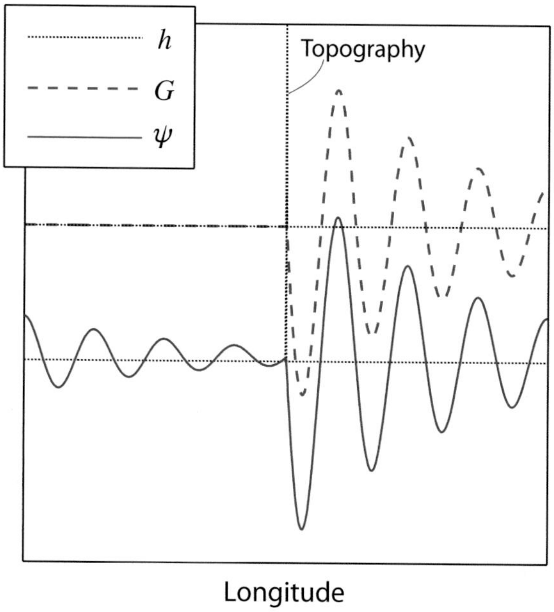

Fig. 16.14 A one-dimensional Rossby wave train excited by uniform eastward flow over a δ-function mountain ridge (h) in the centre of the domain. The upper curve, G, shows the Green's function (16.104), whereas the lower curve shows the exact (linear) response, ψ, in a re-entrant channel calculated numerically using the Fourier method.

The two solutions are both centred around zero and offset for clarity; the only noticeable difference is upstream of the ridge, where there is a finite response in the Fourier case because of the progression of the wave-train around the channel. The stationary wavenumber is 7.5.

16.7.4 The Adequacy of Linear Theory

Having calculated some solutions, we are in a position to estimate, *post facto*, the adequacy of the linear theory by calculating the magnitude of the omitted nonlinear terms. The linear problem here differs in kind from that which arises when using linear theory to evaluate the stability of a flow, as in Chapter 9. In that case, we assume a small initial perturbation and the initial evolution of that perturbation is then accurately described using linear equations. In this case the amplitude of the perturbation is arbitrary, for it may grow exponentially and its size at any given time is proportional to the magnitude of the initial perturbation, which is assumed small but which is otherwise unconstrained. In contrast, when we are calculating the stationary linear response to flow over topography or to a thermal source, the amplitude of the solution is *not* arbitrary; rather, it is determined by the parameters of the problem, including the size of the topography, and represents a real quantity that might be compared to observations. Of course, because the problem *is* linear, the amplitude of the solution is directly proportional to the magnitude of the topography or thermal perturbation.

From (16.98), and recalling that the amplitude of $\widetilde{h}$ is scaled relative to the real topography by the factor $(g/L_d^2 f_0)$, we crudely estimate the amplitude of the response to topography to be

$$|\psi'| \sim \frac{\alpha g h_b}{f_0} \approx \alpha \times 10^8\,\mathrm{m^2\,s^{-1}} = 2\times 10^7\,\mathrm{m^2\,s^{-1}}, \qquad |\eta'| = \frac{f_0|\psi'|}{g} \sim \alpha h_b \approx 0.2\,\mathrm{km}, \tag{16.106}$$

where the nondimensional parameter α accounts for the distance of the response from resonance and the ratio of the length scale to the deformation scale. Choosing $\alpha = 0.2$ and $h_b = 1\,\mathrm{km}$ gives the numerical values above, which are similar to those calculated more carefully, or observed (Fig. 16.12).

If linear theory is to be accurate, we must demand that the self advection of the response is much smaller than the advection by the basic state, and so that

$$|J(\psi', \nabla^2\psi')| \ll |\overline{u}\frac{\partial}{\partial x}\nabla^2\psi'|, \tag{16.107}$$

or, again rather crudely, that $|\psi'/L| \ll \overline{u}$. For $L = 5000\,\mathrm{km}$ we have $\psi'/L = 4\,\mathrm{m\,s^{-1}}$, which is a few times smaller than a typical mid-troposphere zonal flow of 20 m s^{-1}, suggesting that the linear approximation may hold water. However, the inequality is not a large one, especially as a different choice of numerical factors would give a different answer, and the use of a simple barotropic model also implies inaccuracies. Rather, we conclude that we must carefully calculate the linear response, and compare it with the observations and the implied nonlinear terms, before concluding that linear theory is appropriate, although it certainly does give qualitative insight.

16.8 ♦ EFFECTS OF THERMAL FORCING

How does thermal forcing influence the stationary waves? To give an accurate answer for the real atmosphere is a little more difficult than for the orographic case where the forcing can be included reasonably accurately in a quasi-geostrophic model with a term $\overline{u} \cdot \nabla h_b$ at the lower boundary. Anomalous (i.e., variations from a zonal or temporal mean) thermodynamic forcing typically also arises initially at the lower boundary through, for example, variations in the surface temperature. However, such anomalies may be felt throughout the lower troposphere on a relatively short time scale by way of such non-geostrophic phenomena as convection, so that the effective thermodynamic source that should be applied in a quasi-geostrophic calculation has a finite vertical extent. However, an accurate parameterization of this may depend on the structure of the atmospheric boundary layer and this cannot always be represented in a simple way. Because of such uncertainties our treatment concentrates on the fundamental and qualitative aspects of thermal forcing, and the reader should look to the literature for more complete derivations.[9]

The quasi-geostrophic potential vorticity equation, linearized around a uniform zonal flow, is (cf. (16.57))

$$\left(\frac{\partial}{\partial t} + \overline{u}\frac{\partial}{\partial x}\right)\left[\nabla^2\psi' + \frac{f_0^2}{\rho_R}\frac{\partial}{\partial z}\left(\frac{\rho_R}{N^2}\frac{\partial \psi'}{\partial z}\right)\right] + \frac{\partial \psi'}{\partial x}\left[\beta - \frac{f_0^2}{\rho_R}\frac{\partial}{\partial z}\left(\frac{\rho_R}{N^2}\frac{\partial \overline{u}}{\partial z}\right)\right] = \frac{f_0}{N^2}\frac{\partial Q}{\partial z} \equiv T, \tag{16.108}$$

where T is defined for convenience and Q is the source term in the (linear) thermodynamic equation,

$$\frac{\partial}{\partial t}\left(\frac{\partial \psi'}{\partial z}\right) + \overline{u}\frac{\partial}{\partial x}\frac{\partial \psi'}{\partial z} - v'\frac{\partial \overline{u}}{\partial z} + \frac{N^2}{f_0}w' = \frac{Q}{f_0}. \tag{16.109}$$

A particular solution to (16.108) may be constructed if $\overline{u}$ and N^2 are constant, and if Q has a simple vertical structure. If we again write $\psi' = \mathrm{Re}\,\widetilde{\psi}(z)\sin ly\exp(ikx)$ and let $\Phi(z) = \widetilde{\psi}(z)\exp(-z/2H)$ we obtain

$$\frac{\mathrm{d}^2\Phi}{\mathrm{d}z^2} + m^2\Phi = \frac{T}{\mathrm{i}k\overline{u}}\mathrm{e}^{-z/2H}, \qquad \text{where} \quad m^2 = \frac{N^2}{f_0^2}\left(\frac{\beta}{\overline{u}} - K^2 - \gamma^2\right). \tag{16.110}$$

If we let $T = T_0\exp(-z/H_Q)$, so that the heating decays exponentially away from the Earth's surface, then the particular solution to the stationary problem is found to be

$$\widetilde{\psi} = \mathrm{Re}\frac{\mathrm{i}\widehat{T}\mathrm{e}^{-z/H_Q}}{k\overline{u}\left[(N/f_0)^2(K_s^2 - K^2) + H_Q^{-2}(1 + H_Q/H)\right]}, \tag{16.111}$$

where $\widehat{T}$ is a constant proportional to T_0. This solution does not satisfy the boundary condition at $z = 0$, which in the absence of topography and friction is

$$\overline{u}\frac{\partial}{\partial x}\frac{\partial \psi'}{\partial z} - v'\frac{\partial \overline{u}}{\partial z} = \frac{Q(0)}{f_0}. \tag{16.112}$$

A homogeneous solution must therefore be added, and just as in the topographic case this leads to a vertically radiating or a surface trapped response, depending on the sign of m^2. One way to calculate the homogeneous solution is to first use the linearized thermodynamic equation (16.109), or the linearized vorticity equation (16.114), to calculate the vertical velocity at the surface implied by (16.111), $w_p(0)$ say. We then notice that the homogeneous solution is effectively forced by an equivalent topography given by $h_e = -w_p(0)/(\mathrm{i}k\overline{u}(0))$, and so proceed as in the topographic case. The complete solution is rather hard to interpret, and is in any case available only in special cases, so it is useful to take a more qualitative approach.

16.8.1 Thermodynamic Balances

It is the properties of the particular solution that mainly distinguish the response to thermodynamic forcing from that due to topography, because the homogeneous solutions of the two cases are similar. Far from the source region the homogeneous solution will dominate, giving rise to wavetrains as discussed previously.

We can determine many of the properties of the response to thermodynamic forcing by considering the balance of terms in the steady linear thermodynamic equation, which we write as

$$\overline{u}\frac{\partial}{\partial x}\frac{\partial \psi'}{\partial z} - \frac{\partial \psi'}{\partial x}\frac{\partial \overline{u}}{\partial z} + \frac{N^2}{f_0}w' = \frac{Q}{f_0} \equiv R \tag{16.113a}$$

or

$$f_0\overline{u}\frac{\partial v'}{\partial z} - f_0 v'\frac{\partial \overline{u}}{\partial z} + N^2 w' = Q. \tag{16.113b}$$

The vorticity equation is

$$\overline{u}\frac{\partial \zeta'}{\partial x} + \beta v' = \frac{f_0}{\rho_R}\frac{\partial \rho_R w'}{\partial z}. \tag{16.114}$$

Assuming that the diabatic forcing is significant, we may imagine three possible simple balances in the thermodynamic equation:

(i) zonal advection dominates, and $v' = \partial\psi'/\partial x \sim QH_Q/(f_0\overline{u})$;

(ii) meridional advection dominates, and $v' \sim QH_u/(f_0\overline{u})$;

(iii) vertical advection dominates, and $w' \sim Q/N^2$. Then, for large enough horizontal scales the balance in the vorticity equation is $\beta v' \sim f_0 w'_z$ and $v' \sim f_0 Q/(\beta N^2 H_Q)$. For smaller horizontal scales advection of relative vorticity may dominate that of planetary vorticity, and β is replaced by $\overline{u}K^2$.

Here, H_Q is the vertical scale of the source (so that $\partial Q/\partial z \sim Q/H_Q$) and H_u is the vertical scale of the zonal flow (so that $\partial\overline{u}/\partial z \sim \overline{u}/H_u$). We also assume that the vertical scale of the solution is H_Q, so that $\partial v'/\partial z \sim v'/H_Q$. Which of the above three balances is likely to hold? Heuristically, we might suppose that the balance with the smallest v' will dominate, if only because meridional motion is suppressed on the β-plane. Then, zonal advection dominates meridional advection if $H_u > H_Q$, and vice versa. Defining $\widehat{H} = \min(H_u, H_Q)$, then horizontal advection will dominate vertical advection if

$$\mu_1 = \frac{\beta N^2 H_Q \widehat{H}}{\overline{u} f_0^2} \ll 1. \tag{16.115}$$

More systematically, we can proceed in *reductio ad absurdum* fashion by first neglecting the vertical advection term in (16.113), and seeing if we can construct a self-consistent solution. If $\psi' = \mathrm{Re}\,\widetilde{\psi}_p(z)\mathrm{e}^{\mathrm{i}kx}$, and noting that $\overline{u}\partial\widetilde{\psi}_p/\partial z - \widetilde{\psi}_p\partial\overline{u}/\partial z = \overline{u}^2(\partial/\partial z)(\widetilde{\psi}_p/\overline{u})$, we obtain

$$\widetilde{\psi}_p = \frac{\mathrm{i}\overline{u}}{kf_0}\int_z^\infty \frac{\widetilde{Q}}{\overline{u}^2}\,\mathrm{d}z, \tag{16.116}$$

where $\widetilde{Q}$ denotes the Fourier amplitude of Q. Then, from the vorticity equation (16.114), we obtain the (Fourier amplitude of the) vertical velocity

$$\widetilde{w}_p = \frac{-ik}{f_0 \rho_R} \int_z^\infty \rho_R \overline{u}(K_s^2 - K^2)\widetilde{\psi}_p \, dz. \tag{16.117}$$

Using this one may, at least in principle, check whether the vertical advection in (16.113) is indeed negligible. If $\overline{u}$ is uniform (and so $H_u \gg H_Q$), then we find

$$\widetilde{\psi}_p \propto \frac{iQH_Q}{kf_0\overline{u}} \quad \text{and} \quad \widetilde{w}_p \propto \frac{QH_Q^2(K_s^2 - K^2)}{f_0^2}. \tag{16.118a,b}$$

Using this, vertical advection indeed makes a small contribution to the thermodynamic equation provided that

$$\mu_2 = \frac{N^2 H_Q^2 |K_s^2 - K^2|}{f_0^2} \ll 1. \tag{16.119}$$

If $K_s^2 \gg K^2$ and $\widehat{H} = H_Q$ then (16.119) is equivalent to (16.115). If $\overline{u}$ is not constant and if $H_u \ll H_Q$ then H_u replaces H_Q and the criterion for the dominance of horizontal advection becomes

$$\mu = \frac{N^2 \widehat{H} H_Q |K_s^2 - K^2|}{f_0^2} \ll 1. \tag{16.120}$$

This is the condition that the first term in the denominator of (16.111) is negligible compared with the second. For a typical tropospheric value of $N^2 = 10^{-4}\,\mathrm{s}^{-1}$ and for $K > K_S$ we find that $\mu \approx (H_Q/7\,\mathrm{km})^2$, and so we can expect $\mu < 1$ in extra-equatorial regions where the heating is shallow. At low latitudes f_0 is smaller, β is bigger and $\mu \approx (H_Q/1\,\mathrm{km})^2$, and we can expect $\mu > 1$. However, there is both uncertainty and variation in these values.

Equivalent topography

In the case in which zonal advection dominates, the equivalent topography is given by

$$h_e = \frac{-\widetilde{w}_p(0)}{ik\overline{u(0)}} = \frac{1}{\overline{u(0)} f_0 \rho_R(0)} \int_0^\infty \rho_R \overline{u}(K_s^2 - K^2)\widetilde{\psi}_p \, dz, \tag{16.121}$$

where $\widetilde{\psi}_p$ is given by (16.116). The point to notice here is that if $K < K_s$ the equivalent topography is in phase with ψ_p.

16.8.2 Properties of the Solution

In the tropics μ may be large for H_Q greater than a kilometre or so. Heating close to the surface cannot produce a large vertical velocity and will therefore produce a meridional velocity. However, away from the surface the heat source will be balanced by vertical advection. For scales such that $K < K_S$, a criterion that might apply at low latitudes for wavelengths longer than a few thousand kilometres, the associated vortex stretching $f\partial w/\partial z > 0$ is balanced by βv and a poleward meridional motion occurs. This implies a trough west of the heating and/or a ridge east of the heating, although the use of quasi-geostrophic theory to draw tropical inferences may be a little suspect.

In mid-latitudes μ is typically small and horizontal advection locally balances diabatic heating. In this case there is a trough a quarter-wavelength downstream from the heating, and equatorward motion at the longitude of the source. (To see this, note that if the heating has a structure like $\cos kx$ then from either (16.111) or (16.116) the solution goes like $\psi_p \propto -\sin kx$.) The trough may be warm or cold, but is often warm. If $H_Q \ll H_u$, as is assumed in obtaining (16.111), then θ is

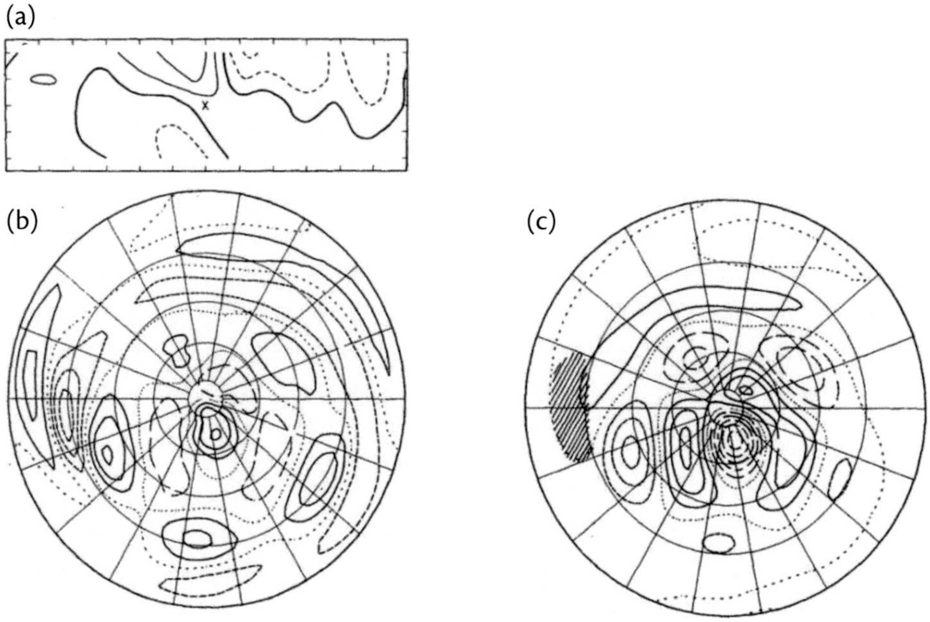

Fig. 16.15 Numerical solution of a baroclinic primitive equation model with a deep heat source at 15° N and a zonal flow similar to that of northern hemisphere winter. (a) Height field in a longitude height at 18° N (the tick marks on the vertical axis are at 100, 300, 500, 700 and 900 hPa); (b) 300 hPa vorticity field; (c) 300 hPa height field. The cross in (a) and the hatched region in (c) indicate the location of the heating.[10]

positive and warm. This is because zonal advection dominates and so the effect of the heating is advected downstream. If $H_Q \gg H_u$ and meridional advection is dominant, then the trough is still warm provided Q decreases with height. The vertical velocity can be inferred from the vorticity balance. If $f_0\partial w/\partial z \approx \beta v$ and if $w = 0$ at the surface (in the absence of Ekman pumping and any topographic effects) there is *descent* in the neighbourhood of a heat source. This counter-intuitive result arises because it is the horizontal advection that is balancing the diabatic heating. (This result cannot be inferred from the particular solution alone.) If the advection of relative vorticity balances vortex stretching, the opposite may hold.

The homogeneous solution can be inferred from (16.121) and (16.75). Consider, for example, waves that are trapped ($m^2 < 0$) but still have $K < K_S$; that is $K^2 < K_S^2 < K^2 + \gamma^2$. The homogeneous solution forced by the equivalent topography is out of phase with that topography, and so out of phase with ψ_p, using (16.121). For still shorter waves, $K > K_s$, the homogeneous solution is in phase with the equivalent topography, and so again out of phase with ψ_p. Thermal sources produced by large-scale continental land masses may have $K^2 < K_s^2$ and, if $K^2 + \gamma^2 < K_S^2$ they will produce waves that penetrate up into the stratosphere and typically these solutions will dominate far from the source. Evidently though, the precise relationship between the particular and homogeneous solution is best dealt with on a case-by-case basis. A few more general points are summarized in the box on page 619.

16.8.3 Numerical Solutions

The numerically calculated response to an isolated heat source is illustrated in Figs. 16.15 and 16.16. The first figure shows the response to a 'deep' heating at 15° N. As the reasoning above would suggest, the vertical velocity field (not shown) is upwards in the vicinity of the source. Away from

Thermal Forcing of Stationary Waves

(i) The solution is composed of a particular solution and a homogeneous solution.

(ii) The homogeneous solution may be thought of as being forced by an 'equivalent topography', chosen so that the complete solution satisfies the boundary condition on vertical velocity at the surface.

(iii) For a localized source, the far field is dominated by the homogeneous solution. This solution has the same properties as a solution forced by real topography. Thus, it may include waves that penetrate vertically into the stratosphere as well as wavetrains propagating around the globe with an equivalent barotropic structure.

(iv) In the extratropics a heating is typically balanced by horizontal advection, producing a trough a quarter wavelength east (downstream) of a localized heat source. The heat source is balanced by advection of cooler air from higher latitudes, and there may be sinking air over the heat source. This can occur when $\mu \ll 1$; see (16.120).

(v) In the tropics, a heat source may be locally balanced by vertical advection, that is adiabatic cooling as air ascends. This can occur when $\mu \gg 1$.

(vi) In the real atmosphere, the stationary solutions must coexist with the chaos of time-dependent, nonlinear flows. Thus, they are likely to manifest themselves only in time averaged fields and in a modified form.

the source, the solution is dominated by the homogeneous solutions in the form of wavetrains, as described in Section 16.7.3, with a simple vertical structure. (In fact, the pattern is quite similar to that obtained with a suitably forced barotropic model, as was found in the topographically forced case.)

Figure 16.16 shows the response to a perturbation at 45° N, and again the solutions are qualitatively in agreement with the reasoning above. The local heating is balanced by an equatorward wind, and there is a surface trough about 20° east of the source, and an upper-level pressure maximum, or ridge, about 60° east. The scale height of the wind field, H_u is about 8 km, greater than that of the source, and the balance in the thermodynamic equation is between the zonal advection of the temperature anomaly $\overline{u}\theta'_x$ and the heat source, so producing a temperature maximum downstream. Again, the far field is dominated by the wavetrain of the homogeneous solution.

Finally, we show a calculation (Fig. 16.17) that, although linear, includes realistic forcing from topography, heat sources and observed transient eddy flux convergences, and uses a realistic zonally averaged zonal flow, although some physical parameters representing friction and diffusion in the calculation must be changed in order that a steady solution can be achieved. Such a calculation is likely to be the most accurate achievable by a linear model, and discrepancies from observations indicate the presence of nonlinearities that are neglected in the calculation. In fact, a generally good agreement with the observed fields is found, and provides some *post facto* justification for the use of linear, stationary wave models.[13]

In such realistic calculations it is virtually impossible to see the wavetrains emerging from iso-

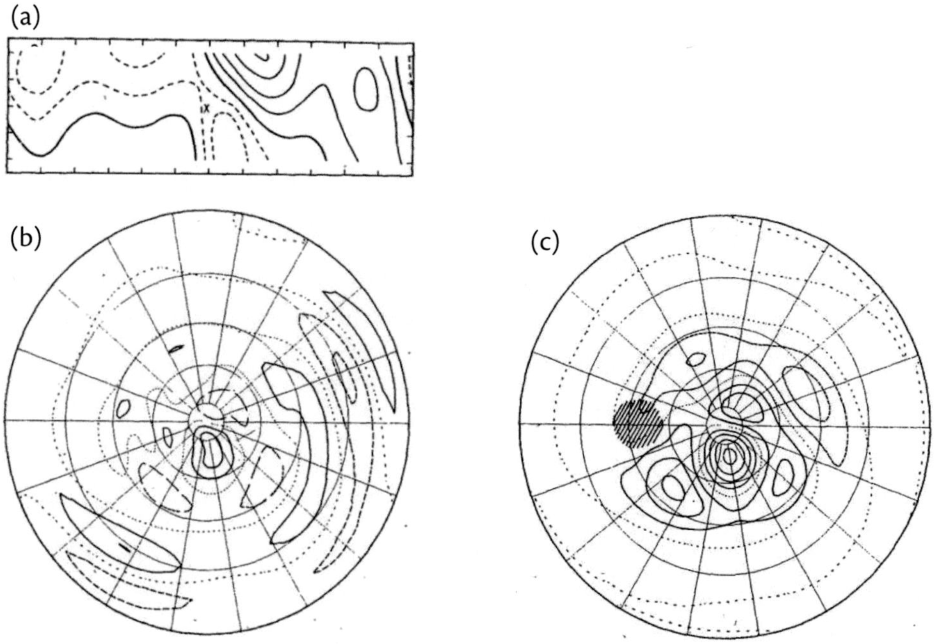

Fig. 16.16 As for Fig. 16.15, but now the solution of a baroclinic primitive equation model with a deep heat source at 45° N. (a) Height field in a longitude height at 18° N; (b) 300 hPa vorticity field; (c) 300 hPa height field. The cross in (a) and the hatched region in (c) indicate the location of the heating.[11]

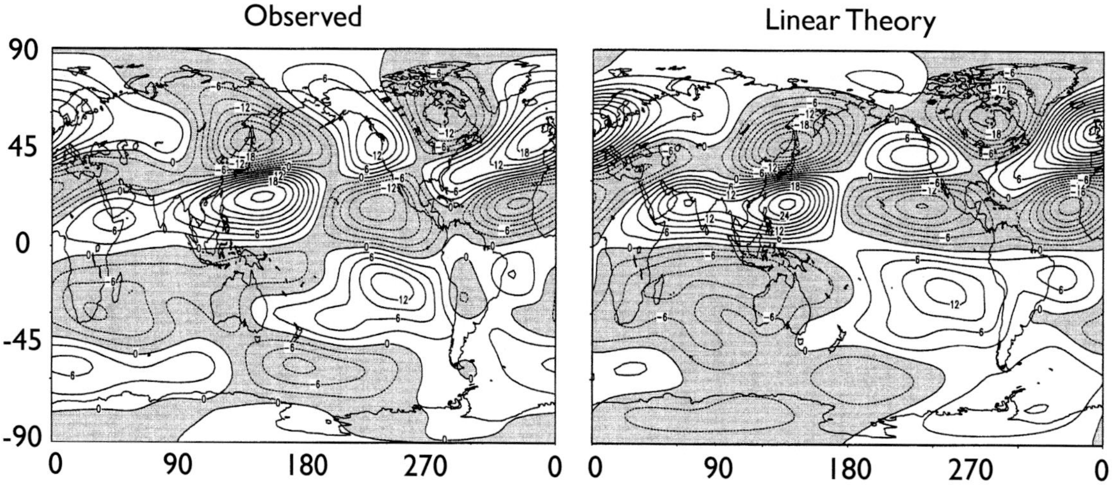

Fig. 16.17 Left: the observed stationary (i.e., time-averaged) streamfunction at 300 hPa (about 0.5 km altitude) in northern hemisphere winter. Right: the steady, linear response to forcing by orography, heat sources and transient eddy flux convergences, calculated using a linear model with the observed height-varying zonally averaged zonal wind. Contour interval is $3\times10^6\ m^2\ s^{-1}$, and negative values are shaded. Note the generally good agreement, and also the much weaker zonal asymmetries in the southern hemisphere.[12]

lated features like the Rockies or Himalayas, because they are combined with the responses from all the other sources included in the calculation. Breaking up the forcing into separate contributions from orographic forcing, heating, and the time averaged momentum and heat fluxes from transient eddies reveals that all of these separate contributions have a non-neglible influence. We should also remember that the effects of the fluxes from the transient eddies are not explained by such a calculation, merely included in a diagnostic sense. Nevertheless, the agreement does reveal the extent to which we might understand the steady zonally asymmetric circulation of the real atmosphere as the response due to the interaction of a zonally uniform zonal wind with the asymmetric features of the Earth's geography and transient eddy field. The quasi-stationary response of the planetary waves to surface anomalies, and the interaction of transient eddies with the large-scale planetary wave field, are important factors in the natural variability of climate, and their understanding remains a challenge for dynamical meteorologists.

16.9 ♦ WAVE PROPAGATION USING RAY THEORY

Catch a wave and you're sitting on top of the world.
Brian Wilson and Mike Love (The Beach Boys), *Catch a Wave*, 1963.

Rossby waves propagate meridionally as well as zonally, and we can expect the major mountain ranges on Earth, as well as thermal anomalies, to generate Rossby waves that propagate both zonally and meridionally. The coefficients of the linear equations that determine this propagation will vary with space: on the sphere β is a function of latitude and in general topography is a function of both latitude and longitude; thus calculating the trajectories of the waves will be difficult, and we cannot expect to solve the full problem except numerically, but a few ideas from wave tracing illustrate many of the features of the response, and indeed of the stationary wave pattern in the Earth's atmosphere.[14]

Let us first recall a few results about rays and ray tracing that we encountered in Section 6.3. Most of the important properties of a wave, such as the energy (if conserved) and the wave activity, propagate along rays at the group velocity. Rays themselves are lines that are parallel to the group velocity, generally emanating from some wave source. A ray is perpendicular to the local wave front, and in a homogeneous medium a wave propagates in a straight line. In non-homogeneous media the group velocity varies with position; however, if the medium varies only slowly, on a scale much larger than that of the wavelength of the waves, the wave activity still propagates along rays at the group velocity.

The variation of wavenumber and frequency vary according to, respectively,

$$\frac{\partial k_i}{\partial t} + c_{gj}\frac{\partial k_i}{\partial x_j} = -\frac{\partial \Omega}{\partial x_i}, \qquad \frac{\partial \omega}{\partial t} + c_{gj}\frac{\partial \omega}{\partial x_j} = \frac{\partial \Omega}{\partial t}. \tag{16.122}$$

If the frequency is not an explicit function of space then the wavenumber is constant along a ray, and similarly for time and frequency. Thus, in problems of the form

$$\frac{\partial}{\partial t}\nabla^2\psi + \beta(y)\frac{\partial \psi}{\partial x} = 0, \tag{16.123}$$

then the frequency and the x-wavenumber, but not the y-wavenumber, are constant along a ray. The wavenumber is not constant in the y-direction because the frequency is a function of y.

16.9.1 Rossby Waves and Rossby Rays

If the wave source is localized, then ray theory provides a useful way of calculating and interpreting the response. On the β-plane and away from the orographic source the steady linear response to a zonally uniform but meridionally varying zonal wind will obey

$$\overline{u}(y)\frac{\partial}{\partial x}\left(\frac{\partial^2}{\partial x^2}+\frac{\partial^2}{\partial y^2}\right)\psi+\beta\frac{\partial\psi'}{\partial x}=0. \tag{16.124}$$

In fact, an equation of this form applies on the sphere. To see this, we transform the spherical coordinates (λ, ϑ) into Mercator coordinates with the mapping[15]

$$x'=a\lambda,\quad \frac{1}{a}\frac{\partial}{\partial\lambda}=\frac{\partial}{\partial x'},\qquad y'=\frac{a}{2}\ln\left(\frac{1+\sin\vartheta}{1-\sin\vartheta}\right),\quad \frac{1}{a}\frac{\partial}{\partial\vartheta}=\frac{1}{\cos\vartheta}\frac{\partial}{\partial y'}. \tag{16.125}$$

The spherical-coordinate vorticity equation then becomes

$$\overline{u}_M\frac{\partial}{\partial x}\left(\frac{\partial^2}{\partial x'^2}+\frac{\partial^2}{\partial y'^2}\right)\psi+\beta_M\frac{\partial\psi}{\partial x'}=0, \tag{16.126}$$

where $\overline{u}_M=\overline{u}/\cos\vartheta$ and

$$\beta_M=\frac{2\Omega}{a}\cos^2\vartheta-\frac{\mathrm{d}}{\mathrm{d}y'}\left[\frac{1}{\cos^2\vartheta}\frac{\mathrm{d}}{\mathrm{d}y}(\overline{u}_M\cos^2\vartheta)\right]=\cos\vartheta\left(\beta_s+\frac{1}{a}\frac{\partial\overline{\zeta}}{\partial\vartheta}\right), \tag{16.127}$$

where $\beta_s=2a^{-1}\Omega\cos\vartheta$. Thus, β_M is the meridional gradient of the absolute vorticity, multiplied by the cosine of latitude. An advantage of Mercator coordinates over their spherical counterparts is that (16.126) has a Cartesian flavour to it, with the metric coefficients being absorbed into the parameters $\overline{u}_M$ and β_M. Of course, unlike the case on the true β-plane, the parameter β_M is not a constant, but this is not a particular disadvantage if $\overline{u}_{yy}$ is also varying with y.

Having noted the spherical relevance we revert to the Cartesian β-plane and seek solutions of (16.124) with the form $\psi'=\widetilde{\psi}(y)\exp(ikx)$, whence

$$\frac{\mathrm{d}^2\widetilde{\psi}}{\mathrm{d}y^2}=\left(k^2-\frac{\beta}{\overline{u}}\right)\widetilde{\psi}=(k^2-K_s^2)\,\widetilde{\psi}, \tag{16.128}$$

where $K_s=(\beta/\overline{u})^{1/2}$. From this equation it is apparent that if $k<K_s$ the solution is harmonic in y and Rossby waves may propagate away from their source. On the other hand, wavenumbers $k>K_s$ are trapped near their source; that is, short waves are meridionally trapped by eastward flow. Without solving (16.128), we can expect an isolated mountain to produce two wavetrains, one for each meridional wavenumber $l=\pm(K_s^2-k^2)^{1/2}$. These wavetrains will then propagate along a ray, and given the dispersion relation this trajectory can be calculated (usually numerically) using the expressions of the previous section. The local dispersion relation of Rossby waves is

$$\omega=\overline{u}k-\frac{\beta k}{k^2+l^2}, \tag{16.129}$$

so that their group velocity is

$$c_g^x=\frac{\partial\omega}{\partial k}=\overline{u}-\frac{\beta(l^2-k^2)}{(k^2+l^2)^2}=\frac{\omega}{k}+\frac{2\beta k^2}{(k^2+l^2)^2},\qquad c_g^y=\frac{\partial\omega}{\partial l}=\frac{2\beta kl}{(k^2+l^2)^2}. \tag{16.130a,b}$$

The sign of the meridional wavenumber thus determines whether the waves propagate polewards (positive l) or equatorwards (negative l). Also, because the dispersion relation (16.130) is independent of x and t, the zonal wavenumber and frequency in the wave group are constant along the ray, and the *meridional wavenumber* must adjust to satisfy the local dispersion relation (16.129). Thus, from (16.128), the meridional scale becomes larger as K_s approaches k from above and an

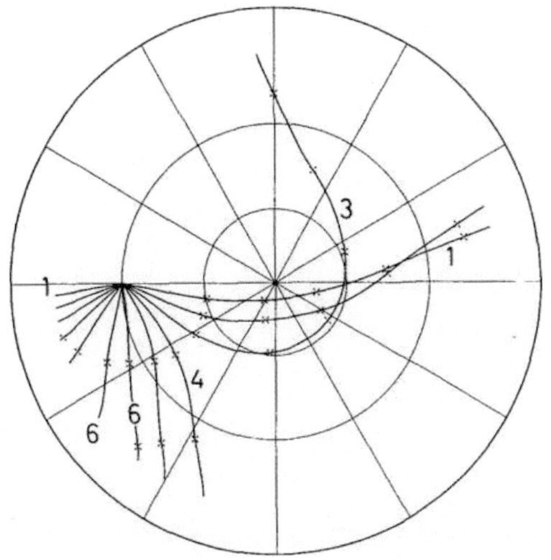

Fig. 16.18 The rays emanating from a point source at 30° N and 180° (nine o'clock), calculated using the observed value of the wind at 300 hPa.[16] The crosses mark every 180° of phase, and mark the positions of successive positive and negative extrema. The numbers indicate the zonal wavenumber of the ray. The ray paths may be compared with the full linear calculation shown in Fig. 16.19.

incident wavetrain is reflected, its meridional wavenumber changes sign, and it continues to propagate eastwards.

Stationary waves have $\omega = 0$, and the trajectory of a ray is parameterized by

$$\frac{\mathrm{d}y}{\mathrm{d}x} = \frac{c_g^y}{c_g^x} = \frac{l}{k}. \tag{16.131}$$

For a given zonal wavenumber the trajectory is then fully determined by this condition and that for the local meridional wavenumber, which from (16.129) is

$$l^2 = K_s^2 - k^2. \tag{16.132}$$

Finally, from (16.130) the magnitude of the group velocity is

$$|c_g| = [(c_g^x)^2 + (c_g^y)^2]^{1/2} = 2\frac{k}{K_s}\overline{u}, \tag{16.133}$$

which is double the speed of the projection of the basic flow, $\overline{u}$, onto the wave direction. Given the above relations, and the zonal wind field, we can compute rays emanating from a given source, although the calculation must still be done numerically. One example is given in Fig. 16.18.

✦ *A WKB solution*

The wave amplitudes along a ray can be obtained using a WKB approach. We write (16.128) as

$$\frac{\mathrm{d}^2\widetilde{\psi}}{\mathrm{d}y^2} + l^2(y)\widetilde{\psi} = 0, \qquad \text{where} \qquad l^2(y) = K_s^2 - k^2, \tag{16.134a,b}$$

and if $l(y)$ is varying sufficiently slowly in y then the WKB solution for the stationary streamfunction is

$$\psi(x, y) = Al^{-1/2} \exp\left[\mathrm{i}\left(kx + \int^y l(y)\,\mathrm{d}y\right)\right], \tag{16.135}$$

where A is a constant. Consider, for example, the disturbance excited by an isolated low-latitude peak, with $\overline{u}$ increasing, and so K_s decreasing, polewards of the source. Assuming that initially there exists a zonal wavenumber k less than K_s then two eastward propagating wavetrains are excited. The meridional wavenumber of the poleward wavetrain diminishes according to (16.134b),

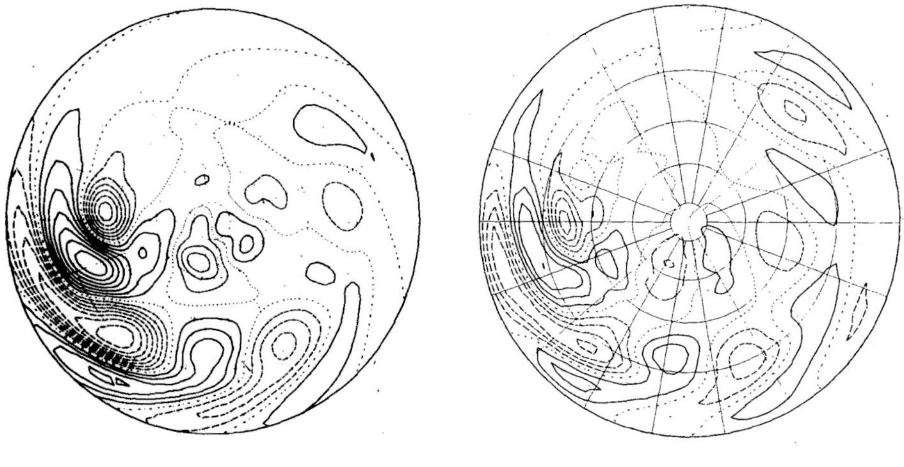

Fig. 16.19 The linear stationary response induced by a circular mountain at 30° N and at 180° longitude (nine o'clock). The figure on the left uses a barotropic model, whereas the figure on the right uses a multi-layer baroclinic model.[18] In both cases the mountain excites a low-wavenumber polar wavetrain and a higher-wavenumber subtropical train.

so that, using (16.131), the ray becomes more zonal. At the latitude where $k = K_s$, the 'turning latitude' the wave is reflected but continues propagating eastwards. The southward propagating wavetrain is propagating into a medium with smaller $\overline{u}$ and larger K_s. At the critical latitude, where $\overline{u} = 0$, $l \to \infty$ but c_g^x and c_g^y both tend to zero, but [using (16.130)] in such a way that $c_g^x/c_g^y \to 0$. That is, the rays become meridionally oriented and their speed tends to zero, and the waves may be absorbed.[17] Finally, we mention without derivation that for zonal flows with constant angular velocity the trajectories are great circles.

16.9.2 Application to an Idealized Atmosphere

Given the complexity of the real atmosphere, and the availability of computers, it is probably best to think of the remarks above as helping us interpret more complete numerical, but still linear, calculations of stationary Rossby waves — for example, numerical solutions of the stationary barotropic vorticity equation in spherical coordinates,

$$\frac{\overline{u}}{a\cos\vartheta}\frac{\partial\zeta'}{\partial\lambda} + v'\left(\frac{1}{a}\frac{\partial\overline{\zeta}}{\partial\vartheta} + \beta\right) = -\frac{\overline{u}f_0}{aH\cos\vartheta}\frac{\partial h_b}{\partial\lambda} - r\zeta', \tag{16.136}$$

where $[u, v] = a^{-1}[-\partial\psi/\partial\vartheta, (\partial\psi/\partial\lambda)/\cos\vartheta]$, $\beta = 2\Omega a^{-1}\cos\vartheta$ and $\zeta = \nabla^2\psi$. The last term in (16.136) crudely represents the effects of friction and generally reduces the sensitivity of the solutions to resonances. Solutions to (16.136) may be obtained first by discretizing and then numerically inverting a matrix, and although the actual procedure is quite involved it is analogous to the Fourier methods used earlier for the simpler one-dimensional problem. Such linear calculations, in turn, help us interpret the stationary wave pattern from more comprehensive models and in the Earth's atmosphere.

Figure 16.19 shows the stationary solution to the problem with a realistic northern hemisphere zonal flow and an isolated, circular mountain at 30° N. The topography excites two wavetrains, both of which slowly decay downstream because of frictional effects, rather like the one-dimensional wavetrain in Fig. 16.14. The polewards propagating wavetrain develops a more meridional orientation, corresponding to a smaller meridional wavenumber l, before moving southwards again, developing a much more zonal orientation eventually to decay completely as it meets the equatorial westward flow. The equatorially propagating train decays a little more rapidly than its polewards moving counterpart because of its proximity to the critical latitude. More complicated patterns naturally result if a realistic distribution of topography is used, as we see in Fig. 16.17. We can see wavetrains emanating from both the Rockies and the Himalayas, but distinct poleward and equatorward wavetrains are hard to discern.

Notes

1 Thanks to Isaac Held and Peter Haynes for various comments on Rossby waves and critical lines.

2 Early discussions of Rossby wave critical layers include those of Warn & Warn (1976), Stewartson (1977) and Killworth & McIntyre (1985). Booker & Bretherton (1967) consider gravity wave critical layers. Haynes (2015) provides an accessible review of both.

3 Modelled after a figure in Haynes (2015).

4 After Charney & Drazin (1961).

5 A quite extensive discussion of the thermal (and the topographically forced) problem is given by Pedlosky (1987a).

6 Much of our basic understanding in this area stems from conceptual and numerical work on forced Rossby waves by Charney & Eliassen (1949), who looked at the response to orography using a barotropic model. This was followed by a study by Smagorinsky (1953) on the response to thermodynamic forcing using a baroclinic, quasi-geostrophic model. Seeking more realism later studies have employed the primitive equations and spherical coordinates in studies that are at least partly numerical (e.g., Egger 1976, and a host of others), although most theoretical studies perforce still use the quasi-geostrophic equations. We also draw from various review articles, among them Smith (1979), Dickinson (1980), Held (1983), particularly for Sections 16.7.3 and 16.8, and Wallace (1983). See also the collection in the *Journal of Climate*, vol. 15, no. 16, 2002.

7 To obtain the solutions shown in Fig. 16.11 and Fig. 16.12, the topography is first specified in physical space. Its Fourier transform is taken and the streamfunction in wavenumber space is calculated using (16.98). The inverse Fourier transform of this gives the streamfunction in physical space.

8 The difference between wavetrains emanating from an isolated topographic feature and a global resonant response is relevant for intra-seasonal variability, which might be considered a quasi-stationary response to slowly changing boundary conditions like the sea-surface temperature. If resonance is important, we might expect to see global-scale anomalies, whereas the viewpoint of damped wave-trains is more local. This whole area is one of continuing, active, research with deep roots going back to Namias (1959) and Bjerknes (1959) and beyond.

A different point of view, one that we do not explore in this book, is that the zonally asymmetric features of the Earth's atmosphere are predominantly due to *nonlinear* effects. One possibility is that eddies might significantly modify (and perhaps amplify and sustain) stationary patterns through their large-scale turbulent transfers; see, for example, Green (1977) and Shutts (1983). We could incorporate such effects into a linear model by including the eddy effects as a forcing term on the right-hand side of a linear equation such as (16.90), or its two- or three-dimensional analogue, although the forcing term would have to be calculated using a nonlinear theory or taken from observations. Different again is the notion, inspired by models of low-order dynamical systems, that the atmosphere might have *regimes* of behaviour, and that the zonally asymmetric patterns are manifestations of the time spent in a particular regime before transiting to another. See for example Kimoto & Ghil (1993) and Palmer (1997).

9 See endnote 6 above for references. Because of these difficulties, understanding the effects of sea-surface temperature anomalies on the atmosphere has become largely the subject of GCM experiments, and one plagued with ambiguous results that depend in part on the particular configuration of the GCM. Some of the modelling issues are reviewed by Kushnir *et al.* (2002).

10 From Hoskins & Karoly (1981).

11 From Hoskins & Karoly (1981).

12 Adapted from Held *et al.* (2002).

13 Such solutions are nearly always most easily obtained numerically. One way is to use a Fourier method described earlier. A related method is to write the equations in finite difference form, schematically as $\mathbf{A}\boldsymbol{X} = \boldsymbol{F}$, where $\boldsymbol{X}$ is the vector of all the model fields, $\boldsymbol{F}$ represents the known forcing and $\mathbf{A}$ is a matrix obtained from the equations of motion and boundary conditions, and solve for $\boldsymbol{X}$. A quite different method is to use a nonlinear time-dependent model, such as a GCM: prescribe or hold steady the zonally averaged zonal flow as well as all the zonally asymmetric forcing terms, but multiply the asymmetric terms by a small number (e.g., 0.01) to ensure the response is linear; then calculate the steady response by forward time integration, and then divide that solution by the small number to obtain the final solution.

14 The description of the stationary waves in terms of wavetrains comes from Hoskins & Karoly (1981), with some earlier theoretical results having been derived by Longuet-Higgins (1964).

15 Steers (1962) and Phillips (1973).

16 From Hoskins & Karoly (1981).

17 At the critical latitude the WKB analysis fails and both dissipative and nonlinear effects are likely to play a role, as discussed by Dickinson (1968), Tung (1979) and others.

18 From Grose & Hoskins (1979) and Hoskins & Karoly (1981).

And beyond it, the deep blue air, that shows
Nothing, and is nowhere, and is endless.
Philip Larkin, *High Windows*, 1974.

CHAPTER 17

The Stratosphere

THE *STRATOSPHERE* IS THE REGION OF THE ATMOSPHERE above the troposphere and below the mesosphere; thus, it extends from the tropopause at a height of about 8–15 km, or a pressure of around 200–300 hPa, to the stratopause at about 50 km or about 1 hPa (see Fig. 15.24 on page 574). The *middle atmosphere* is the somewhat larger region that also includes the mesosphere, and so that extends up to the mesopause at about 90 km or 2×10^{-3} hPa, but we won't consider the mesosphere here. Our goal in this chapter is to provide an introduction to the dynamics giving rise to the structure and variability of the stratosphere.[1]

The outline of this chapter is roughly as follows. We begin with a rather descriptive overview of the stratosphere as a whole. Then, starting in Section 17.2, we discuss the Rossby and gravity waves that in many ways serve to drive the circulation. We come back to the circulation itself in Section 17.4, focusing mainly on the generation of zonal flows and the meridional residual overturning circulation. We round out the chapter with discussions of two striking examples of stratospheric variability, namely the quasi-biennial oscillation in Section 17.6, and extratropical variability and sudden warmings in Section 17.7, with these terms to be defined in the sections ahead.

17.1 A DESCRIPTIVE OVERVIEW

In the troposphere the stratification is determined by dynamical processes — largely by convection at low latitudes and additionally by baroclinic instability at high latitudes — and the tropopause is the height to which the dynamical activity reaches, as discussed in Chapter 15. In contrast, in the stratosphere the temperature is determined to a much greater degree by radiative processes and the dynamics are, compared to those in the tropopause, slow. Over much of the stratosphere the temperature actually increases with height, and this is due to a layer of ozone that absorbs solar radiation in the mid-stratosphere between about 20 and 30 km. If there were no ozone we would certainly have a tropopause and a stratosphere, but the temperature in the stratosphere would increase much less with height than it in fact does.

The radiative-equilibrium temperature for January is illustrated in Fig. 17.1. This temperature is that which would putatively ensue without any stratospheric fluid motion, although we take the distribution of absorbers (such as ozone) to be those present in the actual, moving, atmosphere, and the calculation involves a linearization around the observed temperature.[2] There is quite a strong lateral gradient in the winter hemisphere and a weaker reversed temperature in the summer hemisphere, and in fact the part of the stratosphere with the highest radiative equilibrium

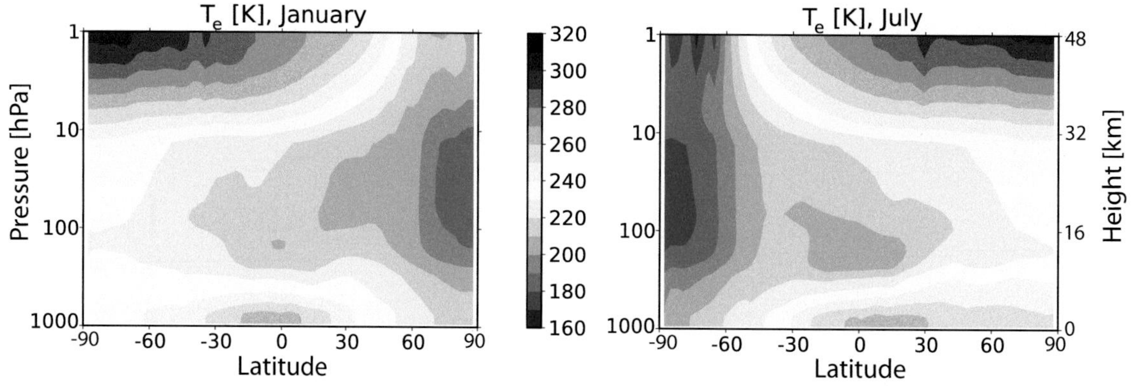

Fig. 17.1 The zonally averaged radiative-equilibrium temperature in January and July; that is, the temperature that would nominally arise in the absence of fluid motion in the stratosphere, but with the actual distribution of radiative absorbers. Above about 50 km the equilibrium temperature generally diminishes with height. The ordinate is pressure, and the height values are approximate.

temperature is the upper-stratosphere summer pole, at around 1 hPa. The actual observed zonally averaged temperature and zonal-wind structure are plotted in Fig. 17.2. From these figures we infer the following:

- The stratosphere is very stably stratified, with a typical lapse rate corresponding to $N \approx 2\times10^{-2}$ s, about twice that of the troposphere on average. This is in part due to the absorption of solar radiation by ozone between 20 and 50 km.
- In the summer the solar absorption at high latitudes leads to a reversed temperature gradient (warmer pole than equator) and, by thermal wind balance, a negative vertical shear of the zonal wind. The temperature distribution is not far from the radiative equilibrium distribution, and over much of the summer stratosphere the mean winds are negative (westward).
- In winter high latitudes receive very little solar radiation and there is a strong meridional temperature gradient and consequently a strong vertical shear in the zonal wind. Nevertheless, this temperature gradient is significantly weaker than the radiative equilibrium temperature gradient, implying a poleward heat transfer by the fluid motions.

How do the dynamics of the stratosphere differ from the troposphere? One way is that there is little, if any, baroclinic instability in the stratosphere — for various reasons. Suppose we first think of the stratosphere in isolation. There is no clear reversal of the potential vorticity gradient and no real opportunity for counter-propagating edge waves or Rossby waves to interact in the stratosphere, and hence stratosphere alone may simply be baroclinically stable. If the stratosphere were baroclinically unstable the instability would be much weaker, because of its higher stratification. A typical value of the static stability in the stratosphere is $N \approx 2\times10^{-2}\,\mathrm{s}^{-1}$, and using a height scale of 20 km gives a value of the deformation radius NH/f of about 4000 km, as opposed to the canonical value of 1000 km in the troposphere. (The stratospheric estimate is very approximate because the scale height, H_s is much less than 20 km, and a relevant deformation radius is then $N\sqrt{HH_s}/f$. But on the other hand one could also take $H > 20$ km, so 4000 km may be a fair estimate.) Thus, even with the same horizontal temperature gradient as the troposphere, a typical instability scale (of the stratosphere in isolation) would be large, perhaps at wavenumber 2 rather than wavenumber 8. The stratospheric growth rate would then be much less than in the troposphere: the Eady growth rate is given by $\sigma_E \equiv 0.31\Lambda H/L_d = 0.31U/L_d$, where Λ is the shear, giving the growth rate that is several times smaller than its tropospheric counterpart. Of course, if baroclinic instability has a modal form then the instability has the same horizontal scale and grows at the same rate in

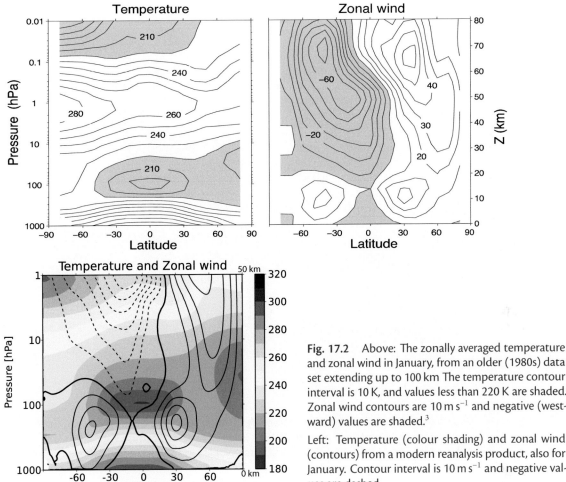

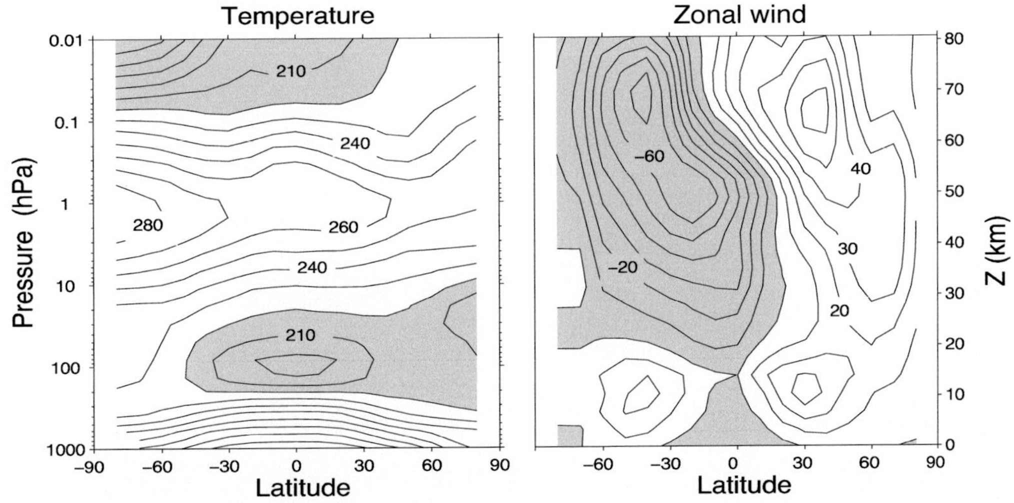

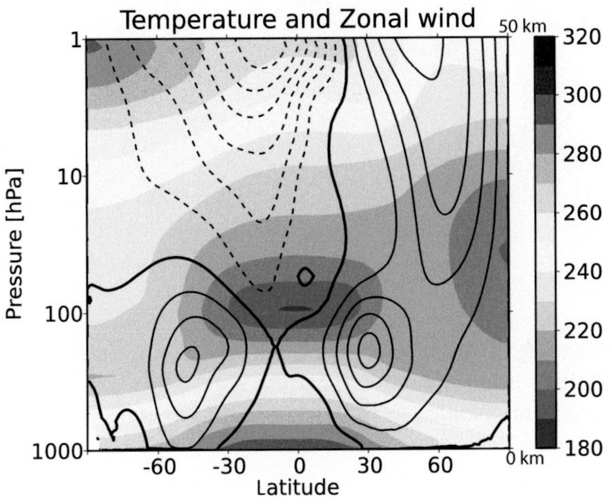

Fig. 17.2 Above: The zonally averaged temperature and zonal wind in January, from an older (1980s) data set extending up to 100 km The temperature contour interval is 10 K, and values less than 220 K are shaded. Zonal wind contours are 10 m s^{-1} and negative (westward) values are shaded.[3]

Left: Temperature (colour shading) and zonal wind (contours) from a modern reanalysis product, also for January. Contour interval is 10 m s^{-1} and negative values are dashed.

the stratosphere as the tropospheric one — it is the same mode! But in this case the higher lapse rate suppresses the amplitude of the stratospheric instability, as shown in Fig. 9.21.

For all these reasons, baroclinic instability is not the main process leading to a circulation in the stratosphere — the main process is the propagation and subsequent breaking and dissipation of gravity and Rossby waves from the troposphere to the stratosphere. This breaking will produce an acceleration of the zonal flow, and/or a meridional overturning circulation, and a good fraction of this chapter will be devoted to describing that process. But first we'll provide a little more description about the circulation itself, and it is convenient to divide that into two parts:

(i) a quasi-horizontal circulation;

(ii) a meridional overturning circulation (MOC) that is most usefully described as a residual circulation (the residual meridional circulation, or RMOC) using the TEM formalism.

17.1.1 The Quasi-Horizontal Circulation

In the extra-tropics the stratification is high and the Rossby number small and, at least to the extent that the scales of motion are not truly hemispheric the circulation is well described by the quasi-geostrophic equations. Now, not only does any stratospheric baroclinic instability tend to occur on a large scale, but so does any wave activity that arises from the propagation of Rossby waves up from

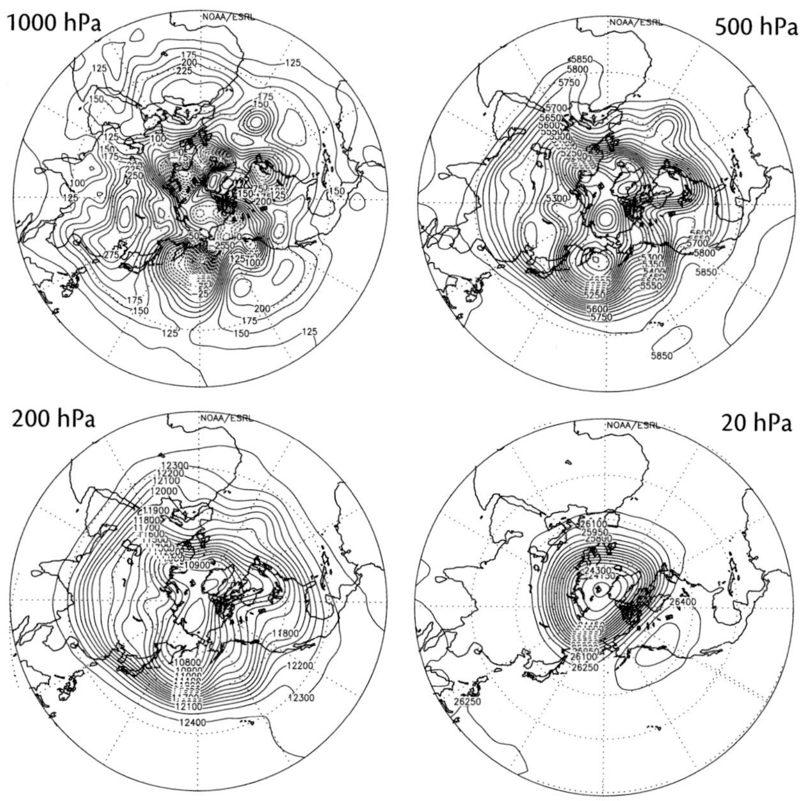

Fig. 17.3 The geopotential height on 1 February, 2000, at various levels in the atmosphere — 1000 and 500 hPa are in the troposphere, 200 hPa is around the tropopause and 20 hPa is in the mid-stratosphere, at about 30 km. Note the general increase in the scale of the variations of the geopotential with height.

the troposphere. This is because of Charney–Drazin filtering, summarized in Fig. 16.6: the smaller the wavelength the smaller is the range of zonal winds through which the waves can propagate. If the wind is too high the waves encounter a turning surface, whereas if the wind is too low they encounter a critical layer. Thus, we would expect that the general horizontal scale of motion is larger in the stratosphere than in the troposphere, and this is borne out by inspection of Fig. 17.3, which shows geopotential height at various levels. The complex patterns of the lower and mid-troposphere are well filtered, and in mid-stratosphere the pattern is dominated by wavenumbers 1 and 2. Indeed it seems from the figure (which is typical) that much of the motion is concentrated around a *polar vortex*.

Looking at geopotential (which roughly corresponds to a streamfunction) gives a somewhat misleading impression of the lack of activity away from the poles. Here, because diabatic effects occur on a rather longer time scale than advective processes, the flow may be characterized by the advection of potential vorticity on more slowly evolving isentropic surfaces, as illustrated in Figs. 17.4 and 17.5. Both the potential vorticity and the tracer are evocative of two-dimensional turbulence. We see Rossby waves breaking and vortices stretched into filaments and tendrils, the features of an enstrophy cascade. We also perceive some idea of the spectral non-locality of the enstrophy transfer — a single large vortex overturns and breaks and there is little sense of a spectrally-local cascade of enstrophy to dissipative scales. For this reason, the mid-latitude region is sometimes known as the *surf zone*. It is precisely this wave breaking that gives rise to the enstrophy flux to

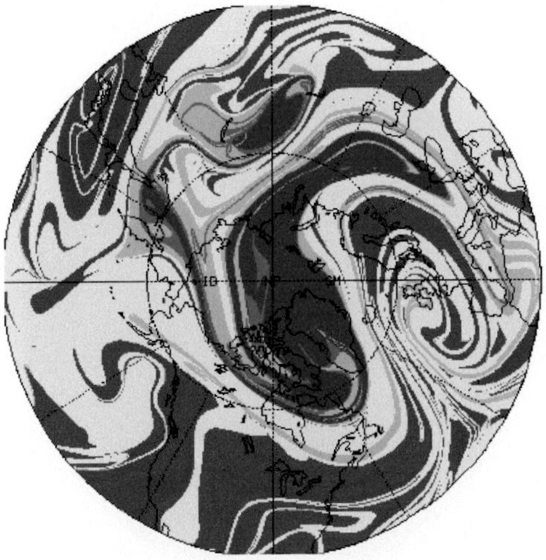

Fig. 17.4 The tracer distribution in the northern hemisphere lower stratosphere on 28 January 1992. The tracer was initialized on 16 January by setting it equal to the potential vorticity field calculated from an observational analysis, and then advected for 12 days by the observed wind fields.[4]

small scales and its dissipation, and which in turn gives rise to the overturning circulation that we discuss below.

The surf-zone does not usually extend to the pole, and in winter dense cold air over the pole forms itself into a cyclonic vortex, apparent in both Fig. 17.3 and 17.5. Although the vortex is ultimately the result of diabatic forcing, and has a preferred location, the tendency of quasi-two-dimensional flow to organize itself into vortices (as we see in Figs. 9.6 and 11.8) contributes to its coherence and isolation from the rest of the hemisphere. The boundary of the vortex, as measured by the value of the potential vorticity or of the tracer, is quite sharp with the value of PV often jumping by a factor of 2 or so, and the vortex is quite persistent — in fact it is a near-permanent feature of the winter hemisphere. Within the vortex potential vorticity tends to homogenize, and once formed the main communication that the vortex has with the surf zone is via occasional wave breaking at its boundary. It is interesting that, although the potential vorticity gradient is strong at the edge of the vortex, the exchange of properties is weak, implying a failure of notions of diffusion, or at least diffusion with a constant value of diffusivity; the edge of the vortex is a *mixing barrier*. We saw this property before, in our discussion of potential vorticity staircases in Section 12.1.3.

Stable as it is, the polar vortex is nevertheless sometimes disrupted by wave activity from below; this tends to occur when the wave activity itself is quite strong, and when the mean conditions are such as to steer that wave activity polewards. Occasionally, this activity is sufficiently strong so as to cause the vortex to break down, or to split into two smaller vortices, and so allow warm mid-latitude air to reach polar latitudes — an event known as a *stratospheric sudden warming*, and one such is illustrated in Fig. 17.22. We come back to the mechanism of such warmings later in the chapter.

17.1.2 The Overturning Circulation

That there is a meridional overturning circulation in the stratosphere was inferred by A. Brewer and G. Dobson based on observations of water vapour and chemical transport, and it is now often called the *Brewer–Dobson circulation*.[7] Brewer and Dobson both inferred the circulation on the basis of tracer transports, rather than performing an Eulerian average of velocity measurements (which would have been impossible then, and is still difficult now). Thus, the circulation they inferred was, in modern parlance, a residual circulation and some modern observations of this circulation are shown in Fig. 17.6. The figure actually shows the observed thickness-weighted circulation, which is almost equivalent to the residual circulation (Section 10.3.3), and which represents both the Eulerian mean and eddy-contributed components. We see a single, equator-to-pole cell

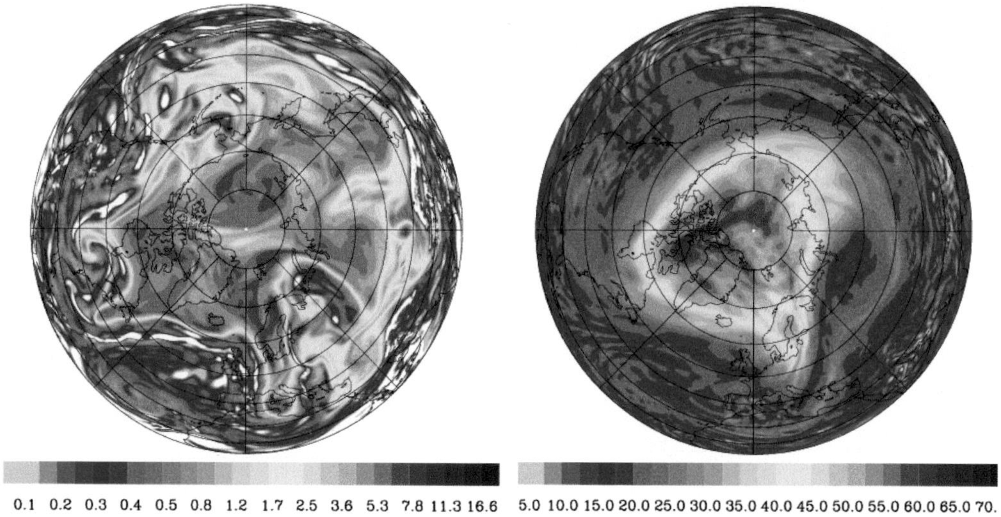

Fig. 17.5 The potential vorticity on two isentropic surfaces, the 310 K surface (left) and the 475 K surface (right), on 19 January, 2005. The shaded bar is in PV units. The 310 K surface is mainly in the troposphere (see Fig. 15.16) where baroclinic instability is abundant. The 475 K surface is at about 20 km altitude, and on it we see a polar stratospheric vortex with a fairly sharp boundary where the PV gradient is high, and a mid-latitude region of smaller-scale features and wave breaking.[5]

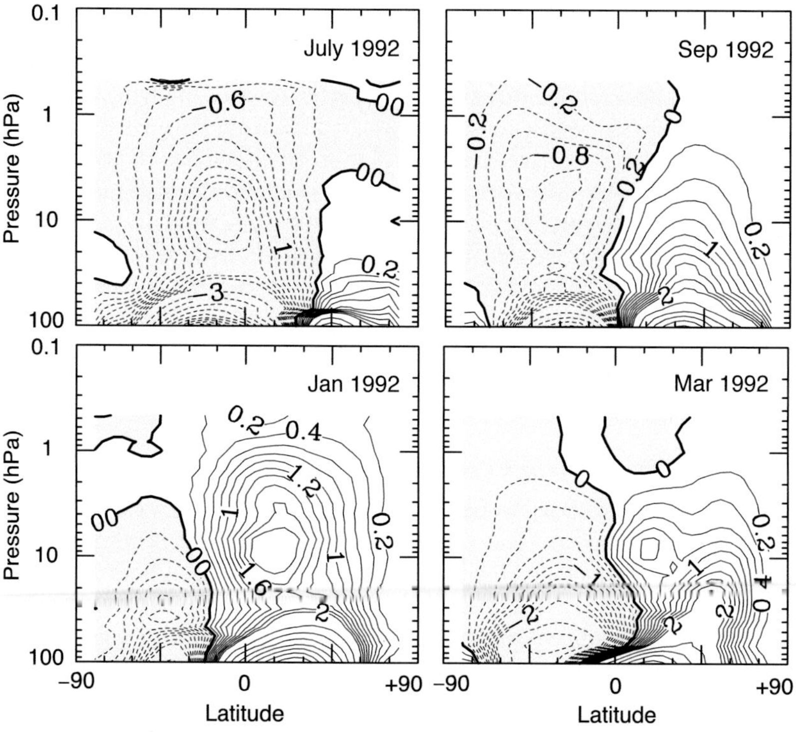

Fig. 17.6 The observed thickness-weighted (residual) streamfunction in the stratosphere, in Sverdrups ($10^9\,\mathrm{kg\,s^{-1}}$). The circulation is clockwise where the contours are solid.

The circulation is stronger in the winter hemispheres whereas the equinoctial circulations (September, March) are more inter-hemispherically symmetric.[6]

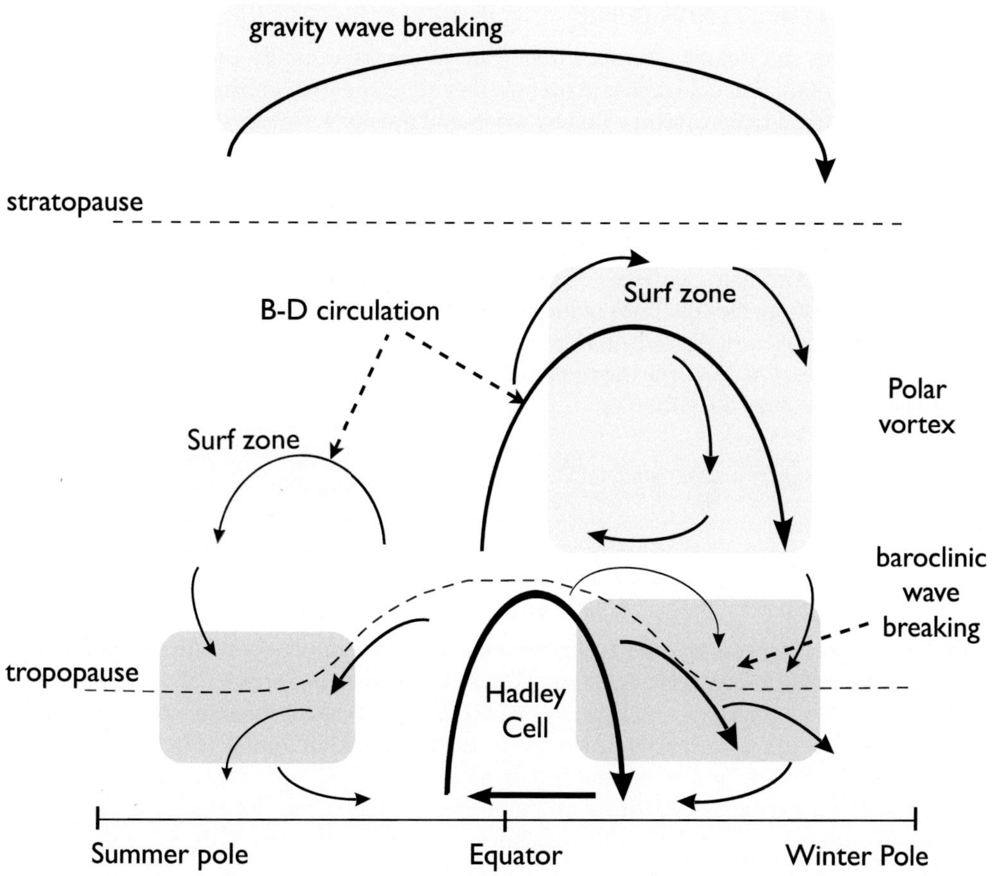

Fig. 17.7 A sketch of the residual mean meridional circulation of the atmosphere. The solid arrows indicate the residual circulation (B-D for Brewer–Dobson) and the shaded areas the main regions of wave breaking (i.e., enstrophy dissipation) associated with the circulation. In the surf zone the breaking is mainly that of planetary Rossby waves, and in the troposphere and lower stratosphere the breaking is that of baroclinic eddies. The surf zone and residual flow are much weaker in the summer hemisphere. Only in the Hadley Cell does the residual circulation consist mainly of the Eulerian mean; elsewhere the eddy component dominates.

in each hemisphere, stronger in the winter hemisphere where it goes high into the stratosphere. There is also a distinct lower branch to the circulation, present in all seasons although strongest in winter, that is confined to the lower stratosphere and is in some ways a vertical extension of (the residual circulation of) the tropospheric Ferrel Cell. Not all the upper circulation is ventilated by the troposphere — some of it recirculates within the stratosphere. This circulation and some of the associated dynamics are illustrated schematically in Fig. 17.7,[8] and three regions may usefully be delineated: (i) a tropical region; (ii) a mid-latitude region; (iii) the polar vortex. The tropical region is relatively quiescent, an area of upward motion where air is drawn up from the troposphere. In mid-latitudes the residual flow is generally directed poleward before sinking at high latitudes. In winter the extreme cold leads to the formation of the polar vortex, a strong cyclonic vortex that appears quite isolated from mid-latitudes although, especially in the Northern Hemisphere, it is not always centred over the pole.

Let us now turn to the dynamics of the circulation, and since this is in large measure dependent on the waves that exist in the stratosphere we first discuss them.

17.2 WAVES IN THE STRATOSPHERE

Both gravity waves and Rossby waves are important in the stratosphere and we have already discussed aspects of both. Our goal now is to see how they affect the stratosphere, first in mid-latitudes and then in equatorial regions, where Rossby waves and gravity waves are intertwined.

17.2.1 Linear Equations of Motion

Because we are dealing explicitly with compressible atmosphere we will use the ideal gas equations with log-pressure coordinates, as discussed in Section 2.6.3. (Nevertheless we will often find that, to a decent approximation, the equations reduce to the Boussinesq form, with compressibility having only a small effect.) We will restrict attention to the hydrostatic case, thereby limiting ourselves to relatively large scales. On a β-plane the equations of motion in log-pressure coordinates, linearized about a resting state, may be written as

$$\frac{\partial u}{\partial t} - fv = -\frac{\partial \Phi}{\partial x}, \qquad \frac{\partial v}{\partial t} + fu = -\frac{\partial \Phi}{\partial y}, \tag{17.1a,b}$$

$$\frac{\partial u}{\partial x} + \frac{\partial v}{\partial y} + \frac{1}{\rho_R}\frac{\partial(\rho_R w)}{\partial z} = 0, \qquad \frac{\partial}{\partial t}\frac{\partial \Phi}{\partial z} + wN_*^2 = 0. \tag{17.1c,d}$$

The equations are, respectively, the two horizontal momentum equations, the mass continuity equation and the thermodynamic equation. The notation is as in Section 2.6.3 except we use z, not Z, and w, not ω; thus, $z = -H\ln(p/p_R)$ where p_R is a constant reference pressure and H is a reference height, and $w = \mathrm{D}z/\mathrm{D}t$. As usual u and v are the horizontal velocities and Φ is the geopotential. The density profile ρ_R is an exponential, $\rho_R(z) = \rho_0\exp(-z/H)$, where we may take $\rho_0 = 1$, and N_*^2 is a reference stratification parameter similar to but not exactly the same as the buoyancy frequency; we will take it to be constant and drop the subscript $*$. The Coriolis parameter f varies as $f = f_0 + \beta y$; when we consider equatorial waves we will take $f_0 = 0$, and when we consider gravity waves in mid-latitudes we will take $\beta = 0$.

As in Section 16.5 it is convenient to extract that part of the solution that grows exponentially with height, and so seek wave solutions of the form

$$[u, v, w, \Phi] = [\tilde{u}(y), \tilde{v}(y), \tilde{w}(y), \tilde{\Phi}(y)]\mathrm{e}^{z/2H}\mathrm{e}^{\mathrm{i}(kx+mz-\omega t)}. \tag{17.2}$$

We cannot assume a simple harmonic form in the y-direction because the equations of motion have coefficients (i.e., f) that depend on y. Substituting (17.2) into (17.1) yields

$$-\mathrm{i}\omega\tilde{u} - f\tilde{v} = -\mathrm{i}k\tilde{\Phi}, \qquad -\mathrm{i}\omega\tilde{v} + f\tilde{u} = -\frac{\partial\tilde{\Phi}}{\partial y}, \tag{17.3a}$$

$$\mathrm{i}k\tilde{u} + \frac{\partial \tilde{v}}{\partial y} + \mathrm{i}\left(m + \frac{\mathrm{i}}{2H}\right)\tilde{w} = 0, \qquad -\mathrm{i}\omega\left(\frac{1}{2H} + \mathrm{i}m\right)\tilde{\Phi} + \tilde{w}N^2 = 0. \tag{17.3b}$$

Perhaps surprisingly, in many situations we can ignore the factor $1/2H$ in this system. Many observed stratospheric waves have a vertical wavelength, λ, that is of order a few kilometres and usually less than 10 km. Also, $T_0 = 240\,\mathrm{K}$ then $H \approx 7\,\mathrm{km}$. The $1/2H$ factor is small when $m \gg 1/2H$ or $4\pi H/\lambda \gg 1$. In fact, it is the square of this ratio that needs to be large, and this is true for all but the deepest stratospheric waves. Compressibility remains in the system because, using (17.2), all the perturbation variables grow exponentially with height, albeit slowly.

In the sections that follow we look at some of the waves supported by this system. The analysis is more complicated for equatorial regions, where gravity and Rossby waves are intertwined, so we begin with the mid-latitudes.

17.2.2 Waves in Mid-Latitudes

In mid-latitudes there is a good frequency separation between Rossby waves and gravity waves so they can be treated separately, and since we have already treated both in Chapters 7 and 16 our discussion focuses on their stratospheric relevance.

Rossby waves

If we neglect the factor of $1/H^2$ the x- and z-components of the group velocity are

$$c_g^x = \frac{(k^2 - l^2 - f_0^2 m^2/N^2)\beta}{(k^2 + l^2 + f_0^2 m^2/N^2)^2}, \qquad c_g^z = \frac{2km\beta f_0^2/N^2}{(k^2 + l^2 + f_0^2 m^2/N^2)^2}. \tag{17.4a,b}$$

Since $k < 0$ for Rossby waves then, in order for the waves to be upwardly propagating, (17.4b) requires that $m < 0$. Thus, the lines of constant phase tilt westward with height. The ratio of the vertical to the horizontal components of group velocity is not, unlike the case with gravity waves, a simple function of the wavenumbers and it is not possible to determine wither c_g^x is positive or negative without knowing the value of l, the meridional wavenumber. To obtain a typical value of the vertical group velocity in the atmosphere we may take $k^{-1} = 1000\,\text{km}$, $m^{-1} = 10\,\text{km}$, $f_0/N = 10^{-2}$, $\beta = 10^{-11}\,\text{m}^{-1}\,\text{s}^{-1}$, giving $c_g^z \sim 0.1\,\text{m}\,\text{s}^{-1} \approx 10\,\text{km/day}$.

Because Rossby waves grow in amplitude as they ascend the linearity assumption will eventually fail and the waves may break and dissipate, and in doing so they will deposit momentum. They may also break and/or dissipate if they encounter a critical line. This deposition is responsible for the production of the stratospheric meridional overturning circulation that we discuss later.

Gravity waves

Gravity waves may also propagate up into the stratosphere from the troposphere. If we take f to be a constant, f_0, then, except for the factor of ρ_R the set (17.1) is the same as the f-plane hydrostatic Boussinesq equations, namely (7.144) on page 280. It is a straightforward matter to show that the dispersion relation is

$$\omega^2 = \frac{f^2 m'^2 + (k^2 + l^2)N^2}{m'^2} = f^2 + \frac{N^2(k^2 + l^2)}{m'^2}, \tag{17.5}$$

where $m'^2 = m^2 + 1/4H^2$. As noted above the factor of $1/4H^2$ is often small and we shall ignore it. If we suppose that the horizontal component of the wave vector is aligned with the x-axis (i.e. $l = 0$) then the group velocity components are

$$c_g^x = \frac{N^2 k}{\omega m^2} = \frac{N^2}{\omega m}\cos\vartheta, \qquad c_g^z = -\frac{N^2(k^2 + l^2)m}{\omega m^4} = \frac{-N^2}{\omega m}\cos^2\vartheta, \tag{17.6a,b}$$

where $\cos^2\vartheta = k^2/(k^2 + m^2) \approx k^2/m^2 \ll 1$. The above expressions are most easily derived directly from (17.5) but are also the hydrostatic limit of the full expression (7.140). The directional aspects of these expressions are the same as those given for the non-rotating case in (7.74) with $\sin\vartheta = 1$, consistent with the hydrostatic limit — indeed we can obtain (17.6) from (7.74) by setting $\cos\lambda = 1$, $\sin\vartheta = 1$, $\omega = N\cos\vartheta$ and $\kappa = m$. Thus, the relation of the group velocity to the phase speed is much the same as for the gravity waves considered in Chapter 7, and in particular we have $c_g^z/c_g^x = -k/m$. If the waves are generated in the troposphere then c_g^z must be positive and so m must be negative. In mid-latitudes there is however no requirement that the horizontal propagation be in any particular direction. The distributions of velocity, pressure and temperature are illustrated in the two panels of Fig. 17.8 for waves with a positive and negative horizontal wavenumber and a negative vertical wavenumber.

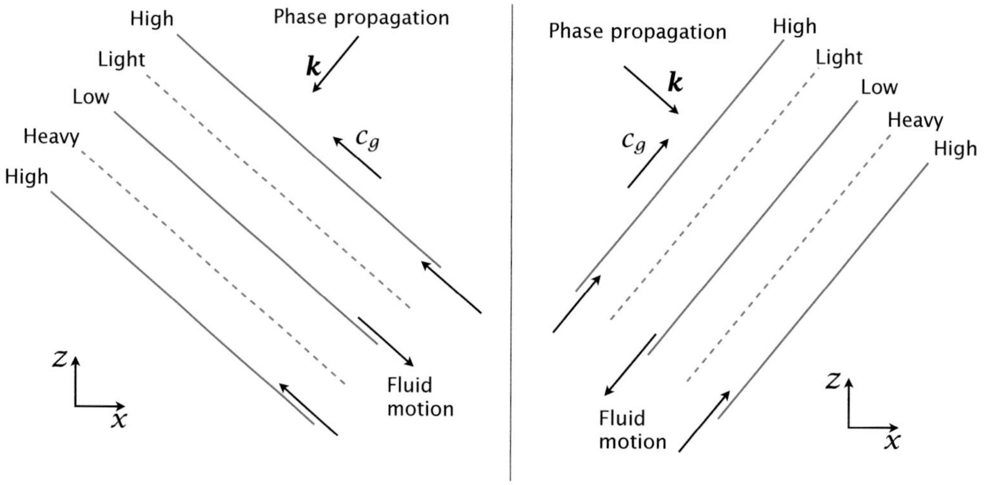

Fig. 17.8 Phase relationships for two examples of upwardly propagating gravity waves. The sketch on the left shows waves propagating to the left, with $k < 0$, and the one on the right shows waves with $k > 0$. The solid and dashed lines are contours of constant phase: 'high' and 'low' refer to pressure and 'light' and 'heavy' refer to density and correspond to warm and cold, respectively. In both sketches m is negative and the group velocity is directed upward and phase propagates downward. For hydrostatic flow the phase lines would be nearly horizontal. The figure may be compared with Fig. 7.3.

17.2.3 Waves in the Equatorial Stratosphere

Equatorial waves in the stratosphere — Kelvin waves and Rossby waves — turn out to be particularly important in generating stratospheric variability. It transpires that there can be vertically propagating waves with both eastward and westward phase speeds, even at relatively low frequencies, and this difference in phase speed has important consequences. We first look at Kelvin waves, for these provide a gentle introduction via a special treatment, and follow this by a more general treatment that includes Rossby and gravity waves.

Kelvin waves

We obtain the Kelvin wave solution by setting $\widetilde{v} = 0$ everywhere in (17.3), whence, after eliminating $\widetilde{w}$, (17.3) straightforwardly becomes

$$\omega\widetilde{u} = k\widetilde{\Phi}, \qquad f\widetilde{u} = -\frac{\partial\widetilde{\Phi}}{\partial y}, \qquad \omega\left(m^2 + \frac{1}{4H^2}\right)\widetilde{\Phi} - N^2 k\widetilde{u} = 0. \tag{17.7a,b,c}$$

Equations (17.7a,b) give $\omega\partial\widetilde{u}/\partial y + kf\widetilde{u} = 0$, which upon integration and with $f = \beta y$ yields

$$\widetilde{u}(y) = \widetilde{u}_0 \mathrm{e}^{-\beta y^2/2c_p}, \tag{17.8}$$

where $c_p = \omega/k$ and $\widetilde{u}_0$ is the value of $\widetilde{u}$ at the equator. The exponential fall-off is familiar from our earlier studies of Kelvin waves in Chapter 8 and requires that $c_p > 0$, meaning that the phase speed of the waves is eastward. Also, since $\omega > 0$ by convention, the x-wavenumber is positive (i.e, $k > 0$). The dispersion relation for Kelvin waves follows easily from (17.7a,c) and is

$$\omega^2 = \frac{N^2k^2}{m^2 + 1/4H^2}. \tag{17.9}$$

Aside from the factor of $1/4H^2$, which in any case is often small compared to m^2, (17.9) is essentially the same as the dispersion relation for hydrostatic gravity waves, namely (7.60) on page 261. The zonal and vertical components of the group velocity are

$$c_g^x = \frac{N}{(m^2 + 1/4H^2)^{1/2}}, \qquad c_g^z = \frac{\partial \omega}{\partial m} = \frac{-Nkm}{(m^2 + 1/4H^2)^{3/2}}. \tag{17.10}$$

Now, for upwardly propagating waves (and so for waves that emanate from the troposphere) we require $c_g^z > 0$ and therefore (because $k > 0$) $m < 0$. The combined conditions of $k > 0$ and $m < 0$ mean that the phase lines tilt eastward with height, as in the right panel of Fig. 17.8. Finally, note that the frequency of Kelvin waves, unlike inertia-gravity waves, is uninfluenced by rotation and thus, as seen in Fig. 8.6, can extend over a broad range.

A more general treatment of equatorial waves

For simplicity let us assume that the scale height H is very large compared to the vertical wavelengths of interest; that is, $m^2 \gg 1/H^2$ and $\exp(-z/H) = 1$. Eqs. (17.1b) and (17.1c) then combine to give

$$\frac{\partial}{\partial t}\frac{\partial^2 \Phi}{\partial z^2} - N^2\left(\frac{\partial u}{\partial x} + \frac{\partial v}{\partial y}\right) = 0. \tag{17.11}$$

If we assume a vertical structure of the form $\Phi(x, y, z, t) = \widetilde{\Phi}(x, y, t)\exp(imz)$, and similarly for u and v, then we obtain

$$\frac{\partial \widetilde{\Phi}}{\partial t} + \frac{N^2}{m^2}\left(\frac{\partial \tilde{u}}{\partial x} + \frac{\partial \tilde{v}}{\partial y}\right) = 0, \tag{17.12a}$$

with corresponding momentum equations

$$\frac{\partial \tilde{u}}{\partial t} - f\tilde{v} = -\frac{\partial \widetilde{\Phi}}{\partial x}, \qquad \frac{\partial \tilde{v}}{\partial t} + f\tilde{u} = -\frac{\partial \widetilde{\Phi}}{\partial y}. \tag{17.12b,c}$$

Evidently, (17.12) are isomorphic to the linear shallow water equations (8.20) on page 304 with the replacement $c^2 = N^2/m^2$, where in (8.20) $c^2 \equiv gH_e$ where H_e is the equivalent depth. (Note that c is not necessarily the phase speed in this problem; we will denote that as c_p.) The correspondence between the continuously-stratified and shallow water equations is a general one, as we found in Section 3.4. All of the machinery following (8.20) can now be applied to (17.12), with Rossby-gravity waves and Kelvin waves emerging in the same way (and the Kelvin waves identified above emerge as a special case); thus, in what follows we draw directly from Section 8.2. We will use m to denote the vertical wavenumber and n to denote the order of the Hermite function, akin to a meridional wavenumber.

Rossby-gravity waves

The dispersion relation that emerges from (17.12) is, by analogy with (8.37b), (8.53) and (8.63),

$$\omega^2 - c^2k^2 - \beta\frac{kc^2}{\omega} = (2n+1)\beta c, \qquad n > 0, \tag{17.13a}$$

$$\omega^2 - \omega k c - \beta c = 0, \qquad n = 0, \tag{17.13b}$$

$$\omega = ck, \qquad n = -1, \tag{17.13c}$$

where $c = N/m$. The case with $n = 0$ is the Yanai wave (a Rossby-gravity wave) and the cases with $n \geq 1$ are planetary waves or gravity waves. The '$n = -1$' case is the Kelvin wave, and all cases are illustrated in Fig. 8.6.

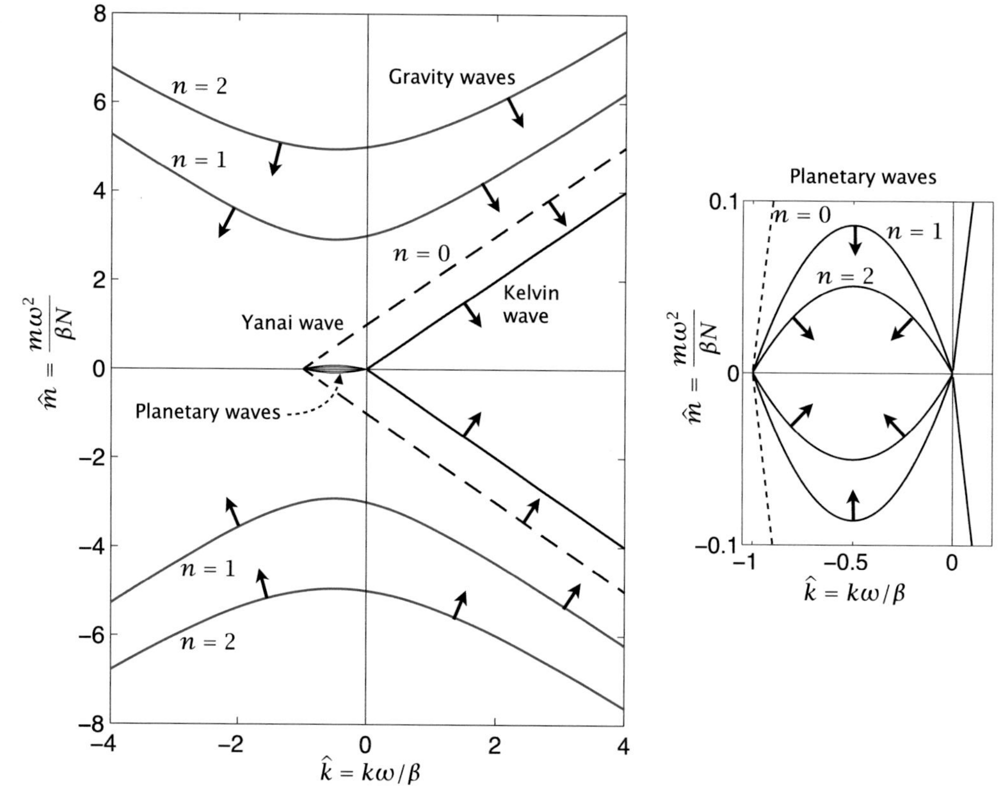

Fig. 17.9 Dispersion curves plotted in $\widehat{m}$–$\widehat{k}$ space, using (17.15). Shown are gravity waves, the Yanai wave and the Kelvin wave for positive and negative $\widehat{m}$, with the plot at the right showing a magnification of the region near the origin. The arrows in the figure indicate the group velocity, which, being the gradient of the frequency, is perpendicular to the curves. Upward propagating waves occur for negative m. Compare with Fig. 8.6.

For waves whose origin is in the troposphere we may think of the frequency as being given and (17.13) then provides a condition on the vertical wavenumber. This in turn suggests that we nondimensionalize the wavenumbers by defining

$$\widehat{m} = \frac{\omega^2}{\beta N} m = \frac{\omega^2}{\beta c}, \qquad \widehat{k} = \frac{\omega}{\beta} k, \tag{17.14}$$

with the hats denoting nondimensional variables. Equations (17.13) become

$$\widehat{m}^2 - (2n+1)\widehat{m} - \widehat{k}^2 - \widehat{k} = 0, \qquad n > 0, \tag{17.15a}$$

$$\widehat{m} - \widehat{k} - 1 = 0, \qquad n = 0 \tag{17.15b}$$

$$\widehat{m} = \widehat{k}, \qquad n = -1. \tag{17.15c}$$

These equations define a set of curves in $\widehat{m}$–$\widehat{k}$ space that are similar to the curves in ω–k space defined by (17.13), although (17.15) are lower order. Equation (17.15a) has the solution

$$\widehat{m} = \left(n + \tfrac{1}{2}\right) \pm \left[\left(\widehat{k} + \tfrac{1}{2}\right)^2 + n(n+1)\right]^{1/2}, \tag{17.16}$$

and the complete set of curves is plotted in Fig. 17.9. The curves at the top and bottom are gravity waves, corresponding to the positive sign in (17.16), and the planetary waves are the curves just to the left of the origin, corresponding to the negative sign. The $n = 0$ curve (the Yanai wave) and the $n = -1$ curve (the Kelvin wave) are labelled.

We can infer the group velocity from the figure by noting that, since β and N are constant, and using (17.14), the curves are contours of constant frequency. The group velocity is the gradient of frequency in wavenumber space and so is at right angles to these curves and directed toward a higher frequency, and is marked by short arrows. For waves propagating up from the troposphere the group velocity must be upward, and therefore have negative m, as can be seen from (17.6) and (17.10).

17.3 WAVE MOMENTUM TRANSPORT AND DEPOSITION

Steady waves by themselves don't affect the mean flow. Rather, they affect the mean flow when they are generated or dissipated, and typically they are generated in the troposphere and dissipated in the stratosphere. Let's look into that, beginning with Rossby waves.

17.3.1 Rossby Waves

The vertical transport of zonal momentum by eddies is given by $\overline{w'u'}$, where an overbar denotes a zonal average. However, directly evaluating this expression from the quasi-geostrophic equations is not particularly simple because w is not a first order variable — it results from the divergence of the ageostrophic horizontal velocities.[9] In fact, under quasi-geostrophic scaling we neglect the vertical eddy flux divergences, but nevertheless the eddy fluxes may certainly make themselves felt aloft, among other things by generating a form stress that acts to transfer momentum vertically and generating a meridional overturning circulation, as discussed in Section 10.4.3.

To proceed we will use the transformed Eulerian mean (TEM) framework of Section 10.3, for which the inviscid and adiabatic zonally-averaged momentum and thermodynamic equations are

$$\frac{\partial \overline{u}}{\partial t} - f_0 \overline{v}^* = \overline{v'q'}, \qquad \frac{\partial \overline{b}}{\partial t} + N^2 \overline{w}^* = 0, \tag{17.17a,b}$$

where potential vorticity flux is related to the EP flux, $\boldsymbol{\mathcal{F}}$, by

$$\overline{v'q'} = \nabla \cdot \boldsymbol{\mathcal{F}}, \qquad \boldsymbol{\mathcal{F}} = -\overline{u'v'}\,\mathbf{j} + \frac{f_0}{N^2}\overline{v'b'}\,\mathbf{k}, \tag{17.18}$$

and now for simplicity we use the Boussinesq equations with b as buoyancy. If the eddy fluxes are due to the presence of waves that satisfy a dispersion relation then, as shown in Section 10.2.2, the EP flux is related to the group velocity by

$$\boldsymbol{\mathcal{F}} = (\mathcal{F}^y, \mathcal{F}^z) = \mathbf{c}_g \mathcal{P}. \tag{17.19}$$

Combining the above equations we obtain

$$\frac{\partial \overline{u}}{\partial t} - f_0 \overline{v}^* = \frac{\partial}{\partial y}(c_g^y \mathcal{P}) + \frac{\partial}{\partial z}(c_g^z \mathcal{P}) = \nabla \cdot \boldsymbol{\mathcal{F}}. \tag{17.20}$$

Now, reprising (10.29a), the wave activity, specifically the pseudomomentum, $\mathcal{P}$, satisfies a conservation law of the form

$$\frac{\partial \mathcal{P}}{\partial t} + \nabla \cdot \boldsymbol{\mathcal{F}} = \mathcal{D}, \tag{17.21}$$

where $\mathcal{P} = \overline{q'^2}/2\beta$, which is a positive quantity, and $\mathcal{D}$ represents dissipation. If $\mathcal{D} = 0$ and the waves are steady then $\nabla \cdot \boldsymbol{\mathcal{F}} = 0$ so that the left-hand side of (17.20) is zero. (This is the non-acceleration result of Section 10.4.2.) Evidently in order to get a zonal flow acceleration or an MOC we need to invoke some dissipative or time-dependent processes.

Consider the case in which Rossby waves propagate up from the troposphere, with $c_g^z > 0$. Suppose that there is some dissipation in the system (and/or that Rossby waves break as they ascend) and that wave activity $\mathcal{P}$ falls with height, and suppose further that we are in a statistically steady state. In this case $\nabla \cdot \boldsymbol{\mathcal{F}} < 0$ and from (17.17a) this will produce a mean flow deceleration (i.e., a westward acceleration) and/or a polewards residual flow. The balance between these two possibilities is discussed later, but one may intuit that close to the equator the more likely outcome is a zonal acceleration rather than a meridional circulation. Why is there a preferred sense of acceleration when the waves break? Ultimately it is because of the beta effect which distinguishes east from west, and the pseudomomentum $\mathcal{P}$ is proportional to beta. The beta effect leads to a particular orientation of the phase of Rossby waves and this carries westward momentum away from the source region. When the waves break, be it in the tropospheric subtropics or in the stratosphere, a westward momentum is deposited.

17.3.2 Gravity and Kelvin waves

Now consider the vertical momentum transport in gravity waves. If these are uninfluenced by rotation, or if they reside on the f-plane, then there is no preferred horizontal direction of propagation. Kelvin waves, on the other hand, propagate their phase eastward only. The upward transport of momentum from waves originating in the troposphere can occur only for waves with a positive group velocity, and thus for either of the examples illustrated in Fig. 17.8. Those waves that propagate phase westward (i.e., have $k < 0$) have $\overline{u'w'} < 0$, and those that propagate phase eastward, such as equatorial Kelvin waves, have $\overline{u'w'} > 0$. The contribution to the zonal flow acceleration by the wave transport is given by

$$\frac{\partial \overline{u}}{\partial t} = -\frac{\partial}{\partial z}\overline{u'w'} + \text{other terms.} \tag{17.22}$$

and if the amplitude of the waves stays constant with height then no mean flow acceleration is induced. However, if the amplitude diminishes with height, because of dissipative processes, then the waves that have a westward (eastward) phase propagation will cause the zonal flow to accelerate westward (eastward). Thus *the dissipation of Kelvin waves as they propagate vertically will cause an eastward acceleration of the zonal flow.*

17.3.3 ✦ Processes of Wave Attenuation

We now explicitly consider the dissipation of waves and the associated momentum deposition as gravity waves propagate vertically. (The reader may wish to skim Section 16.3 before proceeding, for there we consider similar but algebraically simpler problems.) To keep the algebra manageable we will consider the propagation of two-dimensional (x–z) gravity waves in a Boussinesq fluid uninfluenced by rotation. The momentum and buoyancy equations, linearized about a zonal flow $U(z)$ and constant stratification N^2, are

$$\frac{\partial u}{\partial t} + U\frac{\partial u}{\partial x} = -\frac{\partial \phi}{\partial x}, \qquad \frac{\partial w}{\partial t} + U\frac{\partial w}{\partial x} = -\frac{\partial \phi}{\partial z} + b, \tag{17.23a}$$

$$\frac{\partial b}{\partial t} + U\frac{\partial b}{\partial x} + wN^2 = -\alpha b. \tag{17.23b}$$

We include a damping term, $-\alpha b$, where α is a constant, in the buoyancy equation but neglect viscous effects in the momentum equation. If we cross-differentiate the momentum equation and use the mass continuity equation ($\partial u/\partial x + \partial w/\partial z = 0$) we obtain the linear vorticity equation

$$\left(\frac{\partial}{\partial t} + \frac{\partial}{\partial x}\right)\nabla^2\psi + \frac{\partial \psi}{\partial x}\frac{\mathrm{d}^2 U}{\mathrm{d}z^2} = \frac{\partial b}{\partial x}, \tag{17.23c}$$

where ψ is such that $w = \partial\psi/\partial x$ and $u = -\partial\psi/\partial z$. We seek solutions of (17.23) in the form

$$[\psi, b] = [\widetilde{\psi}(z), \tilde{b}(z)]\mathrm{e}^{\mathrm{i}k(x-ct)}, \tag{17.24}$$

and a little algebra reveals

$$\mathrm{i}(-kc + Uk)\left(-k^2\psi + \frac{\mathrm{d}^2\widetilde{\psi}}{\mathrm{d}z^2}\right) + \mathrm{i}k\psi\frac{\mathrm{d}^2U}{\mathrm{d}z^2} = \mathrm{i}k\tilde{b}, \tag{17.25a}$$

$$\mathrm{i}(-kc + Uk)\tilde{b} + \mathrm{i}kN^2\widetilde{\psi} = -\alpha\tilde{b}. \tag{17.25b}$$

These last two equations combine to give

$$\frac{\mathrm{d}^2\widetilde{\psi}}{\mathrm{d}z^2} + m^2(z)\widetilde{\psi} = 0, \qquad m^2(z) = \left[\frac{N^2[1 + \mathrm{i}\alpha/k(U-c)]}{(U-c)^2 + \alpha^2/k^2} - k^2 - \frac{\mathrm{d}^2U/\mathrm{d}z^2}{(U-c)}\right]. \tag{17.26}$$

This is an equation for the vertical structure of the streamfunction. If U and N^2 were constant in (17.26) then m would be constant and its real part would be the vertical wavenumber. There is an imaginary component to m which can be expected to cause the solution to decay in the vertical. In most circumstances the decay is slow but in the neighbourhood of a critical line where $c = U$ then the decay will be rapid. The wave can be expected to deposit momentum and accelerate or decelerate the mean flow, depending on the direction of the phase propagation of the wave. When $\alpha = 0$ there is no dissipation and no deposition (except possibly at the critical line itself). Although the equation seems complicated, we can proceed to a solution if we make some reasonable simplifying assumptions:

(i) We consider (as is realistic) low aspect ratio flows ($L_z/L_x \ll 1$) so that the factor of k^2 is small compared to m^2 and may be neglected.

(ii) We assume that the variation of the mean flow occurs on a long vertical scale compared to m, and so neglect the term in $\mathrm{d}^2U/\mathrm{d}z^2$. This assumption also allows us to use WKB methods.

(iii) We assume dissipation is small, and in particular that $\alpha/k(U-c) \ll 1$.

With these approximations (17.26b) becomes

$$m(z) \approx \left[\frac{N^2[1 + \mathrm{i}\alpha/k(U-c)]}{(U-c)^2 + \alpha^2/k^2}\right]^{1/2} \approx \frac{N}{U-c}\left[1 + \frac{\mathrm{i}\alpha}{2k(U-c)}\right], \tag{17.27}$$

and we can proceed with a WKB solution.

WKB solution and momentum flux

The WKB solution to (17.26) is

$$\widetilde{\psi}(z) = Am^{-1/2}\exp\left(\pm\mathrm{i}\int^z m\,\mathrm{d}z'\right), \tag{17.28}$$

where A is a constant. The wave momentum flux, F, associated with the wave is

$$F_k(z) = \overline{u'w'} = -\mathrm{i}k\left(\widetilde{\psi}\frac{\partial\widetilde{\psi}^*}{\partial z} - \widetilde{\psi}^*\frac{\partial\widetilde{\psi}}{\partial z}\right), \tag{17.29}$$

where the overbar denotes a zonal average and the right-hand side is always real, and the subscript on F indicates we are considering the effects of a single wave of wavenumber k.

Now, the fact that m varies only slowly with z (specifically $m^2 \gg |\mathrm{d}m/\mathrm{d}z|$) means that when we take the vertical derivative of $\widetilde{\psi}$ we can ignore the derivative of the vertical derivative of the amplitude, $Am^{-1/2}$. Given this, and using (17.28) and (17.27) in (17.29) we obtain

$$F_k(z) = F_0\exp\left(\mathrm{i}\int_0^z (m - m^*)\,\mathrm{d}z'\right) = F_0\exp\left(\int_0^z \frac{-N\alpha}{k(U-c)^2}\,\mathrm{d}z'\right). \tag{17.30}$$

where F_0 is the value of the flux at $z = 0$ and we have chosen the sign in the exponent to be appropriate for upwardly propagating gravity waves. The integrand in the right-most expression is the attenuation rate of the wave and, referring to (7.80) on page 266, it can be written as

$$\text{Attenuation rate} = \frac{\alpha}{k(U-c)^2/N} = \frac{\text{Dissipation rate}}{\text{Vertical group velocity}}. \tag{17.31}$$

As $U - c$ diminishes the group velocity falls, giving the dissipative processes more time to act. The result of (17.31) is a general one; we found an almost identical result when looking at the absorption of Rossby waves in Section 16.3 — see (16.49). (The dissipation rate in expressions like (16.49) and (17.31) is that of wave activity, which in the gravity wave case here equals the thermal dissipation rate.)

Effect on the mean flow

If a wave propagating upward is attenuated there will be a divergence in the eddy momentum flux associated with that wave. In particular, if α in (17.31) is non-zero then momentum flux deposition will increase rapidly as a critical layer is approached, $\partial F_k/\partial z$ will be non-zero and the zonal mean flow will be accelerated or decelerated. For definiteness, consider a wave propagating upward with a positive phase speed (so $m < 0$ and $k > 0$). From (17.30) F_k diminishes with height and the mean flow is accelerated eastward. Similarly, absorption of a wave of negative phase speed leads to a negative, or westward mean-flow acceleration. It is not inconceivable to imagine that the wave deposition will affect the mean flow to an extent that the position of the deposition is significantly altered, leading to interesting dynamical behaviour. Indeed this is precisely what happens in the quasi-biennial oscillation of the equatorial stratosphere. But before we discuss variability let us discuss the maintenance of the mean state.

17.4 PHENOMENOLOGY OF THE RESIDUAL OVERTURNING CIRCULATION

We now return to a discussion of the general circulation of the stratosphere and in particular the maintenance of the residual meridional overturning circulation (RMOC), or the Brewer–Dobson circulation. We expect the circulation to be a consequence of waves coming up from the troposphere and breaking, with both tropospheric baroclinic instability and flow over thermal and topographic zonal asymmetries being sources of wave activity. We may naturally ask such questions as whether wavebreaking can give rise to a circulation of the right strength and the right sense, whether it will give the correct seasonal variability, and what determines vertical extent of the circulation. We begin with some elementary theory and phenomenology, for we will find that it will explain a number of the main features of the RMOC, and the advanced or confident reader may skip ahead to Section 17.5.

17.4.1 Wave Breaking and Residual Flow

The equations of motion governing the mean fields are the zonally averaged momentum and thermodynamic equations, which with quasi-geostrophic scaling and in residual form (Section 10.3.1) may be written as

$$\frac{\partial \overline{u}}{\partial t} - f_0 \overline{v}^* = \nabla \cdot \boldsymbol{\mathcal{F}} + \overline{F}, \qquad \frac{\partial \overline{\theta}}{\partial t} + \frac{\partial \overline{\theta}}{\partial z}\overline{w}^* = \overline{J}, \tag{17.32a,b}$$

where $\overline{F}$ represents frictional effects and $\overline{J}$ represents heating, and on the β-plane the residual velocities are related to the Eulerian means by

$$\overline{v}^* = \overline{v} - \frac{1}{\rho_R}\frac{\partial}{\partial z}\left(\rho_R \frac{\overline{v'\theta'}}{\partial_z \overline{\theta}}\right), \qquad \overline{w}^* = \overline{w} + \frac{\partial}{\partial y}\left(\frac{\overline{v'\theta'}}{\partial_z \overline{\theta}}\right). \tag{17.33}$$

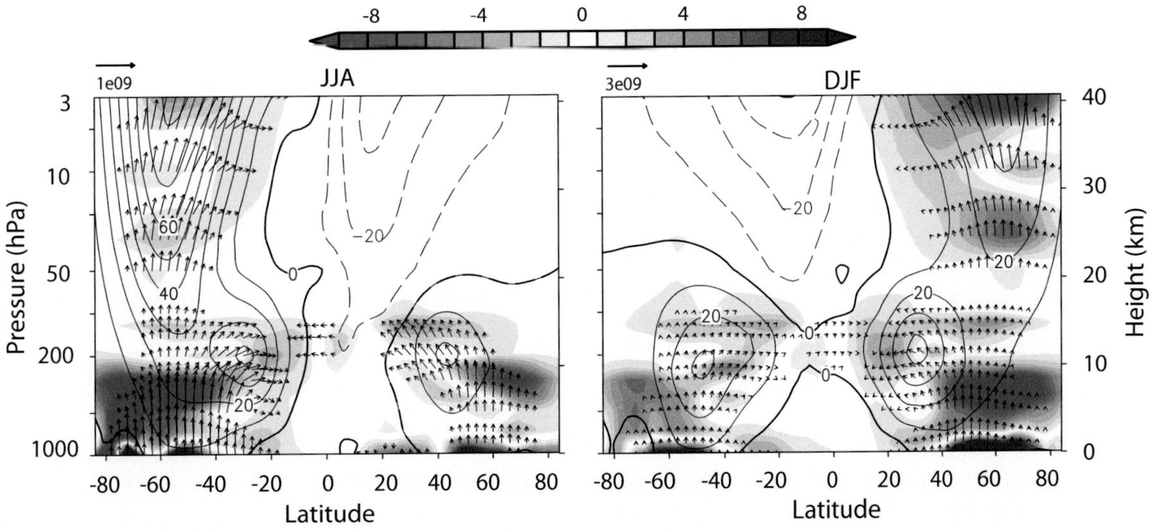

Fig. 17.10 The Eliassen–Palm (EP) flux vectors (arrows), the EP flux divergence (shading) and the zonally averaged zonal wind (contours) for northern hemisphere summer and winter. The tropospheric EP fluxes are of similar magnitudes in the summer and winter hemispheres, but are almost zero in the summer in the stratosphere. Note also strong convergence at high latitudes in the stratospheric winter hemispheres, leading to poleward residual flow and/or zonal flow acceleration.[10]

The vector $\boldsymbol{\mathcal{F}}$ is the Eliassen–Palm flux, and this is related to the meridional flux of potential vorticity by $\nabla \cdot \boldsymbol{\mathcal{F}} = \overline{v'q'}$. The wave activity (or pseudomomentum) obeys the Eliassen–Palm relation

$$\frac{\partial \mathcal{P}}{\partial t} + \nabla \cdot \boldsymbol{\mathcal{F}} = \mathcal{D}, \tag{17.34}$$

where $\mathcal{P}$ is the pseudomomentum, $\boldsymbol{\mathcal{F}}$ is its flux and $\mathcal{D}$ is its dissipation.

From the autumn to the spring, the zonal wind in the stratosphere is generally receptive to planetary-scale Rossby waves propagating up from the troposphere (Fig. 16.6), although at high latitudes in winter there may be a period when the eastward zonal winds are too strong for waves to propagate. If these waves break in the stratosphere then there will be an enstrophy flux to small scales and dissipation. In a statistically-steady state and with small frictional effects the dominant balance in the zonal momentum equation (17.32a) is

$$- f_0 \overline{v}^* \approx \overline{v'q'}, \tag{17.35}$$

where $\overline{v}^*$ is the residual velocity and the potential vorticity flux on the right-hand side is induced by the Rossby wave breaking. In dissipative regions the zonally averaged potential vorticity flux will tend to be down its mean gradient and, if the potential vorticity gradient is polewards (largely because of the β-effect), the residual velocity will be positive if f_0 is positive. That is, the residual flow will be *polewards,* in both hemispheres, and the mechanism giving rise to this is called the 'Rossby wave pump'. Put another way, Rossby waves propagating up from the troposphere break and deposit westward momentum in the stratosphere, and in the mean this wave drag is largely balanced by the Coriolis force on the polewards residual meridional circulation.

This meridional circulation is weakest in summer mainly because linear Rossby waves cannot propagate upward through the westward mean winds, as illustrated in Fig. 17.10. It is quite striking how the EP vectors avoid the region of westward winds in the summer hemisphere, even though the level of wave activity at low elevations is relatively similar in the summer and winter hemispheres (look between 10 and 15 km in the figure). We can interpret this by noting that for nearly plane waves the EP flux obeys the group velocity property, meaning that $\boldsymbol{\mathcal{F}} = \mathbf{c}_g \mathcal{P}$; however, as discussed in Section 16.5, if the mean winds are westward the waves evanesce and do not propagate, and thus almost the entire summer hemisphere is shielded from upwardly propagating waves, leaving it in a near-radiative equilibrium state. In the other seasons, the EP flux is able to propagate into the stratosphere and a circulation is generated. This acts to weaken the pole–equator temperature gradient, as we see by inspection of the thermodynamic equation: if the heating is represented by a simple relaxation to a radiative equilibrium state, θ_E, then in a steady state we have

$$N^2 \overline{w}^* = \frac{\theta_E - \theta}{\tau}. \tag{17.36}$$

Poleward flow in mid-latitudes must be supplied by rising air at low latitudes, and sinking air at high latitudes. Thus, from autumn to spring, at low latitudes we have $\theta < \theta_E$ and at high latitudes $\theta > \theta_E$.

Although cause and effect can be very difficult to disentangle in fluid dynamical problems, and the ultimate cause of nearly all fluid motions in the atmosphere is the differential heating from the Sun, it is important to realize that the meridional overturning in the stratosphere is not a direct response to differential solar heating: note that the most intense solar heating is over the summer pole, yet here there is little or no ascent. Rather, the circulation is more usefully thought of as a response to potential vorticity fluxes which in turn are determined by the upward propagation of Rossby waves from the troposphere combined with the poleward gradient of potential vorticity in the stratosphere. It is salutary to note that without motion we have $\theta = \theta_E$, so there is no net heating at all — the heating is a consequence of the wave forcing.

17.5 ♦ DYNAMICS OF THE RESIDUAL OVERTURNING CIRCULATION

We now discuss the dynamics of the residual meridional overturning circulation, the RMOC, in rather more detail than in the previous section and with a little repetition of important matters.[11] The dynamics of the RMOC can be usefully couched as a quasi-linear problem. That is, the governing equations can be written with the linear terms on the left-hand side and the nonlinear terms as forcing terms on the right-hand side. This cannot be regarded as a full solution, but if the right-hand sides can be determined, if only approximately, by independent means then the equations can be solved fairly straightforwardly and the structure of the RMOC so determined. Such a procedure is likely to be more successful in the stratosphere than in the troposphere; in the latter, the nonlinear terms are a truly essential part of the solution and cannot properly be separated from the linear dynamics (although we might choose to separate the terms to *diagnose* what forces the RMOC). In the stratosphere the nonlinear terms represent the effects of waves and wave breaking on the mean flow. These waves — both gravity waves and Rossby waves — often have their origins in the troposphere, and although the propagation and breaking of the waves does depend strongly on the background flow the basic features of the forcing of the RMOC can still usefully be considered independently of the RMOC itself.

17.5.1 Equations of Motion

Away from the equator the Rossby number is small and the equations governing the large scale flow are in good geostrophic balance. The equations of motion governing the mean fields are the zonally averaged momentum and thermodynamic equations, along with the thermal wind equation and the mass continuity equations. We write the equations in their full form in spherical coordinates using the ideal gas equations in log-pressure coordinates, since both sphericity and compressibility are important, using the TEM formalism, giving

$$\frac{\partial \overline{u}}{\partial t} - f\overline{v}^* = \mathcal{G} + \mathcal{D} = \nabla \cdot \boldsymbol{\mathcal{F}} - \gamma \overline{u}, \qquad f\frac{\partial \overline{u}}{\partial z} + \frac{R}{aH}\frac{\partial \overline{T}}{\partial \vartheta} = 0, \tag{17.37a,b}$$

$$\frac{\partial \overline{T}}{\partial t} + \overline{w}^* S = Q_s + Q_l = \mu(T_R - \overline{T}), \qquad \frac{1}{a\cos\vartheta}\frac{\partial}{\partial \vartheta}(\overline{v}^* \cos\vartheta) + \frac{1}{\rho_R}\frac{\partial}{\partial z}(\rho_R \overline{w}^*) = 0. \qquad (17.37\text{c,d})$$

These equations are, respectively, the zonal momentum equation, the thermal wind equation, the thermodynamic equation and the mass continuity equation, with an overbar denoting a zonal average. The vertical coordinate, z is log pressure and $S = H_\rho N^2/R$ where R is the gas constant and H_ρ is the scale height used to define z; thus, z has dimensions of height and $\rho_R = \exp(-z/H_\rho)$. The velocity components $\overline{v}^*$ and $\overline{w}^*$ are the residual, or transformed Eulerian mean, meridional and vertical velocities. The equations have a very similar form if written in height coordinates using the anelastic approximation (Section 2.5); in that case, ρ_R is a reference profile of density and the thermodynamic equation is written using potential temperature or buoyancy as the thermodynamic variable, the factor H_ρ/R no longer appears, and z really is physical height.

The right-hand side of (17.37a) represents wave forcing and friction: $\mathcal{G} = \nabla\cdot\boldsymbol{\mathcal{F}}$ is the divergence of the Eliassen–Palm flux and $\mathcal{D}$ is the frictional force, which we take to be a simple linear drag. Such a drag is a little arbitrary but its form greatly simplifies the ensuing analysis. On the right-hand side of the thermodynamic equation Q_s and Q_l represent the forcing due to solar and long wave radiation; we take $Q_l = -\mu\overline{T}$ where μ is a constant thermal damping rate, and we may write $Q_s = \mu T_r(\vartheta, z)$ where T_R is a radiative equilibrium temperature, assumed known. There are no fluid-dynamic wave-forcing terms in the TEM form of the thermodynamic equation. Typically, the momentum dissipation is small and $\gamma \ll \mu$, and indeed we may take $\gamma = 0$ without much loss of realism, except close to the ground.

17.5.2 An Equation for the RMOC

If the right-hand sides are known, (17.37) constitutes a closed set of equations for the response of the temperature and the three components of the velocity to an applied wave force $\mathcal{G}$ and solar heating Q_s. Although nominally the equations have two time derivatives, the zonal wind and temperature are related through the thermal wind relation and the equations are balanced and no gravity waves are present. Our interest here is in the RMOC, and it is possible to derive a single equation for either $\overline{v}^*$ and $\overline{w}^*$, and we will focus on $\overline{w}^*$.

The procedure is similar to that used in Section 14.5.3. Essentially, we differentiate (17.37a) with respect to z and (17.37c) with respect to ϑ and then use (17.37c) to eliminate the time derivatives. We then use the mass continuity equation to obtain a single equation in $\overline{w}^*$. We are particularly interested in the dependence of the RMOC on the spatial structure and time dependence of $\mathcal{G}$ and Q_s, and to this end it is instructive to consider the case in which the time dependence is harmonic; that is, $\mathcal{G} = \widetilde{\mathcal{G}}e^{i\omega t}$, $Q_s = \widetilde{Q}_S e^{i\omega t}$ and $\widetilde{w} = \overline{w}^* e^{i\omega t}$. After a little algebra, we obtain

$$\frac{\partial}{\partial z}\left[\frac{1}{\rho_0}\frac{\partial(\rho_0 \widetilde{w})}{\partial z}\right] + \left(\frac{i\omega + \gamma}{i\omega + \mu}\right)\frac{N^2}{4\Omega^2 a^2 \cos\vartheta}\frac{\partial}{\partial\vartheta}\left[\frac{\cos\vartheta}{\sin^2\vartheta}\frac{\partial \widetilde{w}}{\partial \vartheta}\right]$$
$$= \frac{1}{2\Omega a\cos\vartheta}\frac{\partial}{\partial\vartheta}\left[\frac{\cos\vartheta}{\sin\vartheta}\frac{\partial\widetilde{\mathcal{G}}}{\partial z}\right] + \left(\frac{i\omega + \gamma}{i\omega + \mu}\right)\frac{R}{4H\Omega^2 a^2\cos\vartheta}\frac{\partial}{\partial\vartheta}\left[\frac{\cos\vartheta}{\sin^2\vartheta}\frac{\partial \widetilde{Q}_s}{\partial\vartheta}\right]. \qquad (17.38\text{a})$$

The equation is quite a handful, but it is useful to realize that, schematically and without all the metric factors, it is of the form

$$\frac{\partial^2 \widetilde{w}}{\partial z^2} + A\frac{N^2}{f^2}\frac{\partial^2 \widetilde{w}}{\partial y^2} \sim \frac{1}{f}\frac{\partial}{\partial y}\frac{\partial \widetilde{\mathcal{G}}}{\partial z} + \frac{A}{f^2}\frac{\partial^2 \widetilde{Q}_s}{\partial y^2}, \qquad (17.38\text{b})$$

where

$$A = \frac{i\omega + \gamma}{i\omega + \mu}.$$

Equation (17.38b) is similar to (10.63), since with the addition of diabatic terms and a slight change in notation (10.63) is

$$f_0^2 \frac{\partial^2 \widetilde{\psi}}{\partial z^2} + AN^2 \frac{\partial^2 \widetilde{\psi}}{\partial y^2} = f_0 \frac{\partial \widetilde{\mathcal{G}}}{\partial z} + A \frac{\partial \widetilde{Q}_s}{\partial y}, \tag{17.38c}$$

where $\widetilde{\psi} = \psi^* e^{i\omega t}$ is the amplitude of the residual streamfunction of the overturning circulation. Since $\overline{w}^* = \partial \psi^* / \partial y$, (17.38b) is almost the y-derivative of (17.38c). (In (10.63) we took $\gamma = \mu$ so that $A = 1$.) Much of the physical interpretation in what follows comes from (17.38b) and (17.38c), although we will allow f to vary spatially — that is, we use f and not f_0.

With quasi-geostrophic scaling the wave forcing term in (17.38) is

$$\mathcal{G} = \overline{v'q'} = -\frac{\partial}{\partial y}\overline{u'v'} + \frac{\partial}{\partial z}\left(\frac{f_0}{N^2}\overline{v'b'}\right) = \nabla \cdot \boldsymbol{\mathcal{F}}, \quad \text{where} \quad \boldsymbol{\mathcal{F}} = -\overline{u'v'}\,\mathbf{j} + \frac{f_0}{N^2}\overline{v'b'}\,\mathbf{k}, \tag{17.39}$$

and $\boldsymbol{\mathcal{F}}$ is the Eliassen–Palm flux.

17.5.3 The Nature of the Response

The operator on the left-hand side of (17.38) is elliptic, similar to a Poisson equation. Thus, the response will be much less localized than the forcing itself, a concept that is familiar from potential vorticity inversion. To understand the equation better it is useful to take a heuristic look at some special cases, as follows. The cases are not all 'orthogonal' to each other — thus, for example, the low-latitude limit could be either low frequency or high frequency.

(i) The aspect ratio of the response

From (17.38c) the natural aspect ratio of the response, α_r say, is given by

$$\alpha_r = \frac{H_r}{L_r} = \frac{1}{A^{1/2}}\frac{f}{N} = \left(\frac{i\omega + \mu}{i\omega + \gamma}\right)^{1/2}\frac{f}{N}, \tag{17.40}$$

where H_R and L_R are the vertical and horizontal scales of the response. If the thermal and mechanical dissipation are zero, or have the same time scale, then $A = 1$, but more generally the presence of dissipation can alter the aspect ratio considerably. Also, the thermal dissipation can be expected to be much stronger than the mechanical dissipation, meaning that $\mu \gg \gamma$. The high- and low-frequency limits then have somewhat similar behaviour.

(ii) The high-frequency limit

In this case the thermal and mechanical damping are negligible and $A = (i\omega+\gamma)/(i\omega+\mu) \approx 1$. Since μ typically varies between 1/(20 days) in the lower stratosphere and 1/(5 days) in the upper stratosphere, and $1/\gamma$ is an even longer time, phenomena of order a few days fall into this category. Sudden stratospheric warmings are one example, although since the timescale of warmings is of order days thermal effects are not wholly negligible.

Using (17.40) we see that the aspect ratio of the response is simply of the order of Prandtl's ratio. That is, $\alpha_r = H/L \sim f/N$ and since $f \sim 10^{-4}\,\mathrm{s}^{-1}$ and $N \sim 10^{-2}\,\mathrm{s}^{-1}$ or larger, the response to rapid forcing is typically quite shallow. Of course, although the Prandtl ratio is small it is a natural scaling of vertical to horizontal scales in atmospheric dynamics and shallowness should be interpreted in that context. Still, as we approach the equator the response shallows still further, although at the equator itself (17.38) ceases to be valid. A quantitative analysis of the right-hand side of (17.38a) further suggests that both waves (the $\mathcal{G}$ term) and solar forcing act to drive the overturning circulation.

(iii) The low-frequency limit

In this case the frequency is less than the thermal damping rate; that is, $\omega \ll \mu$, with one obvious example being the annual cycle. If as is realistic, $\gamma \ll \mu$ then $|A|$ becomes small. The effect of the solar heating thus also becomes small in (17.38c). The response generally deepens, with the ratio of the vertical to the horizontal scales being given by

$$\alpha_r = \frac{H_r}{L_r} \approx \left(\frac{1}{A^{1/2}}\right)\frac{f}{N} = \left(\frac{\mu}{i\omega + \gamma}\right)^{1/2}\frac{f}{N} \gg \frac{f}{N}, \tag{17.41}$$

if $\mu \gg \gamma$. In the steady-state limit the solar heating is balanced by the thermal relaxation and the right-hand side of (17.37c) nearly vanishes. We discuss this limit in more detail below, in the section on downward control.

(iv) Deep and shallow forces

A force may be regarded as deep or shallow depending on whether its aspect ratio (vertical to horizontal, α_F say) is greater or less than f/N, and it turns out that deep force tends to gives rise to an acceleration of the zonal wind (a non-zero $\partial\overline{u}/\partial t$) whereas a shallow force tends to give rise to a meridional circulation. To see this we will consider the simplified forms of the momentum equation

$$\frac{\partial \overline{u}}{\partial t} - f_0\overline{v}^* = \mathcal{G}, \tag{17.42}$$

along with the MOC equation, (17.38c) which, if f_0 and N are both constant, is a Poisson equation with a right-hand side equal to $\partial\mathcal{G}/\partial z$. Solutions can be obtained by Fourier series methods with terms having the form

$$\widetilde{\psi} = \Psi \cosh k_z z \sin k_y y. \tag{17.43}$$

Let us suppose the forcing is also of this form, so that by a deep forcing we mean that $k_y/k_z \gg f_0/N$ and shallow means $k_y/k_z \gg f_0/N$. We'll also suppose that $A = \mathcal{O}(1)$. From (17.38c) we see that

$$(f_0^2 k_z^2 - N^2 k_y^2)\widetilde{\psi} \sim f_0 k_z G, \tag{17.44}$$

where G is the forcing amplitude. From this we can informally infer the form of the solution for deep and shallow forcing, as follows:

- *Deep forcing:* The dominant balance in (17.44) is between the second term on the left-hand side and the right-hand side and therefore $\widetilde{\psi} \sim f_0 k_z G/(N^2 k_y^2)$ and

$$\widetilde{w} \sim \frac{f_0 k_z G}{N^2 k_y}, \qquad \widetilde{v} \sim \frac{f_0 k_z^2 G}{N^2 k_y^2}. \tag{17.45}$$

  Now look at the momentum equation (17.42). The ratio of the Coriolis force to the forcing on the right-hand side is given by

$$\frac{|f_0 v|}{|\mathcal{G}|} \sim \frac{f_0^2 k_z^2}{N^2 k_y^2} \ll 1, \tag{17.46}$$

  where the inequality follows by definition of what is deep. The dominant balance in the momentum equation must then be between the wave forcing and the acceleration.

- *Shallow forcing:* The dominant balance in (17.44) is between the first term on the left-hand side and the right-hand side and therefore $\widetilde{\psi} \sim G/(f_0 k_z)$ and

$$\widetilde{w} \sim \frac{k_y G}{f_0 k_z}, \qquad \widetilde{v} \sim \frac{G}{f_0}. \tag{17.47}$$

  The ratio of the Coriolis term to the forcing term in the momentum equation is now $\mathcal{O}(1)$, and therefore the response to a shallow forcing appears in the meridional circulation rather than as an acceleration.

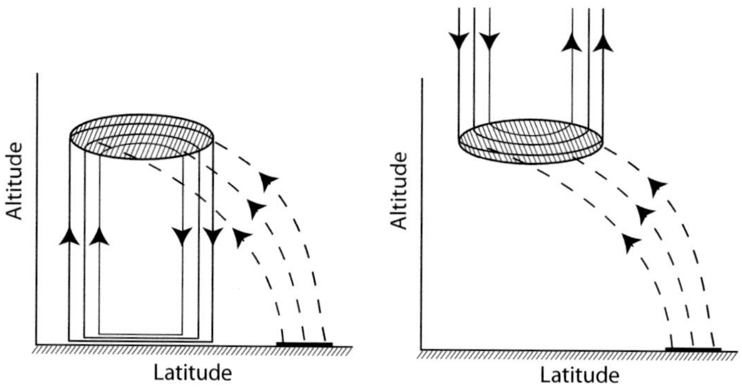

Fig. 17.11 Downward control. Left panel: wave activity propagates upward (dashed lines) from the troposphere, breaking and depositing zonal momentum in the shaded region. This induces an overturning circulation (solid lines) connecting the wave-breaking region with a bottom frictional boundary layer. Right panel: putative 'upward control', requiring friction above the wave breaking region for a steady response.

The underlying reason for the two different responses arises from the need to satisfy the thermal wind equation. A shallow force means that a shear is generated, and hence via thermal wind the temperature must change, but this can only be accomplished if there is an RMOC to affect temperature. If the force is deep then the response can and will be in the form of an acceleration. Close to the equator all forces are essentially deep because f is small.

(v) *Deep and shallow heating*
A similar analysis may be applied to a heating field, but given the intuition we have just developed about the response to a mechanical forcing there is no need to go through the details (although these are straightforward). A shallow heating (perhaps more usefully thought of as a broad heating) can and will produce a direct response in the temperature field itself. However, if the heating is deep (or latitudinally confined) then any direct response in the temperature field would have to be associated with a response in the zonal wind. Instead, a latitudinally confined heating tends to produce a response in the meridional circulation.

(vi) *The low-latitude limit*
At low-latitudes most forces become deep because f is small. More precisely, the criterion for deepness, that $H_f/L_f \gg N/f$ where H_f and L_f are the vertical and horizontal scales of the forcing, becomes easier to satisfy. Thus, wavebreaking at low-latitudes is more likely to induce an acceleration than a similar wavebreaking in mid-latitudes, which will tend to induce an overturning circulation. By the same token, a heating source at low-latitudes has a greater tendency to induce an overturning circulation than a similar heat source in mid-latitudes. The above statements are rules of thumb and not necessarily quantitative.

17.5.4 The Steady-state Limit and Downward Control

Let us now consider in a little more detail the steady-state response in which $\omega/\mu \to 0$, and we also take $\gamma = 0$, so that there is no momentum forcing. Although (17.38) of course still holds, it is also useful to look directly at the momentum equation and thermodynamic equations. The momentum equation, (17.37a) reduces to a balance between the Coriolis force and the wave driving, namely

$$- f\overline{v}^* = \mathcal{G}, \tag{17.48}$$

and the thermodynamic equation becomes

$$\overline{w}^* S = Q_s + Q_l = \mu(T_R - \overline{T}). \tag{17.49}$$

The thermodynamic equation gives us very little information about the vertical velocity because the right-hand side contains the unknown temperature, $\overline{T}$; rather, we can glean much of the information we want from (17.48).

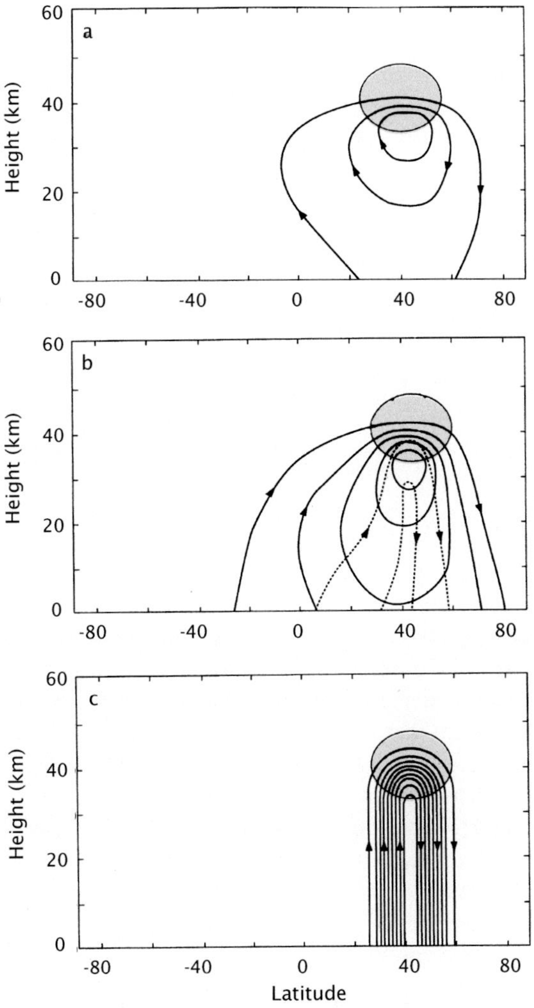

Fig. 17.12 The numerically-computed response of the meridional overturning circulation to a longitudinally symmetric westward force, with the frequency of the forcing decreasing from top to bottom.[12] Contours are streamlines of the residual circulation, with the same uniform interval in all panels, and the shading denotes the forcing region.

(a) Response to high-frequency forcing, $\omega/\mu \gg 1$, $\omega \gg \gamma$. The response is adiabatic and weakly spreads into the opposite hemisphere.
(b) A lower frequency case with $\omega/\mu = 0.34$, corresponding to an annual cycle and a 20-day thermal relaxation timescale. The solid and dashed lines show the response that is in phase and out of phase with the forcing, respectively.
(c) Steady state response, $\omega/\mu \ll 1$. The circulation increases in magnitude and narrows as the frequency decreases, and in panel (c) it is given using the downward control expression (17.51a).

If we differentiate (17.48) with respect to y (or ϑ) and use the mass continuity equation we obtain

$$\frac{1}{\rho_0}\frac{\partial \rho_0 w}{\partial z} = \frac{1}{a\cos\vartheta}\frac{\partial}{\partial\vartheta}\left(\frac{\mathcal{G}\cos\vartheta}{f}\right). \tag{17.50}$$

This is a first-order partial differential equation for the vertical velocity, and we can obtain the vertical velocity itself by a vertical integration, using a single boundary condition. If we require that vertical velocity stays finite at $z \to \infty$ and that $\rho_0 w = 0$,s then we obtain

$$\overline{w}^*(z) = \frac{1}{a\rho_0(z)\cos\vartheta}\frac{\partial}{\partial\vartheta}\int_z^\infty \left(\frac{\rho_0(z')\mathcal{G}(\vartheta,z')\cos\vartheta}{f}\right)\mathrm{d}z'. \tag{17.51a}$$

The Cartesian quasi-geostrophic version of this equation is just

$$\overline{w}^*(z) = -\frac{1}{\rho_R}\frac{\partial}{\partial y}\int_z^\infty \rho_R(z')\frac{\mathcal{G}(\vartheta,z')}{f_0}\,\mathrm{d}z'. \tag{17.51b}$$

Equation (17.51) implies that, in the steady-state limit, the vertical velocity at a given height is determined by the wave forcing *above* that height. The physical situation is illustrated in the left panel of Fig. 17.11. Here, a wave source in the troposphere propagates upward and breaks in the middle atmosphere depositing momentum. This induces a meridional circulation as illustrated in the left panel, with a response *below* the momentum source. The numerically-computed response to an imposed force is illustrated in Fig. 17.12, showing how the response changes depending on the time-scales of the forcing and damping, with the steady-state response illustrated in the bottom panel.

The above derivation may seem a little disingenuous, for surely we might just as well have assumed $w = 0$ at $z = 0$, leading (in the quasi-geostrophic case) to

$$\overline{w}^*(z) = \frac{1}{\rho_R}\frac{\partial}{\partial y}\int_0^z \rho_R \frac{\mathcal{G}+\mathcal{D}}{f_0}\,\mathrm{d}z'. \tag{17.52}$$

This might appear to give 'upward control', as illustrated in the right panel of Fig. 17.11. However, if we are integrating from the ground up then frictional effects are important near the surface and must be included, as represented by the frictional term $\mathcal{D}$ in (17.52). Furthermore, mass conservation demands that

$$\int_0^\infty \rho_R \overline{v}^*\,\mathrm{d}z = 0, \qquad \text{implying} \qquad \int_0^\infty \rho_R(\mathcal{G}+\mathcal{D})\,\mathrm{d}z = 0, \tag{17.53a,b}$$

using (17.37a) for steady state conditions. Thus, above the level of the momentum source $\mathcal{G}$, (17.52) also in fact implies that the vertical velocity is zero, because $\mathcal{G}$ and $\mathcal{D}$ have cancelling effects. Thus, the location of the frictional boundary layer where the momentum is removed is one way to distinguish up from down. Equation (17.53b) tells us that the frictional boundary layer at the bottom must adjust to remove the same amount of momentum that is deposited by wave breaking higher up, if there is to be a steady state. If there were a momentum sink above the momentum deposition region there would be no justification for downward control, for we would have to include that frictional term in (17.51). However, it is hard to envision how such a sink could exist without violating angular momentum conservation. If there were no frictional sink at the ground the disturbance would initially propagate down, but on reaching the ground would then propagate up.

From the point of view of the diagnostic equation for the meridional overturning circulation, in the steady state limit we have $A = 0$, and the solar forcing on the right-hand side and the y-derivative on the left-hand side of (17.38) both vanish, and the equation for the RMOC becomes

$$\frac{\partial}{\partial z}\left[\frac{1}{\rho_0}\frac{\partial(\rho_0\overline{w})}{\partial z}\right] = \frac{1}{2\Omega a\cos\vartheta}\frac{\partial}{\partial\vartheta}\left[\frac{\cos\vartheta}{\sin\vartheta}\frac{\partial\widetilde{\mathcal{G}}}{\partial z}\right], \tag{17.54a}$$

or, in the quasi-geostrophic limit,

$$f_0\frac{\partial^2\widetilde{\psi}}{\partial z^2} = \frac{\partial\widetilde{\mathcal{G}}}{\partial z}. \tag{17.54b}$$

That is to say, in a steady state the solar forcing provides no input to the meridional circulation! This may seem a little counter-intuitive, but if there is no wave forcing and $\mathcal{G} = 0$ then the vertical velocity is zero and the temperature adjusts to the radiative equilibrium temperature, so that the diabatic forcing is zero.

The temperature field

Given that in a statistically steady state the vertical velocity field is determined by the wave forcing, the temperature field can be determined diagnostically from the thermodynamic equation. Thus, using (17.37c) with no time-dependence and a vertical velocity given by (17.51), we obtain in the quasi-geostrophic case

$$\overline{T}(z) - T_r(z) = \frac{S}{\mu\rho_R}\frac{\partial}{\partial y}\int_z^\infty \rho_R\frac{\mathcal{G}(\vartheta,z')}{f_0}\,\mathrm{d}z', \tag{17.55}$$

Parameter	Background	Rossby-gravity waves	Kelvin waves
Static stability, N	$2.2 \times 10^{-2}\,\mathrm{s}^{-1}$		
Beta at equator, β	$2.3 \times 10^{-11}\,\mathrm{m}^{-1}\,\mathrm{s}^{-1}$		
Coriolis parameter, f, at 5°	$1.27 \times 10^{-5}\,\mathrm{s}^{-1}$		
Wave period		4–5 days	10–20 days
Zonal wavelength		10,000 km	20,000–40,000 km
Zonal wavenumber		4	1–2
dimensionally		$6.3 \times 10^{-7}\,\mathrm{m}^{-1}$	1.6–$3.2 \times 10^{-7}\,\mathrm{m}^{-1}$
Meridional scale		1,200 km	1,500 km
Vertical wavelength		4–8 km	6–10 km
Phase speed, relative to ground		20–$25\,\mathrm{m\,s}^{-1}$ (westward)	$25\,\mathrm{m\,s}^{-1}$ (eastward)
Amplitudes:			
zonal velocity		2–$3\,\mathrm{m\,s}^{-1}$	4–$8\,\mathrm{m\,s}^{-1}$
meridional velocity		2–$3\,\mathrm{m\,s}^{-1}$	0
vertical velocity		1–$2\,\mathrm{mm\,s}^{-1}$	1–$2\,\mathrm{mm\,s}^{-1}$
temperature		1 K	2–3 K
geopotential height		4 m	30 m
F_0, wave forcing at 17 km, see (17.59).		3–$6 \times 10^{-3}\,\mathrm{m}^2\,\mathrm{s}^{-2}$	4–$10 \times 10^{-3}\,\mathrm{m}^2\,\mathrm{s}^{-2}$
Thermal damping rate, α		0.5–$1.5 \times 10^{-6}\,\mathrm{s}^{-1}$	0.5–$1.5 \times 10^{-6}\,\mathrm{s}^{-1}$

Table 17.1 Typical, approximate, values of parameters appropriate for waves and background flow in the equatorial lower stratosphere.[13]

with a similar but more complicated expression in the full case. The temperature field at a given height is determined purely by the momentum forcing, being given by the meridional gradient of the zonal force *above* that height.

† An oceanic comparison

It is instructive to compare downward control with the Stommel problem in oceanography (Section 19.1.1). In TEM (residual) form, the approximate zonally averaged zonal momentum equation may be written, as in (17.32a), as

$$\frac{\partial \overline{u}}{\partial t} - f_0 \overline{v}^* = \mathcal{F} + \mathcal{D}, \tag{17.56}$$

The steady version of (17.56) and the equation for the streamfunction for the *horizontal* flow, ψ, in the ocean, (19.6), may thus respectively be written

$$f_0 \frac{\partial \psi^*}{\partial z} = \mathcal{F} + \mathcal{D}, \qquad \beta \frac{\partial \psi}{\partial x} = \mathcal{F}_w - r\nabla^2 \psi, \tag{17.57a,b}$$

where $\mathcal{F}_w$ represents the wind forcing at the ocean surface, and the second term on the right-hand side of (17.57b) represents friction. In (17.57a), $\overline{v}^* = -\partial \psi^*/\partial z$ and in (17.57b), $v = \partial \psi/\partial x$. The two equations have a formal similarity — is there more?

In the ocean interior, the frictional term is negligible, and in solving the resulting first-order equation ($\beta \partial \psi/\partial x = \mathcal{F}_w$) we may apply the boundary condition of $\psi = 0$ only at one meridional boundary. The natural choice is to choose the eastern boundary for this, and then invoke frictional processes to bring ψ to zero on the west. It is a natural choice because Rossby waves propagate westward ('westward control', as in Fig. 19.14); thus, the boundary current (e.g., the Gulf Stream) is on

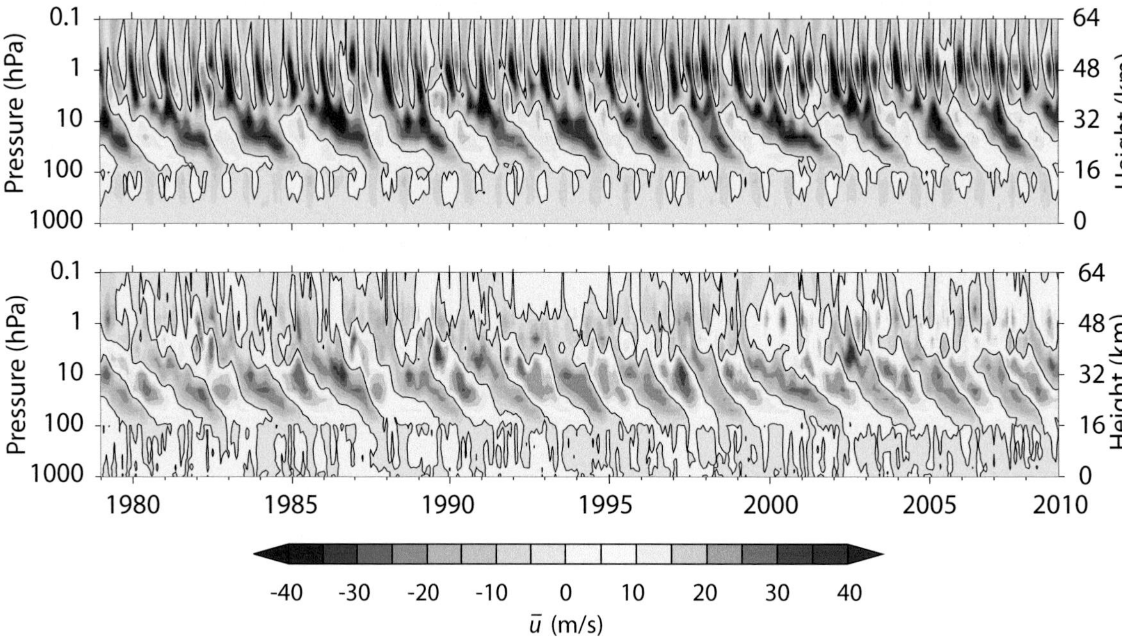

Fig. 17.13 Time-height sections of the observed zonal-mean zonal wind for 1979–2010, averaged from 5° S to 5° N. The contour is the zero wind line, and in the bottom panel the seasonal cycle is removed. The quasi-biennial oscillation, or QBO, is clearly visible between 20 and 40 km, or about 60 and 3 hPa.[15]

the west *because* the wind's influence is carried westward by Rossby waves, not vice versa. The consequence is that westward control is an enormously robust effect that pervades almost every aspect of large-scale physical oceanography. In the atmospheric case there is no similar mechanism that demands that the influence of the momentum source be propagated *only* downward. However, even without friction there is an up-down asymmetry in the atmospheric case because of (a) thermal damping, and (b) density variations. If we imagine a case with no boundaries at either top or bottom then the forcing creates meridional cells that propagate both up and down, growing (it turns out) like $t^{1/2}$. However, the solution also indicates that the upward propagating cell would eventually disappear, essentially to satisfy a boundary condition of boundedness at positive infinity (where density vanishes), leaving only a downward influence.[14] In the case with boundaries a finite distance from the source, the final steady state does depend on the location of the frictional layers, and in particular a steady solution with only a downward influence results because the frictional boundary layer is at the bottom, not vice versa.

As we mentioned, the mechanism of downward control is related to that which gives rise to the Ferrel Cell in the troposphere, and that is certainly a strong and robust effect. Whether the downward control effect following wavebreaking *in the stratosphere* is strong enough to influence circulation in the troposphere, or the structure of the tropopause, remains an open question.

17.6 THE QUASI-BIENNIAL OSCILLATION

17.6.1 A Brief Review of the Observations

The *quasi-biennial oscillation,* or QBO, is a nearly periodic reversal of the zonal wind in the equatorial stratosphere, as illustrated in Fig. 17.13 and Fig. 17.14. It is the most dominant variability of that region, and the following lists some of the main features of the phenomenon:[16]

- The zonal winds in the equatorial region between about 5 and 100 hPa (about 40 and 18 km)

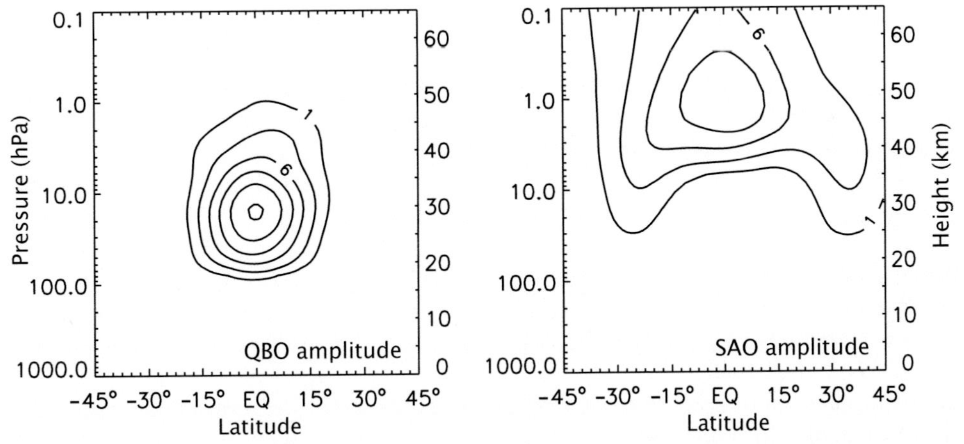

Fig. 17.14 The amplitudes (the root-mean square of the zonally-averaged wind of the oscillations after temporal filtering, with contours at 1, 3, 6, 9, 12 and 15 m s^{-1}) of the QBO and the SAO. The QBO evidently extends roughly from 15° S to 15° N and from 100 hPa to 1 hPa (about 17 km to 50 km). The SAO is broader, higher, weaker and faster.[17]

alternate between being eastward and westward with an average period of about 28 months, with the period varying between 22 and 34 months.

- The QBO is almost latitudinally symmetric about the equator. The amplitude is approximately Gaussian with a half width of about 12°.
- The phenomenon is approximately zonal symmetric; that is, the longitudinal variation is small.
- The maximum amplitude of the oscillation is about 30 m s^{-1}(±15 m s^{-1}) at about 20 hPa on the equator. The westward winds are slightly stronger than the eastward winds, after removing the annual cycle.
- The wind pattern descends at about 1 km per month with little loss of amplitude until it reaches 100 hPa, and the cycle begins again. (This does not mean that information propagates downward, as we discuss later.)
- The QBO is mildly synchronized to the annual cycle, with transitions between eastward and westward flow having a tendency to occur more commonly in March–June than in the other months.

See also the summary on page 655. Although we will not discuss it here, another oscillatory phenomenon occurs above the QBO known as the semi-annual oscillation, or SAO. The SAO is an oscillation in the zonal wind with an approximate period of six months (it should perhaps be called the quasi semi-annual oscillation) and it occurs between 1 hPa and 0.1 hPa and extends from about 30° N to 30° S (Fig. 17.14).

17.6.2 A Qualitative Discussion of Mechanisms

Candidate mechanisms

We first note that the QBO *must* involve zonally asymmetric motions. Without such asymmetric or eddying motions there can be no maximum of angular momentum within the fluid interior, and therefore no eastward winds at the equator, as explained in Sections 13.5.1 and 14.2.8. Given this, let us ponder for a moment what *might* be the mechanism of the QBO. One might suppose that horizontally propagating planetary waves would be a likely mechanism, for the transport of momentum by Rossby waves is known to be an important mechanism for the maintenance of jets in mid-latitudes. However, the descent of the wind pattern with no loss of amplitude cannot be easily explained by such a mechanism.[18] Other proposed mechanisms have involved interactions with the annual cycle or its harmonics (natural enough given the period of the QBO) or have invoked external forcing or some nonlinear feedback. However, no candidate mechanism was able to explain all the features noted above, until a mechanism involving the vertical propagation and absorption of gravity waves was proposed, as we now describe.

A gravity wave mechanism

The mechanism for the QBO that is now generally accepted involves the upward propagation and absorption of gravity waves and their effect on zonal flow.[19] We first describe the basic mechanism rather roughly and qualitatively.

A broad spectrum of gravity waves, with phase speeds in both eastward and westward directions, is generated in the upper equatorial troposphere by deep convection and various other instabilities. The waves will in general have a component of the group velocity that is directed upward, and if these waves are dissipated via mechanical or thermal damping then they will force mean flow accelerations (steady, non-dissipative waves cannot force a mean flow acceleration). A critical level, where the phase speed of the waves equals the speed of the mean flow (i.e., where $c = \overline{u}$) is one place where wave absorption and mean-flow acceleration will be particularly effective, because as a wave approaches a critical level it slows, giving more time for dissipation to act. However, it is not necessary for there to be an actual critical level; indeed, waves approaching a critical level will often be largely dissipated before reaching it.

Let us suppose that initially there is a westward shear (that is $\partial\overline{u}/\partial z > 0$) and that there are upward propagating gravity waves with positive phase speed c. These will be very efficiently absorbed as they approach the critical level, depositing momentum and causing the mean flow to accelerate. As pictured in Fig. 17.15, this causes the critical level to descend and hence the subsequent absorption of gravity waves and acceleration of the zonal flow will be at a lower level. The wind anomaly thus descends, and so on. A similar effect will still occur even if there is no critical level, provided there is enough dissipation for the gravity wave to be absorbed somewhere, causing the mean flow to accelerate. Even if this acceleration is insufficient to induce a critical level, the difference between the wave speed and the zonal fluid speed will be reduced (i.e., $c - \overline{u}$ diminishes) and gravity wave absorption is enhanced. Gravity waves are thus absorbed at a lower level than previously and the anomaly in the zonal wind descends, as before.

Eventually, in the above models, the wind anomaly descends to the level of the gravity wave source. Depending on the strength of the dissipation one might imagine that dissipative processes could then wipe out the wind anomaly completely and the whole process would start all over again, or perhaps a low level westward wind anomaly would persist, so redefining the mean flow. However, in either case the zonal wind anomaly would not change sign (i.e., become westward), as is observed in the real QBO. For that to occur we may invoke a second wave in conjunction with an instability, as we now explain.

A two-wave model

Suppose now there are two upward propagating gravity waves with speeds $+c$ and $-c$ (where c itself is positive), each of which will slowly be dissipated as it propagates, with the dissipation enhanced for smaller values of $|\overline{u} - c|$ or $|\overline{u} + c|$, respectively. (There will of course be very large dissipation if there is a true critical level.) Suppose that the mean flow has no shear, then simply by symmetry that state can persist, with the eastward and westward waves being dissipated equally as they ascend with no zonal flow generation. However, that symmetric state is unstable; to see this suppose that there is a small eastward perturbation to the zonal wind, as illustrated in the left panel of Fig. 17.15. The eastward propagating wave will then be preferentially dissipated, because $\overline{u} - c$ is smaller for it than for the westward wave. The eastward anomaly in the zonal wind will therefore grow and

Essentials of the QBO

What is the Quasi-Biennial Oscillation?

- The QBO is a quasi-periodic reversal of the zonal-mean zonal winds between about 20 km and 45 km altitude and 15° S to 15° N, with an irregular period of about 28 months. It is the dominant pattern of variability in the equatorial stratosphere and it is the clearest manifestation of a non-directly forced nearly-periodic phenomenon in the atmosphere.
- The eastward and westward zonal winds appear to propagate downward at about 1 km per month, reversing at the end of each half cycle.
- The half-amplitude of the zonal wind cycle is about 15 m s^{-1}, with the westward winds being slightly stronger.

What is the mechanism?

- The oscillation is caused by the upward propagation and absorption of Kelvin waves and Rossby-gravity waves at the equator. If a wave has an eastward phase speed then, on absorption, it will cause the mean zonal flow to accelerate eastward. Furthermore, the absorption is strongest near a critical layer, where the mean zonal wind speed equals the phase speed. An upward propagating Kelvin wave thus causes the mean flow aloft to accelerate eastward, and then the maximum in eastward winds to move downward and eventually to be dissipated. An upward propagating Rossby wave then generates a westward zonal wind anomaly aloft, which similarly propagates down. In this way the zonal wind oscillates between positive and negative values, as illustrated in Fig. 17.16.
- The waves are generally considered to be primarily excited by moist convection in the upper tropical troposphere.
- The period is determined by a combination of parameters involving the wave and mean flow. In the simplest model of two upwardly propagating gravity waves the period is given by

$$P = \frac{Akc^3}{\alpha N_0 F_0} \qquad \text{(QBO.1)}$$

where A is a nondimensional number weakly dependent on viscosity, and the other parameters, properties of the waves and mean flow, are defined in the text. The period is not proportional to the period of the waves; rather it is inversely proportional to their strength, F_0, because stronger waves cause more mean flow acceleration and a faster descent of the pattern.

Why is the phenomenon equatorially confined?

- The mechanism requires there to be upwardly propagating waves with very different phase speeds in order that the mean flow can oscillate between the two values. In equatorial regions such a forcing can be provided by Rossby waves and Kelvin waves.
- In mid-latitudes the tropospheric flow is largely balanced and it is primarily long Rossby waves that reach the stratosphere with a spectrum of phase speeds. If and when they break they would provide only a westward acceleration. Furthermore, in mid- and high latitudes an imposed force tends to induce a mean meridional circulation, not a mean flow acceleration (Section 17.5).

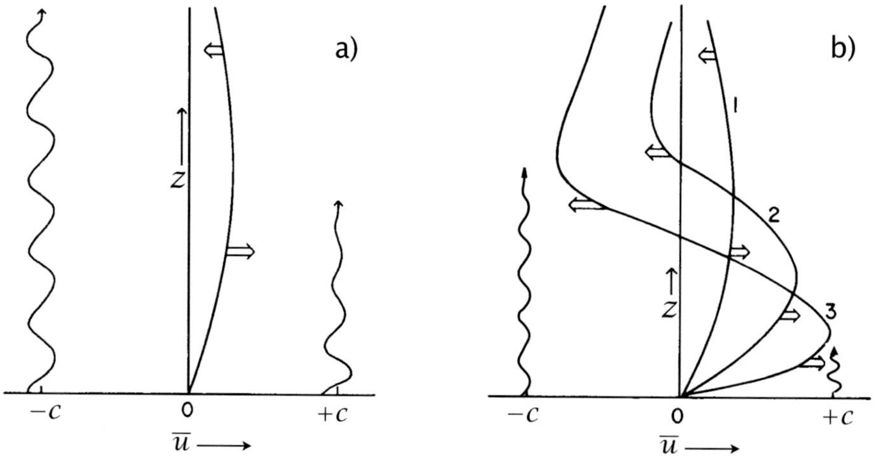

Fig. 17.15 Schema of the initial instability leading to the QBO.[20] The solid lines show the zonal flow, the wavy arrows indicate the gravity-wave penetration from below and the double arrows indicate wave-induced acceleration. Initially, as in the left panel, a small eastward perturbation is added to a stationary mean flow. The eastward moving wave is preferentially absorbed and the perturbation is amplified and then descends, with the right panel showing the zonal flow at the successive times indicated, with the gravity wave penetration illustrated at $t = 3$. After the flow develops an eastward component the westward wave penetrates higher before being absorbed, inducing a westward flow aloft that then itself descends, making the eastward anomaly thinner. Subsequent stages and the development of a periodic oscillation are illustrated in Fig. 17.16.

descend, just as described above. The upward propagation of the eastward wave is then limited, but the westward wave is unconstrained so it reaches higher levels before eventually being absorbed, providing a westward acceleration to the zonal flow, as illustrated in the right panel of Fig. 17.15.

As the the westward anomaly descends it squeezes the eastward anomaly which becomes thinner and thinner. Dissipative processes then become more efficient and can erode the eastward anomaly completely, with the flow becoming entirely westward, as illustrated in panel (b) of Fig. 17.16. A high level eastward anomaly is then created (panel (c) of Fig. 17.16), descending and squeezing the westward anomaly, and a mirror image of the first stage takes place. The entire cycle repeats itself and an oscillation is born, with the period of the oscillation being determined by the strength of the gravity waves and the rate of dissipation: stronger gravity waves lead to a faster acceleration of the mean flow and so a greater rate of descent and so a shorter period. Finally, note that the waves need not have speeds symmetric on either side of zero, $+c$ and $-c$. Suppose, for example, the wave speeds were both positive, a and b say. The mean flow could accelerate to an average value of $(a + b)/2$, with the flow then oscillating between a and b in a fashion similar to the symmetric case.

17.6.3 A Quantitative Model of the QBO

We now consider the above wave–mean-flow interaction model a little more quantitatively, and our first goal will be to obtain equations of motion for the interaction. To this end we will parameterize the vertical propagation and absorption of gravity waves by simple expressions resulting from gravity wave theory described in Section 17.2.3. The absorption leads to a zonal flow acceleration, which in turn affects the wave absorption, and so on.

Let us consider a semi-infinite (no top), non-rotating, stratified fluid subject to a standing wave

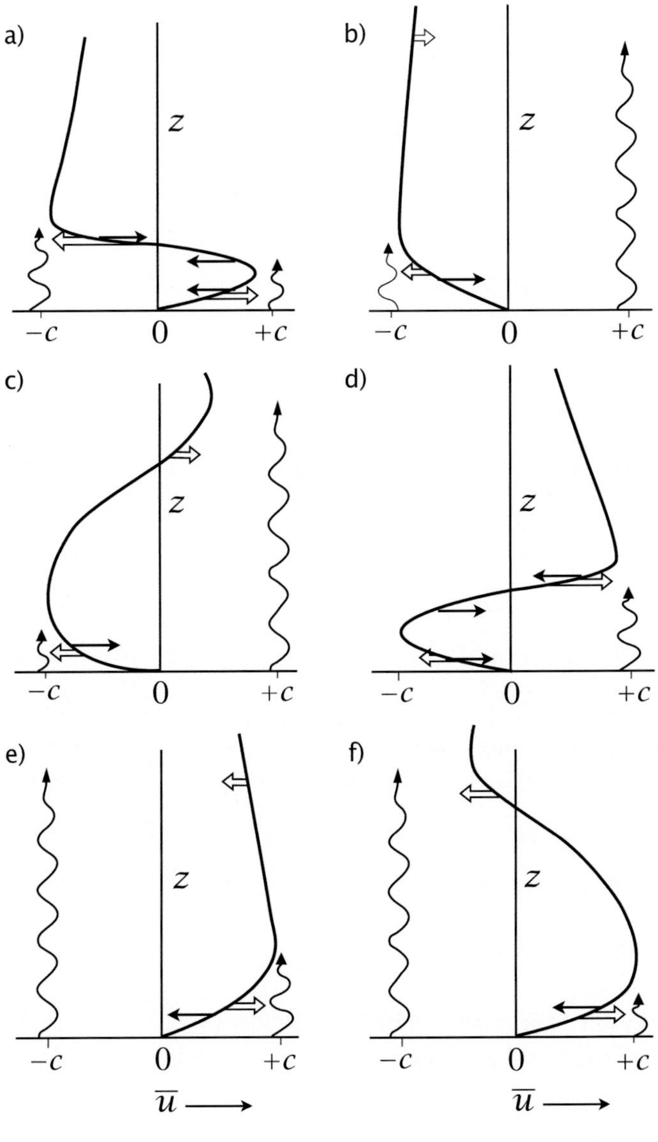

Fig. 17.16 Schema of the evolution of the QBO with gravity-wave forcing from below, following an initial perturbation illustrated in Fig. 17.15.[21] The solid lines show the mean flow and the wavy lines indicate the propagation of gravity waves. Horizontal double arrows indicate wave forcing and single arrows indicate viscous relaxation.

The panels are at successive times, with the top four panels showing a half cycle, and panels (d), (e) and (f) are mirror images of (a), (b) and (c). Wave-induced acceleration of the mean flow occurs preferentially near critical levels where $\overline{u} = c$.

forcing at the lower boundary. Specifically, the waves are of the form

$$w = \operatorname{Re} \widetilde{w}_1(z)e^{ik(x-ct)} + \widetilde{w}_2(z)e^{ik(x+ct)}. \tag{17.58}$$

The waves have a dispersion relation as discussed in Section 17.2.3, a positive (upward) group velocity, and we will take $\widetilde{w}_1 = \widetilde{w}_2$. If there is a source of gravity waves such as convection there is no difficulty in exciting waves with either an eastward or westward phase speed: a Kelvin wave has a purely eastward phase speed ($c_p > 0$), a Rossby-gravity wave has a westward phase speed, and gravity waves completely uninfluenced by rotation can have a phase speed in either direction. The Kelvin and Rossby-gravity waves, probably the most important waves for the QBO, typically have zonal wavenumbers 1–4, and so zonal wavelengths greater than 10 000 km, and periods of 3 days or longer.

As the waves propagate up they are dissipated, primarily by thermal rather than viscous dissipation, and their amplitude diminishes in the vertical and consequently they deposit momentum into the mean flow. From the WKB calculation of Section 17.3.3 the wave momentum flux,

$F_k(z) = \overline{u'w'}$, of a given upward-propagating wave is of the form

$$\overline{F}_k(z) = \overline{F}_k(0)\exp\left[-\int_0^z g_k(z')\,dz'\right], \tag{17.59}$$

where the subscript k indicates the zonal wave number and the attenuation rate, $g_k(z)$, for a given upward-propagating internal wave is given by

$$g_k(z) = \frac{\text{damping rate}}{\text{vertical group velocity}} = \frac{\alpha}{k(\overline{u}-c)^2/N}, \tag{17.60}$$

where c is the phase speed of the waves. The mean flow, $\overline{u}(z,t)$, is influenced by many such waves and so evolves according to

$$\frac{\partial \overline{u}}{\partial t} = -\sum_k \frac{\partial \overline{F}_k}{\partial z} + \nu \frac{\partial^2 \overline{u}}{\partial z^2}. \tag{17.61}$$

In writing (17.61) we include a dissipative term but neglect terms representing advection by the mean flow (such as $w\partial\overline{u}/\partial z$) and the Coriolis force fv. Equation (17.61), with (17.59) and (17.60), is a closed partial differential equation in a single unknown for the mean flow. If the forcing consists of two waves, one with a phase speed c that is positive and one with a negative phase speed, the model produces behaviour that is quite similar to that of the QBO, as we see shortly.

Direction of influence

Although both the observations and the schematic solutions illustrated in Fig. 17.16 suggest that influence is somehow propagating downward, this is in fact not the case when the gravity waves are propagating upward. From (17.59) and (17.61) the wave-driven acceleration of the mean flow, A_w say, is given by

$$A_w = -\frac{\partial \overline{F}}{\partial z} = +\overline{F}(0)g(z)\exp\left[\int_0^z g(z')\,dz'\right] = +g(z)\overline{F}(z). \tag{17.62}$$

That is, the acceleration is a function only of the profile of g in the region from 0 to z, that is the region through which the wave has propagated. Furthermore, attenuation rate $g(z)$ is itself, from (17.60), a function only of the local value of $\overline{u}(z)$ and not of the derivatives of $\overline{u}$. Thus, the wave forcing at some level z is a function only of the profile of $\overline{u}(z')$ for $z' < z$ and independent of the profile at higher altitudes. In other words, and in so far as the diffusivity term in (17.61) is negligible, there is no downward propagation of influence of the mean flow and the mean flow evolution is independent of what takes place above. The physical origin of this result is simply that waves are propagating upward and are absorbed by the mean profile as they ascend. If there were a *source* of waves at very high altitude, or if waves were reflected within the fluid (in which case the first-order WKB approximation is incomplete) then there could be a downward propagation of influence.

17.6.4 Scaling and Numerical Solutions

Scaling the equations — nondimensionalizing in an intelligent way — not only makes numerical integration easier but also indicates what the natural height and time scales are for the problem. Important external parameters that determine the problem are the stratification N (which has units of inverse time, T^{-1}), the damping rate α (also units of inverse time) and the strength of the wave forcing, $\overline{F}$ (units of $(L/T)^2$). A natural horizontal scale is the inverse of the wavenumber k.

Denoting nondimensional quantities with a hat, let

$$\widehat{F} = \overline{F}/F_0, \qquad \widehat{N} = N/N_0, \tag{17.63}$$

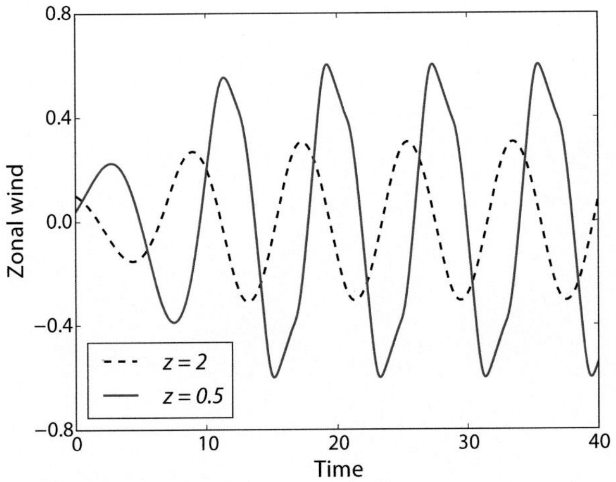

Fig. 17.17 The evolution of the mean zonal wind at two different levels in a numerical solution of (17.67). All the variables are nondimensional, with the only parameter in the problem being viscosity, and here $\widehat{\nu} = 0.15$. A small perturbation is added to $\widehat{u}$, and relatively quickly the solution becomes periodic.

where $F_0 = \overline{F}(0)$ and N_0 is a typical value of N. If N were uniform then we would simply choose $N_0 = N$ whence $\widehat{N} = 1$. To obtain sensible nondimensional quantities we note that the attenuation rate, g, has dimensions of inverse height, so using (17.60) we choose a scaling height H as

$$H = \frac{kc^2}{\alpha N_0}. \tag{17.64}$$

This might suggest a time scaling of $T = kc/\alpha N$. However, because we have a forced problem, the form of (17.61) indicates that we choose

$$T = \frac{cH}{F_0} = \frac{kc^3}{\alpha N_0 F_0}, \tag{17.65}$$

with a velocity scaling of $U = F_0 T/H = c$ (note that this is not an advective scaling). The nondimensional coefficient of viscosity and thermal damping coefficients are then

$$\widehat{\nu} = \nu \frac{T}{H^2} = \nu \frac{\alpha N_0}{F_0 kc}, \qquad \widehat{\alpha} = \alpha T = \frac{kc^3}{N_0 F_0}. \tag{17.66}$$

The nondimensional equation for the mean flow evolution is then, for a single wave,

$$\frac{\partial \widehat{u}}{\partial \widehat{t}} = -\frac{\partial \widehat{F}}{\partial \widehat{z}} + \widehat{\nu}\frac{\partial^2 \widehat{u}}{\partial \widehat{z}^2}, \qquad \text{where} \qquad \widehat{F}(z) = \exp\left[-\int_0^{\widehat{z}} \frac{1}{(\widehat{u}-1)^2}\,\mathrm{d}\widehat{z}'\right], \tag{17.67a,b}$$

with the hats indicating nondimensional quantities. The great simplification that (17.67) offers over (17.59)–(17.61) is that in (17.67) there are no parameters, save for the viscosity, and so the time and vertical scales of the problem are laid bare. In particular, if viscosity is small the only significant timescale in the system is (17.65) and the period of the oscillation must be proportional to that, and the vertical scale of the oscillation must be given by (17.64). Evidently the period of the oscillation is inversely proportion to the strength of the waves, but the vertical extent and the amplitude of the oscillation are both independent of the wave strength.

A numerical solution

Equation (17.67) may readily be numerically integrated[22] and solutions are illustrated in Figs. 17.17, 17.18 and 17.19. The simulations show many of the qualitative features of the observed QBO, including the decay of the pattern with height and its apparent downward propagation. The simulations

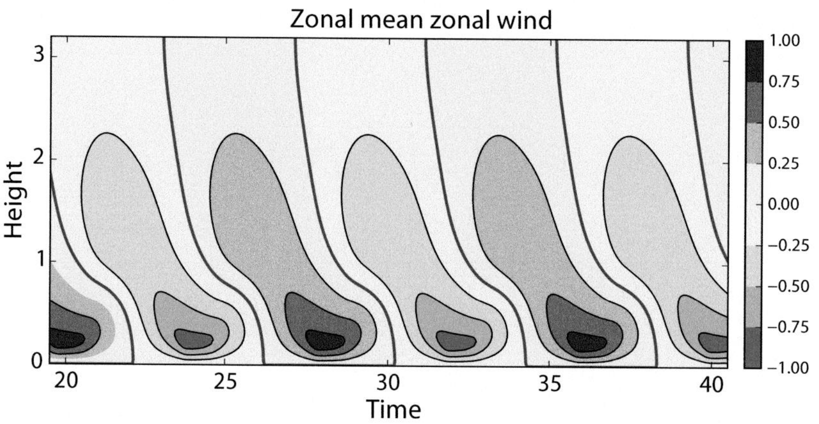

Fig. 17.18 Time height section of nondimensional zonal wind in a numerical solution of (17.67), showing the last 20 time units of the same integration as Fig. 17.17. The zero contour is thicker.

are dependent on the viscosity $\widehat{\nu}$ in that, if $\widehat{\nu} = 0$, the jet at the bottom of the domain cannot be dissipated and the system in fact evolves to a steady state. On the other hand, if the viscosity is large then the jets become too broad, and the boundary layer near $z = 0$ that is evident in Fig. 17.18 is thicker. Still, if the viscosity is small but non-zero then over the bulk of the cycle it plays little role and the oscillation period is only weakly dependent on its value. Thus, for example, in Fig. 17.16 viscosity is needed to wipe out the low level westward jet between panels (d) and (e), but has little role in the rest of the half cycle and so only a small effect on the period. If viscosity is unimportant then the only timescale in the problem is that given by (17.65) and the period is proportional to it, and from numerical integrations we find that it is given by

$$P = A\frac{kc^3}{\alpha N_0 F_0} \tag{17.68}$$

where $A \approx 8$. Note that the period of the QBO is not directly dependent on the period of oscillation of the waves themselves, but is dependent on their strength.

17.6.5 The Roles of Rossby and Kelvin Waves

The waves that propagate into the stratosphere in equatorial regions are of two main types, Kelvin waves and Rossby waves. Kelvin waves are a form of gravity wave but have only an eastward phase propagation, whereas Rossby waves are balanced waves with a westward phase propagation. The theoretical development paralleling Section 17.6.3 is naturally more complex, in part because the problem is now, in principle, a three-dimensional one. However, it is much simplified if we consider motions at the equator and if we take note that, in general, the attenuation rate of a wave is equal to its damping rate divided by its group velocity, as in (17.60). The corresponding attenuation rates for Kelvin and Rossby waves are then given by

$$\text{Kelvin wave:} \qquad g_K(z) = \frac{\alpha}{k_K(\overline{u} - c_K)^2/N}, \tag{17.69a}$$

$$\text{Rossby wave:} \qquad g_R(z) = \frac{\alpha}{k_R(\overline{u} - c_R)^2/N}\left(\frac{\beta}{k_R^2(\overline{u} - c_R)} - 1\right). \tag{17.69b}$$

The Kelvin wave attenuation rate is just the same as that for a non-rotating gravity wave, although the wave speed, c_K, is strictly positive. The Rossby wave attenuation rate (whose derivation requires a little work) involves the equatorial beta parameter and a negative phase speed, c_R. The full problem is defined by (17.61) and (17.59), now with $g(z)$ given by (17.69). It is evident that the problem is no longer east–west symmetric, but the essential structure of the problem remains.

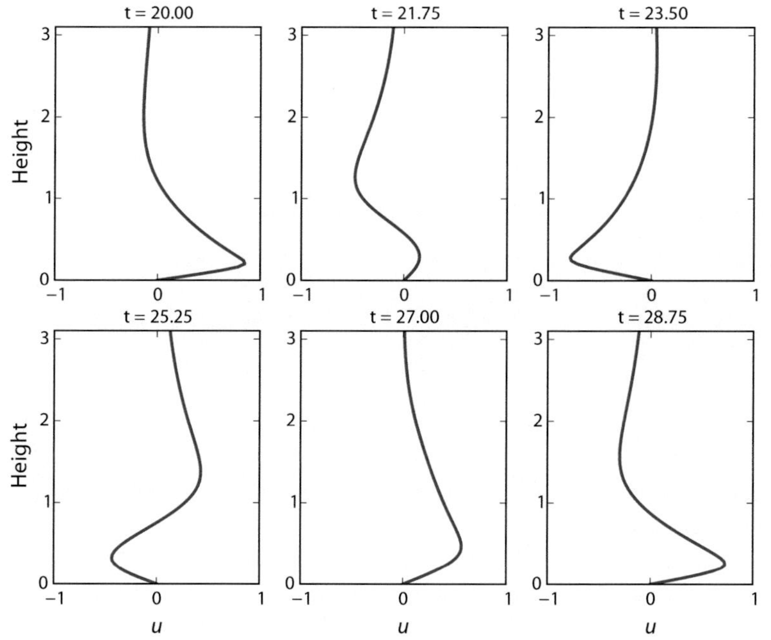

Fig. 17.19 Profiles of zonal wind at the times indicated from the numerical solution of (17.67), shown in Fig. 17.18.

Rossby wave absorption is enhanced near a critical layer where $\overline{u} = c_R$, and Kelvin wave absorption still occurs near $\overline{u} = c_K$, so we expect to see an oscillation contained between these two values. Further, just as in the gravity wave problem, influence propagates upward with the waves.

The equation set (17.61), (17.59) and (17.69) may be numerically integrated and solutions are illustrated in Fig. 17.20. It is the factor g_R that is responsible for the westward acceleration of the mean flow, for 'dragging' $\overline{u}$ toward the value c_R, which is negative. But g_R is zero when $\overline{u} - c_R = \beta/k_R^2$ which, for our numerical simulation, occurs when $\overline{u} = 1$. Thus, when $\overline{u}$ is close to its eastward (Kelvin-wave induced) peak the westward acceleration is small. Another source of east–west asymmetry is the likelihood that Rossby waves and Kelvin waves may have different amplitudes. If Kelvin waves were stronger, for example, then the eastward acceleration would be stronger than the westward and that part of the cycle would be faster.

17.6.6 General Discussion

The above sections have described a couple of relatively simple models that seem to capture the essence of the QBO. The simple model using both Rossby waves and Kelvin waves is not noticeably more realistic in its predictions; rather, it is attractive because Rossby waves and Kelvin waves are observed in the equatorial stratosphere and may be more realistic in its assumptions. The observed east–west asymmetry in the observed QBO is not obviously caused by the differences between Rossby waves and Kelvin waves; other possibilities include the effects of a mean circulation and possible differences in the strength of the eastward and westward forcing.

The model with two non-rotating gravity waves is attractive because it allows a more complete analysis of its properties. In particular, the upward propagation of waves leading to a downward propagation of the zonal wind pattern, and the factors determining the period of the oscillation, are made transparent. The period of the problem is given by (17.68). Using Table 17.1 as a guide, let us take the following dimensional values of the parameters: $k = 2 \times 10^{-7}\,\mathrm{m}^{-1}$, $\alpha = 1 \times 10^{-6}\,\mathrm{s}^{-1}$, $c = 25\,\mathrm{m\,s}^{-1}$, $N_0 = 2.2 \times 10^{-2}\,\mathrm{s}^{-1}$, $F_0 = 5 \times 10^{1}0^{-3}\,\mathrm{m}^2\,\mathrm{s}^{-2}$. We obtain a timescale of $T = (kc^3/\alpha N_0 F_0) \approx 160$ days or about 5 months and so, using (17.68), a period of 40 months. Obviously there is considerable uncertainty in the parameters chosen and it would not be difficult

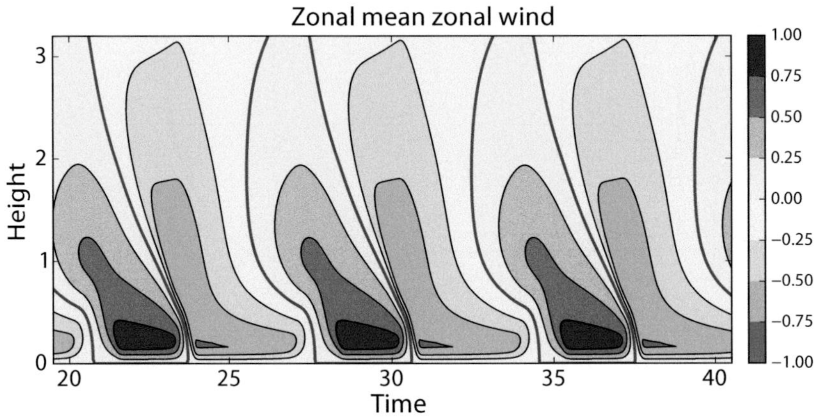

Fig. 17.20 Time height section of nondimensional zonal wind in a numerical solution similar to that of Fig. 17.18, but now using (17.69) and so with both Rossby and Kelvin waves. The zero contour is thicker.

to choose a set of parameters giving the observed value of about 26 months — or for that matter, to choose a set that gave a still longer period.

The vertical scale of the oscillation is given by (17.64) and with the above set of parameters we obtain $H = (kc^2/\alpha N_0) \approx 6\,\text{km}$. From the numerical simulations we see that the vertical penetration of the phenomena is 2 or 3 times this, so about 15 km. This value again is reasonably close to the observed value of, from Fig. 17.13, about 20 km, but again we should be wary of too close an agreement, especially using a one-dimensional quasi-Boussinesq model. It is interesting that the period and vertical extent of the observed oscillation vary only a little, implying only a little interannual variability in the forcing strength and other parameters of the problem.

The actual waves themselves are primarily generated by convection in the tropical troposphere, they then propagate up into the stratosphere. It is difficult to numerically simulate a QBO with an explicit representation of gravity waves because of the large range of scales involved in the problem, although three-dimensional simulations with parameterized gravity waves have been quite successful. Still, a striking demonstration of the mechanism above came from laboratory experiments, using an annulus of stratified water subject to a standing wave forced by a flexible lower boundary. Given a strong enough forcing an oscillating mean flow was generated whose structure was found to be in very good agreement with the two-wave theory. Thus, at the very least, the mechanism does describe a real physical phenomenon.[23]

There are a great many aspects of the QBO that we have not discussed, including its latitudinal structure, the effects of a mean circulation, and three-dimensional numerical simulations, and a few references that may serve as an introduction to these topics are given here.[24] The QBO is also not completely regular, as we see from Fig. 17.13, and one somewhat unusual example occurred in early in 2016 (Fig. 17.21). The QBO normally shows a fairly steady downward propagation of the westerly phase, but in January 2016 the westerly winds in the lower stratosphere switched back to easterlies after only about 6 months, the shortest period of westerlies above 20 km in a record going back to 1953. This behaviour may have been caused by anomalous horizontal propagation of Eliassen–Palm fluxes from mid-latitudes because of the absence of a subtropical critical line, although we cannot be definitive. Still, it is the regularity of the QBO rather than the occasional anomaly that is most striking.

Finally, to make a personal remark, the QBO is both a curiosity and a triumph. The former because its relationship to and influence on tropospheric circulation, and the climate and weather that affect humankind, is not obvious to the casual or even expert observer; it does not have the impact of an El Niño event or a cold winter, for example. Yet, excepting directly forced oscillations like the diurnal and seasonal cycle, it is the clearest example of a nearly periodic phenomenon in the atmosphere and its simple and beautiful explanation must rank as a major achievement in geophysical fluid dynamics.

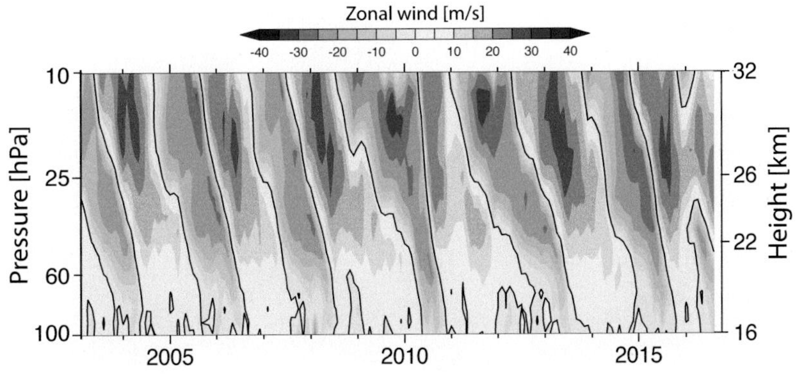

Fig. 17.21 Equatorial zonal wind, similar to Fig. 17.13, but now with cycles of the QBO from 2003 to 2016 and showing the interruption of the eastward winds below about 25 km in early 2016.[25]

17.7 VARIABILITY AND EXTRA-TROPICAL WAVE–MEAN-FLOW INTERACTION

As noted in Section 17.1, the stratosphere has only mild, if any, baroclinic instability and because of its high stratification the amplitude of a baroclinic mode reaching up from the troposphere thus tends to decay rapidly in the stratosphere, as in Fig. 9.21. If there were to be a baroclinic instability confined to the stratosphere it would be at large scales, perhaps at wavenumbers 1, 2 or 3, as opposed to wavenumber 8 or so in the troposphere. But this is not to say there is no variability in the stratosphere, with the variability arising in two main ways:

(i) From waves propagating up from the troposphere, with the stratospheric variability arising from the variability of the troposphere.

(ii) From oscillatory or even chaotic flow arising from the interaction, within the stratosphere, of large-scale planetary waves with themselves and with the mean flow. The forcing may still come from the troposphere but, even when this is steady, intra-stratospheric interactions may give rise to unsteadiness.

In either case the variability tends to be relatively slow (compared to the troposphere) and at a large scale — waves from the troposphere undergo Charney–Drazin filtering and tend to occur at wavenumbers 1 and 2, and as noted any baroclinic instability is also at large scale. It is therefore useful to think of the variability as a wave–mean-flow problem rather than as a problem in geostrophic turbulence.

17.7.1 Upward Propagating Disturbances and Sudden Warmings

Consider planetary waves that are excited in the troposphere and propagate upward, as described in Chapter 16, with this occurring predominantly in winter when the tropospheric forcing is strongest. The wave activity obeys

$$\frac{\partial \mathcal{P}}{\partial t} + \nabla \cdot \boldsymbol{\mathcal{F}} = D, \tag{17.70}$$

where $\mathcal{P} = \overline{q'^2}/2\overline{q}_y$ and $\nabla \cdot \boldsymbol{\mathcal{F}} = \overline{v'q'}$. Thus, if dissipation is small and $\partial \mathcal{P}/\partial t$ is positive, $\overline{v'q'}$ is negative. Now consider the zonal momentum equation in quasi-geostrophic TEM form, namely

$$\frac{\partial \overline{u}}{\partial t} - f_0 \overline{v}^* = \overline{v'q'}. \tag{17.71}$$

Rossby waves propagating into the stratosphere or, by the same token, dissipating Rossby waves in a statistically steady state, thus induce a deceleration (i.e., a westward tendency) of the zonal mean flow and/or a poleward meridional flow, and only for a very deep forcing is the residual circulation response negligible. Also, the zonal-wind response to a wave forcing will tend to be of larger scale

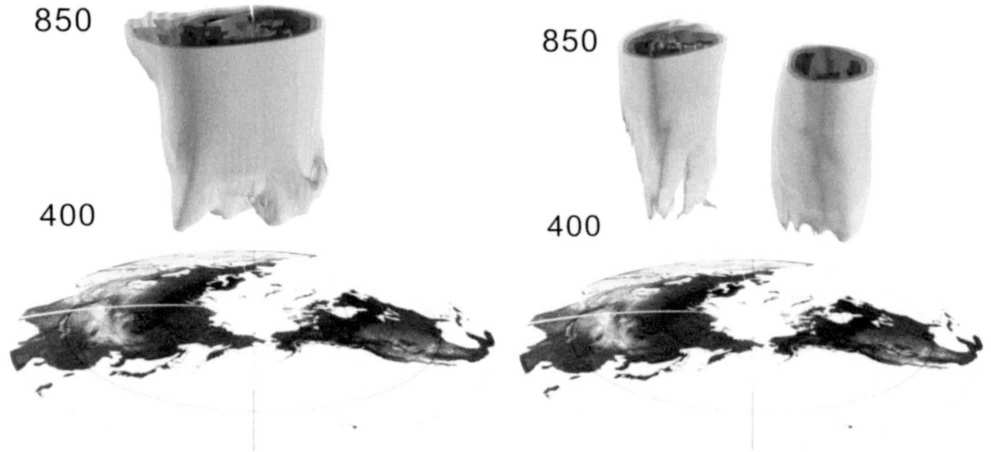

Fig. 17.22 The edge of the stratospheric polar vortex in December 1984. Plotted is the 35 PVU isosurface of $Q^* = Q(\theta/\theta_0)^{-4.5}$, where Q is Ertel PV and $\theta_0 = 475\,\mathrm{K}$. The vertical coordinate is potential temperature. Like Q, Q^* is materially conserved in adiabatic flow, and roughly compensates for the change in density with height that affects the Ertel PV. The left panel (14 December) shows the vortex in a fairly usual state, and the right panel (30 December) shows a split vortex following a stratospheric sudden warming.[26]

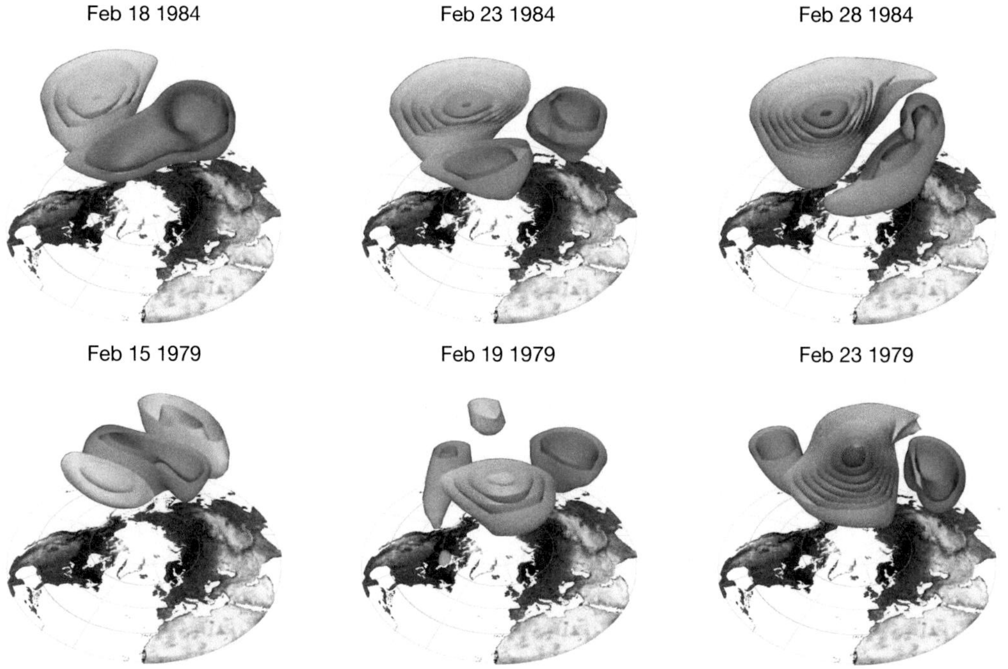

Fig. 17.23 Time sequence of two stratospheric warmings, with the top row showing a displacement of the initial (blue) polar vortex and the bottom showing a split, with the dates marked. Contours are anomalous geopotential height larger (smaller) than 4km (-4km) in red (blue), between 200 and 10 hPa, spaced at 2km intervals. In both cases the initial polar vortex is cold (blue) with an anomalously warm end state (red).[27]

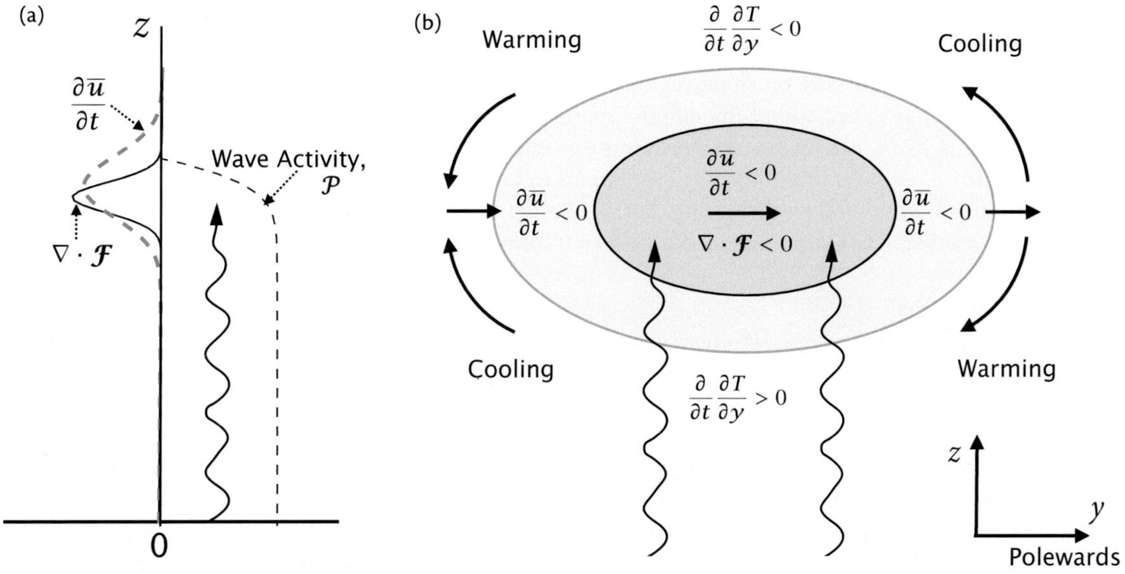

Fig. 17.24 The wave–mean-flow dynamics of a stratospheric warming. (a) Upward propagating Rossby waves (wavy lines)break in the stratosphere, and the wave activity (dashed line) diminishes. The EP flux divergence is negative ($\nabla \cdot \mathcal{F} < 0$), inducing a westward acceleration ($\partial \overline{u}/\partial t < 0$) over a broader region because of the ellipticity in (17.72). (b) The circulation and temperature response induced. The negative EP flux divergence is shaded dark, inducing westward acceleration over the broader region (light shading). Assuming there is no acceleration far away from the wavebreaking, the temperature response can be inferred from thermal wind, with a warming at the lower poleward end of the breaking and an induced residual circulation as shown by the arrows.

than the forcing itself, as can be seen from (10.87),

$$\left[\frac{\partial^2}{\partial y^2} + \frac{\partial}{\partial z}\left(\frac{f_0^2}{N^2}\frac{\partial}{\partial z}\right)\right]\frac{\partial \overline{u}}{\partial t} = \frac{\partial^2}{\partial y^2}\overline{v'q'}. \tag{17.72}$$

The elliptic nature of the operator acting on $\partial \overline{u}/\partial t$ produces a response on a larger scale than the right-hand side.

Suppose, then, that Rossby waves propagate upward from the troposphere and break in the stratosphere. The mean eastward flow (sometimes called the polar night jet) will be weakened, so allowing more waves to propagate up, since strong eastward flow inhibits propagation (Fig. 16.6). If the process continues the winds will eventually reverse, forming a critical layer (as described in Section 16.6.1) where $\overline{u} = 0$. This completely inhibits further upward propagation and wave breaking is intensified, inducing a westward flow at the level to which the propagation reaches. There is a rapid changeover to westward flow and the critical layer descends. This sequence has an obvious similarity with the westward acceleration phase of the QBO, but in the extra-tropics there is no eastward counterpart as there are no Kelvin waves, and thus no oscillation. Rather, the eastward winds of the polar night jet are gradually restored by radiative effects.

A reduced (or reversed) vertical shear is, by thermal wind, associated with a reduced (or reversed) meridional temperature gradient, so that the polar night jet is replaced by a warmer westward flow. Put simply, the deposition of westward wave momentum leads to a warming of the high-latitude stratosphere. Such an event can at times be strong enough to split asunder the cold polar vortex, as illustrated in Figs. 17.22 and 17.23, and when it does the event is known as a *sudden stratospheric warming.*[28]

The interaction of the waves and mean-flow is sketched in Fig. 17.24. In the left panel we see a wave propagating up from the troposphere and breaking, with wave activity then falling. The EP flux is negative in the breaking region causing a deceleration of the zonal flow over a somewhat broader region because of the elliptic operator in (17.72). The temperature response, shown in the right panel, can be inferred from thermal wind balance, noting that above the breaking region $\partial_t(\partial u/\partial z) > 0$ so that $\partial_t(\partial T/\partial y) < 0$, and oppositely for below. The direction of the residual circulation then follows by noting that adiabatic warming (cooling) results from descent (ascent). The residual circulation may also be inferred from (17.38b) or (17.38c).

Numerical simulations

To illustrate the above mechanism in a more realistic setting we show some results from a primitive equation simulation that mimics the broad features of the observations quite well. The advantage over showing the observations is that a great many events are simulated and full diagnostics can be obtained.[29] To obtain the results, the model (which had a well-resolved stratosphere) was integrated for many decades, during which time many sudden warmings occurred. Composites of these events are shown in Fig. 17.25.

The results show an anomalous upward flux of wave activity (EP flux) that, on dissipating in the stratosphere, induces a westward acceleration of the zonal flow, a weakening of the polar vortex and a warming of the polar regions, extending equatorward as far as 60°. It seems to be the condition of the stratosphere in filtering, or not, upward propagating waves that determines whether or not a warming occurs, rather than anomalous bursts originating in the troposphere. Vortex dynamics also play a role: a vortex, once formed, is rather stable and has a natural tendency to persist rather than break up; in two-dimensional turbulence vortices tend to merge and not split. This stability prevents warmings from occurring too frequently, since the wave activity must be strong enough to overcome the elastic properties of the vortex edge.

17.7.2 ♦ Wave–Mean-Flow Interaction and Stratospheric Variability

Stratospheric variability need not arise solely from waves propagating up from the tropopause, and we can illustrate this with a simple numerical model of wave–mean-flow interaction, similar to those discussed in Section 10.1.3. Specifically the model consists of the following quasi-geostrophic ideal-gas equations.[30] The zonally-averaged fields obey

$$\frac{\partial \overline{q}}{\partial t} = \overline{F} - \frac{\partial}{\partial y}\overline{v'q'}, \tag{17.73a}$$

where

$$\overline{q}(y,z,t) - \beta y = \left[\frac{1}{\rho_R}\frac{\partial}{\partial z}\left(\rho_R \frac{f_0^2}{N^2}\frac{\partial \Psi}{\partial z}\right) + \frac{\partial^2 \Psi}{\partial y^2}\right], \qquad \left(\overline{u}, \frac{R}{Hf_0}\overline{T}\right) = \left(-\frac{\partial \Psi}{\partial y}, \frac{\partial \Psi}{\partial z}\right). \tag{17.73b,c}$$

The eddies obey

$$\frac{\partial q'}{\partial t} + \overline{u}\frac{\partial q'}{\partial x} + v'\frac{\partial \overline{q}}{\partial y} = F', \tag{17.74a}$$

where

$$q'(x,y,z,t) = \nabla^2\psi' + \frac{1}{\rho_R}\frac{\partial}{\partial z}\left(\rho_R\frac{f_0^2}{N^2}\frac{\partial \psi'}{\partial x}\right), \qquad (u',v') = \left(-\frac{\partial \psi'}{\partial y}, \frac{\partial \psi'}{\partial x}\right) \tag{17.74b}$$

and

$$\frac{\partial \overline{q}}{\partial y} = \beta - \frac{\partial^2 \overline{u}}{\partial y^2} - \frac{1}{\rho_R}\frac{\partial}{\partial z}\left(\rho_R\frac{f_0^2}{N^2}\frac{\partial \overline{u}}{\partial z}\right). \tag{17.74c}$$

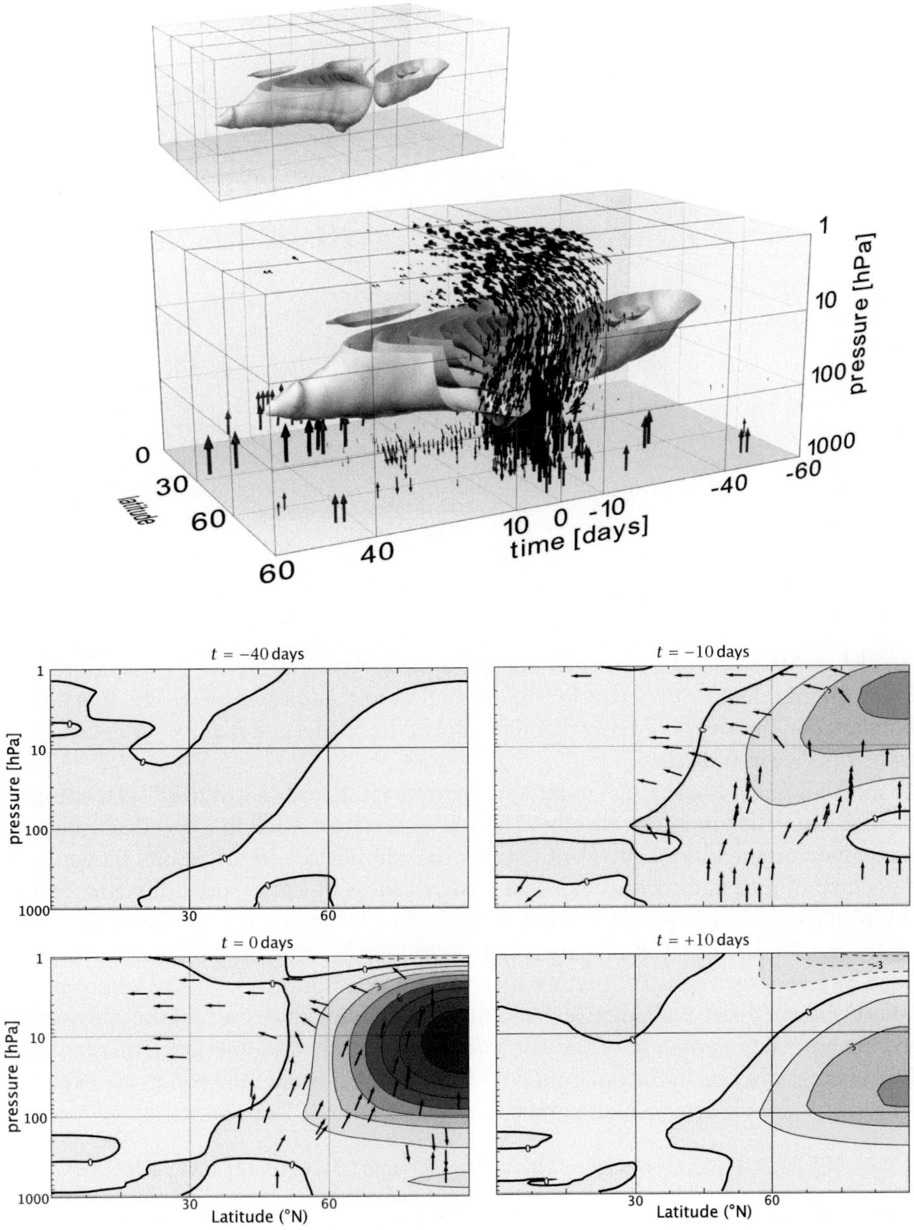

Fig. 17.25 The evolution of a stratospheric warming obtained with a general circulation model, compositing several events together. The top plots show anomalous zonal mean zonal wind, with surface intervals of 2 m/s, with time running from right to left. The bottom plots show time slices of anomalous temperature in colour contours (interval 1.5 K, red is warm), with times relative to the peak warming. In both plots the black arrows show anomalous EP flux.

There is a strengthening and slight northward propagation of the polar vortex prior to the onset, and a strong weakening during and after the onset. At forty days prior to the onset (upper left panel in lower set of plots), anomalies are very small. Ten days before onset (upper right), warming has appeared in the polar upper stratosphere, and anomalously strong EP fluxes appear throughout the atmosphere. At the onset (lower left), the warming and the EP flux anomalies are strongest. Ten days later (lower right), the temperature anomalies are weaker and confined to the lower stratosphere, and the EP fluxes are very weak.[31]

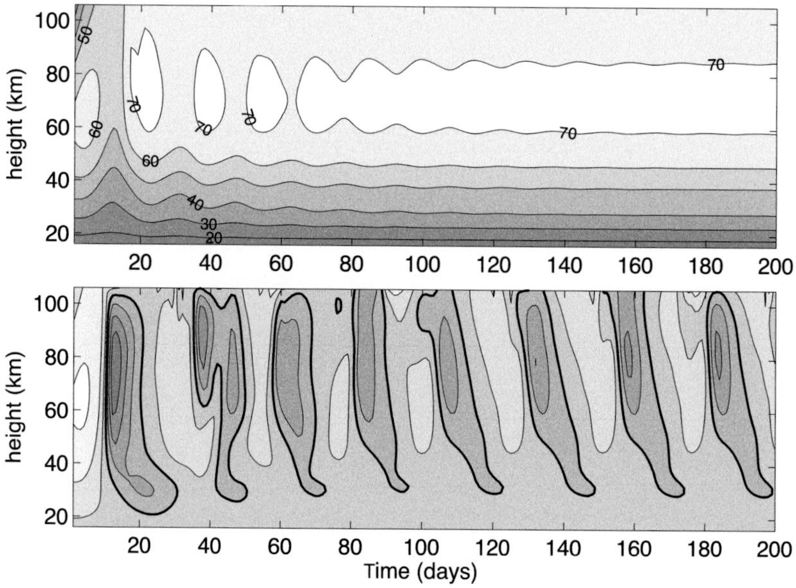

Fig. 17.26 Evolution of the zonal mean zonal wind in a case with steady wave forcing with a value of 200 in the top panel and 300 in the bottom panel. In the bottom panel the contours are every 20 m s^{-1}, positive values have lighter shades and the zero contour is heavy.

The notation follows our usual conventions, with $\rho_R(z)$ being a reference density profile and $\overline{F}$ and F' the forcing/dissipation terms for the mean flow and eddies, respectively. It is the domain, the boundary conditions and forcing that distinguish the model and make it representative of the stratosphere, as we now discuss.

The model domain is a channel nominally centred at 60° and of width 60°, extending upward to about 100 km. The forcing on the zonal flow is a relaxation back to a specified radiative equilibrium temperature field (or equivalently a thermal wind field). In the results shown below this is independent of time and corresponds to a constant shear of 1 m s^{-1} per kilometre, or a temperature difference of about 15 K across the domain, with a relaxation timescale that varies from 20 days at 20 km to 4 days at 50 km. There is also a weak linear drag on the mean flow. The eddies are forced by imposing a constant perturbation at the lower boundary, with wavenumber 2 in the simulations shown. There is a radiative damping on the eddies ensuring that the eddies are mostly damped before reaching the top of the domain. The vertical variations are represented using finite differencing, whereas in the horizontal both the mean flow and the eddies are expanded in a Fourier series with only a very small number of terms retained. Thus, we write

$$[\overline{q}, \Psi] = [Q_0(z,t), \Psi_0(z,t)] \cos ly, \quad [q', \psi'] = \mathrm{Re}\left[q_0(z,t), \psi_0(z,t)\right] \sin ly \exp(\mathrm{i}kx), \qquad (17.75)$$

and after some manipulation we can obtain evolution equations for Q_0 and q_0 with diagnostic equations for Ψ_0 and ψ_0. The quadratic terms in the equations of motion create higher order terms that are projected back onto the retained terms. (Aside from the severe horizontal truncation the numerical method used to find results is not a key aspect of the model.)

Some numerical results and interpretation

Results of two numerical integrations are shown in Fig. 17.26 and Fig. 17.27. In one integration the geopotential forcing at the lower boundary has an amplitude of 200 m, whereas in the other it has an amplitude of 300 m. In the first case the flow evolves into an absolute steady state, whereas in the second the mean flow and the waves oscillate with a period of about 25 days, with the mean flow actually becoming negative over half the cycle. The streamfunction in the unsteady case is tilting into the mean shear (the right panel of Fig. 17.27), evocative of baroclinic instability. The oscillations are in some way redolent of stratospheric warmings. The climatological eastward winds transition

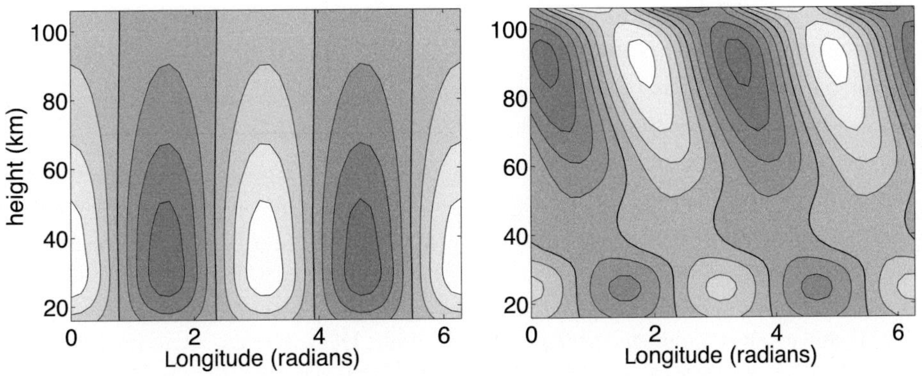

Fig. 17.27 Snapshots of the wave streamfunction in the steady case (forcing of 200) in the left panel and in the oscillatory case (forcing of 300) in the right panel. Zero contour is heavy.

rather quickly to westward winds (darker shading in Fig. 17.26), with the westward winds then descending and with a slower recovery back to climatology. There are two points to be made:

(i) Interactions that are internal to the stratosphere can give rise to oscillatory motion.

(ii) The source of energy for the waves may in part arise from a baroclinic instability and in part from tropospheric forcing.

To summarize, stratospheric warmings are not solely a response to tropospheric forcing; the internal dynamics of the stratosphere, and stratosphere-troposphere interactions, can also play a role, albeit a secondary one.

Notes

1 To read more about the middle atmosphere see, for example, the review by Hamilton (1998), the collection of articles in *Journal of the Meteorological Society of Japan*, vol. 80, no. 4B, 2002, the book by Andrews *et al.* (1987) and the review by Haynes (2005). I am also grateful to Nili Harnik and Peter Haynes for many comments on this chapter.

2 The calculation is described in Jucker *et al.* (2013).

3 Upper figure courtesy of J. Wilson of GFDL, using data from Fleming *et al.* (1988). Lower figure uses data from the ECMWF ERA-interim reanalysis.

4 Courtesy of D. Waugh.

5 Courtesy of A. Dörnbrack.

6 Adapted from Eluszkiewicz *et al.* (1997).

7 Brewer (1949) and Dobson (1956). Brewer deduced upward motion into the stratosphere at low latitudes based on the water vapour distribution, while Dobson deduced a poleward transport within the stratosphere based on the ozone distribution — the circulation takes ozone from the low latitudes toward the poles. Although originally the Brewer–Dobson circulation was taken to mean the chemical transport circulation, it is now usually taken to mean the residual (i.e., thickness weighted) overturning circulation. The two may differ if there is a mixing of chemicals without a mixing of mass, and the chemical transport may differ among chemicals.

8 See also Plumb (2002), which motivated this figure.

9 It turns out that $\overline{u'w'} > 0$ for upward propagating Rossby waves and thus, if the waves were to be dissipated, an eastward acceleration would seemingly be implied. In fact it is the form stress that is the most important aspect of vertical momentum transport in such waves, and when the waves are dissipated a westward acceleration ensues.

10 Blanca Ayarzaguena Porras kindly made this figure using an ERA-interim reanalysis.

11 Drawing from Garcia (1987) and Haynes (2005).

12 Adapted from Holton *et al.* (1995).

13 Values are taken from Wallace (1973), Plumb (1984) and Andrews *et al.* (1987).

14 The solution with no boundaries is not obvious from the description given here and the reader should consult the solution given in Haynes *et al.* (1991), as well as exploring the analogy further.

15 I am grateful to Verena Schenzinger who kindly made this plot using winds from the ERA-interim reanalysis. See also Gray (2010).

16 A broad overview of QBO is provided by Baldwin *et al.* (2001), with updates and additions by Gray (2010), and a more theoretical review is given by Plumb (1984). The term quasi-biennial oscillation seems to have been coined by Angell & Korshover (1964), although the discovery of the QBO is generally credited to R. J. Reed and R. A. Ebdon, independently and at about the same time (Ebdon 1960, Veryard & Ebdon 1961, Reed 1960, Reed *et al.* 1961).

17 Figure adapted from Gray (2010), who used the method of Pascoe *et al.* (2005).

18 Wallace & Holton (1968), Baldwin *et al.* (2001).

19 The first theory of the QBO along these lines was put forward by Lindzen & Holton (1968) and Holton & Lindzen (1972), with clarifications and simplifications by Plumb (1977), and it is these models that we draw from. Prior to Lindzen & Holton's work, Booker & Bretherton (1967) had shown how the momentum deposition by gravity waves can be enhanced near critical lines and this was a key theoretical advance. A host of papers elaborating on the basic mechanism have since appeared, discussing such thing as the particular type of gravity waves involved, the role of the Coriolis force and the meridional confinement of the QBO, the possible influence of the solar cycle and El Niño, the impact on tracer transport and so on.

20 Adapted from Plumb (1984).

21 Adapted from Plumb (1984).

22 Code is available to the reader.

23 Plumb & McEwan (1978). Regarding atmospheric relevance, and as with some other scientific theories of complex phenomena, it is hard to be absolutely certain that the theory is correct, for the sceptic can always point to observational disagreements or say that another theory might be the correct one. Sometimes the analogue of deciphering a complex communication may be apt: if an encrypted signal is deciphered to reveal a meaningful message, it may seem perverse to ask whether the deciphering is unique, and whether some other message might have emerged from a different decryption.

24 Examples of additional theoretical development are Dunkerton (1982, 1997), Boyd (1978) and Plumb & Bell (1982). Simulations of a QBO in an atmospheric GCM have been achieved by Takahashi (1996), Hamilton *et al.* (2001), Scaife *et al.* (2002), Giorgetta *et al.* (2002) and others. The possible effects of the QBO on the extra-tropical circulation are discussed by Holton & Tan (1980, 1982), Jones *et al.* (1998), Kushner (2010), Labitzke *et al.* (2006), Randel *et al.* (1999), Scott & Haynes (1998), Dunkerton *et al.* (1988) and others. The anomalous QBO of 2015–2016 is documented by Newman *et al.* (2016) and Osprey *et al.* (2016).

25 This plot was kindly made by Varena Schenzinger using data from the Singapore radiosonde. For a description of QBO datasets see http://www.geo.fu-berlin.de/en/met/ag/strat/produkte/qbo/.

26 Figure kindly prepared by M. Jucker using ERA-interim reanalysis. See Lait (1994) for a discussion of the alternative PV.

27 Data from ECMWF ERA-Interim, visualization with the software 'pv-atmos', described in Jucker (2014).

28 The model described here was proposed by Matsuno (1971) and although nonlinear effects (and, to a lesser degree, non-geostrophic effects) play a quantitative role, Matsuno's model remains the

foundation of our understanding. An early review is that of Schoeberl (1978) and there have been numerous studies since. To name but a few, Dunkerton *et al.* (1981) and Palmer (1981) explored the phenomenon from a TEM perspective, Limpasuvan *et al.* (2000) and Charlton & Polvani (2007) provide a comprehensive view of the observations of warmings using reanalysis datasets, Charlton *et al.* (2007) look at various simulations with GCMs, and Gray *et al.* (2001) look at external influences on the timing of warmings.

29 These simulations were kindly performed by Martin Jucker. See also Jucker *et al.* (2014) and Jucker (2016).

30 This model was introduced by Holton & Mass (1976) and the numerical results we show use a code adapted from one by J. Holton. Plumb (1981), Yoden (1987, 1990) and others have explored the model further with a view to better understanding the parameters for which steady, oscillatory or chaotic motion was present. Christiansen (1999, 2000), Scaife & James (2000), Scott & Haynes (2000), Sjoberg & Birner (2012), Jucker *et al.* (2014) and others have explored related behaviour using various types of models and observations, for example with the primitive equations and/or including eddy-eddy interactions and with different boundary conditions.

31 Figure created using the software 'pv-atmos', described in Jucker (2014).

I'm singing in the rain, Just singing in the rain,
What a glorious feelin', I'm happy again.
I'm laughing at clouds, So dark up above,
The sun's in my heart, And I'm ready for love .

Lyrics by Arthur Freed, music by Nacio Herb Brown, *Singin' in the Rain*, 1929.

CHAPTER 18

Water Vapour and the Tropical Atmosphere

WATER IS AN ORDINARY SUBSTANCE WITH EXTRAORDINARY EFFECTS. The most obvious is that oceans themselves are made of water, and if our planet were dry this book would perforce be much shorter (if only). Leaving aside the dynamical effects of the oceans, water covers over two-thirds of Earth's surface and because it is warm in some places and cold in others, and because the atmosphere is in motion, water evaporates into the atmosphere in one place and condenses from it elsewhere. The condensation leads to rain, one of the most talked-about aspects of weather and climate. Water also freezes to form ice, so that at any given time water exists on Earth in all three phases. Radiatively, water vapour is a greenhouse gas, meaning that it absorbs infrared radiation that might otherwise be lost to space and so maintains the surface of the planet at a temperature over 20 K higher than an equivalent dry planet. Dynamically, the condensation of water vapour in the atmosphere releases energy, warming the air and tending to make it more unstable than otherwise and leading to convection. Further, the net transport of water vapour from low to high latitudes is effectively a meridional transport of energy.

In this chapter we focus on a small number of these issues, mainly on the kinematics and dynamics of water vapour itself and on some aspects of the dynamics of the tropical atmosphere, where the effects of water vapour are most manifest. The tropics would certainly differ from the mid-latitudes even if the atmosphere were dry — its Coriolis parameter is small among other things — so our attention there is by no means confined to the effects of water vapour. Nevertheless, tropical convection and the attendant 'radiative-convective equilibrium' are greatly influenced by the presence of water. We begin with a discussion of the thermodynamic properties of water vapour itself. We then move on to an essentially kinematic description of the factors determining the large scale distribution of relative humidity, before finally looking at convection and at tropical dynamics more generally.[1]

18.1 A MOIST IDEAL GAS

Water is the compound of hydrogen and oxygen with the chemical formula H_2O, although in informal conversation water is often understood to mean only the liquid form of the compound. *Water vapour* is a gas made up of molecules of H_2O, and *ice* is the solid form of water. *Steam*, in common parlance, is a mixture of air, water vapour and suspended droplets of water, usually at a very high temperature. Steam is formed when water vapour at temperatures above boiling point cools under contact with air and some of the water vapour condenses, forming a fine mist. Cloud and fog are also mixtures of dry air, water vapour and water droplets, but need not be at high temperature. Our focus will be on water vapour which, as we will see, can exist over a wide range of temperatures; let us first say how we quantify it and how it affects the equation of state. A number of thermodynamic derivations are also given in Appendix A (page 720). Those derivations are more systematic but less pedagogical than those below, and may appeal to some.

18.1.1 Ideal Gas Equation of State

The thermal equation of state for an ideal gas is conventionally written in the form

$$pV = Nk_BT = nR^*T, \tag{18.1}$$

where N is the total number of molecules in the volume V, n is the number of moles in that volume, and $N = nN_A$ where N_A is Avogadro's number. A mole is the amount of a substance that contains the same number of elementary entities, usually atoms or molecules, as there are atoms in 12 grams of carbon-12, that number being Avogadro's number ($N_A \approx 6.02 \times 10^{23}$). Two moles of a substance contains two times Avogadro's number of elementary units. The constants in the above equation are Boltzmann's constant, k_B, and the universal gas constant, R^*, where $R^* \equiv N_Ak_B = 8.314\,\mathrm{J\,mol^{-1}\,K^{-1}}$. As noted in Chapter 1, for any particular gas it is convenient to define the specific gas constant by $R = R^*/\mu$ where μ is the molar mass (mean molecular weight in kg/mol). For a single component gas we then divide (18.1) by the total mass $M = n\mu$ to obtain

$$p = \rho RT. \tag{18.2}$$

Throughout this chapter we will be concerned only with 'simple ideal gases', or 'perfect gases', for which the gas constants at constant composition are, in fact, constant.

For a multi-component ideal gas the partial pressure of each component is independent of the presence of the other components (because the volume of the molecules is negligible) and so is equal to the hypothetical pressure of that gas if it alone occupied the volume of the mixture. The total pressure is therefore the sum of the partial pressures of each gas, a dictum known as Dalton's law of partial pressures. The partial pressure of each constituent in a mixture is proportional to the number of molecules of that constituent, and therefore also proportional to the number of moles. Because of Dalton's dictum we can obtain a simple expression for the equation of state of a mixture, as follows. Denoting the constituents by subscript i, the total pressure is given by

$$p = \sum_i p_i = \sum_i \frac{1}{V} n_i R^* T = \sum_i \left(\frac{M}{V}\right) \frac{n_i\mu_i}{M\mu_i} R^* T. \tag{18.3}$$

Let us define the effective molar mass, μ_e, by

$$\frac{1}{\mu_e} = \sum_i \frac{n_i\mu_i}{M\mu_i} = \sum_i \frac{\varphi_i}{\mu_i}, \tag{18.4}$$

where $\varphi_i = (n_i\mu_i)/M$ is the mass fraction of the i-th constituent. We then have

$$p = \rho RT, \qquad \text{where} \qquad R = \frac{R^*}{\mu_e} = \sum_i \varphi_i R_i, \tag{18.5a,b}$$

and $R_i = R^*/\mu_i$. The effective gas constant of the mixture is thus the mass-weighted mean of the specific gas constants of its constituents. Any given gas has a specific gas constant that is inversely proportional to its molecular weight. Thus, for a given fluid density and temperature a gas with a higher molecular weight will exert a lower pressure than one that has a smaller molecular weight, because it will have fewer molecules per unit mass. Similar expressions apply to the heat capacities c_p and c_v, so that a heavier gas (higher molecular weight) has a smaller specific heat capacity.

18.1.2 Application to Moist Air

Dry air has virtually constant composition and its mean molar mass is $\mu^d = 29.0 \times 10^{-3}\,\mathrm{kg\,mol^{-1}}$, giving $R^d = R^*/\mu^d = 287\,\mathrm{J\,kg^{-1}\,K^{-1}}$. Water vapour has a molar mass of $\mu^v = 18.014\times10^{-3}\,\mathrm{kg\,mol^{-1}}$ giving $R^v = 461.5\,\mathrm{J\,kg^{-1}\,K^{-1}}$. The two gas constants are related by

$$\frac{R^v}{R^d} = \frac{\mu^d}{\mu^v} \equiv \frac{1}{\epsilon} \approx 1.608. \tag{18.6}$$

Now consider a mixture of dry air and water vapour.

Measures of moisture

When mixtures are present we use superscripts d, v and l to denote thermodynamic quantities associated with dry air, water vapour and liquid water. The *absolute humidity* is the amount of water vapour per unit volume, with units of $\mathrm{kg\,m^{-3}}$, or informally $\mathrm{g\,m^{-3}}$. The *mixing ratio,* w, is the ratio of the mass of water vapour, m^v, to that of dry air, m^d, in some volume of air and is thus

$$w \equiv \frac{m^v}{m^d} = \frac{\rho^v}{\rho^d}. \tag{18.7}$$

It is a nondimensional measure but it is often expressed in terms of grams per kilogram. In the atmosphere values range from close to zero to about $20\,\mathrm{g\,kg^{-1}}$ (2×10^{-2}) in the tropics on a humid day.

The *specific humidity,* q, is the ratio of the mass of water vapour to the total mass of air — dry air plus water vapour — and so is

$$q \equiv \frac{m^v}{m^d + m^v} = \frac{w}{1+w} \quad \text{and} \quad w = \frac{q}{1-q}. \tag{18.8a,b}$$

The specific humidity is just the mass concentration of water vapour in air. In most circumstances in Earth's atmosphere $m^v \ll m^d$ so that $q \approx w$, usually to an accuracy of about one percent. In most of this chapter we will ignore the differences between w and q, but this is not appropriate for all planetary atmospheres.

The partial pressure of water vapour in air, e, is the pressure exerted by water molecules and is proportional to the number of moles of water vapour in the volume. It is given by

$$e = \frac{n^v}{n^d + n^v} p = \frac{m^v/\mu^v}{m^d/\mu^d + m^v/\mu^v} p, \tag{18.9}$$

where n^v and n^d are the number of moles of water vapour and dry air in the mixture and p is the total pressure. Using (18.7) we can write (18.9) as

$$e = \frac{wp}{w+\epsilon} \quad \text{or} \quad w = \frac{\epsilon e}{p-e}. \tag{18.10}$$

In terms of q instead of w these expressions are

$$e = \frac{qp}{q+\epsilon(1-q)} \quad \text{and} \quad q = \frac{\epsilon e}{p - e(1-\epsilon)}, \tag{18.11}$$

In Earth's atmosphere $w \ll 1$ so that

$$e \approx w\frac{p}{\epsilon} = 1.61wp \qquad \text{and} \qquad q \approx w \approx \epsilon\frac{e}{p}. \tag{18.12}$$

If the mixing ratio of water vapour is $10\,\mathrm{g\,kg^{-1}}$ (a typical tropical value) and $p = 1000\,\mathrm{hPa}$ then $e \approx 16\,\mathrm{hPa}$.

The *relative humidity*, $\mathcal{H}$, is the ratio of the actual vapour pressure to the saturation vapour pressure, e_s, which is the maximum vapour pressure that can occur at a given temperature before condensation occurs, as will be discussed in Section 18.1.4. Thus, $\mathcal{H} = e/e_s \approx q/q_s$ where q_s is the specific humidity at saturation.

18.1.3 Equation of State and Virtual Temperature

Using (18.5b) the effective gas constant of moist air varies with humidity according to

$$R = \frac{m^d R^d + m^v R^v}{m^d + m^v} = (1-q)R^d + qR^v = R^d\left[1 + q\left(\frac{1}{\epsilon} - 1\right)\right], \tag{18.13}$$

with similar expressions for c_p and c_v. In humid air, with $R^d = 287\,\mathrm{J\,kg^{-1}\,K^{-1}}$ and $q = 0.02$ say, we have $R = R^d(1 + 0.02 \times 0.61) = 290.5\,\mathrm{J\,kg^{-1}\,K^{-1}}$,

The heat capacity of water vapour can be estimated from its molecular properties. Water vapour is a triatomic molecule with three translational and three rotational degrees of freedom. If these were the only degrees of freedom then the internal energy would be given by $I = 6R^vT/2$, whence $c_v^v \approx 3R^v = 1384\,\mathrm{W\,m^{-2}}$ and $c_p^v = R^v + c_v^v = 1846\,\mathrm{J\,kg^{-1}\,K^{-1}}$, where c_v^v and c_p^v are the specific heat capacities for water vapour at constant volume and pressure, respectively. In fact vibrational degrees of freedom can sometimes be excited and the measured values are a little higher, namely $c_v^v = 1397\,\mathrm{J\,kg^{-1}\,K^{-1}}$ and $c_p^v = 1859\,\mathrm{J\,kg^{-1}\,K^{-1}}$ (at 273 K, increasing very slightly with temperature). The heat capacity of moist air is thus slightly higher than that of dry air, but since values of q are small the difference is only about 1%.

The variation of gas constant with humidity can be inconvenient in numerical models. A workaround is to define a so-called virtual temperature, T_v, which is the temperature that dry air would need to be in order to have the same density and pressure as moist air. That is, by definition,

$$p = \rho RT = \rho R^d T_v, \tag{18.14}$$

where R is given by (18.13). Using (18.13) we obtain

$$p = \rho R^d T_v, \quad \text{where} \quad T_v = T\left[1 + q\left(\frac{1}{\epsilon} - 1\right)\right] \approx T(1 + 0.61q). \tag{18.15}$$

The virtual temperature, T_v, increases with specific humidity and if $q = 20\,\mathrm{g\,kg^{-1}}$ then T_v is about 12%, or 3 K, larger than the actual temperature. Such a temperature is often used in numerical models of the atmosphere because it enables various thermodynamic equations to keep their original form, with gas constants that actually are constant.

Because the concentration of water vapour in Earth's atmosphere is so small, the variations of the heat capacities are small and constant values are often used to calculate quantities such as the potential temperature and the adiabatic lapse rate. This is not always appropriate, and Appendix A of this chapter indicates how, in principle, more accurate calculations could be made.

18.1.4 Saturation Vapour Pressure

Vapour pressure is the partial pressure of water vapour in the atmosphere. At any given temperature, there is a maximum value of that vapour pressure beyond which condensation normally occurs and this is known as the *saturation vapour pressure*. Why should this be so, and why don't other atmospheric gases, such as oxygen or carbon dioxide, also condense? To understand this, we have to understand the thermodynamic equilibrium between a liquid and a gas.

Equilibration of the Gibbs function

Consider a system that consists of an enclosed, insulated container partially filled with liquid, and with vapour above it. The two subsystems can exchange mass and energy, with liquid potentially evaporating into vapour and vapour condensing into the liquid. Energy is required to evaporate the liquid into a vapour to overcome the molecular forces in the liquid and this is given by $M(h^v - h^l)$, where M is the mass that has evaporated, h^v is the specific enthalpy of the vapour and h^l is the specific enthalpy of the liquid. The enthalpy of vaporization, more commonly called the *latent heat of evaporation,* is defined by the difference between the two enthalpies at a temperature T, namely,

$$L(T) \equiv h^v - h^l, \tag{18.16}$$

with L having units of $\mathrm{J\,kg^{-1}}$. It is a function of temperature, because the enthalpies of liquid water and water vapour are both functions of temperature, and a very weak function of pressure — see Appendix A for details. For water, L diminishes almost linearly by about 10% going from 0° C to 100° C, from 2.5×10^6 to $2.26 \times 10^6\,\mathrm{J\,kg^{-1}}$.

Now suppose we leave the container alone for a long time so that the liquid and vapour come into equilibrium at a temperature T. If a mass M is to evaporate from liquid into vapour then the energy required, E, can be related to the entropy difference between the liquid and vapour phases,

$$E = ML = M(h^v - h^l) = MT(\eta^v - \eta^l), \tag{18.17}$$

where η^v and η^l are the specific entropies of the vapour and liquid, and the temperature is fixed because all the energy put into the liquid is used for evaporation. Re-arranging we find

$$h^v - T\eta^v = h^l - T\eta^l \qquad \text{or} \qquad g^v = g^l, \tag{18.18a,b}$$

where $g^l \equiv h^l - T\eta^l$ and $g^v \equiv h^v - T\eta^v$ are the specific Gibbs functions for the liquid and vapour (Section 1.5.2). That is, *the specific Gibbs functions for the liquid and water phases of a substance are the same at equilibrium.* The result follows directly from (18.17): the energy required to evaporate a mass of liquid, which is equal to the mass times the specific enthalpy difference between the vapour and the liquid, is also equal to the mass times the specific entropy difference between the vapour and liquid. The equality is true only at equilibrium, when temperature remains fixed, so that (18.18b) is an equation, not an identity.

Another way to derive the above result is to begin with the fact that the total Gibbs function, for the entire system, must remain constant. That is, if M_l and M^v are the masses of liquid and water, then

$$G = M_l g^l + M^v g^v \tag{18.19}$$

and

$$\delta G = M_l\,\delta g^l + M^v\,\delta g^v + (g^v - g^l)\,\delta M = 0, \tag{18.20}$$

where δM is the mass exchanged between liquid and vapour arising from a small fluctuation. Now, from (1.78) changes in Gibbs functions arise because of changes in temperature and pressure; that is, in general,

$$\delta g = -\eta\,\delta T + \alpha\,\delta p, \tag{18.21}$$

and given this, (18.20) becomes

$$\delta G = M_l(-\eta^l\,\delta T + \alpha^l\,\delta p) + M^v(-\eta^v\,\delta T + \alpha^v\,\delta p) + (g^v - g^l)\,\delta M = 0, \tag{18.22}$$

where α_l and α^v are the specific volumes (the inverse density) of the liquid and vapour, respectively. But the temperature and pressure are fixed, and thus, in order that $\delta G = 0$ we must have that $g^v = g^l$. The reason for equality of the two Gibbs functions — as opposed to the equality of some other thermodynamic potential — stems from the fact that the Gibbs function is the only potential for which the natural variables are intensive, namely temperature and pressure. The derivation we have given exploits this directly, for we kept p and T fixed.

18.1.5 Clausius–Clapeyron Equation

Now suppose that the temperature of the liquid-vapour system changes by an amount δT, leading to a change in the vapour pressure and in the specific Gibbs functions for the liquid and water vapour. Using (18.21) the change in the two Gibbs functions is given by

$$\delta g^l = -\eta^l \, \delta T + \alpha^l \, \delta p, \qquad \delta g^v = -\eta^v \, \delta T + \alpha^v \, \delta p, \tag{18.23}$$

and, since the two changes must be the same, $\delta g^l = \delta g^v$. Re-arranging (18.23) and taking the limit of small changes then gives

$$\frac{\mathrm{d}p}{\mathrm{d}T} = \frac{\eta^l - \eta^v}{\alpha^v - \alpha^l}. \tag{18.24}$$

Using (18.16) and (18.17) this equation can be written

$$\frac{\mathrm{d}p}{\mathrm{d}T} = \frac{h^l - h^v}{T(\alpha^v - \alpha^l)} = \frac{L}{T(\alpha^v - \alpha^l)}, \tag{18.25}$$

where to obtain the rightmost expression we use the definition of L. The quantity p is the vapour pressure of the vapour above the liquid, which we are denoting e. Furthermore, it is the *saturation* vapour pressure, e_s, because the vapour is in equilibrium with the liquid: if more vapour were added it would immediately condense. Using this notation, the *saturation vapour pressure* of a condensible gas above a liquid is given by

$$\frac{\mathrm{d}e_s}{\mathrm{d}T} = \frac{L}{T(\alpha^v - \alpha^l)}. \tag{18.26}$$

This is the *Clausius–Clapeyron* equation, and it tells us how the pressure of a vapour that is in thermodynamic equilibrium with an adjacent liquid varies with temperature. If for some reason the vapour pressure is higher than this value, and if there is an adjacent surface of liquid water, then the vapour will condense into a liquid — a common manifestation of which is the formation of clouds and rain. Evidently, since $L > 0$ and $\alpha^v > \alpha^l$, the saturation vapour pressure increases with temperature so that a reduction in temperature can lead to saturation. At the temperature at which the saturation vapour pressure of a substance equals that of the ambient pressure then any liquid present will *boil*. For water at a pressure of 1000 hPa this occurs at about 100° C, with a lower temperature needed at a lower pressure, which is why it takes longer to properly boil an egg at high altitude.

The presence of a liquid surface is crucial to the derivation, and it means that the equation only applies to a condensible. For gases such as carbon dioxide or oxygen at temperatures encountered on Earth, the saturation vapour pressure is very much higher than the actual pressure at the Earth's surface and the gas never condenses; the partial pressure of the gas is then determined by the ideal gas relation and not by (18.26). On Mars, temperatures are sufficiently low that carbon dioxide (the main constituent of the Martian atmosphere) is a condensate and as much as 25% of the Martian atmosphere will condense in winter. On Titan temperatures are even lower and methane is a condensate, and methane lakes are scattered over the dystopian surface.

On Earth, the partial pressure of water vapour, however, is constrained by (18.26), meaning that the partial pressure will often reach the saturated value and the vapour will then normally condense. Condensation is not, however, guaranteed, and to see this imagine a container that contains unsaturated water vapour and no liquid water, and suppose its temperature is then lowered; the pressure of the vapour will then fall following the ideal gas law. The saturation vapour pressure falls more quickly than this and so at some temperature the vapour pressure will exceed the saturation vapour pressure, but the vapour will not automatically condense since there is no liquid present and (18.26) does not apply. The vapour is then said to be *supersaturated.* A supersaturated state is unstable and condensation will eventually occur and liquid water will form, and subsequent changes in temperature induce pressure changes that satisfy the Clausius–Clapeyron equation. Supersaturated water vapour is fairly rare in Earth's atmosphere because there is usually no shortage of condensation nuclei.

At the other end of the temperature scale, liquid water can exist at temperatures well below freezing when there is insufficient water for the molecules to become organized into a crystalline structure, and super-cooled water droplets rather than ice then result — a common situation in clouds. If the liquid that is present is in the form of small spherical droplets then the saturation vapour pressure will differ slightly from that when the vapour is over a flat surface, because surface tension will affect the energy required for a molecule to escape from the droplet and so the latent heat of vaporization will differ. If the vapour is in contact with ice instead of water then its saturation vapour pressure will differ again, because the specific enthalpy of ice is different from that of liquid water.

Application to an ideal gas

In a mixture of ideal gases the partial pressure of one gas is unaffected by the presence of the other gases (because the volume of the gas molecules is assumed to be negligible) and in particular the saturation vapour pressure for a particular component is independent of the presence of other components. We can then, at least approximately, integrate (18.26) to see how the saturation vapour pressure of a particular component varies with temperature. Let us assume that density of the vapour is much less than that of the liquid, so that $\alpha^v \gg \alpha^l$. Given that the partial pressure of the vapour satisfies the ideal gas law, namely $e^v\alpha^v = R^vT$, where R^v is the specific gas constant for the vapour, the Clausius–Clapeyron equation becomes

$$\frac{\mathrm{d}e_s}{\mathrm{d}T} = \frac{Le_s}{R^vT^2} . \tag{18.27}$$

This is the form of Clausius–Clapeyron equation that is normally used in atmospheric applications. If we further assume that L is a constant then (18.27) can be integrated to give

$$e_s = e_0 \exp\left[\frac{L}{R^v}\left(\frac{1}{T_0} - \frac{1}{T}\right)\right], \tag{18.28}$$

where e_0 and T_0 are constants, for example $T_0 = 273\,\mathrm{K}$ and $e_0 = 6.12\,\mathrm{hPa}$. Equation (18.28) is a good approximation if the temperature range is not too wide, and as seen in Fig. 18.1 the saturation vapour pressure of water increases approximately exponentially over commonly encountered terrestrial temperatures. Note that the expression for saturation vapour pressure does not depend on the presence or otherwise of dry air.

The fact that water vapour content cannot normally exceed the saturation value distinguishes the distribution of water from other tracers in the atmosphere, even without taking the heating effects of condensation into account. Note finally that if the atmosphere were motionless it would everywhere be in thermodynamic equilibrium with the moist surface and the surface layers would be saturated, and diffusion of water vapour upwards would then saturate the rest of the atmosphere. Thus, *the relative humidity of the atmosphere is determined by its circulation,* as we now discuss.

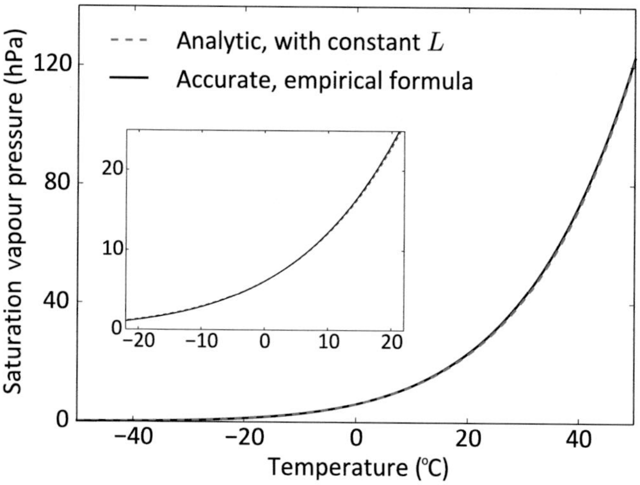

Fig. 18.1 The saturation vapour pressure of water vapour, calculated using the analytic formula (18.28), which assumes that L is constant (dashed line), and using a more accurate, semi-empirical formula (solid line).[2] The inset is the same plot over a smaller range.
The parameters used in the analytic formula are $T_0 = 273\,\text{K}$, $e_0 = 6.12\,\text{hPa}$, $L = 2.44 \times 10^6\,\text{J kg}^{-1}$ and $R^v = 462\,\text{J kg}^{-1}\,\text{K}^{-1}$.

18.2 THE DISTRIBUTION OF RELATIVE HUMIDITY

To a first approximation, the distribution of water in the atmosphere is determined by the distribution of temperature. This is because of the near-exponential dependence of absolute humidity on temperature through the Clausius–Clapeyron equation, so that variations of relative humidity of even an order of magnitude, from 10% to 100% say, are barely noticeable in the specific humidity distribution, as seen in Fig. 18.2. It is, however, the relative humidity that determines such basic quantities as rainfall, and its distribution (Fig. 18.3 and Fig. 18.4) shows a quite different picture, with the following features evident:

- High, near-saturated values close to the ground.
- Low relative humidity in the subtropics at latitudes between 15° to 40° (depending on season) in both hemispheres.
- High values near the equator extending up to the tropopause.
- Vertically near-uniform values in mid- and high latitudes, increasing with latitude close to the pole in some cases.
- Very low values over much of the stratosphere.

The gross distribution of zonally-averaged temperature can be understood, at least in a rough way, using fairly simple arguments. The incoming solar radiation at the top of the atmosphere would lead, in the absence of atmospheric motion, to a strong meridional radiative equilibrium temperature gradient, as in Fig. 14.1. The meridional transport of heat by the Hadley Cell and by mid-latitude baroclinic eddies flattens that temperature gradient, and one might crudely model this transport as a diffusion. In the vertical heat is transported upwards by convection and baroclinic eddies, which might be modelled as a relaxation back to some specified neutrally stable profile or specified isentropic slope. However, arguments of this type cannot capture some basic features of the relative humidity distribution, and diffusive arguments in particular can be quite misleading. The effects of advection, either explicitly or as represented by a stochastic process, are crucial, and in this section we consider advection-diffusion-condensation models of the general form

$$\frac{\partial q}{\partial t} + \boldsymbol{v} \cdot \nabla q = \nabla \cdot \kappa \nabla q - S, \tag{18.29}$$

where q is the specific humidity, $\boldsymbol{v}$ is a specified velocity field, κ is a diffusion coefficient and S is the condensational sink.[4] Condensation in the atmosphere involves complicated microphysical processes but the basic effect is to remove liquid water once the volume becomes saturated and to

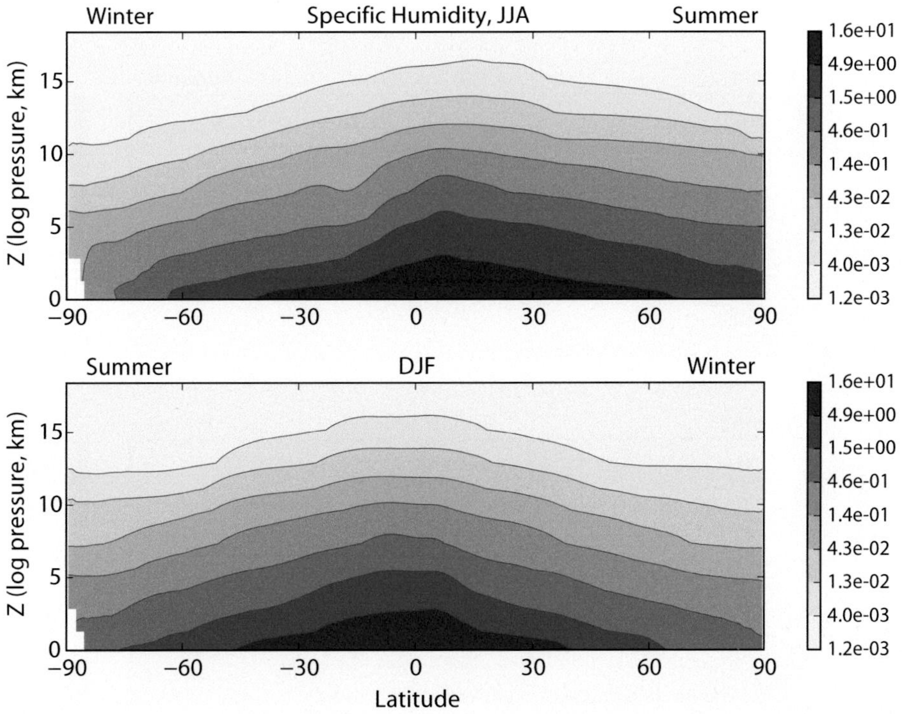

Fig. 18.2 Zonally-averaged specific humidity distribution (g/kg) in the atmosphere, as retrieved from a microwave satellite, for boreal summer in 2008 (top) and boreal winter 2008–2009 (bottom).[3] Note the logarithmic scale.

largely prevent relative humidity from exceeding 100% (although supersaturation can occur locally if there are no particulates in the air onto which the water vapour may condense). A simple analytic way to represent such a condensation process is to let

$$S = \begin{cases} 0, & q \le q_s, \\ (q - q_s)/\tau, & q > q_s, \end{cases} \tag{18.30}$$

where the time τ is much smaller than any large scale diffusion time, L^2/κ, where L is a characteristic length. Such a sink effectively prevents q from exceeding q_s except by a tiny amount. In (18.29) we may consider the advection and diffusion terms as representing larger scale processes and the sink as representing small-scale microphysical processes. We begin our exploration by omitting advection, and a summary can be found on page 683.

18.2.1 A Diffusion-Condensation Model

If we omit advection in (18.29) we have a simple diffusion-condensation model

$$\frac{\partial q}{\partial t} = \nabla \cdot \kappa \nabla q - S. \tag{18.31}$$

Although superficially plausible, such a model is too-often unable to reproduce locally unsaturated regions. For simplicity consider the one-dimensional case satisfying

$$\frac{\partial q}{\partial t} = \kappa \frac{\partial^2 q}{\partial x^2} - S, \tag{18.32}$$

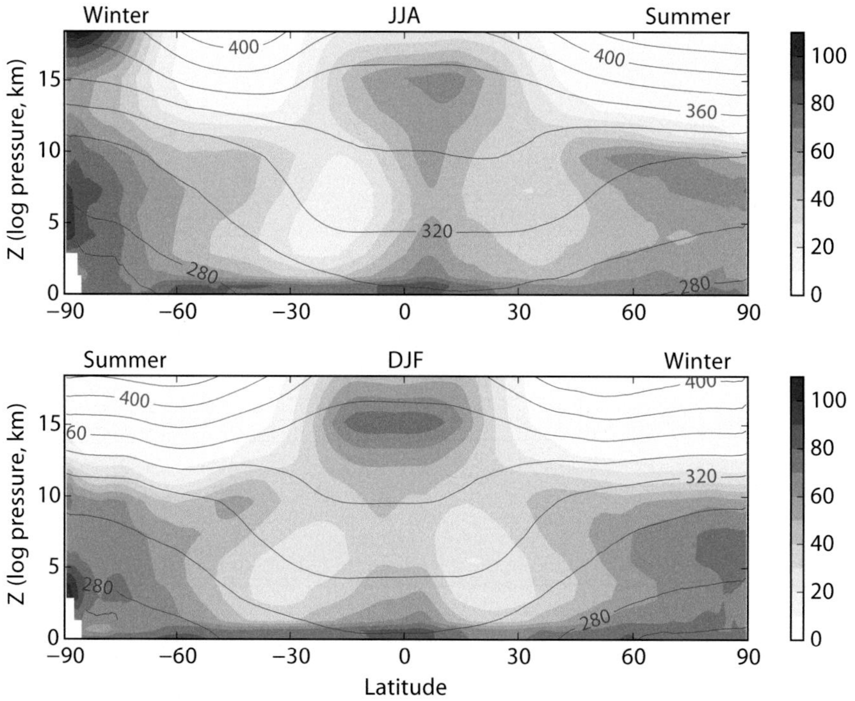

Fig. 18.3 Zonally-averaged relative humidity distribution in the atmosphere (in percent, shading), and isolines of equivalent potential temperature (contours), inferred from satellite as in Fig. 18.2. (Equivalent potential temperature is a modification of potential temperature to account for water vapour, and is a approximately an adiabat; see Section 18.3.2.)

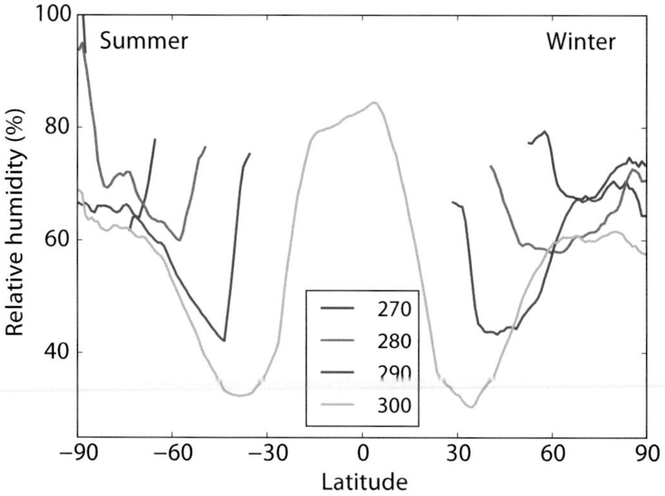

Fig. 18.4 Values of zonally-averaged relative humidity in boreal winter plotted along isolines of equivalent potential temperature (red contours in Fig. 18.3), with values of θ_{eq} as indicated in the legend.

Water Vapour Transport and Relative Humidity

- Water vapour in the atmosphere is primarily transported by advection, much of this on large, near planetary scales but also by convection and smaller scale turbulence. Specific humidity is materially conserved in the absence of condensation and diffusion.
- If the vapour pressure exceeds the saturated value (as given by the Clausius–Clapeyron relation) condensation will occur provided condensation nuclei or liquid water are present. The condensation normally occurs much more quickly than large-scale advective processes, and this situation is known as the *fast condensation limit.*
- When dealing with large-scale flows in Earth's atmosphere the limit is a good approximation. Levels of relative humidity are then determined mainly by advective processes rather than the microphysical details of the condensation process. Specifically, the relative humidity of a parcel is determined by the temperature at the location of last saturation, as in (18.35).
- If the advection is not fully resolved — for example if there is some small-scale turbulence in the flow — then introducing some diffusion seems natural, as is commonly done for tracers, but a large diffusivity can give unrealistic results because of the irreversible nature of condensation. Diffusion is then not a good representation of small-scale quasi-random flow and is overly prone to produce saturation.
- In Earth's atmosphere some of the large-scale features of the relative humidity distribution may be explained as follows:
    - High levels of relative humidity close to the surface. These are due to transport from a saturated surface, especially over the ocean and moist ground.
    - High levels of relative humidity in the ascending branch of the Hadley Cell. These are due to upward advection from a nearly saturated surface. The branch is, however, not saturated on the zonal-average, because of the presence of smaller scale motion such as downdrafts that unsaturate the air.
    - A subtropical minimum of relative humidity. This largely arises because of the mean descending motion, advecting water vapour into a warmer region and decreasing its relative humidity.
    - Variable relative humidity in mid- and high latitudes, with locally strong gradients. Chaotic advection by baroclinic eddies takes moisture upwards and polewards into cooler regions where it becomes saturated, but also downwards and equatorwards so reducing relative humidity.
    - Very low levels of relative humidity in the stratosphere. The tropopause is a cold trap and so relative humidity is very low beyond it. The cold-trap effect occurs in both advective and diffusive models. In Earth's atmosphere, the little water vapour that is in the stratosphere mainly enters advectively through the tropical tropopause.

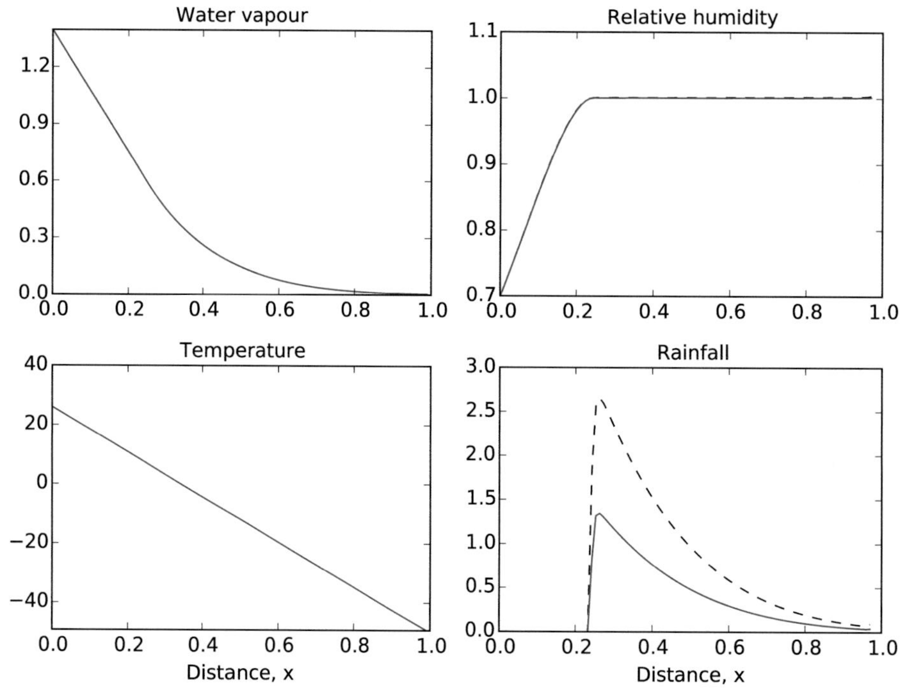

Fig. 18.5 Steady solution of the diffusion-condensation model (18.32) and (18.33), with $\mathcal{H}_b = 0.7$ at $x = 0$ and temperature (°C) falling linearly from $x = 0$ as shown. Water vapour falls linearly away from the boundary at $x = 0$ until it becomes saturated at $x = 0.23$, after which the region is saturated. Two solutions are plotted, the one with dashed lines showing the solution with twice the diffusivity as that with solid lines. The two solutions are nearly identical except for the sink term, the rainfall.

with constant κ. In this model, any interior minimum of q_s will lead to saturation in that neighbourhood. To see this, note that in any region where moisture is present and that has $\partial^2 q/\partial x^2 > 0$ and $q < q_s$, there will be a net flux of water vapour into that region. Saturation must eventually occur, at which point q remains very close to the value of q_s. If $\partial^2 q_s/\partial x^2 > 0$ then the diffusive flux will maintain the saturated state. Given the monotonic dependence of q_s on temperature this result means that, in a moist atmosphere in which water vapour is transported diffusively, the neighbourhood of an interior minimum of temperature will become saturated.

A corollary of this result is that, unless there is a source of moisture at a boundary, a region with $\partial^2 q_s/\partial x^2 > 0$ everywhere will under many conditions eventually lose *nearly all* its moisture. Suppose that $\partial q/\partial x = 0$ at $x = 0$ and that q_s has a maximum at $x = 0$, and that the region extends to infinity and is initially saturated. (Envision a semi-infinite domain with temperature decreasing linearly away from a no-flux boundary, and therefore with no source of moisture, at $x = 0$.) Moisture is transported to higher values of x where condensation occurs and moisture is removed. As time progresses the region of saturation moves to higher and higher vales of x, but nevertheless water is continuously removed. In a finite domain, with no flux boundary conditions at either end, a small amount of moisture will remain in the system for all time because a finite amount of water vapour is needed for condensation to occur.

Now consider the more atmospherically relevant situation with a moisture source at $x = 0$, with q_s decreasing monotonically away and with the other boundary either extending to infinity or being a no-flux boundary at finite x. Such a situation might represent an atmosphere sitting

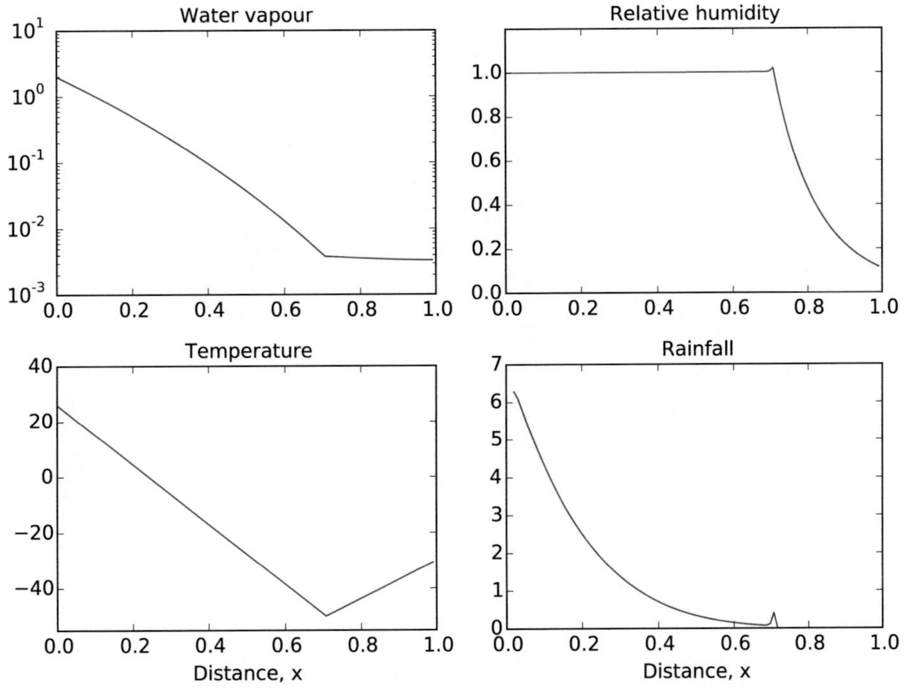

Fig. 18.6 As for Fig. 18.5, but now with $\mathcal{H}_b = 1$ and an interior temperature minimum, a 'cold trap', at $x = 0.7$ (see temperature panel), and water vapour has a log scale. Relative humidity falls rapidly beyond the cold trap and the rainfall there is zero.

atop a moist surface with temperature decreasing with height. Consider the case

$$\begin{aligned} q &= \mathcal{H}_b q_s(T_0), \qquad & x &= 0, \\ \frac{\partial q}{\partial x} &= 0, & x &= 1, \end{aligned} \tag{18.33}$$

where $\mathcal{H}_b$ is a parameter such that if the boundary is effectively saturated then $\mathcal{H}_b = 1$, and $\mathcal{H}_b < 1$ otherwise, and we suppose that T decreases between T_0 at $x = 0$ and T_1 at $x = x_1$. Consider first the case with $H_b = 1$. If T falls linearly then the value of q_s falls approximately exponentially between $x = 0$ and $x = 1$, with $\partial^2 q_s/\partial x^2 > 0$, and the steady solution of this problem is that the domain is saturated everywhere. To see this, suppose that $q < q_s$ is some region so that $S = 0$. Water vapour will then diffuse into that region until condensation begins, maintaining $q \approx q_s$ everywhere, with, if (18.30) applies, q in fact exceeding q_s by a very small amount so that the diffusion into the region is balanced by condensation. If $H_b < 1$ then the region next to the surface will not be saturated and in steady state the water vapour content will decrease linearly, to satisfy $\partial^2 q/\partial x^2 = 0$, until at some value of x, x_s say, the atmosphere becomes saturated and remains so for $x > x_s$. The actual solution is

$$q(x) = \begin{cases} \mathcal{H}_b q_s(0) + x\Delta q/x_s, & x < x_s, \\ q_s(x), & x \geq x_s, \end{cases} \tag{18.34}$$

where $\Delta q = (q_s(x_s) - \mathcal{H}_b q_s(0))$ and x_s is such that the flux is continuous there. A moment's thought reveals that $x_s = \Delta q/(\partial q_s/\partial x)_{x=x_s}$ and $x_s = 0$ if $\mathcal{H}_b = 1$. It is interesting that the values of water vapour in the solution (plotted in Fig. 18.5) do not depend upon κ and only very weakly on τ. In the fast condensation limit (in which τ is small compared to the diffusion time) variations of τ determine only the tiny amount by which q exceeds q_s. The amount of condensation is actually

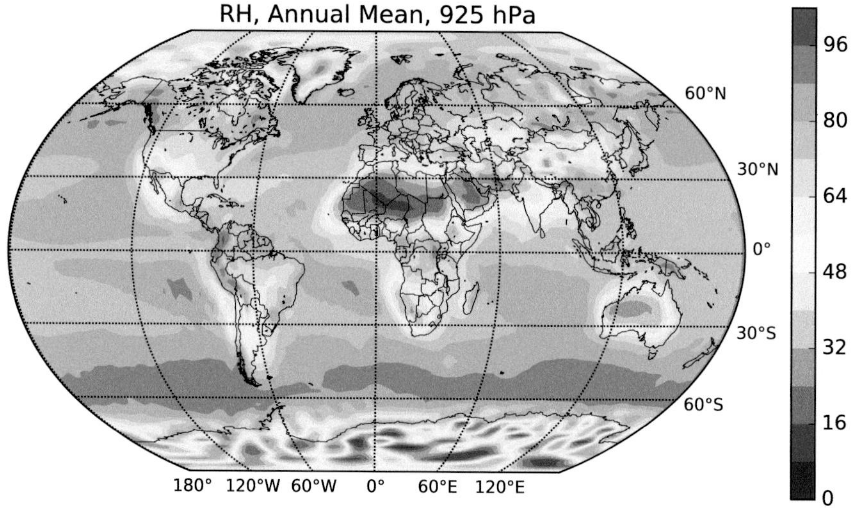

Fig. 18.7 Annually-averaged relative humidity (in percent) at 925 hPa, about 750 m above sea level. The contrast over land and ocean is apparent especially in the subtropics. (In regions of high topography values are interpolated.)

largely determined by the value of κ: large values of κ lead to a stronger diffusion of water vapour into dryer regions where it is almost immediately removed by condensation. This result illustrates the tenet that on large scales precipitation is at leading order determined by the motion of the fluid (here represented by diffusion), with variations in τ (crudely representing complex microphysical processes) being of less import. Microphysical processes are nevertheless important in many ways — a weather forecast model with poor microphysics would likely have little skill in forecasting the onset of precipitation, even if the climatology of the model were good.

Although the above model is over-simple in some respects, the dependence of the solution on $\mathcal{H}_b$ does capture the dependence of relative humidity in the lower atmosphere on the nature of the surface beneath, as seen in Fig. 18.7 showing relative humidity at 925 hPa. Over the desert regions the relative humidity is unsurprisingly low. Perhaps what is surprising is that the general dryness of the subtropics cannot be seen over the oceans — the surface moisture source simply overwhelms the drying effects of descending air (discussed more below).

A variation on the above theme introduces a temperature minimum, or 'cold trap', in the interior of the domain, as at $x = 0.7$ in Fig. 18.6. This configuration is a crude model of the tropopause, with temperatures increasing in the stratosphere beyond. Water vapour has to pass through the cold trap and so, since the specific humidity cannot be higher than the saturated value at the cold trap, the atmosphere will be unsaturated beyond it with relative humidity decreasing rapidly, as is seen in the real atmosphere in Fig. 18.3.

Although informative, diffusive-condensation models are fundamentally limited in what they can achieve, because of the deficiencies of diffusion in parameterizing the motion of a tracer in the presence of condensation. In particular, in the absence of a cold trap, diffusive models are prone to produce saturation everywhere. If the atmosphere obeyed (18.31) with a saturated surface, then the atmosphere would become saturated everywhere up to the tropopause, which from Fig. 18.3 is manifestly not the case. To remedy this we turn our attention to the effects of advection.

18.2.2 Advection-Diffusion-Condensation models

Consider now the effects of advection. Neglecting diffusion, the specific humidity of a parcel is conserved unless condensation occurs. If a moist parcel travels into a region of decreasing temperature

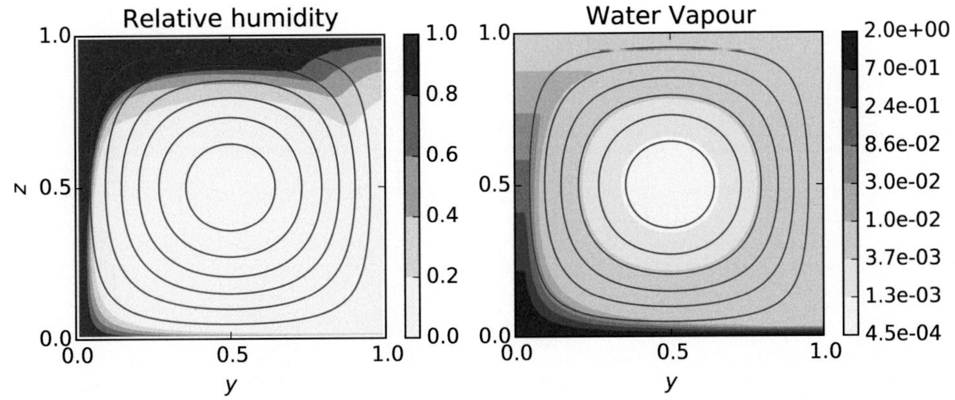

Fig. 18.8 Steady-state distributions of relative humidity (filled contours, left) and water vapour (right) in a single cell as defined by the streamfunction (red contours, clockwise flow. so air rising at the 'equator' at $y = 0$) in a closed domain. The domain boundary is saturated at the bottom, and temperature decreases linearly with height. The diffusivity $\kappa = 0.001$ and so the Peclet number, $Pe \equiv UL/\kappa \sim 1000$.

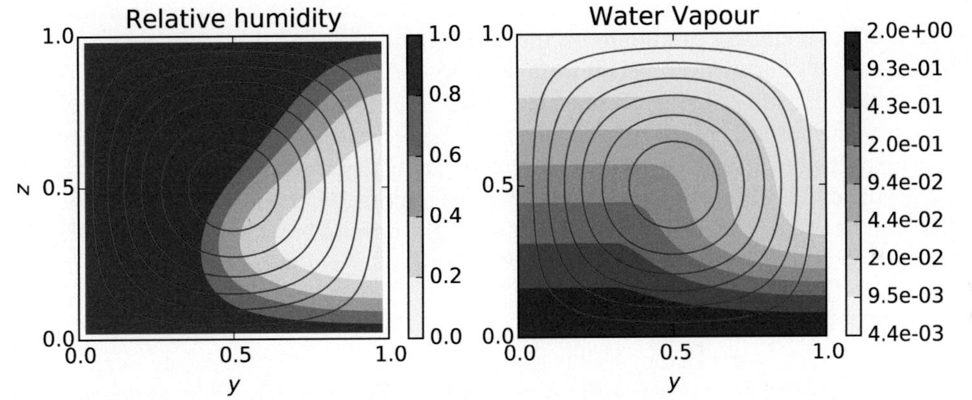

Fig. 18.9 As for Fig. 18.8, but with bigger diffusivity, $\kappa = 0.1$ and $Pe \sim 10$. Most of the domain is now much wetter.

it will eventually become saturated; however, if it passes into a region of increasing temperature then its relative humidity drops. Consider, for example, a cold trap with a temperature minimum at $x = x_{ct}$, as in the previous section. A parcel advected through the trap becomes saturated at x_{ct}, beyond which its specific humidity is constant, so that relative humidity is given by

$$\mathcal{H}(x) = \frac{q_s(T(x_{ct}))}{q_s(T(x))} . \tag{18.35}$$

That is to say, the relative humidity of a parcel is given by the value of the saturated vapour pressure at the point of last saturation, divided by the saturated vapour pressure at its current location.

Let us see the extent to which a simple advection-diffusion-condensation model can explain some of the features of the zonally-averaged relative humidity seen in Fig. 18.3. Consider a two-dimensional model in the y-z plane obeying (18.29), with a divergence-free advecting velocity, a saturated surface at $z = 0$ and no flux boundary conditions elsewhere. A simple model representing some of the features of the Hadley Cell is that of a single cell in a unit-sized square domain with a

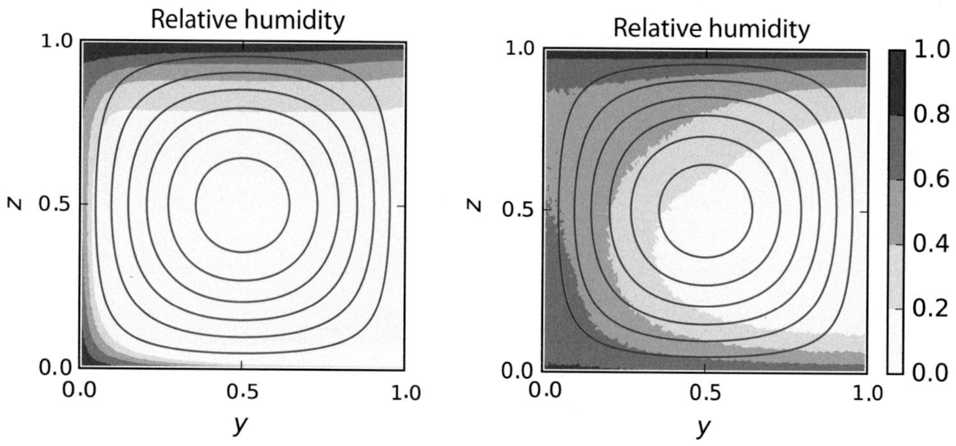

Fig. 18.10 The relative humidity produced using a stochastic advection-condensation model with the same imposed temperature, mean streamfunction (red contours) and boundary conditions as Fig. 18.8 and Fig. 18.9, with a larger stochasticity in the right panel. Specifically, the stochastic components of the left and right panels correspond to diffusivities of Figs. 18.8 and 18.9 respectively.

streamfunction of the form $\psi = \sin \pi y \sin \pi z$. We suppose the lower boundary is saturated, that there are no-flux boundary conditions on the other boundaries, and that temperature is uniform in latitude and falls linearly with height from 26° C to -50° C.

Solutions are illustrated in Fig. 18.8 and Fig. 18.9, the latter having diffusivity κ that is a hundred times larger. The upward branch in both cases is completely saturated. In the low diffusivity case the interior of the cell is quite dry, with a relative humidity that drops almost to zero in the centre. Increasing the diffusivity simply tends to moisten the interior, but does little to reduce the relative humidity in the upward branch of the cell. However, the upward branch of the real Hadley Cell is *not* saturated on the zonal mean (Fig. 18.3). Let us try to remedy this model failing.

Stochastic effects

The upwards branch of the Hadley Cell is not saturated because it is not steady — there is both ascending and descending motion, with the relative humidity of parcels falling as they descend so reducing the average value. We can mimic this effect by introducing a stochastic component into a Lagrangian model and performing 'Monte Carlo' simulations.[5] The model advects infinitesimal parcels of water vapour — one million in the simulations shown in Fig. 18.10 — by a prescribed mean field plus a random component. Whenever the parcels touch the ground they become saturated and whenever their relative humidity exceeds one condensation occurs. We show results with two levels of stochastic motion, one comparable to the diffusivity used in Fig. 18.8 and the other comparable to that of Fig. 18.9, all with the same temperature, boundary condition and mean velocity fields, and the results shown in Fig. 18.10 are coarse-grained time averages. Although the results with weak stochasticity do resemble those with small diffusion (compare the left-hand panel of Fig. 18.10 with Fig. 18.8), those with large stochasticity have less resemblance to those with large diffusivity. Most notably, the explicit randomness reduces the relative humidity in the ascending branch without saturating the interior, a property that diffusion is unable to capture.

Relative humidity in mid-latitudes

The diffusive steady overturning model fails qualitatively when applied to an entire hemisphere, even with more realistic overturning circulation and temperature fields. To see this we construct a model in which the overturning streamfunction (representing a residual circulation) resembles that of the bottom panel of Fig. 15.22, as illustrated (red contours) in Fig. 18.11. To this we add

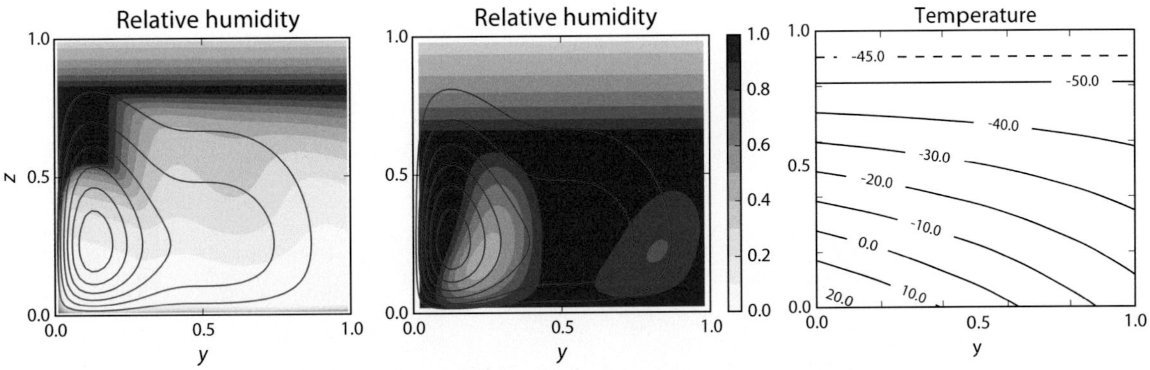

Fig. 18.11 Relative humidity (shading) for small diffusivity (left, $\kappa = 0.001$, $Pe = 1000$) and larger diffusivity (centre, $\kappa = 0.05$, $Pe = 20$), and a clockwise overturning circulation as contoured. The imposed temperature (right) falls linearly with height to a tropopause at $z = 0.8$ with $T = -50°$ C, then rises again to $-40°$ C, and diminishes meridionally at the surface from 26° C at the equator ($y = 0$) to -15° C at the pole ($y = 1$).

a temperature distribution that decreases polewards and upwards until it reaches a tropopause at $z = 0.8$, then increases upwards again in the stratosphere, and we assume that the surface is saturated. With a small diffusivity (which is more realistic) the lower tropopause dries out too readily because of the sinking motion, and with larger diffusivity the atmosphere becomes more saturated but with little of the observed structure. Two features that the model *can* reproduce are the dry stratosphere, because of the tropospheric cold trap, and the subtropical minimum, a robust feature of the downwelling in the Hadley Cell.

The main failings of this model arise from the fact that in mid-latitudes water vapour is not primarily advected by the overturning circulation, and diffusion is a poor model of the meridional transport. Rather, moisture is transported from the boundary layer into the free atmosphere by convection and polewars by larger-scale baroclinic eddies in a quasi-horizontal fashion, roughly along moist isentropic surfaces. Mid-latitude relative humidity is thus highly variable, as can be seen in Fig. 18.12, with sharp gradients between high and low values and with variations being strongly correlated with the parcel trajectory — note for example the swath of high relative humidity just west of the UK associated with moisture moving poleward and to a lower temperature. As previously note, on the zonal average relative humidity is high in the boundary layer, diminishes as one moves upwards, then stays roughly constant as one moves polewards along a moist isentrope before increasing again at very high latitudes (Fig. 18.3 and Fig. 18.4).

We can understand some of these features using a simple advective-diffusive-condensation model, with the unsteadiness explicitly incorporated into the advection, and with no other stochastic component. Consider (18.29) in two horizontal dimensions in a channel of size (L_x, L_y), periodic in x, with a saturated boundary at $y = 0$ and no flux at $y = L + y$, and with imposed meridional temperature gradient (° C) and advecting streamfunction of the forms

$$T(y) = 20 - 30y/L_y, \qquad \psi(x, y, t) = -Uy + \Psi_0 \sin(\pi y/L_y)\, e^{i(kx-\omega t)}. \tag{18.36}$$

We will choose $|\Psi_0| = 1$, $U = 10$, $k = 4\pi/L_x$, $\kappa = 0.005$ (so $Pe \sim 200$ on the domain scale), and a solution is shown in Fig. 18.13. The details of the solution depend on the parameters chosen, but there are a couple of quite robust results:

(i) Relative humidity at a point depends on where a parcel has come from, and hence is correlated with the streamlines, and is generally higher for parcels moving to colder regions.

(ii) Relative humidity is close to one (i.e, 100%) next to the saturated boundary, then decreases

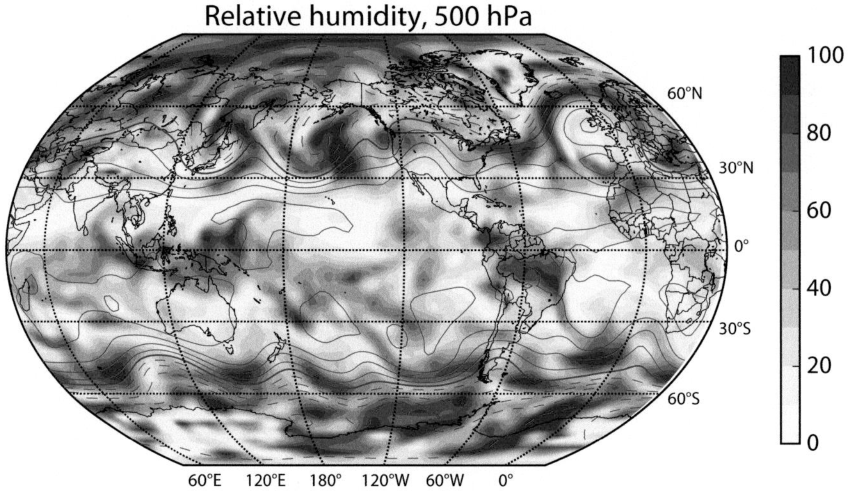

Fig. 18.12 Snapshot of relative humidity at 500 hPa on 9 February, 2015 (shading, in percent), with red contours of geopotential height, from reanalysis.

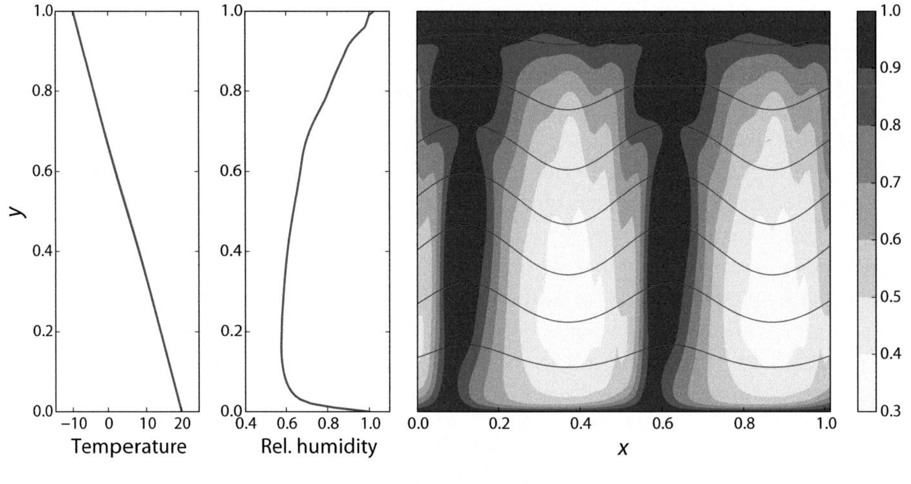

Fig. 18.13 A snapshot of relative humidity (right panel, shading) obtained by a numerical solution of the advective-diffusive-condensation model (18.29) and (18.36) in the horizontal (x–y) plane. The contours are the streamfunction. The middle panel is the zonally-averaged relative humidity and the left panel is the imposed temperature gradient (° C).

to a minimum in the interior, rising again at high latitudes (cf., the observations shown in Fig. 18.4). The width of the saturated region near the saturated boundary diminishes as the Peclet number increases, and the mid-domain minimum arises because of the drying effects of advection, which diminish near either boundary.

There are many limitations to such a model, one being that in reality temperature itself is advected by the flow and another being that advection is three-dimensional, but exploring these effects is beyond the scope of our story.[6] A summary of relative humidity transport is provided in the box on page 683, but we now move on to the dynamical effects of water vapour, and in particular convection.

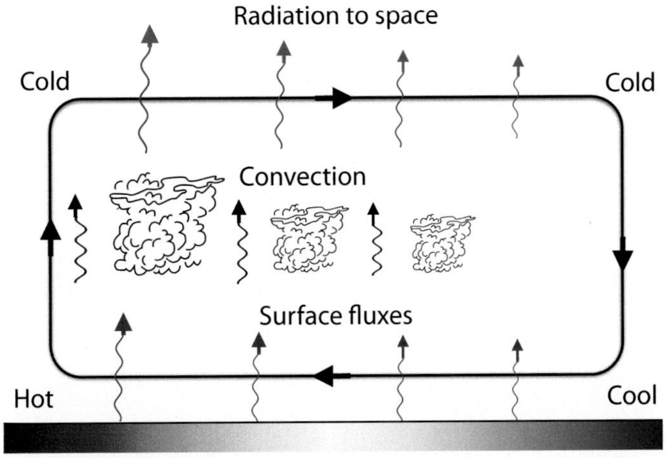

Fig. 18.14 Idealization of small-scale convection embedded within a large scale circulation. The fluid is in contact with a warm surface below, and cools by radiation to space from above. The horizontal temperature gradient drives a large-scale meridional overturning circulation, embedded in which convection forms, especially at the warmer end.

18.3 ATMOSPHERIC CONVECTION

As introduced in Section 2.10, convection occurs when a parcel that is displaced upwards (downwards) finds itself lighter (heavier) than its surroundings, and hence becomes subject to a buoyancy force that amplifies the initial displacement. Convection is particularly prevalent in the tropics for reasons that are, at lowest order, independent of the presence of water vapour; rather, it is simply that the surface is warmer there, as sketched in Fig. 18.14. Consider, rather heuristically, a fluid that is heated by a warm surface from below and cools by radiation from above. If the surface is warmer at one end, as illustrated, a large-scale circulation will arise, transporting energy from warmer regions to cooler ones. In the absence of convection, there will be a larger vertical temperature gradient where the surface temperature is warm, and so these regions are more prone to convection. Having said this, moisture *does* affect convection in rather profound ways, one being that moisture is a *destabilizing* influence, as we now discover.

18.3.1 Generalized Enthalpy and Adiabatic Lapse Rates

A column of air that is convectively stable when dry can be made convectively unstable by the presence of water vapour, because when water vapour condenses heat is released, warming the parcel and potentially making it more buoyant than its surroundings. To explore this we begin with a derivation of the conditions under which a moist column of air may be unstable — essentially an extension of Section 2.10 to include the effects of condensation.

An adiabatic (and convectively neutral) profile is one in which the entropy is constant. Recalling the arguments in Sections 1.6 and 1.10.3, the entropy may be related to the enthalpy by writing the fundamental relation as

$$dh = T d\eta + \alpha \, dp, \tag{18.37}$$

and where, if the column is hydrostatically balanced, $\alpha \, dp = -g \, dz$. Thus, the condition that $d\eta = 0$ is equivalent to

$$d(h + gz) = 0. \tag{18.38}$$

That is, the dry static stability or generalized enthalpy, $h^* = h + gz$, is constant in an adiabatic, hydrostatic, profile. The enthalpy of a dry parcel of ideal gas is given by $h^d = c_p T$ and thus,

$$h^* \equiv h^d + gz = c_p T + gz. \tag{18.39}$$

Since an adiabatic profile has $dh^* = 0$ we recover the dry adiabatic lapse rate (Section 2.10.2), to wit

$$\Gamma^d = -\left.\frac{dT}{dz}\right|_{ad} = \frac{g}{c_p}. \tag{18.40}$$

The adiabatic lapse rate for a moist parcel is given using similar reasoning, but we need an appropriate expression for the enthalpy. A parcel of total mass M^t composed of dry air, water vapour and liquid water such that $M^t = M^v + M^l$ has an enthalpy given by

$$H = h^d M^d + h^v M^v + h^l M^l, \tag{18.41}$$

where h denotes specific enthalpy and the superscripts denote dry air, water vapour and liquid water respectively. Dividing by the total mass, the specific enthalpy of the parcel is thus

$$h = (1 - q^w)h^d + q^v h^v + q^l h^l, \tag{18.42}$$

where

$$q^w = q^v + q^l, \quad q^v = \frac{M^v}{M^t}, \quad q^l = \frac{M^l}{M^t}. \tag{18.43}$$

As in (18.16) the latent heat of evaporation is defined to be $L = h^v - h^l$ so that (18.42) becomes

$$h = (1 - q^w)h^d + q^w h^l + Lq^v. \tag{18.44}$$

The generalized enthalpy is equal to this quantity plus a potential ϕ, and if that potential is equal to gz the quantity is the *moist static energy*,

$$h_m^* = (1 - q^w)h^d + q^w h^l + Lq^v + gz. \tag{18.45}$$

Since $h^d = c_p^d T$ and $h^l = c^l T$ (to a very good approximation and where c^l is the heat capacity of liquid water) we have

$$h_m^* = c_p^{dl} T + Lq^v + gz, \tag{18.46}$$

where $c_p^{dl} = (1 - q^w)c_p^d + q^w c^l$. This quantity varies with liquid water content but in Earth's atmosphere the variation is small and in the derivation below we will take c_p^{dl} to be a constant and denote it c_p (its value is very similar to that of c_p^d). For adiabatic motion, the moist static energy is a constant, whether or not water is evaporating or condensing: the latent heat of evaporation or condensation is merely exchanged with the dry generalized enthalpy.

The saturated adiabatic lapse rate

In a moist atmosphere the moist static energy is conserved as a parcel ascends and $\mathrm{d}h_m^*/\mathrm{d}z = 0$. Using (18.46), an ascending parcel then has a lapse rate given by

$$c_p \frac{\mathrm{d}T}{\mathrm{d}z} = -L\frac{\mathrm{d}q^v}{\mathrm{d}z} - g. \tag{18.47}$$

If the air is moist but not saturated then an ascending parcel will follow the dry adiabatic lapse rate (because $\mathrm{d}q^v/\mathrm{d}z = 0$) but if it is saturated (and $q^v = q_s$) then as a parcel ascends it will cool and some water vapour will condense. Since $q_s \approx \epsilon e_s/p$, and so is a function of temperature and pressure, we have

$$\frac{\mathrm{d}q_s}{\mathrm{d}z} = \left(\frac{\partial q_s}{\partial T}\right)_p \frac{\mathrm{d}T}{\mathrm{d}z} + \left(\frac{\partial q_s}{\partial p}\right)_T \frac{\mathrm{d}p}{\mathrm{d}z} = \left(\frac{\partial q_s}{\partial T}\right)_p \frac{\mathrm{d}T}{\mathrm{d}z} + \left(\frac{q_s}{p}\right)\rho g, \tag{18.48}$$

using hydrostasy. Using the Clausius–Clapeyron equation and the ideal gas equation of state gives

$$\frac{\mathrm{d}q_s}{\mathrm{d}z} = \frac{Lq_s}{R^v T^2}\frac{\mathrm{d}T}{\mathrm{d}z} + \frac{gq_s}{R^d T}. \tag{18.49}$$

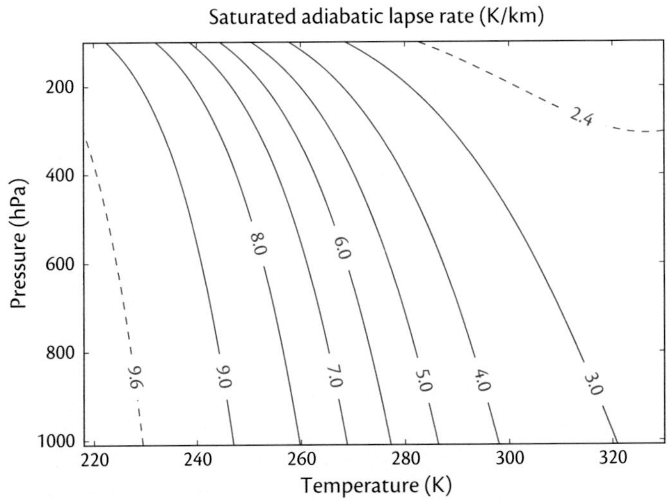

Fig. 18.15 Contours of saturated adiabatic lapse rate (K/km) as a function of pressure and temperature, calculated using (18.50) with $q_s = \epsilon e_s/p$.

Using (18.47) and (18.49) we obtain an expression for the lapse rate of an adiabatically ascending saturated parcel,

$$\Gamma_s = -\left.\frac{\mathrm{d}T}{\mathrm{d}z}\right|_{ad} = \frac{g}{c_p}\frac{1 + Lq_s/(R^d T)}{1 + L^2 q_s/(c_p R^v T^2)} . \tag{18.50}$$

The quantity Γ_s is the *saturated adiabatic* lapse rate, plotted in Fig. 18.15. (This quantity is often called the moist adiabatic lapse rate, but that name is better given to the lapse rate of moist air that does not condense, and which differs slightly from the dry adiabatic lapse rate. In fact Γ_s is properly a *pseudo-adiabatic* lapse rate, because liquid water is not accounted for — see Appendix A.) The lapse rate is a function of temperature and pressure because $q_s = \epsilon e_s/p$ and e_s is given by the solution of the Clausius–Clapeyron equation, (18.28). Values of Γ_s are typically around 6 K km^{-1} in the lower atmosphere although since $\mathrm{d}q_s/\mathrm{d}T$ is an increasing function of T, Γ_s decreases with increasing temperature and can be as low as 3 K km^{-1}. The second term in the numerator of (18.50) is usually quite small (around 0.1, although it can become large at very high temperatures) but the second term in the denominator is positive and order unity. Since g/c_p is the dry adiabatic lapse rate, the saturated adiabatic lapse rate is smaller than the dry adiabatic lapse rate, as can be seen directly from (18.47) since $\mathrm{d}q^v/\mathrm{d}z < 0$.

The saturated adiabatic lapse rate determines the stability of a saturated profile. Using parcel theory, just as in Section 2.10, a profile will be stable or unstable depending on whether the lapse rate is less than or greater than (18.50); that is

$$\text{Stability}: \quad -\frac{\partial \widetilde{T}}{\partial z} < \Gamma_s, \qquad\qquad \text{Instability}: \quad -\frac{\partial \widetilde{T}}{\partial z} > \Gamma_s, \tag{18.51a,b}$$

where $\widetilde{T}$ is the environmental temperature. The lapse of the atmosphere rarely exceeds the saturated adiabatic lapse rate, as was seen in Fig. 15.25. In the tropics and subtropics the lapse is, on average, very close to the saturated adiabat up to about 300 hPa (about 9 km), whereas in mid-latitudes it is considerably less, and so more stable, because of the upward transport of heat by baroclinic eddies. Convection only directly determines the lapse rate over a small fraction of the tropics where convection actually occurs, but nevertheless the average lapse rate is close to moist neutral. This is because the gravity waves emanating from convective regions adjust the tropics to have weak horizontal temperature gradient, in a process akin to geostrophic adjustment, hence maintaining approximately the same vertical profile even away from regions of active convection, as discussed more in Sections 18.8 and 18.9.

18.3.2 Equivalent Potential Temperature

In a dry atmosphere adiabatic motion is characterized by the material conservation of potential temperature, θ, a surrogate for entropy. Thus in adiabatic flow $\mathrm{D}\theta/\mathrm{D}t = 0$, and the dry adiabatic lapse rate can be characterized equivalently either as $\mathrm{d}T/\mathrm{d}z = -g/c_p$ or $\mathrm{d}\theta/\mathrm{d}z = 0$. We can construct a similar quantity for moist air. Suppose that a parcel is lifted, and so cooled, until all its moisture condenses, and that all the latent heat released goes into heating the parcel. The *equivalent potential temperature*, θ_{eq}, is the potential temperature that the parcel then achieves.[7] If the parcel is then moved along a dry adiabat to a reference pressure p_R (commonly 1000 hPa) the actual temperature it will then have is θ_{eq}. As a consequence of the near adiabatic nature of the process, θ_{eq} is an approximate measure of the entropy of the parcel, as we will show (see also Appendix A).

We may obtain an approximate analytic expression for θ_{eq} by noting that the first law of thermodynamics, $đQ = T\,\mathrm{d}\eta$, implies, by definition of potential temperature,

$$-L\,\mathrm{d}q = c_p T\,\mathrm{d}\ln\theta, \tag{18.52}$$

during the condensation process, where $\mathrm{d}q$ is the change in water vapour content. Integrating gives, by definition of equivalent potential temperature,

$$-\int_q^0 \frac{L}{c_p T}\,\mathrm{d}q = \int_\theta^{\theta_{eq}} \mathrm{d}\ln\theta. \tag{18.53}$$

Here, q is the initial amount of water vapour contained in the parcel and T is the temperature at which condensation occurs. We might imagine lifting a parcel from its initial position to a temperature T at which saturation first occurs. To remove *all* the water vapour the parcel must be lifted to great height, because all of the water vapour will only condense if the final temperature is very low, with condensation occurring continuously and so with varying temperature along the way. But if we assume that temperature is constant during condensation, and that L and c_p are also constants, then (18.53) gives

$$\theta_{eq} = \theta\exp\left(\frac{Lq}{c_p T}\right) = T\left(\frac{p_R}{p}\right)^{R/c_p}\exp\left(\frac{Lq}{c_p T}\right). \tag{18.54}$$

The equivalent potential temperature so defined is approximately conserved during condensation, the approximation arising going from (18.53) to (18.54). It is a useful expression for diagnostic purposes and in constructing theories of convection, but it is not accurate enough to use as a primary prognostic variable in a numerical model that aims to be realistic. In saturated adiabatic flow a parcel will follow a 'moist isentrope' (an isoline of equivalent potential temperature) more closely than a dry isentrope.

To see that the equivalent potential temperature is an approximate measure of the entropy of a saturated parcel we begin with the entropy of dry air, which referring to (1.107) is

$$\eta^d = c_p\ln\theta = c_p\ln(T/T_0) - R\ln(p/p_0), \tag{18.55}$$

where T_0 and p_0 are constants. If we add to this a contribution from liquid water (see Appendix A on page 720 for a more complete tratement) then the saturation entropy, η_s, is approximately given by

$$\eta_s \approx \eta^d + \frac{Lq_s}{T} = c_p\ln(T/T_0) - R\ln(p/p_0) + \frac{Lq_s}{T}, \tag{18.56}$$

where q_s is the saturation specific humidity and final liquid water content. If we now define θ_{eq} by

$$c_p\ln(\theta_{eq}/T_0) \equiv \eta_s, \tag{18.57}$$

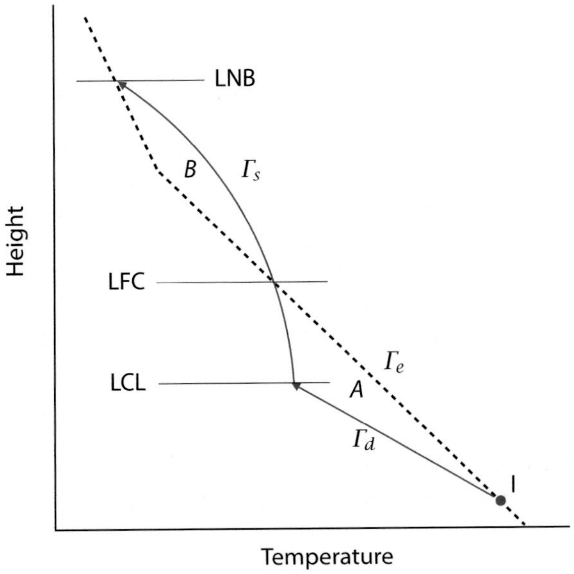

Fig. 18.16 Schematic of a conditional instability in an atmosphere with environmental lapse rate Γ_e (dashed line). A parcel at I is forced to rise, and it does so along the dry adiabat, Γ_d, until it is saturated at the lifting condensation level, LCL. It will then rise along the saturated adiabat, Γ_s, and after reaching the level of free convection (LFC) it is convectively unstable. The parcel continues to rise along the saturated adiabat, without any external forcing, until it reaches the level of neutral buoyancy (LNB). Not to scale.

then using (18.56) and (18.57) we recover (18.54) with $q = q_s$ — meaning that the logarithm of the equivalent potential temperature is, approximately, the entropy. To relate this result to the lapse rate, consider a parcel ascending into a colder region and condensing. The saturation moist entropy will stay the same: the last term on the right-hand side of (18.56) will fall, but the ensuing heating causes the temperature to increase. If we differentiate (18.56) with respect to z and set $\partial\eta_s/\partial z = 0$ then, using hydrostasy and a little algebra, we recover the saturated adiabatic lapse rate given by (18.50), which, therefore, has constant θ_{eq} and constant η_s.

18.4 CONVECTION IN A MOIST ATMOSPHERE

Over a wide range of temperatures and pressures the saturated adiabatic lapse rate is considerably less than the dry lapse rate (Fig. 18.15), and thus a moist atmosphere can be stable to dry convection but unstable to moist convection. If an environmental profile is unstable then convection ensues, transporting energy upward, until the atmosphere nearly stabilizes. But if the atmosphere is not saturated it is the dry adiabatic lapse rate that is the relevant one (or, strictly, the adiabatic lapse rate of moist, non-condensing air, which is very slightly different). The observed environmental profile in convecting situations may thus be a combination of the dry adiabatic and saturated adiabatic profiles. An unsaturated parcel that is unstable by the dry criterion will rise and cool following a dry adiabat, Γ_d, until it becomes saturated, above which it will rise following a saturated adiabat, Γ_s. The temperature then does not fall as rapidly as the dry case because the latent heat release warms the parcel as it rises.

These facts give moist convection a particular flavour and lead to the notion of *conditional stability,* whereby a parcel is stable to an infinitesimal perturbation but unstable to a finite perturbation. Consider an environment in which the lapse rate lies between the moist and dry rates, as in Fig. 18.16, and consider a parcel near the surface at position I. Suppose the parcel is adiabatically lifted (perhaps mechanically) then its temperature profile follows the dry adiabat until it is saturated at, by definition, the *lifting condensation level* (LCL). However, the parcel is actually negatively buoyant and so would sink unless the parcel is forced to continue rising, but if it does continue to rise it will be along a saturated adiabat and eventually, at the *level of free convection* (LFC) is will become buoyant and convectively unstable. It will continue to rise until its buoyancy no longer exceeds that of the environment, at the *level of neutral buoyancy,* which may well be at or close to the tropopause where the temperature starts to increase again. (The height of the tropopause is not independent of the convection itself, a matter we discuss in Section 18.6.) Evidently, to trigger such an instability a *finite* perturbation is needed (for example a large scale flow over a hill forcing a parcel upwards) and this is known as *conditional instability*. A parcel that is initially sitting at the LCL will only be unstable if it is saturated, which is by no means usually the case: the water has to get there somehow.[8]

Convection in the atmosphere need not be 'conditional', in the sense described above, nor is the atmosphere necessarily always in a conditionally unstable state. Indeed, on average the tropical atmosphere is closer to a moist profile than a dry profile. It is also important to remember that convection ultimately arises because the radiative forcing produces a vertical profile that is convectively unstable, with large-scale horizontal temperature gradients confining the convection to warmer regions, as in Fig. 18.14.

18.4.1 Energetics of Convection

How much energy is available for a parcel in convection? One way is to calculate the work released by, or required for, a parcel as it moves through an environment that has a different density. The upward force per unit mass on a parcel of density ρ_p in an environment with density ρ_e is just the familiar buoyancy force,

$$F_b = g\frac{\rho_e - \rho_p}{\rho_p} = g\frac{\alpha_p - \alpha_e}{\alpha_e}. \tag{18.58}$$

Thus, the energy released as a parcel ascends from z_1 to z_2 is

$$\text{Energy} = \int_{z_1}^{z_2} g\frac{\alpha_p - \alpha_e}{\alpha_e}\,\mathrm{d}z = -\int_{p_1}^{p_2} (\alpha_p - \alpha_e)\,\mathrm{d}p, \tag{18.59}$$

if the environment is in hydrostatic balance. If we take the limits of integration to be the level of free convection and the level on neutral buoyancy, the energy is known as the *convective available potential energy*, or CAPE. Using the ideal gas relation, and assuming that the parcel has the same pressure as the environment, (18.59) becomes

$$\text{CAPE} = -R\int_{\text{LFC}}^{\text{LNB}} (T_p - T_e)\frac{\mathrm{d}p}{p}. \tag{18.60}$$

If the height axis in Fig. 18.16 were log pressure then the CAPE would be proportional to the area *B*. If the limits were taken from the initial height to the LFC then the integral would be proportional to the area *A* and would be negative, and is known as the 'convective inhibition,' or CIN. That is

$$\text{CIN} = -R\int_{p_{bot}}^{\text{LFC}} (T_p - T_e)\frac{\mathrm{d}p}{p}, \tag{18.61}$$

where p_{bot} is the pressure at the bottom or in a boundary layer, and CIN is the amount of energy that must be supplied to initiate convection. In some accounts the definition of CAPE includes the CIN, in which case CAPE may be negative (but this is unusual). It is often the case that observed profiles of temperature in the tropics exhibit both CIN and CAPE, and a parcel in the lower atmosphere is stable and the instability is only conditional. That a finite perturbation may be needed to initiate convection is a distinguishing feature of moist convection.

Given the initiation of convection, one may envision an unstable parcel of air experiencing a buoyant force and accelerating though the unstable region, gaining kinetic energy, and this energy can be quite significant. Suppose that $T_p - T_e = 1°\,\mathrm{C}$ and that the limits of integration are 250 hPa and 750 hPa. Then we estimate

$$\mathrm{CAPE} \sim R(T_p - T_e)\frac{\Delta p}{p} = R \times 1 \times \frac{750 - 250}{500} = 286\,\mathrm{J\,kg^{-1}}. \tag{18.62}$$

This is actually an underestimate, by a factor of a few, of the value of CAPE often found in tropical atmospheres, but nevertheless if we translate it to a vertical velocity using $w = 2\sqrt{\mathrm{CAPE}}$ then we find $w \approx 24\,\mathrm{m\,s^{-1}}$! In fact, CAPE is a significant overestimate of the kinetic energy that is imparted to a buoyant parcel: a parcel will not keep its identity because of mixing or entrainment with its surroundings and the environmental profile itself becomes altered, and even if the parcel did keep its identity it would give up some of its kinetic energy to the environment in pushing it out of the way. Nevertheless, the estimate gives the sense that atmospheric convection is a vigorous process, occurring on a fast timescale compared to large-scale dynamics. Let us suppose that the above estimate is too big by a factor of 10. The time taken for a parcel to travel half the height of the troposphere in the tropics is then approximately $7500\,\mathrm{m}/2.4\,\mathrm{m\,s^{-1}} = 3000\,\mathrm{s}$, or less than one hour. *Thus, the timescales of atmospheric convection are measured in hours, not days.*

18.4.2 Effects of Convection

We have seen that, with even small differences in between an environmental profile and a stable profile, convection will be vigorous and act on rather fast timescales. What will the result of that convection be? Parcels will ascend until they are no longer unstable, at which point they can mix irreversibly with their environment, warming the environment aloft. (Indeed the parcels will mix as they ascend, but they have more time to mix as the ascent slows.) The convection and the mixing will proceed until the environmental profile is no longer unstable, a process that, if it occurs on a very short timescale, is called *convective adjustment* or, if on a longer timescale, *convective relaxation*. One might suppose that the profile to which the environment will adjust will be the saturated adiabatic or dry adiabatic lapse rates, depending on whether the atmosphere is saturated or not, although this is something of an oversimplification.[9] The convection will also transfer moisture vertically, and if this transfer causes the profile to become saturated then condensation and precipitation may occur.

Let us denote the adjusted reference profiles of temperature and specific humidity as $T_r(z)$ and $q_r(z)$ (or equivalents in pressure). If the vertical structure of these profiles is given, the magnitude of the profile will be determined by the fact that convection will conserve enthalpy. If, for example, atmosphere is far from saturated then the reference profiles will satisfy

$$\int_{p_B}^{p_T} c_p(T_r - T_i)\,\mathrm{d}p = 0 \quad \text{and} \quad \int_{p_B}^{p_T} (q_r - q_i)\,\mathrm{d}p = 0, \tag{18.63}$$

where the subscript i denotes the initial environmental profile and p_B and p_T denote the levels at the base and top of the convection. If precipitation does occur then only the total enthalpy is conserved and we have

$$\int_{p_B}^{p_T} (h_r - h_i)\,\mathrm{d}p = 0, \tag{18.64}$$

where $h = c_pT + Lq$ is the (moist) specific enthalpy. One then needs a second constraint to determine the profiles of T_r and q_r separately.

The specification of these profiles, and the levels of the base and the top of the convection, is not an exact science since the notion of adjustment is an approximation. As a first estimate in a moist atmosphere one might suppose that T_r is the dry adiabat below saturation and the saturated adiabatic lapse rate above, and that the reference profile of q_r is such that the atmosphere remains saturated, or nearly so, after adjustment, and that p_B and p_T are the levels of free convection and of

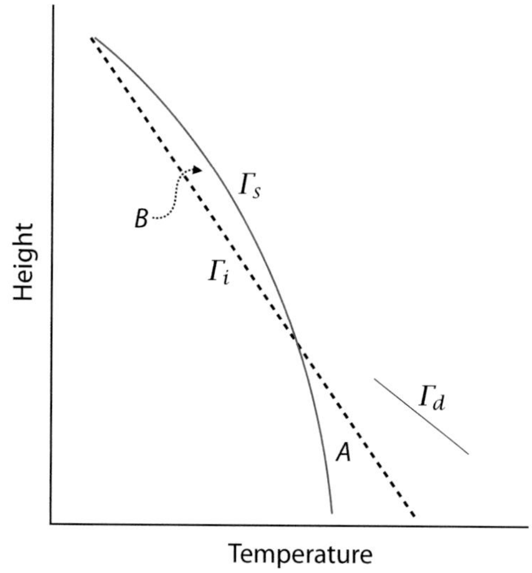

Fig. 18.17 Schematic of convective adjustment in a saturated atmosphere. The initial environment, with profile Γ_i, is unstable to moist convection and adjusts to the saturated adiabatic lapse rate Γ_s, with the dry adiabat Γ_d shown for reference. The average temperature of the final profile is such that the enthalpy is conserved, which provides a relationship between the areas *A* and *B*.

neutral buoyancy, respectively. Application of (18.64) then uniquely determines the final profiles of temperature and humidity, as schematically illustrated for a saturated atmosphere in Fig. 18.17. Details apart, two robust, key points should be emphasized:

(i) To a first approximation, convection determines the *profiles* of temperature and humidity rather than the actual heating rate or the precipitation. The heating and precipitation released by convection are a function of the environmental profile that in turn is largely determined by the larger-scale processes, although convection feeds back onto this profile.

(ii) After convection has occurred the local profile is, by construction, stable. However, this profile is not necessarily in equilibrium with the large-scale motion, or with the radiative processes that initially gave rise to the unstable profile. This disequilibrium may cause the profile to evolve further, possibly giving rise to more convection.

The above state of affairs — a putative initial profile set up by radiation and/or large-scale dynamics, modified quickly by convection, leading to a further slow evolution and convection — is called *quasi-equilibrium.*

18.4.3 † Convective Quasi-Equilibrium

Convective quasi-equilibrium is a posited state in which the forcing of a convectively unstable profile by large-scale dynamics and/or radiation is statistically balanced by convection.[10] The large-scale forcing may change, and the system may evolve, but in quasi-equilibrium the convection is assumed to occur on a faster timescale than the other processes and the fluid quickly enters into a statistical equilibrium state. Even as the large-scale evolves, the vertical profile at each step is largely determined by convection. That such a state exists is an assumption, but in Earth's tropical atmosphere it seems a fairly good one when dealing with timescales of longer than a few days, because convective timescales are relatively short and so convection is able to significantly alter the stratification. The assumption of quasi-equilibrium is marginally satisfied over the diurnal cycle.

The most obvious consequence of quasi-equilibrium is that the temperature profile is constrained to be close to being neutrally stable, which generally means a saturated-adiabatic profile except possibly in some regions in the lower atmosphere where the appropriate profile may be dry adiabatic, and in the boundary layer where properties are well mixed. A second important consequence concerns CAPE, convective available potential energy. Quasi-equilibrium implies that the

Convection, Quasi-Equilibrium and Radiative-Convective Equilibrium

- Atmospheric convection ultimately arises because the radiative forcing tends to produce temperature profiles, and hence buoyancy profiles, that are statically unstable. Large-scale advection, moving cold air over warm air, also leads to convection.
- Convection is most prevalent in the tropics because it is here that the surface is warmest, and the radiative-equilibrium lapse rate would be most unstable (Section 18.5). This would be true even in the absence of moisture, but moisture is most important in the tropics because it is warmest, and because water is imported via the Hadley Cell.
- The presence of moisture may cause the critical lapse rate for convection to be lower than that for dry air, because of the release of heat when water vapour condenses. The effect only arises if a parcel is saturated. Thus, a column of air may be *conditionally unstable* if it requires a finite perturbation to raise a parcel to a level where it saturates (Fig. 18.16).
- Atmospheric convection tends to occur on much faster timescales than those associated with the large-scale circulation. Once convection occurs the lapse rate is closely constrained to the critical lapse rate in the regions of convection.
- In a process analogous to geostrophic adjustment, internal waves propagating away from convection maintain weak horizontal temperature gradients and ensure that the lapse rate does not deviate too far from saturated adiabatic anywhere in the tropics (Fig. 15.25). In mid-latitudes, in contrast, energy is also transported upward by large-scale baroclinic eddies and the lapse rate is lower (more stable).
- When the large-scale dynamical or radiative forcing changes, the convective flux changes to constrain the lapse rate, and the production of CAPE by the large scale is *approximately* balanced by the relaxation of CAPE by convection. This state of affairs is called quasi-equilibrium, and its occurrence is really a hypothesis that appears to be fairly well-satisfied, if the larger timescales are sufficiently long, in Earth's tropical atmosphere.
- Quasi-equilibrium, if and where closely satisfied, leads to considerable simplifications regarding the dynamics of large-scale flow for it de-emphasizes the need to explicitly calculate vertical energy fluxes because the lapse rate is constrained. Equations resembling the shallow-water equations, with a small number of vertical modes, then become a decent approximation for large-scale baroclinic flow, with diabatic effects forcing the first baroclinic mode.
- A radiative-convective equilibrium state arises when the radiative-equilibrium state is convectively unstable. Convection then balances the radiative forcing, producing at first approximation a convectively active state up to some finite height, with a radiative equilibrium state beyond (Fig. 18.18).
- The height to which the constrained lapse rate extends is determined by the need to maintain an overall radiative balance — the incoming solar radiation must equal the outgoing infrared radiation — and this is the leading-order determinant of the tropopause height (Section 18.6.2). A similar calculation could be applied in mid-latitudes, with a lapse rate determined as much or more by baroclinic instability as by convection.
- Quasi-equilibrium is a useful idealization when convection occurs quickly and efficiently, but it does not mean that the large scale always controls the convection. There are many situations when there is no causal separation between convection and the large scale — indeed the convection may create its own large scale flow, the MJO being one example.

CAPE is released by convection at about the same rate as it is generated by the forcing. Thus, at the same time as radiative forcing may be creating an unstable profile with a finite amount of CAPE, convection is destroying that potential energy, largely by converting it to kinetic energy where it is dissipated. The amount of CAPE in a profile may (and does) vary on slow timescales, the variation being produced by the weak imbalance between production and destruction; observations indicate it can vary by a factor of a few over the course of many days, and the environmental profile may also depart from the saturated adiabat away from convectively active regions. However, it turns out that the *average* stratification over the tropics will not deviate too much from moist neutral because internal waves extend the moist-neutral profile in the convective regions to the cloud-free regions, and horizontal pressure and temperature gradients are constrained to be relatively small. This 'weak temperature gradient' effect is described in Section 18.8.

Finally, albeit a little less tangibly, quasi-equilibrium gives rise to a *point of view* of tropical dynamics, that convection does not act as a primary determinant of the heat source for the large-scale circulation. Rather, convection controls the temperature and water vapour profiles, with the heating associated with the convection being determined by the requirement that such a profile be maintained, and thus ultimately arising from the larger scale forcing. (This process is analogous to the production of condensation in the fast-condensation models of Section 18.2, in which the condensation criteria determine the profile of water vapour but not the amount of condensation.) The resulting environmental profile then lies on the edge of stability, with a non-zero CAPE because of the finite time-scale and finite efficiency of convection and with the intensity of convection being just what is required to maintain that profile. The value of CAPE then evolves slowly, compared to convective timescales, if and as external factors change.

Having said this, quasi-equilibrium should not be thought of as being exact or even especially profound — it is a way to gain an understanding of the structure of the tropics but is not a universal recipe. There are many situations — the diurnal cycle, aspects of boundary layer development, weak convection — in which the quasi-equilibrium assumptions do not hold. Furthermore, the recipe may mislead if misapplied: even when quasi-equilibrium holds in so far as the buoyancy profile is constrained to be moist neutral, it should not be thought that convection and the associated condensation is always a quasi-passive process. That is, the scale separation between convective events and the large-scale does not mean that the large-scale always 'controls' the convection and the convection merely 'feeds back' on the large scale. For example, a localised source of convection near the equator may itself create a large-scale flow pattern similar to that of the Matsuno–Gill problem (Section 8.5), and this effect may be at the heart of the Madden–Julian oscillation (Section 18.10). Here one might say the convection controls the large-scale and the large-scale feeds back on the convection! Also, in the moist model of the Hadley Cell (Section 14.2.7) the release of latent heat in the upward branch serves to change the horizontal distribution of heating and greatly intensify the overturning circulation. The convection should not be thought of as the primary driver of the Hadley Cell, but it has a lowest-order effect.

Caveats aside, quasi-equilibrium is a very useful concept and in following sections we look at two particularly important consequences. First (albeit after an introduction to radiative equilibrium) we discuss radiative-convective equilibrium, and we then use quasi- equilibrium to simplify the equations of motion for the larger-scale circulation.

18.5 RADIATIVE EQUILIBRIUM

In order to understand the effects of convection we must first determine what the profile of temperature would be in its absence — the *radiative-equilibrium* state. The electromagnetic radiation in the Earth's atmosphere may usefully be divided into two types, solar (or shortwave) radiation and infrared (or longwave) radiation. We will assume the atmosphere is semi-grey (i.e., grey in the infrared) and, in some instances, transparent to solar radiation. Neither assumption is quantitatively good, but they capture the essence. In Appendix B to this chapter we show that the upward,

U, and downward, D, streams of longwave radiation then satisfy the radiative-transfer equations,

$$\frac{\mathrm{d}D_L}{\mathrm{d}\tau} = B - D_L, \qquad \frac{\mathrm{d}U_L}{\mathrm{d}\tau} = U_L - B. \tag{18.65a,b}$$

Here, $B = \sigma T^4$ is the radiative flux emitted by a black body, τ is the optical depth and σ is Stefan's constant. The optical depth is related to the geometric height z by $\mathrm{d}\tau = -e_L\,\mathrm{d}z$ where e_L is the longwave emissivity. The net flux of longwave radiation is $N_L = U_L - D_L$ and the longwave heating is proportional to the net flux divergence, $-\partial N_L/\partial z$.

18.5.1 Solutions

Formal solution to radiative transfer equations

If the temperature profile is known then B is known and (18.65) is a pair of first order differential equations in the two unknowns U and D. The solution requires two boundary conditions and in atmospheric problems these might be provided at the top of the atmosphere, for example by requiring that the downward infrared radiation be zero (i.e., $D_L(\tau = 0)$), and by specifying the upward radiation (i.e., let $U_L(\tau = 0) = U_0$ where U_0 is given). Alternatively, the upward radiation at the bottom of the atmosphere could be specified if the ground temperature were known. The solution in physical space is found by specifying the form of $\tau(z)$; that is, by specifying the emissivity. To obtain the solution we multiply (18.65) by the integrating factors $\exp(\tau)$ and $\exp(-\tau)$ to give

$$\frac{\mathrm{d}}{\mathrm{d}\tau}(D_L \mathrm{e}^{\tau}) = B\mathrm{e}^{\tau}, \qquad \frac{\mathrm{d}}{\mathrm{d}\tau}(U_L \mathrm{e}^{-\tau}) = -B\mathrm{e}^{-\tau}. \tag{18.66a,b}$$

Integrating between 0 and τ' straightforwardly gives

$$D_L(\tau') = \mathrm{e}^{-\tau'}\left[D_L(0) - \int_0^{\tau'} B(\tau)\mathrm{e}^{\tau}\,\mathrm{d}\tau\right], \qquad U_L(0) = U_L(\tau')\mathrm{e}^{-\tau'} + \int_0^{\tau'} B(\tau)\mathrm{e}^{-\tau}\,\mathrm{d}\tau. \tag{18.67a,b}$$

There are other ways to write the solutions that may be appropriate depending on the boundary conditions, but in any case the solutions are in general *non-local*, for they depend on the temperature along the path. The terms in (18.67) represent the attenuation of radiation as it travels along its path, as well as the cumulative emission. However, in some important special cases we can get a local solution, as we now see.

Radiative equilibrium solution

A *radiative equilibrium* state has, by definition, no radiative heating. If the atmosphere is transparent to solar radiation then the condition implies that the vertical divergence of the longwave radiation is zero:

$$\frac{\partial(U_L - D_L)}{\partial z} = 0 \quad \text{implying} \quad \frac{\partial(U_L - D_L)}{\partial \tau} = 0. \tag{18.68a,b}$$

This condition is normally *not* satisfied in the atmosphere because the air is in motion. If it were satisfied then (18.65) and (18.68b) form three equations in three unknowns, U_L, D_L, and B, and a solution can be found as follows.

Consider an atmosphere with net incoming solar radiation S_{net} and suppose the planet is in radiative equilibrium with the incoming solar radiation balanced by outgoing infrared radiation. That is, $U_{Lt} \equiv U_L(\tau = 0) = S_{net}$ where U_{Lt} is the net outgoing longwave radiation (OLR) at the top of the atmosphere. The downward infrared radiation at the top of the atmosphere is zero, so that the boundary conditions on the radiative transfer equations at the top of the atmosphere are

$$D_L = 0, \quad U_L = U_{Lt} \quad \text{at} \quad \tau = 0. \tag{18.69}$$

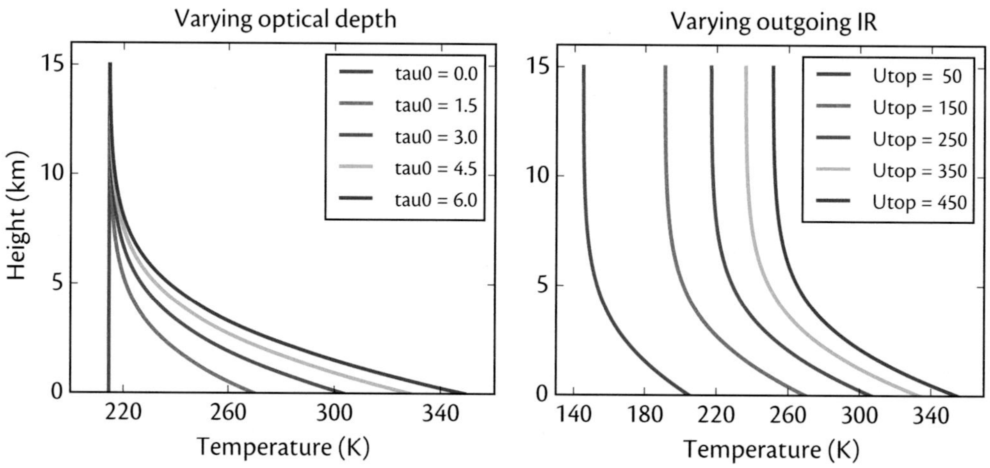

Fig. 18.18 Radiative equilibrium temperature calculated using (18.73), with $H_a = 2$ km. Left: outgoing IR (and incoming solar) radiation is 240 W m^{-2} and surface optical depth varies from 0 to 6. Right: the optical depth is 3.0 and outgoing IR radiation varies from 50 to 450 W m^{-2}.

To obtain a solution we rewrite (18.65) as

$$\frac{\partial}{\partial \tau}(U_L - D_L) = U_L + D_L - 2B, \qquad \frac{\partial}{\partial \tau}(U_L + D_L) = U_L - D_L. \tag{18.70a,b}$$

and a little algebra reveals that a solution of these equations that satisfies (18.68b) is

$$D_L = \frac{\tau}{2} U_{Lt}, \qquad U_L = \left(1 + \frac{\tau}{2}\right) U_{Lt}, \qquad B = \left(\frac{1+\tau}{2}\right) U_{Lt}, \tag{18.71a,b,c}$$

as can be easily verified by substitution back into the equations.

It remains to explicitly relate τ to z, and one approximate recipe is to suppose that τ has an exponential profile,

$$\tau(z) = \tau_0 \exp(-z/H_a), \tag{18.72}$$

where τ_0 is the optical depth at $z = 0$ and H_a is the scale height of the absorber. In the Earth's atmosphere the optical depth is determined by the concentrations of water vapour (primarily) and carbon dioxide (secondarily) and τ_0 (the scaled optical depth) typically varies between 2 and 4, depending on the water vapour content of the atmosphere, and $H_a \approx 2$ km, this being a typical scale height for water vapour. From (18.71c) the temperature then varies as

$$T^4 = U_{Lt} \left(\frac{1 + \tau_0 e^{-z/H_a}}{2\sigma} \right), \tag{18.73}$$

which is illustrated in Fig. 18.18. The following aspects of the solution deserve mention:

(i) Temperature increases rapidly with height near the ground.

(ii) The upper atmosphere, where τ is small, is nearly isothermal.

(iii) The temperature at the top of the atmosphere, T_t is given by

$$\sigma T_t^4 = \frac{U_{Lt}}{2}. \tag{18.74}$$

Thus, if we define the emitting temperature, T_e, to be such that $\sigma T_e^4 = U_{Lt}$, then $T_t = T_e/2^{1/4} < T_e$; that is, the temperature at the top of the atmosphere is *lower* than the emitting temperature.

(iv) Related to the previous point, $B_t/U_{Lt} = 1/2$. That is, the upwards long wave flux at the top of the atmosphere is twice that which would arise if there were a black surface at a temperature T_t. The reason is that there is radiation coming from all heights in the atmosphere.

Interestingly, in obtaining a solution we have not imposed a boundary condition at the ground — in fact there is no ground at all in this problem! What happens if we add one? That is, suppose that we declare that there is a black surface at some height, say $z = 0$, and we require that the atmosphere remain in radiative equilibrium with the same temperature profile. What temperature does the ground take? From (18.71) the upward irradiance and temperature at any height z are related by

$$U_L(z) = \left(\frac{2+\tau(z)}{1+\tau(z)}\right)\sigma T^4(z). \tag{18.75}$$

At $z = 0$ the ground will have to supply upwards radiation equal to that given by (18.75), and therefore its temperature, T_g is given by

$$\sigma T_g^4 = \left(\frac{2+\tau_0}{1+\tau_0}\right)\sigma T_s^4, \tag{18.76}$$

where T_s is the temperature of the fluid adjacent to the ground (the 'surface temperature'). That is, $T_g > T_s$ and there is a temperature discontinuity at the ground, especially if optical depth is small. In fact, in very still and clear conditions a very rapid change of temperature near the ground can sometimes be observed, but usually the presence of conduction and convection, as we discuss below, ensures that T_g and T_s are equal.

We note that in the limit in which $\tau = 0$ in the upper atmosphere we have

$$D_L = 0, \qquad U_L = U_{Lt}, \qquad B = \frac{U_{Lt}}{2}. \tag{18.77}$$

That is, the upper atmosphere is isothermal, there is no downwelling irradiance and the upward flux is constant. The upper-atmosphere temperature, T_{ua} say, and the emitting temperature are related by $T_{ua} = T_e/2^{1/4}$.

The above results imply that at a given optical depth, the temperature difference between the surface and the ground increases with temperature. From (18.73) we have that

$$T_t^4 = \frac{U_{Lt}}{2\sigma}, \qquad T_s^4 = \frac{U_{Lt}(1+\tau_0)}{2\sigma}, \tag{18.78}$$

whence

$$T_s^4 - T_t^4 = T_t^4\tau_0 \qquad \text{or} \qquad \Delta T \approx \frac{\tau_0}{4}T_t, \tag{18.79}$$

where $\Delta T = T_s - T_t$ and we assume $\Delta T \ll T_s, T_t$. Thus, in the absence of other effects, higher temperatures lead to larger average lapse rates, and so are potentially more conducive to convection.

18.6 RADIATIVE-CONVECTIVE EQUILIBRIUM

In the radiative equilibrium solution the temperature gradient near the ground varies so rapidly that $-\partial T/\partial z$ may exceed even the dry adiabatic lapse rate. If so, the radiative equilibrium solution is convectively unstable and convection will be triggered and energy (and moisture) will be redistributed throughout the column. What is the resulting temperature profile? The answer is given in Fig. 18.19, which we now explain.

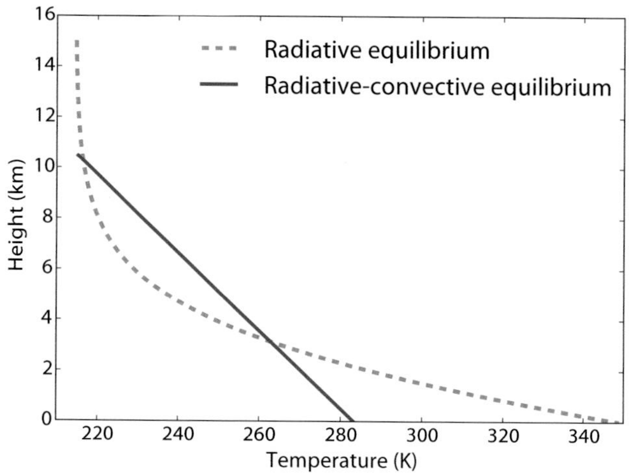

Fig. 18.19 Radiative and radiative-convective equilibrium profiles. The initial radiative equilibrium temperature adjusts to a specified profile (in this instance a constant lapse rate) that extends to a finite height, beyond which radiative equilibrium holds. This height (the tropopause) is determined by the requirement of radiative balance, and the overall adjustment process is not adiabatic. The curves are results of numerical calculations with $\tau_0 = 6$, $\Gamma = 6.5\,\mathrm{K\ km^{-1}}$ and outgoing radiation of $240\,\mathrm{W\ m^{-2}}$ (see Section 18.6.2).

18.6.1 ✦ The General Case

We first approach the problem without making any simplifying quasi-equilibrium assumptions, supposing only that we have a balance between longwave and solar radiation and the effects of convection. Let us assume that upward and downward shortwave fluxes, U_S and D_S, obey the two-stream radiative equations with no thermal emission and no scattering, known as the Schwarzchild equations, namely

$$\frac{\mathrm{d}D_S}{\mathrm{d}\tau_S} = -D_S, \qquad \frac{\mathrm{d}U_S}{\mathrm{d}\tau_S} = U_S. \tag{18.80a,b}$$

Here, τ_S is the shortwave optical depth, which is related to the physical height by $\mathrm{d}\tau_S = -e_S\,\mathrm{d}z$, where e_S is the shortwave emissivity. and we can then write the radiative equations as

$$\frac{\mathrm{d}D_S}{\mathrm{d}\tau_L} = -\frac{e_S}{e_L}D_S, \qquad \frac{\mathrm{d}U_S}{\mathrm{d}\tau} = \frac{e_S}{e_L}U_S. \tag{18.81a,b}$$

The solar heating is proportional to the convergence of the net solar fluxes, $-\partial N_S/\partial z$ where $N_S = U_S - D_S$. The convection also provides a local heating, and this is proportional to the convergence of the upwards enthalpy flux. Thus, in a steady state,

$$\frac{\partial}{\partial z}(N_S + N_L + \mathcal{H}) = 0 \quad \text{implying} \quad \frac{\partial}{\partial \tau}(N_S + N_L + \mathcal{H}) = 0, \tag{18.82a,b}$$

where $\mathcal{H}$ is the enthalpy flux. Using (18.70a) we write (18.82b) as

$$U_L + D_L - 2B + \frac{1}{e_L}Q_S + \frac{\partial \mathcal{H}}{\partial \tau} = 0, \tag{18.83}$$

where Q_S is the solar heating given by $Q_S = -\partial N_S/\partial z = e_L^{-1}\partial N_S/\partial \tau$. If we had a theory for the enthalpy fluxes, $\mathcal{H}$, then (18.83), in conjunction with the Schwarzchild equations and knowledge of the solar radiation and atmospheric emissivity, would determine the temperature profile with height. We have no such theory, but we can still make progress.

Effects of solar fluxes

Suppose first (and taking a diversion from convection) that the enthalpy fluxes are zero, much as in the stratosphere. We then have a radiative equilibrium profile that satisfies

$$\frac{\partial}{\partial z}(N_L + N_S) = 0, \tag{18.84}$$

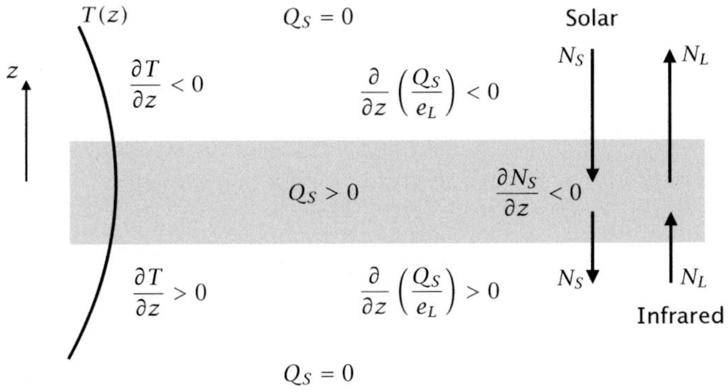

Fig. 18.20 Effects of a band of solar radiation absorption. Shading marks the absorbing region where $\partial N_s/\partial z < 0$, with consequences for the heating, Q_S, and temperature profile, $T(z)$, determined by (18.86b), as indicated.

implying

$$\frac{\partial N_L}{\partial z} = Q_S \qquad \text{or} \qquad \frac{\partial N_L}{\partial \tau} = -\frac{Q_S}{e_L}. \tag{18.85a,b}$$

If we differentiate (18.85b) with respect to τ and use (18.70a,b) we obtain

$$2\frac{\partial B}{\partial \tau} = N_L + \frac{\partial}{\partial \tau}\left(\frac{Q_S}{e_L}\right) \qquad \text{or} \qquad 8\sigma T^3\frac{\partial T}{\partial z} = -e_L N_L + \frac{\partial}{\partial z}\left(\frac{Q_S}{e_L}\right). \tag{18.86a,b}$$

The middle of the stratosphere in Earth's atmosphere contains a layer of ozone that absorbs solar radiation. Thus, $\partial N_s/\partial z < 0$ and Q_S is positive. The net longwave radiation N_L is positive in this region, as there is much more upwelling radiation than downwelling, and this tends to produce a negative vertical temperature gradient, but the effect is weak because e_L is small. However, the solar radiation can have a strong effect because e_L appears in the *denominator* of the term involving Q_S. Just below the heating region $\partial Q_S/\partial z$ is positive and this produces a positive value of $\partial T/\partial z$, with the converse above the ozone layer. Thus, the region of solar heating corresponds to a maximum of temperature, or at least a region of small vertical temperature gradient, with negative values of $\partial^2 T/\partial z^2$, as schematically illustrated in Fig. 18.20. The result is intuitively reasonable, but (18.86) provides the explicit solution. A non-intuitive result that follows from (18.86b) concerns the effects of an increasing infrared emissivity, such as happens with global warming. Suppose that $\partial T/\partial z > 0$, as in the stratosphere because of the presence of a layer of ozone, and suppose further that e_L then increases, because of an increased concentration of greenhouse gases. Changes in the longwave term are small, because e_L is small, but changes in the solar term are large, and these cause $\partial T/\partial z$ to diminish. That is, an increase in the concentration of greenhouse gases will cause stratospheric temperatures to *fall*.[11] The result arises because of the balance, in this case, between solar heating and infrared cooling, as in (18.85a) and Fig. 18.20. If the emissivity increases there will still be the same infrared loss to space, to achieve radiative balance, and if the emissivity is higher the loss can be achieved at a lower temperature.

Effects of enthalpy fluxes

Now consider the effects of an enthalpy flux, supposing that the atmosphere is transparent to solar radiation. The equilibrium condition, (18.83), becomes

$$U_L + D_L - 2B + \frac{\partial \mathcal{H}}{\partial \tau} = 0. \tag{18.87}$$

Differentiating with respect to τ and using (18.70b) gives

$$2\frac{\partial B}{\partial \tau} = N_L + \frac{\partial^2 \mathcal{H}}{\partial \tau^2}. \tag{18.88}$$

Now, from (18.82a) we must have $N_S + N_L + \mathcal{H} = 0$ for all z, if the atmosphere is in an overall radiative balance and if $\mathcal{H} = 0$ at the atmosphere's top. If the solar flux is constant with height then, also for all z,

$$N_L = S_{net} - \mathcal{H}, \tag{18.89}$$

where S_{net} is the net incoming solar radiation, a positive quantity. Using this expression (18.88) becomes

$$2\frac{\partial B}{\partial \tau} = S_{net} - \mathcal{H} + \frac{\partial^2 \mathcal{H}}{\partial \tau^2} \quad \text{or} \quad \frac{8\sigma T^3}{e_L}\frac{\partial T}{\partial z} = -S_{net} + \mathcal{H} - \frac{1}{e_L}\frac{\partial}{\partial z}\left(\frac{1}{e_L}\frac{\partial \mathcal{H}}{\partial z}\right). \tag{18.90a,b}$$

These expressions explicitly tell us how the lapse rate is affected by enthalpy fluxes. If we knew $\mathcal{H}$ we could integrate (18.90) to give us the temperature at every level, and if $\mathcal{H} = 0$ the equations are equivalent to the derivative of (18.71c). Enthalpy fluxes are usually positive (i.e., upwards) in the atmosphere and so they tend to increase $\partial T/\partial z$ and reduce the lapse rate; that is, they tend to warm the upper atmosphere and cool the lower atmosphere and the surface. Since we do not have a quantitative theory for these fluxes let us invoke some ideas of quasi-equilibrium and specify the resulting lapse rate rather than the fluxes themselves.

18.6.2 † Convective Adjustment and the Height of the Tropopause

Let us assume that convection occurs with sufficient efficiency so that it will establish a convectively neutral lapse rate, Γ say, which for simplicity we here assume is constant. We may also assume that the convection equalizes the ground temperature and the surface temperature (the temperature of the layer of air immediately above the ground). Such a model simplifies the problem enormously, for we avoid completely the problem of solving the radiative transfer equations. However, radiation is still present, and we must still satisfy an overall radiative balance, and this determines the height to which the convection extends — which is, effectively, the height of the tropopause. In the first instance we can suppose that the convection occurs quickly and adiabatically, so that the re-arrangement of the temperature profile conserves energy. However, such a profile will not necessarily be a solution of the radiative transfer equations (18.67) and the profile must adjust until a solution is found. If the lapse rate is fixed, the only degree of freedom is the height to which convection reaches — that is, the height of the tropopause — above which radiative equilibrium holds.

The computation of this height follows a straightforward algorithm. If the top of the convecting layer occurs at a height, H_T, at which the optical depth is small, then outgoing long-wave radiation there will be approximately equal to the outgoing radiation at the top of the atmosphere, $U_L(\tau = 0)$, which is equal to the incoming solar radiation and therefore known. This gives the tropopause temperature (from (18.76), namely $T_T^4 = U_L(\tau = 0)/2\sigma$) and therefore, if we know H_T and the lapse rate, the temperature at all heights. We can then calculate the upwards radiative flux using a variant of (18.67b), which we may write as

$$U_L(z = 0) = U_L(z = H_T)e^{\tau_s} + \int_0^{H_T} B(\tau)e^{\tau_s - \tau}\frac{d\tau}{dz}\,dz, \tag{18.91}$$

where τ_s is the optical depth at the surface. However, for an arbitrary tropopause height, the upwelling radiation at the bottom, $U_L(z = 0)$, as given by (18.91), will *not* equal σT_s^4, which it must do if radiative balance is to be satisfied. The height of the tropopause must therefore adjust, and an iterative algorithm for finding the equilibrium solution goes as follows:

(*i*) Obtain the radiative equilibrium temperature profile.

(*ii*) Make a guess for the height of the tropopause, and using the given lapse rate obtain the temperature all the way down to the ground.

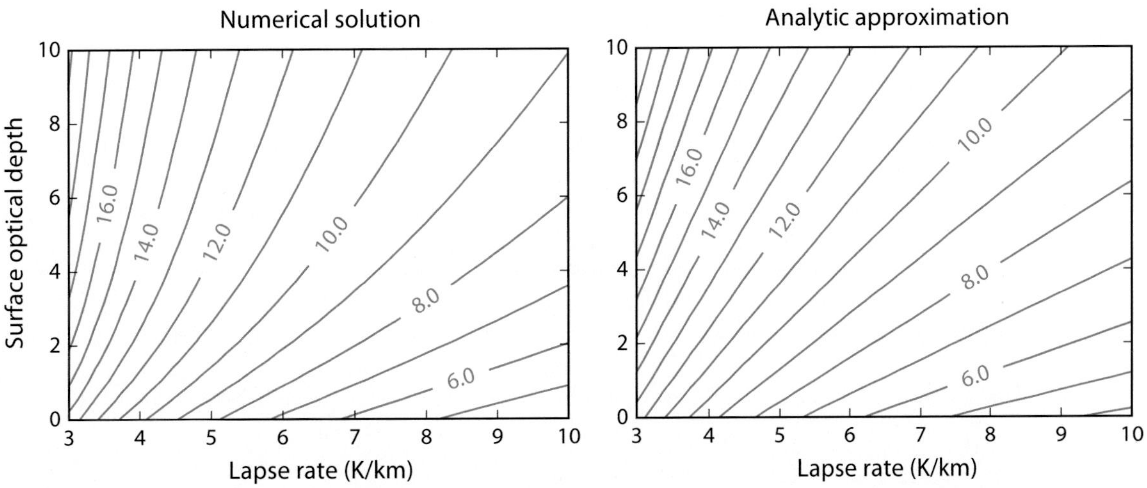

Fig. 18.21 Contours of tropopause height (km) as a function of lapse rate and surface optical depth in a grey atmosphere, calculated numerically by iterating the Schwarzchild equations (left) and using an analytic approximation (right), with $H_a = 2$ km and outgoing longwave radiation of 242 W m^{-2}, or tropopause temperature of 215 K.

(iii) Integrate the radiative transfer equations, (18.91), down from the top. The outgoing radiative balance is achieved this way but there is no balance at the surface if temperature is continuous. That is, $\sigma T_s^4 \neq U_L(z=0)$.

(iv) Change the height of the tropopause, find another solution, and iterate until the surface radiative balance is properly achieved.

A profile of temperature so calculated, along with the radiative-equilibrium temperature, is shown in Fig. 18.19, and solutions as a function of lapse rate and temperature are given in Fig. 18.21.

18.6.3 Approximate Analytic Solution

In Appendix C (page 725) we show that an approximate analytic solution of the above algorithm for the height of the tropopause, H_T, is

$$8\Gamma H_T^2 - C H_T T_T - \tau_s H_a T_T = 0, \tag{18.92}$$

or

$$H_T = \frac{1}{16\Gamma}\left(CT_T + \sqrt{C^2 T_T^2 + 32\Gamma\tau_s H_a T_T}\right), \tag{18.93}$$

where $C = \log 4 \approx 1.4$, Γ is the lapse rate, T_T is the temperature at the tropopause, τ_s is the surface optical depth and H_a is the scale height of the main infrared absorber. For Earth's atmosphere, $H_a \approx 2$ km, $\tau_s \approx 5$ and $\Gamma \approx 6.5$ K km^{-1}. All three terms in (18.92) are then approximately the same size and (18.93) gives $H_T \approx 11$ km. We can now be a little more precise about what it means for an atmosphere to be optically thin or thick. Using (18.93) and approximating $C^2 = 2$ we see that the optically thick limit arises when

$$\tau_s H_a \gg \frac{T_T}{16\Gamma} \quad \text{whence} \quad H_T \approx \sqrt{\frac{T_T \tau_s H_a}{8\Gamma}}. \tag{18.94}$$

The optically thin case has

$$\tau_s H_a \ll \frac{T_T}{16\Gamma} \quad \text{whence} \quad H_T \approx \frac{CT_T}{8\Gamma}. \tag{18.95}$$

With parameters appropriate for Earth's atmosphere both of the above limits give estimates in the range 6–15 km, and they are additive effects. Plots of the tropopause height as a function of lapse rate and optical depth, calculated using (18.93), as well as numerical solutions using the algorithm of the previous subsection, are given in Fig. 18.21, and the agreement is fairly good in Earth's parameter regime. A robust result is that as the lapse rate diminishes ($-\partial T/\partial z$ becomes smaller) the tropopause height increases, essentially to maintain the same tropopause temperature.

The main quantitative deficiency of this argument is that the atmosphere is not grey — the absorption of radiation is a function of wavelength. A less severe approximation is to suppose that the infrared radiation occurs in two bands — a 'window' region and the remainder. In the window region the atmosphere is fairly transparent, meaning that a fraction ($\sim 1/4$ or $1/3$) of the radiation emitted by the surface goes straight to space. This has the effect of slightly decoupling the tropopause temperature from the effective emitting temperature. Nonetheless, the general arguments leading to (18.93) remain valid and the way in which the tropopause height varies with optical depth and stratification will carry through in the more realistic case.[12]

Effects of lateral energy transport and application to mid-latitudes.

Although we have set the above calculation in the context of convection, it applies in any situation where the lapse rate can be specified; the argument would equally well apply if the lapse rate were set by baroclinic instability. We can also apply a modified argument even when there is a horizontal transport of energy, for such a transport effectively just acts to change the radiative emitting temperature. If the lateral convergence of energy and the incoming solar radiation are known then the amount of radiation that the column must emit to space in order to maintain energy balance is easily calculated. For example, if there is a convergence of energy at high latitudes because of the transport of energy by baroclinic eddies, then the column needs to emit less infrared radiation to space than it would otherwise in order to maintain radiative equilibrium. In actuality, in mid-latitudes the lapse rate cannot be regarded as being specified independently of the lateral heat transport, and the determination of the lapse rate and the tropopause height become intertwined. The reader is referred back to Section 15.5 to follow this argument.

18.7 VERTICALLY-CONSTRAINED EQUATIONS OF MOTION FOR LARGE SCALES

Let us now segue toward the large scale dynamics of the tropics. We first show how having a nearly constant lapse rate constrains the vertical degrees of freedom, so reducing the equations of motion to something akin to the shallow water equations, and readers may wish to skim Section 3.4 before continuing.

18.7.1 Reduction of Vertical Degrees of Freedom

The assumption that convection maintains a moist-adiabatic profile everywhere constrains the vertical structure of the horizontal temperature gradient and, at least in so far as the momentum dynamics are linear, reduces the equations of motion governing the large-scale to a set similar to the shallow-water equations.[13] To see this we begin with hydrostasy in pressure coordinates applied to the fluctuating fields,

$$\frac{\partial \phi'}{\partial p} = -\alpha', \tag{18.96}$$

where ϕ' and α' are variations in the geopotential and the specific volume. The latter can be taken to be a function of the state variables pressure and saturation entropy, η (dropping the subscript s). Variations in α at constant pressure thus obey

$$\alpha' = \left(\frac{\partial \alpha}{\partial \eta}\right)_p \eta' = \left(\frac{\partial T}{\partial p}\right)_\eta \eta', \tag{18.97}$$

using one of Maxwell's equations (page 19). If we integrate (18.96) in the vertical and use (18.97) we obtain

$$\phi'(x,y,p,t) = -\int \alpha \, dp = \int \left(\frac{\partial T}{\partial p}\right)_\eta \eta' \, dp = \eta' \int dT, \tag{18.98}$$

where we can take η' out of the integral *because it is constant if the atmosphere has a saturated adiabatic profile.* Integrating gives

$$\phi(x,y,p,t) - \overline{\phi}(x,y,t) = \left(\overline{T}(x,y,t) - T(x,y,p,t)\right)\eta(x,y,t), \tag{18.99}$$

where we have written the constant of integration such that $\overline{\phi}$ and $\overline{T}$ are the vertical means, in pressure coordinates, of ϕ and T. Following conventional usage, we call the vertically integrated components 'barotropic' and the deviations 'baroclinic'.

The temperature difference $\left(\overline{T}(x,y,t) - T(x,y,p,t)\right)$ in (18.99) is determined by the saturated adiabatic lapse rate and in the horizontal this varies quite weakly with the temperature variations found in the tropics. Thus, the horizontal variations of the terms on the right-hand side are dominated by variations in entropy, and to a good approximation we can write

$$\nabla\phi(x,y,p,t) = \nabla\overline{\phi}(x,y,t) + M(p)\nabla\eta(x,y,t), \tag{18.100}$$

where $M(p) = \overline{T}(x,y,t) - T(x,y,p,t)$ and which *we take to be a function of pressure alone.* The profile of $M(p)$ has a single node in the vertical, and this constrains the horizontal velocity to have a similar variation in the vertical, and the vertical integral of $M(p)$ vanishes.

Velocity decomposition

First consider the vertical velocity. In pressure coordinates the mass continuity equation is

$$\frac{\partial \omega}{\partial p} = -\nabla_p \cdot \boldsymbol{u}, \tag{18.101}$$

where $\overline{\boldsymbol{u}} = (u, v)$. Let $\boldsymbol{u} = \overline{\boldsymbol{u}} + \boldsymbol{u}^*$, where $\overline{\boldsymbol{u}}$ and $\boldsymbol{u}^*$ are the barotropic and baroclinic components of $\boldsymbol{u}$, and ω is the vertical (pressure) velocity. We suppose that the fluid is confined between the ground at $p = p_g$ and a rigid tropopause at p_t, with $\omega = 0$ at both. The baroclinic components of the velocity will, as a consequence of their definition, vanish when integrated over the troposphere. An integration of (18.101) over the troposphere implies the barotropic flow is divergence-free:

$$\nabla_p \cdot \overline{\boldsymbol{u}} = 0. \tag{18.102}$$

The vertical velocity is thus related to the baroclinic flow by,

$$\frac{\partial \omega}{\partial p} = -\nabla_p \cdot \boldsymbol{u}^*. \tag{18.103}$$

The horizontal flow itself is given using the momentum equation which, using (18.100), we write as

$$\frac{\partial \boldsymbol{u}}{\partial t} + \boldsymbol{v}\cdot\nabla\boldsymbol{u} + \boldsymbol{f}\times\boldsymbol{u} = -\nabla\overline{\phi} + M(p)\nabla_p\eta, \tag{18.104}$$

where $\boldsymbol{v}$ is the three-dimensional velocity and we omit forcing and viscous terms, and ∇ without a subscript denotes horizontal derivative. The above equation suggests that it will be a good approximation to suppose that the horizontal velocity has the same vertical structure as $M(p)$ and we let

$$\boldsymbol{u}(x,y,p,t) = \boldsymbol{u}_0(x,y,t) + m(p)\boldsymbol{u}_1(x,y,t), \tag{18.105}$$

where $\boldsymbol{u}_0$ ($= \overline{\boldsymbol{u}}$) and $\boldsymbol{u}_1$ are the 'barotropic' and 'first baroclinic' components of the flow, and $m(p) = M(p)/N$ where N is a normalizing coefficient, say $\left[\int M^2\,\mathrm{d}p/(p_t - p_g)\right]^{1/2}$, so that $m(p)$ is a dimensionless basis function. The reader will appreciate the similarity of this approach to that of Section 3.4, now with the quasi-equilibrium constraints determining the form of the vertical eigenfunctions for us.

Momentum equations

Neglecting the advective term the horizontal momentum equation decomposes exactly into separate barotropic and baroclinic equations which, omitting frictional terms, are

$$\frac{\partial \boldsymbol{u}_0}{\partial t} + \boldsymbol{f} \times \boldsymbol{u}_0 = -\nabla\phi_0, \qquad \frac{\partial \boldsymbol{u}_1}{\partial t} + \boldsymbol{f} \times \boldsymbol{u}_1 = -\nabla\phi_1, \tag{18.106a,b}$$

where $\phi_0 = \overline{\phi}$ and $\phi_1 = N\eta$. These have the same form as the linearized shallow water equations. We may obtain a diagnostic equation for the barotropic pressure by taking the divergence of (18.106a), whence the time derivatives disappear because $\nabla\cdot\boldsymbol{u}_0 = 0$. Alternatively, take the curl of (18.106a) to give the linear barotropic vorticity equation,

$$\frac{\partial \zeta_0}{\partial t} + \beta v_0 = 0. \tag{18.107}$$

This equation is closed because, since the flow is divergence-free, there exists a streamfunction ψ such that $(u_0, v_0) = (-\partial\psi/\partial y, \partial\psi/\partial x)$ and $\zeta_0 = \nabla^2\psi$.

If we add nonlinearity back in then the baroclinic and barotropic modes interact. Thus, the terms $\boldsymbol{u}_1\cdot\nabla\boldsymbol{u}_1$ and $\boldsymbol{u}_0\cdot\nabla\boldsymbol{u}_0$ both project onto the barotropic flow, and the terms $\boldsymbol{u}_1\cdot\nabla\boldsymbol{u}_0$ and $\boldsymbol{u}_0\cdot\nabla\boldsymbol{u}_1$ project onto the baroclinic flow, in a way that is analogous to that occurring in two-layer quasi-geostrophic flow (Section 12.2.2). A baroclinic–baroclinic interaction also arises because of vertical advection and we obtain

$$\frac{\partial \boldsymbol{u}_0}{\partial t} + \boldsymbol{u}_0\cdot\nabla\boldsymbol{u}_0 + D_0(\boldsymbol{u}_1, \boldsymbol{u}_1) + \boldsymbol{f}\times\boldsymbol{u}_0 = -\nabla\phi_0, \tag{18.108a}$$

$$\frac{\partial \boldsymbol{u}_1}{\partial t} + \boldsymbol{u}_1\cdot\nabla\boldsymbol{u}_0 + \boldsymbol{u}_0\cdot\nabla\boldsymbol{u}_1 + D_1(\boldsymbol{u}_1, \boldsymbol{u}_1) + \boldsymbol{f}\times\boldsymbol{u}_1 = -\nabla\phi_1, \tag{18.108b}$$

where D_0 and D_1 are nonlinear (largely advective) operators whose exact form is not of concern here. The barotropic flow is divergence free and this determines the pressure ϕ_0, but ϕ_1 is as yet undetermined.

Thermodynamics

To close the equation for baroclinic flow we use an equation for the entropy or for the equivalent potential temperature in conjunction with (18.100). Since entropy is assumed invariant with height there is no vertical dependence and the equation might be written in the general form,

$$\frac{\partial \eta}{\partial t} + \boldsymbol{u}\cdot\nabla\eta = S, \tag{18.109}$$

where S are various source and sink terms of both heat and moisture. This term hides a multitude of sins, for it potentially includes radiative, condensational, evaporative and turbulent flux terms with complex flow-dependent specifications, possibly needing separate equations for dry air and moisture. However, since entropy is (it is assumed) constant with height, we need a thermodynamic equation only at one level, which we might take to be near the surface, or we may use a vertical average.

In summary, the constraints on the vertical temperature structure imposed by convection lead to a relatively simple vertical structure of all the dynamical fields, and as a consequence a heat source will primarily force the first baroclinic mode. In the linear approximation the baroclinic flow is uncoupled from the barotropic flow. The procedure provides some justification for using shallow water-like equations in tropical dynamics, as in Chapter 8, with the velocity fields to be interpreted as being those of the first baroclinic mode, an approach that is especially useful if the dynamics are predominantly linear. More generally, it is perhaps useful to think of this type of quasi-equilibrium model in a similar light as quasi-geostrophy: neither is quantitatively accurate for flow evolution (we would use neither model for an accurate weather forecast) but they may be able to provide insight where the full equations are too complex.

The equations derived above are still unbalanced, and are simpler than the primitive equations only in their constrained vertical structure. Let us try to go a little further and see if we can incorporate any balances in the horizontal that might simplify matters.

18.8 SCALING AND BALANCED DYNAMICS FOR LARGE-SCALE FLOW IN THE TROPICS

In earlier chapters we looked at flow with small Rossby number and, by performing a scale analysis, determined what the dominant balance of terms was in the equations of motion. This allowed us to derive the quasi-geostrophic equations, which are the basis for much of the theory of mid-latitude motion. Is such a program possible for the tropics? The answer is, 'well, in part'. It *is* possible to derive some reduced sets of equations for the tropics, and that will be the main topic of this section. However, these reduced sets have not proven nearly as useful for the tropics as quasi-geostrophy has been for the mid-latitudes, in part because it is harder to make relevant equations that are simple, or simple equations that are relevant. In any case, let us begin with the stratified primitive equations, without explicitly invoking a constrained vertical structure.

18.8.1 Balanced, Adiabatic Flow

We will present a scaling for tropical flow side-by-side with the corresponding scaling for mid-latitude flow.[14] We begin with the hydrostatic primitive equations for adiabatic, frictionless flow, which, reprising (5.15b), may be written as,

$$\frac{\mathrm{D}\boldsymbol{u}}{\mathrm{D}t} + \boldsymbol{f} \times \boldsymbol{u} = -\nabla_z \phi, \qquad \frac{\partial \phi}{\partial z} = b, \tag{18.110a,b}$$

$$\frac{\mathrm{D}b}{\mathrm{D}t} + N^2 w = 0, \qquad \nabla \cdot (\tilde{\rho} \boldsymbol{v}) = 0. \tag{18.110c,d}$$

These are nominally the anelastic equations in height coordinates in our standard notation, but an entirely equivalent derivation could use pressure coordinates as these have similar form. The reader will recall that $b = g\,\delta\theta/\theta_0$ is the buoyancy and $N^2 = \mathrm{d}\tilde{b}/\mathrm{d}z$ where $\tilde{b}(z)$ is a reference stratification. We will suppose that the basic variables scale according to

$$(x, y) \sim L, \quad z \sim H, \quad (u, v) \sim U, \quad w \sim W, \quad t \sim \frac{L}{U}, \quad \phi \sim \Phi, \quad b \sim B, \quad f \sim f_0. \tag{18.111}$$

The quantity B is representative of horizontal variations in buoyancy. Vertical variations scale differently, hence their separate representation in (18.110c). By choosing the time t to scale advectively we are implicitly eliminating gravity waves. The nondimensional numbers that will arise are the Rossby, Burger and Richardson numbers,

$$Ro = \frac{U}{f_0 L}, \quad Bu = \left(\frac{L_d}{L}\right)^2 = \left(\frac{NH}{f_0 L}\right)^2, \quad Ri = \left(\frac{NH}{U}\right)^2, \tag{18.112}$$

and evidently

$$Bu = Ri \times Ro^2. \tag{18.113}$$

The Rossby number Ro is generally small in mid-latitudes for large-scale flow, but in the tropics it is $\mathcal{O}(1)$ or larger. The Richardson number (which is the inverse square of a Froude number) is usually large in both mid-latitudes and tropics, *except* in regions of active convection where N is very small. If we take $N = 10^{-2}\,\mathrm{s}^{-1}$, $H = 10^4\,\mathrm{m}$ and $U = 10\,\mathrm{m\,s}^{-1}$ then $Ri = 100$. In fact, for large-scale flow the Richardson number is usually sufficiently large that $1/(Ro\;Ri)$ is small in both mid-latitudes and tropics.

The difference between the tropics and mid-latitudes is apparent from the dominant balance in the momentum equation. At small Rossby number we have the familiar geostrophic balance and with hydrostatic balance we obtain the scaling:

$$f \times \boldsymbol{u} \approx -\nabla_z \phi, \quad \frac{\partial \phi}{\partial z} = b, \quad \Longrightarrow \quad \Phi = f_0 UL, \quad B = \frac{f_0 UL}{H}. \tag{18.114}$$

In the tropics the advective term, or the advectively-scaled time derivative, balances the pressure gradient meaning that $\mathrm{D}\boldsymbol{u}/\mathrm{D}t \sim \nabla_z \phi$ and, since we still have $\partial\phi/\partial z = b$ we find

$$\Phi = U^2, \qquad B = \frac{U^2}{H}. \tag{18.115}$$

If U is of similar magnitude in the tropics and mid-latitudes (and in the absence of a dynamical analysis this is an assumption), then *variations of pressure and temperature are smaller in the tropics than in mid-latitudes.* This is an important and not-quite obvious result and it is the essence of the *weak temperature gradient approximation*, discussed further in the next section. If we were to carry through the derivation in pressure co-ordinates (see the shaded box on page 81) with geopotential and temperature as the variables we would find

$$\text{Mid-latitudes:} \quad \Phi = f_0 UL, \quad T_s = \frac{f_0 UL}{R}, \tag{18.116a}$$

$$\text{Tropics:} \quad \Phi = U^2, \qquad T_s = \frac{U^2}{R}, \tag{18.116b}$$

where Φ is now the scaling for geopotential, T_s is the scaling for temperature and R is the ideal gas constant. For $U = 10\,\mathrm{m\,s}^{-1}$, $L = 10^6\,\mathrm{m}$ and, for mid-latitudes only, $f_0 = 10^{-4}\,\mathrm{s}^{-1}$, we obtain

$$\text{Mid-latitudes:} \quad \Phi \sim 1000\,\mathrm{m}^2\,\mathrm{s}^{-2}, \quad T_s \sim 3\,\mathrm{K}, \tag{18.117a}$$

$$\text{Tropics:} \quad \Phi \sim 100\,\mathrm{m}^2\,\mathrm{s}^{-2}, \quad T_s \sim 0.3\,\mathrm{K}. \tag{18.117b}$$

The implications of these results are illustrated in Fig. 18.22, which shows a snapshot of observed contours of geopotential height, temperature and zonal wind; the large-scale variability of geopotential and temperature is evidently much smaller in the tropics than mid-latitudes.

Vertical velocity

The obvious scaling for the vertical velocity is that suggested by the mass continuity equation, namely $W = UH/L$. However, as we know from Chapter 5, the vertical velocity may be much less than this estimate in a stratified, rotating fluid, and here we start with the thermodynamic equation for adiabatic flow,

$$\boldsymbol{u} \cdot \nabla b + wN^2 = 0, \qquad \Longrightarrow \qquad W = \frac{UB}{LN^2}. \tag{18.118}$$

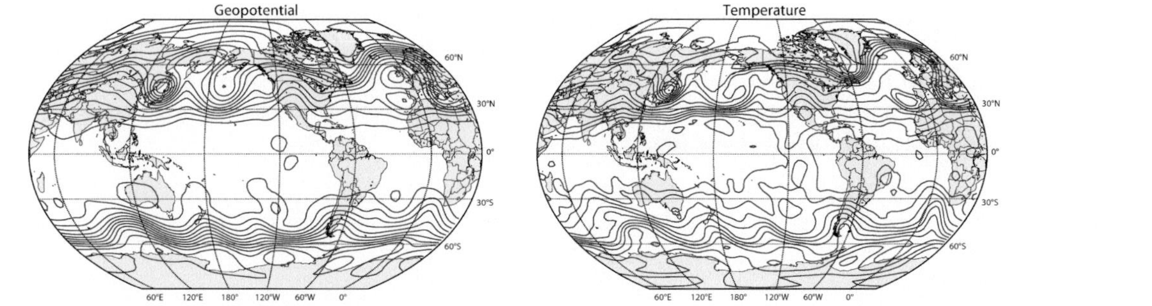

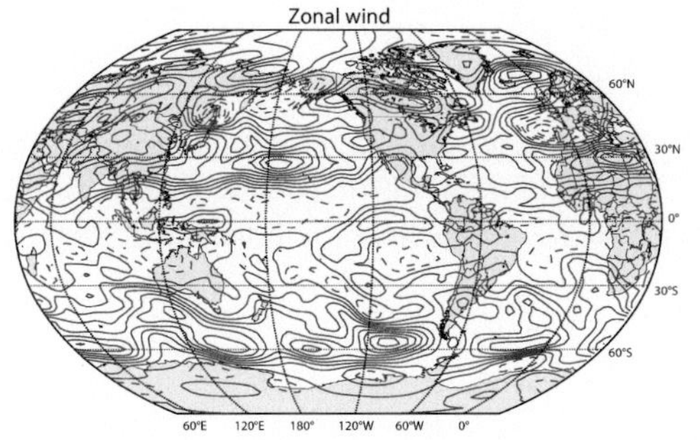

Fig. 18.22 Geopotential height, temperature and zonal wind at 500 hPa on 9 February 2015. The contour interval is uniform in each plot, with same number of contours for each field, and negative values for u are dashed. Noticeable is the lack of variability of the geopotential and temperature fields in the tropics.

Using (18.114) and (18.115) gives

$$\text{Mid-latitudes:} \qquad W = \frac{f_0 U^2}{HN^2} = \frac{f_0 L}{U}\frac{U^2}{N^2H^2}\frac{UH}{L} = (Ro\ Ri)^{-1}\left(\frac{UH}{L}\right), \tag{18.119a}$$

$$\text{Tropics:} \qquad W = \frac{U^3}{LHN^2} = \frac{U^2}{N^2H^2}\frac{UH}{L} = Ri^{-1}\left(\frac{UH}{L}\right), \tag{18.119b}$$

where $Ri \equiv N^2H^2/U^2$ is the Richardson number, with typical values of $\mathcal{O}(10$–$\mathcal{O}(100$ for large-scale flow. Thus, again perhaps non-intuitively, the vertical velocity is, for adiabatic flow, *smaller* in the tropics than in mid-latitudes, by order of a mid-latitude Rossby number. In mid-latitudes the scaling for W is more commonly written as

$$W = \frac{f_0 U^2}{HN^2} = Ro\,\frac{L^2}{L_d^2}\frac{UH}{L} = \frac{Ro}{Bu}\frac{UH}{L}, \tag{18.120}$$

so that for scales comparable to the mid-latitude deformation radius (i.e., with $Bu \sim 1$) the vertical velocity is order Rossby-number smaller than mass-continuity scaling suggests.

Vorticity

Cross-differentiating the horizontal momentum equation gives, as in (4.66) but without a baroclinic term, the vertical component of the vorticity equation with associated scalings,

$$\frac{\mathrm{D}}{\mathrm{D}t}(\zeta + f) = -(\zeta + f)\left(\frac{\partial u}{\partial x} + \frac{\partial v}{\partial y}\right) + \left(\frac{\partial u}{\partial z}\frac{\partial w}{\partial y} - \frac{\partial v}{\partial z}\frac{\partial w}{\partial x}\right), \tag{18.121a}$$

$$\text{Tropics:} \qquad \left(\frac{U}{L}\right)^2 \sim \left(\frac{U}{L} + \frac{1}{Ro}\frac{U}{L}\right)\left(\frac{1}{Ri}\frac{U}{L}\right) \quad \left(\frac{U}{H}\right)\left(\frac{1}{Ri}\frac{UH}{L^2}\right), \tag{18.121b}$$

$$\text{Mid-latitudes:} \qquad \left(\frac{U}{L}\right)^2 \sim \left(\frac{U}{L} + \frac{1}{Ro}\frac{U}{L}\right)\left(\frac{Ro}{Bu}\frac{U}{L}\right) \quad \left(\frac{U}{H}\right)\left(\frac{Ro}{Bu}\frac{UH}{L^2}\right). \tag{18.121c}$$

In the tropical case with $Ro = \mathcal{O}(1)$ or larger *all* the terms on the right-hand side are much smaller than the terms on the left-hand side, whereas in the mid-latitude case vortex stretching by the Coriolis term, $f(\partial u/\partial x + \partial v/\partial y)$, is the same order (because f is big and the divergence is small). Thus, in the tropical case the vorticity equation simplifies severely and at lowest order becomes the two-dimensional vorticity equation,

$$\frac{\mathrm{D}}{\mathrm{D}t}(\zeta + f) = 0. \tag{18.122}$$

The large-scale velocity is, at this order, purely rotational and is given by a streamfunction ψ such that

$$\nabla^2\psi = \zeta, \qquad (u, v) = \left(-\frac{\partial\psi}{\partial u}, \frac{\partial\psi}{\partial x}\right). \tag{18.123}$$

This equation nominally holds independently at each vertical level, although the assumptions that give rise to it assume that the depth scale is large. The pressure is not needed to step forward (18.123) but it may be obtained diagnostically. To do so we go back to the horizontal velocity equation written in the vector-invariant form,

$$\frac{\partial \boldsymbol{u}}{\partial t} - \boldsymbol{u} \times (\zeta + \boldsymbol{f}) = -\nabla\left(\phi + \frac{1}{2}\boldsymbol{u}^2\right), \tag{18.124}$$

neglecting the vertical and divergent velocities. Taking the divergence gives the so-called nonlinear balance or gradient-wind equation,

$$\nabla^2\phi = \nabla \cdot \left[(f + \zeta)\nabla\psi - \frac{1}{2}\nabla(\nabla\psi)^2\right], \tag{18.125}$$

which is similar to (2.238) and expresses a balance between the pressure gradient, Coriolis and centrifugal forces.

18.8.2 A few Remarks

The above derivations and results require some comment:

- The relative weakness of large-scale horizontal gradients of pressure (or geopotential) and temperature in the tropics, for a given velocity field, is a robust result of the scaling analysis and borne out in observations.
- The smallness of the vertical velocity requires that the Richardson number, N^2H^2/U^2 be large; that is, stratification is strong. This is only true in regions that are not actively convecting; in convective regions N may be small.
- Relatedly, the scaling does not take into account diabatic sources, which may be expected to be particularly important in tropical regions.
- Equally tellingly, (18.123) tells us nothing about the vertical structure and so, unlike quasi-geostrophy in mid-latitudes, is not sufficiently complete to be a useful prognostic, or even diagnostic, equation for tropical motion.

Let us now look at how diabatic effects might affect large-scale motion.

18.9 † SCALING AND BALANCE FOR LARGE-SCALE FLOW WITH DIABATIC SOURCES

In the tropics we might expect that heat sources, for example condensational heating, would be important to the extent that they should be explicitly included in any development of reduced equations. Let us see if and how this is possible, using the shallow water equations for illustration.

Using the standard nonlinear shallow water equations for the tropical atmosphere is rather ad hoc (and not justified by quasi-equilibrium except in the linear case) but our intention here is primarily illustrative. The underlying physical assumption is that in an air column adiabatic cooling associated with vertical motion balances diabatic heating, which in the shallow water equations becomes a balance between the heating and the divergence.

18.9.1 Diabatic Balanced Shallow Water Equations

On an f-plane and in conventional notation the equations may be written in vorticity-divergence form as

$$\frac{\partial h}{\partial t} + \nabla\cdot(\boldsymbol{u}h) = Q, \tag{18.126a}$$

$$\frac{\partial \zeta}{\partial t} + \nabla\cdot\left[\boldsymbol{u}(\zeta + f_0)\right] = -r\zeta, \tag{18.126b}$$

$$\frac{\partial \delta}{\partial t} + \nabla^2\left(\frac{1}{2}\boldsymbol{u}^2 + gh\right) - \mathbf{k}\cdot\nabla\times\left[\boldsymbol{u}(\zeta + f_0)\right] = -r\delta, \tag{18.126c}$$

where $\delta = \partial u/\partial x + \partial v/\partial y$ is the divergence, Q is the mass or heating source, r is a frictional coefficient and other notation is standard. The height field is a proxy for both pressure and temperature and we will assume that horizontal gradients are weak, by which we mean that the dominant balance in (18.126a) is characterized by the scaling

$$H\Delta = Q_0, \tag{18.127}$$

where H is the mean thickness of the layer, Δ is a scaling for the divergence and Q_0 is the magnitude of the heating. We then choose the velocity scale, U, and the vorticity scale, $\mathcal{Z}$, to be

$$U = \frac{Q_0 L}{H}, \qquad \mathcal{Z} = \Delta = \frac{Q_0}{H}. \tag{18.128}$$

Finally, the magnitude of horizontal deviations in the height field, $\mathcal{H}$, are determined from the divergence equation. Depending on whether rotation is or is not important we deduce

$$\mathcal{H} = \frac{f_0 UL}{g} = \frac{Q_0 f_0 L^2}{gH}, \quad \text{or} \quad \mathcal{H} = \frac{U^2}{g} = \frac{Q_0^2 L^2}{gH^2}. \tag{18.129a,b}$$

The height field may then be separated into a mean and deviation, $h = H + \eta$, where $\eta = \mathcal{H}\hat{\eta}$. These scalings involve the heating in an essential way and so are fundamentally different from the adiabatic scaling of the previous section.

Using the above scalings, with rotation, (18.126) may be written in nondimensional form as,

$$Bu^{-1}\left[\frac{1}{f_0 T}\frac{\partial \hat{\eta}}{\partial \hat{t}} + Ro\,\nabla\cdot(\hat{\boldsymbol{u}}\hat{\eta})\right] + \hat{\delta} = \widehat{Q}, \tag{18.130a}$$

$$\frac{1}{f_0 T}\frac{\partial \hat{\zeta}}{\partial \hat{t}} + \nabla\cdot\left[\boldsymbol{u}(\hat{\zeta} + \hat{f}_0)\right] = -\frac{r}{f_0}\hat{\zeta}, \tag{18.130b}$$

$$\frac{1}{f_0 T}\frac{\partial \hat{\delta}}{\partial \hat{t}} + \nabla^2\left(\frac{1}{2}\hat{\boldsymbol{u}}^2 + \hat{\eta}\right) - \mathbf{k}\cdot\nabla\times\left[\hat{\boldsymbol{u}}(\hat{\zeta} + \hat{f}_0)\right] = -\frac{r}{\hat{f}_0}\hat{\delta}, \tag{18.130c}$$

where T is the scaling for time, $Bu = (L_d/L)^2$ with $L_d = \sqrt{gH}/f_0$, and $\hat{f}_0 = 1$. The Rossby number is given by $Ro = Q_0/f_0 H$ and is not necessarily small. In the non-rotating case a similar set of equations can be derived but with different coefficients.

Reduced equations

Let us suppose that the mass source determines the divergence in (18.130a). The condition for this is that

$$\max\left(\frac{1}{f_0 T}, Ro\right) Bu^{-1} \ll 1, \tag{18.131}$$

which means that the scale of motion cannot be too large and the time scale cannot be too short. If (18.131) is satisfied then the dimensional height equation, (18.126a), becomes

$$\nabla \cdot \boldsymbol{u} = \frac{Q}{H}. \tag{18.132a}$$

This value of the divergence is used in (18.126b) and (18.126c) which, retaining all terms since none are obviously small, become

$$\frac{\partial \zeta}{\partial t} + \boldsymbol{u} \cdot \nabla(\zeta + f_0) + (\zeta + f)\frac{Q}{H} = -r\zeta, \tag{18.132b}$$

$$g\nabla^2 h = \mathbf{k} \cdot \nabla \times [\boldsymbol{u}(\zeta + f_0)] - \frac{1}{H}\frac{\partial Q}{\partial t} - r\delta - \nabla^2 \frac{\boldsymbol{u}^2}{2}. \tag{18.132c}$$

The equation set (18.132) has but one prognostic equation, namely (18.132b), and so is truly balanced and may be thought of as a generalization of (18.122) and (18.125) to the case with non-zero heating. The divergence equation is a nonlinear balance equation, similar to (18.125), except now with a diabatic term on the right-hand side. The divergent flow itself is computed using the height equation, by an assumed balance between adiabatic cooling and diabatic heating. The relationship between velocity and geopotential (or pressure) is the same as in the adiabatic case, because this arises through the momentum equation. Thus, even in the presence of a heating, gradients of geopotential and temperature remain relatively weak, a result that ultimately arises from the smallness of the Coriolis parameter. The arguments that lead to (18.132), along with the scaling of Section 18.8, provide us with the *weak temperature-gradient approximation*.[15] The importance of the result lies in what it implies about the response of the atmosphere to a localized heating: without making any linear approximation, the equations provide a scaling for the response of the velocity, and suggest that the response may become spread out over a sufficient area to keep the temperature gradients small.

Gravity wave adjustment and weak temperature gradients

Although convection certainly constrains the lapse rate where it is occuring, it does not directly do so elsewhere. Nevertheless, the observed tropical atmosphere has an *average* lapse rate close to the saturated adiabat. Why should this be so? The reason is that a process akin to geostrophic adjustment (Section 3.9) brings the atmosphere close to a state with weak horizontal temperature gradients. Suppose that a particular column undergoes moist convection and, perhaps in a matter of hours, adjusts to a moist neutral profile. Away from the cloud the buoyancy profile will in general be different from that, and so there will be unbalanced horizontal pressure (and hence temperature) gradients. Gravity waves, initiated by the motion surrounding the convection ('compensating subsidence'), will spread from the cloud and will adjust the environmental buoyancy to the same profile as that of the convecting region. The timescale for the adjustment is determined by the time that gravity waves take to propagate horizontally between convecting regions. Internal gravity wave speeds are typically about 10 m s^{-1} or somewhat faster and so will spread a distance 50 km from a cloud in a couple of hours, and the adjustment time on this scale will be of this order. This buoyancy adjustment time is much less than the time it would take a passive tracer to homogenize between clouds by advective or mixing processes. The actual process of gravity wave initiation and subsequent adjustment is complex and beyond our scope, and the reader is referred to the literature.[16]

18.9.2 ♦ Weak Temperature Gradient and Stratified Flow

The weak temparature gradient approximation is formally independent of any quasi-equilibrium argument, so in this section we briefly discuss its application to the stratified, three-dimensional equation of motion. The thermodynamic equation in pressure coordinates (see Equation (P.5) on page 81) may be written,

$$\frac{\partial T}{\partial t} + \boldsymbol{u} \cdot \nabla T + \omega \frac{\partial s}{\partial p} = Q, \tag{18.133}$$

where $s = T + gz/c_p$ is the dry static energy divided by c_p and Q represents heating terms. In the weak temperature gradient approximation this equation becomes

$$\omega \frac{\partial s}{\partial p} = Q. \tag{18.134}$$

If the stratification s is known (e.g., moist neutral) the above equation becomes a diagnostic for the vertical velocity, namely $\omega(p) = Q/\partial_p s$. Even with this approximation the remaining stratified equations (e.g., momentum equation) are somewhat complex, with both nonlinearity and continuous vertical structure.

If we are willing in addition to make the quasi-equilibrium assumption then further simplifications are possible because all the fields are assumed to have a simple vertical structure and the flow may be described, as before, by a system similar to the shallow water equations. The vertical and horizontal velocities are related by the mass continuity equation, (18.101). The barotropic flow has zero horizontal divergence so that the vertical (pressure) velocity and the horizontal are related by

$$\omega(x, y, p, t) = -\int_{p_s}^{p} \nabla \cdot \boldsymbol{u} \, \mathrm{d}p = -n(p) \nabla \cdot \boldsymbol{u}_1, \tag{18.135}$$

where $n(p) = \int m(p) \, \mathrm{d}p$ and we take $\omega = 0$ at the bottom boundary. Thus, we may write $\omega = n(p)\omega_1(x, y, t)$ and $\omega_1 = -\nabla \cdot \boldsymbol{u}_1$. If we divide (18.134) and vertically integrate we obtain

$$\overline{s} \nabla \cdot \boldsymbol{u}_1 = \overline{Q}, \tag{18.136}$$

where $\overline{Q}$ is a vertically integrated heating term, weighted by $1/n(p)$. This equation is the direct analogue of (18.132a) and it provides a predictive equation for the divergence of the baroclinic flow if the heating and stratification are known. Moisture may be added to the mix, in which case a moist static energy appears, and the effects of evaporation and condensation must be included in Q. The precise form of (18.136) and other particulars of implementation will depend on the basis functions chosen for the vertical structure, but the form is generically $\nabla \cdot \boldsymbol{u}_1 = [Q]/[s]$, where $[Q]$ is a measure of heating and $[s]$ a measure of stratification.[17] Equation (18.136) may be used to close the momentum equation (18.108b). If we take the curl of that equation we obtain a vorticity equation analogous to (18.132b), and if we take its divergence we obtain a diagnostic equation for $\nabla^2 \phi_1$ analogous to (18.132c). The combination of a weak temperature gradient approximation and shallow-water-like equations arising from a constrained vertical stratification may be as close as the tropical atmosphere allows us to get to tractable and understandable equations of motion.

18.10 † CONVECTIVELY COUPLED GRAVITY WAVES AND THE MJO

We now look at some of the manifestations of the theoretical development of this chapter and Chapter 8, and three phenonema suggest themselves, namely the Walker circulation, monsoons and the Madden–Julian Oscillations (MJO). We leave monsoons for another day, in part because the subject is large and its theoretical development is a moving target. And we defer discussion of the Walker circulation — an atmospheric overturning circulation largely in the zonal (i.e., y–z)

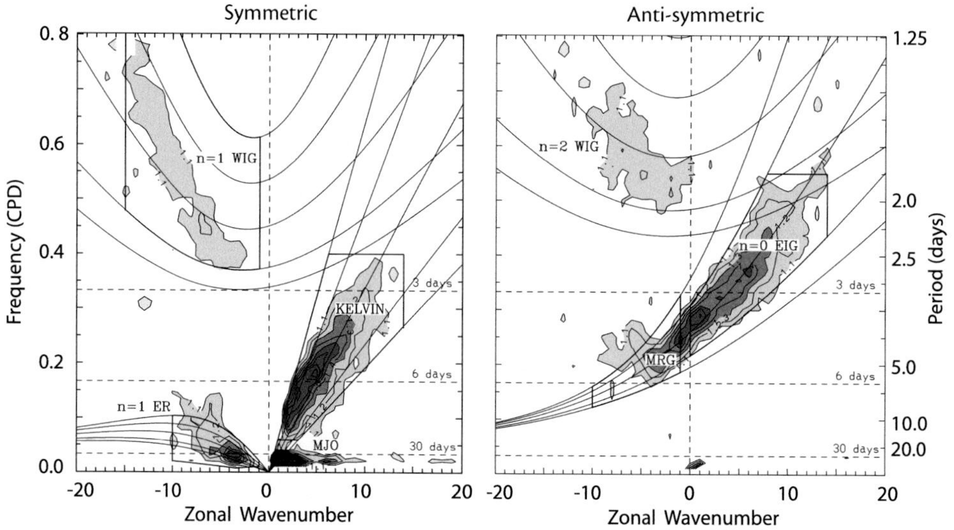

Fig. 18.23 A power spectrum of cloud brightness (colouring) from 15° S to 15° N, measured from satellite and Fourier-transformed into frequency-wavenumber space. Left panel is the symmetric spectrum (Northern plus Southern Hemispheres) and right panel the anti-symmetric spectrum. The solid lines are the corresponding dispersion relations with multiple equivalent depths ranging from 8 m to 90 m, with the best fit at about 25 m. Compare with Fig. 8.6 and Fig. 8.7.[20]

plane — to Chapter 22 because the phenomenon is closely tied to the equatorial ocean. So let us briefly discuss the MJO, although we cannot give it a wholly crisp explanation because at the time of writing none exists.[18]

What is the MJO? It is a pattern of precipitation and winds that takes shape across the western tropical Indian Ocean and drifts eastward at about 4–8 m s^{-1} into the western Pacific before dying out over the cooler waters of the eastern tropical Pacific. It often recurs roughly every 30–60 days or so, although it does not oscillate like a conventional wave. Rather, it is more like a somewhat coherent, drifting pattern several thousand kilometres across, consisting of a wet, rainy region of ascending air flanked by dryer regions on either side.

18.10.1 The Observations

The MJO has a characteristic spectral signature that can be obtained from satellite measurements of cloud brightness. Figure 18.23 shows the symmetric (Northern Hemisphere plus Southern Hemisphere) and antisymmetric power spectrum (that is, the intensity of the field, Fourier analysed in zonal wavenumber and frequency) of tropical satellite brightness temperature, after filtering away some background noisiness.[19] Power spectra of other fields, including velocity fields from re-analysis, show similar features. The main shaded regions in the figure fall nicely on the theoretical dispersion relations for Rossby, Kelvin and mixed Rossby-gravity waves as derived in Section 8.2. This is perhaps somewhat startling to see, but evidently equatorial waves do exist!

The theoretical curves match the observations best if the unadorned gravity wave speed, c, is in the region of 10 to 20 m s^{-1}. This speed is given by $c = \sqrt{gH_e}$ where H_e is the equivalent depth (Section 3.4.2), so theoretical values match the observations with $H_e \approx$ 10–50 m, perhaps with the best match around 25 m, somewhat smaller than the first equivalent depth computed for the atmosphere in Section 3.4.[21] Part of the difference may come from the fact that the tropopause is not a rigid lid, and part from the fact that the presence of moisture may reduce the effective static stability of the atmosphere below that implied by the value of N^2 computed using dry potential temperature $(g/\theta \partial\theta/\partial z)$. In a saturated atmosphere the (smaller) value of $g/\theta_e \partial\theta_e/\partial z$ may be

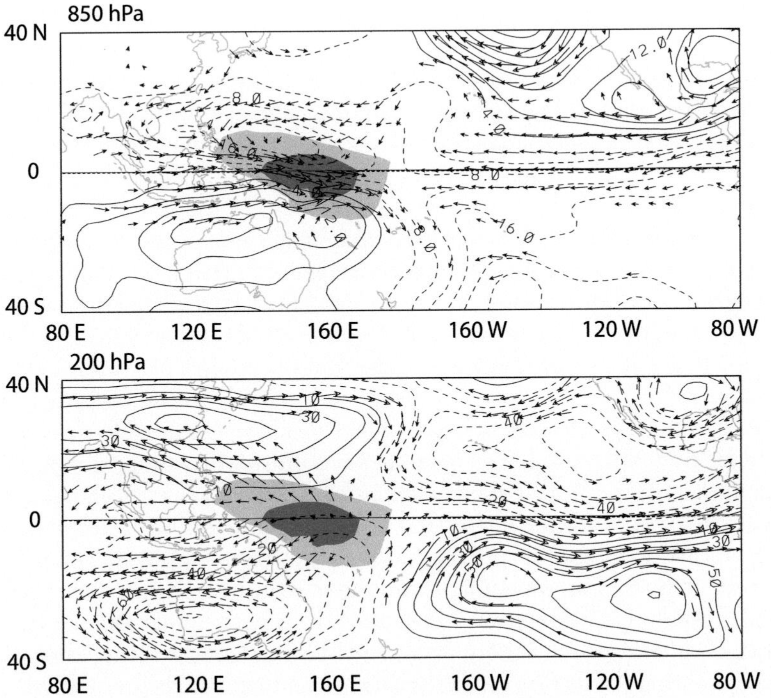

Fig. 18.24 Composite of an observed pattern during an MJO period. Shading indicates anomalously low outgoing IR radiation (less than 16 and 32 W m^{-2} below normal) and indicates the presence of high clouds and precipitation. The solid lines are streamfunctions and the arrows indicate velocities.[22]

more relevant for the generation and propagation of gravity waves. The Rossby wave speed is given by $c_R \approx -\beta k/[(2m+1)\beta/c + k^2]$, so it too depends, albeit more weakly, on c.

In addition to the relatively well understood gravity and Rossby waves, there appears to be a prominent spectral signature in Fig. 18.23 at a small positive zonal wavenumber (between 1 and 5 in the left-hand panel) and a timescale of about 40 days, and this is a signature of the MJO. If we look in physical space, we get the sense that the MJO resembles a drifting Matsuno–Gill pattern more than an oscillating wave — compare Fig. 18.24 with Fig. 8.11 on page 325. Consistently, the observed vertical structure largely has a first baroclinic mode structure, as expected if the lapse rate is constrained to be nearly constant by convection — note the oppositely directed velocity fields in Fig. 18.24, just as in Fig. 8.12. The shaded region corresponds to a region of heating in the Matsuno–Gill model and without too much imagination one may see a Kelvin wave response to its east, along the equator, and off-equatorial Rossby lobes to the west over Australia and South East Asia. A defining characteristic of the MJO is that this pattern drifts eastward at a speed of a few metres per second, much lower than the Kelvin wave speed at that equivalent depth.

18.10.2 † The Mechanism

Precisely why this pattern should move eastward remains unclear, and there is little certainty as to the underlying mechanism except in so far as it likely has to do with the interaction of moist convection with the large-scale circulation — perhaps moist convection *producing* a large-scale circulation that then modulates the convection. Thus, a locally warm region near the equator (initiated by a sea-surface temperature anomaly, for example) will produce a pattern that evolves towards that of the Matsuno–Gill solution (Fig. 8.11), with a low-level convergence of moisture and condensation amplifying the initial pattern. Furthermore, the convergence will produce a dryer region on either side of the heat source (as observed). However, the timescales of condensation are short and the process is unsteady, and it is unlikely that the condensation can produce a self-sustained stationary pattern — in the initial value problem it takes several days for a heating anomaly to settle into a steady Matsuno–Gill pattern in the tropical atmosphere.[23]

Thus, instead of a truly steady pattern, the convection will initiate an eastward moving Kelvin wave, along with slower westward moving Rossby waves trapped nearer the source. Now, the initial moisture convergence arises because the Kelvin wave draws air in from the west, and the Rossby wave similarly draws air from the east. If the Kelvin waves moves east the region of low-level moisture convergence, and thus the region of condensational heating, will also move eastward, and the process begins again. However, the convergence region will move eastward much more slowly than the Kelvin wave, since it is the advective convergence that provides the heat source that is the source of both the Kelvin and Rossby waves. A Kelvin wave that breaks free of the convergence will eventually decay since it will not produce the moisture convergence to feed itself. Rather, the MJO is a self-sustained interaction between moisture convergence and the forced-dissipative Kelvin and Rossby waves it produces, with the requirement for moisture convergence greatly slowing the Kelvin wave. Radiative effects tied to the variable cloud field also almost certainly play a role in the way the pattern is damped.

Although wavelike, in this picture the MJO is not a conventional linear wave with a dispersion relation; rather it is related to the translation of a forced quasi-steady pattern with a longer timescale than either of those waves themselves, with the forcing *maintained by the pattern it itself produces*. Evidently, this will be a delicate balance for the interaction of moist convection with Kelvin and Rossby waves is a complex process, so it is not surprising that even the most sophisticated numerical models have trouble reproducing the phenomenon properly, even if the underlying mechanism can seemingly be described in relatively simple terms. Note also that it is not helpful to think of the convection as being in quasi-equilibrium with a more slowly and independently evolving external environment — the convection helps produce its own external environment and the two evolve in synchrony.

APPENDIX A: MOIST THERMODYNAMICS FROM THE GIBBS FUNCTION

Many of the thermodynamic quantities of interest for moist air can be obtained from knowledge of the Gibbs function, in a manner analogous to that followed for dry air in the appendix on page 47. The benefits are that the approach is systematic, explicit formulae for thermodynamic variables can be given, and approximations can be made consistently as needed. Here we outline the methodology and provide some examples.[24] We use superscripts d, v and l to denote dry air, water vapour, and liquid water, respectively, and a subscript 0 always denotes a constant, for any quantity. Quantities with multiple superscripts are mixtures of the quantities, but the superscripts are occasionally omitted. A subscript s denotes the saturated value.

Given (1.217), the specific Gibbs functions for dry air and water vapour are, respectively,

$$g^d(p,T) = R^d T \ln(p^d/p_0) + c_p^d T[1 - \ln(T/T_0)], \tag{18.137a}$$

$$g^v(p,T) = R^v T \ln(p^v/p_0) + c_p^v T[1 - \ln(T/T_0)] - L_0^v \frac{T}{T_0} + L_0^v, \tag{18.137b}$$

where p^d and p^v are the partial pressures of dry air and water vapour. A constant, L_0^v, appears in the expression for the Gibbs function of water vapour because we will be concerned with water in its liquid phase. We take the analogous constant to be zero for dry air.

The two fluids are at the same temperature, T, and the total pressure, p, is given by the sum of the partial pressures. Let q be the mass fraction of moist air in the mixture (i.e., the specific humidity), and let ϵ be the ratio of the molecular weights of water vapour and dry air, μ^v/μ^d. The two partial pressures are given by

$$p^v = \frac{qp}{q + \epsilon(1-q)}, \qquad p^d = \frac{\epsilon(1-q)p}{q + \epsilon(1-q)}, \tag{18.138}$$

satisfying $p = p^d + p^v$, and p^v is often denoted e in the literature. The Gibbs function for the mixture is

$$g^{dv}(p,T,q) = (1-q)g^d + qg^v, \tag{18.139}$$

whence

$$\begin{aligned} g^{dv}(p,T,q) = {} & \left(qc_p^v + (1-q)c_p^d\right)T[1-\ln(T/T_0)] \\ & + T\left[qR^v \ln\left(\frac{qp/p_0}{q+\epsilon(1-q)}\right) + (1-q)R^d \ln\left(\frac{\epsilon(1-q)p/p_0}{q+\epsilon(1-q)}\right)\right] \\ & + qL_0^v\left(1-\frac{T}{T_0}\right). \end{aligned} \tag{18.140}$$

This expression is symmetric in dry air and water vapour, and valid for any mass fraction of moisture in the air (given that $q \ln q \to 0$ as $q \to 0$). We will also need the Gibbs function for liquid water, which for our purposes may be written

$$g^l(p,T) = c^l T[1-\ln(T/T_0)] - \eta_0^l T + \alpha^l p + g_0^l. \tag{18.141}$$

This expression is a simplified version of (1.146) with slightly different notation, with c^l being the heat capacity and α^l the inverse density.

Density and the thermal equation of state

The inverse density of the gas mixture is given by

$$\alpha^{dv} = \left(\frac{\partial g}{\partial p}\right)_{q,T} = \frac{T}{p}\left[qR^v + (1-q)R^d\right], \tag{18.142}$$

or equivalently $p = \rho RT$ where $R = qR^v + (1-q)R^d$ is the specific gas 'constant' for the mixture (which of course varies with q).

Entropy

The entropy is given by $\eta = -(\partial g/\partial T)_{p,q}$, giving

$$\begin{aligned} \eta^{dv} = {} & \left(qc_p^v + (1-q)c_p^d\right)\ln(T/T_0) \\ & - \left[qR^v \ln\left(\frac{qp/p_0}{q+\epsilon(1-q)}\right) + (1-q)R^d \ln\left(\frac{\epsilon(1-q)p/p_0}{q+\epsilon(1-q)}\right)\right] + \frac{qL_0^v}{T_0}. \end{aligned} \tag{18.143}$$

The entropy of the mixture may also be written as

$$\eta^{dv} = (1-q)\eta^d + q\eta^v, \tag{18.144}$$

where the specific entropies of dry air and water vapour are

$$\begin{aligned} \eta^d &= c_p^d \ln(T/T_0) - R^d \ln(p/p_0) - R^d \ln(1-p^v/p), \\ \eta^v &= c_p^v \ln(T/T_0) - R^v \ln(p/p_0) - R^v \ln(1-p^d/p) + L_0^v/T_0. \end{aligned} \tag{18.145}$$

The third terms on the right-hand sides are called the entropies of mixing.

The entropy of liquid water is given by

$$\eta^l = \eta_0^l + c^l \ln(T/T_0). \tag{18.146}$$

The entropy of a mixture of dry air, liquid water and water vapour is given by

$$\eta^{dvl} = (1-q^{vl})\eta^d + q^v\eta^v + q^l\eta^l, \tag{18.147}$$

where q^v and q^l are the mass concentrations of vapour and liquid in the total mixture and $q^{vl} = q^v + q^l$.

Heat capacities

The heat capacity at constant pressure of moist air (i.e., dry air and water vapour, with no liquid content) is given by

$$c_p^{dv} \equiv T\left(\frac{\partial\eta}{\partial T}\right)_{p,q} = qc_p^v + (1-q)c_p^d. \tag{18.148}$$

That is, it is the mass weighted heat capacity of water vapour and dry air, as expected. The heat capacity at constant volume is given by

$$c_v^{dv} \equiv T\left(\frac{\partial\eta}{\partial T}\right)_{\alpha,q} = \frac{T(g_{pT}^2 - g_{pp}g_{TT})}{g_{pp}} = qc_v^v + (1-q)c_v^d, \tag{18.149}$$

where $c_v^v = c_p^v - R^v$ and $c_v^d = c_p^d - R^d$, and details are left to the reader. Alternatively, eliminate p in favour of α in (18.143) using (18.142) and then directly evaluate $T(\partial\eta/\partial T)_{\alpha,q}$.

Potential temperature

The potential temperature is the temperature that a parcel has if moved adiabatically and at constant composition to a reference pressure, and so satisfies

$$\eta(p_R, \theta, q) = \eta(p, T, q). \tag{18.150}$$

For moist air we use (18.143) for the entropy and (18.150) becomes

$$\begin{aligned} c_p \ln\theta - \left[qR^v \ln\left(\frac{qp_R/p_0}{q+\epsilon(1-q)}\right) + (1-q)R^d \ln\left(\frac{\epsilon(1-q)p_R/p_0}{q+\epsilon(1-q)}\right)\right] \\ = c_p \ln T - \left[qR^v \ln\left(\frac{qp/p_0}{q+\epsilon(1-q)}\right) + (1-q)R^d \ln\left(\frac{\epsilon(1-q)p/p_0}{q+\epsilon(1-q)}\right)\right]. \end{aligned} \tag{18.151}$$

With a couple of lines of algebra this expression simplifies to

$$c_p \ln\theta - [qR^v + (1-q)R^d]\ln(p_R/p) = c_p \ln T \qquad \text{or} \qquad \theta = T\left(\frac{p_R}{p}\right)^{R/c_p}, \tag{18.152}$$

where $R = qR^v + (1-q)R^d$ and c_p is given by (18.148).

Latent heat of evaporation and condensation

The latent heat of evaporation (or condensation) is the amount of energy that must be supplied to evaporate a unit mass of liquid to a vapour, or equivalently the amount of energy released when water vapour condenses. It is therefore equal to the difference between the specific enthalpy of water vapour and liquid water at a given temperature and also known as the enthalpy of vaporization. Using (18.141) the entropy and enthalpy of liquid water are given by

$$\eta^l = -\left(\frac{\partial g^l}{\partial T}\right)_p = \eta_0^l + c^l \ln(T/T_0), \qquad h^l = g^l + T\eta = g_0^l + \alpha^l p + c^l T. \tag{18.153}$$

Using (18.137b) the enthalpy of water vapour is given by

$$h^v = g - T\left(\frac{\partial g^v}{\partial T}\right)_p = c_p^v T + L_0^v. \tag{18.154}$$

The enthalpy of vaporization, L, is therefore given by

$$L = h^v - h^l = L_0^v - (g_0^l + \alpha^l p) + (c_p^v - c^l)T. \tag{18.155}$$

The pressure term is negligibly small: $\alpha^l p \sim 100\,\mathrm{J\,kg^{-1}}$ whereas, by experiment, $L = 2.501 \times 10^6\,\mathrm{J\,kg^{-1}}$ at $0°\,\mathrm{C}$. Using this value for L_0^v (and setting $g_0^l = 0$), and with $c_p^v = 1859\,\mathrm{J\,kg^{-1}\,K^{-1}}$, $c^l = 4218\,\mathrm{J\,kg^{-1}\,K^{-1}}$, we obtain $L \approx (2.501 \times 10^6 - 2359T)\,\mathrm{J\,kg^{-1}}$, with T in Celsius. The temperature dependence of L arises because we have to expend the same amount of energy evaporating water at (say) $10°\,\mathrm{C}$ and then raising the vapour temperature to $20°\,\mathrm{C}$ as we do in raising the water temperature from $10°$ to $20°$ and then evaporating it.

Using the latent heat provides a convenient expression for the entropy of a saturated mixture of dry air, water vapour and liquid water. If the mixture is in equilibrium then $q^v = q_s^v$ where q_s^v is given by the Clausius–Clapeyron relation and is a function of p and T. Furthermore, the differences in entropy between vapour and liquid water are then related to the latent heat of vaporization by

$$T(\eta^v - \eta^l) = h^v - h^l = L, \tag{18.156}$$

and (18.147) becomes

$$\eta_s^{dvl} = (1 - q^{vl})\eta^d + q^{vl}\eta^l + \frac{Lq_s^v}{T}. \tag{18.157}$$

Equivalent potential temperature

There are various definitions of equivalent potential temperature, θ_{eq}, with different names and small quantitative differences.[7] For qualitative uses the differences don't really matter as they are small, and for an exact calculation one rarely needs to use equivalent potential temperature (entropy is often better). Nevertheless, it is a commonly used thermodynamic variable. The definition used in the main text was that equivalent potential temperature, θ_{eq}, is the potential temperature a parcel reaches after it is lifted adiabatically and at constant composition to a level at which it becomes saturated, and then all the water vapour is condensed and all of the latent heat released is used to heat the parcel. To obtain an analytic expression for θ_{eq} so defined we assume the condensation occurs at constant temperature and so obtain (18.54), namely

$$\theta_{eq}^p = \theta \exp\left(\frac{Lq}{c_p T}\right) = T\left(\frac{p_R}{p}\right)^{R/c_p} \exp\left(\frac{Lq}{c_p T}\right). \tag{18.158}$$

Here we give it a superscript p and call it the pseudo-adiabatic equivalent potential temperature. It is related to entropy, but it is not a true measure of it. One problem is that the above process is not physically realizable. In order to condense all the water vapour the temperature must be taken to absolute zero (or in practice we must lift the parcel to a great height where the temperature is very low), but we cannot simply condense all the water at a constant temperature. Second, we have neglected the contribution of liquid water. To see this, begin with an expression for the entropy of a mixture of dry air, water vapour and liquid water in equilibrium, with an entropy given by (18.157). An equivalent potential temperature, θ_{eq}, may then be defined by

$$(1 - q^{vl})c_p^d \ln(\theta_{eq}/T_0) = (1 - q^{vl})\eta^d + q^{vl}\eta^l + \frac{Lq_s^v}{T}. \tag{18.159}$$

Using the expressions for η^d and η^l given in (18.145) the above expression becomes

$$(1 - q^{vl})c_p^d \ln\left(\frac{\theta_{eq}}{T_0}\right) = (1 - q^{vl})c_p^d\left[\ln\left(\frac{\theta^d}{T_0}\right) - \frac{R^d}{c_p^d}\ln\left(1 - \frac{p^v}{p}\right)\right] + q^{vl}c^l \ln\left(\frac{T}{T_0}\right) + \frac{Lq_s^v}{T}. \tag{18.160}$$

Solving this for θ_{eq} gives

$$\theta_{eq} = \theta^d \exp\left(\frac{Lw_s}{c_p^d T}\right)\left(\frac{T}{T_0}\right)^{w^{vl}c^l/c_p^d}\left(1 - \frac{p^v}{p}\right)^{-R^d/c_p^d}, \tag{18.161}$$

where $w = q/(1-q^{vl})$ and $\theta^d = T(p_R/p)^{R^d/c_p^d}$. Equation (18.161) is a true measure of the entropy of a moist parcel, albeit a convoluted one. It differs slightly from (18.54) because θ_{eq}^p is not obtained by an adiabatic process, since the entropy of the liquid water is lost during condensation. If the water content is small ($q \ll 1,\ p^v/p \ll 1$) then (18.161) reduces to the more commonly used expression (18.158), with $q = q_s^v$.

Various other temperature-like quantities may be defined, notably the dew-point temperature and the wet-bulb temperature. The dew-point temperature is the temperature at which moist air, when cooled at constant pressure, becomes saturated. The wet-bulb temperature is the temperature that a parcel of air would have if cooled to saturation by the evaporation of water into it, with the energy required being supplied by the air parcel itself; wet-bulb temperature is directly measurable, for it is the temperature that a thermometer shows when wrapped in a wet cloth in a breeze. These quantities are both useful and have instinctive and human appeal. Potential temperature also has an intuitive attraction, but there are rarely objective reasons to use it in a quantitative calculation — entropy itself is usually a more straightforward alternative. This comment also applies to seawater.

APPENDIX B: EQUATIONS OF RADIATIVE TRANSFER

Consider a beam of radiation propagating through a thin slab of gas. Some of the incoming radiation may be absorbed, some may be scattered, and the slab may emit radiation of its own. Scattering is the change in direction of the radiation, so that it may reduce or — if radiation from other directions is scattered into the beam – amplify the beam's intensity. The difference between the incoming and outgoing radiation is

$$\mathrm{d}I_\nu \equiv I_\nu^{\text{in}} - I_\nu^{\text{out}} = -\mathrm{d}\tau_\nu I_\nu + \mathrm{d}J_\nu. \tag{18.162}$$

In this expression I_ν is the spectral radiance (power per unit area, per unit solid angle, per unit frequency interval) of the radiation, the term $\mathrm{d}\tau_\nu I_\nu$ is the extinction (the absorption plus the radiation scattered away) and $\mathrm{d}J_\nu$ is the emission plus the scattering into the beam. The quantity $\mathrm{d}\tau_\nu$ is the nondimensional *optical depth;* it may be written as $\mathrm{d}\tau_\nu = k_\nu \rho \mathrm{d}s$ where $\mathrm{d}s$ is the slab thickness, ρ is its density and k_ν is the extinction coefficient, a property of the gas in question. The minus sign on $\mathrm{d}\tau_\nu$ is appropriate when τ increases in the direction of the beam, and all of the above quantities depend on the frequency, ν, of the radiation.

Suppose there is no scattering, which is a good approximation for infrared radiation. The emission of radiation is, in thermal equilibrium, then given by the Planck function, B_ν, multiplied by the optical depth. Equation (18.162) becomes

$$\mathrm{d}I_\nu = -\mathrm{d}\tau_\nu (I_\nu - B_\nu) \quad \text{or} \quad \frac{\mathrm{d}I_\nu}{\mathrm{d}\tau_\nu} = -(I_\nu - B_\nu). \tag{18.163}$$

This equation is the foundation of much of radiative transfer. If radiation is propagating in all directions we must integrate over solid angle to obtain the upward and downward spectral irradiances (power per unit area per unit frequency interval). This is a complicated procedure in general but, to a good approximation for infrared radiation in Earth's atmosphere, the simple upshot is the multiplication of the optical depth by a geometric, order-one (for example, 5/3) factor of γ — because most of the radiation is passing slantwise through the medium — and the multiplication of B by π because of the integration over a hemisphere, giving the two-stream approximation.[25] Equation (18.163) becomes

$$\frac{\mathrm{d}F_\nu}{\mathrm{d}\tau_\nu^*} = -F_\nu - \pi B_\nu, \tag{18.164}$$

where F_ν is the spectral irradiance along a vertical path along which τ increases and τ_ν^* is the 'scaled' optical depth given by $\tau^* = \gamma\tau_\nu$. We will drop the asterisk on τ_ν^* and we will absorb the factor π into the definition of B_ν.

In Earth's atmosphere it is common to choose τ increasing downwards, from 0 at the top of the atmosphere, although no physical result depends on this choice. The downwards (D_ν) and upwards (U_ν) irradiances are then

$$\frac{\mathrm{d}D_\nu}{\mathrm{d}\tau_\nu} = B_\nu - D_\nu, \qquad \frac{\mathrm{d}U_\nu}{\mathrm{d}\tau_\nu} = U_\nu - B_\nu. \tag{18.165a,b}$$

These are the *two-stream equations,* without scattering, commonly known as the *Schwarzchild equations.* The downward and upward fluxes are uncoupled, because of the absence of scattering.

If there is no dependence of the optical depth on frequency then the medium is said to be *grey* and we may integrate (18.165) over frequency to give

$$\frac{\mathrm{d}D}{\mathrm{d}\tau} = B - D, \qquad \frac{\mathrm{d}U}{\mathrm{d}\tau} = U - B, \tag{18.166a,b}$$

where U and D are the upward and downward (total) irradiances, $B = \sigma T^4$ (all three with units of $\mathrm{W\,m^{-2}}$), and $\sigma = 5.6704\times10^{-8}\,\mathrm{W\,m^{-2}\,K^{-4}}$ is Stefan's constant. The grey assumption is not accurate for Earth's atmosphere. Nevertheless, for conceptual or approximate calculations it is often useful to suppose that the atmosphere is grey in the infrared, in which case (18.166) applies to infrared radiation with separate equations (that might include scattering or reflection, but that might also be grey) for solar radiation; this is the *semi-grey* approximation.

APPENDIX C: ANALYTIC APPROXIMATION OF TROPOPAUSE HEIGHT

Here we provide an approximate, analytic, expression for the height of the tropopause, given the optical depth and lapse rate. The idea is to solve for a self-consistent radiative-convective state, with a specified lapse rate extending upward to a tropopause and then transitioning to a radiative-equilibrium state, with the tropopause height being determined by the requirement of overall radiative balance.[26]

Instead of trying to use the formal solutions to the Schwarzchild equations, it is easier to approximately solve the radiative-transfer equations for the upward long wave irradiance, U_L, *ab initio.* We make one other approximation, that the value of B/U_L varies linearly from the tropopause (where its value is 0.5) to its value at the surface (where $B/U_L = 1$). Thus,

$$\frac{B}{U_L} = 1 - \frac{z}{2H_T}. \tag{18.167}$$

Numerical calculations suggest this is a decent approximation. Proceeding, we write (18.65b) as

$$\frac{\mathrm{d}\log U_L}{\mathrm{d}\tau} = 1 - \frac{B}{U_L} = \frac{z}{2H_T}. \tag{18.168}$$

Using $\tau(z) = \tau_s \exp(-z/H_a)$ we obtain

$$\frac{\mathrm{d}\log U_L}{\mathrm{d}z} = -\frac{z}{2H_T H_a}\tau_s \exp(-z/H_a). \tag{18.169}$$

We can integrate this expression by parts to obtain a value of the upwelling radiation at the tropopause $U_L(H_T)$, namely

$$\log\left(\frac{U_L(H_T)}{U_L(0)}\right) = -\frac{\tau_s}{2H_T}\int_0^{H_T} \exp(-z/H_a)\,\mathrm{d}z \approx -\frac{\tau_s H_a}{2H_T}, \tag{18.170}$$

for $H_T \gg H_a$. This is an expression for the upwellling longwave radiation, $U_L(H_T)$, and the unknowns in the equation are $U_L(0)$, the upwelling radiation at the surface, and the tropopause height,

H_T. The value of $U_L(0)$ is given by the surface temperature, which is a function of the tropopause temperature, T_T, the specified lapse rate, Γ, and the tropopause temperature, T_T. Inverting the argument, if we are given $U_L(H_T)$ and T_T then we can calculate H_T.

Let us assume that the stratosphere is optically thin, in which case $U_L(H_T)$ is the outgoing longwave radiation, equal to $2\sigma T_T^4$, and this is known because it equals the incoming solar radiation. Thus, we can in principle now solve (18.170) for H_T. To do this note that the upwelling radiation at the surface is given by $U_L(0) = \sigma T_g^4 = \sigma T_s^4$, where $T_s = T_T + \Gamma H_T$. The left-hand side of (18.170) then becomes

$$\log\left(\frac{2\sigma T_T^4}{\sigma T_s^4}\right) = \log 2 + 4\log\frac{T_T}{T_s} = \log 2 + 4\log\left(\frac{T_T}{T_T + \Gamma H_T}\right) \approx \log 2 - \frac{4\Gamma H_T}{T_T}. \tag{18.171}$$

The rightmost terms in (18.171) and (18.170) are approximately equal so that

$$\log 2 - \frac{4\Gamma H_T}{T_T} = -\frac{\tau_s H_a}{2H_T} \qquad \text{or} \qquad 8\Gamma H_T^2 - \mathrm{C} H_T T_T - \tau_s H_a T_T = 0, \tag{18.172}$$

where $\mathrm{C} = 2\log 2 \approx 1.39$. The solution of this equation is

$$H_T = \frac{1}{16\Gamma}\left(\mathrm{C}T_T + \sqrt{\mathrm{C}^2 T_T^2 + 32\Gamma\tau_s H_a T_T}\right). \tag{18.173}$$

The tropopause height given by this equation is fairly close to the actual solution of the radiative-convective equations, obtained by numerically integrating the Schwarzchild equations and iterating to obtain the correct tropopause height following the algorithm of Section 18.6.2, as seen in Fig. 18.21. The analytic approximation may be further improved with a bit of effort, but even in the form above it captures the essential aspects of the true solution.

Notes

1 Many thanks to Adam Sobel for a number of conversations and notes that informed the sections on convection and quasi-equilibrium, and to Will Beeson and his colleagues in Chicago for many useful comments. I am also grateful to Brian Mapes for a detailed critique of this chapter; I was unable to address all of his concerns but his point of view was salutary.

2 I use a formula given by Bolton (1980), which is a variant on the original Magnus formula, *aka* the August–Roche–Magnus formula or the Tetens formula — see Lawrence (2005) for discussion.

3 Measurements come from the hybrid advanced microwave sounding unit, Atmospheric Infrared Sounder (AIRS), as in Sherwood *et al.* (2010).

4 To read more about models of this form see Pierrehumbert *et al.* (2007), with extensions and applications by O'Gorman *et al.* (2011), Sukhatme & Young (2011), Tsang & Vanneste (2016) and others.

5 The simulations here used a model developed by Dr. Yue-Kin Tsang, and I am grateful to him for discussions and help.

6 The interested reader might start with Sherwood *et al.* (2010) or Schneider *et al.* (2010) and go forward and back from there.

7 This definition of θ_{eq} is sometimes called the 'pseudo-equivalent potential temperature', because the condensation product, liquid water, is assumed to fall out of the air parcel and the process is 'pseudo-adiabatic', not adiabatic. Other names with slightly different definitions exist (Betts 1973, Emanuel 1994, Ambaum 2010).

8 The notion of conditional instability has been with us for many years — it appears in Haurwitz (1941) for example — and an influential form was introduced Ooyama (1963) (see Ooyama 1982) and Charney & Eliassen (1964). They proposed models of a cooperative mechanism between the

convection and larger scale flow, with the convection producing a convergence at low levels, leading to ascent, more latent heat release and convection and so on. This mechanism and variations about it became known as 'Conditional Instability of the Second Kind', or CISK, to distinguish it from more conventional conditional instability ('of the first kind') which does not involve such a feedback with the large scale. The CISK mechanism tends to produce vortical updraughts and may be important for hurricane growth, but this remains a topic of some debate (e.g., Raymond 1995, Emanuel 1994, Smith 1997). Other theories of hurricanes tend to de-emphasize the CISK mechanism in favour of model involving a feedback between wind speed and evaporation, called wind-induced surface heat exchange or WISHE, with stronger winds giving more evaporation, leading to saturation and thence convection. The convection then mainly serves to establish a moist adiabatic lapse rate (as in the quasi-equilibrium ideas of Section 18.4.3) which ties the boundary layer to a warm core extending upward (e.g., Craig & Gray 1996).

9 Convective adjustment is a great simplification over what actually occurs, but nevertheless it is a useful concept and the basis of many early parameterization schemes for numerical models of the atmosphere. Convective adjustment was introduced into modelling by Manabe & Strickler (1964) and a theoretical and observational discussion of the general problem was given by Ludlam (1966). A popular variation (a relaxation rather than an adjustment) was proposed by Betts (1986) and Betts & Miller (1986). These days (c. 2017) GCMs rarely use simple convective adjustment or relaxation schemes, although the underlying ideas endure.

10 The nature of a 'quasi-equilibrium' between convection and large-scale forcing was made explicit by Betts (1973) and Arakawa & Schubert (1974), with precedents to be found in Scorer & Ludlam (1951) and Ludlam (1966). Many papers have since followed, with, for example, an extended discussion in Emanuel *et al.* (1994), a counterpoint, examples and further discussion in Mapes (1997, 1998, 2000), and application to the boundary layer in Raymond (1997). More references can be found in the reviews by Arakawa (2004) and Emanuel (2007), and debates continue about applicability and efficacy.

11 This behaviour is seen in many comprehensive General Circulation Models of the atmosphere, going back to Manabe & Wetherald (1980) and beyond. The explanation given here follows Vallis *et al.* (2015) but has much earlier roots.

12 Numerical calculations of the radiative constraint with more realistic treatments of radiation were carried out by Thuburn & Craig (2000).

13 We roughly follow Emanuel (1987). Neelin & Zeng (2000) and Zeng *et al.* (2000) give details of how a functional reduced model, including diabatic and frictional terms, may be constructed. Lindzen & Nigam (1987) have another, albeit related, take on the problem.

14 The tropical scaling was presented by Charney (1963).

15 The derivation of the weak temperature-gradient approximation given here is more-or-less that of Sobel *et al.* (2001), and may be regarded as an extension of Charney's (1963) ideas to include diabatic effects. A number of authors previously used the approximation, implicitly or explicitly, in one form or another (e.g., Neelin 1988, Browning *et al.* 2000), and various extensions and rigour have been added by Majda & Klein (2003) and others.

16 See Bretherton & Smolarkiewicz (1989) and Mapes (1997) and go from there.

17 Details are described in Bretherton & Sobel (2002) and Neelin & Zeng (2000).

18 A review of convectively coupled equatorial waves is provided by Kiladis *et al.* (2009). The MJO was first described by Madden & Julian (1971, 1972), and a review may be found in Zhang (2005). Schubert & Masarik (2006) discuss aspects of a moving Matsuno–Gill pattern and its relation to the MJO, and Raymond & Fuchs (2009), Majda & Stechmann (2009) and Sobel & Maloney (2013), among others, offer theoretical models of the MJO.

19 Diagrams such as these are known as Wheeler–Kiladis diagrams, after Wheeler & Kiladis (1999). Filtering the noise is required to obtain clean plots and requires some attention.

20 Adapted from Kiladis *et al.* (2009).

21 Dias & Kiladis (2014) discuss reasons why the value is smaller than might be expected.

22 Adapted from Kiladis *et al.* (2005).

23 Heckley & Gill (1984).

24 Feistel *et al.* (2010) and Thuburn (2017) give more details. I am grateful to John Thuburn for a number of very useful conversations on this matter.

25 There are various versions of the two-stream approximation; see Goody & Yung (1995) or Pierrehumbert (2010).

26 The contents of this appendix are the results of joint work with Pablo Zurita-Gotor. A seemingly casual question to me by Rich Kerswell led to the theoretical development.

Part IV

LARGE-SCALE OCEANIC CIRCULATION

As I ebb'd with the ocean of life,
As I wended the shores I know,
As I walk'd where the ripples continually wash you Paumanok ...
As the ocean so mysterious rolls toward me closer and closer ...
I perceive I have not really understood any thing,
not a single object, and that no man ever can,
Nature here in sight of the sea taking advantage of me
to dart upon me and sting me,
Because I have dared to open my mouth to sing at all.

Walt Whitman, *As I Ebb'd with the Ocean of Life*, from *Leaves of Grass*, 1881.

CHAPTER 19

Wind-Driven Gyres

UNDERSTANDING THE CIRCULATION OF THE OCEAN involves a combination of observations, comprehensive numerical modelling, and more conceptual modelling or theory.[1] All are essential, but in this chapter and the ones following our emphasis is on the last of the triad. Its (continuing) role is not to explain every feature of the observed ocean circulation, nor to necessarily describe details best left to numerical simulations. Rather, it is to provide a conceptual and theoretical framework for understanding the circulation of the ocean, for interpreting observations and suggesting how new observations may best be made, and to aid the development and interpretation of numerical models.

The aspect of the ocean that most affects the climate is the sea-surface temperature (SST), as illustrated in Fig. 19.1, and aside from the expected latitudinal variation there is significant zonal variation too — the western tropical Pacific is particularly warm, and the western Atlantic is warmer than the corresponding latitude in the east. These variations owe their existence to ocean currents, and the main ones are sketched — in a highly schematic and non-quantitative fashion — in Fig. 19.2. Over most of the ocean, the vertically averaged currents have a similar sense to the surface currents, one exception being at the equator where the surface currents are mainly westward but the vertical integral is dominated by the eastward undercurrent. Two dichotomous aspects of this picture stand out: (i) the complexity of the currents as they interact with topography and the geography of the continents; (ii) the simplicity and commonality of the large-scale structures in the major ocean basins, and in particular the ubiquity of subtropical and subpolar gyres. Indeed these gyres, sweeping across the great oceans carrying vast quantities of water and heat, are perhaps the single most conspicuous feature of the circulation. The subtropical gyres are anticyclonic, extending polewards to about 45°, and the subpolar gyres are cyclonic and polewards of this, primarily in the Northern Hemisphere. The existence of the great gyres, and that they are strongest in the west, has been known for centuries; this *western intensification* leads to such well-known currents as the Gulf Stream in the Atlantic (charted by Benjamin Franklin), the Kuroshio in the Pacific, and the Brazil Current in the South Atlantic.

For much of this chapter we consider a model, and variations about it, that explains the large-scale features of ocean gyres and that lies at the core of ocean circulation theory — the steady, forced-dissipative, homogeneous model of the ocean circulation first formulated by Stommel.[2] In all of the geosciences there is perhaps no other model that combines elegance and as relevance as much as this one.

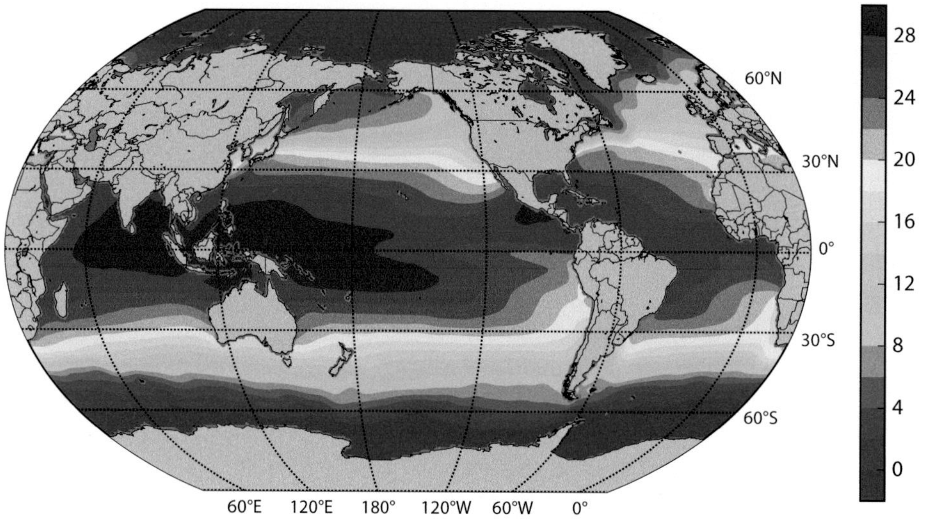

Fig. 19.1 The sea-surface temperature (SST, ° C) of the world's ocean, as determined from a great many observations, combined in the World Ocean Circulation Experiment (WOCE).

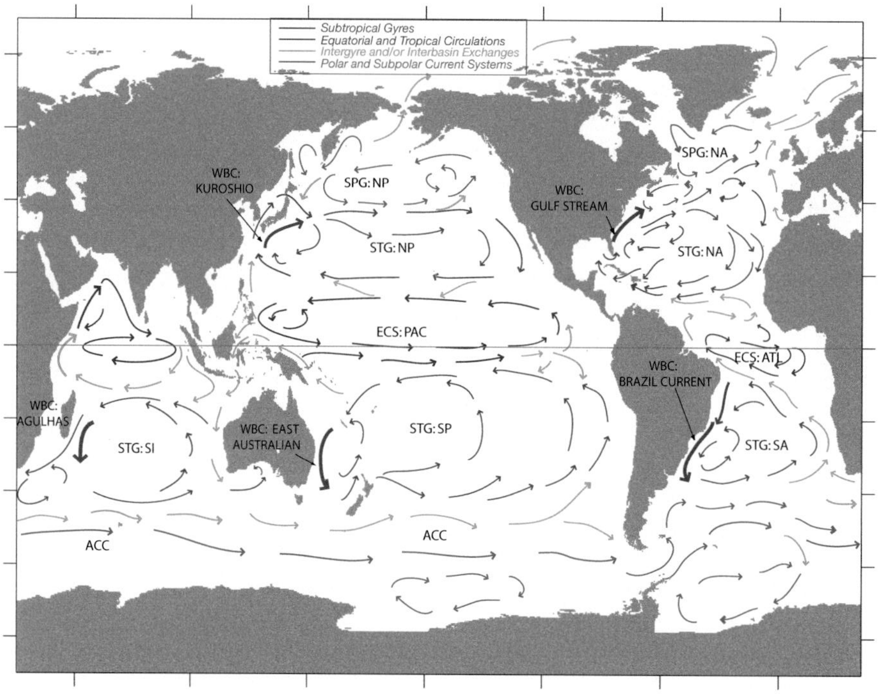

Fig. 19.2 Idealization of the main currents of the global ocean. Key: STG – Subtropical Gyre; SPG – Subpolar Gyre; WBC – Western Boundary Current; ECS – Equatorial Current System; NA – North Atlantic; SA – South Atlantic; NP – North Pacific; SP – South Pacific; SI – South Indian; ACC – Antarctic Circumpolar Current; ATL – Atlantic; PAC – Pacific. The figure is only a qualitative representation of the actual flow. Most of the currents are manifested at the surface, but near the equator the figure shows the undercurrent which flows in the opposite way to the surface current.

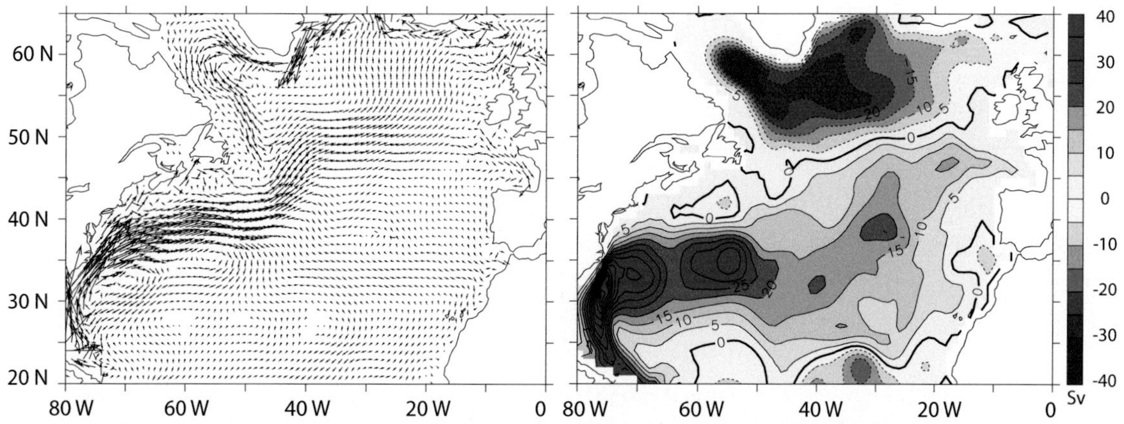

Fig. 19.3 Left: the time averaged velocity field at a depth of 75 m in the North Atlantic, obtained by constraining a numerical model to observations (so giving a 'state estimate'). Right: the streamfunction of the vertically integrated flow, in Sverdrups ($1\,\mathrm{Sv} = 10^9\,\mathrm{kg\,s^{-1}}$). Note the presence of an anticyclonic subtropical gyre (clockwise circulation, shaded red), a cyclonic subpolar gyre (anticlockwise, blue), and intense western boundary currents.[4]

19.1 THE DEPTH INTEGRATED WIND-DRIVEN CIRCULATION

Although even today we barely have sufficient observations to produce a detailed synoptic map of the ocean currents, except at the surface, the large-scale mean currents are fairly well mapped and Fig. 19.3 illustrates the average current pattern of the North Atlantic using a combination of observations and a numerical model, and the Gulf Stream is clearly visible. Similar features are seen in all the major ocean basins (Fig. 19.4) where we see subtropical and subpolar gyres, all of them intensified in the west.[3] Our goal in this chapter is to explain the main features seen in these figures in as simple and straightforward a manner as is possible.

The equations that govern the large-scale flow in the oceans are the planetary-geostrophic equations. Greatly simplified as these are compared to the Navier–Stokes equations, or even the hydrostatic Boussinesq equations, they are still quite daunting: a prognostic equation for buoyancy is coupled to the advecting velocity via hydrostatic and geostrophic balance, and the resulting problem is formidably nonlinear. However, it turns out that thermodynamic effects can effectively be eliminated by the simple device of vertical integration; the resulting equations are linear, and the only external forcing is that due to the wind stress. The resulting model then, at the price of some comprehensiveness, gives a useful picture of the *wind-driven* circulation of the ocean. We will consider the vertical structure of this flow in the next chapter.

19.1.1 The Stommel Model

The planetary-geostrophic equations for a Boussinesq fluid are:

$$\frac{\mathrm{D}b}{\mathrm{D}t} = \dot{b}, \qquad \nabla_3 \cdot \boldsymbol{v} = 0, \tag{19.1a,b}$$

$$\boldsymbol{f} \times \boldsymbol{u} = -\nabla\phi + \frac{1}{\rho_0}\frac{\partial \boldsymbol{\tau}}{\partial z}, \qquad \frac{\partial \phi}{\partial z} = b. \tag{19.2a,b}$$

These equations are, respectively, the thermodynamic equation (19.1a), the mass continuity equation (19.1b), the horizontal momentum equation (19.2a), (i.e., geostrophic balance, plus a stress term), and the vertical momentum equation (19.2b) — that is, hydrostatic balance. These equations are derived more fully in Chapter 5, but they are essentially the Boussinesq primitive equations with the advection terms omitted from the horizontal momentum equation, on the basis of small Rossby number. In this chapter we will henceforth absorb the factor of ρ_0 into the $\boldsymbol{\tau}$, so that $\boldsymbol{\tau}$ denotes the kinematic stress, and the gradient operator will be two dimensional, in the x-y plane, unless noted.

Take the curl of (19.2a) (that is, cross differentiate its x and y components) and integrate over the depth of the ocean to give

$$\int f\nabla\cdot\boldsymbol{u}\,\mathrm{d}z + \frac{\partial f}{\partial y}\int v\,\mathrm{d}z = \mathrm{curl}_z(\boldsymbol{\tau}_T - \boldsymbol{\tau}_B), \tag{19.3}$$

where the operator curl_z is defined by $\mathrm{curl}_z\boldsymbol{A} \equiv \partial A^y/\partial x - \partial A^x/\partial y = \mathbf{k}\cdot\nabla\times\boldsymbol{A}$, and the subscripts T and B are for top and bottom. The divergence term vanishes if the vertical velocity is zero at the top and bottom of the ocean. Strictly, at the top of the ocean the vertical velocity is given by the material derivative of height of the ocean's surface, $\mathrm{D}h/\mathrm{D}t$, but on the large-scales this has a negligible effect and we may make the rigid-lid approximation and set it to zero. At the bottom of the ocean the vertical velocity is only zero if the ocean is flat-bottomed; otherwise it is $\boldsymbol{u}\cdot\nabla\eta_B$, where η_B is the orographic height at the ocean floor. The neglect of this topographic term is probably the most restrictive single approximation in the model. Given this neglect, (19.3) becomes

$$\beta\overline{v} = \mathrm{curl}_z(\boldsymbol{\tau}_T - \boldsymbol{\tau}_B), \tag{19.4}$$

where henceforth, in this section, quantities with an overbar are understood to be the vertical integral over the depth of the ocean. If the stresses depend only on the velocity fields then thermodynamic fields do not affect the vertically integrated flow.

At the top of the ocean, the stress is given by the wind. At the bottom, in the absence of topography we assume that the stress may be parameterized by a linear drag, or Rayleigh friction, as might be generated by an Ekman layer; it is this assumption that particularly characterizes this model as being due to Stommel. (Note that we parameterize the friction by a drag acting on the vertically integrated velocity. Using the velocity at the bottom of the ocean would be more realistic, but this wrinkle is beyond the scope of vertically integrated models.) Equation (19.4) then becomes

$$\beta\overline{v} = -r\overline{\zeta} + F_\tau(x, y), \tag{19.5}$$

where $F_\tau = \mathrm{curl}_z\boldsymbol{\tau}_T$ is the wind-stress curl at the top of the ocean and is a known function. Because the velocity is divergence-free, we can define a streamfunction ψ such that $\overline{u} = -\partial\psi/\partial y$ and $\overline{v} = \partial\psi/\partial x$. Equation (19.5) then becomes

$$r\nabla^2\psi + \beta\frac{\partial\psi}{\partial x} = F_\tau(x, y). \tag{19.6}$$

This equation is often referred to as the *Stommel problem* or the *Stommel model,* and may be posed in a variety of two dimensional domains.

19.1.2 Alternative Formulations

A number of alternative formulations leading to (19.6) are possible. None are perhaps as well justified as the derivation via the planetary-geostrophic equations but the differences in the specific assumptions made give some indication of the robustness of the derivation, and show how the model might be extended to include topographic or nonlinear effects.

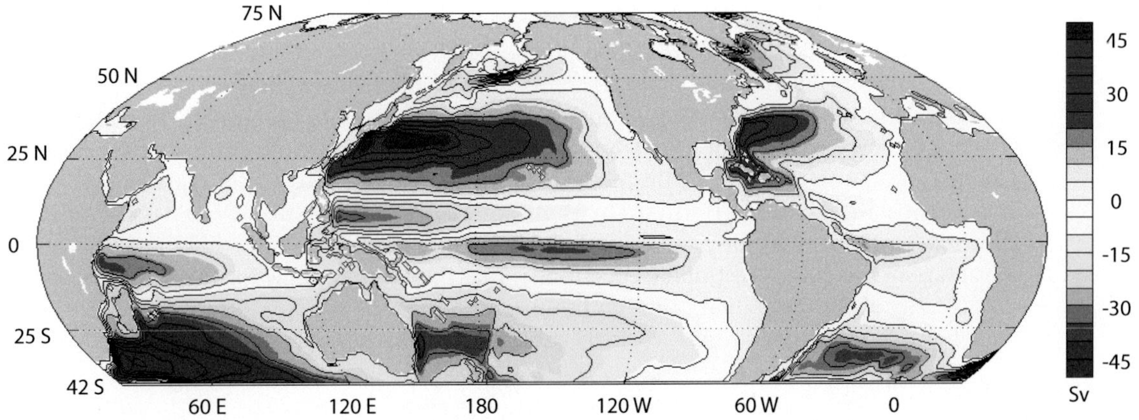

Fig. 19.4 A state estimate of the streamfunction of the vertically integrated flow for the near global ocean. Red shading indicates clockwise flow, and blue shading anticlockwise, but in both hemispheres the subtropical (subpolar) gyres are anticyclonic (cyclonic).[5]

I. A homogeneous model

Rather than vertically integrating, we may suppose that the ocean is a homogeneous fluid obeying the shallow water equations (Chapter 3). The potential vorticity equation (cf. (3.96) on page 121) is:

$$\frac{\mathrm{D}}{\mathrm{D}t}\left(\frac{\zeta + f}{h}\right) = \frac{F}{h}, \tag{19.7}$$

where F represents friction and forcing. In an ocean with a rigid-lid and flat bottom (19.7) gives the barotropic vorticity equation,

$$\frac{\mathrm{D}\zeta}{\mathrm{D}t} + \beta v = F, \tag{19.8}$$

where the term F again represents the wind-stress curl and a linear drag. Further, since the horizontal velocity is divergence-free (because of the flat-bottom and rigid-lid) we may represent it as a streamfunction, whence we obtain the closed equation

$$\frac{\mathrm{D}}{\mathrm{D}t}\nabla^2\psi + \beta\frac{\partial\psi}{\partial x} = F_\tau(x, y) - r\nabla^2\psi, \tag{19.9}$$

where F_τ again represents the wind forcing. This equation is the 'time-dependent nonlinear Stommel problem'. The steady nonlinear problem is sometimes of interest, too, and this is just

$$J(\psi, \nabla^2\psi) + \beta\frac{\partial\psi}{\partial x} = F_\tau(x, y) - r\nabla^2\psi. \tag{19.10}$$

To obtain the original Stommel model we just ignore the advective derivative, which will be valid if $|\zeta| \sim Z \ll \beta L$ where $Z = U/L$ is a representative value of vorticity. This condition is equivalent to

$$R_\beta \equiv \frac{U}{\beta L^2} \ll 1. \tag{19.11}$$

R_β is called the *beta-Rossby number*. On sufficiently large scales, $\beta \sim f/L$ and (19.11) is similar to a small Rossby number assumption.

II. Quasi-geostrophic formulation

In the planetary-geostrophic formulation, the horizontal velocity is divergence-free only because we have vertically integrated. If the scales of motion are not too large the horizontal flow at every level is divergence-free for another reason: because it is in geostrophic balance. In reality, over a single oceanic gyre (say from 15° to 40° latitude), variations in Coriolis parameter are not large, and this prompts us to formulate the model in terms of the quasi-geostrophic equations. Formally, such a model would then be restricted to length scales, L, of no more than $\mathcal{O}(Ro^{-1})$ larger than the deformation radius, and for gyre scales this criterion is marginally satisfied if $L_d = 100\,\mathrm{km}$. An advantage of the quasi-geostrophic equations is that they readily allow for the inclusion of both nonlinearity and stratification. (For an informal summary of quasi-geostrophy, see the box on page 193.) The quasi-geostrophic vorticity equation for a Boussinesq system is:

$$\frac{\mathrm{D}\zeta}{\mathrm{D}t} + \beta v = f_0 \frac{\partial w}{\partial z} + \mathrm{curl}_z \frac{\partial \boldsymbol{\tau}}{\partial z}. \tag{19.12}$$

If we neglect the advective derivative, and vertically integrate, we obtain

$$\beta \int_B^T v \, \mathrm{d}z = f_0 [w]_B^T + \mathrm{curl}_z [\boldsymbol{\tau}]_B^T. \tag{19.13}$$

where T denotes the ocean top and B the bottom. We now make one of two virtually equivalent choices:

(i) We suppose that the integration is over the entire depth of the ocean, in which case the term $[w]_B^T$ vanishes given a rigid lid and a flat bottom. If the stress at the top of the ocean is given by the wind stress, and at the bottom of the ocean it is parameterized by a linear drag, then we obtain

$$\beta \overline{v} = F_\tau(x, y) - r\overline{\zeta} \tag{19.14}$$

just as in (19.5), and where an overbar denotes a vertical integral. Writing $\overline{v} = \partial\psi/\partial x$ and $\overline{\zeta} = \nabla^2\psi$ then gives (19.6).

(ii) We suppose that the integration is between two thin Ekman layers at the top and bottom of the ocean. The stress is zero at the interior edge of these layers, but the vertical velocity is not. At the base of the upper Ekman layer, at $z = -\delta_T$, the vertical velocity is given by:

$$w(x, y, -\delta_T) = \mathrm{curl}_z(\boldsymbol{\tau}_T / f_0), \tag{19.15}$$

where the top of the ocean is at $z = 0$ and δ_T is the thickness of the upper Ekman layer. Similarly, at the top of the lower Ekman layer, the vertical velocity is:

$$w(x, y, -H + \delta_B) = \delta_B \zeta, \tag{19.16}$$

where $z = -H$ at the ocean bottom and δ_B is the thickness of the bottom Ekman layer. Neglecting the advective derivative, and integrating over the ocean between the two Ekman layers, (19.12) becomes

$$\beta \overline{v} = \mathrm{curl}_z \boldsymbol{\tau}_T - f_0 \delta_B \zeta = \mathrm{curl}_z \boldsymbol{\tau}_T - f_0 \delta_B \overline{\zeta}/H, \tag{19.17}$$

where, to obtain the second equality, we assume that the bottom drag may be parameterized using the vertically integrated vorticity, $\overline{\zeta}$. Defining the drag coefficient r by $r = f_0\delta_B/H$, and introducing a streamfunction gives

$$r\nabla^2\psi + \beta\frac{\partial \psi}{\partial x} = \mathrm{curl}_z \boldsymbol{\tau}_T, \tag{19.18}$$

as before.

19.1.3 Approximate Solution of Stommel Model

Sverdrup balance

Equation (19.6) is linear and it is possible to obtain an exact, analytic solution. However, it is more insightful to approach the problem perturbatively, by supposing that the frictional term is small, meaning there is an approximate balance between wind stress and the β-effect.[6] Friction is small if $|r\zeta| \ll |\beta v|$ or

$$\frac{r}{L} = \frac{f\delta_B}{HL} \ll \beta \tag{19.19}$$

using $r = f\delta_B/H$, and where L is the horizontal scale of the motion, and generally speaking this inequality is well satisfied for large-scale flow. The vorticity equation becomes

$$\beta\overline{v} \approx \mathrm{curl}_z \boldsymbol{\tau}_T, \tag{19.20}$$

which is known as *Sverdrup balance*.[7] (Sometimes Sverdrup balance is taken to mean the linear geostrophic vorticity balance $\beta v = f\partial w/\partial z$, but we will restrict its use to mean a balance between the beta effect and wind stress curl.) The observational support for Sverdrup balance is rather mixed, discrepancies arising not so much from the failure of (19.19), but from the presence of small-scale eddying motion with concomitantly large nonlinear terms, and the presence of non-negligible vertical velocities induced by the interaction with bottom topography.[8] Nevertheless, Sverdrup balance provides a useful, if not impregnable, foundation on which to build.

Boundary-layer solution

For simplicity, consider a square domain of side a and rescale the variables by setting

$$x = a\widehat{x}, \qquad y = a\widehat{y}, \qquad \tau = \tau_0\widehat{\tau}, \qquad \psi = \widehat{\psi}\frac{\tau_0}{\beta}, \tag{19.21}$$

where τ_0 is the amplitude of the wind stress. The hatted variables are nondimensional and, assuming our scaling to be sensible, these are $\mathcal{O}(1)$ quantities in the interior. Equation (19.18) becomes

$$\frac{\partial\widehat{\psi}}{\partial\widehat{x}} + \epsilon_S\nabla^2\widehat{\psi} = \mathrm{curl}_z\widehat{\boldsymbol{\tau}}_T, \tag{19.22}$$

where $\epsilon_S = (r/a\beta) \ll 1$, in accord with (19.19). For the rest of this section we will drop the hats over nondimensional quantities. Over the interior of the domain, away from boundaries, the frictional term in (19.22) is small. We can take advantage of this by writing

$$\psi(x,y) = \psi_I(x,y) + \phi(x,y), \tag{19.23}$$

where ψ_I is the interior streamfunction and ϕ is a boundary layer correction. Away from boundaries ψ_I is presumed to dominate the flow, and this satisfies

$$\frac{\partial\psi_I}{\partial x} = \mathrm{curl}_z\boldsymbol{\tau}_T. \tag{19.24}$$

The solution of this equation (called the 'Sverdrup interior') is

$$\psi_I(x,y) = \int_0^x \mathrm{curl}_z\boldsymbol{\tau}(x',y)\,\mathrm{d}x' + g(y), \tag{19.25}$$

where $g(y)$ is an arbitrary function of integration that gives rise to an arbitrary zonal flow. The corresponding velocities are

$$v_I = \mathrm{curl}_z\boldsymbol{\tau}, \qquad u_I = -\frac{\partial}{\partial y}\int_0^x \mathrm{curl}_z\boldsymbol{\tau}(x',y)\,\mathrm{d}x' - \frac{\mathrm{d}g(y)}{\mathrm{d}y}. \tag{19.26}$$

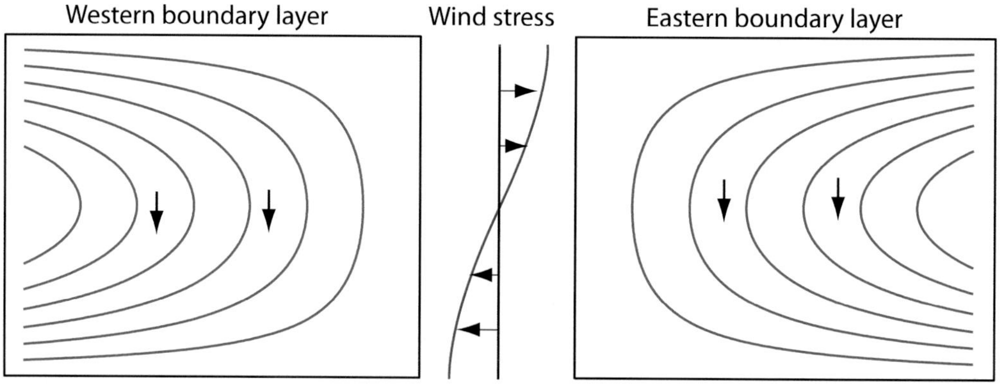

Fig. 19.5 Two possible Sverdrup flows, ψ_I, for the wind stress shown in the centre. Each solution satisfies the no-flow condition at either the eastern or western boundary, and a boundary layer is therefore required at the other boundary. Both flows have the same, equatorward, meridional flow in the interior. Only the flow with the western boundary current is physically realizable, however, because only then can friction produce a curl that opposes that of the wind stress, so allowing the flow to equilibrate.

The dynamics is most clearly illustrated if we now restrict our attention to a wind-stress curl that is zonally uniform, and that vanishes at two latitudes, $y = 0$ and $y = 1$. An example is

$$\tau_T^y = 0, \qquad \tau_T^x = -\cos(\pi y), \tag{19.27}$$

for which $\mathrm{curl}_z \boldsymbol{\tau}_T = -\pi \sin(\pi y)$. The Sverdrup (interior) flow may then be written as

$$\psi_I(x, y) = [x - C(y)]\mathrm{curl}_z \boldsymbol{\tau}_T = \pi[C(y) - x] \sin \pi y, \tag{19.28}$$

where $C(y)$ is the arbitrary function of integration $[C(y) = -g(y)/\mathrm{curl}_z \boldsymbol{\tau}]$. If we choose C to be a constant, the zonal flow associated with it is $C\,\mathrm{curl}_z \boldsymbol{\tau}_T$. We can then satisfy $\psi = 0$ at *either* $x = 0$ (if $C = 0$) *or* $x = 1$ (if $C = 1$). These solutions are illustrated in Fig. 19.5 for the particular stress (19.27).

Regardless of our choice of C we cannot satisfy $\psi = 0$ at both zonal boundaries. We must choose one, and then construct a *boundary layer* solution (i.e., we determine ϕ) to satisfy the other condition. Which choice do we make? On intuitive grounds it seems that we should choose the solution that satisfies $\psi = 0$ at $x = 1$ (the solution on the left in Fig. 19.5), for the interior flow then goes round in the same direction as the wind: the wind is supplying a clockwise torque, and to achieve an angular momentum balance anticlockwise angular momentum must be supplied by friction. We can imagine that this would be provided by the frictional forces at the western boundary layer if the interior flow is clockwise, but not by friction at an eastern boundary layer when the interior flow is anticlockwise. Note that this argument is not dependent on the sign of the wind-stress curl: if the wind blew the other way a similar argument still implies that a western boundary layer is needed. We will now see if and how the mathematics reflects this intuitive but non-rigorous argument.

Asymptotic matching

Near the walls of the domain the boundary layer correction $\phi(x, y)$ must become important in order that the boundary conditions may be satisfied, and the flow, and in particular $\phi(x, y)$, will vary rapidly with x. To reflect this, let us *stretch* the x-coordinate near this point of failure (i.e., at either $x = 0$ or $x = 1$, but we do not know at which yet) and let

$$x = \epsilon\alpha \qquad \text{or} \qquad x - 1 = \epsilon\alpha. \tag{19.29a,b}$$

Here, α is the stretched coordinate, which has values $\mathcal{O}(1)$ in the boundary layer, and ϵ is a small parameter, as yet undetermined. We then suppose that $\phi = \phi(\alpha, y)$, and using (19.23) in (19.22), we obtain

$$\epsilon_S(\nabla^2\psi_I + \nabla^2\phi) + \frac{\partial \psi_I}{\partial x} + \frac{1}{\epsilon}\frac{\partial \phi}{\partial \alpha} = \operatorname{curl}_z \boldsymbol{\tau}_T, \tag{19.30}$$

where $\phi = \phi(\alpha, y)$ and $\nabla^2\phi = \epsilon^{-2}\partial^2\phi/\partial\alpha^2 + \partial^2\phi/\partial y^2$. Now, by choice, ψ_I exactly satisfies Sverdrup balance, and so (19.30) becomes

$$\epsilon_S\left(\nabla^2\psi_I + \frac{1}{\epsilon^2}\frac{\partial^2\phi}{\partial\alpha^2} + \frac{\partial^2\phi}{\partial y^2}\right) + \frac{1}{\epsilon}\frac{\partial\phi}{\partial\alpha} = 0. \tag{19.31}$$

We now choose ϵ to obtain a physically meaningful solution. An obvious choice is $\epsilon = \epsilon_S$, for then the leading-order balance in (19.31) is

$$\frac{\partial^2\phi}{\partial\alpha^2} + \frac{\partial\phi}{\partial\alpha} = 0, \tag{19.32}$$

the solution of which is

$$\phi = A(y) + B(y)\mathrm{e}^{-\alpha}. \tag{19.33}$$

Evidently, ϕ grows exponentially in the negative α direction. If this were allowed, it would violate our assumption that solutions are small in the interior, and we must eliminate this possibility by allowing α to take only positive values in the interior of the domain, and by setting $A(y) = 0$. We therefore choose $x = \epsilon\alpha$ so that $\alpha > 0$ for $x > 0$; the boundary layer is then at $x = 0$, that is, it is a *western boundary*, and it decays eastwards in the direction of increasing α — that is, into the ocean interior. We now choose $C = 1$ in (19.28) to make $\psi_I = 0$ at $x = 1$ in (19.28) and then, for the wind stress (19.27), the interior solution is given by

$$\psi_I = \pi(1 - x)\sin \pi y. \tag{19.34}$$

This alone satisfies the boundary condition at the eastern boundary. The function $B(y)$ is chosen to satisfy the additional condition that

$$\psi = \psi_I + \phi = 0 \qquad \text{at} \quad x = 0, \tag{19.35}$$

and using (19.34) this gives

$$\pi \sin \pi y + B(y) = 0. \tag{19.36}$$

Using this in (19.33), with $A(y) = 0$, then gives the boundary layer solution

$$\phi = -\pi \sin \pi y \mathrm{e}^{-x/\epsilon_S}. \tag{19.37}$$

The composite (boundary layer plus interior) solution is the sum of (19.34) and (19.37), namely

$$\psi = (1 - x - \mathrm{e}^{-x/\epsilon_S})\pi \sin \pi y. \tag{19.38}$$

With dimensional variables this is

$$\psi = \frac{\tau_0 \pi}{\beta}\left(1 - \frac{x}{a} - \mathrm{e}^{-x/(a\epsilon_S)}\right)\sin\frac{\pi y}{a}. \tag{19.39}$$

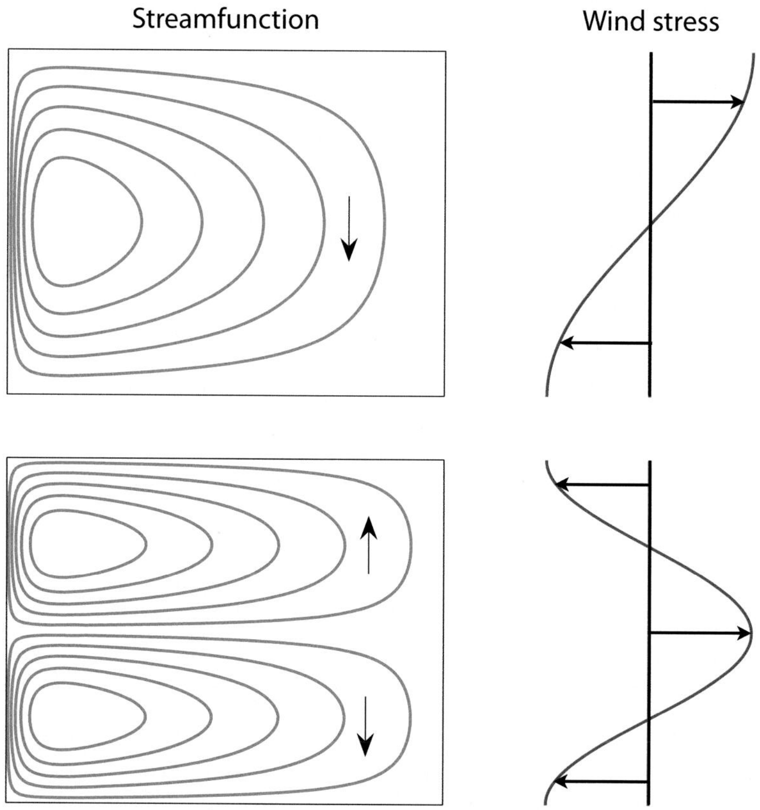

Fig. 19.6 Two solutions of the Stommel model. Upper panel shows the streamfunction of a single-gyre solution, with a wind stress proportional to $-\cos(\pi y/a)$ (in a domain of side a), and the lower panel shows a two-gyre solution, with wind stress proportional to $\cos(2\pi y/a)$. In both cases $\epsilon_S = 0.04$.

This is a 'single gyre' solution. Two or more gyres can be obtained with a different wind forcing, such as $\tau^x = -\tau_0 \cos(2\pi y)$, as in Fig. 19.6.

It is a relatively straightforward matter to generalize to other wind stresses, provided these also vanish at the two latitudes between which the solution is desired. It is left as a problem to show that in general

$$\psi_I = \int_{x_E}^{x} \mathrm{curl}_z \boldsymbol{\tau}(x', y)\,\mathrm{d}x', \tag{19.40}$$

and that the composite solution is

$$\psi = \psi_I - \psi_I(0, y)\mathrm{e}^{-x/(x_E \epsilon_S)}. \tag{19.41}$$

19.2 USING VISCOSITY INSTEAD OF DRAG

A natural variation on the Stommel problem is to use a harmonic viscosity, $\nu\nabla^2\zeta$, in place of the drag term $-r\zeta$ in the vorticity equation, the argument being that the wind-driven circulation does not reach all the way to the ocean bottom so that an Ekman drag is not appropriate. This variation is called the 'Munk problem' or 'Munk model',[9] and if both drag and viscosity are present we have the 'Stommel–Munk' model. The particular form of the lateral friction used in the Munk problem is still somewhat hard to justify because it relies on an ill-founded eddy diffusion of relative vorticity (Chapter 13). Our treatment is brief, focusing on aspects that differ from the Stommel problem. The problem is to find and understand the solution to the (dimensional) equation

$$\beta\frac{\partial\psi}{\partial x} = \mathrm{curl}_z \boldsymbol{\tau}_T + \nu\nabla^2\zeta = \mathrm{curl}_z \boldsymbol{\tau}_T + \nu\nabla^4\psi \tag{19.42}$$

The Stommel and Munk models of the Wind-Driven Circulation

Formulation

- Vertically integrated planetary-geostrophic equations, or a homogeneous fluid with nonlinearity neglected.
- Friction parameterized by a linear drag (Stommel model) or a harmonic Newtonian viscosity (Munk model) or both (Stommel–Munk model).
- Flat bottomed ocean.

Variations on the theme include allowing nonlinearity in the vorticity equation, posing the problem in domains of various shapes, and allowing bottom topography (in particular sloping sidewalls). The solutions are most usefully calculated in the boundary-layer approximation, but some exact solutions exist.

Properties

- The transport in the Sverdrup interior is equatorwards for an anti-cyclonic wind-stress curl. This transport is exactly balanced by the poleward transport in the western boundary layer.
- There must be a boundary layer to satisfy mass conservation, and this must be a *western* boundary layer if the friction acts to provide a force of opposite sign to the motion itself. As there is a balance between friction and the β-effect, it is a 'frictional boundary layer'. The western location does not depend on the sign of the wind stress, nor on the sign of the Coriolis parameter, but it does depend on the sign of β, and so on the direction of rotation of the Earth.
- In the Stommel model the boundary layer width arises by noting that the terms $r\nabla^2\psi$ and $\beta\partial\psi/\partial x$ are in approximate balance in the western boundary layer, implying boundary-layer scale of $L_S = (r/\beta)$. If r, the inverse frictional time, is chosen to be $1/20\,\text{days}^{-1}$ then $L_S \approx 60\,\text{km}$, similar to the width of the Gulf Stream. Unless the wind has a special form the Sverdrup flow is non-zero on the zonal walls and there must also be boundary layers there, but they are weaker and less visible.
- In the Munk model the balance in western boundary layer is between $\nu\nabla^4\psi$ and $\beta\partial\psi/\partial x$, implying a scale of $L_M = (\nu/\beta)^{1/3}$. There are also weak boundary layers on the eastern and zonal walls, to satisfy the no-slip (or free slip) condition.

in a given domain, for example a square of side a. We need two boundary conditions at each wall to solve the problem uniquely, and as before for one of them we choose $\psi = 0$ to satisfy the no-normal-flow condition. For the other condition, two possibilities present themselves:

(i) Zero vorticity, or $\zeta = 0$. Since $\psi = 0$ along the boundary, this possibility is equivalent to $\partial^2\psi/\partial n^2 = 0$ where $\partial/\partial n$ denotes a derivative normal to the boundary. This is known as the 'free-slip' condition. At $x = 0$, for example, the condition becomes $\partial v/\partial x = 0$; that is, there is no horizontal shear at the boundary.

(ii) No flow along the boundary (the no-slip condition). That is $\psi_n = 0$ where the subscript denotes the normal derivative of the streamfunction. At $x = 0$ we have $v = 0$.

There is little a-priori justification for choosing either of these. The second choice would be demanded if ν were a molecular viscosity, but then we would have to resolve a molecular boundary layer which perhaps would be a few millimetres thick. Instead, ν must be interpreted as some form of eddy viscosity, as we discuss in Chapter 13. In that case, one might argue that the free-slip condition should be preferred, but in the absence of a proper theoretical basis for such eddy viscosities there is no truly rational way to make the choice. We will solve the no-slip problem.

Let the wind stress be the canonical $\tau^x = -\cos(\pi y/a)$. Then the interior (Sverdrup) flow is given by (19.34), as for the Stommel problem. This satisfies the no-slip boundary conditions at $y = 0, 1$, namely $\partial_y \psi = 0$, automatically. However, we need boundary layers at both the western and eastern boundaries, because the interior solution cannot satisfy all four boundary conditions required by (19.42). The eastern boundary layer will be relatively weak, and needed only to satisfy the no-slip condition but, as in the Stommel problem, there will be a strong western boundary layer, needed to satisfy the no-normal flow condition. How thick will this be? Inspection of (19.42) suggests that the frictional term and the β term will balance in a boundary layer of thickness of order L_M where

$$L_M = \left(\frac{\nu}{\beta}\right)^{1/3}. \tag{19.43}$$

nondimensionalizing (19.42) in a similar way to the Stommel problem yields

$$-\epsilon_M \nabla^4 \widehat{\psi} + \frac{\partial \widehat{\psi}}{\partial \widehat{x}} = \mathrm{curl}_z \widehat{\boldsymbol{\tau}}_T, \tag{19.44}$$

where $\epsilon_M = (\nu/\beta a^3)$. Considering only the western boundary layer correction, we let the solution be the sum of an interior (Sverdrup) streamfunction plus a boundary layer correction:

$$\widehat{\psi} = \psi_I + \phi_W(\alpha, \widehat{y}), \tag{19.45}$$

where α is a stretched coordinate such that $\widehat{x} = \epsilon\alpha$ where ϵ is some small parameter. Substituting (19.45) into (19.44) and subtracting the Sverdrup balance gives

$$-\epsilon_M \left(\nabla^4 \psi_I + \frac{1}{\epsilon^4}\frac{\partial^4 \phi_W}{\partial \alpha^4}\right) + \frac{1}{\epsilon}\frac{\partial \phi_W}{\partial \alpha} = 0. \tag{19.46}$$

A non-trivial balance is obtained when $\epsilon = \epsilon_M^{1/3}$, implying a dimensional western boundary thickness of $\epsilon a = (\nu/\beta)^{1/3}$, consistent with (19.43). Equation (19.46) then becomes, at leading order,

$$-\frac{\partial^4 \phi_W}{\partial \alpha^4} + \frac{\partial \phi_W}{\partial \alpha} = 0. \tag{19.47}$$

The boundary conditions on (19.47) are that:

(i) $\phi_W \to 0$ as $\alpha \to \infty$: this states that the perturbation decays as it extends into the interior;

(ii) $\phi_W = -\psi_I$ at $\widehat{x} = \alpha = 0$: this is the no-normal-flow condition on the meridional boundary.

(iii) $\partial \phi_W/\partial \widehat{x} = -\partial \psi_I/\partial \widehat{x}$ at $\widehat{x} = \alpha = 0$: this is the no-slip condition.

In addition, zonal boundary layers must exist at $\widehat{y} = 0, 1$ and another meridional boundary layer must exist at $\widehat{x} = 1$, in order to satisfy the no slip condition. Solving the full problem is a straightforward albeit non-trivial algebraic exercise and, omitting the weak zonal boundary layers at $\widehat{y} = 0, 1$ but including the eastern boundary layer correction, we eventually find the solution

$$\begin{aligned}\psi = \psi_I &- \mathrm{e}^{-\widehat{x}/(2\epsilon)}\left\{\psi_I(0, y)\left[\cos\left(\frac{\sqrt{3}\widehat{x}}{2\epsilon}\right) + \frac{1}{\sqrt{3}}\sin\left(\frac{\sqrt{3}\widehat{x}}{2\epsilon}\right)\right] + \frac{2\epsilon}{\sqrt{3}}\sin\left(\frac{\sqrt{3}\widehat{x}}{2\epsilon}\right)\frac{\partial \psi_I}{\partial \widehat{x}}\bigg|_{x=0}\right\} \\ &- \epsilon \mathrm{e}^{(\widehat{x}-1)/\epsilon}\frac{\partial \psi_I}{\partial \widehat{x}}\bigg|_{x=1},\end{aligned} \tag{19.48}$$

where $\epsilon = (\nu/\beta a^3)^{1/3}$. With the canonical wind stress, (19.27), the interior solution is given by

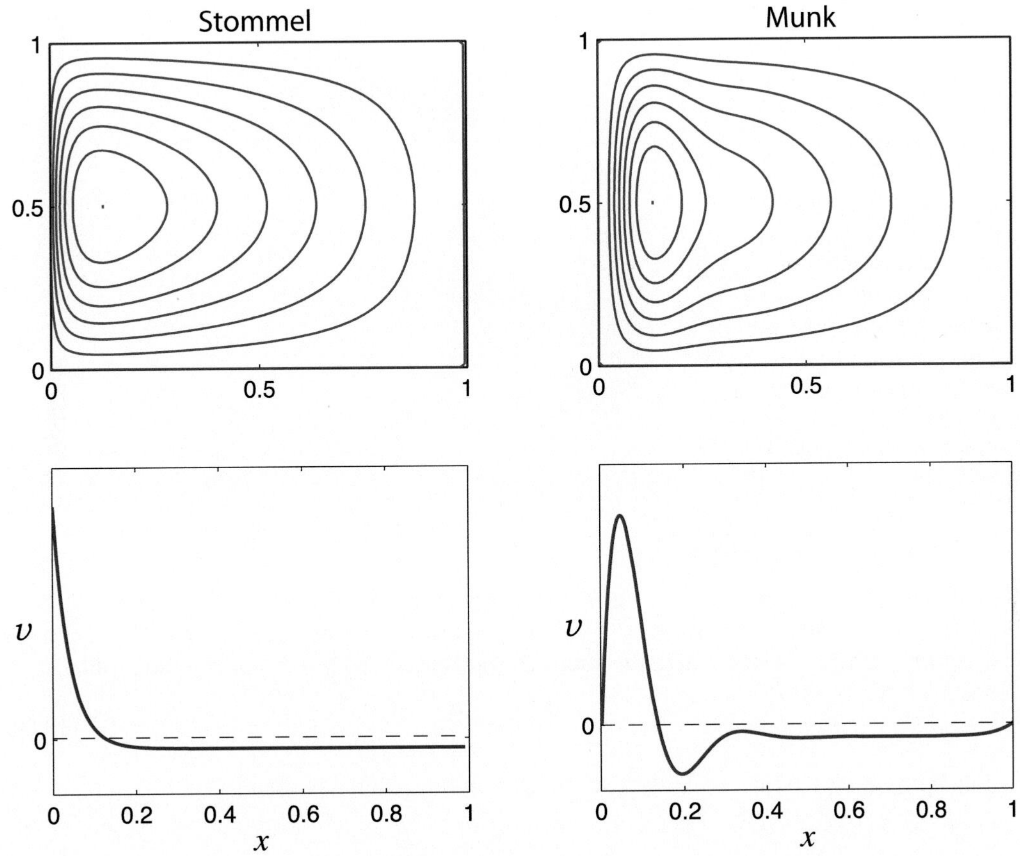

Fig. 19.7 The Stommel and Munk solutions, (19.49) with $\epsilon_S = \epsilon_M^{1/3} = 0.04$, with the wind stress $\tau = -\cos \pi y$, for $x, y \in (0, 1)$. Upper panels are contours of streamfunction in the x-y plane, and the flow is clockwise. The lower panels are plots of meridional velocity, v, as a function of x, in the centre of the domain ($y = 0.5$). The Munk solution can satisfy both no-normal flow and one other boundary condition at each wall, here chosen to be no-slip.

$\psi_I = \pi(1 - x) \sin \pi y$ and the above solution becomes

$$\widehat{\psi} = \pi \sin(\pi \widehat{y}) \left\{ 1 - \widehat{x} - e^{-\widehat{x}/(2\epsilon)} \left[\cos\left(\frac{\sqrt{3}\widehat{x}}{2\epsilon} \right) + \frac{1 - 2\epsilon}{\sqrt{3}} \sin\left(\frac{\sqrt{3}\widehat{x}}{2\epsilon} \right) \right] + \epsilon e^{(\widehat{x}-1)/\epsilon} \right\}. \qquad (19.49)$$

The solutions of this are plotted in Fig. 19.7. Note how the Munk layers bring the tangential as well as the normal velocity to zero. The eastern boundary layer has a similar thickness to the western boundary layer, but is not as dynamically important since its *raison d'etre* is to enable the no-slip condition to be satisfied, a relatively weak frictional constraint that manifests itself by a boundary layer in which the flow parallel to the boundary is slowed down. On the other hand the western boundary layer exists in order that the no-normal flow condition can be satisfied, which causes a qualitative change in the flow pattern. It should be emphasized that neither the Stommel nor the Munk models are accurate descriptors of the real ocean, but taken together the similarities of their solutions are a powerful argument for the relative insensitivity of the qualitative form of the solution to the detailed form of the friction. The models do in fact produce reasonably realistic patterns of large-scale flow in the major basins of the world, as illustrated in Fig. 19.8.

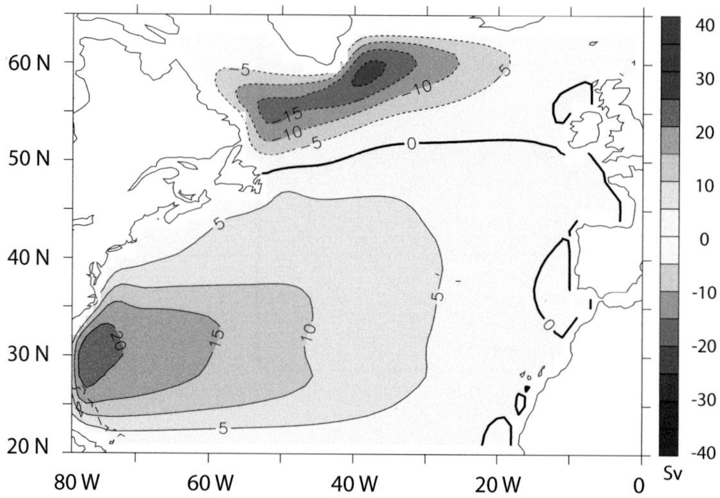

Fig. 19.8 The solution (streamfunction, in Sverdrups) to the Stommel–Munk problem numerically calculated for the North Atlantic, using the observed wind field. The model has realistic geometry, but is flat-bottomed.

The calculation reproduces the observed large-scale patterns, but the Gulf Stream and its extension are too diffuse, and its separation from the coast is a little too far north. Compare with Figs. 19.3 and 19.4.

19.3 ZONAL BOUNDARY LAYERS

The canonical wind stress $[\tau^x = -\tau_0 \cos(\pi y/a)]$ is special because its curl vanishes at $y = 0$ and a, and so the interior solution satisfies $\psi_I = 0$ at $y = 0$ and a. We cannot expect the real wind to be so accommodating. Consider, then, the Stommel problem forced by the linear wind profile,

$$\tau^x = \frac{\tau_0}{a}\left(y - \frac{a}{2}\right). \tag{19.50}$$

Scaling the variables in the usual way leads to the nondimensional problem

$$\epsilon_S \nabla^2 \widehat{\psi} + \frac{\partial \widehat{\psi}}{\partial \widehat{x}} = -1. \tag{19.51}$$

The interior flow obeys $\partial\widehat{\psi}/\partial\widehat{x} = -1$ everywhere, and evidently does not satisfy $\widehat{\psi} = 0$ at either $\widehat{y} = 0$ or $\widehat{y} = 1$; thus, boundary layers are needed at the meridional and zonal boundaries and we let the solution be the sum of five parts,

$$\widehat{\psi} = \psi_I + \phi_W + \phi_E + \phi_N + \phi_S, \tag{19.52}$$

with self-explanatory notation. The interior solution, ψ_I, is the solution of $\partial\psi_I/\partial\widehat{x} = -1$, so that a solution satisfying $\psi_I = 0$ at $\widehat{x} = 1$ is $\psi_I = (1 - \widehat{x})$ and there is no need for an eastern boundary layer. By the same methods we used in Section 19.1.3 the western boundary layer correction is easily found to be

$$\phi_W = -\mathrm{e}^{-\widehat{x}/\epsilon_S}. \tag{19.53}$$

It remains to find ϕ_N and ϕ_S, the boundary layer corrections at the northern and southern boundaries.

Consider the boundary layer correction at $y = 1$, and introduce the stretched coordinate α where $\epsilon'\alpha = y - 1$ and where ϵ' is a small parameter. Thus, we let $\phi_N = \phi_N(x, \alpha)$, and on substituting (19.52) into (19.51) we find

$$\epsilon_S\left(\nabla^2\psi_I + \frac{\partial^2\phi_N}{\partial\widehat{x}^2} + \frac{1}{\epsilon'^2}\frac{\partial^2\phi_N}{\partial\alpha^2}\right) + \frac{\partial\psi_I}{\partial\widehat{x}} + \frac{\partial\phi_N}{\partial\widehat{x}} = -1, \tag{19.54}$$

having neglected the small contributions from the other boundary layer streamfunctions (such as ϕ_S). To obtain a non-trivial balance we choose $\epsilon'^2 = \epsilon_S$ and obtain the dominant balance

$$\frac{\partial^2\phi_N}{\partial\alpha^2} + \frac{\partial\phi_N}{\partial\widehat{x}} = 0. \tag{19.55}$$

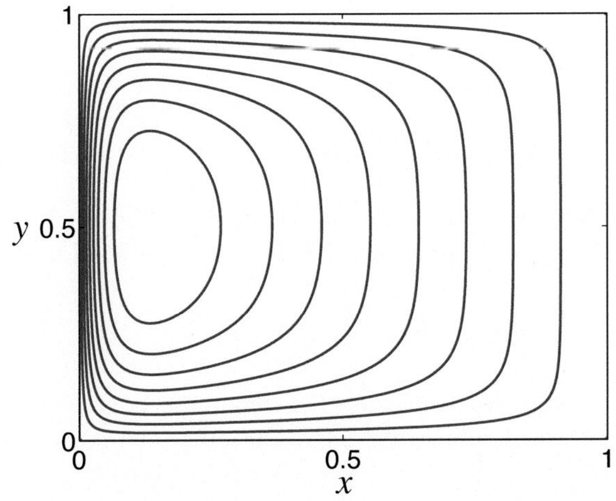

Fig. 19.9 Solutions to the Stommel problem with a wind stress that increases linearly from $y = 0$ to $y = 1$, as in (19.50). The interior solution is $\psi_I = (1 - x)$, or $v_I = -1$, necessitating zonal boundary layers at $y = 0$ and $y = 1$, as well as a western boundary layer at $x = 0$.

The boundary conditions necessary to complete the solution are:

(i) The total streamfunction (interior plus boundary correction) must vanish at the northern boundary; that is, $\phi_N(\hat{x}, \alpha = 0) = -\psi_I(\hat{x}, \hat{y} = 1) = -(1 - \hat{x})$.

(ii) The boundary solution should match to the interior streamfunction far from the boundary; that is $\phi_N(\hat{x}, \alpha \to -\infty) = 0$.

(iii) At the eastern boundary ϕ_N should also vanish, for otherwise it would provide a velocity into the eastern wall; that is $\phi_N(1, \alpha) = 0$.

The solution at the southern boundary is obtained in an analogous way, and the complete solution is illustrated in Fig. 19.9 (obtaining the solution analytically is quite algebraically tedious and is easier done numerically). The nondimensional thickness of the zonal (northern and southern) boundary layers is $\epsilon_S^{1/2}$, which, because it scales like the half power of a small number, is much thicker than the western boundary current. The thickness arises from the dimensional equations that dominate at the boundary, namely

$$r\frac{\partial^2 \psi}{\partial y^2} + \beta\frac{\partial \psi}{\partial x} = 0. \tag{19.56}$$

This follows from (19.18) by noting that the Laplacian operator must be dominated by the derivatives in the y-direction, and the β term has an (interior) component that annihilates the wind-stress curl, plus a boundary layer correction to balance the Laplacian. Inspection of (19.56) yields a dimensional thickness $L_Z \sim \sqrt{ra/\beta} = \epsilon_S^{1/2} a$, where a is the length scale in the x-direction.

19.4 ✦ THE NONLINEAR PROBLEM

In the nonlinear problem we seek solutions to

$$\frac{\partial \zeta}{\partial t} + J(\psi, \zeta) + \beta\frac{\partial \psi}{\partial x} = -r\nabla^2\psi + \mathrm{curl}_z \boldsymbol{\tau}_T + \nu\nabla^2\zeta, \tag{19.57}$$

which we have written in dimensional form. In the Stommel problem we set $\nu = 0$ and in the Munk problem we set $r = 0$. In general, solutions will be time-dependent and turbulent and this will create motion on small scales, so that ν cannot be neglected. The 'steady nonlinear Stommel–Munk problem' is

$$J(\psi, \zeta) + \beta\frac{\partial \psi}{\partial x} = -r\nabla^2\psi + \mathrm{curl}_z \boldsymbol{\tau}_T + \nu\nabla^2\zeta. \tag{19.58}$$

We can scale this by first supposing that the leading order balance is Sverdrupian (i.e., $\beta \partial\psi/\partial x \sim \mathrm{curl}_z \boldsymbol{\tau}_T$), from which we obtain the scales $\Psi = |\tau|/\beta$ and $U = |\tau|/(\beta L)$. Equation (19.58) may then be nondimensionalized to yield

$$R_\beta J(\widehat{\psi}, \widehat{\zeta}) + \frac{\partial \widehat{\psi}}{\partial \widehat{x}} = -\epsilon_S \nabla^2 \widehat{\psi} + \mathrm{curl}_z \widehat{\boldsymbol{\tau}}_T + \epsilon_M \nabla^2 \widehat{\zeta}, \tag{19.59}$$

where $R_\beta = U/\beta L^2 = |\tau|/(\beta^2 L^3)$, the β-Rossby number for this problem, is a measure of the nonlinearity. Evidently, the nonlinear term increases in importance with increasing wind stress and for a smaller domain.

19.4.1 A Perturbative Approach

A direct attack on the full nonlinear problem (19.58) is possible only through numerical methods, so first we shall explore the problem perturbatively, assuming the nonlinear term to be small; the analysis is straightforward, albeit messy. We begin with the Stommel problem, (19.59) with $\epsilon_M = 0$, and expand the streamfunction in terms of R_β,

$$\widehat{\psi} = \psi_0 + R_\beta \psi_1 + \dots . \tag{19.60}$$

Now substitute this into (19.59) and equate powers of R_β. The lowest-order problem is simply

$$\epsilon_S \nabla^2 \psi_0 + \frac{\partial \psi_0}{\partial \widehat{x}} = \mathrm{curl}_z, \widehat{\boldsymbol{\tau}} \tag{19.61}$$

which is the Stommel problem we have already solved. At the next order,

$$\epsilon_S \nabla^2 \psi_1 + \frac{\partial \psi_1}{\partial \widehat{x}} = J(\psi_0, \zeta_0). \tag{19.62}$$

This equation has precisely the same form as the Stommel problem, with the known nonlinear term on the right-hand side playing the part of the wind stress. The algebra to obtain the solution is rather tedious, because the right-hand side varies with both $\widehat{x}$ and $\widehat{y}$, but this is much ameliorated by the use of computer algebraic manipulation languages. For the canonical wind stress $\tau^x = -\tau_0 \cos(\pi \widehat{y})$ the corrected solution, in the boundary layer approximation and ignoring any corrections at the zonal boundaries, is found to be[10]

$$\widehat{\psi} \approx \sin(\pi \widehat{y})(1 - \widehat{x} - \mathrm{e}^{-\widehat{x}/\epsilon_S}) - \frac{R_\beta \pi^3}{2\epsilon_S^3} \sin(2\pi \widehat{y}) \widehat{x} \mathrm{e}^{-\widehat{x}/\epsilon_S}. \tag{19.63}$$

The solution is illustrated in Fig. 19.10. The perturbation is antisymmetric about $\widehat{y} = 1/2$, being positive for $\widehat{y} > 1/2$ and negative for $\widehat{y} < 1/2$. This tends to move the centre of the gyre polewards, narrowing and intensifying the flow in the poleward half of the western boundary current, whereas the western boundary current equatorwards of $\widehat{y} = 1/2$ is broadened and weakened. The net effect is that the centre of the gyre is pushed poleward — essentially because the western boundary current is advecting the vorticity of the gyre poleward. In the perturbation solution the advection is both by and of the linear Stommel solution; thus, negative vorticity is advected polewards, intensifying the gyre in its poleward half, weakening it in its equatorward half. The solution illustrated in Fig. 19.10 has $\epsilon = 0.04$ and $R_\beta = 10^{-4}$; for larger values of R_β the perturbation itself starts to dominate.

The problem with this perturbative approach is that a boundary layer solution to the Stommel problem does not calculate derivatives accurately, so that the nonlinear term $J(\psi, \nabla^2\psi)$ is poorly approximated in the western boundary layer; however, in the interior where the errors are small the perturbative correction is negligible. A more accurate perturbative approach begins with the *exact* solution to the Stommel problem, and then proceeds in the same way. However, the analytic effort is considerable, and the intuitive sense of the way nonlinearities affect the solution is not apparent.

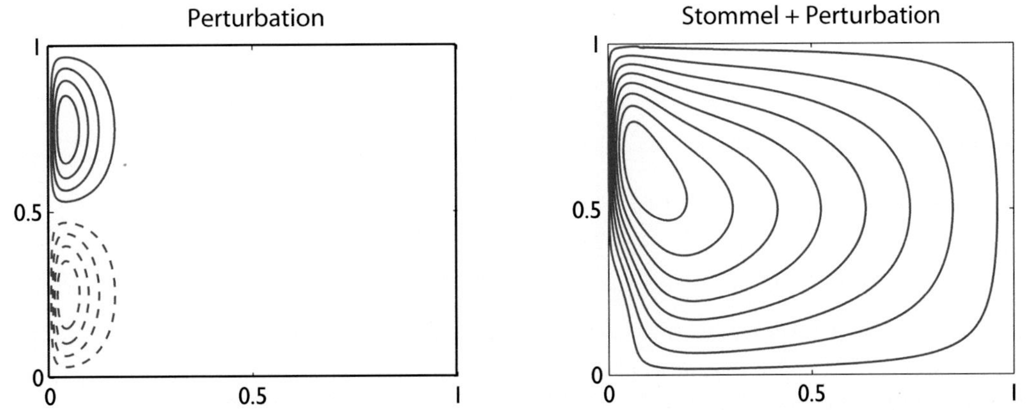

Fig. 19.10 The nonlinear perturbation solution of the Stommel problem, calculated according to (19.63). On the left is the perturbation, $-R_\beta \pi^3/(2\epsilon_S^2)\sin(2\pi y)xe^{-x/\epsilon_S}$, and on the right is the reconstituted solution, using $R_\beta = 10^{-4}$ and $\epsilon = 0.04$. Dashed contours are negative.

19.4.2 A Numerical Approach

Fully nonlinear solutions show qualitatively similar effects to those seen in the perturbative solutions, as we see in Fig. 19.11, where the solutions to (19.59) for the Stommel and Munk problems are obtained numerically by Newton's method.[11]

Just as with the perturbative procedure, for both the Stommel and Munk problems small values of nonlinearity lead to the poleward advection of the gyre's anticyclonic vorticity in the western boundary current, strengthening and intensifying the boundary current in the northwest corner. A higher level of nonlinearity results in a strong recirculating regime in the upper westward quadrant, and ultimately much of the gyre's transport is confined to this regime. The western boundary current itself becomes less noticeable as nonlinearity increases, the more nonlinear solutions have a much greater degree of east-west symmetry than the linear ones, just as the fully nonlinear Fofonoff solutions (Section 19.5.3).

The qualitative effects above do not depend on the precise formulation of the model, but the boundary conditions do play an important role in the detailed solution. For example, for a given value of R_β, nonlinearity has a stronger effect in the Munk problem with slip boundary conditions than with no-slip, because in the latter the velocity is reduced to zero at the boundary with a corresponding reduction in the advection term. However, these solutions themselves are unlikely to be relevant for larger values of nonlinearity, because then the flow becomes hydrodynamically unstable

19.5 ✦ INERTIAL SOLUTIONS

In this section we further explore inertial effects and ask: might purely inertial effects be sufficient to satisfy boundary conditions at the western boundary? Can we envision a purely inertial gyre circulation? Our question is motivated by the steady wind-driven, Rayleigh-damped quasi-geostrophic equation, namely

$$J(\psi, \nabla^2\psi) + \beta\frac{\partial \psi}{\partial x} = F_\tau - r\nabla^2\psi, \tag{19.64}$$

where F_τ is the wind forcing. Since the inertial (advective) terms are of a higher order than the linear terms, indeed they are of a higher order than the Rayleigh drag, it is natural to wonder if they themselves might serve to satisfy the no-normal flow condition on the western boundary, without

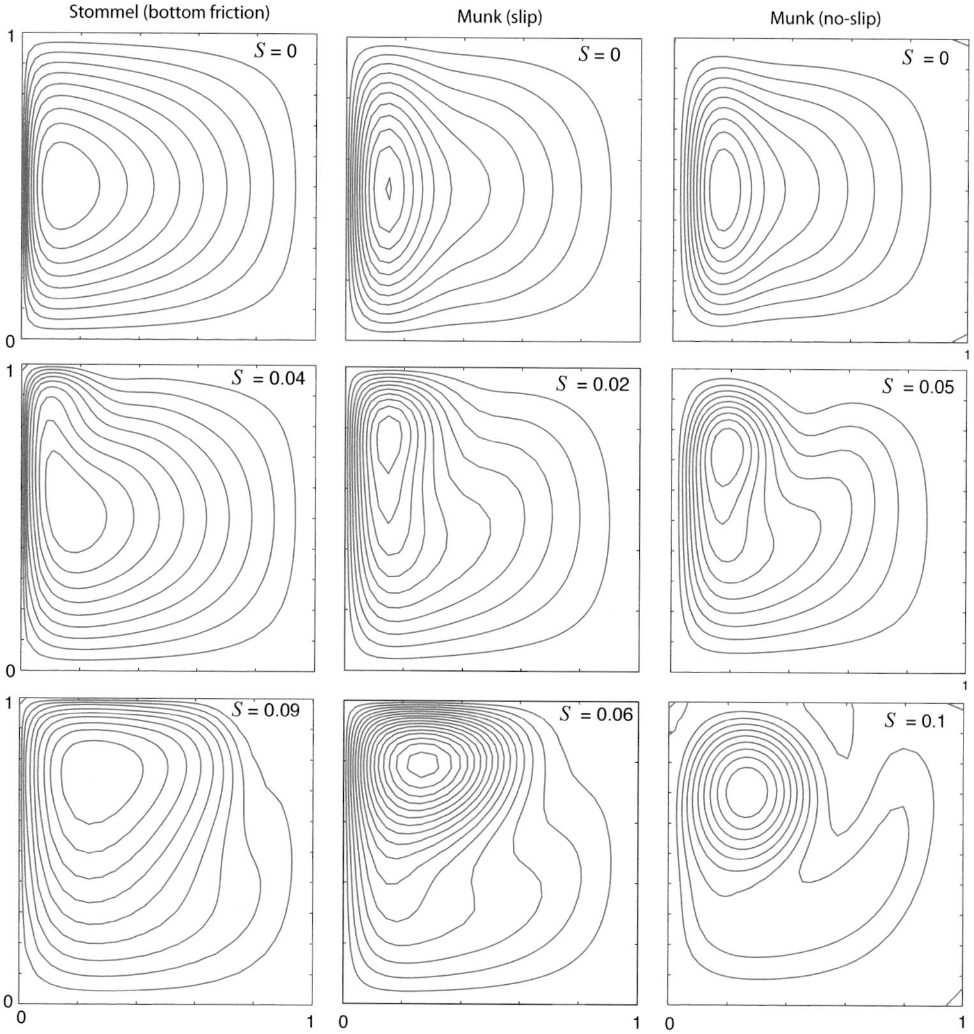

Fig. 19.11 Streamfunctions in solutions of the nonlinear Stommel and Munk problems, obtained numerically with a Newton's method, for various values of the nonlinearity parameter $S = R_\beta^{1/2}$. As in the perturbation solution, for small values of nonlinearity the centre of the gyre moves polewards, strengthening the boundary current in the north-western quadrant (for a northern-hemisphere solution). As nonlinearity increases, the recirculation of the gyre dominates, and the solutions become increasingly inertial.[12]

recourse to rather ill-defined frictional terms. The answer is no, as we see below, but nevertheless nonlinear effects may be important in the western boundary layer even if they are small in the interior.

19.5.1 The Need for Friction

Consider the steady barotropic flow satisfying

$$\boldsymbol{u} \cdot \nabla q = \mathrm{curl}_z \boldsymbol{\tau}_T + \mathrm{Fr}, \tag{19.65}$$

where $q = \nabla^2\psi + \beta y$ and Fr represents frictional effects. Noting that $\boldsymbol{u}$ is divergence-free and integrating the left-hand side of (19.65) over the area, A, between two closed streamlines (ψ_1 and

ψ_2, say) and using the divergence theorem we find

$$\int_A \nabla \cdot (\boldsymbol{u}q)\,\mathrm{d}A = \oint_{\psi_1} \boldsymbol{u}q \cdot \boldsymbol{n}\,\mathrm{d}l - \oint_{\psi_2} \boldsymbol{u}q \cdot \boldsymbol{n}\,\mathrm{d}l = 0. \tag{19.66}$$

Here, $\boldsymbol{n}$ is the unit vector normal to the streamline so that $\boldsymbol{u} \cdot \boldsymbol{n} = 0$. The integral of the wind-stress curl over the same area will not, in general, be zero. Now, we can take these two streamlines as close together as we wish; thus, a balance in (19.65) can only be achieved if *every* closed contour passes through a region where frictional effects are non-zero. This does not mean that nonlinear terms may not locally dominate the friction, just that friction must be somewhere important. In the Stommel and Munk problems, it means that every streamline must pass through the frictional western boundary layer.

Frictional and inertial scales

The ratio of the size of the nonlinear terms to the linear terms is given by the β-Rossby number, $R_\beta = U/(\beta L^2)$. In the western boundary layer the length scales can be expected to be much smaller than the basin scale, and if the balance in the western boundary layer were between the nonlinear and beta term, as in

$$u\frac{\partial}{\partial x}\nabla^2\psi \sim \beta\frac{\partial \psi}{\partial x}, \tag{19.67}$$

then the *inertial boundary layer thickness*, δ_I, is given by

$$\delta_I = \left(\frac{U}{\beta}\right)^{1/2}. \tag{19.68}$$

This, of course, gives $R_\beta = 1$ if $L = \delta_I$. The more energetic the flow, the wider the region where non-linearity is important, and the corresponding scale is sometimes called the Charney thickness.[13]

The linearized Stommel equation has a boundary layer of dimensional thickness of order $\delta_S = (r/\beta)$, obtained by equating $\beta\,\partial\psi/\partial x$ and $r\nabla^2\psi$. This thickness is equal to the inertial boundary layer thickness when $U = (r^2/\beta)$. If $\delta_I > \delta_S$, i.e., if $U > (r^2/\beta)$, then nonlinearity *must* be important in the western boundary layer, because the nonlinear terms are at least as important as the beta term in (19.64). On the other hand, if the Stommel boundary layer is wider than the inertial boundary layer, there is no obvious need for nonlinearity to be important, since the Stommel boundary layer generates no length scales smaller than δ_I and R_β remains small.

19.5.2 Attempting an Inertial Western Boundary Solution

Although friction must be important, it is nevertheless instructive to try to find a purely inertial solution for the western boundary layer, and to see if and how the attempt fails. Let us suppose that the interior solution has purely zonal flow, with flow either towards or away from the western boundary current, and consider the dimensional equation of motion

$$J(\psi, \nabla^2\psi + \beta y) = 0. \tag{19.69}$$

This has the general solution

$$\nabla^2\psi + \beta y = G(\psi), \tag{19.70}$$

where G is an arbitrary function of its argument. Consider first the case with flow entering the boundary layer with a local velocity $-U$ (i.e., westward); that is, $\psi = Uy$. The potential vorticity in this region (just outside our putative boundary layer) is $Q_I = \beta y$. Thus, the relation between potential vorticity and streamfunction is $Q_I = \beta\psi_I/U$ and so

$$G(\psi) = \frac{\beta\psi}{U}. \tag{19.71}$$

Because we are assuming the flow is inviscid, the fluid will preserve this relationship between potential vorticity and streamfunction *even as it moves through the western boundary layer.* Thus, using (19.70) and (19.71), the flow in the interior *and* in the boundary layer is given by the solution of

$$\nabla^2\psi - \frac{\beta\psi}{U} = -\beta y, \tag{19.72}$$

with $\psi = 0$ on the boundary. To obtain a solution, we let $\psi = \psi_I + \phi$, where $\psi_I = Uy$ is the particular solution to (19.72) (and, of course, the interior flow). The boundary layer correction then obeys

$$\nabla^2\phi - \frac{\beta\phi}{U} = 0, \tag{19.73}$$

with $\phi = -\psi_I$ at $x = 0$. Now, in the boundary layer, length scales in the x-direction are much smaller than length scales in the y-direction, and so (19.73) becomes approximately

$$\frac{\partial^2\phi}{\partial x^2} - \frac{\beta\phi}{U} = 0. \tag{19.74}$$

(We could formalize this procedure by nondimensionalizing and introducing a stretched coordinate, as we did for the Stommel problem.) Solutions of (19.74) are $\phi = -\psi_I e^{-x/\delta_I}$, where $\delta_I = (U/\beta)^{1/2}$, and so the full solution is

$$\psi = \psi_I(1 - e^{-x/\delta_I}). \tag{19.75}$$

Clearly, this solution smoothly transitions into the interior solution for large x.

What about solutions exiting the boundary layer? We might attempt a similar procedure, but now the interior boundary condition that we must match is that of eastward flow, namely $\psi_I = -Uy$. Thus, analogously to (19.72), we seek solutions to the problem,

$$\nabla^2\psi + \frac{\beta\psi}{U} = -\beta y, \tag{19.76}$$

for which the homogeneous (boundary layer) equation is

$$\frac{\partial^2\phi}{\partial x^2} + \frac{\beta\phi}{U} = 0. \tag{19.77}$$

Solutions of (19.77) are

$$\psi = \psi_I(1 - e^{-x/\delta_I^*}), \tag{19.78}$$

where $\delta_I^* = i\,\delta_I = i\,(U/\beta)^{1/2}$, and may be compared with (19.75). The solution is therefore wave-like, and does not transition smoothly to the interior flow. This suggests that friction must be important in allowing the boundary layer to smoothly connect to the interior for reasons connected to the conservation of potential vorticity, as explained heuristically in the next subsection.[14]

In addition to this transition problem, we note that (19.75) and (19.78) together do not constitute a globally acceptable inviscid solution, for the simple reason that the flow in the westward flowing interior region has a different potential vorticity–streamfunction (q–ψ) relationship than does the eastwards flowing interior. If these two regions are connected by a boundary layer, the flow in that boundary layer must be viscous or unsteady since the value of potential vorticity on the streamlines has changed, whereas inviscid flow conserves potential vorticity.

The connection between the boundary layer and the interior

We have seen that solutions with a boundary layer character that blend smoothly to a flow interior exist only for westward interior flow ($u_I = -U < 0$): an inertial western boundary layer on a beta-plane evidently cannot easily release fluid into the interior. The underlying reason for this stems from the conservation of potential vorticity, as we now explain (and see Fig. 19.12).

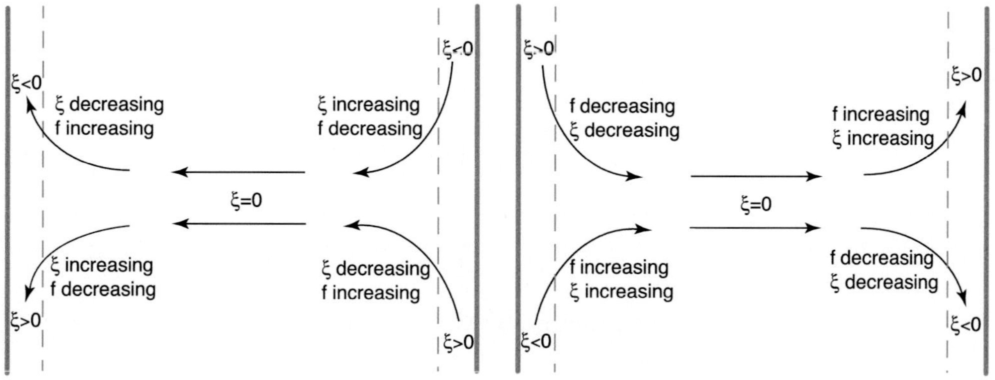

Fig. 19.12 Putative inertial boundary layers connected to a westward flowing interior flow (left panel) or eastwards flowing interior flow (right panel), in the Northern Hemisphere. Westward flow into the western boundary layer, or flow emerging from an eastern boundary layer, is able to conserve its potential vorticity through a balance between changes in relative vorticity and Coriolis parameter. But flow cannot emerge smoothly from a western boundary layer into an eastwards flowing interior and still conserve its potential vorticity. The right panel thus has inconsistent dynamics.

Conservation of potential vorticity demands that, in barotropic flow, $v_x - u_y + \beta y$ is a constant on streamlines. In the interior, relative vorticity, ζ, is zero, and in the meridional boundary layers it is effectively v_x. Consider first the case with westward flow in the interior (left panel of Fig. 19.12). Fluid from the irrotational interior approaches the western boundary where it is deflected either polewards or equatorwards. In the former case its relative vorticity falls, so allowing potential vorticity to be conserved because the reduction of ζ can be balanced by an increase in the planetary vorticity, f. Similarly, flow deflected southwards produces positive relative vorticity, which is compensated for by the reduced value of f. In the eastern boundary layer, southwards (northwards) moving flow has negative (positive) relative vorticity. As it emerges into the interior its relative vorticity increases (decreases), this being balanced by a fall (rise) in the value of f. Thus, we see that the solution with a westward flowing interior can indeed conserve potential vorticity, at both eastern and western boundaries.

On the other hand suppose the interior flow were eastwards (right panel of Fig. 19.12). Flow moving polewards in the western boundary layer has negative relative vorticity. It cannot be freely released into an irrotational interior because f and ζ would then both need to increase, violating potential vorticity conservation. Flow moving southwards with positive relative vorticity similarly is trapped within the western boundary current, unless it meets a zonal boundary which allows an eastward moving boundary current with positive relative vorticity. Similar arguments show that an eastern boundary current cannot entrain fluid from an eastward flowing irrotational interior.

One might ask, why cannot we simply reverse the trajectory of all the fluid parcels in an inviscid flow and thereby obtain a solution with an eastward-flowing interior? The answer is that such a flow will only be a solution if we also reverse the direction of the Earth's rotation, and so reverse the sign of β, and hence the location of the frictional boundary current.

19.5.3 A Fully Inertial Approach: the Fofonoff Model

Rather than attempt to match an inertial boundary layer with an interior Sverdrup flow, we may look for a purely inertial solution that holds basinwide, and such a construction is known as the *Fofonoff model*.[15] That is, we seek global solutions to the inviscid, unforced problem,

$$J(\psi, \nabla^2\psi + \beta y) = 0. \tag{19.79}$$

We should not regard this problem as representing even a very idealized wind-driven ocean; rather, we may hope to learn about the properties of purely inertial solutions and this might, in turn, tell us something about the ocean circulation.

The general solution to (19.79) is

$$\nabla^2\psi + \beta y = Q(\psi), \tag{19.80}$$

where $Q(\psi)$ is an arbitrary function of its argument. For simplicity we choose the linear form,

$$Q(\psi) = A\psi + B, \tag{19.81}$$

where $A = \beta/U$ and $B = \beta y_0$, where U and y_0 are arbitrary constants. Thus, (19.80) becomes

$$\left(\nabla^2 - \frac{\beta}{U}\right)\psi = \beta(y_0 - y). \tag{19.82}$$

We will further choose $\beta/U > 0$, which we anticipate will provide a westward-flowing interior flow, and which (from our experience in the previous section) is more likely to provide a meaningful solution than an eastward interior, and we will use boundary-layer methods to find a solution. A natural scaling for ψ is UL, where L is the domain size, and with this the nondimensional problem is

$$(\epsilon_F\nabla^2 - 1)\widehat{\psi} = \widehat{y}_0 - \widehat{y}, \tag{19.83}$$

where $\epsilon_F = U/(\beta L^2)$ and $\widehat{y} = y/L$. If we take ϵ_F to be small (note that $\epsilon_F = R_\beta$) we can find a solution by boundary-layer methods similar to those used in Section 19.1.3 for the Stommel problem. Thus, we write $\widehat{\psi} = \widehat{\psi}_I + \widehat{\phi}$ where $\widehat{\psi}_I = \widehat{y} - \widehat{y}_0$ (dimensionally, $\psi_I = U(y - y_0)$) and $\widehat{\phi}$ is the boundary layer correction, to be calculated separately for each boundary using

$$\epsilon_F\nabla^2\widehat{\phi} - \widehat{\phi} = 0 \tag{19.84}$$

and the boundary condition that $\widehat{\phi} + \widehat{\psi}_I = 0$. For example, at the northern boundary, $\widehat{y} = \widehat{y}_N$, the y-derivatives will dominate and (19.84) may be approximated by

$$\epsilon_F\frac{\partial^2\widehat{\phi}}{\partial\widehat{y}^2} - \widehat{\phi} = 0, \tag{19.85}$$

with solution

$$\widehat{\phi} = B\exp[-(\widehat{y}_N - \widehat{y})/\epsilon_F^{1/2}], \tag{19.86}$$

where $B = \widehat{y}_0 - \widehat{y}_N$, hence satisfying the boundary condition that $\widehat{\phi}(\widehat{x}, \widehat{y}_N) = -\widehat{\psi}_I(\widehat{x}, \widehat{y}_N)$. We follow a similar procedure at the other boundaries to obtain the full solution, and in dimensional form this is

$$\psi = U(y - y_0)\left[1 - e^{-x/\delta_I} - e^{-(x_E - x)/\delta_I}\right] + U(y_0 - y_N)e^{-(y_N - y)/\delta_I} + U y_0 e^{-y/\delta_I}, \tag{19.87}$$

where $\delta_I = (U/\beta)^{1/2}$ is the boundary layer thickness. Evidently, only positive values of U corresponding to a westward interior flow give boundary-layer solutions that decay into the interior.

A typical solution is illustrated in Fig. 19.13. On approaching the western boundary layer, the interior flow bifurcates at $y = y_0$. The western boundary layer, of width δ_I, accelerates away from this point, being constantly fed by the interior flow. The westward return flow occurs in zonal boundary layers at the northern and southern edges, also of width δ_I. Flow along the eastern boundary layers is constantly being decelerated, because it is feeding the interior. If one of the zonal boundaries corresponds to y_0 (e.g., if $y_N = y_0$) there would be no boundary layer along it, since ψ is already zero at $y = y_0$. Rather, there would be westward flow along it, just as in the interior. Indeed, a slippery wall placed at $y = 0.5$ would have no effect on the solution illustrated in Fig. 19.13.

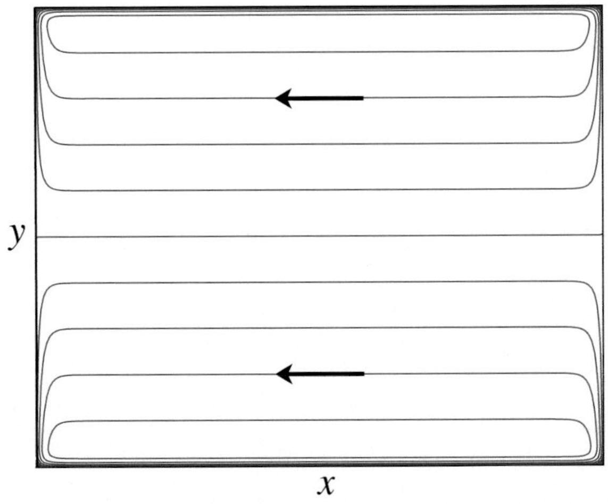

Fig. 19.13 The Fofonoff solution. Plotted are contours (streamlines) of (19.87) in the plane $0 < x < x_E$, $0 < y < y_N$ with $U = 1$, $y_N = 1$, $x_E = y_N = 1$, $y_0 = 0.5$ and $\delta_I = 0.05$. The interior flow is westward everywhere, and $\psi = 0$ at $y = y_0$. In addition, boundary layers of thickness $\delta_I = \sqrt{U/\beta}$ bring the solution to zero at $x = (0, x_E)$ and $y = (0, y_N)$, excepting small regions at the corners.

19.6 ♦ TOPOGRAPHIC EFFECTS ON WESTERN BOUNDARY CURRENTS

The above sections have emphasized the role of friction in satisfying the boundary conditions in the west. However, we should certainly not think of friction as being the *cause* of the western boundary layer and in this section we shall show that if there are sloping sidewalls the role of friction is significantly different, and the western boundary current may even be largely inviscid![16] The key point is that the flow may be inviscid if it is able to follow potential vorticity contours. In a flat-bottomed western boundary layer the flow is moving to larger values of f (and not as a direct response to the wind) and so the flow *must* be frictional. However, if the sidewalls are sloping, then the flow may preserve its potential vorticity (approximately, its value of f/h) if it moves offshore as it moves polewards. In the treatment below, we focus on homogeneous fluids, noting that the interaction of topography and stratification is a subtle and rather complex problem.

19.6.1 Homogeneous Model

The potential vorticity evolution equation for a homogeneous model with topography and a rigid lid may be written as:

$$\frac{\mathrm{D}q}{\mathrm{D}t} = \frac{F}{h}, \tag{19.88}$$

where $q = (\zeta + f)/h$ where $h = h(x, y)$ is the time-independent depth of the fluid, and F represents vorticity forcing and frictional terms. The advecting velocity is determined by noting that the mass conservation equation is just $\nabla \cdot [\boldsymbol{u}h(x, y)] = 0$, which allows us to define the mass-transport streamfunction ψ such that

$$u = -\frac{1}{h}\frac{\partial \psi}{\partial y}, \qquad v = \frac{1}{h}\frac{\partial \psi}{\partial x}. \tag{19.89}$$

The streamfunction itself is obtained from the vorticity by solving the elliptic equation,

$$\nabla \cdot \left(\frac{1}{h} \nabla \psi \right) = \zeta = qh - f. \tag{19.90}$$

Equations (19.88), (19.90) and (19.89) form a closed system. Unlike the quasi-geostrophic case, neither the Rossby number nor the topography need be small for the model to be valid. Including finite size topography but not stratification is not especially realistic *vis-à-vis* the real ocean, but nevertheless this model is physically realizable and a useful tool.

The 'topographic Stommel problem' is obtained by neglecting relative vorticity in the potential vorticity in (19.88) (i.e., let $q = f/h$) and using a linear drag acting on the vertically integrated fields

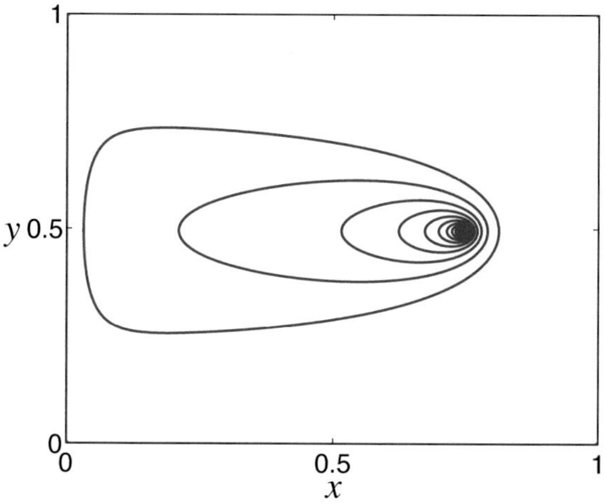

Fig. 19.14 The β-plume, namely the Green function for the Stommel problem. Specifically we plot the solution of (19.93) with $\psi = 0$ at the walls, and a delta-function source at $x = 0.75$, $y = 0.5$.

The streamfunction trails westward from the source, as if it were a tracer being diffused while being advected westward along lines of constant f.

for friction, and a wind-stress curl forcing. Multiplying (19.88) by h, expanding the advective term and omitting time dependence gives

$$J\left(\psi, \frac{f}{h}\right) = -r\nabla^2\psi + \mathrm{curl}_z(\boldsymbol{\tau}_T/h), \tag{19.91}$$

where the boundary conditions are $\psi = 0$ at the domain edges and $\boldsymbol{\tau}_T$ represents the wind stress at the top.

19.6.2 Advective Dynamics

An illuminating way to begin to study the problem is to write (19.91) in the form[17]

$$J(\Psi, \psi) = +r\nabla^2\psi - \mathrm{curl}_z(\boldsymbol{\tau}_T/h), \tag{19.92}$$

where $\Psi \equiv f/h$. Thus, noting that $J(\Psi, \psi) = \boldsymbol{U} \cdot \nabla\psi$ where $\boldsymbol{U} = \mathbf{k} \times \nabla(f/h)$, we regard ψ as being advected by the pseudovelocity $\boldsymbol{U}$. This advection is along f/h contours and is quasi-westward, meaning that high values of potential vorticity lie to the right. Equation (19.92) is then an advection-diffusion equation for the tracer ψ, with the 'source' of ψ being the wind stress curl with ψ being diffused by the first term on the right-hand side of (19.92), and advected by $\boldsymbol{U}$. This same interpretation applies to the original Stommel problem, of course, where the pseudovelocity, $-\beta\mathbf{i}$, is purely westward, and it is useful to first revisit this problem.

Consider, then, a flat-bottomed ocean, where the wind-stress curl is just a point source at $\boldsymbol{x}_0$. With $\Psi = f$, (19.92) becomes

$$r\nabla^2\psi + \beta\frac{\partial\psi}{\partial x} = \delta(\boldsymbol{x} - \boldsymbol{x}_0). \tag{19.93}$$

This can be transformed to a Helmholtz equation by writing $\phi = \psi\exp(\beta x/2r)$, giving $\nabla^2\phi - [\beta/(2r)]^2\phi = \delta(\boldsymbol{x} - \boldsymbol{x}_0)$. This may then be solved exactly, and the solution (for ψ) is illustrated in Fig. 19.14 — this is the Green function for the Stommel problem. The tracer ψ is 'advected' westward along f/h contours — lines of latitude in this case — spreading diffusively as it goes; the resulting structure is called a β-*plume*. The western boundary layer results as a consequence of the f/h contours colliding with the western boundary, along with the need to satisfy the boundary condition $\psi = 0$. If there were no diffusion at all, ψ would just propagate westward from the source, and if in addition the source were spatially distributed, for example as $\sin \pi y$, the solution streamfunction would represent Sverdrup interior flow. This case is illustrated in Fig. 19.15(a), which

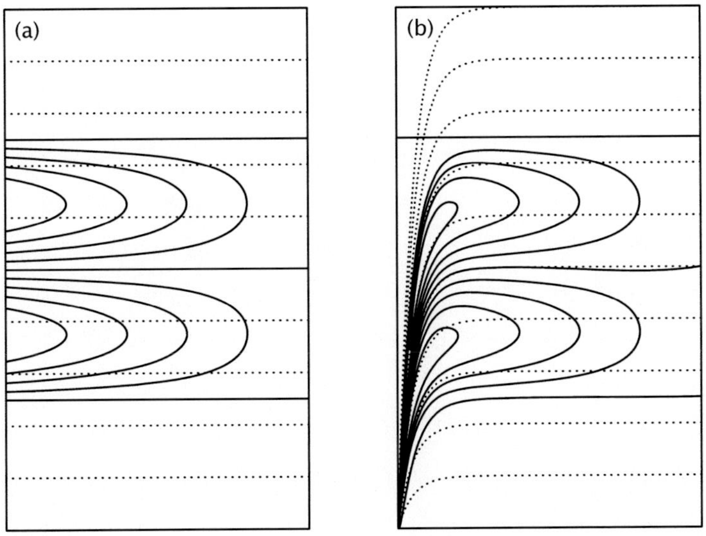

Fig. 19.15 The two-gyre Sverdrup flow (solid contours) for (a) a flat-bottomed domain, and (b) a domain with sloping sidewalls. The f/h contours are dotted.[18]

shows the solution to (19.93) with $r = 0$ and with the right-hand side replaced by a conventional 'two-gyre' wind stress.

Now consider the case with a sloping sidewall on the western boundary. The f/h contours (the dotted lines in Fig. 19.15b) tend to converge at the southwest corner of the domain, and only where f/h contours intersect the boundary is a diffusive boundary layer required. In terms of the interpretation above, wind stress provides a source for the streamfunction ψ and the latter is advected *pseudowestward* — i.e., along potential vorticity contours, with higher values to the right. The source in this case is distributed over the entire domain, but the contours all converge in the southwest corner. The (numerically obtained) solution to the associated Stommel problem is illustrated in Fig. 19.16, and the western boundary current in this case is no longer a frictional boundary layer. Friction is necessarily important where the flow crosses f/h contours; linear theory suggests that this will occur at the southwest corner. It also occurs on the western boundary where the f/h contours are densely packed and the vorticity in the topographic Sverdrup flow is large, and the friction enables the flow to move across the f/h contours. (In the flat bottomed case the f/h contours are zonal and friction allows the flow to move meridionally.)

19.6.3 Bottom Pressure Stress and Form Drag

In the homogeneous problem f and h appear only in the combination f/h, and we may solve the problem entirely without considering pressure effects. It is nevertheless informative to think about how the pressure interacts with the topography to produce meridional flow, and as a way toward addressing the effects of stratification. The geostrophic momentum equation is

$$\boldsymbol{f} \times \boldsymbol{u} = -\nabla\phi + \boldsymbol{F}, \tag{19.94}$$

where $\boldsymbol{F}$ represents both wind forcing and frictional terms. Integrating this over the depth of the ocean (with $z = 0$ at the top), and using the Leibnitz rule ($\nabla \int_{\eta_B}^{0} \phi \, \mathrm{d}z = \int_{\eta_B}^{0} \nabla\phi \, \mathrm{d}z - \phi_B \nabla \eta_B$) to evaluate the pressure term, gives

$$\boldsymbol{f} \times \overline{\boldsymbol{u}} = -\nabla\overline{\phi} - \phi_B \nabla \eta_B + \overline{\boldsymbol{F}}. \tag{19.95}$$

Here, the overbar denotes a vertical integral (e.g., $\overline{\boldsymbol{u}} = \int_{\eta_B}^{0} \boldsymbol{u} \, \mathrm{d}z$), ϕ_B is the pressure at $z = \eta_B$ and η_B is the z-coordinate of the bottom topography. We take the top of the ocean at $z = 0$, and note

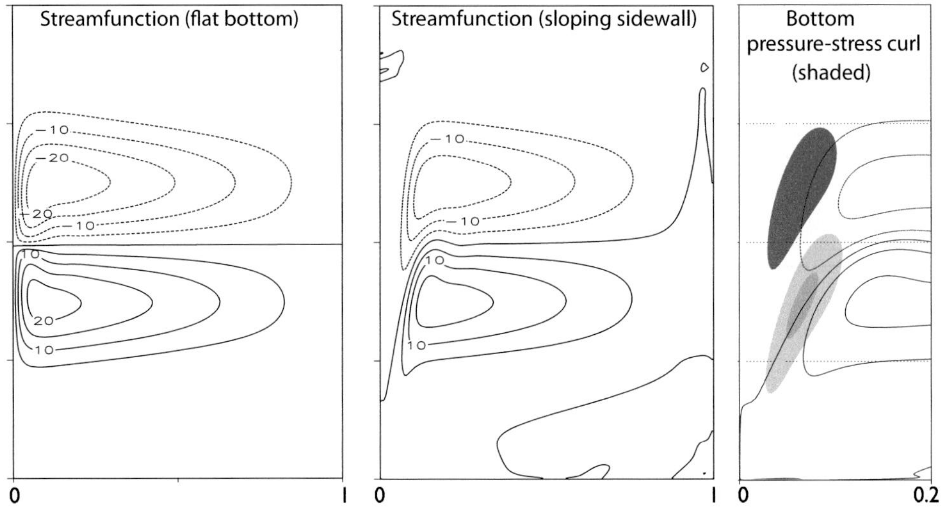

Fig. 19.16 The numerically obtained steady-state solution to the homogeneous problem with a two-gyre forcing and friction, for a flat-bottomed domain and a domain with sloping western sidewall. The shaded regions in the right panel show the regions where bottom pressure-stress curl is important in the meridional flow of the western boundary currents.[19]

that $\nabla\eta_B = -\nabla h$ where h is the fluid thickness. The second term on the right-hand side of (19.95) is the stress on the fluid due to a correlation between the pressure gradient and the topography at the ocean bottom — it is the bottom *form drag*, first encountered in Section 3.6.

Taking the curl of (19.95) gives

$$\beta\overline{v} = -\mathrm{curl}_z(\phi_B\nabla\eta_B) + \mathrm{curl}_z\overline{\boldsymbol{F}} = -J(\phi_B, \eta_B) + \mathrm{curl}_z\overline{\boldsymbol{F}}. \tag{19.96}$$

This equation holds for both a stratified and a homogeneous fluid. The first term on the right-hand side is the bottom pressure-stress curl, or the form-drag curl. (It is also sometimes informally referred to as the bottom pressure torque.)

Equation (19.96) is similar to (19.91), in that both arise from (19.95). To derive an equation with the same form as (19.91) but valid for a stratified fluid, we write the vertical integral of the pressure as

$$\int_{-h}^{0}\phi\,\mathrm{d}z = \int_{-h}^{0}\left[\mathrm{d}(\phi z) - z(\partial\phi/\partial z)\,\mathrm{d}z\right] = \int_{-h}^{0}\left[\mathrm{d}(\phi z) - zb\,\mathrm{d}z\right] = h\phi_B + \varGamma, \tag{19.97}$$

using hydrostatic balance, and where $\varGamma \equiv -\int_{-h}^{0} zb\,\mathrm{d}z$. Using (19.97) in (19.95) we obtain:

$$\boldsymbol{f}\times\overline{\boldsymbol{u}} = -h\nabla\phi_B + \nabla\varGamma + \overline{\boldsymbol{F}}. \tag{19.98}$$

The curl of this equation just gives back (19.96), but if we divide by h before taking the curl we obtain

$$J(\psi, f/h) + J(h^{-1}, \varGamma) = \mathrm{curl}_z(\overline{\boldsymbol{F}}/h). \tag{19.99}$$

The second term in (19.99) is known as the JEBAR TERM — joint effect of baroclinicity and relief — and it couples the depth integrated flow with the baroclinic flow.[20] Thus, if the bottom is not flat, the Stommel–Munk models are not solutions for the vertically integrated flow. If the stratification vanishes (i.e., b is constant) then $\varGamma$ is a function of h alone and $J(h^{-1}, \varGamma) = 0$, and (19.99) reprises (19.91), given an appropriate choice of $\overline{\boldsymbol{F}}$.

Bottom pressure stress in a homogeneous gyre

For the remainder of this section we restrict attention to a homogeneous (i.e., unstratified) gyre. If there is no forcing or friction, then from (19.96) and (19.98) we see that the flow simultaneously satisfies

$$\beta v = -\nabla\phi_B \times \nabla\eta_B \equiv -J(\phi_b, \eta_B), \qquad \text{and} \qquad J(\psi, f/h) = 0. \tag{19.100a,b}$$

The right-hand side of (19.100a), the form-drag curl, is non-zero whenever the pressure gradient has a component parallel to the topographic contour (i.e., when the isobars are not aligned with the topographic contours). From (19.100b) we may conclude that if there *is* meridional flow in an unforced, inviscid fluid it must be along f/h contours, and this meridional flow may be thought of as being driven by the curl of the form drag. If the domain is flat-bottomed then the form drag is zero, and in that case all meridional flow is forced or viscous.

Real flows are both forced and viscous. Bottom pressure stresses may — and likely do — locally dominate viscous stresses. However, the bottom pressure-stress curl cannot balance the wind-stress curl when integrated over the whole domain, or indeed when integrated over an area bounded by a line of constant ϕ_B or constant η_B, because its integral over such an area vanishes and (19.91) cannot be balanced if $r = 0$. In the numerical simulations shown in Fig. 19.16, it is the bottom pressure-stress curl term that largely balances the poleward flow term (βv) in the vorticity equation in parts of the western boundary current, with friction acting in the opposite sense. That is, over some regions where the flow is crossing f/h contours we have the balance

$$[\beta v] \approx [\text{bottom pressure-stress curl}] - [\text{friction}], \tag{19.101}$$

where the terms in square brackets are positive, and friction is small. In contrast, in the flat-bottomed case in the western boundary layer we have the classical balance $[\beta v] \approx +[\text{friction}]$, with both terms positive.

Now consider the balance of momentum, integrated zonally across the domain. We write the vorticity equation (19.96) in the form

$$\nabla\cdot(f\overline{\boldsymbol{u}}) = -\text{curl}_z(\phi_B\nabla\eta_B) + \text{curl}_z\boldsymbol{\tau}_T - \text{curl}_z\boldsymbol{\tau}_B, \tag{19.102}$$

where $\boldsymbol{u} = (u, v)$, $\boldsymbol{\tau}_T$ is the wind stress at the top and $\boldsymbol{\tau}_B$ the frictional stress at the bottom. Integrate (19.102) over the area of a zonal strip bounded by two nearby lines of latitude, y_1 and y_2, and the coastlines at either end. The term on the left-hand side vanishes by mass conservation and using Stokes' theorem we obtain:

$$\int_{y_1} \phi_B \frac{\partial \eta_B}{\partial x}\,\mathrm{d}x - \int_{y_2} \phi_B \frac{\partial \eta_B}{\partial x}\,\mathrm{d}x = \int_{y_1} (\tau_T^x - \tau_F^x)\,\mathrm{d}x - \int_{y_2} (\tau_T^x - \tau_F^x)\,\mathrm{d}x. \tag{19.103}$$

If the topography is non-zero, there is nothing in this equation to prevent the wind stress being balanced by the form stress terms, with the friction being a negligible contribution. If, for example, friction were to be confined to the southwest corner, then bottom pressure stress is the proximate driver of fluid polewards in the western boundary current. This may hold only if the scale of the sloping sidewall is greater than the thickness of the Stommel layer; if the converse holds then the sidewalls appear to be essentially vertical to the flow. If the sidewalls are truly vertical, then the form stress is confined to delta-functions at the walls. Friction must then be important even in the zonal balance, because if we restrict the integral in (19.103) to a strip that does not quite reach the sidewalls, the left-hand side vanishes identically and the wind stress can only be balanced by friction.

To conclude this discussion, we note that the effects of topography are likely greater in homogeneous fluids than in stratified fluids, because the stratification will partially shield the wind-driven upper ocean from feeling the topography, but we leave the exploration of that topic for another day.

Notes

1 Paumanok is the Native American name for Long Island in New York state. Studying the natural world is a humbling experience, and Whitman's beautiful writing reflects my feelings. But to try to understand and convey that understanding is what we are drawn to do, is perhaps what we *have* to do.

2 Henry Stommel (1920–1992) was one of the most creative and influential physical oceanographers of the twentieth century. Spending most of his career at Woods Hole Institute of Oceanography, his enduring contributions include the first essentially correct theory of western intensification (and so of the Gulf Stream), some of the first models of abyssal flow and the thermohaline circulation (Chapter 21), and his foundational work on the thermocline. His *forté* was in constructing elegantly simple models of complex phenomena — often models that were physically realizable in the laboratory — while at the same time testing and encouraging others to test the models against observations. This chapter might have been entitled 'Variations on a theme of Stommel'.

3 Both Fig. 19.3 and Fig. 19.4 are 'state estimates' — outputs of a model constrained by or combined with observations in such a way as to produce an approximation of the ocean state, hopefully more accurate than either models or data can separately produce. An atmospheric 'reanalysis' is also a state estimate, but for historical reasons meteorologists use a non-standard terminology. Because of the dearth of data in the ocean compared to the atmosphere, oceanic state estimates are less accurate and more model dependant than their atmospheric counterparts.

4 Courtesy of R. Zhang. See also Zhang & Vallis (2007).

5 I'm grateful to P. Heimbach for this figure, obtained using the ECCO state estimation system.

6 The asymptotic solution to this boundary value problem was obtained by Wasow (1944), a few years prior to Stommel's work, and further investigated by Levinson (1950). However, it seems unlikely these two investigators were motivated by the oceanographic problem.

7 Harald Sverdrup (1888–1957) was a Norwegian meteorologist/oceanographer who is most famous for the balance that now bears his name, but he also played a leadership role in scientific policy and was the director of Scripps Institution of Oceanography from 1936–1948. The Sverdrup unit is also named for him. Originally defined as a measure of volume transport, with $1\,\mathrm{Sv} \equiv 10^6\,\mathrm{m^3\,s^{-1}}$, it is more generally thought of as a mass transport with $1\,\mathrm{Sv} \equiv 10^9\,\mathrm{kg\,s^{-1}}$, in which case it can also be used as a measure of transport in the atmosphere. The Hadley Cell, for example, has an average transport of about 100 Sv (Figs. 14.3 and 15.22).

8 Leetmaa *et al.* (1977) and Wunsch & Roemmich (1985) offer complementary views on the matter.

9 After Munk (1950). Many thanks to Manuel Lopez Mariscal of CICESE for a number of useful comments and corrections.

10 Hendershott (1987), Veronis (1966a).

11 Veronis (1966a,b) was one of the first to investigate nonlinear effects in wind-driven gyres. See also Fox-Kemper & Pedlosky (2004) and references therein.

12 Solutions kindly provided by B. Fox-Kemper.

13 After Charney (1955).

14 See also Greenspan (1962). Il'in & Kamenkovich (1964) and Ierley & Ruehr (1986) show numerically that the friction must be sufficiently strong for steady boundary-layer solutions to exist.

15 After Fofonoff (1954).

16 Hughes & de Cuevas (2001).

17 Welander (1968).

18 Figure kindly provided by Laura Jackson.

19 Adapted from Jackson *et al.* (2006).

20 Sarkisyan & Ivanov (1971).

Further Reading

Mainly theory and modelling

Abarbanel H. D. I. & Young, W. R., Eds., 1987. *General Circulation of the Ocean.*
Contains several useful review articles on the oceanic general circulation as it was then understood.

In increasing order of size, the following books cover a variety of topics in ocean circulation, all at about the graduate student level:

Samelson, R. M., 2011. *The Theory of Large-Scale Ocean Circulation.*

Dijkstra, H. A. 2008. *Dynamical Oceanography.*

Pedlosky, J., 1996. *Ocean Circulation Theory.*

Huang, R. X., 2010. *Oceean Circulation.*

Olbers, D., Willebrand, J. & Eden, C., 2012. *Ocean Dynamics.*
These last two books are both hefty treatises on the topic.

Treatments of El Niño are to be found in

Clarke, A. J., 2008. *An Introduction to the Dynamics of El Niño and the Southern Oscillation.*

Sarachik, E. and Cane, M., 2010. *The El Niño–Southern Oscillation Phenomenon.*

Description and observation

Talley, L. D., Pickard, G.L., W. J. Emery, W. J. & Swift, J. H. 2011. *Descriptive Physical Oceanography: An Introduction.*
This text gives a sense of the big picture, as well as being full of useful maps of ocean properties.

Wunsch, C. 2015. *Modern Observational Physical Oceanography.*
Shows how modern observing systems (floats, satellites, etc.) can be used alongside numerical models to provide a more complete view of the ocean.

There is a tide in the affairs of men,
which, taken at the flood, leads on to fortune.
Omitted, all the voyage of their life is bound in shallows and in miseries.
On such a full sea are we now afloat,
and we must take the current when it serves, or lose our ventures.
William Shakespeare, *Julius Caesar*, c. 1599.

CHAPTER 20

Structure of the Upper Ocean

IN THE PREVIOUS CHAPTER we developed an understanding of the vertically integrated flow of the worlds oceans. If we are to proceed further we must develop an understanding of the *vertical structure* of the oceans, and that is the subject of this chapter. Our main focus will be on the upper ocean and we will proceed as follows:

1. We first explore the vertical structure of the wind-driven circulation, largely as a continuation of the investigation of the previous chapter. We use the quasi-geostrophic equations to understand why the subsurface ocean moves at all, and we introduce the notion of potential vorticity homogenization.
2. A limitation of the quasi-geostrophic approach is that these equations take the stratification, $N(z)$, as a given and therefore cannot provide a answer to the question as to what produces the density structure itself. Thus, beginning in Section 20.4, we relax the quasi-geostrophic restriction and, using the *planetary-geostrophic* equations, we try to understand the dynamics that give rise to the vertical structure of density itself. We focus on the *main thermocline*, the region of the upper ocean in which temperature and density vary most rapidly, in all seasons, and we discuss the structure of both the internal thermocline and the ventilated thermocline, the meaning of which will become apparent later.

As with many fluid problems, the dynamics becomes intertwined with the thermodynamics, and the mean flow becomes intertwined with the smaller, turbulent, baroclinic eddies, in rather subtle ways that, to this day, are not fully understood and that large numerical models are only beginning to properly simulate. We begin by looking at the vertical structure of the wind-driven gyres, and if and how the influence of the wind can be communicated to the subsurface ocean.

20.1 VERTICAL STRUCTURE OF THE WIND-DRIVEN CIRCULATION

20.1.1 A Two-layer Quasi-Geostrophic Model

We pose the problem using the quasi-geostrophic equations, taking the background stratification of the ocean as a given.[1] The simplest system that has vertical structure is a two-layer model and that is where we start. We don't yet wish to consider the effects of mesoscale eddies, so we'll limit ourselves to motion larger than the deformation scale, although not so large that the quasi-geostrophic system itself does not hold.

Scales of motion

On scales that are sufficiently larger than the deformation radius we can ignore the relative vorticity compared to planetary vortex stretching and the β-effect. Since quasi- geostrophic scaling itself applies only to scales that are not significantly larger than the deformation scale, our analysis will be formally valid under the following set of inequalities:

$$\begin{aligned}
\beta L &\ll f_0 && \text{(small variations in Coriolis parameter),}\\
\beta L &> U/L && \text{(to ignore relative vorticity compared to planetary vorticity),}\\
L^2 &> L_d^2 && \text{(to ignore relative vorticity compared to vortex stretching),}\\
Ro\, L^2 &\ll L_d^2 && \text{(to keep the variations in stratification small),}
\end{aligned}$$

where L_d is the deformation radius and L the scale of the motion. The first and last of the above inequalities are standard quasi-geostrophic requirements, with the '$\gg$' symbol denoting the asymptotic ordering. The middle two inequalities are taken within the quasi-geostrophic dynamics, and are needed in order to ignore relative vorticity and give a balance between the β-effect and vortex stretching. The simultaneous satisfaction of all these conditions may seem restrictive, but the plangent dynamics contained within the quasi-geostrophic equations and the generality of the method employed below will suggest that the principal results obtained may transcend the limitations of the equations used. In the mid-latitude ocean $L_d \approx 10^5$ m and the above inequalities are reasonably well satisfied for $L \approx 10^6$ m and $U \approx 0.1\,\mathrm{m\,s^{-1}}$ with $\beta = 10^{-11}\,\mathrm{m^{-1}\,s^{-1}}$ and $f_0 = 10^{-4}\,\mathrm{s^{-1}}$.

Constructing the model

We now make the following simplifications for our model ocean:

(i) We use the two-layer quasi-geostrophic equations, with layers of equal thickness.

(ii) We seek only statistically-steady solutions.

(iii) We include a frictional term coming from a downgradient flux of potential vorticity. Given the neglect of relative vorticity, this is equivalent to an interfacial drag.

(iv) We neglect the western boundary layer.

Because of the equal-layer-thickness assumption, which makes the algebra simpler, it is best considered as a model for the upper ocean above a level where the vertical velocity is approximately zero. The equations of motion are then

$$J(\psi_1, q_1) = \frac{1}{H_0}\mathrm{curl}_z \boldsymbol{\tau}_T - \nabla\cdot\boldsymbol{T}_1, \qquad J(\psi_2, q_2) = -\nabla\cdot\boldsymbol{T}_2 \tag{20.1a,b}$$

where

$$q_1 = \beta y + F(\psi_2 - \psi_1), \qquad q_2 = \beta y + F(\psi_1 - \psi_2). \tag{20.2a,b}$$

Here, $F = f_0^2/(g' H_0) = 1/L_d^2$ is a measure of the stratification, where H_0 is the thickness of either layer, and the $\nabla\cdot\boldsymbol{T}$ terms represent interfacial eddy stresses, which, if needed, we will parameterize by a downgradient flux of potential vorticity,

$$\boldsymbol{T}_1 = -\kappa\nabla q_1 = -\kappa(F\nabla(\psi_2 - \psi_1) + \beta\mathbf{j}), \qquad \boldsymbol{T}_2 = -\kappa\nabla q_2 = -\kappa(F\nabla(\psi_1 - \psi_2) + \beta\mathbf{j}), \tag{20.3}$$

where κ is a constant. We will mostly be interested in the limit of small κ, or more specifically $UL/\kappa \gg 1$, which is a large Péclet number condition. (The Péclet number is similar to a Reynolds number, but with the diffusivity replacing the kinematic viscosity.) So first consider the case when κ is identically zero. An *exact* solution to (20.1) has $\psi_2 = 0$, so that (20.1a) becomes $\beta\partial\psi_1/\partial x = H_0^{-1}\mathrm{curl}_z\boldsymbol{\tau}_T$, with solution

$$\psi_1 = -\frac{1}{H_0\beta}\int_x^{x_E} \mathrm{curl}_z\boldsymbol{\tau}_T\,\mathrm{d}x. \tag{20.4}$$

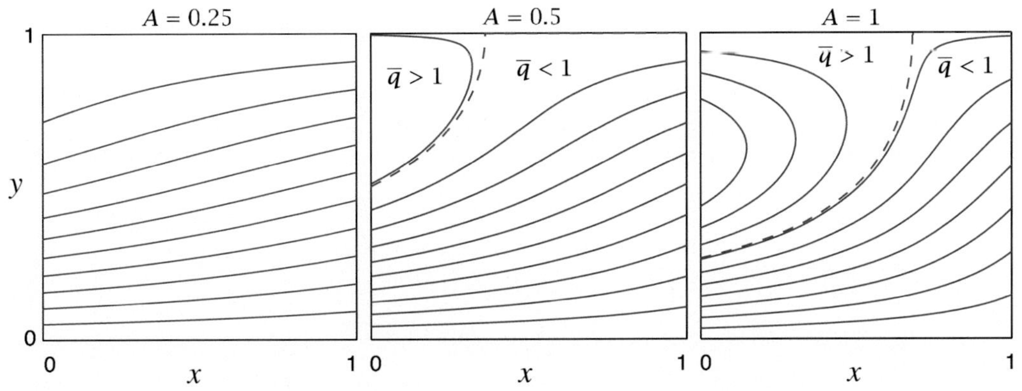

Fig. 20.1 Contours of $\overline{q} = \beta y + A \sin \pi y(1 - x)$, with $\beta = 1$, for three values of A. The red dashed line is $\overline{q} = 1$, which separates the blocked region to the east ($\overline{q} < 1$) from the closed region to the west ($\overline{q} > 1$). See Fig. 20.2 for plots of the other fields.

That is, *there is no flow in the lower layer*, and the upper layer solution is given by Sverdrup balance. The solution satisfies $\psi_1 = 0$ at $x = x_E$ and, because $\psi_2 = 0$, the nonlinear term on the left-hand side of (20.1a) vanishes identically. This is both counter-intuitive and counter-observations, for we know the subsurface ocean is not quiescent. Is there another solution?

A general solution

We now construct the solution without assuming $\psi_2 = 0$. Although the equations are nonlinear, we can obtain a linear equation for a streamfunction by adding (20.1a) and (20.1b), giving

$$J(\psi_1, \beta y + F(\psi_2 - \psi_1)) + J(\psi_2, \beta y + F(\psi_1 - \psi_2)) = \frac{1}{H_0}\operatorname{curl}_z \boldsymbol{\tau}_T. \tag{20.5}$$

The nonlinear terms cancel leaving

$$J(\overline{\psi}, \beta y) = \frac{1}{H_0}\operatorname{curl}_z \boldsymbol{\tau}_T, \qquad \text{where} \quad \overline{\psi} = \psi_1 + \psi_2, \tag{20.6a,b}$$

with solution, as in (20.4),

$$\overline{\psi} = -\frac{1}{H_0 \beta}\int_x^{x_E} \operatorname{curl}_z \boldsymbol{\tau}_T \, dx'. \tag{20.7}$$

This simply says that the vertically integrated flow obeys Sverdrup balance. For the canonical wind stress

$$\boldsymbol{\tau}_T = -\tau_0 \cos \pi y \, \mathbf{i}, \tag{20.8}$$

and we obtain $\overline{\psi} = (\pi\tau_0/\beta H_0)(x_E - x)\sin \pi y$. It is useful to define

$$\overline{q} \equiv (\beta y + F\overline{\psi}), \tag{20.9}$$

and then $\overline{q} = \beta[y + A(1 - x)\sin \pi y]$, where $A = \pi\tau_0/(\beta H_0)$ parameterizes the wind strength, and this is plotted in Fig. 20.1. For $\overline{q} < 1$ (below and to the right of the dashed line) all the geostrophic contours intersect the eastern boundary and the flow is 'blocked'. For $\overline{q} > 1$ the flow is 'closed'.

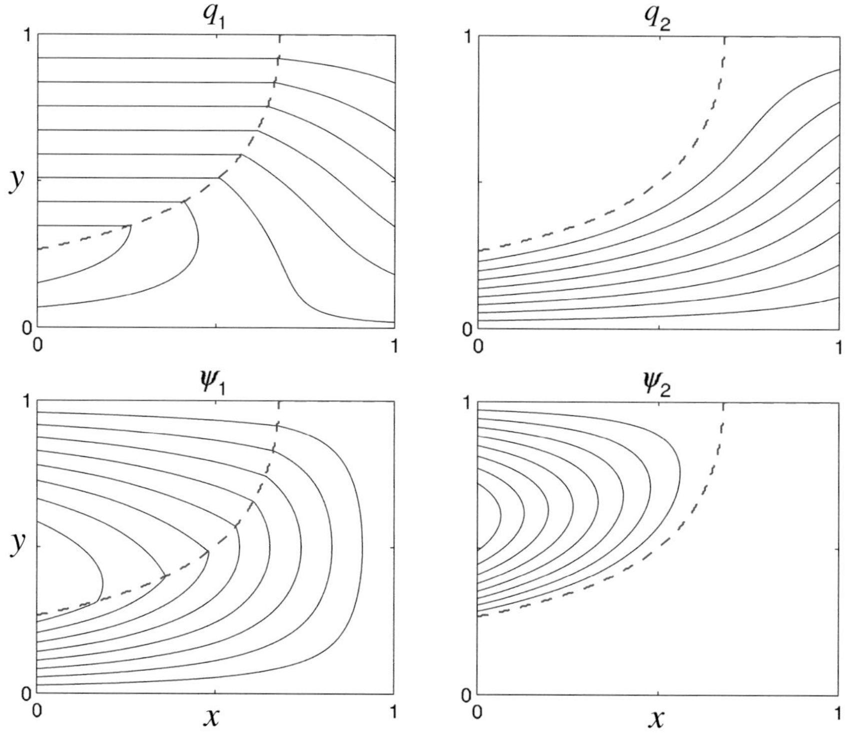

Fig. 20.2 Upper- and lower-level potential vorticity and streamfunction for the canonical wind stress (20.8). The field of $\overline{q}$ is that of Fig. 20.1 with $A = 1$. The dashed line divides the blocked region from the closed region. The lower layer streamfunction ψ_2 is non-zero only in the closed region, and here $q_2 = \beta L$ and $q_1 = 2\beta y - \beta L$. In the blocked region the upper layer carries all of the Sverdrup transport. Both the streamfunction and potential vorticity are continuous at the divide: $\psi_2 = 0$ and $q_2 = \overline{q} = \beta L$.

Lower layer

Although the full equations are nonlinear, using (20.9) we can obtain a linear equation for the lower layer. Because the Jacobian of a field with itself vanishes, (20.1b) and (20.2b) imply that

$$J(\psi_2, \overline{q}) = -\nabla \cdot \boldsymbol{T}_2, \tag{20.10}$$

and this is useful because $\overline{q}$ is a function of the wind, using (20.9). If $\nabla \cdot \boldsymbol{T}_2 = 0$ then

$$J(\psi_2, \overline{q}) = 0. \tag{20.11}$$

As well as the possibility that $\psi_2 = 0$ we now have the more general solution

$$\psi_2 = G(\overline{q}), \tag{20.12}$$

where G is an *arbitrary* function of its argument. Isolines of ψ_2 and $\overline{q}$ are then coincident. (Contours that are isolines of both streamfunction and potential vorticity are known as geostrophic contours.)

Consider a blocked isoline of $\overline{q}$; that is, one that intersects the eastern boundary (see Fig. 20.1). The ψ_2 contour coincident with this has a value of zero at the eastern boundary (by the no-normal flow condition). Thus $\psi_2 = 0$ *everywhere* in the blocked region, and $q_2 = \overline{q}$. In this region the Sverdrup transport is carried everywhere by the upper layer, and the lower layer is at rest. This

Summary of Wind-Driven, Two-layer Solution

The vertically integrated flow in a wind-driven two-layer quasi-geostrophic model is determined by Sverdrup balance. The effects of eddies may be crudely parameterized by a downgradient diffusion of potential vorticity. If this is identically zero, then the lower-layer flow is identically zero and the upper-layer flow carries all the transport. If the diffusion is small but non-zero, the lower-layer streamfunction approximately satisfies $J(\psi_2, \overline{q})$, where $\overline{q}$ is given by (20.9), and therefore ψ_2 is a function of $\overline{q}$ — that is, $\psi_2 \approx G(\overline{q})$. For a typical subtropical wind, contours of $\overline{q}$, and therefore contours of ψ_2, are naturally divided into two regions (Fig. 20.1):

(i) A blocked region (the shadow zone), in which contours of $\overline{q}$ intersect the eastern boundary, the lower layer flow is zero and the upper layer carries all the Sverdrup transport.

(ii) A closed region in which (if we envision a nearly inviscid western boundary current) the flow recirculates. In this region we posit that the lower layer potential vorticity becomes homogeneous, with a value determined by the value at the region's boundary, and this in turn is determined by tracing $\overline{q}$ back to the domain boundary.

To satisfy a circulation constraint the function $G(\overline{q})$ must be a linear function, and given this, the entire solution may be determined. If, for example, the wind is zonal and a function of y only, and $\text{curl}_z \boldsymbol{\tau}_T = g(y)$ then, in both regions:

$$\overline{\psi} \equiv \psi_1 + \psi_2 = -\frac{1}{\beta H_0} g(y)(x_E - x), \qquad \overline{q} \equiv \beta y + F\overline{\psi}. \tag{OC.1a,b}$$

In the blocked region:

$$\psi_2 = 0, \qquad \psi_1 = -\frac{1}{\beta H_0}(x_E - x) g(y), \tag{OC.2a}$$

$$q_1 = \beta y + F(\psi_1 - \psi_2), \qquad q_2 = \beta y + F(\psi_2 - \psi_1). \tag{OC.2b}$$

In the closed region:

$$q_2 = \beta L \quad \text{(by homogenization)}, \tag{OC.3a}$$

$$\psi_2 = \frac{1}{2F}(\overline{q} - \beta L), \qquad \psi_1 = \overline{\psi} - \psi_2, \tag{OC.3b}$$

$$q_1 = \beta y + F(\psi_2 - \psi_1) = 2\beta y - \beta L. \tag{OC.3c}$$

For $g(y) = -\sin \pi y$ these solutions are illustrated in Figs. 20.1 and 20.2.

This approach provides a solution to the conundrum of what drives the subsurface (quasi-geostrophic) ocean, for if there are no eddy effects at all (i.e., in (20.3) $\boldsymbol{T}_1 = \boldsymbol{T}_2 = 0$), then the lower layer flow is stationary. This solution is not wholly realistic, for the upper layer flow could be made quite shallow. Another solution to this issue is provided in Section 20.7, wherein it is assumed that the lower layers may outcrop and so feel the wind directly.

region is called a 'shadow zone', for the fluid is in the shadow of the eastern boundary, and it will re-appear in a model of the ventilated thermocline later on in this chapter. In the region of closed contours, ψ_2 cannot be given by this argument. But if κ is sufficiently small, we can expect (20.11) to approximately hold, and that the presence of a small amount of dissipation will determine the functional relationship between ψ_2 and $\overline{q}$. Thus, in summary, there are two regions of flow:

(i) the blocked region in which $\psi_1 \approx \overline{\psi} \gg \psi_2$ and ψ_1 is approximately given by (20.4);

(ii) a closed region in which $\psi_2 = G(\overline{q}) + O(\kappa)$.

20.1.2 Relation Between Streamfunction and Potential Vorticity

A general argument

In Chapter 13 we showed that, within a region of closed contours, the values of a tracer that is materially conserved except for the effects of a small diffusion would become *homogeneous*. In the case at hand, potential vorticity is that tracer, so that within potential vorticity contours or closed streamlines potential vorticity will become homogenized. If we can determine the value of q_2 within the region of closed contours, then from (20.2) ψ_2 is given by

$$\psi_2 = (1/2F)(\overline{q} - q_2), \tag{20.13}$$

and the solution would be complete. Now, outside the closed region $\psi_2 \ll \psi_1$, so that the outermost contour of the closed region must be characterized by $q_2 \approx \overline{q}$, for this makes ψ_2 continuous between closed and blocked regions. Thus, the value of q_2 within the closed homogeneous region is that of $\overline{q}$ (i.e., $\beta y + F\overline{\psi}$) on its boundary. Since this contour intersects the poleward edge of the domain, where $\overline{\psi}$ is zero, the value of this contour is just βy at $y = L$; that is, βL. Thus, within the closed region,

$$q_2 = \beta L. \tag{20.14}$$

A specific calculation

Now consider the steady, lower-layer potential vorticity equation (20.1b); noting that $J(\psi_2, F(\psi_1 - \psi_2)) = J(\psi_2, F(\psi_1 + \psi_2))$, (20.1b) may be written as

$$J(\psi_2, \overline{q}) = -\nabla \cdot \boldsymbol{T}_2. \tag{20.15}$$

Integrating around a closed contour of $\overline{q}$ the left-hand side vanishes and

$$R\int(\nabla\psi_1 - \nabla\psi_2)\cdot \boldsymbol{n}\,\mathrm{d}l = 0 \qquad \text{or} \qquad \oint \boldsymbol{u}_1 \cdot \mathrm{d}\boldsymbol{l} = \oint \boldsymbol{u}_2 \cdot \mathrm{d}\boldsymbol{l}. \tag{20.16a,b}$$

Thus, the deep circulation around a mean geostrophic contour (i.e., isoline of $\overline{q}$) is equal to the upper-level circulation.

Previously we argued that

$$\psi_2 = G(\overline{q}) = G(\beta y + F(\psi_1 + \psi_2)), \tag{20.17}$$

where G is an arbitrary function of its argument. In order to satisfy (20.16) (a linear relation between $\boldsymbol{u}_1$ and $\boldsymbol{u}_2$) G must be a linear function, and so we write

$$\psi_2 = C\left[\frac{\beta y}{F} + (\psi_1 + \psi_2)\right] + B, \tag{20.18}$$

where C and B are constants. This may be rearranged to give

$$\psi_1 = -C\frac{\beta y}{F} + (\psi_1 + \psi_2)(1 - C) - B. \tag{20.19}$$

The above two equations are consistent with (20.16) if $C = 1/2$. With this, (20.18) gives

$$\overline{q} = 2F(\psi_2 - B), \tag{20.20}$$

and the potential vorticity in the closed contour region of the lower layer is

$$q_2 = \beta y + F\overline{\psi} - 2F\psi_2 = -2FB. \tag{20.21}$$

That is, it is constant. Outside the closed contours $\psi_2 \ll \psi_1$ so that $q_2 \approx \overline{q} = \beta y + F\psi_1$. If we trace this contour to the edge of the domain where $\psi_1 = 0$ and $y = L$ then we see that the value of $\overline{q}$ on the contour, and hence q_2 in the closed region, is βL, as in (20.14), and $B = -\beta L/(2F)$. Using (20.20) then gives

$$\psi_2 = (2F)^{-1}(\overline{q} - \beta L). \tag{20.22}$$

Given ψ_2 and q_2, from (20.14), we obtain q_1 and ψ_1 using (20.2) and (20.6b), giving

$$q_1 = 2\beta y - \beta L, \qquad \psi_1 = \overline{\psi} - \psi_2. \tag{20.23}$$

All these fields are illustrated in Fig. 20.2, and see the shaded box on page 765 for a summary.

20.2 ✦ A MODEL WITH CONTINUOUS STRATIFICATION

We now look at the dynamics of the continuously stratified circulation, largely by way of an extension of our two-layer procedure. Let us first consider how deep the wind's influence is.

20.2.1 Depth of the Wind's Influence

The thermal wind relationship in the form $f\,\partial u/\partial z = \partial b/\partial y$ implies a vertical scale H given by

$$H = \frac{fUL}{\Delta b}, \tag{20.24}$$

where Δb is a typical magnitude of the horizontal variation of the buoyancy. We can relate this to the Ekman pumping velocity W_E using the linear geostrophic vorticity equation, $\beta v = f\partial w/\partial z$, which, (assuming that the horizontal components of velocity are roughly similar, i.e., $V = U$), implies that

$$U = \frac{fW_E}{\beta H}. \tag{20.25}$$

Equations (20.24) and (20.25) may be combined to give an estimate of the depth of the wind-driven circulation, namely

$$H = \left(\frac{f^2 W_E L}{\beta\,\Delta b}\right)^{1/2}, \tag{20.26}$$

where L may be interpreted as the gyre scale. We now use quasi-geostrophic scaling to relate the horizontal temperature gradient to the stratification using the thermodynamic equation,

$$\frac{\mathrm{D}b}{\mathrm{D}t} + wN^2 = 0, \tag{20.27}$$

with implied scaling

$$\Delta b = \frac{W_E N^2 L}{U} = \frac{N^2 \beta H L}{f_0}, \tag{20.28}$$

where the second equality uses (20.25). Using (20.26) and (20.28) gives

$$H = \left(\frac{W_E f^3}{\beta^2 N^2}\right)^{1/3}. \tag{20.29}$$

Potential vorticity interpretation

The estimate (20.29) can be obtained and interpreted more directly: *the wind-driven circulation penetrates as far as it can alter the potential vorticity q from its planetary value βy.* Recall that, ignoring relative vorticity,

$$q = \beta y + \frac{\partial}{\partial z}\left(\frac{f_0^2}{N^2}\frac{\partial \psi}{\partial z}\right). \tag{20.30}$$

The two terms are comparable if

$$\frac{f_0^2}{N^2 H^2} UL \approx \beta L \qquad \text{or} \qquad H^2 \approx \frac{f_0^2 U}{N^2 \beta}. \tag{20.31}$$

Using (20.25) to eliminate U in favour of W_E recovers (20.29). Thus, for a given stratification, we have an estimate of the depth of the wind-driven circulation, or at least a scaling for depth of the vertical influence of the wind.

20.2.2 The Complete Solution

Armed with an estimate for the depth of the wind's influence, we can obtain a solution for the continuously stratified case analogous to that found in the two-layer case in Section 20.1. Our assumptions are as follows:

(i) In the limit of small dissipation, streamfunction and potential vorticity have a functional relationship with each other.

(ii) Potential vorticity is homogenized within closed isolines of q or ψ. The value of q within the homogenized pool is that of the outermost contour, which here is the value of q at the poleward edge of the barotropic gyre.

(iii) Outside of the pool region, (i.e., below the depth of the wind's influence) the streamfunction is zero, and the potential vorticity is given by the planetary value, i.e., βy.

Given these, finding a solution is not difficult. If N^2 is constant and neglecting relative vorticity, the expression for potential vorticity is

$$q = \frac{\partial^2}{\partial z^2}\left(\frac{f_0^2}{N^2}\psi\right) + \beta y. \tag{20.32}$$

We nondimensionalize by writing

$$z = \left(\frac{f_0^2 U}{N^2 \beta}\right)^{1/2} \widehat{z}, \quad q = \beta L \widehat{q}, \quad \psi = \widehat{\psi} UL, \quad y = L\widehat{y}, \quad w = \frac{U^2 f_0}{N^2 H}\widehat{w}, \tag{20.33}$$

where the hatted variables are nondimensional, and the scaling for w arises from the thermodynamic equation $N^2 w \sim J(\psi, f_0 \psi_z)$. With this, (20.32) becomes

$$\widehat{q} = \frac{\partial^2 \widehat{\psi}}{\partial \widehat{z}^2} + \widehat{y}. \tag{20.34}$$

The flow is then given by solving the following equations:

$$\psi_{zz} + y = y_0, \qquad -D(x,y) < z < 0, \tag{20.35a}$$

$$\psi = 0, \qquad z \le -D(x,y), \tag{20.35b}$$

where D is the (to be determined) depth of the bowl, y_0 is a constant, and we have dropped the hats over the nondimensional variables. The solution in $D(x, y) < z < 0$ corresponds to the closed

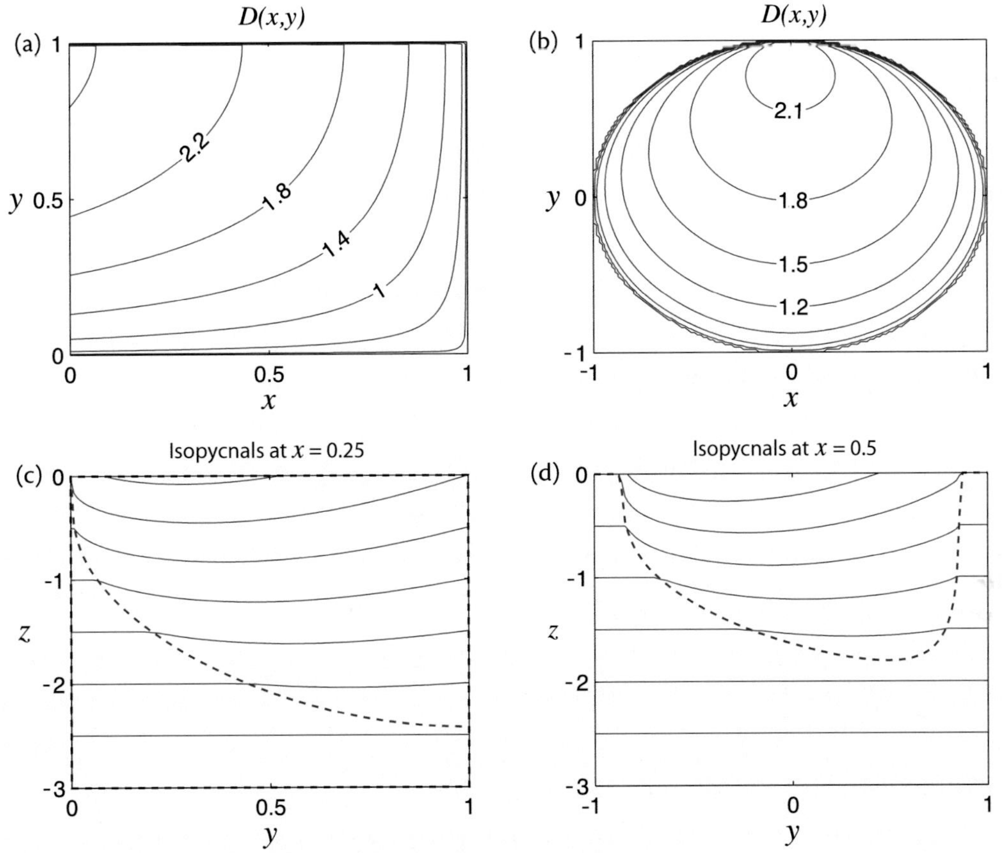

Fig. 20.3 Solutions of (20.9) for two different barotropic streamfunctions. On the left $\psi_B = (1 - x)\sin \pi y$ and on the right $\psi_B = 1 - (x^2 + y^2)$ for $x^2 + y^2 < 1$, zero elsewhere. The upper panels show contours of the depth of the wind-influenced region [solutions of (20.40)]. The depth increases to the northwest in the left panel, and to the north in the right panel, so that in both cases the area of the bowl shrinks with depth. The lower panels are contours of $z + (\beta L/f_0)\psi_z/2$, with $\beta L/f_0 = 1/2$, obtained from (20.36) or (20.41), at $x = 0.25$ and $x = 0.5$ in the two cases. These are isopycnal surfaces, with a rather large value of $\beta L/f_0$ to exaggerate the displacement in the bowl region. The dashed lines indicate the boundary of the bowl region, outside of which the isopycnals are flat.

region of the two-layer model, and the solution $z \le -D(x, y)$ corresponds to the blocked region of zero lower-layer flow. The constant y_0 is the nondimensional value of potential vorticity within the pool region, and following our reasoning in the two-layer case this is the value of the potential vorticity at the northern boundary. Dimensionally this is βL, so that in nondimensional units $y_0 = 1$.

The lower boundary condition on (20.35a) is that $\psi = \psi_z = 0$ at $z = -D$, because in the abyss $\psi = \partial\psi/\partial z = 0$ and we require that both ψ and $\partial\psi/\partial z$ be continuous (note that the buoyancy perturbation is proportional to $\partial\psi/\partial z$). The solution that satisfies this is

$$\psi = \frac{1}{2}(z + D)^2(y_0 - y), \tag{20.36}$$

and $\psi = 0$ for $z < -D$.

To obtain an expression for D we first note that the nondimensional vertical velocity at $z = 0$

is given by

$$w = -J(\psi, \psi_z), \tag{20.37}$$

which, using (20.36), gives

$$w = \frac{1}{2}(z + D)^2 (y_0 - y)\frac{\partial D}{\partial x}. \tag{20.38}$$

At $z = 0$ the vertical velocity is the Ekman pumping velocity and (20.38) becomes

$$D^2 \frac{\partial D}{\partial x} = \frac{2w_E}{(y_0 - y)}. \tag{20.39}$$

But the Ekman pumping velocity is related to the barotropic streamfunction, ψ_B, by the Sverdrup relationship, so that integrating (20.39) gives

$$D^3 = \frac{6\psi_B}{(y_0 - y)} = -\frac{6(x_E - x)w_E}{(y_0 - y)}, \tag{20.40}$$

where the second equality holds if w_E is not a function of x. This is a solution for the depth of moving region, the bowl in which potential vorticity is homogenized. An expression for the streamfunction is then obtained by using (20.40) in (20.36), and is found to be

$$\psi = \begin{cases} \frac{1}{2}\left[z(y_0 - y)^{1/2} + (6\psi_B)^{1/3}(y_0 - y)^{1/6}\right]^2 & -D < z < 0, \\ 0 & z < -D. \end{cases} \tag{20.41}$$

The potential vorticity corresponding to this solution is

$$q = \begin{cases} y_0 & -D < z < 0, \\ y & z < -D. \end{cases} \tag{20.42}$$

Solutions are illustrated in Figs. 20.3 for cases with two different barotropic streamfunctions.

It is possible to heuristically extend models such as the one described above by appending a western boundary layer, and indeed the homogenization of potential vorticity depends upon the presence of such a region to allowing the flow to recirculate. However, as we saw in Section 19.5.3, it is difficult for flow to leave a western boundary layer without the help of friction, and a neutrally stable, damped, stationary Rossby wave typically forms. The critical issue then is whether the presence of dissipation in the western boundary layer affects the homogenization of potential vorticity in the gyre itself. This problem is the province of observation and numerical simulation, and solutions with both quasi-geostrophic and primitive equation models do in fact show that potential vorticity is able to homogenize under many circumstances.[3] Let us now take a brief look at some observations.

20.3 OBSERVATIONS OF POTENTIAL VORTICITY

Homogenization of potential vorticity in the real ocean has been observed in both Pacific and Atlantic Oceans, in both hemispheres, and to a lesser degree in the Indian Ocean, and various maps are shown in Fig. 20.4 through Fig. 20.7.[5] In all of the plots we see that the near-equatorial variation of potential vorticity is dominated by the beta effect, more so at depth where the potential vorticity isolines are more-or-less along latitude lines until almost 20°, but in the subtropical gyres there are large regions of homogeneous potential vorticity.

Looking first at the Pacific, the upper two plots in Fig. 20.4 show the potential vorticity (here defined as $f\partial\rho/\partial z$) on potential density surfaces in the main thermocline. These surfaces slope

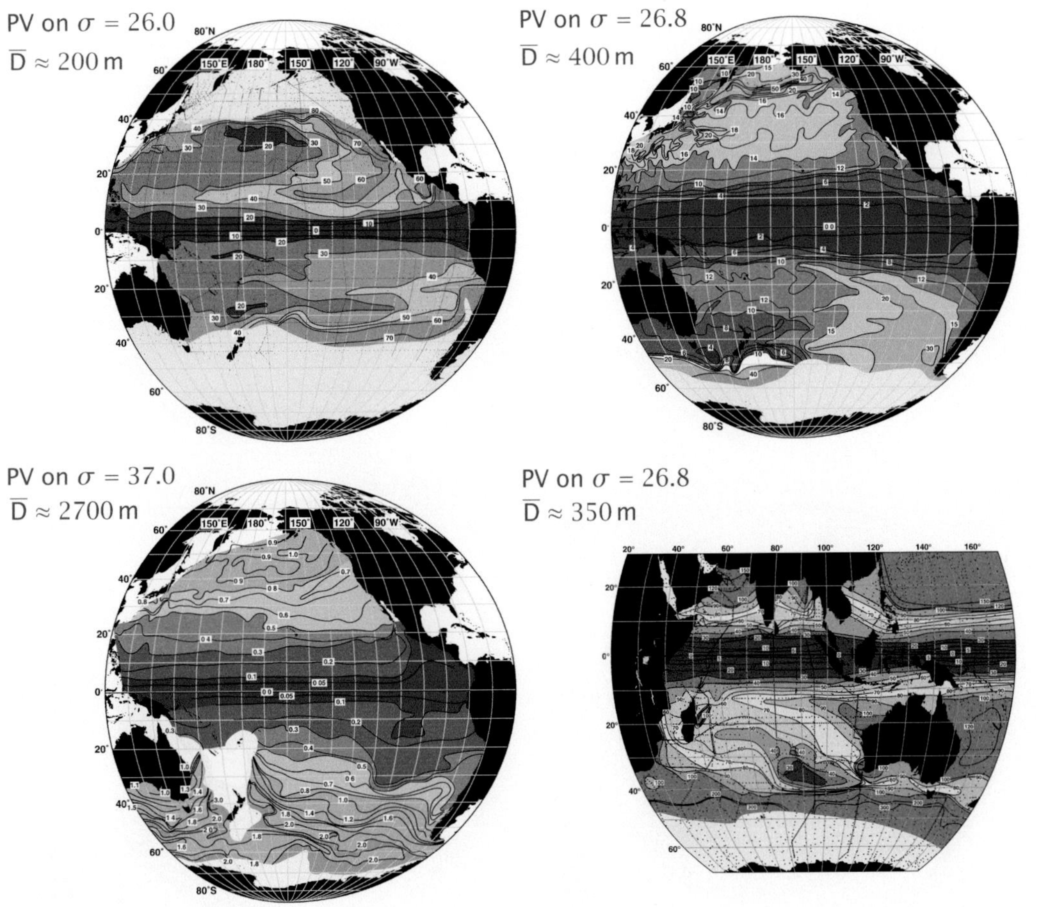

Fig. 20.4 Potential vorticity (i.e., $f(\partial\rho/\partial z)$) in the Pacific and Indian Oceans on the potential density surfaces labelled, which each have approximate average depths $\overline{\mathrm{D}}$. Some potential vorticity homogenization can be seen in the subtropical gyre in the Pacific (the upper two plots) but less so at depth in the Pacific (lower left) and less so at all depths in the Indian Ocean (just the upper ocean is shown, lower right), which has a less pronounced gyre structure.[2]

up toward the pole, and the $\sigma = 26$ surface (i.e., a surface with a potential density of approximately $1026\,\mathrm{kg\,m^{-3}}$) outcrops at about 40°, a little equatorward of the boundary between the subpolar and subtropical gyre. On these surfaces there are large swathes of near-uniform potential vorticity in the subtropical gyre, perhaps a little more obviously so in the Northern Hemisphere, with strong gradients quite noticeable at the gyre edge at about 50° N. Tongues of high potential vorticity are advected by the gyre itself, sweeping equatorward and westward along the $\sigma = 26$ surface in the Northern Hemisphere.

Moving into the deep Pacific there is less homogenization, with isolines of potential vorticity generally crossing the entire Pacific at all latitudes, with a just the odd pool of closed contours. The Indian Ocean has less potential vorticity homogenization at all depths, most likely because the subtropical gyre itself is less pronounced in the Indian Ocean, and the subpolar gyre is largely replaced by the eastward flow of the Antarctic Circumpolar Current system.

The Atlantic also shows large regions of homogenization in the upper ocean as seen in Fig. 20.5 and Fig. 20.6. These maps were constructed from a different set of observations, and using a different method, than those of Fig. 20.4, but show similar features — homogenization in the upper gyre

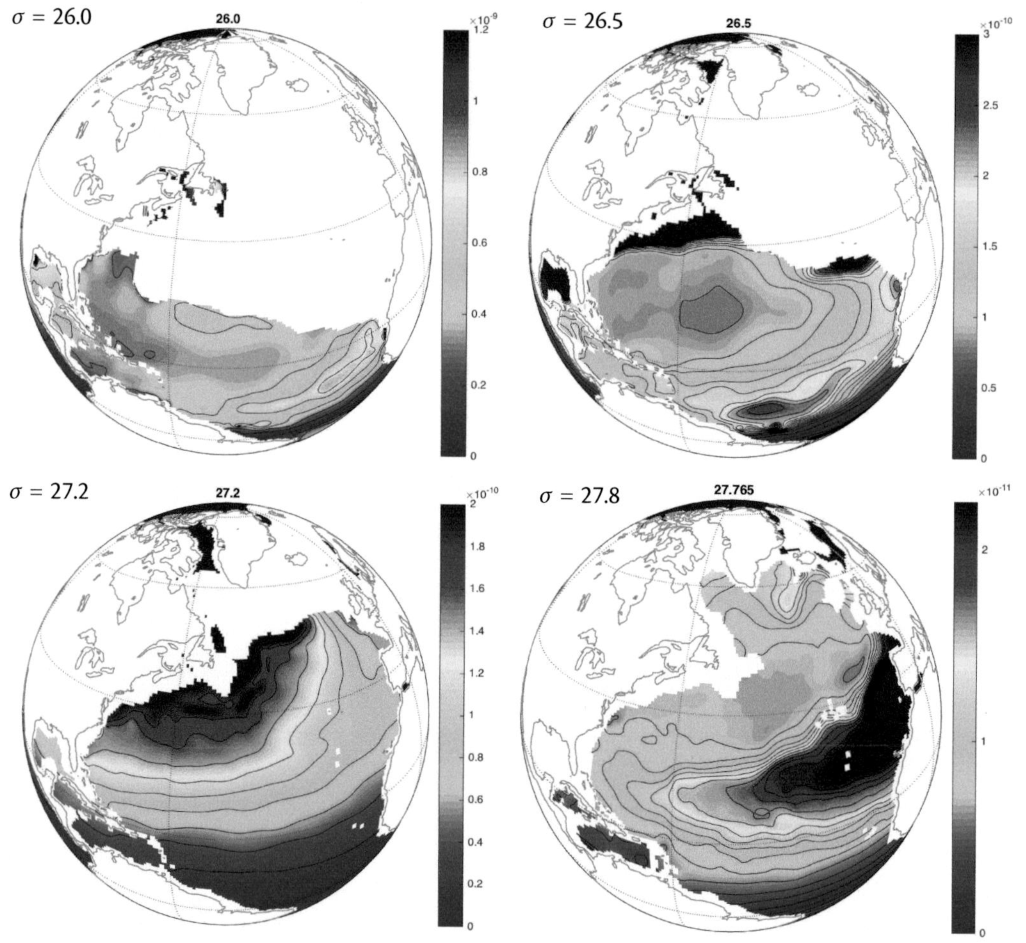

Fig. 20.5 Potential vorticity in the Atlantic Ocean on the potential density surfaces labelled, averaged over several Januaries using the MIMOC climatology. Potential vorticity is a normalized version of f/h. Specifically it is $|f|\delta\sigma/(\rho h)$ where $h(x, y)$ is the isopycnal layer thickness and $\delta\sigma$ is a fixed difference between layer interface potential densities, so here PV has units of $\mathrm{m^{-1}s^{-1}}$. The PV is homogenized over much of the subtropical gyre around $\sigma = 26.5$. At deeper levels the potential vorticity is more dominated by the beta effect and an influx of Mediterranean water.[4]

but planetary values (and now a Mediterranean influence) dominating at depth. Consider Fig. 20.5 layer-by-layer, from the top down, where a close inspection of Fig. 20.8 will reveal the depths of each layer. The shallow, $\sigma = 26.0$ layer (typically tens of metres deep) outcrops in the middle of the subtropical gyre, receiving most of its fluid directly by Ekman-pumping from the mixed layer, and has a relatively small pool of homogenized potential vorticity. The deeper, $\sigma = 26.5$ level (with typical depths of a few hundred metres over much of the gyre) outcrops much further poleward and consequently has an extensive recirculating regime that homogenizes the potential vorticity. We also see a region between about 10° N and 25° N where potential vorticity increases moving southward — that is, $\partial Q/\partial y < 0$ — so enabling baroclinic instability. Going deeper, at $\sigma = 27.24$ (with typical depth of several hundred to a thousand or so metres) the planetary influence begins to dominate, with the subtropical gyre shrinking and a smaller region of potential vorticity homogenization further north. Finally, at $\sigma = 27.76$, or about 1500 m depth, the circulation is dominated by low potential vorticity Labrador Sea Water to the north and high potential vorticity

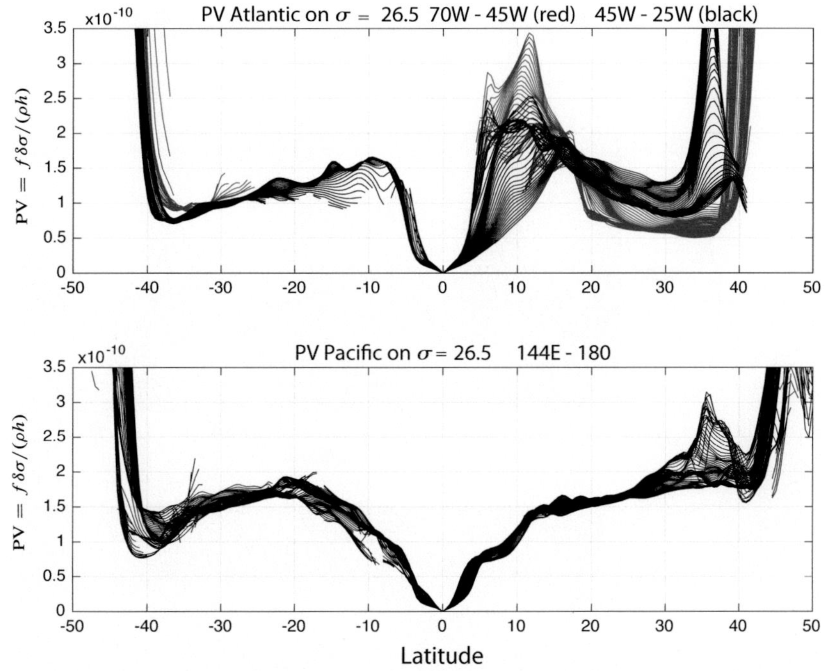

Fig. 20.6 Potential vorticity in the Atlantic and Pacific at the $\sigma_0 = 26.5$ level, using the same data as in Fig. 20.5.[4] The Coriolis parameter is latitudinally varying but taken as positive in both hemispheres for graphical convenience. See text for more discussion.

from Mediterranean Salt Tongue to the south — again with a potentially baroclinically unstable flow where potential vorticity increases equatorward. The homogenization in both Pacific and Atlantic is very strikingly displayed in Fig. 20.6 and Fig. 20.7, with broad plateaus of near-constant potential vorticity reaching to the poleward edge of the gyres, where there is a sudden leap that acts as a mixing barrier.

The story of potential vorticity is a rich one. The mesoscale eddies provide the stirring that leads to the homogenization, but competing processes complicate the picture. The eddies weaken with depth and planetary effects begin to dominate, and in the upper ocean the varied influences of the western boundary current, topographic effects and the stripping of potential vorticity sheets from solid boundaries turn the ocean into a complex tapestry.

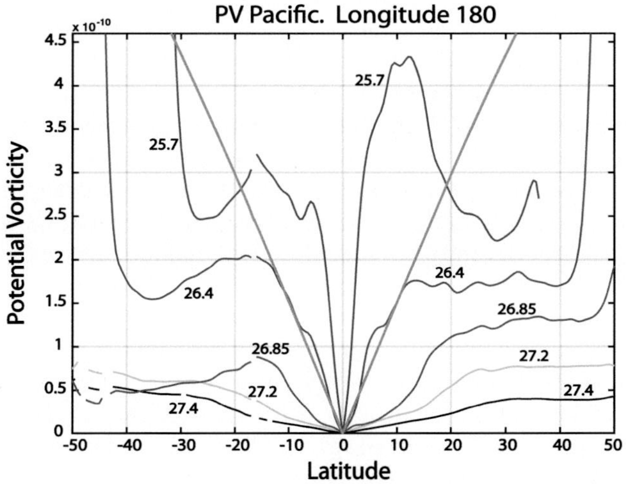

Fig. 20.7 As for Fig. 20.6, but now showing the potential vorticity at various sigma levels in the Pacific, on the date line at longitude 180. The v-shaped grey lines show the reference variation due to βy for the $\sigma = 26.4$ level. For the other levels, the reference variation is similar to the actual variation between about 0° and 10° latitude.

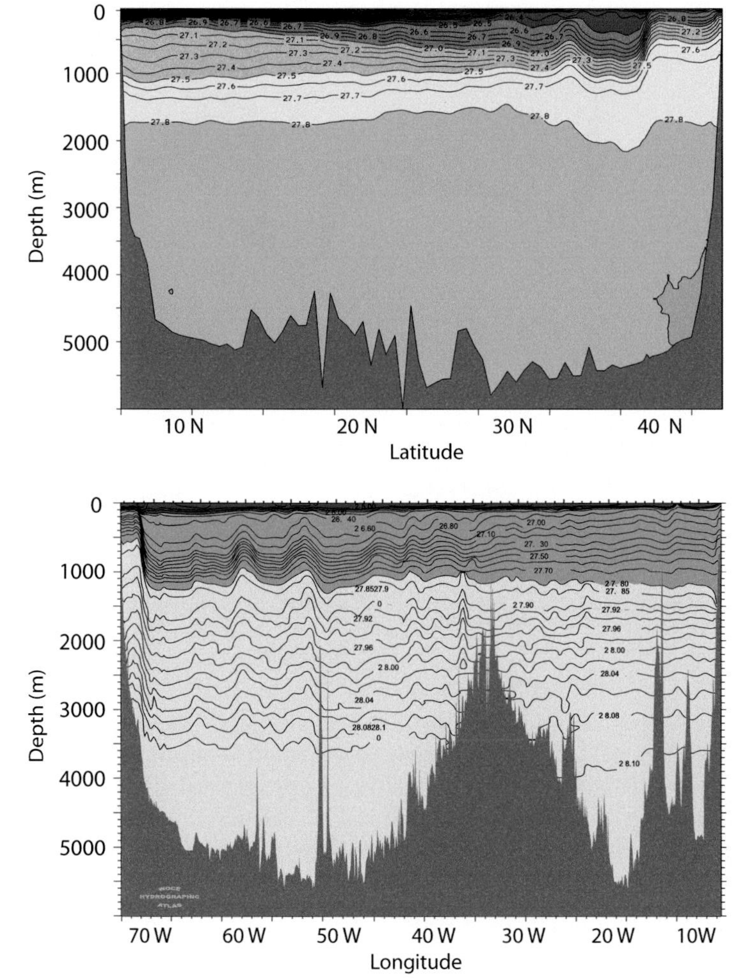

Fig. 20.8 Sections of density in the North Atlantic.

Upper panel: meridional section of potential density at 53° W, from 5° N to 45° N, with a uniform contour interval. In the upper northwestern region of the subtropical thermocline there is a region of low stratification known as MODE water: isopycnals above this outcrop in the subtropical gyre and are 'ventilated'; isopycnals below the MODE water outcrop in the subpolar gyre, north of about 45°, and/or in ACC.

Lower panel: zonal section of neutral density at 36° N, from about 75° W to 10° W. Contour interval changes where the colour changes. Note the front associated with the western boundary current at about 70° W.[6]

20.4 THE MAIN THERMOCLINE

We now approach the dynamics of the upper ocean from another angle and address the mechanisms that give rise to the actual density structure of the upper ocean, and in particular to the *main thermocline,* which is the region of the upper ocean, about 1 km deep, in which the density and temperature change most rapidly, as illustrated in Fig. 20.8 and Fig. 20.9.

We will consider the circulation in a closed, single hemispheric basin, and suppose that there is a net surface heating at low latitudes and a net cooling at high latitudes that maintains a meridional temperature gradient at the surface. Let us presume, *ab initio,* that there is a single overturning cell, with water rising at low latitudes before returning to polar regions, as illustrated schematically in Fig. 20.10. We will investigate the dynamics of this meridional overturning circulation (MOC) in much more detail in the next chapter, but here our interest is mainly in why and how affects the density structure in the upper ocean and this is less affected by interhemispheric effects. (Readers for whom the MOC is of primary interest may wish to read Chapter 21 before proceeding.) We will also, by and large, omit considerations of saline effects and assume a linear equation of state, so that the thermocline is synonymous with the *pycnocline,* the region where density changes rapidly.

The physical picture we have is the following. Cold, dense water at high latitudes sinks, so that dense water extends all the way to the ocean floor. By hydrostasy the pressure in the deep ocean is then higher at high latitudes than at low, where the water is warmer. Thus the water moves

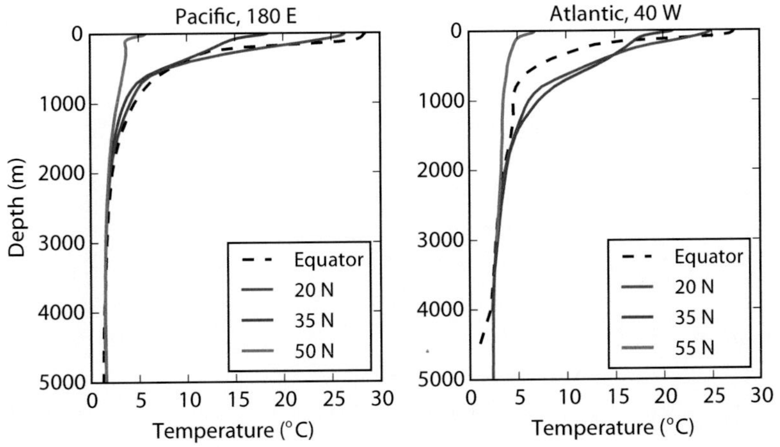

Fig. 20.9 Profiles of mean temperature in the North Pacific and Atlantic, from WOCE, at the longitudes and latitudes indicated. The profiles are considerably smoother than instantaneous ones.

Note the shallowness of the equatorial thermoclines (especially in the Atlantic), and weakness of the subpolar thermoclines.

equatorward filling the abyss. This water is also slowly *warmed* by heat diffusion down from above, and it is this diffusion that enables the circulation to persist: if diffusion were zero, the entire ocean would eventually fill with the densest available water and the circulation would cease. The water that fills the interior from the cold pole is colder and denser than the surface waters at lower latitude so there must be a vertical temperature gradient, except at the highest latitudes where the water is sinking, and we indeed see in Fig. 20.9 how the vertical temperature profile varies with latitude. However, without considering the dynamics it is hard to see what form the temperature profile will take; for example, it is conceivable that the polar waters might fill up the abyss nearly all the way to the surface, leaving a thermocline only a few metres thick. Or there might be a uniform temperature gradient from the surface to the ocean floor.

Complicating matters, the thermocline is also the region where the gyre circulation is most prominent, so that the potential vorticity dynamics of the previous few sections must play a role. Putting that complication aside for now, let us first look at a simple kinematic model.

20.4.1 A Simple Kinematic Model

The fact that cold water with polar origins upwells into a region of warmer water suggests that we consider the simple one-dimensional advective–diffusive balance,

$$w\frac{\partial T}{\partial z} = \kappa\frac{\partial^2 T}{\partial z^2}, \tag{20.43}$$

where w is the vertical velocity, κ is a diffusivity and T is temperature. In mid-latitudes, where this might hold, w is positive and the equation represents a balance between the upwelling of cold water and the downward diffusion of heat. If w and κ are given constants, and if T is specified at the top ($T = T_T$ at $z = 0$) and if $\partial T/\partial z = 0$ at great depth ($z = -\infty$) then the temperature falls exponentially away from the surface according to

$$T = (T_T - T_B)e^{wz/\kappa} + T_B, \tag{20.44}$$

where T_B is the temperature at depth. Temperature decays exponentially away from its surface value with the scale

$$\delta = \frac{\kappa}{w}, \tag{20.45}$$

and this is an estimate of the thermocline thickness. It is not particularly useful, because the magnitude of w depends on κ, as we will see. However, it is reasonable to see if the observed ocean is

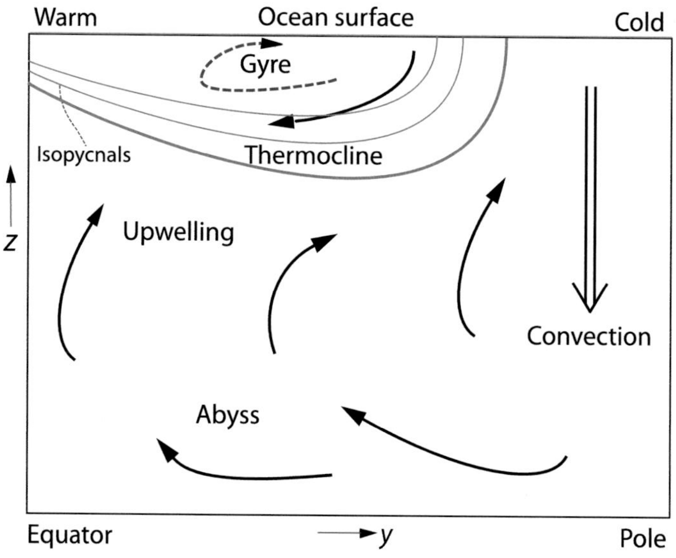

Fig. 20.10 Cartoon of a single-celled meridional overturning circulation, with a wall at the equator.

Sinking is concentrated at high latitudes and upwelling spread out over lower latitudes. The thermocline is the boundary between the cold abyssal waters, with polar origins, and the warmer near-surface subtropical water. Wind forcing in the subtropics pushes the warm surface water into the fluid interior, deepening the thermocline as well as circulating as a gyre.

broadly consistent with this expression. The diffusivity κ can be measured; it is an eddy diffusivity, maintained by small-scale turbulence, and measurements produce values that range between $10^{-5}\,\mathrm{m^2\,s^{-1}}$ in the main thermocline and $10^{-4}\,\mathrm{m^2\,s^{-1}}$ in abyssal regions over rough topography and in and near continental margins, with still higher values locally.[7] The vertical velocity is too small to be measured directly, but various estimates based on deep water production suggest a value of about $10^{-7}\,\mathrm{m\,s^{-1}}$. Using this and the smaller value of κ in (20.44) gives an e-folding vertical scale, κ/w, of just 100 m, beneath which the stratification is predicted to be very small (i.e., nearly uniform potential density). Using the larger value of κ increases the vertical scale to 1000 m, which is probably closer to the observed value for the total thickness of the thermocline (Fig. 20.9), but using such a large value of κ in the main thermocline is not supported by the observations. Similarly, the deep stratification of the ocean is rather larger than that given by (20.43), except with values of diffusivity on the large side of those observed, a topic we return to in Chapter 21.[8]

Aside from diffusion, mechanical forcing, and in particular the wind, will deepen the thermocline, as Fig. 20.10 suggests. The wind-stress curl forces water to converge in the subtropical Ekman layer, thereby forcing relatively warm water to downwell and meet the upwelling colder abyssal water at some finite depth, thus deepening the thermocline from its purely diffusive value. Indeed, in so far as we can separate the two effects of wind and diffusion, we can say that the strength of the wind influences the *depth* at which the thermocline occurs, whereas the strength of the diffusivity influences the *thickness* of the thermocline.

20.5 SCALING AND SIMPLE DYNAMICS OF THE MAIN THERMOCLINE

We now begin to consider the dynamics that produce an overturning circulation and a thermocline. The Rossby number of the large-scale circulation is small and the scale of the motion large, and the flow obeys the planetary-geostrophic equations,

$$\boldsymbol{f} \times \boldsymbol{u} = -\nabla\phi, \qquad \frac{\partial \phi}{\partial z} = b, \qquad \nabla \cdot \boldsymbol{v} = 0, \qquad \frac{\mathrm{D}b}{\mathrm{D}t} = \kappa \frac{\partial^2 b}{\partial z^2}, \tag{20.46a,b,c,d}$$

in our standard notation, in which $\boldsymbol{u}$ and $\boldsymbol{v}$ refer to the two- and three-dimensional velocities. We suppose that these equations hold below an Ekman layer, so that the effects of a wind stress may be included by specifying a vertical velocity, w_E, at the top of the domain. The diapycnal diffusivity, κ, is some kind of eddy diffusivity, but since its precise form and magnitude are uncertain we proceed with due caution, and a useful practical philosophy is to try to ignore dissipation and viscosity where possible, and to invoke them only if there is no other way out. Let us therefore scale the equations in two ways, with and without diffusion; these scalings will be central to our theory.

20.5.1 A Diffusive Scale

Suppose that the circulation is steady and resembles that of Fig. 20.10, but with no wind forcing. Can we estimate how deep the diffusive layer will be in the subtropical gyre? We will suppose that, as in the kinematic model, the thermodynamic equation reduces to the advective-diffusive balance of (20.43), but we will use the other equations in (20.46) to give an estimate of the vertical velocity. If we take the curl of (i.e., cross differentiate) the momentum equation (20.46a) and use mass continuity we obtain the linear vorticity equation, $\beta v = f\partial w/\partial z$, and if we take the vertical derivative of the momentum equation and use hydrostasy we obtain thermal wind, $\partial \boldsymbol{u}/\partial z = \mathbf{k}\times\nabla b$. Collecting these equations together we have

$$w\frac{\partial b}{\partial z} = \kappa\frac{\partial^2 b}{\partial z^2}, \qquad \beta v = f\frac{\partial w}{\partial z}, \qquad f\frac{\partial \boldsymbol{u}}{\partial z} = \mathbf{k}\times\nabla b, \tag{20.47a,b,c}$$

with corresponding scales

$$\frac{W}{\delta} = \frac{\kappa}{\delta^2}, \qquad \beta V = \frac{fW}{\delta}, \qquad \frac{U}{\delta} = \frac{\Delta b}{fL}, \tag{20.48a,b,c}$$

where δ is the vertical scale and other scaling values are denoted with capital letters. We suppose that $V \sim U$, where U is the zonal velocity scale, and henceforth we will denote both by U, and L is the horizontal scale of the motion, which we take as the gyre or basin scale. Typical values for the subtropical gyre are $\Delta b = g\Delta\rho/\rho_0 = g\beta_T\Delta T \sim 10^{-2}\,\mathrm{m\,s^{-2}}$, $L = 5000\,\mathrm{km}$, $f = 10^{-4}\,\mathrm{s^{-1}}$ and $\kappa = 10^{-5}\,\mathrm{m^2\,s^{-2}}$.

Equation (20.48a) is the same as (20.45), as expected, but we can now use (20.48b,c) to obtain an estimate for the vertical velocity, namely

$$W = \frac{\beta\delta^2\Delta b}{f^2 L}. \tag{20.49}$$

Using this and (20.48a) gives the diffusive vertical scale, and the estimates

$$\delta = \left(\frac{\kappa f^2 L}{\beta\,\Delta b}\right)^{1/3}, \qquad W = \left(\frac{\kappa^2\beta\,\Delta b}{f^2 L}\right)^{1/3}. \tag{20.50}$$

With values of the parameters as above, (20.50) gives $\delta \approx 150\,\mathrm{m}$ and $W \approx 10^{-7}\,\mathrm{m\,s^{-1}}$.

20.5.2 An Advective Scale

The value of the vertical velocity obtained above is very small, much smaller than the Ekman pumping velocity at the top of the ocean, which is of order 10^{-6}–$10^{-5}\,\mathrm{m\,s^{-1}}$. This difference suggests that we might ignore the diffusive term in (20.47a) — indeed, ignore the thermodynamic term completely — and construct an adiabatic scaling estimate for the depth of the wind's influence. Further, in subtropical gyres the Ekman pumping is downward, whereas the diffusive velocity is upward, meaning that at some level, D_a, we expect the vertical velocity to be zero.

The equations of motion are just the thermal wind balance and the linear geostrophic vorticity equation, namely

$$\beta v = f\frac{\partial w}{\partial z}, \qquad \boldsymbol{f}\times\frac{\partial \boldsymbol{u}}{\partial z} = -\nabla b, \tag{20.51}$$

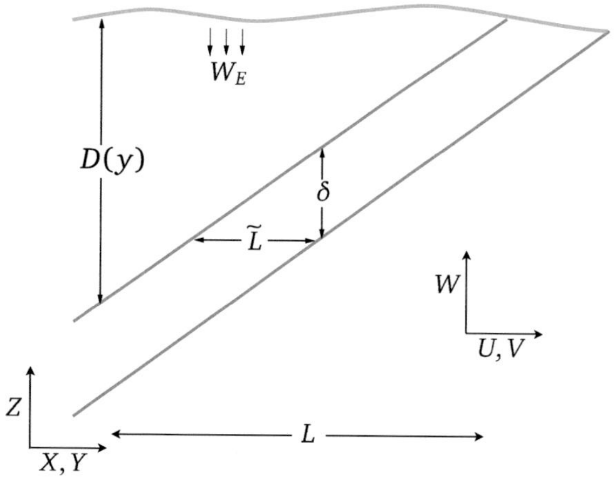

Fig. 20.11 Scaling the thermocline. The diagonal lines mark the diffusive thermocline of thickness δ and depth $D(y)$. The advective scaling for $D(y)$, i.e., D_a, is given by (20.53), and the diffusive scaling for δ is given by (20.55).

with corresponding scales

$$\beta U = f\frac{W}{D_a}, \qquad \frac{U}{D_a} = \frac{1}{f}\frac{\Delta b}{L}, \tag{20.52}$$

recalling that $V \sim U$.

The thermodynamic equation does not enter, but we take the vertical velocity to be that due to Ekman pumping, W_E. From (20.52) we immediately obtain

$$D_a = W_E^{1/2}\left(\frac{f^2 L}{\beta\,\Delta b}\right)^{1/2}, \tag{20.53}$$

which may be compared with the estimate of (20.26). If we relate U and W_E using mass conservation, $U/L = W_E/D_a$, instead of using (20.51a), then we write L in place of f/β and (20.53) becomes $D_a = (W_E f L^2/\Delta b)^{1/2}$, which is not qualitatively different from (20.53) for large scales.

The important aspect of the above estimate is that the depth of the wind-influenced region increases with the magnitude of the wind stress (because $W_E \propto \mathrm{curl}_z\tau$) and decreases with the meridional temperature gradient. The former dependence is reasonably intuitive, and the latter arises because as the temperature gradient increases the associated thermal wind-shear U/D_a correspondingly increases. But the horizontal transport (the product UD_a) is fixed by mass conservation; the only way that these two can remain consistent is for the vertical scale to decrease. Taking $W_E = 10^{-6}\,\mathrm{m\,s^{-1}}$, and other values as before, gives $D_a = 500\,\mathrm{m}$, and $W_E = 10^{-5}\,\mathrm{m\,s^{-1}}$ gives $D_a = 500\,\mathrm{m}$. Such a scaling argument cannot be expected to give more than an estimate of the depth of the wind-influenced region; nevertheless, because D_a is much less than the ocean depth, the estimate does suggest that the wind-driven circulation is predominantly an upper-ocean phenomenon.

♦ *A wind-influenced diffusive scaling*

The scalings above assume that the length scale over which thermal wind balance holds is the gyre scale itself. In fact, there is another length scale that is more appropriate, and this leads to a slightly different diffusive scaling for the thickness of the thermocline. To obtain this scaling, we first note that the depth of the subtropical thermocline is not constant: it shoals up to the east because of Sverdrup balance, and it may shoal up polewards as the curl of the wind stress falls (and is zero at the poleward edge of the gyre). Thus, referring to Fig. 20.11, the appropriate horizontal length scale $\widetilde{L}$ is given by

$$\widetilde{L} = \delta \frac{L}{D_a}. \tag{20.54}$$

This is no longer an externally imposed parameter, but must be determined as part of the solution. Using $\widetilde{L}$ instead of L as the length scale in the thermal wind equation (20.48c) gives, using (20.53), the modified diffusive scale

$$\delta = \kappa^{1/2} \left(\frac{f^2 L}{\Delta b \beta D_a} \right)^{1/2} = \kappa^{1/2} \left(\frac{f^2 L}{\Delta b \beta W_E} \right)^{1/4}. \tag{20.55}$$

Substituting values of the various parameters results in a thickness of about 100–200 m. The thermocline thickness now scales as $\kappa^{1/2}$. The interpretation of this scale and that of (20.50) is that the thickness of the thermocline scales as $\kappa^{1/3}$ in the absence of a wind stress, but scales as $\kappa^{1/2}$ if a wind stress is present that can provide a finite slope to the base of the thermocline that is independent of κ, and this is confirmed by numerical simulations.[9] From (20.47a) the vertical velocity, and hence the meridional overturning circulation, no longer scale as $\kappa^{2/3}$ but as

$$W = \frac{\kappa}{\delta} \propto \kappa^{1/2}. \tag{20.56}$$

20.5.3 Summary of the Physical Picture

What do the vertical scales derived above represent? The wind-influenced scaling, D_a, is the depth to which the directly wind-driven circulation can be expected to penetrate. Thus, over this depth we can expect to see wind-driven gyres and associated phenomena. At greater depths lies the abyssal circulation, and this is not wind-driven in the same sense. Now, in general, the water at the base of the wind-driven layer will not have the same thermodynamic properties as the upwelling abyssal water — this being cold and dense, whereas the water in the wind-driven layer is warm and subtropical (look again at Fig. 20.10). The thickness δ characterizes the diffusive transition region between these two water masses and in the limit of very small diffusivity this becomes a *front*. One might say that D_a is the *depth* of the thermocline, while δ is the *thickness* of the thermocline. In the diffusive region, no matter how small the diffusivity κ is in the thermodynamic equation, the diffusive term is important. Of course if the diffusion is sufficiently large, the thickness will be as large or larger than the depth, and the two regions will blur into each other, and this may indeed be the case in the real ocean. Nevertheless, these scales are a useful foundation on which to build.[10]

20.6 THE INTERNAL THERMOCLINE

We now try to go beyond simple scaling arguments and investigate in more detail the dynamics of the thermocline. In this section we consider the diffusive, or internal, thermocline and in Section 20.7 we consider the advective, or ventilated, thermocline. The advective term in the thermodynamic equation makes such an investigation difficult, and prevents us from constructing exact analytic models, but not from constructing informative models. We begin by expressing the planetary-geostrophic equations as an equation in a single unknown.

20.6.1 The *M* Equation

The planetary-geostrophic equations can be written as a single partial differential equation in a single variable, although the resulting equation is of quite high order and is nonlinear. We write the equations of motion as

$$-fv = -\frac{\partial \phi}{\partial x}, \qquad fu = -\frac{\partial \phi}{\partial y}, \qquad b = \frac{\partial \phi}{\partial z}, \tag{20.57a,b,c}$$

$$\nabla \cdot \boldsymbol{v} = 0, \qquad \frac{\partial b}{\partial t} + \boldsymbol{v} \cdot \nabla b = \kappa \nabla^2 b, \tag{20.58a,b}$$

where we take $f = \beta y$. Cross-differentiating the horizontal momentum equations and using (20.58a) gives the linear geostrophic vorticity relation $\beta v = f \partial w / \partial z$ which, using (20.57a) again, may be written as

$$\frac{\partial \phi}{\partial x} + \frac{\partial}{\partial z}\left(-\frac{f^2}{\beta} w\right) = 0. \tag{20.59}$$

This equation is the divergence in (x, z) of $(\phi, -f^2 w/\beta)$ and is automatically satisfied if

$$\phi = M_z \qquad \text{and} \qquad \frac{f^2 w}{\beta} = M_x, \tag{20.60a,b}$$

where the subscripts on M denote derivatives. Then straightforwardly

$$u = -\frac{\partial_y \phi}{f} = -\frac{M_{zy}}{f}, \qquad v = \frac{\partial_x \phi}{f} = \frac{M_{zx}}{f}, \qquad b = \partial_z \phi = M_{zz}. \tag{20.61a,b,c}$$

The thermodynamic equation, (20.58b) becomes

$$\frac{\partial M_{zz}}{\partial t} + \left(\frac{-M_{zy}}{f} M_{zzx} + \frac{M_{zx}}{f} M_{zzy}\right) + \frac{\beta}{f^2} M_x M_{zzz} = \kappa M_{zzzz} \tag{20.62}$$

or

$$\frac{\partial M_{zz}}{\partial t} + \frac{1}{f} J(M_z, M_{zz}) + \frac{\beta}{f^2} M_x M_{zzz} = \kappa M_{zzzz}, \tag{20.63}$$

where J is the usual horizontal Jacobian. This is the *M equation,*[11] somewhat analogous to the potential vorticity equation in quasi-geostrophic theory in that it expresses the entire dynamics of the system in a single, nonlinear, advective–diffusive partial differential equation, although M_{zz} is materially conserved (in the absence of diabatic effects) by the three-dimensional flow. Because of the high differential order and nonlinearity of the system, analytic solutions of (20.63) are hard to find, and from a numerical perspective it is easier to integrate the equations in the form (20.57) and (20.58) than in the form (20.63). Nevertheless, it is possible to move forward by approximating (20.63) to one or two dimensions, or by a priori assuming a boundary-layer structure.

A one-dimensional model

Let us consider an illustrative one-dimensional model (in z) of the thermocline.[12] Merely setting all horizontal derivatives in (20.63) to zero is not very useful, for then all the advective terms on the left-hand side vanish. Rather, we look for steady solutions of the form $M = M(x, z)$, and the M equation then becomes

$$\frac{\beta}{f^2} M_x M_{zzz} = \kappa M_{zzzz}, \tag{20.64}$$

which represents the advective–diffusive balance

$$w \frac{\partial b}{\partial z} = \kappa \frac{\partial^2 b}{\partial z^2}. \tag{20.65}$$

In proceeding in this way we have assumed that the value of κ varies meridionally in the same manner as does β / f^2; without this technicality M would be a function of y, violating our premise.

If the ocean surface is warm and the abyss is cold, then (20.64) represents a balance between the upward advection of cold water and the downward diffusion of warm water. The horizontal advection terms vanish because the zonal velocity, u, and the meridional buoyancy gradient, b_y, are each zero. Let us further consider the special case

$$M = (x - x_e)W(z), \tag{20.66}$$

where the domain extends from $0 \le x \le x_e$, so satisfying $M = 0$ on the eastern boundary. Equation (20.64) becomes the ordinary differential equation

$$\frac{\beta}{f^2} W W_{zzz} = \kappa W_{zzzz}, \tag{20.67}$$

where W has the dimensions of velocity squared. We nondimensionalize this by setting

$$z = H\hat{z}, \qquad \kappa = \hat{\kappa}(HW_S), \qquad W = \left(\frac{f^2 W_S}{\beta}\right)\widehat{W}, \tag{20.68a,b,c}$$

where the hatted variables are nondimensional and W_S is a scaling value of the dimensional vertical velocity, w (e.g., the magnitude of the Ekman pumping velocity W_E). Equation (20.67) becomes

$$\widehat{W}\widehat{W}_{\hat{z}\hat{z}\hat{z}} = \hat{\kappa}\widehat{W}_{\hat{z}\hat{z}\hat{z}\hat{z}}. \tag{20.69}$$

The parameter $\hat{\kappa}$ is a nondimensional measure of the strength of diffusion in the interior, and the interesting case occurs when $\hat{\kappa} \ll 1$; in the ocean, typical values are $H = 1$ km, $\kappa = 10^{-5}\,\mathrm{m\,s^{-2}}$ and $W_S = W_E = 10^{-6}\,\mathrm{m\,s^{-1}}$ so that $\hat{\kappa} \approx 10^{-2}$, which is indeed small. (It might appear that we could completely scale away the value of κ in (20.67) by scaling W appropriately, and if so there would be no meaningful way that one could say that κ was small. However, this is a chimera, because the value of κ would still appear in the boundary conditions.

The time-dependent form of (20.69), namely $\widehat{W}_{\hat{z}\hat{z}t} + \widehat{W}\widehat{W}_{\hat{z}\hat{z}\hat{z}} = \hat{\kappa}\widehat{W}_{\hat{z}\hat{z}\hat{z}\hat{z}}$, is similar to Burger's equation, $V_t + VV_z = \nu V_{zz}$, which is known to develop fronts. (In the inviscid Burger's equation, $DV/Dt = 0$, where the advective derivative is one-dimensional, and therefore the velocity of a given fluid parcel is preserved on the line. Suppose that the velocity of the fluid is positive but diminishes in the positive z-direction, so that a fluid parcel will catch-up with the fluid parcel in front of it. But since the velocity of a fluid parcel is fixed, there are two values of velocity at the same point, so a singularity must form. In the presence of viscosity, the singularity is tamed to a front.) Thus, we might similarly expect (20.69) to produce a front, but because of the extra derivatives the argument is not as straightforward and it is simplest to obtain solutions numerically.

Equation (20.69) is fourth order, so four boundary conditions are needed, two at each boundary. Appropriate ones are a prescribed buoyancy and a prescribed vertical velocity at each boundary, for example

$$\begin{aligned} &\widehat{W} = \widehat{W}_E, \quad -\widehat{W}_{\hat{z}\hat{z}} = B_0, \qquad \text{at top},\\ &\widehat{W} = 0, \qquad -\widehat{W}_{\hat{z}\hat{z}} = 0, \qquad \text{at bottom}, \end{aligned} \tag{20.70}$$

where $\widehat{W}_E$ is the (nondimensional) vertical velocity at the base of the top Ekman layer, which is negative for Ekman pumping in the subtropical gyre, and B_0 is a constant, proportional to the buoyancy difference across the domain. We obtain solutions numerically by Newton's method,[13] and these are shown in Figs. 20.12 and 20.13. The solutions do indeed display fronts, or boundary layers, for small diffusivity. If the wind forcing is zero (Fig. 20.13), the boundary layer is at the top of the fluid. If the wind forcing is non-zero, an internal boundary layer — a front — forms in the fluid interior with an adiabatic layer above and below. In the real ocean, where wind forcing is of course non-zero, the frontal region is known as the *internal thermocline*.

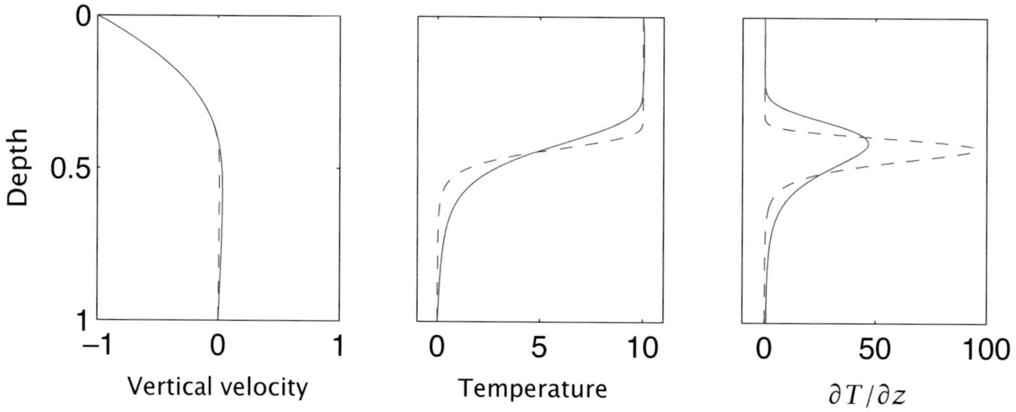

Fig. 20.12 Solution of the one-dimensional thermocline equation, (20.69), with boundary conditions (20.70), for two different values of the diffusivity: $\widehat{\kappa} = 3.2 \times 10^{-3}$ (solid line) and $\widehat{\kappa} = 0.4 \times 10^{-3}$ (dashed line), in the domain $0 \le \widehat{z} \le -1$. 'Vertical velocity' is W, 'temperature' is $-W_{\widehat{z}\widehat{z}}$, and all units are the nondimensional ones of the equation itself. A negative vertical velocity, $\widehat{W_E} = -1$, is imposed at the surface (representing Ekman pumping) and $B_0 = 10$.

The internal boundary layer thickness increases as $\widehat{\kappa}^{1/3}$, so doubling in thickness for an eightfold increase in $\widehat{\kappa}$. The upwelling velocity also increases with $\widehat{\kappa}$ (as $\widehat{\kappa}^{2/3}$), but this is barely noticeable on the graph because the downwelling velocity, above the internal boundary layer, is much larger and almost independent of $\widehat{\kappa}$. The depth of the boundary layer increases as $\widehat{W_E}^{1/2}$, so if $\widehat{W_E} = 0$ the boundary layer is at the surface, as in Fig. 20.13.

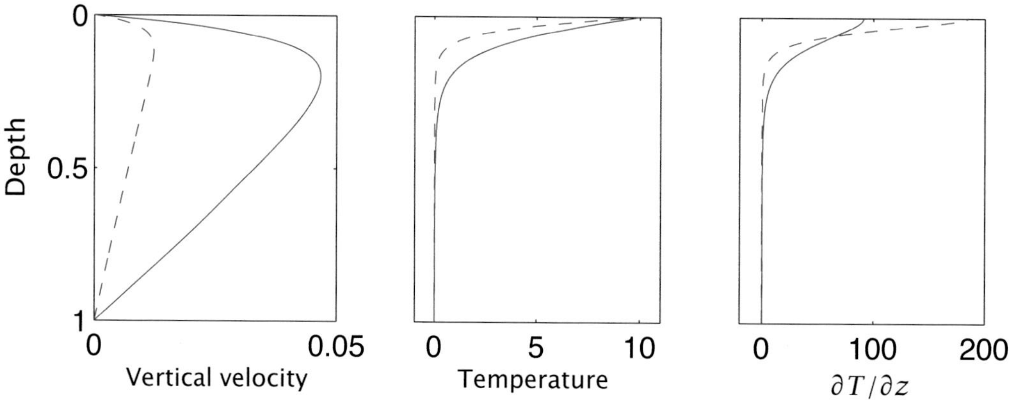

Fig. 20.13 As for Fig. 20.12, but with no imposed Ekman pumping velocity at the upper boundary ($\widehat{W_E} = 0$), again for two different values of the diffusivity: $\widehat{\kappa} = 3.2 \times 10^{-3}$ (solid line) and $\widehat{\kappa} = 0.4 \times 10^{-3}$ (dashed line). The boundary layer now forms at the upper surface. The boundary thickness again increases with diffusivity and, even more noticeably, so does the upwelling velocity — this scales as $\widehat{\kappa}^{2/3}$, and so increases fourfold for an eightfold increase in $\widehat{\kappa}$.

20.6.2 ♦ Boundary-layer Analysis

The reasoning and the numerical solutions of the above sections suggest that the internal thermocline has a boundary-layer structure whose thickness decreases with κ. If the Ekman pumping at the top of the ocean is non-zero, the boundary layer is internal to the fluid. To learn more, let us perform a boundary layer analysis, much as we did when investigating western boundary currents in Section 19.1.3. The nonlinearity precludes a complete solution of the equation, but we can nevertheless obtain some useful information.

One-dimensional model

Let us now *assume* a steady two-layer structure of the form illustrated in Fig. 20.14, and that the dynamics are governed by (20.69) in a domain that extends from 0 to −1. The buoyancy thus varies rapidly only in an internal boundary layer of nondimensional thickness $\widehat{\delta}$ located at $\widehat{z} = -h$; above and below this the buoyancy is assumed to be only very slowly varying. Following standard boundary layer procedure we introduce a stretched boundary layer coordinate ζ where

$$\widehat{\delta}\zeta = \widehat{z} + h. \tag{20.71}$$

That is, ζ is the distance from $\widehat{z} = -h$, scaled by the boundary layer thickness $\widehat{\delta}$, and within the boundary layer ζ is an order-one quantity. We also let

$$\widehat{W}(\widehat{z}) = \widehat{W}_I(\widehat{z}) + \widetilde{W}(\zeta), \tag{20.72}$$

where $\widehat{W}_I$ is the solution away from the boundary layer and $\widetilde{W}$ is the boundary layer correction. Because the boundary layer is presumptively thin, $\widehat{W}_I$ is effectively constant through it and, furthermore, for $\widehat{z} < -h$, $\widehat{W}$ vanishes in the limit as $\kappa = 0$. We thus take $\widehat{W}_I = 0$ throughout the boundary layer. (The small diffusively-driven upwelling below the boundary layer is part of the boundary layer solution, not the interior solution.) Now, buoyancy varies rapidly in the boundary layer but it remains an order-one quantity throughout. To satisfy this we explicitly scale $\widetilde{W}$ in the boundary layer by writing

$$\widetilde{W}(\zeta) = \widehat{\delta}^2 B_0 A(\zeta), \tag{20.73}$$

where B_0 is defined by (20.70) and A is an order-one field. The derivatives of W are

$$\frac{\partial \widehat{W}}{\partial \widehat{z}} = \frac{1}{\widehat{\delta}}\frac{\partial \widetilde{W}}{\partial \zeta} = \widehat{\delta} B_0 \frac{\partial A}{\partial \zeta}, \qquad \frac{\partial^2 \widehat{W}}{\partial \widehat{z}^2} = B_0 \frac{\partial^2 A}{\partial \zeta^2}, \tag{20.74}$$

so that $\widehat{W}_{\widehat{z}\widehat{z}}$ is an order-one quantity. Far from the boundary layer the solution must be able to match the external conditions on temperature and velocity, (20.70); the buoyancy condition on $W_{\widehat{z}\widehat{z}}$ is satisfied if

$$A_{\zeta\zeta} \to \begin{cases} 1 & \text{as } \zeta \to +\infty \\ 0 & \text{as } \zeta \to -\infty. \end{cases} \tag{20.75}$$

On vertical velocity we require that $W \to (\widehat{z}/h + 1)W_E$ as $\zeta \to +\infty$, and $W \to$ constant as $\zeta \to -\infty$. The first matches the Ekman pumping velocity above the boundary layer, and the second condition produces the abyssal upwelling velocity, which as noted vanishes for $\kappa \to 0$.

Substituting (20.72) and (20.73) into (20.69) we obtain

$$B_0 A A_{\zeta\zeta\zeta} = \frac{\widehat{\kappa}}{\widehat{\delta}^3} A_{\zeta\zeta\zeta\zeta}. \tag{20.76}$$

Because all quantities are presumptively $\mathcal{O}(1)$, (20.76) implies that $\widehat{\delta} \sim (\widehat{\kappa}/B_0)^{1/3}$. We restore the dimensions of δ by using $\kappa = \widehat{\kappa}(HW_S)$ and $\Delta b = B_0 L f^2 W_S/(\beta H^2)$, where Δb is the dimensional buoyancy difference across the boundary layer — note that $b = M_{zz} = (x-1)W_{zz} \sim LW_{zz} \sim LB_0 f^2 W_S/(\beta H^2)$ using (20.68). The dimensional boundary layer thickness, δ, is then given by

$$\delta \sim \left(\frac{\kappa f^2 L}{\Delta b\, \beta}\right)^{1/3}, \tag{20.77}$$

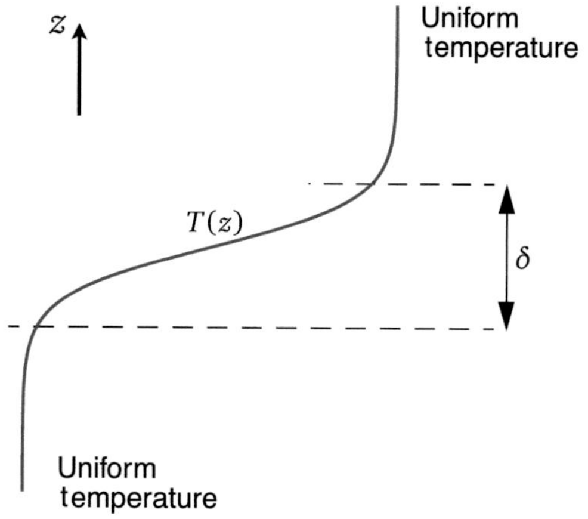

Fig. 20.14 The simplified boundary-layer structure of the internal thermocline. In the limit of small diffusivity the internal thermocline forms a boundary layer, of thickness δ in the figure, in which the temperature and buoyancy change rapidly.

which is the same as the heuristic estimate (20.50). The dimensional vertical velocity scales as

$$W \sim \frac{\kappa}{\delta} \sim \kappa^{2/3} \left(\frac{\Delta b \, \beta}{f^2 L} \right)^{1/3}, \tag{20.78}$$

this being an estimate of strength of the upwelling velocity at the base of the thermocline and, more generally, the strength of the diffusively-driven component of meridional overturning circulation of the ocean.

Although the detailed properties of such one-dimensional thermocline models depend on the details of model construction, two significant features are robust:

(i) The thickness of the internal thermocline increases with increasing diffusivity, and decreases with increasing buoyancy difference across it, and as the diffusivity tends to zero the thickness of the internal thermocline tends to zero.

(ii) The strength of the upwelling velocity, and hence the strength of the meridional overturning circulation, increase with increasing diffusivity and increasing buoyancy difference.

♦ *The three-dimensional equations*

We now apply boundary layer techniques to the three-dimensional M equation.[14] The main difference is that the depth of the boundary layer is now a function of x and y, so that the stretched coordinate ζ is given by

$$\widehat{\delta}\zeta = z + h(x, y). \tag{20.79}$$

(The coordinates (x, y, z) in this subsection are nondimensional, but we omit their hats to avoid too cluttered a notation.) Just as in the one-dimensional case we rescale M in the boundary layer and write

$$M = B_0 \widehat{\delta}^2 \widehat{A}(x, y, \zeta), \tag{20.80}$$

where the scaling factor $\widehat{\delta}^2$ again ensures that the temperature remains an order-one quantity. In the boundary layer the derivatives of M become

$$\frac{\partial M}{\partial z} = \frac{1}{\widehat{\delta}} \frac{\partial A}{\partial \zeta}, \tag{20.81}$$

and

$$\frac{\partial M}{\partial x} = \widehat{\delta}^2 B_0 \left(\frac{\partial A}{\partial \zeta} \frac{\partial \zeta}{\partial x} + \frac{\partial A}{\partial x} \right) = \widehat{\delta}^2 B_0 \left(\frac{\partial A}{\partial \zeta} \frac{1}{\widehat{\delta}} \frac{\partial h}{\partial x} + \frac{\partial A}{\partial x} \right). \tag{20.82}$$

Substituting these into (20.62) we obtain, omitting the time-derivative,

$$\begin{aligned}\widehat{\delta}\left[\frac{1}{f}\left(A_{\zeta x}A_{\zeta\zeta y}-A_{\zeta y}A_{\zeta\zeta x}\right)+\frac{\beta}{f^2}A_x A_{\zeta\zeta\zeta}\right]+\frac{\beta}{f^2}h_x A_\zeta A_{\zeta\zeta\zeta}&\\ +\frac{1}{f}\left[h_x\left(A_{\zeta\zeta}A_{\zeta\zeta y}-A_{\zeta y}A_{\zeta\zeta\zeta}\right)+h_y\left(A_{\zeta x}A_{\zeta\zeta\zeta}-A_{\zeta\zeta}A_{\zeta\zeta x}\right)\right]&=\frac{\kappa}{B_0\widehat{\delta}^2}A_{\zeta\zeta\zeta\zeta},\end{aligned} \tag{20.83}$$

where the subscripts on A and h denote derivatives. If $h_x = h_y = 0$, that is if the base of the thermocline is flat, then (20.83) becomes

$$\frac{1}{f}\left[A_{\zeta x}A_{\zeta\zeta y}-A_{\zeta y}A_{\zeta\zeta x}\right]+\frac{\beta}{f^2}A_x A_{\zeta\zeta\zeta}=\frac{\kappa}{B_0\widehat{\delta}^3}A_{\zeta\zeta\zeta\zeta}. \tag{20.84}$$

Since all the terms in this equation are, by construction, order one, we immediately see that the nondimensional boundary layer thickness $\widehat{\delta}$ scales as

$$\widehat{\delta}\sim\left(\frac{\kappa}{B_0}\right)^{1/3}, \tag{20.85}$$

just as in the one-dimensional model. On the other hand, if h_x and h_y are order-one quantities then the dominant balance in (20.83) is

$$\frac{1}{f}\left[h_x(A_{\zeta\zeta}A_{\zeta\zeta y}-A_{\zeta y}A_{\zeta\zeta\zeta})+h_y(A_{\zeta x}A_{\zeta\zeta\zeta}-A_{\zeta\zeta}A_{\zeta\zeta x})\right]=\frac{\kappa}{B_0\widehat{\delta}^2}A_{\zeta\zeta\zeta\zeta} \tag{20.86}$$

and

$$\widehat{\delta}\sim\left(\frac{\kappa}{B_0}\right)^{1/2}, \tag{20.87}$$

confirming the heuristic scaling arguments. Thus, if the isotherm slopes are fixed independently of κ (for example, by the wind stress), then as $\kappa \to 0$ an internal boundary layer will form whose thickness is proportional to $\kappa^{1/2}$. We expect this to occur at the base of the main thermocline, with purely advective dynamics being dominant in the upper part of the thermocline, and determining the slope of the isotherms (i.e., the form of h_x and h_y), as in Fig. 20.11. Interestingly, the balance in the three-dimensional boundary layer equation does not in general correspond locally to $wT_z \approx \kappa T_{zz}$. Both at $\mathcal{O}(1)$ and $\mathcal{O}(\delta)$ the horizontal advective terms in (20.83) are of the same asymptotic size as the vertical advection terms. In the boundary layer the thermodynamic balance is thus $\boldsymbol{u}\cdot\nabla_z T + wT_z \approx \kappa T_{zz}$, whether the isotherms are sloping or flat. We might have anticipated this, because the vertical velocity passes through zero within the boundary layer.

What are the dynamics above the diffusive layer, presuming that it does not extend all the way to the surface? Answering this leads us into our next topic, the 'ventilated thermocline'.

20.7 THE VENTILATED THERMOCLINE

We now consider the nature of the dynamics *above* the diffusive layer, presuming that the diffusivity is sufficiently small that there is a meaningful separation of the internal boundary layer and the advective dynamics above. In the advective region there is no general reason that the temperature profile should be uniform, and we envision an essentially adiabatic region that is both wind-driven and stratified. This region of the thermocline has become known, for reasons that will become apparent, as the *ventilated thermocline*. The main thermocline is composed of the internal thermocline plus a ventilated region, and to set our bearings it may be useful to refer now to the overall picture sketched in Fig. 20.15 and the shaded box on page 790.

To elucidate the structure of the ventilated thermocline we will assume the following:[15]

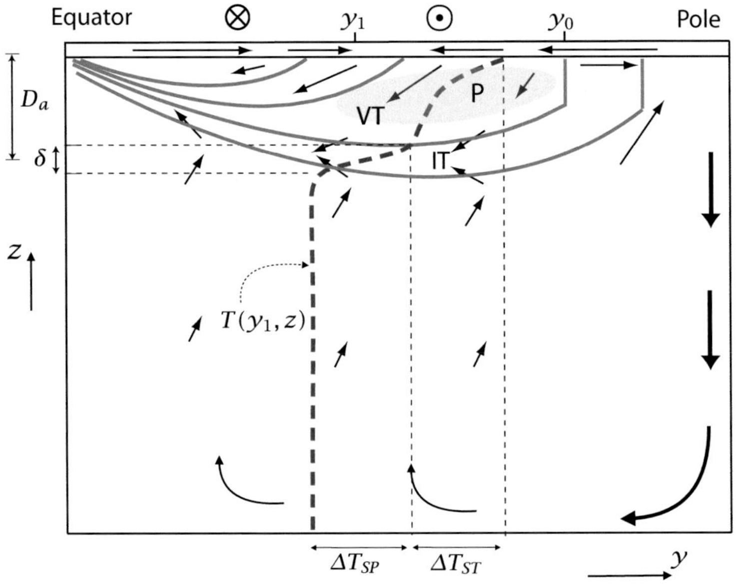

Fig. 20.15 Idealization of the large-scale circulation and structure of the main thermocline, in a single-hemisphere ocean driven by wind stress (⊙ and ⊗) and a surface temperature that decreases monotonically from equator to pole.

The thin arrows indicate the meridional overturning circulation and the flow in the Ekman layer near the ocean surface. The thick dashed line is a temperature profile, $T(y_1, z)$ at the subtropical latitude y_1, where the horizontal axis is temperature. The solid blue lines are isotherms and the homogenized western pool region is shaded grey.

Key: VT – ventilated thermocline. IT – internal thermocline. P – pool region in west. δ – thickness of the internal thermocline. Da – depth of the ventilated thermocline. ΔT_{ST} – temperature drop across subtropical gyre and across ventilated thermocline. ΔT_{SP} – temperature drop across the subpolar gyre and internal thermocline. y_0 – subtropical-subpolar gyre boundary. y_1 – a latitude in the subtropical gyre.

- The motion satisfies the ideal, steady, planetary-geostrophic equations.
- The surface temperature, and the vertical velocity due to Ekman pumping, are given. (These surface conditions are, in reality, influenced by the ocean's dynamics, but we assume that we can calculate a solution with specified surface conditions.) At the base of the wind-influenced region we will impose $w = 0$.
- Rather than use the continuously stratified equations, we will assume that the solution can be adequately represented by a small number of layers, each of constant density. The abyss is represented by a single stationary layer.
- We will not take into account the possible effects of a western boundary current. In that sense the model is an extension of the Sverdrup interior of homogeneous models.

The model is thus not a complete one, yet we may hope that it is revealing about the structure of the real ocean.

20.7.1 A Reduced Gravity, Single-layer Model

The simplest possible model along these lines is to suppose the ocean is composed of just two layers, and only one moving layer, as illustrated in Fig. 20.16. The upper layer of density ρ_1 is wind-driven, whereas the lower layer of density ρ_2 is assumed to be stationary; this is called a 'one-and-a-half-layer' model or a 'reduced gravity single-layer' model. Pertinent questions are, how deep is the upper layer? What is the velocity field in it?

In the planetary-geostrophic approximation, the momentum and mass conservation equations

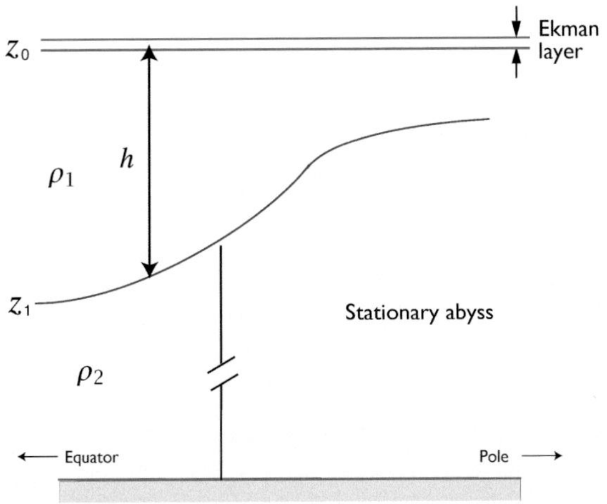

Fig. 20.16 A reduced gravity, single-layer model. A single moving layer lies above a deep, stationary layer of higher density. The upper surface is rigid. A thin Ekman layer may be envisioned to lie on top of the moving layer, providing a vertical velocity boundary condition.

of the reduced gravity shallow water model may be written as:

$$\boldsymbol{f} \times \boldsymbol{u} = -g'\nabla h, \qquad \nabla \cdot \boldsymbol{u} = -\frac{\partial w}{\partial z}, \tag{20.88a,b}$$

where ∇ is a two-dimensional operator (as it will be for the rest of this section) and $g' = g(\rho_2 - \rho_1)/\rho_0$ is the *reduced gravity*. Taking the curl of (20.88a) gives the geostrophic vorticity equation, $\beta v + f\nabla \cdot \boldsymbol{u} = 0$, and integrating this over the depth of the layer and using mass conservation gives

$$h\beta v = f(w_E - w_b), \tag{20.89}$$

where w_E is the velocity at the top of the layer, due mainly to Ekman pumping, and w_b is the vertical velocity at the layer base. If the flow is steady, w_b is zero for then

$$w_b = \boldsymbol{u} \cdot \nabla h = -\frac{g'}{f}\frac{\partial h}{\partial y}\frac{\partial h}{\partial x} + \frac{g'}{f}\frac{\partial h}{\partial x}\frac{\partial h}{\partial y} = 0. \tag{20.90}$$

Using this result and geostrophic balance, (20.89) becomes

$$\frac{g'}{f}\beta h\frac{\partial h}{\partial x} = f w_E, \tag{20.91}$$

which integrates to

$$h^2 = -2\frac{f^2}{g'\beta}\int_x^{x_e} w_E \, dx' + H_e^2, \tag{20.92}$$

where H_e is the (unknown) value of h at the eastern boundary x_e, and it is a constant to satisfy the no-normal flow condition. This apart, the equation contains complete information about the solution. We note that:

- the depth of the moving layer scales as the magnitude of the wind stress (or Ekman pumping velocity) to the one-half power;
- the horizontal solution is similar to the simpler Sverdrup interior solution previously obtained in Section 19.1.3;
- there is no solution if w_E is positive; that is, if there is Ekman upwelling;
- the solution depends on the unknown parameter H_e, the layer depth at the eastern boundary. (That the eastern boundary depth is undetermined is perhaps the main incomplete aspect of the theory as presented here.[16])

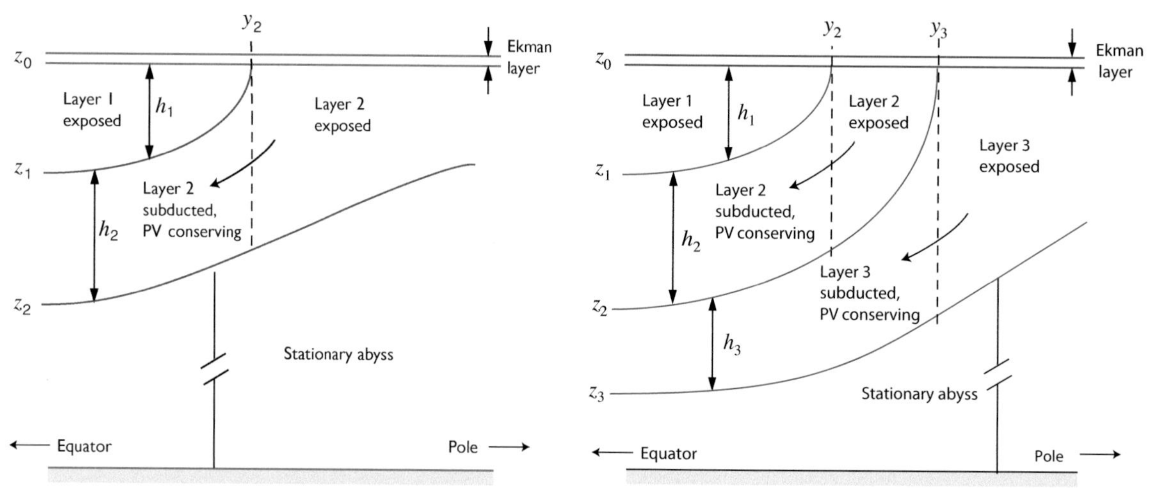

Fig. 20.17 Two-layer (left) and three-layer (right) schematics of the ventilated thermocline, each with a stationary abyss. Models with still more moving layers may be constructed, at least in principle, by extension.

20.7.2 Two-layer and Three-layer Models

Imagine now there are two moving layers above a stationary abyss, as in Fig. 20.17. If there is a meridional buoyancy gradient at the surface then isopycnals *outcrop*, or intersect the surface. Thus, at some latitude (say $y = y_2$, which for simplicity we assume not to be a function of longitude) layer 2 passes underneath layer 1, which is of lower density, as sketched in Fig. 20.17.

Thus, polewards of y_2 the dynamics are just those of a single layer discussed above, whereas equatorward of y_2 layer 2 does not feel the wind directly, and its dynamics are governed by two principles:

(i) Sverdrup balance. This still applies to the vertically integrated motion, and thus to the sum of layer 1 and layer 2.

(ii) Conservation of potential vorticity. The motion in layer 2 is shielded from the wind forcing, and the effects of dissipation are assumed to be negligible. Thus, the fluid parcels in the layer will conserve their potential vorticity.

We first use potential vorticity conservation to obtain an expression for the depth of each layer in terms of the total depth of the moving fluid, h, and then use Sverdrup balance to obtain h.

We can apply exactly the same procedure if there are three (or more) moving layers, as in the right-hand panel of Fig. 20.17, and in principle go to the limit of continuous stratification. However, the algrebra becomes considerably more complicated and most of the essential dynamics are contained in two layers, so that is our focus.

Potential vorticity conservation

Conservation of potential vorticity in the region equatorward of y_2 is, for steady flow,

$$\boldsymbol{u}_2 \cdot \nabla q_2 = 0 \qquad \text{for} \quad y < y_2, \tag{20.93}$$

where $q_2 = f/h_2$. Now, the velocity field in layer 2 is given by $\boldsymbol{u}_2 = (g_2'/f)\mathbf{k}\times\nabla h$, where $h = h_1 + h_2$ is the total depth of the moving fluid (see the appendix to this chapter). Thus, (20.93) becomes

$$-\frac{g_2'}{f}\frac{\partial h}{\partial y}\frac{\partial}{\partial x}\left(\frac{f}{h_2}\right) + \frac{g_2'}{f}\frac{\partial h}{\partial x}\frac{\partial}{\partial y}\left(\frac{f}{h_2}\right) = \frac{g_2'}{f}J\left(\frac{f}{h_2}, h\right) = 0. \tag{20.94}$$

This is an equation relating h and h_2 and it has the general solution

$$q_2 \equiv \frac{f}{h_2} = G_2(h), \tag{20.95}$$

where G_2 is an *arbitrary* function of its argument. However, we *know* what the potential vorticity of layer 2 is at the moment it is subducted; it is just

$$q_2(y_2) \equiv \frac{f(y_2)}{h_2} = \frac{f_2}{h}, \tag{20.96}$$

where $f_2 \equiv f(y_2)$, and $h_2 = h$ because $h_1 = 0$. This relationship must therefore hold everywhere in layer 2, equatorwards of y_2; that is,

$$G_2(h) = \frac{f_2}{h}. \tag{20.97}$$

Thus, in the subducted region, and taking $z_0 = 0$,

$$\frac{f}{h_2} = \frac{f_2}{h} \qquad \text{or} \qquad \frac{f}{z_1 - z_2} = -\frac{f_2}{z_2}. \tag{20.98}$$

From this we easily obtain expressions for the depth of each layer as a function of the total depth, h, namely

$$h_2 = z_1 - z_2 = \frac{f}{f_2}h \qquad \text{and} \qquad h_1 = -z_1 = \left(1 - \frac{f}{f_2}\right)h. \tag{20.99}$$

It remains only to find an expression for the total depth of the moving fluid, h, and this we do using Sverdrup balance. Note that because potential vorticity, f/h_2, is conserved, as the subducted fluid column moves equatorward its thickness must decrease.

Using Sverdrup balance to find the total depth

Equations (20.99) contain the unknown total depth h, and we now use Sverdrup balance to find this and close the problem. The linear vorticity equation is $\beta v = f\partial w/\partial z$, where the velocity at the top of layer one is that due to the Ekman layer and the velocity at the base of layer two is zero. Given this, we may write the Sverdrup balance as

$$\beta(h_1 v_1 + h_2 v_2) = f w_E, \tag{20.100}$$

where, using (20.128), the velocities in each layer are given by

$$f v_1 = \frac{\partial}{\partial x}(g_2' h + g_1' h_1) \qquad \text{and} \qquad f v_2 = \frac{\partial}{\partial x}(g_2' h). \tag{20.101}$$

Using these, Sverdrup balance becomes

$$\beta h_1 \frac{\partial}{\partial x}(g_2' h + g_1' h_1) + \beta(h - h_1) g_2' \frac{\partial h}{\partial x} = f^2 w_E, \tag{20.102}$$

or

$$\frac{\partial}{\partial x}(g_2' h^2 + g_1' h_1^2) = \frac{2f^2}{\beta} w_E. \tag{20.103}$$

On integrating, the above equation becomes

$$\left(h^2 + \frac{g_1'}{g_2'} h_1^2\right) = D_0^2 + C, \qquad \text{where} \qquad D_0^2(x, y) = -\frac{2f^2}{\beta g_2'} \int_x^{x_e} w_E(x', y)\, dx', \tag{20.104a,b}$$

Thermocline Dynamics — an Overview

The model of the main thermocline that we have constructed in sections 20.4–20.7 is illustrated schematically in Fig. 20.15. Some of the features, and limitations, of this model are listed below:

- The main subtropical thermocline consists of an advective upper region overlying a diffusive base.
  - The diffusive base forms the *internal thermocline,* and in the limit of small diffusivity this is an internal boundary layer. The advective region forms the *ventilated thermocline.* The separation of the two regions may, in reality, not be sharp.
  - The relative thickness of these layers is a function of various parameters, notably the strength of the wind and the magnitude of the diffusivity.
- Above the ventilated thermocline there may be a mixed layer with a seasonally varying depth. In certain regions, for example at the poleward edge of the subtropical gyre, convection may deepen the mixed layer as far as the base of the thermocline.
- In the thermocline theories we have presented there is no explicit western boundary layer. Such a boundary layer is needed to close the circulation and the heat budget.
- The single-hemisphere model assumes that the water that sinks at high latitude either upwells through the main thermocline or returns to the subpolar gyre beneath the main thermocline. In reality some of this water may cross into the other hemisphere before upwelling — more so in the Atlantic than Pacific.
  - In this case, the diffusion-dependent overturning circulation represents only part of the overall meridional overturning circulation.
  - Nevertheless, there would remain a diffusive internal thermocline (and a ventilated thermocline above it) because there is still a boundary between the warm subtropical water and cold abyssal water.
- Within the ventilated thermocline there are two regions — the shadow zone and the western pool — whose dynamics are not determined without additional assumptions. Plausible assumptions for the western pool are:
  - All the water within it is ventilated, leading to a model of mode water.
  - The potential vorticity within the pool is homogenized through the action of mesoscale eddies.

  Some combination of these might also apply. The size of the pool region increases as the poleward boundary of the pool region approaches the latitude of the outcrop (Fig. 20.21) and can extend almost across the entire gyre.
- Topics of research include questions of how potential vorticity homogenization is affected by a western boundary current, if and how the non-passive nature of potential vorticity affects the pool region, and if and how all these concepts apply in the real ocean.

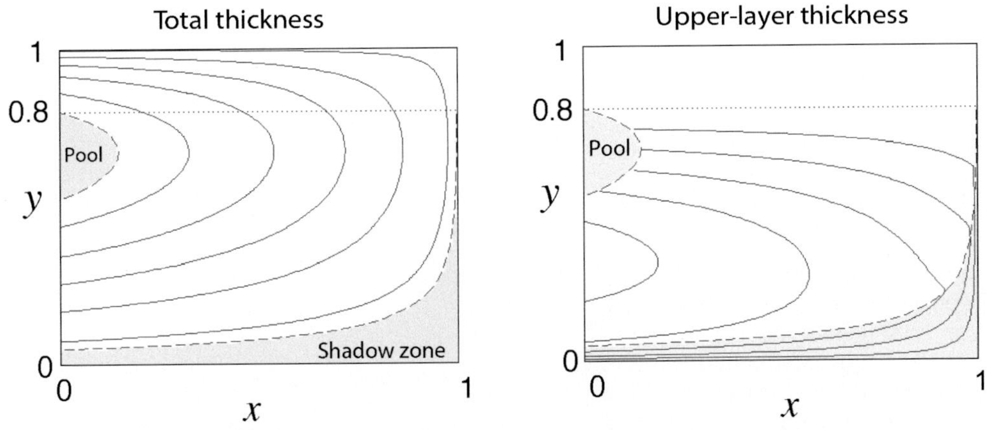

Fig. 20.18 Contour plots of total thickness and upper layer thickness in a two-layer model of the ventilated thermocline. The thickness generally increases westwards, and the flow is clockwise. The shadow zone and the western pool are shaded, and no contours are drawn in the latter. The outcrop latitude, $y_2 = 0.8$, is marked with a dotted line. The parameters used are $g_1' = g_2' = 1, \beta = 1$, $f_0 = 0.5, H_e = 0.5$, and $w_E = -\sin(\pi y)$.

which by construction vanishes at the eastern wall ($x = x_e$). The constant of integration C may be interpreted as follows. Let us write $C = H_e^2 + (g_1'/g_2')H_1^2$ where H_e is the (unknown) total depth of layers 1 and 2 at the eastern boundary, and H_1 is the depth of layer 1. These must both be constants in order to satisfy the no-normal flow condition. However, H_1 must be zero, because at the outcrop line $h_1 = 0$. Thus, H_1 is zero at $y = y_2$, and therefore zero everywhere, and $C = H_e^2$.

Using (20.99) and (20.104) we obtain a closed expression for h, namely

$$h = -z_2 = \frac{(D_0^2 + H_e^2)^{1/2}}{[1 + (g_1'/g_2')(1 - f/f_2)^2]^{1/2}} . \tag{20.105}$$

Using (20.99) the depths in each layer, and the corresponding geostrophic velocities, can readily be obtained.

A typical solution is shown in Fig. 20.18. The upper layer exists only equatorward of the outcrop latitude, $y_2 = 0.8$, and isolines of total thickness correspond to streamlines of the lower layer. We see, as expected, the overall shape of a subtropical gyre, with the circulation being closed by an implicit western boundary current that is not part of the calculation. Two regions are shaded in the figure, the 'pool' region in the west and the 'shadow zone' in the south-east. The solutions above do not apply to these, and they require some special attention.

20.7.3 The Shadow Zone

In the fluid interior the potential vorticity of a parcel in layer 2 is determined by tracing its trajectory back to its outcrop latitude where the potential vorticity is given. That trajectory is determined by its velocity, and this in turn is determined by inverting the potential vorticity. Now, parcels subducted at y_2 sweep equatorward and westward, so that a parcel, labelled '***a***' say, subducted at the eastern boundary will in general leave the eastern boundary tracing a southwestern trajectory. Consider another parcel, '***b***' say, in the interior that lies eastward of the subducted position of ***a***, in the shaded region of Fig. 20.19. It is impossible to trace ***b*** back to the outcrop line without

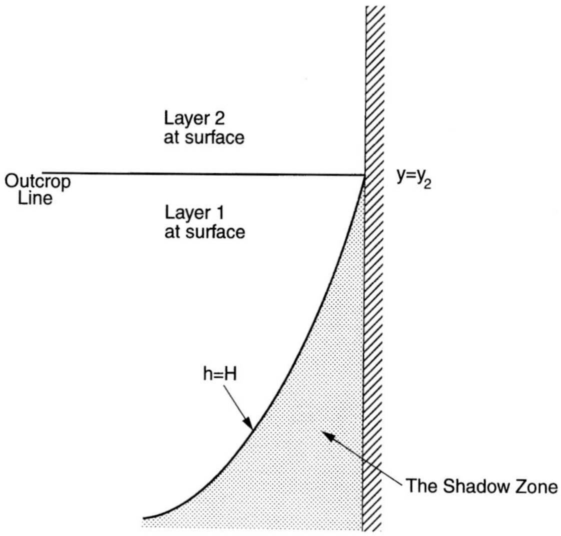

Fig. 20.19 The shadow zone in the ventilated thermocline. Layer 2 outcrops at $y = y_2$. A column moving equatorward along the eastern boundary in layer 2 is subducted at y_2.

The column cannot remain against the eastern wall and both preserve its potential vorticity, which implies the column shrinks, at the same time that the no-normal flow condition is satisfied, as by geostrophy this implies the layer depth is constant. Thus, the column must move westward, along the boundary of a 'shadow zone' within which there is no motion. The streamline it follows is the one of constant total thickness of the two moving layers — see (20.128) or (20.130c).

trajectories crossing, and this is forbidden in steady flow. Rather, it seems as if the trajectory of $\boldsymbol{b}$ would emanate from the eastern wall. What is the potential vorticity there?

At the eastern boundary the condition of no normal flow at the boundary demands that h be constant (so that $u_2 = 0$), and h_1 be constant (so that $u_1 = 0$). But if a parcel in layer 2 moves *along* the boundary potential vorticity conservation demands that f/h_2 is constant, and therefore h_2 must change, contradicting the no-normal flow requirement. Thus, the velocity at the boundary can have neither a normal component nor a tangential component, and so we cannot trace parcels in the shaded region back to the wall. Rather, in the absence of closed trajectories (for example, eddying motion), we may assume that the shaded region is stagnant, and h is constant. Of course, potential vorticity is everywhere given by f/h_2, which varies spatially, but since there is no motion potential vorticity is still, rather trivially, conserved along trajectories. This region is aptly called the *shadow zone,* since the region falls under the shadow of the eastern boundary; an analogous region arose in the quasi-geostrophic discussion of Section 20.1.

To obtain an expression for the fields within the shadow zone, first note that because h is constant, its value is equal to that on the eastern wall; that is, $h = H_e$. The wind forcing must then all be taken up by the upper layer, and Sverdrup balance then implies

$$\beta v_1 h_1 = f w_E, \tag{20.106}$$

and using (20.101) we obtain an expression for h_1, to wit

$$h_1^2 = -\frac{2f^2}{\beta g_1'} \int_x^{x_e} w_E(x', y)\,\mathrm{d}x' = \frac{g_2'}{g_1'} D_0^2, \tag{20.107}$$

which is zero at the eastern wall. In the lower layer the thickness is just $h_2 = H_e - h_1$. The boundary of the shadow zone is given by the trajectory of a fluid parcel in layer 2 that emanates from the eastern boundary at the outcrop line where $h_1 = 0$ and $h = h_2 = H_e$. Since the flow is steady, the trajectory is an isoline of h. Thus, from (20.105) we have

$$h^2 = \frac{(D_0^2(x_s, y_s) + H_e^2)}{[1 + g_1'/g_2'(1 - f/f_2)^2]} = H_e^2, \tag{20.108}$$

where (x_s, y_s) denotes the boundary of the shadow zone. (Note that $x_s = x_e$ at $y = y_2$.) The above equation yields

$$D_0^2(x_s, y_s) = H_e^2 \left[\frac{g_1'}{g_2'} \left(1 - \frac{f}{f_2} \right)^2 \right], \tag{20.109}$$

which, given the wind stress, determines the shadow zone boundary x_s as a function of y.

20.7.4 † The Western Pool

Polewards of the outcrop latitude the fluid of layer 2 feels the wind directly and the layer thickness is determined by Sverdrup balance. Equatorward of the outcrop latitude the properties of this layer are determined by potential vorticity conservation, with the potential vorticity being determined by the layer thickness at the outcrop. However, just as there is a region in the east where trajectories cannot be traced back to the outcrop, there is a 'pool' region in the west that is bounded by the trajectory that emerges from the western boundary at the outcrop latitude Within the pool, trajectories cannot be traced back to the outcrop (Fig. 20.18), and one might suppose that they emerge from the western boundary current. There are two plausible hypotheses for determining the layer depths within this region:

(i) Within layer 2, potential vorticity is homogenized.

(ii) Because there is no source for layer-2 water, layer-2 water does not exist and the pool consists solely of ventilated, layer-1 water.

Neither of the above can be derived from the governing equations of motion without making additional physical assumptions that are neither a priori true nor obvious. We discuss both hypotheses briefly below, followed by a more general discussion of how the pool region fits together with the earlier discussions about potential vorticity homogenization in the quasi- geostrophic equations.

(i) Potential vorticity homogenization

The pool region is a region of recirculation, receiving water from and depositing water into the western boundary current. Thus, following the ideas described in Chapter 13 and employed in Section 20.1, we hypothesize that the potential vorticity within this region becomes homogenized. The value of potential vorticity within the pool is just the value of potential vorticity at its boundary, and this is given by $f_2/h_2(w)$, where $h_2(w)$ is the thickness of layer 2 at the western boundary at the outcrop latitude. This is given using (20.92) with $f = f_2$ and $g' = g_2'$, and thus the potential vorticity in the pool is given by

$$q_{pool} = \frac{f_2}{D_w^2 + H_e^2}, \tag{20.110}$$

where $D_w^2 = -2(f_2^2/g_2'\beta) \int_{x_w}^{x_e} w_E(x', y_2)\,dx'$. The thickness of layer 2 in the pool must be consistent with this, and so is given by

$$h_2 = \frac{f}{q_{pool}}. \tag{20.111}$$

The thickness of layer 1 is determined by using Sverdrup balance, (20.100), which, given h_2 and geostrophy, reduces to an equation for h_1, and solutions are shown in Fig. 20.20.

The extent of the pool region is dependent upon the outcrop latitude of the moving layer, since the boundary of the pool is a thickness contour. As the outcrop latitude moves poleward toward the gyre boundary then the pool region expands, as seen in Fig. 20.21. By the same token, if we have more moving layers then, since the deeper layers outcrop further poleward (Fig. 20.17) those deeper layers will have a more extensive pool region. That is, the pool expands with depth and the layer that outcrops just equatorward of the gyre boundary may have a pool region reaching across the gyre, as illustrated in Fig. 20.22, which shows the pool boundaries in a calculation (not shown here) with a three-layer model.

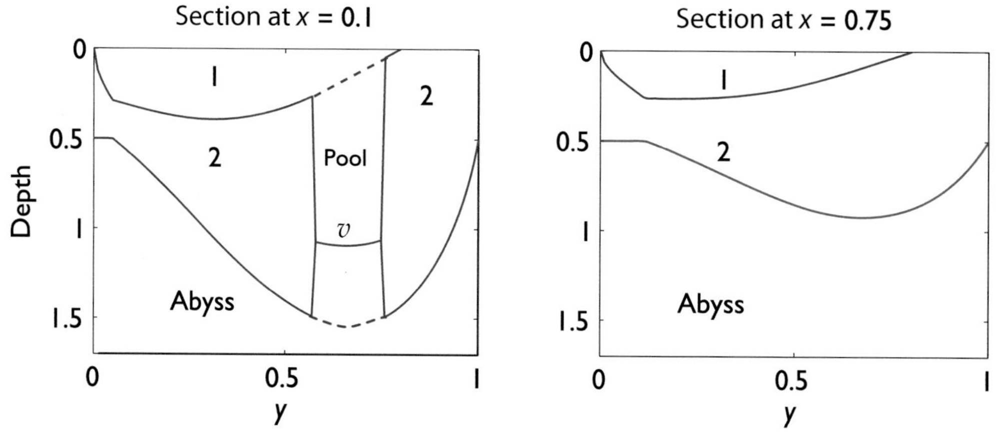

Fig. 20.20 Two north-south sections of layer thickness, at different longitudes, from the same solution as Fig. 20.18. The numbers refer to the fluid layer. The section on the left passes through the western pool region. In the homogenized PV model of the pool, the dashed lines should be solid and the line labelled v should be removed. In the ventilated pool model, the dashed lines should be removed. The region near $y = 0$ in both plots where the total depth of the thermocline is constant is the shadow zone.

(ii) The ventilated pool

'I came for the waters.' 'What waters?' 'I was misinformed.'
From *Casablanca* (1942).

The homogenization hypothesis, although entirely plausible, depends on the assumption of down-gradient diffusion of potential vorticity by eddies. Also, because there is no source of layer-2 water in the pool, we must suppose that it is ventilated by eddy pathways that meander down from the surface. An alternative hypothesis, and one that does not rely on the properties of mesoscale eddies, is to suppose that the western pool is filled with water that is directly ventilated from the surface. That is, if there is no surface source for a water mass, we simply suppose that that water mass does not exist. In the two-layer model, this means that the western pool is filled entirely with layer-1 fluid. If a non-ventilated (e.g., layer-2) fluid is present initially, then we hypothesize that it is slowly expunged by the continuous downwards Ekman pumping of layer-1 water into the pool.[17]

Because the layer-2 fluid is absent, the layer-1 fluid extends all the way down to the stagnant abyss; it takes up all the Sverdrup transport, and this determines the depth of the ventilated pool. Thus, rounding up the usual equations, we set $h = h_1$ in (20.104) to give

$$h_1^2 = D_1^2 + g_2' H_e^2, \tag{20.112}$$

where

$$D_1^2(x, y) = -\frac{2f^2}{\beta g_{1a}'} \int_x^{x_e} w_E(x', y)\, dx', \tag{20.113}$$

with $g_{1a}' = g_1' + g_2'$ being the reduced gravity between layer 1 and the abyss, and H_e, as before, being the thickness of layer 2 at the eastern boundary. Because $g_{1a}' > g_2'$ this pool will generally be shallower than the total depth of the moving fluid ($h_1 + h_2$) just outside, but the depth of layer-1 fluid alone will be much greater; that is, there will be discontinuities in layer depths at the pool boundary. A section through the pool region is shown in Fig. 20.20. The figure also shows the configuration of the pool if the homogenized potential vorticity hypothesis is used.

Although we may be shocked by the appearance of discontinuities in layer depths in a fluid model, the model does provide a simple mechanism for the appearance of *mode water*. This is a

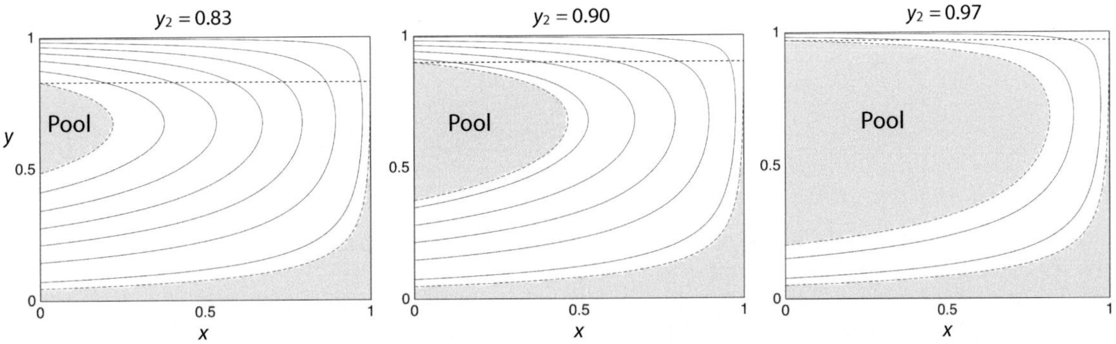

Fig. 20.21 Contour plots of total thickness in a two-layer model of the ventilated thermocline, with parameters as in Fig. 20.18, except for the outcrop latitude, y_2, that takes values as indicated. The pool and the shadow zone are shaded and contours within are not shown. The pool expands considerably as the outcrop latitude approaches the gyre boundary at $y = 1$.

distinct mass of weakly stratified, low potential vorticity water appearing in the north-west corner of the North Atlantic subtropical gyre (where it is sometimes called '18 degree water'), with analogues in the other gyres of the world's oceans; it is so-called because it appears as a distinct mode in a census of water properties. The proximate mechanism for mode water formation is convection in winter, but for such convection to occur the large-scale ocean circulation must maintain a weakly stratified region, and it is the ventilated pool that enables this, and sets the formation in the context of thermocline structure. In reality, the vertical isopycnals predicted by the simple model will be highly baroclinically unstable, and the ensuing mesoscale eddies will erode the pool interface and cause the isopycnals to slump, so that the discontinuities in layer depths will be manifest only as rapid changes or fronts.

Observations show that mode water exists only over a small region in the northeastern corner of the subtropical gyre, as the pool region in Fig. 20.18 might suggest. Regions of potential vorticity homogenization can extend much further, as in Fig. 20.4, with a similar degree of homogenization in the Atlantic. It may be that the effects of the wind are only able to homogenize the water masses in the upper regions of the pool. At greater depths the effects of baroclinic instability and the consequent mesoscale eddies may dominate the wind effects and produce regions of homogenized potential vorticity of greater horizontal extent.

20.7.5 † Remarks on Thermocline Structure

In the first two sections of this chapter we discussed a quasi-geostrophic, wind-driven view of the upper thermocline and found potentially large regions in which the flow re-circulated and potential vorticity was homogenized. One might call this an *elliptic* view of the upper ocean circulation, with the value of potential vorticity being determined by a diffusion problem, with potential vorticity diffusion along isopycnals. We then discussed the thermocline in a rather different way: in the ventilated thermocline the potential vorticity is subducted in from the surface, more along the lines of a *hyperbolic* problem or a problem in characteristics. The ventilated region has an unavoidably *diffusive* internal boundary layer at its base, with diffusion across isopycnals connecting the thermocline to the abyss.

These are not necessarily competing theories; rather, the three regions (an elliptic region of potential vorticity homogenization, the hyperbolic ventilated thermocline and a diffusive internal thermocline) can co-exist in the upper ocean, as sketched in Fig. 20.15. Depending on the value of the diapycnal diffusivity, the diffusive and internal thermoclines may be almost indistinguishable, observationally if not dynamically, and, depending on the stratification and wind structure the

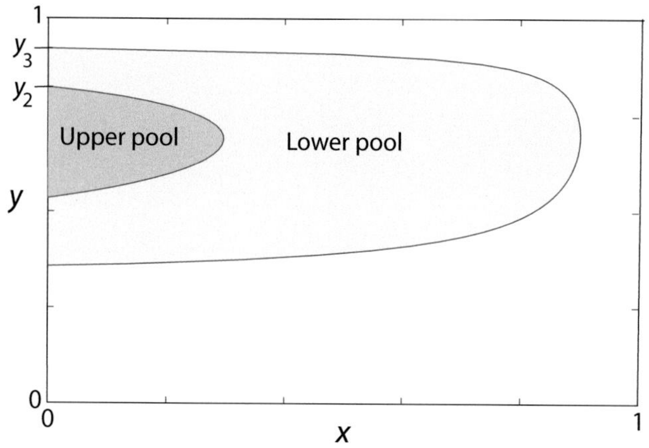

Fig. 20.22 The two pool regions in a calculation of the ventilated thermocline with three moving layers. The upper pool refers to the western pool in layer 2, which is the upper subducted layer that outcrops at y_2, and the lower pool is that in layer 3, which outcrops at y_3 (see right-hand panel of Fig. 20.17).

regions of potential vorticity homogenization may be larger or smaller.

Finally we remark that potential vorticity is not a passive scalar, which means that the boundary of the pool region will be affected by the homogenization process. This process may entrain additional water into the pool, which then grows and may prevent subducted water from entering into it, and the edge of the homogenized region will then become a mixing barrier. However, the picture of if and how the homogenization of potential vorticity precisely affects the structure of the ventilated thermocline is a little murky, and transparency will require high resolution numerical simulations and more complete observations to guide us to the truth, or an approximation of it.

APPENDIX A: MISCELLANEOUS RELATIONSHIPS IN A LAYERED MODEL

Here we collect various expressions relating pressure, density and velocity in a geostrophic and Boussinesq layered model. The layers and the interfaces are numbered, increasing downwards, as in Fig. 20.23, and the bottom layer is stationary.

A.1 Hydrostatic Balance

Hydrostatic balance is $\partial p/\partial z = -\rho g$. We can integrate this to give, in layers n and $n-1$,

$$p_n = -\rho_n g z + p_n'(x, y), \qquad p_{n-1} = -\rho_{n-1} g z + p_{n-1}'(x, y). \tag{20.114}$$

Since pressure is continuous, at $z = z_{n-1}$ these two expressions are equal so that

$$-\rho_n g z_{n-1} + p_n' = p_{n-1} = -\rho_{n-1} g z_{n-1} + p_{n-1}' \tag{20.115}$$

whence

$$g_{n-1}' z_{n-1} = \frac{1}{\rho_0}(p_n' - p_{n-1}') \tag{20.116}$$

where $g_n' = g(\rho_{n+1} - \rho_n)/\rho_0$ is the *reduced gravity*, and ρ_0 is the constant, reference, value of the density used in the Boussinesq approximation, taken to be equal to ρ_1.

A.2 Geostrophic and Thermal Wind Balance

In the Boussinesq approximation, geostrophic balance is:

$$\rho_0 f \boldsymbol{u}_n = \mathbf{k} \times \nabla p_n. \tag{20.117}$$

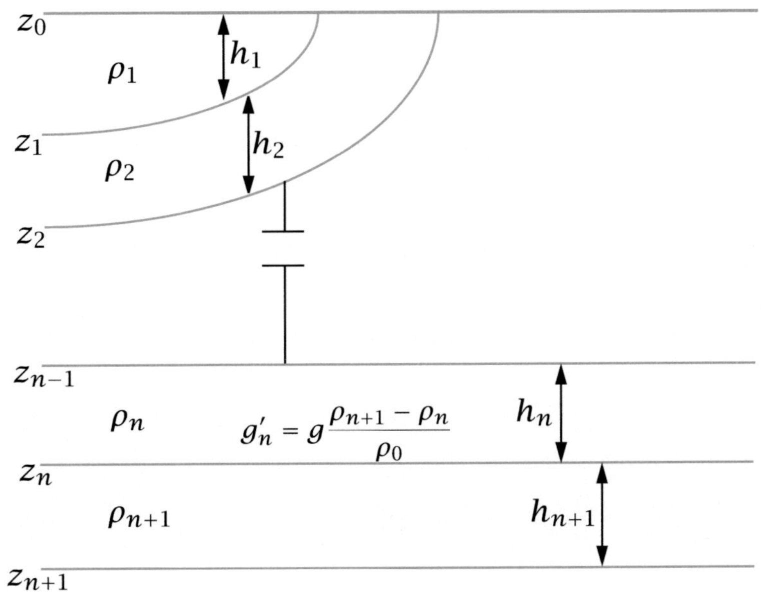

Fig. 20.23 Structure and notational conventions used for a multi-layered model.

Using (20.117) with (20.116) gives

$$\boldsymbol{u}_{n+1} - \boldsymbol{u}_n = \frac{g'_n}{f}\mathbf{k} \times \nabla z_n, \tag{20.118}$$

which is the appropriate form of thermal wind balance for this system. Let us suppose that at sufficient depth there is no motion, and in particular that layer $N+1$ is stationary and contains no pressure gradients. That is, $p'_{N+1} = 0$ and so, from (20.115) and (20.117)

$$p'_N = -(\rho_{N+1} - \rho_N) g z_N = -g'_N \rho_0 z_N. \tag{20.119}$$

Integrating upwards we obtain the pressure in each layer,

$$p'_n = -\rho_0 \sum_{i=n}^{i=N} g'_i z_i \tag{20.120}$$

where $n \le N$. Thus, the geostrophic velocities in each layer are given by

$$f\boldsymbol{u}_n = -\mathbf{k} \times \nabla \left(\sum_{i=n}^{i=N} g'_i z_i \right). \tag{20.121}$$

The quantity in brackets on the right-hand side is not a velocity streamfunction because it is $f\boldsymbol{u}_n$, and not $\boldsymbol{u}_n$ that is given by its curl. Nevertheless, the velocity is normal to its gradient, and therefore its isolines define streamlines.

The upper surface of the ocean is assumed to be fixed; this is the 'rigid-lid' approximation. Thus, $z_0 = 0$ and $h_1 = -z_1$. More generally, the layer thicknesses and the interfaces between the layers are related by

$$z_n = -\sum_{i=M}^{i=n} h_i \tag{20.122}$$

where M is the index of the uppermost layer, and $n \ge M$. If there is no outcropping, then $M = 1$.

The geostrophic velocity in the lowest moving layer is given by

$$f\boldsymbol{u}_N = -g'_N \mathbf{k} \times \nabla z_N = g'_N \mathbf{k} \times \nabla h. \tag{20.123}$$

This means that lines of constant depth of the lowest layer are also streamlines; the velocity moves parallel to the depth contours. The vertical velocity at the base of the lowest layer is given by, for steady flow,

$$w(z = -h) = \boldsymbol{u}_N \cdot \nabla h = \frac{1}{f} g'_N (\mathbf{k} \times \nabla h) \cdot \nabla h = 0. \tag{20.124}$$

That is, there is no vertical motion at the base of the moving layers.

A.3 Explicit Cases

A one-layer reduced-gravity model

The perturbation pressure in the moving layer (layer 1) is

$$p'_1 = -\rho_0 g'_1 z_1 = \rho_0 g'_1 h_1. \tag{20.125}$$

The geostrophic velocity is given by

$$f\boldsymbol{u}_1 = \frac{1}{\rho_0} \mathbf{k} \times \nabla p'_1 = g'_1 \mathbf{k} \times \nabla h_1. \tag{20.126}$$

(In a single-layer model, the subscripts are often omitted.)

A two-layer model

The perturbation pressures in the upper and lower moving layers are given by

$$p'_1 = -\rho_0 (g'_2 z_2 + g'_1 z_1) = \rho_0 (g'_2 h + g'_1 h_1), \tag{20.127a}$$
$$p'_2 = -\rho_0 g'_2 z_2 = \rho_0 g'_2 (h_1 + h_2) = \rho_0 g'_2 h, \tag{20.127b}$$

where $h = h_1 + h_2 = -z_2$.

The corresponding geostrophic velocities are

$$f\boldsymbol{u}_1 = \mathbf{k} \times \nabla (g'_2 h + g'_1 h_1), \tag{20.128a}$$
$$f\boldsymbol{u}_2 = \mathbf{k} \times \nabla (g'_2 h). \tag{20.128b}$$

A three-layer model

The perturbation pressures in the three moving layers are

$$p_1 = -\rho_0 [g'_3 z_3 + g'_2 z_2 + g'_1 z_1] = \rho_0 [g'_3 h + g'_2 (h_1 + h_2) + g'_1 h_1], \tag{20.129a}$$
$$p_2 = -\rho_0 [g'_2 z_2 + g'_3 z_3] = \rho_0 [g'_2 (h_1 + h_2) + g'_3 h], \tag{20.129b}$$
$$p_3 = -\rho_0 g'_3 z_3 = \rho_0 g'_3 h, \tag{20.129c}$$

where $h = h_1 + h_2 + h_3 = -z_3$. The corresponding geostrophic velocities are:

$$f\boldsymbol{u}_1 = \mathbf{k} \times \nabla [g'_3 h + g'_2 (h_1 + h_2) + g'_1 h_1] \tag{20.130a}$$
$$f\boldsymbol{u}_2 = \mathbf{k} \times \nabla [g'_2 (h_1 + h_2) + g'_3 h] \tag{20.130b}$$
$$f\boldsymbol{u}_3 = \mathbf{k} \times \nabla [g'_3 h]. \tag{20.130c}$$

Notes

1 Drawing from Rhines & Young (1982b). Young & Rhines (1982) also considered the problem of a western boundary layer.

2 Indian Ocean results are from McCarthy & Talley (1999), and I am very grateful to Lynne Talley for the Pacific plots. The data is mainly from CTD profiles and bottle casts. The potential density is σ_0 in the upper ocean plots, and σ_2 for the deep ocean.

3 See Rhines & Young (1982a) and Holland *et al.* (1984) for numerical examples of PV homogenization and Xu *et al.* (2015) for some related high-resolution simulations of the Atlantic.

4 I am very grateful to Peter Rhines for constructing Figs. 20.5, 20.6 and 20.7 from MIMOC data, and for his insightful interpretations. The MIMOC climatology is described at http://www.pmel.noaa.gov/mimoc/ and by Schmidtko *et al.* (2013). The date comes mainly from Argo CTDs, supplemented by shipboard and ice-tethered profiler CTDs, and is put onto a 0.5° grid.

5 See Keffer (1985), Talley (1988), Lozier *et al.* (1996) and McCarthy & Talley (1999) for some earlier observations of potential vorticity.

6 Sections courtesy of L. Talley and the WOCE hydrographic atlas. Globally continuous, unique, neutral-density surfaces cannot be constructed exactly because of the presence of salinity and the thermobaric term in the equation of state for seawater, and a parcel will not necessarily return to its level of departure when displaced to the same (x, y) position. Neutral density is nonetheless useful, and for the purposes of this figure it is the same as potential density.

7 See, among others, Toole *et al.* (1994), Polzin *et al.* (1997), Gregg (1998) and Ledwell *et al.* (1998).

8 For estimates of the strength of the overturning circulation in the ocean, and its relation to diapycnal diffusivity and the observed stratification, see Munk (1966), revisited by Munk & Wunsch (1998) and Wunsch & Ferrari (2004). If the abyssal flow is along rather than across isopycnals, smaller values of diffusivity suffice to maintain deep stratification — see Section 21.6.

9 Vallis (2000).

10 The modern development of the theory of the main thermocline began with two back-to-back papers in 1959 in the journal *Tellus*. Welander (1959) suggested an adiabatic model, based on the ideal-fluid thermocline equations (i.e., the planetary-geostrophic equations, with no diffusion terms in the buoyancy equation), whereas Robinson & Stommel (1959) proposed a model that is intrinsically diffusive. In this model (developed further by Stommel & Webster (1963), Salmon (1990), and others) the thermocline is an internal boundary layer or front that forms at the convergence of two different homogeneous water types, warm surface fluid above and cold abyssal fluid below. Meanwhile, the adiabatic model continued its own development (see Veronis 1969), culminating in the ventilated thermocline model of Luyten *et al.* (1983) and its continuous extensions (e.g., Killworth 1987). Signs that the two classes of theory might not be wholly incompatible came from Welander (1971b) and Colin de Verdière (1989) who noted that the diffusion might become important below an adiabatic near-surface flow, and Samelson & Vallis (1997) eventually suggested a model in which the upper thermocline is adiabatic, as in the ventilated thermocline model, but has a diffusive base, constituting an internal boundary layer. Mesoscale eddies play a role in homogenizing potential vorticity in the pool regions (e.g. Henning & Vallis 2004, Cessi & Fantini 2004, Maze & Marshall 2011). Extensions of the ventilated-style of model to the subpolar gyre are provided by Bell (2015b,a).

11 Welander (1971a).

12 Drawing from Salmon (1990).

13 Newton's method is an iterative way to numerically solve certain types of differential equations. The solutions here are obtained using about 1000 uniformly spaced grid points to span the domain, taking just a few seconds of computer time. Because of the boundary layer structure of the solutions employing a non-uniform grid would be even more efficient for this problem, but there is little point in designing a streamlined hat to reduce the effort of walking.

14 Following Samelson (1999b).

15 Following Luyten *et al.* (1983).

16 It may be that the eastern boundary depth is determined by global thermodynamic and/or mass constraints; see, for example, Boccaletti *et al.* (2004). There must also be a poleward transport across the boundary of the subtropical–subpolar gyre to balance the equatorward transport in the Ekman layer and that of the meridional overturning circulation, and this balance requirement may influence the boundary depth.

17 Dewar *et al.* (2005). This paper also discusses the nature of discontinuities at the pool boundaries, and their treatment via shock conditions.

In the shadow zone, layer-2 fluid also has no direct surface source, so one may wonder why this is also not expunged. However, the shadow zone is not a recirculating regime and the Ekman induced displacement will be much less efficient. More directly, the eastern boundary condition $h_1(x = x_e) = 0$ precludes the vanishing of layer-2 water there, and this boundary condition is propagated westwards into the interior.

Our heads are round so our thoughts may change direction.
Francis Picabia (1879–1953).

Paradigms Lost.
Apologies to John Milton.

CHAPTER 21

The Meridional Overturning Circulation and the Antarctic Circumpolar Current

THE *MERIDIONAL OVERTURNING CIRCULATION*, or the MOC, of the ocean is the circulation associated with sinking mostly at high latitudes and upwelling elsewhere, with much of the meridional transport taking place below the main thermocline. Understanding this circulation is one of the main goals of this chapter. The theory explaining the MOC is not nearly as settled as that of the quasi-horizontal wind-driven circulation discussed in Chapter 19, but considerable progress has been made, in particular with a significant re-thinking of the fundamentals occurring in the late 20th and early 21st century, as we will discover. Our other main goal is to glean an understanding of the Antarctic Circumpolar Current, or ACC, the theory of which has also undergone a transformation over that same period. The ACC is important not only in its own right, but because it mediates the MOCs of the individual ocean basins, connecting them into a global circulation.

That there *is* a deep circulation has been known for a long time, largely from observations of tracers such as temperature, salinity, and constituents such as dissolved oxygen and silica.[1] We can also take advantage of numerical models that are able to assimilate observations (from hydrographic measurements, floats and satellites) and produce a state estimate of the overturning circulation that is consistent with both the observations and the equations of motion, and one such estimate is illustrated in Fig. 21.1. We see that the water does not all upwell in the subtropics as we tacitly assumed in the previous chapter. In fact, much of the mid-depth circulation more-or-less follows the isopycnals that span the two hemispheres (Fig. 21.2), sinking in the North Atlantic and upwelling in the Southern Ocean, with the transport in between being, at least in part, adiabatic.

The MOC used to be known as the 'thermohaline' circulation, reflecting the belief that it was primarily driven[2] by buoyancy forcing arising from gradients in temperature and salinity. Such a circulation requires that the diapycnal mixing must be sufficiently large, but many measurements have suggested this is not the case and that has led to a more recent view that the MOC is at least partially, and perhaps primarily, mechanically driven, mostly by winds, and so *along* isopycnals instead of across them. However, the situation is not wholly settled, and it is almost certain that both buoyancy and wind forcing, as well as diapycnal diffusion, all contribute. The possible role of multiple basins (Atlantic, Pacific, etc.) on the MOC is likewise not fully understood.

In the first half of the chapter we mainly discuss somewhat classical topics associated with the buoyancy forcing. Then, beginning in Section 21.6, we discuss the role of wind forcing in

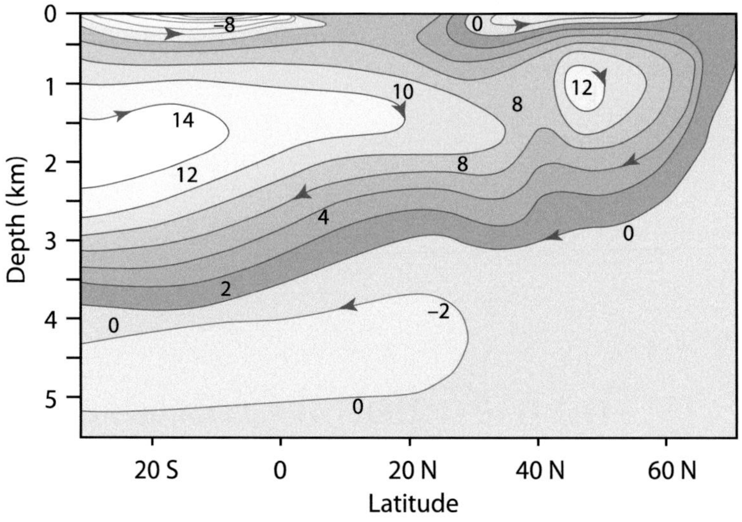

Fig. 21.1 An estimate of the mean meridional overturning circulation of the Atlantic (i.e., the streamfunction of the zonally averaged meridional flow) in Sverdrups.[3]

producing a MOC. This forces us to take an extended diversion into the dynamics of the ACC in Section 21.7 and then, in the last two sections, we present a theory of the MOC that incorporates both wind and buoyancy effects. We start by considering a simple but revealing fluid model of buoyancy forcing at the surface in a very idealised setting.

21.1 SIDEWAYS CONVECTION

Perhaps the simplest and most obvious fluid dynamical model of the overturning circulation is that of *sideways convection*. The physical situation is sketched in Fig. 21.3. A fluid (two- or three-dimensional) is held in a container that is insulated on all of its sides and its bottom, but its upper surface is non-uniformly heated and cooled. In the purest fluid dynamical problem the heat enters the fluid solely by conduction at the upper surface, and one may suppose that here the temperature is imposed. Thus, for a simple Boussinesq fluid the equations of motion are

$$\frac{\mathrm{D}\boldsymbol{v}}{\mathrm{D}t} + \boldsymbol{f} \times \boldsymbol{v} = -\nabla\phi + b\mathbf{k} + \nu\nabla^2\boldsymbol{v}, \qquad \frac{\mathrm{D}b}{\mathrm{D}t} = \kappa\nabla^2 b, \qquad \nabla\cdot\boldsymbol{v} = 0, \tag{21.1a,b,c}$$

where $\boldsymbol{f}\, f\mathbf{k}$, and with boundary conditions

$$b(x, y, 0, t) = g(x, y), \tag{21.2}$$

where $g(x, y)$ is a specified field, and $\partial_n b = 0$ on the other boundaries, meaning that the derivative normal to the boundary, and so the buoyancy flux, is zero. The oceanographic relevance of (21.1) and (21.2) should be clear: the ocean is heated *and* cooled from above, and although the thermal forcing in the real ocean may differ in detail (being in part a radiative flux, and in part a sensible and latent heat transfer from the atmosphere), (21.2) is a useful idealization. An alternative upper boundary condition would be to impose a flux condition whereby

$$\text{flux} = \kappa\frac{\partial}{\partial z}b(x, y, 0, t) = h(x, y), \tag{21.3a}$$

where $h(x, y)$ is given. In some numerical models of the ocean, the heat input at the top is parameterized by way of a relaxation to some specified temperature. This is a form of flux condition in which

$$\kappa\frac{\partial b}{\partial z} = C(b^*(x, y) - b), \tag{21.3b}$$

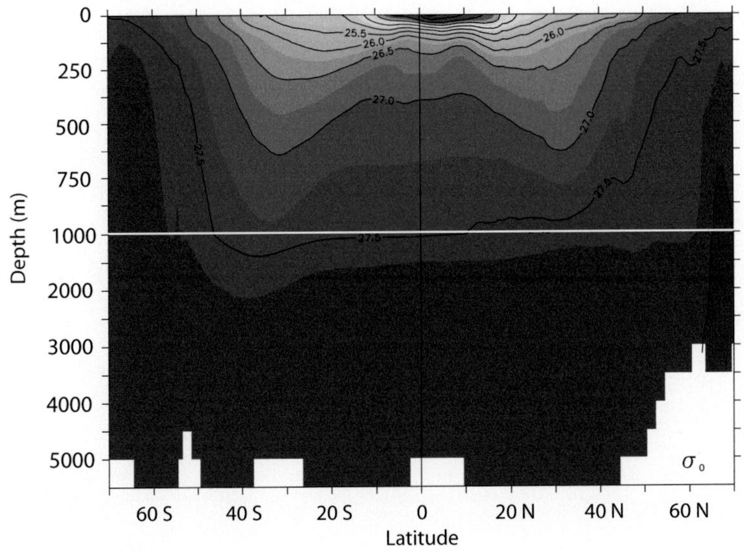

Fig. 21.2 The climatological zonally-averaged potential density (σ_θ) in the Atlantic ocean. Note the break in the vertical scale at 1000 m. The region of rapid change of density (and temperature) is concentrated in the upper kilometre, in the main thermocline, below which the density is more uniform. The flow of the MOC is largely, but not exactly, parallel to the isopycnals. [4]

and C is an empirical constant and $b^*(x, y)$ is given.[5] Although this may be a little more relevant than (21.2) for the real ocean, which of the three boundary conditions is chosen will not affect the arguments below, and we use (21.2).

21.1.1 Two-dimensional Convection

We may usefully restrict attention to the two-dimensional problem, in latitude and height. This is a poor model of the actual overturning circulation of the ocean, but the results do not depend on this idealization. The incompressibility of the flow then allows one to define a streamfunction such that

$$v = -\frac{\partial \psi}{\partial z}, \qquad w = \frac{\partial \psi}{\partial y}, \qquad \zeta = \nabla_x^2 \psi = \left(\frac{\partial^2 \psi}{\partial y^2} + \frac{\partial^2 \psi}{\partial z^2} \right), \tag{21.4}$$

where ζ is the vorticity in the meridional plane. We will omit the subscript x on the Laplacian operator where there is no ambiguity.

Taking the curl of Boussinesq equations of motion (21.1) then gives

$$\frac{\partial \nabla^2 \psi}{\partial t} + J(\psi, \nabla^2 \psi) = \frac{\partial b}{\partial y} + \nu \nabla^4 \psi, \tag{21.5a}$$

$$\frac{\partial b}{\partial t} + J(\psi, b) = \kappa \nabla^2 b, \tag{21.5b}$$

where $J(a, b) \equiv (\partial_y a)(\partial_z b) - (\partial_z a)(\partial_y b)$.

Nondimensionalization and scaling

We nondimensionalize (21.5) by formally setting

$$b = \Delta b\, \hat{b}, \qquad \psi = \Psi \hat{\psi}, \qquad y = L\hat{y}, \qquad z = H\hat{z}, \qquad t = \frac{LH}{\Psi}\hat{t}, \tag{21.6}$$

where the hatted variables are nondimensional, Δb is the temperature difference across the surface, L is the horizontal size of the domain, and Ψ, and ultimately the vertical scale H, are to be

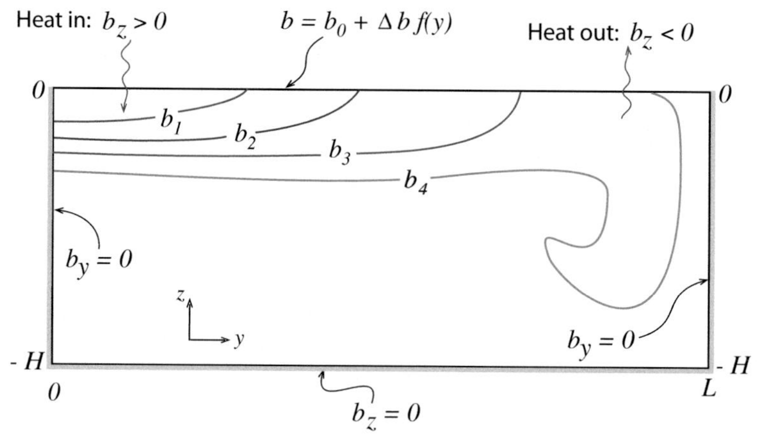

Fig. 21.3 Sketch of sideways convection. The fluid is differentially heated and cooled along its top surface, whereas all the other walls are insulating.

The result is, typically, a small region of convective instability and sinking near the coldest boundary, with generally upwards motion elsewhere.[6]

determined. Substituting (21.6) into (21.5) gives

$$\frac{\partial \hat{\nabla}^2 \hat{\psi}}{\partial \hat{t}} + \hat{J}(\hat{\psi}, \nabla^2 \hat{\psi}) = \frac{H^3 \Delta b}{\Psi^2} \frac{\partial \hat{b}}{\partial \hat{y}} + \frac{\nu L}{\Psi H} \hat{\nabla}^4 \hat{\psi}, \tag{21.7a}$$

$$\frac{\partial \hat{b}}{\partial \hat{t}} + \hat{J}(\hat{\psi}, \hat{b}) = \frac{\kappa L}{\Psi H} \hat{\nabla}^2 \hat{b}, \tag{21.7b}$$

where $\hat{\nabla}^2 = (H/L)^2 \partial^2/\partial \hat{y}^2 + \partial^2/\partial \hat{z}^2$ and the Jacobian operator is similarly nondimensional. If we now use (21.7b) to choose Ψ as

$$\Psi = \frac{\kappa L}{H}, \tag{21.8}$$

so that $t = H^2 \hat{t}/\kappa$, then (21.7) becomes

$$\frac{\partial \hat{\nabla}^2 \hat{\psi}}{\partial \hat{t}} + \hat{J}(\hat{\psi}, \hat{\nabla}^2 \hat{\psi}) = Ra\sigma\alpha^5 \frac{\partial \hat{b}}{\partial \hat{y}} + \sigma \hat{\nabla}^4 \hat{\psi}, \tag{21.9}$$

$$\frac{\partial \hat{b}}{\partial \hat{t}} + \hat{J}(\hat{\psi}, \hat{b}) = \hat{\nabla}^2 \hat{b}. \tag{21.10}$$

It is possible to make different scaling choices, but they all lead to the appearance of the same non dimensional parameters, or combinations thereof, and the three that govern the behaviour of the system are

$$Ra = \left(\frac{\Delta b L^3}{\nu\kappa}\right), \qquad \text{(the Rayleigh number)}, \tag{21.11a}$$

$$\sigma = \frac{\nu}{\kappa}, \qquad \text{(the Prandtl number)}, \tag{21.11b}$$

$$\alpha = \frac{H}{L}, \qquad \text{(the aspect ratio)}. \tag{21.11c}$$

Sometimes H is used instead of L in the Rayleigh number definition; we use L here because it is an external parameter. The Rayleigh number is a measure of the strength of the buoyancy forcing relative to the viscous term, and in the ocean it will be very large indeed, perhaps $\sim 10^{24}$ if molecular values are used.

For steady non-turbulent flows, and also perhaps for statistically steady flows, we can demand that the buoyancy term in (21.9) is $\mathcal{O}(1)$. If it is smaller then the flow is not buoyancy driven, and

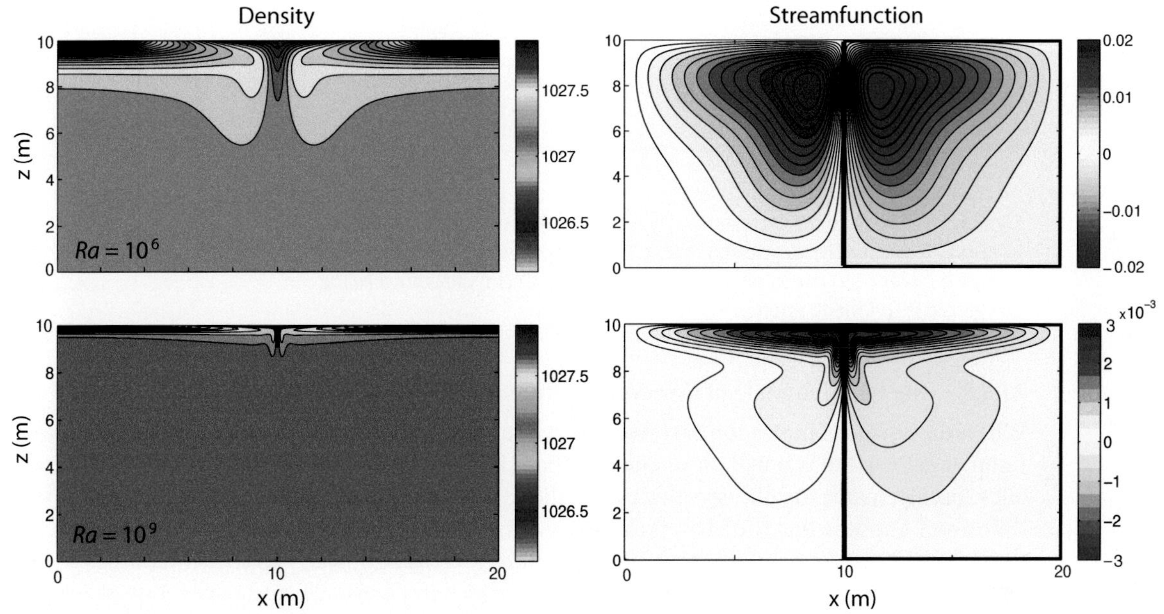

Fig. 21.4 The density and streamfunction in two numerical simulations of two-dimensional sideways convection, with Rayleigh numbers of 10^6 (top) and 10^9 (bottom). The imposed temperature at the top linearly decreases from the centre outward, the side and bottom walls are insulating, and the Prandtl number is 10. The two density plots use the same colourmap, but the streamfunction plots do not. There is a sinking plume at the centre, with a weaker circulation and a thinner thermocline at the higher Rayleigh number.[8]

if it is larger there is nothing to balance it. Our demand can be satisfied only if the vertical scale of the motion adjusts appropriately and, for $\sigma = \mathcal{O}(1)$, this suggests the scalings:[7]

$$H = L\sigma^{-1/5}Ra^{-1/5} = \left(\frac{\kappa^2 L^2}{\Delta b}\right)^{1/5}, \qquad \Psi = Ra^{1/5}\sigma^{-4/5}\nu = (\kappa^3 L^3 \Delta b)^{1/5}. \tag{21.12a,b}$$

The vertical scale H arises as a consequence of the analysis, and the vertical size of the domain plays no direct role. [For $\sigma \gg 1$ we might expect the nonlinear terms to be small and if the buoyancy term balances the viscous term in (21.9) the right-hand sides of (21.12) are multiplied by $\sigma^{1/5}$ and $\sigma^{-1/5}$. For seawater, $\sigma \approx 7$ using the molecular values of κ and ν. If small scale turbulence exists, then the eddy viscosity will likely be similar to the eddy diffusivity and $\sigma \approx 1$.] Numerical experiments (an example is shown in Fig. 21.4) provide support for the scaling of (21.12), and a few simple and robust points that have relevance to the real ocean emerge:

- Most of the box fills up with the densest available fluid, with a boundary layer in temperature near the surface required in order to satisfy the top boundary condition. The boundary gets thinner with decreasing diffusivity, consistent with (21.12). This is a diffusive prototype of the oceanic thermocline.
- The horizontal scale of the overturning circulation is large, nearly the scale of the box.
- The downwelling regions (the regions of convection) are of smaller horizontal scale than the upwelling regions, especially as the Rayleigh number increases.

Let us now try to explain some of the features in a simple and heuristic way, beginning with the scale of the motion.

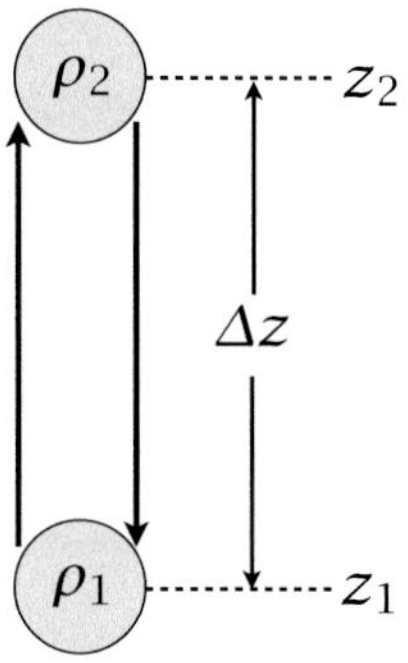

Fig. 21.5 Two fluid parcels, of density ρ_1 and ρ_2 and initially at positions z_1 and z_2 respectively, are interchanged. If $\rho_2 > \rho_1$ then the final potential energy is lower than the initial potential energy, with the difference being converted into kinetic energy.

21.1.2 The Relative Scale of Convective Plumes and Diffusive Upwelling

Why is the downwelling region narrower than the upwelling? The short answer is that high Rayleigh number convection is much more efficient than diffusive upwelling, so that the convective buoyancy flux can match the diffusive flux only if the convective plumes cover a much smaller area than diffusion.[9] Suppose that the basin is initially filled with water of an intermediate temperature, and that surface boundary conditions of a temperature decreasing linearly from low latitudes to high latitudes are imposed. The deep water will be convectively unstable, and convection at high latitudes (where the surface is coldest) will occur, quickly filling the abyss with dense water. After this initial adjustment the deep, dense water at lower latitudes will be slowly warmed by diffusion, but at the same time surface forcing will maintain a cold high latitude surface, thus leading to high latitude convection. A steady state or statistically steady state is eventually reached with the deep water having a slightly lower potential density than the surface water at the highest latitudes, and so maintaining continual convection, but convection that takes place only at the highest latitudes.

To see this more quantitatively consider the respective efficiencies of the convective heat flux and the diffusive heat flux. Consider an idealized re-arrangement of two parcels, initially with the heavier one on top as illustrated in Fig. 21.5. The potential energy released by the re-arrangement, ΔP is given by

$$\Delta P = P_{\text{final}} - P_{\text{initial}} \tag{21.13}$$
$$= g\left[(\rho_1 z_2 + \rho_2 z_1) - (\rho_1 z_1 + \rho_2 z_2)\right] \tag{21.14}$$
$$= g(z_2 - z_1)(\rho_1 - \rho_2) = \rho_0 \Delta b \Delta z, \tag{21.15}$$

where $\Delta z = z_2 - z_1$ and $\Delta b = g(\rho_1 - \rho_2)/\rho_0$.

The kinetic energy gained by this re-arrangement, ΔK is given by $\Delta K = \rho_0 w^2$ and equating this to (21.13) gives

$$w^2 = -\Delta b \Delta z. \tag{21.16}$$

If the heavier fluid is initially on top then $\rho_2 > \rho_1$ and, as defined, $\Delta b < 0$. The vertical convective buoyancy flux per unit area, B_c, is given by $B_c = w\Delta b$ and using (21.16) we find

$$B_c = (-\Delta b)^{3/2}(\Delta z)^{1/2}. \tag{21.17}$$

The upwards diffusive flux, B_d, per unit area is given by

$$B_d = \kappa \frac{\Delta b}{H}, \tag{21.18}$$

where H is the thickness of the layer over which the flux occurs. In a steady state the total diffusive flux must equal the convective flux so that, from (21.17) and (21.18),

$$(-\Delta b)^{3/2}(\Delta z)^{1/2}\delta = \kappa \frac{\Delta b}{H}, \tag{21.19}$$

where δ is the fractional area over which convection occurs. If we set $\Delta z = H$, and use (21.12a) we find

$$(-\Delta b)^{3/2} \left(\frac{\kappa^2 L^2}{\Delta b} \right)^{1/10} \delta = \kappa \frac{\Delta b}{(\kappa^2 L^2 / \Delta b)^{1/5}}, \tag{21.20}$$

giving

$$\delta = \left(\frac{\kappa^2}{\Delta b L^3} \right)^{1/5} = (Ra\, \sigma)^{-1/5}. \tag{21.21}$$

For geophysically relevant situations this is a very small number, usually smaller than 10^{-5}. Although the details of the above calculation may be questioned (for example, the use of the same buoyancy difference and vertical scale in the convection and the diffusion), the physical basis for the result is transcendent: for realistic choices of the diffusivity the convection is much more efficient than the diffusion and so will occur over a much smaller area.

21.1.3 Phenomenology of the Overturning Circulation

No water can be denser (or, more accurately, have a greater potential density) than the densest water at the surface. If the surface water is denser than the water at depth then it will be convectively unstable and sink in a plume.[10] The plume slowly entrains the warmer water that surrounds it, and then spreads horizontally when it reaches the bottom or when its density becomes similar to that of its surroundings. The presence of water denser than its surroundings creates a horizontal pressure gradient, and the ensuing flow will displace any adjacent lighter fluid, and so the domain fills with the densest available fluid. This process is a continuous one: the plumes take cold water into the interior, where the water slowly warms by diffusion, and the source of cold water at the surface is continuously replenished. If diffusion is small, the end result is that the potential density of the fluid in the interior will be slightly less than that of the densest fluid formed at the surface. (Because diffusion can act only to reduce extrema, no fluid in the interior can be colder than the coldest fluid formed at the surface.)

However, the value at the surface is given by the boundary condition $b(x, y, z = 0) = f(x, y)$. Thus, the interior cannot fill all the way to the surface with this cold water and there must be a boundary layer connecting the cold, dense interior with the surface; its thickness δ is given by the height scale of (21.12a); that is $\delta \sim H = (\kappa^2 L^2 / \Delta b)^{1/5}$. Such a strong boundary layer will not necessarily be manifest in the velocity field, however, because the no-normal flow boundary condition on the velocity field is satisfied by setting $\psi = 0$ as a boundary condition to the elliptic problem $\nabla^2 \psi = \zeta$, where ζ is the prognostic variable in (21.5a), and this boundary condition has a global effect on the velocity field.

Why is the horizontal scale of the circulation large? The circulation transfers heat meridionally, and it is far more efficient to do this by a single overturning cell than by a multitude of small cells; hence, although we cannot entirely eliminate the possibility that some instability will produce such small scales of motion, it seems likely the horizontal scale of the mean circulation will be determined by the domain scale. (At low Rayleigh number we can in fact explicitly calculate an approximate analytic solution for the flow, demonstrating this.) For higher Rayleigh number perturbation approaches fail and we must resort to numerical solutions; these (e.g., Fig. 21.4), do show the circulation dominated by a single overturning circulation rather than many small convective cells over a large range of Rayleigh number.

Finally, it is important to realize that *even for large diffusion and viscosity there is no stationary solution:* as soon as we impose a temperature gradient at the top the fluid begins to circulate, a manifestation of the dictum that a baroclinic fluid is a moving fluid, encountered in Section 4.2. Put simply, a temperature gradient leads a density gradient, which in turn leads to a pressure gradient.

The pressure gradient leads to motion: viscosity cannot prevent that, for it can have an effect only if the velocity is non-zero.

21.2 THE MAINTENANCE OF SIDEWAYS CONVECTION

In most conventional convection problems the fluid is heated from below, becomes buoyant and rises, and is cooled at the top. In contrast, in sideways convection the heating and cooling occur at the same level, and the conditions under which a circulation can be maintained are by no means clear; our purpose here is to make them clearer. The energetic derivations of this section are but an extension of Section 2.4.3, but now with starring roles for friction, diffusion and the boundary conditions, and the reader may wish to review that section first. The derivations below are not difficult, but they lead to powerful and perhaps counter-intuitive results that provide important information about the overturning circulation of the real ocean. To begin, we rewrite the equations of motion, (21.1), in a slightly different way, namely

$$\frac{\partial \boldsymbol{v}}{\partial t} + (\boldsymbol{f} + 2\boldsymbol{\omega}) \times \boldsymbol{v} = -\nabla B + b\mathbf{k} + \nu\nabla^2\boldsymbol{v}, \tag{21.22a}$$

$$\frac{\partial b}{\partial t} + \nabla\cdot(b\boldsymbol{v}) = \dot{Q} = J + \kappa\nabla^2 b, \tag{21.22b}$$

$$\nabla\cdot\boldsymbol{v} = 0, \tag{21.22c}$$

where $B = \boldsymbol{v}^2/2 + \phi$ is the Bernoulli function for Boussinesq flow and $\dot{Q}$ $(= \dot{b})$ is the total rate of heating (including diffusion, and absorbing constant factors such as heat capacity into its definition) with J its non-diffusive component. The fluid occupies a finite volume, and in a steady state $\langle\dot{Q}\rangle = 0$, where the angle brackets denote a volume and time integration.

21.2.1 The Energy Budget

To obtain an energy budget we follow the procedure of Section 2.4.3. First take the dot product of (21.22a) with $\boldsymbol{v}$ to give

$$\frac{1}{2}\frac{\partial \boldsymbol{v}^2}{\partial t} = -\nabla\cdot(\boldsymbol{v}B) + wb + \nu\boldsymbol{v}\cdot\nabla^2\boldsymbol{v}. \tag{21.23}$$

Integrating over a domain bounded by stress-free rigid walls gives the kinetic energy equation

$$\frac{\mathrm{d}}{\mathrm{d}t}\left\langle\frac{1}{2}\boldsymbol{v}^2\right\rangle = \langle wb\rangle - \varepsilon, \tag{21.24}$$

where angle brackets denote (for the moment) just a volume average and ε is the average dissipation of kinetic energy ($\varepsilon = -\nu\langle\boldsymbol{v}\cdot\nabla^2\boldsymbol{v}\rangle = \nu\langle\boldsymbol{\omega}^2\rangle$), a positive definite quantity. Thus, in a statistically steady state in which the left-hand side vanishes after time averaging, the dissipation of kinetic energy is maintained by the buoyancy flux; that is, by a release of potential energy with light fluid ascending and dense fluid descending.

We obtain a potential energy budget by using (21.22b) to write

$$\frac{\mathrm{D}bz}{\mathrm{D}t} = z\frac{\mathrm{D}b}{\mathrm{D}t} + b\frac{\mathrm{D}z}{\mathrm{D}t} = z\dot{Q} + bw, \tag{21.25}$$

and integrating this over the domain gives the potential energy equation

$$\frac{\mathrm{d}}{\mathrm{d}t}\langle bz\rangle = \langle z\dot{Q}\rangle + \langle bw\rangle. \tag{21.26}$$

Subtracting (21.26) from (21.24) gives the energy equation

$$\frac{\mathrm{d}}{\mathrm{d}t}\left\langle\frac{1}{2}\boldsymbol{v}^2 - bz\right\rangle = -\langle z\dot{Q}\rangle - \varepsilon. \tag{21.27}$$

21.2.2 Conditions for Maintaining a Thermally-driven Circulation

In a statistically steady state the left-hand side of (21.27) vanishes and the kinetic energy dissipation is balanced by the buoyancy source terms; that is

$$\langle z\dot{Q}\rangle = -\varepsilon < 0. \tag{21.28}$$

The right-hand side (the term $-\varepsilon$) is negative definite, and to balance this the heating must be negatively correlated with height. (Recall that $\langle\dot{Q}\rangle = 0$, and the origin of the z-coordinate is then immaterial.) Thus, *in order to maintain a circulation in which kinetic energy is dissipated, the heating (including diffusive heating) must occur, on average, at lower levels than the cooling.* This result is related to, but not quite the same as, a postulate due to Sandström, discussed below.

In the ocean the non-diffusive heating occurs predominantly at the surface, except for the negligible effects of hydrothermal vents. Thus, $\langle Jz\rangle \approx 0$ and a kinetic-energy-dissipating circulation can *only* be maintained, in the absence of mechanical forcing, if the diffusion is non-zero — in that case heat may be diffused from the surface to depth, so effectively providing a deep heat source. In the atmosphere, the heating is mostly at the surface and cooling is mostly in the mid-troposphere so that $\langle z\dot{Q}\rangle < 0$; thus, (21.28) is readily satisfied and the circulation is not restricted.

✦ *Maintaining a steady baroclinic circulation*

By a rather different method — and one closer to a suggestion of Sandström dating from 1908 — we can obtain a result that is different from but related to (21.28).[11] Using (4.37) and (21.22a), the circulation in a Boussinesq system obeys

$$C = \oint \boldsymbol{v}\cdot \mathrm{d}\boldsymbol{r}, \qquad \frac{\mathrm{D}C}{\mathrm{D}t} = \oint b\mathbf{k}\cdot \mathrm{d}\boldsymbol{r} + \oint \boldsymbol{F}\cdot \mathrm{d}\boldsymbol{r}, \tag{21.29a,b}$$

where C is the circulation, $\mathrm{d}\boldsymbol{r}$ is a path element, and $\boldsymbol{F}$ represents the frictional terms. (The Coriolis parameter plays no role in this argument because the Coriolis force does no work, and f may be set to zero without loss of generality.) Now we may write the rate of change of circulation in the form

$$\frac{\mathrm{D}C}{\mathrm{D}t} = \oint\left(\frac{\partial \boldsymbol{v}}{\partial t} + \boldsymbol{v}\cdot\nabla\boldsymbol{v}\right)\cdot \mathrm{d}\boldsymbol{r} = \oint\left(\frac{\partial \boldsymbol{v}}{\partial t} + \boldsymbol{\omega}\times\boldsymbol{v}\right)\cdot \mathrm{d}\boldsymbol{r}, \tag{21.30}$$

because the integral of the potential term that arises when going to the last expression vanishes. Let us assume the flow is steady, so that $\partial\boldsymbol{v}/\partial t$ vanishes. Let us further choose the path of integration to be a streamline, which since the flow is steady is also a parcel trajectory. The second term on the right-most expression of (21.30) then also vanishes and (21.29b) becomes

$$\oint b\,\mathrm{d}z = -\oint \boldsymbol{F}\cdot \mathrm{d}\boldsymbol{r} = -\oint \frac{\boldsymbol{F}}{|\boldsymbol{v}|}\cdot \boldsymbol{v}\,\mathrm{d}r, \tag{21.31}$$

where the last equality follows because the path is everywhere parallel to the velocity. Let us now assume that the friction retards the flow, and that $\oint \boldsymbol{F}\cdot\boldsymbol{v}/|\boldsymbol{v}|\,\mathrm{d}r < 0$. (One form of friction that has this property is linear drag, $\boldsymbol{F} = -C\boldsymbol{v}$ where C is a constant. The property is similar to, but not the same as, the property that the friction dissipates kinetic energy over the circuit.) Making this assumption, if we integrate the term on the left-hand side by parts we obtain

$$\oint z\,\mathrm{d}b < 0. \tag{21.32}$$

Now, because the integration circuit in (21.32) is a fluid trajectory, the change in buoyancy $\mathrm{d}b$ is proportional to the heating of a fluid element as it travels the circuit; in the notation of (21.22b),

$db = \dot{Q}\,dt$, where the heating, $\dot{Q}$, includes diffusive effects. Thus, the inequality implies that the net heating must be negatively correlated with height: that is, *the heating must occur, on average, at a lower level than the cooling in order that a steady circulation may be maintained against the retarding effects of friction.*

A similar result can be obtained for a compressible fluid. From Equations (4.43)–(4.45) on page 151 we write the baroclinic circulation theorem as

$$\frac{DC}{Dt} = \oint p\,d\alpha + \oint \boldsymbol{F}\cdot d\boldsymbol{r} = \oint T\,d\eta + \oint \boldsymbol{F}\cdot d\boldsymbol{r}, \tag{21.33}$$

where η is the specific entropy. Then, by precisely the same arguments as led to (21.32), we are led to the requirements that

$$\oint T\,d\eta > 0 \qquad \text{or equivalently} \qquad \oint p\,d\alpha > 0. \tag{21.34a,b}$$

Equation (21.34a) means that parcels must gain entropy at high temperatures and lose entropy at low temperatures; similarly, from (21.32b), a parcel must expand ($d\alpha > 0$) at high pressures and contract at low pressures.

For an ideal gas we can put these statements into a form analogous to (21.32) by noting that $d\eta = c_p(d\theta/\theta)$, where θ is potential temperature, and using the definition of potential temperature, (1.105). With these we have

$$\oint T\,d\eta = \oint c_p \frac{T}{\theta}\,d\theta = \oint c_p \left(\frac{p}{p_R}\right)^{\kappa} d\theta, \tag{21.35}$$

and (21.34a) becomes

$$\oint c_p \left(\frac{p}{p_R}\right)^{\kappa} d\theta > 0. \tag{21.36}$$

Because the path of integration is a fluid trajectory, $d\theta$ is proportional to the heating of a fluid element. Thus, as with the Boussinesq result, (21.36) implies that *the heating (the potential temperature increase) must occur at a higher pressure than the cooling in order that a steady circulation may be maintained against the retarding effects of friction.*

These results may be understood by noting that the heating must occur at a higher pressure than the cooling in order that work may be done, the work being necessary to convert potential energy into kinetic energy to maintain a circulation against friction. Intuitively, if the heating is below the cooling, then the heated fluid will expand and become buoyant and rise, and a steady circulation between heat source and heat sink can readily be imagined. On the other hand, if the heating is above the cooling there is no obvious pathway between source and sink.

Intuitive as these results may be, the conditions required to prove (21.32) and (21.36) are much more restrictive than those needed to prove (21.28). To prove the former, we must *assume* that the flow is absolutely steady, *and* that streamlines form a closed path, *and* that the friction has retarding properties. The second of these conditions is not generally satisfied in three dimensions, even when the flow is steady. Furthermore, one cannot prove that Newtonian viscosity ($\nu\nabla^2\boldsymbol{v}$) will always act to retard the flow. On the other hand, (21.28) provides a condition for the maintenance of a statistically steady circulation, assuming only that the friction acts to dissipate kinetic energy. In any case, it is clear from all of the above results that the overturning circulation is greatly affected by the relative pressures at the locations of the heating and cooling, and this is called *Sandström's effect.* In all of these cases, the heating must be taken to include diffusive effects; if the molecular diffusivity is small and the heating is at the surface we can further constrain the flow, as we now see.

21.2.3 Surface Fluxes and Non-turbulent Flow at Small Diffusivities

Suppose that the only heating to the fluid is via diffusion through the upper surface; that is $J = 0$ in (21.22b). Is a circulation possible? If the diffusivity is finite then heat can diffuse into the fluid and thereby potentially provide a difference in altitude between the heating and the cooling. However, as $\kappa \to 0$ this mechanism ceases to operate and we therefore expect that the left-hand sides of (21.28) and (21.32) will go to zero, and the circulation will cease. In what follows we put this argument on a more rigorous footing: we will show that as $\kappa \to 0$ the kinetic energy dissipation also goes to zero, and therefore the flow is 'non-turbulent', the meaning of which will be made clearer below.

Assuming a statistically steady state, integrating (21.22b) horizontally gives

$$\frac{\partial \overline{bw}}{\partial z} = \kappa \frac{\partial^2 \overline{b}}{\partial z^2}, \tag{21.37}$$

where an overbar indicates a horizontal average. Integrating this equation up from the bottom (where there is no flux) to a level z gives

$$\overline{wb} - \kappa \overline{b}_z = 0 \tag{21.38}$$

at every level in the fluid. The two terms on the left-hand side together give rise to the total buoyancy flux through the level z, and the flux must vanish because there is no buoyancy input except at the surface. If we integrate this vertically we have

$$\langle wb \rangle = H^{-1}\kappa \left[\overline{b}(0) - \overline{b}(-H) \right], \tag{21.39}$$

where the angle brackets denote an average over the entire volume. In the limit $\kappa \to 0$, the integrated advective buoyancy flux will vanish, because the term $\overline{b}(0) - \overline{b}(-H)$ remains finite. This follows because b is conserved on parcels, except for the effects of diffusion, which can only act to reduce the value of extrema in the fluid. Thus, $\overline{b}(0) - \overline{b}(-H)$ can only be as large as the temperature difference at the surface, which is set by the boundary conditions.

Now consider the kinetic energy budget. Using (21.24) and (21.39) we have in a statistically steady state

$$\varepsilon = H^{-1}\kappa \left[\overline{b}(0) - \overline{b}(-H) \right]. \tag{21.40}$$

Because, as noted above, the buoyancy difference on the right-hand side is bounded, the kinetic energy dissipation must go to zero if the thermal diffusivity goes to zero; that is, $\varepsilon \to 0$ as $\kappa \to 0$ and in particular $\varepsilon < \kappa \Delta b / H$ where Δb is the maximum buoyancy difference at the surface. We may also consider the limit $(\kappa, \nu) \to 0$ with a fixed Prandtl number, $\sigma \equiv \nu/\kappa$, and in this limit the energy dissipation also vanishes with κ.

Finally, let us see how the surface buoyancy is related to the buoyancy flux, for any value of κ. Multiplying (21.22b) (with $J = 0$) by b and integrating over the domain gives the buoyancy variance equation

$$\frac{1}{2}\frac{\mathrm{d}\langle b^2 \rangle}{\mathrm{d}t} = \kappa \left[\overline{b \frac{\partial b}{\partial z}}\bigg|_{z=0} - \langle |\nabla b|^2 \rangle \right]. \tag{21.41}$$

We have assumed that the normal derivative of b vanishes on all surfaces except the top one ($z = 0$) and an overbar denotes a horizontal average. In a statistically steady state,

$$\overline{b \frac{\partial b}{\partial z}}\bigg|_{z=0} = \langle |\nabla b|^2 \rangle, \tag{21.42}$$

where the overbar and angle brackets now also imply a time average. The right-hand side is positive definite, and thus there must be a positive correlation between b and $\partial b/\partial z$, meaning that there is a heat flux into the fluid where it is hot, and a heat flux out of the fluid where it is cold. This result holds no matter whether the upper boundary condition is a condition on b or on $\partial b/\partial z$.

Interpretation

The result encapsulated by (21.40) means that, for a fluid forced only at the surface by buoyancy forcing, as the diffusivity goes to zero so does the energy dissipation. For a fluid of finite viscosity the vorticity in the fluid must then go to zero, because $\varepsilon = \nu \langle \boldsymbol{\omega}^2 \rangle$; this in turn means that the flow cannot be baroclinic, because baroclinicity generates vorticity, even in the presence of viscosity (Section 4.2). An even more interesting result follows for a fluid with small viscosity. In turbulent flow, the energy dissipation at high Reynolds number is not a function of the viscosity; if the viscosity is reduced, the cascade of energy to smaller scales merely continues to still smaller scale, generating vorticity at these smaller scales, and the energy dissipation is unaltered, remaining finite even in the limit $\nu \to 0$. In contrast, for a fluid heated and cooled only at the upper surface, the energy dissipation *tends to zero* as $\kappa \to 0$, whether or not one is in the high-Reynolds number limit. This means that vorticity cannot be generated at the viscous scales by the action of a turbulent cascade, as that would lead to energy dissipation. Effectively, the result prohibits an ocean that is forced only at the surface by a buoyancy flux from having an 'eddy viscosity' that would enable the fluid to efficiently dissipate energy, and if there is no small scale motion producing an eddy viscosity there can be no eddy diffusivity either. Thus, such an ocean is *non-turbulent.* This is a rather different picture from that of the real ocean, where there is some dissipation of energy in the interior because of breaking gravity waves, and dissipation at the boundary in Ekman layers, and the eddy diffusivity is needed for there to be a non-negligible buoyancy-driven meridional overturning circulation.

Of course, thermal forcing in the ocean is in part an imposed flux, coming from radiation among other things, and this penetrates below the surface. However, this makes little physical difference to the argument, provided that this forcing remains confined to the upper ocean. If so, then for any level below this forcing we still have the result (21.38), and the final result (21.40) holds, assuming that the range of temperatures produced by the forcing is still finite.

21.2.4 The Importance of Mechanical Forcing

The results of (21.28) and (21.40) do not, strictly speaking, prohibit there from being a thermal circulation, with fluid sinking at high latitudes and rising at low, even for zero diffusivity. However, in the absence of any mechanical forcing, this circulation must be laminar, even at high Rayleigh number, meaning that flow is not allowed to break in such a way that energy can be dissipated — a very severe constraint that most flows cannot satisfy. The scalings (21.12) further suggest that the magnitude of the circulation in fact scales (albeit nonlinearly) with the size of the molecular diffusivity, and if these scalings are correct the circulation will in fact diminish as $\kappa \to 0$. For small diffusivity, the solution most likely to be adopted by the fluid is for the flow to become confined to a very thin layer at the surface, with no abyssal motion at all, which is completely unrealistic vis-à-vis the observed ocean. Thus, the deep circulation of the ocean cannot be considered to be wholly forced by buoyancy gradients at the surface.

Suppose we add a mechanical forcing, $\boldsymbol{F}$, to the right-hand side of (21.22a); this might represent wind forcing at the surface, or tides. The kinetic energy budget becomes

$$\varepsilon = \langle wb \rangle + \langle \boldsymbol{F} \cdot \boldsymbol{v} \rangle = H^{-1}\kappa[\overline{b}(0) - \overline{b}(-H)] + \langle \boldsymbol{F} \cdot \boldsymbol{v} \rangle . \tag{21.43}$$

In this case, even for $\kappa = 0$, there is a source of energy and therefore turbulence (i.e., a dissipative circulation) can be maintained. The turbulent motion at small scales then provides a mechanism of mixing and so can effectively generate an 'eddy diffusivity' of buoyancy. *Given* such an eddy diffusivity, wind forcing is no longer necessary for there to be an overturning circulation. Therefore, it is useful to think of mechanical forcing as having two distinct effects:

(i) The wind provides a stress on the surface that may directly drive the large-scale circulation, including the overturning circulation. (An example of this is discussed in Section 21.6.)

(ii) Both tides and the wind provide a mechanical source of energy to the system that allows the flow to become turbulent and so provides a source for an eddy diffusivity and eddy viscosity.

In either case, we may conclude that the presence of mechanical forcing is necessary for there to be an overturning circulation in the world's oceans of the kind observed. Let us first suppose that the most important effect of the wind is that it enables there to be an eddy diffusivity that is much larger than the molecular one; the eddy diffusivity enables large volumes of the ocean to become mixed, so allowing a buoyancy-driven overturning circulation (a 'thermohaline circulation') to exist.

21.2.5 The Mixing-driven Ocean

Before moving on to other matters, let's make a connection to the scaling for the overturning circulation discussed in Section 20.5.1. In that section we considered the planetary-geostrophic equations,

$$\boldsymbol{v} \cdot \nabla b = \kappa \nabla^2 b, \qquad \nabla \cdot \boldsymbol{u} + \frac{\partial w}{\partial z} = 0, \qquad \boldsymbol{f} \times \boldsymbol{u} = -\nabla \phi, \qquad b = \frac{\partial \phi}{\partial z}, \tag{21.44a,b,c,d}$$

and obtained, after a little algebra, the scalings for vertical velocity and thermocline thickness,

$$W = \kappa^{2/3} \left(\frac{\beta \, \Delta b}{f^2 L} \right)^{1/3}, \qquad \delta = \kappa^{1/3} \left(\frac{f^2 L}{\beta \, \Delta b} \right)^{1/3}. \tag{21.45}$$

These are different in detail from the Rossby scalings, (21.12), but they also require a finite diffusivity to produce a circulation and a thermocline. The Sandström effect applies, as it must, to oceanically relevant equations.

In order for the scales given in (21.45) to be representative of those observed in the real ocean, we must use an eddy diffusivity for κ. Using $f = 10^{-4}\,\mathrm{s}^{-1}$, $\beta = 10^{-11}\,\mathrm{m}^{-1}\,\mathrm{s}^{-1}$, $L = 5 \times 10^6\,\mathrm{m}$, $g = 10\,\mathrm{m\,s}^{-2}$, $\kappa = 10^{-5}\,\mathrm{m}^2\,\mathrm{s}^{-1}$, $\Delta b = -g\Delta\rho/\rho_0 = g\beta_T \Delta T$ and $\Delta T = 10\,\mathrm{K}$ we find the not unreasonable values of $\delta \approx 150\,\mathrm{m}$ and $W = 10^{-7}\,\mathrm{m\,s}^{-1}$, albeit δ is rather smaller than the thickness of the observed thermocline. However, if we take the molecular value of $\kappa \approx 10^{-7}\,\mathrm{m}^2$ the values of W and δ are unrealistically small (although still non-zero). Evidently, if the deep circulation of the ocean is buoyancy (or mixing) driven, it must take advantage of turbulence that enhances the small scale mixing and produces an eddy diffusivity.

21.3 ♦ SIMPLE BOX MODELS

This section is marked with a black diamond not because it is advanced; rather, it is a little peripheral to our main development. The purist may consider this section a diversion away from a consideration of the fluid dynamical properties of the ocean, and the content implied by the title of this book, but such box models have been quite fecund and an evident source of qualitative understanding, and thus find a place in our discussion, if not in our canon. Readers may skim this section without fear of disapprobation.

Even though they are far simpler than the real ocean, the fluid dynamical models of the previous sections are still quite daunting. The analysis that can be performed is either very specific and of little generality, for example the construction of solutions at low Rayleigh number, or it is a very general form such as scaling or energetic arguments. Models based on the fluid dynamical equations do not easily allow for the construction of explicit solutions in the parameter regime — high

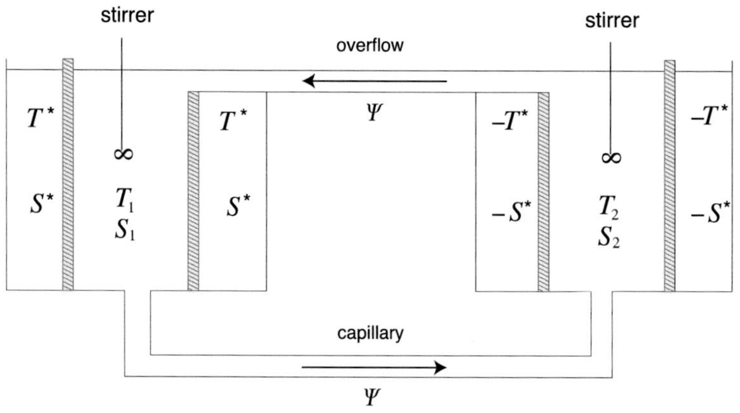

Fig. 21.6 A two-box model of the overturning circulation of the ocean. The shaded walls are porous, and each box is well mixed by its stirrer. Temperature and salinity evolve by way of fluid exchange between the boxes via the capillary tube and the overflow, and by way of relaxation with the two infinite reservoirs at $(+T^*, +S^*)$ and $(-T^*, -S^*)$.

Rayleigh and Reynolds numbers — of interest. It is therefore useful to consider an extreme simplification of the overturning circulation, namely *box models*. These are caricatures of the circulation, constructed by dividing the ocean into a small number of boxes with simple rules determining the transport of fluid properties between them.[12]

21.3.1 A Two-box Model

Consider two boxes as illustrated in Fig. 21.6. Each box is well-mixed and has a uniform temperature and salinity, T_1, T_2 and S_1, S_2. The boxes are connected with a capillary tube at the bottom along which the flow is viscous, obeying Stokes' Law. That is, the flow along the tube is proportional to the pressure gradient which, because the flow is hydrostatic, is proportional to the density difference between the two boxes. An overflow at the top keeps the upper surfaces of the two boxes at the same level. Thus, the circulation, Ψ, is given by

$$\Psi = A(\rho_1 - \rho_2), \tag{21.46}$$

where ρ_1 and ρ_2 are the densities of the fluids in the two boxes and A is a constant. The boxes are enclosed by porous walls beyond which are reservoirs of constant temperature and salinity, and we are at liberty to choose the origin of the temperature scale such that the two reservoirs are at $+T^*$ and $-T^*$, and similarly for salinity. Thus, heat and salt are transferred into and out of the boxes as represented by simple linear laws and we have

$$\begin{aligned} \frac{dT_1}{dt} &= c(T^* - T_1) - |\Psi|(T_1 - T_2), & \frac{dT_2}{dt} &= c(-T^* - T_2) - |\Psi|(T_2 - T_1), \\ \frac{dS_1}{dt} &= d(S^* - S_1) - |\Psi|(S_1 - S_2), & \frac{dS_2}{dt} &= d(-S^* - S_2) - |\Psi|(S_2 - S_1). \end{aligned} \tag{21.47}$$

The advective transfer is independent of the sign of the circulation, because it occurs through both the capillary tube and the overflow. From these equations it is easy to show that the sum of the temperatures, $T_1 + T_2$ decays to zero and is uncoupled from the difference, and similarly for salinity. Defining $\widehat{T} = (T_1 - T_2)/(2T^*)$ and $\widehat{S} = (S_1 - S_2)/(2S^*)$, we obtain

$$\frac{d\widehat{T}}{dt} = c(1 - \widehat{T}) - 2|\Psi|\widehat{T}, \qquad \frac{d\widehat{S}}{dt} = d(1 - \widehat{S}) - 2|\Psi|\widehat{S}. \tag{21.48a,b}$$

Using a linear equation of state of the form $\rho = \rho_0(1 - \beta_T T + \beta_S S)$ (where the variables are dimensional) the circulation (21.46) becomes

$$\Psi = 2\rho_0 T^* \beta_T A \left(-\widehat{T} + \frac{\beta_S S^*}{\beta_T T^*} \widehat{S} \right). \tag{21.49}$$

Finally, nondimensionalizing time using $\tau = ct$, the equations of motion become

$$\frac{d\widehat{T}}{d\tau} = (1 - \widehat{T}) - |\Phi|\widehat{T}, \qquad \frac{d\widehat{S}}{d\tau} = \delta(1 - \widehat{S}) - |\Phi|\widehat{S}, \qquad \Phi = -\gamma(\widehat{T} - \mu\widehat{S}), \tag{21.50a,b,c}$$

where $\Phi = 2\Psi/c$ and the three parameters that determine the behaviour of the system are

$$\gamma = \frac{4\rho_0 T^* \beta_T A}{c}, \qquad \delta = \frac{d}{c}, \qquad \mu = \frac{\beta_S S^*}{\beta_T T^*}. \tag{21.51}$$

The parameter γ measures the overall strength of the forcing in determining the strength of the circulation, and is the ratio of a relaxation time scale to an advective time scale. The parameter δ is the ratio of the reciprocal time constants of temperature and salinity relaxation, and μ is a measure of the ratio of the effect of the salinity and temperature forcings on the density. Salinity transfer will normally be much slower than heat transfer so that $\delta \ll 1$, whereas if salinity and temperature are both to play a role in the dynamics we need $\mu = \mathcal{O}(1)$. We might also expect both advection and relaxation to be important if $\gamma = \mathcal{O}(1)$, and this will depend on the properties of the capillary tube.

Interpretation

Although the above model describes a potentially real system, one that might be constructed in the laboratory, it is the analogy to aspects of the ocean circulation that interests us here. To make the analogy, we suppose that one box represents the entire high-latitude ocean and the other the entire low-latitude ocean, and the capillary tube and the overflow carry the overturning circulation between them. The reservoirs at $\pm T^*$ and $\pm S^*$ represent the atmosphere. Typically, we would choose the low latitudes to be both heated and salted (the latter because of the low rainfall and high evaporation in the subtropics) and the high latitudes to be cooled and freshened by rainfall. Thus, T^* and S^* have the same sign, and they force the circulation in opposite directions.

It is a common fluid-dynamical experience that the behaviour of a highly-truncated system has little resemblance to that of the corresponding continuous system, and so we expect the model to be only a cartoon of the ocean circulation. For example, we have restricted the circulation to be of basin scale, and the parameterization of the intensity of the overturning circulation by (21.50c) must be regarded with caution, because it represents a frictionally controlled flow rather than a nearly inviscid geostrophic flow. Nevertheless, observations and numerical simulations do indicate that the overturning circulation does have a relatively simple vertical and horizontal structure: the circulation in the North Atlantic is similar to that of a single cell, for example, indicating that an appropriate low-order model may be useful.

One might also question the oceanic appropriateness of the linear relaxation terms. For temperature, the bulk aerodynamic formulae often used to parameterize air–sea fluxes do have a similar form, but the freshening of seawater by rainfall is more akin to an imposed negative flux of salinity, and evaporation is a function of temperature. An alternative might be to impose an effective salt flux so that

$$\frac{d}{dt}(S_1 - S_2) = 2E - 2|\Psi|(S_1 - S_2), \tag{21.52}$$

where E is an imposed, constant, effective rate of salt exchange with the atmosphere. After nondimensionalization, using E/c to nondimensionalize salt, (21.50b) is replaced by

$$\frac{dS}{d\tau} = 1 - |\Phi|S. \tag{21.53}$$

Another aspect of the model that is oceanographically questionable is that the model assumes that the water masses can be mixed below the surface. Thus, when water enters one box from the

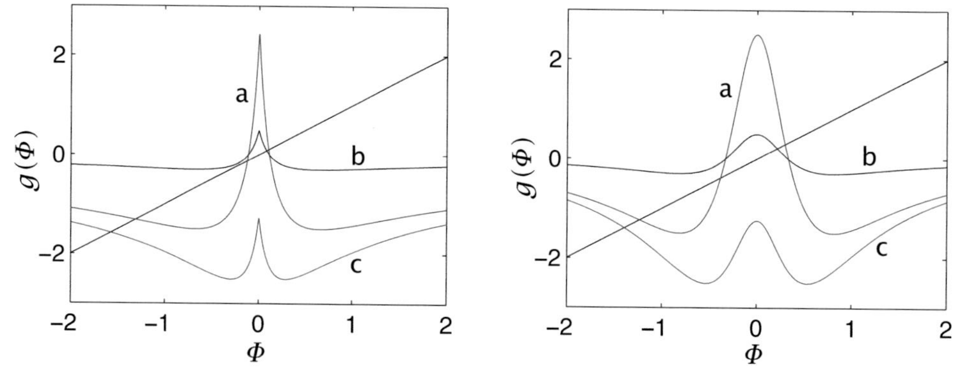

Fig. 21.7 Left panel: graphical solution of the two-box model. The straight line has unit slope and passes through the origin, and is therefore Φ itself. The curved lines plot the function $f(\Phi)$ as given by the right-hand side of (21.54). The intercepts of the two are solutions to the equation. The parameters for the three curves are: a, $\gamma = 5$, $\delta = 1/6$, $\mu = 1.5$; b, $\gamma = 1$, $\delta = 1/6$, $\mu = 1.5$; c, $\gamma = 5$, $\delta = 1/6$, $\mu = 0.75$. Right panel: the same except with Φ^2 in place of $|\Phi|$ on the rhs of (21.54).

other it immediately mixes with its surroundings. Without the stirrer this would not occur and the equations of the box model would not represent a real system. In the real ocean, most of the mixing of water masses happens near the surface (in the mixed layer) and near lateral boundaries or regions of steep topography. Elsewhere in the ocean, mixing is quite small, and probably far from sufficient to mix a large volume of water in the advective or relaxation times of the box model. Having noted all these objections, we will put them aside and continue with an analysis of the model.

Solutions

Equilibria occur when the time-derivatives vanish, and the circulation then satisfies

$$\Phi = g(\Phi) \equiv \gamma\left(\frac{-1}{1+|\Phi|} + \frac{\mu}{1+|\Phi|/\delta}\right). \tag{21.54}$$

A graphical solution of this is obtained as the intercept of the right-hand side with the left-hand side, the latter being a straight line through the origin at an angle of 45°, and this is plotted in Fig. 21.7. Perhaps the most interesting aspect of the solutions is that they exhibit *multiple equilibria;* that is, there are multiple steady solutions with the same parameters.

Evidently, for a range of parameters three solutions are possible, whereas for others only one solution exists. Although a fairly complete analysis of the nature of the steady solutions is possible, it is instructive to consider the special case with $\gamma \gg 1$ and $\delta \ll 1$ This corresponds to the situation in which the advective time scale is shorter than the diffusive one and temperature relaxation is much faster than salt relaxation. Using the graphical solution as a guide, two of the solutions are then close to the origin, with $\Phi \ll 1$, and satisfy

$$\Phi \approx \gamma\left(-1 + \frac{\mu\delta}{\delta + |\Phi|}\right), \tag{21.55}$$

giving, for small $|\Phi|$ and $\mu > 1$,

$$\Phi \approx \pm[\delta(\mu - 1)]. \tag{21.56}$$

The positive solution, with flow in the capillary tube from box 1 to box 2, is salinity driven — driven by the density gradient of the same sign as that caused by the salinity, with the density gradient due to temperature opposing the motion. That is, box 1 is denser than box 2 because it

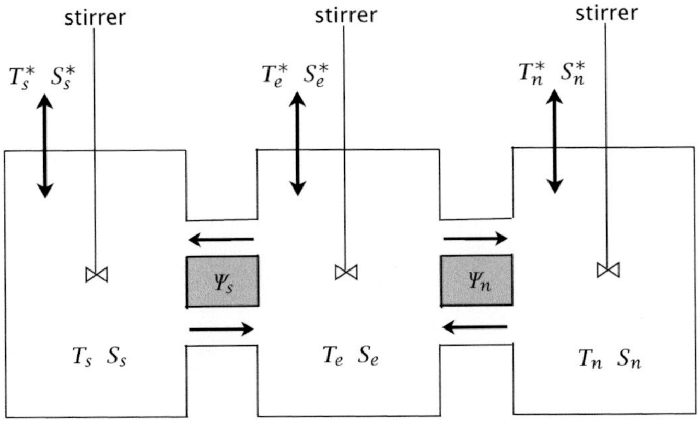

Fig. 21.8 A three-box model. Each box contains fluid with uniform values of temperature and salinity, each exchanges fluid with its neighbour, and in each the temperature and salinity are relaxed toward fixed values.

is more saline, even though it is also warmer. The negative solution is thermally driven, with the flow in the capillary tube going from the denser (cold and fresh) box 2 to the lighter (warm and salty) box 1. However, this solution is unstable, and any small perturbation will amplify and the system will move away from this solution. Solving for temperature and salinity we find that $T \approx 1$ (i.e., it is close to its relaxation value and hardly altered by advection), and $S \approx 1/\mu$.

The third solution has a circulation far from the origin, and the balance in (21.54) is between the left-hand side and the first term on the right. In the limiting case we find

$$\Phi \approx -\sqrt{\gamma}. \tag{21.57}$$

This solution has a density gradient dominated by the temperature effect: the temperature difference is $T \approx 1/\sqrt{\gamma}$ whereas the salinity difference is $S \approx \delta/\sqrt{\gamma}$, and thus its effect on density is much smaller.

21.3.2 ♦ More Boxes

More boxes can be added in a variety of ways and, now forgoing an easy relevance to a laboratory apparatus, one such is illustrated in Fig. 21.8. The three boxes represent the mid- and high-latitude Northern Hemisphere, the mid- and high-latitude Southern Hemisphere, and the equatorial regions. Each of the three boxes can exchange fluid with its neighbour, and each is also in contact with a reservoir and subject to a relaxation to a fixed value of temperature and salinity, (T_s^*, S_s^*), (T_e^*, S_e^*) and (T_n^*, S_n^*). Then, with obvious notation, we infer the equations of motion for temperature

$$\frac{\mathrm{d}T_s}{\mathrm{d}t} = c(T_s^* - T_s) - |\Psi_s|(T_s - T_e), \qquad \frac{\mathrm{d}T_n}{\mathrm{d}t} = c(T_n^* - T_n) - |\Psi_n|(T_n - T_e),$$
$$\frac{\mathrm{d}T_e}{\mathrm{d}t} = c(T_e^* - T_e) - |\Psi_s|(T_e - T_s) - |\Psi_n|(T_e - T_n), \tag{21.58}$$

and similarly for salt, with flow rates given by the density differences

$$\Psi_s = A\rho_0[-\beta_T(T_s - T_e) + \beta_S(S_s - S_e)], \qquad \Psi_n = A\rho_0[-\beta_T(T_n - T_e) + \beta_S(S_n - S_e)]. \tag{21.59}$$

These equations may be nondimensionalized and reduced to four prognostic equations for the quantities $T_e - T_n$, $T_e - T_s$, $S_e - S_n$, $S_e - S_s$. Not surprisingly, multiple equilibria can again be found. One interesting aspect is that stable asymmetric solutions arise with symmetric forcing ($T_s^* = T_n^*$, $S_s^* = S_n^*$). These effectively have a pole-to-pole circulation, illustrated in the upper row of Fig. 21.9. Such a circulation can be thought of as the superposition of a thermal circulation in one hemisphere and a salinity-driven circulation in the other.

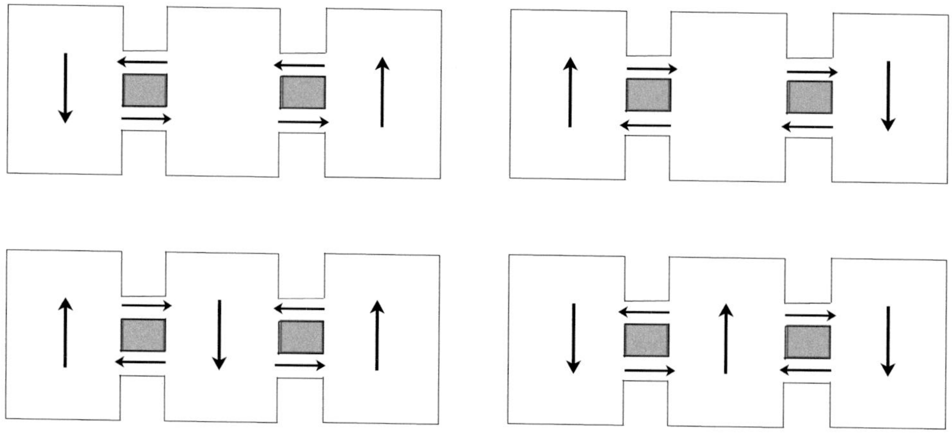

Fig. 21.9 Four solutions to the three-box model with the symmetric forcing $S_s^* = S_n^*$ and $T_n^* = T_s^*$. The two solutions on the top row have an asymmetric, 'pole-to-pole', circulation whereas the solutions on the bottom row are symmetric.[13]

The box models are useful because they are suggestive of behaviour that might occur in real fluid systems, and because they provide a means of interpreting behaviour that does occur in more complete numerical models and perhaps in the real world. However, they are by no means good approximations of the real equations of motion and without other supporting evidence the solutions found in box models should not be regarded as representing real solutions of the fluid equations for the world's oceans.[14] Indeed, the mechanism for the observed pole-to-pole circulation in the real ocean may be quite different from that of the box models — see the sections beginning with sec:windmoc.

21.4 A LABORATORY MODEL OF THE ABYSSAL CIRCULATION

We now return to a more fluid dynamical description of the deep ocean circulation, and consider two simple, closely related, models that are relevant to aspects of the deep circulation, still assuming it to be buoyancy- or mixing-driven. The first, which we consider in this section, is a laboratory model, originally envisioned as being a prototype for the deep circulation. The second model, considered in the following sections, is explicitly a model of the deep circulation. Both models are severe idealizations that describe only limited aspects of the circulation, but they are both very helpful tools that enable us to understand more complete models and, in part, the real circulation itself.

21.4.1 Set-up of the Laboratory Model

Let us consider flow in a rotating tank, as illustrated in Fig. 21.10. The fluid is confined by vertical walls to occupy a sector, and the entire tank rotates anticlockwise when viewed from above, like the Northern Hemisphere. When the fluid is stationary in the rotating frame, the fluid slopes up toward the outer edge of the tank and the balance of forces in the rotating frame is between a centrifugal force pointing outwards and the pressure gradient due to the sloping fluid pointing inwards. In the inertial frame of the laboratory itself, the pressure gradient pointing inwards provides a centripetal force that causes the fluid to accelerate toward the centre of the tank, resulting in a circular motion. (Recall that steady circular motion is always accompanied by an acceleration toward the centre of the circle.)

This set-up, and the accompanying theory, have become known as the *Stommel–Arons–Faller* model.[15] The motivation of the construct is clear, in that the sector represents an ocean basin.

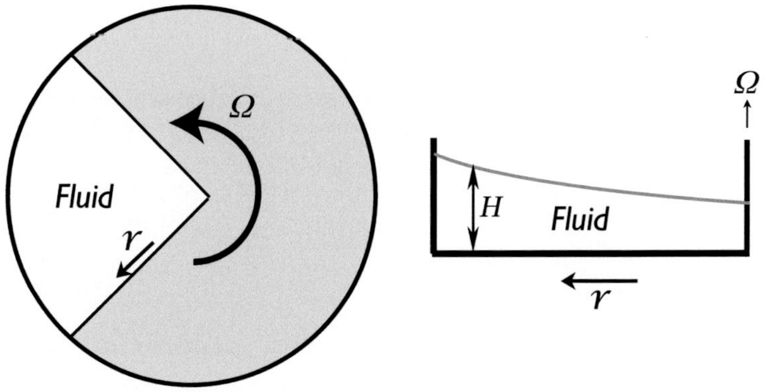

Fig. 21.10 Stommel–Arons–Faller rotating tank experiment. Left: A plan view, with the fluid in the sector at left. Right: Side view. The free surface of the fluid slopes up with increasing radius, giving a balance between the centrifugal force pointing outwards and the pressure force pointing inwards. Small pipes (not shown) provide mass sources and sinks.

However, rather than driving the fluid with wind or by differential heating, we drive it with localized mass sources and sinks, for example from small pipes inserted into the tank. If an oceanic analogy is desired, the mass source might be thought of as representing a sources of deep abyssal water due to deep convection. The oceanic analogy is not perfect but it helps build intuition about the real ocean.

21.4.2 Dynamics of Flow in the Tank

Let us assume that the motion of the fluid in the tank is sufficiently weak that its Rossby number is small, and that it obeys the shallow water planetary-geostrophic equations, namely

$$f_0 \times \boldsymbol{u} = -g\nabla h + \Omega^2 r \hat{\boldsymbol{r}} + \boldsymbol{F}, \tag{21.60a}$$

$$\frac{\partial h}{\partial t} + \nabla \cdot (\boldsymbol{u}h) = S, \tag{21.60b}$$

where $\boldsymbol{u} = (v^r, v^\theta)$ is the horizontal velocity in cylindrical (r, θ) coordinates, $\hat{\boldsymbol{r}}$ is a unit vector in the direction of increasing r, $\boldsymbol{F}$ represents frictional terms (which we will suppose are small except in boundary layers) and S represents mass sources. These two equations yield the potential vorticity equation,

$$\frac{\mathrm{D}}{\mathrm{D}t}\left(\frac{f_0}{h}\right) = \frac{\mathrm{curl}_z \boldsymbol{F}}{h} - \frac{f_0 S}{h^2}. \tag{21.61}$$

Let us write the height field as

$$h = H(r, t) + \eta(r, \theta, t), \tag{21.62}$$

where $H(r, t)$ is the height field corresponding to the rest state of the fluid (in the rotating frame) and η the perturbation. Thus, from (21.60a)

$$0 = -g\nabla H + \Omega^2 r \hat{\boldsymbol{r}}, \tag{21.63}$$

which gives

$$H = \frac{\Omega^2 r^2}{2g} + \widehat{H}(t), \tag{21.64}$$

where $\widehat{H}$ is a measure of the overall mass of the fluid. Its rate of change is determined by the mass source

$$\frac{\mathrm{d}\widehat{H}}{\mathrm{d}t} = \langle S \rangle, \tag{21.65}$$

the angle brackets indicating a domain average. The equations of motion (21.60) become

$$f_0 \times \boldsymbol{u} = -g\nabla\eta + \boldsymbol{F}, \tag{21.66a}$$

$$\frac{\partial}{\partial t}(\eta + H) + \nabla \cdot [\boldsymbol{u}(\eta + H)] = 0. \tag{21.66b}$$

Equation (21.66a) tells us that, away from frictional regions, the velocity is in geostrophic balance with the pressure field due to the perturbation height η.

Let us now suppose $|\eta| \ll H$ and $|\partial\eta/\partial t| \ll |\partial H/\partial t|$, which holds if the mass source is small and gentle enough. Then (21.66b) may be written

$$\frac{\partial H}{\partial t} + \nabla \cdot (\boldsymbol{u}H) = 0. \tag{21.67}$$

In this approximation, the potential vorticity equation (21.61) becomes, away from friction and mass sources,

$$\frac{\mathrm{D}}{\mathrm{D}t}\left(\frac{f_0}{H}\right) = 0 \qquad \text{or} \qquad \frac{\mathrm{D}H}{\mathrm{D}t} = 0, \tag{21.68a,b}$$

where the second equation follows because f_0 is a constant. (This equation also follows directly from (21.67), because the velocity is geostrophic and divergence-free where friction is absent; however, it is better thought of as a potential vorticity equation, not a mass conservation equation.) Equation (21.68b) means that fluid columns change position in order to keep the same value of H. Further, because H only varies with r, (21.68b) becomes

$$\frac{\partial H}{\partial t} + v^r \frac{\partial H}{\partial r} = 0, \tag{21.69}$$

where the superscript r indicates the radial component of velocity. Using (21.64) and (21.65) then gives

$$v^r = -\frac{g}{\Omega^2 r} \langle S \rangle . \tag{21.70}$$

This is a remarkable result, for it implies that, if $\langle S \rangle$ is positive, the flow is *toward* the apex of the dish, except at the location of the mass sources and in frictional boundary layers, *no matter where the mass source is actually located.* The explanation of this counter-intuitive result is simple enough. If $\langle S \rangle > 0$ the overall height of the fluid increases with time; thus, in order that a given material column of fluid keep its height fixed, it must move toward the apex of the dish. The full velocity may be obtained, away from the frictional regions, using the divergence-free nature of the velocity:

$$\nabla \cdot \boldsymbol{u} = \frac{1}{r}\left[\frac{\partial (r v^r)}{\partial r} + \frac{\partial v^\theta}{\partial \theta}\right] = 0. \tag{21.71}$$

Then, using (21.70), $\partial v^\theta/\partial\theta = 0$ except at a source or sink, or in a frictional boundary layer. Assuming there is only one frictional boundary layer, $v^\theta = 0$ except at those latitudes (i.e., values of r) that contain a mass source or sink.

To balance the flow toward the apex there must, then, be a *boundary layer* in which the flow has the opposite sense, and therefore in which frictional effects are important. To determine where the boundary layer is — on the east or west side of the domain — we need some vorticity dynamics. Away from the mass source, but including friction, the potential vorticity equation is

$$\frac{\mathrm{D}}{\mathrm{D}t}\left(\frac{f_0}{H}\right) = \frac{\mathrm{curl}_z \boldsymbol{F}}{H} \qquad \text{or} \qquad -\frac{f_0}{H^2}\frac{\mathrm{D}H}{\mathrm{D}t} = \mathrm{curl}_z \boldsymbol{F}, \tag{21.72a,b}$$

and the free surface of the water slopes downwards toward the apex, as illustrated in Fig. 21.10. Now, suppose that there are a mass source and a sink of equal magnitudes, with the source further from the apex than the sink, as in the panel at the bottom right of Fig. 21.11. The flow from source

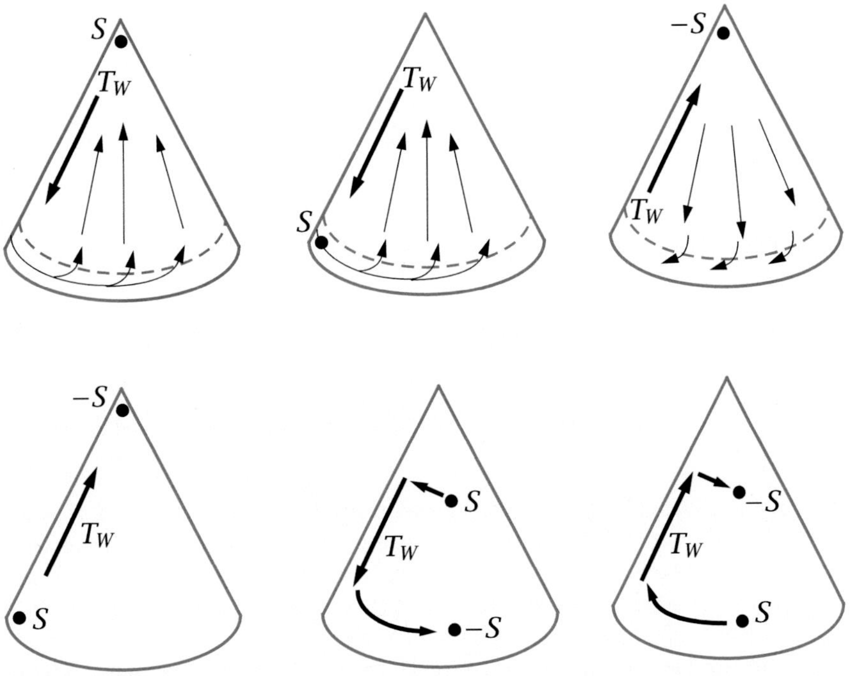

Fig. 21.11 Idealized examples of the flow in the rotating sector experiments, with various locations of a source (S) or sink ($-S$) of mass.

to sink must be along either the left or right boundary of the container. To see which, note that the flow is toward smaller values of H, and therefore the left-hand side of (21.72a) is positive. To balance this, the friction in the boundary current must impart a positive vorticity to the flow (i.e., $\text{curl}_z \boldsymbol{F} > 0$), which means in general that the flow itself must have negative vorticity, and the flow is clockwise. (For example, if $\boldsymbol{F} = -\lambda \boldsymbol{u}$ the right-hand side of (21.72) is $-(\lambda/H)\text{curl}_z \boldsymbol{u}$ and this is positive if the flow is clockwise.) Clockwise flow implies a *western* boundary layer, on the left of the container. A western boundary layer is a general feature, not dependent on the placement of mass sources or sinks. For suppose there is a single source of mass, as for example in the upper left example of Fig. 21.11; the interior mass flow will then be toward the apex and the flow in the boundary layer away from the apex. The left-hand side of (21.72a) is then negative, and so $\text{curl}_z \boldsymbol{F}$ must be negative. The flow must then have an anticlockwise sense, again requiring a western boundary layer to achieve a balance in the potential vorticity equation.

The flow is in some ways analogous to flow on the β-plane, and in particular:

(i) the r-dependence of the height field provides a background potential vorticity gradient, analogous to the β-effect;

(ii) the time-dependence of H is analogous to a wind curl, since it is this that ultimately drives the fluid motion.

The analogies are drawn out explicitly in the shaded box on page 823; the box also includes a column for abyssal flow in the ocean, discussed in the next two sections.

21.5 A MODEL FOR OCEANIC ABYSSAL FLOW

We will now extend the reasoning applied to the rotating tank to the rotating sphere, and so construct a model — the *Stommel–Arons model* — of the abyssal flow in the ocean.[16] The basic idea is simple: we model the deep ocean as a single layer of homogeneous fluid in which there is a

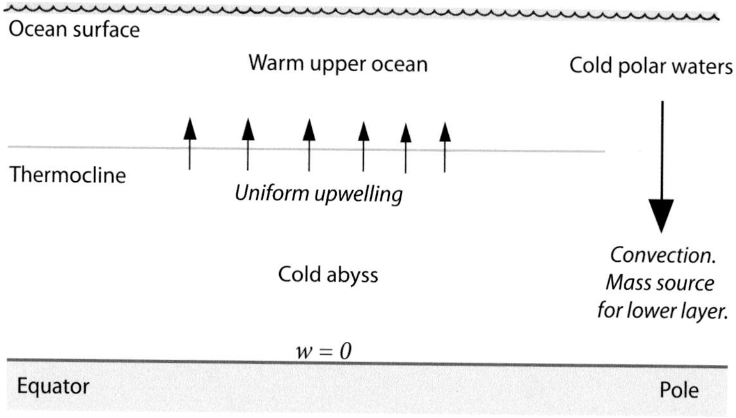

Fig. 21.12 The structure of a Stommel–Arons ocean model of the abyssal circulation. Convection at high latitudes provides a localized mass-source to the lower layer, and upwelling through the thermocline provides a more uniform mass sink.

localized injection of mass at high latitudes, representing convection (Fig. 21.12). However, unlike the rotating dish, mass is extracted from this layer by upwelling into the warmer waters above it, keeping the average thickness of the abyssal layer constant. We assume that this upwelling is nearly uniform, that the ocean is flat-bottomed, and that a passive western boundary current may be invoked to satisfy mass conservation, and which does not affect the interior flow. Obviously, these assumptions are very severe and the model can at best be a conceptual model of the real ocean. Given that, we will work in Cartesian coordinates on the β-plane, and use the planetary-geostrophic approximation.

The momentum and mass continuity equations are

$$\boldsymbol{f} \times \boldsymbol{u} = -\nabla_z \phi \quad \text{and} \quad \nabla_z \cdot \boldsymbol{u} = -\frac{\partial w}{\partial z}, \tag{21.73a,b}$$

where $\boldsymbol{f} = (f_0 + \beta y)\mathbf{k}$. On elimination of ϕ, (21.73) yields the now-familiar balance,

$$\beta v = f \frac{\partial w}{\partial z}. \tag{21.74}$$

Except in the localized regions of convection, the vertical velocity is, by assumption, positive and uniform at the top of the lower layer, and zero at the bottom. Thus (21.74) becomes

$$v = \frac{f}{\beta} \frac{w_0}{H}, \tag{21.75}$$

where w_0 is the uniform upwelling velocity and H the layer thickness. Thus, the flow is *polewards* everywhere (including the Southern Hemisphere), vanishing at the equator.

21.5.1 Completing the Solution

Since $v = f^{-1}(\partial \phi / \partial x)$, the pressure is given by

$$\phi = \int_{x_0}^{x} \left(\frac{f^2 w_0}{\beta H} \right) \mathrm{d}x', \tag{21.76}$$

where x_0 is a constant of integration, to be determined by the boundary conditions. Because there is no flow into the eastern boundary, x_E, we set ϕ = constant at $x = x_E$, and because this is a one-layer model we are at liberty to set that constant equal to zero. Thus,

$$\phi(x) = -\int_{x}^{x_E} \left(\frac{f^2 w_0}{\beta H} \right) \mathrm{d}x' = -\frac{f^2}{\beta H} w_0 (x_E - x). \tag{21.77}$$

Analogies Between a Rotating Dish, Wind-Driven and Abyssal Flows

Consider homogeneous models of: (i) a rotating dish; (ii) wind-driven flow on the β-plane; and (iii) abyssal flow on the β-plane. We model them all with a single layer of homogeneous fluid satisfying the planetary geostrophic equations. In (i) the mass source, $\langle S \rangle$, is localized and the total depth of the fluid layer changes with time; fluid columns move to keep their depth constant. In (ii) there is no mass source and the depth of the fluid layer is constant; the fluid motion is determined by the wind-stress curl, $\mathrm{curl}_z \boldsymbol{\tau}$, and by β. In (iii) the fluid source (convection) is localized at high latitudes and exactly balanced by a mass loss, S_u, due to upwelling everywhere else, so that layer depth is constant and S_u is uniform and negative nearly everywhere. The equations below then apply away from frictional boundary layers and localized mass sources.

	(i) Rotating dish	*(ii) Wind-driven flow*	*(iii) Abyssal flow*
PV conservation	$\dfrac{\mathrm{D}}{\mathrm{D}t}\left(\dfrac{f_0}{H}\right) = 0$	$\dfrac{\mathrm{D}}{\mathrm{D}t}\left(\dfrac{f}{H_0}\right) = \dfrac{1}{H_0}\mathrm{curl}_z \boldsymbol{\tau}$	$\dfrac{\mathrm{D}}{\mathrm{D}t}\left(\dfrac{f}{h}\right) = -\dfrac{fS_u}{h^2}$
This leads to	$v^r \dfrac{\partial H}{\partial r} = -\dfrac{\partial H}{\partial t}$	$\dfrac{v}{H_0}\dfrac{\partial f}{\partial y} = \dfrac{1}{H_0}\mathrm{curl}_z \boldsymbol{\tau}$	$\dfrac{v}{h}\dfrac{\partial f}{\partial y} = -\dfrac{fS_u}{h^2}$
and	$v^r = -\dfrac{g}{\Omega^2 r}\langle S \rangle$	$v = \dfrac{1}{\beta}\mathrm{curl}_z \boldsymbol{\tau}$	$v = -\dfrac{fS_u h}{\beta}$
	$\langle S \rangle$ is localized mass source	$\mathrm{curl}_z \boldsymbol{\tau}$ is wind stress curl	S_u is upwelling mass loss

Meridional mass flow away from boundaries is thus determined by:

(i)	(ii)	(iii)
sign (and not location) of localized mass source, $\langle S \rangle$,	sign of wind-stress curl, $\mathrm{curl}_z \boldsymbol{\tau}$,	upwelling and sign of f, so polewards if $S_u < 0$ (upwelling).

The zonal velocity follows using geostrophic balance,

$$u = \frac{1}{f}\frac{\partial \phi}{\partial y} = \frac{2}{H} w_0 (x_E - x), \tag{21.78}$$

where we have also used $\partial f/\partial y = \beta$ and $\partial \beta/\partial y = 0$. Thus the velocity is eastwards in the interior, and independent of f and latitude, provided x_E is not a function of y.

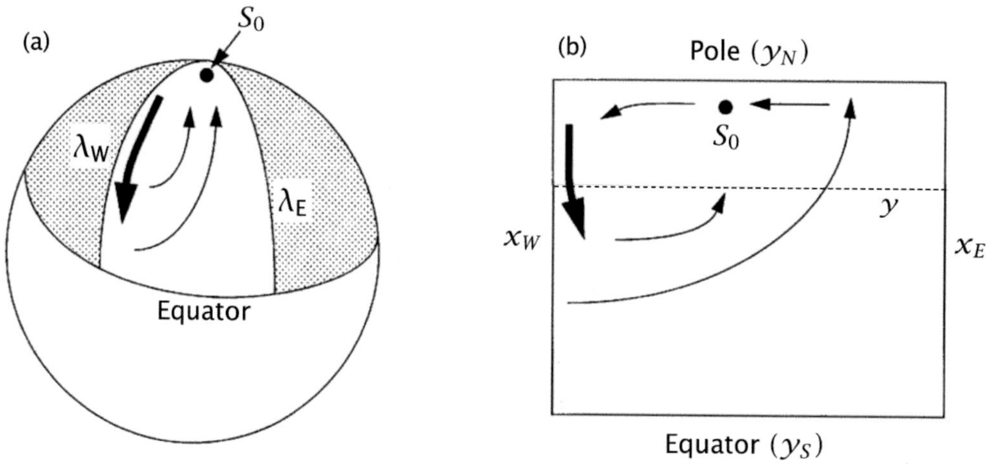

Fig. 21.13 Abyssal circulation in a spherical sector (left) and in a corresponding Cartesian rectangle (right).

Using (21.75) and (21.78) we can confirm mass conservation is indeed satisfied:

$$\frac{\partial u}{\partial x} + \frac{\partial v}{\partial y} + \frac{\partial w}{\partial z} = -\frac{2w_0}{H} + \frac{w_0}{H} + \frac{w_0}{H} = 0. \tag{21.79}$$

21.5.2 Application to the Ocean

Let us consider a rectangular ocean with a mass source at the northern boundary, balanced by uniform upwelling (see Figs. 21.13 and 21.14). Since the interior flow will be northwards, we anticipate a southwards flowing western boundary current to balance mass. Conservation of mass in the area polewards of the latitude y demands that

$$S_0 + T_I(y) = T_W(y) + U(y), \tag{21.80}$$

where S_0 is the strength of the source, T_W the equatorwards transport in the western boundary current, T_I the poleward transport in the interior, and U is the integrated loss due to upwelling polewards of y. Then, using (21.75),

$$T_I = \int_{x_W}^{x_E} vH \,\mathrm{d}x = \int_{x_W}^{x_E} \frac{fw_0}{\beta} \,\mathrm{d}x = \frac{f}{\beta} w(x_E - x_W). \tag{21.81}$$

The upwelling loss is given by

$$U = \int_{x_W}^{x_E} \int_{y}^{y_N} w \,\mathrm{d}x = w_0(x_E - x_W)(y_N - y), \tag{21.82}$$

where y_N denotes the northern (polar) boundary. Assuming the source term is known, then using (21.80) we obtain the strength of the western boundary current,

$$T_W(y) = S_0 + T_I - U = S_0 + \frac{f}{\beta} w(x_E - x_W) - w_0(x_E - x_W)(y_N - y). \tag{21.83}$$

To close the problem we use the fact that over the entire basin mass must be balanced, which gives a relationship between w_0 and S_0,

$$S_0 = w_0 \Delta x \Delta y, \tag{21.84}$$

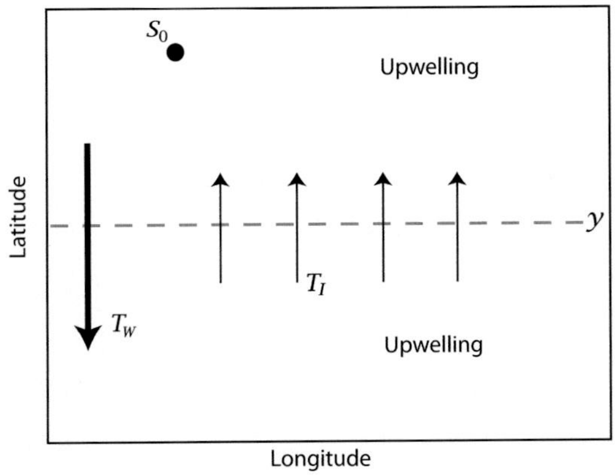

Fig. 21.14 Mass budget in an idealized abyssal ocean. Polewards of some latitude y, the mass source (S_0) plus the poleward mass flux across y (T_I) are equal to the sum of the equatorward mass flux in the western boundary current (T_W) and the integrated loss due to upwelling (U) polewards of y. See (21.80).

where $\Delta x = x_E - x_W$ and $\Delta y = y_N - y_S$, with y_S being the southern boundary of the domain. The strength of the circulation (i.e., the magnitude of S_0 or w_0) is in reality determined by the diffusivity, κ, as previously discussed, and here we take it as a given.

Using (21.84), (21.83) becomes

$$T_W(y) = -w_0\left(\Delta x(y_N - y) - \frac{f}{\beta}\Delta x - \Delta x\Delta y\right) = w_0\Delta x\left(y - y_S + \frac{f}{\beta}\right). \tag{21.85}$$

With no loss of generality we will take $y_S = 0$ and $f = f_0 + \beta y$. Then (21.85) becomes

$$T_W(y) = w_0\Delta x\,(2y + f_0/\beta)\,, \tag{21.86}$$

or, using $S_0 = w_0\Delta x y_N$,

$$T_W(y) = \frac{S_0}{y_N}\left(2y - \frac{f_0}{\beta}\right). \tag{21.87}$$

With a slight loss of generality (but consistent with the spirit of the planetary-geostrophic approximation) we take $f_0 = 0$, which is equivalent to supposing that the equatorial boundary of the domain is at the equator, and finally obtain

$$T_W(y) = 2S_0\frac{y}{y_N}\,. \tag{21.88}$$

At the northern boundary this becomes

$$T_W(y) = 2S_0, \tag{21.89}$$

which means that the flow southwards from the source is *twice* the strength of the source itself! We also see that:

(i) the western boundary current is equatorward everywhere;
(ii) at the northern boundary the equatorward transport in the western boundary current is equal to *twice* the strength of the source;
(iii) the northward mass flux at the northern boundary is equal to the strength of the source itself.

We may check this last point directly: from (21.81)

$$T_I(y_N) = \frac{\beta y_N}{\beta}w_0\Delta x = S. \tag{21.90}$$

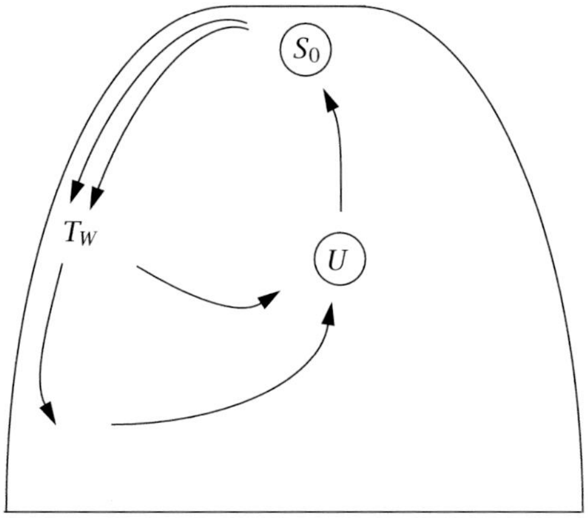

Fig. 21.15 Schematic of a Stommel–Arons circulation in a single sector. The transport of the western boundary current is greater than that provided by the source at the apex, illustrating the property of *recirculation.* The transport in the western boundary current T_W decreases in intensity equatorward, as it loses mass to the polewards interior flow, and thence to upwelling. The integrated sink, due to upwelling, U, exactly matches the strength of the source, S_0.

The fact that convergence at the pole balances T_W and S_0 does not of course depend on the particular choice we made for f and y_S.

The flow pattern evidently has the property of *recirculation* (see Fig. 21.15): this is one of the most important properties of the solution, and one that is likely to transcend all the limitations inherent in the model. This single-hemisphere model may be thought of as a crude model for aspects of the abyssal circulation in the North Atlantic, in which convection at high latitudes near Greenland is at least partially associated with the abyssal circulation. In the North Pacific there is, in contrast, little if any deep convection to act as a mass source. Rather, the deep circulation is driven by mass sources in the opposite hemisphere, and we now consider a simple model of this.

21.5.3 A Two-hemisphere Model

Our treatment now is even more obviously heuristic, since our domain crosses the equator yet we continue to use the planetary-geostrophic equations, invalid at the equator. We also persist with Cartesian geometry, even for these global-scale flows. In our defence, we remark that the value of the solutions lies in their qualitative structure, not in their quantitative predictions. Let us consider a situation with a source in the Southern Hemisphere but none in the Northern Hemisphere. For later convenience we take the Southern Hemisphere source to be of strength $2S_0$, and we suppose the two hemispheres have equal area. As before, the upwelling is uniform, so that to satisfy global mass balance S_0 and w_0 are related by

$$S_0 = w_0 \, \Delta x \, \Delta y, \tag{21.91}$$

where $\Delta x \Delta y$ is the area of each hemisphere. Then, for a given w_0, the zonally integrated poleward interior flow in each hemisphere, away from the equator, follows from Sverdrup balance,

$$T_I(y) = \frac{f}{\beta} w_0 (x_E - x_W) = S_0 \frac{y}{y_p}, \tag{21.92}$$

where y_p is either y_N (the northern boundary) or y_S. The western boundary current is assumed to 'take up the slack', that is to be able to adjust its strength to satisfy mass conservation. Thus, since $T_I(y_N) = S_0$, where S_0 is half the strength of the source in the *Southern* Hemisphere, it is plain that there must be a southwards flowing western boundary current near the northern end of the Northern Hemisphere, even in the absence of any deep water formation there!

In the northern hemisphere, the total loss due to upwelling polewards of a latitude y is given by

$$U(y) = w_0 \Delta x |y_N - y|. \tag{21.93}$$

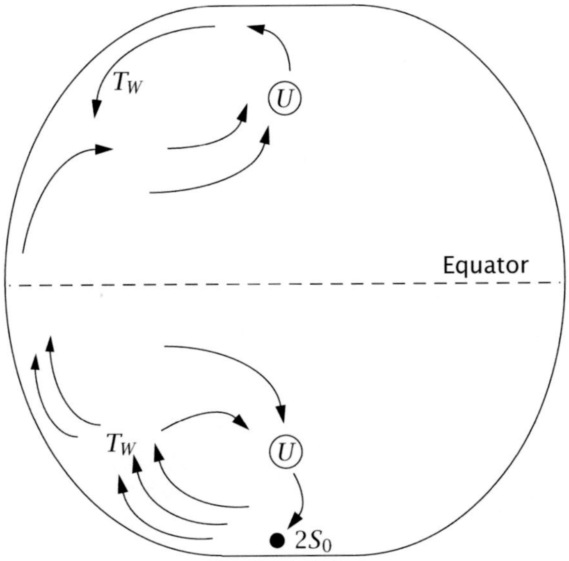

Fig. 21.16 Schematic of a Stommel–Arons circulation in a two-hemisphere basin. There is only one mass source, and this is in the Southern Hemisphere and for convenience it has a strength of 2. Although there is no source in the Northern Hemisphere, there is still a western boundary current and a recirculation. The integrated sinks due to upwelling exactly match the strength of the source.

The strength of the western boundary current is then given by, with southward flow positive,

$$T_W(y) = T_I - U = \frac{f}{\beta} w_0 \Delta x - w_0 \Delta x (y_N - y) = -w_0 \Delta x (y_N - 2y), \tag{21.94}$$

using $f = \beta y$. The boundary current thus changes sign halfway between equator and North Pole, at $y = y_N/2$. [In spherical coordinates, the analogous latitude turns out to be at $\theta = \sin^{-1}(1/2)$.] At the North Pole $y = y_N$ and we have

$$T_W(y_N) = w_0 \Delta x y = S. \tag{21.95}$$

The solution is illustrated schematically in Fig. 21.16. We can (rather fancifully) imagine this to represent the abyssal circulation in the Pacific Ocean, with no source of deep water at high northern latitudes.[17]

21.5.4 Summary Remarks on the Stommel–Arons Model

If we were given the location and strength of the sources of deep water in the real ocean, the Stommel–Arons model could give us a global solution for the abyssal circulation. The solution for the Atlantic, for example, resembles a superposition of Fig. 21.15 and Fig. 21.16 (with deep water sources in the Weddell Sea and near Greenland), and that for the Pacific resembles Fig. 21.16 (with a deep water source emanating from the Antarctic Circumpolar Current). Perhaps the greatest success of the model is that it introduces the notions of deep western boundary currents and recirculation — enduring concepts of the deep circulation that remain with us today. For example, the North Atlantic ocean does have a well-defined deep western boundary current running south along the eastern seaboard of Canada and the United States, as seen in Fig. 21.17. However, in other important aspects the model is found to be in error, in particular it is found that there is little upwelling through the main thermocline — much of the water formed by deep convection in the North Atlantic in reality upwells in the Southern Hemisphere.[18] Are there fundamental problems with the model, or just discrepancies in details that might be corrected with a slight reformulation? To help answer that we summarize the assumptions and corresponding predictions of the model, and distinguish the essential aspects from what is merely convenient:

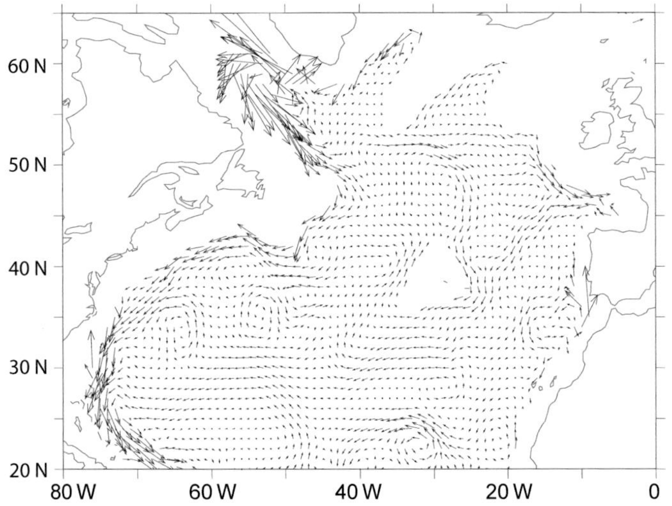

Fig. 21.17 The ocean currents at a depth of 2500 m in the North Atlantic, obtained using a combination of observations and model (as in Fig. 19.3). Note the southwards flowing *deep western boundary current.*

(i) A foundational assumption is that of linear geostrophic vorticity balance in the ocean abyss, represented by $\beta v \approx f \partial w / \partial z$, or its shallow water analogue.

- The effects of mesoscale eddies are thereby neglected. As discussed in Chapter 12, in their mature phase mesoscale eddies seek to barotropize the flow, and so create deep eddying motion that might dominate the deep flow.

(ii) A second important assumption is that of uniform upwelling, across isopycnals, into the upper ocean, and that $w = 0$ at the ocean bottom. When combined with *(i)* this gives rise to a poleward interior flow, and by mass conservation a deep western boundary current. The upwelling is a consequence of a finite diffusion, which in turn leads to deep convection as in the model of sideways convection of Section 21.1.

- The uniform-upwelling assumption might be partially relaxed, while remaining in the Stommel–Arons framework, by supposing (for example) that the upwelling occurs near boundaries, or intermittently, with corresponding detailed changes to the interior flow.
- If bottom topography is important, then $w \neq 0$ at the ocean bottom. This effect may be most important if mesoscale eddies are present, for then in an attempt to maintain its value of potential vorticity the abyssal flow will have a tendency to meander nearly inviscidly along contours of constant topography. In the presence of a mid-ocean ridge, some of the deep western boundary current might travel meridionally along the eastern edge of the ridge instead of along the coast.
- The deep water might not upwell across isopycnals, but might move along isopycnals that intersect the surface (or are connected to the surface by convection). If so, then in the presence of mechanical forcing a deep circulation could be maintained even in the absence of a diapycnal diffusivity. The circulation might then be qualitatively different from the Stommel–Arons model, although a linear vorticity balance might still hold, with deep western boundary currents. This is discussed in Section 21.6.

Even if the Stommel–Arons picture were to be essentially correct, we should not consider the deep flow as being driven by deep convection at the source regions. It is a *convenience* to specify the strength of the source term in these regions for the calculations but, just as in the models of sideways convection considered in Section 21.1, the overall strength of the circulation (insofar as it is buoyancy driven) is a function of the size of the diffusivity and the meridional buoyancy gradient at the surface.

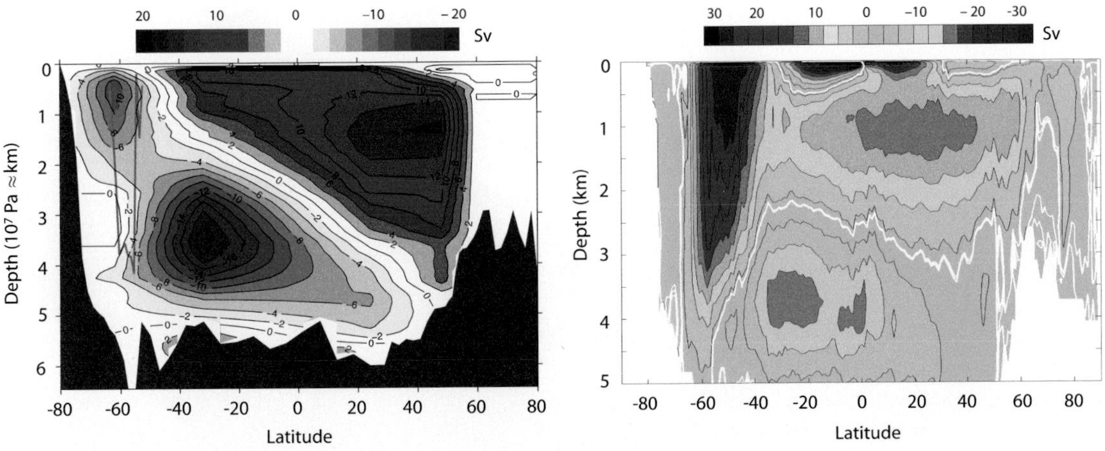

Fig. 21.18 Two independent estimates of the zonally-averaged overturning circulation of the world's ocean. The left panel is from an inverse model that mainly uses hydrographic observations, and shows the residual circulation. The right panel is a state estimation that makes explicit use of a numerical model, and shows the Eulerian circulation. [20]

21.6 † A MODEL OF DEEP WIND-DRIVEN OVERTURNING

There is no need to ask the question 'Is the model true?'. If 'truth' is to be the 'whole truth' the answer must be 'No'. The only question of interest is 'Is the model illuminating and useful?'

George E. Box, *Robustness in the strategy of scientific model building*, 1979.

We previously noted that, with values of the diapycnal diffusivity that are measured in the main thermocline, the theoretical predictions of the MOC are rather weaker than observations suggest. There are two possible resolutions to this problem. One is that the measured diapycnal diffusivity is in fact large in some parts of the ocean (e.g., in the abyss over steep topography), and if this were sufficient to produce the measured overturning and stratification the issue would be resolved. However, such a calculation would likely be fraught with uncertainty. A second and more straightforward resolution would arise if a deep circulation, and deep stratification, could be maintained by a mechanism that was *independent* of the diapycnal diffusivity. This second approach is the one we shall take in for much of the rest of this chapter. Specifically, our goal is to construct and explore models of the overturning circulation of the ocean, and the concomitant deep stratification, that have a 'wind-driven' component that persists even as the diffusivity goes to zero.[19]

21.6.1 Observations and Physical Principles

We are motivated by the observation that the MOC is in large part *interhemispheric*, with water sinking at high northern latitudes and upwelling in the Antarctic Circumpolar Current (ACC), as seen for the global circulation in Fig. 21.18 (where the left panel better shows the trajectory of water parcels) and in the Atlantic in Fig. 21.29. The Atlantic MOC (which is the dominant contributor to the global MOC) is dominated by two cells, an upper cell of North Atlantic Deep Water (NADW) with water sinking at about 60° N, moving southwards largely along isopycnals and upwelling in the south. Beneath this cell lies Antarctic Bottom Water (AABW), with sinking at high southern latitudes followed by a deep cross hemispheric circulation and upwelling again in the ACC. We would like to construct a purely wind-driven model that shows these features as simply as possible.

In the absence of a diapycnal diffusivity no upwelling can occur *through* the stratification, because that is a diabatic process. Rather, if there is deep stratification, the deep water must be directly connected to the surface along isopycnals or via a convective pathway, for convection, although

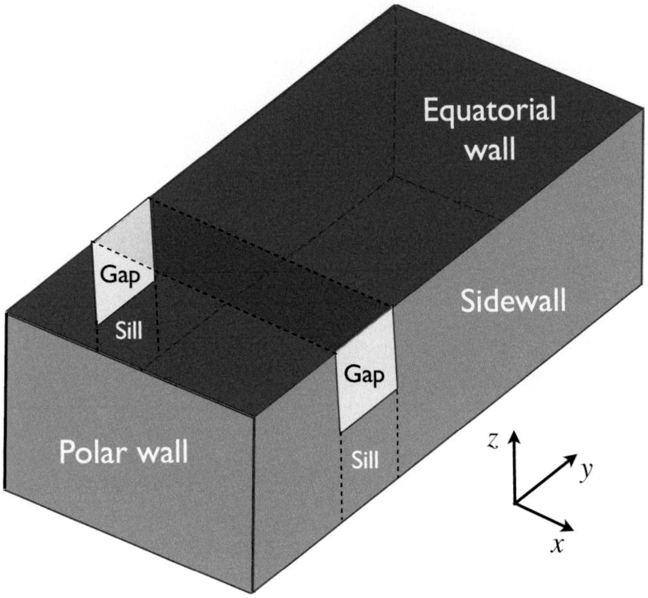

Fig. 21.19 Idealized geometry of the Southern Ocean: a re-entrant channel, partially blocked by a sill, is embedded within a closed rectangular basin; thus, the channel has periodic boundary conditions. The channel is a crude model of the Antarctic Circumpolar Current, with the area over the sill analogous to the Drake Passage.

diabatic, does not rely on a finite eddy diapycnal diffusivity. Let us recall two *de facto* principles about deep circulation:

(i) A basin will, in the absence of mechanical forcing, tend to fill with the densest available fluid.

(ii) Light fluid forced down by wind may displace the cold fluid, so producing stratification.

A completely closed ocean thus fills completely with dense polar water, except in the upper several hundred metres where the main thermocline forms. However, suppose that the polewards part of the basin is not fully enclosed but is periodic, as illustrated in Fig. 21.19, with a sill across it at mid-depth, and suppose too that the surface temperature decreases monotonically polewards. A fully enclosed basin exists only beneath the level of the sill, and we may expect the densest water in the basin, formed at the polewards edge of the domain, to fill the basin only below the level of the sill, and that above this may lie warmer water with origins at lower latitudes. Furthermore, suppose that an eastward wind blows over the channel that produces an equatorial flow in the Ekman layer. Then mass conservation demands that there must exist a *subsurface* return flow, and thus a meridional overturning circulation is set up. Note the essential role of the channel in this: if the gap were closed, then the return flow could take place at the surface via a western boundary current, as in a conventional subpolar gyre, and no overturning circulation need be set up. But in a zonally-periodic channel, an eastward wind produces a northward Ekman flow that can only be balanced by a return flow at depth — that is, a meridional overturning circulation.

21.6.2 A Single-hemisphere Model

Let us first a single-hemisphere basin with a periodic channel near its poleward edge. We suppose the basin to be in the southern hemisphere, so the channel represents the Antarctic Circumpolar Current (ACC), and that the dynamics are Boussinesq and planetary-geostrophic. We will choose extremely simple forms of wind and buoyancy forcing to allow us to obtain an analytic solution, and then later discuss how the qualitative forms of these solutions might more apply generally.

Wind and buoyancy forcing

Thermodynamic forcing is imposed by fixing the surface buoyancy, b_s. (In the discussion following salinity is absent, and buoyancy is virtually equivalent to temperature.) South of the gap we

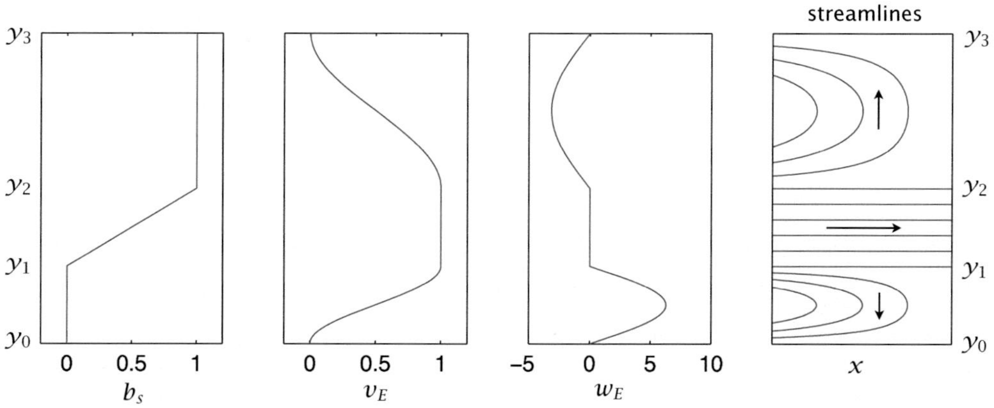

Fig. 21.20 The surface buoyancy b_s, meridional Ekman velocity v_E, vertical Ekman velocity w_E and the solution streamlines for the geostrophic horizontal flow, omitting the western boundary currents. The ordinate in all plots is latitude, with the pole at the bottom, and the four fields are given by, respectively, (21.96), (21.97a), (21.97b) and (21.98), with purely zonal flow given by (21.101) in the channel.

suppose the buoyancy to be constant, then that it increases linearly across the gap, and is constant again polewards of the gap. Thus, there is no temperature gradient across the subtropical gyre, focusing attention on the influence of the channel. Thus, referring to Fig. 21.20 or Fig. 21.21 for the definitions of the geometric factors,

$$b_s = \begin{cases} b_1, & y_0 \le y \le y_1, \\ b_1 + \dfrac{(b_2 - b_1)(y - y_1)}{y_2 - y_1}, & y_1 \le y \le y_2, \\ b_2, & y \ge y_2, \end{cases} \tag{21.96}$$

where $b_2 > b_1$, and both are constants, and we may take $b_1 = 0$ and $y_0 = 0$.

The wind forcing is purely zonal, and it is convenient to express this in terms of the Ekman transport and associated pumping (refer to section 5.7). In the channel the Ekman transport is chosen to be (realistically) equatorward and (less realistically) constant, a simplification that avoids complications of wind-driven upwelling in the channel. South (polewards) of the channel there is a conventional subpolar gyre, with an Ekman upwelling and an equatorward Ekman transport that joins smoothly to that of the channel. Equatorwards of the channel there is a conventional subtropical gyre, with Ekman downwelling. All this may be achieved by specifying:

$$v_E = \begin{cases} \dfrac{V}{2}\left[1 - \cos\left(\dfrac{\pi y}{\Delta y_1}\right)\right] \\ V \\ \dfrac{V}{2}\left[1 + \cos\left(\dfrac{\pi(y - y_2)}{\Delta y_2}\right)\right] \end{cases} \qquad w_E = \begin{cases} W_1 \sin\left(\dfrac{\pi y}{\Delta y_1}\right) & 0 \le y < y_1 \\ 0 & y_1 \le y < y_2 \\ -W_2 \sin\left(\dfrac{\pi(y - y_2)}{\Delta y_2}\right) & y_2 \le y < y_3, \end{cases} \tag{21.97a,b}$$

where $\Delta y_1 = y_1, \Delta y_2 = y_3 - y_2$, and V is a constant that determines the magnitude of the meridional Ekman flow. The meridional Ekman transport, v_E, is related to the Ekman pumping by $w_E/\delta_E = \partial v_E/\partial y$, so that $W_i = \delta_E \pi V/(2\Delta y_i)$, where δ_E is the Ekman layer thickness. If f were constant, the wind-stress curl would be proportional to the w_E field above. The precise details of the forcing do not affect the qualitative form of the solution — they merely allow an analytic solution to be obtained — but there are two essential aspects to it:

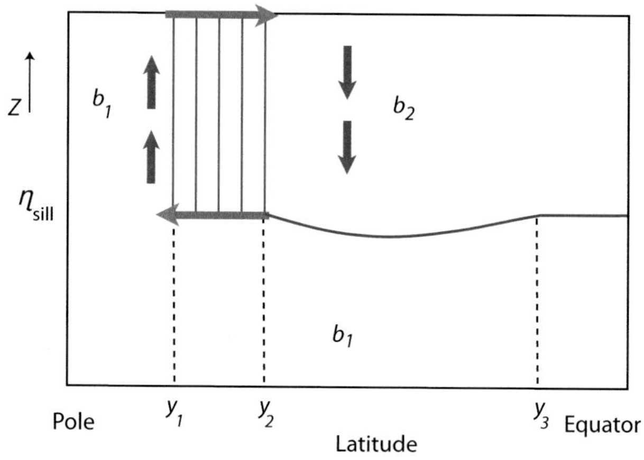

Fig. 21.21 Cross-section of the structure of the single-hemisphere ocean model described in Section 21.6.2. The domain is zonally closed equatorwards of y_2 and polewards of y_1, with a zonally periodic channel between latitudes y_1 and y_2 and above the sill, which has height η_{sill}. The arrows indicate the fluid flow driven by the equatorward Ekman transport in the channel, and the solid lines are isopycnals.

(i) The surface is cold south of the channel, warm north of the channel, and there is a temperature gradient across the channel.

(ii) The Ekman flow is equatorwards within the channel, with conventional gyres to either side.

The meridional extent of the region south of the channel and the wind forcing within it are relatively unimportant, and the region could be shrunk to nearly zero.

Solution in the gyres

Below the depth of the sill the basin is fully enclosed, and therefore up to that level the basin will fill with the densest available water (much as described in Section 21.1), except where it may be displaced by warmer fluid equatorward of the gap that is pumped down below the level of the sill by the wind (Fig. 21.21). Thus, all of the domain south of the channel, and nearly everywhere below the sill, the water has buoyancy b_1. Polewards of the channel, then, the fluid is barotropic and its vertically integrated horizontal circulation is given by Sverdrup balance, $\beta V = f w_E/H$, where V is the vertically integrated flow. With the wind stress of (21.97) we get a conventional barotropic subpolar gyre (and associated western boundary current) by the same methods that we employed in Chapter 19.

Above the sill, net meridional geostrophic transfer is forbidden in the channel region, because by geostrophic balance $f\overline{v}_g = \partial\overline{\phi}/\partial x = 0$, where ϕ is the pressure and the overbar denotes a zonal average. Equatorward of the channel the region above the sill will therefore tend to fill with the densest water available to it, and this is water with buoyancy equal to b_2 (which is the buoyancy of the water as it emerges from the channel). However, because of the presence of wind forcing, the base of this layer is not flat; rather, this fluid obeys the dynamics of the reduced-gravity single-layer ventilated thermocline model discussed in Section 20.7.1. In such a model the depth of the fluid on the eastern boundary is constant, and this must be specified. Here, this is given by the height of the sill, and therefore $h(x = x_e, y) = h_e = H - \eta_{sill}$, where H is the total depth of the basin and η_{sill} is the sill height. Then, using (20.92) and (21.97), the thickness of the moving layer equatorward of the sill is given by, for $y_2 < y < y_3$,

$$h^2 = D^2(x, y) + h_e^2, \tag{21.98}$$

where

$$D^2 = -\frac{2f^2}{g'\beta}\int_x^{x_e} w_E \,\mathrm{d}x' = \frac{2f^2}{g'\beta} W_2(x_e - x)\sin\left(\frac{\pi(y - y_2)}{\Delta y_2}\right), \tag{21.99}$$

with $g' = b_2 - b_1$. The solution is closed by the addition of a western boundary current. Note that because $h > h_e$, the light fluid is pushed below the level of the sill in the subtropical gyre.

Solution in the channel

In the channel, the fluid in the Ekman layer flows equatorward, and therefore there must be a compensating poleward flow at depth. This will occur just below the level of the sill: it cannot be deeper, because here the basin is full of denser, b_1 fluid, and in the absence of eddying or ageostrophic flow it cannot be shallower because of the geostrophic constraint. Now, because of the temperature gradient across the channel the polewards flowing fluid is warmer than the fluid at the surface, and therefore convectively unstable. Convection ensues, the result of which is the entire column of fluid between the top of the sill and the surface mixes and takes on the temperature of the surface. Thermal wind demands that there be a zonal flow associated with this meridional temperature gradient, so this temperature distribution is advected eastwards into the interior of the channel. Because the interior is presumed to be adiabatic, this temperature field extends zonally throughout the channel. Thus, in steady state, the temperature *everywhere* in the channel above the level of the sill is given by

$$b(x, y, z) = b_s(y) = b_1 + \frac{(b_2 - b_1)(y - y_1)}{y_2 - y_1}, \qquad y_1 \le y \le y_2, \;\; z > \eta_{sill}. \tag{21.100}$$

Convective mixing does not rely on a diapycnal diffusivity other than a molecular one: convective plumes are generally turbulent, generating small scales in the fluid interior where mixing and entrainment may occur; failing that, the lighter fluid is displaced to the surface where it cools by way of interaction with the atmosphere. The zonal velocity within the channel is then given by thermal wind balance, so that

$$u(x, y, z) = -\frac{1}{f}\left(\frac{b_2 - b_1}{y_2 - y_1}\right)(z - \eta_{sill}), \tag{21.101}$$

and since $f < 0$ the shear, $\partial u/\partial z$, is positive.

Regarding the depth-integrated zonal momentum budget, the wind stress at the surface is balanced by a pressure force against the sill walls. This pressure gradient arises through the meridional circulation, as the southward return flow just below the level of the sill is associated with a zonal pressure gradient that is exactly equal, but opposite, to the stress exerted by the wind. That is to say, in the Ekman layer the wind stress is balanced by the Coriolis force on the equatorward flow in the Ekman layer, which by mass conservation is equal and opposite to the Coriolis force on the deep poleward flow, which by geostrophy is equal to the net pressure force on the sill walls. The wind stress plays no role in determining the zonal transport of the channel: if the wind increases, the meridional overturning and the pressure force increase but with no change to the transport. This is a somewhat unrealistic feature of the model, for in reality the form stress induced by the flow over bottom topography (and that balances the wind stress) is likely to be a function of the zonal transport as well as the meridional transport.

A qualitative summary

The circulation of the model may be described as follows. The entire basin polewards of the channel fills with dense, b_1, water. Below the sill this fluid extends equatorward, filling the lower part of the channel and subtropical basin, up to the level of the sill. Now, Ekman pumping in the channel forces near-surface fluid equatorward, which warms as it goes, entering the subtropical basin with buoyancy b_2. This fluid fills the basin down to the level of the sill, where it encounters the dense, b_1, fluid. The subtropical basin is wind-driven, and it forms a subtropical gyre with a single moving layer. Its dynamics are completely determined by specifying the wind, the reduced gravity ($g' = b_2 - b_1$), and the depth of the fluid at the eastern boundary (the sill depth). Because of the requirements of mass conservation, there must be a poleward return flow at depth, and so at the level of the sill warm water flows polewards. This flow is convectively unstable (because the water is lighter than that at the surface), and so the entire column of fluid mixes and its density takes on the value at the surface. The meridional temperature gradient gives rise to an eastward flow, and this temperature field is advected zonally, and in steady state the temperature distribution is zonally symmetric and given by (21.100). The overturning circulation within the ACC is known as the Deacon Cell, and this is a crude model of it. It is considered further in Section 21.7.

If the diapycnal diffusivity were non-zero, the sharp boundary between the two fluid masses at the sill height would be diffused to a front of finite thickness, with some upwelling and water mass transformation occurring across the front. This diffusive loss of dense fluid would be compensated by water-mass formation at the surface, polewards of the channel, leading to a deep, diffusively-driven circulation. That is, the deep water mass of b_1 fluid would circulate: this is a crude model of the 'Antarctic Bottom Water' cell.

Suppose now that the wind were everywhere zero, and the diffusivity small but non-zero. The cold, b_1 fluid would quickly completely fill the basin polewards of the channel, and would also fill the basin equatorward of the channel up to the level of the sill. However, with no wind to drive an overturning circulation dense b_1 water would slowly drift ageostrophically across the channel, displacing any warmer water until the *entire basin* were filled with the dense, b_1, fluid, except for a thin boundary layer at the top needed to satisfy the upper boundary condition. The final state would be one of no motion, and no stratification, below this boundary layer.

The important overall conclusion to be drawn is the following: *a deep meridional circulation and a deep stratification can be maintained, even as the diapycnal diffusivity goes to zero, in the presence of a wind forcing and a circumpolar channel.* Of course there are a number of idealized or unrealistic aspects to this model, perhaps the most egregious being:

- The vertical isopycnals in the channel will be highly baroclinically unstable. This will cause the isopycnals to slump and will potentially set up an eddy-induced circulation. We consider this at length later on.
- This model has no surface temperature gradient across the subtropical gyre. If one were present, it would lead to the formation of a 'main' subtropical thermocline, a full treatment of which would require determining its eastern boundary conditions. This would not qualitatively affect the presence of a deep, wind-driven overturning circulation.
- The wind stress in the model channel is chosen so that the meridional Ekman transport is constant. (This means the wind stress is chosen to vary in the same fashion as the Coriolis parameter, and if f were constant, the wind-stress curl would vanish.) Thus, there is no wind-driven downwelling or upwelling in the channel, and this simplifies the solution. Numerical simulations suggest that this choice does not affect the qualitative nature of the overturning circulation or temperature distribution.

21.6.3 A Cross-equatorial Wind-driven Deep Circulation

We qualitatively and heuristically extend the above model to consider flow across the equator. Thus, we suppose that the ocean basin extends to high northern latitudes, where there is, potentially, another source of cold deep water. To keep the model simple and tractable we will assume a very simple buoyancy structure:

$$b_s = \begin{cases} b_1, & 0 \le y \le y_1, \\ b_1 + \dfrac{(b_2 - b_1)(y - y_1)}{y_2 - y_1}, & y_1 \le y \le y_2, \\ b_2, & y_2 \le y \le y_4, \\ b_3 & y > y_4, \end{cases} \tag{21.102}$$

where the geometry is illustrated in Fig. 21.22. Given that $b_2 > b_1$, there are three cases to consider:

(*i*) $b_3 > b_2$. This is not oceanographically relevant to today's climate, nor does it provide another potential deep water source.

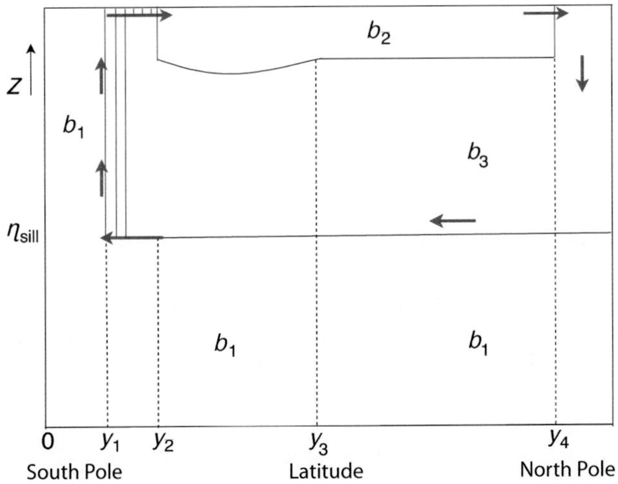

Fig. 21.22 As for Fig. 21.21, but now for a two-hemisphere ocean with a source of dense water, b_3, at high northern latitudes. The solid lines are isopycnals, and here the wind is zero in the Northern Hemisphere.

(ii) $b_3 < b_1$. The northern water is now the densest in the ocean, and would fill up the entire basin north of the channel (except near the surface in regions where some b_2 water is pushed down by the wind), and so provide no mid-depth stratification.

(iii) $b_1 < b_3 < b_2$. This is the most interesting and relevant case, and the only one we explore further.

As regards the wind, we will assume that south of the equator this is given by (21.97). North of the equator the wind forcing does not affect the qualitative nature of the overturning circulation, and may be taken to be zero.

Descriptive solution

In case *(iii)*, the entire basin below the sill fills with b_1 water, except where wind forcing forces warmer fluid below the sill level, as before. However, unlike the earlier case, the fluid above the sill is predominantly b_3 water from high northern latitudes. This forms in high polar latitudes and fills most of the basin above the sill, from the basin boundary in the north to the channel in the south (as discussed more below). However, except at latitudes where the b_3 is formed, it does not reach the surface because of the presence of b_2 water. That water is pushed down by the wind in the southern hemisphere to some as yet undetermined depth (discussed below), the boundary between b_2 and b_3 water then forming the upper ocean thermocline.

These water masses circulate because of the wind forcing in the channel. As in the single-hemisphere case, northwards flowing water emerges from the channel with buoyancy b_2. This emerges into a region of Ekman downwelling, with a northward transport carried by a western boundary current. This transport crosses the equator finally reaching the latitudes where b_3 water is formed where it sinks and returns equatorward, again in a western boundary current. (Away from the western boundary layer there is no meridional flow in the absence of diffusion, because the flow satisfies $\beta v = f\partial w/\partial z$ and there is no upwelling.) This water then crosses the sill. However, unlike the single-hemisphere case, in the northern part of the sill this water is denser than the surface water; no convection occurs and so the b_3 water extends upwards to the surface, where it warms by contact with the atmosphere and is advected equatorward to become b_2 water. Further south the surface buoyancy in the channel is less than b_3, and the column now mixes convectively, much as in the single-hemisphere case. The solution is completed by specifying the thickness of the layer of b_2 water at the surface. Now, if the circulation is in steady state, the meridional transport between the gyres must equal that of the northward Ekman flow at the northern edge of the circumpolar channel, and given the wind forcing, this is determined by the depth of the layer at the eastern boundary, a constant. Thus, in this model, global constraints determine the depth of the eastern boundary of the thermocline.

Suppose that the wind were everywhere zero. Then, as in the single-hemisphere case, the circulation would eventually die. Again, though, slow ageostrophic motion across the channel would first allow the entire basin, within and on both sides of the channel, to fill with the densest available water, and in the final steady state there would be no stratification (and no motion) below a thin surface layer.

Suppose, on the other hand, that a small amount of diffusion were added to the wind-forced model above. Then there would be mass exchange between the layers and, in particular, the deep cell of b_1 water would begin to circulate diffusively. In addition, the mid-depth cell would begin to upwell through the b_2–b_3 interface, and develop a diffusively driven circulation, in much the same way as is illustrated in Fig. 20.15.

Summary remarks

The key result of this model is that, even as the diffusivity falls and the interior of the ocean becomes more and more adiabatic, a meridional cross-hemispheric circulation can be maintained, provided that the wind across the circumpolar channel remains finite. The diabatic water mass transformations all occur at the surface or in convection: these processes require a non-zero diffusivity, but this can be the molecular diffusivity because the associated mixing involves turbulence, which can generate arbitrarily small scales. (Note also that the convection that occurs in the circumpolar channel reduces the potential energy of the column, and requires no mechanical input of energy.) Aside from the region of the ACC, the meridional transport will occur (in this model) in western boundary layers. Indeed, we may still expect to see a southwards flowing deep western boundary current south of y_4 and below the b_2 water in Fig. 21.22, just as in the implicitly diffusive Stommel–Arons model. In the ACC itself, the meridional transport occurs in a subsurface current, nestled against the sill. Although the overturning circulation in this model is 'wind-driven', the possibility that it may be cross-equatorial depends upon the thermodynamic forcing; in particular, if there is no source of dense water in the northern hemisphere, then the basin above the sill simply fills with b_2 water, as in the model of Section 21.6.2, and there need be little or no interhemispheric flow. We emphasize, too, that our model of interhemispheric flow is quite heuristic: we have essentially *posited* that b_2 water may continuously flow across the equator, possibly in a western boundary current but without examining the equatorial dynamics at all.

The ACC plays a key role in the above description but we have grossly oversimplified it. In particular, the nearly vertical isopycnals of the model will be highly baroclinically unstable, and this provides a convenient segue into our next topic.

21.7 THE ANTARCTIC CIRCUMPOLAR CURRENT

We now take a closer look at the Antarctic Circumpolar Current (ACC) itself, with a focus on its own internal dynamics; we come back to the connection with the rest of the world's oceans in Section 21.8. The ACC system, sketched in Fig. 21.23, differs from other oceanic regimes primarily in that the flow is, like that of the atmosphere, predominantly zonal and re-entrant. The two obvious influences on the circulation are the strong, eastward winds (the 'roaring forties' and the 'furious fifties') and the buoyancy forcing associated with the meridional gradient of atmospheric temperature and radiative effects that cause ocean cooling at high latitudes and warming at low ones. Providing a detailed description of the resulting flow is properly the province of numerical models, and here our goals are much more modest, namely to describe and understand some of the fundamental dynamical mechanisms that determine the structure and transport of the system, with a view to then connecting the ACC to the rest of the world's oceans.[22]

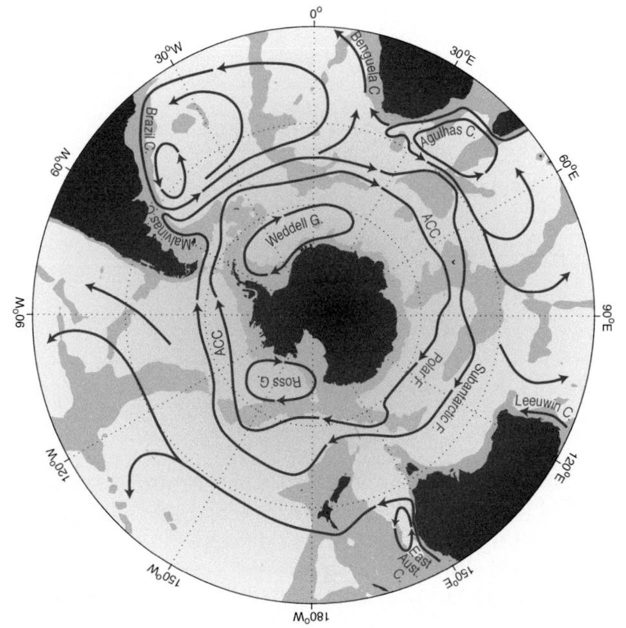

Fig. 21.23 The major currents in the Southern Ocean. Notable are the South Atlantic subtropic gyre and the two main cores of the ACC, associated with the Polar front and the sub-Antarctic front.[21]

21.7.1 Steady and Eddying Flow

Consider again the simplified geometry of the Southern Ocean as sketched in Fig. 21.19. The ocean floor is flat, except for a ridge (or 'sill') at the same longitude as the gyre walls; this is a crude representation of the topography across the Drake Passage, that part of the ACC between the tip of South America and the Antarctic Peninsula. In the planetary-geostrophic approximation, the steady response is that of nearly vertical isopycnals in the area above the sill, as illustrated in Fig. 21.21. Below the sill a meridional flow can be supported and the isotherms spread polewards, as illustrated in the left panel of numerical solutions using the primitive equations (Fig. 21.24).[23]

The stratification of the non-eddying simulation is similar to that predicted by the idealized model illustrated in Fig. 21.21. However, the steep isotherms within the channel contain a huge amount of available potential energy (APE), and the flow is highly baroclinically unstable. If baroclinic eddies are allowed to form, the solution is dramatically different: the isotherms slump, releasing that APE and generating mesoscale eddies that exercise control over much of the circulation. An important conclusion is that *baroclinic eddies are of leading-order importance in the dynamics of the ACC*. A dynamical description of the ACC without eddies would be *qualitatively* in error, in much the same way as would a similar description of the mid-latitude troposphere (i.e., the Ferrel Cell). These eddies transfer both heat and momentum, and much of the rest of our description will focus on their effects.

21.7.2 Vertically Integrated Momentum Balance

The momentum supplied by the strong eastward winds must somehow be removed. Presuming that lateral transfers of momentum are small the momentum must be removed by fluid contact with the solid Earth at the bottom of the channel. Thus, let us first consider the vertically integrated momentum balance in a channel, without regard to how the momentum might be vertically transferred. We begin with the frictional–geostrophic balance, namely

$$\boldsymbol{f} \times \boldsymbol{u} = -\nabla\phi + \frac{\partial \widetilde{\boldsymbol{\tau}}}{\partial z}, \tag{21.103}$$

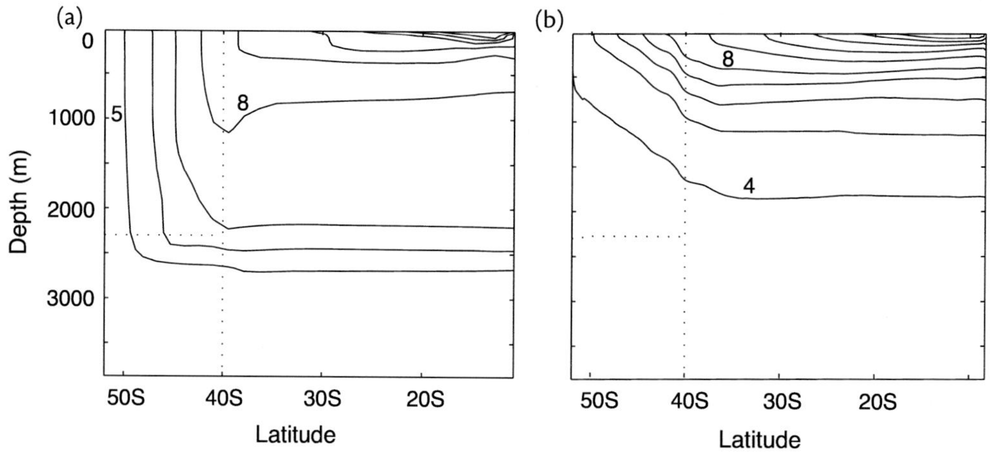

Fig. 21.24 The zonally averaged temperature field in numerical solutions of the primitive equations in a domain similar to that of Fig. 21.19 (except that here the channel and sill are nestled against the poleward boundary). Panel (a) shows the steady solution of a diffusive model with no baroclinic eddies, and (b) shows the time averaged solution in a higher-resolution model that allows baroclinic eddies to develop. Two contour values in each panel are labelled. The dotted lines show the channel boundaries and the sill.[24]

where $\widetilde{\boldsymbol{\tau}}$ is the kinematic stress, $\boldsymbol{\tau}/\rho_0$, and $\phi = p/\rho_0$. Integrating over the depth of the ocean and using Leibniz's rule (that is, $\nabla \int_{\eta_B}^{0} \phi \, \mathrm{d}z = \int_{\eta_B}^{0} \nabla\phi \, \mathrm{d}z - \phi_B \nabla \eta_B$), gives

$$\boldsymbol{f} \times \widehat{\boldsymbol{u}} = -\nabla\widehat{\phi} - \phi_b \nabla \eta_b + \widetilde{\boldsymbol{\tau}}_w - \widetilde{\boldsymbol{\tau}}_f, \tag{21.104}$$

where $\widetilde{\boldsymbol{\tau}}_w$ is the stress at the surface (due mainly to the wind) and $\widetilde{\boldsymbol{\tau}}_f$ is the frictional stress at the bottom, a hat denotes a vertical integral and ϕ_b is the pressure at $z = \eta_b$, where η_b is the z-coordinate of the bottom topography. The x-component of (21.104) is

$$-f\widehat{v} = -\frac{\partial \widehat{\phi}}{\partial x} - \phi_b \frac{\partial \eta_b}{\partial x} + \widetilde{\tau}^x_w - \widetilde{\tau}^x_f, \tag{21.105}$$

and on integrating around a line of latitude the term on the left-hand side vanishes by mass conservation and we are left with

$$-\overline{\phi_b \frac{\partial \eta_b}{\partial x}} + \overline{\widetilde{\tau}^x_w} - \overline{\widetilde{\tau}^x_f} = 0, \tag{21.106}$$

where overbars denote zonal averages. The first term is the bottom, or topographic, form drag, encountered in Sections 3.6 and 19.6, and observations and numerical simulations indicate that it is this, rather than the frictional term $\widetilde{\tau}^x_f$, that predominantly balances the wind stress.[25] We address the question of *why* this should be so in section 21.7.5.

The vorticity balance is similarly dominated by a balance between bottom form-stress curl and wind-stress curl. Taking the curl of (21.104), noting that $\nabla \cdot \widehat{\boldsymbol{u}} = 0$, gives

$$\beta\widehat{v} = -\mathbf{k} \cdot \nabla\phi_b \times \nabla\eta_b + \mathrm{curl}_z \widetilde{\boldsymbol{\tau}}_w - \mathrm{curl}_z \widetilde{\boldsymbol{\tau}}_f. \tag{21.107}$$

Now, on integrating over an area bounded by two latitude circles and applying Stokes' theorem the β-term vanishes by mass conservation and we regain (21.106). This means that *Sverdrup balance, in the usual sense of* $\beta v \approx \mathrm{curl}_z \boldsymbol{\tau}_w$, *cannot hold in the zonal average:* the left-hand side vanishes but the right-hand side does not. The same could be said for the zonal integral of (21.107) across a gyre, but the two cases do differ: In a gyre Sverdrup balance can (in principle) hold over most of the interior, with mass balance being satisfied by the presence of an intense western boundary current. In contrast, in a channel where the dynamics are zonally homogeneous then v must be, on average, zero at *all* longitudes and form drag and/or frictional terms must balance the wind-stress curl in a given water column. Sverdrup balance is thus a less useful foundation for channel dynamics — at least zonally homogeneous ones — than it is for gyres. Of course, the real ACC is *not* zonally homogeneous, and may contain regions of poleward Sverdrup flow balanced by equatorward flow in boundary currents along the eastern edges of sills and continents, and the extent to which Sverdrup flow is a leading-order descriptor of its dynamics is a matter of geography (and debate!). See also Section 21.7.6.

Even though topographic drag may be dominant in removing momentum, non-conservative frictional terms cannot be neglected, for two reasons. First, they are the means whereby kinetic energy is dissipated. Second, if there is a contour of constant orographic height encircling the domain (i.e., encircling Antarctica) then the form drag will vanish when integrated along it. However, the same integral of the wind stress will not vanish, and therefore must be balanced by something else. To see this explicitly, write the vertically integrated vorticity equation, (21.107), in the form

$$\beta\widehat{v} + J(\phi_b, \eta_b) = \mathrm{curl}_z\widetilde{\boldsymbol{\tau}}_w - \mathrm{curl}_z\widetilde{\boldsymbol{\tau}}_f. \tag{21.108}$$

If we integrate over an area bounded by a contour of constant orographic height (i.e., constant η_b) then both terms on the left-hand side vanish, and the wind stress along that line must be balanced by friction. In the real ocean there may be no such contour that is confined to the ACC— rather, any such contour would meander through the rest of the ocean; indeed, no such confined contour exists in the idealized geometry of Fig. 21.19.

21.7.3 Form Drag and Baroclinic Eddies

How does the momentum put in at the surface by the wind stress make its way to the bottom of the ocean where it may be removed by form drag? We saw in Section 21.6.2 that one mechanism is by way of a mean meridional overturning circulation, with an upper branch in the Ferrel Cell and a lower branch at the level of the sill, with no meridional flow between. However, the presence of baroclinic eddies allows an eddy form drag to pass momentum vertically within the fluid. Let's see how that works.

We model the channel as a finite number of fluid layers, each of constant density and lying one on top of the other — a 'stacked shallow water' model, and one equivalent to a model expressed in isopycnal coordinates. The wind provides a stress on the upper layer, which sets it into motion, and this in turn, via the mechanism of form drag, provides a stress to the layer below, and so on until the bottom is reached. The lowest layer then equilibrates via form drag with the bottom topography or via Ekman friction, and the general mechanism is illustrated in Fig. 21.25.

Recalling the results of Section 3.6, the zonal form drag at a layer interface is given by

$$\tau_i = -\overline{\eta_i \frac{\partial p_i}{\partial x}} = -\rho_0 f \overline{\eta_i v_i}, \tag{21.109}$$

where p_i is the pressure and η_i is the displacement at the i-th interface (i.e., between the i-th and $(i+1)$-th layer as in Fig. 21.26), and the overbar denotes a zonal average. If we define the averaged meridional transport in each layer by

$$V_i = \int_{\eta_i}^{\eta_{i-1}} \rho_0 v \,\mathrm{d}z, \tag{21.110}$$

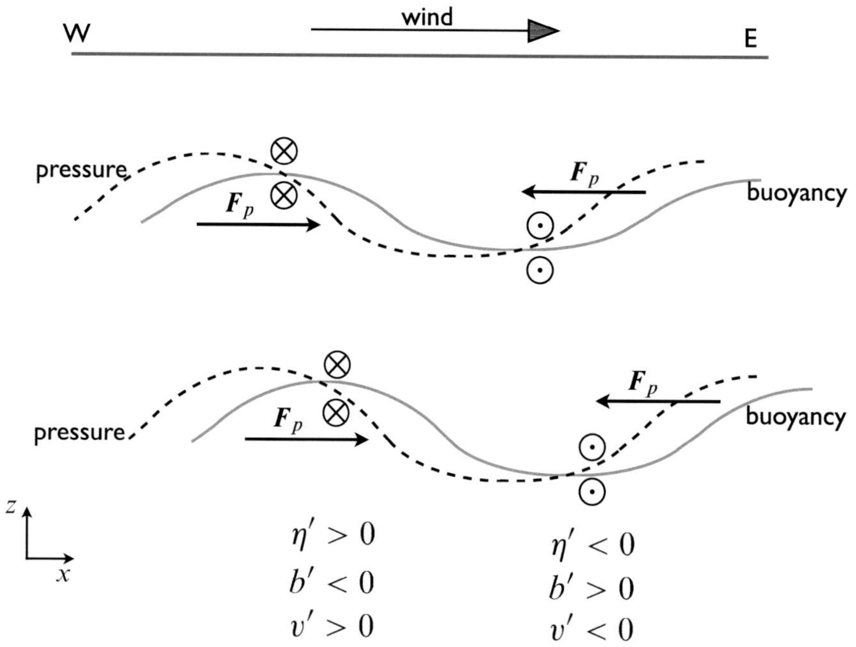

Fig. 21.25 Eddy fluxes and form drag in a Southern Hemisphere channel, viewed from the south. Cold (less buoyant) water flows equatorwards and warm water poleward, so that $\overline{v'b'} < 0$. The pressure field (dashed lines) provides a form drag on the successive layers, F_p, shown. At the ocean bottom the westward form drag on the fluid arising through its interaction with the orography is equal and opposite to that of the eastward wind stress at the top. The mass fluxes in each layer are given by $\overline{v'h'} \approx -\partial_z(\overline{v'b'}/N^2)$. If the magnitude of buoyancy displacement increases with depth then $\overline{v'h'} < 0$.

then, neglecting the meridional momentum flux divergence (for reasons given in the next subsection), the time and zonally averaged zonal momentum balance for each layer of fluid are:

$$-f\overline{V}_1 = \tau_w - \tau_1 = \overline{\eta_1 \frac{\partial p_1}{\partial x}} + \tau_w, \tag{21.111a}$$

$$-f\overline{V}_i = \tau_{i-1} - \tau_i = -\overline{\eta_{i-1} \frac{\partial p_{i-1}}{\partial x}} + \overline{\eta_i \frac{\partial p_i}{\partial x}}, \tag{21.111b}$$

$$-f\overline{V}_N = \tau_{N-1} - \tau_N = -\overline{\eta_{N-1} \frac{\partial p_{N-1}}{\partial x}} + \overline{\eta_b \frac{\partial p_b}{\partial x}} - \tau_f, \tag{21.111c}$$

where the subscripts 1, i and N refer to the top layer, an interior layer, and the bottom layer, respectively. Also, η_b is the height of the bottom topography and τ_w is the zonal stress imparted by the wind which, we assume, is confined to the uppermost layer. The term τ_f represents drag at the bottom due to Ekman friction, but we have neglected any other viscous terms or friction between the layers.

The vertically integrated meridional mass transport must vanish, and thus summing over all the layers (21.111) becomes

$$0 = \tau_w - \tau_f - \tau_N, \tag{21.112}$$

or, noting that $\tau_N = -\overline{\eta_b \partial p_b / \partial x}$,

$$\tau_w = \tau_f - \overline{\eta_b \frac{\partial p_b}{\partial x}}. \tag{21.113}$$

Thus, the stress imparted by the wind (τ_w) may be communicated vertically through the fluid by form drag, and ultimately balanced by the sum of the bottom form stress (τ_N) and the bottom friction (τ_f).

Momentum dynamics in height coordinates

We now look at these same dynamics in height coordinates, using the quasi-geostrophic TEM formalism, and it may be helpful to review Section 10.3 before proceeding. As in (10.61), we write the zonally averaged momentum equation in the form

$$-f_0\overline{v}^* = \nabla_m \cdot \boldsymbol{\mathcal{F}} + \frac{\partial \overline{\tau}}{\partial z}, \tag{21.114}$$

where $\overline{v}^* = \overline{v} - \partial_z(\overline{v'b'}/\overline{b}_z)$ is the residual meridional velocity, $\overline{\tau}$ is the zonal component of the kinematic stress (wind-induced and frictional, and typically important only in an Ekman layer at the surface and in a frictional layer at the bottom) and $\boldsymbol{\mathcal{F}}$ is the Eliassen–Palm flux, which satisfies

$$\nabla_m \cdot \boldsymbol{\mathcal{F}} = -\frac{\partial}{\partial y}\overline{u'v'} + \frac{\partial}{\partial z}\left(\frac{f_0}{N^2}\overline{v'b'}\right) = \overline{v'q'}. \tag{21.115}$$

Now, if the horizontal velocity and buoyancy perturbations are related by $v' \sim b'/N$ (meaning available potential energy and kinetic energy are roughly similar, see also Section 12.4), then the two terms comprising the potential vorticity flux scale as

$$\frac{\partial}{\partial y}\overline{u'v'} \sim \frac{v'^2}{L_e}, \qquad \frac{\partial}{\partial z}\left(f_0\frac{\overline{v'b'}}{\overline{b}_z}\right) \sim \frac{v'^2}{L_d}, \tag{21.116}$$

where L_e is the scale of the eddies and L_d is the deformation radius. If the former is much larger than the latter, as we might expect in a field of developed geostrophic turbulence (and as is observed in the ACC), then the potential vorticity flux is dominated by the buoyancy flux and (21.114) becomes

$$-f_0\overline{v}^* \approx \frac{\partial \overline{\tau}}{\partial z} + \frac{\partial}{\partial z}\left(f_0\frac{\overline{v'b'}}{\overline{b}_z}\right). \tag{21.117}$$

In the ocean interior the frictional terms, $\partial\overline{\tau}/\partial z$, are small, and (21.117) represents a balance between the Coriolis force on the residual flow and the form stress associated with the vertical component of the EP flux (an association further explained in Section 10.4.3).

If we integrate (21.117) over the depth of the channel the term on the left-hand side vanishes and we have

$$\widetilde{\tau}_w = \widetilde{\tau}_f - \left[f_0\frac{\overline{v'b'}}{\overline{b}_z}\right]_{-H}^{0}, \tag{21.118}$$

where $\widetilde{\tau}_w$ is the wind stress and $\widetilde{\tau}_f$ is the frictional stress at the bottom (both divided by ρ_0). Equation (21.118) expresses essentially the same momentum balance as (21.113). Thus, the EP flux expresses the passage of momentum vertically through the water column, the momentum being removed at the bottom through frictional stresses and/or form drag with the orography.

Mass fluxes and thermodynamics

Associated with the form drag is a meridional mass flux in each layer, which in the layered model appears as V_i (a thickness flux) in each layer. The satisfaction of the momentum balance at a particular latitude goes hand-in-hand with the satisfaction of the mass balance. Above any topography the Eulerian mean momentum equation is, with quasi-geostrophic scaling and neglecting eddy momentum fluxes,

$$f_0\overline{v} = \frac{\partial \overline{\tau}}{\partial z}, \tag{21.119}$$

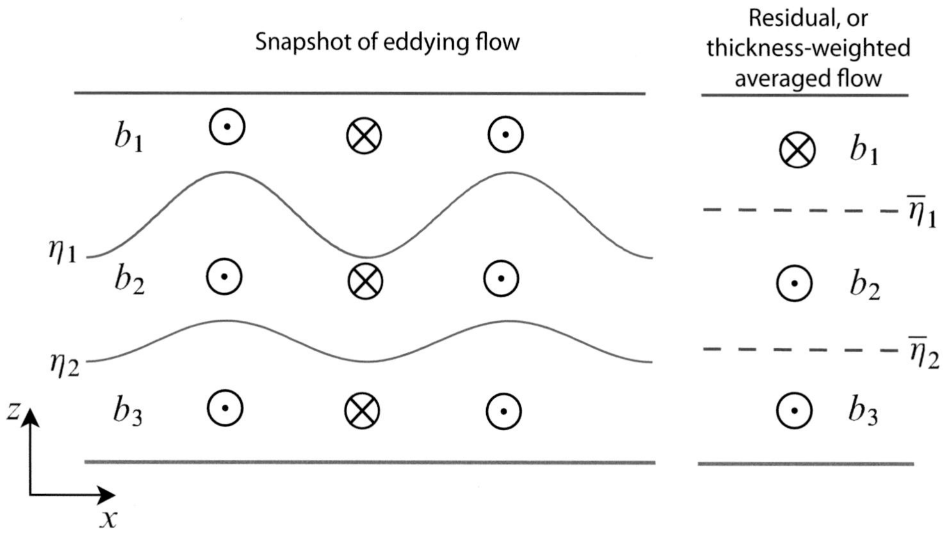

Fig. 21.26 An example of the meridional flow in an eddying channel. The eddying flow may be organized such that, even though at any given level the Eulerian meridional flow may be small, there is a net flow in a given isopycnal layer. The residual ($\overline{v}^*$) and Eulerian ($\overline{v}$) flows are related by $\overline{v}^* = \overline{v} + \overline{v'h'}/\overline{h}$; thus, the thickness-weighted average of the eddying flow on the left gives rise to the residual flow on the right, where $\overline{\eta}_i$ denotes the mean elevation of the isopycnal interface η_i.

where $\overline{v}$ is the zonally averaged meridional velocity and $\widetilde{\tau}$ is the zonal component of the kinematic stress. The zonally averaged meridional flow is thus purely ageostrophic and since the stress, $\widetilde{\tau}$, is fairly constant in the interior, the mean meridional flow is non-zero only near the surface (i.e., equatorward Ekman flow) and near the ocean bottom, where the flow can be supported by friction and/or form drag. Even in an eddying flow, the Eulerian circulation is primarily confined to the upper Ekman layer and a frictional or topographically interrupted layer at the bottom, as sketched in Fig. 21.27. This is a perfectly acceptable description of the flow, and is not an artifact in any way.

However, and analogously to the atmospheric Ferrel Cell (Sections 14.7 and 15.2.2), if the flow is unsteady this circulation does not necessarily represent the flow of water parcels, nor does it imply that water parcels cross isopycnals, as might be suggested by the dark blue circulation (ψ_{Euler}) in Fig. 21.27. The flow of parcels is better represented by the *residual,* or *thickness-weighted,* flow, and as sketched in Fig. 21.26 and Fig. 21.27 there can be a net meridional residual flow in a given layer (i.e., of a given water mass type) *even when the net meridional Eulerian flow at the level of mean height of the layer is zero.*

The vertically integrated residual mass flux must vanish, and even though one component of this — the equatorward Ekman flow — is determined mechanically, the overall sense of the residual circulation is not determined by the momentum balance alone: thermodynamic effects play a role. The zonally averaged thermodynamic equation may be written in TEM form as

$$\frac{\partial \overline{b}}{\partial t} + J(\psi^*, \overline{b}) = Q_{[b]}, \tag{21.120}$$

where $J(\psi^*, \overline{b}) = (\partial_y \psi^*)(\partial_z b) - (\partial_z \psi^*)(\partial_y b) = \overline{v}^* \partial_y b + \overline{w}^* \partial_z b$, ψ^* is the streamfunction of the residual flow and $Q_{[b]}$ represents heating and cooling, which occur mainly at the surface. In the ocean interior and in a statistically steady state we therefore have

$$J(\psi^*, \overline{b}) = 0, \tag{21.121}$$

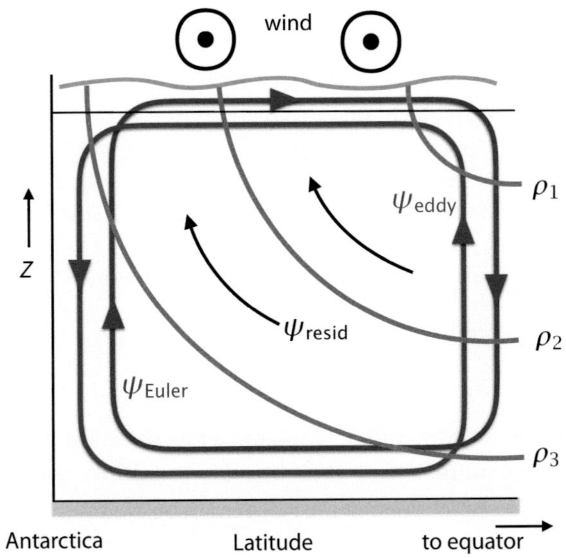

Fig. 21.27 Idealization of the Eulerian, eddy-induced ('bolus'), and residual streamfunctions in a circumpolar channel. The clockwise Eulerian circulation is forced by the eastward winds, the bolus circulation opposes it, and the net, or residual, circulation is nearly along isopycnals.[26]

the general solution of which is $\psi^* = G(\overline{b})$, where G is an arbitrary function. That is, the interior residual flow is along isopycnals (Fig. 21.27). At the surface, however, the flow is generally not adiabatic, because of heat exchange with the atmosphere, and so the residual flow can be across isopycnals. The sense of the subsurface circulation determines how the form drag varies with depth; if the residual flow were zero, for example, then, either from (21.111) or from (21.117), we see that the form drag must be constant with depth.

21.7.4 † An Idealized Adiabatic Model

We finally consider a simple but rather illuminating model of the ACC.[27] The simplifying assumption we make is that the flow is adiabatic everywhere; it then follows that the net overturning, as given by the residual circulation, is zero. We can see this by first noting that in a statistically steady state the flow satisfies (21.121), implying that the residual flow is along isopycnals. However, if there is a meridional buoyancy gradient at the surface (where isopycnals outcrop) there can be no surface residual flow (because this would be cross-isopycnal); it then follows that there can be no net flow along isopycnals in the interior, because if these outcrop there would be a net fluid source, and hence diapycnal flow, at the surface. This idealized limit has thus led to the 'vanishing of the Deacon Cell'. In reality the flow is not adiabatic near the surface and the residual flow will not vanish, but it is likely to be weaker than either the Eulerian or the eddy-induced flow (as sketched in Fig. 21.27).

The zonal momentum equation in this limit follows from (21.117), which with $\overline{v}^* = 0$ gives

$$\frac{\partial \tilde{\tau}}{\partial z} \approx -\frac{\partial}{\partial z}\left(f_0 \frac{\overline{v'b'}}{\overline{b}_z}\right). \tag{21.122}$$

The equivalent balance for the Eulerian flow is, using the definition of $\overline{v}^*$,

$$-f_0\overline{v} = \frac{\partial \tilde{\tau}}{\partial z} \qquad \rightarrow \qquad f_0\overline{v} = f_0\frac{\partial}{\partial z}\left(\frac{\overline{v'b'}}{\overline{b}_z}\right). \tag{21.123a,b}$$

These equations represent *dynamical* balances; they do not follow from the momentum equation without making additional assumptions, in this case that $\overline{v}^* = 0$. In the residual equation, (21.122), the wind stress is balanced by the divergence of the Eliassen–Palm flux, which is dominated by the contribution from the buoyancy flux, and in the ocean interior where the stress is small the meridional buoyancy flux will be constant with height. Equation (21.123b) represents a balance between the Coriolis force on the equatorward flow and form drag. If the frictional stress is small in the interior then the form drag does not vary in the vertical and the right-hand side of (21.123) is small. The zonally averaged meridional flow in the interior is then also small, and the equatorial flow in the top Ekman layer is balanced by a return flow at the bottom of the ocean involving topographic form stress or a bottom Ekman layer.

Integrating (21.122) from the surface (where $\widetilde{\tau} = \widetilde{\tau}_w$) to a stress-free level in the interior (where $\widetilde{\tau} = 0$) gives

$$\widetilde{\tau}_w = f_0 \frac{\overline{v'b'}}{\overline{b}_z}, \tag{21.124}$$

if the buoyancy flux at the surface is small. If we are now willing to parameterize the eddy fluxes in terms of the mean flow, then we can predict the stratification. Thus, let $\overline{v'b'} = -\kappa\, \partial\overline{b}/\partial y$, where κ is an eddy diffusivity, and noting that $s = -\overline{b}_y/\overline{b}_z$ is the slope of the isopycnals, we find

$$\widetilde{\tau}_w = \kappa f_0 s = \kappa \frac{f_0^2}{\overline{b}_z} \frac{\partial \overline{u}}{\partial z}, \tag{21.125}$$

where the second equality uses thermal wind balance. Thus, given κ, we can predict the isopycnal slope $[s = \widetilde{\tau}_w/(\kappa f_0)]$ and, potentially, the total baroclinic transport of the ACC as a function of the wind stress. The sense of the residual circulation can be inferred if the diabatic fluxes at the surface are known, but at the same time these fluxes depend in a complicated way on both the lateral eddy fluxes and the general circulation itself. We come back to this in Sections 21.8 and 21.9.

21.7.5 Form Stress and Ekman Stress at the Ocean Bottom

Earlier, we noted that the stress at the ocean bottom is observed to be dominated by form stress, rather than Ekman friction, in the ACC. A simple scaling argument helps understand why this should be. The form stress scales like

$$\tau_{form} \sim \eta_b \frac{\partial p_b}{\partial x} \sim \eta_b \rho_0 U f, \tag{21.126}$$

where we have used geostrophic balance and U is a scaling for the horizontal velocity. The frictional stress due to an Ekman layer (Section 5.7) scales like

$$\tau_{Ekman} \sim \rho_0 A \frac{\partial u}{\partial z} \sim \frac{\rho_0 A U}{d} \sim \rho_0 d U f, \tag{21.127}$$

where A is the eddy kinematic viscosity and $d = \sqrt{A/f}$ is the Ekman layer thickness. The ratio of these two stresses thus scales as

$$\frac{\tau_{form}}{\tau_{Ekman}} \sim \frac{\eta_b}{d}. \tag{21.128}$$

We therefore expect the form stress to dominate the Ekman stress if the variations in topography are greater than the Ekman layer thickness, and if the flow goes *over* the topography rather than around it. In the ACC the topography is hundreds or even thousands of metres high whereas the bottom Ekman layer may be of order tens of metres, and furthermore the predominantly eastward flow must (unlike the situation in gyre circulations) go *over* the topography. Thus, form stress dominates the frictional, Ekman layer, stress at the bottom of the ACC.

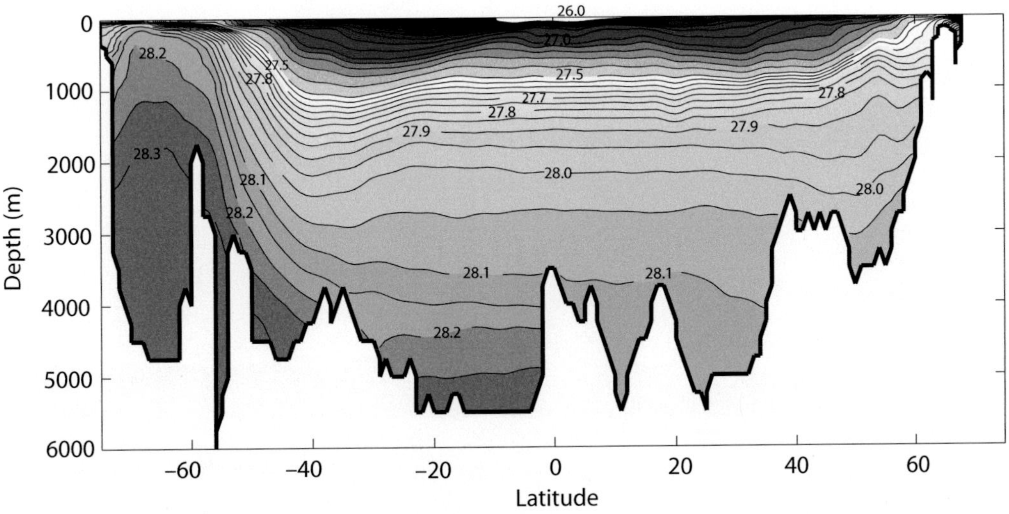

Fig. 21.28 Neutral density in the Atlantic at 25° W from WOCE. A weakly stratified water mass at mid-depth, roughly between the 28 and 28.1 isopycnals, is associated with the inflow of NADW in the Atlantic Ocean. The contour intervals are 0.1 and 0.05 kg m^{-3} for isopycnals greater and lower than 27.5 kg m^{-3}, coloured green and red, respectively.

21.7.6 Differences Between Gyres and Channels

In the dynamics of the ACC, the wind stress itself seems to play an important role, whereas in our discussion of gyres in chapter 19 the wind stress *curl* was dominant. What is the root of this difference?[28] Suppose that we change the wind stress, but not its curl, in a closed basin. The vertically integrated gyral flow, as given for example by the Stommel solution (19.39) or its two-gyre counterpart, does not change at all. However, the vertical structure of this flow will in general change; for example, if the wind is made uniformly more eastward, there will be a corresponding increase in the equatorward flux in the Ekman layer that must return polewards at depth (assuming that the western boundary current balances only the Sverdrup flow). At the same time, the added force from the wind must be balanced by an increased pressure difference between the western and eastern boundaries. This may be achieved if the sea-surface tilts upwards to the east, so producing a net (vertically integrated) poleward geostrophic flow. The subsurface isopycnal slopes may then adjust in order to reduce this flow to near zero in the abyss. The added force provided by the basin walls on the fluid in the basin is a kind of form drag (rather like the force provided by the sill in Section 21.6), and integrated around the basin this force must be equal and opposite to the force supplied by the wind. In contrast, in a channel adding a constant wind produces a direct change in its zonal transport. This is because the wind stress is balanced by form drag and bottom friction, and both of these depend on the zonal flow at the channel bottom.

21.8 † A DYNAMICAL MODEL OF THE RESIDUAL OVERTURNING CIRCULATION

In the last section it became clear that the ACC is a region of strongly eddying activity, and one effect of these eddies is to reduce the slope of the isopycnals, so reducing the available potential energy of the flow. Thus, the sketches of Fig. 21.21 and Fig. 21.22 do not properly represent the state of the channel region: not only do the isopycnals slope, but the southward flowing water parcels can enter the channel region above any sill. That is, the zonally averaged *residual* meridional flow can be non-zero, and the deep stratification can be non-zero, even without topography. In this section we seek to build a model that combines our view of the ACC, as described in Section 21.7, with

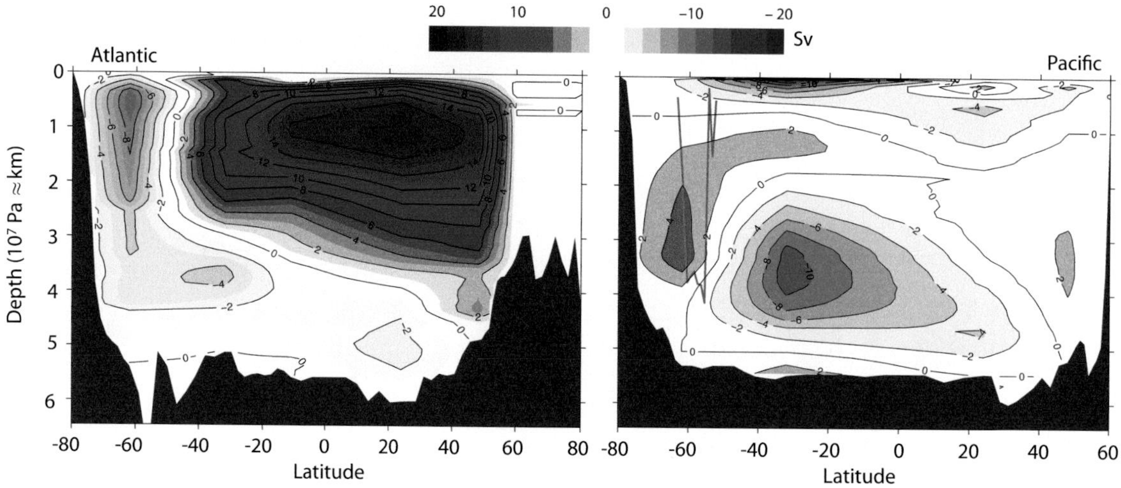

Fig. 21.29 Overturning circulations in the Atlantic and Pacific Oceans as determined by an inverse calculation. South of about 35° the circulation is not a true streamfunction, because of the open boundaries, and this may lead to errors, especially in the Pacific. (See Fig. 21.1 for another Atlantic estimate.)

the view of the partially wind-driven overturning circulation described in Section 21.6. In spite of these additions the model is still incomplete, for it treats only one basin, and it has ad hoc aspects in its treatment of eddy effects.

Our model is partially motivated by the plot of the stratification shown in Fig. 21.28 and the overturning circulations of Fig. 21.29, as abstracted in Fig. 21.30. Although nominally a sketch of the global MOC, of the individual basins it most resembles the circulation of the Atlantic where there are two main circulating masses of water, North Atlantic Deep Water (NADW) and Antarctic Bottom Water (AABW), as seen in Fig. 21.29. The NADW outcrops in high northern latitudes and high southern latitudes, and AABW just at high Southern latitudes. The Pacific overturning circulation (Fig. 21.29) is rather different, for here there is really no mid-depth cell corresponding to NADW; there is essentially *only* a bottom cell of Antarctic bottom water spreading northward. Finally, we note that isopycnals are flat over most of the ocean, but have a fairly uniform slope in the Southern Ocean. (Figure 21.28 shows the Atlantic; the situation is similar for the Pacific.) We will now construct a dynamical model that attempts to describe these features.[30]

Let us first imagine there is a wall at the equator, and make a model of the circulation in the Southern Hemisphere, that is, essentially of AABW. There's an obvious connection to the Indian Ocean and, if a little less obvious at the moment, to the Pacific.

21.8.1 Model Phenomenology

We divide the basin into two regions, a *Southern Channel* and a *basin*, as in Fig. 21.31 (see also the shaded box on page 850). In the channel the isopycnals slope, and we anticipate a balance between wind effects and baroclinic instability: in the absence of eddies the isopycnals are vertical, as in Fig. 21.21, and baroclinic activity causes the isopycnals to slump. In the basin region we invoke an ansatz that the isopycnals are flat — the model applies below the thermocline where wind effects cause stratification. Wind over the channel induces a northwards Ekman flux, and the return flow occurs at the bottom of the channel, as in the thick arrows in Fig. 21.31, because in the interior the flow is nearly geostrophic and the zonally-averaged geostrophic meridional flow is zero. However, it is the residual circulation that carries water properties and that will connect to the basin flow, and thermodynamic considerations suggest that the flow will circulate along the dashed lines in

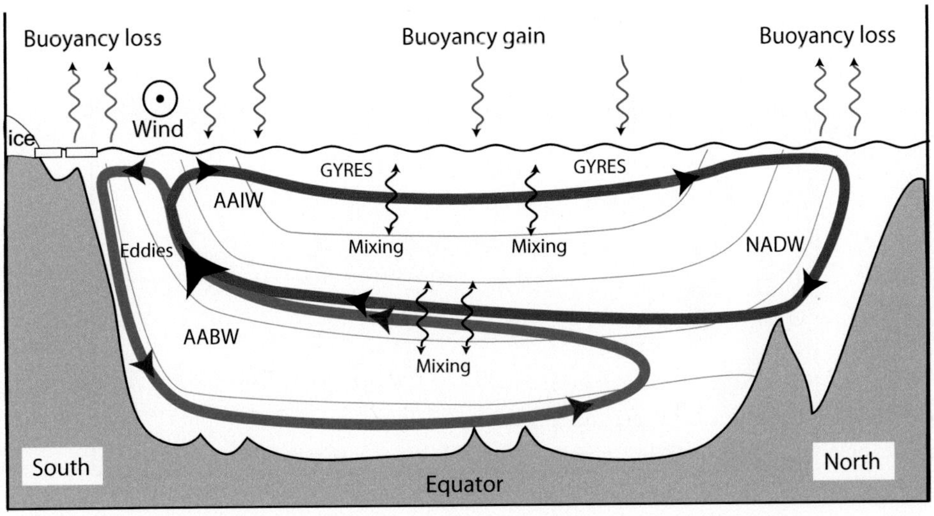

Fig. 21.30 The overturning circulation of the ocean and the main processes that produce it — winds, mixing, baroclinic eddies and surface buoyancy fluxes.[29] The sketch is most representative of the Atlantic, which is the major contributor to the global average. Observational views are given in Fig. 21.18 and Fig. 21.29.

Fig. 21.31. In the basin there will be an advective-diffusive balance in the vertical, and the flow will be non-zero only if the diffusivity is non- zero. This flow should connect smoothly to the more adiabatic flow in the channel. Let us see how the equations allow this to be accomplished, and if we can obtain estimates for the strength and structure of the flow.

21.8.2 Equations of Motion

We will use zonally-averaged equations of motion and write them in residual, or TEM, form because the treatment of mesoscale eddies is more convenient and the equations directly predict the velocities that advect the tracers. Thus, following the methodology of Section 10.3, we define a *residual flow* such that

$$\overline{v}^* = \overline{v} - \frac{\partial}{\partial z}\left(\frac{1}{N^2}\overline{v'b'}\right), \qquad \overline{w}^* = \overline{w} + \frac{\partial}{\partial y}\left(\frac{1}{N^2}\overline{v'b'}\right), \tag{21.129}$$

where $N^2 = \partial\overline{b}/\partial z$, which is assumed to vary only very slowly. The residual velocities $\overline{v}^*$ and $\overline{w}^*$ more nearly represent the trajectories of fluid parcels than the Eulerian velocities, $\overline{v}$ and $\overline{w}$. There are no fluxes in the buoyancy equation and only the potential vorticity flux, $\overline{v'q'}$, need be parameterized.

We will further suppose that the large scale flow satisfies planetary-geostrophic scaling, and so we drop the time derivative in the momentum equation and assume the zonal flow is in geostrophic wind balance. Including forcing and dissipation terms, our equations of motion become

$$-f\overline{v}^* = \overline{v'q'} + \frac{\partial \tau}{\partial z}, \qquad \frac{\partial \overline{b}}{\partial t} + \overline{v}^*\frac{\partial \overline{b}}{\partial y} + \overline{w}^*\frac{\partial \overline{b}}{\partial z} = \kappa_v \frac{\partial^2 \overline{b}}{\partial z^2}. \tag{21.130a,b}$$

The velocities are non-divergent and may be represented by a streamfunction so that $(\overline{v}^*, \overline{w}^*) = (-\partial\psi/\partial z, \partial\psi/\partial y)$, and we will assume that the residual velocities themselves satisfy the boundary

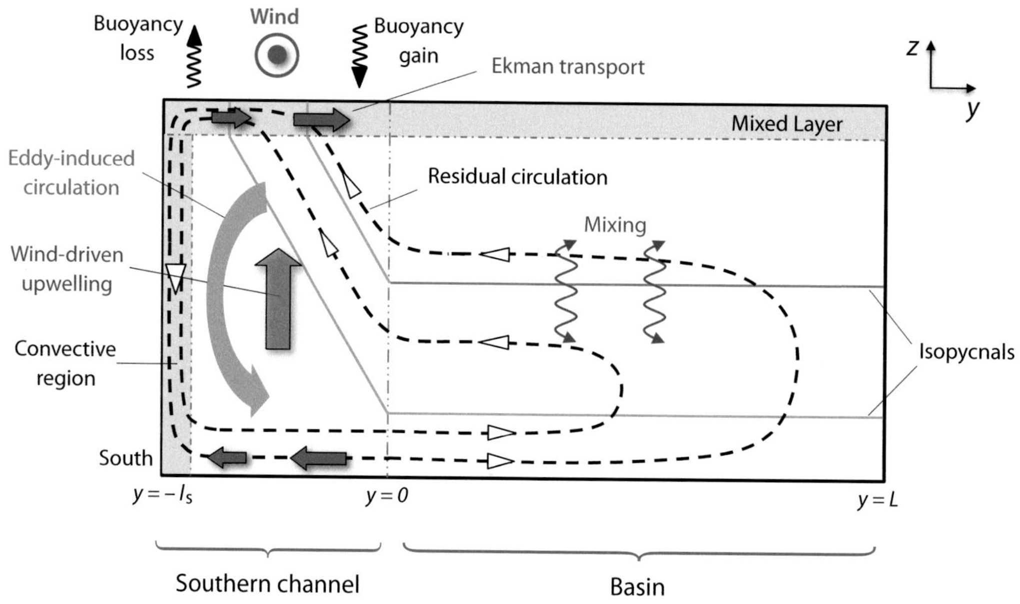

Fig. 21.31 A model of the single hemisphere meridional overturning circulation, crudely representing an idealized Antarctic Bottom Water (AABW) cell. In the Atlantic this cell sits below the interhemispheric North Atlantic Deep Water Cell, but it sits alone in the Pacific and India Oceans. Thin solid lines are the isopycnals, the dashed black line is a residual overturning streamfunction. The thick dark blue arrows are the Eulerian circulation, namely the top and bottom Ekman transport and the wind-driven upwelling.

conditions of no normal flow. The zonal wind, $\overline{u}$, may be obtained from thermal wind balance, $f\partial\overline{u}/\partial z = -\partial\overline{b}/\partial y$, and the stress, τ, is only non-zero near the top (wind-stress) and bottom (Ekman drag). We will henceforth drop the * notation, and all variables are understood to be residuals and zonal averages. These equations apply in both the channel and basin regions, but with different dominant balances.

Equations in the channel

The right-hand side of (21.130a) contains the eddy flux of potential vorticity which we parameterize using an eddy diffusivity,

$$\overline{v'q'} = -K_e\frac{\partial\overline{q}}{\partial y}, \tag{21.131}$$

where K_e is the eddy diffusivity. (It is more-or-less a 'Gent–McWilliams' coefficient, as in Section 13.6.) The Coriolis parameter is almost constant in the channel, and we denote it f_S. For the large-scale ocean the potential vorticity is given by

$$\overline{q} \approx f_S\frac{\partial}{\partial z}\left(\frac{\overline{b}}{\overline{b}_z}\right), \quad \text{so that} \quad \frac{\partial\overline{q}}{\partial y} \approx f_S\frac{\partial}{\partial z}\left(\frac{\overline{b}_y}{\overline{b}_z}\right) = -f_S\frac{\partial S}{\partial z}, \tag{21.132}$$

where $S = -\overline{b}_y/\overline{b}_z$ is the slope of the isopycnals (and the similarity with the Gent–McWilliams scheme is now clear). The potential vorticity flux is then given by

$$\overline{v'q'} \approx f_S K_e\frac{\partial S}{\partial z}, \tag{21.133}$$

and the momentum equation becomes

$$-f_S\overline{v} = f_S K_e \frac{\partial S_b}{\partial z} + \frac{\partial \tau}{\partial z}. \tag{21.134}$$

Since $\overline{v}^* = -\partial\psi/\partial z$ we integrate this from the top to a level z and obtain

$$\psi = -\frac{\tau_w}{f_S} + K_e S, \tag{21.135}$$

where both f_S and S are negative and τ_S is the surface kinematic stress in the channel. We have assumed $\psi = 0$, $S = 0$ at the surface (the base of the mixed layer) and $\tau = 0$ in the interior.

The buoyancy equation in terms of streamfunction is

$$\overline{\boldsymbol{v}} \cdot \nabla\overline{b} = \kappa_v \frac{\partial^2 \overline{b}}{\partial z^2} \qquad \text{or} \qquad \frac{\partial\psi}{\partial y}\frac{\partial\overline{b}}{\partial z} - \frac{\partial\psi}{\partial z}\frac{\partial\overline{b}}{\partial y} = \kappa_v \frac{\partial^2 \overline{b}}{\partial z^2}, \tag{21.136}$$

which can be written as

$$\frac{\partial\psi}{\partial y} + S\frac{\partial\psi}{\partial z} = \kappa_v \frac{\partial_z^2 \overline{b}}{\partial_z \overline{b}}. \tag{21.137}$$

The boundary condition on ψ for this equation will be supplied by the basin! The other boundary condition we will need is the buoyancy distribution at the top, and so we specify

$$\overline{b}(y, z = 0) = b_0(y). \tag{21.138}$$

Equations in the basin

In the basin the slope of the isopycnals is assumed zero and (21.137) becomes the conventional upwelling diffusive balance,

$$w\frac{\partial\overline{b}}{\partial z} = \kappa\frac{\partial^2\overline{b}}{\partial z^2} \qquad \text{or} \qquad \frac{\partial\psi}{\partial y}\frac{\partial\overline{b}}{\partial z} = \kappa_v \frac{\partial^2\overline{b}}{\partial z^2}. \tag{21.139}$$

If we integrate this from the edge of the channel, $y = 0$, to the northern edge, $y = L$, we obtain

$$\psi|_{y=0} = -\kappa_v L \frac{\overline{b}_{zz}}{\overline{b}_z}. \tag{21.140}$$

This equation then becomes the needed boundary condition for the equations in the channel.

21.8.3 Scaling

The above equations do not give up analytic solutions, but we can use them to obtain estimates of the flow strength and structure. Let us scale the equations by letting

$$z = h\widehat{z}, \qquad y = l\widehat{y}, \qquad \tau_S = \tau_0\widehat{\tau}_S, \qquad f_S = \overline{f}_S\widehat{f}, \qquad \psi = \frac{\tau_0}{\overline{f}_S}\widehat{\psi}, \qquad S = \frac{h}{l}\widehat{S}, \tag{21.141}$$

where $\overline{f}_S = |f_S|$, a hat denotes a nondimensional value and h is a characteristic vertical scale such that $S \sim h/l$, and *this will emerge as part of the solution.*

A Model of the Meridional Overturning Circulation

The essential features of the model of the MOC of Sections 21.8 and 21.9 are:

Formulation

- The model is zonally averaged, in a single basin, with simple geometry: a zonally re-entrant channel at high latitudes, with an enclosed basin between it and the northern boundary. The effects of wind-driven gyres are neglected; the dynamics in the enclosed basin can be regarded as being below the main thermocline.
- The equations solved are the planetary geostrophic equations, in a transformed Eulerian mean form, with the effects of mesoscale eddies being parameterized with a simple down-gradient buoyancy flux scheme in the momentum equation.
- The ocean is divided into three regions: a northern convective region, a cross-equatorial basin region, and a southern channel.
- The dynamics are treated separately in these three regions, and the solutions matched at the boundaries:
  - *(i)* In the southern channel there is a balance between wind stress (causing isopycnals to steepen) and the mesoscale eddies, which have a flattening effect. Buoyancy satisfies the full nonlinear advective-diffusive equation.
  - *(ii)* In the basin region the isopycnals are assumed flat, and by integrating meridionally over the basin this region essentially becomes a boundary condition for the southern channel. The buoyancy equation reduces to a vertical advective-diffusive balance ($w\,\partial b/\partial z = \kappa\,\partial^2 b/\partial z^2$).
  - *(iii)* In the northern convective region the isopycnals are vertical, descending sufficiently far down to connect to the corresponding horizontal isopycnals of the basin. If the northern region is too warm to allow this, the basin isopycnals will extend all the way to the northern boundary.

Properties and Predictions

- If the surface boundary conditions on buoyancy permit, there is an isopycnal pathway from the northern convective region to the southern channel. For small values of diffusivity flow can then circulate, largely adiabatically, from high northern latitudes to high southern latitudes, mechanically pumped by the wind over the southern channel. This may roughly correspond to flow in the Atlantic.
- If the boundary conditions are such that the surface of the northern region is too buoyant, then there is no interhemispheric wind-driven mid-depth circulation and no northern convection, roughly corresponding to flow in the Pacific and Indian Ocean.
- For large values of diffusivity, the flow sinks at high latitudes and upwells in low latitudes, as in a conventional buoyancy/mixing-driven circulation.
- Beneath the wind-driven mid-depth cell, a diffusive cell corresponding to Antarctic Bottom Water forms. Its strength is determined by the diapycnal diffusivity and surface meridional buoyancy gradients.
- The southern channel there is convection at the southern end and elsewhere the isopycnal slope is determined by a balance between wind forcing and eddy effects.
- The effects of mesoscale eddies are parameterized by an eddy diffusivity, but the overall model framework is not fundamentally dependent on that.
- The model cannot account for inter-basin pathways of water, for example between the Atlantic and Pacific Oceans.

If we have scaled properly then variables with hats on are of order one. The nondimensional equations of motion are then

$$\text{Buoyancy evolution:} \qquad \partial_{\hat{y}}\hat{\Psi} + \hat{S}\,\partial_{\hat{z}}\hat{\Psi} = \epsilon\left(\frac{l}{L}\right)\frac{\partial_{\hat{z}\hat{z}}\hat{b}}{\partial_{\hat{z}}\hat{b}}, \tag{21.142a}$$

$$\text{Momentum balance:} \qquad \hat{\Psi} = -\frac{\hat{\tau}_S}{\hat{f}} + \Lambda\hat{S}, \tag{21.142b}$$

$$\text{Boundary condition:} \qquad \hat{\Psi}|_{\hat{y}=0} = -\epsilon\frac{\partial_{\hat{z}\hat{z}}\hat{b}}{\partial_{\hat{z}}\hat{b}}, \tag{21.142c}$$

where

$$\Lambda = \frac{\text{Eddies}}{\text{Wind}} = \frac{K_e}{\tau_0/\overline{f}_S}\frac{h}{l}, \qquad \epsilon = \frac{\text{Mixing}}{\text{Wind}} = \frac{\kappa_v}{\tau_0/\overline{f}_S}\frac{L}{h}. \tag{21.143a,b}$$

These are two important nondimensional numbers, and we can obtain estimates of their values by using some observed values for the other parameters. Let us take

$$\begin{gathered} h = 1\,\text{km},\ \kappa_v = 10^{-5}\,\text{m}^2\,\text{s}^{-1},\ K_e = 10^3\,\text{m}^2\,\text{s}^{-1},\ \rho_0 = 10^3\,\text{kg}\,\text{m}^{-3}, \\ \tau_0 = 0.1\,\text{N}\,\text{m}^{-2}/\rho_0 = 10^{-4}\,\text{N}\,\text{m}\,\text{kg}^{-1},\ \overline{f}_S = 10^{-4}\,\text{s}^{-1},\ L = 10\,000\,\text{km},\ l_s = 1000\,\text{km}, \end{gathered} \tag{21.144}$$

and we find

$$\Lambda \approx 1, \qquad \epsilon \approx 0.1. \tag{21.145}$$

These values come with large error bars: the diffusivity, κ_v, may be much larger in the abyss, and the eddy coefficient K_e is very poorly constrained (indeed, it is a property of the flow itself, not the fluid). Finally, note that Λ and ϵ are not independent of each other for they both depend on the vertical scale of stratification, h, which is a part of the solution. To obtain some theoretical estimates of h we look at some limiting cases.

The small diffusion limit

Suppose that mixing is small and that $\epsilon \ll 1$. We can then *require* that $\Lambda = 1$ in order that the eddy-induced circulation nearly balance the wind-driven circulation (because the diffusive term is small), whence the vertical scale h is given by

$$\frac{h}{l} = \frac{\tau_0/f_S}{K_e}. \tag{21.146}$$

As K_e diminishes h becomes larger, meaning that the isopycnals are near vertical. Using (21.146) in (21.143b) gives

$$\epsilon = \frac{\kappa_v K_e}{(\tau_0/f_S)^2}\frac{L}{l}. \tag{21.147}$$

This is an appropriate nondimensional measure of the strength of the diapycnal diffusion in the ocean. Using (21.142c) we see that $\hat{\Psi} \sim \epsilon$ so that the dimensional strength of the circulation goes as

$$\Psi = \epsilon\frac{\tau_0}{f_0} = \kappa_v\frac{K_e}{\tau_0/f_S}\frac{L}{l}. \tag{21.148}$$

Another way to obtain this is to use the fact that for weak diffusion the balance in the dimensional momentum equation is between wind forcing and eddy effects (because they must nearly cancel) so that

$$\frac{\tau_w}{f} \sim K_e S, \qquad \text{or equivalently} \qquad \frac{h}{l} \sim \frac{\tau_w}{K_e f_S}. \tag{21.149a,b}$$

Advective-diffusive balance in the basin gives

$$\frac{\partial \psi}{\partial y}\frac{\partial \overline{b}}{\partial z} = \kappa_v \frac{\partial^2 \overline{b}}{\partial z^2} \qquad \text{whence} \qquad \Psi = \frac{\kappa_v L}{h} \tag{21.150a,b}$$

and (21.149b) and (21.150b) together give (21.148).

The high diffusion limit

To explore the high diffusion limit we take $\epsilon \gg 1$. The nondimensional strength of the circulation is given by

$$\widehat{\psi} = \mathcal{O}(\epsilon) \gg 1. \tag{21.151}$$

The circulation is now 'strong', since $\widehat{\psi} \neq \mathcal{O}(1)$. Dimensionally we still have that

$$\Psi = \epsilon \frac{\tau_0}{f_S} \qquad \text{or} \qquad \Psi = \frac{\kappa_v L}{h} \tag{21.152a,b}$$

but h and ϵ will be different than in the low diffusion limit. Now, if $\widehat{\psi} \sim \epsilon \gg 1$ the diffusion driven circulation in the basin cannot be matched by a purely wind-driven circulation in the channel, since the latter is $\mathcal{O}(1)$. Put more physically, as we increase diffusivity the circulation increases in strength, but this cannot connect smoothly to the flow in the channel unless the eddy-driven circulation changes, because the wind-driven circulation is externally fixed. We thus match the basin circulation to an eddy-driven channel circulation and require $\Lambda = \mathcal{O}(\epsilon)$. In particular, if we set $\Lambda = \epsilon$ then

$$\epsilon = \Lambda = \sqrt{\frac{K_e \kappa_v L}{(\tau_0/f_S)^2 l}}. \tag{21.153}$$

This is the square root of the expression for ϵ in the weak diffusion limit. Using (21.153) and (21.152a) we find

$$\frac{h}{l} = \sqrt{\frac{\kappa_v L}{K_e l}}, \qquad \Psi = \sqrt{\frac{K_e \kappa_v L}{l}}. \tag{21.154a,b}$$

These are expressions for the characteristic depth and strength of the circulation in a strong mixing regime.

Meaning of the limits

If diffusion is weak the stratification is set by a trade-off between the eddies and wind and this determines h, as in (21.146), and this does not involve diapycnal diffusion at all. However, the strength of the circulation is determined by an upwelling-diffusion balance which gives the estimate $\Psi \sim \kappa_v L/h$. Since h is independent of diffusivity we obtain a circulation strength that is linearly proportional to diffusivity, as in (21.148). Since diffusion is, in this limit, small then the circulation is weak, even in the southern channel. The weakness arises because there is a cancellation between the wind stress and eddy terms in (21.135), and a total cancellation would lead to the so-called 'vanishing of the Deacon Cell — the Deacon Cell here being the residual overturning in the southern channel. It is interesting that the circulation gets weaker as the wind gets stronger;

this counter-intuitive effect arises because the wind steepens the isopycnals and deepens the stratification, so that the diffusive term ($\kappa_v \overline{b}_{zz}$) in the basin gets smaller. From an asymptotic perspective, in the small ϵ limit the residual circulation is zero to lowest order, and at the next order the flow is parallel to the isopycnals in the channel (except in the mixed layer). We will see in the next section that the Deacon cell need not vanish, even in the limit of weak diffusion, if there is a northern source of water.

In the strong diffusion case the diffusivity itself directly affects the stratification, and consequently we get a weaker dependence of the circulation strength on κ_v. In this limit diapycnal mixing deepens the isopycnals in the basin away from the channel, and this deepening in turn means that the diffusion has a weaker effect. Thus, although the circulation is stronger than in the weak diffusion case it has a weaker dependence on diffusivity, to the one half power in fact (21.154b). The second consequence of the deepening is that the isopycnals are steeper in the channel, with the steepening being balanced by the enhanced slumping effects of baroclinic instability, with the wind then only having a secondary effect.

Finally, instead of varying diffusivity we can think of the wind changing. In the weak wind limit the circulation is diffusively driven and independent of the wind strength, as in (21.154b). In the strong wind limit the circulation, as noted above, actually decreases as the wind increases, but still remains proportional to the diapycnal diffusivity κ_v.

21.9 † A MODEL OF THE INTERHEMISPHERIC CIRCULATION

We now introduce another 'water mass' into the mix — *North Atlantic Deep Water*, or NADW. We thus divide the ocean into three regions as sketched in Fig. 21.32, namely:

(i) a southern channel (south of about 50° S) where, as before, we expect a balance between eddy effects and wind effects;

(ii) a basin region (from about 50° S to, say, 60° N), where the isopycnals are fairly flat;

(iii) a northern convective region (north of 60° N) in which convection produces vertical isopycnals that connect with those in the basin.

Although the dynamics of all three regions are locally different, they must act in concert to produce a dynamically consistent circulation. The main difference, and it is an important one, between this model and the previous one is the presence of an interhemispheric cell that, we will find, is primarily wind driven, and that (for realistic parameter values) sits on top of the lower cell. In the presentation that follows we focus our description on the northern convective region and the upper cell, for the dynamics of the lower cell are very similar to those of the previous section. Further, we seek only scaling relations rather than full analytic or semi-analytic solutions.[31] We use lower case letters (e.g., h, ψ) to denote field variables and upper case symbols (e.g., H, Ψ) for representative values.

21.9.1 Model Phenomenology

The upper cell has similar characteristics to the wind-driven cell sketched in Fig. 21.22, but we now require the flow in the basin to connect smoothly to an eddy-rich southern channel region, in a similar manner to that described in the previous section. We also suppose that in the northern region the interior flow connects to the surface by way of convection. To see how this occurs, consider a given isopycnal, b_0 say, that outcrops at the surface in the southern channel, slopes down in the channel and becomes horizontal in the basin. If the surface values of b in the northern convective region are all larger (warmer) than b_0 then the isopycnal never outcrops in the north; rather, it continues northward until it intersects the northern wall. If, on the other hand, at some latitude there is a latitude, y_n say, at which the surface values become lower than b_0 convection will occur and the b_0 isopycnal becomes vertical. There is then an isopycnal pathway from the surface

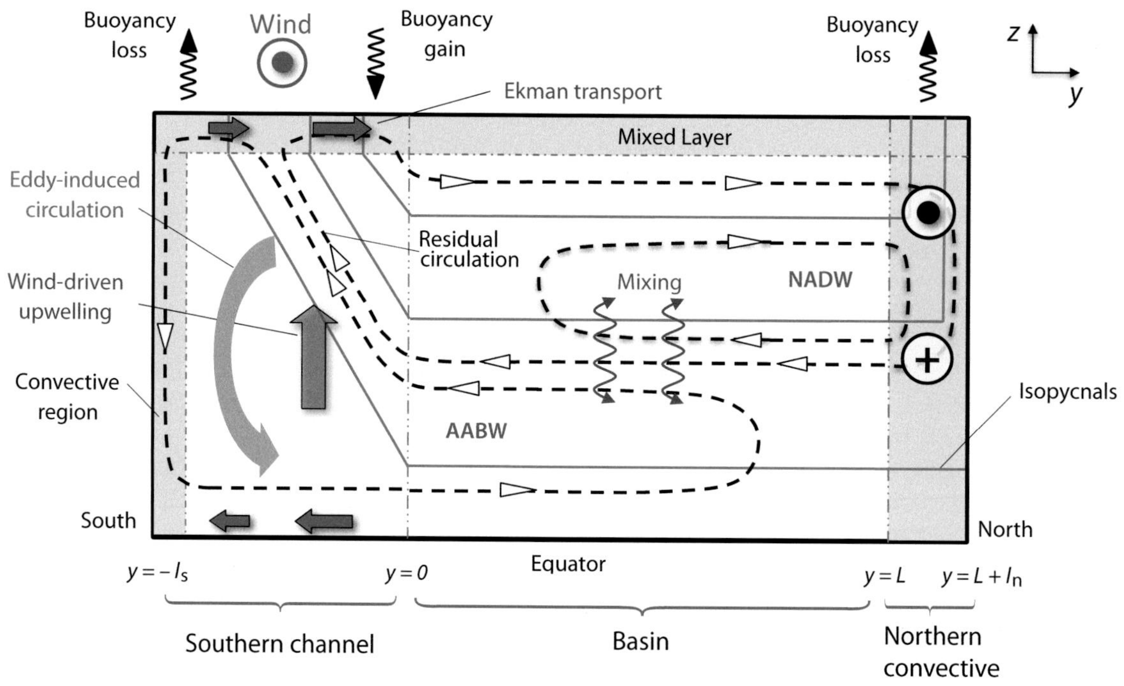

Fig. 21.32 An idealized interhemispheric MOC in a single basin, crudely representing a zonally-averaged Atlantic circulation. The solid blue lines are the isopycnals, the dashed lines with arrows are the streamlines, the dashed vertical lines are the boundaries between adjacent regions, shaded grey areas are the convective regions at high latitudes and the surface mixed layer, and the red curly arrows represent mixing giving a downward diffusive heat flux.

at y_n through the interior to the southern channel. A parcel of water may move along this pathway even in the absence of diffusion; that is, there can be an interhemispheric mid-depth adiabatic circulation.[32] Below this cell (which we associate with NADW) there can be a bottom cell of AABW that is diffusively driven, as in the previous section.

21.9.2 Dynamics in the Northern Convective Region

In the northern region (denoted with a subscript N) the values of buoyancy at the surface (i.e., $b_N(y, z = 0)$) are mapped on to the flat isopycnals, $b_B(z)$ in the interior basin region (denoted with a subscript B), and the simplest assumption to make is that the matching occurs by convection. That is, the surface waters convect downward to the level of neutral buoyancy, producing vertical isopycnals ($\partial b_N/\partial z = 0$), and then flow meridionally. By thermal wind the vertical isopycnals give rise to a zonal flow, with the total zonal transport being determined by the meridional temperature gradient and the depth, h, to which flow convects. The zonal flow is thus

$$u_N(y,z) = -\frac{1}{f}\int_{-h}^{z} \frac{\partial b_N}{\partial y}\,\mathrm{d}z' + \text{constant}, \tag{21.155}$$

where the constant is determined by the requirement that $\int_{-h}^{0} u_N \,\mathrm{d}z = 0$, and there are boundary layers in both east and west to bring the flow to zero. When the relatively shallow eastward moving zonal flow collides with the eastern wall it subducts and returns, as sketched in Fig. 21.33, and when the flow reaches the western wall it then moves equatorward in the deep western boundary current. Similarly, it is the upper, northward moving branch of the western boundary current that feeds the

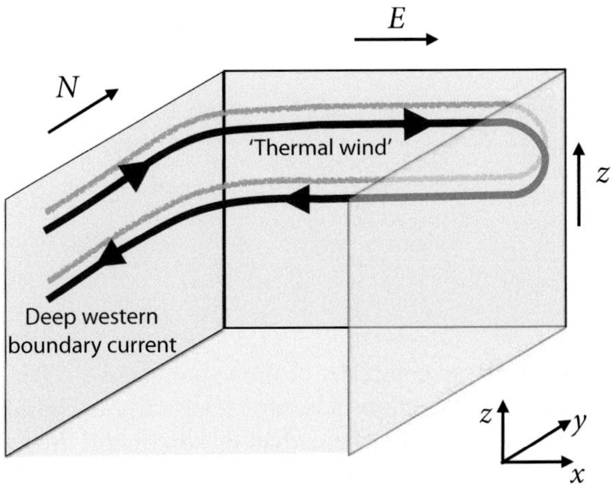

Fig. 21.33 The envisioned mean flow in the northern region of the model of Fig. 21.32. The north–south temperature gradient induces a zonal 'thermal wind', which is supplied by and feeds the deep western boundary current, as shown.

eastward moving flow. The total volume transport ($m^3\,s^{-1}$) in these thermally-induced zonal flows thus translates to a meridional streamfunction given by

$$\int_{x_W}^{x_E} \psi_N \,\mathrm{d}x = \int_{-h}^{z} \mathrm{d}z' \int_{L}^{L_n} u_N \,\mathrm{d}y, \tag{21.156}$$

where L_x is the zonal extent of the region, L is the latitude of the southern edge of the convecting region and $L_n = L + l_n$ is the northern edge of the domain (see Fig. 21.32). Using (21.155) gives an estimate for the value of this streamfunction as

$$\Psi_N = \frac{\Delta b\, H^2}{L_x f_N}, \tag{21.157}$$

where Δb is the surface buoyancy difference across the northern convective region, which in the theory we are describing is an external parameter, and f_N is the Coriolis parameter in the northern region. The streamfunction ψ_N is function of space and Ψ_N is a representative value of it, and H is a representative value of h, the depth to which the convection reaches and so of the stratification.

21.9.3 Connection to Other Regions

In the basin region we posit flat isopycnals and an upwelling diffusive balance, whence

$$w_B \frac{\partial b}{\partial z} = \kappa_v \frac{\partial^2 b}{\partial z^2} \qquad \text{giving} \qquad \frac{\Psi_B}{L_y} = \frac{\Psi_N - \Psi_S}{L_y} = \frac{\kappa_v}{H}. \tag{21.158}$$

In the channel region the residual circulation arises from a balance between the wind and eddy effects and is given by (21.135), which we can write as

$$\Psi_S = \left(\frac{\tau_0}{\overline{f}_S} - K_e \frac{H}{l_s} \right), \tag{21.159}$$

with $\overline{f}_S = |f_S|$ as before. Collecting the various expressions above for streamfunction we have

$$\Psi_S = \left(\frac{\tau_0}{\overline{f}_S} - K_e \frac{H}{l_s} \right), \qquad \Psi_N - \Psi_S = \frac{\kappa_v}{H} L_y, \qquad \Psi_N = \frac{\Delta b\, H^2}{f_N L_x}, \tag{21.160a,b,c}$$

with unknowns Ψ_S, Ψ_N and H. The various parameters have approximate values as given in (21.144) as well as

$$L_x = 5000\,\text{km}, \quad L_y = 10\,000\,\text{km}, \quad \Delta b = 10^{-2}\,\text{m}\,\text{s}^{-2}, \quad f_N = 10^{-4}\,\text{s}^{-1}. \tag{21.161}$$

Equation (21.160) may be reduced to

$$\frac{\Delta b H^2}{f_N} - \left(\frac{\overline{\tau_0}}{f_S} - K_e \frac{H}{l_s}\right) L_x = \frac{\kappa_v}{H} L_x L_y, \tag{21.162}$$

which is a cubic equation for the characteristic depth, H, of the upper, NADW, cell in Fig. 21.32. Although there are analytic solutions to cubic equations it is more instructive to consider limiting cases, first with either the northern region or the southern channel absent and then with low or high diffusivity.

No northern source

Suppose that $\Delta b = 0$ and that there is no deep water formation in the north. If κ_v is small then we obtain $H/l_s = (\tau_0/\overline{f_S})/K_e$ a, equivalent to (21.146), and $\Psi_S = \Psi_B = 0$. If κ_v is large then we find

$$H^2 = \frac{\kappa_v L l_s}{K_e}, \tag{21.163}$$

so recovering (21.154a). Thus, the dynamics are essentially those of Section 21.8, and there is a single deep, and rather weak, diffusively-driven, cell. The same situation arises if the northern region is too buoyant (e.g., too warm) for then there is no isopycnal pathway between high northern hemispheres and the southern channel, and deep convection does not occur. This case may have relevance to the Pacific Ocean, where the surface at high northern latitudes is insufficiently dense and there is no Pacific equivalent of NADW.

No southern channel

If there is no southern channel then $\Psi_S = 0$ and we have

$$\frac{\Delta b H^2}{f_N} = \frac{\kappa_v}{H} L_x L_y, \tag{21.164}$$

giving

$$H^3 = \kappa_v \left(\frac{f_N L_x L_y}{\Delta b}\right) \quad \text{and} \quad \Psi_N = \Psi_B = (\kappa_v L_y)^{2/3} \left(\frac{\Delta b}{L_x f_N}\right)^{1/3}. \tag{21.165}$$

These are the classical expressions for the thickness of a diffusive thermocline and the strength of a diffusively-driven overturning circulation, essentially as obtained in Sections 21.2.5 and 20.5.1.

Let us now look at the case with all three regions, in the limits of weak and strong diffusivity.

Low diffusivity limit

In this case the upwelling is weak and $|\Psi_N| \approx |\Psi_S|$ and

$$\frac{\Delta b H^2}{f_N} - \left(\frac{\overline{\tau_0}}{f_S} - K_e \frac{H}{l_s}\right) L_x = 0. \tag{21.166}$$

In this case the basin is just a 'pass-through' region: water formed in the North Atlantic just passes through the basin without change, and upwells in the Southern Ocean. For the above expression to be physical, τ_0 must be non-zero; this requirement is a manifestation of Sandström's effect, that in the absence of diffusivity a mechanical forcing is needed, and if $\tau_0 = 0$ in the above then the only physical solution is $H = 0$.

If we assume that K_e is small then we obtain

$$H = \left(\frac{\tau_0 f_N L_x}{\overline{f}_S \Delta b}\right)^{1/2}, \qquad \Psi_S = \Psi_N = \frac{\tau_0 L_x}{\overline{f}_S}, \tag{21.167}$$

which with the parameters chosen earlier gives $H \sim 320$ m and $\Psi \sim 10$ Sv. The residual circulation *does not vanish in the limit of small diffusivity;* rather, it is wind driven, adiabatic, interhemispheric and independent of diffusivity. This is to be contrasted with the case in which deep waters are not produced in the north, as observed in the Pacific Ocean, where in the low diffusivity limit the eddy-induced circulation nearly cancels the wind-driven circulation resulting in small residual circulation, dependent on that diffusivity.

In the more general case we solve (21.166) to give

$$H = \left(\frac{\tau_0 f_N L_x}{\overline{f}_S \Delta b}\right)^{1/2} \left(-\alpha + \sqrt{1+\alpha^2}\right), \tag{21.168}$$

where α is a nondimensional number giving the ratio of eddy to wind effects,

$$\alpha = \frac{1}{2}\frac{K_e}{l_s}\left(\frac{L_x \overline{f}_S f_N}{\tau_0 \Delta b}\right)^{1/2} = \frac{1}{2}\frac{\Psi_{eddy}}{\Psi_{wind}}, \tag{21.169}$$

where

$$\Psi_{wind} = \frac{\tau_0}{\overline{f}_S} \quad \text{and} \quad \Psi_{eddy} = \frac{K_e}{l_s}\left(\frac{\tau_0 f_N L_x}{\overline{f}_S \Delta b}\right)^{1/2}. \tag{21.170}$$

Putting in values from (21.144) and (21.161) gives $\alpha \sim 0.1$, $\Psi_{eddy} \sim 1.6$ Sv and $\Psi_{wind} \sim 10$ Sv, suggesting that wind effects are dominant, but there is considerable uncertainty because K_e is ill-defined and does not have a definitive value.

High diffusivity limit

In the high diffusivity limit the wind-driven upwelling in the Southern Hemisphere is small compared to the mixing-driven upwelling in the ocean basin, and although it seems to be not relevant for today's circulation it may have been important in glacial climates. Equation (21.162) simply becomes (21.164), which as already noted gives us the classical scaling for a diffusively-driven circulation. The upper cell thus fades out before reaching the southern channel but there remains a lower, AABW, cell that connects to the flow in the southern channel as described in Section 21.8.

21.9.4 Final Remarks and Relevance to the Ocean

Over the last several pages we've described a conceptual, but quantitative, theoretical model of the overturning circulation in the ocean. Aside from its idealizations (e.g, simplified geometry), the model has two main shortcomings: it uses an eddy-diffusivity parameterization for the effects of mesoscale eddies, and it treats the ocean one basin at a time. Putting these aside, what does the model tell us?

In the limit of weak diapycnal mixing, which seems relevant to the present mid-depth ocean, and with a northern source of deep water, then the model produces a circulation relevant to the

Atlantic — Fig. 21.32 is an idealization of the Atlantic panel in Fig. 21.29. The strength of the mid-depth overturning circulation is then largely determined by the Ekman transport in the Southern Ocean and, secondarily, eddy effects. The rest of the ocean is essentially forced to adjust and produce the amount of deep water demanded by the Ekman transport and the associated wind-driven upwelling in the Southern Ocean. Beneath this mechanically-forced mid-depth cell lies a diffusively-driven deep cell, and this is the model representation of AABW.

The Pacific Ocean is insufficiently dense at high northern latitudes to produce deep water — compared to the Atlantic it is relatively fresh. That is, there is no isopycnal pathway from the surface waters at high latitudes to the southern channel; and consequently there is no wind-driven mid-depth cell comparable to that of the Atlantic. The model then produces a circulation similar to that of Fig. 21.31, where the northern wall is at a high latitudes in the Northern Hemisphere, and this is an idealization of the Pacific panel in Fig. 21.29. Since the diffusivity is weak the circulation is weak, particularly in the Northern Hemisphere. The flow in the world's ocean is much more interconnected than these simple ideas suggest, and in reality the flow travels from basin-to-basin on what has been metaphorically called a conveyor belt.[33] But our own story ends here, for now. How the models above might be extended to produce a global flow is a chapter for another day.

Notes

1 Warren (1981) provides a review and historical background and Schmitz (1995) surveys the observations and provides an interpretation of the deep global circulation. Marshall & Speer (2012) review the role of the Southern Ocean.

2 The word 'driven' is fraught with ambiguity, even when the subject matter is well understood. Does it refer to the proximate mechanical forces producing the motion, or to the controlling device? The former (which is quite common in physical science) suggests that an engine drives a car, for that is what makes the wheels go round, whereas the latter suggests that, in fact, the driver drives the car. For the less well-understood ocean there is scope for still more confusion, and the context in which the word is used becomes important. What we in this chapter sometimes call buoyancy-driven might be better called mixing-driven, since it is the mixing of fluid parcels that makes potential energy available for the circulation. *Caveat lector.*

3 Adapted from Wunsch (2002). The figure shows a 'state estimate' — a combination of models and observation, similar to an atmospheric reanalysis.

4 Figure kindly prepared by Neven Stjepan Fučkar, using the climatology of Conkright *et al.* (2001).

5 As in Haney (1971). The value of C, which is not necessarily related to that of κ, is often taken to be such that the heat flux is of order $30\,\mathrm{W\,m^{-2}\,K^{-1}}$, but it is certainly not a universal constant.

6 Adapted from Paparella & Young (2002).

7 Rossby (1965).

8 Adapted from Ilicak & Vallis (2012).

9 I am grateful to Tom Haine for pointing out this argument. See also Haine & Marshall (1998) and Hughes & Griffiths (2008).

10 Ocean convection is also reviewed by Marshall & Schott (1999).

11 Sandström (1908, 1916). Sandström's discussion was rather qualitative and generally thermodynamic in nature, with friction playing only an implicit role. Since then a number of related statements with varying degrees of generality and preciseness have been given (e.g., Dutton 1986, Huang 1999, Paparella & Young 2002). Section 21.2.3 follows Paparella and Young.

12 The original box model is due to Stommel (1961), and many studies with variations around this have followed. Rooth (1982) developed the idea of a buoyancy-driven pole-to-pole overturning circulation, and Welander (1986) discussed, among other things, the role of boundary conditions on temperature and salinity at the ocean surface. Thual & McWilliams (1992) systematically explored how box models compare with two-dimensional fluid models of sideways convection, Quon & Ghil

(1992) explored how multiple equilibria arise in related fluid models, and Dewar & Huang (1995) discussed the problem of flow in loops. Cessi & Young (1992) tried to derive simple models systematically from the equations of motion, obtaining various nonlinear amplitude equations. Our discussion is just a fraction of all this — see also Whitehead (1995) and Cessi (2001) for reviews.

13 Adapted from Welander (1986).

14 Having said this, Bryan (1986), Manabe & Stouffer (1988) and Marotzke (1989) did find evidence of multiple equilibria in various three-dimensional numerical models, motivated in part by the solutions of box models.

15 After Stommel *et al.* (1958).

16 Following Stommel & Arons (1960).

17 A global Stommel–Arons-like solution was presented by Stommel (1958). The discovery of deep western boundary currents by Swallow & Worthington (1961) was motivated by the theoretical model. Using neutrally-buoyant floats underneath the Gulf Stream they found an equatorward-flowing undercurrent with typical speeds of 10–20 cm s^{-1}. Some relevant observations of the deep circulation are summarized by Hogg (2001).

18 For example, Toggweiler & Samuels (1995).

19 Drawing from the various numerical, conceptual and analytic models of Toggweiler & Samuels (1995, 1998), Döös & Coward (1997), Gnanadesikan (1999), Vallis (2000), Webb & Suginohara (2001), Nof (2003), Samelson (1999a, 2004), Wolfe & Cessi (2011), and Nikurashin & Vallis (2011, 2012). The notion of a deep interhemispheric circulation driven by winds in the ACC was earlier proposed by Eady (1957), albeit rather sketchily. None of our models of the MOC (including the ones in later sections) are complete, never mind true, but some may be useful.

20 Loic Jullion graciously provided the inverse calculations, which are similar to those described in Lumpkin & Speer (2007). Patrick Heimback kindly provided the state estimates, which are from the ECCO suite of calculations.

21 From Rintoul *et al.* (2001).

22 See Rintoul *et al.* (2001) and Olbers *et al.* (2004) for ACC reviews.

23 These simulations, described in Henning & Vallis (2005), solve the primitive equations in a domain similar to Fig. 21.19. The wind forcing produces a poleward Ekman drift across the channel, as well as a subtropical gyre, and there is a meridional temperature gradient across the whole domain, so giving rise to a subtropical thermocline.

24 Adapted from Henning & Vallis (2005).

25 Munk & Palmén (1951), Gille (1997) and Stevens & Ivchenko (1997).

26 Adapted from a figure in Burke *et al.* (2015).

27 Models of this ilk stem from Johnson & Bryden (1989). Straub (1993), Hallberg & Gnanadesikan (2001) Karsten *et al.* (2002), Marshall & Radko (2003), Henning & Vallis (2005), consider related issues and extensions.

28 See also Munk & Palmén (1951), Warren *et al.* (1996), Olbers (1998) and Hughes (2002).

29 Similar to a figure in Watson *et al.* (2015).

30 Largely following Nikurashin & Vallis (2011, 2012). For related numerical simulations see Vallis (2000) and Wolfe & Cessi (2010, 2011).

31 A full description may be found in Nikurashin & Vallis (2012). In the form described here the model becomes similar to the one of Gnanadesikan (1999).

32 A continuous, unique, pole-to-pole isoneutral pathway is a chimera, because of the nonlinear dependence of density on pressure, temperature and salinity in the seawater equation of state. But at the level of our theory there is an approximate one. See also endnote 9 on page 53.

33 For variations involving multiple basins see Ferrari *et al.* (2014) and Thompson *et al.* (2016).

In the afternoon they came unto a land
In which it seemed always afternoon.
All round the coast the languid air did swoon,
Breathing like one that hath a weary dream.
Alfred Tennyson, *The Lotus Eaters*, 1832.

CHAPTER 22

Equatorial Circulation and El Niño

EQUATORIAL OCEANOGRAPHY DECEIVES US, hiding fascinating, non-intuitive dynamics beneath the languorous tropical air. The mid-latitudes give us the great gyres with their intense western boundary currents and mesoscale eddies, and by comparison the equatorial currents may seem, on the surface, featureless and vapid. Yet the equatorial regions are home to the resolute equatorial undercurrents that tunnel across the basins, opposite in bearing to the winds that drive them. And the equatorial ocean and atmosphere — in a collaboration that is more tango than waltz — give rise to the marvellous phenomenon that is El Niño, the most dramatic example of climate variability on human timescales that this planet has to offer. Such phenomena are the subjects of this chapter.

The defining feature of equatorial dynamics is that the Coriolis parameter becomes small, at least by comparison with the mid-latitudes, and balanced and unbalanced dynamics become intertwined, as we encountered in Chapter 8. Yet if we move more than a few degrees away from the equator the Rossby number again becomes quite small, suggesting that familiar ways of investigating the dynamics — Sverdrup balance for example — might yet play a role. Let's first see what we are trying to understand and if the observations can give us some intuition.

22.1 OBSERVATIONAL PRELIMINARIES

In mid-latitudes the gyres are very robust features, existing in all the basins, and may be understood as the direct response to the curl of the wind stress. In the equatorial regions the currents also display some robust and distinctive features, illustrated in Fig. 22.1 and the top panel of Fig. 22.2, but their relation to the winds is less obvious. The main features are as follows:

1. A shallow westward[1] flowing surface current, typically confined to the upper 50 m or less, strongest within a few degrees of the equator, although not always symmetric about the equator. Its speed is typically a few tens of centimetres per second.

2. A strong coherent eastward undercurrent extending to about 200 m depth, confined to within a few degrees of the equator. Its speed is up to a metre per second or a little more, and it is this current that dominates the vertically integrated transport at the equator. Beneath the undercurrent the flow is relatively weak.

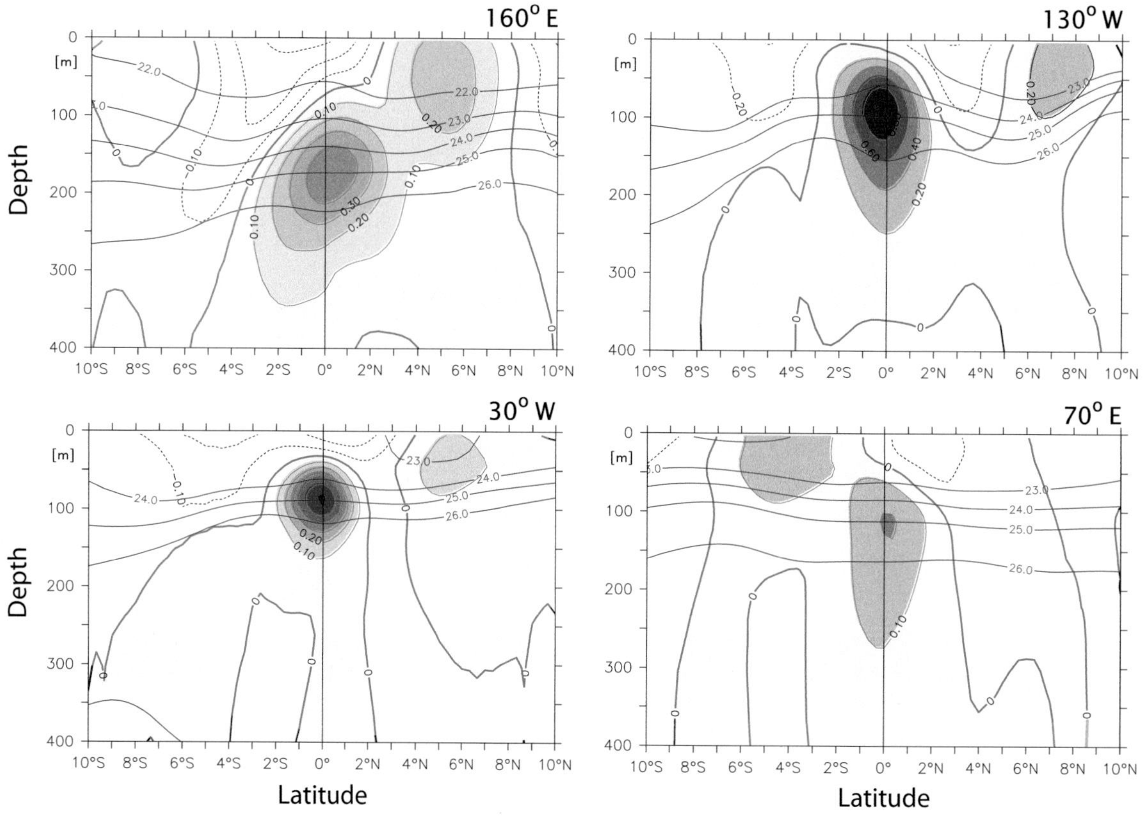

Fig. 22.1 Sections of the observed mean zonal current (shading and associated contours) at two longitudes in the Pacific (upper panels), in the Atlantic (lower left) and in the Indian Ocean (lower right). The contours are every $20\,\mathrm{cm\,s^{-1}}$ in the upper two panels and every $10\,\mathrm{cm\,s^{-1}}$ in the lower panels. Note the well-defined eastward undercurrent at the equator in all panels, and a weaker eastward countercurrent at about 6° N and/or 6° S. The red, more horizontal, lines are isolines of potential density.[2]

3. Westward flow on either side of the undercurrent, with eastward countercurrents poleward of this. In the Pacific the countercurrent is strongest in the Northern Hemisphere where it reaches the surface.

22.2 DYNAMICAL PRELIMINARIES

In mid-latitudes the large scale currents system may be understood using the planetary geostrophic equations of motion. Applying these allows us to understand formation of the great wind-driven gyres, with Sverdrup balance providing a solid foundation on which to build. As we approach lower latitudes the Coriolis parameter, f, decreases and the Rossby number increases and one might expect that dynamics based on geostrophic balance will ultimately fail. A little surprisingly, it is only very close to the equator that the Rossby number exceeds unity: if we take a velocity of $0.5\ \mathrm{m\ s^{-1}}$ and a length scale of 500 km then the Rossby number at 5° latitude is 0.08, at 2°, 0.2 and at 1°, 0.4. These numbers suggest that until we are virtually at the equator (where the Rossby number is infinite) we can use some of the familiar tools from the mid-latitude dynamics. At the equator the Coriolis parameter switches sign and this leads to some interesting features. The vertical structure is also a little complex so let us first see the extent to which the familiar Sverdrup balance can

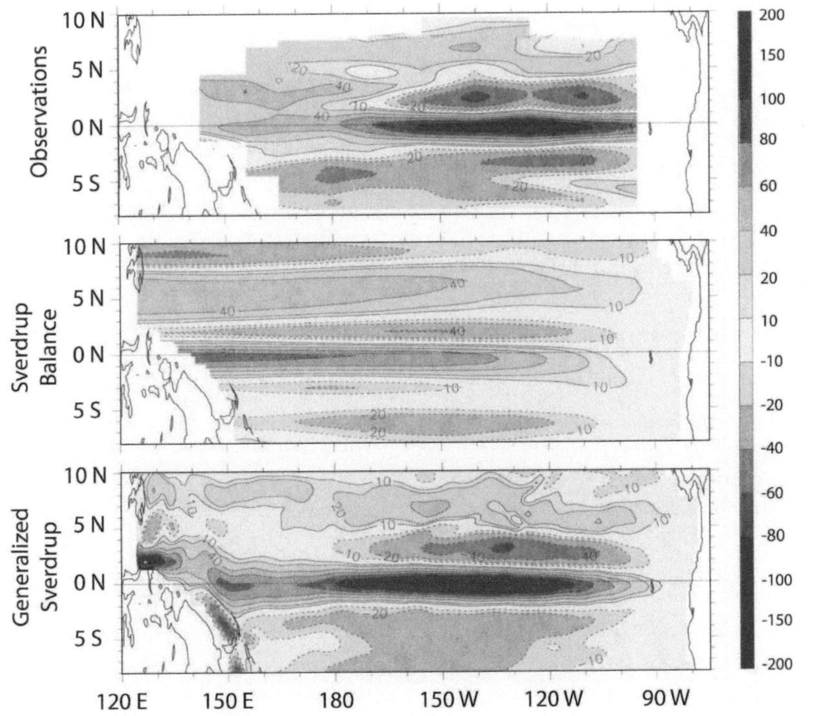

Fig. 22.2 Vertically integrated zonal transport in the Pacific. Red colours indicate eastward flow, blue colours westward. The top panel shows the observed flow, the middle panel shows the flow calculated using Sverdrup balance with the observed wind, and the bottom panel shows the flow calculated with a 'generalized' Sverdrup balance that includes the nonlinear terms in a diagnostic way.[3]

explain the vertically integrated flow.

22.2.1 The Vertically Integrated Flow and Sverdrup Balance

The horizontal momentum may be written

$$\frac{\partial \boldsymbol{u}}{\partial t} + \boldsymbol{u} \cdot \nabla \boldsymbol{u} + \boldsymbol{f} \times \boldsymbol{u} = -\nabla \phi + \frac{1}{\rho_0} \frac{\partial \boldsymbol{\tau}}{\partial z}, \tag{22.1}$$

where $\boldsymbol{\tau}$ is the stress on the fluid. The mass conservation equation is

$$\frac{\partial u}{\partial x} + \frac{\partial v}{\partial y} + \frac{\partial w}{\partial z} = 0, \tag{22.2}$$

which, on vertical integration over the depth of the ocean, gives

$$\frac{\partial U}{\partial x} + \frac{\partial V}{\partial y} = 0, \tag{22.3}$$

where U and V are the vertically integrated zonal and meridional velocities (e.g., $U = \int u \, \mathrm{d}z$) and we assume the ocean has a flat bottom and a rigid lid at the top. If we assume the flow is steady and integrate (22.1) vertically, then take the curl and use (22.3), we obtain

$$\beta V = \mathrm{curl}_z(\widetilde{\boldsymbol{\tau}}_T - \widetilde{\boldsymbol{\tau}}_B) + \mathrm{curl}_z \boldsymbol{N}, \tag{22.4}$$

where $\widetilde{\boldsymbol{\tau}}$ is the kinematic stress ($\widetilde{\boldsymbol{\tau}} = \boldsymbol{\tau}/\rho_0$ where ρ_0 is the reference density of seawater) with the subscripts T and B denoting top and bottom, $\boldsymbol{N}$ represents all the nonlinear terms and curl_z is defined by $\mathrm{curl}_z \boldsymbol{A} \equiv \partial A^y/\partial x - \partial A^x/\partial y = \mathbf{k} \cdot \nabla_3 \times \boldsymbol{A}$. Equations (22.4) and (22.3) are closed equations for the vertically averaged flow. In oceanography we very often deal with the kinematic

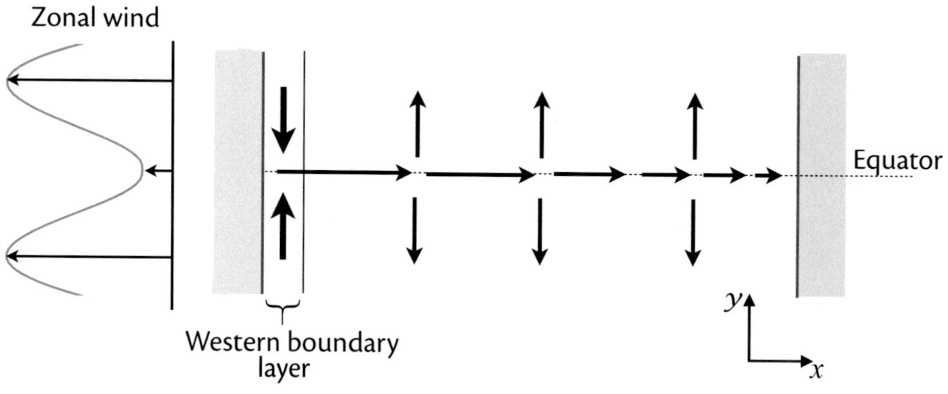

Fig. 22.3 Schema of Sverdrup flow at the equator between two meridional boundaries. The mean winds are all westward, but with a minimum in magnitude at the equator. By Sverdrup balance, (22.5), the wind stress produces the divergent meridional flow shown, which in turn induces an eastward equatorial zonal flow, strongest in the western part of the basin.

stress rather than the stress itself, so henceforth we will drop the tilde over the τ symbol as well as the adjective 'kinematic'. In the cases that we need to refer to the actual stress we will denote this by τ^*; thus, $\tau = \tau^*/\rho_0$.

If we neglect the nonlinear terms and the stress at the bottom (we'll come back to these terms later) then (22.4) becomes

$$\beta V = \text{curl}_z \boldsymbol{\tau}_T. \tag{22.5}$$

This is just Sverdrup balance, familiar from Chapter 19. The zonal transport is obtained by differentiating (22.5) with respect to y, using (22.3) to replace $\partial_y V$ with $\partial_x U$, and then integrating from the eastern boundary (x_E). This procedure gives

$$U = -\frac{1}{\beta}\int_{x_E}^{x} \frac{\partial}{\partial y}\text{curl}_z \boldsymbol{\tau}_T \, \mathrm{d}x' + U(x_E, y). \tag{22.6}$$

We don't integrate from the western boundary because a boundary layer can be expected there, whereas the value of U at the eastern boundary will be small.

If $U(x_E, y) = 0$ and the stress is zonal and uniform, then (22.6) becomes

$$U(x, y) = \frac{1}{\beta}(x - x_E)\frac{\partial^2 \tau_T^x}{\partial y^2}. \tag{22.7}$$

That is, the depth integrated flow is proportional to the second derivative of the zonal wind stress, and because $x < x_E$ we have $U \propto -\partial^2 \tau_T^x / \partial y^2$. Evidently, the result will depend rather sensitively on the wind pattern. Although the zonal wind is generally westward in the tropics there is a minimum in the magnitude of that wind near the equator (that is, a local maximum as schematized in Fig. 22.3) so that $\partial^2 \tau_T^x / \partial y^2$ is negative. Thus, using (22.7), U will generally be positive at the equator. Using the observed wind field the Sverdrup flow — that is, the solution of (22.6) with $U(x_E, y) = 0$ — can be calculated and this is plotted in the middle panel of Fig. 22.2. There is a good but not perfect agreement with the observations: the observed flow has its maximum further east. Further, in the western equatorial Pacific the observed eastward flow is quite broad whereas the eastward Sverdrup flow is narrow, flanked on either side by westward flow. Some of the discrepancy can be attributed to the role of the nonlinear and frictional terms, as illustrated in the bottom panel of Fig. 22.2. To obtain the flow illustrated, the calculation proceeds from (22.4) in the same way as before, but now includes the nonlinear terms and a representation of frictional effects in

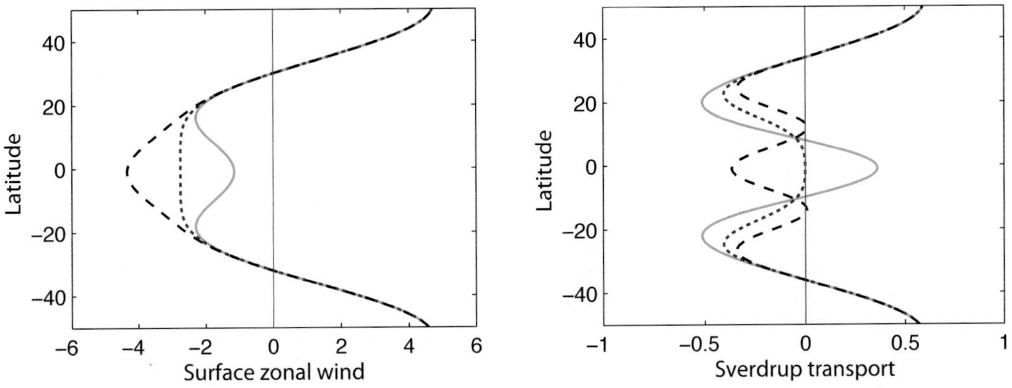

Fig. 22.4 The left panel shows three putative surface zonal (atmospheric) winds, u, all with westward winds in the tropics and with the solid line being the most realistic. The right panel shows the corresponding negative of the second derivative, $-\partial^2 u/\partial y^2$, proportional to the (oceanic) Sverdrup transport, in arbitrary units. The wind represented by solid (blue) line gives an eastward transport at the equator, as is observed, with the others differing markedly.

a diagnostic fashion. Thus, for example, the nonlinear terms of the form $\mathrm{curl}_z(\int \boldsymbol{u} \cdot \nabla \boldsymbol{u}\, \mathrm{d}z)$ are evaluated and used to calculate a generalized Sverdrup flow, where the velocities are taken from a nonlinear model forced by the observed winds. Of the nonlinear terms, the largest ones involve the meridional derivatives of the zonal flow, for example $\partial_y(uu_x)$. The effect of the nonlinear terms is to decelerate the eastward flow in the eastern Pacific, with friction tending to damp the flow especially in the central Pacific, and the resulting flow is evidently closer to the observations than is linear Sverdrup balance. Of course the full solution (22.4) must give a vertically integrated flow that closely resembles the observations, because there are only very weak approximations made in deriving it. The success of the Sverdrup theory lies in the extent to which the vertically integrated flow can be satisfied by the simple linear balance (22.5), and then improved by adding nonlinear and dissipative terms in a diagnostic fashion.

22.2.2 Sensitivity of the Sverdrup Flow

Although the calculations of Sverdrup flow do show good agreement with observations, the calculation — and, most likely, the observed flow — is rather sensitive to the precise form of the winds, as in Fig. 22.4. The figure shows three surface zonal wind distributions, with the 'w' shaped solid line having a minimum in the westward flow (i.e., a minimum in the trade winds) at the equator and so being the most realistic. The right-hand panel shows the negative of the second derivative of the winds which is proportional to the zonal Sverdrup flow. Only in the one case does the wind produce an eastward Sverdrup flow. In fact, in the case illustrated with the dashed lines, the small changes in the meridional gradient of the wind between 15° and 20° produce large variations in the Sverdrup transport. Given this delicacy, the small difference in the latitudinal variation of the Sverdrup flow and the observed flow, illustrated in the top and middle panels of Fig. 22.2, is not surprising and cannot be considered a major failure of the theory. However, the difference in the longitudinal structure of the two fields is indicative of the importance of other terms in the vorticity balance.

22.3 A LOCAL MODEL OF THE EQUATORIAL UNDERCURRENT

The most conspicuous feature of the ocean current system at low latitudes is the equatorial undercurrent, and we now consider its dynamics.[4] The physical picture that we first discuss is a 'local' one, and is essentially the following.[5] The mean winds are westward and provide a stress on the upper ocean, pushing the near-surface waters westward. Given that there is a boundary in the west, the water piles up there so creating a pressure-gradient force that pushes fluid eastward. To some degree the pressure gradient and the wind stress compensate each other leading to a state of no motion. However, the compensation is not perfect. Close to the surface the stress is dominant and a westward surface current results. Below the surface the pressure gradient dominates, resulting in an eastward flowing undercurrent, as in the observations in Fig. 22.1.

The above description makes no mention of the Coriolis parameter or Sverdrup balance or the wind-stress curl. On the one hand this suggests that the dynamics are likely to be robust and will not depend in a sensitive way on the wind pattern (as the Sverdrup flow does). On the other hand, given the usefulness of the Sverdrupian concept, such a description is also likely to be incomplete. To proceed further we'll construct a small hierarchy of mathematical models of the equatorial current system, beginning with the very simplest model of a homogeneous fluid subject to a uniform westward stress at the surface. Following that we will discuss a more inertial and non-local physical picture, in which the undercurrent may be thought of as being pushed by a pressure head that begins in extra-equatorial regions. In the extreme limiting case of this picture, the winds at the equator have no effect on the undercurrent. The real equatorial undercurrent likely involves a combination of local and inertial dynamics, and remains a topic of research.

22.3.1 Response of a Homogeneous Layer to a Uniform Zonal Wind

Let us first consider the simple case of the response of a layer of homogeneous fluid to a steady zonal wind that is uniform in the y-direction. With our usual notation the equations of motion in the presence of momentum and mass forcing are

$$\frac{\mathrm{D}u}{\mathrm{D}t} - fv = -g'\frac{\partial \eta}{\partial x} + \frac{\tau^x}{H}, \tag{22.8a}$$

$$\frac{\mathrm{D}v}{\mathrm{D}t} + fu = -g'\frac{\partial \eta}{\partial y} + \frac{\tau^y}{H}, \tag{22.8b}$$

$$\frac{\mathrm{D}h}{\mathrm{D}t} + h\left(\frac{\partial u}{\partial x} + \frac{\partial v}{\partial y}\right) = M, \tag{22.8c}$$

where (τ^x, τ^y) are the zonal and meridional kinematic stresses on the fluid, H is the depth of the fluid and M is a mass source, which for now we take to be zero. For steady flow and neglecting the nonlinear terms the equations become

$$-fv = -g'\frac{\partial \eta}{\partial x} + \frac{\tau^x}{H}, \qquad +fu = -g'\frac{\partial \eta}{\partial y}, \qquad H\left(\frac{\partial u}{\partial x} + \frac{\partial v}{\partial y}\right) = 0. \tag{22.9a,b,c}$$

If we take the y-derivative of (22.9a) and subtract it from the x-derivative of (22.9b), and noting that $\partial \tau^x/\partial y = 0$, we obtain

$$\beta v = 0. \tag{22.10}$$

Thus, using the continuity equation (22.9c), we have $\partial u/\partial x = 0$. That is, the zonal velocity is uniform. If there is a zonal boundary at which $u = 0$ then the zonal flow is zero everywhere and the complete solution is

$$u = 0, \qquad v = 0, \qquad g\frac{\partial \eta}{\partial x} = \frac{\tau^x}{H}, \qquad \frac{\partial \eta}{\partial y} = 0. \tag{22.11}$$

That is to say, the ocean is motionless and the wind stress is balanced by a pressure gradient. If the wind is westward, as it is on the equator, then $\partial \eta/\partial x < 0$ and the thermocline slopes down and deepens toward the west. The fact that there is no flow might have been anticipated from

Sverdrup balance in the absence of a wind-stress curl. Although the real ocean is not as simple as our model of it, the analysis exposes a truth with some generality: *the wind stress is largely opposed by a pressure gradient* rather than inducing a large westward acceleration that is halted by friction.

22.3.2 An Unstratified Local Model

Let us now consider a model with some vertical structure, thereby allowing the wind stress to be taken up in the upper ocean. The wind will still push near-surface water westward and create a zonal pressure gradient. The deeper water will feel the pressure-gradient force — because the pressure is hydrostatic — but not the wind stress, and so flows eastward. A simple model that can capture these effects begins with the three-dimensional momentum equations, namely

$$-fv = -\frac{\partial\phi}{\partial x} + \nu_z\frac{\partial^2 u}{\partial z^2} + \nu_h\nabla^2 u, \tag{22.12a}$$

$$fu = -\frac{\partial\phi}{\partial y} + \nu_z\frac{\partial^2 v}{\partial z^2} + \nu_h\nabla^2 v. \tag{22.12b}$$

The parameters ν_z and ν_h are eddy viscosities acting on vertical and horizontal shear, respectively, and the ∇ operator is purely horizontal, so that $\nabla^2 u = \partial^2 u/\partial x^2 + \partial^2 u/\partial y^2$. Dealing with a horizontal viscosity requires a more mathematically cumbersome treatment that we defer to Section 22.3.3; instead, we will first invoke a linear drag whence the momentum equations, along with the mass continuity equation, become

$$-fv = -\frac{\partial\phi}{\partial x} + \nu_z\frac{\partial^2 u}{\partial z^2} - ru, \qquad fu = -\frac{\partial\phi}{\partial y} + \nu_z\frac{\partial^2 v}{\partial z^2} - rv, \tag{22.13a,b}$$

$$\frac{\partial u}{\partial x} + \frac{\partial v}{\partial y} + \frac{\partial w}{\partial z} = 0. \tag{22.13c}$$

The drag terms are presumed to act throughout the depth of the fluid and so are a little arbitrary, but their presence enables us to construct a simple and very illuminating model. We should also remember that almost *any* frictional terms in a model of the large-scale circulation are to some degree ad hoc: the viscosities (ν_z, ν_h) are certainly not molecular viscosities and there is no proper justification for the use of eddy viscosities on momentum.

The vertical friction terms $(\partial^2 u/\partial z^2, \partial^2 v/\partial z^2)$ enable the wind's influence to be felt in the upper ocean via the boundary conditions, namely

$$\nu_z\frac{\partial u}{\partial z} = \tau^x, \qquad \nu_z\frac{\partial v}{\partial z} = \tau^y, \qquad \text{at } z = 0, \tag{22.14a}$$

$$\nu_z\frac{\partial u}{\partial z} = 0, \qquad \nu_z\frac{\partial v}{\partial z} = 0, \qquad \text{at } z = -H, \tag{22.14b}$$

where (τ^x, τ^y) is the kinematic wind stress. With boundary conditions of $w = 0$ at top and bottom the vertical integral of (22.13c) is

$$\frac{\partial U}{\partial x} + \frac{\partial V}{\partial y} = 0, \tag{22.14c}$$

where $(U, V) = \int (u, v)\,\mathrm{d}z$ is the vertically integrated flow. Equation (22.14c) allows for the introduction of a streamfunction ψ such that $U = -\partial\psi/\partial y$ and $V = \partial\psi/\partial x$. Cross-differentiating (22.13a,b) and vertically integrating then gives

$$r\nabla^2\psi + \beta\frac{\partial\psi}{\partial x} = \mathrm{curl}_z\tau. \tag{22.15}$$

This is the equation of Stommel's model, as in (19.6), and in the absence of the frictional term the vertically integrated flow is given by Sverdrup balance. If the wind has no curl the vertically integrated flow is zero, as before. However, the flow is not zero at each vertical level, as we now see.

Let us now assume the flow is unstratified and that the buoyancy b is a constant, which we take to be zero. The hydrostatic relation is $\partial\phi/\partial z = b = 0$ so that ϕ is uniform with height. From (22.13a,b) the vertically integrated momentum equations are then

$$H\frac{\partial\phi}{\partial x} = \tau^x - rU + fV, \qquad H\frac{\partial\phi}{\partial y} = \tau^y - rV - fU. \tag{22.16a,b}$$

Let us further suppose that the stress (i.e., $\boldsymbol{\tau} = \nu\partial\boldsymbol{u}/\partial z$) is non-zero only in a shallow layer — an Ekman layer — in the upper ocean. Below this layer we have, from (22.13a,b),

$$-fv = -\frac{\partial\phi}{\partial x} - ru, \qquad fu = -\frac{\partial\phi}{\partial y} - rv. \tag{22.17a,b}$$

and using (22.16) we obtain

$$-fv' = -\frac{\tau^x}{H} - ru', \qquad fu' = -\frac{\tau^y}{H} - rv'. \tag{22.18a,b}$$

where $u' \equiv u - U/H$ and $v' \equiv v - V/H$ is the deviation of the flow from the vertical average (i.e., the deviation from Sverdrup balance). That is to say, we may solve the equations assuming the Sverdrup flow is zero, and add it back in at the end of the day, noting also that the presence of a Sverdrup flow makes no difference to the vertical velocity. Given this, we'll drop the prime on u' and v' unless ambiguity would arise. Solving for u and v gives the expressions for the deep flow, namely

$$u = \frac{-\tau^x r - \tau^y f}{H(r^2 + f^2)}, \qquad v = \frac{\tau^x f - \tau^y r}{H(r^2 + f^2)}. \tag{22.19a,b}$$

The transport in the Ekman layer at the surface is in the opposite direction to the deep flow, in order to satisfy the integral constraints that $\int u\,\mathrm{d}z = \int v\,\mathrm{d}z = 0$. To complete the solution we use the mass continuity equation, (22.13c), to obtain w, giving

$$w = -\frac{(z + H)}{H}\,\frac{\beta(r^2 - \beta^2 y^2)\tau^x + 2r\beta^2 y\tau^y}{(r^2 + \beta^2 y^2)^2}. \tag{22.20}$$

To better understand these solutions it is useful to look at the nondimensional form, and we obtain that by setting

$$(u, v) = (\widehat{u}, \widehat{v})\frac{\tau}{2\Omega H}, \quad y = \widehat{y}a, \quad (\tau^x, \tau^y) = (\widehat{\tau}^x, \widehat{\tau}^y)\tau, \quad \beta = \frac{2\Omega}{a}, \tag{22.21}$$

where a hat denotes a nondimensional quantity and a is the radius of Earth. The nondimensional versions of (22.18) are then

$$-\widehat{y}\widehat{v} = -E_r\widehat{u} - \widehat{\tau}^x, \qquad \widehat{y}\widehat{u} = -E_r\widehat{v} - \widehat{\tau}^y, \tag{22.22}$$

where $E_r = r/(2\Omega)$ is a horizontal Ekman number and if, for example, the wind is zonal and westward then $\widehat{\tau}^x = -1$ and $\tau^y = 0$. The nondimensional versions of (22.19) are

$$\widehat{u} = \frac{-E_r\widehat{\tau}^x - \widehat{\tau}^y\widehat{y}}{E_r^2 + \widehat{y}^2}, \qquad \widehat{v} = \frac{\widehat{\tau}^x\widehat{y} - \widehat{\tau}^y E_r}{E_r^2 + \widehat{y}^2}. \tag{22.23a,b}$$

The overall strength of the undercurrent scales, unsurprisingly given the nature of the model, with the wind stress, and the Ekman number determines the width and height of the profile. A

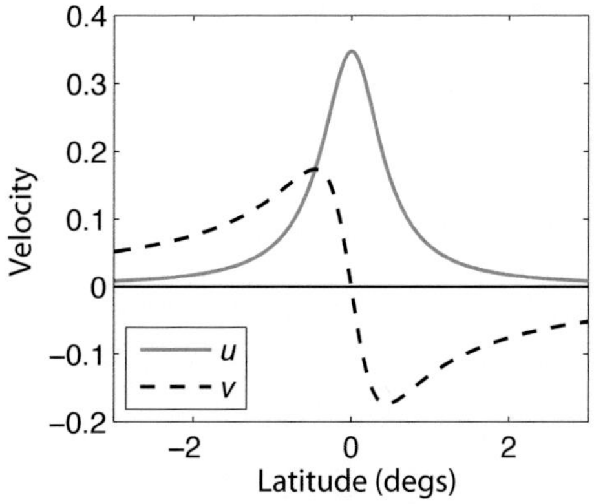

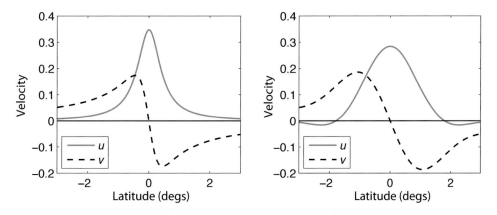

Fig. 22.5 Horizontal profiles of the undercurrent with friction represented by a linear drag [left, using (22.19)] and by a harmonic viscosity [right, using (22.43)], nominally in dimensional units (metres/second and degrees).

typical solution is plotted in Fig. 22.5 (along with a solution using harmonic friction that we discuss later). The parameters are $\widehat{\tau}^x = -1, \widehat{\tau}^y = 0$ and $E_r = 8 \times 10^{-3}$, which corresponds to a purely westward wind and a frictional decay timescale of about 10 days. If we further suppose that the dimensional value of the stress is about $4 \times 10^{-2}\,\mathrm{N/m^2}$ and take $H = 100$ m we obtain the dimensional values shown in the plot. The zonal flow as given by (22.19a) is then *eastward*, for the reason we have mentioned before, namely that, overall, the wind is balanced by an opposing pressure gradient and the deep ocean feels the pressure gradient but not the wind stress; thus, the deep zonal flow is in the opposite direction to the surface wind. The deep meridional flow is zero at the equator, where $f = 0$, but is toward the equator in both hemispheres and so induces equatorial upwelling. The shallow Ekman-layer flow is directed away from the equator, in order that the vertically integrated flow is zero. A consequence of this is that the vertical velocity is positive — that is, there is *upwelling* at the equator, as can be seen directly from (22.20) when $\tau^y = 0$, $y = 0$ and $\tau^x < 0$.

The zonal undercurrent falls with latitude with a width proportional to E_r. The peak value at the equator is proportional to E_r^{-1}, so that by reducing the drag we make the equatorial peak sharper. However, and as that scaling suggests, the overall transport is independent of E_r (at least for a constant, zonal stress). To see this we integrate (22.23) with $\widehat{\tau}^x = -1$ and $\widehat{\tau}^y = 0$:

$$\widehat{U}_T = \int_{-\infty}^{\infty} \widehat{u}\,\mathrm{d}\widehat{y} = \int_{-\infty}^{\infty} \frac{E_r}{E_r^2 + \widehat{y}^2}\,\mathrm{d}\widehat{y} = \left[\tan^{-1}\frac{\widehat{y}}{E_r}\right]_{-\infty}^{\infty} = \pi. \tag{22.24}$$

Dimensionally, this translates to

$$U_T = H\int_{-\infty}^{\infty} u\,\mathrm{d}y = \frac{-\pi a \tau^x}{2\Omega}. \tag{22.25}$$

It is pleasing that the total transport of the undercurrent does not depend on the rather poorly-constrained frictional coefficient, although the transport as given by (22.25) is somewhat smaller than observed. This can be guessed from Fig. 22.5 where the parameters are such that the width of the undercurrent is similar to that observed but its magnitude is too low (compare with Fig. 22.1). If we take $\tau^x = 4\times10^{-2}\,\mathrm{N/m^2}$ then using (22.25) we obtain a transport of about $5\times10^6\,\mathrm{m^3\,s^{-1}}$ or 5 Sv whereas the observed transport, with the vertical average (i.e., the Sverdrup flow) removed is 10–15 Sverdrups. Part of the discrepancy may come from the neglect of nonlinearity and stratification, and part of it from there being an inertial component to the equatorial undercurrent that is not a local response to the wind field, as we discuss in Section 22.4.

The expressions are also useful when the wind is not purely zonal. In the somewhat less realistic situation in which the wind is northward ($\tau^y > 0, \tau^x = 0$), the deep flow is southward. If the wind blows toward the northwest the undercurrent flows down the pressure gradient to the southeast.

Vertical structure at the equator

Because there is no lateral friction the solution at the equator is independent of the solution elsewhere and an analytic form for the vertical profile may easily be obtained. The Coriolis parameter is zero and so, from (22.13), the equations of motion become

$$0 = -\frac{\partial \phi}{\partial x} + \nu \frac{\partial^2 u}{\partial z^2} - ru, \qquad 0 = -\frac{\partial \phi}{\partial y} + \nu \frac{\partial^2 v}{\partial z^2} - rv. \tag{22.26a,b}$$

If the meridional wind stress at the surface is zero (i.e., $\nu_z \partial v/\partial z = 0$ at $z = 0$) then $v = 0$ everywhere. The zonal pressure gradient is given by (22.16a) and the zonal flow is then given by the solution of

$$\nu_z \frac{\partial^2 u}{\partial z^2} - ru = \frac{\tau^x}{H}, \tag{22.27}$$

with boundary conditions

$$\nu_z \frac{\partial u}{\partial z} = \begin{cases} \tau^x & \text{at } z = 0, \\ 0 & \text{at } z = -H. \end{cases} \tag{22.28}$$

The solution is easily found to be

$$u = A\mathrm{e}^{\alpha z} + B\mathrm{e}^{-\alpha z} - \frac{\tau^x}{Hr}, \tag{22.29}$$

where $\alpha = \sqrt{r/\nu_z}$ and A and B are obtained from the boundary conditions. We find

$$A = \frac{\tau^x}{\sqrt{\nu_z r}} \left(\frac{\mathrm{e}^{\alpha H}}{\mathrm{e}^{\alpha H} - \mathrm{e}^{-\alpha H}} \right), \qquad B = \frac{\tau^x}{\sqrt{\nu_z r}} \left(\frac{\mathrm{e}^{-\alpha H}}{\mathrm{e}^{\alpha H} - \mathrm{e}^{-\alpha H}} \right). \tag{22.30a,b}$$

A key parameter is the depth scale $d = \alpha^{-1} = \sqrt{\nu_z/r}$ that determines the depth to which the surface flow extends: if ν_z/r is small, the flow in the direction of the wind is confined to a shallow layer near the surface with the undercurrent beneath. A few example solutions are illustrated in Fig. 22.6.

These solutions indicate one failing of this simple model: the undercurrent is too deep and in fact extends all the way to the bottom of the ocean; evidently the model fails to reproduce a coherent, focused eastward flowing jet of finite vertical extent such as is seen in Fig. 22.1. The main ingredient that can overcome this limitation is stratification, with nonlinearity an important secondary effect, and we'll consider how this constrains the vertical extent in Section 22.3.4.

A note on the undercurrent in the presence of a Sverdrup flow

The zonal winds in the tropics have a minimum in the westward flow, that is a local maximum in u, at the equator and produce an eastward vertically integrated (Sverdrup) flow, as sketched by the solid line in Fig. 22.4. It seems natural to associate this flow with the eastward undercurrent but this can be misleading. The Sverdrup flow is produced by the wind-stress curl whereas the undercurrent is a consequence of the wind itself, and the two are not necessarily in the same direction. If, for example, the meridional variation of the wind differed, and were more akin to the dashed line in Fig. 22.4, then the Sverdrup flow would be westward. Whether the undercurrent would be eastward or westward now depends on the relative strength of the Sverdrup flow as well as other parameters, as the following very simple calculation shows.

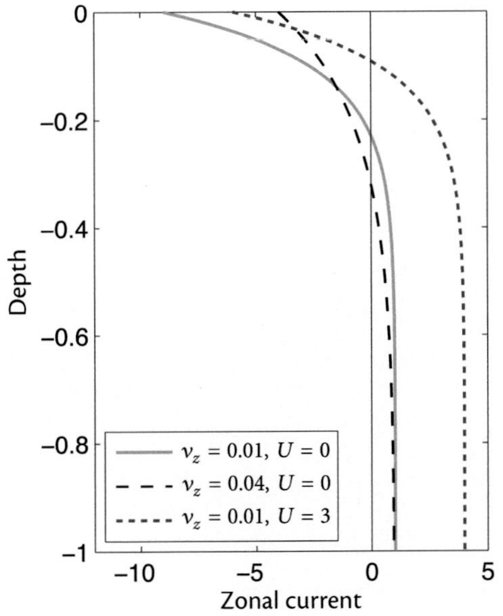

Fig. 22.6 Vertical profile of the zonal current at the equator, obtained using the analytic solutions (22.29) and (22.30), with $H = 1, r = 1, \tau^x = -1$ and the values of ν_z and U indicated in the legend.

The deep flow is the superposition of the Sverdrup flow, U, and the vertically varying flow, so that at the equator and with $\tau^y = 0$ the deep flow is approximately given by

$$u = \frac{-\tau^x + rU}{Hr}, \qquad v = V/H. \tag{22.31}$$

If the magnitude of U is sufficiently large then the zonal undercurrent u will take the sign of U, rather than automatically opposing the direction of the wind stress. In this simple linear model, the deep flow is just the sum of two components, one proportional to and opposing the surface wind stress, and one in the direction of the Sverdrup flow. With the wind as it is today, the two effects reinforce each other and for that reason the undercurrent is significantly stronger than the surface flow, but this is not a general rule.

22.3.3 ♦ Effect of Horizontal Viscosity

We now explore the effects of using a horizontal viscosity in place of a linear drag. Neither horizontal viscosity nor linear drag are wholly defensible representations of frictional effects, so that one purpose of this exercise is to see what aspects of the solution are robust to the choice made.

Formulating the problem

As we see from Fig. 22.2, meridional variations tend to occur on a smaller scale than zonal variations, so we'll neglect the zonal derivatives in the lateral friction. Our equations of motion then become

$$-fv = -\frac{\partial \phi}{\partial x} + \nu_z \frac{\partial^2 u}{\partial z^2} + \nu_h \frac{\partial^2 u}{\partial y^2}, \tag{22.32a}$$

$$fu = -\frac{\partial \phi}{\partial y} + \nu_z \frac{\partial^2 v}{\partial z^2} + \nu_h \frac{\partial^2 v}{\partial y^2}, \tag{22.32b}$$

$$\frac{\partial u}{\partial x} + \frac{\partial v}{\partial y} + \frac{\partial w}{\partial z} = 0, \tag{22.32c}$$

The Mean Equatorial Currents

The main observed features of tropical currents are:

- A vertically integrated flow that is in approximate Sverdrup balance, but with non-negligible contributions from nonlinearity and friction. At the equator this flow is eastward, flanked by narrow westward and eastward moving strips, transitioning to broader westward flow polewards of about 10° latitude that is part of the main sub-tropical gyres.
- A shallow westward flow at the equator, no more than a few tens of meters deep and a few degrees of latitude wide, with speeds of a few tens of centimetres per second.
- A strong eastward flowing undercurrent, typically from about 50 m to 200 m depth and a few degrees of latitude wide, with velocities up to a metre per second.

The leading order dynamics of this flow is roughly as follows:

- The zonal Sverdrup flow is proportional to the meridional derivative of the wind-stress curl, and so roughly to $-\partial^2 u_s/\partial y^2$ where u_s is the surface zonal wind. The vertically integrated eastward flow at the equator is thus a response to the *minimum* in the westward trade winds at the equator.
- The shallow surface westward flow, and the strong eastward undercurrent, are primarily a response to the westward winds themselves, rather than the curl of the winds, and so are very robust features. (See the shaded box on page 877 for more about the equatorial undercurrent.)
- If the surface zonal winds were uniformly westward in the tropics, or had a westward maximum at the equator, the vertically integrated flow would be quite different and might be westward because of the dependence of the zonal flow on the second derivative of the wind stress in Sverdrup theory. However, there might well still be an eastward equatorial undercurrent, depending on the strength of the Sverdrup flow.

where $f = \beta y$ and with boundary conditions given by (22.14), as before. The vertically integrated horizontal flow, (U, V), satisfies

$$-fV = -H\frac{\partial \phi}{\partial x} + \tau^x + \nu_h \frac{\partial^2 U}{\partial y^2}, \qquad (22.33a)$$

$$fU = -H\frac{\partial \phi}{\partial y} + \tau^y + \nu_h \frac{\partial^2 V}{\partial y^2}, \qquad (22.33b)$$

$$\frac{\partial U}{\partial x} + \frac{\partial V}{\partial y} = 0, \qquad (22.33c)$$

and cross differentiating leads to an equation similar to (22.15), namely

$$\nu_z \nabla^2 \frac{\partial^2 \psi}{\partial y^2} + \beta \frac{\partial \psi}{\partial x} = \mathrm{curl}_z \boldsymbol{\tau}. \qquad (22.34)$$

Once again, in the absence of a wind-stress curl, the vertically integrated flow is zero and the wind stress is balanced by a pressure gradient. The flow relative to the vertical average, (u', v'), is given by subtracting (22.33) (divided by H) from (22.32) giving

$$-fv' = \nu_z \frac{\partial^2 u'}{\partial z^2} + \nu_h \frac{\partial^2 u'}{\partial y^2} - \frac{\tau^x}{H}, \tag{22.35a}$$

$$fu' = \nu_z \frac{\partial^2 v'}{\partial z^2} + \nu_h \frac{\partial^2 v'}{\partial y^2} - \frac{\tau^y}{H}, \tag{22.35b}$$

$$\frac{\partial u'}{\partial x} + \frac{\partial v'}{\partial y} + \frac{\partial w}{\partial z} = 0. \tag{22.35c}$$

This is independent of the vertical average itself (as was the case with the linear drag) and henceforth, we'll take the vertical averaged flow to be zero and drop the prime on the velocity, with the understanding that it may be added back as needed. A full solution of (22.35) is both difficult to obtain and uninformative, so we will concentrate on various special cases, as follows.

Solution away from the equator

Away from the equator we neglect the horizontal friction terms and (22.35) becomes

$$- fv = \nu_z \frac{\partial^2 u}{\partial z^2} - \frac{\tau^x}{H}, \qquad fu = \nu_z \frac{\partial^2 v}{\partial z^2} - \frac{\tau^y}{H}, \tag{22.36a,b}$$

The particular solution to this is the depth independent flow,

$$v_p = \frac{\tau^x}{fH}, \qquad u_p = \frac{-\tau^y}{fH}. \tag{22.37a,b}$$

To this we must add the solution of the homogeneous equation

$$\nu_z \frac{\partial^2 u}{\partial z^2} + fv = 0, \qquad \nu_z \frac{\partial^2 v}{\partial z^2} - fu = 0. \tag{22.38a,b}$$

These are the equations for an Ekman layer, as encountered in Section 5.7. As there, the solution spirals down from the surface while decaying exponentially with an e-folding depth of $\sqrt{2\nu_z/f}$. The transport in the Ekman layer, $(\tau^y/f, -\tau^x/f)$, is equal and opposite to the transport of the particular solution so that the total transport, relative to the vertical average, is indeed zero.

Solution below the Ekman layer

When f is small the lateral friction terms cannot be ignored and we are left with the full problem again. However, below the surface layer (which is the Ekman layer itself except very close to the equator) the vertical friction may be neglected and we can obtain a solution analogous to (22.19). The flow in this deep layer satisfies

$$- fv = \nu_h \frac{\partial^2 u}{\partial y^2} - \frac{\tau^x}{H}, \qquad fu = \nu_h \frac{\partial^2 v}{\partial y^2} - \frac{\tau^y}{H}, \tag{22.39}$$

where $f = \beta y$, and these equations are very similar to (22.18). We nondimensionalize by setting

$$(u, v) = (\widehat{u}, \widehat{v})\frac{\tau}{2\Omega H}, \quad y = \widehat{y}a, \quad (\tau^x, \tau^y) = (\widehat{\tau}^x, \widehat{\tau}^y)\tau, \quad \beta = \frac{2\Omega}{a}, \tag{22.40}$$

where a hat denotes a nondimensional quantity and a is the radius of Earth. The nondimensional versions of (22.39) are then

$$- \widehat{y}\widehat{v} = E_h \frac{\partial^2 \widehat{u}}{\partial \widehat{y}^2} - \widehat{\tau}^x, \qquad \widehat{y}\widehat{u} = E_h \frac{\partial^2 \widehat{v}}{\partial \widehat{y}^2} - \widehat{\tau}^y, \tag{22.41}$$

where $E_h = \nu_h/(2\Omega a^2)$ is a horizontal Ekman number.

The easiest way to obtain a solution is to multiply the second equation by i (i.e., $\sqrt{-1}$) and add to the first, to give

$$E_h \frac{\partial^2 Z}{\partial \hat{y}^2} - \mathrm{i}\hat{y}Z = T, \tag{22.42}$$

where $Z \equiv \hat{u} + \mathrm{i}\hat{v}$ and $T = \hat{\tau}^x + \mathrm{i}\hat{\tau}^y$, which we henceforth take to be equal to -1 (i.e., a purely westward stress). Equation (22.42) is a particular form of Airy's equation and its solution is given by[6]

$$Z(\hat{y}) = \int_0^\infty \exp\left[-E_h\alpha^3/3 - \mathrm{i}y\alpha\right] \mathrm{d}\alpha. \tag{22.43}$$

This solution asymptotes to the geostrophic balance $Z = 1/(\mathrm{i}\hat{y})$ (i.e., $u = 0, v = \partial\phi/\partial x = \tau^x/fH$) for large $|\hat{y}|$. The solution, just like the one obtained using a linear drag, has total transport that is independent of the frictional coefficient; that is

$$\int_{-\infty}^{\infty} \hat{u}\,\mathrm{d}\hat{y} = \pi \qquad \text{or} \qquad H\int_{-\infty}^{\infty} u\,\mathrm{d}y = -\tau^x \frac{\pi a}{2\Omega}. \tag{22.44}$$

The mathematical derivation of this is left as a (tricky) exercise for the reader or a literature search. The integral is in fact exactly the same as that obtained using a linear drag, so that the quantitative underestimate of the magnitude of the undercurrent remains. The lack of dependence of the total transport on the viscosity arises because the width of the undercurrent increases with the (one third power of the) horizontal viscosity but the peak value diminishes with the (one third power of the) viscosity. The dependence on the one third power follows from a simple scaling of (22.42): at large $\hat{y}$ the flow is geostrophic and lateral friction unimportant, whereas at small $\hat{y}$ the lateral friction is required to remove the equatorial singularity. Thus, the nondimensional width of the undercurrent, $\hat{L}$ say, is determined by the requirement that the terms on the left-hand side of (22.44) are both important and so that

$$\frac{E_h}{\hat{L}^2} \sim \hat{L}. \tag{22.45}$$

Dimensionally, this translates to

$$L \sim E_h^{1/3} a = \left(\frac{\nu_h a}{2\Omega}\right)^{1/3} \sim 100\,\mathrm{km}, \tag{22.46}$$

if $E_h \approx 10^{-6}$ (which implies $\nu_h \approx 10^4\,\mathrm{m^2\,s^{-1}}$, but this value should not be seen as fundamental).

Horizontal profiles of u and v obtained from (22.43) are plotted in the right-hand panel of Fig. 22.5, and may be compared with the corresponding solutions obtained with a linear drag in the left-hand panel. The results shown are obtained with $E_h = 2 \times 10^{-6}$ but otherwise the same values as were used with a linear drag. Evidently, the results with the two frictional schemes display the same qualitative features, with a peak at the equator and a decay away, and a meridional velocity directed toward the equator in both hemispheres, which gives rise to equatorial upwelling. Thus, as with the Stommel (linear drag) and Munk (harmonic friction) solutions for an ocean-gyre, the similarity of solutions with two different forms of solution gives confidence in the insensitivity of the solution to the form of the frictional terms.

22.3.4 Effects of Stratification: A Layered Model of the Undercurrent

One unrealistic aspect of the models described above is that the undercurrent appears to extend all the way to the bottom of the ocean, whereas in reality it is confined to the upper few hundred meters of the ocean, with the deeper fluid being almost quiescent. A potential reason for this discrepancy is that we have neglected stratification, which tends to limit vertical communication

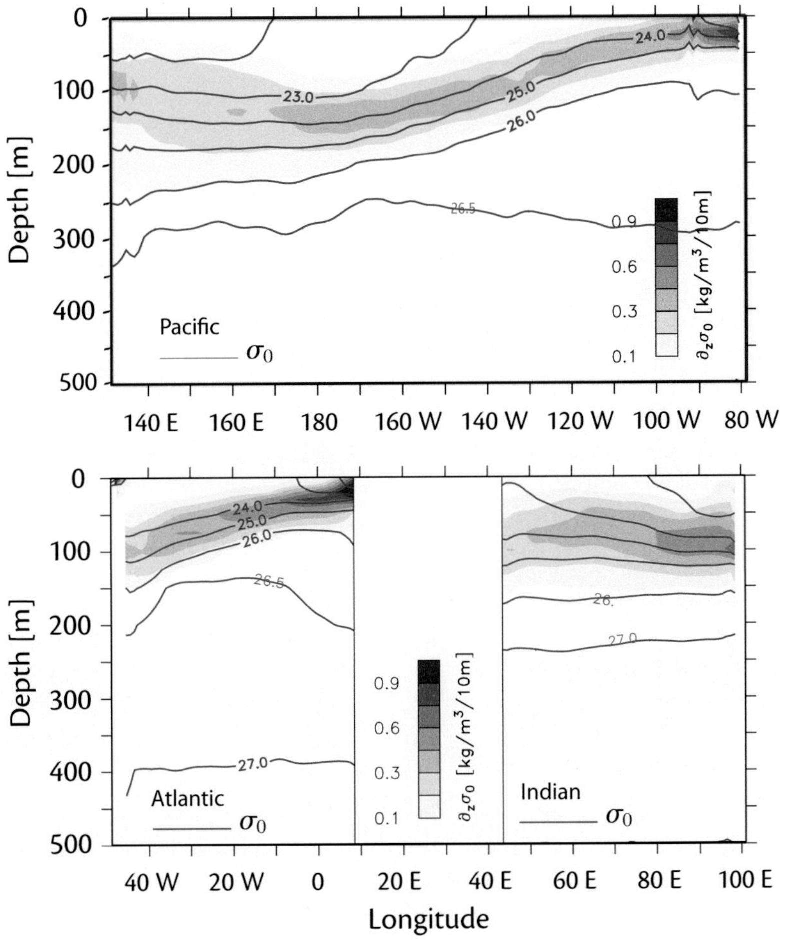

Fig. 22.7 Zonal sections of annual mean density at the equator in the Pacific (top) and Atlantic and Indian oceans. The contours are of potential density and the shading is the vertical derivative of potential density, with data from the World Ocean Atlas.

within an ocean column. Let's try to model this with a simple layered model, reverting to the use of linear drag.

Let us suppose that the ocean consists of two homogeneous layers. The continuous, homogeneous model described above describes the solution in the upper layer while the lower layer, of slightly greater density, represents the abyssal ocean and is assumed stationary. The pressure gradient must therefore be zero in the lower level, and we will see that this requires that the interface between the layers must slope, and indeed will usually slope upwards toward the east. The interface is, of course, a crude representation of the equatorial thermocline.

The zonal pressure gradient at the equator at the base of the upper layer is given by (22.16a), namely $\partial\phi/\partial x = (\tau^x - rU)/H$, and this is usually negative. If the upper layer has a density ρ_1 and the lower layer has a density ρ_2 then, in order for there to be no pressure gradient in the lower layer the interface must slope by an amount

$$s \equiv \frac{\partial z}{\partial x} = -\frac{1}{g'}\frac{\partial \phi}{\partial x}, \tag{22.47}$$

where $g' \equiv g\Delta\rho/\rho_1 \equiv g(\rho_2 - \rho_1)/\rho_1$ is the reduced gravity and, as we recall, $\phi \equiv p/\rho_1$. Thus, an estimate of the slope of the thermocline is

$$s \approx \frac{1}{g'H}(\tau^x - rU). \tag{22.48}$$

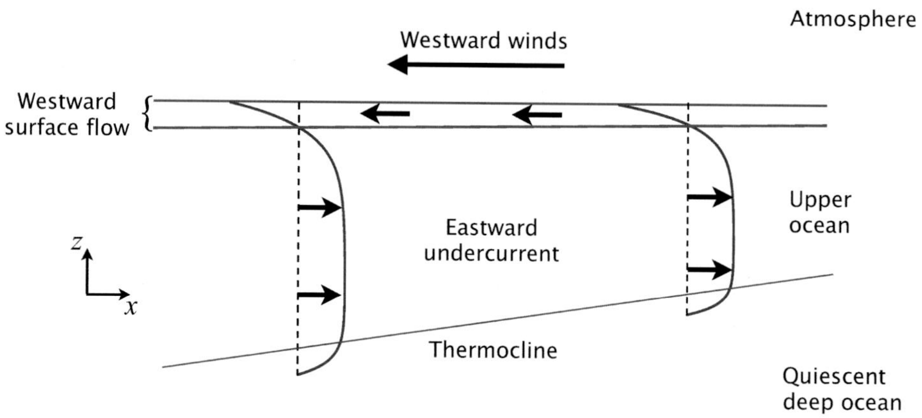

Fig. 22.8 A zonal section of the local model of the undercurrent, showing a near-surface flow in the direction of the wind, a counter-flowing undercurrent beneath and a quiescent deep ocean below the thermocline.

The quantitative effects of the Sverdrup flow are hard to gauge because of the rather ill-constrained frictional coefficient r. The mean wind stress at the equator is westward and about $0.04\,\mathrm{N\,m^{-2}}$, and with $\Delta\rho = \rho_2 - \rho_1 = 3\,\mathrm{kg\,m^{-3}}$ and $H = 200\,\mathrm{m}$ we find

$$s \approx \frac{\tau^{*x}}{g\Delta\rho H} \approx \frac{0.04}{10 \times 3 \times 200} = 6.7 \times 10^{-6}. \tag{22.49}$$

This suggests that over the 15 000 km extent of the equatorial Pacific we might expect the thermocline to shoal upwards toward the east by about 100 m. This slope is comparable to that observed (see Fig. 22.7), although considering the simplicity of the model the agreement is perhaps a little fortuitous. The thermocline slopes up toward the east in both the Atlantic and Pacific, where the prevailing winds are westward, but not in the Indian Ocean where the prevailing winds are seasonally variable because of the monsoons. The undercurrent itself is also a seasonal phenomenon in the Indian Ocean.

Except for the presence of the frictional coefficient, (22.48) is fairly insensitive to the details of the model; a virtually identical expression results if we model the ocean as two immiscible layers of fluid using the shallow water equations. The parameters determining the thermocline slope are just the thickness, H, of the upper layer and the density difference, $\Delta\rho$, between the upper and lower layers. A schematic of the flow is given in Fig. 22.8.

22.4 AN IDEAL FLUID MODEL OF THE EQUATORIAL UNDERCURRENT

The model of the equatorial undercurrent presented in the previous sections is physically appealing and transparent, but it has two potential shortcomings:

(i) The detailed results depend on the frictional parameters chosen. This is unsatisfactory because we have no well-founded basis on which to choose those parameters.

(ii) The model makes no connection to the extra-tropical circulation of the ocean; that is, all the dynamics are essentially local.

The second shortcoming is of no import in itself — the real ocean might be that way. However, observations suggest that at least some of the water in the equatorial undercurrent has its origins in the subtropical gyre: temperatures in the core of the undercurrent are mostly in the range of 16° C to 22° C, rather lower than the surface temperatures in the equatorial region except at the

The Equatorial Undercurrent

What is it?

The equatorial undercurrent is the single most striking feature of the low latitude ocean circulation. It is an eastward flowing subsurface current, mostly confined to depths between about 50 m and 250 m and to latitudes within 2° of the equator, with speeds of up to $1\ \mathrm{m\,s^{-1}}$ (Fig. 22.1). It is sometimes connected to an eastward flowing current a few degrees north or south of the equator. The undercurrent is a permanent feature of the Atlantic and Pacific Oceans, but varies with season in the Indian Ocean because of the monsoon winds.

What are its dynamics?

Most models of the equatorial undercurrent lie between two idealized end members that we refer to as the *local theory* and the *inertial theory*.[5]

- The local theory regards the undercurrent as a direct response to the westward winds at the equator. The winds push water westward and create a balancing eastward pressure gradient force. Below a frictional surface layer the influence of the wind stress is small and the pressure gradient leads to an eastward undercurrent.
  - In the frictional surface layer the flow is away from the equator and there is upwelling at the equator. The circulation is closed in the equatorial region. Continuous stratification may be included in the theory, although if there is upwelling through stratified water the diapycnal diffusivity must be non-zero.
  - The dynamics of the simplest models of this ilk are linear, but their quantification relies on the use of somewhat poorly constrained frictional and mixing coefficients.

- In the inertial theory, the equatorial current system is connected to the extra-equatorial region. A subsurface current moves inertially from higher latitudes, conserving its potential vorticity (which includes, crucially, a relative vorticity component) and Bernoulli function into the equatorial region. A pressure head is created in the western equatorial basin, which then pushes the undercurrent along.
  - Even if there were no wind at the equator the theory, in its simplest form, would still predict the presence of an undercurrent.
  - The theory, which is unavoidably nonlinear, contains parameters that must be specified somewhat arbitrarily but to which results are not especially sensitive.

- In reality, the undercurrent contains aspects of both theories, and more. Neither theory can be entirely correct; the local theories do not properly take into account distant effects and, in contrast to the inertial theory, numerical experiments show that the undercurrent *does* depend on the wind at the equator.
  - Part of the undercurrent is closed within the equatorial region, and part connects to higher latitudes. A more complete model involves treating the equatorial undercurrent as one branch of a more complex tropical current system.
  - It would be hard, perhaps impossible, to construct a theory of the whole system that is elegant, complete and correct. Understanding arises via careful treatments of special cases along with numerical and conceptual models of the areas in between.

eastern end of the ocean basins — that is, at the end of the undercurrent. Furthermore, as we see from the upper two panels Fig. 22.1, the current gains strength as it moves eastward, implying that it is drawing water from higher latitudes as it moves.

Modern observational analyses of the equatorial ocean indeed suggest that the equatorial current system is a three-dimensional beast, connecting smoothly with the subtropical current system described in earlier chapters.[7] As subsurface water approaches the equator it largely rises along isopycnal surfaces as it moves eastward, with the cross-isopycnal velocity being only a small fraction of the total vertical velocity. This is in some contrast to the more local picture imagined in Section 22.3 in which there is overturning in the vertical-meridional plane, and hence (to the extent that the water is stratified) with cross-isopycnal upwelling at the equator.

The above discussion suggests that it would be useful to construct a model that both connects to the subtropics and does not depend in any essential way on dissipative processes. That is, we should try to construct an ideal fluid model of the equatorial ocean. We'll do this in a way that is analogous to our treatment of the ventilated thermocline in Chapter 20. That is, we'll represent the vertical structure of the ocean with a small number (one or two) of immiscible layers, and we'll assume that the subsurface layer conserves its potential vorticity.[8]

22.4.1 A Simple Barotropic Model

Suppose that a fluid parcel at some latitude moves toward the equator, preserving its potential vorticity in a shallow-water system. If the fluid parcel originates from a latitude y_0 where, we suppose, its relative vorticity is negligible then, as it moves its vorticity, ζ, is determined by

$$\frac{f+\zeta}{h} = \frac{f_0}{h_0} \tag{22.50}$$

where, on the equatorial beta plane, $f = \beta y$, $f_0 = f(y_0) = \beta y_0$ and h_0 is the depth of the fluid column at y_0. If, simplifying still further, the depth of the fluid column is assumed constant and meridional derivatives are much larger than zonal derivatives so that $\zeta = \partial v/\partial x - \partial u/\partial y \approx -\partial u/\partial y$, we have

$$\beta y - \frac{\partial u}{\partial y} = \beta y_0. \tag{22.51}$$

Integrating this expression, with $u = 0$ at $y = y_0$, gives

$$u = \frac{\beta}{2}(y - y_0)^2. \tag{22.52}$$

Interestingly, at $y = 0$, $u = \beta y_0^2/2$, which is positive. That is, conservation of absolute vorticity has, virtually by itself, produced an eastward flowing current at the equator (Fig. 22.9). The solution resembles the angular momentum conserving solution to the equinoctial Hadley Cell discussed in Section 14.3, specifically Equation (14.38) but with a different constant of integration: essentially, in the atmospheric case $y_0 = 0$, because the meridionally moving air in the upper branch of the Hadley Cell originates at the equator in the equinoctial case. However, in the oceanic case we do not expect angular momentum to be conserved because of the presence of a zonal pressure gradient, absent in the zonally-averaged atmospheric case. Rather, it is absolute vorticity conservation, in its simplest form, that leads to (22.52).

However, from a quantitative standpoint the solution is not very satisfactory. It depends heavily on the value of y_0, and for y_0 greater than a few degrees the value of the zonal flow at the equator as predicted by the model is far too large, as can be inferred from Fig. 22.9. Also, the model eastward flow at the equator is not as jetlike as the undercurrent in the real ocean (Fig. 22.1). Nevertheless, the qualitative success suggests that it might be useful to proceed with a more complete model, in particular one in which the value of h does vary with latitude, perhaps accounting for a good fraction of the variation of the potential vorticity.

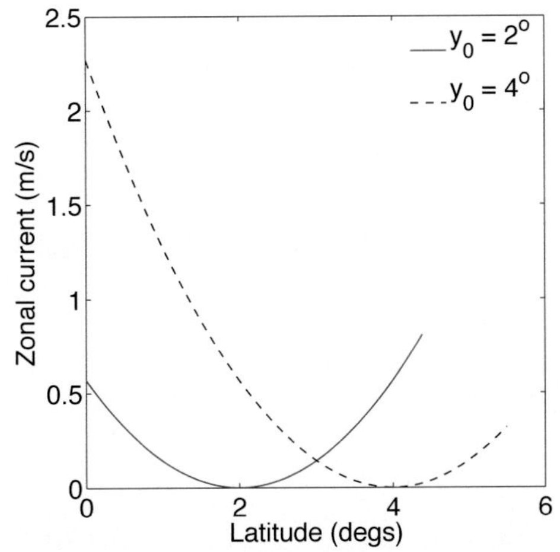

Fig. 22.9 Zonal current as produced by the absolute vorticity conserving model. Specifically, solutions are plotted of (22.52) with $y_0 = 2°$ and $y_0 = 4°$ ($\times 2\pi a/360$) and $\beta = 2\Omega/a = 2.27 \times 10^{-11}\,\mathrm{m^{-1}\,s^{-1}}$.

22.4.2 A Two-layer Model of the Inertial Undercurrent

We now extend the barotropic model to two moving layers. We presume that the flow in the lower level conserves potential vorticity, with a height field h determined in a self-consistent fashion rather than being fixed. Thus, the features of the model are as follows:

1. The use of the ideal form (i.e., inviscid, no dissipation) of the two-layer shallow water equations, with the lower level shielded from the wind's influence and conserving potential vorticity and giving rise to the equatorial undercurrent.
2. At low latitudes the equations are solved in a boundary-layer approximation, with variations in y being of smaller scale than variations in x. Unlike our treatment of the ventilated thermocline in mid-latitudes, no assumption is made that the flow satisfies the planetary geostrophic equations. It is the inertial terms that prevent the solution from becoming singular at the equator.
3. At higher latitudes the solutions are constructed to blend in with the solution of a mid-latitude ventilated thermocline model, described in Section 20.7. Put another way, the ventilated thermocline provides a high-latitude boundary condition for the model.

To get a sense of the overall flow the reader may look ahead to Fig. 22.10.

Equations of motion

Our primary concern will be the lower layer (layer 2) for which the momentum and mass continuity equations are, respectively,

$$\frac{\mathrm{D}\boldsymbol{u}_2}{\mathrm{D}t} + \boldsymbol{f} \times \boldsymbol{u}_2 = -\frac{1}{\rho_0}\nabla p_2 = -g_2'\nabla h \tag{22.53a}$$

$$\frac{\partial h_2}{\partial t} + \nabla\cdot(h_2\boldsymbol{u}_2) = 0, \tag{22.53b}$$

where h_2 is the thickness of the layer and $\boldsymbol{u}_2$ the horizontal velocity within it, and $g_2' = g(\rho_3 - \rho_2)/\rho_0$. We remain on the equatorial beta plane so that $\boldsymbol{f} = f\mathbf{k} = \beta y\,\mathbf{k}$, where $\mathbf{k}$ is the unit vector in the vertical direction, and we will consider only the steady versions of these equations. We may also write the momentum equation in terms of the Bernoulli function,

$$\frac{\partial u_2}{\partial t} + (f + \zeta_2)v_2 = -\frac{\partial B_2}{\partial x}, \qquad \frac{\partial v_2}{\partial t} + (f + \zeta_2)u_2 = -\frac{\partial B_2}{\partial y}, \tag{22.54}$$

where $B_2 = g_2' h + \boldsymbol{u}_2^2/2$.

The above equations conserve potential vorticity, $Q_2 = (f + \zeta)/h_2$ and, because the flow is presumed steady, the Bernoulli function. That is,

$$\boldsymbol{u}_2 \cdot \nabla Q_2 = 0, \qquad \boldsymbol{u}_2 \cdot \nabla B_2 = 0. \tag{22.55a,b}$$

Also, because of the form of the mass continuity equation in the steady state, namely $\nabla \cdot (h_2 \boldsymbol{u}_2) = 0$, we can define a streamfunction, ψ, such that

$$h_2 \boldsymbol{u}_2 = \mathbf{k} \times \nabla \psi \qquad \text{or} \qquad h u_2 = -\frac{\partial \psi}{\partial y}, \quad h v_2 = \frac{\partial \psi}{\partial x}, \tag{22.56}$$

Using the streamfunction, conservation of potential vorticity and Bernoulli function may be written as

$$J(\psi, Q_2) = 0, \qquad J(\psi, B_2) = 0, \tag{22.57a,b}$$

where $J(a,b) = \partial_x a \, \partial_y b - \partial_y a \, \partial_x b$. Equations (22.57a) and (22.57b) imply, respectively that isolines of Q_2 and ψ, as well as isolines of B_2 and ψ, are everywhere parallel to each other. Thus, in general, Q_2 is a function of B; that is,

$$Q_2 = F(B_2), \tag{22.58}$$

where the function, F, is as yet unknown. It is also the case that Q_2 is a function of ψ; that is, $Q_2 = G(\psi)$ where G is some other function. However, it is not the case that, in general, Q_2 is a function of the height field h, because h is not proportional to the streamfunction for the flow. This is in contrast to the mid-latitude case in which geostrophic balance may be written as $\boldsymbol{u}_2 = (g_2'/f)\mathbf{k} \times \nabla h$, and so the relation $\boldsymbol{u}_2 \cdot \nabla Q_2$ implies that Q_2 is a function of h itself. We do not assume geostrophic balance in the equatorial region.

The equations and the properties of the equations so far discussed are quite general (save for the restriction to the beta plane). Let us now consider the equatorial region, and then how it connects to the subtropics.

Equatorial dynamics

Let us first derive some elementary scaling relations between the variables. Consider motion within a narrow strip of distance no more than L_y from the equator where L_y is the characteristic meridional scale of the undercurrent, as yet undetermined. If L_x is the characteristic zonal scale, typically the scale of the ocean basin itself, then $L_y \ll L_x$. We expect that L_y will be the scale over which the relative vorticity becomes comparable to the planetary vorticity, or equivalently the scale such that the beta Rossby number is $\mathcal{O}(1)$. If the scale of the zonal velocity is U then this requirement is $U/(\beta L_y^2) = 1$ or

$$L_y = \left(\frac{U}{\beta}\right)^{1/2} \qquad \text{or} \qquad U = \beta L_y^2. \tag{22.59}$$

The disparity between zonal and meridional scales implies that there will also be a disparity between the zonal and meridional velocities, and in particular from the mass continuity equation we expect that

$$V = \frac{U L_y}{L_x}, \tag{22.60}$$

and so $V \ll U$, where V is the scale of the meridional velocity.

At a (non-zero) distance L_y from the equator the relevant Rossby number in the meridional momentum equation is given by $U/(\beta L_x^2)$, and this remains small. Thus, essentially because U is so much larger than V, even very close to the equator the *zonal* flow will be in near geostrophic balance with the meridional pressure gradient. The meridional momentum equation then becomes

$$\beta y u_2 = -g_2' \frac{\partial h}{\partial y}, \tag{22.61}$$

implying the scaling

$$H = \frac{\beta L_y^2 U}{g_2'} = \frac{\beta^2 L_y^4}{g_2'}, \tag{22.62}$$

using (22.59), where H is the scale of the variation of thickness in layer 2.

Now let's consider the equations themselves. As we noted the flow conserves potential vorticity, Q_2. Close to the equator $Q_2 \approx (f - \partial u_2/\partial y)/h_2$ so that, using (22.58),

$$\frac{\beta y - \partial u_2/\partial y}{h_2} = F(B_2), \tag{22.63}$$

where $B_2 = g_2'h + u_2^2/2$, noting that $|u_2| \gg |v_2|$ in the equatorial region. There is an obvious similarity between (22.63) and (22.51). Note also that (22.63) and (22.61) are ordinary differential equations, although of course u_2 and h do vary in x. If we knew the function $F(B_2)$, and we knew the upper layer thickness h_1, then the equations would be closed and we could find a solution. For this we turn to the dynamics in the subtropics.

Extra-equatorial dynamics

The role of the extra-equatorial region in out treatment is to provide a boundary condition for the equatorial dynamics, and to determine the functional relationship between potential vorticity and the Bernoulli function, $F(B_2)$. We will suppose that the fluid obeys the dynamics of the two-layer model of the ventilated thermocline discussed in Section 20.7, in which the fluid obey the planetary geostrophic equations. The total depth of the moving fluid, h, is given by

$$h^2 = -z_2 = \frac{D_0^2}{[1 + (g_1'/g_2')(1 - f/f_2)^2]}, \tag{22.64}$$

where

$$D_0^2(x, y) = -\frac{2f^2}{\beta g_2'} \int_x^{x_e} w_E(x', y)\, dx', \tag{22.65}$$

with $w_E = \mathrm{curl}_z(\boldsymbol{\tau}/f)$ being the vertical velocity at the base of the Ekman layer. Assuming the stress is zonal then, at low latitudes, $w_E \approx \beta\tau^x/f^2 = \tau^x/(\beta y^2)$. If the stress is also independent of longitude then we find

$$D_0^2 = \frac{-2(x_e - x)\tau^x}{g_2'} \tag{22.66}$$

so that

$$h^2 = \frac{-2(x_e - x)\tau^x}{g_2'[1 + (g_1'/g_2')(1 - f/f_2)^2]}. \tag{22.67}$$

The extra-equatorial solution is completed by noting the expressions of the depths of each layer as a function of the total depth, as in (20.99)

$$h_2 = z_1 - z_2 = \frac{f}{f_2}h \quad \text{and} \quad h_1 = -z_1 = \left(1 - \frac{f}{f_2}\right)h. \tag{22.68a,b}$$

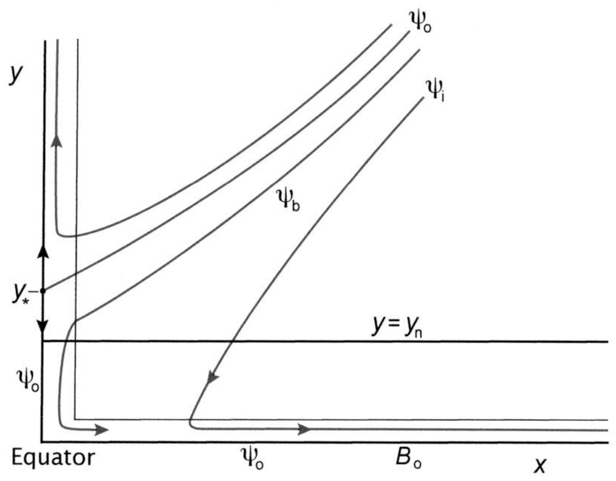

Fig. 22.10 Schematic of the flow streamlines leading to an equatorial undercurrent. Subsurface flow from the subtropics heads westward and equatorward, before veering equatorward and back eastward to form the equatorial undercurrent. The critical streamline ψ_0 hits the stagnation point on the western boundary at y_*, and the value of the Bernoulli function at the equator is equal to the value on this streamline.

Connections

We now start to connect the extra-equatorial solution to the tropical one. We first note that (22.67) provides a scaling relation for h; that is $h = \mathcal{O}(H)$ where

$$H = \left(\frac{L_x \tau}{g_2'}\right)^{1/2}. \tag{22.69}$$

Now, we are assuming that the equatorial dynamics transition smoothly to the extra-equatorial solution, so that (22.69) must be consistent with (22.62). Taken together they give us estimates for the meridional scale, the zonal velocity and the depth of the moving fluid purely in terms of external parameters, to wit:

$$L_y = \left(\frac{L_x \tau g_2'}{\beta^4}\right)^{1/8}, \qquad H = \left(\frac{L_x \tau}{g_2'}\right)^{1/2}, \qquad U = (g_2' \tau L_x)^{1/4}. \tag{22.70a,b,c}$$

These scalings are important results of the model, just as much as the precise form of the solution discussed below. Note that the scaling for zonal velocity is qualitatively different from that derived earlier using the frictional model — compare (22.70c) with (22.19a) or (22.25), for example. The dependence of U on the wind stress in (22.70c) is perhaps surprisingly weak, although both the layer thickness and the horizontal scale also increase with the wind so that the total transport increases almost linearly with wind stress. To the extent that the thickness of the upper layer stays constant, the transport of the lower layer scales as

$$HU = \left(\frac{L_x^3 \tau^3}{g_2'}\right)^{1/4} \quad \text{and} \quad HUL = \left(\frac{L_x^7 \tau^7}{g_2' \beta^4}\right)^{1/8}. \tag{22.71}$$

We now obtain the functional connection between Q_2 and B_2 necessary to close (22.63). In the extra-equatorial region, the horizontal shear becomes small compared to the Coriolis term so that (22.63) becomes

$$Q_2 = \frac{\beta y}{h_2} = F(B_2), \tag{22.72}$$

and the Bernoulli function itself, $B_2 = g_2' h + \boldsymbol{u}^2/2$, may be approximated by $B_2 = g_2' h$. Therefore, at the edge of the equatorial region,

$$Q_2 = \frac{f}{h_2} = \frac{f_2}{h}, \tag{22.73}$$

using (22.68a). This functional form holds throughout the equatorial region, and therefore $Q_2(h) = f_2/h$. More generally, $Q_2(\phi) = f_2/\phi$ for any variable ϕ and in particular,

$$Q_2(B_2) = \frac{f_2}{B_2} = \frac{f_2}{g_2'h + u_2^2/2}. \tag{22.74}$$

Our quest for the solution is now all over bar the shouting, in the sense that we can write down the equations of motion and the boundary conditions. Using geostrophic balance, (22.74), and (22.63), we write down the equations determining the subsurface flow in the equatorial region, namely

$$\frac{\beta y - \partial u_2/\partial y}{h_2} = \frac{f_2}{g_2'h + u_2^2/2}, \tag{22.75}$$

$$\beta y u_2 = -g_2'\frac{\partial h}{\partial y}, \tag{22.76}$$

$$h_2 = h - h_1. \tag{22.77}$$

In addition to specifying the value of the upper layer field, h_1, we need to specify the boundary conditions. At large values of y ($y/L_y \gg 1$) the value of h should be that given by (22.67). A second boundary condition may be applied at the equator, and if we suppose that the flow is hemispherically symmetric we have

$$v_2 = 0 \qquad \text{at } y = 0. \tag{22.78}$$

Taken with (22.54) this equation implies that, for steady flow, $\partial B_2/\partial x = 0$ and so that

$$B_2 = g_2'h + \frac{u_2^2}{2} = B_0 \qquad \text{at } y = 0, \tag{22.79}$$

where B_0 is a constant. That is to say, the equator is a streamline of the flow. (If there is flow across the equator the problem becomes more complicated, but we leave that for another day.) The value of B_0 is plausibly given by supposing it to be the value of B_2 at the western edge of the basin just outside the equatorial region but other choices might be made. Finally, we need to specify the field h_1, and there are a number of reasonable ways to proceed, although no obviously correct one. One choice would be to suppose that, just as in the extra-tropics, the total thickness of the moving layers is given by Sverdrup balance. If we were to do this we would essentially be extending the ventilated thermocline model all the way to the equator, with the addition of inertial terms. Although Sverdrup balance is qualitatively reasonable in equatorial regions (Fig. 22.2), quantitatively it is not particularly good and a simpler recipe is appropriate. One option is to choose h_1 to be a function of x only, such that the value of h_1 is equal to the value that it has at the high latitude edge of the equatorial region, at $y = y_n \gg L_y$. Using (22.68b) and (22.64) this gives

$$h_1^2 = \frac{D_0^2(1 - y_n/y_2)^2}{[1 + (g_1'/g_2')(1 - y_n/y_2)^2]} = \frac{-2(x_e - x)\tau^x(1 - y_n/y_2)^2}{g_2'\,[1 + (g_1'/g_2')(1 - y_n/y_2)^2]}, \tag{22.80}$$

using (22.66). The choice is simple albeit a little special, but it turns out that the solution is not especially sensitive to it. That it is a reasonable choice can be seen by noting that for $y_n \ll y_2$, $h \to h_1$ so that

$$h \to h_1 = \left[\frac{-2(x_e - x)\tau^x}{g_2'\,[1 + (g_1'/g_2')]}\right]^{1/2}. \tag{22.81}$$

Re-arranging and differentiating this expression with respect to x we obtain

$$\frac{\partial}{\partial x}\left(1 + \frac{g_1'}{g_2'}\right)h_1 = \frac{\tau^x}{g_2'h_1}. \tag{22.82}$$

That is, there is a balance between the applied wind stress and the pressure gradient force in the upper layer.

Equations of Motion for the Inertial Undercurrent

The dimensional forms of the equations of motion are

$$\frac{\partial u_2}{\partial y} - \beta y = \frac{f_2 h_2}{h + u_2^2/2}, \qquad (h_2 = h - h_1) \tag{U.1a}$$

$$\beta y u_2 = -g_2' \frac{\partial h}{\partial y}, \tag{U.1b}$$

where h_1 is a specified function of x. One plausible choice is given by choosing it to be the value of h_1 just outside the equatorial region, as given by (22.80).

Two boundary conditions are needed. The first is that in the extra-equatorial region h approaches the value given by the ventilated thermocline model, for example

$$h^2 = \frac{-2(x_e - x)\tau^x}{g_2' \left[1 + (g_1'/g_2')(1 - f/f_2)^2\right]}. \tag{U.2}$$

At the equator the boundary condition on h is obtained by setting $v_2 = 0$ and specifying the Bernoulli function, B_2, there. That is, we specify

$$B_2 = g_2' h + \frac{1}{2} u_2^2 = B_0 \qquad \text{at } y = 0. \tag{U.3}$$

Here, B_0 is a constant, chosen to be equal to the value of the Bernoulli function on the western edge of the basin just outside the equatorial region (that is, using (U.2) at $x = 0$ and $y = y_n$). There is then a pressure head at the western edge of the equatorial region, and the flow accelerates zonally along the equator preserving its Bernoulli function.

The nondimensional form of (U.1) is

$$\frac{\partial \hat{u}_2}{\partial \hat{y}} - \hat{y} = \frac{-y_2 \hat{h}_2}{\hat{h} + \hat{u}_2^2/2}, \qquad y\hat{u}_2 = -\frac{\partial \hat{h}}{\partial \hat{y}}, \tag{U.4}$$

where $y_2 = f_2/(\beta L_y)$.

The solution and its properties

The equations of motion and the boundary conditions for this model are summarized in the shaded box above. They are nonlinear and rather complex, and solutions must in general be obtained numerically by an iterative method. However, some qualitative properties (as in Fig. 22.10) may be deduced from the form of the equations, as describe below. We use Northern Hemisphere terminology, but the ideas apply equally to the Southern Hemisphere.

The ventilated thermocline gives rise to fluid that, at low latitudes, flows southward and westward. As it flows equatorward, potential vorticity conservation must lead to, in the absence of changes in layer thickness, an increase in relative vorticity — an anticlockwise or cyclonic turning — and the flow veers more southward and then eastward, so giving rise to the equatorial undercurrent. This property is present in the barotropic model of absolute vorticity conservation discussed in Section 22.4.1. However, the two-layer model differs from the barotropic model in two important regards. First, the layer thickness is allowed to change in a self-consistent fashion. Second, the

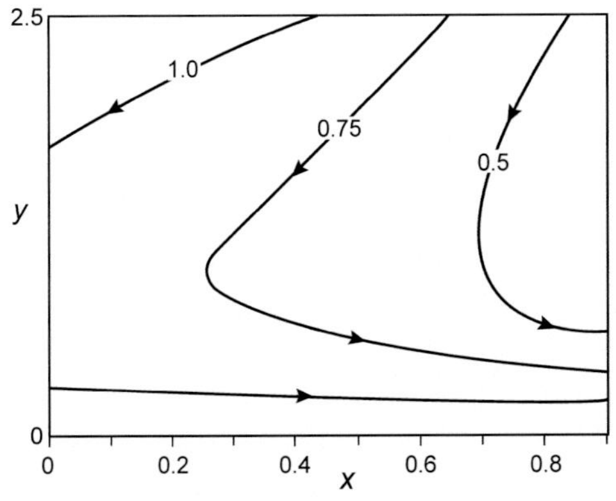

Fig. 22.11 The streamlines in a solution of the equations for an inertial equatorial undercurrent. The wind stress is constant, $g_1'/g_2' = 1$, $y_2 = 5$ and $B_0 = 1.26$.

flow is not particularly sensitive to the matching latitude at which we connect the equatorial equations of motion to the extra-equatorial region. This is because the ventilated thermocline model is itself based on conservation of potential vorticity, so that changing the matching latitude will have little effect on the potential vorticity entering the equatorial region. The two layer model does have some parameters that cannot be deduced a priori, in particular the thickness of the top layer and the value of the Bernoulli parameter, but the solutions are not especially sensitive to it. That is, and in common with some other models in geophysical fluid dynamics (for example, the Stommel model of western intensification in a gyre), the behaviour of the solutions is quite robust and transcends the detailed limitations of the model itself.

A numerically obtained solution is illustrated in Fig. 22.11. We see the streamlines sweeping westward and equatorward before taking a sharp equatorward and then eastward turn, with the flow being purely eastward at the equator. The solutions of u_2 and h are shown in Fig. 22.12, illustrating the formation of the undercurrent and its intensification as it moves eastward at the equator. Note also the latitudinal variation of the layer depth, h_2 ($h_2 = h - h_1$). In the extra-tropical ventilated thermocline, the thickness of layer 2 diminishes as we move equatorward, and if the limit of $y \to 0$ were taken the thickness of layer 2 would go to zero (a consequence of f going to zero and f/h being preserved). However, in the equatorial boundary layer the layer thickness actually increases as the equator is approached, and so the relative vorticity must increase to compensate, because now $(f + \zeta)/h$ is preserved. That is, the cyclonic intensification of the flow is somewhat more pronounced than in the barotropic model. Note also that the layer depth diminishes eastward; that is, the thermocline slopes up toward the east.

22.4.3 Relation of Inertial and Frictional Undercurrents

In the above sections we have discussed two different conceptual models of the equatorial thermocline. The first one is frictional and local, the second one is inertial and remote. In the local model, the westward winds set up a compensating pressure gradient, and below the frictional layer near the surface the pressure gradient dominates leading to an eastward flow. The only cross-latitudinal effects come from a lateral friction, and if this is replaced with a linear drag the zonal flow at the equator is wholly independent of the dynamics at other latitudes. In contrast, in the inertial model the undercurrent arises as a consequence of potential vorticity conservation of the subsurface flow, with the value of the potential vorticity set in the extra-equatorial region. The undercurrent is fed by extra-equatorial waters at all longitudes and so builds up as it moves eastward (as is observed). There is a pressure head at the western edge of the equatorial basin, so that the flow accelerates eastward *without the need for any winds at all at the equator.* It is the link with the geostrophically balanced motion in the extra-equatorial region that determines the structure of the equatorial undercurrent, not the local winds.

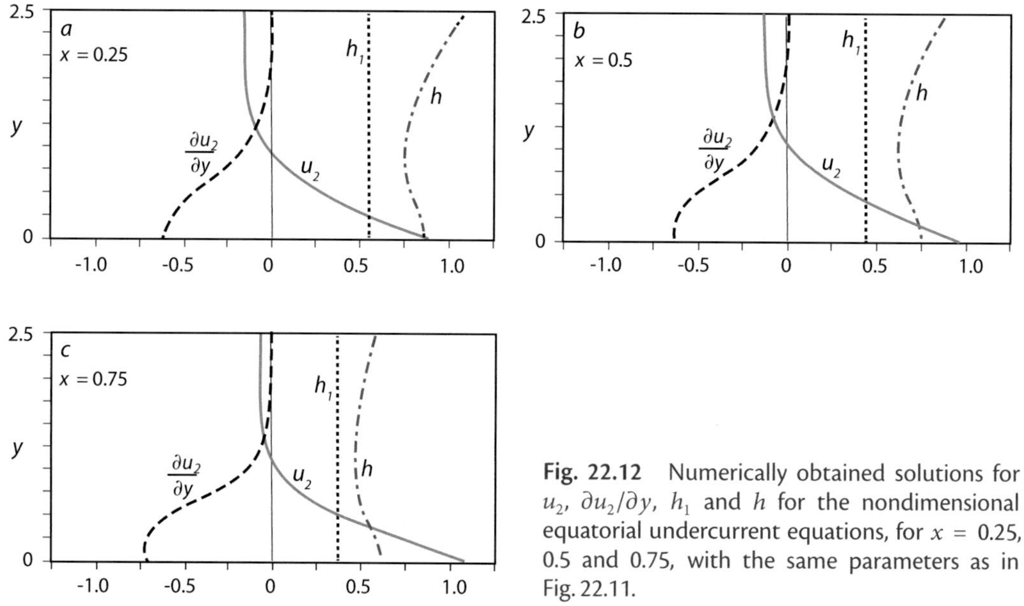

Fig. 22.12 Numerically obtained solutions for u_2, $\partial u_2/\partial y$, h_1 and h for the nondimensional equatorial undercurrent equations, for $x = 0.25$, 0.5 and 0.75, with the same parameters as in Fig. 22.11.

Are these two views of the equatorial undercurrent in complete opposition, to the extent that only one can be true? In fact, the real ocean may have elements of both, and the two models are best thought of as end-members, or limiting cases, of the a model of the true system, as summarized in the shaded box on page 877 and discussed more in the literature.[9]

22.5 AN INTRODUCTION TO EL NIÑO AND THE SOUTHERN OSCILLATION

El Niño! One of the most famous phenomena in the climate sciences, and certainly one with an enormous impact on humankind. It is an anomalous warming of the surface waters in the eastern equatorial Pacific, and its appealing name (el niño, without capitalization, is Spanish for male infant, el Niño refers to the Christ Child and El Niño to the oceanic phenomenon[10]) belies its enormous power and global effects, bringing heavy rains to California and Northern Argentina and anomalously dry weather to South East Asia and Northern and Eastern Australia; it also raises the global average surface temperature by about half a degree Celsius. Taken with the associated changes in the atmosphere, in which case the whole phenomenon is known as the El Niño–Southern Oscillation (ENSO), it is the largest and most important source of global climate variability on interannual timescales, and it is one of only two unequivocal 'oscillations' in the atmosphere–ocean system that is not directly forced by an external agent (the other one being the Quasi-Biennial Oscillation; however, neither phenomena are modes of a linear system). Our focus will be on the phenomenon itself, not its global effects.[11]

22.5.1 A Descriptive Overview

Every few years the temperature of the surface waters in the eastern tropical Pacific rises quite significantly. The strongest warming takes place between about 5° S to 5° N, and from the west coast of Peru (a longitude of about 80° W) almost to the dateline, at 180° W, as illustrated in Fig. 22.13. The warming is large, with a difference in temperature up to 6° C from an El Niño year to a non-El Niño year. The warmings occur rather irregularly, with typical intervals between warmings of 3 to 7 years, as seen in Fig. 22.14, and with particularly large events in 1887–88, 1982–83, 1997–98 and 2015–2016; the development of the last event is illustrated in Fig. 22.15.

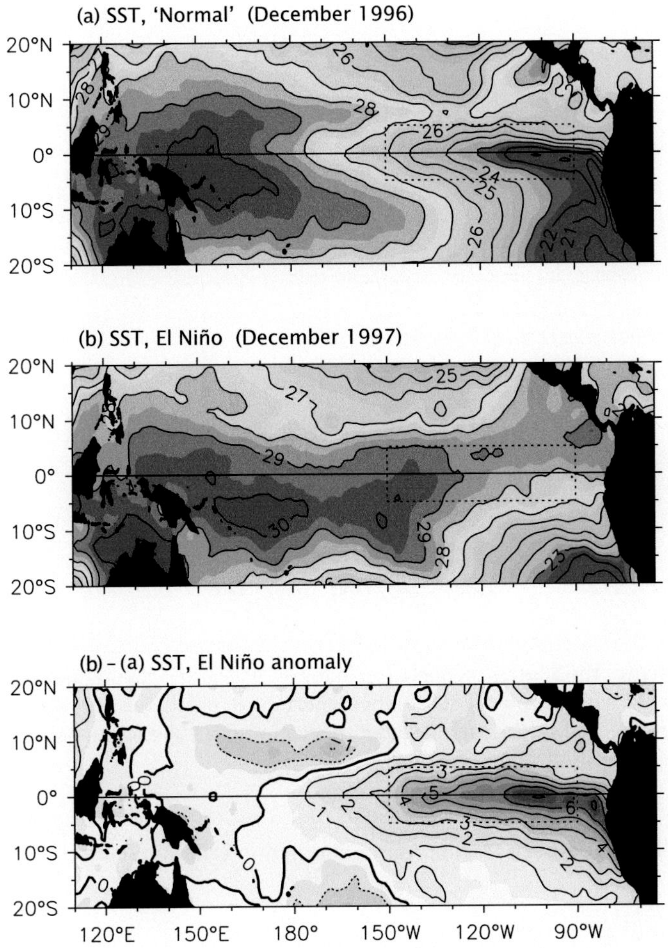

Fig. 22.13 The sea-surface temperature in December of a non-El Niño year (December 1996, top panel), a strong El Niño year (December 1997, middle panel) and their difference (bottom panel). An El Niño year is typically characterized by an anomalously warm tongue of water in the eastern tropical Pacific.[12]

The warmings have become known as El Niño events or even (abusing the Spanish language) El Niños. The warmings typically last for several months, occasionally up to two years, and appear as an enhancement to the seasonal cycle with high temperatures appearing at a time when the waters are already warming. If definiteness is desired, an event may be said to occur when there is a warming of at least 0.5° C averaged over the eastern tropical Pacific lasting for six months or more.[14] Ocean temperatures tend to fluctuate between warm El Niño years and those years in which the equatorial ocean temperatures are colder in the east and warmer in the west, and a *La Niña* is said to occur when it is particularly warm in the west (la niña being Spanish for young girl). We have direct observational evidence — temperature measurements from ships and buoys — of El Niño events for over a century, but the events have almost certainly gone on for a much longer period of time, probably many millennia, judging from proxy records of tree rings and coral growth.[15]

The overlying atmospheric winds and the surface pressure in the equatorial Pacific tend to co-vary with the sea-surface temperature (SST) and during El Niño events the equatorial Pacific trade winds, become much weaker and may even reverse. A convenient measure of the atmospheric signal is the pressure difference between Darwin (12° S, 130° E) and Tahiti (17° S, 150° W), and the normalized record of this signal is known as the *Southern Oscillation.* (Other similar measures of the subtropical high pressure zone of the South Pacific are sometimes used instead of the pressure at Tahiti.[18]) The Southern Oscillation and the SST record of El Niño are highly correlated

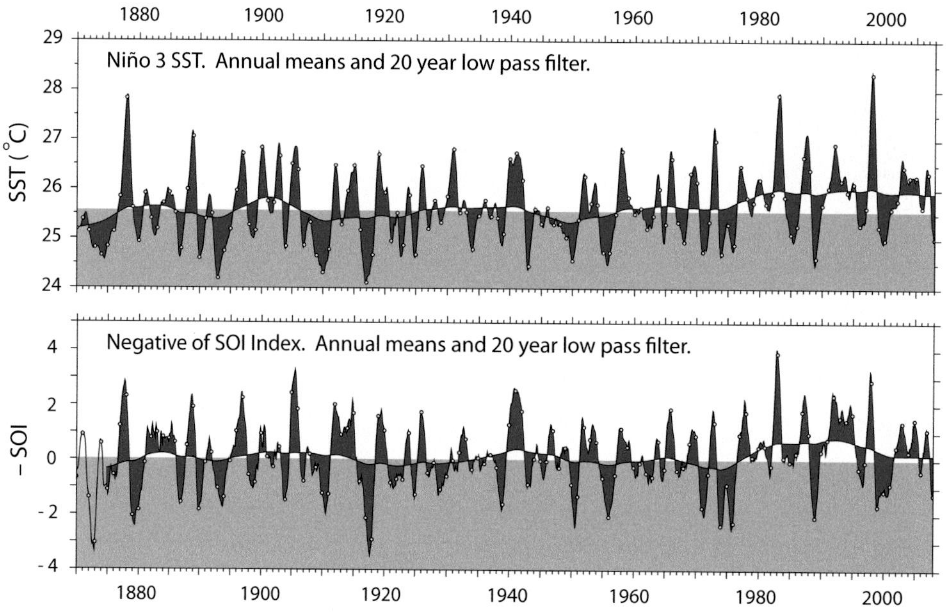

Fig. 22.14 Upper: Times series of the sea-surface temperature (SST) in the Eastern Equatorial Pacific (Niño 3) region. Lower: Negative of the Southern Oscillation Index (SOI), the anomalous pressure difference between Tahiti and Darwin. The pointy curves show the annual means, the dots are values in December, the smoother curves show the variable after application of a 20 year low-pass filter, and the top of the grey shading is the 1876–1975 mean.[13]

(Fig. 22.14), and the combined El Niño–Southern Oscillation phenomenon is denoted *ENSO*.

22.5.2 A Qualitative View of the Mechanism

ENSO is the only unambiguous example in the climate system of dynamical coupling between the atmosphere and ocean on timescales of months to years, and the essential mechanism is sketched in Fig. 22.16. First consider the mean state. The trade winds blow predominantly from higher latitudes toward the equator, and from the east to the west. This leads to a current system as described earlier in this chapter, with equatorial divergence of surface waters and upwelling — particularly in the east because here it must also replenish the surface waters moving westward away from the South American continent. The upwelling water here is cold, because much of it comes from below the thermocline, so that the SST of the eastern equatorial Pacific is relatively low, and the surface waters warm as they move westward. Furthermore, the thermocline deepens further west and so the upwelling does not bring as much cold abyssal water to the surface. The result of all this is that the SST is high in the western Pacific, up to about 30° C, and low in the eastern Pacific, about 21° C (see Fig. 22.13, top panel).

The strong zonal gradient of SST affects the atmosphere. The warm western Pacific region becomes more convectively unstable with respect to convection, the eastern Pacific is correspondingly cool and, as sketched in Fig. 22.16, an east-west overturning circulation is set up in the atmosphere — the *Walker circulation*, which we discuss more in Section 22.6. Evidently, the oceanic and the atmospheric states reinforce each other: the westward trade winds give rise to an east–west SST gradient in the ocean which generates the Walker Cell, so the surface winds are stronger than they would be if there were no ocean or if the ocean extended all the way round the globe.

By their nature positive feedbacks reinforce initial tendencies, whatever those tendencies may be, and such a feedback is at the core of El Niño. If the trade winds weaken then so will the zonal

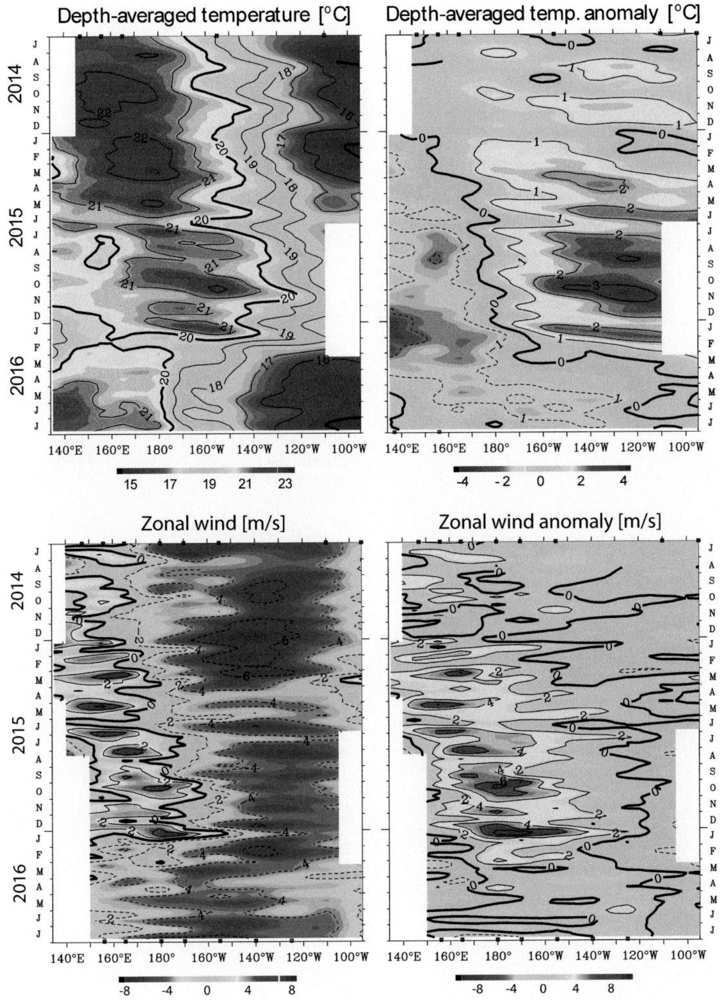

Fig. 22.15 The development of the 2015–2016 ENSO event. The top panels show the evolution of the depth-averaged temperature over the upper 300 m from mid-2014 to mid-2016, and the bottom panels show the surface zonal wind, both over a strip from 2° S to 2° N.

An eastward propagation in both fields can be seen over the second half of 2015, ceasing in early 2016 just after the temperature anomalies in the eastern tropical Pacific reach their maximum in the east.[16]

SST gradient, and the trades will weaken further and soon we have a full fledged El Niño event. The feedback cannot amplify without bound and, in fact, there are are natural damping mechanisms for El Niño, one of them being that the warm pool in the east is able to leak out of the tropics along the coast of the Americas, both north and south, through the mechanism of coastal Kelvin waves. No such leakage can occur in the western boundary, and this may be one reason that there is an asymmetry between El Niño events and La Niña events (after subtracting out the basic state asymmetry between east and west because of the direction of the trade winds). Another damping mechanism is simply that a warm sea-surface will give up heat to the atmosphere. The El Niño event then begins to decay and the whole feedback then occurs in the opposite sense and eventually the system reverts to its normal state. The mechanism that gives rise to the ENSO cycle is known as the *Bjerknes feedback*,[19] and the feedback is generally regarded as being stronger in the eastern ocean because there the thermocline is shallower. Thermocline depth and SST are correlated, with a shallow thermocline corresponding to low SST because cold water can then upwell through the thermocline, with the correlation getting weaker as the thermocline thickens.

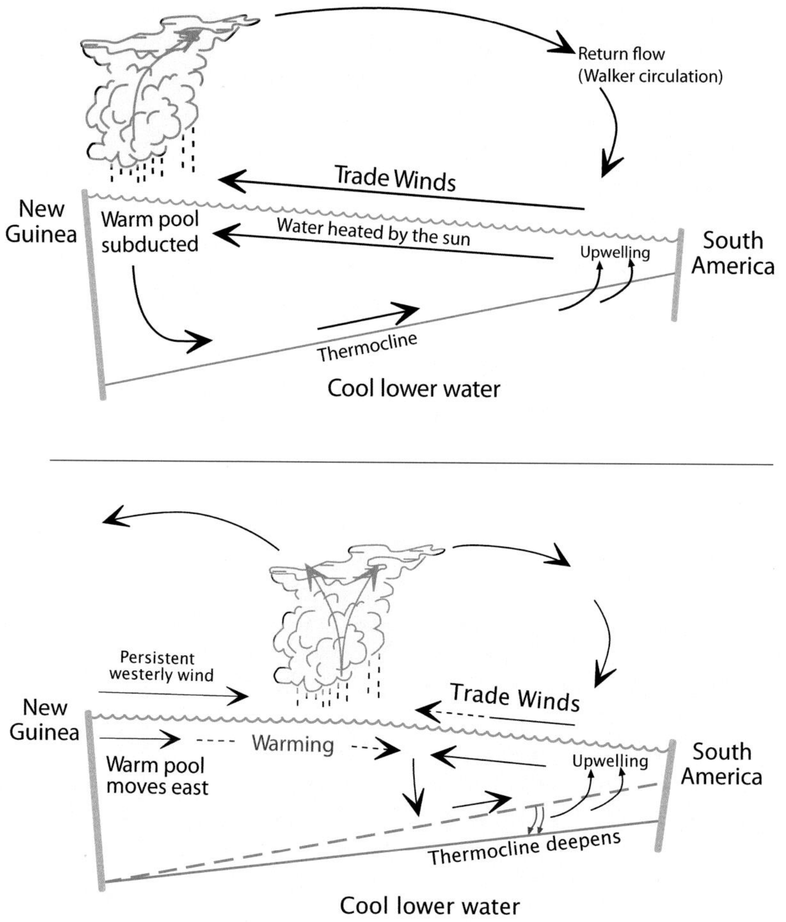

Fig. 22.16 Sketch of the end members of the ENSO cycle. The top panel shows a cross-section along the equator during non-El Niño, years, and the bottom panel at the peak of El Niño. Not to scale, and slopes are exaggerated.[17]

22.5.3 Why the Tropics?

Why do these dynamics occur in the tropics, and not in mid-latitudes? The tropics are different because there is a close and reasonably direct connection between SST and the winds, so enabling a feedback to occur that reinforces initial tendencies, and that arises because of a combination of the following factors:

(i) Equatorial sea surface temperatures are generally high, around 27° C (Fig. 22.13), and convection is readily triggered if the temperature further increases (although there is nothing magical about 27° C). Ascending motion occurs over warm regions with associated low-level convergence, and the surface winds thus directly respond to a changing SST gradient, as in the Matsuno–Gill problem.

(ii) The equatorial thermocline is quite shallow, varying from 200 m in the west to 50 m in the east, and is thus sensitive to changing wind patterns. Furthermore, because the thermocline is shallow upwelling can occur *through* it (Fig. 22.16), leading to a close connection between thermocline depth and SST, especially in the east.

(iii) Equatorial Rossby and Kelvin waves are quite fast, certainly compared to advective motion, so allowing cross-basin communication to occur on short, but not immediate, timescales. The slight delay (of order months) may contribute to the ability of the system to oscillate.

In mid-latitudes none of the above are as effective. Here the atmosphere is internally highly variable and its response to an SST anomaly has small signal to noise ratio and the induced winds may have little immediate local correspondence with the anomaly. Adding to this, the mid-latitude thermocline is deep and anomalous winds do not necessarily reinforce an existing anomaly. For all of these reasons mid-latitude ocean–atmosphere coupling is weaker and slower than its tropical counterpart.[20] None of this discussion, of course, is to imply that mid-latitude SST anomalies do not affect the atmosphere; indeed, such anomalies may be main cause of seasonal anomalies in the weather.

When described in the terms above the El Niño phenomenon seems readily understandable, but it glosses over a host of issues. If the above feedback is robust why doesn't it occur in the tropical Atlantic? Why doesn't the system find a stable fixed point, or oscillate in a regular manner? What determines the interval between events, and the magnitude of the events? We can't answer all these questions, but to begin let us consider the dynamics in a little more detail. We first consider the atmosphere where the response is, essentially, to modify the Walker circulation.

22.6 THE WALKER CIRCULATION

The *Walker circulation*[21] is an overturning cell in the tropical atmosphere predominantly in the zonal, or x–z plane, as schematically illustrated in the top panel of Fig. 22.16, and as a solution of the Matsuno–Gill problem in the top panel of Fig. 8.12. Convection is embedded within the cell, especially at the warm western end, as in Fig. 18.14 on page 691, but the Walker Cell is not 'driven' by convection any more than the meridional overturning circulation in the ocean is driven by convection in the North Atlantic. Rather, if there is a driver, it is the warm surface waters in the western equatorial Pacific that arise because of the westward surface flow and that are heated by the Sun as they move (see also the shaded box on page 872). These winds are part of the trade wind system and at leading order are independent of the presence of the ocean — they arise by the action of the Coriolis force on the equatorward surface branch of the Hadley Cell. There are other zonal overturning cells in the tropics but the one in the Pacific — the Walker Cell — is the strongest.

22.6.1 A Matsuno–Gill Model

Perhaps the simplest dynamical description of the Walker circulation is the Matsuno–Gill model.[22] The model (described in Section 8.5) finds the steady solutions of the linear shallow water equations on an equatorial beta-plane forced with a mass source. As discussed in Section 18.7, the solution of these equations may be interpreted as being the first baroclinic mode of a continuously stratified convecting atmosphere in which the static stability is maintained by convection, with the mass source representing thermal forcing and coming in this instance from a warm ocean.

What does such a solution look like? Figure 22.17 shows the solution to the Matsuno–Gill problem in which the thermal forcing linearly increases with longitude, crudely representing a sea-surface temperature gradient increasing westward, falling of exponentially away from the equator with an e-folding scale of an equatorial deformation radius. The solution shows a westward surface velocity between the cooling in the east and heating in the west, with ascent over the heating and descent over the cooling. Ascent is largest over the heating region, and so the trades weaken and contract as the heating moves east, as in the schematic (Fig. 22.16).

We can make some rough estimates of the importance of the above effects. Suppose that an SST anomaly of ΔT produces a heat flux of $Q = \lambda \Delta T$, where λ has units $\mathrm{m^2\,s^{-1}\,K^{-1}}$. The nondimensional parameter, γ^* say, that determines the importance of the heating is, using the expression involving Q in (8.115),

$$\gamma^* = \frac{\lambda\, \Delta T\, L_{eq}}{c_a^3}, \tag{22.83}$$

El Niño and the Southern Oscillation

- *El Niño* is the name given to the aperiodic warming of the ocean surface in the eastern equatorial Pacific. The interval between warm events is typically from two to seven years but is quite irregular (Fig. 22.14).
- El Niño events are associated with a weakening of the trade winds and an eastward shift of the region of convection, conveniently measured by a pressure difference between Tahiti and Darwin and known as the *Southern Oscillation*. The combined phenomenon is known as the El Niño–Southern Oscillation, or ENSO.
- El Niño, and its complement La Niña (an anomalous, weaker, warming in the western equatorial Pacific) are caused by the mutual interaction between the atmosphere and ocean in the equatorial Pacific. The key ingredients are:
  - *(i)* A close correlation between the thermocline thickness and surface temperature, especially in the east, with a shallow thermocline associated with a cool surface. A shallow thermocline allows water to upwell through it, bringing cold abyssal water to the surface (Fig. 22.16).
  - *(ii)* A positive feedback between the winds and the SST. Surface warming in the eastern Pacific leads to convergence of the winds (Fig. 22.17), a deepening of the thermocline, and a further warming of the surface.
  - *(iii)* The back-and-forth 'sloshing' of the thermocline depth anomalies, mediated by Kelvin and Rossby waves propagating quickly eastward and more slowly westward, respectively, with corresponding basin crossing times of about 70 and 200 days.
  - *(iv)* The interaction of these timescales with the annual cycle of the trade winds and thermodynamic forcing, and with the natural ('stochastic') variability of the atmosphere on shorter timescales.
  - *(v)* Damping by way of leakage in the west via coastal Kelvin waves and loss of heat to the atmosphere and deep ocean, and a delayed negative feedback because of the finite crossing time of the waves.
- The above factors combine to produce irregular oscillations, partially phase-locked to the annual cycle.
- It seems that the crossing times in the Atlantic are too short for such phase-locking and amplification, and there is no equivalent phenomenon there.

where c_a is the gravity wave speed in the atmosphere. Using (8.136a) a rough estimate for the magnitude of the wind response is then

$$u \sim \frac{Q}{c_a r_a} = \frac{\lambda \Delta T}{c_a r_a}, \tag{22.84}$$

where r_a is the linear drag on the wind (denoted α in Section 8.5, and with units of $\mathrm{s^{-1}}$) in the Matsuno–Gill problem. If $r_a \approx 1/10\,\mathrm{days^{-1}}$ and $c_a = 50\,\mathrm{m\,s^{-1}}$ then u increases by $3 \times 10^4 \lambda$ per degree Celsius. Observations show that $\lambda \sim 10^{-4}\,\mathrm{m^2\,s^{-1}\,K^{-1}}$, suggesting that sea-surface temperature

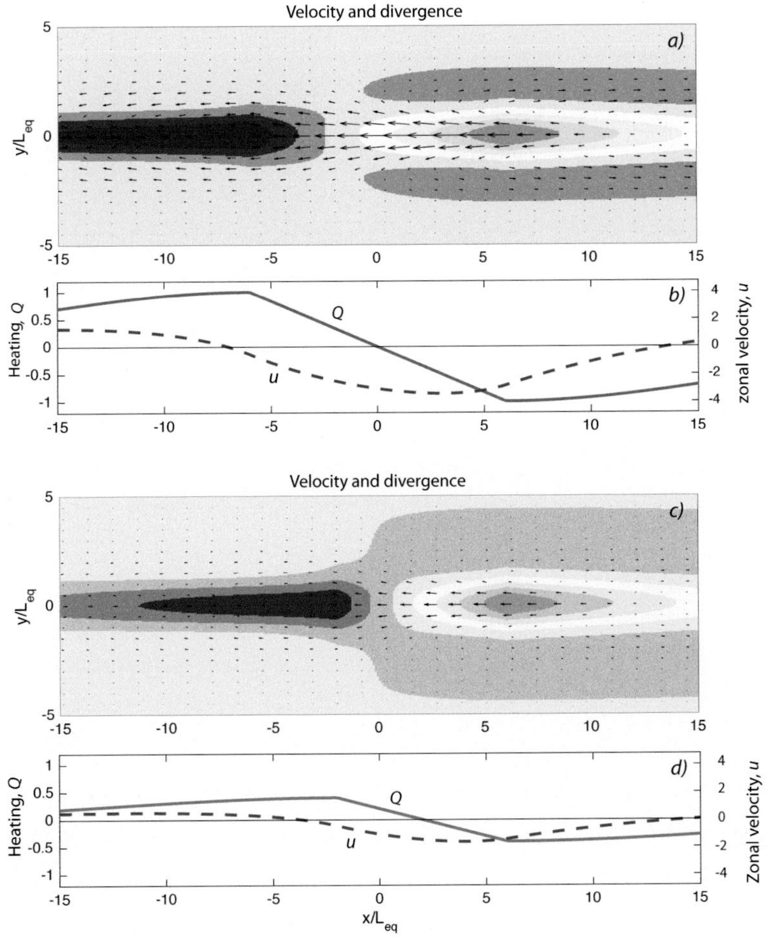

Fig. 22.17 Two solutions to the Matsuno–Gill model for the atmosphere, each with different distributions of heating, corresponding to a normal year (panels (a) and (b), top) and El Niño (panels (c) and (d), bottom).

Panels (a) and (c) show the low-level velocity (arrows) and velocity divergence (shading) and (b) and (d) show the heating, Q, and zonal velocity, u, along $y = 0$. (a) and (b) have a maximum heating at $x = -6$ and a cooling at $x = 6$, and (c) and (d) have heating and cooling at $x = -2$ and $x = 6$ with a much reduced amplitude.

In both cases, there is descent (blue shading) in the cooling region, westward flow toward the heating where air ascends (red shading), but the westward winds are weaker and extend less in (c) and (d).

anomalies can have a measurable impact on surface winds in the tropics.

22.7 THE OCEANIC RESPONSE

The oceanic response during an ENSO cycle is mediated by Kelvin and Rossby waves propagating westward and eastward respectively, as discussed in Chapter 8. These waves do not intrinsically preferentially cause upwelling or downwelling, nor do they cause a systematic change in the thermocline thickness or sea-surface temperature. Rather, the waves are the means by which the ocean transitions from one state to another, just as gravity waves effect an adjustment to geostrophic balance.

The linearized equations of motion for the equatorial thermocline are

$$\frac{\partial u}{\partial t} - \beta y v = -g' \frac{\partial h}{\partial x} - ru + \widetilde{\tau}^x, \qquad \frac{\partial v}{\partial t} + \beta y u = -g' \frac{\partial h}{\partial y} - ru + \widetilde{\tau}^y, \tag{22.85a,b}$$

$$\frac{\partial h}{\partial t} + H\left(\frac{\partial u}{\partial x} + \frac{\partial v}{\partial y}\right) = 0, \tag{22.85c}$$

where $\widetilde{\tau} = \tau/\rho_0 H$ and H is the undisturbed mean thickness of the thermocline.

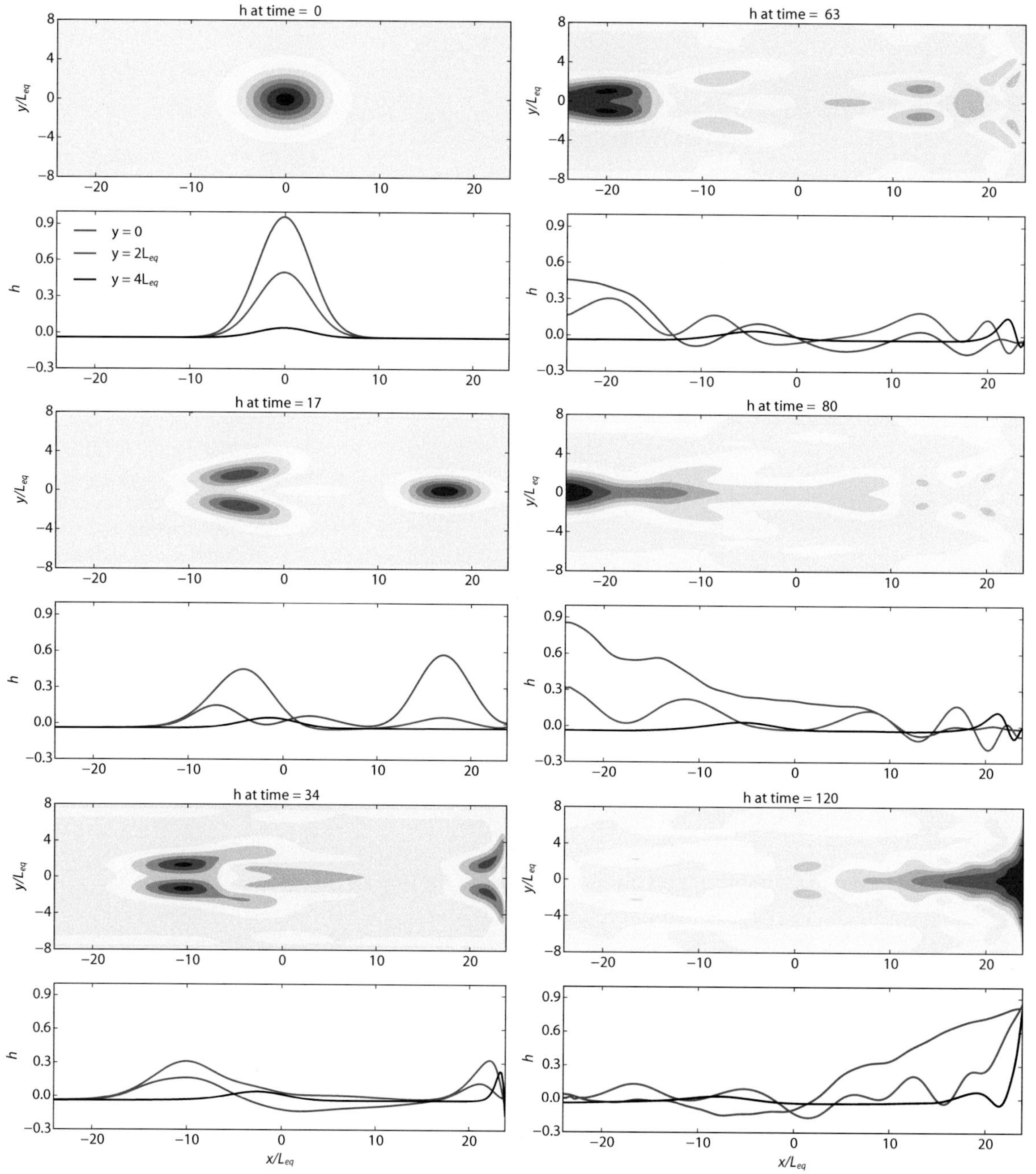

Fig. 22.18 Numerical evolution of the height field on an equatorial beta plane shown at the times indicated, starting from a Gaussian hump centred at the equator. The blue curve shows the height field along the equator and the red and black curves at two and four deformation radii poleward. All quantities are nondimensional, so the Kelvin wave (the hump propagating eastward) moves at a speed of one with off-equatorial Rossby waves moving more slowly westward. At time 34 we see the equatorial Kelvin wave partially reflected back as Rossby waves and partially propagating polewards as coastal Kelvin waves. At times 63 and 80 we see the Rossby waves reflected back as an equatorial Kelvin wave.

If the trade winds blow steadily westward, with no y-variation, then there is a stationary, steady state situation with

$$u = 0, \quad v = 0, \quad g\frac{\partial h}{\partial x} = \frac{\widetilde{\tau}^x}{H}, \quad \frac{\partial h}{\partial y} = 0, \tag{22.86}$$

where h is the perturbation thickness of the upper layer (a positive number) and H is its mean thickness. For negative τ_x (westward winds) the thermocline thickness deepens going west. If the trade winds slacken (as during an El Niño) then the slope of the thermocline will, in its equilibrium state, diminish (as in Fig. 22.16) but the new equilibrium is not achieved instantly; rather, it is mediated by Kelvin and Rossby waves, just as the passage to geostrophic balance is mediated by gravity waves. This means that any feedback that the ocean may have on the atmosphere will be delayed by a time that is of the same order as, and perhaps somewhat more than, the time it takes Rossby and Kelvin waves to traverse the Pacific.

To see these waves, we consider how the ocean responds to a perturbation on the equator in the middle of the ocean, and specifically we set up the initial value problem with a perturbation height field, and no velocity perturbation, as illustrated in Fig. 22.18. This perturbation will potentially project onto multiple modes and at the equator we expect there to be a significant Kelvin wave response, decaying away from the equator as $\exp(-y^2/L_{eq}^2)$ as in Section 8.2.2. The Kelvin wave moves east at a speed $(g'H)^{1/2}$, eventually colliding with the eastern boundary. Here it generates Kelvin coastal waves that move along the coast away from the boundary and are eventually dissipated, and a weak reflected Rossby wave.

Rossby waves will also be generated by the perturbation, the gravest of which has an $m = 1$ structure in the v field, where m is the index of the Hermite polynomial (Section 8.2). This mode contains both $m = 0$ and $m = 2$ components in the height field (see (8.161) on page 332) represented in physical space as lobes slanted away from the equator, initially rather like the lobes in the stationary problem illustrated in Fig. 8.11. These lobes propagate westward (second panel of Fig. 22.18), but at about a third the speed of the Kelvin wave, before they eventually collide with the western boundary. Now, in mid-latitudes, when Rossby waves hit a western wall they can only reflect as short Rossby waves, in order to have an eastward directed group velocity (Section 6.6) and in practice they often dissipate in the western boundary current. But here at the equator the wave energy can (and does) return eastward as an equatorial Kelvin wave, because the frequencies of Kelvin and Rossby waves can be the same — see Fig. 8.6 — as required for reflection. The whole sequence is illustrated numerically in Fig. 22.18 and schematically in Fig. 22.19. The initial conditions for the numerical simulation are a single positive depth anomaly (top left panel), whereas the wording on the schema of assumes that the wind anomaly produces a deepening of the thermocline on the equator and a shallowing off the equator (generated by the *curl* of the wind stress) which are propagated away by Kelvin and Rossby waves respectively.

22.8 COUPLED MODELS AND UNSTABLE INTERACTIONS

22.8.1 Equations of Motion

The equatorial atmosphere and ocean are governed by very similar sets of shallow-water equations, although the scales of the two systems, and how the variables are to be interpreted, differ. In the ocean the equations represent the water above the equatorial thermocline and should be interpreted as reduced gravity equations, with the variable h being the thermocline depth and g' being a direct measure of the density difference between the upper and deep ocean. In the atmosphere the equations represent the first baroclinic mode of the atmosphere, and the deformation radius is about 1000 km, several times that of the ocean.

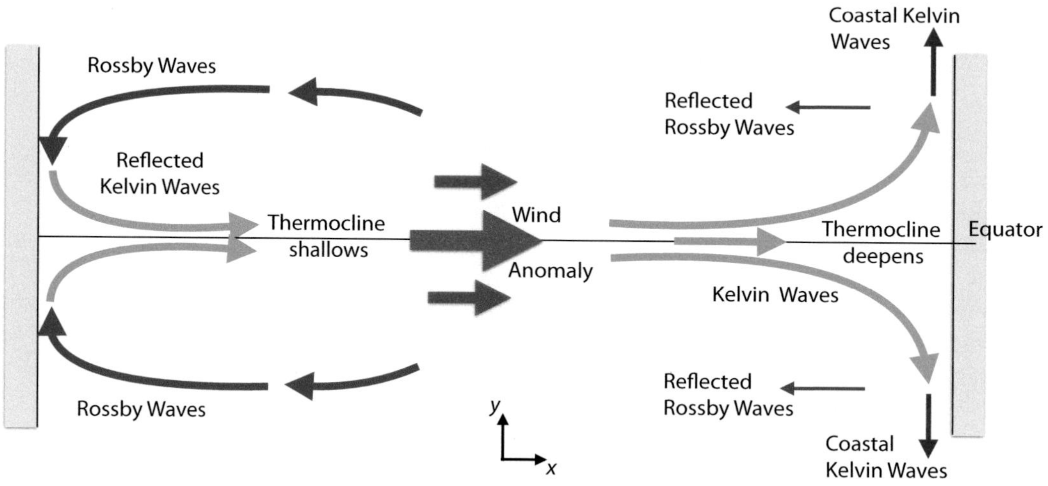

Fig. 22.19 Oceanic waves induced by a wind anomaly on the equator, leading to a local thickening of the thermocline and a propagating Kelvin wave that carries the anomaly westward, where it is partially propagated away as coastal Kelvin waves. The wind-stress curl may generate an off-equatorial shallow anomaly that is carried westward by Rossby waves and then reflected back as Kelvin waves.

Ocean

In the longwave approximation (discussed in Section 8.2) the zonal scales are much greater than meridional scales and (22.85) reduces to geostrophic balance and the above equations become

$$\frac{\partial u}{\partial t} - \beta y v = -\frac{\partial \phi}{\partial x} - ru + \widetilde{\tau}^x, \qquad \beta y u = -\frac{\partial \phi}{\partial y}, \qquad \frac{\partial \phi}{\partial t} + c_o^2\left(\frac{\partial u}{\partial x} + \frac{\partial v}{\partial y}\right) = 0, \qquad (22.87a,b,c)$$

where we have also written $\phi = g'h$ and $c_o^2 = g'H$; in the equatorial ocean c_o has a value of 1.5–2 m s^{-1}. The above equations are complete, but do not provide a prediction for the surface temperature, which is the field that the atmosphere actually cares about. We may heuristically add a mixed-layer temperature variable by supposing that the temperature is passive and advected by the flow, and is heated or cooled from above. Then, linearizing around a mean temperature $\overline{T}$, we write

$$\frac{\partial T}{\partial t} + u\frac{\partial \overline{T}}{\partial x} + v\frac{\partial \overline{T}}{\partial y} + \gamma_1 \frac{\overline{w}}{H'}(T - T_d) = -\lambda_1(T - T_a). \qquad (22.88)$$

The term $\gamma_1 \overline{w}(T - T_d)/H'$ represents upwelling from some depth H' where the temperature is T_d, and the term on the right-hand side represents exchange with the atmosphere, and γ_1 and λ_1 are constants. The derivations of equations like this are all somewhat empirical,[23] so it is appropriate to maximally simplify. The most important processes are exchange with the atmosphere and upwelling, and the efficacy of the upwelling is largely dependent on the depth of the thermocline, h, since this determines the temperature of the upwelling water. Thus, simplifying further, we have

$$\frac{\partial T}{\partial t} = +\gamma_2 h - \lambda T \quad \text{or, with still more approximation,} \quad T = Ah, \qquad (22.89a,b)$$

where γ_2 is a constant and $A = \gamma_2/\lambda$ and $h = \phi'/g'$. Equation (22.89b) assumes that equatorial upwelling is the dominant factor in determining surface temperature, and states that the anomalous temperature is proportional to the anomalous thickness of the upper layer.

Atmosphere

On the atmospheric side the equations are almost the same, and using capital letters for the variables and making the longwave approximation we have

$$\frac{\partial U}{\partial t} - \beta y V = -\frac{\partial \Phi}{\partial x} - r_a U, \qquad \beta y U = -\frac{\partial \Phi}{\partial y}, \qquad \frac{\partial \Phi}{\partial t} + c_a^2 \left(\frac{\partial U}{\partial x} + \frac{\partial V}{\partial y} \right) = Q_a, \tag{22.90a,b,c}$$

where the constant r_a is a drag coefficient and c_a is the gravity wave speed of the first baroclinic mode, around $30\,\mathrm{m\,s^{-1}}$. The term on the right-hand side of (22.90) represents a thermodynamic forcing by SST anomalies, and the simplest parameterization of this is $Q_a = \alpha' T$ where α' is a constant. Then, using (22.89), we obtain $Q_a = \alpha h$, where $\alpha = A\alpha'$. Similarly, the stress on the ocean, represented by $\widetilde{\tau}^x$ in (22.87), arises from the linear drag term $r_a U$ in (22.90), since stress is continuous across the atmosphere-ocean interface. Thus, we write $\widetilde{\tau}^x = -\gamma U$ where γ is a constant — the negative sign arising because (22.90) describes the first baroclinic mode and the surface wind is proportional to $-U$. Coupling between the atmosphere and ocean in (22.90) and (22.87) is thus, in the simplest case, effected by

$$\widetilde{\tau}^x = -\gamma U, \qquad Q_a = \alpha \phi. \tag{22.91a,b}$$

Equations (22.85), (22.90) and (22.91), are a complete set of equations for a coupled, equatorial, ocean–atmosphere system. Nearly all extant theories of ENSO are essentially theories of the behaviour of these or related sets of equations. The most ad hoc aspects of these equations are the relationships in (22.91), and the coefficients γ and α should be regarded as semi-empirical — a problem that would remain even if we were to use (22.89a). The model could be further extended by a more complete treatment of diabatic effects and the coupling terms — for example, one could add an evolution equation for the sea-surface temperature with advection and boundary layer processes, and decouple it from the thermocline depth. But even with a comprehensive numerical model (e.g., a coupled GCM) a degree of empiricism is unavoidable.

22.8.2 Unstable Air–Sea Interactions

We can expose some of the essential dynamics of the coupled system if we severely approximate (22.87), (22.90) and (22.91) by omitting y-derivatives, Coriolis terms and meridional velocities. We also neglect time derivatives in the atmosphere, since it achieves a quasi-equilibrium with the state of the ocean on a shorter timescale than that of the ocean. The coupled set then reduces to[24]

$$\frac{\partial u}{\partial t} = -\frac{\partial \phi}{\partial x} - r_o u - \gamma U, \qquad \frac{\partial \phi}{\partial t} + c_o^2 \frac{\partial u}{\partial x} = 0, \tag{22.92a,b}$$

$$0 = -\frac{\partial \Phi}{\partial x} - r_a U, \qquad c_a^2 \frac{\partial U}{\partial x} = \alpha \phi. \tag{22.92c,d}$$

These equations combine into a single equation for ϕ (a proxy for ocean surface temperature), namely

$$\frac{\partial^2 \phi}{\partial t^2} - c_o^2 \frac{\partial^2 \phi}{\partial x^2} + r_o \frac{\partial \phi}{\partial t} - \frac{c_o^2}{c_a^2} \gamma \alpha \phi = 0. \tag{22.93}$$

Seeking a harmonic solution of the form $\mathrm{e}^{\mathrm{i}(kx - \omega t)}$ yields the dispersion relation

$$\omega^2 = -\mathrm{i} r_o \omega + c_0^2 k^2 - \gamma \alpha \frac{c_o^2}{c_a^2}. \tag{22.94}$$

The terms on the right-hand side give rise to damping (the term involving r_o, the drag on the oceanic flow), waves (the term $c_o^2 k^2$), and, potentially, exponential growth (the term $\gamma\alpha$, which is the feedback between atmosphere and ocean). If this feedback is large enough the flow will be unstable, and this is a simple mathematical representation of the feedback described verbally in Section 22.5.2. The solution of (22.94) is

$$\omega = -\frac{ir_o}{2} \pm \sqrt{c_o^2 k^2 - \frac{r_o^2}{4} - \gamma\alpha\frac{c_o^2}{c_a^2}} \qquad \text{or} \qquad \sigma = -\frac{r_o}{2} \pm \sqrt{\gamma\alpha\frac{c_o^2}{c_a^2} + \frac{r_o^2}{4} - c_o^2 k^2}\,, \tag{22.95a,b}$$

where $\sigma \equiv -i\omega$ is the growth rate, and depending on the size of $\alpha\gamma$ the solution may be decaying or a growing, and if decaying it may be oscillatory. Note that the frequency of the oscillations is reduced (from its pure oceanic value, $c_o k$) by the interaction with the atmosphere.

The above analysis is very instructive in revealing the basic source of the instability, but it has some unrealistic aspects. For example, (22.95) has both westward and eastward propagating gravity waves; however, since these are at the equator they should really represent Kelvin waves that propagate only eastward; that is, the westward propagating solution in (22.95) is artifactual. Also, the analysis does not yield steady self-sustained oscillations and to obtain these we need to delve further; we will do so by way of numerical integrations and toy models.

22.9 SIMPLE CONCEPTUAL AND NUMERICAL MODELS OF ENSO

We now discuss two conceptual, or toy, models of ENSO. Toy models are *ad hoc* models, meaning that their governing equations cannot be derived in a rigorous or even a systematic fashion from the fundamental physics or dynamics. Rather, the models often arise using verbal reasoning and/or severe approximation, and their purpose is to illustrate mechanisms and suggest developments rather than to be accurate predictive tools or proper reductions of the full equations. However, the models may be very useful and the behaviour they describe may be quite robust.[25]

We first construct and analyse a *delayed-oscillator* model, which provides a basis for understanding how the finite propagation time of equatorial waves in conjunction with the Bjerknes feedback can give rise to oscillatory behaviour. We then look at the perhaps more realistic *recharge-discharge* model and finally show some numerical simulations.

22.9.1 A Delayed-oscillator Model

The mechanisms involved in the delayed-oscillator model are illustrated in Fig. 22.19.[26] To begin, suppose there is an eastward wind anomaly at the equator, and that this develops a positive (i.e., deeper) thermocline anomaly in mid-basin at the equator. This anomaly propagates eastward as a Kelvin wave, thickening the thermocline, warming the sea surface and generating anomalous eastward winds, that further thicken the thermocline (the Bjerknes feedback). At the same time a negative thermocline anomaly is generated off the equator, which is propagated westward as Rossby waves before being reflected back as a Kelvin wave. When this reaches the eastern Pacific it lowers the sea-surface temperature, so providing a delayed negative feedback. Appendix A at the end of the chapter attempts to derive a simple equation governing such a process from the equations of motion and the result is

$$\frac{\partial T}{\partial t} = aT(t) - bT(t-\delta) - rT^3(t), \tag{22.96}$$

where a, b, c and δ are constants and T is the SST anomaly, nominally in the eastern equatorial Pacific. The three terms on the right-hand side of this equation may be interpreted as follows:

(*i*) The term $aT(t)$ represents a local wind–surface-temperature positive feedback. A positive SST anomaly in the East Pacific produces an anomalous eastward wind stress and eastward propagating downwelling Kelvin wave, deepening the thermocline, further warming the sea surface.

(ii) The term $bT(t-\delta)$ represents a delayed negative feedback. A wind-stress curl anomaly leads to off-equatorial thermocline shallowing and westward propagating Rossby waves. These reflect back as a Kelvin wave that, after a time δ, cools the eastern equatorial ocean.

(iii) The cubic nonlinear damping term, rT^3, represents all other dissipative processes in the system.

In some ways the verbal reasoning above is as convincing as the derivation in Appendix A.

Analysis of the oscillator

Equation (22.96) simplifies if we write $\hat{t} = ta$ and $\hat{T} = T\sqrt{r/a}$, giving the nondimensional equation

$$\frac{\mathrm{d}\hat{T}}{\mathrm{d}\hat{t}} = \hat{T} - \alpha\hat{T}(\hat{t} - \hat{\delta}) - \hat{T}^3, \tag{22.97}$$

where $\hat{\delta} = \delta a$ is the nondimensional delay and $\alpha = b/a$. Equations such as (22.97) are not 'simple'. Formally, the equation has an infinite number of degrees of freedom — note that we need to specify initial conditions not just at a single time but at all times from $t = -\delta$ to $t = 0$ — and corresponds to an infinite number of ordinary differential equations. Nevertheless, we can obtain useful information about the solution by way of a linear analysis. Dropping the hats from (22.97), equilibria occur at $T = T_0$ where

$$T_0 = 0 \quad \text{and} \quad T_0 = \pm\sqrt{1-\alpha}. \tag{22.98}$$

We set $T = T_0 + T'$ and linearize, and then (22.97) becomes

$$\frac{\mathrm{d}T'}{\mathrm{d}t} = T'(1 - 3T_0^2) - \alpha T'(t - \delta). \tag{22.99}$$

In the usual manner we now let $T' = \widetilde{T}\mathrm{e}^{\sigma t}$, where $\sigma = \sigma_r + \mathrm{i}\sigma_i$ and obtain

$$\sigma = 1 - 3T_0^2 - \alpha\mathrm{e}^{-\sigma\delta}, \qquad \sigma_r = 1 - 3T_0^2 - \alpha\mathrm{e}^{-\sigma_r\delta}\cos\sigma_i\delta, \quad \sigma_i = \alpha\mathrm{e}^{-\sigma_r\delta}\sin\sigma_i\delta. \tag{22.100}$$

These are transcendental equations, since σ appears on the right-hand sides in the sine and cosine terms. For $T_0 = 0$ we get

$$\sigma_r = 1 - \alpha\mathrm{e}^{-\sigma_r\delta}\cos\sigma_i\delta, \qquad \sigma_i = \alpha\mathrm{e}^{-\sigma_r\delta}\sin\sigma_i\delta. \tag{22.101a,b}$$

A solution of this satisfies $\sigma_i = 0$ and $\sigma_r = 1 - \alpha\mathrm{e}^{-\sigma_r\delta}$. A value for σ_r can easily be obtained graphically, but even without doing that we can see that σ_r will generally be positive, because the term $\alpha\mathrm{e}^{-\sigma_r\delta}$ is positive and less than unity. The solution is therefore unstable.

For the case in which $T_0 = \pm\sqrt{1-\alpha}$ the stability equation is

$$\sigma_r = 3\alpha - 2 - \alpha\mathrm{e}^{-\sigma_r\delta}\cos\sigma_i\delta, \qquad \sigma_i = \alpha\mathrm{e}^{-\sigma_r\delta}\sin\sigma_i\delta. \tag{22.102a,b}$$

The neutral curves, where $\sigma_r = 0$, are given by

$$\delta = \frac{1}{\sigma_i}\cos^{-1}[(3\alpha - 2)/\alpha], \qquad \sigma_i = [\alpha^2 - (3\alpha - 2)^2]^{1/2}. \tag{22.103a,b}$$

These are plotted in Fig. 22.20 and the lower curve corresponds to a boundary between stable and unstable regions: for the shaded regions the solution to (22.97) converges to a fixed point, whereas in the unshaded region the fixed points are unstable and the solutions are oscillatory. (Because the

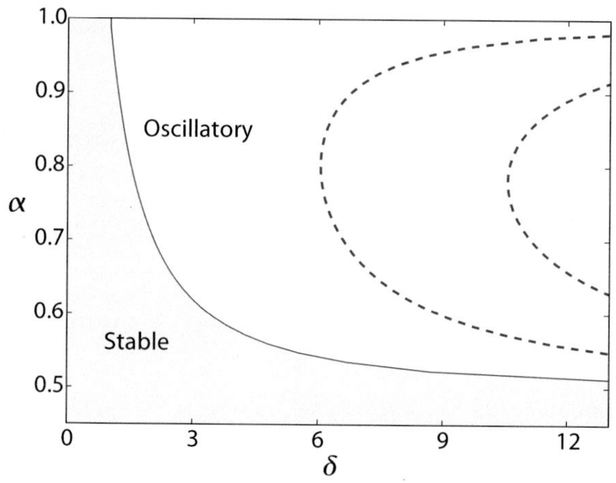

Fig. 22.20 Stability regime for a delayed-oscillator model. Plotted are values of α vs δ as given by (22.103).

signal will diminish as the Rossby waves propagate westward and return as Kelvin waves, we might expect that the negative feedback in the delay term will be weaker than the positive feedback in the linear term and therefore $\alpha < 1$.)

Unstable modes arise more readily for *larger* values of the negative feedback parameter α and for *larger* values of the delay time δ. To help understand this consider the case with zero delay, for which the equation of motion is just

$$\frac{d\widehat{T}}{d\hat{t}} = (1 - \alpha)\widehat{T} - \widehat{T}^3. \tag{22.104}$$

This equation has three stationary states: an unstable state at $T = 0$ and stable states at $T = \pm\sqrt{1 - \alpha}$, as in the case with finite delay, but no oscillatory modes are present. As α diminishes the effect of the delay also diminishes, and for $\alpha = 0$ stable states occur at $T = \pm 1$ and no oscillatory modes are present. The period of the oscillations is not easily obtained except on the neutral curves, and for that we turn to numerical solutions.

Numerical solutions

Numerical solutions of (22.97) may be obtained easily and some typical solutions are shown in Fig. 22.21. The actual periods close the neutral line are in fact similar to those given by the linear analysis, (22.103b). The period gets longer for a longer delay, although the ratio of the period to the delay itself diminishes slightly as the delay increases. Typically, the period is two or three times the delay, although for small values of δ the period can be five times the delay. What is that delay time, dimensionally? Given the speed of Rossby waves and the size of the Pacific, a typical value is about eight months, with a maximum possible value of about 12 months. This gives a period of about 24 months, which is rather less than the typical observed interval between El Niño events. However, it is hard to be precise because we need to determine what nondimensional value of δ corresponds to a dimensional value of 8 months: we have scaled time by the value of the amplifying parameter a in (22.96), and this is uncertain to within a factor of a few. Unless the parameter regime of the Pacific ocean is such that a dimensional delay of 8 months corresponds to a nondimensional delay of about 2, which would give a period of four or five times 8 months, then the period delayed oscillator model is somewhat shorter than that of the real El Niño.

The delayed-oscillator model illustrates two properties that may be important in the real system, namely the positive feedback by the wind (the underlying Bjerknes hypothesis) and the importance of a delayed damping, so enabling oscillations to occur. Still, it has notably ad hoc aspects, especially its treatment of the negative feedback from the Rossby waves and its introduction of a cubic damping.

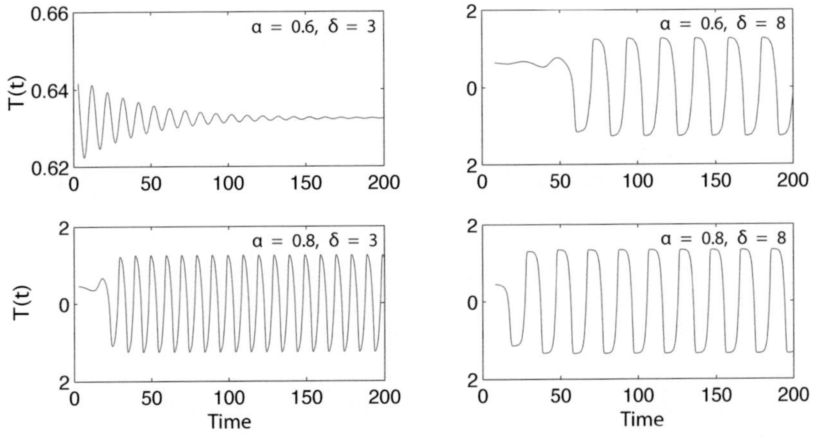

Fig. 22.21 Various time series obtained by numerically integrating (22.97) for the parameters indicated. The initial conditions are the value at the fixed point, $T = \sqrt{1-\alpha}$, plus a small perturbation.

22.9.2 The Recharge-Discharge Oscillator

A basin-wide mechanism that has become known as the *recharge-discharge* hypothesis seems to capture many of the salient features of the ENSO phenomenon in an appealingly simple way, roughly as follows.[27]

The prevalent westward trade winds push warm water westward, depressing the thermocline in the west and building up a significant amount of warm water. At some point a fluctuation in the atmospheric circulation will occur that causes the trade winds to relax and give rise to an eastward Kelvin wave surge, warming the central and Eastern equatorial ocean, further weakening the trades and allowing El Niño to develop fully. Then, along with the wave reflection that occurs at the eastern boundary, some of that warm water drains away polewards along the coast in the form of coastal Kelvin waves, meaning that some of the warm pool is irretrievably lost to the equatorial ocean, and this discharge is the demise of the event. The trade winds then resume, slowly building up the warm pool in the west and priming the system for the onset of another El Niño. The timescale of the oscillation — that is, the time between events — is then determined by the time taken to build up the warm water, rather than being a multiple of the wave propagation time across the basin. The two are related, since Kelvin and Rossby waves are still the mechanisms that determine the timescale of propagation of the signal across the ocean basin, and so determine the timescales of the El Niño genesis and recharge. In this model, the positive feedback is (still) the Bjerknes mechanism, the growth rate (or, perhaps more relevantly, the onset time) is determined by the passage of the equatorial ocean waves across the basin, the damping is the drainage of warm water into coastal Kelvin waves as well as heat loss to the atmosphere and deeper ocean, and the interval between events is determined by the time it takes to refill the basin with warm water, along with the interaction with the seasonal cycle.

In some ways the recharge-discharge mechanism is more physically grounded and straightforward and than that of the delayed-oscillator, although the two are not wholly in opposition — waves propagate back and forth, leading to delayed negative feedbacks. Similarly to the delayed-oscillator case, we can write down, or try to derive more systematically, simple ODE models mimicking the recharge-discharge behaviour, and readers are invited to construct their own.[28] Such models can be transparent and simple but the derivations have unavoidably ad hoc aspects. Rather than proceeding down that path let us illustrate the ENSO phenomenon with a simple but fluid-dynamical numerical model, namely the shallow water equations of Section 22.8 for the ocean coupled to a simplified atmosphere.

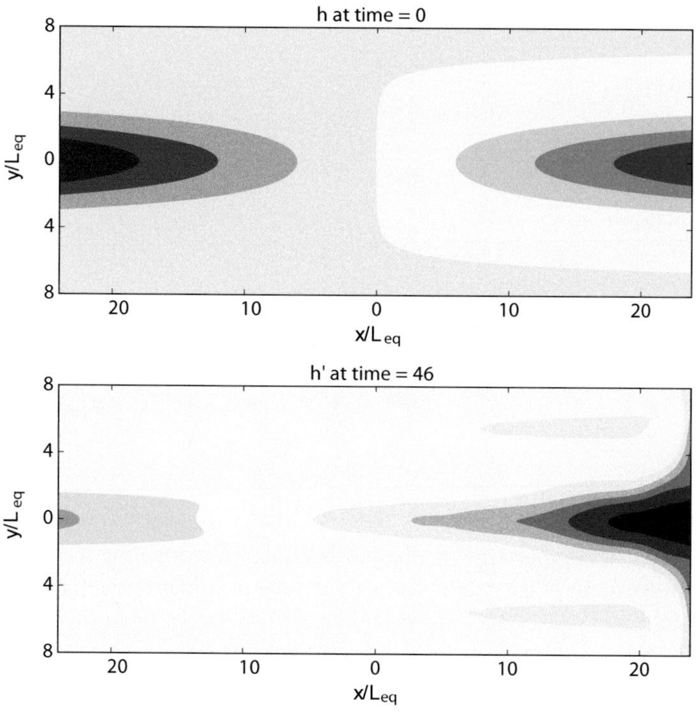

Fig. 22.22 A numerical solution of the shallow water equations showing the initial thermocline depth and the perturbation depth (i.e., final minus initial) at $t = 46/\sqrt{\beta c} \approx 70$ days, approximately the Kelvin wave crossing time for the basin. Red shading indicates deeper, warmer, water.

The initial field increases linearly toward the west, and the later state has a tongue of warm water in the east.

22.10 NUMERICAL SOLUTIONS OF THE SHALLOW WATER EQUATIONS

We now show some numerical solutions of the time-dependent shallow water equations on an equatorial beta-plane, (22.85). The model is in a regime in which the linear equations are a very good approximation, and have a mean thermocline depth of 50 m and $g' = 0.05\,\mathrm{m\,s^{-2}}$ (so that $c = (g'H)^{1/2} \approx 2.2\,\mathrm{m\,s^{-1}}$) in a domain about 15 000 km across. The equatorial deformation radius, $L_{eq} = (c/\beta)^{1/2}$ is then about 300 km, the nondimensional time $T_{eq} = 1/(c\beta)^{1/2}$ is 1.6 days and the basin width is about 50 L_{eq}. The model also has a linear damping in all fields with a timescale of 16 months.

The simplest illustration of El Niño arises from the initial value problem illustrated in Fig. 22.22. The initial thermocline thickness is set to deepen linearly toward the west and falls off exponentially away from the equator over a few deformation radii, as if held by a westward wind. The flow is released (i.e., the nominal wind is 'turned off') and a Kelvin wave propagates eastward, colliding with the eastern wall some time later, generating poleward propagating coastal Kelvin waves and, most obviously, a warm tongue that appears to spread from the east — this is the model analogue of an El Niño event. The disturbance is partially damped by those coastal waves but also propagates back across the basin as a Rossby wave, and then back again as a Kelvin wave, producing another El Niño event, and so forth, with the oscillations eventually dying out.

Self-sustained oscillations can be generated by adding a wind forcing that depends on the ocean state. Although it is straightforward to couple the ocean to a corresponding shallow water atmosphere, the mechanism is seen transparently if we reduce (22.90) and (22.91) to the simple equation

$$\widetilde{\tau}^x = \gamma'(h_E - h_W) - \widetilde{\tau}^x_0, \tag{22.105}$$

where the constant $\widetilde{\tau}^x_0$ represents the trade winds and $\overline{h}_E$ and $\overline{h}_W$ are the averaged thermocline thicknesses in the eastern and western parts of the domain, so representing the Bjerknes feedback in its most basic form. The results are quite sensitive to the value of the coupling coefficient γ' but much less so to the way $\overline{h}_E$ and $\overline{h}_W$ are calculated.

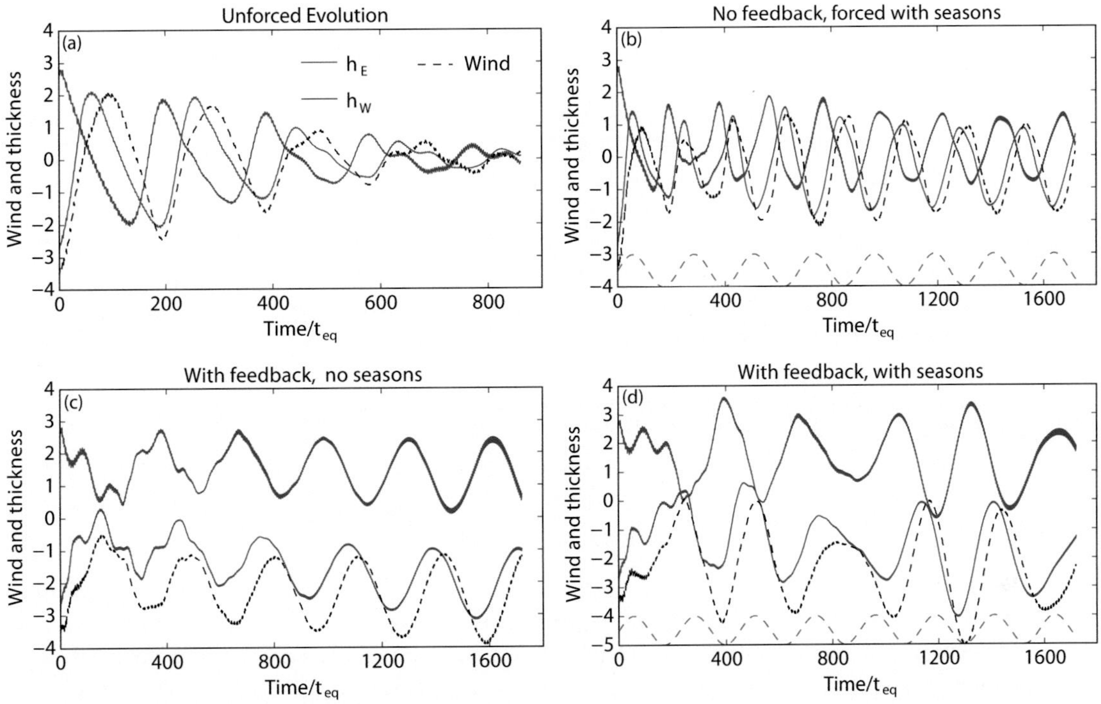

Fig. 22.23 Time series of thermocline thickness in the east and west, h_E and h_W, and the wind given by (22.105), which has no effect in the top two panels. Time is in units of $\sqrt{1/(c\beta)}$ or 1.6 days. The panels variously show the effects of a wind feedback and a seasonal cycle: (a) shows the unforced evolution, (b) includes an annual cycle of winds but no feedback on those winds, (c) includes a wind feedback as in (22.105), and (d) includes both a wind feedback and annual cycle. A year is 226 time units ($226/\sqrt{c\beta}$) and its cycle is shown in the dashed lines at bottom. A Kelvin wave crossing time is 48 time units and a Rossby wave crossing time is about three times that.

Various results are shown in Fig. 22.23, all with the initial condition of Fig. 22.22. The plots all show time-series of the thermocline thickness at the eastern and western edges, h_E and h_W, and the wind stress, in rather arbitrary, order one, units. Panel (a) shows the continuation of the integration of Fig. 22.22, showing a damped oscillator (the wind stress has no effect) with a period of approximately the sum of Kelvin wave and Rossby wave crossing times, or about 190 time units, somewhat less than one year. The difference in the two crossing times is apparent in the saw-tooth nature of the thickness time series. If we now force the model with seasonal cycle in winds, but do not allow the thermocline thickness to feedback on the wind, then the oscillations become phase-locked to the seasonal cycle and eventually become quite steady, as in panel (b). If we allow a feedback using (22.105), but remove the seasonal cycle, the system develops self-sustained, nearly periodic, oscillations with a period that is significantly longer than the unforced period, as in panel (c). The combined effects of a seasonal cycle and a wind feedback produce more irregular oscillations, partially locked to the seasonal cycle but sometimes skipping a year. Even more irregular behaviour can be obtained if some stochasticity is added to the forcing. On the other hand, if the feedback is too strong (i.e., γ' too large in (22.105)) the model goes to a fixed point — the wind feedback is too strong to allow the model to escape from either a permanent El Niño or La Niña.

Although still some distance from reality, the model illustrates the most important and robust features of the ENSO cycle (see also the shaded box on page 892):

(i) The back-and-forth of Kelvin and Rossby waves, illustrated even in the unforced simulation

of panel (a);

(ii) The sustaining effects of the feedback of the thermocline thickness on the wind, which tends to lengthen the period of oscillation;

(iii) The damping that arises through leakage at the western boundary and the explicit damping in the model reflecting loss to the atmosphere and the deeper ocean, and the delayed damping because of the finite propagation speed of the waves;

(iv) The irregularity arising from interactions with another timescale, in this case the seasonal cycle.

The quantitative importance of these various effects — for example the relative importance of western boundary leakage versus loss to the atmosphere — remains a matter of investigation, as is the exact mechanism that produces the great irregularity in the observed El Niño events — the role of the seasonal cycle, chaos, stochasticity, and so forth.[29] In these ways El Niño phenomenon reflects the puzzles, challenges and triumphs of geophysical fluid dynamics itself, and on that note we close this book. Thank you for reading.

APPENDIX A: DERIVATION OF A DELAYED-OSCILLATOR MODEL

Our goal here is to give a sense of how equations for toy-like models might be derived, supplementing verbal reasoning. Although the derivation is mathematical it not rigorous and it may overstate the relevance of the final model. We begin with (22.87) and include a damping on the height field to give

$$\frac{\partial u}{\partial t} - \beta y v = -\frac{\partial \phi}{\partial x} - r_o u + \widetilde{\tau}^x, \qquad \beta y u = -\frac{\partial \phi}{\partial y}, \qquad \frac{\partial \phi}{\partial t} + c_o^2\left(\frac{\partial u}{\partial x} + \frac{\partial v}{\partial y}\right) = -r_o\phi. \tag{22.106a,b,c}$$

The properties of this equation were explored in Section 8.4, but we proceed *ab initio*. We are interested in thermocline depth variations so we combine the above into a single equation for ϕ,

$$\beta y^2\left(\frac{\partial}{\partial t} + r_o\right)\phi + \frac{c_o^2}{\beta}\left[\frac{2}{y} - \frac{\partial^2}{\partial y^2}\right]\left(\frac{\partial}{\partial t} + r_o\right)\phi - c_o^2\frac{\partial \phi}{\partial x} + \left[\widetilde{\tau}^x - y\frac{\partial \widetilde{\tau}^x}{\partial y}\right] = 0. \tag{22.107}$$

Solutions of these equations will have the meridional form discussed in Sections 8.2 and 8.4, namely Hermite functions multiplied by a Gaussian, and we focus on the gravest of these modes.

Equatorial response

If the wind perturbation $\widetilde{\tau}^x$ is peaked at the equator then it will excite a Kelvin wave that decays as $\exp(-y^2/L_{eq}^2)$ away from the equator (as in (8.62) on page 311). Thus motivated, we seek a solution to (22.107) of the form

$$\phi(x, y, t) = \Phi(x, t)\exp(-y^2/2L_{eq}^2), \tag{22.108}$$

where $L_{eq}^2 = c_o/\beta$. Substituting into (22.108) gives, after a few lines of algebra,

$$\left(\frac{\partial}{\partial t} + r_o\right)\Phi + c_o\frac{\partial \Phi}{\partial x} = \frac{1}{c_o}\left[\widetilde{\tau}^x - y\frac{\partial \tau^x}{\partial y}\right]. \tag{22.109}$$

At the equator this is the equation for a forced-damped Kelvin wave, propagating eastward at speed c_o, forced by the wind (and not its curl). We will also excite Rossby waves at the equator, but as seen in the previous section the Kelvin wave response will normally be dominant.

Off equatorial response

Because of the fast meridional decay of the Kelvin wave, off the equator we expect the response to wind perturbations to be largely in the form of Rossby waves. To see this explicitly, note that the relative size of the first and second terms in (22.107) is $\beta y^2 : (c_o^2/\beta y^2) = y^4 : L_{eq}^4$. Thus, if $y \gg L_{eq}$ then the second term will be small and if we neglect it we have

$$\left(\frac{\partial}{\partial t} + r_o - \frac{c_o^2}{\beta y^2}\frac{\partial}{\partial x}\right)\phi = \frac{1}{\beta}\frac{\partial}{\partial y}\left(\frac{\widetilde{\tau}^x}{y}\right). \tag{22.110}$$

This is an equation for forced-dissipative equatorial Rossby waves travelling westward at speed $c_o^2/\beta y^2$, and it is in fact nothing other than a forced-dissipative version of the planetary-geostrophic wave equation (8.16). It may seem odd that the speed decreases as beta increases, but in the mid-latitude case the speed is $c_o^2\beta/f_0^2$, which is the same as $c_o^2/\beta y^2$ if we replace f_0 by βy. Note that the wave is forced by the wind-stress curl (divided by f) rather than the wind itself, and typically the two have the opposite sign as the reader may verify with a few examples.

Wave characteristics and solution

By following the Rossby and Kelvin waves as they propagate we may see their remote effects on sea-surface height as time progresses. Suppose that a Rossby wave takes a time t_R to cross the basin, and a Kelvin wave takes t_K, and that we are interested in the depth of the thermocline at time t in the eastern equatorial Pacific (where the thermocline is shallowest and its variations have most effect on the SST and thence the wind). The Kelvin wave influence at the east, Φ_E, is given by the solution to (22.109), which may be obtained using the method of characteristics — we introduce a new variable $t' = t + c_o x$, whence (22.109) becomes an integrable ordinary differential equation from which we obtain the solution

$$\Phi_E(t) = \Phi_W(t - t_K)\mathrm{e}^{-r_o t_K} + \frac{1}{c_o}\widetilde{\tau}^x(t - t_K/2)\mathrm{e}^{-r_o t_K/2}. \tag{22.111}$$

Here, Φ_W is the thermocline thickness in the west, to be evaluated at time $t - t_K$, and the wind, assumed to be confined to the centre of the domain, is evaluated at $t - t_K/2$. Variations in Φ_W arise from Rossby waves propagating westward from the wind anomaly and being reflected back at the equator as a Kelvin wave, and are thus given by the solution of (22.110), namely:

$$\Phi_W(t - t_K) = -\frac{1}{\beta}\frac{\partial}{\partial y}\left(\frac{\widetilde{\tau}^x(t - t_K - t_R/2)}{y}\right)\mathrm{e}^{-r_o t_R/2}. \tag{22.112}$$

Combining these last two expressions gives

$$\Phi_E(t) = \frac{1}{c_o}\widetilde{\tau}^x(t - t_K/2)\mathrm{e}^{-r_o t_K/2} - \frac{1}{\beta}\frac{\partial}{\partial y}\left(\frac{\widetilde{\tau}^x(t - t_K - t_R/2)}{y}\right)\mathrm{e}^{-r_o(t_K + t_R/2)}. \tag{22.113}$$

There may be additional contributions from successive wave crossings but these will be weak, especially since reflection does not occur without loss. Although the derivation of (22.113) is far from rigorous, the essential aspect is that *it is the wind at some time in the past that affects the thermocline thickness in the present.* If we now use (22.89a) to relate the SST to the thermocline thickness we have

$$\frac{\partial T_t}{\partial t} = A\widetilde{\tau}^x(t - t_K/2) - B\widetilde{\tau}^x(t - t_K/2 - t_R/2) - \lambda T, \tag{22.114}$$

where A and B are dimensional coefficients that incorporate the constants in (22.113) and account for the decay of the waves as they propagate. The delay in the first term on the right-hand side, due to Kelvin waves, is much smaller than that in the second term due to Rossby waves and Kelvin waves, and we will neglect it.

It remains to relate the wind stress to the SST. Using the Matsuno–Gill model of Section 8.5 the atmosphere responds approximately to an SST anomaly with a structure of the form

$$\tau^x \sim \mu A(T_e, x, y) \mathrm{e}^{-y^2/2L_a^2}, \tag{22.115}$$

where T_e is the SST in the east, L_a is the atmospheric equatorial deformation radius and A is a function that depends linearly on the SST and only slowly on y. (More generally the response is a Green function multiplied by $T_e(x, y)$ and integrated over the domain.) The wind-stress curl, $\partial(\tau^x/y)/\partial y$ is then proportional to the negative of τ^x, which means that in (22.113) the Rossby waves will carry a height anomaly of *opposite* sign to that of the Kelvin waves. Also, the atmospheric response is much faster than that of the ocean, and occurs over larger zonal scales and thus the wind-stress in the ocean centre can be approximately related to the SST in the East by

$$\widetilde{\tau}^x(t) = CT_e(t), \qquad \frac{\partial}{\partial y}\left(\frac{\widetilde{\tau}^x(t)}{y}\right) = -DT_e(t), \tag{22.116}$$

where C and D are positive constants. Using (22.116) in (22.114) finally gives

$$\frac{\partial T}{\partial t} = aT(t) - bT(t-\delta) - rT^3(t), \tag{22.117}$$

where we have neglected the delay in the Kelvin wave term, we introduced a cubic damping term for stability, and $a = AC + \lambda$, $b = BD$ and $\delta = t_K + t_R/2$.

Notes

1 *'I'm getting youngerly every day* — Paul Kushner, Gordon Conference, Santorini, 2007. Meteorologists tend to talk about westerly winds — the winds that come from the west — because it is where the winds come from that determines the weather. Oceanographers tend to talk about eastward currents, because this is where the currents will take things (or perhaps oceanographers are just a more forward looking crowd). We follow the lead of the oceanographers and talk about eastward and westward flow, for both currents and winds.

2 Figure kindly made by Neven Fučkar, using a state estimation from NCEP/GODAS (http://www.esrl.noaa.gov/psd/data/gridded/data.godas.html). Many of the observations themselves are made with acoustic Doppler current profilers (ADCP), which measure the currents by measuring the Doppler shift from a sonar.

3 Adapted from Kessler *et al.* (2003).

4 The undercurrent itself seems to have been first discovered in the Atlantic by J. Y. Buchanan in the 1880s. He measured a southeastward flowing current with speeds of more than 1 knot (about 0.5 m s^{-1}) at depths around 30 fathoms (55 metres) at the equator and 13° W from the steamship *Buccaneer,* which was chartered to do a survey prior to the laying of a telegraph cable (Buchanan 1886). The discovery of the undercurrent in the Pacific is often credited to Townsend Cromwell (1922–1958) in the early 1950s, and there the current is called the Cromwell Current. Cromwell also provided the first credible theoretical model of the undercurrent, as noted below. He tragically died in 1958 in a plane crash while en route to an oceanography expedition.

5 The local theories began with a description by Cromwell (1953) of the currents produced by a westward wind at the equator and were extended and put into mathematical form by Stommel (1960) with thermal effects added by Veronis (1960) and with a later variation by Robinson (1966). This class of model, which at its core is essentially linear and unavoidably dissipative, was further developed and clarified by Gill (1971), McKee (1973) and Gill (1975) and we mostly follow their treatment. The effects of nonlinearities were looked at first by Charney (1960) and then by McKee (1973) and Cane (1979a,b). The linear model was significantly extended by McCreary (1981) to include the effects of continuous stratification.

The inertial theories of the undercurrent, described in Section 22.4, have their seeds in Fofonoff & Montgomery (1955) and were developed by Pedlosky (1987b). This viewpoint was extended and, in part, reconciled with the local viewpoint by McCreary & Lu (1994) who considered the equatorial undercurrent as part of a larger and more complex subtropical current system, with both local and inertial aspects. A complete description of the integrated dynamics is perforce largely numerical.

6 The canonical Airy equation is $\partial^2 y/\partial x^2 - xy = 0$. The solution, the Airy function, is discussed in many books on ordinary differential equations and special functions (e.g., Jeffreys & Jeffreys 1946, Abramowitz & Stegun 1965) and, perhaps of more relevance to the modern reader, may be calculated using mathematical software such as Maple and Python. The form of solution we use was presented by McKee (1973).

7 One of the first observational analyses to unambiguously link the equatorial ocean to higher latitudes was Bryden & Brady (1985).

8 Much of our discussion follows Pedlosky (1987b).

9 Read McCreary & Lu (1994) and work backward and forward.

10 The name El Niño was originally used by fishermen along the coasts of Ecuador and Peru to refer to a warm ocean current that often appears in December (i.e., around Christmas) between Paita and Pascamayo, and lasts for several months. An early, perhaps the first, written account of El Niño is to be found in Carillo (1892), a report by a Navy captain on a warm ocean current know to the fishermen as *Corriente del Niño*, or Current of the Christ Child. These days, the name is applied only when the warming is particularly strong and to the warming over the whole eastern tropical Pacific.

11 My thanks to Eli Tziperman for a number of conversations on the topic, and for kindly providing some unpublished lecture notes. Thanks also to Malte Jansen for some helpful remarks about, among other things, unstable air-sea interactions.

12 Figure kindly constructed by A. Wittenberg by passing satellite observations through an optimal interpolation analysis.

13 Data from http://www.cgd.ucar.edu/cas/catalog/climind/soi.html. See also Wittenberg (2009). The Niño 3 region is the rectangular region outlined by dotted lines in Fig. 22.13.

14 Trenberth (1997) provides more detail.

15 See, for example, Tudhope *et al.* (2001) and Wittenberg (2009).

16 Data from the Tropical Atmosphere Ocean (TAO) project, http://www.pmel.noaa.gov/tao/.

17 Adapted from drawings by Dr. Billy Kessler.

18 The pressure at Easter Island was used by Quinn (1974). For more methodology on calculating the Southern Ocean index see https://www.ncdc.noaa.gov/teleconnections/enso/indicators/soi/.

19 Jacob Bjerknes (1897–1975), son of Vilhelm Bjerknes (see note 3 on page 169), was a leading player in the Bergen school of meteorology. He was responsible for the now-famous frontal model of cyclones (Bjerknes 1919), and was one of the first to seriously discuss the role of cyclones in the general circulation of the atmosphere. With Halvor Solberg and Tor Bergeron the frontal model led to a picture of the lifecycle of extratropical cyclones (see Chapter 12), in which a wave grows initially on the polar front (akin to baroclinic instability with the meridional temperature gradient compressed to a front, but baroclinic instability theory was not then developed), develops into a mature cyclone, occludes and decays. In 1939 Bjerknes moved to the US and, largely because of World War II, stayed, joining UCLA and heading its Department of Meteorology after its formation in 1945. He developed an interest in air-sea interactions, and notably proposed the feedback between sea-surface temperatures and the strength of the trade winds (Bjerknes 1969). See also Friedman (1989), Cressman (1996), Shapiro & Grønas (1999), and the memoir by Arnt Eliassen at http://www.nap.edu/readingroom/books/biomems/jbjerknes.html.

20 Thus, for example, Latif & Barnett (1996) used a coupled model to identify a plausible coupling between atmosphere and ocean in the mid-latitudes that might give rise to decadal variability, but the mechanism is not robustly seen in other models.

21 Gilbert Walker (1868–1958) was a British meteorologist who described this eponymous circulation in the 1920s. He has also been credited with discovery of the Southern Oscillation and the North Atlantic Oscillation. Walker began his career as an applied mathematician but was afforded greater renown by his analysis of meteorological observations.

22 Studies of the Walker circulation using models that make quasi-equilibium assumptions and/or use variations of the Matsuno–Gill model include Bretherton & Sobel (2002), Stechmann & Ogrosky (2014) and, in the first instance, Gill (1980). Stechman and Ogrosky argue that the Matsuno–Gill model can in fact give quantitatively accurate results if the forcing and damping are properly chosen. An exploration of the Matsuno–Gill problem in the weak-temperature-gradient (WTG) limit is given by Bretherton & Sobel (2003).

23 See, for example, Hirst (1986) and Zebiak & Cane (1987).

24 A related system of equations was considered by Lau (1981).

25 Oxymoronically, so-called toy models have a rather distinguished role in geophysical fluid dynamics in general, and in ENSO dynamics in particular. Toy models are neither a theory (such as the theory of the western boundary current in Chapter 19) nor a realistic model (such as a GCM might aim to be), and they thus occupy rather questionable territory. Still, the models can sometimes spawn great insight and be very useful. The Stommel box model of the deep overturning circulation, energy balance models of the global climate, and the Lorenz model of chaotic convection are important toy models in related areas.

26 Such models began with Suarez & Schopf (1988) and Battisti (1988), and we also draw from Galanti & Tziperman (2000).

27 The recharge, or recharge-discharge, or refill, hypothesis was suggested by Wyrtki (1985) and Cane & Zebiak (1985). Wyrtki had for some time been of the view that El Niño was a basinwide phenomenon resembling seiches in smaller enclosed seas in which winds cause water to pile up at one end of the basin before sloshing back if and when the winds relax (Wyrtki 1952, 1975). A historical perspective is provided by McPhaden *et al.* (2015).

28 Versions of the mechanism have been put in the form of toy models by Jin (1997a,b) and Clarke *et al.* (2007).

29 Papers exploring stochastic, chaotic and seasonal effects include Vallis (1988a), Tziperman *et al.* (1994), Chang *et al.* (1996) and Samelson & Tziperman (2001). See also the book by Clarke (2008).

References

Abarbanel, H. D. I. & Young, W. R., Eds., 1987. *General Circulation of the Ocean*. Springer-Verlag, 291 pp.

Abbe, C., 1901. The physical basis of long-range weather forecasts. *Mon. Wea. Rev.*, **29**, 551–561.

Ablowitz, M. J., 2011. *Nonlinear dispersive waves: asymptotic analysis and solitons*. Vol. 47, Cambridge University Press, 311 pp.

Abramowitz, M. & Stegun, I. A., 1965. *Handbook of Mathematical Functions*. Dover Publications, 1046 pp.

Allen, J. S., 1993. Iterated geostrophic intermediate models. *J. Phys. Oceanogr.*, **23**, 2447–2461.

Allen, J. S., Barth, J. A. & Newberger, P. A., 1990a. On intermediate models for barotropic continental shelf and slope flow fields. Part I: Formulation and comparison of exact solutions. *J. Phys. Oceanogr.*, **20**, 1017–1042.

Allen, J. S., Barth, J. A. & Newberger, P. A., 1990b. On intermediate models for barotropic continental shelf and slope flow fields. Part II: Comparison of numerical model solutions in doubly periodic domains. *J. Phys. Oceanogr.*, **20**, 1043–1082.

Allen, J. S., Holm, D. D. & Newberger, P. A., 2002. Extended-geostrophic Euler–Poincaré models for mesoscale oceanographic flow. In J. Norbury & I. Roulstone, Eds., *Large-scale Atmosphere–Ocean Dynamics I*. Cambridge University Press.

Ambaum, M. H. P., 2010. *Thermal Physics of the Atmosphere*. Wiley, 239 pp.

Andrews, D. G., 1987. On the interpretation of the Eliassen–Palm flux divergence. *Quart. J. Roy. Meteor. Soc.*, **113**, 323–338.

Andrews, D. G., 2010. *An Introduction to Atmospheric Physics*. Cambridge University Press, 237 pp.

Andrews, D. G., Holton, J. R. & Leovy, C. B., 1987. *Middle Atmosphere Dynamics*. Academic Press, 489 pp.

Andrews, D. G. & McIntyre, M. E., 1976. Planetary waves in horizontal and vertical shear: the generalized Eliassen–Palm relation and the mean zonal acceleration. *J. Atmos. Sci.*, **33**, 2031–2048.

Andrews, D. G. & McIntyre, M. E., 1978. Generalized Eliassen–Palm and Charney–Drazin theorems for waves on axisymmetric mean flows in compressible atmospheres. *J. Atmos. Sci.*, **35**, 175–185.

Angell, J. K. & Korshover, J., 1964. Quasi-biennial variations in temperature, total ozone, and tropopause height. *J. Atmos. Sci.*, **21**, 479–492.

Arakawa, A., 2004. The cumulus parameterization problem: Past, present, and future. *J. Climate*, **17**, 13, 2493–2525.

Arakawa, A. & Schubert, W. H., 1974. Interaction of a cumulus cloud ensemble with the large-scale environment, part i. *J. Atmos. Sci.*, **31**, 3, 674–701.

Arbic, B. K., Flierl, G. R. & Scott, R. B., 2007. Cascade inequalities for forced-dissipated geostrophic turbulence. *J. Phys. Oceanogr.*, **37**, 1470–1487.

Arnold, V. I., 1965. Conditions for nonlinear stability of stationary plane curvilinear flows of an ideal fluid. *Dokl. Akad. Nauk SSSR*, **162**, 975–978. Engl. transl.: *Sov. Math.* **6**, 773–777 (1965).

Arnold, V. I., 1966. On an a priori estimate in the theory of hydrodynamic stability. *Izv. Vyssh. Uchebn. Zaved. Math.*, **54**, 3–5. Engl. transl.: *Am Mat. Soc. Transl. Ser.*, 2, **79**, 267–289 (1969).

Assmann, R., 1902. Über die Existenz eines wärmeren Luftstromes in der Höhe von 10 bis 15 km. (On the existence of a warmer airflow at heights from 10 to 15 km). *Sitzber. Königl. Preuss. Akad. Wiss. Berlin*, **24**, 495–504.

Aubin, D. & Dahan Dalmedico, A., 2002. Writing the history of dynamical systems and chaos: *longue durée* and revolution, disciplines and cultures. *Historia Mathematica*, **29**, 1–67.

Baldwin, M. P., Gray, L. J., Dunkerton, T. J., Hamilton, K. *et al.*, 2001. The quasi-biennial oscillation. *Rev. Geophys.*, **39**, 179–229.

Ball, J. M. & James, R. D., 2002. The scientific life and influence of Clifford Ambrose Truesdell III. *Arch. Rational Mech. Anal.*, **161**, 1–26.

Bannon, P. R., 1995. Potential vorticity conservation, hydrostatic adjustment, and the anelastic approximation. *J. Atmos. Sci.*, **52**, 2301–2312.

Bannon, P. R., 1996. On the anelastic equation for a compressible atmosphere. *J. Atmos. Sci.*, **53**, 3618–3628.

Bartello, P. & Warn, T., 1996. Self-similarity of decaying two-dimensional turbulence. *J. Fluid Mech.*, **326**, 357–372.

Batchelor, G. K., 1953a. The conditions for dynamical similarity of motions of a frictionless perfect-gas atmosphere. *Quart. J. Roy. Meteor. Soc.*, **79**, 224–235.

Batchelor, G. K., 1953b. *The Theory of Homogeneous Turbulence*. Cambridge University Press, 197 pp.

Batchelor, G. K., 1959. Small-scale variation of convected quantities like temperature in turbulent fluid. Part 1: General discussion and the case of small conductivity. *J. Fluid Mech.*, **5**, 113–133.

Batchelor, G. K., 1967. *An Introduction to Fluid Dynamics*. Cambridge University Press, 615 pp.

Batchelor, G. K., 1969. Computation of the energy spectrum in homogeneous two-dimensional turbulence. *Phys. Fluids Suppl.*, **12**, II–233—II–239.

Battisti, D. S., 1988. Dynamics and thermodynamics of a warming event in a coupled tropical atmosphere–ocean model. *J. Atmos. Sci.*, **45**, 2889–2919.

Baumert, H. Z., Simpson, J. & Sündermann, J., Eds., 2005. *Marine Turbulence: Theories, Observations and Models*. Cambridge University Press, 630 pp.

Bell, E. T., 1937. *Men of Mathematics*. Simon & Schuster, 590 pp.

Bell, M. J., 2015a. Meridional overturning circulations driven by surface wind and buoyancy forcing. *J. Phys. Oceanogr.*, **45**, 2701–2714.

Bell, M. J., 2015b. Water mass transformations driven by Ekman upwelling and surface warming in subpolar gyres. *J. Phys. Oceanogr.*, **45**, 2356–2380.

Bender, C. M. & Orszag, S. A., 1978. *Advanced Mathematical Methods for Scientists and Engineers*. McGraw-Hill, 593 pp.

Berrisford, P., Marshall, J. C. & White, A. A., 1993. Quasi-geostrophic potential vorticity in isentropic coordinates. *Quart. J. Roy. Meteor. Soc.*, **119**, 778–782.

Betts, A. K., 1973. Non-precipitating convection and its parameterization. *Quart. J. Roy. Meteor. Soc.*, **99**, 178–196.

Betts, A. K., 1986. A new convective adjustment scheme. Part I: Observational and theoretical basis. *Quart. J. Roy. Meteor. Soc.*, **112**, 677–691.

Betts, A. K. & Miller, M. J., 1986. A new convective adjustment scheme. Part II: Single column test using GATE wave, BOMEX, ATEX and Arctic air mass data sets. *Quart. J. Roy. Meteor. Soc.*, **112**, 693–709.

Birner, T., 2006. Fine-scale structure of the extratropical tropopause region. *J. Geophys. Res.*, **111**, D04104.

Birner, T., Dörnbrack, A. & Schumann, U., 2002. How sharp is the tropopause at midlatitudes? *Geophys. Res. Lett.*, **29**, 1700.

Bjerknes, J., 1919. On the structure of moving cyclones. *Geofys. Publ*, **1** (2), 1–8.

Bjerknes, J., 1937. Die Theorie der aussertropischen Zyklonenbildung (The theory of extra tropical cyclone formation). *Meteor. Zeitschr*, **12**, 460–466.

Bjerknes, J., 1959. Atlantic air–sea interaction. In *Advances in Geophysics*, Vol. 10, pp. 1–82. Academic Press.

Bjerknes, J., 1969. Atmospheric teleconnections from the equatorial Pacific. *Mon. Wea. Rev.*, **97**, 163–172.

Bjerknes, V., 1898a. Über die Bildung von Cirkulationsbewegungen und Wirbeln in reibunglosen Flüssigkeiten (On the generation of circulation and vortices in inviscid fluids). *Skr. Nor. Vidensk.-Akad. 1: Mat.-Naturvidensk. Kl.*, **5**, 3–29.

Bjerknes, V., 1898b. Über einen hydrodynamischen Fundamentalsatz und seine Anwendung besonders auf die Mechanik der Atmosphäre und des Weltmeeres (On a fundamental principle of hydrodynamics and its application particularly to the mechanics of the atmosphere and the world's oceans). *Kongl. Sven. Vetensk. Akad. Handlingar*, **31**, 1–35.

Bjerknes, V., 1902. Cirkulation relativ zu der Erde (Circulation relative to the Earth). *Meteor. Z.*, **37**, 97–108.

Bjerknes, V., 1904. Das Problem der Wettervorhersage, betrachtet vom Standpunkte der Mechanik und der Physic (The problem of weather forecasting as a problem in mathematics and physics). *Meteor. Z.*, January, 1–7. Engl. transl.: Y. Mintz, in Shapiro and Grønas (1999), pp. 1–7.

Blumen, W., 1968. On the stability of quasi-geostrophic flow. *J. Atmos. Sci.*, **25**, 929–933.

Boccaletti, G., Pacanowski, R. C., Philander, S. G. H. & Fedorov, A. V., 2004. The thermal structure of the upper ocean. *J. Phys. Oceanogr.*, **34**, 888–902.

Boer, G. J. & Shepherd, T. G., 1983. Large-scale two-dimensional turbulence in the atmosphere. *J. Atmos. Sci.*, **40**, 164–184.

Boffetta, G. & Ecke, R. E., 2012. Two-dimensional turbulence. *Ann. Rev. Fluid Mech.*, **44**, 427–451.

Bohren, C. F. & Albrecht, B. A., 1998. *Atmospheric thermodynamics*. Oxford University Press, 416 pp.

Bolton, D., 1980. The computation of equivalent potential temperature. *Mon. Wea. Rev.*, **108**, 1046–1053.

Booker, J. R. & Bretherton, F. P., 1967. The critical layer for internal gravity waves in a shear flow. *J. Fluid Mech.*, **27**, 513–539.

Boussinesq, J., 1903. Théorie analytique de la chaleur (Analytic theory of heat). *Tome, Paris, Gauthier-Villars*, **II**, 170–172.

Box, G. E. P., 1976. Science and statistics. *J. Am. Stat. Assoc*, **71**, 791–799.

Box, G. E. P., 1979. Robustness in the strategy of scientific model building. In R. L. Launer & G. N. Wilkinson, Eds., *Robustness in Statistics*, pp. 201–236. Academic Press.

Boyd, J. P., 1976. The noninteraction of waves with the zonally averaged flow on a spherical Earth and the interrelationships of eddy fluxes of energy, heat and momentum. *J. Atmos. Sci.*, **33**, 2285–2291.

Boyd, J. P., 1978. The effects of latitudinal shear on equatorial waves. Part 1. Theory and method. *J. Atmos. Sci.*, **35**, 2236–2258.

Boyd, J. P., 1980. The nonlinear equatorial Kelvin wave. *J. Phys. Oceanogr.*, **10**, 1–11.

Branscome, L. E., 1983. The Charney baroclinic stability problem: approximate solutions and modal structures. *J. Atmos. Sci.*, **40**, 1393–1409.

Bretherton, C. S. & Schär, C., 1993. Flux of potential vorticity substance: a simple derivation and a uniqueness property. *J. Atmos. Sci.*, **50**, 1834–1836.

Bretherton, C. S. & Smolarkiewicz, P. K., 1989. Gravity waves, compensating subsidence and detrainment around cumulus clouds. *J. Atmos. Sci.*, **46**, 740–759.

Bretherton, C. S. & Sobel, A. H., 2002. A simple model of a convectively coupled Walker circulation using the weak temperature gradient approximation. *J. Climate*, **15**, 2907–2920.

Bretherton, C. S. & Sobel, A. H., 2003. The Gill model and the weak temperature gradient approximation. *J. Atmos. Sci.*, **60**, 451–460.

Bretherton, F. P., 1964. Low frequency oscillations trapped near the equator. *Tellus*, **16**, 181–185.

Bretherton, F. P., 1966a. Baroclinic instability and the short wavelength cut-off in terms of potential vorticity. *Quart. J. Roy. Meteor. Soc.*, **92**, 335–345.

Bretherton, F. P., 1966b. Critical layer instability in baroclinic flows. *Quart. J. Roy. Meteor. Soc.*, **92**, 325–334.

Bretherton, F. P., 1969. Momentum transport by gravity waves. *Quart. J. Roy. Meteor. Soc.*, **95**, 213–243.

Brewer, A. W., 1949. Evidence for a world circulation provided by the measurements of helium and water vapour distribution in the stratosphere. *Quart. J. Roy. Meteor. Soc.*, **75**, 251–363.

Brillouin, L., 1926. La mécanique ondulatoire de Schrödinger; une méthode générale de resolution par approximations successives (The wave mechanics of Schrödinger: a general method of solution by successive approximation). *Comptes Rendus*, **183**, 24–26.

Browning, G., Kreiss, H. & Schubert, W., 2000. The role of gravity waves in slowly varying in time tropospheric motions near the equator. *J. Atmos. Sci.*, **57**, 4008–4019.

Bryan, F., 1986. High-latitude salinity effects and interhemispheric thermohaline circulations. *Nature*, **323**, 301–304.

Bryden, H. & Brady, E. C., 1985. Diagnostic study of the three-dimensional circulation of the upper equatorialPacific Ocean. *J. Phys. Oceanogr.*, **15**, 1255–1273.

Buchanan, J. Y., 1886. On the similarities in the physical geography of the great oceans. *Proc. Roy. Geogr. Soc.*, **8**, 753–770.

Bühler, O., 2009. *Waves and Mean Flows*. Cambridge University Press, 370 pp.

Burger, A., 1958. Scale considerations of planetary motions of the atmosphere. *Tellus*, **10**, 195–205.

Burke, A., Stewart, A. L., Adkins, J. F., Ferrari, R. *et al.*, 2015. The glacial mid-depth radiocarbon bulge and its implications for the overturning circulation. *Paleoceanog.*, **30**, 1021–1039.

Caballero, R., 2014. *Physics of the Atmosphere*. IOP Publishing, 132 pp.

Callen, H. B., 1985. *Thermodynamics and an Introduction to Themostatistics*. John Wiley & Sons, 493 pp.

Cane, M. A., 1979a. The response of an equatorial ocean to simple wind stress patterns: I. Model formulation and analytic results. *J. Mar. Res.*, **37**, 232–252.

Cane, M. A., 1979b. The response of an equatorial ocean to simple wind stress patterns: II. Numerical results. *J. Mar. Res.*, **37**, 355–398.

Cane, M. A. & Zebiak, S. E., 1985. A theory for El Niño and the Southern Oscillation. *Science*, **228**, 1084–1087.

Carillo, C. N., 1892. Desertacion sobre las corrientes y estudios de la corriente Peruana de Humboldt. (Dissertation on currents and studies of the Peruvian Humboldt current). *Bol. Soc. Geogr. Lima*, **11**, 72–110.

Carlini, F., 1837. Ricerche sulla convergenza della serie che serva alla soluzione del problema di Keplero (Research on the convergence of series for the solution of Kepler's problem). Milan.

Cessi, P., 2001. Thermohaline circulation variability. In *Conceptual Models of the Climate, Woods Hole Program in Geophysical Fluid Dynamics* (2001). Also available from http://gfd.whoi.edu/proceedings/2001/PDFvol2001.html.

Cessi, P. & Fantini, M., 2004. The eddy-driven thermocline. *J. Phys. Oceanogr.*, **34**, 2642–2658.

Cessi, P. & Young, W. R., 1992. Multiple equilibria in two-dimensional thermohaline circulation. *J. Fluid Mech.*, **241**, 291–309.

Chai, J., 2016. Understanding geostrophic turbulence in a hierarchy of models. Ph. D thesis, Princeton University.

Chandrasekhar, S., 1961. *Hydrodynamic and Hydromagnetic Stability*. Oxford University Press, 652 pp. Reprinted by Dover Publications, 1981.

Chang, E. K. M. & Orlanski, I., 1994. On energy flux and group velocity of waves in baroclinic flows. *J. Atmos. Sci.*, **51**, 3823–3828.

Chang, P., Ji, L., Li, H. & Flügel, M., 1996. Chaotic dynamics versus stochastic processes in El Niño–Southern Oscillation in coupled ocean–atmosphere models. *Physica D*, **98**, 301–320.

Chapman, D. C., Malanotte-Rizzoli, P. & Hendershott, M., 1989. Wave motions in the ocean. Unpublished notes based on lectures by Myrl Hendershott.

Chapman, S. & Lindzen, R. S., 1970. *Atmospheric Tides*. Gordon and Breach, 200 pp.

Charlton, A. J. & Polvani, L. M., 2007. A new look at stratospheric sudden warmings. Part I: Climatology and modeling benchmarks. *J. Climate*, **20**, 449–469.

Charlton, A. J., Polvani, L. M., Perlwitz, J., Sassi, F. *et al.*, 2007. A new look at stratospheric sudden warmings. Part II: Evaluation of numerical model simulations. *J. Climate*, **20**, 470–488.

Charney, J. G., 1947. The dynamics of long waves in a baroclinic westerly current. *J. Meteor.*, **4**, 135–162.

Charney, J. G., 1948. On the scale of atmospheric motion. *Geofys. Publ. Oslo*, **17** (2), 1–17.

Charney, J. G., 1955. The Gulf Stream as an inertial boundary layer. *Proc. Nat. Acad. Sci.*, **41**, 731–740.

Charney, J. G., 1960. Non-linear theory of a wind-driven homogeneous layer near the equator. *Deep-Sea Res.*, **6**, 303–310.

Charney, J. G., 1963. A note on large-scale motions in the tropics. *J. Atmos. Sci.*, **20**, 607–609.

Charney, J. G., 1971. Geostrophic turbulence. *J. Atmos. Sci.*, **28**, 1087–1095.

Charney, J. G. & Drazin, P. G., 1961. Propagation of planetary scale disturbances from the lower into the upper atmosphere. *J. Geophys. Res.*, **66**, 83–109.

Charney, J. G. & Eliassen, A., 1949. A numerical method for predicting the perturbations of the mid-latitude westerlies. *Tellus*, **1**, 38–54.

Charney, J. G. & Eliassen, A., 1964. On the growth of the hurricane depression. *J. Atmos. Sci.*, **21**, 68–75.

Charney, J. G., Fjørtoft, R. & Neumann, J. V., 1950. Numerical integration of the barotropic vorticity equation. *Tellus*, **2**, 237–254.

Charney, J. G. & Stern, M. E., 1962. On the stability of internal baroclinic jets in a rotating atmosphere. *J. Atmos. Sci.*, **19**, 159–172.

Charnock, H., Green, J., Ludlam, F., Scorer, R. & Sheppard, P., 1966. Dr. E. T. Eady, B. A. (Obituary). *Quart. J. Roy. Meteor. Soc.*, **92**, 591–592.

Chasnov, J. R., 1991. Simulation of the inertial-conductive subrange. *Phys. Fluids A*, **3**, 1164–1168.

Chelton, D. B., de Szoeke, R. A., Schlax, M. G., Naggar, K. E. & Siwertz, N., 1998. Geographical variability of the first-baroclinic Rossby radius of deformation. *J. Phys. Oceanogr.*, **28**, 433–460.

Christiansen, B., 1999. Stratospheric vacillations in a general circulation model. *J. Atmos. Sci.*, **56**, 1858–1872.

Christiansen, B., 2000. Chaos, quasiperiodicity, and interannual variability: studies of a stratospheric vacillation model. *J. Atmos. Sci.*, **57**, 3161–3173.

Clarke, A. J., 2008. *An Introduction to the Dynamics of El Niño and the Southern Oscillation*. Elsevier, 308 pp.

Clarke, A. J., Van Gorder, S. & Colantuono, G., 2007. Wind stress curl and ENSO discharge/recharge in the equatorial Pacific. *J. Phys. Oceanogr.*, **37**, 1077–1091.

Colin de Verdière, A., 1980. Quasi-geostrophic turbulence in a rotating homogeneous fluid. *Geophys. Astrophys. Fluid Dyn.*, **15**, 213–251.

Colin de Verdière, A., 1989. On the interaction of wind and buoyancy driven gyres. *J. Mar. Res.*, **47**, 595–633.

Conkright, M. E., Antonov, J., Baranova, O., Boyer, T. P. *et al.*, 2001. World ocean database 2001, vol. 1. In S. Levitus, Ed., *NOAA Atlas NESDIS 42*, pp. 167. US Government Printing Office, Washington DC.

Coriolis, G. G., 1832. Mémoire sur le principe des forces vives dans les mouvements relatifs des machines (On the principle of kinetic energy in the relative movement of machines). *J. Ec. Polytech*, **13**, 268–301.

Coriolis, G. G., 1835. Mémoire sur les équations du mouvement relatif des systèmes de corps (On the equations of relative motion of a system of bodies). *J. Ec. Polytech*, **15**, 142–154.

Corrsin, S., 1951. On the spectrum of isotropic temperature fluctuations in an isotropic turbulence. *J. Appl. Phys.*, **22**, 469–473. Erratum: *J. Appl. Phys.* **22,** 1292, (1951).

Craig, G. C. & Gray, S. L., 1996. CISK or WISHE as the mechanism for tropical cyclone intensification. *J. Atmos. Sci.*, **53**, 3528–3540.

Cressman, G. P., 1996. The origin and rise of numerical weather prediction. In J. R. Fleming, Ed., *Historical Essays on Meteorology 1919–1995*, pp. 617. American Meteorological Society.

Cromwell, T., 1953. Circulation in a meridional plane in the central equatorial Pacific. *J. Mar. Res.*, **12**, 196–213.

Da Vinci, L., 1500. *The notebooks of Leonardo da Vinci*, Vol. 2. J. P. Richter, Ed. Dover Publications, 1970.

Danielsen, E. F., 1990. In defense of Ertel's potential vorticity and its general applicability as a meteorological tracer. *J. Atmos. Sci.*, **47**, 2353–2361.

Danilov, S. & Gryanik, V., 2002. Rhines scale and spectra of the β-plane turbulence with bottom drag. *Phys. Rev. E*, **65**, 067301–1–067301–3.

Danilov, S. & Gurarie, D., 2001. Quasi-two-dimensional turbulence. *Usp. Fiz. Nauk.*, **170**, 921–968.

Davidson, P., 2015. *Turbulence: An Introduction for Scientists and Engineers*. Oxford University Press, 688 pp.

Davies-Jones, R., 2003. Comments on "A generalization of Bernoulli's theorem". *J. Atmos. Sci.*, **60**, 2039–2041.

Davis, R. E., de Szoeke, R., Halpern, D. & Niiler, P., 1981. Variability in the upper ocean during MILE. Part I: The heat and momentum balances. *Deep-Sea Res.*, **28**, 1427–1452.

De Morgan, A., 1872. *A Budget of Paradoxes*. Thoemmmes Continuum, 814 pp.

de Szoeke, R. & Bennett, A. F., 1993. Microstructure fluxes across density surfaces. *J. Phys. Oceanogr.*, **24**, 2254–2264.

de Szoeke, R. A., 2000. Equations of motion using thermodynamic coordinates. *J. Phys. Oceanogr.*, **30**, 2814–2829.

de Szoeke, R. A., 2004. An effect of the thermobaric nonlinearity of the equation of state: A mechanism for sustaining solitary Rossby waves. *J. Phys. Oceanogr.*, **34**, 2042–2056.

Dee, D., Uppala, S., Simmons, A., Berrisford, P. *et al.*, 2011. The ERA-interim reanalysis: Configuration and performance of the data assimilation system. *Quart. J. Roy. Meteor. Soc.*, **137**, 553–597.

Defant, A., 1921. Die Zirkulation der Atmosphäre in den gemäßigten Breiten der Erde. Grundzüge einer Theorie der Klimaschwankungen (The circulation of the atmosphere in the Earth's mid-latitudes. Basic features of a theory of climate fluctuations). *Geograf. Ann.*, **3**, 209–266.

Dellar, P. J., 2011. Variations on a beta-plane: derivation of non-traditional beta-plane equations from Hamilton's principle on a sphere. *J. Fluid Mech.*, **674**, 174–195.

Dewar, W. K. & Huang, R. X., 1995. Fluid flow in loops driven by freshwater and heat fluxes. *J. Fluid Mech.*, **297**, 153–191.

Dewar, W. K., Samelson, R. S. & Vallis, G. K., 2005. The ventilated pool: a model of subtropical mode water. *J. Phys. Oceanogr.*, **35**, 137–150.

Dias, J. & Kiladis, G. N., 2014. Influence of the basic state zonal flow on convectively coupled equatorial waves. *Geophys. Res. Lett.*, **41**, 6904–6913.

Dickinson, R. E., 1968. Planetary Rossby waves propagating vertically through weak westerly wind wave guides. *J. Atmos. Sci.*, **25**, 984–1002.

Dickinson, R. E., 1969. Theory of planetary wave–zonal flow interaction. *J. Atmos. Sci.*, **26**, 73–81.

Dickinson, R. E., 1980. Planetary waves: theory and observation. In *Orographic Effects on Planetary Flows*, Number 23 in GARP Publication Series. World Meteorological Organization.

Dijkstra, H. A., 2008. *Dynamical Oceanography*. Springer, 407 pp.

Dima, I. & Wallace, J. M., 2003. On the seasonality of the Hadley Cell. *J. Atmos. Sci.*, **60**, 1522–1527.

Dobson, G. M. B., 1956. Origin and distribution of the polyatomic molecules in the atmosphere. *Proc. Roy. Soc. Lond. A*, **236**, 187–193.

Döös, K. & Coward, A., 1997. The Southern Ocean as the major upwelling zone of the North Atlantic. *Int. WOCE Newsletter* 27, 3–4.

Drazin, P. G. & Reid, W. H., 1981. *Hydrodynamic Stability*. Cambridge University Press, 527 pp.

Drijfhout, S. S. & Hazeleger, W., 2001. Eddy mixing of potential vorticity versus temperature in an isopycnic ocean model. *J. Phys. Oceanogr.*, **31**, 481–505.

Dritschel, D. & McIntyre, M., 2008. Multiple jets as PV staircases: The Phillips effect and the resilience of eddy-transport barriers. *J. Atmos. Sci.*, **65**, 855–874.

Dunkerton, T., Hsu, C.-P. & McIntrye, M. E., 1981. Some Eulerian and Lagrangian diagnostics for a model stratosphere warming. *JAS*, **38**, 819–843.

Dunkerton, T. J., 1980. A Lagrangian-mean theory of wave, mean-flow interaction with applications to non-acceleration and its breakdown. *Rev. Geophys. Space Phys.*, **18**, 387–400.

Dunkerton, T. J., 1982. Shear zone asymmetry in the observed and simulated quasi-biennial oscillation. *J. Atmos. Sci.*, **38**, 461–469.

Dunkerton, T. J., 1997. The role of gravity waves in the quasi-biennial oscillation. *JGR*, **102**, 26053–26076.

Dunkerton, T. J., Delisi, D. P. & Baldwin, M. P., 1988. Distribution of major stratospheric warmings in relation to the quasi-biennial oscillation. *Geophys. Res. Lett.*, **115**, 136–139.

Durran, D. R., 1989. Improving the anelastic approximation. *J. Atmos. Sci.*, **46**, 1453–1461.

Durran, D. R., 1990. Mountain waves and downslope winds. *Meteor. Monogr.*, **23**, 59–81.

Durran, D. R., 1993. Is the Coriolis force really responsible for the inertial oscillations? *Bull. Am. Meteor. Soc.*, **74**, 2179–2184.

Durran, D. R., 2015. Lee waves and mountain waves. *Encycl. Atmos. Sci.*, 2nd edn, **4**, 95–102.

Durst, C. S. & Sutcliffe, R. C., 1938. The effect of vertical motion on the "geostrophic departure" of the wind. *Quart. J. Roy. Meteor. Soc.*, **64**, 240.

Dutton, J. A., 1986. *The Ceaseless Wind: An Introduction to the Theory of Atmospheric Motion*. Dover Publications, 617 pp.

Eady, E. T., 1949. Long waves and cyclone waves. *Tellus*, **1**, 33–52.

Eady, E. T., 1950. The cause of the general circulation of the atmosphere. In *Cent. Proc. Roy. Meteor. Soc.* (1950), pp. 156–172.

Eady, E. T., 1954. The maintenance of the mean zonal surface currents. *Proc. Toronto Meteor. Conf. 1953*, **138**, 124–128. Royal Meteorological Society.

Eady, E. T., 1957. The general circulation of the atmosphere and oceans. In D. R. Bates, Ed., *The Earth and its Atmosphere*, pp. 130–151. New York, Basic Books.

Eady, E. T. & Sawyer, J. S., 1951. Dynamics of flow patterns in extra-tropical regions. *Quart. J. Roy. Meteor. Soc.*, **77**, 531–551.

Ebdon, R. A., 1960. Notes on the wind flow at 50 mb in tropical and subtropical regions in January 1957 and in 1960. *Quart. J. Roy. Meteor. Soc.*, **86**, 540–542.

Eden, C. & Willebrand, J., 1999. Neutral density revisited. *Deep Sea Res., Part II*, **46**, 33–54.

Edmon, H. J., Hoskins, B. J. & McIntyre, M. E., 1980. Eliassen–Palm cross sections for the troposphere. *J. Atmos. Sci.*, **37**, 2600–2616.

Egger, J., 1976. Linear response of a two-level primitive equation model to forcing by topography. *Mon. Wea. Rev.*, **104**, 351–364.

Ekman, V. W., 1905. On the influence of the Earth's rotation on ocean currents. *Arch. Math. Astron. Phys.*, **2**, 1–52.

Eliassen, A. & Palm, E., 1961. On the transfer of energy in stationary mountain waves. *Geofys. Publ.*, **22**, 1–23.

Eluszkiewicz, J., Crisp, D., Grainger, R. G., Lambert, A. *et al.*, 1997. Sensitivity of the residual circulation diagnosed from the UARS data to the uncertainties in the input fields and to the inclusion of aerosols. *J. Atmos. Sci.*, **54**, 1739–1757.

Emanuel, K., 2007. Quasi-equilibrium dynamics of the tropical atmosphere. In T. Schneider & A. Sobel, Eds., *The Global Circulation of the Atmosphere: Phenomena, Theory, Challenges*, pp. 143–185. Princeton University Press.

Emanuel, K. A., 1987. An air-sea interaction model of intraseasonal oscillations in the tropics. *J. Atmos. Sci.*, **44**, 2324–2340.

Emanuel, K. A., 1994. *Atmospheric Convection*. Oxford University Press, 580 pp.

Emanuel, K. A., 2011. Edward Norton Lorenz, 1917–2008. In *Biog. Memoirs*, pp. 1–28. Nat'l. Acad. Sci.

Emanuel, K. A., Neelin, J. D. & Bretherton, C. S., 1994. On large-scale circulations in convecting atmospheres. *Quart. J. Roy. Meteor. Soc.*, **120**, 1111–1143.

Er-El, J. & Peskin, R., 1981. Relative diffusion of constant-level balloons in the Southern Hemisphere. *J. Atmos. Sci.*, **38**, 2264–2274.

Ertel, H., 1942a. Ein neuer hydrodynamischer Wirbelsatz (A new hydrodynamic eddy theorem). *Meteorol. Z.*, **59**, 277–281.

Ertel, H., 1942b. Über des Verhältnis des neuen hydrodynamischen Wirbelsatzes zum Zirculationssatz von V. Bjerknes (On the relationship of the new hydrodynamic eddy theorem to the circulation theorem of V. Bjerknes). *Meteorol. Z.*, **59**, 385–387.

Ertel, H. & Rossby, C.-G., 1949a. Ein neuer Erhaltungs-satz der Hydrodynamik (A new conservation theorem of hydrodynamics). *Sitzungsber. d. Deutschen Akad. Wissenschaften Berlin*, **1**, 3–11.

Ertel, H. & Rossby, C.-G., 1949b. A new conservation theorem of hydrodynamics. *Geofis. Pura Appl.*, **14**, 189–193.

Fang, M. & Tung, K. K., 1996. A simple model of nonlinear Hadley circulation with an ITCZ: analytic and numerical solutions. *J. Atmos. Sci.*, **53**, 1241–1261.

Fang, M. & Tung, K. K., 1999. Time-dependent nonlinear Hadley circulation. *J. Atmos. Sci.*, **56**, 1797–1807.

Farrell, B., 1984. Modal and non-modal baroclinic waves. *J. Atmos. Sci.*, **41**, 668–673.

Farrell, B. & Ioannou, P. J., 1995. Stochastic dynamics of the midlatitude atmospheric jet. *J. Atmos. Sci.*, **52**, 1642–1656.

Farrell, B. F. & Ioannou, P. J., 1996. Generalized stability theory. Part I: autonomous operators. *J. Atmos. Sci.*, **53**, 2025–2040.

Farrell, B. F. & Ioannou, P. J., 2008. Formation of jets by baroclinic turbulence. *J. Atmos. Sci.*, **65**, 3353–3375.

Feistel, R., 2008. A Gibbs function for seawater thermodynamics for −6° C to 80° C and salinity up to 120 g kg^{-1}. *Deep Sea Res.*, **55**, 1639–1671.

Feistel, R., Wright, D., Kretzschmar, H.-J., Hagen, E. *et al.*, 2010. Thermodynamic properties of sea air. *Oce. Sci.*, **6**, 91–141.

Ferrari, R., Griffies, S. M., Nurser, G. & Vallis, G. K., 2010. A boundary-value problem for the parameterized mesoscale eddy transport. *Oce. Model.*, **32**, 143–156.

Ferrari, R., Jansen, M. F., Adkins, J. F., Burke, A. *et al.*, 2014. Antarctic sea ice control on ocean circulation in present and glacial climates. *Proceedings of the National Academy of Sciences*, **111**, 8753–8758.

Ferrel, W., 1856a. An essay on the winds and currents of the ocean. *Nashville J. Med. & Surg.*, **11**, 287–301.

Ferrel, W., 1856b. The problem of the tides. *Astron. J*, **4**, 173–176.

Ferrel, W., 1858. The influence of the Earth's rotation upon the relative motion of bodies near its surface. *Astron. J.*, **V**, No. 13 (109), 97–100.

Ferrel, W., 1859. The motion of fluids and solids relative to the Earth's surface. *Math. Monthly*, **1**, 140–148, 210–216, 300–307, 366–373, 397–406.

Fjørtoft, R., 1950. Application of integral theorems in deriving criteria for laminar flows and for the baroclinic circular vortex. *Geophys. Publ.*, **17**, 1–52.

Fjørtoft, R., 1953. On the changes in the spectral distribution of kinetic energy for two-dimensional nondivergent flow. *Tellus*, **5**, 225–230.

Fleming, E. L., Chandra, S., Schoeberl, M. R. & Barnett, J. J., 1988. Monthly mean global climatology of temperature, wind, geopotential height, and pressure for 0-120 km. Technical report, NASA/Goddard Space Flight Center, Greenbelt, MD. NASA Tech. Memo. 100697.

Fofonoff, N. P., 1954. Steady flow in a frictionless homogeneous ocean. *J. Mar. Res.*, **13**, 254–262.

Fofonoff, N. P., 1959. Interpretation of oceanographic measurements – thermodynamics. In *Physical and Chemical Properties of Sea Water*, Vol. 600. Nat. Acad. Sci., Nat. Res. Counc., Publ.

Fofonoff, N. P. & Montgomery, R. B., 1955. The equatorial undercurrent in the light of the vorticity equation. *Tellus*, **7**, 518–521.

Fox-Kemper, B. & Pedlosky, J., 2004. Wind-driven barotropic gyre I: Circulation control by eddy vorticity fluxes to an enhanced removal region. *J. Mar. Res.*, **62**, 169–193.

Franklin, W. S., 1898. Review of P. Duhem, *Traité Elementaire de Méchanique Chimique fondée sur la Thermodynamique*, Two volumes. Paris, 1897. *Phys. Rev.*, **6**, 170–175.

Friedman, R. M., 1989. *Appropriating the Weather: Vilhelm Bjerknes and the Construction of a Modern Meteorology*. Cornell University Press, 251 pp.

Frierson, D. M. W., Lu, J. & Chen, G., 2007. Width of the Hadley cell in simple and comprehensive general circulation models. *Geophys. Res. Lett.*, **34**, L18804.

Fu, L. L. & Flierl, G. R., 1980. Nonlinear energy and enstrophy transfers in a realistically stratified ocean. *Dyn. Atmos. Oceans*, **4**, 219–246.

Gage, K. S. & Nastrom, G. D., 1986. Theoretical interpretation of atmospheric wavenumber spectra of wind and temperature observed by commercial aircraft during GASP. *J. Atmos. Sci.*, **43**, 729–740.

Galanti, E. & Tziperman, E., 2000. ENSO's phase locking to the seasonal cycle in the fast-SST, fast-wave, and mixed-mode regimes. *J. Atmos. Sci.*, **57**, 2936–2950.

Galperin, B. & Read, P. L., Eds., 2017. *Zonal Jets: Phenomenology, Genesis, Physics*. Cambridge University Press, 431 pp.

Galperin, B., Sukoriansky, S. & Dikovskaya, N., 2010. Geophysical flows with anisotropic turbulence and dispersive waves: flows with a β-effect. *Ocean Dynamics*, **60**, 427–441.

Galperin, B., Sukoriansky, S., Dikovskaya, N., Read, P. *et al.*, 2006. Anisotropic turbulence and zonal jets in rotating flows with a β-effect. *Nonlinear Proc. Geophys.*, **13**, 83–98.

Garcia, R. R., 1987. On the mean meridional circulation of the stratosphere. *J. Atmos. Sci.*, **44**, 2599–2609.

Gardiner, C. W., 1985. *Handbook of Stochastic Methods*. Springer-Verlag, 442 pp.

Gent, P. R. & McWilliams, J. C., 1990. Isopycnal mixing in ocean circulation models. *J. Phys. Oceanogr.*, **20**, 150–155.

Gent, P. R., Willebrand, J., McDougall, T. J. & McWilliams, J. C., 1995. Parameterizing eddy-induced transports in ocean circulation models. *J. Phys. Oceanogr.*, **25**, 463–474.

Gierasch, P. J., 1975. Meridional circulation and the maintenance of the Venus atmospheric rotation. *J. Atmos. Sci.*, **32**, 1038–1044.

Gill, A. E., 1971. The equatorial current in a homogeneous ocean. *Deep-Sea Res.*, **18**, 421–431.

Gill, A. E., 1975. Models of equatorial currents. In *Proceedings of Numerical Models of Ocean Circulation*, pp. 181–203. National Academy of Science.

Gill, A. E., 1980. Some simple solutions for heat induced tropical circulation. *Quart. J. Roy. Meteor. Soc.*, **106**, 447–462.

Gill, A. E., 1982. *Atmosphere–Ocean Dynamics*. Academic Press, 662 pp.

Gill, A. E. & Clarke, A. J., 1974. Wind-induced upwelling, coastal currents and sea-level changes. *Deep-Sea Res.*, **21**, 325–345.

Gill, A. E., Green, J. S. A. & Simmons, A. J., 1974. Energy partition in the large-scale ocean circulation and the production of mid-ocean eddies. *Deep-Sea Res.*, **21**, 499–528.

Gille, S. T., 1997. The Southern Ocean momentum balance: evidence for topographic effects from numerical model output and altimeter data. *J. Phys. Oceanogr.*, **27**, 2219–2232.

Gilman, P. A. & Glatzmaier, G. A., 1981. Compressible convection in a rotating spherical shell. I. Anelastic equations. *Astrophys. J. Suppl. Ser.*, **45**, 335–349.

Giorgetta, M. A., Manzini, E. & Roeckner, E., 2002. Forcing of the quasi-biennial oscillation from a broad spectrum of atmospheric waves. *Geophys. Res. Lett.*, **29**, 1245.

Gnanadesikan, A., 1999. A simple predictive model for the structure of the oceanic pycnocline. *Science*, **283**, 2077–2079.

Godske, C. L., Bergeron, T., Bjerknes, J. & Budgaard, R. C., 1957. *Dynamic Meteorology and Weather Forecasting.* American Meteorological Society, 864 pp.

Goody, R. M. & Yung, Y. L., 1995. *Atmospheric radiation: theoretical basis.* Oxford University Press, 544 pp.

Gough, D. O., 1969. The anelastic approximation for thermal convection. *J. Atmos. Sci.*, **216**, 448–456.

Graham, F. S. & McDougall, T. J., 2013. Quantifying the nonconservative production of conservative temperature, potential temperature and entropy. *J. Phys. Oceanogr.*, **43**, 838–862.

Grant, H. L., Hughes, B. A., Vogel, W. M. & Moilliet, A., 1968. The spectrum of temperaure fluctuation in turbulent flow. *J. Fluid Mech.*, **344**, 423–442.

Grant, H. L., Stewart, R. W. & Moilliet, A., 1962. Turbulent spectra from a tidal channel. *J. Fluid Mech.*, **12**, 241–268.

Gray, D. D. & Giorgini, A., 1976. The validity of the Boussinesq approximation for liquids and gases. *Int. J. Heat and Mass Transfer*, **19**, 545–551.

Gray, L. J., 2010. Stratospheric equatorial dynamics. In *The Stratosphere: Dynamics, Transport, and Chemistry*, Geophys. Monogr. Ser, Vol. 190 (2010).

Gray, L. J., Crooks, S., Pascoe, C. & Palmer, M., 2001. Solar and QBO influences on the timing of stratospheric sudden warmings. *J. Atmos. Sci.*, **61**, 2777–2796.

Gray, L. J., Phipps, S. J., Dunkerton, T. J., Baldwin, M. P. *et al.*, 2001. A data study of the influence of the upper stratosphere on northern hemisphere stratospheric warmings. *Quart. J. Roy. Meteor. Soc.*, **127**, 1985–2003.

Greatbatch, R. J., 1998. Exploring the relationship between eddy-induced transport velocity, vertical momentum transfer, and the isopycnal flux of potential vorticity. *J. Phys. Oceanogr.*, **28**, 422–432.

Green, G., 1837. On the motion of waves in a variable canal of small depth and width. *Trans. Camb. Phil. Soc.*, **6**, 457–462.

Green, J. S. A., 1960. A problem in baroclinic stability. *Quart. J. Roy. Meteor. Soc.*, **86**, 237–251.

Green, J. S. A., 1970. Transfer properties of the large-scale eddies and the general circulation of the atmosphere. *Quart. J. Roy. Meteor. Soc.*, **96**, 157–185.

Green, J. S. A., 1977. The weather during July 1976: some dynamical considerations of the drought. *Weather*, **32**, 120–128.

Green, J. S. A., 1999. *Atmospheric Dynamics.* Cambridge University Press, 213 pp.

Greenspan, H., 1962. A criterion for the existence of inertial boundary layers in oceanic circulation. *Proc. Nat. Acad. Sci.*, **48**, 2034–2039.

Gregg, M. C., 1998. Estimation and geography of diapycnal mixing in the stratified ocean. In J. Imberger, Ed., *Physical Processes in Lakes and Oceans*, pp. 305–338. American Geophysical Union.

Griffies, S. M., 1998. The Gent–McWilliams skew flux. *J. Phys. Oceanogr.*, **28**, 831–841.

Griffies, S. M., 2004. *Fundamentals of Ocean Climate Models.* Princeton University Press, 518 pp.

Grose, W. & Hoskins, B., 1979. On the influence of orography on large-scale atmospheric flow. *J. Atmos. Sci.*, **36**, 223–234.

Hadley, G., 1735. Concerning the cause of the general trade-winds. *Phil. Trans. Roy. Soc.*, **29**, 58–62.

Haine, T. W. N. & Marshall, J., 1998. Gravitational, symmetric, and baroclinic instability of the ocean mixed layer. *J. Phys. Oceanogr.*, **28**, 634–658.

Hallberg, R. & Gnanadesikan, A., 2001. An exploration of the role of transient eddies in determining the transport of a zonally reentrant current. *J. Phys. Oceanogr.*, **31**, 3312–3330.

Hamilton, K., Wilson, R. J. & Hemler, R. S., 2001. Spontaneous QBO-like oscillations simulated by the GFDL SKYHI general circulation mode. *J. Atmos. Sci.*, **58**, 3271–3292.

Hamilton, K. P., 1998. Dynamics of the tropical middle atmosphere: a tutorial review. *Atmosphere–Ocean*, **36**, 319–354.

Haney, R. L., 1971. Surface thermal boundary condition for ocean circulation models. *J. Phys. Oceanogr.*, **1**, 241–248.

Harnik, N. & Heifetz, E., 2007. Relating overreflection and wave geometry to the counterpropagating Rossby wave perspective: Toward a deeper mechanistic understanding of shear instability. *J. Atmos. Sci.*, **64**, 7, 2238–2261.

Haurwitz, B., 1941. *Dynamic meteorology*. Haurwitz press, 380 pp. Reprinted in 2007.

Hayes, M., 1977. A note on group velocity. *Proc. Roy. Soc. Lond. A*, **354**, 533–535.

Haynes, P., 2005. Stratospheric dynamics. *Ann. Rev. Fluid Mech.*, **37**, 263–293.

Haynes, P. H., 2015. Critical layers. *Encycl. Atmos. Sci.*, 2nd edn, **2**, 317–323.

Haynes, P. H., Marks, C. J., McIntyre, M. E., Shepherd, T. G. & Shine, K. P., 1991. On the "downward control" of extratropical diabatic circulations by eddy-induced mean zonal forces. *J. Atmos. Sci.*, **48**, 651–678.

Haynes, P. H. & McIntyre, M. E., 1987. On the evolution of vorticity and potential vorticity in the presence of diabatic heating and frictional or other forces. *J. Atmos. Sci.*, **44**, 828–841.

Haynes, P. H. & McIntyre, M. E., 1990. On the conservation and impermeability theorem for potential vorticity. *J. Atmos. Sci.*, **47**, 2021–2031.

Heckley, W. & Gill, A., 1984. Some simple analytical solutions to the problem of forced equatorial long waves. *Quart. J. Roy. Meteor. Soc.*, **110**, 203–217.

Heifetz, E. & Caballero, R., 2014. An alternative view on the role of the β-effect in the Rossby wave propagation mechanism. *Tellus A*, **66**, 1–6.

Held, I. M., 1982. On the height of the tropopause and the static stability of the troposphere. *J. Atmos. Sci.*, **39**, 412–417.

Held, I. M., 1983. Stationary and quasi-stationary eddies in the extratropical troposphere: theory. In B. Hoskins & R. P. Pearce, Eds., *Large-Scale Dynamical Processes in the Atmosphere*, pp. 127–168. Academic Press.

Held, I. M., 1985. Pseudomomentum and the orthogonality of modes in shear flows. *J. Atmos. Sci.*, **42**, 2280–2288.

Held, I. M., 2000. The general circulation of the atmosphere. In *Woods Hole Program in Geophysical Fluid Dynamics* (2000), pp. 66.

Held, I. M. & Hou, A. Y., 1980. Nonlinear axially symmetric circulations in a nearly inviscid atmosphere. *J. Atmos. Sci.*, **37**, 515–533.

Held, I. M. & Larichev, V. D., 1996. A scaling theory for horizontally homogeneous, baroclinically unstable flow on a beta-plane. *J. Atmos. Sci.*, **53**, 946–952.

Held, I. M., Tang, M. & Wang, H., 2002. Northern winter stationary waves: theory and modeling. *J. Climate*, **15**, 2125–2144.

Helmholtz, H., 1858. Über Integrale der hydrodynamischen Gleichungen welche den Wirbelbewegungen entsprechen (On the integrals of the hydrodynamic equations that correspond to eddy motion). *J. Reine Angew. Math*, **25**, 25–55. Engl. transl.: C. Abbe, *Smithson. Misc. Collect.*, no. 34, pp. 78–93, Smithsonian Institution, Washington DC, 1893.

Helmholtz, H., 1868. Über discontinuirliche Flüssigkeitsbewegungen (On discontinuous liquid motion). *Monats. Königl. Preuss. Akad. Wiss. Berlin*, **23**, 215–228. Engl. trans.: F. Guthrie: On discontinuous movements of fluids. *Phil. Mag*, **36**, 337–346 (1868).

Hendershott, M., 1987. Single layer models of the general circulation. In H. Abarbanel & W. R. Young, Eds., *General Circulation of the Ocean*, pp. 202–267. Springer-Verlag.

Henning, C. C. & Vallis, G. K., 2004. The effect of mesoscale eddies on the main subtropical thermocline. *J. Phys. Oceanogr.*, **34**, 2428–2443.

Henning, C. C. & Vallis, G. K., 2005. The effects of mesoscale eddies on the stratification and transport of an ocean with a circumpolar channel. *J. Phys. Oceanogr.*, **35**, 880–896.

Hide, R., 1969. Dynamics of the atmospheres of major planets with an appendix on the viscous boundary layer at the rigid boundary surface of an electrically conducting rotating fluid in the presence of a magnetic field. *J. Atmos. Sci.*, **26**, 841–853.

Hilborn, R. C., 2004. Sea-gulls, butterflies, and grasshoppers: a brief history of the butterfly effect in nonlinear dynamics. *Am. J. Phys.*, **72**, 425–427.

Hirst, A. C., 1986. Unstable and damped equatorial modes in simple coupled ocean–atmosphere models. *J. Atmos. Sci.*, **43**, 606–632.

Hockney, R., 1970. The potential calculation and some applications. In *Methods of Computational Physics*, Vol. 9, pp. 135–211. Academic Press.

Hogg, N., 2001. Quantification of the deep circulation. In G. Siedler, J. Church, & J. Gould, Eds., *Ocean Circulation and Climate: Observing and Modelling the Global Ocean*, pp. 259–270. Academic Press.

Hoinka, K. P., 1997. The tropopause: discovery, definition and demarcation. *Meteorol. Z.*, **6**, 281–303.

Holland, W. R., Keffer, T. & Rhines, P. B., 1984. Dynamics of the oceanic circulation: The potential vorticity field. *Nature*, **308**, 698–705.

Holloway, G. & Hendershott, M. C., 1977. Stochastic closure for nonlinear Rossby waves. *J. Fluid Mech.*, **82**, 747–765.

Holm, D. D., Marsden, J. E., Ratiu, T. & Weinstein, A., 1985. Nonlinear stability of fluid and plasma equilibria. *Phys. Rep.*, **123**, 1–116.

Holmes, M. H., 2013. *Introduction to Perturbation Methods*. 2nd edn. Springer, 436 pp.

Holton, J. R., 1974. Forcing of mean flows by stationary waves. *J. Atmos. Sci.*, **31**, 942–945.

Holton, J. R., 1992. *An Introduction to Dynamic Meteorology*. 3rd edn. Academic Press, 507 pp.

Holton, J. R. & Hakim, G., 2012. *An Introduction to Dynamic Meteorology*. 5th edn. Academic Press, 552 pp.

Holton, J. R., Haynes, P. R., McIntyre, M. E., Douglass, A. R. *et al.*, 1995. Stratosphere-troposphere exchange. *Rev. Geophys.*, **33**, 403–439.

Holton, J. R. & Lindzen, R. S., 1972. An updated theory for the quasi-biennial cycle of the tropical stratosphere. *J. Atmos. Sci.*, **29**, 1076–1080.

Holton, J. R. & Mass, C., 1976. Stratospheric vacillation cycles. *J. Atmos. Sci.*, **33**, 2218–2215.

Holton, J. R. & Tan, H.-C., 1980. The influence of the equatorial quasi-biennial oscillation on the global circulation at 50 mb. *J. Atmos. Sci.*, **37**, 2200–2208.

Holton, J. R. & Tan, H.-C., 1982. The quasi-biennial oscillation in the northern hemisphere lower stratosphere. *J. Meteor. Soc. Japan*, **60**, 140–158.

Hoskins, B. J. & Karoly, D. J., 1981. The steady linear response of a spherical atmosphere to thermal and orographic forcing. *J. Atmos. Sci.*, **38**, 1179–1196.

Hough, S. S., 1897. On the application of harmonic analysis to the dynamical theory of the tides. Part I: On Laplace's "Oscillations of the first species", and on the dynamics of ocean currents. *Phil. Trans. (A)*, **189 (IX)**, 201–258.

Hough, S. S., 1898. On the application of harmonic analysis to the dynamical theory of the tides. Part II: On the general integration of Laplace's dynamical equations. *Phil. Trans. (A)*, **191 (V)**, 139–186.

Huang, R. X., 1998. Mixing and available potential energy in a Boussinesq ocean. *J. Phys. Oceanogr.*, **28**, 669–678.

Huang, R. X., 1999. Mixing and energetics of the oceanic thermohaline circulation. *J. Phys. Oceanogr.*, **29**, 727–746.

Huang, R. X., 2010. *Ocean Circulation*. Cambridge University Press, 791 pp.

Hughes, C. W., 2002. Sverdrup-like theories of the Antarctic Circumpolar Current. *J. Mar. Res.*, **60**, 1–17.

Hughes, C. W. & de Cuevas, B., 2001. Why western boundary currents in realistic oceans are inviscid: a link between form stress and bottom pressure torques. *J. Phys. Oceanogr.*, **31**, 2871–2885.

Hughes, G. O. & Griffiths, R. W., 2008. Horizontal convection. *Ann. Rev. Fluid Mech.*, **185–2008**, 40.

Ierley, G. R. & Ruehr, O. G., 1986. Analytic and numerical solutions of a nonlinear boundary value problem. *Stud. Appl. Math.*, **75**, 1–36.

Ilicak, M. & Vallis, G. K., 2012. Simulations and scaling of horizontal convection. *Tellus A*, **64**, 1–17.

Il'in, A. M. & Kamenkovich, V. M., 1964. The structure of the boundary layer in the two-dimensional theory of ocean currents (in Russian). *Okeanologiya*, **4** (5), 756–769.

Ingersoll, A. P., 2005. Boussinesq and anelastic approximations revisited: potential energy release during thermobaric instability. *J. Phys. Oceanogr.*, **35**, 1359–1369.

IOC, SCOR & IAPSO, 2010. The international thermodynamic equation of seawater – 2010: Calculation and use of thermodynamic properties. Technical report, Intergovernmental Oceanographic Commission, Manuals and Guides No. 56, UNESCO (English).

Iwayama, T., Shepherd, T. G. & Watanabe, T., 2002. An 'ideal' form of decaying two-dimensional turbulence. *J. Fluid Mech.*, **456**, 183–198.

Jackett, D. R. & McDougall, T. J., 1997. A neutral density variable for the world's oceans. *J. Phys. Oceanogr.*, **28**, 237–263.

Jackson, L., Hughes, C. W. & Williams, R. G., 2006. Topographic control of basin and channel flows: the role of bottom pressure torques and friction. *J. Phys. Oceanogr.*, **36**, 1786–1805.

Jansen, M. & Ferrari, R., 2012. Macroturbulent equilibration in a thermally forced primitive equation system. *J. Atmos. Sci.*, **69**, 695–713.

Jansen, M. & Ferrari, R., 2013. Equilibration of an atmosphere by adiabatic eddy fluxes. *J. Atmos. Sci.*, **70**, 2948–2962.

Jeffreys, H., 1924. On certain approximate solutions of linear differential equations of the second order. *Proc. London Math. Soc.*, **23**, 428–436.

Jeffreys, H., 1926. On the dynamics of geostrophic winds. *Quart. J. Roy. Meteor. Soc.*, **51**, 85–104.

Jeffreys, H. & Jeffreys, B. S., 1946. *Methods of Mathematical Physics*. Cambridge University Press, 728 pp.

Jin, F.-F., 1997a. An equatorial ocean recharge paradigm for ENSO. Part I: Conceptual model. *J. Atmos. Sci.*, **54**, 811–829.

Jin, F.-F., 1997b. An equatorial ocean recharge paradigm for ENSO. Part II: A stripped-down coupled model. *J. Atmos. Sci.*, **54**, 830–847.

Johnson, G. C. & Bryden, H. L., 1989. On the size of the Antarctic Circumpolar Current. *Deep-Sea Res.*, **36**, 39–53.

Jones, D. B. A., Schneider, H. R. & McElroy, M. B., 1998. Effects of the quasi-biennial oscillation on the zonally averaged transport of tracer. *J. Geophys. Res.*, **103**, 11235–11249.

Jones, W. L., 1967. Propagation of internal gravity waves in fluids with shear flow and rotation. *J. Fluid Mech.*, **30**, 439–448.

Jucker, M., 2014. Scientific visualisation of atmospheric data with ParaView. *J. Open Res. Software*, **2**, e4.

Jucker, M., Fueglistaler, S. & Vallis, G. K., 2013. Maintenance of stratospheric structure in an idealized general circulation model. *J. Atmos. Sci.*, **70**, 3341–3358.

Jucker, M., Fueglistaler, S. & Vallis, G. K., 2014. Stratospheric sudden warmings in an idealized GCM. *J. Geophys. Res. (Atmospheres)*, **119**, 11054–11064.

Juckes, M. N., 2000. The static stability of the midlatitude troposphere: the relevance of moisture. *J. Atmos. Sci.*, **57**, 3050–3057.

Juckes, M. N., 2001. A generalization of the transformed Eulerian-mean meridional circulation. *Quart. J. Roy. Meteor. Soc.*, **127**, 147–160.

Kalnay, E., 1996. The NCEP/NCAR 40-year reanalysis project. *Bull. Amer. Meteor. Soc.*, **77**, 437–471.

Karsten, R., Jones, H. & Marshall, J., 2002. The role of eddy transfer in setting the stratification and transport of a circumpolar current. *J. Phys. Oceanogr.*, **32**, 39–54.

Keffer, T., 1985. The ventilation of the world's oceans: maps of potential vorticity. *J. Phys. Oceanogr.*, **15**, 509–523.

Kessler, W. S., Johnson, G. C. & Moore, D. W., 2003. Sverdrup and nonlinear dynamics of the Pacific equatorial currents. *J. Phys. Oceanogr.*, **33**, 994–1008.

Kevorkian, J. & Cole, J. D., 2011. *Multiple Scale and Singular Perturbation Methods*. Springer-Verlag, 648 pp.

Kibel, I., 1940. Priloozhenie k meteorogi uravnenii mekhaniki baroklinnoi zhidkosti (Application of baroclinic fluid dynamic equations to meteorology). *SSSR Ser. Geogr. Geofiz.*, **5**, 627–637.

Kiladis, G. N., Straub, K. H. & Haertel, P. T., 2005. Zonal and vertical structure of the Madden–Julian oscillation. *J. Atmos. Sci.*, **62**, 2790–2809.

Kiladis, G. N., Wheeler, M. C., Haertel, P. T., Straub, K. H. & Roundy, P. E., 2009. Convectively coupled equatorial waves. *Rev. Geophys.*, **47**, RG2003.

Killworth, P. D., 1987. A continuously stratified nonlinear ventilated thermocline. *J. Phys. Oceanogr.*, **17**, 1925–1943.

Killworth, P. D., 1997. On the parameterization of eddy transfer. Part I: theory. *J. Marine Res.*, **55**, 1171–1197.

Killworth, P. D. & McIntyre, M. E., 1985. Do Rossby-wave critical layers absorb, reflect, or over-reflect? *J. Fluid Mech.*, **161**, 449–492.

Kim, H.-K. & Lee, S., 2001. Hadley cell dynamics in a primitive equation model. Part II: Nonaxisymmetric flow. *J. Atmos. Sci.*, **58**, 19, 2859–2871.

Kimoto, M. & Ghil, M., 1993. Multiple flow regimes in the northern hemisphere winter. Part 1: Methodology and hemispheric regimes. *J. Atmos. Sci.*, **50**, 2625–2643.

Kolmogorov, A. N., 1941. The local structure of turbulence in incompressible viscous fluid for very large Reynolds numbers. *Dokl. Acad. Sci. USSR*, **30**, 299–303.

Kolmogorov, A. N., 1962. A refinement of previous hypotheses concerning the local structure of turbulence in a viscous incompressible fluid at high Reynolds numbers. *J. Fluid Mech.*, **13**, 82–85.

Kraichnan, R., 1967. Inertial ranges in two-dimensional turbulence. *Phys. Fluids*, **10**, 1417–1423.

Kraichnan, R. & Montgomery, D., 1980. Two-dimensional turbulence. *Rep. Prog. Phys.*, **43**, 547–619.

Kramers, H. A., 1926. Wellenmechanik und halbzahlige Quantisierung (Wave mechanics and semi-integral quantization). *Zeit. fur Physik A*, **39**, 828–840.

Kundu, P., Allen, J. S. & Smith, R. L., 1975. Modal decomposition of the velocity field near the Oregon coast. *J. Phys. Oceanogr.*, **5**, 683–704.

Kundu, P., Cohen, I. M. & Dowling, D. R., 2015. *Fluid Mechanics.* Academic Press, 928 pp.

Kuo, H.-I., 1949. Dynamic instability of two-dimensional nondivergent flow in a barotropic atmosphere. *J. Meteorol.*, **6**, 105–122.

Kuo, H.-I., 1951. Vorticity transfer as related to the development of the general circulation. *J. Meteorol.*, **8**, 307–315.

Kushner, P. J., 2010. Annular modes of the troposphere and stratosphere. In *The Stratosphere: Dynamics, Transport, and Chemistry*, Geophys. Monogr. Ser (2010).

Kushnir, Y., Robinson, W. A., Bladé, I., Hall, N. M. J. *et al.*, 2002. Atmospheric GCM response to extratropical SST anomalies: synthesis and evaluation. *J. Climate*, **15**, 2233–2256.

Labitzke, K., Kunze, M. & Bronnimann, S., 2006. Sunspots, the QBO and the stratosphere in north polar regions — 20 years later. *Meteor. Z.*, **15**, 355–363.

LaCasce, J. H. & Ohlmann, C., 2003. Relative dispersion at the surface of the Gulf of Mexico. *J. Mar. Res.*, **65**, 285–312.

Lait, L. R., 1994. An alternative form for potential vorticity. *J. Atmos. Sci.*, **51**, 1754–1759.

Lamb, H., 1932. *Hydrodynamics.* Cambridge University Press, reissued by Dover Publications 1945, 768 pp.

Lanczos, C., 1970. *The Variational Principles of Mechanics.* University of Toronto Press, Reprinted by Dover Publications 1980, 418 pp.

Landau, L. D., 1944. On the problem of turbulence. *Dokl. Akad. Nauk SSSR*, **44**, 311–314.

Landau, L. D. & Lifshitz, E. M., 1987. *Fluid Mechanics* (Course of Theoretical Physics, v. 6). 2nd edn. Pergamon Press, 539 pp.

Larichev, V. D. & Held, I. M., 1995. Eddy amplitudes and fluxes in a homogeneous model of fully developed baroclinic instability. *J. Phys. Oceanogr.*, **25**, 2285–2297.

Latif, M. & Barnett, T., 1996. Decadal climate variability over the North Pacific and North America: dynamics and predictability. *J. Climate*, **10**, 219–239.

Lau, K.-M., 1981. Oscillations in a simple equatorial climate system. *J. Atmos. Sci.*, **38**, 248–261.

Lawrence, M. G., 2005. The relationship between relative humidity and the dewpoint temperature in moist air: A simple conversion and applications. *Bull. Am. Meteor. Soc.*, **86**, 225–233.

LeBlond, P. H. & Mysak, L. A., 1980. *Waves in the Ocean.* Elsevier, 616 pp.

Ledwell, J., Watson, A. & Law, C., 1998. Mixing of a tracer released in the pycnocline. *J. Geophys. Res.*, **103**, 21499–21529.

Lee, M.-M., Marshall, D. P. & Williams, R. G., 1997. On the eddy transfer of tracers: advective or diffusive? *J. Mar. Res.*, **55**, 483–595.

Lee, T. D., 1951. Difference between turbulence in a two-dimensional fluid and in a three-dimensional fluid. *J. Appl. Phys.*, **22**, 524.

Leetmaa, A., Niiler, P. & Stommel, H., 1977. Does the Sverdrup relation account for the mid-Atlantic circulation? *J. Mar. Res.*, **35**, 1–10.

Leith, C. E., 1968. Diffusion approximation for two-dimensional turbulence. *Phys. Fluids*, **11**, 671–672.

Lesieur, M., 1997. *Turbulence in Fluids: Third Revised and Enlarged Edition*. Kluwer, 515 pp.

Levinson, N., 1950. The 1st boundary value problem for $\epsilon \Delta U + A(x,y)U_x + B(x,y)U_y + C(x,y)U = D(x,t)$ for small epsilon. *Ann. Math.*, **51**, 429–445.

Lewis, R., Ed., 1991. *Meteorological Glossary*. 6th edn. Her Majesty's Stationery Office, 335 pp.

Lighthill, J., 1978. *Waves in Fluids*. Cambridge University Press, 504 pp.

Lighthill, M. J., 1965. Group velocity. *J. Inst. Math. Appl.*, **1**, 1–28.

Lighthill, M. J., 1969. Dynamic response of the Indian Ocean to onset of the southwest monsoon. *Phil. Trans. Roy. Soc. Lond. A*, **265**, 45–92.

Lilly, D. K., 1969. Numerical simulation of two-dimensional turbulence. *Phys. Fluid Suppl. II*, **12**, 240–249.

Lilly, D. K., 1996. A comparison of incompressible, anelastic and Boussinesq dynamics. *Atmos. Res*, **40**, 143–151.

Limpasuvan, V., Thompson, D. W. J. & Hartmann, D. L., 2000. The life cycle of the Northern Hemisphere sudden stratospheric warmings. *J. Climate*, **17**, 2584–2596.

Lindborg, E., 1999. Can the atmospheric kinetic energy spectrum be explained by two-dimensional turbulence? *J. Fluid Mech.*, **388**, 259–288.

Lindborg, E. & Alvelius, K., 2000. The kinetic energy spectrum of the two-dimensional enstrophy turbulence cascade. *Phys. Fluids*, **12**, 945–947.

Lindzen, R. S. & Farrell, B., 1980. The role of the polar regions in global climate, and a new parameterization of global heat transport. *Mon. Wea. Rev.*, **108**, 2064–2079.

Lindzen, R. S. & Holton, J. R., 1968. A theory of the quasi-biennial oscillation. *J. Atmos. Sci.*, **25**, 1095–1107.

Lindzen, R. S. & Hou, A. Y., 1988. Hadley circulation for zonally averaged heating centered off the equator. *J. Atmos. Sci.*, **45**, 2416–2427.

Lindzen, R. S., Lorenz, E. N. & Plazman, G. W., Eds., 1990. *The Atmosphere — a Challenge: the Science of Jule Gregory Charney*. American Meteorological Society, 321 pp.

Lindzen, R. S. & Nigam, S., 1987. On the role of sea surface temperature gradients in forcing low-level winds and convergence in the tropics. *J. Atmos. Sci.*, **44**, 2418–2436.

Liouville, J., 1837. Sur le développement des fonction ou parties de fonction en séries (On the development of functions of parts of functions in series). *J. Math. Pures Appl.*, **2**, 16–35.

Lipps, F. B. & Hemler, R. S., 1982. A scale analysis of deep moist convection and some related numerical calculations. *J. Atmos. Sci.*, **39**, 2192–2210.

Longuet-Higgins, M. S., 1964. Planetary waves on a rotating sphere, I. *Proc. Roy. Soc. Lond. A*, **279**, 446–473.

Longuet-Higgins, M. S., 1968. The eigenfunctions of Laplace's tidal equations over a sphere. *Proc. Roy. Soc. Lond. A*, **262**, 511–607.

Lorenz, E. N., 1955. Available potential energy and the maintenance of the general circulation. *Tellus*, **7**, 157–167.

Lorenz, E. N., 1963. Deterministic nonperiodic flow. *J. Atmos. Sci.*, **20**, 130–141.

Lorenz, E. N., 1967. *The Nature and the Theory of the General Circulation of the Atmosphere*. WMO Publications, Vol. 218, World Meteorological Organization.

Lozier, S., Owens, W. B. & Curry, R. G., 1996. The climatology of the North Atlantic. *Prog. Oceanog.*, **36**, 1–44.

Ludlam, F. H., 1966. Cumulus and cumulonimbus convection. *Tellus*, **18**, 687–698.

Lumpkin, R. & Speer, K., 2007. Global ocean meridional overturning. *J. Phys. Oceanogr.*, **37**, 2550–2562.

Luyten, J. R., Pedlosky, J. & Stommel, H., 1983. The ventilated thermocline. *J. Phys. Oceanogr.*, **13**, 292–309.

Madden, R. A. & Julian, P. R., 1971. Detection of a 40–50 day oscillation in the zonal wind in the tropical Pacific. *J. Atmos. Sci.*, **28**, 702–708.

Madden, R. A. & Julian, P. R., 1972. Description of global-scale circulation cells in the tropics with a 40–50 day period. *J. Atmos. Sci.*, **29**, 1109–1123.

Majda, A. J. & Klein, R., 2003. Systematic multiscale models for the tropics. *J. Atmos. Sci.*, **60**, 393–408.

Majda, A. J. & Stechmann, S. N., 2009. The skeleton of tropical intraseasonal oscillations. *Proc. Nat. Acad. Sci.*, **106**, 8417–8422.

Maltrud, M. E. & Vallis, G. K., 1991. Energy spectra and coherent structures in forced two-dimensional and beta-plane turbulence. *J. Fluid Mech.*, **228**, 321–342.

Manabe, S. & Stouffer, R. J., 1988. Two stable equilibria of a coupled ocean–atmosphere model. *J. Climate*, **1**, 841–866.

Manabe, S. & Strickler, R. F., 1964. Thermal equilibrium of the atmosphere with a convective adjustment. *J. Atmos. Sci.*, **21**, 361–385.

Manabe, S. & Wetherald, R. T., 1980. On the distribution of climate change resulting from an increase in CO_2 content of the atmosphere. *J. Atmos. Sci.*, **37**, 99–118.

Mapes, B. E., 1997. Equilibrium vs. activation control of large-scale variations of tropical deep convection. In R. K. Smith, Ed., *The Physics and Parameterization of Moist Atmospheric Convection*, pp. 321–358. Springer.

Mapes, B. E., 1998. The large-scale part of tropical mesoscale convective system circulations: A linear vertical spectral band model. *J. Meteor. Soc. Japan*, **76**, 29–55.

Mapes, B. E., 2000. Convective inhibition, subgrid-scale triggering energy, and stratiform instability in a toy tropical wave model. *J. Atmos. Sci.*, **57**, 1515–1535.

Marcus, P. S., 1993. Jupiter's Great Red Spot and other vortices. *Ann. Rev. Astron. Astrophys*, **31**, 523–573.

Margules, M., 1903. Über die Energie der Stürme (On the energy of storms). *Jahrb. Kais.-kön Zent. für Met. und Geodynamik, Vienna*, 26 pp. Engl. transl.: C. Abbe, *Smithson. Misc. Collect.*, no. 51, pp. 533-595, Smithsonian Institution, Washington D. C., 1910.

Marotzke, J., 1989. Instabilities and multiple steady states of the thermohaline circulation. In D. L. T. Anderson & J. Willebrand, Eds., *Oceanic Circulation Models: Combining Data and Dynamics*, pp. 501–511. NATO ASI Series, Kluwer.

Marshall, D. P., Maddison, J. R. & Berloff, P. S., 2012. A framework for parameterizing eddy potential vorticity fluxes. *J. Phys. Oceanogr.*, **42**, 539–557.

Marshall, D. P., Williams, R. G. & Lee, M.-M., 1999. The relation between eddy-induced transport and isopycnic gradients of potential vorticity. *J. Phys. Oceanogr.*, **29**, 1571–1578.

Marshall, J. & Plumb, R. A., 2008. *Atmosphere, Ocean and Climate Dynamics: An Introductory Text*. Academic Press, 344 pp.

Marshall, J. & Speer, K., 2012. Closure of the meridional overturning circulation through southern ocean upwelling. *Nature Geosciences*, **5**, 171–180.

Marshall, J. C., 1981. On the parameterization of geostrophic eddies in the ocean. *J. Phys. Oceanogr.*, **11**, 257–271.

Marshall, J. C. & Nurser, A. J. G., 1992. Fluid dynamics of oceanic thermocline ventilation. *J. Phys. Oceanogr.*, **22**, 583–595.

Marshall, J. C. & Radko, T., 2003. Residual-mean solutions for the Antarctic Circumpolar Current and its associated overturning circulation. *J. Phys. Oceanogr.*, **22**, 2341–2354.

Marshall, J. C. & Schott, F., 1999. Open-ocean convection: observations, theory, and models. *Rev. Geophys.*, **37**, 1–64.

Matsuno, T., 1966. Quasi-geostrophic motions in the equatorial area. *J. Meteor. Soc. Japan*, **44**, 25–43.

Matsuno, T., 1971. A dynamical model of the sudden stratospheric warming. *J. Atmos. Sci.*, **28**, 1479–1494.

Maze, G. & Marshall, J., 2011. Diagnosing the observed seasonal cycle of Atlantic subtropical mode water using potential vorticity and its attendant theorems. *J. Phys. Oceanogr.*, **41**, 1986–1999.

McCarthy, M. C. & Talley, L. D., 1999. Three-dimensional isoneutral potential vorticity structure in the Indian Ocean. *J. Geophys. Res. (Oceans)*, **104**, 13251–13267.

McCreary, J. P., 1981. A linear stratified ocean model of the equatorial undercurrent. *Phil. Trans. Roy. Soc. Lond.*, **A298**, 603–635.

McCreary, J. P., 1985. Modeling equatorial ocean circulation. *Ann. Rev. Fluid Mech.*, **17**, 359–407.

McCreary, J. P. & Lu, P., 1994. Interaction between the subtropical and equatorial ocean circulations: the subtropical cell. *J. Phys. Oceanogr.*, **24**, 466–497.

McDougall, T. J., 1987. Neutral surfaces. *J. Phys. Oceanogr.*, **17**, 1950–1964.

McDougall, T. J., 1998. Three-dimensional residual mean theory. In E. P. Chassignet & J. Verron, Eds., *Ocean Modeling and Parameterization*, pp. 269–302. Kluwer Academic.

McDougall, T. J., 2003. Potential enthalphy: a conservative oceanic variable for evaluating heat content and heat fluxes. *J. Phys. Oceanogr.*, **33**, 945–963.

McIntosh, P. C. & McDougall, T. J., 1996. Isopycnal averaging and the residual mean circulation. *J. Phys. Oceanogr.*, **26**, 1655–1660.

McIntyre, M. E. & Norton, W. A., 1990. Dissipative wave–mean interactions and the transport of vorticity or potential vorticity. *J. Fluid Mech.*, **212**, 403–435.

McIntyre, M. E. & Norton, W. A., 2000. Potential vorticity inversion on a hemisphere. *J. Atmos. Sci.*, **57**, 1214–1235.

McIntyre, M. E. & Shepherd, T. G., 1987. An exact local conservation theorem for finite-amplitude disturbances to nonparallel shear flows, with remarks on Hamiltonian structure and on Arnol'd's stability theorems. *J. Fluid Mech.*, **181**, 527–565.

McKee, W. D., 1973. The wind-driven equatorial circulation in a homogeneous ocean. *Deep-Sea Res.*, **20**, 889–899.

McPhaden, M. J., Timmermann, A., Widlansky, M. J., Balmaseda, M. A. & Stockdale, T. N., 2015. The curious case of the El Niño that never happened: A perspective from 40 years of progress in climate research and forecasting. *Bull. Am. Meteor. Soc.*, **96**, 1647–1665.

McWilliams, J. C., 1984. The emergence of isolated coherent vortices in turbulent flow. *J. Fluid Mech.*, **146**, 21–43.

Mihaljan, J. M., 1962. A rigorous exposition of the Boussinesq approximations applicable to a thin layer of fluid. *Astrophysical J.*, **136**, 1126–1133.

Millero, F. J., Feistel, R., Wright, D. G. & McDougall, T. J., 2008. The composition of standard seawater and the definition of the reference-composition salinity scale. *Deep Sea Res., Part I*, **55**, 50–72.

Moffatt, H. K., 1983. Transport effects associated with turbulence with particular attention to the influence of helicity. *Rep. Progress Phys.*, **46**, 621–664.

Monin, A. S. & Yaglom, A. M., 1971. *Statistical Fluid Mechanics: Mechanics of Turbulence, Vols. 1 and 2*. MIT Press and Dover Publications, 1680 pp.

Morel, P. & Larcheveque, M., 1974. Relative dispersion of constant-level balloons in the 200 mb general circulation. *J. Atmos. Sci.*, **31**, 2189–2196.

Mundt, M., Vallis, G. K. & Wang, J., 1997. Balanced models for the large- and meso-scale circulation. *J. Phys. Oceanogr.*, **27**, 1133–1152.

Munk, W. H., 1950. On the wind-driven ocean circulation. *J. Meteorol*, **7**, 79–93.

Munk, W. H., 1966. Abyssal recipes. *Deep-Sea Res.*, **13**, 707–730.

Munk, W. H. & Palmén, E., 1951. Note on dynamics of the Antarctic Circumpolar Current. *Tellus*, **3**, 53–55.

Munk, W. H. & Wunsch, C., 1998. Abyssal recipes II: energetics of tidal and wind mixing. *Deep-Sea Res.*, **45**, 1976–2009.

Namias, J., 1959. Recent seasonal interaction between North Pacific waters and the overlying atmospheric circulation. *J. Geophys. Res.*, **64**, 631–646.

Neelin, J. D., 1988. A simple model for surface stress and low-level flow in the tropical atmosphere driven by prescribed heating. *Quart. J. Roy. Meteor. Soc.*, **114**, 747–770.

Neelin, J. D. & Zeng, N., 2000. A quasi-equilibrium tropical circulation model—formulation. *J. Atmos. Sci.*, **57**, 1741–1766.

Newell, A. C., 1969. Rossby wave packet interactions. *J. Fluid Mech.*, **35**, 255–271.

Newman, P. A., Coy, L., Pawson, S. & Lait, L. R., 2016. The anomalous change in the QBO in 2015–2016. *Geophys. Res. Lett.*, **43**, 8791–8797, doi:10.1002/2016GL070373.

Nicholls, S., 1985. Aircraft observations of the Ekman layer during the joint air-sea interaction experiment. *Quart. J. Roy. Meteor. Soc.*, **111**, 391–426.

Nikurashin, M. & Vallis, G. K., 2011. A theory of deep stratification and overturning circulation in the ocean. *J. Phys. Oceanogr.*, **41**, 485–502.

Nikurashin, M. & Vallis, G. K., 2012. A theory of the interhemispheric meridional overturning circulation and associated stratification. *J. Phys. Oceanogr.*, **42**, 1652–1667.

Nof, D., 2003. The Southern Ocean's grip on the northward meridional flow. In *Progress in Oceanography*, Vol. 56, pp. 223–247. Pergamon.

Novikov, E. A., 1959. Contributions to the problem of the predictability of synoptic processes. *Izv. An. SSSR Ser. Geophys.*, **11**, 1721. Eng. transl.: *Am. Geophys. U. Transl.*, 1209-1211.

Nycander, J., 2011. Energy conversion, mixing energy, and neutral surfaces with a nonlinear equation of state. *J. Phys. Oceanogr.*, **41**, 28–41.

Nycander, J. & Roquet, F., 2015. The nonlinear equation of state of sea water and the global water mass distribution. *Geophys. Res. Lett.*, **42**, 7714–7721. In press.

Oberbeck, A., 1879. Über die Wärmeleitung der Flüssigkeiten bei Berücksichtigung der Strömungen infolge vor Temperaturdifferenzen (On the thermal conduction of liquids taking into account flows due to temperature differences). *Ann. Phys. Chem., Neue Folge*, **7**, 271–292.

Oberbeck, A., 1888. Über die Bewegungserscheinungen der Atmosphäre (On the phenomena of motion in the atmosphere). *Sitzb. K. Preuss. Akad. Wiss*, **7**, 383–395 and 1129–1138. Engl trans.: in B. Saltzman, Ed.,*Theory of Thermal Convection*, Dover, 162–183.

Obukhov, A. M., 1941. Energy distribution in the spectrum of turbulent flow. *Izv. Akad. Nauk. SSR, Ser. Geogr. Geofiz.*, **5**, 453–466.

Obukhov, A. M., 1949. Structure of the temperature field in turbulent flows. *Izv. Akad. Nauk. SSR, Ser. Geogr. Geofiz.*, **13**, 58–63.

Obukhov, A. M., 1962. On the dynamics of a stratified liquid. *Dokl. Akad. Nauk SSSR*, **145**, 1239–1242. Engl. transl.: *Soviet Physics–Dokl.* 7, 682–684.

Oetzel, K. & Vallis, G. K., 1997. Strain, vortices, and the enstrophy inertial range in two-dimensional turbulence. *Phys. Fluids*, **9**, 2991–3004.

O'Gorman, P. A., Lamquin, N., Schneider, T. & Singh, M. S., 2011. The relative humidity in an isentropic advection–condensation model: Limited poleward influence and properties of subtropical minima. *J. Atmos. Sci.*, **68**, 3079–3093.

O'Gorman, P. A. & Pullin, D. I., 2005. Effect of Schmidt number of the velocity-scaler cospectrum in isotropic turbulence with a mean scalar gradient. *J. Fluid Mech.*, **532**, 111–140.

Ogura, Y. & Phillips, N. A., 1962. Scale analysis of deep and shallow convection in the atmosphere. *J. Atmos. Sci.*, **19**, 173–179.

Olbers, D., 1998. Comments on "On the obscurantist physics of 'form drag' in theorizing about the Circumpolar Current". *J. Phys. Oceanogr.*, **28**, 1647–1654.

Olbers, D., Borowski, D., Völker, C. & Wolff, J.-O., 2004. The dynamical balance, transport and circulation of the Antarctic Circumpolar Current. *Antarctic Science*, **16**, 439–470.

Olbers, D., Willebrand, J. & Eden, C., 2012. *Ocean Dynamics*. Springer, 704 pp.

Ollitrault, M., Gabillet, C. & Colin de Verdière, A., 2005. Open ocean regimes of relative dispersion. *J. Fluid Mech.*, **533**, 381–407.

Onsager, L., 1931. Reciprocal relations in irreversible processes. II. *Phys. Rev.*, **38**, 2265–2279.

Onsager, L., 1949. Statistical hydrodynamics. *Nuovo Cim. (Suppl.)*, **6**, 279–287.

Ooyama, K. ., 1963. A dynamical model for the study of tropical cyclone development. New York University, 26pp, unpublished manuscript.

Ooyama, K. V., 1982. Conceptual evolution of the theory and modeling of the tropical cyclone. *J. Meteor. Soc. Japan*, **60**, 369–380.

Orlanski, I. & Sheldon, J. P., 1995. Stages in the energetics of baroclinic systems. *Tellus A*, **47**, 605–628.

Osprey, S. M., Butchart, N., Knight, J. R., Scaife, A. *et al.*, 2016. An unexpected disruption of the atmospheric quasi-biennial oscillation. *Science*, **353**, 1424–1427.

Paldor, N., Rubin, S. & Mariono, A. J., 2007. A consistent theory for linear waves of the shallow-water equations on a rotating plane in midlatitudes. *J. Atmos. Sci.*, **37**, 115–128.

Paldor, N. & Sigalov, A., 2011. An invariant theory of the linearized shallow water equations with rotation and its application to a sphere and a plane. *Dyn. Atmos. Oceans*, **51**, 26–44.

Palmer, T. N., 1981. Diagnostic study of a wavenumber-2 stratospheric sudden warming in a transformed Eulerian-mean formalism. *J. Atmos. Sci.*, **38**, 844–855.

Palmer, T. N., 1997. A nonlinear dynamical perspective on climate prediction. *J. Climate*, **12**, 575–591.

Palmer, T. N., 2009. Edward Lorenz, 1917–2008. *Biogr. Mems Fell. R. Soc.*, **55**, 139–155.

Paparella, F. & Young, W. R., 2002. Horizontal convection is non-turbulent. *J. Fluid Mech.*, **466**, 205–214.

Pascoe, C. L., Gray, L. J., Crooks, S. A., Juckes, M. N. & Baldwin, M. P., 2005. The quasi-biennial oscillation: Analysis using ERA-40 data. *J. Geophys. Res.*, **110**, D08105.

Pauluis, O., 2007. Sources and sinks of available potential energy in a moist atmosphere. *J. Atmos. Sci.*, **64**, 2627–2641.

Pauluis, O., 2008. Thermodynamic consistency of the anelastic approximation for a moist atmosphere. *J. Atmos. Sci.*, **65**, 2719–2729.

Pedlosky, J., 1964. The stability of currents in the atmosphere and ocean. Part I. *J. Atmos. Sci.*, **21**, 201–219.

Pedlosky, J., 1987a. *Geophysical Fluid Dynamics*. 2nd edn. Springer-Verlag, 710 pp.

Pedlosky, J., 1987b. An inertial theory of the equatorial undercurrent. *J. Phys. Oceanogr.*, **17**, 1978–1985.

Pedlosky, J., 1996. *Ocean Circulation Theory*. Springer-Verlag, 453 pp.

Pedlosky, J., 2003. *Waves in the Ocean and Atmosphere: Introduction to Wave Dynamics*. Springer-Verlag, 260 pp.

Peixoto, J. P. & Oort, A. H., 1992. *Physics of Climate*. American Institute of Physics, 520 pp.

Peltier, W. R. & Stuhne, G., 2002. The upscale turbulence cascade: shear layers, cyclones and gas giant bands. In R. P. Pearce, Ed., *Meteorology at the Millennium*, pp. 43–61. Academic Press.

Persson, A., 1998. How do we understand the Coriolis force? *Bull. Am. Meteor. Soc.*, **79**, 1373–1385.

Philander, S. G., 1990. *El Niño, La Niña, and the Southern Oscillation*. Academic Press, 289 pp.

Phillips, N. A., 1954. Energy transformations and meridional circulations associated with simple baroclinic waves in a two-level, quasi-geostrophic model. *Tellus*, **6**, 273–286.

Phillips, N. A., 1956. The general circulation of the atmosphere: a numerical experiment. *Quart. J. Roy. Meteor. Soc.*, **82**, 123–164.

Phillips, N. A., 1963. Geostrophic motion. *Rev. Geophys.*, **1**, 123–176.

Phillips, N. A., 1966. The equations of motion for a shallow rotating atmosphere and the traditional approximation. *J. Atmos. Sci.*, **23**, 626–630.

Phillips, N. A., 1973. Principles of large-scale numerical weather prediction. In P. Morel, Ed., *Dynamic Meteorology*, pp. 1–96. Riedel.

Phillips, O., 1972. Turbulence in a strongly stratified fluid—is it unstable? In *Deep Sea Research and Oceanographic Abstracts*, Vol. 19 (1972), pp. 79–81. Elsevier.

Pierini, S. & Vulpiani, A., 1981. Nonlinear stability analysis in multi-layer quasigeostrophic system. *J. Phys. A*, **14**, L203–L207.

Pierrehumbert, R. T., 2010. *Principles of Planetary Climate*. Cambridge University Press, 652 pp.

Pierrehumbert, R. T., Brogniez, H. & Roca, R., 2007. On the relative humidity of the atmosphere. In T. Schneider & A. Sobel, Eds., *The Global Circulation of the Atmosphere: Phenomena, Theory, Challenges*, pp. 143–185. Princeton University Press.

Pierrehumbert, R. T. & Swanson, K. L., 1995. Baroclinic instability. *Ann. Rev. Fluid Mech.*, **27**, 419–467.

Plumb, R., 1981. Instability of the distorted polar night vortex: A theory of stratospheric warmings. *J. Atmos. Sci.*, **38**, 2514–2531.

Plumb, R. & McEwan, A., 1978. The instability of a forced standing wave in a viscous stratified fluid: A laboratory analogue of. the Quasi-Biennial Oscillation. *J. Atmos. Sci.*, **35**, 1827–1839.

Plumb, R. A., 1977. The interaction of two internal waves with the mean flow: Implications for the theory of the Quasi-Biennial Oscillation. *J. Atmos. Sci.*, **34**, 1847–1858.

Plumb, R. A., 1979. Eddy fluxes of conserved quantities by small-amplitude waves. *J. Atmos. Sci.*, **36**, 1699–1704.

Plumb, R. A., 1984. The quasi-biennial oscillation. In J. R. Holton & T. Matsuno, Eds., *Dynamics of the Middle Atmosphere*, pp. 217–251. Terra Scientific Publishing.

Plumb, R. A., 1990. A nonacceleration theorem for transient quasi-geostrophic eddies on a three-dimensional time-mean flow. *J. Atmos. Sci.*, **47**, 1825–1836.

Plumb, R. A., 2002. Stratospheric transport. *J. Meteor. Soc. Japan*, **80**, 793–809.

Plumb, R. A. & Bell, R. C., 1982. An analysis of the quasi-biennial oscillation on an equatorial beta-plane. *Quart. J. Roy. Meteor. Soc.*, **108**, 335–352.

Poincaré, H., 1893. *Théorie des Tourbillons (Theory of Vortices [literally, Swirls])*. Georges Carré, Éditeur. Reprinted by Éditions Jacques Gabay, 1990, 211 pp.

Poincaré, H., 1908. *Science and Method*. T. Nelson and Sons. Engl. transl.: F. Maitland. Reprinted in *The Value of Science: Essential Writings of Henri Poincaré*, Ed. S. J. Gould, Random House, 584 pp.

Polzin, K. L., Toole, J. M., Ledwell, J. R. & Schmidt, R. W., 1997. Spatial variability of turbulent mixing in the abyssal ocean. *Science*, **276**, 93–96.

Pope, S. B., 2000. *Turbulent Flows*. Cambridge University Press, 754 pp.

Price, J. F., Weller, R. A. & Schudlich, R. R., 1987. Wind-driven ocean currents and Ekman transport. *Science*, **238**, 1534–1538.

Proudman, J., 1916. On the motion of solids in liquids. *Proc. Roy. Soc. Lond. A*, **92**, 408–424.

Queney, P., 1948. The problem of air flow over mountains: A summary of theoretical results. *Bull. Am. Meteor. Soc.*, **29**, 16–26.

Quinn, W. H., 1974. Monitoring and predicting El Niño invasions. *J. Appl. Meteor.*, **13**, 825–830.

Quon, C. & Ghil, M., 1992. Multiple equilibria in thermosolutal convection due to salt-flux boundary conditions. *J. Fluid Mech.*, **245**, 449–484.

Randall, D. A., 2015. *An Introduction to the Global Circulation of the Atmosphere*. Princeton University Press, 456 pp.

Randel, W. J., Wu, F., Swinbank, R., Nash, J. & O'Neill, A., 1999. Global QBO circulation derived from UKMO stratospheric analyse. *J. Atmos. Sci.*, **56**, 457–474.

Rayleigh, Lord, 1880. On the stability, or instability, of certain fluid motions. *Proc. London Math. Soc.*, **11**, 57–70.

Rayleigh, Lord, 1894. *The Theory of Sound, Volume II*. 2nd edn. Macmillan, 522 pp. Reprinted by Dover Publications, 1945.

Rayleigh, Lord, 1912. On the propagation of waves through a stratified medium, with special reference to the question of reflection. *Proc. Roy. Soc. Lond. A*, **86**, 207–226.

Raymond, D. J., 1995. Regulation of moist convection over the west Pacific warm pool. *J. Atmos. Sci.*, **52**, 3945–3959.

Raymond, D. J., 1997. Boundary layer quasi-equilibrium. In R. K. Smith, Ed., *The Physics and Parameterization of Moist Atmospheric Convection*, pp. 387–397. Springer.

Raymond, D. J. & Fuchs, Ž., 2009. Moisture modes and the Madden–Julian oscillation. *J. Climate*, **22**, 3031–3046.

Read, P. L., 2001. Transition to geostrophic turbulence in the laboratory, and as a paradigm in atmospheres and oceans. *Surveys Geophys.*, **33**, 265–317.

Reed, R. J., 1960. The structure and dynamics of the 26-month oscillation. Paper presented at the 40th anniversary meeting of the Am. Meter. Soc., Boston.

Reed, R. J., Campbell, W. J., Rasmussen, L. A. & Rogers, D. G., 1961. Evidence of a downward-propagating annual wind reversal in the equatorial stratosphere. *J. Geophys. Res.*, **66**, 813–818.

Reif, F., 1965. *Fundamentals of Statistical and Thermal Physics*. McGraw-Hill, 651 pp.

Rhines, P. B., 1975. Waves and turbulence on a β-plane. *J. Fluid. Mech.*, **69**, 417–443.

Rhines, P. B., 1977. The dynamics of unsteady currents. In E. A. Goldberg, I. N. McCane, J. J. O'Brien, & J. H. Steele, Eds., *The Sea*, Vol. 6, pp. 189–318. J. Wiley and Sons.

Rhines, P. B. & Holland, W. R., 1979. A theoretical discussion of eddy-driven mean flows. *Dyn. Atmos. Oceans*, **3**, 289–325.

Rhines, P. B. & Young, W. R., 1982a. Homogenization of potential vorticity in planetary gyres. *J. Fluid Mech.*, **122**, 347–367.

Rhines, P. B. & Young, W. R., 1982b. A theory of wind-driven circulation. I. Mid-ocean gyres. *J. Mar. Res. (Suppl)*, **40**, 559–596.

Richardson, L. F., 1920. The supply of energy from and to atmospheric eddies. *Proc. Roy. Soc. Lond. A*, **97**, 354–373.

Richardson, L. F., 1922. *Weather Prediction by Numerical Process*. Cambridge University Press, 236 pp. Reprinted by Dover Publications.

Richardson, L. F., 1926. Atmospheric diffusion on a distance-neighbour graph. *Proc. Roy. Soc. Lond. A*, **110**, 709–737.

Richardson, P. L., 1983. Eddy kinetic-energy in the North Atlantic from surface drifters. *J. Geophys. Res.*, **88**, 4355–4367.

Riehl, H. & Fultz, D., 1957. Jet stream and long waves in a steady rotating-dishpan experiment: structure of the circulation. *Quart. J. Roy. Meteor. Soc.*, **82**, 215–231.

Rintoul, S. R., Hughes, C. & Olbers, D., 2001. The Antarctic Circumpolar Current system. In G. Siedler, J. Church, & J. Gould, Eds., *Ocean Circulation and Climate*, pp. 271–302. Academic Press.

Ripa, P., 1981. On the theory of nonlinear wave–wave interactions among geophysical waves. *J. Fluid Mech.*, **103**, 87–115.

Robinson, A. R., 1966. An investigation into the wind as the cause of the equatorial undercurrent. *J. Mar. Res.*, **24**, 179–204.

Robinson, A. R., Ed., 1984. *Eddies in Marine Science*. Springer-Verlag, 609 pp.

Robinson, A. R. & McWilliams, J. C., 1974. The baroclinic instability of the open ocean. *J. Phys. Oceanogr.*, **4**, 281–294.

Robinson, A. R. & Stommel, H., 1959. The oceanic thermocline and the associated thermohaline circulation. *Tellus*, **11**, 295–308.

Rooth, C., 1982. Hydrology and ocean circulation. *Prog. Oceanogr.*, **11**, 131–149.

Roquet, F., Madec, G., Brodeau, L. & Nycander, J., 2015. Defining a simplified yet 'realistic' equation of state for seawater. *J. Phys. Oceanogr.*, **45**, 2564–2579.

Roquet, F., Madec, G., McDougall, T. J. & Barker, P. M., 2015. Accurate polynomial expressions for the density and specific volume of seawater using the TEOS-10 standard. *Oce. Model.*, **90**, 29–43.

Rossby, C.-G., 1936. Dynamics of steady ocean currents in the light of experimental fluid dynamics. *Papers Phys. Oceanog. Meteor.*, **5**, 1–43.

Rossby, C.-G., 1938. On the mutual adjustment of pressure and velocity distributions in certain simple current systems, II. *J. Mar. Res.*, **5**, 239–263.

Rossby, C.-G., 1939. Relations between variation in the intensity of the zonal circulation and the displacements of the semi-permanent centers of action. *J. Marine Res.*, **2**, 38–55.

Rossby, C.-G., 1940. Planetary flow patterns in the atmosphere. *Quart. J. Roy. Meteor. Soc.*, **66,** suppl., 68–87.

Rossby, C.-G., 1949. On the nature of the general circulation of the lower atmosphere. In G. P. Kuiper, Ed., *The Atmospheres of the Earth and Planets*, pp. 16–48. University of Chicago Press.

Rossby, H. T., 1965. On thermal convection driven by non-uniform heating from below: an experimental study. *Deep-Sea Res.*, **12**, 9–16.

Ruddick, B., McDougall, T. & Turner, J., 1989. The formation of layers in a uniformly stirred density gradient. *Deep-Sea Res.*, **36**, 597–609.

Rudnick, D. L. & Weller, R. A., 1993. Observations of superinertial and near-inertial wind-driven flow. *J. Phys. Oceanogr.*, **23**, 2351–2359.

Ruelle, D. & Takens, F., 1971. On the nature of turbulence. *Commun. Math. Phys.*, **20**, 167–192.

Salmon, R., 1980. Baroclinic instability and geostrophic turbulence. *Geophys. Astrophys. Fluid Dyn.*, **10**, 25–52.

Salmon, R., 1983. Practical use of Hamilton's principle. *J. Fluid Mech.*, **132**, 431–444.

Salmon, R., 1990. The thermocline as an internal boundary layer. *J. Mar. Res.*, **48**, 437–469.

Salmon, R., 1998. *Lectures on Geophysical Fluid Dynamics*. Oxford University Press, 378 pp.

Saltzman, B., 1962. Finite amplitude free convection as an initial value problem. *J. Atmos. Sci.*, **19**, 329–341.

Samelson, R. M., 1999a. Geostrophic circulation in a rectangular basin with a circumpolar connection. *J. Phys. Oceanogr.*, **29**, 3175–3184.

Samelson, R. M., 1999b. Internal boundary layer scaling in 'two-layer' solutions of the thermocline equations. *J. Phys. Oceanogr.*, **29**, 2099–2102.

Samelson, R. M., 2004. Simple mechanistic models of middepth meridional overturning. *J. Phys. Oceanogr.*, **34**, 2096–2103.

Samelson, R. M., 2011. *The Theory of Large-Scale Ocean Circulation*. Cambridge University Press, 193 pp.

Samelson, R. M. & Tziperman, E., 2001. Instability of the chaotic ENSO: the growth-phase predictability barrier. *J. Atmos. Sci.*, **58**, 3613–3625.

Samelson, R. M. & Vallis, G. K., 1997. Large-scale circulation with small diapycnal diffusion: the two-thermocline limit. *J. Mar. Res.*, **55**, 223–275.

Sandström, J. W., 1908. Dynamische Versuche mit Meerwasser (Dynamical experiments with seawater). *Annal. Hydrogr. Marit. Meteorol.*, **36**, 6–23.

Sandström, J. W., 1916. Meteorologische Studien im Schwedischen Hochgebirge (Meteorological studies in the Swedish high mountains). *Goteborgs Kungl. Vetenskaps-och Vitterhets-Samhalles, Handingar*, **27**, 1–48.

Sarachik, E. S. & Cane, M. A., 2010. *The El Niño-Southern Oscillation Phenomenon.* Cambridge University Press, 369 pp.

Sarkisyan, A. & Ivanov, I., 1971. Joint effect of baroclinicity and relief as an important factor in the dynamics of sea currents. *Izv. Akad. Nauk Atmos. Ocean. Phys.*, **7**, 116–124.

Scaife, A., Butchart, N., Warner, C. D. & Swinbank, R., 2002. Impact of a spectral gravity wave parameterization on the stratosphere in the Met Office Unified Model. *J. Atmos. Sci.*, **59**, 1473–1489.

Scaife, A. & James, I. N., 2000. Response of the stratosphere to interannual variability of tropospheric planetary waves. *Quart. J. Roy. Meteor. Soc.*, **126**, 275–297.

Schär, C., 1993. A generalization of Bernoulli's theorem. *J. Atmos. Sci.*, **50**, 1437–1443.

Schmidtko, S., Johnson, G. C. & Lyman, J. M., 2013. MIMOC: a global monthly isopycnal upper-ocean climatology with mixed layers. *J. Geophys. Res. (Oceans)*, **118**, 4, 1658–1672.

Schmitz, W. J., 1995. On the interbasin-scale thermohaline circulation. *Rev. Geophysics*, **33**, 151–173.

Schneider, E. K., 1977. Axially symmetric steady-state models of the basic state for instability and climate studies. Part II: nonlinear calculations. *J. Atmos. Sci.*, **34**, 280–297.

Schneider, T., Held, I. & Garner, S. T., 2003. Boundary effects in potential vorticity dynamics. *J. Atmos. Sci.*, **60**, 1024–1040.

Schneider, T., O'Gorman, P. A. & Levine, X. J., 2010. Water vapor and the dynamics of climate changes. *Rev. Geophys.*, **48**, RG3001.

Schneider, T. & Sobel, A., Eds., 2007. *The Global Circulation of the Atmosphere.* Princeton University Press.

Schneider, T. & Walker, C. C., 2006. Self-organization of atmospheric macroturbulence into critical states of weak nonlinearity. *J. Atmos. Sci.*, **63**, 1569–1586.

Schoeberl, M. R., 1978. Stratospheric warmings: observations and theory. *Revs. Geophys. Space Phys.*, **16**, 521–538.

Schubert, W. H., Hausman, S. A., Garcia, M., Ooyama, K. V. & Kuo, H.-C., 2001. Potential vorticity in a moist atmosphere. *J. Atmos. Sci.*, **58**, 3148–3157.

Schubert, W. H. & Masarik, M. T., 2006. Potential vorticity aspects of the MJO. *Dyn. Atmos. Oceans*, **42**, 127–151.

Schubert, W. H., Ruprecht, E., Hertenstein, R., Nieto Ferreira, R. *et al.*, 2004. English translations of twenty-one of Ertel's papers on geophysical fluid dynamics. *Meteor. Z.*, **13**, 527–576.

Scinocca, J. F. & Shepherd, T. G., 1992. Nonlinear wave-activity conservation laws and Hamiltonian structure for the two-dimensional anelastic equations. *J. Atmos. Sci.*, **49**, 5–28.

Scorer, R. S. & Ludlam, F. H., 1951. Bubble theory of penetrative convection. *Q. J. Mech. Appl. Maths*, **3**, 107–112.

Scott, R. B., 2001. Evolution of energy and enstrophy containing scales in decaying, two-dimensional turbulence with friction. *Phys. Fluids*, **13**, 2739–2742.

Scott, R. K. & Dritschel, D. G., 2012. The structure of zonal jets in geostrophic turbulence. *J. Fluid Mech.*, **711**, 576–598.

Scott, R. K. & Haynes, P. H., 1998. Internal interannual variability of the extratropical stratospheric circulation: The low latitude flywheel. *Quart. J. Roy. Meteor. Soc.*, **124**, 2149–2173.

Scott, R. K. & Haynes, P. H., 2000. Internal vacillations in stratosphere-only models. *J. Atmos. Sci.*, **57**, 2333–2350.

Shapiro, M. & Grønas, S., Eds., 1999. *The Life Cycles of Extratropical Cyclones.* American Meteorological Society, 359 pp.

Shepherd, T. G., 1983. Mean motions induced by baroclinic instability in a jet. *Geophys. Astrophys. Fluid Dyn.*, **27**, 35–72.

Shepherd, T. G., 1987. A spectral view of nonlinear fluxes and stationary-transient interaction in the atmosphere. *J. Atmos. Sci.*, **44**, 1166–1179.

Shepherd, T. G., 1990. Symmetries, conservation laws, and Hamiltonian structure in geophysical fluid dynamics. *Adv. Geophys.*, **32**, 287–338.

Shepherd, T. G., 1993. A unified theory of available potential energy. *Atmosphere–Ocean*, **31**, 1–26.

Sherwood, S. C., Roca, R., Weckwerth, T. M. & Andronova, N. G., 2010. Tropospheric water vapor, convection and climate. *Rev. Geophys.*, **48**, RG2001.

Shutts, G. J., 1983. Propagation of eddies in diffluent jet streams: eddy vorticity forcing of blocking flow fields. *Quart. J. Roy. Meteor. Soc.*, **109**, 737–761.

Silberstein, L., 1896. O tworzeniu sie wirow, w plynie doskonalym (On the creation of eddies in an ideal fluid). *W Krakaowie Nakladem Akademii Umiejetnosci (Proc. Cracow Acad. Sci.)*, **31**, 325–335.

Simmonds, J. G. & Mann, J. E., 1998. *A First Look at Perturbation Theory*. Dover Publications, 139 pp.

Simmons, A. & Hoskins, B., 1978. The life-cycles of some nonlinear baroclinic waves. *J. Atmos. Sci.*, **35**, 414–432.

Sjoberg, J. P. & Birner, T., 2012. Transient tropospheric forcing of sudden stratospheric warmings. *J. Atmos. Sci.*, **69**, 3420–3432.

Smagorinsky, J., 1953. The dynamical influences of large-scale heat sources and sinks on the quasi-stationary mean motions of the atmosphere. *Quart. J. Roy. Meteor. Soc.*, **79**, 342–366.

Smagorinsky, J., 1969. Problems and promises of deterministic extended range forecasting. *Bull. Am. Meteor. Soc.*, **50**, 286–311.

Smith, K. S., Boccaletti, G., Henning, C. C., Marinov, I. *et al.*, 2002. Turbulent diffusion in the geostrophic inverse cascade. *J. Fluid Mech.*, **469**, 13–48.

Smith, K. S. & Vallis, G. K., 1998. Linear wave and instability properties of extended range geostrophic models. *J. Atmos. Sci.*, **56**, 1579–1593.

Smith, K. S. & Vallis, G. K., 2001. The scales and equilibration of mid-ocean eddies: freely evolving flow. *J. Phys. Oceanogr.*, **31**, 554–571.

Smith, K. S. & Vallis, G. K., 2002. The scales and equilibration of mid-ocean eddies: forced-dissipative flow. *J. Phys. Oceanogr.*, **32**, 1669–1721.

Smith, L. M. & Waleffe, F., 1999. Transfer of energy to two-dimensional large scales in forced, rotating three-dimensional turbulence. *Phys. Fluids*, **11**, 1608–1622.

Smith, R. B., 1979. The influence of mountains on the atmosphere. In B. Saltzman, Ed., *Advances in Geophysics*, vol. 21, pp. 87–230. Academic Press.

Smith, R. K., 1997. On the theory of CISK. *Quart. J. Roy. Meteor. Soc.*, **123**, 407–418.

Sobel, A. & Maloney, E., 2013. Moisture modes and the eastward propagation of the MJO. *J. Atmos. Sci.*, **70**, 187–192.

Sobel, A. H., Nilsson, J. & Polvani, L., 2001. The weak temperature gradient approximation and balanced tropical moisture waves. *J. Atmos. Sci.*, **58**, 3650–3665.

Spall, M. A., 2000. Generation of strong mesoscale eddies by weak ocean gyres. *J. Mar. Res.*, **58**, 97–116.

Spiegel, E. A. & Veronis, G., 1960. On the Boussinesq approximation for a compressible fluid. *Astrophys. J.*, **131**, 442–447. (Correction: *Astrophys. J.*, **135**, 655–656).

Squire, H., 1933. On the stability of three-dimensional disturbances of viscous flow between parallel walls. *Proc. Roy. Soc. Lond. A*, **142**, 621–628.

Srinivasan, K. & Young, W., 2012. Zonostrophic instability. *J. Atmos. Sci.*, **69**, 1633–1656.

Stammer, D., 1997. Global characteristics of ocean variability estimated from regional TOPEX/Poseidon altimeter measurements. *J. Phys. Oceanogr.*, **27**, 1743–1769.

Starr, V. P., 1948. An essay on the general circulation of the Earth's atmosphere. *J. Meteor.*, **78**, 39–43.

Starr, V. P., 1968. *Physics of Negative Viscosity Phenomena*. McGraw Hill, 256 pp.

Stechmann, S. N. & Ogrosky, H. R., 2014. The Walker circulation, diabatic heating, and outgoing longwave radiation. *Geophys. Res. Lett.*, **41**, 9097–9105.

Steers, J. A., 1962. *An Introduction to the Study of Map Projections*. Univ. of London Press, 288 pp.

Stern, M. E., 1963. Trapping of low frequency oscillations in an equatorial boundary layer. *Tellus*, **15**, 246–250.

Stevens, D. P. & Ivchenko, V. O., 1997. The zonal momentum balance in an eddy-resolving general-circulation model of the southern ocean. *Quart. J. Roy. Meteor. Soc.*, **123**, 929–951.

Stewart, G. R., 1941. *Storm*. Random House, 349 pp.

Stewartson, K., 1977. The evolution of the critical layer of a Rossby wave. *Geophys. Astrophys. Fluid Dyn.*, **9**, 185–200.

Stips, A., 2005. Dissipation measurement: theory. In H. Z. Baumert, J. Simpson, & J. Sündermann, Eds., *Marine Turbulence*. Cambridge University Press.

Stommel, H., 1958. The abyssal circulation. *Deep-Sea Res.*, **5**, 80–82.

Stommel, H., 1960. Wind-drift near the equator. *Deep-Sea Res.*, **6**, 298–302.

Stommel, H., 1961. Thermohaline convection with two stable regimes of flow. *Tellus*, **13**, 224–230.

Stommel, H. & Arons, A. B., 1960. On the abyssal circulation of the world ocean—I. Stationary planetary flow patterns on a sphere. *Deep-Sea Res.*, **6**, 140–154.

Stommel, H., Arons, A. B. & Faller, A. J., 1958. Some examples of stationary planetary flow patterns in bounded basins. *Tellus*, **10**, 179–187.

Stommel, H. & Moore, D. W., 1989. *An Introduction to the Coriolis Force*. Columbia University Press, 297 pp.

Stommel, H. & Webster, J., 1963. Some properties of the thermocline equations in a subtropical gyre. *J. Mar. Res.*, **44**, 695–711.

Stone, P. H., 1972. A simplified radiative-dynamical model for the static stability of rotating atmospheres. *J. Atmos. Sci.*, **29**, 405–418.

Stone, P. H., 1978. Baroclinic adjustment. *J. Atmos. Sci.*, **35**, 561–571.

Stone, P. H. & Nemet, B., 1996. Baroclinic adjustment: a comparison between theory, observations, and models. *J. Atmos. Sci.*, **53**, 1663–1674.

Straub, D. N., 1993. On the transport and angular momentum balance of channel models of the Antarctic Circumpolar Current. *J. Phys. Oceanogr.*, **23**, 776–782.

Straub, D. N., 1999. On thermobaric production of potential vorticity in the ocean. *Tellus A*, **51**, 314–325.

Suarez, M. & Schopf, P., 1988. A delayed action oscillator for ENSO. *J. Atmos. Sci.*, **45**, 323–3287.

Sukhatme, J. & Young, W. R., 2011. The advection–condensation model and water-vapour probability density functions. *Quart. J. Roy. Meteor. Soc.*, **137**, 1561–1572.

Sukoriansky, S., Dikovskaya, N. & Galperin, B., 2007. On the arrest of inverse energy cascade and the Rhines scale. *J. Atmos. Sci.*, **64**, 3312–3327.

Sutcliffe, R. C., 1939. Cyclonic and anticylonic development. *Quart. J. Roy. Meteor. Soc.*, **65**, 518–524.

Sutcliffe, R. C., 1947. A contribution to the problem of development. *Quart. J. Roy. Meteor. Soc.*, **73**, 370–383.

Sutherland, B., 2010. *Internal Gravity Waves*. Cambridge University Press, 394 pp.

Swallow, J. C. & Worthington, V., 1961. An observation of a deep countercurrent in the western North Atlantic. *Deep-Sea Res.*, **8**, 1–19.

Tailleux, R., 2016. Neutrality versus materiality: A thermodynamic theory of neutral surfaces. *Fluids*, **1**, 32.

Takahashi, M., 1996. Simulation of the stratospheric quasi-biennial oscillation using a general circulation model. *Geophys. Res. Lett.*, **23**, 661–664.

Talley, L., 1988. Potential vorticity distribution in the North Pacific. *J. Phys. Oceanogr.*, **18**, 89–106.

Talley, L. D., Pickard, G., Emery, W. J. & Swift, J. H., 2011. *Descriptive Physical Oceanography: An Introduction*. Academic press, 555 pp.

Taylor, G. I., 1921a. Diffusion by continuous movements. *Proc. London Math. Soc.*, **2** (20), 196–211.

Taylor, G. I., 1921b. Experiments with rotating fluids. *Proc. Roy. Soc. Lond. A*, **100**, 114–121.

Tennekes, H. & Lumley, J. L., 1972. *A First Course in Turbulence*. The MIT Press, 330 pp.

Tesserenc De Bort, L. P., 1902. Variations de la température de l'air libre dans la zone comprise 8 km et 13 km d'altitude (Variations in the temperature of the free air in the zone between 8 km and 13 km of altitude). *C. R. Hebd. Séances Acad. Sci.*, **134**, 987–989.

Thompson, A. F., Stewart, A. L. & Bischoff, T., 2016. A multibasin residual-mean model for the global overturning circulation. *J. Phys. Oceanogr.*, **46**, 2583–2604.

Thompson, A. F. & Young, W. R., 2006. Scaling baroclinic eddy fluxes: vortices and energy balance. *J. Phys. Oceanogr.*, **36**, 720–738.

Thompson, P. D., 1957. Uncertainty of initial state as a factor in the predictability of large scale atmospheric flow patterns. *Tellus*, **9**, 275–295.

Thompson, R. O. R. Y., 1971. Why there is an intense eastward current in the North Atlantic but not in the South Atlantic. *J. Phys. Oceanogr.*, **1**, 235–237.

Thompson, R. O. R. Y., 1980. A prograde jet driven by Rossby waves. *J. Atmos. Sci.*, **37**, 1216–1226.

Thomson, J., 1892. Bakerian lecture. On the grand currents of atmospheric circulation. *Phil. Trans. Roy. Soc. Lond. A*, **183**, 653–684.

Thomson, W. (Lord Kelvin), 1869. On vortex motion. *Trans. Roy. Soc. Edinburgh*, **25**, 217–260.

Thomson, W. (Lord Kelvin), 1871. Hydrokinetic solutions and observations. *Phil. Mag. and J. Science*, **42**, 362–377.

Thomson, W. (Lord Kelvin), 1879. On gravitational oscillations of rotating water. *Proc. Roy. Soc. Edinburgh*, **10**, 92–100.

Thorncroft, C. D., Hoskins, B. J. & McIntyre, M. E., 1993. Two paradigms of baroclinic-wave life-cycle behaviour. *Quart. J. Roy. Meteor. Soc.*, **119**, 17–55.

Thorpe, A. J., Volkert, H. & Ziemianski, M. J., 2003. The Bjerknes' circulation theorem: a historical perspective. *Bull. Am. Meteor. Soc.*, **84**, 471–480.

Thual, O. & McWilliams, J. C., 1992. The catastrophe structure of thermohaline convection in a two-dimensional fluid model and a comparison with low-order box models. *Geophys. Astrophys. Fluid Dyn.*, **64**, 67–95.

Thuburn, J., 2017. Use of the Gibbs thermodynamic potential to express the equation of state in atmospheric models. *Quart. J. Roy. Meteor. Soc.*. submitted.

Thuburn, J. & Craig, G. C., 1997. GCM tests of theories for the height of the tropopause. *J. Atmos. Sci.*, **54**, 869–882.

Thuburn, J. & Craig, G. C., 2000. Stratospheric influence on tropopause height: the radiative constraint. *J. Atmos. Sci.*, **57**, 17–28.

Tobias, S. & Marston, J., 2013. Direct statistical simulation of out-of-equilibrium jets. *Phys. Rev. Lett.*, **110**, 10, 104502.

Toggweiler, J. R. & Samuels, B., 1995. Effect of Drake Passage on the global thermohaline circulation. *Deep-Sea Res.*, **42**, 477–500.

Toggweiler, J. R. & Samuels, B., 1998. On the ocean's large-scale circulation in the limit of no vertical mixing. *J. Phys. Oceanogr.*, **28**, 1832–1852.

Toole, J. M., Polzin, K. L. & Schmitt, R. W., 1994. Estimates of diapycnal mixing in the abyssal ocean. *Science*, **264**, 1120–1123.

Tort, M. & Dubos, T., 2014. Dynamically consistent shallow-atmosphere equations with a complete Coriolis force. *Quart. J. Roy. Meteor. Soc.*, **140**, 684, 2388–2392.

Tréguier, A. M., Held, I. M. & Larichev, V. D., 1997. Parameterization of quasi-geostrophic eddies in primitive equation ocean models. *J. Phys. Oceanogr.*, **29**, 567–580.

Trenberth, K. E., 1997. The definition of El Niño. *Bull. Am. Meteor. Soc.*, **78**, 2771–2777.

Trenberth, K. E. & Caron, J. M., 2001. Estimates of meridional atmosphere and ocean heat transports. *J. Climate*, **14**, 3433–3443.

Tritton, D. J., 1988. *Physical Fluid Dynamics*. Oxford University Press, 519 pp.

Truesdell, C., 1951. Proof that Ertel's vorticity theorem holds in average for any medium suffering no tangential acceleration on the boundary. *Geofis Pura Appl.*, **19**, 167–169.

Truesdell, C., 1954. *The Kinematics of Vorticity*. Indiana University Press, 232 pp.

Truesdell, C., 1969. *Rational Thermodynamics*. McGraw Hill, 208 pp.

Tsang, Y.-K. & Vanneste, J., 2016. Advection-condensation of water vapor in a model of coherent stirring. Submitted to *Proc. Roy. Soc. A*.

Tudhope, A. W., Chilcott, C. P., McCulloch, M. T., Cook, E. R. *et al.*, 2001. Variability in the El Niño Southern Oscillation through a glacial-interglacial cycle. *Science*, **291**, 1511–1517.

Tung, K. K., 1979. A theory of stationary long waves. Part III: quasi-normal modes in a singular wave guide. *Mon. Wea. Rev.*, **107**, 751–774.

Tziperman, E., Stone, L., Cane, M. A. & Jarosh, H., 1994. El Niño chaos: Overlapping of resonances between the seasonal cycle and the Pacific ocean-atmosphere oscillator. *Science*, **264**, 72–73.

Valdes, P. J. & Hoskins, B. J., 1988. Baroclinic instability of the zonally averaged flow with boundary layer damping. *J. Atmos. Sci.*, **45**, 1584–1593.

Vallis, G. K., 1982. A statistical dynamical climate model with a simple hydrology cycle. *Tellus*, **34**, 211–227.

Vallis, G. K., 1985. Instability and flow over topography. *Geophys. Astrophys. Fluid Dyn.*, **34**, 1–38.

Vallis, G. K., 1988a. Conceptual models of El Niño and the Southern Oscillation. *J. Geophys. Res.*, **93**, 13979–13991.

Vallis, G. K., 1988b. Numerical studies of eddy transport properties in eddy-resolving and parameterized models. *Quart. J. Roy. Meteor. Soc.*, **114**, 183–204.

Vallis, G. K., 1996. Potential vorticity and balanced equations of motion for rotating and stratified flows. *Quart. J. Roy. Meteor. Soc.*, **122**, 291–322.

Vallis, G. K., 2000. Large-scale circulation and production of stratification: effects of wind, geometry and diffusion. *J. Phys. Oceanogr.*, **30**, 933–954.

Vallis, G. K. & Maltrud, M. E., 1993. Generation of mean flows and jets on a beta plane and over topography. *J. Phys. Oceanogr.*, **23**, 1346–1362.

Vallis, G. K., Zurita-Gotor, P., Cairns, C. & Kidston, J., 2015. The response of the large-scale structure of the atmosphere to global warming. *Quart. J. Roy. Meteor. Soc.*, **141**, 1479–1501.

Vanneste, J. & Shepherd, T. G., 1998. On the group-velocity property for wave-activity conservation laws. *J. Atmos. Sci.*, **55**, 1063–1068.

Verkley, W. T. M. & van der Velde, I. R., 2010. Balanced dynamics in the tropics. *Quart. J. Roy. Meteor. Soc.*, **136**, 41–49.

Veronis, G., 1960. An approximate theoretical analysis of the equatorial undercurrent. *Deep-Sea Res.*, **6**, 318–327.

Veronis, G., 1966a. Wind-driven ocean circulation – Part 1: Linear theory and perturbation analysis. *Deep-Sea Res.*, **13**, 17–29.

Veronis, G., 1966b. Wind-driven ocean circulation – Part 2: Numerical solutions of the non-linear problem. *Deep-Sea Res.*, **13**, 30–55.

Veronis, G., 1969. On theoretical models of the thermocline circulation. *Deep-Sea Res.*, **31** Suppl., 301–323.

Veryard, R. G. & Ebdon, R. A., 1961. Fluctuations in tropical stratospheric winds. *Meteor. Mag*, **90**, 125–143.

Visbeck, M., Marshall, J., Haine, T. & Spall, M., 1997. Specification of eddy transfer coefficients in coarse-resolution ocean circulation models. *J. Phys. Oceanogr.*, **27**, 381–402.

Von Neumann, J., 1955. Methods in the physical sciences. In L. G. Leary, Ed., *The Unity of Knowledge*. Doubleday.

Walker, C. & Schneider, T., 2005. Response of idealized Hadley circulations to seasonally varying heating. *Geophys. Res. Lett.*, **32**, L06813. doi:10.1029/2004GL022304.

Wallace, J. M., 1973. General circulation of the tropical lower stratosphere. *Rev. Geophys.*, **11**, 191–222.

Wallace, J. M., 1983. The climatological mean stationary waves: observational evidence. In B. Hoskins & R. P. Pearce, Eds., *Large-Scale Dynamical Processes in the Atmosphere*, pp. 27–63. Academic Press.

Wallace, J. M. & Hobbs, P. V., 2006. *Atmospheric Science: An Introductory Survey*. 2nd edn. Elsevier, 483 pp.

Wallace, J. M. & Holton, J. R., 1968. A diagnostic numerical model of the quasi-biennial oscillation. *J. Atmos. Sci.*, **25**, 280–292.

Warn, T., Bokhove, O., Shepherd, T. G. & Vallis, G. K., 1995. Rossby number expansions, slaving principles, and balance dynamics. *Quart. J. Roy. Meteor. Soc.*, **121**, 723–739.

Warn, T. & Warn, H., 1976. On the development of a Rossby wave critical level. *J. Atmos. Sci.*, **33**, 2021–2024.

Warren, B. A., 1981. Deep circulation of the world ocean. In B. A. Warren & C. Wunsch, Eds., *Evolution of Physical Oceanography*, pp. 6–41. The MIT Press.

Warren, B. A., 1999. Approximating the energy transport across oceanic sections. *J. Geophys. Res.*, **104**, 7915–7920.

Warren, B. A., 2006. The first law of thermodynamics in a salty ocean. *Progress in Oceanography*, **70**, 2, 149–167.

Warren, B. A., LaCasce, J. H. & Robbins, P. E., 1996. On the obscurantist physics of form drag in theorizing about the Circumpolar Current. *J. Phys. Oceanogr.*, **26**, 2297–2301.

Wasow, W., 1944. Asymptotic solution of boundary value problems for the differential equation $\Delta U + \lambda(\partial/\partial x)U = \lambda f(x, y)$. *Duke Math J.*, **11**, 405–415.

Watson, A., Vallis, G. K. & Nikurashin, M., 2015. Southern Ocean buoyancy forcing of ocean ventilation and glacial atmospheric CO_2. *Nature Geosciences*, **8**, 861–864. doi:10.1038/ngeo2538.

Webb, D. J. & Suginohara, N., 2001. Vertical mixing in the ocean. *Nature*, **409**, 37.

Weinstock, R., 1952. *Calculus of Variations*. McGraw-Hill. Reprinted by Dover Publications, 1980, 328 pp.

Welander, P., 1959. An advective model of the ocean thermocline. *Tellus*, **11**, 309–318.

Welander, P., 1968. Wind-driven circulation in one- and two-layer oceans of variable depth. *Tellus*, **20**, 1–15.

Welander, P., 1971a. Some exact solutions to the equations describing an ideal-fluid thermocline. *J. Mar. Res.*, **29**, 60–68.

Welander, P., 1971b. The thermocline problem. *Phil. Trans. Roy. Soc. Lond. A*, **270**, 415–421.

Welander, P., 1973. Lateral friction in the ocean as an effect of potential vorticity mixing. *Geophys. Fluid Dyn.*, **5**, 101–120.

Welander, P., 1986. Thermohaline effects in the ocean circulation and related simple models. In J. Willebrand & D. L. T. Anderson, Eds., *Large-scale Transport Processes in Oceans and Atmospheres*, pp. 163–200. Reidel.

Wentzel, G., 1926. Eine Verallgemeinerung der Quantenbedingungen für die Zwecke der Wellenmechanik (A generalization of the quantum conditions for the purposes of wave mechanics). *Zeit. fur Physic A*, **38**, 518–529.

Wheeler, M. & Kiladis, G. N., 1999. Convectively coupled equatorial waves: Analysis of clouds and temperature in the wavenumber-frequency domain. *J. Atmos. Sci.*, **56**, 374–399.

White, A. A., 1977. Modified quasi-geostrophic equations using geometric height as vertical co-ordinate. *Quart. J. Roy. Meteor. Soc.*, **103**, 383–396.

White, A. A., 2002. A view of the equations of meteorological dynamics and various approximations. In J. Norbury & I. Roulstone, Eds., *Large-Scale Atmosphere-Ocean Dynamics I*, pp. 1–100. Cambridge University Press.

White, A. A., 2003. The primitive equations. In J. Holton, J. Pyle, & J. Curry, Eds., *Encyclopedia of Atmospheric Science*, pp. 694–702. Academic Press.

White, A. A., Hoskins, B. J., Roulstone, I. & Staniforth, A., 2005. Consistent approximate models of the global atmosphere: shallow, deep, hydrostatic, quasi-hydrostatic and non-hydrostatic. *Quart. J. Roy. Meteor. Soc.*, **131**, 609, 2081–2107.

Whitehead, J. A., 1975. Mean flow generated by circulation on a beta-plane: An analogy with the moving flame experiment. *Tellus*, **27**, 358–364.

Whitehead, J. A., 1995. Thermohaline ocean processes and models. *Ann. Rev. Fluid Mech.*, **27**, 89–113.

Whitham, G. B., 1974. *Linear and Nonlinear Waves*. Wiley-Interscience, 656 pp.

Williams, G. P., 1978. Planetary circulations: 1. Barotropic representation of Jovian and terrestrial turbulence. *J. Atmos. Sci.*, **35**, 1399–1426.

Wittenberg, A. T., 2009. Are historical records sufficient to constrain ENSO simulations? *Geophys. Res. Lett.*, **36**, L12702. doi:10.1029/2009GL038710.

Wolfe, C. L. & Cessi, P., 2010. What sets the middepth stratification of eddying ocean models? *J. Phys. Oceanogr.*, **40**, 1520–1538.

Wolfe, C. L. & Cessi, P., 2011. The adiabatic pole-to-pole overturning circulation. *J. Phys. Oceanogr.*, **41**, 1795–1810.

World Meteorological Organization, 1957. Definition of the tropopause. *WMO Bulletin*, **6**, 136.

Wunsch, C., 2002. What is the thermohaline circulation? *Science*, **298**, 1179–1180.

Wunsch, C., 2015. *Modern Observational Physical Oceanography*. Princeton University Press, 481 pp.

Wunsch, C. & Ferrari, R., 2004. Vertical mixing, energy, and the general circulation of the oceans. *Ann. Rev. Fluid Mech.*, **36**, 281–314.

Wunsch, C. & Roemmich, D., 1985. Is the North Atlantic in Sverdrup balance? *J. Phys. Oceanogr.*, **15**, 1876–1880.

Wyrtki, K., 1952. Der Einfluss des Windes auf den mittleren Wasserstand Der Nordsee und ihren Wasserhausalt (the influence of the wind on the mean sea level and water budget of the north sea). *Dtsch.. Hydrogr. Z.*, **5**, 21–27.

Wyrtki, K., 1975. El Niño—The dynamic response of the equatorial Pacific ocean to atmospheric forcing. *J. Phys. Oceanogr.*, **5**, 572–584.

Wyrtki, K., 1985. Water displacements in the Pacific and the genesis of El Niño cycles. *J. Geophys. Res.*, **90**, 7129–7132.

Wyrtki, K., Magaard, L. & Hager, J., 1976. Eddy energy in oceans. *J. Geophys. Res.*, **81**, 2641–2646.

Xu, X., Rhines, P. B., Chassignet, E. P. & Schmitz, W. J., 2015. Spreading of Denmark Strait overflow water in the western subpolar North Atlantic: insights from eddy-resolving simulations with a passive tracer. *J. Phys. Oceanogr.*, **45**, 2913–2932.

Yaglom, A. M., 1994. A. N. Kolmogorov as a fluid mechanician and founder of a school in turbulence research. *Ann. Rev. Fluid Mech.*, **26**, 1–22.

Yanai, M. & Maruyama, T., 1966. Stratospheric wave disturbances in the tropical stratosphere. *J. Meteor. Soc. Japan*, **44**, 291–294.

Yoden, S., 1987. Bifurcation properties of a stratospheric vacillation model. *J. Atmos. Sci.*, **44**, 1723–1733.

Yoden, S., 1990. An illustrative model of seasonal and interannual variations of the stratospheric circulation. *J. Atmos. Sci.*, **47**, 1845–1853.

Young, W. R., 2010. Dynamic enthalpy, conservative temperature, and the seawater Boussinesq approximation. *J. Phys. Oceanogr.*, **40**, 394–400.

Young, W. R., 2012. An exact thickness-weighted average formulation of the Boussinesq equations. *J. Phys. Oceanogr.*, **42**, 692–707.

Young, W. R. & Rhines, P. B., 1982. A theory of the wind-driven circulation II. Gyres with western boundary layers. *J. Mar. Res.*, **40**, 849–872.

Zebiak, S. E. & Cane, M. A., 1987. A model El Niño Southern Oscillation. *Mon. Wea. Rev.*, **115**, 2262–2278.

Zeng, N., Neelin, J. D. & Chou, C., 2000. A quasi-equilibrium tropical circulation model – implementation and simulation. *J. Atmos. Sci.*, **57**, 1767–1796.

Zhang, C., 2005. Madden–Julian oscillation. *Rev. Geophys.*, **43**, 1–36.

Zhang, R. & Vallis, G. K., 2007. The role of the bottom vortex stretching on the path of the North Atlantic western boundary current and on the northern recirculation gyre. *J. Phys. Oceanogr.*, **37**, 2053–2080.

Zurita-Gotor, P. & Lindzen, R., 2007. Theories of baroclinic adjustment and eddy equilibration. In T. Schneider & A. Sobel, Eds., *The Global Circulation of the Atmosphere: Phenomena, Theory, Challenges*. Princeton University Press.

Zurita-Gotor, P. & Vallis, G. K., 2009. Equilibration of baroclinic turbulence in primitive equations and quasi-geostrophic models. *J. Atmos. Sci.*, **66**, 837–863.

Zurita-Gotor, P. & Vallis, G. K., 2011. Dynamics of mid-latitude tropopause height in an idealized model. *J. Atmos. Sci.*, **68**, 823–838.

Index

Bold face denotes a primary entry or an extended discussion.

Abbe, Cleveland, 442
Absolute humidity, 675
Abyssal ocean circulation, **818–828**
    wind driven, 829
ACC, **836–845**
    adiabatic model of, 843
    and mesoscale eddies, 837
    form drag in, 844
    momentum balance, 837, 839, 841
Acoustic-gravity waves, **293–296**
Adiabatic lapse rate
    dry, 99
    moist, 692
Adiadbatic lapse rate, 48
Advection-diffusion-condensation models, 686
Advective derivative, 4
Anelastic approximation, **75–78**
Anelastic equations, 76, 78
    energetics of, 78
Angular momentum, 67
    spherical coordinates, 67
Antarctic Circumpolar Current, **836–845**
Antisymmetric turbulent diffusivity, 491
APE, 137
Arnold stability conditions, 406
Asymptotic models
    conservation properties of, 187
    quasi-geostrophy, 188
Atmospheric stratification, **572–581**
Auto-barotropic fluid, 14
Available potential energy, **137–140**
    Boussinesq fluid, 138
    ideal gas, 139

Balanced dynamics
    tropics, 711
Baroclinic adjustment, 576
Baroclinic circulation theorem, 151
Baroclinic eddies, **464–471**
    effect on Hadley Cell, 528
    in atmosphere, 465
    in ocean, 468
    magnitude and scale, 464
Baroclinic eddy diffusivities, 494
Baroclinic fluid, 14
Baroclinic instability, **335**, **347–367**
    beta effect in continuous model, 369
    beta effect in two-layer model, 360
    Eady problem, 351
    effect of stratosphere, 372
    energetics of, 367
    high-wavenumber cut-off, 343, 360, 410, 459
    in ocean, 373
    interacting edge waves, 363
    linear QG equations, 349
    mechanism of, 347
    minimum shear, 361
    necessary conditions for, 351, 408, 410
    neutral curve in two-layer problem, 362
    non-uniform shear and stratification, 372
    sloping convection, 347
    two-layer problem, 356
Baroclinic lifecycle, 466
Baroclinic lifecycles
    in atmosphere, 465
    in ocean, 468, 470
Baroclinic term, 145
Baroclinic triads, 457
Barotropic fluid, 14, 20
Barotropic instability, **335**
Barotropic jet, **540–549**
    and Rossby waves, 542
    and the EP flux, 547

numerical example, 548
Barotropic triads, 457
Batchelor scale, 439
Batchelor spectrum, 439
Bernoulli function, 44
Bernoulli's theorem, 44
and potential vorticity flux, 167
Beta effect, 154, 155
in two-dimensional turbulence, 445
Beta plane vorticity equation, 156
Beta scale, 446, 447
Beta-plane approximation, **69**
Beta-Rossby number, 559, 735
Bjerknes, Jacob, 907
Bjerknes, Vilhelm, 169
Bjerknes-Silberstein circulation theorem, 151
Boiling point, 678
Bolus velocity, 500, 503
Bottom pressure stress, 755
Boundary layers
Ekman, 201
Boussinesq approximation, **70–75**
Boussinesq equations, 71
asymptotic derivation, 101
energetics of, 74
potential vorticity conservation, 163
relation to pressure coordinates, 82
strong and weak versions, 78
summary, 74
Box ocean models, **813–818**
many boxes, 817
two boxes, 814
Breaking waves, 599
Bretherton's boundary layer, 191
Brewer–Dobson circulation, **631–643**, 644, 669
Brunt–Väisälä frequency, 97
Buoyancy frequency, 73, 97
ideal gas, 99
ocean, 99
Buoyancy-driven ocean circulation, 801, 813
Burger number, 173

Cabbeling, 39
CAPE, 695
Centrifugal force, 57
Chaos, 433, 435
a brief history, 443
Charney problem, 369
Charney, Jule, 376
Charney–Drazin condition, 599, 601, 603
Charney–Eliassen problem, 609
Charney–Green number, 370
Charney–Stern–Pedlosky criterion for instability, 351
Chemical potential, 16, 17
CIN, 695
Circular reference, *see* Reference, circular
Circulation, **143–153**
Circulation theorem, 150, **147–156**
baroclinic, 151
barotropic fluid, 150
beta effect, 154
hydrostatic flow, 152
rotating flow, 152
CISK, 727
Clausius–Clapeyron equation, **678–679**
ideal gas, 679
Closure problem of turbulence, 413
Compensating subsidence, 716
Compressible flow, 41
Concentration and mixing ratio, 10
Condensation-diffusion models, 681
Condensation-diffusion-advection models, 686
Conditional instability, 695, 696
Conservative temperature, 36, 39
Conservative tracers, 10
Convection, 699
energetics of, 696
moist, 695
Convective adjustment, 697, 706
Convective instability, 261
Convective plumes, 805
Convective quasi-equilibrium, 698
Conventional equation of state, 13
Coriolis acceleration, 57
Coriolis force, **57–58**
Coriolis, Gaspard Gustave de, 103
Critical layers, 594
gravity wave, 597
Rossy wave, 594
Critical levels, 634
Critical line, 593, **594**, 607
Cyclostrophic balance, 91, 95, 96

Deacon Cell, 833, 841, 843, 852
Deacon cell, 834
Deformation radius, 124, 125
Delayed oscillator model of El Niño, 904
Density
neutral, 32
potential, 24
Dew point, 724
Diffusion
equation of, 473
potential vorticity, 505
turbulent, 473
Diffusion-condensation models, 681
Diffusive fluxes, 490
Diffusive thermocline, **779–785**, 805
Diffusive transport, 473
Diffusivity tensors, 490

Dispersion relation, 215, 216, **219**
  Rossby waves, 228
Divergence equation, 96
Downward control, 648
Dry adiabatic lapse rate, 29, 48, 99
Dumbbell in beta-plane turbulence, 448

Eady problem, **351–356**, 357
  eddy effect on mean-flow, 399
  secondary circulation, 399
  with beta, 371
Eady, Eric, 376
Eddy diffusion, 475
  two-dimensional, 483
Eddy transport
  and the TEM, 506
  velocity, 500
Eddy vicoscity, 202
Edge waves, 340, 341, 343
  Eady problem, 365
  in shear flows, 340
Effective gravity, 59
Egg, boiling, 678
Ekman layers, **201–211**
  integral properties of, 204
  momentum balance, 203
  observed, 209
  stress in, 201
Ekman number, 202
Ekman spiral, 206, 209
El Niño, **886–904**
  toy models, 898
Elephant, 610
Eliassen–Palm flux, **383–387**, 568
  and barotropic jets, 547
  and form drag, 397
  observed, 568
  primitive equations, 581
  spherical coordinates, 581
Eliassen–Palm relation, 384
Energetics
  of quasi-geostrophic equations, 198
Energy budget, 42
  constant density fluid, 42
  variable density fluid, 43
  viscous effects, 45
Energy conservation
  Boussinesq equations, 74
  primitive equations, 65
  shallow water equations, 122
Energy flux, 234
  Bernoulli function, 44
  Rossby waves, 234–236
Energy inertial range
  in two-dimensional turbulence, 429
Energy transfer in two-dimensional flow, 424
ENSO, 886
Enstrophy inertial range, 424, 429
  passive tracer in, 439
Enstrophy transfer in two-dimensional flow, 424
Enthalpy, 17, 45, 49
  dynamic, 37
  fluxes in convection, 705
  generalized, 45
  ideal gas, 21
  of vaporization, 677
  potential, 36, 37, 49
  potential, generalized, 46
Enthalpy of vaporization, 722
Entropy, 14, 24, 48
  moist air, 721
Equation of State, **14**
Equation of state, **13**
  conventional or thermal, 13
  fundamental, 14, 15, 20
  ideal gas, 13
  moist air, 674
  seawater, 13, 33
Equations of motion, **3**
  for tropics, 711
  rotating frame, 55
  tropical atmosphere, 708
Equatorial undercurrent, **865**, 867, 874
  ideal fluid model, **876–885**
  local model, **865–876**
Equatorial waves
  stratosphere, 634
Equivalent potential temperature, 680, **694**, 723
Equivalent topography, 617
Euler, Leonard, 52
Eulerian and Lagrangian, 3
Eulerian derivative, 4
Eulerian viewpoint, 4
Exner function, 136
Explication, 156, 170

f-plane approximation, 69
Ferrel Cell, **534–536**, 571
  eddy fluxes in, 536
  surface flow in, 535
Ferrel, William, 537
Field or Eulerian viewpoint, 4
First law of thermodynamics, 15, **49**
Fjørtoft's criterion for instability, 347, 408
Fluid element, 4
Fofonoff model, 751
Force
  centrifugal, 57
  Coriolis, 58
Form drag, 119
  and Eliassen Palm flux, 397

at ocean bottom, 755
in ACC, 839, 844
Form stress, **119–120**
Four-thirds law, 482
Free energy, 17
Free-slip condition, 741
Frequency, 216
Frictional–geostrophic balance, 202
Froude number, 86, 173, 494
Frozen in property of vorticity, 147
Fundamental equation of state, 14, 15, 20
ideal gas, 25, **47–49**
Fundamental postulate of thermodynamics, 14
Fundamental thermodynamic relation, 16

Gas constant, 674
Generalized enthalpy, 46
Gent–McWilliams scheme, 499
Geopotential surfaces, 60
Geostrophic adjustment, **127–134**
energetics of, 130
Rossby problem, 128
Geostrophic balance, **87–93**, 95, 118
a variational perspective, 133
frictional, 202
in shallow water equations, 118
pressure coordinates, 91
Geostrophic contours, 764
Geostrophic scaling, 171
in continuously stratified equations, 174
in shallow water equations, 171
Geostrophic turbulence, **445**, 460
Larichev–Held model, 460
stratified, 454
two layers, 455
two-dimensional, beta-plane, 445
Gibbs function, 17, 18
equilibration, 677
for seawater, 33
ideal gas, 25, **47**
moist air, 720
Gradient wind balance, 91, 96, **94–97**
in Eulerian equations, 96
Gravity waves, 100, 101, **251**
acoustic, 293
critical line, 594
hydrostatic, 261
stratosphere, 634
Green and Stone turbulent transport, 494
Group velocity, **220–224**, **240**
internal waves, 264
property for wave activity, 384
Group velocity property, **242–246**
Gyres, 731

Hadley Cell, **516–534**
angular-momentum-conserving model, 516
effects of eddies on, 528, 532
effects of moisture on, 522
poleward extent, 517
radiative equilibrium solution, 523
seasonal effects and hemispheric asymmetry, 525
shallow water model of, 524
strength of, 517
Hadley, George, 537
Haney boundary condition, 802, 858
Heat capacity, 21
Held–Hou model of Hadley Cell, 516
Helmholtz function, 18
Hermite polynomials, **330**
Hide's theorem, 523, 537
Holton, Jim, 104
Homentropic fluid, 14, 20, 53
Homogenization of a tracer, **487–489**
Horizontal convection, **802–813**
maintenance of, 808
Humidty
measures of, 675
Hurricanes , 727
Hydrostasy, 12
accuracy, 86
scaling for, 83, 84
Hydrostatic approximation
accuracy, 86
in deriving primitive equations, 64
Hydrostatic balance, 12, **87**
effects of rotation, 92
effects of stratification, 85
scaling for, 84, 93, **83–93**
Hydrostatic equations
potential vorticity conservation, 164
Hydrostatic internal waves, 261

Ideal gas, 20
buoyancy frequency, 99
enthalpy, 21
equation of state, 13
fundamental equation of state, 25, 47
heat capacity, 21
simple and general, 20
thermodynamics of, 23
Impermeability of potential vorticity, 165
Incompressible flow, 41, **41–42**
conditions for, 41
Inertial flow, 96
Inertial oscillations, 126
Inertial range, **418**, 423
3D, 418
energy in 2D, 429
energy in 3D, 419

enstrophy, 429
two-dimensional turbulence, 427
Inertial western boundary currents, **747**
Inertial-diffusive range, 441
Inflection point criterion, 346
Instability
baroclinic, 335, 347
barotropic, 335
Kelvin–Helmholtz, 335
necessary conditions in baroclinic flow, 351
necessary conditions in shear flow, 345
parallel shear flow, 337
Intermediate models, 186
Intermittency, 423
Internal energy, 20, 48
Internal thermocline, **779–785**
Internal waves, 100, **259–261**
energetics, 267
group velocity, 264, 268
polarization properties, 261
polarization relations, 261
rays, 273
reflection, 268
stratosphere, 634
topographic generation, 283
Inverse cascade, **424**, 429
energy-enstrophy argument, 425
similarity theory, 426
vorticity elongation, 424
Inversion, 147
of vorticity, 147
Inviscid western boundary currents, **753–757**
Isentropic coordinates, **134–137**
and quasi-geostrophy, 196
Boussinesq fluid, 135
ideal gas, 136
Isopycnal coordinates, 135

JEBAR, joint effect of baroclinicity and relief, 756
Jets, **448–453**, **540–549**
and the pseudomomentum budget, 544
and the vorticity budget, 541
atmospheric, 540
eddy-driven, 540
in beta-plane turbulence, 448
numerical simulation of, 449
Joint effect of baroclinicity and relief, 756
Joint effect of beta and friction, 449
Jump conditions, 339
JWKB approximation, 247, 623

K41 theory, 416, 418
Kelvin cat's eye, 597
Kelvin waves, 126, 636
Kelvin's circulation theorem, **150**, 156
Kelvin–Helmholtz instability, **335**, 340
Kinematic stress, 202
Kinematic viscosity, 12
Kinematics
of waves, 215
Kolmogorov scale, 420
Kolmogorov theory, **416–422**
Kolmogorov theory of turbulence, 418
Kolmogorov, A. N., 441

Lagrange, Joseph-Louis, 52
Lagrangian derivative, 4
Lagrangian viewpoint, 3, 4
Lamb waves, 295
Lapse rate, **99**
adiabatic, of density, 27
adiabatic, of temperature, 29
dry adiabatic, 29, 99
dry ideal gas, 99
of seawater, 34
saturated, 692
Latent heat, 722
Latent heat of evaporation, 677
Level of free convection, 695
Lifecycle of baroclinic waves, 466
in atmosphere, 465
in ocean, 468, 470
Lifting condensation level, 695
Liouville–Green approximation, 247
Locality in turbulence, 423
Log-pressure coordinates, 82
Lorenz equations, 433
Lorenz, Edward, 141, 142
LPS model, 785
Luyten–Pedlosky–Stommel model, 785

M equation, 779
one-dimensional model, 780
Mach number, 42
Macro-turbulence, 445
Madden–Julian oscillation, 717
Main thermocline, 774
Margules relation, 119
Mass continuity, **7–10**
Eulerian derivation, 7
in a rotating frame, 58
in Boussinesq equations, 72
Lagrangian derivation, 9
Mass continuity equation
shallow water, 107
Material derivative, **4–7**
finite volume, 5
fluid property, 5
Material viewpoint, 4
Maxwell relations, **17**, **19**
Mercator coordinates, 621
Meridional overturning circulation, 801

atmospheric, Eulerian, 514
of atmosphere, 514
of ocean, 801
wind-driven, ocean, 829
Mid-latitude atmospheric circulation, **549–571**
Minimum shear for baroclinic instability, 361
Mixing length theory, **484–487**
Mixing ratio, 675
Mixing ratio and concentration, 10
MJO, 717
MOC, 514, 801
stratosphere, 644
wind-driven, ocean, 829
Moist convection, **695–700**
Moist thermodynamics, 720
Moisture
effect on potential vorticity, 160
effects on Hadley Cell, 522
Momentum equation, **11–13**
in a rotating frame, 58
shallow water, 106
vector invariant form, 66
monsoons, 717
Montgomery potential, 136
Mountain waves, 283
Multi-layer QG equations, 185
Munk wind-driven model, 740
properties of, 741

Natural coodinates, 94
Navier, Claude, 53
Necessary conditions for baroclinic instability, 408, 410
Necessary conditions for instability, **403–411**
baroclinic flow, 351
Charney–Stern–Pedlosky criterion, 351, 404
Fjørtoft's criterion, 347, 408
Rayleigh–Kuo criterion, 346, 404
relation to eddy fluxes, 564
shear flow, 345
use of pseudoenergy, 406
use of pseudomomentum, 403
Neutral density, 32, 33, 53, 774, 799
No-slip condition, 741
Non-acceleration result, 379, 639
Non-acceleration theorem, **394–399**
Non-homentropic term, 145
Nondimensionalization
in rotating flow, 171
nondimensionalization, **46–47**

Oblate spheroid, 59, 60
Observations
MJO, 718
abyssal ocean, 829
Atlantic Ocean, 803
atmospheric meridional overturning circulation, 514, 572
atmospheric stratification, mean, 572
atmospheric wind and temperature, 513
deep ocean circulation, 801
deep western boundary current, 827
Ekman layers, 209
Eliassen–Palm flux, 568, 570
Eliassen–Palm flux divergence, 570
equatorial ocean currents, 861
Global Ocean, 735
global ocean currents, 732
main thermocline, 774
North Atlantic, 827
North Atlantic currents, 733
ocean stratification, 802
oceanic meridional overturning circulation, 801
of the atmosphere, 511
potential vorticity, ocean, 770
reanalysis, 537
relative humidity, 682
surface winds, 514
zonally-averaged atmosphere, 515
zonally-averaged zonal wind, 570
Ocean circulation
abyssal, 821
laboratory model of, 818
scaling for buoyancy-driven, 813
wind-driven, 733
wind-driven abyssal, 829
Ocean currents, 732
Ocean gyres, 731
Omega equation, 192
Outcropping, 788

Parabolic cylinder functions, 330
Parcel method, **97–99**, 695
Passive tracer, **437–441**
in three dimensions, 439
in two dimensions, 439
spectra of, 437
Perfect gas, 47
Phase speed, 217, **216–219**
Phase velocity, 219
Phillips instability problem, 356
Piecewise linear flows, 338
Plane waves, 216
Planetary waves, 585
Planetary-geostrophic equations, **176–180**
for shallow water flow, 176
for stratified flow, 178
Planetary-geostrophic potential vorticity equation, 178, 179
shallow water, 178

stratified, 179
Poincaré waves, 124, 125, 244
Poincaré, Henri, 141
Polar vortex, 629
Polarization properties of internal waves, 261
Polarization relations, 261
Polytropic fluid, 14
Potential density, 24, 25, 27, 30, **32**
and static instability, 97
of seawater, 27, 32
Potential enthalpy, 36, 46, 49
Potential temperature, 24, 30, 49
equivalent, 694
ideal gas, 25
moist air, 722
of liquids, 28
seawater, 28, 34
Potential vorticity, 143, **156–168**
and Bernoulli's theorem, 167
and the frozen-in property, 158
Boussinesq equations, 163
concentration, 165
conservation of, 156
diffusion of, 505
for baroclinic fluids, 157
for barotropic fluids, 156
homogenization of, 793
hydrostatic equations, 164
impermeability of isentropes, 165
mixing, 575, 576
moisture effect, 160
ocean observations, 770
on isentropic surfaces, 164
planetary-geostrophic, 178
quasi-geostrophic, 190
relation to circulation, 156
salinity effect, 160
shallow water, 120, 162, 182
staircase, 451–453
substance, 165
Potential vorticity flux, 566
Potential vorticity fluxes, atmospheric, 566
Potential vorticity homogenization, 487
Potential vorticity transport
and tropospheric stratification, 574
Prandtl number, 439, 804
Predictability, **433–437**
of Lorenz equations, 433
of turbulence, 435
of weather, 437
Pressure, 11, 17
Pressure coordinates, **79**, 81
and quasi-geostrophy, 192
relation to Boussinesq equations, 82
Primitive equations, **64**
potential vorticity conservation, 164
vector form, 65
Pseudoenergy, 406
and hydrodynamic instability, 406
and wave activity, 407
Pseudomomentum, 384
and hydrodynamic stability, 403
and zonal jets, 544

QBO, **652–662**
essentials, 655
Quasi-biennial oscillation, **652–662**
Quasi-equilibrium, **698**, 699
Quasi-geostrophic
wave–mean-flow interaction, 380
Quasi-geostrophic potential vorticity
equation, 190
relation to Ertel PV, 195
Quasi-geostrophic turbulence, 454
Quasi-geostrophy, **180–195**
asymptotic derivation, 188
buoyancy advection at surface, 191
continuously stratified, 187
energetics, 198
in isentropic coordinates, 196
informal derivation, 193
multi-layer, 185–186
pressure coordinates, 192
shallow water, 180
sheet at boundary, 191
single layer, 180
stratified equations, **187–194**
two-layer, 184–185
two-level, 194

Radiation condition, 543
Radiative equilibrium temperature, 511
Radiative equililibrium, 700–703
Radiative transfer, 724
Radiative-convective equililibrium, 703–706
Radius of deformation, 124, 125
Random walk, 476
Ray theory, **224–226**, 621
Ray tracing, 621
Rayleigh criterion for instability, 345
Rayleigh equation, 338
Rayleigh number, 804
Rayleigh's equation, 338
Rayleigh–Kuo criterion, 346, 404
Rayleigh–Kuo equation, 338
Rays, 226, 273
equatorial, 314
in internal waves, 273
Reanalysis, **537**, 537
Recharge-discharge oscillator, 901
Reduced gravity equations, **110–112**

Reference, circular, *see* Circular reference
Reflection
internal waves, 268
Rossby waves, 237
Refractive index, 587, 599
Relative humidity, 675, 676, **680–690**
in mid-latitudes, 688
Relative vorticity, 152
Residual circulation, 388
and thickness-weighted circulation, 391
atmospheric, mid-latitude, 571
atmospheric, observations of, 572
stratospheric, 642
Resonance of stationary waves, 610
Reynolds number, 46
Reynolds stress, 415
Reynolds, Osborne, 53
Rhines length, 447
Rhines scale, 446
Rhines–Young model, 761
Richardson number, 494
Richardson's four-thirds law, 482
Richardson, Lewis Fry, 442
Rigid body rotation, 144
Rigid lid, 108, 111
Rossby number, **87**
Rossby wave trains, 611
Rossby waves, **226–240**, 585
and barotropic jets, 542
and ray tracing, 621
and turbulence, 446
barotropic, 227
breaking, 599, 642
continuously stratified, 231, 599
critical layers, 594
dispersion relation, 228
energy flux, **234–236**
finite deformation radius, 229
group velocity property, 384
horizontal propagation, 588
mechanism of, 229
meridional propagation, 621
momentum transport in, 542
planetary geostrophic, 302
propagation, 585
reflection, 237
topographic, 602
two layers, 230
vertical propagation, 599, 603, 605, 606
Rossby, Carl-Gustav, 212
Rotating frame, **55–59**

Salinity, 13, 33
effect on potential vorticity, 160
in box models, 814
Salt, 13
Sandström's effect, 809
Saturated adiabatic lapse rate, 692, 695
Saturation vapour pressure, 676
Scale height
atmosphere, 42
density, 27
temperature, 28
Scale height, atmosphere, 83
Scaling, **46–47**
geostrophic, 171
in rotating shallow water equations, 171
in rotating stratified equations, 174
Schwarzchild equations, 724
Seawater, 13, **33**
adiabatic lapse rate, 34
equation of state, 13, 33, 34
heat capacity, 34
potential temperature, 34
thermodynamic properties, 33
Shadow zone, 791
Shallow water
quasi-geostrophic equations, 180
Shallow water equations
multi-layer, 112
potential vorticity conservation, 162
reduced gravity, 110
rotation effects, 121
Shallow water model of Hadley Cell, 524
Shallow water systems, **105–123**
conservation properties of, 120
potential vorticity in, 120
Shallow water waves, **123–127**
Shallow-fluid approximation, 65
Sideways convection, **802–813**
conditions for maintenance, 809
energy budget, 808
limit of small diffusivity, 811
maintenance of, 808
mechanical forcing of, 812
phenomenology, 807
Sigma coordinates, 82
Singing in the rain, 673
Single-particle diffusivity, 478
Skew diffusion, 491
Skew flux, 491
Skew fluxes, 490
Sloping convection, 347
Solenoidal term, 145
Solenoids, 146, 151
Sound waves, 40
Southern Ocean, 836
Southern Oscillation, 888
Specific heat capacities, 20
Specific humidity, 675
Spectra of passive tracers, 437

Spherical coordinates, **59–68**
centrifugal force in, 59
Squire's theorem, 375
Stacked shallow water equations, 112
Staircases of potential vorticity, 451–453
Standard atmosphere, 572
State estimate, 537
Static instability, **97–99**, 695
Stationary phase, 223
Stationary waves, **609–625**
adequacy of linear theory, 614
and ray tracing, 621
Green function, 613
in a single-layer, 609
meridional propagation, 621
one-dimensional wave trains, 611
resonant response, 610
thermal forcing of, 615
Stokes, George, 53
Stommel box models, 813
Stommel wind-driven model, **733**
boundary layer solution, 737
properties of, 741
quasi-geostrophic formulation, 736
the nonlinear problem, 745
Stommel, Henry, 758
Stommel–Arons model, **821**
single-hemisphere, 821
two-hemisphere, 826
Stommel–Arons–Faller model, 818
Stratification
in mid-latitudes, 579
of the atmosphere, 572
Stratified geostrophic turbulence, 454
Stratosphere, 514, 572, **627**
polar vortex, 629
sudden warming of, 631
Stratospheric dynamics, **627–669**
Stratospheric sudden warmings, 663
Stress
Ekman layer, 201
kinematic, 202
Stretching, 149
Sudden warming, 631
Sudden warmings, 663
Super-rotation, 523
Surf zone, 629
Surface drifters, 483
Surface westerlies, 540
Surface winds, 567
observed, 514
Sverdrup balance, 737
near the equator, 863
Sverdrup interior flow, 738
Symmetric diffusivity tensor, 490
Tangent plane, 69
Taylor–Goldstein equation, 597
Taylor–Proudman effect, 90
TEM, 379, 387, 389, 392
TEM equations, **387**
for primitive equations, 581
Temperature, **17**, 724
dew point, 724
potential, 24, 724
wet-bulb, 724
Thermal equation of state, 13, 47
moist air, 721
Thermal wind, 87
in shallow water equations, 118, 119
Thermal wind balance, **87–93**
pressure coordinates, 91
Thermobaric effect, 14, 33
on potential density, 36
on potential vorticity, 161
Thermocline, 761, **774–798**
advective scaling, 777
boundary-layer analysis, 782
diffusive, 805
diffusive scaling, 777
internal, 779
kinematic model, 775
main, 774
one-dimensional model, 780
reduced-gravity, single-layer model, 786
scaling for, 776
summary and overview, 790
ventilated, 785
wind-influenced diffusive scaling, 778
Thermodynamic equation, **21–30**
Boussinesq equations, 72
for liquids, 25, 30
summary table, 26
Thermodynamic equilibrium, 21
Thermodynamic potentials, 17–19
Thermodynamic relations, **14–21**
fundamental, 16
Maxwell, 17
Thermodynamics
first law, 15
fundamental postulate, 14
moist, 720
Thermohaline circulation, 801
Thickness, 83
Thickness diffusion, **502**, 504
Thomas, Dylan, 3
Tilting and tipping, 149
Topographic effects
atmospheric stationary waves, 609
JEBAR, 756
oceanic western boundary current, 753

Toy models, 908
  El Niño, 898
Tracer continuity equation, 10
Tracer homogenization, 487
Traditional approximation, 65
Transformed Eulerian Mean, 379, 392, **387–392**
  and eddy transport, 506
  isentropic coordinates, 389
  primitive equations, 581
  quasi-geostrophic form, 387
  spherical coordinates, 581
Triad interactions, 415
  two-layer geostrophic turbulence, 457
Tropics, 673
Tropopause, 572–581, **706–708**
  definition, 572
Tropopause height
  in mid-latitudes, 579
  theory of, 706, 707, 725
Troposphere, 514, 572
  and potential vorticity transport, 574
  stratification, 572, 574, 578
  ventilation and moist convection, 577
Truesdell, Clifford, 52
Turbulence, **413**
  beta-plane, 448
  closure problem, 413
  degrees of freedom, 422
  fundamental problem, 413
  predictability of, 433
  three-dimensional, 416
  two-dimensional, 423
Turbulent diffusion, **473**, **475**, 490
  and the TEM, 506
  in the atmosphere and ocean, 493
  macroscopic perspective, 487
  requirements for, 485
  thickness, 502
  two-dimensional, 483
Turbulent diffusivity, 478
Turning line, 589, 607
Two-box model, 814
Two-dimensional turbulence, **423–433**
  beta effect, 445
  eddy diffusion in, 483
  energy and enstrophy transfer, 424
  numerical solutions, 432
Two-dimensional vorticity equation, 147
Two-layer instability problem, 356
Two-layer model
  of atmospheric mid-latitudes, 554
Two-layer QG equations, 184
Two-level QG equations, 194
Two-particle diffusivity, 480, 482

Under Milk Wood, 3
Unit vectors
  rate of change on sphere, 62

Vapour pressure
  saturation, 676
Vector invariant momentum equation, 66
Ventilated pool, 794
Ventilated thermocline, **785–798**
  reduced-gravity, single-layer model, 786
  two-layer model, 788
Vertical coordinates, **79–83**
Vertical vorticity equation, 155
Virtual temperature, 674, 676
Viscosity, 12
  effect on energy budget, 45
Viscous scale, 420
Viscous-advective range, 439
Vorticity, 96, **143–153**
  equation for a barotropic fluid, 146
  equation on beta plane, 156
  evolution equation, **145**
  evolution in a rotating frame, 153
  frozen-in property, 147
  in two dimensional fluids, 147
  stretching, 151
  stretching and tilting, 149
  vertical component, 155
Vorticity equation, 96
Vorticity, relative, 152
vr vortex, 144

Walker Cell, 717
Walker circulation, 328, 888, **891**
Water, 673
Water vapour, 673–676
  measures of, 675
Wave activity, 384
  and pseudomomentum, 384
  group velocity property, 384
  orthogonality of modes, 385
Wave breaking, 642
Wave packet, 221
Wave propagation, 585
Wave trains, 611
Wave–mean-flow interaction, **379**, 382
  quasi-geostrophic, 380
Wave–turbulence cross-over, 446
Waveguides
  for equatorial waves, 314
  for internal waves, 274
Wavelength, 217
Waves, **215**
  acoustic-gravity, 293
  barotropic Rossby, 227
  breaking, 642

- frequency, 216
- group velocity property, 240
- hydrostatic gravity, 261
- inertial, 126
- Kelvin, 126, 636
- kinematics, 215
- Lamb, 295
- Poincaré, 124, 125, 244
- Rossby, 226
- Rossby dispersion relation, 228
- Rossby wave mechanism, 229
- Rossby, continuously stratified, 231
- Rossby, single-layer, 227
- Rossby, two-layer, 230
- rotating shallow water, 124
- shallow water, 123
- sound, 40
- wavevector, 216

Wavevector, 216
Weak temperature gradient approximation, 712, 714
- adjustment to, 716
- for stratified flow, 717
- in shallow water equations, 716

Weather predictability, 437
West, Mae, 156
Western boundary currents
- topographic and inviscid, 753

Western boundary layer, 739
- frictional, 738
- inertial, 747

Western intensification, 731
Western pool, 793
Wet-bulb temperature, 724
Wind-driven gyres, 731
Wind-driven ocean circulation, **733–770**
- continuously stratified, 767
- homogeneous model, 733
- two-layer model, 761
- vertical structure, 761

WKB approximation, **247–249**, 623
- internal waves, 271
- Rossby waves, 589, 590, 598

Zonal boundary layers in ocean gyres, 744
Zonal flow in turbulence, 446
Zonal flows in beta-plane turbulence, 448
Zonally-averaged atmospheric circulation, 539

Printed in the United States
by Baker & Taylor Publisher Services